2025 IEEE Applied Power Electronics Conference and Exposition (APEC 2025)

Atlanta, Georgia, USA
16-20 March 2025

Pages 2654-3339

IEEE Catalog Number: CFP25APE-POD
ISBN: 979-8-3315-1612-3

**Copyright © 2025 by the Institute of Electrical and Electronics Engineers, Inc.
All Rights Reserved**

Copyright and Reprint Permissions: Abstracting is permitted with credit to the source. Libraries are permitted to photocopy beyond the limit of U.S. copyright law for private use of patrons those articles in this volume that carry a code at the bottom of the first page, provided the per-copy fee indicated in the code is paid through Copyright Clearance Center, 222 Rosewood Drive, Danvers, MA 01923.

For other copying, reprint or republication permission, write to IEEE Copyrights Manager, IEEE Service Center, 445 Hoes Lane, Piscataway, NJ 08854. All rights reserved.

*** This is a print representation of what appears in the IEEE Digital Library. Some format issues inherent in the e-media version may also appear in this print version.

IEEE Catalog Number:	CFP25APE-POD
ISBN (Print-On-Demand):	979-8-3315-1612-3
ISBN (Online):	979-8-3315-1611-6
ISSN:	1048-2334

Additional Copies of This Publication Are Available From:

Curran Associates, Inc
57 Morehouse Lane
Red Hook, NY 12571 USA
Phone: (845) 758-0400
Fax: (845) 758-2633
E-mail: curran@proceedings.com
Web: www.proceedings.com

TABLE OF CONTENTS

Versatile Controller Architecture for a Universal DC Fast Charging Front-End 1
Anurag Singh, Sayan Paul, Tejas Bhuse, Trent Martin, Hien Nguyen, Inder Vedula, Nikola Milivojeviæ, Dragan Maksimoviæ, Luca Corradini

A 10 kV SiC MOSFET based Three-Phase Single-Stage Isolated MVAC/LVDC Converter for Solid State Transformer Applications 9
Anup Anurag, Chi Zhang, Rudy Wang, Peter Barbosa

Direct Digital Control Applied to T-Type Vienna Rectifiers for Power Factor Correction 16
Jun-Yang Chang, Tsai-Fu Wu, Chien-Chih Hung, Jui-Yang Chiu

Active Power Decoupling Method based on Dual Active Bridge Converter without Additional Components 21
Kosuke Takeuchi, Takashi Ohno, Hiroki Watanabe, Yuki Nakata, Jun-Ichi Itoh

An ANPC-Based Building Block for Medium-Voltage Applications 27
Ahmed Rahouma, Hui Cao, David A. Porras, Zhuxuan Ma, Yue Zhao, Juan C. Balda

Analog Control of a 2.5 kW GaN based CRM PFC with Input Filter Optimization 34
Naveed Ishraq, Ayan Mallik

An iTHD and Efficiency Optimized Control Method for Triangular Conduction Mode Totem Pole Bridgeless PFC with Zero Current Detection 42
Brent McDonald, Sheng-Yang Yu

Resonance Current Suppression for AC-DC Active-Clamp Flyback Converter by Triangular Current Mode 48
Yasuo Uchida, Hiroki Watanabe, Jun-Ichi Itoh

A Universal DC Fast Charging Front-End with Optimized Film Capacitor Design 54
Sayan Paul, Anurag Singh, Tejas Bhuse, Trent Martin, Hien Nguyen, Inder Vedula, Nikola Milivojeviæ, Dragan Maksimoviæ, Luca Corradini

Power Characterization of a 1200-V/800-V 22-kVA 30-kHz Unity-Gain Dual-Active-Bridge Converter Prototype 62
Radhika Sarda, Abishek Sethupandi, Madasamy Palavesha Thevar, Howe Li Yeo, Praveenkumar Palani, Vaisambhayana B. Sriram, Anshuman Tripathi

Design of Fully Soft-Switched Semi-Dual Active DC-DC Converter for Battery Charging Application 69
Siva Prabhakar, Shiladri Chakraborty, Sandeep Anand

A ZCS-ZVS Strategy for Low Impedance Dual Active Bridges in MHz Range 77
Pushkar Saraf, Michael Solomentsev, Alex Hanson

A 6.6 kW Highly Efficient Reconfigurable Dual Active Bridge Converter Designed using Planar Transformer, SiC-Fets and Monolithic Bidirectional Devices 90
Reza Barzegarkhoo, Fabian Groon, Arkadeb Sengupta, Marco Liserre

Interleaved Switched-Inductor-Based SIPO Partial Power Converter Module for Battery Management Systems 98
Fengwang Lu, Henry Shu-Hung Chung

Single Sensor-Based Fault Localization and Detection in GaN Three-Phase Dual Active Bridge Converters 103

Satyam Sa, Yi Han, Cheng Feng Wang, Olivier Trescases

Enhanced Cocharge Operation Scheme in Bidirectional PhaseShift Full-Bridge Converters with Eliminated Voltage Overshoot and Reduced Freewheeling Current.................. 111

Tien-Sheng Li, Minh Ngo, Rolando Burgos, Dong Dong

DC Bias Elimination in Isolated DC-DC Converters using Fundamental-Frequency Ripple 118

Arkadeb Sengupta, Thiago Antonio Pereira, Marco Liserre

Tunable Matching Network with Dual Phase-Switched Impedance Modulation Actuators.................. 124

Alexander Jurkov, David Perreault

Soft-Switched Pulsed Bias Plasma Supply System 132

Julia Estrin, Alexander Jurkov, David J. Perreault

Analysis and Design of a Cyclo-Active-Bridge Inverter for Single-Stage Three-Phase Grid Interface.................. 139

Mian Liao, Tanuj Sen, Yang Wu, Minjie Chen

Modular Nanosecond Pulse Generator Leveraging GaN and SiC for Versatility and Performance 147

P. Briz, H. Sarnago, O. Lucía

A Variable Frequency Technique for EMI and Efficiency Improvements in High-Level Count Flying Capacitor Multilevel Converters 151

Francesca Giardine, Sahana Krishnan, Logan Horowitz, Robert C. N. Pilawa-Podgurski

Analysis and Implementation of Minimum-Sensor Capacitor Voltage Estimators for Flying Capacitor Multilevel Converters 157

S. Tahmid Mahbub, Rahul K. Iyer, Ivan Z. Petriæ, Robert C. N. Pilawa-Podgurski

Single-Stage Bidirectional High-Frequency Link DC to Three-Phase AC (4-Wire) Grid-Tied Microinverter.................. 164

Aniruddh Marellapudi, Satish Belkhode, Joseph Benzaquen Sune, Deepak Divan

Analysis and Design of a Constant Current LCC Class-E Inverter 171

Ju Gao, Ziheng Liu, Jiayin He, Hongjie Peng, Chengkang Ao, Jinyan Wang

Series Connected Class-E Push-Pull Converters using GaN HEMT for High-Efficiency RF Generators in Float Zone Silicon Production 175

Faheem Ahmad, Thore Stig Aunsborg, Jannick Kjær Jørgensen, Stig Munk-Nielsen

State of the Art 1.7kV Lateral GaN HEMTs, an Alternative to SiC 180

Karthick Murukesan, Robert Yang, Kamal Varadarajan, Sorin Georgescu, Doug Kang

Modeling and Characterization of Current and Future 1.2 kV Wide Bandgap Semiconductor-Based MOSFETs.................. 185

Sushanta Gautam, Austin M. Szczublewski, Samuel K. Atwimah, Aidan P. Fox, William M. Collings, Tolen Nelson, Daniel G. Georgiev, Raghav Khanna, Andrew D. Koehler, Karl D. Hobart

2.5-kV 6.4-ns 100-kHz Repetitive GaN Marx Generator.................. 192

Ruize Sun, Ci Pan, Wanjun Chen, Bo Zhang

Novel Dual Output LDO Architecture in 650-V GaN Technology for Power ICs 195

Plinio Bau, Thanh Hai Phung, Deniz Aygun, Bart Coomans, Mike Wens

Impact of Substrate Bias on the Stability of Bidirectional GaN HEMT in Hard- and Soft-Switching 202
Qihao Song, Hongchang Cui, Qiang Li, Yuhao Zhang

Characterization of LED Driven GaN-Based Photoconductive Switches 207
Samuel K. Atwimah, Tolen M. Nelson, Geoffrey M. Foster, Daniel G. Georgiev, Andrew D. Koehler, Alan G. Jacobs, Karl D. Hobart, Micheal R. Hontz, Raghav Khanna

Development and Validation of Repetitive Transient Gate Overvoltage Rating for GaN HEMTs 214
Ricardo Garcia, Angel Espinoza, Siddhesh Gajare, Shengke Zhang

Junction Temperature Monitoring of GaN HEMT by using On-Resistance with Voltage Clamp and
Current Shunt 219
Xiao Wang, Mingrui Zou, Jiakun Gong, Yulei Wang, Zheng Zeng

False Turn-On Failure and Protection of p-Gate GaN HEMT in MHz Class-E Resonant Inverter 225
Ziheng Liu, Ju Gao, Hongjie Peng, Jiayin He, Jinyan Wang, Maojun Wang

Heat Extraction from Ferrite Cores using Metallic Laminations 231
Alyssa Brown, Duy T. Nguyen, Alex J. Hanson

Folded Flex-PCB Winding Planar Transformer for High-Frequency Isolated DC-DC Converters 238
Soundhariya G. Soundararajan, Hans Wouters, Wout Vanderwegen, Wilmar Martinez

Winding Strategy Analysis and Optimization for High-Current Matrix Transformer 246
Bima Nugraha Sanusi, Pinhe Wang, Michael A. E. Andersen, Ziwei Ouyang

Investigation on Impact of Transformer Parasitic Capacitance on Standby Power Consumption in
Power Converters 252
Kamran Kamran, Andrea Russo, Federica Cammarata, Claudia Malannino, S. Yuri Ciardo, Ziwei Ouyang

PCB-Winding Integrated Transformer for 800-V Dual Active Bridge Converter using 1.2-kV GaN
Devices 258
Hans Wouters, Wei-Ren Lin, Nicolas Pirson, Thomas Jochmans, Yu Zuo, Wilmar Martinez

Comparative Assessment of Inductance Modeling for PCB-Based Circular Spiral Coils in Inductive
Power Transfer Systems 266
Gaia Petrillo, Drazen Dujic

Compact Air-Core Inductors for Variable Frequency Soft-Switching in 3 Phase Inverters 272
Youssef A. Fahmy, Matthias Preindl

Simulation and Experimental Research on Cooling Performance of Fully-Immersed Evaporative
Cooling High-Frequency Transformer 278
Zhanlei Liu, Lingyu Zhu, Yuntian Gao, Yongliang Dang, Cao Zhan, Shengchang Ji

High-Efficiency PCB-Embeddable Inductor for Vertical Power IVR Applications 285
Youssef Kandeel, Liang Ye, John Flannery, Cian Ó Mathúna, Ranajit Sai, Seamus O'Driscoll, Takayuki Tsuchida, Naoya Terauchi, Sumiaki Kishimoto, Toshio Hiraoka, Masanori Nagano

An Adaptive Zero Current Switching Control Technique for Multi-Resonant Switched-Capacitor
Converters 291
Haifah B. Sambo, Rose A. Abramson, Sahana Krishnan, Robert C. N. Pilawa-Podgurski

Small-Signal Analysis and External Ramp Design for Multiphase Current-Mode Constant On-Time
Control with Phase Overlapping 299
Sundaramoorthy Sridhar, Qiang Li

Multiphase Constant-on-Time Minimum-Deviation Controller for Modern Processors 307
    Duo Li, Gianluca Roberts, Aleksandar Prodić, Alan Wu

Closed-Loop Control of a Dual-Side Series/Parallel Piezoelectric-Resonator-Based DC-DC
Converter 315
    Wen-Chin B. Liu, Gaël Pillonnet, Patrick P. Mercier

High-Bandwidth Embedded Rogowski Coil on Multilayer Substrate with Minimal Contribution to
Power Loop Inductance 321
    Takahiro Okamoto, Masataka Ishihara, Kazuhiro Umetani, Eiji Hiraki

Operating and Switching Frequency Circulating Current Control in Paralleled High Power
Adjustable Speed Drives with Common DC Link 327
    Kevin Lee, Zhihao Song, Wenxi Yao, Bo Wei

Mixed-Signal Sliding Mode Controller for Non-Inverting Buck-Boost Photovoltaic DC Optimizers 334
    Anurag Singh, Sayan Paul, Dragan Maksimović, Luca Corradini

A Current Sensorless Output Voltage Tracking Controller-Observer for a Boost Inverter using
Feedback Linearization 342
    Ion Leandro Dos Santos, Tailan Orlando, Yohannes Amilcar Tekle Scherer, Telles Brunelli
    Lazzarin, Hector Bessa Silveira

Modeling and Control of a Cyclo-Active-Bridge Inverter for Single-Stage Three-Phase Grid
Interface 349
    Tanuj Sen, Mian Liao, Yang Wu, Minjie Chen

Turn-On Transient Modeling of 10 kV SiC MOSFET Half-Bridge Power Module in LTspice 357
    Nianzun Qi, Jannick Kjær Jørgensen, Gao Liu, Zhixing Yan, Morten Rahr Nielsen, Asger
    Bjørn Jørgensen, Hongbo Zhao, Stig Munk-Nielsen

A Compact, Automated Sawyer-Tower System for Characterization of the High-Frequency, Soft-
Switching C_{oss} Loss of Wide Bandgap Devices 363
    Katherine Liang, Malachi Hornbuckle, Juan Rivas-Davila

Enhancing Behind-the-Meter Visibility of Grid Edge PV Systems and Electric Vehicle Charging
Loads Through Integration of Compact Low-Cost Sensors 370
    Mehrnaz Madadi, Paul Ohodnicki, Subhashish Bhattacharya

Supercapacitor based TMS Pulse Generator Design-Experimental Results Versus MATLAB
MOSFET Simulation Model 378
    Soniya Raju, Nihal Kularatna, Marcus Wilson, Alistair Steyn-Ross

Application of Artificial Intelligence for Modeling SiC Power MOSFETs 385
    Fredo Chavez, Danial Bavi, Sourabh Khandelwal

Multi-Objective Design Automation in Power Electronics using Bayesian Optimization Techniques 389
    Tung-Tan Nguyen, Man-Hay Pong, Huang-Jen Chiu

Reduced Order Thermal Modelling of Multi-Chip Silicon Carbide Power Modules 395
    Aamir Rafiq, Blake Nelson, Marshal Olimmah

Design and Evaluation of Dual-Resolver Emulation for Control System Verification in Aerospace
Actuation Applications 401
    Tomas Sadilek, Julian Opificius, Jason Wright, Alec Leslie, Jeremie Tuzizila, Cesar Alzate,
    Hunter Burnett, Joshua Atkinson, Justin Stricula

Un-Terminated Blackbox Modeling for Electric Machines...409
    Xinliang Yang, Vladimir Mitrovic, Qing Lin, Rolando Burgos

7.2 kW GaN-Based DAB Converter with 37 kW/L Power Density and High Efficiency...............416
    Esmaeil Jalalabadi, Xiaoyu Wang, Jaksa Rubinic, Yang Jiao, Lucas Lu

A Novel Interleaving Method for High Power Integrated Electric Vehicle Charger with Three-Phase
Permanent Magnet Synchronous Motor 423
    *Ryota Tanaka, Toshihiro Kai, Kenta Takishima, Yoshiyuki Nagai, Tetsuya Hayashi, Kantaro
    Yoshimoto*

A Three-Phase CLLC Resonant Converter with Integrated Planar Magnetics for 22-kW On-Board
Chargers...429
    *Tianlong Yuan, Zhangwei Xiang, Abdelrahman Ali, Feng Jin, Qiang Li, Wendell Da-Cunha-
    Alves, Xiaoshan Liu*

Reconfigurable LLC Resonant Converter for Wide Voltage Range and Reduced Voltage Stress in
DC-Connected EV Charging Stations 436
    Yu Zuo, Xiaobing Shen, Bangli Du, Qingcheng Sui, Tim Geboers, Wilmar Martinez

Design and Control of GaN based Three-Phase / Single-Phase Combo Three-Level Flying
Capacitor PFC for OBC Applications...442
    Nidhi Haryani, Laszlo Huber, Anup Anurag, Juan Ruiz, Peter Barbosa

Optimization Strategy for Battery Electric Vehicle (BEV) DC Fast Charging (FC) in Cold
Environments..449
    Seif Sarofim, Cheng Feng Wang, Satyam Sa, Avram Kachura, Isaac Muscat, Olivier Trescases

DC-Link Voltage Reduction with Synergetic Common-Mode Voltage Control of Single-Phase Two-
Stage Non-Isolated EV Chargers...457
    Dongsu Lee, Juwon Lee, Jung-Ik Ha

DC-DC Converter Architecture for Fast Electric Vehicle (EV) Battery Charging Applications 464
    Shibaji Basu, Arjun Ivimey, Praveen Jain

Fast Simulator for the Estimation of Inverter DC-Link Temperature in e-Drives Subjected to Highly
Variable Working Cycles 472
    Simone Giuffrida, Fabio Mandrile, Radu Bojoi

A Monolithic Regulated 160 MHz Resonant DC-DC Converter 479
    *Giacomo Ripamonti, Stefano Michelis, Georgios Bantemits, Pablo Daniel Antoszczuk, Khalil
    Khalife, Nils Hans Van Der Blij, Sokratis Koseoglou, Mattia Balutto, Francesco Driussi,
    Stefano Saggini*

Reconfigurable Trans-Inductor Voltage Regulator with Improved Light Load Efficiency in Data
Center Applications 485
    Ziyao Wang, Zehui Li, Haoyu Wang

Fully Integrated Voltage Regulators (FIVRs) with Package In-Situ Coupled CoaxMIL Inductor for
High Power Density Microprocessor Applications 491
    *Jaeil Baek, Beomseok Choi, Siddharth Kulasekaran, Huong Do, Brandon Marin, Jose
    Chavarria, Leigh Wojewoda, Kaladhar Radhakrishnan*

Multiphase Lateral Flux Indirect Coupled Inductor for Vertical Power Delivery Voltage Regulator
Module 498
    Adhistira M. Naradhipa, Qiong Wang, Qiang Li

A High Density Three-Level Quadratic Buck Hybrid Converter for 48V-to-PoL Conversion 505
Kejia Wang, Si Yuan Sim, Yin Quen Choong, Xin Zhang, Sriharsh Pakala, Cheng Huang

Air-LEGO: A Magnetic-Free Ultra-Thin 24V-to-1V 120A VRM with Air-Coupled Inductors 510
Haoran Li, Wenliang Zeng, Youssef Elasser, Minjie Chen

A 15A 48V-Input Dual-Path Hybrid Dickson Converter with 6 mm³ Low Saturation Current
Inductors for Point-of-Load Conversion 518
Hua Chen, Young-Seok Noh, Minxiang Gong, Vivek De, Arijit Raychowdhury

An Ultra-Fast Control Strategy and Pre-Current-Balancing Measures Prepared for Rapid Transients
in Constant On-Time Controllers 524
Yijie Qian, Yuan Gao, Wenze Shu, Lingyun Li, Shen Xu, Weifeng Sun

Loosely Coupled Trans-Inductor Voltage Regulator (LC-TLVR) Inductor as Compensation Inductor
(Lc) 530
Pavan Kumar, Arturo Sanchez Hernandez

Novel Complex Permeability Model of Powder Magnetic Materials 538
Lukas Mueller, James Cox, Jun Wang, Enrique Garcia

Design Study Evaluating Impact of Gap Loss on Nanocrystalline Inductor Cores with Experimental
Validation 544
Maurice Sturdivant, Brandon Grainger, Christopher Bracken, Paul R. Ohodnicki

A Permanent Magnet Variable Inductor for DC Fault Current Limiting Applications 552
Mark Nations, Subhashish Bhattacharya

Design-Oriented Modeling and Multi-Objective Optimization of Two-Phase Coupled Inductors in
Multiphase PWM Converters 558
Yicheng Zhu, Jiarui Zou, Robert C. N. Pilawa-Podgurski

MagNetX: Extending the Magnet Database for Modeling Power Magnetics in Transient 566
*Hyukjae Kwon, Shukai Wang, Haoran Li, Youssef Elasser, Gyeong-Gu Kang, Daniel Zhou,
Davit Grigoryan, Minjie Chen*

Non-Monotonic Influence of DC Bias on Ferrite Core Loss Up to 10 MHz with Sine Wave
Excitation 573
Bohua Zhang, Martin Pfost

Comprehensive SPICE Model for Inductors Considering Magnetic Losses Under DC Bias Current 579
Yuki Sato, Hirokazu Matsumoto, Junichi Kotani, Shohei Tomioka, Kenichiro Tanaka

Indented Core to Reduce and Desensitize Inductor's Fringing Losses without Increasing Volume 586
Rajaie Nassar, Promit Datta, Guo-Quan Lu, Christina DiMarino, Khai Ngo

Coupled Inductor Analysis and Finite Element Modeling Assisted Design for Boost Extender
Topology 594
Vikas Kumar Rathore, Michael Evzelman, Mor Mordechai Peretz

Stability Analysis of Current-Limited Grid-Forming Inverters with Frequency Stabilization: An
Equivalent Impedance Approach 602
Bowen Yang, Gab-Su Seo

Revisit Active Power Oscillation in Multi-Virtual Synchronous Generators Gride 609
Junjie Xiao, Pavol Bauer, Zian Qin

A Novel Current Control Technique for Off-Grid Single-Phase Inverters 616
Arpan Laha, Abirami Kalathy, Praveen Jain, Majid Pahlevani

Intelligent Low-Bandwidth Frequency Controller for VSGs at Economic Dispatch in Islanded
Microgrid.. 622
Shraf Eldin Sati, Ahmed Al-Durra, Hatem H. Zeineldin, Tarek H.M. El-Fouly, Ehab F. El-Saadany

Hardware-in-the-Loop of a Grid Forming Control Strategy Applied to a DC Off-Grid Green
Hydrogen Production System... 629
*Diego Montoya-Acevedo, René Contreras-Barrios, Ángel Maureira-Riquelme, Esteban
Ibáñez-Muñoz, Catalina Gonzalez-Castaño, Carlos Restrepo*

Experimental Validation of a 40kW, 480V Point-to-Point DC Interlinks for Controller-Agnostic,
Interoperable Networked Microgrids 637
Maximiliano Ferrari, Michael Starke, John Smith, Joao Pereira, Misael Montejano

Andronov-Hopf Oscillator-Based Grid-Forming Converters with Embedded Disturbance Rejection
for Non-Ideal Loading Condition 645
Vikram Roy Chowdhury, Gab-Su Seo, Barry Mather

Estimation of Rectifier Output Current of the LLC Converter.. 651
Xin Wu, Yi Zhou, Haihong Long, Dehong Xu

A 100kHz Digitally Controlled 10kW, 2-Channel Solar MPPT Converter using 3-Level Topology
with >75W/in³ Power Density and >98.5% Peak Efficiency.. 658
Ranajay Mallik, Akshat Jain

A Bootstrapless KY-S-Hybrid Buck-Boost Converter with Full Range iLs Reduction and 400%
Line Transient Response Acceleration for AI-Mobile Application... 664
Chuan-En Chang, Cheng-Ta Chuang, Hao-Ran Huang, Chieh-Ju Tsai, Ching-Jan Chen

Digital Control of a 600-V to 28-V 20-kW Two-Stage DC-DC Converter 670
*Shreyas B. Shah, Rachit Pradhan, Jiaqi Yuan, Mohamed Ibrahim, Ahmed Elezab, Samuel
Hemming, Giorgio Pietrini, Piranavan Suntharalingam, Mario F. Cruz, Ali Emadi*

Self-Calibrated Digital Current Emulation for High-Frequency Hysteretic Current-Mode Control in
GaN PFC Converters 676
Mohammad Shawkat Zaman, Olivier Trescases

High-Frequency Flying Capacitor Four-Level Drain Supply Modulator 682
Audrey Cheshire, Paul Flaten, Zoya Popoviœ, Dragan Maksimoviœ

Discontinuous Modulation Strategy for Voltage and Temperature Balancing of MMCs 689
Davide D'Amato, Stayner Nóbrega Barros, Jun-Hyung Jung, Marco Liserre

Damping Control and Improvement of Grid-Forming Inverter from a Wideband Stability
Perspective.. 696
Rui Kong, Subham Sahoo, Yubo Song, Frede Blaabjerg

A Grid-Forming Split-Phase Three-Leg Inverter with Unbalanced Loading and Active Power
Decoupling 703
Namwon Kim, Renata Kimpara, Michael Starke

Completely Decentralized Active and Reactive Power Control of Grid-Connected Cascaded H-
Bridge Inverters with Integrated Battery Storage 711
Soham Dutta, Brian Johnson

Small-Signal Modeling and Damping Design of Unfolding-Based Single Stage AC-DC Converter using the Extra Element Theorem 719

Dakota Goodrich, Aditya Zade, Shubhangi Gurudiwan, Mahmoud Mansour, Regan Zane, Hongjie Wang

Methods to Enhance Cybersecurity of Multiple Inverters in Large Grid Connected PV / Battery Energy Storage Systems 727

Hasan Ibrahim, Jaewon Kim, Peng-Hao Huang, Vishwam Raval, Prasad Enjeti

Optimal DC-DC Converter Topology and Control Algorithm for Fuel Cell Electric Vehicle with Series-Connected Supercapacitor 733

Hyeon Soo Kim, Yun Seong Hwang, Seung Hyun Kang, Man Jae Kwon, Byoung Kuk Lee

Reliability-Constrained Design of a High-Gain Power Optimizer based on a Real Mission Profile 738

Stefano Cerutti, Francesco Iannuzzo, Ariya Sangwongwanich, Tamás Kerekes, Mario Giuseppe Pavone, Francesco Gennaro, Natale Aiello, Francesco Musolino, Paolo Stefano Crovetti

Submodule Voltage Balancing Technique of Solar MMC for Firing the Switches using Integrated PWM Modules 746

Ahmed Elsanabary, Saad Mekhilef, Mokhtar Aly, José Rodriguez

Single-Stage High-Frequency-Link Split-Phase Microinverter with High Voltage Gain based on Buck-Boost AC Chopper 751

Xuewen Li, Jia Liu, Jinjun Liu

Fault Diagnosis and Tolerant Strategy for Triple-Port Hydrogen Converter using SSA-Optimized Random Forest Algorithm 757

Shiqi Zhang, Yiyina Teng, Naizhe Diao, Xiaoqiang Guo, Vladimir Terzija, Lichong Wang

Resilient Operation for Grid-Connected Cascaded H-Bridge Multilevel Inverter with Improving PV Source Stress 761

Jinli Zhu, Yuan Li, Hector Akuta, Jeonghun Kim, Uthandi Selvarasu, Shumeng Wang, Vikram Roy Chowdhury, Brad Lehman, Fang Z. Peng

A Medium Voltage Grid-Connected PV Inverter with a New Modular High Voltage Gain Converter Featuring Internal Modified Voltage Doubling Balancers 768

Kajanan Kanathipan, Muhammad Ali Masood Cheema, John Lam

Split-Source Common-Ground Inverter for Photovoltaic Applications 775

Mahmoud A. Gaafar, Mohamed Orabi, Samir Kouro, Ahmed Ibrahim, Eltaib Abdeen D. Ibrahim

Comprehensive Investigation and Proposal of a New Wireless Charging Road Structure using Low-Environmental-Impact Magnetic Concrete 782

Shuntaro Inoue, Yuko Kano, Shin Tajima

Design of a Bidirectional High Power Inductive Power Transfer System with Auxiliary Winding for Automotive Applications 788

Luis Ruiz Chamorro, Nikola Mirkoviœ, Alberto Delgado Expósito, Pedro Alou Cervera, Miroslav Vasiœ

Mutual Inductance and Load Identification Method based on the Voltage Transients of WPT Systems 795

Xiaosheng Wang, C.Q. Jiang, Yibo Wang, Liping Mo

Digitally Controlled Misalignment-Tolerant Inductive Power Transfer System with Adaptive Hybrid Compensation for CC/CV Charging of E-Scooter 801
Niranjan Shrestha, V.S.R.Varaprasad Oruganti, Sheldon Williamson

On/Off Control of Modular Inductive Power Transfer System 809
Kunxiao Zhou, Guangdong Ning, Heyuan Li, Xinlin Wang, Minfan Fu

Receiver Side Regulation of LCC Wireless Power Transfer System with Variable Notch Filter 815
Hsin-Che Hsieh, Jih-Sheng Lai

84.7 Percent Peak Efficiency Stress Tolerant DC DC Buck Converter for Li Ion Battery Driven Standby Circuits in 18nm FDSOI 821
Gautam Dey Kanungo, Pijush Kanti Panja, Vikas Bugade, Kallol Chatterjee

Leveraging Ultrasound and Neural Networks for Non-Invasive Power Converter Efficiency Estimation 828
Youssof Fassi, Vincent Heiries, Jérôme Boutet, Julien Marianne, Sébastien Martin, Mathilde Chareyron, Clément Chambon, Sébastien Boisseau

A Load-Independent Multi-Relays Wireless Power Transfer with Self-Regulation and Single Compensation Network 834
Jong-Hun Kim, Najam Ul Hassan, Seogyong Jeong, Myeong-Ho Kim, Min-Sik Kim, Jee-Hoon Jung, Byunghun Lee, Se-Un Shin

A GaN-Based Single-Stage Solid-State Transformer Replacement for 40 VA Class 2 Line-Frequency Transformers 840
Allen T. Nguyen, Charles R. Sullivan

Survey of Components and Topologies for High-Efficiency and High-Power Density 48V DC-DC Converters 848
Joseph Winkler, Niklas Deneke, Bernhard Wicht

A Novel Solid-State Circuit Breaker using B-TRAN™ 854
Mudit Khanna, Ruiyang Yu, Milad Tayebi, Jiankang Bu, Jeffrey Knapp

Development of a Supercritical Fluid-Insulated Fast Mechanical Switch for MVDC Hybrid Circuit Breakers 860
Zhiyang Jin, Qichen Yang, Alfonso Cruz, Lukas Graber

Dynamic Impedance Matching for a Variable Reluctance Energy Harvesting Application with Constrained Space 868
Fernando Pérez, Alejandro Redondo, Airán Francés, Gabriel Mujica

Renewable Energy-Powered DC-Converted Refrigerator based on a Supercapacitor-Assisted Technique 874
Nirashi Polwaththa Gallage, Nihal Kularatna, Alistair Steyn-Ross, Dulsha Kularatna-Abeywardana

Design and Evaluation of Flexible Inductors for Wearable Power Electronics 880
Sean Logi, F. Selin Bagci, Katherine A. Kim

Design of Boost Power Factor Corrector and Asymmetrical Half-Bridge Flyback Converter for USB-PD Applications 887
Yun-Keng Cheng, Tsorng-Juu Liang, Kai-Hui Chen, Ming-Chang Tsou

Computationally Efficient Current Sensorless Predictive Control for PMSM Drive Fed by a Matrix Converter with CMV-Free Operation 895
Ali Sarajian, Ibrahim Harbi, Quanxue Guan, Davood Arab Khaburi, Ralph Kennel, José Rodriguez, Patrick Wheeler, Mokhtar Aly

PMSM Motor Drive with Current Direct Digital Control and Near 1st-Order Speed Control 900
Po-Chang Lee, Tsai-Fu Wu, Han Ku, Chien-Chih Hung, Jui-Yang Chiu

Fault-Tolerant Multilevel Converter for Multiphase Switched Reluctance Motor Drives based on q+2 Converter 906
Mahmoud A. Gaafar, Mohamed Orabi, Hao Chen, Mostafa Dardeer

Uncertainty-Aware Artificial Intelligence for Gear Fault Diagnosis in Motor Drives 912
Subham Sahoo, Huai Wang, Frede Blaabjerg

Neural Network based Digital Twin Health Monitoring of BLDC Motor Drives for Robots 919
Mohamed Y. Metwly, Benjamin Luckett, Landon Clark, JiangBiao He, Biyun Xie

MTPA Control using Predictive P&O Method for Dual Parallel Surface-Mounted Permanent Magnet Synchronous Motor Drives Fed by a Single Inverter 925
Jae-Seong Kim, Kyo-Beum Lee

A Novel I-f Startup Strategy with Smooth Transition to Sensorless Control for CSI-Fed PMSM Drives used in Submersible Pumps 930
Milad Bahrami-Fard, Majid Ghasemi Korrani, Babak Fahimi

Simulation-Assisted Design and Implementation of an Electrically Excited Synchronous Motor Drive System 938
Shih-Gang Chen, Jun-Ming Hsu, Chun-Yen Chen, Ming-Shi Huang

Implementation and Analysis of Direct Torque Control on High-Speed PMSMs: A Comparative Study of Commercial and Laboratory-Developed Motors 943
Md Moniruzzaman, Kishor Joshi, Md Rashedur Rahman, Md Khurshedul Islam, Seungdeog Choi, Masoud Karimi Ghartemani

A Ferrite based Carbon Reinforced Composite Wrapped IPM Rotor Design for High-Speed Traction Applications 951
Md Rashedur Rahman, Md Khurshedul Islam, Md Moniruzzaman, Seungdeog Choi, Han-Gyu Kim, Andrew Walters

A Novel Phase-Mode Controller for Resonant Converters 958
Claudio Adragna, Daniele Cazzaniga, Stefano Manzoni

A Regulated 36V-60V-Input VIN-Insensitive Resonant Switched-Capacitor Converter with Large Voltage Conversion Ratio 966
Yichao Ji, Jingyi Yuan, Lin Cheng

A Hybrid Switched Capacitor Converter Enabling Capacitive-Based Wireless Power Transfer for Battery Charging Applications 971
Jade Sund, Samantha Coday

A 48V to 50-110V Resonant Power-Bus Charger with Reduced Conduction Loss for MHz-Frequency Long-Range LiDAR Driver 978
Hangxiao Ma, Xuchu Mu, Yang Jiang, Weihang Zhang, Jincheng Zhang, Rui P. Martins, Pui-In Mak

A Trajectory Controlled 48-to-24 V Resonant Switched Capacitor Converter with 98.7% Efficiency and Ultrafast Dynamic Response 983
   Hélène T.W. Ma Yang, Liang Wang, Haoyu Wang, Wai Tung Ng

Low Power, Non-Isolated, Extremely-High Step-Up, Quasi-Resonant Hybrid DC–DC Converter 990
   Kumar Joy Nag, Aleksandar Prodić

Isolated Soft-Switching Flying-Capacitor based Quasi-Resonant Step-Up Converter.................................... 997
   Kumar Joy Nag, Aleksandar Prodić

Accurate Small-Signal Phasor Transformation-Based Modeling of Secondary-Side Diode-Bridge Rectifiers for Battery Charging Applications 1004
   Aditya Zade, Regan Zane

High-Efficiency Isolated Piezoelectric Transformers for Magnetic-Less DC-DC Power Conversion 1012
   Sourav Naval, Wentao Xu, Mustapha Touhami, Jessica D. Boles

First Characterization of GaN Power Device and IC at Deep Cryogenic Temperatures Down to 100 mK 1020
   Xin Yang, Matthew Porter, Zineng Yang, Zichen Xi, Liyang Jin, Liyan Zhu, Linbo Shao, Yuhao Zhang

Dynamic Environment-Aware Lifetime Prediction of SiC MOSFET Modules Through LSTM 1026
   Md Zakir Hasan, Seungdeog Choi, Youssef Aider, Prashant Singh, Chun-Hung Liu

Guarding-Based C-V Characterization of 10 kV SiC MOSFET in Half-Bridge Module Configuration.. 1034
   Nianzun Qi, Gao Liu, Zhixing Yan, Shaokang Luan, Pawel Piotr Kubulus, Yuan Gao, Stefan Meyer, Hongbo Zhao, Asger Bjørn Jørgensen, Stig Munk-Nielsen

Automated Characterization Platform for Comprehensive Dynamic R_{dson} Assessment of GaN HEMTs from 50 K to 400 K.. 1040
   Tian Qiu, Zheyu Zhang, Purushottam Khadka, Ahmed Siraj, Dilip Rana

A Gate Driving Scheme for GaN Git with Enhanced Short Circuit Capability for Motor Drive Application 1047
   Zongjie Zhou, Yan Cheng, Kevin J. Chen

Online Detection and Reduction of the Influence of Parameter Tolerance of Paralleled SiC MOSFETs in an EV Inverter Environment.. 1051
   Hadiuzzaman Syed, Jochen Streit, Robert Kragl, Muhammad Muneeb Alam, Alberto Martinez-Limia, Karl Oberdieck, Ertuðrul Sönmez

Dynamic Current Sharing Issues with Paralleling SiC Power MOSFETs 1058
   Ching-Yao Liu, Chen-Chan Lee, Jih-Sheng Lai

Integrated Short-Circuit Protection Design based on Dual-Channel Gate Driver for Series Connected Medium-Voltage SiC MOSFETs 1063
   Rui Wang, Drazen Dujic

Long-Term High-Temperature Dynamic Gate Stress Reliability of a Last-Generation, Automotive-Grade, Planar 1200 V SiC MOSFET.. 1070
   Giuseppe Mauromicale, Alessandro Sitta, Michele Fiore, Michele Calabretta, Francesco Iannuzzo

Innovative Gate Driver Structure Achieving Low Time Skew Across Isolation Barrier for Parallel Connected SiC Modules 1076
Louison Gouy, Anne-Sophie Descamps, Nicolas Ginot, Christophe Batard

Fully Integrated Closed-Loop Active Gate Driver IC with Real-Time Control of Gate Current Change Timing by Gate Current Sensing 1084
Yaogan Liang, Katsuhiro Hata, Makoto Takamiya

Analyze and Design of Digitally Load Current Modulated Active Gate Driver for GaN HEMTs based Buck DC-DC 1090
Wentao Liu, Zhina Lian, Taotao Wu, Xiaochuan Peng, Hao Min

Impact of Real-Time Variable Gate-Drive Strength on Drive Cycle Efficiency in SiC Inverter-Fed PMSM Traction Drives 1096
Matteo Pizzuto, Aiswarya Balamurali, Aniket Anand, Narayan C. Kar

Demonstration of Efficiency Increase of 350 V-to-13.3 V Isolated DC-DC Converters for Electric Vehicles by Active Gate Driving 1102
Yohei Sukita, Katsuhiro Hata, Hiroki Kondo, Kenichi Watanabe, Kenichi Nagayoshi, Makoto Takamiya

A Multi-Level Active Gate Driver for Achieving Thermal Balance in Parallel Connected Power MOSFETs 1108
Jingyuan Liang, Lingwei Sun, Wen Tao Cui, Wai Tung Ng, Motomitsu Iwamoto, Haruhiko Nishio

A Fast Short-Circuit Protection Method for Ohmic Gate P-GaN HEMT based on Gate Charge 1114
Yue Wu, Xi Jiang, Song Yuan, Xiaowu Gong, Zhaoheng Yan, Jiahong Chen, Yun Xu, Jinjie Liu

Comparison of Ultrafast-Rise-Time Gate Drivers for Wide-Bandgap Devices in Sub-Microsecond Pulsed Power Applications 1121
Soham Roy, Duy T. Nguyen, Neeraj Anantha, Alex J. Hanson

A Discrete Multilevel Active Gate Driver for GaN HEMTs to Optimize the Switching Behavior 1129
Celine Lawniczak, Martin Pfost

Attenuation of Fundamental Component of Differential Mode Noise using Active EMI Filter 1135
Guru Abhilash Mulumudi, Naveed Ishraq, Ayan Mallik

Graph Neural Network based Performance Modeling for the Dual Active Bridge Converter with Operational Generalization 1143
Weihao Lei, Fanfan Lin, Xinze Li, Xiaokun Bao, Xin Zhang

An Augmented State Space Modelling Approach for DC-DC Converter Start-Up in Closed Loop 1148
Waseah Anjum, Arkadeb Sengupta, Marco Liserre

The Utilization of a Parallel Computing Algorithm for Accelerating Switching-Level Modeling of Power Electronics Simulations in a T-Type PV Inverter 1153
Buck F. Brown III, Liwei Wang, Zheyu Zhang, Johan Enslin, Yi Li

A New Reduced Order Analytical Switching Model for eGaN HEMTs 1159
Ruqi Li, Douglas Arduini, Phen Lumod, Shobhana Punjabi, River Lin, Harold Gutierrez

Proposal of an Alternative Reverse Recovery Calculation Method 1167
Brian Deboi, Blake Nelson, Austin Curbow

Improvement of CM EMI Attenuation Ability of Transformer with Negative Capacitor 1173
  Qinghui Huang, Yiming Li, Yirui Yang, Shuo Wang, Yanwen Lai, Zhedong Ma

Damping Factor based PCB Parasitic Inductance Value Optimization to Minimize Voltage
Overshoot and Settling Time of Semiconductors 1179
  Reza Shahbazi, Yunting Liu

Hardware Implementation of Virtual Resistance based FRT Logic in Programmable 3-Level ANPC
Inverters...1184
  Mohammad Safayet Hossain, Shuvangkar Chandra Das, Paychuda Kritprajun, Amin Banaie,
  Tapas Barik, Deepak Ramasubramanian, Aboutaleb Haddadi, Evangelos Farantatos, Ulrich
  Muenz

Rad-Hard PSFB Controller for High-Voltage Space Applications 1190
  Reynaldo S. Gonzalez, Robert E. Bolaños

Modeling, Control and Digital Implementation of a Buck Converter Operating in Triangular
Current Mode for a Wide Output Voltage Range Space Application..1197
  Regina Ramos, Sara Pérez, Guillermo Núñez, Pedro Alou, Javier Torres

Thermal Model and Optimization of a Multi-Winding Transformer for Lunar Surface Power
Transmission... 1203
  Zhining Zhang, Yuzhou Yao, Junchong Fan, Juchen Yang, Robert Guenther, Pengyu Fu, Jin
  Wang

Active Gate Driver Power Supply for High-Reliability Applications 1211
  Joseph P. Kozak, Juan Ramirez, Jesse Lin, Allison Orr, Alexander Martin, Hala Tomey

A Hybrid Energy Storage System for eVTOL Unmanned Aerial Vehicles using Supercapacitors................ 1217
  Ali Alenezi, PengHao Huang, Prasad Enjeti

Evaluation of Retired Lithium-Ion Batteries for Second-Life Applications Through Electrochemical
Impedance Spectroscopy 1224
  Latha Anekal, Sheldon Williamson

Uninterruptable Non-Isolated Integrated Power Electronics Converter (UNIPEC) for Commercial
Truck Auxiliary Power Unit 1230
  Pouya Zolfi, Ahmad Alzahrani, Ayman El-Refaie

Investigation of Electrical Safety for Non-Isolated Single-Phase On-Board Chargers used in
BEV/PHEV 1237
  Soya Kataoka, Shohei Funatsu, Hiroaki Matsumori, Takashi Kosaka, Keisuke Nakamura,
  Subrata Saha

An 8-Level Flying Capacitor Multilevel Converter for Electric Aircraft Pulse Deicing 1242
  Nicole Stokowski, Andrew Freeman, Aidan Rodgers, Aria Delmar, Jonathan Sengstock, Alex
  Solecki, Andrew Stillwell

Impact of Position Measurement Delay Angle on Performance of PMSM Drives for Electric Power
Steering in a Wide Speed Range... 1248
  Yingzhe Wu, Hengbin Zhang, Yuxiang Xue, Lisheng Wang, Hui Li, Shan Yin

Physical Parameter Estimation for a Two-Level VSI Three-Phase PMSM Electric Drivetrain 1255
  Bernard Steyaert, Ananda Tjakra Adisurja, Matthias Preindl

A Novel Two-Dimensional Random Switching Frequency PWM Method for Variable Frequency Drives 1261
Mostafa Abarzadeh, Kevin Lee

Optimized Maximum Torque and Minimum Loss Fault-Tolerant Control Schemes for Dual Three-Phase PMSM 1267
Syed Mohammad Maaz, Dong-Choon Lee

Wireless Actuation of Magnetic Robots with a Modular 60 mT 3-D Helmholtz Coil System 1274
Konstantinos Manos, Yifan Rao, Tuo Zhao, Kevin Liu, Daniel Zhou, Calvin Nguyen, Eric Chen, Glaucio H. Paulino, Minjie Chen

A Versatile PHIL based Motor Emulator Testbench using a High-Performance Power Amplifier Testbench.. 1279
Seyedeh Nazanin Afrasiabi, Rajendra Thike, Mathews Boby, K. S. Amitkumar

A 450V Three Phase GaN IPM Achieving 99.1% Efficiency in Smallest 12mm x12mm Package for 250W Power Delivery without Heatsink 1286
Maik Kaufmann, Manu Balakrishnan, Stefan Herzer, Anand Chellamuthu, Hely Zhang

FEA based High-Frequency Synthesis for the Design and Optimization of GaN-Based Dual Three Phase Motor Drive System 1294
Syed Imam Hasan, Alper Uzum, Ashraf Siddiquee, Yilmaz Sozer, Krishna Namburi

Evaluation of Passive Common-Noise Canceller Considering Both of Thermal Equilibrium and Common-Mode Noise Cancellation 1299
Koji Mitsui, Kenshiro Katsura, Koki Notake, Koji Yamaguchi

Performance Evaluation of Isolated DC/DC Converters in Modularized Bridge Rectifier Solid-State Transformer 1305
Zhenchao Li, Andrea Cervone, Drazen Dujic

Active and Reactive Power Flow Control of the Dual Active Bridge Converter 1311
Lauryn Morris, Thomas W. Francois, Jonathan Saelens, Oroghene Oboreh-Snapps, Arnold Fernandes, Praneeth Uddarraju, Sophia A. Strathman, Jonathan W. Kimball

Comparative Analysis of Carbon Footprints and Material Usage of Solid-State Transformers and Low-Frequency-Transformer-Based MVac-LVdc Interfaces for High-Power EV Charging 1318
Luc Imperiali, Rudy Wang, Anup Anurag, Peter Barbosa, Johann W. Kolar, Jonas Huber

Trade Study of Isolation Requirements and Magnetic Core Selection for Medium Frequency-Medium Voltage Transformers 1326
Mohendro Kumar Ghosh, Mark A. Juds, Brandon Grainger, Ahmad El Shafei, Bogdan S. Borowy, Paul Ohodnicki

Comparative Evaluation of a Multilevel LLC Resonant Converter for a Modular DC/DC Stage in a Electrolyzer Power Supply 1334
Samuel S. Queiroz, Levy F. Costa

Cost-Effectiveness Assessment of SiC MOSFET and Si IGBT Semiconductors in a Three-Level Resonant Converter for Solid-State Transformer 1341
Samuel S. Queiroz, Levy F. Costa

Comparative Performance Analysis of Medium Voltage 3L-ANPC and 3L-DNPC Pole Enabled by Series-Connection of 10kV SiC MOSFETs and 10kV SiC JBS Diodes for Sine Triangle PWM Operation 1347
  Sanket Parashar, Shubham Rawat, Nithin Kolli, Raj Kumar Kokkonda, Subhashish Bhattacharya

A Zero Harmonic Distortion Master Converter for Medium Voltage Microgrids 1355
  Gabriel V. Ramos, Dener A. de L. Brandão, Thiago M. Parreiras, Danilo I. Brandão, Braz de J.C. Filho

An MILP Approach for Modeling and Analyzing the BESS for Smoothing Renewable Fluctuations Considering BESS Capacity Attenuation in the Bulk Power System with High Inverter-Based Resource Penetration 1363
  Hualong Liu, Wenyuan Tang

Thermal and Efficiency Characterization of Immersion Cooled SiC Traction Inverter 1368
  Yiju Wang, Reza Ilka, JiangBiao He

FPGA-Based Hybrid Simulator for Real-Time 3-D Temperature Monitoring of Power Converters 1375
  Xianghao Mo, Daniel Ríos Linares, Regina Ramos, Miroslav Vasiœ

A New Subassembly Concept for Enhanced Heat Dissipation and Reliability of Power Module 1383
  Yosuke Nakata, Yuji Sato, Shin Uegaki, Jun Fujita, Akihiko Furukawa, Masayoshi Tarutani

Stand-Alone R_{DS-ON} Sensor for In-Situ Prognostic, Protection and Reliability Enhancement of Power Converters 1388
  Zaheen Mustakin, Qiang Mu, Lucas Pereira, Jiale Zhou, Tiefu Zhao, Babak Parkhideh

Electrical Evaluation of a Modular High Voltage 3D Power Module using Direct Dielectric Liquid Cooling 1396
  Omar Sanjakdar, Yvan Avenas, Rachelle Hanna, Guillaume Piquet Boisson, Emmanuel Marcault, Antoine Philippe

Board Level Reliability of Gull-Wing, Micro-Leaded and Lead-Less Packaged MOSFETs in Automotive Environments 1403
  Christopher Liu, Vijayakrishna Satyamsetti, Xuanjing Wei, Christian Radici, Peter Vines, Wayne Lawson

Cost Effective and High Noise Immunity Methodology for Aging Evaluation of DC-Link Capacitors in Traction Inverters 1408
  Seyed Hossein Aleyasin, Fausto Stella, Radu Bojoi, Enrico Vico

A 3D Structure of Single-Sided Cooling Power Module with Low Thermal Resistance and Low Inductance 1414
  Hirofumi Hisamochi, Koki Notake, Yoshiaki Takahashi, Koji Yamaguchi

Aging of Y-Capacitor in an EMI Filter and Its Impact on Common-Mode Noises 1420
  Tahmid Ibne Mannan, Seungdeog Choi, Subarto Kumar Ghosh, Md Moniruzzaman

2200A/48V-to-1V Low-Profile Direct Power Converter with Standard PCB Transformer 1427
  Alejandro Figueroa, Pablo Mazariegos, Álvaro Cobos, Javier Goicoechea, Alejandro Castro, José A. Cobos

Single-Stage 48V-to-1V Regulator with a Half-Turn Transformer and Current-Doubler Rectifier 1433
  Xinmiao Xu, Qiang Li

Ultra-Low-Profile Single-Stage Voltage Regulator Module (VRM) for Next-Generation AI Accelerators 1439
    Xufu Ren, Jinfeng Zhang, Zhenshuai Rong, Borong Hu, Teng Long

Novel TLVR Operation in Multi-Stage Voltage Regulator Module with Current Multipliers 1444
    Kevin Zufferli, Roberto Rizzolatti, Mario Ursino, Simone Mazzer, Gerald Deboy, Stefano Saggini

Interphase LC-Oscillation Suppression with Fast Line-Transient Response in 48-V Series-Capacitor Buck Converters for Automotive Applications 1451
    W.L. Jiang, Y. Liu, N. Khan, J. Pigott, H.J. Bergveld, V. Chaturvedi, O. Trescases

An Approach to Compensate for Low Frequency DC-Link Voltage Ripple in High Power ANPC Inverter 1459
    Shaozhe Wang, Ankit Vivek Deshpande, Rolando Sandoval, Erick Pool-Mazun, Enrique Garza-Arias, Prasad Enjeti

A Cascaded Multilevel Inverter System with Hot-Swapping and Fault Isolation Capability for Improved Resiliency 1465
    Uthandi Selvarasu, Vikram Roy Chowdhury, Shumeng Wang, Jinli Zhu, Mahshid Amirabadi, Yuan Li, Brad Lehman

Layout Optimization for Parasitic Inductance Reduction of GaN-Based NPL.X Multilevel Inverter 1473
    Ali Halawa, Jinyeong Moon, Woongkul Lee

Topology Selection and Design Methodology for SiC based Solar Photovoltaic Medium Voltage Direct Grid Connect Inverters 1481
    Jenson Joseph C. Attukadavil, Baylon G. Fernandes

EMI Modeling of PCB-Based Three-Level Active Neutral-Point-Clamped GaN Converter 1489
    Mohammad Hassan Adeli, Necmi Altin, Erkan Deniz, Adel Nasiri

A Novel Layout for Improving Current Sharing of Paralleled SiC MOSFETs with TO-247 Package 1495
    Che-Wei Chang, Matthias Spieler, Rolando Burgos, Ayman El-Refaie, Renato Amorim Torres, Dong Dong

A Sensor-Less IGBT On-State Voltage Estimation Method using Inverter Control Variables 1501
    Shuyu Ou, Subham Sahoo, Ariya Sangwongwanich, Yongjie Liu, Frede Blaabjerg

A Novel Non-Intrusive Online Monitoring Method for Diagnosing the Lift-Off of Bonding Wires in SiC MOSFETs 1507
    Keqi Song, Henry Shu-Hung Chung, Ho-Tin Tang

Optimizing MOSFET Selection for EMC-Critical Automotive Applications 1512
    Sacha J. Cazzitti, Christian Radici, Andrew J. Forsyth, Cheng Zhang, Peter Vines

Improving Dynamic Current Sharing Between Parallel MOSFETs by Optimizing Device Parameters 1519
    Kunal Jha, Kapil Kelkar, Marina Hedenik, David Penof

A 21.6 kW/L Two-Phase Immersion-Cooled Isolated DC-DC Converter 1529
    Aleksandar Ristic-Smith, Kawsar Ali, Daniel Rogers

Extraction of Common Mode Parasitic Capacitance in Balance Filter for the Prediction of EMI Noise Suppression 1537
    Qiuzhe Yang, Xingyu Chen, Zijian Wang, Qiang Li

A 660W, 96% Efficiency 3D Heterogeneously Integrated Digital DC/DC Power Module for Vertical Power Delivery 1544
Haoyu Wang, Xuliang Wang, Yan Wang, Xiaosen Liu

Planar Rogowski Coil-Based Switch Current Measurement for a 1.2 kV SiC MOSFET Embedded Die PCB.............. 1551
Matthias Spieler, Che-Wei Chang, Ayman M. El-Refaie, Dong Dong, Rolando Burgos

Effect of Magnetic Couplings on Conducted EMI of GaN-Based PFC Converter 1557
Tyler McGrew, Qiang Li

Optically-Controlled 3.3 kV SiC MOSFET with Fast Switching Speed and Low Optical Power 1564
Xin Yang, Guannan Shi, Liyang Jin, Yuan Qin, Matthew Porter, Che-Wei Chang, Xiaoting Jia, Dong Dong, Linbo Shao, Yuhao Zhang

Optimization Techniques for Parallel-Connected Devices in IPMs for Consumer Use 1569
Keisuke Kawamoto, Haruhiko Murakami, Teruaki Nagahara, Michael Rogers, Akiko Goto, Shoji Saito, Koichiro Noguchi

Investigating the Temperature Dependency and Operating Parameters of a Self-Driving Active Gate Driver 1576
Vin Loong Choo, Martin Pfost

Use of Switched-Capacitor Circuit to Generate Negative Gate-Source Voltage Pulses 1582
Ho-Tin Tang, Henry Shu-Hung Chung

An Optically Isolated Gate Driver with Simultaneous Data and Power Transmission Through a Miniaturized, Efficient Photonic Platform.............. 1590
Jiajun Li, Mariia Klymenko, Yanqiao Li, William Scheideler, Jason T. Stauth

Optimal Shared Energy Storage Capacity Configuration in Multi-Energy Microgrids Considering Battery Lifetime Loss based on Relaxation Techniques.............. 1597
Hualong Liu, Wenyuan Tang

Virtual Resistance Control for an Active Battery Management System 1602
Alastair P. Thurlbeck, Ashraf Siddiquee, Mithat John Kisacikoglu, Yilmaz Sozer

Internal Voltage Source Saturation Impact on Stability Limits of Grid Forming Converter 1610
Divyanshu Bansal, Aravind G., L. Umanand

A Zero Harmonic Distortion Grid-Connected Grid-Forming Converter for Battery Energy Storage System Applications 1615
Gabriel V. Ramos, Thiago M. Parreiras, Fangzhou Zhao, Xiongfei Wang, Braz de J.C. Filho

Single Cell Energy Router Justification for Three Phase Near Zero Energy Buildings 1622
Hossein Nourollahi Hokmabad, Tala Hemmati Shahsavar, Oleksandr Matiushkin, Tanel Jalakas, Oleksandr Husev, Juri Belikov

A Multi-UAV Charging Station Enabling Free Landing by Grid Pattern Transmitter.............. 1629
Jungho Kim, Hyunkyeong Jo, Seoktae Seo, Bonyoung Lee, Hyungki Min, Franklin Bien

Capacitor Design for Self-Resonant Coils for Long-Distance Wireless Power Transfer System.............. 1635
Mostak Mohammad, Vandana Rallabandi, Omer C. Onar, Gui-Jia Su

A 10.4-kW High-Power-Transfer-Density Multi-MHz Capacitive Wireless Power Transfer System for EV Charging Utilizing Stacked-Inverter Stacked-Rectifier Architecture 1640
    Dheeraj Etta, Miguel Alvarez Dominguez, Sounak Maji, Syed Saeed Rashid, Khurram K. Afridi

Reduced-Fringing-Field Multi-MHz Capacitive Wireless Power Transfer System using Metasurface-Based Couplers with Active Field Cancellation 1646
    Syed Saeed Rashid, Dheeraj Etta, Matteo Ciabattoni, Francesco Monticone, Khurram K. Afridi

Living Object Detection in Wireless Power Transfer Systems using Remote Capacitive Bio-Signals Monitoring.. 1653
    Bruno M.G. Rosa, Paul D. Mitcheson

Modified N:1 Switched Capacitor Converter with Reduced Capacitor DC Bias Voltage for High Power Density 1659
    Taewoo Lee, Dam Yun, Sunghyuk Choi, Jung-Ik Ha

Wide Range Digital Control for Three-Level Buck Converters with Sensorless Flying-Cap Voltage Balancing... 1666
    Hossein Hajisadeghian, Giovanni Bonanno

A Comparative Investigation of a New Continuous Voltage Conversion Ratio Approach in a Zero-Inductor Voltage Converter... 1673
    Sina Salehi Dobakhshari, Aamna Nasir Hameed, Binghui He, Mojtaba Forouzesh, Yan-Fei Liu

A 96.1% Peak Efficiency, 6.8 kW/in³, 48V-to-6V On-Package Intermediate Bus Converter with LV-GaN Power Transistors.. 1681
    Mausamjeet Khatua, Nachiket Desai, Harish Krishnamurthy, Sheldon Weng, Jingshu Yu, Huong Do, Samuel Bader, Han Wui Then, Krishnan Ravichandran, James Tschanz, Kaladhar Radhakrishnan, Vivek De

A 48V to 2.4V-5V 95.8%-Peak-Efficiency 869W/in³-Power-Density Fibonacci Dual-Path Hybrid DC-DC Converter with Inductor Current Reduction and Low Output Resistance...................................... 1687
    Yichao Ji, Zeguo Liu, Lin Cheng

An Ultra-Fast Very Large Scale Interleaved Li-Fi Transmitter 1693
    Daniel H. Zhou, Konstantinos Manos, Minjie Chen

Isolated PWM DC-DC Converter with Single Magnetic Component, ZVS and Self-Balanced Switched-Capacitor Voltage 1701
    Pablo M. Gil, Juan Rodríguez, Diego G. Lamar

Analysis and Design of a Low-Complexity ZVS Buck-Boost Converter 1707
    Burkhard Ulrich

A High Conversion-Ratio Hybrid Series-Parallel DC-DC Converter with Pseudo-Soft-Charging and Inductor Current Frequency Multiplication... 1715
    Avinash Maddela, Kishalay Datta, Jason T. Stauth

A Real-Time Variation Control of Deadtime in GaN-Based Bidirectional Buck-Boost Converter for Lithium-Ion Battery Formation System... 1723
    Jong-Hun Lim, Go Woon Heo, Je-Yeong Lim, Dong Hwan Kim, Byoung Kuk Lee

A Space Vector PWM Strategy for Charging of Bootstrap Capacitor in Three-Level Neutral-Point-Clamped Inverter 1728

Anantha Hegde, Asamira Suzuki, Hirokazu Nakamura, Takamune Kabashima, Koji Higashiyama, Keiji Akamatsu

A Complementary Carrier based PWM Strategy for Average Current Sampling of Three-Phase Inverter using Single Current Sensor 1734

Byeong-Il Kim, Joon-Seok Kim, Yeongsu Bak, June-Seok Lee

Short-Circuit Ride-Through for a CRM-Based Soft-Switching Three-Phase Inverter 1741

Xingyu Chen, Gibong Son, Qiang Li

Modified Space Vector Modulation with Low Bandwidth Sensor to Reduce Losses in Soft Switching Three-Phase Inverters 1746

Md Didarul Alam, Nazmul Hassan, Iqbal Husain, Liming Liu, Hongrae Kim

A Feedforward Ripple Reduction Control Strategy based on a Hybrid GaN/Si Interleaved Inverter 1754

Mowei Lu, Jurgis Reinotas, Xiaoyang Tian, Stefan M. Goetz

IGBT Comparison for Optimized Switching Behavior in the SiC/Si-Hybrid Switch 1759

Adrian Amler, Thomas Heckel, Daniel Ruppert, Cornelius Rettner, Martin März

Forward Recovery and its Mitigation in Hybrid Si/SiC-Based DC–AC Converters 1767

Yan Zhou, Thomas Lehmeier, Adrian Amler, Martin März

Real-Time IGBT Module Ageing Characterization Through Temperature Monitoring 1774

Quirc Perez-Farre, Luis F. Gomez-Rivera, Carlos Lopez-Torres, Kai Dannehl, Antoni García-Espinosa, Alejandro Paredes-Camacho

Experimental Validation of Triangular SOA via Infrared Thermography of a MOSFET Die Operating in the Thermally Unstable Linear-Mode for Automotive Applications 1781

Yacine Ayachi Amor, Christian Radici, Kerry J. Abrams, Philip Ellis, Peter Vines, Wayne Lawson

Feasibility Study of the SuperIGBT: A Series-Connected High Voltage IGBT with a Single Gate 1786

Junhong Tong, Alex Q. Huang, Huanghaohe Zou, Zhiyuan Ma

Low Profile, Laminated Nife Transformers for Flyback Converters 1791

Xuan Wang, Reza Mounesi, Matthew Catanoso, Matthew Fox, Adel Nasiri, Mark G. Allen

Comprehensive Demonstration of New Magnetic Designs Utilizing Magnetic Anisotropy of the Cores for Integrated Magnetics 1797

Yota Takamura, Honami Nitta, Tatsuya Miyazaki, Kimito Yamanaka, Ryosuke Ishido, Akira Namba, Keisuke Fujisaki, Shigeki Nakagawa

A Two – Stage Artificial Neural Network (ANN) – based Design and Optimization of High Frequency Transformers for Dual Active Bridge Converter 1803

Lufan Zhou, Alberto Delgado Expósito, Adam Ruszczyk, Simon Round, Miroslav Vasiœ

Modeling and Optimizing Winding Arrangement for Gapped Planar Magnetics based on Artificial Neural Network 1810

Hanqing Cao, Bima Nugraha Sanusi, Ziwei Ouyang

Free-Shape Optimization of VHF Air-Core Inductors using a Constraint-Aware Genetic Algorithm 1816

Thomas Guillod, Charles R. Sullivan

Organic Direct Bonded Copper-Based Rapid Prototyping for Silicon Carbide Power Module Packaging 1824

Shuofeng Zhao, Joshua Major, Douglas DeVoto, Sarwar Islam, Xiaoling Li, Mike Tant, Faisal Khan, Sreekant Narumanchi

Discrete Power Device Packaging with Integrated Direct Two-Phase Cooling 1832

Jinpeng Cheng, Jinxiao Wei, Hao Feng, Li Ran

Investigation of Die Top-Side Re-Metallization for SiC-Based Double-Side Cooled Power Modules 1836

Narayanan Rajagopal, Christina DiMarino

Design of Low Parasitic Inductance GaN HEMT Flip-Chip Power Module 1844

Mohammad Dehan Rahman, Tanzila Akter, Abu Shahir Md Khalid Hasan, H. Alan Mantooth, Xiaoqing Song

A Scalable Dual-Orthogonal-Cooling Packaging Concept for Parallel-Series SiC Chips 1850

Ekaterina Muravleva, Youssef Abotaleb, Blake Anderson, Zichen Zhang, Boyi Zhang, Jerry Hudgins, Jun Wang

Parasitic Impact Analysis and Design of Hybrid EMI Filter for Active Clamp Flyback SMPS 1858

Tahmid Ibne Mannan, Seungdeog Choi, Masoud Karimi-Ghartemani

Overview of Dynamic Characterization of Switches for Three Phase Voltage Source, Current Source, and Matrix Converter Applications 1866

Sneha Narasimhan, Sathya Rupan Thirumoorthi, Subhashish Bhattacharya

Advanced Modeling Technique of Class-E Inverter Considering Low R_{on} of eGaN FETs and Different Design Procedures 1874

Manas Palmal, Jungwon Choi

PiezoNet and Data-Driven Models for Time-Domain Characterization of Piezoelectric Resonators 1882

Davit Grigoryan, Mian Liao, Haoran Li, Shukai Wang, Tanuj Sen, Matthew Tan, Minjie Chen

A New Gate Charge De-Embedding Method for Accurate On-Wafer Characterization of HV MOSFET Devices 1889

João R.R.O. Martins, Rachid Hamani, Vincent Quenette, Joerg Gessner

4 kW Auxiliary Power Module for Electric Vehicles Utilizing a Dual-Phase LLC DC-DC Converter 1892

Mojtaba Forouzesh, Xiang Yu, Yan-Fei Liu, Paresh C. Sen

New Reverse Mode Control Method of Phase-Shift Full-Bridge Converter for Bidirectional Auxiliary Power Module 1899

Jongyoon Chae, Dongmin Kim, Dongmin Choi, Gun-Woo Moon

In-Situ EV EIS with a High-Density Flying Capacitor Multi-Level Converter Supercapacitor System 1905

Avram Kachura, Gaël Vergès, Samantha K. Murray, Olivier Trescases

A Novel 500-kHz LLC-T Resonant Converter with Wide Output Range 1913

Zhengming Hou, Dong Jiao, Jih-Sheng Lai

High Efficiency Traction Drive Operation with a Partial Load Three-Phase Triangular Current Mode Modulation Concept 1919

Bhaskar Chatterjee, Jan Allgeier, Thomas Plum, Marc Hiller

Analysis of Maximum Power Transfer Limit for Linear Operation of Dual-Active-Bridge Converters 1927

Radhika Sarda, Ezequiel Ramos Rodriguez, Gaowen Liang, Glen G. Farivar, Josep Pou, Vaisambhayana B. Sriram, Anshuman Tripathi

Enhanced Control for Integrated Active Power Decoupling in Single-Phase Three-Level Flying Capacitor PFC Converter 1935

Gleisson Balen, Cristian Blanco, Ángel Navarro-Rodríguez, Pablo García, Rafael Peña-Alzola

Improving Transient Stability of PLL-Synchronized Grid-Following Inverters 1940

Surya Prakash, Kalpana Beura, Mohamed Alkhatib, Omar Al Zaabi, Khalifa Al Hosani, Utkal Ranjan Muduli

Online Impedance-Based Analysis for Power System Stability Assessment using Transformer-Less and Filter-Less Switch-Mode Perturbation Generator 1946

Tomoya Ide, Yuko Hirase, Cheng Huang, Takanori Isobe

PIR-R Control for Three-Phase Grid-Connected Inverter with Unbalanced Grid Current Correction 1953

Haneen Ghanayem, Xingyu Yang, Mohammad Alathamneh, R.M. Nelms

Design and Placement of a Passive Clamp Snubber for Isolated SEPIC and Cuk Converters Working as Automatic Power Factor Correctors 1959

Abraham López, Juan Rodríguez, Duberney Murillo-Yarce, Javier Sebastián, Diego G. Lamar

Current Sensorless Control Strategy for Single-Phase T-Type PFC Converter 1967

Che-Yu Lu, Jia-En Zeng

Three-Phase Single-Stage Multiport AC-DC Converter with Integrated DC-DC Conversion Stages 1972

Asad Hameed, Gerry Moschopoulos

High Efficiency AC-Adapter Realized by Voltage-Clamper with Mid-Voltage AHB Converter using Synchronous Rectification 1977

Shuichiro Motoori, Akihiro Kawano, Toshiyuki Zaitsu, Riku Tatetsu, Kohei Sebata, Kazuki Miyanjou, Kimihiro Nishijima

Active Soft Switching Technique for Single Phase Series Capacitive Link Universal Rectifier 1983

Anran Wei, Brad Lehman, Mahshid Amirabadi

A Multi Mode Control Algorithm for Totem-Pole Bridgeless PFC 1990

Bosheng Sun, Sheng-Yang Yu, Amir Hussain

Protection Strategy for Flying Capacitor Totem-Pole PFC Under the AC Drop Transient 1995

Yanqing Wu, Wending Zhao, Zhenhai Zhu, Xinke Wu

Three-Phase with Three Single-Phase Single-Stage Isolated AC-DC Converters for EV Charging Station Applications 2002

Misha Kumar, Peter M. Barbosa, Juan M. Ruiz

400V SiC in Next-Generation 3-Level Flying Capacitor Bridgeless Totem-Pole PFC 2009

Rytis Beinarys, Seamus O'Driscoll

Extended Smart-Link Quasi-Single-Stage 3-Phase AC-DC Power Supply Module for AI-Driving Data Centers 2014

D. Biadene, J. Huber, J.W. Kolar, P. Mattavelli

A New Three-Phase Multi-Mode AC/DC LLC Converter with Output-Controlled Active Rectifier (with V2G and G2V Functions) for Fast DC Charging Application.. 2022
    Xiaoyi Xia, John Lam

Capacitorless Notch Resonant Converters for Miniaturized LLCLC Resonant Converters in Electric Vehicle Charging Applications 2029
    Haitham Kanakri, Euzeli Cipriano Dos Santos Jr., Maher Rizkalla

Multiple-Core Transformer Design based on Half-Turn Structure in Two-Stage DC-DC Converter for Battery Storage System... 2035
    Yilei Li, Bima Nugraha Sanusi, Pinhe Wang, Tianming Luo

Bidirectional DC-DC Converter Utilizing Coupled Inductors for Energy Storage System........................... 2043
    Wen-Hsuan Lee, Jiann-Fuh Chen, Hsuan Liao, Kuo Fu Liao

Comparison of 2-Level and Quasi-2-Level Topologies in a Bidirectional Isolated DC-DC Converter for MVDC Networks 2051
    José Andrés Aguilar Croston, Jean-Yves Gauthier, Cyril Buttay, Maryam Saeedifard, Besar Asllani, Piotr Dworakowski

Sling Forward Converter for Offline Operation: Achieving High Efficiency and Wide Voltage Range Performance 2059
    Nasherul Islam, Guozhu Chen, Honglei Miao, Fuxing Zhang

A Pulse Width Alternating Modulation Strategy for Three-Level Buck-Boost Converter 2066
    Xinlong He, Caifeng Liu, Xudong Zou, Jiaao Zou, Tianyi Zhang, Yong Kang

ZVT Circuit Applied for Wide Input Range Isolated Converters 2070
    Linguo Wang, Zhongyin Guo, Junjie Zhu, Bing Zhang, Zhiling Zuo, Xiaoguang Gao, Guangji Ma

Impact of Asymmetrical Leakage Inductance on a 380 V-12 V LLC Converter with Synchronous Rectifier for DCX Application 2075
    Jinshu Lin, Shan Yin, Chen Song, Honglang Zhang, Minhai Dong, Limei Xu, Hui Li

Start-Up Techniques and Universal Closed Loop Control of Immittance Network based Resonant Converter.. 2082
    Ripun Phukan, Misha Kumar, Randy Beckemeyer, Juan Ruiz, Peter Barbosa

Multi-Objective Efficiency-Oriented Optimization for DAB Converters Minimizing Current Stress and Backflow Power with Soft-Switching Assurance 2088
    Kun Wang, Ian Laird, Jun Wang

An ISOP-PSFB PWM Converter based on Coupled Output Inductors and Phase-Shifted Modulation with Full ZVZCS Range 2096
    Kang Hong, Guo Xu, Guangfu Ning, Mei Su

Design and Implementation of a GaN-Based Soft-Switched Series-Capacitor Buck Converter Operating at the CCM-DCM Boundary for High-Performance Computing Systems 2101
    Ramin Rahimzadeh Khorasani, Kolman Puterman Ghitelman, Madhavan Swaminathan

Intrinsic Feedback Model for Coupled-Damped Self-Balancing of General Multiphase Hybrid Converters 2109
    Haoran Xu, Weijia Hao, Desheng Zhang, Run Min, Qiaoling Tong, Xuecheng Zou

A High-Efficiency Switching Oscillation Suppression Strategy based on Damped Oscillation for Synchronous DC-DC Converter 2117
Hao Yuan, Chuan Ni, Zhengyu Ye, Wei Lu, Hui Xue, Ting Qian

Efficient and Streamlined Demodulation Strategy for High-Frequency Talkative DC-DC Converters 2125
Abdelmoumin Allioua, Hendrik Gockel, Gerd Griepentrog

A 90.9% Peak Efficiency KY Single-Inductor Bipolar-Output Converter with Conductance Modulation Controller for Active-Matrix Organic Light Emitting Diode Power Supply 2131
Sheng-Han Yu, Chieh-Ju Tsai, Hao-Ran Huang, Ching-Jan Chen

Constant-on-Time Control for Zero-Bias Trans-Inductor Voltage Regulators.................................... 2138
Hank Zeng, Justin Lee, Rixin Lai, Hang Shao

An Improved PFM Control Scheme for Three-Level Buck Converter based on Ton Extension Achieving an 810% Frequency Reduction 2143
Yi-Chun Chang, Chieh-Ju Tsai, Ting-Lun Lee, Ching-Jan Chen

A Concept for Current Ripple or Transient Improvements in Multiphase Converters 2149
Alexandr Ikriannikov, Alex Gao

System Solutions and Design Trade-Offs to the Input Filter Interactions with Battery Chargers 2157
Xigen Zhou, Dan Mavencamp, Kuang-Yao Cheng

Modeling and Implementation of a Zero Bias TLVR 2162
Lei Wang, Travis Guthrie, Peyman Asadi, Mark Alexander, Kunrong Wang, Brandon Howell

cGANET-Enhanced Voltage Gain Modeling: Elevating CLLC Converter Accuracy.................................... 2167
Yu Zuo, Xiaobing Shen, Fanghao Tian, Jiaze Kong, Hans Wouters, Wilmar Martinez

Capacitive vs Inductive Coupling based DC-DC Converter Operating in MHz Switching Frequency Range... 2173
Saeid Pourjafar, Parham Mohseni, Oleksandr Husev, Ryszard Strezelecki, Oleksandr Matiushkin

LLC Converter Main Transformer Losses: Eliminating Air Gaps and Integrating Parallel External Inductors... 2179
Yu-Chen Liu, Shang-Syun Wu

Small-Signal Phasor Modeling of T-Type Bridge-Based Single-Sided and Double-Sided LCC Resonant Converters for WPT Applications... 2194
Aditya Zade, Shubhangi Gurudiwan, Regan Zane

A Hybrid Three-Port Topology for Urban Charging Stations... 2202
Mohammadreza Khodaparast Klidbari, Naser Souri, Zahra Sadat Habibolahi, Hamid Montazeri Hedeshi

Reconfigurable H5-Bridge based LLC-DAB Sigma Converter for EV Fast Charging Stations 2207
Huangsheng Xu, Mingde Zhou, Qishan Pan, Haoyu Wang

A Resonant Reset Forward Converter with Ultra-High Conversion Gain using Differential Transformation Technique (DTT)... 2213
Shubham Srivastava, Mandeep S. Rana, Santanu K. Mishra

Full-Range ZVS Modulation of Switched Capacitor Converter for Sensorless Voltage Balancing 2220
Md Tanvir Ahammed, Wensong Yu

Dimensional Parasitics Absorption in Capacitively-Isolated Ćuk Converter for Medium-Voltage High Step-Down Converters 2228

    Aakash Kamalapur, Jung-Soo Bae, Mark Cairnie, Rajaie Nassar, Jack Knoll, Dushan Boroyevich, Guo-Quan Lu, Christina DiMarino, Qiang Li, Khai D. T. Ngo

A 36-to-60V Input Dual-Phase 2MHz 93%-Efficiency ZVT Series-Parallel Hybrid Buck Converter using Single Auxiliary Inductor and Adaptive Time Multiplexing Control 2236

    Qi Cheng, Hoi Lee

Improved Efficiency in a 10 W Class-Φ_2 Converter Utilizing a Resonant Gate Drive 2241

    Malachi Hornbuckle, Katherine Liang, Juan Rivas-Davila

The Analysis and Design of a Resonant Capacitively-Isolated Cockcroft-Walton Converter 2249

    Elizabeth Rabenold, Raiphy Jerez, Samantha Coday

SHSC: Non-Isolated High-Density 4:1 IBC for 48 V Applications 2254

    Mario Ursino, Roberto Rizzolatti, Simone Mazzer

High-Performance Current Multiplier: A Hybrid Switched Capacitor Solution for High-Current Applications............................. 2260

    Kevin Zufferli, Roberto Rizzolatti, Mario Ursino, Simone Mazzer, Gerald Deboy, Stefano Saggini

Representation and Design Methodology for Generalized Switched-Capacitor Converter Topologies 2268

    Seokwon Choi, Dam Yun, Jung-Ik Ha

A 48-V-to-1-V Gallium Nitride Switching Bus Converter for Processor Vertical Power Delivery with 2.7 mm Thickness and 3048 W/in³ Power Density 2276

    Jiarui Zou, Yicheng Zhu, Nathan M. Ellis, Logan Horowitz, Robert C. N. Pilawa-Podgurski

Ripple Reduction and Efficiency Improvement of Always-Dual-Path Hybrid DC-DC Converter based on Phase Shift Operation 2284

    Katsuhiro Hata, Shinsaku Tanaka, Toru Ashikaga, Yasuhiro Rikiishi

Ultralocal PQ Theory: A New Approach for Model-Free Predictive Direct Power Control of Shunt Active Power Filters 2290

    Mahdi S. Mousavi, Abolfazl Nassaji, Ibrahim Harbi, Behnam Nikmaram, S. Alireza Davari, Mokhtar Aly, José Rodriguez

Symmetrical Balanced Circuit for Common-Mode Noise Mitigation in LCL-T Resonant Converter 2296

    Ripun Phukan, Boyi Zhang, Juan Ruiz, Peter Barbosa

A Single-Phase Soft-Switching Buck-Boost Inverter 2303

    Lukas Wipprecht, Burkhard Ulrich

Low-Complexity Model Predictive Control Method for Driving Dual Induction Motors Fed by Five-Leg Inverter............................. 2311

    Jun Young Lee, Eun Woo Lee, Dongho Choi, June-Seok Lee

Overvoltage Mitigation Filter using High-Frequency Cable Modeling in Long Transmission Lines for Silicon Carbide Inverter Systems............................. 2317

    Yun-Jin Lee, Kyo-Beum Lee

Power Delivery Network (PDN) Design and Analysis to Achieve Low Impedance in Fast Edge Rate DC-DC Converters for EMI Compliance............................. 2322

    Manraj Singh Ladhar, Sheldon Williamson

Enhancing the Performance of Dual Input Split Source Inverters using an Advanced Modulation Strategy.. 2327

Mustafa Abu-Zaher, Fang Zhuo, Mokhtar Aly, Mahmoud A. Gaafar, Mohamed Orabi, José Rodriguez, Alaaeldien Hassan, Jiachen Tian, Samir Kouro

A Novel GaN-HEMT Single-Phase Single-Stage Buck-Boost Micro-Inverter Topology for PV Applications... 2332

Pengwei Li, Uiliam Kutrolli, Ali Bazzi

A Dynamic Current Sharing Method using Novel Clip Considering Mutual Inductance Coupling.............. 2343

Zexiang Zheng, Jianwei Lv, Yiyang Yan, Baihan Liu, Yifan Zhang, Linhao Ren, Jiaxin Liu, Cai Chen, Yong Kang, Xiong Zhang, Hao Yu, Wei Jiang

Application-Oriented Test Setup for Measuring Dynamic Output and Transfer Characteristics of GaN-HEMTs 2348

Philipp Swoboda, Martin Fein, Simon Frank, Andreas Liske, Marc Hiller

Mitigating Gate Voltage Oscillation in Parallel SiC Power Modules for xEV 2356

Hideo Komo, Michael Rogers, Mark Steiner, Eric Motto, Koichi Taguchi, Chihiro Kawahara, Junichi Nakashima, Yasushige Mukunoki, Seiichiro Inokuchi, Rei Yoneyama

Switching Performance Comparison of Low-Voltage GaN and Si Devices....................................... 2361

Tianxiao Chen, Haoyang Liu, Pedro A.M. Bezerra, Eckart Hoene, Sibylle Dieckerhoff

Modeling of Switching Transients for Frequency-Domain CM EMI Analysis in Double Sided Cooling Power Modules 2369

Sijia Liu, Liu Yang, Heng Zhang, Yifan Zhang, Zexiang Zheng, Jianwei Lv, Jiaxin Liu, Cai Chen, Yong Kang, Yuebin Zhou, Daming Wang, Shuang Zhao

Leakage Current Detection Scheme for Aging Test of 10kV SiC MOSFET Power Module 2375

Peiyang Ding, Hong Zhang, Tianshu Yuan, Qiling Chen, Jiacheng Guo, Dingkun Ma, Peiyuan Sun, Ting Hou, Laili Wang

Physics-Informed Neural Network Approach for Early Degradation Trajectory Prediction of Power Semiconductor Modules 2380

Jie Kong, Yi Zhang, Yichi Zhang, Lukas Wick, Frederik Lillebæk Hansen, Dao Zhou, Huai Wang

Nonlinear Output Capacitance of Bidirectional Gallium Nitride Power Switches 2387

Michael Bosch, Jeremy Nuzzo, Dominik Koch, Mathias C.J. Weiser, Ingmar Kallfass

Novel Approach of Determining and Predicting SiC MOSFET's on Resistance from Device Case Temperature using Machine Learning 2393

Paul Bradford, Conner Deppe, Hongjie Wang

Comparison of Static Characteristics in GaN HEMTs Across 50K to 400K Considering Diverse Techniques and Statistical Variation 2400

Purushottam Khadka, Saumil Shivdikar, Zheyu Zhang, Tian Qiu, Ahmed Siraj

Compact Model of β-Ga$_2$O$_3$ Schottky Barrier Diode 2407

Abu Shahir Md Khalid Hasan, Mohammad Dehan Rahman, Tanzila Akter, Md Majharul Islam, Md Maksudul Hossain, Xiaoqing Song, H. Alan Mantooth

DC-Link Capacitor Board Design for Low Parasitic Inductance 2413

Mikayla Benson, Lifang Yi, Kangbeen Lee, Jinyeong Moon, Woongkul Lee

First Demonstration of a Gallium Oxide Power Converter 2419
Joshua J. Piel, Elizabeth A. Sowers, Daniel M. Dryden, Thaddeus J. Asel, Adam T. Neal,
Brenton A. Noesges, Shin Mou, Andrew J. Green

Optimized Integrated EMI Filter Design in SiC Power Modules with Terminal Inductor for Better
High-Frequency EMI Suppression 2426
Yifan Zhang, Wenzhe Xu, Jianwei Lv, Yiyang Yan, Baihan Liu, Sijia Liu, Jiaxin Liu, Cai Chen,
Yong Kang, Xiong Zhang, Hao Yu, Wei Jiang

Balanced Technique using Integrated Winding Coupled Inductor for High-Power Density Two-
Phase Interleaved Boost Converter.. 2431
Yuta Imaeda, Jun Imaoka, Masayoshi Yamamoto, Hiroyuki Onishi

MagNetX: Foundation Neural Network Models for Simulating Power Magnetics in Transient................... 2438
Shukai Wang, Hyukjae Kwon, Haoran Li, Youssef Elasser, Gyeong-Gu Kang, Daniel Zhou,
Davit Grigoryan, Minjie Chen

Revisiting Models of Common Mode Inductors to Include the Magnetized Capacitance Effect................... 2446
Rafael Bogo Portal Chagas, Marcelo Lobo Heldwein

A High Frequency Coupled Inductor with Distributed Air Gap for High Power DC-DC Converters............ 2453
Muhammad Fasih Uddin, Ahmed H. Ismail, Peyman Darvish, Baher Abu Sba, Yue Zhao

High-Power Planar Transformer Design for Four-Port Converters 2461
Arya Sadasivan, Behrooz Mirafzal

Optimal Design of Inductors with Aluminum Litz Wire for Inductive Power Transfer Systems 2468
Jesús Acero, Claudio Carretero, Ignacio Lope, Óscar Lahuerta, José-Miguel Burdío

Analytic Design of Flat-Wire Inductors for High-Current and Compact DC-DC Converters....................... 2474
Sajjad Mohammadi, James L. Kirtley, Alireza Namadmalan

Insulation Dielectric Loss of High-Frequency Transformer Under Square Voltage Excitation with
Edge Oscillation 2482
Zhanlei Liu, Lingyu Zhu, Yuntian Gao, Yongliang Dang, Cao Zhan, Shengchang Ji

Improved High-Speed Thermal Analysis based on Two-Step Simulation for High-Frequency
Transformers... 2488
Zheyuan Yi, Kai Sun, Qiang Li, Zengyang Liu

Core Material Characterization Under DC Bias Conditions.. 2495
Jonas Mühlethaler, Fabrice Locher, Frédéric Mathieu, Edward Herbert

A Low-Cost Setup and Procedure for Measuring Losses in Inductors 2502
Burkhard Ulrich

Effect of Temperature of Additively Manufactured Cores 2510
Ken Johnson, Ali Bazzi

Extreme Temperature Permeability Engineered Soft Magnetics 2516
Tyler W. Paplham, Alex M. Leary, Paul R. Ohodnicki Jr.

An Isolated RF Power Combining Approach with Multiple Decoupled Input Coils 2521
Ziyang Xu, Yifan Zhao, Zhan Liu, Alex J. Hanson, Ming Liu

Simulation of a Custom Core, 15kV Isolated Gap Transformer Optimized for High Power Density 2527
Andrew Galamb, Fei Teng, Srdjan Lukic

Low Interwinding Capacitance Design for PCB-Winding based Transformer in Self-Powered Gate Drive Power Supply for High-Voltage SiC MOSFET 2535
Yuan Zhou, Li Zhang, Yilun Chen, Tianxiang Yin, Lei Lin

Integrated 4-Level Dual-Phase Superimposed Quadratic Power Converter for High-Density Direct 48V/1V Conversion 2541
Prosenjit Ghosh, Jin Woong Kwak, Fei Zhou, D. Brian Ma

Compensation Method for Unbalance of the Multi-Channel Class E Power Amplifier using the Closed Loop Frequency Control 2547
Kyungmin Lee, Sungku Yeo

High Temperature Operation of Digital Gate Driver Integrated Into a Power Module 2551
Kazuma Saiga, Shohei Zaizen, Satoshi Nakano, Shigeru Kusunoki, Kiyoto Watabe, Katsuhiro Hata, Makoto Takamiya, Shin-Ichi Nishizawa, Wataru Saito

Evaluation Index-Based Multiphysics Coupling Model and Analysis Methodology for High-Reliable Power Supply Module 2556
Haoyu Wang, Xuliang Wang, Yan Wang, Xiaosen Liu

Electrical Characterization of Modular 3D Packaging Assembled with Compressed Metal Foams 2562
Paul Bruyere, Alexis Derbey, Betina Zynger-Capaverde, Yvan Avenas, Eric Vagnon, Jean-Luc Schanen, Jean-Michel Guichon, Omar Sanjakdar

Improvement in Short-Circuit Robustness of SiC-MOSFETs based Power Modules using Two-Level Turn-On (2LTO) 2569
Muhammad Muneeb Alam, Saad Khalid, Nisar Ahmed Khan, Ngoc Ho Tran, Sebastian Strache

GaN-Based Two Stage Point-of-Load (PoL) Converter with 2.5D Embedded Substrate Implementation 2576
Samuel Defaz, Yang Li, Fang Luo

Near-Field Coupling Mitigation of the Noise from High Voltage DC-Link Decoupling Capacitors in Voltage Source Converters 2582
Yuxuan Wu, Kushan Choksi, Samuel Defaz, Fang Luo

Advantages of Paralleling SiC MOSFETs in High-Performance Power Modules 2589
Steffen Beushausen, Dominik Alexander Ruoff, Wenqi Zhou, Karl Oberdieck

A SiC Half-Bridge Power Module based on Liquid Metal Packaging for High Performance and Low Thermal Stress 2597
Wei Mu, Ameer Janabi, Luke Shillaber, Borong Hu, Teng Long

Analysis and Modeling of Radiated EMI Considering Coupling Between Power Converter and Power Cable with LC-Type EMI Filter 2603
Qinghui Huang, Yingjie Zhang, Shuo Wang, Yirui Yang, Zhedong Ma, Yanwen Lai

Simple Prediction Method for Impacts of Switching Characteristics on EMI Noise of a Three-Phase PWM Inverter 2610
Shinobu Nagasawa, Toshiya Tadakuma, Keita Takahashi

Coaxially Nested 3.3 kV SiC MOSFET Packages with Uniform Interpackage Electric Field Distribution 2616
Jack Knoll, Mark Cairnie, Christina DiMarino

Thermal Modeling and Performance of a Bare-Die Embedded PCB for High Power Density Converters Design 2624
Shahid Aziz Khan, Feng Zhou, Mengqi Wang, DucDung Le, Shivam Chaturvedi

Research on the Voltage Fluctuation Suppression Strategy in Weak Grid Under Pulsed Power Load Integration 2628
Xi Chen, Jiazheng Zhang, Mingjun Bao

An Optimized Firmware-Based Cycle-by-Cycle Current Limiting Method for Power Electronic Converters in UPS 2634
Teng Wu, Hong Liu

Frequency Stop-Band Management System for DC-DC Converters 2640
Alessandro Bertolini, Alberto Cattani, Claudio Luise, Alessandro Gasparini

Multi-Stage Model Predictive Control with Enhanced Discrete-Time Models for Multilevel Inverters 2647
Hoang Le, Apparao Dekka, Deepak Ronanki, Abdul R. Beig

Direct Effective Power Control (D-EPC) for LLC Resonant Converters Operating in Boost Mode using Event-Driven-Timer based Digital Controller 2654
Yuto Yoshimura, Kenji Funatani, Kazuhiro Umetani, Toshiyuki Zaitsu, Akito Nakagaki, Masataka Ishihara, Eiji Hiraki

Mitigation Method of Resonance Between Paralleled On-Line UPS 2660
Teng Wu, Zhenguo Huo, Shangxian Ning

An Extra-Element Small-Signal Model for a Current-Fed Resonant Dual-Active-Bridge Converter 2667
Paolo Sbabo, Paolo Mattavelli, Giorgio Spiazzi, Andrea Petucco

Concurrent Charge Distribution and Time-Optimal Control for Unordered Single-Inductor Dual-Output Converter 2675
Xuliang Wang, Haoyu Wang, Yang Liu, Yunxin Wang, Boran Zhang, Hongru Liu, Yan Wang, Xiaosen Liu

Circulating Current Control with Loss Reduction for Parallel Connected Inverters 2681
Shun Endo, Takae Shimada, Masato Ando, Yuuichi Mabuchi, Masaki Miyamae, Naoki Takayama, Yohei Matsumoto, Naoto Onuma

Analysis of Power and Power Spectral Density for Quaternary Random Pulse Position Modulation 2687
Hung-Chi Chen, Hsiang-Kai Wu, Chih-Chiang Wu

Bidirectional CLLC Converter using a Hybrid Control Method for Wide Voltage Range Applications 2692
Jhih-Cheng Hu, Hong-Xuan Liao, Chien-Lung Liu, Wei Wang, Ming-Shi Huang

Design and Control of a High-Bandwidth Dual Active Bridge DC-DC Converter 2698
Alper Uzum, Syed Imam Hasan, Yilmaz Sozer, Kenneth A. Loparo

Unified Model Predictive Control for DC-DC Buck Converters: From Start-Up to Steady-State Operation 2703
Zhengchen Guo, R.M. Nelms

A Novel IPPC Method for Precise Overload Protection and Burst Mode Operation in LLC Resonant Converters 2708
Manikanta Pallantla, Ramkumar S

An Improved Current-Sensorless Model Predictive Voltage Control for Four-Leg Voltage Source Inverters 2713
Heng Guo, Yuxin Wei, Mengmeng Jing, Wenlong Ding, Bin Duan, Chenghui Zhang

A Highly Integrable, Modular and Multi-Functional Fault Monitoring Active Gate Driver with Parallel Buffers for a Global Enhanced Reliability of Gen. 3 SiC Power MOSFETs 2718
Mathis Picot-Digoix, Léo Seugnet, Frédéric Richardeau, Jean-Marc Blaquière, Sébastien Vinnac, Thanh-Long Le, Stéphane Azzopardi

A 24 – 16 V to 0.8 – 1.2 V Merged 4-Stage Hybrid-SC-SL Converter with 96.5% Peak Efficiency and Larger Than 50% iL Reduction 2725
Chien-Hao Tseng, Cheng-Ta Chuang, Chieh-Ju Tsai, Ching-Jan Chen

Innovation Active Gate Drive Method (Named TriC3™) for MOSFET Heat Reduction and EMI 2730
Hisashi Sugie

A KY Buck-Boost Converter with Extended Ramp Control Achieving 1500% Output Variation Reduction for Smooth Mode Transition 2735
Yu-Ting Hung, Chieh-Ju Tsai, Ching-Jan Chen, Chun-Yu Hsieh

An USB Cable based Extended Conversion Range L-First Hybrid-Converter using Valley-Virtual-Inductor-Current-Mode Control with Auto-Tracking Slope Compensation Against ±50% Inductance Variation 2741
Chun-I Li, Chieh-Ju Tsai, Ching-Jan Chen

Impact of Gate Resistor Configurations on Current Balancing in Paralleled SiC MOSFETs 2746
Yifu Zhang, Shashank Karanth, Emanuel Eni

Exploring the Potential of FPGA in High-Frequency Switching DC-DC Boost Converters using Model Predictive Control 2752
Qingcheng Sui, Bangli Du, Yu Zuo, Wilmar Martinez

A 7 Bit 5A 6.7 GHz Gate-Shaping Digital Gate Driver with Burst-Sampling ADC for Iterative Switching Optimization of SiC Power MOSFETs 2757
Tobias Zekorn, Kenny Vohl, Erik Wehr, Leon Weihs, Michael Hanhart, Ralf Wunderlich, Stefan Heinen

Decentralized Interleaving of Series-Stacked DC-DC Converters via Extremum-Seeking Control 2764
Ivan Petrić, Vignesh Iyer, Shoudong Hu, Chirayu Rajpurohit, Bailey Sauter, Milan Ilić, Luca Corradini, Dragan Maksimović

Online Dead-Time Control for Half Bridges without Preliminary Training based on Switching Transient Steepness 2772
Lukas Knappstein, Niklas Falkenberg, Martin Pfost

Impedance-Based State-of-Health Estimation for Lithium-Ion Battery Management Systems 2779
Mohammad K. Al-Smadi, Jaber A. Abu Qahouq

Stability Analysis and Resonance Damping of LC Filter-Based Voltage Source Converter with Single-Loop Voltage Control 2785
Aravind G., Divyanshu Bansal, L. Umanand

Finite Control Set Model Predictive Control Combined with Online Junction Temperature Estimation for Reliability Enhancement of Voltage Source Inverters 2790
Qiang Mu, Jiale Zhou, Zaheen Mustakin, Lucas Pereira, Babak Parkhideh, Tiefu Zhao

Framework for Dynamic Control and Operation of Power Electronics Interfaces 2797
Radha Sree Krishna Moorthy, Steven Campbell

Achieving Soft-Charging and Over 20% Input Current Ripple Reduction in a 48-to-6 V Dickson
Converter using 3-Phase Split-Phase Control 2805
Nagesh Patle, Rose A. Abramson, Sahana Krishnan, Jiarui Zou, Robert C. N. Pilawa-
Podgurski

Experimental Verification of Circuit-Losses Analysis-Model of DC-Output Converter Developed
using Approximated Equations from Measurement Data and Datasheet Data 2813
Ryota Kondo, Tsuyoshi Funaki

Scattering Parameter Measurement System using Probes for Surface Mount Devices Operating in
the Frequency Range from 50 kHz to 1 GHz 2821
Ryoko Kishikawa, Masahiro Horibe, Tomokazu Shoji, Shigenori Yabuta, Toshi Ohi, Ryo
Takeda, Takamasa Arai

Optical Transformer Design with Additional Common-Mode Noise Reduction Winding for Flyback
DC-DC Converters 2828
Yusuke Irie, Shinichiro Eguchi, Yoichi Ishizuka, Toshiro Takeuchi, Akio Iwabuchi, Takahiro
Koga, Toshiyuki Tanaka

Enhanced Bus Voltage Stability Through Digital Twin-Enabled Adaptive Controller Tuning 2833
Matthew Belanger, Andy Wong, Kerry Sado, Enrico Santi

Modeling and Performance Characterization of Lithium-Ion Capacitor at Different Temperature
and Voltage Values 2840
Mohammad K. Al-Smadi, Jaber A. Abu Qahouq, Sajad Saberi

Conveniently Identify Coils in Inductive Power Transfer System using Machine Learning 2846
Yifan Zhao, Mowei Lu, Ting Chen, Heyuan Li, Xiang Gao, Zhenbin Zhang, Minfan Fu, Stefan
M. Goetz

Accurate Modeling of LLC Resonant Converters with Enhanced Analytical Approach Considering
of Parasitic Capacitance 2851
Dong Jiao, Zhengming Hou, Jih-Sheng Lai

High-Frequency Conditioning Circuits for Power-Related Information Extraction in Non-
Sinusoidal Power Electronic Systems 2857
Haoyu Wang, Yuanxin Zhang, Di Mou, Alex Hanson, Shiqi Ji

Transconductance Model of the Dual Active Bridge Converter Under Single and Dual Phase Shift
Control 2865
Jared Cronin, Andrew Wunderlich, Enrico Santi

Lumped Parameter Modeling for Real-Time Thermal Regulation of Li-Ion Battery Packs 2871
Utkal Ranjan Muduli, Mohamed Shawky El Moursi, Khalifa Al Hosani, Ahmed Al-Durra

A Physics-Based Temperature Dependent Analytical Model for 2DEG Density in AlGaN/GaN
HEMT Devices 2877
Kashfia Tajmim Nabila, Jerry L. Hudgins

Comparative Analysis of Stator-PM Machines: Design Optimization and Electromagnetic
Performance Evaluation 2883
Maryam Salehi, Madhav Manjrekar

Elimination of Deadtime Effect on Resolver Offset Estimation using the Pulsating Current Command for Electric Vehicle Application 2889
Yingfeng Ji, Nurani Chandrasekhar

A Generic Load Emulator for Testing Motor Drives of E-Mobility 2894
Qingzheng Zhang, Kaiyuan Feng, Changsheng Hu, Dehong Xu

Design and Implementation of Power Assisted Control System for E-Bikes 2900
Che-Yu Lu, Tzu-Ping Cheng

A Hybrid PWM Strategy with Reduced Common-Mode Voltage and Extended Output Voltage Linearity for Adjustable Speed Drives 2907
Zhe Zhang, Kevin Lee

Single-Phase Open-Circuit Fault-Tolerant Control of Three-Phase PMSM Drives 2913
Yuichiro Minato, Yuki Nakata, Jun-Ichi Itoh

Multi-Vendor Encoder Position Sensing Interface using Programmable IP based Solution........................... 2920
Rajul Bhambay, Dhaval Khandla, Pratheesh Gangadhar, Thomas Leyrer, Achala Ram, Manoj Koppolu, Archit Dev

Sensorless Control Method at Low-Speed Range using High-Frequency Voltage Injection for Synchronous Reluctance Motors Considering to Nonlinear Characteristic Due to Magnetic Saturation 2924
Sota Takizawa, Sari Maekawa

Hybrid Control Scheme for Permanent Magnet Gear Motor.. 2932
Bing Li, Takayoshi Matsuo, Ahmed Sayed-Ahmed, Yujia Cui, Jiangang Hu

Cost-Effective Fault Diagnosis for Motor and Inverter using Bootstrap Charging and Single DC Link Current Sensor 2937
Gyu Cheol Lim, Won Hyo Jeong, Kahyun Lee, Jung-Ik Ha

Improved PWM to Suppress Motor Overvoltage Caused by Voltage Reflection............................. 2943
Sung-Oh Kim, Kyo-Beum Lee

Analysis of Double Pulsing Effect in Motor Drives based on Vector Diagram............................ 2948
Byeong-Woo Kang, Kyo-Beum Lee

A Novel Speed Sensor-Less Control of a Solar-Powered PMSM Drive 2953
Abirami Kalathy, Arpan Laha, Praveen Jain, Majid Pahlevani

Design of a Compact Low-Loss MMC Double Submodule for MVDC and HVDC Applications 2960
Ali Sharaf Addin, Rainer Marquardt, Thomas Brückner

A Series-Type Dynamic Voltage Restorer Control Strategy to Cope with Voltage Swell.............................. 2968
Jiazheng Zhang, Hongyu Chen, Xi Chen, Mingjun Bao

Machine Learning Approach for Accurate Lithium-Ion Battery Temperature Prediction using Electrochemical Features Independent of Battery SOC and SOH.. 2973
Vincent Masabiar Tingbari, Oluwaseun Isaiah Ekuewa, Anshul Nagar, Asad Abbas, Jamil Umar, Yuxin Zhang, Woonki Na, Jonghoon Kim

A Battery Strings Circulating Current Blocking Method for Battery Energy Storage Systems 2981
Haihong Long, Ziang Sun, Yucheng Fan, Xin Wu, Dehong Xu

A Hybrid Multilevel Converter-Based High-Gain Isolated DC/DC Converter for Grid-Tied Energy Storage Applications 2986
Pengyu Fu, Yizhou Cong, Jin Wang, Anant Agarwal

LCL Filter Parameter Selection using Graphical Method for a 13.8 kV ac 1.1 MVA 7-Level Flying Capacitor Grid-Connected Converter Utilizing Variable Switching Frequency 2992
Arthur Mendes, David Nam, Mingze Gao, Thimothy Thacker, Dong Dong, Rolando Burgos

Online Extraction of Electrochemical Impedance Spectroscopy Pattern based on EV Load Profile and Short Time Fourier Transform for Diagnosis of Lithium-Ion Battery Safety 3000
Miyoung Lee, Dongcheol Lee, Youngmin Bae, Jongchan An, Garam Yang, Woonki Na, Jonghoon Kim

Enhanced Incremental Capacity Analysis for Evaluating Battery Degradation Mechanisms of Optimized Fast Charging Methods 3006
Taehyeon Gong, Jaehyeong Lee, Sungjun Lee, Yura Kim, Bomyeong Ko, Woonki Na, Sungjin Choi, Jonghoon Kim

Co-Estimation of SOC and SOT in Lithium-Ion Batteries using an RLS-Based Heat Generation Model 3012
Seongkyu Lee, Eunjin Kang, Minhyeok Kim, Seunghyun Lee, Minwoo Song, Jaea Lee, Woonki Na, Jonghoon Kim

Three-Stage Adaptive Control Strategy for Stability Improvement of Grid-Connected Inverter in Weak Grid 3018
Longxiang You, Sicong Jin, Xin Zhang, Zuoshuai Wang, Sunqing Wang

Degradation Analysis of Offshore Bifacial PV Modules Under Multiple Climatic Stressors 3024
Aidha Muhammad Ajmal, Yongheng Yang

A Flexible Energy Management System for Solar Powered Electric-Bus Charging Stations 3030
Supun Amarathunga, Pasan Gunawardena, Xiaoting Wang, Yunwei Li

A Vienna Rectifier based Grid-Connected Powertrain for Hydrokinetic Turbine Systems 3036
Peidong Li, Md Tariquzzaman, Yue Cao

Condition Monitoring for DC-Link Capacitors and PV Arrays based on the Start-Up Process of the PV System 3042
Yongjie Liu, Ariya Sangwongwanich, Chen Liu, Xing Wei, Shuyu Ou, Tamás Kerekes, Jiahong Liu, Huai Wang

Electrically and Thermally Efficient Reliable Power Converter Design for Micro–Hydrokinetic Turbine 3048
Md Tariquzzaman, Peidong Li, Yue Cao

Comprehensive Evaluation of Cyber Attacks on Grid-Connected Smart Inverters 3054
Rishabh Singla, Vishwam Raval, Hasan Ibrahim, Jaewon Kim, Prasad Enjeti, Narsimha Reddy

Parallel Operation of Grid-Forming Converters based on Kuramoto Oscillators with Virtual Cable Emulation for Improved Power Sharing 3059
Vikram Roy Chowdhury, Gab-Su Seo, Barry Mather

Enhancing Hydrogen Production in Hybrid Standalone Microgrids 3064
Utkal Ranjan Muduli, Mohamed Shawky El Moursi, Khalifa Al Hosani, Ahmed Al-Durra

LSTM-Based Sub-Synchronous Oscillation Detection Scheme for Type 4 Wind Farm Interfaced with Weak AC Grid 3071

Omar Abu-Rub, Muhammad F. Umar, Jana A. Sheikh Ali, Yazan Qiblawey, Abdulrahman Alassi, Maryam Saeedifard, Mohammad B. Shadmand

A Study of Module Design Method to Suppress the Oscillation Occurs Between Parallel-Connected Power Devices 3077

Shinji Yato, Hiroto Sakai, Hideo Araki, Shumei Shimosako

A High-Efficient Hybrid Traction Inverter in Electric Vehicle Applications 3083

Yousefreza Jafarian, Omid Salari, Praveen Jain, Alireza Bakhshai, Mohamed Z. Youssef

Dual-Use of Onboard Chargers to Achieve Controllable DC Bus Voltage for Electric Vehicles 3089

Anuj Maheshwari, Elie Libbos, Arijit Banerjee

Isolated Single-Phase Onboard Chargers for BEV/PHEV using Active Power Decoupling Technology 3096

Yoshiki Amano, Keigo Nishimura, Hiroaki Matsumori, Takashi Kosaka, Kenichi Nagayoshi, Kenichi Watanabe

A Practical Use of xEVCap: The Modular and Standard DC-Link Capacitor Solution for the Main EV Powertrain Inverter 3100

David Olalla, Tomas Wagner, Fernando Rodriguez, Alberto Espinar

Optimized Bidirectional On-Board Charger using a Novel Unfolder-DAB Topology 3109

Héctor Sarnago, Ignacio Álvarez, Pablo Briz, Óscar Lucía

Critical Thermal Characterization of Next-Generation Solid-State Batteries for Automotive Battery Management Systems 3114

Chandan Chetri, Sheldon Williamson

Nanocrystalline CMC Inductors for EV Charging: Trade Studies and Testing Standardization 3119

Christopher Bracken, Mark A. Juds, Paul R. Ohodnicki, Bharadwaj Reddy Andapally, Jose Gato

Predicting Efficiency of On-Board and Off-Board EV Charging Systems using Machine Learning 3124

Mohamed Yasko, Fanghao Tian, Wilmar Martinez, Johan Driesen

High-Power and High-Speed Multi-Channel VCSEL Arrays with GaN Driver for Automotive LiDAR 3129

Yifu Liu, Sichao Li, Junlei He, Changyu Hu, Bill He, Karthik Krishnamurthy, Andy Shen

Double Pulse Test Platform for Hybrid SiC-IGBT Switch Characterization and Optimal Gate Control Strategy for EV Traction Inverters 3133

Rosario Attanasio, Harsha Ademane, Ryan Satterlee, Gianni Vitale

Critical Role of Individual Cell Temperature Monitoring in Mitigating Thermal Runaway and Reducing Accelerated Degradation in Lithium-Ion Batteries 3141

Mohit Sharma, Akash Samanta, William Locke, Sheldon Williamson

Loss-Optimized Design of a Triple Active Bridge DC-DC Converter for an Electric Vehicle Application 3147

Sreejith Chakkalakkal, Kyle Kozielski, Wesam Taha, Yicheng Wang, Aniket Anand, Ali Emadi

A Magnetic-Less DC/DC Converter with Pulse Charging for 800 V Powertrains from 400 V DC Fast Chargers 3155

Duc Dung Le, Shivam Chaturvedi, Shahid Aziz Khan, Mengqi Wang, Mohamed Elshaer

Boosting Charger Efficiency: A GaN-Based Flyback Converter with Energy Recycling 3160
Ahmad Nabizadah, Majid Ghasemi Korrani, Babak Fahimi

A Hybrid Three-Level Buck Converter with Flying Supercapacitor for High Load Current Surge
Capability using Peak Current Mode Control 3167
Finlay Lodge, Rafael Peña-Alzola, Martin MacFadyen, Patrick Norman, Mark Sweet,
Graeme Burt

Supercritical Carbon Dioxide (sCO.)-Cooled Current Source Inverter-based Integrated Motor Drive
for MW-Scale Electric Aviation Applications 3174
Hang Dai, John Yagielski, Thomas Jahns, Kum-Kang Huh, Vandana Rallabandi, Libing
Wang, Tarak Saha, Wenda Feng, Bulent Sarlioglu

The Challenge of Thermal Runaway in Soft Magnetic Materials for Inductive Power Transfer 3181
Yibo Wang, Ben Zhang, Weisheng Guo, Tianlu Ma, Sheng Ren, C.Q. Jiang

A Capacitively Coupled Alternative Electric Field Control for Freeze-Free based High Quality
Food Preservation 3187
Jaeyong Cho, Junhyeong Park, Sung-Bum Park, Daehyun Kim, Jinsoo Choi

The Characteristics of the Long Length Primary Loop and the Power Supply for the SCMaglev's
DWPT System 3194
Keisuke Yamamoto, Jun Enomoto, Shunsaku Koga, Junichi Kitano

A Wireless EV Charging System with a Double-Sided LCC Network using Variable Switching
Frequency and DC-Link Voltage Control 3200
Chae-Lyn Kim, Hyeonu Jo, Ju-A Lee, Dong Hyeon Sim, Byoung Kuk Lee

Class E/EF Inductive Power Transfer to Achieve Stable Output Under Variable Low Coupling 3206
Yifan Zhao, Mowei Lu, Heyuan Li, Zhenbin Zhang, Minfan Fu, Stefan M. Goetz

A Motorized Air-Core Variable Inductance Winding Structure 3212
Xindong Li, Sampath Jayalath, Cheng Zhang

Wireless Power Transfer System with Automatic Tuning Capability in Metallic Environment 3220
Renjie Zhang, Yue Wu, Delin Zhao, Yaohua Li, Yongbin Jiang, Yi Tang, Huan Yuan, Xiaohua
Wang, Mingzhe Rong

Design of Wireless Power Transmitters for Enhanced Transmission Distance and Output Power 3227
Kaiyuan Wang, Shuang Zhao, Shuye Shang, Eric Ka-Wai Cheng, Siew-Chong Tan, Yun Yang

Optimization of Wireless Power Transfer Waveforms and In-Vivo Receivers for Implantable
Medical Devices 3232
Hanbing Liu, Xin Zan

Comparison of Compact Power Amplifier Designs for High Frequency Resonant Wireless Power
Transfer Systems at 6.78 MHz using High-Q Resonators 3241
Manuel Rueß, Kilian Müller, Mathias C.J. Weiser, Ingmar Kallfass

Analysis and Design of Capacitive Coupling Wireless Power Transfer System using Load-
Independent Class-EF Inverter 3248
Takumi Kobayashi, Yutaro Komiyama, Akihiro Konishi, Hiroaki Ota, Yuki Ito, Taichi Mishima,
Takeshi Uematsu, Kien Nguyen, Hiroo Sekiya

Design and Optimization of a 600 W Wireless Drone Charger for High Gravimetric Power Density 3253
Arka Basu, Daniel Costinett

Stabilization Method for DC-Bus Oscillation in Dynamic Wireless Power Transfer Systems...... 3261
Yuki Ochiai, Keisuke Kusaka

Unveiling Aliasing Effect on Resonant Pole Locations in Wireless Battery Chargers 3267
Anwesha Mukhopadhyay, Daniel Costinett

Integrated Hybrid Inductive and Capacitive Power Transfer System with Asymmetrical PCB Self-Resonator........ 3275
Yao Wang, Zhen Sun, Xiangrong Zhang, Yun Yang, Shu Yuen Ron Hui

High Frequency Noise Reduction Method of the Class E Power Amplifier.................... 3281
Kyungmin Lee, Sungku Yeo

Single-Stage Three-Phase Buck-Matrix Rectifier with Series-Parallel Connected Transformers for High-Power 48 V Data Center Power Supplies........ 3285
Yuki Ishikura, Chinmay Bhagat

Sector Transition PWM Modulation Scheme for a Three-Phase Isolated Buck-Matrix Rectifier.......... 3291
Chinmay Bhagat, Yuki Ishikura

Adaptive Capacitance Circuit for Optimal Dynamic Impedance Matching in Variable Reluctance Energy Harvesting Applications........ 3298
Alejandro Redondo, Fernando Pérez, Sofia García, Gabriel Mujica, Airán Francés

Gallium Nitride (GaN) based Topology Comparison for Low Power Battery Charging Applications 3304
Jai Aditya Chaudhary, Rosario Attanasio, Gianni Vitale

Server Motherboard Power Performance Study Under Immersion Cooling Environment.................... 3312
Meng Wang, Haiyan Wang, Pavan Kumar, Haijin Zhang, Xiang Li, Fengwei Bian, Jianting Deng, Jiaqi Zhu, Yiming Lei

Practical PCB Design Considerations for GaN HEMTs based Isolated DC-DC Converter 3316
Gaureej Gauttam, Harish S. Krishnamoorthy, Sai Sushma Pasupuleti

Data-Driven Characterization and Forecasting of Metal-Oxide Varistor Degradation in DC Circuit Breakers........ 3321
Zhi Jin Zhang, Yang Liu, Lukas Graber, Maryam Saeedifard

A Thyristor-Based Fault Current Bypass Solid-State Circuit Breaker for DC Microgrid Applications 3328
Jiale Zhou, Xiuhu Sun, Qiang Mu, Tiefu Zhao

Single-Stage Three-Phase AC-AC Isolated Inertialess Converter (IIC) for Industrial Drives.................... 3334
Brad Houska, Decheng Yan, Aniruddh Marellapudi, Satish Belkhode, Joseph Benzaquen Sune, Deepak Divan

Author Index

Direct Effective Power Control (D-EPC) for LLC Resonant Converters Operating in Boost Mode Using Event-Driven-Timer Based Digital Controller

Yuto Yoshimura
Graduate school of environmental, life, natural science and technology
Okayama University
Okayana, Japan
pqx88wuy@s.okayama-u.ac.jp

Kenji Funatani
System Solutions Engineering HQ
ROHM Co., Ltd.
Yokohama, Japan
Kenji.Funatani@dsn.rohm.co.jp

Kazuhiro Umetani
Graduate school of environmental, life, natural science and technology
Okayama University
Okayana, Japan
umetani@okayama-u.ac.jp

Toshiyuki Zaitsu
LSI Business Unit
ROHM Co., Ltd.
Yokohama, Japan
toshiyuki.zaitsu@rohm.co.jp

Akito Nakagaki
Graduate school of environmental, life, natural science and technology
Okayama University
Okayana, Japan
pr3g4r2e@s.okayama-u.ac.jp

Masataka Ishihara
Graduate school of environmental, life, natural science and technology
Okayama University
Okayana, Japan
masataka.ishihara@s.okayama-u.ac.jp

Eiji Hiraki
Graduate school of environmental, life, natural science and technology
Okayama University
Okayana, Japan
hiraki@s.okayama-u.ac.jp

Abstract—The LLC converter is a topology using resonance and is effective for miniaturization and high efficiency. Generally, LLC converters use a direct frequency control (DFC) method, which has the issue of slow transient response characteristics. Therefore, power factor control (called previously), which directly controls the effective power (D-EPC) of the resonant circuit in LLC, has been proposed previously. However, it has been adopted only in series resonant operation (buck mode) because the magnetizing inductance was not considered. In this paper, a small-signal characteristics considering the effect of magnetizing inductance is derived, showed that D-EPC is effective even in boost mode, and confirm the characteristics by experiments

Keywords—LLC converter, power factor, stability, digital control

I. INTRODUCTION

The LLC converter is a topology that uses resonance, enabling high efficiency operation at high frequencies. The high-frequency drive also enables miniaturization of passive components. Because of these features, research to replace conventional buck converters with LLC converters was promoted [1]-[5]. However, it has become difficult to have apply LLC converter to point-of-load (PoL) converter which, require stability against sudden load fluctuations [6][7]. This is due to the LLC converter is conventionally controlled by direct frequency control (DFC) using a voltage-controlled-oscillator (VCO). In the case of DFC, the dynamic characteristics of the LLC converter vary significantly depending on the load condition [8][9]. As a result, fast response to load changing is difficult. As previous study, There are many challenges to overcome this drawback of DFC [12]-[19].

Among them, power factor control (called previously) [12]-[14] is promising for fast response of LLC converters because small-signal characteristics with little load dependence have been revealed. This method controls the phase difference between the voltage and current waveform in the resonant circuit. Therefore, it is a method of direct-effective power control (D-EPC) that adjusts the power factor of the resonant circuit. However, the derivation of the small-signal characteristics of D-EPC ignores the effect of the magnetizing inductance in the previous study [12]-[14]. Thus, it could not be used in the boost mode that is commonly used in the LLC converter applications.

In this paper, the small-signal characteristics of D-EPC considering magnetizing inductance are derived from the power-focused equation of state to show the effectiveness of the control characteristics and the adaptability of D-EPC in boost mode. The remainder of this paper consists of five sections. Section II briefly describes the shortcomings of conventional control and the control method of D-EPC. Next, in Section III, we propose a method to analyze the small-signal characteristics of the D-EPC considering the magnetizing inductance. In Section IV, we confirm the validity of the proposed analysis method by conducting actual measurements of the control characteristics. Finally, conclusions are presented in Section V.

II. PRINCIPLE OF D-EPC

A. Review of Convetinaol contorol (DFC)

Fig.1 shows LLC converter with direct frequency control DFC using VCO. Here, Q_1 is the high-side switch, Q_2 is the low-

979-8-3315-1612-3/25 $31.00 © 2025 IEEE

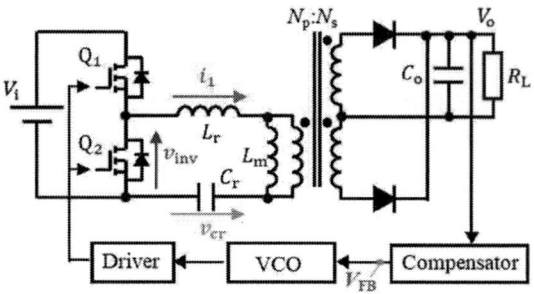

Fig.1. Block diagram of LLC using VCO

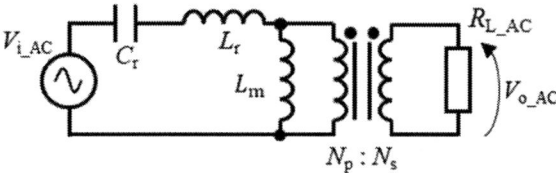

Fig.2. Equivalent circuit of LLC converter using FHA.

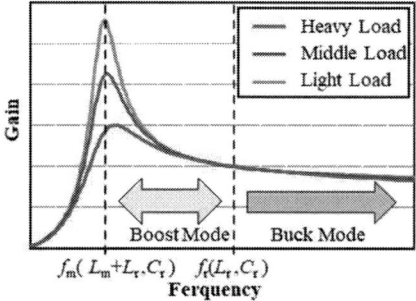

Fig.3. Frequency vs gain characteristic LLC

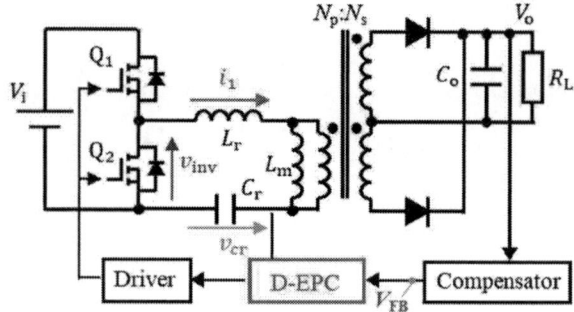

(a) Block diagram of LLC using D-EPC

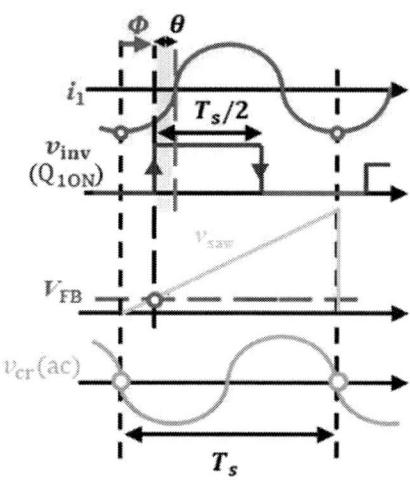

(b) Timing chart of D-EPC

Fig.4. The principle of operation of D-EPC for LLC converter

side switch, L_m of the transformer is magnetizing inductance, L_r is the resonant inductance, C_r is the resonant capacitor, C_o is the smoothing capacitor and R_L is the load resistance. f_s is the operating frequency.Fig.2 shows the equivalent circuit of the LLC converter based on the FHA (Fundamental Harmonics Analysis) method. Using Fig.2, the frequency versus gain characteristic is expressed by (1)

$$\frac{V_{o\_AC}}{V_{i\_AC}} = \frac{1}{\sqrt{\left(1 + \frac{1}{S} - \frac{1}{SF^2}\right) + Q^2\left(F - \frac{1}{F}\right)^2}} \quad (1)$$

$$S = \frac{L_m}{L_r}, f_r = \frac{1}{2\pi\sqrt{L_r C_r}}, F = \frac{f_s}{f_r}, Z_r = \sqrt{\frac{L_r}{C_r}}$$

$$N = \frac{N_p}{N_s}, R_{L\_AC} = \frac{8}{\pi^2} N^2 R_L, Q = \frac{Z_r}{R_{L\_AC}}$$

As shown in (1), the LLC converter can control the output voltage by adjusting the operation frequency f_s. Fig.3 shows the gain versus frequency characteristics of the LLC converter. Fig.3 shows that the characteristic of gain and frequency depends on the load. Therefore, conventional control (DFC)

using VCO is difficult to control design because the transfer function of the plant is highly load-dependent. On the other hand, D-EPC has a small load dependence and is easy to control design.

B. Operation of D-EPC[13][14]

D-EPC (previously called "power factor control") method regulates the output voltage by adjusting the angle θ which is defined as the phase difference between the inverter voltage v_{inv} and the primary resonant current i_1. Fig 4(a) shows the block diagram of D-EPC and Fig.4 (b) the timing chart. Here, V_{saw} is the saw wave of the control circuit, V_{FB} is the feedback voltage, Q_{1on} is the turn-on signal of the high-side switch Q_1, and ϕ is the phase difference from zero crossing of the resonant capacitor to turning on the high-side switch. D-EPC algorithm generates V_{saw} with respect to the zero crossing point of the resonant capacitor voltage V_{cr} instead of detecting the current phase, as shown in Fig.4. Then, the switch Q_1 is turned on at the timing, when it crosses V_{FB} to control the phase difference ϕ between the inverter voltage v_{inv} and the primary resonant current i_1. The phase difference ϕ is related to the power factor $\cos\theta$ by the (2).

$$\cos\theta = \cos(90 - \phi) \quad (2)$$

From the relationship in(2), the effective power output by the inverter is controlled based on the power factor $\cos\theta$.

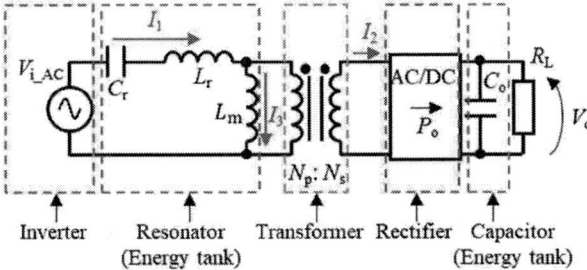

Fig.5. Proposed analysis model for LLC converter

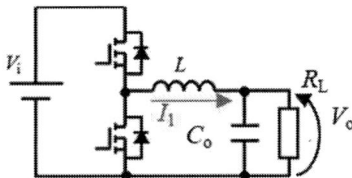

Fig.6. Buck converter

C. Small signal analysis in buck mode[12]

The small-signal characteristic analysis establishes equations for the effective-power flow from the inverter to the load resistor in reference [12]. Fig.5 shows an analysis model of LLC converter. In this model, the input voltage is approximated as only Fundamental wave and the rectifier circuit is considered as a lossless power conversion circuit. V and I is root-mean-square value of the voltage and current in each part. V_{i_AC} and V_{o_AC} values are expressed as follows.

$$V_{i\_AC} = \frac{\sqrt{2}}{\pi} V_i \tag{3}$$

$$V_{o\_AC} = \frac{2\sqrt{2}}{\pi} V_o \tag{4}$$

In this case, the effective input power P_i, output power P_{tr} from the secondary winding of the transformer per unit time can be expressed as follows

$$P_i = I_1 V_{i\_AC} \cos\theta \tag{5}$$

$$P_{tr} = I_2 \cdot V_{o\_AC} \tag{6}$$

In the case of the LLC converter in buck mode the LLC converter is used in buck mode, the magnetizing inductance L_m should be large. Therefore, the value of the magnetizing current I_3 is smaller than the secondary current I_2, thus the secondary current I_2 is expressed as

$$I_2 = NI_1 \tag{7}$$

Assuming that the LC resonator is driven at its resonant frequency f_r, the stored energy in the resonator E_{res} during the period of AC voltage is considered to be almost constant. Here, Note that the instantaneous AC voltage of C_r is zero at the instant of peak current in L_r, E_{res} can be formulated as

$$E_{res} = \frac{1}{2} L_r \left(\sqrt{2} I_1 \right)^2 \tag{8}$$

The difference between P_{in} and P_{tr} indicates the energy stored in the LC resonator per unit time. Therefore, from (5)-(8), the equation of state for I_1 is expressed as

$$\frac{dE_{res}}{dt} = P_i - P_{tr} \tag{9}$$

$$\therefore \frac{d}{dt} \left\{ \frac{1}{2} L_r \left(\sqrt{2} I_1 \right)^2 \right\} = I_1 V_{i\_AC} \cos\theta - NI_1 \cdot V_{o\_AC} \tag{10}$$

Next, deriving the expression for V_o from V_{o_ac}, the dc output power per unit time P_o of the LLC converter is expressed as

$$P_o = \frac{V_o^2}{R_L} \tag{11}$$

Noting that P_{tr} indicates the ac power per unit time supplied to the rectifier, $P_{tr}-P_o$ can be interpreted as the energy charged in the output smoothing capacitor per unit time. If E_{cap} denotes the total energy stored in the output smoothing capacitor, E_{cap} can be expressed as

$$E_{cap} = \frac{1}{2} C_o V_o^2 \tag{12}$$

Therefore, from (6), (11) and (12), the equation of state for V_{out} is expressed as

$$\frac{dE_{cap}}{dt} = P_{tr} - P_o \tag{13}$$

$$\therefore \frac{d}{dt} \left(\frac{1}{2} C_o V_o^2 \right) = NI_1 \cdot V_{o\_AC} - \frac{V_o^2}{R_L} \tag{14}$$

Finally, from (10) and (14), "state-space-equation" of LLC Power Factor Control is derived as follows.

$$\frac{d}{dt} \begin{bmatrix} I_1 \\ V_o \end{bmatrix} = \begin{bmatrix} 0 & -\frac{\sqrt{2}N}{\pi L_r} \\ \frac{2\sqrt{2}N}{\pi C_o} & -\frac{1}{C_o R_L} \end{bmatrix} \begin{bmatrix} I_1 \\ V_o \end{bmatrix} + \begin{bmatrix} \frac{\sqrt{2}N}{\pi L_r} \\ 0 \end{bmatrix} \frac{V_i}{2N} \cos\theta \tag{14}$$

On the other hand, Back converter (Fig.6) is well known "state-space equation" is expressed as

$$\frac{d}{dt} \begin{bmatrix} I_1 \\ V_o \end{bmatrix} = \begin{bmatrix} 0 & -\frac{1}{L} \\ \frac{1}{C_o} & -\frac{1}{C_o R_L} \end{bmatrix} \begin{bmatrix} I_1 \\ V_o \end{bmatrix} + \begin{bmatrix} \frac{1}{L} \\ 0 \end{bmatrix} V_i D \tag{15}$$

Comparing (14) and (15), it is found that they have the same form. Therefore, the method that directly controls the effective power based on the power factor $\cos\theta$ (D-EPC) has the same control characteristics as the buck converter. This allows the compensator design method of the buck converter to be adapted and enables a faster response.

III. SMALL SIGNAL CHARACTERISTICS IN BOOST MODE

A. Derivation of Small-Signal Model

Similar to the buck mode, the small-signal characteristics in the boost mode are also derived by using (9) and (13), which focus on the effective power. However, in the boost mode, the equation is derived by considering the effect of the magnetizing current I_3, which was ignored in (7) in the buck mode.

Since the rectifier circuit is a power load, it can be considered a resistive load. Therefore, the magnetizing current I_3 and the transformer secondary current I_2 are orthogonal. Therefore, the transformer secondary current I_2 is expressed as

$$I_2 = \sqrt{I_1^2 - I_3^2} = \sqrt{I_1^2 - \left(\frac{V_{o\_AC}}{\omega L_m}\right)^2} \quad (16)$$

From (16), which considering the magnetizing current, the equations of state (9) and (13) are expressed once again

$$\frac{d}{dt}\left\{\frac{1}{2}L_r\left(\sqrt{2}I_1\right)^2\right\} =$$

$$I_1 V_{i\_AC}\cos\theta - N\sqrt{I_1^2 - \left(\frac{V_{o\_AC}}{\omega L_m}\right)^2} \cdot V_{o\_AC} \quad (17)$$

$$\frac{d}{dt}\left(\frac{1}{2}C_o V_o^2\right) = N\sqrt{I_1^2 - \left(\frac{V_{o\_AC}}{\omega L_m}\right)^2}\cdot V_{o\_AC} - \frac{V_o^2}{R_L} \quad (18)$$

Then the small signal transfer function from $\cos\theta$ to V_o derive from (17)(18) to (19).

$$\frac{\delta V_o}{\delta \cos\theta} = \frac{\dfrac{1}{\eta N}V_i}{2L_r C_o s^2 + \left(C_o\left(1-\eta^2\right)R_L + \dfrac{2L_r}{R_L\eta^2}\right)s + 1} \quad (19)$$

where η is the ratio of transformer primary current I_1 and secondary current I_2 in steady state as shown in (20)

$$\eta = \frac{NI_2}{I_1} = \frac{\omega L_m}{\sqrt{R_{L\_AC}^2 + \omega^2 L_m^2}}$$

$$\eta = \frac{\omega L_m}{\sqrt{R_{L\_AC}^2 + \omega^2 L_m^2}} \quad (20)$$

On the other hand, it is well known that the buck converter (Fig.6) can be linearly controlled and the small-signal characteristic is expressed by (21).

$$\frac{\delta V_o}{\delta D} = \frac{V_i}{LC_o s^2 + \dfrac{L}{R_L}s + 1} \quad (21)$$

Comparing (19) and (21), we can see that both are second-order delay systems and that the corner frequency is constant. It also shows that the damping factor depends on the load. Therefore, the LLC converter in boost mode has the same control characteristics as the buck converter by controlling the effective power based on $\cos\theta$.

B. Calculating plant transfer function

Fig. 7 shows the calculation characteristic the plant (control to output) transfer function at $R_o=1.6\Omega$ and 4.7Ω based on (19), and the values of TABLE.I were used for each parameter. Fig.7 shows that there is no increase in gain at the resonance frequency regardless of the load condition. This is due to the effect of the

magnetizing inductance, which tends to increase the damping coefficient. This effect causes the separation of the poles, resulting in a Bode plot similar to that of a first-order system. It can also be seen that the pole separation increases with light load. This model is considered to be a light load increases the attenuation coefficient by increasing the value of η in (19).

IV. IMPRIMENTATION AND EXPERIMENTS

A. Imprimentation and Operating Mechanism

Fig.8 shown circuit diagram. This circuit consists of an LLC converter main circuit, a capacitor voltage detection circuit, a saw wave circuit, and a digital controller. The digital controller is the ML62Q203x/4x (16-bit, 16 MHz clock, 64 MHz timer)[20]. This controller can realize event-driven operation

TABLE.I Calculation parameters

Parameter	Symbol	Value
DC input Voltage	V_{in_DC}	48[V]
Regulated Output Voltage	V_o	12[V]
Transformer turn-ratio	N	2
Series resonant inductance	L_r	2[μH]
Magnetizing inductance	L_m	7[μH]
Resonant capacitance	C_r	500[nF]
Output capacitance	C_o	350[μF]

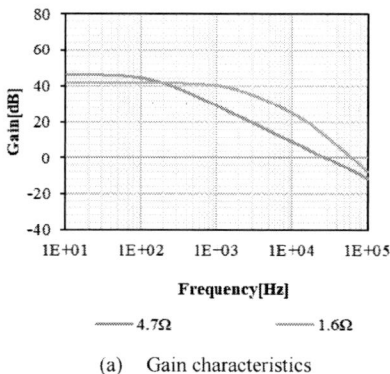

(a) Gain characteristics

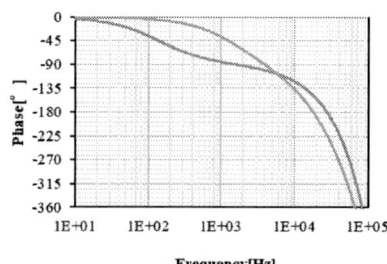

(b) Phase characteristics

Fig.7. Calculation result (control to output bode plot)

without CPU control, and the timer has a frequency tracking function with an internal comparator.

Fig.9 shows the timing chart. The control method is based on the saw wave as in the D-EPC operation. The ON period of the switch is determined by a timer in the digital controller. In this circuit, the rising edge of the internal comparator CMP_{tr} is used as an event, and frequency tracking is performed from the previous resonance period to the triangular wave timer for phase delay timing adjustment.

B. Experiments

Fig.10 shows the prototype board. The experiment also used Table.I for each parameter. This circuit is stepped down from 48 V to 12V. The maximum output current is 8 A and the maximum output power is 100 W. The series resonance frequency is 150 kHz. In the experiment, the feedback voltage V_{FB} is VFB is applied externally, and the circuit is operated in open loop. Here, the feedback voltage is adjusted so that the output voltage is 12 V.

Fig.11 shows actual operating waveform with load resistance of 4.7Ω. The current waveforms in Fig.11 show that the LLC converter is operating at a frequency lower than the resonant frequency, to operate in boost mode. The high-side switch Q1 turns on at the point where the feedback voltage and the saw wave intersect. Therefore, it was found that D-EPC circuit operates even in the boost mode and can be adapted to all modes of operation.

C. Measuring plant transfer function

Fig. 12 shows the measured plant transfer function at $Ro=1.6Ω$ and 4.7Ω. From Fig. 12, the transfer function shows the same characteristics as calculation result of proposed model. From this result, it can be said that the small-signal characteristics in (19) sufficiently predict damping factor. The characteristics of LLC converters using D-EPC are similar to those of buck converters.. This indicates that it is possible to adapt the compensator design of the buck converter to the LLC converter using D-EPC.

V. CONCULTION

Direct effective power control(D-EPC) method for LLC control was proposed. A small-signal characteristic of the plant transfer function (control to output) considering magnetizing inductance for the boost mode operation was derived from "power -focused equation of state". Test board was implemented and evaluated at Vin=48V, V_{out}=12V, R_o=1.6 Ω and 4.7Ω. As a result, D-EPC LLC

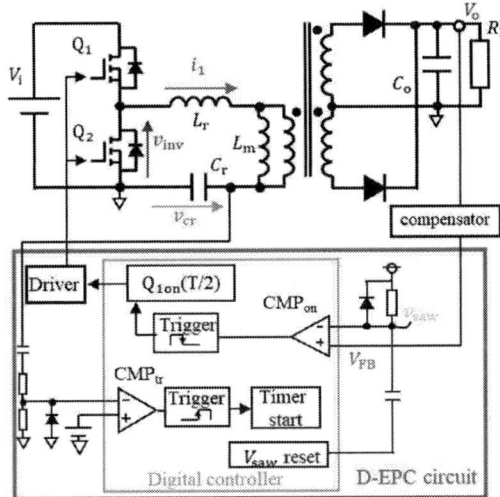

Fig.8. Circuit Diagram of LLC D-EPC

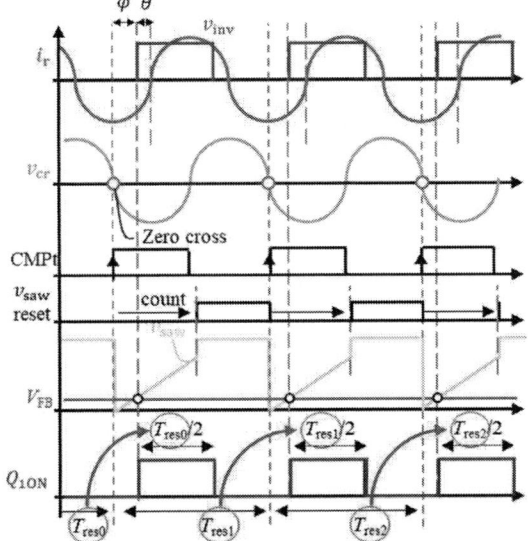

Fig.9. Timing chart of LLC D-EPC

Fig.10. Prototype board of LLC D-EPC

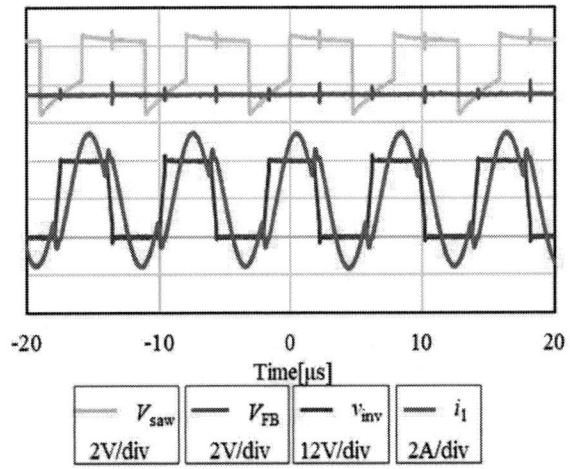

Fig.11. Operating wave form

converter in boost mode was working properly and "the plant transfer function" was in good agreement between calculated and measured bode-plot. It is confirmed that D-EPC method is effective even in the boost mode.

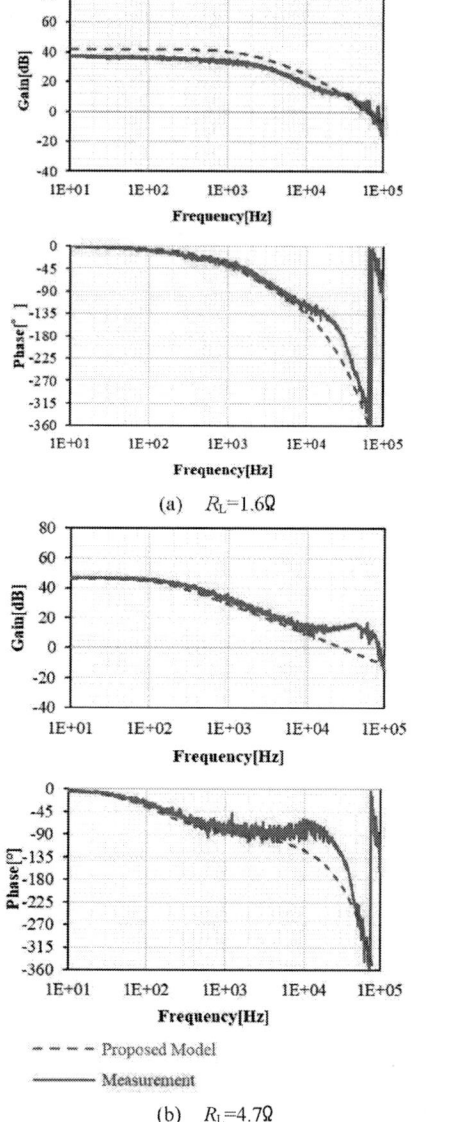

(a) R_L=1.6Ω

- - - - Proposed Model

———— Measurement

(b) R_L=4.7Ω

Fig. 12. Experimental result (control to output bode plot)

REFERENCES

[1] G. Yang, P. Dubus, and D. Sadarnac, "Double-phase highefficiency,wide load range high- voltage/low-voltage LLC DC/DC converter forelectric/hybrid vehicles," IEEE Trans. Power Electron., vol. 30, no. 4,pp. 1876-1886, Apr. 2015.

[2] C. Duan, H. Bai, W. Guo, and Z. Nie, "Design of a 2.5-kW 400/12Vhigh-efficiency DC/DC converter using a novel synchronous rectification control for electric vehicles," IEEE Trans. Transportation Electrification, vol. 1, no. 1, pp. 106-114, Jun. 2015.

[3] J.-B. Lee, J.-K. Kim, J.-H. Kim, J.-I. Baek, and G.-W. Moon, "A high efficiency PFM half-bridge converter utilizing a half-bridge LLC converter under light load conditions," IEEE Trans. Power Electron., vol. 30, no. 9, pp. 4931-4942, Sept. 2015.

[4] C. Fei, R. Gadelrab, Q. Li, and F. C. Lee, "High-frequency three-phase interleaved LLC resonant converter with GaN devices and integrated planar magnetics," IEEE J. Emerg. Select. Topics Power Electron., vol. 7, no. 2, pp. 653-663, Jun. 2019.

[5] R. Beiranvand, B. Rashidian, M. R. Zolghadri, and S. M. H. Alavi, "A design procedure for optimizing the LLC resonant converter as a wide output range voltage source," IEEE Trans. Power Electron., vol. 27, no. 8, pp. 3749-3763, Aug. 2012.

[6] C.-O Yeon, J.-W. Kim, M.-H. Park, I.-O. Lee, and G.-W. Moon, " Improving the light-load regulation capability of LLC series resonant converter using impedance analysis," IEEE Trans. Power Electron., vol. 32, no. 9, pp. 7056-7067, Sept. 2017.

[7] Z. Fang, J. Wang, R. Liu, L. Xiao, J. Zhang, G. Hu, and Q. Liu, "Energy feedback control of light-load voltage regulation for LLC resonant converter," IEEE Trans. Power Electron., vol. 34, no. 5, pp. 4807-4819, May 2019.

[8] H.-P. Park and J.-H. Jung, "Power stage and feedback loop design for LLC resonant converter in high-switching-frequency operation," IEEE Trans. Power Electron., vol. 32, no. 10, pp. 7770-7782, Oct. 2017.

[9] M. F. Menke, A. R. Seidel, and R. V. Tambara, "LLC LED driver small-signal modeling and digital control design for active ripple compensation," IEEE Trans. Ind. Electron., vol. 66, no. 1, pp. 387-396, Jan. 2019.

[10] Z. Fang, J. Wang, S. Duan, K. Liu, and T. Cai, "Control of an LLC resonant converter using load feedback linearization," IEEE Power Electron., vol. 33, no. 1, pp. 887-898, Jan. 2018.

[11] S. Tian, F. C. Lee, Q. Li, "Equivalent Circuit Modeling of LLC Resonant Converter," IEEE Transactions on Power Electronics, Vol. 35, No.8, pp.8833-8845, Aug. 2020.

[12] K.Umetani, K. Shimomura, K. Yamada, T. Kawakami, M. Ishihara, and E. Hiraki, "A Control Method Based on Power Factor for Improving Output Voltage Stability and Efficiency of LLC Converter in Wide Range of Output Voltage and Load Impedance," IEEE ECCE 2021, pp.3436-3443.

[13] K. Yamada, K. Umetani, E. Hiraki, and M. Ishihara, "Phase-Shift Based on Power Factor Control for LLC Converter with High Output Stability Against Load Fluctuation," IEEE Southern Power Electronics Conference (SPEC), 2022.

[14] T. Zaitsu, Y. Yoshimura, K. Umetani, M. Ishihara, E. Hiraki and K. Horii, "Compact Hardware Implementation of Power Factor Control for LLC converter with Event-Driven-Timer Based Digital Controller," 2024 IEEE Applied Power Electronics Conference and Exposition (APEC), Long Beach, CA, USA, 2024, pp. 2015-2020

[15] C. Adragna, "Time-shift control of LLC resonant converters" PCIM Europe 2010 Proceedings, Paper #113, Page(s) 661- 666, May 2010.

[16] C. Adragna, D. Ciambellotti, M. Dell'Oro and F. Gallenda, "Digital implementation and performance evaluation of a time-shift-controlled LLC resonant half-bridge converter," 2014 IEEE Applied Power Electronics Conference and Exposition - APEC 2014, Fort Worth, TX, USA, 2014, pp. 2074-2080

[17] Z. Hu, Y. -F. Liu and P. C. Sen, "Bang-Bang Charge Control for LLC Resonant Converters," in IEEE Transactions on Power Electronics, vol. 30, no. 2, pp. 1093-1108, Feb. 2015

[18] B. McDonald and Y. Li, "A novel LLC resonant controller with best-in-class transient performance and low standby power consumption," 2018 IEEE Applied Power Electronics Conference and Exposition (APEC), San Antonio, TX, USA, 2018, pp. 489-493

[19] S. S. Shah, U. Raheja and S. Bhattacharya, "Input Impedance Analyses of Charge Controlled and Frequency Controlled LLC Resonant Converter," 2018 IEEE Energy Conversion Congress and Exposition (ECCE), Portland, OR, USA, 2018, pp. 1-5

[20] https://www.rohm.com/products/micon/logicoa/ml62q20xx-series/ml62q2035-nnngd_taping_-product#productDetail

Mitigation Method of Resonance between Paralleled On-line UPS

Teng Wu, Zhenguo Huo, and Shangxian Ning
Secure Power Division
Schneider Electric (China) Co., Ltd. Shanghai Pudong Branch
Shanghai 200120, China
warren.wu@se.com, zhenguo.huo@se.com, shangxian.ning@se.com

Abstract—On-line uninterrupted power supply (UPS) is used to guarantee stable power supply in case of grid faults. It is widely applied in industry, commerce, medical and transportation. In many application scenarios, several on-line UPSs are required to work in parallel due to high load demand. However, high-frequency resonance might occur in the output current of paralleled UPSs. This paper provides impedance model analysis of the resonance and proposes a mitigation method based on digital notch filter. Through this method, the resonance can be mitigated significantly, and voltage dynamic performance is not deteriorated. The effectiveness of the proposed method is verified through experimental tests.

Keywords—*uninterrupted power supply, parallel, resonance, impedance model, notch filter*

I. INTRODUCTION

Uninterruptible Power Supply (UPS) is a power protection device used to provide temporary power to loads when main power failure. It can protect critical load equipment from the impact of power issues, such as voltage fluctuations, power outages, frequency changes, etc. High power UPS, as a key component of the power supply system, has been widely used in data centers, industrial control, medical, transportation and other fields [1-6].

According to the working principle and design structure, high-power UPS is generally divided into three types: backup, online, and online interactive [7-8]. Both backup UPS and online interactive UPS have switching time issues, that is, when the input mains power is abnormal, the UPS will experience short-term interruption of load power supply during the working mode switching process, and its adaptability to the input mains power is low. The distortion and interference of the mains power will be transmitted to the load. Therefore, both types of UPS are not suitable for application in critical power supply areas where power cannot be interrupted. The working mode of online UPS is different from the previous two types of UPS. Its inverter is always in working state, so there is no switching time problem for online UPS. Moreover, it has stronger adaptability to input mains power and can effectively isolate distortion and interference on the input side. Therefore, online UPS is more suitable for occasions with strict requirements for power supply quality and continuity, especially in data centers.

As mentioned above, high-power online UPS is an important component of data center power supply system. In recent years, with the rapid development of artificial intelligence, the electricity consumption of data centers has been increasing year by year. In order to meet the power supply needs of large data centers, multiple high-power UPS are often required to operate in parallel. When the inverter operates in parallel, it may cause voltage or current resonance at the output terminal. The reason for resonance may be the switch sub high frequency resonance caused by the asynchronous carrier of the parallel converter, the LC resonance caused by the unreasonable design of the inverter output filter parameters, the resonance caused by the mismatch between the inverter output impedance and the grid impedance, or the instability phenomenon during load fluctuations caused by the unreasonable design of droop control or virtual synchronous machine controller parameters.

According to the causes of resonance, different methods need to be used to suppress it. The resonance problem caused by carrier asynchrony needs to be addressed by introducing precise carrier synchronization strategies [9]. The instability phenomenon under droop control or virtual synchronous machine control requires optimizing controller parameters to reduce voltage fluctuations during load fluctuations (essentially reducing inverter output impedance) [10]. Resonance problems caused by impedance mismatch or output filter parameters can be suppressed through passive damping or active damping methods. Passive damping refers to increasing system damping by series or parallel resistors on the inductance or capacitance of the output filter [11]. This method is simple to implement and does not require additional control steps, but it increases the power loss and cost of the system, and when external conditions change, series or parallel resistors may not be suitable and resonance cannot be eliminated. There are many ways to implement the active damping method, which can suppress resonance by providing feedback on the output filter capacitor current [12] and simulating virtual capacitor impedance. However, this method requires an additional capacitor current sensor; It is also possible to suppress resonance by increasing the output impedance of the inverter through feedback output current [13] and corresponding algorithms, but increasing the output impedance across the entire frequency band may affect the response characteristics beyond the resonance frequency.

This article analyzes the root causes of parallel resonance in high-power UPS and selects a notch filter to suppress parallel resonance. This method does not increase system losses and costs, and does not affect the response characteristics of the fundamental frequency while suppressing high-frequency resonance. The effectiveness of this method has been verified through experiments on mass-produced products.

II. PROBLEM DESCRIPTION

In the two modular online UPS parallel systems shown in Figure 1, each consisting of 1-10 Power Modules, short for PM, each PM includes a Power Factor Corrector (PFC), an inverter, and a DC-DC converter connected to the battery (since DC-DC conversion is not within the scope of this article, the battery and

979-8-3315-1612-3/25 $31.00 © 2025 IEEE

DC-DC converter are omitted in Figure 1) [14]. The rated power of each PM is 50kVA, and the rated power of UPS is the sum of the rated power of the PM equipped with it. The output filter of each UPS consists of two parts:

- An LCL filter composed of L1, C1 and L2 at the output terminal of each PM.
- EMI filtering capacitor C2 on the UPS framework.

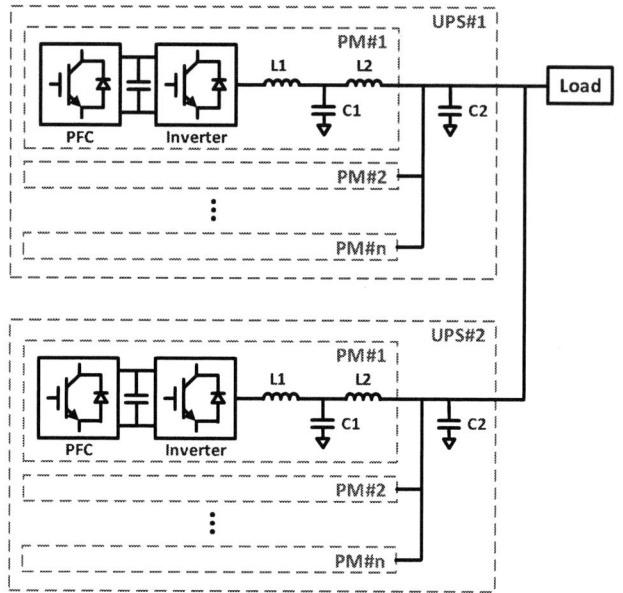

Figure 1. Equivalent circuit of paralleled modular on-line UPS

When the inverters of two UPS PMs are started and the UPS runs in parallel without load, high-frequency resonant current can be observed on the output current of each UPS. The characteristics of this high-frequency resonant current are as follows:

- High frequency current resonates between two UPS systems and does not flow to the load. If add load, the high-frequency resonance disappears.
- The resonant frequency of high-frequency current is between 3.5kHz and 7kHz, and the frequency decreases with the increase of PM number.
- The resonance amplitude of high-frequency current is relatively high, and its peak value may even exceed the PM rated current peak value.
- When the output power cable length of two UPS units in parallel is less than 3 meters, high-frequency resonance disappears.

The risks brought by high-frequency resonance mainly include the following three aspects:

- High frequency current causes the filter capacitor to charge and discharge more frequently, affecting its lifespan and potentially causing device damage.
- High frequency current can cause inductance whistling, bringing noise that users can perceive.
- High frequency current affects the calculation of UPS output power, resulting in deviations in the power values displayed on the HMI interface.

III. Root Cause Analysis

A. Resonant Path

In order to eliminate or suppress the high-frequency resonance current caused by UPS parallel connection, this chapter analyzes the root causes of high-frequency resonance.

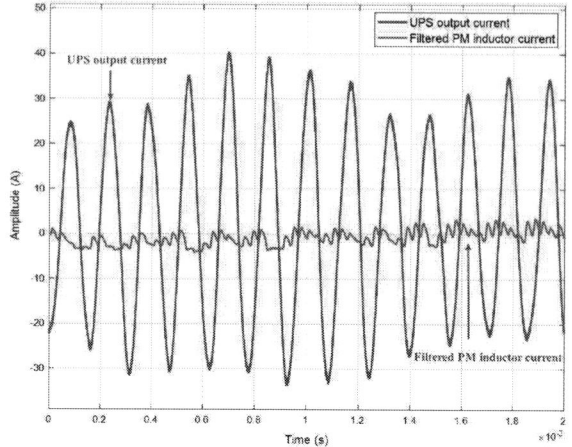

Figure 2. Comparison of UPS output current and inverter inductor current

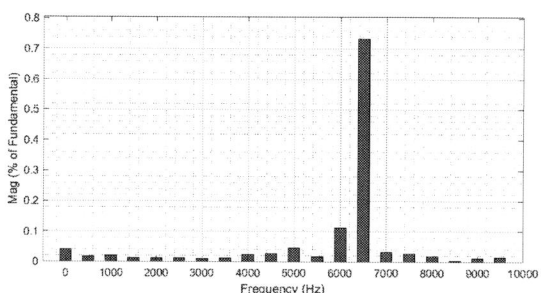

(a) Harmonic content of UPS output current

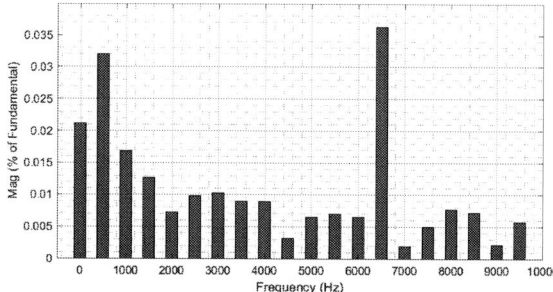

(b) Harmonic content of inverter inductor current

Figure 3. THD of UPS output current and inverter inductor current

Firstly, the resonant path of high-frequency current was detected, and the UPS output current and PM inverter output inductor current were measured simultaneously. After filtering out the switching ripple of inductor current using a low-pass filter with a cutoff frequency of 10kHz, the waveform was drawn as shown in Figure 2. The blue waveform represents the UPS output current, and the orange waveform represents the PM output inductor current after filtering out the switching ripple. FFT analysis was conducted on the current data with a resolution of 500Hz, as shown in Figure 3. It can be seen that

979-8-3315-1612-3/25 $31.00 © 2025 IEEE   2661

although the high-frequency resonant component is high in the UPS output current, the high-frequency resonant component in the PM output inductor current is very low.

Secondly, the UPS output current, PM output filter capacitor current, and EMI filter capacitor current were measured simultaneously, and their effective values are shown in Table I. It can be seen that the sum of the effective values of the capacitor currents has exceeded the effective value of the UPS output current. The above phenomenon indicates that the main component of high-frequency current is resonance between the output capacitors of two UPS systems, and a small component flows into the DC bus through the inverter of the PM.

TABLE I. RMS VALUE OF UPS OUTPUT CURRENT AND FILTER CAPACITOR CURRENT

Measure current	Effective current value/A
UPS output current	15.73
PM output filtering capacitor current	18.67
UPS frame EMI filtering capacitor current	3.12

In summary, it is reasonable to infer that the resonant frequency is the LC resonant frequency formed by the UPS output filtering capacitor, filtering inductor, and line inductance. The output impedance of the inverter exhibits an underdamped state at this resonant frequency and continuously provides excitation to the resonant system, resulting in this high-frequency resonance.

B. Model Validation

The control block diagram of the power module inverter is shown in Figure 4, where:

- V_{ref}, output voltage reference value of the inverter.
- V_{dc}, DC bus voltage,
- V_{inv}, inverter bridge arm voltage,
- V_{out}, inverter output voltage,
- i_L, inverter output inductor current,
- $G_v(s)$, voltage loop controller
- $G_i(s)$, current loop controller
- $G_p(s)$, PWM modulation link gain
- $G_{sv}(s)$, voltage sampling zero order hold

- $G_{si}(s)$, current sampling zero order hold.

Derive the transfer function $Z_{INV}(s)$ from V_{out} to i_L as shown in (1):

$$Z_{INV}(s) = \frac{G_i(s)G_p(s)G_{si}(s)V_{dc}}{1+s^2L_1C_1+G_v(s)G_i(s)G_p(s)G_{sv}(s)V_{dc}-G_p(s)G_{sv}(s)} \quad (1)$$

The equivalent output impedance $Z_{PM}(s)$ of a single PM is derived as shown in (2):

$$Z_{PM}(s) = \frac{Z_{INV}(s)}{sC_1(Z_{INV}(s)+1/sC_1)} + sL_2 \quad (2)$$

The equivalent output impedance $Z_{UPS}(s)$ of the entire UPS is derived as shown in (3):

$$Z_{UPS}(s) = \frac{Z_{PM}(s)}{sC_2N_{PM}(Z_{PM}(s)/N_{PM}+1/sC_2)} \quad (3)$$

where N_{PM} refers to the number of PMs equipped on a single UPS.

The impedance $Z_{pas}(s)$ of the output passive filter of another UPS in the parallel system is shown in (4):

$$Z_{pas}(s) = \frac{1/sC_1+sL_2}{N_{PM}+C_2/C_1+s^2L_2C_2} \quad (4)$$

The system equivalent impedance $Z_{sys}(s)$ composed of the equivalent output impedance of one UPS, the passive filter of another UPS, and the line impedance is shown in (5):

$$Z_{sys}(s) = Z_{UPS}(s) + Z_{pas}(s) + Z_{line}(s) \quad (5)$$

Among them, $Z_{line}(s)$ is the line inductance, calculated at 0.6 μH per meter.

Draw Bode plots of the equivalent impedance of the system corresponding to different PM quantities, as shown in Figure 5. It can be seen that the system has resonance points in the range of 3.5 kHz – 7 kHz, and the phase crosses from -180° to +180°, indicating that the system is in an underdamped state in this frequency band. Additionally, it can be observed that as the number of PM increases, the resonant frequency decreases.

Draw the Bode plot of the equivalent impedance of the system under different line impedances, as shown in Figure 6. It can be seen that when the parallel cable is too long or too short, it will not cause high-frequency resonance.

The above modeling analysis results are consistent with the measured phenomena, verifying that the fundamental cause of parallel high-frequency resonance is the LC resonance

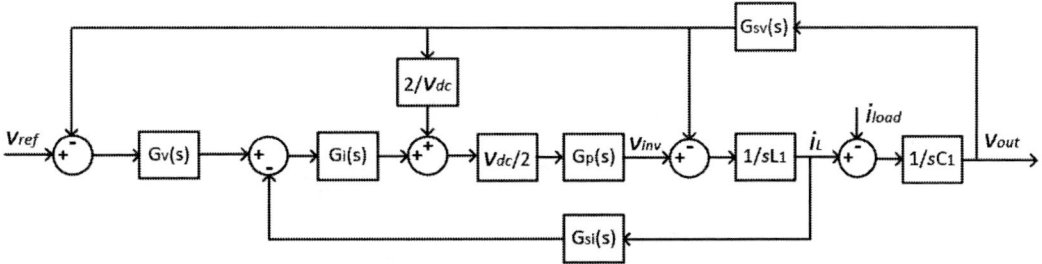

Figure 4. Control block diagram of UPS inverter

frequency formed by the UPS output filtering capacitor, filtering inductor, and line inductance. The output impedance of the inverter shows an underdamped state at this resonance frequency and continuously provides excitation for the resonant system, thus generating this high-frequency resonance.

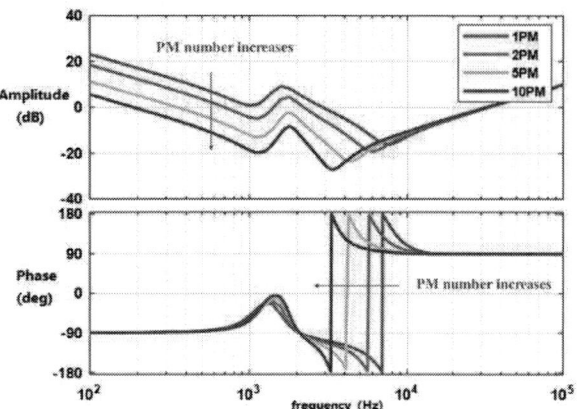

Figure 5. Bode diagram of system impedance with different amount of PM

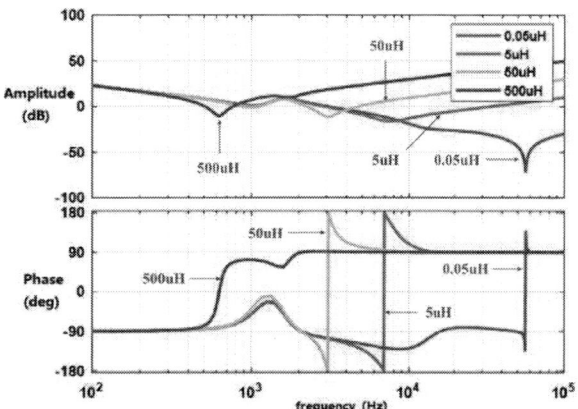

Figure 6. Bode diagram of system impedance with different line impedance

IV. SOLUTION AND VERIFICATION

A. Solution Design

The general approach to suppressing high-frequency resonance is to increase the equivalent impedance of the system, and reducing the gain of the voltage loop controller is one way to lower the equivalent output impedance of the inverter. However, if the overall gain of the voltage loop controller is reduced, it will decrease the bandwidth of the inverter and affect its response characteristics at the fundamental frequency. Therefore, this article adopts a scheme of connecting a voltage loop controller in series with a digital notch filter to achieve the effect of improving the equivalent output impedance of the inverter resonant frequency band.

From the impedance Bode plot, it can also be seen that when the cable length changes (represented by the inductance value in the Bode plot), its line impedance characteristics also change, from underdamped status to overdamped status or zero damped status. At this time, due to the lack of excitation from under damping on the resonant signal, it attenuates.

Therefore, the notch filter designed in this paper only needs to process signals within a certain frequency range, and signals beyond this frequency range will not resonate.

As shown in Figure 7, a notch filter (dashed block) is added before the inverter voltage loop controller to reduce the transmission of voltage components in the resonant frequency band to the current loop, thereby improving the equivalent output impedance of the resonant frequency band. The center frequency of the notch filter is selected as 4.5 kHz, with a depth of -3.68 dB. This design ensures that the notch depth is less than -0.17 dB at 3 kHz, which means the impact on the original controller gain is less than 2%. The transfer function of the notch filter is shown in (6):

$$G_{notch}(s) = \frac{s^2 + 1300\pi \cdot s + (2\pi \cdot 4500)^2}{s^2 + 2000\pi \cdot s + (2\pi \cdot 4500)^2} \tag{6}$$

The Bode plot of the notch filter is shown in Figure 8, where the blue curve represents the open-loop gain of the original voltage loop controller, the orange curve represents the gain curve of the added notch filter, and the yellow curve represents the open-loop gain of the voltage loop controller after connecting the notch filters in series. It can be seen that the notch filter reduces the gain of the voltage loop controller in the 3 k-7 kHz range but has very little effect on the gain of the frequency band below 3 kHz, has no effect on the gain and phase of the power frequency range.

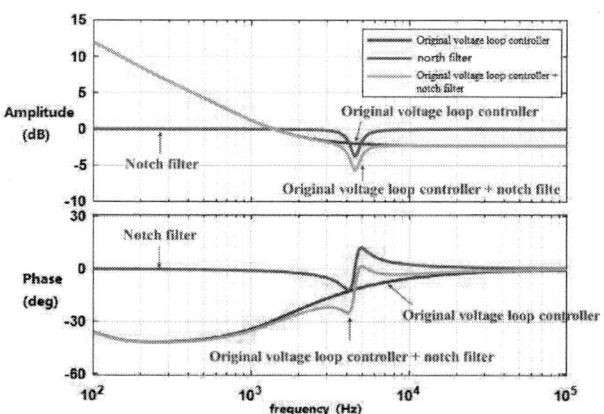

Figure 8. Bode diagram of open loop gain of voltage regulator with and without notch filter

B. Experimental Verification

This article conducted experimental verification of the proposed solution on mass-produced high-power UPS products. Due to concerns that the introduction of notch filters would affect the voltage dynamic response speed, the experiment was divided into two parts:

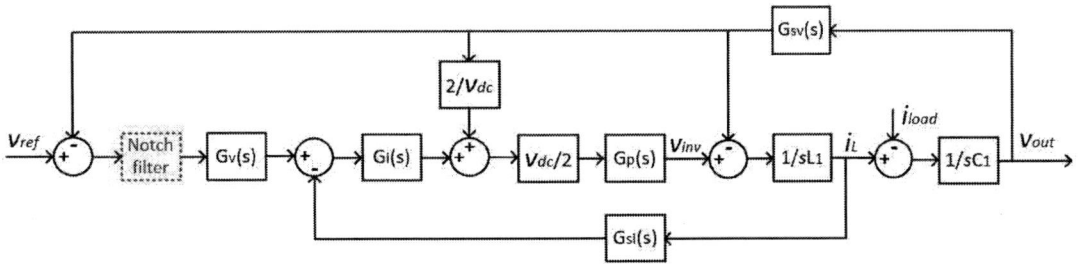

Figure 7. Control block diagram of UPS inverter with notch filter

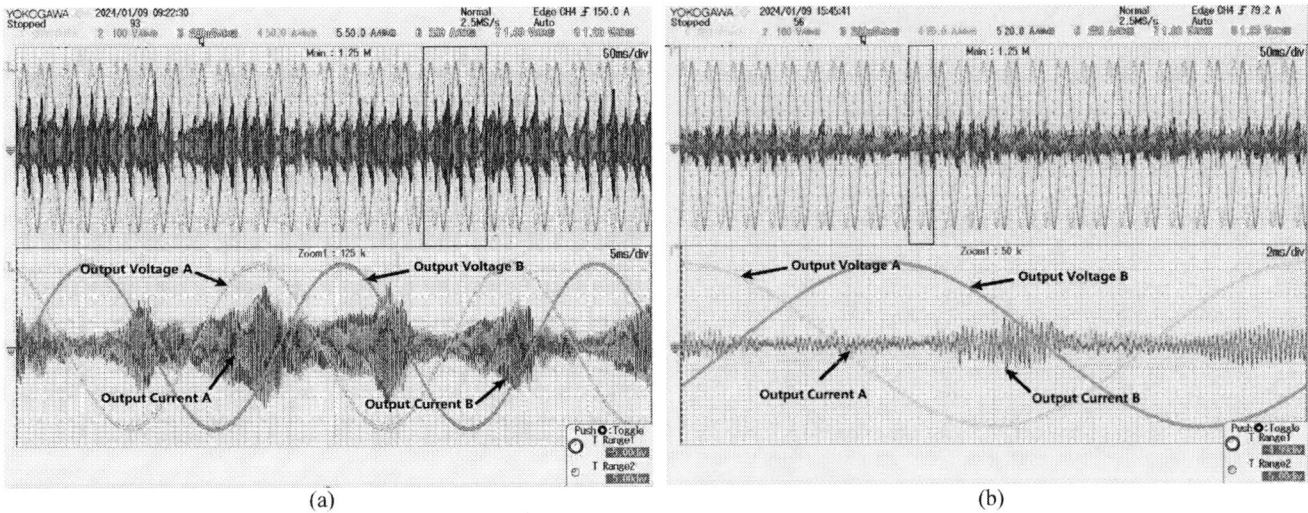

Figure 9. Experimental waveforms of output voltage and current of one of the paralleled UPSs with 1 PM for each (a) before, and (b) after applying the method described in this article

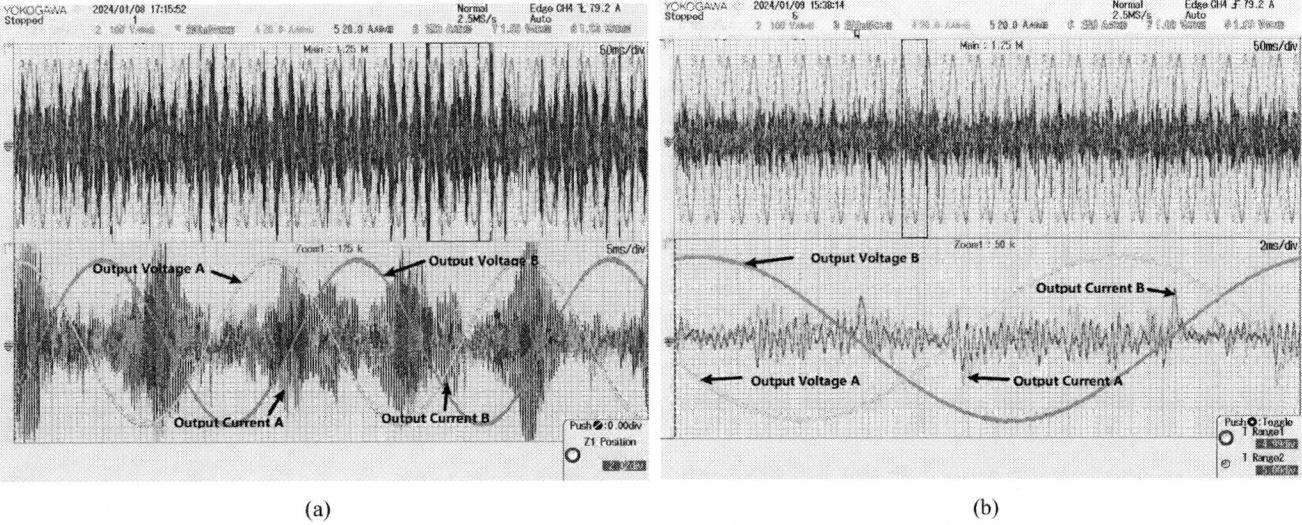

Figure 10. Experimental waveforms of output voltage and current of one of the paralleled UPSs with 5 PMs for each (a) before, and (b) after applying the method described in this article

- Static testing, which tested the output current of two UPS units each equipped with one PM and five PMs.

- Dynamic testing to test whether the output voltage deviation during sudden loading of parallel UPS meets the requirements of IEC standards.

1) Static test

Firstly, each of the two UPS used in the experiment is equipped with one PM. The output voltage and current waveform of one UPS is shown in Figure 9, where Figure 9 (a) shows the output voltage and current waveform of the UPS before applying the proposed scheme. It can be seen that the UPS output current has obvious high-frequency resonance, with a resonance frequency of 6.54 kHz and an effective resonance value of up to 37.36 A; Figure 9 (b) shows the waveform of the UPS output voltage and current after applying the scheme

resonance frequency of 4.75 kHz and an effective resonance value of up to 57.29 A; Figure 10 (b) shows the waveform of the UPS output voltage and current after applying the scheme proposed in this article. It can be seen that the high-frequency resonance is significantly reduced, with an effective value of only 8.45 A.

The UPS output current data from the two tests are shown in Table II. It can be seen that although the current resonance frequency is different due to different PM quantity configurations, the method described in this article has a significant suppression effect on both.

TABLE II. UPS OUTPUT CURRENT DATA BEFORE AND AFTER APPLYING THE PROPOSED METHOD

PM quantity configuration	Current resonance frequency/kHz		Before applying the method described in this article Effective current value/A		After applying the method described in this article Effective current value/A	
	Phase A	B phase	Phase A	B phase	Phase A	B phase
1PM + 1PM	6.6	6.54	29.81	37.36	5.49	5.62
5PM + 5PM	4.8	4.75	45.67	57.29	10.23	8.45

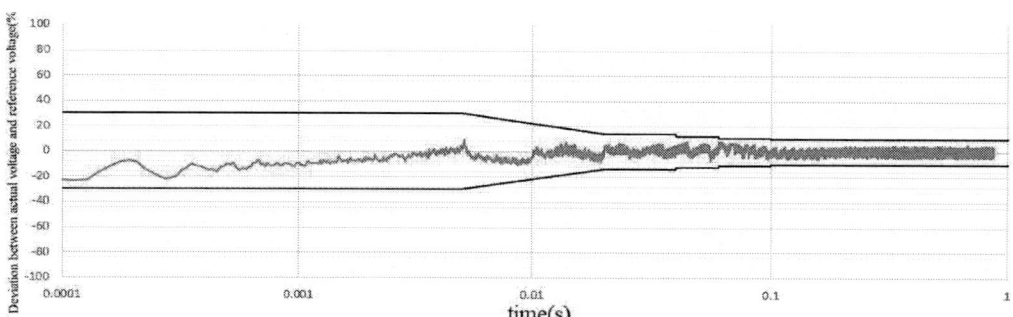

(a) Dynamic fluctuation of output voltage before applying the method described in this article

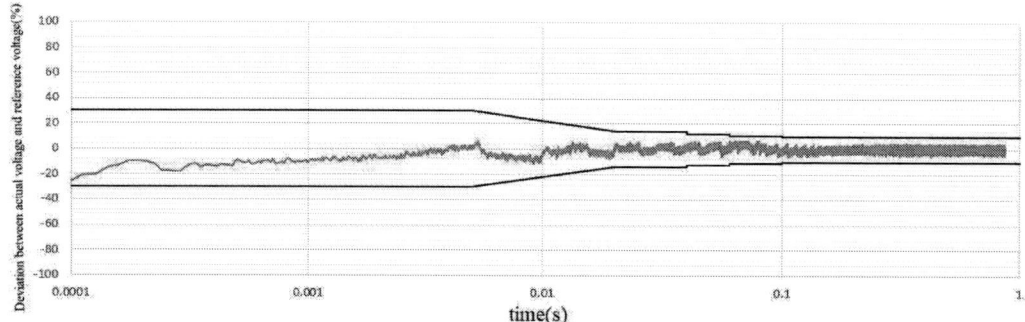

(b) Dynamic fluctuation of output voltage after applying the method described in this article

Figure 11 Experimental waveforms of output voltage dynamics of paralleled UPS with load steps

proposed in this article. It can be seen that the high-frequency resonance is significantly reduced, with an effective value of only 5.62 A.

Then, the two UPS systems used in the experiment were each equipped with 5 PMs. The output voltage and current waveform of one UPS is shown in Figure 10, where Figure 10 (a) shows the output voltage and current waveform of the UPS before applying the proposed scheme. It can be seen that the UPS output current has obvious high-frequency resonance, with a

2) Dynamic test

In order to verify the effect of the proposed scheme on voltage dynamic response, this paper conducted a sudden load test on the scheme, with a load rate step from 0 to 100%. The deviation between the measured voltage and the reference voltage is shown in Figure 11, where Figure 11 (a) reflects the dynamic fluctuation of the UPS output voltage before applying the proposed scheme, and Figure 11 (b) reflects the dynamic fluctuation of the UPS output voltage after applying the proposed scheme. It can be seen that the proposed scheme has little effect on voltage dynamic response, and the voltage fluctuation during the sudden load process always remains within the Class I range specified in the IEC 62040-3 standard [15] (within the black line boundary shown in Figure 11).

V. CONCLUSION

This article establishes and analyzes an impedance model for the high-frequency resonance problem generated during parallel operation of online UPS and proposes a method to suppress it by applying a digital notch filter in series with the voltage loop controller. The experiment results prove that this method effectively suppresses high-frequency resonance without affecting the voltage dynamic response. The solution described in this article can provide reference for similar problems encountered in the development of paralleled power electronic converters.

REFERENCES

[1] Y. Chen, K. Shi, M. Chen, and D. Xu, "Data Center Power Supply Systems from Grid Edge to Point-of-Load," IEEE Journal of Emerging and Selected Topics in Power Electronics, vol. 11, no. 3, pp. 2441–2456, 2023.

[2] A. Lahyani, P. Venet, and A. Guermazi, "Battery/supercapacitors combination in uninterruptible power supply (UPS)," IEEE Transactions on Power Electronics, vol. 28, no. 4, pp. 1509–1522, 2013.

[3] Y. Zhang, M. Yu, F. Liu, and Y. Kang, "Instantaneous current-sharing control strategy for parallel operation of UPS modules using virtual impedance," IEEE Transactions on Power Electronics, vol. 28, no. 1, pp. 432–440, 2013.

[4] B. Zhao, Q. Song, W. Liu, and Y. Xiao, "Next-generation multi-functional modular intelligent UPS system for smart grid ," IEEE Transactions on Industrial Electronics, vol. 60, no. 9, pp. 3602–3618, 2013.

[5] C. G. C. Branco, R. P. Torrico-Bascope, C. M. T. Cruz, and F. K. D. A. Lima, "Proposal of three-phase high-frequency transformer isolation UPS topologies for distributed generation applications," IEEE Transactions on Industrial Electronics, vol. 60, no. 4, pp. 1520–1531, 2013.

[6] W. Solter, "A new international UPS classification by IEC 62040-3," In Proceedings of IEEE Telecommunications Energy Conference, Montreal, QC, Canada, 2018, pp. 541–545.

[7] S. Karve, "Three of a kind [UPS topologies, IEC standard],". IEE Review, vol. 46, no. 2, pp. 27–31, 2000.

[8] S. B. Bekiarov, and A. Emadi, "Uninterruptible power supplies: Classification, operation, dynamics, and control," In Proceedings of IEEE Applied Power Electronics Conference and Exposition, Dallas, TX, 2002, pp. 597–604.

[9] J. Hu, and H. Ma, "Synchronization of the carrier wave of parallel three-phase inverters with virtual oscillator control," IEEE Transactions on Power Electronics, vol. 32, no. 10, pp. 7998–8007, 2017.

[10] T. Wu, Z. Liu, J. Liu, S. Wang, and Z. You, "A Unified Virtual Power Decoupling Method for Droop-Controlled Parallel Inverters in Microgrids," IEEE Transactions on Power Electronics, vol. 31, no. 8, pp. 5587–5603, 2016.

[11] C. Xie, K. Li, J. Zou, and J. M. Guerrero, "Passivity-based stabilization of lcl-type grid-connected inverters via a general admittance model," IEEE Transactions on Power Electronics, vol. 35, no. 6, pp. 6636–6648, 2020.

[12] X. Dou, Y. Jiao, K. Yang, Z. Wu, W. Gu, and H. Li, "An optimal grid current control strategy with grid voltage observer (gvo) for LCL-filtered grid-connected inverters," IEEJ Transactions on Electrical and Electronic Engineering, vol. 13, no. 5, pp. 777–784, 2018.

[13] J. R. Massing, M. Stefanello, H. A. Grundling, and H. Pinheiro, "Adaptive current control for grid-connected converters with LCL filter," IEEE Transactions on Industrial Electronics, vol. 59, no. 12, pp. 4681–4693, 2012.

[14] J. H. Choi, J. M. Kwon, J. H. Jung, and B. H. Kwon, "High-performance online UPS using three-leg type converter," IEEE Transactions on Industrial Electronics, vol. 52, no. 13, pp. 889–897, 2005.

[15] International Standard IEC 62040-3, Uninterruptible Power Systems (UPS) Part 3: Method of Specifying the Performance and Test Requirements.

An Extra-Element Small-Signal Model for a Current-Fed Resonant Dual-Active-Bridge Converter

Paolo Sbabo*, Paolo Mattavelli†, Giorgio Spiazzi*, and Andrea Petucco†
*Department of Information Engineering (DEI) – University of Padova, Italy
†Department of Management and Engineering (DTG) – University of Padova, Italy

Abstract—This paper proposes a novel small-signal modeling method for the Current-Fed Resonant Dual-Active-Bridge (CF-RDAB) converter, which enables to strongly simplify the design of the regulators in the system. Such an approach divides the Small-Signal Model (SSM) of the CF-RDAB into the SSM of an Interleaved Boost (IB) converter and the SSM of a Resonant Dual-Active-Bridge (RDAB) converter, enabling to control them separately. Additionally, this article exposes a specific operating point in which the SSM of the CF-RDAB reduces to the SSM of the cascaded connection of the IB converter and the RDAB converter, as they were fully independent. In this operating condition, the stability of the whole CF-RDAB converter can be studied on the IB+RDAB cascaded connection with the Impedance-Based Stability Criterion (IBSC) and the Extra-Element Theorem (EET). Afterwards, this paper presents a control strategy that utilizes both the input duty-cycle and the phase-shift of the CF-RDAB to guarantee the stability of the system. The converter's model and control are experimentally verified on a 500W CF-RDAB prototype, displaying an excellent agreement between the theory and the measurements.

Index Terms—Power Electronics, Switching Converters, Resonant Converters, Dual Active Bridge, Stability, Small-Signal Modeling

I. INTRODUCTION

IN the transition towards a more sustainable power generation, a pivotal role is played by smartgrids [1], which integrates power generation, energy storage systems, residential houses, telecommunication infrastructures and different types of loads [2]. The evolution of smart grids is supported by the latest advancements and the high integration of power electronic devices [3]. One of these advancements is the use of high-frequency transformers for the realization of bidirectional power converters [4], enabling to efficiently exchange energy between the input port and the output port of the converter.

One example of bidirectional power converter is the single-phase Dual-Active-Bridge (DAB) converter used as On-Board Charger (OBC) [5]. Such a topology is one of the most popular bidirectional power converters because of its high efficiency at high power and simple control. Another example of OBC is the LLC resonant converter present in the Kostal 4th Gen. [6]. Even though the LLC is a unidirectional converter, such a topology is widely used for its Zero-Voltage Switching (ZVS) capabilities, thus increasing the efficiency of the system.

This study was carried out within the MOST–Sustainable Mobility Center and received funding from the European Union Next- GenerationEU (PIANO NAZIONALE DI RIPRESA E RESILIENZA (PNRR)–MISSIONE 4 COMPONENTE 2, INVESTIMENTO 1.4 – D.D. 1033 17/06/2022, CN00000023).

The bidirectionality of the DAB converter and the extended ZVS capabilities of the LLC resonant converter are both present in the RDAB converter, which achieves high efficiency and power density [7].

In contrast, when charging/discharging a battery, a continuous and controlled battery current is highly desirable, reducing the battery stress and optimizing its charge/discharge. This is achieved by means of current-fed topologies, like the Current-Fed Dual-Active-Bridge (CF-DAB) [8], the Bidirectional Isolated Boost with Coupled Inductors (BIBCI) [9], [10] and the CF-RDAB [11], [12].

This paper delves into the CF-RDAB converter by developing its comprehensive SSM with the generalized average method [13]. Such an approach performs the first-harmonic approximation to include the dynamics of the resonant tank and has been widely employed for the analysis of power converters' dynamics, like in the series and parallel resonant converters [14], [15], the LLC resonant converter [16], [17], and the RDAB converter [18]–[20].

In [21], the authors proposed a generalized average SSM of the CF-RDAB by neglecting the transformer's magnetizing inductance and the control of the duty-cycles. Such a model is here extended to include the magnetizing inductance and by considering all the available degrees of freedom: the input inverter's duty-cycle, the output inverter's duty-cycle, the phase-shift, and the switching frequency.

After having obtained the SSM of the CF-RDAB converter, this article demonstrates that, when the input inverter operates with 50% duty-cycle, the SSM of the CF-RDAB reduces to the SSM of the cascaded connection of an IB converter and an RDAB converter. In this specific condition the design of the regulators in the system is strongly simplified, with separate controls for the IB and the RDAB converters. Additionally, in this operating point, the stability of the IB+RDAB connection can be studied with the IBSC [22] and the EET [23], [24], despite them being integrated inside the CF-RDAB converter.

This paper is organized as follows: in Section II the SSM of the IB and the RDAB converters are derived. Section III combines the SSMs of the IB and the RDAB converters to build the SSM of the CF-RDAB converter. In Section IV the stability of the CF-RDAB converter is studied with the IBSC and the EET. Section V provides the experimental validation of the proposed circuit model and its control. Finally, Section VI summarizes the presented theory and draws the conclusions.

979-8-3315-1612-3/25 $31.00 © 2025 IEEE

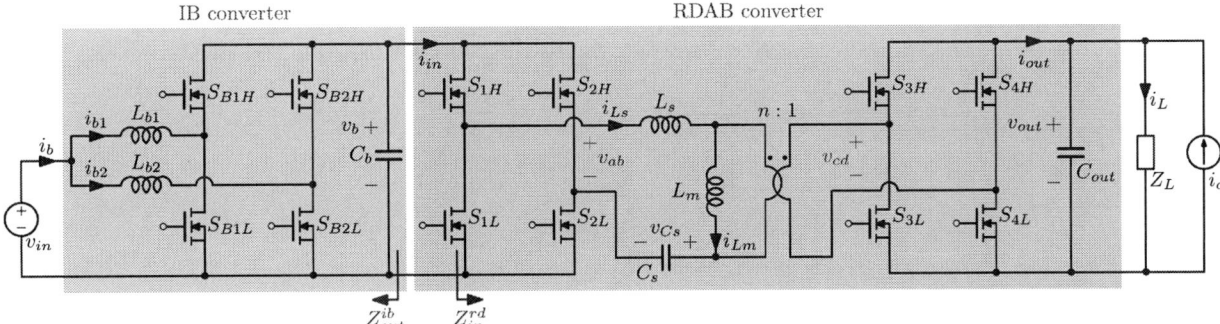

Fig. 1: Cascaded connection of an IB converter and an RDAB converter. In the figure are also present with arrows the sections for the calculation of the output impedance of the IB Z_{out}^{ib}, and the input impedance of the RDAB Z_{in}^{rd}.

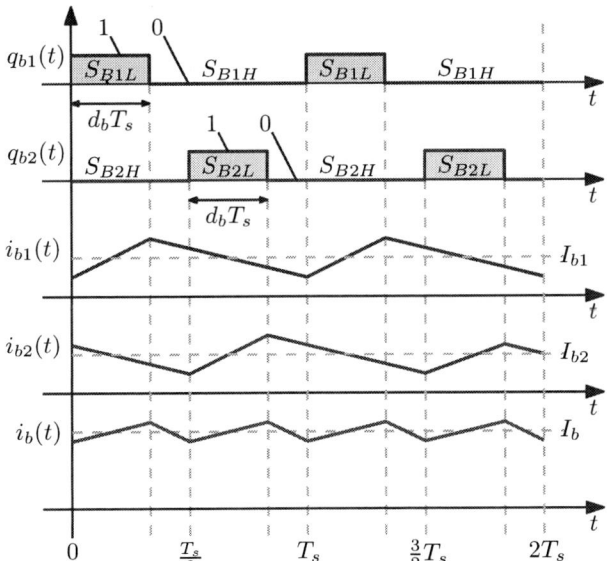

Fig. 2: Waveforms of the IB converter.

II. IB CONVERTER + RDAB CONVERTER

Fig. 1 shows the cascaded connection of an IB converter and an RDAB converter. Being the two converters fully independent, they can be analyzed separately. The aim of this section is to obtain their SSMs, with the goal of addressing the stability of the connection.

A. Small-signal modeling of the IB converter

Typical waveforms of the IB converter are displayed in **Fig. 2**, where d_b is the duty-cycle of the boost converter and the switching functions q_{b1} and q_{b2} are defined as:

$$
q_{b1} = \begin{cases} 1 & \text{if } S_{B1L} = 1 \text{ and } S_{B1H} = 0 \\ 0 & \text{if } S_{B1L} = 0 \text{ and } S_{B1H} = 1 \end{cases}
$$
$$
q_{b2} = \begin{cases} 1 & \text{if } S_{B2L} = 1 \text{ and } S_{B2H} = 0 \\ 0 & \text{if } S_{B2L} = 0 \text{ and } S_{B2H} = 1. \end{cases}
\tag{1}
$$

The differential equations of the boost are derived from **Fig. 1** as:

$$
L_{b1}\frac{di_{b1}}{dt} = v_{in} - (1 - q_{b1})v_b
$$
$$
L_{b2}\frac{di_{b2}}{dt} = v_{in} - (1 - q_{b2})v_b
\tag{2}
$$
$$
C_b\frac{dv_b}{dt} = i_{b1}(1 - q_{b1}) + i_{b2}(1 - q_{b2}) - i_{in},
$$

where $i_{in} = i_{Ls}(v_{ab}/v_b)$ is the input current of the RDAB.

By averaging (2) over a switching period, the average model of the IB converter is obtained as:

$$
L_{b1}\frac{d\bar{i}_{b1}}{dt} = \bar{v}_{in} - (1 - d_b)\bar{v}_b
$$
$$
L_{b2}\frac{d\bar{i}_{b2}}{dt} = \bar{v}_{in} - (1 - d_b)\bar{v}_b
\tag{3}
$$
$$
C_b\frac{d\bar{v}_b}{dt} = \bar{i}_{b1}(1 - d_b) + \bar{i}_{b2}(1 - d_b) - \bar{i}_{in},
$$

where d_b is the average duty-cycle of the boost, $\bar{v}_{in}$ is the average input voltage, $\bar{v}_b$ is the average output voltage, $\bar{i}_{b1}$ and $\bar{i}_{b2}$ are the average currents of the inductors and $\bar{i}_{in}$ is the average input current of the RDAB.

The SSM of the IB converter is obtained from (3) by applying the perturbation and linearization method [25], where each average value $\bar{x}$ is divided into the steady-state value of the average X and the small-signal variation of the average $\hat{x}$. The resulting state-space model is:

$$
\begin{cases} \dot{\hat{x}}_{ib} = A_{ib} \cdot \hat{x}_{ib} + B_{ib} \cdot \hat{u}_{ib} \\ \hat{y}_{ib} = C_{ib} \cdot \hat{x}_{ib} + D_{ib} \cdot \hat{u}_{ib}, \end{cases}
\tag{4}
$$

where the state vector is $\hat{x}_{ib} = (\hat{i}_b \; \hat{v}_b)^T$, the input vector is $\hat{u}_{ib} = (\hat{v}_{in} \; \hat{d}_b \; \hat{i}_{in})^T$ and the output vector is $\hat{y}_{ib} = (\hat{i}_b \; \hat{v}_b)^T$. While the matrices present in (4) are defined as:

$$
A_{ib} = \begin{pmatrix} 0 & -\frac{1-D_b}{L_b} \\ \frac{1-D_b}{C_b} & 0 \end{pmatrix}
$$
$$
B_{ib} = \begin{pmatrix} \frac{1}{L_b} & \frac{V_b}{L_b} & 0 \\ 0 & -\frac{I_b}{C_b} & -\frac{1}{C_b} \end{pmatrix}
\tag{5}
$$
$$
C_{ib} = \begin{pmatrix} 1 & 0 \\ 0 & 1 \end{pmatrix}
$$
$$
D_{ib} = 0_{2\times 3}.
$$

979-8-3315-1612-3/25 $31.00 © 2025 IEEE 2668

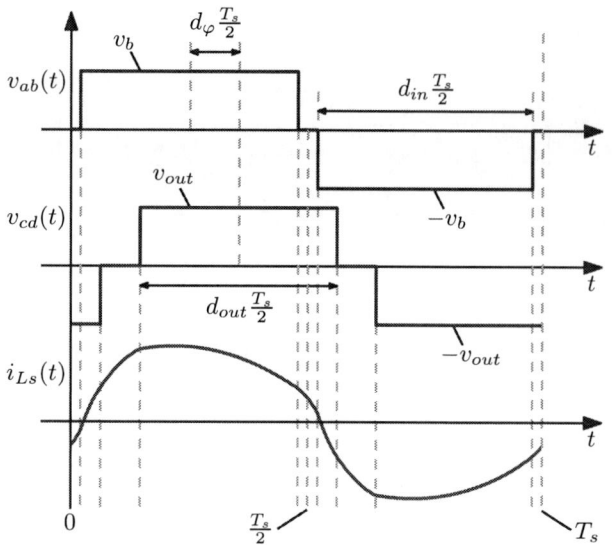

Fig. 3: Waveforms of the RDAB converter.

As can be observed in (4) and (5), the average state-space description of the IB converter considers the current $i_b = i_{b1} + i_{b2}$ as state variable of the system. This is achieved by defining the equivalent boost inductance $L_b = L_{b1}/2 = L_{b2}/2$.

Finally, the open-loop transfer functions of the boost converter are derived as:

$$
\frac{Y_{ib}(s)}{U_{ib}(s)} = \begin{pmatrix} G_{11}^{ib} & G_{12}^{ib} & G_{13}^{ib} \\ G_{21}^{ib} & G_{22}^{ib} & G_{23}^{ib} \end{pmatrix} = \tag{6}
$$
$$
= C_{ib} \cdot (s \cdot I_2 - A_{ib})^{-1} \cdot B_{ib} + D_{ib},
$$

where $Y_{ib}(s)$ and $U_{ib}(s)$ are the Laplace transformations of the outputs and the inputs of the state-space model of the IB converter, and I_2 is the 2×2 identity matrix.

B. Small-signal modeling of the RDAB converter

The SSM of the RDAB converter is derived by using the generalized average method [13]. In such an approach the variables in the resonant tank (v_{ab}, v_{cd}, i_{Ls}, v_{Cs} and i_{Lm}) are approximated with their fundamental harmonics, while outside the resonant tank the variables i_{in}, i_{out} and v_{out} are approximated with their average values over a switching period.

Typical waveforms of the RDAB converter are displayed in **Fig. 3**, where the input duty-ratio d_{in}, the output duty-ratio d_{out} and the normalized phase-shift d_φ are defined.

The differential equations of the RDAB converter are derived from **Fig. 1** as:

$$
L_s \frac{di_{Ls}}{dt} = v_{ab} - n\, v_{cd} - v_{Cs}
$$
$$
C_s \frac{dv_{Cs}}{dt} = i_{Ls}
$$
$$
L_m \frac{di_{Lm}}{dt} = n\, v_{cd} \tag{7}
$$
$$
C_{out} \frac{dv_{out}}{dt} = i_{out} - i_L + i_o.
$$

Then, in (7), the terms v_{ab}, v_{cd}, i_{Ls}, v_{Cs} and i_{Lm} are approximated with their fundamental harmonics as:

$$
v_{ab} \simeq v_b \frac{4}{\pi} \sin\left(\frac{\pi}{2} d_{in}\right) \sin(\omega_s t)
$$
$$
v_{cd} \simeq v_{out} \frac{4}{\pi} \sin\left(\frac{\pi}{2} d_{out}\right) [\sin(\omega_s t) \cos(\pi d_\varphi)
$$
$$
- \cos(\omega_s t) \sin(\pi d_\varphi)] \tag{8}
$$
$$
i_{Ls} \simeq i_{Ls,\alpha} \cos(\omega_s t) + i_{Ls,\beta} \sin(\omega_s t)
$$
$$
v_{Cs} \simeq v_{Cs,\alpha} \cos(\omega_s t) + v_{Cs,\beta} \sin(\omega_s t)
$$
$$
i_{Lm} \simeq i_{Lm,\alpha} \cos(\omega_s t) + i_{Lm,\beta} \sin(\omega_s t),
$$

where ω_s is the switching frequency of the RDAB converter (in rad/s), while $i_{Ls,\alpha}$, $i_{Ls,\beta}$, $v_{Cs,\alpha}$, $v_{Cs,\beta}$, $i_{Lm,\alpha}$ and $i_{Lm,\beta}$ are the cosine (alpha) and sine (beta) components of the state variables i_{Ls}, v_{Cs} and i_{Lm}.

Additionally, in (7), the current i_{out} is approximated with its average value over a switching period as:

$$
i_{out} \simeq \frac{2}{\pi} n \sin\left(\frac{\pi}{2} d_{out}\right) \left[(i_{Ls,\beta} - i_{Lm,\beta}) \cos(\pi d_\varphi) \right.
$$
$$
\left. - (i_{Ls,\alpha} - i_{Lm,\alpha}) \sin(\pi d_\varphi) \right]. \tag{9}
$$

Then, after inserting (8) and (9) in (7), it is possible to equate cosine/sine terms, thus exploiting the concept of harmonic balance. This process yields to the following non-linear differential equations:

$$
\frac{di_{Ls,\alpha}}{dt} = \frac{1}{L_s} \left[-v_{Cs,\alpha} + n\, v_{out} \frac{4}{\pi} \sin\left(\frac{\pi}{2} d_{out}\right) \sin(\pi d_\varphi) \right]
$$
$$
- \omega_s i_{Ls,\beta}
$$
$$
\frac{di_{Ls,\beta}}{dt} = \frac{1}{L_s} \left[-v_{Cs,\beta} - n\, v_{out} \frac{4}{\pi} \sin\left(\frac{\pi}{2} d_{out}\right) \cos(\pi d_\varphi) \right.
$$
$$
\left. + v_b \frac{4}{\pi} \sin\left(\frac{\pi}{2} d_{in}\right) \right] + \omega_s i_{Ls,\alpha}
$$
$$
\frac{dv_{Cs,\alpha}}{dt} = \frac{i_{Ls,\alpha}}{C_s} - \omega_s v_{Cs,\beta}
$$
$$
\frac{dv_{Cs,\beta}}{dt} = \frac{i_{Ls,\beta}}{C_s} + \omega_s v_{Cs,\alpha}
$$
$$
\frac{di_{Lm,\alpha}}{dt} = -\frac{1}{L_m} n\, v_{out} \frac{4}{\pi} \sin\left(\frac{\pi}{2} d_{out}\right) \sin(\pi d_\varphi) - \omega_s i_{Lm,\beta}
$$
$$
\frac{di_{Lm,\beta}}{dt} = \frac{1}{L_m} n\, v_{out} \frac{4}{\pi} \sin\left(\frac{\pi}{2} d_{out}\right) \cos(\pi d_\varphi) + \omega_s i_{Lm,\alpha}
$$
$$
\frac{dv_{out}}{dt} = \frac{1}{C_{out}} \left\{ \frac{2}{\pi} n \sin\left(\frac{\pi}{2} d_{out}\right) \left[(i_{Ls,\beta} - i_{Lm,\beta}) \cos(\pi d_\varphi) \right. \right.
$$
$$
\left. \left. - (i_{Ls,\alpha} - i_{Lm,\alpha}) \sin(\pi d_\varphi) \right] - i_L + i_o \right\}. \tag{10}
$$

Additionally, for input-port characterization, the input current i_{in} is approximated with its average value over a switching period:

979-8-3315-1612-3/25 $31.00 © 2025 IEEE

$$i_{in} \simeq i_{Ls,\beta} \frac{2}{\pi} \sin\left(\frac{\pi}{2} d_{in}\right). \tag{11}$$

Finally, by applying the perturbation-and-linearization process, the SSM of the RDAB converter is obtained as:

$$\begin{cases} \dot{\hat{x}}_{rd} = A_{rd} \cdot \hat{x}_{rd} + B_{rd} \cdot \hat{u}_{rd} \\ \hat{y}_{rd} = C_{rd} \cdot \hat{x}_{rd} + D_{rd} \cdot \hat{u}_{rd}, \end{cases} \tag{12}$$

where the state vector is $\hat{x}_{rd} = (\hat{i}_{Ls,\alpha} \; \hat{i}_{Ls,\beta} \; \hat{v}_{Cs,\alpha} \; \hat{v}_{Cs,\beta} \; \hat{i}_{Lm,\alpha} \; \hat{i}_{Lm,\beta} \; \hat{v}_{out})^T$, the input vector is $\hat{u}_{rd} = (\hat{d}_{in} \; \hat{d}_{out} \; \hat{d}_\varphi \; \hat{\omega}_s \; \hat{v}_b \; \hat{i}_o)^T$ and the output vector is $\hat{y}_{rd} = (\hat{v}_{out} \; \hat{i}_{in})^T$. While the matrices present in (12) are showcased in (14) where the variables $I_{Ls,\alpha}$, $I_{Ls,\beta}$, $V_{Cs,\alpha}$, $V_{Cs,\beta}$, $I_{Lm,\alpha}$, $I_{Lm,\beta}$, V_{out}, D_{in}, D_{out}, D_φ, Ω_s and V_b are

the steady-state values of the quantities present in (10), and the equivalent resistance R_L is the differential resistance of the load (V_{out}/I_L for resistive load, 0 for constant current load, and $-V_{out}/I_L$ for constant power load).

From **Fig. 12**, the transfer functions of the RDAB are obtained as:

$$\frac{Y_{rd}(s)}{U_{rd}(s)} = \begin{pmatrix} G_{11}^{rd} & G_{12}^{rd} & G_{13}^{rd} & G_{14}^{rd} & G_{15}^{rd} & G_{16}^{rd} \\ G_{21}^{rd} & G_{22}^{rd} & G_{23}^{rd} & G_{24}^{rd} & G_{25}^{rd} & G_{26}^{rd} \end{pmatrix} = \tag{13}$$
$$= C_{rd} \cdot (s \cdot I_7 - A_{rd})^{-1} \cdot B_{rd} + D_{rd},$$

where $Y_{rd}(s)$ and $U_{rd}(s)$ are the Laplace transformations of the outputs and the inputs of the state-space model of the RDAB converter, and I_7 is the 7×7 identity matrix.

$$A_{rd} = \begin{pmatrix}
0 & -\Omega_s & \frac{-1}{L_s} & 0 & 0 & 0 & \frac{4n\sin\left(\frac{\pi}{2}D_{out}\right)\sin(\pi D_\varphi)}{\pi L_s} \\
\Omega_s & 0 & 0 & \frac{-1}{L_s} & 0 & 0 & \frac{4n\sin\left(\frac{\pi}{2}D_{out}\right)\cos(\pi D_\varphi)}{-\pi L_s} \\
\frac{1}{C_s} & 0 & 0 & -\Omega_s & 0 & 0 & 0 \\
0 & \frac{1}{C_s} & \Omega_s & 0 & 0 & 0 & 0 \\
0 & 0 & 0 & 0 & 0 & -\Omega_s & \frac{4n\sin\left(\frac{\pi}{2}D_{out}\right)\sin(\pi D_\varphi)}{-\pi L_m} \\
0 & 0 & 0 & 0 & \Omega_s & 0 & \frac{4n\sin\left(\frac{\pi}{2}D_{out}\right)\cos(\pi D_\varphi)}{\pi L_m} \\
\frac{2n\sin(\frac{\pi}{2}D_{out})\sin(\pi D_\varphi)}{-\pi C_{out}} & \frac{2n\sin(\frac{\pi}{2}D_{out})\cos(\pi D_\varphi)}{\pi C_{out}} & 0 & 0 & \frac{2n\sin(\frac{\pi}{2}D_{out})\sin(\pi D_\varphi)}{\pi C_{out}} & \frac{2n\sin(\frac{\pi}{2}D_{out})\cos(\pi D_\varphi)}{-\pi C_{out}} & \frac{-1}{R_L C_{out}}
\end{pmatrix}$$

$$B_{rd} = \begin{pmatrix}
0 & \frac{2nV_{out}\cos(\frac{\pi}{2}D_{out})\sin(\pi D_\varphi)}{L_s} & \frac{4nV_{out}\sin(\frac{\pi}{2}D_{out})\cos(\pi D_\varphi)}{L_s} & -I_{Ls,\beta} & 0 & 0 \\
\frac{2V_b\cos(\frac{\pi}{2}D_{in})}{L_s} & \frac{2nV_{out}\cos(\frac{\pi}{2}D_{out})\cos(\pi D_\varphi)}{-L_s} & \frac{4nV_{out}\sin(\frac{\pi}{2}D_{out})\sin(\pi D_\varphi)}{L_s} & I_{Ls,\alpha} & \frac{4\sin(\frac{\pi}{2}D_{in})}{\pi L_s} & 0 \\
0 & 0 & 0 & -V_{Cs,\beta} & 0 & 0 \\
0 & 0 & 0 & V_{Cs,\alpha} & 0 & 0 \\
0 & \frac{2nV_{out}\cos(\frac{\pi}{2}D_{out})\sin(\pi D_\varphi)}{-L_m} & \frac{4nV_{out}\sin(\frac{\pi}{2}D_{out})\cos(\pi D_\varphi)}{-L_m} & -I_{Lm,\beta} & 0 & 0 \\
0 & \frac{2nV_{out}\cos(\frac{\pi}{2}D_{out})\cos(\pi D_\varphi)}{L_m} & \frac{4nV_{out}\sin(\frac{\pi}{2}D_{out})\sin(\pi D_\varphi)}{-L_m} & I_{Lm,\alpha} & 0 & 0 \\
0 & \frac{n\cos(\frac{\pi}{2}D_{out})(I_{Ls,\beta}-I_{Lm,\beta})\cos(\pi D_\varphi)}{-C_{out}} - \frac{n\cos(\frac{\pi}{2}D_{out})(I_{Ls,\alpha}-I_{Lm,\alpha})\sin(\pi D_\varphi)}{C_{out}} & \frac{2n\sin(\frac{\pi}{2}D_{out})(I_{Ls,\beta}-I_{Lm,\beta})\sin(\pi D_\varphi)}{-C_{out}} + \frac{2n\sin(\frac{\pi}{2}D_{out})(I_{Ls,\alpha}-I_{Lm,\alpha})\cos(\pi D_\varphi)}{-C_{out}} & 0 & 0 & \frac{1}{C_{out}}
\end{pmatrix}$$

$$C_{rd} = \begin{pmatrix} 0 & 0 & 0 & 0 & 0 & 0 & 1 \\ 0 & \frac{2}{\pi}\sin(\frac{\pi}{2}D_{in}) & 0 & 0 & 0 & 0 & 0 \end{pmatrix} \qquad D_{rd} = \begin{pmatrix} 0 & 0 & 0 & 0 & 0 & 0 \\ I_{Ls,\beta}\cos(\frac{\pi}{2}D_{in}) & 0 & 0 & 0 & 0 & 0 \end{pmatrix}.$$

$$\tag{14}$$

III. SSM OF THE CF-RDAB CONVERTER

Fig. 4 shows the CF-RDAB converter obtained by the integration of the IB converter inside the RDAB converter. Considering the new switching functions q_1 and q_2 as:

$$q_1 = \begin{cases} 1 & \text{if } S_{1L} = 1 \text{ and } S_{1H} = 0 \\ 0 & \text{if } S_{1L} = 0 \text{ and } S_{1H} = 1 \end{cases}$$
$$q_2 = \begin{cases} 1 & \text{if } S_{2L} = 1 \text{ and } S_{2H} = 0 \\ 0 & \text{if } S_{2L} = 0 \text{ and } S_{2H} = 1, \end{cases} \tag{15}$$

the differential equations of the circuit in **Fig. 4** are:

$$L_{b1}\frac{di_{b1}}{dt} = v_{in} - (1-q_1)v_b$$
$$L_{b2}\frac{di_{b2}}{dt} = v_{in} - (1-q_2)v_b$$
$$C_b\frac{dv_b}{dt} = i_{b1}(1-q_1) + i_{b2}(1-q_2) - i_{in}$$
$$L_s\frac{di_{Ls}}{dt} = v_{ab} - n\,v_{cd} - v_{Cs}$$
$$C_s\frac{dv_{Cs}}{dt} = i_{Ls} \tag{16}$$
$$L_m\frac{di_{Lm}}{dt} = n\,v_{cd}$$
$$C_{out}\frac{dv_{out}}{dt} = i_{out} - i_L + i_o,$$

where $i_{in} = i_{Ls}(v_{ab}/v_b)$.

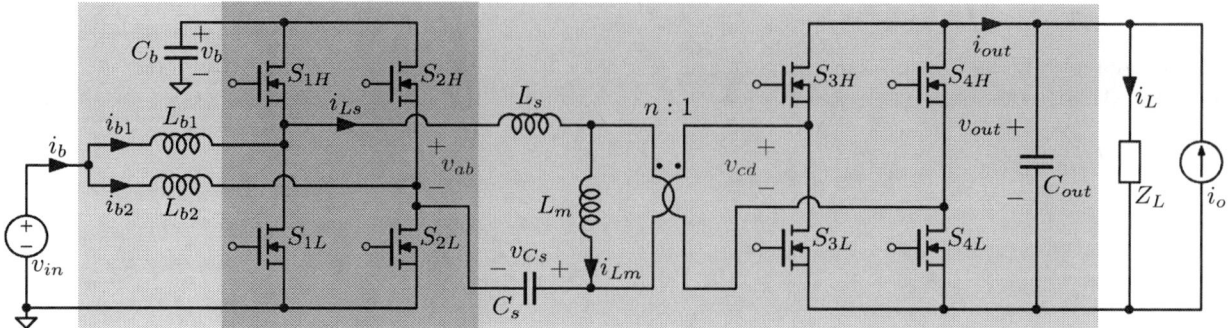

Fig. 4: CF-RDAB converter obtained by the integration of an IB converter within a RDAB converter.

Comparing (16) with (2) and (7), it is evident that the differential equations of the RDAB converter mirror the differential equations of the IB+RDAB cascaded connection. The only difference between the two models is the input voltage of the resonant tank v_{ab} which depends now on the boost's switching functions q_1 and q_2:

$$v_{ab} = (q_2 - q_1)v_b. \tag{17}$$

Repeating the averaging process and the perturbation and linearization process, the SSM of the CF-RDAB results the same as the IB+RDAB cascaded connection, where the duty-cycle d_b of the IB also determines the duty-ratio d_{in} of the input inverter of the RDAB. Such a concept is summarized in **Fig. 5**, where a representation of the CF-RDAB SSM is shown.

In **Fig. 5** the transfer function G_d from $\hat{d}_b$ to $\hat{d}_{in}$ can be derived by considering the relationship between d_b and d_{in}:

$$d_{in} = \begin{cases} 2\,d_b & \text{if } d_b < 0.5 \\ 2\,(1 - d_b) & \text{if } d_b > 0.5. \end{cases} \tag{18}$$

As depicted in **Fig. 5**, the CF-RDAB SSM differs from the SSM of the IB+RDAB cascaded connection by means of a feedforward path that links the duty-cycle of the IB to the duty-ratio of the RDAB.

However, by deriving the transfer functions in (13), it is found that, **when the boost's steady-state duty-cycle D_b is 50%, the transfer functions G_{11}^{rd} and G_{21}^{rd} are zero. This can be easily demonstrated by observing that any partial derivative of (10) and (11) with respect to d_{in} becomes zero when $D_{in} = 1$, or equivalently when $D_b = 50\%$. As a result, this eliminates the additional feedforward path, making the SSM of the CF-RDAB identical to the SSM of the IB+RDAB cascaded connection.**

In this specific condition, the SSM of the IB and the SSM of the RDAB are independent, enabling the development of two separate control schemes: one for the IB and one for the RDAB. Afterwards, the stability of the connection can be addressed by exploiting the IBSC [22] and the EET [23], [24] to get insight about the voltage loop gain of the RDAB converter.

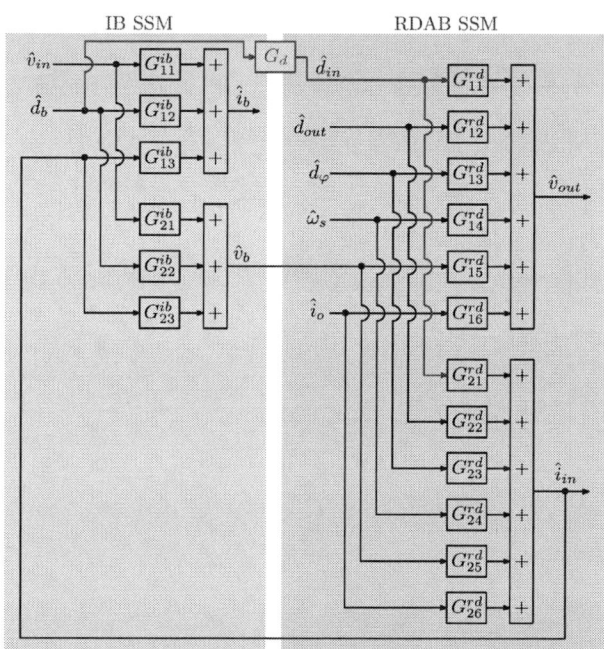

Fig. 5: SSM of the CF-RDAB converter, obtained by linking the SSMs of an IB converter (red), and the SSM of a RDAB converter (blue). Notice how the connection between the two SSMs are the output voltage of the IB (v_b), the input current of the RDAB (i_{in}), and an additional feedforward path (in green), that connects the duty-cycle of the IB d_b to the input duty-ratio of the RDAB d_{in} (with the transfer function G_d).

The key concept is that, the IBSC and the EET are here used on the IB and the RDAB converters despite them being integrated inside the CF-RDAB converter. This is possible because, if $D_b = 50\%$, the only connection between the IB SSM and the RDAB SSM is the output voltage of the IB converter and the input current of the RDAB converter, as the two systems were in a typical cascaded connection.

IV. EET Applied to the CF-RDAB

The proposed control scheme for the CF-RDAB is illustrated in **Fig. 6**. In this scheme, the IB control utilizes an inner current loop to control the input current i_b, and an outer voltage loop to control voltage v_b. On the other hand, the RDAB control employs a single-voltage loop to adjust

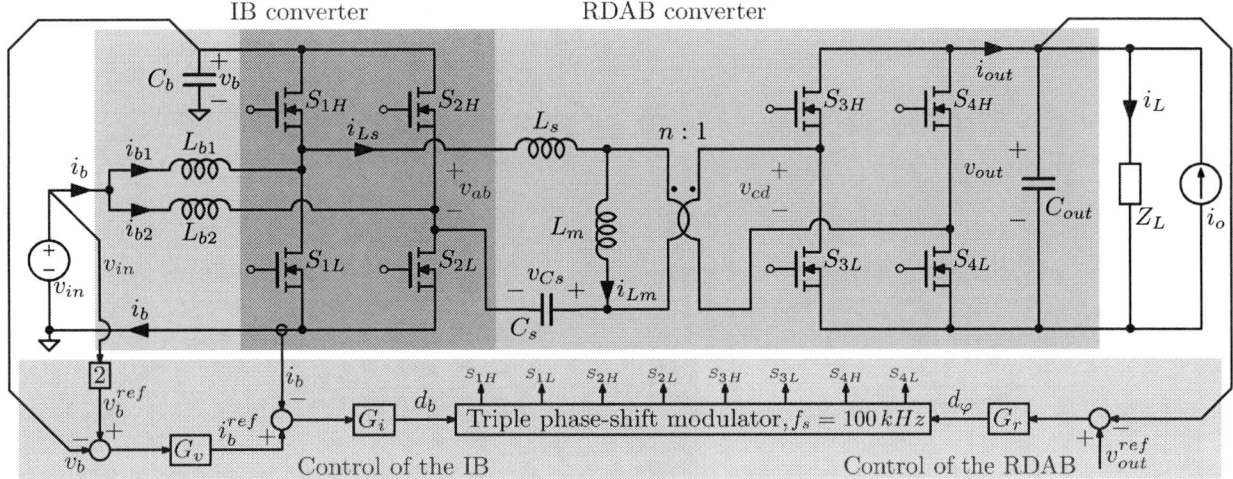

Fig. 6: Proposed control scheme for the CF-RDAB converter. The control strategy considers a double loop control for the IB, where an external voltage controller G_v regulates the voltage v_b to $2\,v_{in}$, by adjusting the boost current reference i_b^{ref}. Then, the current regulator G_i provides the boost duty-cycle d_b to follow the current reference with the boost current i_b. On the other hand, the RDAB uses a single output voltage loop (with the regulator G_r), to bring the output voltage v_{out} to its reference v_{out}^{ref}, by adjusting the phase-shift d_φ.

the phase-shift d_φ for regulating the output voltage v_{out}. Additionally, the boost's voltage reference v_b^{ref} is set to $2\,v_{in}$, allowing the boost to operate at a $50\,\%$ duty-cycle. This configuration, not only **decouples the SSMs of the IB and the RDAB as discussed in Section III, but also enables maximum power transfer and eliminates the ripple in the input current i_b.**

Finally, in this analysis, the switching frequency is kept constant, while the power transfer is maximized by setting the output inverter duty-ratio to $d_{out} = 1$.

To study the stability of the overall system, the IBSC is used referring to **Fig. 1**, where the output impedance of the IB converter Z_{out}^{ib} and the input impedance of the RDAB converter Z_{in}^{rd} are defined. Those impedances can be computed as:

$$Z_{out}^{ib} = -\frac{G_{23}^{ib}(1 + G_i G_{12}^{ib}) - G_i G_{13}^{ib} G_{22}^{ib}}{1 + G_i(G_{12}^{ib} + G_v G_{22}^{ib})}$$

$$Z_{in}^{rd} = \frac{1 + G_r G_{13}^{rd}}{G_{25}^{rd}(1 + G_r G_{13}^{rd}) - G_r G_{23}^{rd} G_{15}^{rd}}. \qquad (19)$$

Considering the parameters in **Tab. I**, the plots of Z_{out}^{ib} and Z_{in}^{rd} are reported in **Fig. 7a** (in **Tab. I**: f_s is the switching frequency, $f_{cr,i}$ is the crossover frequency of the current loop of the IB converter, $f_{cr,v}$ is the crossover frequency of the voltage loop of the IB converter, $f_{cr,d}$ is the crossover frequency of the voltage loop of the RDAB converter, and I_L is the current of the constant-current load. Additionally all the controllers G_i, G_v and G_r are designed considering a constant current load for both the IB and the RDAB. Finally, the test generator i_o has an average current $I_o = 0\,\text{A}$ since it is used for the only purpose of output impedance measurement).

As depicted in **Fig. 7a**, the impedances Z_{out}^{ib} and Z_{in}^{rd} do not cross. This guarantees the stability of the system in **Fig. 1** for the IBSC [22].

TABLE I: Circuit parameters

Parameter			
f_s	50 kHz	V_{out}	100 V
V_{in}	50 V	L_s	28.5 µH
V_b	100 V	C_s	730 nF
n	1	L_m	275 µH
L_{b1} , L_{b2}	350 µH	C_{out}	705 µF
C_b	470 µF	I_L	5 A
$f_{cr,i}$	1 kHz	I_o	0 A
$f_{cr,v}$	15 Hz	$f_{cr,d}$	500 Hz

Finally, the voltage loop gain T of the RDAB converter can be computed with the EET [23], [24] as:

$$T = T_0 \frac{1 + \frac{Z_{out}^{ib}}{Z_N}}{1 + \frac{Z_{out}^{ib}}{Z_D}}, \qquad (20)$$

where:

$$T = T_0 \frac{1 + \frac{Z_{out}^{ib}}{Z_N}}{1 + \frac{Z_{out}^{ib}}{Z_D}},$$

$$T_0 = G_r G_{13}^{rd},$$

$$Z_N = \frac{1}{G_{25}^{rd} - G_{23}^{rd} \frac{G_{15}^{rd}}{G_{13}^{rd}}},$$

$$Z_D = \frac{1}{G_{25}^{rd}}. \qquad (21)$$

Fig. 7b shows the voltage loop gain T of the RDAB converter computed with the EET. In the same figure is also present the original loop gain T_0 of the RDAB converter without the IB converter. As depicted in such a figure, the presence of the IB reduces the loop gain of the RDAB in the frequency range where the output impedance of the IB

979-8-3315-1612-3/25 $31.00 © 2025 IEEE

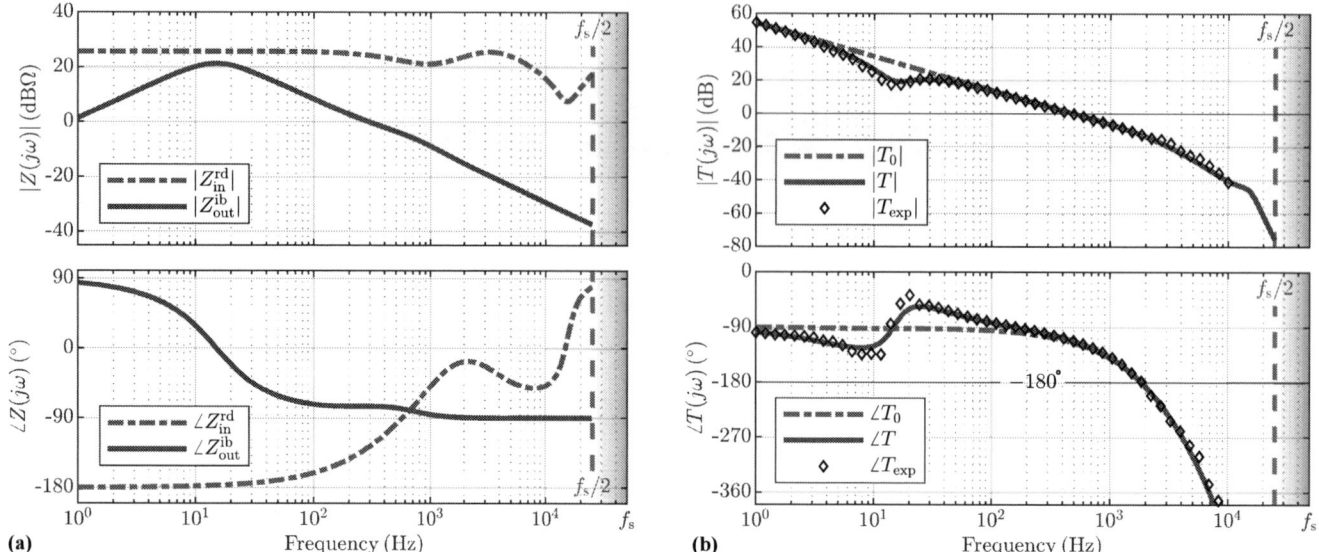

Fig. 7: (a) Impedances of the system in **Fig. 1**: in continuous line the output impedance of the IB converter, and in dashed line the input impedance of the RDAB converter. (b) Loop gain of the RDAB converter: in dashed line the original loop gain without the IB converter, in continuous line the loop gain with the IB converter, and in markers the experimentally measured output voltage loop gain of the CF-RDAB converter displayed in **Fig. 8a**.

is at its peak. Nevertheless, the overall system remains stable as predicted with the IBSC. In **Fig. 7b** is also present the output voltage loop gain of the CF-RDAB measured on the prototype in **Fig. 8a**. Such a measure mirrors the loop gain of the IB+RDAB cascaded connection, thus validating the model and the proposed control strategy for the CF-RDAB converter.

V. EXPERIMENTAL VALIDATION

To fully validate the presented theory, a CF-RDAB prototype is built with the Imperix Power Rack and the Imperix PEB8024 power modules. The CF-RDAB prototype is shown in **Fig. 8a**, and is digitally controlled by the Imperix B-Box 3.0. Referring to **Fig. 6**, the circuit parameters are reported in **Tab. I**. As a first measure, the steady-state waveforms of the CF-RDAB are presented in **Fig. 8b**, illustrating a comprehensive view of the electrical quantities of the CF-RDAB, and proving the converter's stability under steady-state conditions. A second measurement is shown in **Fig. 7b**, where the theoretical loop gain of the CF-RDAB, denoted as T (continuous line), is compared with the experimentally measured loop gain, T_{exp} (markers) [22]. This figure reveals that the measured loop gain closely follows its theoretical counterpart, confirming both the SSM of the CF-RDAB and the effectiveness of the proposed control strategy. Additionally, the matching between the experimentally measured loop gain T_{exp} and the theoretical loop gain T computed with the EET on the IB+RDAB cascaded connection, confirms the validity of the proposed approach for the stability analysis of current-fed topologies.

VI. CONCLUSION

This paper has presented a generalized-average SSM of the CF-RDAB converter as the combination of an IB an RDAB. Such a SSM simplifies the control of the topology,

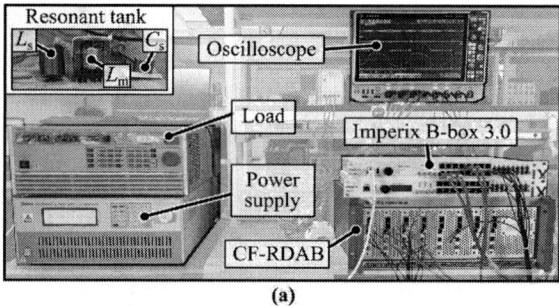

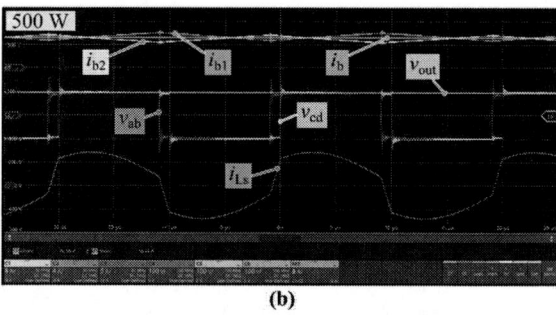

Fig. 8: Experimental validation of the CF-RDAB: (a) Prototype of the CF-RDAB with the parameters reported in **Tab. I**. (b) Steady-state waveforms of the CF-RDAB.

with separate regulators for the IB and the RDAB. This articles has derived that, when the input inverter operates with 50% duty-cycle, the SSM of the CF-RDAB reduces to the IB+RDAB cascaded connection. In this specific operating point, the stability of the whole CF-RDAB can be studied with the IBSC and the EET, as the IB and the RDAB were fully independent.

Afterwards, this paper has proposed a control strategy for the CF-RDAB that uses both the input duty-cycle and the

979-8-3315-1612-3/25 $31.00 © 2025 IEEE

phase-shift to regulate the output voltage of the converter. The proposed model and control have been experimentally validated on a 500 W CF-RDAB converter, showing an excellent alignment between the theory and the measurements.

While the focus of this analysis has been on the CF-RDAB, the small-signal-modeling approach used in this article can be applied to different current-fed topologies, facilitating the stability analysis and the development of effective control strategies in a wide range of applications.

REFERENCES

[1] P. Sanjeev, N. P. Padhy, and P. Agarwal, "Peak energy management using renewable integrated dc microgrid," *IEEE Transactions on Smart Grid*, vol. 9, no. 5, pp. 4906–4917, 2018.

[2] V. C. Gungor, D. Sahin, T. Kocak, S. Ergut, C. Buccella, C. Cecati, and G. P. Hancke, "Smart grid technologies: Communication technologies and standards," *IEEE Transactions on Industrial Informatics*, vol. 7, no. 4, pp. 529–539, 2011.

[3] C. Kang, D. Kirschen, and T. C. Green, "The evolution of smart grids," *Proceedings of the IEEE*, vol. 111, no. 7, pp. 691–693, 2023.

[4] R. Panigrahi, S. K. Mishra, S. C. Srivastava, and P. Enjeti, "Microgrid integration in smart low-voltage distribution systems," *IEEE Power Electronics Magazine*, vol. 9, no. 2, pp. 61–66, 2022.

[5] J. Yuan, L. Dorn-Gomba, A. D. Callegaro, J. Reimers, and A. Emadi, "A review of bidirectional on-board chargers for electric vehicles," *IEEE Access*, vol. 9, pp. 51 501–51 518, 2021.

[6] J. Rodriguez, F. Blaabjerg, and M. P. Kazmierkowski, "Energy transition technology: The role of power electronics," *Proceedings of the IEEE*, vol. 111, no. 4, pp. 329–334, 2023.

[7] T. Chen, R. Yu, and A. Q. Huang, "A bidirectional isolated dual-phase-shift variable-frequency series resonant dual-active-bridge gan ac–dc converter," *IEEE Transactions on Industrial Electronics*, vol. 70, no. 4, pp. 3315–3325, 2023.

[8] Y. Shi, R. Li, Y. Xue, and H. Li, "Optimized operation of current-fed dual active bridge dc–dc converter for pv applications," *IEEE Transactions on Industrial Electronics*, vol. 62, no. 11, pp. 6986–6995, 2015.

[9] G. Spiazzi, S. Buso, D. Biadene, G. Rossetto, and F. Mela, "High efficiency battery charger for photovoltaic inverters," in *2017 IEEE Southern Power Electronics Conference (SPEC)*, 2017, pp. 1–6.

[10] F. Toniolo, Q. Liu, T. Caldognetto, S. Buso, and G. Spiazzi, "Digital current control for a bidirectional interleaved boost converter with coupled inductors," in *2019 IEEE 15th Brazilian Power Electronics Conference and 5th IEEE Southern Power Electronics Conference (COBEP/SPEC)*, 2019, pp. 1–6.

[11] Y. Li, Y. Xing, Y. Lu, H. Wu, and P. Xu, "Performance analysis of a current-fed bidirectional llc resonant converter," in *IECON 2016 - 42nd Annual Conference of the IEEE Industrial Electronics Society*, 2016, pp. 2486–2491.

[12] X. Pan, H. Li, Y. Liu, T. Zhao, C. Ju, and A. K. Rathore, "An overview and comprehensive comparative evaluation of current-fed-isolated-bidirectional dc/dc converter," *IEEE Transactions on Power Electronics*, vol. 35, no. 3, pp. 2737–2763, 2020.

[13] S. Sanders, J. Noworolski, X. Liu, and G. Verghese, "Generalized averaging method for power conversion circuits," *IEEE Transactions on Power Electronics*, vol. 6, no. 2, pp. 251–259, 1991.

[14] E. Yang, F. Lee, and M. Jovanovic, "Small-signal modeling of series and parallel resonant converters," in *[Proceedings] APEC '92 Seventh Annual Applied Power Electronics Conference and Exposition*, 1992, pp. 785–792.

[15] S. Tian, F. C. Lee, and Q. Li, "A simplified equivalent circuit model of series resonant converter," *IEEE Transactions on Power Electronics*, vol. 31, no. 5, pp. 3922–3931, 2016.

[16] C.-H. Chang, E.-C. Chang, C.-A. Cheng, H.-L. Cheng, and S.-C. Lin, "Small signal modeling of llc resonant converters based on extended describing function," in *2012 International Symposium on Computer, Consumer and Control*, 2012, pp. 365–368.

[17] S. Tian, F. C. Lee, and Q. Li, "Equivalent circuit modeling of llc resonant converter," in *2016 IEEE Applied Power Electronics Conference and Exposition (APEC)*, 2016, pp. 1608–1615.

[18] A. K. Dubey and N. Lakshminarasamma, "Modeling of series resonant dual active bridge converter for dc microgrid application," in *2022 IEEE International Conference on Environment and Electrical Engineering and 2022 IEEE Industrial and Commercial Power Systems Europe (EEEIC / I&CPS Europe)*, 2022, pp. 1–6.

[19] R. Nitheesh, N. Dandare, A. K. Dubey, N. Lakshminarasamma, and A. K. B., "Modeling of series resonant dual active bridge with grid-connected inverter for vehicle-to-everything applications," in *2023 IEEE IAS Global Conference on Emerging Technologies (GlobConET)*, 2023, pp. 1–6.

[20] A. K. Dubey and N. Lakshminarasamma, "Generalized model for lc resonant dual active bridge dc-dc converter for varied control techniques," in *2022 IEEE International Conference on Power Electronics, Drives and Energy Systems (PEDES)*, 2022, pp. 1–7.

[21] R. Nitheesh, A. K. Dubey, N. Lakshminarasamma, and B. A. Karuppaswamy, "Modeling and analysis of high gain pole point inductor-based series resonant dual active bridge converter," in *2023 25th European Conference on Power Electronics and Applications (EPE'23 ECCE Europe)*, 2023, pp. 1–8.

[22] R. W. Erickson and D. Maksimović, *Fundamentals of Power Electronics*, 3rd ed. Springer, 2020.

[23] R. Middlebrook, "Null double injection and the extra element theorem," *IEEE Transactions on Education*, vol. 32, no. 3, pp. 167–180, 1989.

[24] R. Middlebrook, V. Vorperian, and J. Lindal, "The n extra element theorem," *IEEE Transactions on Circuits and Systems I: Fundamental Theory and Applications*, vol. 45, no. 9, pp. 919–935, 1998.

[25] R. D. Middlebrook and S. Cuk, "A general unified approach to modelling switching-converter power stages," in *1976 IEEE Power Electronics Specialists Conference*, 1976, pp. 18–34.

Concurrent Charge Distribution and Time-Optimal Control for Unordered Single-Inductor Dual-Output Converter

Xuliang Wang[1], Haoyu Wang[1], Yang Liu[2], Yunxin Wang[1], Boran Zhang[1], Hongru Liu[1], Yan Wang[1], Xiaosen Liu[1]

[1]Tsinghua University, Beijing, China
[2]Hong Kong University of Science and Technology, Hong Kong
xwangef@tsinghua.edu.cn

Abstract—Single-inductor-multiple-output (SIMO) converters are pivotal for providing versatile V-F domains and fine-grained DVFS for many-core XPUs under stringent cost and size constraints. However, acceptable cross-regulation (CR) can only be promised in scenarios with less than 50% current discrepancies between output channels, making previous work unfeasible for frequent sleep/wake-up demands. Charging sequence reordering has emerged as a promising technique for mitigating last-channel CR in SIMO converters. However, conventional unordered control techniques introduce unwanted preceding-channel cross-regulation issues. The root cause is the mismatch in the loop response between the peak current (PC) control loop for the half-bridge and the channel-wise charge-driven loop. Although response mismatches can be alleviated through real-time calibration, significant CRs still occur during load transients with extreme channel discrepancies and hinder practical applications. To address these limitations, we introduce a calibration-free concurrent charge distribution and time-optimal control (CCD-TOC) technique. Moreover, a comparative study has explored the stability boundary and the benefits of an unfixed charging sequence. The experimental results of the single-inductor dual-output (SIDO) prototype demonstrate a transient CR of 0.176 mV/mA and a 92-mV (3.1%) undershoot against a 500-mA load step and a 400-mA (66.7%) channel discrepancy.

Index Terms—SIMO Buck converters, cross-regulation, sequence re-ordering, charge expectation, digital control, time-optimal control

I. INTRODUCTION

In typical many-core XPU designs, fine-grained DVFS and sleep/wake-up are enabled using per-core fully integrated VRs (FIVRs) or linear VRs and PLLs, thus posing major challenges to integrating many high-Q inductors within the thin package or degraded efficiency in high dropout channels. Single-inductor-multiple-output (SIMO) converters are naturally favorable for providing versatile V-F domains under stringent cost and size constraints. However, acceptable cross-regulation can only be promised in scenarios with less than 50% workload discrepancies among output channels, making prior work unfeasible for frequent sleep/wake-up demands. Fig. 2 illustrates the last-channel cross-regulation (CR) issue in ordered SIDO converters and the preceding-channel CR issue in unordered SIDO converters. In a conventional SIDO converter with a fixed charging sequence, the last channel may receive insufficient or surplus energy [1]. Therefore, the victim-last

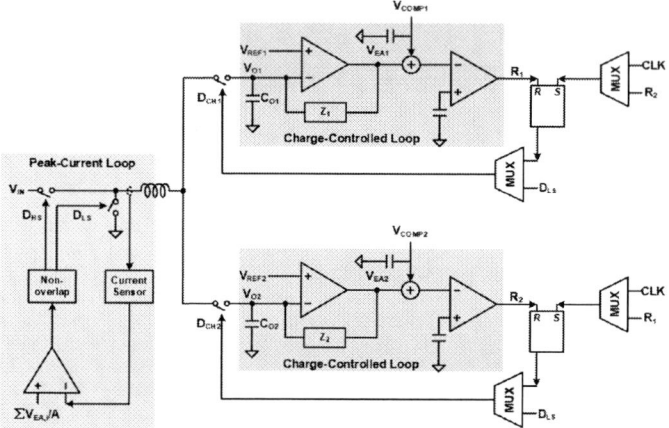

Fig. 1. General diagram of ordered/unordered SIDO Converter.

control senses which channel has the transient load and then dispatches the victim to the last place in the charging sequence [2]. Although the victim-last control addresses the last-channel CR, it introduces CR to the preceding channel, caused by the mismatch between the PC control loop and the channel-wise charge-driven loop. The duty stretching/shrinking strength of the half-bridge is intentionally bounded in time-multiplexing SIMO converters to prevent conflicts with the charged-based domain duty calculation. This conflict limits the cycle time for load transient recovery, especially when a large channel discrepancy exists. In this paper, we present an all-digital solution to supply independent V-F domains with reduced cross-regulation issues, even with a 400-mA (66.7%) channel discrepancy, via a SIMO Buck converter.

The rest of this article is organized as follows. The operation principle of the proposed concurrent charge distribution and time-optimal control is presented in Section II. The instability phenomena of ordered SIDO Buck converters are revealed by bifurcation analysis, including the effect of load parameters on stability boundaries. In Section IV, the working principle of the inductor current estimator is discussed, and the parameters are investigated for calibration. Section V gives the experimental results of the studied converter to verify the theoretical anal-

979-8-3315-1612-3/25 $31.00 © 2025 IEEE

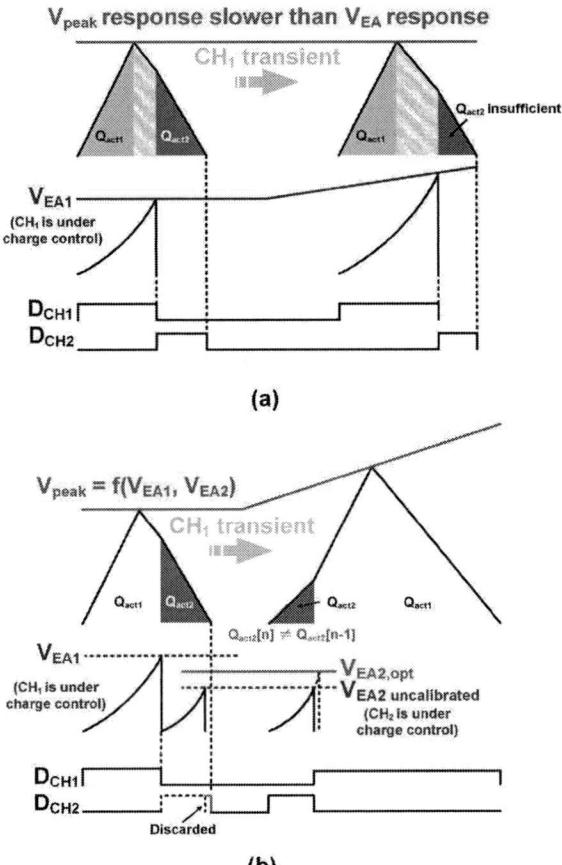

(a)

(b)

Fig. 2. Cross-regulation issues of (a) ordered and (b) unordered SIDO converters.

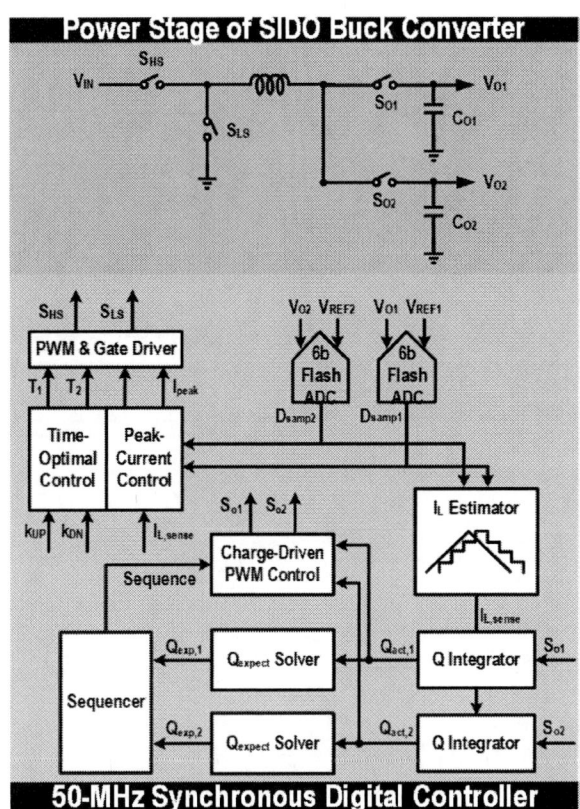

Fig. 3. Diagram of proposed SIDO architecture.

ysis, followed by the conclusion in Section VI.

II. PROPOSED CONCURRENT CHARGE DISTRIBUTION AND TIME-OPTIMAL CONTROL

To address the above problems and unleash the full benefits of an unfixed charging sequence, we propose a concurrent charge distribution and time-optimal control (CCD-TOC) strategy with all-digital implementation. Independent control loops are assigned to the half-bridge and load switch in each channel [5]. For the half-bridge, a peak current control loop compares the estimated instantaneous inductor current and the PID-regulated reference code $I_{L,REF}$ to update the duty ratio.

$$
\begin{cases}
I_{L,REF}[z] = K_P \sum_{i=1}^{2} V_{err,i}[z] + \frac{K_I \sum_{i=1}^{2} V_{err,i}[z]}{z-1} \\
V_{err,i}[z] = K_q(V_{REF,i} - v_{OUT,i}[z]) \\
K_q = \frac{1}{V_{res}}
\end{cases} \quad (1)
$$

where V_{res} is the resolution of the ADC.

The design and calibration of the digital inductor current estimator follow [6]-[9] and will be discussed in Section IV. At each load point, a digital integrator records the charge received from the given channel in each period, which is used for

calculating the charge expectation [10][11]. Unlike the victim-last control strategy [1][2], the charge expectations (Eq. 1) are resequenced at the beginning of the new period to determine the charging sequence.

$$
Q_{exp}[n] = Q_{act}[n-1] + 2C_L V_{err}[n] - C_L V_{err}[n-1] \quad (2)
$$

where Q_{act} and Q_{exp} are the per-channel received charge and charge expectation, respectively. For the preceding channel under charge iteration, Q_{act} is the same as Q_{exp}. When load transients between low load and rated load cause voltage undershoots/overshoots at any channel, the time-optimal control (TOC) loop takes over the regulation. During the TOC phase, the inductor current is directed to the unloaded channel at the beginning of the inductor charging/discharging sub-phases. The transient channel is always placed in the last order of voltage regulation to minimize CR in light-load channels. The charge iteration equation of the new preceding channel is derived from Eq. 1 and rewritten as shown in Eq. 2.

$$
Q_{exp} = \frac{T_2}{T_1} Q_{act} + (1 + \frac{T_2}{T_1}) C_L V_{err}[n] - \frac{T_2}{T_1} C_L V_{err}[n-1] \quad (3)
$$

For a large load step, the proposed TOC response involves (1) maximally slewing I_L beyond I_{load} to stop and reverse the voltage droop until T_1; (2) slewing I_L back toward I_{load}

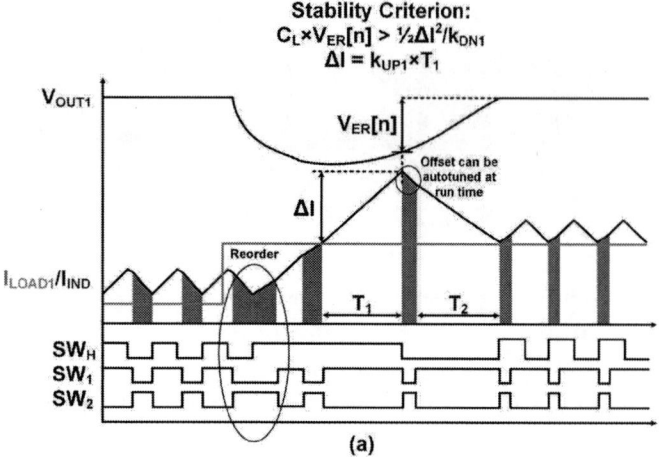

Stability Criterion:
$$C_L \times V_{ER}[n] > \frac{1}{2}\Delta I^2/k_{DN1}$$
$$\Delta I = k_{UP1} \times T_1$$

Fig. 4. Proposed CCD-TOC technique.

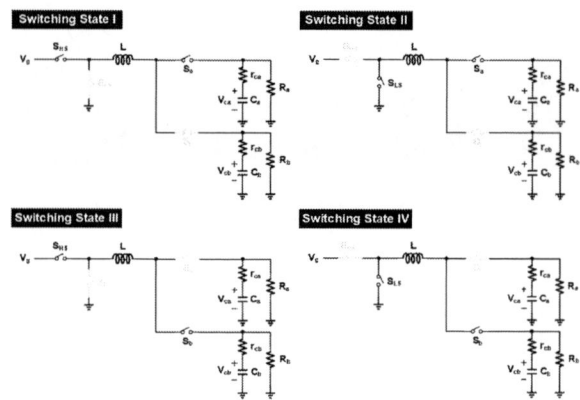

Fig. 5. Switching states of ordered SIDO converter..

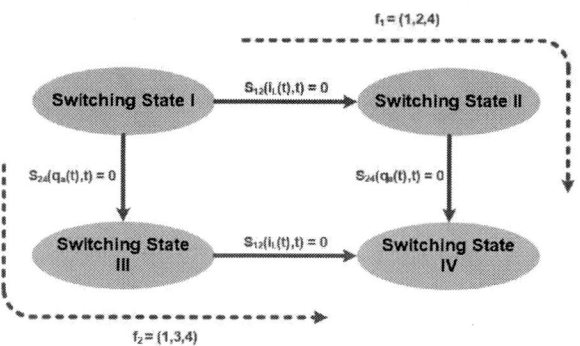

Fig. 6. Two switching state sequences.

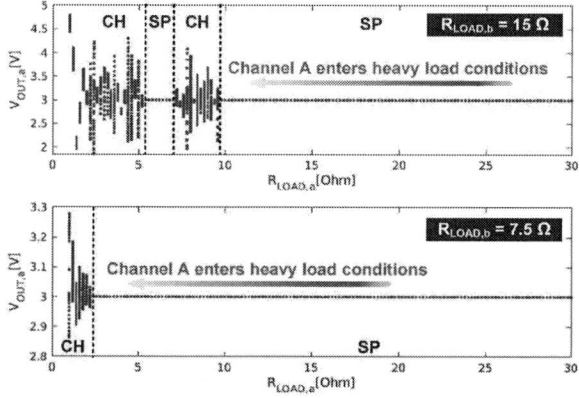

Fig. 7. Bifurcation diagrams of ordered SIDO converter.

until T_2; (3) I_L tracks I_{load} after T_2, and the linear controller regains the overall regulation with a preset duty constant for smooth transition between the transient state and the steady state. The time intervals (denoted as T_1 and T_2 in Fig. 4) of the inductor charging/discharging subphases are recorded starting from the current crossing point where the voltage recovery begins [12][13]. Timing T_1 and T_2 precisely is pivotal for optimal response and is achieved by tracking the surplus charge $Q_{surplus}$:

$$Q_{surplus} = 1/2(k_{UP}T_1)^2/k_{DN} \qquad (4)$$

where k_{UP} and k_{DN} are the slew rate of the inductor charging/discharge subphases. An optimal response is achieved when $Q_{surplus}$ provides just the right quantity of charge to restore V_{OUT} without overshooting, and it is guaranteed by continuous evaluation of the stability constraint:

$$Q_{surplus} <= C_L V_{err} \qquad (5)$$

Our proposed CCD-TOC technique relaxes the matching requirement between the PC control and the channel-wise charge control, allowing the former to operate at a moderate bandwidth and achieve a wider load range without stability issues. Despite the vast adoption of time-optimal control (TOC) in single-channel digital Buck or LDO converters, this is the first work to use TOC for a SIMO converter. The details of the stability analysis will be discussed in the following section.

III. DISCRETE-TIME MAP MODEL AND STABILITY BOUNDARIES

In this section, bifurcation diagrams are utilized to explore the stability limits of conventional ordered SIDO converters. The derivation of the discrete-time map model follows [14][15]. An ordered SIDO converter with two output channels has four available switching states, which is defined by Eq. 6-9 and shown in Fig. 5.

$$\begin{cases} \dfrac{di_L}{dt} = \dfrac{V_g}{L} - \dfrac{R_a r_{ca} i_L}{L(R_a+r_{ca})} - \dfrac{R_a v_{ca}}{L(R_a+r_{ca})} \\ \dfrac{dv_{ca}}{dt} = \dfrac{R_a i_L}{C_a(R_a+r_{ca})} - \dfrac{v_{ca}}{C_a(R_a+r_{ca})} \\ \dfrac{dv_{cb}}{dt} = -\dfrac{v_{cb}}{C_b(R_b+r_{cb})}. \end{cases} \qquad (6)$$

$$\begin{cases} \dfrac{di_L}{dt} = -\dfrac{R_a r_{ca} i_L}{L(R_a+r_{ca})} - \dfrac{R_a v_{ca}}{L(R_a+r_{ca})} \\ \dfrac{dv_{ca}}{dt} = \dfrac{R_a i_L}{C_a(R_a+r_{ca})} - \dfrac{v_{ca}}{C_a(R_a+r_{ca})} \\ \dfrac{dv_{cb}}{dt} = -\dfrac{v_{cb}}{C_b(R_b+r_{cb})}. \end{cases} \qquad (7)$$

$$\begin{cases} \frac{di_L}{dt} = \frac{V_g}{L} - \frac{R_b r_{cb} i_L}{L(R_b+r_{cb})} - \frac{R_b v_{cb}}{L(R_b+r_{cb})} \\ \frac{dv_{ca}}{dt} = -\frac{v_{ca}}{C_a(R_a+r_{ca})} \\ \frac{dv_{cb}}{dt} = \frac{R_b i_L}{C_b(R_b+r_{cb})} - \frac{v_{cb}}{C_b(R_b+r_{cb})}. \end{cases} \quad (8)$$

$$\begin{cases} \frac{di_L}{dt} = -\frac{R_b r_{cb} i_L}{L(R_b+r_{cb})} - \frac{R_b v_{cb}}{L(R_b+r_{cb})} \\ \frac{dv_{ca}}{dt} = -\frac{v_{ca}}{C_a(R_a+r_{ca})} \\ \frac{dv_{cb}}{dt} = \frac{R_b i_L}{C_b(R_b+r_{cb})} - \frac{v_{cb}}{C_b(R_b+r_{cb})}. \end{cases} \quad (9)$$

As shown in Fig. 6, the independent control of the half-bridge and channel-wise conduction time leads to two potential switching sequences. The transition between adjacent switching states occurs when meeting Eq. 10 or Eq. 11.

$$\begin{cases} S_{12}(i_L(t),t) = i_L(t) - i_{L,REF} = 0 \\ S_{24}(q_a(t),t) = q_a(t) - q_{a,REF} = 0. \end{cases} \quad (10)$$

$$\begin{cases} S_{13}(q_a(t),t) = q_a(t) - q_{a,REF} = 0 \\ S_{34}(i_L(t),t) = i_L(t) - i_{L,REF} = 0. \end{cases} \quad (11)$$

TABLE I
RATED CIRCUIT PARAMETERS OF SIDO BUCK CONVERTER

Variable	Definition	Value
V_g	Input Voltage	8 V
R_a, R_b	Load resistors of channel a and channel b	30 Ω, 15 Ω
V_{ca}, V_{cb}	Output voltages of channel a and channel b	3.3 V, 5 V
L	Inductor	3.9 μH
C_a, C_b	Capacitors of channel a and channel b	20 μF
r_{ca}, r_{cb}	ESR resistors	2.5 mΩ
K_p, K_I	Proportional and integral parameters	0.01, 0.001
K_q	Quantization gain	1

Bifurcation diagrams with the variation of load parameters are presented based on the established discrete iterative map model. The rated parameters are listed in Table. I, extracted from the prototype in Section IV. The iterative calculation begins from $x_1 = 0$ and stops at the 5000th period with a 2-ns step. Finally, the last 50 points at the end of the simulation are used to draw the bifurcation diagram. As shown in Fig. 7, the motion orbit mutates from stable period-1 (SP) to chaos (CH) via border-collision bifurcation as the leading channel enters heavy-load conditions. When the trailing channel has a 15-Ω resistive load (333 mA), the collision point is around 300 mA. When the trailing channel has a 7.5-Ω resistive load (666 mA), the collision point is around 1.2 A. In conclusion, the border collision occurs as the channel discrepancy increases. This transition hinders the dynamic range of ordered SIDO converters and highlights the superiority of the proposed charge-expectation-driven reordering technique.

IV. DIGITAL INDUCTOR CURRENT ESTIMATION

DCR current sensing is one of the main methods of analog current sensing, and the working principle expression is shown as Eq. 12:

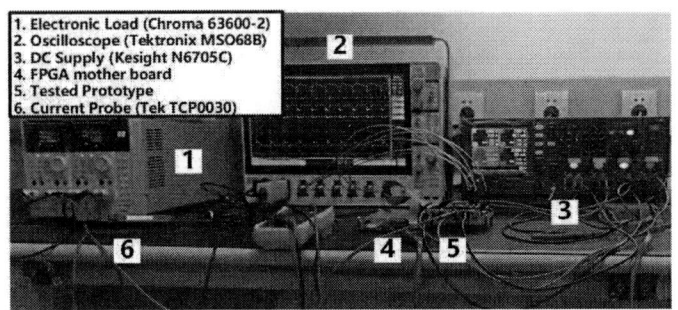

Fig. 8. Experimental setup.

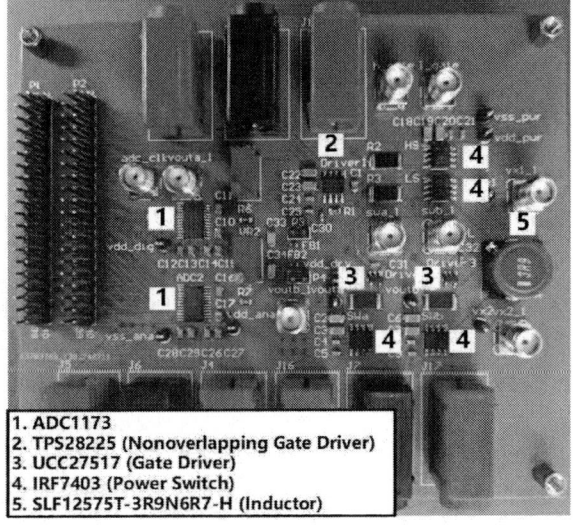

1. ADC1173
2. TPS28225 (Nonoverlapping Gate Driver)
3. UCC27517 (Gate Driver)
4. IRF7403 (Power Switch)
5. SLF12575T-3R9N6R7-H (Inductor)

Fig. 9. Tested Prototype.

$$I_{sense} = \frac{V_L(s)}{sL_e + R_{Le}} \frac{I_L(s)R_L}{R_{Le}} \frac{1 + s\frac{L}{R_L}}{1 + s\frac{L_e}{R_{Le}}} \quad (12)$$

where L_e and R_{Le} are the estimated inductance and DCR value set in the filter, respectively. By manipulating and applying the bilinear transformation, the digital equivalence of Eq. 12 can be obtained as given in [8]. The calibration procedure of the parameters (L_e and R_{Le}) consists of two steps: (1) gain calibration and (2) time constant calibration. Similar to [7], a 50-MHz sampling clock (100 times the switching frequency) with low-pass filtering is used to eliminate the estimation error of the averaged duty cycle D for accurate calculation of V_L:

$$V_L[n] = D \times V_g[n] - V_o[n] \quad (13)$$

To track the channel-wise R_{ds} of the power switches, while a test current sink circuit with known load change ΔI_{test} is applied in one channel, the other channel will be disabled and disconnected from the half-bridge. The online calibration procedure will be repeated twice and the estimated equivalent resistance R_{eq} of the entire converter becomes:

$$R_{eq} = R_L + D_{CH1}[n]R_{ds,1} + D_{CH2}[n]R_{ds,2} \quad (14)$$

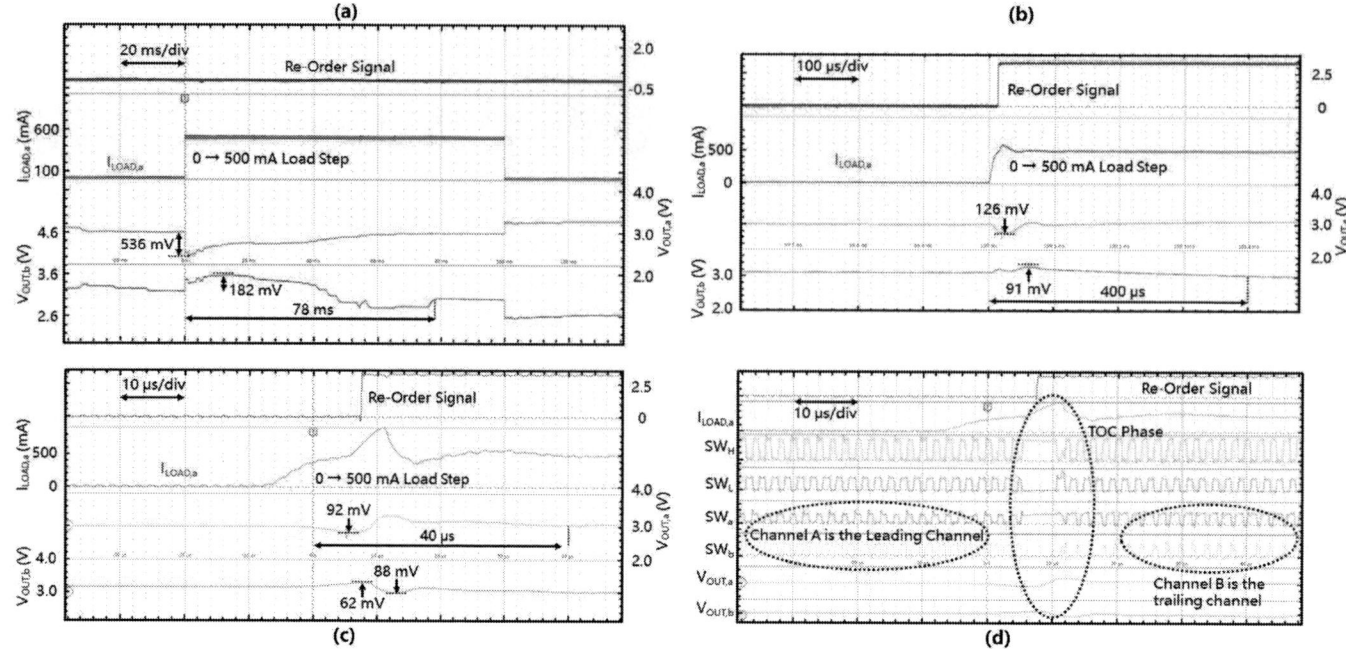

Fig. 10. Waveforms of load transient responses (a) without sequence re-ordering or time-optimal control (TOC); (b) with sequence reordering and well-tuned PID control; (c) with sequence reordering and TOC and (d) its detailed gate driving signals (right).

where D_{CH1} and D_{CH2} are the average conduction time of each channel within a switching period. Eq. 14 is used to replace R_L in Eq. 12 and regard R_{ds} of the power switch as a part of the DCR resistance.

V. EXPERIMENT VERIFICATION

An experimental prototype of a SIDO CCM buck converter is implemented with IRF7403 MOSFET as power switches, TPS28225 and UCC27517 as gate drivers. The 500-kHz gate driving signals are generated by FPGA with a 6.25-MHz ADC sampling frequency. Figs. 10 (a)-(d) present the experimental waveforms of output voltages, current loads, and gate driving signals of the tested prototype. The resistive loads of each channel are 30 Ω and 15 Ω, respectively. Compared Fig. 3(b) with the baseline in Fig. 3(a), channel sequence reordering prevents the response mismatches between the peak-current control loop and channel-wise charge iteration loops. When applying a 500-mA load step to the original preceding channel A, the voltage undershoot (V_{US}), the cross-regulation (CR), and the settling time (T_{SETTLE}) are reduced to 126 mV, 91 mV, and 400 μs by dispatching the unloaded channel B as the new leading channel. As shown in Fig. 3(c)-(d), concurrent sequence reordering and time-optimal control further reduce V_{US} and T_{SETTLE} to 92 mV and 40 μs without deterioration in CR. Table II lists the parameters of previous works with the standalone charge control, the standalone unordered control, and the standalone time-optimal control. Owing to the concurrent utilization of unordered charge control and time-optimal control, this work achieved one of the best undershoots and CR against one of the largest load steps and channel discrepancies.

VI. CONCLUSION AND FUTURE WORK

The major contributions of our work are concluded here:

- We proposed a calibration-free concurrent charge distribution and time-optimal control (CCD-TOC) technique to address both the last-channel CR issue of ordered SIDO converters and the preceding-channel CR problem introduced by conventional reordering techniques.
- We rendered a charge-expectation-driven reordering technique and conducted analytical models to explore the stability boundaries.
- We introduced channel discrepancy as a supplement to cross-regulation to give a comprehensive performance evaluation of SIDO converters for real-world applications.

Although the proposed CCD-TOC technique was verified by a simple Buck prototype with only two output channels, neither the converter topology nor the number of output channels is a fundamental constraint. The following methodologies are proposed for future investigation to exploit the positive impact of out-of-order control on performance and power efficiency in practical applications.

- To support concurrent load changes in more than one channel, the victim channels should be bundled during the time-optimal control phase. The reordering within the bundle will be executed at a higher frequency to guarantee the channel-wise crossing points are close to each other for precise timing of T_1 and T_2.
- With a proper decomposition of the I_L estimation error [6], iterative correction of the current estimator can be achieved without interrupting the converter operation.

TABLE II
COMPARISON TABLE WITH PRIOR WORKS

Parameter	TPE'12 [6]	ISSCC'21 [10]	TIE'21 [1]	APEC'24① [16]	TCAS-I'24 [2]	*This work*
No. of channels	2 Boost	4 Buck	2 Buck	2 Boost	4 Buck	**2 Buck**
Control Scheme	Ordered Charge Control (Digital)	Ordered Charge Control (Digital)	Unordered TOC Control (Digital)	Ordered Charge Control (Analog)	Unordered Charge Control (Analog)	**Unordered Charge & TOC Control (Digital)**
Supply Voltage	2.5 V	1.8 V	3.3 V	2.7 V – 4.5 V	5 V	**5 V – 12 V**
C_{OUT}/L_{IND}	10 µF / 4.7 µH	1 µF / 10 µH	10 µF / 4.7 µH	22 µF / 1 µH	22 µF / 1 µH	**20 µF / 3.9 µH**
F_{SW}	390 kHz	2.5 MHz	1 MHz	2 MHz	1.5 MHz	**500 kHz**
Load Step / Undershoot	350 mA / 500 mV (12.5 %)	78 mA / 33 mV (3.3 %)	190 mA / 95 mV (5.3 %)	600 mA / 69 mV (1.1 %)	225 mA / 145 mV (9.7 %)	**500 mA / 92 mV (3.1 %)**
Channel Discrepancy②	200 mA / 430 mA (46.5 %)	106 mA / 101 mA (105 %)	N/A	350 mA / 600 mA (58.3 %)	380 mA / 375 mA (101.3 %)	**200 mA / 600 mA (33.3 %)**
Cross Regulation	0.429 mV/mA	0.359 mV/mA	0.05 mV/mA	0.060 mV/mA	0.057 mV/mA	**0.176 mV/mA**
FoM③	0.199 mV/mA	0.377 mV/mA	N/A	0.035 mV/mA	0.058 mV/mA	**0.058 mV/mA**

[1]SIMPLIS simulation results; [2]Channel Discrepancy = The current sum of unloaded channels / The maximum load of the victim channel; [3]FoM = Cross Regulation × Channel Discrepancy.

ACKNOWLEDGMENT

This work was supported by a grant from the National Key RD Program of China (Project No. 2022YFB4401100) and the National Natural Science Foundation of China (Grant No. 62374100).

REFERENCES

[1] C.-W. Liu, L.-R. Chang-Chien, "Area efficient high-performance digitally controlled power management unit," *IEEE Trans. Ind. Electron.*, vol. 68, no. 3, pp. 2437-2446, Mar. 2021.

[2] Y. Li, M. Huang, R. P. Martins, and Y. Lu, "A single-inductor multiple-output DC–DC converter with fixed-frequency victim-last charge control for reduced Cross Regulation," *IEEE Trans. Circuits and Syst. I*, vol. 71, no. 8, pp. 3904-3914, Aug. 2024.

[3] J. Haj-Yahya et al., "FlexWatts: A power- and workload-aware hybrid power delivery network for energy-efficient microprocessors," in *IEEE/ACM International Symposium on Microarchitecture (MICRO)*, Oct. 2020, pp. 1051-1066.

[4] S. C. Lee et al., "A hybrid converter with dual outputs for low cross regulation and improved current balance," *IEEE Trans. on Power Electron.*, vol. 39, no. 8, pp. 9591-9601, Aug. 2024.

[5] Y. Jiang and A. Fayed, "A 1 A, dual-inductor 4-output Buck converter with 20 MHz/100 MHz dual-frequency switching and integrated output filters in 65 nm CMOS," *IEEE J. Solid-State Circuits*, vol. 51, no. 10, pp. 2485-2500, Oct. 2016.

[6] Z. Shen, X. Chang, W. Wang, X. Tan, N. Yan, and H. Min, "Predictive digital current control of single-inductor multiple-output converters in CCM with low cross regulation," *IEEE Trans. Power Electron.*, vol. 27, no. 4, pp. 1917-1925, Apr. 2012.

[7] L. Yu, S. Xu, C. Yang, Y. Wu, L. Shi, and W. Sun, " A high-bandwidth current estimator with self tunning for digital Buck controller," *IEEE Trans. Ind. Electron.*, vol. 70, no. 11, pp. 11598-11607, Nov. 2023.

[8] Z. Lukic, S. M. Ahsanuzzaman, Z. Zhao, and A. Prodic, " Sensorless self-tuning digital CPM controller with multiple parameter estimation and thermal stress equalization," *IEEE Trans. Power Electron.*, vol. 26, no. 12, pp. 3948-3963, Dec. 2011.

[9] M. P. Chan and P. K. T. Mok, "A monolithic digital ripple-based adaptive-off-time DC-DC converter with a digital inductor current sensor," *IEEE J. Solid-State Circuits*, vol. 49, no. 8, pp. 1837-1847, Aug. 2014.

[10] C.-H. Huang et al., "A single-inductor 4-output SoC with dynamic droop allocation and adaptive clocking for enhanced performance and energy efficiency in 65nm CMOS," in *IEEE Int. Solid-State Circuits Conf. Dig. Tech. Papers (ISSCC)*, Feb. 2021, pp. 416–417.

[11] B. Wang, X. Zhang, J. Ye and H. B. Gooi, "Deadbeat control for a single-inductor multiple-input multiple-output DC–DC converter," *IEEE Trans. Power Electron.*, vol. 34, no. 2, pp. 1914-1924, Feb. 2019.

[12] S. Kim et al., "A 1.8W high-frequency SIMO converter featuring digital sensor-less computational zero-current operation and non-linear duty-boost," in *IEEE Int. Solid-State Circuits Conf. Dig. Tech. Papers (ISSCC)*, Feb. 2023, pp. 10-12.

[13] X. Sun, A. Boora, R. Pamula, C. -H. Huang, D. Peña-Colaiocco and V. S. Sathe, "Model predictive control of an integrated buck converter for digital SoC domains in 65nm CMOS," in *IEEE Symp. VLSI Circuits*, Jun. 2020, pp. 1-2.

[14] Y. Wang, J. Xu, F. Qin and D. Mou, "A capacitor current and capacitor voltage ripple controlled SIDO CCM buck converter with wide load range and reduced cross regulation," *IEEE Trans. Ind. Electron.*, vol. 69, no. 1, pp. 270-281, Jan. 2022.

[15] Y. Wang, L. Xu, L. Chen and J. Zhou, "Discrete iterative map model-based stability analysis of capacitor current ripple-controlled SIDO CCM Buck converter," *IEEE J. Emerg. Sel. Top. Power Electron.*, vol. 8, no. 4, pp. 3272-3280, Dec. 2020.

[16] H. A V, P. Patra, M. Parmar and N. Chen, "Cumulative charge balanced single-inductor dual-output converter for improved transient and cross regulation," in *IEEE Applied Power Electron. Conf. and Exposition (APEC)*, Feb. 2024, pp. 712-718.

Circulating Current Control with Loss Reduction for Parallel Connected Inverters

Shun Endo
Research & Development Group, Hitachi, Ltd.
Hitachi, Ibaraki, Japan
shun.endo.aw@hitachi.com

Takae Shimada
Research & Development Group, Hitachi, Ltd.
Hitachi, Ibaraki, Japan
takae.shimada.sv@hitachi.com

Masato Ando
Research & Development Group, Hitachi, Ltd.
Hitachi, Ibaraki, Japan
masato.ando.ur@hitachi.com

Yuichi Mabuchi
Research & Development Group, Hitachi, Ltd.
Hitachi, Ibaraki, Japan
yuichi.mabuchi.wz@hitachi.com

Masaki Miyamae
Building Systems Business Unit, Hitachi, Ltd.
Hitachinaka, Ibaraki, Japan
masaki.miyamae.qw@hitachi.com

Naoki Takayama
Building Systems Business Unit, Hitachi, Ltd.
Hitachinaka, Ibaraki, Japan
naoki.takayama.we@hitachi.com

Yohei Matsumoto
Hitachi Building Systems Co., Ltd.
Hitachinaka, Ibaraki, Japan
yohei.matsumoto.xb@hitachi.com

Naoto Onuma
Hitachi Building Systems Co., Ltd.
Hitachinaka, Ibaraki, Japan
naoto.onuma.gv@hitachi.com

Abstract—Connecting inverters in parallel is a common method for increasing current capacity. Due to the difference in the delay time and on-voltage of the gate circuit and the switching element between the parallel inverters, a circulating current is generated between the inverters connected in parallel. In this paper, we propose a method to suppress the circulating current of a parallel inverter and to reduce the loss. By adjusting the turn-on and turn-off timing, the voltage difference between the inverters is reduced. As a result, the circulation current is suppressed. The effect of the proposed method of adjusting the turn-on delay time by matching the turn-off delay time was experimentally verified. By adjusting the turn-on and turn-off timing, the voltage difference between the inverters is eliminated, the circulating current is suppressed, and the current imbalance rate is less than 1%. It was experimentally verified that increasing the turn-on delay time difference from 39 ns to 76 ns and narrowing the turn-off delay time difference from 43 ns to 3 ns could reduce total switching losses by 3.5%.

Keywords— Parallel, Circulating current, Loss reduction, inverter.

I. INTRODUCTION

Parallel inverters are an effective way of solving the problem of high-capacity output requirements by connecting inverter units in parallel. In addition, there are advantages such as economic efficiency and expandability of design work due to modular design and easy installation of inverters because the inverter can carry around each module[1]. Between inverter units connected in parallel, a circulation current is generated due to the difference in delay time and on voltage between the gate circuit and the switching element, and the impedance variation of the inverter output. The circulating current causes current to concentrate on some switching elements, causing problems such as shortened product life and failure due to increased loss. Therefore, Parallel inverters are required to "suppress circulating current " and "reduce loss" [2] [3]. The circulating

current is divided into a DC component and an instantaneous component. The DC component is caused by the difference in the voltage and time of the switching element. The instantaneous component is caused by the difference in turn-on or turn-off timing due to the variation in the delay time of the switching element or gate circuit [4]. A common solution for suppressing circulation current is to apply the reactor between the outputs of the parallel inverters [5] [6] [7]. However, adding a reactor increases cost, weight, and volume. As a method of suppressing circulating current without using a reactor, there is a method of detecting circulating current between inverter outputs and adjusting to the switching timing of the gate driver [8] [9] [10] [11]. However, in order to detect the current, it is necessary to add a current sensor and a control system, and a sensorless method of suppressing the circulating current is desired. In addition no method has been proposed in which both "circulating current suppression" and "loss reduction" achieved by the above method. This paper proposes a method to realize "circulating current suppression" and "loss reduction" without using a reactor or a sensor for circulating current detection.

II. CAUSES OF CIRCULATING CURRENT

A. Circuit Method

Fig. 1 shows a circuit diagram of the Parallel inverter. The parallel inverter connects each module in parallel, and connects each module output to the motor via a harmonic suppression reactor that suppresses the harmonics of the motor. For example, Fig. 1 (a) shows a typical circuit configuration in which a harmonic suppression reactor is connected to each module output, but the number of reactors increases, and the cost, weight, and volume increase. Conversely, Fig. 1 (b) shows a proposed circuit configuration in which harmonic suppression reactors are integrated on the motor side, and the inverter can reduce the number of reactors. A delay adjustment circuit is built in to suppress the circulating current. In this paper, the circuit shown in Fig. 1 (b) is targeted. Fig. 2 shows the circuit for one phase of the inverter. The main causes of circulating current are the variation in the switching delay time of the control circuit, the variation in the $V_{ce(sat)}$ of the IGBT, and the variation in the forward voltage of the FRD. Here, conventionally, there is a method of suppressing the variation in switching delay time by adjusting the switching timing of the gate driver [5] [9] [10] [11] [12]. However, it is not clear whether the turn-on or turn-off delay should be adjusted, and the method of adjusting by pulse width is common.

B. Factors That Generate Circulating Current

In Fig. 2, the circulation current ΔI flows between the module outputs. The main cause of the circulation current is the voltage difference ΔV at the module output, and the component variation factors that cause the voltage difference are shown below.

(1) Variation in delay time during switching between gate circuit and IGBT

(2) $V_{ce(sat)}$ variation of IGBTs

(3) Variation of FRD forward voltage V_F

When there is a difference in the impedance value of the inverter output wiring of Module 1 and Module 2, a voltage drop occurs in the impedance of the difference, and the same phenomenon occurs as when $V_{ce(sat)}$ variation or V_F variation occurs. In this paper, the average value of the circulation current is the DC component, and the circulating current that occurs instantaneously during one period of the carrier frequency is the instantaneous component. Due to the delay time variation, there is a timing when one module outputs voltage and the other module does not. When an output voltage difference occurs at that timing, a circulation current is generated. The instantaneous component of the circulation current occurs when the output voltage occurs.

C. How to Calculate the Circulating Current

The circulation current generated by component variation is described separately for each element. As shown in Fig. 3 (a), the difference in pulse width Δt_p of the module output voltage occurs due to the delay time variation, and the voltage difference ΔV occurs. This voltage difference occurs at the switching frequency of the inverter f_S times in 1 second and is applied to the resistor R between the outputs of the modules connected in parallel. As a result, the circulating current I_{cc_P} due to pulse width variation is shown by equation (1).

$$I_{cc\_P} = \frac{\Delta t_p \cdot V_{DC} \cdot f_S}{R} \quad \text{......................................} (1)$$

As shown in Fig. 3 (b), a voltage difference occurs between the outputs of inverters connected in parallel by $\Delta V_{ce(sat)}$. From this voltage difference and resistance, the I_{cc_ce} of the circulating current is shown by Equation (2).

$$I_{cc\_ce} = \frac{\Delta V_{ce(sat)}}{2R} \quad \text{...} (2)$$

As shown in Fig. 3 (c), a voltage difference occurs between the outputs of inverters connected in parallel by ΔV_F. From this voltage difference and resistance, the circulating current I_{cc_F} is shown by Equation (3).

$$I_{cc\_F} = \frac{\Delta V_F}{2R} \quad \text{..} (3)$$

Equations (2) and (3) calculate the duty of one pulse as 0.5. From (1)-(3), the DC component I_{cc_DC} of the circulating current

between the two parallel modules is shown by equation (4).

$$I_{cc\_DC} = \frac{(\Delta t_p \cdot V_{DC} \cdot f_S) + \frac{1}{2}(\Delta V_{ce(sat)} + \Delta V_F)}{R} \quad \text{..................................} (4)$$

Here, the differences in pulse width Δt_p, $\Delta V_{ce(sat)}$, and ΔV_F are due to manufacturing variations of components constituting the gate circuit and IGBTs.

As shown in Fig. 3 (a), the instantaneous component of the circulation current is generated due to the delay time variation. This occurs equally for both the turn-on difference Δt_{on} and the turn-off difference Δt_{off}. This instantaneous component is applied to the inductance L between the modules and is represented by equations (5) and (6).

$$I_{cc\_ton} = \frac{\Delta t_{on} \cdot V_{DC}}{L} \quad \text{.......................................} (5)$$

$$I_{cc\_toff} = \frac{\Delta t_{off} \cdot V_{DC}}{L} \quad \text{.......................................} (6)$$

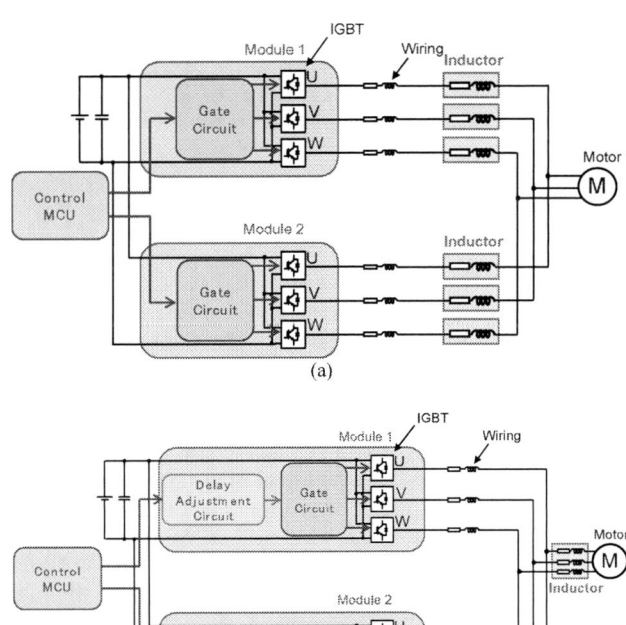

(a)

(b)

Fig. 1. Schematic of parallel inverter circuits. (a)Connect the reactor to each module. (b)Consolidate reactors on the motor side.

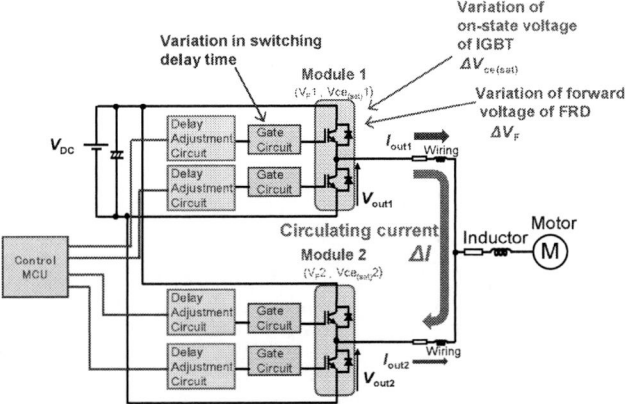

Fig. 2. Causes of circulating current for one phase.

979-8-3315-1612-3/25 $31.00 © 2025 IEEE   2682

III. Turn-on Delay Reduces Circulating Current and Switching Losses

A. Delay Adjustment Circuit

As shown in equations (1), (5), and (6), the circulating current is caused by the delay time difference. As shown in Equation (4), the DC component of lateral flow can be reduced by adjusting Δt_p to cancel the voltage difference due to $\Delta V_{ce(sat)}$ and ΔV_F. Fig. 4 shows a delay adjustment circuit. The RC circuit is composed of a variable resistor and a capacitor, and a threshold voltage is set by a comparator. By adjusting the variable resistance, the time constant of the RC circuit changes, and the time until the threshold voltage is reached changes. In addition, by providing a turn-on and turn-off path by a diode, the turn-on delay time and turn-off delay time can be adjusted by adjusting the variable resistance. A delay adjustment circuit is provided between the MCU and the gate circuit to adjust the switching timing. Since the problem is that the output current of one module increases due to the unbalance of the output current of the module due to the circulating current, $\Delta V_{ce(sat)}$ and ΔV_F used in the calculation need to be considered as the value of the maximum output current of the inverter. In this paper, the inverter output current suppresses the circulation current at DC 315 A output.

B. Waveform During Turn-on Delay Adjustment

Fig. 5 shows the waveform of the proposed method during turn-on delay adjustment. As an example, the conditions with the difference between $V_{ce(sat)}$ and V_F will be described. The voltage difference is suppressed by delaying the turn-on of the module with a high voltage by eliminating the turn-off delay time difference. As a result, the voltage component of equation (4) disappears, and the DC component of the circulating current decreases. As a result, the voltage component of equation (4) is eliminated, and the DC component of the circulation current becomes zero. The inverter can also reduce the turn-on timing current of the Modules 1 and Module 2. Thereby, switching loss can be suppressed.

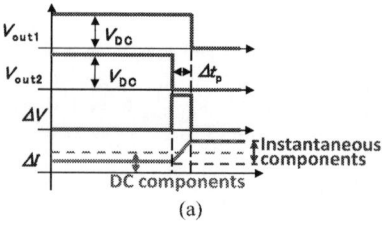

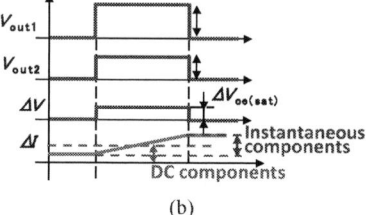

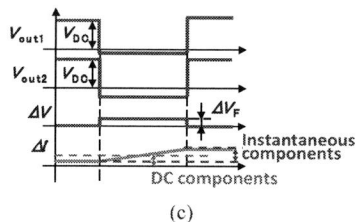

Fig. 3. Circulating current between modules. (a)Circulating current caused by delay time variation. (b)Circulating current caused by $\Delta V_{ce(sat)}$. (c) Circulating current caused by ΔV_F.

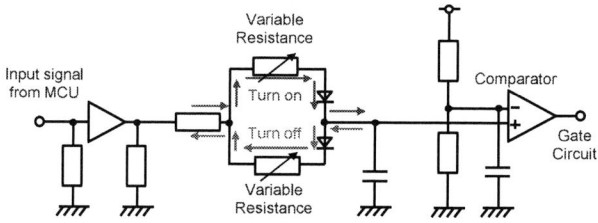

Fig. 4. Delay adjustment circuit.

TABLE I. SIMULATION CONDITIONS

Item		Value
Inverter output current		315 A (DC output)
Module 1 wiring	Resistance	1.7 mΩ
	Inductance	2.4 uH
Module 1 add wiring	Resistance	2.9 mΩ (Equivalent to $\Delta V_{ce(sat)}$, ΔV_F 0.46V)
	Inductance	2.4 uH
Module 2 wiring	Resistance	1.6 mΩ
	Inductance	2.2 uH
V_{out} delay time difference	Turn-on delay time Δt_{on}	89 ns
	Turn-off delay time Δt_{off}	0 ns
	Difference in pulse width Δt_p	89 ns

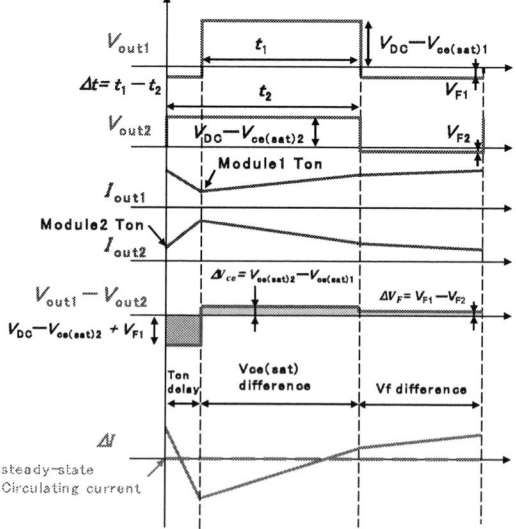

Fig. 5. Proposed circulating current suppression method.

C. Circuit Simulation

TABLE 1 shows the simulation conditions. Ansys simplore was used for the simulation. As analysis conditions, output wiring of 1.7 mΩ, 1.6 mΩ and 2.4 uH, 2.2 uH is set for the output of Module 1 and Module 2, and additional wiring of 2.9 mΩ and 2.4 uH is used for Module 1. The additional wiring of 2.9 mΩ causes a voltage drop of 0.46 V to flow through about 158 A, which is half of the inverter output current of 315 A. This corresponds to $\Delta V_{ce(sat)} = 0.46$ V, $\Delta V_F = 0.46$ V. Fig. 6 shows the analysis results. From equation (4), it can be seen that Δt_p is -89 ns, and the DC component of the circulating current is within a small value of 1% or less of the output current. Equation (5) indicates that the instantaneous component of the circulation current is 8.6 A, and the same result was obtained in the analysis. Furthermore, it was confirmed by analysis that the turn-on of Module 1 was 153.3 A and the turn-on of Module 2 was 153.3 A, and that switching was performed at low current due to the instantaneous circulating current generated by the turn-on delay. In the next chapter, we will conduct a test on the actual device and confirm the effect.

IV. EXPERIMENTAL RESULTS

TABLE 2 shows the experimental conditions. The wire impedance is equal to the analysis. This paper compares the following two methods.

(I). A method in which Δt_p is provided with the same difference between turn-on delay and turn-off delay.

(II). Compared with the case where Δt_p is provided due to the turn-on delay.

Fig. 7 shows the experimental waveform. In the method (I) shown in Fig. 7 (a), when V_{out} is 340 V, which is half of the DC link voltage, there is a turn-on delay difference of 39 ns and a turn-off delay time difference of 43 ns. As a result, Δt_p is set to 82 ns. In the method (II) shown in Fig. 7 (b), the turn-on delay time difference is 76 ns and the turn-off delay time difference is 3 ns. As a result, Δt_p is set to 79 ns. Fig. 8 shows the output current and circulation current of each module. It can be seen that the lateral flow of the delay time adjustment is suppressed. The circulation current of the experiment is calculated by Equation (7).

$$\Delta I = \frac{I_{out1} - I_{out2}}{2} \quad \text{(7)}$$

Fig. 8 (a) is the module output current of the (I) method. Twice in one switching cycle, a sudden current change occurs due to instantaneous circulation current. This occurs during turn-on timing and turn-off timing. Because the noise in the switching timing was large, the waveform was approximated as a straight line to estimate the switching timing current. The turn-on current for Module 1 was 154 A, and for Module 2 it was 158 A. The average current for one cycle was 162 A for Module 1 and 159 A for Module 2, and the current imbalance $\Delta I_{imbance}$ was within 1%. Fig. 8 (b) is the module output current of the (II) method. Since the difference in turn-on delay time is large, the instantaneous component of turn-on is large, and the difference between the current of Module 1 and Module 2 is large, but the instantaneous component is small at the turn-

off timing, and it can be seen that the difference between the current of Module1 and Module 2 is not large.

TABLE II. EXPERIMENTAL CONDITIONS

Item		Value
Motor current		322 A
Equivalent on-voltage difference		0.46 V
(a) Pulse width adjustment	Turn-on delay time Δt_{on}	39 ns
	Turn-off delay time Δt_{off}	43 ns
	Difference in pulse width Δt_p	82 ns
	Simple diagram of the Vout Waveform	Module1≫Module2 Turn-on 39ns Turn-off 43ns
(b) Turn-on delay time adjustment	Turn-on delay time Δt_{on}	76 ns
	Turn-off delay time Δt_{off}	3 ns
	Difference in pulse width Δt_p	79 ns
	Simple diagram of the V_{out} Waveform	Module1 ≫Module2 Turn-on 76ns Turn-off 3ns
Measurement point		

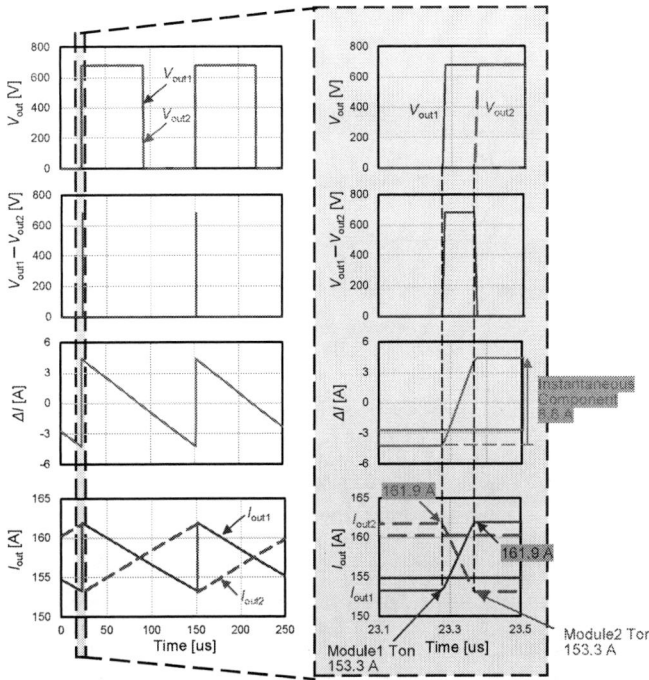

Fig. 6. Circuit simulation waveform.

979-8-3315-1612-3/25 $31.00 © 2025 IEEE

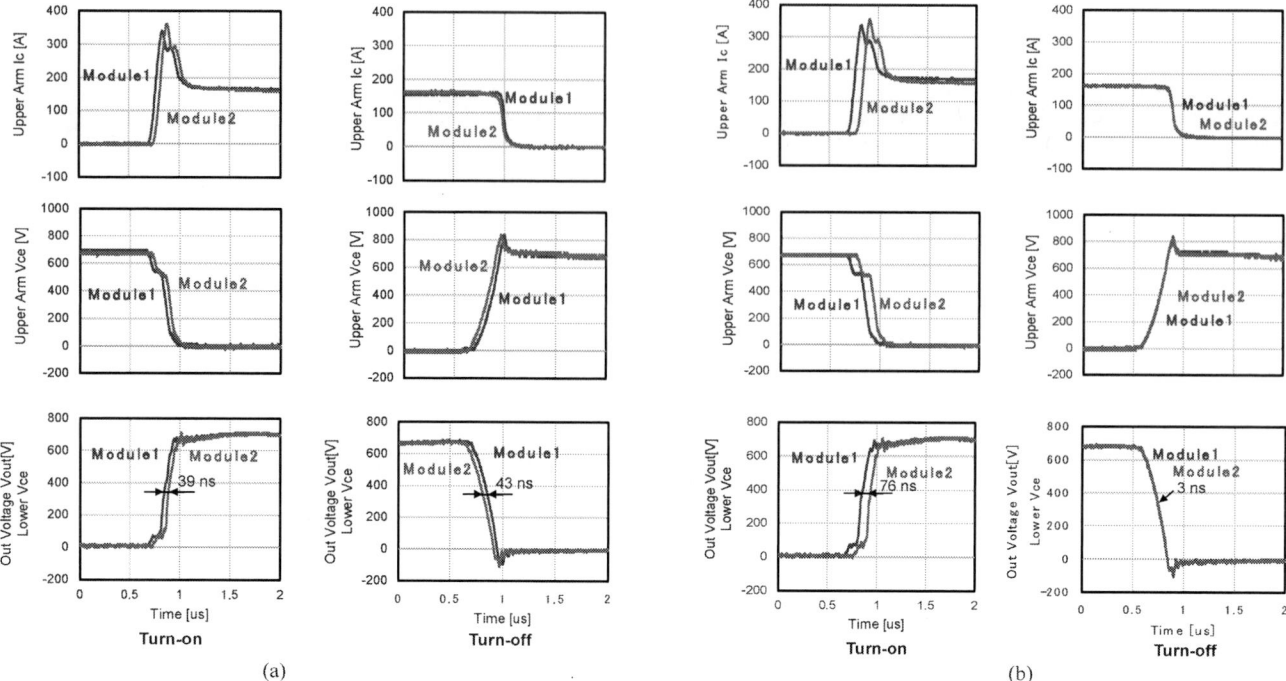

Fig. 7. Experimental waveform. (a) Turn-on delay of 39 ns and turn-off delay of 43 ns in Pulse width adjustment. (b) Turn-on delay of 76 ns and turn-off delay of 3 ns in Turn-on delay time adjustment.

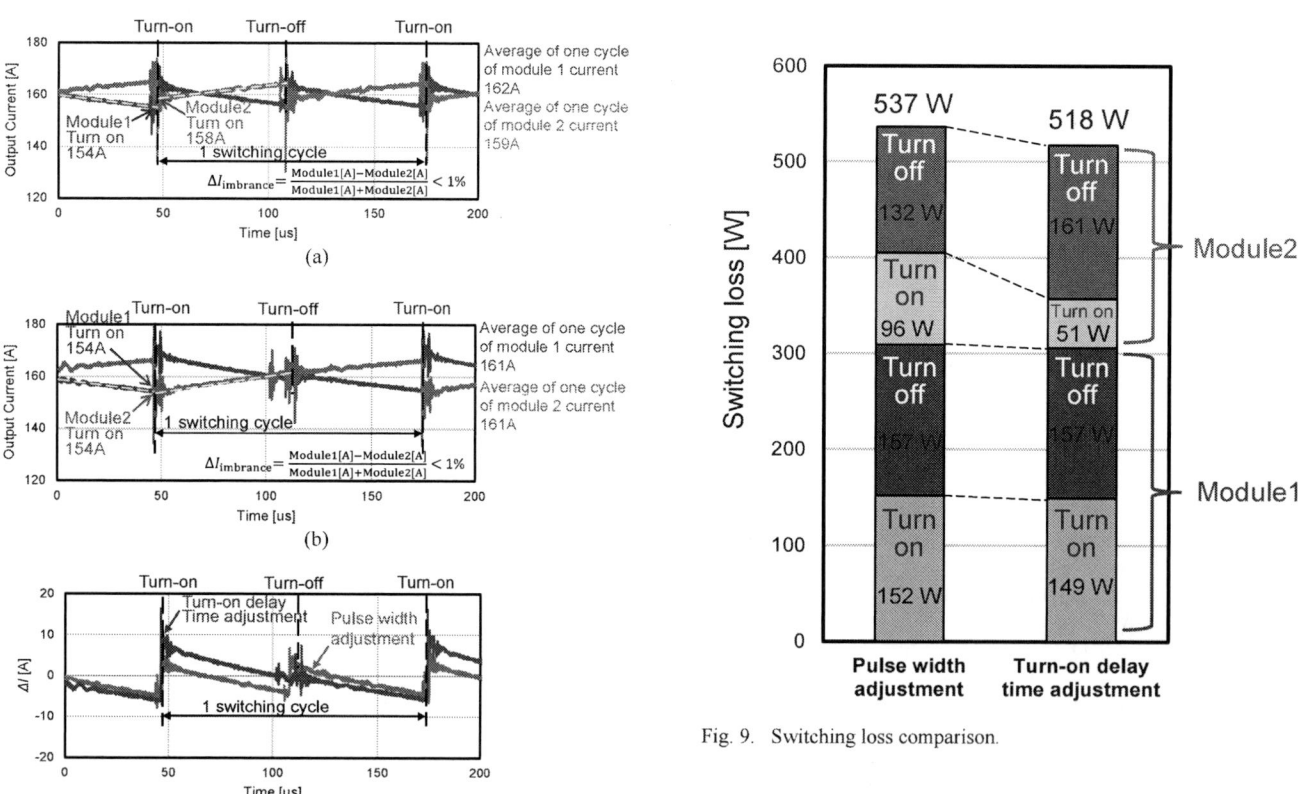

Fig. 8. Current waveform. (a) Pulse width adjustment output current. (b) Turn-on delay time adjustment output current. (c) Circulating Current.

Fig. 9. Switching loss comparison.

The current during switching was 154 A for Module 1 and 154 A for Module 2. The average current per cycle was 161 A for Module 1 and 161 A for Module 2, and the current imbalance $\Delta I_{imbrance}$ was within 1%. Fig. 8 (c) indicates the circulating current. Both the turn-on delay method and the pulse width adjustment method are methods that suppress circulating current, so the current imbalance $\Delta I_{imbrance}$ of the two is less than 1%, as shown in Fig. 8 (a) and (b). Fig. 9 shows the difference in switching loss. By widening the turn-on delay time difference from 39 ns to 76 ns, the turn-on loss was reduced from 248 W to 200 W. By narrowing the turn-off delay time difference from 43 ns to 3 ns, the turn-off loss increases from 289 W to 318 W. The total switching loss can be reduced by 3.5% from 537 W to 518 W. Therefore, it was clarified that the circulating current due to $\Delta V_{ce(sat)}$ and ΔV_F can be reduced by changing the size of Δt_p by delay time adjustment, and switching loss can be reduced by making Δt_p with the turn-on delay time difference.

V. CONCLUSIONS AND FUTURE WORK

In this paper, we propose a method that can suppress the circulation current of a parallel inverter without a sensor or inductor and suppress switching loss by adjusting the turn-on and turn-off timing based on the delay time change of the inverter component and the output voltage fluctuation of the switching element. The causes of the circulation current, such as the circulation current generated by the switching delay time variation of the gate circuit and the IGBT, the $V_{ce(sat)}$ variation of the IGBT, and the V_F variation of the FRD, were formulated, and the required switching timing adjustment amount was calculated. In addition, we experimentally verified that the turn-on and turn-off delay can suppress the circulation current and suppress the switching loss. In the experiment, by adjusting the delay time, the circulating current could be suppressed to current imbalance ratio of less than 1%, which is sufficiently small compared to the inverter output current of 315 A. In addition, we experimentally verified that increasing the turn-on delay time difference from 39 ns to 76 ns and narrowing the turn-off delay time difference from 43 ns to 3 ns could reduce total switching losses by 3.5%. This gives it an advantage in suppressing the circulating current by the proposed turn-on delay method. The circulating current during switching and the phenomenon that occurs during turn-off will be future issues.

REFERENCES

[1] Fei Xiao, Wei Chen, Jilong Liu, Hengli Wang,"Parallel connected three phase inverters based on modular design and distributed control ," International Power Electronics Conference (IPEC), 2014.

[2] Nasser Talebi, M.A. Sadrnia, S.M.R. Rafiei,"Current and voltage control of paralleled multi-module inverter systems," Mediterranean Conference on Control and Automation,2009.

[3] J. Hu, O. Alatise, J. A. O. Gonzalez, R. Bonyadi, L. Ran and P. A. Mawby, "The Effect of Electrothermal Nonuniformities on Parallel Connected SiC Power Devices Under Unclamped and Clamped Inductive Switching", IEEE Trans. Power Electron, vol. 31, no. 6, pp. 4526-4535, 2016.

[4] Nan Lin, Yuheng Wu, Mohammad Mahmud, Yue Zhao, Alan Mantooth,"Current Balancing Methods for a High Power Silicon Carbide Inverter with Paralleled Modules," IEEE Applied Power Electronics Conference and Exposition (APEC), 2022.

[5] Peng Zhang, Guorong Zhang, Haibo Du,"Circulating Currents Control for Parallel Grid-Connected Three-Phase Inverters," IEEE Transactions on Circuits and Systems ,Vol.65, Issue.9,2018.

[6] T. Kawabata and S. Higashino, "Parallel Operation of Voltage Source Inverters", IEEE Transactions on Industry Applications, vol. 24, no. 2, pp. 281-287, March/April 1988.

[7] J.W. Dixon and B.T. Ooi, "Series an Parallel Operation of Hysteresis Current-Controlled PWM Rectifiers", IEEE Transactions on Industry Applications, vol. 25, no. 6, pp. 1817-1823, November/December 1988.

[8] Toni Itkonen, Kimmo Rauma, Hannu Saren, Ossi Laakkonen, Olli Pyrhonen, Pertti Silventoinen,"Parallel Connected Voltage Source Inverters without Intermodule Reactors," International Power Electronics and Motion Control Conference, 2006.

[9] S. Ogasawara, J. Takagaki, H. Akagi and A. Nabae, "A Novel Control Scheme of a Parallel Current-Controlled PWM inverter", IEEE Transactions on Industry Applications, vol. 28, no. 5, pp. 1023-1030, September/October 1982.

[10] N. Kawakami, M. Mitsuyuki, T. Ikimi, A. Ueda, J. Takahashi and K. Kamiyama, "Quick Response and Low-Distortion Current Control for Multiple Inverter-Fed Induction Motor Drives", IEEE Transactions on Industry Applications, vol. 9, no. 2, pp. 240-247, March 1994.

[11] Sungjoon Cho, Yun Jang, Sebong Jeon, Kyo-Beum Lee, "A Reliable Suppression Method of High Frequency Circulating Current in Parallel Grid Connected Inverters", IEEE Energy Conversion Congress and Exposition (ECCE), 2019.

[12] Mason Parker, Sebastian Neira, Philip Waite, Edward L. Horsley, Stephen Finney, Paul D. Judge, "Current Balancing of Parallel High Current SiC Half Bridge Modules using Delay Based Active Gate Driving with Inter-Device Inductances", 2024 IEEE Applied Power Electronics Conference and Exposition (APEC), 2024.

Analysis of Power and Power Spectral Density for Quaternary Random Pulse Position Modulation

Hung-Chi Chen
Department of Electronics and
Electrical Engineering
National Yang Ming Chiao Tung
University (NYCU)
Hsinchu, Taiwan, (R.O.C.)
hcchen @nycu.edu.tw

Hsiang-Kai, Wu
Institute of Electrical and
Control Engineering
National Yang Ming Chiao Tung
University (NYCU)
Hsinchu, Taiwan (R.O.C.)
hcchen @nycu.edu.tw

Chih-Chiang Wu
Mechanical and Mechatronics Systems
Research Laboratories,
Industrial Technology Research
Institute (ITRI),
Hsinchu, Taiwan (R.O.C.)
John.Wu@itri.org.tw

Abstract—**In this paper, the quaternary random pulse position modulation (QRPPM) is proposed to further reduce the power of dominant harmonic clusters. From the analysis of the power and power spectral density, the minimum integer-harmonics power dispersion rate (HPDR) is increased from 50.0% of the conventional random center/edge alignment PWM (RCEA-PWM) to 75.0% of the proposed QRPPM. Some results are provided to validate the proposed method.**

Keywords—harmonic spread, power spectral density, random pulse position modulation

I. INTRODUCTION

The conventional PWM strategy is widely used in DC-DC converters due to its simplicity. However, this strategy generates the deterministic switching signals and contributes to the concentration of harmonic energy at the specific frequency. The dominant harmonic clusters may lead to increased EMI [1-3].

Some random pulse position modulation (RPPM) techniques had been developed to disperse some integer-harmonic energy. Fig. 1. shows two common RPPMs.

In Fig. 1(a), two deterministic signals with the same duty ratio are generated from the comparisons of the control signal and two saw-teeth signals [4]. Since one of two deterministic signals always begins with "High" and the other always begins with "Low" within every time slots, it is also named random lead-lag (RLL) PWM.

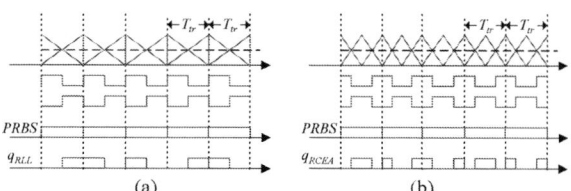

(a)   (b)

Fig. 1 Two common random pulse position modulations
(a) random lead-lag PWM (RLL-PWM) [4-5]
(b) random center/edge aligned PWM (RCEA-PWM)[6-9].

In Fig. 1(b), two deterministic signals with the same duty ratio are generated from the comparisons of one control signal and two triangular signals [5-8]. Since one of two deterministic signals always begins and ends with "High" and the other always begins and ends with "Low" within every time slots, it is also named random center/edge-aligned PWM (RCEA).

It is noted that RCEA-PWM possesses better integer-harmonics power dispersion performance than RLL-PWM. Thus, RCEA-PWM often cooperates with other random schemes to enhance the power dispersion performance.

In [7], random carrier frequency modulation (RCFM) was cooperated with RCEA-PWM. In [8], random pulse number modulation (RPNM) was combined with RCEA-PWM. Table I tabulates the comparison of various RPPM. The integer-harmonics power dispersion rate is defined to evaluate the power dispersion performance. For the conventional PWM, PWM's power is concentrated at the integer multiples of the switching frequency. Thus, PWM's HPDR is 0% in Table I.

Due to randomness, some concentrated power may be dispersed, and thus HPDR of random modulation may be larger than 0%. It is noted that power dispersion rate is dependent on the duty ratio, and RLL and RCEA yield the peak HPDR 100% at the duty ratio 0.5.

Table I. Summary of various RPPM

	PWM	RLL	RCEA	QRPPM
Triangular Frequency	f_{tr}	f_{tr}	f_{tr}	f_{tr}
Average Switching Frequency	f_{tr}	$0.75f_{tr}$	$1.25f_{tr}$	$1.1875f_{tr}$
1st nonzero harmonic Frequency	f_{tr}	f_{tr}	$2f_{tr}$	$4f_{tr}$
HPDR	0%	50%~100%	50%~100%	75%~100%

Since the switching signals of the random modulation are not deterministic, the power spectrum may include the continuous and discrete parts [10-12].

Normally, the popular Fourier transform (FT) can be used to obtain the deterministic signal's spectral characteristics and its frequency domain distribution. However, the FT may not clearly describe the frequency domain characteristics of non-deterministic signals.

For random signals, analyzing their spectrum with the FT is possible, but the result is not comprehensive since the amplitude at each frequency represents just one sample within the signal's sample space.

Thus, power spectral density (PSD) are widely used to evaluate the spectral distribution of a random signal [6, 13]. However, PSD can also be used to evaluate the spectral harmonics of a deterministic signal.

In this paper, the quaternary random pulse position modulation (QRPPM) is proposed and its integer-harmonics power dispersion effect is evaluated by the analysis of power and power spectral density. The minimum power dispersion rate (PDR) is increased from 50% of the conventional random center/edge alignment PWM (RCEA-PWM) to 75.0% of the proposed QRPPM without the decrease of the switching number and switching loss. Some simulation and experimental results are provided to demonstrate the proposed QRPPM.

979-8-3315-1612-3/25 $31.00 © 2025 IEEE

II. THE PROPOSED QRPPMs

The block diagram of the proposed QRPPM is plotted in Fig. 2(a) where the random signal *PRBS* changes within every fixed time period T_{tr} and selects one of four pulse signals to generate the proposed pulse signal $q_{QRPPM}(t)$.

Four deterministic pulse signals $q_1(t)$, $q_2(t)$, $q_3(t)$ and $q_4(t)$ are generated from the comparisons of the control signal $v_{cont}(t)$ and four interleaved triangular signals $v_{tr1}(t)$, $v_{tr2}(t)$, $v_{tr3}(t)$ and $v_{tr4}(t)$ with the unified amplitude $\hat{v}_{tr}$. Thus, all four pulse signals yield the same duty ratio d. The duty ratio d of the pulse signals can be expressed by

$$d = \frac{v_{cont}}{\hat{v}_{tr} = \hat{v}_{tr1} = \hat{v}_{tr2} = \hat{v}_{tr3} = \hat{v}_{tr4}} \tag{1}$$

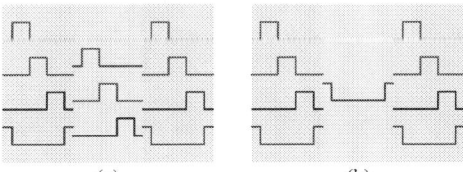

(a)　　　　　　　　　(b)
Fig. 2 (a) Block diagram of the proposed QRPPM;
(b) illustrated waveforms of the proposed QRPPM.

From the illustrated waveforms in Fig. 2(b), four pulse signals during $nT_{tr} \leq t \leq (n+1)T_{tr}$ can be expressed by

$$q_1(t) = u\left(t - nT_{tr}\right) - u\left(t - nT_{tr} - \frac{dT_{tr}}{2}\right)$$
$$+ u\left(t - (n+1)T_{tr} + \frac{dT_{tr}}{2}\right) - u\left(t - (n+1)T_{tr}\right) \tag{2}$$

$$q_2(t) = \begin{cases} u\left(t - nT_{tr} - (\frac{T_{tr}}{4} - \frac{dT_{tr}}{2})\right) \\ \quad - u\left(t - nT_{tr} - (\frac{T_{tr}}{4} + \frac{dT_{tr}}{2})\right), & \text{when } 0 \leq d \leq \frac{1}{2} \\ u\left(t - nT_{tr}\right) - u\left(\begin{matrix} t - nT_{tr} \\ -(\frac{T_{tr}}{4} + \frac{dT_{tr}}{2}) \end{matrix}\right) \\ + u\left(\begin{matrix} t - (n+1)T_{tr} \\ -(\frac{T_{tr}}{4} - \frac{dT_{tr}}{2}) \end{matrix}\right) - u\left(t - (n+1)T_{tr}\right), & \text{when } \frac{1}{2} \leq d \leq 1 \end{cases} \tag{3}$$

$$q_3(t) = u\left(t - nT_{tr} - (\frac{T_{tr}}{2} - \frac{dT_{tr}}{2})\right) - u\left(t - nT_{tr} - (\frac{T_{tr}}{2} + \frac{dT_{tr}}{2})\right) \tag{4}$$

$$q_4(t) = \begin{cases} u\left(t - nT_{tr} - (\frac{3T_{tr}}{4} - \frac{dT_{tr}}{2})\right) \\ \quad - u\left(t - nT_{tr} - (\frac{3T_{tr}}{4} + \frac{dT_{tr}}{2})\right), & \text{when } 0 \leq d \leq \frac{1}{2} \\ u\left(t - nT_{tr}\right) - u\left(\begin{matrix} t - nT_{tr} \\ +(\frac{T_{tr}}{4} - \frac{dT_{tr}}{2}) \end{matrix}\right) \\ + u\left(\begin{matrix} t - nT_{tr} \\ -(\frac{3T_{tr}}{4} - \frac{dT_{tr}}{2}) \end{matrix}\right) - u\left(t - (n+1)T_{tr}\right), & \text{when } \frac{1}{2} \leq d \leq 1 \end{cases} \tag{5}$$

The random signal PRBS may be 00, 01, 10 and 11 with equal possibility 1/4. The pulse signal $q_{QRPPM}(t)$ can be expressed by

$$q_{QRPPM}(t) = \begin{cases} q_1, & when \ \text{PRBS} = 00 \\ q_2, & when \ \text{PRBS} = 01 \\ q_3, & when \ \text{PRBS} = 10 \\ q_4, & when \ \text{PRBS} = 11 \end{cases} \tag{6}$$

To calculate the average switching number per second, the illustrated classifications with duty ratio <0.5 are plotted in Fig. 3. From Fig. 3(a), when the PRBS is 01, 10 and 11, the selected pulses begin and end with "Low", and thus, the switching number is counted by 1.

From Fig. 3(b), when the PRBS is 00, the pulse signal $q_1(t)$ is selected and it begins with "High", goes to "Low", returns to "High" and ends with "High", and the switching number may be counted by more than 1.

If the previous pulse ends with "Low", the switching number should be additionally counted by 0.5. If the next pulse begins with "Low", the switching number should be additionally counted by 0.5.

With considering the possibility, the average switching number is calculated by 1+1/4*3/4*1/2+1/4*3/4*1/2=19/16.

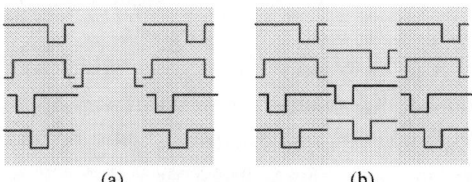

(a)　　　　　　　　　(b)
Fig. 3 Classifications of the proposed pulse signal with duty ratio d<0.5
(a) the cases of the switching number counted by 1;
(b) the case of the switching number counted by more than 1.

The illustrated classifications with duty ratio >0.5 are plotted in Fig. 4. From Fig. 4(a), when the PRBS is 10, the selected pulse begins and ends with "Low", and thus, the switching number is counted by 1.

From Fig. 4(b), the selected pulses begin with "High", go to "Low", return to "High" and end with "High", and the switching number may be counted by more than 1.

If the previous pulse ends with "Low", the switching number should be additionally counted by 0.5. If the next pulse begins with "Low", the switching number should be additionally counted by 0.5.

With considering the possibility, the average switching number is calculated by 1+3/4*1/4*1/2+3/4*1/4*1/2=19/16.

It follows that the proposed QRPPM yields the average switching frequency $1.1875 f_{tr}$.

(a)　　　　　　　　　(b)
Fig. 4 Classifications of the proposed pulse signal with duty ratio d>0.5
(a) the cases of the switching number counted by 1;
(b) the case of the switching number counted by more than 1.

III. Power Spectral Density Anaysis

In dc/dc converter, the output voltage is dependent on the switching signal. Thus, the analysis of power spectral density of the pulse signal $q_{QRPPM}(t)$ is able to evaluate the power distribution of the proposed strategie.

The power spectral density PSD $S_q(f)$ of the combined pulse signal $q_{QRPPM}(t)$ consists of the dc PSD component $S_q^{dc}(f=0)$, the integer-harmonics PSD $S_q^h(f=nf_{tr})$ and the other continuous PSD component $S_q^c(f)$.

$$S_q(f) = S_q^{dc}(f=0) + S_q^h(f=nf_{tr}) + S_q^c(f) \quad (7)$$

$$S_q^{dc}(0) = \frac{\int_0^{T_{tr}} q(t)dt}{T_{tr}} \delta(0) \quad (8)$$

$$S_q^h(f=nf_{tr}) = \sum_{n=1}^{\infty} \left(2\left|E\left[Q(nf_{tr})\right]\right|^2 (f_{tr})^2 \delta(f-nf_{tr}) \right) \quad (9)$$

$$S_q^c(f) = (f_{tr}) \left(E\left[|Q(f)|^2\right] - \left|E[Q(f)]\right|^2 \right) \quad (10)$$

where $f_{tr} = 1/T_{tr}$ is the triangle frequency, and $E[.]$ represents the operator for taking the expected value across the entire FT ensemble of the possible deterministic pulse signals. For a pulse signal $q_{QRPPM}(t)$ with fixed duty ratio d within every period T_{tr}, the dc PSD component is $S_q^{dc}(f=0) = d \cdot \delta(0)$ and thus, the dc power is d^2.

The integer-harmonics power P_q^h can be calculated from the integer-harmonics PSD $S_q^h(f)$.

$$P_q^h = \sum_{\mathrm{int}\,n=1}^{\infty} \left(2\left|E\left[Q(nf_{tr})\right]\right|^2 (f_{tr})^2 \right) \quad (11)$$

Similarly, the power P_q^c can be obtained from the continuous PSD component.

$$P_q^c = 2\int_{0^+}^{\infty} S_q^c(f)df = P_q - P_q^{dc} - P_q^h \quad (12)$$

where P_q is the total power of the pulse signal $q_{QRPPM}(t)$, and it is equal to the sum $P_q = P_q^{dc} + P_q^h + P_q^c$.

For any comparison signal $q_{QRPPM}(t)$ varying between 0 and 1 with the fixed duty ratio d within each triangle period T_{tr}, the rms value of $q_{QRPPM}(t)$ is $q_{rms} = \sqrt{d}$, and thus, the total power of the signal $q_{QRPPM}(t)$ is $P_q = (\sqrt{d})^2 = d$. Because that the dc value of the signal $q_{QRPPM}(t)$ is d, and thus, the dc power P_q^{dc} of the signal $q_{QRPPM}(t)$ is $P_q^{dc} = d^2$. From (12), the sum of the integer-harmonics power P_q^h and the power P_q^c is constant $P_q^c + P_q^h = d - d^2$ [13].

If the pulse signal is deterministic, the continuous PSD component $S_q^c(f)$ must be zero, and thus, the power P_q^c is zero. It follows that the integer-harmonics power P_q^h is $P_q^h = P_q - P_q^{dc} = d - d^2$ for a deterministic signal and there is no integer-harmonics power dispersion.

For a nondeterministic comparison signal with fixed duty ratio d, the continuous PSD component $S_q^c(f)$ must be larger than zero, and the power P_q^c must be larger than zero. That is, the integer-harmonics power is $P_q^h = P_q - P_q^{dc} - P_q^c < P_q - P_q^{dc} = d - d^2$.

That is, the strategy generating the nondeterministic comparison signal may disperse some integer-harmonics power P_q^h to the power P_q^c due to the constant $P_q^c + P_q^h = d - d^2$. The power dispersion effect is evaluated by the defined integer-harmonics power dispersion ratio ξ (HPDR).

$$\xi = \frac{P_q^c}{P_q^h + P_q^c} = \frac{P_q^c}{d - d^2} * 100\% \quad (13)$$

The Fourier transform (FT) $Q_1(f)$, $Q_2(f)$, $Q_3(f)$ and $Q_4(f)$ of the deterministic pulse signals can be expressed by

$$Q_1(f) = \frac{\sin(\pi f d T_{tr})}{\pi f} \left[\cos(2\pi f T_{tr}) - j\sin(2\pi f T_{tr})\right] \quad (14)$$

$$Q_2(f) = \int_{\frac{T_{tr}}{4} - \frac{dT_{tr}}{2}}^{\frac{T_{tr}}{4} + \frac{dT_{tr}}{2}} e^{-j2\pi ft} dt = \int_{\frac{T_{tr}}{4} - \frac{dT_{tr}}{2}}^{\frac{T_{tr}}{4} + \frac{dT_{tr}}{2}} \frac{de^{-j2\pi ft}}{(-j2\pi f)} = \frac{\sin(\pi f d T_{tr})}{\pi f} \begin{bmatrix} \cos\left(\frac{\pi f T_{tr}}{2}\right) \\ -j\sin\left(\frac{\pi f T_{tr}}{2}\right) \end{bmatrix} \quad (15)$$

$$Q_3(f) = \frac{\sin(\pi f d T_{tr})}{\pi f} \left[\cos(\pi f T_{tr}) - j\sin(\pi f T_{tr})\right] \quad (16)$$

$$Q_4(f) = \int_{\frac{3T_{tr}}{4} - \frac{dT_{tr}}{2}}^{\frac{3T_{tr}}{4} + \frac{dT_{tr}}{2}} e^{-j2\pi ft} dt = \int_{\frac{3T_{tr}}{4} - \frac{dT_{tr}}{2}}^{\frac{3T_{tr}}{4} + \frac{dT_{tr}}{2}} \frac{de^{-j2\pi ft}}{(-j2\pi f)} = \frac{\sin(\pi f d T_{tr})}{\pi f} \begin{bmatrix} \cos\left(\frac{3\pi f T_{tr}}{2}\right) \\ -j\sin\left(\frac{3\pi f T_{tr}}{2}\right) \end{bmatrix} \quad (17)$$

The expected value of the square amplitude of FT can be expressed by

$$E\left[|Q(f)|^2\right] = \frac{1}{4}|Q_1(f)|^2 + \frac{1}{4}|Q_2(f)|^2 + \frac{1}{4}|Q_3(f)|^2 + \frac{1}{4}|Q_4(f)|^2 \quad (18)$$

The square $\left|E[Q(f)]\right|^2$ of the expected FT $E[Q(f)]$ can be expressed by

$$\left|E[Q(f)]\right|^2 = \left|\frac{1}{4}Q_1(f) + \frac{1}{4}Q_2(f) + \frac{1}{4}Q_3(f) + \frac{1}{4}Q_4(f)\right|^2 \quad (19)$$

From (9), the integer-harmonics PSD component $S_{TRPPM}^h(f)$ can be derived by

$$S_{QRPPM}^h(f) = \sum_{n=1}^{\infty} \left(\frac{2\sin^2(dn4\pi)}{(4n\pi)^2} \delta(f - 4nf_{tr}) \right) \quad (20)$$

From (20), the integer-harmonics power P_{QRPPM}^h can be simplified to

$$P_{QRPPM}^h(d) = \begin{cases} -(d - \frac{1}{8})^2 + \frac{1}{64}, & \text{when } 0 \le d \le \frac{1}{4} \\ -(d - \frac{3}{8})^2 + \frac{1}{64}, & \text{when } \frac{1}{4} \le d \le \frac{2}{4} \\ -(d - \frac{5}{8})^2 + \frac{1}{64}, & \text{when } \frac{2}{4} \le d \le \frac{3}{4} \\ -(d - \frac{7}{8})^2 + \frac{1}{64}, & \text{when } \frac{3}{4} \le d \le 1 \end{cases} \quad (21)$$

The dispersion power P_{QRPPM}^c is equal to $P_{QRPPM}^c = d - d^2 - P_{QRPPM}^h$.

For reference, the integer-harmonics power for RCEA-PWM is included here.

$$P_{RCEA}^h(d) = \sum_{n=1}^{\infty} \frac{\left[1 + (-1)^n\right]\sin^2(dn\pi)}{(n\pi)^2}$$

$$= \begin{cases} -(d - \frac{1}{4})^2 + \frac{1}{16}, & \text{when } 0 \le d \le 0.5 \\ -(d - \frac{3}{4})^2 + \frac{1}{16}, & \text{when } 0.5 \le d \le 1.0 \end{cases} \quad (22)$$

The derived power distribution are plotted in Fig. 5(a) and Fig. 5(b) where the **blue line** indicates the function $d - d^2 = P_{QRPPM}^h + P_{QRPPM}^c$, the gray line indicates the integer-harmonic power P_{QRPPM}^h and the orange line is the dispersion power P_{QRPPM}^c . The proposed QRPPM yields larger dispersion power than the conventional RCEA-PWM. Fig. 3 plots the power dispersion ratio curves of the proposed QRPPM (**red line**) and the conventional RCEA-PWM (**black line**). Obviously, the minimum power dispersion ratio PDR is increased from 50% to 75%.

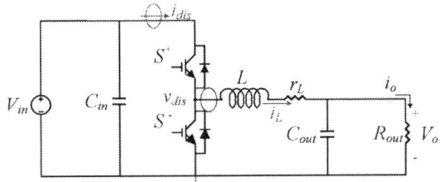

Fig. 7 Buck converter for demonstration.

TABLE II CIRCUIT PARAMETERS

PWM Schemes	Conventional PWM	RCEA-PWM	Proposed QRPPM
Triamgular Frequency f_{tr}	40kHz	40kHz	40kHz
Input Voltage V_{in}	200V		
Inductor L	2.24mH		
Output Capacitance C_{out}	300μF		
Load R_{out}	25Ω		

From Fig. 6 and (13), the curve at duty ratio d=0.25 shows that the HPDR of the proposed QRPPM is 100% and the HPDR of the RCEA-PWM is 66.7%.

It is obvious that from Fig. 8(b), some integer-harmonic power is dispersed by RCEA-PWM and all integer-harmonic power are dispersed by the proposed QRPPM.

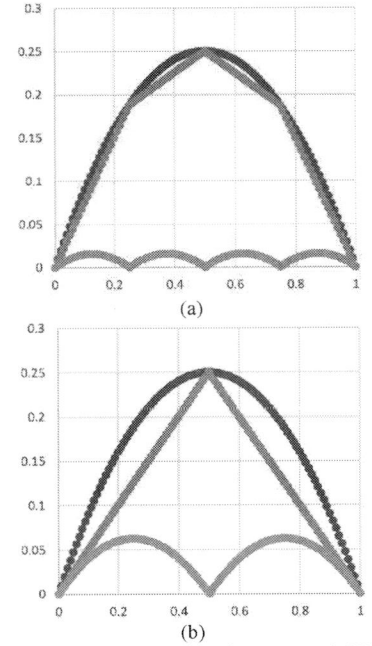

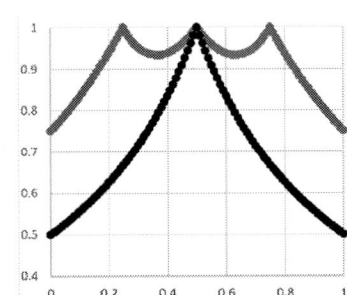

Fig. 5 Power distribution (a) the proposed QRPPM; (b) the conventional RCEA-PWM;

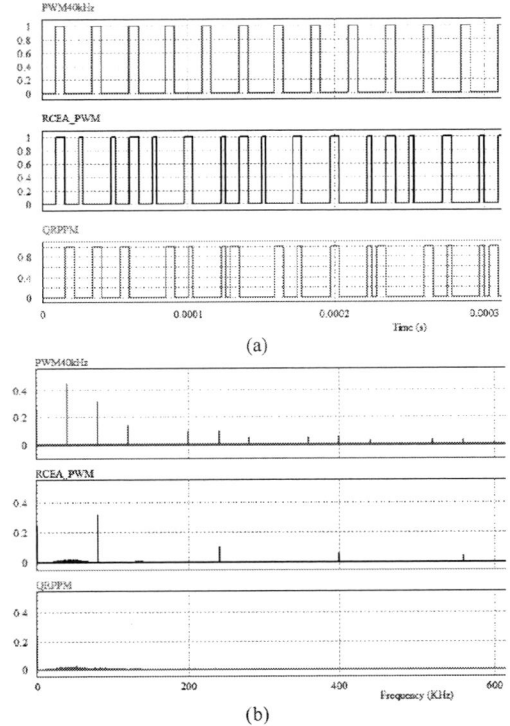

Fig. 8 Simulation results (d=1/4) of various PWM strategies: (a) switching signals; (b) FFT.

Fig. 6 HPDR of the proposed QRPPM (red line) and RCEA-PWM (black line).

IV. SIMULATION RESULTS

To demonstrate the proposed QRPPM, the buck converter in Fig. 7 is setup, and its circuit parameters are tabulated in Table II.

The simulation results of the duty ratio d=1/4 with the conventional PWM, RCEA-PWM and the proposed QRPPM are plotted in Fig. 8 where Fig. 8(a) plots their switching signals and Fig. 8(b) plots FFT of the switching signals in Fig. 8(a).

The simulation results of the duty ratio d=3/8 with PWM, RCEA-PWM and the proposed QRPPM are plotted in Fig. 9. In Fig. 6, the curve at duty ratio d=0.375 indicates that the HPDR of the proposed QRPPM is 93.3% and the HPDR of the RCEA-PWM is 80%. From Fig. 9(b), it is obvious that the proposed QRPPM disperses more integer-harmonic power than RCEA-PWM, and the first dominant integer-harmonic is located at 160kHz (i.e. $4f_{tr}$).

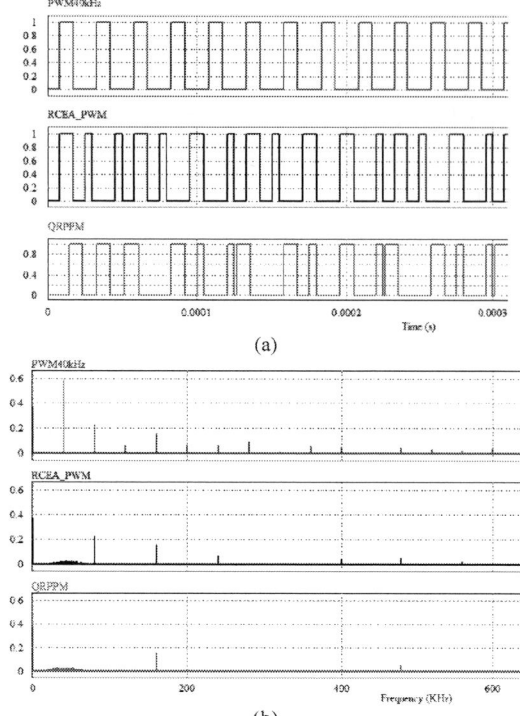

(a)

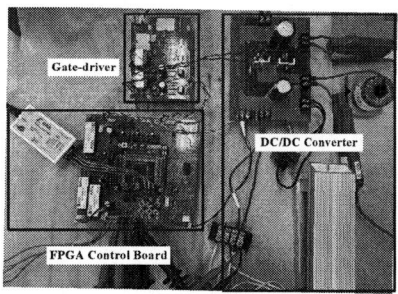

(b)

Fig. 9 Simulation results (*d*=3/8) of various PWM strategies:
(a) switching signals; (b) FFT.

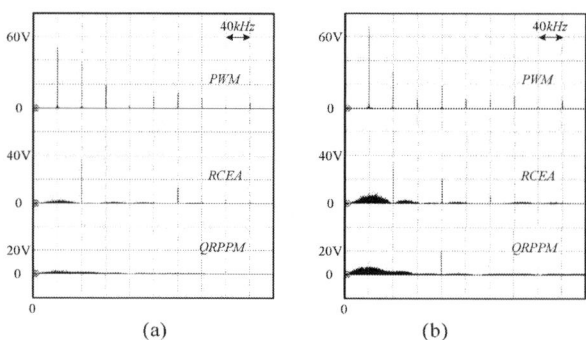

(a)   (b)

Fig. 11 FFT of the switching voltage v_{dis} in buck converter
(a) *d*=1/4; (b) *d*=3/8.

V. EXPERIMENTAL RESULTS

Fig. 10 shows the implementation hardware in this paper, featuring a bidirectional DC-DC converter. The control circuit, developed on a Xilinx Spartan-6 series XC6SLX9 FPGA board, includes the controller, PWM strategy designs, and Analog-to-Digital conversion circuits.

Fig. 10. Experimental circuit architecture.

Fig. 11 plots the FFTs of the switching voltage v_{dis} in buck converter by the proposed QRPPM where the digital oscilloscope R/S is used.

The proposed QRPPM shows superior power dispersion performance than the conventional PWM and RCEA especially at duty ratio *d*=0.25. From Fig. 11(b), the first dominant peak of QRPPM is located at 160kHz (i.e. $4f_{tr}$) larger than those of PWM and RCEA-PWM.

VI. CONCLUSION

In this paper, the proposed QRPPM shows superior power dispersion performance among the other RPPMs. In addition, the DC/DC converter would also benefit from the QRPPM's first dominant peak located at four time the triangular frequency.

REFERENCES

[1] A. Vedde, M. Neuburger and H. Reuss, "An optimized high-frequency EMI filter design for an automotive DC/DC-converter," 2021 National Power Electronics Conference (NPEC), Bhubaneswar, India, 2021, pp. 01-06, doi: 10.1109/NPEC52100.2021.9672512.

[2] F. Mihali and D. Kos, "Reduced Conductive EMI in Switched-Mode DC–DC Power Converters Without EMI Filters: PWM Versus Randomized PWM," in *IEEE Transactions on Power Electronics*, vol. 21, no. 6, pp. 1783-1794, Nov. 2006, doi: 10.1109/TPEL.2006.882910.

[3] S. Kaboli, J. Mahdavi, and A. Agah, "Application of random PWM technique for reducing the conducted electromagnetic emissions in active filters," in *IEEE Transactions on Industrial Electronics*, vol. 54, no. 4, pp. 2333–2343, Aug. 2007.

[4] Y. -C. Lim, S. -O. Wi, J. -N. Kim and Y. -G. Jung, "A Pseudorandom Carrier Modulation Scheme," in *IEEE Transactions on Power Electronics*, vol. 25, no. 4, pp. 797-805, April 2010, doi: 10.1109/TPEL.2009.2035699.

[5] M. M. Bech, F. Blaabjerg and J. K. Pedersen, "Random modulation techniques with fixed switching frequency for three-phase power converters," in *IEEE Transactions on Power Electronics*, vol. 15, no. 4, pp. 753-761, July 2000, doi: 10.1109/63.849046.

[6] Bor-Ren Lin, "Implementation of nondeterministic pulse width modulation for inverter drives," in *IEEE Transactions on Aerospace and Electronic Systems*, vol. 36, no. 2, pp. 482-490, April 2000, doi: 10.1109/7.845228.

[7] H. C. Chen, Y. F. Tsai and W. C. Tsai, "Analysis of Random Center/Edge Alignment PWM and its Application to Digital Current Control for Boost DC-DC Converters," 16th IEEE Energy Conversion Congress and Exposition (ECCE), Pheonix, USA, 2024.

[8] K. -S. Kim, Y. -G. Jung and Y. -C. Lim, "A New Hybrid Random PWM Scheme," in *IEEE Transactions on Power Electronics*, vol. 24, no. 1, pp. 192-200, Jan. 2009, doi: 10.1109/TPEL.2008.2006613.

[9] Y. -S. Lai, Y. -T. Chang and B. -Y. Chen, "Novel Random-Switching PWM Technique With Constant Sampling Frequency and Constant Inductor Average Current for Digitally Controlled Converter," in *IEEE Transactions on Industrial Electronics*, vol. 60, no. 8, pp. 3126-3135, Aug. 2013, doi: 10.1109/TIE.2012.2201436.

[10] A. M. Stankovic, G. E. Verghese and D. J. Perreault, "Analysis and synthesis of randomized modulation schemes for power converters," in *IEEE Transactions on Power Electronics*, vol. 10, no. 6, pp. 680-693, Nov. 1995, doi: 10.1109/63.471288.

[11] K. K. Tse, H. S. -H. Chung, S. Y. R. Hui and H. C. So, "A comparative investigation on the use of random modulation schemes for DC/DC converters," in *IEEE Transactions on Industrial Electronics*, vol. 47, no. 2, pp. 253-263, April 2000, doi: 10.1109/41.836340.

[12] Y. Shrivastava, S. Sathiakumar and V. G. Agelidis, "Analysis and Verification of Two-Level Random Aperiodic PWM Schemes for DC–DC Converters," in *IEEE Transactions on Power Electronics*, vol. 24, no. 9, pp. 2138-2147, Sept. 2009, doi: 10.1109/TPEL.2009.2021762.

979-8-3315-1612-3/25 $31.00 © 2025 IEEE

Bidirectional CLLC Converter Using A Hybrid Control Method for Wide Voltage Range Applications

Jhih-Cheng Hu
Department of Electrical Engineering
National Taipei University of Technology
Taipei, Taiwan
t111319010@ntut.org.tw

Hong-Xuan Liao
Department of Electrical Engineering
National Taipei University of Technology
Taipei, Taiwan
t112318013@ntut.org.tw

Chien-Lung Liu
Delta Electronics Inc.
Taoyuan, Taiwan
CL.LIU@deltaww.com

Wei Wang
Department of Electrical Engineering
National Taipei University of Technology
Taipei, Taiwan
t111318126@ntut.org.tw

Ming-Shi Huang
Department of Electrical Engineering
National Taipei University of Technology
Taipei, Taiwan
simonh@ntut.edu.tw

Abstract—To achieve a wide voltage range for battery applications, this paper presents a hybrid control strategy combining DC-link voltage (DCV) control, pulse-frequency modulation (PFM), and phase-shift modulation (PSM) for a bidirectional CLLC resonant converter. Additionally, a constant turn-on delay (CTOD) method is proposed for synchronous rectifiers (SRs), reducing control complexity. In this paper, a digital signal processor (DSP)-based bidirectional full-bridge CLLC resonant converter with SRs was constructed, featuring a wide voltage range of 360 V to 400 V on the high-voltage side and 36 V to 60 V on the low-voltage side. Operating in constant current (CC) charging mode, the CLLC converter achieves a peak efficiency of 97.9% at 2.5 kW and 60 V on the low-voltage side, demonstrating the effectiveness of the proposed hybrid control method. Furthermore, the bidirectional performance and smooth mode transition between different control modes are verified through the experimental results.

Keywords—CLLC resonant converter, bidirectional, wide range, hybrid control

I. INTRODUCTION

Due to global warming and air pollution concerns, electrical vehicles (EVs) and renewable energy sources have gained significant attention in recent years. These applications require both AC-DC converter and DC-DC converters. For instance, EV batteries support bidirectional functions, enabling grid-to-vehicle (G2V) and vehicle-to-grid (V2G) operations [1]. This has increased demand for bidirectional DC-DC converters with galvanic isolation [2]-[3]. Among common topologies are the dual active bridge (DAB) converter and CLLC resonant converters. DAB converters provide a wide voltage gain range across varying loads but only achieve zero-voltage switching (ZVS) at full load [4]-[5], resulting in higher switching loss and frequency limitation. Conversely, CLLC resonant converter can achieve ZVS on the primary-side and zero-current switching

(ZCS) on the secondary-side switches at the resonant frequency, making them well-suited for high frequency applications with wide-band-gap power semiconductor devices. These benefits lead to high efficiency, frequency, and power density [6]-[8].

Pulse-frequency modulation (PFM) control is a common control strategy for CLLC converters, allowing a wide output voltage range through frequency adjustment. However, this can increase copper and core losses and exacerbate electromagnetic interference (EMI) due to a broad frequency range. To mitigate these issues, phase-shift modulation (PSM) was adopted to adjust the phase angle between two switching legs at a fixed frequency, minimizing frequency variation [9]-[10]. In addition, [11] introduced variable DC-link voltage (DCV) control for the CLLC converter, enabling operation at the resonant frequency for optimized efficiency over a wide voltage range.

This paper proposes a hybrid control method for a full-bridge CLLC resonant converter to achieve a wide output voltage range, combining DCV control, PFM, and PSM. The paper is organized as follows: Section II analyzes the voltage gains under various control modes and compares traditional PFM with the proposed hybrid methods. Section III presents a simulation-based strategy for synchronous rectifiers to reduce control complexity. Finally, Section IV demonstrates a 2.5 kW DSP-based bidirectional CLLC converter achieving an output voltage range from 36 V to 60 V under constant current (CC) charging mode, verifying the effectiveness of the proposed method, mode transition, and bidirectional performance.

II. HYBRID CONTROL METHOD FOR CLLC CONVERTER

As depicted in Fig. 1, L_{rp} and C_{rp} are the resonant inductor and capacitor on the high-voltage side, respectively, while L_{rs} and C_{rs} are the resonant inductor and capacitor on the low-voltage side, respectively. L_m is the magnetizing inductance of the transformer. Using the first harmonic approximation (FHA), the voltage gains G_1 and G_2 for the charging and discharging modes under PFM control, as shown in Fig. 2 (a), can be expressed as follows:

$$|G_1| = \frac{V_{LV}}{V_{HV}} = \frac{1}{n} \cdot \frac{1}{\sqrt{A_1^2 + (B_1 - C_1)^2}} \qquad (1)$$

$$|G_2| = \frac{V_{HV}}{V_{LV}} = n \cdot \frac{1}{\sqrt{A_2^2 + (B_2 - C_2)^2}} \qquad (2)$$

Where

$$\begin{cases} A_1 = \dfrac{1}{K_p} - \dfrac{1}{K_p \cdot f_{np}^2} + 1 \\[2mm] B_1 = \dfrac{Q_p}{f_{np}} \cdot \left(\dfrac{\alpha_p}{K_p} + \dfrac{1}{K_p \cdot \beta_p} + \dfrac{1}{\beta_p} + 1 \right) \\[2mm] C_1 = Q_p \cdot f_{np} \cdot \left(\dfrac{\alpha_p}{K_p} + \alpha_p + 1 \right) - \dfrac{Q_p}{K_p \cdot \beta_p \cdot f_{np}^3} \end{cases} \qquad (3)$$

$$\begin{cases} A_2 = \dfrac{1}{K_s} - \dfrac{1}{K_s \cdot f_{ns}^2} + 1 \\[2mm] B_2 = \dfrac{Q_s}{f_{ns}} \cdot \left(\dfrac{\alpha_s}{K_s} + \dfrac{1}{K_s \cdot \beta_s} + \dfrac{1}{\beta_s} + 1 \right) \\[2mm] C_2 = Q_s \cdot f_{ns} \cdot \left(\dfrac{\alpha_s}{K_s} + \alpha_s + 1 \right) - \dfrac{Q_s}{K_s \cdot \beta_s \cdot f_{ns}^3} \end{cases} \qquad (4)$$

Moreover, the turn ratio n, the normalized switching frequency f_n, the quality factor Q, the equivalent ac load resistance R_{ac}, and inductance ratio K are defined as follows.

$$\begin{cases} n = \dfrac{N_p}{N_s}, \; K_p = \dfrac{L_m}{L_{rp}}, \; K_s = \dfrac{L_m}{n^2 \cdot L_{rs}} \\[2mm] Q_p = \dfrac{1}{R_{acp}} \sqrt{\dfrac{L_{rp}}{C_{rp}}}, \; Q_s = \dfrac{1}{R_{acs}} \sqrt{\dfrac{L_{rs}}{C_{rs}}} \\[2mm] R_{acp} = \dfrac{8 \cdot n^2 \cdot R_L}{\pi^2}, \; R_{acs} = \dfrac{8 \cdot R_L}{\pi^2 \cdot n^2} \\[2mm] \alpha_p = \dfrac{n^2 \cdot L_{rs}}{L_{rp}}, \; \alpha_s = \dfrac{L_{rp}}{n^2 \cdot L_{rs}} \\[2mm] \beta_p = \dfrac{C_{rs}}{n^2 \cdot C_{rp}}, \; \beta_s = \dfrac{n^2 \cdot C_{rp}}{C_{rs}} \\[2mm] f_{np} = \dfrac{f_s}{f_{rp1}}, \; f_{rp1} = \dfrac{1}{2\pi \cdot \sqrt{L_{rp} \cdot C_{rp}}} \\[2mm] f_{ns} = \dfrac{f_s}{f_{rs1}}, \; f_{rs1} = \dfrac{1}{2\pi \cdot \sqrt{L_{rs} \cdot C_{rs}}} \end{cases} \qquad (5)$$

The subscript p and s represent primary-side and secondary-side, respectively. If the resonant tanks on both sides are symmetrical, G_1 and G_2 will be nearly the same. In addition, the output voltage can be adjusted through PSM control, as shown in Fig. 2 (b), while the voltage gain G_3 can be expressed as below:

$$|G_3| = |G_1| \cdot \sin\left(\frac{\pi - \varphi_s}{2} \right) \qquad (6)$$

Where φ_s is the phase shift angle between S_1 and S_4 shown in Fig. 2 (b).

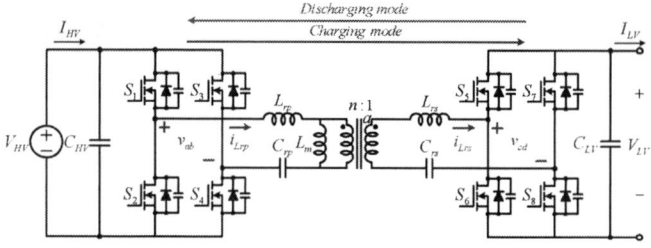

Fig. 1. Full-bridge CLLC resonant converter.

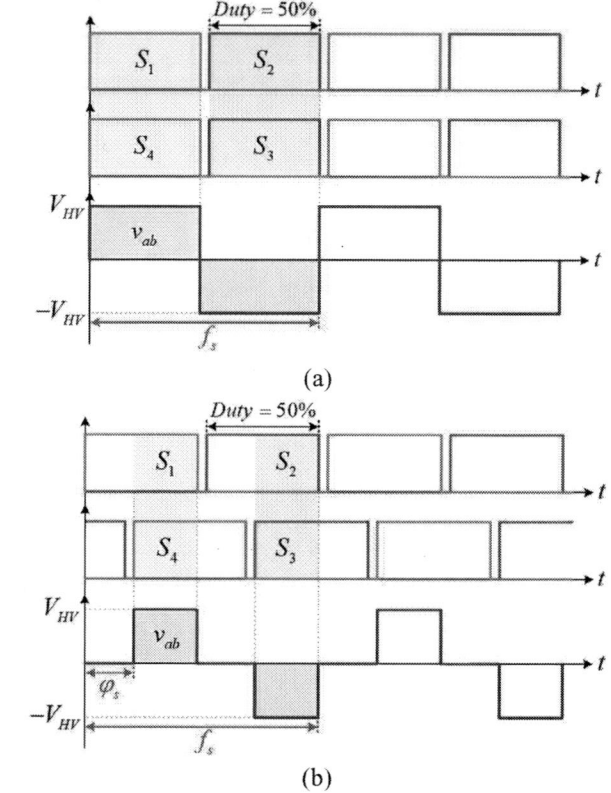

Fig. 2. Key waveforms under PFM and PSM in the charging mode. (a) PFM. (b) PSM.

Based on the aforementioned analysis, this paper designs a bidirectional CLLC converter using a hybrid control method to achieve a wide voltage range. The control curve under CC charging mode is shown in Fig. 3. In this paper, the mode transition points are chosen based on the optimal measured efficiency presented in Chapter IV. Comparison between traditional PFM and proposed hybrid control under different V_{LV} values is listed in Table I. The key concepts for proposed control in Fig. 3 (b) are noted as follows:

(i) The converter operates at f_{rp1} under DCV control, achieving the optimized efficiency at low-voltage side voltage V_{LV}=54-60 V and high-voltage side voltage V_{HV}=360-400 V.

(ii) PFM control is utilized for V_{LV}=42-54 V at V_{HV}=360 V, with the f_s ranging from f_{rp1} to 1.56 f_{rp1}.

(iii) PSM control is used to provide lower voltages for V_{LV}=36-42 V at V_{HV}=360 V, reducing the maximum f_s from 1.91 f_{rp1} to 1.56 f_{rp1} and decreasing losses compared to Fig. 3 (a).

(iv) Additionally, as shown in Table I, it can be observed that f_s is significantly reduced using the proposed method.

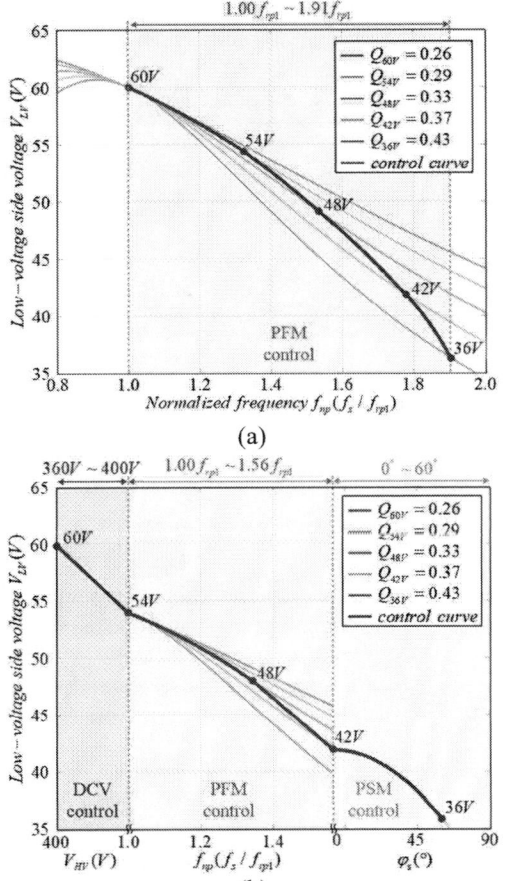

(a)

(b)

Fig. 3. The voltage curves under CC charging mode (V_{HV}=360 V, I_{LV}=41.7 A). (a) Traditional PFM control. (b) Proposed hybrid control method.

TABLE I
COMPARISON BETWEEN TRADITIONAL AND PROPOSED
CONTROL METHOD UNDER DIFFERENT V_{LV} VALUES

Control method		V_{LV} (V)	Control variable
Traditional PFM		36-60	f_s
Proposed	DCV control	54-60	V_{HV}
	PFM	42-54	f_s
	PSM	36-42	φ_s

Then, the proposed hybrid control system is depicted in Fig. 4. Some descriptions of the approach are illustrated as follows:

(i) The converter operates in PFM control when V_{LV}=42-54 V, with feedforward control implemented based on (1).

(ii) When V_{LV}=36-42 V, the PSM control is activated at a fixed f_s, utilizing feedforward control as defined in (6).

(iii) As shown in Fig. 5, φ_s gradually decreases to zero during the transition from PSM to PFM control. In addition, hysteresis control is employed to prevent chattering effects during mode transitions.

(iv) The DCV control is excluded in this paper, as it is managed by the frond-end AC-DC converter, and the CLLC resonant converter always operates at the resonant frequency.

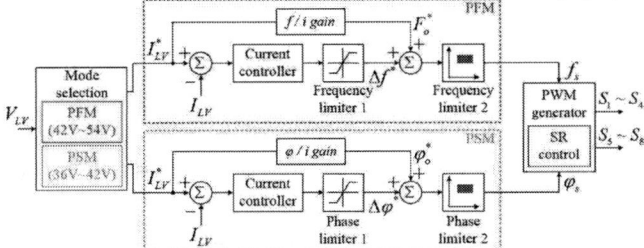

Fig. 4. Proposed hybrid control system for CLLC resonant converter in charging mode.

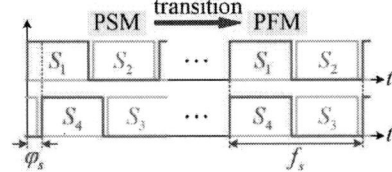

Fig. 5. Reference waveform of mode transition from PSM to PFM in charging mode.

III. PROPOSED SIMULATION-BASED STRATEGY FOR SYNCHRONOUS RECTIFICATION

The SRs are commonly used to improve efficiency, but their driving signals depend on the control method and operating conditions [12]-[13], which increases control complexity. To address this, a simulation-assisted approach described in [14] is adopted in this paper. Fig. 6 displays the simulation results for the required SR driving signals v_{gs5} under various V_{LV} levels. Regardless of the control method, the SR switch S_5 can be turned on after a slight delay $t_{d,on}$ following the main switch S_4, and turned off synchronously with S_4. Based on simulation analysis, this paper proposes a constant turn-on delay (CTOD) method for SR driving signals under the hybrid control strategy, selecting 560 ns for $t_{d,on}$.

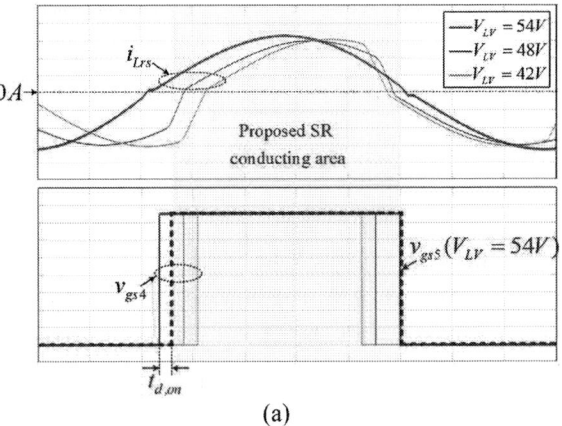

(a)

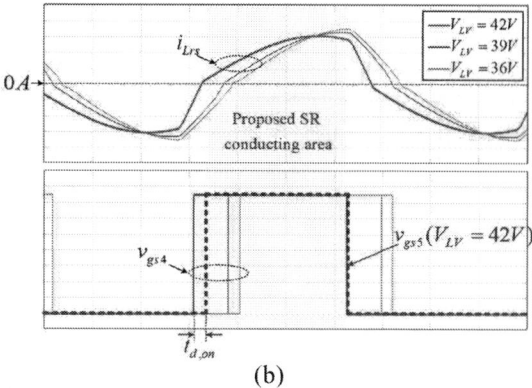

(b)

Fig. 6. Proposed CTOD method for SR switches under different control methods (CC mode, I_{LV}=41.7 A). (a) PFM. (b) PSM.

IV. EXPERIMENTAL VERIFICATION

The proposed 2.5 kW DSP-based CLLC converter setup is shown in Fig. 7, with system specifications detailed in Table II. The high-voltage side uses four SiC MOSFETs (Infineon IMT65R022M1H), while the low-voltage side employ four Si MOSFETs (Infineon IPB027N10N5). Furthermore, waveforms and efficiency measurements were obtained using a digital oscilloscope (R&S RTE 1034) and a power analyzer (Keysight PA2201A).

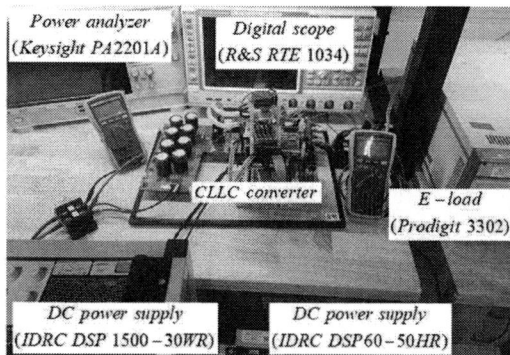

Fig. 7. The experiment setup of the proposed DSP-based CLLC resonant converter.

TABLE II
THE SPECIFICATIONS OF THE PROPOSED CLLC CONVERTER

Symbol	Description	Value
V_{HV}, V_{LV}	High/Low-voltage side voltage	400 V, 36-60 V
P_o	Rated power	2.5 kW
f_r	Resonant frequency	100 kHz
f_s	Switching frequency	100-142 kHz
n	Turns ratio	20 : 3
L_m	Magnetizing inductor	189.0 μH
L_{rp}, L_{rs}	High/Low-voltage side resonant inductor	21.5 μH, 484.0 nH
C_{rp}, C_{rs}	High/Low-voltage side resonant capacitor	118.0 nF, 5.2 μF

Fig. 8 displays the measured efficiencies in CC charging mode for various V_{LV} values, with the efficiency curve for traditional PFM in dashed lines and the proposed hybrid control in solid lines. Key observations are as follows:

(i) Both control methods allow the proposed converter to reach a maximum efficiency of 97.9% at V_{LV}=60 V and f_s=f_{rp1}.

(ii) In traditional PFM, efficiency decreases as V_{LV} is reduced due to the increase in f_s.

(iii) The proposed hybrid control, integrating DCV control, PFM, and PSM, improves efficiency across various V_{LV} conditions compared to traditional PFM.

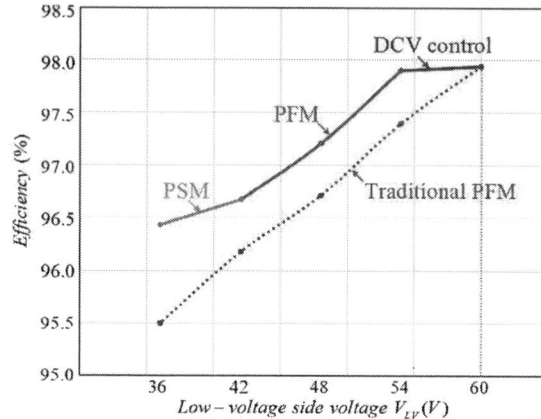

Fig. 8. Measured efficiencies in CC charging mode (I_{LV}=41.7A).

Key parameters based on these observations are summarized in Table III, and the advantages of the proposed method are as follows:

(i) Using DCV control, V_{HV} decreases from 400 V to 360 V while f_s is fixed, resulting in maintained efficiency of 97.9% at V_{LV}=54 V.

(ii) DCV control transitions to PFM when V_{LV} falls below 54 V, allowing for reduced f_s at same V_{LV}. For instance, f_s decreases to 142 kHz compared to the traditional method.

(iii) PSM is activated when V_{LV} is below 42 V. φ_s increases as V_{LV} decreases, while f_s remains at 142 kHz, resulting in a 0.9% efficiency improvement at V_{LV}=36 V.

TABLE III
COMPARISON OF MEASURED RESULTS BETWEEN TRADITIONAL AND PROPOSED CONTROL METHOD

Control method	V_{LV} (V)	η (%)	V_{HV} (V)	f_s (kHz)	φ_s (°)
Traditional	60	97.9	400	100	0
	54	97.4	400	127	0
	42	96.1	400	160	0
	36	95.5	400	170	0
Proposed	60	97.9	400	100	0
	54	97.9	360	100	0
	42	97.2	360	142	0
	36	96.4	360	142	49

The experimental waveforms at V_{LV}=60 V are displayed in Fig. 9, while Fig. 10 presents the waveforms at V_{LV}=36 V. It can be observed that both high-voltage and low-voltage side switches have soft-switching characteristics whether V_{LV} equals to 60 V or 36 V.

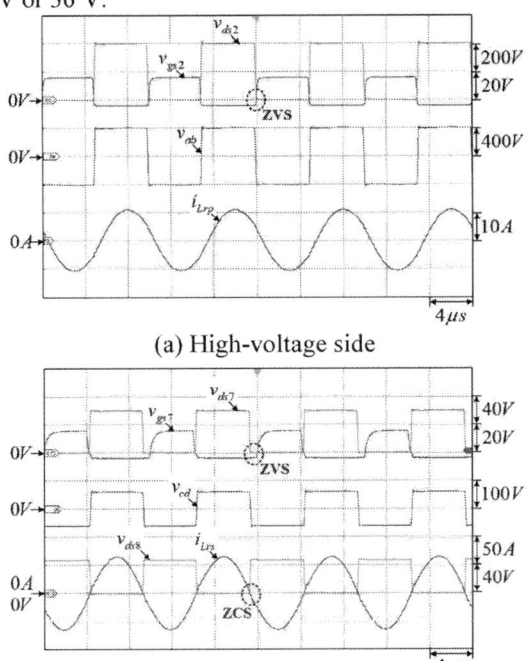

(a) High-voltage side

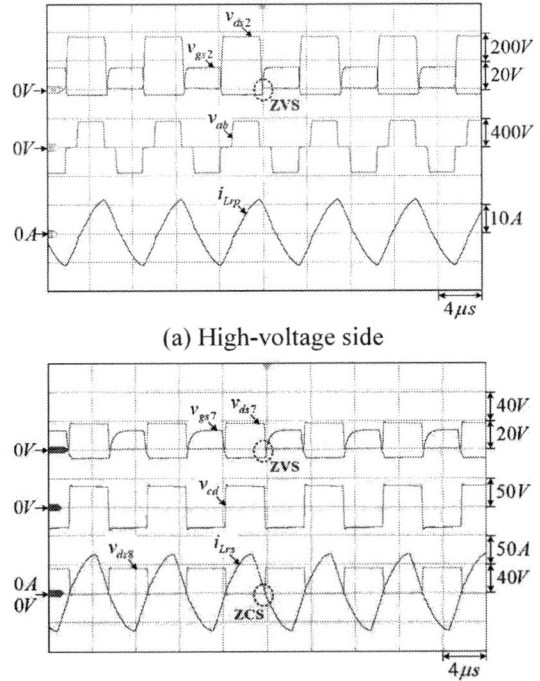

(b) Low-voltage side

Fig. 9. The experimental waveforms at V_{LV}=60 V using DCV control (V_{HV}=400 V, I_{LV}=41.7 A, f_s=100 kHz).

(a) High-voltage side

(b) Low-voltage side

Fig. 10. The experimental waveforms at V_{LV}=36 V using PSM (V_{HV}=360 V, I_{LV}=41.7 A, f_s=142 kHz, φ_s=49°).

Fig. 11 presents the mode transition from PSM to PFM, based on the method described in Fig. 5. Notably, the proposed converter achieves a seamless transition, maintaining stability without any resonant current spike between control methods, which verifies the effectiveness of the proposed approach.

Additionally, Fig. 12 demonstrates bidirectional operation at rated load, confirming that the proposed converter can successfully handle transitions between charging and discharging modes. This bidirectional capability supports efficient energy transfer in both directions, further enhancing the converter's adaptability across various operating conditions.

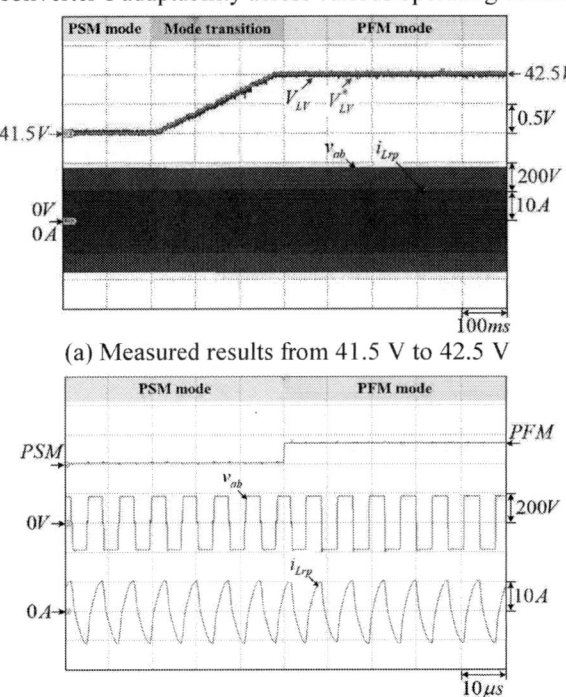

(a) Measured results from 41.5 V to 42.5 V

(b) zoom-in waveform at mode transition

Fig. 11. The experimental waveforms of control modes transition (V_{HV}=360 V, I_{LV}=41.7 A).

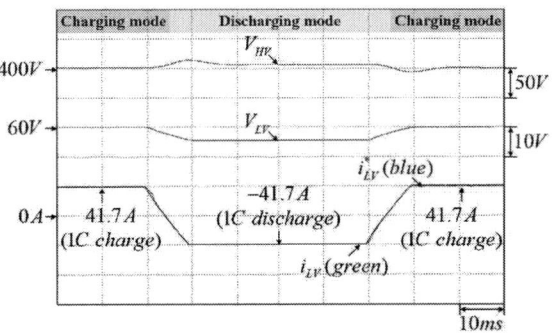

Fig. 12. The experimental waveforms of the bidirectional operation (V_{HV}=400 V, I_{LV}=41.7 A).

V. CONCLUSIONS

A CLLC resonant converter was built to verify the wide voltage range capability using the proposed hybrid control strategy, which incorporates DCV control, PFM, and PSM. Additionally, the control complexity of the SRs is decreased through the DSP implementation of a CTOD method. Key contributions of this paper are summarized as follows:

(i) The proposed converter achieves a maximum efficiency of 97.9% at V_{LV}=60 V. Furthermore, both high-voltage and low-voltage side switches demonstrate soft-switching behavior across a wide voltage range.

(ii) Experimental results verify that the converter's good performance during mode transition and bidirectional operation, highlighting the effectiveness of the proposed method.

ACKNOWLEDGMENT

This work was partially supported by a project from Delta Electronics. The authors would like to extend their gratitude to all R&D engineers who contributed to this work.

REFERENCES

[1] A. B. N. Lingaiah and N. R. Tummuru, "A PV-utility integrated cascaded interleaved configuration-based IPT charging system for residential V2G and G2V applications of EV," *IEEE Trans. Transp. Electrif.*, vol. 10, no. 3, pp. 6583-6595, Sep. 2024.

[2] B. K. Lee, J. P. Kim, S. G. Kim, and J. Y. Lee, "An isolated/bidirectional PWM resonant converter for V2G(H) EV on-Board charger," *IEEE Trans. Veh. Technol.*, vol. 66, no. 9, pp. 7741–7750, Sep. 2017.

[3] Z. U. Zahid, Z. M. Dalala, R. Chen, B. Chen, and J. S. Lai, "Design of bidirectional DC-DC resonant converter for vehicle-to-grid (V2G) applications," *IEEE Trans. Transp. Electrif.*, vol. 1, no. 3, pp. 232–244, Oct. 2015.

[4] J. Lee, S. Choi, and J. -I. Ha, "Design and four-degree-of-freedom modulation method of dual active bridge converter," *IEEE J. Emerg. Sel. Top. Power Electron.*, vol. 12, no. 5, pp. 4478-4493, Oct. 2024.

[5] G. Arena, A. Chub, M. Lukianov, R. Strzelecki, D. Vinnikov, and G. De Carne, "A comprehensive review on DC fast charging stations for electric vehicles: Standards, power conversion technologies, architectures, energy management, and cybersecurity," *IEEE Open J. Power Electron.*, vol. 5, pp. 1573-1611, Sep. 2024.

[6] P. He and A. Khaligh, "Comprehensive analyses and comparison of 1 kW isolated DC–DC converters for bidirectional EV charging systems," *IEEE Trans. Transp. Electrif.*, vol. 3, no. 1, pp. 147-156, Mar. 2017.

[7] B. Li, L. Jing, X. Wang, N. Chen, B. Liu, and M. Chen, "A smooth mode-switching strategy for bidirectional OBC base on V2G technology," in *Proc. IEEE Appl. Power Electron. Conf. (APEC)*, Anaheim, CA, USA, Mar. 2019, pp. 3320-3324.

[8] Z. Bai, J. Shao, J. Gu, H. Liu, X. Zhang, and D. Xu, "Design and modeling of CLLC converter for bidirectional on-board charger," *IEEE Trans. Ind. Appl.*, vol. 59, no. 5, pp. 6095-6102, Sep./Oct. 2023.

[9] T. Zhu, F. Zhuo, F. Zhao, F. Wang, H. Yi, and T. Zhao, "Optimization of extended phase-shift control for full-bridge CLLC resonant converter with improved light-load efficiency," *IEEE Trans. Power Electron.*, vol. 35, no. 10, pp. 11129-11142, Oct. 2020.

[10] I. Kim, W. -Y. Jang, S. -J. Lee, and J. -W. Park, "Enhanced PFM-PSM hybrid control of CLLC resonant converter for electric vehicles: Comprehensive study and verification," *IEEE Trans. Power Electron.*, vol. 39, no. 10, pp. 12978-12990, Oct. 2024.

[11] B. Li, F. C. Lee, Q. Li, and Z. Liu, "Bi-directional on-board charger architecture and control for achieving ultra-high efficiency with wide battery voltage range," in *IEEE Proc. Appl. Power Electron. Conf. Expo.*, Tampa, FL, USA, Mar. 2017, pp. 3688–3694.

[12] W. Feng, F. C. Lee, P. Mattavelli, and D. Huang, "A universal adaptive driving scheme for synchronous rectification in LLC resonant converters," *IEEE Trans. Power Electron.*, vol. 27, no. 8, pp. 3775-3781, Aug. 2012.

[13] Y. Wei, Q. Luo, and H. A. Mantooth, "Synchronous rectification for LLC resonant converter: An overview," *IEEE Trans. Power Electron.*, vol. 36, no. 6, pp. 7264-7280, Jun 2021.

[14] C. -C. Liao, M. -S. Huang, Z. -F. Li, F. -J. Lin, and W. -T. Wu, "Simulation-assisted design of a bidirectional wireless power transfer with circular sandwich coils for E-bike sharing system," *IEEE Access*, vol. 8, pp. 110003-110017, Jun. 2020.

Design and Control of a High-Bandwidth Dual Active Bridge DC-DC Converter

Alper Uzum
Dept. of Electrical and Computer Engineering
The University of Akron
Akron, OH, USA
au25@uakron.edu

Syed Imam Hasan
Dept. of Electrical and Computer Engineering
The University of Akron
Akron, OH, USA
sh328@uakron.edu

Yilmaz Sozer
Dept. of Electrical and Computer Engineering
The University of Akron
Akron, OH, USA
ys@uakron.edu

Kenneth A. Loparo
Dept. of Electrical, Computer, and Systems Engineering
Case Western Reserve University
Cleveland, OH, USA
kal4@case.edu

Abstract—**For the stable operation of sensitive systems, achieving high-bandwidth control in power converters is critical, as the current and voltage ripple directly reflects the quality of the synthesized waveform. The selection of an appropriate DC-link capacitor, which significantly influences the system's bandwidth, plays a pivotal role in ensuring performance. This paper explores the intricate relationship between input and output DC-link capacitance and its effect on the system bandwidth in Dual Active Bridge (DAB) converters, a topic that remains relatively unexplored despite numerous research on high-bandwidth control for other power converters. To address the challenges posed by the bulky DC-link capacitor typically required to manage ripple current, the proposed approach integrates a harmonic trap filter to effectively mitigate these limitations. Using state-space modeling and state-variable feedback control, the design and control strategy is validated on a 5 kW DAB converter, demonstrating its ability to enhance system bandwidth without compromising ripple performance.**

Keywords—High bandwidth control, DAB, State space modeling, State variable feedback control, THD

I. INTRODUCTION

High bandwidth DC-DC converters are vital for ensuring the stable operation of sensitive systems through their fast response to changes in load or input conditions specially in aerospace applications. These converters provide a fast transient response, improved power supply regulation, and enhanced efficiency. Besides, the capability of the high bandwidth current synthesizing enables rapid performance evaluation, testing and interaction with the rest of electrical system.

The Dual Active Bridge (DAB) converter is critical in DC microgrids due to its capability to handle bidirectional power flow, wide voltage conversion range, and high efficiency achieved inherently by zero-voltage switching (ZVS) [1]. The DAB is particularly well-suited for interfacing energy storage systems (ESSs) such as batteries and supercapacitors, as well as for use in solid-state transformers (SSTs) that manage power flow between DC microgrids and distribution networks [2]. While numerous studies have addressed high bandwidth controller design for DC-DC converters [3], the application of DAB in this context remains relatively novel. Various modeling techniques and control efforts are discussed in [4]. The relationship between system bandwidth and filter design remains unexplored for Dual Active Bridge (DAB) converters.

While a full-order continuous-time model for the DAB has been developed using generalized average modeling, focusing on the purely AC transformer current [5], this study neglects dynamic effects. Instead, it prioritizes an intuitive representation of the bandwidth relationship with system parameters. Additionally, it examines the potential for volumetric downsizing by incorporating an extra filter stage without compromising system bandwidth, offering insights into optimizing system design which is crucial for aerospace applications [6-8].

In this paper, a reduced-order state-space model is derived, and a close-loop controller is designed to analyze the impact of the DC-link capacitor selection on the system bandwidth. The model's validity is verified by comparing the open-loop step response of the state-space model with that of the circuit model. Following the introduction of closed-loop control, the system bandwidth is assessed through the frequency response of the closed-loop system. By employing a harmonic trap filter, the ripple current is attenuated, allowing for a reduction in the DC-link capacitor requirement and ensuring higher achievable bandwidth. The desired 5 kHz bandwidth is achieved while the total harmonic distortion (THD) is represented as a metric for the ripple current estimation.

II. DUAL ACTIVE BRIDGE CONVERTER MODELING

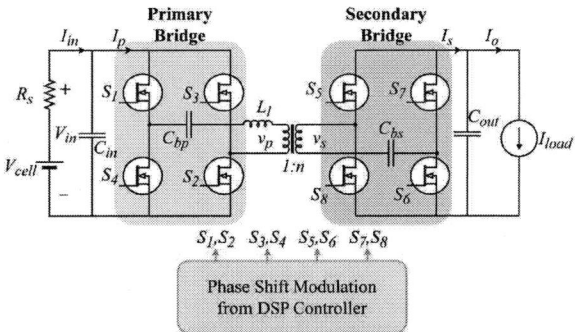

Fig. 1. Generic DAB converter [1]

Incorporating an overall system model of the DAB converter is crucial for achieving the desired system bandwidth and ensuring optimal performance. To facilitate this, a simplified reduced-order modeling approach is employed, which involves neglecting the detailed dynamics of the transformer current. This

979-8-3315-1612-3/25 $31.00 © 2025 IEEE

reduction in complexity allows for a more tractable model while still capturing the essential behavior of the system, as outlined in [4]. In this approach, the primary and secondary bridges of the DAB converter are represented as two dependent current sources, which are crucial for accurately modeling the power transfer between the two sides of the converter. These dependent sources are mathematically defined in Equations (1) and (2), respectively. The use of dependent current sources simplifies the representation of the power flow and enables a more efficient analysis of the system's response to various operating conditions. The schematic of generic DAB converter is shown in Fig. 1. In Fig. 2, the reduced order model of DAB converter is presented where the relationships between the primary and secondary circuits are clearly depicted, providing a foundation for further analysis and design of the control system.

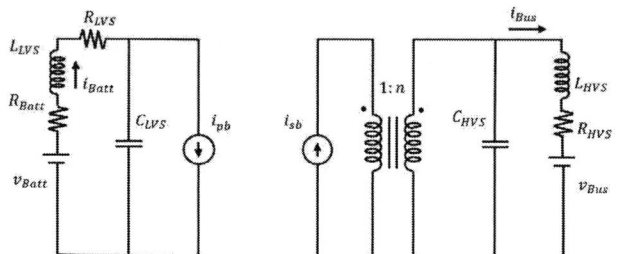

Fig. 2. DAB converter reduced-order model

$$i_{pb} = \frac{d(1-d)v_{BUS}}{2f_{sw}nL_t} \tag{1}$$

$$i_{sb} = \frac{d(1-d)v_{Batt}}{2f_{sw}nL_t} \tag{2}$$

Where, i_{pb} and i_{sb} are the primary and secondary bridge average current respectively, single-phase shift (SPS) modulation is employed between the primary and secondary bridges, denoted as d. The switching frequency is f_{sw} [Hz], the transformer leakage inductance is L_t [H], and the transformer turns ratio is n. The state equations are derived below,

$$L_{LVS}\frac{di_{Batt}}{dt} = v_{Batt} - i_{Batt}(R_{Batt} + R_{LVS}) - v_{C_{LVS}} \tag{3}$$

$$C_{LVS}\frac{dv_{C_{LVS}}}{dt} = i_{Batt} - \frac{d(1-d)v_{Bus}}{2nf_{sw}L_t} \tag{4}$$

$$L_{HVS}\frac{di_{Bus}}{dt} = v_{C_{HVS}} - R_{HVS}i_{Bus} - v_{Bus} \tag{5}$$

$$C_{HVS}\frac{dv_{C_{HVS}}}{dt} = \frac{d(1-d)v_{Batt}}{2nf_{sw}L_t} - i_{Bus} \tag{6}$$

$$\dot{x} = $$
$$\begin{bmatrix} -(R_{Batt}+R_{LVS})/L_{LVS} & -1/L_{LVS} & 0 & 0 \\ 1/C_{LVS} & 0 & 0 & 0 \\ 0 & 0 & -R_{HVS}/L_{HVS} & 1/L_{HVS} \\ 0 & 0 & -1/C_{HVS} & 0 \end{bmatrix} +$$
$$\begin{bmatrix} 1/L_{LVS} & 0 \\ 0 & -\frac{d(1-d)}{2nf_{sw}L_tC_{LVS}} \\ 0 & -1/L_{HVS} \\ \frac{d(1-d)}{2nf_{sw}L_tC_{HVS}} & 0 \end{bmatrix} u \tag{7}$$

The large-signal state-space model of the system, presented in Equation (7), is derived from the state equations (3–6). This model exhibits nonlinearity with respect to the control input phase shift, d. To facilitate analysis and control design, the model is linearized using the small-signal perturbation method. In this approach, the system states and inputs are expressed as the sum of their DC (steady-state) and AC (perturbation) components, as described in [6]. During the linearization process, DC components and higher-order terms are omitted, yielding the simplified small-signal model shown in Equation (8).

$$\tilde{x} = $$
$$\begin{bmatrix} -(R_{Batt}+R_{LVS})/L_{LVS} & -1/L_{LVS} & 0 & 0 \\ 1/C_{LVS} & 0 & 0 & 0 \\ 0 & 0 & -R_{HVS}/L_{HVS} & 1/L_{HVS} \\ 0 & 0 & -1/C_{HVS} & 0 \end{bmatrix} \tilde{x}$$
$$+ \begin{bmatrix} 0 \\ \frac{-(1-2D)V_{Bus}}{2nf_{sw}L_tC_{LVS}} \\ 0 \\ \frac{(1-2D)V_{Batt}}{2nf_{sw}L_tC_{HVS}} \end{bmatrix} \tilde{d} \tag{8}$$

The derived small-signal model is designed to facilitate control of both the battery current and the bus current. However, since the primary control objective is the regulation of the battery current, the system can be effectively decoupled. This decoupling simplifies the system dynamics, enabling a reduction in system order. The dynamic behavior of the battery current in response to a step change in phase shift has been verified through simulation. The results, illustrated in Fig. 3, demonstrate the transient response of the system, with simulation parameters outlined in Table I. These findings confirm the accuracy of the derived model in capturing the system's dynamic behavior, reinforcing its applicability for achieving the desired control objectives.

In Fig. 3, the step response of the circuit model has ripple (400 kHz) due to the switching frequency of 200 kHz. Besides, the settling time and the overshoot observed in Fig. 3 are 70 us and 5% respectively. The step response of the circuit and the small signal model are close to each other.

979-8-3315-1612-3/25 $31.00 © 2025 IEEE

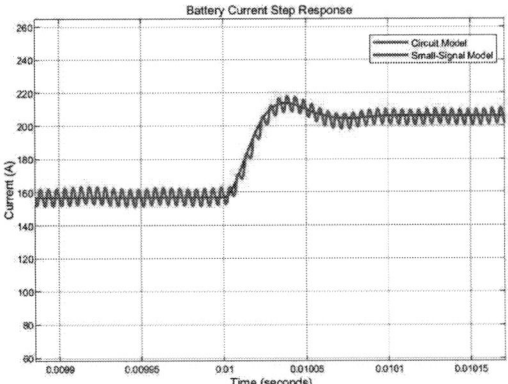

Fig. 3. Battery current step response circuit model vs small-signal model

TABLE I. Converter Specification

Parameter	Value
Nominal Battery Voltage ($V_{Batt-nom}$)	28 V
DC Bus Voltage (V_{Bus})	400 V
Transformer Turns Ratio (n)	1:14
Switching Frequency (f_{sw})	200 kHz
Rated Power (P_{rated})	5 kW
HVS DC-Link Capacitance (C_{HVS})	560 μF
HVS Leakage Inductance (L_t)	14.55 μH
LVS Cable Resistance (R_{LVS})	0.213 mΩ
LVS Cable Inductance (L_{LVS})	105.33 nH
Battery Internal Resistance (R_{Batt})	10 mΩ
HVS Cable Resistance (R_{HVS})	1.6 mΩ
HVS Cable Inductance (L_{HVS})	344 nH

III. STATE VARIABLE FEEDBACK CONTROLLER DESIGN

Controller and hardware design aim to achieve multiple objectives, including attaining the desired bandwidth of 5 kHz and maintaining THD below 1% for the battery current. To enhance the system's dynamic behavior, the state-variable feedback control method is employed. However, due to the system type, an integrator is incorporated to compensate for steady-state error. The closed-loop block diagram is illustrated in Fig. 4. The frequency response analysis for the closed-loop system, considering only state-variable feedback, reveals that increasing the DC-link capacitance significantly impacts the system's bandwidth.

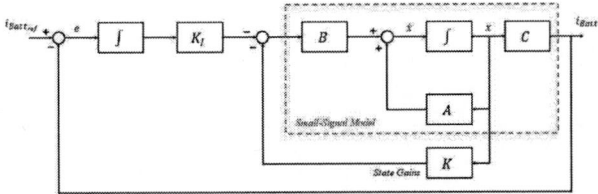

Fig. 4. Closed-loop system block diagram

The results in Fig. 5 highlight a key trade-off in system performance. Fig. 5 shows that increasing the DC-link capacitance reduces system bandwidth, due to the capacitor's greater energy storage, which slows the system's dynamic response.

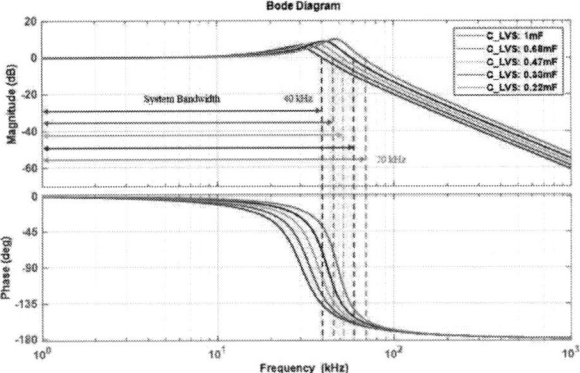

Fig. 5. Closed-loop system frequency response for varying DC-link capacitor values

To address the issue of ripple current in the battery, this study proposes the design of a harmonic trap filter. The primary aim of this filter is to reduce the required DC-link capacitance by attenuating the dominant harmonic content in the battery current. This solution not only minimizes the need for large capacitance values but also facilitates the achievement of the desired system bandwidth, while concurrently reducing current ripple. The battery current exhibits a 400 kHz ripple, which results from the 200 kHz switching frequency of the converter. To mitigate this high-frequency ripple, a harmonic trap filter is employed. The filter is designed as an LC resonance tank circuit, as depicted in Fig. 6. This circuit is specifically tuned to target and suppress the 400 kHz harmonic component, thereby improving the overall performance of the system by enhancing current quality and minimizing harmonic distortion.

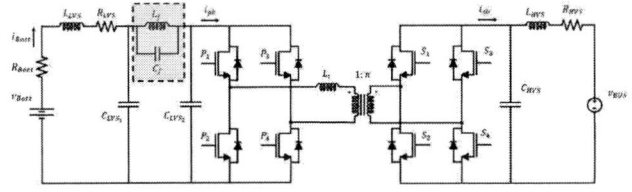

Fig. 6. DAB with LC trap filter at the battery side

The relation between trap filter resonance frequency, f_r and the filter capacitance, C_f and inductance, L_f can be expressed using the following equation (9). From (9) the required capacitance and inductance to attenuate 400 kHz harmonic is calculated.

$$f_r = \frac{1}{2\pi\sqrt{L_f C_f}} \tag{9}$$

The implementation of the proposed harmonic trap filter has demonstrated a substantial improvement in the harmonic performance of the system. Specifically, the THD in the battery current represents a significant enhancement in current quality. This improvement is achieved while maintaining the same DC-link capacitor values ensuring that no additional burden is placed on the system's capacitive components. Furthermore, it is observed that the voltage ripple at the LVS remains unaffected, as depicted in Fig. 7 (a) and (b), thereby preserving the stability of the output voltage under the new filtering conditions.

In addition, the analysis of the open-loop system's pole-zero map, presented in Fig. 8, provides further insight into the system's dynamic behavior. The inclusion of the additional filter stage does not merely suppress harmonic distortion; it also contributes to an overall enhancement of the system's dynamic response. This is evidenced by the modified pole and zero placements, which indicate improved stability margins and faster transient response. Consequently, the proposed filter design not only achieves its primary objective of harmonic reduction but also reinforces the robustness and efficiency of the overall system.

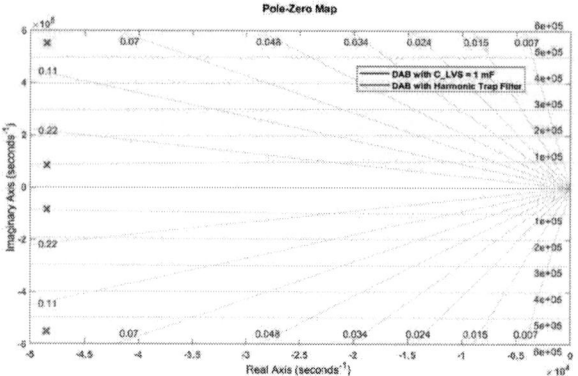

Fig. 8. Open-loop system's pole-zero map with trap filter (red) and without trap filter (blue)

IV. SIMULATION RESULTS

The simulation utilized specific filter parameters, which are detailed in Table II. These parameters were selected to achieve harmonic suppression and maintain system stability. As illustrated in Fig. 9, the battery current demonstrates excellent tracking performance, accurately following the sinusoidal reference current at a frequency of 5 kHz. This precise tracking underscores the effectiveness of the control strategy and the filter design in ensuring high fidelity in current regulation.

At time t=0.01s, a step change is applied to the reference current, wherein its peak amplitude is increased from 89.3 A to 178.6 A. This sudden transition introduces an overshoot of 3.4% in the battery current, which is promptly mitigated, with the current settling to the desired amplitude within half a cycle. Such rapid settling behavior highlights the robustness of the system's dynamic response and its ability to handle abrupt changes in operating conditions without compromising performance.

The implementation of the designed harmonic trap filter further enhances system performance by achieving a THD of merely 0.5% in the battery current relative to its fundamental amplitude. This exceptionally low THD level demonstrates the filter's capability to suppress harmonic components, thereby improving the overall quality of the current waveform. The significant reduction in harmonic content not only ensures compliance with stringent power quality standards but also contributes to the enhanced efficiency and reliability of the system.

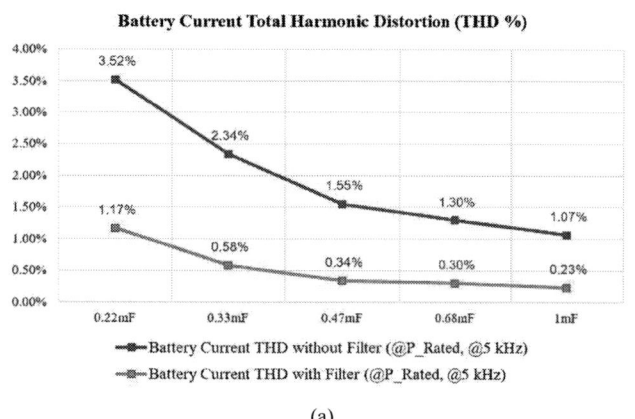

(a)

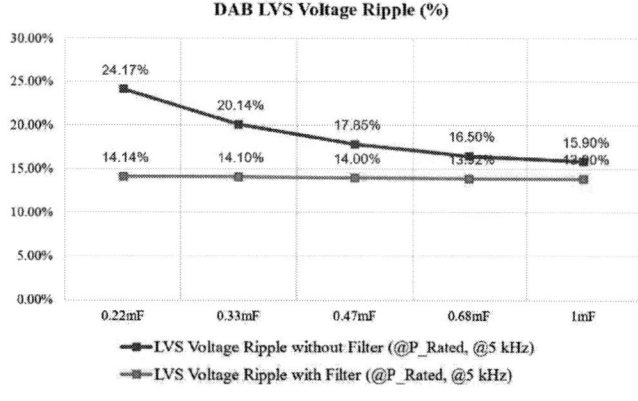

(b)

Fig. 7. (a) Battery current THD; (b) DAB LVS voltage ripple -for varying DC-link capacitor values with and without trap filter

TABLE II. FILTER PARAMETERS

Parameter	Value
HVS DC-Link Capacitance (C_{HVS})	80 μF
LVS DC-Link Capacitance (C_{LVS})	350 μF
Trap Filter Inductance (L_f)	171 nH
Trap Filer Capacitance (C_f)	1 μF

979-8-3315-1612-3/25 $31.00 © 2025 IEEE 2701

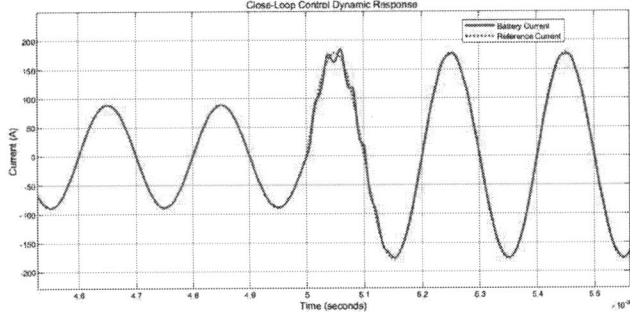

Fig. 9. Battery current dynamic response

V. CONCLUSIONS

In conclusion, this study successfully demonstrates the implementation of high-bandwidth control in a DAB converter using state-space modeling and state-variable feedback control. The research highlights the critical relationship between the DC-link filter capacitor and system bandwidth, emphasizing the necessity of precise design considerations, particularly in applications with stringent performance requirements such as aerospace systems. To address the challenges of achieving the desired bandwidth while meeting the bulky DC-link capacitor requirements, a harmonic trap filter approach is incorporated into the system design. By integrating the harmonic trap filter into the overall system model, notable improvements in system response are achieved. The proposed strategy, which includes reduced-order continuous-time state-space modeling and state variable feedback control, is validated through the simulation of a 5 kW DAB converter. The results confirm the efficacy of this approach, demonstrating its potential for practical, high-performance applications.

REFERENCES

[1] S. I. Hasan, A. Uzum, M. E. Haque, M. Arifur Rahman, A. Siddiquee, M. J. Kisacikoglu, and Y. Sozer, "Design, Optimization, and Validation of GaN-Based DAB Converter for Active Cell Balancing in BTMS Applications," *2023 IEEE Energy Conversion Congress and Exposition (ECCE)*, Nashville, TN, USA, 2023, pp. 434-441.

[2] B. Zhao, Q. Song, W. Liu, and Y. Sun, "Overview of dual-active-bridge isolated bidirectional DC-DC converter for high-frequency-link power-conversion system", *IEEE Trans. Power Electron.*, vol. 29, no. 8, pp. 4091-4106, Aug. 2014.

[3] S. Buso and T. Caldognetto, "A non-linear wide bandwidth digital current controller for DC-DC and DC-AC converters," *IECON 2014 - 40th Annual Conference of the IEEE Industrial Electronics Society*, Dallas, TX, USA, 2014, pp. 1090-1096.

[4] S. Shao, L. Chen, Z. Shan, F. Gao, H. Chen, D. Sha, and T. Dragičević, "Modeling and Advanced Control of Dual-Active-Bridge DC–DC Converters: A Review," in *IEEE Transactions on Power Electronics*, vol. 37, no. 2, pp. 1524-1547, Feb. 2022.

[5] H. Qin and J. W. Kimball, "Generalized Average Modeling of Dual Active Bridge DC–DC Converter," in *IEEE Transactions on Power Electronics*, vol. 27, no. 4, pp. 2078-2084, April 2012, doi: 10.1109/TPEL.2011.2165734.

[6] K. Zhang, Z. Shan, and J. Jatskevich, "Large- and Small-Signal Average-Value Modeling of Dual-Active-Bridge DC–DC Converter Considering Power Losses," in *IEEE Transactions on Power Electronics*, vol. 32, no. 3, pp. 1964-1974, March 2017, doi: 10.1109/TPEL.2016.2555929.

[7] Bash, M., Pekarek, S., and Zumberge, J., "The Utility of Wide-Bandwidth Emulation to Evaluate Aircraft Power System Performance," SAE Int. J. Aerosp. 9(1):14-22, 2016

[8] J. -M. M. Choe, P. Channegowda, and X. Wu, "Development of High Bandwidth Power Hardware-in-Loop Test Platform for Aircraft Hybrid Electric Propulsion," *2021 AIAA/IEEE Electric Aircraft Technologies Symposium (EATS)*, Denver, CO, USA, 2021, pp. 1-5, doi: 10.23919/EATS52162.2021.9704849.

Unified Model Predictive Control for DC-DC Buck Converters: From Start-up to Steady-State Operation

Zhengchen Guo
Dept. of Electrical and Computer Cngineering
Auburn University
Auburn, USA
zzg0024@auburn.edu

R. M. Nelms
Dept. of Electrical and Computer Engineering
Auburn University
Auburn, USA
nelmsrm@auburn.edu

Abstract—**Presented in this paper is a unified model predictive control (MPC) method for DC-DC buck converters, designed to offer stable output voltage while simultaneously regulating inductor current. Utilizing only one control loop, this method simplifies the design and regulation process and enhances dynamic performance by avoiding additional control loops. The strategy supports both start-up and steady-state conditions, providing a soft start-up process and desired stable steady-state without extra procedures. Furthermore, the algorithm is suitable for buck converters operating in both continuous (CCM) and discontinuous conduction modes (DCM). The effectiveness of this approach is validated via simulation and experimental results.**

Index Terms—**model predictive control, dc-dc buck converter, start-up process, CCM, DCM**

I. INTRODUCTION

The rapid advancement of renewable energy and electronic technology have made power electronic converters ubiquitous. As a result, performance requirements for these converters have increased significantly [1] - [3]. Modern applications demand broader stability ranges, higher output voltage accuracy, faster load transient response, and robust over-current protection [4].

Traditional voltage-mode control for DC-DC converters is often limited by poor input and load transient responses, especially when dealing with the non-linearities and different operation modes [5]. Current-mode control for a dc-dc converter typically utilizes the inductor current in an inner loop, with an outer loop generating its reference based on voltage regulation [6]. To simplify controller design, the inner and outer loops are typically decoupled by restricting the outer controller's bandwidth to a fraction of the inner loop's bandwidth. However, this approach limits the dynamic performance of higher control levels. Additionally, tuning multiple control parameters in multi-loop systems presents significant challenges [7].

Model Predictive Control (MPC) is well-suited to meet these demands and has been successfully implemented across various power electronic topologies, including DC-DC converters [8]. And Finite Control Set Model Predictive Control (FCS-MPC) has gained widespread adoption in power converters and drives in recent years [9]. Its capability to handle multiple objectives and nonlinear control makes it particularly suitable for power electronic converters [10]. For instance, researchers have successfully applied FCS-MPC to control the output voltage of DC-DC converters [11].

Furthermore, current-mode MPC for a dc-dc converter has been developed to enhance their dynamic performance of the converter [12]. While researchers first proposed model predictive current control for boost converters [13], this implementation maintained a multi-loop architecture, merely replacing the PID module in the inner loop. This approach required a Kalman filter for the outer loop, increasing both system complexity and implementation challenges. Similarly, another proposed MPC strategy regulated inductor current but required two additional loops for indirect output voltage control [14]. A more recent NPI-MPC algorithm addressed the non-minimum phase behavior in DC-DC boost converters [15]. However, this approach utilized continuous-control set MPC and focused solely on steady-state performance.

Previous studies have not fully leveraged the multi-objective and nonlinear control capabilities of MPC. Instead, these implementations often merely substitute PI modules in the inner current control loop or require additional control modules. This multi-loop architecture increases the number of control parameters requiring tuning, thereby adding complexity to both design and implementation. Moreover, the inclusion of multiple control modules compromises the system's overall dynamic performance. Additionally, these controllers generally aim only at steady-state design and usually have a large current overshoot during the start-up.

This paper proposes an MPC method for DC-DC buck converters that synchronously controls both voltage and current. By employing MPC as a single loop, the method enhances transient performance, directly controlling output voltage and inductor current from start-up to steady-state operation in both CCM and DCM without changing the control structure. Furthermore, the integrated current regulation functionality eliminates the need for separate over-current protection and soft-start processes, thereby simplifying the hardware design. The use of weighting factors enables fully customizable control performance from start-up to steady-state.

This paper is organized as follows: Section II presents the continuous and discrete-time models of the converter and proposes the cost function design. Section III analyzes control performance through simulation and compares results with

979-8-3315-1612-3/25 $31.00 © 2025 IEEE

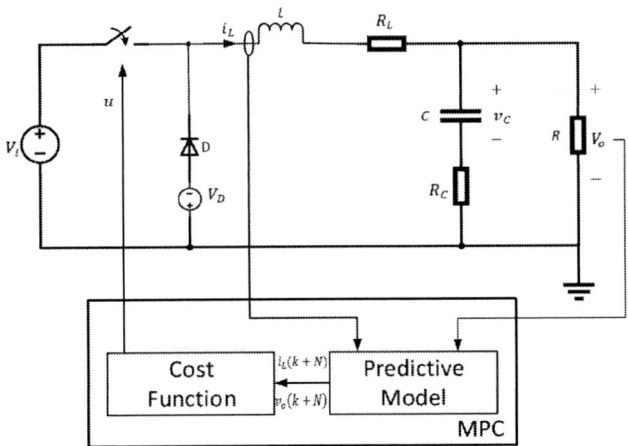

Fig. 1: Block diagram of MPC on buck converters.

other controllers. Section IV presents the experimental results, and Section V concludes the paper.

II. WORKING PRINCIPLES OF PROPOSED CONTROL STRATEGY

A. Predictive Model for Buck Converter

Fig. 1 shows the circuit topology of a dc-dc buck converter, where L is the inductance and C is the capacitance, R_L and R_C are their parasitic resistances, respectively. R is the load resistance. V_i and v_o are input and output voltage respectively. The switching state, denoted by u, assumes a value of 1 for the switch-on state and 0 for the switch-off state. The predictive model and cost function are included in the MPC block. The independent states of the converter are the inductor current and output voltage. The state vector is defined as $x(t) = [i_L(t) \ v_o(t)]^T$. Suppose that the input voltage is constant and the controlled switching state $u(t)$ is a system input, the continuous system is described by the following equations:

$$\dot{x}(t) = Ax(t) + Bu(t), \tag{1}$$

where the matrices A and B are given by

$$A = \begin{bmatrix} -\frac{R_L}{L} - (\frac{RR_C}{R+R_C}) & \frac{1}{L}(\frac{R_C}{R+R_C} - 1) \\ \frac{R}{C(R+R_C)} & -\frac{R}{C(R+R_C)} \end{bmatrix}, B = \begin{bmatrix} V_i \\ 0 \end{bmatrix}.$$

The dc-dc buck converter can operate in CCM and DCM, depending on the value of inductor current $i_L(t)$. Specifically, CCM refers to the case where the current $i_L(t)$ is always positive and DCM means that the current $i_L(t)$ reaches zero during the switching cycles and at that time $u = 0$. Since only the value of inductor current changes operation in DCM, the model (1) can also be applied in DCM. Besides, (1) is a nonlinear model rather than the small-signal model. Hence it can be applied to all running processes from start-up to steady state.

A discrete-time model is required for MPC implementation as an internal prediction model. For simplicity, the forward Euler's method is used, resulting in the following discrete-time model:

$$x(k+1) = A_d x(k) + B_d u(k), \tag{2}$$

where, $A_d = I + T_s A, B_d = T_s B$, I is the identity matrix, and T_s is the sampling period.

B. Cost Function Design

The cost function $J(k)$, detailed in (3), facilitates the optimal selection of switching state sequences for extended prediction horizons.

$$J(k) = \alpha < i_{L,err} > (k) + v_{o,err}(k) + \beta \Delta u(k). \tag{3}$$

In this specific case, the cost function encapsulates three control objectives: the average inductor current error, output voltage error, and switching frequency constraint. The averaged current error objective is represented as the average value of the inductor current over the prediction horizon, given by

$$< i_{L,err} > (k) = \frac{1}{N} \sum_{l=k}^{k+N-1} ||i_L(l) - I_L||,$$

where N is the prediction horizon and I_L is the average current reference. The output voltage objective and switching constraint are defined as

$$v_{o,err}(k) = \sum_{l=k}^{k+N-1} ||v_o(l) - V_o||,$$

$$\Delta u(k) = ||u(k) - u(k-1)||,$$

in which V_o is output voltage reference and $V_o = I_L/R$, R is the load resistance. The constraint term, the third term in (3), reduces the switching frequency to avoid excessive switching. The parameters α and β are weighting factors for inductor current and switching frequency constraints respectively.

III. ANALYSIS AND SIMULATION RESULTS

A. Weighting Factors

Unlike conventional cost functions that typically have one or two control objectives, this system incorporates three objectives, making the design of the weighting factors critical to the system's control performance.

Different combinations of weighting factors will influence the control performance relative to these three objectives. Specifically, increasing the value of α enhances the emphasis on inductor current control, while the other objectives become less significant. For the DC-DC buck converter, the primary objective is to achieve smooth and stable output voltage with over-current protection. The secondary objective is to maintain good transient performance under varying conditions. The tertiary objective is to lower the switching frequency to reduce switching losses. Consequently, the voltage error is prioritized, the current error is secondary, and switching frequency is last. Since switching loss is not the primary concern, it should be adjusted after establishing a balance between current and voltage control.

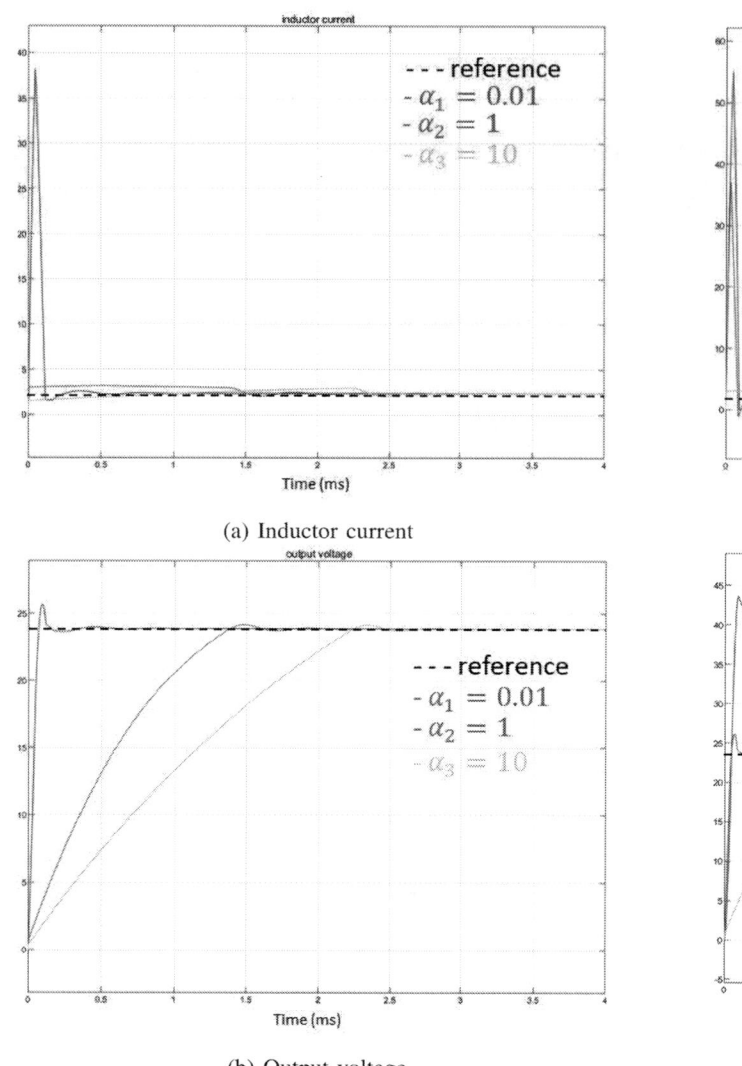

(a) Inductor current

(b) Output voltage

Fig. 2: Simulation results for start-up process with different weighting factors.

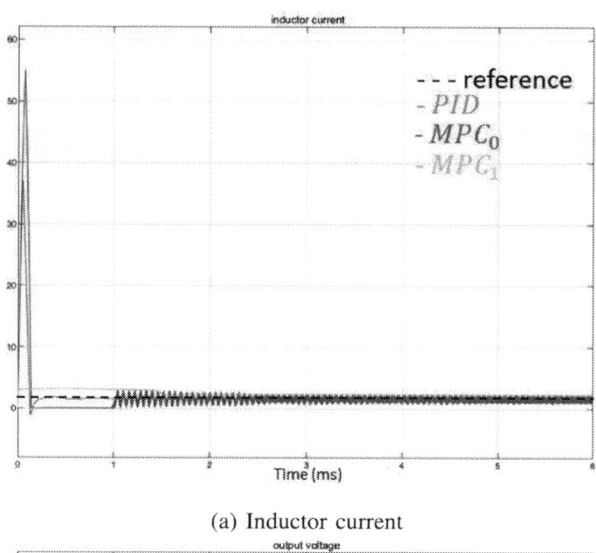

(a) Inductor current

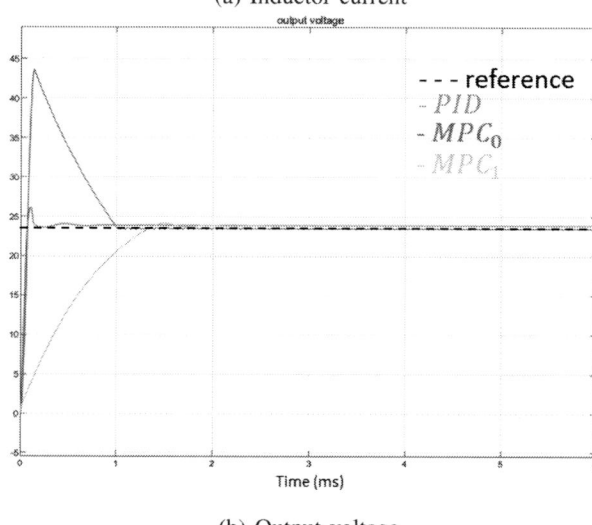

(b) Output voltage

Fig. 3: Simulation results for start-up process with different controllers.

Fig. 2 presents the Matlab/Simulink simulation results, demonstrating the control effects on inductor current and output voltage using three different sets of weighting factors (all $\beta = 0.1$). The simulation parameters are shown in Table I. All three sets successfully reach the desired reference values, with the voltage error normalized at steady state. As shown in Fig. 2, when α_1 is set to a small value (0.1), the current control becomes limited, and the system prioritizes voltage control. This prioritization results in a large current spike—approximately 20 times the nominal current value—which could lead to inductor saturation. Such extreme current conditions would necessitate additional soft-start and over-current protection mechanisms.

However, as the current weighting increases with α_2 and α_3 (same and 10 times weighted as voltage control), the current overshoot is considerably reduced to a safe range. This adjustment, however, leads to an increased voltage settling

time, due to the lower emphasis on voltage control.

This control algorithm effectively regulates transient performance during start-up while maintaining stable steady-state operation and offering high customizability to meet various performance requirements. Designers can achieve trade-offs between voltage and current control based on specific hardware specifications, enabling faster voltage dynamic response while maintaining inductor current within safe operating limits. And these does not require additional control modules.

B. Comparison with Other Controllers

Fig. 3 demonstrates the control performances of three different control: PID controller, conventional MPC and the proposed MPC.

As shown in Fig. 3, the PID controller (blue lines) exhibits significant overshoot in both output voltage and inductor current. This larger overshoot necessitates longer settling time to reach the desired reference values. The conventional MPC

TABLE I: Simulation and Experimental Setup Parameters

Parameters	Values (CCM / DCM)
Input voltage	48 V
Output voltage	24 V / 9 V
Load	10 Ω / 25 Ω
Inductance	47 μH
Capacitance	94 μF
Switching frequency (PI controller)	200 kHz
PI gains	$K_p = 0.2$, $K_i = 10$
Prediction horizons	$N = 4$

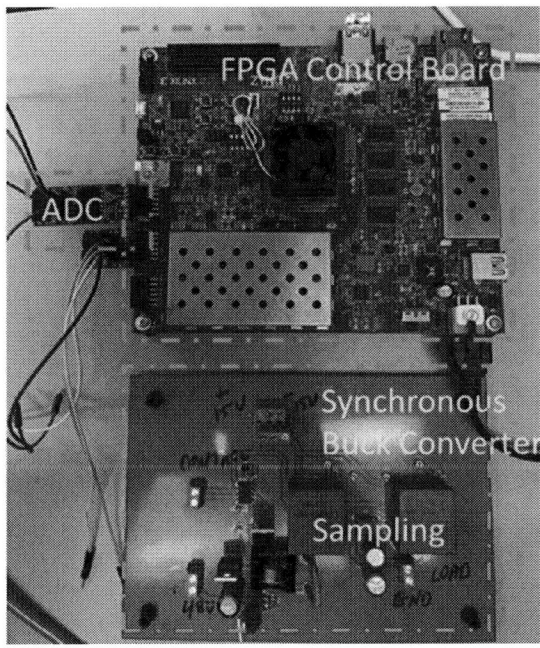

Fig. 4: Experimental setup.

approach (red lines), which considers only output voltage control in its cost function, achieves smaller voltage overshoot and reduced settling time. However, since its control objective is limited to voltage regulation, the inductor current rises excessively as the controller attempts to reach the reference voltage as quickly as possible. The proposed MPC strategy (yellow lines) incorporates both voltage and current control objectives. This dual-objective approach maintains the inductor current within safe operating limits while effectively regulating output voltage. The settling time can be optimized by adjusting the weighting factor based on hardware-specific current overshoot tolerances.

IV. EXPERIMENTAL RESULTS

To verify the proposed MPC algorithm's ability controlling voltage and current, a buck converter was built with MPC implemented on a field-programmable gate array (FPGA). The FPGA-based MPC controller uses a hardware acceleration method as described in [16]. The FPGA board is Xilinx MPSoC ZCU104, the ADC is AD7476a with 1 Msps sampling rate and the MOSFET is IRF540N. Fig. 4 is the experimental setup. Table I shows the system's parameters.

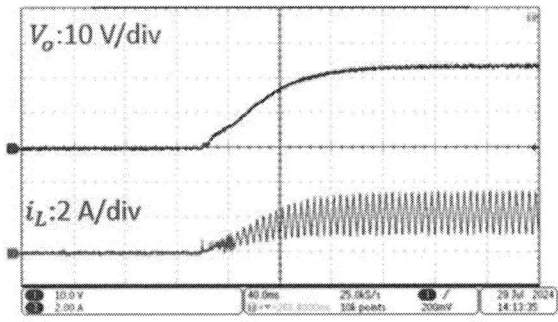

(a) proposed MPC

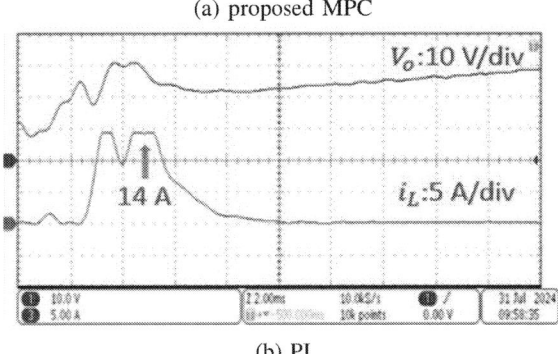

(b) PI

Fig. 5: Experimental results for start-up process (a) proposed MPC; (b) PI.

Fig. 5(a) illustrates the start-up process using the proposed MPC method ($\alpha = 0.5$, $\beta = 0.1$), where the output voltage and inductor current gradually reach their reference values without any overshoot or current spike. In contrast, Fig. 5(b) shows the start-up process with a PI controller ($K_p = 0.2$, $K_i = 10$), which exhibits a large current spike of around 14 A—about six times the current reference—along with a corresponding voltage overshoot. Consequently, additional processes or hardware are required to limit the over-current.

Fig. 6(a) and 6(b) demonstrate the steady-state performance in CCM and DCM, respectively. The proposed control algorithm successfully regulates the DC-DC buck converter in both operating modes using mode-specific parameters. As shown in Fig. 6, this MPC method achieves stable operation in both CCM and DCM without altering control modules.

V. CONCLUSIONS

This paper presents a novel MPC strategy for DC-DC buck converters, utilizing a single control loop to regulate both voltage and current directly. This approach not only enhances transient performance but also simplifies the hardware with stable output voltage from start-up to steady-state operation in both CCM and DCM.

REFERENCES

[1] J. M. Guerrero, M. Chandorkar, T.-L. Lee, and P. C. Loh, "Advanced-control architectures for intelligent microgrids—Part I: Decentralized and hierarchical control," IEEE Trans. Ind. Electron., vol. 60, no. 4, pp. 1254–1262, Apr. 2013.

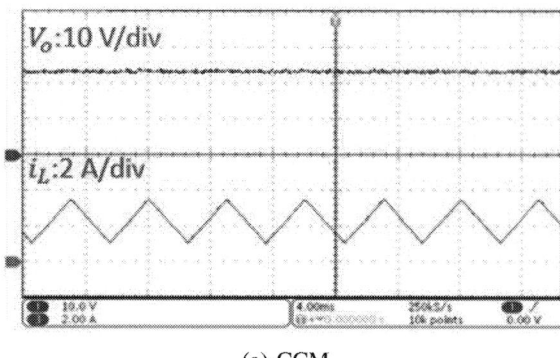

(a) CCM

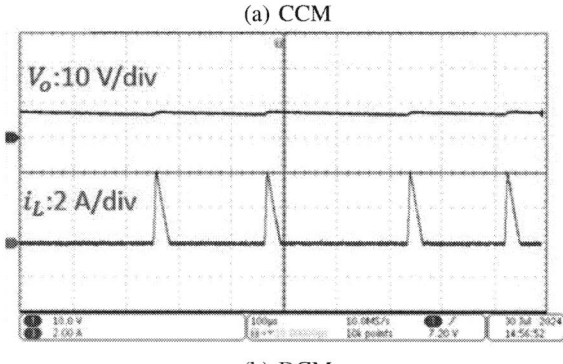

(b) DCM

Fig. 6: Experimental results with proposed MPC in steady state (a) CCM; (b) DCM.

[2] M. Su, Z. Liu, Y. Sun, H. Han and X. Hou, "Stability Analysis and Stabilization Methods of DC Microgrid With Multiple Parallel-Connected DC–DC Converters Loaded by CPLs," in IEEE Transactions on Smart Grid, vol. 9, no. 1, pp. 132-142, Jan. 2018, doi: 10.1109/TSG.2016.2546551.

[3] Z. Guo, X. Zou, Y. Huang, Y. Kang and K. Zou, "Full-State Feedback Based Active Damping Control Design for LCL-type Grid-Connected Converter under Weak Grid," 2018 IEEE International Power Electronics and Application Conference and Exposition (PEAC), Shenzhen, China, 2018, pp. 1-6, doi: 10.1109/PEAC.2018.8590676.

[4] T. Dragičević, S. Vazquez and P. Wheeler, "Advanced Control Methods for Power Converters in DG Systems and Microgrids," in IEEE Transactions on Industrial Electronics, vol. 68, no. 7, pp. 5847-5862, July 2021.

[5] Erickson, Robert W., and Dragan Maksimovic. Fundamentals of power electronics. Springer Science & Business Media, 2007.

[6] S. Chattopadhyay and S. Das, "A Digital Current-Mode Control Technique for DC–DC Converters," in IEEE Transactions on Power Electronics, vol. 21, no. 6, pp. 1718-1726, Nov. 2006, doi: 10.1109/TPEL.2006.882929.

[7] R. Heydari et al., "Model-Free Predictive Control of Grid-Forming Inverters With LCL Filters," in IEEE Transactions on Power Electronics, vol. 37, no. 8, pp. 9200-9211, Aug. 2022, doi: 10.1109/TPEL.2022.3159730.

[8] P. Karamanakos, E. Liegmann, T. Geyer, and R. Kennel, "Model Predictive Control of Power Electronic Systems: Methods, Results, and Challenges," IEEE Open J. Ind. Appl., vol. 1, pp. 95–114, 2020.

[9] M. Narimani, Bin Wu, V. Yaramasu, Zhongyuan Cheng and N. R. Zargari, "Finite Control-Set Model Predictive Control (FCS-MPC) of Nested Neutral Point-Clamped (NNPC) Converter," in IEEE Transactions on Power Electronics, vol. 30, no. 12, pp. 7262-7269, Dec. 2015, doi: 10.1109/TPEL.2015.2396033.

[10] E. Zerdali, M. Rivera and P. Wheeler, "A Review on Weighting Factor Design of Finite Control Set Model Predictive Control Strategies for AC Electric Drives," in IEEE Transactions on Power Electronics, vol. 39, no. 8, pp. 9967-9981, Aug. 2024, doi: 10.1109/TPEL.2024.3370550.

[11] K. Z. Liu and Y. Yokozawa, "An MPC-PI approach for buck DC-DC converters and its implementation," 2012 IEEE International Symposium on Industrial Electronics, Hangzhou, China, 2012, pp. 171-176, doi: 10.1109/ISIE.2012.6237079.

[12] Bonanno, Giovanni, and Luca Corradini. "Digital predictive current-mode control of three-level flying capacitor buck converters." IEEE Transactions on Power Electronics 36.4 (2020): 4697-4710.

[13] Karamanakos, Petros, Tobias Geyer, and Stefanos Manias. "Direct model predictive current control strategy of DC–DC boost converters." IEEE Journal of Emerging and Selected Topics in Power Electronics 1.4 (2013): 337-346.

[14] Cheng, Long, et al. "Model predictive control for DC–DC boost converters with reduced-prediction horizon and constant switching frequency." IEEE Transactions on Power Electronics 33.10 (2017): 9064-9075.

[15] Li, Yuan, et al. "Stability-oriented design of model predictive control for DC/DC boost converter." IEEE Transactions on Industrial Electronics 71.1 (2023): 922-932.

[16] Z. Guo, M. Nelms, "Fpga-based Hardware Acceleration for Model Predictive Control of Power Electronic Converters," IECON 2024- 50th Annual Conference of the IEEE Industrial Electronics Society, Chicago, IL, 2024.

A Novel IPPC Method for Precise Overload Protection and Burst Mode Operation in LLC Resonant Converters

Manikanta Pallantla, Ramkumar S

Texas Instruments

Emails: m-pallantla@ti.com, ramkumar.s@ti.com

Abstract—**In this paper, we provide a new method for regulating resonant converters that compares the input power to a control signal and uses that comparison to change the switching frequency. This approach reduces sensitivity to power stage parameters and tolerances by directly managing input power via the feedback loop that controls output voltage or current. It is perfect for a variety of uses, including LED drivers and battery chargers, due to its inherent power output limits and its ability to effectively monitor loads regardless of input or output voltage fluctuations. Additionally, the method guarantees top-notch responsiveness in real-time. This paper provides a theoretical analysis and employs an LLC resonant converter evaluation module to demonstrate the effectiveness of the method.**

Index Terms—**LLC, Resonant Converter, LCC, IPPC, Input Power Proportional Control, HHC Control**

I. INTRODUCTION

LLC resonant DC-DC converters are becoming more prevalent in off-line DC-DC converters for LED drivers, adapters, chargers, and televisions, covering power ranges from 100W to 1kW. This trend is due to the adoption of soft-switching techniques, extensive implementation of anti-capacitive mode protection, and the use of burst mode control to improve efficiency under light load conditions.

Recent advancements in control methods have significantly enhanced the dynamic properties of LLC resonant converters, leading to their incorporation into commercial products. These innovative techniques manipulate a distinct physical parameter, diverging from traditional feedback loops that regulate the converter's switching frequency. Consequently, the control action autonomously adjusts the switching frequency to align with the load requirements and the converter's input-output power flow. Charge-Mode Control forms the cornerstone of many of these state-of-the-art methods.

Charge-Mode Control (CMC), as its name suggests, controls the electric charge per cycle drawn from the input source by the converter, requiring a form of current integration. CMC methods for resonant converters fall into two categories: those that integrate by processing the signal indicative of the resonant current, and those that employ the resonant capacitor's natural current integration. Regardless of the method, CMC results in a single-pole dynamic system, ensuring superior dynamic performance under all operating conditions, including input voltage (Vin) and output power (Pout). Compared to

Direct Frequency Control (DFC), CMC requires significantly less design effort and fine-tuning to compensate the feedback loop that regulates the output voltage or current due to its integrative nature [1]. However, challenges remain, especially in repeatability, since the control signal in the feedback loop is affected by the switching frequency, which complicates the use of the control signal as a load monitor for initiating burst-mode operation at light loads and implementing overload protection at heavy loads, practices common in PWM converters [2].

Traditionally, protecting LLC converters from overload and deciding when to enter burst mode during light load conditions involves measuring the current through the load. However, charge mode control uses the "charge" control variable, which includes power information of the load, offering an alternative to sensing load current. The challenge arises from the nonlinear relationship between charge and load power, where different load powers may correspond to the same charge, complicating load monitoring without direct current sensing. The proposed Input Power Proportional Control (IPPC) establishes a nearly linear relationship between the "normalized charge" feedback control variable and the load current by normalizing the charge with the switching frequency. Moreover, IPPC aligns this control variable with the input power using input voltage feed-forward, facilitating a linear relationship regardless of input and output voltage changes, between the control parameter and the load power. This method accurately reflects the load current, enhancing precise detection.

This paper presents a concise discussion on the working principles of Hybrid Hysteretic Control (HHC, incorporating CMC) [3] and IPPC. Experimental validations were carried out on a 240W, 24V/10A LLC resonant DC-DC converter prototype board to confirm the theoretical concepts of these methods.

II. HYBRID HYSTERETIC CONTROL FOR LLC RESONANT CONVERTER

Fig. 1 illustrates the HHC operating principle [3]. The resonant capacitor voltage (V_r) is monitored by a capacitor divider (C_{up}, C_{dw}) to manage cycle-by-cycle charge control. This is because the net input charge to the resonant tank over a half cycle correlates with the change in resonant capacitor voltage. HHC modulates the ΔV_{CR} value by comparing the monitored resonant capacitor voltage with the threshold

979-8-3315-1612-3/25 $31.00 © 2025 IEEE

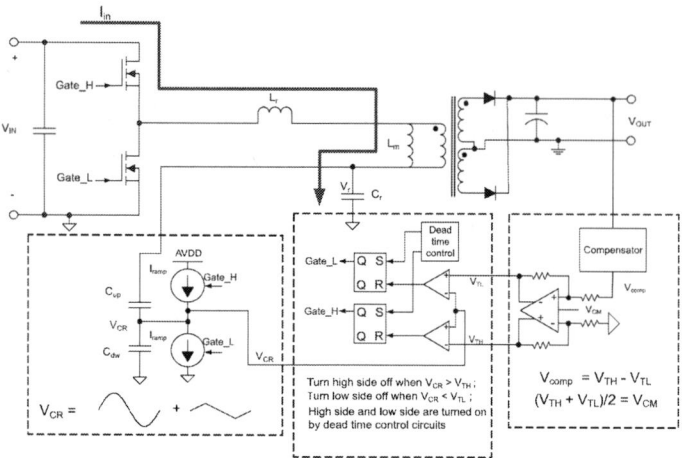

Figure 1: HHC operating principle

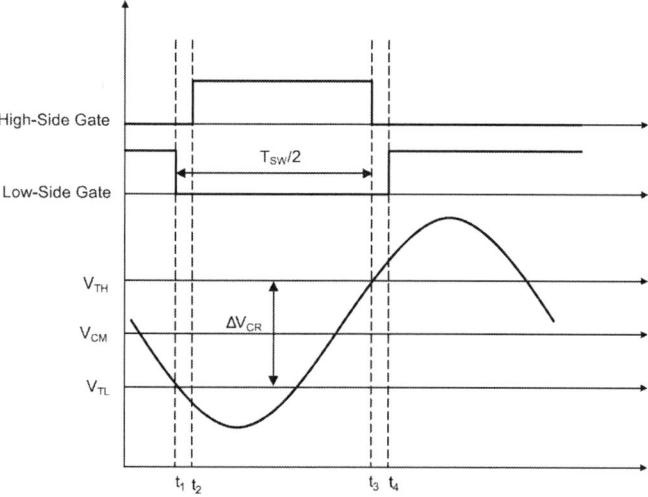

Figure 2: HHC Gate on/off control

voltages V_{TH} and V_{TL}, as depicted in Fig. 2. Additionally, a ramp voltage, created by I_{ramp} current sources, is combined with the monitored resonant capacitor voltage V_{CR}. This ramp voltage is crucial for preventing instability during light load conditions.

$$V_{TH} = V_{CM} + \frac{V_{comp}}{2} \tag{1}$$

$$V_{TL} = V_{CM} - \frac{V_{comp}}{2} \tag{2}$$

$$V_{TH} - V_{TL} = V_{comp} \tag{3}$$

$$\Delta V_{CR} = \frac{C_{up}}{C_{up} + C_{dw}} \cdot \Delta V_r + \frac{1}{C_{up} + C_{dw}} \cdot I_{ramp} \cdot \frac{T_{sw}}{2} \tag{4}$$

$$\Delta V_{CR} = V_{comp} = \frac{C_{up}}{C_{up} + C_{dw}} \cdot \frac{1}{C_r} \cdot T_{sw} \cdot I_{in(avg)} + \frac{1}{C_{up} + C_{dw}} \cdot I_{ramp} \cdot \frac{T_{sw}}{2} \tag{5}$$

According to equation 5, it is evident that in the HHC, the control variable V_{comp} depends on both the input current and the switching frequency. Consequently, the comp signal cannot be relied upon to determine precise load power information. This is because a change in input voltage for the same load power will result in adjustments to both the switching frequency and the input DC average current to maintain the same output voltage.

III. INPUT POWER PROPORTIONAL CONTROL FOR LLC RESONANT CONVERTER

Fig. 3 illustrates the internal workings of the IPPC control [4]. This block measures the resonant tank current through an external differentiator consisting of a resistor R_{ISNS} and a capacitor C_{ISNS}. Unlike HHC, the V_{CR} signal here is synthesized from the sensed resonant current signal V_{ISNS} using an integrator. The V_{CR} synthesizer block produces the compensated internal V_{CR} signal by applying ramp compensation and feed-forward gain, reflecting the sensed DC input voltage V_{BLK}. This V_{CR} signal is then compared with two threshold sets to determine the switching off moments for the high-side and low-side switches, using V_{TH} and V_{TL} as thresholds. These thresholds, V_{TH} and V_{TL}, are set by the compensator output signal $FBReplica$ and the on-time of the high-side and low-side switches from the preceding half-switching cycle. As depicted in Fig. 4, when the V_{CR} voltage surpasses the V_{TH} threshold, the high-side switch is turned off. Similarly, when the V_{CR} voltage drops below the V_{TL} threshold, the low-side switch is turned off.

$$FBReplica \cong \frac{V_{CM}}{R_{RAMP} \cdot C_{VCR} \cdot k} + \left(\frac{2}{k} \cdot \frac{k_{FF}}{R_{VCR} \cdot C_{VCR}} \cdot \frac{R_{ISNS} \cdot C_{ISNS}}{C_r} \right) \cdot \left(\frac{P_{out}}{\eta} \right) \tag{6}$$

$$\Delta V_{CR} = V_{TH} - V_{TL} = k \cdot FBReplica \cdot \frac{T_{sw}}{2} \tag{7}$$

In Equation 6 [5], it can be observed that the control output signal, $FBReplica$, depends solely on the output power (P_{out}) and is independent of the input voltage (V_{in}), switching period (T_{sw}), and the sensitivity of resonant tank parameters. Consequently, it can serve as a load monitor. This feature is particularly beneficial for achieving consistent burst mode and overload protection in scenarios with a wide input range (PFC off condition where LLC is supplied by rectified universal AC input) and wide output applications, such as LED and battery chargers.

IV. HARDWARE RESULTS

A 240W, 24V, 10A LLC resonant half-bridge converter evaluation module shown in Fig. 5, is used to show the differences between $FBReplica$ when the converter is operated using IPPC and HHC control. Also, the same EVM is used to extract the over load power limit (OLP) across the input voltage range. Finally, the EVM load transient response result is provided.

979-8-3315-1612-3/25 $31.00 © 2025 IEEE

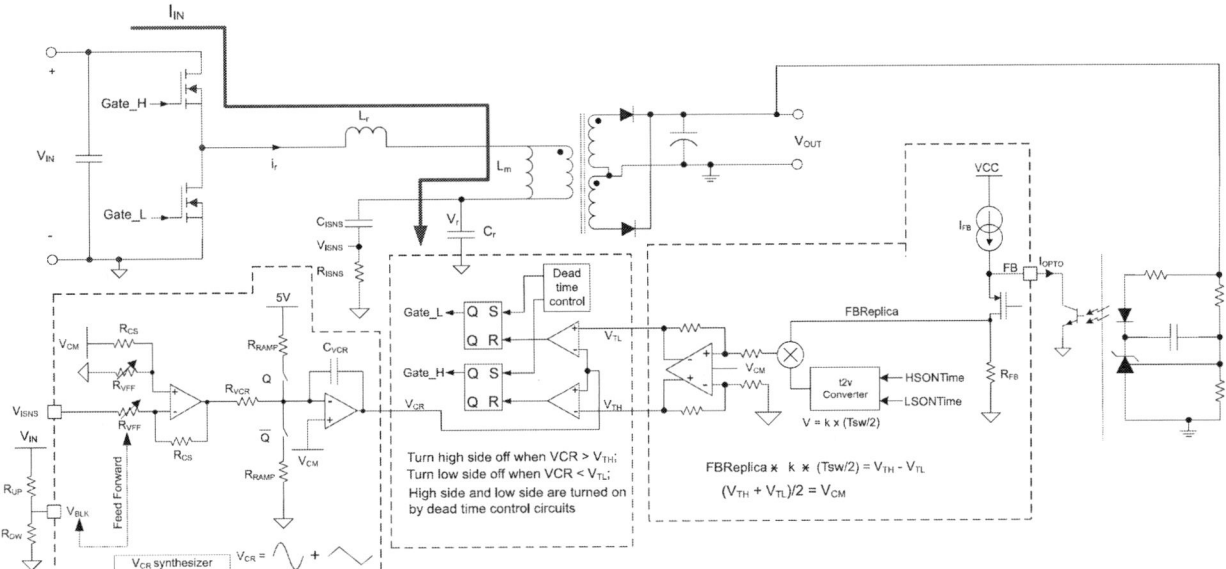

Figure 3: IPPC operating principle

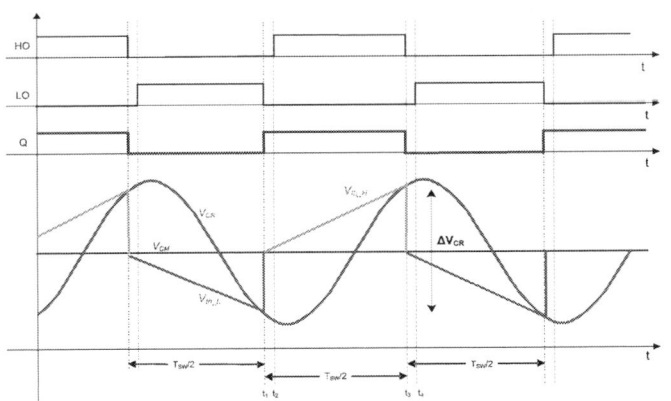

Figure 4: IPPC waveforms

The LLC power stage elements values of this EVM are given in Table I.

Parameters	Value
Magnetizing Inductance	69 uH
Resonant Inductance	23 uH
Resonant Capacitor	12 nF
Resonant Frequency	300 kHz
Primary-to-secondary turns ratio	8:1
Output capacitance	2.82 mF

Table I: LLC EVM power stage element values

A. *FBReplica variation as a function of Output Power and Input Voltage:*

Figures 6 and 7 illustrate the variation in *FBReplica* as the load power ranges from 24W to 240W and the input voltage varies from 230V to 410V. In case of IPPC control, the *FBReplica* variation at any given power is less than 0.35V across the input voltage variation whereas in case of HHC Control (similar to charge mode control), the variation

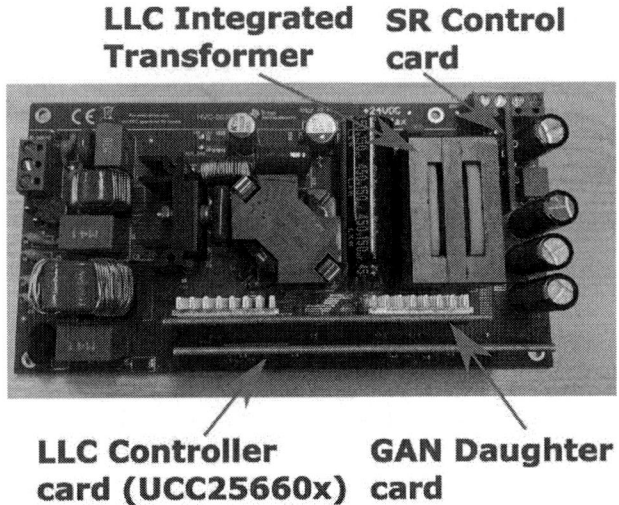

Figure 5: LLC EVM

of *FBReplica* varies as the input voltage changes. At high load, variation of *FBReplica* is close to 1.5V across the input voltage, whereas at light load, variation is 0.8V.

B. *FBReplica variation as a function of Output Power and Wide Output Voltage:*

Figures 8 and 9 depict the variation in *FBReplica* at a constant output power of 100W and 120W, with an output voltage regulated from 15V to 30V and at a 390V DC input. Under HHC control, the *FBReplica* variation at any given power remains below 1.5V. Conversely, with IPPC control, the *FBReplica* variation is considerably smaller, staying under 0.3V across the output voltage range.

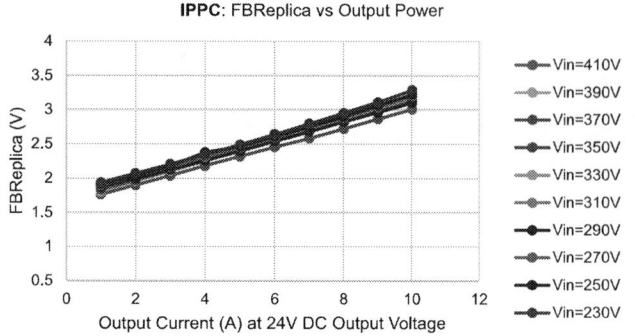

Figure 6: $FBReplica$ variation in case of IPPC Control

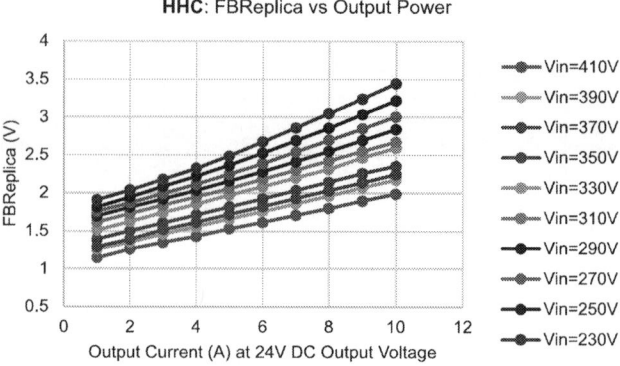

Figure 7: $FBReplica$ variation in case of HHC Control

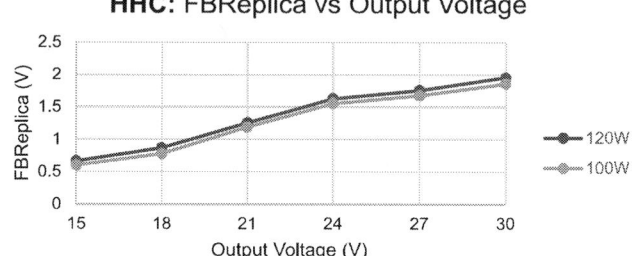

Figure 8: $FBReplica$ variation in case of IPPC Control

Figure 9: $FBReplica$ variation in case of HHC Control

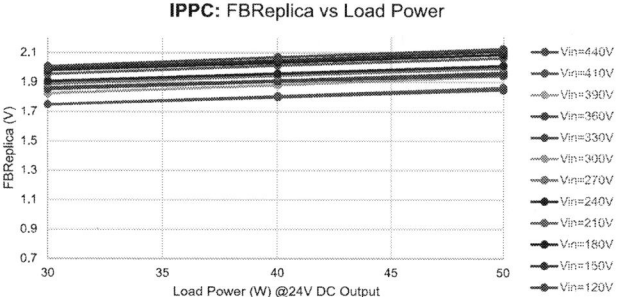

Figure 10: $FBReplica$ variation in case of IPPC Control

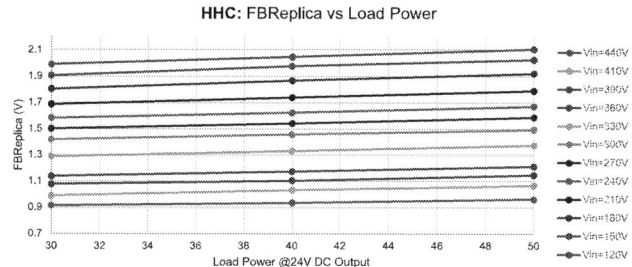

Figure 11: $FBReplica$ variation in case of HHC Control

C. $FBReplica$ variation as a function of Output Power and Wide Input Voltage:

Figures 10 and 11 shows how $FBReplica$ varies across much wider input voltage range (120V to 440V) while the load power spans from 30W to 50W. With HHC control, the $FBReplica$ variation at any given power stays below 1.1V. In contrast, with IPPC control, the $FBReplica$ variation is significantly smaller, remaining under 0.35V across the input voltage.

D. Variation of Over Load Power Limit vs Input Voltage in case of IPPC Control:

From Table II, it can be observed that the variation of OLP with the variation in input voltage is very small. This is due to the fact that $FBReplica$ variation is small in case of IPPC control.

Input Voltage (V)	Over Load Power Limit in Amps @ 24V Regulated Output Voltage
410	12
390	11.56
370	11.22
350	11
330	10.94
310	10.76
290	10.56
270	10.31
250	10.1

Table II: OLP variation vs Input Voltage incase of IPPC Control

E. Load Transient Results:

Figure 12 illustrates the IPPC control's load transient performance. In this captured waveform load current periodically changed from 0.1A to 10A every 40ms at a slew rate of 5A/us. The output voltage ripple during each load transient is approximately 200mV.

979-8-3315-1612-3/25 $31.00 © 2025 IEEE   2711

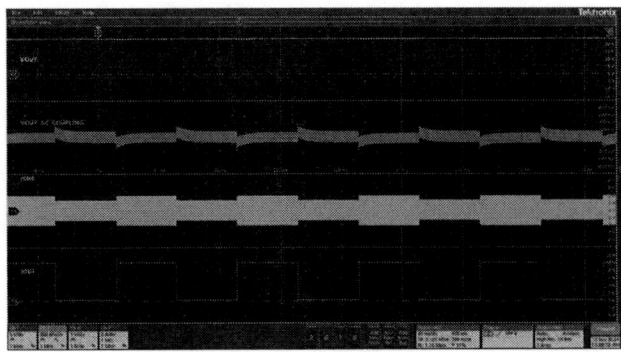

Figure 12: Load Transient Waveform: Ch1: Output Voltage; Ch3: Output Voltage with AC Coupling; Ch4: Sensed Resonant Current; Ch5: Output Current

V. CONCLUSION

The Input Power Proportional Control (IPPC) for resonant converters has been proposed and its main advantages compared to charge mode control theoretically analyzed and validated by the hardware results. Its been shown that the control signal in case of the IPPC control can be used as a load monitor irrespective of changes in input voltage and output voltages of the LLC due to which consistent over load power limit and consistent burst mode entries can be obtained.

REFERENCES

[1] C. Adragna, "Llc resonant converters: An overview of modeling, control and design methods and challenges," *Foundations and Trends® in Electric Energy Systems*, vol. 5, no. 2–4, pp. 75–491, 2022. [Online]. Available: http://dx.doi.org/10.1561/3100000029

[2] J. Chen, T. Sato, K. Yano, H. Shiroyama, M. Owa, and M. Yamadaya, "An average input current sensing method of llc resonant converters for precise overload protection," in *2017 IEEE Applied Power Electronics Conference and Exposition (APEC)*, 2017, pp. 267–272.

[3] B. McDonald and Y. Li, "A novel llc resonant controller with best-in-class transient performance and low standby power consumption," in *2018 IEEE Applied Power Electronics Conference and Exposition (APEC)*, 2018, pp. 489–493.

[4] "Ucc25660 750-khz wide vin /vout range llc controller optimized for light-load efficiency," 2023. [Online]. Available: https://www.ti.com/lit/gpn/ucc25660

[5] M. Pallantla and R. S, "Input power proportional control of llc resonant converter," in *2024 IEEE Transportation Electrification Conference and Expo (ITEC)*, 2024, pp. 1–6.

An Improved Current-Sensorless Model Predictive Voltage Control for Four-Leg Voltage Source Inverters

Heng Guo
School of
Control Science and Engineering
Shandong University
Jinan, China
guoheng@mail.sdu.edu.cn

Yuxin Wei
School of
Control Science and Engineering
Shandong University
Jinan, China
202434888@mail.sdu.edu.cn

Mengmeng Jing
School of
Control Science and Engineering
Shandong University
Jinan, China
202014057@mail.sdu.edu.cn

Wenlong Ding
School of
Control Science and Engineering
Shandong University
Jinan, China
wlding@sdu.edu.cn

Bin Duan
School of
Control Science and Engineering
Shandong University
Jinan, China
duanbin@sdu.edu.cn

Chenghui Zhang*
School of
Control Science and Engineering
Shandong University
Jinan, China
zchui@sdu.edu.cn

Abstract—Four-leg voltage source inverters (VSIs) with LC output filter are commonly utilized to stabilize three-phase sinusoidal voltages under unbalanced loads conditions. However, conventional model predictive voltage control (MMPVC) for four-leg VSIs relies heavily on current sensors, leading to increased costs and suboptimal output voltage performance. To address these issues, this paper proposes an improved current-sensorless modulated model predictive voltage control (MMPVC) for four-leg VSIs to effectively reduce voltage tracking errors and total harmonic distortion (THDv). First, a readily implementable capacitor current extended state observer is constructed to replace current sensors and a corresponding discrete model is reconstructed. Then, four-vector MMPVC for four-leg VSIs is carefully designed. Especially, the ineffective voltage vectors are found, and their duration time is set as zero to optimize the output voltage performance in edge regions of sectors and over-modulation regions. Finally, experimental results validate the proposed approach, demonstrating substantial improvements in output voltage quality and tracking accuracy in over-modulation regions.

Index Terms—model predictive voltage control, current-sensorless, four-leg voltage source inverters.

I. Introduction

Three-phase four-leg voltage source inverters (VSIs) with LC output filter stand out as one of the best solutions to provide an ideal three-phase sinusoidal output voltage, even under severe unbalanced loads, thanks to high DC-link voltage utilization and flexible control of zero-sequence component

This paper was supported by the National Key R&D Plan of China, Grant/AwardNumber: 2022YFF0712700, National Natural Science Foundation of China (No. 62203275, 62203265, 62133007, 62333013), Shandong Province Science Foundation (No. 2022HYYQ-022, ZR2022QF028, ZR2024QF101), Fellowship of China National Postdoctoral Program (No. BX20230209, 2024M751809).

[1], [2]. Therefore, it has been used in many independent power supply systems, including the distributed generation (DG) systems, uninterruptible power supply (UPS) systems, energy storage (ES) systems, and stand-alone AC applications systems [3].

Remarkably, model predictive voltage control (MPVC) is favorable for four-leg VSIs due to the merits of simplicity, fast dynamic response, and ability to handle multivariate constraints [4]. To reduce costs, many implementations replace physical current sensors with current estimators [5]–[8]. However, conventional MPVC suffers from degraded output voltage quality, leading to inaccuracy in current estimation and difficulty in voltage tracking.

To overcome these issues, researchers have proposed modulated model predictive voltage control (MMPVC), which employs multiple voltage vectors within a single cycle and incorporates modulation stages to improve output performance. In MMPVC, the duration time of each voltage vector is determined based on its cost function value [9]. Yet, MMPVC usually combines two voltage vectors or three voltage vectors in one sampling period [10]. While this approach generally enhances voltage quality compared to traditional MPVC, it typically combines only two or three voltage vectors per sampling period, limiting its effectiveness for four-leg VSIs. The additional zero-sequence component in four-leg VSIs demands the use of four voltage vectors for optimal performance. Based on this, in [11], four voltage vectors were selected in one cycle, and the performance in the over-modulation region is also optimized by removing the zero vector. The article realized high performance output voltage both in the linear region and in the over-modulation region, but exhibited poor tracking

979-8-3315-1612-3/25 $31.00 © 2025 IEEE

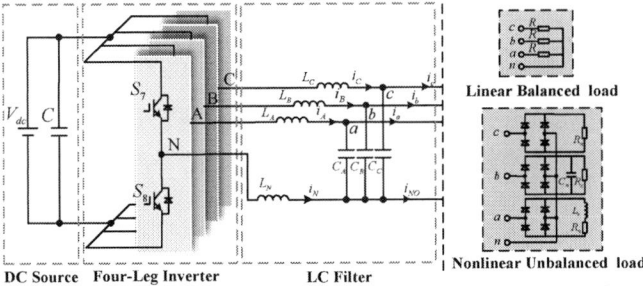

Fig. 1. Topology of four-leg VSI

performance.

To overcome these shortcomings, this paper proposed an improved current-sensorless MMPVC for four-leg VSIs. The key contributions are:

1) A capacitor current extended state observer is constructed to replace current sensors, enhancing estimation accuracy.
2) An improved MMPVC strategy is proposed to reduce voltage tracking errors and voltage total harmonic distortion (THDv), especially in edge regions of sectors and over-modulation regions.

The remaining parts of this article are arranged as follows. In Section II, the state variable is reconstructed as capacitor current, and the change in output current is used as an error for the extend state observer design. Based on this, a simple current-sensorless mathematical model is developed. In Section III, the four-vector MMPVC is briefly extended to four-leg VSIs, the duration time of ineffective voltage vectors are set to zero. In Section IV, the effectiveness of the proposed MMPVC strategy is verified by comparative experimental studies. Conclusions are drawn in section V.

II. SYSTEM DESCRIPTION

The topology of the four-leg VSI studied in this article is shown in Fig. 1, where V_{dc} is the DC-link voltage, L_a, L_b and L_c are three-phase filter inductor, i_{La}, i_{Lb}, i_{Lc} are three-phase inductor currents, i_{ca}, i_{cb}, i_{cc} are three-phase capacitive currents, i_{oa}, i_{ob}, i_{oc} are three-phase output currents, L_n is the zero-sequence inductor.

With the addition of the fourth leg, the coupling between the three phases is more severe. To better handle the coupling introduced by the fourth leg, the three-phase variables are transformed into the $\alpha\beta\gamma$ reference frame, enabling independent control of decoupled state variables [12]. The mathematical model in $\alpha\beta\gamma$ reference frame is described by equation (1).

$$\begin{cases} \dfrac{di_{Lx}}{dt} = \dfrac{1}{L_x}(v_{ix} - v_{cx}) \\ \dfrac{dv_{cx}}{dt} = \dfrac{1}{C_x}(i_{Lx} - i_{ox}) \end{cases} \tag{1}$$

where $x = \alpha$, β, γ, $L_{\alpha,\beta} = L$, $L_\gamma = 3L_n + L$, $C_{\alpha,\beta,\gamma} = C$. As can be seen from (1), the output current needs to be measured to observe the inductor current. If only the capacitor voltage is used, it is impossible to directly observe

two currents at the same time. Therefore, the state variable is firstly converted from inductive current to capacitive current, as described in equation (2).

$$\begin{cases} \dfrac{di_{cx}}{dt} = \dfrac{di_{Lx}}{dt} - \dfrac{di_{ox}}{dt} = \dfrac{1}{L_x}(v_{ix} - v_{cx}) - \dfrac{di_{ox}}{dt} \\ \dfrac{dv_{cx}}{dt} = \dfrac{1}{C}(i_{Lx} - i_{ox}) = \dfrac{i_{cx}}{C} \end{cases} \tag{2}$$

To simplify the design of the observer, it is often assumed that change in output current is constant ($di_{ox}/dt = 0$). However, this approximation results in inaccurate current estimation and poor immunity. Therefore, denoting di_{ox}/dt as f_x, the capacitor current extended state observer can be designed as (3).

$$\begin{cases} \dfrac{d\hat{v}_{xn}}{dt} = \dfrac{1}{C}\hat{i}_{cx} + k_1(v_{cx} - \hat{v}_{cx}) \\ \dfrac{d\hat{i}_{cx}}{dt} = \dfrac{1}{L_x}(v_{ix} - \hat{v}_{cx}) - \hat{f}_x + k_2(v_{cx} - \hat{v}_{cx}) \\ \dfrac{d\hat{f}_x}{dt} = k_3(v_{cx} - \hat{v}_{cx}) \end{cases} \tag{3}$$

where the notation " $\wedge$ " represents the estimated value, while k_1, k_2 and k_3 are the estimator gains.

Furthermore, based on the forward Euler method, the discrete model of the proposed capacitor current extended state estimator is derived as

$$\begin{cases} \hat{v}_{cx}(k+1) = \hat{v}_{cx}(k) + T_s\left(\dfrac{1}{C_f}\hat{i}_{cx}(k) + k_1 e_x(k)\right) \\ \hat{i}_{cx}(k+1) = \hat{i}_{cx}(k) \\ \quad\quad + T_s\left(\dfrac{1}{L_x}(v_{ix}(k) - \hat{v}_{cx}(k)) - f_x(k) + k_2 e_x(k)\right) \\ \hat{f}_x(k+1) = \hat{f}_x(k) + T_s k_3 e_x(k) \end{cases} \tag{4}$$

where T_s is the sampling time, $e_x(k) = v_{cx}(k) - \hat{v}_{cx}(k)$.

The estimator gains k_1, k_2 and k_3 in (4) can be selected using the direct pole assignment strategy. Besides, the proposed capacitor current extend estimator is rewritten as follows:

$$\underbrace{\begin{bmatrix} \hat{V}_{Cx}(k+1) \\ \hat{i}_{c\alpha}(k+1) \\ \hat{F}(k+1) \end{bmatrix}}_{\hat{\mathbf{x}}(k+1)} = \underbrace{\begin{bmatrix} 1 & \frac{T_s}{C_f} & 0 \\ -\frac{T_s}{L_f} & 1 & T_s \\ 0 & 0 & 1 \end{bmatrix}}_{\mathbf{G}} \underbrace{\begin{bmatrix} \hat{v}_{f\alpha}(k) \\ \hat{i}_{c\alpha}(k) \\ \hat{F}(k) \end{bmatrix}}_{\mathbf{x}(k)} + \underbrace{\begin{bmatrix} 0 \\ \frac{T_s}{L_f} \\ 0 \end{bmatrix}}_{\mathbf{H}} v_{i\alpha}(k)$$
$$+ \underbrace{\begin{bmatrix} k_1 T_s & 0 & 0 \\ k_2 T_s & 0 & 0 \\ k_3 T_s & 0 & 0 \end{bmatrix}}_{\mathbf{K}} [\mathbf{x}(k) - \hat{\mathbf{x}}(k)] \tag{5}$$

Accordingly, the eigenpolynomial of (5) is expressed as

$$\det(z\mathbf{I} - \mathbf{G} + \mathbf{K}) = z^3 + (k_1 T_s - 3)z^2$$
$$+ \left(\dfrac{T_s^2}{CL_x} + 3 + \dfrac{T_s^2 k_2}{C} - 2T_s k_1\right)z$$
$$- \dfrac{T_s}{C_f} - 1 - \dfrac{T_s^2 k_2}{C} + \dfrac{T_s^3 k_3}{C} + T_s k_1 \tag{6}$$

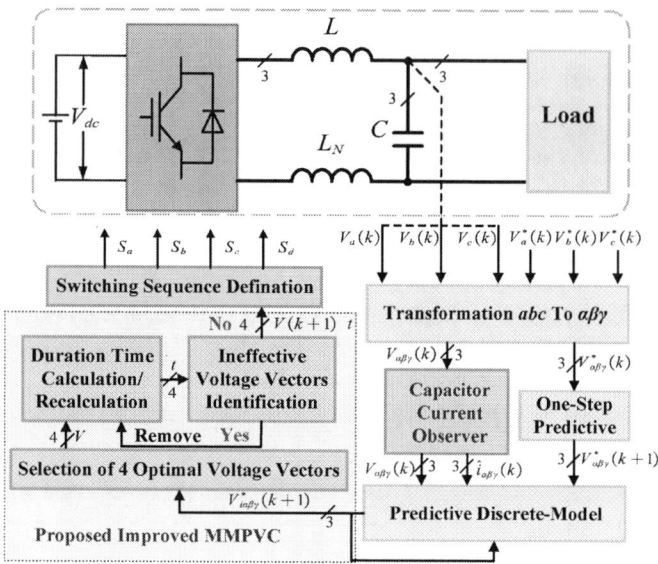

Fig. 2. Control block diagram of the proposed method.

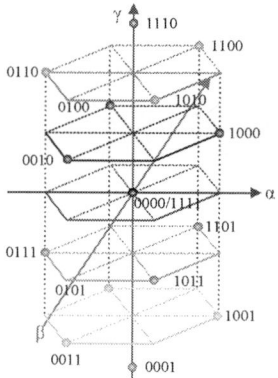

Fig. 3. Voltage vectors of four-leg VSIs in 3-D space.

where **I** is the three-dimensional unit matrix. To make the roots of the characteristic equation both fall at $-\omega_0$, k_1, k_2 and k_3 can be obtained by

$$\det\left(z\mathbf{I} - \mathbf{G} + \mathbf{K}\right) = \left(z - \left(1 - \omega_0 T_s\right)\right)^2 \quad (7)$$

By comparing the coefficients of (7), the estimator gains k_1, k_2 and k_3 can be obtained as

$$\begin{cases} k_1 = 3\omega_0 \\ k_2 = 3C_x\omega_0^2 - \dfrac{1}{L_x} \\ k_3 = C_x\omega_0^3 \end{cases} \quad (8)$$

Based on deadbeat control proposed in [12], voltage reference can be obtained by

$$u_{ix}(ref) = \frac{L_x}{T_s}\left(\frac{C}{T_s}\left(v_{cx}(ref) - v_{cx}(k)\right) - \hat{i}_{cx}(k)\right) + \hat{f}_x(k)) + v_{cx}(ref) \quad (9)$$

where i_{cx} and f are replaced by $\hat{i}_{cx}$ and $\hat{f}$ because they can not be directly measured. where $v_{cx}(ref)$ is the voltage reference and it is synthesized by proposed MMPVC to generate the final gating pulses in this article.

III. PROPOSED IMPROVED MMPVC FOR FOUR-LEG VSIS

The control block diagram of the proposed method is shown in Fig. 2. Fig. 3 shows voltage vectors of four-leg VSIs in 3-D space. To obtain the optimal voltage vectors, all basic voltage vectors are evaluated by the cost function for voltage tracking [11].

$$g(v) = \sqrt{\sum_{x=\alpha,\beta,\gamma}\left(v_x(ref) - v_x\right)^2} \quad (10)$$

Then, the optimal four voltage vectors are applied in one cycle (recorded as u_0, u_1, u_2 and u_3), and their duration time is usually calculated by (11) [9]. However, equation (11) inaccurately calculates the duration time of voltage vectors, resulting in low voltage tracking accuracy in the edge regions of sectors and the overmodulation regions. Specifically, in these regions, the duration time of some voltage vectors should be close to zero, but the results calculated using equation (11) are contrary. Therefore, in the proposed method, such voltage vectors are recognized as ineffective voltage vectors, and their duration time is set to zero and reallocated to the remaining voltage vectors.

$$\begin{cases} t_{1,u0} = \dfrac{G_{u2}G_{u3}G_{u4}}{G_u}T_s \\[2mm] t_{1,u1} = \dfrac{G_{u1}G_{u3}G_{u4}}{G_u}T_s \\[2mm] t_{1,u2} = \dfrac{G_{u1}G_{u2}G_{u4}}{G_u}T_s \\[2mm] t_{1,u3} = \dfrac{G_{u1}G_{u2}G_{u3}}{G_u}T_s \end{cases} \quad (11)$$

where T_s is the duration time of one cycle, $G_u = G_{u1}G_{u2}G_{u3} + G_{u1}G_{u2}G_{u4} + G_{u1}G_{u2}G_{u4} + G_{u2}G_{u3}G_{u4}$.

The simplified execution steps of the proposed MMPVC are shown below:

Step 1: Based on (4) and (9), the capacitor current is estimated, and the reference voltage can be calculated.

Step 2: According to the location of the reference voltage, four candidate voltage vectors are preselected, and their duration time is calculated by (11).

Step 3: Among all candidate voltage vectors, the duration time of the voltage vector with the largest cost function value is set to zero. Assuming that the duration time of u_3 is set to zero, the duration time of the remaining vector is allocated according to (12).

Step 4: If the remaining voltage vectors synthesize the reference voltage vector more accurately, u_3 is identified as an ineffective voltage vector. Step 3 is repeated to further determine whether there are still ineffective voltage vectors. If not, it indicates no ineffective vector among the candidate voltage vectors. Then, four voltage vectors and their reallocated time is fed into the modulator.

$$\begin{cases} t'_{u0} = t_{u0} * t_{u3} \dfrac{t_{u0}}{T_s} \\[2mm] t'_{u1} = t_{u1} * t_{u3} \dfrac{t_{u1}}{T_s} \\[2mm] t'_{u2} = t_{u2} * t_{u3} \dfrac{t_{u2}}{T_s} \end{cases} \quad (12)$$

Compared with traditional methods, this approach can eliminate ineffective voltage vectors, and improve tracking accuracy in overmodulation regions.

IV. EXPERIMENTAL RESULTS

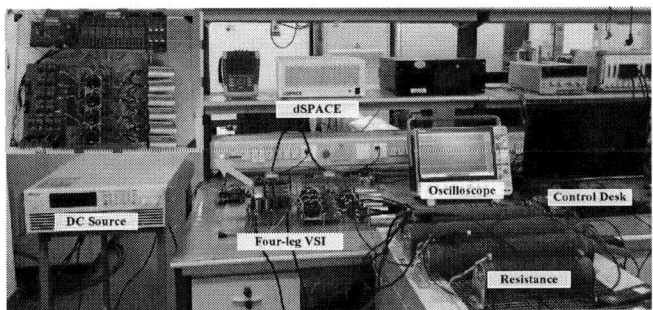

Fig. 4. Experimental setup for verification of the proposed scheme.

TABLE I
GENERAL PARAMETERS OF THE IMPLEMENTED SYSTEMS

Parameter	Symbol	Value
Input Voltage	V_{dc}	240V/200V
Filter Capacitor	C_A, C_B,C_C	$60\mu F$
Filter Inductor	L_A, L_B, L_C, L_N	$1.5mH$
Resistive load	R_a,R_b,R_c	10Ω

In order to verify the effectiveness and superiority of the proposed MMPVC, an experimental platform of the four-leg VSI was established, as shown in Fig. 4. The main parameters are listed in Table I. The FS-MPVC [4], four-vector MM-PVC and proposed MMPVC are implemented on a dSPACE DS2004 board. The experiment comprises the following two cases:

1) **In the linear region**, Fig. 5 shows the voltage waveforms and current waveforms of three methods under the condition that the peak of phase reference voltage at 120V and DC-link voltage at 240V. The comparison results of THDv and tracking error are recorded in Table II.

2) **In the overmodulation region**, Fig. 6 shows the voltage and current waveforms of the three methods under the

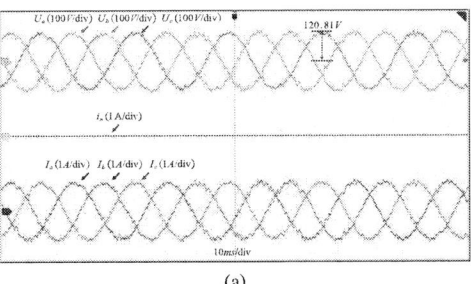

(a)

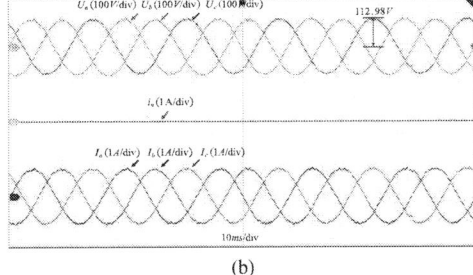

(b)

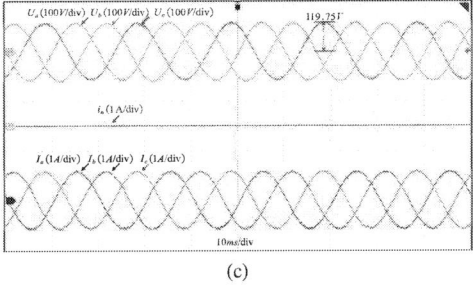

(c)

Fig. 5. Voltage and current waveforms in the linear region. (a) MPVC. (b) Four-vector MMPVC. (c) Proposed MMPVC.

condition that the peak of the phase reference voltage at 120V and the DC-link voltage at 200V. The comparison results of THDv and tracking error are recorded in Table III.

TABLE II
COMPARISON RESULTS OF THDv AND TRACKING ERROR
IN THE LINEAR REGION.

Method	THDv	Tracking error
FS-MPVC	7.46%	0.81V
Four-Vector MMPVC	2.22%	7.02V
Proposed MMPVC	2.18%	0.25V

Form the experimental results, the following conclusions can be drawn:

1) All three methods can effectively track the three-phase voltage reference, both in the linear region and in the over-modulation region. Among them, MPVC and the proposed method have better tracking performance than the four-vector MMPVC.

2) While the four-vector MMPVC can demonstrate improved output THDv in the linear region region, it works

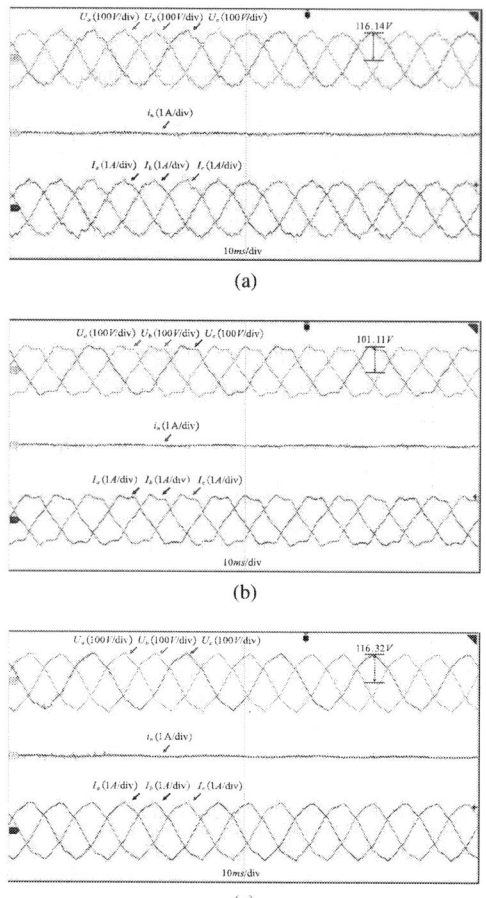

Fig. 6. Voltage and current waveforms in the overmodulation region. (a) MPVC. (b) Four-vector MMPVC. (c) Proposed MMPVC.

TABLE III
COMPARISON RESULTS OF THDv AND TRACKING ERROR
IN THE OVER-MODULATION REGION.

Method	THDv	Tracking error
FS-MPVC	5.38%	3.86V
Four-Vector MMPVC	8.35%	18.89V
Proposed MMPVC	4.18%	3.68V

very poorly in the over-modulation region, which is attributed to the difficulty in maintaining the output voltage. Whereas, MPVC exhibits poor output THDv performance both in the linear region and in the over-modulation region, which is higher than the proposed method.

In summary, the experimental results show that although all three methods can track the reference voltage, but the proposed MMPVC achieves better output voltage performance and tracking capability.

V. CONCLUSIONS

This paper proposes a novel current-sensorless MMPVC strategy for four-leg VSIs with LC output filter. By replacing current sensors with a capacitor current extended state observer

and optimizing the duration time of voltage vectors, the proposed method significantly enhances voltage tracking accuracy and reduces THDv. Experimental results verified its superiority over conventional methods in both linear and over-modulation regions, highlighting its potential for high-performance and cost-effective power conversion.

REFERENCES

[1] R Zhang, VH Prasad, D Boroyevich, and FC Lee. Three-dimensional space vector modulation for four-leg voltage-source converters. *IEEE TRANSACTIONS ON POWER ELECTRONICS*, 17(3):314–326, MAY 2002.

[2] Roberto Cardenas, Carlos Juri, Ruben Pena, Patrick Wheeler, and Jon Clare. The application of resonant controllers to four-leg matrix converters feeding unbalanced or nonlinear loads. *IEEE TRANSACTIONS ON POWER ELECTRONICS*, 27(3):1120–1129, MAR 2012.

[3] Joan Rocabert, Alvaro Luna, Frede Blaabjerg, and Pedro Rodriguez. Control of power converters in ac microgrids. *IEEE TRANSACTIONS ON POWER ELECTRONICS*, 27(11, SI):4734–4749, NOV 2012.

[4] Changming Zheng, Tomislav Dragicevic, and Frede Blaabjerg. Current-sensorless finite-set model predictive control for LC-filtered voltage source inverters. *IEEE TRANSACTIONS ON POWER ELECTRONICS*, 35(1):1086–1095, JAN 2020.

[5] Sebastian Gomez Jorge, Jorge A. Solsona, and Claudio A. Busada. Control scheme for a single-phase grid-tied voltage source converter with reduced number of sensors. *IEEE TRANSACTIONS ON POWER ELECTRONICS*, 29(7):3758–3765, JUL 2014.

[6] Yasser Abdel-Rady Ibrahim Mohamed, Ehab F. El-Saadany, and Magdy M. A. Salama. Adaptive grid-voltage sensorless control scheme for inverter-based distributed generation. *IEEE TRANSACTIONS ON ENERGY CONVERSION*, 24(3):683–694, SEP 2009.

[7] Yasser Abdel-Rady I. Mohamed, M. A. Rahman, and R. Seethapathy. Robust line-voltage sensorless control and synchronization of LCL-filtered distributed generation inverters for high power quality grid connection. *IEEE TRANSACTIONS ON POWER ELECTRONICS*, 27(1), JAN 2012.

[8] Roberto A. Fantino, Claudio A. Busada, and Jorge A. Solsona. Observer-based grid-voltage sensorless synchronization and control of a VSI-LCL tied to an unbalanced grid. *IEEE TRANSACTIONS ON INDUSTRIAL ELECTRONICS*, 66(7):4972–4981, JUL 2019.

[9] Leilei Guo, Mo Chen, Yanyan Li, Pengshuai Wang, Nan Jin, and Jie Wu. Hybrid multi-vector modulated model predictive control strategy for voltage source inverters based on a new visualization analysis method. *IEEE TRANSACTIONS ON TRANSPORTATION ELECTRIFICATION*, 9(1):8–21, MAR 2023.

[10] Tong Liu, Alian Chen, Wei Wang, Xi Liu, Qicai Ren, Guanguan Zhang, and Chenghui Zhang. An improved model predictive control to enhance voltage performance for LC filtered three-level inverters with voltage feedback only. *IEEE TRANSACTIONS ON INDUSTRIAL INFORMATICS*, 19(9):9809–9820, SEP 2023.

[11] Dan Xiao, Kazi Saiful Alam, and Muhammed Fazlur Rahman. Predictive duty cycle control for four-leg inverters with LC output filter. *IEEE TRANSACTIONS ON INDUSTRIAL ELECTRONICS*, 68(5):4259–4268, MAY 2021.

[12] Mohammad Pichan, Hasan Rastegar, and Mohammad Monfared. Dead-beat control of the stand-alone four-leg inverter considering the effect of the neutral line inductor. *IEEE TRANSACTIONS ON INDUSTRIAL ELECTRONICS*, 64(4):2592–2601, APR 2017.

A Highly Integrable, Modular and Multi-Functional Fault Monitoring Active Gate Driver with Parallel Buffers for a Global Enhanced Reliability of Gen. 3 SiC Power MOSFETs

Mathis Picot-Digoix
Electrical & Electronic Systems
Safran Tech
Châteaufort, France
mathis.picot-digoix@safrangroup.com

2nd Léo Seugnet
LAPLACE Laboratory
Univ. de Toulouse, CNRS, INPT, UPS
Toulouse, France
leo.seugnet@etu.toulouse-inp.fr

3rd Frédéric Richardeau
LAPLACE Laboratory
Univ. de Toulouse, CNRS, INPT, UPS
Toulouse, France
frederic.richardeau@laplace.univ-tlse.fr

4th Jean-Marc Blaquière
LAPLACE Laboratory
Univ. de Toulouse, CNRS, INPT, UPS
Toulouse, France
blaquiere@laplace.univ-tlse.fr

5th Sébastien Vinnac
LAPLACE Laboratory
Univ. de Toulouse, CNRS, INPT, UPS
Toulouse, France
sebastien.vinnac@laplace.univ-tlse.fr

6th Thanh-Long Le
Electrical & Electronic Systems
Safran Tech
Châteaufort, France
thanh-long.le@safrangroup.com

7th Stéphane Azzopardi
Electrical & Electronic Systems
Safran Tech
Châteaufort, France
stephane.azzopardi@safrangroup.com

Abstract—**To address their fragility and limitations in high-reliability, safety-critical applications, especially in future aeronautic on-board power systems, SiC MOSFETs are increasingly paired with smart gate drivers. These drivers provide health diagnostics, dv/dt control to reduce losses and EMI, and integrated fault detection. With newer generations, SiC MOSFETs demonstrate increasingly shorter short-circuit withstand times. A multi-channel parallel gate driver with extensive fault detection and gate-oxide integrity monitoring, based on gate-source voltage monitoring is proposed. On commercial 1200 V Gen. 3 SiC MOSFETs, Hard Switching Faults and Faults Under Load are detected within 890 ns and 235 ns, respectively. Gate-oxide cracks are identified before controllability is compromised, and the robustness of this detection method is experimentally validated.**

Index Terms—**Power MOSFET, driver circuits, electrical fault detection, power system reliability, fault currents, power transistors**

I. INTRODUCTION

Smart gate drivers (SGDs) play a key role in improving the reliability and efficiency of power systems in More Electric Aircraft, which are essential for sustainable aviation [1] [2] [3]. By enabling proactive monitoring, SGDs help prevent failures and extend power semiconductor devices lifespan, supporting the shift toward fully electric aircraft. Dual-channel gate drivers, such as the Infineon 2EDi product family, Texas Instruments UCC21222 [4] and UCC27444-Q1 [5], and Skyworks Si827x [6], can control high and low-side switches alternately

or be paralleled for increased output capability. Parallel-buffer gate-driving architectures can be used for switching speed dynamic optimization [7] and for health monitoring and aging compensation [8]. Such a parallel architecture is leveraged for electrical fault detection in a previous generation of SiC MOSFETs (Gen. 2), as detailed in [9]. In this study, short-circuit detection relies on the transistor's physical properties, including the chip's active area (linked to the Schottky emission gate leakage current), the internal capacitance (C_{GD}), and the source stray inductance (L_S). These parameters are reduced in newer generations of SiC MOSFETs, as shown in Tab. I. Consequently, this article focuses on evaluating the performance of this fault detection method on Gen. 3 SiC MOSFETs.

TABLE I
COMPARISON OF MOSFET DESIGN PARAMETERS ACROSS VARIOUS
GENERATIONS WITH SIMILAR RATINGS, FROM [10] [11].

Device Under Test (DUT)	C2M0080120D Gen. 2 1200 V 80 mΩ	C3M0075120K Gen. 3 1200 V 75 mΩ
Package	TO-247-3	TO-247-4
t_{SC} (V_{BUS} = 600 V, 25 °C)	8.5 µs	5 µs (-41 %)
Critical energy (E_{CRIT})	720 mJ	500 mJ (-30 %)
C_{GD} (V_{DS} = 600 V, V_{GS} = 0 V)	10 pF	2 pF (-80 %)
Chip active area (A_0)	6 mm²	4.6 mm² (-23 %)

These changes result in a decreased short-circuit withstand time (t_{SC}) and potentially lower detection sensitivity. Current short-circuit detection typically employs the desaturation (DeSat) method, which monitors the MOSFET's drain-source voltage for a rise above a threshold, indicating a short-circuit. This method requires a blanking time and costly and footprint-intensive high-voltage diodes [12]. The considered gate driver detects all types of short-circuits, permanent gate-oxide damage, and over-currents. It is versatile, modular, and compatible with recent SiC MOSFET health monitoring work [13] that relies on a punctual ultra-slow turn-on to monitor the plateau (V_P) and threshold (V_{TH}) voltages, as shown in dashed lines in Fig. 1 and in Fig. 2 as an optional plug-in solution.

The applicability of this fault detection method to Gen. 3 SiC MOSFETs is evaluated and its robustness to bus voltage and temperature variations is characterized.

II. DESCRIPTION OF THE PROPOSED ARCHITECTURE

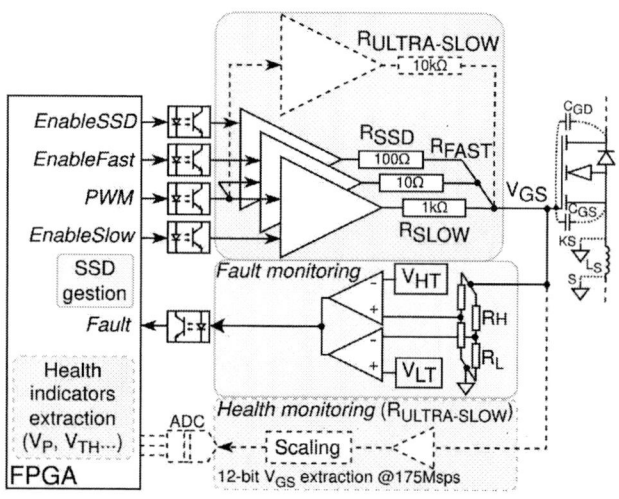

Fig. 1. Detailed schematic of the proposed gate driving architecture (in blue) and fault detection scheme (in green). Compatibility with digital health monitoring is highlighted in purple.

Monitoring faults via the gate voltage (V_{GS}) avoids the need for high-voltage sensing, making it ASIC-compatible.
Every fault impacts the MOSFET's gate voltage.

Hard Switching Faults (HSF) (Fig. 3, red line) occur due to an erratic turn-on command that drives the transistor into saturation under high drain voltage, causing a rapid junction temperature rise. HSF initially manifest on the gate through an internal C_{GD} current [14] (Fig. 6(a), area '1') that pulls charge from gate to drain, reducing V_{GS}. This is followed by a thermally triggered Schottky emission gate leakage current [15] via C_{GS} (Fig. 6(a), area '2'), which further reduces V_{GS} by drawing charge from the gate, creating an inflection point on V_{GS}.

Faults Under Load (FUL) can arise from various sources, including the breakdown of the complementary antiparallel diode (depicted in brown line in Fig. 3), an erratic turn-on of

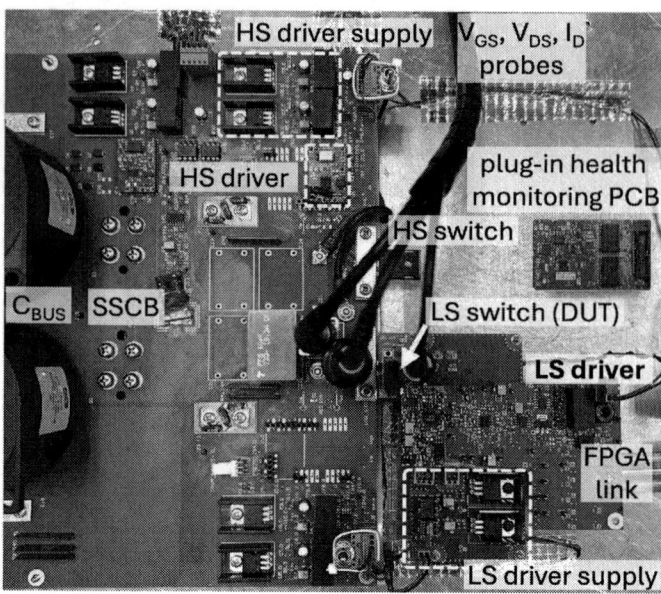

Fig. 2. 600 V inverter leg short-circuit test bench with proposed gate driver on the low-side.

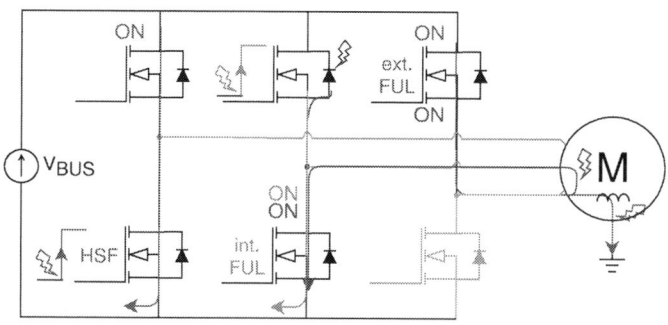

Fig. 3. Overview of potential short-circuits in a three-phase inverter.

the complementary switch (orange line), insulation breakdown between a motor phase and ground (purple line), or an inter-phase insulation breakdown (blue line), among others. During FUL, the fault current flows through the channel of an already-on transistor, leading to an ohmic drain voltage rise. This

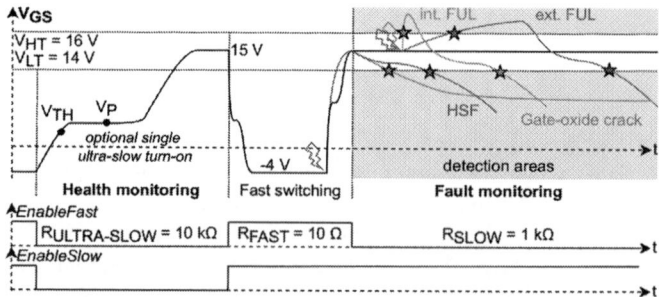

Fig. 4. SiC MOSFET gate control chronograms, ultra-slow turn-on for health monitoring, faults impact on V_{GS} and detection principle.

injects charge back to the gate via C_{GD} [16], increasing V_{GS}. When the transistor operates in the ohmic region at low V_{DS}, C_{GD} is high, further enhancing the fault current's effect on V_{GS}, which improves fault detectability. Additionally, in TO-247-3 packages, stray source inductance (L_S) also helps raise V_{GS}.

However, with traditional low gate resistance values of a few ohms, these effects on V_{GS} are minimal or even indiscernible. To make use of these phenomena for fault detection, the proposed gate driver in Fig. 1 significantly increases gate resistance (R_{SLOW}) during the on-state, amplifying fault signatures on V_{GS} (*EnableFast* = 0). Low gate resistance (R_{FAST}) is used during switching events for high dynamics (*EnableFast* = 1), and during off-state, reducing V_{GS} sensitivity to dv/dt switching phases to prevent false turn-ons and excessive gate-oxide stress. This is achieved with commercial gate drivers that have an *Enable* function for selectively setting high impedance or activating specific channels.

During the on-state, V_{GS} is compared to fixed thresholds (V_{HT} and V_{LT}) set around the nominal gate voltage of 15 V (Fig. 4), adjusted manually with multi-turn potentiometers. Although implemented on the low side (LS) of the 600 V inverter leg (Fig. 2), the gate driver design accommodates high-side use, with an isolation barrier as shown in Fig. 1. Additionally, the voltage dividers sensing V_{GS} (R_H and R_L in Fig. 1) serve as gate-source pull-down resistors, critical for meeting aeronautic safety standards.

As illustrated in Fig. 4, when V_{GS} enters a detection area marked by the end of *Fast* mode and the detection thresholds, the Soft Shut-Down (SSD) protection is triggered, as shown in the algorithm in Fig. 5.

III. EXPERIMENTAL RESULTS

Short-circuit detection time (t_{SSD}) characterization campaigns were conducted on the presented 600 V inverter leg to evaluate the performance of the proposed fault detection method.

A. Hard Switching Faults

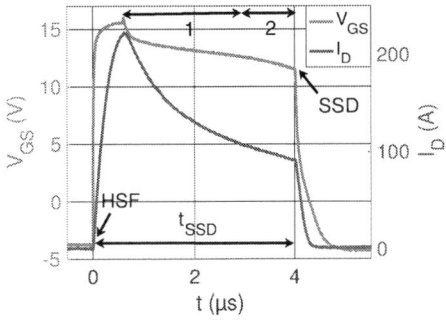

(a) Experimental HSF V_{GS} waveform. Area '1': V_{GS} is decreased by an internal C_{GD} current ; Area '2': V_{GS} is decreased by the Schottky gate leakage current (V_{LT} = 12 V, R_{SLOW} = 10 kΩ, V_{BUS} = 600 V).

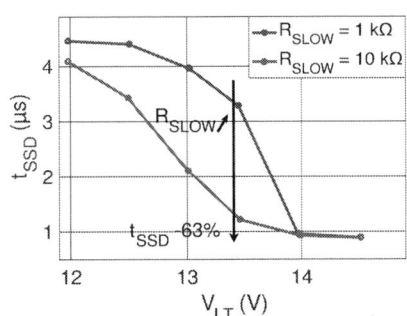

(b) Influence of R_{SLOW} and low detection threshold (V_{LT}) on HSF detection time (t_{SSD}).

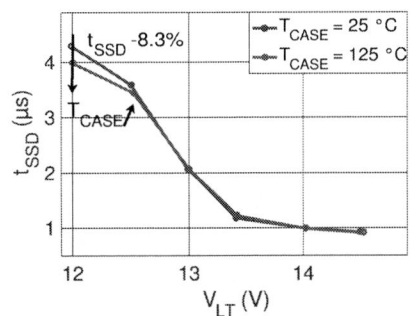

(c) Influence of operating temperature (T_{CASE}) and low detection threshold (V_{LT}) on HSF detection time (t_{SSD}).

Fig. 6. HSF experimental results on C3M0075120D.

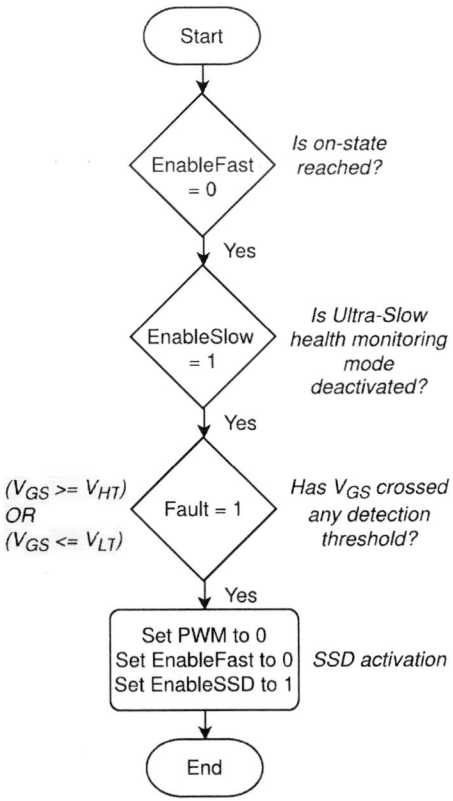

Fig. 5. SSD activation flowchart.

979-8-3315-1612-3/25 $31.00 © 2025 IEEE

In HSF tests, the source inductance L_S has minimal impact on V_{GS} and fault detectability. As a result, the fault detection method performs equivalently in MOSFETs with TO-247-3 and TO-247-4 packages. Therefore, the HSF tests were conducted on a C3M0075120D DUT with a TO-247-3 package.

Fig. 6(b) outlines that increasing R_{SLOW} more strongly depolarizes the gate, which shortens t_{SSD}. This allows the detection threshold V_{LT} to be set further from the nominal drive voltage (15 V), reducing the risk of false triggers. Despite a 23 % reduction in chip active area between Gen. 2 and Gen. 3 MOSFETs (see Tab. I), HSF detection is achieved in only 890 ns (18 % of t_{SC}).

Fig. 6(c) illustrates that higher DUT operating temperature (T_{CASE}) improves detection, as the Schottky emission gate leakage current appears sooner, resulting in up to 8.3 % faster detection.

Then, Fig. 7 reveals that a 50 % decrease in the DC bus voltage (V_{BUS}) increases the dissipated energy (E_{dissip}) by only 13 %, remaining well below the critical threshold (E_{CRIT} = 500 mJ), validating robustness against V_{BUS} fluctuations.

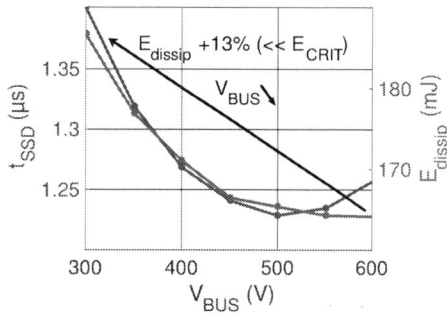

Fig. 7. Influence of V_{BUS} variations on t_{SSD} and E_{dissip} in HSF condition (R_{SLOW} = 10 kΩ, V_{LT} = 14 V, DUT: C3M0075120D).

B. Gate-Oxide Crack Detection

Under repeated thermo-mechanical stress, irreversible cracks can form in the DUT gate-oxide [17]. These cracks create leakage paths between gate and source detectable by

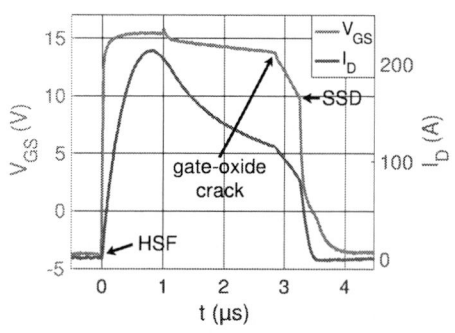

Fig. 8. Gate-oxide crack manifestation on V_{GS} during an HSF pulse (V_{LT} = 12.5 V, R_{SLOW} = 10 kΩ, DUT: C3M0075120D).

the proposed gate driver architecture, which amplifies their impact on V_{GS} as shown in Fig. 8.

Early detection of these cracks is crucial, as they are permanent and progressively worsen, ultimately leading to a loss of MOSFET controllability [18].

C. Faults Under Load

FUL detection relies on electrical couplings through C_{GD} and L_S, which are inherently more dynamic than HSF thermal signatures, resulting in shorter t_{SSD} (Fig. 9). Fig. 10(a) shows FUL waveforms for a TO-247-3 package MOSFET, marked by a sharp increase in V_{GS} as soon as the fault appears due to C_{GD} and L_S couplings.

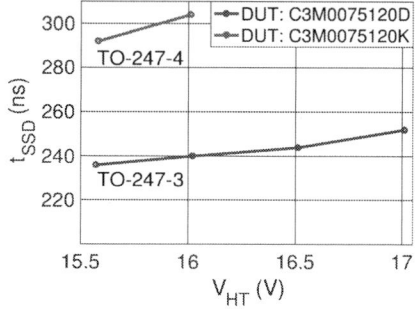

Fig. 9. Influence of Kelvin source contact on FUL detection times (t_{SSD}) (V_{BUS} = 600 V, R_{SLOW} = 1 kΩ).

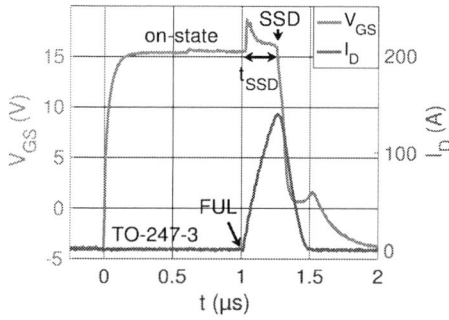

(a) DUT: C3M0075120D (TO-247-3 package).

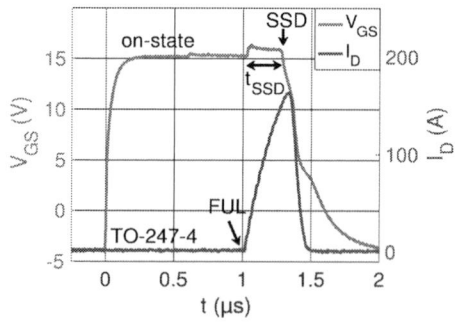

(b) DUT: C3M0075120K (TO-247-4 package).

Fig. 10. FUL experimental results on Gen. 3 SiC MOSFETs, influence of Kelvin source contact (V_{HT} = 15.5 V, R_{SLOW} = 1 kΩ, V_{BUS} = 600 V).

In contrast, TO-247-4 package MOSFETs with a Kelvin Source (KS) contact eliminate the L_S effect on V_{GS} (Fig. 10(b)). Despite reduced V_{GS} overpolarization, FUL detection occurs in just 290 ns. The influence of T_{CASE} on t_{SSD} is negligible at low V_{HT} (Fig. 9). Although FUL detection time is slightly longer on the Kelvin Source device, Fig. 11 shows it experiences lower energy dissipation. Referencing the gate driver to KS eliminates the voltage drop across L_S in the gate loop, enabling faster MOSFET turn-off [19] and reducing dissipated energy. FUL can be detected in as little as 235 ns on the TO-247-3 device, corresponding to 8.5 mJ dissipation (1.7 % of E_{CRIT}), and in 290 ns on the TO-247-4 device, with 6.5 mJ dissipation (1.3 % of E_{CRIT}). This ensures the chips can withstand repeated FUL events.

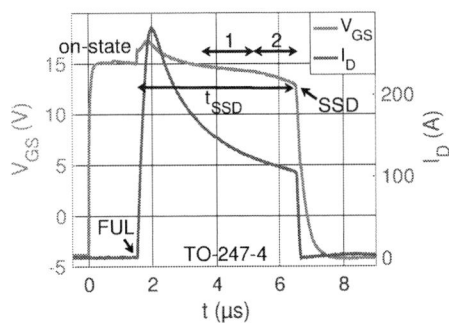

Fig. 12. Experimental waveform of a FUL detection by an intersection of V_{GS} with the low threshold V_{LT} (V_{BUS} = 600 V, R_{SLOW} = 1 kΩ, V_{HT} = 21 V, V_{LT} = 13 V, DUT: C3M0075120K).

faster detection. In this configuration, detection occurs within 5.0 µs, corresponding to a dissipation of 360 mJ or 72 % of E_{CRIT}. This redundancy is essential for ensuring fault detection reliability.

E. Robustness Against False Detections

The shortest detection times are achieved with detection thresholds set close to the nominal drive voltage. Therefore, it is crucial to evaluate the system's robustness against false detections under this configuration using double pulse tests,

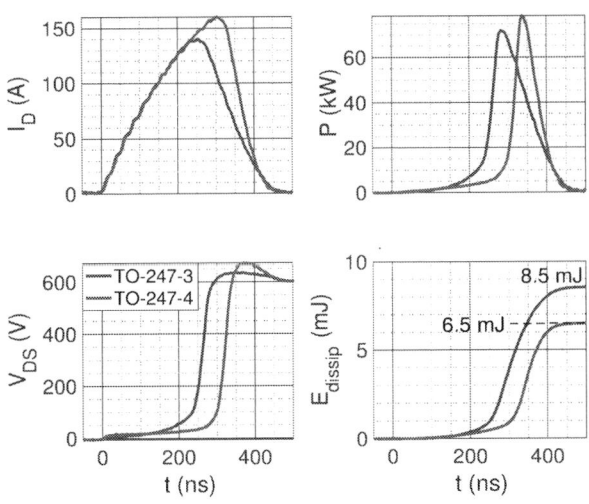

Fig. 11. Influence of the Kelvin source contact on power and energy dissipation in Fig. 10 FUL tests (DUTs: C3M0075120D (TO-247-3 package) and C3M0075120K (TO-247-4 package)).

The limiting factor for the detection delay t_{SSD} is the propagation delay of the different integrated circuits, which currently stands at 220 ns. Ongoing work aims to reduce this delay to enable even faster detection.

Tab. II summarizes the performance of the proposed fault detection method for both Gen. 2 and Gen. 3 DUTs.

D. Fault Under Load Detection Redundancy

During FUL events, if the high detection threshold (V_{HT}) system malfunctions or if the fault is too inductive and the faulty di/dt is too low, detection may not occur during the rising phase of the drain current. In this case, the MOSFET will eventually saturate, causing its drain voltage and junction temperature to rise. As previously shown in Fig. 4, this leads to the appearance of HSF signatures, which can serve as a fallback detection method. Thus, FUL can be detected not through the high threshold V_{HT} but rather through the low threshold V_{LT}, as illustrated in Fig. 12. In Fig. 12, the threshold is set as low as 13 V to illustrate the full phenomenon on V_{GS}. In practice, it can be set closer to the drive voltage of 15 V for

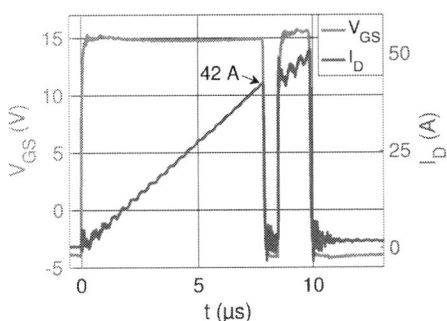

(a) DUT: C3M0075120D (TO-247-3 package, dV_{DS}/dt = 24 kV/µs).

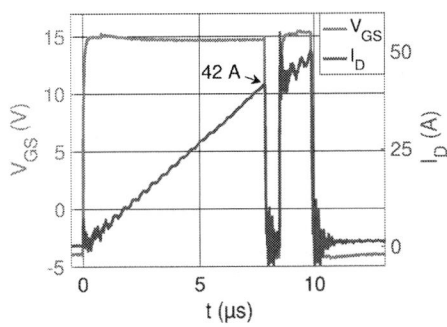

(b) DUT: C3M0075120K (TO-247-4 package, dV_{DS}/dt = 26 kV/µs).

Fig. 13. Double pulse tests at 140 % the rated current of the DUTs, (L = 114 µH, R_{SLOW} = 1 kΩ, V_{BUS} = 600 V, V_{LT} = 14.5 V, V_{HT} = 15.5 V).

which enable the MOSFET to switch at specific drain current levels, mimicking real operating conditions [20]. Fig. 13 shows the experimental waveforms obtained at a worst-case load current of 42 A, or 140 % of the 30 A current rating of the DUTs.

During these tests, the detection thresholds were set only 0.5 V apart from the drive voltage of 15 V, and no false detections were observed, validating the robustness of the concept against false triggering.

As shown in Fig. 13, in normal operation during the on-state, V_{GS} slightly decreases from the nominal drive voltage. This drop is due to the gate-source pull-down resistors (R_H and R_L) used to sense V_{GS}, as illustrated in Fig. 1. When the driver operates in *Slow* mode (*EnableFast* = 0), the gate is in a quasi-flying state and is highly sensitive to such leakage currents. From this observation, a criterion emerges for the values of the sensing resistors. Specifically, this normal voltage drop should not cause V_{GS} to cross the low detection threshold V_{LT}, thereby avoiding false triggering. Assuming the on-state V_{GS} should not drop more than 0.4 V below the nominal drive voltage (V_{DD}), this yields the first condition for designing R_H and R_L as shown in eq. (1).

$$R_H + R_L = 2R_{SLOW} \left(\frac{V_{DD} - 0.4}{0.4} \right) = 73 \; k\Omega \qquad (1)$$

Another criterion arises from the input limitations of the comparators, which cannot exceed 5 V, while the maximum value of V_{GS} is assumed to be 25 V. The attenuation of the voltage dividers is therefore set as shown in eq. (2).

$$R_H + R_L = 5R_H \qquad (2)$$

Combining both criteria gives R_H = 58.4 kΩ and R_L = 14.6 kΩ, resulting in a global value of 36.5 kΩ, which is within the typical range for a gate-source pull-down resistance.

To further enhance robustness against false detections, particularly from noise on sensed V_{GS} caused by switching in other inverter legs within a three-phase converter, a short blanking period of a few nanoseconds can be introduced. The

fault signal should maintain a value of '1' for longer than this blanking period to prevent false triggering due to parasitic oscillations on V_{GS}.

IV. CONCLUSION

The proposed parallel gate driver architecture, featuring a dynamic high on-state gate resistance value, compensates for reduced design parameters in Gen. 3 SiC MOSFETs, achieving reliable Hard Switching Fault and Fault Under Load detection within 890 ns and 235 ns, respectively. This rapid response supports repeated short-circuit endurance and enables robust gate-oxide integrity monitoring. Designed to be modular and integrable, the gate driver architecture also allows for additional SiC MOSFET health monitoring through a simple buffer addition. The proposed gate driver is experimentally validated on TO-247-3 and TO-247-4 Gen. 3 SiC MOSFETs on a 600 V inverter leg. The architecture's reliability is further validated under variations in key parameters such as temperature and DC bus voltage, with minimal impact on detection time and energy dissipation. Future work will adapt this gate driver to a multi-chip SiC MOSFET power module.

REFERENCES

[1] G. Buticchi, P. Wheeler, and D. Boroyevich, "The More-Electric Aircraft and Beyond," *Proceedings of the IEEE*, vol. 111, no. 4, pp. 356–370, 2023.

[2] T. C. Cano, I. Castro, A. Rodríguez, D. G. Lamar, Y. F. Khalil, L. Albiol-Tendillo, and P. Kshirsagar, "Future of Electrical Aircraft Energy Power Systems: An Architecture Review," *IEEE Transactions on Transportation Electrification*, vol. 7, no. 3, pp. 1915–1929, 2021.

[3] V. Madonna, P. Giangrande, and M. Galea, "Electrical Power Generation in Aircraft: Review, Challenges, and Opportunities," *IEEE Transactions on Transportation Electrification*, vol. 4, no. 3, pp. 646–659, 2018.

[4] *UCC21222 4A, 6A, 3kVRMS Isolated Dual-Channel Gate Driver*, Texas Instruments, 2018, rev. C. [Online]. Available: https://www.ti.com/lit/ds/symlink/ucc21222.pdf?ts=1733668688754

[5] *UCC27444-Q1 20V, 4A Dual-Channel Low-Side Gate Driver with −5V Input Capability For Automotive Applications*, Texas Instruments, 2022, rev. A.

[6] *Si827x 4 Amp ISOdriver with High Transient (dV/dt) Immunity*, Skyworks, 2022. [Online]. Available: https://www.skyworksinc.com/-/media/Skyworks/SL/documents/public/data-sheets/Si827x.pdf

[7] W. T. Cui, W. J. Zhang, J. Y. Liang, H. Nishio, H. Sumida, H. Nakajima, Y.-T. Hsieh, H.-H. Tsai, Y.-Z. Juang, W.-K. Yeh, and W. T. Ng, "A Dynamic Gate Driver IC with Automated Pattern Optimization for SiC Power MOSFETs," in *IEEE 34th International Symposium on Power Semiconductor Devices and ICs (ISPSD)*, 2022, pp. 33–36.

[8] M. Wang, W. J. Zhang, J. Liang, W. T. Cui, W. Tung Ng, H. Nishio, H. Sumida, and H. Nakajima, "A Smart Gate Driver for SiC Power MOSFETs with Aging Compensation and Ringing Suppression," in *33rd International Symposium on Power Semiconductor Devices and ICs (ISPSD)*, 2021, pp. 67–70.

[9] M. Picot-Digoix, F. Richardeau, J.-M. Blaquière, S. Vinnac, S. Azzopardi, and T.-L. Le, "Quasi-Flying Gate Concept Used for Short-Circuit Detection on SiC Power MOSFETs Based on a Dual-Port Gate Driver," *IEEE Transactions on Power Electronics*, vol. 38, no. 6, pp. 6934–6938, 2023.

[10] B. Kakarla, "Short Circuit Behavior of SiC MOSFETs," Doctoral Thesis, ETH Zurich, Zurich, 2021.

[11] A. O. Adan, D. Tanaka, L. Burgyan, and Y. Kakizaki, "The Current Status and Trends of 1,200-V Commercial Silicon-Carbide MOSFETs: Deep Physical Analysis of Power Transistors From a Designer's Perspective," *IEEE Power Electronics Magazine*, vol. 6, no. 2, pp. 36–47, 2019.

TABLE II
SYNTHESIS OF THE PERFORMANCE OF THE PROPOSED FAULT DETECTION METHOD ON GEN. 2 [9] AND GEN. 3 SiC MOSFETs.

DUT		C2M0080120D Gen. 2 80 mΩ TO-247-3	C3M0075120D Gen. 3 75 mΩ TO-247-3	C3M0075120K Gen. 3 75 mΩ TO-247-4
HSF	t_{SSD}	1.2 µs	890 ns	
	% t_{SC}	15 %	18 %	
	E_{dissip}	140 mJ	123 mJ	
	% E_{CRIT}	20 %	25 %	
FUL	t_{SSD}	250 ns	235 ns	290 ns
	E_{dissip}	20 mJ	8.5 mJ	6.5 mJ
	% E_{CRIT}	3 %	1.7 %	1.3 %

[12] S. Mocevic, J. Wang, R. Burgos, D. Boroyevich, C. Stancu, M. Jaksic, and B. Peaslee, "Comparison between desaturation sensing and Rogowski coil current sensing for shortcircuit protection of 1.2 kV, 300 A SiC MOSFET module," in *2018 IEEE Applied Power Electronics Conference and Exposition (APEC)*, 2018, pp. 2666–2672.

[13] M. Picot-Digoix, T.-L. Le, S. Azzopardi, S. Vinnac, F. Richardeau, and J.-M. Blaquière, "On-Line Digital Fine Monitoring of SiC MOSFET Gate-Oxide Health: A Dual-Channel Gate Driving Approach," in *2024 36th International Symposium on Power Semiconductor Devices and ICs (ISPSD)*, 2024, pp. 84–87.

[14] S. Kochoska, J. R. Guitart, L. Richert, and B. Vlachakis, "Gate Current Peaks Due to CGD Overcharge in SiC MOSFETs Under Short-Circuit Test," in *2023 35th International Symposium on Power Semiconductor Devices and ICs (ISPSD)*, 2023, pp. 246–249.

[15] F. Boige, D. Trémouilles, and F. Richardeau, "Physical Origin of the Gate Current Surge During Short-Circuit Operation of SiC MOSFET," *IEEE Electron Device Letters*, vol. 40, no. 5, pp. 666–669, 2019.

[16] M. Cui, J. Li, Y. Du, and Z. Zhao, "Behavior of SiC MOSFET under Short-Circuit during the On-State," *IOP Conference Series: Materials Science and Engineering*, vol. 439, no. 2, p. 022026, nov 2018.

[17] P. D. Reigosa, F. Iannuzzo, and L. Ceccarelli, "Effect of short-circuit stress on the degradation of the SiO2 dielectric in SiC power MOS-FETs," *Microelectronics Reliability*, vol. 88-90, pp. 577–583, 2018, 29th European Symposium on Reliability of Electron Devices, Failure Physics and Analysis (ESREF 2018).

[18] P.-D. Reigosa, F. Iannuzzo, and L. Ceccarelli, "Failure Analysis of a Degraded 1.2 kV SiC MOSFET after Short Circuit at High Temperature," *2018 IEEE International Symposium on the Physical and Failure Analysis of Integrated Circuits (IPFA)*, pp. 1–5, 2018.

[19] C. Bödeker, E. Ayerbe, and N. Kaminski, "Impact of a Kelvin Source Connection on Discrete High Power SiC-MOSFETs," in *Silicon Carbide and Related Materials 2017*, ser. Materials Science Forum, vol. 924. Trans Tech Publications Ltd, 7 2018, pp. 723–726.

[20] B. Mondal, R. T. Pogulaguntla, and A. K. B, "Double Pulse Test Setup: Hardware Design and Measurement Guidelines," in *2022 IEEE International Conference on Power Electronics, Drives and Energy Systems (PEDES)*, 2022, pp. 1–6.

A 24 – 16 V to 0.8 – 1.2 V Merged 4-Stage Hybrid-SC-SL Converter with 96.5% Peak Efficiency and larger than 50% i_L reduction

Chien-Hao Tseng
Graduate Institute of Electronics Engineering
National Taiwan University
Taipei, Taiwan
chienhaotseng@gmail.com

Cheng-Ta Chuang
Graduate Institute of Electrical Engineering
National Taiwan University
Taipei, Taiwan
darcyboogi286@gmail.com

Chieh-Ju Tsai
Graduate Institute of Electrical Engineering
National Taiwan University
Taipei, Taiwan
f04943123@ntu.edu.tw

Ching-Jan Chen
Graduate Institute of Electrical Engineering
National Taiwan University
Taipei, Taiwan
chenjim@ntu.edu.tw

Abstract—This paper introduces a merged 4-stage hybrid switched-capacitor-switched-inductor converter for 24 – 16 V AI-notebook applications. The topology is based on a cascaded switched-capacitor and switched-inductor step-down converter with merged switching and energy transfer. This design ensures that the front-end switched-capacitor stage achieves complete soft-charging and automatic balance, enabling the use of mixed voltage domain power switch devices to optimize on-chip area and efficiency. Additionally, the converter achieves a greater than 50% reduction in inductor current, significantly reducing the need for bulky low DCR inductors. Implemented in a 180 nm BCD process, the proposed converter achieves a peak efficiency of 96% for 24 V to 1 V conversion and 95.52% efficiency for 24 V to 0.8V conversion. Furthermore, a peak-VIC-control with adaptive slope compensation is employed, resulting in an undershoot/overshoot of 56 mV/58 mV for a 2 A/100 ns load step-up/-down.

Keywords—*Merged 4-stage hybrid converter, buck converter, peak-VIC control, load transient response*

I. INTRODUCTION

Artificial intelligence (AI) is transforming modern personal technology, particularly through its integration into the latest notebook designs. The addition of AI capabilities requires significantly enhanced power management systems to support high-performance processing, often necessitating an extended input voltage range from 24 V to 16 V. This range helps minimize copper line transmission losses and enables a large conversion ratio with high efficiency, as noted in recent studies [1]–[12]. Conventional buck converters, however, face challenges in meeting these demands due to the large inductor current that induces considerable conduction losses when achieving high conversion ratios. Additionally, the voltage stress across the switches requires the use of large power MOSFETs [13], which increases both the area and cost of the circuitry, making the design less suitable for compact AI-driven notebooks.

To address the high efficiency requirements of demanding AI applications, this paper presents a novel hybrid converter architecture based on a merged 4-stage hybrid-switched-capacitor-switched-inductor (SC-SL) approach. This design achieves a high conversion ratio while maintaining high efficiency tailored for AI-intensive devices. The core concept of the proposed converter stems from the voltage-dividing characteristic of the switched-capacitor (SC) voltage divider. Simple SC voltage dividers effectively reduce the voltage in stages but have limitations in terms of efficiency and output range. By contrast, switched-inductor (SL) converters, known for their higher efficiency and broader output range, offer advantages over SC converters but are constrained by the impact of inductor voltage on output efficiency, which also imposes critical design limitations.

The proposed merged 4-stage hybrid-SC-SL converter combines the benefits of both SC and SL technologies, creating a complementary structure that reduces each other's limitations. This hybrid approach leverages the SC voltage divider's efficiency in handling high voltage while taking advantage of the SL converter's capacity for efficient current management and wider output regulation range. Together, the 4-stage hybrid-SC-SL configuration achieves an optimized balance, providing both high conversion ratios and reduced component stress.

The detailed architecture of the proposed converter with the derivation steps of its design will be discussed in Section II. Additionally, a peak-VIC-control feedback controller is introduced to regulate the 4-stage hybrid-SC-SL converter, the control method will be covered in Section III. Measurement results and comparative analyses are presented in Section IV. Finally, the conclusions are summarized in Section V.

979-8-3315-1612-3/25 $31.00 © 2025 IEEE

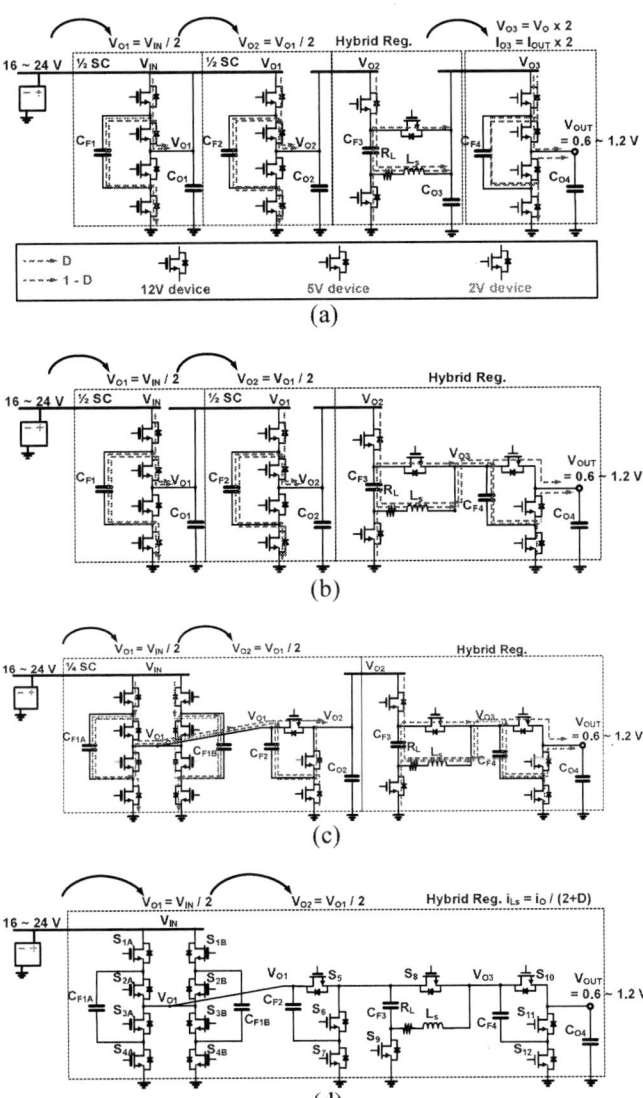

Fig. 1 The evolution of proposed topology: (a) original 4 stage SC-SL converter with current paths, (b) after merging last two stage with current paths, (c) after merging first and second stage with current paths, and (d) final architecture after merging all.

II. MERGED 4-STAGE HYBRID-SC-SL CONVERTER

A. Architecture

The full concept and evolution of the proposed topology are illustrated in Fig. 1. The initial design of this topology incorporates two 1/2 SC voltage dividers that first reduce the high input voltage by a factor of four. This is followed by a cascaded hybrid SC-SL converter, which further regulates the voltage and reduces inductor current stress. Additionally, a final 1/2 SC stage after the hybrid SC-SL converter serves as a current doubler, enhancing the step-down capability. The full topology with the current paths during duty cycle on (D) and duty cycle off ($(1-D)$) are shown in Fig. 1(a). The topology enables the use of mixed-voltage domain devices: 12V for the first stage, 5V for the second, and 2V for both the third and fourth stages. This arrangement optimizes on-chip area utilization and improves efficiency. However, it requires large power switches and numerous active and bypass capacitors, which occupy additional space. Specifically, C_{O1} and C_{O2} serve as output bypass capacitors for the first and second stages of the SC divider, respectively, while C_{O3} is the bypass capacitor for the hybrid SC-SL converter.

To minimize the use of active components and bypass capacitors (C_{O1}, C_{O2}, C_{O3}) between stages, we combined several circuit elements to reduce the overall size. First, we merged the last two stages by removing C_{O3} and the first switch of the final 1/2 SC divider. This combination created a hybrid SC-SL dual-path converter for the last two stages. Additionally, C_{F4} accumulates charge from C_{F3} and L_s during the $(1-D)$ time interval and discharges it to the output. As a result, the DC stress on the inductor current is reduced to $I_O / (2+D)$ [2]. This integrated 3rd and 4th stage configuration significantly lowers inductor current stress, allowing for a smaller inductor (2.2 µH with 56 mΩ SMD 0805 DCR) and thus increases power density.

Second, the first two 1/2 SC stages can be combined using a similar approach by removing the first switch of the second 1/2 SC divider. Additionally, the first 1/2 SC divider is divided into two interleaved 1/2 SC dividers [3], as shown in Fig. 1(c). This configuration ensures auto-balancing in the first stage and reduces the area needed for passive and active devices.

Finally, the 1st-2nd stages and the 3rd-4th stages can be further combined into a final version. The final architecture, shown in Fig. 1(d), not only removes C_{O1}, C_{O2}, and a power switch but also enables fully soft-charging and discharging of the 1st-2nd SC stages via inductor current, eliminating hard-charging losses. As a result, this topology reduces two power switches and three output capacitors in total compared to the original design. Although it adds an additional capacitor, C_{F1B}, the topology still achieves both high efficiency and high power density due to its compact size.

B. Operating Principle

To be more clearly, we simply the actual switches and introduce the operating principle with short circuit when power MOS is conducting and open circuit when power MOS is turned off.

Fig. 2 illustrates the complete operation of the proposed converter, which follows an 8-phase cycle. The eight phases are defined as Φ_{A1} through Φ_{A3} and Φ_{B1}, with Φ_{A2} serving as a reset phase that recurs after each main phase. The sequence progresses in the order of Φ_{A1}, Φ_{A2}, Φ_{A3}, Φ_{A2}, and Φ_{B1}, repeating every cycle.

In Φ_{A1}, current initially flows through C_{F1A} and then divides into two paths: one path flows to ground through C_{F1B}, while the other flows to the output through C_{F2}, C_{F3}, and L_s. Simultaneously, C_{F4} provides an additional current path to the output, having been charged during the preceding reset phase Φ_{A2}. This dual-path configuration increases the converter's current-handling capacity and reduces the stress on individual components.

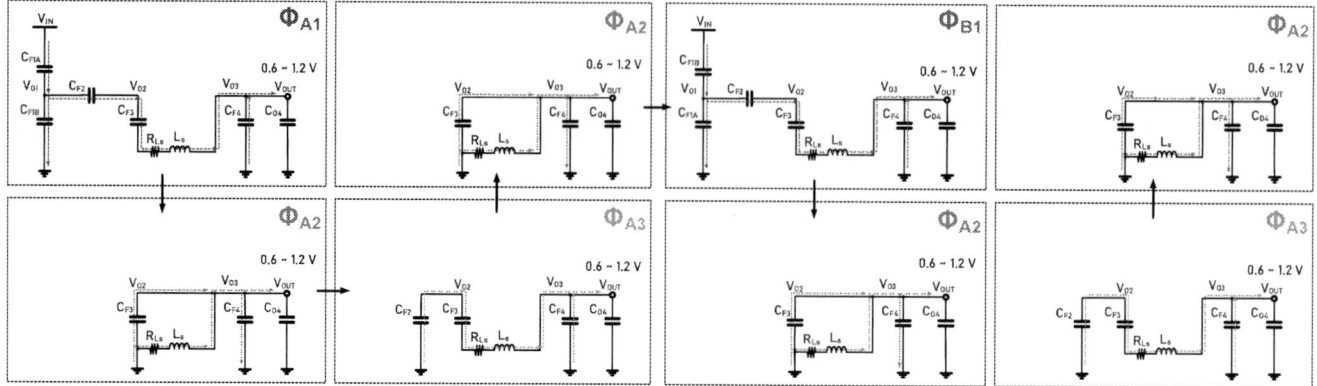

Fig. 2 Operation principle.

During the next phase, Φ_{A2}, CF3 and Ls provide two separate current paths to the output without connections to the components from previous phases, further facilitating charge balancing and reducing the impact on any single capacitor. In Φ_{A3}, C_{F2} supplies energy to charge C_{F3}, L_s, and the output in preparation for later phases. Meanwhile, C_{F4} offers an additional pathway to the output, which helps achieve overall charge balance across the circuit.

After another reset phase, Φ_{A2}, the converter proceeds to Φ_{B1}. In Φ_{B1}, the current flows through C_{F1B}, reversing the initial phase flow, and divides into two paths: one going to ground through C_{F1A} and the other directed to the output via C_{F2}, C_{F3}, and L_s. C_{F4} continues to provide an additional current path to the output, ensuring stable and balanced output performance.

The cycle completes with three additional phases in the sequence Φ_{A2}, Φ_{A3}, and another Φ_{A2}. After that, the operation returns to Φ_{A1}. This multi-phase operation provides robust charge redistribution and stabilizes the output. The strategic interleaving of paths not only ensures efficient energy transfer but also mitigates current stress on individual components, enhancing the converter's overall reliability and efficiency.

III. PEAK-VIC CONTROL FEEDBACK CONTROL FOR THE MERGED4-STAGE HYBRID CONVERTER

In order to stabilize the output voltage while minimizing sensitivity to variations in passive components, we propose a peak-virtual-inductor-current (peak-VIC) modulation method. This control not only eliminates the need for current-sensing circuitry, which is difficult implemented in hybrid converters, but also preserve the single pole system dynamic as peak-current-mode control [14]—[16]. The overall feedback control design is illustrated in Fig. 3. Instead of relying on traditional current-sensing circuits to detect inductor current, this method employs a virtual inductor current for modulation [14]—[16]. By connecting an R-C network, composed of R_{VIC} and C_{VIC}, to V_{LX} and V_{OUT}, the circuit generates an i_{Ls}-like ramp signal, V_{VIC}. Then, the modulation logic is the same as peak current mode control with the modulation ramp replaced by this VIC ramp.

Moreover, we incorporate an adaptive slope compensation circuit with a voltage-controlled current source, G_m, whose output is proportional to the DV_{LX} signal [4]. This feature ensures avoiding subharmonic oscillations while keeping wider inner loop bandwidth without affecting the loop bandwidth. To further enhance this control, a capacitor C_{SLP} is stacked on V_{VIC}, and G_m charges this capacitor to produce the modulation signal V_{RAMP}. A switch, controlled by the inverse duty cycle (D bar), is placed in parallel with C_{SLP} to dynamically adjust for duty cycle changes. The resulting ramp signal V_{RAMP} can be expressed by Equation (1). T_s is the period of a full cycle.

$$V_{RMP} = V_{VIC} + \frac{G_m DV_{LX1}}{C_{SLP} DT_S} \tag{1}$$

The control loop compares V_{RAMP} with V_C, the output of the error amplifier, which takes the reference voltage V_{REF} and feedback voltage V_{FB} as inputs. This comparison produces the signal V_{CMP}, which synchronizes with the clock (F_{clk}) to modulate the duty cycle. Fig. 3 also illustrates the waveforms of V_{RAMP}, V_{VIC}, V_C and D. This careful balance between reduced inductor current stress, minimized voltage stress, and high efficiency highlights the effectiveness of the proposed topology in high-performance applications.

IV. MEASUREMENT RESULT

The proposed design demonstrates high efficiency based on the measurement results. Fig. 4 shows the pre-layout simulation results. Owing to the converter's hybrid architecture, which optimizes power transfer and minimizes losses across components, the converter achieves peak efficiencies of 96% and 95.52% with an input voltage V_{IN} of 24 V and output voltages V_{OUT} of 1 V and 0.8 V, respectively. The efficiency performance of conventional buck converters is also overlaid on the same figure, highlighting the superiority of the proposed converter in comparison.

Table I summarizes the voltage and current stresses for each switch in the converter. Notably, all switch current stresses are reduced by a factor of *(2+D)*, due to the additional current path to the output provided by C_{F4}, which significantly lowers power dissipation and enhances reliability. The comparison of inductor current stress reduction ratio with previous works are shown in Fig. 5 [4] − [11]. Voltage stresses across the

979-8-3315-1612-3/25 $31.00 © 2025 IEEE

MOSFETs are also minimized by utilizing flying capacitors, allowing the use of smaller, lower-rated components and contributing to an overall reduction in converter size and cost [13].

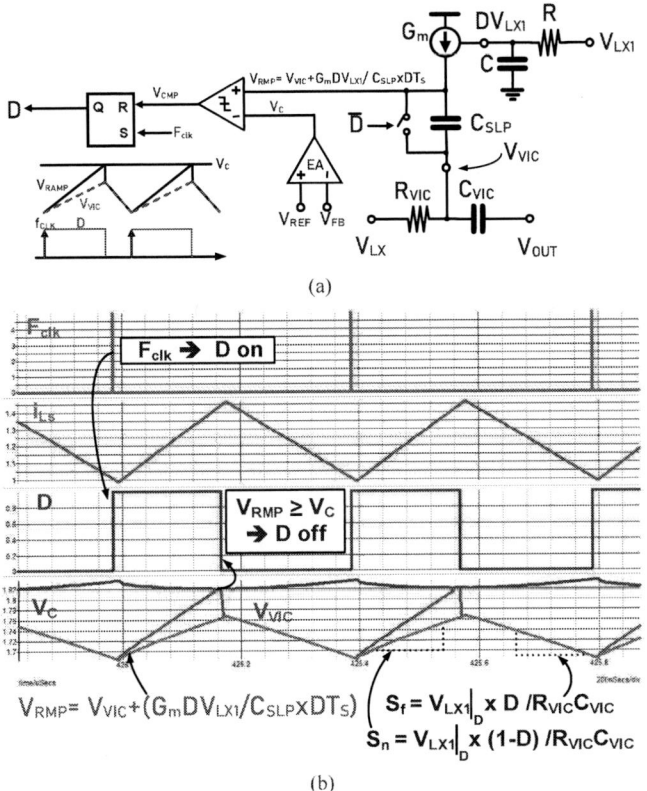

(a)

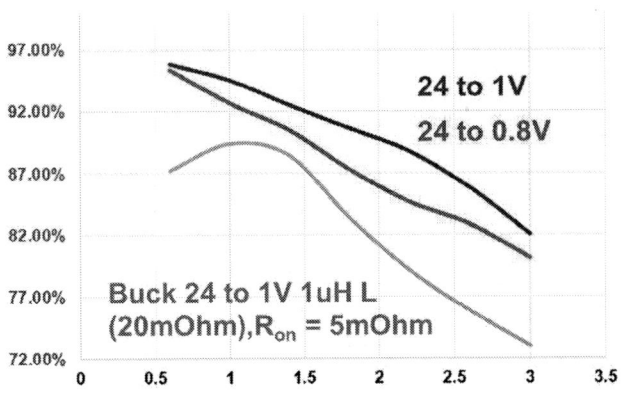

(b)

Fig. 3 (a) Control signal and architecture of peak-VIC control and (b) detailed modulation waveform.

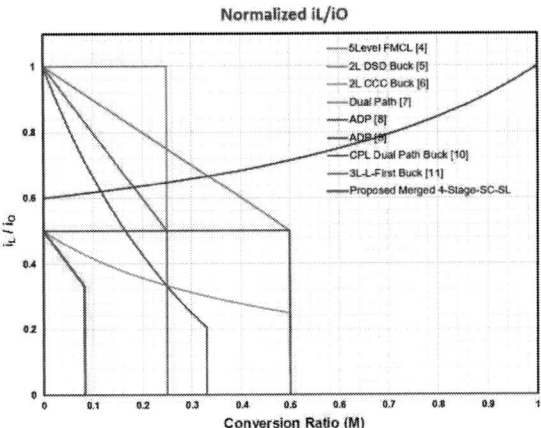

Fig. 4 Efficiency comparison of proposed converter with 24 V input to 1 V and 0.8 V output and conventional buck.

TABLE I. SWITCHES' VOLTAGE AND CURRENT STRESS

Switch	V Stress	I Stress
$S_{1A} \sim S_{4B}$	$V_{IN}/2$	$D/(16+4D) I_O$
$S_5 \sim S_7$	$V_{IN}/4$	D

Switch	V Stress	I Stress
S_8	$V_{IN}/4 - V_{OUT}$	$(1-D)/(2+D) I_O$
S_9	$V_{IN}/4 - 2V_{OUT}$	$I_O/(2+D)$
$S_{10} \sim S_{12}$	V_{OUT}	$I_O/(2+D)$

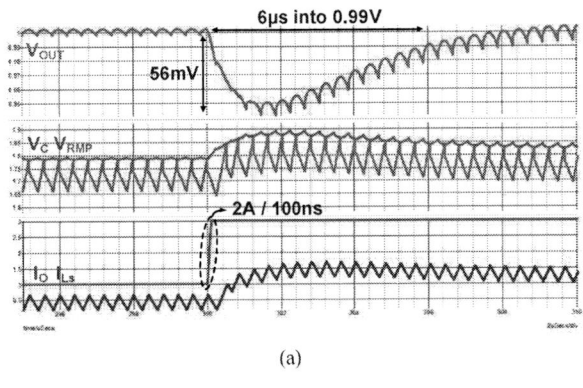

Fig. 5 The comparison of inductor current stress reduction ratio.

The proposed peak-VIC-control modulation simplifies real circuit implementation and perform a fast load transient response time. Simulated load transient response tests show a 56 mV undershoot for a 2 A/100 ns load step-up and a 58 mV overshoot for a 2 A/100 ns load step-down. Fig. 6(a) and Fig. 6 (b) demonstrates the load transient response for step-up and step-down condition respectively.

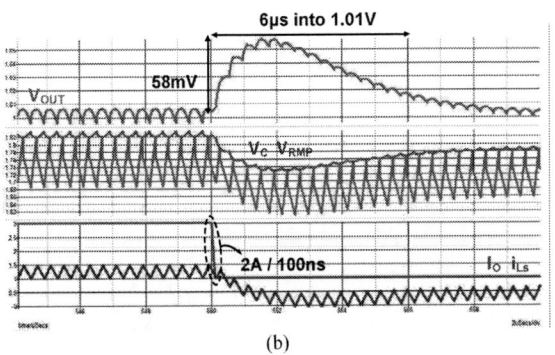

(a)

(b)

Fig. 6 (a) Simulated load transient response under 2 A / 100 ns step-up and (b) 2A / 100 ns step-down.

979-8-3315-1612-3/25 $31.00 © 2025 IEEE

V. CONCLUSION

This paper introduces a merged 4-stage switched-capacitor-switched-inductor (SC-SL) hybrid converter designed to achieve over 50% reduction in inductor current stress, while maintaining low current and voltage stress on the power switches. This advanced topology supports a mix of 12 V, 5 V, and 2 V devices, optimizing the use of on-chip power switches and allowing for fully soft switching to eliminate hard-charging losses and improve overall reliability. The converter achieves peak efficiencies of 96% and 95.52% when the input voltage V_{IN} is 24 V, with output voltages V_{OUT} of 1 V and 0.8 V, respectively, showcasing its effectiveness in high-efficiency applications. Moreover, the regulation of the output voltage is managed through a peak-virtual-inductor-current-control (peak-VIC-control) modulation, ensuring stable performance across various loads, which is a 56 mV undershoot for a 2 A/100 ns load step-up and a 58 mV overshoot for a 2 A/100 ns load step-down, without additional current-sensing circuits. The converter is fabricated using a 180 nm BCD process, with ongoing chip fabrication and measurement to verify its performance and efficiency. This innovative design holds promise for applications requiring high power density, reduced component stress, and efficient voltage conversion in power-sensitive environments.

ACKNOWLEDGMENT

The authors sincerely appreciate the support provided by the National Science and Technology Council (NSTC), Taiwan, along with the educational sponsorship for chip fabrication from TSRI, Taiwan. The authors also extend their gratitude to SIMPLIS Technologies for offering access to the SIMPLIS simulation tool.

REFERENCES

[1] L. Chilumba, A. T. Mushi and B. M. M. Mwinyiwiwa, "Developing a Laptop Power Adaptor for 12 V and 24 V Solar PV Source," in *Proc. 2023 ECCE Asia*, Jeju Island, Korea, Republic of, 2023, pp. 2157-2162.

[2] Q. Ma et al., "A 10.5 W, 93% Efficient Dual-Path Hybrid (DPH)-Based DC–DC Converter Incorporating a Continuous-Current-Input Switched-Capacitor Stage and Enhanced IL Reduction for 12 V/24 V Inputs," *IEEE Trans. on Circuits and Systems I: Regular Papers*, vol. 70, no. 12, pp. 5482-5495, Dec. 2023.

[3] C. Schaef et al., "8.1 A 93.8% Peak Efficiency, 5V-Input, 10A Max ILOAD Flying Capacitor Multilevel Converter in 22nm CMOS Featuring Wide Output Voltage Range and Flying Capacitor Precharging," in *Proc. 2019 ISSCC*, San Francisco, CA, USA, 2019, pp. 146-148.

[4] K. Nishijima, K. Harada, T. Nakano, T. Nabeshima and T. Sato, "A double step-down two-phase buck converter for VRM," in *Proc. 2005 European Conference on Power Electronics and Applications*, Dresden, Germany, 2005, pp. 8 pp.-P.8.

[5] H. Shin et al., "A 96.6%-Efficiency Continuous-Input-Current Hybrid Dual-Path Buck-Boost Converter with Single-Mode Operation and Non-Stopping Output Current Delivery," in *Proc. IEEE 2021 Symposium on VLSI*, Kyoto, Japan, 2021, pp. 1-2.

[6] Tingxu Hu, Mo Huang, Yan Lu and R. P. Martins, "A Capacitor-Cross-Connected Boost Converter With Duty Cycle < 0.5 Control for Extended Conversion-Ratio and Soft Start-Up," *IEEE Transactions on Circuits and Systems I: Regular Papers*. vol. 69, no. 10, pp. 4272–4283, Oct-2022.

[7] Y. Huh, S. -W. Hong and G. -H. Cho, "A Hybrid Structure Dual-Path Step-Down Converter With 96.2% Peak Efficiency Using 250-m Ω Large-DCR Inductor," *IEEE Journal of Solid-State Circuits*, vol. 54, no. 4, pp. 959-967, April 2019..

[8] Dae-Hyeon Kim and Hyun-Sik Kim, "A 96.9%-Peak-Efficiency Bilaterally-Symmetrical Hybrid Buck-Boost Converter Featuring Seamless Single-Mode Operation, Always- Reduced Inductor Current, and the Use of All CMOS Switches," in *Proc. 2024 IEEE ISSCC*, pp. 146-147, Feb. 2024.

[9] K. Hata, Y. Jiang, M. -K. Law and M. Takamiya, "Always-Dual-Path Hybrid DC-DC Converter Achieving High Efficiency at Around 2:1 Step-Down Ratio," in *Proc. 2021 IEEE Applied Power Electronics Conference and Exposition (APEC)*, Phoenix, AZ, USA, 2021, pp. 1302-1307.

[10] J. -Y. Ko, Y. Huh, M. -W. Ko, G. -G. Kang, G. -H. Cho and H. -S. Kim, "A 4.5V-Input 0.3-to-1.7V-Output Step-Down Always-Dual-Path DC-DC Converter Achieving 91.5%- Efficiency with 250mΩ-DCR Inductor for Low-Voltage SoCs," in *Proc. 2021 Symposium on VLSI Circuits*, Kyoto, Japan, 2021, pp. 1-2.

[11] C. Hardy and H. -P. Le, "11.5 A 21W 94.8%-Efficient Reconfigurable Single-Inductor Multi-Stage Hybrid DC-DC Converter," in *Proc. 2023 IEEE International Solid-State Circuits Conference (ISSCC)*, San Francisco, CA, USA, 2023, pp. 190-192.

[12] R. W. Erickson and D. Maksimovic, Fundamental of Power Electronics, Norwell, MA: Kluwer, 2001.

[13] B. Razavi, Design of Analog CMOS Integrated Circuits. Boston, MA: McGFraw-Hill, 2001.

[14] Cheung Fai Lee and P. K. T. Mok, "A monolithic current-mode CMOS DC-DC converter with on-chip current-sensing technique," in *IEEE J. Solid-State Circuits*, vol. 39, no. 1, pp. 3-14, Jan. 2004.

[15] C. Deisch, "Simple switching control method changes power converter into a current source," in *Proc. IEEE Power Electronics Specialists Conf.*, 1978, pp. 300-306.

[16] S. Hsu, A. Brown, L, Rensink, and R. Middlebrook, "Modeling and analysis of switching Dc-to-Dc converters in constant frequency current programmed mode," in *Proc. IEEE Power Electronics Specialists Conf.*, 1979, pp. 284-301.

Innovation Active Gate Drive method (named TriC3™) for MOSFET heat reduction and EMI

Hisashi Sugie
ROHM Co., Ltd.
Yokohama, Japan
hisashi.sugie@dsn.rohm.co.jp

Abstract— **An innovative Active Gate Drive method (TriC3™) that suppresses EMI and reduces heat generation of MOSFET power devices in motor drive systems has been developed. TriC3™ achieves these functions by monitoring various voltages around the MOSFET power device and using this voltages information to intelligently control the gate drive current of the MOSFET power device. On an actual motor system, TriC3™ has been proven to lower EMI and measurements have shown that heat generated by the MOSFET power devices can be reduced by 35% compared to conventional constant current drive with the same EMI level.**

Keywords—EMI, Power Loss, Motor, Active Gate Driver, MOSFET, Reverse recovery, Gate Drive

I. Introduction

Methods for MOSFET heat reduction and EMI suppression by way of switching gate current using a timer [1]-[13] or sensing the slew rate of output voltage [14], [15] or sensing of output current information [16]-[20] are currently being actively researched. Reference [21] presents TriC3™ that uses direct sensing of output and gate voltages. This paper presents a quantitative comparison evaluation carried out on an actual motor system.

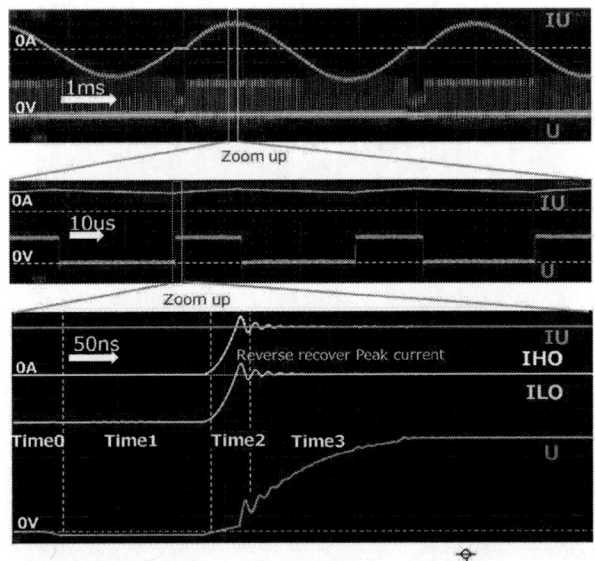

Fig.1(a) Driving three phase motor waveform

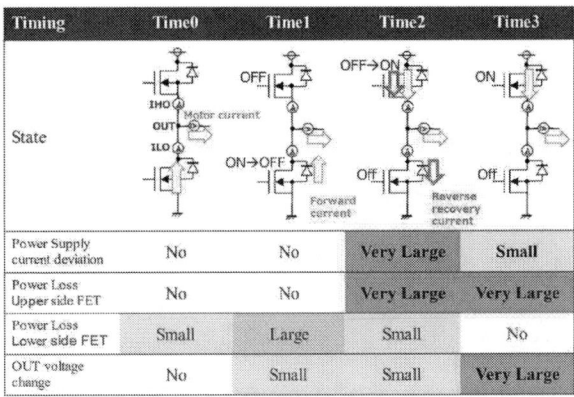

Timing	Time0	Time1	Time2	Time3
Power Supply current deviation	No	No	Very Large	Small
Power Loss Upper side FET	No	No	Very Large	Very Large
Power Loss Lower side FET	Small	Large	Small	No
OUT voltage change	No	Small	Small	Very Large

Fig.1(b) MOSFET current at PWM switching(source rising)

Fig.1 PWM switching during motor drive

Motor systems typically employ Pulse Width Modulation (PWM) to drive large motor currents as shown in Figure 1(a).

Figure 1(b) shows the current flowing through the upper and lower MOSFET of PWM operation specifically when OUT node is transitioning from low to high (rising). This operation is divided into different "Time" periods. At "Time1" period, the low side is transitioning from ON to OFF state and the motor current then flows through the body diode of the lower side MOSFET. "Time2" is the period where EMI is of great concern. The power supply current deviation causes EMI. One of the elements is the peak current of reverse recovery. Thus the gate current during "Time2" should not be increased. At "Time3" the upper side MOSFET is finishing turning ON and the power supply current does not change much. This period is what causes switching loss on the MOSFET and thus this "Time3" must be reduced.

II. Conventional method(Constant current drive)

Constant current drive is shown as an example of a conventional system in Figure 2(a).

In order to reduce ringing and change in power supply current, the gate current must be reduced uniformly for constant current drive. Therefore, output slew rate is slower, increasing switching loss. The reverse recovery peak current and change in power supply current depends on the magnitude of the gate current at Time2. Especially at Time3, the timing with almost no change in power supply current does not affect the power supply EMI, but the power consumption of the upper side MOSFETs is high in Figure 2(b).

979-8-3315-1612-3/25 $31.00 © 2025 IEEE

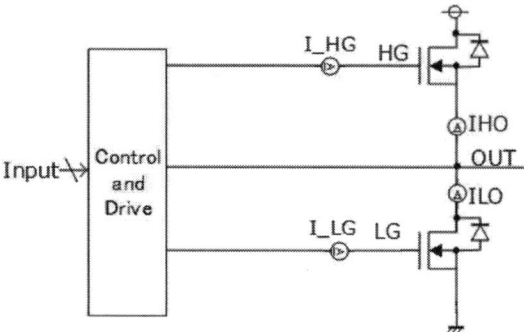

Fig.2(a) Circuit configuration of Constant current

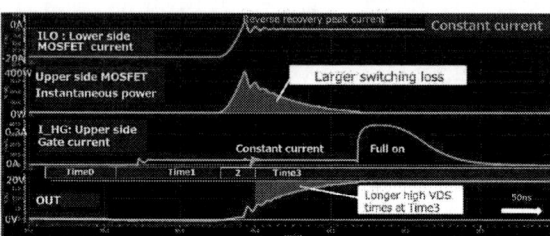

Fig.2(b) Constant current drive waveform at 20A source rising in simulation

Si MOSFET:
Ron=1mohm
Total Gate Charge (Qg)=50nC@(VGS=0 to 10V)

Fig.2 Constant current drive

III. INNOVATIVE ACTIVE GATE DRIVE METHOD (TRiC3™)

Figure 3(a) shows the circuit configuration of TriC3™. Four sensors are used to actively monitor various voltages and information is then used to control the gate drive voltage. The four sensors are the Gate-Source (GS) sensor of the upper MOSFET, the GS sensor of the lower MOSFET, the lower potential sensor of the OUT voltage and the upper potential sensor of the OUT voltage. These sensors are required to perform at high speed, because this speed of operation has a significant impact to their efficiency. To reduce the overall power consumption of the gate driver, these sensors are configured to keep their current consumption low.

Figure 3(b) shows Micrograph of Three phase gate driver (BD67871MWV) chip with TriC3™ integrated. The BD67871MWV has three half bridges, each equipped with high speed and low consumption sensors. The area of sensors on BD67871MWV is indicated by the blue frame in Figure 3(b). All those sensors are implemented in the BD67871MWV chip.

Figure 3(c) shows circuit configuration of TriC3™ Drive block. This block is divided into two parts to ensure smooth active gate drive. One is for the 2nd step, and the rest is for the other steps.

The specific gate current switching operation of the TriC3™ is shown in Fig. 3(d).

The GS sensor of the upper MOSFET is used from 1st step to 2nd step and the OUT voltage sensor is used from 2nd step to 3rd step. This configuration is one of the features of TriC3™.

At Time 2, the gate drive current is reduced because of the timing of large power supply current deviation. This has the effect of suppressing power supply current deviation and avoiding excessive ringing during body diode reverse recovery.

At Time3, the gate drive current is increased because the timing of the power supply current change is small. This has the effect of reducing the switching losses of the MOSFETs.

Figure 3(e) compares conventional Constant Current drive (Figure 2(b)) versus the TriC3™ in simulation. Thereby TriC3™ accomplishes both suppression of EMI and reduction of heat generated on the MOSFET.

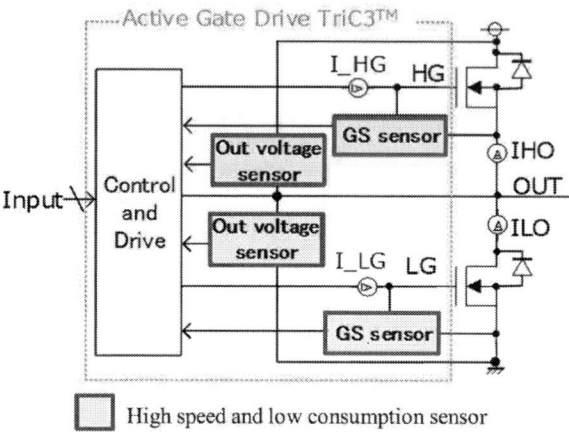

High speed and low consumption sensor

Fig.3(a) Circuit configuration of TriC3 ™

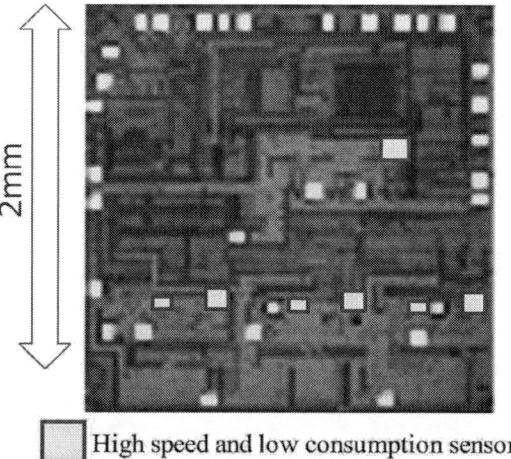

High speed and low consumption sensor

Fig.3(b) Micrograph of Three phase gate driver (BD67871MWV) chip with TriC3 ™ integrated

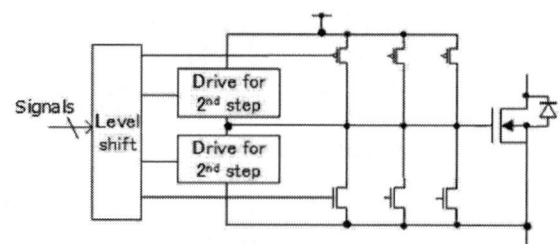

Fig.3(c) Circuit configuration of TriC3™ Drive block

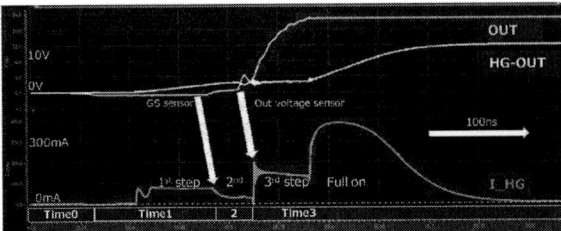

Fig.3(d) The specific gate current switching of TriC3™ in simulation

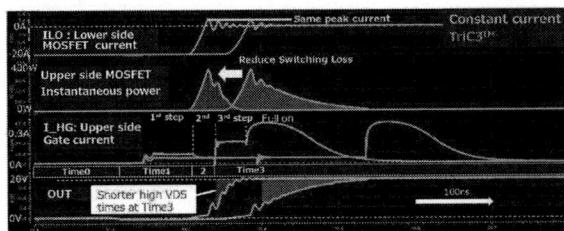

Fig.3(e) Comparison of Constant current drive(Fig.2(b)) versus TriC3™ in simulation

Fig.3 TriC3™ drive

IV. MEASUREMENT RESULT

Experiments are carried out to compare device implementing TriC3™ with Constant Current drive.

Figure 4(a) shows the measurement conditions, the settings are the same except for gate driver setting. Gate driver uses BD67871MWV which can change TriC3 and constant current mode by serial command.

Figure 4(b) shows waveform of 2.TriC3™ during Three phase motor drive in actual motor system.

Figure 4(c) shows comparison of source rising voltage waveforms for 1.CC, 2.TriC3™, and Reference Large Constant Current (Ref. Large CC). Ref. Large CC has significant ringing due to the large gate current at Time2. On the other hand, 1.CC with a small gate current causes the output voltage to rise slowly at time 3, resulting in higher power consumption of the MOSFETs. The gate current of 2.TriC3™ at Time2 is the same as 1.CC. Therefore 1.CC and 2.TriC3™ have similar waveforms at Time2. 2.TriC3™ has lower power consumption in the MOSFETs at Time3 due to the increased gate current.

EMI result is shown in Figure 4(e). EMI of 2. TriC3™ are almost the same as 1. CC.

Reduction of heat generated by 35% using TriC3™ method compared to "1.CC" with the same EMI level is seen in Figure 4(d).

While Figure 4(f) shows the overall comparison of both EMI and heat reduction performance.

"Reference Large Constant Current" of Figure.4 is listed for comparison purposes only.

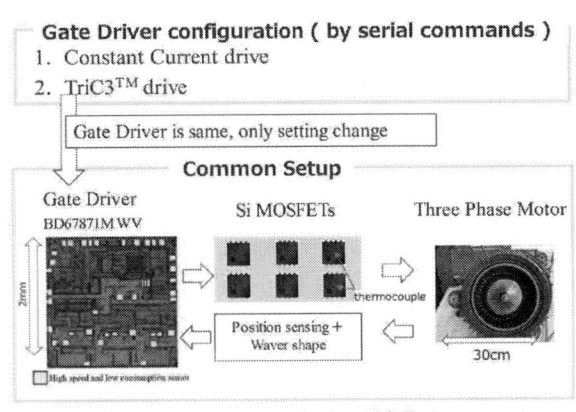

Fig.4(a) Measurement condition in **Actual Motor system**

Three Phase Motor:
    Surface Permanent Magnet Synchronous Motor (SPMSM)
Power supply:
    Voltage = 15V
    Current(average) = 5A at 3500rpm(revolutions per minute)
Si MOSFETs:
    Ron = 1mohm
    Total Gate Charge (Qg) = 50nC@(VGS=0 to 10V)
Gate Driver:
    BD67871MWV
PWM frequency:
    20kHz

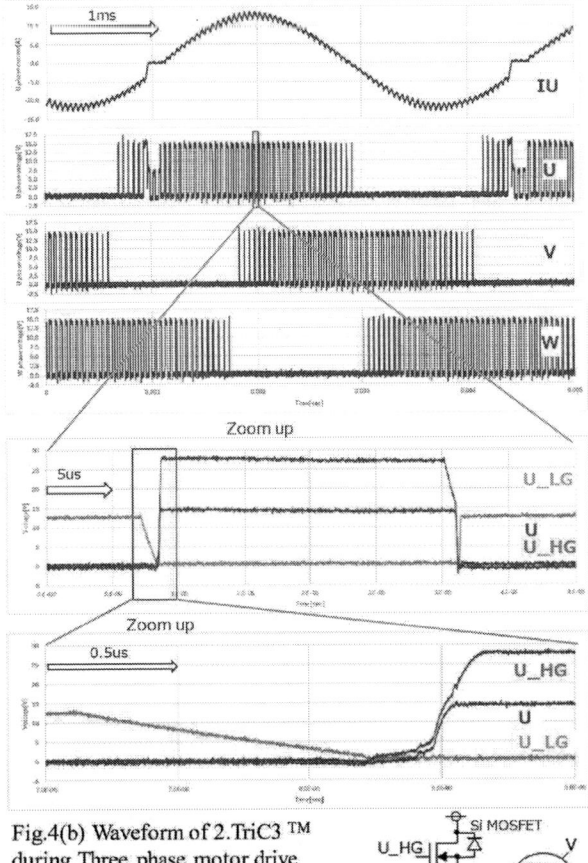

Fig.4(b) Waveform of 2.TriC3™ during Three phase motor drive

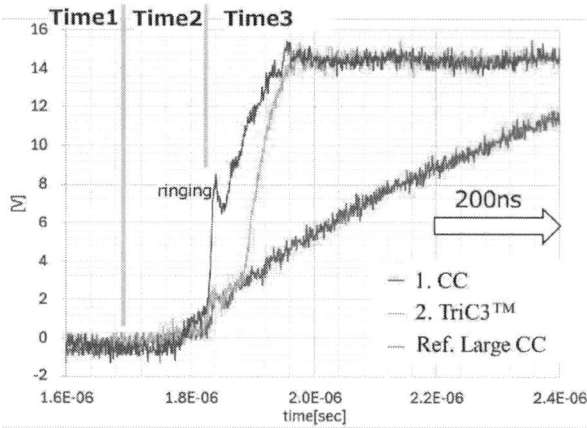

Fig.4(c) Comparison of source rising voltage waveforms of 1.CC, 2.TriC3 ™, and Reference Large CC

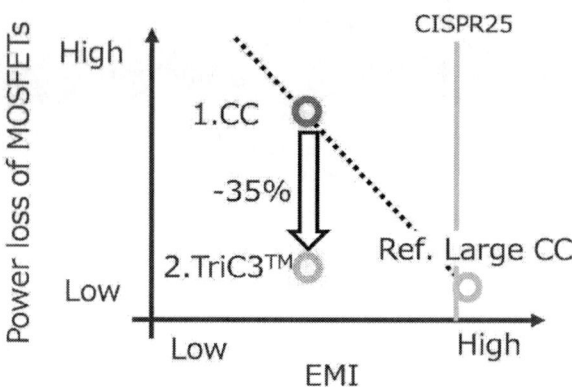

Fig.4(f) EMI vs. Power loss of MOSFET Image

Fig.4 Measurement result in Actual Motor system

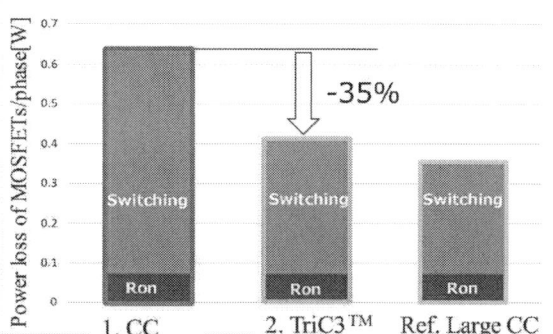

Fig.4(d) Comparison of Power Loss

"Calculate Switching" = "Measured Power Loss" – "Calculate Ron"

V. CONCLUSIONS

It has been shown through verification on an actual motor system that the Innovative Active Gate Drive Method (TriC3™) by sensing output and gate voltages directly and switching gate drive current strength is effective in both suppressing EMI and reducing heat generated on MOSFET.

Simply changing from an existing gate driver to TriC3™ contributes to energy saving. This reduction in energy consumption minimizes environmental impact, making TriC3™ a valuable asset in promoting a sustainable society.

REFERENCES

[1] H.Zhou,T.Inuma,D.Zhang,K.Hata,andM.Takamiya,"Variable Gate Current Range Digital Gate Driver IC Always Providing 6bit Controllability in Various IGBTs"IEEE Applied Power Electronics Conference and Exposition(APEC), LongBeach, USA, pp. 1125-1129, Feb.2024.

[2] R. Katada, K. Hata, Y. Yamauchi, T. -W. Wang, R. Morikawa, C. -H. Wu, T. Sai, P. -H. Chen, and M. Takamiya, "5 V, 300 MSa/s, 6-bit digital gate driver IC for GaN achieving 69 % reduction of switching loss and 60 % reduction of current overshoot," in Proc. IEEE Int. Symp. Power Semicond. Devices ICs, May 2021, pp. 55-58.

[3] Kohei Horii; Katsuhiro Hata; Ruizhi Wang; Wataru Saito; Makoto Takamiya," Large Current Output Digital Gate Driver Using Half-Bridge Digital-to-Analog Converter IC and Two Power MOSFETs" 2022 IEEE 34th International Symposium on Power Semiconductor Devices and ICs (ISPSD), pp.293-296

[4] W. J. Zhang, J. Yu, W. T. Cui, Y. Leng, J. Liang, Y.-T. Hsieh, H.-H. Tsai, Y.-Z. Juang, W.-K. Yeh, and W. T. Ng, "A smart gate driver IC for GaN power HEMTs with dynamic ringing suppression," IEEE Trans. on Power Electronics, vol. 36, no. 12, pp. 14119-14132, Dec. 2021.

[5] K. Horii, R. Morikawa, K. Hata, K. Morokuma, Y. Wada, Y. Obiraki, Y. Mukunoki, and M. Takamiya, "Sub-0.5 ns step, 10-bit time domain digital gate driver IC for reducing radiated EMI and switching loss of SiC MOSFETs," inProc. IEEE Energy Conversion Congress & Exposition, Oct. 2022, pp. 1-8.

[6] T. Inuma, K. Hata, T. Sai, W. Saito, and M. Takamiya, "Two Stop-and-Go Gate Driving to Reduce Switching Loss and Switching Noise in Automotive IGBT Modules," IEEE Southern Power Electronics Conference (SPEC), Nadi, Fiji, pp.1-7, Dec. 2022.

[7] D. Yamaguchi, Y. S. Cheng, T. Mannen, H. Obara, K. Wada, T. Sai, M. Takamiya, and T. Sakurai, "Digital Active Gate Control for a Three-Phase Inverter Circuit for a Surge Voltage Suppression and Switching

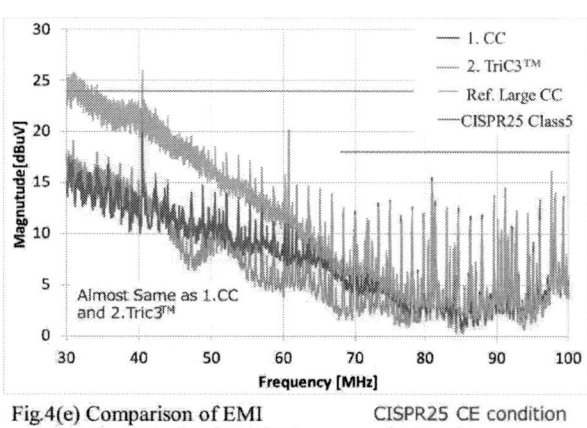

Fig.4(e) Comparison of EMI

CISPR25 CE(Conducted Emissions)
Average Measurement

CISPR25 CE condition

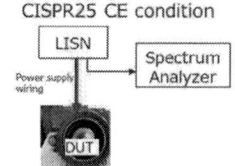

979-8-3315-1612-3/25 $31.00 © 2025 IEEE

Loss Reduction," IEEE Energy Conversion Congress & Exposition (ECCE), Detroit, USA, pp. 3782-3787, Oct. 2020.

[8] Y. Cheng, D. Yamaguchi, T. Mannen, K. Wada, T. Sai, K. Miyazaki, M. Takamiya, and T. Sakurai, "Digital Active Gate Drive with Optimal Switching Patterns to Adapt to Sinusoidal Output Current in a Full Bridge Inverter Circuit," 45th Annual Conference of the IEEE Industrial Electronics Society (IECON), Lisbon, Portugal, pp. 1684-1689, Oct. 2019.

[9] H. Obara, T. Mannen, K. Wada, K. Miyazaki, T. Sai, M. Takamiya, and T. Sakurai, "Design and Implementation of Digital Active Gate Control with Variable 63-level Drivability Controlled by Serial 4-bit Signals," IEEE Energy Conversion Congress & Exposition (ECCE), Baltimore, USA, pp. 2408-2412, Sep. 2019.

[10] Y. Cheng, T. Mannen, K. Wada, K. Miyazaki, M. Takamiya, and T. Sakurai, "Optimization Platform to Find a Switching Pattern of Digital Active Gate Drive for Reducing Both Switching Loss and Surge Voltage," IEEE Transactions on Industry Applications, Vol.55, No.5, pp. 5023 - 5031, Sep./Oct. 2019.

[11] Y. S. Cheng, T. Mannen, K. Wada, K. Miyazaki, M. Takamiya, and T. Sakurai, "High-Speed Searching of Optimum Switching Pattern for Digital Active Gate Drive Circuit of Full Bridge Inverter Circuit," IEEE Applied Power Electronics Conference and Exposition (APEC), Anaheim, USA, pp. 2740-2745, March 2019.

[12] [12] M. Kumar, Z. Feng, S. Wang, M. Sandell and W. Ming, "Closed Loop Digital Design of Active Gate Driver based Power Converter," 2024 IEEE Applied Power Electronics Conference and Exposition (APEC), Long Beach, CA, USA, 2024, pp. 1135-1140

[13] T.-W. Wang, T. Inuma, P.-H. Chen, and M. Takamiya, "Search Method of Robust Gate Driving Vectors for Digital Gate Drivers With Low Test Cost Against Load Current and Temperature Variations in IGBTs," IEEE Transactions on Power Electronics, Vol.38, No.9, pp. 10669-10679, Sep. 2023.

[14] S. Kawai, T. Ueno, H. Ishikuro and K. Onizuka, "An Active Slew Rate Control Gate Driver IC With Robust Discrete-Time Feedback

Technique for 600-V Superjunction MOSFETs," in IEEE Journal of Solid-State Circuits, vol. 58, no. 2, pp. 428-438, Feb. 2023.

[15] M. Blank, T. Glück, A. Kugi, and H.-P. Kreuter, "Digital slew rate and s-shape control for smart power switches to reduce EMI generation," IEEE Trans. Power Electron., vol. 30, no. 9, pp. 5170–5180, Sep. 2015.

[16] S. Kawai et al., "A Load Adaptive Digital Gate Driver IC With Integrated 500 ksps ADC for Drive Pattern Selection and Functional Safety Targeting Dependable SiC Application," in IEEE Transactions on Power Electronics, vol. 38, no. 6, pp. 7079-7091.

[17] D. Zhang, K. Horii, K. Hata, and M. Takamiya, "Digital gate driver IC with fully integrated automatic timing control function in stop-and-go gate drive for IGBTs," inProc. IEEE Applied Power Electronics Conf. and Expo., March 2023, pp. 1225-1231.

[18] D. Zhang, K. Horii, K. Hata, and M. Takamiya, "Digital Gate Driver IC with Real-Time Gate Current Change by Sensing Drain Current to Cope with Operating Condition Variations of SiC MOSFET," 11th International Conference on Power Electronics ECCE Asia (ICPE 2023-ECCE Asia), Jeju, Korea, pp. 374-380, May 2023.

[19] D. Zhang, K. Horii, K. Hata, and M. Takamiya, "Digital Gate ICs for Driving and Sensing Power Devices to Achieve Low-Loss, Low-Noise, and Highly Reliable Power Electronic Systems," IEEE Custom Integrated Circuits Conference (CICC), San Antonio, USA, pp. 1-7, April 2023. (Invited)

[20] B. Hyon, J. -S. Park and J. -H. Kim, "The Active Gate Driver for Switching Loss Reduction of Inverter," 2020 IEEE Energy Conversion Congress and Exposition (ECCE), Detroit, MI, USA, 2020, pp. 2219-2223

[21] H. Sugie, "Innovative Gate Drive method TriC3^{TM}(TM) for Motor," PCIM Europe 2024; International Exhibition and Conference for Power Electronics, Intelligent Motion, Renewable Energy and Energy Management, Nürnberg, Germany, 2024, pp. 1760-1764, doi: 10.30420/566262240.

A KY Buck-Boost Converter with Extended Ramp Control Achieving 1500% Output Variation Reduction for Smooth Mode Transition

Yu-Ting Hung
Graduate Institute of Electronics Engineering
National Taiwan University
Taipei, Taiwan
d09943004@ntu.edu.tw

Chieh-Ju Tsai
Department of Electrical Engineering
National Taiwan University
Taipei, Taiwan
f04943123@ntu.edu.tw

Ching-Jan Chen
Department of Electrical Engineering
National Taiwan University
Taipei, Taiwan
chenjim@ntu.edu.tw

Chun-Yu Hsieh
NovaTek Corporation
Taipei, Taiwan
azure_hsieh@novatek.com.tw

Abstract—**In this paper, a KY buck-boost converter with extended ramp control for the Li-ion battery in the mobile applications is presented. Conventional non-inverting buck-boost converter suffers from several issues, including high conduction loss due to the excessive average inductor current and right-half-plane (RHP) zero existence which degrades the system stability and the transient performance. A KY buck-boost converter is a good candidate to relieve these problems. Unlike the conventional triple mode control utilized in the buck-boost converters, the proposed extended ramp control can achieve smaller output voltage fluctuation when the input voltage is close to the output voltage. The chip prototype is designed using a TSMC 0.18um BCD process. Simulation results show that compared with triple mode non-inverting buck-boost converter, 1500% output voltage fluctuation reduction (less than 10mV) during mode transition is achieved in the proposed control. The peak efficiency of this work can achieve 95.1%.**

Keywords—*Buck-Boost converter, KY Buck-Boost converter, Li-ion battery management, Smooth mode transition*

I. INTRODUCTION

In the recent years, the Li-ion battery has been widely used for mobile devices [1]. As time goes by, the voltage of Li-ion battery will gradually decrease from 4.2V to 2.7V. However, most functional blocks in the mobile devices require a stable 3.4V supply voltage. The common method is implemented by a non-inverting step up/down converter [2] – [4]. The structure and operation of the conventional non-inverting buck-boost converter is shown in Fig. 1(a). A large conversion range can be achieved to meet the requirement for the Li-ion battery in the mobile devices. Nevertheless, as depicted in Fig. 1(b), the RHP zero attributed to the discontinuous power delivery to output not only degrades the system stability but also slows down the load transient response of the converter. Besides, large average inductor current leads to significant conduction

loss and thus deteriorates the power efficiency at heavy load which is also demonstrated in Fig. 1(b).

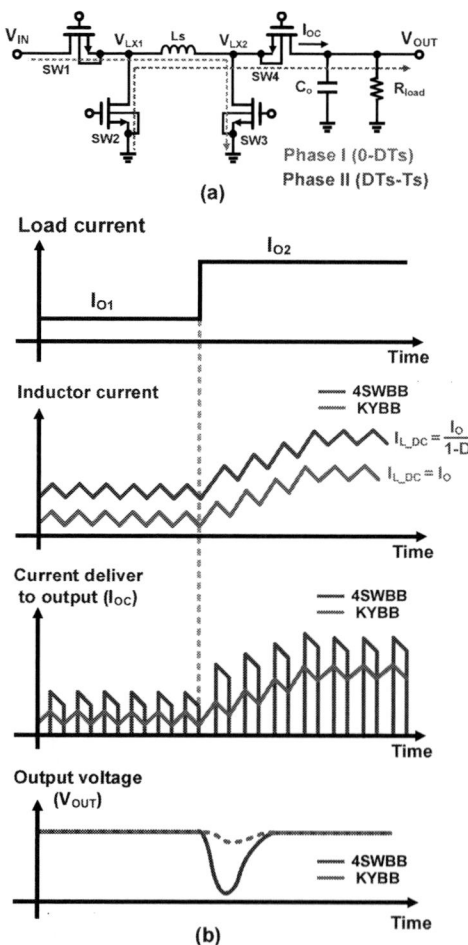

Fig. 1. (a) Structure & operation of conventional non-inverting buck-boost converter. (b) Load transient waveform comparisons of two buck-boost converter.

A hybrid topology named as KY buck-boost converter [5] has the potential to relieve the issues which are mentioned above. Fig. 2 shows the structure and operation of the KY buck-boost converter. The buck mode is identical to the conventional buck converter which is composed of phase I and phase II. For the boost mode operation that is the same as the KY boost converter [6], during phase III, V_{LX} is pumped to $2V_{IN}$ causing the inductor L_s to be magnetized and the flying capacitor C_{fly} to be discharged. On the other hand, during phase IV, C_{fly} is charged by the input voltage. The ideal conversion ratio of the boost mode can be derived as 1+D which D represents the duty cycle of the switch control signal. Since the inductor is placed at the output as the buck converter, the KY buck-boost converter delivers a continuous output current and a faster load transient response is achieved compared to the conventional non-inverting buck-boost converter. Furthermore, the average inductor current is independent to the duty cycle of the switch control signal, thereby causing a smaller conduction loss when the input voltage is close to the output voltage.

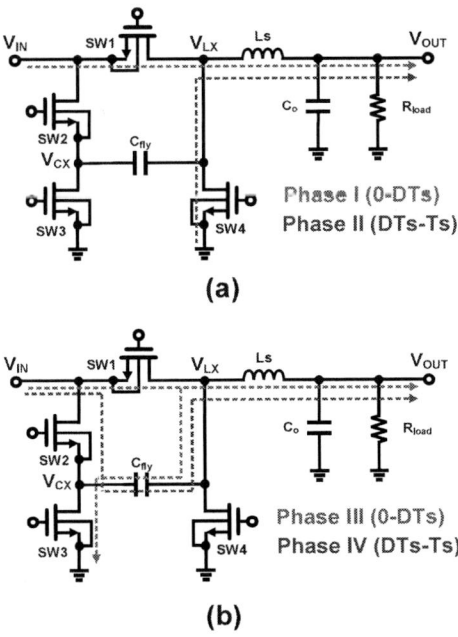

Fig. 2. Structure & operation of KY buck-boost converter : (a) buck mode (b) boost mode.

To improve the power efficiency of the converter, triple mode buck-boost converter has been proposed in [7] – [8]. The operation of the triple mode buck-boost converter is depicted in Fig. 3(a). Buck mode and boost mode are added. The original operation mode is equal to the buck-boost mode. As shown in Fig. 3(b), the average inductor current (I_L) is reduced in both buck mode and boost mode. However, there is still no reduction in the buck-boost mode. Furthermore, some challenges including large output fluctuation during mode transition as demonstrated in Fig. 3(c) and RHP zero existence in the boost and buck-boost mode occur in the triple mode buck-boost converter. Although some prior works [9] – [11] insert a buffer region to replace the buck-boost mode to achieve smooth mode transition, circuit design complexity is also increased. Therefore, an extended ramp control without

additional mode control is proposed in this paper not only to achieve smooth mode transition but also to reduce the design complexity.

The paper is organized as follows. System architecture and the introduction of the proposed extended ramp control are presented in section II. The chip simulation results of the proposed KY buck-boost converter are demonstrated in section III. Finally, a conclusion will be given in section IV.

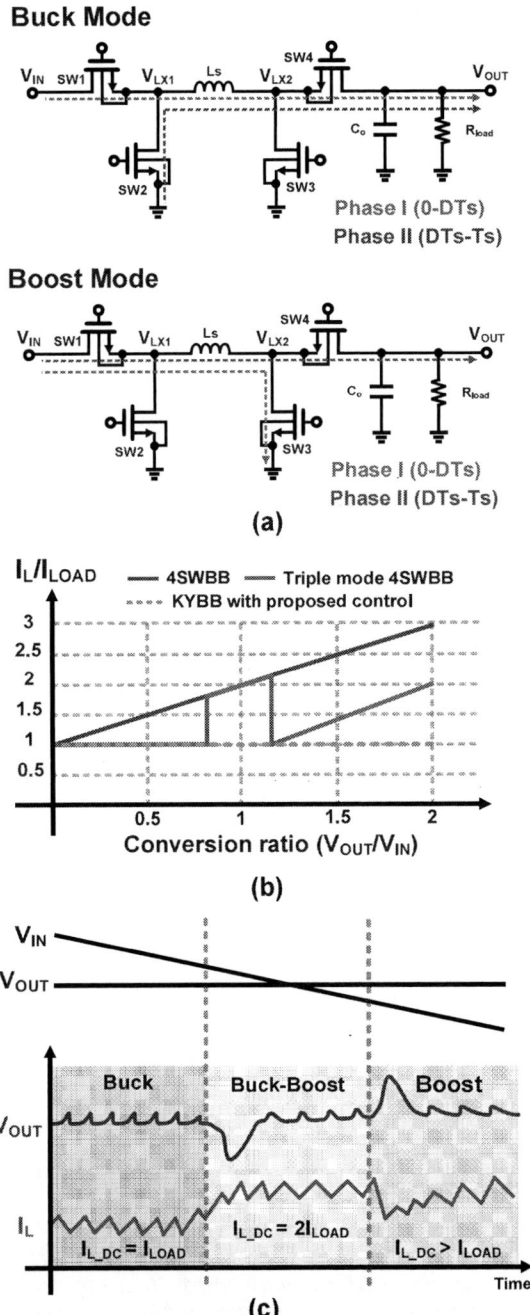

Fig. 3. (a) Operation of the triple mode buck-boost converter. (b) Ratio of inductor current versus load current among three converters. (c) Mode transient waveforms of the triple mode buck-boost converter.

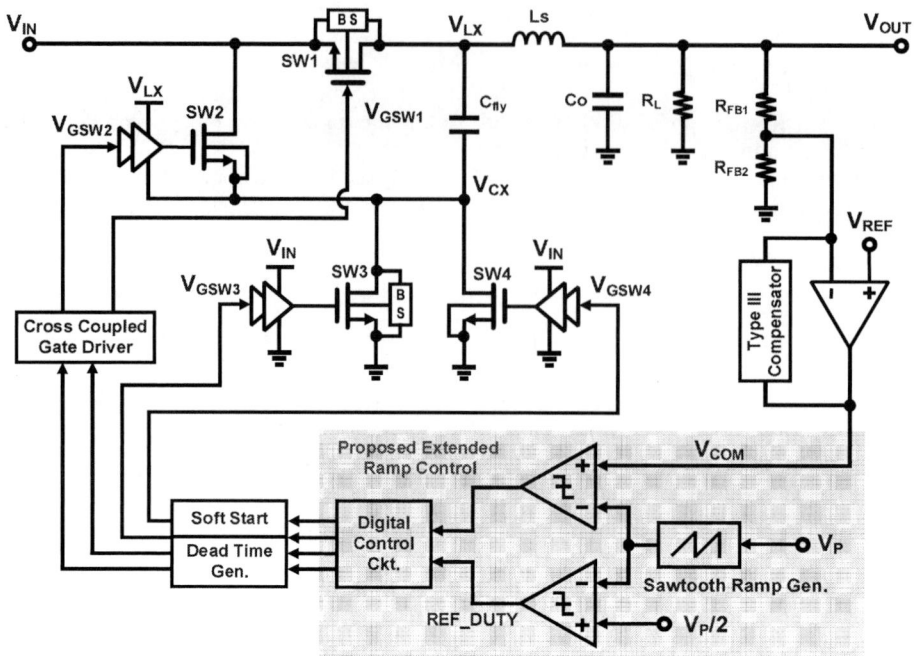

Fig. 4. Architecture of the proposed KY buck-boost converter with the extended ramp control.

II. THE PROPOSED KY BUCK-BOOST CONVERTER WITH EXTENDED RAMP CONTROL

A. The proposed system architecture

Fig. 4 demonstrates the architecture of the proposed KY buck-boost converter with the extended ramp control. The converter is mainly composed of a KY buck-boost power stage, a sawtooth waveform generator and digital control circuits including pulse-width modulation (PWM) control circuits. The switch control signal ($V_{GSW1} \sim V_{GSW4}$) is generated through the digital control circuits and the cross coupled gate driver. A proportional integral derivative (PID) compensator is applied in this work to stabilize the converter. Besides, this converter employs all NMOS power switches (SW1 ~ SW4) for the power stage and the body switching technique [12] is used for SW1 and SW3. Smaller chip size and high efficiency of the proposed converter can be achieved simultaneously. In order to drive the NMOS power switch properly, we usually utilize a bootstrap circuit to generate the driving signal. However, it will increase the chip size due to a large bootstrap capacitor. In this work, the gate driver of SW2 reuses the flying capacitor C_{fly} as bootstrap capacitor which significantly reduces the chip area [13].

B. The proposed extended ramp control

The architecture and operation of the proposed extended ramp control are shown in Fig. 4 and Fig. 5, respectively. The controller consists of two comparators, one sawtooth ramp generator and digital control circuits. The reference signal REF_DUTY is generated to help determine the operation mode. The error amplifier output signal V_{COM} is compared with a sawtooth ramp signal. As depicted in Fig. 5(a), the sawtooth ramp signal is divided into two parts by a DC signal V_{mid} which is half of the amplitude of the sawtooth ramp generator. When V_{COM} is larger than V_{mid}, the converter enters the boost mode. On the contrary, when V_{COM} is smaller than V_{mid}, the converter will enter the buck mode. The complete switching diagram at two operation modes are shown in Fig. 5(b) and Fig. 5(c), respectively.

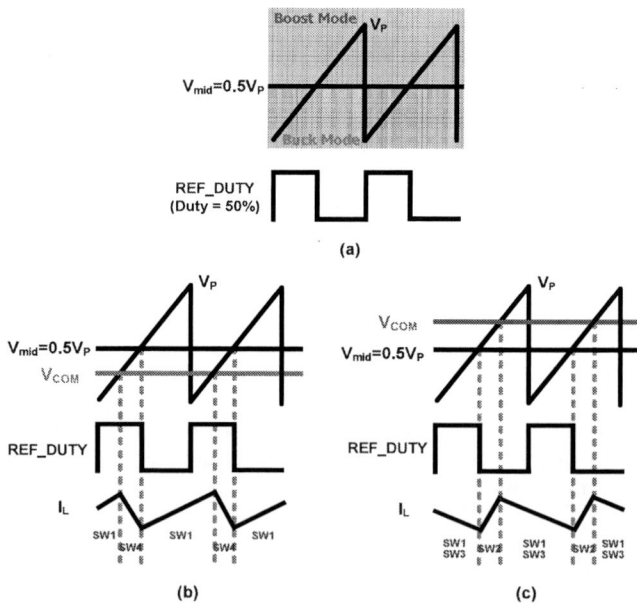

Fig. 5. Proposed extended ramp control : (a) Conceptual diagram (b) Buck mode operation (c) Boost mode opration.

For the conventional control utilized in the conventional non-inverting buck-boost converter, PWM nonlinearity [14] issue usually occurs at the interface of the buck mode and boost mode. As demonstrated in Fig. 6, when the input and output voltages are similar and the V_{COM} is approaching zero or one, PWM discontinuity will cause pulse-skipping and then result in significant fluctuation at the output voltage. Previous works have proposed various methods to avoid this problem. [9] – [11] add a buffer region and [15] – [17] implement other controls to make the improvements for smooth mode transition. In the proposed extended ramp control, when the input voltage is close to the output voltage, V_{COM} is roughly located at the middle of the sawtooth ramp. The PWM nonlinearity issue is naturally resolved without inserting extra buffer region and the smooth mode transition can be attained at the same time. The line transient waveform of the proposed control is depicted in Fig. 7. Fig. 8 shows the SIMPLIS simulated output voltage waveforms of two buck-boost converters with their corresponding controllers during line transition. It can be observed that 1500% output deviation reduction during mode transition is achieved when the proposed extended ramp control is applied.

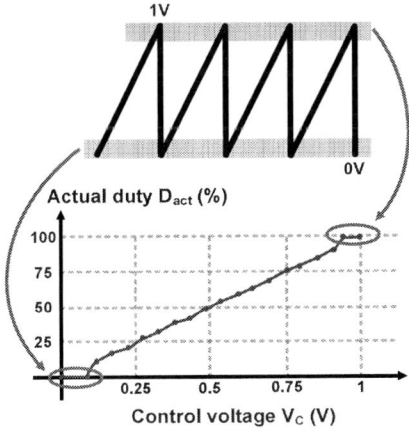

Fig. 6. Issue of PWM nonlinearity.

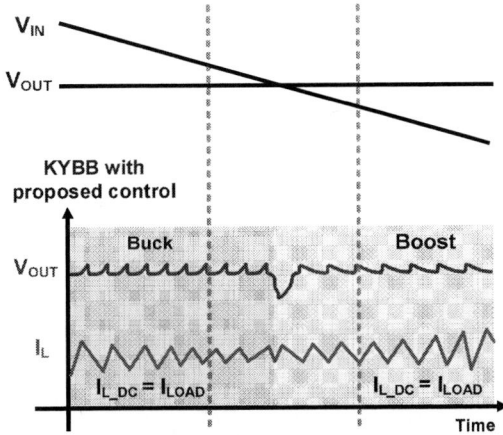

Fig. 7. Line transient waveform of the proposed KY buck-boost converter.

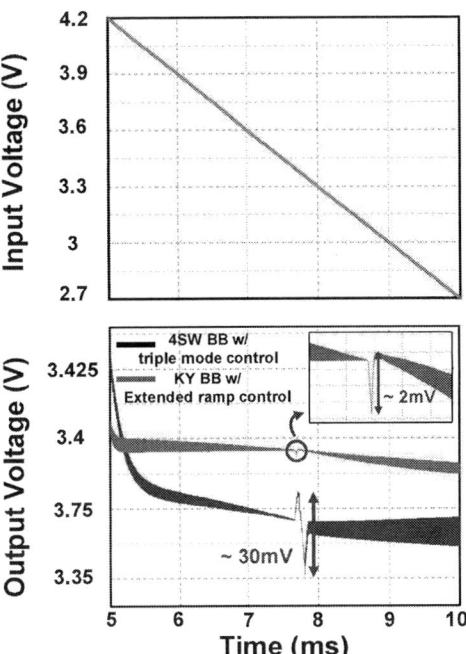

Fig. 8. Simulated line transient waveform of two buck-boost converters with their corresponding controllers.

III. CHIP SIMULATION RESULT

The proposed KY buck-boost converter with extended ramp control has been implemented in TSMC 0.18um BCD process to verify the proposed design. Fig. 9 depicts the overall chip layout. The total chip area is about 1.4 mm x 1.5 mm including pads. This converter operates with an input voltage ranging from 2.7V to 4.2V and the nominal output voltage is 3.4V. The switching frequency in this work is designed as 1.5 MHz at continuous conduction mode (CCM). Besides, for the off-chip components, this work includes a 2.2 μH inductor, a 4.7 μF flying capacitor and a 4.7 μF output capacitor.

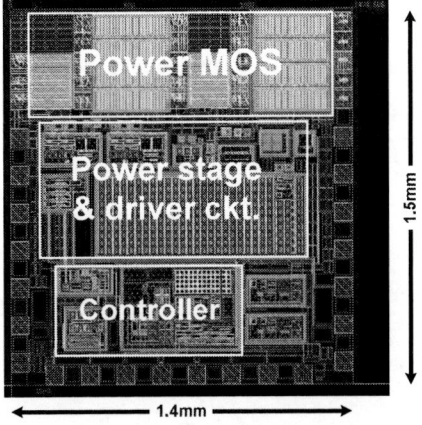

Fig. 9. Chip layout of the proposed KY buck-boost converter.

Fig. 10(a) and (b) demonstrate the simulated steady-state results of the proposed KY buck-boost converter when the input voltage is 2.7V and 4.2V, respectively. The proposed converter can operate properly at both buck mode and boost mode. Fig. 11 depicts the line transient response when the load current is 100mA. The input voltage varies from 4.2V to 2.7V within 500us. It is shown that when the proposed extended ramp control is applied, the output voltage fluctuation during mode transition can be reduced less than 10mV. The simulated power efficiency of the proposed converter is shown in Fig. 12. The peak efficiency of this work is 95.1% at 200mA load current.

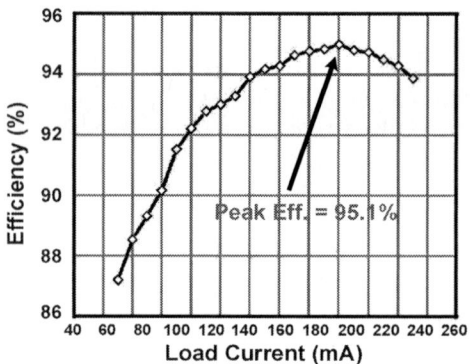

Fig. 12. Simulated power efficiency versus load current.

IV. CONCLUSION

The paper presents a KY buck-boost converter with extended ramp control. The proposed converter achieves a seamless mode transition without applying extra buffer region. When the proposed extended ramp control is applied, the output voltage fluctuation is less than 10mV. 1500% output variation reduction during mode transition and 95.1% peak efficiency is realized in the proposed converter. The chip fabrication and measurement are in-progress.

ACKNOWLEDGMENT

The authors would like to thank SIMPLIS Technologies Corporation, USA, for providing SIMPLIS simulation tool and the educational subsidy for chip fabrication from Taiwan Semiconductor Research Institute, Taiwan,.

REFERENCES

[1] B. Sahu and G. A. Rincon-Mora, "A low voltage, dynamic, noninverting, synchronous buck-boost converter for portable applications," *IEEE Trans. Power Electron.*, vol. 19, no. 2, pp. 443–452, March 2004.

[2] Y. -K. Cho and K. C. Lee, "Noninverting Buck–Boost DC–DC Converter Using a Duobinary-Encoded Single-Bit Delta-Sigma Modulator," *IEEE Trans. Power Electron.*, vol. 35, no. 1, pp. 484–495, Jan. 2020.

[3] J. -J. Chen, P. -N. Shen and Y. -S. Hwang, "A High-Efficiency Positive Buck–Boost Converter With Mode-Select Circuit and Feed-Forward Techniques," *IEEE Trans. Power Electron.*, vol. 28, no. 9, pp. 4240–4247, Sept. 2013.

[4] P. Malcovati, M. Belloni, F. Gozzini, C. Bazzani and A. Baschirotto, "A 0.18μm CMOS 91%-efficiency 0.1-to-2A scalable buck-boost DC-DC converter for LED drivers," in *IEEE Int. Solid-State Circuits Conf. (ISSCC) Dig. Tech. Papers*, Feb. 2012, pp. 280–282.

[5] Y.-A. Lin et al., "A right-half-plane zero-free buck-boost DC–DC converter with 97.46% high efficiency and low output voltage ripple," in *Proc. Symp. VLSI Circuits*, Jun. 2019, pp. C174–C175.

[6] Y. -T. Hung, C. -J. Tsai, C. -J. Chen, C. -H. Hsu and C. -Y. Hsieh, "An All NMOS KY-Boost Converter With Double Injection Control for Fast Line and Load Transient Response," *IEEE Trans. Circuits Syst. I, Reg. Papers*, vol. 71, no. 11, pp. 5005–5016, Nov. 2024.

[7] C. Zheng and D. Ma, "A 10 MHz 92.1%-efficiency green-mode automatic reconfigurable switching converter with adaptively compensated single-bound hysteresis control," in *IEEE Int. Solid-State Circuits Conf. (ISSCC) Dig. Tech. Papers*, Feb. 2010, pp. 204–205.

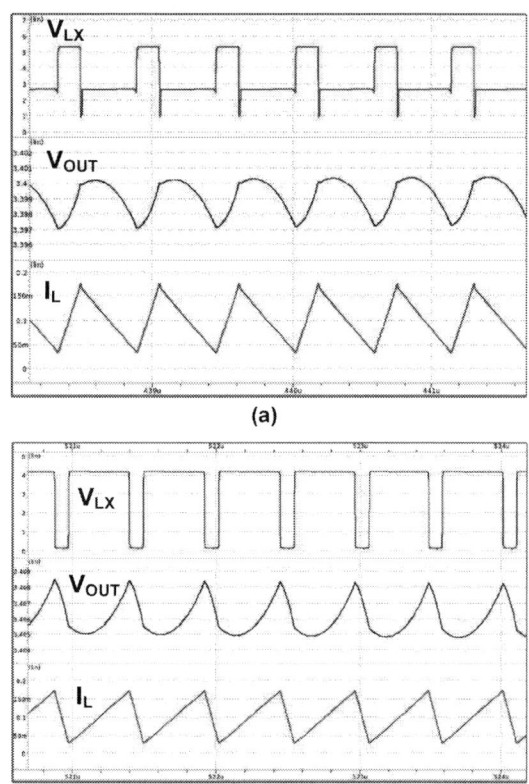

(a)

(b)

Fig. 10. Simulated steady-state waveform when V_{IN} is (a) 2.7V (b) 4.2V.

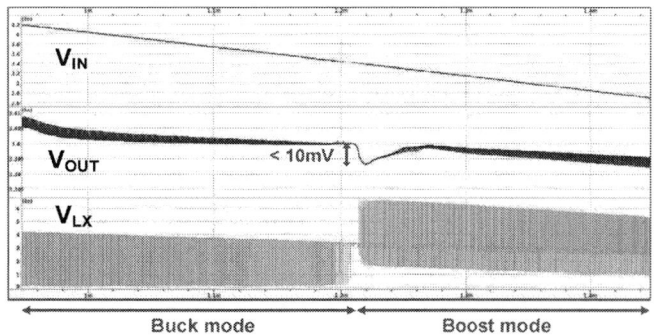

Fig. 11. Simulated line transient response when V_{IN} decreases from 4.2V to 2.7V.

[8] S. Rao et al., "A 1.2-A buck-boost LED driver with on-chip error averaged SenseFET-based current sensing technique," *IEEE J. Solid-State Circuits*, vol. 46, no. 12, pp. 2772–2783, Dec. 2011.

[9] P.-C. Huang, W.-Q. Wu, H.-H. Ho, and K.-H. Chen, "Hybrid buck-boost feedforward and reduced average inductor current techniques in fast line transient and high-efficiency buck-boost converter," *IEEE Trans. Power Electron.*, vol. 25, no. 3, pp. 719–730, Mar. 2010.

[10] L. Wang, X. Wu and J. Lou, "A Multi-Mode Four-Switch Buck-Boost DC/DC Converter," *2009 Asia-Pacific Power and Energy Engineering Conference*, Wuhan, China, 2009, pp. 1–4.

[11] S. Zhao, C. Zhan and Y. Lu, "A Battery-Input Three-Mode Buck–Boost Hybrid DC–DC Converter With 97.6% Peak Efficiency," *IEEE J. Solid-State Circuits*, vol. 59, no. 5, pp. 1567–1577, May 2024.

[12] P. Favrat, P. Deval and M. J. Declercq, "A high-efficiency CMOS voltage doubler," *IEEE J. Solid-State Circuits*, vol. 33, no. 3, pp. 410–416, March 1998

[13] C.-J. Tsai, C.-H. Hsu, C.-J. Chen, Y.-T. Hung, C.-Y. Hsieh, "A Monolithic All-1.8V-Thin-Gate-NMOS KY-Boost Converter With Reused Flying-Capacitor Bootstrap Gate Driver Achieving 94.42% Peak Efficiency," *IEEE Trans. Power Electron.*, vol. 39, no. 6, pp. 7238–7251, Jun. 2024.

[14] P. Rajarshi and D. Maksimovic, "Analysis of PWM nonlinearity in non-inverting buck-boost power converters," in *Proc. IEEE Power Electron. Spec. Conf.*, 2008, pp. 3741–3747.

[15] K. -D. Kim et al., "A Noninverting Buck-Boost Converter With State-Based Current Control for Li-ion Battery Management inMobile Applications," *IEEE Trans. Ind. Electron.*, vol. 66, no. 12, pp. 9623–9627, Dec. 2019.

[16] X. -E. Hong et al., "98.1%-Efficiency Hysteretic-Current-Mode Noninverting Buck-Boost DC-DC Converter With Smooth Mode Transition," *IEEE Trans. Power Electron.*, vol. 32, no. 3, pp. 2008–2017, March 2017.

[17] C. -L. Wei, C. -H. Chen, K. -C. Wu and I. -T. Ko, "Design of an Average-Current-Mode Noninverting Buck–Boost DC–DC Converter With Reduced Switching and Conduction Losses," *IEEE Trans. Power Electron.*, vol. 27, no. 12, pp. 4934–4943, Dec. 2012.

An USB Cable Based Extended Conversion Range L-First Hybrid-Converter using Valley-Virtual-Inductor-Current-Mode Control with Auto-Tracking Slope Compensation against ±50% Inductance Variation

Chun-I Li
Graduate Institute of Electrical Engineering
National Taiwan University
Taipei, Taiwan
jordan891119@gmail.com

Chieh-Ju Tsai
Graduate Institute of Electrical Engineering
National Taiwan University
Taipei, Taiwan
f04943123@ntu.edu.tw

Ching-Jan Chen
Graduate Institute of Electrical Engineering
National Taiwan University
Taipei, Taiwan
chenjim@ntu.edu.tw

Abstract—This paper introduces an L-first hybrid converter with extended conversion ratio that uses an USB cable as the power inductor to maximize power density both on-chip and on-PCB. The inductor current, reduced by the conversion ratio, naturally equals the input current, enabling the USB cable to function effectively as an inductor despite its parasitic resistance. The converter achieves an estimated power density of 3.3 W/3mm³ with a peak efficiency of 95%. To accommodate varying USB cable lengths, a valley-virtual-inductor-current mode control with auto-tracking slope compensation is proposed. This scheme eliminates the need for inductor current sensing and adjusts slope compensation based on changes in voltage and inductance. A 5 V input and 1.8–3.3 V output prototype, implemented in a 180 nm BCD process, covers inductances from 500 nH to 1 µH, corresponding to USB cable lengths of 1 to 2 meters.

Keywords—Hybrid converter, USB cable, auto-tracking slope compensation, valley-virtual-inductor-current

I. INTRODUCTION

Modern mobile and IoT devices require efficient, compact power management solutions that can operate directly from a USB cable (5 V) for charging and discharging batteries. Conventional buck converters struggle to achieve both high power density and efficiency due to increased DCR and conduction losses from miniaturized inductors, particularly on the high-current side. To mitigate these losses, inductor-first topologies [1]–[3] place the inductor on the input side, where it handles lower current, reducing conduction losses. Some L-first designs even use the USB cable as the inductor to boost power density [1]. However, these traditional L-first topologies face challenges, including a limited conversion ratio of 0.5–1 and controller design difficulties for varying USB cable inductances. This digest extends the traditional L-first converter [1] from a dual-path to a triple-path design, expanding the conversion ratio to 1/3–1 by adding a flying capacitor and switch network. A virtual-inductor-current (VIC) technique is employed to enable valley current mode modulation, eliminating the need for current sensing circuitry, which is challenging in L-first converters. Additionally, an auto-tracking slope compensation technique is developed to handle large inductance variations.

II. USB-CABLE BASED L-FIRST CONVERTER WITH EXTENDED CONVERSION RANGE

Fig. 1(a) illustrates the basic concept of a traditional L-first hybrid converter, which uses an output switching network to transfer the switching node V_{LX} toggling between V_O and $2V_O$ for inductor charging and discharging. This approach limits the conversion ratio to a range of 0.5 to 1, as determined by the volt-second balance. By simply modifying the switching network, following the concept in Fig. 1(a), a new L-first converter that operates with can be developed, extending the conversion range to 1/3 to 1, making it suitable for mobile devices powered by a USB cable. The actual implementation of the converter and its operating states are shown in Fig. 2. Notably, the inductor current is equal to the input current, which reduces the inductor current stress by the conversion ratio M ($= 1/(3-2D)$). A comparison of inductor current stress, inductor current ripple, estimated inductor volume, additional on-board capacitor requirements, and total volume for various converters, including buck, 3-level, and double-step-down converters, is provided in Table I. Since this architecture uses the USB cable as a parasitic inductance along with appropriate control methods, there's no need for an external inductor. Consequently, compared to other converters, the power density can be significantly increased. An estimated volume of 3 mm³ by accounting passive devices and power density of 3.3 W / 3 mm³ is achieved and gradually superior to conventional converter!!

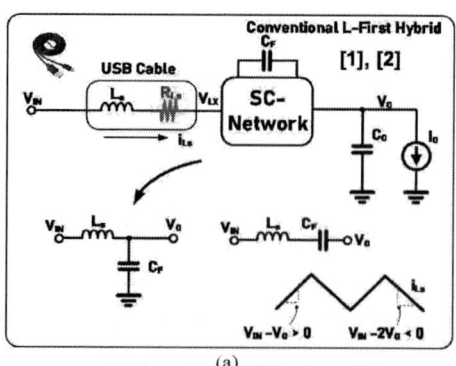

(a)

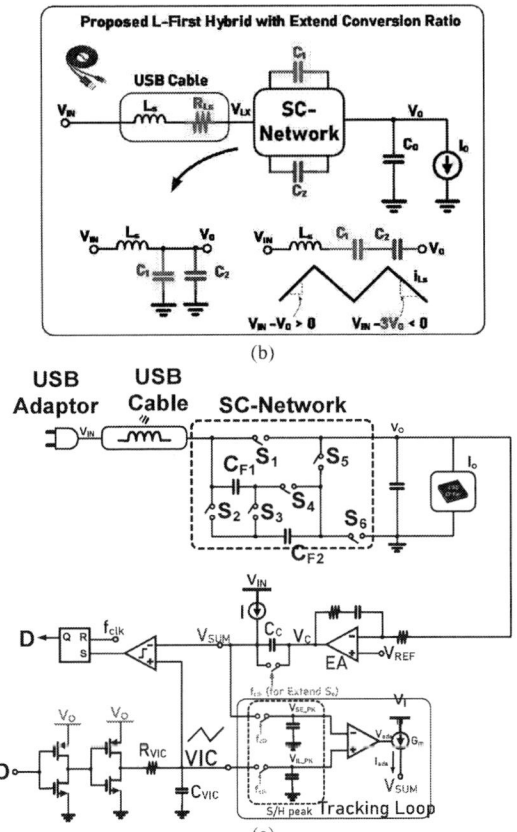

Fig. 1. concept of (a): conventional L-first converter, (b): proposed L-first with extended conversion ratio, (c): architecture of the proposed converter

TABLE I.
COMPARISON OF TOPOLOGY

Topology	i_{Ls}/I_O	Δi_{Ls}	Estimated L	C_F / C_O	Total Volume
Buck	1	$\frac{V_{IN}(1-M)M}{L_s F_{SW}}$	500nH, 4mΩ DCR 100 mm³	0 / 1 mm³	100 mm³
3-L Buck [4]	1	$\frac{V_{IN}(1-M)M}{2L_s F_{SW}}$	250nH, 4mΩ DCR, 25 mm³	1 / 1 mm³	26 mm³
DSD Buck [5]	1/2	$\frac{V_{IN}(1-2M)M}{L_s F_{SW}}$	250nH X 2, 4mΩ DCR, 50 mm³	1 / 1 mm³	52 mm³
Proposed	M	$\frac{V_{IN}(1-M)M}{L_s F_{SW}}$	500nH – 1000nH, 50mΩ DCR (USB Cable)	2 / 1 mm³	3 mm³

III. THE CONCEPT OF THE PROPOSED CONVERTER

The proposed triple path converter is illustrated in Figure. 1(c). Since the inductor is connected to Vin, the inductance can be implemented using the parasitic inductance of a USB cable, leading to a significant improvement in power density. It is evident that the parasitic inductance value is related to the cable length and independent of the cable shape [5], indicating the stability of the parasitic inductance value. We measured one testing cable's inductance using an Impedance Analyzer and found it to be 750nH with a DC resistance of 10mΩ. Subsequent derivations of L are based on this set of data.

A. Operating principle

Fig. 2 depicts the two states of the circuit in normal mode. As shown in Fig. 2(a), In state A, capacitors C_1 and C_2 are connected in parallel with V_o. From this, it can be inferred that when the switching frequency is sufficiently high, C_1 and C_2 will be approximately equal to V_o and can achieve auto balance. Inductor and capacitors supply I_o through the parallel connection of V_o. And in state B, I_{Ls} charges C_1 and C_2 to maintain the principle of capacitor charge balance in one cycle.

The voltage conversion ratio can be calculated based on charge flow analysis [9]. The deduced result is consistent with using voltage second balance. First, we assume output capacitor is very large. Then, calculate the conversion ratio by dividing the total input charge by the total output charge. It follows that:

$$q_{in}^A = I_{Ls}DT \quad , \quad q_{in}^B = I_{Ls}(1-D)T \quad (1)$$

$$q_o^A = q_{in}^A + 2q_{in}^B \quad , \quad q_o^B = q_{in}^B \quad (2)$$

$$\text{Conversion ratio (M)} = \frac{q_{in}^A + q_{in}^B}{q_o^A + q_o^B} = \frac{1}{3-2D} \quad (3)$$

By using the formula in ideal case $P_{in} = P_{out}$ and $I_{in} = I_{Ls,DC}$, we can derive:

$$I_{in} = I_{Ls,DC} = MI_o \quad (4)$$

$$\Delta I_{Ls} = \frac{1}{L}(V_{in} - V_O)DT = \frac{V_{in}}{L}T(1-M)\left(\frac{3}{2} - \frac{1}{2M}\right) \quad (5)$$

With the derivations provided above, we can easily observe that $I_{Ls,DC}$ is reduced by a factor of M when compared to the conventional buck converter and 3-level buck converter, and this reduction is directly related to DCR loss.

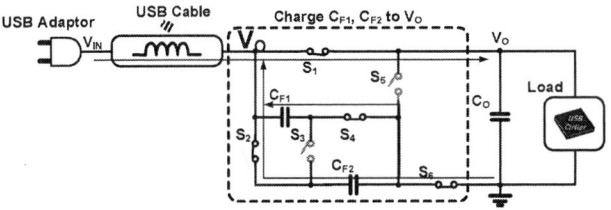

(a)

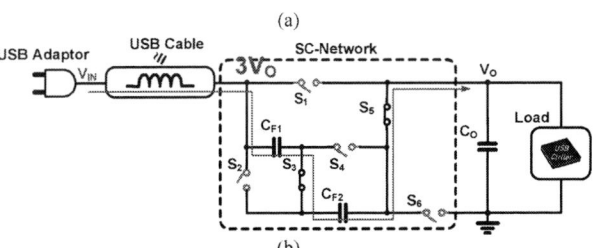

(b)

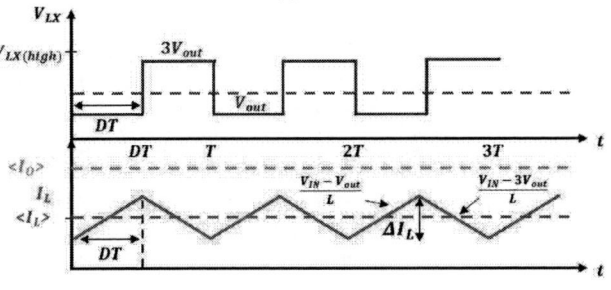

(c)

Fig. 2. The proposed converter: (a)State A (b)State B (c)Steady State Waveform

B. Conduction Loss Analysis and Cflly hard charging Loss

Because of the hard-charging issue with C_1 and C_2, it is difficult to analyze Ro using conventional methods. Thus, charge flow analysis is applied to estimate the hard-charging average current and determine Ro. A charge-flow analysis framework is presented as follows: From Fig .2 , we derive the charge flows of the two states at (1), encompassing input (q_{in}^A, q_{in}^B). Next, by utilizing the charge flowing through the switch and the conduction time, we can calculate the average current during each switch conduction. Substituting this into the energy formula $E = I^2 R * t_{on}$, we can obtain the losses on each switch. Then, by using the relationship between I_L and I_o, we can determine the equivalent output resistance $R_{o,cond}$. Fig. 3 presents the graph of Ro corresponding to f_{sw}, serving as a reference for selecting the switching frequency.

$$R_{o,cond} = R_{o,FSL} = \left(\frac{1}{3-2D}\right)^2 (\frac{(2-D)^2}{D}R_{SW1} + \frac{(1-D)^2}{D}R_{SW2}$$
$$+ (1-D)R_{SW3} + \frac{(1-D)^2}{D}R_{SW4} + (1-D)R_{SW5}$$
$$+ \frac{(1-D)^2 * 4}{D}R_{SW6})$$

(6)

The SSL loss analysis:

$$\frac{\Delta q^2}{2CT} = I_O^2 R_{o,SSL} , \Delta q = I_L(1-D)T$$

(7)

By using (3),(4) and (7):

$$R_{o,SSL} = \frac{(1-D)^2}{C_{fly}F_{sw}(3-2D)^2}$$

(8)

$$R_{O,total} = \sqrt{R_{o,cond}^2 + R_{o,SSL}^2}$$

(9)

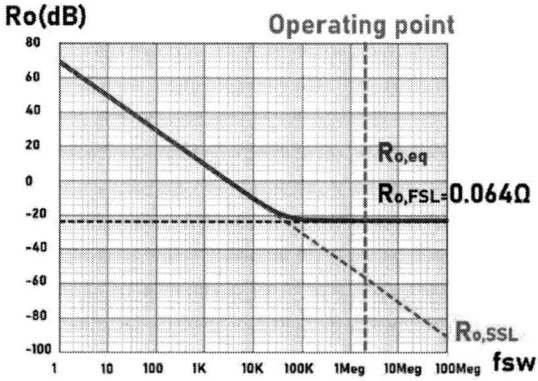

Fig. 3 Conduction loss contribution of Switches and C_1, C_2 charging loss.

IV. VALLEY VIC CONTROL WITH AUTO-TRACKING SLOPE COMPENSATION

To address the challenges of inductance variation when using a USB cable, we propose a valley virtual-inductor-current (VIC) mode control with auto-tracking slope compensation. Fig. 3 illustrates the control loop of the proposed scheme, which is a modification of conventional valley current mode control [6]. Instead of using the actual inductor current, a "virtual-inductor-current" [7] is generated

by passing the duty cycle through two inverters, sourced by the output voltage V_O, to create an inductor current-shaped signal. An error amplifier senses the output voltage and compares it to a reference voltage V_{REF}, generating V_C for modulation. Slope compensation is applied directly to V_C using a stacked capacitor C_C and a charging current source, eliminating the need for a current sensor while maintaining the control dynamics and type-II compensation. This approach is particularly suitable for L-first hybrid converters, where inductor current sensing is challenging. The slope compensation is implemented at the output of the type-II compensation with a stacked current source charging the capacitor, reducing the need for an additional analog adder and enabling adaptive slope compensation.

The adaptive slope compensation tracks the modulation waveform to ensure stability across varying inductance and V_{IN}/V_O conditions. Based on the modulation waveform shown in Fig. 4(a):

$$VIC_{max} = V_c + S_e * (1-D)T + S_n * DT$$

(10)

If $S_e = S_n$:

$$V_{SUM,max} = V_c + S_e * T = VIC_{max}$$

(11)

Based on the (11), the automatic tracking loop samples the peak value of V_{SUM} and VIC at the clock edge and adjusts the slope compensation to match the modulation current ramp's rising slope S_n. If $V_{SUM,max} > VIC_{max}$, V_{ada} will decrease, so the current flow into C_c is decreased which cause $V_{SUM,max}$ decrease. By the same concept if $V_{SUM,max} < VIC_{max}$, V_{ada} and $V_{SUM,max}$ will increase.

This tracking loop ensuring $S_e = S_n$ [8], which prevents subharmonic oscillation under all operating conditions, even with ±50% inductance variation. The simulated control-to-output transfer function G_{vc} in Fig. 4 confirms the absence of subharmonic instability, as indicated by the lack of $f_{sw}/2$ complex poles.

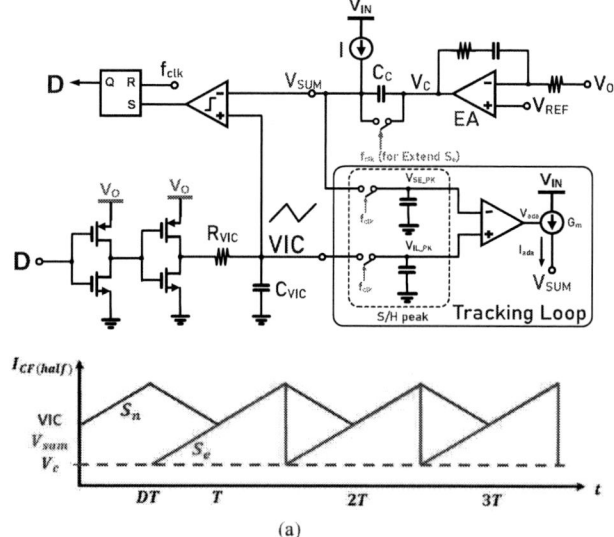

(a)

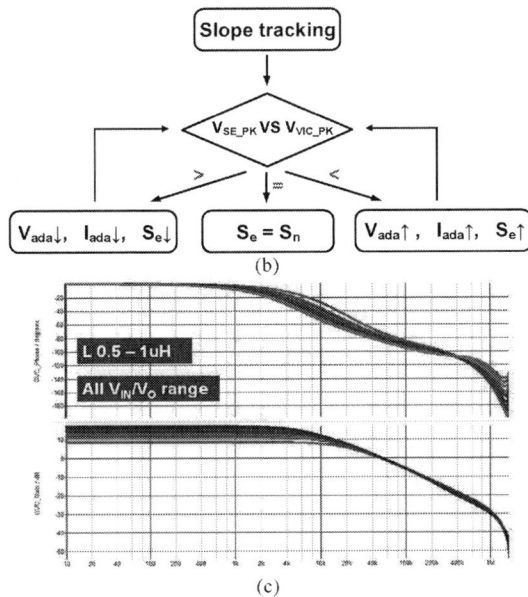

(b)

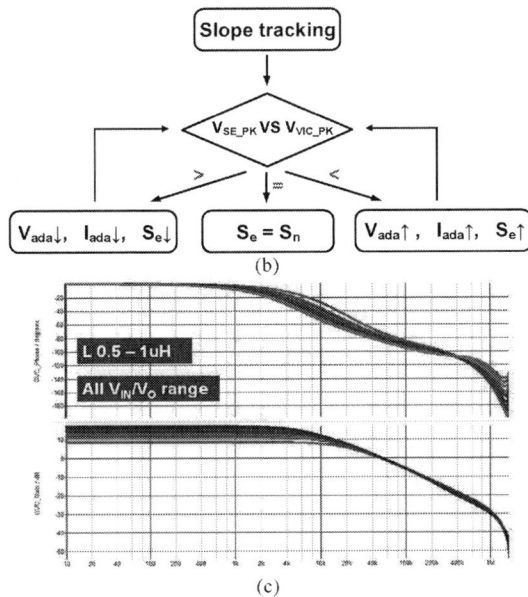

(c)

Fig. 4 The Valley VIC control with auto-slope compensation tracking technique

V. SIMULATION RESULTS

Fig. 5 (a) shows the load transient response of valley VIC control. During load transients, V_{sum} decreases due to the feedforward action of V_o. At the same time, V_c rises as the error between V_o and V_{REF} becomes large, especially when V_o experiences a significant drop under heavy load. This causes the duty cycle to be activated quickly.

From Fig. 5(b), it shows the waveform with an inductance variation of 50%. This waveform shows that the system remains stable in every condition. Besides, when the output current (Io) ranges from 0 to 1 A, the settling time is 12 us in the worst-case scenario.

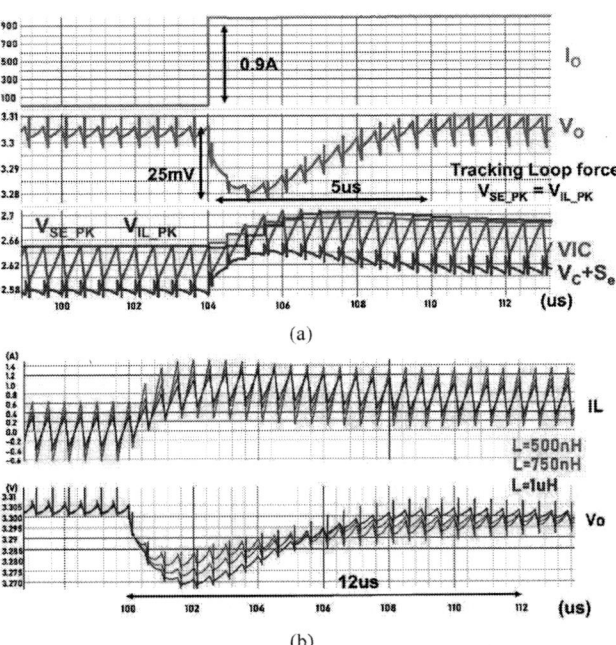

(a)

(b)

Fig. 5. Simulation waveform: (a)load transient response (b)load transient response with inductance variation

We first build up a PCB prototype design with a measured efficiency of 5 V to 3.3 V after whole chip tape out shown in Fig. 6 to verify the converter. The peak efficiency is 95.7% of this prototype PCB. The on-chip version will increase the maximum loading current to 3 A and the power density will be increased to around 10 W / 3mm³ with flying capacitor and output capacitor mounted on die.

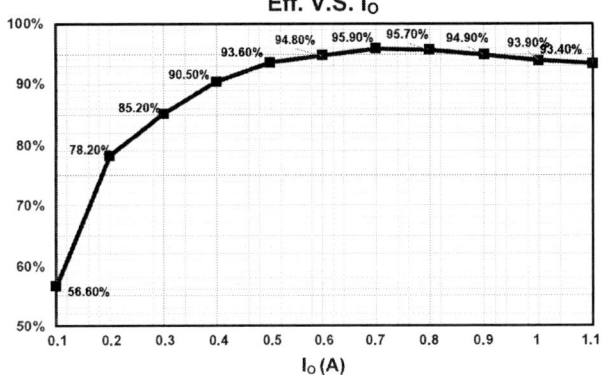

Fig. 6 Efficiency measured on prototype PCB version proposed converter

VI. CONCLUSION

This paper presents an L-first hybrid converter with extended conversion ratio and generalized valley-VIC-control scheme with auto-tracking slope compensation against USB cable ±50% inductance variation. The converter has been implemented a 180 nm BCD process, achieving a peak efficiency of 95.7% and 3.3 W / 1mm³ power density. The chip fabrication process and measurement are in progress.

	[1] TIE 2020	[3] JSSC 19	[11] ISSCC 23	[10] JSSC 24	This work
Technology	DSP Ctrller	130nm	180nm	180nm	180nm
Topology	L-1st 1C_F	L-1st 1C_F 3C_o	L-1st 1C_F	2Phase L-1st 2C_F	L-1st 2C_F
Control scheme	V-mode	V-mode	V-mode	V-mode	V-VIC
$V_O(V)$/ $V_{IN}(V)$	3.3 / 5	3 - 4.2 / 9	8 - 12.6 / 12-20	0.8 - 1.8 / 1.5 - 3.3	1.6–3.3 /5
f_{sw}(MHZ)	2	2	5	0.5	2
Ls(uH) / Co(uF)	Cable (1) / 13.2	Cable (0.5)/4.4	Cable (1) / N.A.	1*2 /4*4.7	Cable (0.75) / 10
Inductance Variation Tolerance	X	X	X	X	50%
Peak-Eff(%)	90	96	95.1	97.3	95.9
$I_{O,MAX}$(A)	3.9	3	3.3	3.3	4

ACKNOWLEDGMENT

The authors sincerely appreciate the support provided by the National Science and Technology Council (NSTC), Taiwan, along with the educational sponsorship for chip fabrication from TSRI, Taiwan. The authors also extend their gratitude to SIMPLIS Technologies for offering access to the SIMPLIS simulation tool.

REFERENCES

979-8-3315-1612-3/25 $31.00 © 2025 IEEE   2744

[1] G. -S. Seo and H. -P. Le, "S-Hybrid Step-Down DC–DC Converter—Analysis of Operation and Design Considerations," *IEEE Tran.s on Industrial Electronics*, vol. 67, no. 1, pp. 265-275, Jan. 2020.

[2] A. Abdulslam and P. P. Mercier, "A Battery-Connected Inductor-First Flying Capacitor Multilevel Converter Achieving 0.77W/mm2 and 97.1% Peak Efficiency," in *Proc. 2021 IEEE Custom Integrated Circuits Conference (CICC)*, Austin, TX, USA, 2021, pp. 1-2.

[3] Casey Hardy, Yogesh Ramadass, Kevin Scoones, Hanh-Phuc Le, "A Flying-Inductor Hybrid DC–DC Converter for 1-Cell and 2-Cell Smart-Cable Battery Chargers", *IEEE Journal of Solid-State Circuits*, vol.54, no.12, pp.3292-3305, 2019.

[4] X. Liu, P. K. T. Mok, J. Jiang and W. -H. Ki, "Analysis and Design Considerations of Integrated 3-Level Buck Converters," *IEEE Transactions on Circuits and Systems I: Regular Papers*, vol. 63, no. 5, pp. 671-682, May 2016.

[5] J. Yuan, Z. Liu, F. Wu and L. Cheng, "A 12V/24V-to-1V DSD Power Converter with 56mV Droop and $0.9\mu \mathrm{S}$ 1% Settling Time for a 3A/20ns Load Transient," in *Proc. 2022 IEEE International Solid-State Circuits Conference (ISSCC)*, San Francisco, CA, USA, 2022, pp. 1-3.

[6] J. Li and F. C. Lee, "Modeling of V2 Current-Mode Control." in *Proc. 2009 Twenty-Fourth Annual IEEE Applied Power Electronics Conference and Exposition*, Washington, DC, USA, 2009, pp. 298-304.

[7] Y.-C. Lin, C.-J. Chen, D. Chen, B. Wang, "A Ripple-Based Constant On-Time Control With Virtual Inductor Current and Offset Cancellation for DC Power Converters," *IEEE Transactions on Power Electronics*, Vol. 27, No.10, pp. 4301 - 4310, 2012.

[8] P. -H. Liu, Y. Yan, F. C. Lee and P. Mattavelli, "External ramp autotuning for current mode control of switching converters," in *Proc. 2013 Twenty-Eighth Annual IEEE Applied Power Electronics Conference and Exposition (APEC)*, Long Beach, CA, USA, 2013, pp. 276-280.

[9] M. D. Seeman and S. R. Sanders, "Analysis and Optimization of Switched-Capacitor DC–DC Converters," in IEEE *Trans. on Power Electronics*. vol. 23, no. 2, pp. 841-851, March 2008.

[10] X. Zhang et al., "An Outphase-Interleaved Switched-Capacitor Hybrid Buck Converter With Relieved Capacitor Inrush Current and COUT-Free Operations," *IEEE Journal of Solid-State Circuits*, vol. 59, no. 4, pp. 1078-1092, April 2024.

[11] Z. Tong, J. Huang, Y. Lu and R. P. Martins, "A 42W Reconfigurable Bidirectional Power Delivery Voltage-Regulating Cable," in *Proc. 2023 IEEE International Solid-State Circuits Conference (ISSCC)*, San Francisco, CA, USA, 2023, pp. 192-194.

Impact of Gate Resistor Configurations on Current Balancing in Paralleled SiC MOSFETs

Yifu Zhang
Infineon Technologies AG
Neubiberg, Germany
Yifu.Zhang@infineon.com

Shashank Karanth
Infineon Technologies AG
Neubiberg, Germany
Shashank.Karanth@infineon.com

Emanuel Eni
Infineon Technologies AG
Neubiberg, Germany
Emanuel.Eni@infineon.com

Abstract—To ensure availability and larger 2nd sourcing capabilities, paralleling discrete SiC MOSFETS is becoming more popular as the application's current requirements are increasing rapidly. However, the dynamic and static drain currents need to be well balanced when driving SiC MOSFETs in parallel in order to ensure power loss balancing. Apart from the device characteristics and the power circuit layout, the gate driver circuits also play a significant role, especially when dealing with high slew rates during switching. In this work, the switching behavior of paralleled discrete SiC MOSFETs are evaluated with different gate resistor configurations under various mismatch conditions. The switching loss imbalance and oscillations are analyzed for each case. In the end, based on the measurement and analysis results, a design principle is proposed to determine the optimal gate resistor configuration to drive paralleled devices.

Keywords—gate resistor, paralleling, SiC MOSFET, gate driver, switching loss balancing, DM oscillations

I. INTRODUCTION

The transition to renewable energy, especially the growth of energy storage systems (ESS) and electric vehicles (EV), which demand a very high efficiency, is pushing the demand for SiC MOSFETs forward. The power rating of these applications is increasing drastically, which in turn requires higher current ratings from the power devices. To achieve these requirements, a common approach is to parallel multiple chips, which usually includes two methods: building a power module or paralleling discrete devices [1]. The latter solution is becoming more and more prevalent, as it provides more second sourcing flexibility, which is especially beneficial during semiconductor shortages. Therefore, this paper will focus on paralleling discrete SiC MOSFETs.

When paralleling SiC MOSFETs, the most critical consideration is the drain-source current balancing, which generally includes two aspects: static current balancing and dynamic current balancing [1]. Unbalanced current sharing can reduce the reliability of the devices and could require derating of the devices [2]. Devices with higher current might get overloaded and fail due to overcurrent [3]. Furthermore, an unbalanced current will lead to unbalanced power loss. This will further cause an imbalance in junction temperature, which further results in a difference in the lifetime of the paralleled device [1]. On the other hand, in real power applications, devices are switching continuously, which means there will be a temperature fluctuation. Different power losses will cause different temperature fluctuations, which means different cycling stresses. The device with higher temperature fluctuation

will age earlier than other devices. As a result, the degradation of the whole system, where devices are paralleled, will accelerate [1]. Fortunately, the positive temperature coefficient (PTC) of the on-state resistance (R_{dson}) enables the self-balancing of static current [1-3]. However, dynamic current is more challenging due to its strong dependence on parameters like threshold voltage (V_{th}), which typically exhibits a negative temperature coefficient (NTC). Therefore, the dynamic currents need to be actively balanced. Various investigations have been conducted to achieve a balanced dynamic current, such as novel methods to select and classify the devices [4][5], optimizing the layout [6-8], introducing extra passive components like differential mode choke in the power loops [9] or common mode choke in the gate loops [10]. In other studies [11-13], feedback loops are used to balance the dynamic current or switching loss in paralleled SiC MOSFETs. While these methods have shown promising results, it is generally impractical or impossible to implement such approaches into the market due to manufacturing constraints or extensive calibration needs. Moreover, apart from all these methods, the fundamental role of gate resistor configurations is often overlooked and not yet fully understood. Driving paralleled devices with a single gate resistor on the common path might lead to totally different results, compared to setting the gate resistor only on the individual paths for each device. Fig. 1 shows the basic concept of the gate resistor configuration, where two types of resistors are involved: common gate resistor (R_c) and individual gate resistor (R_i).

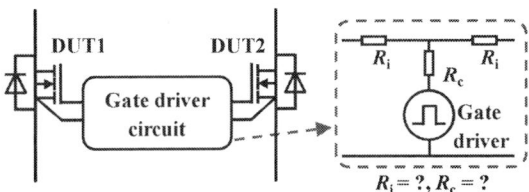

Fig. 1. Concept of the gate resistor configuration.

In [14], the authors mention that higher R_i helps to suppress the oscillations while higher R_c helps to balance the currents. But there are no supporting waveforms or data. Paper [15] also argues that lower R_i helps to decrease the switching loss imbalance caused by the mismatch in gate capacitance. However, the impact on switching loss imbalance caused by other factors such as mismatched PCB layout are not discussed. Moreover, the measurement was based on 3-pin devices, where a Kelvin source connection is not available. Therefore, more investigations are needed to gain a further understanding of the

979-8-3315-1612-3/25 $31.00 © 2025 IEEE

influence of different gate resistor configurations on the switching behavior of paralleled devices, which will be the topic of this work.

Apart from the switching loss balancing, behavior of the oscillations during switching in paralleled devices is another important aspect, which generally can be divided into common-mode (CM) oscillations and differential-mode (DM) oscillations [16][17]. CM oscillations are namely the switching ringing, which also happens for a single device due to the device parasitic capacitance and loop stray inductance [18]. DM oscillations are oscillations that occur between paralleled devices, which are also known as inter-chip oscillations [19][20] or parallel oscillations [21]. Paper [20] studied the inter-chip oscillations in paralleled SiC MOSFETs and concluded that higher individual gate resistance helps to mitigate the inter-chip oscillations. Therefore, different gate resistor configurations should also have an influence on the DM oscillations. In this work, the impact of gate resistor configurations will be experimentally investigated with various mismatch conditions. The measurement results will be evaluated in terms of the switching loss imbalance and the oscillations. Section II introduces the measurement setup and analysis of the stray inductance. Section III introduces the methods used to sort the devices to control the impact of the mismatch in device properties. Section IV presents the measurement results and the analysis, where conclusions and the design recommendation are drawn. Section V summarizes this paper.

II. MEASUREMENT SETUP

A. Test setup description

To evaluate the switching performance of paralleled devices, a custom-designed double pulse test platform was developed. Fig. 2 shows an overview of the measurement setup in this work. The test PCBs are located inside a safety box to provide

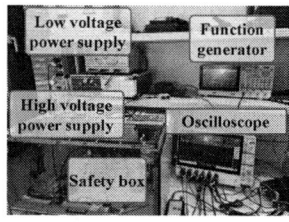
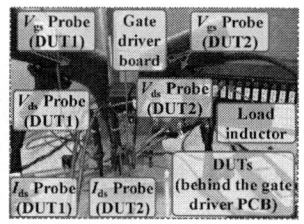

Fig. 2. An overview of the test setup.

protection from the high voltages. DUTs (devices-under-test) are mounted on the test boards. The pulse signals from the function generator are fed into the input leads of the gate driver. Low voltage (LV) power supply provides the power for gate driver circuits, and high voltage (HV) power supply provides the power for the power circuit. All the waveforms are measured via a 6-channel oscilloscope with 1 GHz bandwidth for each channel. To measure the gate-source voltage waveform, optically isolated probes are used, which can provide a high CMRR (common-mode rejection ratio). The test PCBs used the same design principles as in [22]. The power circuitry and gate driver circuitry are separately designed on different boards. The gate driver board is vertically inserted on the power board to minimize the possible capacitive coupling between the boards.

To exclude the limitation of the gate current on the switching performance, the gate driver used is 1ED3124MU12F, which can provide more than 14 A peak current. On the power PCB, the paralleled devices are placed symmetrically so that the stray inductance is matched, which is also verified through FEM analysis. The gate driver PCB also uses symmetric routing, which means the gate loops of both switches are identical. An accurate measurement of the transient current is challenging, as the switching speed of SiC MOSFETs is quite high. For this, SMD shunt resistors are connected in parallel to reduce the stray inductance and improve the measurement accuracy.

B. Analysis of the stray inductances

The PCBs are analyzed with Ansys Q3D to gain a better understanding of the stray inductances. Fig. 3 depicts the extracted values on the schematic of paralleled devices, where L_{d1} and L_{d2} indicate the stray inductances on the drain path of DUT1 and DUT2. L_{s1}, L_{s2} are the stray inductances on the source path of DUT1 and DUT2. L_{d1}, L_{d2}, L_{ks1}, and L_{ks2} are the stray inductance on the gate path and Kelvin source path of both DUTs, respectively. It can be seen that for DUT1 and DUT2, the stray inductances are quite similar, which indicates a matched layout.

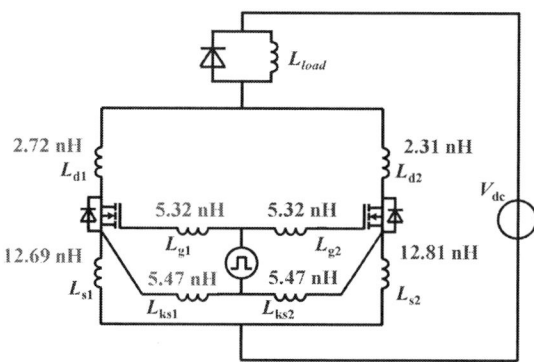

Fig. 3. Extracted stay inductance values with Ansys Q3D

III. SORTING OF THE DEVICES

The purpose of this work is to investigate the influence of gate resistor configurations under different mismatch conditions. Therefore, the switches need to be sorted to control the mismatch on the devices. The device used in the investigation is IMZA120R030M1H, which is a SiC MOSFET in TO247-4 package. Table I shows the relevant parameters from the datasheet [23]. It is worth noting that the internal gate resistance is 2.1 Ω, which means there is an inherent 2.1 Ω resistance on the gate path.

TABLE I. RELEVANT PARAMETERS OF THE SELECTED SiC MOSFET

Parameter	Symbol	Value	Unit
Rated drain-source voltage	V_{DS}	1200	V
Continuous DC drain current	I_{DC}	70	A
On-state resistance	R_{dson}	30	mΩ
Gate threshold voltage	V_{th}	4.2	V
Internal gate resistance	$R_{G,int}$	2.1	Ω

In total 100 samples were sourced from the market and then characterized by the curve tracer. Fig. 4. shows the spread of

979-8-3315-1612-3/25 $31.00 © 2025 IEEE

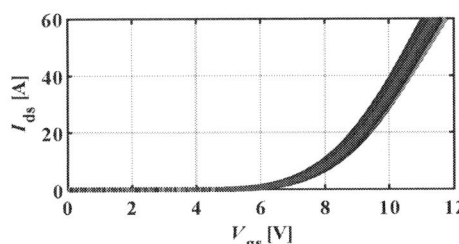

Fig. 4. Extracted stay inductance values with Ansys Q3D

the measured transfer curves, which are used as the reference to sort the devices. The parallelization number is 2, which generally means that two devices are to be selected from 100 samples for paralleling. Therefore, there will be $C_{100}^2 = 4950$ possible combinations. For each combination, the transfer curve difference (TCD) is calculated with (1), which used the method in [22] as reference.

$$ \text{TCD} = \sum_{nn=N_0}^{N} |f_{\text{TC1}}(nn) - f_{\text{TC2}}(nn)| \qquad (1) $$

where N_0 is the nearest index to the smaller V_{th} value of in the paralleled DUTs. The lower TCD the combination exhibits, the better the devices in this combination are matched. The TCD values of all the combinations fall between 0.0030 and 2.457. Combinations with the TCD value lower than 0.5 are preselected and considered matched. Then, the selected combinations are experimentally verified through the double-pulse-test platform mentioned in the above section.

IV. MEASUREMENT RESULTS AND ANALYSIS

After the devices are sorted, experimental measurements are conducted with different gate resistor configurations under different mismatch conditions.

A. Experimental measurements

1) Different gate resistor configurations

To achieve a high dv/dt (~30 kV/μs), the equivalent gate resistance per switch is selected as 10 Ω, which is kept constant for all the measurements. Note that the equivalent gate resistance per switch consists of two parts, namely:

$$ R_{\text{eq}} = R_{\text{i}} + 2R_{\text{c}} \qquad (2) $$

R_{i} is swept from 0 Ω to 10 Ω with a 2 Ω step. Therefore, the configurations are shown in Table II.

TABLE II. RELEVANT GATE RESISTOR CONFIGURATIONS

Configuration	#1	#2	#3	#4	#5	#6
R_{i}	0 Ω	2 Ω	4 Ω	6 Ω	8 Ω	10 Ω
R_{c}	5 Ω	4 Ω	3 Ω	2 Ω	1 Ω	0 Ω
R_{eq}	10 Ω	10 Ω	10 Ω	10 Ω	10 Ω	10 Ω

2) Creating various mismatch conditions:

Various mismatch conditions are created based on the above-mentioned PCB and selected devices. The mismatch conditions are divided into different aspects: mismatch in stray inducatance, mismatch

in deivce transfer characteristics, and mismatch in the parasitic capacitance. According to the characterization data, the gate-drain capacitance and drain-source capacitance does not exhibit a large difference. Therefore, only the mismatch in gate-source capacitance was considered.

a) Creating mismatch in stray inductance: As shown earlier, the test PCB is symmetrical. Therefore, to introduce asymmetry of stray inductance in the power loop as well as the gate loop, the leads of the DUTs are bent individually. Fig. 5 shows an example of this method, where the drain lead is bent.

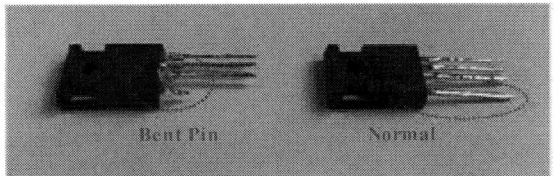

Fig. 5. An example of the bent device lead

The structure is also analyzed with Q3D analysis. The results show that for drain and source, around 4.75 nH additional inductance is introduced. For gate or Kelvin source, this value is 5.77 nH, which is slightly larger than that of drain or source leads, as they have a smaller cross-sectional area.

b) Creating mismatch in transfer characteristics of the MOSFETs: Mismatch on transfer characteristics of the devices can be achieved by simply selecting the devices with a large difference in their transfer curves.

c) Creating mismatch in C_{gs} of the MOSFETs: Mismatch in C_{gs} of the devices is achieved based on the matched devices. External capacitors are soldered directly between the gate and Kelvin source on one of the devices.

For each mismatched condition, all the gate resistor configurations defined in Table II are tested. For each measurement, waveforms of switching transients are captured with the setup mentioned in Section II.

B. Switching loss imbalance vs. gate resistor configurations

After the measurement, all the measured waveforms are analyzed to evaluate the switching losses and oscillations. For all the measurements, the overall switching losses are similar and are around 0.75 mJ during turn-off and 2.65 mJ during turn-on, which indicates that gate resistor configurations do not affect the overall switching losses. This is because the equivalent gate resistances of all the measurements are the same. Furthermore, as mentioned in Section I, in paralleled devices, the switching loss imbalance plays a significant role, which is defined as:

$$ \Delta \text{Loss} = \frac{|E_{\text{Loss1}} - E_{\text{Loss2}}|}{E_{\text{Loss1}}} \times 100\% \qquad (3) $$

where E_{Loss1} and E_{Loss2} are the switching loss of DUT1 and DUT2, respectively, and ΔLoss represents the calculated switching loss imbalance as a percentage. This calculation is applied for all the measurements. Fig. 6 shows the switching loss imbalance with different gate resistor configurations when

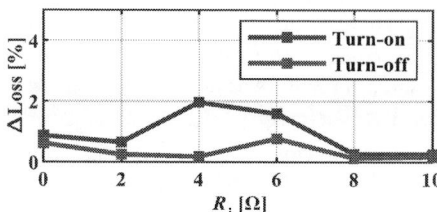

Fig. 6. Extracted switching loss imbalance vs. R_i, with matched devices and symmetrical layout

both the DUTs and stray inductance are matched. It can be seen that the switching loss imbalances fall below 2% for all the cases. Based on this, it can be concluded that no essential correlation between the switching loss imbalance and R_i can be found. Fig. 7a-d shows the switching loss imbalance with different gate resistor configurations when there is an asymmetry on the drain, source, gate, and Kelvin source, respectively. The asymmetry was created with the method mentioned in Fig. 5. It can be seen that the switching loss imbalances don't vary with R_i, which means there is no essential

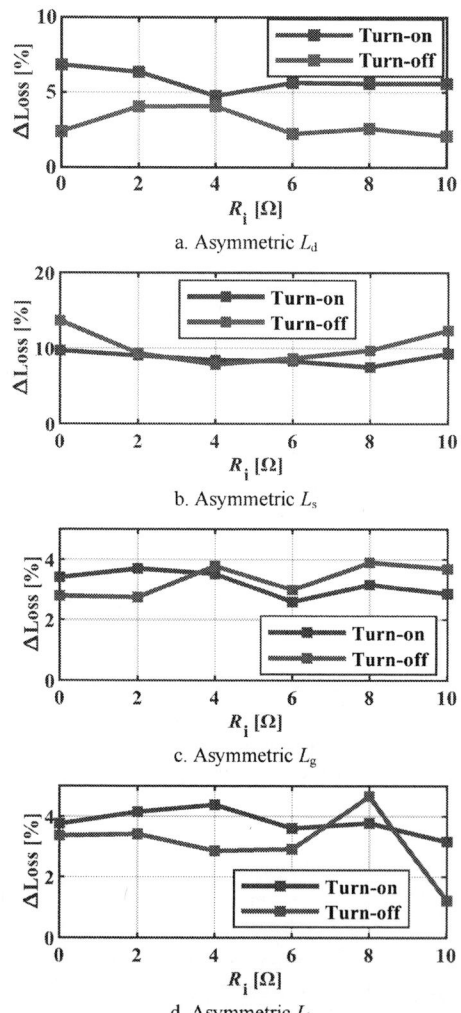

a. Asymmetric L_d

b. Asymmetric L_s

c. Asymmetric L_g

d. Asymmetric L_{ks}

Fig. 7. Extracted switching loss imbalance vs. R_i, with matched devices and asymmetrical layout

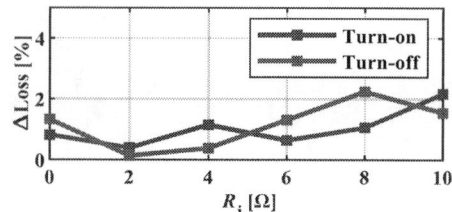

Fig. 8. Extracted switching loss imbalance vs. R_i, with unmatched devices transfer characteristics and symmetrical layout

correlation between the switching loss imbalance caused by mismatch in stray inductances and gate resistor configurations. Comparing Fig. 7a and Fig. 7b, one can see that asymmetry on the source path has a higher impact than asymmetry on the drain path.

Fig. 8 shows the switching loss imbalance with different gate resistor configurations when the stray inductances are symmetric, but there is a difference in the device transfer characteristics. The TCD of the DUTs is around 1.7, which is around 70% of the maximum value in all the combinations. Generally, a rising tendency is visible. However, the value only changes from around 1% to 2%. Therefore, it is concluded as not-correlated.

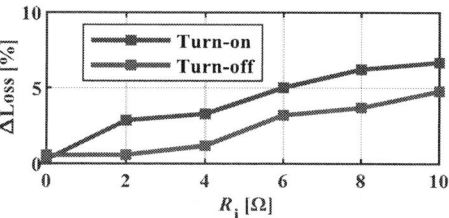

Fig. 9. Extracted switching loss imbalance vs. R_i, with unmatched C_{gs} and symmetrical layout

Fig. 9 shows the switching loss imbalance with different gate resistor configurations when the stray inductances are symmetric, but there is a mismatch on the gate-source capacitance. In this case, it is around 300 pF, which is about 10%. One can see that the switching loss imbalance increases with R_i. An obvious positive correlation can be found, which is also in line with the conclusion of [15], where the 3-pin devices were studied.

Fig. 10a shows the model of driving paralleled devices. Before the Miller plateau, the drain-source voltage is almost constant.

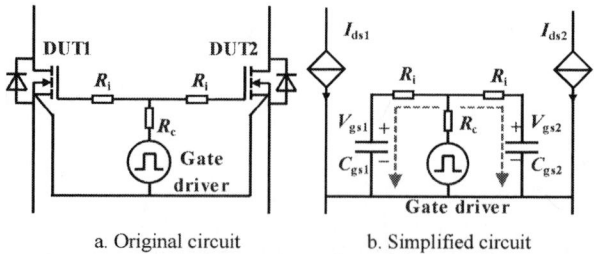

a. Original circuit   b. Simplified circuit

Fig. 10. Circuit model to explain the correlation of switching loss imbalance and gate resistor configurations

979-8-3315-1612-3/25 $31.00 © 2025 IEEE

Therefore, the C_{gd} and C_{ds} can be neglected. The circuit can be simplified into two fully decoupled stages: charging or discharging the gate-source capacitance and mapping V_{gs} to I_{ds} through the transfer characteristics. Note that for this case, the transfer characteristics of the DUTs are matched. Therefore, the dynamic current imbalance or switching loss imbalance originates from the gate-source voltage imbalance. The gate voltages of the devices can be obtained by solving the RC transient shown in Fig. 10b. The circuit can be represented in the s-domain. The gate voltage of DUT1 is:

$$V_{gs1}(s) = V_{dr}(s) \frac{\frac{R_i}{sC_{gs1}} + \frac{1}{s^2 C_{gs1} C_{gs2}}}{R_i^2 + 2R_c R_i + (R_c + R_i)\left(\frac{1}{sC_{gs1}} + \frac{1}{sC_{gs2}}\right) + \frac{1}{s^2 C_{gs1} C_{gs2}}} \quad (4)$$

Similarly, the gate voltage of DUT2 is:

$$V_{gs2}(s) = V_{dr}(s) \frac{\frac{R_i}{sC_{gs2}} + \frac{1}{s^2 C_{gs1} C_{gs2}}}{R_i^2 + 2R_c R_i + (R_c + R_i)\left(\frac{1}{sC_{gs1}} + \frac{1}{sC_{gs2}}\right) + \frac{1}{s^2 C_{gs1} C_{gs2}}} \quad (5)$$

The gate voltage difference can be obtained by subtracting V_{gs1} and V_{gs2}, namely,

$$V_{gs\_diff} = V_{dr}(s) \frac{R_i(C_{gs2} - C_{gs1})s}{(R_i^2 + 2R_c R_i)C_{gs1}C_{gs2}s^2 + (R_c + R_i)(C_{gs1} + C_{gs2})s + 1} \quad (6)$$

For a given gate capacitance difference, the lower R_i is, the closer the gate voltages of the paralleled devices are, which further means the lower switching loss imbalance there is.

Considering on the analysis in this section, it can be concluded that the switching loss caused by the stray inductance mismatch and transfer characteristic mismatch cannot be reduced or mitigated by gate resistor configurations. Only the switching loss caused by C_{gs} mismatch can be reduced by lowering R_i.

C. Oscillations vs. gate resistor configurations

As mentioned in section I, apart from the switching loss imbalance, oscillations are another factor to be considered when driving paralleled SiC MOSFETs. Since in this paper, only two devices are connected in parallel, the DM and CM current can simply be obtained by:

$$I_{CM} = \frac{I_{ds1} + I_{ds2}}{2}, I_{DM} = I_{ds1} - I_{ds2} \quad (7)$$

With (7), more insights can be gained in terms of the oscillation behaviors. Fig. 11a-d show the decomposed waveforms measured with four different gate resistor configurations, where there is an asymmetry on the gate path. For other mismatch conditions, the behaviors are similar and will be skipped. It is worth noting that DM oscillation mainly occurs during turn-off events in our cases. Hence, we will focus on those transients. Comparing Fig. 11a, Fig. 11b, and Fig. 11c, one can see that increasing R_i helps to decrease the DM oscillations, while on the other hand, the CM oscillations are rarely affected by gate resistor configurations. Furthermore, comparing the waveform shown in Fig. 11c and Fig. 11d, it can be observed that if the gate resistors are placed on the individual Kelvin source path, the oscillations in DM current become much

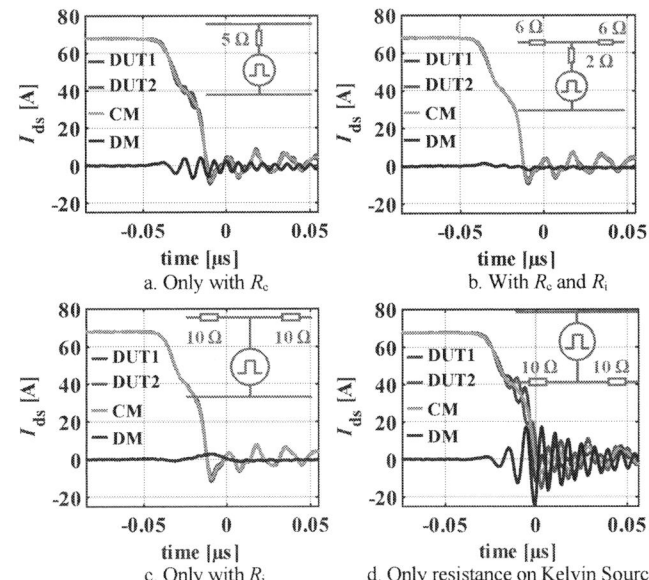

a. Only with R_c b. With R_c and R_i

c. Only with R_i d. Only resistance on Kelvin Source

Fig. 11. Measured and decomposed drain-source waveforms for different gate resistor configurations with asymmetric L_g

worse than that on the gate path, which is also in line with [20].

D. Trade-off when determining the gate resistor configuration

Based on the above-presented measurements and analysis, a trade-off can be concluded in terms of the gate resistor configuration. When driving paralleled devices, the individual gate resistor is critical. A smaller R_i helps to reduce the influence of the C_{gs} mismatch on the dynamic imbalance, as a smaller R_i decreases the RC time constant difference. However, if R_i is too small, the DM oscillations could appear. On the other hand, increasing R_i helps in mitigating the DM oscillation, but makes the switching loss imbalance more sensitive to the mismatch in C_{gs}.

Therefore, a design principle can be proposed accordingly: for a given R_{eq}, the value of R_i required to damp the DM oscillation should be as low as possible, which can be verified and fine-tuned in bench measurements. The remaining gate resistance should be assigned to R_c.

V. SUMMARY AND CONCLUSIONS

This work experimentally investigated the influence of gate resistor configurations on the switching loss imbalance and oscillations in paralleled SiC MOSFETs. First a test setup is designed in order to enable the measurements and evaluations of the switching transients. Afterwards, the devices are sorted in order to ensure that the devices are matched and any mismatch can be easily controlled. From measurements, a trade-off regarding the gate resistor combination is identified. Increasing the individual gate resistance generally means lower DM oscillation but amplifies the switching loss imbalance caused by the mismatch in C_{gs}. In the end, based on this finding, a design principle is proposed. For a given equivalent gate resistance, the lowest value of individual gate resistance required to damp the DM oscillation should be used. The rest of the gate resistance should be assigned to the common gate resistance.

REFERENCES

[1] H. Li, S. Zhao, X. Wang, L. Ding and H. A. Mantooth, "Parallel Connection of Silicon Carbide MOSFETs—Challenges, Mechanism, and Solutions," in IEEE Transactions on Power Electronics, vol. 38, no. 8, pp. 9731-9749, Aug. 2023.

[2] T. Bertelshofer, A. Maerz and M. Bakran, "Derating of Parallel SiC MOSFETs Considering Switching Imbalances," PCIM Europe 2018; International Exhibition and Conference for Power Electronics, Intelligent Motion, Renewable Energy and Energy Management, Nuremberg, Germany, 2018, pp. 1-8.

[3] Wu, R.; Mendy, S.; Agbo, N.; Gonzalez, J.O.; Jahdi, S.; Alatise, O. Performance of Parallel Connected SiC MOSFETs under Short Circuits Conditions. Energies 2021.

[4] J. Ke, Z. Zhao, P. Sun, H. Huang, J. Abuogo and X. Cui, "Chips Classification for Suppressing Transient Current Imbalance of Parallel-Connected Silicon Carbide MOSFETs," in IEEE Transactions on Power Electronics, vol. 35, no. 4, pp. 3963-3972, April 2020.

[5] Abuogo, James Opondo, and Zhibin Zhao. "Machine learning approach for sorting SiC MOSFET devices for paralleling." Journal of Power Electronics 20 (2020): 329-340.

[6] C. Zhao, L. Wang, F. Zhang and F. Yang, "A Method to Balance Dynamic Current of Paralleled SiC MOSFETs With Kelvin Connection Based on Response Surface Model and Nonlinear Optimization," in IEEE Transactions on Power Electronics, vol. 36, no. 2, pp. 2068-2079, Feb. 2021.

[7] J. Qu, Q. Zhang, X. Yuan and S. Cui, "Design of a Paralleled SiC MOSFET Half-Bridge Unit With Distributed Arrangement of DC Capacitors," in IEEE Transactions on Power Electronics, vol. 35, no. 10, pp. 10879-10891, Oct. 2020.

[8] M. Zhang, H. Li, Z. Yang, S. Zhao, X. Wang and L. Ding, "Layout Design Principle for Optimization of Transient Current Distribution among Paralleled SiC MOSFETs in Multichip Modules," 2023 25th European Conference on Power Electronics and Applications (EPE'23 ECCE Europe), Aalborg, Denmark, 2023, pp. 1-6.

[9] Z. Zeng, X. Zhang and Z. Zhang, "Imbalance Current Analysis and Its Suppression Methodology for Parallel SiC MOSFETs with Aid of a Differential Mode Choke," in IEEE Transactions on Industrial Electronics, vol. 67, no. 2, pp. 1508-1519, Feb. 2020.

[10] J. Liu and Z. Zheng, "Switching Current Imbalance Mitigation for Paralleled SiC MOSFETs Using Common-mode Choke in Gate Loop," 2020 IEEE Energy Conversion Congress and Exposition (ECCE), Detroit, MI, USA, 2020, pp. 705-710.

[11] Y. He, X. Wang, S. Shao and J. Zhang, "Active Gate Driver for Dynamic Current Balancing of Parallel-Connected SiC MOSFETs," in IEEE Transactions on Power Electronics, vol. 38, no. 5, pp. 6116-6127, May 2023.

[12] C. Luedecke, F. Krichel, M. Laumen and R. W. De Doncker, "Balancing the Switching Losses of Paralleled SiC MOSFETs Using an Intelligent Gate Driver," PCIM Asia 2020; International Exhibition and Conference for Power Electronics, Intelligent Motion, Renewable Energy and Energy Management, Shanghai, China, 2020, pp. 1-7.

[13] Zhang, Y., Song, Q., Tang, X. and Zhang, Y., "Gate driver for parallel connection SiC MOSFETs with over-current protection and dynamic current balancing scheme." Journal of Power Electronics 20 (2020): 319-328.

[14] S. La Mantia, L. Abbatelli, C. Brusca, M. Melito and M. Nania, "Design Rules for Paralleling of Silicon Carbide Power MOSFETs," PCIM Europe 2017; International Exhibition and Conference for Power Electronics, Intelligent Motion, Renewable Energy and Energy Management, Nuremberg, Germany, 2017, pp. 1-6.

[15] J. Ke, H. Huang, P. Sun, J. Abuogo, Z. Zhao and X. Cui, "Influence of Parasitic Capacitances on Transient Current Distribution of Paralleled SiC MOSFETs," 2018 1st Workshop on Wide Bandgap Power Devices and Applications in Asia (WiPDA Asia), Xi'an, China, 2018, pp. 1-6.

[16] Y. Shen, X. Dong, T. Schuetz and R. Roesner, "Stability Modeling for Multichip SiC MOSFET Power Modules," CIPS 2022; 12th International Conference on Integrated Power Electronics Systems, Berlin, Germany, 2022, pp. 1-6.

[17] Z. Dong, R. Ren, W. Zhang, F. F. Wang and L. M. Tolbert, "Instability Issue of Paralleled Dies in an SiC Power Module in Solid-State Circuit Breaker Applications," in IEEE Transactions on Power Electronics, vol. 36, no. 10, pp. 11763-11773, Oct. 2021.

[18] T. Liu, R. Ning, T. T. Y. Wong and Z. J. Shen, "Modeling and Analysis of SiC MOSFET Switching Oscillations," in IEEE Journal of Emerging and Selected Topics in Power Electronics, vol. 4, no. 3, pp. 747-756, Sept. 2016.

[19] F. Sawallich and H. -G. Eckel, "Inter-chip Oscillation of paralleled SiC MOSFETs," PCIM Europe 2023; International Exhibition and Conference for Power Electronics, Intelligent Motion, Renewable Energy and Energy Management, Nuremberg, Germany, 2023, pp. 1-7.

[20] F. Sawallich and H. -G. Eckel, "Mitigating Inter-Chip Oscillation of paralleled SiC MOSFETs," 2023 25th European Conference on Power Electronics and Applications (EPE'23 ECCE Europe), Aalborg, Denmark, 2023, pp. 1-11.

[21] S. Takeda, E. Miyake, H. Kono, T. Ohashi, T. Iguchi and K. Kodani, "Novel Approach to Mitigate Parasitic Oscillation of Power Modules with Parallel Connected SiC-MOSFETs," 2024 36th International Symposium on Power Semiconductor Devices and ICs (ISPSD), Bremen, Germany, 2024, pp. 514-517.

[22] A. Piccioni and N. Seltner, "Impact of Parameter Spread in Parallel-Operated SiC MOSFETs for Hard-Switching Conversion," PCIM Europe 2024; International Exhibition and Conference for Power Electronics, Intelligent Motion, Renewable Energy and Energy Management, Nürnberg, Germany, 2024, pp. 2675-2680.

[23] Infineon Technologies AG, "CoolSiC™ 1200 V SiC Trench MOSFET," IMZA120R030M1H datasheet, May 2023.

Exploring the Potential of FPGA in High-Frequency Switching DC-DC Boost Converters Using Model Predictive Control

Qingcheng Sui
Dept. of Electrical Engineering
KU Leuven - EnergyVille
Diepenbeek - Genk, Belgium
qingcheng.sui@kuleuven.be

Bangli Du
Dept. of Electrical Engineering
KU Leuven - EnergyVille
Diepenbeek - Genk, Belgium
bangli.du@kuleuven.be

Yu Zuo
Dept. of Electrical Engineering
KU Leuven - EnergyVille
Diepenbeek - Genk, Belgium
yu.zuo@kuleuven.be

Wilmar Martinez
Dept. of Electrical Engineering
KU Leuven - EnergyVille
Diepenbeek - Genk, Belgium
wilmar.martinez@kuleuven.be

Abstract—**Model Predictive Control (MPC) has demonstrated significant potential in the control of power electronics; however, the trade-off between the online computational demands and the storage requirements for offline explicit control laws presents a persistent challenge for industrial controller implementation. This work introduces an innovative FPGA-based neural network controller tailored for high-switching frequency DC/DC Boost converters. Multi-Layer Perceptron (MLP) is used to approximate parametrized Explicit MPC controller capable of adapting to fluctuations in the systems variables and parameters. The high-speed implementation of the MLP using High-Level Synthethis (HLS) on the FPGA enables operation at higher switching frequencies and allows for higher power density in power converters design with wide bandgap (WBG) devices such as GaN. This method offers substantial improvements over conventional techniques by delivering low-latency computation. Experimental validation on a 500W, 5MHz GaN-based Boost converter prototype demonstrates that the MLP-based controller closely replicates the performance of the original Explicit MPC (EMPC) controller, achieving superior transient response.**

Index Terms—**FPGA, HLS, MPC, MLP, Boost Converter**

I. INTRODUCTION

Model Predictive Control (MPC) has become a powerful tool in power electronics control due to its ability to handle system constraints and provide fast dynamic responses[1]. By predicting system behavior over a finite horizon and optimizing control actions accordingly, MPC explicitly accounts for constraints, making it more suitable for practical applications than traditional controllers like Linear Quadratic Regulator (LQR) or Linear Quadratic Gaussian (LQG). However, the real-time implementation of online MPC is often hindered by the significant computational and memory demands required to solve quadratic programming (QP) problems, especially in industrial settings that require high-speed solutions[2], [3].

Explicit MPC (EMPC) mitigates some of these computational challenges by pre-storing control laws for various input states, eliminating the need for online computation[4]. While EMPC reduces computation time, it suffers from substantial memory requirements, particularly for systems with long prediction horizons, limiting its implementation on microcontroller units (MCUs)[4][5]. Approximation methods that reduce the number of stored regions can decrease memory usage but may compromise control accuracy.

Recent advancements in machine learning (ML) offer new avenues for approximating MPC control laws, specifically through the use of multilayer perceptron (MLP) neural networks. Learning-based MPC (LB-MPC) leverages data-driven models to approximate optimal control actions, significantly reducing online computation time[6]. MLPs, known for their universal function approximation capabilities, can handle the complexity of MPC control laws without incurring substantial memory overhead on conventional DSP[5].

Field-Programmable Gate Arrays (FPGAs) present an ideal platform for implementing these MLP-approximated MPC controllers. FPGAs offer parallel processing capabilities and hardware-level configurability, enabling them to execute complex algorithms with low latency and high data-throughput[7]. Traditionally, the adoption of FPGAs in industrial control has been limited due to the complexity of hardware design and long time-to-market, with Digital Signal Processors (DSPs) and MCUs being preferred for their maturity and ease of programming in C/C++[8]. However, the introduction of High-Level Synthesis (HLS) tools has lowered the barrier to FPGA programming, allowing developers to design hardware using high-level languages and significantly reducing development time[9], [10].

By combining MLP-approximated MPC with FPGA acceleration, we can implement more complex control algorithms with less programming effort and achieve shorter computation times. This approach not only overcomes the computational limitations of traditional MPC implementations but also leverages the parallelism of FPGAs to attain higher control loop rates. This is particularly crucial for modern power converters employing wide bandgap (WBG) devices like silicon carbide (SiC) and gallium nitride (GaN), which operate at higher switching frequencies[10].

In this research, we propose a data-driven method to approximate a parameterized MPC controller using an MLP that adapts to changes in both plant state variables and parameters. The MLP is efficiently implemented on an FPGA using HLS tools, supporting high-frequency switching and enhancing power density in power converter designs. Our approach enables the execution of more complex algorithms with less programming effort and shorter computation times, addressing the challenges of real-time MPC implementation in industrial applications.

The main contributions of this research are as follows:

- **Development of a Data-Driven Approach**: We develop a data-driven method to approximate a parameterized MPC controller using an MLP, capable of adapting to variations in plant state variables and parameters.
- **Efficient FPGA Implementation**: We achieve fast implementation of MLP computations on an FPGA using HLS tools, supporting high-frequency switching operations and enhancing power density in power converter designs.

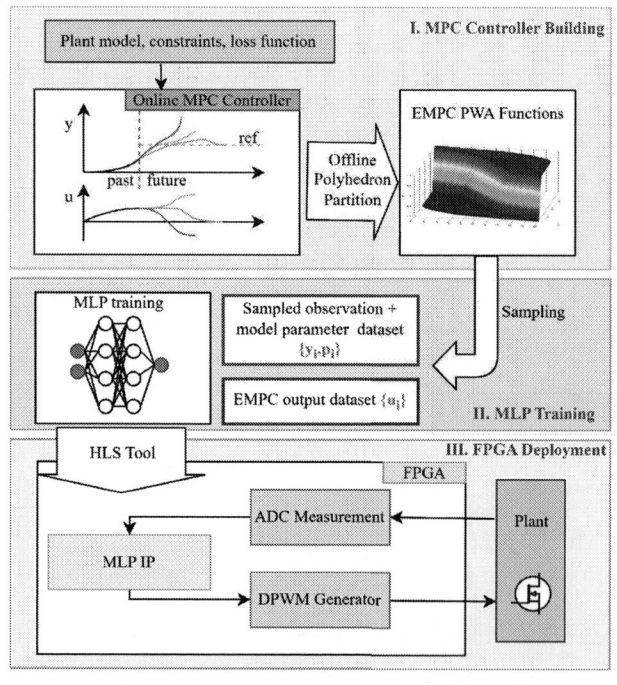

Fig. 1: Overview of proposed approach

II. METHODOLOGY

The design process of our proposed FPGA-based MPC controller for high-switching frequency Boost converter consists of following steps. First, the Boost converter model is built and the MPC problem is formulated. Second, the paired state-control dataset are constructed by sampling input state variables and output control variables of EMPC controller, and quantization-aware training of MLP. Third, MLP are packaged as IP and deployed in the FPGA design with other necessary modules connected.

A. EMPC controller formulation for Boost Converter

Fig.X shows the DC/DC Boost Converter. Under assumption that sampling frequency is sufficiently high, using the Euler approximation, the discretized form of the state-space average equations can be derived as:

$$
\begin{aligned}
V_o(k+1) &= V_o(k) + \frac{1-d}{C} i_L(k) T_S - \frac{1}{RC} V_o(k) T_S \\
i_L(k+1) &= i_L(k) - \frac{1-d}{L} V_o(k) T_S + \frac{1}{L} V_g T_S
\end{aligned}
\tag{1}
$$

where i_L is the current in the inductor of the Boost converter, V_o is the output voltage on the load, V_g is the input source voltage, d denotes the duty cycle, T_s represents the sampling interval, and R, C, L are the load resistance, output capacitance, and inductance, respectively.

We derive the linearized model around the operating point $x_{\text{ref}} = (V_{o,\text{ref}}, i_{L,\text{ref}})$, $u_{\text{ref}} = d_{\text{ref}}$.

$$
\Delta x_{k+1} = A' \Delta x_k + B' \Delta u_k
\tag{2}
$$

$$
A' = \begin{bmatrix} 1 - \frac{T_S}{RC} & \frac{(1-d_0)T_S}{C} \\ -\frac{(1-d_0)T_S}{L} & 1 \end{bmatrix}, \quad B' = \begin{bmatrix} -\frac{i_{L0}T_S}{C} \\ \frac{V_{o0}T_S}{L} \end{bmatrix}
\tag{3}
$$

The subscript 0 denotes constants under a specific equilibrium and reference operating condition for the MPC tracking problem. The reference values of the output voltage are given as $V_{o,\text{ref}}$, the reference values of the other key variables can be derived as:

$$
d_{\text{ref}} = 1 - \frac{V_{o,\text{ref}}}{V_g}, \quad I_{L,\text{ref}} = \frac{V_{o,\text{ref}}}{R_{\text{load}}} \left(1 - \frac{V_{o,\text{ref}}}{V_g} \right)
$$

.

The discretized linear state-space form of the Boost converter can be derived and applied to the MPC problem formulation. The MPC problem can be established as follows:

$$
u_{\text{mpc}}(\Delta x'_k) = u_{\text{ref}} + [1, 0, \ldots, 0] \arg \min_{\Delta u'_i, \ldots, \Delta u'_{k+N-1}}
$$

$$
\sum_{i=k}^{k+N} \left(\Delta x_i'^{T} Q \Delta x'_i + \Delta u_i'^{T} R \Delta u'_i \right)
\tag{4}
$$

subject to:

$$
\Delta x'_{k+1} = A' \Delta x'_k + B' \Delta u'_k, \ldots
$$

$$
\Delta x'_{k,1} + V_{o,\text{ref}} \leq V_{o,\text{max}}, \quad \Delta x'_{k+1,1} + V_{o,\text{ref}} \leq V_{o,\text{max}}, \ldots
$$

$$0 \leq u_k \leq 1, \quad 0 \leq u_{k+1} \leq 1, \ldots$$

where Q and R denote the penalty and control regularization coefficient matrices in the cost function, and k denotes the time step. After formulating the MPC problem, we use multiparametric optimization tools, such as MPT3 [11], to obtain an explicit solution.

By identifying the active constraints and collecting primal and dual optimality conditions, the algorithm of EMPC identifies different regions or polyhedrons in state space, which may lead to piecewise affine (PWA) local optimal solutions. The algorithm repeatedly samples states within the predefined state-space limits until whole coverage of the predefined state-space by the PWA control function. Multiparametric optimization tools such as MPT3 [11] can be utilized to obtain an explicit solution. The mapping of two input variables (that is, V_o, I_L) to the control output of the MLP is illustrated in Fig. 2.

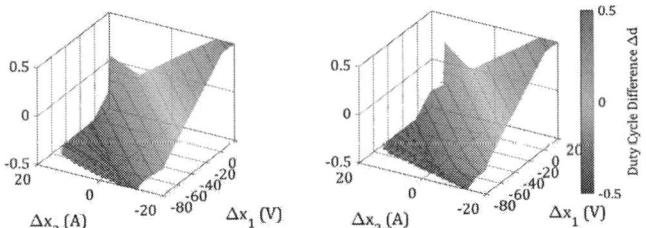

Fig. 2: EMPC state/control mapping visualization under different R_{load} (5Ω and 10Ω)

B. Quantization-aware MLP-based Controller Training

By considering ML algorithms together with heterogeneous computing, we can greatly benefit from developments outside the field of power electronics, while maintaining the excellent control characteristics of advanced control algorithms such as MPC. Compared with online MPC controllers and explicit model predictive control (EMPC), MLP-based MPC controllers can operate in real-time on specialized hardware (e.g., FPGA), thanks to the parallelized and quantized neural network inference computation.

To fully exploit the capabilities of FPGA hardware and achieve efficient real-time control, it is essential to consider quantization during the training of the MLP-based controller. Quantization reduces the numerical precision of the neural network's weights and activations, allowing for faster computations on hardware with limited resources. However, naive quantization after training can lead to significant performance degradation. Therefore, we employ quantization-aware training to mitigate this issue.

In our approach, we focus on accelerating machine learning inference by performing quantization-aware training of the MLP for real-time power electronics controllers. This involves incorporating the quantization effects into the training process so that the neural network learns to compensate for the reduced precision. Specifically, we simulate the finite precision of the FPGA's arithmetic units during training by quantizing the weights and activations to fixed-point representations that match the target hardware.

To prepare the MLP for quantization-aware training, we first generate a dataset required for training and validation. Within the permitted state-space range, we uniformly sample the system states $x^{(i)} = (v_o, i_L, i_o)$ as input for the neural network model. The corresponding output of EMPC $y^{(i)} = (u_i)$ is also collected to form a paired data set $D = \{(x^{(i)}, y^{(i)}) : 1 \ldots N\}$. The dataset is then divided into three parts for training, validation, and testing separately. A MLP-type neural network model, consisting of more than three fully connected feedforward layers, is built and trained on the dataset. ReLU activation functions are used for the sake of simplicity in the FPGA implementation.

C. FPGA Implementation of MLP-based controller

In contrast to conventional DSPs used for power electronics control, FPGAs offer the advantage of custom parallel computation when running, which can greatly reduce computational latency.

However, due to the limited computational resources available for consideration of cost-effectiveness in industrial control, a trade-off between latency and resource utilization is required. In the context of neural network implementations on FPGAs, the most critical resources are often digital signal processing (DSP) slices. In deploying the quantized MLP on the FPGA, we map the fixed-point representations directly onto the hardware's DSP slices and lookup tables (LUTs), optimizing for parallel execution by using tools like HLS. HLS tool (e.g. Vitis HLS) can help to easily find an RTL design that realizes the desired behavior in four simple steps: C-simulation, synthesis, C/RTL co-simulation, IP exporting[8]. The IP block can then be integrated to a block design together with other modules like ADC reading, DPWM control signal generation, DMA data storage, industrial communications, etc.

III. EXPERIMENTS

To demonstrate the effectiveness of the proposed MLP-based MPC controller, a 500W, 5MHz GaN-based Boost converter simulation is built. The system operates with an output voltage of 48V, input voltage of 24V. The load resistance varies between 5Ω and 10Ω. The inductor 4μH, and the capacitor is 20μF. The prediction horizon is 50, the output voltage limit is 100V, and the inductor current is limited to 60A.

The MLP model presented here consists of four fully connected layers with decreasing dimensions, specifically configured as follows: an input layer of 3 units, two hidden layers each containing 10 units, and two subsequent layers with 2 and 1 unit, respectively. Activation is applied using the ReLU function for the initial three layers, while the final output layer employs a clamping function to restrict output values within the [0, 1] range. PI controller is tuned optimally using the same lost function as the MPC controller under the given transient testing cases.

979-8-3315-1612-3/25 $31.00 © 2025 IEEE

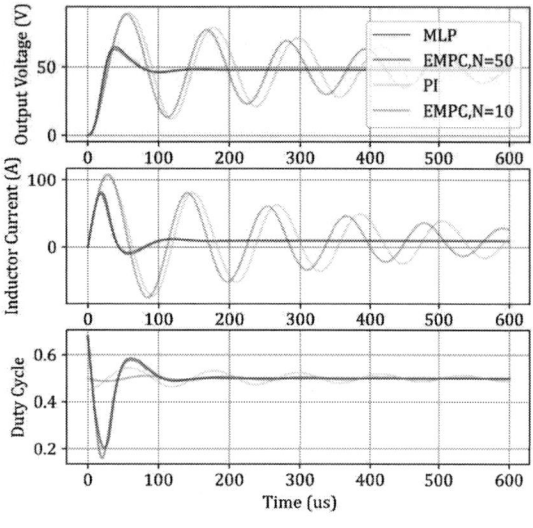

(a) $V_{o,init} = 0V$, $I_{L,init} = 0A$

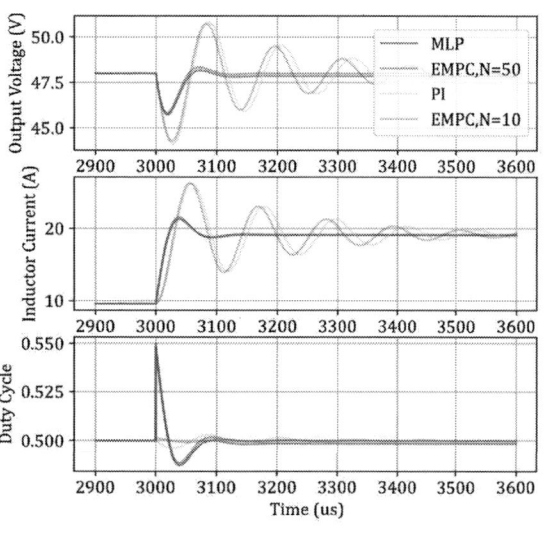

(b) $V_{o,init} = 48V$, $I_{L,init} = 9.6A$

Fig. 3: Simulation results of boost DC-DC Boost converters using PI, EMPC controller, proposed MLP-based MPC controller under cold-start and sudden R_{load} changes

Fig. 3 illustrates the system response under rated converter working conditions with various controllers. The results demonstrate that the MLP-based controller closely approximates the performance of the original EMPC controller, with only minor deviations, while both controllers effectively reduce output voltage overshoot. As load resistance varies, which typically necessitates a linear model parameter update in the online MPC scheme for controller adaptability, the offline EMPC requires an additional transformation to an explicit form to accommodate these changes. Without this adjustment,

model inaccuracies can lead to significant deviations from the reference values. In contrast, because the MLP-based controller is directly trained offline with load current as an input parameter, it adapts seamlessly to changes without requiring controller updates; the load current serves as an additional input feature for the MLP model, ensuring robustness across continuously varying operating conditions.

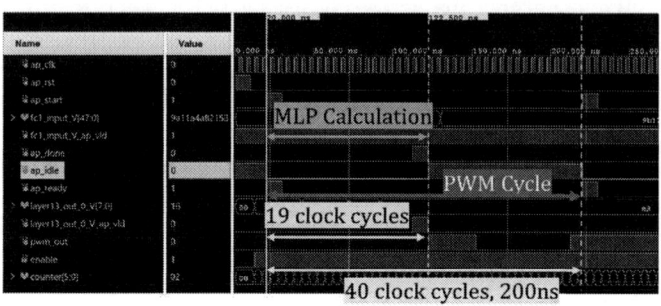

Fig. 4: Waveform Diagram of our RTL implementation on Zynq7020

TABLE I: FPGA Logic and Memory Resource Requirements for MLP IP

FPGA Resources	Req.	Available	Used %
DSP48E	26	220	11.8%
FF	5801	106K	5.5%
BRAM_18K	0	280	0%
LUT	6747	53K	12.7%

Conventional DSPs typically use a sequential execution style, where tasks like signal sensing, conditioning, control algorithm computation, and ePWM register updates must be completed within a tight control time window, often resulting in limited determinism. In contrast, the prototype MLP-based controller was implemented on a Zynq-7020 FPGA (xc7z020clg400-2) with a 200 MHz clock frequency. As shown in Fig. 4, the digital PWM cycle requires 40 clock cycles, while the MLP computation takes only 19 clock cycles, utilizing DSP48E, Flip-Flops (FFs), and LUTs. As shown in Table I, this efficient MLP implementation leaves ample FPGA resources available for additional tasks such as health monitoring and industrial communication. A comparison of various MPC controllers on different platforms (FPGA, DSP, Lab-box) is provided in Table II. Our model demonstrates superior performance with minimal computation delay and an extended prediction horizon.

IV. CONCLUSION

This work presents a novel FPGA-based neural network controller tailored for high-switching frequency DC/DC Boost converters. By leveraging a dataset to approximate a parameterized Model Predictive Control (MPC) controller, the proposed approach effectively adapts to variations in both state variables and model parameters. The FPGA implementation is resource-efficient, enabling high loop rates and low latency,

979-8-3315-1612-3/25 $31.00 © 2025 IEEE

TABLE II: Comparison of Different Approaches on NN-Enabled MPC Computation Time

Approach	Hardware	Prediction Horizon	Computation Delay
Ours	Zynq7020	50	$19 \times 5ns$
MPC-NPI [12]	dSpace DS1202	1	N/A
DualReLU [5]	DSP28335	30	$4.3\mu s$
MPC-RL [13]	Altera Cyclone IV-E	15	$3 \times 250ns$

which are crucial for exploiting the potential of wide bandgap (WBG) devices like GaN in power converters. Experimental validation on a 500W, 5 MHz GaN-based Boost converter prototype demonstrates that the proposed MLP-based controller not only closely mimics the performance of an Explicit MPC (EMPC) controller but also provides superior transient response and steady-state accuracy compared to traditional PI controllers. This solution paves the way for more efficient and robust control strategies in high-performance power electronics systems.

This proof-of-concept work demonstrates the feasibility of integrating machine learning (ML) and heterogeneous computing into power electronics control for high-frequency switching power converters. However, there are several avenues for future research and industrial development to fully mature this technology:

- Enhancing Robustness: A more robust control scheme should be developed to handle uncertainties in parameter estimation, especially for high-switching frequency applications where precise control relies on accurate and robust model parameter estimation.
- Exploration of Model-Free Techniques: The use of model-free RL controllers should be investigated as a potential alternative, offering the ability to adapt dynamically without requiring an accurate system model.
- Hardware Co-Design Optimization: Further optimization of the hardware-software co-design could enable even faster computation and open up opportunities for integrating additional functionalities such as fault diagnosis and predictive maintenance.

REFERENCES

[1] D. E. Quevedo, R. P. Aguilera, and T. Geyer, "Model predictive control for power electronics applications," *Handbook of Model Predictive Control*, pp. 551–580, 2019.

[2] B. Stellato, T. Geyer, and P. J. Goulart, "High-Speed Finite Control Set Model Predictive Control for Power Electronics," *IEEE Transactions on Power Electronics*, vol. 32, no. 5, pp. 4007–4020, May 2017. DOI: 10.1109/TPEL.2016.2584678. (visited on 01/31/2024).

[3] J. Chen, Y. Chen, L. Tong, L. Peng, and Y. Kang, "A Backpropagation Neural Network-Based Explicit Model Predictive Control for DC–DC Converters With High Switching Frequency," *IEEE Journal of Emerging and Selected Topics in Power Electronics*, vol. 8, no. 3, 2020.

[4] F. Borrelli, A. Bemporad, and M. Morari, *Predictive control for linear and hybrid systems*. 2017. (visited on 11/13/2024).

[5] Y. Xiang, H. S.-H. Chung, and H. Lin, "On the Use of DualReLU ANN for Approximating Explicit Model Predictive Control for Buck Converters," in *2024 IEEE Applied Power Electronics Conference and Exposition (APEC)*, Feb. 2024, pp. 2822–2827. DOI: 10.1109/APEC48139.2024.10509403. (visited on 07/19/2024).

[6] J. M. Manzano, D. Limon, D. Muñoz de la Peña, and J.-P. Calliess, "Robust learning-based MPC for nonlinear constrained systems," *Automatica*, vol. 117, p. 108948, Jul. 2020. DOI: 10.1016/j.automatica.2020.108948. (visited on 11/13/2024).

[7] J. Duarte, P. Harris, S. Hauck, *et al.*, "FPGA-Accelerated Machine Learning Inference as a Service for Particle Physics Computing," *Computing and Software for Big Science*, vol. 3, no. 1, p. 13, Oct. 2019. DOI: 10.1007/s41781-019-0027-2. (visited on 10/23/2024).

[8] S. Sinha and T. Srikanthan, "High-Level Synthesis: Boosting Designer Productivity and Reducing Time to Market," *IEEE Potentials*, vol. 34, no. 4, pp. 31–35, Jul. 2015. DOI: 10.1109/MPOT.2013.2292957. (visited on 10/31/2024).

[9] M. Levental, A. Khan, R. Chard, K. Yoshii, K. Chard, and I. Foster, *OpenHLS: High-Level Synthesis for Low-Latency Deep Neural Networks for Experimental Science*, Mar. 2023. (visited on 07/03/2024).

[10] J. Liu, T. Wei, N. Chen, W. Liu, J. Wu, and P. Xiao, "Data-Mining-Based Hardware-Efficient Neural Network Controller for DC–DC Switching Converters," *IEEE Journal of Emerging and Selected Topics in Power Electronics*, vol. 11, no. 4, pp. 4222–4232, Aug. 2023. DOI: 10.1109/JESTPE.2022.3192354. (visited on 10/29/2024).

[11] M. Herceg, M. Kvasnica, C. N. Jones, and M. Morari, "Multi-parametric toolbox 3.0," in *2013 European control conference (ECC)*, 2013, pp. 502–510. (visited on 08/09/2024).

[12] Y. Li, S. Sahoo, T. Dragičević, Y. Zhang, and F. Blaabjerg, "Stability-Oriented Design of Model Predictive Control for DC/DC Boost Converter," *IEEE Transactions on Industrial Electronics*, vol. 71, no. 1, pp. 922–932, Jan. 2024. DOI: 10.1109/TIE.2023.3247785. (visited on 08/09/2024).

[13] D. Alfred, D. Czarkowski, and J. Teng, *Reinforcement Learning-Based Control of a Power Electronic Converter*, Jan. 2024. DOI: 10.3390/math12050671. (visited on 05/27/2024).

A 7 bit 5 A 6.7 GHz Gate-Shaping Digital Gate Driver with Burst-Sampling ADC for Iterative Switching Optimization of SiC Power MOSFETs

Tobias Zekorn, Kenny Vohl, Erik Wehr, Leon Weihs, Michael Hanhart, Ralf Wunderlich, and Stefan Heinen

Integrated Analog Circuits and RF Systems
RWTH Aachen University
Kopernikusstr. 16, D-52074 Aachen, Germany
E-mail: tobias.zekorn@ias.rwth-aachen.de

Abstract— **Wide bandgap semiconductors provide the potential to significantly increase the efficiency of today's power electronics while simultaneously their superior characteristics regarding dV/dt and dI/dt pose a challenge to EMI and EMC. Through shaping the waveform driving the HV power semiconductor the challenging figures of EMI and EMC can be significantly improved while achieving a high efficiency. This work presents a gate-shaping digital gate driver for reducing EMI and optimizing the switching behavior of HV SiC power-MOSFETs. It includes a 7 bit 5 A 6.7 GHz driver unit with a tunable current resolution of up to 40 mA and larger steps of 160 mA and a time resolution of up to 150 ps. For iteratively closing the loop driving the SiC power-MOSFET it contains a burst-sampling ADC with a sampling rate of up to 1.66 MS s^{-1}. Further, it includes a separate dead-time control circuit, an 8-phase 832 MHz clock generation and a 0.4 W PMU supplying the digital gate driver with the appropriate clock frequency phases and voltage levels of 5 V and 1.8 V.**

Index Terms—**Digital gate driver, Gate-Shaping, Burst-Sampling, Iterative Switching Optimization, SiC, power MOSFET**

I. INTRODUCTION

Today's power electronics are essentially contributing to improving the power efficiency of industrial electronics, consumer electronics and electric vehicles. Wide bandgap semiconductors including Silicon-Carbide (SiC) power-MOSFETs provide the potential to further increase their efficiency as they provide a number of superior characteristics compared to their silicon counterparts. These superior characteristics include a higher breakdown voltage, higher operating frequency, higher drift velocity, higher operating temperature and higher thermal conductivity. Thus, they are increasingly being used in power electronic converters in various applications like power supplies, inverters in renewable energy technologies and electric vehicles (EV). A faster and therefore more efficient switching of power-MOSFETs employing their superior characteristics regarding dV/dt and dI/dt simultaneaously poses a challenge to preserve electromagnetic interference (EMI) and electromagnetic compatibility (EMC) for power electronic products especially in unshielded environments [1] [2]. Shaping the switching waveform driving the SiC power-MOSFET can solve this challenge while still employing the superior char-

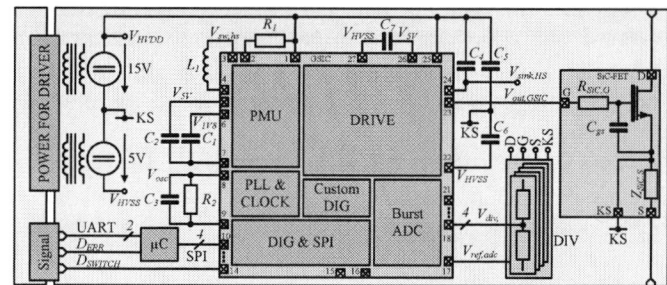

Fig. 1: System overview of the digital gate driver IC.

acteristics of SiC [2] [3]. The capacitive behavior of the gate of the power-MOSFET and thus, shaping the gate voltage by modifying the charging current while switching is an intuitive and promising approach [1] [4].

This work presents a highly integrated gate driver for Gate-Shaping of high-voltage (HV) Silicon-Carbide (SiC)-Power MOSFETs. A system overview of the digital gate driver with its periphery is shown in Fig. 1.

For driving the SiC power-MOSFET the digital gate driver offers 34 current sources for NMOS and PMOS each with a maximum total output current of up to 5 A at a resolution of 40 mA and 150 ps from a maximum supply voltage of up to 23 V referred to Kelvin Source (KS). The supply voltage of the digital gate can be chosen in the design with the digital gate driver being operational in a range of 15 V to 23 V referred to Kelvin Source. For iteratively controlling the charging profile of the SiC power-MOSFET, the digital gate driver includes a burst-mode analog-to-digital converter (ADC) for sampling the switching voltages of the power-MOSFET. This allows sampling of the drain and gate voltage as well as drain current with up to 30 samples and an adjustable sampling time. A new charging and discharging profile needs to be calculated externally as shown in [11] and can be written to a lookup table (LUT) in the digital part of the digital gate driver via a Serial Peripheral Interface (SPI).

Further, a power management unit (PMU) is included in the digital gate driver supplying the internal supply rails of 5 V for analog processing and driving power-MOSFET and 1.8 V

979-8-3315-1612-3/25 $31.00 © 2025 IEEE

for digital processing. A multiphase clock generation circuit supplies the digital gate driver with the required frequencies of 104 MHz for digital processing, and an 832 MHz 8 phase clock for processing the charging profile in the digital driver when switching. The digital dead-time control circuit allows to employ soft switching of a power-MOSFET by determining the transition of the switching edge.

II. POWER MANAGEMENT UNIT

Since there is only a limited budget for power, die area and number of pins available at the package of the digital gate driver, the power management needs to be small and efficient while supplying a load current of at least 50 mA on both voltage rails $V_{5V} = 5$ V and $V_{1V8} = 1.8$ V. With the supply voltage V_{HVDD} being in the range of 15 V to 23 V and generally, single-inductor multiple-output (SIMO) architectures being employed in various applications for supplying multiple voltages and currents using only one inductor [5] [6] [7], a SIMO buck converter is supplying the voltage rails of the digital gate driver. A system level block diagram of the PMU is shown in Fig. 2.

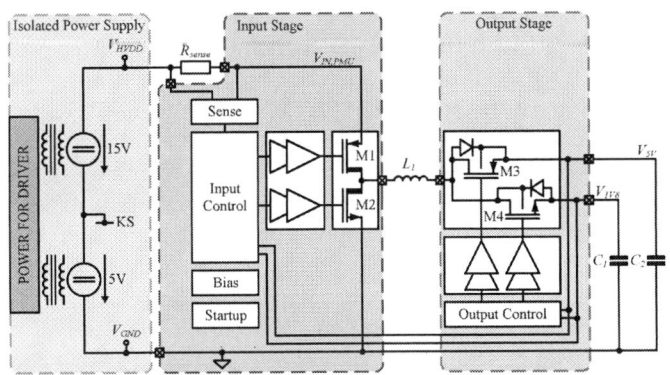

Fig. 2: System overview of the power management unit of the GSIC.

The control of the SIMO buck converter is split into the input control and the output control with the input control switching the power-MOSFETs M1 and M2, which are charging and discharging the inductor L_1 to provide the total amount of power needed for supplying the voltage rails V_{5V} and V_{1V8}. The total amount of power needed at the output of the PMU is represented as a summarized error voltage which sets the target for the input current of the buck converter. The input current of the buck converter is sensed at the high-side (HS) sense resistor $R_{sense} = 100$ mΩ and compared to the target current. This control scheme determines the duty cycle of the buck converter as shown in [10]. As the input current is calculated accordingly to the required power of both output voltage rails, the output control determines the appropriate power share to the output voltage rails V_{5V} and V_{1V8}. Due to the larger voltage difference from V_{HVDD} to V_{1V8} the voltage rail of V_{1V8} is charged first with the inductor current and second the voltage rail of V_{5V} is charged with the remaining current in

the inductor. Both voltage rails have a large external capacitor C_1 and C_2 for storing the charge.

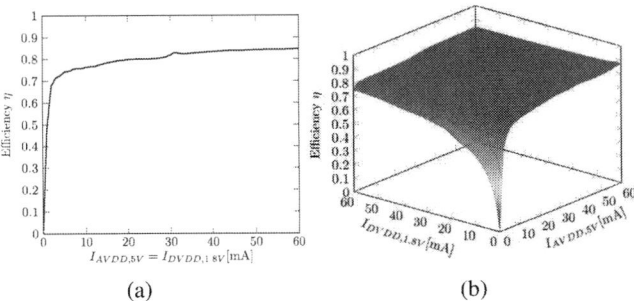

(a)   (b)

Fig. 3: Measured efficiency of the SIMO buck converter as a function of the output currents I_{5V} and $I_{1.8V}$ in a 2-dimensional figure (a) and a 3-dimensional figure (b).

The measured efficiency of the SIMO buck converter supplying the two voltage rails of I_{5V} and $I_{1.8V}$ is shown in a three-dimensional plot over the load currents of both outputs in Fig. 3 (b). With a maximum current of 60 mA for each of the two output voltage rails 5 V and 1.8 V a total output power of 0.4 W is generated with one external inductor of 47 μH at a peak efficiency of 87 %. Further, a two-dimensional plot of the efficiency for the same load current on both outputs $I_{AVDD,5V} = I_{DVDD,1.8V}$ is shown in Fig. 3 (a).

III. HIGH-FREQUENCY CLOCK GENERATION

To achieve a time resolution of 150 ps in the used 180 nm BCD process, the high-frequency clock generator implements a multi-phase clock architecture. Choosing eight clock phases and a clock frequency of 832 MHz represents an adequate trade-off between the number of clock phases and feasible clock speed in the process technology at hand.

The generation of eight clock phases is performed by a voltage-controlled ring oscillator as illustrated in Fig. 4. It consists of four differential delay stages and is regulated to the target clock frequency of 832 MHz in a phase-locked loop (PLL). By tapping both outputs of each delay stage, an eight-phase clock signal is derived with uniform phase shift between the clock phases. In conjunction with a frequency divider in the feedback path of the PLL, this architecture combines the multi-phase generation and frequency up-conversion from a fixed 8 MHz reference frequency.

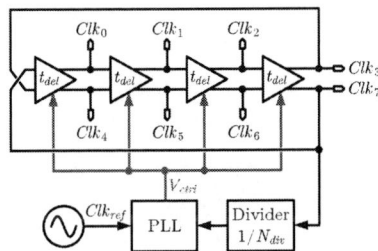

Fig. 4: Illustration of the eight-phase clock generation.

Fig. 5 shows a detailed overview of entire the clock generation unit. The PLL for regulation of the eight-phase voltage-controlled ring oscillator is implemented as an integer-N charge-pump PLL. A programmable frequency divider allows a clock frequency tuning with 16 MHz step size. Following the eight-phase ring VCO, a background phase spacing calibration circuit is employed to correct phase spacing mismatches due to process variations and layout asymmetries. The calibration is based on a digitally-controlled variable delay buffer for each clock phase, which allows an individual time delay adjustment of each clock phase with a resolution of approximately 1.5 ps.

The 8 MHz reference frequency generation for the PLL is performed by a relaxation oscillator. As indicated in Fig. 5, it derives its frequency from the time constant of a discrete off-chip resistor R_o and capacitor C_o. Simultaneously, resistor R_o is utilized for a precise reference current generation for the gate driver IC. As presented in [9], this reference generation circuit based on off-chip passive components achieves an accurate reference frequency and current independent from on-chip process variations and thus does not require any post-fabrication trimming.

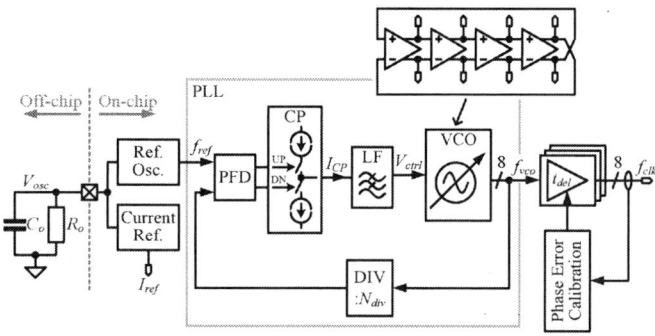

Fig. 5: System overview of the clock generation unit.

Fig. 6 presents an oscilloscope measurement of the eight clock phases at the output of the phase error calibration circuit. The residual phase spacing errors are within $+2.1\,\text{ps}/-1.3\,\text{ps}$, demonstrating the timing accuracy of the clock system for the proposed digital gate driver.

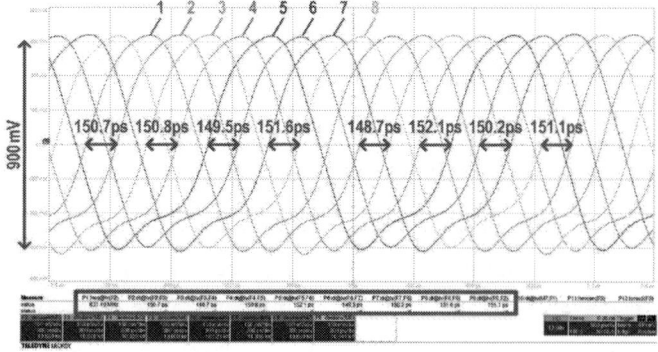

Fig. 6: Oscilloscope measurement of the eight clock phases.

IV. BURST-MODE ADC

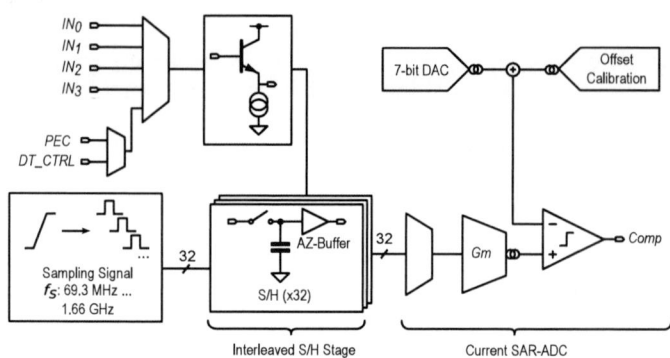

Fig. 7: System overview of the burst-mode ADC.

In order to iteratively adjust the gate-shaping profile of the digital gate driver the burst-mode ADC provides measurements of the drain voltage, gate voltage or drain current of the SiC power-MOSFET. The system level block diagram of the burst-mode ADC is shown in Fig. 7. The input signal of the burst-mode ADC is buffered using an emitter-follower and fed to the interleaved sampling stage. Additionally, for shifting the common mode level of the input voltage a buffered voltage of 1.6 V is provided. Because the switching of the SiC power-MOSFET is a very short event compared to the following on- or off-time of the powerMOSFET, the burst-mode ADC was designed with interleaved high-speed sample-and-hold (S&H) stages. During a measurement, there are 30 interleaved S&H stages following the input signal until the respective sampling signal activates the sampling of the respective sampling stage. Then, the samples taken during the switching event are multiplexed to a single successive-approximation (SAR)-ADC, thus achieving a high sampling rate during the interval of interest while operating the conversion circuit at a lower speed relaxing the requirements. The SAR-ADC itself consists of an adjustable V-to-I stage, a 7 bit current-steering DAC and a 6 bit DAC for offset calibration. The sampling signals for each

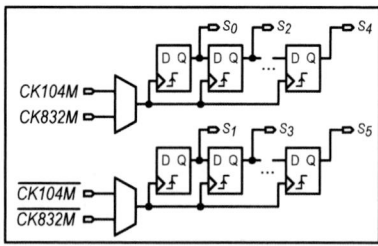

Fig. 8: Trigger generator of the burst-mode ADC.

S&H circuit of the burst-mode ADC are generated by two shift registers as shown in Fig. 8, one clocked by the rising and one by the falling edge of either a 104 MHz or an 832 MHz clock. By multiplexing the shift register's outputs to the S&H circuits, sampling rates of 69.3 MHz, 104 MHz, 208 MHz, 554.7 MHz, 832 MHz, and 1664 MHz are realized.

Measurement results of the signal-to-noise and distortion ratio (SINAD) and effective number of bits (ENOB) of the

979-8-3315-1612-3/25 $31.00 © 2025 IEEE

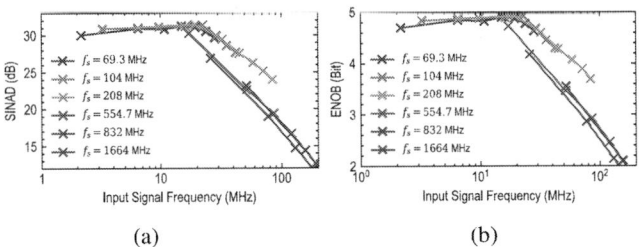

(a)   (b)

Fig. 9: Measured signal-to-noise and distortion ratio (SINAD) (a) and measured effective number of bits (ENOB) of the burst-mode ADC.

burst-mode ADC for the range of sampling frequencies are shown in Fig. 9.

V. DEAD-TIME REGULATION

Figure 10 illustrates the dead-time control mechanism implemented in this gate driver. The ability follows the concept proposed in [8], making use of the non-linearity of the reverse transfer capacitance to infer the voltage across the device without requiring further external components. It improves on its implementation through the added incorporation within the high-speed burst-mode ADC. The discharge current $I_{c,gd}$ is measured throughout the transition of the driven SiC-FET, with the peak-current being targeted for zero-crossing of the voltage. This current is measured resistively using resistor R_1 within the driver stage, benefiting from the current-mirror topology while not impacting driver performance and requiring negligible additional area. However, instead of relying on iterative approaching of the peak value to determine ZVS, the burst-mode stage is used to achieve a single-shot capture and evaluation of the peak current timing index. This in turn allows for a much faster achievement of ZVS, reducing the number of cycles the transistor is exposed to hard switching, resulting in improved application efficiency and device longevity. To ensure ZVS is maintained throughout load and line changes, periodic reacquisition is done automatically after a (tunable) set amount of switching cycles, and can further be triggered

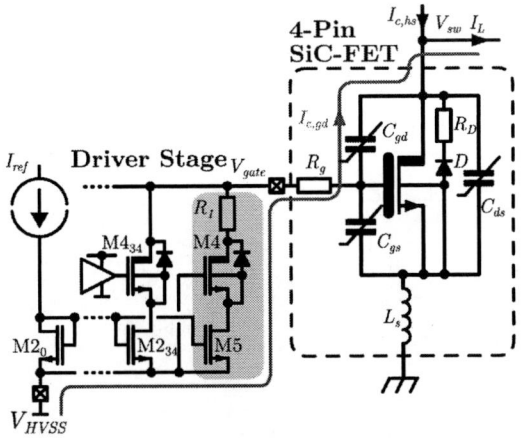

Fig. 10: System overview of the dead-time control implementation within the driver stage.

by an external signal from the application control algorithm over SPI.

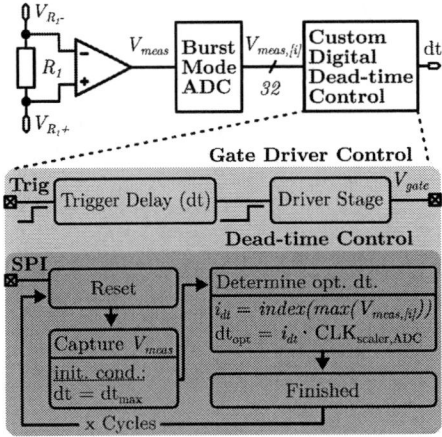

Fig. 11: Control algorithm of the dead-time control implementation.

VI. DRIVER

In order to shape the switching behavior of the SiC power-MOSFET an optimized switching profile is transferred to the lookup table (LUT) in the digital gate driver via SPI. This switching profile is applied to the gate of the SiC power-MOSFET when the digital gate driver is triggered. To accurately control the ringing and EMI of the switching event the output current of the digital gate driver is defined with 31 current sources of $4 \cdot I_{FRR}$ and 3 current sources of I_{FRR} which can be turned on for a pulse width $\Delta t_o n$. This pulse width $\Delta t_o n$ can be adjusted with a resolution of up to $150\,\text{ps}$. The system level schematic of the driver is shown in Fig. 12. The reference current I_{FRR} of the driver is programmable with a default configuration of $32\,\text{mA}$ and a maximum current of $40\,\text{mA}$.

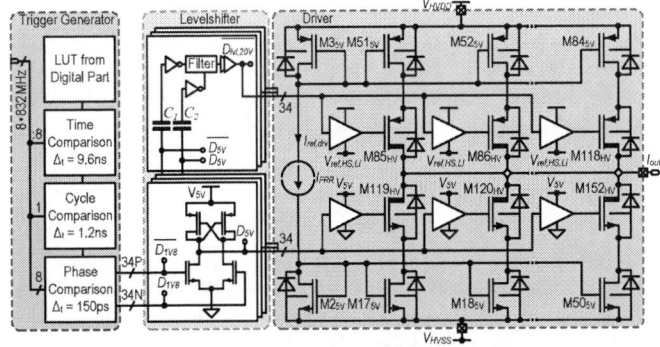

Fig. 12: Driver stage of the digital gate driver.

The driver consists of a digital trigger generator, cross-coupled pair low-voltage levelshifters, capacitive high-voltage levelshifters, an accurate high-voltage reference supplying $V_{ref,HS,LI}$, a fast response current source supplying I_{FRR}, the current mirrors $M17_{5V}$ to M_{84}, and the power-MOSFETs

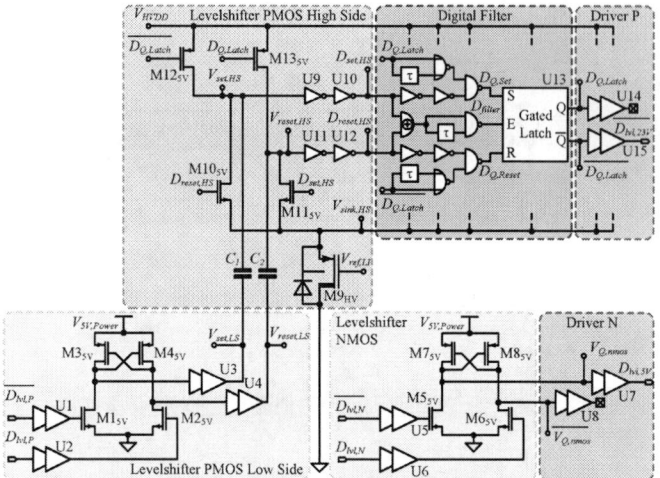

Fig. 13: Capacitive levelshifter in the driver for minimum time delay levelshift.

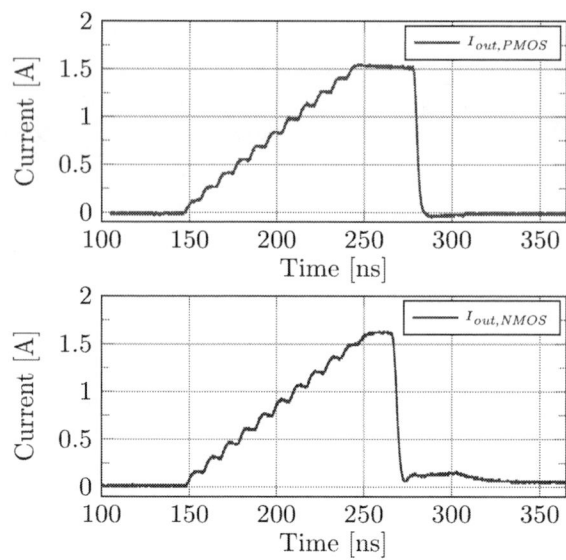

Fig. 14: Measured incremental increase of the digital gate driver output current driving a $10\,\Omega$ resistor.

$M85_{HV}$ to $M152_{HV}$. After the trigger signal is received by the digital gate driver, the driving stage ramps up the driver within less than $50\,\mathrm{ns}$. After the ramp up the switching cycle is started. In the digital trigger generator, the entries in the LUT are compared to the time counted from the start of the switching cycle. Due to the minimum step width of $150\,\mathrm{ps}$ the comparison is split into 3 comparisons with the time comparison $\Delta t_1 = 9.6\,\mathrm{ns}$, cycle comparison $\Delta t_2 = 1.2\,\mathrm{ns}$ and phase comparison $\Delta t_3 = 150\,\mathrm{ps}$. For every power-MOSFET in the driver there is a trigger generator that activates only the levelshifter of this power-MOSFET. Thus, the timing information for turn-on and turn-off of this respective power-MOSFET is loaded to the trigger generator. Then the trigger generator compares the target time with the counted time since the beginning of the switching cycle and drives the levelshifter once the time comparison $\Delta t_1 = 9.6\,\mathrm{ns}$, cycle comparison $\Delta t_2 = 1.2\,\mathrm{ns}$ and phase comparison $\Delta t_3 = 150\,\mathrm{ps}$ are equal. For the NMOS-transistors the voltage is shifted to a driving voltage of $0\,\mathrm{V}$ and $5\,\mathrm{V}$ and for the PMOS-transistors to V_{HVDD} and $V_{HVDD} - 5\,\mathrm{V}$ respectively. Usually, the digital gate driver is used with a supply voltage of $V_{HVDD} = 20\,\mathrm{V}$. In the output stage, power transistors are in series with $5\,\mathrm{V}$ current mirrors, which are switched on and off through the power transistors. The voltage low-impedance high-side reference voltage $V_{ref,HS,LI}$ is provided from an accurate high-voltage bandgap reference. Further, the current mirrors are biased with a reference current of I_{FRR} from a reference current source with a startup time of less than $45\,\mathrm{ns}$ in order to start switching within $50\,\mathrm{ns}$ after the trigger signal.

The schematic of the capacitive high-side levelshifter is shown in Fig. 13. The digital signals $D_{lvl,N}$ and $D_{lvl,P}$ as well as the inverse signals are received from the trigger generator with a voltage level of $1.8\,\mathrm{V}$ and shifted to a voltage level of $5\,\mathrm{V}$. The driver U7 drives the signal driving the NMOS-transistors in the driver stage. For driving the PMOS-transistors the drivers U3 and U4 drive the capacitors C_1 and C_2 capacitively, which is detected and amplified in U9 to U12

at the HS of the levelshifter. The digital filter protects the high-side driver (HS driver) from triggering in the case of common mode noise on the supply voltage rails V_{HVDD} or $V_{sink,HS}$. In case a differential signal is received at the high-side of the levelshifter the gated latch is set and the driver U15 drives the PMOS power-transistor in the driver stage.

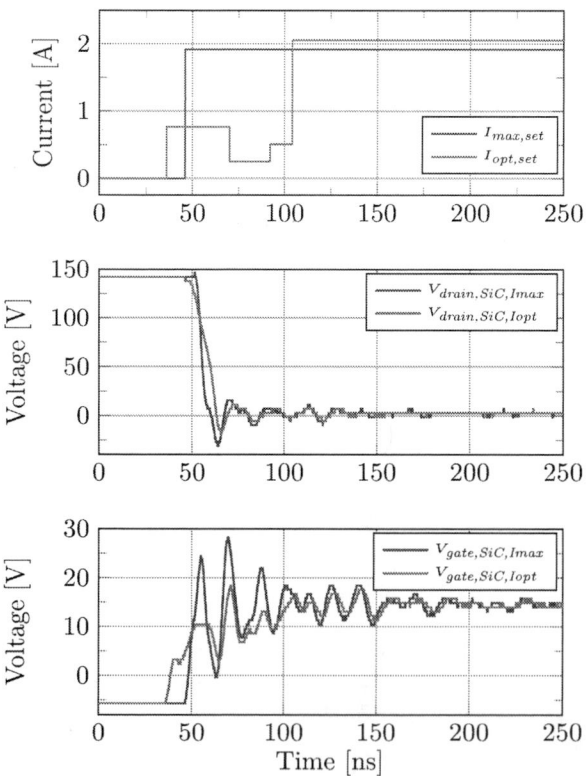

Fig. 15: Measurement of a double pulse test with a hard switching SiC power-MOSFET and a simple optimization.

Fig. 14 shows a transient ramp up of the output current of the digital gate driver characterizing the driver stage. In the measurement, the current of the output stage driving a $10\,\Omega$ resistor is increased in 11 steps by $128\,\text{mA}$ every $10\,\text{ns}$ up to a current of roughly $1.5\,\text{A}$ for PMOS and $1.6\,\text{A}$ for NMOS. Fig. 15 shows $V_{ds,SiC}$, $V_{gs,SiC}$, and the given current profile $I_{max,set}$ and $I_{opt,set}$ in a double pulse test with a hard switching SiC power-MOSFET at a drain voltage of $140\,\text{V}$ and a current of $2\,\text{A}$ with a simple optimization.

Further, the measurements of the individual current sources against each other show a maximum deviation of $7.5\,\text{mA}$ for $128\,\text{mA}$ NMOS-transistors and $6\,\text{mA}$ for $128\,\text{mA}$ PMOS-transistors. The jitter of two consecutive edges of the driver has a standard deviation of roughly $0.2\,\%$ to $0.3\,\%$ of the pulse width which results in $220\,\text{ps}$ for a pulse width of $100\,\text{ns}$ to $30\,\text{ps}$ for a pulse width of $10\,\text{ns}$ and is independent of NMOS or PMOS sources being switched.

VII. COMPARISON

TABLE I: Comparison table with previous active gate drivers

	TPE'23 [12]	TPE'21 [13]	ISSCC'19 [14]	**This work**
Process	0.5 µm CMOS	0.18 µm CMOS	0.13 µm CMOS	**0.18 µm BCD**
Output current	3.6 A	N/A	N/A	**5 A**
Output voltage	+18 V/0 V	+5 V/0 V	+10 V/0 V	**+18 V/-5 V**
Control strategy	Digital	Digital	Analog	**Digital**
Timer architecture	Starved inverter	Tapped delay line	N/A	**Multiphase PLL**
Time step	1 ns	100 ps	continuous	**150 ps**
LUT	8	1	N/A	**1**
Integrated ADC	500 kS SAR	No	No	**Burst-mode**
Power supply	Yes	Yes	No	**Yes**
Simultaneous PMOS & NMOS conduction	Yes	Yes	No	**Yes**

Table I shows a comparison of the gate-shaping digital gate driver presented in this with other reported state-of-the-art active gate driver designs, that manipulate the gate of the power-MOSFET during the switching cycle. With a current of $5\,\text{A}$ and an output voltage range of up to $23\,\text{V}$ the proposed active gate driver is targeting a slightly higher range of power-MOSFETs, which makes it suitable for driving larger SiC devices. Compared to the proposed design and [12] the reported active gate driver in [13] has an even higher time step resolution targeting to drive power-MOSFETs with a lower gate voltage more accurately. From the system architecture the reported active gate driver in [12] is most similar, as it also includes a power management unit, clock generation unit and an integrated ADC for providing feedback. Further, most active gate driver architectures share the feature to simultaneously turn on NMOS and PMOS drivers. The presented digital gate driver offers a significantly wider output voltage and driving current range and thus at a precise regulated timing resolution of $150\,\text{ps}$ and thus allows for more applications while also enabling sensing of the SiC power-MOSFET. Furthermore,

the integration of the power supply as well as additional functionalities such as integrated dead-time regulation offers further possibilities in system design, reducing the overall system footprint and cost.

VIII. CONCLUSION

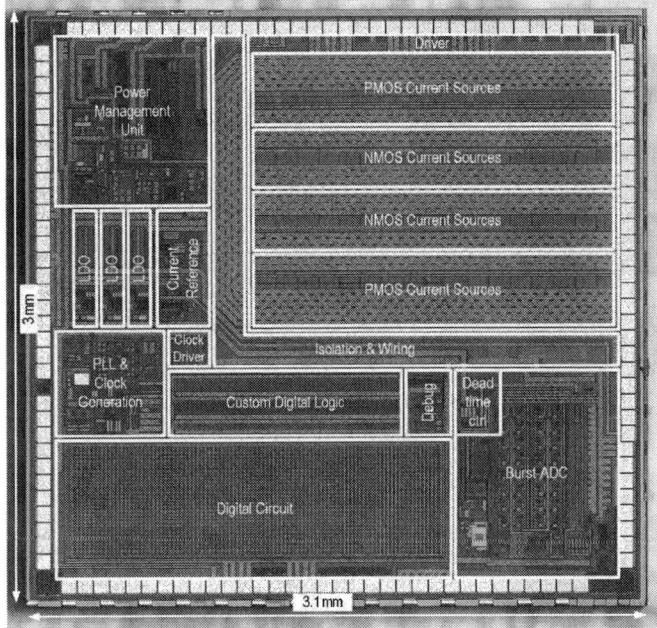

Fig. 16: Micrograph of the fabricated digital gate driver die.

A gate-shaping digital gate driver for driving SiC power-MOSFETs in power electronic applications has been implemented. The designed gate driver has been simulated and fabricated in a $0.18\,\mu\text{m}$ HV BCD technology. Further, the device functions have been characterized in the laboratory and the digital gate driver has been measured in a double pulse test and a buck converter. The system architecture of the digital gate driver and the individual blocks of the design have been presented and the performance has been evaluated in simulations and measurements. By driving the SiC power-MOSFET with a dedicated switching profile the EMI and EMC can be significantly improved while still employing the superior characteristics of the wide bandgap semiconductor power MOSFET. By sampling the switching behavior of the SiC power-MOSFET during the switching cycle, the switching profile of the digital gate driver can be adjusted with an appropriate algorithm, which have been evaluated in earlier works. The micrograph of the die in Fig. 16 shows the fabricated die in an $0.18\,\mu\text{m}$ BCD technology with an area of $9.3\,\text{mm}^2$. A comparison with the state of the art active gate drivers is presented.

REFERENCES

[1] J. Biela, M. Schweizer, S. Waffler and J. W. Kolar, "SiC versus Si—Evaluation of Potentials for Performance Improvement of Inverter and DC–DC Converter Systems by SiC Power Semiconductors," in IEEE Transactions on Industrial Electronics, vol. 58, no. 7, pp. 2872-2882, July 2011, doi: 10.1109/TIE.2010.2072896.

979-8-3315-1612-3/25 $31.00 © 2025 IEEE

[2] A. Schindler, B. Koeppl and B. Wicht, "EMC and switching loss improvement for fast switching power stages by di/dt, dv/dt optimization with 10ns variable current source gate driver," 2015 10th International Workshop on the Electromagnetic Compatibility of Integrated Circuits (EMC Compo), Edinburgh, UK, 2015, pp. 18-23, doi: 10.1109/EMC-Compo.2015.7358323.

[3] G. Engelmann, T. Senoner and R. W. De Doncker, "Experimental investigation on the transient switching behavior of SiC MOSFETs using a stage-wise gate driver," in CPSS Transactions on Power Electronics and Applications, vol. 3, no. 1, pp. 77-87, March 2018, doi: 10.24295/CPSSTPEA.2018.00008.

[4] J. Henn et al., "Intelligent Gate Drivers for Future Power Converters," in IEEE Transactions on Power Electronics, vol. 37, no. 3, pp. 3484-3503, March 2022, doi: 10.1109/TPEL.2021.3112337.

[5] G. Chen, Z. Jin, Y. Deng, X. He and X. Qing, "Principle and Topology Synthesis of Integrated Single-Input Dual-Output and Dual-Input Single-Output DC–DC Converters," in IEEE Transactions on Industrial Electronics, vol. 65, no. 5, pp. 3815-3825, May 2018, doi: 10.1109/TIE.2017.2760856.

[6] Z. Dong, X. L. Li and C. K. Tse, "Single-Inductor Multiple-Output Current-Source Converter With Improved Cross Regulation and Simple Control Strategy," 2018 International Power Electronics Conference (IPEC-Niigata 2018 -ECCE Asia), Niigata, Japan, 2018, pp. 3768-3772, doi: 10.23919/IPEC.2018.8507396.

[7] W. Xu, Y. Li, Z. Hong and D. Killat, "A 90% peak efficiency single-inductor dual-output buck-boost converter with extended-PWM control," 2011 IEEE International Solid-State Circuits Conference, San Francisco, CA, USA, 2011, pp. 394-396, doi: 10.1109/ISSCC.2011.5746367.

[8] L. Weihs, E. Wehr, K. Vohl, T. Zekorn, R. Wunderlich and S. Heinen, "A Fully Integrated Adaptive Dead-Time Controlling Gate Driver Enabling ZVS in HV Converters," ESSCIRC 2023- IEEE 49th European Solid State Circuits Conference (ESSCIRC), Lisbon, Portugal, 2023, pp. 201-204, doi: 10.1109/ESSCIRC59616.2023.10268753.

[9] E. Wehr, T. Zekorn, M. Hanhart, K. Vohl, L. Weihs, R. Wunderlich, and S. Heinen, "A Trimming-Less External-RC Relaxation Oscillator With Self-Calibrating Current Reference for a SiC Active Gate Driver Application," 2024 IEEE International Symposium on Circuits and Systems (ISCAS), Singapore, 2024, pp. 1-5, doi: 10.1109/IS-CAS58744.2024.10558102.

[10] T. Zekorn, F. Schimkat, K. Vohl, E. Wehr, R. Wunderlich and S. Heinen, "A High-Voltage Single-Inductor Multiple-Output DC-DC Buck Converter for the Power Management Unit of a Gate-Shaping Digital Gate Driver," 2024 IEEE International Symposium on Circuits and Systems (ISCAS), Singapore, Singapore, 2024, pp. 1-5, doi: 10.1109/IS-CAS58744.2024.10557929.

[11] J. Kuhn, T. Zekorn, K. Vohl, R. Wunderlich and S. Heinen, "An Active Gate Driver for Iteratively Optimizing the Switching Characteristics of SiC MOSFETs," 2023 25th European Conference on Power Electronics and Applications (EPE'23 ECCE Europe), Aalborg, Denmark, 2023, pp. 1-8, doi: 10.23919/EPE23ECCEEurope58414.2023.10264429.

[12] S. Kawai et al., "A Load Adaptive Digital Gate Driver IC With Integrated 500 ksps ADC for Drive Pattern Selection and Functional Safety Targeting Dependable SiC Application," in IEEE Transactions on Power Electronics, vol. 38, no. 6, pp. 7079-7091, June 2023, doi: 10.1109/TPEL.2023.3244200.

[13] D. Liu et al., "Full Custom Design of an Arbitrary Waveform Gate Driver With 10-GHz Waypoint Rates for GaN FETs," in IEEE Transactions on Power Electronics, vol. 36, no. 7, pp. 8267-8279, July 2021, doi: 10.1109/TPEL.2020.3044874.

[14] S. Kawai, T. Ueno and K. Onizuka, "15.8 A 4.5V/ns Active Slew-Rate-Controlling Gate Driver with Robust Discrete-Time Feedback Technique for 600V Superjunction MOSFETs," 2019 IEEE International Solid-State Circuits Conference - (ISSCC), San Francisco, CA, USA, 2019, pp. 252-254, doi: 10.1109/ISSCC.2019.8662534.

Decentralized Interleaving of Series-Stacked DC-DC Converters via Extremum-Seeking Control

Ivan Petrić[a], Vignesh Iyer[b], Shoudong Hu[b], Chirayu Rajpurohit[b], Bailey Sauter[b], Milan Ilić[a], Luca Corradini[b], and Dragan Maksimović[b]

[a]Future Energy Technology Center, Hanwha Qcells Technologies, Santa Clara, CA 95054, USA
[b]Department of Electrical, Computer, and Energy Engineering, University of Colorado, Boulder, CO 80309, USA

Abstract—This paper introduces a fully decentralized solution for achieving optimal interleaving of series-stacked power converters based on distributed, module-level extremum-seeking controllers. This approach ensures switching-cycle-level synchronization, effectively minimizing switching ripple and significantly reducing the size of passive filters. The proposed control system is model-free, straightforward to design and implement, and offers continuous synchronization to mitigate clock drifts while also supporting asymmetric module operation. To guide system design, dynamic modeling is presented for a series-stacked architecture with two modules. The proposed approach is validated through simulations involving up to seven modules and by experiments with up to three modules.

Index Terms—Decentralized interleaving, Ripple minimization, Series-stacked dc-dc converters

I. INTRODUCTION

SERIES-stacked (cascaded) converter modules allow multiport power conversion with reduced voltage-rated semiconductors in applications such as photovoltaic (PV) dc optimizers [1], [2], as illustrated in Fig. 1. With a proper switching-cycle-level synchronization (interleaving), the cascaded architecture offers additional system-level improvements, in terms of significantly faster dynamic performance and reduction of passive filters, due to harmonic cancellation effects [3].

While interleaving is widely implemented across the field, it is most often set fixed using the symmetric phase-shifted pulsewidth modulation (PS-PWM), which is optimal for the balanced module operation [3]. For applications where mismatches between the modules are significant, asymmetric interleaving should be imposed instead [4]. However, calculating the optimal asymmetric interleaving angles represents a difficult optimization problem that brings a high computational complexity, unsuitable for real-time implementation [5]. Additionally, in many applications, fully decentralized control of the modules within a multi-cell architecture is preferred [6], [7], as it avoids non-standard wiring, reduces cost, and increases reliability. There, module interleaving becomes very challenging due to the lack of a synchronizing signal and the appearance of clock drifts due to crystal oscillator tolerances. All this has resulted in an expanded research interest in adaptive and decentralized interleaving, without any fast inter-module communication [5], [8]–[12].

In [8], the output ripple of paralleled converters is minimized using digital nonlinear oscillators, which requires a

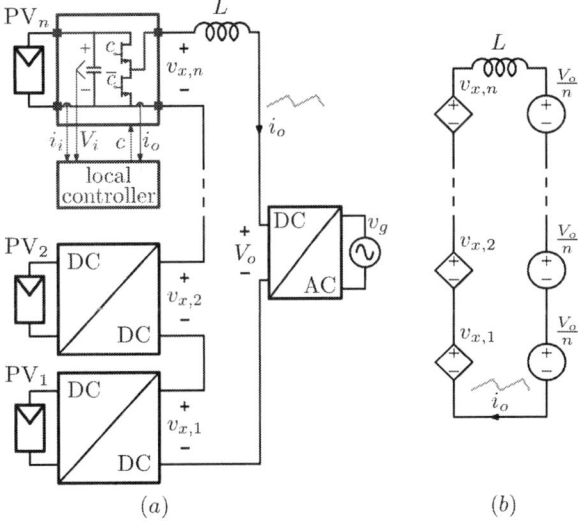

Fig. 1: (a) A series-stacked dc converter architecture employed in a PV system based on dc optimizers, and (b) an equivalent circuit for calculating the current ripple.

very fast FPGA implementation and is limited to symmetric paralleled units. In [9], it is assumed that all units operate with equal duty cycles, and the method depends on a single sample taken at the switching instants, which inherently increases sensitivity to noise. The works [5], [11] offer two decentralized methods based on gradient descent. However, both approaches necessitate an iterative, sequential optimization of individual modules, rather than enabling continuous and simultaneous corrective actions across all modules. This limitation may be problematic in the presence of clock drifts. Furthermore, both methods are computationally intensive and rely on model-based approaches, making them susceptible to unmodeled dynamics and parameter tolerances. The methods from [10], [12] feature several approximative assumptions and are also based on the gradient descent, where the gradient estimation relies on precisely setting the single-sampling instant within a switching cycle, depending on the system parameters, including the sensor bandwidth. All the above-mentioned methods are strictly model-based and have several disadvantages in terms of implementation robustness and complexity.

To overcome these limitations, this paper proposes a model-free, adaptive algorithm for decentralized interleaving, based on extremum seeking control (ESC) [13]. The algorithm has a simple design procedure and is capable of minimizing the current ripple robustly, regardless of asymmetries between the modules, while rejecting the impact of clock drifts. It is computationally efficient, suitable for implementation in standard digital signal processors (DSPs), and compatible with oversampling techniques that are often implemented in state-of-the-art feedback control loops [14]. As the first investigation of the ESC capabilities for decentralized interleaving of power converter modules, this paper focuses its analyses and demonstrations on minimizing the output current ripple of series-stacked dc architectures, such as the system shown in Fig. 1. For system design guidelines and convergence analysis, dynamic modeling is presented and validated in simulations for two series-stacked modules. Results showing decentralized interleaving are provided for up to seven modules in simulations and up to three modules experimentally.

The paper is organized as follows. A series-stacked dc architecture is introduced and the principles of harmonic cancellation by interleaving are briefly summarized in Section II. In Section III, the proposed ESC approach is analyzed and a small-signal dynamic model is derived for a system with two modules. Verifications of the algorithm operation and the proposed model are presented in Section IV. Finally, conclusions are presented in Section V.

II. SERIES-STACKED DC-DC ARCHITECTURE

Fig. 1(a) shows a series-stacked dc-dc converter architecture that can be found in, for example, residential PV systems based on dc optimizers [1], [2]. It consists of n low-voltage (LV) dc-dc converter modules that are connected in series to interface to the high-voltage (HV) dc link via a single inductive filter L. The grid connection is then established using a string inverter, which regulates the dc bus voltage V_o. Aside from enabling improved energy capture via module-level maximum power point tracking, the main motivation for the series-stacked architecture comes from being able to interface to the HV dc link without isolation and relying only on LV components. This enables ultra-high efficiency and power density while using cost-effective, high figure-of-merit low-voltage Si or GaN semiconductors. An important target feature is the fully decentralized control of LV modules, which obviates the need for fast inter-module communication. This brings plug-and-play operability and, hence, increases reliability, simplifies installation and wiring, and reduces the overall cost.

In this paper, each LV module's output stage utilizes a simple half-bridge buck-type switching cell; however, the method can be readily applied to other topologies. The plug-and-play LV modules feature only local controllers [2] that generate the switching signals $c(t)$, using PWM with the nominal frequency f_{sw}. The instantaneous switched node voltages are equal to $v_x(t) = c(t)V_i$ and their average values are equal to $V_x = DV_i$, where V_i is the module's input voltage and D is the duty cycle of c. The output voltage of the series-stacked architecture is equal to $V_o = \sum_{k=1}^{n} D_k V_{i,k}$, which, for

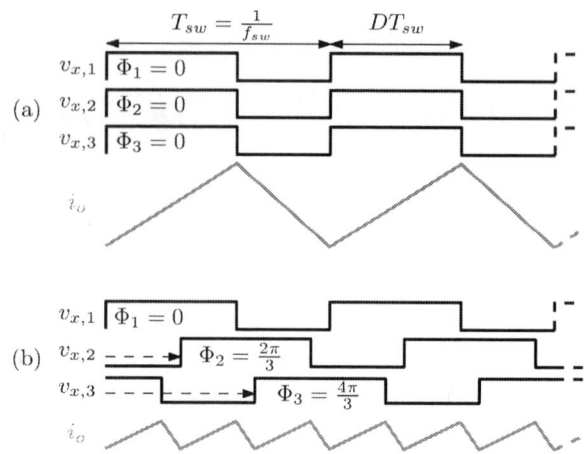

Fig. 2: Illustration of the worst case in-phase operation and the optimal symmetric PS-PWM for $n = 3$ balanced LV modules.

balanced LV modules ($\forall k \in \{1, ..., n\}$ $D_k = D$ and $V_{i,k} = V_i$) results in $V_o = nDV_i$.

The output current spectrum, which depends on the sum of all LV modules' switched node voltages, can be calculated by applying the superposition principle to each pair of $v_{x,k}$ and V_o/n, as shown in Fig. 1(b). For the considered half-bridge topology, k^{th} module's contribution to the current ripple, represented by phasors at multiples of the switching frequency, can be found as

$$\underline{I}_{o,k}(jl\omega_{sw}) = \frac{V_{i,k}}{\pi\omega_{sw}L}\frac{1}{l^2}\left[e^{-j2\pi l D_k} - 1\right]e^{-jl\Phi_k}, \quad (1)$$

where $\omega_{sw} = 2\pi f_{sw}$, l is the switching harmonic order, j is the imaginary coefficient, and Φ_k is the phase angle (in radians) of the switched-node voltage $v_{x,k}$. The total current ripple phasors are found as a sum of (1) for all $k \in \{1, ..., n\}$, and are a function of modules' input voltages, duty cycles, and switched nodes' phase angles. A simplification is obtained for balanced LV modules, for which the output current harmonics are equal to

$$\underline{I}_{o}(jl\omega_{sw}) = \frac{V_{i}}{\pi\omega_{sw}L}\frac{1}{l^2}\left[e^{-j2\pi l D} - 1\right]\left[1 + \sum_{k=2}^{n} e^{-jl\Delta\Phi_k}\right]. \tag{2}$$

Note that the first LV module is taken as a reference for other modules' phase shifts, i.e., $\Phi_1 = 0$ and $\Delta\Phi_k = \Phi_k$. As illustrated in Fig. 2, for a balanced series-stacked architecture, current harmonics are maximized with the in-phase operation (zero phase shifts) and minimized with the symmetric PS-PWM, for which the switched node voltages are phase-shifted $2\pi/n$ apart. Besides reducing the magnitude of the voltage pulse applied to the filter inductor, with balanced modules the symmetric PS-PWM results in the first $n - 1$ current ripple harmonics being equal to 0, thus pushing its spectral content to higher frequencies. This enables a drastic reduction of the output inductance and RMS ratings of the HV dc capacitors without having to increase the switching frequency of the individual modules, hence, reducing losses and electromagnetic interference.

In balanced series-stacked systems with centralized (or distributed) control, imposing symmetric PS-PWM to minimize (2) is straightforward and widely implemented. Complexity is significantly increased for unbalanced systems, where the minimal current ripple may no longer correspond to the symmetric PS-PWM. There, finding the optimal phase shift distribution represents a non-convex optimization problem. Calculating a solution in real time is impractical, especially for a large number of modules [4], [5]. Finally, achieving optimal interleaving for decentralized LV modules is highly challenging due to the absence of a synchronization signal and the fact that each local controller generates its own PWM clock. This leads to clock drifts between units, stemming from the tolerances of the crystal oscillators. As a result, a continuously-applied synchronizing action is necessary, rather than relying on synchronization at specific moments, such as during start-up. To address these challenges, this work proposes a new method for optimal decentralized interleaving, based on distributed adaptive extremum-seeking controllers.

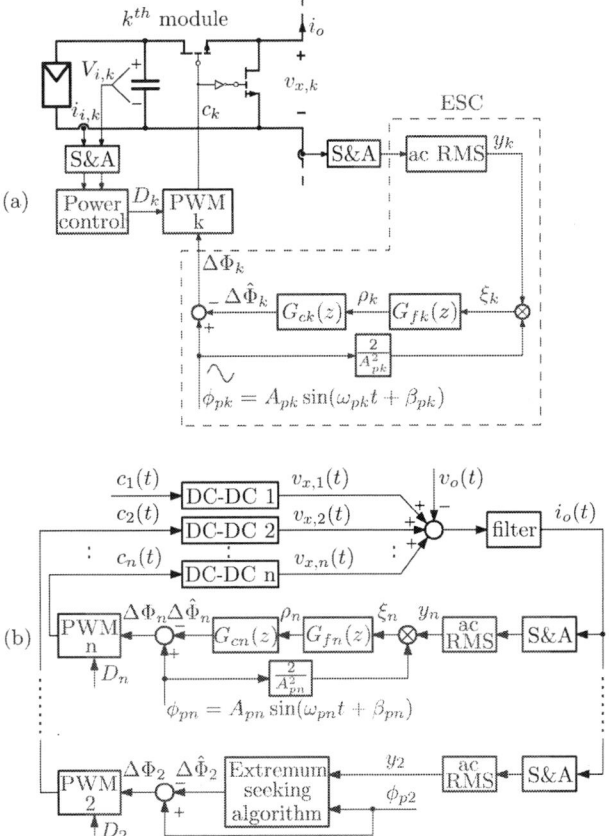

Fig. 3: (a) The proposed module-level ESC system for decentralized interleaving. (b) Block diagram of the series-stacked system forming the multi-variable ESC.

III. RIPPLE MINIMIZATION VIA EXTREMUM SEEKING CONTROL

The goal of the method proposed in this paper is to bring the series-stacked system from Fig. 1 to its optimal point in terms of minimizing the output current ripple, using decentralized interleaving. To achieve this, a decentralized multi-variable ESC algorithm illustrated in Fig. 3 is introduced in Section III-A, followed by a dynamic model development in Section III-B.

A. System overview

A block diagram of the ESC system for the kth module is shown in Fig. 3(a). The ESC determines a trajectory of the optimization variables by perturbing them to measure the local gradient of the chosen cost function. In the considered multiple-input single-output system, the optimization variables are the LV modules' phase shift angles $\Delta\Phi_k$ and the cost function y_k quantifies the output current ripple. Due to the series connection, the output current i_o is shared, meaning that all LV modules have access to the same cost function. The output current acts as a system link, creating a multi-variable ESC system from the distributed controllers, as shown in the equivalent system block diagram in Fig. 3(b). For n modules there are $n-1$ phase shifts. Hence, one module can be employed without running the ESC algorithm (e.g. the module 1 sets a fixed $\Phi_1 = 0$ and serves as a reference for other phase shifts).

First, it is necessary to perform the sensing and acquisition (S&A) of the output current. In the digital implementation considered in this paper, the current is oversampled $N_s = 32$ times over the switching period and the cost function is calculated using the standard deviation (AC *root mean square* (RMS)) of the obtained samples. Other measures can be used as well, for example, the variance, the peak-to-peak ripple value, and similar. Once every switching period, after N_s samples are stored in the memory (e.g. using a *direct memory access* (DMA) module), the standard control interrupt is used

for the ESC algorithm. Referring to the kth module, a sinusoidal phase-shift perturbation signal ϕ_{pk}, with the frequency $f_{pk} = \omega_{pk}/2\pi$ and the magnitude A_{pk}, is calculated. A sinusoidal signal is chosen, although other waveforms can be used as well [13]. To allow independent perturbations, all $n-1$ modules have different perturbation frequencies, which can be set during the system start-up.

The cost function y_k is demodulated to allow the kth module to independently react to its perturbation. A scaling factor $2/A_{pk}^2$ used for the demodulation allows the convergence dynamics to be invariant to the perturbation magnitude, as explained in the following subsection. The demodulated signal ξ_k contains a dc component, indicative of the local partial derivative of the cost function, and higher-order components at frequencies that are, at least, linear combinations of all the perturbation frequencies. Those components do not contribute to the convergence and are, therefore, attenuated using a digital filter $G_{fk}(z)$, where z is the variable of the Z-transform. The obtained signal ρ_k is processed by a compensator $G_{ck}(z)$ (in this work selected as an integrator with the gain k_i), which ensures the system's convergence. The compensator output $\Delta\hat{\Phi}_k$ determines the slowly-varying phase shift value, which is subtracted from the perturbation to calculate the total phase shift of the kth PWM module. Due to the decentralized nature of the system, it is not possible to directly apply phase shifts

979-8-3315-1612-3/25 $31.00 © 2025 IEEE

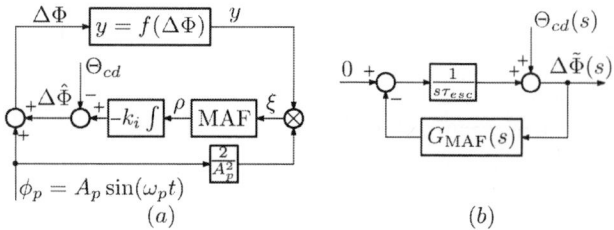

Fig. 4: Block diagrams of (a) the single-input, single-output ESC loop for a system with two modules where the clock drift is analyzed using a phase disturbance signal Θ_{cd} and (b) a small-signal system model for operation around the optimal phase shift $\Delta\Phi^* = \pi$, which can be used to analyze clock-drift disturbance rejection.

between the modules. Rather, the module PWM frequency is adjusted to achieve the target phase increment between the two consecutive ESC executions. Besides the phase shift command, the PWM receives a duty cycle command D_k from the power flow controller and outputs the switching signal $c_k(t)$.

An important design step concerns selecting the perturbation frequencies and magnitudes. Considerations on the necessary time-scale separation between the ESC dynamics, perturbation frequencies, and the optimization plant dynamics can be found in [13]. In particular, a simplification is obtained for applications where the cost function can be considered to be a static map. In the considered application, this imposes the need to limit the ESC dynamics to frequencies well below the switching harmonics. Therefore, perturbation frequencies are set well below the switching frequency, and the compensator gain is set significantly lower than the perturbation angular frequency, i.e., $k_{ik} \ll \omega_{pk} \ll \omega_{sw}$. As for distributing the perturbation frequencies among the LV modules, the demodulation process causes frequency components at their linear combinations to appear in the system. In [13], to ensure convergence without additional filters, it is recommended that the multi-variable ESC is implemented such that $\omega_{pp} + \omega_{pq} \neq \omega_{pr}$, for any triplet of the used perturbation frequencies. This approach is adopted here. Moreover, having very low differences between the perturbation frequencies will result in low-frequency components that must be rejected by the loop filters, thus limiting the achievable dynamics. Hence, the desired system dynamics must be considered when choosing the minimal difference between any two perturbation frequencies.

B. Convergence analysis and small-signal modeling for $n = 2$

Besides ensuring the system's convergence, dynamics of the ESC loop employed for decentralized interleaving must be fast enough to compensate the clock drifts between the LV modules. Following the approach from [13], a simple small-signal model is derived here to gain an insight into the system dynamics and, consequently, parameter design guidelines. As a starting point, this paper analyzes the simplest case of two LV modules, which results in a single-input single-output (SISO) ESC.

Consider the system from Fig. 4(a), which shows the ESC loop and introduces a signal Θ_{cd}, which models the phase

disturbance caused by the clock drift between the modules. For $n = 2$, the cost function y has one optimization parameter, which is the phase shift $\Delta\Phi = \Delta\Phi_2$ between the two modules. The optimal value of the phase shift is denoted as $\Delta\Phi^*$. The cost function is nonlinear, which is why the convergence analysis is performed using a Taylor expansion around the optimal point determined by $\Delta\Phi^*$ and y^*,

$$y = f(\Delta\Phi) = y^* + \frac{y''}{2}\left(\Delta\Phi - \Delta\Phi^*\right)^2. \tag{3}$$

The ESC error $\Delta\tilde{\Phi}$ is defined as

$$\Delta\tilde{\Phi} = \Delta\Phi^* - \Delta\hat{\Phi}. \tag{4}$$

Based on the block diagram in Fig. 4(a), the cost function can be expanded as

$$
\begin{aligned}
y &= y^* + \frac{y''}{2}\left(\Delta\hat{\Phi} + \phi_p - \Delta\Phi^*\right)^2 = y^* + \frac{y''}{2}\left(\phi_p - \Delta\tilde{\Phi}\right)^2 \\
&= y^* + \frac{y''}{2}\left(A_p\sin(\omega_p t) - \Delta\tilde{\Phi}\right)^2.
\end{aligned} \tag{5}
$$

The cost function is then demodulated, which results in the signal ξ,

$$
\begin{aligned}
\xi &= -\Delta\tilde{\Phi}y'' + \Delta\tilde{\Phi}y''\cos(2\omega_p t) \\
&+ \left(\frac{2}{A_p}y^* + \frac{1}{A_p}y''\Delta\tilde{\Phi}^2\right)\sin(\omega_p t) + A_p y''\sin^3(\omega_p t).
\end{aligned} \tag{6}
$$

The demodulation factor $2/A_p^2$ is chosen to make the dc component of ξ independent from the perturbation magnitude. Observing the spectral content of ξ in (6), the filter $G_f(z)$ is selected as a moving average filter (MAF) across the perturbation period window. Such filtering completely removes all non-dc frequency components from (6). Note that, for $n > 2$, such filtering does not ensure cancellation of all non-dc components, due to perturbations coming from the other LV modules. However, as subsequently demonstrated, the low-pass response of the ESC loop helps suppress the remaining impact.

The output of the MAF is the signal ρ, which measures the local gradient of the cost function

$$\rho = \mathrm{MAF}\{-\Delta\tilde{\Phi}y''\} = -y''\mathrm{MAF}\{\Delta\tilde{\Phi}\}, \tag{7}$$

where $\mathrm{MAF}\{\}$ represents the filtering action, which does not impact the static term y''. Although the system is implemented digitally, the remaining analysis is performed in the s-domain, which is adequate considering that dynamics of interest are well below the algorithm execution rate. The output of the extremum-seeking algorithm is obtained by integrating the signal ρ and adding the disturbance signal Θ_{cd},

$$
\begin{aligned}
\Delta\hat{\Phi}(s) &= -\frac{k_i}{s}\rho(s) - \Theta_{cd}(s) \\
s\Delta\hat{\Phi}(s) &= k_i y'' G_{\mathrm{MAF}}(s)\Delta\tilde{\Phi}(s) - s\Theta_{cd}(s),
\end{aligned} \tag{8}
$$

where $G_{\mathrm{MAF}}(s)$ is the s-domain transfer function of the averaging filter. From (4), it follows that

$$s\Delta\hat{\Phi} = -s\Delta\tilde{\Phi} \tag{9}$$

which yields

$$
\begin{aligned}
-s\Delta\tilde{\Phi}(s) &= k_i y'' G_{\mathrm{MAF}}(s)\Delta\tilde{\Phi}(s) - s\Theta_{cd}(s) \\
&= \frac{1}{\tau_{esc}} G_{\mathrm{MAF}}(s)\Delta\tilde{\Phi}(s) - s\Theta_{cd}(s),
\end{aligned}
\tag{10}
$$

where $\tau_{esc} = 1/k_i y''$ is the time constant of the ESC system, which describes the first-order dynamics around the optimal point. The resulting small-signal closed-loop system is shown in Fig. 4(b), and the disturbance-rejection transfer function is found as

$$
\frac{\Delta\tilde{\Phi}(s)}{\Theta_{cd}(s)} = \frac{s}{s + G_{\mathrm{MAF}}(s)\frac{1}{\tau_{esc}}}.
\tag{11}
$$

The ESC small-signal stability is limited by the perturbation frequency due to the phase lag introduced by the MAF.

As expected, the transfer function (11) has a high-pass property and fully rejects constant disturbances (e.g. a step change of the phase shift), whereas the suppression of ac disturbances (such as the one imposed by clock drifts) depends on the loop bandwidth, $\omega_{esc} = 1/\tau_{esc}$. For analyzing the clock drift rejection, the phase disturbance can be represented as $\Theta_{cd}(s) = \omega_{cd}/s^2$, where the ramp slope is determined by the clock drift angular frequency ω_{cd}. Since the ESC loop has a first-order structure, it results in a non-zero steady-state response to the ramp waveform $\Theta_{cd}(s)$,

$$
\Delta\tilde{\Phi}(t \to \infty) = \omega_{cd}\tau_{esc}.
\tag{12}
$$

A design criterion can be to limit this steady-state error to, for example, 1%. As an example, for a clock drift equal to $5\,\mu\mathrm{s/s}$ and $f_{sw} = 20\,\mathrm{kHz}$, $\omega_{cd} = 2\pi/10\,\mathrm{rad/s}$ and $\tau_{esc} < 0.1\,\mathrm{s}$.

From (11), it is clear that the system dynamics are determined by the integrator gain, k_i, and the second derivative at the optimal point, y'', which depends on all system parameters. This means that the information on y'' is necessary for determining dynamics. For a preliminary gain selection, or to narrow-down the search-space, it is possible to find off-line the worst-case (lowest) value of y'', such that a satisfactory clock drift rejection is achieved.

As an example, the cost function y is numerically calculated, based on (2), and is shown in Fig. 5, for the balanced case of two LV modules with $V_i = 60\,\mathrm{V}$, $D = 0.4$, $f_{sw} = 20\,\mathrm{kHz}$, and $L = 200\,\mu\mathrm{H}$. The traces show the impact of using different number of harmonics in (2), highlighting the minor contribution of the ones above the nth. The second derivative at the optimal point, y'', as a function of D, is calculated for the first 1000 harmonics and plotted in Fig. 6. It is noteworthy that the AC RMS cost function is not smooth at $D = 0.5$; however, it was verified that this did not pose issues in the implemented ESC scheme.

Some additional remarks concerning the proposed modeling approach are provided here. First, the derivations are only valid around the optimal point y''. Considering operation far from the optimal interleaving, e.g., to analyze convergence speed starting from the worst-case in-phase operation, the first-order Taylor-expansion term (i.e. the cost-function gradient) of y can be added to the model. Regarding extensions to $n > 2$, the convergence and small-signal analyses become dependent on the multi-variable cost function matrix partial derivatives

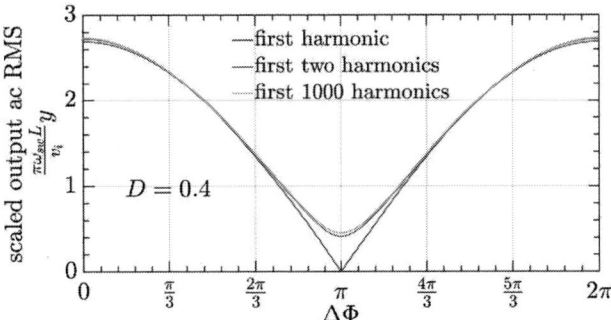

Fig. 5: Normalized cost function y for $n = 2$ balanced modules with $V_i = 60\,\mathrm{V}$, $D = 0.4$, $f_{sw} = 20\,\mathrm{kHz}$, and $L = 200\,\mu\mathrm{H}$, using (2) with different number of considered harmonics.

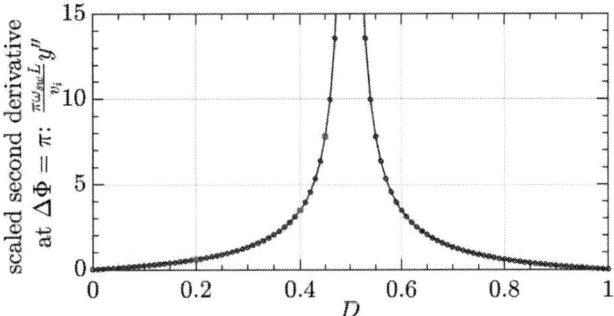

Fig. 6: Normalized second derivative y'' at the optimal point $\Delta\Phi = \pi$ as a function of duty cycle D. Operation at points marked with red are subsequently validated by simulations shown in Fig. 8.

(elements of the gradient and the *Hessian*). For a larger number of modules and, moreover, considering their possible unbalanced operation, the optimization space becomes very large. These topics are left for future work.

IV. Verifications

This section presents simulations and experimental results that validate the derived model and demonstrate the proposed approach's ability to achieve optimal interleaving in the presence of clock drifts and asymmetries among the LV modules. For all results, the switching frequency is $f_{sw} = 20\,\mathrm{kHz}$ and the ESC algorithm runs at the same rate. The PWM is implemented using trailing-edge carriers and all results are obtained for open-loop dc modulation.

A. Simulation results

The simulation model, implemented in MATLAB/Simulink, features a series-stacked system from Fig. 1(a). The load for the series-stacked system is formed using a passive parallel RC network, with $C = 33\,\mu\mathrm{F}$ and R determined based on target output power for different values of n and D. Simulation results are shown for balanced LV modules with input voltages equal to $V_i = 60\,\mathrm{V}$ and the total output filter inductance $L = 200\,\mu\mathrm{H}$. Sampling, modulation, and the ESC algorithm faithfully emulate the implementation in the DSPs used for the experimental results. The current measurements are simulated using $200\,\mathrm{kHz}$ bandwidth sensors with the ADC resolution that results in $LSB = 10.5\,\mathrm{mA}$.

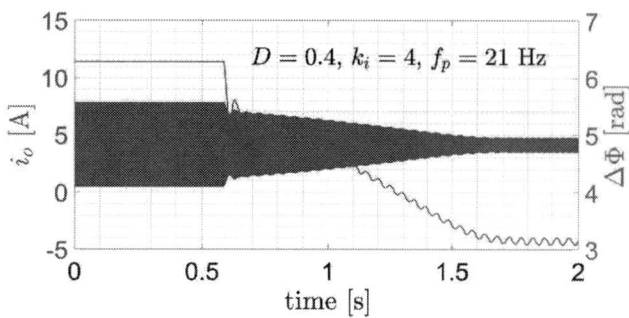

Fig. 7: Simulated ESC convergence from the worst case in-phase condition for $n = 2$ LV modules. The blue trace represents the output current and the red trace represents the phase shift.

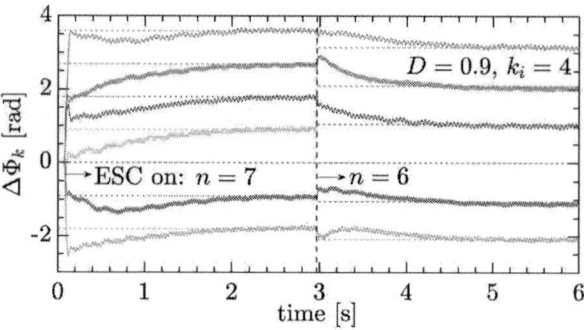

Fig. 9: ESC convergence for $n = 7$ LV modules, followed by a shut-down of one of them and the convergence to the optimal interleaving for $n = 6$.

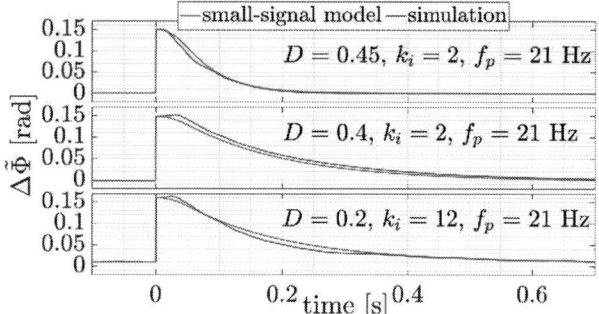

Fig. 8: Simulated validation of the disturbance rejection dynamics for $n = 2$ modules with different values of k_i and D corresponding to the red markers in Fig. 6. A step phase disturbance of $0.15\,rad$ is imposed at $t = 0$. The simulated traces closely align with the small-signal model (11).

The first set of results is given for $n = 2$ balanced LV modules and the total output power of $200\,\text{W}$. The perturbation frequency and magnitude are set to $f_p = 21\,\text{Hz}$ and $A_p = 2\pi/100$, and several values of k_i are tested to affirm the dynamic modeling. For these results, the clock drift is not included in the simulation. In Fig. 7, the convergence of the ESC is shown for $D = 0.4$ and $k_i = 4$. The system starts from the in-phase condition, which corresponds to the worst-case i_o ripple magnitude. After the ESC algorithm is enabled, the phase shift angle $\Delta\Phi$ converges to the optimal value, π, which minimizes the current ripple. Note that the small-signal modeling of Section III-B is not sufficient to estimate the total convergence time starting from the in-phase condition, as the starting point is far away from the optimum. to address this issue, the model can be extended based on the first-order term in the Taylor expansion of y.

To verify the first-order dynamics, a step phase disturbance $\Theta_{cd} = 0.15\ rad$ is imposed to the LV module operating with the optimal interleaving angle. The results, comparing the simulated response of $\Delta\hat{\Phi} - \pi$ with $\Delta\tilde{\Phi}$ modeled by (11), are given in Fig. 8 for the three values of D marked in Fig. 6. First, the top subplot is given for $D = 0.45$ and $k_i = 2$, which results in $\tau_{esc} = 84.4\,\text{ms}$. For the middle subplot, the duty cycle is reduced to $D = 0.4$ and the gain is kept equal to $k_i = 2$. This increases the time constant to $\tau_{esc} = 144.3\,\text{ms}$, which is predicted well by the model. Finally, the third subplot shows the case of $D = 0.2$ and the

gain $k_i = 12$, which is chosen to result in the same time constant $\tau_{esc} = 144.3\,\text{ms}$. As shown in the figure, the small-signal model accurately predicts the first-order dynamics. The presence of some unmodeled dynamics is anticipated due to the nonlinear characteristics of the ESC. Small steady-state errors can be seen, with values always bounded below the peak-to-peak value of the perturbation signal ($2A_p = \pi/100$). Similar verifications were repeated for various other operating conditions and, qualitatively, results did not change in terms of the dominant first-order response. However, it was noticed that the response does feature some dependency on the time at which the transient is imposed, which may be caused by different starting angles of the perturbation signal or the PWM carrier.

Subsequently, the ESC is tested for 6 and 7 balanced LV modules. The output load resistance is set to obtain $3.8\,\text{kW}$ with $V_o = 380\,\text{V}$ and the duty cycles are fixed to $D = 0.9$. The perturbation frequencies are set as $\{21, 24, 28, 32, 36, 40\}$ Hz, while the magnitude $A_p = 2\pi/100$ and the compensator gain $k_i = 4$ are set for all the modules. The clock drifts are simulated by injecting ramp signals to the PWM phases of all LV modules employing the ESC algorithm. The drift values are set equal to $\{2.44, -2.38, 2.27, -2.22, 2.78, -2.63\}\ \mu\text{s/s}$, which corresponds to an arbitrary distribution around the maximal value measured between the microcontrollers used in the experimental setup.

The results shown in Fig. 9 first demonstrate the convergence of a system with 7 balanced LV modules, starting from the in-phase operation. The full-line traces show simulated values of phase shifts $\Delta\Phi_k$ and the dashed traces show their optimal values corresponding to the symmetric PS-PWM. Following the initial ESC convergence, around $t = 3\,\text{s}$, one of the modules is shut down, to which the system adapts and converges to the new optimal point, determined by $n = 6$, without modifying anything in the ESC loops. This is an important result, as it demonstrates a true plug-and-play nature of the ESC-based decentralized interleaving.

B. Experimental results

A photo of the experimental set-up is shown in Fig. 10(a), while the developed LV module prototype is shown in Fig. 10(b). Intended to serve as dc optimizers or as dc-ac units, the LV modules feature a soft-switched boost front-

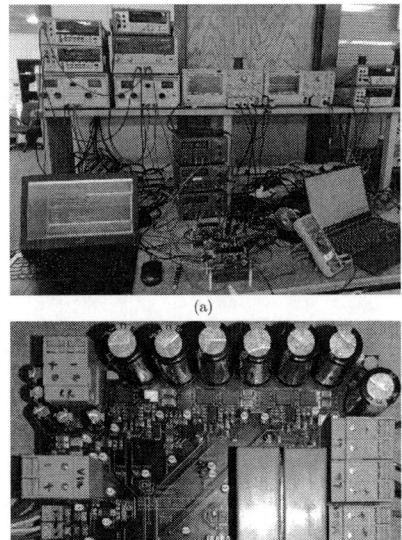

(a)

(b)

Fig. 10: (a) Photo of the experimental set-up with 3 LV modules; (b) photo of the LV module hardware prototype.

end that regulates the module dc-link voltage, and a full-bridge output that is, in this paper, configured to operate as a half-bridge stage. Each LV module is rated for 450 W and is controlled using its own controller based on the Texas Instruments F28379D. The load resistor is $R = 22\ \Omega$, the output capacitor is $C = 33\ \mu\text{F}$ and the filter inductor is $L = 180\ \mu\text{F}$.

Fig. 11(a) shows operation of $n = 2$ balanced LV modules with $V_{i,1} = V_{i,2} = 58\ \text{V}$, $D_1 = D_2 = 0.5$, $f_p = 20\ \text{Hz}$, $A_p = 2\pi/100$, and $k_i = 4$, including an illustration of the convergence to ideal interleaving after the ESC algorithm is enabled, as well as steady-state switched-node waveforms showing ideal π interleaving of the two balanced stacked modules. The phase shift is measured by outputting the ESC result using the digital-to-analog converter (DAC). One may also observe continuous small perturbations in the switched-node waveform of the second module. The second result, shown in Fig. 11(b), is given for $n = 3$ balanced LV modules, with $V_{i,1} = V_{i,2} = 56\ \text{V}$, $D_1 = D_2 = 0.4$, $f_{p1} = 109\ \text{Hz}$, $f_{p1} = 117\ \text{Hz}$, $A_p = 2\pi/160$, and $k_i = 0.2$. It can again be seen that the ESC algorithm results in the optimal interleaving. For both results, the modules are initially synchronized using an external signal, to allow starting from the in-phase condition.

Finally, the capability of ESC to synchronize unbalanced LV modules was validated for $n = 2$, by lowering the input voltage of the second module. Steady-state waveforms comparing the operation of symmetric and asymmetric interleaving are shown in Fig. 12. The ESC parameters are $f_p = 30\ \text{Hz}$ and $A_p = 2\pi/100$. The asymmetry is achieved by setting $V_{i,1} = 58\ \text{V}$ and $V_{i,2} = 40\ \text{V}$, while the duty cycles are kept equal, $D_1 = D_2 = 0.8$. It is verified analytically that the optimal phase shift for equal duty cycles and mismatched input voltages remains equal to π, which is achieved experimentally.

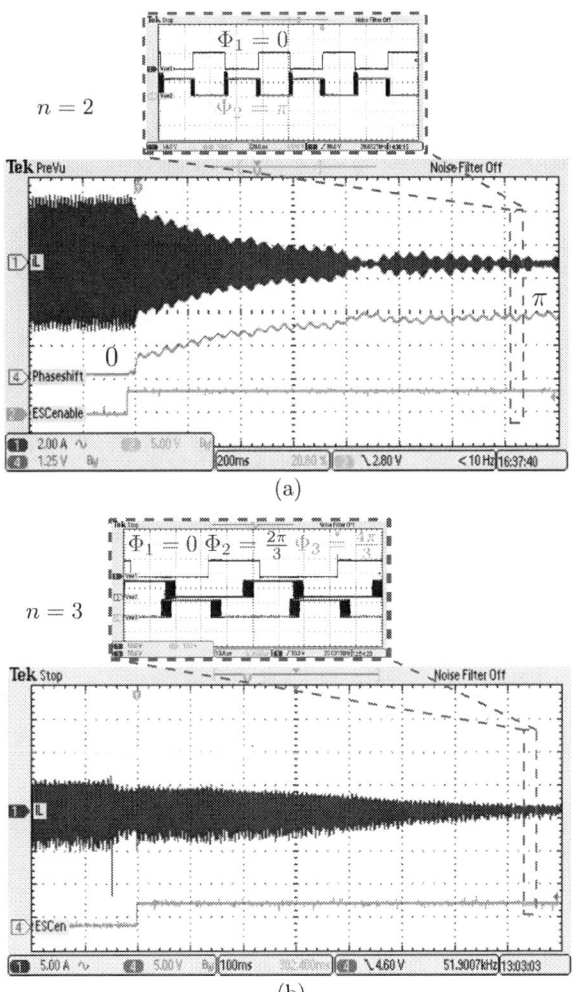

(a)

(b)

Fig. 11: (a) ESC convergence for: (a) $n = 2$ LV modules with $V_{i,1} = V_{i,2} = 58\ \text{V}$, $D_1 = D_2 = 0.5$, $f_p = 20\ \text{Hz}$, $A_p = 2\pi/100$, and $k_i = 4$; (b) $n = 3$ LV modules with $V_{i,1} = V_{i,2} = 56\ \text{V}$, $D_1 = D_2 = 0.4$, $f_{p1} = 117\ \text{Hz}$, $A_p = 2\pi/160$, and $k_i = 0.2$. The ESC enable signal is triggered when the third module algorithm starts.

The corresponding theoretical AC RMS value is equal to $0.52\ \text{A}$, which is close to the experimentally measured value of $0.55\ \text{A}$.

It is important to note that the presented results primarily focus on verifying the ESC's ability to compensate for clock drifts among the module microcontrollers in the prototype hardware setup. Faster dynamic performance of the ESC may be necessary to mitigate increased component stresses during start-up transients, which is an area for future work.

V. CONCLUSIONS

This paper presents a decentralized solution for achieving optimal interleaving of series-stacked dc-dc converters using a module-level extremum-seeking control (ESC) approach. The developed method is capable of minimizing the current ripple without any communication between the stacked modules, while compensating for the impact of clock drifts among module microcontrollers. Compared to existing methods, advantages include the model-free nature of the ESC, plug-and-play

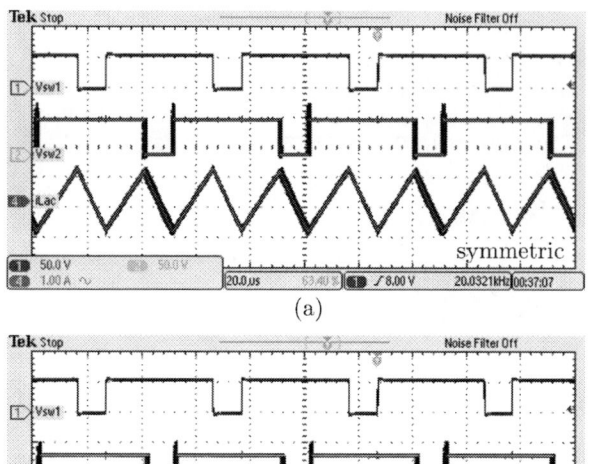

(a)

(b)

Fig. 12: Steady-state waveforms showing interleaving of $n = 2$: (a) balanced LV modules with $V_{i,1} = V_{i,2} = 58\,\text{V}$, $D_1 = D_2 = 0.8$ and (b) unbalanced LV modules with $V_{i,1} = 58\,\text{V}$, $V_{i,2} = 40\,\text{V}$, $D_1 = D_2 = 0.8$.

operability with asymmetric modules, and simple design and digital implementation without relying on sophisticated sampling schemes. To facilitate a preliminary analysis of system dynamics and clock drift disturbance rejection, a small-signal dynamic model is proposed and validated for the case when the system consists of two modules. Decentralized interleaving is validated by simulations for up to seven series-connected modules, also showing the capability to ride-through a module shut-down. Experimental results are provided for a system comprising up to three modules, including unbalanced module operation.

REFERENCES

[1] G. Walker and P. Sernia, "Cascaded dc-dc converter connection of photovoltaic modules," *IEEE Transactions on Power Electronics*, vol. 19,

no. 4, pp. 1130–1139, 2004.

[2] L. Linares, R. W. Erickson, S. MacAlpine, and M. Brandemuehl, "Improved energy capture in series string photovoltaics via smart distributed power electronics," in *2009 Twenty-Fourth Annual IEEE Applied Power Electronics Conference and Exposition*, 2009, pp. 904–910.

[3] D. Holmes and B. McGrath, "Opportunities for harmonic cancellation with carrier-based PWM for a two-level and multilevel cascaded inverters," *IEEE Transactions on Industry Applications*, vol. 37, no. 2, pp. 574–582, 2001.

[4] M. Schuck and R. C. N. Pilawa-Podgurski, "Ripple minimization through harmonic elimination in asymmetric interleaved multiphase dc–dc converters," *IEEE Transactions on Power Electronics*, vol. 30, no. 12, pp. 7202–7214, 2015.

[5] J. Poon, B. B. Johnson, S. V. Dhople, and S. R. Sanders, "Minimum distortion point tracking," *IEEE Transactions on Power Electronics*, vol. 35, no. 10, pp. 11 013–11 025, 2020.

[6] B. P. McGrath, D. G. Holmes, and W. Y. Kong, "A decentralized controller architecture for a cascaded H-bridge multilevel converter," *IEEE Transactions on Industrial Electronics*, vol. 61, no. 3, pp. 1169–1178, 2014.

[7] P. K. Achanta, D. Maksimovic, and M. Ilic, "Decentralized control of series stacked bidirectional dc-ac modules," in *2018 IEEE Applied Power Electronics Conference and Exposition (APEC)*, 2018, pp. 1008–1013.

[8] M. Sinha, J. Poon, B. B. Johnson, M. Rodriguez, and S. V. Dhople, "Decentralized interleaving of parallel-connected buck converters," *IEEE Transactions on Power Electronics*, vol. 34, no. 5, pp. 4993–5006, 2019.

[9] S. Dutta, R. Mallik, B. Majmunovic, S. Mukherjee, G.-S. Seo, D. Maksimovic, and B. Johnson, "Decentralized carrier interleaving in cascaded multilevel dc-ac converters," in *2019 20th Workshop on Control and Modeling for Power Electronics (COMPEL)*, 2019, pp. 1–6.

[10] S. Dutta, B. Majmunovic, S. Mukherjee, R. Mallik, G.-S. Seo, D. Maksimovic, and B. Johnson, "A novel decentralized PWM interleaving technique for ripple minimization in series-stacked dc-dc converters," in *2021 IEEE Applied Power Electronics Conference and Exposition (APEC)*, 2021, pp. 487–493.

[11] J. Poon, B. Johnson, S. V. Dhople, and J. Rivas-Davila, "Decentralized carrier phase shifting for optimal harmonic minimization in asymmetric parallel-connected inverters," *IEEE Transactions on Power Electronics*, vol. 36, no. 5, pp. 5915–5925, 2021.

[12] S. Dutta and B. Johnson, "A practical digital implementation of completely decentralized ripple minimization in parallel-connected dc–dc converters," *IEEE Transactions on Power Electronics*, vol. 37, no. 12, pp. 14 422–14 433, 2022.

[13] K. B. Ariyur and M. Krstic, *Real-time optimization by extremum-seeking control*. John Wiley & Sons, 2003.

[14] S. Buso and P. Mattavelli, "Digital control in power electronics, 2nd edition," *Synthesis Lectures on Power Electronics*, Morgan & Claypool Publishers, USA, 2015.

979-8-3315-1612-3/25 $31.00 © 2025 IEEE

Online Dead-Time Control for Half Bridges Without Preliminary Training Based on Switching Transient Steepness

1st Lukas Knappstein
Chair of Energy Conversion
TU Dortmund University
Dortmund, Germany
lukas.knappstein@tu-dortmund.de

2nd Niklas Falkenberg
Chair of Energy Conversion
TU Dortmund University
Dortmund, Germany
niklas.falkenberg@tu-dortmund.de

3rd Martin Pfost
Chair of Energy Conversion
TU Dortmund University
Dortmund, Germany
martin.pfost@tu-dortmund.de

Abstract— **Half bridges operating at high switching frequencies allow power electronics to achieve higher energy densities. However, they also suffer from proportionally high switching losses. Soft-switching techniques can mitigate these losses, but achieving optimal switching dead times is crucial, as shorter or longer dead times lead to increased switching losses. This study proposes an online dead-time controller that utilizes in-situ measurements. To achieve this, the controller evaluates the steepness of the switch node voltage, leveraging the fact that the maximum steepness increases during hard switching. The proposed algorithm utilizes this dependency independently of system characteristics, thereby eliminating the need for preliminary training.**

Index Terms—**Dead-Time Control, Switching Slope Derivative, Switching Losses, Peak Detection**

I. INTRODUCTION

Modern power electronic applications utilize higher switching frequencies, thus allowing for higher energy densities. However, operating at high frequencies can lead to significant switching losses. These losses can be mitigated through soft switching, which typically necessitates longer dead times to avoid hard switching. Conversely, excessively long dead times can cause substantial reverse-conduction losses [1][2]. Therefore, the dead time should be optimized to be as close as possible to the threshold that ensures soft switching.

Recent studies have introduced dead-time controllers that utilize look-up tables, demonstrating a significant impact on switching efficiency [3]. However, these controllers are strictly limited to a specific set of operational points and require extensive training procedures. More advanced dead-time controllers use analytical transistor models, which perform online calculations of the switching dead time instead of relying on look-up tables [4][5]. Despite these advancements, such controllers lack adaptability to changes of system parameters.

This limitation poses practical challenges, as modifications in system design necessitate retraining. Furthermore, these studies rely on DC current and voltage measurements taken outside the converters. These measurements are influenced not only by the transistors but also by passive system components, which can change over time due to factors like aging and temperature. As a result, they can significantly impact the controller performance.

To address these issues, a gate driver based on the switching slope is being developed in [6]. The design process reveals that a circuit that consists of 18 transistors is required, which necessitates a very sensitive calibration. Moreover, a tightly constrained calibration could lead to critical detection errors in practical applications.

In [7], a dead-time controller is introduced that also relies on measuring the switching slope. The key difference is that it utilizes the simple relationship between hard and soft switching. This controller can operate without a preliminary training process and has been tested in real-time on a running synchronous converter. However, to evaluate the full switching slope, the approach requires an expensive FPGA and a separate ADC to implement the metric, making it less attractive for industrial applications.

This study introduces a dead-time controller that automatically adapts on the switch-node voltage steepness by using a simple analog measurement. This approach eliminates the need for static reference values by leveraging the strong correlation between steepness and dead time during hard switching. This dependency remains valid across various systems, allowing the controller to function effectively without extensive calibration or reliance on specific transistor models. The proposed method does not have special requirements and works even with a cost-effective microcontroller with an integrated ADC.

II. METHODOLOGY FOR DEAD-TIME CONTROL

The proposed method is used in a synchronous converter utilizing a half bridge operating GaN Systems GS66516B at $V_{in} = 400\,\mathrm{V}$. During a switching event, the output capacitances C_{OSS} of the transistors have to be recharged. In a hard switching case, this is done by the opposite transistor during its turn on, resulting in an excessive switching loss. In a DC-to-DC conversion, a half bridge is connected to an inductor. The

979-8-3315-1612-3/25 $31.00 © 2025 IEEE

inductor current i_q can be assumed constant during a switching event. For soft switching, C_{OSS} is recharged by i_q. This asks for a dead time t_D after a transistor gets turned off before the opposite transistor will be turned on. A too short t_D will result in partial hard switching, while a too long t_D results in reverse conduction mode and increased losses. This study proposes a dead-time controller which yields an optimal t_D preventing hard switching, but also large reverse conduction losses.

For this, the strong dependency of the switch-node voltage steepness dv_{sw}/dt on t_D during hard switching is being used. During the brief duration of hard switching, the current that these transistors will handle is significantly higher than i_q. This results in a consistently higher dv_{sw}/dt during hard switching compared to soft-switching operation. During soft switching where i_q fully discharges C_{OSS}, dv_{sw}/dt remains unaffected by t_D even when considering excessively long values of t_D. The dead-time controller utilizes the dependency that if t_D is increased during hard switching, dv_{sw}/dt will continuously decrease until soft switching is achieved. To determine this, the peak value of dv_{sw}/dt during a switching event is captured.

This study uses a a first-order high-pass filter to obtain dv_{sw}/dt. For a better understanding of the method, the circuit diagram of an LTSpice simulation is shown in Fig. 1. Note that the operation of the controller is exemplified for negative switching transients resulting from the transition from the high-side transistor (HS) to the low-side transistor (LS).

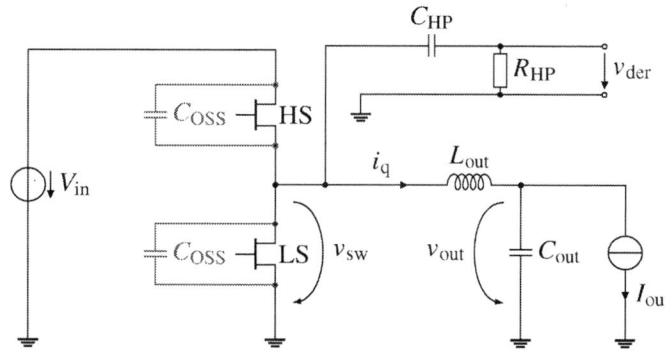

Fig. 1. Overview of the simulation setup with a half bridge in a synchronous converter.

Using this simulation, the impact of t_D on the switching performance is investigated with an outflowing inductor current of $i_q = 2\,\mathrm{A}$ during a negative switching event, i.e. transferring i_q from the HS to the LS. The simulation results in Fig. 2 show the switch node voltage and the high-pass filter output voltage v_{der} for different values of t_D. These simulation results show how the output of the high-pass filter, v_{der}, can clearly represent the derivative dv_{sw}/dt. Furthermore, it becomes evident that the dependency of the minimum v_{der} value, $v_{der,min}$, strongly depends on t_D during partial hard switching, which is the basic requirement for the dead-time controller algorithm.

Dependencies between $v_{der,min}$ and t_D are shown in Fig. 3 for output currents from $i_q = 2\,\mathrm{A}$ to $i_q = 10\,\mathrm{A}$. This shows a decreasing behavior of $v_{der,min}$ with respect to t_D during partial

hard switching, as the portion of v_{sw} where hard switching occurs diminishes. Once soft switching is reached, $v_{der,min}$ remains constant.

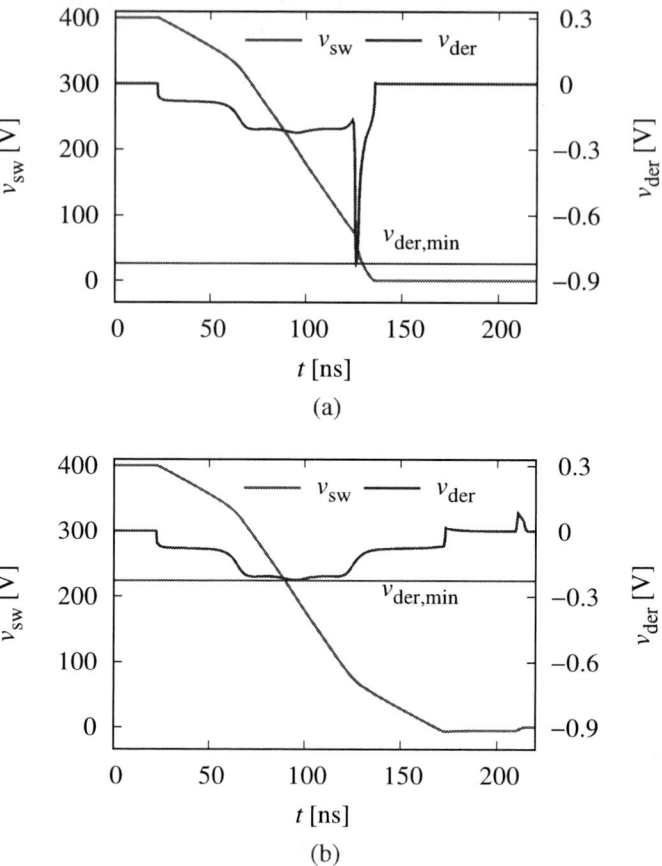

(a)

(b)

Fig. 2. Simulated switch node voltage v_{sw} and its derivative v_{der} with (a) a too short dead time and (b) a too long dead time for a positive current $i_q = 2\,\mathrm{A}$.

The dead-time controller algorithm is shown in Fig. 4. It adjusts t_D by monitoring and utilizing changes in $v_{der,min}$. If an increase in t_D results in a lower dv_{sw}/dt and therefore a lower value of $v_{der,min}$, this indicates a hard-switching operation, which suggests a further increase of t_D. To avoid an excessive t_D, which leads to reverse conduction losses, it should be decreased to identify the precise t_D where $v_{der,min}$ stabilizes. Similarly, when t_D is decreased, a negative change in $v_{der,min}$ indicates a hard switching, while a steady or positive change in $v_{der,min}$ a soft-switching operation, cf. Fig. 3.

III. Test Bench

To demonstrate the functionality of the dead-time controller, it is run online in a test bench while its resulting dead times are recorded as $t_{D,ctrl}$. This test bench consists of a synchronous converter utilizing GaN Systems GS66516B in a half bridge configuration operating at a switching frequency of $f_s = 500\,\mathrm{kHz}$. The converter is powered by a power source (PS) that is set to a constant input voltage of $V_{in} = 400\,\mathrm{V}$. Additionally, a computer runs an operating-point controller that manages an electrical load (EL) using measurements of v_{sw} and i_q. This allows for the benchmarking of the dead-time controller with various inductor currents i_q. On this computer, the dead-time controller operates by using only the measurement of the maximum switch-node voltage steepness $v_{der,rect}$, which is a rectified version of v_{der} and thus corresponds to $v_{der,min}$.

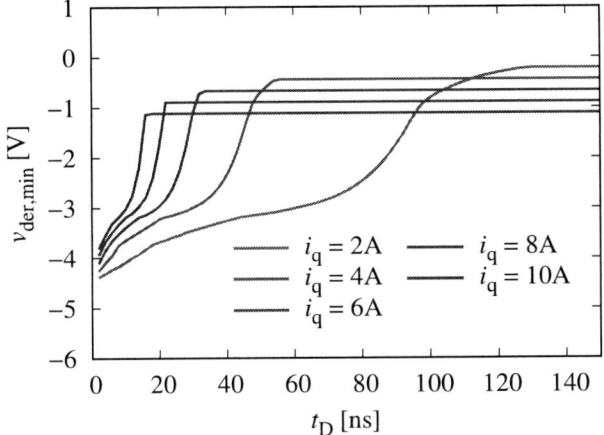

Fig. 3. Dependency of $v_{der,min}$ on t_D with i_q from 2 A to 10 A during a switching event.

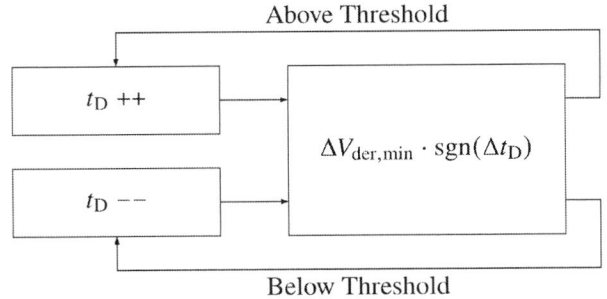

Fig. 4. Dead-time controller using the dependency of $v_{der,min}$ on t_D.

An overview of the test bench is provided by Fig. 5. Additionally, Fig. 6 presents a photograph of the synchronous converter. A more detailed description of the converter circuit is shown in Fig. 7. It also includes the measurement functionality for gathering $v_{der,rect}$, which is explained in the following section.

Centerpiece of the proposed method is the measurement of the maximum switch node voltage steepness: This involves a derivative measurement circuit utilizing first-order high-pass filter, which outputs v_{der}. This high-pass filter consists of a capacitor $C_{HP} = 1$ pF, which is loaded by a resistor $R_{HP} = 50\,\Omega$. The capacitor C_{HP} is implemented directly on the PCB using two copper planes with an area of $4\,\text{mm}^2$, positioned in the first two PCB layers with a separation of $140\,\mu\text{m}$. To compensate for the high impedance of the high-pass filter, it is followed by an operational amplifier, which is configured as a non-inverting amplifier circuit with a gain of $G = 3$ to achieve maximum output swing during hard switching events.

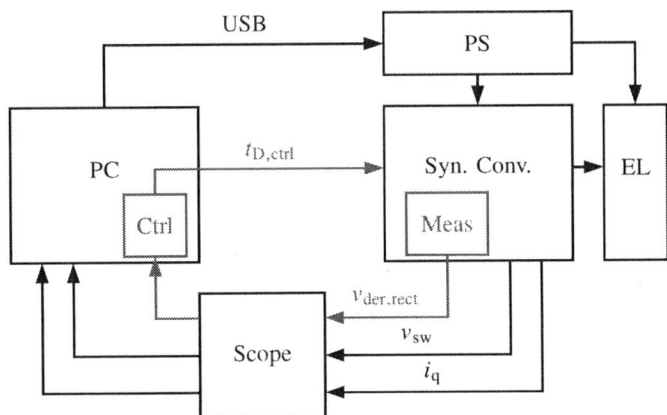

Fig. 5. Overview of the test bench, in which the dead-time controller was tested. All signals used by the dead-time controller are marked red, other signals are used for setting the operational points.

Fig. 6. Photograph of the synchronous converter.

In this example where negative switching transients are considered, hard switching is defined by the negative peak values of v_{der}, seen in $v_{der,min}$. To facilitate continuous measurement of $v_{der,min}$, a peak detection of v_{der} is implemented by a rectifier. That rectifier is composed of a diode D_{rect} and a capacitor $C_{rect} = 1\,\mu\text{F}$, cf. Fig. 7, and outputs $v_{der,rect}$. Given that changes of $v_{der,rect}$ are expected within a few switching cycles while the controller adjusts t_D, the capacitor C_{rect} must be continuously discharged. Here, a discharge resistor $R_{rect} = 100\,\text{k}\Omega$ is connected to the positive supply voltage of the operational amplifier, $V_{CC} = 12$ V. A photograph of the half-bridge along with the associated measurement setup is shown in Fig. 8.

This configuration ensures proper rectification of all negative peak values of v_{der} below 0 V, accounting for the forward voltage drop across the diode. Due to this voltage drop, $v_{der,rect}$ will not accurately represent the peak value of v_{der}. However, this voltage shift does not impact the functionality of the dead-time controller, as it primarily responds to relative changes of $v_{der,rect}$.

To evaluate the performance of the proposed dead-time controller on the test bench, it is essential that i_q can be set

979-8-3315-1612-3/25 $31.00 © 2025 IEEE

to a specific level precisely at the moment of a switching event. This cannot be guaranteed by directly setting I_{out} on the EL, as i_q is also influenced by the output filter ripple current. Therefore, i_q is measured using a current probe directly before a switching event, denoted as $i_{q,sw}$ in Fig. 9. This allows the operating-point controller to regulate $i_{q,sw}$ to a desired value by successively adjusting the electrical load. Any measurements of i_q and v_{sw} are only used for testing the dead-time controller and are not required for the operation of the dead-time controller itself.

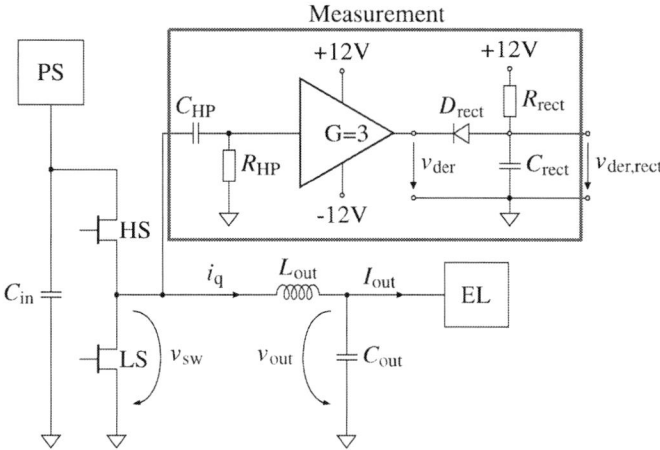

Fig. 7. Circuit of the test bench consisting of a synchronous converter and the derivative measurement circuit. $v_{der,rect}$ is the negative rectification v_{der} and therefore represents $v_{der,min}$ for the dead-time controller.

IV. ONLINE DEAD-TIME CONTROL ON THE TEST BENCH

Before the dead-time controller can be operated, it is essential to verify that the derivative measurement v_{der} yields the expected results. This specifically means that v_{der} shows a distinct negative swing during a hard switching event and that $v_{der,rect}$ shows a significant reaction to that swing. Tests were conducted for an outflowing current of $i_{q,sw} = 2\,\text{A}$, using both a too-short and a too-long dead time. The results are presented in Fig. 10.

These measurements demonstrate the expected dependence of $v_{der,rect}$ on the hard switching of v_{sw}. However, they also reveal that $v_{der,rect}$ does not completely reach the minimum of v_{der} due to the forward voltage drop of the rectification diode D_{rect}. Therefore, it is necessary to further investigate whether the dependency of $v_{der,rect}$ on t_D is sufficient for the algorithm to operate effectively. For this, the dependency of $v_{der,rect}$ requires a constant increase during the hard-switching range until the optimal level of t_D is reached, where they remain independent of t_D. Consequently, $v_{der,rect}$ was measured with $i_{q,sw}$ ranging from 2 A to 10 A, each with t_D varying from 0 ns to 150 ns, cf. Fig. 11. The measurements confirm that this requirement is fulfilled.

After those measurements have shown the sufficient behavior of $v_{der,rect}$, the dead-time controller was operated online. To evaluate its performance, it was enabled on the running converter for 200 switching cycles. For improved visibility of

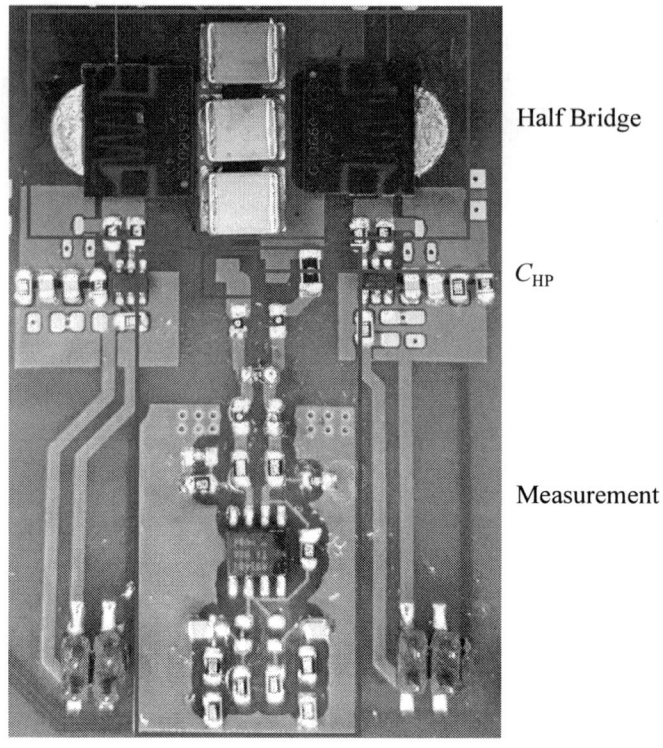

Fig. 8. Detailed photograph of the half bridge and the measurement. Note that C_{HP} is implemented in PCB instead of using a discrete component.

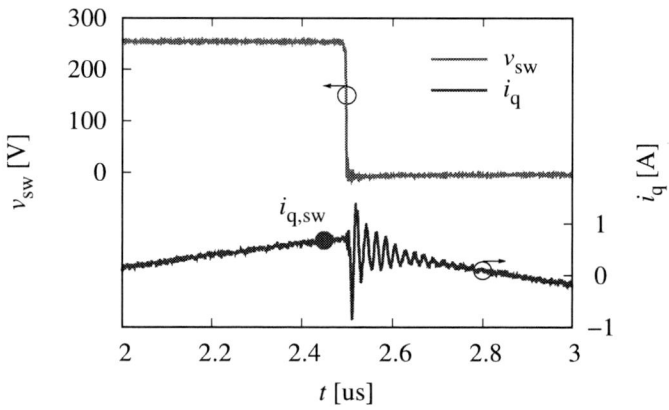

Fig. 9. During operation the measurement of i_q shows a large current ripple. Therefore i_q will be captured directly before the switching event as basis for dead-time control.

the dead-time controller behavior, a distribution of the last 100 cycles, after the dead-time controller had settled, is presented in Fig. 12. The values for $t_{D,ctrl}$ over the 200 switching cycles are shown in Fig. 13.

To evaluate the performance of the dead-time controller, it is necessary to measure the switching losses of the low-side transistor (LS) separately. An investigation of the efficiency of the test bench would encompass the entire synchronous converter, including the output filters and the high-side (HS) transistor. Since only the switching losses of LS are relevant

979-8-3315-1612-3/25 $31.00 © 2025 IEEE

for assessing the dead-time controller, temperature differences between LS and the heat sink, denoted as ΔT_{LS}, are measured as a metric for its losses, see Fig. 14. The minimum of each temperature curve is referenced as $t_{\mathrm{D,opt}}$.

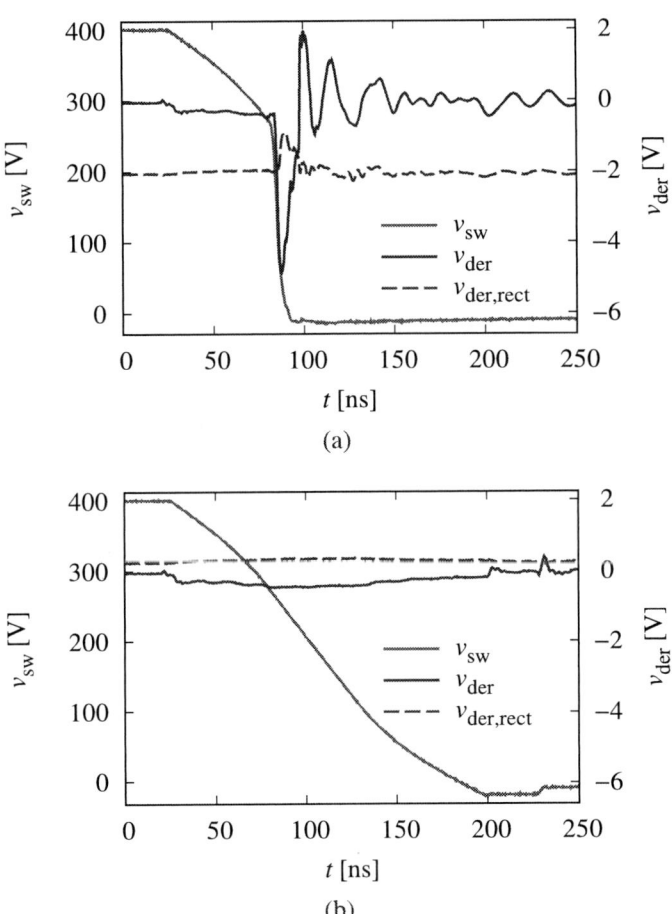

(a)

(b)

Fig. 10. Measurement results showing v_{sw}, v_{der}, and the rectified $v_{\mathrm{der,rect}}$ for an outflowing current of $i_{\mathrm{q,sw}} = 2\,\mathrm{A}$ during a switching event. (a) regards a switching dead time of $t_{\mathrm{D}} = 50\,\mathrm{ns}$ (too short), (b) of $t_{\mathrm{D}} = 200\,\mathrm{ns}$ (too long).

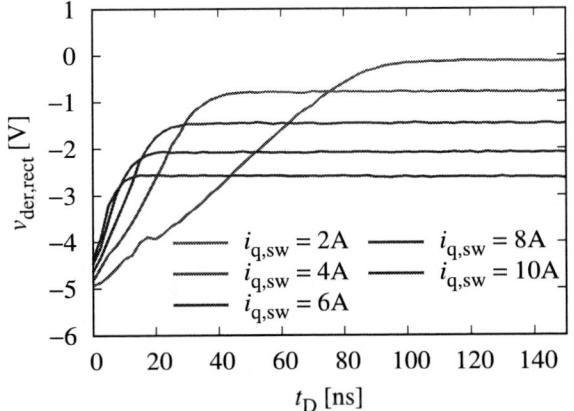

Fig. 11. Measured $v_{\mathrm{der,rect}}$ while running the converter in static conditions with $i_{\mathrm{q,sw}}$ from 2 A to 10 A and t_{D} from 0 ns to 250 ns.

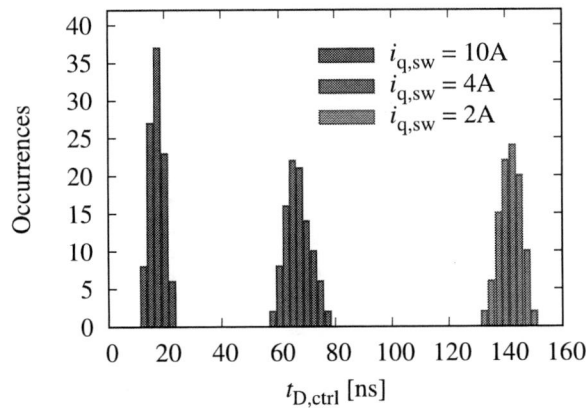

Fig. 12. Distribution of $t_{\mathrm{D,ctrl}}$ during a 200 switch cycle testrun regarding the last 100 switching cycles, where $t_{\mathrm{D,ctrl}}$ has settled.

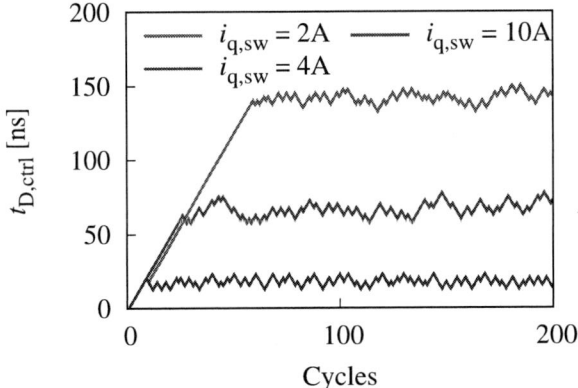

Fig. 13. Values of $t_{\mathrm{D,ctrl}}$ during an online run of the dead-time controller for 200 switching cycles.

A comparison of the resulting controller dead times $t_{\mathrm{D,ctrl}}$ to the empirically determined $t_{\mathrm{D,opt}}$ is presented in Fig. 15 for currents ranging from $i_{\mathrm{q,sw}} = 2\,\mathrm{A}$ to $i_{\mathrm{q,sw}} = 10\,\mathrm{A}$ in increments of 2 A. To enhance the visibility of the controller results, only the values between the first and third quartiles of the distributions for $t_{\mathrm{D,ctrl}}$ are displayed.

These values show that the dead-time controller, on average, matches the $t_{\mathrm{D,opt}}$ values, but maintains a small variation in t_{D}, fluctuating around them. It is important to investigate how significant these fluctuations are. The losses of LS, represented by ΔT_{LS}, are therefore tested for three examples of static t_{D} values and compared to the operation of the dead-time controller in Fig. 16. It can be observed that a short static t_{D} causes ΔT_{LS} to rise to a high level due to hard switching behavior. Conversely, a high t_{D} results in significant reverse conduction losses, particularly at larger currents. Additionally, a comparison of the losses between $t_{\mathrm{D,opt}}$ and $t_{\mathrm{D,ctrl}}$ is illustrated in Fig. 17. An operation with the dead time controller results in a ΔT_{LS}, which is only 2 °C higher than with $t_{\mathrm{D,opt}}$. These tests demonstrate that the effects of the small varying deviation of $t_{\mathrm{D,ctrl}}$ against $t_{\mathrm{D,opt}}$ on the losses remain within a reasonable range.

979-8-3315-1612-3/25 $31.00 © 2025 IEEE

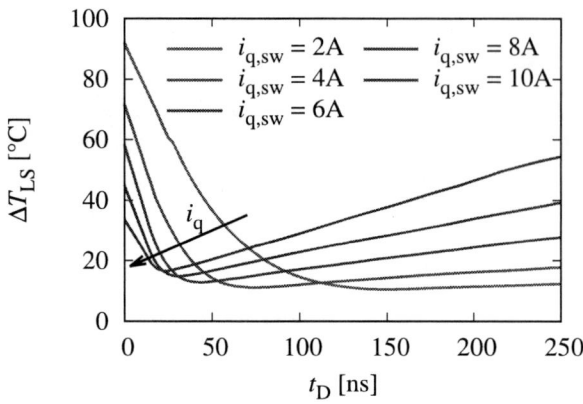

Fig. 14. Losses of LS, represented by ΔT_{LS} for $i_{q,sw}$ from 2 A to 10 A and t_D from 0 ns to 250 ns.

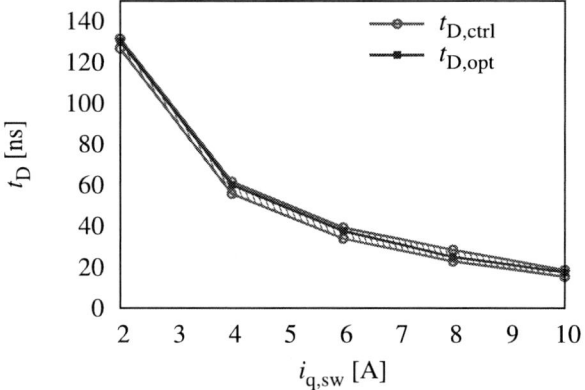

Fig. 15. Resulting $t_{D,ctrl}$ of the dead-time controller compared to $t_{D,opt}$. The dead-time controller, on average, matches the ideal dead times, but remains within a varying deviation.

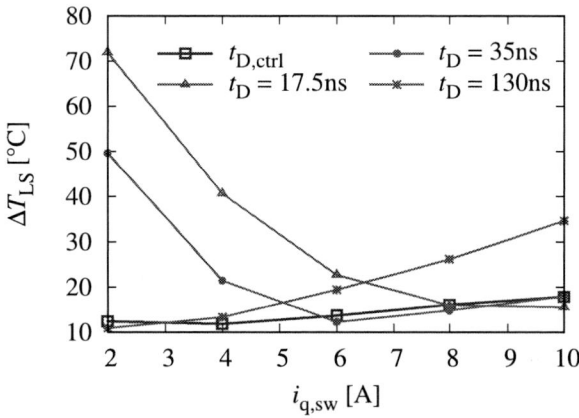

Fig. 16. Losses represented by ΔT_{LS} when operating with three static values for t_D and $t_{D,ctrl}$. The dead-time controller shows reasonably low losses compared to the static dead times.

V. CONCLUSION

This study proposes an active dead-time controller designed to significantly reduce switching losses by dynamically adjusting the dead times to an optimal level during the operation

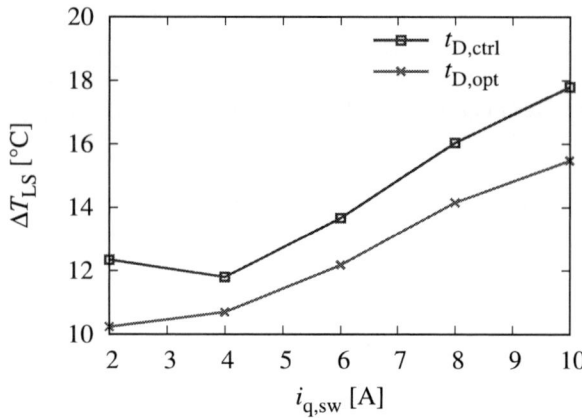

Fig. 17. Losses represented by ΔT_{LS} when operating with the dead-time controller agains operating with $t_{D,opt}$.

of power electronic applications. The controller leverages the relationship between switching transient steepness and hard switching behavior.

It can be implemented on a cost-effective microcontroller with an integrated ADC, eliminating the need for high sample rates. The advantages of this approach include simplicity, cost efficiency, and adaptability. Unlike other methods that rely on pre-trained system models, this approach leverages the intrinsic relationship between switch-node voltage steepness and hard switching behavior, which is common for many systems.

VI. ACKNOWLEDGMENT

This work is the result of activities within the "Automotive Intelligence for Connected Shared Mobility" (AI4CSM) project, which has received funding from the ECSEL Joint Undertaking (JU) under grant agreement No. 101007326. The ECSEL-JU is supported by the European Union's Horizon 2020 research and innovation programme, as well as by Germany, Austria, Norway, Belgium, Italy, the Netherlands, the Czech Republic, Latvia, and India.

REFERENCES

[1] R. Reiner, P. Waltereit, B. Weiss, R. Quay, and O. Ambacher, "Investigation of GaN-HEMTs in reverse conduction," in *PCIM Europe 2017; International Exhibition and Conference for Power Electronics, Intelligent Motion, Renewable Energy and Energy Management*, May 2017, pp. 1–8.

[2] M. Asad, A. K. Singha, and R. M. S. Rao, "Dead time optimization in a GaN-based buck converter," *IEEE Transactions on Power Electronics*, vol. 37, no. 3, pp. 2830–2844, March 2022.

[3] B. Kohlhepp, D. Kübrich, M. Tannhäuser, and T. Duerbaum, "Adaptive dead time in high frequency GaN-inverters with LC output filter," in *The 10th International Conference on Power Electronics, Machines and Drives (PEMD 2020)*, vol. 2020, Dec 2020, pp. 372–377.

[4] Y. Zhang, H. Peng, C. Chen, Y. Xie, T. Liu, and Y. Kang, "A high-efficiency dynamic inverter dead-time adjustment method based on an improved GaN HEMTs switching model," *IEEE Transactions on Power Electronics*, vol. 37, no. 3, pp. 2667–2683, March 2022.

[5] Q. Huang, A. Q. Huang, W. Yu, and R. Yu, "Adaptive zero-voltage-switching control and hybrid current control for high efficiency GaN-based mhz totem-pole PFC rectifier," in *2017 IEEE Applied Power Electronics Conference and Exposition (APEC)*, March 2017, pp. 1763–1770.

[6] W. Zhu, A. Mishra, A. Karmakar, and V. D. Smedt, "A 400V gate driver with complementary slope sensing ZVS detector and high voltage level shifter for full-bridge phase-shifted converters," in *2022 20th IEEE Interregional NEWCAS Conference (NEWCAS)*, June 2022, pp. 193–197.

[7] L. Knappstein, N. Falkenberg, and M. Pfost, "An adaptive dead time control based on the switch-node-voltage derivative," in *PCIM Europe 2024; International Exhibition and Conference for Power Electronics, Intelligent Motion, Renewable Energy and Energy Management*, June 2024, pp. 1740–1745.

Impedance-Based State-of-Health Estimation for Lithium-Ion Battery Management Systems

Mohammad K. Al-Smadi, *Student Member, IEEE* and Jaber A. Abu Qahouq, *Senior Member, IEEE*

The University of Alabama
Department of Electrical and Computer Engineering
Tuscaloosa, Alabama 35487, USA

Abstract— This paper discusses an impedance-based state-of-health (SOH) estimation for lithium-ion battery management systems. Features obtained from the Nyquist plot of the electrochemical impedance spectroscopy (EIS) are utilized to estimate the SOH. Two commercial lithium-ion batteries were aged (cycled) in the laboratory to collect aging and impedance data. An algorithm for SOH estimation is presented and evaluated. The performance evaluation results show that the SOH features extracted from the Nyquist plot can be utilized for SOH estimation. Estimated SOH values can be then utilized in various battery management systems (BMS) functions such as for calibrating the available capacity, adjusting charging/discharging strategies, and protection.

Keywords—Lithium-ion battery, impedance, electrochemical impedance spectroscopy (EIS), Nyquist, state of health estimation.

I. INTRODUCTION

Lithium-ion batteries are widely used in a range of applications such as electric vehicles [1-3], portable electronics [4], and grid energy storage [5]. Developing advanced battery management systems (BMS) has become crucial in meeting the high-performance requirements of battery systems [6]. BMS is essential to maintain safe and efficient operation of battery packs. BMS functions include, but are not limited to, battery balancing, thermal management, protection (e.g., over-current, over-voltage, under-voltage, over-temperature, etc.), and state estimation (e.g., state of charge "SOC", state of health "SOH", state of energy "SOE", and state of power "SOP")[7, 8].

SOH estimation is a critical function of BMS to maintain safe operation of the battery pack and predict potential issues. SOH estimation is also utilized in adjusting charging/discharging strategies [9]. Developing accurate SOH estimation have also become critical in adopting second-life batteries in various applications such as stationary energy storage and electric vehicle charging stations [2, 10]. Battery aging yields power and capacity fade. In terms of capacity fade, SOH can be expressed as the ratio between the available capacity Q_{ava} and the nominal capacity when the battery is new Q_{nom} as in (1). The available capacity Q_{ava} is the charge amount a battery can discharge before reaching its minimum voltage.

$$SOH = Q_{ava}/Q_{nom} \qquad (1)$$

SOH estimation techniques in the literature can classified into experimental and model-based techniques as shown in Fig. 1 [11]. Experimental SOH estimation techniques are based on direct and indirect measurements. Coulomb counting [12] and

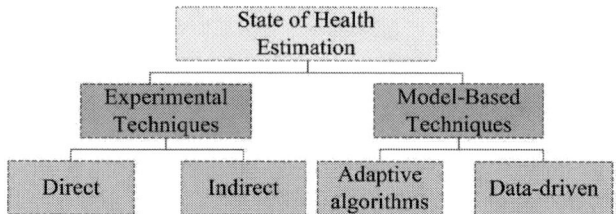

Fig. 1: SOH estimation techniques.

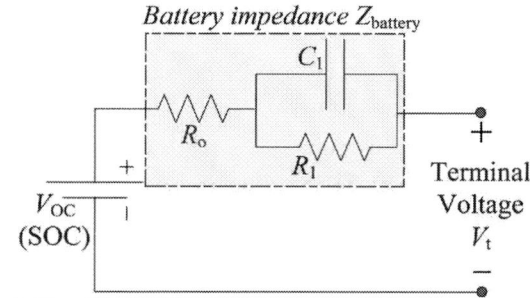

Fig. 2: Simplified equivalent circuit model for a battery.

internal resistance measurement [13] are two popular examples of direct measurement techniques. Coulomb counting is based on counting the amount of charge (Ah) the battery discharges. The internal resistance increases as the battery ages and is usually measured by current pulse method [14, 15]. Indirect experimental techniques are based on analyzing some battery measurements such as differential capacity and voltage analyses[16-18] and ultrasonic methods [19]. Model-based SOH estimation techniques are categorized into adaptive algorithms and data-driven techniques. Examples on adaptive algorithms include particle filter [20] and Extended Kalman Filter (EKF) [21]. Data-driven techniques have attracted more interest to be utilized for SOH estimation as there is no need to fully realize the aging mechanisms [22]. Data-driven techniques utilize SOH features obtained from battery data such charging/discharging curves and impedance curves. For example, the charging and discharging times decrease as the battery capacity deteriorates, and the rate of change of the battery voltage during charging/discharging changes as the battery ages [22]. As the battery ages, its impedance curve changes and that can be noticed in impedance magnitude, phase, and Nyquist plots [23]. Fig. 2 shows a simple equivalent circuit model (ECM) for a battery (referred to as Thevenin model [24]). The shown ECM consists of a variable voltage source (i.e., open circuit voltage or

979-8-3315-1612-3/25 $31.00 © 2025 IEEE

V_{oc}) that indicates the SOC of the battery in addition to resistive and capacitive elements. The parameter R_o represents the internal ohmic resistance of the battery where the parallel RC branch (R_1 and C_1) models the nonlinear behavior of the battery. More RC branches (higher order) can be considered to model the nonlinear behavior of the battery [24]. Both R_o and the parallel RC branch represent the battery impedance. The impedance spectrum is usually measured by electrochemical impedance spectroscopy (EIS) using two methods, namely, potentiostatic EIS, and galvanostatic EIS. In potentiostatic mode, an AC voltage perturbation $v_{ac\text{-}fp}$ is applied to the battery and the AC current response $i_{ac\text{-}fp}$ is recorded. In galvanostatic mode, an AC current perturbation is applied where the AC voltage response is recorded. In both modes, the AC impedance is calculated as expressed in (2), where f_p is the perturbation frequency. The terms $v_{ac\text{-}fp}$ and $i_{ac\text{-}fp}$ are the AC peak voltage and peak current at the perturbation frequency f_p, respectively. The impedance phase is represented by $\theta_{z\text{-}fp}$ and is equal to the phase difference between $v_{ac\text{-}fp}$ and $i_{ac\text{-}fp}$. The voltage or current perturbations are applied over a range of frequency to obtain the EIS spectrum.

$$Z_{fp} = \frac{v_{ac-fp}}{i_{ac-fp}} e^{j\theta_{z-fp}} \tag{2}$$

The battery impedance shows a direct relationship with the SOH value of the battery, therefore, it can be used for SOH estimation [25]. Some examples on utilizing the impedance for SOH estimation include using the impedance value at 316 Hz [26], 0.1 Hz [27], minimum impedance magnitude [9], the impedance magnitude when the impedance phase is equal to 0° [28], and the peak point (both real and imaginary parts) of the Nyquist plot [29]. Some SOH estimation algorithms utilize the entire impedance curve (all real and all imaginary impedance parts) such as the one presented in [30] for 45-mAh coin cell batteries. Also, the EIS curve can be utilized to estimate the SOH by fitting it to find the parameters of the ECM. For example, the EIS curve can be utilized to calculate values for the constant phase elements (CPE) [31], charge transfer resistance [32], and solid electrolyte interphase (SEI) such as in [33].

This paper presents a data-driven impedance based SOH estimator based on some features obtained from the Nyquist plot. The SOH estimation process is explained and the SOH estimation model performance is evaluated. Also, one possible use for the presented SOH estimator is presented for online BMS functions.

II. BATTERY AGING AND SOH INDICATORS

Two commercial lithium-ion battery cells (LG INR18650 3500 mAh [34] with main specifications listed in Table I) were cycled in the laboratory by following the aging protocol shown in Fig. 3. Fig. 4 shows the SOH value of the two aged battery cells Every aging cycle consists of a full discharge at ~1 C (3.4 A) and a full charge. The battery is discharged to its minimum voltage of 2.5 V. The charging process consists of two modes, namely, constant current (CC) charging mode and constant voltage (CV) charging mode. During CC charging mode, the battery charging current is regulated at a constant value (~1 C or 3.4 A in this paper) until the battery cell voltage reaches the maximum value of 4.2 V. Then, CV charging mode is started by regulating the battery cell voltage at 4.2 V until the battery

Table I: Main manufacturer's specifications for LG INR18650MJ1-3.5Ah battery cell [34]

Nominal capacity C	2.5 Ah
Nominal voltage V_{nom}	3.635 V
Minimum voltage V_{min}	2.75 V
Maximum voltage V_{max}	4.2 V

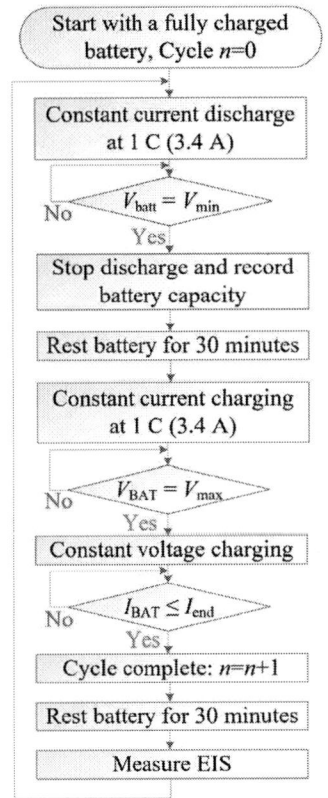

Fig. 3: Battery aging and data collection protocol.

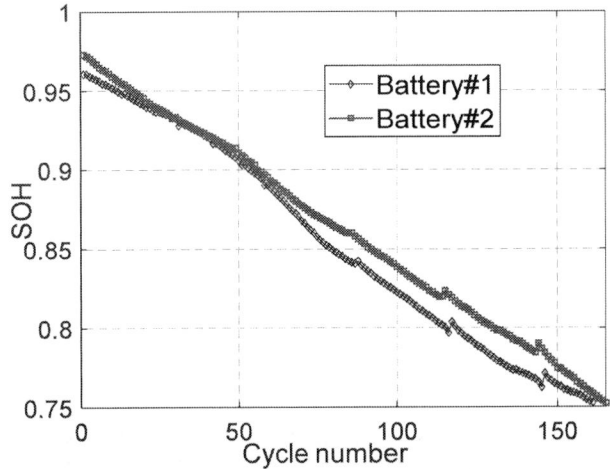

Fig. 4: SOH for two batteries at different aging cycles.

charging current drops below the charging end current I_{end} (50 mA for the cells used in this paper). The battery is rested for 30 minutes after each discharge cycle and after each charge cycle

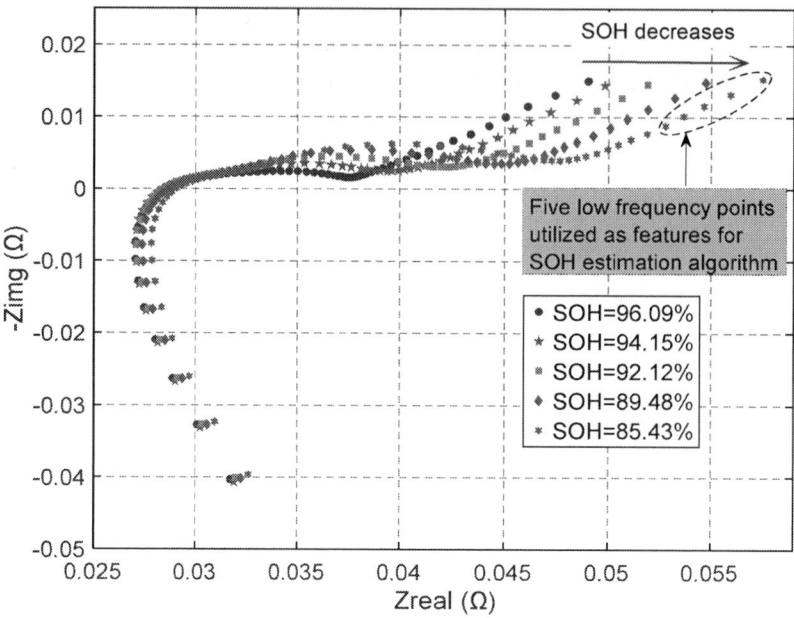

Fig. 5: Nyquist plots at different SOH values.

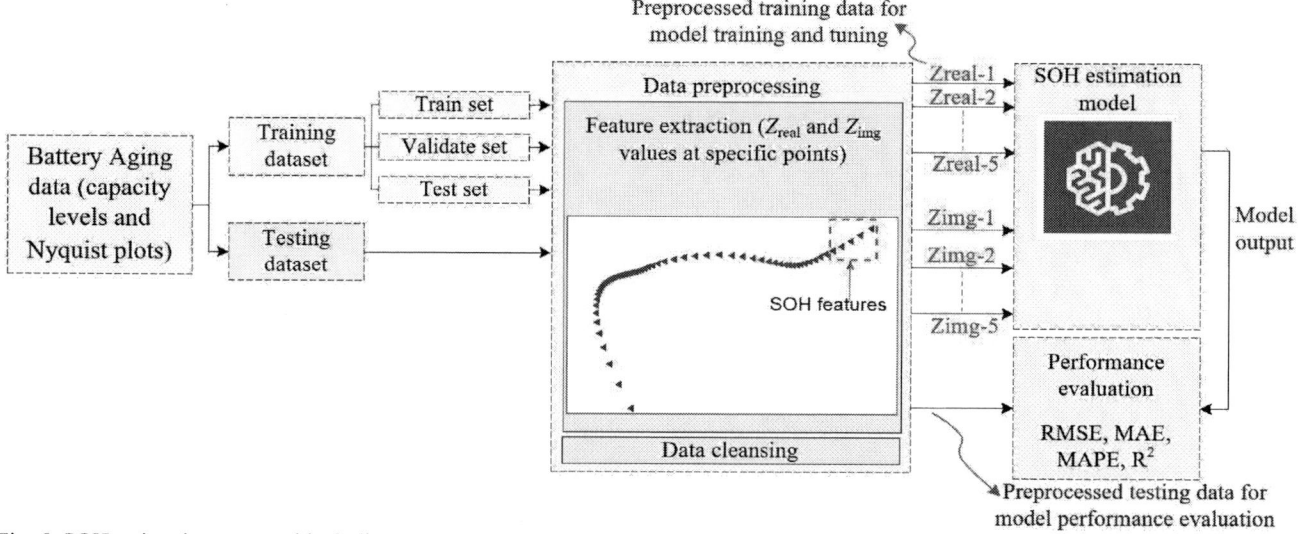

Fig. 6: SOH estimation process block diagram.

After each cycle, the impedance spectrum is measured over a frequency range of 10 kHz to 0.01 Hz with 10 points per decade. Fig. 5 shows sample Nyquist plots at different SOH levels. As can be observed from Fig. 5, the Nyquist plot has notable changes within the very low frequency range (right side of the Nyquist plot) as SOH values decrease. Therefore, the changes within this low frequency range can be utilized for SOH estimation. In this paper, the first five points (0.01 Hz to 0.251 Hz) in the Nyquist plot are utilized for SOH estimation. Each complex impedance point consists of a real part Z_{real} and an imaginary part Z_{img}. Therefore, the total number of points utilized for the SOH estimation is 10 ($Z_{real\text{-}1}$ through $Z_{real\text{-}5}$ and $Z_{img\text{-}1}$ through $Z_{img\text{-}5}$). Although more points can be used for SOH estimation to obtain better accuracy in general, there is a tradeoff

between the number of input features for the SOH estimation algorithm and the computational burden.

III. SOH ESTIMATION ALGORITHM

The SOH estimation process in this paper utilizes the SOH features explained in the previous section, namely, the magnitudes of the real and imaginary impedance parts of the first five complex impedance points in the low frequency range of the Nyquist plot. The SOH estimation process utilizes a neural network (NN) to map the relationship between the SOH indicators and the SOH values. The main block diagram for the presented SOH estimation process is illustrated in Fig. 6. Battery aging data and EIS curves are collected according to the aging and data collection protocol presented in the previous section

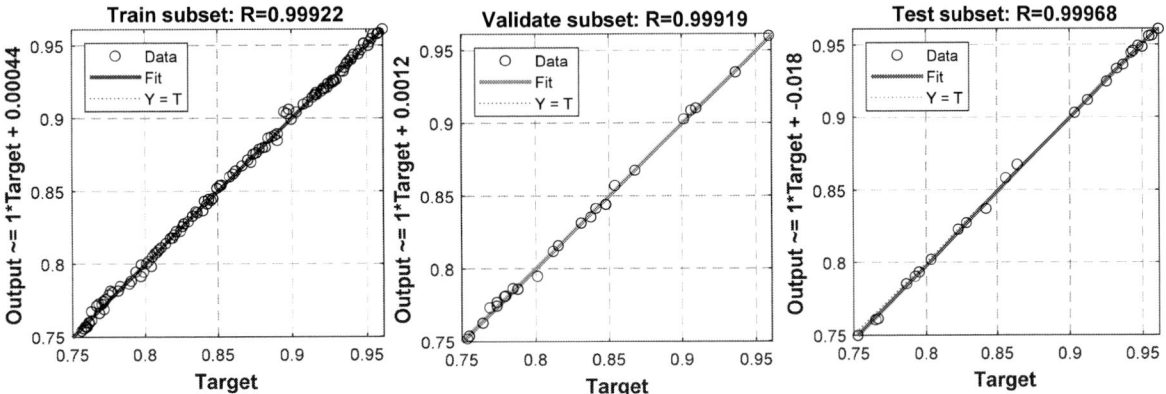

Fig. 7: SOH algorithm (neural network) training session in Matlab showing model training using three subsets of the training dataset.

and shown in Fig. 3. Since aging data is available for two battery cells, the data of one battery cell is used as a training set whereas the data of the second battery cell is used as a testing set. Then, both training and testing datasets undergo data preprocessing in which the battery health features are extracted from the Nyquist plots and the data is cleansed from any outliers. Then, the preprocessed training set is used to train the SOH estimation model. Once the SOH estimation algorithm performance using the training dataset is acceptable, its performance is evaluated on the testing dataset. If the model performance using the testing dataset is not acceptable, the hyper parameters of the neural network (e.gs., number of hidden layers, number of neurons in each hidden layer, activation functions, etc.) are adjusted and the process is repeated until a satisfactory performance is achieved. The loss function used in the neural network training is the mean square error (MSE) which is expressed in (3) where $y(i)$ represents the actual value (true value) of the SOH and $y'(i)$ is the predicted value of the SOH for the ith sample. The SOH estimation algorithm training is carried out using deep learning toolbox in Matlab, and the training algorithm is Levenberg-Marquardt [35].

$$MSE = \frac{1}{n}\sum_{i=1}^{n}|y(i) - \acute{y}(i)| \tag{3}$$

The performance (accuracy) of the SOH estimation algorithm is evaluated through multiple error indices expressed in (4) to (7) where RMSE is the root mean square error, MAE is the mean absolute error, MAPE is the mean absolute percentage error, and R^2 is the coefficient of determination. In (4) to (7), $A(i)$ is the true SOH value and $P(i)$ is the predicted SOH value. In (7), SS_{res} is the sum of squared residuals (differences) between the true SOH values and the predicted SOH values, and SS_{tot} is the sum of squared differences between the true SOH values and their mean.

$$RMSE = \sqrt{\frac{1}{n}\sum_{i=1}^{n}\big(A(i) - P(i)\big)^2} \tag{4}$$

$$MAE = \frac{1}{n}\sum_{i=1}^{n}|A(i) - P(i)| \tag{5}$$

$$MAPE = \frac{100}{n}\sum_{i=1}^{n}\frac{|A(i)-P(i)|}{A(i)} \tag{6}$$

$$R^2 = 1 - \frac{SS_{res}}{SS_{tot}} = 1 - \frac{\sum_{i=1}^{n}\big(A(i)-P(i)\big)^2}{\sum_{i=1}^{n}\big(A(i)-\overline{A(i)}\big)^2} \tag{7}$$

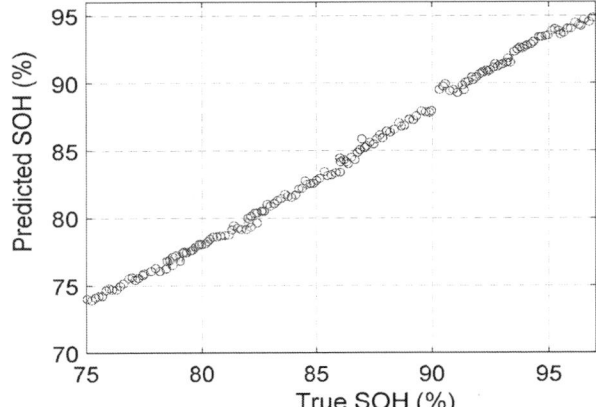

Fig. 8: True SOH vs predicted SOH.

Table II: Performance evaluation results of the SOH estimation algorithm

Error index	Value
RMSE	0.01831
MAE	0.01788
MAPE	0.02090
R^2	0.9172

The neural network utilized in the SOH estimation algorithm consists of one input layer that has 10 inputs (the SOH features), one hidden layer of 10 neurons, and one output layer that gives the predicted SOH value. The training dataset itself is divided into three subsets, namely, train, validate, and test (different from the main testing dataset). The training subset is used to train the algorithm whereas the validate subset is used to evaluate the algorithm's performance during the training phase. The test subset is used to evaluate the final model before testing on the main testing dataset that the algorithm has never seen before. Fig. 7 shows the neural network training session in Matlab.

After training the SOH estimation algorithm, the parameters of the trained neural network were used to predict the SOH value

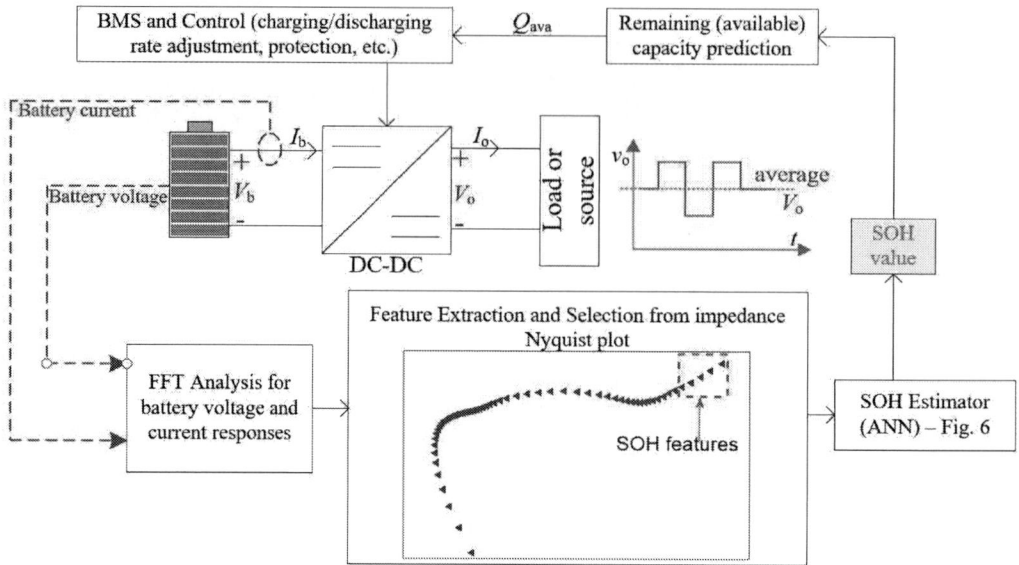

Fig. 9: SOH estimation for real-time BMS functions.

using the testing dataset. The performance (accuracy) indices for the SOH estimation algorithm on the testing dataset are summarized in Table II. The predicted SOH value is plotted against the true SOH value in Fig. 8. As shown, the presented SOH estimation algorithm is capable of accurately predicting the SOH values by utilizing the SOH features presented earlier.

IV. SOH Estimator Unitizatiuon in BMS and Control

The presented SOH estimator can be utilized in various battery powered applications for several functions including, but not limited to, protection, monitoring the SOH of the battery, and estimating the remaining capacity of the battery for user information or for State-Of-Charge (SOC) balancing. For example, for a battery with a nominal capacity Q_{nom} of 5 Ah, the available capacity Q_{ava} when the battery is fully charged (SOC=100%) is equal to 5 Ah when the battery is new (the battery is not aged and therefore $Q_{ava} = Q_{nom}$). As the battery ages, the available battery capacity Q_{ava} when SOC=100% is no longer equal to the nominal capacity ($Q_{ava} \neq Q_{nom}$). For example, if the SOH is estimated to be equal to 80%, then the available battery capacity is equal to 80%×5 Ah=4 Ah. Estimating the available battery capacity can be utilized in different BMS functions such as protection and battery balancing. For a system of multiple batteries where each battery is connected to a DC-DC converter [9], battery balancing can be carried out based on the estimated capacity of each battery. For example, the BMS can utilize the estimated batteries' capacities to draw less current from the battery of low capacity (low SOH value) compared with batteries of higher capacities (high SOH value). Also, BMS can adjust the charge/discharge strategies of the battery based on the estimated remaining capacity (e.g., charging at a lower rate) to prolong the battery lifetime.

Fig. 9 shows one possible use of the presented SOH estimator for online BMS functions. A battery is connected to a load or a charging source through a DC-DC converter. The DC-DC converter control can be utilized to measure the battery impedance [36]. The output voltage reference of the DC-DC converter is perturbed around its average value to generate both voltage and current responses at the battery side (converter's input). The battery voltage and current responses are analyzed by Fast Fourier Transform (FFT) to obtain battery voltage and current harmonics. The harmonics are then utilized to find the battery impedance at the frequency at which the converter's output voltage is perturbed. After obtaining the battery impedance curve, features are extracted from the Nyquist plot and fed into the SOH estimator presented earlier in this paper. The SOH estimator gives a prediction about the battery SOH which can be used to estimate the remaining/available capacity Q_{ava}. Then, the BMS can command changes in charging/discharging rates or trigger protection functions.

V. Conclusion

This paper presents a SOH estimator for lithium-ion batteries by utilizing SOH features obtained from the battery impedance Nyquist plot within the low frequency range. The battery SOH shows a direct connection with the real and imaginary parts of the battery impedance at very low frequency. The performance of the presented SOH estimator is evaluated, and the results show that it can be utilized for remaining capacity calibration, protection, and other BMS functions.

Acknowledgement

This material is based upon work supported in part by the U.S. Department of Energy's Office of Energy Efficiency and Renewable Energy (EERE) under the Vehicle Technologies Program Award Number DE-EE0010402. The views expressed herein do not necessarily represent the views of the U.S. Department of Energy or the United States Government.

References

[1] A. G. Olabi *et al.*, "Battery electric vehicles: Progress, power electronic converters, strength (S), weakness (W), opportunity (O), and threats (T),"

International Journal of Thermofluids, vol. 16, p. 100212, 2022/11/01/ 2022, doi: https://doi.org/10.1016/j.ijft.2022.100212.

[2] Y. Zhao and J. A. Abu Qahouq, "Second-Use Battery System for EV Charging Stations," in *2023 IEEE Energy Conversion Congress and Exposition (ECCE),* 2023: IEEE, pp. 2122-2126.

[3] X. Chen, W. Shen, T. T. Vo, Z. Cao, and A. Kapoor, "An overview of lithium-ion batteries for electric vehicles," in *2012 10th International Power & Energy Conference (IPEC),* 2012: IEEE, pp. 230-235.

[4] Y. Liang *et al.,* "A review of rechargeable batteries for portable electronic devices," *InfoMat,* vol. 1, no. 1, pp. 6-32, 2019.

[5] H. Zhu *et al.,* "Energy storage in high variable renewable energy penetration power systems: technologies and applications," *CSEE Journal of Power and Energy Systems,* vol. 9, no. 6, pp. 2099-2108, 2020.

[6] S. Nyamathulla and C. Dhanamjayulu, "A review of battery energy storage systems and advanced battery management system for different applications: Challenges and recommendations," *Journal of Energy Storage,* vol. 86, p. 111179, 2024.

[7] R. R. Kumar, C. Bharatiraja, K. Udhayakumar, S. Devakirubakaran, S. Sekar, and L. Mihet-Popa, "Advances in batteries, battery modeling, battery management system, battery thermal management, SOC, SOH, and charge/discharge characteristics in EV applications," *Ieee Access,* 2023.

[8] M. K. Al-Smadi and J. A. Abu Qahouq, "Evaluation of current-mode controller for active battery cells balancing with peak efficiency operation," *IEEE Transactions on Power Electronics,* vol. 38, no. 2, pp. 1610-1621, 2022, doi: 10.1109/TPEL.2022.3211905

[9] Z. Xia and J. A. Abu Qahouq, "State-of-charge balancing of lithium-ion batteries with state-of-health awareness capability," *IEEE Transactions on Industry Applications,* vol. 57, no. 1, pp. 673-684, 2020.

[10] V. S *et al.,* "State of Health (SoH) estimation methods for second life lithium-ion battery—Review and challenges," *Applied Energy,* vol. 369, p. 123542, 2024/09/01/ 2024, doi: https://doi.org/10.1016/j.apenergy.2024.123542.

[11] O. Demirci, S. Taskin, E. Schaltz, and B. A. Demirci, "Review of battery state estimation methods for electric vehicles-Part II: SOH estimation," *Journal of Energy Storage,* vol. 96, p. 112703, 2024.

[12] K. S. Ng, C.-S. Moo, Y.-P. Chen, and Y.-C. Hsieh, "Enhanced coulomb counting method for estimating state-of-charge and state-of-health of lithium-ion batteries," *Applied Energy,* vol. 86, no. 9, pp. 1506-1511, 2009/09/01/ 2009, doi: https://doi.org/10.1016/j.apenergy.2008.11.021.

[13] L. Chen, Z. Lü, W. Lin, J. Li, and H. Pan, "A new state-of-health estimation method for lithium-ion batteries through the intrinsic relationship between ohmic internal resistance and capacity," *Measurement,* vol. 116, pp. 586-595, 2018/02/01/ 2018.

[14] Y. Bao, W. Dong, and D. Wang, "Online internal resistance measurement application in lithium ion battery capacity and state of charge estimation," *Energies,* vol. 11, no. 5, p. 1073, 2018.

[15] C. Zhang, J. Liu, and S. Sharkh, "Identification of dynamic model parameters for lithium-ion batteries used in hybrid electric vehicles," 2009.

[16] J. He, Z. Wei, X. Bian, and F. Yan, "State-of-health estimation of lithium-ion batteries using incremental capacity analysis based on voltage–capacity model," *IEEE Transactions on Transportation Electrification,* vol. 6, no. 2, pp. 417-426, 2020.

[17] D.-I. Stroe and E. Schaltz, "Lithium-ion battery state-of-health estimation using the incremental capacity analysis technique," *IEEE Transactions on Industry Applications,* vol. 56, no. 1, pp. 678-685, 2019.

[18] E. Schaltz, D.-I. Stroe, K. Nørregaard, L. S. Ingvardsen, and A. Christensen, "Incremental capacity analysis applied on electric vehicles for battery state-of-health estimation," *IEEE Transactions on Industry Applications,* vol. 57, no. 2, pp. 1810-1817, 2021.

[19] K. Liu, Y. Liu, S. Zhao, X. Li, and Q. Peng, "An Ultrasonic Wave-Based Method for Efficient State-of-Health Estimation of Li-Ion Batteries," *IEEE Transactions on Industrial Electronics,* 2024.

[20] T. Wu, S. Liu, Z. Wang, and Y. Huang, "SOC and SOH joint estimation of lithium-ion battery based on improved particle filter algorithm," *Journal of Electrical Engineering & Technology,* vol. 17, no. 1, pp. 307-317, 2022.

[21] C. Hu, B. D. Youn, and J. Chung, "A multiscale framework with extended Kalman filter for lithium-ion battery SOC and capacity estimation," *Applied Energy,* vol. 92, pp. 694-704, 2012.

[22] Z. Xia and J. A. Abu Qahouq, "Lithium-ion battery ageing behavior pattern characterization and state-of-health estimation using data-driven method," *Ieee Access,* vol. 9, pp. 98287-98304, 2021.

[23] K. Mc Carthy, H. Gullapalli, K. M. Ryan, and T. Kennedy, "Use of impedance spectroscopy for the estimation of Li-ion battery state of charge, state of health and internal temperature," *Journal of the Electrochemical Society,* vol. 168, no. 8, p. 080517, 2021.

[24] X. Ding, D. Zhang, J. Cheng, B. Wang, and P. C. K. Luk, "An improved Thevenin model of lithium-ion battery with high accuracy for electric vehicles," *Applied Energy,* vol. 254, p. 113615, 2019/11/15/ 2019, doi: https://doi.org/10.1016/j.apenergy.2019.113615.

[25] J. A. Abu Qahouq, "An Electrochemical Impedance Spectrum-Based State of Health Differential Indicator with Reduced Sensitivity to Measurement Errors for Lithium–Ion Batteries," *Batteries,* vol. 10, no. 10, p. 368, 2024., doi: https://doi.org/10.3390/batteries10100368

[26] C. T. Love *et al.,* "Lithium-ion cell fault detection by single-point impedance diagnostic and degradation mechanism validation for series-wired batteries cycled at 0 C," *Energies,* vol. 11, no. 4, p. 834, 2018.

[27] A. Eddahech, O. Briat, E. Woirgard, and J.-M. Vinassa, "Remaining useful life prediction of lithium batteries in calendar ageing for automotive applications," *Microelectronics Reliability,* vol. 52, no. 9-10, pp. 2438-2442, 2012.

[28] Z. Xia and J. A. Abu Qahouq, "Adaptive and fast state of health estimation method for lithium-ion batteries using online complex impedance and artificial neural network," in *2019 IEEE applied power electronics conference and exposition (APEC),* 2019: IEEE, pp. 3361-3365.

[29] Y. Fu, J. Xu, M. Shi, and X. Mei, "A fast impedance calculation-based battery state-of-health estimation method," *IEEE Transactions on Industrial Electronics,* vol. 69, no. 7, pp. 7019-7028, 2021.

[30] Y. Zhang, Q. Tang, Y. Zhang, J. Wang, U. Stimming, and A. A. Lee, "Identifying degradation patterns of lithium ion batteries from impedance spectroscopy using machine learning," *Nature communications,* vol. 11, no. 1, p. 1706, 2020.

[31] F. Luo, H. Huang, L. Ni, and T. Li, "Rapid prediction of the state of health of retired power batteries based on electrochemical impedance spectroscopy," *Journal of Energy Storage,* vol. 41, p. 102866, 2021.

[32] X. Wang, X. Wei, and H. Dai, "Estimation of state of health of lithium-ion batteries based on charge transfer resistance considering different temperature and state of charge," *Journal of Energy Storage,* vol. 21, pp. 618-631, 2019.

[33] R. Xiong, J. Tian, H. Mu, and C. Wang, "A systematic model-based degradation behavior recognition and health monitoring method for lithium-ion batteries," *Applied energy,* vol. 207, pp. 372-383, 2017.

[34] LGChem, Product specification: Rechargeable Lithium Ion Battery, Model: INR18650 MJ1 3500mAh.

[35] J. J. Moré, "The Levenberg-Marquardt algorithm: implementation and theory," in *Numerical analysis: proceedings of the biennial Conference held at Dundee, June 28–July 1, 1977,* 2006: Springer, pp. 105-116.

[36] J. A. Abu Qahouq and Z. Xia, "Single-perturbation-cycle online battery impedance spectrum measurement method with closed-loop control of power converter," *IEEE Transactions on Industrial Electronics,* vol. 64, no. 9, pp. 7019-7029, 2017.

Stability Analysis and Resonance Damping of LC Filter-Based Voltage Source Converter With Single-Loop Voltage Control

Aravind G, Divyanshu Bansal, L Umanand
Department of Electronics Systems Engineering
Indian Institute of Science
Bengaluru, India
aravindg@iisc.ac.in, divyanshub@iisc.ac.in, lums@iisc.ac.in

Abstract—**This paper presents a lead compensator-based voltage control and resonance damping of a single-loop LC filter-based Voltage Source Inverter (VSI). Passive damping approaches lead to additional losses in the system; hence active methods to damp the filter resonances are widely utilized. Active methods either need additional current sensors or current estimator. In this paper, the effect of an undamped LC resonance on the system transfer functions is discussed and a lead compensator-based control system is designed to improve the system stability, thereby consequently damping the resonance in the output impedance transfer function as well. A detailed mathematical analysis is provided along with simulation and experimental results to validate the concept.**

Index Terms—**Filter resonance, Active damping (AD), stability, output impedance, Voltage Source Inverter (VSI).**

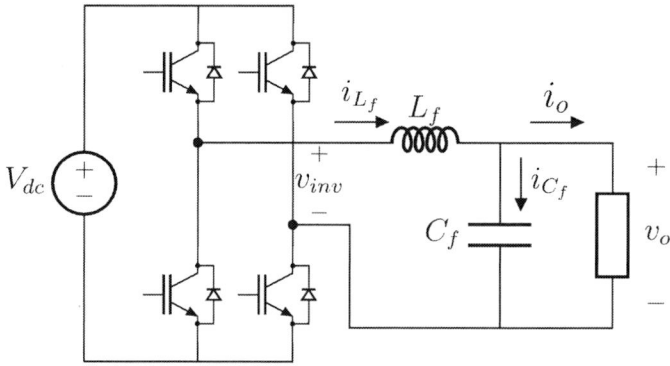

Fig. 1. LC filter-based 1-ϕ VSI with load.

I. INTRODUCTION

Voltage Source Inverter (VSI) systems with LC filter-based voltage control are widely used in applications where output voltage quality is crucial, such as Uninterruptible Power Supplies (UPS), Dynamic Voltage Restores (DVR), grid emulators, Grid Forming Inverters (GFM) [1] [2]. The second-order filtering provides more attenuation to the harmonic voltages present due to the inverter switching as well as a better output impedance profile, thus providing much better output quality when compared to a first-order filter. However, the issue of resonance in the second-order filter must be properly addressed in order to have stable operation of the control system. Passive damping of filter resonance leads to additional power loss in the system and hence damping techniques that combine resistors with capacitors or inductors (and their variants) have been explored to reduce the power dissipation [3]. Active damping methods involves the synthesis of virtual resistors [4] to achieve a virtual damping effect. Like in the passive case, synthesis of combinations of the vitual elements have also been analyzed [5]. However, they typically require additional current sensors in the case of inductor or load current sensing which increases the system cost, or a current estimator for capacitor current-based active damping.

For the output voltage control, both single and multi-loop voltage control methods are discussed in literature. Multi-loop control involves the typical inner current and outer voltage control loops [6]. The outer voltage control loop provides the reference for the inner current loop in which either the filter inductor or filter capacitor current is controlled. Since tight current control is typically not required, a proportional controller-based control is often utilized in practice. A rearrangement of the proportional gain of the current loop into the feedback and reference path reveals that it actually emulates either an inductor or capacitor current-based AD, depending on which current is being sensed [7].

In this paper, a single-loop control structure is realized using a Proportional-Resonant (PR) controller in cascade with a lead compensator. PR controller helps achieve zero steady-state error at its resonant frequency, which will be the required fundamental frequency, while the lead compensator provides the necessary phase gain at the gain crossover frequency (ω_{gc}) of the open-loop gain transfer function, thus improving the phase margin and relative stablity of the system. The phase improvement consequently provides sufficient damping required in the output impedance transfer function at and around the filter resonant frequency. Simulation and experimental results presented shows the effectiveness of the lead compensator in damping the filter resonance. The rest of the paper organization is as follows. Section II gives a brief system description along with the mathematical model. The issues associated with the open-loop operation is discussed. Section III describes the

979-8-3315-1612-3/25 $31.00 © 2025 IEEE

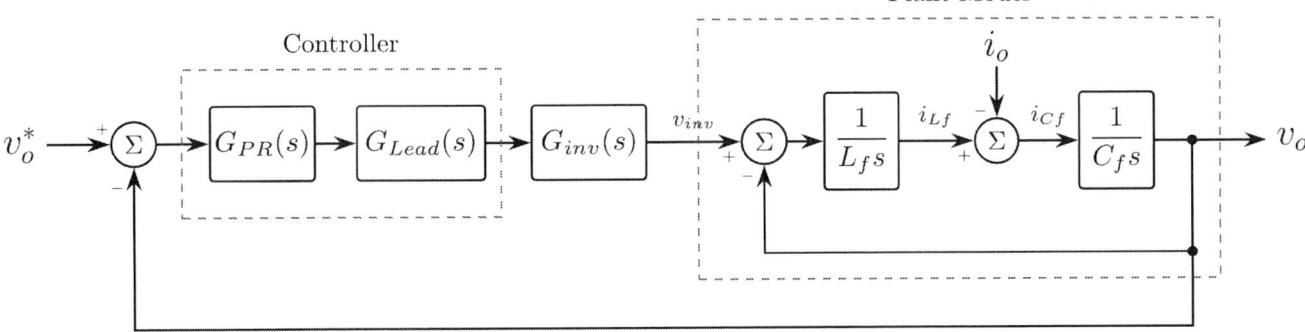

Fig. 2. Lead compensator-based single-loop voltage control block diagram.

controller structure along with the relevant frequency domain plots. The simulation and experimental results are discussed in section IV. Section V concludes the document.

II. SYSTEM DESCRIPTION

The circuit schematic of the system is shown in Fig. 1. It consists of a 1-ϕ VSI cascaded with a second order LC filter. Refering to Fig. 1, the expression for the output voltage (v_o) in Laplace domain can be written as:

$$v_o(s) = \frac{1}{s^2/\omega_f^2 + 1} v_{inv}(s) - \frac{L_f s}{s^2/\omega_f^2 + 1} i_o(s)$$

$$= G_p(s) \times v_o^*(s) + Z_{out_{OL}}(s) \times [-i_o(s)] \quad (1)$$

where,

$G_p(s)$ - Plant transfer function

$Z_{out_{OL}}$ - Open-loop output impedance transfer function

It can be observed from the above equation that the undamped filter resonance causes resonant poles to appear in both the transfer functions. The sharp transition of the phase of the plant from 0° to -180° at the resonant frequency degrades the phase margin of the overall control system, thus leading to an oscillatory behaviour during closed-loop operation theoretically, but can lead to instability because of the delays present in a practical system. On the other hand, the infinite resonant gain in $Z_{out}(s)$ can cause even the smallest disturbance input (persistent or otherwise) to get amplified. Thus, the above problems need to be addressed for the stable operation of the closed-loop control system. Hence, the controller needs to improve the phase in the open-loop transfer function (loop gain) and damp the resonance in the output impedance.

III. CONTROLLER STRUCTURE AND CLOSED-LOOP TRANSFER FUNCTIONS

The block diagram of the closed-loop control system is depicted in Fig. 2. The controller part, as indicated in Fig. 2 uses a Proportional-Resonant (PR) controller cascaded with a lead compensator.

A. PR Controller

The PR controller can be tuned to provide a very large gain at the resonant frequency (50 Hz), thereby resulting in zero steady-state error for a 50 Hz reference input. The transfer function of the PR controller is given as

$$G_{PR}(s) = K_{pr} + \frac{K_{ir}s}{s^2 + \omega_r^2}$$

$$= K_{pr} \frac{s^2 + \frac{K_{ir}}{K_{pr}}s + \omega_r^2}{s^2 + \omega_r^2} \quad (2)$$

The controller gain K_{pr} affects the overall gain while the ratio of $\frac{K_{ir}}{K_{pr}}$ affect the damping of the zero and hence widens the gain around the resonant frequency and also increases the frequency range upto which the phase lag of the PR controller persists. Here, the PR controller gains are chosen in such a way that a desired minimum gain is obtained in a ± 2 Hz band around the resonant frequency of 50 Hz. This limits the influence of the phase of the PR block to a limited region, which is desirable since that phase lag is not desired at and around the unity gain crossover frequency of the loop gain.

B. Lead compensator

The lead compensator is cascaded with the PR controller in order to provide additional phase lead at and around the gain crossover frequency of the loop gain . The transfer function of the lead compensator is given as

$$G_{Lead}(s) = G_o \frac{1 + s/\omega_z}{1 + s/\omega_p}, \quad \omega_p > \omega_z$$

The maximum phase lead for a lead compensator occurs at the geometric mean of the pole and zero frequencies and hence the by selecting the gain crossover frequency as the geometric mean, maximum phase lead can be achieved. However, this method requires the gain (G_o) of the compensator to be less than one since the placement of the zero and pole to the left and right respectively, of the required crossover frequency causes it to shift to the right. Since further reduction in gain is not desired in this system, the zero of the lead compensator is placed exaclty at the filter pole location, which helps to achieve a phase margin of about 50° at ω_{gc} (which is typically chosen

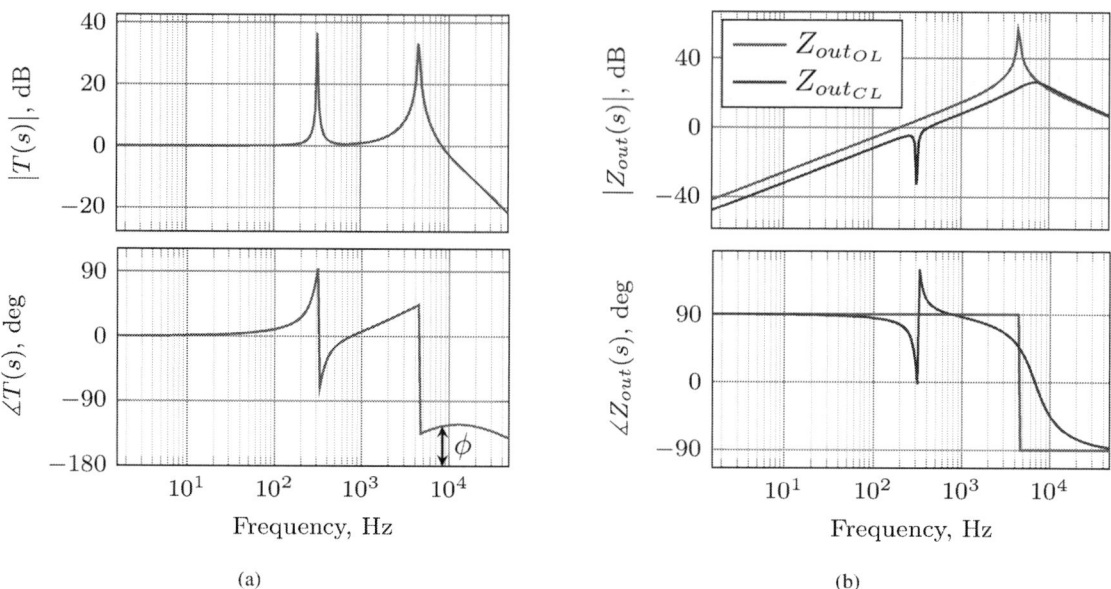

Fig. 3. Frequency response plots of (a) Compensated Loop gain, $L(s)$, (b) Output impedance, $Z_{out}(s)$

very close to the filter resonant frequency). Also, considering a $\pm 15\%$ variation in the pole location to the change in the filter parameters, a reasonable phase margin can still be achieved without too much deviation in ω_{gc}.

C. Power amplifier block

The power amplifier, which is the VSI, can be modelled as a gain with a delay period of one sampling interval (T_s). Double sampling, where the computation and PWM updation is done twice in a carrier cycle, is utilized. Thus, $G_{inv}(s) = V_{dc}e^{-sT_s}$. This block adds an additional phase lag of $36°$ and and $18°$ at $\left(\frac{1}{10}\right)^{th}$ and $\left(\frac{1}{20}\right)^{th}$ of the sampling frequency (F_s) respectively. This additional phase lag is taken into consideration for checking the final available phase margin for the loop gain.

D. Closed-loop system

The expression for the output voltage of the closed-loop system as derived from the system block diagram in Fig. 2, is

TABLE I
SYSTEM PARAMETERS

Parameters	Values
Filter inductor, L_f	5mH
Filter capacitor, C_f	10 μF
Filter resonant frequency, ω_f	711 Hz
Switching frequency, F_{sw}	10 kHz
Sampling time, Ts	20 kHz

given as:

$$v_o(s) = \frac{L(s)}{1 + L(s)} v_o^*(s) + \frac{Z_{out_{OL}}(s)}{1 + L(s)} [-i_o(s)]$$

$$= T(s) \times v_o^*(s) + Z_{out_{CL}}(s) \times [-i_o(s)] \quad (3)$$

where,

$L(s) = G_{PR} \times G_{Lead} \times G_{inv} \times G_P$, is the system loop gain (open-loop transfer function)

$T(s)$ - Transmission function (Reference-to-output transfer function)

$Z_{out_{CL}}$ - Closed-loop output impedance transfer function

The frequency response plots for the loop gain ($L(s)$) along with the uncompensated and compensated output impedance transfer fucntions for the design parameters of Table. I is shown in Fig. 3. The magnitude plot in Fig. 3a exhibits high gain at both resonant integrator frequency (ω_o) and the filter resonant frequency (ω_f). It seems like the resonance is undamped even in the compensated open-loop gain. However, it is evident from the phase plot that due to the addition of the lead compensator, the sharp transition in the phase at the filter resonant frequency does not lead to instability. Also, sufficient phase margin is achieved.

Also, it can be observed from 3b that the resonant peak in the output impedance is effectively suppressed after the addition of the lead compensator. The high gain at the resonant frequency of the numerator and denominator polynomial of $Z_{o_{CL}}(s)$ gets subtracted, thereby limiting the gain to the value of the inductive reactance at the filter resonant frequency and producing a resistor-like effect in the gain plot at ω_f.

In addition, proper choice of the value of the filter components (for the same ω_f) can help bring down the gain further. The regulation performance of the system under harmonic disturbances can be addressed by using multi-resonant PR controllers [8] [9]. This however, not being the focus of this paper will, not be discussed.

IV. RESULTS AND DISCUSSION

The system parameters used for simulation are shown in Table. I. The resonant frequency of the LC filter is selected close to $\left(\frac{1}{10}\right)^{th}$ of the switching frequency of the inverter. The simulation is performed using the switching model of the inverter with all the filter components being ideal. The simulation results for no-load condition is shown in Fig. 4a. The damping provided by external elements during the no-load condition is virtually zero. It can be seen that the lead compensator is stabilizing the system and the reference voltage is getting perfectly tracked. At a point in time (see Fig. 4a), the lead compensator is disables and then re-enabled again. It can be observed that due to the degradation of the relative stability of the system, the oscillations kick in immediately and starts to grow. After re-enabling the lead compensator, oscillations subside and the system stabilizes again in little more than one line cycle. Fig. 4b shows the simulation results for the steady state operation of the inverter when loaded with an inductive load of 2.5 kVA. It can be seen that the controller perfectly tracks the reference voltage, thus providing a good quality output voltage. Experimental results for no-load condition for a voltage reference peak of 150 V is shown in Fig. 5. The DC bus voltage is 200 V. It is evident that the lead compensator is able to stabilize the system and the reference voltage is getting perfectly tracked. Lead compensator is then disabled for 5 ms duration (shown by the bottom trace) and then re-enabled. It can be observed that oscillations at the filter resonant frequency is triggered immediately. However, because of the damping present in the practical system and also due to the reduced DC bus voltage, the peak voltage rise is not as high as in the simulation. The re-enabling of the compensator causes the oscillation to die down immediately. Experimental results thus show the effectiveness of the lead compensator in damping the system.

V. CONCLUSION

A single-loop voltage controlled LC filter-based VSI is presented. The issues caused by an undamped second order filter is discussed and a simple solution involving a lead compensator in cascade with a resonant controller is presented, which improves the phase margin of the system loop gain transfer function and consequently damps the resonant peak in the output impedance transfer function, thereby ensuring stable operation. Additional current sensor, which is typically used for active damping (in case of inductor and load current-based methods) can be done away with, thereby reducing the overall system cost. Simulation studies performed using the inverter switching model along with the experimental results validates the concept.

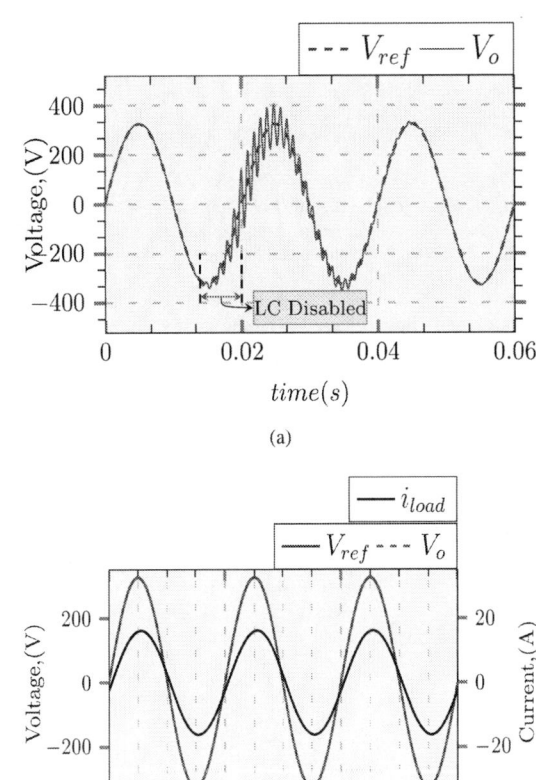

(a)

(b)

Fig. 4. Simulation results. (a) Lead compensator (LC) disabled for a short interval during no-load condition, (b) Steady-state performance for 2.5 kVA load

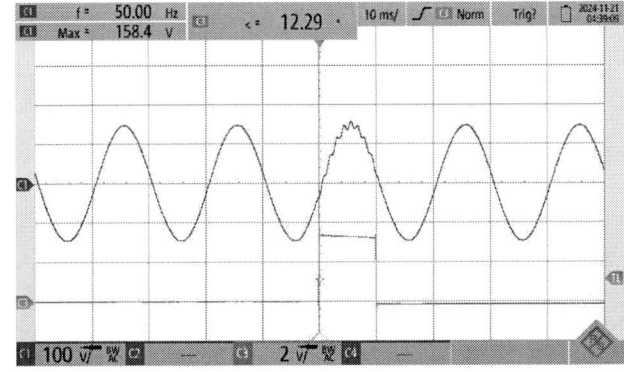

Fig. 5. Experimental results showing output voltage (C1), lead compensator disable and re-enable (C3) during no-load condition.

REFERENCES

[1] Y. W. Li, "Control and resonance damping of voltage-source and current-source converters with *lc* filters," *IEEE Transactions on Industrial Electronics*, vol. 56, no. 5, pp. 1511–1521, 2009.

[2] X. Wang, Y. W. Li, F. Blaabjerg, and P. C. Loh, "Virtual-impedance-based control for voltage-source and current-source converters," *IEEE Transactions on Power Electronics*, vol. 30, no. 12, pp. 7019–7037, 2015.

[3] A. K. Balasubramanian and V. John, "Analysis and design of split-capacitor resistive-inductive passive damping for lcl filters in grid-connected inverters," *IET Power Electronics*, vol. 6, no. 9, pp.

979-8-3315-1612-3/25 $31.00 © 2025 IEEE

1822–1832, 2013. [Online]. Available: https://ietresearch.onlinelibrary. wiley.com/doi/abs/10.1049/iet-pel.2012.0679

[4] P. Dahono, Y. Bahar, Y. Sato, and T. Kataoka, "Damping of transient oscillations on the output lc filter of pwm inverters by using a virtual resistor," in *4th IEEE International Conference on Power Electronics and Drive Systems. IEEE PEDS 2001 - Indonesia. Proceedings (Cat. No.01TH8594)*, vol. 1, 2001, pp. 403–407 vol.1.

[5] X. Wang, F. Blaabjerg, and P. C. Loh, "Virtual rc damping of lcl-filtered voltage source converters with extended selective harmonic compensation," *IEEE Transactions on Power Electronics*, vol. 30, no. 9, pp. 4726–4737, 2015.

[6] P. C. Loh, M. Newman, D. Zmood, and D. Holmes, "A comparative analysis of multiloop voltage regulation strategies for single and three-phase ups systems," *IEEE Transactions on Power Electronics*, vol. 18, no. 5, pp. 1176–1185, 2003.

[7] V. Blasko and V. Kaura, "A novel control to actively damp resonance in input lc filter of a three-phase voltage source converter," *IEEE Transactions on Industry Applications*, vol. 33, no. 2, pp. 542–550, 1997.

[8] L. F. A. Pereira, J. V. Flores, G. Bonan, D. F. Coutinho, and J. M. G. da Silva, "Multiple resonant controllers for uninterruptible power supplies—a systematic robust control design approach," *IEEE Transactions on Industrial Electronics*, vol. 61, no. 3, pp. 1528–1538, 2014.

[9] F. Hans, W. Schumacher, S.-F. Chou, and X. Wang, "Design of multifrequency proportional–resonant current controllers for voltage-source converters," *IEEE Transactions on Power Electronics*, vol. 35, no. 12, pp. 13 573–13 589, 2020.

979-8-3315-1612-3/25 $31.00 © 2025 IEEE

Finite Control Set Model Predictive Control Combined with Online Junction Temperature Estimation for Reliability Enhancement of Voltage Source Inverters

Qiang Mu, Jiale Zhou, Zaheen Mustakin, Lucas Pereira, Babak Parkhideh, Tiefu Zhao
Dept. of Electrical and Computer Engineering
Energy Production and Infrastructure Center (EPIC)
University of North Carolina at Charlotte
Charlotte, NC, USA
qmu1@charlotte.edu; jzhou20@charlotte.edu;
zmustaki@charlotte.edu; lpereira@charlotte.edu;
bparkhideh@charlotte.edu; tzhao5@charlotte.edu

Abstract—As the efficiency of the power converter has significantly improved, its long-term reliable operation has become a critical factor in most power electronics applications. Among these factors, monitoring real-time temperature variations in the power converter during operation is a key priority for assessing reliability and predicting its lifetime. However, limited research has been conducted on the online monitoring of junction temperatures in power semiconductors using on-resistance measurements. Online temperature monitoring can help prevent individual device failure caused by transient thermal overstress, which could otherwise undermine the reliability of the entire converter. The power semiconductor junction temperature prediction method discussed in this paper still relies on the less commonly used direct measurement of on-resistance. Building on this approach, a combined method is proposed that integrates finite control set model predictive control (FCS-MPC) to minimize power loss and thermal stress in power semiconductors, thereby improving the overall reliability of the converter. Above all, acknowledging the variations in parameter distribution among identical power semiconductor devices, this method aims to estimate the online junction temperatures of all SiC power MOSFETs in a three-phase, two-level voltage source inverter. The simulation results demonstrate that the proposed method enables online estimation of the junction temperature of all SiC power MOSFETs in a three-phase inverter, and even allows closed-loop regulation of the junction temperature of the SiC power MOSFETs. Simultaneously, the reliability of the power MOSFET is notably enhanced due to the considerable reduction in junction temperature achieved by the proposed method. In addition, the single-vector MPC under finite set model predictive control, along with the double-vector and triple-vector approaches under continuous set model predictive control, are used for comparative analysis with the proposed method. After comparison, the excellence of the proposed method was finally verified through simulation.

Keywords—junction temperature, on-resistance, voltage source inverter, model predictive control (MPC), single-vector, double-vector, triple-vector, reliability assessment.

I. INTRODUCTION

The voltage source inverter is widely used in renewable energy generation, energy storage, uninterruptible power supply and electric vehicle drive systems, mainly due to its simple design [1-2]. With significant improvements in the efficiency of the voltage source inverter, its long-term reliable operation has become a crucial factor in most power electronics applications. Power semiconductors are recognized as one of the most vulnerable components in power converters, with their failure rate closely correlated to the junction temperature during operation. Thus, it is essential to keep these temperatures within safe limits. Despite considerable efforts by the scientific community and power module manufacturers, accurately measuring the junction temperature of power semiconductor devices continues to be a challenge. Effective thermal management of SiC devices is especially challenging due to their low thermal inertia and high power density. One way to prevent temperature-related failures is by estimating junction temperatures, which allows for precise thermal management and improves converter reliability. Over time, several methods have been developed to measure or estimate the junction temperature of power semiconductors. As noted in [3], current methods for estimating the temperature of power semiconductor devices can be broadly classified into optical, physical contact, and electrical methods, each offering distinct advantages and disadvantages. Optical methods involve detecting the temperature-dependent optical properties of semiconductors, such as through thermal imaging of the semiconductor die with infrared cameras [4-8]. These methods offer high accuracy and produce detailed thermal images of the semiconductor die, allowing for easy identification of the maximum temperature point and the temperature gradient across the die. However, these methods require visual access to the chip, removal of the dielectric gel, and additional computational resources to analyze the thermal images. Physical contact methods involve making direct contact between the die and a heat-sensitive material. This approach necessitates mechanical contact with the die inside the module and typically provides limited accuracy and dynamic response.

979-8-3315-1612-3/25 $31.00 © 2025 IEEE

Electrical methods, especially those utilizing thermosensitive electrical parameters (TSEPs), have emerged as the most promising approach for estimating junction temperature in power semiconductors. These methods use the power semiconductor device itself as a temperature sensor, converting its junction temperature into electrical variables. TSEPs in SiC MOSFETs can be categorized into transient and steady-state groups. Transient TSEPs include parameters such as turn-on saturation current [9], turn-on or turn-off delay [10-11], internal gate resistance [12], and gate threshold voltage (V_{th}) [13-15]. Steady-state TSEPs mainly refer to parameters like on-state voltage (V_{on}) [16-17] and on-resistance (R_{dson}) [18-20]. Additionally, some transient TSEPs have been proposed, including turn-on drain-source current overshoot [21], turn-off drain-source voltage undershoot [22], and gate current [23]. Notably, V_{dson} and the on-state voltage drop of the body diode (V_{bdon}) are often combined as indicators of junction temperature (T_j) [24]. Among the various TSEPs, the relationship between on-resistance (R_{dson}) and junction temperature in MOSFETs is especially reliable and practical. This correlation enables direct estimation of T_j, facilitating real-time temperature monitoring to prevent individual device failures caused by transient thermal overstress, thus improving the reliability of the entire converter. Therefore, this paper uses on-state resistance (R_{dson}) as an indicator for the online junction temperature of power semiconductor devices.

At the same time, based on the successful online monitoring of junction temperature, it is still necessary to combine some advanced control methods to improve the reliability of the converter. Thermal stress has been identified as one of the main causes of failure in power modules. The thermal stress caused by power loss can accelerate the degradation of semiconductor devices and reduce system reliability. Adjusting the switching frequency is the most straightforward method for achieving thermal control, as it directly affects power switching losses. MPC, a nonlinear control technology, has received more attention due to its advantages of flexible control and simple implementation [25-26]. A converter topology with a specific level has a limited number of switching states; therefore, MPC heuristically selects the best switching state from the possibilities by minimizing an objective function. At present, few scholars have studied the method of using MPC to improve reliability [27]. Although FCS-MPC was used in combination with on-site junction temperature to improve reliability in [28], it did not take into account the parameter differences of semiconductor devices that lead to junction temperature differences. Therefore, acknowledging the variations in parameter distribution among identical power semiconductor devices, this study aims to estimate the online junction temperatures of all SiC power MOSFETs in a three-phase, two-level voltage source inverter. Additionally, while monitoring the online junction temperature of power semiconductor devices, the finite control set model predictive control method can be applied further by incorporating power loss as a secondary objective within the control function. This approach helps reduce power loss in semiconductor devices, thereby minimizing their thermal stress.

This paper presents an FCS-MPC method integrated with online junction temperature prediction to limit the rise in

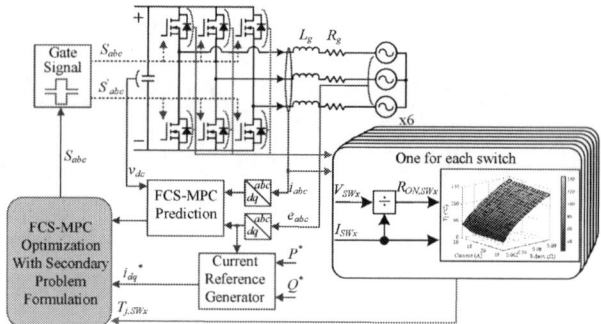

Fig. 1. Finite control set model predictive control diagram with online junction temperature estimation.

junction temperature of power devices, thereby improving the reliability of the entire converter. The proposed method enables online estimation of the junction temperatures for all silicon carbide (SiC) power MOSFETs in a three-phase inverter and even allows closed-loop regulation of the junction temperature of the SiC power MOSFETs.

II. THE PROPOSED FINITE CONTROL PREDICTIVE CONTROL METHOD WITH ONLINE JUNCTION TEMPERATURE ESTIMATION

Fig. 1 illustrates finite control set model predictive control diagram combined with online junction temperature estimation. It mainly includes two parts: FCS-MPC and online junction temperature estimation. In the FCS-MPC framework, the primary objective is to implement power flow control, while the secondary objective focuses on minimizing power loss, which is determined using the power loss model of semiconductor devices. The secondary problem formulation, which employs a similar l_2 norm-2 least squares objective function for power loss, is presented as follows [29]:

$$J_s = \sum_{j=1}^{H} ||0 - E_{abc}[k+j]||_2^2 \qquad (1)$$

Here, 0 represents the reference power loss, E_{abc} represents the power loss calculated using the loss model, and J_s corresponds to the secondary objective. Thus, the combined objective function, which integrates power flow control and power loss reduction, is expressed as follows:

$$\min_{S_{abc}} \lambda_p \cdot J_p + \lambda_s \cdot J_s \qquad (2)$$

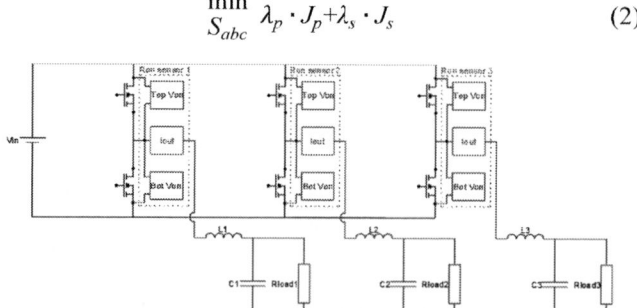

Fig. 2. Three-phase voltage source inverters architecture with proposed $R_{ds\text{-}on}$ estimation architecture and package.

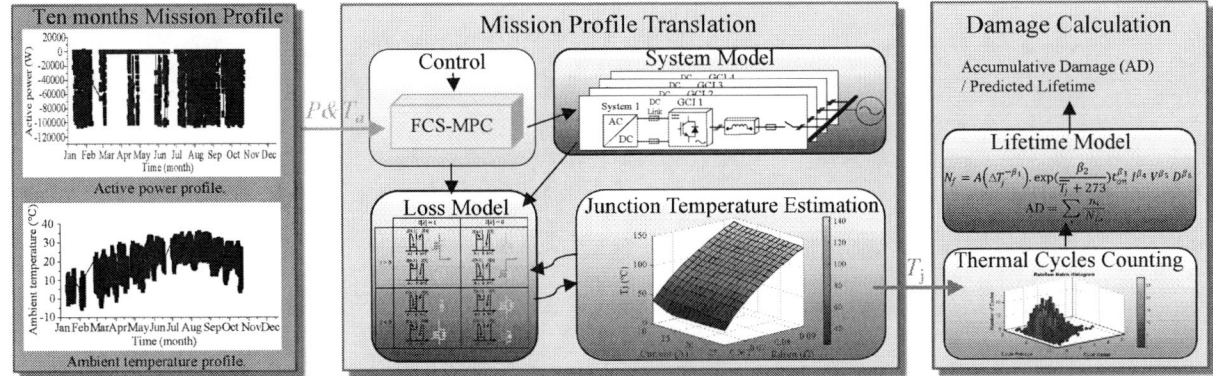

Fig. 3. Reliability assessment framework of semiconductor for voltage source inverter inverters using the proposed method [30].

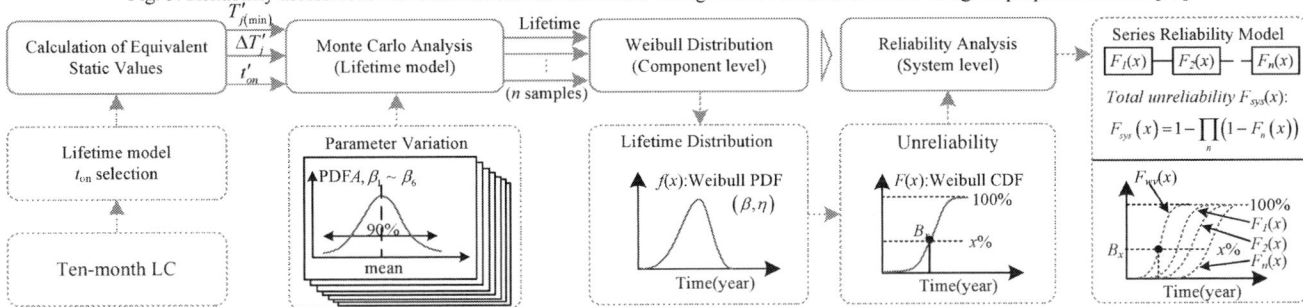

Fig. 4. Flow diagram of the reliability assessment of SiC inverters with the Monte Carlo analysis and reliability block diagram. LC: lifetime consumption, t_{on}: heating time, T_j(min): equivalent minimum junction temperature, ΔT_j: equivalent cycle amplitude, t'on: equivalent heating time, $f(x)$: Weibull Probability density function (PDF), β: sharp parameter, η: scale parameter, $F(x)$: cumulative density function (CDF), $F_i(x)$: unreliability function of the ith device in the system, $F_{sys}(x)$: total unreliability of the system, and B_x: operation time when $x\%$ of the populations fail [31].

Here, S_{abc} represents the switching state of the power semiconductor devices, J_p denotes the primary objective value for power flow control, and λ_p and λ_s are the weighting factors for the primary and secondary objective functions, respectively. λ_p is set to 1 by default here. The value of λ_s will be selected based on the requirements of the specific application.

For the online junction temperature estimation part, to measure the junction temperature online, ultrafast sensors are used to monitor the on-state current and voltage of all power devices. In [32], the traditional method of measuring $R_{ds\text{-}on}$ requires each phase bridge arm's upper and lower power devices to have their own voltage sensors, while the current of the power device is determined using the output current sensor. However, this approach lacks better sensor integration and involves more complex isolation of power supplies. Figure 2 illustrates the design of three-phase voltage source inverters architecture with proposed $R_{ds\text{-}on}$ estimation architecture and package. The proposed $R_{ds\text{-}on}$ estimation architecture forms a standalone package by positioning the current sensor between the voltage sensors of the upper and lower power devices, effectively addressing the integration and isolation power supply challenges associated with the traditional method. These measurements are then fed into a lookup table, created from double pulse tests in [28], which defines on-resistance as a function of junction temperature and conduction current. If all semiconductor devices were to be tested, the workload required in practice would be too large. This paper aims to verify the proposed control method; therefore, only a single power semiconductor is tested, and the corresponding mapping is obtained. The

remaining mappings are derived from the obtained data by making minor adjustments to generate all the mappings. The junction temperature is subsequently determined online during operation by referencing this lookup table, based on the real-time sampled conduction current and voltage of the device. Finally, the junction temperature of all SiC power MOSFETs, derived from the lookup table, is integrated into the previously analyzed secondary objective function of the FCS-MPC, enabling the closed-loop regulation of their junction temperatures.

III. RELIABILITY ASSESSMENT BASED ON THE PROPOSED METHOD

The reliability assessment framework is illustrated in Fig. 3, is based on the real-world mission profiles of the PV system in North Carolina. The field data has been processed to create the mission profile, which serves as the input for the reliability assessment framework. To translate the mission profile between the input and output of the framework, control methods, system models, loss models, and junction temperature estimation are required. The framework outputs the accumulative damage or predicted lifetime of the semiconductors. The quantified

TABLE I - EQUIVALENT STATIC VALUES OF THE STRESS PARAMETERS

Mean junction temperature T'_{jm}	21.8°C
Cycle amplitude $\Delta T'_j$	0.1741°C
Cycle period t'_{on}	0.01s
Number of cycles per ten-month n'_i	$(10\times30\times24\times60\times60)\times50$

TABLE II - SIMULATION PARAMETERS

Symbol	Parameters	Values
P	Rated power	20 kW
V_{dc}	DC-link voltage	1200 V
e_{abc}	Grid voltage	277/480 V
f	Grid frequency	60 Hz
i_{abc}	Grid current	24 A
L_g	Line inductance	20 mH
R_g	Line resistance	10 mΩ
MOSFET	Semiconductor device	C2M0080120D
T_a	Air temperature	25 °C
$R_{\theta AH}$	Thermal resistance from air to heatsink	0.5 °C/W
$R_{\theta HC}$	Thermal resistance from heatsink to case	0.5 °C/W

accumulative damage is closely associated with the semiconductor specifications, PV inverter topology, and control methods. The results are representative and will be further refined upon receiving the detailed design and specifications from the inverter vendor. The framework serves as a universal tool for quantitatively comparing the reliability of PV inverters from different vendors employing various control methods. Based on the ten-month lifetime of the power device calculated above, it can be concluded that the lifetime of the corresponding power device is a fixed value. This often deviates from reality because variations in device parameters and thermal stresses are not accounted for [33]. In practice, these uncertainties can cause the lifetime of power devices to vary within a specific range [34]. Therefore, lifetime predictions are typically expressed as statistical values rather than fixed values. Building on the analysis above, a statistical approach utilizing Monte Carlo analysis is applied, as illustrated in Fig. 4. Table I summarizes the equivalent static values derived from the mission profile.

IV. SIMULATION RESULTS

To validate the effectiveness of the proposed finite predictive control method with online junction temperature estimation, simulations are performed comparing the proposed method with single-vector MPC under finite set model predictive control, the double-vector and triple-vector approaches under continuous set model predictive control, using the MATLAB and PLECS environments. The three traditional modulated MPC strategies mentioned above are not the focus of

TABLE III - SIMULATION PARAMETERS OF THE LIFETIME MODEL

V	1200	Blocking voltage
A	$9.34*10^{14}$	Technology factor
I	300	Current per wire bond
D	$150*10^{-5}$	Diameter of bonding wire
β_1	-4.416	Contribution of Coffin-Manson law
β_2	1285	Contribution of Arrhenius law
β_3	-0.463	Influence of transient thermal response on the chip
β_4	-0.716	Contribution of accelerated wire bonds failure close to end of life
β_5	-0.761	Accounted correlation between blocking voltage and chip thickness
β_6	-0.19	Considered impact of wire diameter on bond interface and thermal expansion

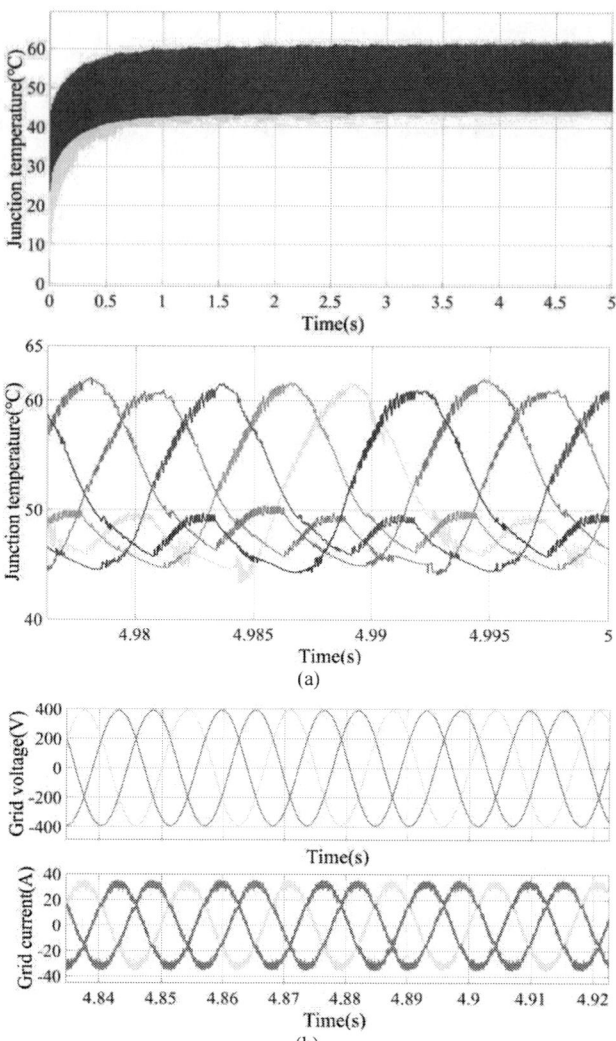

Fig. 5. (a) The junction temperature of all MOSFETs and corresponding zoom-in details (b) Grid voltage and current (Sampling period T_s=33μs, weighting factor $\lambda_s = 250*10^4$).

this paper; further details on these methods can be found in [35]. The primary simulation parameters are listed in Table II. Furthermore, to minimize the data volume in the mission profile, the ambient temperature profile is kept constant, the ambient temperature profile is kept constant, while the active power profile is scaled down to a maximum limit of 20 kW. Simultaneously, it is assumed that the mission profile repeats annually. The simulation parameters of the lifetime model are selected from Table III.

As illustrated in Figure 5, when the grid voltage and current operate under normal conditions, the six power MOSFETs with varying on-resistances exhibit nearly identical junction temperatures, with a maximum value of approximately 60°C. A closer examination of their junction temperature details reveals that the maximum temperature difference among them does not exceed 1°C. Therefore, the proposed method enables the online estimation of junction temperatures for all SiC power MOSFETs in a three-phase voltage source inverter and facilitates the closed-loop regulation of their junction temperatures by

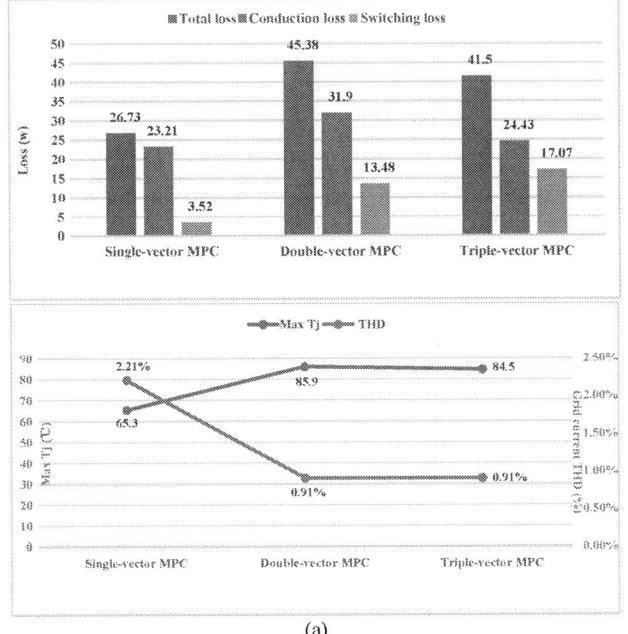

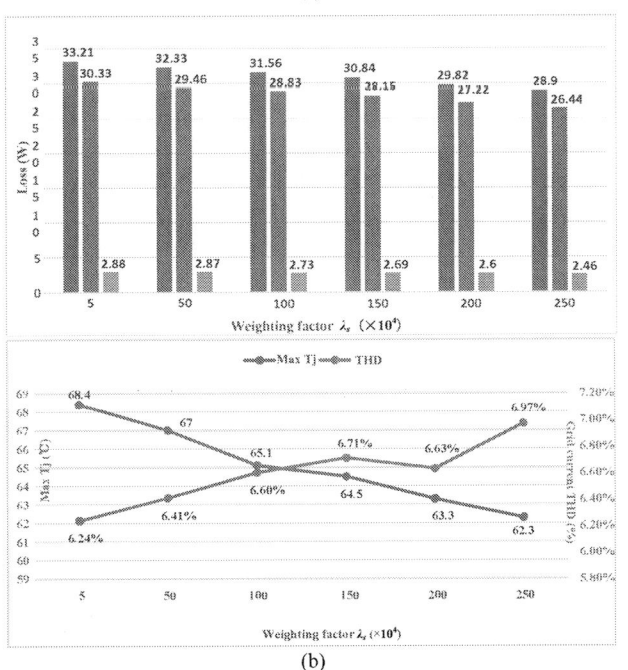

Fig. 6. Quantitative comparison of different control strategies (a) Other modulated MPC methods (b) Proposed method (T_s=33μs).

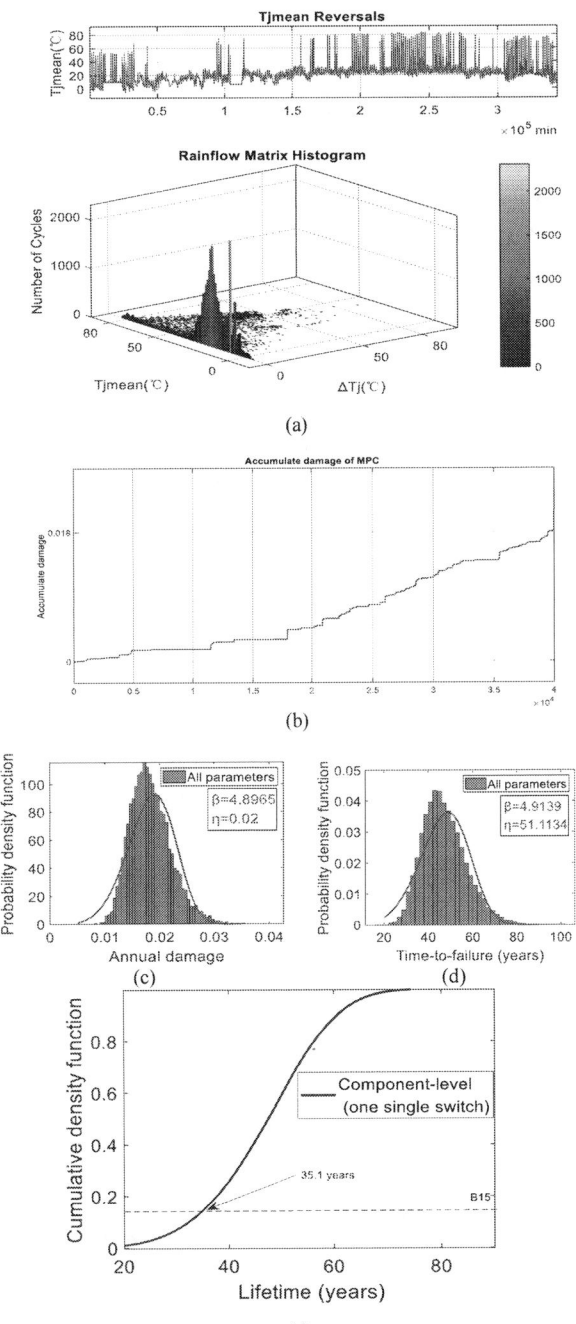

Fig. 7. Reliability assessment based on the proposed method (a) Rainflow counting results (b) Accumulative damage results (c) Annual damage; (d) Time-to-failure distribution; (e) Cumulative density function along with the lifetime (T_s=25μs, $\lambda s = 5*10^4$).

incorporating the junction temperature into the previously analyzed secondary objective function of the FCS-MPC. A comparison between the proposed method and conventional modulated MPC methods is shown in Fig. 6. It is evident that, with the proposed method, as the weighting factor increases, the maximum junction temperature of the power semiconductor decreases, achieving lower values compared to other methods. For instance, when the weighting factor λ_s is set to 250, the maximum junction temperature of the power MOSFETs is 3°C lower than that achieved by the traditional single-vector MPC.

Therefore, the reliability of the converter is significantly enhanced. However, the THD of grid current correspondingly increases. Nevertheless, if reliability improvement is prioritized, the proposed method demonstrates superior performance. Simultaneously, it is observed that although the double-vector and triple-vector MPC methods are compared at the same fixed switching frequency, theoretical analysis indicates that the triple-vector MPC incurs higher switching losses due to its use of additional vectors. However, its conduction loss is lower than that of the corresponding double -vector MPC. Additionally, while the theoretical THD value for the triple-vector method is expected to be higher than that of the double-vector method, at higher switching frequencies, the change in THD becomes negligible or remains unchanged.

Based on the previous reliability analysis, the component-level unreliability functions of the converters are presented in Fig. 7, along with their corresponding component-level B_{15} lifetimes. FCS-MPC, as a nonlinear control algorithm, selects the optimal switching state from a finite set of candidates by minimizing the objective function. As a result, the system-level B_{15} lifetime does not reflect uniform unreliability across all power devices. Specifically, the component-level B_{15} lifetime for the MOSFET under the proposed method is 35.1 years.

V. CONCLUSION

In this article, a finite control set model predictive control (FCS-MPC) scheme combined with online junction temperature estimation is proposed. The simulation results indicate that the proposed method facilitates the online estimation of junction temperatures for all SiC power MOSFETs in a three-phase inverter, enabling the closed-loop regulation of their junction temperatures. Additionally, the method significantly enhances the reliability of power MOSFETs by substantially reducing their junction temperatures, but it sacrifices the corresponding THD of the grid current. Finally, the superiority of the proposed method is verified by comparing it with the single-vector MPC under finite set model predictive control, and the double-vector and triple-vector MPC methods under continuous set model predictive control. This paper lays the foundation for better applying MPC algorithm to reliability improvement projects in the future.

REFERENCES

[1] J. Teng, X. Sun, M. Zhang, W. Zhao and X. Li, "Low-Capacitance CHB-Based SST Based on Resonant Push–Pull Decoupling Channel," in *IEEE Transactions on Industrial Electronics*, vol. 71, no. 3, pp. 2477-2488, March 2024.

[2] Z. Wang *et al.*, "Common-Mode Voltage Suppression Strategy for CHB-Based Motor Drive Based on Topology and Modulation Optimization," in *IEEE Transactions on Power Electronics*, vol. 40, no. 1, pp. 1697-1716, Jan. 2025.

[3] Y. Avenas, L. Dupont and Z. Khatir, "Temperature Measurement of Power Semiconductor Devices by Thermo-Sensitive Electrical Parameters—A Review," in IEEE Transactions on Power Electronics, vol. 27, no. 6, pp. 3081-3092, June 2012.

[4] C. Fang, T. An, F. Qin, X. Bie and J. Zhao, "Study on temperature distribution of IGBT module," 2017 18th International Conference on Electronic Packaging Technology (ICEPT), Harbin, China, 2017.

[5] E. Valor et al., "Comparison of in Situ Land Surface Temperatures Measured with Radiometers and Pyrgeometers: Consequences for Calibration and Validation of Thermal Infrared Sensors," IGARSS 2018 - 2018 IEEE International Geoscience and Remote Sensing Symposium, Valencia, Spain, 2018, pp. 7961-7964.

[6] L. Rossi, G. Breglio, A. Irace and P. Spirito, "Thermal Mapping of Power Devices with a Completely Automated Thermoreflectance Measurement System," 2006 Ph.D. Research in Microelectronics and Electronics, Otranto, Italy, 2006, pp. 41-44.

[7] O.Olanrewaju, Z. Yang, N. Evans, A. Fayyaz, T. Lagier and A. Castellazzi, "Investigation of Temperature Distribution in SIC Power Module Prototype in Transient Conditions," 2019 20th International Symposium on Power Electronics (Ee), Novi Sad, Serbia, 2019, pp. 1-5.

[8] L. Dupont, Y. Avenas and P. -O. Jeannin, "Comparison of Junction Temperature Evaluations in a Power IGBT Module Using an IR Camera and Three Thermosensitive Electrical Parameters," in IEEE Transactions on Industry Applications, vol. 49, no. 4, pp. 1599-1608, July-Aug. 2013.

[9] H. -C. Yang, R. Simanjorang and K. Y. See, "A Method for Junction Temperature Estimation Utilizing Turn-on Saturation Current for SiC MOSFET," 2018 International Power Electronics Conference (IPEC-Niigata 2018 -ECCE Asia), Niigata, Japan, 2018, pp. 2296-2300.

[10] F. Yang, S. Pu, C. Xu, and B. Akin, "Turn-on Delay Based Real-Time Junction Temperature Measurement for SiC MOSFETs With Aging Compensation," in IEEE Transactions on Power Electronics, vol. 36, no. 2, pp. 1280-1294, Feb. 2021.

[11] Z. Zhang et al., "Online Junction Temperature Monitoring Using Intelligent Gate Drive for SiC Power Devices," in IEEE Transactions on Power Electronics, vol. 34, no. 8, pp. 7922-7932, Aug. 2019.

[12] H. Niu and R. D. Lorenz, "Sensing Power MOSFET Junction Temperature Using Gate Drive Turn-On Current Transient Properties," in IEEE Transactions on Industry Applications, vol. 52, no. 2, pp. 1677-1687, March-April 2016.

[13] X. Jiang et al., "Online Junction Temperature Measurement for SiC MOSFET Based on Dynamic Threshold Voltage Extraction," in IEEE Transactions on Power Electronics, vol. 36, no. 4, pp. 3757-3768, April 2021.

[14] B. Strauss and A. Lindemann, "Measuring the junction temperature of an IGBT using its threshold voltage as a temperature sensitive electrical parameter (TSEP)," 2016 13th International Multi-Conference on Systems, Signals & Devices (SSD), Leipzig, Germany, 2016, pp. 459-467.

[15] M. Du, Y. Tang, M. Gao, Z. Ouyang, K. Wei and W. G. Hurley, "Online Estimation of the Junction Temperature Based on the Gate Pre-Threshold Voltage in High-Power IGBT Modules," in IEEE Transactions on Device and Materials Reliability, vol. 19, no. 3, pp. 501-508, Sept. 2019.

[16] M. Guacci, D. Bortis and J. W. Kolar, "On-state voltage measurement of fast switching power semiconductors," in CPSS Transactions on Power Electronics and Applications, vol. 3, no. 2, pp. 163-176, June 2018.

[17] Y. Zhang and Y. C. Liang, "A simple approach on junction temperature estimation for SiC MOSFET dynamic operation within safe operating area," 2015 IEEE Energy Conversion Congress and Exposition (ECCE), Montreal, QC, Canada, 2015, pp. 5704-5707.

[18] Q. Zhang, G. Lu, Y. Yang and P. Zhang, "A High-Frequency Online Junction Temperature Monitoring Method for SiC mosfets Based on on-State Resistance With Aging Compensation," in IEEE Transactions on Industrial Electronics, vol. 70, no. 7, pp. 7393-7405, July 2023.

[19] A. Koenig, T. Plum, P. Fidler and R. W. De Doncker, "On-line Junction Temperature Measurement of CoolMOS Devices," 2007 7th International Conference on Power Electronics and Drive Systems, Bangkok, Thailand, 2007, pp. 90-95.

[20] J. O. Gonzalez, O. Alatise, J. Hu, L. Ran and P. A. Mawby, "An Investigation of Temperature-Sensitive Electrical Parameters for SiC Power MOSFETs," in IEEE Transactions on Power Electronics, vol. 32, no. 10, pp. 7954-7966, Oct. 2017.

[21] Q. Zhang, G. Lu and P. Zhang, "A High-Sensitivity Online Junction Temperature Monitoring Method for SiC mosfets Based on the Turn-on Drain–Source Current Overshoot," in IEEE Transactions on Power Electronics, vol. 37, no. 12, pp. 15505-15516, Dec. 2022.

[22] Y. Yang, Y. Wu, X. Ding and P. Zhang, "Online Junction Temperature Estimation Method for SiC MOSFETs Based on the DC Bus Voltage Undershoot," in IEEE Transactions on Power Electronics, vol. 38, no. 4, pp. 5422-5431, April 2023.

[23] H. Niu and R. D. Lorenz, "Sensing Power MOSFET Junction Temperature Using Gate Drive Turn-On Current Transient Properties," in IEEE Transactions on Industry Applications, vol. 52, no. 2, pp. 1677-1687, March-April 2016.

[24] Q. Zhang, W. Li and P. Zhang, "An Online Junction Temperature Estimating Method for SiC MOSFETs Based on Steady-State Features and GPR," in IEEE Transactions on Industrial Electronics, vol. 71, no. 10, pp. 13299-13309, Jan. 2024.

[25] Y. Zhang *et al.*, "A Robust Predictive Current Control of T-Type Three-Level Power Converters Based on Adaptive Linear Neural

Network," *2023 IEEE 2nd International Power Electronics and Application Symposium (PEAS)*, Guangzhou, China, 2023, pp. 1122-1127.

[26] H. Chen, Z. Zhang, Z. Li, P. Zhang and M. Zhang, "Data-Driven Predictive Current Control for Active Front Ends with Neural Networks," *2022 IEEE 17th Conference on Industrial Electronics and Applications (ICIEA)*, Chengdu, China, 2022, pp. 201-206.

[27] M. -H. Nguyen, s. kwak and S. Choi, "Model Predictive Control Algorithm for Prolonging Lifetime of Three-Phase Voltage Source Converters," in IEEE Access, vol. 11, pp. 72781-72802, 2023.

[28] J. Zhou *et al.*, "Finite Control Set Model Predictive Control Based on In-Situ Junction Temperature for Reliability Enhancement of Power Converters," *2023 IEEE 10th Workshop on Wide Bandgap Power Devices & Applications (WiPDA)*, Charlotte, NC, USA, 2023, pp. 1-6.

[29] L. Wang, J. He, T. Han, and T. Zhao, "Finite Control Set Model Predictive Control With Secondary Problem Formulation for Power Loss and Thermal Stress Reductions," in *IEEE Transactions on Industry Applications*, vol. 56, no. 4, pp. 4028-4039, July-Aug. 2020.

[30] Q. Mu, J. Zhou, L. Wang, and T. Zhao. "Universal Reliability Assessment of Inverters in Photovoltaic Systems Based on Real-Field Mission Profiles," 2024 IEEE Energy Conversion Congress and Exposition (ECCE), Phoenix, AZ, USA, 2024, in press.

[31] J. He, A. Sangwongwanich, Y. Yang and F. Iannuzzo, "Lifetime Evaluation of Three-Level Inverters for 1500-V Photovoltaic Systems," in IEEE Journal of Emerging and Selected Topics in Power Electronics, vol. 9, no. 4, pp. 4285-4298, Aug. 2021.

[32] C. Roy, N. Kim, J. Gafford and B. Parkhideh, "On-State Voltage Measurement of High-Side Power Transistors in Three-Phase Four-Leg Inverter for In-Situ Prognostics," *2021 IEEE Energy Conversion Congress and Exposition (ECCE)*, Vancouver, BC, Canada, 2021, pp. 2770-2776.

[33] K. Kuang, X. Guo, C. Li, X. Li, X. Xi and H. Fang, "A Novel Multiscale Perspective Based Hotspot Temperature Assessment Method for Film Capacitor in DC-Link Applications," in *IEEE Transactions on Industry Applications*, vol. 60, no. 6, pp. 9111-9122, Nov.-Dec. 2024.

[34] K. Kuang, X. Guo, C. Li and X. Li, "A Novel Lifetime Estimation Method and Structural Optimization Design for Film Capacitors in EVs Considering Material Aging and Power Losses," in *IEEE Transactions on Device and Materials Reliability*, vol. 24, no. 3, pp. 365-379, Sept. 2024.

[35] L. Guo, M. Chen, Y. Li, P. Wang, N. Jin and J. Wu, "Hybrid Multi-Vector Modulated Model Predictive Control Strategy for Voltage Source Inverters Based on a New Visualization Analysis Method," in *IEEE Transactions on Transportation Electrification*, vol. 9, no. 1, pp. 8-21, March 2023.

Framework for Dynamic Control and Operation of Power Electronics Interfaces

Radha Sree Krishna Moorthy
Grid Systems Architecture Group
Oak Ridge National Laboratory
Knoxville, TN, USA
krishnamoorr@ornl.gov

Steven Campbell
Grid Systems Architecture Group
Oak Ridge National Laboratory
Knoxville, TN, USA
campbellsl@ornl.gov

Abstract— **The paper introduces a dynamic control framework for power electronics (PE) interfaces, aimed at improving resource utilization, flexibility, and system reliability. The proposed strategy enables PE interfaces to respond effectively to multiple abnormalities at the point of connection (POC) while delivering multiple ancillary services from a single entity. Key features include seamless and interlaced autonomous and non-autonomous transitions with a response time of less than 10 switching cycles. Validation of the approach is demonstrated using a controller hardware-in-the-loop (CHIL) testbed, showcasing its effectiveness for grid applications. The paper also highlights the system-level benefits achievable through this strategy, emphasizing its potential to enhance operational efficiency and reliability.**

Keywords—ancillary services, autonomous response, dynamic control, grid architectures, multimode-transition control, solid state power substations, reliability.

I. INTRODUCTION

Distribution networks are facing new challenges owing to the growing integration of distributed energy resources (DERs). The need for system support from DERs has become imperative with the evolution of modern power system and the increasing demand for greater system reliability and efficiency. To enhance reliability, DERs can help maintain the system voltage and frequency by dynamically adjusting their output during abnormalities [1].

Traditionally, support from DERs evolved to ease interconnection around the globe. Grid support services like Volt-Var capability were initially mandated to allow for higher PV penetration without impacting the hosting capacity of the network. Ride through capabilities were later recommended to maintain system stability and reliability [1]. Additionally, other functions like Freq-Watt were pivotal to overcome high frequency events as in the "German 50.2 Hz problem" [2]. In the evolving renewable energy landscape, entities including distribution utilities and independent system operators are mandating coordinated and improved dynamic response from DERs. A few examples of this can be seen throughout several sources [3-5].

Besides interconnections, new aggregation and coordination architectures for distribution—such as solid-state power substations (SSPS) [6], dynamic virtual power plants [7], vehicle to grid integration solutions [8], and microgrids [9] —are also propelling dynamic response from DERs to enhance the reliability and resilience of both the grid and downstream assets/systems. As dynamic response from DERs becomes more structured the reliability of the overall electrical system can be ensured over a short period of time to an entire day.

The integrated or aggregated DERs can support the system in the event of internal and external abnormalities with an array of functions or services [10], [1] as shown in Fig. 1. These functions/services also allow for dynamic modification of system constraints/limits including stability & operating limits amidst contingencies and intermittencies [11]. The evolving energy landscape mandates advanced capabilities at the power electronics (PE) interface layer of the DERs, including the following to allow for more flexibility in asset utilization and to improve reliability.

a) Hybrid operation
- Autonomous response – respond to local voltage and frequency changes [12].
- Non-autonomous response – can be triggered to respond to an event or abnormality through communication channels in anticipation of an event.

b) Versatile functionalities
- Dynamic change of setpoints, settings, limits and regulation curves pertaining to each control mode.
- Ability to turn on/off the control modes in real-time.
- Multimode operation: PE interfaces should be able to switch between different control modes (e.g., voltage regulation, frequency control) to meet system demands without negatively impacting system stability.
- Allow seamless transitions between control modes.
- Be able to respond flexibly to the requirements of aggregators or utilities.

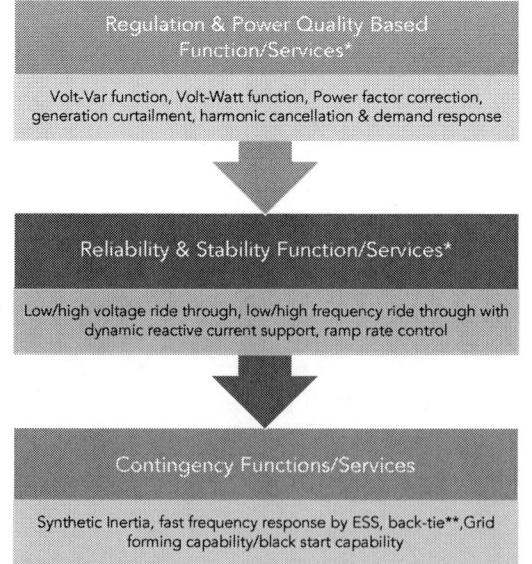

Fig. 1. Categorization of few DER control functions based on system level benefits [10].

Notice: This manuscript has been authored by UT-Battelle, LLC under Contract No. DE-AC05-00OR22725 with the U.S. Department of Energy. The United States Government retains and the publisher, by accepting the article for publication, acknowledges that the United States Government retains a non-exclusive, paid-up, irrevocable, world-wide license to publish or reproduce the published form of this manuscript, or allow others to do so, for United States Government purposes. The Department of Energy will provide public access to these results of federally sponsored research in accordance with the DOE Public Access Plan (http://energy.gov/downloads/doe-public-access-plan).

Dynamic response from DERs with an array of functions can be orchestrated by controlling the PE interface through control loops, sometimes referred to as control modes. These functions or control mode responses can either be instigated via local measurements (autonomously) or through communications (non-autonomously) depending on the system configuration and approach.

With the requirements for dynamic response established, state of the art control approaches, available to handle the requirements, were investigated [13-20]. Literature reports several centralized and decentralized control philosophies and the associated PE interface functionalities required to handle system abnormalities with varied responses. In the reported decentralized approaches, the PE interfaces are traditionally designed to respond to abnormalities in local measurements thus permitting autonomous response. While centralized approaches utilize a secondary control layer for decision making to instigate a non-autonomous response from a PE interface in anticipation or in the event of an abnormality.

A decentralized communication-less control approach for mode transition in multiple inverters was demonstrated in [13], where the system autonomously transitioned between grid-tied and standalone modes to ensure voltage and frequency stability. In this, a state machine in the PE system has been realized to autonomously transition between standalone and grid-tied modes with voltage and frequency control capabilities. A similar communication less decentralized approach for mode transition in multiple inverters is detailed in [14]. In [15] an alternate approach, using lookup tables has been demonstrated using a hardware in loop (HIL) testbed.

Several centralized control solutions also exist [16-20]. One example for a PV inverter system contained an event management layer as discussed in [16], where the PE system responds to triggers from the management layer to change the operating conditions. Another example uses an agent-based interface for power electronics systems that hosts a decision tree for control modes and transitions as described in [17]. A similar structure with a microgrid state transition framework in the microgrid controller layer is reported in [18]. Additionally, a supervisory control theory based methodology with a management layer is proposed in [19]. The management layer is designed to control the grid functions in a PV inverter. Finally, a framework for adaptive droop in which the droop/regulation curves can be changed dynamically or on the fly is reported in [20].

In general, trade-offs are seen with the different control philosophies. In decentralized approach, the speed at which the controller is able to respond to an event is very fast. However, the controller is limited to the specific task it was preprogrammed to do and lacks the flexibility for further optimization within a system. In the centralized approach, this flexibility is achieved via optimization using predictive models and forecasts. The main limitation of centralized controllers can be seen when unexpected behavior occurs, such as a grid fault, and fast responses are required. In these cases the converter would likely fault and be required to restart.

The growing integration of DERs highlights the need for PE interfaces capable of hybrid operations to effectively combine decentralized and centralized control philosophies.

These PE interfaces must be versatile enough to dynamically manage and switch between multiple functions (in an autonomous, non-autonomous or hybrid fashion) based on the system needs. The implementation of multiple PE interface controls also comes with increased setpoints, computational burden, novel and adaptive protection schemes and configurations and stability constraints that are not part of standard controllers today [21].

Realizing this need, this work proposes a dynamic control for PE interfaces with attributes including multimode operation, dynamic limit adjustments and capability for hybrid transitions (i.e., interlaced autonomous and non-autonomous transitions) to handle system abnormalities. As these attributes are highlighted throughout the paper, the proposed dynamic control philosophy will be referred to as "dynamic multimode-transition control" henceforth. The proposed framework maximizes services from a PE interface or an associated node and enhances system reliability and resiliency. The strategy also allows for larger flexibility and more effective resource utilization in any application. Given that the proposed strategy supports hybrid operation and other versatile functions, it can be easily adopted by any application.

The manuscript is organized as follows: Section – II explains the dynamic multimode-transition control in the context of a system. Section III – elaborates on the strategy for dynamic multimode-transition control at the PE interface level. Section – IV highlights the controller hardware in the loop (CHIL) testbed developed to demonstrate the concept. Finally Section – V highlights the experimental results for various case scenarios pertaining to multimode-transition operation for grid applications.

II. MULTIMODE – TRANSITION CONTROL – A SYSTEMS APPROACH

A PE interface that is capable of dynamic response attributes can be differentiated from commercially available solutions by the software layer that governs its advanced functionality [22]. Traditional PE interfaces are typically not designed to handle multiple functions or transitions simultaneously. As an example, in conventional systems, settings for autonomous responses are often preprogrammed as lookup tables [23-25]. Generally, the preprogrammed responses are limited to one or two functions. Provisions to modify these settings remotely via communication links may be available as shown in Fig. 2.

The introduction of the dynamic multimode-transition control layer (Fig. 2), enabled by an additional algorithm in the PE interface controller, allows for the simultaneous execution and seamless transition between various functions. This algorithm facilitates dynamic adjustments to control parameters such as ramp limits and fault thresholds, ensuring different types of transitions while maintaining system stability.

Key settings, including regulation curves and enaction thresholds for defining the type of transition, can be communicated from the node controller as needed. Effective communication is therefore essential to enable these advanced features at the node level. For nodes with distributed agents (as described in [26]), both autonomous and non-autonomous responses can be activated based on parameters sent from the node controller via a resource

979-8-3315-1612-3/25 $31.00 © 2025 IEEE

integration layer. These parameters include enaction thresholds and other transition attributes. For example, setting enaction thresholds to zero and defining the transition as permanent would classify the PE system response as non-autonomous.

At the system level, the proposed dynamic multimode-transition control strategy enables, system support based objectives and other control objectives to now be an integral part of the optimization in the upstream controller. Traditionally, the upstream controllers were designed only for power and energy management. As an example, PE systems in microgrids or distribution nodes [27] can manage both internal and external harmonics while addressing voltage events when equipped with the proposed dynamic multi-mode transition control. Mitigating these issues at the POC helps reduce system losses and operational costs while improving node reliability, all without imposing penalties on customers.

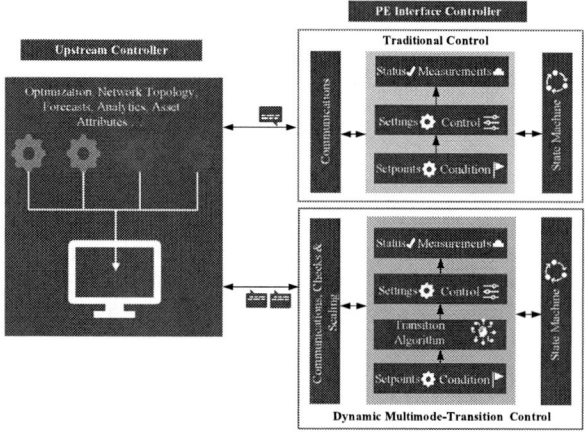

Fig. 2. Variations between traditional and proposed dynamic multimode-transition control to enable advanced features upstream and locally.

III. METHODOLOGY FOR DYNMAMIC OPERATION OF PE INTERFACES

a. Categoraization of control modes to maximize services or functions from a PE interface

At the PE interface level, the control modes can be classified as primary, secondary, and simultaneous modes based on their placement/position in the control loop. The available control modes and hence the categorization can vary depending on the type of PE interface as in [28]. Furthermore, transition modes for primary and secondary modes can be defined based on the interface type. Such a grouping is mandatory to establish modes that can operate in unison without leading to conflicts or stability concerns. The categorization also informs the upstream controller of the available modes and associated transitions. The categorization and mode definitions are as follows:

a) **Primary Mode -** Primary modes are predominantly real-power based modes. These are modes defining the fundamental behavior and operation of the system. Examples include dc bus regulation, active power (P) control etc.

b) **Secondary Modes -** Secondary modes are associated with reactive power (Q) and can work in unison with

the primary modes. Examples include constant reactive power (Q) control, Volt – Var etc.

c) **Simultaneous Modes -** These modes augment the primary and secondary control modes. An example of a simultaneous mode is harmonic compensation in case of a AC interface

d) **Transition Modes -** A set of primary or secondary modes that can be transitioned to/from based on the operation/functionality required by the system. The transitions can be autonomous or non-autonomous. Ex: P control is a primary mode that can be transitioned to Volt-Watt control. Transition modes have three additional options upper and lower enaction thresholds and a temporary/permanent flag. The enaction thresholds determine when the autonomous transitions occurs. When the temporary/permanent flag is set to temporary, this allows two different modes to transition back and forth autonomously based on the enaction thresholds. If the temporary/permanent flag is set to permanent, and enaction threshold values are set, then an autonomous transition will occur when the enaction thresholds are crossed. In such a case, a user or upstream controller will need to decide when to change the mode again.

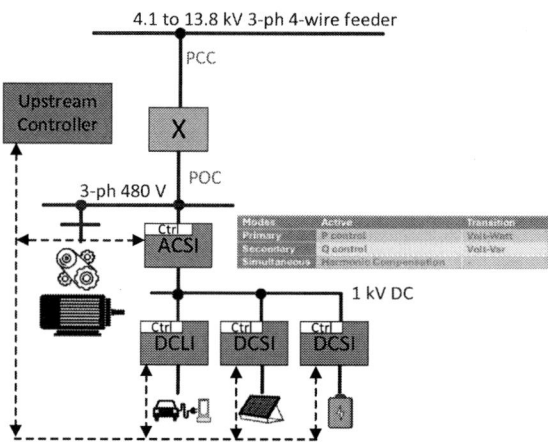

Fig. 3. Illustration of dynamic multimode-transition control in a distribution node as an example. Note: ACSI – AC source interface, DCLI – DC load interface and DCSI – DC source interface [28].

TABLE I. GROUPING OF SOME CONTROL MODES IN PE INTERFACES

Operation Type	Primary modes (Can Transition)	Secondary Modes (Can transition)	Simultaneous Modes (Can be turned ON/OFF)
Grid Following	Constant P, Volt-Watt, Freq-Watt, Ride through, DC-regulation, direct current control	Constant Q, constant PF, Volt-Var, quadrature current control	Harmonic compensation (i3, i5, i7, i9, i11….)
Grid Forming	PF droop, virtual inertia	QV Droop	Balancing control, frequency/voltage restoration etc.
DC	DC bus regulation, IV droop, PV droop, current and P control, MPPT, LPPT	-	Curtailment

The mode categorization is elaborated using a grid-tied PE interface i.e., an AC source interface (ACSI) as an example. An ACSI in a node can operate with multiple control modes including P control (primary) and Q control (secondary) while compensating for grid harmonics (simultaneous) as shown in Fig. 3. With multiple modes enabled at given instant, the proposed dynamic multimode-transition control allows for simultaneous transitions of the modes. Hence, in this case the primary mode can transition from P control to Volt-Watt during a grid event as shown in Fig. 3. The secondary mode can transition to Volt-Watt in response to grid abnormalities as well. Note that each of the individual transition can be autonomous or non-autonomous. Additionally, harmonic compensation can be turned on/off as required while dynamically changing the associated fault limits, control gains, etc. The multimode operation and the hybrid transitions can be handled together and therein lies the complexity of the dynamic multimode-transition control introduced in this paper. Such a combination of modes in ACSI i.e., harmonic control in addition to voltage support can increase the reliability of the grid and the connected assets by managing the events at the POC. Such control strategies are crucial to handle case scenarios in industrial systems with power quality issues during a voltage event in the network. Table-I, categorizes few possible control modes for DC and AC systems (grid following or grid forming).

b. Software framework for dynamic multimode-transition control

Following the control mode categorization, any PE interface controller can be designed with dedicated mode transition state machine as shown in Fig. 4, to realize dynamic response. The mode transition state machine, operating at switching frequency, ties to the main state machine of PE interface, operating in the order of ms, and works in unison. Further, the mode transition state machine executes a state during one control loop ISR cycle and thus completes any transition in < 10 switching cycles unless a zero crossing is to be detected. The mode transition state machine is designed to,

- enable/disable the primary and secondary modes and the associated simultaneous and transitions modes.

- Permit dynamic change of setpoints or regulation curves, check for operation and protection limits for all control modes.

- Allow hybrid operation and thus support autonomous, non-autonomous or hybrid responses.

Although the transition occurs quickly the communication does not need to be as fast. The communications validate the data transferred and then updates the transition modes in preparation of an event or when a change of modes is desired. The details on each state in the state machine is described below:

a) **Trigger check**: If there is a primary or secondary transition mode setup that is not the same as the current primary or secondary modes then a set of predefined trigger conditions are checked for specific modes. Typically this consists of a measurement that is greater than or less than either the upper or lower enaction threshold. This can be setup generically but will require a specific measurement to compare against for a particular transition mode. If one is found then the upper and lower enaction thresholds, as well as the temporary/permanent flag, are updated in the current mode options to match the transition mode options. Only one mode can be transitioned at a time, primary or secondary. Once complete then the state machine moves to the next state.

b) **Transition setup:** Based on the current mode and the transition mode, certain simultaneous modes may need to be disabled, fault limits modified, etc. to setup the converter for the mode transition. Disabling simultaneous modes could take some time, depending on the simultaneous modes enabled, so a hand shake was developed using two variables. One to identify that the simultaneous modes need to be disabled and another that identifies it has been disabled. Once completed the first variable is cleared, fault limit variables are updated, and the state machine moves to the next state.

c) **Transition check:** If the transition mode is identified as a temporary transition then the current mode is stored. If all specific operational conditions are met for a smooth transition then the current mode transitions to the new mode and the state machine moves to the next state.

d) **Start:** If the simultaneous modes still apply to the new mode then they are reenabled one at a time. Similarly to the disable handshake an enable variable is reserved to identify that the simultaneous modes should be reenabled. Once complete the disabled variable is cleared and the state machine moves to the next state.

e) **Finish:** The limits and other settings are restored to default values and the state machine returns to the Trigger check state.

To achieve a non-autonomous change one would simply set the upper and lower enaction thresholds both to 0. This is a predefined trigger condition in the "Trigger check" state and immediately proceeds through the state machine just the same as an autonomous transition. Note that while transitioning to a new primary control mode, the simultaneous modes associated with the primary mode is disabled before the transition commences. This is owing to the following reasons: a) disabling the simultaneous modes ensures that the system stability is not impacted or any interferences between the modes can be avoided and b) this also ensures that no conflicts arise between the simultaneous modes of the primary and the transition.

For a system with architecture as discussed in [26], the node controller decides on the operating mode and the type of transition in any given instant. While the settings, mode of operation etc. is decided by the node controller, the resource integration layer can help translate these settings including enaction thresholds etc. to the PE interface. Thus, communication variables like enaction thresholds, setpoints, and modes can be communicated from the resource integration controller to the PE interface controller. With this updated every 10 ms, the PE interface responds to the transition requests or requests for enabling and disabling control modes via the mode transition state machine. As mentioned earlier, the mode transition state machine is coupled to the main state machine and the communication module in the PE interface with dependencies as highlighted in Fig. 4.

979-8-3315-1612-3/25 $31.00 © 2025 IEEE

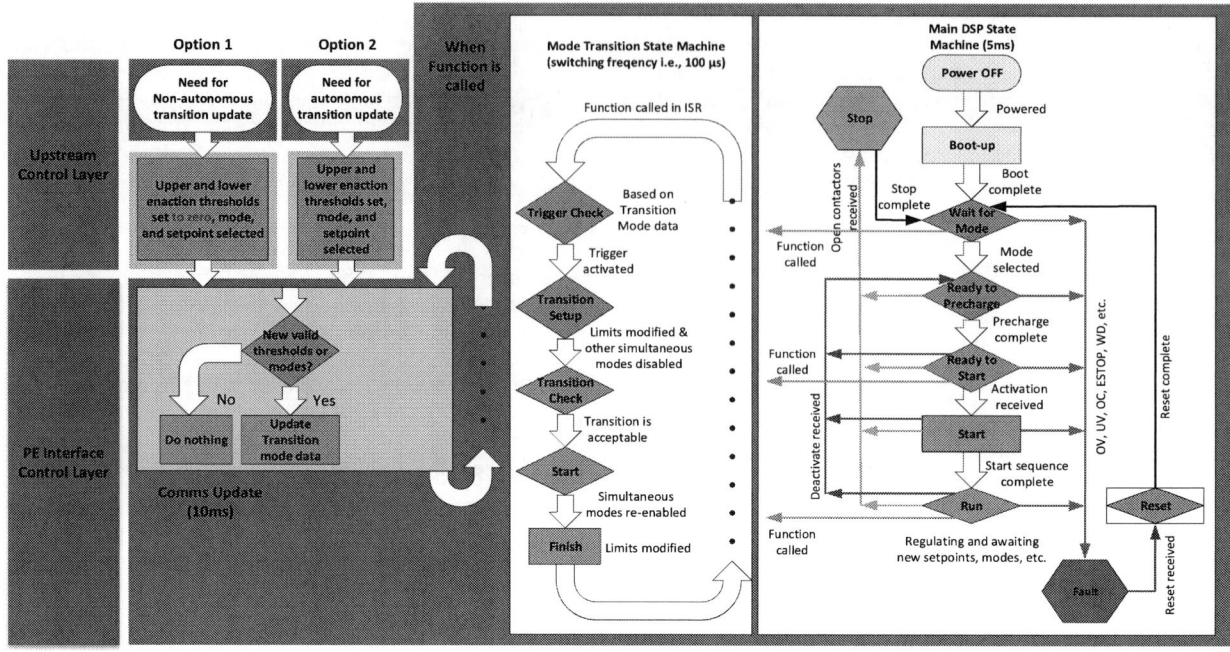

Fig. 4. Illustration of the different modules in the PE interface layer required for the dynamic multimode-transition control strategy.

IV. CONTROLLER HARDWARE IN LOOP (CHIL) TESTBED FOR VALIDATING THE DYNAMIC MULITIMODE-TRANSITION CONTROL

The CHIL testbed as shown in Fig. 5, was used to demonstrate scenarios pertaining to the dynamic multimode-transition control of PE interfaces for grid applications. The CHIL testbed is vital to verify the functionality of the introduced software layer and the PE interface controller. This validation ensures that the timing and computational requirements associated with the software modules in the PE interface controller can be handled in real-time without overruns or errors.

The testbed encompasses a real-time simulator (OP5707) and associated expansion units for modeling the SSPS 1.0 node. The PE interface control layer in the node is hosted in a TI F28379D based customized digital signal processor (DSP) board. The PE interface layer communicates with the upstream controller hosted in a computational platform. For example, in architectures as in [26], the PE interface layer communicates to the resource integration layer hosted in the Raspberry Pi using user datagram protocol. The resource integration layer in turn communicates with the centralized node controller using message queuing telemetry transport (MQTT) protocol. The node controller is hosted on a DELL OptiPlex system. However, the focus of this research was to validate the dynamic multimode-transition capability at the PE interface controller before upgrading the higher-level controller's communication. The upstream control layer is only mentioned and shown to make the concept easier to understand.

Only the grid side converter i.e., ACSI as discussed in the previous section, was used for the initial demonstration of the dynamic control logic reported in this manuscript. Also, the communication from the upstream controller was mimicked on the PE interface controller for initial concept

validation. The grid-side converter was modelled as a 480V, 75 kW system with a 1 kV dc-link. The grid-side converter was operating with a frequency of 10 kHz with only an inductive filter. The CHIL model parameters are laid out in Table-II. The converter was programmed with four control modes namely P, Volt-Watt control (primary modes) and Q, Volt-Var control (secondary modes). The simultaneous modes were not included in this case. The inverter was connected to a 4.16 kV distribution network via a transformer. The POC measurements from the real-time simulator, i.e., phase voltages and currents measured on the low side of the transformer, was fed to a DSP board which generated the switching signals of the inverter in response to programmed control modes.

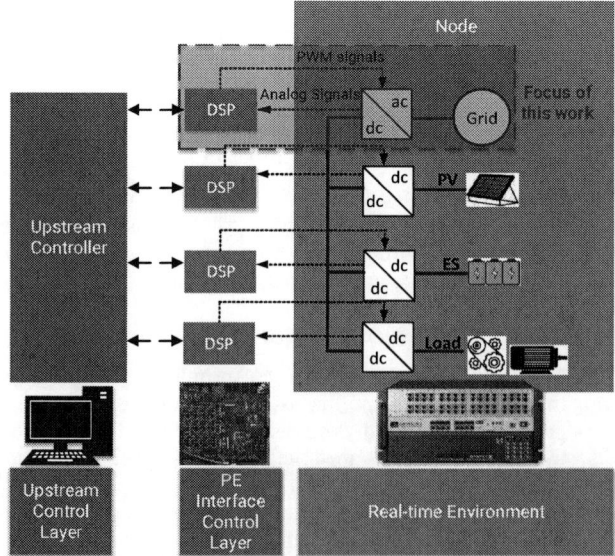

Fig. 5. CHIL testbed established to verify a SSPS node with assets capable of multimode.

TABLE II. PARAMETER OF THE PE SYSTEM USED TO DEMONSTARTE THE MULTIMODE-TRANSITION CONTROL SCENARIOS

Parameters	Values
Rated power	75 kVA
DC-link voltage	1 kV
AC voltage	480 V (L-L)
Switching frequency	10 kHz
DC-link capacitance	880 µF
Filter inductance	1 mH per phase

The CHIL testbed was used to demonstrate three different operating scenarios pertaining to the dynamic multimode-transition control for grid applications. a) Multimode operation with non-autonomous transition of primary and secondary modes, b) multimode operation with autonomous transition of primary and secondary modes and c) multi-mode operation with hybrid transitions i.e., non-autonomous transition of primary mode and autonomous transition of secondary modes. The communication variables used to communicate the regulation curves from the upstream controller was preprogrammed in the DSP.

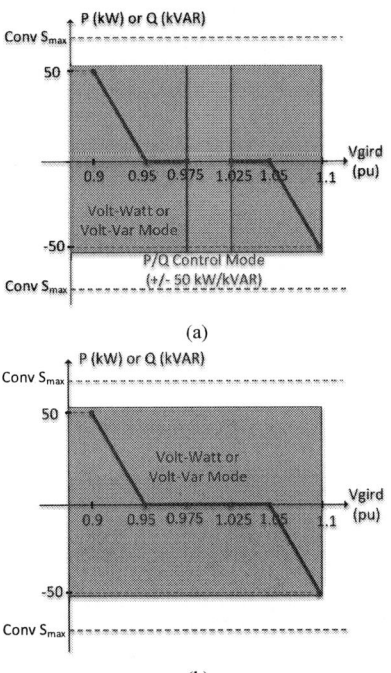

Fig. 6. Regulation curves used to demonstrate the multimode transition control. (a) Autonomous operation and (b) Non-autonomous operation.

The regulation curves used for the scenario validation for both autonomous and non-autonomous operation is shown in Fig. 6. For autonomous operation that is setup as a temporary transition, the mode autonomously switches back and forth between P/Q control and voltage droop using the enaction thresholds to initiate the mode transition as shown in Fig. 6(a). The lower and the upper enaction thresholds were set to be 0.975 and 1.025 pu respectively and between these values the P/Q control follows the fixed setpoint provided. The non-autonomous transition to voltage droop mode sets the enaction thresholds to 0 as they are only used to request an immediate mode transition by the upstream controller. The three segment droop curve has a slope of -3.61 for $v_{grid} <$ 0.95 pu, a slope of 0 between 0.95 and 1.05, and a slope of -3.61 for $v_{grid} >$ 1.05 pu. Anything beyond +/-50 kW causes

the curve to saturate on power. In case of non-autonomous operation, when the grid voltage is within the deadband (0.95 to 1.05 pu), the PE interface doesn't inject/absorb real/reactive power as shown in Fig. 6(b). Checks and conditions are in place to ensure that the PE interfaces' apparent power limits are not exceeded at any instant. The experimental results from the aforementioned scenarios are presented below.

V. CASE SCENARIOS FOR MULTIMODE OPERATION WITH MODE TRANSITION – EXPERIMENTAL RESULTS

a. Scenario – 1: Multimode Operation with Non-autonomous Transition of Primary and Secondary Modes in Response to a Grid Voltage Event

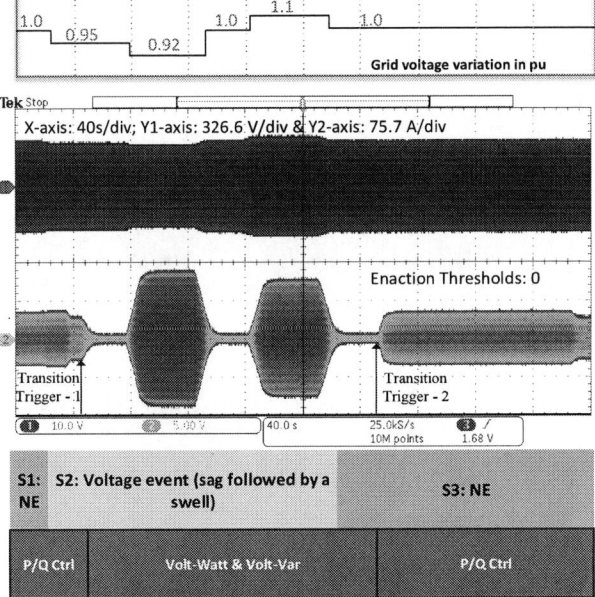

Fig. 7. Experimental result highlighting non-autonomous transition of primary and secondary modes.

Scenario 1 demonstrates multimode operation with non-autonomous transitions using four control modes: constant P, constant Q, Volt-Var, and Volt-Watt. The regulation curves shown in Fig. 6(b) were applied to both primary and secondary modes.

Under normal grid conditions (no event, NE) in segment S1, the inverter operated with constant P and constant Q as the primary and secondary modes, respectively, with setpoints of 20 kW and 20 kVAR. During S1, when the grid voltage dropped to 0.95 pu, the current increased to maintain constant power. Because this is a non-autonomous response to a grid event, there is a delay before the control modes gets switched. At this instance labelled "Transition Trigger 1" in Fig. 7, the control modes were switched to Volt-Watt and Volt-Var for primary and secondary modes, respectively, via communication variables. With 0.95 pu falling within the dead-band of the regulation curve, the inverter's output was reduced to zero power.

When the grid voltage further decreased to 0.92 pu, the inverter injected P and Q into the grid according to the voltage droop regulation curve shown in Fig. 6(b). A similar response occurred during voltage swell events since the control modes remained set to Volt-Watt and Volt-Var.

After the grid voltage stabilized at 1.0 pu, communication variables were used to switch the PE interface back to P and Q control modes, as indicated by "Transition Trigger 2" in Fig. 7. The system resumed normal operation, injecting 20 kW and 20 kVAR into the grid.

It is important to note that the "transition triggers" occurred when the enaction thresholds were set to 0, via communication variables, initiating control mode updates using the dynamic multimode-transition algorithm. Additionally, the measurements for the results in Fig. 7 have a scaling and attenuation factor from the CHIL testbed. The appropriate scaling is noted in Fig. 7 for the grid voltage and current measurements.

b. Scenario – 2: Multimode Operation with Autonomous Transition of Primary and Secondary Modes in Repsonse to a Grid Voltage Event

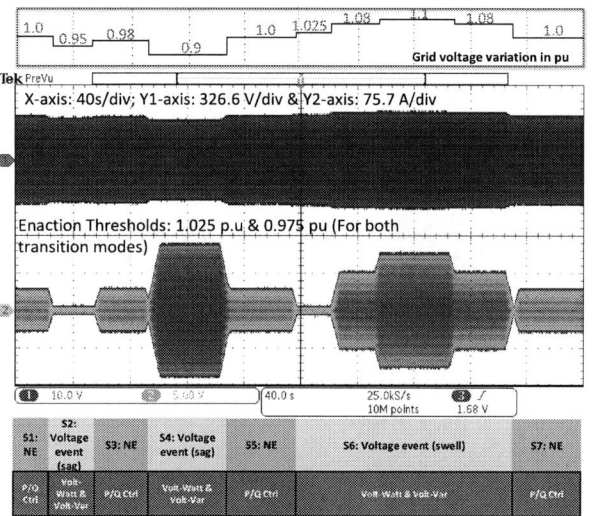

Fig. 8. Experimental result highlighting autonomous transition of primary modes and secondary modes.

The same setup was utilized to demonstrate multimode operation with autonomous transitions among four control modes: constant P, constant Q, Volt-Var, and Volt-Watt. Initially, P and Q control were configured using non-autonomous control mode updates. Subsequently, Volt-Watt and Volt-Var modes were established based on activation thresholds of 0.975 pu and 1.025 pu, respectively, with the transition set as temporary.

Under normal conditions (S1), the inverter operated in constant P and constant Q modes as the primary and secondary modes, respectively. During a voltage event (sag or swell), as illustrated in S2, S4, and S6 of Fig. 8, the primary and secondary modes autonomously transitioned to Volt-Watt and Volt-Var modes, respectively, in response to the grid disturbance, following the regulation curve in Fig. 6(a).

As shown in Fig. 8, all transitions occurred swiftly without delays caused by external intervention, adhering to ramp rates of 5 kW/s or 5 kVAR/s programmed into the state machine. The seamless transitions, free from overshoots, demonstrate that the system stability was maintained even during consecutive voltage sag and swell events.

c. Scenario – 3: Multimode Operation with Non-autonomous Transition of Primary Modes and Autonomous Transitions of Secondary Modes in Repsonse to a Grid Voltage Event

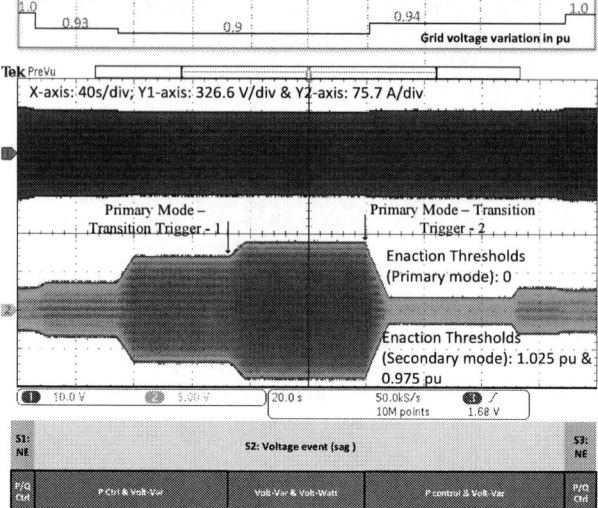

Fig. 9. Experimental result highlighting non-autonomous transition of primary modes and autonomous transition of secondary modes.

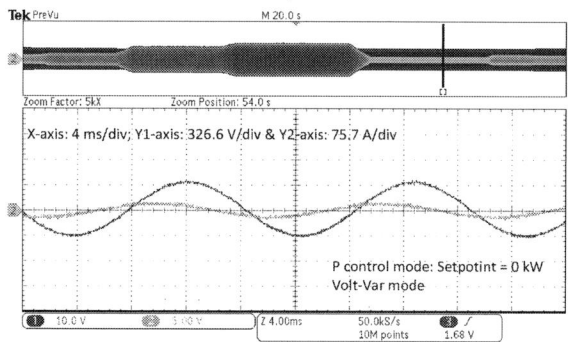

Fig. 10. Zoomed out image capture highlighting the system P control mode (primary) and Volt-Var mode (secondary).

Scenario 3 illustrates the non-autonomous operation of the primary mode (triggered externally) and the autonomous operation of the secondary modes. Like the previous case, the system starts in normal conditions (S1) while the inverter operated in P and Q control modes as the primary and secondary modes respectively. Upon detecting a grid voltage sag, identified as S2 in Fig. 9, the secondary mode autonomously transitioned to Volt-Var control, while the primary mode continued operating in P control. Subsequently, the voltage sag dropped from 0.93 pu to 0.9 pu and the inverter responded to the event based on the Volt-Var curve, while the primary control mode was held at P control with a fixed setpoint. Then, for additional voltage compensation, the primary mode was transitioned non-autonomously to Volt-Watt control, noted as "Transition Trigger 1" in Fig. 9. The primary mode remained in Volt-Watt mode until the non-autonomous transition request to P control with 0 kW setpoint noted as "Transition Trigger 2" in Fig. 9. A zoomed-out waveform depicting this operating condition in shown in Fig. 10. Near the end of (S2) voltage sag event, the P setpoint was set back to 20 kW. Finally, when the grid voltage returned to the nominal value of 1 pu, the primary mode remained at 20 kW and the secondary mode autonomously transitioned back to Q control mode with a reactive power setpoint of 0 kVAR as depicted in S3 of Fig. 9.

VI. Conclusion

The paper presents a strategy for dynamic control for PE interfaces with attributes like multimode operation, hybrid transitions, etc. for enhancing system reliability. The proposed dynamic control enables multiple control modes and associated parameter changes on the fly. The software layer necessary at the PE interface level to realize the proposed dynamic functionality has been elaborated. Validation using a CHIL testbed demonstrates the effectiveness of the proposed strategy in managing grid voltage events. Future work will focus on integrating the dynamic control enabled PE interfaces into a distribution node, validating optimization strategies centered on control at the node level and quantifying the system-level benefits.

Acknowledgment

This project is supported by the U.S. Department of Energy (DOE) – Office of Electricity's Transformer Resilience and Advanced Components (TRAC) program, led by the program manager Andre Pereira. The authors are grateful to Madhu Chinthavali, Oak Ridge National Laboratory for their review and feedback.

References

[1] J. Johnson, "Changing grid codes around the world," SunSpec Alliance Members Meeting, San Francisco, Mar. 2015.

[2] https://www.nsenergybusiness.com/analysis/featuredealing-with-the-50-2-hz-problem/.

[3] "Default New England bulk system area setting requirement," PJM

[4] MISO guideline for IEEE Std 1547-2018 implementation", Nov. 2019.

[5] https://www.cleanenergycouncil.org.au/news/new-inverter-standard-to-improve-grid-stability.

[6] U.S Department of Energy, Office of Electricity, "Solid State Power Substation Technology Roadmap," Transformer Resilience and Advanced Components (TRAC) Progam, Jun. 2020.

[7] B. Marinescu, O.G.-Bellmunt, F. Dorfler, H. Schulte and A. L. Sigrist, "Dynamic virtual power plant: a new concept for grid integraion of renewable energy sources," in *IEEE Access*, vol. 10, pp. 104980-104995, Sept. 2022.

[8] K. L.-Healey, L. Jones, m. M. Haque and B. Sturmberg, "Electric vehicles and the grid – analysis, gaps and recommendations," Battery storage and integration program, The Australian National University, Canberra, Australia, Jan. 2022.

[9] A. Maitra and Spirae, "grid interactive microgrid controller for resilient communities," *Final Scientific/Technical Report*, Electric Power Research Institute, 2018.

[10] T. Lowder and K. Xu, "The evolving US distribution system: technologies, architectures and regulations for realizing a transactive energy marketplace," *Technical report, NREL/TP-7A40-74412*, May 2020.

[11] "IEEE Standard for interconnection and interoperability of distributed energy resources with associated electric power system interfaces," *IEEE 1547*, 2018.

[12] California Public Utilities Commission, "Electric Rule 21: Generating facilities interconnection," 2020.

[13] S. Ghosh and A. Chattopadhyay, "Three-loop based universal control architecture for decentralized operation of multiple inverters in an autonomous grid-interactive microgrid," in *IEEE Trans. Ind. Appls.* vol. 56, no. 2, pp. 1966-1979, March. 2020.

[14] O. V. Kulkarni, S. Doolla and B. G. Fernandes, "Mode transition control stratgey for multiple inverter-based distributed generators operating in grid-coinnected and standalone mode," in *IEEE Trans. Ind. Appls.* Vol. 53, no. 6, pp. 5927-5939, Nov./Dec. 2017.

[15] K. Prabakar, M. Shirazai, A. Singh and S. Chakraborthy, "Advanced photovoltaic inverter control development and validation in controller-hardware-in-the-loop test bed," in *Proc. IEEE Energy Conversion Congress and Expsoition (ECCE)*, Nov. 2017, Cincinnati, OH, USA.

[16] V. E. S. Barbosa, E. G. Carati, J. P. da Coata, R. Cardosa, C. M. O. Stein and Z. L. I. Nadal, "Event management layer for distributed generation photovoltaic inverters," in *Proc. IEEE International Conference on Industrial Technology*, Apr. 2020, Buenos Aires, Argentina.

[17] M. Starke, M. Chinthavali, C. Winstead, Z . Sheng, S. Campbell, R. Zeng, T. Kuruganti, Y. Xue and C. Thomas, "Networked control and optimization for widescale integration of power electronics in residential homes," in *Proc. IEEE Energy Conversion Congress and Exposition (ECCE)*, Nov. 2019, Baltimore, MD, USA.

[18] J. Wang, A. Pratt and M. Baggu, "Design of a state machine for smooth microgrid transition operation," in *Proc. Innovative Smart Grids Technologies Conference (ISGT)*, 2018, Washington, DC, USA.

[19] E.G. Carati, V. E. S. Barbosa, R. Cardoso, C. M. O. Stein and J. P. Costa, "Supervisory layer for improved intercativity of distributed generation inverters in smart grids," in *IEEE Journal of Sensors and Actuator Networks*, vol. 10, no. 64, 2021.

[20] M. Starke, N. Kim, B. Dean, S. Campbell, P. Kandula, R. S. K. Moorthy and M. Chinthavali, "Framework for supporting adaptive droop based power electronics systems," in *Proc. Innovative Smart Grid Technologies Conference (ISGT)*, Feb. 2024, Washington, DC, USA.

[21] S. Xu, Y. Xue and L. Chang, "Review of power system support functions for inverter-based distributed energy resources – standards, control algorithms and trends," in *IEEE Open Journal of Power Electronics*, vol. 2, pp. 88-105, 2021.

[22] E. Reiter, K. Ardani, R. Margolis and R. Edge, "Industry perspective on advanced inverters for US solar photovoltaic systems: grid benefits, deployment challenges and emerging solutions," *Technical report*, NREL/TP-7A40-65063, Sept. 2015.

[23] F. Ding, A. Nagarajan, M. Baggu, S. Chakraborthy, A. Hguyen, S. Walinga, M. Mc Carty and F. Bell "Application of autonomous inverter Volt-Var function for voltage reduction energy savings and power quality in electric distribution systems," in *Proc. Eighth Conference on Innovative Smart Grid Technologies (ISGT 2017)*, Apr. 2017, Washington DC, USA.

[24] R. K. Varma and E. M. Siavashi, "PV-STATCOM: A New Smart Inverter for Voltage Control in Distribution Systems," *IEEE Transactions on Sustainable Energy*, vol. 9, no. 4, pp. 1681-1691, Oct 2018.

[25] K. Rahimi, A. Tbaileh, R. Broadwater, J. Woyak and M. Dilek, "Voltage regulation performance of smart inverters: Power factor versus volt-VAR control," in *Proc. 2017 North American Power Symposium*, Sept. 2017, Morgantown, WV, USA.

[26] M. Starke, B. Xiao, P. K. Bhowmik, S. L.Campbell, M. Chinthavali, B. R. Dean, R. S. K. Moorthy, M. Smith and A. Thapa, "Agent-Based Distributed Energy Resources for Supporting Intelligence at the Grid Edge," in *IEEE Journal of Emerging and Selected Topics in Power Electronics*, vol. 3, no. 1, pp. 69-78, Jan. 2022.

[27] M. Chinthavali, R. S. K. Moorthy, M. Starke and B. Dean, "Solid state power substations (SSPS): a multi-hierarchical architecture from substation to grid edge," in *Proc. International Symposium on Power Electronics for Distributed Generation Systems (PEDG)*, Jun. 2022, Kiel, Germany.

[28] R. S. K. Moorthy, G. Liu, J. Choi, A. Adib and M. Starke, "Architecture of an residential solid state power substation (SSPS) node," in *Proc. Innovative Smart Grid Technologies Conference (ISGT)*, Feb. 2021, Washington DC, USA.

Achieving Soft-Charging and Over 20% Input Current Ripple Reduction in a 48-to-6 V Dickson Converter Using 3-Phase Split-Phase Control

Nagesh Patle, Rose A. Abramson, Sahana Krishnan, Jiarui Zou, and Robert C. N. Pilawa-Podgurski

Department of Electrical Engineering and Computer Sciences
University of California, Berkeley
Email: {nageshpatle, rose_abramson, sahana_krishnan, jiarui.zou, pilawa}@berkeley.edu

Abstract—Hybrid switched-capacitor (HSC) converters can outperform conventional switched-capacitor and switched-inductor converters by incorporating inductors within capacitor charging/discharging loops, thereby eliminating hard-charging losses. Among these, the Dickson HSC topology achieves the minimum possible switch VA stress, resulting in superior efficiency compared to other HSC topologies. While odd-conversion-ratio Dickson topologies can achieve complete soft-charging using two-phase operation, even-conversion-ratio topologies cannot. Previous work has introduced a 4-phase split-phase control scheme to enable soft-charging operation. However, the 4-phase scheme increases the input current RMS due to reduced conduction time of the input switch. This poses challenges in applications where line losses are significant, buffer capacitor volume is critical, or minimal input noise and electromagnetic interference (EMI) are required. This work presents a 3-phase split-phase control scheme that maintains input current conduction for a constant 50% of the switching period, regardless of the conversion ratio. The method is also extendable to any even-ratio Dickson topology. Hardware validation of the proposed scheme on a 48-to-6 V Dickson converter confirms significant reductions in input current ripple compared to prior techniques.

I. INTRODUCTION

Hybrid switched-capacitor (HSC) converters [1] are increasingly utilized in applications such as data centers [2]–[4] and automotive power trains [5]–[8], driven by the growing demand for high-efficiency and high-power-density converters. HSC converters often outperform conventional switched-inductor topologies by utilizing energy-dense capacitors as the primary energy storage elements [9], in combination with small-valued inductors that help mitigate hard-charging losses between capacitor loops [10], [11].

The efficiency of a typical Intermediate-Bus Converter (IBC) is also influenced by line losses in the system external to the converter. With increasing power demands, there is a shift from traditional 12 V systems to 48 V systems [12]–[14], as these systems enable higher power delivery at lower line currents, thereby reducing I^2R losses and improving overall system efficiency. Fig. 1 illustrates such a system, where R represents the input path parasitic resistance that incurs line losses. Bulk capacitors with low equivalent series resistance (ESR) are typically added to provide a low-impedance path for high-frequency ripple current, thus reducing the input RMS current and input ripple voltage [15]–[18]. However, these capacitors increase system volume and reduce power density. Therefore, minimizing input

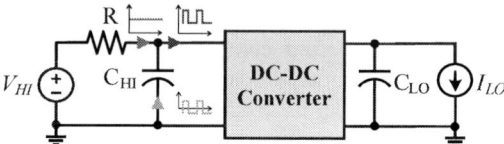

Fig. 1: Intermediate-Bus Converter (IBC) system illustrating converter current distribution through input path parasitic resistance and buffer capacitors.

RMS currents is critical in 48 V systems to maximize both efficiency and power density.

Among the various HSC topologies for 48 V to Intermediate-Bus power conversion, the Dickson topology (shown in Fig. 2) is particularly noteworthy. It achieves the theoretical minimum switch stress [19], which often translates to reduced converter losses. Consequently, numerous works [20]–[23] have been inspired by the Dickson structure. For odd-conversion ratios, the Dickson converter achieves soft-charging of all flying capacitors using a two-phase operation scheme. However, for even-conversion ratios—commonly required for converting 48 V to standard IBC voltages such as 12 V, 8 V, or 6 V—the converter cannot achieve soft-charging with two-phase operation. Instead, a split-phase control scheme is necessary, which introduces two additional sub-phases within the two main phases [24]–[26]. Despite its advantages, the 4-phase split-phase scheme presents three significant challenges for step-down applications:

1) *Reduced input current conduction time:* This increases the input RMS current needed to deliver the same average current (and power), posing a disadvantage for applications where line losses due to parasitic resistance are significant.

2) *Increased bulk capacitor requirement:* Higher input peak-to-peak ripple currents necessitate larger capacitors, reducing overall system power density.

3) *Electromagnetic interference (EMI):* Higher rates of voltage and current change during phase switching exacerbate the conducted EMI [23], [27]. While interleaved solutions [28] mitigate EMI by reducing the input RMS current, they do not address the significant non-conduction periods.

This work proposes a novel 3-phase split-phase control scheme that introduces only one additional sub-phase to

979-8-3315-1612-3/25 $31.00 © 2025 IEEE

two-phase operation. This scheme achieves soft-charging of all flying capacitors for even-conversion ratio Dickson converters while maintaining a consistent 50% input current conduction time, regardless of the conversion ratio. This results in a significant reduction in input RMS current and ripple voltage, minimizing line losses and bulk capacitor requirements. Additionally, the scheme simplifies interleaving techniques for noise and EMI mitigation.

The presented switching scheme is demonstrated on an 8-to-1 step-down Dickson converter, as shown in Fig. 2, and can be extended to other Dickson-based topologies, such as the dual inductor hybrid (DIH) [29] and symmetric dual inductor hybrid (SDIH) [26]. Section II revisits the conventional 2-phase and 4-phase split-phase techniques. Section III discusses the capacitor sizing and control timing for the proposed 3-phase technique. Section IV presents a theoretical comparison of the 4-phase and 3-phase split-phase schemes, while Section V provides hardware validation and experimental results.

II. OVERVIEW OF CONVENTIONAL OPERATION

A. 2-Phase Operation

Fig. 3(a) illustrates the equivalent circuits for the basic 2-phase scheme for the 8-to-1 Dickson converter (Fig. 2). In this scheme, all capacitors deliver or absorb equal magnitudes of charge q, in both phases. The charge delivered to the output remains constant at $4q$ for each phase. For a constant current load, the phase durations are chosen to be proportional to the charge delivered to the output, resulting in a 50% duty cycle. This implies that the input voltage source remains connected for half of the period duration. As described in [11], the capacitor charge flow vectors and voltage loops are analyzed to formulate the system of equations that define the capacitance relations for each phase, which are then used to derive the required flying capacitor sizing for complete soft charging. In Phase 1, the capacitor charge flow is given by: $q_{C1}^{\phi_1} : q_{C2}^{\phi_1} : q_{C3}^{\phi_1} : q_{C4}^{\phi_1} : q_{C5}^{\phi_1} : q_{C6}^{\phi_1} : q_{C7}^{\phi_1} = [1 : -1 : 1 : -1 : 1 : -1 : 1]$. Three capacitor loops are analyzed using Kirchhoff's Voltage Law (KVL), expressed in terms of incremental capacitor voltage ripple. The voltage ripple in each phase, Φ_k, is related to the capacitance by $\Delta V_C^{\Phi_k} = q_C^{\Phi_k}/C$. The capacitance relations for Phase 1 are derived as follows:

$$\Delta V_{C2}^{\phi_1} - \Delta V_{C1}^{\phi_1} = \Delta V_{C4}^{\phi_1} - \Delta V_{C3}^{\phi_1} = \Delta V_{C6}^{\phi_1} - \Delta V_{C5}^{\phi_1} = \Delta V_{in}^{\phi_1} - \Delta V_{C7}^{\phi_1}$$

$$\Rightarrow \frac{q_{C2}^{\phi_1}}{C_2} - \frac{q_{C1}^{\phi_1}}{C_1} = \frac{q_{C4}^{\phi_1}}{C_4} - \frac{q_{C3}^{\phi_1}}{C_3} = \frac{q_{C6}^{\phi_1}}{C_6} - \frac{q_{C5}^{\phi_1}}{C_5} = 0 - \frac{q_{C7}^{\phi_1}}{C_7}$$

$$\Rightarrow \frac{1}{C_1} + \frac{1}{C_2} = \frac{1}{C_3} + \frac{1}{C_4} = \frac{1}{C_5} + \frac{1}{C_6} = \frac{1}{C_7} \quad (1)$$

In Phase 2, the charge flow relation is: $q_{C1}^{\phi_2} : q_{C2}^{\phi_2} : q_{C3}^{\phi_2} : q_{C4}^{\phi_2} : q_{C5}^{\phi_2} : q_{C6}^{\phi_2} : q_{C7}^{\phi_2} = [-1 : 1 : -1 : 1 : -1 : 1 : -1]$. The capacitance relations are derived similarly:

$$\Delta V_{C1}^{\phi_2} = \Delta V_{C3}^{\phi_2} - \Delta V_{C2}^{\phi_2} = \Delta V_{C5}^{\phi_2} - \Delta V_{C4}^{\phi_2} = \Delta V_{C7}^{\phi_2} - \Delta V_{C6}^{\phi_2}$$

$$\Rightarrow \frac{q_{C1}^{\phi_2}}{C_1} = \frac{q_{C3}^{\phi_2}}{C_3} - \frac{q_{C2}^{\phi_2}}{C_2} = \frac{q_{C5}^{\phi_2}}{C_5} - \frac{q_{C4}^{\phi_2}}{C_4} = \frac{q_{C7}^{\phi_2}}{C_7} - \frac{q_{C6}^{\phi_2}}{C_6}$$

$$\Rightarrow \frac{1}{C_1} = \frac{1}{C_2} + \frac{1}{C_3} = \frac{1}{C_4} + \frac{1}{C_5} = \frac{1}{C_6} + \frac{1}{C_7} \quad (2)$$

From (1) and (2),

$$\frac{1}{C_1} : \frac{1}{C_2} : \frac{1}{C_3} : \frac{1}{C_4} : \frac{1}{C_5} : \frac{1}{C_6} : \frac{1}{C_7} = [1 : 0 : 1 : 0 : 1 : 0 : 1]$$

$$C_1 : C_2 : C_3 : C_4 : C_5 : C_6 : C_7 = [1 : \infty : 1 : \infty : 1 : \infty : 1] \quad (3)$$

From (3), it is concluded that the converter cannot achieve complete soft-charging with finite capacitor values. Consequently, the input current experiences sharp transients, which degrades the quality of the RMS current despite being able to conduct for a longer duration of the period.

B. 4-Phase Split-Phase Operation

The work in [24] proposed a 4-phase split-phase control scheme to enable complete soft-charging, with the equivalent circuits illustrated in Fig. 3(b). This approach involves splitting the two original phases into four sub-phases, i.e. $1 \rightarrow [1a, 1b]$ and $2 \rightarrow [2a, 2b]$. In each of the two additional sub-phases (Phase 1b and Phase 2b), one capacitor branch from the original phases is disconnected. This is implemented by shortening the duty cycles of the switches S_5 and S_{12}. In Phase 1a, the capacitor charge flow relation is: $q_{C1}^{\phi_{1a}} : q_{C2}^{\phi_{1a}} : q_{C3}^{\phi_{1a}} : q_{C4}^{\phi_{1a}} : q_{C5}^{\phi_{1a}} : q_{C6}^{\phi_{1a}} : q_{C7}^{\phi_{1a}} = [0.5 : -0.5 : 0.5 : -0.5 : 0.5 : -0.5 : 1]$. The capacitance relation is derived as follows:

$$\Delta V_{C2}^{\phi_{1a}} - \Delta V_{C1}^{\phi_{1a}} = \Delta V_{C4}^{\phi_{1a}} - \Delta V_{C3}^{\phi_{1a}} = \Delta V_{C6}^{\phi_{1a}} - \Delta V_{C5}^{\phi_{1a}} = -\Delta V_{C7}^{\phi_{1a}}$$

$$\Rightarrow \frac{q_{C2}^{\phi_{1a}}}{C_2} - \frac{q_{C1}^{\phi_{1a}}}{C_1} = \frac{q_{C4}^{\phi_{1a}}}{C_4} - \frac{q_{C3}^{\phi_{1a}}}{C_3} = \frac{q_{C6}^{\phi_{1a}}}{C_6} - \frac{q_{C5}^{\phi_{1a}}}{C_5} = 0 - \frac{q_{C7}^{\phi_{1a}}}{C_7}$$

$$\Rightarrow \frac{1}{C_1} + \frac{1}{C_2} = \frac{1}{C_3} + \frac{1}{C_4} = \frac{1}{C_5} + \frac{1}{C_6} = \frac{1}{0.5 C_7} \quad (4)$$

Phase 1b is constrained by a subset of the equations in (4) and is therefore omitted for brevity. In Phase 2a, the capacitor charge flow relation is: $q_{C1}^{\phi_{2a}} : q_{C2}^{\phi_{2a}} : q_{C3}^{\phi_{2a}} : q_{C4}^{\phi_{2a}} : q_{C5}^{\phi_{2a}} : q_{C6}^{\phi_{2a}} : q_{C7}^{\phi_{2a}} = [-1 : 0.5 : -0.5 : 0.5 : -0.5 : 0.5 : -0.5]$. The capacitance relation is derived in a similar manner:

$$\Delta V_{C1}^{\phi_{2a}} = \Delta V_{C3}^{\phi_{2a}} - \Delta V_{C2}^{\phi_{2a}} = \Delta V_{C5}^{\phi_{2a}} - \Delta V_{C4}^{\phi_{2a}} = \Delta V_{C7}^{\phi_{2a}} - \Delta V_{C6}^{\phi_2}$$

$$\Rightarrow \frac{q_{C1}^{\phi_{2a}}}{C_1} = \frac{q_{C3}^{\phi_{2a}}}{C_3} - \frac{q_{C2}^{\phi_{2a}}}{C_2} = \frac{q_{C5}^{\phi_{2a}}}{C_5} - \frac{q_{C4}^{\phi_{2a}}}{C_4} = \frac{q_{C7}^{\phi_{2a}}}{C_7} - \frac{q_{C6}^{\phi_{2a}}}{C_6}$$

$$\Rightarrow \frac{1}{0.5 C_1} = \frac{1}{C_2} + \frac{1}{C_3} = \frac{1}{C_4} + \frac{1}{C_5} = \frac{1}{C_6} + \frac{1}{C_7} \quad (5)$$

Phase 2b is similarly constrained by a subset of the equations in (5) and is omitted for redundancy. Solving (4) and (5) yields:

$$\frac{1}{C_1} : \frac{1}{C_2} : \frac{1}{C_3} : \frac{1}{C_4} : \frac{1}{C_5} : \frac{1}{C_6} : \frac{1}{C_7} = [1 : 1 : 1 : 1 : 1 : 1 : 1]$$

$$C_1 : C_2 : C_3 : C_4 : C_5 : C_6 : C_7 = [1 : 1 : 1 : 1 : 1 : 1 : 1] \quad (6)$$

979-8-3315-1612-3/25 $31.00 © 2025 IEEE

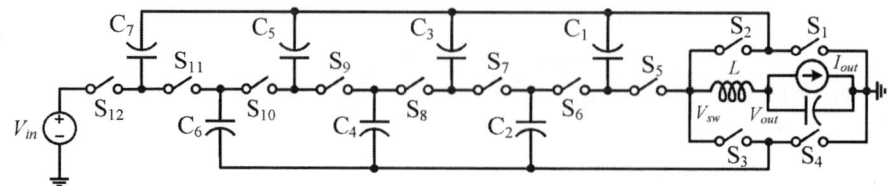

Fig. 2: 8-to-1 Dickson topology.

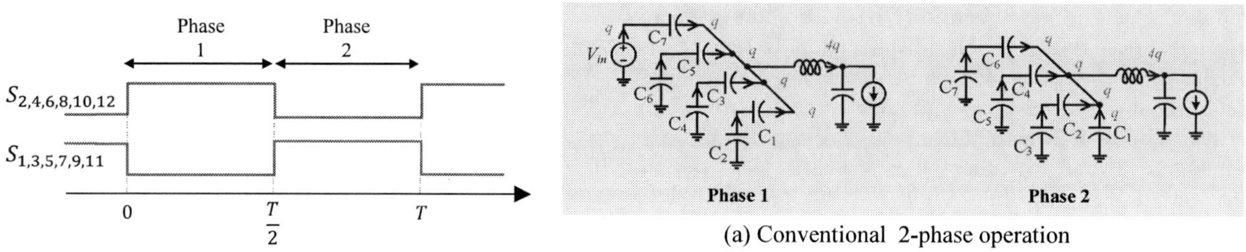

(a) Conventional 2-phase operation

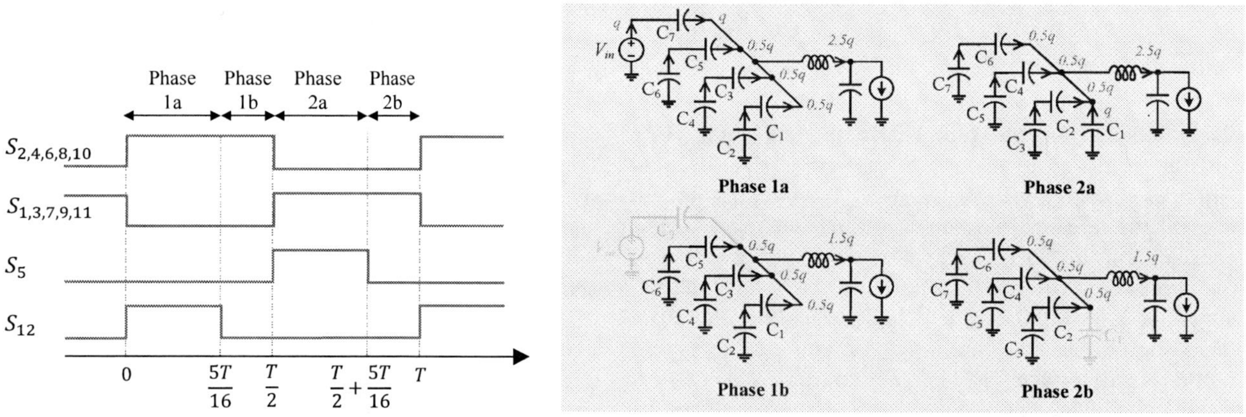

(b) Conventional 4-phase split-based operation

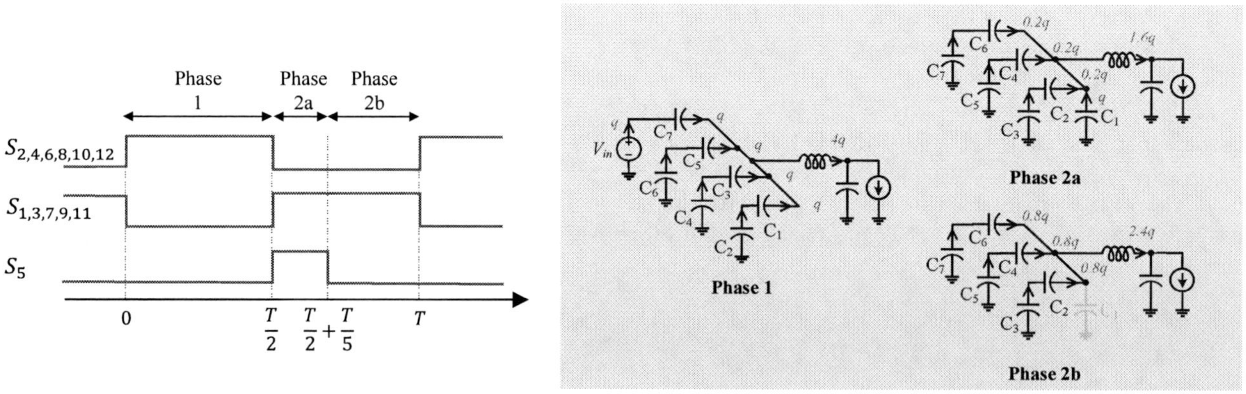

(c) Proposed 3-phase split-based operation

Fig. 3: Gate signals of switches and corresponding phase-equivalent circuits for 8-to-1 Dickson topology. (a) 2-phase scheme (hard-charged), (b) conventional 4-phase split-based scheme (soft-charged), and (c) proposed 3-phase split-based scheme (soft-charged).

Thus, this scheme achieves soft-charging with all capacitors sized equally. The charge delivered to the output is proportional to $[2.5q : 1.5q : 2.5q : 1.5q]$. For a constant current load, the phase durations are chosen in proportion to $\left[\frac{5}{16} : \frac{3}{16} : \frac{5}{16} : \frac{3}{16}\right]$. However, the input current conduction duration is reduced from $\frac{1}{2}$ to $\frac{5}{16}$ of the switching period, as it is tied to the duty cycle of switch S_{12}. This reduction

increases the input RMS current and necessitates larger bulk capacitors. Interleaving is a common solution to improve the input current conduction time [23], but it introduces greater complexity and requires additional components. This raises the question of whether a non-interleaved solution could be developed to improve the input current RMS.

III. Proposed 3-Phase Split-Phase Scheme

The proposed 3-phase control scheme no longer symmetrically splits both phases, but splits only one (i.e. $2 \rightarrow [2a, 2b]$). As the phase that is split corresponds to zero conduction from the input, this preserves the 50% conduction time of the input current. Fig. 3(c) illustrates the equivalent circuits for the 3-phase split-phase control scheme. In Phase 1, the capacitor charge flow remains the same as in 2-phase operation, as described by the vector: $q_{C1}^{\phi_1} : q_{C2}^{\phi_1} : q_{C3}^{\phi_1} : q_{C4}^{\phi_1} : q_{C5}^{\phi_1} : q_{C6}^{\phi_1} : q_{C7}^{\phi_1} = [1 : -1 : 1 : -1 : 1 : -1 : 1]$. The equation for the capacitance relation in Phase 1 is derived similarly to that in 2-phase operation:

$$\Delta V_{C2}^{\phi_1} - \Delta V_{C1}^{\phi_1} = \Delta V_{C4}^{\phi_1} - \Delta V_{C3}^{\phi_1} = \Delta V_{C6}^{\phi_1} - \Delta V_{C5}^{\phi_1} = \Delta V_{in}^{\phi_1} - \Delta V_{C7}^{\phi_1}$$
$$\Rightarrow \frac{q_{C2}^{\phi_1}}{C_2} - \frac{q_{C1}^{\phi_1}}{C_1} = \frac{q_{C4}^{\phi_1}}{C_4} - \frac{q_{C3}^{\phi_1}}{C_3} = \frac{q_{C6}^{\phi_1}}{C_6} - \frac{q_{C5}^{\phi_1}}{C_5} = 0 - \frac{q_{C7}^{\phi_1}}{C_7}$$
$$\Rightarrow \frac{1}{C_1} + \frac{1}{C_2} = \frac{1}{C_3} + \frac{1}{C_4} = \frac{1}{C_5} + \frac{1}{C_6} = \frac{1}{C_7} \quad (7)$$

Across the Phases 2a and 2b, the total charge delivered through the capacitor branches will be the same as the original Phase 2. Since C_1 only participates in Phase 2a, it delivers the entire q charge in that phase. For the other capacitors, let us assume they deliver xq charge in Phase 2a and $q - xq$ charge in Phase 2b, where $0 < x < 1$. The capacitor charge flow vector in Phase 2a can then be written as: $q_{C1}^{\phi_{2a}} : q_{C2}^{\phi_{2a}} : q_{C3}^{\phi_{2a}} : q_{C4}^{\phi_{2a}} : q_{C5}^{\phi_{2a}} : q_{C6}^{\phi_{2a}} : q_{C7}^{\phi_{2a}} = [-1 : x : -x : x : -x : x : -x]$. The capacitance relation for Phase 2a is then given by:

$$\Delta V_{C1}^{\phi_{2a}} = \Delta V_{C3}^{\phi_{2a}} - \Delta V_{C2}^{\phi_{2a}} = \Delta V_{C5}^{\phi_{2a}} - \Delta V_{C4}^{\phi_{2a}} = \Delta V_{C7}^{\phi_{2a}} - \Delta V_{C6}^{\phi_2}$$
$$\Rightarrow \frac{q_{C1}^{\phi_{2a}}}{C_1} = \frac{q_{C3}^{\phi_{2a}}}{C_3} - \frac{q_{C2}^{\phi_{2a}}}{C_2} = \frac{q_{C5}^{\phi_{2a}}}{C_5} - \frac{q_{C4}^{\phi_{2a}}}{C_4} = \frac{q_{C7}^{\phi_{2a}}}{C_7} - \frac{q_{C6}^{\phi_{2a}}}{C_6}$$
$$\Rightarrow \frac{1}{xC_1} = \frac{1}{C_2} + \frac{1}{C_3} = \frac{1}{C_4} + \frac{1}{C_5} = \frac{1}{C_6} + \frac{1}{C_7} \quad (8)$$

Phase 2b produces a subset of the equations in (8), and are thus omitted for simplicity. The general capacitor sizing solution can be derived from equations (7) and (8) as:

$$C_1 : C_2 : C_3 : C_4 : C_5 : C_6 : C_7$$
$$= \left[1 : \frac{4x}{3-3x} : \frac{4x}{1+3x} : \frac{2x}{1-x} : \frac{2x}{1+x} : \frac{4x}{1-x} : \frac{4x}{3+x}\right] \quad (9)$$

The most practical capacitor sizing ratio is achieved when $x = 0.2$, as the proportions can be simplified into small integers. These values effectively represent the number of capacitors required to be connected in parallel for realization. This is derived as follows:

$$\frac{1}{C_1} : \frac{1}{C_2} : \frac{1}{C_3} : \frac{1}{C_4} : \frac{1}{C_5} : \frac{1}{C_6} : \frac{1}{C_7} = [1 : 3 : 2 : 2 : 3 : 1 : 4]$$
$$C_1 : C_2 : C_3 : C_4 : C_5 : C_6 : C_7 = \left[1 : \frac{1}{3} : \frac{1}{2} : \frac{1}{2} : \frac{1}{3} : 1 : \frac{1}{4}\right]$$
$$= [12 : 4 : 6 : 6 : 4 : 12 : 3] \quad (10)$$

The charge delivered to the output is in the proportion $[4q : (1 + 3x)q : (3 - 3x)q] = [4q : 1.6q : 2.4q]$. For a constant current load, the phase durations are therefore chosen to be in the ratio $\left[\frac{1}{2} : \frac{1}{5} : \frac{3}{10}\right]$. The input current conduction duration is increased to 50% since the switch S_{12} operates with a 50% duty cycle. The following sections present a comparison of the proposed scheme with the conventional scheme across various aspects.

IV. Theoretical Comparison of Split-Phase Schemes

A. Capacitor Sizing

The 3-phase scheme requires capacitors to be sized unequally in a specific ratio (similar to the design choice in [29]). As shown in (9), equal capacitance sizing is not part of the solution space for the 3-phase scheme. For applications focused on power density, it is a common practice to use Class II capacitors, which derate with voltage, temperature, and age. Additionally, since capacitors are rated for different voltages and may use different dielectrics (e.g., [25]), capacitance derating can vary across capacitors. This means that even the uniform sizing in the 4-phase scheme demands meticulous capacitor selection. Therefore, neither scheme offers a clear advantage in terms of straightforward design.

B. Switch RMS Current

Fig. 4 compares the RMS current through all switches between the two split-phase schemes. In principle, a switch's RMS current is minimized when the duration of its conduction is maximized, and the current through the switch remains constant for the longest period of time. For the 3-phase scheme, switches with an extended duty cycle and flat current in Phase 1 experience reduced RMS current, while most switches in Phase 2 see an increased RMS as more charge must be supplied over a shorter time duration. Switches S_1 and S_2 maintain a 50% duty cycle in both phases, so their RMS current remains unchanged. Furthermore, switch S_{12} connects the input voltage source to the rest of the converter. Let the duty cycle of this switch be D_{in}, which corresponds to the input current conduction duration.

$$I_{in,pk-pk} \propto \frac{1}{D_{in}}$$
$$I_{in,RMS} \propto \frac{1}{\sqrt{D_{in}}} \quad (11)$$

D_{in} improves from $5/16 = 0.3125$ in the 4-phase scheme to 0.5 in the 3-phase scheme, reducing I^2R line losses by 37.5% (assuming no input bulk capacitor is connected and all current ripple is seen by the source). This can result in improved overall system efficiency, particularly in applications where line losses are dominant. Fig. 5 shows that as D_{in} decreases for higher even-conversion ratios in the 4-phase scheme, it remains at 50% in the 3-phase scheme. This indicates that the 3-phase scheme provides more significant RMS current benefits at higher conversion ratios.

C. Equivalent Output Resistance

In a 48-V-to-6-V Dickson converter, the switches typically block voltages between 6 V and 13 V for both 4-phase and 3-phase schemes. However, since optimized switches are not always available with such granularity (in practice low voltage switches are only available in 15 V, 25 V, 30 V, and 40 V), it is common practice to use switches with the same blocking voltage. In this case, the equivalent output resistance is a useful metric for comparing the topologies. If all switches have an on-resistance of R_{on}, the equivalent output resistance is given by $R_{on}\frac{\sum I_k^2}{I_{out}^2}$, where I_k represents the current through the kth switch, and I_{out} is the output current. The equivalent output resistance for the 4-phase and 3-phase schemes is computed to be $1.9R_{on}$ and $1.92R_{on}$, respectively. This suggests that the 3-phase scheme has a marginally worse equivalent output resistance, and thus efficiency, at full load.

In applications where safe startup is critical, over-rated switches are often used (e.g., for switch S_{12}) to provide an extra margin of safety, albeit with performance penalties (e.g., higher R_{on}). In such cases, the 3-phase scheme can offer better equivalent output resistance if losses due to switch S_{12} are dominant.

D. Bulk Capacitor Requirement

A large input buffer capacitor, C_{HI}, can be used to reduce the input current RMS. Assuming that all the ripple current is attributed to the input capacitor and only dc current flows through the input voltage source, we can establish the following relation:

$$C_{HI}\Delta V_{C,HI}f_{sw} = I_{in}(1 - D_{in}). \qquad (12)$$

Assuming constant I_{in} and fixed input voltage ripple [15]–[18], it follows that $C_{HI} \propto 1 - D_{in}$. This implies that the 3-phase scheme can reduce the input capacitor C_{HI} by 25% compared to the 4-phase scheme, thereby improving the overall system power density and power quality at the input.

E. Simplified Control Signal

The 3-phase scheme requires one fewer control signal to be generated, simplifying the control circuitry. The other two signals have a 50% duty cycle and are 180-degree out of phase, which is straightforward to design. Furthermore, the 3-phase scheme is particularly useful for implementing active split-phase control [25], as there is only one flying capacitor voltage that must be sensed to ensure soft-charging.

Table I presents a summary of the key design and performance differences between the 3-phase and 4-phase split-phase schemes.

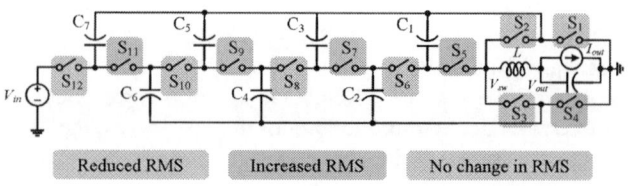

(a) Visual illustration of the switch RMS current change in 3-phase scheme relative to the 4-phase split-phase scheme.

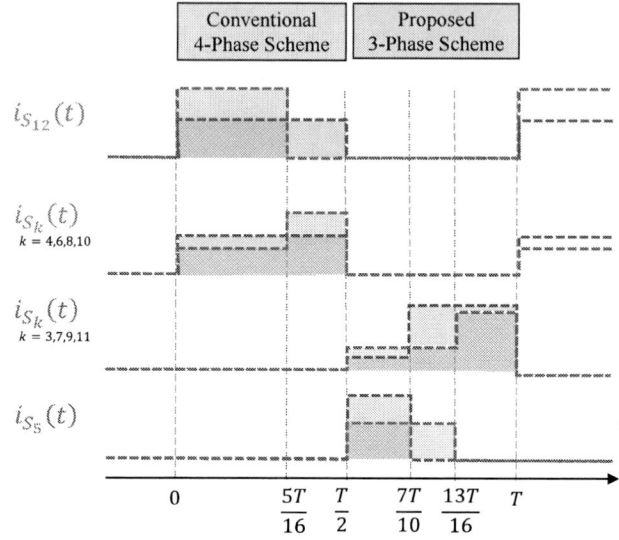

(b) Time domain plot of switch current waveforms with relative magnitudes, not to scale. Switches S_1 and S_2 are omitted due to identical waveforms.

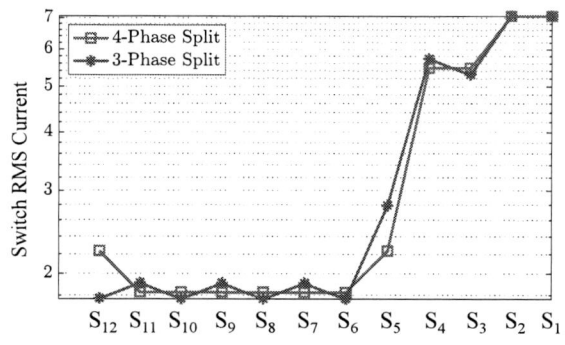

(c) Plot of switch RMS currents.

Fig. 4: Comparison between the conventional 4-phase and the proposed 3-phase split-phase schemes in an 8-to-1 Dickson converter.

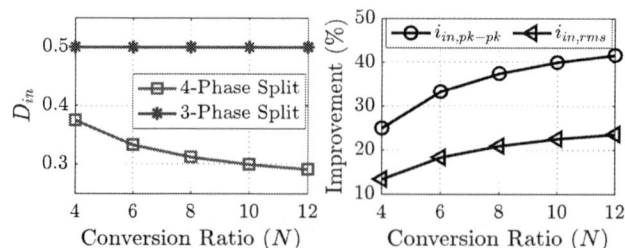

Fig. 5: Improvement in input current conduction time and current RMS of the proposed 3-phase scheme compared to the conventional 4-phase split-phase scheme as a function of the conversion ratio.

Table I: Summary of Key Points for Conventional 4-Phase vs Proposed 3-Phase Split-Phase Schemes for 8-to-1 Conversion.

Aspect	4-Phase Scheme	3-Phase Scheme
Capacitor Sizing	Equal ratio	Unequal ratio
Input Switch Duty	31.25% duty	50% duty
Ripple at $V_{C,HI}$	Higher	27.3% reduction
Input Current RMS	Higher	20.9% reduction
Equivalent R_{out}	1.9 R_{on}	1.92 R_{on}
Bulk Capacitor	Larger C_{HI}	25% reduction
Control Complexity	4 signals	3 signals
Phase Symmetry	Symmetric	Asymmetric

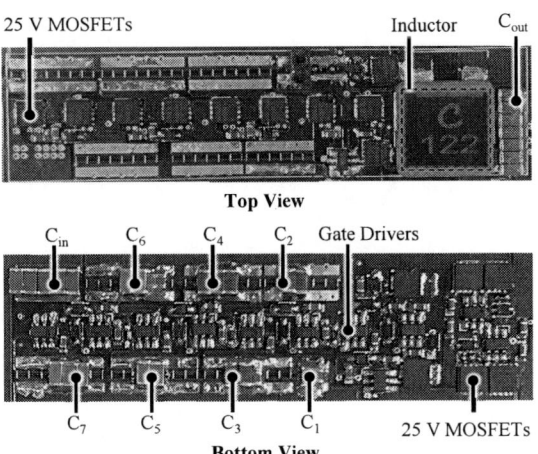

Fig. 6: Annotated photograph of the 8-to-1 Dickson converter with main components labeled.

V. CONTROL SCHEME VALIDATION IN HARDWARE

The proposed 3-phase control scheme for the 48-to-6 V Dickson converter was validated in hardware using the prototype shown in the annotated photograph in Fig. 6. The operating parameters of the prototype and list of the main power stage components are provided in Table II and III. The switches theoretically block voltages in the range of 6-13 V. In the hardware, all switches are selected to block 25 V, providing a margin of safety. The capacitor passive volume ($\sum_k C_k V_{C_k}^2$) was kept the same across both the 4-phase and 3-phase schemes to ensure a fair comparison. Class I MLCC capacitors were selected as they do not derate with DC bias, making it easier to validate the soft-charging capability of the proposed split-phase scheme. Class II capacitors can still be used in applications where high power density is required, with slight adjustments in timing [25]. The converter was also operated high above resonance to mimic operation with an ideal constant current load. This results in the phase timings being exactly proportional to the charge delivered to the output.

Table II: Operating Parameters

Description	Value
Input Voltage	48 V
Output Voltage	6 V
Max. Load Current	10 A
Switching Frequency	100-300 kHz

Table III: Component listing of the converter prototype.

Component	Part Number	Parameters
C_{1-7}	GRM31C5C1H224GE02L	$k^\dagger \times$ 0.22 μF
Inductor	XGL3530-122MEC	1.2 μH
MOSFET	IQE006NE2LM5	25 V, 0.65 $m\Omega$
Gate Driver	LTC4440-5	80 V, high-side
Bootstrap Diode	PMEG6002EJ,115	60 V, Schottky

† k represents the number of capacitors connected in parallel.
For 4-Phase scheme, k = 12:12:12:12:12:12:12
For 3-Phase scheme, k = 24:8:12:12:8:24:6

Fig. 7 presents the experimental waveforms obtained for a load current of 10 A, showing the switch node voltage (V_{sw}), capacitor C_1 voltage (V_{C1}), and inductor current (i_L). The primary phases are shaded in dark yellow and dark blue for clarity, while the introduced phases are shaded in light yellow and light blue. The 4-phase scheme exhibits a

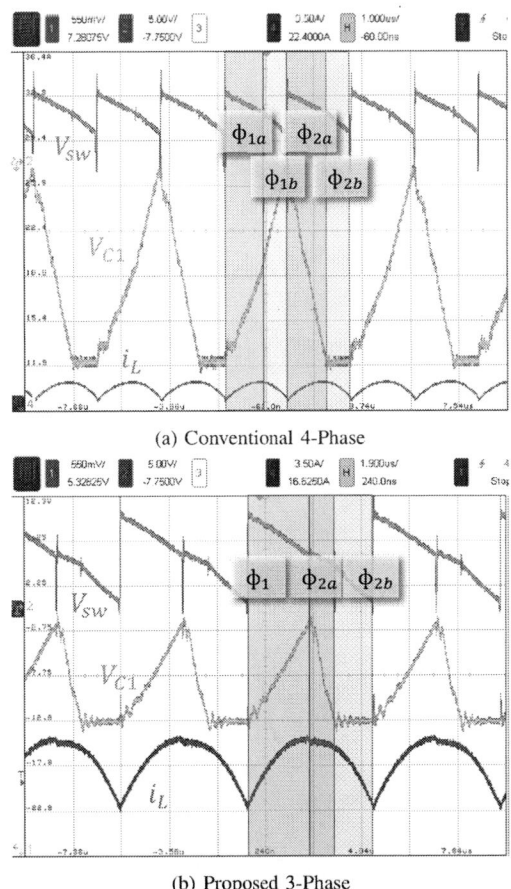

(a) Conventional 4-Phase

(b) Proposed 3-Phase

Fig. 7: Experimental results for 48 V input, 6 V output, and a switching frequency of 200 kHz at I_{out} = 10 A with capacitor sizing in Table III.

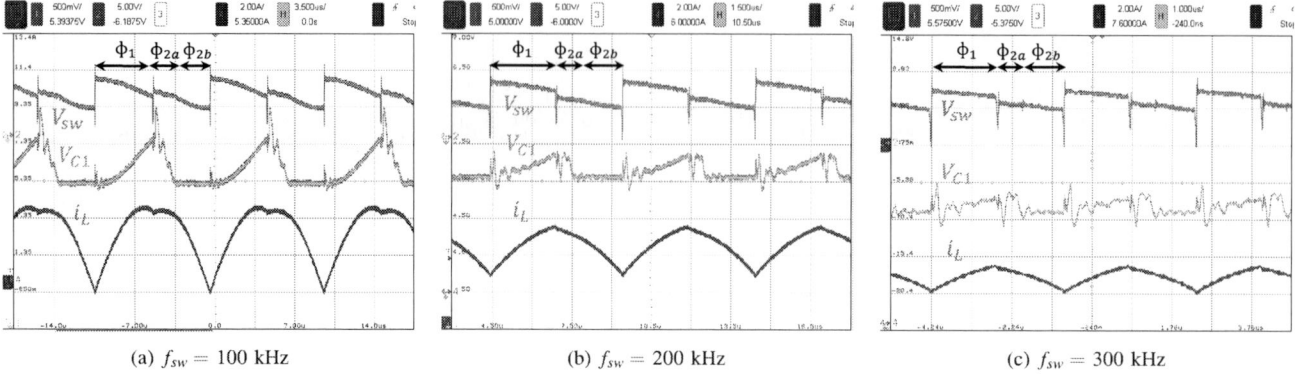

(a) $f_{sw} = 100$ kHz (b) $f_{sw} = 200$ kHz (c) $f_{sw} = 300$ kHz

Fig. 8: Experimental results for 48 V input, 6 V output for 3-phase scheme for different frequencies at a load current of 2.5 A.

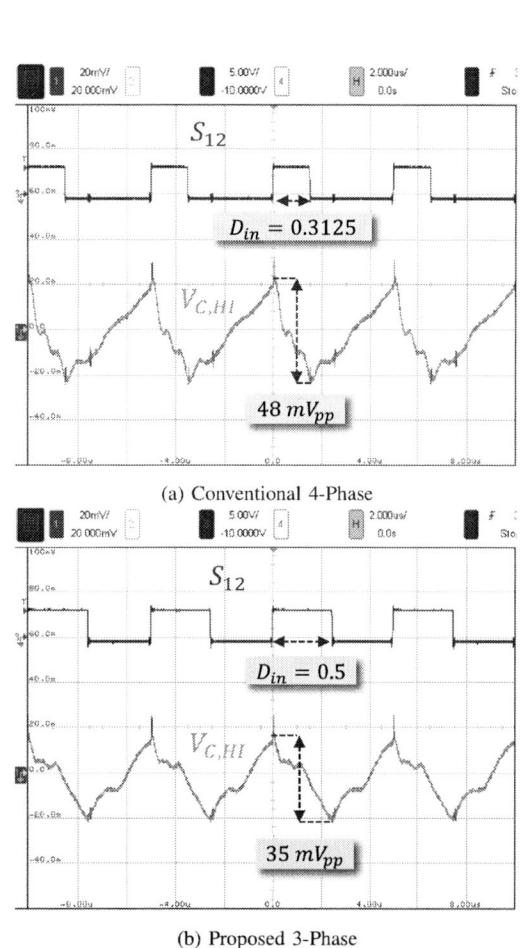

(a) Conventional 4-Phase

(b) Proposed 3-Phase

Fig. 9: Experimental results showing the increase in input current conduction duration (as indicated by the S_{12} control signal) and the corresponding reduction in input bulk capacitor voltage ripple for a 48 V input, 6 V output, 10 A load current at a switching frequency of 200 kHz. An approximate 27% reduction in voltage ripple is observed, in agreement with theoretical calculations.

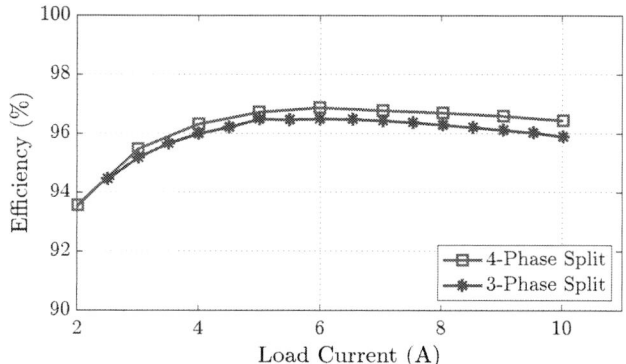

Fig. 10: Converter efficiency plot versus load current (excluding gate driver losses).

2x frequency multiplication effect on the inductor current due to symmetrical split-phase operation, reducing inductor current ripple. Capacitor C_1 experiences higher voltage ripple in the 4-phase scheme, primarily because C_1 is half the size compared to the 3-phase configuration. Fig. 7(b) shows that capacitor C_1 absorbs a charge of q during Phase 1 and delivers q in Phase 2a, while remaining unconnected in Phase 2b, as described in Fig. 3(c)

Fig. 8 shows the variation in key waveforms (switch node voltage, capacitor voltage, and inductor current) for the 3-phase scheme at switching frequencies of 100 kHz, 200 kHz, and 300 kHz. As frequency increases, the inductor current magnitude decreases, and the current waveform becomes more linear.

Fig. 9 compares the voltage ripple observed at the input bulk capacitor alongside the switch S_{12} control signal for both split-phase schemes. Based on (12), it can be concluded that $\Delta V_{C,HI} \propto (1 - D_{in})$ for a given C_{HI}, f_{sw}, and I_{in}. An increase in D_{in} from 0.3125 to 0.5 between 4-phase and 3-phase schemes results in a theoretical 27.3% reduction in the input voltage ripple. Experimentally, the peak-to-peak ripple decreases from $48mV_{pp}$ in the 4-phase scheme to $35mV_{pp}$ in the 3-phase scheme, reflecting a reduction of approximately 27.1% and thereby confirming the advantage of the proposed scheme.

Fig. 10 shows efficiency variation with load current for both split-phase schemes. Both schemes exhibit high efficiency, demonstrating that the converter successfully achieves complete soft-charging. The 3-phase scheme shows a slightly lower efficiency due to its higher output impedance compared to the 4-phase scheme, as discussed in Section IV. However, for power converter systems where line losses are significant, the 3-phase scheme holds potential for improved overall system efficiency.

VI. CONCLUSION

This work introduces a novel split-phase control scheme for the Dickson converter operating at an even-conversion ratio. Unlike the conventional approach, which adds two sub-phases, the proposed scheme introduces only one additional sub-phase. This modification significantly reduces the ripple of the input current, thereby simplifying the design of the input filter capacitor for noise/EMI management, reducing control complexity, and facilitating easier interleaving strategies. The scheme was successfully validated on a 48-to-6 V Dickson converter prototype.

ACKNOWLEDGEMENT

The authors would like to acknowledge the Berkeley Power and Energy Center (BPEC) for its financial support.

REFERENCES

[1] R. C. N. Pilawa-Podgurski, D. M. Giuliano, and D. J. Perreault, "Merged two-stage power converter architecture with soft charging switched-capacitor energy transfer," in *2008 IEEE Power Electronics Specialists Conference*, June 2008, pp. 4008–4015.

[2] Y. Li, X. Lyu, D. Cao, S. Jiang, and C. Nan, "A 98.55% Efficiency Switched-Tank Converter for Data Center Application," *IEEE Transactions on Industry Applications*, vol. 54, no. 6, pp. 6205–6222, 2018.

[3] A. Lidow, "Powering graphics processors from a 48-v bus," *Power Electronic Tips*, 2019.

[4] Q. Ouyang, R. Samsi, and J. Zhou, "Increasing the power density and efficacy of datacenters using a two-stage solution for 48v power distribution," *Monolithic Power Syst., Kirkland, WA, USA*, 2017.

[5] Y. Zhang, Q. Liu, Y. Gao, J. Li, and M. Sumner, "Hybrid switched-capacitor/switched-quasi-z-source bidirectional dc–dc converter with a wide voltage gain range for hybrid energy sources evs," *IEEE Transactions on Industrial Electronics*, vol. 66, no. 4, pp. 2680–2690, 2019.

[6] M. Ashourloo, V. R. Namburi, G. V. Piqué, J. Pigott, H. J. Bergveld, A. El Sherif, and O. Trescases, "Decentralized quasi-fixed-frequency control of multiphase interleaved hybrid dickson converters for fault-tolerant automotive applications," *IEEE Transactions on Power Electronics*, vol. 35, no. 7, pp. 7653–7663, 2019.

[7] M. Ashourloo, V. R. Namburi, G. V. Pique, J. Pigott, H. J. Bergveld, A. El Sherif, and O. Trescases, "An automotive-grade monolithic masterless fault-tolerant hybrid dickson dc–dc converter for 48-v multi-phase applications," *IEEE Journal of Solid-State Circuits*, vol. 56, no. 12, pp. 3608–3618, 2021.

[8] J. W. Kwak and D. B. Ma, "An automotive-use dual-f sw-zone hybrid switching power converter with vo-jitter-immune spread-spectrum modulation," in *ESSCIRC 2022-IEEE 48th European Solid State Circuits Conference (ESSCIRC)*. IEEE, 2022, pp. 281–284.

[9] N. C. Brooks, J. Zou, S. Coday, T. Ge, N. M. Ellis, and R. C. Pilawa-Podgurski, "On the size and weight of passive components: Scaling trends for high-density power converter designs," *IEEE Transactions on Power Electronics*, 2024.

[10] Y. Lei and R. C. N. Pilawa-Podgurski, "A general method for analyzing resonant and soft-charging operation of switched-capacitor converters," *IEEE Transactions on Power Electronics*, vol. 30, no. 10, pp. 5650–5664, Oct 2015.

[11] N. Patle, R. A. Abramson, R. K. Iyer, and R. C. Pilawa-Podgurski, "A ripple-equivalent circuit-based method for analyzing soft-charging operation in hybrid switched-capacitor converters," in *2024 IEEE Energy Conversion Congress and Exposition (ECCE)*. IEEE, Oct 2024, pp. 1–8.

[12] Infineon Technologies, "48 v power distribution for hyperscale computing," accessed: 2024-11-16. [Online]. Available: https://www.infineon.com/cms/en/applications/information-communication-technology/hyperscale-computing/48v-power-distribution

[13] Vicor Corporation, "Efficiently bridging 48v and 12v systems," 2024, accessed: 2024-11-16. [Online]. Available: https://www.vicorpower.com/industries-and-innovations/bridging-48v-and-12v

[14] S. Lovati, "Reasons for migrating power distribution design to 48v bus," Power Electronics News, 2019, accessed: 2024-11-16. [Online]. Available: https://www.powerelectronicsnews.com/reasons-for-migrating-power-distribution-design-to-48v-bus/

[15] Analog Devices, "Simple methods reduce input ripple for all charge pumps," Analog Devices, Inc., 2024, accessed: 2024-11-17. [Online]. Available: https://www.analog.com/en/resources/technical-articles/simple-methods-reduce-input-ripple-for-all-charge-pumps.html

[16] R. Rizzolatti, M. Ursino, and C. Rainer, "Hybrid switched capacitor converter using source-down mosfets," Infineon Technologies, Application Note AN_2305_PL15_2305_112212, 2024, accessed: 2024-11-17. [Online]. Available: https://www.infineon.com/dgdl/Infineon-AN_2305_PL15_2305_112212_Hybrid_switched_capacitor_converter_using_source-down_MOSFETs-ApplicationNotes-v01_00-EN.pdf

[17] A. Elkhateb, N. Abd Rahim, J. Selvaraj, and B. W. Williams, "Dc-to-dc converter with low input current ripple for maximum photovoltaic power extraction," *IEEE Transactions on Industrial Electronics*, vol. 62, no. 4, pp. 2246–2256, 2014.

[18] Texas Instruments, "Input and output capacitor selection," Texas Instruments, Application Report SLTA055, 2024, accessed: 2024-11-17. [Online]. Available: https://www.ti.com/lit/an/slta055/slta055.pdf

[19] Z. Ye, S. R. Sanders, and R. C. N. Pilawa-Podgurski, "Modeling and comparison of passive component volume of hybrid resonant switched-capacitor converters," in *2019 20th Workshop on Control and Modeling for Power Electronics (COMPEL)*, June 2019, pp. 1–8.

[20] S. Jiang, C. Nan, X. Li, C. Chung, and M. Yazdani, "Switched tank converters," in *2018 IEEE Applied Power Electronics Conference and Exposition (APEC)*, March 2018, pp. 81–90.

[21] P. H. McLaughlin, P. A. Kyaw, M. H. Kiani, C. R. Sullivan, and J. T. Stauth, "A 48-v: 16-v, 180-w resonant switched-capacitor converter with high-q merged multiphase lc resonator," *IEEE Journal of Emerging and Selected Topics in Power Electronics*, vol. 8, no. 3, pp. 2255–2267, 2019.

[22] N. M. Ellis, *Hybrid switched-capacitor power converter techniques*. University of California, Davis, 2020.

[23] S. Krishnan, M. E. Blackwell, and R. C. Pilawa-Podgurski, "An emi-compliant and automotive-rated 48v to point-of-load dickson-based hybrid switched-capacitor dc-dc converter," in *2023 IEEE Transportation Electrification Conference & Expo (ITEC)*. IEEE, 2023, pp. 1–7.

[24] Y. Lei, R. May, and R. C. N. Pilawa-Podgurski, "Split-phase control: Achieving complete soft-charging operation of a dickson switched-capacitor converter," *IEEE Transactions on Power Electronics*, vol. 31, no. 1, pp. 770–782, Jan 2016.

[25] R. A. Abramson, S. Krishnan, M. E. Blackwell, and R. C. Pilawa-Podgurski, "An active split-phase control technique for hybrid switched-capacitor converters using capacitor voltage discontinuity detection," in *2023 IEEE 24th Workshop on Control and Modeling for Power Electronics (COMPEL)*. IEEE, 2023, pp. 1–7.

[26] N. M. Ellis and R. Amirtharajah, "Large signal analysis on variations of the hybridized dickson switched-capacitor converter," *IEEE Transactions on Power Electronics*, vol. 37, no. 12, pp. 15 005–15 019, 2022.

[27] L. Yuan, J. Zhang, Z. Liang, M. Hu, G. Chen, and W. Lu, "Emi challenges in modern power electronic-based converters: recent advances and mitigation techniques," *Frontiers in Electronics*, vol. 4, p. 1274258, 2023.

[28] M. E. Blackwell, *Switching Schemes for Hybrid Switched-Capacitor DC-DC Power Converters*. University of California, Berkeley, 2023.

[29] R. Das, G.-S. Seo, and H.-P. Le, "Analysis of dual-inductor hybrid converters for extreme conversion ratios," *IEEE Journal of Emerging and Selected Topics in Power Electronics*, vol. 9, no. 5, pp. 5249–5260, 2020.

Experimental Verification of Circuit-Losses Analysis-Model of DC-Output Converter developed using Approximated Equations from Measurement Data and Datasheet Data

Ryota Kondo
Advanced Technology R&D Center
Mitsubishi Electric Corporation
Hyogo, Japan
Kondo.Ryota@db.MitsubishiElectric.co.jp

Tsuyoshi Funaki
Division of Electric Electronics and Information Engineering
Osaka Univercity
Osaka, Japan
funaki@eei.eng.osaka-u.ac.jp

Abstract— This paper proposes a losses-analysis model of a DC-output converter based on the approximated equations from measurements and a datasheet. Target verification circuit consists of a DC-voltage source, a smooth capacitor, two half bridge legs, and a load inductor connected to the midpoints of the half bridge legs, which controls the continuous average DC current of the load inductor. The use of self-measured data, compared with the use of datasheet values, enables expanding the ranges of the considerable target parameters. Hence, lesser order approximated equations can deal with more target parameters with less calculation period than that in a conventional time domain simulation. In this study, calculation flow of losses from power devices losses and inductor losses to circuit losses was developed. Subsequently, a power losses model and inductor losses model were developed via the linear least square method. Finally, a 400 V experimental system was developed to verify the accuracy of the circuit losses at 16 specified operating conditions by varying three verification parameters, switching frequency, gate resistance, and inductance. The result shows that the losses calculated using the proposed model has good accuracy with an error of 1.9 % to 10.6 % to the measured result.

Keywords—Power Converter, Power devices losses, inductor losses, loss analysis model

I. INTRODUCTION

The market of DC/DC converter installed in an electric mobility and IoT (Internet of Things) devices is expanding. The converters, with basic functions fulfilled at early design stage, requires performance of higher efficiency for saving energy and higher-power density for easy installation in spatially constrained area[1]-[3]. A common requirement of these requirements is lower losses.

The compatibility with loss target values is considered by not only developing circuit configurations, control schemes, power devices, and inductors[4]-[10] but also optimizing circuit and design parameters[11]-[16]. A development of these core technologies and parameter optimization at early design stage needs to make specified requirements to sub-component designs within a constrained duration. Therefore, an accurate circuit-losses analysis model with an instant calculation is necessary.

Time domain circuit simulation[17][18] and finite element analysis for inductor losses[19][20] can simulate operating waveforms with a given analysis parameters, and calculate accurate circuit losses. However, it makes too heavy effort of losses calculation to set many simulation patterns. On the other hand, a scheme to calculate circuit losses by model equations of inductor and power device formulated from datasheet values[21], makes not only calculation effort lighter but also can explore more analysis conditions than that in a conventional time domain simulation. However, the switching characteristics of semiconductor devices described in the data sheet are limited to the values measured in specific circuits and operating conditions, and the range to which this method can be applied is limited.

The authors have proposed a losses-analysis model of power devices and of inductor based on the approximated equations from measurements and datasheet[22]. The use of self-measured data, compared with the conventional one of datasheet values, enables expanding the ranges of the considerable target parameters. Hence, reduced-order approximated equations can deal with more target parameters with less calculation effort than that in a conventional time domain simulation.

This paper proposes to integrate each loss equations to one circuit-losses equation for further analysis. It is expected to analyze circuit losses, not parts losses, by varying parameters from multiple viewpoints with shorter calculation duration. In this paper, a concept and a modeling flow of the proposed circuit-losses analysis model are explained, and then a 400 V experimental system is built to verify accuracy of the circuit-losses analysis model.

II. VERIFICATION CONFIGURATION

A. Circuit configuration

Fig. 1 shows the circuit configuration for losses verification. It consists of a DC-voltage source V_{dc}, a smooth capacitor C_{dc}, power devices Q_i and a load inductor L. The gate drive circuit GD_i outputs the on-voltage V_{g+} or the off-voltage V_{g-}, according to the gate signal S_{gi}, to a gate of Q_i via a gate resistor R_g. The switching frequency is f_{sw}. Table 1 shows the operating specifications and circuit parameters. Fig. 2 shows a 400 V – 5 A prototype used for storing measured losses characteristics of power devices and an inductor, and for verifying accuracy of the circuit loss. It is noted that "Gate drive board" in Fig.2 consists of two GD_i shown in Fig.1.

B. Circuit losses defenition

Fig. 3 is a diagram showing the inductor current i_L with reference to the gate signal S_{gi}. In period (1), the DC voltage

979-8-3315-1612-3/25 $31.00 © 2025 IEEE

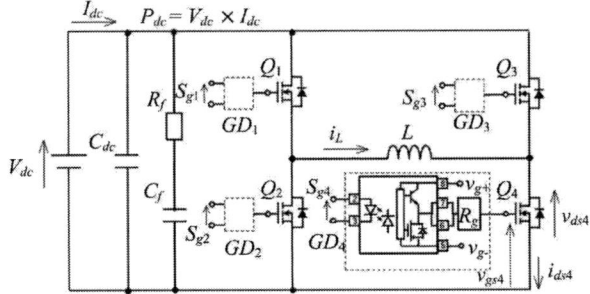

Fig.1 Circuit configuration for losses verification of power devices, $Q_1 \sim Q_4$, and inductor, L.

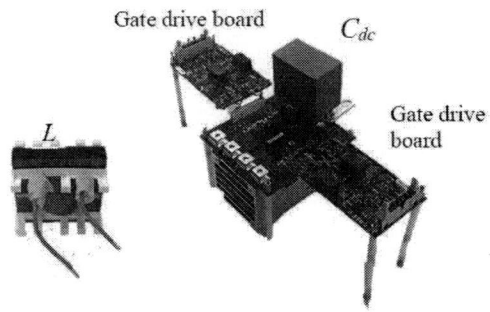

Fig.2 A 400 V – 7.5A prototype for losses verification with a circuit configuration indicated in Fig.1

TABLE I. OPERATING SPECIFICATIONS AND CIRCUIT PARAMETERS OF THE CIRCUIT CONFIGURATION DESCRIBED IN FIG.1

Items	Unit	Value
DC Input voltage, V_{dc}	V	150 ~ 400
Load current, I_c	A	~7.5
Smooth capacitance, C_{dc}	μF	1
Snubber capacitance, C_f	μF	10
Snubber resistance, R_f	Ω	6.8
Switching frequency, f_{sw}	kHz	75 ~ 125
Dead time, T_d	μs	1
Gate on voltage, V_{g+}	V	20
Gate off voltage, V_{g-}	V	-5
Core material of L		PC40
Core shape of L		EC70
Cross section Ae	mm²	280
Core volume, V_{core}	mm³	40420

V_{dc} is applied to the inductor L via Q_1 and Q_4. In period (2), the DC voltage V_{dc} is applied to the inductor L via Q_2 and Q_3. If the length of period (1) and period (2) are equal, capacitance of the smoothing capacitor C_{dc} is sufficiently large, and equivalent series resistance (ESR) of the capacitor C_{dc} is negligible, the DC voltage source V_{dc} supplies only loss generated by the power devices Q_i and the inductor L as a DC current I_{dc}. Therefore, the measured loss P_{dc} is evaluated by multiplying the DC current I_{dc} and the DC voltage V_{dc}. In this paper, the period (1) is adjusted by feedback control of an average current I_L to reference value I_{Lref}. The circuit loss generated in the circuit shown in Fig.1 is sum of losses of the power devices Q_i and the inductor L. The losses of the power device Q_i consists of conduction loss E_{ds} when Q_i is conductive, parasitic diode conduction loss E_{sd} when Q_i is non-conductive, turn-on loss E_{on}, and turn-off loss E_{off}. The losses of the inductor L consists of core loss P_{core} and winding loss P_{wire}. Table 2 shows loss definition of the power device Q_i in the circuit operation of one switching period in Fig. 3. It is assumed that i_L is always positive in this paper. When Q_1 and Q_4 are turned on at period of T_d, forward current changes from 0 to I_{LP-} and turn-on loss E_{on} occurs. When Q_1 and Q_4 are turned off at period of $T_{sw}/2 + T_d$, forward current changes from I_{LP+} to 0 and turn-off loss E_{off} occurs. At period of $T_{sw}/2 + 2T_d$, Q_2 and Q_3 are ZVS (Zero-Voltage Switching) operation because body diodes are conducting with the reverse current I_{LPTd+}, and turn-on losses E_{on} of Q_2 and Q_3 are zero. Similarly, at period of T_{sw} when the channels of Q_2 and Q_3 are turned off, the body diodes continue to conduct reverse current I_{LPTd-}. So turn-off losses E_{off} are zero. During the period when Q_1 and Q_4 are turn on($T_d \sim T_{sw}/2 + T_d$) and period when Q_2 and Q_3 are turn on($T_{sw}/2 + 2T_d \sim T_{sw}$), Q_i has forward conduction loss E_{ds}. Hence, during the dead-time period ($T_{sw}/2 + T_d \sim T_{sw}/2 + 2 T_d$ and $T_{sw} \sim T_{sw} + T_d$),

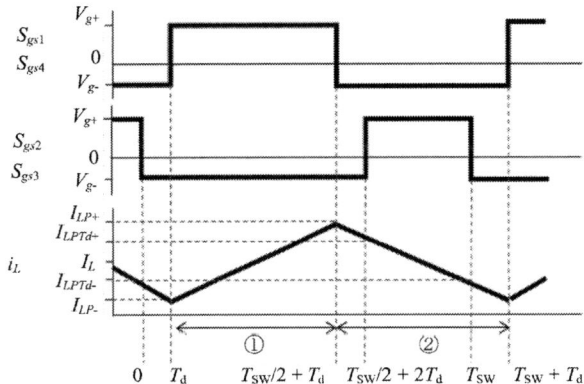

Fig.3 Reference operating waveforms of gate signals, $v_{gs1} \sim v_{gs4}$, and an inductor current, i_L, in Fig.1

TABLE II. LOSSES DEFINITION OF POWER DEVICES, $Q_1 \sim Q_4$, DURING A SWITCHING PERIOD FROM T_D TO $T_s + T_D$ IN FIG. 2

Period	Losses of Q_1 and Q_4	Losses of Q_2 and Q_3
T_d	Turn on loss, E_{on}	(Recovery loss)
$T_d \sim T_{sw}/2+T_d$	Forward conduction loss, E_{ds}	
$T_{sw}/2+T_d$	Turn off loss, E_{off}	
$T_{sw}/2+T_d \sim T_{sw}/2+2T_d$		Reverse conduction loss, E_{sd}
$T_{sw}/2+2T_d$		Zero voltage switch, $E_{on} = 0$
$T_{sw}/2+2T_d \sim T_{sw}$		Forward conduction loss, E_{ds}
T_{sw}		Zero voltage switch, $E_{off} = 0$
$T_{sw} \sim T_d$		Reverse conduction loss, E_{sd}

Q_i has reverse conduction loss E_{sd}. Recovery loss of Q_i is ignored because the loss is sufficiently small compared to turn-on loss E_{on} and turn-off loss E_{off} due to use of SiC-MOSFETs (C2M0080120D).

III. LOSSES EQUATION MODELS

Fig. 4 shows a calculation flow of the circuit losses. A work box of "Circuit parameters" sets circuit design parameters and gate drive parameters, then operating parameters are

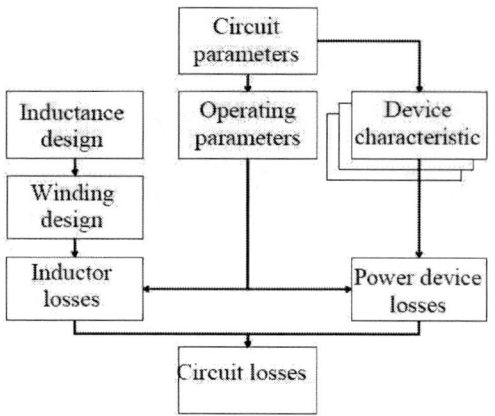

Fig. 4 Calculation flow of circuit losses, consisting of power device losses and inductor losses, in the circuit configuration described in Fig.1

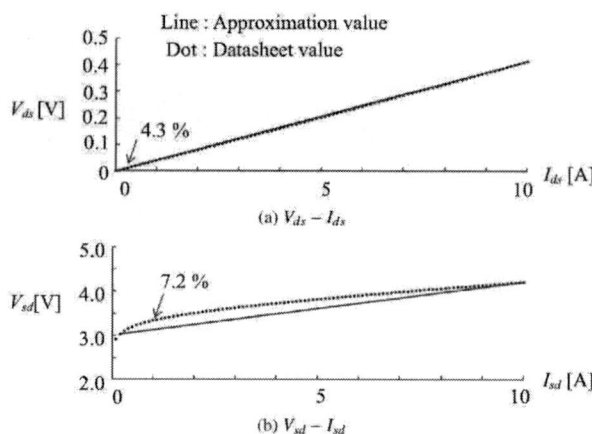

Fig.5 $V_{ds} - I_{ds}$ and $V_{sd} - I_{sd}$ characteristic of C2M0080120D in conduction mode

calculated from these parameters subordinately in "operating parameters".

Parameters for the inductance design, turn numbers and core shapes, are adjusted in "Inductance design". Parameters for the winding design are calculated in "Winding design" subordinately from "Inductance design". Then, "Inductor losses" calculates core and winding losses subordinately. "Device characteristic" selects a target device in which losses characteristics of the selected power device is modeled by approximated equations from the measured data and the data sheet. Then "Power device losses" calculates the power device loss subordinately. Finally, "circuit losses" calculate total circuit losses from inductor losses in "Inductor losses" and power device losses in "power device losses".

A. Conduction loss of power device

In Fig.4, "Circuit parameters" sets circuit conditions (C_{dc}, C_f, R_f, and L) and operating conditions(V_{dc}, I_L, and f_{sw}) shown in Table 1. An inductor current i_L and a conduction current i_c of the selected power device are approximated by a first-order linear equation, assuming that an inductance L is a constant value and ignoring the voltage drop of Q_1. In this case, conduction current i_c of the power device is expressed by (1), assuming that the current at start point of conduction period ($t = 0$) is I_{c1} and the current at end point of conduction period ($t = \Delta T_c$) is I_{c2}.

$$i_c = I_{c1} + \left(\frac{I_{c2} - I_{c1}}{\Delta T_c}\right) t \qquad (1)$$

Hence, I_{LP+}, I_{LPTd+}, I_{LP-}, and I_{LPTd-} of the inductor current i_L shown in Fig.3 are expressed as following equations.

$$I_{LP\pm} = I_L \pm \frac{1}{2} \cdot V_{dc} \cdot \frac{1}{2f_{sw}} \cdot \frac{1}{L} \qquad (2)$$

$$I_{LPTd\pm} = I_{LP\pm} \mp V_{dc} \cdot T_d \cdot \frac{1}{L} \qquad (3)$$

"Device characteristic" selects SiC-MOSFET (C2M0080120 D) and set drive parameters (R_g, V_{g+}, V_{g-} and T_d). A junction temperature T_j of the power device is set to room temperature of 25 °C, which is obtained from the data sheet of the C2M0080120D. Fig. 5 shows the V_{ds}-I_{ds} characteristic and the

V_{sd}-I_{sd} characteristics of the SiC- MOSFET (C2M0080120D) selected by first-order linear equations when V_{g+} is 20 V and V_{g-} is -5 V. An approximate equation for a voltage drop V_c as a first-order linear equation of time is expressed as follows.

$$V_c = V_c(I_{c1}) + \left\{\frac{V_c(I_{c2}) - V_c(I_{c1})}{\Delta T_c}\right\} \cdot t \qquad (4)$$

Therefore, a conduction loss P_{ds} is calculated by I_c in (1), V_c in (4), ΔT_c and switching frequency f_{sw} as follows.

$$P_{ds} = f_{sw} \cdot \int_0^{\Delta T_c} I_c \cdot V_c \, dt$$
$$= \left\{\begin{array}{c} \dfrac{I_{c1} \cdot V_c(I_{c1}) + I_{c2} \cdot V_c(I_{c2})}{3} \\ + \dfrac{I_{c1} \cdot V_c(I_{c2}) + I_{c2} \cdot V_c(I_{c1})}{6} \end{array}\right\} \cdot \Delta T_c \cdot f_{sw} \qquad (5)$$

The conduction losses of Q_1 and Q_4 (P_{dsQ1} and P_{dsQ4}) are occurred in the period from T_d to $T_{sw}/2 + T_d$ in Fig.3. V_c, I_{C1}, I_{C2}, and ΔT_c in (5) are calculated as V_{ds}, I_{LP+}, I_{LP-}, and $T_{sw}/2$. The conduction losses of Q_2 and Q_3 (P_{dsQ2} and P_{dsQ3}) are occurred in the period from $T_{sw}/2 + 2T_d$ to T_{sw} in Fig.3. V_c, I_{C1}, I_{C2}, and ΔT_c in (5) are calculated as V_{ds}, I_{LPTd+}, I_{LPTd-}, and $T_{sw}/2$ - $2T_d$. The parasitic diode conduction losses of Q_2 and Q_3 (P_{sdQ2} and P_{sdQ3}) are occurred in the period from $T_{sw}/2 + T_d$ to $T_{sw}/2 + 2 T_d$ and from T_{sw} to $T_{sw} + T_d$ in Fig. 3. In the period from $T_{sw}/2 + T_d$ to $T_{sw}/2 + 2T_d$, V_c, I_{C1}, I_{C2}, and ΔT_c in (5) are V_{sd}, I_{LP+}, I_{LPTd+}, and T_d. In the period from T_{sw} to $T_{sw} + T_d$, V_c, I_{C1}, I_{C2}, and ΔT_c in (5) are V_{sd}, I_{LPTd-}, I_{LP-}, and T_d.

B. Switching losses of power device

It is assumed that turn-off loss E_{off} and turn-on loss E_{on} are characterized by R_g, I_{sw}, and V_{dc}. Therefore, turn on loss E_{on} and turn off loss E_{off} of the SiC-MOSFET(C2M0080120D) are measured by a double pulse measurement at total 72 conditions with R_g of 10 Ω, 20 Ω, 30 Ω, V_{dc} of 150 V, 200 V, 300 V, 400 V, and I_{sw} of 0 A, 5 A, 10 A, 15 A, 20 A, 25 A. Fig.6 shows measured characteristics of E_{off} and E_{on} to R_g, V_{dc}, and I_{sw} at the 72 conditions. Fig.6(a)(b) show E_{off} - R_g characteristic and E_{on} - R_g one when I_{sw} and V_{dc} are parameters, Fig.7(c)(d) show E_{off} - V_{dc} characteristic and E_{on} - V_{dc} one when I_{sw} and R_g are parameters, and Fig.7(e)(f) show E_{off} - I_{sw} characteristic and E_{on} - I_{sw} one when R_g and V_{dc} are parameters. Due to a symmetrical approximate process between E_{on} and

979-8-3315-1612-3/25 $31.00 © 2025 IEEE

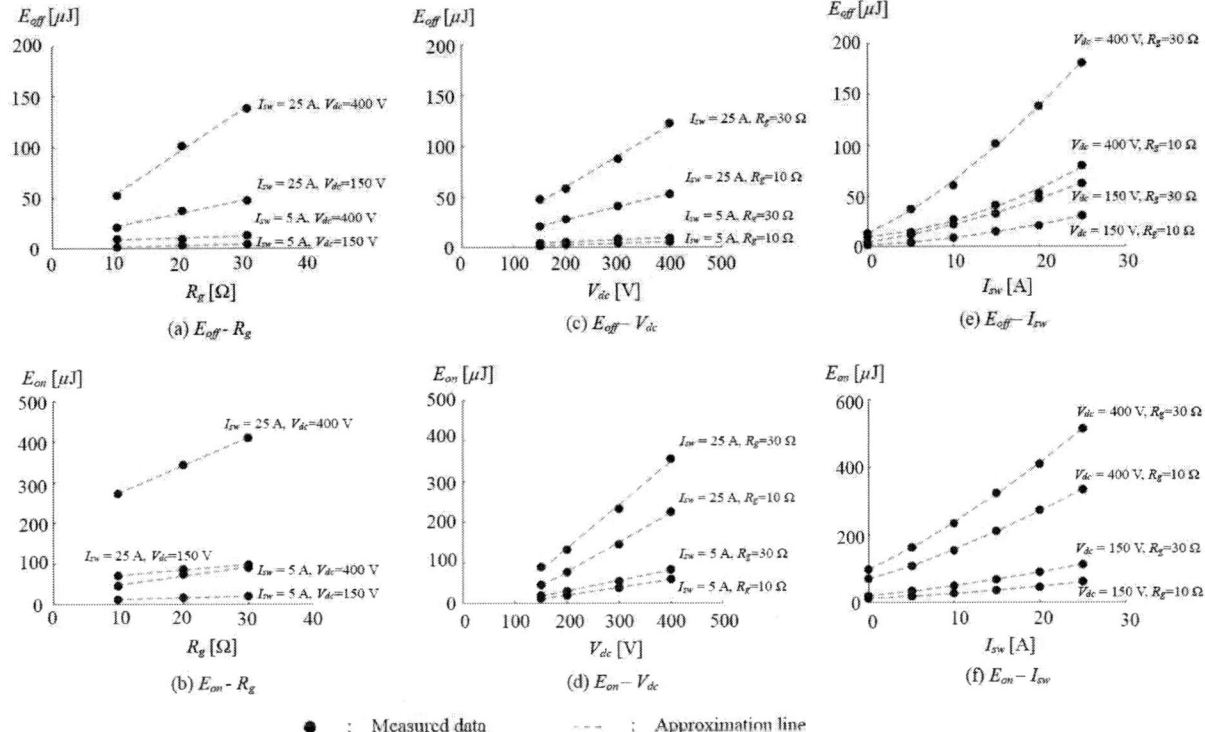

Fig.6 Measured characteristic of E_{off} and E_{on} versus R_g, V_{dc}, I_{sw}

● : Measured data   - - - : Approximation line

E_{off}, E_{sw} is set to a common symbol. E_{sw} -R_g characteristics, E_{sw} - V_{dc} characteristics are approximated using a first-order linear equation as shown in (6) and (7). Hence, E_{sw}-I_{sw} characteristics are approximated using a second-order linear equation as shown in (8).

TABLE III.   EXTRACTED COEFFICIENTS FOR EQ.(9) IN CASE OF E_{OFF} AND E_{ON}

	E_{off}		E_{on}
A_{on1}	2.52×10^{-5}	A_{off1}	1.47×10^{-5}
A_{on2}	-9.57×10^{-13}	A_{off2}	1.81×10^{-12}
A_{on3}	2.6×10^{-3}	A_{off3}	9.03×10^{-5}
A_{on4}	-5.5×10^{-3}	A_{off4}	-2.30×10^{-3}
A_{on5}	1.28×10^{-1}	A_{off5}	4.93×10^{-2}
A_{on6}	-6.36×10^{-1}	A_{off6}	-5.41×10^{-2}
A_{on7}	-1.97×10^{-4}	A_{off7}	-1.13×10^{-4}
A_{on8}	2.4×10^{-2}	A_{off8}	3.90×10^{-3}
A_{on9}	4.24×10^{-2}	A_{off9}	-2.4×10^{-3}
A_{on10}	9.58×10^{-2}	A_{off10}	5.63×10^{-2}
A_{on11}	$- 4.62$	A_{off11}	1.47×10^{-5}
A_{on12}	4.97	A_{off12}	1.81×10^{-12}

$$E_{sw} = A_{Rg} \cdot R_g + B_{Rg} \qquad (6)$$

$$E_{sw} = A_{Vdc} \cdot V_{dc} + B_{Vdc} \qquad (7)$$

$$E_{sw} = A_{Isw} \cdot I_{sw}^2 + B_{Isw} \cdot I_{sw} + C_{Isw} \qquad (8)$$

By substituting the approximate equation (7) and (8) into the slope A_{Rg} and intercept B_{Rg}, an integrated approximate equation of E_{sw} for R_g, V_{dc}, and I_{sw} is expressed as shown in (9). The coefficients $A_{sw1} \sim A_{sw12}$ is $A_{on1} \sim A_{on12}$ in case of calculating E_{on} and $A_{off1} \sim A_{off12}$ in case calculating E_{off} as

shown in Table 3. These coefficients are calculated by least squares method under the 72 conditions.

$$\begin{aligned}
E_{sw} = {} & A_1 \cdot I_{SW}^2 \cdot V_{dc} \cdot R_g + A_2 \cdot I_{SW} \cdot V_{dc} \cdot R_g + A_3 \cdot V_{dc} \cdot R_g \\
& + A_4 \cdot I_{SW}^2 \cdot R_g + A_5 \cdot I_{sw} \cdot R_g + A_6 \cdot R_g \\
& + A_7 \cdot I_{SW}^2 \cdot V_{dc} + A_8 \cdot I_{sw} \cdot V_{dc} + A_9 \cdot V_{dc} \\
& + A_{10} \cdot I_{SW}^2 + A_{11} \cdot I_{sw} + A_{12} \qquad (9)
\end{aligned}$$

C. Inductor losses

"Inductor design" in Fig.4 selects the core shape (EC70) and the core material (PC40) as precondition, and design inductance, which is specified from circuit parameters, by adjusting the gap length L_g and the turn number N.

The gap length L_g is obtained that satisfies the saturation magnetic flux density B_{max} of the core material (PC40). The lower limit of L must satisfy the condition equation expressed in (10) from "Circuit Parameters".

$$L > V_{dc} \cdot \frac{1}{2 f_{sw}} \cdot \frac{1}{i_L} \qquad (10)$$

The designed inductance L_{des} is defined by (11).

$$L_{des} = \frac{\mu \cdot \mu_0 \cdot Ae}{\mu_0 \cdot Lc + \mu \cdot Lg} \cdot N^2 \qquad (11)$$

The magnetic permeability μ_0, the relative permeability μ, and the average magnetic path length L_c of the selected core(PC40) are $\mu_0 = 1.26 \times 10^{-6}$ H/m, $\mu = 2532$, and $L_c = 0.14$

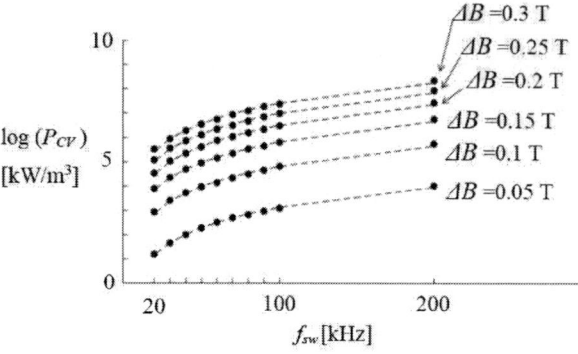

Fig. 7. Core - loss density, P_{cv}, characteristics of PC40

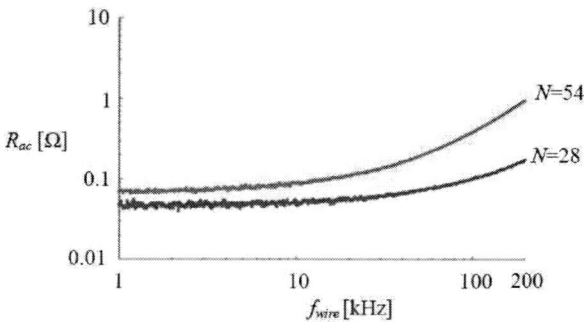

Fig. 8. Frequency characteristic of measured ac-winding resistance(R_{ac}) for inductor L when N is 28 and 54

TABLE IV. EXTRACTED PARAMETERS FOR EQ.(9) IN CASE OF E_{OFF} AND E_{ON}

	N=28	N=54
R_{w1}	-4.84×10^{-18}	-2.91×10^{-19}
R_{w2}	1.29×10^{-11}	1.22×10^{-12}
R_{w3}	2.17×10^{-5}	3.9×10^{-7}
R_{w4}	3.45×10^{-2}	4.37×10^{-2}

m from the datasheet. The condition equation from (10) and (11) is expressed in (12).

$$\frac{\mu \cdot \mu_0 \cdot Ae}{\mu_0 \cdot Lc + \mu \cdot Lg} \cdot N^2 > V_{dc} \cdot \frac{1}{2f_{sw}} \cdot \frac{1}{i_L} \tag{12}$$

ΔB is constrained by (13) from the saturated magnetic flux density B_{max}.

$$\Delta B = \frac{V_{dc}}{Ae \cdot N \cdot 2f_{sw}} < B_{max} \tag{13}$$

In "Winding design" described in Fig.4, the turn number N satisfying (12) and (13) is 28 at V_{dc} < 300 V, and 54 at V_{dc} > 300 V. Then, from (11), the L_g is calculated to satisfy the inductance L.

In "Inductor losses" in Fig.4, the effective cross-sectional area Ae and the core volume V_{core} are specified from the selected core shape (EC70). Hence, the core loss density characteristic P_{cv} of the selected core material is specified from the datasheet. The core loss density characteristic P_{cv} obtained by sinusoidal excitation of the core in the datasheet is approximated by the steinmetz equation[22] shown in (14).

$$P_{cv} = K_1 \cdot f_{sw}^{K_2} \cdot \Delta B^{K_3} \tag{14}$$

For (15), which is logarithm of both sides of (14), the P_{cv}-f_{sw}-ΔB characteristics read from the datasheet of the PC40 material shown in Fig.7 are substituted. K_1, K_2, and K_3 are calculated as K_1 = 0.0235, K_2 = 1.22, and K_3 = 2.50 by a least squares method.

$$\log(P_{cv1}) = \log(K_1) + K_2 \cdot \log(f_{sw}) + K_3 \cdot \log(\Delta B) \tag{15}$$

The core loss P_{core} is calculated as following (16) from the core loss density P_{cv} and the core volume V_{core}

$$P_{core} = P_{cv} \cdot V_{core} \tag{16}$$

The winding loss P_{wire} is calculated from the measured winding resistance R_{ac} and the inductor current i_L. R_{ac} is frequency-dependent due to skin effect and proximity effect. The DC component R_{dc} measured without the core is 0.048 Ω at N = 28 and 0.069 Ω at N = 54.

Fig.8 shows the measured frequency characteristics of winding resistor R_{ac} with a frequency(f_{wire}) from 1 kHz to 200 kHz by an impedance analyzer. The measured value of R_{ac} is approximated by a third-order polynomial shown in (17) and the approximate coefficients $R_1 \sim R_4$ are shown in Table 4.

$$R_{ac} = R_{w1} \cdot f_{wire}^3 + R_{w2} \cdot f_{wire}^2 + R_{w3} \cdot f_{wire} + R_{w4} \tag{17}$$

The DC component I_{dc} of the winding current i_L is shown in Table 1. Since i_L is a triangular wave from Fig. 3, the fundamental frequency component I_{ac1}, the third harmonic component I_{ac3}, and the fifth harmonic component I_{ac5} are calculated as follows, by Fourier series expansion from the lower limit of the peak current I_{LP-} and the upper limit of the peak current I_{LP+} described in (2) and (3).

$$I_{acn} = \frac{8}{n^2 \cdot \pi^2} \cdot \left(\frac{I_{LP+} - I_{LP-}}{2} \right) \tag{18}$$

Since the DC component P_{dc} and the high-frequency component P_{ac} can be superimposed, the winding loss is calculated by considering the DC components P_{dc} and I_{dc} of the winding resistance and the flow current, the fundamental frequency components R_{ac1} and I_{ac1}, and the third-order harmonic components R_{ac3} and I_{ac3} and the fifth harmonic components R_{ac5} and I_{ac5} as the following equation.

$$P_{wire} = R_{dc} \cdot I_{dc}^2 + R_{ac1} \cdot I_{ac1}^2 + R_{ac3} \cdot I_{ac3}^2 + R_{ac5} \cdot I_{ac5}^2 \tag{19}$$

D. Circuit losses

The above component losses, consisting of the power device losses defined in section B and the inductor losses defined in section C, are integrated to a circuit losses $P_{circuit}$ in "Circuit losses". Adjustable parameters are set to R_g and f_{sw} as an example in this paper. Moreover, unit of circuit losses is W.

The conduction-loss characteristic $V_{ds} - I_{ds}$ described in Fig.5 is approximated by a following first-order linear equation with coefficients of a_{ds} and b_{ds}.

$$V_{ds} = a_{ds} \cdot I_{ds} + b_{ds} \tag{20}$$

TABLE V. EXTRACTED PARAMETERS FOR EQ.(24) IN CASE OF E_{off} AND E_{on}

	E_{off}		E_{on}
K	$\dfrac{1}{2} \cdot \dfrac{V_{dc} \cdot D}{L}$	K	$\dfrac{1}{2} \cdot \dfrac{V_{dc} \cdot D}{L}$
R_{off1}	$K^2 \cdot (A_{off1} \cdot V_{dc} + A_{off4})$	R_{on1}	$K^2 \cdot (A_{on1} \cdot V_{dc} + A_{on4})$
R_{off2}	$K^2 \cdot (A_{off7} \cdot V_{dc} + A_{off10})$	R_{on2}	$K^2 \cdot (A_{on7} \cdot V_{dc} + A_{on10})$
R_{off3}	$(A_{off1} \cdot V_{dc} + A_{off4}) \cdot I_L^2$ $+(A_{off2} \cdot V_{dc} + A_{off5}) \cdot I_L$ $+(A_{off3} \cdot V_{dc} + A_{off6})$	R_{on3}	$(A_{on1} \cdot V_{dc} + A_{on4}) \cdot I_L^2$ $+(A_{on2} \cdot V_{dc} + A_{on5}) \cdot I_L$ $+(A_{on3} \cdot V_{dc} + A_{on6})$
R_{off4}	$(A_{off7} \cdot V_{dc} + A_{off10}) \cdot I_L^2$ $+(A_{off8} \cdot V_{dc} + A_{off11}) \cdot I_L$ $+(A_{off9} \cdot V_{dc} + A_{off12})$	R_{on4}	$(A_{on7} \cdot V_{dc} + A_{on10}) \cdot I_L^2$ $+(A_{on8} \cdot V_{dc} + A_{on11}) \cdot I_L$ $+(A_{on9} \cdot V_{dc} + A_{on12})$
R_{off5}	$-2K \cdot (A_{off1} \cdot V_{dc} + A_{off4}) \cdot I_L$ $-K \cdot (A_{off2} \cdot V_{dc} + A_{off5})$	R_{on5}	$-2K \cdot (A_{on1} \cdot V_{dc} + A_{on4}) \cdot I_L$ $-K \cdot (A_{on2} \cdot V_{dc} + A_{on5})$
R_{off6}	$-2K \cdot (A_{off7} \cdot V_{dc} + A_{off10}) \cdot I_L$ $-K \cdot (A_{off8} \cdot V_{dc} + A_{off11})$	R_{on6}	$-2K \cdot (A_{on7} \cdot V_{dc} + A_{on10}) \cdot I_L$ $-K \cdot (A_{on8} \cdot V_{dc} + A_{on11})$

The reverse characteristic is also approximated by a first-order linear equation with coefficients of a_{sd} and b_{sd} as following equation.

$$V_{sd} = a_{sd} \cdot I_{sd} + b_{sd} \tag{21}$$

The forward conduction losses of Q_1 and Q_4, which are defined at period from T_d to $T_{sw}/2 + T_d$ in Table 2, are expressed to a following equation by inserting I_{LP-} to I_{c1} and I_{LP+} to I_{c2} in (5).

$$P_{dsQ14} = \left\{ a_{ds} \cdot I_L^2 + b_{ds} \cdot I_L + \frac{a_{ds}}{3} \cdot \left(\frac{V_{dc}}{L}\right)^2 \cdot \left(\frac{D}{2f_{sw}}\right)^2 \right\} \cdot D \tag{22}$$

The forward conduction losses of Q_2 and Q_3 at period from $T_{sw}/2 + 2T_d$ to T_{sw} are also obtained as a following equation by inseting I_{LTd-} to I_{c1} and I_{LTd+} to I_{c2} in (5).

$$P_{dsQ23} = \left\{ a_{sd} \cdot I_L^2 + b_{sd} \cdot I_L + \frac{a_{sd}}{3} \cdot \left(\frac{V_{dc}}{L}\right)^2 \cdot \left(\frac{D}{2f_{sw}} - \frac{T_d}{2}\right)^2 \right\} \cdot (D - 2T_d \cdot f_{sw}) \tag{23}$$

The reverse conduction losses of Q_2 and Q_3 at period from $T_{sw}/2 + T_d$ to $T_{sw}/2 + 2T_d$ are obtained by inserting I_{LTd-} to I_{c1} and I_{LP+} to I_{c2} in (5). Hence, the reverse conduction losses of Q_2 and Q_3 at period from 0 to T_d are obtained by inserting I_{LP-} to I_{c1} and I_{LTd-} to I_{c2}. The total reverse conduction losses are expressed as a following equation by summing up these two reverse losses.

$$P_{sdQ23} = \left\{ 2a_{sd} \cdot I_L^2 + 2b_{sd} \cdot I_L + \frac{2a_{sd}}{3} \cdot \left(\frac{V_{dc}}{2L}\right)^2 \right. \\ \left. \cdot \left\{ 3 \cdot \left(\frac{D}{f_{sw}}\right)^2 - \frac{3DT_d}{2f_{sw}} + T_d^2 \right\} \right\} \cdot T_d \cdot f_{sw} \tag{24}$$

For the turn on losses of Q_1 and Q_4 at T_d, E_{on} is expressed to (25) by inserting I_{LP-} defined in (2) to I_{sw} in (9). Also for the turn off losses of Q_1 and Q_4 at $T_{sw}/2$, E_{off} is expressed to (26) by inserting I_{LP+} defined in (2) to I_{sw} in (9). Coefficients of R_{off1} through R_{off6} and R_{on1} through R_{on6} are indicated in Table 5.

$$E_{on} = R_{on1} \cdot \frac{R_g}{f_{sw}} + R_{on2} \cdot \frac{1}{f_{sw}} + R_{on3} \cdot R_g \cdot f_{sw} \\ + R_{on4} \cdot f_{sw} + R_{on5} \cdot R_g + R_{on6} \tag{25}$$

$$E_{off} = R_{off1} \cdot \frac{R_g}{f_{sw}} + R_{off2} \cdot \frac{1}{f_{sw}} + R_{off3} \cdot R_g \cdot f_{sw} \\ + R_{off4} \cdot f_{sw} + R_{off5} \cdot R_g + R_{off6} \tag{26}$$

The core losses P_{core} is obtained by inserting (13) to (14) as a following equation.

$$P_{core} = K_1 \cdot \left(\frac{V_{dc}}{Ae \cdot N \cdot 2f_{sw}}\right)^{K_2} \cdot f_{sw}^{K_3} \cdot V_{core} \tag{27}$$

The winding loss of P_{wire} is calculated by inserting (2) to (18), and then by inserting (17) and (18) to (19). As a result, P_{wire} is expressed as a following equation.

$$P_{wire} = \left(R_{f1} \cdot f_{sw}^3 + R_{f2} \cdot f_{sw}^2 + R_{f3} \cdot f_{sw} + R_{f4}\right) \\ \cdot \left(\frac{1}{\sqrt{2}} \cdot \frac{V_{dc} \cdot D}{2 \cdot L} \cdot \frac{1}{f_{sw}} \cdot \frac{8}{\pi^2}\right)^2 \\ + \left(R_{f1} \cdot (3f_{sw})^3 + R_{f2} \cdot (3f_{sw})^2 + R_{f3} \cdot (3f_{sw}) + R_{f4}\right) \\ \cdot \left(\frac{1}{\sqrt{2}} \cdot \frac{V_{dc} \cdot D}{2 \cdot L} \cdot \frac{1}{f_{sw}} \cdot \frac{8}{9\pi^2}\right)^2 \\ + \left(R_{f1} \cdot (5f_{sw})^3 + R_{f2} \cdot (5f_{sw})^2 + R_{f3} \cdot (5f_{sw}) + R_{f4}\right) \\ \cdot \left(\frac{1}{\sqrt{2}} \cdot \frac{V_{dc} \cdot D}{2 \cdot L} \cdot \frac{1}{f_{sw}} \cdot \frac{8}{25\pi^2}\right)^2 \tag{28}$$

From the above calculation process, total circuit losses $P_{circuit}$ is calculated by integrating equation(22), (23), (24), (25), (26), (27) and (28) with parameters of switching frequency f_{sw} and gate resistance R_g as follows.

$$P_{circuit} = P_1 \cdot f_{sw}^{K_2 - K_3} + P_2 \cdot f_{sw}^2 + P_3 \cdot f_{sw} + P_4 \cdot \frac{R_g}{f_{sw}} \\ + P_5 \cdot \frac{1}{f_{sw}} + P_6 \cdot \frac{1}{f_{sw}^2} + P_7 \cdot R_g \cdot f_{sw} + P_8 \cdot R_g + P_9 \tag{29}$$

$$P_1 = K_1 \cdot \left(\frac{V_{dc} \cdot D}{Ae \cdot N}\right)^{K_3} \cdot V_{core} \tag{30}$$

$$P_2 = 0 \tag{31}$$

$$P_3 = 2 \cdot \left\{ \frac{a_{sd}}{6} \cdot \left(\frac{V_{dc}}{L}\right)^2 \cdot T_d^3 + 2 \cdot (a_{sd} \cdot I_L^2 + b_{sd} \cdot I_L) \cdot T_d \right\} \\ - 2 \cdot \left\{ \frac{a_{ds}}{3} \cdot \left(\frac{V_{dc}}{L}\right)^2 \cdot \left(\frac{T_d^3}{2}\right) + 2 \cdot (a_{ds} \cdot I_L^2 + b_{ds} \cdot I_L) \cdot T_d \right\} \\ + 2 \cdot (R_{on4} + R_{off4}) + \frac{23}{15} R_{w1} \cdot \left(\frac{1}{\sqrt{2}} \cdot \frac{V_{dc} \cdot D}{2 \cdot L} \cdot \frac{8}{\pi^2}\right)^2 \tag{32}$$

$$P_4 = 2 \cdot (R_{on1} + R_{off1}) \tag{33}$$

$$P_5 = -\frac{2a_{ds}}{3} \cdot \left(\frac{V_{dc}}{L}\right)^2 \cdot T_d \cdot D^2 + a_{sd} \cdot \left(\frac{V_{dc}}{L}\right)^2 \cdot D^2 \cdot T_d \\ + (R_{on2} + R_{off2}) + \frac{56505}{50625} R_{w3} \cdot \left(\frac{1}{\sqrt{2}} \cdot \frac{V_{dc} \cdot D}{2 \cdot L} \cdot \frac{8}{\pi^2}\right)^2 \tag{34}$$

979-8-3315-1612-3/25 $31.00 © 2025 IEEE

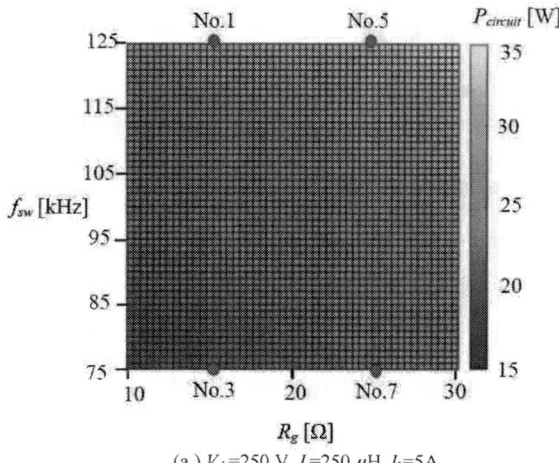

(a) V_{dc}=250 V, L=250 μH, I_L=5A

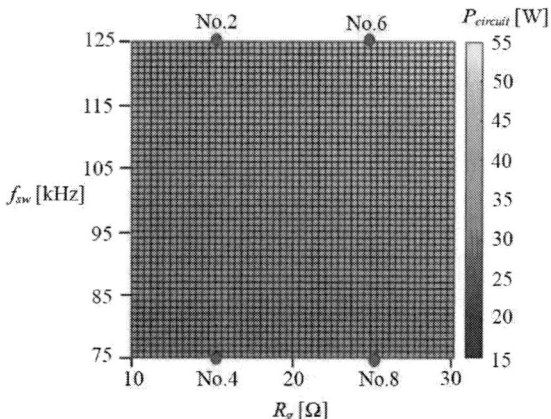

(b) V_{dc}=350 V, L=590 μH, I_L=5A

Fig. 9. Calculated characteristic of circuit losses ($P_{circuit}$)

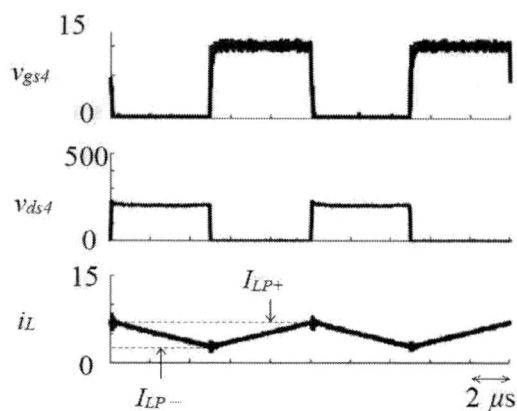

Fig.10 Operating waveforms at No.7 in Fig. 9(a)

TABLE VI. MEASURED LOSSES WITH DIFFERENT CIRCUIT / OPERATING CONDITIONS (V_{DC}, I_L, f_{sw}, L, R_g, Lg, AND N) AND ERROR OF CIRCUIT LOSS ($P_{circuit}$)

No	Vdc	IL	fsw	Rg	L	N	Lg	Error
	V	A	kHz	Ω	μH		mm	%
1	250	5	125	15	250	28	1.1	2.7
2	350	5	125	15	250	28	1.1	5.0
3	250	5	75	15	590	54	2.9	4.8
4	350	5	75	15	590	54	2.9	10.6
5	250	5	125	25	590	54	2.9	5.2
6	350	5	125	25	590	54	2.9	1.9
7	250	5	75	25	590	54	2.9	2.1
8	350	5	75	25	590	54	2.9	7.6
9	250	7.5	125	15	250	28	1.1	1.6
10	350	7.5	125	15	250	28	1.1	0.8
11	250	7.5	75	15	590	54	2.9	3.6
12	350	7.5	75	15	590	54	2.9	2.3
13	250	7.5	125	25	590	54	2.9	9.1
14	350	7.5	125	25	590	54	2.9	3.1
15	250	7.5	75	25	590	54	2.9	3.7
16	350	7.5	75	25	590	54	2.9	0.6

$$P_6 = \frac{2a_{ds}}{3} \cdot \left(\frac{V_{dc}}{L}\right)^2 \cdot \frac{D^3}{2}$$
$$+ \frac{51331}{50625} R_{w4} \cdot \left(\frac{1}{\sqrt{2}} \cdot \frac{V_{dc} \cdot D}{2 \cdot L} \cdot \frac{8}{\pi^2}\right)^2 \quad (35)$$

$$P_7 = 2 \cdot (R_{on3} + R_{off3}) \quad (36)$$

$$P_8 = 2 \cdot (R_{on5} + R_{off5}) \quad (37)$$

$$P_9 = 2 \cdot \left\{ \frac{5 \cdot a_{ds}}{12} \cdot \left(\frac{V_{dc}}{L}\right)^2 \cdot D \cdot T_d^2 + 2 \cdot (a_{ds} \cdot I_L^2 + b_{ds} \cdot I_L) \cdot D \right\}$$
$$- \frac{a_{sd}}{2} \cdot \left(\frac{V_{dc}}{L}\right)^2 \cdot D \cdot T_d^2 + \frac{259}{225} R_{w2} \cdot \left(\frac{1}{\sqrt{2}} \cdot \frac{V_{dc} \cdot D}{2 \cdot L} \cdot \frac{8}{\pi^2}\right)^2$$
$$+ 2 \cdot (R_{on6} + R_{off6}) \quad (38)$$

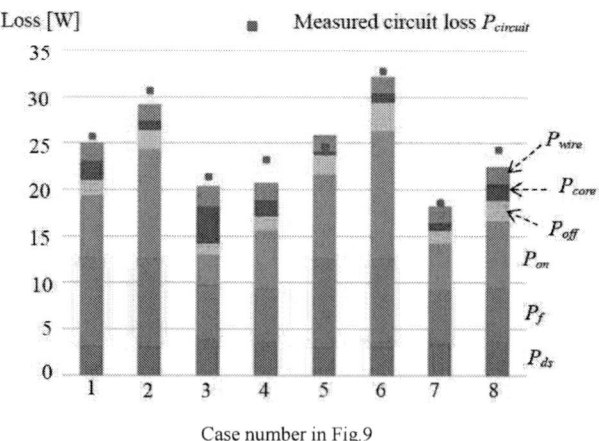

Fig.11 Losses breakdown of P_{ds}, P_{sd}, P_{on}, P_{off}, P_{core} and P_{wire}, and measured circuit loss of $P_{circuit}$

IV. EXPERIMENTAL VERIFICATION

An experimental system is constructed with the circuit configuration shown in Fig.1 and Table 1, and the error of the calculated the circuit loss $P_{circuit}$, expressed in (28) through (37) with respect to the switching frequency f_{sw} and the gate resistance R_g at power supply voltages V_{dc} and the inductance value L was verified. Fig.9 shows $P_{circuit}$ characteristics

calculated for f_{sw} and R_g when L is 250 μH (N = 24, L_g = 1.1 mm) and V_{dc} is 250 V at (a) and L is 590 μH (N = 54, L_g = 1.9 mm) and V_{dc} is 350 V at (b). Calculation software is MATLAB(Online) and the calculation time is less than 1s at total calculation numbers of 2601 points described in Fig.9. At

979-8-3315-1612-3/25 $31.00 © 2025 IEEE

No.7 point in Fig.9(a), the prototype described in Fig.2 realized a stable operation as shown in Fig.10. Table.6 verify errors between the calculated $P_{circuit}$ and measured one of at total 16 conditions. Hence, Fig.11 shows losses breakdown of P_{ds}, P_{sd}, P_{on}, P_{off}, P_{core} and P_{wire}, and the measured circuit loss of $P_{circuit}$ at No.1 through No.8 in Table 6. For measurement of $P_{circuit}$, P_{dc} was measured with the power meter (WT1600 manufactured by YOKOGAWA). Since the measured voltage range is 600 V, the current range is 1 A, and the power accuracy is up to 0.2%, ±1.2 W is the maximum measurement error. Under the conditions shown in Fig.9(a), V_{dc} = 250 V, L=250 μH, I_L=5A (No.1, No.3, No.5, No.7), the error between the calculated and measured values of the circuit loss $P_{circuit}$ ranges from 2.1 % to 5.7 %.

Hence, the error is in the range of 1.9 % to 10.6 % at the conditions shown in Fig.9(b), V_{dc} = 350 V, L = 590 μH, I_L=5 A (No.2, No.4, No.6, No.8).

Furthermore, the results of verifying the error, when only the load current I_L at the conditions No. 1 through No. 8 of Table 6 are changed from 5 A to 7.5 A, are shown at No. 9 through 16 in Table6. Even when I_L was increased to 7.5 A, the maximum error is 9.1 %.

V. CONCLUSION

This paper proposes a losses-analysis model of a DC-output converter based on the approximated equations from measurements and a datasheet. Modeling component loss characteristic by measurements at extended the conditions of the data sheet enables to expand number of analysis parameters and its variable range with less calculation effort.

In this paper, circuit losses calculation flow based on the loss of the power devices and inductor is built, and model the loss of the power devices, of inductor and of circuit as simplified approximated equations. Then, a 400V experimental system is built to verify an error of the circuit losses between measured losses and calculated losses. The result reveals that the calculated losses has good accuracy with an error of 0.8 % to 10.6 % with respect to the measured losses.

REFERENCES

[1] A. Bindra : "Driven by Density, Efficiency at Low Cost, Power Integration Reaches New Heights: PSiPs and PwrSoCs are growing faster than the traditional power supplies", IEEE Power Electronics Magazine, Vol. 9, No. 1, pp. 14-19, Mar. 2022

[2] G.Liu, D.Li,, Y.Jang and J. Zhang : "Over 300kHz GaN Device Based Resonant Bi-directional DCDC Converter With Integrated Magnetics", Conference paper, APEC2016, 20-24 Mar. 2016, Long Beach USA

[3] G-J.Su, C.White and Z.Liang : "Design and evaluation of a 6.6 kW GaN converter for onboard charger applications", Conference paper, 2017 IEEE 18th Workshop on Control and Modeling for Power Electronics (COMPEL), 9-12 July 2017, Stanford, USA

[4] R.Kondo, Y.Higaki, and M.Yamada : "Experimental Verification of Reducing Power Loss under Light Load Condition of a Bi-Directional Isolated DC/DC Converter for a Battery Charger-Discharger of Electric Vehicle", IEEJ Transaction on Industry Applications, Vol.137, No.8, pp.673-680, 2017 (in Japanese)

[5] Y.Cao, M.Ngo, N.Yan, Y.Bai, D,Dong, R.Burgos and I.Agirman : "Design and Implementation of High-density Isolated Bi-directional Soft-switching Resonant DC-DC Converter with Partial Power Processing", Conference paper, 2017 IEEE Applied Power Electronics Conference and Exposition (APEC , 14-17 June. 2021, Phoenix, AZ, USA

[6] U. Badst uebner, J. Biela, D. Christen and J.W.Kolar, "Optimization of a 5-kW Telecom Phase-Shift DC-DC Converter With Magnetically Integrated Current Doubler" IEEE Transaction on Industrial Electronics. Vol.58, No. 10, pp.4736-

4745, Oct. 2011

[7] L.Xue, M.Mu, D.Boroyevich and P.Mattaveli : "The optimal design of GaN-based Dual Active Bridge for bi-directional Plug-IN Hybrid Electric Vehicle (PHEV) charger", Conference paper, 2015 IEEE Applied Power Electronics Conference and Exposition (APEC), 15-19 Mar. 2015, Charlotte USA

[8] F.Krismer and J.W.Kolar, "Efficiency-Optimized High-Current Dual Active Bridge Converter for Automotive Applications" IEEE Transaction on Industrial Electronics. Vol.59, No. 7, pp.2745-2760, Jul. 2012

[9] M. Guacci, D.Bortis and J.W.Kolar, "High-Efficiency Weight-Optimized Fault-Tolerant Modular Multi-Cell Three-Phase GaN Inverter for Next Generation Aerospace Applications", Conference paper , 2018 IEEE Energy Conversion Congress and Exposition (ECCE), 23-27 Sept. 2018

[10] D.Zhang, M.Guacci, M.Haider, D.Bortis, J.W.Kolar and J.Everts, "Three-Phase Bidirectional Buck-Boost Current DC-Link EV Battery Charger Featuring a Wide Output Voltage Range of 200 to 1000V", 2020 IEEE Energy Conversion Congress and Exposition (ECCE), 11-15 Oct. 2020, Detroit USA

[11] T.M.Evans, S.Mukherjee, Y.Peng and H.A.Mantooth, "Electronic Design Automation (EDA) Tools and Considerations for Electro-Thermo-Mechanical Co-Design of High Voltage Power Modules", Conference paper, 2020 IEEE Energy Conversion Congress and Exposition (ECCE), 15-18 Oct.2020, Online

[12] Quang Le, Imam Al Razi, Yarui Peng, H. Alan Mantooth, "Fast and Accurate Inductance Extraction for Power Module Layout Optimization Using Loop-Based Method", 2021 IEEE Energy Conversion Congress and Exposition (ECCE), 10-14 Oct.2021, Online

[13] C.Fei, F.C.Lee and Q.Li : "High-efficiency high-power-density 380V/12V DC/DC converter with a novel matrix transformer", Conference paper, 2017 IEEE Applied Power Electronics Conference and Exposition (APEC), 26-30 Mar. 2017, Tampa USA

[14] G.Ortiz, J. Biela and J.W.Kolar : "Optimized design of medium frequency transformers with high isolation requirements", Conference Paper, IECON 2010 - 36th Annual Conference on IEEE Industrial Electronics Society, 7-10 Nov. 2010, Glendale, AZ, USA

[15] A.Melkonyan : "Optimized analsysis of inductance and winding for high efficiency", Conference Paper, ANSYS Conference & 30th CADFEM User's Meeting 2012, 24-26 Oct. 2012

[16] R.Kondo, P.Shuelting, A.H.Wienhausen and R. W. De Doncker "An Automated Component-Based Hardware Design of a Three-Phase Dual-Active Bridge Converter for a Bidirectional On-Board Charger", Conference Paper, 2020 IEEE Energy Conversion Congress and Exposition (ECCE), 15-20 Oct 2020, Online

[17] Y.Mukunoki, Y.Nakamura, T.Horiguchi,S.Kinouchi, Y.Nakayama, T, Terashima, M.Kuzumoto and H.Akagi, "Characterization and Modeling of a 1.2-kV 30-A Silicon-Carbide MOSFET", IEEE Transaction on Power Electronics, Vol.63, No. 11, pp.4339-4345, Nov.2016

[18] S.Muff, A.Shih, L.Eichinger, B.Holzinger and H.Tanigawa, " From device modeling to characterization: a complete end to end design flow for SiC devices half bridge design", PCIM Europe digital days 2020; International Exhibition and Conference for Power Electronics, Intelligent Motion, Renewable Energy and Energy Management, 7-8 July 2020, Germany

[19] Y.Park, S.Chakraborty and A.Khaligh : "DAB Converter for EV Onboard Chargers Using Bare-Die SiC MOSFETs and Leakage-Integrated Planar Transformer", IEEE Transaction on Transportation Electrification, Vol. 8, No. 1, March 2022

[20] M.Mu and F.C.Lee : "Design and Optimization of a 380–12 V High-Frequency, High-Current LLC Converter With GaN Devices and Planar Matrix Transformers", IEEE JOURNAL OF EMERGING AND SELECTED TOPICS IN POWER ELECTRONICS ,Vol.4, No.3, pp.854-862, 2016

[21] D. Zhang, M.Guacci, M.Haider, D.Bortis, J.W.Kolar and J. Everts, "Three-Phase Bidirectional Buck-Boost Current DC-Link EV Battery Charger Featuring a Wide Output Voltage Range of 200 to 1000V", 2020 IEEE Energy Conversion Congress and Exposition (ECCE), 11-15 Oct. 2020, Detroit, USA

[22] Venkatachalam. K, Sullivan. C.R, Abdallah. T., Tacca. H, "Accurate prediction of ferrite core loss with non-sinusoidal waveforms using only Steinmetz parameters", Conference Paper, 2002 IEEE Workshop on Computers in Power Electronics, 3-4 June 2002, USA

[23] R.Kondo and T.Funaki, "Experimental Verification of Losses-Analysis Model of DC-Output Converter developed using Approximated Equations from Measurement Data and Datasheet Data", IEEJ Transaction on Industry Applications, Vol.143, No.4, pp.1-10, 2023

Scattering Parameter Measurement System Using Probes for Surface Mount Devices Operating in the Frequency Range from 50 kHz to 1 GHz

Ryoko Kishikawa
Research Institute for Physical Measurement
National Institute of Advanced Industrial Science and Technology
Tsukuba, Japan
ryoko-kishikawa@aist.go.jp

Masahiro Horibe
Research Strategic Planning Office
National Institute of Advanced Industrial Science and Technology
Tsukuba, Japan
masahiro-horibe@aist.go.jp

Tomokazu Shoji
R&D
T Plus Co. Ltd.
Chiba, Japan
shoji@technoprobe.co.jp

Shigenori Yabuta
R&D
T Plus Co. Ltd.
Chiba, Japan
yabuta@technoprobe.co.jp

Toshi Ohi
R&D
T Plus Co. Ltd.
Chiba, Japan
toshi-ohi@tplus-co.com

Ryo Takeda
Automotive Energy Solutions
Keysight Technologies International Japan G.K.
Hachioji, Japan
ryo_takeda@keysight.com

Takamasa Arai
Automotive Energy Solutions
Keysight Technologies Japan K.K.
Hachioji, Japan
takamasa_arai@keysight.com

Abstract—This paper presents a new probe system for measuring the scattering parameters of surface mount devices (SMDs) in the frequency range from 50 kHz to 1 GHz. This system addresses the high demand for power devices operating at high-switching frequencies; thereby facilitating the downsizing of power electronic circuits. The measurement of the scattering parameters of power devices, including the parasitic elements of the packages, is a prerequisite step in realizing the high-switching frequency operations. Therefore, probes with a pitch of 3.7 mm between the signal and ground tips and a probe station for controlling the probes were manufactured as original components. Using a commercially available vector network analyzer, a source/measure unit, and bias-tees in addition to the developed components, the scattering parameters with a dc bias applied are conveniently measured. The advantage of the proposed system is that the probes can directly contact an SMD under test, and no test fixture is required to establish a connection between a device under test and the test port. Moreover, a single type of probe can contact multiple types of SMDs under test. This is a pioneering measurement system for SMDs in power electronics operated at radio frequencies.

Keywords—Surface mount device, scattering parameter, radio frequency, vector network analyzer

I. INTRODUCTION

In recent years, the operating frequencies of power devices have been gradually increasing to reduce the size of power electronic circuits and consequently improve the energy utilization efficiency. To address these requirements, various types of compound semiconductors, such as silicon carbide (SiC) and gallium nitride (GaN), have attracted considerable attention as potential candidates for next-generation power electronic semiconductors [1]-[2]. Considering that the electron

mobility of such compound semiconductors exceeds that of silicon (Si), SiC and GaN are presumably suitable for devices operated at high frequencies. To maximize the advantageous characteristics of such compound semiconductors and consequently realize high-performance power modules, peripheral circuits that match the characteristics of power devices at radio frequencies must be designed. To achieve this design, the characteristics of SiC- and GaN-based power devices must be precisely measured at analog frequencies above 10 MHz.

The scattering parameters (S-parameters) describe the device characteristics in the high frequency region, and a vector network analyzer (VNA) is used to measure the S-parameters of power devices [3]-[18]. The conventional current-voltage (IV) and capacitance-voltage (CV) measurements of power transistors only yield the resistance and capacitance components. However, the S-parameters measured using a VNA convey the resistance, capacitance, and inductance elements. Device modeling, including device packages, exemplifies the importance of inductance components. Generally, wires in a package generate inductance components. These components distort the current waveform, generate noise, and reduce the power conversion efficiency of power modules. Since the inductance component exerts a particularly large effect in the high frequency range, VNA measurements are essential in the latest research and development of power electronics.

A major challenge encountered when measuring the S-parameters of packaged power devices with a VNA is that the electrode shape of the device under test, typically a flat metal with various shapes, is different from that of the test port of a

VNA, typically a coaxial connector. These incompatible ports cannot be connected directly; therefore, operators must design and create a test fixture that converts the shape of a metal plate into a coaxial connector. Electrode shapes vary; thus it is challenging to create various types of test fixtures customized to various electrode shapes [19]. In addition, the effects of a test fixture must be calibrated because the measured data indicate the combination of the characteristics of the device under test and test fixture. This must be resolved to popularize the application of S-parameter measurements for packaged power devices operating at high frequencies.

We have proposed an S-parameter measurement system with probes for surface mount devices (SMDs), a type of packaged power device [20]. The probe pitch between the signal and ground tips is 3.7 mm, and the probes can directly contact an SMD under test. This simplifies the calibration procedure because this probe system does not require a customized test fixture. Moreover, the proposed system enables the measurements of SMDs under test with various electrode shapes using a single type of probe.

This study presents a novel probe system for measuring the S-parameters of SMDs operating in the frequency range from 50 kHz to 1 GHz. The rest of this paper is organized as follows. Section II presents the system configuration, including the developed probes and probe station. Section III discusses calibration of the system. Section IV presents measurement examples of a commercially available power transistor using the developed probe system. In addition, comparisons between the developed system and a conventional system are shown. Finally, Section V presents the conclusions of the study.

II. PROBE MEASUREMENT SYSTEM

A. Scattering Parameters

The S-parameters are quantities used to describe the characteristics of a linear network circuit [21][22]. The S-parameters are useful because they are convenient for analysis and simple for measurements at high frequencies compared with the impedance parameters (Z-parameters) and admittance parameters (Y-parameters). The S-parameters can be mathematically converted into the Z- or Y-parameters [21].

The S-parameters are defined where the reference impedance is 50 Ω. For a 2-port device, the S-parameters, S_{ij} (i, j = 1, 2), are expressed by the following two equations (Fig. 1(a)):

$$b_1 = S_{11}a_1 + S_{12}a_2 \qquad (1)$$

$$b_2 = S_{21}a_1 + S_{22}a_2 \qquad (2)$$

Here, a_i is the incident pseudo wave at port i, which is defined as the ratio of the incident voltage wave at port i and the square root of the reference impedance. Similarly, b_i is the reflected pseudo wave at port i, defined as the ratio of the reflected voltage wave at port i and the square root of the reference impedance.

The S-parameters are typically measured with a VNA (Fig. 1(b)). For forward configuration, a signal source provides power

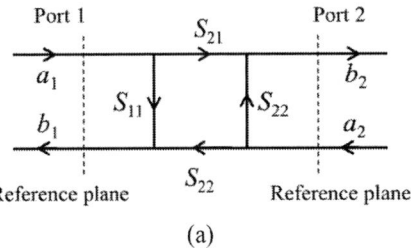

(a)

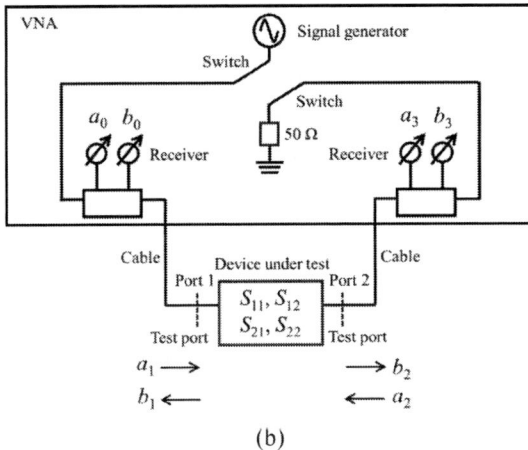

(b)

Fig. 1. S-parameters and pseudo waves for a 2-port device (a). Block diagram of a VNA for forward configuration (b).

to a device under test through port 1. The incident and reflected signals, a_0 and b_0, are measured using two receivers at port 1, and the transmitted signal, b_3, is measured using a receiver at port 2. The measured data, a_0, b_0, and b_3, are mathematically converted into a_1, b_1, and b_2 through a calibration process (Section III), and finally provide S_{11} and S_{21} in the frequency domain. For reverse configuration, the switches connect in the other direction, and S_{22} and S_{12} in the frequency domain are obtained.

B. Probe System for Scattering Parameter Measurements

We have developed a probe system for measuring the S-parameters of SMDs operating in the frequency range from 50 kHz to 1 GHz, as shown in Fig. 2 [20]. This system comprises a VNA for measuring the S-parameters, a source/measure unit (SMU) and two bias-tees (bias-Ts) for applying a dc bias, two probes for contacting an SMD under test, a probe station for controlling the probes, and coaxial cables for connecting the components. The probes and probe station were originally designed and developed as part of this study by National Institute of Advanced Industrial Science and Technology (AIST) and T Plus. The VNA, SMU, and bias-Ts are commercially available products manufactured by Keysight Technologies.

Ground-signal (GS) probes were originally manufactured to electrically connect coaxial cables and an SMD under test with various forms of electrodes. The probe possesses a ground tip and a signal tip; the pitch between both tips is 3.7 mm. The

979-8-3315-1612-3/25 $31.00 © 2025 IEEE

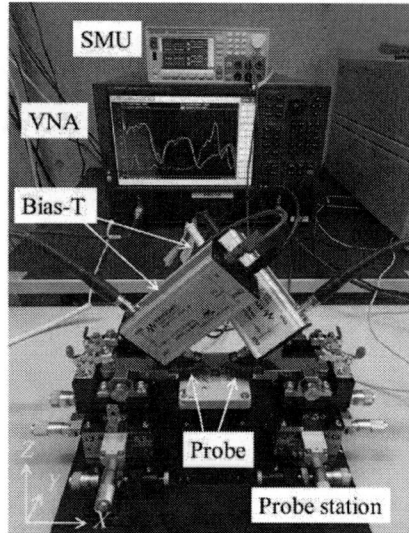

(a)

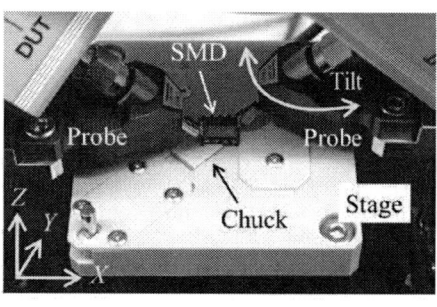

(b)

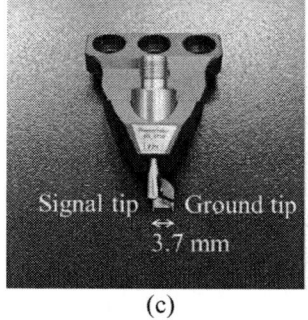

(c)

Fig. 2. Photographs of the developed probe system (a), the stage on the probe station (b), and the GS probe (c).

signal and ground tips contact the signal and ground lines of SMDs under test, respectively. These probes were manufactured to enable high power operations while maintaining measurement accuracy at high frequencies. The probe material was selected for its resistance to high voltage and current. A cantilever structure provides a long overdrive. The springiness of the cantilever absorbs the microscopic irregular flatness of the

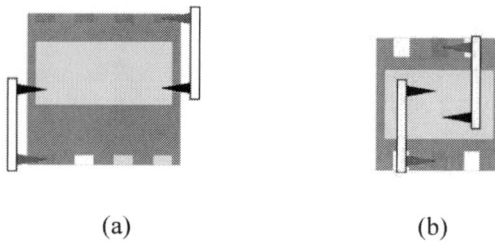

(a)   (b)

Fig. 3. Examples of contact configuration between the developed GS probes and commercially available power transistors. The source, drain, and gate electrodes of the power transistors are illustrated in yellow, green, and blue, respectively. The signal and ground tip of the probe are illustrated in red and black, respectively. The signal tip on ports 1 and 2 contact the gate and drain electrodes, respectively. The ground tips on ports 1 and 2 contact the source electrode. The transistors reported in [23]-[28] belong to the type depicted in (a), whereas those reported in [29]-[36] belong to the type presented in (b).

electrode surfaces of an SMD under test, and the ground and signal tips contact the electrodes firmly and stably. Furthermore, because the probe tips are clearly visible to an operator, the contact status between the probe tips and the electrodes of an SMD under test can be observed. These structures suppress sparks caused by poor contact between the probe tip and an electrode.

The probe station controls the movement of the probes in the X, Y, Z, and tilt directions. To improve the contact reproducibility between the probes and an SMD under test, the magnitude of movement is quantified using the scales on the operation rods. Therefore, probe control is achieved with a pitch of approximately 10 μm along the X, Y, and Z axes and one degree in the tilt direction. The probe station is equipped with a stage, which holds an SMD under test with a movable chuck that fits various lengths of rectangle-form devices. These structures enable stable measurements at high powers and frequencies.

The developed probe system offers two distinct advantages. First, the probe can directly contact the electrode of SMDs under test. Conventionally, a test fixture is required to connect the electrode of an SMD under test to the coaxial test port of a VNA [19]. Considering that the structure of the electrode is unique depending on SMDs, operators must design and create various types of test fixtures customized to the various electrode structures. However, with the proposed probe, which enables direct contact with an SMD under test, whose length between the electrodes matches the probe pitch, no such test fixture is required. This advantage enables simpler calibration, i.e., operators need not design and create calibration devices for customized test fixtures.

Then, a single type of probe can contact multiple types of SMDs under test. Fig. 3 shows examples of contact configurations between the probes and commercially available

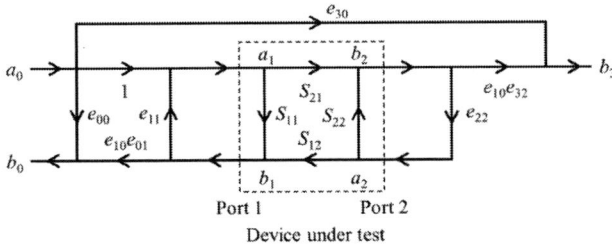

Fig. 4. Error model for OSLT calibration. The signal flow graph is for forward configuration. The six error terms, e_{00}, e_{11}, $e_{10}e_{01}$, $e_{10}e_{32}$, e_{22}, and e_{30}, indicate the directivity, port-1 matching, reflection tracking, transmission tracking, port-2 matching, and leakage, respectively. a_0, b_0, and b_3 are the measured incident, measured reflected, and measured transmitted pseudo waves with the receivers of a VNA, respectively. The reverse model is a mirror image of the forward model, including additional six error terms. Once the twelve error terms are determined, the S-parameters of a device under test, S_{11}, S_{21}, S_{12}, and S_{22}, can be calculated using the measured pseudo waves and error terms. [37][38].

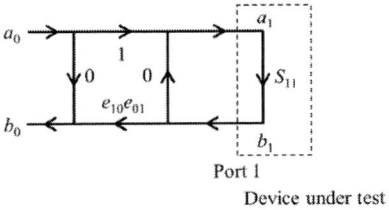

Fig. 5. Error model for port extension calibration. The signal flow graph is for port 1. The error term, $e_{10}e_{01}$, indicates a phase shift between the receivers of a VNA and the reference plane. a_0 and b_0 are the measured incident and measured reflected pseudo waves with the VNA receivers, respectively. Once the error term is determined, the S-parameters of a device under test, S_{11}, can be calculated using the measured pseudo waves and error term.

power transistors. With these configurations, the S-parameters can be measured with a dc bias supplying the drain and gate.

III. CALIBRATION

The calibration of a VNA is essential for correcting the loss and phase shift of electromagnetic waves between the reference plane of the test port and VNA receivers and calculating the S-parameters of a device under test [37] [38]. The phase shift is a simple example that warrants the importance of calibration. Within a low frequency range, in which the wavelength of the electromagnetic waves exceeds the size of the measurement system, the phase shift between the test port and VNA receivers can be neglected. In contrast, within a high frequency range, in

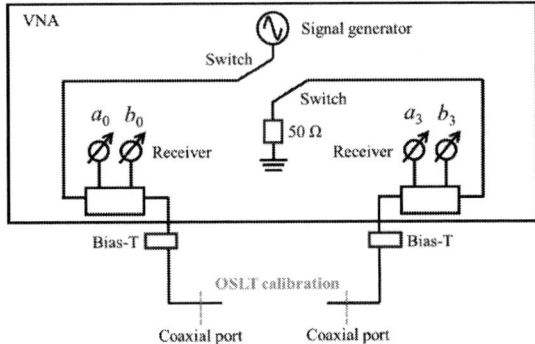

Fig. 6. Schematic drawings of OSLT calibration procedure of the developed system. The calibration was performed at the coaxial test ports to correct the errors from the receivers to the coaxial ports.

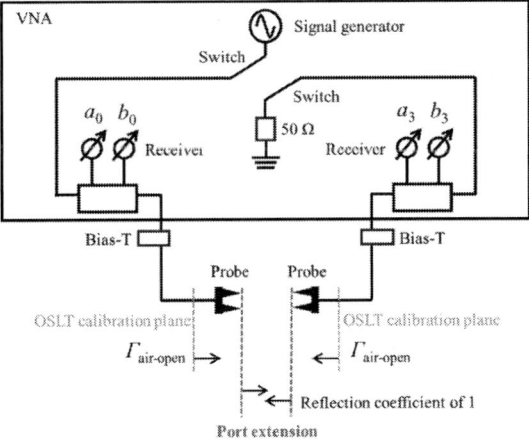

Fig. 7. Schematic drawings of port extension calibration procedure of the developed system. The calibration was performed to move the calibration planes from the coaxial ports to the edges of the probes.

which the wavelength is equal to or shorter than the size of the measurement system, the phase shift is not negligible. Certain models that express the loss and phase shift using a few error terms and methods of calculating the terms using known calibration devices have been proposed [37] [38]. For example, open-short-load-thru (OSLT) calibration is a familiar method. This method includes twelve error terms, which represent the propagation of electromagnetic waves between VNA receivers and reference planes (Fig. 4). The error terms are determined using an open device, a short device, a load device, and a thru connection whose S-parameters are known. Once all the error terms are determined, calculating the S-parameters of a device connected to the reference plane is possible. Port extension is another method for obtaining the characteristics of a device under test. This calibration method removes the phase shift of

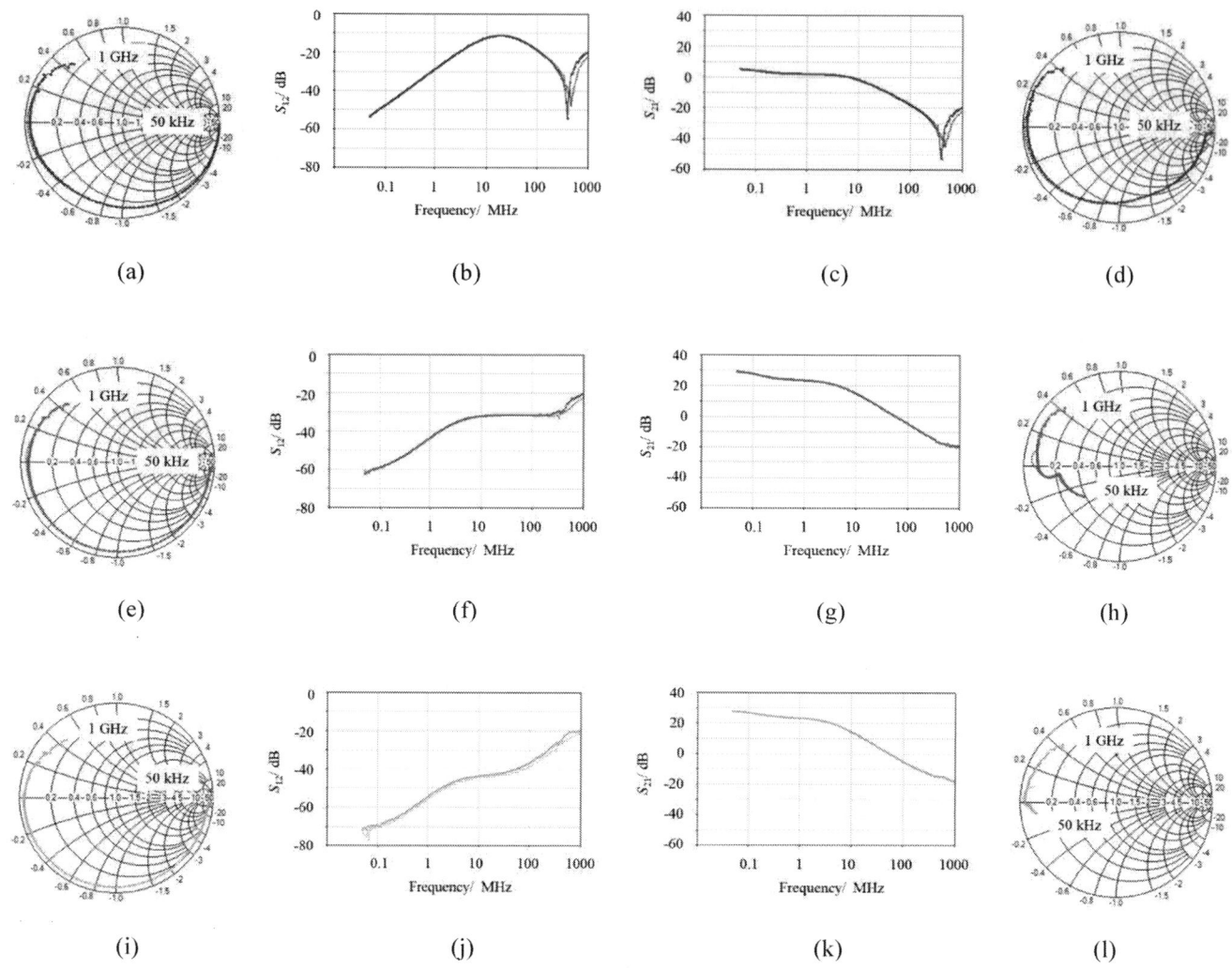

Fig. 8. Measurement results of the S-parameters of the power transistor, PGA26E19BA, from 50 kHz to 1 GHz. S_{11} with V_{ds} of 1 V and V_{gs} of 1.4 V (a), $|S_{12}|$ with V_{ds} of 1 V and V_{gs} of 1.4 V (b), $|S_{21}|$ with V_{ds} of 1 V and V_{gs} of 1.4 V (c), S_{22} with V_{ds} of 1 V and V_{gs} of 1.4 V (d), S_{11} with V_{ds} of 1 V and V_{gs} of 1.6 V (e), $|S_{12}|$ with V_{ds} of 1 V and V_{gs} of 1.6 V (f), $|S_{21}|$ with V_{ds} of 1 V and V_{gs} of 1.6 V (g), S_{22} with V_{ds} of 1 V and V_{gs} of 1.6 V (h), S_{11} with V_{ds} of 1 V and V_{gs} of 1.8 V (i), $|S_{12}|$ with V_{ds} of 1 V and V_{gs} of 1.8 V (j), $|S_{21}|$ with V_{ds} of 1 V and V_{gs} of 1.8 V (k), and S_{22} with V_{ds} of 1 V and V_{gs} of 1.8 V (l). The solid and dashed lines indicate the results using the developed probe system and the test fixture system, respectively.

the electromagnetic wave between the VNA receivers and reference plane (Fig. 5).

To calibrate the developed measurement system at the probe tip, two conventional methods, OSLT and port extension, were applied. The former was performed to correct the error terms between the coaxial ports at the bias-T and VNA receivers. The latter was conducted to move the calibrated plane from the coaxial ports to the ends of the probe tips.

First, we performed OSLT calibration at the coaxial ports of the bias-Ts, as illustrated in Fig. 6, to correct the error from the receivers to the coaxial ports. Here, the calibration at the coaxial

port was performed using a commercially available calibration kit, 85052D by Keysight Technologies.

Thereafter, port extension calibration was conducted to compensate for the error term from the coaxial ports to the probe tips (Fig. 7). The probes were connected at the calibrated coaxial ports, and the reflection coefficient, $\Gamma_{\text{air-open}}$, was measured by setting the probes in the air (air-open). This process assesses the phase shift errors of the electromagnetic waves between the calibrated coaxial port and the probe tips, and extends the calibrated coaxial planes to the probe tips. For example, the measured phase of $\Gamma_{\text{air-open}}$ was $-93.47°$ at 1 GHz. When the reflection components of the S-parameters of the probes were

979-8-3315-1612-3/25 $31.00 © 2025 IEEE

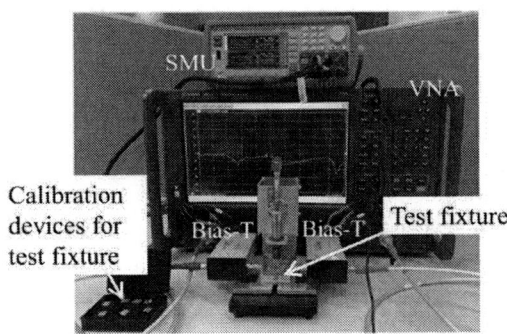

Fig. 9. Photographs of the conventional test fixture measurement system and calibration devices.

assumed to be zero and the reflection coefficient of the air-open was assumed to be 1, the probes presumably caused a phase shift of 46.735° at 1 GHz. The S-parameters at the probe tips were obtained by subtracting the estimated phase shift from the measured values at the calibrated coaxial ports. By combining OSLT and port extension calibrations, the errors between the receivers and probe tips can be eliminated, and the S-parameters of an SMD under test can be measured.

IV. EXAMPLE OF POWER TRANSISTOR MEASUREMENTS

This section reports examples of the S-parameter measurements of a commercially available power transistor, PGA26E19BA by Panasonic [23], using the developed GS probes and probe station. The measurements were performed with the port power of −20 dBm, the IF bandwidth of 100 Hz, and the average of 1. Fig. 3(a) shows a schematic representation of the measurements.

As represented by the solid lines in Fig. 8 (a) to (l), the S-parameters, S_{11}, $|S_{21}|$, $|S_{12}|$, and S_{22}, in the frequency range from 50 kHz to 1 GHz were measured by supplying the drain-source voltage, V_{ds}, of 1 V and gate-source voltage, V_{gs}, of 1.4 V, 1.6 V, and 1.8 V, respectively. The results imply that the proposed system captures the change in the characteristics of the power transistor under test from the off state to the on state by applying the bias voltages.

Figure 8 validates the measurement results obtained using the developed probe system, represented by the solid lines, in comparison with the results obtained using a conventional test fixture system (Fig. 9), represented by the dashed lines. The conventional test fixture system comprised the same VNA, SMU, two bias-Ts as the developed probe system, and a commercially available test fixture, PD1000A, by Keysight Technologies [19][39]. OSLT method was performed to calibrate the test fixture system using the commercially available calibration devices, PD1000A. The results were essentially identical.

V. CONCLUSION AND FUTURE WORK

This study presents an innovative probe system for measuring the S-parameters of SMDs operating in the frequency range from 50 kHz to 1 GHz. This measurement system was developed to increase the operating frequencies of power transistors in the research and development of power electronics. The system is advantageous because the probes are in direct contact with an SMD under test, and thus, test fixtures are not required. In addition, multiple types of SMDs can be measured using a single type of probe. To the best of our knowledge, a measurement system using probes with a 3.7 mm pitch for power devices, as presented in this study, is a pioneering development in this field. Notably, this system further simplifies the measurements of S-parameters for SMDs. Consequently, the research and development of power electronics is expected to rapidly progress. This paper presents the case of the 3.7 mm probe pitch, however, the method discussed herein is not limited to the probe form. Many SMDs can be measured by fabricating various probe pitches according to the electrode shape, installing them in the probe station, and applying the calibration method discussed in this paper.

In the future, the performance of the proposed system will be characterized and improved. For instance, the contact reproducibility may be further improved by applying a force-feedback structure on probe tips or grinding the contact area of a device to be measured. Extending the frequency range of the system may also be explored. In addition, we will modify the measurement system such that the two probes can approach a device under test at the angle of 90 °. Through this modification, the proposed system can be used to measure more commercially available power devices [40]-[47].

REFERENCES

[1] T. Kimoto, "Material science and device physics in SiC technology for high-voltage power devices," Jpn. J. Appl., vol.54, Mar. 2015, Art. No.040103.

[2] R. Quay, "Gallium Nitride Electronics" Springer Series in Material Science 96, Springer.

[3] D. D. Mahajan, S. A. Albahrani, R. Sodhi, T. Eguchi, and S. Khandewal, "Physics-Oriented Device Model for Packaged GaN Devices" IEEE Transactions on Power Electronics, vol.35, no.6, pp.6332-6339, Jun. 2020.

[4] T. Liu, Y. Feng, R. Ning, W. Wang, T. T. Y. Wong, and Z. J. Shen, "Extraction of Parasitic Inductances of SiC MOSFET Power Modules Based on Two-Port S-Parameters Measurement," 2017 IEEE Energy Conversion Congress and Exposition, Oct. 2017.

[5] L. Pace, N. Defrance, A. Videt, N. Idir, J. C. De Jaeger, and V. Avramovic, "Extraction of Packaged GaN Power Transistors Parasitics Using S-Parameters," IEEE Transactions on Electron devices, vol.66, no.6, pp.2583-2588, Jun. 2019.

[6] H. Sakairi, T. Yanagi, H. Otake, N. Kuroda, H. Tanigawa, and K. Nakahara, "Measurement Methodology for Accurate Modeling of SiC MOSFET Switching Behavior Over Wide Voltage and Current Ranges," IEEE Transactions on Power electronics, vol.33, no.9, pp.7314-7325, Sept. 2018.

[7] G. Crupi, D. M. M. P. Schreurs, and A. Caddemi, "On the Small Signal Modeling of Advanced Microwave FETs: A Comparative Study," International Journal of RF and Microwave Computer-aided Engineering, vol.18, no.5, pp.417-425, May. 2008.

[8] K. Li, P. Evans, and M. Johnson, "SiC/GaN power semiconductor devices: a theoretical comparison and experimental evaluation under different switching conditions," IET Electrical systems in Transportation, vol.8, no.1, pp.3-11, Mar. 2018.

[9] T. Hirano, K. Okada, J. Hirosawa, and M. Ando, "Accuracy investigation of de-embedding techniques based on electromagnetic simulation for on-wafer RF measurements," InTech Open Access Book, Numerical Simulation – From Theory to Industry, pp.233-258, Sept. 2012.

979-8-3315-1612-3/25 $31.00 © 2025 IEEE

[10] T. Liu, T. T. Y. Wong, and Z. J. Shen, "A new characterization technique for extracting parasitic inductances of SiC power MOSFETs in discrete and module packages based on two-port S-parameters," IEEE Transactions on Power Electronics, vol.33, no.11, pp.9819-9833, Nov. 2018.

[11] T. Liu, R. Ning, T. T. Y. Wong, and Z. J. Shen, "A new characterization technique for extracting parasitic inductances of fast switching power MOSFETs using two-port vector network analyzer," 2017 29th International Symposium on Power Semiconductor Devices and IC's, May. 2017.

[12] T. Liu, R. Ning, T. T. Y. Wong, and Z. J. Shen, "Equivalent circuit models and model validation of SiC MOSFET oscillation phenomenon," 2016 IEEE Energy Conversion Congress and Exposition, Sept. 2016.

[13] A. M. Bouchour, P. Dherbecourt, A. E. Oualkadi, and O. Latry, "Parasitic Elements Extraction of the GaN HEMT Packaged Power Transistor based on S-parameter Measurements," 2020 International Symposium on Advanced Elecrical and Communication Technologies, Nov. 2020.

[14] Yanagi, H. Sakairi, H. Otake, N. Kuroda, S. Kitagawa, N. Hashimoto, R. Takeda, and K. Nakahara, "Circuit simulation of a silicon-carbide MOSFET considering the effect of the parasitic elements on circuit boards by using S-parameters," 2018 IEEE Applies Power Electronics Conference and Exposition, Mar. 2018.

[15] Z. Wang, Z. Yuan, and Y. Zhao, "Parasitic Inductances Extraction for SiC Power Modules Using An Enhanced Two-Port S-parameter Approach," 2021 IEEE Applied Power Electronics Conference and Exposition, Jun. 2021.

[16] E. McShane and K. Shenai, "RF de-embedding technique for extracting power MOSFET package parasitics," IWIPP 2000. International Workshop on Integrated Power Packaging, Aug. 2002.

[17] J. Saijets and M. Aberg, "MOSFET RF Extraction Uncertainties Due to S Parameter Measurement Errors," Physical Scripta, vol.2004, no.T114, 2004.

[18] M. Horibe and I. Hirano, "Vector Network Analyzer Calibration for Characterization of Packaged Power MOSFET Device at RF Frequency," Proceeding of 2020 95th ARFTG Microwave Measurement Conference, Aug 2020.

[19] "Keysight PD1000A Power Device Test Fixtures for S-Parameter Measurements", Operation Guide, Keysight Technologies. [Online]. Available: https://www.keysight.com/jp/ja/assets/9018-04658/user-manuals/9018-04658.pdf?success=true

[20] R. Kishikawa, M. Horibe, T. Ohi, A. Yamamoto, N. Hashimoto, and R. Takeda, "Probe Measurement System for Surface Mount Devices at Radio Frequencies," Proceeding of 2021 98th ARFTG Microwave Measurement Conference, Jan. 2021.

[21] D. M. Pozar, "MICROWAVE ENGINEERING fourth edition," Wiley.

[22] V. Teppati, A. Ferrero, and M. Sayed, "Modern RF and Microwave Measurement Techniques," The Cambridge RF and Microwave Engineering Series, Cambridge.

[23] PGA26E19BA, Panasonic. [Online]. Available: https://industrial.panasonic.com/content/data/SC/ds/ds4/PGA26E19BA_J.pdf

[24] PGA26E06BA, Panasonic. [Online]. Available: https://industrial.panasonic.com/content/data/SC/ds/ds4/PGA26E06BA_J.pdf

[25] PGA26E07BA, Panasonic. [Online]. Available: https://industrial.panasonic.com/content/data/SC/ds/ds4/PGA26E07BA_J.pdf

[26] PGA26E17BA, Panasonic. [Online]. Available: https://industrial.panasonic.com/content/data/SC/ds/ds4/PGA26E17BA_J.pdf

[27] IGLD60R070D1, Infineon. [Online]. Available: https://www.infineon.com/dgdl/Infineon-IGOT60R070D1-DataSheet-v02_12-EN.pdf?fileId=5546d46265f064ff016685fa65066523

[28] IGLD60R190D1, Infineon. [Online]. Available: https://www.infineon.com/dgdl/Infineon-IGLD60R190D1-DataSheet-v02_13-EN.pdf?fileId=5546d46269e1c019016a6d78ff5e2aba

[29] SG26F30S-D, SUMITOMO ELECTRIC. [Online]. Available: https://www.sedi.co.jp/file.jsp?/pdf/SG26F30S-D_ED1-2.pdf

[30] SG36F30S-D, SUMITOMO ELECTRIC. [Online]. Available: https://www.sedi.co.jp/file.jsp?/pdf/SG36F30S-D_ED1-2.pdf

[31] SGFCF10S-D, SUMITOMO ELECTRIC. [Online]. Available: https://www.sedi.co.jp/file.jsp?/pdf/SGFCF10S-D_ED1-2.pdf

[32] SGFCF15S-D, SUMITOMO ELECTRIC. [Online]. Available: https://www.sedi.co.jp/file.jsp?/pdf/SGFCF15S-D_ED1-2.pdf

[33] SGFCF20S-D, SUMITOMO ELECTRIC. [Online]. Available: https://www.sedi.co.jp/file.jsp?/pdf/SGFCF20S-D_ED1-2.pdf

[34] SGFCF20T-D, SUMITOMO ELECTRIC. [Online]. Available: https://www.sedi.co.jp/file.jsp?/pdf/SGFCF20T-D_ED1-2.pdf

[35] SGFCF30T-D, SUMITOMO ELECTRIC. [Online]. Available: https://www.sedi.co.jp/file.jsp?/pdf/SGFCF30T-D_ED1-2.pdf

[36] SGFCF40T-D, SUMITOMO ELECTRIC. [Online]. Available: https://www.sedi.co.jp/file.jsp?/pdf/SGFCF40T-D_ED1-2.pdf

[37] D. Rytting, "Network Analyzer Error Models and Calibration Methods," IEEE MTT/ED Seminar: Calibration and Error Correction Techniques for Network Analysis, Sept. 2004.

[38] D. Rytting, "An Analysis of Vector Measurement Accuracy Enhancement Techniques," IEEE MTT/ED Seminar: Calibration and Error Correction Techniques for Network Analysis, Sept. 2004.

[39] PD1000A Power Device Measurement System for Advanced Modeling, Keysight Technologies. [Online]. https://www.keysight.com/us/en/product/PD1000A/power-device-measurement-system-advanced-modeling.html

[40] GS66506T, GaN systems. [Online]. Available: https://www.mouser.jp/datasheet/2/692/GS66506T_DS_Rev_200402-1837976.pdf

[41] GS-065-004-1-L, GaN systems. [Online]. Available: https://www.mouser.jp/datasheet/2/692/GS_065_004_1_L_DS_Rev_21 0104_1-2300385.pdf

[42] GS-065-008-1-L, GaN systems. [Online]. Available: https://www.mouser.jp/datasheet/2/692/GS_065_008_1_L_DS_Rev_21 0104-2887158.pdf

[43] GS-065-011-1-L, GaN systems. [Online]. Available: https://gansystems.com/wp-content/uploads/2022/07/GS-065-011-1-L-DS-Rev-220708.pdf

[44] GS66502B, GaN systems. [Online]. Available: https://www.mouser.jp/datasheet/2/692/GS66502B-DS-Rev-180213-1314162.pdf

[45] GS66504B, GaN systems. [Online]. Available: https://www.mouser.jp/datasheet/2/692/GS66504B_DS_Rev_200402-1838010.pdf

[46] GS66508B, GaN systems. [Online]. Available: https://gansystems.com/wp-content/uploads/2020/04/GS66508B-DS-Rev-200402.pdf

[47] GS66516B, GaN systems. [Online]. Available: https://www.mouser.jp/datasheet/2/692/GS66516B_DS_Rev_211025-2887030.pdf G. Eason, B. Nobl.

979-8-3315-1612-3/25 $31.00 © 2025 IEEE

Optical Transformer Design with Additional Common-Mode Noise Reduction Winding for Flyback DC-DC Converters

Yusuke Irie
Graduate School of Engineering
Nagasaki University
Nagasaki, Japan
bb521123208@ms.nagasaki-u.ac.jp

Shinichiro Eguchi
Graduate School of Engineering
Nagasaki University
Nagasaki, Japan
long.student1333@gmail.com

Yoichi Ishizuka
Graduate School of Engineering
Nagasaki University
Nagasaki, Japan
isy2@nagasaki-u.ac.jp

Toshiro Takeuchi
Sanken Electric Co., Ltd.
Saitama, Japan
ttake@ms3.sanken-ele.co.jp

Akio Iwabuchi
Smart Power Semi
Tokyo, Japan
aiwabuchi@smartpowersemi.com

Takehiro Koga
Ansys Japan K. K
Tokyo, Japan
takehiro.koga@ansys.com

Toshiyuki Tanaka
Graduate School of Engineering
Nagasaki University
Nagasaki, Japan
t-tanaka@nagasaki-u.ac.jp

Abstract—This paper proposes a method and theory for reducing conducted noise in the frequency range of 150 kHz to 30 MHz in accordance with CISPR16 for flyback converter circuits, which are the most used topologies in isolated DC/DC converters. Among conducted noise, we focus on common-mode noise, which is the most important to counteract. Choke coils and Y-capacitors are commonly used as countermeasures, but they have the disadvantage of increasing the circuit volume. Therefore, we focused on transformers used in switching power supplies and confirmed that by adding shielded windings between the primary and secondary windings and designing the optimal number of turns, the choke coil commonly used as a countermeasure for common mode noise is no longer necessary and the volume can be reduced by 6%. While other studies have theorized a simple transformer structure in their analysis, this study presents a design that considers transformer variations due to transformer design errors by theorizing so that the analysis can be performed even with complex structures, based on practical applications. A 98.3% reduction of common-mode noise was confirmed by simulation using the proposed common-mode noise reduction method. The number of shield windings optimal for common-mode noise suppression is also demonstrated through measurements on actual equipment in accordance with CISPR standards.

Keywords—Flyback DC-DC Converters, Common-Mode Noise, Transformer

I. INTRODUCTION

In recent years, the stable supply of electric power and increasing energy efficiency have demanded smaller and lighter switching power supplies with higher efficiency and higher frequency [1,2]. For this reason, materials for switching element MOSFETs are shifting from conventional silicon to next-generation semiconductor elements such as GaN and SiC. While the use of next-generation semiconductors enables higher switching frequencies and smaller noise filters, the higher switching frequencies generate high-frequency noise, which

has become a serious EMI (Electromagnetic Interference) problem. In general, conducted EMI noise can be classified into common mode (CM) noise and normal mode (DM) noise. CM noise is particularly difficult to counteract and is considered important in EMI[3,4]. Shielding techniques between transformer windings include the use of shield foil for countermeasures [9-12] and shielded windings [13-15]. In order to counter CM noise, it is important to identify the CM noise transmission path. Figure 1 shows the CM noise transfer path in a flyback converter equivalent circuit. The circuit diagram in Fig. 1 is based on other research [16]. The difference is that the secondary MOSFET is connected to the Line side; CM noise is evaluated by the current flowing through the resistance of the pseudo power supply network (LISN).

C_{p1s}, C_{p2s}, C_{p3s} and C_{psls} are the parasitic capacitance between the transformer windings. Those parasitic capacitances and voltage variations generated by the MOSFET (dV_P/dt) generate CM currents and form the CM noise transmission path [5,17-20]. I_{p1s}, i_{p2s} and i_{p3s} are the CM current propagating from the primary winding to the secondary winding. One side of the shield winding is open, and the other side is connected to the GND of the primary winding. The shield winding is wound in the same way as the secondary winding to produce a current

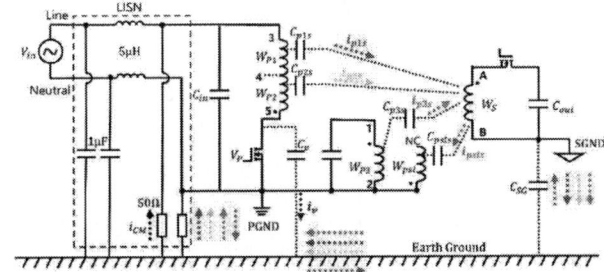

Fig. 1. CM noise path in a flyback converter

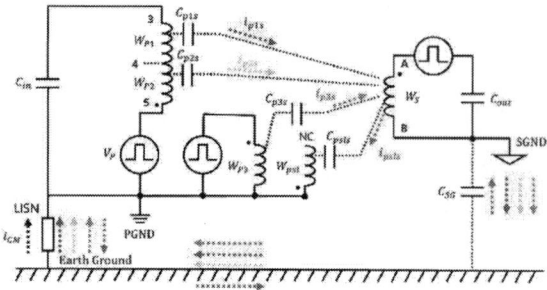

Fig. 2. Equivalent circuit of CM noise path of flyback converter

that is in phase with the CM current i_{psls}. This suppresses CM noise. The CM current transmitted from the primary MOSFET to the ground through the parasitic capacitance C_v with LISN i_v can be eliminated by using a heat sink [16,21]. In this paper, the parasitic capacitance between windings in a complex transformer is obtained by analysis, the current equation of CM noise is calculated by numerical analysis, and a design method for the optimal number of shield winding turns is proposed. Chapter 2 describes the theoretical analysis of CM noise. Chapter 3 presents the analysis results of the proposed method using the electromagnetic field simulator Maxwell, and Chapter 4 verifies the validity of the simulation analysis results by verifying the actual equipment. In Chapter 5, the winding variation of transformers and CM noise effects caused by the manufacturing process are verified by analysis using Maxwell. Finally, Chapter 6 summarizes this paper.

II. THEORETICAL ANALYSIS OF CM NOISE

Figure 2 shows the equivalent circuit of the CM noise path. The CM noise transmitted through the circuit is evaluated by the current flowing through the LISN resistors. Therefore, to evaluate the current in the analysis using Maxwell, the relationship between the current and parasitic capacitance is expressed as an equation. The total CM current i_{CM} observed in LISN is expressed as follows

$$i_{CM} = i_{p1s} + i_{p2s} + i_{p3s} - i_{spsl} \quad (1)$$

TABLE I.   Transformer winding specifications structure

name	number of turns	Wire	Diameter[mm]	coating[mm]
Primary winding P1	15	S UEW	0.516	0.065
Primary winding P2	20	S UEW	0.516	0.04
Secondary winding S1_1, S1_2, S1_3	6 (2 parallel winding)	TIW-3	0.6	0.085
Shielded winding PSL1, PSL2, PSL3	12~17 (2 parallel winding)	2UEW	0.12×2	0.006
Primary auxiliary winding P3	6	2UEW	0.26	0.085
Synchronous rectifier winding S2	4	TIW-3	0.32	0.1

The sign of each current is positive in the direction of flow from the primary winding to the secondary winding shown in Fig. 2. i_{p1s} can be expressed by the following equation

$$i_{p1s} = \frac{C_{p1s}}{2} \cdot \frac{N_{p1} - N_s}{N_p} \cdot \frac{dV_p}{dt} \quad (2)$$

N_{p1} is the number of turns of the primary winding P1, N_p is the total number of primary windings (P1, P2), N_s is the secondary winding, V_p is the voltage supplied to the primary winding and primary MOSFET. To accurately determine the parasitic capacitance and minimize calculation errors during design, adjacent inter-winding paths and minor coupling paths must be considered [16]. Minor couplings are CM current paths formed by the boundaries between electric and magnetic fields between non-adjacent windings due to fringing effects. Considering them, i_{p1s} can also be expressed by the following equation

$$i_{p1s} = \sum_{i=1}^{N_p} \sum_{j=1}^{N_s} C_{pisj} \frac{N_p - N_s - i + j}{N_p} \frac{dV_p}{dt} \quad (3)$$

It can be seen that (2) and (3) have general expressions that are functions of parasitic capacitance and winding. Therefore, we express them as follows.

$$i_{p1s} = (\frac{C_{p1s}}{2} \cdot N_{p1} - N_s + \sum_{i=1}^{N_p} \sum_{j=1}^{N_s} C_{pisj} \cdot (N_p - N_s - i + j)) \cdot \frac{1}{N_p} \frac{dV_p}{dt}$$
$$= k_{c1}(C, N) \cdot \frac{1}{N_p} \cdot \frac{dV_p}{dt} \quad (4)$$

The coefficient shown in (4) $k_{c1}(C, N)$ is defined as an indicator of the magnitude of CM noise (CM noise coefficient). Similarly for the other windings, the CM noise coefficient is defined as $k_{c2}(C, N)$, $k_{c3}(C, N)$, $k_{c4}(C, N)$. From (1), the total CM mode noise coefficient can be expressed as the sum of the CM noise coefficients for each winding, which is k_{CM}. Therefore, the expression for the CM noise current can be expressed as follows

$$i_{CM} = (k_{c1}(C, N) + k_{c2}(C, N) + k_{c3}(C, N) - k_{c4}(C, N)) \cdot \frac{1}{N_p} \frac{dV_p}{dt}$$
$$= k_{CM}(C, N) \cdot \frac{1}{N_p} \cdot \frac{dV_p}{dt} \quad (5)$$

From (5), to evaluate the total CM current i_{CM} for different numbers of shield windings, it is better to compare the k_{CM}. k_{CM} is close to zero, i_{CM} is also close to zero, it is thought that CM noise can be suppressed.

III. SIMULATION RESULT

Maxwell was used for electromagnetic field analysis of parasitic capacitance. Fig. 3 shows the winding structure of the

979-8-3315-1612-3/25 $31.00 © 2025 IEEE

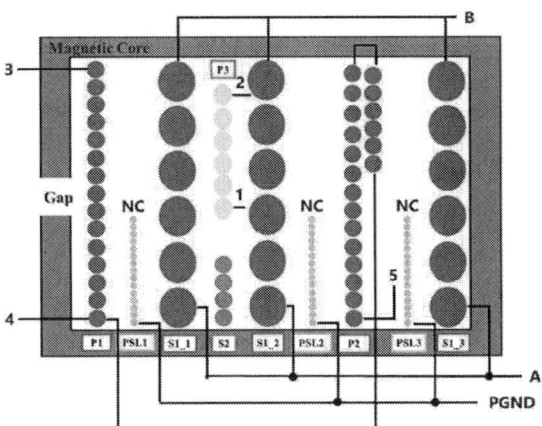

Fig. 3. Transformer winding structure

transformer used for the Maxwell electromagnetic field analysis, and TABLE I shows the specifications of the transformer windings.

Unlike other studies [16], we believe that the number of shield winding turns can be optimized by increasing the parameters of the current equation for CM noise and allowing for the design error of the transformer by considering the transformer winding folds in the analysis as shown in Figure 3.

Figure 4 shows the relationship between the number of shield winding turns and the coefficient representing the magnitude of CM noise calculated from the parasitic capacitance in the electromagnetic field analysis. 15 turns shows a large attenuation of the coefficient, indicating that the optimal number of

TABLE II. Power circuit specifications

Input Voltage	100 V
Output Voltage	24V
Output Current	4.2
Maximum Output Power	100.8W
Switching Frequency	65kHz

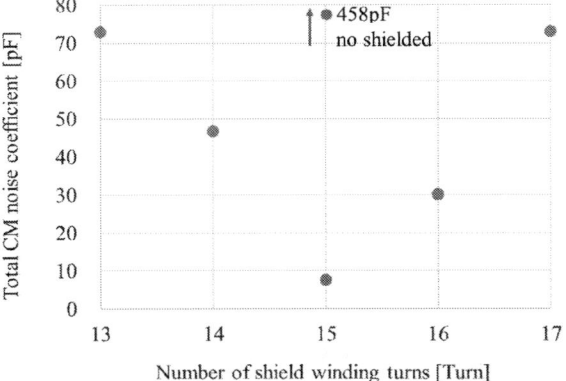

Fig. 4. Number of shield winding turns and CM noise coefficient

shield winding turns is 15Turns. It was also confirmed that the total CM noise coefficient can be reduced by 98.3% by using shielded windings. In other words, it can be said that designing the shield winding with 15 turns is expected to be more effective in suppressing CM noise.

IV. EXPERIMENT RESULT

To validate the analysis, actual equipment measurements were performed in accordance with the CISPR16 standard. The circuit specifications are shown in TABLE II and a photograph of the transformer compared to the flyback converter is shown in Fig. 5.

Figure 6 shows the measurement results of the shield winding and CM noise level at 195 KHz. 195 KHz was chosen because of the frequency characteristic at 195 KHz, which is the third harmonic of the switching frequency in the CISPR16 frequency band, was found to have a significant impact on CM noise. This is because Similar to the Fig. 6 simulation analysis, the CM noise was most attenuated at 15 turns, confirming the validity of the simulation analysis results. However, when several transformers with 15 turns of shielded windings were measured, there were deviations in the CM noise level measurements as shown in Fig. 7.

Figure 7 shows the results of CM noise level measurements

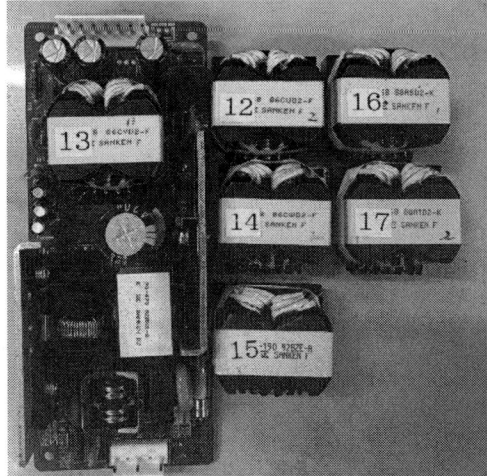

Fig. 5. Power circuit and transformers

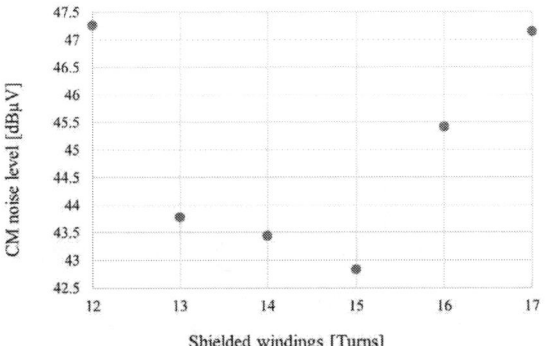

Fig. 6. Shielded windings and CM noise level

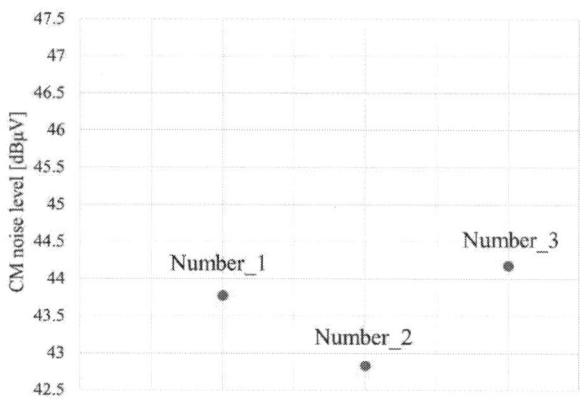

Fig. 7. Transformer with 15 turns of shielded windings and CM noise level

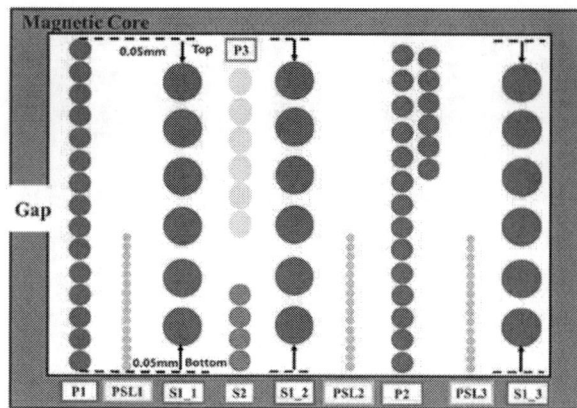

Fig. 9. Transformer winding structure when the secondary winding is shifted up and down by 0.05 mm

using transformers with multiple shield windings of 15 turns. It can be seen that the CM noise level varies even for the same number of shield winding turns. Furthermore, the error range of the CM noise suppression effect in the 15-turn shield winding is lower than the CM noise levels of the transformers with 12 and 17 turns shown in Fig. 6. In other words, the CM noise suppression effect due to the difference in the number of shield winding turns can be confirmed. Considering the above, it can be equivalently considered that CM noise is suppressed in the range of the number of shield winding turns from 13 to 15 in Fig. 6. As shown in Fig. 7, the reason for the difference in CM noise levels even for transformers with the same number of shield winding turns is considered to be the misalignment of the windings due to transformer design errors. Unlike the transformer model that was analyzed, it can be said that the actual winding misalignment occurs at the design stage and the parasitic capacitance between the windings changes, causing the CM noise level to vary depending on the individual transformer.

Figure 8 shows an image of a transformer with 15 turns of shielded windings analyzed by Computed Tomography. The left side is the analyzed image of the Number_1 transformer shown in Fig. 7, and the right side is the analyzed image of the Number_2 transformer. It can be seen that even though the transformers have the same specifications, there is a misalignment of the windings during the transformer manufacturing process. Therefore, the effect of the winding misalignment on the CM noise is verified by simulation analysis using Maxwell.

Fig. 8. Comparison of windings misalignment in 15 turns of shielded windings analyzed by computed tomography

V. ANALYSIS OF PARASITIC CAPACITANCE AND CM NOISE DUE TO WINDING MISALIGNMENT

When analyzing the parasitic capacitance between the windings, the effect of minor paths must be considered, as described in Chapter 2. Therefore, we verified how the parasitic capacitance values and CM noise coefficients change when the secondary winding is shifted for two windings, the primary and secondary windings, in a 15 turns transformer, which had the highest CM noise suppression effect.

Figure 9 shows the cross-section of the transformer with the secondary winding shifted inward by 0.05 mm (top and bottom). 0.05 mm was shifted because the image analyzed by Computed Tomography and the cross-section of the transformer when it was cut showed that the secondary winding was shifted by approximately 0.05 mm. In addition, when transformers are wound in layers, the position of the beginning of the winding is shifted. The misalignment of the secondary winding was verified by taking this effect into account. TABLE III shows the change in parasitic capacitance values between the primary and secondary windings.

TABLE III shows that when the position of the secondary winding is shifted inward, the parasitic capacitance value between the primary and secondary winding becomes smaller. This can be attributed to the fact that shifting the secondary winding inward increases the effect of the shield winding, which reduces the influence of the minor path between the pri

TABLE III. Parasitic capacitance values between the primary and secondary windings.

	Top [pF]	Bottom [pF]
No Positional Displacement	0.708	0.081
Positional Displacement of 0.05 mm	0.684	0.073

TABLE IV. Comparison results of the total CM noise coefficient

	Total CM noise coefficient [pF]
No Positional Displacement	7.62
Positional Displacement of 0.05 mm	7.22

mary and secondary windings. TABLE IV shows the change in the total CM noise coefficient.

By shifting the position of the secondary winding, the total CM noise coefficient decreased by approximately 5.2%. In other words, the CM noise suppression is considered to be more effective even if design errors occur due to the misalignment of the start of the secondary winding.

VI. CONCLUSION

In this paper, CM noise was theorized so that it could be analyzed even with complex transformer structures based on practical applications, and the results of transformer analysis and actual machine verification were described. 98.3% reduction of the total CM noise coefficient was achieved by using shielded windings in the transformer as a CM noise countermeasure, and the CM noise reduction effect was confirmed. The results showed that the total CM noise coefficient was reduced by 98.3% by using shielded windings. Furthermore, the optimal number of turns of the shielded winding was shown to be 15 turns by analysis and measurement of actual equipment. Considering the winding variation during transformer manufacturing, the optimal number of shield windings can be equivalently considered to be 13 to 15 turns. Therefore, we calculated the current equation of CM noise by considering the deviation of the position of the secondary winding for practical use and showed that the variation of the transformer winding during manufacturing can be considered in the analysis and actual measurement. In the future, we plan to establish an optimal transformer design method for CM noise suppression by showing the relationship between the variation of transformer windings and the change in parasitic capacitance between transformer windings.

REFERENCES

[1] M. Joung, H. Kim and J. Baek, "Dynamic analysis and optimal design of high efficiency full bridge LLC resonant converter for server power system,"2012 Twenty-Seventh Annual IEEE Applied Power Electronics Conference and Exposition (APEC), 2012, pp. 1292-1297, doi: 10.1109/APEC.2012.6165985.

[2] H. Park, H. Choi and J. Jung, "Design and implementation of high switching frequency LLC resonant converter for high power density," 2015 9th International Conference on Power Electronics and ECCE Asia (ICPE-ECCE Asia), 2015, pp. 502-507, doi: 10.1109/ICPE.2015.7167832.

[3] Mao Xinkui and Chen Wei, "More precise model for parasitic capacitances in high-frequency transformer," 2002 IEEE 33rd Annual IEEE Power Electronics Specialists Conference. Proceedings (Cat. No.02CH37289), 2002, pp. 1054-1057 vol.2, doi: 10.1109/PSEC.2002.1022595.

[4] Shuo Wang, Lee, F.C. and Odendaal, W.G. "Characterization and parasitic extraction of EMI filters using scattering parameters," IEEE Transactions on Power Electronics, vol. 20, no. 2, 2005, pp. 502-510, doi: 10.1109/TPEL.2004.842949.

[5] J. -L. Kotny, X. Margueron and N. Idir, "High-Frequency Model of the Coupled Inductors Used in EMI Filters," IEEE Transactions on Power Electronics, vol. 27, no. 6, pp. 2805-2812, 2012, doi: 10.1109/TPEL.2011.2175452.

[6] Y. P. Chan, M. H. Pong, N. K. Poon and C. P. Liu, "Effective switching mode power supplies common mode noise cancellation technique with zero equipotential transformer models," 2010 Twenty-Fifth Annual IEEE

Applied Power Electronics Conference and Exposition (APEC), 2010, pp. 571-574, doi: 10.1109/APEC.2010.5433613.

[7] Pentti, L.; Hyvonen, O. "Electrically Decoupled Integrated Transformer Having at Least One Grounded Electric Shield," U.S. Patent 7733205 B2, 8 June 2010.

[8] M. A. Saket, M. Ordonez and N. Shafiei, "Planar Transformers with Near-Zero Common-Mode Noise for Flyback and Forward Converters," IEEE Transactions on Power Electronics, vol. 33, no. 2, pp. 1554-1571, 2018, doi: 10.1109/TPEL.2017.2679717.

[9] H. Chen, J. Xiao, "Determination of Transformer Shielding Foil Structure for Suppressing Common-Mode Noise in Flyback Converters," IEEE Transactions on Magnetics, vol. 52, no. 12, 2016, pp. 1-9, doi: 10.1109/TMAG.2016.2594047.

[10] Y. Yang, D. Huang, F. C. Lee, and Q. Li, "Transformer shielding technique for common mode noise reduction in isolated converters," in Proc. IEEE Energy Convers. Congr. Expo., 2013, pp. 4149–4153, doi: 10.1109/ECCE.2013.6647252.

[11] L. Xie, X. Ruan, Q. Ji, and Z. Ye, "Shielding-cancelation technique for suppressing common-mode EMI in isolated power converters," IEEE Energy Conversion Congress and Exposition (ECCE), 2014, pp. 4769-4776, doi: 10.1109/ECCE.2014.6954054.

[12] Zengyi Lu and Wei Chen, "Common mode EMI noise reduction technique by noise path configuration of high frequency power transformer," IEEE 6th International Power Electronics and Motion Control Conference, 2009, pp. 954-956, doi: 10.1109/IPEMC.2009.5157521.

[13] Y. P. Chan, M. H. Pong, N. K. Poon and C. P. Liu, "Effective switching mode power supplies common mode noise cancellation technique with zero equipotential trans-former models," Twenty-Fifth Annual IEEE Applied Power Electronics Conference and Exposition (APEC), 2010, pp. 571-574, doi: 10.1109/APEC.2010.5433613.

[14] H. Chen, Z. Zheng, and J. Xiao, "Determining the number of transformer shielding winding turns for suppressing common-mode noise in flyback converters," IEEE Transactions on Electromagnetic Compatibility, vol. 60, no. 5, 2018, pp. 1606-1609, doi: 10.1109/TEMC.2017.2777258.

[15] K. Fu and W. Chen, "Evaluation Method of Flyback Converter Behaviors on Common-Mode Noise," IEEE Access, vol. 7, 2019, pp. 28019-28030, doi: 10.1109/ACCESS.2019.2902462.

[16] Z. C. Xu, "Quantitative analysis of common-mode current in flyback converter with balance winding," IET Power Electronics, vol. 16, issue 5, pp.740-751, 2022, doi: 10.1049/pel2.12419.

[17] M. R. Yazdani, H. Farzanehfard and J. Faiz, "EMI Analysis and Evaluation of an Improved ZCT Flyback Converter," IEEE Transactions on Power Electronics, vol. 26, no. 8, 2011, pp. 2326-2334, doi: 10.1109/TPEL.2010.2095884.

[18] M. R. Yazdani, H. Farzanehfard and J. Faiz, "Conducted EMI modeling and reduction in a flyback switched mode power supply," 2011 2nd Power Electronics, Drive Systems and Technologies Conference, 2011, pp. 620-624, doi: 10.1109/PEDSTC.2011.5742494.

[19] P. Kong and F. C. Lee, "Transformer structure and its effects on common mode EMI noise in isolated power converters," 2010 Twenty-Fifth Annual IEEE Applied Power Electronics Conference and Exposition (APEC), 2010, pp. 1424-1429, doi: 10.1109/APEC.2010.5433416.

[20] D. Fu, S. Wang, P. Kong, F. C. Lee and D. Huang, "Novel Techniques to Suppress the Common-Mode EMI Noise Caused by Transformer Parasitic Capacitances in DC–DC Converters," IEEE Transactions on Industrial Electronics, vol. 60, no. 11, 2013, pp. 4968-4977, doi: 10.1109/TIE.2012.2224071.

[21] Qingbin Chen, Wei Chen, Qingliang Song and Yongfa, Zhu."An Evaluation Method of Transformer Behaviors on Common-mode Conduction Noise in SMPS," IEEE Ninth International Conference on Power Electronics and Drive Systems, 2011, pp782-786, doi: 10.1109/PEDS.2011.6147342.

979-8-3315-1612-3/25 $31.00 © 2025 IEEE

Enhanced Bus Voltage Stability through Digital Twin-Enabled Adaptive Controller Tuning

Matthew Belanger*, Andy Wong[†], Kerry Sado, and Enrico Santi

Dept. of Electrical Engineering
University of South Carolina
Columbia, USA
*belangem@email.sc.edu and [†]andyw@email.sc.edu

Abstract—In this paper, a digital twin approach for adaptive controller tuning is presented. For an anticipated load profile, step transitions are fed into a simulation-based optimizer which utilizes a state-space averaged model to determine optimal control parameters for the future operating point after the next upcoming load step. This process aims to minimize voltage overshoot, settling time, and oscillation. Control parameters are adjusted preemptively before each load step by the simulation-based optimizer and subsequently validated through a comprehensive switching model simulation. This guarantees good dynamic performance throughout the load profile. Experimental results are provided for a simple example involving a two-loop PI controlled bidirectional DC/DC converter connected to a pulse load. Compared to the static PI controller, the digital twin tuning approach reduced voltage overshoot by ∼50% and settling time by ∼50% for the single converter. Simulation results are presented for a multi-converter system with a complex load profile, showing significant improvement, with up to 96% improvement in voltage settling time and up to 95% improvement in voltage overshoot.

Index Terms—Digital Twin, Adaptive Tuning, Simulation Based Optimization

I. INTRODUCTION

In shipboard power systems, maintaining stability in the presence of high-power pulsed loads such as advanced radar and weapon systems is a significant challenge. These systems can impose sudden and substantial energy demands on the power system, leading to fluctuations in bus voltage and potentially compromising the operation of sensitive equipment. For instance, the rapid variations of power drawn from high-energy radars and electromagnetic weapons can cause voltage sags, impacting other critical onboard systems and overall power quality [1], [2]. To address these challenges, adaptive control tuning and adjustment mechanisms are essential. The Digital Twin (DT) approach offers a powerful solution by enabling real-time predictive analytics. A DT can be defined as a collection of high accuracy dynamic digital models that incorporate sensor and historical data to represent an existing physical system or subsystem [3]. The DT allows for the preemptive tuning of control parameters based on anticipated load conditions. For instance, the adaptive control algorithm can adjust the gains of a voltage controller to maintain desired performance metrics, such as stability and response time,

This work was supported by the Office of Naval Research under contracts No. N00014-22-C-1003 and N00014-23-C-1012.

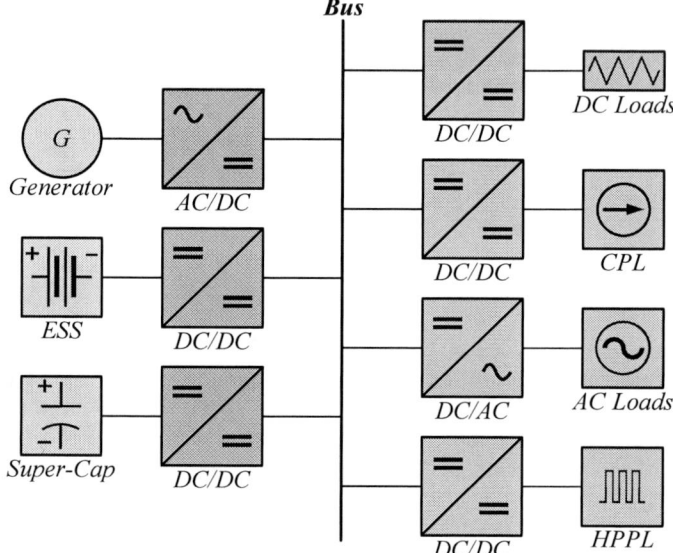

Fig. 1. Islanded DC microgrid

during transient events. This is particularly important for shipboard power systems, where highly dynamic, high-power pulsed loads require swift and precise control adjustments to prevent instability. By employing DTs and advanced control strategies, shipboard power systems can achieve a more resilient and robust performance, ensuring that, even during high-energy operations, systems remain stable and efficient.

In Section II, the proposed DT-enabled control strategy is presented in detail as a framework for real time adaptive tuning. In Section III, the simulation-based optimization procedure is discussed, as well as the process of meta-heuristic optimization. In Section IV, the approach is demonstrated using a single converter system with a pulse load; simulation and experimental results are presented. In Section V, a more complex, multi-converter system is optimized for an arbitrary load profile using three different optimization algorithms; simulation results are examined. Lastly, in Section VI, the overall performance of the proposed approach is reviewed, and future work is explored.

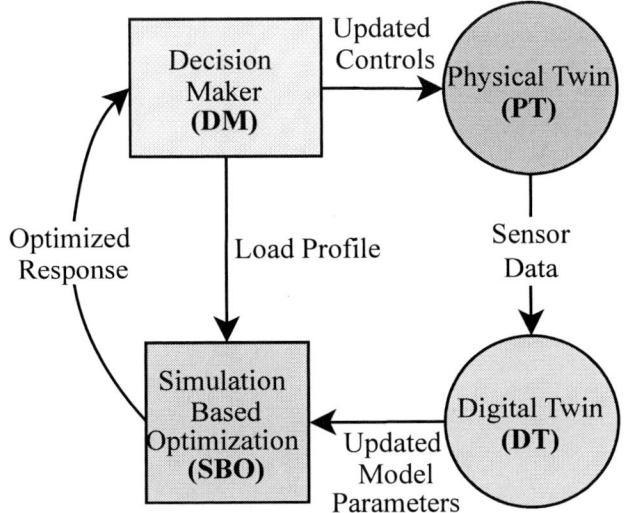

Fig. 2. Closed loop cyber-physical connection.

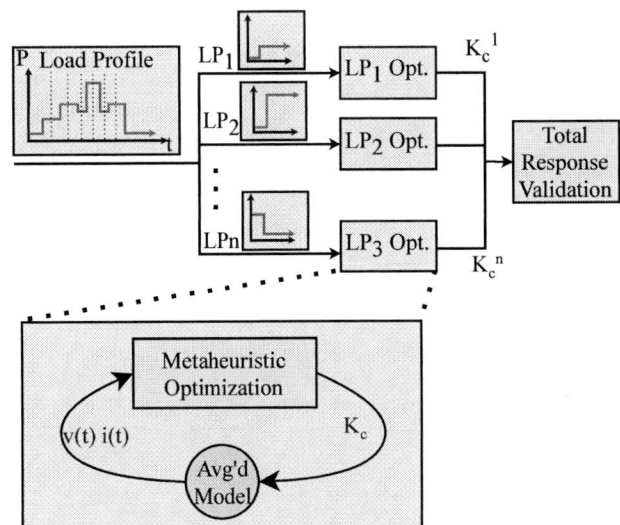

Fig. 3. Simulation based optimization

II. DIGITAL TWIN APPROACH

Using predictive analysis and real-time updates to the virtual model of the system provided by DT technology enables the optimization of the system response to anticipated loads and disturbances. A diagram of the DT tuning approach is shown in Fig. 2. As depicted, a Decision Maker (DM) specifies an anticipated load profile, which may come from an operator or a higher level DT. Once a load profile is specified, Simulation Based Optimization (SBO) is performed to find optimal control parameters for each load step utilizing look-head simulations. The optimized response is sent to the DM to confirm and make updates to the controls of the Physical Twin (PT). By gain scheduling, the control parameters are updated preemptively to achieve the desired response. Models used by the SBO are updated by the DT which corrects for any discrepancies between the physical and digital twins.

The proposed SBO is outlined in Fig. 3. First, the load profile, specified by the DM, is partitioned into step loads. Then, a meta-heuristic optimization, such as Particle Swarm Optimization (PSO), is performed on each load step using an averaged model of the system to obtain optimal controller gains, K_c. The total response of the complex load profile is simulated using a higher fidelity model to validate and evaluate the dynamic performance using the optimized control parameters specific to each step in the load profile.

III. SIMULATION BASED OPTIMIZATION

A. Converter Modeling

Simulation Based Optimization, as described earlier, is often difficult to implement in real time due to the complexity of accurately modeling dynamic systems and the significant computational resources required to run simulations and optimization algorithms quickly. To make this approach realizable for complex systems such as DC/DC converters, it is necessary to choose a model that balances accuracy and simulation time. Thus, the State Space Averaged Model (SSAM) [4] was chosen for the optimizer due to its ability to simplify the dynamic representation of converters. By averaging the

states of the system over a single switching period, the model effectively captures the essential behavior of the converter, reducing system complexity. Paralleling converters can reduce the accuracy of a standard SSAM; however, paralleled converters can be modeled with first order harmonics to improve the accuracy [5].

Fig. 4 shows a block diagram representation of the converter with controls. A standard two-loop PI control scheme, consisting of an inner current loop and outer voltage loop, was considered. In addition, droop control was utilized for multi-converter power sharing. Here, the PI controller blocks, G_{ci} and G_{cv}, represent the transfer functions of the inner current controller and outer voltage controller, respectively. R_d is the droop resistance of the converter, and $K(d)$ is the duty-related gain from the inductor current to the output current.

B. Feature Selection and Cost Function Definition

For the SBO, the inner current loop controller gains, K_{pi} and K_{ii}, and outer voltage loop controller gains, K_{pv} and K_{iv}, of the converters were considered as inputs for the optimization. The first subscript letters, p and i, denote the proportional and integral gains, respectively. After simulating a load step with the selected control parameters, only one feature was extracted from the simulated output response, the integral-time of absolute voltage error (ITAE) is

$$ITAE = \int t|v(t) - V_{ss}|\, dt, \tag{1}$$

where V_{ss} is the steady state voltage.

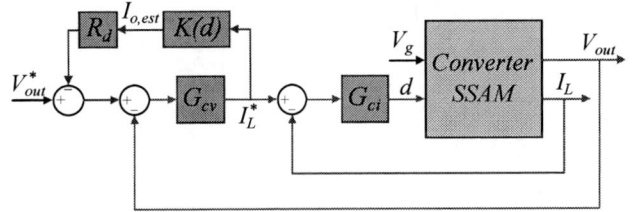

Fig. 4. SSAM with double loop PI control

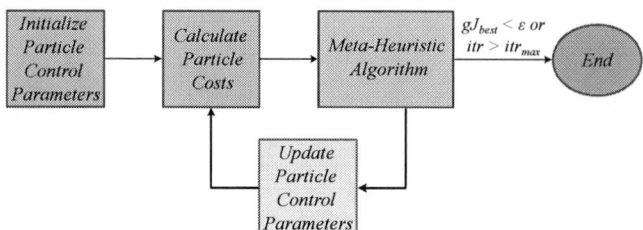

Fig. 5. Meta-heuristic optimization process

From this feature, the objective variable,

$$\theta = \frac{ITAE - ITAE_{des}}{ITAE_{max}}, \quad (2)$$

was defined where subscripts des and max are desired and maximum values, respectively. Soft saturation was applied for normalization and convergence. From this objective variable, a linear cost function was defined, given by

$$J(\theta) = \theta. \quad (3)$$

C. Meta-Heuristic Algorithms for SBO

Another difficulty comes with the optimization itself of complex nonlinear systems. For such systems, there can be multiple local minima, making it hard to find the global optimum. Additionally, the parameters of the system can be highly sensitive to initial conditions, leading to vastly different results from small changes. In order to tackle these problems, meta-heuristic algorithms were explored for control parameter tuning given their global search capability and stochasticity. In addition, these algorithms allow for simulation parallelization which can decrease overall optimization time.

However, implementation of such algorithms comes at a cost of computational complexity and memory intensity. Appropriate hyper-parameter tuning and balance of exploration and exploitation need to be considered for fast convergence and computation time.

Fig. 5 shows the general process for the SBO. The process is as follows. First, n particles are initialized in R^m search space, where m is the number of control parameters being tuned in the system. Given this search space, each particle contains a set of control parameters to be optimized for the system. Next, for each iteration, simulations are run for each particle utilizing its respective control parameters to extract feature values and calculate the cost. Then, as a function of cost and specified hyper-parameter values, the meta-heuristic algorithm updates the position of each particle before another iteration of simulations is performed. This process of simulation, feature extraction, cost calculation, and position update, is performed until a specified cost tolerance or maximum iteration is reached.

For the SBO, three algorithms were considered. Namely: Particle Swarm Optimization (PSO) [6], Crow Search Algorithm (CSA) [7], and Whale Optimization Algorithm (WOA) [8]. The position update equations, parameters, and modified variable definitions are defined in Table I and Table II.

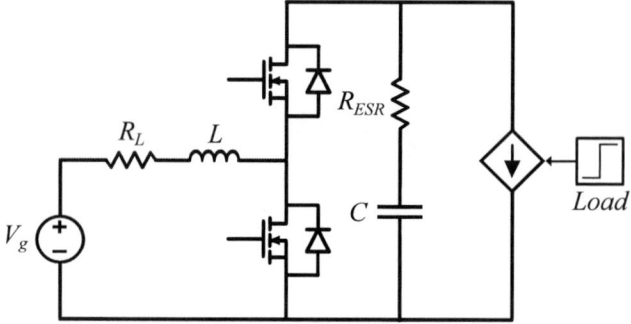

Fig. 6. Bidirectional DC/DC converter circuit

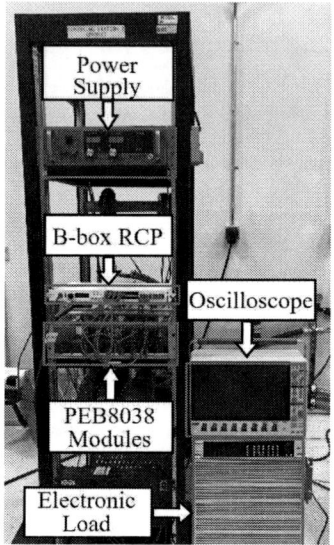

Fig. 7. Experimental setup

D. Total Response Validation

Although the meta-heuristic optimization may produce optimal or close to optimal gains for the lower fidelity model, the response of the system may not follow in practice. The additional complexity inherent in a physical system adds unmodeled dynamics and disturbances to the system that can cause dramatic differences in simulated and experimental results. To achieve a better tuned response and more accurate simulation results, a local optimization can be performed using the higher fidelity switching model, searching around the solution obtained from the SSAM. A high fidelity model, such as a switching model, may require dramatically longer computation time per simulation; therefore, it is imperative that the local search is tuned precisely so that optimal results are achieved in minimum time.

IV. SINGLE CONVERTER TUNING

A. Methods

To demonstrate DT-enabled controller tuning on a simple system, this work utilized a bidirectional DC/DC converter connected to an electronic load as shown in the circuit schematic in Fig. 6, and was implemented in hardware, as shown in Fig. 7. The circuit parameters used are found in

979-8-3315-1612-3/25 $31.00 © 2025 IEEE

TABLE I
POSITION UPDATE EQUATIONS FOR ALGORITHMS

Algorithm	Position Update Equations
Particle Swarm (PSO)	$$\vec{P}_i(t+1) = \vec{P}_i(t) + \chi \cdot \vec{V}_i(t)$$ where $$\vec{V}_i(t+1) = w \cdot \vec{V}_i(t) + c_1 p_1 \cdot (\vec{P}_{\text{best,i}}(t) - \vec{P}_i(t)) + c_2 p_2 \cdot (\vec{G}_{\text{best}}(t) - \vec{P}_i(t))$$
Crow Search (CSA)	$$\vec{P}_i(t+1) = \begin{cases} \vec{P}_i(t) - f_l \cdot (\vec{P}_{\text{best,i}}(t) - \vec{P}_i(t)) & p > AP \\ \vec{P}_{\text{rand}}(t) \cdot f_l \cdot (C \cdot \vec{P}_{\text{rand}}(t) - \vec{G}_{\text{best}}(t)) & \text{otherwise} \end{cases}$$
Whale Optimization (WOA)	$$\vec{P}_i(t+1) = \begin{cases} \vec{G}_{\text{best}}(t) - A \cdot |C \cdot \vec{G}_{\text{best}}(t) - \vec{P}_i(t)| & p < 0.5, A < 1 \\ \vec{G}_{\text{best}}(t) + \cos(2\pi l)e^{bl} \cdot |\vec{G}_{\text{best}}(t) - \vec{P}_i(t)| & p < 0.5, A > 1 \\ \vec{P}_{\text{rand}}(t) - A \cdot |\vec{P}_{\text{rand}}(t) - \vec{G}_{\text{best}}(t)| & p > 0.5 \end{cases}$$

TABLE II
VARIABLE AND PARAMETER DEFINITIONS

Variable or Parameter	Definition or Equation
f_l	$$f_l = (f_{l\max} - f_{l\min}) \cdot \left(1 - \exp\left(-\frac{\text{itr}^2}{\text{itr}_{\max}^2} \cdot 2\pi\right)\right) + f_{l\min}$$
AP	$$AP = AP_{\max} - \frac{AP_{\max} + AP_{\min}}{2} \cdot \left(2 \cdot \frac{\text{itr}}{\text{itr}_{\max}} - \frac{\text{itr}^2}{\text{itr}_{\max}^2}\right)$$
A	$$A = \left(\frac{A_{\max} - A_{\min}}{\text{itr}_{\max}}\right) \cdot l$$
b	$$b = b_{\min} \cdot \left(1 - \exp\left(-\frac{\text{itr}}{\text{itr}_{\max}}\right)\right)$$
w	$$w = \exp\left(-\frac{\text{itr}}{\text{itr}_{\max}} \cdot w_{\text{decay}}\right)$$
p	Random number, $p \in [0, 1]$
l	Random number, $l \in [-1, 1]$
C	Random number, $C \in [0, 2]$

Table III. From this schematic, the following state space averaged equations,

$$\frac{di_L}{dt} = \frac{1}{L}\left[(v_g - ri_L) - (1-d)\left(v_c + R_{ESR}C\frac{dv_c}{dt}\right)\right] \quad (4)$$

and

$$\frac{dv_c}{dt} = \frac{1}{C}\left[(1-d)i_L - i_{\text{Load}}\right], \quad (5)$$

were used to build a model in MATLAB Simulink. In (4), $r = R_L + R_{ds}$.

A standard two-loop PI control scheme was implemented for the converter, producing a search space of 4-dimensions: K_{pv}, K_{iv}, K_{pi}, and K_{ii}. Using the converter to step up the voltage from 250 V to 650 V, results were obtained for the optimized response to a single pulse load. First, the load was partitioned into two steps—one from 1 A to 15 A and another from 15 A to 1 A; then, SBO was performed to achieve the optimal system response with respect to ITAE of voltage deviation from the steady state value: 650 V.

To optimize the controls for bus voltage stability during load transitions, PSO was utilized for the SBO because of its parallelizability and efficiency in finding optimal solutions [9]. Fig. 8a shows 2 dimensions of the search space, representing the voltage loop PI parameters, depicting the particles converging as the epochs increase. Fig. 8b shows the corresponding best transient response at each epoch.

The voltage response in epoch 1 shows highly oscillatory behavior; however, as the particles converge, the oscillations die out. This is due to the cost function, ITAE, which heavily penalize oscillatory behavior around the steady state voltage. Once the control parameters for the step up and step down were optimized, the total response of the system was then validated with a switching model simulation, created in Simulink, using the optimized control parameters. Results were then compared to an "unoptimized" static PI case. The "unoptimized" PI controller was tuned to have a current loop bandwidth of 1200Hz, a current loop phase margin of 75°, a voltage loop bandwidth of 150Hz, and a voltage loop phase margin of 54°.

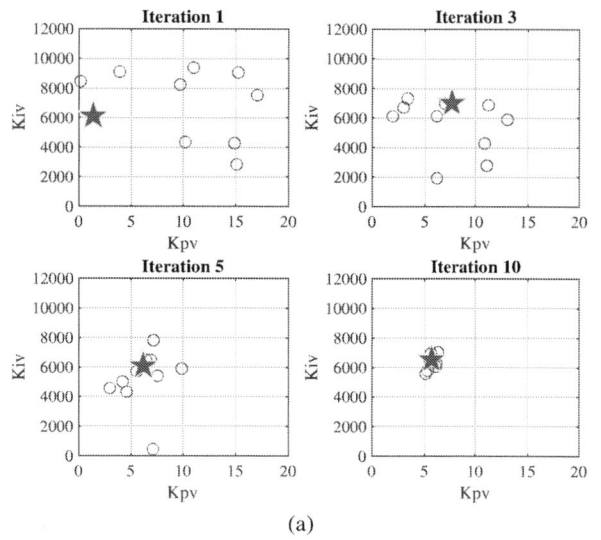
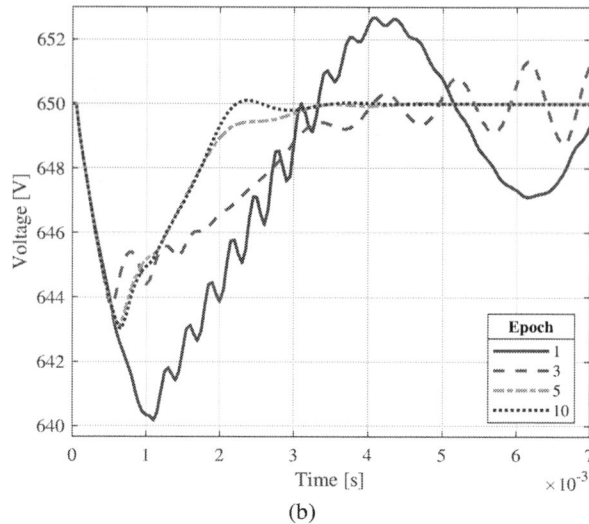

Fig. 8. Comparison of (a) particle convergence and (b) voltage transient improvement with increased iterations.

B. Hardware Results

The control portion of the switching model used to obtain high fidelity simulation results was compiled for experimental use on Imperix PEB8038 modules. Experimental results were obtained for both the optimized and unoptimized case using the hardware shown in Fig. 7. As shown in Fig. 9, the optimized controller improves the voltage settling time by 51 % and 55 % for the step up and step down, respectively, in load current. In addition, the voltage overshoot was decreased by 45 % and 49 %.

V. MULTI-CONVERTER TUNING

A. Modeling

To optimize a more realistic case for a general DC microgrid, or the shipboard power system described in Section I, it is necessary to consider systems with paralleled converters, as shown in Fig. 10. The multi-converter problem is complicated by the additional dynamics created through the coupling of

TABLE III
CONVERTER PARAMETER VALUES

Parameter	Single Converter	Parameter	Multi-Converter
V_g	250 V	$V_{g1,2}$	250 V , 200 V
V_{out}^*	650 V	V_{out}^*	650 V
L	1.5 mH	$L_{1,2}$	1.5 mH
C	500 µF	C_1	1000 µF
R_{ESR}	21 mΩ	R_{ESR}	10 mΩ
R_L	20 mΩ	$R_{L,1,2}$	20 mΩ
R_{ds}	21 mΩ	$R_{ds,1,2}$	21 mΩ

output voltage. This added complexity demands a model of the full system for successful optimization, rather than optimizing each converter subsystem individually. In addition to added model complexity, the search space of controller gains increases in dimension; for each additional double loop PI, 4 dimensions are added to the search space, increasing the minimum number of particles and iterations to achieve the desired response through SBO. To command power sharing for paralleled converters, droop control is employed for each converter, as shown in Fig. 4. The droop resistance, or impedance in some cases, is an additional gain that can be tuned, as it has some effect on the voltage response; however, for this work, the droop gains are kept constant through the optimization.

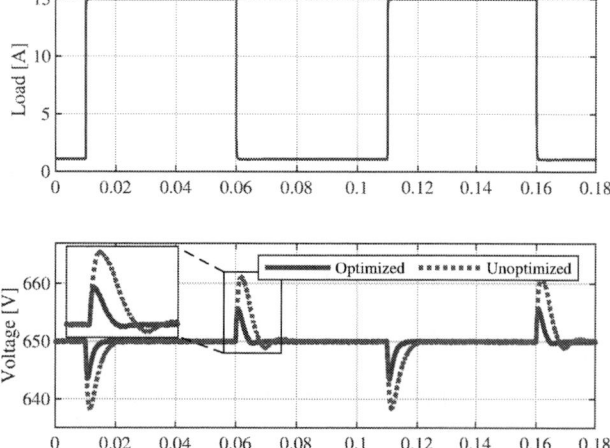

Fig. 9. Experimental results for single converter

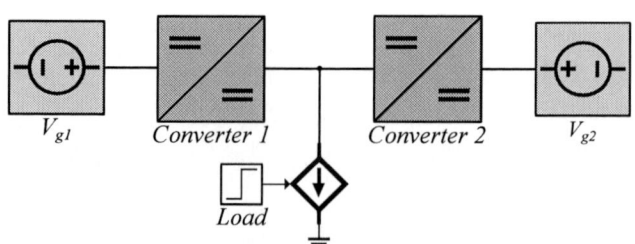

Fig. 10. Droop controlled multi converter system

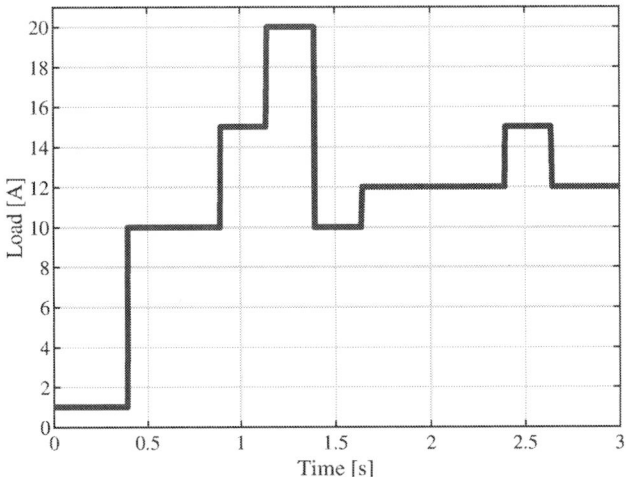

Fig. 11. Load profile for multi-converter system optimization

TABLE IV
PARAMETER VALUES FOR PSO, CSA, AND WOA

PSO		CSA		WOA	
n	100	n	100	n	100
itr_{max}	20	itr_{max}	20	itr_{max}	20
ϵ	1.4e-5	ϵ	1.4e-5	ϵ	1.4e-5
C_1	1.49	flmax	1	b_{min}	0.1
C_2	1.49	flmin	0.4	Amax	2
W_{decay}	15	APmax	0.75	Amin	0
χ	1	APmin	0		

B. Methods

The state space averaged model used for the paralleled converters follows the following state equations:

$$\frac{di_{L1}}{dt} = \frac{1}{L_1}\left[v_{g1} - r_1 i_{L1} - (1-d_1)(v_c + R_{esr}C\frac{dv_c}{dt})\right], \quad (6)$$

$$\frac{di_{L2}}{dt} = \frac{1}{L_2}\left[v_{g2} - r_2 i_{L2} - (1-d_2)(v_c + R_{esr}C\frac{dv_c}{dt})\right], \quad (7)$$

and

$$\frac{dv_c}{dt} = \frac{1}{C}\left[(1-d_1)i_{L1} + (1-d_2)i_{L2} - i_{\text{Load}}\right]. \quad (8)$$

The circuit parameters used are found in Table III. The droop resistance chosen for the system was $R_1 = 1\,\Omega$ and $R_2 = 2\,\Omega$. The resistances chosen effectively command converter one to supply $\frac{2}{3}$ of the load current and converter two to supply the complementary $\frac{1}{3}$.

The multi-converter optimizer was tested using the current load profile, arbitrarily chosen, shown in Fig. 11. PSO, CSA and WOA were all used to optimize the performance of the response; Table IV contains the hyper-parameters used for each algorithm. The "unoptimized" PI controllers were tuned for each step to have a current loop bandwidth of 1400Hz, a current loop phase margin of $75°$, a voltage loop bandwidth of 120Hz, and a voltage loop phase margin of $70°$. These less aggressive tuning requirements allowed the controllers to be tuned separately and still remain stable when paralleled.

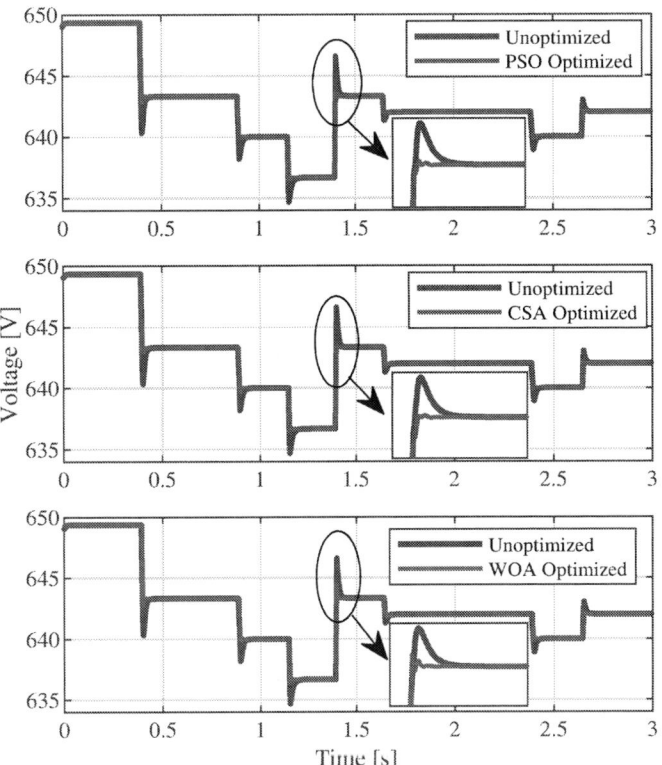

Fig. 12. Results for PSO, CSA, WOA

C. Simulated Results

Fig. 12 shows the SBO successfully decreased the cost function, voltage ITAE, for each step in load current, as compared to the unoptimized case. The different meta-heuristics did not find the same controller gains, as shown by their distinct transient responses, despite the global cost being similar for each. The three optimization algorithms improved the voltage response of the system by a number of metrics, as shown in Table V. The voltage overshoot is reduced for every step by at least 67.9 %. It should be noted that the voltage settling time was not reduced in some cases; the WOA increased the settling time of step 2 by more than 350 %. This discrepancy is to be expected, as the cost function worked to minimize only ITAE, and not necessary voltage settling time or overshoot.

VI. CONCLUSIONS

In this paper, a DT based approach to converter tuning is presented. The simulation based optimization, key to the approach, is preformed for both single and multi-converter systems operating under double loop PI control. Experimental results are obtained for the single converter system and simulation results are obtained for the multi-converter system.

Overall, the DT-enabled preemptive control improves system performance when compared with traditional methods. The computational requirements to run parallel simulations may make real time tuning unrealizable for more complex systems; however, the proposed method is still valid for offline tuning for known load profiles following degradation or a change in the system configuration.

TABLE V
OPTIMIZATION ALGORITHM PERFORMANCE COMPARISON

Metric	Algorithm	Step 1	Step 2	Step 3	Step 4	Step 5	Step 6	Step7	Elapsed Time (s)
Iterations per Load	WOA	6	3	3	11	3	3	4	544.856
	PSO	20	6	6	14	9	5	3	930.691
	CSA	20	20	8	20	8	2	9	1399.986
ITAE ($\times 10^{-5}$)	WOA	142.7	136.9	137.1	149.0	68.5	98.8	69.5	
	PSO	162.3	113.1	91.3	149.1	121.1	123.7	104.5	
	CSA	314.8	138.2	113.9	172.5	139.6	81.7	114.9	
VST Reduction (%)	WOA	95.05	-357.3	96.27	95.53	$\sim$0	$\sim$0	$\sim$0	
	PSO	94.01	3.42	95.84	94.90	$\sim$0	$\sim$0	$\sim$0	
	CSA	81.78	0.72	95.95	94.72	$\sim$0	$\sim$0	$\sim$0	
VOS Reduction (%)	WOA	85.04	92.70	95.37	89.91	84.06	84.77	92.66	
	PSO	86.88	86.47	90.82	94.68	67.93	88.32	82.56	
	CSA	77.09	95.18	82.54	68.25	75.58	81.87	91.45	

Thus far, the simulation based optimization functions well to minimize the cost function defined; however, future work is needed to make the necessary improvements to the SBO algorithms and hyperparameter tuning to allow the DM to specify ideal response characteristics. Other future works includes optimization of different control methods, such as sliding mode control and synergetic control using the framework presented in this work.

ACKNOWLEDGMENTS

This work was supported by the Office of Naval Research under contracts No. N00014-22-C-1003 and N00014-23-C-1012.

REFERENCES

[1] W. Zhu, C. Jin, and Z. Liang, "Hybrid modeling and simulation for shipboard power system considering high-power pulse loads integration," *Journal of Marine Science and Engineering*, vol. 10, no. 10, 2022. [Online]. Available: https://www.mdpi.com/2077-1312/10/10/1507

[2] Y. Zhang, F. Ji, X. Gao, F. Ma, and Q. Hu, "Optimal operation schedule strategy of high-power pulsed loads in shipboard power system," *Journal of Electrical Engineering & Technology*, vol. 19, no. 3, pp. 2089–2101, 2024. [Online]. Available: https://link.springer.com/article/10.1007/s42835-023-01706-6

[3] K. Sado, J. Hannum, E. Skinner, H. L. Ginn, and K. Booth, "Hierarchical digital twin of a naval power system," in *2023 IEEE Energy Conversion Congress and Exposition (ECCE)*, 2023, pp. 1514–1521.

[4] W. M. Polivka, P. R. K. Chetty, and R. D. Middlebrook, "State-space average modelling of converters with parasitics and storage-time modulation," in *1980 IEEE Power Electronics Specialists Conference*, 1980, pp. 119–143.

[5] H. Alrajhi, "A generalized state space average model for parallel dc-to-dc converters," *Computer Systems Science and Engineering*, vol. 41, no. 2, pp. 717–734, 2022. [Online]. Available: http://www.techscience.com/csse/v41n2/45201

[6] B. Li and K. Wada, "Parallelizing particle swarm optimization," pp. 288–291, 2005.

[7] A. Askarzadeh, "A novel metaheuristic method for solving constrained engineering optimization problems: Crow search algorithm," *Computers & Structures*, vol. 169, pp. 1–12, 2016. [Online]. Available: https://www.sciencedirect.com/science/article/pii/S0045794916300475

[8] S. Mirjalili and A. Lewis, "The whale optimization algorithm," *Advances in Engineering Software*, vol. 95, pp. 51–67, 2016. [Online]. Available: https://www.sciencedirect.com/science/article/pii/S0965997816300163

[9] A. G. Gad, "Particle swarm optimization algorithm and its applications: A systematic review," *Archives of Computational Methods in Engineering*, vol. 29, no. 5, pp. 2531–2561, 2022. [Online]. Available: https://doi.org/10.1007/s11831-021-09694-4

Modeling and Performance Characterization of Lithium-Ion Capacitor at Different Temperature and Voltage Values

Mohammad K. Al-Smadi, *Student Member, IEEE*, Jaber A. Abu Qahouq, *Senior Member, IEEE*, and Sajad Saberi

The University of Alabama
Department of Electrical and Computer Engineering
Tuscaloosa, Alabama 35487, USA

Abstract—This paper presents modeling and performance characterization analysis for lithium-ion capacitor (LiC) at different voltage and temperature values. The presented analysis includes electrochemical impedance spectroscopy (EIS), an equivalent circuit model (ECM), and DC performance in terms of equivalent series resistance (ESR) and discharge capacity. Performance characterization at different temperature and voltage values helps in understanding the behavior and performance of LiC, setting operation limits, design, and achieving LiC cell management functions.

Keywords—Lithium-ion capacitor, hybrid capacitor, electrochemical impedance spectroscopy, capacity, energy storage, equivalent series resistance (ESR), pulse current test, equivalent circuit model (ECM).

I. INTRODUCTION

Energy storage has become essential in the progress of transportation electrification and renewable energy integration. Various energy storage devices have been introduced and investigated such as lithium-ion batteries, fuel cells, and lithium-ion capacitors (LiC). LiC has a higher energy density compared to super capacitors and a lower energy density compared to lithium-ion batteries [1-3]. Fig. 1 illustrates the relative differences in energy and power densities between different energy storage devices. LiC are proposed in applications such as renewable energy (e.g., power smoothing) [1, 4], railway transportation [5], and electric vehicles [2, 3, 6]. In general, LiC are well-suited for applications where short-term high-power supply is needed. For example, a system which includes a combination of lithium-ion batteries and LiC can be utilized as shown in Fig. 2 [7]. In such a system, LiC and battery packs supply different shares of power/current by controlling their own DC-DC converters. The LiC can supply short-term high current demand instead of drawing high current from the battery pack which could affect the battery pack lifetime.

Various studies have been reported in the literature for LiC including modeling [8-12], design [13, 14], and assessment [2]. This paper presents modeling and performance characterization for LiC at different voltage and temperature values. The characterization analysis includes electrochemical impedance spectroscopy (EIS), an equivalent circuit model (ECM), equivalent series resistance (ESR), and discharge capacity.

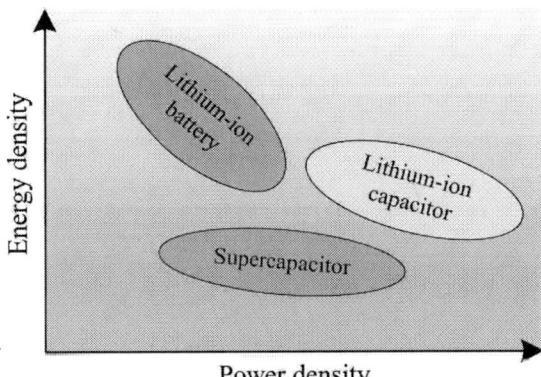

Fig. 1: Ragone plot for different energy storage devices.

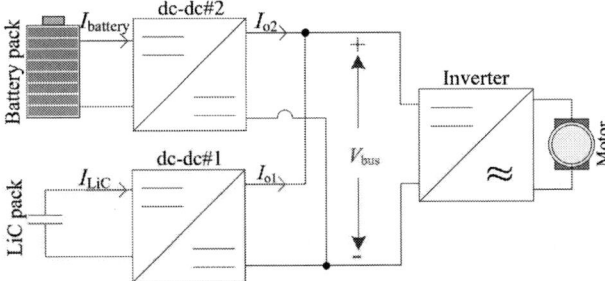

Fig. 2: Example on utilizing LiC in battery-powered applications.

II. ELECTROCHEMICAL IMPEDANCE SPECTROSCOPY

EIS is utilized for characterization and state estimation (e.gs., state of charge "SOC" and state of health "SOH") for various energy storage devices [15]. Also, information obtained from EIS curves can be utilized in LiC management functions. Potentiostatic and galvanostatic modes are usually utilized to obtain EIS data [16]. Potentiostatic EIS mode refers to the process of applying an AC voltage waveform (voltage perturbation) with a magnitude of $v_{\text{ac-fp}}$ where f_{p} is the perturbation frequency. Then, the AC current response $i_{\text{ac-fp}}$ due to the voltage perturbation is observed. In galvanostatic mode, an AC current waveform (current perturbation) is applied where the AC voltage response is observed. In both modes, the impedance is calculated at each perturbation frequency as expressed in (1) where Z_{fp} is the impedance at the perturbation

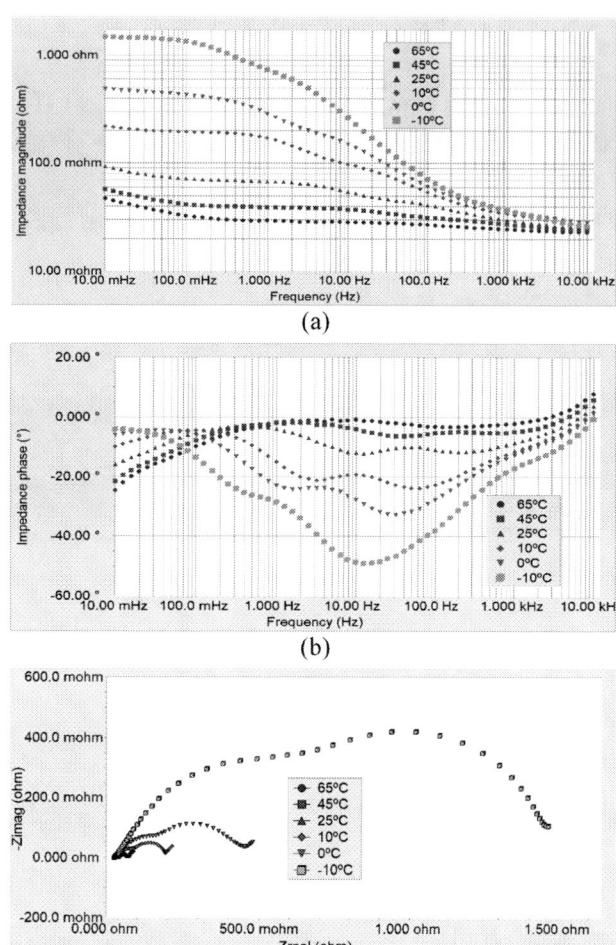

Fig. 3: EIS plots for LiC at 3.8V. (a) Impedance magnitude. (b) Impedance phase. (c) Nyquist plot.

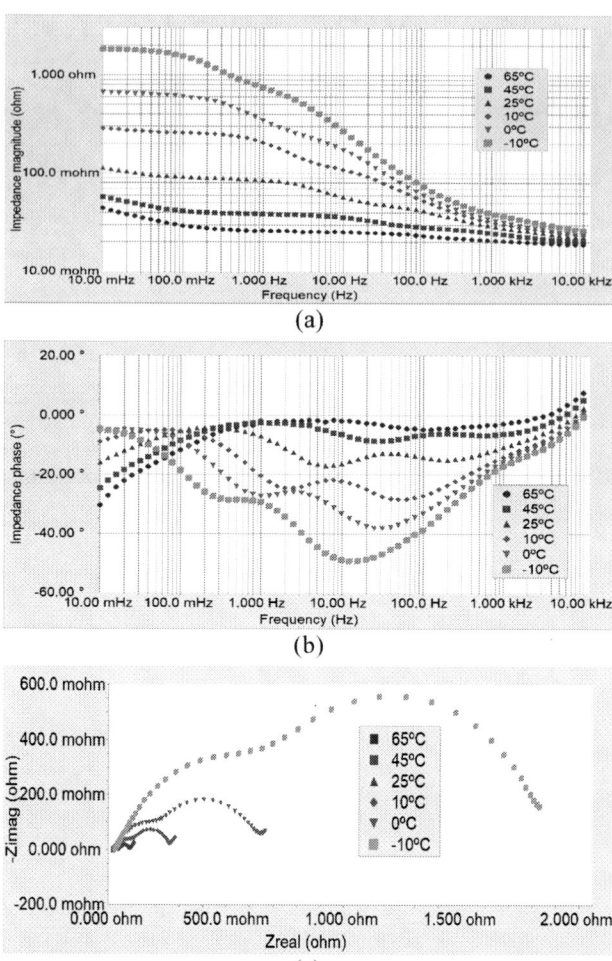

Fig. 4: EIS plots for LiC at 3.4V. (a) Impedance magnitude. (b) Impedance phase. (c) Nyquist plot.

frequency f_p, and $\theta_{z\text{-}fp}$ is the phase difference between $v_{ac\text{-}fp}$ and $i_{ac\text{-}fp}$. The process of perturbing the voltage/current and observing the current/voltage is repeated over a range of frequencies to get the impedance spectrum.

$$Z_{fp} = \frac{v_{ac\text{-}fp}}{i_{ac\text{-}fp}} e^{j\theta_{z\text{-}fp}} \qquad (1)$$

In this paper, the EIS data is collected for a 450 F-LiC cell (capacitance: 450 F, maximum voltage: 4.2 V, and minimum voltage: 2.2 V [17]) under galvanostatic mode. The EIS data is collected at different voltage values of 3.8 V, 3.4 V, 3.0 V, and 2.6 V. At each voltage value, the EIS data is collected at different temperature values of 65 °C, 45 °C, 25 °C, 10 °C, 0 °C, and -10 °C. The EIS was collected from 10kHz to 0.01 Hz using Gamry Instruments Interface 5000E™ [18]. An environmental chamber [19] was utilized to control the temperature. EIS curves (impedance magnitude, impedance phase, and Nyquist plot) at different voltage and temperature values are shown in Fig. 3 through Fig. 6. From these plots, the following can be observed:

- At the same voltage value, the lower the temperature, the higher the impedance of the LiC, especially in the low frequency range. At low frequency, the LiC impedance is

represented by the slow diffusion process (Warburg impedance W and the resistance R_D shown later in Fig. 8). The smaller the impedance, the faster the diffusion process. The impedance magnitude within the low frequency range can be utilized for state estimation (e.g., state of charge "SOC") due to the large differences between the impedance magnitude curves at different temperature values.

- From Nyquist plots, it can be noticed that at the same voltage value, the radius of the semi-circle is inversely proportional to the temperature. The semi-circle represents the charge transfer dynamics of the LiC (the impedance modeled by the parallel RC branches as shown later in Fig. 8). At low temperature, the movement of ions is slower, due to low kinetic energy [20], which indicates a large impedance that is represented by a large semi-circle. The Nyquist plot at 65°C is significantly smaller than other plots, and thus overlaid by other plots.

- The EIS impedance plots (magnitude and phase) are almost identical at high frequency. This is because the impedance becomes more inductive, i.e., the impedance represents the cables and current collectors and is modeled by L_{HF}.

979-8-3315-1612-3/25 $31.00 © 2025 IEEE   2841

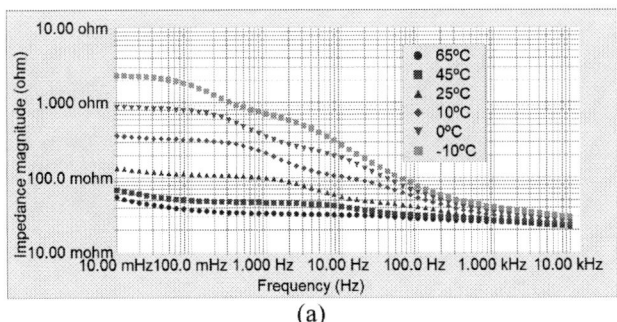

(a)

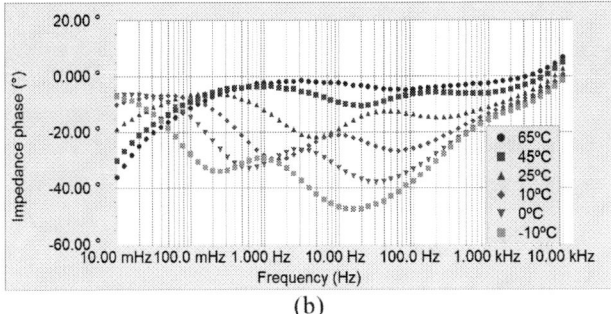

(b)

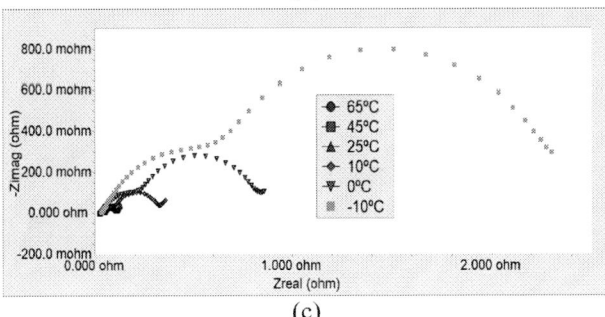

(c)

Fig. 5. EIS plots at 3.0V. (a) Impedance magnitude. (b) Impedance phase. (c) Nyquist plot.

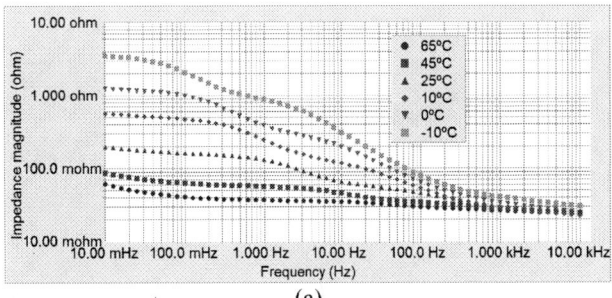

(a)

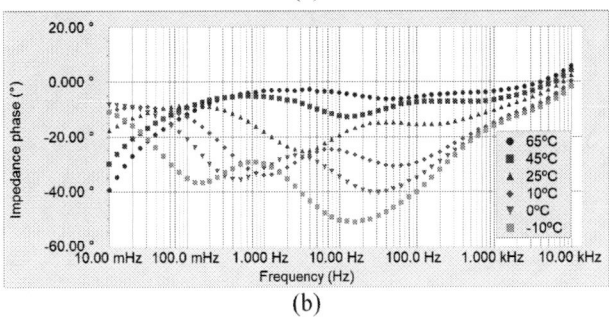

(b)

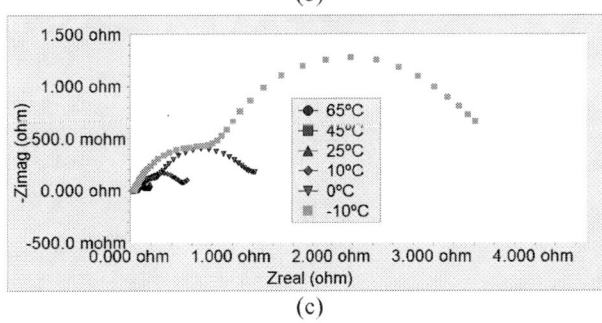

(c)

Fig. 6. EIS plots at 2.6V. (a) Impedance magnitude. (b) Impedance phase. (c) Nyquist plot.

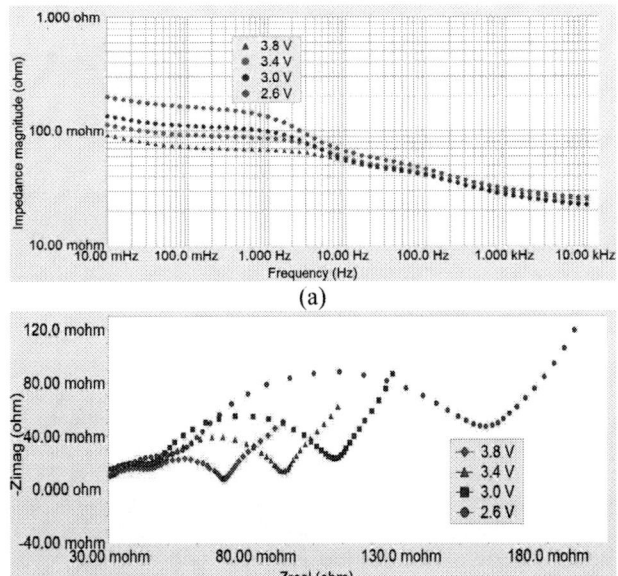

Fig. 7. EIS plots 25 °C. (a) Impedance magnitude. (b) Nyquist plot.

By comparing the EIS curves at different voltage values (same temperature value) similar insights can be obtained. The impedance magnitude and Nyquist plots are shown at 25 °C and different voltage values in Fig. 7. The impedance magnitude increases as the voltage of the LiC decreases, especially in the low frequency range which is because lower voltage value results in lower energy at the anode which results in a slow diffusion [10]. Also, the radius of the semi-circle in the Nyquist plot becomes larger at lower voltage values which indicates high charge transfer impedance. In addition, at high frequency range, the impedance magnitude curves, and phase curves are almost identical due to the inductive effects.

III. EQUIVALENT CIRCUIT MODEL

The change in the EIS curves at different temperature and voltage values can be described through the parameters of the equivalent circuit model (ECM). Fig. 8 shows the LiC ECM considered in this paper. The electrolyte resistance is represented by R_{Con} whose effect appears in the medium frequency range. The cable and connections are modeled by the inductive part L_{HF}. The parallel RC branches model the charge

979-8-3315-1612-3/25 $31.00 © 2025 IEEE

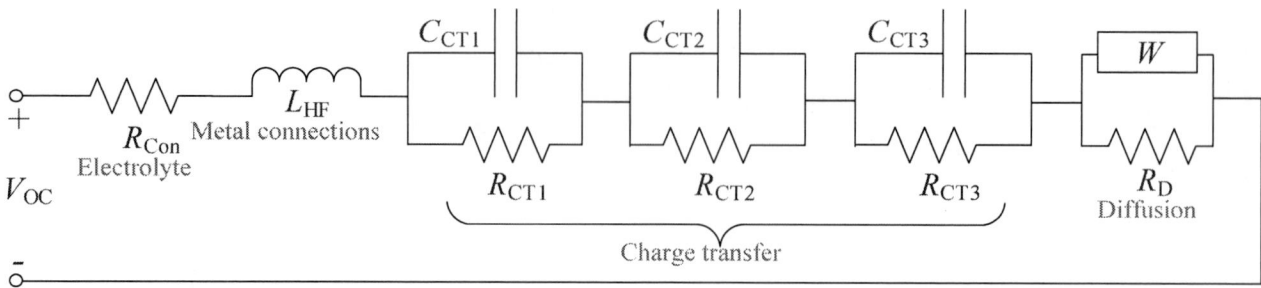

Fig. 8: Equivalent circuit model for LiC.

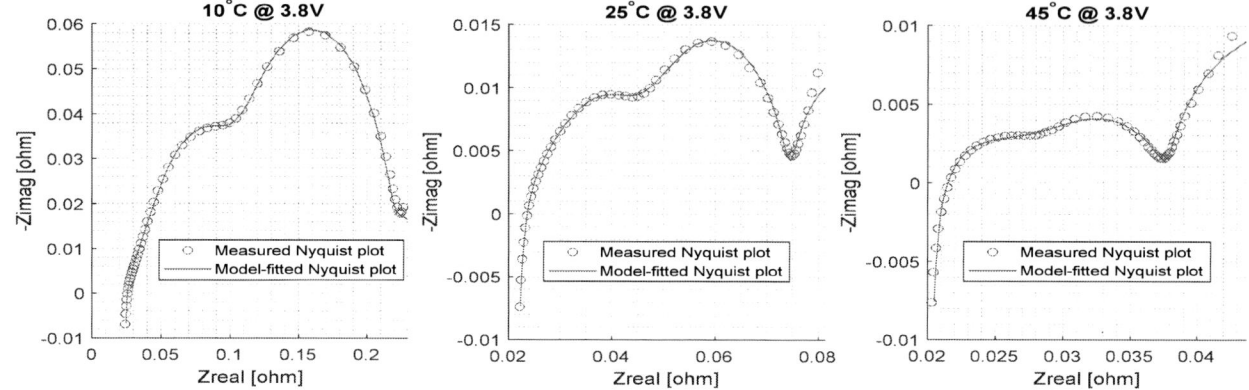

Fig. 9: Measured and model-fitted impedance Nyquist plots at 3.8 V.

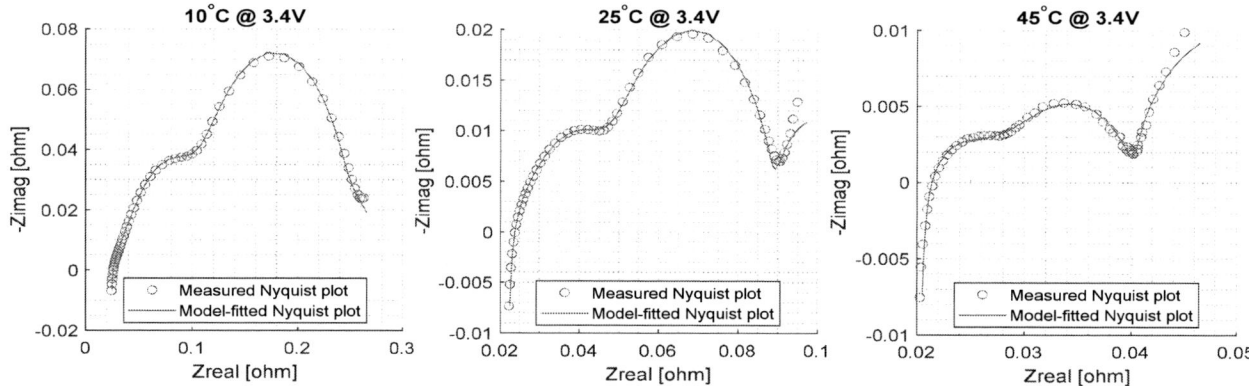

Fig. 10: Measured and model-fitted impedance Nyquist plots at 3.4 V.

transfer or polarization impedance within the low and medium frequency ranges (R_{CT1} in parallel with C_{CT1}, R_{CT2} in parallel with C_{CT2}, and R_{CT3} in parallel with C_{CT3}). The very low frequency range is modeled by the diffusion process and is represented by the parallel connection of Warburg impedance W and the resistance R_D. The open circuit voltage of the LiC is represented by V_{OC}. To fit the EIS curves, the simplex algorithm (used to solve linear programming problems) is utilized [21]. Fig. 9 and Fig. 10 show sample different impedance Nyquist plots against their model-fitted plots for the LiC at 3.8 V and 3.4 V under different temperature values. As shown, the model-fitted Nyquist plots are in good agreement with the measured plots. Model fitting can be utilized to obtain values for the ECM parameters as summarized in Table I.

Table I: ECM parameters obtained from model fitting.

Parameter	3.8 V			3.4 V		
	10°C	25°C	45°C	10 °C	25 °C	45 °C
R_{Con} (Ω)	0.023	0.02	0.01	0.023	0.021	0.020
L_{HF} (μH)	0.066	0.06	0.07	0.066	0.069	0.064
R_{CT1} (Ω)	0.033	0.00	0.00	0.031	0.005	0.003
C_{CT1} (F)	0.183	0.23	1.61	0.190	0.223	0.035
R_{CT2} (Ω)	0.036	0.02	0.02	0.028	0.033	0.006
C_{CT2} (F)	181	239	308	107	1.30	1.80
R_{CT3} (Ω)	0.097	0.02	0.00	0.122	0.029	0.025
C_{CT3} (F)	1.199	1.49	435	1.190	206.0	255.2
W (ΩS$^{-0.5}$)	1.062	1.23	1.30	1.066	1.208	3.616
R_D (Ω)	0.072	0.03	0.01	0.078	0.031	0.012

979-8-3315-1612-3/25 $31.00 © 2025 IEEE

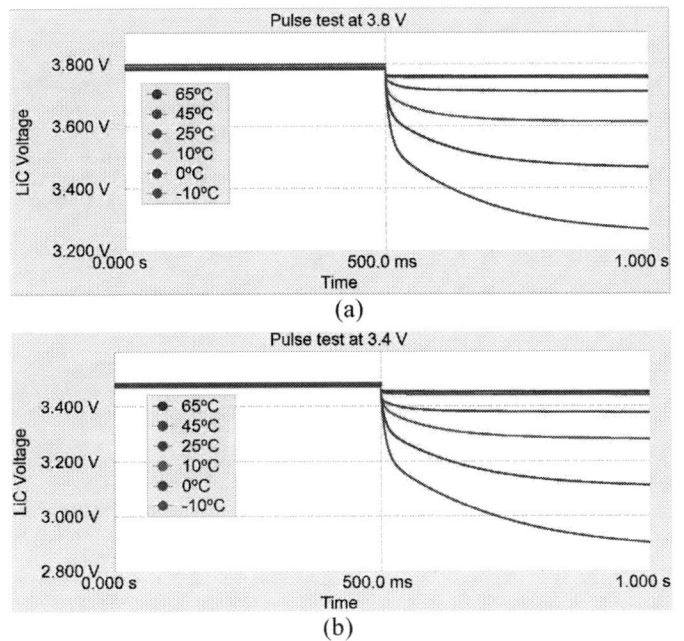

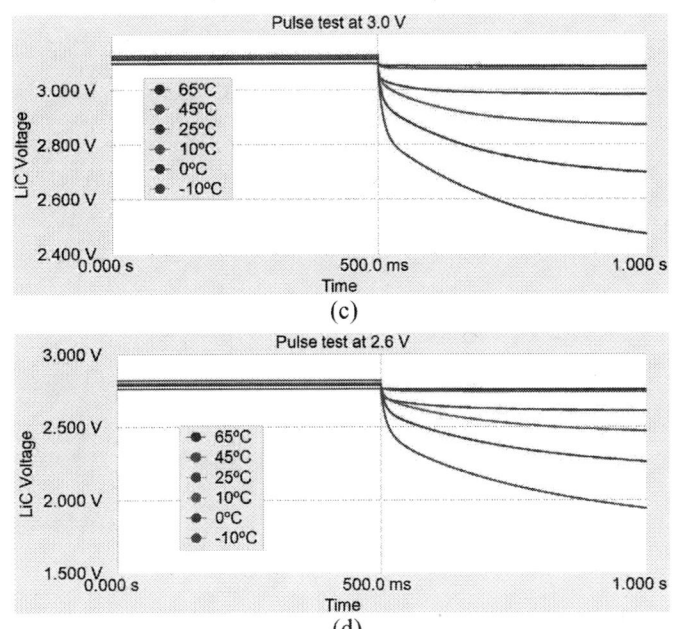

Fig. 11: Pulse tests at (a) 3.8V, (b) 3.4 V, (c) 3.0V, (d) 2.6 V.

IV. PULSE CURRENT RESPONSE AND EQUIVALENT SERIES RESISTANCE (ESR)

Pulse current tests are utilized to characterize the ESR at different conditions (voltage and temperature in this paper) which gives information about the power capability of the LiC. By applying a fixed load current pulse and observing the voltage response of the LiC, the ESR can be calculated [22]. Pulse current tests were carried out at different voltage values (3.8 V, 3.4 V, 3.0 V, and 2.6 V). At each voltage, the pulse current test is carried out at different temperature values (65 °C, 45 °C, 25 °C, 10 °C, 0 °C, and -10 °C). The magnitude of the current pulse was set to be 0.5 A. The voltage response waveforms obtained from the current pulse tests are shown in Fig. 11. From the current pulse tests, the ESR can be calculated by dividing the voltage difference ΔV by the current pulse magnitude I_{pulse} in the linear region of the voltage response curve [19] as expressed in (2). It should be mentioned that the LiC voltage in Fig. 11 appears to be at a value that is not exactly equal to the testing point (for example, in Fig. 11 (d), the LiC voltage appears to be around 2.8 V instead of 2.6 V) which is due to the relaxation in the LiC voltage after discharging/charging to the voltage value of interest. From Fig. 11, it can be noticed that the higher the temperature of the LiC, the smaller the voltage change is due to the pulse current.

$$ESR = \frac{\Delta V}{I_{pulse}} \qquad (2)$$

The curves in Fig. 11 are utilized to calculate the ESR values at different voltage and temperature values using (2). The calculated ESR values are plotted at each voltage with different temperature values in Fig. 12. As shown, at the same voltage value, the ESR increases as the LiC temperature decreases which means less power capability at lower temperature values.

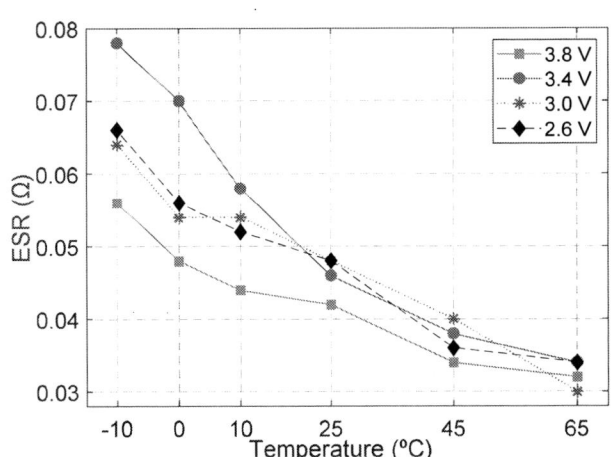

Fig. 12: ESR at different voltage and temperature values.

V. DISCHARGE CAPACITY

LiC capacity refers to the amount of charges that can be supplied by the LiC cell before reaching the minimum voltage. To evaluate the LiC capacity, the LiC is discharged from the maximum voltage (3.8 V) to the minimum voltage (2.2 V). The LiC discharge is performed at different temperature values to study the effect of high/low temperatures on the LiC capacity. Fig. 13 shows the discharging curves for the LiC at different temperature values. The LiC is discharged at 0.5A. As shown, the lower the temperature, the lower the capacity. Since the electrode resistance is higher at low temperature, the voltage drop is higher which justifies the reduction in the LiC capacity at low temperature values. Fig. 14 illustrates the variation of the LiC capacity with temperature.

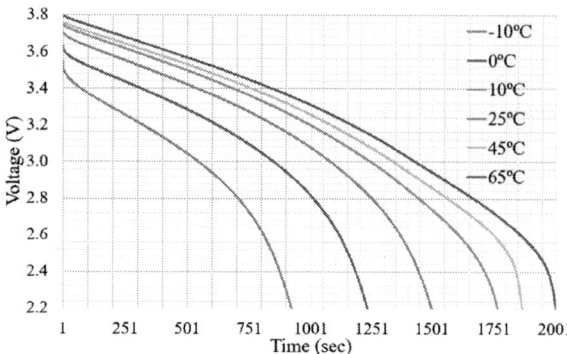

Fig. 13: Discharge curves at different temperature values (discharge current = 0.5 A).

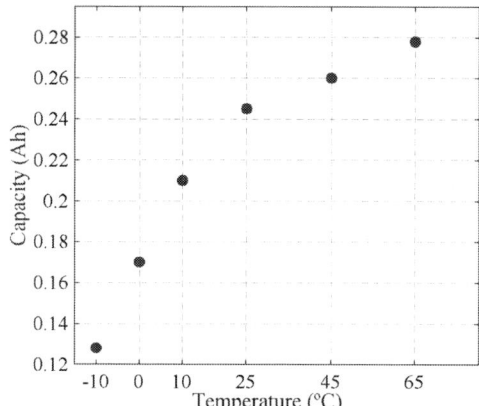

Fig. 14: LiC Discharge capacity at different temperature values.

VI. CONCLUSION

This paper presents modeling and performance characterization for lithium-ion capacitor at different voltage and temperature values. Impedance curves at different voltage and temperature values are presented and discussed. An equivalent circuit model is presented. Also, the equivalent series resistance at different voltage and temperature values is studied in addition to the discharge capacity. It was shown that the impedance magnitude is higher at lower temperature and voltage values. Also, the power capability and discharge capacity of the LiC decrease at low temperature values.

ACKNOWLEDGEMENT

This material is based upon work supported in part by the National Science Foundation under Grant No. 2213918. Any opinions, findings and conclusions or recommendations expressed in this material are those of the author(s) and do not necessarily reflect the views of the National Science Foundation.

REFERENCES

[1] S. A. Hamidi, E. Manla, and A. Nasiri, "Li-ion batteries and Li-ion ultracapacitors: Characteristics, modeling and grid applications," in *2015 IEEE Energy Conversion Congress and Exposition (ECCE)*, 20-24 Sept. 2015 2015, pp. 4973-4979, doi: 10.1109/ECCE.2015.7310361.

[2] N. Omar *et al.*, "Assessment of lithium-ion capacitor for using in battery electric vehicle and hybrid electric vehicle applications," *Electrochimica Acta*, vol. 86, pp. 305-315, 2012/12/30/ 2012.

[3] M. Soltani *et al.*, "Hybrid Battery/Lithium-Ion Capacitor Energy Storage System for a Pure Electric Bus for an Urban Transportation Application," *Applied Sciences*, vol. 8, no. 7, doi: 10.3390/app8071176.

[4] K. Koyanagi *et al.*, "A Smart Photovoltaic Generation System Integrated with Lithium-ion Capacitor Storage," in *2011 46th International Universities' Power Engineering Conference (UPEC)*, 5-8 Sept. 2011 2011, pp. 1-6.

[5] F. Ciccarelli, A. D. Pizzo, and D. Iannuzzi, "Improvement of Energy Efficiency in Light Railway Vehicles Based on Power Management Control of Wayside Lithium-Ion Capacitor Storage," *IEEE Transactions on Power Electronics*, vol. 29, no. 1, pp. 275-286, 2014, doi: 10.1109/TPEL.2013.2253492.

[6] J. Cao and A. Emadi, "A New Battery/UltraCapacitor Hybrid Energy Storage System for Electric, Hybrid, and Plug-In Hybrid Electric Vehicles," *IEEE Transactions on Power Electronics*, vol. 27, no. 1, pp. 122-132, 2012, doi: 10.1109/TPEL.2011.2151206.

[7] A. Di Napoli, F. Crescimbini, F. G. Capponi, and L. Solero, "Control strategy for multiple input DC-DC power converters devoted to hybrid vehicle propulsion systems," in *Industrial electronics, 2002. isie 2002. proceedings of the 2002 ieee international symposium on*, 2002, vol. 3: IEEE, pp. 1036-1041.

[8] M. K. Al-Smadi and J. A. Abu Qahouq, "Parameter Variations of Equivalent Circuit Model of Lithium-ion Capacitor," in *2023 IEEE Energy Conversion Congress and Exposition (ECCE)*, 2023: IEEE, pp. 454-458.

[9] S. Barcellona and L. Piegari, "A lithium-ion capacitor model working on a wide temperature range," *Journal of Power Sources*, vol. 342, pp. 241-251, 2017.

[10] Y. Firouz, N. Omar, J.-M. Timmermans, P. Van Den Bossche, and J. Van Mierlo, "Lithium-ion capacitor–Characterization and development of new electrical model," *Energy*, vol. 83, pp. 597-613, 2015.

[11] D. Moye, P. Moss, X. Chen, W. Cao, and S. Foo, "A design-based predictive model for lithium-ion capacitors," *Journal of Power Sources*, vol. 435, p. 226694, 2019.

[12] S. Song *et al.*, "Equivalent circuit models and parameter identification methods for lithium-ion capacitors," *Journal of Energy Storage*, vol. 24, p. 100762, 2019.

[13] A. Jagadale, X. Zhou, R. Xiong, D. P. Dubal, J. Xu, and S. Yang, "Lithium ion capacitors (LICs): Development of the materials," *Energy Storage Materials*, vol. 19, pp. 314-329, 2019/05/01/ 2019, doi: https://doi.org/10.1016/j.ensm.2019.02.031.

[14] J. J. Lamb and O. S. Burheim, "Lithium-Ion Capacitors: A Review of Design and Active Materials," *Energies*, vol. 14, no. 4, p. 979, 20+21. [Online]. Available: https://www.mdpi.com/1996-1073/14/4/979.

[15] J. A. Abu Qahouq, "An Electrochemical Impedance Spectrum-Based State of Health Differential Indicator with Reduced Sensitivity to Measurement Errors for Lithium–Ion Batteries," *Batteries*, vol. 10, no. 10, p. 368, 2024, doi: https://doi.org/10.3390/batteries10100368

[16] S. Wang, J. Zhang, O. Gharbi, V. Vivier, M. Gao, and M. E. Orazem, "Electrochemical impedance spectroscopy," *Nature Reviews Methods Primers*, vol. 1, no. 1, p. 41, 2021.

[17] Tecate 450 F (EDLC) Supercapacitor, 2022. Accessed: Aug. 19, 2022. [Online]. Available: https://www.tecategroup.com/products/data_sheet.php?t=SERIES&i=TPLC

[18] The Interface 5000E™ potentiostat, Gamry Instruments. [Online]. Available: https://www.gamry.com/potentiostats/interface-5000epotentiostat/.

[19] TestEquity, "Model 101H Temperature/Humidity Chamber," Aug. 2022. Accessed: June.20, 2023. [Online]. Available: https://d3fnwqmod42ein.cloudfront.net/userfiles/documents/pdfs/101h_manual.pdf

[20] Y. Firouz, N. Omar, J. M. Timmermans, P. Van den Bossche, and J. Van Mierlo, "Lithium-ion capacitor – Characterization and development of new electrical model," *Energy*, vol. 83, pp. 597-613, 2015/04/01/ 2015.

[21] Y.-S. Kim, "Refined simplex method for data fitting," in *Astronomical Data Analysis Software and Systems VI*, 1997, vol. 125, p. 206.

[22] M. K. Al-Smadi and J. A. Abu Qahouq, "Evaluation of current-mode controller for active battery cells balancing with peak efficiency operation," *IEEE Transactions on Power Electronics*, vol. 38, no. 2, pp. 1610-1621, 2022, doi: 10.1109/TPEL.2022.3211905

Conveniently Identify Coils in Inductive Power Transfer System Using Machine Learning

Yifan Zhao[1], Mowei Lu[1], Ting Chen[2], Heyuan Li[3], Xiang Gao[3], Zhenbin Zhang[2], Minfan Fu[3] and Stefan M. Goetz[1*]

[1] *Department of Engineering, University of Cambridge, Cambridge, United Kingdom.*
[2] *School of Electrical Engineering, Shandong University, Jinan, China.*
[3] *School of Information Science and Technology, ShanghaiTech University, Shanghai, China*
Email: smg84@cam.ac.uk

Abstract—High-frequency inductive power transfer (IPT) has garnered significant attention in recent years due to its long transmission distance and high efficiency. The inductance values (*L*) and quality factors (*Q*) of the transmitting and receiving coils greatly influence the system's operation. Traditional methods involved impedance analyzers or network analyzers for measurement, which required bulky and costly equipment. Moreover, disassembling it for re-measurement is impractical once the product is packaged. Alternatively, simulation software such as HYSS can serve for the identification. Nevertheless, in the case of very high frequencies, the simulation process consumes a significant amount of time due to the skin and proximity effects. More importantly, obtaining parameters through simulation software becomes impractical when the coil design is more complex. This paper firstly employs a machine learning approach for the identification task. We simply input images of the coils and operating frequency into a well-trained model. This method enables rapid identification of the coil's *L* and *Q* values anytime and anywhere, without the need for expensive machinery or coil disassembly.

Index Terms—Very-High-frequency IPT, machine learning, parameter identification

I. INTRODUCTION

In recent years, IPT has garnered widespread attention due to its ability to overcome the limitations of physical connections. It offers a promising solution for future charging techniques [1]–[10]. By enabling the wireless transfer of electrical energy, there is a consensus that operating them at very-high frequencies can significantly enhance the transmission distance, reduce system size, and improve overall performance. Increasing the operating frequency of IPT systems has therefore become a widely recognized objective.

The performance of IPT systems is heavily influenced by the *L* and *Q* of the transmitting and receiving coils used, especially in high-frequency cases. These parameters are critical for the efficient operation of the system, i.e., System efficiency and ZVS [11]–[25]. Also other fields of electronics and power require coils with often intricate design [26]–[32]. However, measuring these parameters with traditional methods such as impedance or network analyzers can be challenging, as they require expensive and often bulky equipment. Additionally, it is not feasible to disassemble the system for re-measurement once it is manufactured and sealed. Another approach to measure the parameter is to conduct a simulation in software, e.g., HYSS. But in high-frequency cases and when the coil is complex, the simulation will take long time. Therefore, this method becomes impractical.

To solve these issues, this work first proposes a novel approach using machine learning [33]–[35] to identify the *L* and *Q* of coils. By only inputting images of the coils along with their operating frequency into a trained model, we can quickly and accurately determine the necessary parameters without the need for complex and costly measurement devices. This method is portable, easy to use and does not require the coil to be disassembled. Therefore providing a practical solution for measuring coil parameters in various applications. The model uses a convolutional neural network (CNN) architecture which we trained on a diverse dataset of coils. This dataset includes coils with and without a ferromagnetic core, excitation wires of different thicknesses, and coils of different shapes. Such a comprehensive dataset enables the model to adapt to a wide range of identification tasks. Experimental results have demonstrated that the established model has an identification error rate of only 21.6%.[1]

II. HIGH-FREQUENCY IPT SYSTEM

This section describes the basic topology of IPT system and then analyze the influence of *L* and *Q* on operation.

A. Basic Topology of the IPT System

Fig. 1 illustrates the typical topology of an IPT system. The AC current generated by the inverter is introduced into the primary compensation network. Then it traverses the transmitter coil (Ltx), which transfers energy to the receiver coil (Lrx). Subsequently, the energy is delivered to the load via the secondary compensation network.

B. L and Q Influence

An IPT system prototype operating at 6.78 MHz has been established in Fig. 2. The inverter, compensation network, and

---

This work was supported by National Natural Science Foundation of China under Grant 52477013 and Lingang Laboratory under Grant NO. LG-GG-202402-06-10.

[1]Please note that this work just proposes a novel identification frame using machine learning, the identification error will be reduced by increasing the data set.

979-8-3315-1612-3/25 $31.00 © 2025 IEEE

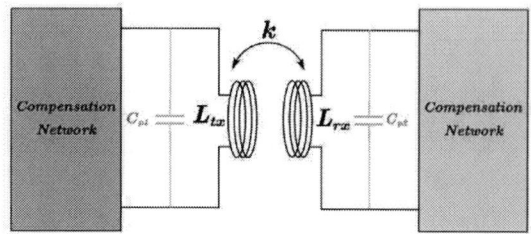

Fig. 1: Basic IPT system

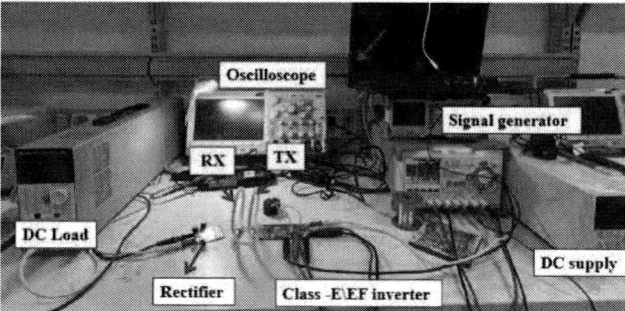

Fig. 2: Experimental setup

rectifier respectively use a Class-E inverter, series compensation, and a full bridge. The system's efficiency is measured by altering L and quality factor Q values of the coils. Fig. 3 presents the experimental results. It has demonstrated that as Q gradually increases, the efficiency also rises. Systems with different L values exhibit distinct efficiency curves, thereby confirming the significance of identifying the L and Q of coils in high-frequency IPT systems.

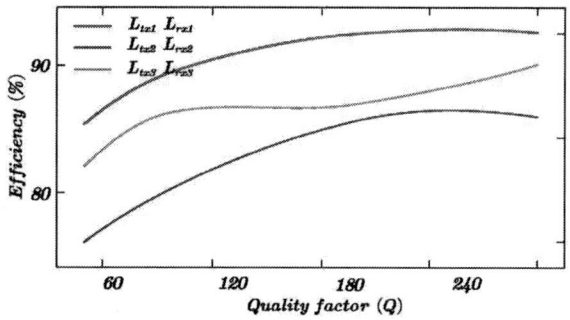

Fig. 3: Experimental efficiency

III. Identification System Based on CNN Model

A. CNN Model

CNNs use convolutional layers that compute only on local pixels and thereby capture local features of the image. Various features can be extracted through multiple convolutional kernels to perform sliding window operations on the input image. These convolutional kernels can detect local information within the image, such as edges, corners, textures, and other

distinctive patterns. The convolution formula can be expressed as

$$O(i, j) = \sum_{m=0}^{M-1} \sum_{n=0}^{N-1} I(i + m, j + n) \cdot K(m, n), \quad (1)$$

where $O(i, j)$ represents the output feature map at position (i, j), M and N are respectively the height and width of the convolutional kernel.

A convolutional layer is typically followed by an activation function, with ReLU being the most common one. ReLU aims to introduce nonlinearity, which allows the network to learn more complex patterns. The ReLU function sets negative input values to zero while it keeps positive values unchanged, which accelerates the training and convergence of the network.

The pooling layer is used to down-sample the feature maps produced by the convolutional layer, reduces computational complexity, and imparts a certain degree of invariance to the features (e.g., translation invariance). Common pooling methods include max pooling and average pooling. Taking max pooling as an example, the pooling layer selects the maximum value within a small window on the feature map. It forms a down-sampled feature map, which could be formulated as

$$O(i, j) = Max_{m=0}^{M-1} Max_{n=0}^{N-1} I(i + m, j + n), \quad (2)$$

$$O(i, j) = \frac{1}{M \times N} \sum_{m=0}^{M-1} \sum_{n=0}^{N-1} I(i + m, j + n), \quad (3)$$

where $M \times N$ is the size of the pooling window.

Fig. 4 illustrates our decoder module. To ensure the model's efficiency and effectiveness, we use two fully connected layers with respective dimensions of 8192 and 128. The features predicted by the CNN are fed into these fully connected layers, after which the L and Q parameters we aim to predict are directly output.

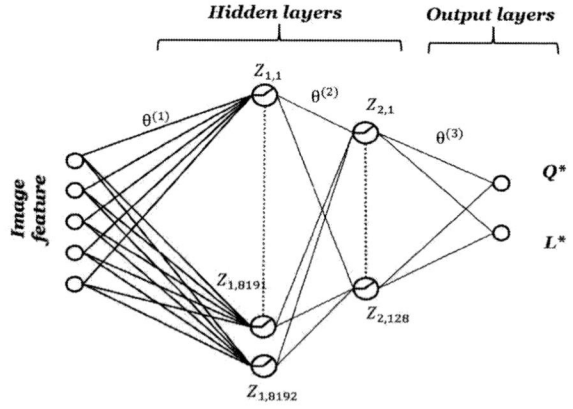

Fig. 4: CNN Network

B. Proposed Architecture

The proposed architecture is shown in Fig. 5. By performing convolution operations only on local parts of the image, CNNs are more adept at capturing detailed features such as diameter and coil turns in the coil parameter identification task and

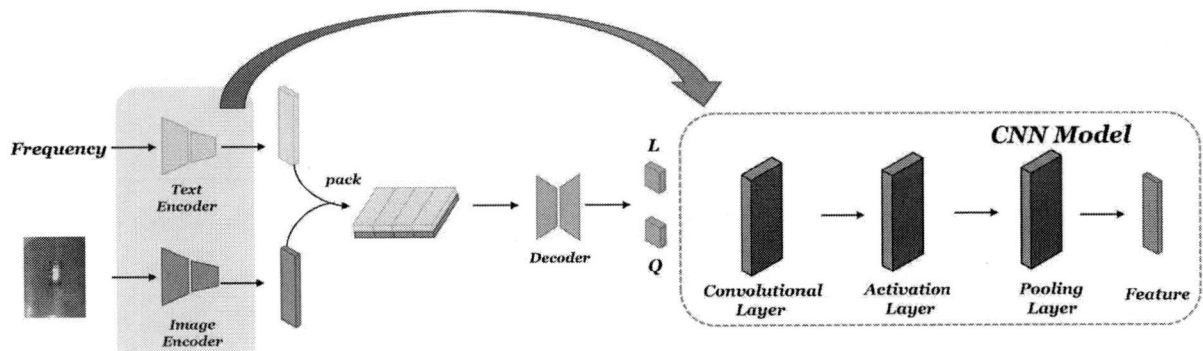

Fig. 5: Proposed architecture based ML

improve recognition accuracy. On the other hand, our design incorporates a multi-modal AI representation, as the L and Q values of the coil are related not only to the coil's parameters but also to the operating frequency of the circuit. Therefore, we simultaneously consider the effect of the operating frequency, as demonstrated in our model diagram. After obtaining the input frequency, we fuse it with the image features to create a multi-modal hybrid feature. Then it is processed by a decoder composed of two fully connected layers to output the resulting L and Q values. We use the commonly used mean squared error (MSE) as the loss function per

$$MSE = \frac{1}{2}\left[(Q_i - Q_i^*)^2 + (L_i - L_i^*)^2\right], \qquad (4)$$

where Q_i represents the identified value of the quality factor, Q_i^* the actual value of the quality factor, L_i the identified value of the inductance value, and L_i^* the actual value of the inductance variable.

C. Data Collection

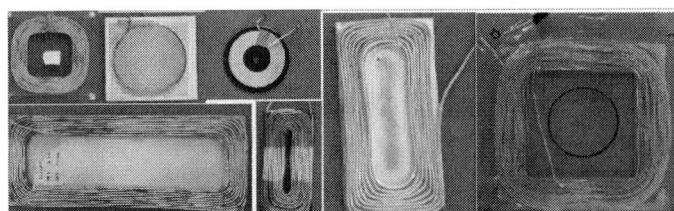

Fig. 6: Example images from the dataset

As shown in Fig. 6, our dataset is a comprehensive collection that encompasses 20 distinct sets of coils. This diverse assortment includes coils with and without a ferromagnetic core and thereby a broad spectrum of magnetic properties. Furthermore, the dataset incorporates coils with excitation wires of varying thicknesses, which allows the model to discern subtle differences in wire gauge that could impact performance. Additionally, it features coils of diverse shapes, such as circular, rectangular, and irregular forms, which further enhances the model's capability to generalize across different physical configurations.

The meticulously curated and sizable dataset enables the machine learning model to adapt effectively to a wide range of identification tasks, ensuring robustness and versatility in its applications. To gather comprehensive data, we collected a total of 100 sets of measurements with five different frequencies for each of these 20 sets of coils. This approach ensures that the model is exposed to a variety of operational conditions and frequencies. It fosters its ability to accurately identify and classify coils under diverse scenarios.

To streamline the processing of the original images and ensure consistency in analysis, we uniformly resized all twenty images to a standard resolution of 64×64 pixels. This normalization step maintains uniformity in the input data.

D. Evaluation metrics

To evaluate the output quality, we ues two different evaluation metrics, i.e., mean square error (MSE) and error, which are defined as

$$MSE = \frac{1}{n}\sum_{i=1}^{n}(y_i - \hat{y}_i)^2, \qquad (5)$$

$$error = \frac{1}{2}\left[(Q_i - Q_i^*)^2 + (L_i - L_i^*)^2\right]. \qquad (6)$$

IV. EXPERIMENTAL RESULTS

A. Model Training

Algorithm 1 Coil Parameter Identify

1: **Data:** $image,\ f,\ Q_{label},\ \mathcal{I}_{label}$
2: **Result:** $Q,\ \mathcal{I}$
3: Initialize $Q = 0,\ \mathcal{I} = 0$
4: **for** $p \in image$ and $N \in f$ **do**
5:      $Q \leftarrow \text{CNN}(image, f_s)$
6:      $\mathcal{I} \leftarrow \text{CNN}(image, f_s)$
7:      **Calculate MSE Loss:**
8:      $L_Q \leftarrow \text{MSELoss}(Q, Q_{label})$
9:      $L_{\mathcal{I}} \leftarrow \text{MSELoss}(\mathcal{I}, \mathcal{I}_{label})$
10:      **Update Parameters:**
11:      $\theta_Q \leftarrow \theta_Q - \eta\frac{\partial L_Q}{\partial \theta_Q}$
12:      $\theta_{\mathcal{I}} \leftarrow \theta_{\mathcal{I}} - \eta\frac{\partial L_{\mathcal{I}}}{\partial \theta_{\mathcal{I}}}$
13: **end for**

All experiments were implemented with the PyTorch framework on a single GPU (i.e, Nvidia GeForce RTX 4090, 24 GB). The model was trained on M = 16 images randomly selected from the 20 sets of coil images we obtained. Algorithm 1 details the method. The input to our algorithm consists of an image and a frequency, with the output being L and Q. First, the input passes through an encoder to extract features, The resulting features are then fed into the decoder, where we employ fully connected layers. The identified values are then compared with the target labels to compute the loss, which is used to update the model parameters and complete one epoch. The training loss we obtained is shown in Fig. 7. As

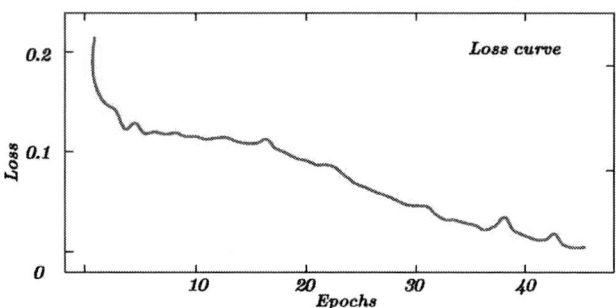

Fig. 7: Loss curve

can be seen, with the increase in the number of epochs, our loss steadily decreases. It has demonstrated the effectiveness of the proposed method.

B. Identification Based on the Proposed Model

Fig. 8 shows the experimental test coil, while Fig. 9 presents the experimental results, which indicate that at 85 kHz, both the MSE loss and error rate reached their lowest values of 0.011% and 21.6%, respectively. As the frequency increases, metrics remain at a commendably high standard. The current study only input twenty sets of coils as the training dataset; if the dataset be expanded, the model's accuracy is expected to significantly improve.

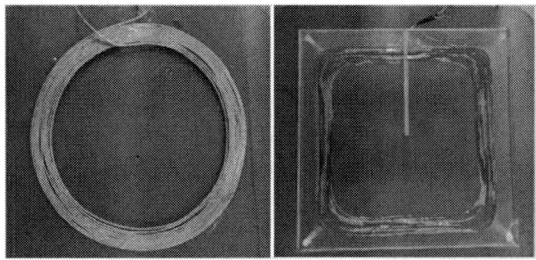

Fig. 8: Test coil

V. Conclusions

This paper first introduces a rapid coil parameter recognition technology based on machine learning. Traditional methods based on impedance analyzers and network analyzers are limited by equipment constraints. Moreover, once the entire

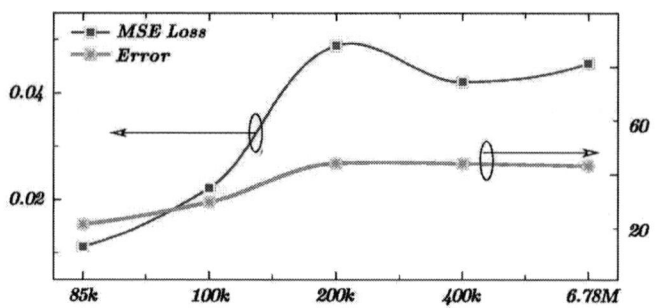

Fig. 9: Recognition results

system is packaged, it is impossible to disassemble it for re-identification. This paper simply recognizes parameters by photographing the coil. Then providing a new paradigm for coil recognition. Experiments have proven that the established model has an identification error rate of only 21.6%. With the addition of more datasets in the future, the accuracy of the model will be further increased.

References

[1] H. Wang, N. Tashakor, W. Jiang, W. Liu, C. Q. Jiang, and S. M. Goetz, "Hacking encrypted frequency-varying wireless power: Cyber-security of dynamic charging," *IEEE Transactions on Energy Conversion*, vol. 39, no. 3, pp. 1947–1957, 2024.

[2] S. Liu, Y. Wu, L. Zhou, R. Mai, Z. He, and S. M. Goetz, "A two-dimensional misalignment-tolerant ipt system based on three-arm voltage doubler rectifier," in *2022 IEEE Energy Conversion Congress and Exposition (ECCE)*, 2022, pp. 1–7.

[3] X. Tian, J. Zhang, H. Wang, and S. M. Goetz, "Design and analysis of automatic modulation and demodulation strategy in wireless power and drive transfer system," *IEEE Transactions on Industrial Electronics*, 2024.

[4] P. Zhao, J. Liang, H. Wang, and M. Fu, "Detuned lcc/ss compensation for stable-output inductive power transfer system under ultra-wide coupling variation," *IEEE Transactions on Power Electronics*, 2023.

[5] H. Feng, T. Cai, S. Duan, X. Zhang, H. Hu, and J. Niu, "A dual-side-detuned series–series compensated resonant converter for wide charging region in a wireless power transfer system," *IEEE Transactions on Industrial Electronics*, vol. 65, no. 3, pp. 2177–2188, 2017.

[6] L. Gu and J. Rivas-Davila, "1.7 kW 6.78 MHz wireless power transfer with air-core coils at 95.7

[7] L. Gu, W. Liang, and J. R. Davila, "Design of very-high-frequency synchronous resonant dc-dc converter for variable load operation," in *2017 IEEE Energy Conversion Congress and Exposition (ECCE)*, 2017, pp. 3447–3454.

[8] M. Liu and M. Chen, "Dual-band wireless power transfer with reactance steering network and reconfigurable receivers," *IEEE Transactions on Power Electronics*, vol. 35, no. 1, pp. 496–507, 2020.

[9] M. Lu, W. Mu, M. Qin, A. Koehler, J. Fang, and S. M. Goetz, "Differential detection of feeder and mesh impedances through a series–parallel direct-injection soft open point," *IEEE Transactions on Power Electronics*, pp. 1–10, 2024.

[10] M. Lu, M. Qin, W. Mu, J. Fang, and S. M. Goetz, "A hybrid gallium-nitride–silicon direct-injection universal power flow and quality control circuit with reduced magnetics," *IEEE Transactions on Industrial Electronics*, vol. 71, no. 11, pp. 14 161–14 174, 2024.

[11] S. Aldhaher, D. C. Yates, and P. D. Mitcheson, "Load-independent class e/ef inverters and rectifiers for MHz-switching applications," *IEEE Transactions on Power Electronics*, vol. 33, no. 10, pp. 8270–8287, 2018.

[12] M. Kim, M. Jeong, M. Cardone, and J. Choi, "Design of a spiral coil for high-frequency wireless power transfer systems using machine learning," *IEEE Journal of Emerging and Selected Topics in Industrial Electronics*, 2023.

979-8-3315-1612-3/25 $31.00 © 2025 IEEE

[13] J. Xu, Z. Tong, and J. Rivas-Davila, "1 kW MHz wideband class e power amplifier," *IEEE Open Journal of Power Electronics*, vol. 3, pp. 84–92, 2022.

[14] K. Surakitbovorn and J. M. Rivas-Davila, "A simple method to combine the output power from multiple class-e power amplifiers," *IEEE Journal of Emerging and Selected Topics in Power Electronics*, vol. 10, no. 2, pp. 2245–2253, 2022.

[15] S. Liu, M. Liu, S. Yang, C. Ma, and X. Zhu, "A novel design methodology for high-efficiency current-mode and voltage-mode class-e power amplifiers in wireless power transfer systems," *IEEE Transactions on Power Electronics*, vol. 32, no. 6, pp. 4514–4523, 2017.

[16] X. Huang, Y. Kong, Z. Ouyang, W. Chen, and S. Lin, "Analysis and comparison of push–pull class-e inverters with magnetic integration for megahertz wireless power transfer," *IEEE Transactions on Power Electronics*, vol. 35, no. 1, pp. 565–577, 2020.

[17] H. Sekiya, K. Tokano, W. Zhu, Y. Komiyama, and K. Nguyen, "Design procedure of load-independent class-e wpt systems and its application in robot arm," *IEEE Transactions on Industrial Electronics*, vol. 70, no. 10, pp. 10 014–10 023, 2023.

[18] C. H. Lee, G. Jung, K. A. Hosani, B. Song, D.-k. Seo, and D. Cho, "Wireless power transfer system for an autonomous electric vehicle," in *2020 IEEE Wireless Power Transfer Conference (WPTC)*, 2020, pp. 467–470.

[19] Z. Yue, Q. Zhang, Z. Yang, R. Bian, D. Zhao, and B.-Z. Wang, "Wall-meshed cavity resonator-based wireless power transfer without blocking wireless communications with outside world," *IEEE Transactions on Industrial Electronics*, vol. 69, no. 7, pp. 7481–7490, 2022.

[20] M. Venkatesan, R. Narayanamoorthi, K. M. AboRas, and A. Emara, "Efficient bidirectional wireless power transfer system control using dual phase shift pwm technique for electric vehicle applications," *IEEE Access*, vol. 12, pp. 27 739–27 755, 2024.

[21] F. Lu, H. Zhang, H. Hofmann, and C. C. Mi, "An inductive and capacitive combined wireless power transfer system with lc-compensated topology," *IEEE Transactions on Power Electronics*, vol. 31, no. 12, pp. 8471–8482, 2016.

[22] D. Ahn and P. P. Mercier, "Wireless power transfer with concurrent 200-khz and 6.78-mhz operation in a single-transmitter device," *IEEE Transactions on Power Electronics*, vol. 31, no. 7, pp. 5018–5029, 2016.

[23] T. Mishima and C.-M. Lai, "Zero-phase-angle load-independent and -adaptable dual-side lcc inductive wireless power transfer system," *IEEE Transactions on Transportation Electrification*, vol. 10, no. 2, pp. 3492–3503, 2024.

[24] D.-W. Seo, J.-H. Lee, and H.-S. Lee, "Optimal coupling to achieve maximum output power in a wpt system," *IEEE Transactions on Power Electronics*, vol. 31, no. 6, pp. 3994–3998, 2016.

[25] T. Fujita, T. Yasuda, and H. Akagi, "A dynamic wireless power transfer system applicable to a stationary system," *IEEE Transactions on Industry Applications*, vol. 53, no. 4, pp. 3748–3757, 2017.

[26] M.-S. Huang, C.-C. Liao, Z.-F. Li, Z.-R. Shih, and H.-W. Hsueh, "Quantitative design and implementation of an induction cooker for a copper pan," *IEEE Access*, vol. 9, pp. 5105–5118, 2020.

[27] O. Lucia, P. Maussion, E. J. Dede, and J. M. Burdío, "Induction heating technology and its applications: past developments, current technology, and future challenges," *IEEE Transactions on industrial electronics*, vol. 61, no. 5, pp. 2509–2520, 2013.

[28] S. Shen, N. Koonjoo, X. Kong, M. S. Rosen, and Z. Xu, "Gradient coil design and optimization for an ultra-low-field mri system," *Applied Magnetic Resonance*, vol. 53, no. 6, pp. 895–914, 2022.

[29] V. Pasku, A. De Angelis, M. Dionigi, G. De Angelis, A. Moschitta, and P. Carbone, "A positioning system based on low-frequency magnetic fields," *IEEE Transactions on Industrial Electronics*, vol. 63, no. 4, pp. 2457–2468, 2015.

[30] J. Zhou, P. Zhang, J. Han, L. Li, and Y. Huang, "Metamaterials and metasurfaces for wireless power transfer and energy harvesting," *Proceedings of the IEEE*, vol. 110, no. 1, pp. 31–55, 2022.

[31] M. Koehler and S. M. Goetz, "A closed formalism for anatomy-independent projection and optimization of magnetic stimulation coils on arbitrarily shaped surfaces," *IEEE Transactions on Biomedical Engineering*, vol. 71, no. 6, pp. 1745–1755, 2024.

[32] Koehler, Max and Goetz, Stefan M., "Simplified H1 coil with a single layer of surface conductors," *IEEE Access*, vol. 12, pp. 59 861–59 867, 2024.

[33] L. Jiang and S. Goetz, "Artificial intelligence exploring the patent field," *arXiv preprint arXiv:2403.04105*, 2024.

[34] G. Li, H. Cao, M. Liu, C. Jiang, and J. Yang, "Unirit: Towards few-shot non-rigid point cloud registration," *arXiv preprint arXiv:2410.22909*, 2024.

[35] A. Hashemi-Zadeh, N. Tashakor, M. Rahnama, C. T. Touko Sieyadji, M. Amirrezai, and S. Goetz, "Comparative analysis of model-free deep reinforcement learning controllers for reconfigurable battery systems output voltage regulation," in *2024 Energy Conversion Congress Expo Europe (ECCE Europe)*, 2024, pp. 1–6.

Accurate Modeling of LLC Resonant Converters with Enhanced Analytical Approach Considering Parasitic Capacitance

Dong Jiao
Future Energy Electronics Center
Virginia Tech
Blacksburg, VA, USA
jdong@vt.edu

Zhengming Hou
Future Energy Electronics Center
Virginia Tech
Blacksburg, VA, USA
hzhengm@vt.edu

Jih-Sheng Lai
Future Energy Electronics Center
Virginia Tech
Blacksburg, VA, USA
laijs@vt.edu

Abstract—This study introduces a novel analytical model designed to improve the accuracy of key parameter predictions in LLC resonant converters. The model addresses limitations in existing approaches that often lose accuracy with varying frequencies and load conditions and do not account for parasitic capacitance. By incorporating the parasitic capacitance of power devices, the model more accurately predicts current and voltage waveforms, voltage gain, and the necessary deadtime for achieving soft switching. Validation through simulations across various operating conditions shows a significant improvement in prediction accuracy, reducing errors in gain prediction and ZVS deadtime calculation. This improved accuracy has the potential to lead to more efficient and reliable LLC converter designs.

Index Terms—modeling, LLC resonant converter, time-domain analysis, parasitic capacitance, deadtime calculation

I. INTRODUCTION

Power electronics systems are integral to various applications, ranging from renewable energy to electric vehicles and data centers [1]–[3]. Among converter topologies, LLC resonant converters have emerged as a preferred choice for applications that demand high efficiency and soft-switching characteristics, particularly where galvanic isolation is necessary [4]–[9]. Despite the extensive application of LLC resonant converters, accurately predicting key parameters, such as resonant current waveforms, voltage gain, and optimal ZVS deadtime, remains a significant challenge. Existing methods often fall short, particularly under varying load and frequency conditions, leading to suboptimal designs and reduced efficiency. Traditional large-signal analysis methods include the fundamental harmonic approximation (FHA) or the extended harmonic approximation (EHA) and time-domain analysis (TDA) or time-interval analysis (TIA) [10]–[14]. FHA, known for its simplicity and effectiveness, is most accurate near the resonant frequency under ideal conditions. However, its precision diminishes when the operating frequency deviates from resonance or when parasitic effects are significant. Moreover, FHA overlooks critical details such as deadtime and voltage switching transients [15], [16]. Time-domain analysis, while offering higher accuracy and versatility, comes with increased complexity and computational demands. Notably, many mod-

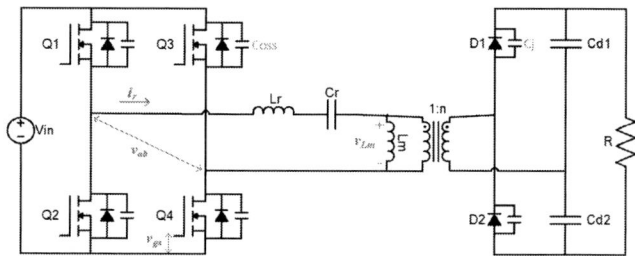

Fig. 1: LLC resonant converter circuit diagram.

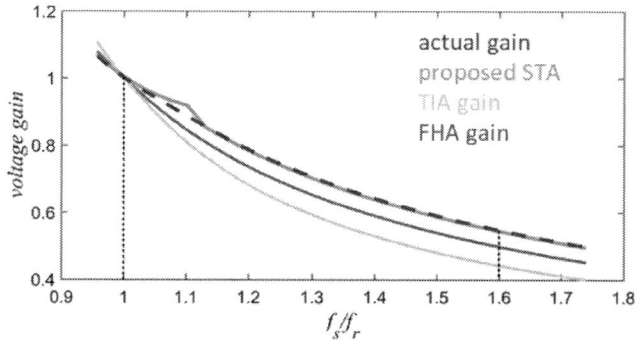

Fig. 2: Gain curves comparison among FHA model, TIA model, STA model, and actual results.

els based on time-domain analysis fail to incorporate the parasitic capacitance of power devices and rectifiers, which are crucial for accurate ZVS performance and overall efficiency [17]–[19].

This study addresses these challenges by proposing a novel switching-transient time-domain analysis (STA) model for LLC resonant converters. Unlike previous approaches, the model incorporates the parasitic capacitance of power devices, such as C_{oss} and C_j, to enhance the accuracy of predictions for key parameters, including current waveforms, voltage gain, and ZVS deadtime. This work contributes to a more comprehensive understanding of resonant converter performance and design optimization.

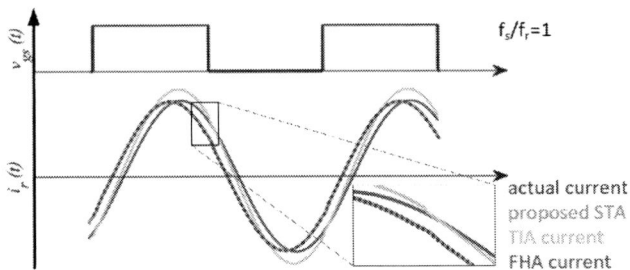

Fig. 3: Resonant currents comparison among different techniques at $f_s = f_r$.

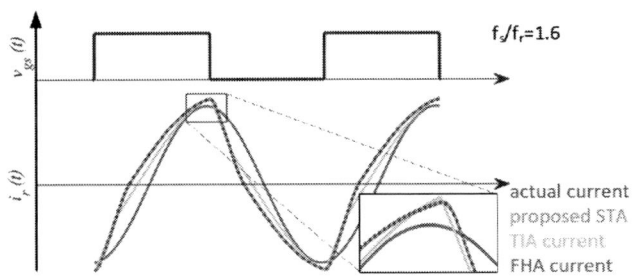

Fig. 4: Resonant currents comparison among different techniques at $f_s > f_r$.

II. THE PROPOSED ACCURATE MODEL FOR LLC CONVERTERS

Fig. 1 shows the conventional LLC resonant converter circuit diagram. Fig. 2 compares the gain curves of the proposed model with existing techniques, illustrating the improved accuracy in gain-frequency curves and waveform behavior of conventional LLC converters. The following sections discuss the implications of these findings for the design and control of LLC converters, as well as their broader applications in power electronics.

A. Assumption and principle of the proposed method

In this study, the parasitic capacitance of power devices is incorporated into the analysis of switching transients. Previous works have considered parasitic capacitance [20]–[22] and diode reverse recovery effect [23] during deadtime to estimate the ZVS performance; however, they often require initial voltage and current conditions or make many assumptions. The proposed STA method improves upon this by accounting for voltage transients at switching nodes, enhancing the accuracy of waveform predictions and gain curves. The behavior of the waveforms can be classified into six conditions, with three basic conditions illustrated in Fig. 5. The estimations for conditions below and above resonance are similar, with the main difference being the polarity changes in v_{Lm} and v_{ab}. Near resonance, there are four distinct conditions where the charging times of C_{oss} (t_0-t_1) and C_j (t_2-t_3) overlap, making the analysis more complex. This paper focuses on the above resonance condition. The following assumptions are made for the analysis:

1) The nonlinear parasitic capacitance C_{oss} and C_j are replaced with the equivalent linear values.
2) The converter can achieve ZVS at the primary side power devices without the need for setting a deadtime T_d.
3) The charging of C_{oss} occurs at $t_0 < t < t_1$, while the charging of C_j takes place at $t_2 < t < t_3$. And the sequence of events is $t_0 < t_1 < t_2 < t_3 < T_s/2$.

If the voltage transients of v_{Lm} and v_{ab} overlap, the results from the proposed approach will introduce errors in gain estimation, as seen in the bulge of the gain curve in Fig. 2, nevertheless, the gain curve should remain continuous, allowing the curve to be accurately extended across the entire frequency range, both above and below resonance.

B. Time-domain equations based on STA for LLC converters

From Fig. 5, the analysis of the waveform behavior above resonance can be performed in four intervals in a half switching cycle. In the first interval $t_0 = 0 < t < t_1$, voltage across primary side of transformer v_{Lm} is constant and input v_{ab} includes the charging and discharging of parasitic capacitance in the primary side devices. The differential equations are shown in (1).

$$
\begin{cases}
v_{Cr,a}(t) = v_{ab,a}(t) + \dfrac{V_{out}}{2n} - L_r \dfrac{di_{r,a}(t)}{dt} \\[2mm]
i_{r,a}(t) = C_r \dfrac{dv_{Cr,a}(t)}{dt} = C_{oss}\dfrac{dv_{ab}(t)}{dt} \\[2mm]
-\dfrac{V_{out}}{2n} = L_m \dfrac{di_{Lm,a}(t)}{dt} \\[2mm]
v_{ab}(t) = -V_{in} - \dfrac{1}{C_{oss}}\displaystyle\int_0^t i_{r,a}(t)\,dt
\end{cases}
\tag{1}
$$

In the interval of $t_1 < t < t_2$, v_{Lm} and v_{ab} are constant, the differential equations are shown as,

$$
\begin{cases}
v_{Cr,b}(t) = V_{in} + \dfrac{V_{out}}{2n} - L_r \dfrac{di_{r,b}(t - t_1)}{dt} \\[2mm]
i_{r,b}(t) = C_r \dfrac{dv_{Cr,b}(t - t_1)}{dt} \\[2mm]
-\dfrac{V_{out}}{2n} = L_m \dfrac{di_{Lm,b}(t - t_1)}{dt}
\end{cases}
\tag{2}
$$

During the interval $t_2 < t < t_3$, v_{ab} is constant and v_{Lm} includes the charging transient of secondary side parasitic capacitance.

$$
\begin{cases}
v_{Cr,c}(t) = V_{in} - v_{Lm,c}(t - t_2) - L_r \dfrac{di_{r,c}(t - t_2)}{dt} \\[2mm]
i_{r,c}(t) = C_r \dfrac{dv_{Cr,c}(t - t_2)}{dt} \\[2mm]
i_{r,c}(t) = i_{Lm,c}(t - t_2) + 2n^2 C_j \dfrac{dv_{Lm,c}(t - t_2)}{dt} \\[2mm]
v_{Lm,c}(t) = L_m \dfrac{di_{Lm,c}(t - t_2)}{dt}
\end{cases}
\tag{3}
$$

In the interval $t_3 < t < \frac{1}{2f_s}$, v_{Lm} and v_{ab} are constant. And the differential equations are presented in (4).

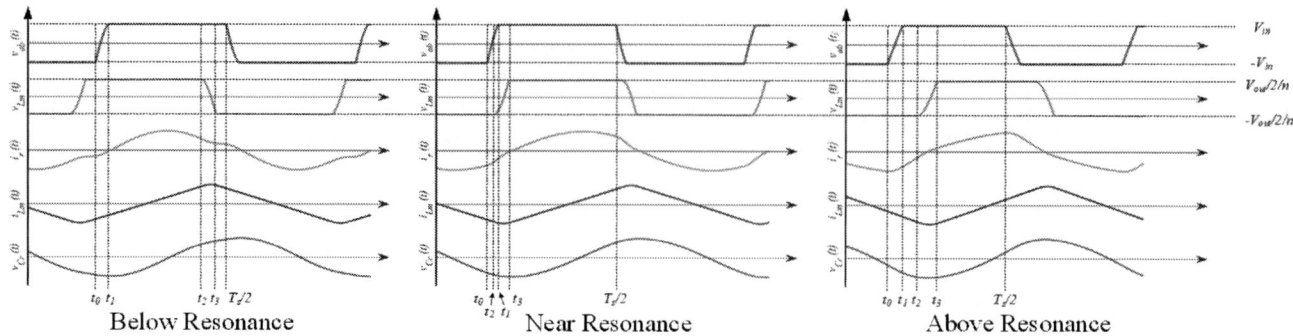

Fig. 5: Waveforms of the behaviors of voltages and currents of the LLC resonant converter under different operating conditions with the consideration of C_{oss} and C_j.

$$\begin{cases} v_{Cr,d}(t) = V_{in} - \dfrac{V_{out}}{2n} - L_r \dfrac{di_{r,d}(t-t_3)}{dt} \\[2mm] i_{r,d}(t) = C_r \dfrac{dv_{Cr,d}(t-t_3)}{dt} \\[2mm] \dfrac{V_{out}}{2n} = L_m \dfrac{di_{Lm,d}(t-t_3)}{dt} \end{cases} \quad (4)$$

The second and fourth interval, $t_1 < t < t_2$ and $t_3 < t < \frac{1}{2f_s}$, are the same as the TIA method [14]. Thus, the expressions of the voltage and current waveforms are presented in (5)-(8). And in these equations, $\omega = 1/\sqrt{L_r C_r}$ and $\omega_a = \sqrt{(C_{oss} + C_r)/(L_r C_r C_{oss})}$ which represent the resonant frequency in different intervals. (7) was simplified and

$$\begin{aligned} K_1 &= C_r(L_m + L_r), \\ K_2 &= C_r(L_m - L_r), \\ K_3 &= 2n^2 C_j L_m, \\ K_4 &= K_1^2 + K_3^2 + 2K_2 K_3, \\ K_5 &= K_1 - K_3 + \sqrt{K_4}, \\ K_6 &= K_3 - K_1 + \sqrt{K_4}, \\ K_7 &= K_1 + K_3 + \sqrt{K_4}, \\ K_8 &= K_1 + K_3 - \sqrt{K_4}, \\ K_9 &= -K_2(V_2 - V_{in}) + \frac{K_1 V_{out}}{2n} \end{aligned}$$

The equations require specific circuit parameters, such as L_r and C_r, before analysis, and the unknown parameters, V_{out}, I_0, V_0, t_1, t_2, t_3, can be solved by boundary conditions:

$$v_{Cr,a}(t_0) = v_{Cr,d}(\frac{1}{2f_s}) = V_0,$$

$$i_{r,a}(t_0) = i_{r,d}(\frac{1}{2f_s}) = I_0,$$

$$v_{Cr,b}(t_1) = v_{Cr,a}(t_1) = V_1,$$

$$i_{r,b}(t_1) = i_{r,a}(t_1) = I_1,$$

$$v_{Cr,c}(t_2) = v_{Cr,b}(t_2) = V_2,$$

$$i_{r,c}(t_2) = i_{r,b}(t_2) = I_2,$$

$$v_{Cr,d}(t_3) = v_{Cr,c}(t_3) = V_3,$$

$$i_{r,d}(t_3) = i_{r,c}(t_3) = I_3$$

And charge-balance equations:

$$-I_0 = i_{r,d}(\frac{1}{2f_s}),$$

$$-V_0 = v_{Cr,d}(\frac{1}{2f_s}),$$

$$-I_2 = i_{Lm,c}(t_3) + \frac{V_{out}}{2nL_m}\left(\frac{1}{2f_s} - t_3 + t_2\right),$$

$$-2V_{in}C_{oss} = \int_0^{t_1} i_{r,a}(t)dt,$$

$$4n^2 C_j \frac{V_{out}}{2n} = \int_{t_2}^{t_3}(i_{r,c}(t) - i_{Lm,c}(t))\,dt,$$

$$\frac{V_{out}^2}{R} = 2f_s\left(\int_0^{t_1} i_{r,a}(t)v_{ab,a}(t)dt + \int_{t_1}^{t_2} V_{in} i_{r,b}(t)dt\right.$$
$$\left. + \int_{t_2}^{t_3} V_{in} i_{r,c}(t)dt + \int_{t_3}^{1/2f_s} V_{in} i_{r,d}(t)dt\right)$$

Then the unknown parameters can be determined numerically, and the key parameters are achieved like the gain and resonant waveforms which would be used to determine the ZVS deadtime range.

III. MATHEMATICAL PROCESSING TO IMPROVE ACCURACY OF STA NEAR RESONANCE

The interval of $[t_0, t_1]$ represents the minimum deadtime required to achieve ZVS at primary side, while $[t_2, t_3]$ is the minimum deadtime for secondary side if synchronous rectification (SR) control is applied to secondary side [15], [18], [24]. However, this study focuses on rectification using diodes. The zero-crossing point of the resonant current defines the maximum deadtime limit, as the current begins to reverse beyond this point. Simplified equations for output voltage and minimum deadtime in the above-resonance case are provided in (9) and (10). Due to their complexity, other key parameters are omitted in this paper. The analysis assumes no overlap between the charging periods of the parasitic capacitance, which can lead to mismatches near resonance. For example, the numerical solution might yield $t_2 < t_1$ or $t_3 > T_s/2$. Fig. 6 shows an example at around $f_s/f_r = 1$, where the STA calculation (depicted by the orange-yellow-red curve) shows $t_3 > T_s/2$. To address this, a simple mathematical correction can be applied within the interval $[t_0, t_3 - T_s/2]$. By adjusting

the red curve with the sum of the orange curve and subtracting the yellow curve, the new result (green curve) aligns closely with the actual performance. This correction allows the waveform behavior of LLC converters to be accurately modeled across the entire frequency range. Consequently, the deadtime estimation should be derived from the updated current waveform by solving $-2V_{in}C_{oss} = \int_0^{t_{d,min}} i_r(t)dt$.

IV. MODEL VALIDATION

In many applications, the deadtime of an LLC converter is typically estimated using the formula $T_d \geq 8L_m f_s C_{oss}$. However, this estimation is independent of both load and C_j, which makes it prone to inaccuracies, particularly the loss of ZVS in real cases. Fig. 7 and Fig. 8 demonstrate that the proposed STA method provides a more precise prediction of the deadtime range across various operating conditions.

$$
\begin{cases}
i_{r,a}(t) = I_0 \cos(\omega_a t) + \dfrac{-V_0 - V_{in} + \frac{V_{out}}{2n}}{\omega_a L_r} \sin(\omega_a t), \\[3mm]
v_{cr,a}(t) = \dfrac{-V_{in} + \frac{V_{out}}{2n} + \frac{V_0 C_r}{C_{oss}}}{L_r C_r \omega_a^2} (1 - \cos(\omega_a t)) + V_0 \cos(\omega_a t) + \dfrac{I_0}{C_r \omega_a} \sin(\omega_a t), \\[3mm]
v_{ab,a}(t) = -V_{in} - \dfrac{1}{C_{oss}}\left(\dfrac{I_0}{\omega_a} \sin(\omega_a t) - \dfrac{-V_0 - V_{in} + \frac{V_{out}}{2n}}{\omega_a^2 L_r} \cos(\omega_a t) + \dfrac{-V_0 - V_{in} + \frac{V_{out}}{2n}}{\omega_a^2 L_r} \right)
\end{cases}
\tag{5}
$$

$$
\begin{cases}
i_{r,b}(t) = I_1 \cos(\omega(t - t_1)) - \left(V_1 - V_{in} - \dfrac{V_{out}}{2n} \right) C_r \omega \sin(\omega(t - t_1)), \\[3mm]
v_{cr,b}(t) = V_{in} + \dfrac{V_{out}}{2n} + I_1 L_r \omega \sin(\omega(t - t_1)) + \left(V_1 - V_{in} - \dfrac{V_{out}}{2n} \right) \cos(\omega(t - t_1))
\end{cases}
\tag{6}
$$

$$
\begin{cases}
i_{r,c}(t) = \dfrac{I_2 K_6}{2\sqrt{K_4}} \cos\left(\dfrac{\sqrt{2}}{\sqrt{K_8}}(t - t_2) \right) + \dfrac{\sqrt{2}\sqrt{K_8}}{4L_r\sqrt{K_4}} \left[(K_3 + \sqrt{K_4})\left(V_{in} - V_2 + \dfrac{V_{out}}{2n} \right) + K_9 \right] \sin\left(\dfrac{\sqrt{2}}{\sqrt{K_8}}(t - t_2) \right) \\[3mm]
\quad + \dfrac{I_2 K_5}{2\sqrt{K_4}} \cos\left(\dfrac{\sqrt{2}}{\sqrt{K_7}}(t - t_2) \right) - \dfrac{\sqrt{2}\sqrt{K_7}}{4L_r\sqrt{K_4}} \left[(K_3 - \sqrt{K_4})\left(V_{in} - V_2 + \dfrac{V_{out}}{2n} \right) + K_9 \right] \sin\left(\dfrac{\sqrt{2}}{\sqrt{K_7}}(t - t_2) \right), \\[3mm]
v_{Cr,c}(t) = V_{in} + \dfrac{K_1(V_{in} - V_2) + \sqrt{K_4}(V_2 - V_{in}) - K_3\left(V_{in} - V_2 + \frac{V_{out}}{2n} \right)}{2\sqrt{K_4}} \cos\left(\dfrac{\sqrt{2}}{\sqrt{K_8}}(t - t_2) \right) \\[3mm]
\quad + \dfrac{I_2\sqrt{2}\sqrt{K_8}(K_8 - 2K_3)}{4C_r\sqrt{K_4}} \sin\left(\dfrac{\sqrt{2}}{\sqrt{K_8}}(t - t_2) \right) - \dfrac{I_2\sqrt{K_8}\left[K_1\sqrt{K_4} + K_1^2 + K_2 K_3 \right]}{2C_r\sqrt{K_4}\sqrt{K_3(K_1 - K_2)}} \sin\left(\dfrac{\sqrt{2}}{\sqrt{K_7}}(t - t_2) \right) \\[3mm]
\quad + \dfrac{K_1(V_2 - V_{in}) + \sqrt{K_4}(V_2 - V_{in}) + K_3\left(V_{in} - V_2 + \frac{V_{out}}{2n} \right)}{2\sqrt{K_4}} \cos\left(\dfrac{\sqrt{2}}{\sqrt{K_7}}(t - t_2) \right), \\[3mm]
i_{Lm,c}(t) = \dfrac{I_2 K_7}{2\sqrt{K_4}} \cos\left(\dfrac{\sqrt{2}}{\sqrt{K_7}}(t - t_2) \right) + \dfrac{\sqrt{2}}{4L_m\sqrt{K_4}} \left[\sqrt{K_7} \left(2C_r L_m(V_{in} - V_2) + V_{Lm0} K_6 \right) \right] \sin\left(\dfrac{\sqrt{2}}{\sqrt{K_7}}(t - t_2) \right) \\[3mm]
\quad - \dfrac{I_2 K_8}{2\sqrt{K_4}} \cos\left(\dfrac{\sqrt{2}}{\sqrt{K_8}}(t - t_2) \right) + \dfrac{\sqrt{2}}{4L_m\sqrt{K_4}} \left[\sqrt{K_8} \left(2C_r L_m(V_2 - V_{in}) + V_{Lm0} K_5 \right) \right] \sin\left(\dfrac{\sqrt{2}}{\sqrt{K_8}}(t - t_2) \right)
\end{cases}
\tag{7}
$$

$$
\begin{cases}
i_{r,d}(t) = I_3 \cos(\omega(t - t_3)) - \left(V_3 - V_{in} + \dfrac{V_{out}}{2n} \right) C_r \omega \sin(\omega(t - t_3)), \\[3mm]
v_{cr,d}(t) = V_{in} - \dfrac{V_{out}}{2n} + I_3 L_r \omega \sin(\omega(t - t_3)) + \left(V_3 - V_{in} + \dfrac{V_{out}}{2n} \right) \cos(\omega(t - t_3))
\end{cases}
\tag{8}
$$

979-8-3315-1612-3/25 $31.00 © 2025 IEEE

$$V_{out} = \frac{2n\left[(V_0 + V_{in})(1 - \cos(\omega_a t_1)) - 2V_{in}C_{oss}\omega_a^2 L_r - I_0\omega_a L_r \sin(\omega_a t_1)\right]}{1 - \cos(\omega_a t_1)} \tag{9}$$

$$t_{d,\min} = t_1 = \frac{1}{\omega_a}\ln\left(-\frac{2\omega_a\sqrt{-nL_r\left(nI_0^2 L_r - 2C_{oss}V_{in}V_{out} + 4nC_{oss}V_{in}^2 + 4nC_{oss}V_0 V_{in} - 4nC_{oss}^2 L_r V_{in}^2 \omega_a^2\right)}}{V_{out} - 2nV_{in} - 2nV_0 + 2nI_0 L_r \omega_a i}\right.$$
$$\left. - \frac{2nV_0 - V_{out} + 2nV_{in} - 4nC_{oss}L_r V_{in}\omega_a^2}{V_{out} - 2nV_{in} - 2nV_0 + 2nI_0 L_r \omega_a i}\right) \tag{10}$$

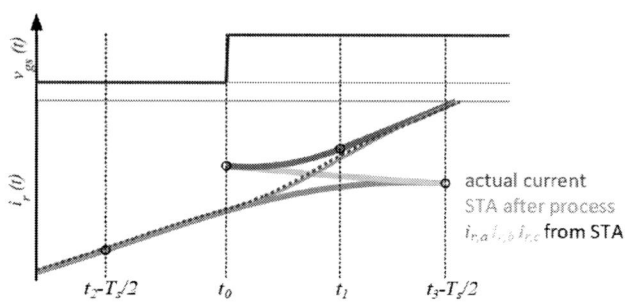

Fig. 6: Mathematical process to eliminate the error near resonant frequency point.

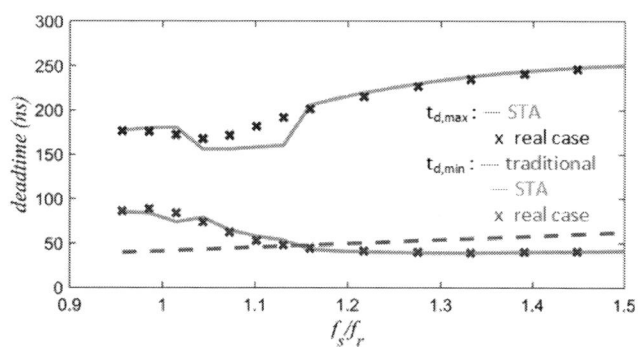

Fig. 7: Deadtime estimation at different magnetizing inductance with $L_m/L_r = 1.5$

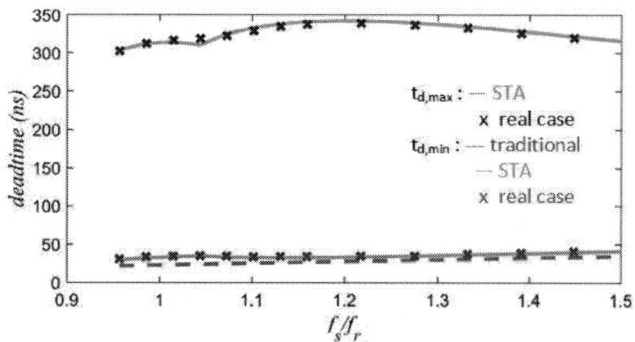

Fig. 8: Deadtime estimation at different magnetizing inductance with $L_m/L_r = 0.83$

CONCLUSIONS

In this paper, an advanced switching-transient time-domain analysis (STA) model is presented for LLC resonant converters, incorporating parasitic capacitance to significantly enhance the accuracy of key parameter predictions, including voltage gain, current waveforms, and the ZVS deadtime. Through a comparison with conventional methods, the paper demonstrated the proposed model overcomes the limitations of existing approaches, particularly under varying load and frequency conditions where traditional models often fail. Future work will focus on further extending its applicability to other types of resonant converters. In future work, experimental validation of the STA model in various real-world cases will be conducted to further confirm its accuracy and utility.

REFERENCES

[1] L. A. D. Ta, N. D. Dao, and D.-C. Lee, "High-efficiency hybrid llc resonant converter for on-board chargers of plug-in electric vehicles," *IEEE Transactions on Power Electronics*, vol. 35, no. 8, pp. 8324–8334, 2020.

[2] Z. Hou, S. C. Kao, D. Jiao, and J.-S. Lai, "Variable turns-ratio matrix transformer based llc converter for two-stage electric vehicle auxiliary power module applications," in *2023 IEEE Energy Conversion Congress and Exposition (ECCE)*, pp. 5859–5865, 2023.

[3] J.-S. Lai, B. Gutierrez, M. Lee, C.-S. Yeh, H. Wen, D. Jiao, Z. Hou, and H.-C. Hsieh, "A hybrid binary multilevel cascaded inverter for medium-voltage applications," in *2021 IEEE Energy Conversion Congress and Exposition (ECCE)*, pp. 1939–1945, 2021.

[4] S. Khan, D. Sha, X. Jia, and S. Wang, "Resonant llc dc–dc converter employing fixed switching frequency based on dual-transformer with wide input-voltage range," *IEEE Transactions on Power Electronics*, vol. 36, no. 1, pp. 607–616, 2021.

[5] X. Wu, R. Li, and X. Cai, "A wide output voltage range llc resonant converter based on topology reconfiguration method," *IEEE Journal of Emerging and Selected Topics in Power Electronics*, vol. 10, no. 1, pp. 969–983, 2022.

[6] Y. Wei, Q. Luo, and H. Alan Mantooth, "A novel llc converter with topology morphing control for wide input voltage range application," *IEEE Journal of Emerging and Selected Topics in Power Electronics*, vol. 10, no. 2, pp. 1563–1574, 2022.

[7] H. Wen, D. Jiao, J.-S. Lai, J. Strydom, and B. Lu, "A mhz llc converter based single-stage soft-switching isolated inverter with hybrid modulation method," in *2021 IEEE Energy Conversion Congress and Exposition (ECCE)*, pp. 1882–1888, 2021.

[8] D. Jiao, Z. Hou, and J.-S. Lai, "Llc type resonant converter adopting peak current shaving with third harmonics injection for wide output voltage range application," in *2024 IEEE Applied Power Electronics Conference and Exposition (APEC)*, pp. 2232–2238, 2024.

[9] C.-Y. Pan, T.-J. Liang, W.-J. Tseng, K.-F. Liao, J.-W. Yeh, and S.-J. Zhan, "Design and implementation of sic-based wide input voltage range full-bridge dc-to-dc resonant converter," in *2019 IEEE 4th International Future Energy Electronics Conference (IFEEC)*, pp. 1–8, 2019.

979-8-3315-1612-3/25 $31.00 © 2025 IEEE

[10] H. Huang, "Fha-based voltage gain function with harmonic compensation for llc resonant converter," in *2010 Twenty-Fifth Annual IEEE Applied Power Electronics Conference and Exposition (APEC)*, pp. 1770–1777, 2010.

[11] H. Wang, Y. Chen, Y.-F. Liu, J. Afsharian, and Z. Yang, "A passive current sharing method with common inductor multiphase llc resonant converter," *IEEE Transactions on Power Electronics*, vol. 32, no. 9, pp. 6994–7010, 2017.

[12] J. Niu, Y. Tong, Q. Ding, X. Wu, X. Xin, and X. Wang, "Time domain simplified equations and its iterative calculation model for llc resonant converter," *IEEE Access*, vol. 8, pp. 151195–151207, 2020.

[13] A. Sankar, A. Mallik, and A. Khaligh, "Extended harmonics based phase tracking for synchronous rectification in cllc converters," *IEEE Transactions on Industrial Electronics*, vol. 66, no. 8, pp. 6592–6603, 2019.

[14] E. S. Glitz and M. Ordonez, "Mosfet power loss estimation in llc resonant converters: Time interval analysis," *IEEE Transactions on Power Electronics*, vol. 34, no. 12, pp. 11964–11980, 2019.

[15] H. Wen, D. Jiao, J.-S. Lai, J. Strydom, and B. Lu, "A mhz lclcl resonant converter based single-stage soft-switching isolated inverter with variable frequency modulation," *IEEE Transactions on Power Electronics*, vol. 37, no. 9, pp. 10797–10807, 2022.

[16] D. Jiao, H. Wen, and J.-S. Lai, "Llc resonant converter based single-stage inverter with multi-resonant branches using variable frequency modulation," in *2023 IEEE Applied Power Electronics Conference and Exposition (APEC)*, pp. 263–270, 2023.

[17] R. Wei, L. Ding, R. Liu, and Y. Li, "An intuitive and noniterative design methodology for cllc chargers employing simplified operation modes model," *IEEE Transactions on Power Electronics*, vol. 38, no. 6, pp. 7771–7784, 2023.

[18] B. Li, M. Chen, X. Wang, N. Chen, X. Sun, and D. Zhang, "An optimized digital synchronous rectification scheme based on time-domain model of resonant cllc circuit," *IEEE Transactions on Power Electronics*, vol. 36, no. 9, pp. 10933–10948, 2021.

[19] J. Sun, L. Yuan, Q. Gu, R. Duan, Z. Lu, and Z. Zhao, "Design-oriented comprehensive time-domain model for cllc class isolated bidirectional dc-dc converter for various operation modes," *IEEE Transactions on Power Electronics*, vol. 35, no. 4, pp. 3491–3505, 2020.

[20] P. Jiang, H. Feng, and L. Ran, "Zvs analysis and a design method for unidirectional medium-voltage llc-dcx with high step-up ratio," *IEEE Transactions on Power Electronics*, vol. 39, no. 3, pp. 2948–2953, 2024.

[21] Y. Sun, Z. Deng, G. Xu, G. Deng, Q. Ouyang, and M. Su, "Zvs analysis and design for half bridge bidirectional llc-dcx converter with consideration of nonlinear capacitance and different load under synchronous turn-on and turn-off modulation," *IEEE Transactions on Transportation Electrification*, vol. 8, no. 2, pp. 2429–2443, 2022.

[22] P. Jiang, H. Feng, and L. Ran, "Zvs analysis and a design method for unidirectional medium-voltage llc-dcx with high step-up ratio," *IEEE Transactions on Power Electronics*, vol. 39, no. 3, pp. 2948–2953, 2024.

[23] H. Wen, D. Jiao, J.-S. Lai, J. Strydom, and L. Zhang, "An accurate voltage gain model considering diode effect for llc resonant converter in wide gain range applications," in *2020 IEEE Energy Conversion Congress and Exposition (ECCE)*, pp. 5436–5441, 2020.

[24] N. Chen, M. Chen, B. Li, X. Wang, X. Sun, D. Zhang, F. Jiang, and J. Han, "Synchronous rectification based on resonant inductor voltage for cllc bidirectional converter," *IEEE Transactions on Power Electronics*, vol. 37, no. 1, pp. 547–561, 2022.

High-Frequency Conditioning Circuits for Power-Related Information Extraction in Non-Sinusoidal Power Electronic Systems

Haoyu Wang
Dept. of Electrical and Computer Eng.
The University of Texas at Austin
wanghaoyu@utexas.edu

Yuanxin Zhang
Department of Electrical Engineering
Tsinghua University
zyx22@mails.tsinghua.edu.cn

Di Mou
Department of Electrical Engineering
Tsinghua University
dimou428@mail.tsinghua.edu.cn

Alex Hanson
Dept. of Electrical and Computer Eng.
The University of Texas at Austin
ajhanson@utexas.edu

Shiqi Ji
Department of Electrical Engineering
Tsinghua University
chic2020@tsinghua.edu.cn

Abstract—**Power electronic converters often have non-sinusoidal voltages and/or currents but must perform calculations on individual harmonics of those waveforms, particularly the fundamental. If these calculations must be done online, then rapidly sampling the waveform and performing a Fast Fourier Transform is often unfeasible in terms of cost and power consumption.**

One example scenario is the control of multi-port DC-DC converters to achieve low rms current, soft switching, and regulation, which is an analytically and computationally difficult task. Prior work has emphasized offline optimization because computational burden of online computations can be high. To address this challenge, control based on real-time measurements of port power at the fundamental frequency has been developed, which uses closed-loop control to minimize rms current with a very low computational burden. The key to this approach, and other design scenarios, is a circuit to compute the power at the fundamental frequency alone.

This paper presents high-frequency conditioning circuits to measure port real and reactive power at the fundamental switching frequency, despite highly non-sinusoidal waveforms, to enable the online control technique mentioned and other scenarios with similar requirements. Test results show that the proposed circuit functions fast and produces accurate measurements. Generally, the conditioning circuits can be implemented in any non-sinusoidal power electronic systems where magnitude or phase information is required at frequencies up to several MHz.

Index Terms—**high frequency, hardware, magnitude and phase, power electronic system**

I. INTRODUCTION

Because power electronic converters use MOSFETs with large-signal excitations, most voltages and currents in them are severely non-sinusoidal. Nevertheless, optimization and control often rely on knowledge of specific harmonics of those waveforms, particularly the fundamental. One example

This work was supported by the Program of National Natural Science Foundation of China under Grant 52277189.

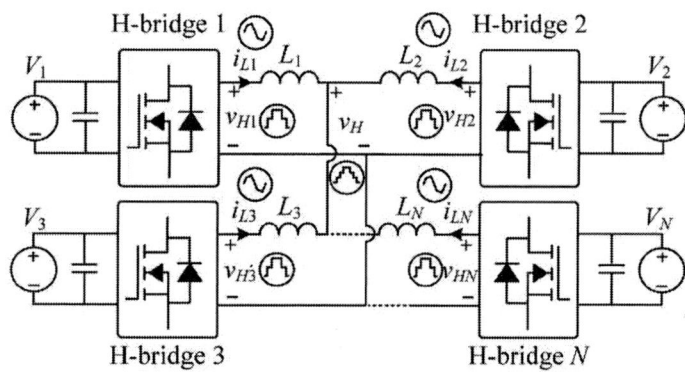

Fig. 1. A generalized multiport DC-DC system with square wave modulation.

is resonant converters and wireless power transfer, in which currents may be largely sinusoidal but switching port voltages are not. Another example is the measurement of magnetic losses at high frequency, which uses power amplifiers to drive resonant circuits and results in the power amplifier producing distorted waveforms.

Of particular interest to this paper is the modular multi-active bridge (MMAB) converter, in which existing control faces a difficult tradeoff between complexity and inefficiency. A useful control that is simple and maintains high efficiency over wide operating ranges is proposed in [1], but it requires information about the fundamental components of the quasi-square-wave voltage at the high-frequency link and the quasi-triangular inductor currents.

This paper will give a detailed design of a high-frequency conditioning circuit capable of performing such sensing at high frequencies (typically dozens of kHz but up to several MHz). The proposed circuit has applications beyond multi-port dc-dc converters, including for more sinusoid-based optimal control

979-8-3315-1612-3/25 $31.00 © 2025 IEEE

strategies [2] to be implemented in a non-sinusoidal power electronic system. The rest of the paper is organized as follows. The background of the MMAB control is given in Section II. Section III will introduce the basic theory of power-relation information extraction. The selection of the components of the conditioning circuit will be given in detail in Section IV. Section V shows the PCB design of the circuit and its corresponding control frame together with other controllers, and Section VI verifies the feasibility and accuracy of the proposed conditioning circuit.

II. BACKGROUND ON MODULAR MULTI-ACTIVE BRIDGE CONTROL

The increasing penetration of renewable energy (e.g., wind power and photovoltaic), storage devices and electric vehicles in distribution grids requires the distribution network to be safer, smarter, and more flexible. In the modern integration of power grids, distributed energy resources, and local loads, high-frequency DC-DC converters have been widely adopted for flexible power conversion [3], [4]. Particularly, dual active bridge (DAB) converters [5], [6], with advantages of galvanic isolation and zero-voltage-switching (ZVS) operation, have proven adept for such applications and play an important role as the hub in power distribution. With the emerging demands of multiport systems to deal with local loads with multiple voltage forms and levels, multi-active bridge (MAB) [7], [8] and MMAB DC-DC converters [9], [10] (shown in Fig. 1) have been developed into the DAB family. These multiport converters can be integrated into high-frequency-link-based intelligent power electronic transformers [11], flexibly connecting different ac and dc ports with different voltage and power ratings.

Typically, the converters in the DAB family, especially the multiport ones, adopt single phase-shift control due to its simple implementation and inherent ZVS potentials [12], [13]. However, this leads to high rms current and loss of soft switching at non-nominal voltages and powers, thus reducing the converter efficiency [14], [15].

More complex control schemes are used to achieve higher efficiency of multiport converters. Some research managed to optimize the switching sequence of the bridges in the multiport converters based on time-domain models, but only in an offline optimization way due to the inherent coupling between ports and the expense of high online computational burden. Particularly, by adding one additional control degree of freedom (DOF) into the converter, [16] achieves the power loss optimization of MMAB converters. Following this principle, [17] further involves more DOFs, thus achieving ZVS and low rms current across wide voltage and power ranges for each port. However, the offline methods relying on look-up tables are limited to applications. Therefore, [18] proposed a time-division multiplexing modulation method to online optimize MMAB converters, although the efficiency improvement is relatively low when compared to offline methods.

Other research in the literature has established frequency-domain models and optimization approaches to minimize reac-

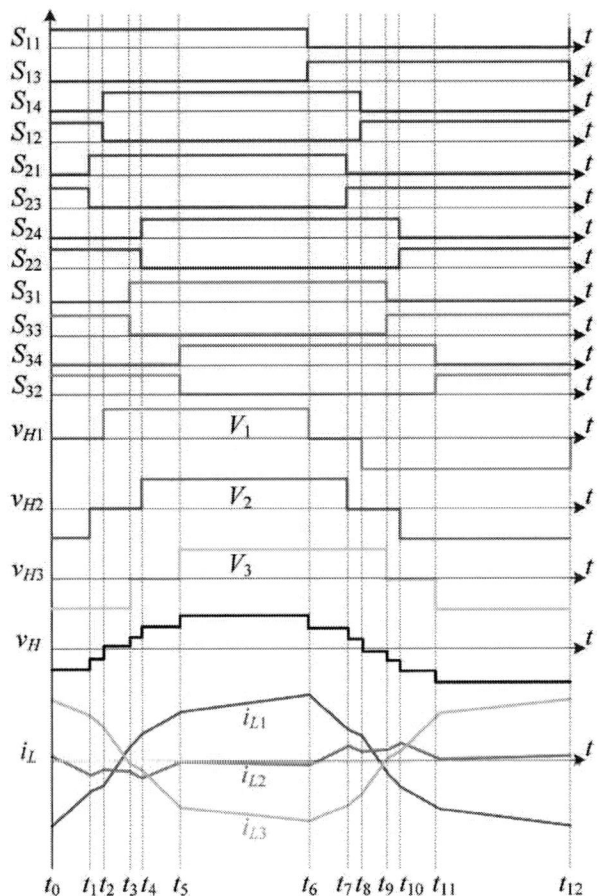

Fig. 2. Key waveforms of a three-port MMAB converter. $S_{i1} - S_{i4}$ are four active switches in the H-bridge of port i.

tive power at the fundamental switching frequency, including [19], [20] for DAB converters and [21], [22] for MMAB converters. Although simple and real-time, these model-based strategies depend on accurate model parameters. The optimization effects are highly determined by the parameter accuracy.

Considering the above model-based methods are typically based on traditional voltage or power loop control [23], [24], another effective way to improve the converter performance is to develop different control strategies. Accordingly, [1] proposed a closed-loop optimal control to eliminate the parameter dependence and achieve independent port control. However, unlike resonant dc-dc converters [25], [26] or wireless power transfer systems [27], [28] where DAB converters are used but indutcor currents are pretty sinusoidal, enabling control approaches like that of [1] in multiport converters requires a high-frequency conditioning circuit to isolate and sense the magnitude-phase information of the fundamental components with highly non-sinusoidal signals (i.e., high-frequency-link voltage and inductor currents). The high-frequency conditioning circuit can save controllers in the converter from large computational burden, such as high-frequency waveform sampling and Fast Fourier Transform (FFT). The extracted fundamental magnitude and phase are related to the fundamental active-reactive powers of each port, therefore can be used to achieve further control strategies (i.e., power-curernt dual loop control) or other fundamental-oriented applications. Additionally, the power-related information of harmonics extracted in similar ways can also be used for system monitoring and optimization.

III. THEORY OF POWER-RELATED INFORMATION EXTRACTION

DC-DC systems generally adopt square-wave modulation, which leads to quasi-square voltages and quasi-triangular (e.g., trapezoidal) inductor currents, as shown in Fig. 2. These intermediate ac bridge voltages and inductor currents can be decomposed into fundamental frequency components and higher-order harmonics through Fourier Series decomposition. If the switching frequency is $f_s = \omega/2\pi$, any quantity x in the system can be expressed as

$$x(t) = X + \sum_{n=1}^{\infty} A_n \sin(n\omega t + \varphi_n) \tag{1}$$

where X is the dc offset of the quantity x, A_n and φ_n are the magnitude and phase of the n-th order harmonic of x.

For example, the bridge voltages and inductor currents together determine the active and reactive powers of certain ports in the converter in steady state. According to (1), the bridge voltage v_{Hi} and inductor current i_{Li} in the port i can be expressed as

$$\begin{cases} v_{Hi}(t) = \sum_{n=1}^{\infty} A_{vi,n} \sin(n\omega t + \varphi_{vi,n}) \\ i_{Li}(t) = \sum_{n=1}^{\infty} A_{ii,n} \sin(n\omega t + \varphi_{ii,n}) \end{cases} \tag{2}$$

where $A_{vi,n}$ and $A_{ii,n}$ represent the magnitudes of the n-th order harmonics of v_{Hi} and i_{Li}, respectively, while $\varphi_{vi,n}$ and $\varphi_{ii,n}$ represent the phase information. Therefore, the active and reactive powers $P_{i,n}$ and $Q_{i,n}$ of the n-th order harmonics in port i can be obtained by

$$\begin{cases} P_{i,n} = A_{vi,n} A_{ii,n} \cos(\varphi_{vi,n} - \varphi_{ii,n}) \\ Q_{i,n} = A_{vi,n} A_{ii,n} \sin(\varphi_{vi,n} - \varphi_{ii,n}) \end{cases} \tag{3}$$

Therefore, with magnitude and phase information, the powers delivered by each harmonics can be easily derived. Typically, the power-related information of the fundamental component is desired, which can be useful in fundamental-optimal strategies. The information of certain harmonics (e.g., the third order) can also be used to monitor the converter or provide other benefits. However, in non-sinusoidal systems, the magnitude-phase information of a given harmonic needs to be extracted from the complete waveform.

It would be conceptually straightforward to sample the signals and compute the FFT, from which the fundamental or other components could be extracted. However, it is hard for an analog-to-digital converter (ADC) to obtain the full-cycle digital waveform of a high-frequency quantity and for the controller to compute the FFT in one switching cycle. The

required sampling speeds and computational burden of such an approach are prohibitive for any but the highest power systems.

Another method is to pre-filter the information and reduce the computational burden. The detailed process is illustrated by

1) generating sinusoids at, for example, the fundamental frequency $\sin(\omega t)$ and $\cos(\omega t)$,
2) multiplying the sinusoids by the signal of interest $x(t)$, resulting in analog signals with dc components related to the fundamental magnitude and phase,
3) low-pass filtering the analog outputs to obtain the dc components,
4) performing the light computations to identify the magnitude and phase.

Mathematically, the multiplication of the signal of interest $x(t)$ and the generated sine waves $\sin(\omega t)$ and $\cos(\omega t)$ lead to

$$x(t)\sin(\omega t) = X\sin(\omega t) + \tag{4}$$
$$\sum_{n=1}^{\infty} \frac{A_n}{2}\left[\cos((n-1)\omega t + \varphi_n) - \cos((n+1)\omega t + \varphi_n)\right]$$

$$x(t)\cos(\omega t) = X\cos(\omega t) + \tag{5}$$
$$\sum_{n=1}^{\infty} \frac{A_n}{2}\left[\sin((n-1)\omega t + \varphi_n) + \sin((n+1)\omega t + \varphi_n)\right]$$

When the products in (4) are filtered by low-pass filters, we have the $\alpha - \beta$ components of the fundamental component of x, which can be expressed by

$$\begin{cases} x_{\alpha,1} = \frac{A_1}{2}\cos\varphi_1 \\ x_{\beta,1} = \frac{A_1}{2}\sin\varphi_1 \end{cases} \tag{6}$$

With the $\alpha - \beta$ components, we have the magnitude-phase information as

$$\begin{cases} A_1 = 2\sqrt{x_{\alpha,1}^2 + x_{\beta,1}^2} \\ \varphi_1 = \arctan\frac{x_{\beta,1}}{x_{\alpha,1}} \end{cases} \tag{7}$$

IV. COMPONENTS OF HIGH-FREQUENCY CONDITIONING CIRCUIT

To process the high-frequency signals according to the above theory, analog components are needed to achieve high-speed signal processing at acceptable power levels. The above theory indicates that the proposed circuit therefore requires: 1) sinusoid generators, 2) analog multipliers, 3) lowpass filters, and 4) ADCs. A digital controller, such as an FPGA, can then manipulate the final data at lower speed, as shown in Fig. 3.

A. High-frequency Sinusoid Generator

Direct Digital Frequency Synthesis (DDS) technology [29] is currently the main way to generate high-frequency waveforms. DDS pre-stores waveform data such as sine waves, triangle waves, and sawtooth waves in its internal memory, and generates waveforms of specific frequencies and phases

979-8-3315-1612-3/25 $31.00 © 2025 IEEE

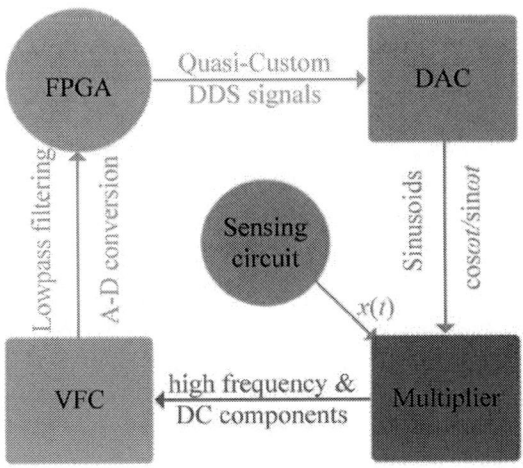

Fig. 3. The diagram of the proposed high-frequency conditioning circuit.

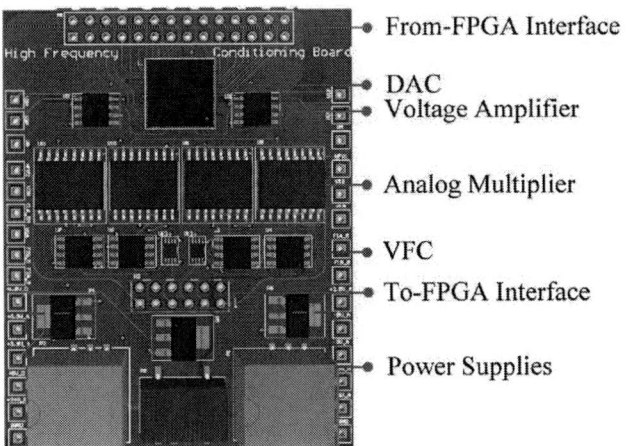

Fig. 4. The designed PCB of the proposed high-frequency conditioning circuit.

through digital control. There are currently two commonly used DDS generation methods:

1) integrate a DDS chip and use protocols such as serial peripheral interfaces to control the output of its internally integrated digital-to-analog converter (DAC);
2) use the collaborative frame of FPGA and DAC. The structure uses the DDS function of the FPGA to generate a waveform data file, and then the FPGA sends serial or parallel data instructions to the DAC, commanding the DAC to output a waveform.

The proposed approach requires high-frequency reference sinusoids to be generated at exactly the switching frequency or its harmonics. Therefore, it is necessary to precisely synchronize the DDS-generated waveform and the switching frequency, which is most easily done by using the same clock to produce both the gate-drive waveforms and the generated sine waves. This corresponds to the second option above. Therefore, in the proposed DDS design, the collaborative frame of FPGA and DAC is used and the data file with the corresponding DAC bits is generated through the manually pre-stored sine waveform. Since the FPGA is controlling both the switching frequency and the sinusoid generation frequency, the frequency error will be naturally eliminated.

The sine wave generator must contain a large number of points in one cycle to reduce the harmonics of the generated waveform. If we need, for example, 500 data points per cycle, then the sinusoid generator should have a generation rate of at least 10 MSPS to obtain a 20 kHz sinusoid. This minimum generation rate increases with the switching frequency or selected harmonic order. Besides, the input data of the DAC at the beginning of the switching cycle determines the phase of the generated sinusoid.

For the DAC, this design selects AD9743, a parallel input DAC with 10-bit accuracy and 250 MSPS sampling rate. The DAC is capable of generating sinusoids up to 200 kHz with 500 data points at a 100 MHz input differential clock frequency. AD9743 has two output channels and can generate

two sinusoids with different phases to meet the design requirements. It receives 10-bit parallel digital input signals from the FPGA and outputs analog differential currents. Therefore, its output needs to be terminated with a resistor to generate a differential voltage with an output range of ±1 V. The differential voltage then passes through the differential amplifier AD8130 with variable amplification coefficient to generate a single-ended sinusoidal voltage with variable amplitudes, which makes it easier to meet the input requirements of subsequent analog components. Specifically, the AD8130 can generate a distortion-free sinusoid with the magnitude of 4 V with a 5 V power supply.

B. Analog Multiplier

An analog multiplier is the key component in the high-frequency conditioning circuit. Its main function is to multiply the generated sinusoid with the high-frequency quantity (e.g., the high-frequency-link voltage or inductor current), so that their product yields a dc component which carries the power-related information. The high-speed real-time multiplying process greatly reduces computational burden.

The analog multiplier used in this design is MPY634, whose power supply voltage is ±15 V, which can meet a wide range of analog voltage inputs. MPY634 has two pairs of differential inputs $X_1 - X_2$ and $Y_1 - Y_2$, and a reference voltage input $Z_1 - Z_2$. Its typical output formula is

$$V_{out} = \frac{(X_1 - X_2)(Y_1 - Y_2)}{10} + Z_1 - Z_2 \qquad (8)$$

Let $X_1 - X_2$ be the generated sinusoid with a 4 V magnitude, $Y_1 - Y_2$ be the sensed high-frequency quantity, whose amplitude is controlled within ±5 V, and $Z_1 - Z_2$ be 2.5 V. Therefore, the output of MPY634 will be between 0.5 V and 4.5 V, which sastifies the subsequent input voltage requirements. Note that when the average output of MPY634 is less than 2.5 V, it indicates the phase might be negative or more than 90 degrees according to (6).

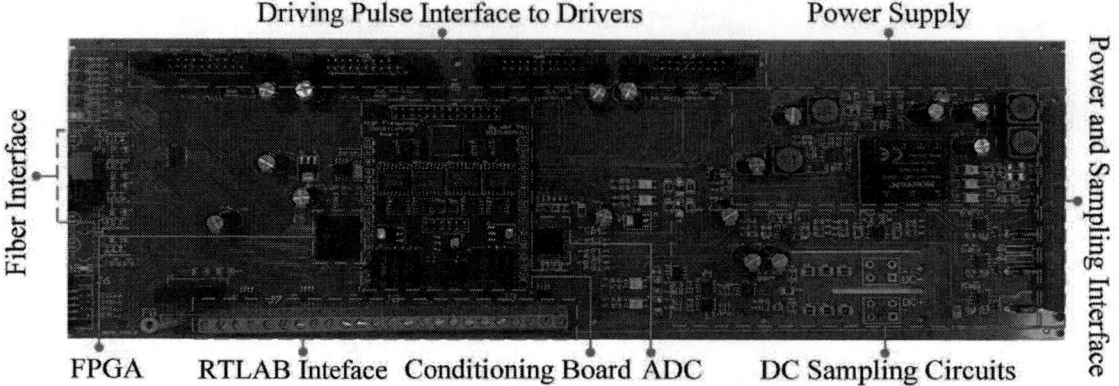

Fig. 5. The PCB of the corresponding FPGA control board.

C. Voltage-to-Frequency Converter

The output of the analog multiplier has a dc component and high-frequency components, where the dc component contains the power-related information of the sensed high-frequency quantity. Therefore, the dc component should be preserved while the other components should be suppressed.

This could be accomplished through an analog lowpass filter followed by an ADC. As an alternative, we propose a circuit that uses a voltage-to-frequency converter (VFC) to simultaneously provide filtering and analog-to-digital conversion functions. The pulses generated by the VFC are transferred to the FPGA. In order to obtain the dc component, The FPGA counts the rising edge of the pulse sequence with variable frequency within a switching cycle. The count value corresponds to the average pulse frequency $f_{out,avg}$ of the VFC within a switching period. If the count value of FPGA in one switching cycle is CNT, then

$$f_{out,avg} = CNT \times f_s \qquad (9)$$

The output frequency is then related to the average voltage through the transfer characteristics of the VFC.

AD7741 is selected as the VFC, which can convert the voltage within the input range into a pulse sequence with a linear relationship between frequency and input voltage. Its supply voltage is 5 V and reference voltage V_{REF} is the supply voltage, so that the input voltage range of AD7741 is 0 to V_{REF}. Its input clock frequency f_{CLKIN} can be up to 6 MHz and is chosen as 5 MHz in this design. The relationship between the output frequency and the input voltage V_{IN} satisfies

$$f_{out} = \frac{2V_{IN}}{5V_{REF}} f_{CLKIN} + \frac{f_{CLKIN}}{20} \qquad (10)$$

It can be seen from (10) that AD7741 can generate a minimum pulse sequence of 250 kHz and a maximum of 2.25 MHz within an input range of 0 V to 5 V. The high level of the pulse is 5 V and the low level is 0 V and the pulse width is half of the input clock period (i.e., $1/f_{CLKIN}/2$=100 ns).

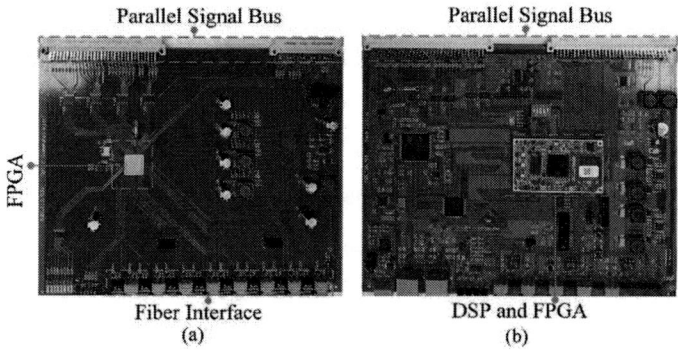

Fig. 6. (a) Signal interfacing board. (b) Main control Board.

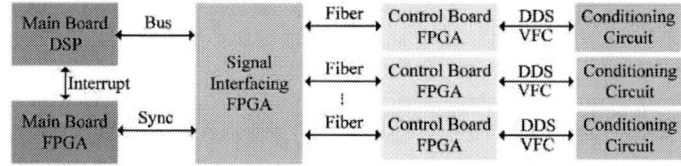

Fig. 7. The collaborative control frame.

According to (10), the dc component $V_{IN,avg}$ has a linear relationship with $f_{out,avg}$.

$$V_{IN,avg} = V_{REF}\left(\frac{5f_{out,avg}}{2f_{CLKIN}} - \frac{1}{8}\right) = V_{REF}\left(\frac{5f_sCNT}{2f_{CLKIN}} - \frac{1}{8}\right) \qquad (11)$$

Taking $v \times 4\sin(\omega t)$ as an example, assume that v is sensed through a k_{div}-times voltage divider. According to (8) and (11), the fundamental magnitude-phase information of v satisfies

$$k_{div}V_1 \cos\varphi_1 = \frac{125f_sCNT}{2f_{CLKIN}} - 15.625 \qquad (12)$$

V. System Implementation

The conditioning circuits discussed are integrated onto a PCB daughter board which interfaces with a motherboard

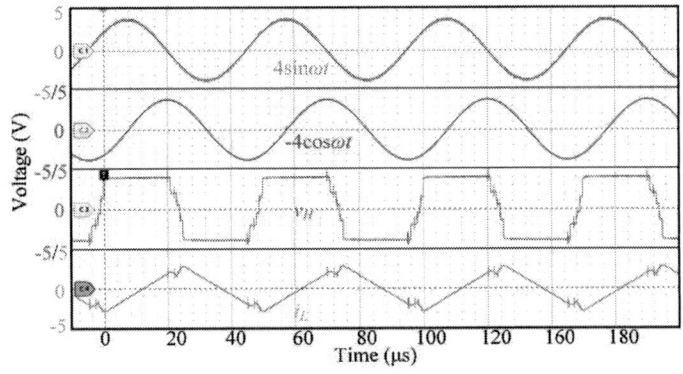

Fig. 8. Generated sinusoids and sensed quantities.

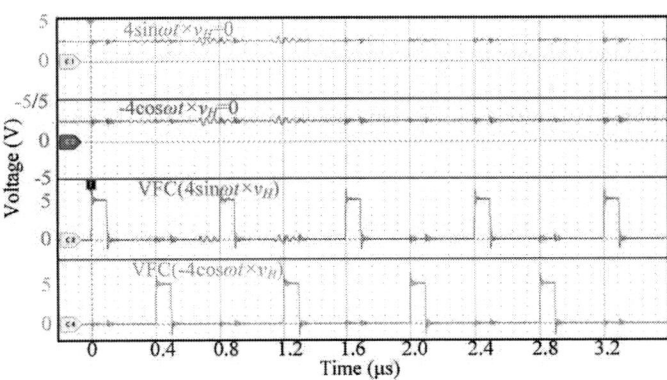

Fig. 10. Output pulses of VFCs with no inputs.

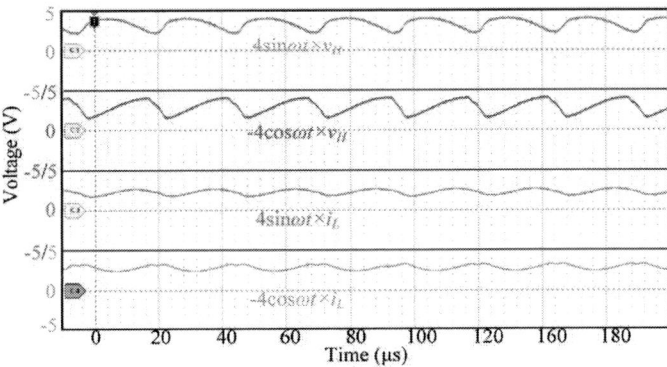

Fig. 9. Outputs of analog multipliers.

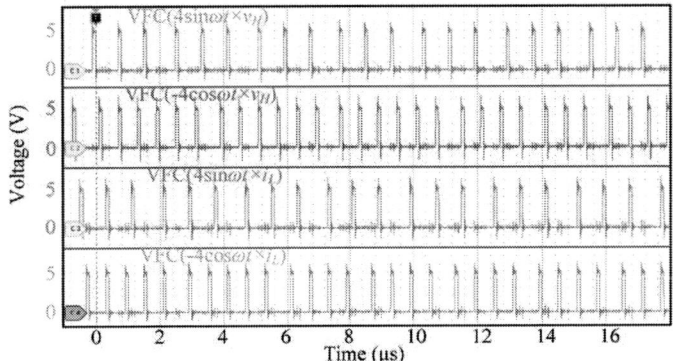

Fig. 11. Output pulses of VFCs with the multiplies.

containing an FPGA and connections to gate drivers as well as an RTLAB hardware-in-the-loop (HIL) system [30].

A. PCB Design of Conditioning Circuit

Integrating all the components introduced in Section IV, the designed PCB of the proposed high-frequency conditioning circuit is shown in Fig. 4. It has 1 DAC (2 outputs) and 2 sensing inputs, and can obtain 4 VFC outputs to meet the conditioning demands of two high-frequency quantities. Note that the from-FPGA interface receives the digital signals from the external FPGA to control the DAC outputs, while the to-FPGA interface offers the pulse sequences generated by VFCs to the FPGA for counting.

B. Collaborative Frame of Controllers and Circuits

The FPGA control board shown in Fig. 5 contains FPGA EG4X20BG256, which sends two channels of DDS sinusoidal signals with a total of 20 bits to the high-frequency conditioning board, and receives the 4-channel VFC pulse sequences returned by the high-frequency conditioning board. In addition, the board has 4 driving signal outputs, which can drive 4 power semiconductor half bridges. It also contains 6 driving signal interfaces that can interact with RTLAB real-time simulators and drive 6 virtual half bridges.

The FPGA control board can also exchange serial data with the higher-level signal interfacing board through the serial optical fiber interface in a large power electronics system. The signal interfacing board is shown in Fig. 6(a), which consists of FPGA PH1A100GCG324. The signal interfacing board sends synchronization, control, protection, driving and other signals to multiple FPGA control boards, and receives signals returned from them. In addition, the signal interfacing board interacts with the DSP controller (e.g., TMS320C6748) of the higher-level main control board (as shown in Fig. 6(b)) through a parallel signal bus and receives the global synchronization signal generated by the FPGA of the main control board. The main control panel is shown in Fig. 7.

VI. Test of High-Frequency Conditioning Circuit

A. Implementation and Test

A multiport DC-DC converter implemented in HIL [1] is used in this case. Its switching frequency is 20 kHz. Therefore, in this test, two orthogonal 20 kHz sinusoids are generated by the conditioning board, as shown in Fig. 8. Note that the initial phase of the sinusoids can be arbitrary, and in this case they are aligned with the initial switching pulse of the slack port in the converter. Then the magnitudes of the sinusoids are amplified by voltage amplifiers AD8130 to 4 V and have little distortion. The high-frequency-link voltage and the inductor current, both non-sinusoidal, are sensed by the FPGA control board.

979-8-3315-1612-3/25 $31.00 © 2025 IEEE

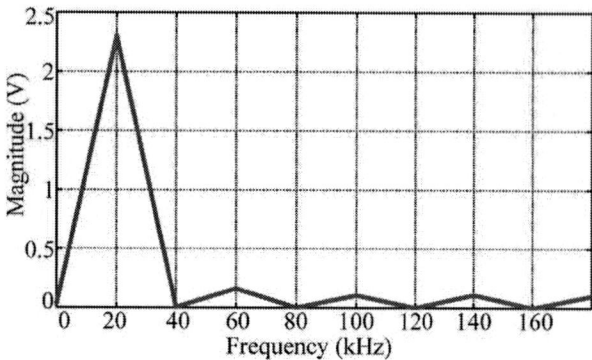

Fig. 12. Magnitude spectrum of the sensed inductor current i_L.

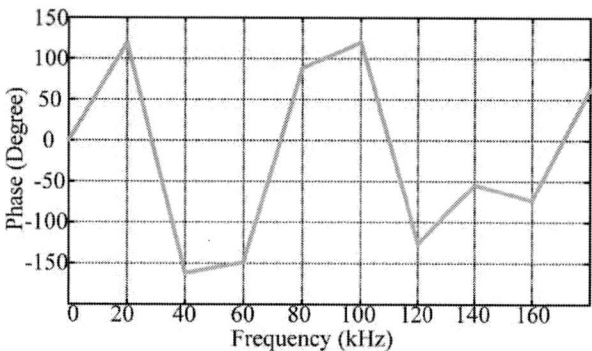

Fig. 13. Phase spectrum of the sensed inductor current i_L.

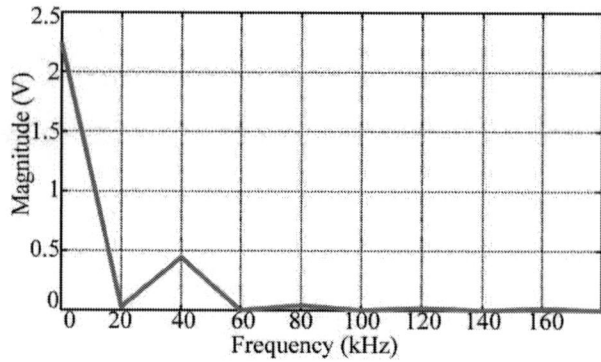

Fig. 14. Magnitude spectrum of the multiplier output of $4\sin\omega t \times i_L$.

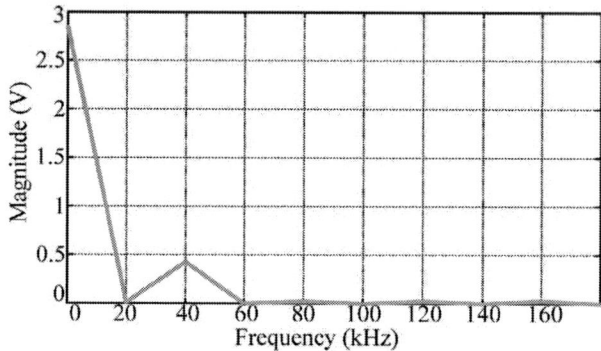

Fig. 15. Magnitude spectrum of the multiplier output of $-4\cos\omega t \times i_L$.

The generated signals and sensed quantities shown in Fig. 8 are multiplied by 4 analog multipliers MPY634. The outputs of the multipliers are shown in Fig. 9. It can be observed that dc components and high-frequency components (double line frequency and higher) are obtained for each output, showing the test results are consistent with (4).

These signals are converted into a pulse train through the VFCs which, when counted and averaged, perform both low-pass filtering and analog-to-digital conversion. For precision, the VFCs must be calibrated when the multipliers have no inputs and therefore the inputs to the VFCs are about 2.5 Vdc. Typical waveforms of the VFCs are shown in Fig. 10. We can see that VFCs output 1.25 MHz pulse sequences with the pulse width as 100 ns. The pulse number in a switching period is the reference value for calibration, which is around 62 in this 20 kHz case. If the VFC is linear, then the total pulse count will be within 0-125 as the inputs vary from 0 V to 5 V, making the resolution as 0.04 V per count. Lower switching frequencies may permit higher resolution, while higher switching frequencies can have fewer pulses per cycle, reducing resolution. Therefore, it is advised that traditional RC filters and high-speed ADCs be used in higher-frequency cases, which can lead to the same results.

When the signals in Fig. 9 are received by their corresponding VFCs, different pulse sequences with variable frequencies will be generated. Fig. 11 shows the VFC pulses of the signals.

By counting the pulses in one switching cycle, the FPGA obtains the dc components that lead to magnitude and phase information. Thus, the proposed circuit can quickly extract the power-related information of non-sinusoidal signals, enabling the low-computation online optimizer approaches discussed in the introduction.

B. Accuracy Verification

In order to verify the accuracy of the proposed conditioning circuit, FFTs of key waveforms have been conducted and compared with the counting results in the FPGA. Take the inductor current i_L in Fig. 8 for example. The frequency spectrums of this sensed quantity are shown in Fig. 12 and Fig. 13. It can be seen that the magnitude of the fundamental component is about 2.22 V while the phase is about 120 degrees.

We conducted FFTs for the two multiplies shown in Fig. 9 (i.e., $4\sin\omega t \times i_L$ and $-4\cos\omega t \times i_L$), as shown in Fig. 14 and Fig. 15, respectively. As can be seen from the figures, there are dc but not fundamental components, and double line frequency is dominant in harmonics, which is consistent with (4). The two dc components are 2.23 V and 2.84 V. When subtracted by the actual dc offset $Z_1 - Z_2 = 2.47$ V, the effective dc values are -0.24 V and 0.37 V, which indicates the equivalent fundamental magnitude is 0.88 V and the phase is 122 degrees according to (7). Because the magnitude of the generated sinusoid is 4 V and the gain of multipliers is 1/10, the actual fundamental magnitude derived by the results should be 2.20 V according to (8), which is exactly the same as the FFT result shown in Fig. 12. The phase under test also aligns with the phase analysis by FFT. Therefore, the effectiveness of using analog multipliers to extract magnitude and phase information is verified. At the VFC stage, the count numbers for the two multiplies are 56 and 71, referring to -0.24 V and 0.36 V, respectively, which is consistent with the test results and verifies the whole circuit.

VII. Conclusion

In this paper, a high-frequency conditioning circuit is designed to sense and extract the magnitude and phase of individual harmonic components of non-sinusoidal waveforms. The basic extraction theory is introduced and corresponding hardware components are chosen for simplicity and at low cost. With the magnitude-phase information, advanced control and optimization strategies are available. The functions of the designed circuit have been verified by step-by-step tests and in an optimal control application [1]. The circuit has potential to be adopted in many other cases where the magnitude and phase of individual harmonic components need to be extracted from non-sinusoidal waveforms as are common in switch-mode power conversion.

References

[1] H. Wang *et al.*, "High-Frequency-Link-Based Reactive-Power Optimal Control for Modular Multi-Active Bridge Converter," in *IEEE Journal of Emerging and Selected Topics in Industrial Electronics*, vol. 5, no. 3, pp. 1333-1337, July 2024.

[2] J. Rocabert *et al.*, "Control of Power Converters in AC Microgrids," in *IEEE Transactions on Power Electronics*, vol. 27, no. 11, pp. 4734-4749, Nov. 2012.

[3] N. Tan *et al.*, "Design and performance of a bidirectional isolated DC–DC converter for a battery energy storage system," in *IEEE Transactions on Power Electronics*, vol. 27, no. 3, pp. 1237–1248, Mar. 2012.

[4] N. Hou *et al.*, "Overview and Comparison of Modulation and Control Strategies for a Nonresonant Single-Phase Dual-Active-Bridge DC–DC Converter," in *IEEE Transactions on Power Electronics*, vol. 35, no. 3, pp. 3148-3172, March 2020.

[5] H. Qin *et al.*, "Generalized Average Modeling of Dual Active Bridge DC–DC Converter," in *IEEE Transactions on Power Electronics*, vol. 27, no. 4, pp. 2078-2084, April 2012.

[6] M. Zhang *et al.*, "New Single-Stage Single-Phase Isolated Bidirectional AC-DC PFC Converter," *2024 IEEE Applied Power Electronics Conference and Exposition (APEC)*, Long Beach, CA, USA, 2024, pp. 1962-1967.

[7] C. Gu *et al.*, "Modeling and Control of a Multiport Power Electronic Transformer (PET) for Electric Traction Applications," in *IEEE Transactions on Power Electronics*, vol. 31, no. 2, pp. 915-927, Feb. 2016.

[8] Y. Chen *et al.*, "Power Flow Control in Multi-Active-Bridge Converters: Theories and Applications," *2019 IEEE Applied Power Electronics Conference and Exposition (APEC)*, Anaheim, CA, USA, 2019, pp. 1500-1507.

[9] H. Wang *et al.*, "ZVS Soft Switching Operation Region Analysis of Modular Multi Active Bridge Converter Under Single Phase Shift Control," in *IEEE Transactions on Industrial Electronics*, vol. 70, no. 7, pp. 6865–6875, Jul. 2023.

[10] P. Zumel *et al.*, "Overall analysis of a modular multi active bridge converter," *2014 IEEE 15th Workshop on Control and Modeling for Power Electronics (COMPEL)*, Santander, Spain, 2014, pp. 1-9.

[11] S. Falcones *et al.*, "A DC–DC Multiport-Converter-Based Solid-State Transformer Integrating Distributed Generation and Storage," in *IEEE Transactions on Power Electronics*, vol. 28, no. 5, pp. 2192-2203, May 2013.

[12] P. Purgat *et al.*, "Zero Voltage Switching Criteria of Triple Active Bridge Converter," in *IEEE Transactions on Power Electronics*, vol. 36, no. 5, pp. 5425-5439, May 2021.

[13] J. Riedel *et al.*, "ZVS soft switching boundaries for dual active bridge DC–DC converters using frequency domain analysis," in *IEEE Transactions on Power Electronics*, vol. 32, no. 4, pp. 3166–3179, Apr. 2017.

[14] H. Wang *et al.*, "Comparison and Improvement of ZVS Operation Under Different Modulation Strategies for Modular Multi Active Bridge Converters," *2023 IEEE Energy Conversion Congress and Exposition (ECCE)*, Nashville, TN, USA, 2023, pp. 2556-2563.

[15] A. Rodríguez *et al.*, "Different purpose design strategies and techniques to improve the performance of a dual active bridge with phase-shift control," in *IEEE Transactions on Power Electronics*, vol. 30, no. 2, pp. 790–804, Feb. 2015.

[16] H. Wang *et al.*, "Switching Characterization and Power Loss Optimization for Modular Multiactive Bridge Converter Under Common Phase Shift Control," in *IEEE Journal of Emerging and Selected Topics in Power Electronics*, vol. 11, no. 4, pp. 3924-3936, Aug. 2023.

[17] H. Wang *et al.*, "Universal Phase-Shift Modulation Scheme and Efficiency Optimization for Modular Multiactive Bridge Converter," in *IEEE Transactions on Industrial Electronics*, vol. 71, no. 7, pp. 7312-7321, July 2024.

[18] D. Mou *et al.*, "High-Efficiency Time-Division Multiplexing Modulation Technology for Modular Multiactive Bridge Converters," in *IEEE Transactions on Industrial Electronics*, vol. 71, no. 6, pp. 5745-5754, June 2024.

[19] B. Zhao *et al.*, "Universal High-Frequency-Link Characterization and Practical Fundamental-Optimal Strategy for Dual-Active-Bridge DC-DC Converter Under PWM Plus Phase-Shift Control," in *IEEE Transactions on Power Electronics*, vol. 30, no. 12, pp. 6488-6494, Dec. 2015.

[20] W. Choi *et al.*, "Fundamental duty modulation of dual-active-bridge converter for wide-range operation," in *IEEE Transactions on Power Electronics*, vol. 31, no. 6, pp. 4048–4064, Jun. 2016.

[21] D. Mou *et al.*, "Reactive Power Minimization for Modular Multi-Active-Bridge Converter With Whole Operating Range," in *IEEE Transactions on Power Electronics*, vol. 38, no. 7, pp. 8011-8015, July 2023.

[22] H. Wang *et al.*, "Model-Free Approximate Fundamental Reactive Power Minimization for Modular Multi-Active Bridge Converter," *2024 IEEE 7th International Electrical and Energy Conference (CIEEC)*, Harbin, China, 2024, pp. 2068-2073.

[23] O. M. Hebala *et al.*, "Generalized active power flow controller for multiactive bridge DC–DC converters with minimum-current-point-tracking algorithm," in *IEEE Transactions on Industrial Electronics*, vol. 69, no. 4, pp. 3764–3775, Apr. 2022.

[24] L. F. Costa *et al.*, "Optimum design of a multiple active-bridge DC–DC converter for smart transformer," in *IEEE Transactions on Power Electronics*, vol. 33, no. 12, pp. 10112–10121, Dec. 2018.

[25] F. Musavi *et al.*, "An LLC Resonant DC–DC Converter for Wide Output Voltage Range Battery Charging Applications," in *IEEE Transactions on Power Electronics*, vol. 28, no. 12, pp. 5437-5445, Dec. 2013.

[26] W. Chen *et al.*, "Snubberless Bidirectional DC–DC Converter With New CLLC Resonant Tank Featuring Minimized Switching Loss," in *IEEE Transactions on Industrial Electronics*, vol. 57, no. 9, pp. 3075-3086, Sept. 2010.

[27] A. P. Sample *et al.*, "Analysis, Experimental Results, and Range Adaptation of Magnetically Coupled Resonators for Wireless Power Transfer," in *IEEE Transactions on Industrial Electronics*, vol. 58, no. 2, pp. 544-554, Feb. 2011.

[28] Y. Wu *et al.*, "A Highly-Efficient and Cost-Effective Reconfigurable IPT Topology for Constant-Current and Constant-Voltage Battery Charging," *2021 IEEE Applied Power Electronics Conference and Exposition (APEC)*, Phoenix, AZ, USA, 2021, pp. 451-455.

[29] D. De Caro *et al.*, "High-performance direct digital frequency synthesizers using piecewise-polynomial approximation," in *IEEE Transactions on Circuits and Systems I: Regular Papers*, vol. 52, no. 2, pp. 324-337, Feb. 2005.

[30] F. Li *et al.*, "Review of Real-time Simulation of Power Electronics," in *Journal of Modern Power Systems and Clean Energy*, vol. 8, no. 4, pp. 796–808, 2020.

Transconductance Model of the Dual Active Bridge Converter Under Single and Dual Phase Shift Control

Jared Cronin
Dept. of Electrical Engineering
University of South Carolina
Columbia, SC, U.S.A.
jared.cronin@sc.edu

Andrew Wunderlich
Integer Technologies LLC
Columbia, SC, U.S.A.
andrew.wunderlich@integer-tech.com

Enrico Santi
Dept. of Electrical Engineering
University of South Carolina
Columbia, SC, U.S.A.
santi@cec.sc.edu

Abstract—**This work proposes a model for the Dual Active Bridge (DAB) converter that recognizes the dominant dynamics and simplifies analysis. The DAB converter is a complex topology with eight semiconductor switches and a high frequency transformer. The DAB is widely applied in literature based on a myriad range of benefits including low losses, galvanic isolation, and excellent power density. The proposed model is based on the recognition of the first order system dynamics that dictate the control-to-output-voltage response. This model utilizes a transconductance term and simplifies analysis significantly compared to previous approaches reported in the literature. This simplification however maintains high accuracy at the terminals of the DAB converter compared to other modeling approaches. This modeling approach benefits the practicing engineer by allowing more efficient simulation of the converter and an intuitive understanding of the dominant features of the DAB. The DAB transconductance model is presented along with time and frequency domain simulations. The model is then validated against a physical DAB converter with resistive load. In recognition of the limitations of the single-phase shift modulation scheme, the transconductance modeling approach is extended to the dual phase shift modulation control.**

Index Terms—**dual active bridge, converter modeling, real-time modeling**

I. INTRODUCTION

The Dual Active Bridge converter, proposed and patented in 1991, has become quite popular over the last decade [1]. The DAB is gaining traction because it offers high power density, high efficiency, guaranteed soft switching on all devices for some operating conditions, galvanic isolation, and seamless bidirectional power flow due to complete input/output symmetry [2]. The topology of the standard DAB converter is shown in Figure 1. The DAB is formed from two H-bridge converter stages connected with an inductive link, this is often a transformer as it allows for voltage scaling and provides desirable galvanic isolation. The DAB is relevant in a range of application fields including electric vehicle chargers and vehicle to load [3], electric aircraft [4], modular marine power converters [5], or grid tied photovoltaic systems [6]. With

This work was supported by the Department of Defense National Defense Science and Engineering Graduate Fellowship Program (NDSEG).

this widespread use, efficient modeling is desired to speed up design and analysis of large or intricate systems. An

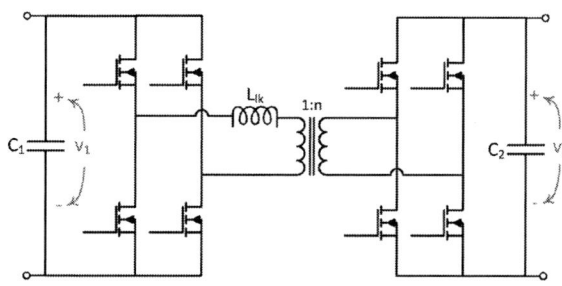

Fig. 1. Dual Active Bridge topology with input and output capacitance and leakage inductance of transformer referred to transformer primary side.

obvious problem with the DAB converter is the complexity that arises when modeling the converter for simulation. With eight semiconductor switches, input and output capacitors, and transformer leakage inductance the DAB presents multiple challenges to the engineer modeling the converter. This paper will explore a transconductance model of the DAB converter and provide a comparison with a more traditional model. The proposed model captures critical terminal characteristics that are important for engineers deploying DABs in systems.

II. DISCUSSION OF PREVIOUS LITERATURE

The challenge in accurately modeling the DAB converter arises from a combination of several factors. The DAB, unlike many other DC to DC converters, ideally transfers power entirely via high frequency alternating current across the central transformer. The lack of a DC component precludes the use of the "small ripple approximation" on the leakage inductance current that is a widely accepted technique in the analysis of other DC converters. This led previous researchers to use advanced nontraditional techniques such as generalized averaged models (GAM), which result in complex and unintuitive models [7]–[9]. The GAM approach involves higher order terms, 6th order, along with inclusion of harmonics of the switching signal. GAM is based upon the application of the

Fourier Series for each state variable. Yet, upon examination of the several model variations proposed in the papers, the result can be considered a first-order response. This result of the generalized averaged modeling approach is a recognition of the nature of the dynamic behavior of the converter. The control-to-output frequency plot is an almost perfect first-order response. This results in the dynamics of the leakage inductor not being significant below the switching frequency because the current is reset to zero each switching period; therefore, the inductor current does not act as a state in the switch-averaged model of the DAB as shown in [10]. This leads the authors to propose the transconductance model of the DAB converter proposed in this work. The model follows the equivalent circuit modeling approach used in [11] to obtain two-port switch models for converters operating in discontinuous conduction mode and current programmed mode. This approach separates the switches and time invariant reactive elements of the circuit, replacing the switches with controlled voltage or current sources.

III. The Transconductance Model

A. Single-Phase-Shift Control

The DAB converter, shown in Fig. 1, consists of two controlled full bridges with a high frequency transformer in between. In the simplest control scheme, termed Single-Phase-Shift (SPS) modulation, each half bridge is operated at a 50% duty cycle. The control input, ϕ, is the phase shift between primary and secondary full bridges. The creation of square wave voltages at the terminals of the transformer allows for the analysis presented in [2]. The phase shifted square wave voltages across the transformer leakage inductance can be considered similar to power transfer through an inductive line in an AC power system. Thus, the critical parameter determining power flow is the transformer leakage inductance, here it is considered referred to the primary side.

B. Transconductance Model

This model of the converter is significantly simpler than previous attempts yet retains the dominant dynamic features of the DAB. Importantly, this model is as accurate as previously proposed approaches. In a DAB under the most common modulation strategy, single-phase-shift (SPS) modulation, the power transfer across the converter from [1] is given by:

$$P = \frac{v_1 v_2}{2\pi n f_s L_{lk}}\phi\left(1 - \frac{\phi}{\pi}\right) \quad (1)$$

Here v_1 and v_2 are the input and output voltages, n is the transformer turns ratio, ϕ is the phase shift between the primary and secondary full bridges, f_s is the switching frequency, and L_{lk} the leakage inductance of the transformer referred to the primary side. Thus, it clearly follows that the average current drawn from the primary side full-bridge stage is i_1:

$$i_1 = \frac{P}{v_1} = \frac{v_2}{2\pi n f_s L_{lk}}\phi\left(1 - \frac{\phi}{\pi}\right) \quad (2)$$

and the current injected into the output capacitor and load by the secondary side full bridge is i_2:

$$i_2 = \frac{P}{v_2} = \frac{v_1}{2\pi n f_s L_{lk}}\phi\left(1 - \frac{\phi}{\pi}\right) \quad (3)$$

These currents can be imposed directly in the model using two controlled current sources which draw and inject current from input and output sides. This approach is similar to that used in [11] but allows for modeling the input-side dynamics and shows explicitly the relationship between the two controlled current sources. In this model, the two controlled sources replace the entire central portion of the DAB converter—including the transformer and all semiconductor switches. The result, shown in Figure 2, is a fully switch-averaged model of the DAB which can be executed much faster than a traditional model. Note that the function $g_m(\phi)$ referenced in the figure is a nonlinear transconductance given by:

$$g_m(\phi) = \frac{1}{2\pi n f_s L_{lk}}\phi\left(1 - \frac{\phi}{\pi}\right) \quad (4)$$

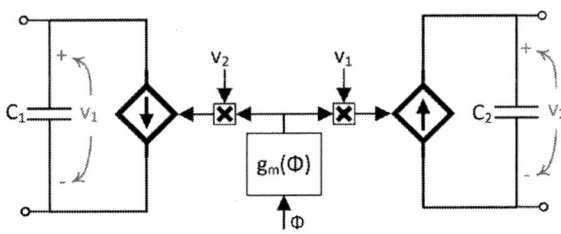

Fig. 2. The equivalent model of the dual active bridge converter replaces the switches and high-frequency transformer with controlled current sources and a nonlinear function of the phase shift between primary and secondary bridges.

IV. Model Comparison Switching and Transconductance

A. Simulation Study - SPS Modulation

The example system for this modeling is a DAB with a resistive load with input voltage and source Thevenin impedance shown in Figure 3. A Thevenin impedance is included on the input to retain the dynamics of the input because an ideal source would render the voltage across the input capacitor fixed. The component values used for the simulation and frequency domain study are shown in Table I.

TABLE I
DAB Component Values for Simulation Study

Component	Value
R_s	0.5Ω
L_s	1 mF
n	1:1
L_{lk}	27.2μH
C_1	500 μF
C_2	100 μF
F_{sw}	50 kHz

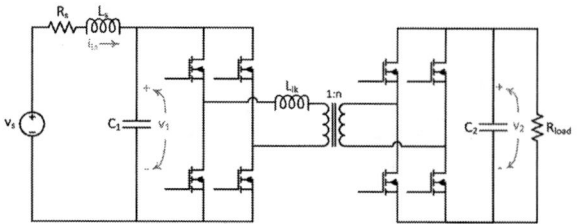

Fig. 3. The schematic of the DAB converter which is used to validate the accuracy of the proposed modeling approaches.

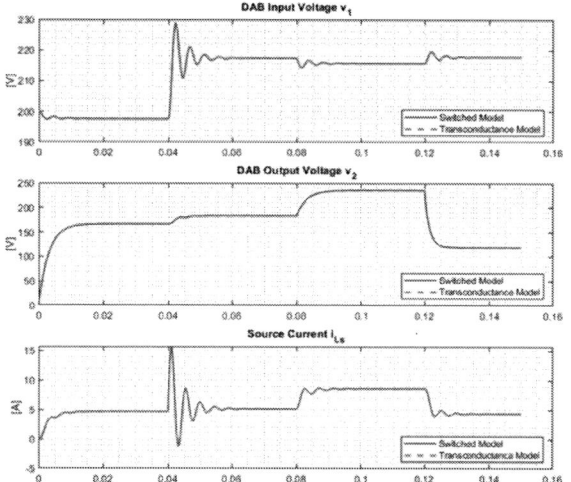

Fig. 4. Simulation results for equivalent DAB converter models using a traditional switching simulation and the transconductance-based average model which show excellent agreement and no loss of dynamic accuracy.

The transconductance model of the system under test is shown in Figure 6. Using MATLAB/Simulink, the transconductance model and a reference traditional switching model are simulated and subjected to identical step function perturbations in the input voltage, load resistance, and phase shift ϕ. The simulation results in Figure 4 from inputs shown in Figure 5 confirm outstanding agreement between both models, showing that no accuracy was lost in the equivalent transconductance-based model.

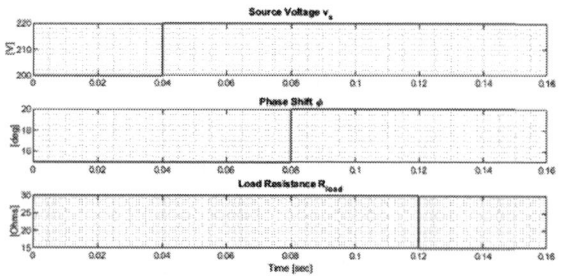

Fig. 5. Simulation inputs for model verification.

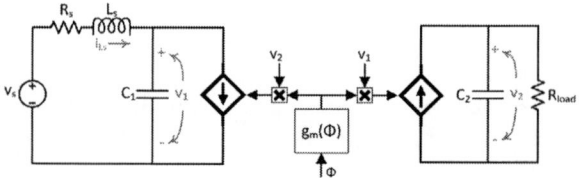

Fig. 6. The schematic of the transconductance-based equivalent model of the DAB converter replaces the switches and transformer with a pair of controlled current sources.

B. Frequency Domain Analysis

To further analyze the suitability of the transconductance model, a brief frequency domain analysis is performed. The control-to-output voltage transfer function is derived analytically, then Bode plots are generated using the MATLAB/Simulink linear analysis tool of the transconductance model. These results are compared against the analysis in [8] which found a first order transfer function for the control-to-output-voltage. While the traditional power electronics analysis fails when considering power transfer across the transformer leakage inductance, the capacitor charge balance can be written for the output capacitor with capacitor current i_c, DAB output current i_o, load resistance R, and output voltage v_{out}. The equivalent circuit used to model the control-to-output-

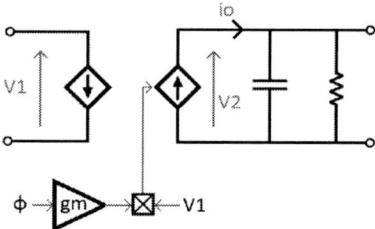

Fig. 7. Equivalent circuit model utilized to generate the analytical frequency domain transfer function $G_{\phi V_{out}}$.

voltage transfer function is shown in Figure 8.

$$i_c = i_o - \frac{v_{out}}{R} \tag{5}$$

The capacitor current can be rewritten as the time derivative of the capacitor voltage, and the output current of the DAB can be represented as the multiplication of the transconductance term $g_m(\phi)$ and DAB input voltage V_1.

$$C\frac{dv}{dt} = g_m(\phi)v_1 - \frac{v_{out}}{R} \tag{6}$$

Performing the small signal analysis from [11], the term related to output current, $g_m(\phi)v_1$, is

$$i_o = g_m(\phi)v_1 = \frac{1}{2\pi n f_s L_{lk}}\phi(1 - \frac{\phi}{\pi}) \tag{7}$$

The coefficients related to the switching frequency and leakage inductance can be written as k.

$$k = \frac{1}{2\pi n f_s L_{lk}} \tag{8}$$

979-8-3315-1612-3/25 $31.00 © 2025 IEEE   2867

The small signal analysis is then performed which results in

$$\hat{i}_o = k\hat{\phi}V_1 - \frac{2}{\pi}\hat{\phi}V_1\Phi + \hat{v_1}\Phi\left(1 - \frac{\Phi}{\pi}\right) \quad (9)$$

The result in 9 can be substituted into 6 and transformed into the s-domain which results in the transfer function

$$G_{\phi v_{out}} = \frac{v_{out}(s)}{\phi(s)} = \frac{R(kV_1 - \frac{2}{\pi}\Phi V_1)}{1 + sRC} \quad (10)$$

As described, the dominant behavior of the DAB under the SPS modulation scheme for control-to-output-voltage is a first order system. Using the MATLAB/Simulink Linear Analysis Toolbox, the transconductance model is linearized and the frequency response is plotted in a Bode plot. The plot is shown in Figure 8 and matches the analytical derivation above which shows a first order response.

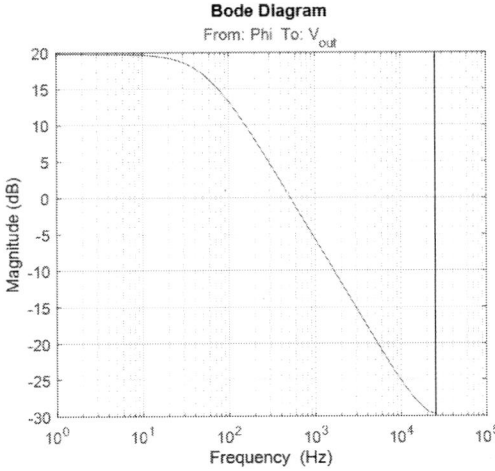

Fig. 8. Bode plot of the control-to-output frequency response for the SPS modulation scheme transconductance model at 50 kHz.

V. EXPERIMENTAL TRANSCONDUCTANCE MODEL VALIDATION

To validate the transconductance model, a hardware study is performed. The DAB transconductance model and the physical hardware are controlled via a PI regulator which regulates output voltage via the phase shift ϕ. The results of the physical converter and transconductance model are comapred for a load step.

A. Hardware Under Study

The DAB converter is built using four Imperix PEB8038 SiC half bridge power modules controlled via an Imperix B-Box controller. The transformer is a planar construction with an external series inductor. Measurement of waveforms is made with a Tektronix oscilloscope. The DAB is connected to a controllable electronic load set in constant resistance mode. The topology of the DAB is shown in Figure 3, component values are displayed in Table II.

TABLE II
DAB COMPONENT VALUES FOR HARDWARE VALIDATION

Component	Value
R_s	0.32Ω
L_s	120 μF
n	1:1
L_{lk}	27μH
C_1, C_2	1000 μF
F_{sw}	30 kHz

B. Comparison of Transconductance and Physical DAB Waveforms

To explore the physical behavior of the DAB under SPS modulation, a step test was conducted. The DAB converter is operating in closed loop control with a PI regulator controlling the output bus voltage. The oscilloscope is utilized to capture the input and output voltages, and the input inductor current. The converter is operating in steady state when a step change to the load resistance is implemented. This steps from a 100 Ω load to a 50 Ω load, which corresponds to a step in the load power from 2.5 kW to 5 kW. The waveforms of the physical

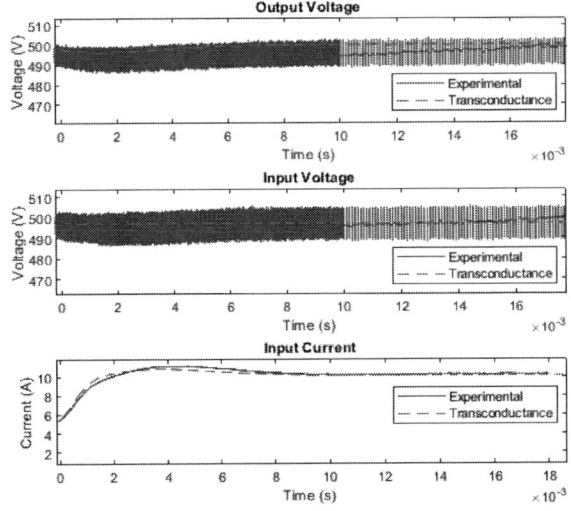

Fig. 9. Simulation and hardware results for a load step in the load resistance from 100 Ω to 50 Ω. The top plot is the output voltage, the middle plot is the input voltage, and the bottom plot is the input current.

DAB converter are plotted in Figure 9 alongside waveforms from a simulation transconductance model of the same DAB. The waveforms show good agreement between the physical and simulation converters. Of note, the tuning and implementation of a closed loop control impact the accuracy of the model when compared to the physical converter. When modeling the DAB, especially when it is in a cascade configuration of converters, it is critical to match the controls and interface setpoints appropriately.

VI. DUAL PHASE SHIFT MODULATION

The transconductance model presented in this paper is an intuitive and useful method for modeling the DAB converter under the SPS modulation scheme. In practice, the SPS modulation scheme has certain limitations relating primarily to the inability to control reactive power in the converter. The Dual Phase Shift (DPS) Modulation introduces a second control input allowing for minimization of reactive power in the converter [12]. This transconductance model is well suited for adaption to the DPS control modulation scheme.

A. DPS Theory

DPS modulation of the DAB converter can be considered a generalization of the SPS control scheme [12], [13]. Similar to SPS modulation, each converter half-bridge leg is switched at a 50% duty cycle. Under DPS modulation, there are two control inputs to the DAB, the inner phase shift, D_1, and the outer phase shift, D_2. These phase shifts are in the range of [0 1], which maps to a [0 π] phase shift over a switching period. The outer phase shift is the same as the phase shift term, ϕ, from SPS modulation. The inner phase shift, D_1, is between the half bridge legs within a full bridge. SPS modulation is thus a specific case of DPS when D_1 is set to 1. From [13], the power is dependent on the relationship between D_1 and D_2 which is termed G_D. The power for the four operating points (OP) is given by

$$P = \frac{V_1 V_2}{4 f_s L_{lk}} \underbrace{\begin{cases} D_2(2D_1 - D_2), & OP\ 1 \\ 2D_2 + 2D_1 - 1 - 2D_2^2 - D_1^2, & OP\ 2 \\ D_1^2, & OP\ 3 \\ 2D_1 - D_2^2 + 2D_2 - 2D_2 D_1 - 1, & OP\ 4 \end{cases}}_{G_D}$$

$$\tag{11}$$

TABLE III
DPS POWER EQUATION OPERATING POINT

OP 1	$D_2 \leq D_1$ and $D_1 + D_2 \leq 1$
OP 2	$D_2 \leq D_1$ and $D_1 + D_2 > 1$
OP 3	$D_2 \geq D_1$ and $D_1 + D_2 \leq 1$
OP 4	$D_2 \geq D_1$ and $D_1 + D_2 > 1$

The operating point is determined by the relationship between D_1 and D_2 according to the equations shown in Table III.

B. DPS Transconductance Model

The creation of the DPS modulation transconductance term is straightforward and follows the same approach as the SPS. The topology of the equivalent circuit is identical to Figure 2. The nonlinear transconductance function is again given by a coefficient term related to f_s and L_{lk}, and a gain term according to the relationship between D_1 and D_2, G_D. G_D is determined for the operating point based on the equations shown in Table III.

$$g_m(\phi) = \frac{1}{4 f_s L_{lk}} G_D \tag{12}$$

C. Simulation Study - DPS Modulation

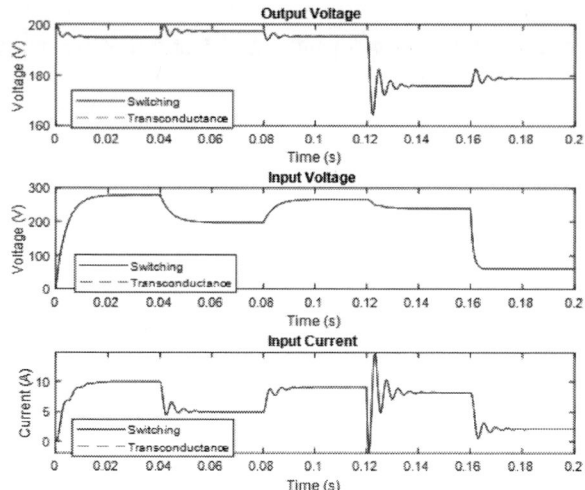

Fig. 10. Simulation results for equivalent DAB converter models using a traditional switching simulation and the DPS transconductance-based average model which show excellent agreement and no loss of dynamic accuracy.

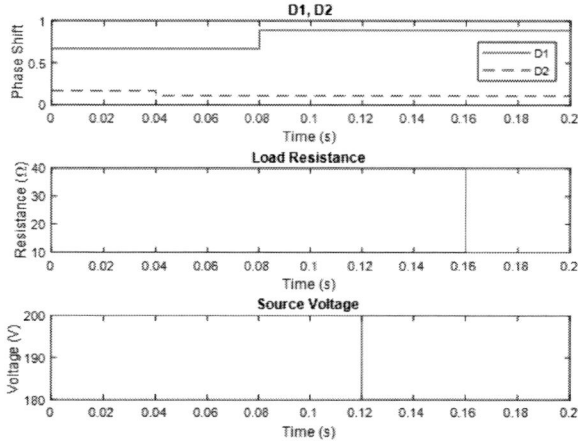

Fig. 11. Simulation inputs for model verification of the DAB converter under DPS modulation.

To validate the performance of the DPS transconductance model, a simulation study is performed. The same parameters from the SPS simulation study are utilized, which are shown in Table I. The results from a switching model and the transconductance model subject to input variations are shown in Figure 10. The inputs to the simulation are shown in Figure 11. As with the SPS case, the model shows an excellent agreement with the switching model.

VII. CONCLUSIONS

In this paper, a computationally efficient and intuitive model of the DAB converter is presented. This model maintains accuracy compared to a traditional switching model yet requires less effort to set up and executes faster. The transconductance model was presented along with time and frequency domain analysis in simulation. The waveforms predicted from the transconductance model were validated with a physical DAB in hardware. This model provides a simple model that captures transient and steady state terminal behavior of interest for engineers implementing DAB across a range of applications. This model is useful for the practicing engineer as it provides an effective method of capturing terminal dynamics of the DAB. This paper introduced the transconductance model along with time and frequency domain analysis of the model compared to a traditional model under SPS modulation. The model was compared experimentally to a DAB operating at $30 \, kHz$, with a unity turns ratio transformer at 500 V. The transconductance approach was then extended in simulation to model the DPS modulation scheme.

REFERENCES

[1] R. De Doncker, D. Divan, and M. Kheraluwala, "A three-phase soft-switched high-power-density dc/dc converter for high-power applications," *IEEE Transactions on Industry Applications*, vol. 27, no. 1, pp. 63–73, 1991.

[2] B. Zhao, Q. Song, W. Liu, and Y. Sun, "Overview of dual-active-bridge isolated bidirectional dc–dc converter for high-frequency-link power-conversion system," *IEEE Transactions on Power Electronics*, vol. 29, no. 8, pp. 4091–4106, 2014.

[3] Y. Liu, X. Wang, W. Qian, A. Janabi, B. Wang, X. Lu, K. Zou, C. Chen, and F. Z. Peng, "Dc voltage control of inverter interfaced dual active bridge converter for v2l applications," in *2019 IEEE 7th Workshop on Wide Bandgap Power Devices and Applications (WiPDA)*, pp. 319–324, 2019.

[4] B. Rahrovi, R. T. Mehrjardi, and M. Ehsani, "On the analysis and design of high-frequency transformers for dual and triple active bridge converters in more electric aircraft," in *2021 IEEE Texas Power and Energy Conference (TPEC)*, pp. 1–6, 2021.

[5] N. Rajagopal, C. DiMarino, R. Burgos, I. Cvetkovic, and M. Shawky, "Design of a high-density integrated power electronics building block (ipebb) based on 1.7 kv sic mosfets on a common substrate," in *2021 IEEE Applied Power Electronics Conference and Exposition (APEC)*, pp. 1–8, 2021.

[6] M. Aguirre and A. Yazdani, "A single-phase dc-ac dual-active-bridge based resonant converter for grid-connected photovoltaic (pv) applications," in *2019 21st European Conference on Power Electronics and Applications (EPE '19 ECCE Europe)*, pp. P.1–P.10, 2019.

[7] H. Qin and J. W. Kimball, "Generalized average modeling of dual active bridge dc–dc converter," *IEEE Transactions on Power Electronics*, vol. 27, no. 4, pp. 2078–2084, 2012.

[8] J. A. Mueller and J. W. Kimball, "An improved generalized average model of dc–dc dual active bridge converters," *IEEE Transactions on Power Electronics*, vol. 33, no. 11, pp. 9975–9988, 2018.

[9] J. A. Mueller and J. W. Kimball, "Modeling dual active bridge converters in dc distribution systems," *IEEE Transactions on Power Electronics*, vol. 34, no. 6, pp. 5867–5879, 2019.

[10] H. Bai, C. Mi, C. Wang, and S. Gargies, "The dynamic model and hybrid phase-shift control of a dual-active-bridge converter," in *2008 34th Annual Conference of IEEE Industrial Electronics*, pp. 2840–2845, 2008.

[11] R. W. Erickson and D. Maksimovic, *Fundamentals of Power Electronics*. No. 3rd, Springer, 2020.

[12] H. Bai and C. Mi, "Eliminate reactive power and increase system efficiency of isolated bidirectional dual-active-bridge dc–dc converters using novel dual-phase-shift control," *IEEE Transactions on Power Electronics*, vol. 23, no. 6, pp. 2905–2914, 2008.

[13] M. Kim, M. Rosekeit, S.-K. Sul, and R. W. A. A. De Doncker, "A dual-phase-shift control strategy for dual-active-bridge dc-dc converter in wide voltage range," in *8th International Conference on Power Electronics - ECCE Asia*, pp. 364–371, 2011.

Lumped Parameter Modeling for Real-Time Thermal Regulation of Li-Ion Battery Packs

Utkal Ranjan Muduli
APEC, EE Department
Khalifa University
Abu Dhabi, UAE
utkal.muduli@ku.ac.ae

Mohamed Shawky El Moursi
APEC, EE Department
Khalifa University
Abu Dhabi, UAE
mohamed.elmoursi@ku.ac.ae

Khalifa Al Hosani
APEC, EE Department
Khalifa University
Abu Dhabi, UAE
khalifa.halhosani@ku.ac.ae

Ahmed Al-Durra
APEC, EE Department
Khalifa University
Abu Dhabi, UAE
ahmed.aldurra@ku.ac.ae

Abstract—Overheating in lithium-ion battery packs can significantly reduce their performance and lifespan. To address this issue, we have developed a lumped parameter thermal model that accurately represents the thermal behavior of these battery packs. This model includes the effects of fluid flow and the channel width between parallel cells, aspects often overlooked in traditional models. By integrating our model with an internal combustion engine cooling method, we form a comprehensive thermal regulation system. We implement this system in real-time using OPAL-RT to evaluate its performance under various operational scenarios. Our results highlight the importance of considering channel width in thermal modeling to enhance accuracy. The real-time implementation demonstrates the practicality of our approach for grid-connected battery energy storage systems. This work advances battery thermal management by providing a model suitable for real-time applications. It offers insights for designing efficient cooling processes and preventing overheating. Our study contributes to improving the safety and efficiency of lithium-ion battery packs.

Index Terms—Lithium-ion Battery Packs; Thermal Management; Numerical Simulation; Real-time Implementation.

I. INTRODUCTION

The surge in demand for high-capacity lithium-ion batteries has spotlighted a pressing issue: excessive heat generation during operation can lead to performance degradation and safety hazards. As electric vehicles (EVs) and grid-connected energy storage systems become more prevalent, efficient thermal management of battery packs is crucial for ensuring reliability and longevity [1]–[4]. Heat accumulation not only reduces the efficiency of batteries but also poses significant risks of thermal runaway [5]. Therefore, understanding and controlling the thermal behavior of lithium-ion battery packs is a vital area of research [6]. Recent studies have focused on various cooling methods to mitigate these thermal challenges [7], [8]. Optimizing airflow configurations has shown promise in improving thermal efficiency [7]. Forced convection cooling strategies have been investigated for their effectiveness in reducing temperature gradients within battery packs [6], [8]. Moreover, innovative placement of air inlets and outlets has demonstrated significant improvements in heat dissipation [9],

[10]. Thermal modeling techniques, such as reduced order models [11] and deep neural networks [12], have been developed to predict temperature distributions and enhance thermal management. These advancements underline the importance of designing thermal management systems that can adapt to the dynamic conditions of battery operation. Addressing thermal issues is not just about safety; it also enhances the overall performance and lifespan of the batteries [5]. Our research contributes to this critical field by proposing a novel thermal regulation system for lithium-ion battery packs.

Despite advancements in cooling techniques, existing thermal management systems often fall short in real-time regulation under varying operational conditions. Traditional models may not accurately capture the complex interactions between fluid flow and heat transfer within battery packs. Moreover, the channel width between parallel battery cells significantly affects the thermal behavior, yet this factor is often overlooked in conventional models [13]. Without precise thermal modeling, batteries are susceptible to uneven temperature distributions, leading to decreased performance and potential safety risks. Real-time thermal regulation requires models that are both accurate and computationally efficient [14]. Current approaches may lack the ability to be implemented in real-time control systems due to their complexity [15]. Therefore, there is a need for a thermal model that accounts for fluid dynamics and can be integrated into real-time applications. By addressing these challenges, we can enhance the safety and efficiency of lithium-ion battery packs in practical applications. Our work aims to fill this gap by developing a lumped parameter model that considers realistic fluid flow and heat processes. This model facilitates real-time thermal regulation, improving the overall performance of battery energy storage systems.

In the quest for effective thermal management, various studies have explored different cooling strategies and modeling techniques for lithium-ion battery packs. Park [7] conducted numerical simulations to enhance thermal efficiency through optimized airflow configurations. Pesaran [8] investigated the impact of cell configurations and forced convection, finding that increased flow rates reduce temperature differences between cells. Mahamud and Park [16] analyzed the effects of alternating airflow rates, demonstrating improvements in thermal performance. Agelin-Chaab et al. [9] studied the placement

"This work was supported in part by Advanced Power and Energy Center (APEC), Khalifa Univeristy, Abu Dhabi, United Arab Emirates; and in part by the Advanced Technology Research Council ASPIRE Virtual Research Institute (VRI) Program, Abu Dhabi, United Arab Emirates, under Award VRI20-07."

of additional air intakes, observing significant decreases in battery temperature and increases in heat uniformity. Wang et al. [6] used three-dimensional models to explore factors influencing forced convection cooling effectiveness. Lu et al. [10] simulated staggered battery packs, showing that top placement of air inlets maximizes cooling performance. Afzal et al. [1] evaluated thermal characteristics of EV battery systems, providing insights into temperature distribution and its impact on performance. Reduced order models have been developed to assess the effects of cell parameter variations in battery pack architectures [11]. Deep neural networks have been employed for battery surface temperature estimation under various conditions [12], enabling accurate predictions essential for thermal management. Fault diagnosis methods have been proposed to detect early-stage faults and enhance safety in battery packs [17], [18]. Optimization of charging strategies from a thermal safety perspective has also been a focus [19]. Iterative learning control designs have been developed for battery electro-thermal management under varying state-of-charge patterns [20]. Physics-informed neural networks have been utilized for temperature field reconstruction in battery packs [14]. Integrated strategies for optimized charging and balancing with thermal management have been proposed [21]. Investigations into the causes of performance degradation and state of health estimation have been conducted [22]. Modeling and inconsistency analysis of battery packs in electric vehicles have been explored [23]. Simulation platforms have been created to analyze aging, temperature, and current inconsistencies in parallel battery packs [24]. Thermal management systems have been modeled for hybrid electric vehicles [4]. Furthermore, methods for abnormality detection using spatiotemporal entropy have been developed [25], and open circuit fault diagnostics have been compared to identify efficient approaches [26]. Phase change material-based thermal management has been explored to enhance battery pack performance [27]. Transient thermal analysis using simulation tools like COMSOL has been conducted to understand voltage and temperature relationships in battery packs [28]. These diverse studies contribute to the understanding of thermal behavior and fault management in battery systems, yet there remains a gap in models suitable for real-time implementation.

Building upon the existing body of research, our study has three primary objectives. First, we aim to develop a lumped parameter thermal model that accurately represents the thermal behavior of lithium-ion battery packs, including the effects of fluid flow and channel width between cells. Second, we intend to integrate this model with an internal combustion engine cooling method to form a comprehensive thermal regulation system. Third, we plan to implement this system in real-time using OPAL-RT to evaluate its performance under various operational scenarios. The key contributions of our work are as follows. We provide a novel thermal model that incorporates realistic fluid flow and heat processes, offering enhanced accuracy over traditional models [29]. We demonstrate the importance of considering channel width in thermal modeling, which has been underrepresented in previous

studies. Additionally, we validate our model through real-time implementation, showcasing its practicality for applications in grid-connected battery energy storage systems [30]. By addressing these areas, our research advances the field of battery thermal management and contributes valuable insights for future developments.

The remainder of this paper is organized as follows. In Section II, we present the detailed development of the lumped parameter thermal model, including the mathematical formulations and assumptions made. Section III describes the integration of the thermal model with the internal combustion engine cooling method to form the thermal regulation system. In Section IV, we discuss the real-time implementation of the proposed system using OPAL-RT and outline the experimental setup. Section V provides the results of our tests under various operational conditions, analyzing the performance of the thermal regulation system. Finally, Section VI concludes the paper by summarizing our findings and suggesting directions for future research.

II. PROPOSED METHODOLOGY

A simulation model was developed to evaluate the thermal performance of a coolant interacting with a battery pack under the assumption of uniform heat generation and laminar fluid flow. The heat generated by the battery cells is removed through forced convection, where the heat transfers from the battery surface to the adjacent cooling fluid. The model relies on steady-state conduction equations and prescribed boundary conditions to assess the temperature distribution and gradients within the battery cells. This approach ensures precise quantification of thermal behavior essential for optimizing battery thermal management systems. The mathematical framework of the model incorporates energy conservation equations as outlined in [1]. The heat generation within the system is governed by the equation:

$$k_a \nabla^2 T + H_g^m = 0 \qquad (1)$$

This equation represents the interplay between thermal conductivity (k_a), the temperature field (T), and volumetric heat generation (H_g^m). It forms the foundation for predicting heat diffusion within the battery cells. To accurately model fluid flow and heat transfer, energy and continuity equations were included, capturing essential variables such as pressure (p), density (ρ), viscosity (μ), and thermal diffusivity (α). The governing equations are expressed as:

$$\nabla p = 0,$$
$$p \nabla p = -\frac{1}{\rho} \nabla \eta + \mu \nabla^2 p, \qquad (2)$$
$$p \nabla T = \alpha \nabla^2 T$$

These equations ensure that the fluid's thermodynamic state and velocity distribution are accurately represented. Dimensionless variables were introduced to normalize spatial and thermal parameters, facilitating scalability and computational efficiency. Relations like

$$L_2^i = \frac{L_2}{L_1}, \quad U = \frac{p}{p_b}, \quad \bar{T} = \frac{T - T_b}{T_1 - T_b} \qquad (3)$$

979-8-3315-1612-3/25 $31.00 © 2025 IEEE

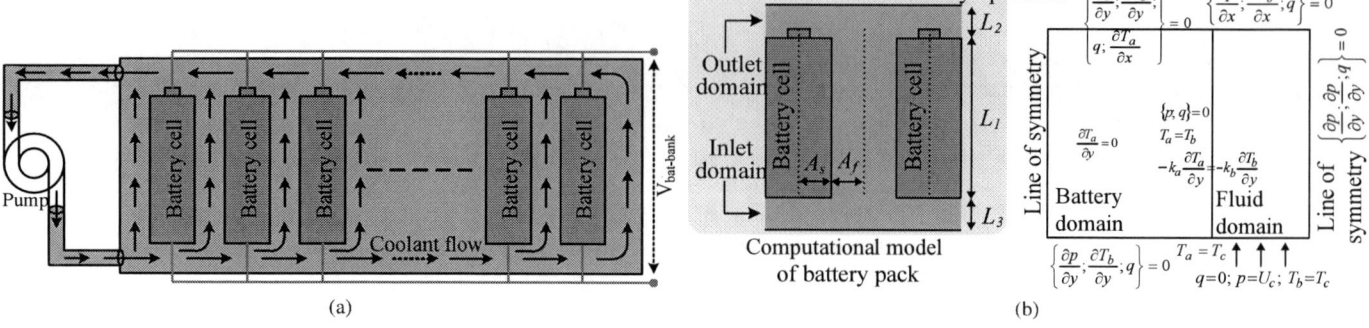

Fig. 1. (a) Thermal process of BPs using battery cells and (b) Boundary conditions of the batter pack during the thermal process.

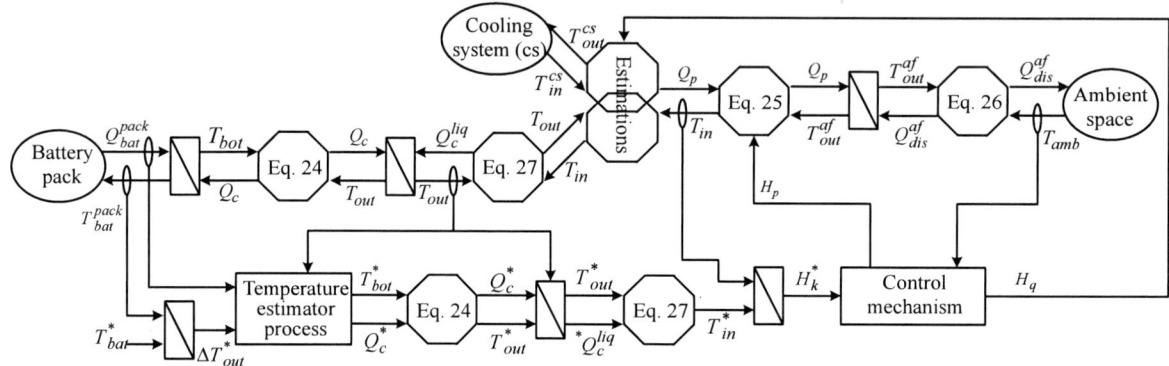

Fig. 2. Energetic Macroscopic Representation for thermal control model of the battery pack

reduce complexity and enhance the interpretability of results. These transformations provide a clearer understanding of the heat transfer process by focusing on non-dimensionalized temperature, pressure, and velocity distributions. Elliptic governing equations dictate the solution domain, necessitating boundary conditions for all sides of the computational region. These conditions are defined mathematically as follows:

$$\frac{\partial \bar{T}_a}{\partial Y} = 0 \qquad \text{for } Y = 0, \ 0 \leq X \leq 1 \tag{4}$$
$$\bar{T}_a = \bar{T}_b \qquad \text{for } Y = 1, \ 0 \leq X \leq 1$$
$$\bar{T}_a = 0 \qquad \text{for } X = 0, \ 0 \leq Y \leq 1$$
$$\frac{\partial \bar{T}_a}{\partial X} \bar{T}_a = 0 \qquad \text{for } X = 1, \ 0 \leq Y \leq 1 \tag{5}$$

These conditions ensure accurate representation of thermal interactions at the boundaries, critical for determining internal and external temperature gradients. The model's robustness, grounded in well-established equations and boundary conditions, supports detailed analysis of thermal behavior in battery cooling systems. By integrating heat conduction, convection, and fluid dynamics, the simulation offers valuable insights into the design of efficient thermal management solutions.

The computational domains, as depicted in Fig. 1, are expanded in the transverse directions to ensure the comprehensive capture of fluid flow and heat transfer constraints. This expansion accounts for the adjustment of vortices during the interaction of the fluid with the lower section and the diffusive effects of low Prandtl liquids. The height was selected as L_3 at the inlet and L_2 at the exit, following a series of initial

studies conducted. These adjustments were made to enforce additional boundary conditions that are physically meaningful [1]. The governing boundary conditions for the computational domain are expressed as:

$$\begin{cases} \dfrac{\partial \bar{T}_b}{\partial Y} = 0; \ V = 0; \ \dfrac{\partial U}{\partial Y} = 0 \\ \qquad \text{for } Y = 1; \ -L_3^i \leq X \leq 0 \text{ and } L_1 \leq X \leq L_2^i, \\[4pt] \dfrac{\partial T_a}{\partial Y} = \dfrac{1}{\xi_b} \dfrac{\partial \bar{T}_b}{\partial Y}; \ V = 0; \ U = 0 \\ \qquad \text{for } Y = 1; \ 0 \leq X \leq L_1, \\[4pt] \dfrac{\partial \bar{T}_b}{\partial Y} = \dfrac{\partial U}{\partial Y} = 0; \ V = 0 \\ \qquad \text{for } Y = \bar{A}_f = 1; \ -L_3^i \leq X \leq (L_1 + L_2^i), \\[4pt] T_a = 0; \ V = 0; \ U = 1 \\ \qquad \text{for } 0 \leq Y \leq 1; \ X = L_3^i, \\[4pt] \dfrac{\partial \bar{T}_a}{\partial X} = \dfrac{\partial U}{\partial Y} = 0; \ V = 0 \\ \qquad \text{for } 0 \leq Y \leq 1; \ X = L_1 + L_2^i \end{cases} \tag{6}$$

The operational temperature of the battery pack is determined using a thermal equivalent electrical circuit, illustrated in Fig. 1. This circuit employs lumped parameters to represent thermal processes. Due to the symmetry of the battery module, only half of the module needs to be considered in the thermal model. In this representation, heat flux is replaced by electrical current, heat capacity by electrical capacitance, thermal resistance by electrical resistance, and temperature differences

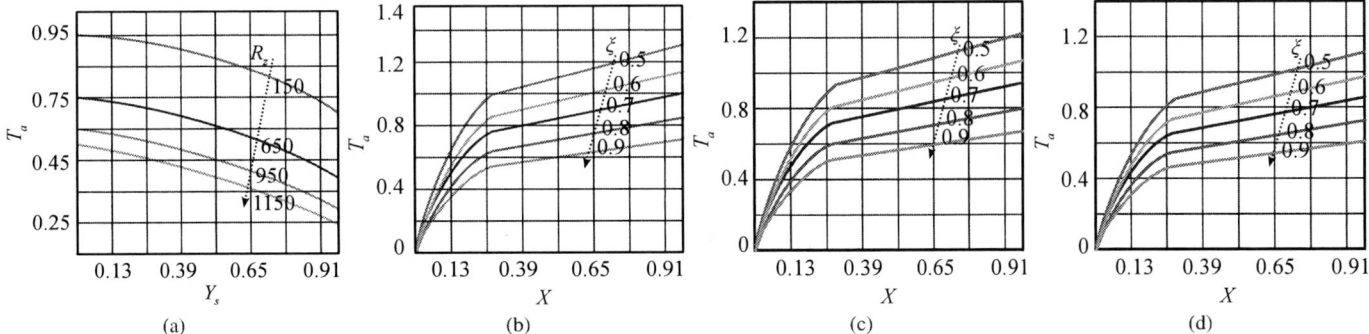

Fig. 3. Battery's temperature distribution profile (a) Effect of R_z, (b) at R_z=150, (c) at R_z=650, (d) at R_z=950

by voltage [4]. The lumped parameters are calculated based on the dimensions and thermal properties of the materials used in the module's heat extraction system. The thermal resistances in the electrical model are determined as follows:

$$R_{th,i} = \frac{(L_1 + L_2 + L_3)}{S_i \lambda_i},$$

$$Q_c = \left(V_{oc} - \frac{T_{cell}\Delta S}{nF} \right),$$

$$\rho V_p \frac{dT_{out}}{dt} = m_c \left(T_{in} - T_{out} \right) + \frac{h_p \left(T_{bot} - T_{out} \right)}{C_{pc}}. \tag{7}$$

Additional expressions governing the dynamics of the thermal system are given as:

$$\frac{dT_{out}^{af}}{dt} = \frac{1}{C_r} \left[H_p C_{pc} m_{af} (T_{in} - T_{out})^{af} - Q_{dis}^{af} \right],$$

$$\frac{dT_{out}}{dt} = \frac{m_p}{V_p \rho} \left(T_{in} - T_{out} \right), \tag{8}$$

where:

$$Q_{dis}^{af} = h^{af} \left(T_{in}^{af} - T_{amb} \right) + \ell H_{fan} C_a m_a \left(T_{out}^{af} - T_{amb} \right). \tag{9}$$

The Energetic Macroscopic Representation (EMR) was developed in 2000 by the L2EP Lab in Lille, France, to model intricate electromechanical and electrochemical systems. The EMR theory proposes that the control system of an object can be derived by inverting the object model. Fig. 2 presents the EMR for the thermal control model of the battery pack.

III. RESULTS AND DISCUSSION

The simulation of the thermal protection system operation took into account five ambient temperature values ranging from 262 to 290K. The standard mean temperature of the cells was selected to be 300 K. The maximum acceptable temperature for the liquid is 340K. The models were compared using different controllers for each ambient temperature, taking into account the average errors between the reference temperature value and the actual values. The errors' mean values were computed following the battery pack heating procedure, once the battery pack's average temperature hit the specified reference value. The Fig. 4 displays the mean temperatures of the cells while the battery pack is in operation for every ambient temperature taken into account. Fig. 4 illustrates the heating time of the battery pack for the reference temperature ranging

from 290 to 1380 seconds. The pack will be positioned within the bus, where the minimum temperature will be around 262K due to the bus heating system, making the proposed heating duration deemed acceptable. It may be necessary to increase the temperature of the liquid in order to reduce the heating time. Proper thermal control of heat-producing Li-ion batteries is crucial to ensure optimal performance, increased efficiency, safe operation, and extended lifespan. The rising need for Lithium-ion batteries in electrical vehicles and electronic devices has motivated numerous researchers to explore the issues related to overheating in lithium ion batteries, often caused by inadequate thermal management systems. An examination of the thermal characteristics in a prismatic battery cell that produces consistent heat throughout its operation is conducted. The conjugate heat transfer condition is taken into account at the interface between the cell and the fluid, ensuring continuity of heat flux and temperatures. Air serves as the cooling agent that transports heat away from the surface of the cell. The assumption is made that the air flow is laminar and in a steady state condition. The parameters' range is considered based on various sources pertaining to modern batteries, and they are normalized using the appropriate terms mentioned previously.

Fig. 3 (a) illustrates the variation in temperature distribution as R_z increases. It is clear from the diagram that higher R_z values lead to a decrease in temperature. At first, a greater decrease in temperature is achieved, which then becomes almost negligible at elevated R_z levels. With an increased R_z, the heat transfer coefficient rises, resulting in greater heat dissipation from the battery's surface and subsequently leading to a decrease in heat. Additionally, an augmented velocity in higher R_z flows leads to a thicker thermal boundary layer and consequently enhances heat transfer. Consequently, it is easy to see that with an increase in R_z, there is an improvement in heat transfer along with a corresponding decrease in the temperature of the cells in pack. However, there is a maximum R_z value beyond which no reduction in temperature can be achieved, leading to an increased requirement for pumping power. If the value of R_z falls below 150, the temperature rise will become more severe, leading to a thermal runaway scenario. Consequently, Re should not be diminished below its lower limit. Likewise, adjusting the spacing in conjunction with R_z will produce a result closely resembling the one depicted in Fig. 3 (b). Higher R_z values will not have any

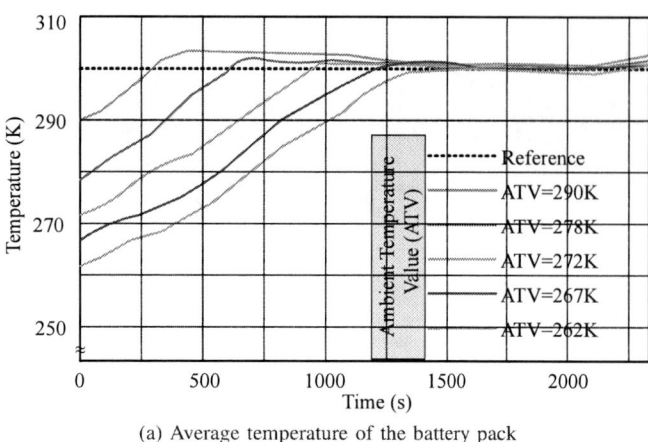

(a) Average temperature of the battery pack

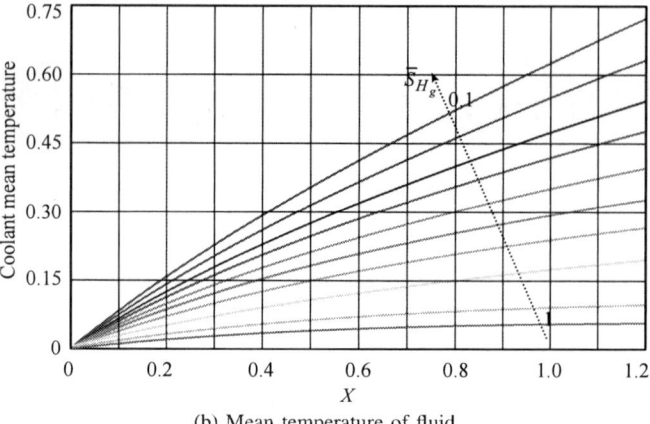

(b) Mean temperature of fluid

Fig. 4. Comparative performance.

spacing observed, whereas lower R_z values will be significantly affected by spacing. The pack temperature at its center is depicted as rising axially from bottom to top in Fig. 3 (b). The analysis here focuses on the impact of R_z and ζ ranging from 0.5 to 0.9. The flow R_z is limited to 150, 650, and 950, as depicted in Figs. 3(b) through 3(d) correspondingly. To begin with, it is crucial to emphasize that as the height increases vertically, the temperature increases sharply and reaches its peak at the highest point. The coolant initially makes contact with the lower side of the cells, resulting in a very low temperature for the pack. This allows for efficient heat dissipation, as there is a significant temperature variation between the coolant and battery. The coolant then flows laterally upwards, as depicted in Figs 1, carrying away the heat. As the coolant conveys greater quantities of heat in the direction of the flow, the temperature differential decreases, it is leading to a reduction in efficiency. Therefore, the temperature is greater at the highest point in comparison to the lowest point. Furthermore, it is important to highlight that as the spacing increases, the cells temperature will be decrease, as illustrated in Fig. 3(b). As the spacing increases, the amount of air being circulated also increases, resulting in more heat being carried away at greater distances. This ultimately enhances the cooling efficiency of the battery system. The thermal conductivity is lower when air is used as the coolant, resulting in reduced heat removal from the battery as it moves towards the trailing edge. Subsequently, as the parameters denoting the conductivity ratio ζ are raised from 0.5 to 0.9, the impact becomes evident in Fig. 3 (a). The rising ζ signifies a rise in the thermal conductivity of the coolant, causing the thermal boundary to expand further due to the increased prominence of thermal diffusivity. Consequently, the heat is readily conducted and subsequently transferred to the coolant. Consequently, the pack's temperature decreases significantly. Furthermore, due to the presence of conjugate conditions at the interface, the temperature within the coolant domain decreases, resulting in a reduction of its average temperature at the outlet. We must search for additional variables that could influence it. Nevertheless, the temperature decrease at R_z 150 is greater than that at R_z 650, which is further reduced at R_z 950, as

illustrated in Figs. 3(c) and 3(d).

Fig. 4(a) displays the mean temperatures of the cells while the battery pack is in operation for every ambient temperature taken into account. It also illustrates the heating time of the battery pack for the reference temperature ranging from 290 to 1380 seconds. The pack will be positioned within the bus, where the minimum temperature will be around 262K due to the bus heating system, making the proposed heating duration deemed acceptable. It may be necessary to increase the temperature of the liquid in order to reduce the heating time. Fig. 4(b) illustrates the change in average coolant fluid temperature and its axial velocity as the parameters vary along the length of the channel. The variation in average temperature of the coolant fluid as shown in Fig. 4(b) can be attributed to the alteration in temperature distribution within the cell. The heat observed by the coolant increases in conjunction with the rise in heat generation $\bar{S}_{H_g}$, maintaining a consistent trend across all $\bar{S}_{H_g}$ values, as illustrated in Fig. 4(b). The increment in temperature of the cells leads to a greater amount of heat being dissipated to the surroundings as the $\bar{S}_{H_g}$ in the cell increases, with all other parameters remaining constant The figure clearly shows that shear effects are most prominent near the cell wall at lower spacing, and as the spacing increases, these effects extend further downstream. It was also noted that the highest velocity occurs near the front edge of the cell wall when the spacing is reduced. Even a minor adjustment in the distance between the cells leads to a substantial change in the maximum velocity's position along the flow direction, moving further away from the front edge.

IV. CONCLUSION

In this work, we aimed to develop a lumped parameter thermal model for lithium-ion battery packs that includes fluid flow effects and the channel width between parallel cells. We integrated this model with an internal combustion engine cooling method to create a comprehensive thermal regulation system. Our real-time implementation using OPAL-RT demonstrated the system's effectiveness under various

979-8-3315-1612-3/25 $31.00 © 2025 IEEE

operational conditions. The results showed that parameters like Reynolds number and spacing between cells significantly affect temperature distribution within the battery pack. Higher Reynolds numbers initially decreased temperatures but reached a point where further increases had little effect. Increasing the spacing between cells improved cooling efficiency by enhancing air circulation. These findings address critical thermal management challenges, contributing to improved safety and performance of battery packs. By accurately modeling thermal behavior, we can design better cooling strategies to prevent overheating. One limitation of our study is the assumption of laminar and steady-state airflow, which may not reflect all real-world conditions. Future research could explore turbulent flow effects and transient conditions to enhance the model's applicability. Investigating different cooling methods or battery configurations may also provide valuable insights. Overall, our work advances battery thermal management by offering a model suitable for real-time applications, contributing to the safety and efficiency of lithium-ion battery packs in various technologies.

REFERENCES

[1] A. Afzal, A. R. Kaladgi, R. Jilte, M. Ibrahim, R. Kumar, M. Mujtaba, S. Alshahrani, and C. A. Saleel, "Thermal modelling and characteristic evaluation of electric vehicle battery system," *Case Studies in Thermal Engineering*, vol. 26, p. 101058, 2021.

[2] M. K. Senapati, O. Al Zaabi, K. Al Hosani, K. Al Jaafari, C. Pradhan, and U. Ranjan Muduli, "Advancing electric vehicle charging ecosystems with intelligent control of dc microgrid stability," *IEEE Transactions on Industry Applications*, vol. 60, no. 5, pp. 7264–7278, 2024.

[3] U. R. Muduli, M. S. E. Moursi, I. P. Nikolakakos, K. A. Hosani, S. A. Mohammad, and T. Ghaoud, "Impedance modeling with stability boundaries for constant power load during line failure," *IEEE Transactions on Industry Applications*, vol. 60, no. 1, pp. 1484–1496, Jan.-Feb. 2024.

[4] K. Murashko, H. Wu, J. Pyrhönen, and L. Laurila, "Modelling of the battery pack thermal management system for hybrid electric vehicles," in *2014 16th European Conference on Power Electronics and Applications*, 2014, pp. 1–10.

[5] A. Belhocine and W. Z. W. Omar, "Analytical solution and numerical simulation of the generalized levèque equation to predict the thermal boundary layer," *Mathematics and Computers in Simulation*, vol. 180, pp. 43–60, 2021.

[6] T. Wang, K. Tseng, J. Zhao, and Z. Wei, "Thermal investigation of lithium-ion battery module with different cell arrangement structures and forced air-cooling strategies," *Applied energy*, vol. 134, pp. 229–238, 2014.

[7] H. Park, "A design of air flow configuration for cooling lithium ion battery in hybrid electric vehicles," *Journal of power sources*, vol. 239, pp. 30–36, 2013.

[8] A. A. Pesaran, "Battery thermal models for hybrid vehicle simulations," *Journal of power sources*, vol. 110, no. 2, pp. 377–382, 2002.

[9] S. Shahid and M. Agelin-Chaab, "Development and analysis of a technique to improve air-cooling and temperature uniformity in a battery pack for cylindrical batteries," *Thermal Science and Engineering Progress*, vol. 5, pp. 351–363, 2018.

[10] Z. Lu, X. Yu, L. Wei, Y. Qiu, L. Zhang, X. Meng, and L. Jin, "Parametric study of forced air cooling strategy for lithium-ion battery pack with staggered arrangement," *Applied Thermal Engineering*, vol. 136, pp. 28–40, 2018.

[11] F. Porpora, M. D'Arpino, Y. Cheng, G. Rizzoni, and G. Tomasso, "Reduced order model of common battery pack architectures for assessment of cell parameter variation propagation," *IEEE Access*, vol. 11, pp. 96 693–96 709, 2023.

[12] M. Naguib, P. Kollmeyer, and A. Emadi, "Application of deep neural networks for lithium-ion battery surface temperature estimation under driving and fast charge conditions," *IEEE Transactions on Transportation Electrification*, vol. 9, no. 1, pp. 1153–1165, March 2023.

[13] K. Sundararaju, P. Priyanka, T. Revanth, K. Niviss, and S. Sabarish, "Optimizing the thermal operation of lithium-ion battery packs using cell to cell variation," in *2023 7th International Conference on Trends in Electronics and Informatics (ICOEI)*, April 2023, pp. 176–180.

[14] Y. Wang, C. Xiong, C. Ju, G. Yang, Y.-w. Chen, and X. Yu, "A deep transfer operator learning method for temperature field reconstruction in a lithium-ion battery pack," *IEEE Transactions on Industrial Informatics*, vol. 20, no. 6, pp. 8089–8101, June 2024.

[15] S. Prakash, M. Alkhatib, O. A. Zaabi, K. A. Hosani, R. K. Behera, and U. R. Muduli, "Adaptive phase synthesizer with in-loop variable gain control for reliable grid integration," *IEEE Journal of Emerging and Selected Topics in Power Electronics*, pp. 1–1, 2024.

[16] R. Mahamud and C. Park, "Reciprocating air flow for li-ion battery thermal management to improve temperature uniformity," *Journal of Power Sources*, vol. 196, no. 13, pp. 5685–5696, 2011.

[17] K. Zhang, L. Jiang, Z. Deng, Y. Xie, J. Couture, X. Lin, J. Zhou, and X. Hu, "An early soft internal short-circuit fault diagnosis method for lithium-ion battery packs in electric vehicles," *IEEE/ASME Transactions on Mechatronics*, vol. 28, no. 2, pp. 644–655, April 2023.

[18] D. Hethu Avinash and A. Rammohan, "Integrating level shift anomaly detection for fault diagnosis of battery management system for lithium-ion batteries," *IEEE Access*, vol. 12, pp. 116 071–116 084, 2024.

[19] L. Wang, Y. Song, C. Lyu, D. Yang, G. Yang, and D. Shen, "Optimization of lithium-ion battery charging strategies from a thermal safety perspective," *IEEE Transactions on Transportation Electrification*, vol. 10, no. 2, pp. 2727–2739, June 2024.

[20] D. H. Nguyen, "Model predictive iterative learning control design for battery optimal electro-thermal management under daily-variant state-of-charge patterns," *IEEE Access*, vol. 12, pp. 155 904–155 914, 2024.

[21] G. D. Astudillo, H. Beiranvand, F. Cecati, C. Werlich, A. Würsig, and M. Liserre, "Integrated strategy for optimized charging and balancing of lithium-ion battery packs," *IEEE Transactions on Transportation Electrification*, pp. 1–1, 2024.

[22] X. Zhang, X. Zhang, and W. Zhang, "Current status of research on factors causing lithium-ion battery packs performance degradation and their state of health estimation," in *2023 5th International Conference on Power and Energy Technology (ICPET)*, July 2023, pp. 306–311.

[23] M. Sarath and K. Shukla, "Parameter estimation and inconsistency analysis of lithium ion battery packs in electric vehicles," in *2023 9th International Conference on Smart Computing and Communications (ICSCC)*, Aug 2023, pp. 724–729.

[24] Y. Zhu, X. Qi, and S. Yang, "Inconsistency analysis of an electrical-thermal-aging coupled model for parallel battery packs in an air cooling system," in *2024 IEEE 10th International Power Electronics and Motion Control Conference (IPEMC2024-ECCE Asia)*, May 2024, pp. 2156–2160.

[25] P. Wei and H.-X. Li, "Spatiotemporal entropy for abnormality detection and localization of li-ion battery packs," *IEEE Transactions on Industrial Electronics*, vol. 70, no. 12, pp. 12 851–12 859, Dec 2023.

[26] S. Zhou, Z. Chen, and T. Lin, "Lithium-ion battery cell open circuit fault diagnostics: Methods, analysis, and comparison," *IEEE Transactions on Power Electronics*, vol. 38, no. 2, pp. 2493–2505, Feb 2023.

[27] M. Amir, A. Sadar, and N. Mohammad, "Enhancing electric vehicle battery pack performance: A pcm-based study on battery thermal management in the energy market," *IEEE Journal of Emerging and Selected Topics in Industrial Electronics*, pp. 1–8, 2024.

[28] S. P. Sanaka, R. Kandula, N. V. S. S. Garikapati, B. Chinatada, A. Mahamad, and G. Gaddam, "Transient thermal analysis on li-ion battery pack used for evs using comsol," in *2023 IEEE 2nd International Conference on Industrial Electronics: Developments & Applications (ICIDeA)*, Sep. 2023, pp. 639–644.

[29] P. Pal, R. K. Behera, and U. R. Muduli, "Eliminating current sensor dependencies in dab converters using a luenberger observer-based hybrid approach," *IEEE Transactions on Industry Applications*, vol. 60, no. 4, pp. 6380–6392, 2024.

[30] S. Prakash, O. A. Zaabi, R. K. Behera, K. A. Jaafari, K. A. Hosani, and U. R. Muduli, "Modeling and dynamic stability analysis of the grid-following inverter integrated with photovoltaics," *IEEE Journal of Emerging and Selected Topics in Power Electronics*, vol. 11, no. 4, pp. 3788–3802, 2023.

A Physics-Based Temperature Dependent Analytical Model For 2DEG Density in AlGaN/GaN HEMT Devices

Kashfia Tajmim Nabila
dept. of Electrical and Computer Engineering
University of Nebraska-Lincoln
Lincoln, USA
knabila2@huskers.unl.edu

Jerry L. Hudgins
dept. of Electrical and Computer Engineering
University of Nebraska-Lincoln
Lincoln, USA
jhudgins2@unl.edu

Abstract—**Accurate modeling of the temperature dependent two-dimensional electron gas (2DEG) is very crucial in applied power electronics, given its critical role in High Electron Mobility Transistor (HEMT) characteristics. In view of this, the temperature dependency of 2 DEG has been analyzed for different gate voltages and modeled for the temperature range from 300K to 500K. Relying on the device physics, the temperature dependence of various parameters including bandgap variation, fermi level energy, barrier height reduction, threshold voltage changes and their effects on electron sheet charge density (2DEG) has been included. This model will also integrate the effects of sub-band energy variations, polarization changes, and material property adjustment due to temperature changes. The goal is to provide a robust analytical framework that can predict device behavior accurately across a spectrum of operational gate voltage and temperature conditions. This anticipated model will facilitate the development of power electronics that are more reliable and efficient, capable of operating effectively in diverse and demanding environments.**

Keywords— *AlGaN/GaN heterostructure, n-type doping, mole fraction, temperature, 2DEG.*

I. INTRODUCTION

In recent years, high electron mobility transistors (HEMTs) have been extensively researched for their wide bandgap, high breakdown voltage, high electron mobility, ability to work under high temperature and high frequency, ensuring high speed and high-power application [1][2]. Previous research has shown that until now, HEMTs are one of the fastest transistor technologies, because of a triangular potential well formed by the lattice mismatch between two wide bandgap materials, facilitating the generation of a two-dimensional electron gas (2DEG) [3]. For successful modeling of AlGaN/GaN single heterostructure devices, understanding the device physics is very important. Accurate modeling of the AlGaN/GaN heterostructure is crucial, given that the 2DEG sheet charge density is integral to the device's overall characterization. As we know, GaN based

HEMTs are expected to be able to work in high temperature environments, and thus it is very important to accurately predict the device characteristics. The device performance is greatly influenced by the density and transportation of the two-dimensional electron gas (2DEG) [4].

Previous studies have proposed various models of HEMTs capturing the electrical behavior. In [5][6], the authors have developed a model focusing on the gate voltage dependence on the 2DEG density calculations. However, those works do not provide insight on the temperature dependency, nor does it give the details on the effects of doping concentration or Al mole fraction. Similarly, in the paper [7], the authors have described the importance of temperature effects on the device but fail to provide the critical connection between the gate voltage and temperature change. There were many empirical data that have been used to determine crucial parameters but fail to help understand the actual physics of real-world devices.

In this research work, we have successfully designed a 2DEG density model that accounts for both active and subthreshold operation regions of a normally on device. It successfully depicts the shift in fermi level, sub-bands in the quantum well, barrier height and polarization sheet charge density that are very critical in modelling the 2DEG density.

II. DEVICE STRUCTURE AND MODELLING APPROACH

A. Device structure

Fig 1., shows the schematic view of the normally-on AlGaN/GaN HEMT considered in this research. The heterostructure is made of $Al_{0.2}Ga_{0.8}N$ with a donor concentration of $1 \times 10^{18} cm^{-3}$ and a unintentionally doped GaN layer. Note that, normally-on AlGaN/GaN heterostructures intrinsically generate a two-dimensional electron gas (2DEG) at the interface due to spontaneous and piezoelectric polarization.

The additional donors in the $Al_{0.2}Ga_{0.8}N$ layer enhances 2DEG density by providing additional free electrons, which increases the carrier concentration in the channel. That is why, there is a very thin layer of an unintentionally doped $Al_{0.2}Ga_{0.8}N$ spacer layer between doped $Al_{0.2}Ga_{0.8}N$ and UID-

979-8-3315-1612-3/25 $31.00 © 2025 IEEE

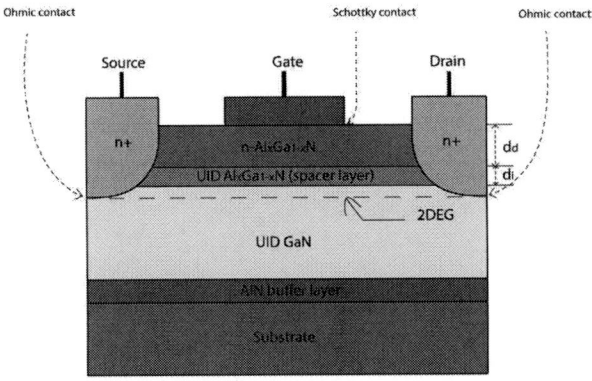

Fig. 1. Cross-sectional view of $n\text{-}Al_xG_{1-x}N/GaN$ HEMT

GaN layers, which reduces the electron scattering between the 2DEG and the impurities in the doped AlGaN layer [8]. This configuration is suitable for high electron density and high mobility, which is very suitable for high-frequency and high-power applications.

TABLE I.   LIST OF SYMBOLS

Symbol	Physical Parameter
$n_s{}^I$	2DEG sheet charge density in subthreshold region (cm^{-2})
$n_s{}^{II}$	2DEG sheet charge density in active region (cm^{-2})
ϕ_b	Barrier height (eV)
V_{off}	Threshold voltage (V)
V_{Th}	Thermal voltage (V)
x	Al mole fraction
D	Density of state (2D) ($eV^{-1}cm^{-2}$)
N_C	Effective density of state (cm^{-3})
N_D	Donor impurity doping concentration (cm^{-3})
m_e^*	Electron effective mass (kg)
$E_f{}^I$	Position of Fermi level in subthreshold region(eV)
$E_f{}^{II}$	Position of Fermi level in active region(eV)
E_0	Position of 1st sub band in the quantum well (eV)
E_1	Position 2nd sub band in the quantum well (eV)
$\varepsilon(x)$	The permittivity of $Al_{0.2}G_{0.8}N$ (Fcm^{-1})
k	Boltzmann Constant (eVK^{-1})
T	Temperature (K)
V_G	Gate voltage (V)
d_d	Thickness of AlGaN layer (nm)
d_i	Thickness of spacer layer (nm)
ΔE_c	Conduction band discontinuity (eV)
$a(0), a(x)$	Lattice constant
$C_{31}(x), C_{33}(x)$	Elastic constant
$e_{31}(x), e_{33}(x)$	Piezoelectric constant
$P_{SP}(0), P_{SP}(x)$	Spontaneous polarization

B. Charge density model

The temperature dependent standard two-dimensional electron gas (2DEG) charge density model is derived from the Schrödinger equation using the density of states and Fermi-Dirac function [9].

$$n_s(T) = DV_{th}\left\{\ln\left[\exp\left(\frac{E_f(T)-E_0(T)}{V_{th}}\right)+1\right]\right.$$
$$\left.+\ln\left[\exp\left(\frac{E_f(T)-E_1(T)}{V_{th}}\right)+1\right]\right\} \quad (1)$$

where, $E_0(T) = \gamma_0 n_s(T)^{\frac{2}{3}}$ (1st lowest sub band)

and, $E_1(T) = \gamma_1 n_s(T)^{\frac{2}{3}}$ (2nd lowest sub band)

Here, γ_0 and γ_1 are the functions of n_s, which can be positive or negative, depending on the position of the sub bands relative to fermi level. In the quantum well electrons are confined and can move only in two directions. Thus, density of states $D=\frac{qm_e^*}{\pi\hbar^2}$, for 2D systems. The change in electron effective mass with temperature is minimal, and therefore can be considered negligible in this analysis. The definitions of the symbols used throughout this paper are given in Table I. In thermal equilibrium the 2DEG density is given by [10], which is modified and written as (2).

$$n_s(T) = \frac{\varepsilon(x)}{qd_d}\left[V_G - V_{off}(T) - E_f(T)\right] \quad (2)$$

Fig 2. illustrates the band diagram of the HEMT of the device. From the diagram we can observe that the position of the fermi level energy (E_f) relative to the conduction band changes depending on which area of the band diagram we are talking about. In the Fig 2., we can see the whole diagram is divided into 2 regions. Region I shows conduction band energy (E_c) is higher than the fermi level energy E_f. However, in the quantum well E_c is lower than E_f, except in a very thin portion of the interface between Region I and Region II. Consequently, as the temperature varies the position of fermi level also varies due to changes in electron distribution. In this research only Region II is taken into account as we are modelling the 2DEG density. To reduce complexity the conduction band minimum, E_C is considered the reference level of energy and is set to zero.

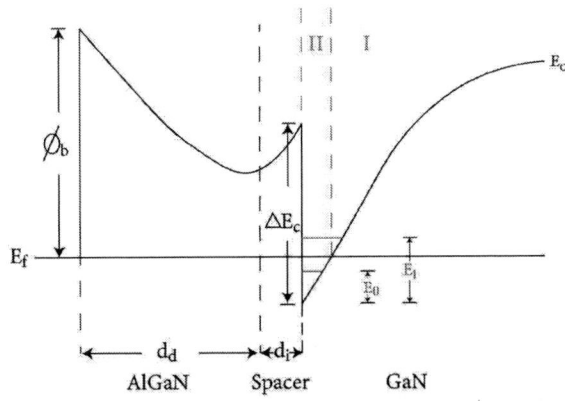

Fig. 2. Band diagram of the *AlGaN/GaN* heterostructure at thermal equilibrium

The position of the fermi level energy is given by [11]

$$E_c - E_f(T) = \pm KTln\left(\frac{N_c(T)}{N_D}\right) \quad (3)$$

The position of the sub bands in eq. (1) changes depending on the gate voltage and the temperature. In the fig 3., different positions of $E_f, E_0, and\ E_1$ are plotted based on numerical calculations using (1), (2), and (3).

$$V_{off}(T) = \phi_b(x, N_d, T) - \Delta E_c(x, T) - \frac{qN_D d_d^2}{2\varepsilon(x)}$$
$$- \frac{\sigma(x)}{\varepsilon(x)}(d_d + d_i) \quad (4)$$

$V_{off}(T)$ is a temperature dependent term and decreases with increasing temperature, shown in fig. 3(d). The cutoff or threshold voltage in heterostructures like AlGaN/GaN is determined by the spontaneous and piezoelectric polarization, the conduction band discontinuity, pinch off voltage and most importantly the Schottky barrier height. The barrier height in the AlGaN/GaN HEMTs is strongly impacted by the Al mole fraction, doping concentration and temperature as they directly affect the conduction band offset, thermal expansion, bandgap narrowing, as well as the electric field and built in potential near the Schottky junction. The relationship is given below [12]

$$\phi_b(x, Nd, T) = \phi_{b0}(x, Nd)\left[1 - \lambda \times exp\frac{\Delta T}{T_f}\right] \quad (5)$$

Where, λ=0.2, and $\Delta T = T_f - T_i$. The initial barrier height, ϕ_{b0} for varying mole fraction and doping concentration is derived using multiple variable linear regression by using the datas from [13][14].

$$\phi_{b0}(x, N_d) = 0.9914 - 1.8559 \times 10^{-21}N_d + 1.9315x \quad (6)$$

The conduction band discontinuity, $\Delta E_c(x, T)$ between the AlGaN and GaN material is given in (7)

$$\Delta E_c(x, T) = 0.7\left[xE_{g_{AlN}}(T) + (1-x)E_{g_{GaN}}(T) - x(1-x)b \right.$$
$$\left. - E_{g_{GaN}}(T)\right] \quad (7)$$

The amount of Al mole fraction in the AlGaN layer contributes to the polarization-induced sheet charge density at the AlGaN/GaN heterostructure interface. We may estimate these polarization characteristics for different compositions of $Al_xGa_{1-x}N$ by performing linear interpolation between the properties of GaN and AlN as these materials have different properties, especially with regard to polarization. The Al composition directly effects the 2DEG density if Al mole fraction is high. The polarization induced sheet charge density as a function of Al mole fraction (x) is given by [15]

$$|\sigma(x)| = \left| 2\frac{a(0) - a(x)}{a(x)}\left\{e_{31}(x) - e_{33}(x)\frac{C_{13}(x)}{C_{33}(x)}\right\} \right.$$
$$\left. - P_{SP}(0) \right| \quad (8)$$

The various parameters that are considered for the device and depend on Al mole fraction are given in Table II.

TABLE II. PARAMETERS FOR $Al_xGa_{1-x}N/GaN$ HEMT

N_d	$1 \times 10^{18}\ cm^{-3}$
x	0.2
d_d	120Å
d_i	30Å
$a(0.2)$	$3.17\times 10^{-8}\ cm$
$a(0)$	$3.189 \times 10^{-8}\ cm$
$e_{31}(0.2)$	$-5.12 \times 10^{-3}\ \frac{C}{cm^2}$
$e_{33}(0.2)$	$8.76 \times 10^{-3}\ \frac{C}{cm^2}$
$C_{13}(0.2)$	$104 \times 10^{11}\ \frac{N}{cm^2}$
$C_{33}(0.2)$	$398.4 \times 10^{11}\ \frac{N}{cm^2}$
$P_{SP}(0)$	$-2.9 \times 10^{-6}\ \frac{C}{cm^2}$
$P_{SP}(0.2)$	$-3.94 \times 10^{-6}\ \frac{C}{cm^2}$

From fig. 3, there are several things to observe. We can see that in (a), when -3.1V gate voltage is applied, which is the cutoff voltage for the device in our case, the 1st and 2nd sub-bands are above the fermi level. In fact, they remain above the fermi level energy position, E_f throughout the various range of temperature and gate voltages. This is identified as a cutoff region, and positioned at the edge of Region I. As we increase the gate voltage the difference between the fermi level, E_f and the 1st sub-band, E_0 decreases but E_f remains still below the 1st sub-band, this region is called subthreshold region. Electrons start accumulating in this region slowly. From fig 3 (b) and (c) we can see after applying certain amount of gate voltage fermi level, E_f begins to rise above the 1st sub-band E_0, allowing the device to fully turn on and populate the quantum well.

1. Subthreshold Region ($E_0 > E_f$)

The subthreshold region occurs at the interface of both regions. It is the region above the cutoff voltage, where the 2DEG is not fully depleted but it is significantly reduced. For our device the subthreshold region continuous till -2.5V. In this region the E_f is close to but still below E_0 at room temperature. But as temperature increases V_{off} decreases, making the E_f go above E_0. The 2DEG density is small in this region. As E_1 is way above the fermi level throughout the range of temperature change between 300K to 500K, we can safely eliminate that. Thus, eliminating the second term of (1) we can write

$$E_f^I = \gamma_0(n_s^I)^{\frac{2}{3}} + V_{th}ln\left[\frac{n_s^I}{DV_{th}}\right] \quad (9)$$

Equation (9) is the fermi level in the subthreshold and cutoff region. We will combine (9) with (3) to develop a temperature dependent fermi level model.

$$E_f^I(T) = \gamma_0\left(n_s^I(T)\right)^{\frac{2}{3}} + V_{th}ln\left[\frac{n_s^I(T)}{DV_{th}}\right] - KT\ln\left[\frac{N_c(T)}{N_d}\right] \quad (10)$$

979-8-3315-1612-3/25 $31.00 © 2025 IEEE

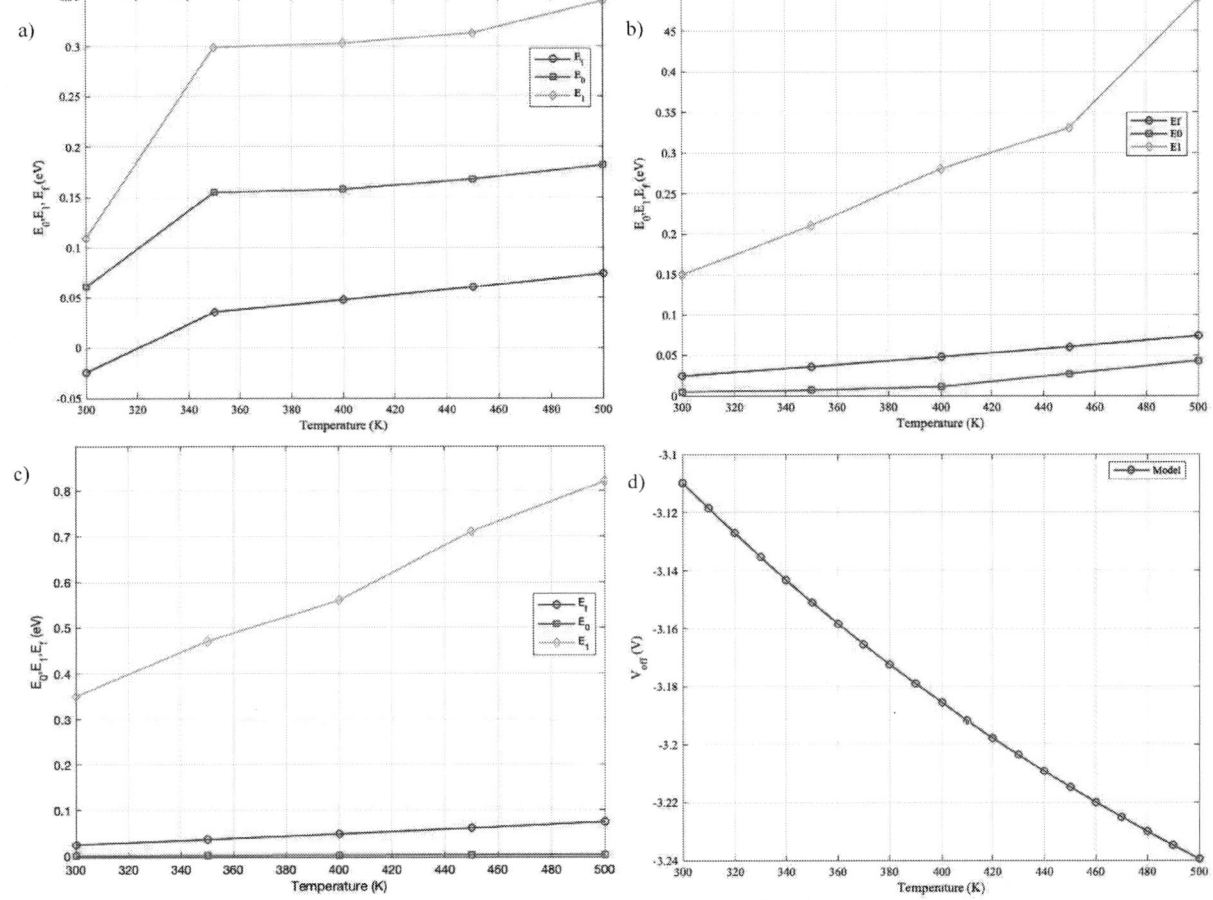

Fig. 3. Numerical calculations of $E_0, E_1,$ and E_f vs gate voltages at a) cutoff voltage ($V_g =-3.1$V), b) V_g -2.1V, and c) $V_g = 1.1$V for temperature ranging from 300K to 500K, d) cutoff voltage variance depending on temperature.

Next, we apply (2) into (10) to eliminate $n_s{}^I$ term from the equation and we obtain the following analytical equation for fermi level in subthreshold and cutoff region as a function of

$$E_f{}^I(T) = V_{Go} \frac{V_{th}\ln(\beta V_{Go}) + \gamma_0\left(\frac{C_g V_{Go}}{q}\right)^{\frac{2}{3}} - KT \ln\left(\frac{N_C(T)}{N_d}\right)}{V_{GO} + \frac{2}{3}\gamma_0\left(\frac{C_g V_{Go}}{q}\right)^{\frac{2}{3}} + V_{th}} \quad (11)$$

where, $V_{Go} = V_G - V_{off}$, $C_g = \frac{\varepsilon(x)}{d_d}$, and $\beta = \frac{C_g}{qDV_{th}}$. Putting (11) into (2), 2DEG density is expressed as (12).

$$n_s{}^I(T)$$
$$= \frac{C_g V_{Go}}{q}\left[\frac{V_{Go} - \frac{\gamma_0}{3}\left(\frac{C_g V_{Go}}{q}\right)^{\frac{2}{3}} + V_{th}(1 - \ln\beta V_{th}) + KT\ln\left(\frac{N_C(T)}{N_d}\right)}{V_{Go} + V_{th} + \frac{2}{3}\gamma_0\left(\frac{C_g V_{Go}}{q}\right)^{\frac{2}{3}}}\right] \quad (12)$$

2. Active Region ($E_0 < E_f$)

The active region is when the device is fully turned on and fermi level energy E_f is greater than E_0. In another word the channel is fully formed. In this region the 2DEG starts populating with electrons at the bottom of the quantum well, allowing efficient conduction of current between the source and the drain. Therefore, the device operates with high density of electrons in the channel.

To calculate the fermi level in active region, we can again combine (1) and (2) and derive the following equation.

$$E_f{}^{II} = \frac{n_s{}^{II}}{D} + \gamma_0(n_s{}^{II})^{\frac{2}{3}} \quad (13)$$

By adding (3) into (13), the temperature dependent fermi level in active region (14) is derived as follows.

$$E_f{}^{II}(T) = \frac{n_s{}^{II}(T)}{D} + \gamma_0(n_s{}^{II}(T))^{\frac{2}{3}} + KT\ln\left[\frac{N_C(T)}{N_d}\right] \quad (14)$$

Putting (12) into (2) we can again eliminate the n_s terms and rewrite the fermi level position in the active region as (15).

$$E_f{}^{II}(T) = V_{Go} \frac{V_{th}\beta V_{Go} + \gamma_0\left(\frac{C_g V_{Go}}{q}\right)^{\frac{2}{3}} - KT\ln\left(\frac{N_C(T)}{N_d}\right)}{(1 + \beta V_{th})V_{Go} + \frac{2}{3}\gamma_0\left(\frac{C_g V_{Go}}{q}\right)^{\frac{2}{3}}} \quad (15)$$

Now, we can take (15) and put it in (2), thus we will have the 2DEG density formula for active region (16).

$$n_s{}^{II}(T) = \frac{C_g V_{Go}}{q}\left[\frac{V_{Go} - \frac{\gamma_0}{3}\left(\frac{C_g V_{Go}}{q}\right)^{\frac{2}{3}} - KT \ln\left(\frac{N_C(T)}{N_d}\right)}{(1 + \beta V_{th})V_{Go} + \frac{2}{3}\gamma_0\left(\frac{C_g V_{Go}}{q}\right)^{\frac{2}{3}}}\right] \quad (16)$$

$$n_s(T) = \frac{C_g V_{Go}}{q} \left[\frac{V_{Go} + V_{th}[1 - (\beta V_{Gon})] + KT \ln\left(\frac{N_C(T)}{N_d}\right) \tanh\left\{\frac{1 - (\alpha_x - V_{Go})}{0.01}\right\} - \frac{\gamma_0}{3}\left(\frac{C_g V_{Go}}{q}\right)^{\frac{2}{3}}}{\left(1 + \frac{V_{th}}{V_{God}}\right) V_{Go} + \frac{2}{3}\gamma_0\left(\frac{C_g V_{Go}}{q}\right)^{\frac{2}{3}}} \right] \quad (17)$$

3. Unified model

To have a smooth transition between the active and subthreshold region it is important to introduce a unified model. Real devices often operate on a range of gate voltages and temperature. For the sake of avoiding the risk of high errors and any kind of discontinuities it is crucial to have a unified model that accurately depicts the 2DEG density in active, subthreshold and cutoff region. The proposed unified model in (17) successfully integrates the physical behaviors of both active and subthreshold regions into a single formula.

In (17), V_{Gon} and V_{God} are the functions of V_{Go} and is given by the interpolation equation below.

$$V_{Gox} = \frac{V_{Gox}\alpha_x}{\sqrt{V_{Gox}^2 + \alpha_x^2}}$$

and $\alpha_n = \frac{q}{\beta}$ and $\alpha_d = \frac{1}{\beta}$.

III. SIMULATION RESULTS AND DISCUSSION

Fig. 4 illustrates the simulation results taken from the unified model and compared with the data from a previously published study that employed a different set of parameters [17]. The objective of selecting the model from [17] is both the models focuses on predicting the 2DEG density at various temperatures and gate voltages. From our model we can see that at cutoff voltage the 2DEG density is zero at room temperature. However, as temperature increases very small amount of two-dimensional electron gas (2DEG) starts to gather in the quantum well. As we raise the gate voltage 2DEG density increases. From the figure we can see for a fixed gate voltage 2DEG reflects temperature dependent characteristics as it shows incremental rise with temperature.

Despite the difference in parameters both models depict a similar trend, that 2DEG electron sheet charge density increases with increasing temperature and gate voltage. This agreement highlights the robustness of our designed model that successfully captures the underlying physics of the n-type doped normally-on AlGaN/GaN HEMT. The proposed model smoothly transitions between subthreshold region and active region based on given gate voltages and temperature.

From the results observed we can surely say that specific device parameters have a strong influence on the number of 2DEG sheet charge density accumulated in the quantum well of the heterostructure. The consistency of the observed trend between the two models is the general validity of our proposed model in successfully predict the device characteristics.

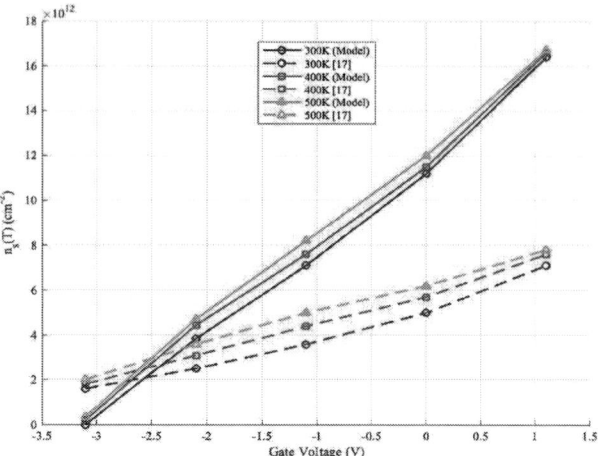

Fig. 4. Comparison of unifies model with [17] where $N_d = 1 \times 10^{15} cm^{-3}$ $Al_{0.3}Ga_{0.7}N/GaN$ HEMT at $V_{DS} = 5.0V$ is used.

IV. CONCLUSION

We have presented a physics-based temperature and gate voltage dependent 2DEG charge density model, that can operate for both subthreshold and active region. This model has very few empirical or fitting parameters. While the proposed theoretical model provides a strong theoretical framework on 2DEG sheet charge density across both subthreshold and active region for various gate voltage and temperature, it is important to note that the model has not been experimentally validated with a real-life device. The future work will focus on comparing the model's predictions with experimental data.

REFERENCES

[1] M. N. A. Aadit, S. G. Kirtania, F. Afrin, M. K. Alam, and Q. D. M. Khosru, "High electron mobility transistors: performance analysis, research trend and applications," in *Different Types of Field-Effect Transistors-Theory and Applications*, pp. 45-64, Jun. 2017.

[2] Y. Liu, D. Chen, K. Dong, H. Lu, R. Zhang, Y. Zheng, ... Z. Lin, "Temperature dependence of the energy band diagram of AlGaN/GaN heterostructure," *Advances in Condensed Matter Physics*, vol. 2018, no. 1, Apr. 2018.

[3] Md. A. Khan, M.A. Alim, and C. Gaquiere, "2DEG transport properties over temperature for AlGaN/GaN HEMT and AlGaN/InGaN/GaN pHEMT," *Microelectronic Engineering*, vol. 238, p. 111508, Feb. 2021.

[4] Wang, M., Shen, B., Xu, F. *et al.*, "High temperature dependence of the density of two-dimensional electron gas in $Al_{0.18}Ga_{0.82}N/GaN$ heterostructures," Applied Physics A, vol. 88, pp. 715–718, Jun 2007.

[5] Khandelwal, S, Goyal, N. and Fjeldly, TA,. "A physics-based analytical model for 2DEG charge density in AlGaN/GaN HEMT devices." IEEE Transactions on Electron Devices, pp. 3622-3625, vol. 5, no. 10, Oct. 2011.

[6] Zhang, J., Binit S., Xing Z., Subramaniam A., and Geok Ing Ng. "A compact model for generic MIS-HEMTs based on the unified 2DEG density expression." *IEEE Transactions on Electron Devices*, vol. 61, no. 2, pp. 314-323, Feb. 2014.

[7] Sahebghalam, N, Shalchian M, "High-temperature HEMT model." IEEE Transactions on Electron Devices, vol. 69, no.9, pp. 4821-4827, Sept. 2022.

[8] Hossain, M. M, Hassan, M.M, and Adnan, M.M.R. "Impact of structural modification by spacer layer inclusion on AlGaN/GaN HEMT performance." In *2021 International Conference on Information and Communication Technology for Sustainable Development (ICICT4SD)*, pp. 76-81. IEEE, Feb. 2021.

[9] Kola, S., Golio, J. M., and Maracas, G. M., "An analytical expression for Fermi level versus sheet carrier concentration for HEMT modeling." *IEEE electron device letters*, vol. 9, no. 3 , pp.136-138, Mar. 1988.

[10] D. Delagebeaudeuf and N. T. Linh, "Metal-(n) AlGaAs-GaAs two-dimensional electron gas FET," in *IEEE Transactions on Electron Devices*, vol. 29, no. 6, pp. 955-960, June 1982.

[11] Bozhkov, V. G., and S. E. Zaitsev. "On the current-voltage characteristic of a tunnel metal-semiconductor Schottky-Barrier contact." *Russian Physics Journal*, vol. 49, no. 3, pp. 251-259, 2006.

[12] M.N.Khan, U.F. Ahmed, M.M. Ahmed, and S. Rehman, "An improved model to assess temperature-dependent DC characteristics of submicron GaN HEMTs," *Journal of Computational Electronics*, vol. 17, no. 2, pp. 653–662, Mar. 2018.

[13] D. Qiao, L. S. Yu, S. S. Lau, J. M. Redwing, J. Y. Lin, and H. X. Jiang, "Dependence of Ni/AlGaN Schottky barrier height on Al mole fraction," vol. 87, no. 2, pp. 801–804, Jan. 2000.

[14] Y. H. Wang, Y. C. Liang, G. S. Samudra, T. F. Chang, C. F. Huang, L. Yuan, and G. Q. Lo, "Modelling temperature dependence on AlGaN/GaN power HEMT device characteristics," *Semiconductor science and technology*, vol. 28, no. 12, pp. 125010–125010, Nov. 2013.

[15] Bellakhdar, A., Soltani, A., "The thermal effect on the output conductance in AlGaN/GaN HEMT's." *24th International Conference on Microelectronics (ICM)*. IEEE, 2012.

[16] M. A. Alim, A. A. Rezazadeh, and C. Gaquiere, "Temperature dependence of the threshold voltage of AlGaN/GaN/SiC high electron mobility transistors," *Semiconductor Science and Technology*, vol. 31, no. 12, p. 125016, Nov. 2016.

[17] Sahebghalam, N., Majid S., Amirali C., and Farzan J., "High-temperature HEMT model." *IEEE Transactions on Electron Devices* vol. 69, no. 9 pp. 4821-4827, Sept. 2022.

Comparative Analysis of Stator-PM Machines: Design Optimization and Electromagnetic Performance Evaluation

Maryam Salehi

Madhav Manjrekar

Department of Electrical and Computer Engineering
University of North Carolina at Charlotte
Charlotte, US

msalehi@charlotte.edu

madhav.manjrekar@charlotte.edu

Abstract—Stator-Permanent Magnet (stator-PM) machines have gained interest over the past years due to their high torque-to-power density, a simple and passive rotor, and better thermal management capabilities. This paper compares the Flux Switching Permanent Magnet Machine (FSPM) with pole winding, Doubly Salient Permanent Magnet Machine (DSPM) with toroidal winding, and Doubly Salient Parallel Path Magnetic Machine (DS-PPMM) with pole winding configurations being investigated in literature. Large-scale optimization is employed for each machine to increase the average torque and reduce torque ripples. 2-D Finite Element Analysis (FEA) is used to analyze electromagnetic performance including flux linkages, back-EMFs, cogging torque, electromagnetic force, and torque vs. speed characteristics.

Keywords—flux switching permanent magnet machines, doubly salient permanent magnet machines, parallel path magnetic technology, torque generation, cogging torque, multi-objective optimization

I. INTRODUCTION

Stator-PM machines have gained interest in high-speed applications due to their robust rotor structure, high torque/power density, and better thermal dissipation [1], [2]. stator-PM machines can be classified in four classes, namely, Doubly Salient PM (DSPM) machine with magnets mounted in the stator back-iron [3], Flux Reversal PM (FRPM) machine with magnets placed on the inner surface of the stator teeth [4], and Flux Switching PM (FSPM) machine with magnets sandwiched in the stator teeth [5], and Parallel Path Magnetic Technology (PPMT) machine with magnets in the stator back iron [6], [7].

FSPM machines has been introduced by Hoang et al., "Switching Flux Permanent Magnet Polyphase Synchronous Machines," circa1997 [8]. Due to their flux concentration effects, FSPMs have advantages over FRPMs and DSPMs, as well as high torque capacity, and more sinusoidal back-EMF compared to FRPM and DSPM machines. Reduced copper area, restricted saturation overload capability, complex stator design, leakage, large magnet volume, more cogging torque due to flux focusing effect are some of the disadvantages of FSPM machines [9], [10]. Additionally, it has been demonstrated that FSPMs can provide some level of fault tolerance [11]. A novel alternate winding has been designed to increase the fault tolerant capacity of the FSPM machines [12]. However, the asymmetrical back-EMF waveforms cause high torque ripples. PPMT is an advanced magnetic control technology to enhance the capability of electric machines.

PPMT is a method of providing greater force because magnetic flux can be controlled and directed within the motor's core. The PPMT not only amplifies the magnetic flux within the core by a factor of four, but also directs the flux, generating greater motive power compared to conventional motors. PPMT method which can be applied to linear actuators, electrical motors, and generators has been introduced by Flynn et al., "Parallel Path Magnetic Technology for High Efficiency Power Generators and Motor Drives," circa 2006 [7]. A comparative analysis between 12/10 FSPM and toroidal wound DSPM has shown that DSPM has better torque and an extended speed range [13]. In [14], a comparison between 6-pole FSPM and PPMT motor has been made, which leads to the conclusion that the PPMT motor has lower cogging torque and better sine back-EMF waveform, making them more suitable for servo applications.

Due to their doubly salient structure, high airgap flux density, and flux-concentrating effect, stator PM machines typically exhibit greater torque ripples than traditional rotor-PM brushless machines. The main cause of torque ripples is cogging torque, and numerous researchers have worked to reduce it using a variety of strategies that can be divided into two categories: machine design based [15-17] and control based [18-20]. These techniques consist of skewing, notching, injecting harmonic current, shaping rotor teeth, etc. In [21], the impacts of the suggested strategies on Unbalanced Magnetic Force (UMF) and acoustic noise are explored after the cogging torque reduction techniques for a 12/10 FSPM machine. Even though multiphase motors have demonstrated superior fault tolerant capabilities, reduced torque ripples, and more flexible control strategies, most of these machines are still only compatible with standard three-phase systems [22], [23]. A new five-phase modular FSPM machine that can address the issues with the current fault tolerant machines has been proposed in [24]. High UMF is another issue in stator PM machines, particularly those with an odd number of rotor poles. A new double rotor FSPM machine without a UMF has been presented in [25].

A comparative study between three stator-PM machines including a pole wound FSPM with magnets between the stator teeth (Fig. 1(a)), a toroidal wound DSPM with magnets between the stator teeth (Fig. 1(b)), and a pole wound Doubly Salient Parallel Path Magnetic Motor (DS-PPMM) with magnets in the stator back iron (Fig. 1(c)) [26] is investigated in this paper. All the above-mentioned machines are three-phase and contain four sectors. A multi objective design optimization based on 2D FEA is presented in Section II. Electromagnetic performance analyses include flux linkages and back-EMF, electromagnetic torque, cogging torque, force analysis, and torque vs speed characteristics are presented in Section III.

979-8-3315-1612-3/25 $31.00 © 2025 IEEE

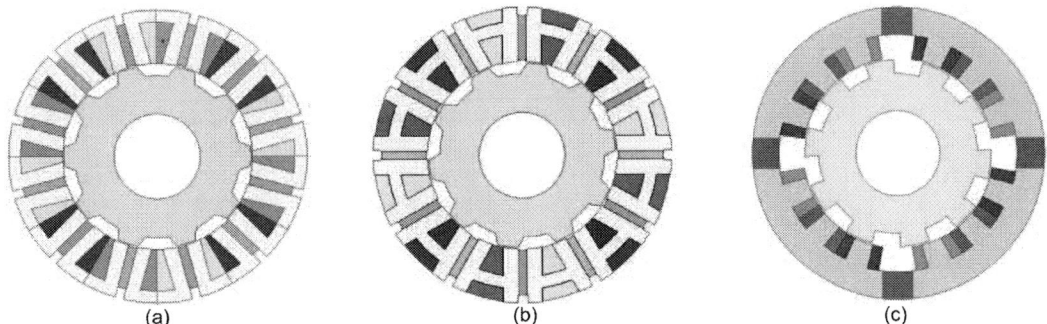

Fig.1. Cross-section of three-phase stator-PM machines, a) FSPM, b) DSPM, c) DS-PPMM.

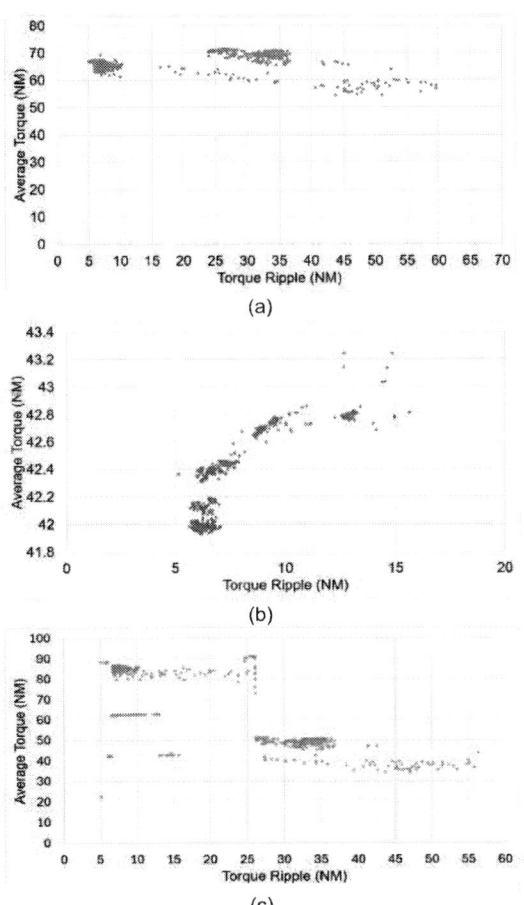

Fig. 2. Optimization results: 2-D pareto front projection with objective of maximum average torque and minimum torque ripple a) FSPM, b) DSPM, and c) DS-PPMM.

II. MULTI-DESIGN OPTIMIZATION

The rated speed for three machines is 5000 r/min, the rated frequency is 833.33 HZ, the rated current is 135 A. First the split ratio (ratio of stator outer diameter to stator inner diameter) and rotor pole width, which have the major effect on average torque and torque ripple are optimized. The initial split ratio for all machines is considered 0.632. The rotor and stator pole width, stator yoke, PM widths, and stack length are identical in the initial design of the FSPM, and DSPM. However, the initial design for DS-PPMM is different from the FSPM and DS-PPMM, in which the outer diameter of DS-PPMM is 39% greater than FSPM and DSPM while the stack length of DS-PPMM is 50% less than FSPM and DS-PPMM to ensure that the machine volume is identical for all three machines. The rotor tooth width varies

in the range of 11-13 mm for FSPM and DSPM, while for DS-PPMM, varies between 22-24 mm. After optimization, the rotor tooth widths for FSPM and DSPM increased slightly to 12.33 and 12.14 mm, respectively. In the case of DS-PPMM, the rotor tooth width reduced slightly from the initial 23 mm to 22.84 mm. The split ratio variation range, for all machines constant at 0.6–0.7, giving the optimized values 0.627 for FSPM, 0.643 for DSPM, and 0.698 for DS-PPMM. The objective of large-scale optimization is to increase average torque and reduce torque ripple. The systematic comparison of three topologies has been done using multi-objective design optimization. As illustrated in Fig. 2, several potential designs were found, with estimated average torques of more than 60NM for FSPM, 40NM for DSPM, and 80NM for DS-PPMM. In terms of increased average torque, the results indicate that DS-PPMM performs better. Additional design criteria for optimized FSPM, DSPM, and DS-PPMM are listed in table II.

III. COMPARISON OF ELECTROMAGNETIC PERFORMANCE

A. No-Load Flux-Linkage and Back-EMF

Fig. 3 depicts the flux-linkage per phase of three machines, with all waveforms appearing sinusoidal. Fig. 4 and 5 depict the phase no-load back-EMF waveform and the back-EMFs harmonic contents, respectively. The fundamental back-EMF in DS-PPMM is approximately 55.6% and 187.65% greater than FSPM and DSPM, respectively. The Total Harmonic Distortion (THD) can be calculated as equation (1)

$$THD = \frac{\sqrt{\sum_{n=2}^{\infty} H_n^2}}{H_1} \qquad (1)$$

where H_1 is the fundamental harmonic and H_n are the other harmonic components. The FSPM machine has the highest THD of 34.84%, which indicates high harmonic content; thus, it may cause noise and vibration and may lead to loss in efficiency. DSPM machine has moderately better THD of 25.11%. The DS-PPMM machine gives the least THD of 17.48%, its back-EMF being smoother which is shown in Fig. 4 (c), and hence is more efficient and reliable. This is also evident from the open circuit field distribution of three machines (Fig. 6) in which the leakage flux outside the stator is larger for DSPM and less for DS-PPMM. However, the leakage fluxes for three machines are still relatively minor compared to the primary flux. By optimizing the other parameters, such as the magnet thickness, the stator tooth width, the harmonic content of the back-EMF can be reduced effectively.

TABLE I. OPTIMIZATION PARAMETERS

Parameter	FSPM	DSPM	DS-PPMM
Rotor Tooth Width Variation Range	11-13mm	11-13mm	22-25
Initial Rotor Tooth Width	11.42 mm	11.42 mm	23
Optimized Rotor Tooth Width	12.33mm	12.14mm	22.84
Split Ratio Variation Range	0.6-0.7	0.6-0.7	0.6-0.7
Initial Split ratio	0.632	0.632	0.632
Optimized Split Ratio	0.627	0.643	0.698

TABLE II. KEY DESIGN PARAMETERS FSPM, DSPM, AND DS-PPMM.

Quantity	FSPM	DSPM	DS-PPMM
Rated speed (rpm)	5000/	5000	5000
Rated current (A)	135	135	135
Coil current density (A/mm²)	4.5	4.5	8
Rated frequency	833.33	833.33	833.33
Number of phases	3	3	3
Number of turns per phase	48	48	30
Magnet remanence	1.23	1.23	1.23
Magnet length (mm)	5	5	21
Magnet Number	12	12	4
Air gap length (mm)	0.74	0.73	0.99
Outer stator Radius (mm)	86.5	86	119
Active axial length (mm)	101	101	50
Inner stator radius (mm)	54.93	54.91	83.062
Outer rotor radius (mm)	54.19	54.18	84.052
Rotor Tooth Width Variation Range (mm)	11-13	11-13	22-24
Optimized Rotor pole width (mm)	11.61	11.47	22.84
Inner rotor radius (mm)	25	25	30
PM volume (mm³)	151500	151500	110000
PM Weight (Kg)	1.21	1.21	0.814
Active Weight	18.65	18.96	12.392

B. Cogging Torque and Output Torque

Comparing the FSPM, DSPM, and DS-PPMM machines, there are differences in their weight, torque, and torque density. The DS-PPMM has the highest average torque of 88 Nm, the lowest weight at 12.392 kg, and the highest torque density at 7.10 Nm/kg. The FSPM with an average torque of 69.2 Nm and weight of 18.65 kg presents a moderate torque density of 3.71 Nm/kg, ensuring decent performance but less efficient than DS-PPMM. On the other hand, DSPM has the minimum average torque equal to 42.1 Nm and a mass of 18.96 kg, yielding the smallest torque density of 2.22 Nm/kg.

Fig. 7 compares the cogging torque of FSPM, DSPM, and DS-PPMM machines. In typical rotor-PM machines, the cogging torque is produced by the interaction between the rotor PMs and the stator slots. However, the cogging torque in doubly salient machines is generated by the interaction of the rotor and stator poles. As shown in Fig. 7, the FSPM has the largest cogging torque variation, about 12 Nm peak to peak (14.5%), indicating large torque ripples and cogging effect. The cogging torque variation is small for the DSPM machine, having a peak-to-peak variation of about 2.5 Nm (5.2%). The DS-PPMM exhibits a moderate cogging torque characteristic, with a peak-to-peak value of approximately 7 Nm (7.9%).

C. Unbalanced Magnetic Force (UMF)

UMF, caused by the asymmetric magnetic distribution in the airgap, refers to the net electromagnetic force operating on the rotor along radial and tangential axes from the stator teeth. The UMF can be caused due to the rotor's eccentricity, manufacturing flaws, and unbalanced phase windings. However, the UMF in rotors with an even number of poles can be disregarded. The precise force/torque computation can be made using the Maxwell stress tensor approach. The following equations can be used to express the radial and tangential electromagnetic force in the airgap can be expressed as equation (2), and (3), respectively:

$$fr(\phi, t) = \frac{Br(\phi,t)^2 - Bt(\phi,t)^2}{\mu_0} \tag{2}$$

$$ft(\phi, t) = \frac{Br(\phi,t)Bt(\phi,t)}{\mu_0} \tag{3}$$

where the radial and tangential air-gap flux densities determined by FEA are $Br(\phi, t)$ and $Bt(\phi, t)$, respectively. The radial and tangential force acting on the stator teeth and rotor protrusions can be achieved by using 2-D FEA. Fig. 8 shows the distributed electromagnetic force on the stator teeth for FSPM, DSPM, and DS-PPMM.

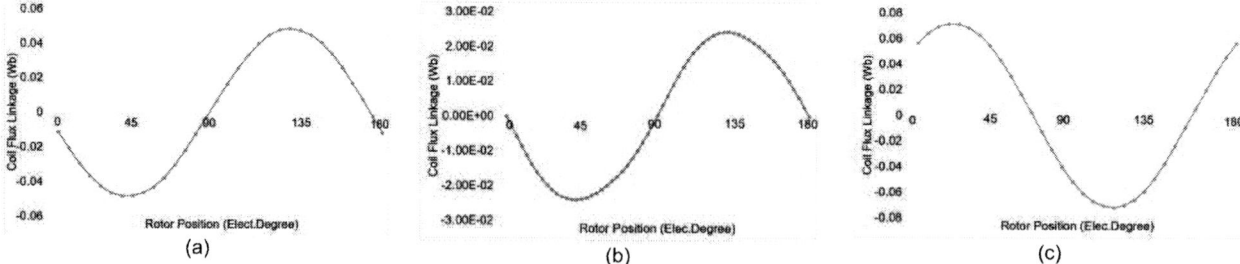

Fig. 3. Flux Linkage per phase of a) FSPM, and b) DSPM, and c) DS-PPMM, 5000rpm.

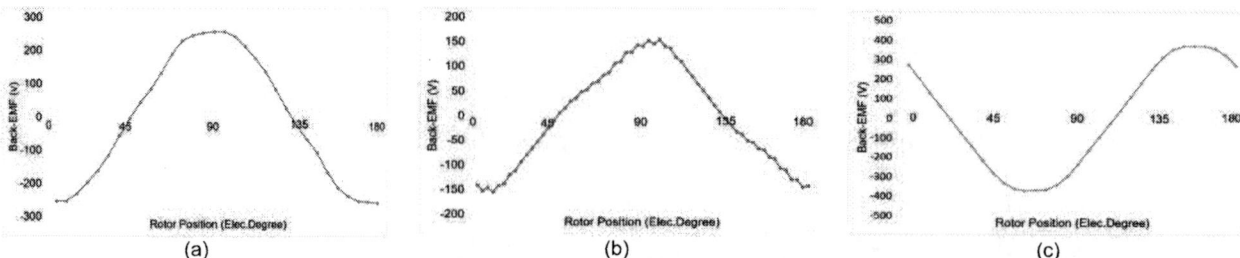

Fig. 4. No-load back-EMF per phase of a) FSPM, and b) DSPM, and c) DS-PPMM, 5000rpm.

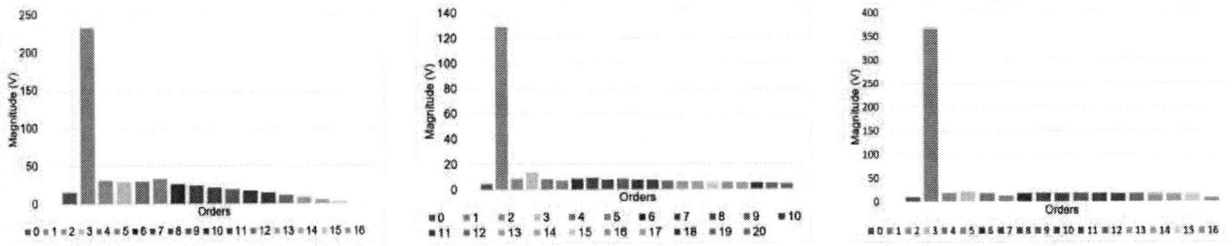

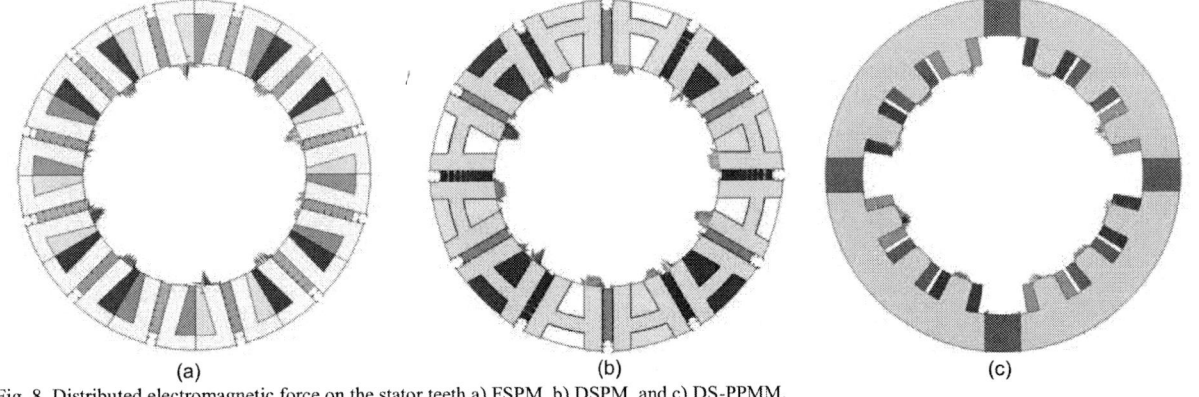

Fig. 5. Spectra of phase back-EMF of a) FSPM, and b) DSPM, and c) DS-PPMM, 5000rpm.

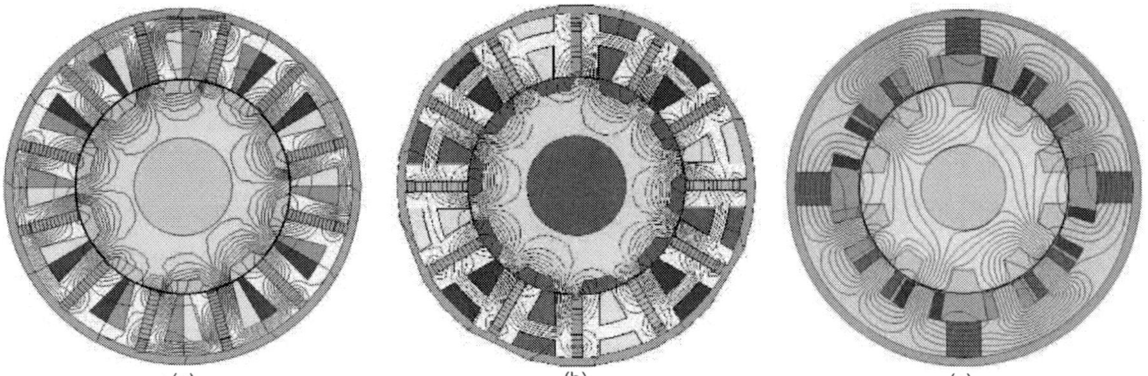

Fig. 6. Magnetic field distribution of a) FSPM, and b) DSPM, and c) DS-PPMM, 5000rpm.

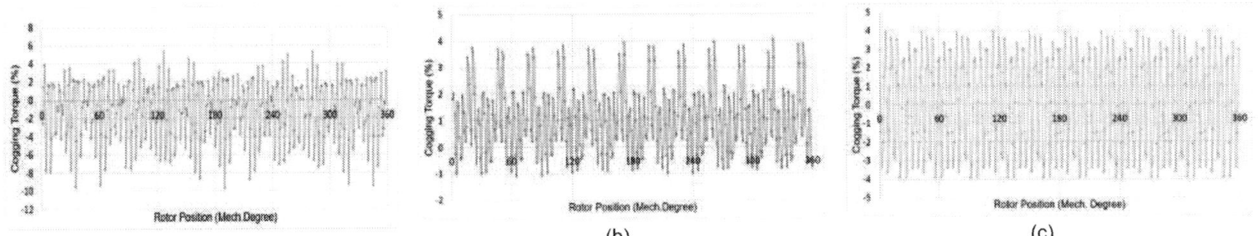

Fig. 7. Cogging torque waveforms of a) FSPM, (b) DSPM, and c) DS-PPMM (Percentage of average torque: 69 NM for FSPM, 42 NM for DSPM, and 88 NM for DS-PPMM.

Fig. 8. Distributed electromagnetic force on the stator teeth a) FSPM, b) DSPM, and c) DS-PPMM.

The force contribution of each stator tooth is different from each other due to the asymmetrical magnetic flux path. Fig. 9 shows the normalized radial and tangential forces at full load on the stator teeth of the machine. As shown, the radial force in DSPM is greater than the tangential force which may result in higher UMF. The radial force in FSPM is approximately equal to the tangential force. DS-PPMM exhibits lower radial force components compared to tangential force, leading to lower mechanical vibration in this machine. This imbalance also can be visualized in the flux lines shown in Fig. 7. The field flux distribution is asymmetrical over 120 mechanical degrees, generating UMF even on open circuit condition.

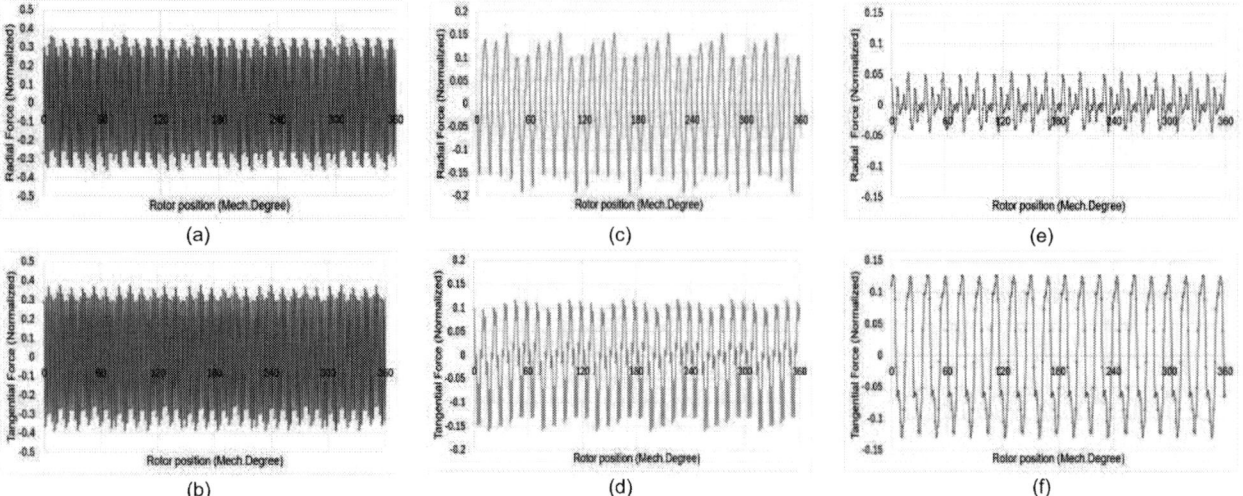

Fig. 9. Force analysis a) Radial force and b) Tangential force of FSPM, c) Radial force and d) Tangential force of DSPM, e) Radial force and f) Tangential force of DS-PPPM (Normalized with absolute force: 996 NM for FSPM, 356 NM for DSPM, and 1132 NM for DS-PPMM).

D. Torque vs Speed Characteristics

The torque vs. speed characteristics show some differences among the machines (Fig. 10): the FSPM has moderate torque (69NM) at low speeds but drops drastically beyond 5,000 RPM, which indicates that the high-speed range is narrow; the DSPM shows a flatter torque profile, with moderate levels of torque (42 NM) up to around 8,000 RPM, after which it drops whit higher slop, which demonstrates better adaptability at lower speed and extended speed ranges in comparison with FSPM. DSPM can operate with a constant power of 40kW at up to 2.5 times the base speed, meanwhile, the DS-PPMM has a high torque (88 NM) at low speeds, and the torque starts to drop smoothly after about 7,000 RPM. This characterizes its smooth torque reduction and broader operational range compared to FSPM and DSPM.

IV. CONCLUSION

This paper compares three stator-PM machines, including FSPM, DSPM, and DS-PPMM, under the same volume. A multi-objective design optimization is conducted to optimize the split ratio and rotor pole width which have a significant impact on average torque and torque ripple. 2-D FEA analyses, including flux linkages, back-EMFs, cogging torque, electromagnetic force, and torque vs. speed characteristics are examined. The DS-PPMM indicates superior performance, with greater fundamental back-EMF and lower THD, demonstrating better sinusoidal back-EMFs. Additionally, DS-PPMM has a lower weight (12.392Kg) compared to FSPM (18.65Kg) and DSPM (18.96Kg) with greater average torque. The cogging torque in DSPM machine is the smallest, whereas the FSPM exhibits higher cogging torque. In the DS-PPMM, the radial force is lower than tangential force, resulting in lower level of UMF. Maximum average torque can be maintained for both DSPM and DS-PPMM up to approximately 8000 RPM and 7000 RPM. however DS-PPMM achieves greater average torque in lower speed with smoother torque reduction over the higher speed range compared to DSPM.

ACKNOWLEDGMENT

The authors would like to thank EPIC at UNC Charlotte for providing us with the generous support of equipment and facilities. We also gratefully appreciate discussions with V. Acharya and JSOL for providing access to its Finite Element Analysis (FEA) program.

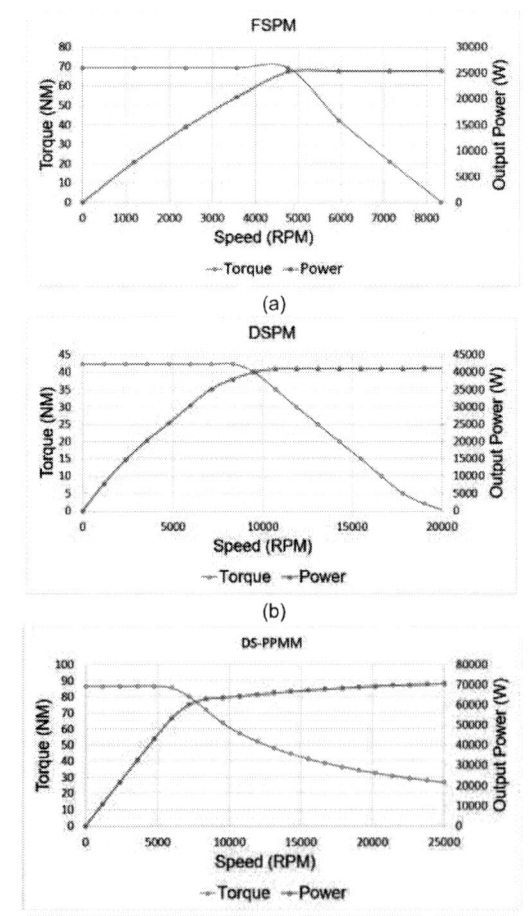

Fig. 10. Torque and power vs. speed characteristics of a) FSPM, b) DSPM, and c) DS-PPMM.

979-8-3315-1612-3/25 $31.00 © 2025 IEEE

REFERENCES

[1] M. Cheng, W. Hua, J. Zhang and W. Zhao, "Overview of Stator-Permanent Magnet Brushless Machines," in IEEE Transactions on Industrial Electronics, vol. 58, no. 11, pp. 5087-5101, Nov. 2011.

[2] K. T. Chau, C. C. Chan, and C. Liu, "Overview of permanent-magnet brushless drives for electric and hybrid electric vehicles," IEEE Trans. Ind. Electron., vol. 55, no. 6, pp. 2246–2257, Jun. 2008.

[3] Y. Liao, F. Liang, and T.A. Lipo, "A Novel Permanent Magnet Motor with Doubly Salient Structure," IEEE Transactions on Industry Applications, vol. 31, no. 5, pp. 1069-1078, 1995.

[4] R. P. Deodhar, S. Andersson, I. Boldea, and T. J. E. Miller, "The flux reversal machine: A new brushless doubly-salient permanent-magnet machine," in Conf. Rec. IEEE IAS Annu. Meeting, pp. 786–793. 1996.

[5] E. Hoang, A. H. Ben-Ahmed, and J. Lucidarme, "Switching flux permanent magnet polyphased machines," in Proc. Eur. Conf. Power Electron. Appl., pp. 903–908, 1997.

[6] C.J. Flynn, N.B. Talsoe, and J.J. Childress, "Parallel path magnetic technology for high-efficiency power generators and motors", in AIP Conference Proceedings, vol. 813, pp. 1205-1212, 2006.

[7] C.J. Flynn (2006), "Parallel Path Magnetic Circuit", US Patent No 0089775 A1.

[8] E. Hoang, A. H. Ben-Ahmed, and J. Lucidarme, "Switching flux permanent magnet polyphased machines," in Proc. Eur. Conf. Power Electron. Appl., pp. 903–908, 1997.

[9] H. Chen, A. M. EL-Refaie and N. A. O. Demerdash, "Flux-Switching Permanent Magnet Machines: A Review of Opportunities and Challenges—Part I: Fundamentals and Topologies," in IEEE Transactions on Energy Conversion, vol. 35, no. 2, pp. 684-698, June 2020.

[10] Z. Q. Zhu, "Switched flux permanent magnet machines — Innovation continues," 2011 International Conference on Electrical Machines and Systems, Beijing, China, pp. 1-10, 2011.

[11] W. Zhao et al., "Stator-flux-oriented fault-tolerant control of flux-switching permanent-magnet motors," IEEE Trans. Magn., vol. 47, no. 10, pp. 4191–4194, Oct. 2011.

[12] W. Zhao, M. Cheng, K. T. Chau, and C. C. Chan, "Control and operation of fault-tolerant flux-switching permanent-magnet motor drive with second harmonic current injection," IET Electr. Power Appl., vol. 6, no. 9, pp. 707–715, Nov. 2012.

[13] C. S. Goli, M. Salehi, S. Essakiappan, D. M. Ionel, J. Gafford and M. Manjrekar, "Comparative Analysis of Doubly Salient Special Machine and Flux Switching Machine with PMs in Stator," 2023 IEEE International Electric Machines & Drives Conference (IEMDC), San Francisco, CA, USA, pp. 1-7, 2023.

[14] Y. Yang, H. Ju, K. Wang, and Z. Guo, "Analysis and Design of a Novel Doubly Salient PM Motor Based on Parallel Path Magnetic Technology," IEEJ Transactions on Electrical and Electronic Engineering, vol. 18, no. 4, pp. 537-545, Apr. 2023.

[15] M. Jin, Y. Wang, J. Shen, P. Luk, W. Fei, and C. Wang, "Cogging torque suppression in a permanent magnet flux switching integrated starter- generator," IET Elect. Power Appl., vol. 4, no. 8, pp. 647–656, Sep. 2010.

[16] K. Abbaszadeh, F. R. Alam, and S. A. Saied, "Cogging torque optimization in surface mounted permanent magnet motors by using design of experiment," Energy Convers. Manage., vol. 52, no. 10, pp. 3075–3082, Sep. 2011.

[17] S. Yousefnejad, H. Heydari, K. Akatsu and J. -S. Ro, "Analysis and Design of Novel Structured High Torque Density Magnetic-Geared Permanent Magnet Machine," in IEEE Access, vol. 9, pp. 64574-64586, 2021.

[18] H. Jia, M. Cheng, W. Hua, Z. Yang, and Y. Zhang, "Compensation of cogging torque for flux-switching permanent magnet motor based on current harmonics injection," in Proc. IEEE Int. Elect. Motor Drives Conf., Miami, FL, USA, May 2009, pp. 286–291.

[19] W. Huang and W. Hua, "A finite-control-set-based model-predictiveflux- control strategy with iterative learning control for torque ripple minimization of flux-switching permanent magnet machines," in Proc. IEEE Veh. Power Propulsion Conf., Hangzhou, China, Oct. 2016, pp. 1–6.

[20] E. Ghanaee, J. I. Pérez-Díaz, D. Fernández-Muñoz, J. Nájera, M. Chazarra and S. Castaño-Solis, "Optimal Scheduling of a Hybrid Wind–Battery Power Plant in the Day-Ahead and Reserve Markets Considering Battery Degradation Cost," 2024 International Conference on Smart Energy Systems and Technologies (SEST), Torino, Italy, pp. 1-6, 2024.

[21] X. Zhu, W. Hua, Z. Wu, W. Huang, H. Zhang and M. Cheng, "Analytical Approach for Cogging Torque Reduction in Flux-Switching Permanent Magnet Machines Based on Magnetomotive Force-Permeance Model," in IEEE Transactions on Industrial Electronics, vol. 65, no. 3, pp. 1965-1979, March 2018.

[22] H. Chen, X. Liu, A. M. EL-Refaie, J. Zhao, N. A. O. Demerdash and J. He, "Comparative Study of Winding Configurations of a Five-Phase Flux-Switching PM Machine," in IEEE Transactions on Energy Conversion, vol. 34, no. 4, pp. 1792-1804, Dec. 2019.

[23] F. Li, W. Hua, M. Tong, G. Zhao and M. Cheng, "Nine-Phase Flux-Switching Permanent Magnet Brushless Machine for Low-Speed and High-Torque Applications," in IEEE Transactions on Magnetics, vol. 51, no. 3, pp. 1-4, March 2015, Art no. 8700204.

[24] X. Xue, W. Zhao, J. Zhu, G. Liu, X. Zhu and M. Cheng, "Design of Five-Phase Modular Flux-Switching Permanent-Magnet Machines for High Reliability Applications," in IEEE Transactions on Magnetics, vol. 49, no. 7, pp. 3941-3944, July 2013.

[25] A. Gandhi and L. Parsa, "Double-Rotor Flux-Switching Permanent Magnet Machine with Yokeless Stator," in IEEE Transactions on Energy Conversion, vol. 31, no. 4, pp. 1267-1277, Dec. 2016.

[26] M. Salehi, and M. Manjrekar, "Doubly Salient Parallel Path Magnetic Motor: a novel doubly salient stator permanent magnet machine," 2024 IEEE Energy Conversion Congress and Exposition (ECCE), Phoenix, Arizona, USA.

Elimination of Deadtime Effect on Resolver Offset Estimation Using the Pulsating Current Command for Electric Vehicle Application

Yingfeng Ji
Ford Motor Company
Dearborn, U.S
yji17@ford.com

Nurani Chandrasekhar
Ford Motor Company
Dearborn, U.S
nchandra@ford.com

Abstract—An accurate position measurement of the rotor is critical for the success of field-oriented control (FOC) for interior permanent magnet synchronous machines (IPMSM). This paper proposes a back electromotive force (BEMF) based approach to determine the resolver offset (RO), commanding a pulsating current to eliminate the effect of inverter dead-time. The results show that the proposed approach successfully eliminates the effects of dead time on the estimation of RO, and estimation accuracy is higher than the method employing zero current command. The simulation results are experimentally verified by dynamometer testing.

Keywords— resolver offset, BEMF, pulsating, FOC

I. INTRODUCTION

The resolver offset (RO) is the angle between the motor's zero reference position, indicated by the zero-crossing point of motor phase V-W line-to-line back electromotive force (BEMF), and the zero-reference of a resolver. In Fig. 1, the zero point of line-to-line BEMF between phase V and W is indicated by point A while the zero position of the resolver is indicated by point B. The difference in angle between point A and B is the resolver offset. In control, at A point the position should be 0 for a perfect FOC alignment. Depending on the relative position of point B, the resolver offset can be positive or negative.

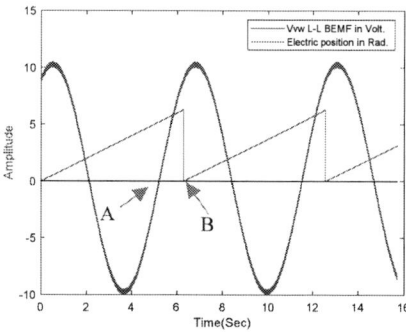

Fig. 1 Illustrative graph for the definition of RO

In field-oriented control (FOC) the RO needs to be compensated to achieve the precise torque control of the permanent magnet synchronous machines (PMSM). However, the RO measurement using the line-to-line voltage involves additional instruments and hence the risk of measurement and manual errors [1]. In some cases, it is not feasible to directly measure the line-to-line voltage, e.g., high-volume production, since the motor terminals are not accessible. In addition, there is a strong desire for a method that does not require extra time to conduct the test. An alternative is to use the inverter voltage commands in place of the measured line-to-line voltage,

allowing the test to conducted much faster without the need for access to the machine terminals. This method has the disadvantage of being dependent on inverter generated distortion between the voltage commands and the applied voltage. Inverter deadtime behavior is one such effect. Another advanced method, called high-frequency voltage injection (HFVI), can estimate the RO while the motor is at standstill [2]. In HFVI a rotating high frequency voltage is injected into stationary reference frame to estimate the rotor position information with or without polarity information. The current response contains the saliency information, as well as the polarity information. The double pulse voltage injection in a synchronous reference frame can be used to detect the polarity information if the current response of HFVI does not contain the polarity information. One of the terms in the current response can be used to extract the spatial information, which is the position offset. However, the estimation accuracy is impacted by the inverter dead-time, and hence is highly dependent on the rotor position and DC-link voltage [2]-[4]. For a surface-mounted PMSM or a motor with low saliency polarity, the difference in inductance is not significant, so the high-frequency injection method's accuracy may decrease [7-8]. HFVI suffers from a low ratio of signal-to-noise as well, therefore, the accuracy is affected and disturbed by environmental conditions for instance EMI/EMC. In [9] a new algorithm was proposed for estimating the RO by injecting an additive voltage at high frequency to simplify the usual HFVI method, but the estimation accuracy was affected significantly by the inverter deadtime.

The method proposed in this paper is based on using the inverter commanded voltages in combination with a pulsating current command, as well as a zero current (Id and Iq) command, to decouple the distortion of current regulator commanded voltage caused by the inverter nonlinearities, e.g., inverter dead time, etc. The RO estimation accuracy obtained using the proposed method is validated using dynamometer testing. The results demonstrate the method is capable of a very high estimation accuracy. Comparing to previous methods, the estimation accuracy has minimal error caused by inverter deadtime.

II. EFFECT OF INVERTER DEADTIME ON RESOLVER OFFSET ESTIMATION

The voltage commands from the current regulator are often employed to estimate the RO, however the output voltage distortion caused by inverter nonlinearities, i.e., the dead time, can introduce errors in the RO estimation. While various references [5-6] have focused on accurately compensating for the voltage distortion at all frequencies, the fundamental component is more important for the accurate estimation of RO.

979-8-3315-1612-3/25 $31.00 © 2025 IEEE

The deadtime introduces error in the commanded voltage compared to the actual voltage according to the current polarity as shown in Fig. 2. The idea proposed in this paper is to command a pulsating q-axis current with alternating positive and negative polarity to cancel the effect of deadtime.

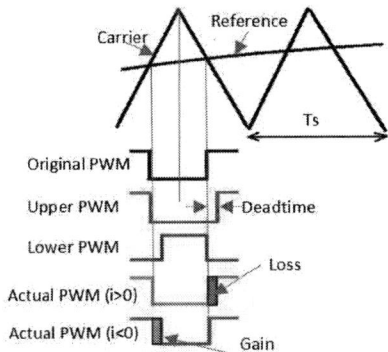

Fig. 2 Voltage distortion due to deadtime for a single phase

The effect of inverter deadtime on voltage commands (Vd and Vq), which are critical for RO estimation, was measured on a dynamometer as shown in Fig. 3. It is obvious that there is voltage offset on Vd_Cmd, and the offset is caused by the inverter dead-time.

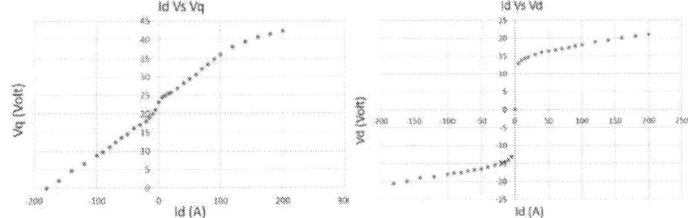

Fig. 3 The relationships between Vd_Cmd and id_Cmd, Vq_Cmd and id_Cmd. Motor parameters: the number of pole pair is 4; the magnet flux is 0.0551Web; the phase resistance @20DegC is 0.0194Ohms; Ld = 0.241mH, Lq = 0.462mH.

III. PROPOSED METHOD

This paper proposes a control method to estimate the RO. The approach eliminates the effect of inverter deadtime by employing specific current commands for the current regulator. To explain the method, it is useful to develop a model of the electric machine. Fig. 4 shows the correlation between the rotor (synchronous) frame and controller frame. The frame with superscript r represents the rotor frame. The frame with superscript c refers to the controller frame. θ_r is the resolver offset. The relationship between the rotor frame and the controller frame can be described by (1). The voltage response signal from the machine in the controller frame will contain offset information θ_r. The general voltage model for an IPMSM machine in a synchronous (d-q) frame is described in (2).

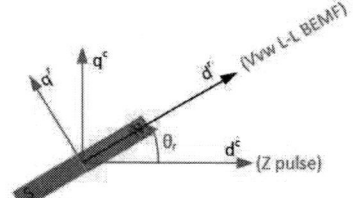

Fig. 4 Definition of reference frames

$$\begin{bmatrix} v_d^r \\ v_q^r \end{bmatrix} = \begin{bmatrix} cos(\theta_r) & sin(\theta_r) \\ -sin(\theta_r) & cos(\theta_r) \end{bmatrix} \begin{bmatrix} v_d^c \\ v_q^c \end{bmatrix} \quad (1)$$

$$\begin{bmatrix} v_d^r \\ v_q^r \end{bmatrix} = \begin{bmatrix} r_s & -\omega_r L_q \\ \omega_r L_d & r_s \end{bmatrix} \begin{bmatrix} i_d^r \\ i_q^r \end{bmatrix} + \frac{d}{dt} \begin{bmatrix} L_d & 0 \\ 0 & L_q \end{bmatrix} \begin{bmatrix} i_d^r \\ i_q^r \end{bmatrix} + \begin{bmatrix} 0 \\ \omega_r \lambda_{PM} \end{bmatrix} \quad (2)$$

Considering steady-state, and assuming the RMS (average) value of i_d^r and i_q^r is zero, (2) can be simplified to (3) and (4).

$$v_d^r = 0 \quad (3)$$

$$v_q^r = \omega_r \lambda_{PM} \quad (4)$$

Transforming (3) and (4) from the rotor synchronous frame to the controller synchronous frame using (1) results in (5) and (6).

$$v_d^c = -\omega_r \lambda_{PM} sin(\theta_r) \quad (5)$$

$$v_q^c = \omega_r \lambda_{PM} cos(\theta_r) \quad (6)$$

By looking at the d-axis, projecting the v_d^c, v_q^c to the d-axis by θ_L, which is the estimated RO by the proposed algorithm, the voltage in the d-axis is shown in (7).

$$v_{d\_projected}^c = v_d^c \cos(\theta_L) - v_q^c \sin(\theta_L) \quad (7)$$

The RO can be determined by regulating the voltage command on d-axis in (9):

$$\theta_L = k_p\left(-v_d^c \cos(\theta_L) + v_q^c \sin(\theta_L)\right) + k_i \int \left(-v_d^c \cos(\theta_L) + v_q^c \sin(\theta_L)\right) dt \quad (9)$$

Based on the proposed estimation method in (9), and submitted (5) and (6), a simplified phase locked loop can be derived, and illustrated by (10).

$$\theta_L = k_p \omega_r \lambda_{PM} sin(\theta_r + \theta_L) + k_i \int \omega_r \lambda_{PM} sin(\theta_r + \theta_L) dt \quad (10)$$

Fig. 5 Illustrative graph of the simplified phase locked loop

Fig. 6 shows the relationship between the main control path of current regulation and the function of RO estimation. In the proposed method the iq_Cmd is a pulsating current command with the frequency of 25 Hz and the amplitude of 75 A, the mean value of iq_Cmd is zero. In the implementation, dead time will add an extra voltage to the actual voltage when the phase current is positive, while an extra voltage will be deducted from the actual voltage due to the dead time if the phase current is negative.

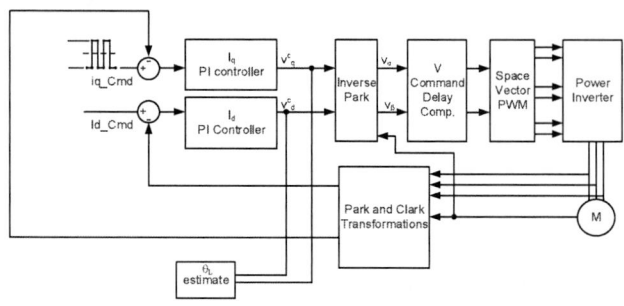

Fig. 6 Control structure of RO estimation, iq_Cmd can be pulsating current or 0

IV. SIMULATION RESULTS

The performance using a pulsating current command on the estimation of the resolver offset was simulated for three cases where the nominal RO was 60, 100, and -30 degrees, respectively. It is obvious to eliminate the estimation deviation of position offset. With respect to the estimation accuracy of the position offset, the comparable simulations are evaluated by comparing with the estimations where the id_Cmd and iq_Cmd are zero.

Command #1: id_Cmd = 0, iq_Cmd = 0

Command #2: id_Cmd = 0, iq_Cmd is a pulsating command with the frequency of 25 Hz and the amplitude of 75 A, the mean value of iq_Cmd is zero.

For the test, the motor was spun at a constant speed of 1000 RPM. The switching frequency was 5 kHz, and the inverter deadtime was 5us. The DC bus voltage was 400 V. The proportional and integral controller gains were 0.1, 15, respectively. The parameters of the electric machine A are listed as TABLE I. The bandwidth of current controller is designed for 50 Hz. The simulation results are shown in Figs. 7-9. It can be seen that the estimated resolver offset converges to the actual resolver offset when a pulsating current command is used but shows an error when the current is regulated to be zero.

TABLE I. The machine parameters of electric machine A

	Value	Unit	Comment
P	4	/	The number of pole pair
λ_m	0.0544	weber	The nominal magnet flux
L_d	0.204	mH	The d-axis inductance
L_q	0.375	mH	The q-axis inductance
R_20C	0.0142	Ohm	The phase resistance @20Deg
R_V	0.071	Ohm	The virtual resistance

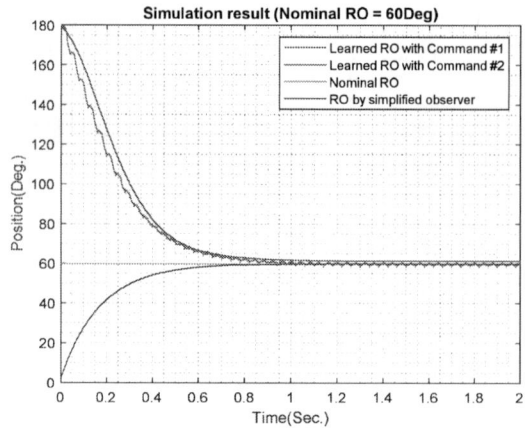

Fig. 7 The estimated RO: 61.35Deg for Command #1, 59.78 for Command #2

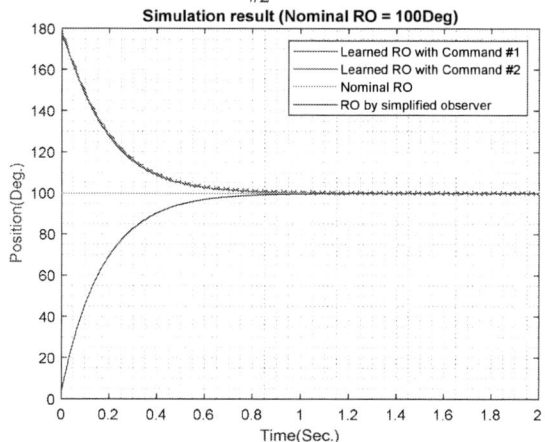

Fig. 8 The estimated RO: 101.41Deg for Command #1, 100.46 for Command #2

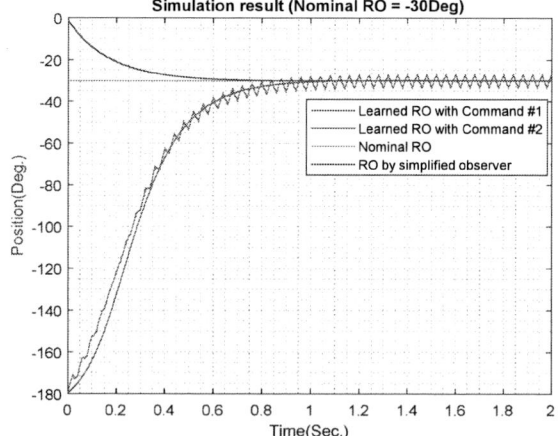

Fig. 9 The estimated RO: -29.66Deg for Command #1, -30.10 for Command #2

V. EXPERIMENTAL RESULTS

The performance of the algorithm was validated experimentally on a dynamometer.. The motor parameters are listed in TABLE II. Fig.10 shows the dynamometer experimental setup.

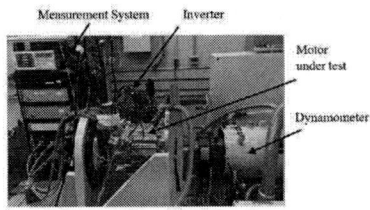

Fig. 10: Dynamometer experimental setup

The validation was performed for both a resolver and an eddy-current inductive position sensor. The reference ROs, which are measured by the traditional method as shown in Fig. 1 in the Introduction. These are deemed to be the accurate measurement of RO since the inverter non-linearities are not present. The measured ROs for the resolver and inductive position sensor are -21.09 and -88.33 degrees, respectively, using this method. Fig. 11 and Fig. 12 are the experimental results to show the estimation performance of RO in time domain. In all cases, the Id_Cmd = 0 while comparing the estimation performance between Iq_Cmd = 0 and a pulsating Iq_Cmd. In the test the amplitude of pulsating Iq_Cmd is 25A and the frequency is 25Hz. For the resolver, the estimated RO is -20.77 degrees with Iq_Cmd = 0, and the estimated RO is -21.10 degrees with Pulsating Iq_Cmd. For the eddy-current inductive position sensor, the estimated RO is 89.46 degrees with Iq_Cmd = 0, and the estimated RO is 87.98 degrees with pulsating Iq_Cmd. The estimation accuracy with pulsating Iq_Cmd is a little bit higher than the one while Iq_Cmd = 0.

The additional test results for other two motors are listed in TABLE III and IV, however only for resolver and under pulsating Iq_Cmd.

TABLE II. The motor parameters for electric motor B

	Value	Unit	Comment
P	4	/	The number of pole pair
λ_m	0.062583	weber	The nominal magnet flux
L_d	9.50E-02	mH	The d-axis inductance
L_q	1.64E-01	mH	The q-axis inductance
R_20C	0.00543	Ohm	The phase resistance @20Deg
R_V	0.034	Ohm	The virtual resistance

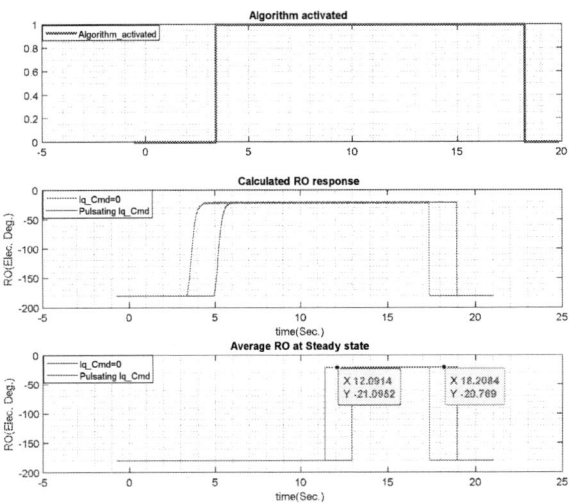

Fig. 11 Experimental result for a resolver

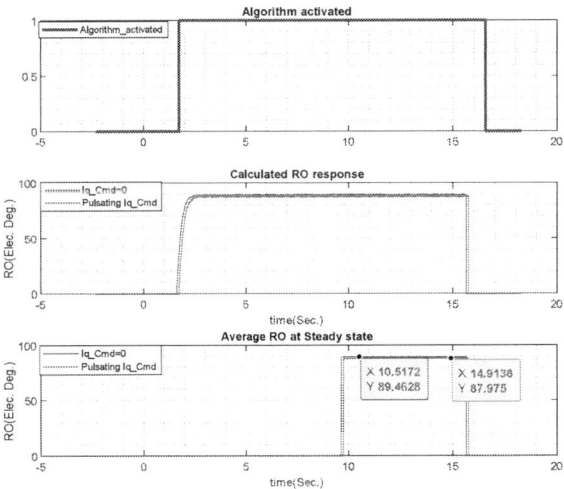

Fig. 12 Experimental result for eddy-current inductive position sensor

TABLE III. The estimated RO for Motor A (The measured RO is 145.55degree)

Test Cases	Motor speed (rpm)	DC-Bus voltage (Volt.)	Estimated RO (degree)
1	1000	346	145.81
2	1000	346	145.46
3	1000	346	144.84
4	1000	346	144.84
5	2000	346	145.90
6	-1000	346	146.25
7	-2000	346	145.81

TABLE IV. The estimated RO for Motor B (The measured RO is 0 degree)

Test Cases	Motor speed (rpm)	DC-Bus voltage (Volt.)	Estimated RO (Counts)
1	1000	270	0.44
2	2000	270	0.53
3	1000	200	0.35

VI. Conclusions

The proposed method requires no extra hardware compared to the measurement method. The proposed method does not need any calibration work since the pulsing current command eliminates the effect of the inverter dead time. The disclosed method improves the estimation accuracy comparing to the case of zero iq and id commands. The results are verified by the dynamometers test.

Acknowledgment

I would like to express my appreciation to Dr. Michael Degner. He spent lots of time on sharing the advice of improving the paper quality. I also thank Dr. Jiyao Wang for the discussion on the idea in this pater and thank Dr. Yang Xu for the technical discussion at the beginning of this topic.

References

[1] Jae Sung Bang and Tae Soo Kim, "Automatic calibration of a resolver offset of permanent magnet synchronous motors for hybrid electric vehicles," 2015 American Control Conference (ACC), pp. 4174-4179.

[2] Hyunbae Kim, " ON-LINE PARAMETER ESTIMATION, CURRENT REGULATION, AND SELF-SENSING FOR IPM SYNCHRONOUS MACHINE DRIVES," Ph.D. Dissertation, UNIVERSITY OF WISCONSIN-MADISON, 2004.

[3] Jong-Woo Choi and S. -K. Sul, "Inverter output voltage synthesis using novel dead time compensation," in *IEEE Transactions on Power Electronics*, vol. 11, no. 2, pp. 221-227, March 1996.

[4] Hag-Wone Kim, Myung-Joong Youn, Kwan-Yuhl Cho and Hyun-Soo Kim, "Nonlinearity estimation and compensation of PWM VSI for PMSM under resistance and flux linkage uncertainty," in *IEEE Transactions on Control Systems Technology*, vol. 14, no. 4, pp. 589-601, July 2006.

[5] [5] Bian Yuanjun, et al., "Initial Rotor Position Estimation of PMSM Based on High Frequency Signal Injection", 2014 IEEE Conference and Expo Transportation Electrification Asia-Pacific (ITEC Asia-Pacific), pp. 1-4, Oct. 2014

[6] [6] Hyung-Do Kang; Seon-Hwan Hwang; Jaesuk Lee, "Initial Rotor Position Estimation of Single-phase Permanent Magnet Synchronous Motor with Asymmetric Air-gap," 2018 IEEE International Transportation Electrification Conference & EXPO Asia-Pacific.

[7] [7] H.-D. Kang, S.-H. Hwang, and J. Lee, "Initial rotor position estimation of single-phase permanent magnet synchronous motor with asymmetric air-gap," in Proc. IEEE Transp. Electric. Conf. Expo, Asia Pacific (ITEC Asia Pacific), Jun. 2018, pp. 1-5.

[8] [8] L. M. Gong and Z. Q. Zhu, "Robust initial rotor position estimation of permanent-magnet brushless AC machines with carrier-signal-injection-based sensor-less control," IEEE Trans. Ind. Appl., vol. 49, no. 6, pp. 2602_2609, Nov. 2013.

[9] [9] S. Medjmadj1, D. Diallo, C. Delpha, G. Yao, "A salient-pole PMSM position and speed estimation at standstill and low speed by a simplified HF injection method," IECON 2017 - 43rd Annual Conference of the IEEE Industrial Electronics Society, pp. 8317-8322, Nov. 2017.

A Generic Load Emulator for Testing Motor Drives of E-mobility

Qingzheng Zhang, Kaiyuan Feng, Changsheng Hu, Dehong Xu
college of Electrial Engineering
Zhejiang University
Hangzhou 310027,China
12410079@zju.edu.cn

Abstract—**A generic load emulator for testing motor drives of e-mobility is studied. The permanent magnet synchronous motor (PMSM) with maximum torque per ampere control (MTPA) is used to emulate the load characteristic, which can emulate different kinds of mechanical loads for the motor drive of e-mobility. Thus, the motor drive under tested can be easily characterized either static and dynamic effectively. Finally, the 20 kW experiment platform is set up to verifying the load emulator.**

Keywords—load emulation, PMSM, MTPA, E-mobility

I. INTRODUCTION

Motor drives are the core of electrified transportation, such as electric vehicles, high-speed trains, electric ships and electric airplanes, etc. Their operating conditions and load characteristics vary significantly in different applications[1]. It is important to characterize the performance of the motor drives in designing stage before they are put into vehicle system test. Conventional motor drive test includes mechanical load emulation, electro-hydraulic load emulation[2] , and electric load emulation systems[3]. Among them, the mechanical load emulation has drawbacks such as bulky equipment, high energy consumption, and not flexible due to the fixed relationship between the torque and speed. Electro-hydraulic load emulation has slower dynamics [2]. The electric load emulation has the advantages of faster response and stepless adjustment of output torque, which makes it widely used. Therefore, motor drives are often used to emulate different types of loads[1], such as rotating paddle-type wave energy converters [4], gas engines[5][6], hydrokinetic turbines[7], wind turbines[8][9], and electric vehicles[10][11][12] .

With regards to the electric load emulation, dynamic dynamometer can emulate linear and nonlinear loads for testing motors drives[13]. PMSM with fuzzy control is used to achieve dynamic emulation of mechanical inertial load[14]. PMSM is used to emulate the cubic relationship between power and rotor speed by using the resistive-capacitive load [15].

A generic load emulator for testing motor drives of e-mobility is studied. The load emulator is implemented by a permanent magnet synchronous motor (PMSM). Its load characteristics can be programmed so that it can emulate arbitrary mechanical load characteristics for the motor drive of e-mobility. PMSM with MTPA control is used in the load emulation[16][17][18][19]. Finally, the 20 kW experiment platform is built to verify the load emulation method.

II. LOAD EMULATION SYSTEM SCHEME

A. Test and load emulation system

Structure of the motor testing system is shown in Figure 1. It consists of DC source, inverter, the motor under test, and the load emulator. The load emulator is composed of a PMSM and a three-phase rectifier. The load emulator is realized by controlling PMSM through the three-phase rectifier.

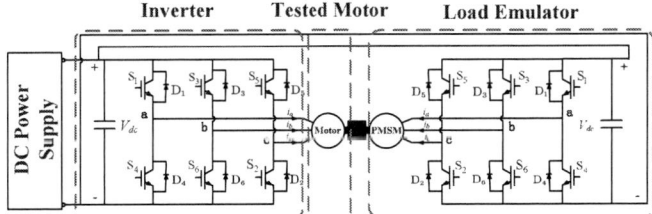

Figure 1. Test and load emulation system

The load emulation motor emulates required mechanical load characteristics with d-q control through the PWM rectifier.

B. Load emulation control

To make the motor operate properly, the amplitude of the stator current cannot exceed the maximum current and the voltage on the stator cannot exceed the maximum phase voltage of the rectifier (voltage limit ellipse) :

$$\begin{cases} \omega_e^2 \left(L_d i_d + \varphi_f\right)^2 + \omega_e^2 \left(L_q i_q\right)^2 \leq U_m^2 \\ i_d^2 + i_q^2 \leq I_m^2 \end{cases} \quad (1)$$

where ω_e is the electrical angular velocity of the motor, U_m is the maximum phase voltage of the rectifier ($\frac{V_{dc}}{\sqrt{3}}$) , I_m is the maximum current of the state motor, i_d and i_q are the d-axis and q-axis currents respectively, L_d and L_q are the d-axis and q-axis inductances, and φ_f is the magnetic flux linkage of the PMSM's permanent magnet.

In Figure 2, the voltage limit ellipse is the restrain to the stator voltage and the current limit circle is restrain to the stator current.

At a certain torque $T_{e\text{-}ref}$, the dq axis current satisfies the relationship:

979-8-3315-1612-3/25 $31.00 © 2025 IEEE

$$i_q = \frac{2T_{e-ref}}{3p\left[(L_d - L_q)i_d + \varphi_f\right]} \quad (2)$$

where p is the number of the pole pair of the PMSM.

In contrast to the case Id=0 when the torque is only controlled only by the q-axis current. Maximum Torque Per Ampere (MTPA) control is used to achieve minimal stator current for the desired torque, which reduces motor losses. Its steady-state operating point at the target torque is shown at point B in Figure 2. In the figure, curve T_{e-ref} shows the relationship between id and iq currents for the given torque.

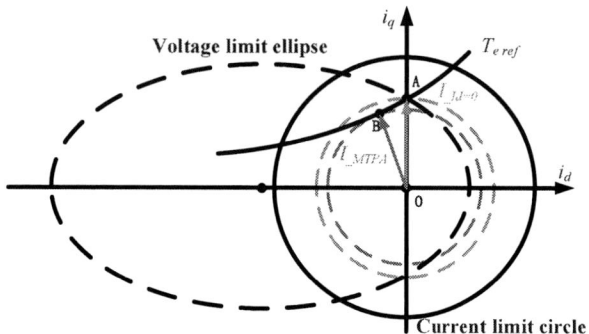

Figure 2. Schematic diagram of MTPA control

The torque of PMSM is represented by:

$$T_e = \frac{3}{2} p\left[(L_d - L_q)i_d + \varphi_f\right]i_q \quad (3)$$

In the dq coordinate, the amplitude of the stator current of the motor is:

$$I_s = \sqrt{i_d^2 + i_q^2} \quad (4)$$

Assuming that the load torque to be emulated is T_{e-ref}, the following constraint is obtained from (1):

$$T_{e-ref} - \frac{3}{2} p\left[(L_d - L_q)i_d + \varphi_f\right]i_q = 0 \quad (5)$$

Following Lagrange function H is introduced:

$$H = \sqrt{i_d^2 + i_q^2} - \lambda\left\{T_{e ref} - \frac{3}{2} p\left[(L_d - L_q)i_d + \varphi_f\right]i_q\right\} \quad (6)$$

To minimize the stator current $I_s = \sqrt{i_d^2 + i_q^2}$, following formulas are obtained:

$$\begin{cases} \frac{\partial H}{\partial i_d} = \frac{i_d}{\sqrt{i_d^2 + i_q^2}} - \frac{3}{2}\lambda p(L_d - L_q)i_q = 0 \\ \frac{\partial H}{\partial i_q} = \frac{i_q}{\sqrt{i_d^2 + i_q^2}} - \frac{3}{2}\lambda p[(L_d - L_q)i_d + \varphi_f] = 0 \\ \frac{\partial H}{\partial \lambda} = T_{e ref} - \frac{3}{2} p[(L_d - L_q)i_d + \varphi_f]i_q = 0 \end{cases} \quad (7)$$

By combining the first and second formula in (7), the dq axis current relationship is obtained:

$$\frac{i_d}{i_q} = \frac{(L_d - L_q)i_q}{(L_d - L_q)i_d + \varphi_f} \quad (8)$$

Combining the third formula in (7) and formula (8), the reference stator currents are derived for the given torque:

$$\begin{cases} i^*_{dref} = -\frac{3k}{4} - \frac{1}{4}\sqrt{k^2 + 4\Delta} - \frac{1}{2}\sqrt{\frac{k^2}{2} - \Delta - \frac{k^3}{2\sqrt{k^2 + 4\Delta}}} \\ i^*_{qref} = \frac{T_{e-ref}}{\frac{3}{2} p\left[(L_d - L_q)i^*_{dref} + \varphi_f\right]} \end{cases} \quad (9)$$

where

$$k = \frac{\varphi_f}{(L_d - L_q)}; \Delta = \frac{\sqrt[3]{2}\Delta_1}{3\sqrt[3]{\Delta_2 + \sqrt{-4\Delta_1^3 + \Delta_2^2}}} + \frac{\sqrt[3]{\Delta_2 + \sqrt{-4\Delta_1^3 + \Delta_2^2}}}{3\sqrt[3]{2}} \quad (10)$$

$$\Delta_1 = -\frac{16T_{e-ref}^2}{3p^2(L_d - L_q)^2}; \Delta_2 = -\frac{12T_{e-ref}^2\varphi_f^2}{p^2(L_d - L_q)^4} \quad (11)$$

According to formula (9), we can estimate reference values of i_{dref}, i_{qref} according to the required torque to emulate certain mechanical load. By controlling the d-axis and q-axis currents i_d and i_q with the rectifier, the output torque of the PMSM motor will follow the given torque. The control diagram is illustrated in Figure 3.

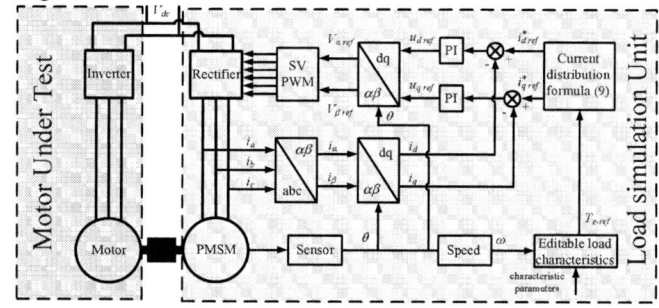

Figure 3. Motor under test system and load emulation unit

III. SIMULATION

The parameters used in the simulation is listed in TABLE I. The DC voltage of the rectifier is 350V. The switching frequency of power devices IGBTs is 10kHz.

TABLE I CIRCUIT PARAMETERS

Parameter	Value	Parameter	Value
DC bus voltage V_{DC}	350V	D-axis inductance L_d	362μH
Switching frequency	10kHz	Q-axis inductance L_q	1.15mH
Number of pole pairs	4	Flux of permanent magnets φ_f	0.0843Wb

In order to verify the effect of the load emulation strategy proposed in this paper, the constant torque load, constant power load, fan load and random load are simulated. The simulation results are shown in Figure 4, 5, 6, 7. Where (a) shows the target speed and the simulation speed of the test motor. (b) shows the torque emulated by the load emulation unit. (c) shows the amplitude of the AC stator current in the load emulation unit under different control strategies, and Id=0 is selected for comparison with MTPA.

Figure 4 shows the simulation result of a 50 N · m constant torque load. Figure 4(a) shows the reference speed and measured speed of the test motor and the load motor. The initial speed is 1000rpm, and increases to 1000rpm in two seconds. In Figure 4(b), the emulated torque is 49.70N.m, which is consistent with the given value. In figure 4(c),Blue line shows the amplitude of the stator current is 82.25A with the MTPA control. Red line shows the amplitude of the stator current is 99.5A by only using iq control with keeping id=0.

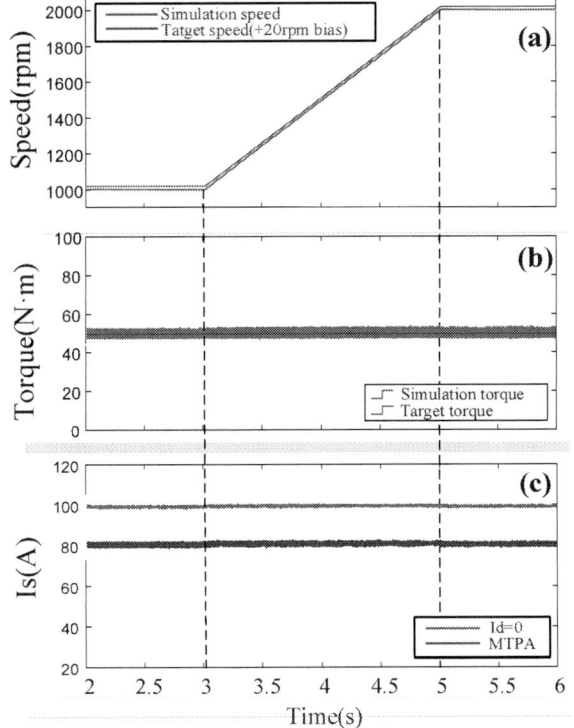

Figure 4. constant torque load simulation. (a) Motor speed (b) Motor emulated torque (c) Motor stator phase current amplitude

Figure 5 shows the simulation of a 15kW constant power load. Figure 5(a) shows the reference speed and measured speed of the test motor and the load motor. In Figure 5(b), the load torque is inversely proportional to the rotational speed, which reduced from 143.24 N · m to 71.62 N · m. The emulated torque deviation is less than 1%. Figure 5(c) show the load emulation with MTPA control can effectively reduce the stator current compared with iq control with id=0.

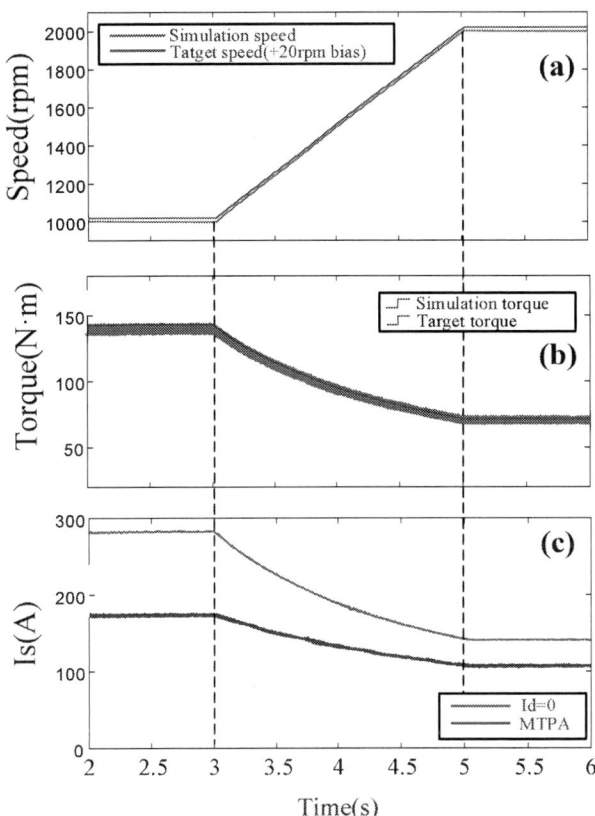

Figure 5. Constant power load simulation. (a) Motor speed (b) Motor emulated torque (c) Motor stator phase current amplitude

Figure 6 shows the simulation of a fan load. Figure 6(a) shows the reference speed and measured speed of the test motor and the load motor. In Figure 6(b), The relationship between torque and rotational speed satisfies the following formula:

$$T_{e_{ref}} = C\omega^2 \qquad (12)$$

where C is a constant, and 0.00002 is taken in the simulation, ω is the speed of motor, the unit is rpm. The emulated torque deviation is less than 1.5%.

Figure 6(c) compares the stator current of MTPA control with that of only iq control.

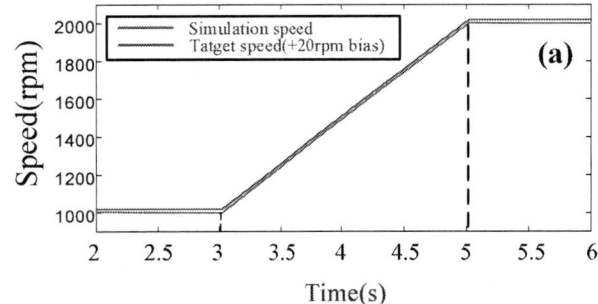

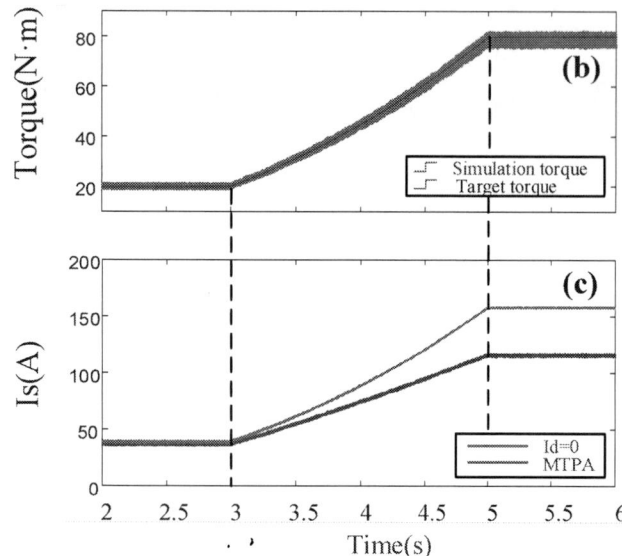

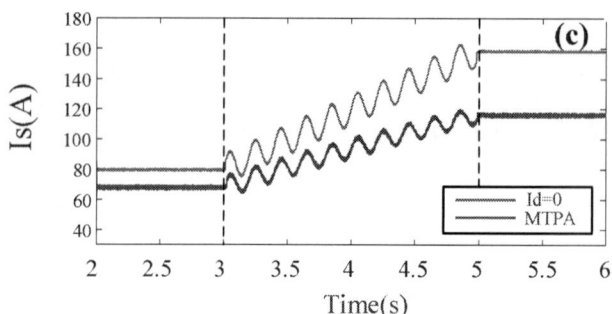

Figure 7. (a) Radom load simulation. (a) Motor speed (b) Motor emulated torque (c) Motor stator phase current amplitude

Figure 6. Fan load simulation. (a) Motor speed (b) Emulated load torque (c) Motor stator phase current amplitude

Figure 7 shows the simulation of a random load, emulate the load that e-mobility might encounter on the road. Figure 7(a) shows the reference speed and measured speed of the test motor and the load motor. In Figure 7(b), the load torque shocks increase as the speed increases, which increased from 40 N · m to 80 N · m. The emulated torque deviation is less than 1%.Figure 7(c) compares the stator current of MTPA control with that of only iq control.

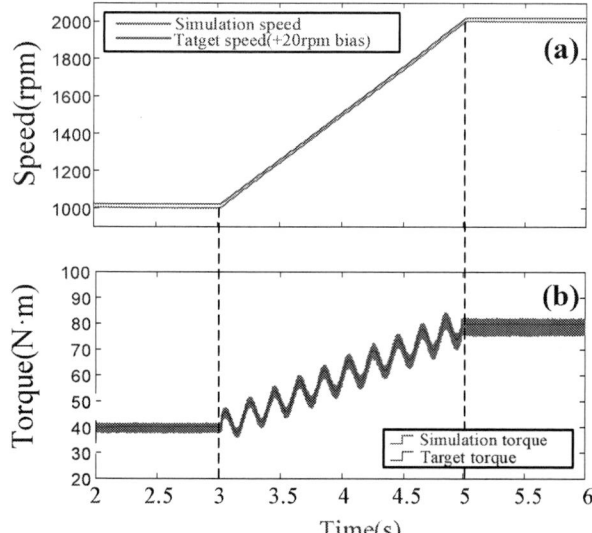

IV. EXPERIMENT

To validate the proposed method, a test platform is set up shown in Figure 8. The platform includes a DC power supply, inverter, a 60 kW permanent magnet synchronous motor under test, a 20 kW load emulator. The supporting system and its system parameters are as follows:

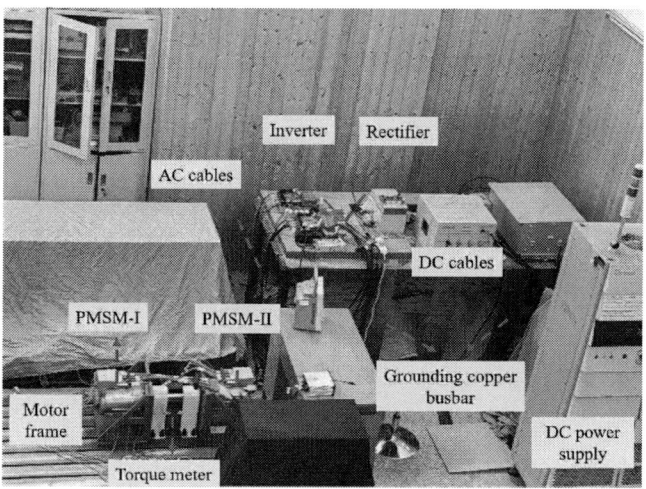

Figure 8.experiment platform

TABLE II SYSTEM PARAMETERS

Parameter	Value	Parameter	Value
DC bus voltage V_{DC}	350V	D-axis inductance L_d	362μH
Switching frequency	10kHz	Q-axis inductance L_q	1.15mH
Rated motor speed	3200RPM	Number of pole pairs	4
Maximum motor speed	9000RPM	Flux of permanent magnets φ_f	0.0843Wb
Rated Motor power P_o	30kW	Rated motor torqueTe_o	90N • m
Motor peak power P_{max}	60kW	Maximum motor torqueTe_{max}	220 N • m

Firstly constant torque load characteristics is emulated. The initial speed is 1000rpm. After 2s, the speed set to be increase linearly for 1s. Then it keeps at 2000rpm. The emulator almost keeps constant torque at about 50 N · m shown in Figure 9. The motor power increases from 5.23 kW to 10.47 kW.

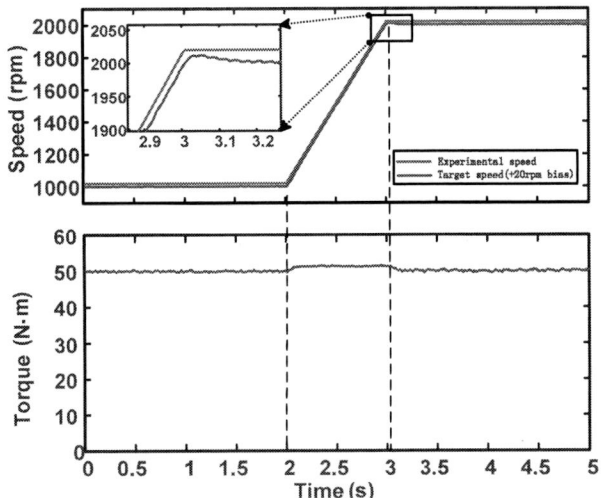

Figure 9. Constatnt torque load experiment (a)Motor speed (b) Motor emulated torque

The emulator is used to emulate the characteristic of the fan load. The emulated torque with speed(rpm) is well matched with theoretical data as shown in Figure 10.

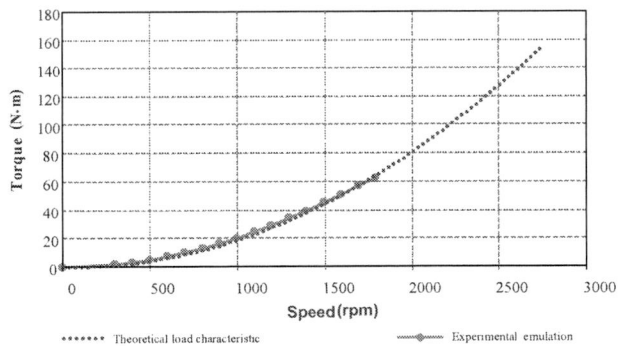

Figure 10. Characteristic curve of fan load emulation experiment

The dynamics performance of the fan load test result are shown in Figure 11. The initial speed is 1000rpm. After 2s, the speed set to be increase linearly for 0.8s. Then it keeps at 1800rpm. When the speed changes, the emulator output torque is proportional to the square of the rotational speed. The motor power increases from 2.09 kW to 12.21 kW.

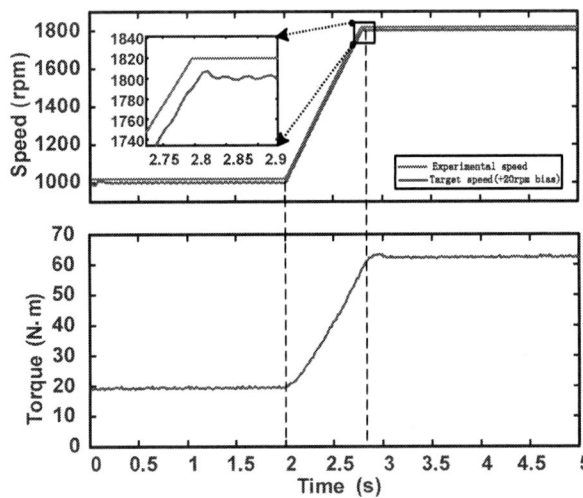

Figure 11. Fan load experiment (a)Motor speed (b) Motor emulated torque

V. CONCLUSIONS

A load emulator with a PMSM and a three-phase rectifier is introduced for testing motor drives of E-mobility. The reference values of i_{dref}, i_{qref} to emulate certain mechanical load are derived based on MTPA. By controlling the d-axis and q-axis currents i_d and i_q with the rectifier, the output torque of the PMSM motor will follow the required torque. load emulation method is verified by experiments with constant torque load and fan load. Further study is needed to verify the emulator in other types of load and its dynamics performance.

REFERENCES

[1] C. M. R. de Oliveira, M. L. de Aguiar, A. G. de Castro, P. R. U. Guazzelli, W. C. d. A. Pereira and J. R. B. d. A. Monteiro, "High-Accuracy Dynamic Load Emulation Method for Electrical Drives," in IEEE Transactions on Industrial Electronics, vol. 67, no. 9, pp. 7239-7249, Sept. 2020

[2] Truong D Q, Kwan A K,Il Yoon J,et al. A Study on Force Control of Electric-Hydraulic Load Simulator Using an Online Tuning Quantitative Feedback Theory[M]. New York:IEEE, 2008; 2272-2277.

[3] Y. NAM, "QFT force loop design for the aerodynamic load simulator," IEEE Transaction on aerospace and electronic systems, vol. 4, no. 37, pp. 1384-1392, 2001.

[4] S. Hazra and S. Bhattacharya, "Modeling and emulation of a rotating paddle type wave energy converter," IEEE Trans. Energy Convers., vol. 33,no. 2, pp. 594–604, Jun. 2018.

[5] C. Gan, R. Todd, and J. M. Apsley, "Drive system dynamics compensator for a mechanical system emulator," IEEE Trans. Ind. Electron., vol. 62,no. 1, pp. 70–78, Jan. 2015

[6] B. A. Correa, Y . Zhang, R. A. Dougal, T. Chiocchio, and K. Schoder, " Mechanical power-hardware-in-the-loop: Emulation of an aeroderivative twin-shaft turbine engine," in Proc. IEEE Elect. Ship Technol. Symp.,Apr. 2013, pp. 464–468.

[7] R. J. Cavagnaro, J. C. Neely, F. Fay, J. L. Mendia, and J. A. Rea, "Evaluation of electromechanical systems dynamically emulating a candidate hydrokinetic turbine," IEEE Trans. Sustain. Energy, vol. 7, no. 1, pp. 390–399, Jan. 2016.

[8] N. R. Averous et al., "Development of a 4 MW full-size wind-turbine test bench," IEEE J. Emerg. Sel. Topics Power Electron., vol. 5, no. 2,pp. 600–609, Jun. 2017.

[9] J. Y an, Y . Feng, and J. Dong, "Study on dynamic characteristic of wind turbine emulator based on PMSM," Renewable Energy, vol. 97, pp. 731–736, Nov. 2016.

[10] Z. Zhang, L. Wang, J. Zhang, and R. Ma, "Study on requirements for load emulation of the vehicle with an electric braking system," IEEE Trans.V eh. Technol., vol. 66, no. 11, pp. 9638–9653, Nov. 2017.

[11] P . Fajri, V . A. K. Prabhala, and M. Ferdowsi, "Emulating on-road operating conditions for electric-drive propulsion systems," IEEE Trans. Energy Convers., vol. 31, no. 1, pp. 1–11, Mar. 2016.

[12] P . Fajri, S. Lee, V . A. K. Prabhala, and M. Ferdowsi, "Modeling and integration of electric vehicle regenerative and friction braking for motor/dynamometer test bench emulation," IEEE Trans. V eh. Technol.vol. 65, no. 6, pp. 4264–4273, Jun. 2016.

[13] P. Sandholdt, E. Ritchie, J. K. Pedersen and R. E. Betz, "A dynamometer performing dynamical emulation of loads with nonlinear friction," Proceedings of IEEE International Symposium on Industrial Electronics, Warsaw, Poland, 1996, pp. 873-878.

[14] Gang Gu, Songjiang Peng, Shijie Su, Yongmei Zhu and Ting Wang, "The experimental research of dynamical load simulation system based on PMSM," 2016 IEEE Advanced Information Management, Communicates, Electronic and Automation Control Conference (IMCEC), Xi'an, China, 2016, pp. 1207-1211

[15] R. Zhou, B. Lu, M. Li, P. Zhu, D. Liu.Control Design and Load Testing of a 65kW 36600rpm High Speed Electric Drive System[A].In: International Power Electronics and Motion Control Conference [C],Chengdu, China.2024.141.

[16] Zhe Wang, Mingyan Wang, Ben Guo and Chai Feng, "Load torque and inertia simulation based on double-stator permanent-magnet synchronous motor," 2014 International Power Electronics Conference (IPEC-Hiroshima 2014 - ECCE ASIA), Hiroshima, 2014, pp. 3129-3133

[17] Chunyan Lai ,Guodong Feng , Jimi Tjong, and Narayan C. Kar,Direct Calculation of Maximum-Torque-Per-Ampere Angle for Interior PMSM Control Using Measured Speed Harmonic,'IEEE TRANSACTIONS ON POWER ELECTRONICS, VOL. 33, NO. 11, NOVEMBER 2018.

[18] G. K. G and M. V . P, "MTPA Based Efficient PMSM Drive for Electric Car," 2023 International Conference on Circuit Power and Computing Technologies (ICCPCT), Kollam, India, 2023, pp. 1130-1135.

[19] M. Alzayed and H. Chaoui, "Efficient Simplified Current Sensorless Dynamic Direct Voltage MTPA of Interior PMSM for Electric Vehicles Operation," in IEEE Transactions on Vehicular Technology, vol. 71, no. 12, pp. 12701-12710, Dec. 2022.

979-8-3315-1612-3/25 $31.00 © 2025 IEEE

Design and Implementation of Power Assisted Control System for E-Bikes

Che-Yu Lu
Department of Electrical Engineering
National United University, Miaoli, Taiwan
cylu@nuu.edu.tw

Tzu-Ping Cheng
Department of Electrical Engineering,
National United University, Miaoli, Taiwan
chengjiang1126@gmail.com

Abstract— This paper aims to develop a power assisted control system for E-bikes. The proposed approach combines a motor drive system with the sliding mode observer and adaptive control strategy to advance E-bike technology. The paper also establishes riding models that account for various environmental factors such as rider weights, road gradients, and wind speeds to ensure that the power assisted modes are well-suited to different riding conditions. In addition, the E-bike motor will utilize an adaptive sliding mode observer control algorithm in conjunction with the encoder and central shaft torque signals. This method enhances control accuracy during low-speed operation and reduces estimated torque ripple errors. Finally, the proposed method significantly improves the competitiveness of power assisted control systems.

I. INTRODUCTION

The power assisted control strategy proposed in this paper utilizes a sensorless control approach, where the rotor position is estimated through motor voltage and current signals to calculate the speed, thus eliminating the need for position feedback sensors, such as encoders [1], [2]. The proposed control system incorporates speed and current control loops to enhance dynamic performance [3]. Specifically, the E-bike motor control employs a sliding mode observer (SMO) for rotor position estimation. To mitigate instability during SMO switching, a low-pass filter (LPF) and a phase-locked loop (PLL) are integrated to address this issue [4].

Traditional SMO control is widely used in brushless DC (BLDC) motors to provide good robustness to the E-bike motor operation system. The signal for SMO phase lock estimation contains high-frequency components. In addition to utilizing LPF and PLL control, it can replace the Sigmoid mapping function instead of the traditional Signum function [5]-[8]; the advantage of the Sigmoid mapping function is that it can eliminate high-frequency signals more linearly during SMO switching. At the same time, the motor may also cause more significant errors in system response and phase due to the function. In addition to the influence of parameter changes of the motor itself and system interference, according to [9], [10], the motor parameter definition model is combined with an adaptive neural network to track the back electromotive force (EMF) voltage and make error compensation with the d-q

axis current to simulate pedaling ripple and eliminate oscillation.

However, the E-bike motor part needs to be controlled more accurately, and the stable output signal of the adaptive control mentioned above needs to be adopted. By integrating SMO and field-oriented control (FOC), along with analyzing magnetic cross-coupling effects in the rotor and stator [11], and enhancing observer circuits in FOC or SMO with adaptive parameter variations [12], [13], the dynamic response and steady-state performance of the system can be substantially improved. The stability of the SMO dynamics is analyzed by monitoring the consistency of measured and estimated rotor position and speed under closed-loop control. This paper utilizes SMO and PLL control to validate the capability of the proposed controller to reduce oscillations [14] and investigates the effects of speed variations on system parameters and overall performance.

II. MATHEMATICAL MODEL OF E-BIKE MOTOR

After the magnetic field generated by the E-bike motor winding is interlinked with the rotor magnet, the magnetic linkage generated after cutting will change with the magnetic linkage of the stator winding according to the structural design. The stator voltage equation can be expressed as

$$
\begin{bmatrix} V_a \\ V_b \\ V_c \end{bmatrix} = \begin{bmatrix} I_{as} \\ I_{bs} \\ I_{cs} \end{bmatrix} \begin{bmatrix} R_{SL} & 0 & 0 \\ 0 & R_{SL} & 0 \\ 0 & 0 & R_{SL} \end{bmatrix} + \frac{d}{dt} [\lambda_{abcs}]^T \tag{1}
$$

where $\begin{bmatrix} V_a & V_b & V_c \end{bmatrix}^T$ represents the stator phase voltage matrix of the E-bike motor, and the unit matrix of R_{SL} is the stator winding matrix; $\begin{bmatrix} I_{as} & I_{bs} & I_{cs} \end{bmatrix}^T$ is the current matrix of the stator winding; $[\lambda_{abs}] = \begin{bmatrix} \lambda_{as} & \lambda_{bs} & \lambda_{cs} \end{bmatrix}$ is the flux linkage generated by the stator winding linkage of phases a, b, and c $\lambda_{abs}(s)$, which is expressed as

$$
\lambda_{abc}(s) = L_S I_{abcs} = \begin{bmatrix} L_{aa} & L_{ab} & L_{ac} \\ L_{ba} & L_{bb} & L_{bc} \\ L_{ca} & L_{cb} & L_{cc} \end{bmatrix} \begin{bmatrix} I_{as} \\ I_{bs} \\ I_{cs} \end{bmatrix} \tag{2}
$$

979-8-3315-1612-3/25 $31.00 © 2025 IEEE

where L_s is the self and mutual inductances of the stator winding. The self-inductance is $L_{aa} = L_{bb} = L_{cc} = L_s$ and the mutual-inductance is $L_{ab} = L_{bc} = L_{ca} = ... = M$. In a three-phase balanced system, $I_{as} + I_{bs} + I_{cs} = .0$, thus, (2) can be written as

$$\frac{d}{dt}[L_{aa}I_{as} + M_{ib} + M_{ic}] = \frac{d}{dt}[L_{aa}I_{as} - M_{ia}] \quad (3)$$

Combining (3) and (1) to get the stator voltage equation as expressed in (4).

$$\begin{bmatrix} V_a \\ V_b \\ V_c \end{bmatrix} = \begin{bmatrix} I_{as} \\ I_{bs} \\ I_{cs} \end{bmatrix} \begin{bmatrix} R_{SL} & 0 & 0 \\ 0 & R_{SL} & 0 \\ 0 & 0 & R_{SL} \end{bmatrix}$$
$$+ \frac{d}{dt} \begin{bmatrix} L_S - M & 0 & 0 \\ 0 & L_S - M & 0 \\ 0 & 0 & L_S - M \end{bmatrix} + \begin{bmatrix} E_{as} \\ E_{bs} \\ E_{cs} \end{bmatrix} \quad (4)$$

The equivalent circuit of the E-bike motor is shown in Fig. 1.

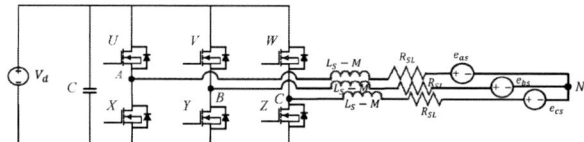

Fig. 1. Equivalent circuit of the E-bike motor.

III. THEORETICAL ANALYSIS OF CONTROL STARTEGIES

A. Field Oriented Control

The motor drive system uses a position sensorless control method to replace the encoder position feedback sensor signal. The circuit structure of this control method can be divided into a speed loop, a current loop, an axis conversion calculation, and a drive circuit that sets the d-axis current to zero in order to fix the stator magnetic field, the q-axis current is determined based on the torque change. The speed loop can be corrected based on the input speed and the actual speed error. Both the current and speed loops sample through current sensing and then calculate through axis conversion to convert the actual current i_d and i_q are compared with the commands i_{dc} and i_{qc}, and then the error of the control quantity is compensated through PI-type compensator.

Moreover, the current loop output is the SVPWM control voltage command. The control voltage is input to the driving circuit after vector modulation and that is composed of the six switching devices $Q_1 \sim Q_6$ and the driving signals $G_1 \sim G_6$ form a complete vector control of the inverter as shown in Fig. 2.

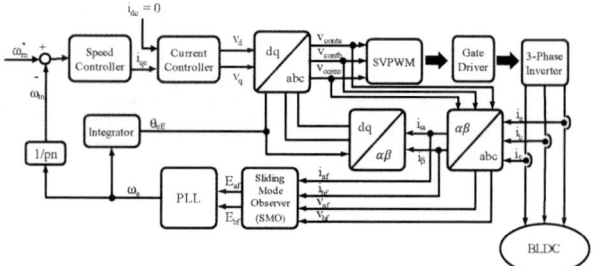

Fig. 2. FOC block diagram of the E-bike motor.

B. Sliding Mode Observer

The E-bike motor control loop adopts the sliding mode observer (SMO) of nonlinear state observation control, which is introduced into the α-β axis voltage equation as presented in (5).

$$\begin{bmatrix} U_\alpha \\ U_\beta \end{bmatrix} = \begin{bmatrix} R_s & 0 \\ 0 & R_s \end{bmatrix} \begin{bmatrix} i_\alpha \\ i_\beta \end{bmatrix} + \frac{d}{dt} \begin{bmatrix} \psi_\alpha \\ \psi_\beta \end{bmatrix} \quad (5)$$

The back electromotive force is expressed as (6).

$$e_\alpha = -\omega_e \psi_f \, sin(\theta_e)$$
$$e_\beta = \omega_e \psi_f \, cos(\theta_e) \quad (6)$$

The magnetic flux is written as in (7).

$$\begin{bmatrix} \psi_\alpha \\ \psi_\beta \end{bmatrix} = \begin{bmatrix} L_\alpha i_\alpha + \psi_f \, cos(\theta_e) \\ L_\beta i_\beta + \psi_f \, sin(\theta_e) \end{bmatrix} \quad (7)$$

The current vary can be observed from (5) to (7) and the observation equation is presented in (8).

$$\frac{d}{dt} \begin{bmatrix} i_\alpha \\ i_\beta \end{bmatrix} = \begin{bmatrix} i_\alpha \\ i_\beta \end{bmatrix} \begin{bmatrix} \dfrac{-R_s}{L_s} & 0 \\ 0 & \dfrac{-R_s}{L_s} \end{bmatrix} + \frac{1}{L_s} \begin{bmatrix} U_\alpha \\ U_\beta \end{bmatrix}$$
$$+ \frac{1}{L_s} \begin{bmatrix} \omega_e \psi_f \, sin(\theta_e) \\ -\omega_e \psi_f \, cos(\theta_e) \end{bmatrix} \quad (7)$$

Eq. (7) is based on the resistance and inductance parameters of the E-bike motor, observes the voltage and current of each α-β axis, and then judges whether the estimation is consistent with the actual current value based on the current error signals, as expressed in (8).

$$\frac{d}{dt} \begin{bmatrix} \widehat{i_\alpha} - i_\alpha \\ \widehat{i_\beta} - i_\beta \end{bmatrix} = \frac{-R_s}{L_s} \begin{bmatrix} \widehat{i_\alpha} - i_\alpha \\ \widehat{i_\beta} - i_\beta \end{bmatrix} - \frac{k}{L_s} \begin{bmatrix} \widehat{i_\alpha} - i_\alpha \\ \widehat{i_\beta} - i_\beta \end{bmatrix}$$
$$+ \frac{1}{L_s} \begin{bmatrix} \widehat{e_\alpha} - e_\alpha \\ \widehat{e_\beta} - e_\beta \end{bmatrix} \quad (8)$$

In order to stabilize the system, it is assumed that the estimated back electromotive force voltage can be equivalent to (9).

$$\begin{bmatrix} \widehat{e_\alpha} \\ \widehat{e_\beta} \end{bmatrix} = \begin{bmatrix} ksign(\widehat{i_\alpha} - i_\alpha) \\ ksign(\widehat{i_\beta} - i_\beta) \end{bmatrix} \quad (9)$$

Assume that the estimated back electromotive force voltage $\begin{bmatrix} \hat{E}_a & \hat{E}_b \end{bmatrix}^T$ and the actual back electromotive force voltage can make the sliding function of the α-β axis current to be σ and the α-β back electromotive force sliding function to be $\dot{\sigma}$.

For α-axis, if the estimated current is greater than the actual value $\hat{i}_a > i_a$, and the functional relationship between the current and back electromotive force of the α-β axis is expressed as (10).

$$(\widehat{i_\alpha} - i_\alpha) \cdot (\dot{\widehat{i_\alpha}} - \ddot{i_\alpha})$$
$$= \frac{-R_s}{L_s}(\widehat{i_\alpha} - i_\alpha)^2 + \frac{1}{L_s}(e_\alpha - k)(\widehat{i_\alpha} - i_\alpha) \quad (10)$$

For contrary, if the estimated current is greater than the actual value $\hat{i}_a < i_a$, and the functional relationship between the current and back electromotive force of the α-β axis is represented as (11).

$$(\widehat{i_\alpha} - i_\alpha) \cdot (\dot{\widehat{i_\alpha}} - \ddot{i_\alpha})$$
$$= \frac{-R_s}{L_s}(\widehat{i_\alpha} - i_\alpha)^2 + \frac{1}{L_s}(e_\alpha + k)(\widehat{i_\alpha} - i_\alpha) \quad (11)$$

In the cases of (10) and (11), if the α-axis is to be stable, it needs to be less than or equal to 0. In order to meet this condition, the control gain k must be greater than e_a to be less than 0, and the same is true for the β-axis. After the sliding mode observation is calculated by sign function, the back electromotive force output by the α-β axis is determined by the current estimation error of the α-β axis and the sliding mode control gain k. The sign calculation adjusts the error of the sliding surface through discontinuous changes in positive and negative functions. The E-bike motor parameters that will be affected is the observer estimated voltage. The final sliding mode observer control system architecture is shown in Fig. 3.

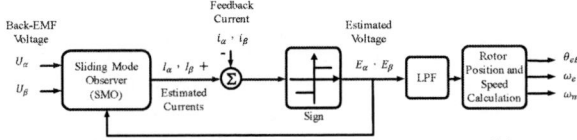

Fig. 3. Sliding mode observer control system architecture.

C. Adaptive Control

A control system operating parameters or conditions will cause uncertain changes in system parameters during operation due to environment, operating factors, or time. Adaptive control can be used to resist parameter changes. The advantage is that it can be controlled according to the self-correcting system with timeliness and uncertainty designed based on the system requirements.

In this paper, the parameter that will be affected by the motor control is the estimated voltage. To stabilize the control voltage during the input phase lock loop, it is necessary to perform adaptive correction of the estimated voltage. The E-bike motor belongs to direct adaptive control and adjusts the controller parameters online. The major purpose is to make the error between system estimation and actual close to zero. Its derivation will be stabilized through Lyapunov criteria to derive the conditions and formulas of the adaptive control law. When the control algorithm operates, there will be a delay time between the transient and steady states. The actual parameter values are tracked and updated with the delay time, giving the adaptive control voltage, current, and observer better tracking timepiece signal and error compensation.

The adaptive control system optimizes original parameters, and adaptive correction emphasizes stability. Uncertain parameters will change depending on the needs of the system. That is, they rely on the adaptive law of parameters for adjustment. Lyapunov stability theory will be used here. The error approaches zero when reaching a steady state, and the external disturbance is minimally suppressed. Equation (13) is the single-input single-output (SISO) reference model of E-bike motor system.

$$\dot{x}_r(t) = a_r x_r + b_r r \quad (12)$$

where x is the actual value of the system output, u is the system control input, and a, b is a constant.

According to the adaptive control designed by the E-bike motor controller, a single output consists of two uncertainty parameters to form an adaptive reference model. From (12), a_r and b_r are the actual parameters of the system but unknown, where $a_r < 0$, $b_r > 0$, then the function will be negative-definite and consistent with the equilibrium state and asymptotic stable state of the system. Moreover, r is the adaptive controller parameter, $\dot{x}_r(t)$ is the actual state value, $\hat{\dot{x}}_r(t)$ is the state estimated value, and the above parameters are used to define the following state equation as (13).

$$\dot{x}_r(t) = a_r x_r + b_r u_r = b(kx + r) \quad (13)$$

The error variable dynamic equation $\dot{e}(t)$ can be expressed as (14).

$$\dot{e}(t) = \dot{x}_r(t) - \hat{\dot{x}}_r(t) = b(kx + r) - \hat{\dot{x}}_r(t) \quad (14)$$

Eventually, the dynamic equations of voltage and speed error variables corresponding to this paper are presented in (15)-(17), respectively.

$$\frac{d}{dt}\hat{E}_\alpha = -\hat{\omega}_e \hat{E}_\beta - I\tilde{E}_\alpha \qquad (15)$$

$$\frac{d}{dt}\hat{E}_\beta = -\hat{\omega}_e \hat{E}_\alpha - I\tilde{E}_\beta \qquad (16)$$

$$\frac{d}{dt}\tilde{\omega}_e = \tilde{E}_\alpha \hat{E}_\beta - \hat{E}_\alpha \tilde{E}_\beta \qquad (17)$$

Above three equations are used to establish the block diagram of adaptive control scheme, as shown in Fig. 4.

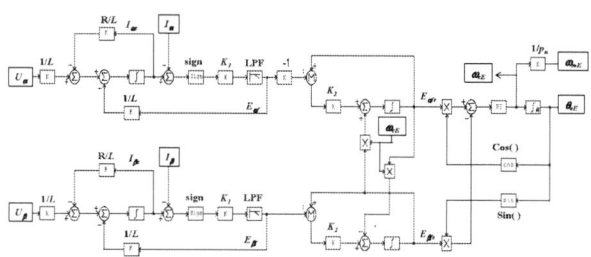

Fig. 4. Block diagram of adaptive control scheme.

IV. POWER ASSISTED CONTROL STRATEGY

The proposed power assisted control system combines the human pedaling force and the torque of the motor. The E-bike motor driving control generally uses the pedaling output or the current speed of the E-bike to determine whether to provide assisted force form E-bike motor, therefore the shaft torque sensing is introduced in the hardware. The current output torque of the wheel motor is transmitted back to the controller in the form of a digital signal as a basis for control parameters. At the same time, the E-bike motor is equipped with a Hall sensing devices, which determines the rotor position through the sensing signal, and determines the speed. This method has higher accuracy of motor control scheme. In order to improve the efficiency of E-bike system, the proposed control strategy adopts FOC and SMO, as shown in Fig. 5.

The torque sensing signal is processed analog to digital by a digital signal processor (DSP) to obtain the current output torque of the motor is shown in Fig. 6. The torque signal is filtered before input to improve the stability of assisted power control loop. After entering the E-bike controller, the torque output ratio is adjusted according to different speeds and functional requirements. The current command from the E-bike controller to the current control loop with the speed change and the assisted mode switching and assisted ratio output by the controller can monitor the E-bike power transmission at any time.

The safety concern of riding an E-bike is the upper speed limit of 25 km/h. The control strategy of the power assisted controller affects the rider's feeling and whether the sensitivity of the assisted force is appropriate. The assisted force can be obtained under conditions that require power. This allows the rider to reduce pedaling force, improve riding time and comfort.

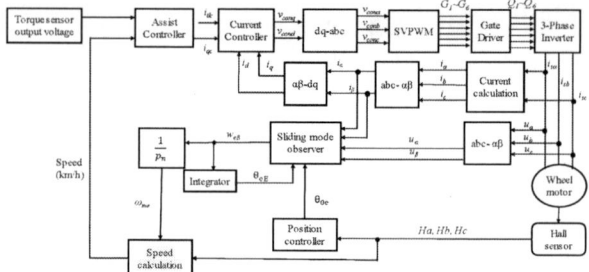

Fig. 5. Block diagram of power assisted control scheme.

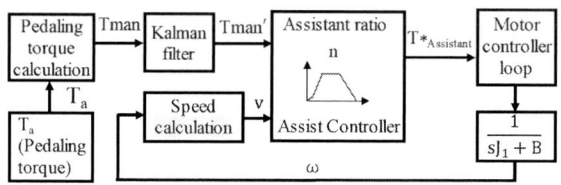

Fig. 6. Power assisted control block diagram of E-bike.

The shaft torque sensor transmits the voltage signal of the human pedaling force back to the controller when the rider pedals. The power assisted controller will turn, and the measured voltage signal of torque is used as the current command of the controller. The E-bike motor has three action stages according to the size of the current command, including starting, fixed assist, and speed reduction. This paper arranges these three stages into 7 modes, as shown in Fig. 7.

➢ **[Mode 1]** This mode mainly determines the start and stop actions. When the starting torque is less than 0.72 Nm, the controller's assisted ratio is 0. At this time, ensuring that it does not start before the starting torque is insufficient. The stop action is mainly determined to prevent the controller from outputting when the pedaling stops. The E-bike motor continues to malfunction due to the current command.

➢ **[Mode 2 & 3]** Both modes at this period are starting and accelerating states. After starting, the E-bike motor used in the experiment is tested to have a minimum speed limit of 14 rpm (1.8 km/h). The speed measured at this time is the speed of the outer rotor because it contains gears inside. The ratio of 12.7 is based on the fact that the lowest stable outer rotor speed falls within the range of 20 rpm to 25 rpm, and the converted speed is between 2.6 km/h and 3.3 km/h. Therefore, the assisted condition of mode 2 is set to the speed before the speed is less than 3 km/h. When the speed command exceeds 3 km/h due to changes in the assisted ratio, it will enter mode 3. The control range of this mode is between 3 km/h and 15 km/h.

979-8-3315-1612-3/25 $31.00 © 2025 IEEE

➤ **[Mode 4]** Generally, the most comfortable speed for an E-bike to ride on flat ground is about 18 km/h to 20 km/h; thus, when the speed is between 15 km/h and 20 km/h, the controller is designed with a fixed assisted ratio, to ensure long-distance riding. It can maintain the speed of the E-bike with stable assisted force for a long time. In the process, the pedaling torque determines the current command output by the controller.

➤ **[Mode 5]** The design speed of 20 km/h is the upper limit of the E-bike. When it is exceeded, it will switch to mode 5 for deceleration. According to the safety regulations of the E-bike, the speed cannot exceed 25 km/h. Thus, the assisted ratio of this mode is designed. It will gradually decrease, and the deceleration range is 20 km/h to 25 km/h.

➤ **[Mode 6]** In this mode, the speed sensed by the controller fails to reach a stable value due to unstable torque input. In mode 6, speed and torque limits are set. The B-bike must meet the requirements of greater than 25 km/h and input to the controller. The torque is less than 3.9 Nm. At this time, the mode can be used to confirm further whether it exceeds 25 km/h, and the torque less than 3.9 Nm is used as the basis for controlling the judgment to enter over-speed stop assistance.

➤ **[Mode 7]** When the speed still exceeds 25 km/h, and the torque is greater than or equal to 4.5 Nm, the current command output by the controller will be quickly reduced by the assisted ratio of mode 7.

Environmental factors are considered to make the riding situation and feeling closer to the practical case. The environmental forces of gravity F_{mg}, tire grip F_{wheel}, and wind F_{wind} are defined as (18) to (20), respectively.

$$F_{mg} = Mg \sin \theta \qquad (18)$$

$$F_{wheel} = C_f Mg \cos \theta \qquad (19)$$

$$F_{wind} = 0.5 C_w \rho_a A (V + U)^2 \qquad (20)$$

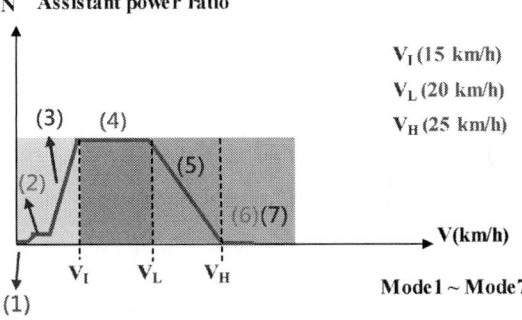

Fig. 7. Power assisted ratio and speed curve.

The rider's weight causes slope resistance on a ramp road. On flat ground, the weight is offset by the positive force on the ground. The most obvious change for the user is uphill. On uphill sections, the tire resistance force must be overcome. The friction force of the ground also needs to face the slope force generated by gravity. Finally, the proposed control system considers various environmental resistance factors and rider habits, combined with the above three equations (18) to (20), and the power assisted control block diagram of E-bike system is plotted in Fig. 8.

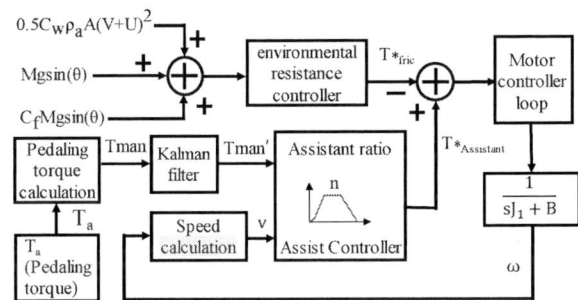

Fig. 8. Power assisted control block diagram of E-bike with environmental resistance factors.

V. VERIFICATION OF EXPERIMENTAL RESULTS

According to the E-bike testing platform shown in Fig. 9, the platform is composed of three motors. The red and blue zones are the servo motors, the green zone is an E-bike motor, and the yellow zone is the machine base for adjusting the permanent magnet motor chain and flywheel. The distance adjustment platform is formed through the screw, stroke direction extension, and base.

Fig. 9. E-bike testing platform.

The control circuit uses TI TMS320F28335PGFA. The voltage, current, and sensor signal GPIO pins in the planned circuit are processed and converted by the analog to digital (ADC) chip. The PWM signal that drives the main power stage circuit can be controlled by logic to determine the GPIO. In order to avoid PWM control signals errors and cause the circuit damage, a complex programmable logic device (CPLD) is used to provide additional processing for interrupt processing or GPIO special control. Finally, Fig. 10 shows the experimental circuit of E-bike system.

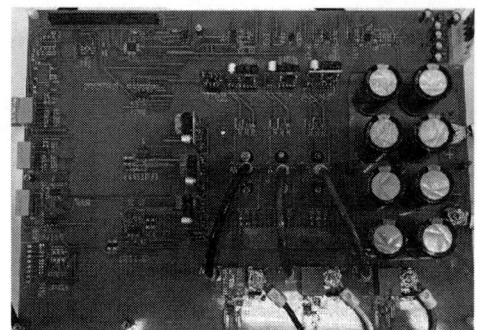

Fig. 10. Experimental circuit of E-bike system.

The permanent magnet servo motor ECMA-F11308RS used in the testing platform is a product manufactured by Delta Electronics. The E-bike motor used in this platform is the AKM100SX series. The factory parameters of the E-bike motor are presented in Table I.

Table. I. Specifications of E-bike motor.

Rated power	250 W
Rated torque	10.6 N-m
Rated speed	225 rpm
Rated current	9 A
Pole number	8
Torque constant (K_t)	0.12 N-m/A
Voltage constant (K_e)	9.26 mV/rpm
Winding resistance	146.4 mΩ
Winding inductance	340.6 uH

At first, the environmental resistance force is generated by the ground, including the type of ground, slope, rolling friction, and the weight of the rider. The second is a wind resistance force whose influencing factors include air resistance coefficient, windward area, E-bike speed, and wind speed, as shown in Table II.

Table. II. Various environmental resistance forces.

Ground friction coefficient (C_f)	0.8
Air friction coefficient (C_w)	0.587
Slope (θ)	30°
Windward area (A)	0.376 m²
Wind speed (U)	5 m/s
Height (H)	1.9 m
Weight (M)	90 kg

When the power assisted mode is without considering environmental resistance forces, the blue line is the average speed close to 22.5 km/h, and the actual speed is a red line; both are shown in Fig. 11. In addition, the average power assisted ratio is 0.0157, and the average pedaling torque is 3.25 Nm. In Fig. 12, the maximum phase current of the E-bike motor is 2.8 A, the practical current is measured as 1.84 A, and the average current is 1.79 A.

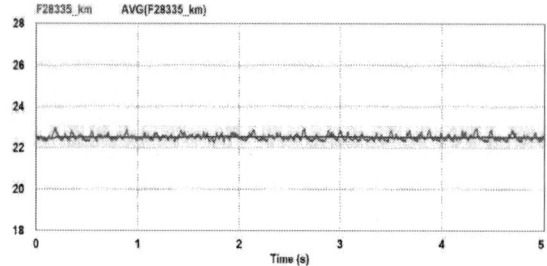

Fig. 11. The experimental speed (22.5 km/h) of the E-bike motor without environmental resistance forces.

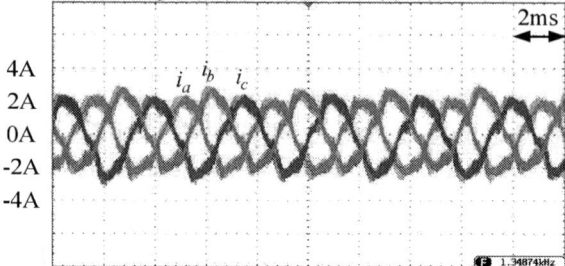

Fig. 12. The experimental three-phase current of the E-bike motor without environmental resistance forces.

After considering environmental resistance forces, the speed dropped from 22.5 km/h to 18.7 km/h, as shown in Fig. 13. Moreover, the environmental resistance forces reduced the current command of the control loop. Thus the output current command of the power-assisted controller will increase from 0.0622 to 0.116. Finally, Fig. 14 shows the actual three-phase current of the E-bike motor at 18.7 km/h. The maximum phase current is 1.83 A, the measured effective current is 1.13 A, and the average current is 1.04 A.

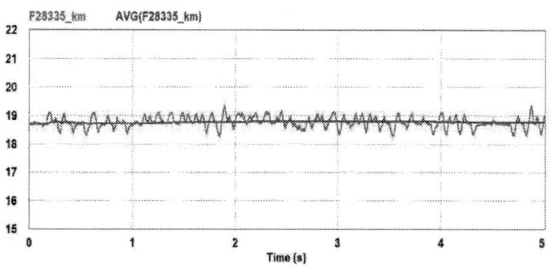

Fig. 13. The experimental speed (18.7 km/h) of the E-bike motor with environmental resistance forces.

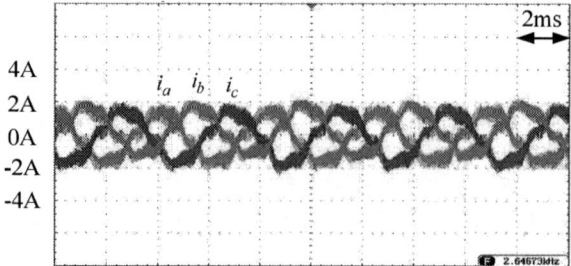

Fig. 14. The experimental three-phase current of the E-bike motor with environmental resistance forces.

979-8-3315-1612-3/25 $31.00 © 2025 IEEE

VI. CONCLUSION

This paper demonstrates the power assisted controller and sensorless control schemes for the E-bikes. The magnetic field-oriented control (FOC) strategy solves the problem of interference while using the encoder. The sliding mode observer (SMO) and adaptive control are adopted to improve the control accuracy after E-bike operation. The torque sensor driving signal is included in the control loop, and the assisted power is determined according to the proposed controller. In the meantime, environmental resistance factors are added to simulate the speed varies of the E-bike system, which is more suitable for practical riding applications.

REFERENCES

[1] S. M.D. et al., "Design and Analysis of E-Bike with Electrical Regeneration and Self-Balancing Assist," *4th International Conference on Trends in Electronics and Informatics (ICOEI)*, Tirunelveli, India, 2020, pp. 119-124.

[2] T. Li, Q. Yang, B. Ren and X. Tu, "A Torque Sensor-less Speed Control Method of Electric Assisted Bicycle," *37th Chinese Control Conference (CCC)*, Wuhan, China, 2018, pp. 3705-3709.

[3] H. Mehta, A. Apte, S. Pawar and V. Joshi, "Vector control of PMSM using TI's launchpad F28069 and MATLAB embedded coder with incremental build approach," *7th International Conference on Power Systems (ICPS)*, Pune, India, 2017, pp. 771-775.

[4] Liu X, Wang Z, Wang W, Lv Y, Yuan B, Wang S, Li W, Li Q, Zhang Q and Chen Q, "SMO-Based Sensorless Control of a Permanent Magnet Synchronous Motor," Front. Energy. Res. 10:839329, 2020.

[5] Z. Zhang, G. Xiong, J. Wang, X. Zhang, S. Wang and X. Wang, "A SMO Based Position Sensorless Permanent Magnet Synchronous Motor Control Strategy," *15th IEEE Conference on Industrial Electronics and Applications (ICIEA)*, Kristiansand, Norway, 2020, pp. 373-379.

[6] H. Kim, J. Son and J. Lee, "A High-Speed Sliding-Mode Observer for the Sensorless Speed Control of a PMSM," *IEEE Trans. on Ind. Electronics*, vol. 58, no. 9, pp. 4069-4077.

[7] Y. Yang, H. Guo and H. Qian, "A sensorless control of SPMSM based on sliding mode observer with linear power drive method," *43rd Annual Conference of the IEEE Industrial Electronics Society (IECON 2017)*, Beijing, China, 2017, pp. 4094-4098.

[8] Z. Qiao, T. Shi, Y. Wang, Y. Yan, C. Xia and X. He, "New Sliding-Mode Observer for Position Sensorless Control of Permanent-Magnet Synchronous Motor," *IEEE Trans. on Industrial Electronics*, vol. 60, no. 2, pp. 710-719, Feb. 2013.

[9] C. C. Wang, C. L. Chen and M. C. Tsai, "Model Reference Impedance Control of Power-Assisted Bike," *21st International Conference on Electrical Machines and Systems (ICEMS)*, Jeju, Korea (South), 2018, pp. 2781-2784.

[10] K. Liu and Z. Q. Zhu, "Parameter estimation of PMSM for aiding PI regulator design of field oriented control," 2014 17th International Conference on Electrical Machines and Systems (ICEMS), Hangzhou, China, 2014, pp. 2705-2711.

[11] A. Podder and D. Pandit, "Study of Sensorless Field-Oriented Control of SPMSM Using Rotor Flux Observer & Disturbance Observer Based Discrete Sliding Mode Observer," 2021 IEEE 22nd Workshop on Control and Modelling of Power Electronics (COMPEL), Cartagena, Colombia, 2021.

[12] V. M. Bida, D. V. Samokhvalov and F. S. Al-Mahturi, "PMSM vector control techniques — A survey," *IEEE Conference of Russian Young Researchers in Electrical and Electronic Engineering (EIConRus)*, Moscow and St. Petersburg, Russia, 2018, pp. 577-581.

[13] Z. Yin, Y. Zhang, X. Cao, D. Yuan and J. Liu, "Estimated Position Error Suppression Using Novel PLL for IPMSM Sensorless Drives Based on Full-Order SMO," *IEEE Transactions on Power Electronics*, vol. 37, no. 4, pp. 4463-4474, Apr. 2022.

[14] G. Li, Y. Yang, Z. Dai, Y. Zhao, B. Li and J. Huang, "Undesirable Oscillations of Phase Locked Loop-Based Position and Speed Estimator for Sensorless Control," *IEEE Trans. on Power Electronics*, vol. 38, no. 7, pp. 8754-8761, July 2023.

A Hybrid PWM Strategy with Reduced Common-mode Voltage and Extended Output Voltage Linearity for Adjustable Speed Drives

Zhe Zhang
Eaton
Menomonee Falls, WI, USA
zhezhang2@eaton.com

Kevin Lee
Eaton
Menomonee Falls, WI, USA
kevinlee@eaton.com

Abstract—Common-mode voltage (CMV), introduced by employing conventional space vector pulse-width modulation (SVPWM) in voltage source inverters, can be detrimental to inverter-fed electric machines electric machines. The utilization of zero voltage vectors produces considerable common-mode voltages, which could result in premature bearing failures. In this paper, a hybrid PWM strategy, eliminating the use of zero voltage vectors, is proposed to reduce CMV within and beyond the rated speed in adjustable speed drives (ASDs). The proposed hybrid PWM strategy features the combination of remote state PWM (RSPWM) and near state PWM (NSPWM) across a wide range of output frequencies with guaranteed output voltage linearity. Simulation and experimental results demonstrate the effectiveness of the proposed hybrid PWM strategy in CMV reduction for ASDs.

Keywords—*Common-mode voltage, output voltage linearity, overmodulation, adjustable speed drive.*

I. INTRODUCTION

Three-phase voltage source inverters are extensively used to drive AC motors because they could offer a faster dynamic response and operate motors without de-rating [1]. Fig. 1 illustrates the typical setup of a two-level adjustable speed drive (ASD). Pulse width modulation (PWM) is the standard technique used to operate inverter switches and generate the desired output voltages [2]. Among various PWM methods, space vector PWM (SVPWM) is the most commonly used in ASDs due to its superior harmonics performance. However, SVPWM generates significant CMV, which could potentially cause premature bearing failures in AC motors.

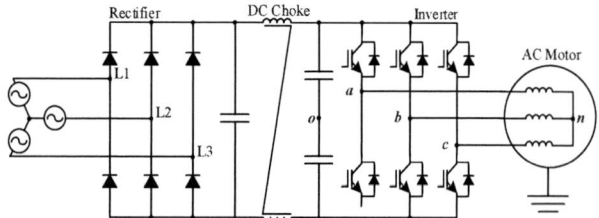

Fig. 1. Standard two-level voltage source inverter based adjustable speed drive system.

$$V_{cm} = V_{no} = \frac{V_{ao} + V_{bo} + V_{co}}{3} \quad (1)$$

The CMV of the inverter, as defined in equation (1), is the average sum of the three-phase voltage outputs. The CMV, when referenced to the midpoint of the DC bus, is never zero. The inverter output voltage V_{inv} can be expressed using space vector notation shown in equation (2), where $a = e^{j2\pi/3}$ denotes the phase shift operator. This transformation yields six active and two zero voltage vectors, forming six equivalent sectors that span a hexagon plane illustrated in Fig. 2. Fig. 3 shows an example of the inverter switches and the corresponding CMV in one switching cycle under conventional SVPWM.

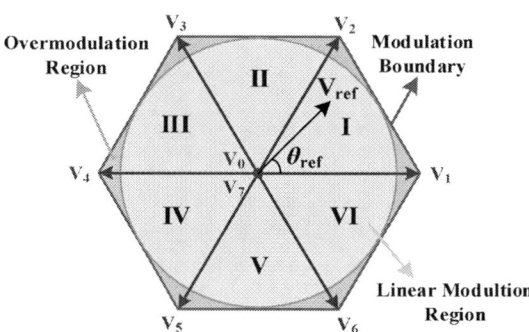

Fig. 2 .Hexagon plane with voltage space vectors.

$$V_{inv} = \frac{2}{3}(V_{ao} + aV_{bo} + a^2 V_{co}) \quad (2)$$

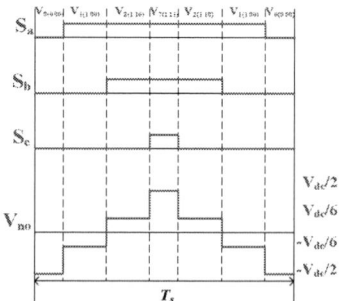

Fig. 3. SVPWM switch states and corresponding CMV in one switching cycle.

The referenced output voltage in ASDs is usually limited within the linear modulation region shown in Fig. 2, featuring a circle inscribed to the hexagon. The overmodulation region is the area between the inscribed circle and the hexagon boundary, which can further increase the utilization of DC bus voltage. The downside of overmodulation is the production of harmonic distortion in the inverter's voltage output, which could deteriorate the motor's torque and speed performance. However, if wider speed range in particular applications is required, overmodulation must be adopted to produce higher motor torque. The modulation index (MI) M_i for a given PWM method is defined in equation (3). Variation in load dynamics cause the output frequency and voltage in an ASD to fluctuate, leading to a wide range of MI values. Fig. 4 shows the typical profile of stator voltage versus output frequency under V/Hz control [3].

$$M_i = V_{1m} / (\frac{2}{\pi} V_{dc}) \qquad (3)$$

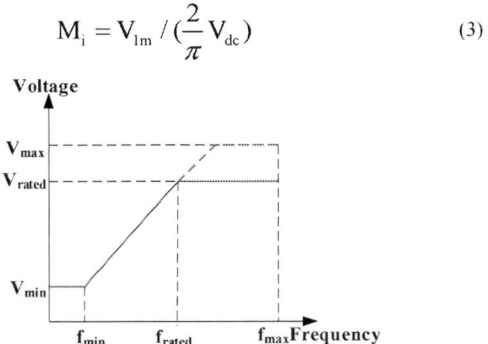

Fig. 4. V/Hz profile for frequency control [3].

The zero voltage vectors will be utilized in synthesizing the referenced output voltage if conventional SVPWM is employed. It's been clearly revealed in [4], [5] that utilizing zero voltage vector produces highest CMV. To combat the adverse CMV effect introduced by conventional SVPWM, PWM methods only relying on non-zero voltage vectors are developed [6]-[13]. Among these methods, AZSPWM [6], RSPWM [7], NSPWM [8] are the most typical representatives. These methods are limited to reducing the CMV in the linear modulation region. Few literatures discuss the CMV reduction in the overmodulation region.

In this paper, a hybrid PWM strategy based on [11] but augmented with extended output voltage linearity is proposed to increase the utilization of the DC bus voltage and facilitate wider operational speed range. The proposed hybrid PWM strategy incorporates RSPWM and NSPWM to ensure output voltage linearity and CMV reduction in the linear modulation region as well as in the overmodulation region.

II. BRIEF REVIEW OF REMOTE STATE PWM AND NEAR STATE PWM

A. Remote State PWM

Three active voltage vectors that are 120° apart from each other are utilized to synthesize the referenced voltage. Either the group of V_1, V_3, V_5 (RSPWM1) or the group of V_2, V_4, V_6 (RSPWM2) can be used to produce the desired output voltage at the same magnitude of CMV. Fig. 5 (a) and (b) depict the synthesis of V_{ref} residing in sector I with RSPWM1 and RSPWM2. The corresponding governing equations for the RSPWM method, with V_{ref} situated in sector I as an example, are summarized in TABLE I.

The output voltage linearity region over a fundamental cycle, where the governing equations provide meaningful solutions for dwelling time, can be determined and shown in Fig. 7 (a). The gray area in Fig. 7 (a) indicates the modulation index range ($0 < M_i < 0.524$) that maintains the output voltage linearity for RSPWM.

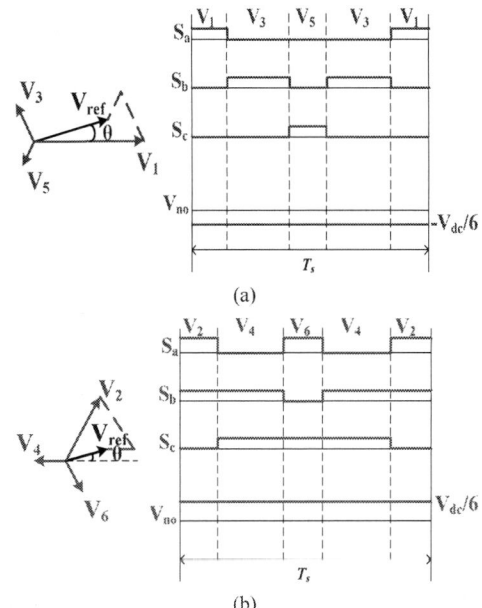

Fig. 5. RSPWM synthesis illustration: (a) Group I (RSPWM1); (b) Group II (RSPWM2).

B. Near State PWM

The two active voltage vectors nearest to V_{ref} are selected, along with a third near-neighbor vector according to V_{ref}'s position. Fig. 6 illustrates the formation of V_{ref} in sector I using NSPWM. The resulting CMV undergoes four transitions in one switching cycle. TABLE I provides the corresponding governing equations for the NSPWM method.

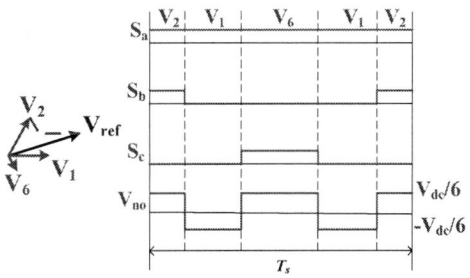

Fig. 6. NSPWM synthesis illustration.

The output voltage linearity region over a fundamental cycle, where the governing equations provide meaningful solutions for dwelling time, can be determined and shown in Fig. 7 (b). The gray area in Fig. 7 (b) indicates the modulation index range ($0.605 < M_i < 0.906$) that maintains the output voltage linearity for NSPWM. Outside the gray region, both RSPWM and NSPWM involve low-frequency harmonics generation in the inverter's output, which could result in significant distortion in motor phase currents. It has been experimentally validated in [14] that comparing to SVPWM,

979-8-3315-1612-3/25 $31.00 © 2025 IEEE

RSPWM and NSPWM do not exhibit significant current total harmonic distortion performance deterioration within their respective voltage linearity regions.

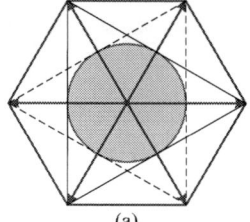 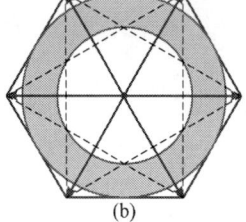

Fig. 7. Output voltage linearity region: (a) RSPWM1/RSPWM2; (b) NSPWM.

TABLE I.   VOLTAGE SYNTHESIS EQUATIONS IN SECTOR I

PWM Method	Dwelling Time Calculation
RSPWM1/RSPWM2	$T_3 = T_s * \left[\dfrac{1}{3} - \dfrac{2}{\pi} M_i \cos(\dfrac{\pi}{3} + \theta) \right] = T_4$ $T_1 = T_s * \left(\dfrac{1}{3} + \dfrac{2}{\pi} M_i \cos\theta \right) = T_2$ $T_5 = 1 - T_3 - T_1$ $T_6 = 1 - T_4 - T_2$
NSPWM	$T_2 = T_s * \left[1 - \dfrac{2\sqrt{3}}{\pi} M_i \cos(\dfrac{\pi}{6} + \theta) \right]$ $T_1 = T_s * \left(\dfrac{6}{\pi} M_i \cos\theta - 1 \right)$ $T_6 = 1 - T_1 - T_2$

III. PORPOSED HYBRID PWM STRATEGY

Given the analysis in Section II, the two reduced CMV PWM methods: NSPWM and RSPWM do not fully cover the entire output voltage linearity region. However, by combining them, it is possible to maintain output voltage linearity in the linear modulation region and extend into the overmodulation region.

A. Single-mode and Dual-mode Opeartion of RSPWM

RSPWM1 or RSPWM2 can each cover the output voltage linearity range of $0 < M_i \le 0.524$, corresponding to single-mode operation. For higher MI values, dual-mode operation can be utilized, where RSPWM1 and RSPWM2 alternate every 60°, as illustrated in Fig. 8. This dual-mode operation extends the voltage linearity region of RSPWM to $0 < M_i \le 0.605$.

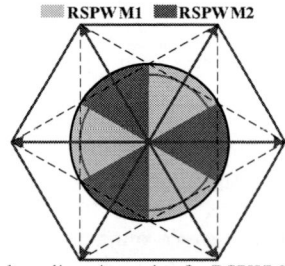

Fig. 8. Output voltage linearity region for RSPWM under dual-mode operation.

B. NSPWM with Extended Output Voltage Linearity

By implementing dual0mode operation, the upper limit of MI for RSPWM can be extended to 0.605. At the same time, the lower limit of MI for NSPWM is also 0.605. This allows for a seamless transition between RSPWM and NSPWMM, ensuring both output voltage linearity and CMV reduction. For a given M_i greater than 0.906, the trajectory of the referenced voltage vector goes beyond the inscribed circle shown in Fig. 9 (a). The dashed blue line delineates the ideal trace of the rotating referenced voltage vector, whereas the solid blue line resembles its actual trace. The generated output voltage in one fundamental cycle under NSPWM will be insufficient to fulfil the desired M_i if no appropriate measure is taken to compensate for the "voltage loss" by the hexagon boundary.

The proposed compensation scheme takes care of the insufficiency in output voltage's fundamental component by extending the magnitude of the referenced voltage vector while keeping its angle unchanged to minimize the phase distortion. The newly developed ideal (dashed red line) and actual (solid red line) trace are shown in Fig. 9 (b), yielding a control parameter α associated with a given M_i. In this way, the fundamental output voltage linearity versus modulation index M_i can be preserved until the trace of the adjusted rotating referenced voltage vector features the circumcircle of the hexagon plane. The magnitude and phase of the adjusted referenced voltage vector V^*_{ref} are given in equations (4) and (5), respectively.

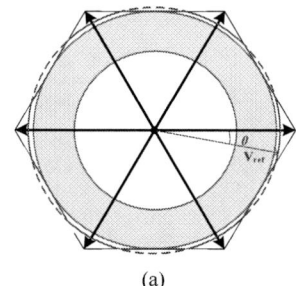

(a)

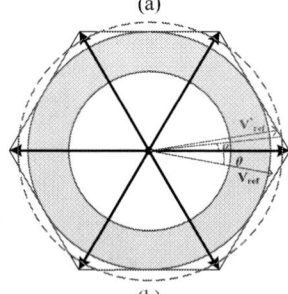

(b)

Fig. 9. Trace of rotating referenced voltage vector: (a) uncompensated; (b) compensated.

$$\left| V^*_{ref} \right| = \frac{\sqrt{3}}{3} V_{dc} / \cos\left(\frac{\pi}{6} - \alpha \right) \tag{4}$$

$$\theta^*_{ref} = \theta_{ref} \tag{5}$$

where $\alpha \in \left[0, \dfrac{\pi}{6} \right]$. To derive the relationship between the control parameter α and a given M_i in the overmodulation region, the equivalent phase voltage waveform in time domain will be investigate [15]-[17]. For a given rotating referenced voltage vector, the corresponding phase voltage equations within the interval of $[0, \pi/2)$, are expresses as:

$$\underbrace{\frac{\sqrt{3}}{3} V_{dc} \cos\theta_{ref} / \cos\left(\frac{\pi}{6} - \alpha\right)}_{f_1}, \theta_{ref} \in \underbrace{[0, \alpha)}_{A} \quad (6)$$

$$\underbrace{\frac{\sqrt{3}}{3} V_{dc} \cos\theta_{ref} / \cos\left(\frac{\pi}{6} - \theta_{ref}\right)}_{f_2}, \theta_{ref} \in \underbrace{\left(\alpha, \frac{\pi}{3} - \alpha\right)}_{B} \quad (7)$$

$$\underbrace{\frac{\sqrt{3}}{3} V_{dc} \cos\theta_{ref} / \cos\left(\frac{\pi}{6} - \alpha\right)}_{f_3}, \theta_{ref} \in \underbrace{\left[\frac{\pi}{3} - \alpha, \frac{\pi}{3} + \alpha\right)}_{C} \quad (8)$$

$$\underbrace{\frac{\sqrt{3}}{3} V_{dc} \cos\theta_{ref} / \cos\left(\frac{\pi}{2} - \theta_{ref}\right)}_{f_4}, \theta_{ref} \in \underbrace{\left[\frac{\pi}{3} + \alpha, \frac{\pi}{2}\right)}_{D} \quad (9)$$

$$F_1(\alpha) = \frac{\pi}{4}\left[\begin{array}{l} \int_A f_1 \cos\theta d\theta + \int_B f_2 \cos\theta d\theta \\ + \int_C f_3 \cos\theta d\theta + \int_D f_4 \cos\theta d\theta \end{array}\right] \quad (10)$$

The resultant fundamental component of phase voltage can be obtained by expanding (6)-(9) in Fourier series and performing numerical integration to get the corresponding Fourier coefficient as expressed in (10), where $F_1(\alpha)$ represents the peak value of the fundamental component of phase voltage. Combining (10) and (3), a relationship between α and M_i is determined. Fig. 10 explicitly illustrates such relationship.

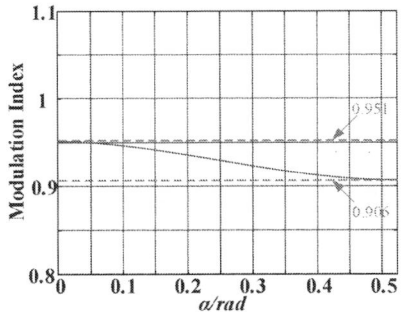

Fig. 10. Relationship between control parameter α and Mi under NSPWM in the overmodulation region.

C. Hybrid PWM Scheme of CMV Reduction in ASDs

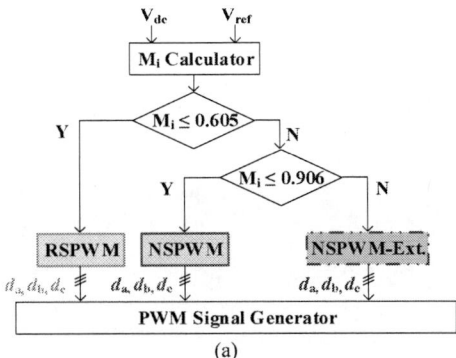

(a)

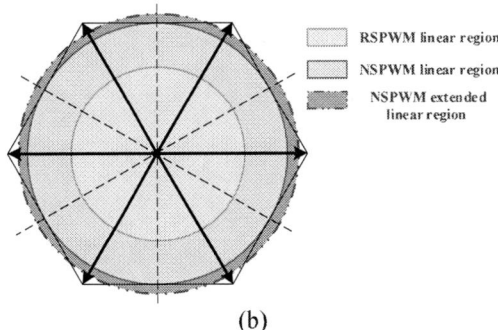

(b)

Fig. 11. (a) High-level flowchart of the proposed hybrid PWM strategy; (b) Output voltage linearity region covered by the proposed reduced-CMV hybrid PWM strategy.

The proposed hybrid PWM strategy can be summarized as follows: The RSPWM will be taking care of the situation where low/medium M_i is desired while the normal NSPWM will be managing the situation where M_i falls within 0.605 and 0.906. The proposed NSPWM with extended output voltage linearity will come into play if even higher M_i is required. Fig. 11 (a) demonstrates the high-level flowchart of the proposed reduced-CMV hybrid PWM strategy. Fig. 11 (b) illustrates the respective output voltage linearity regions associated with the RSPWM, the normal NSPWM, and the proposed NSPWM with extended output voltage linearity. Such hybrid PWM strategy can guarantee the optimal CMV performance in a wide operational range for ASDs.

IV. SIMULATION AND EXPERIMENTATION

A simulation model with a timestep of 1 µs was developed in MATLAB/Simulink to compare the proposed hybrid PWM strategy with the conventional SVPWM. The model features a three-phase voltage source inverter with a 300 V DC bus supplying a three-phase R-L load. The R-L load is set to 10 kW with a power factor of 0.85 at 60 Hz, simulating typical electric motor operation. The inverter switches at 10 kHz in constant V/Hz mode (132 Vrms/60 Hz). Fig. 12 (a) and (b) display the simulated waveforms of 80 Hz ramp operation for the conventional SVPWM and the proposed hybrid PWM, respectively.

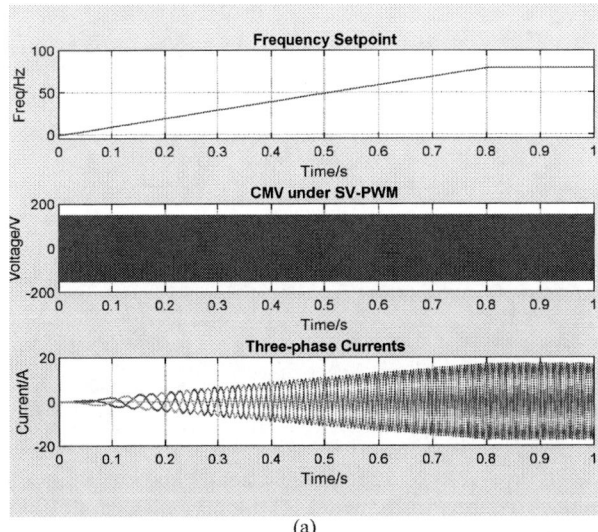

(a)

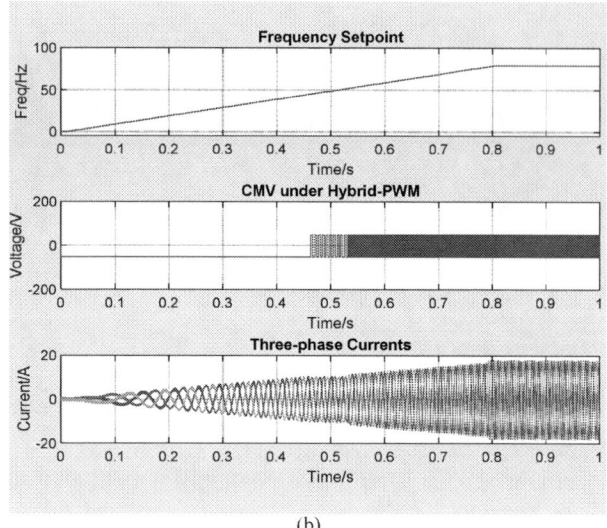

(b)

Fig. 12. Simulated waveforms: (a) conventional SVPWM; (b) proposed reduced-CMV hybrid PWM.

Under SVPWM, the CMV frequently fluctuates within $\pm V_{dc}/2$. In contrast, the proposed hybrid PWM significantly reduces the CMV peak to $\pm V_{dc}/6$. At low speeds, RSPWM1 keeps the CMV constant. At intermediate speeds, RSPWM1 and RSPWM2 alternate, causing the CMV to slowly fluctuate between $V_{dc}/6$ and $-V_{dc}/6$. At high speeds, the CMV begins to exhibit frequent fluctuations under NSPWM, but with a significantly reduced peak-to-peak amplitude.

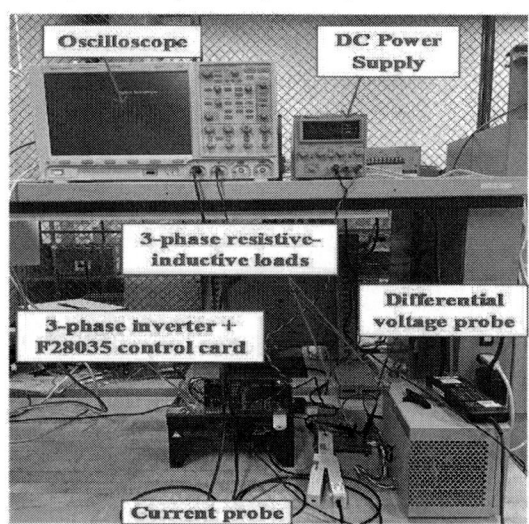

Fig. 13. Experimental setup.

TABLE II. PARAMETERS OF EXPERIMENTAL SETUP

Inverter switching frequency	10 kHz
DC bus voltage	30 V
Fundamental output frequency	60 Hz
Power resistor bank	10 Ω, 3-Ø
Reactor	6.5 mH, 3-Ø
High voltage motor control kit	TMDSHVMTRPFCKIT
Digital signal processor	TMS320F28035

The low-voltage experimental setup, shown in Fig. 13, includes a Texas Instrument (TI) digital motor control kit, a three-phase resistor bank and reactor. The parameters of the setup are detailed in TABLE II. Constant V/Hz control is applied to drive the three-phase R-L load. Fig, 14 and 15 display the measured CMV and phase current under SVPWM and the proposed hybrid PWM, respectively. The output frequency command gradually ramps up to 80 Hz. Compared to SVPWM, the proposed hybrid PWM method significantly reduces both the frequency and peak-to-peak amplitude of CMV across the whole speed range. The experimental results closely match the simulated results.

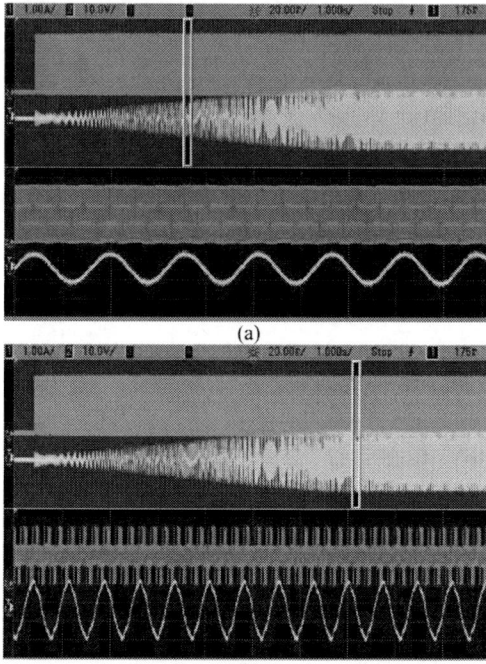

Fig. 14. Experimental waveforms under the conventional SVPWM (Ch. 1: phase current measurement; Ch. 2: CMV measurement).

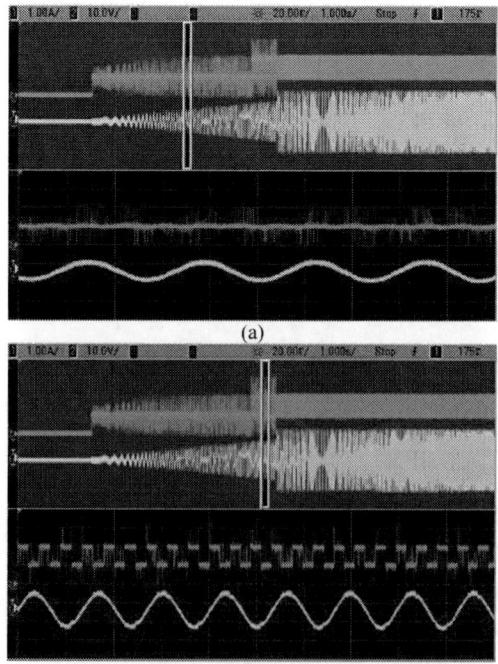

(b)

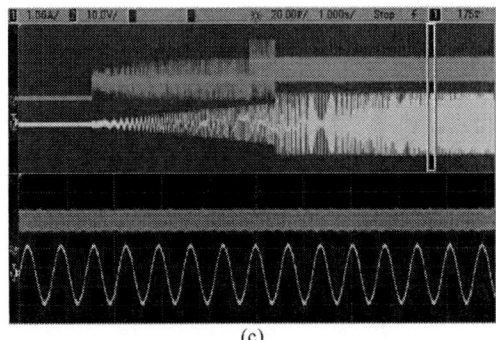

(c)

Fig. 15. Experimental waveforms under the proposed hybrid PWM (Ch. 1: phase current measurement; Ch. 2: CMV measurement).

V. CONCLUSIONS

A hybrid PWM strategy has been developed for ASDs to achieve reduced CMV beyond the rated speed while extending output voltage linearity. The proposed hybrid PWM method, which seamlessly combines RSPWM and NSPWM, achieves CMV reduction as well as maintaining output voltage linearity in the overmodulation region. Both simulation and experiment have verified and validated the effectiveness of the proposed strategy.

REFERENCES

[1] A. VanderMeulen, and J. Maurin, "Current source inverter vs. voltage source inverter topology, White Paper WP020001EN, Eaton, USA, 2014.

[2] A. M. Hava, R. J. Kerkman, and T. A. Lipo, "Simple analytical and graphical methods for carrier-based PWM-VSI drives," *IEEE Trans. Power Electron.*, vol. 14, no. 1, pp. 49–61, Jan. 1999.

[3] B. Akin, and N. Garg, "Scalar (V/f) control of 3-phase induction motors," Texas Instrum., Appl. Rep. SPRABQ8, Jul. 2013.

[4] Z. Zhang, and A. M. Bazzi, "Optimized PWM scheme with minimized common-mode voltage amplitude and frequency in VSI-fed motor drives," *IEEE Energy Conversion Congress and Exposition (ECCE)*, pp. 2819-2824, Vancouver, BC, Canada, 2021.

[5] Z. Zhang, A. M. Bazzi and A. Semin, "An active zero-state switch (AZS) for common mode voltage reduction in voltage source inverter (VSI) drives," *IEEE Applied Power Electronics Conference and Exposition (APEC)*, pp. 711-717, New Orleans, LA, USA, 2020.

[6] Y. S. Lai, P. S. Chen, H. K. Lee, and J. Chou, "Optimal common-mode voltage reduction PWM technique for inverter control with consideration of the dead-time effects—Part II: Applications to IM drives with diode front end," *IEEE Trans. Ind. Appl.*, vol. 40, no. 6, pp. 1613–1620, Nov./Dec. 2004.

[7] M. Cacciato, A. Consoli, G. Scarcella, and A. Testa, "Reduction of common-mode currents in PWM inverter motor drives," *IEEE Trans. Ind. Appl.*, vol. 35, no. 2, pp. 469–476, Mar./Apr. 1999.

[8] E. Un and A. M. Hava, "A near state PWM method with reduced switching frequency and reduced common mode voltage for three-phase voltage source inverters," in *Proc. IEEE-IEMDC*, pp. 235–240, 2007.

[9] L. Du and Q. Fu, "An Improved Nonzero Vector Based Common-Mode Voltage Suppression Strategy for Permanent Magnet Synchronous Motors," *2023 2nd International Symposium on Control Engineering and Robotics (ISCER)*, Hangzhou, China, 2023, pp. 185-190.

[10] C. -C. Hou, C. -C. Shih, P. -T. Cheng and A. M. Hava, "Common-Mode Voltage Reduction Pulsewidth Modulation Techniques for Three-Phase Grid-Connected Converters," in *IEEE Transactions on Power Electronics*, vol. 28, no. 4, pp. 1971-1979, April 2013.

[11] Z. Zhang and K. Lee, "A Hybrid PWM Strategy for Common-mode Voltage Reduction and Voltage Linearity Retainment in Adjustable Speed Drives," *2023 IEEE Energy Conversion Congress and Exposition (ECCE)*, Nashville, TN, USA, 2023, pp. 5229-5234.

[12] T. Kashihara, Y. Araki, H. Yoshida and K. Kobayashi, "Overmodulation Technique on Common Mode Voltage Reduction PWM Inverter using Saw-Wave Carrier Signal," *2022 International Power Electronics Conference (IPEC-Himeji 2022- ECCE Asia)*, Himeji, Japan, 2022, pp. 741-745.

[13] Y. Duan, Z. Li, Y. Guo and X. Zhang, "Synchronized SVPWM Strategy for Common Mode Voltage Reduction of High Power Inverters in the Overmodulation Region," *2019 22nd International Conference on Electrical Machines and Systems (ICEMS)*, Harbin, China, 2019, pp. 1-5.

[14] Z. Zhang and K. Lee, "Voltage and Current Harmonics Evaluation of Reduced Common-mode Voltage PWM Methods in Adjustable Speed Drives," *2023 IEEE International Electric Machines & Drives Conference (IEMDC)*, San Francisco, CA, USA, 2023, pp. 1-7.

[15] J. Holtz, W. Lotzkat, and A. M. Khambadkone, "On continuous control of PWM inverters in the overmodulation range including the six-step mode," *IEEE Trans. Power Electron.*, vol. 8, no. 4, pp. 546–553, 1993.

[16] S. Bolognani, and M. Zigliotto, "Novel digital continuous control of SVM inverters in the overmodulation range," *IEEE Trans. Ind. Applicat.*, vol. 33, no. 2, pp. 525–530, 1997.

[17] D.-C. Lee, and G.-M. Lee, "A novel overmodulation technique for space vector PWM inverters," *IEEE Trans. Power Electron.*, vol. 13, no. 6, pp. 1144–1151, Nov. 1998.

Single-Phase Open-Circuit Fault-Tolerant Control of Three-Phase PMSM Drives

Yuichiro Minato
Technology Development Center, Research and Development Headquarters
Murata Machinery, LTD
Kyoto, Japan
yuuichirou.minato@akl.muratec.co.jp

Yuki Nakata
Dept. of Electrical, Electronics, and Information Engineering
Nagaoka University of Technology
Niigata, Japan
ynakata@vos.nagaoka.ac.jp

Jun-ichi Itoh
Dept. of Electrical, Electronics, and Information Engineering
Nagaoka University of Technology
Niigata, Japan
itoh@vos.nagaokaut.ac.jp

Abstract—**This paper proposes a Fault-tolerant control method for a standard three-phase permanent magnet synchronous motor (PMSM) under a single-phase open-circuit fault condition. The method for driving a three-phase PMSM from a standstill using a three-leg, two-level inverter configuration is proposed. Experimental results show that 65% of the motor's rated speed and 50% of its rated torque maintains.**

Keywords—Degraded operation, Permanent Magnet Motor, Field-oriented Control (FOC) in dq0 control, Open circuit fault

I. INTRODUCTION

In recent years, with the expansion of e-commerce(EC) business, the importance of EC logistics warehouse that handle the storage, picking, and shipping of sold products has been increasing faster speed from order to delivery is required, and the volume of shipments has been increasing as well. In warehouse, mobile systems such as automated guided vehicles (AGVs), autonomous mobile robot (AMR), and robots are moving around freely, contributing to the improvement of order fulfillment efficiency. While sufficient measures are taken, these mobile systems occasionally experience failures. When the drive motor of a mobile unit fails, causing it to stop, it can disrupt the entire logistics operation.

In applications such as electric vehicles (EVs) and aerospace, a four-leg inverter is used to achieve fault-tolerant control in the event of a single-phase open-circuit fault [1-3]. Methods for implementing fault-tolerant control by accessing the neutral point of the stator windings have also been proposed [4]. A method for implementing fault-tolerant control using the modular multi-unit PMSM has been proposed. [5]. However, in general-purpose industrial motors used in applications such as AGVs, there is no neutral line and cost constraints are stringent.

Therefore, fault-tolerant control using a three-leg inverter has been proposed [6]. There have been proposals for generating phase voltage references during fault-tolerant control [7] and switching to fault-tolerant space vectors [8]. Additionally, methods for generating dq-axis current references for fault-tolerant control have also been proposed [9-10]. However, generating accurate phase voltage references requires precise detection of the healthy phase, which can complicate the software due to the need for filtering switching noise and other factors. Furthermore, even in fault-tolerant control that generates dq-axis current references by classifying them into multiple modes, there are regions where physical torque cannot be produced, resulting in a temporary drop to zero torque.

This paper proposes a fault-tolerant control method that maintains a compact program size by simplifying the voltage references calculations, allowing the control of two motors with a single microcontroller (MCU).

The originalities of this paper are in generation dq-axis voltage references based on dq-axis current feedback values, thereby eliminating the need for healthy phase identification, and implementing fault-tolerant control by switching the voltage output to zero in regions where physical torque is not producible. As a result, the proposed method does not require complex processing for fault-tolerant control and is achieved through a simplified software architecture. Furthermore, it avoids excessive current in regions where the torque drop to zero.

II. AMR DRIVE SYSTEM

Fig. 1 shows the motor control board, which is equipped with a single microcontroller and two sets of inverters. The AMR operates with two differential wheels. The motor control scheme implements FOC, incorporating both speed and current control. Each of these controls employs proportional-Integral(PI) controllers, and the target motor is a surface permanent magnet synchronous motor(SPMSM).

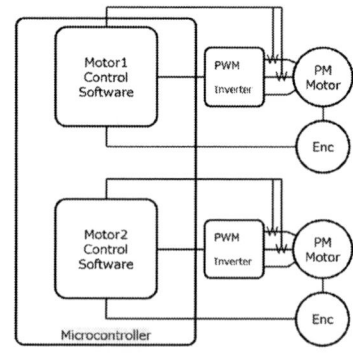

Fig. 1. Block diagram of AMR motor control system.Control of two motors with one MCU.

979-8-3315-1612-3/25 $31.00 © 2025 IEEE

A. Motor Control Unit

Fig. 2 shows the block diagram of the system, which employs a typical software servo utilizing PI control for both speed and current regulation. Although the diagram shows one motor axis, the MCU controls two axes. Speed control is executed at 1kHz, while current control is executed at 10 kHz.

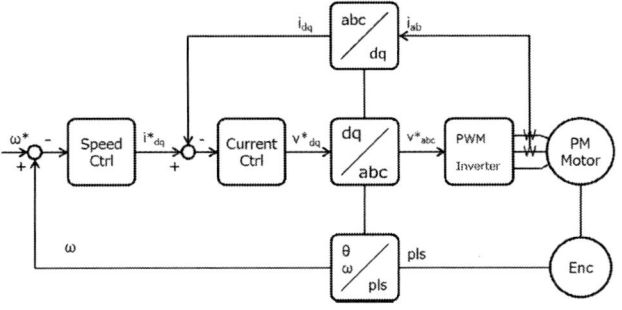

Fig. 2. Block diagram of the motor control unit.

B. Clarke and Park Transformations Equations

The dq-axis current feedback i_d and i_q can be obtained by applying Clark and Park transformations to the phase currents i_a, i_b and i_c.

$$\begin{bmatrix} i_d \\ i_q \end{bmatrix} = \frac{2}{3} \begin{bmatrix} \cos\theta_e & \sin\theta_e \\ -\sin\theta_e & \cos\theta_e \end{bmatrix} \begin{bmatrix} 1 & -\frac{1}{2} & -\frac{1}{2} \\ 0 & \frac{\sqrt{3}}{2} & -\frac{\sqrt{3}}{2} \end{bmatrix} \begin{bmatrix} i_a \\ i_b \\ i_c \end{bmatrix} \quad (1)$$

Where θ_e is rotor electric angle.

C. Current Control Algorithm

The difference between the dq-axis current command input and dq-axis current feedback is fed into a PI control scheme, which generates the dq-axis voltage command.

III. SINE WAVE DRIVE DURING OPEN-PHASE FAULT

Analyze the variations in dq-axis currents under open-phase fault conditions below.

A. In the Case of an Open-Circuit Fault in the C Phase of the Motor

When assuming that an open-circuit fault occurs in phase C of the motor, the current in this phase drops to zero.

$$i_c = 0 \quad (2)$$

The motor under consideration has each phase coil connected at a neutral point inside the motor, and according to Kirchhoff's law, it can be transformed as shown in equation (3).

$$i_a = -i_b \quad (3)$$

By substituting equation (2) and equation (3) into equation (1) and simplifying, equation (4) is obtained.

$$\begin{bmatrix} i_d \\ i_q \end{bmatrix} = \begin{bmatrix} (\frac{1}{\sqrt{3}}\sin\theta_e - \cos\theta_e)i_a \\ (\sin\theta_e + \frac{1}{\sqrt{3}}\cos\theta_e)i_a \end{bmatrix} \quad (4)$$

Using equation (4), Fig 3 shows the waveforms of the phase currents and dq-axis currents over one electrical cycle. As shown in Fig. 3, when a sinusoidal voltage is applied to the motor under open-phase conditions, the dq-axis currents, which were converted to DC quantities during normal operation, oscillate in a sinusoidal manner.

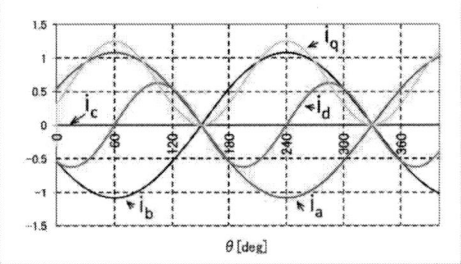

Fig. 3. Currents during post open-C-phase fault.

B. In the Case of an Open-Circuit Fault in the A Phase of the Motor

When assuming that an open-circuit fault occurs in phase A of the motor, the current in this phase drops to zero.

$$i_a = 0 \quad (5)$$

Bring equation (5) into Kirchhoff's laws:

$$i_c = -i_b \quad (6)$$

By substituting equation (5) and equation (6) into equation (1) and simplifying, equation (4) obtained.

$$\begin{bmatrix} i_d \\ i_q \end{bmatrix} = \begin{bmatrix} \frac{2}{\sqrt{3}}\sin\theta_e \\ \frac{2}{\sqrt{3}}\cos\theta_e \end{bmatrix} \quad (7)$$

Using equation (7), fig 4 shows the waveforms of the phase currents dq-axis currents over one electrical cycle.

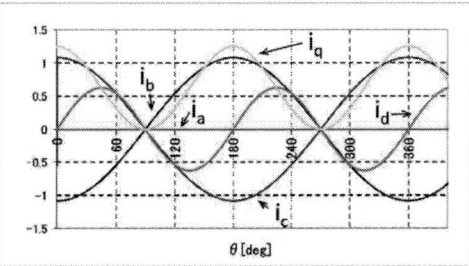

Fig. 4. Currents during post open-A-phase fault.

From Figs. 3 and 4, it can be observed that the oscillation phase of the dq-axis current exhibits specific changes depending on the phase that has experienced the open-phase fault.

IV. FAULT TOLERANT CONTROL

Figs. 3 and 4 show that when a sinusoidal voltage is applied in the same manner as under normal conditions during an open-phase fault, the current in the motor becomes a two-phase AC with a 180-degree phase difference. When a sinusoidal voltage with a 120-degree phase difference is applied during an open-phase fault in phase C, the current flowing through the motor coil flows exclusively from phase A to phase B or vice versa. When an open-phase fault occurs while the motor is rotating, the motor continues to rotate due to inertia even with a two phase AC with a 180-degree phase difference. However, once the motor stops, it is difficult to restart. This is evident from Fig. 3, as the direction of the magnetic field that can be generated in the motor coil is limited to only two opposing directions.

A. Motor Current Applied during an Open-Phase Fault

Since the direction of the magnetic flux is limited to only two opposing directions during an open-phase fault, many papers on open-phase drive systems have devised voltage waveforms generated by two phases. This means switching between the three-phase AC voltage waveform with a 120-degree phase difference under normal conditions and a special waveform of a two-phase AC voltage during an open-phase fault. Additionally, to accommodate any of the three phases experiencing an open-phase fault, three voltage maps for open-phase fault conditions. This raises concerns about insufficient flash memory capacity. In this paper, to ensure that the two-phase voltage waveforms correspond to the open-phase fault condition, the voltage is set to zero to prevent the generation of magnetic flux that would pull the rotor in the reverse direction.

As shown in Fig. 5, in the simplified single-pole pair motor diagram, the direction of the magnetic flux that can be generated by the coil does not rotate in the event of an open-phase fault.

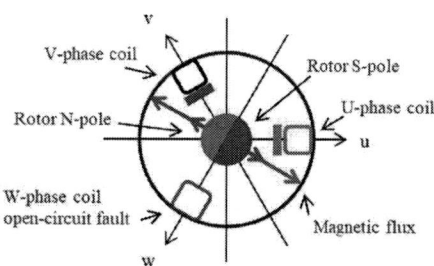

Fig. 5. Direction of magnetic flux in the motor structure daigram.

When the direction of the magnetic flux is superimposed on the current waveforms during open-phase fault condition shown in Fig. 3, it results in Fig. 6. Particularly at electrical angles of 120 degrees and 300 degrees, the direction of the magnetic flux generated in the coils hinders the rotation of the rotor. Therefore, by switching the voltage to zero and not supplying current to the motor at these times, the rotor is allowed to free-run, thereby continuing the rotation of the motor.

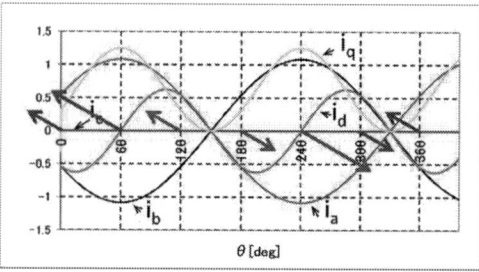

Fig. 6. Phase currents, dq-axis currents, and direction of magnetic flux during open-C-phase fault condition.

Furthermore, since the dq-axis currents during an open-phase fault oscillate with a phase uniquely determined by the faulted phase, the state of the dq-axis currents is monitored to determine whether to output voltage or switch the voltage to zero.

B. Current Control during an Open-Phase Fault

Fig. 7 shows the voltage output, which is switched between the normal voltage output and zero voltage output based on the values of the dq-axis current feedback. The torque is generated in the rotational direction by applying voltage, and then the voltage is set to zero to maintain rotation through inertial force, thus achieving fault-tolerant control. The dq-axis current feedback waveforms uniquely change depending on which phase has experienced an open-circuit fault. In other words, since the dq-axis current feedback contains information about the healthy phases, it eliminates the need for healthy phase identification and prevents motor lock due to misidentification.

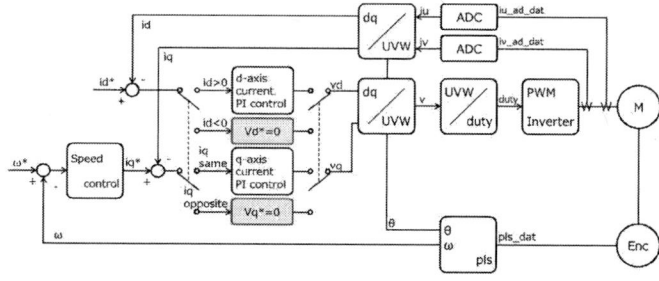

Fig. 7. Block diagram of the fault tolerant control part of the proposed method.

C. Processing Flow for the Dq-axis Voltage Reference

Fig. 8 and Fig. 9 show the processing flow for the d-axis voltage reference and q-axis voltage reference, respectively. The software structures achieves fault-tolerant control by selecting whether to execute PI control using the current deviation or to set the dq-axis voltage references to zero based on the dq-axis current feedback, thereby eliminating the need for a voltage map during an open-circuit fault and simplifying the process. The current references are set similarly to normal operation, with i_d^* set to zero and i_q^* determined by speed control. The q-axis voltage reference is switched according to the direction of rotation (the sign of q-axis current reference).

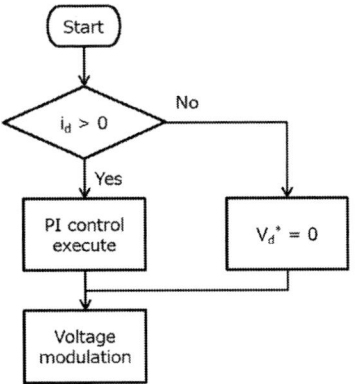

Fig. 8. Flowchart for calculating d-axis voltage reference v_d^*. If i_d is positive, PI control is executed. If it is negative, the d-axis voltage reference is set to zero.

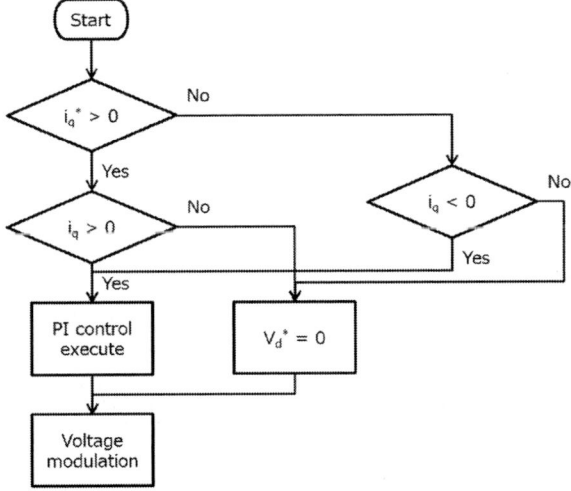

Fig. 9. Flowchart for calculating q-axis voltage reference d_q^*. If i_q has the same sign as the current reference i_q^*, PI control is executed. If it has the opposite sign, the q-axis voltage reference is set to zero.

D. Equations for Calculating the Dq-axis Voltage References

The equations for calculating the dq-axis voltage references are as follows:

The proposed voltage reference calculation formulas for counterclockwise (CCW) operation ($i_q^* > 0$)

When $i_d > 0$:

$$v_d^* = K_c(1 + \frac{1}{T_c s})(i_d^* - i_d) \tag{8}$$

When $i_q > 0$:

$$v_q^* = K_c(1 + \frac{1}{T_c s})(i_q^* - i_q) \tag{9}$$

When $i_d < 0$:

$$v_d^* = 0 \tag{10}$$

When $i_q < 0$:

$$v_q^* = 0 \tag{11}$$

Where v_d^* and v_q^* are dq-axis voltage references, K_c is the current proportional gain, T_c is the current integral time.

The proposed voltage reference calculation formulas for clockwise (CW) operation ($i_q^* < 0$)

When $i_d > 0$:

$$v_d^* = K_c(1 + \frac{1}{T_c s})(i_d^* - i_d) \tag{12}$$

When $i_q < 0$:

$$v_q^* = K_c(1 + \frac{1}{T_c s})(i_q^* - i_q) \tag{13}$$

When $i_d < 0$:

$$v_d^* = 0 \tag{14}$$

When $i_q > 0$:

$$v_q^* = 0 \tag{15}$$

V. EXPERIMENTAL RESULTS

The proposed method was implemented on a motor test rig. Experiments were conducted by inducing an open-circuit fault condition and stopping the motor, demonstrating that the motor could be successfully restated using the proposed fault-tolerant control. Experiments have shown that during a single-phase open-circuit fault, the motor operates at 65% of its rated speed and 50% of its rated torque.

A. A Results from the Experimental Setup

Fig. 10 shows the experimental setup for fault-tolerant control. The test motor is connected to the motor control board, which outputs the motor speed obtained via a digital-to-analog (DA) converter, calibrated such that 1000 rpm corresponds to 0.5V, and it is configured so that a voltage of 2.5 V corresponds to a motor speed of zero. The speed command, motor speed, and torque meter output are then measured using a measuring instrument to obtain the waveforms. This experimental setup was used to conduct open-circuit fault tests on the C-phase.

979-8-3315-1612-3/25 $31.00 © 2025 IEEE

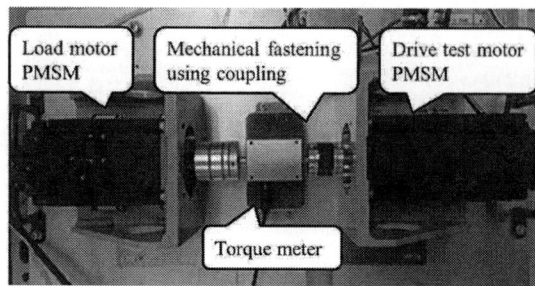

Fig. 10. Experimental setup for fault-tolerant control.

B. Motor Specifications for Experimental Setup for Fault-Tolerant Control

Table 1 shows the specifications of the motors used in the experimental setup for fault-tolerant control.

TABLE I.   Motor Specifications

Motor Test Rig	
Parameter	**Value**
Rated voltage	DC24 V
Rated output	300 W
Rated speed	3800 rpm
Rated torque	0.74 Nm
Rated current	15 A
Inductance	0.0127mH
Resistance	0.08 Ω

C. Speed and Torque Waveforms from Experimental Setup for Fault-Tolerant Control

Fig. 11 shows the speed and torque waveforms when the load is set to 35% of the rated value and the motor is accelerated to 1000 rom from a stop under a C-phase open circuit fault condition.

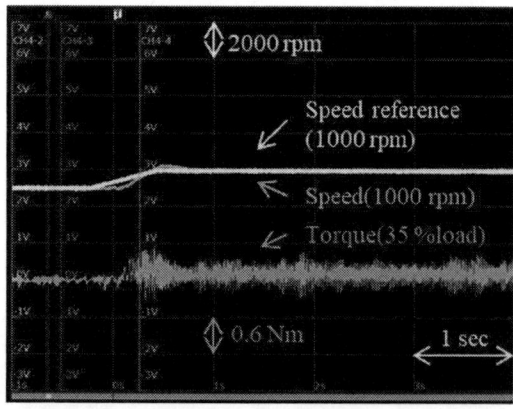

Fig. 11. Speed and torque waveforms during acceleration from standstill to 1000 rpm under C-phase open-circuit fault condition.

Fig. 11 shows that although there is a slight time lag during acceleration, the motor closely follows the speed reference. Although the torque waveform exhibits significant ripple, the average torque is directed in the rotational direction, indicating that fault-tolerant control is being successfully implemented.

D. Phase Current Waveforms from Experimental Setup for Fault-Tolerant Control

Fig. 12 shows that since the C-phase is in an open-Circuit fault condition, i_c is 0 A.

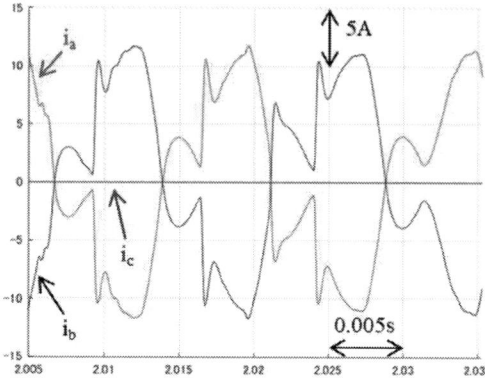

Fig. 12. Current waveforms during acceleration from standstill to 1000 rpm under C-phase open-circuit fault condition. The rated current is 15 A.

Fig. 13 shows the speed and torque waveforms when the load is set to 50% of the rated value and the motor is accelerated to 2500 rpm from a stop under a C-Phase open-circuit fault.

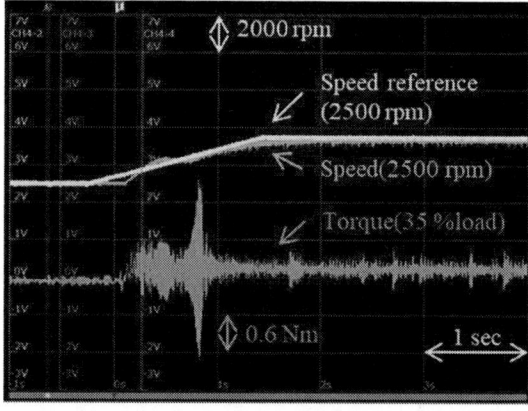

Fig. 13. Speed and torque waveforms during acceleration from standstill to 2500 rpm under c-phase open-circuit fault condition.

E. Configuration of the AMR

Following the experiments on the motor test rig, the feasibility of open-phase fault-control drive using the AMR was verified. Fig. 14 shows the AMR used in the experiments. This AMR uses two motors and is driven by a differential two-wheel drive system. The MCU used is the SH7286, which drives two motors by connecting its two PWM outputs to the inverter. Fig. 1 shows the configuration of the motor control board that drives AMR in this experiment. In the experiments using AMR, an open-circuit fault in the A-phase was induced.

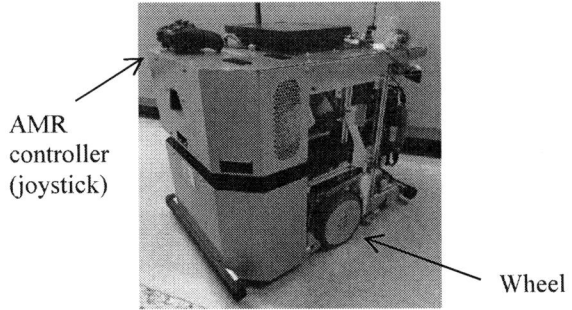

AMR
controller
(joystick)

Wheel

Fig. 14. Overview of the AMR. AMR operated by joystick.

F. Motor Specifications for AMR

Table II shows the specifications of the motor. The motor is a three-phase SPMSM.

TABLE II.   MOTOR SPECIFICATIONS

AMR	
Parameter	**Value**
Rated voltage	DC24 V
Rated output	100 W
Rated speed	3000 rpm
Rated torque	0.32 Nm
Rated current	6 A
Inductance	1.0mH
Resistance	0.67 Ω

G. Waveforms during Nomal Operation Using AMR

Fig 15 shows the speed control waveform under normal conditions. Since the motors are mounted in opposition, the sign of the speed is reversed. Using a DA converter to output the motor speed, a DA converter output of 2.5 V corresponds to a motor speed of 0 rpm. The phase current waveform is measured only from the side conducting the open-phase fault experiment.

H. Waveforms Observed During an Open-Circuit Fault in Normal Operation Using AMR

Fig. 16 shows the speed and current waveforms when an open-circuit-fault occurs from a normal operation under speed control. In this case, an open-circuit-fault occurred in the A phase, resulting in the A phase current dropping to 0 A. The software detects the open-circuit fault and switches to the voltage for fault-tolerant control, thereby maintaining speed control. As a result, the AMR can continue moving straight.

I. Waveforms Under Open-Circuit Fault Using AMR

Fig. 17 shows the speed and current waveforms under open-phase fault-tolerant control. The right-side motor is in normal operation, while the left-side motor is under an open-phase fault. The AMR can move straight, indicating that the open-phase fault-tolerant control is functioning properly.

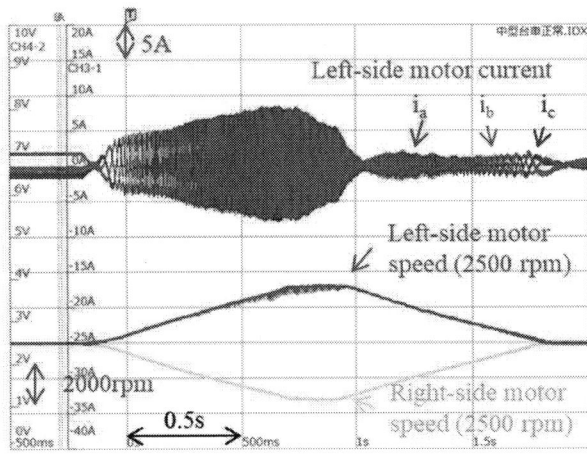

Fig. 15. Waveforms during speed control (normal operation).

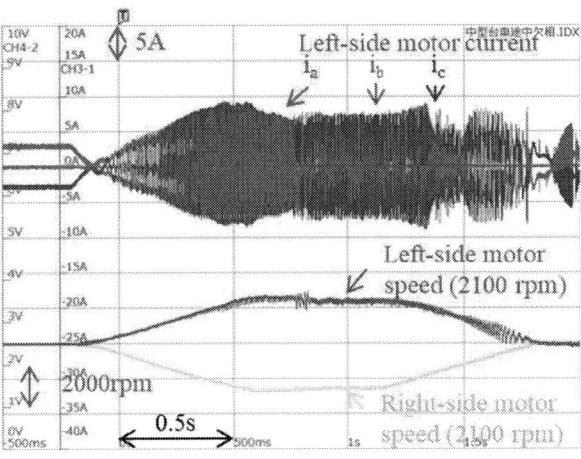

Fig. 16. Waveforms during speed control (an open-circuit fault occurred under normal operation).

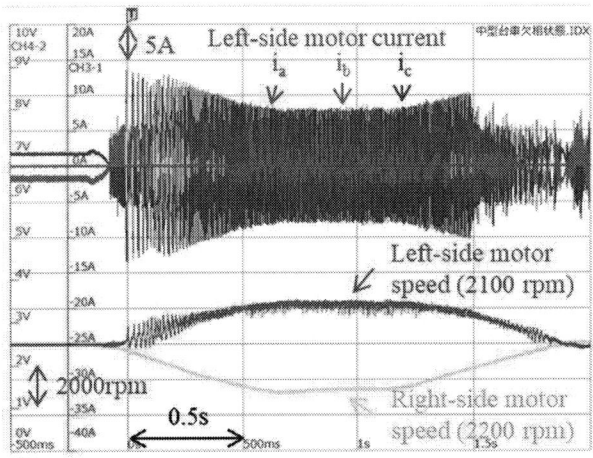

Fig. 17. Waveforms during speed control (under an open-circuit fault occurred).

J. Waveforms When Normal Operation Is Restored During Open-Circuit Fault Control Using AMR

Fig. 18 shows the case where normal operation is restored while running in an open-circuit fault condition. When returning to normal operation from a fault condition, the accumulated speed deviation causes significant current fluctuations. The current fluctuations when returning to normal operation from an open-phase fault condition will be addressed as a future challenge. The AMR can move straight even in an open-phase fault condition and maintains its straight movement upon returning to normal operation. Examining the current waveforms reveals that three-phase AC current flows after recovery from the open-phase fault.

Due to the tendency for torque ripple to increase during open-phase fault-tolerant control, the dq-axis voltages are limited. This results in a limitation of the output during open-phase fault-tolerant control. The voltage limit is not lifted even after returning to normal from an open-phase fault condition. This is necessary there is a possibility of an open-circuit fault occurring again, requiring inspection by a service technician.

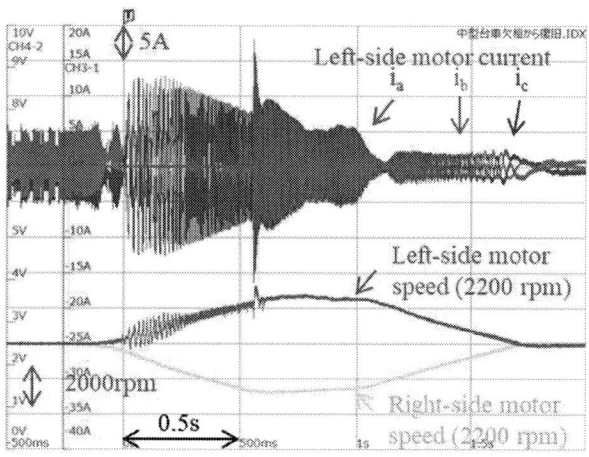

Fig. 18. Waveforms during speed control (normal operation is restored during open-circuit fault-tolerant control).

VI. Conclusion

This paper proposes a fault-tolerant control method for open-circuit fault conditions without increasing program size. The proposed method determines whether to execute PI control based on the dq-axis current feedback values or to set the dq-axis voltage references to zero. This method was implemented on the control board for the two motors driving the AMR, and fault-tolerant control was successfully validated on the motor test rig. Furthermore, with the proposed fault-tolerant control, the motor was able to operate at 65% of the rated speed and 50% of the rated load. These are the limit values, so driving the motor under these conditions for an extended period can cause damage to the bearings and other components due to vibrations. Therefore, in AMR experiments, a larger margin is provided. In experiments using AMR for open-circuit fault-tolerant control, various scenarios were tested, including inducing an open-circuit fault during AMR operation and recovering to normal operation from an open-circuit condition. When an open-circuit failure occurs in an AGV or AMR using this method, a limited from a few minutes to several tens of minutes is planned.

References

[1] B. Sun, J. Fu, P. Cui, Z. Hou, W Wang, X Wang, X Yan, D Ji "Fault-Tolerant Control of Three-Phase Four-Leg PMSM Servo Systems," IEEE 6th International Electrical and Energy Conference (CIEEC) Jul. 2023.

[2] Y. Guo, L. Wu, X. Huang, Y. Fang, J. Liu, "Adaptive Torque Ripple Suppression Methods of Three-Phase PMSM During Single-Phase Open-Circuit Fault-Tolerant Operation," IEEE Transactions on Industry Applications, Vol. 56, Issue. 5, Sep.-Oct.2020.

[3] H. Tang, W. Li, J. Li, H. Gao, Z. Wu, and X. Shen, "Calculation and Analysis of the Electromagnetic Field and Temperature Field of the PMSM Based on Fault-Tolerant Control of Four-Leg Inverters," IEEE Transactions on Energy Conversion, Vol. 35, Issue. 4, Dec. 2020.

[4] A. Gaeta, G Scelba, A Consoli, "Modeling and Control of Three-Phase PMSMs Under Open-Phase Fault," IEEE Transactions on Industry Applications, Vol. 49, Issue. 1, Jan.-Feb. 2013.

[5] Z. Yang, F. Ying, W. Moyang "Fault-tolerant Control Strategy for Modular Multi-Unit PMSM Based on Vector Decoupling and Minimum Copper Loss," 2023 26th International Conference on Electrical Machines and Systems (ICEMS)

[6] A. Kontarˇcek, P. Bajec, M. Nemec, V. Ambroˇziˇc, and D Nedeljkoviˇc, "Cost-Effective Three-Phase PMSM Drive Tolerant to Open-Phase Fault," IEEE Transactions on Industrial Electronics, Vol. 62, Issue. 11, Nov. 2015

[7] N. D. Veeramaneni and R. Kanagaraj, "Post-Fault Operation of Fault-Tolerant Three Phase Surface Mounted PMSM", 2019 National Power Electronics Conference (NPEC), Dec. 2019.

[8] Z. Zhang, Y. Hu, G. Luo, C. Gong, X. Liu, and S. Chen, "An Embedded Fault-Tolerant Control Method for Single Open-Switch Faults in Standard PMSM Drives," IEEE Transactions on Power Electronics, Vol. 37, Issue. 7, Jul. 2022

[9] X. Wang, Z. Wang, M. Gu, B. Wang, W. Wang, M. Cheng, "Current Optimization-Based Fault-Tolerant Control of Standard Three-Phase PMSM Drives," IEEE Transactions on Energy Conversion, Vol. 36, Issue. 2, Jun. 2021

[10] F. Li, K. Zeng, Y. Jiang, Y. Zhao, S. Chen, "Tri-mode Fault-Tolerant Control of Three-Phase Permanent Magnet Synchronous Motor Drive System under Single-Phase Open Fault," IEEE 2023 26th International Conference on Electrical Machines and Systems (ICEMS), Dec. 2023.

Multi-vendor encoder position sensing interface using programmable IP based solution

Rajul Bhambay
Embedded Processors
Texas Instruments Inc.
Bengaluru, India
r-bhambay@ti.com

Dhaval Khandla
Embedded Processors
Texas Instruments Inc.
Bengaluru, India
dhavaljk@ti.com

Pratheesh Gangadhar
Embedded Processors
Texas Instruments Inc.
Bengaluru, India
pratheesh@ti.com

Thomas Leyrer
Embedded Processors
Texas Instruments Inc.
Munich, Germany
t-leyrer@ti.com

Achala Ram
Embedded Processors
Texas Instruments Inc.
Bengaluru, India
a-ram@ti.com

Manoj Koppolu
Embedded Processors
Texas Instruments Inc.
Bengaluru, India
manoj_koppolu@ti.com

Archit Dev
Embedded Processors
Texas Instruments Inc.
Bengaluru, India
a-dev@ti.com

Abstract—The aim of this work is to look at a software programmable solution for multi-vendor encoders protocol in which TI utilizes "Three Channel Peripheral Interface (3CPI)" or general purpose inputs (GPIs)/general purpose outputs (GPOs) in programmable real-time unit subsystem and industrial communication subsystem (PRU-ICSS) on Texas Instruments AM24x/AM26x micro-controllers (MCUs) and AM64x processors [10]–[12]. Existing solutions are FPGA based which come with a few challenges like high power consumption, high cost, and latency in operating control loop. This approach cuts down huge amount of cost by not implementing with FPGA, but using programmable IP and offloading MCUs or processors CPU load via PRU-ICSS. It also has the capability to optimize control loop, by moving the position data after validation and post processing to the tightly coupled memory of CPU running control loop. This makes TI's solution cost effective and low latency in terms of running control loop, highly flexible and scalable in terms of implementing multiple vendor encoder interfaces.

Index Terms—Texas Instruments (TI), Intellectual Property (IP), Programmable IP, Field-Programmable Gate Array (FPGA), Programmable Real-Time Unit and Industrial Communication Subsystem (PRU-ICSS), Texas Instruments Sitara System-on-chips (SoCs), Three Channel Peripheral Interface (3CPI), Industrial Ethernet Peripheral (IEP), Compare Events (CMP), Cyclic Redundancy Check (CRC), Hardware Description languages (HDLs), Broad Side (BS)

I. INTRODUCTION

Encoders are devices used to measure position, speed, or direction of motion in various systems and provide accurate and precise communication of this information. This makes them an indispensable part of applications which require accurate position data and control like robotics, manufacturing processes, industrial automation, medical devices like robotic arm surgery systems, and many more. A major challenge in using encoders in industrial environments is the variety of protocols available to interface the hardware from multiple vendors. Traditionally, Field Programmable Gate Arrays (FPGAs) have been utilized to handle the high-speed and parallel processing demands of multiple encoders. Section II

gives high level overview of how PRU-ICSS based solution is different from FPGA based solution. Section III introduces the AM243x SoC and its internal architecture. Section IV gives implementation overview. Section V shows the results achieved using TI's implementation.

II. MOTIVATION

Position encoders are fundamental components in a wide range of applications, including robotics, motion control, and industrial automation. Traditionally, FPGAs have been utilized to handle the high-speed and parallel processing demands of multiple encoders, but complexity of FPGA is designed specific to protocol, and often requires specialized knowledge in HDLs, making them hard IP. Furthermore, changes in encoder IP specification, the need for high-speed design and testing cycles can lead to delays in the overall development process as integrating FPGAs into embedded systems often requires additional components, such as external memory, or specialized interfaces, which further complicates the design and increases the overall system footprint resulting in higher overall system costs. Texas Instruments (TI) PRU-ICSS is software programmable, thereby offering an innovative alternative to FPGAs by providing highly efficient, fast real-time processing and response capabilities integrated into a Sitara family of devices. The ability of PRU-ICSS to handle time-critical tasks such as signal processing and encoder interfacing, with lower power consumption and simpler integration into wide range of applications, makes it an ideal candidate for replacing FPGAs supporting multiple encoder interfaces.

III. PROPOSED SOLUTION

A. PRU-ICSS Module Overview

The 32-bit RISC CPU Core architecture referred to as PRU-ICSS (Programmable Real time Unit Industrial Communications Sub System) can serve as an intermediary subsystem to enable low-latency real-time data communication, that is essential for developing systems that need time-sensitive interaction between micro-controllers and any external sensors, in this case motor position sensing encoders. This subsystem responds fast in real time due to its low latency access to SoC level pins and other resources, which eventually offloads work from other processor arm cores and handles data in a unique method.

Each PRU-ICSS instance consists of 2 slices (slice 0 and slice 1) and in each slice there are three 32-bit load/store RISC CPU cores called PRUs (Programmable Real-time Units) [16]–[18]. Each core consists of following configurations [16]–[18]:

- Core clock frequency range from 200Mhz to 333Mhz
- 20 Enhanced General-Purpose Inputs (EGPI)
- 20 Enhanced General-Purpose Outputs (EGPO)
- 8 to 12 KB program memory
- 2KB Broadside (BS) RAM
- Hardware Accelerators

Along with cores, there are other components involved such as [16]–[18]:

- 64 KB general purpose memory RAM (Data RAM2), shared across all cores
- Two 8 KB (shared) data memories (Data RAM0 and Data RAM1)
- Two 3CPI Interfaces
- Two Industrial Ethernet Peripheral's (IEP0/IEP1) for time stamping.
- Scratchpad Memory (SPAD) with 8 banks of 30 registers (32-bit).
- One Enhanced Capture Module (ECAP0)
- Interrupt controller (INTC)
- UART along with separate clock
- Two real-time Ethernet ports

Slice 0 contains PRU0, RTU PRU0 and TX PRU0 and Slice 1 contains PRU1, RTU PRU1, and TX PRU1 [16]–[18]. Features listed above are present in TI Sitara AM243x MCUs [10] and TI Sitara AM64x. Some other SoCs like TI Sitara AM263x MCUs [11] and TI Sitara AM263Px MCUs [12] have PRU-ICSS IP with limited features, like two cores only (instead of six), no task manager, etc. More details on features available are present in Technical Reference Manual (TRM) of TI Sitara AM243x MCUs and TI Sitara AM263x MCUs [16]–[18].

B. PRU Core Overview

The PRU turned out to be a optimized processor for embedded systems operations which are necessary to work along with mapped structures of data in compacted memory engaged with subsystems outside of the system-on-chip (SoC) while controlling system-level events with severe real-time restrictions. Whenever it involves organizing various embedded tasks, the PRU appears compact and highly effective.

Quick real-time feedback, concentrated handling of data processes, distinctive peripheral interfaces, and task outsourcing from the device's other cores of processor are all made possible by the programmable capabilities of the PRU cores as well as their proximity to pins, and all other system resources.

Every PRU has the ability to function individually or in a synchronized way with one another, or in association alongside the host processor at the system level. The particular kind of firmware that is programmed into the PRU's memory dictates the way these processors communicate with one another. Moreover, section 6.4.5.1 in AM64x /AM243x Processors TRM summarizes the PRU's fundamental features [17].

The 4-bus topology of processor enable transfer of data to take place simultaneously alongside the retrieval and processing of instructions. Furthermore, an input is included to enable the inbuilt processor status register update by external status data. A schematic representation of the computation component with the corresponding command ROM/RAM, that holds the program that needs to be implemented, can be summarized from section 6.4.5.1 in AM64x /AM243x Processors TRM [17].

C. Three Channel Peripheral Interface (3CPI) Overview

The PRU-ICSS module provide 2 types of internal registers called PRU registers R30 and R31 highlighted in section 6.4.5.2.2 in AM64x /AM243x Processors TRM with their functionality [17]. The R31 register functions as an interface through specific PRU GPI pins and PRU ICSS INTC. Accessing R31 retrieves status details from the GPI pins and PRU ICSS INTC through accessing real time PRU status interface. Event generation triggered by writing data to R31 register using PRU event interface. The R30 register functions as an interface through specific PRU GPO pins.

The 3CPI mode for IO as depicted in fig. 1 enables serial communication implementation for position sensing using multiple vendors encoders makes. This module supports half-duplex communication with wide range of baud rate options using PRU-ICSS UART/Core clock as an input to independent fractional clock dividers and produce a TX clock and of over-sampled Rx clock. The major configurable parameters mentioned in table I and other major functionalities mentioned below [16]–[18].

- Programmable RX termination
- Programmable hardware wire and tst delays
- Programmable TX termination
- Individual channel and all 3 channels simultaneous TX start trigger
- Configurable Start bit polarity (0 or 1)
- Variety of options in hardware assisted clock output generation that allow
  - free running clock
  - stop high and stop low once last RX bit received
  - stop high once last TX bit received

D. Features of Proposed Solution

- Supporting multiple channels:

In each PRU-ICSS instance, there are two 3CPIs available to program encoder interface efficiently. Each 3CPI can implement three channels of one encoder type. One PRU-ICSS instance is capable of interfacing two different vendor encoder

979-8-3315-1612-3/25 $31.00 © 2025 IEEE

S.NO	Parameter	Values
1	No. of channels	3
2	Baud Rate	100 (KHz) to 50 (MHz)
3	PRU-ICSS Core Clock	200, 225, 250, 300, 333 (MHz)
4	PRU-ICSS UART Clock	192 (MHz)
5	TX FIFO Size	32 Bits
6	RX FIFO Size	32 Bits
7	Rx Oversampling rate	Upto 8x

TABLE I

3CPI MAJOR CONFIGURABLE PARAMETERS

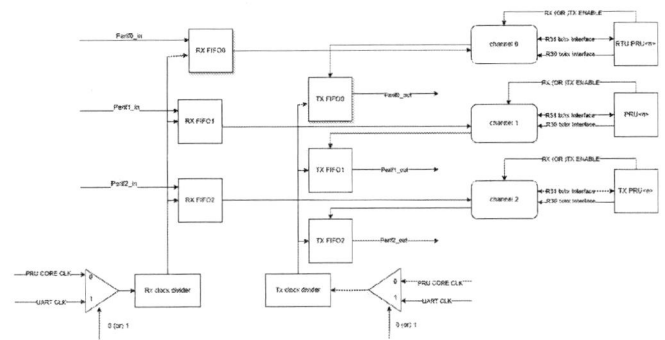

Fig. 1. 3 Channel Peripheral architecture

types and provides six channels in total for six encoders position sensing interface. On devices like Sitara AM243x, which have two PRU-ICSS, TI can implement 12 channels.

- Support daisy-chain mode:

TI's solution supports daisy chaining operation bus for multiple encoders of same type. This saves additional hardware components needed like RS-485 used for adding additional encoders for position sensing.

- Synchronization Capabilities:

PRU-ICSS for the servo motor drive applications has the potential to synchronize the motor control loop with industrial communications, and also precisely trigger the position sense based on encoder and sigma-delta filtering based current sense within the control loop. Additionally, PRU-ICSS PWMs can be used to generate PWM outputs for servo motors and this is achieved using IEP module [9].

- Offer support for high baud rates:

The PRU-ICSS clock can be configured to run at different options like 200 MHz, 225 MHz, 250 MHz, 300 MHz or 333 MHz. With an appropriate clock divider value and up to 8x oversampling, baud rates between 100 KHz-50 MHz are achievable.

- Different sized data frames:

Using the 32-bit Tx FIFO and optional programmable TX termination, a variety of encoders with different sized data frames can be interfaced with this solution. On the fly CRC computation: The solution can support up-to 16 M bauds with on-fly CRC computation using CRC broad side accelerator. This also off-loads the ARM Cortex R5F core in TI Sitara SoCs that acts as the application core

IV. IMPLEMENTATION

A. Hardware Mapping

Fig. 2 shows mapping of multi-vendor encoders connected to TI Sitara AM243x SoC. Here PRU-ICSS's Slice 0 and Slice 1 are used to run two different vendor position encoder processing. Three cores from within a slice are used to handle three position encoders from single vendor. Slice 0 handles three A encoders and Slice 1 handles three B encoders.

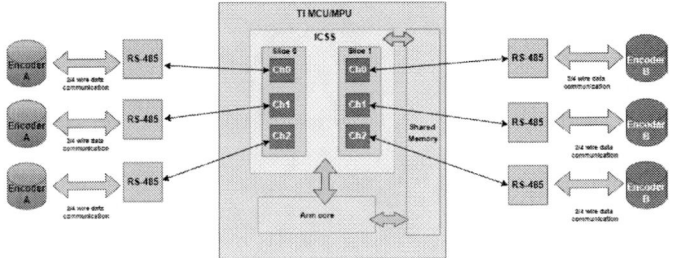

Fig. 2. Proposed Solution Architecture

B. Software Mapping

Fig. 3 represents software flow for a PRU slice, that is used for the PRU-ICSS subsystems. TI sets the PRU-ICSS in 3CPI mode to control Three channel peripheral's TX, RX, TX EN and clock lines. A PRU-ICSS Slice has three cores, which are used to process 3 channels. During initialization, one of the core configures the global level settings such as clock source, encoder baud rate, RX overlap rate, etc. for all the 3 channels. The RTU_PRU core is used to process channel 0, the PRU core is used to process channel 1 and the TX_PRU core is used to process channel 2.

Each channel can be used in either TX or RX mode. In TX mode, the firmware configures TX FIFO (Hardware queue for data to be transmitted) size and PRU-ICSS firmware loads TX data to TX FIFO which will be sent to encoder. At the end of TX mode, PRU-ICSS, the firmware enables communication to send TX data on TX line. After transmission is done, the firmware switches to RX mode. In RX mode, it waits for RX start bit. Once the start bit is received, the PRU-ICSS firmware starts reading the data form RX FIFO (Hardware queue for data to be received) which is sent by the encoder.

On the application side, R5F core is used to enable PRU cores as represented in Fig. 3, and then to create command format of TX data, and parsing of RX data and shared memory to exchange data between R5F core and PRU-ICSS cores.

V. RESULTS

The position sensing interface for multiple vendors [3]–[8] has been implemented on AM243x TI Sitara SoCs with less than 300 KB of MS RAM with the configurations mentioned in the table II.

There are few systems that can be implemented for position sensing interface for different encoders.

979-8-3315-1612-3/25 $31.00 © 2025 IEEE

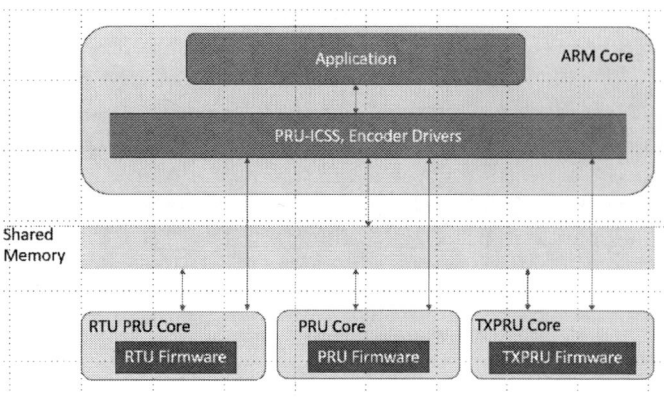

Fig. 3. Software Architecture

S.No	R5F Core Frequency (Mhz)	PRU Core frequency (Mhz)	Encoder vendor	Encoder baud rate (Mbps)
1	800	300	EnDAT 2.2	16
2	800	300	EnDAT 3	25
3	800	225/300	HDSL	9.375
4	800	200	Tamagawa	5
5	800	300	Nikon	16
6	800	200	BISS-C	10

TABLE II

AM243X IMPLEMENTATION SPECIFICATION

A. SoC with an external FPGA (System 1)

In this category, MPU/MCU is used with an external FPGA IC which performs the functionality of position sensing interface as an encoder master IP core. For reference, SoCs with XILINX FPGAs and many more [1], [14], [15].

B. Using Single TI Sitara SoC with integrated ICSS module (System 2)

Using PRU-ICSS module in Sitara SoCs, there is a huge advantage in terms of number of channels or encoder master interfaces as it comes with multiple cores (up to 12), for instance, TI's AM243x/AM64x Sitara SoCs.

With FPGA based implementation, multiple channel interfaces for encoders are possible, but raising number of channels by "n" may impact overall system cost heavily. Using PRU-ICSS module, TI have tested different encoders from multiple vendors without using an external FPGA drive. Table III, as represented below mentions a few results, which explain how using FPGA based system may lead to high cost for supporting multi-vendor positional encoder interface.

System type	FPGA usage	No. Logic elements for n channels (k)	FPGA cost increase factor	Overall System Cost
I	Yes	(6 – 200)*n	n	High
II	No	150	1	Low

TABLE III

FPGA VS PRU-ICSS SYSTEM COST COMPARISON

VI. SUMMARY

To conclude, using the ability of PRU-ICSS to implement a Generic Multi-encoder interface for all UART and customized interfaces benefits real-time position sensing and data processing, with cost effective and flexible solution for maintaining different position sensing encoder interface together. TI's solution caters to varied needs of Motor Control customers by supporting a wide array of position sense encoder protocols on multiple encoders with up to 12 channels and high baud rates. Along with other PRU blocks [e.g., IEP (Industrial Ethernet Peripheral), CMP (Compare event register), CRC (Cyclic redundancy check module), etc.]. TI's solution can also be used to implement features like time-triggered send, and on-fly CRC computation at high speeds, making it a great choice when used together with various industrial communication protocols like EtherCAT, Profinet, etc. Also, a typical Servo Motor Drive receives position reference over industrial Ethernet from a motion controller and uses that reference to close the position/speed/torque loops to actuate the drive.

REFERENCES

[1] AMD Technical Information Portal. (n.d.). Docs.amd.com. https://docs.amd.com/v/u/en-US/7-series-product-selection-guide

[2] Krah, Jens and Schmidt, Tobias and Richter, Rolf. (2021). Multi-Protocol Position Encoder Interface IP for Safety-Related Automation Drives.

[3] The IP-Core "DSL Master" allows implementing the motor-feedback protocol HIPERFACE DSL ® on drives. (n.d.). Retrieved November 14, 2024, from https://cdn.sick.com/media/docs/9/59/759/ Product_information_IP_Core_DSL_Master_V1.07_en_IM0074759.PDF

[4] "EnDat 2 Interface." HEIDENHAIN, endat.heidenhain.com/endat2.

[5] HIPERFACE DSL® - the Digital Evolution — SICK. www.sick.com/us/en/technologies/hiperface-dsl-the-digital-evolution/w/hiperface-dsl.

[6] Rotary Encoders — TAMAGAWA SEIKI CO.,LTD. tamagawa-seiki.com/products/rotaryencoder.

[7] Encoders — Nikon Business. www.nikon.com/business/encoder.

[8] BiSS Interface. "About BiSS - BiSS Interface." BiSS Interface -, 14 Aug. 2022, biss-interface.com/about-biss.

[9] Reeder, Jason, et al. "Single Chip Connected Multi-Axis Servo Drive for Industrial Systems." 2022 IEEE International Conference on Consumer Electronics (ICCE). IEEE, 2022.

[10] AM2434 data sheet, product information and support — TI.com. (n.d.). https://www.ti.com/product/AM2434

[11] AM2634 data sheet, product information and support — TI.com. (n.d.). https://www.ti.com/product/AM2634

[12] AM263P4 data sheet, product information and support — TI.com. (n.d.). https://www.ti.com/product/AM263P4

[13] Wang, H., ECE Department, & Southern Illinois University. (n.d.). Programmable ASIC design (pp. 5-1-5–13). https://www.engr.siu.edu/haibo/ece428/notes/ece428_logcell.pdf

[14] Farooq, U., Marrakchi, Z., & Mehrez, H. (2012). FPGA Architectures: An Overview. Tree-Based Heterogeneous FPGA Architectures, 7–48. https://doi.org/10.1007/978-1-4614-3594-5_2

[15] Cyclone® V FPGA and SoC FPGA Product Table. (2024). Intel. https://www.intel.com/content/www/us/en/content-details/714207/cyclone-v-fpga-and-soc-fpga-product-table.html

[16] Texas Instruments Ins. AM263x SitaraTM Microcontrollers Texas Instruments Families of Products, Texas Instruments Inc, Oct. 2024, www.ti.com/lit/ug/spruj17h/spruj17h.pdf?ts=1729222767400.

[17] AM64x /AM243x Processors Silicon Revision 2.0 Texas Instruments Families of Products Technical Reference Manual, Texas Instruments Inc, Oct. 2023, www.ti.com/lit/ug/spruim2h/spruim2h.pdf?ts=1716401947466.

[18] Texas Instruments Inc. (2019, December). AM437x and AMIC120 ARM® CortexTM-A9 Processors Technical Reference Manual (Rev. I). https://www.ti.com/lit/ug/spruhl7i/spruhl7i.pdf?ts=1733258045202

Sensorless Control Method at Low-Speed Range using High-Frequency Voltage Injection for Synchronous Reluctance Motors considering to NonLinear Characteristic due to Magnetic Saturation

Sota Takizawa
Depertment of Electrical Engineering
Meiji University, Kawasaki, Kanagawa
Email: ce241024@meiji.ac.jp

Sari Maekawa
Depertment of Electrical Engineering
Meiji University, Kawasaki, Kanagawa
Email: sari_maekawa@meiji.ac.jp

Abstract— **Synchronous Reluctance Motor (SynRM) has large magnetic saturation, and it experiences fluctuations in parameters such as flux and inductance because of magnetic saturation as the amount of current passing through it changes. Because of this, when saliency based sensorless control method is used for position estimation, the characteristic for position estimation becomes nonlinear. In this case, the accuracy of position estimation may decrease, or the controlled motor may lose synchronization. To solve this problem, the characteristics arising from the high-frequency square wave injection method is modeled including the cross-coupling effect and a method for combining two characteristics on dq-axes from the model equation is proposed in this paper. Then, by deriving characteristics from the result of electromagnetic field analysis using Finite Element Analysis (FEA) and theoretical equation, it is possible to verify their validity from the comparison results with the characteristics obtained from the actual machine. In the end, the proposed method is verified by experiments on a 2.2-kW SynRM drive platform.**

Keywords—Synchronous reluctance motors, Sensorless control, High-frequency voltage injection, Magnetic saturation, Cross-coupling effect.

I. INTRODUCTION

In recent years, there has been a growing trend across various fields to seek resource conservation and increased efficiency. SynRM is a type of synchronous machine that operates using reluctance torque alone, without employing rare-earth permanent magnets in the rotor. Additionally, its simple and robust structure makes it more cost-effective compared to Permanent Magnet Synchronous Motor (PMSM), allowing for low cost and resource conservation. Then, the torque ripple and vibrations of Switched Reluctance Motor (SRM) can be reduced in SynRM and it is also possible to achieve higher power density and efficiency compared to Induction Motor (IM) because there are no losses associated with rotor conductors. Due to these advantages, its application areas are expanding, and it is

expected to play a significant role as a next-generation motor. However, SynRM is known as a motor with large magnetic saturation, and when using sensorless control method that rely on saliency, the influence of changes in dq-axes inductances on the position estimation error is considerable. In addition, the influence of cross-coupling effect is also present in actual motor drives [1], [2]. These makes high-precision position estimation difficult. For the sensorless Field Oriented Control (FOC) of SynRM, the position estimation error is calculated as the function of dq-axes current, using for direct compensation [3], [4]. By using this compensation table, the sensorless control system is achieved [3]. Then, the flux is mathematically modeled, and a sensorless control configuration is constructed by applying an observer based on this model [4], [5]. By modeling with Recursive Least Squares (RLS), an observer can be constructed that allows online parameter identification. This enables the creation of a sensorless control system [6]. Then, a sensorless control configuration using a Neural Network (NN) is employed [7]. Electromagnetic field analysis results from FEA and Linear Least Squares (LLS) is used for parameter identification of flux [8]-[10], and a control system is constructed using the compensation coefficients derived from the obtained analysis results [11]. However, the addition of compensation table or compensation coefficient [3], [7], applying the observer [4], [5] and the use of NN [7] increases the computation time. Furthermore, there are issues such as compensation delay due to direct compensation control [3]-[5] and increased complexity in the control system caused by the observer for real time parameter identification [6].

As demonstrated by the sensorless control methods outlined above, when considering control strategies that takes magnetic saturation and cross-coupling effect into account, it is essential to understand parameters such as flux and inductance affected by magnetic saturation and cross-coupling effect, and modeling the variations of these parameters is crucial. While various methods have been used to construct mathematical models of

979-8-3315-1612-3/25 $31.00 © 2025 IEEE

parameter variations, there is a lack of research that includes theoretical examination and quantitative evaluation of the magnetic nonlinearity of high-frequency current responses used for position estimation in the high-frequency voltage injection method. Therefore, in examining a sensorless control method for SynRM that considers magnetic saturation and cross-coupling effect, the sensorless control configuration applying the high-frequency square wave voltage injection method is constructed, and detailed theoretical investigation into the two characteristics in dq-axes of high-frequency current responses affected by magnetic saturation and cross-coupling effect is conducted using electromagnetic field analysis result from FEA. Then, the theoretical examination results is compared with the experimental verification results and a quantitative evaluation is conducted. By utilizing the combined characteristics derived from the two properties on the dq-axes, which are designed to improve the effects of magnetic nonlinearity and cross-coupling terms, it becomes possible to achieve high-precision sensorless control for position estimation.

This proposed method has the following features.

i. This method involves theoretical modeling from the voltage equations in the high-frequency injection method, eliminating the need for complex processing. Therefore, the issues of increased complexity in the control configuration can be solved.

ii. If electromagnetic field analysis can be performed, it is believed that modeling can be achieved regardless of the motor type, making it widely applicable to motors with salient pole structures.

iii. It is possible to model transient inductance changes in high-frequency voltage injection, resulting in high accuracy.

iv. It can directly address magnetic nonlinearity characteristics and cross-coupling effects, and it is believed that this approach can solve response issues caused by delay elements in compensatory control methods that take these factors into account. Additionally, since parameters are not used directly in control, it is robust to parameter mismatches.

This paper theoretically derives the characteristics of high-frequency current responses considering the magnetic nonlinearity and cross-coupling effects in sensorless control for SynRM with significant magnetic saturation. Furthermore, it shows the results of analyzing dq-axes inductance variations using electromagnetic field analysis and describes the results of the quantitative evaluation of the high-frequency current responses of SynRM based on the comparison between the theoretical examination and the experimental verification. In the end, it discusses the sensorless drive results on the actual machine using the proposed method, which employs combined characteristic in a saturation-based position estimation approach.

II. SENSARLESS CONTROL SYSTEM AT LOW-SPEED RANGE

This Chapter describes principle of sensorless control methods at low-speed range using high-frequency square wave injection and derivation of theoretical equations for high-frequency current response considering cross-coupling effect.

The letters shown below are used in deriving the theoretical equations used for position estimation.

V_{dh}, V_{qh}: dq-axes high-frequency voltage
I_d, I_q: dq-axes current φ_d, φ_q: dq-axes flux
I_{dh}, I_{qh}: dq-axes high-frequency current
I_{dch}, I_{qch}: Estimated dq-axes high-frequency current
I_{dh_pp}, I_{qh_pp}: High-frequency current response
L'_{dd}, L'_{qq}: dq-axes inductance
L'_{qd}: Effect of q-axis current on d-axis flux
L'_{dq}: Effect of d-axis current on q-axis flux
p: Differential operator $\Delta\theta$: Estimated position error
v_{ing}: Injected high-frequency voltage
V_{ing}: Integral Calculation Result of v_{ing}
V_h: Amplitude of injected high-frequency voltage
f: Frequency of injected high-frequency voltage

A. High-Frequency Square Wave Injection Method

The sensorless control configuration used in this paper is shown in Fig. 1. It shows that high-frequency square wave voltage is injected for d-axis direction. Where, θ_a is current phase angle. Then, motor model of SynRM to be controlled is shown in Fig. 2. SynRM in Fig. 2 is defined as d-axis in the direction of the iron core and q-axis in the direction of the air gap.

When there is no position error between the actual dq-axes and estimated dq-axes, the high-frequency current is generated only in the d-axis. However, when there is position error, it arise not only in the d-axis but also in q-axis. When the peak and bottom value of triangular wave-shaped high-frequency current is detected and differential is taken, it results in a high-frequency current response, I_{qh_pp} characteristic, represented by $sin2\Delta\theta$, as shown in (11) below.

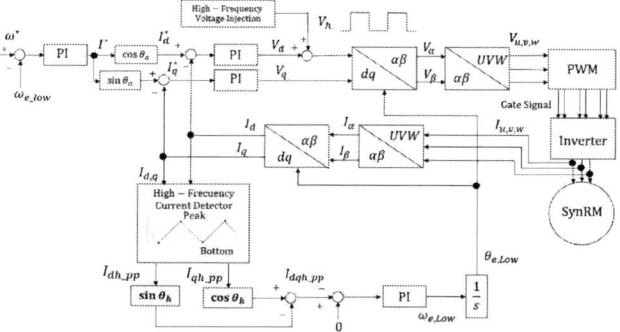

Fig. 1. Sensorless control configuration for SynRM at low-speed range

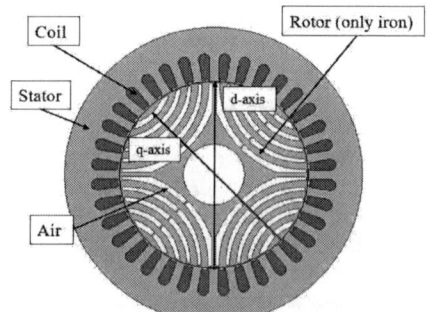

Fig. 2. Motor Model of SynRM used in This Paper.

If the coefficient part remains constant, I_{qh_pp} characteristic is influenced solely by the position estimation error $\Delta\theta$. Unfortunately, I_{qh_pp} characteristic is affected by inductance variation due to magnetic saturation, resulting in nonlinear characteristic with respect to $\Delta\theta$. In position estimation using a PLL configuration that feeds back such position estimation characteristic, achieving high-precision position estimation may become difficult.

B. Derivation of Theoretical Equations

This section describes derivation of theoretical equations of I_{qh_pp} characteristic when high-frequency square wave voltage is injected. In conducting a theoretical investigation of I_{qh_pp} characteristic, the influence of cross-coupling effect typically occurs when driving a motor. Therefore, theoretical equations are derived considering the cross-coupling effect. Voltage equation including cross-coupling effect in the high-frequency voltage injection method is shown in (1).

$$\begin{pmatrix} V_{dh} \\ V_{qh} \end{pmatrix} = \begin{pmatrix} L'_{dd} & L'_{qd} \\ L'_{dq} & L'_{qq} \end{pmatrix} p \begin{pmatrix} I_{dh} \\ I_{qh} \end{pmatrix} \quad (1)$$

To apply the high-frequency voltage injection method, as mentioned earlier, high-frequency current flows along dq-axes due to the injected high-frequency voltage. Therefore, high-frequency current is assumed by using small change in dq-axes current, dI_d, dI_q, and each inductance can be determined from (2) and (3).

$$\begin{pmatrix} L'_{dd} \\ L'_{qq} \end{pmatrix} = \begin{pmatrix} \frac{d\varphi_d}{dI_d} \\ \frac{d\varphi_q}{dI_q} \end{pmatrix} \quad (2)$$

$$\begin{pmatrix} L'_{qd} \\ L'_{dq} \end{pmatrix} = \begin{pmatrix} \frac{d\varphi_d}{dI_q} \\ \frac{d\varphi_q}{dI_d} \end{pmatrix} \quad (3)$$

Then, V_{dh} and V_{qh} with position error is shown in (4) and also v_{ing} is shown in (5).

$$\begin{pmatrix} V_{dh} \\ V_{qh} \end{pmatrix} = \begin{pmatrix} v_{ing} \cos \Delta\theta \\ v_{ing} \sin \Delta\theta \end{pmatrix} \quad (4)$$

$$v_{ing} = \frac{4V_h}{\pi} \sum \frac{\sin 2(2k-1)}{2k-1} \quad (5)$$

Here, the dq-axes high-frequency current on the actual axis can be calculated from (1) and (4), as shown in (6). (for convenience, the detailed calculation process is omitted.)

$$\begin{pmatrix} I_{dh} \\ I_{qh} \end{pmatrix} = \begin{pmatrix} \frac{v_{ing}}{L'_{dd}L'_{qq}-L'_{qd}L'_{dq}} (L'_{qq} \cos \Delta\theta - L'_{qd} \sin \Delta\theta) \\ \frac{v_{ing}}{L'_{dd}L'_{qq}-L'_{qd}L'_{dq}} (L'_{dd} \sin \Delta\theta - L'_{dq} \cos \Delta\theta) \end{pmatrix} \quad (6)$$

Where V_{ing} is shown in (7).

$$V_{ing} = -\frac{2V_h}{\pi^2 f} \sum \frac{\cos 2(2k-1)}{(2k-1)^2} \quad (7)$$

Regarding to (6), the current that can be estimated is the high-frequency current on the estimated axis; therefore, coordinate transformation to the estimated reference frame is performed. Then, I_{dch} and I_{qch} are shown in (8) and (9).

$$I_{dch} = \frac{V_{ing}}{2(L'_{dd}L'_{qq} - L'_{qd}L'_{dq})} \{ (L'_{dd} + L'_{qq}) \\ -(L'_{dd} - L'_{qq}) \cos 2\Delta\theta - (L'_{qd} + L'_{dq}) \sin 2\Delta\theta \} \quad (8)$$

$$I_{qch} = \frac{V_{ing}}{2(L'_{dd}L'_{qq} - L'_{qd}L'_{dq})} \{ (L'_{dd} - L'_{qq}) \sin 2\Delta\theta \\ -(L'_{qd} + L'_{dq}) \cos 2\Delta\theta + (L'_{qd} - L'_{dq}) \} \quad (9)$$

Equation (8) and (9) represent the triangular waveform high-frequency currents, and the characteristics used for position estimation in the high-frequency voltage injection method are the high-frequency current response obtained by taking the difference between the peaks and bottoms. Finally, I_{dh_pp} and I_{qh_pp} are shown in (10) and (11).

$$I_{dh\_pp} = \frac{V_h}{4f(L'_{dd}L'_{qq} - L'_{qd}L'_{dq})} \{ (L'_{dd} + L'_{qq}) \\ -(L'_{dd} - L'_{qq}) \cos 2\Delta\theta - (L'_{qd} + L'_{dq}) \sin 2\Delta\theta \} \quad (10)$$

$$I_{qh\_pp} = \frac{V_h}{4f(L'_{dd}L'_{qq} - L'_{qd}L'_{dq})} \{ (L'_{dd} - L'_{qq}) \sin 2\Delta\theta \\ -(L'_{qd} + L'_{dq}) \cos 2\Delta\theta + (L'_{qd} - L'_{dq}) \} \quad (11)$$

III. CHARACTERISTIC ANALYSIS

In this chapter, the results of electromagnetic field analysis by using JMAG from JSOL are described, and based on these results, The factors contributing to magnetic nonlinearity and cross-coupling effect are identified from the theoretical equations established in the previous chapter. Finally, the proposed method is shown based on the result of the analysis of I_{dh_pp} and I_{qh_pp} characteristics.

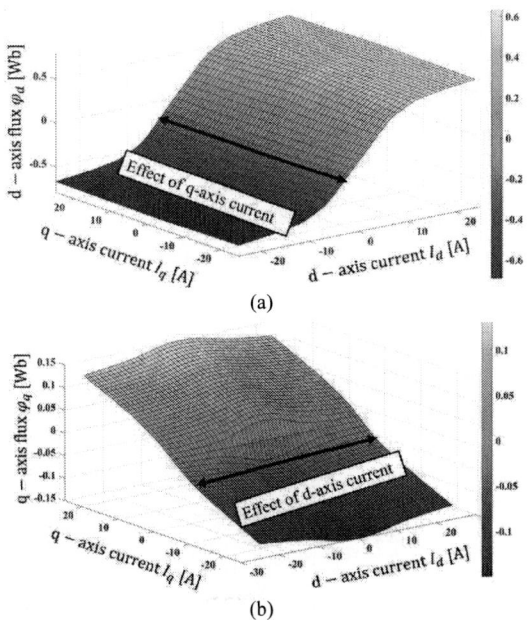

Fig. 3. Characteristics of dq-axes flux to dq-axes current. (a) d-axis flux. (b) q-axis flux.

A. Electromagnetic field analysis results

Fig. 3 shows the results of the analysis of the dq-axes flux using JMAG. During the operation of a conventional motor drive, cross-coupling terms due to cross-coupling effects are present. As a result, the d-axis flux is affected by the q-axis current. Similarly, the q-axis flux is influenced by the d-axis current, and this effect is more pronounced.

Next, the result of L'_{dd} and L'_{qq} derived from (2) in conjunction with the results in Fig. 3 are shown in Fig. 4. Additionally, Fig. 5 illustrates L'_{qd} and L'_{dq} derived from (3) based on the results obtained in Fig. 3. (Where, $dI_d=dI_q=1$.) It is evident that L'_{dd} and L'_{qq} varies significantly with changes in the amount of current due to the effect of magnetic saturation (see Fig. 4). Then, L'_{qd} and L'_{dq} is small compared to L'_{dd} and L'_{qq}, and it is clear that its influence increases with the increase in the amount of current (see Fig. 5).

B. Analysis of Nonlineariy due to Magnetic Saturation

The L'_{qd} and L'_{dq} is considered sufficiently small compared to L'_{dd} and L'_{qq}. Therefore, coefficient part of $sin2\Delta\theta$ in (11), $(L'_{dd}-L'_{qq})/(L'_{dd}L'_{qq}-L'_{qd}L'_{dq})$[1/H] is dominant. $(L'_{dd}-L'_{qq})/(L'_{dd}L'_{qq}-L'_{qd}L'_{dq})$ calculated from L'_{dd}, L'_{qq}, L'_{qd} and L'_{dq} obtained in the previous section is shown in Fig. 6. The unit in this figure is the reciprocal of inductance [1/H] and also this figure shows Maximum per Toruqe Ampere (MTPA) direction. In the current phase of Fig. 6, if a positive position error occurs, the actual current vector changes in a counterclockwise direction. In this case, the amplitude of I_{qh_pp} characteristic dosen't change significantly regardless of the position error or the amount of current. In contrast, when a negative position error occurs, the actual current vector changes in a clock wise direction, and the amplitude of I_{qh_pp} characteristic varies significantly in a decreasing direction, depending on $\Delta\theta$ and the amount of current. Therefore, it is considered to be a source of magnetic nonlinearity of I_{qh_pp} characteristic with respect to $\Delta\theta$ in SynRM with large magnetic saturation.

Equation (11) calculated from L'_{dd}, L'_{qq}, L'_{qd} and L'_{dq} obtained in the previous section is shown in Fig. 7. Equation (10) and (11) changes depending on the voltage and frequency of the injected high-frequency voltage. Where, V_h=50[V], f=1250[Hz] is chosen to obtain characteristics of sufficient magnitude and current phase angle is 50[deg] (from d-axis direction). Firstly, when the amount of current is small, no nonlinearity is observed with respect to $\Delta\theta$. However, as the amount of current increases, the effect of magnetic saturation becomes stronger and the nonlinearity characteristic become more pronounced. Secondly, the second and third terms in (11) contain cross-coupling terms, and zero crossing point shifts, resulting in I_{qh_pp} even if the position error is zero. When the amount of current is small, the effect of cross-coupling terms is minute, about 0.1[A], but as the amount of current is increases, the effect increases to about 0.3[A], which is significant. This is also evident from the characteristics of L'_{qd} and L'_{dq} (see Fig. 5). Finally, when such position estimation characteristic is fed back in the PLL-based position estimation section, I_{qh_pp} characteristic is almost zero when a negative position error occurs.

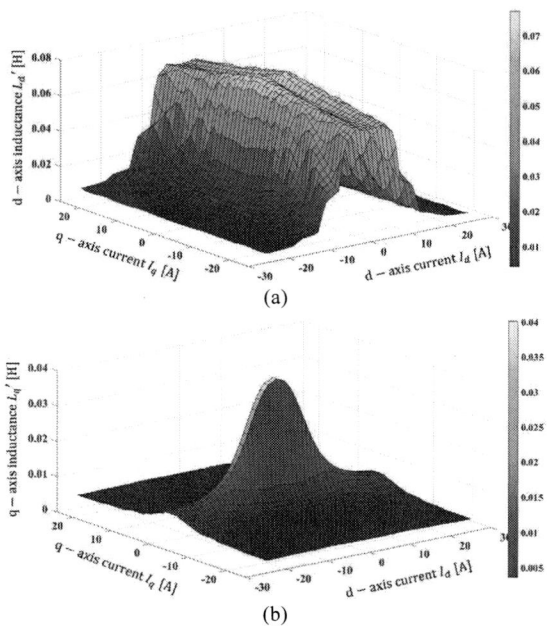

Fig. 4. Characteristics of dq-axes inductance to dq-axes current. (a) d-axis inductance. (b) q-axis inductance.

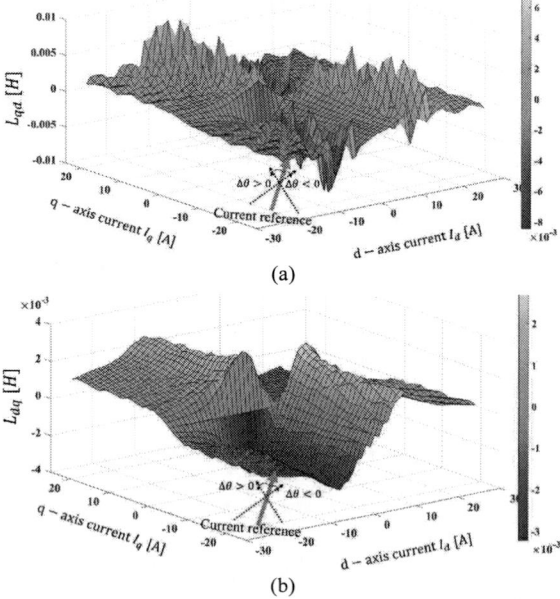

Fig. 5. Characteristics of dq-axes cross-coupling inductance to dq-axes current. (a) cross-coupling inductance L'_{qd}. (b) cross-coupling inductance L'_{dq}

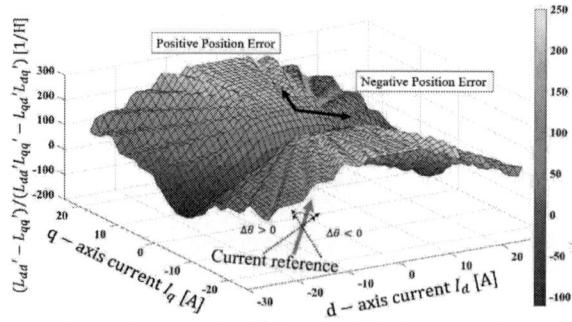

Fig. 6. Characteristic of $(L'_{dd} - L'_{qq})/(L'_{dd}L'_{qq} - L'_{qd}L'_{dq})$.

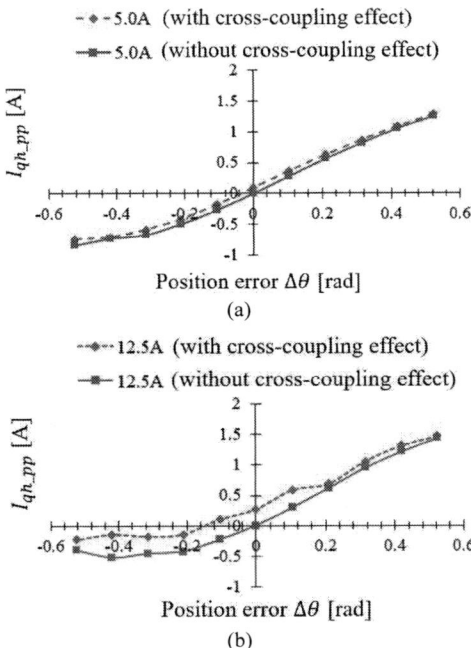

- ◆ - 5.0A (with cross-coupling effect)
- ■ - 5.0A (without cross-coupling effect)

(a)

- ◆ - 12.5A (with cross-coupling effect)
- ■ - 12.5A (without cross-coupling effect)

(b)

Fig. 7. Theoretical result of I_{qh_pp} characteristic to $\Delta\theta$. (The blue line represents the I_{qh_pp} characteristic when the cross-coupling effect is not considered. $L'_{qd} = L'_{dq} = 0$.) (a) Current reference: 5.0 [A]. (b) Current reference: 12.5 [A]

Because of this, position estimation is difficult even if the PI controller tries to make the I_{qh_pp} characteristic to zero.

C. Proposed Sensorless Control Method

In the I_{dh_pp} characteristic shown in (10), the dominant part, $(L'_{dd}+L'_{qq})/(L'_{dd}L'_{qq}-L'_{qd}L'_{dq})$ is shown in Fig. 8. In the current phase of Fig. 8, I_{dh_pp} characteristic increases as the $\Delta\theta$, where I_{qh_pp} characteristic decreases. Then, the result of calculating (10) using L'_{dd}, L'_{qq}, L'_{qd} and L'_{dq} is shown in Fig. 9. From Fig. 7 and 9, the nonlinear characteristic and the problem of the shifted zero-crossing point can be improved by combining I_{dh_pp} characteristic, which can correct I_{qh_pp} characteristic when a negative position error occurs, with the nonlinear I_{qh_pp} characteristic. The combination characteristic I_{dqh_pp} characteristic is shown in (12).

$$I_{dqh\_pp} = I_{qh\_pp} \cos\theta_h - I_{dh_{pp}} \sin\theta_h \qquad (12)$$

Where, θ_h is proportion of the combination. Then, I_{dqh_pp} characteristic can be calculated as shown in Fig. 10 ($\theta_h = 20$[deg]). The combined characteristic shows that the magnetic nonlinearity and the problem of shifted zero-crossing point can be improved. This suggest that I_{dqh_pp} characteristic enables highly accurate sensorless control that takes into account magnetic saturation and cross-coupling effect.

IV. EXPERIMENTAL RESULTS

This chapter describes comparison and evaluation of high-frequency current response characteristics derived from theoretical calculations in the previous chapter with those measured through actual device verification. The experimental setup that was constructed is shown in Fig. 11. (Fig. 11 is

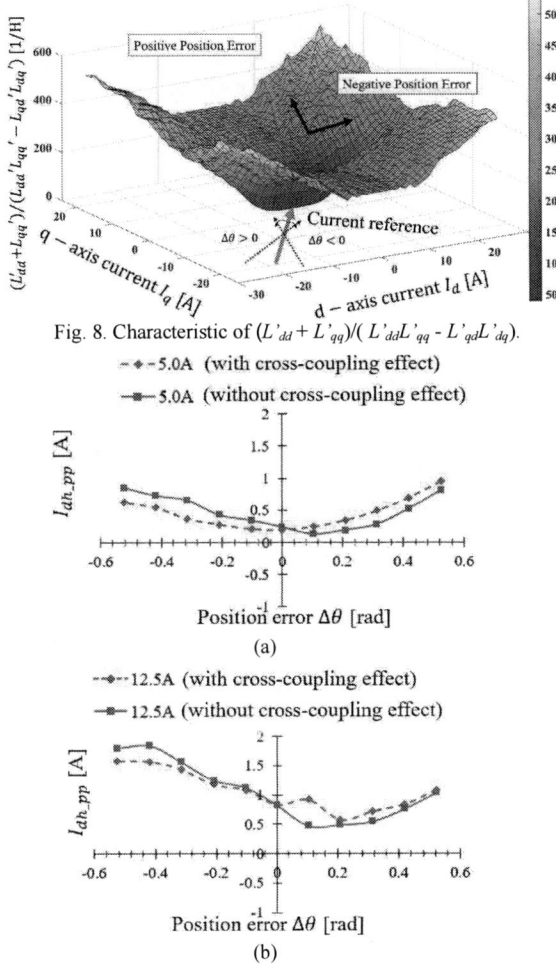

Fig. 8. Characteristic of $(L'_{dd} + L'_{qq})/(L'_{dd}L'_{qq} - L'_{qd}L'_{dq})$.

- ◆ - 5.0A (with cross-coupling effect)
- ■ - 5.0A (without cross-coupling effect)

(a)

- ◆ - 12.5A (with cross-coupling effect)
- ■ - 12.5A (without cross-coupling effect)

(b)

Fig. 9. Theoretical result of I_{dh_pp} characteristic to $\Delta\theta$. (The blue line represents the I_{dh_pp} characteristic when the cross-coupling effect is not considered. $L'_{qd} = L'_{dq} = 0$.) (a) Current reference: 5.0 [A]. (b) Current reference: 12.5 [A]

- ◆ - 12.5A (Theoretical Value)

Fig. 10. Theoretical result of I_{dqh_pp} characteristic to $\Delta\theta$ with cross-coupling effect. (Current reference: 12.5 [A], $\theta_h = 20$[deg])

divided into two because it could not fit into a single figure due to experimental environment. The three-phase output of inverter for SynRM in the upper figure is connected to SynRM in the lower figure.) Then, an IM for the load is connected to the SynRM to be controlled, and the IM is controlled by an inverter for the IM (see Fig. 11). The motor specifications and parameters used in the actual verification are shown in Table 1. Fig. 12 shows the control configuration for verification of the high-frequency current response characteristics.

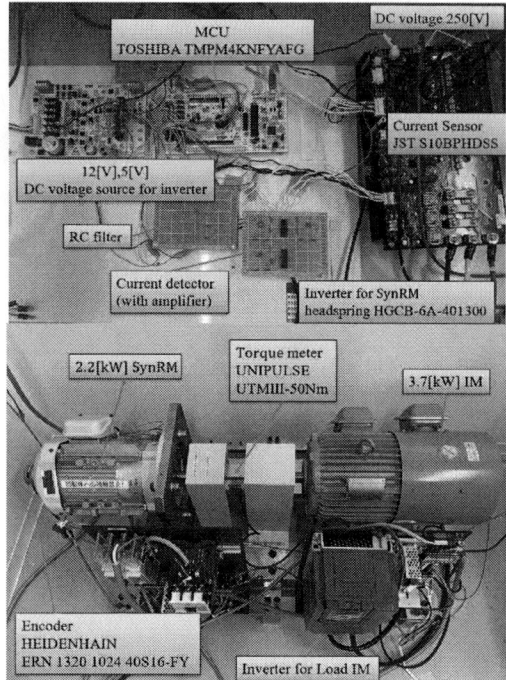

Fig. 11. Experimental environment.

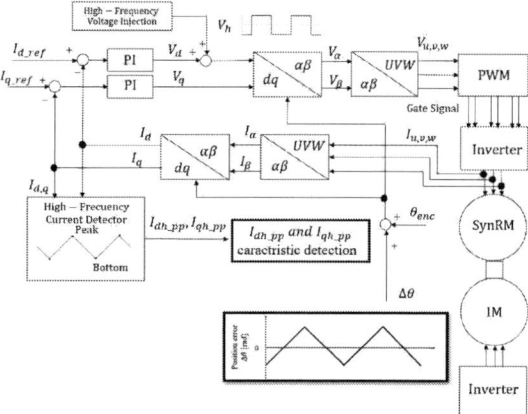

Fig. 12. Control configuration for I_{dh_pp} and I_{qh_pp} characteristics detection.

The SynRM was driven by a constant current, 5.0[A], 12.5[A], and by current control with a sensor. In addition, during the verification, it was performed under constant speed conditions by the load IM. In this case, a range of $\pm\pi/6$ [deg] was given to the detected sensor signal as the magnetic position error to the control system .

A. Results of High-Frequency Current Response

The I_{dh_pp} characteristic and I_{qh_pp} characteristic for $\Delta\theta$ obtained from experimental results is shown in Fig. 13. In the case of current reference 5.0[A], I_{qh_pp} characteristic changes according to $\Delta\theta$ (see Fig. 13(a)). In other words, the effect of magnetic saturation is small and magnetic nonlinearity is not observed. In contrast, in the case of current reference 12.5[A], I_{qh_pp} characteristic hardly changes with respect to the variation of $\Delta\theta$ when a negative position error occurs (see Fig. 13(b)). These indicate that as the amount of current increases, the effect of magnetic saturation becomes stronger and the part insensitive

TABLE I. EXPERIMENTAL CONDITION

Parameters		
Type	*Item*	*Value*
Motor Parameters (SynRM)	Rated Speed	1500 [rpm]
	Rated Current	10.92 [A$_{rms}$]
	Rated Voltage	179.9 [V$_{rms}$]
	Pole Pair Number	2
	Inertia Moment	0.00712 [Kg·m^2]
	Resistance	0.795 [Ω]
Drive Condition	Rotation Speed	150 [rpm]
	Current Phase Angle	50 [deg]
	High-Frequency Voltage	50 [V]
	Injected Voltage Frequency	1250 [Hz]
Control parameter	Current Control gain (d-axis) K_{pcd}, K_{icd}	30, 600
	Current Control gain (q-axis) K_{pcq}, K_{icq}	10, 600
Inverter	Carrier Frequency	10 [kHz]
	Inverter DC Voltage	250 [V]

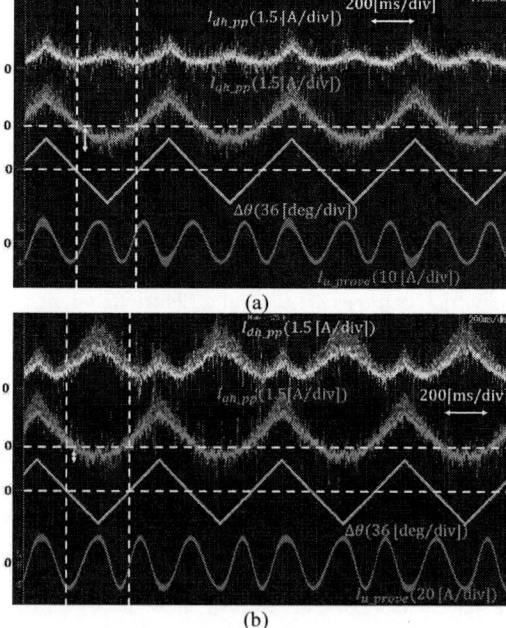

Fig. 13. Experimental result of I_{qh_pp} characteristic to $\Delta\theta$. (a) Current reference: 5.0 [A]. (b) Current reference: 12.5 [A]

to $\Delta\theta$ is also found. Then, as for the I_{dh_pp} characteristic, its features do not change significantly in Fig. 13.

B. Consideration of Validity

The results of the comparison of I_{dh_pp} characteristic and I_{qh_pp} characteristic obtained by theoretical investigation and obtained by actual experiment are shown in Fig. 14 and Fig. 15. In the case of current reference 12.5[A], the maximum error in I_{qh_pp}

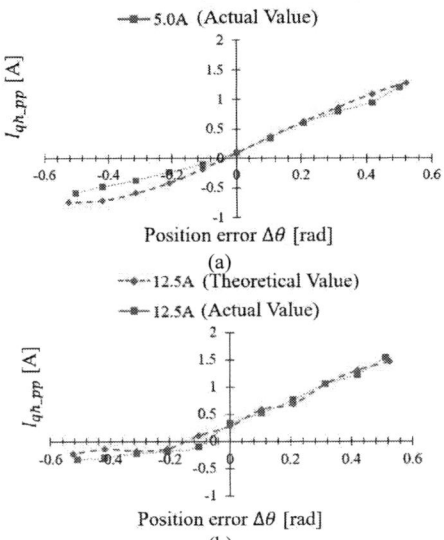

Fig. 14. Comparison of theoretical and actual I_{qh_pp} characteristic to $\varDelta\theta$. (a) Current reference: 5.0 [A]. (b) Current reference: 12.5 [A]

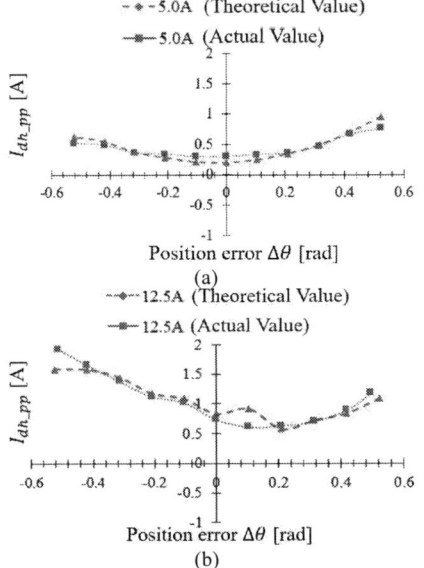

Fig. 15. Comparison of theoretical and actual I_{dh_pp} characteristic to $\varDelta\theta$. (a) Current reference: 5.0 [A]. (b) Current reference: 12.5 [A].

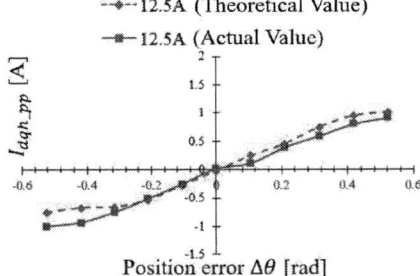

Fig. 16. Theoretical result of I_{dqh_pp} characteristic to $\varDelta\theta$ with cross-coupling effect. (Current reference: 12.5 [A], $\theta_h = 20$[deg])

characteristic between theoretical value and actual value was 0.15[A] and the maximum error in I_{dh_pp} characteristic was 0.18[A]. In the case of current reference 5.0[A], the maximum error in I_{qh_pp} characteristic was 0.23[A] and the maximum error in I_{qh_pp} characteristic was 0.19[A]. From these results, it can be concluded that similar characteristics was obtained, suggesting that the modeling of the high-frequency current response was successful. In addition, the validity of the modeling method for high-frequency current response, which includes the cross-coupling effect used in this paper, was confirmed. Then, the result of comparison of I_{dqh_pp} characteristic is shown in Fig. 16 ($\theta_h = 20$[deg]). Similar characteristics were obtained, and issues such as magnetic nonlinearity and shifted zero-crossing points were improved in the actual verification.

C. Experimental Result of Proposed Method

This section describes experimental result of proposed method that used combination characteristic for position estimation. The motor drive conditions involved speed control using the proposed method. Experimental result is shown in Fig. 17, 18, 19, and speed control gain were $K_{ps}=1.3$, $K_{is}=5.0$, current control gain were $K_{pcd}=10$, $K_{icd}=10$, $K_{pcd}=5$, $K_{icd}=100$, and sensoless gain were $K_{p\_Vh}=15$, $K_{i\_Vh}=90$.

In Fig. 17, SynRM was operated with a speed reference, 300[rpm], and the load was gradually increased to maximum 12[Nm]. Current ripple was occurring, which was due to the effect of the injected high-frequency voltage. The sensorless drive, which used only the I_{qh_pp} characteristic, showed an increase in position estimation error as the load increased, ultimately leading to a loss of synchronization. This occurred because the amount of current increased as the load rose, leading to issue such as the shifted zero-crossing point due to cross-coupling effect and magnetic nonlinearity caused by magnetic saturation.

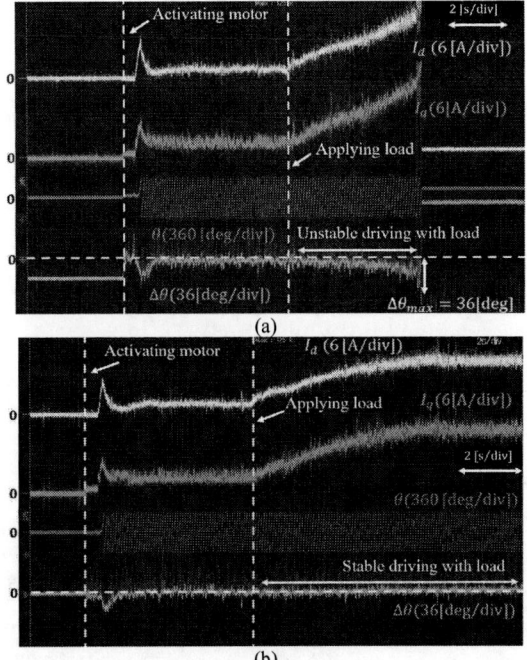

Fig. 17. Operation result using proposed sensorless control method or not. (θ: estimated position.) (a) Operation result using only I_{qh_pp} characteristic (b) Operation result using I_{dqh_pp} characteristic.

In other words, it is difficult to construct a sensorless control scheme that takes magnetic saturation and cross-coupling effect into account using only I_{qh_pp} characteristic. In contrast, when using I_{dqh_pp} characteristic, the position error remained relatively stable even as the load increased, enabling reliable sensorless drive operation .

In Fig. 18, SynRM was operated with a constant speed reference, 300[rpm], and the experimental results were shown for when load was varied between 0[Nm], 12[Nm] and 0[Nm]. $\Delta\theta$ was sufficiently small and stable when a step load was applied and when the load was suddenly removed. Equation (11) showed the estimation accuracy increased when V_h became larger or the frequency decreased. Then, in Fig. 18(b), it showed that $\Delta\theta$ became small when V_h=75[V]. However, increasing the voltage resulted in a larger ripple, which limited stability, so it was necessary to set an optimal value, and the same applied to the frequency.

In Fig. 19, the SynRM was operated with load torque, 12[Nm], and the experimental results were shown for when

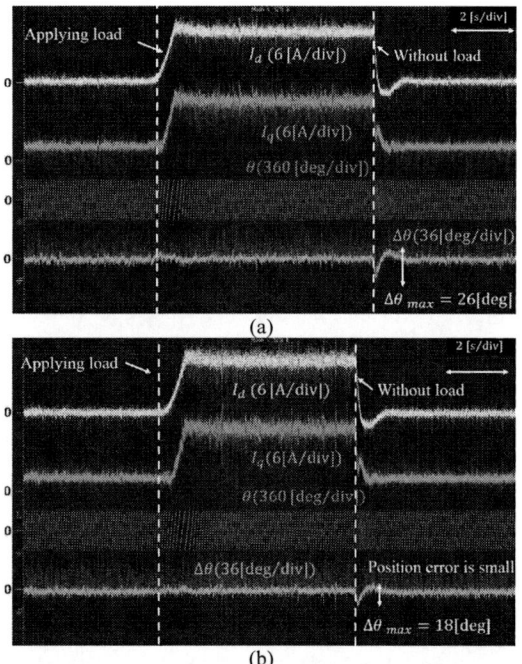

Fig. 18. Operation results at a constant speed of 300[rpm] when the load torque was varied. (θ: estimated position, θ_{max}: maximum position error) (a) V_h =50[V], (b) V_h =75[V]

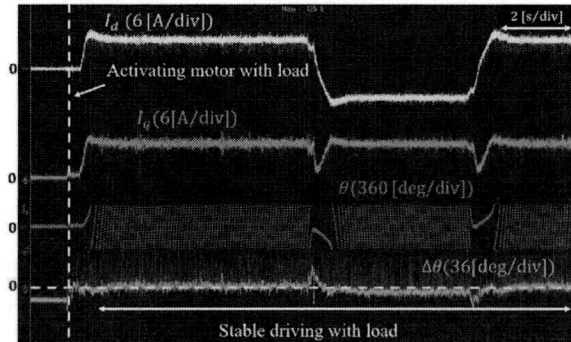

Fig. 19. Operation result at a constant load torque of 12[Nm] when the speed was varied. (θ: estimated position)

speed was varied between 300[rpm], -300[rpm] and 300[rpm]. Even when alternating between forward and reverse rotation under load, $\Delta\theta$ during the transient state was found to be sufficiently small.

V. CONCLUSION

This paper focuses on the development of a sensorless control configuration for SynRM with large magnetic saturation. First, theoretical equations for high-frequency current response considering magnetic saturation and cross-coupling effect in the high-frequency square wave injection method is derived. Second, the high-frequency current response is modeled using the derived theoretical equations and the result of electromagnetic field analysis. Finally, by combining the two modeled high-frequency responses, a sensorless control method for SynRM is proposed to improve the effect of magnetic nonlinearity and cross-coupling terms. The performance of proposed method is validated by experiments on 2.2-kW SynRM drive platform, and the effectiveness of the proposed method is demonstrated.

REFERENCES

[1] Masayuki Nashiki, Yoshimitu Inoue, Youihi Kawai and Shigeru Okuma, "Inductance Calculation and New Modeling of a Synchronous Reluctance Motor Using Flux Linkages", IEEJ Trans. IA, Vol.127, No.2, pp.1212-1219 (2007) (in Japanese)

[2] Shu Yamamoto, Takashi Kano, Yoshihiro Yamaguchi and Takahiro Ara, "A Method to Determine Direct- and Quadrature-Axis Inductances of Permanent Magnet Synchronous Motors", IEEJ Trans. IA, Vol.128, No.7, pp.910-918 (2008) (in Japanese)

[3] G. Wang, D. Xiao, N. Zhao, X. Zhang, W. Wang, and D. Xu, "Low frequency pulse voltage injection scheme-based sensorless control of IPMSM drives for audible noise reduction," IEEE Trans. Ind. Electron., vol. 64, no. 11, pp. 8415–8426, Nov. 2017.

[4] P. Guglielmi, M. Pastorelli, and A. Vagati, "Impact of cross-saturation in sensorless control of transverse-laminated synchronous reluctance motors," IEEE Trans. Ind. Electron., vol. 53, no. 2, pp. 429–439, Apr. 2006.

[5] E. Capecchi, P. Guglielmi, M. Pastorelli, and A. Vagati, "Position sensorless control of the transverse-laminated synchronous reluctance motor," IEEE Trans. Ind. Appl., vol. 37, no. 6, pp. 1768–1776, Nov./Dec. 2001.

[6] Shigeo Morimoto, Masayuki Sanada and Yoji Takeda: "Mechanical Sensorless Drives of IPMSM With Online Parameter Identification", IEEE TRANSACTIONS ON INDUSTRY APPLICATIONS, Vol.42, No.5, pp.1241-1248 (2006)

[7] P. Garcia, D. Reigosa, F. Briz, D. Raca, and R. D. Lorenz, "Automatic self-commissioning for secondary-saliencies decoupling in sensorless controlled ac machines using structured neural networks," in Proc. 2007 IEEE Int. Symp. Ind. Electron., 2007, pp. 2284–2289.

[8] Masayuki Nashiki, Yoshimitu Inoue, Youihi Kawai and Shigeru Okuma: "Inductance Calculation and New Modeling of a Synchronous Reluctance Motor Using Flux Linkages", IEEJ Trans. IA, Vol.127, No.2, pp.1212-1219 (2007) (in Japanese)

[9] Marko Hinkkanen, Paolo Pecetto, Eemeli Mölsä, Seppo E. Saarakkala, Gianmmario Pellegrino and Radu Bojoi: "Sensorless Self-Commissioning of Synchronous Reluctance Motors at Standstill Without Rotor Locking", IEEE TRANSACTIONS ON INDUSTRY APPLICATIONS, Vol.53, No.3, pp.2120-2129 (2017)

[10] Marth Bugsch, Dminik Großmann and Bernhard Piepenbreier: "Voltage Square-Wave-Injection-Based HF Parameter Identification Method for Sensorless Control of a Synchronous Reluctance Machine", EPE'17 ECCE Europe, pp.1-8 (2017)

[11] Chengrui Li, Gaolin Wang, Guoqiang Zhang and Dianxun Xiao: "Saliency-Based Sensorless Control of Position Estimation Error", IEEE TRANSACTIONS ON INDUSTRIAL ELECTRONICS, Vol.66, No.8, pp.5839-5849 (2019)

Hybrid Control Scheme for Permanent Magnet Gear Motor

Bing Li, Takayoshi Matsuo, Ahmed Sayed-Ahmed, Yujia Cui, Jiangang Hu
Intelligent Devices, Rockwell Automation
Mequon, WI 53092, USA

Abstract—A gear motor is a special motor that integrates an AC motor and gearbox within a single housing. Typically, a permanent magnet (PM) motor is used in a gear motor. Due to its special construction, high torque accuracy, minimal torque ripple and a sensorless control scheme has to be considered. Consequently, a hybrid control scheme is proposed in this paper for a PM gear motor. At low speed, an I/F control is proposed to generate high starting torque and stable speed control. For medium to high-speed range, a flux vector (FV) sensorless control with an adaptive observer is proposed to achieve high efficiency and accurate torque control. During the transition between low and high speed, a blending method is used to seamlessly transfer from I/F to FV control during speed ramp-up, and vice versa during ramp-down. Experiment results, obtained from a general variable speed drive with a PM gear motor, validate the robust and high performance of this control scheme.

Index terms—PM gear motor, hybrid control, IF control, flux vector control, adaptive observer, sensorless control, variable speed drive

I. INTRODUCTION

To develop a variable speed AC drive to operate a permanent magnet (PM) gear motor, certain special considerations must be addressed. Firstly, due to the large gear ratio, the inertia perceived by the motor is very low, which requires precise torque control to achieve smooth speed regulation. Secondly, installing an encoder on the motor shaft inside a gear motor is not feasible, making sensorless control a must. Lastly, the backlash phenomena, caused by the small space between two inter-meshing gear teeth, requires the minimal torque ripple to ensure smooth speed control in this application.

Sensorless PM control has been extensively researched over the past several decades. There are two major schemes: flux vector (FV) control and scalar control. For the FV scheme, the rotor position must be known, which can be estimated by means of fundamental excitation method at medium to high speed or high frequency signal injection (HFI) method at low speed. Several fundamental excitation methods are developed in the literature, including extended back electromotive force (BEMF) method [1], [2], adaptive observer method [3], sliding mode observer [4], etc. The HFI method is utilized from zero to medium speed. However, the HFI method's drawbacks include high torque ripple due to signal injection and complex, heavy computation load due to the filters and signal processing involved. The scalar control consists of two methods, V/F and I/F control. V/F control on a PM motor is unstable by nature and the PM can easily lose synchronization under abrupt velocity and load changes. Conversely, I/F directly controls the current of the motor by current regulator, producing stable torque. Based on the analysis above, I/F control at low speed and adaptive observer based FV sensorless scheme at medium and high speed have been selected for PM gear motor control in this paper.

Research focusing specifically on PM gear motor is limited, with more attention on general PM control. A method for adjusting the frequency slope during ramp-up using I/F control is proposed in [5]; however, the addition of filters and torque estimation increase the complexity of the otherwise simple I/F control. In [6], reactive power is monitored to regulate the current reference via a PI controller, enhancing efficiency in I/F control. Another method, detailed in [7], suggests regulating the current reference through a torque angle loop, utilizing instantaneous power estimation to bolster stability. [8] introduces a controller that modulates the electrical frequency in I/F control for high-speed PM motors. The transition method between I/F and FV controls is the key for this hybrid control scheme. [9] proposes maintaining a constant speed reference while linearly decreasing the Q-axis current during the transition. In contrast, [10] advocates for a nonlinear current reduction function to facilitate the transition. Furthermore, [11] utilizes a first-order lag compensator to guarantee a seamless transition of the electrical angle from I/F to FV control. For medium to high speed range, Back-EMF sensorless scheme is used in all above papers.

Different from methods used in previous research, a simple I/F control is utilized in this paper with predefined current amplitude, ensuring sufficient starting torque and stable speed control. In the medium to high-speed range, an adaptive observer based FV sensorless control scheme is first explored. Furthermore, an innovative blending method for the electrical angle is introduced for the transition between I/F and FV control, facilitating smooth speed ramp-up and ramp-down.

979-8-3315-1612-3/25 $31.00 © 2025 IEEE

During these transition, a torque estimation method is proposed within I/F control to reset the integrator of speed regulator, ensuring smooth speed regulation.

II. MODEL OF PM MOTOR

A generalized voltage model for PM synchronous motor is presented in dq reference frame as follows:

$$V_d = R_s I_d + L_d \frac{\mathrm{d}I_d}{\mathrm{d}t} - \omega L_q I_q$$
$$V_q = R_s I_q + L_q \frac{\mathrm{d}I_q}{\mathrm{d}t} + \omega(\psi_p + L_d I_d) \tag{1}$$

where $V_{d,q}$, $I_{d,q}$, $L_{d,q}$ are the components of voltage, current and inductance in dq frame, R_s is the stator resistance, ω is the electrical speed of motor, and ψ_p is the rotor flux linkage. Torque is produced as follows:

$$T = \frac{3p}{2}\psi_p I_q + \frac{3p}{2}(L_d - L_q)I_d I_q \tag{2}$$

where p is the number of pole pairs. The first portion on the right of Eqn. 2 is magnetic torque and the second portion is reluctance torque.

For surface permanent magnet (SPM) synchronous motor, $L_{d,q}$ are equal. For interior permanent magnet (IPM) synchronous motor, L_q is larger than L_d. IPM produces both magnetic and reluctance torque, whereas SPM only produce magnetic torque because $L_{d,q}$ are equal.

III. PROPOSED CONTROL ALGORITHM

A. I/F Control at Low Speed

I/F control scheme, as shown in Fig. 1, operates in two stages. The first stage is rotor alignment, where the electrical angle θ_e is kept at a constant position. The direct current reference, $I_{d_{ref}}$ is set to zero, and the quadrature current reference $I_{q_{ref}}$ is assigned and ramping up to a predefined level sufficient to rotate the shaft. This predefined value varies based on the application; for example, if the rated starting torque is required, $I_{q_{ref}}$ could be set equal to motor rated current. In the second stage, the speed reference will ramp up with a fixed ramping rate, θ_e is generated by integration of ramped speed reference. Concurrently, an adaptive observer, described in the next section, operates in the background to estimate both the rotor position and the torque.

Fig. 2 illustrates the dq axis used in the I/F control and FV control under load condition. θ_L denoted the angle difference between these two reference frames, θ_{IF} represents the angle generated and used during I/F control, and θ_{FV} is the angle generated and used during FV control.

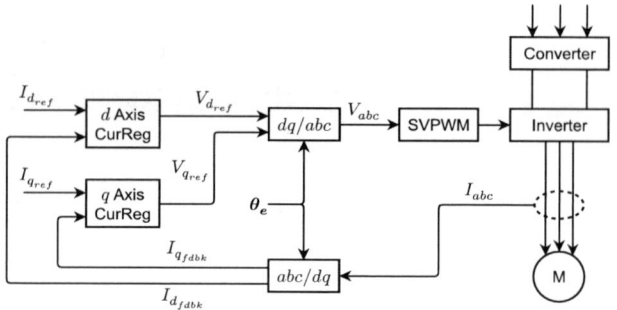

Fig. 1. I/F control scheme

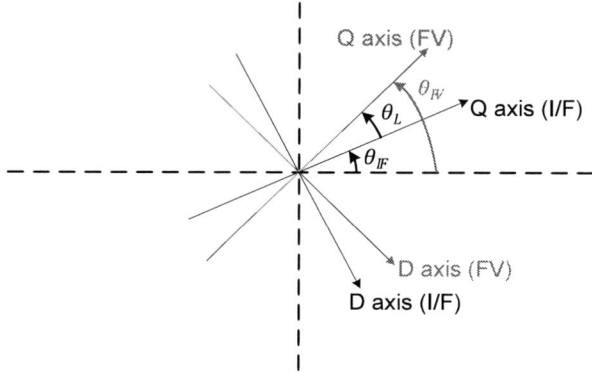

Fig. 2. Angle defined in I/F and FV control

The estimated torque can be calculated as follows:

$$T = \frac{3p}{2}\psi_p I_q \cos(\theta_L) \tag{3}$$

At this stage, a speed regulator is unnecessary and can be deactivated. However, for medium and high speeds in the next stage, the speed regulator must be reset and activated. This estimated torque can be used to reset the integrator of the speed regulator, assuming a simple PI controller is used for speed regulator.

B. FV Control with Adaptive Observer

At medium to high speed range, an adaptive observer is used to estimate rotor position and speed, which are based on the estimation error between the actual motor and observer. The control structure is depicted in Fig. 3 . The PM voltage model, as stated in Eqn. 1, can be expressed in a state space form as

$$\dot{x} = Ax + Bu$$
$$y = Cx \tag{4}$$

where x is state variable, stator flux $\psi_s = \begin{bmatrix} \psi_d & \psi_q \end{bmatrix}^T$; u is input, stator voltage $V_s = \begin{bmatrix} V_d & V_q \end{bmatrix}^T$; y is output, and stator current $I_s = \begin{bmatrix} I_d & I_q \end{bmatrix}^T$.

$$A = \begin{bmatrix} -\frac{R_s}{L_d} & \omega \\ -\omega & \frac{R_s}{L_q} \end{bmatrix}, \quad B = \begin{bmatrix} 1 & 0 \\ 0 & 1 \end{bmatrix}, \quad C = \begin{bmatrix} \frac{1}{L_d} & 0 \\ 0 & \frac{1}{L_q} \end{bmatrix} \tag{5}$$

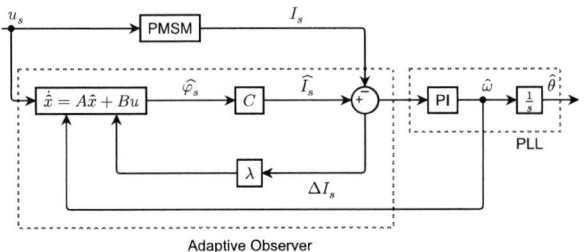

Fig. 3. Adaptive observer

Based on this, an observer can be constructed and presented by

$$\dot{\hat{x}} = A\hat{x} + Bu + \lambda \Delta I_s$$
$$\hat{y} = C\hat{x}$$

(6)

where, ΔI_s represents the stator current error, which is the difference between the actual motor current I_s and the estimated current $\widehat{I}_s$ from observer, while λ denotes the observer gain. Signals estimated by the observer are indicated by $\widehat{}$ and expressed in the estimated rotor reference frame.

There are various ways to select observer gain and choose ΔI_d or ΔI_q to derive the error term from the current error. In this instance, a simple voltage model is selected, only ΔI_q are selected for adaption with $\lambda = R_s$. A stability analysis confirms the system remains stable under these conditions.

The electrical angle and speed estimates are obtained using a phase-locked loop (PLL) as illustrated in Fig. 3. A proportional-integral (PI) regulator generates the estimated speed $\hat{\omega}$ based on the current error. By integrating the estimated speed, the estimated rotor position is determined, and the estimated speed is subsequently fed back to the observer.

C. Transition Between I/F and FV Control

When the motor speed ramps up above a threshold, the transition phase starts. Electrical angle θ_e will smoothly transfer from the angle used in I/F control θ_{IF} to θ_{FV}, which is estimated angle by observer and used in FV control as shown

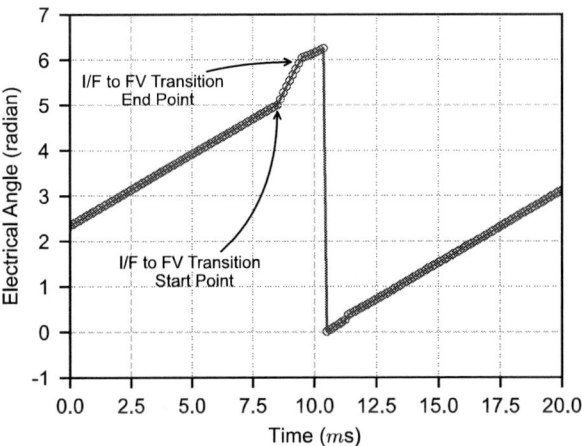

Fig. 4. Angle transition

in Fig. 2. To facilitate a smooth transition, an angle blending method is proposed. Fig. 4 is the angle transition captured in an experiment test.

- $\theta_e = \theta_{IF}$ at low speed I/F control
- $\theta_e = \theta_{FV}$ at high speed FV control
- $\theta_e = K\theta_{IF} + (1 - K)\theta_{FV}$.

During the transition from I/F to FV control, K increases linearly from 0 to 1, and decreases linearly from 1 to 0 during the transition from FV to I/F control.

IV. EXPERIMENT RESULTS

Finally, the proposed method has been implemented and tested in a drive-motor setup, as shown in Fig. 5. The gear motor under test contains a PM motor within its housing, and its nameplate data is listed in TABLE I. The rated speed of the gear motor is 108 RPM, which indicates a gear ratio of 18.5. A 2HP AC drive, with the proposed method implemented in control firmware, controls the gear motor, the output shaft of gear motor is coupled with a DC motor that servers as the load.

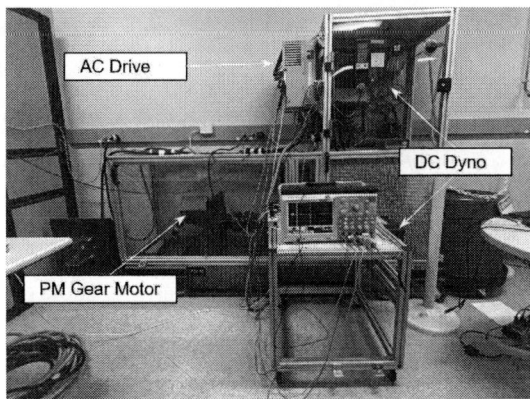

Fig. 5. Experiment setup

TABLE I. PM Motor Nameplate Data

Parameter	Value
Nominal power	1HP
Nominal voltage	400V
Nominal current	1.8A
Nominal Frequency	167Hz
Nominal Speed	2000RPM
Number of Poles	10
Stator Resistance	5.17Ω
Inductance	47.3mH
Inertia	0.000226Kgm^2

Fig. 6 illustrates a test case where the motor is ramped up to its rated speed of 167Hz without any load. The yellow signal

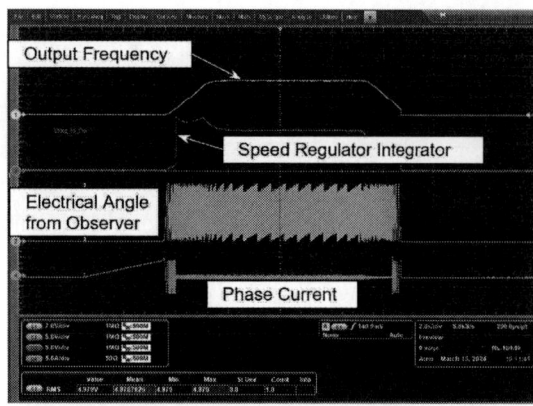

Fig. 6. Ramp up to 167Hz without load

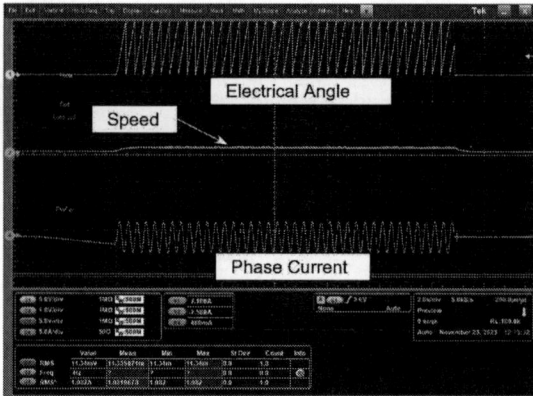

Fig. 8. Ramp up to 3Hz with 100% load

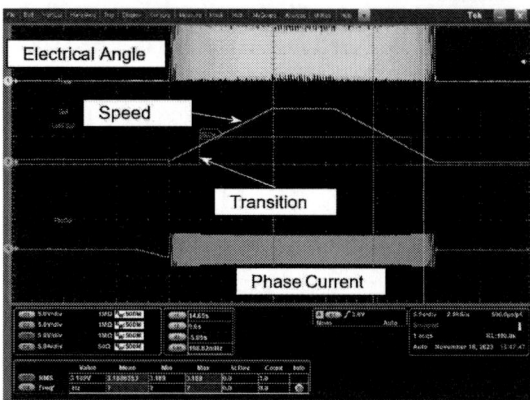

Fig. 7. Ramp up to 167Hz with 100% load

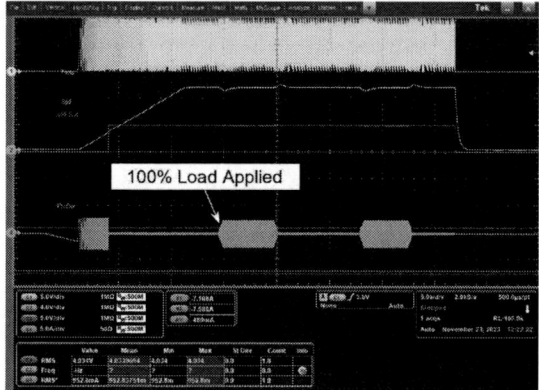

Fig. 9. Step load test at 167Hz

indicates the output frequency, which remains sufficiently smooth during both ramp-up and ramp-down that the transition is almost imperceptible. The pink signal denotes the integrator of speed regulator, which is reset at the moment when the control scheme switches from I/F to FV. The purple signal represents the estimated angle by the observer, while the green signal shows one phase of the motor current. This current ramps up initially to generate sufficient torque and then reduces to a lower level after the transition of control scheme to FV.

Fig. 7 shows a test case of ramping up the motor to the rated speed of 167Hz with 100% load. The yellow signal is the electrical angle, at low speed it is the angle generated in I/F control scheme, while at high speed, it becomes the estimated rotor angle generated by adaptive observer; The cyan signal denotes the estimated motor speed, at low speed it is the ramped speed reference, while at high speed , it becomes the estimated rotor speed from the adaptive observer. The purple signal indicates the transition moment of control scheme from I/F to FV during motor ramp-up, and vice versa during ramp-down. The green signal is the phase current. Initially, the current ramps up to the rated 1.8A, followed by the speed ramp-up. Once the speed exceeds a threshold, the control scheme is switched from I/F to FV observer mode. It demonstrates the capability of starting torque 100%, and stable speed control cross all speed

range and during transition. Fig. 8 shows the low speed (3Hz) performance with 100% load, electrical angle and speed are stable. Fig. 9 is the 100% step load test at the rated speed of 167Hz.

V. CONCLUSIONS

PM gear motors are widely used in industry. By analyzing the characteristics of this type of gear motor, a hybrid control scheme is proposed. At low speed, a simple I/F control is employed, while at high speed, FV control with an adaptive observer method is utilized. Additionally, a blending transition scheme in between I/F and FV control is implemented during ramp-up and ramp-down. The experiment results validated the high performance of this approach - high starting torque, stable and good dynamic speed control across all speed ranges, both with and without load.

REFERENCES

[1] Z. Chen, M. Tomita, S. Doki, and S. Okuma, "An extended electromotive force model for sensorless control of interior permanent-magnet synchronous motors," *IEEE transactions on Industrial Electronics*, vol. 50, no. 2, pp. 288–295, 2003.

[2] S. Morimoto, K. Kawamoto, M. Sanada, and Y. Takeda, "Sensorless control strategy for salient-pole PMSM based on extended EMF in

rotating reference frame," in *Conference Record of the 2001 IEEE Industry Applications Conference. 36th IAS Annual Meeting (Cat. No. 01CH37248)*, 2001, pp. 2637–2644.

[3] A. Piippo, M. Hinkkanen, and J. Luomi, "Analysis of an adaptive observer for sensorless control of interior permanent magnet synchronous motors," *IEEE Transactions on Industrial Electronics*, vol. 55, no. 2, pp. 570–576, 2008.

[4] F. Parasiliti, R. Petrella, and M. Tursini, "Sensorless speed control of a PM synchronous motor by sliding mode observer," in *ISIE'97 Proceeding of the IEEE International Symposium on Industrial Electronics*, 1997, pp. 1106–1111.

[5] S. V. Nair, K. Hatua, N. D. Prasad, and D. K. Reddy, "A smooth and stable open-loop IF control for a surface mount PMSM drive by ensuring controlled starting torque," in *IECON 2018-44th Annual Conference of the IEEE Industrial Electronics Society*, 2018, pp. 355–360.

[6] H. Shen and C. Zhang, "A new efficient sensorless I/f control method for IPMSM drives," in *2017 IEEE 26th International Symposium on Industrial Electronics (ISIE)*, 2017, pp. 209–213.

[7] J. Yang, W. Huang, R. Cao, and X. Jiang, "A closed-loop I/f sensorless control based on current vector orientation for permanent magnet synchronous motors," in *2015 18th International Conference on Electrical Machines and Systems (ICEMS)*, 2015, pp. 1609–1614.

[8] A. Borisavljevic, H. Polinder, and J. Ferreira, "Realization of the I/f control method for a high-speed permanent magnet motor," in *The XIX International Conference on Electrical Machines-ICEM 2010*, 2010, pp. 1–6.

[9] Z. Wang, K. Lu, and F. Blaabjerg, "A simple startup strategy based on current regulation for back-EMF-based sensorless control of PMSM," *IEEE Transactions on Power Electronics*, vol. 27, no. 8, pp. 3817–3825, 2012.

[10] M. Fatu, R. Teodorescu, I. Boldea, G.-D. Andreescu, and F. Blaabjerg, "IF starting method with smooth transition to EMF based motion-sensorless vector control of PM synchronous motor/generator," in *2008 IEEE Power Electronics Specialists Conference*, 2008, pp. 1481–1487.

[11] C. L. Baratieri and H. Pinheiro, "An IF starting method for smooth and fast transition to sensorless control of BLDC motors," in *2013 Brazilian Power Electronics Conference*, 2013, pp. 836–843.

Cost-Effective Fault Diagnosis for Motor and Inverter Using Bootstrap Charging and Single DC Link Current Sensor

Gyu Cheol Lim
Dept. of Electrical and Computer Engineering
Seoul National University
Seoul, Korea
gclim@snu.ac.kr

Won Hyo Jeong
Dept. of Electrical and Computer Engineering
Seoul National University
Seoul, Korea
wonhyo0428 @snu.ac.kr

Kahyun Lee
Dept of Electronic and Electrical Engineering
Ewha Womans University
Seoul, Korea
kh.lee@ewha.ac.kr

Jung-Ik Ha
Dept. of Electrical and Computer Engineering
Seoul National University
Seoul, Korea
jungikha@snu.ac.k

Abstract—**Motor drive systems in home appliances are meticulously designed with minimal sensors and components to maximize reliability, reduce costs, and minimize size. These systems often operate continuously, making fault detection critical to ensure rapid maintenance and extended system longevity. Among the components, the bootstrap configuration—commonly adopted for driving the upper-side switches—requires charging of the bootstrap capacitor during the initial stage of operation. This paper introduces a novel fault diagnosis method for motor drive systems that leverages a single DC link current sensor and the existing bootstrap topology without requiring additional sensors or circuits. The proposed fault diagnosis is performed during the bootstrap charging phase by observing the current flowing through the sensor. The charging impedance and corresponding current under different fault conditions are analyzed and classified, enabling accurate detection and distinction between motor and inverter faults. Experimental validation demonstrates the effectiveness and practicality of the approach.**

Keywords—bootstrap, fault diagnosis, single dc link current sensor, single shunt

I. INTRODUCTION

In many industrial applications, motor drive systems are widely used and have become essential in home appliances such as fans, washing machines, air conditioners, and refrigerators. These systems are typically designed to balance cost, efficiency, and reliability to meet the demands of the consumer electronics market. To reduce manufacturing and maintenance costs while enhancing system reliability, most motor drive systems in home appliances minimize the use of sensors [1]-[3]. One common strategy involves replacing multiple phase current sensors with a single DC link current sensor, as shown in Fig. 1. This approach not only reduces the overall component count but also simplifies the circuit design and improves system robustness.

In addition to reducing sensors, many motor drive systems adopt a bootstrap configuration to drive the high-side switches of inverters as shown in Fig. 1. This widely used configuration eliminates the need for three separate DC-DC converters, significantly reducing component complexity and cost [4]-[7]. The bootstrap capacitor charges during specific intervals of inverter operation, making it an integral part of the switching

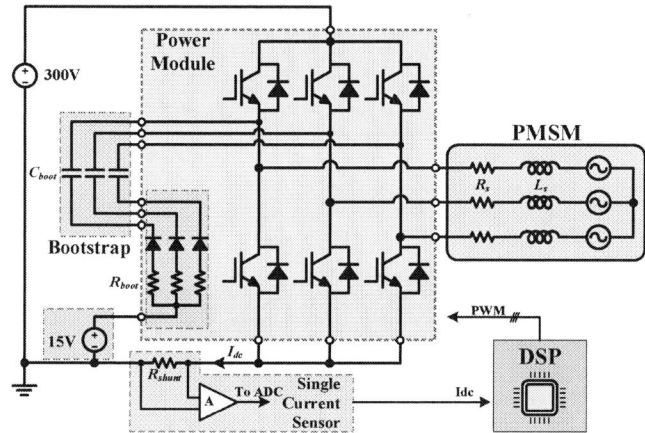

Fig. 1. Motor drive system consisting of a single dc link current sensor and bootstrap configuration.

mechanism. However, this reliance on bootstrap charging introduces diagnostic opportunities that remain underexplored in existing literature.

Home appliances often operate continuously for extended periods, making system reliability and fault detection critical to ensuring consistent performance and longevity [8]-[11]. While a significant body of research focuses on inverter and motor fault detection, most studies rely on multiple current sensors or complex external diagnostic circuits or even machine learning methods [12]-[16]. Few methods have been proposed that leverage a single DC link current sensor for comprehensive fault diagnosis [17]-[20]. Furthermore, most existing techniques primarily address inverter open-circuit faults and are limited in their ability to differentiate between motor and inverter faults. This limitation can lead to unnecessary component replacements, increased downtime, and higher maintenance costs.

Accurate fault identification is essential for targeted maintenance and reducing unnecessary costs. Differentiating between motor and inverter faults using minimal components aligns with the design philosophy of home appliances, which prioritize cost-effectiveness and simplicity. Fault diagnosis

979-8-3315-1612-3/25 $31.00 © 2025 IEEE

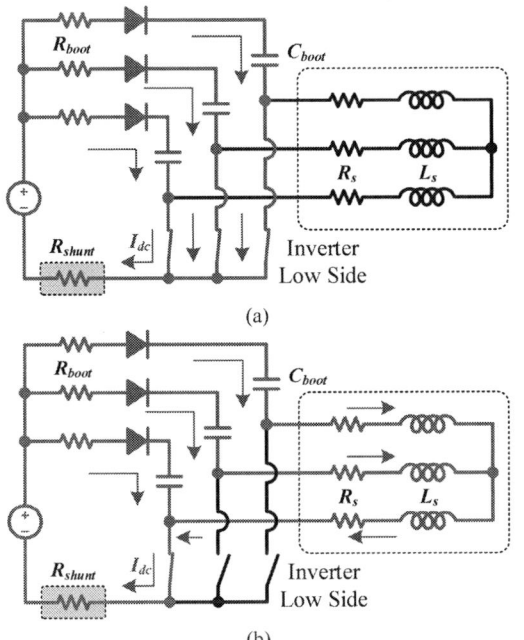

(a)

(b)

Fig. 2. Bootstrap charging circuit (a) conventional procedure, (b) proposed charging procedure one of three steps.

methods that utilize existing components without introducing additional hardware are particularly attractive for their practicality and ease of implementation in consumer electronics [21], [22].

This paper introduces a novel fault diagnosis method for motor drive systems that utilizes a single DC link current sensor in conjunction with the bootstrap configuration. By analyzing the current and impedance characteristics during the bootstrap capacitor charging phase, the proposed method accurately detects and distinguishes between motor and inverter faults. Unlike conventional methods, it does not require additional external circuits or high-voltage diagnostic signals, thereby minimizing the risk of further damage to the system. The proposed method ensures reliability while maintaining simplicity, making it highly suitable for cost-sensitive and space-constrained applications like home appliances. Experimental validation demonstrates the effectiveness of the approach, with results confirming its accuracy and robustness under various fault scenarios.

II. Fault Diagnosis via Bootstrap Charging

A. Bootstrap Charging Impedance

In motor drive systems, the bootstrap configuration is a widely adopted method to drive the high-side switches of inverters. At the beginning of operation, the bootstrap capacitor

must undergo a charging process to ensure proper functioning of the high-side gate driver. Fig. 2(a) illustrates the conventional bootstrap charging circuit, where all three low-side switches are simultaneously turned on. This allows the bootstrap capacitors to charge via the DC link through the low-side switches and a single shunt resistor. In this configuration, the charging current flows through the switches without utilizing the motor windings.

This paper proposes an alternative charging procedure, as shown in Fig. 2(b), where the low-side switches are sequentially activated instead of being operated simultaneously. This sequential activation modifies the current path, causing the charging current to flow not only through the switches but also through the motor windings. This approach introduces a unique opportunity to extract diagnostic information from the bootstrap charging phase. By analyzing the charging impedance in both healthy and faulty conditions, it is possible to distinguish fault scenarios effectively.

The equivalent charging circuits for various conditions are illustrated in Fig. 3. In the proposed method, the charging impedance during this phase can be modeled as follows:

$$Z_{boot} = R_{boot} + 1/sC_{boot} \tag{1}$$

$$Z_s = R_s + sL_s \tag{2}$$

where Z_{boot} represents the impedance of the bootstrap capacitor, Z_s represents the impedance of the switches and motor windings, R_{boot} and C_{boot} are the bootstrap resistance and capacitance, respectively, and R_s and L_s are the series resistance and inductance of the motor windings.

During the bootstrap charging phase, the charging impedance is influenced by the configuration of the circuit and the fault conditions present. To understand these dynamics, the charging impedance is modeled for both healthy and faulty conditions using the equivalent circuit representations. The charging impedances for different conditions are derived as (3) to (6).

$$Z_{charge(healthy)} = R_{shunt} + \left(\frac{1}{Z_s + \left(\frac{2}{Z_{boot} + Z_s}\right)^{-1}} + \frac{1}{Z_{boot}} \right)^{-1} \tag{3}$$

$$Z_{charge(phase\ short)} = R_{shunt} + \left(\frac{1}{Z_{boot} + Z_s + \left(\frac{2}{Z_s}\right)^{-1}} + \frac{2}{Z_{boot}} \right)^{-1} \tag{4}$$

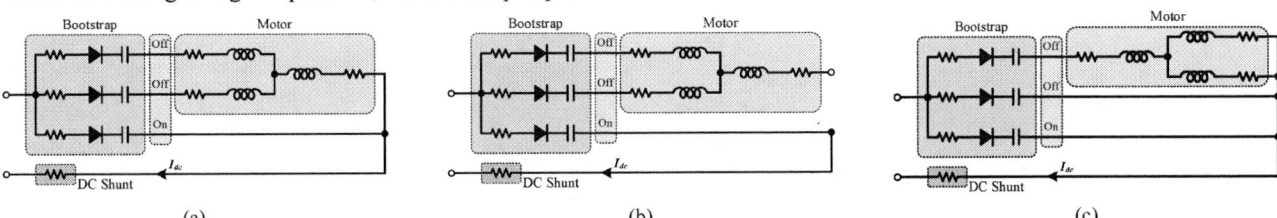

(a)   (b)   (c)

Fig. 3. Equivalent circuit of proposed bootstrap charging, (a) healthy system (b) motor phase open (ON) (c) motor phase short.

979-8-3315-1612-3/25 $31.00 © 2025 IEEE

$$Z_{charge(phase\ open\ ON)}$$
$$= R_{shunt} + \left(\frac{1}{Z_{boot} + 2Z_s} + \frac{1}{Z_{boot}} \right)^{-1} \quad (5)$$

$$Z_{charge(phase\ open\ OFF)}$$
$$= R_{shunt} + Z_{boot} \quad (6)$$

In healthy condition, the charging current flows through all three motor windings and the bootstrap capacitors, resulting in a balanced distribution of impedance as (1). In phase short fault condition, the short circuit significantly reduces the impedance in the affected phase, altering the overall charging dynamics. In phase open fault condition, two cases exist: open phase switch ON or OFF. When a phase is open but its switch is still ON, the open phase completely disconnects from the current path, leaving only the bootstrap impedance and the shunt resistor as the dominant impedance components. In other case, when its switch is OFF, the remaining two phases contribute to the charging current path, resulting in an intermediate impedance.

The impedance plots corresponding to these equations are shown in Fig. 4. The results illustrate how the charging impedance varies significantly across the fault conditions. In the healthy condition, the impedance remains relatively low due to the balanced contribution of all three phases. In contrast, a phase short fault reduces the overall impedance, as the short circuit introduces a low-impedance path. For phase open faults, the impedance depends on whether the switch is ON or OFF, with the ON condition exhibiting the highest impedance due to the lack of current contribution from the disconnected phase.

The variations in impedance under different conditions directly influence the charging current dynamics. As depicted in Fig. 4, the impedance magnitudes change across a wide frequency range, with distinct patterns for each fault scenario. These unique characteristics provide a reliable basis for fault diagnosis, as the charging current waveform reflects the underlying impedance variations. By analyzing the measured charging current during the bootstrap phase, the proposed method can accurately identify the type and location of faults in the system.

B. Bootstrap Charging Current

The proposed diagnostic method relies on analyzing the bootstrap charging current, which is measured using a single DC link current sensor denoted as I_{dc}. This current is influenced by the sequential activation of the low-side switches and reflects the charging behavior of the bootstrap capacitor as well as the system's fault condition.

Simulation results of the bootstrap charging current under healthy and faulty conditions are shown in Fig. 5. The figure illustrates four scenarios: (a) a healthy system, (b) an open fault in the inverter low-side switch (Phase A), (c) a motor open fault in Phase A, and (d) a motor phase-to-phase short between phases A and B. Each case includes the PWM signals driving the switches, the DC link current waveform I_{dc}, and two derived signals: the cumulative sum of the DC current $\sum I_{dc}$ (7) and the sum of absolute difference of the DC current $\sum |\Delta I_{dc}|$ (9).

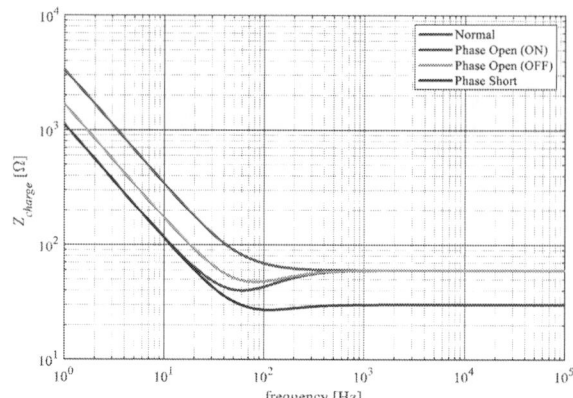

Fig. 4. Charging impedance according to the fault cases

$$\sum I_{dc} = I_{dc}[n] + \sum I_{dc}[n-1] \quad (7)$$

$$\Delta I_{dc} = I_{dc}[n] - I_{dc}[n-1] \quad (8)$$

$$\sum |\Delta I_{dc}| = |\Delta I_{dc}[n]| + \sum |\Delta I_{dc}[n-1]| \quad (9)$$

As shown in Fig. 5, these signals exhibit distinct patterns for each fault condition, providing a reliable basis for fault identification.

In the healthy condition, the charging current follows a consistent pattern, with smooth cumulative and absolute difference signals.

In the inverter low-side open fault (Phase A), the current does not flow when the switch turned ON. Then at the next sequence, the other currents show similar pattern.

In the motor phase open fault (Phase A), the charging current further reduces, with pronounced deviations in the absolute difference signal.

In the motor phase-to-phase short (A-B), the current magnitude increases due to the short circuit, resulting in a higher cumulative current and significant change in the absolute difference signal.

C. Proposed Diagnosis Method

The proposed fault diagnosis algorithm leverages the observed current signals during the bootstrap charging phase to classify fault conditions. The key steps are explained as follows.

Sequential Low-Side Switch Activation: Each low-side switch is sequentially activated to charge the bootstrap capacitors. The DC link current is measured during each activation period using the single current sensor. Then the signals are extracted using (7) to (9). These signals are used to capture the dynamic behavior of the charging current under various conditions. Lastly pattern recognition and fault classification are performed. The extracted signals are compared against reference patterns derived from simulation and experimental data. Distinct deviations in the signals enable classification of the following fault types:

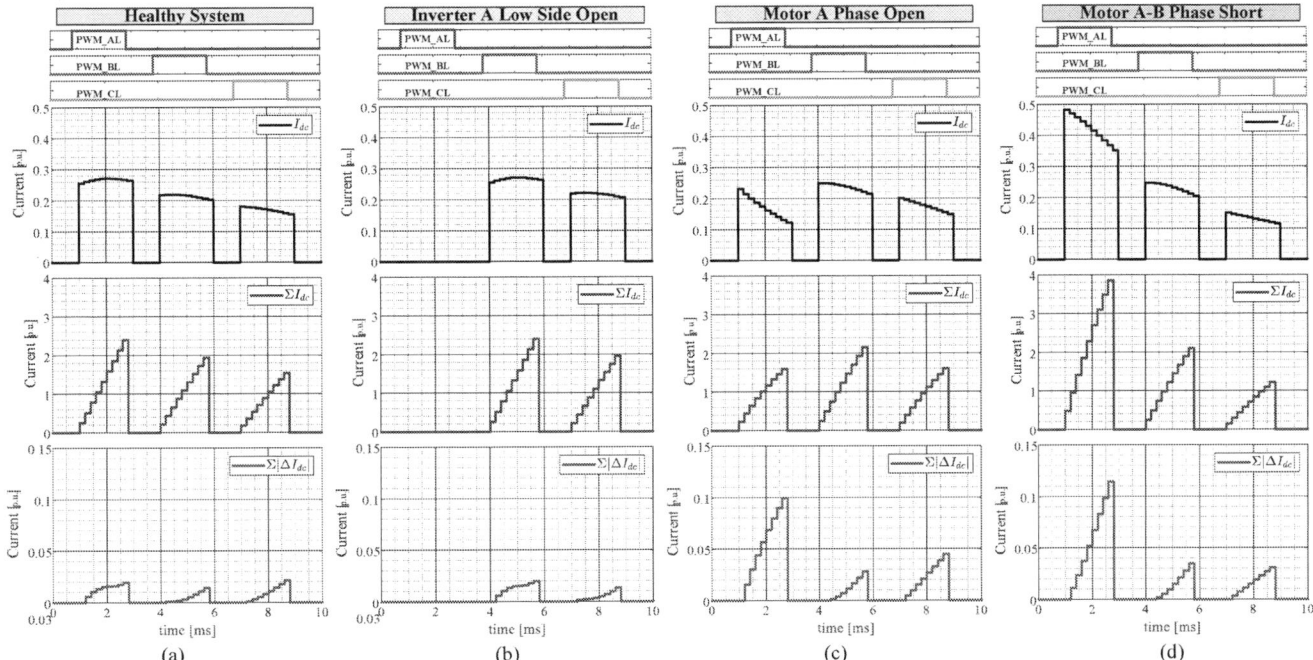

Fig. 6. Simulation results of proposed bootstrap charging with current signals for fault diagnosis method ($I_{dc}, \sum I_{dc}, \sum |\Delta I_{dc}|$), (a) Healthy system, (b) Inverter Phase A low-side switch open, (c) Motor Phase A open, (d) Motor A-B phase short.

- Inverter Faults: Open-circuit faults in the low-side switches (Phases A, B, or C).

- Motor Faults: Open-circuit faults in the motor windings (Phases A, B, or C).

- Motor Short Faults: Phase-to-phase short circuits.

The algorithm reliably distinguishes between eight fault scenarios and the healthy condition, using only the existing bootstrap configuration and a single DC link current sensor. This approach eliminates the need for additional diagnostic circuits, simplifying the system design while maintaining diagnostic accuracy.

III. EXPERIMENTAL RESULTS

The experiment is conducted to validate the effectiveness of the proposed fault diagnosis method under various operating conditions. The experimental setup, depicted in Fig. 6, includes a fan motor for an air conditioner, a single DC link current sensor, and a TI 28377S DSP-based control system. The power stage employs an IGBT-integrated power module (IPM) with an embedded bootstrap diode and resistor, facilitating the sequential bootstrap charging process. The parameters of drive and control system are listed in Table I.

Fig. 7 shows the measured waveforms for the proposed method under different conditions: (a) a healthy system, (b) an inverter A-phase low-side switch open fault, (c) a motor A-phase open fault, and (d) a motor A-B phase short fault. Each subfigure contains the PWM signals for the low-side switches, the bootstrap capacitor voltage, the measured DC link current and the calculated signals of (7) and (9).

The experimental results closely match the simulated waveforms, confirming the validity of the proposed method.

TABLE I   PARAMETERS OF DRIVE SYSTEM

3 Phase IGBT IPM	SanKen SIM6822M
Motor	120 W Fan Motor

Fig. 5. Experimental setup. Fan motor from the outdoor unit of an air-conditioner. 3-phase IGBT inverter with single dc link current sensing

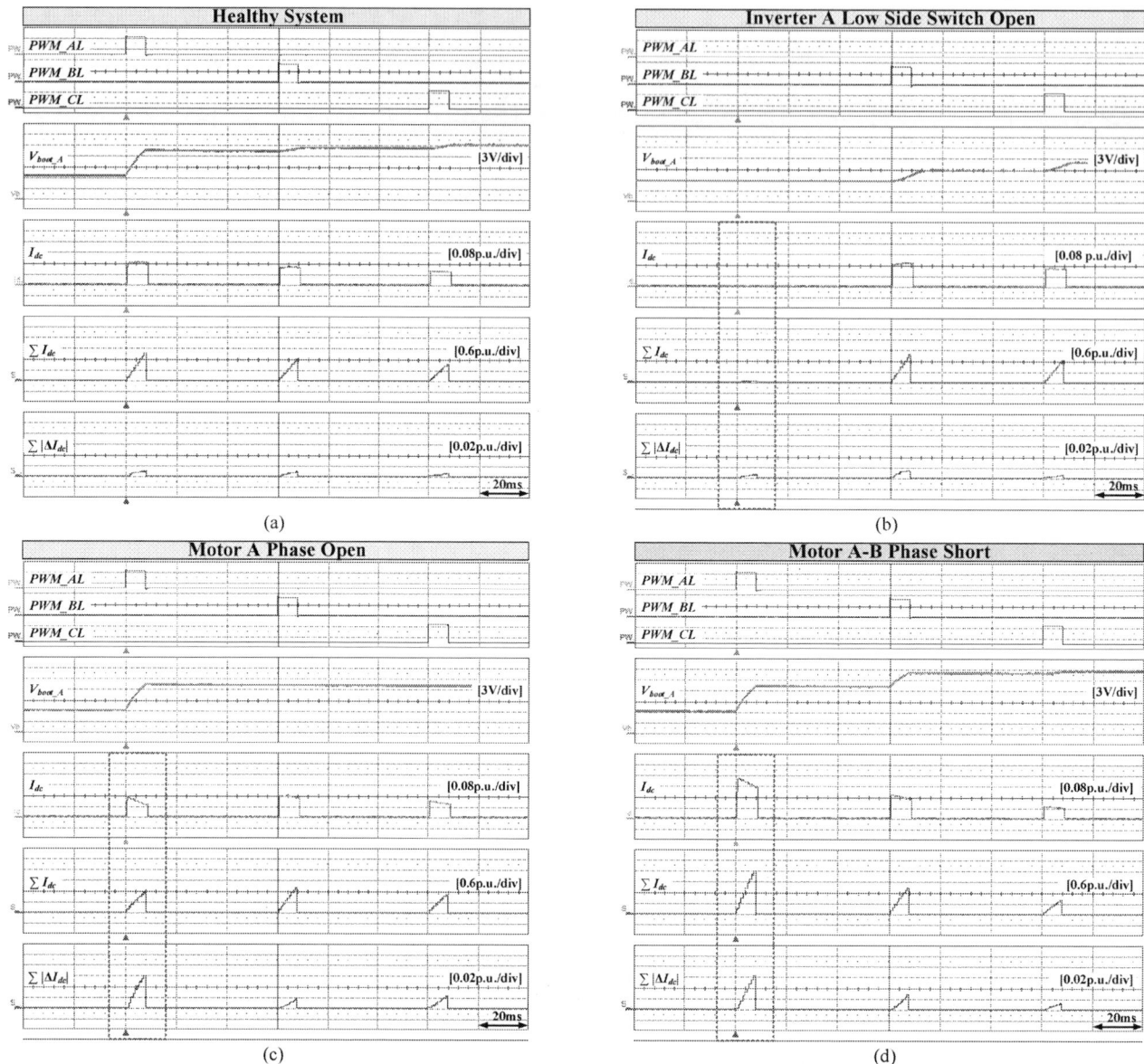

Fig. 7. Experimental results of proposed bootstrap charging with current signals for fault diagnosis method (I_{dc}, $\sum I_{dc}$, $\sum |\Delta I_{dc}|$), (a) healthy system (b) inverter A phase low side switch open (c) motor A phase open (d) motor A-B phase short.

Both sets of results demonstrate clear differences in current signals for each fault condition.

Unlike conventional methods that rely on high-voltage testing, the proposed method operates at the bootstrap low voltage (approximately 15 V). This approach minimizes the risk of overcurrent damage and allows for repeated application without compromising system reliability.

The distinct current signal patterns enable real-time differentiation of all eight fault conditions. The proposed method can accurately classify inverter and motor faults during the initial bootstrap charging phase, providing a practical and efficient solution for fault diagnosis in motor drive systems.

Fig. 8 demonstrates the effectiveness of the proposed method in distinguishing all eight fault scenarios based on the measured and derived current signals. The normalized errors, calculated with respect to the healthy system, reveal distinct patterns for each fault condition across the current signals

Key observations from Fig. 8 include significant deviations in current values for fault cases compared to the healthy system, with unique signal characteristics observed for each fault type. The values indicate the normalized values of error compared the healthy values calculated as (10).

$$normalized\ error\ \% = \left| \frac{X_{healthy} - X_{fault}}{X_{healthy}} \right| \cdot 100 \quad (10)$$

	I_{dc1}	I_{dc2}	I_{dc3}	$\sum I_{dc1}$	$\sum I_{dc2}$	$\sum I_{dc3}$	$\sum |\Delta I_{dc1}|$	$\sum |\Delta I_{dc2}|$	$\sum |\Delta I_{dc3}|$
Healthy Value	0.167	0.129	0.099	1.498	1.174	0.925	0.008	0.008	0.006
Inverter AL Open	98.19	35.22	33.36	97.63	32.12	31.72	37.50	62.50	0.00
Inverter BL Open	1.24	97.65	34.40	1.59	97.16	31.16	12.50	37.50	16.67
Inverter CL Open	3.71	4.00	96.94	3.72	3.43	96.28	25.00	25.00	0.00
Motor A Open	45.08	12.81	2.08	28.41	17.78	9.16	612.50	87.50	250.00
Motor B Open	11.12	31.22	3.13	5.72	11.53	11.17	75.00	600.00	250.00
Motor C Open	13.59	21.61	7.30	6.90	14.43	14.52	112.50	137.50	800.00
Motor AB Short	42.61	14.41	18.76	61.78	24.99	15.86	650.00	225.00	83.33
Motor BC Short	4.94	42.42	13.55	4.62	65.56	26.14	25.00	687.50	433.33
Normallized error with respect to healthy case [%]									

Fig. 8. Current signal representation for motor and inverter fault diagnosis. The values are normalized errors with respect to the healthy case

These distinct signal patterns across the measured and derived parameters enable reliable differentiation of all eight fault conditions.

The results confirm that the proposed method effectively distinguishes between motor and inverter faults, leveraging the unique characteristics of the bootstrap charging current under each fault scenario. This demonstrates the robustness and practicality of the approach for real-time fault detection in motor drive systems.

IV. CONCLUSION

This paper presents a low-cost fault diagnosis method for motor drive systems using bootstrap charging with a single DC link current sensor. The proposed method is especially suitable for motor and inverter systems in home appliances like fans, compressors, and vacuum cleaners, which typically feature a single current sensor and a bootstrap structure to minimize costs and enhance reliability. The method requires no additional circuitry, as fault diagnosis is conducted during the bootstrap capacitor's charging process. Experimental results demonstrate the method's effectiveness in distinguishing various fault conditions.

REFERENCES

[1] K. D. Hurst and T. G. Habetler, "Sensorless Control of Induction Machines," *IEEE Transactions on Industry Applications*, vol. 33, no. 1, pp. 33-47, Jan. 1997.

[2] B. R. Park, G. C. Lim, Y. Han and J. -I. Ha, "Six-Vertex Voltage Injection Strategy for Sensorless IPMSM Control Using Single DC-Link Current Sensor," in *IEEE Transactions on Industrial Electronics*, vol. 71, no. 11, pp. 13879-13889, Nov. 2024

[3] I. Boldea and S. A. Nasar, Electric Drives, Boca Raton, FL, USA: CRC Press, 2005.

[4] B. J. Baliga, "The Role of Bootstrap Capacitors in Gate Drivers for Power MOSFETs," in *Power Electronics Handbook*, 3rd ed. Amsterdam, Netherlands: Elsevier, 2017, pp. 327-333.

[5] N. Mohan, Power Electronics: Converters, Applications, and Design, 3rd ed. Hoboken, NJ, USA: John Wiley & Sons, 2003.

[6] Shihong Park and T. M. Jahns, "A self-boost charge pump topology for a gate drive high-side power supply," in *IEEE Transactions on Power Electronics*, vol. 20, no. 2, pp. 300-307, March 2005.

[7] H. Li, Y. Qian, S. Asgarpoor and H. Sharif, "Machine Current Sensor FDI Strategy in PMSMs," in *IEEE Access*, vol. 7, pp. 158575-158583, 2019.

[8] M. E. H. Benbouzid, "A Review of Induction Motors Signature Analysis as a Medium for Faults Detection," *IEEE Transactions on Industrial Electronics*, vol. 47, no. 5, pp. 984-993, Oct. 2000.

[9] S. Yang, A. Bryant, P. Mawby, D. Xiang, L. Ran, and P. Tavner, "An Industry-Based Survey of Reliability in Power Electronic Converters," *IEEE Transactions on Industry Applications*, vol. 47, no. 3, pp. 1441-1451, May 2011.

[10] Y. Luo, L. Zhang, C. Chen, K. Li and K. Li, "Real-Time Diagnosis of Open Circuit Faults in Three-Phase Voltage Source Inverters," in *IEEE Tran. on Power Electronics*, vol. 39, no. 6, pp. 7572-7585, June 2024.

[11] F. Wu and J. Zhao, "A Real-Time Multiple Open-Circuit Fault Diagnosis Method in Voltage-Source-Inverter Fed Vector Controlled Drives," in *IEEE Trans. on Power Electronics*, vol. 31, no. 2, pp. 1425-1437, Feb. 2016.

[12] M. Riera-Guasp, J. A. Antonino-Daviu, and G. A. Capolino, "Advances in Electrical Machine, Power Electronic, and Drive Condition Monitoring and Fault Detection: State of the Art," *IEEE Transactions on Industrial Electronics*, vol. 62, no. 3, pp. 1746-1759, Mar. 2015.

[13] S. Castellan, A. Zucca, R. Menis, F. Milani, and M. Cipriani, "Current Sensing in Low-Cost Inverters Using a Single DC-Link Sensor," *IEEE Trans. on Power Electronics*, vol. 29, no. 7, pp. 3409-3417, Jul. 2014.

[14] T. Orlowska-Kowalska et al., "Fault Diagnosis and Fault-Tolerant Control of PMSM Drives–State of the Art and Future Challenges," in *IEEE Access*, vol. 10, pp. 59979-60024, 2022.

[15] F. Wu and J. Zhao, "A Real-Time Multiple Open-Circuit Fault Diagnosis Method in Voltage-Source-Inverter Fed Vector Controlled Drives," in *IEEE Trans. on Power Electronics*, vol. 31, no. 2, pp. 1425-1437, Feb. 2016.

[16] J. R. Heredia, F. Perez Hidalgo and J. L. Duran Paz, "Sensorless control of induction motors by artificial neural networks," in *IEEE Transactions on Industrial Electronics*, vol. 48, no. 5, pp. 1038-1040, Oct. 2001.

[17] P. Zhang, Y. Du, T. G. Habetler, and R. G. Harley, "A Survey of Condition Monitoring and Protection Methods for Medium-Voltage Induction Motors," *IEEE Transactions on Industry Applications*, vol. 47, no. 1, pp. 34-46, Jan. 2011.

[18] A. Bellini, F. Filippetti, C. Tassoni, and G. A. Capolino, "Advances in Diagnostic Techniques for Induction Machines," *IEEE Transactions on Industrial Electronics*, vol. 55, no. 12, pp. 4109-4126, Dec. 2008.

[19] H. Yan, Y. Xu, J. Zou, Y. Fang and F. Cai, "A Novel Open-Circuit Fault Diagnosis Method for Voltage Source Inverters With a Single Current Sensor," in *IEEE Transactions on Power Electronics*, vol. 33, no. 10, pp. 8775-8786, Oct. 2018.

[20] Y. Shen, Z. Ma, N. Jin and L. Guo, "Open-circuit Fault Diagnosis Strategy Based on Current Reconstruction with A Single Current Sensor for Voltage Source Inverter," *2023 IEEE 6th International Electrical and Energy Conference (CIEEC)*, Hefei, China, 2023, pp. 3806-3811.

[21] A. Yazidi, R. Bensalah, B. Essouri, and K. Benmessaoud, "Fault Diagnosis and Fault-Tolerant Control in Induction Motor Drives: A Review," *IEEE Transactions on Industrial Electronics*, vol. 67, no. 3, pp. 2042-2053, Mar. 2020.

[22] A. Benyahia and M. E. H. Benbouzid, "Induction Motors Faults Detection and Localization Using Stator Current Advanced Signal Processing Techniques," *IEEE Transactions on Power Electronics*, vol. 30, no. 12, pp. 3474-3482, Dec. 2015.

Improved PWM to Suppress Motor Overvoltage Caused by Voltage Reflection

Sung-Oh Kim
Department of Electrical and Computer Engineering
Ajou University
Suwon, South Korea
kimso7980@ajou.ac.kr

Kyo-Beum Lee
Department of Electrical and Computer Engineering
Ajou University
Suwon, South Korea
kyl@ajou.ac.kr

Abstract—This paper presents an improved PWM to suppress motor overvoltage caused by voltage reflection. The motor drive systems used in industry encounter overvoltage due to voltage reflection when long transmission lines are used. This overvoltage leads to motor damage, making its mitigation essential. One effective method for mitigating overvoltage is adjusting the PWM. The method of adjusting PWM dynamically controls the duty cycle to address the motor overvoltage by utilizing the oscillation frequency associated with motor overvoltage. The proposed method generates a new reference voltage using the new look-up table to suppress motor overvoltage even with adjusting the duty cycle. A flowchart of the proposed method is presented, considering its application to the controller. The validity of the proposed method is verified through simulations.

Keywords—*AC motor, transmission lines, motor overvoltage, voltage reflection, voltage wave*

I. INTRODUCTION

The voltage reflection arises from the impedance mismatch between the transmission line and the motor in motor drive systems with long transmission cables [1]–[2]. This voltage reflection generates overvoltage and oscillation at the motor terminal. The level of overvoltage is determined by the reflection coefficient, while the oscillation frequency depends on the propagation time. A transient state is observed at the motor terminal. If another pulse arrives during this transient phase, a higher peak voltage may occur. This phenomenon is known as the double pulsing effect, where motor overvoltage can exceed twice the inverter output voltage under these circumstances [3]. Such effects can lead to insulation failure in motor windings, ultimately reducing the motor's lifespan. Additionally, these high-voltage stresses can degrade overall system performance, causing efficiency losses and potential downtime in industrial applications. Therefore, addressing the double pulsing effect and voltage reflection is essential for improving system reliability and extending motor life. Various techniques, including improved control strategies and appropriate filtering, can help mitigate these issues and enhance motor drive system performance.

Methods involving additional hardware have been proposed to alleviate motor overvoltage [3]–[6]. One fundamental approach to suppressing overvoltage is adding filters. Various passive filters reduce overvoltage by decreasing dv/dt or achieving impedance matching.

Commonly utilized passive filters include L filters, RLC filters, RC filters [3], and RL filters [4]. While these filters are relatively straightforward to implement within the system, they also tend to increase power loss. To address this issue, filters that account for system losses have been developed, such as the RL-plus-C filter [5] and the inductorless filter [6]. The RL-plus-C filter necessitates the addition of a capacitor on the motor side, making its applicability contingent on the specific requirements of the application. Meanwhile, the inductorless filter requires a separation of 1 to 1.5 meters between the power device and the output terminals for effective implementation. Another method to reduce losses is the addition of switching devices, which was also proposed previously. This approach suppresses motor overvoltage using a soft switching technique. However, these solutions still result in increased system volume and associated costs [7].

This paper proposes a method of adjusting PWM to mitigate overvoltage without additional hardware. This paper analyzed motor overvoltage and ringing caused by the voltage reflection. The proposed method mitigates the double pulsing effect while ensuring that the error with the reference voltage is minimized. A flowchart of the proposed method is presented. The simulation results demonstrate the validity and effectiveness of the proposed method of adjusting PWM.

II. METHOD OF ADJUSTING PWM

The phenomenon of voltage reflection occurring as the transmission line length increases is explained. The overvoltage phenomenon resulting from voltage reflection is analyzed in motor drive systems with long cables. Oscillations in motor overvoltage cause the motor-side voltage to encounter a transient period. The analysis focuses on the principle of larger overvoltage occurring under specific conditions due to the transient period. Based on this analysis, a method for adjusting PWM to alleviate overvoltage is presented.

A. Analysis of the double pulsing effect

Fig. 1 shows a motor drive system with a transmission line. The voltage reflection occurs in motor drive systems due to impedance mismatching at each terminal. The impedance of the motor Z_{mot} is commonly greater than the characteristic impedance of the cable Z_{cab}, and the impedance of the inverter Z_{inv} is smaller than the characteristic impedance of the cable.

979-8-3315-1612-3/25 $31.00 © 2025 IEEE

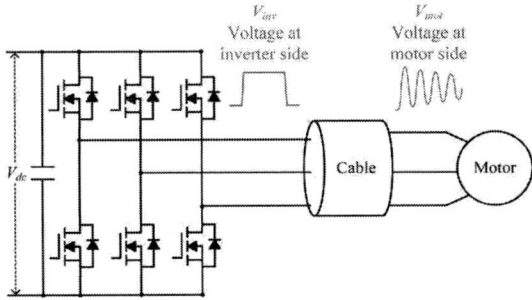

Fig. 1. Motor drive system with long cable.

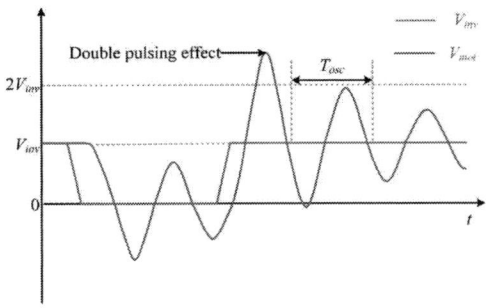

Fig. 2. Double pulsing effect at motor side.

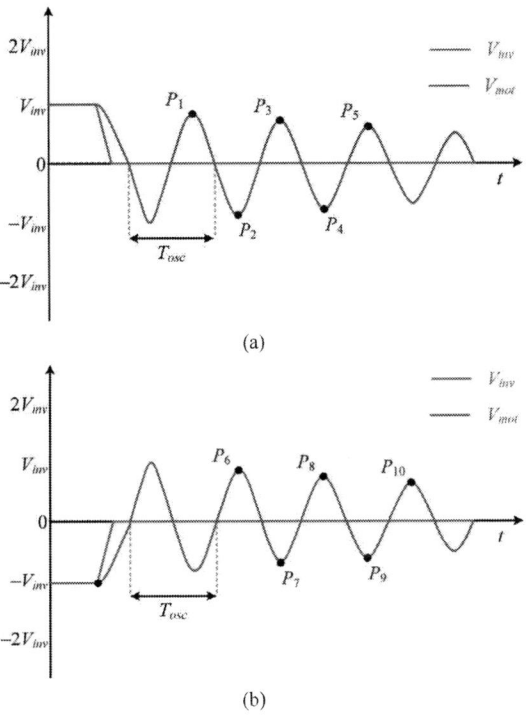

Fig. 3. Oscillation of overvoltage caused by the voltage reflection: (a) switching-off pulse of oscillation voltage and (b) switching-on pulse of oscillation voltage.

The reflection coefficient on the inverter side Γ_{inv} is given by Eq. (1). The reflection coefficient on the motor side Γ_{mot} is given by Eq. (2).

$$\Gamma_{inv} = \frac{Z_{inv} - Z_{cab}}{Z_{inv} + Z_{cab}}, \quad (1)$$

$$\Gamma_{mot} = \frac{Z_{mot} - Z_{cab}}{Z_{mot} + Z_{cab}}. \quad (2)$$

The reflection coefficients cause oscillations of overvoltage on the motor side. The double pulsing effect occurs when the next pulse arrives after the voltage from the previous cycle has fully decayed. Fig. 2 shows the superposition of the oscillation voltage from the previous cycle with the next pulse. This superposition results in overvoltage on the motor side that exceeds twice the inverter voltage.

B. Principle of adjusting PWM

The overvoltage exceeding twice the inverter voltage is generated on the motor side if a switching-on command occurs at points P_2 and P_4 in Fig. 3(a). An overvoltage is generated on the motor side if a switching-off command is applied at points P_6, P_8, and P_{10} in Fig. 3(b), following the same principle. Conversely, overvoltage is prevented when a switching-on command is applied at P_1, P_3, and P_5 in Fig. 3(a). At P_1, the differing directions of the pulses result in their offset. The method of adjusting PWM uses the principle of pulse

offset to suppress overvoltage. The adjusting PWM generates a modified duty cycle if the reference voltage causes a double pulsing effect. The duty cycle is adjusted to operate at either P_1 or P_3 when switching on at P_2. Eq. (3) provides the method for calculating the duty cycle offset d_{off} in regions with large reference voltage magnitudes, which is determined by the switching frequency f_{sw} and the oscillation period of motor overvoltage T_{osc}.

$$d_{off} = 1 - N \cdot T_{osc} \cdot f_{sw}, \ (N = 1, 2, 3 \text{L}), \quad (3)$$

where the T_{osc} is determined through cable modeling. The duty cycle in regions with small reference voltage is defined as follows:

$$d_{off} = N \cdot T_{osc} \cdot f_{sw}, \ (N = 1, 2, 3 \text{L}). \quad (4)$$

The modified voltage using the calculated duty cycle is either higher or lower than the reference voltage. The accumulated error e_{acc} is between the reference voltage and the compensated voltage in all previous compensations. A duty cycle lower than the reference voltage is selected when the accumulated error is greater than zero. Conversely, a higher duty cycle is generated when the accumulated error is lower than zero. The modified reference voltage is generated using the duty cycle based on the accumulated error.

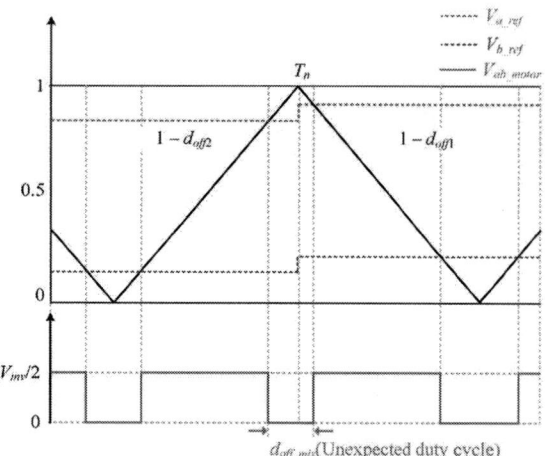

Fig. 4. Phenomenon of the conventional method at high reference voltage in *a*-phase.

III. IMPROVED PWM

This paper analyzes the phenomenon arising from the application of the conventional PWM adjustment method to the controller. A flowchart of the proposed method is provided to address this phenomenon.

Fig. 4 illustrates the phenomenon of the conventional method at high reference voltage in *a*-phase. The magnitude of the reference voltage fluctuates at the peak of the carrier wave when interrupted at the peak of a carrier wave. SVPWM characteristics cause a 1/2 zero vector voltage to be applied at both the start and end of the control period when the *a*-phase reference voltage is large. The conventional method still leads to a double pulsing effect under specific conditions. The previous control period influences the resulting duty cycle, causing it to become d_{off_mis}. The phenomenon occurs with a large reference voltage when the control period aligns with the peak of the carrier wave and with a small reference voltage when the control period aligns with the valley.

An offline look-up table (LUT) for d_{off} is first obtained using the measured value of T_{osc}. The control cycle is selected second, as the compensation method changes based on its point. The LUT generates the duty cycle for the reference voltage if compensation is not needed. The even or odd determination of N is performed if compensation is needed. LUT_{ev} is used to generate the duty cycle if N is even; LUT_{od} is used if N is odd. The compensated duty cycle is selected using an accumulated error e_{acc} from the previous control period. Fig. 5 presents the flowchart of the proposed method. The flowchart for the reference voltage of *a*-phase is shown here; it applies equally to all phases. The variable N_a represents the *a*-phase of N in Eqs. (3) and (4).

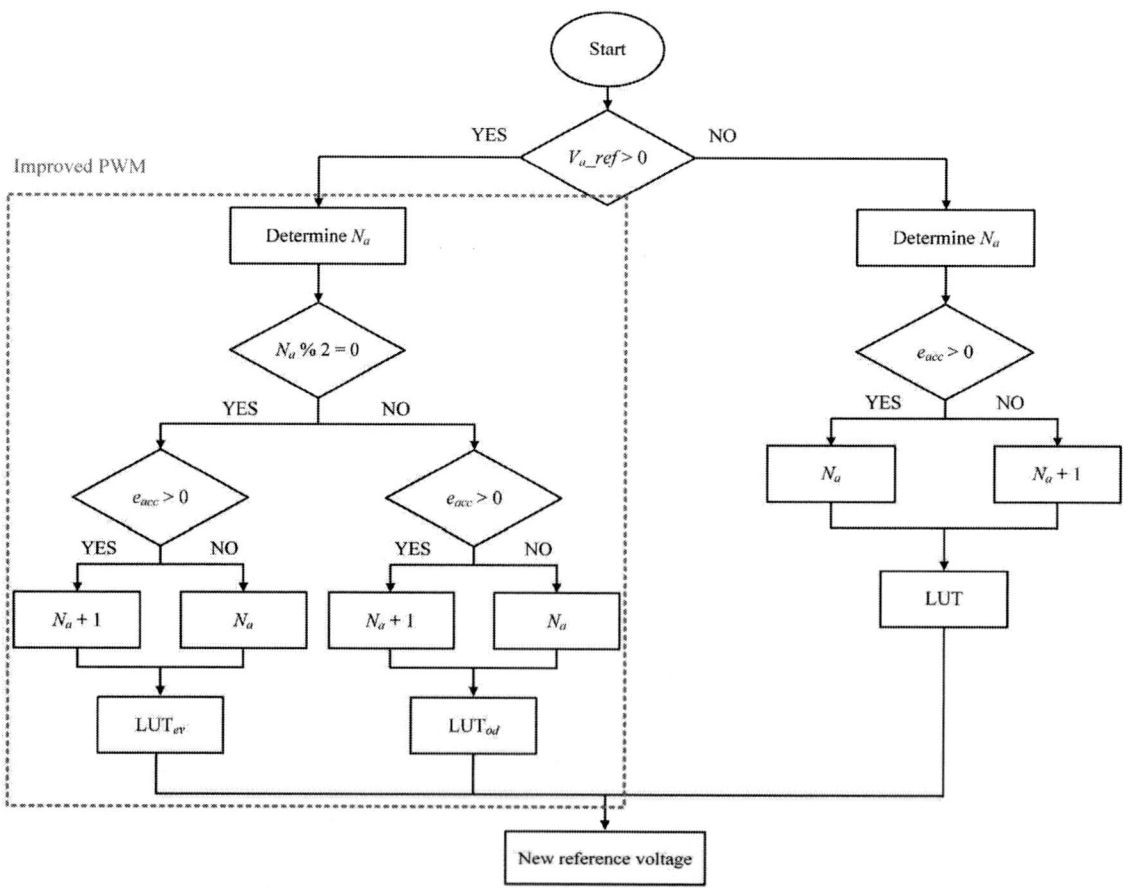

Fig. 5. Flowchart of the proposed method.

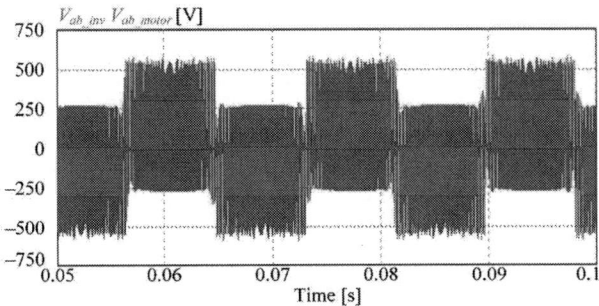

Fig. 6. Simulation results of line-to-line voltage in MI 1.0.

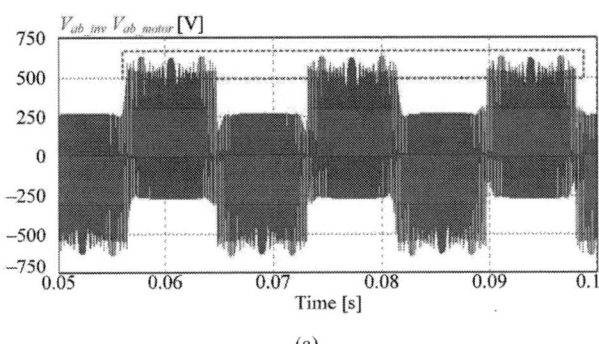

(a)

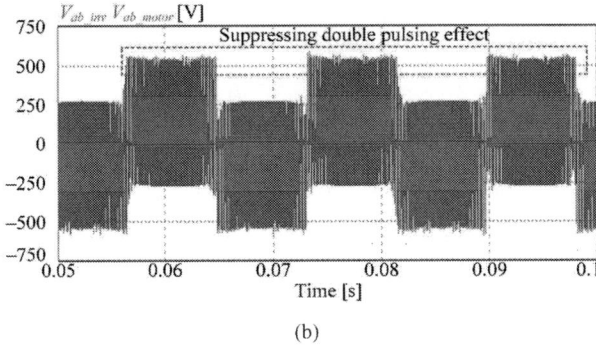

(b)

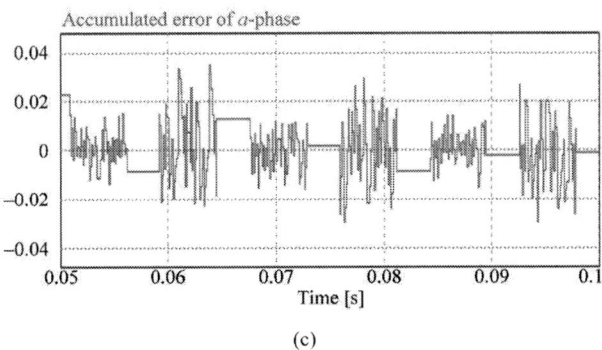

(c)

Fig. 7. Simulation results of proposed method in MI 1.05: (a) line-line voltage, (b) line-to-line voltage with proposed method and (c) accumulated error of a-phase.

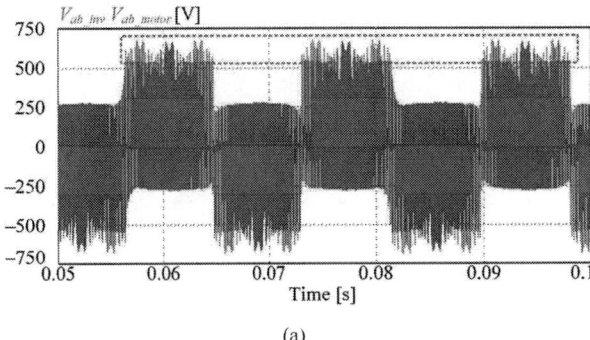

(a)

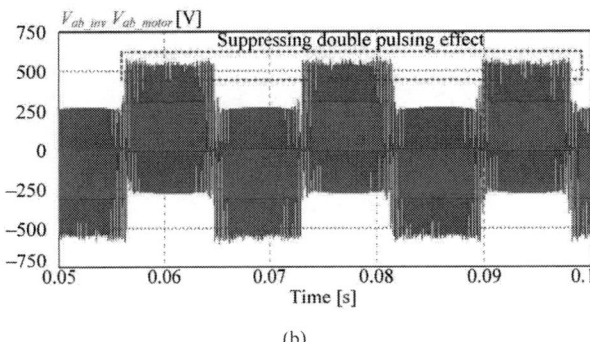

(b)

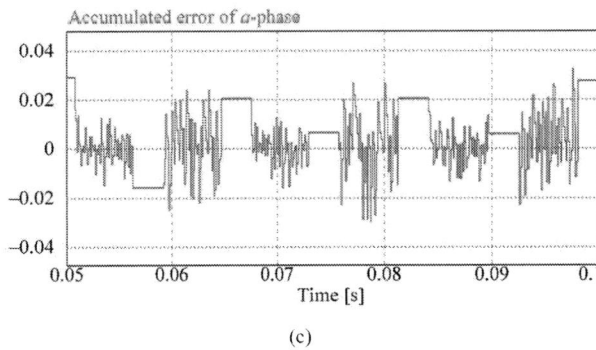

(c)

Fig. 8. Simulation results of proposed method in MI 1.1: (a) line-line voltage, (b) line-to-line voltage with proposed method and (c) accumulated error of a-phase.

IV. SIMULATION RESULTS

The motor drive system with a long cable is modeled using PSIM software, and the performance of the proposed method is validated through simulations. The simulation was conducted using a motor with an 11 kW power, and detailed specifications are provided in Table I. T_{osc} obtained through cable modeling is used to generate an offline LUT, LUT_{ev}, and LUT_{od}. Fig. 6 shows the line-to-line voltage between phase a and phase b when the method is not applied in modulation index (MI) 1.0. The double pulsing effect is not present in regions with a low MI. Fig. 7 shows the line-to-line voltage at an MI of 1.05, the line-to-line voltage with the proposed technique, and the accumulated error of a-phase. A line-to-line voltage of 644 V, more than twice the input voltage of 300 V, was observed on the motor input side in Fig 7 (a). Fig. 7(b) illustrates the suppression of motor overvoltage through the application of the proposed method. The magnitude of the

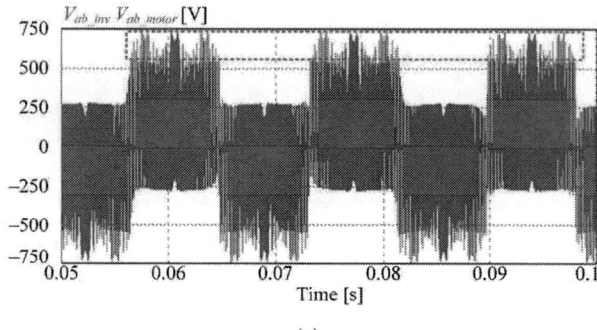

(a)

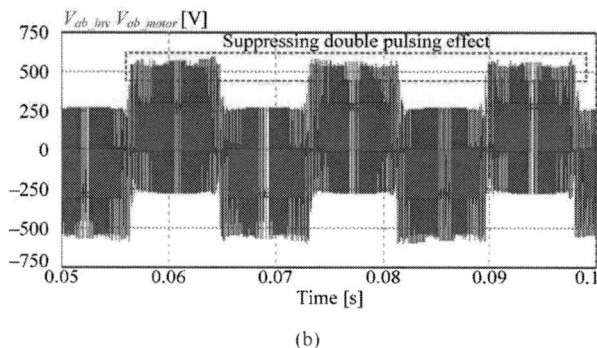

(b)

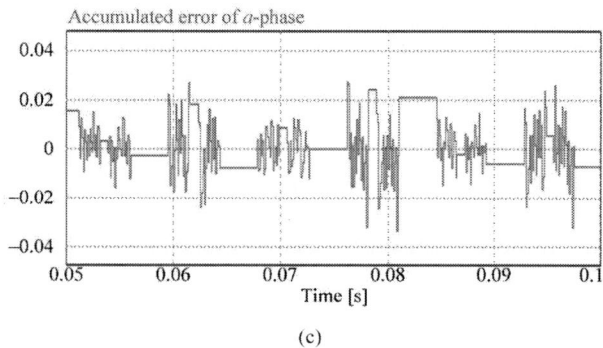

(c)

Fig. 9. Simulation results of proposed method in MI 1.14: (a) line-line voltage, (b) line-to-line voltage with proposed method and (c) accumulated error of *a*-phase.

line-to-line voltage with the proposed method applied is 582 V, resulting in a voltage below 600 V. Fig. 7 (c) presents the accumulated error. The magnitude of the accumulated error is within 0.04, satisfying the volt-sec balance condition. Fig. 8 and Fig. 9 show the line-to-line voltages of phases *a* and *b*, and the accumulated error of *a*-phase, at MI values of 1.1 and 1.14, respectively. The magnitude of the overvoltage rises to 676 V as the MI increases in Fig. 8(a). The magnitude of the overvoltage is 748 V, the largest value across all MIs in Fig. 9(a). The suppression of overvoltage for both MIs using the proposed method is demonstrated in Fig. 8(b) and Fig. 9(b). All accumulated errors satisfy the volt-sec balance condition. As a result, the proposed method demonstrates a high performance in suppressing the double pulsing effect.

TABLE I. SIMULATION PARAMETERS

Parameter	Value
DC-link voltage	300 [V]
Switching frequency	10 [kHz]
Cable length	30 [m]
d-axis inductance	13.17 [mH]
q-axis inductance	15.6 [mH]
Stator resistor	0.349 [Ω]

V. CONCLUSIONS

This paper proposed improved PWM for mitigating the motor overvoltage caused by the voltage reflection. The proposed method was applied to mitigate the double pulsing effect. The motor overvoltage was suppressed to below twice the inverter output voltage by the proposed method. The phenomenon of unexpected duty cycle occurring was analyzed when the adjusting PWM was applied to mitigate overvoltage. The flowchart of the proposed method was presented considering this phenomenon. The validity of the proposed method was demonstrated by the simulations.

REFERENCES

[1] T. Lackie, Y. Jiang, L. Shillaber, and T. Long, "Motor Overvoltage Mitigation by Active Cancellation of Reflections Using Parallel SiC Devices with a Coupled Inductor," *IEEE Trans. Power Electron.*, vol. 38, no. 9, pp. 11368–11384, Sep. 2023.

[2] N. Wang, Z. Wang, Z. Liu, G. Xin, X. Shi, and Y. Kang, "A Band-Stop-Type dv/dt Filter for Terminal Overvoltage Mitigation of SiC Motor Drives," *IEEE Trans. Power Electron.*, vol. 39, no. 12, pp. 16391–16399, Dec. 2024.

[3] B. Narayanasamy, A. S. Sathyanarayanan, F. Luo, and C. Chen, "Reflected Wave Phenomenon in SiC Motor Drives: Consequences, Boundaries, and Mitigation," *IEEE Trans. Power Electron.*, vol. 35, no. 10, pp. 10629–10642, Oct. 2020.

[4] R. Ruffo, P. Guglielmi, and E. Armando, "Inverter Side RL Filter Precise Design for Motor Overvoltage Mitigation in SiC-Based Drives," *IEEE Trans. Ind. Electron.*, vol. 67, no. 2, pp. 863–873, Feb. 2020.

[5] Y. Jiang, W. Wu, Y. He, H. S.-H. Chung, and F. Blaabjerg, "New Passive Filter Design Method for Overvoltage Suppression and Bearing Currents Mitigation in a Long Cable Based PWM Inverter-Fed Motor Drive System," *IEEE Trans. Power Electron.*, vol. 32, no. 10, pp. 7882–7893, Oct. 2017.

[6] E. Velander, G. Bohlin, A. Sandberg, T. Wiik, F. Botling, M. Lindahl, G. Zanuso, and H.-P. Nee, "An Ultralow Loss Inductorless dv/dt Filter Concept for Medium-Power Voltage Source Motor Drive Converters with SiC Devices," *IEEE Trans. Power Electron.*, vol. 33, no. 7, pp. 6072–6081, Jul. 2018.

[7] H. Xiong, J. Zhang, and A. V. Jouanne, "Control of Variable Frequency Drive PWM to Mitigate Motor Overvoltage due to Double Pulsing in Reflected Wave Phenomenon," in *Proc. ECCE Conf.*, pp. 6563–6570, 2018.

Analysis of Double Pulsing Effect in Motor Drives Based on Vector Diagram

Byeong-Woo Kang
Department of Electrical and Computer Engineering
Ajou University
Suwon, South Korea
blackstar11740@ajou.ac.kr

Kyo-Beum Lee
Department of Electrical and Computer Engineering
Ajou University
Suwon, South Korea
kyl@ajou.ac.kr

Abstract— **This paper analyzes the double pulsing effect based on vector diagram in motor drives with variable modulation techniques. The overvoltage caused by voltage reflection is analyzed in a long transmission line between an inverter and a motor. The region where the double pulsing effect occurs is compared and analyzed through voltage vector diagrams of different modulation techniques. A modulation index and an on-time of a zero-voltage vector are considered to generalize the region where the double pulsing effect occurs. The validity of the analysis on the region where the double pulsing effect occurs with variable modulation techniques is verified through simulation results.**

Keywords—The double pulsing effect, vector diagram, long transmission line, modulation techniques

I. INTRODUCTION

The overvoltage from the reflected wave phenomenon deteriorates motor winding insulation in a long cable between a power inverter and a motor [1]. The voltage between an inverter and a motor winding terminal reaches up to twice the DC bus voltage when the rising and falling time of power semiconductor devices is fast [2]–[4]. The overvoltage caused by the double pulsing effect increases up to four times the maximum DC bus voltage and leads to motor insulation failure [5]. The magnitude of overvoltage varies depending on factors such as cable length, impedance mismatch between a cable and motor, and pulse damping time [6]. The method for mitigating the double pulsing effect is based on the switching characteristics of an inverter and a motor overvoltage.

The method for mitigating the double pulsing effect includes adding of filters and modulating PWM patterns. The dv/dt filters used to mitigate overvoltage are large and lead to high power losses [7]–[9]. Applying PWM methods reduces system costs and mitigates the double pulsing effect more effectively than filters [10]–[12]. Analyzing the region where the double pulsing effect occurs with each modulation technique is important for applying methods for mitigating the double pulsing effect. This analysis is the essential component for understanding the mechanism of the method for mitigating the double pulsing effect.

This paper analyzes and generalizes the region where the double pulsing effect occurs in motor drives based on a vector diagram. The overvoltage caused by voltage reflection between a power inverter and a motor is analyzed. The region where the double pulsing effect occurs is compared and analyzed based on vector diagrams for each modulation technique. The analysis generalizes the region where the double pulsing effect occurs by considering a modulation index and an on-time of the zero-voltage vector. The validity of the analysis is demonstrated by various simulation results.

II. OVEROLTAGE BY VOLTAGE REFLECTION

The reflected wave phenomenon is analyzed based on transmission line theory in the transmission line between an inverter and a motor. The principle causing the double pulsing effect is analyzed during the voltage reflection process with switching signals applied to the inverter.

A. Analysis of the region where the Double pulsing effect occurs with modulation techniques

Fig. 1(a) illustrates a circuit composed of infinitesimal segments of a transmission line represented by the interconnection of R, L, G, and C. R represents resistance due to thermal losses when current flows through the cable. L represents the inductance caused by the magnetic field when current flows through the cable. G represents resistance related to leakage current losses. C represents capacitance caused by the electric field when a voltage is applied to the cable. The voltage and current are expressed as shown in equation (1) and equation (2) by applying Kirchhoff's voltage law and current law to the circuit in Fig. 1(a).

$$V(z) = (R + j\omega L)\Delta z I(z) + V(z + \Delta z), \quad (1)$$

$$I(z) = (G + j\omega C)\Delta z V(z + \Delta z) + I(z + \Delta z). \quad (2)$$

The transmission line equations are expressed as shown in equation (3) and equation (4) as Δz approaches zero.

$$\frac{dV(z)}{dz} = \lim_{\Delta z \to 0} \frac{V(z + \Delta z) - V(z)}{\Delta z} = -(R + j\omega L)I(z), \quad (3)$$

$$\frac{dI(z)}{dz} = \lim_{\Delta z \to 0} \frac{I(z + \Delta z) - I(z)}{\Delta z} = -(G + j\omega C)V(z). \quad (4)$$

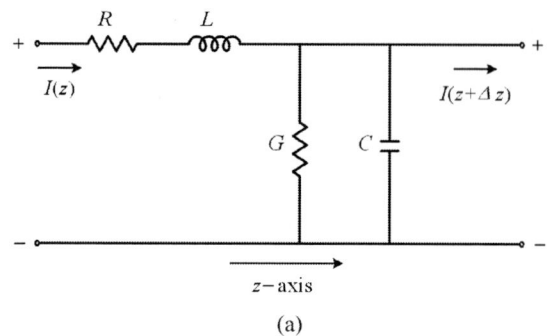
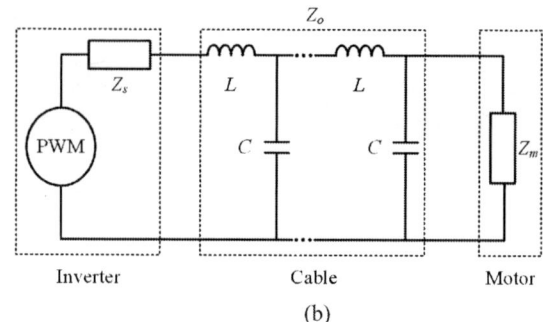

Fig. 1. The cable and motor drive system: (a) The cable in an infinitesimal segment and (b) Equivalent circuit of a motor drive system.

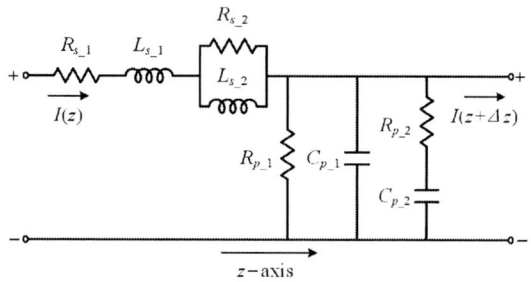

Fig. 2. Equivalent cable model considering high-frequency range.

The voltage and current waves derived from the second-order linear differential equations in equation (1) and equation (2) are expressed in equation (3) and equation (4). The voltage and current waves consist of incident waves and reflected waves as wave components. The $+z$ direction represents the incident wave while the $-z$ direction represents the reflected wave in equations (5) and equation (6).

$$V(z) = V_0^+ e^{-\gamma z} + V_0^- e^{\gamma z}, \tag{5}$$

$$I(z) = I_0^+ e^{-\gamma z} - I_0^- e^{\gamma z}. \tag{6}$$

The equations for the voltage and current waves given in equation (5) and equation (6) are differentiated with respect to z. This differentiation assumes no reflected voltages or currents and results in the equations presented in equation (7) and equation (8).

$$\frac{dV(z)}{dz} = -\gamma V_0^+ e^{-\gamma z}, \tag{7}$$

$$\frac{dI(z)}{dz} = -\gamma I_0^+ e^{-\gamma z}. \tag{8}$$

Fig. 1(b) illustrates a motor drive system utilizing a cable between the inverter and the motor when the cable loss is negligible. The characteristic impedance Z_o is expressed by equation (9) when equation (5) and equation (6) are substituted into equation (3) and (4).

$$Z_o = \frac{V_0^+}{I_0^+} = \sqrt{\frac{R + j\omega L}{G + j\omega C}} \approx \sqrt{\frac{L}{C}}. \tag{9}$$

The impedance mismatch between an inverter and motor terminals generates incident and reflected waves of the inverter and motor. The reflection coefficients Γ_s and Γ_m of the inverter and motor are expressed as shown in equation (10) and equation (11) based on the characteristic impedance Z_o derived in equation (9). The Z_s and Z_m in equation (10) and equation (11) represent the impedance of the inverter and motor, respectively. The voltage reflection does not occur when the inverter and motor impedance matches the characteristic impedance Z_o. Z_s is approximately 0, which results in the reflection coefficient at the inverter output side being approximately -1. Γ_m which is greater than Z_o causes the reflection coefficient at the motor input side to be approximately 1. The reflected voltage at the inverter output is reflected back to the motor side with the reverse phase. The reflected voltage at the motor input is reflected back to the inverter side with the same phase.

$$\Gamma_s = \frac{Z_s - Z_o}{Z_s + Z_o} \approx -1, \tag{10}$$

$$\Gamma_m = \frac{Z_m - Z_o}{Z_m + Z_o} \approx 1. \tag{11}$$

The propagation time t_p is the time required for a voltage pulse to travel from an inverter to motor terminals and is expressed as follows:

979-8-3315-1612-3/25 $31.00 © 2025 IEEE

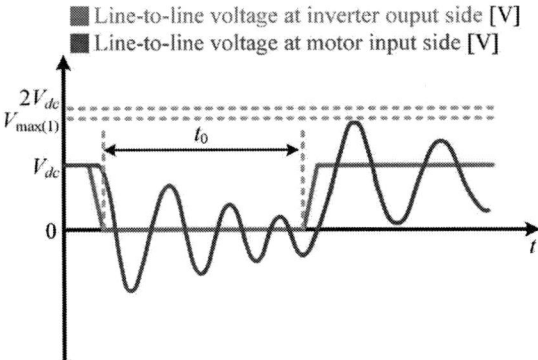

Fig. 3. Line-to-line voltage at the inverter output side and the motor input side when the double pulsing effect does not occur.

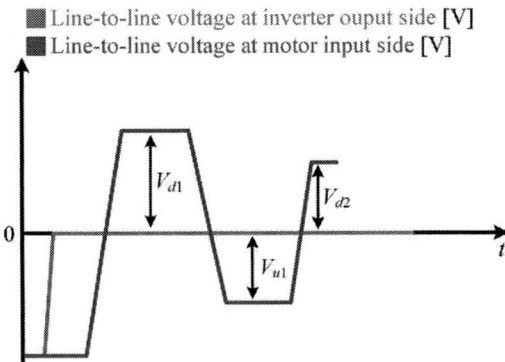

Fig. 5. Amplified waveform of the line-to-line voltage at the inverter output side and the motor input side.

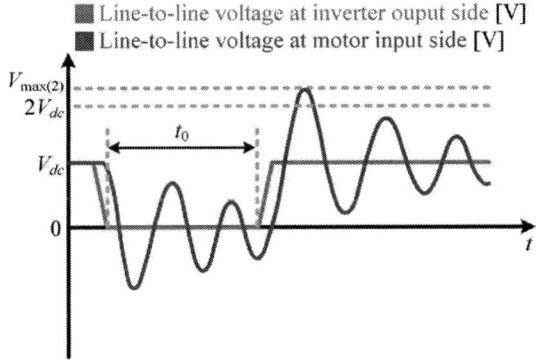

Fig. 4. Line-to-line voltage at the inverter output side and the motor input side when the double pulsing effect occurs.

$$t_p = \frac{l}{v} = l\sqrt{LC}. \tag{12}$$

The propagation speed v is determined by L, C, which constitute the characteristic impedance Z_o.

Fig. 2 illustrates the equivalent cable model used to model the cable between the inverter and the motor. R_{s-2} and L_{s-2} represent parameters considering dielectric losses. R_{p-2} and C_{p-2} represent parameters considering the proximity effect and the skin effect.

Fig. 3 and Fig. 4 illustrate the line-to-line voltages at the inverter output and motor input side as a function of the on-time of a zero-voltage vector t_0. Fig. 5 illustrates the enlarged waveforms of overvoltage at the inverter and motor input sides. The sufficient on-time of the zero-voltage vector ensures adequate decay time for motor overvoltage in Fig. 3. It prevents the occurrence of the double pulsing effect. The waveform where the double pulsing effect occurs due to a reduction in t_0 in Fig. 4. The descending pulse V_{dn} is converted to the ascending pulse V_{un} to facilitate the analysis of the

variation in the line-to-line voltage at the motor input side in Fig. 5. Generalizing the variation in the motor line-to-line voltage based on the calculation results of the descending and ascending pulses, it is expressed as follows:

$$V_{dn} = \left\{ \sum_{i=0}^{n+1} \Gamma_m^i \Gamma_s^i \cdot \left(1 + \Gamma_m\right) \cdot V_{dc} \right\} - V_{dc}, \tag{13}$$

$$V_{u,n} = \sum_{i=0}^{n+1} \Gamma_m^{i+1} \Gamma_s^i \cdot \left(1 + \Gamma_s\right) \cdot V_{dc}. \tag{14}$$

B. The principle of the Double pulsing effect

The occurrence of the double pulsing effect is determined by the on-time of the zero-voltage vector t_0. The double pulsing effect occurs when the motor input overvoltage exceeds 2 times the DC voltage due to overlapping pulses. The risk of insulation breakdown in the motor windings increases if V_{max} exceeds 2 times the DC voltage. V_{u3} overlaps with V_{d3} due to the small t_0, leading to an overvoltage at the motor input in Fig. 4. $V_{max(1)}$ and $V_{max(2)}$ represent the maximum value of the overvoltage at the motor input side, as shown in Fig. 3 and Fig. 4.

III. ANALYSIS OF THE REGION WHERE THE DOUBLE PULSING EFFECT OCCURS WITH MODULATION TECHNIQUES

The region where the double pulsing effect occurs is analyzed and compared based on the vector diagrams of SPWM and SVPWM.

A. Analysis of the region where the double pulsing effect occurs in SPWM

The minimum on-time of the zero-voltage vector t_{DPE} and the maximum modulation index MI_{DPE} where the double pulsing effect does not occur are defined as follows:

$$t_{DPE} = t_p + \left(n-1\right)t_{ocs} + \frac{t_{ocs}}{2}, \tag{15}$$

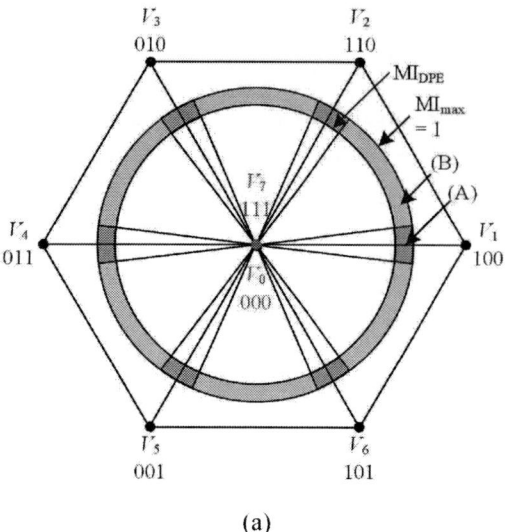
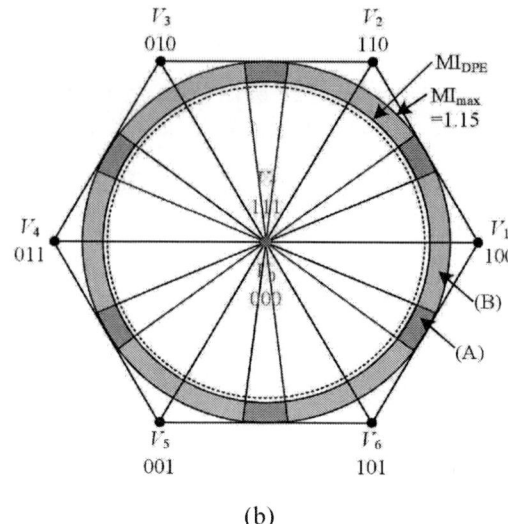

(a)   (b)

Fig. 6. The region where the double pulsing effect occurs in the vector diagram.: (a) SPWM and (b) SVPWM.

TABLE I. SIMULATION PARAMETERS

Parameter	Value
DC-link voltage	300 [V]
Cable length	30 [m]
Resistance load	10 [Ω]
Inductance load	2.5 [mH]
Switching frequency	10 [kHz]

TABLE II. CABLE PARAMETERS

Parameter	Value
R_{s-1}	17.5 [mΩ]
R_{s-2}	650 [mΩ]
L_{s-1}	332 [nH]
L_{s-2}	36.6 [nH]
R_{p-1}	47.9 [MΩ]
R_{p-2}	55.9 [kΩ]
C_{p-1}	43.8 [pF]
C_{p-2}	12.3 [pF]

$$\mathrm{MI_{DPE}} = \frac{T_s - t_{\mathrm{DPE}}}{T_s} = \frac{T_s - \left\{ t_p + (n-1)t_{ocs} + \dfrac{t_{ocs}}{2} \right\}}{T_s}. \quad (16)$$

Fig. 6(a) illustrates the vector diagram of SPWM indicating the region where the double pulsing effect occurs. $\theta_{\mathrm{DPE(SPWM)}}$

is calculated using the inverse cosine function of the duty cycle in the region where the magnitude of the reference voltage for each phase is greater than $\mathrm{MI_{DPE}}$. $\theta_{\mathrm{DPE(SPWM)}}$ is expressed as follows:

$$\theta_{\mathrm{DPE(SPWM)}} = \cos^{-1}\left(\frac{V_{ref}}{V_{dc}/2} \right) = \cos^{-1}\left(\frac{T_s - t_{\mathrm{DPE}}}{T_s} \right). \quad (17)$$

The region where the double pulsing effect occurs is determined based on $\mathrm{MI_{DPE}}$ and $\theta_{\mathrm{DPF(SPWM)}}$. The red area indicates the region where the double pulsing effect occurs, characterized by the magnitude of the reference voltage for each phase being greater than $\mathrm{MI_{DPE}}$. In Fig. 6(a) and Fig. 6(b), the possibility of the double pulsing effect occurrence decreases in the following order: red and blue.

B. *Analysis of the region where the double pulsing effect occurs in SVPWM*

Fig. 6(b) illustrates the vector diagram of SVPWM indicating the region where the double pulsing effect occurs. The analysis is conducted in the same way as the method used to analyze the region where the double pulsing effect occurs in SPWM. The maximum modulation index $\mathrm{MI_{max}}$ in the linear region for SVPWM is 1.15, unlike SPWM. The value of $\theta_{\mathrm{DPE(SVPWM)}}$ for SVPWM differs by 30 degrees compared to SPWM and is expressed as shown in equation (18).

$$\theta_{\mathrm{DPE(SVPWM)}} = \cos^{-1}\left(\frac{T_s - t_{\mathrm{DPE}}}{T_s} \right) - \frac{\pi}{6}. \quad (18)$$

979-8-3315-1612-3/25 $31.00 © 2025 IEEE   2951

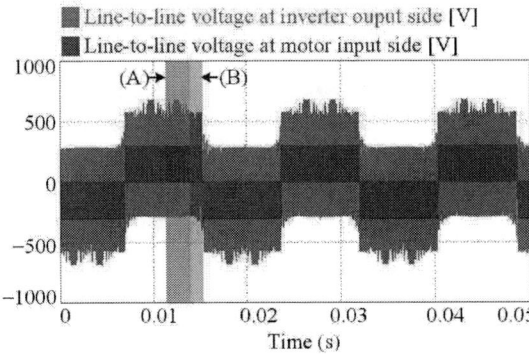

Fig. 7. PISM simulation result of SPWM.

IV. SIMULATION RESULTS

A two-level inverter system is modeled by PSIM software. The analysis of the region where the double pulsing effect occurs is verified through the simulations. The simulation is performed under the conditions shown in Table I and Table II. The simulation result for SPWM indicates the region (A) where the reference voltage magnitude exceeds MI_{DPE} in Fig. 7. The region (B) has a lower probability of the double pulsing effect occurrence than region (A). This is because the magnitude of the reference voltage does not exceed MI_{DPE}. The simulation result for SVPWM shows the region (A) where the double pulsing effect occurs is different from SPWM.

V. CONCLUSION

This paper analyzed the region where the double pulsing effect occurs for SPWM and SVPWM in motor drives. The overvoltage in the motor line-to-line voltage on long transmission lines was analyzed. The probability for the double pulsing effect was analyzed according to the magnitude of the reference voltage. This forms the foundation for analyzing the region in which the double pulsing effect occurs for each modulation technique. The validity of the proposed strategy was demonstrated through PSIM simulations.

REFERENCES

[1] H. Kim, A. Anurag, S. Acharya, and S. Bhattacharya, "Analytical Study of SiC MOSFET Based Inverter Output dv/dt Mitigation and Loss Comparison with a Passive dv/dt Filter for High Frequency Motor Drive Applications," *IEEE Access*, vol. 9, pp. 15228–15238, Jan. 2021.

[2] Y. Xu, X. Yuan, F. Ye, Y. Zhang, M. Diab, and W. Zhou, "Impact of High Switching Speed and High Switching Frequency of Wide-Bandgap Motor Drives on Electric Machines," *IEEE Access*, vol. 9, pp. 82866–82880, Jun. 2021.

[3] W. Zhou and X. Yuan, "Experimental Evaluation of SiC MOSFETs in Comparison to Si IGBTs in a Soft-Switching Converter," *IEEE Trans. Ind. Electron.*, vol 67, no. 5, pp. 5108–5118, Sep./Oct. 2020.

[4] S. D. Caro, S. Foti, T. Scimone, A. Testa, G. Scelba, M. Pulvirenti, and S. Russo, "Motor overvoltage mitigation on sic mosfet drives exploiting an open-end winding configuration," *IEEE Trans. Power Electron.*, vol 34, no. 11, pp. 11128–11138, Nov. 2019.

[5] M. Diab, W. Zhou, C. Emersic, X. Yuan, and I. Cotton, "Impact of

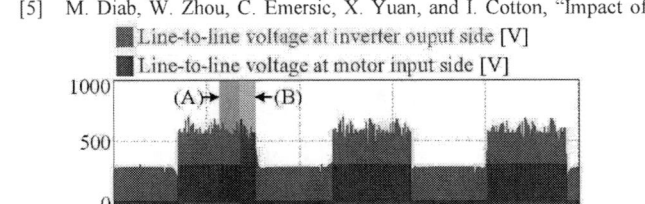
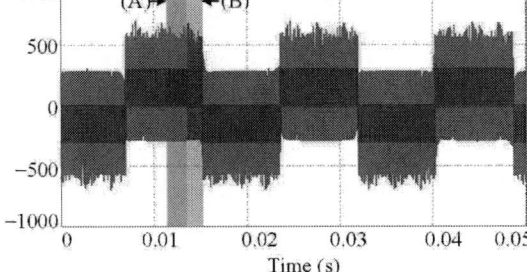

Fig. 8. PISM simulation result of SVPWM.

PWM Voltage Waveforms on Magnet Wire Insulation Partial Discharge in SiC-Based Motor Drives," *IEEE Access*, vol. 9, pp. 156599–156612, Nov. 2021.

[6] L. Wang, C, N. Ho, F. Canals, and J. Jatskevich, "High-Frequency Modeling of the Long-Cable-Fed Induction Motor Drive System Using TLM Approach for Predicting Overvoltage Transients," *IEEE Trans. Power Electron.*, vol. 25, no. 10, pp. 2653–2664, Oct. 2010.

[7] H. Kim, A. Anurag, S. Acharya, and S. Bhattacharya, "Analytical Study of SiC MOSFET Based Inverter Output dv/dt Mitigation and Loss Comparison with a Passive dv/dt Filter for High Frequency Motor Drive Applications," *IEEE Access*, vol. 9, no. 6, pp. 15228–15238, Jan. 2021.

[8] R. Ruffo, P. Guglielmi, and E. Armando, "Inverter Side RL Filter Precise Design for Motor Overvoltage Mitigation in Sic-Based Drives," *IEEE Trans. Ind. Electron.*, vol 67, no. 2, pp. 863–873, Feb. 2020.

[9] Y. Jiang, W. Wu, Y. He, H. S. H. Chung, and F. Blaabjerg, "New Passive Filter Design Method for Overvoltage Suppression and Bearing Currents Mitigation in a Long Cable Based PWM Inverter-Fed Motor Drive System," *IEEE Trans. Power Electron.*, vol. 32, no. 10, pp. 7882–7893, Oct. 2017.

[10] B. Narayanasamy, A. S. Sathyanarayanan, F. Luo, and C. Chen, "Reflected Wave Phenomenon in SiC Motor Drives: Consequences, Boundaries, and Mitigation," *IEEE Trans. Power Electron.*, vol. 35, no. 10, pp. 10629–10632, Oct. 2020.

[11] R. M. Tallam and D. Leggate, "Control of a PWM Voltage-Source Inverter in the Pulse-Dropping Region to Reduce Reflected-Wave Motor Overvoltage," *IEEE Trans. Ind. Appl.*, vol. 49, no. 2, pp. 873–879, March-April. 2013.

[12] H. Xiang, J. Zhang, and A. Jouanne, "Control of Variable Frequency Drive PWM to Mitigate Motor Overvoltage Due to Double Pulsing in Reflected Wave Phenomenon," in *Proc. ECCE conf.*, pp. 6563–6570, 2018.

A Novel Speed Sensor-less Control of a Solar-Powered PMSM Drive

Abirami Kalathy
Dept. of ECE
Queen's University
Kingston, Canada
18ak67@queensu.ca

Arpan Laha
Dept. of ECE
Queen's University
Kingston, Canada
arpan.laha@queensu.ca

Praveen Jain
Dept. of ECE
Queen's University
Kingston, Canada
praveen.jain@queensu.ca

Majid Pahlevani
Dept. of ECE
Queen's University
Kingston, Canada
majid.pahlevani@queensu.ca

Abstract—This paper proposes a novel control methodology for the solar-powered motor drive of a surface-mounted Permanent Magnet Synchronous Motor (SPMSM). Standalone solar-powered motor drives face considerable challenges due to inherent PV power fluctuations and the absence of an external energy storage system. These challenges are further exacerbated with solar microinverters utilizing a low DC link capacitance due to the limited energy storage capacity of the DC link. Reliable operation under such conditions requires a control strategy that optimally utilizes fluctuating solar power while maintaining stability during input power and load transients, despite the low DC link stiffness. The proposed control strategy quickly modulates the drive frequency to follow the Maximum Torque per Ampere (MTPA) trajectory while the current reference is directly set by DC link voltage regulation. Compared to existing observer-based vector control methods, the proposed approach offers fast dynamic performance, simplified implementation, and eliminates the need for a separate startup routine.

Index Terms—PMSM control, vector control, field-oriented control, FOC, PV water pump, solar motor drive.

I. INTRODUCTION

Solar-powered motor drives have gained traction for applications such as water pumps as we look to shift away from conventional energy sources due to environmental concerns. PMSMs are being increasingly considered for these pump drives owing to their high efficiency, high power density, and low maintenance requirements. PMSMs are typically controlled by V/f control, direct torque control (DTC), or vector control methods. Scalar V/f control is an open-loop control method that is simple to implement but suffers from stability issues and may lose synchronism without stabilizing loops [1], [2]. DTC method provides a fast torque response but can induce torque and current ripples in the motor [3], [4]. Vector control methods rely on the rotor position information obtained either from a position sensor or from a sensorless estimation method. For pump drives such as submersible pumps, the use of position sensors is not feasible due to the harsh and noise-prone environment.

Sensorless vector control methods utilize either signal injection [5]–[7] or model-based estimation [8]–[10]. Signal injection based methods are preferred for low speed operation while model-based methods are used for medium to high speed operation. Back-emf observer-based methods are widely used for pump drives as they are generally intended to work in the

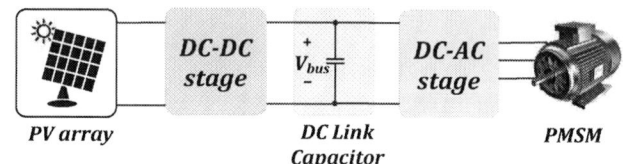

Fig. 1. A typical two-stage solar-powered motor drive.

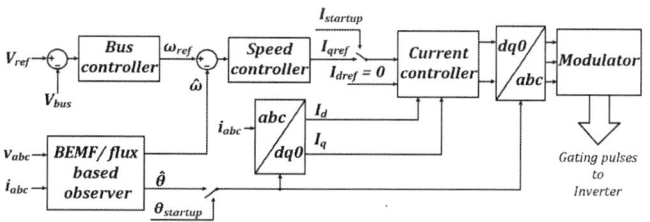

Fig. 2. Conventional observer-based vector control for PV driven PMSM.

medium-to-high speed range. However, these methods suffer from estimation errors at low speed and standstill [11]. Hence, they require a separate start-up routine to accelerate the motor to a minimum speed before the observer-based control can be applied. For this purpose, a startup procedure using I-f control is favored, where the motor is accelerated in open-loop with a constant current reference [12], [13].

Fig. 1 illustratess a two-stage solar-powered PMSM drive where the DC-DC stage performs maximum power point tracking (MPPT) for the photovoltaic (PV) panels and supplies power to the DC-AC stage, which drives the motor. The two stages are interfaced by a DC link capacitor. Fig. 2 depicts a conventional observer-based control scheme using I-f startup. During startup, the bus voltage remains unregulated due to the use of a constant current reference. In solar microinverters with low DC link capacitance, this can result in DC link voltage overshoot or collapse, potentially causing failure of drive startup. Moreover, the low DC link stiffness exacerbates the effects of fluctuating input power, leading to significant voltage oscillations at the DC link and system instability. Previous studies have predominantly focused on string inverters with

979-8-3315-1612-3/25 $31.00 © 2025 IEEE

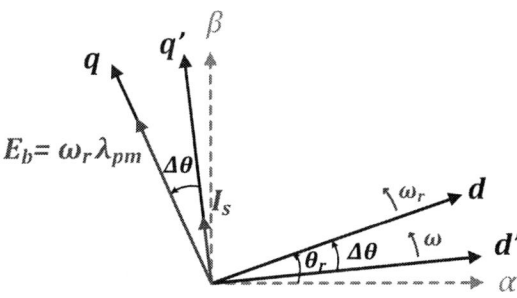

Fig. 3. Illustration of the control reference frame $d'q'$ and rotor dq frame.

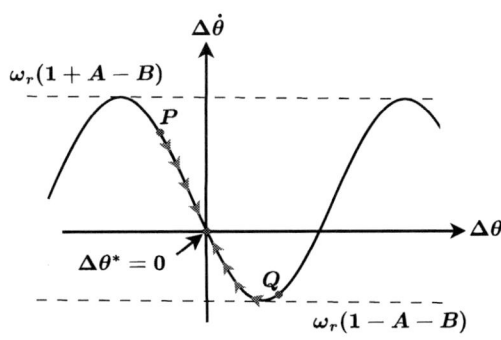

Fig. 4. Visual depiction of equation (9).

large DC link capacitance for PMSM drives [14], [15]. In [14], a PV array-fed water pumping system was proposed using a DC link capacitance of 4700μF compared to an 8μF capacitor used in this research. A low DC link capacitance allows the use of film capacitors enhancing the system's lifespan [16] which is a critical factor influencing the economic viability of PV water pumping applications [17].

This paper proposes a novel control scheme for PMSM drives powered by a solar microinverter with low DC link capacitance. The proposed controller offers fast dynamic performance, eliminates the need for a dedicated startup routine, and is simple to implement compared to observer-based control methods. It ensures reliable drive startup even under low irradiance conditions, while the use of small DC link capacitance enhances the power density and longevity of the system.

II. PROPOSED CONTROL STRATEGY

Let us consider a synchronous control reference frame $d'q'$ such that the q' axis is aligned with the stator current I_s as illustrated in Fig. 3. In this frame, the model of the SPMSM is given as

$$
\frac{d}{dt}\begin{pmatrix} \Delta\theta \\ J\omega_r \\ Li_{d'} \\ Li_{q'} \end{pmatrix} = \begin{pmatrix} \omega_r - \omega \\ \frac{P}{2}(T_{em} - T_L) - B\omega_r \\ V_{d'} - R_s i_{d'} + \omega L i_{q'} + \lambda_{pm}\omega_r \sin\Delta\theta \\ V_{q'} - R_s i_{q'} - \omega L i_{d'} - \lambda_{pm}\omega_r \cos\Delta\theta \end{pmatrix}
\tag{1}
$$

$$
T_{em} = \frac{3}{2} \cdot \frac{P}{2} \cdot \lambda_{pm}\left(i_{q'}\cos\Delta\theta - i_{d'}\sin\Delta\theta\right)
\tag{2}
$$

$$
i_{q'} = I_s, \quad i_{d'} = 0
\tag{3}
$$

where $\Delta\theta = \theta_r - \theta'$ is the angle difference between the rotor dq frame and the $d'q'$ control frame where θ_r and θ' are the angles of the rotor dq frame and $d'q'$ frame from the stationary axis respectively. ω_r is the rotor electrical speed, ω is the speed of the $d'q'$ frame, λ_{pm} is the peak magnetic flux linkage, T_L is the load torque, T_{em} is the developed electromagnetic torque, J is the moment of inertia, B is the viscous friction coefficient, P is the number of poles, R_s is the stator per-phase resistance,

and L is the synchronous inductance. $V_{d'}$, $V_{q'}$, $i_{d'}$, and $i_{q'}$ are the voltages and currents resolved along the d'-q' axes.

From (2), we observe that achieving Maximum Torque per Ampere (MTPA) control requires $\Delta\theta = 0$, meaning that the control frame $d'q'$ must align with the rotor dq frame. Given that the control frame is fixed to the stator current vector, the speed of the $d'q'$ frame is the speed of the current vector. Therefore, in a synchronous machine, both frames rotate at the same speed ($\omega_r = \omega$) under steady-state conditions. However, during transients, the speed of the two frames can differ changing the angle $\Delta\theta$ between them. Thus, by controlling the speed of the stator current vector ω, we can drive $\Delta\theta$ to zero, aligning the two frames.

In the absence of a position sensor, the rotor position information can be deduced from the motor voltages and currents due to the presence of the back-emf. From (1), we have

$$
V_{d'} = -\omega L i_{q'} - \lambda_{pm}\omega_r \sin\Delta\theta
\tag{4}
$$

$$
V_{q'} = L\frac{di_{q'}}{dt} + R_s i_{q'} + \lambda_{pm}\omega_r \cos\Delta\theta
\tag{5}
$$

Let us consider a control law $\omega = k_1 V_{d'} + k_2 V_{q'}$, where k_1 and k_2 are some coefficients, then:

$$
\omega = k_1\left(-\omega L i_{q'} - \lambda_{pm}\omega_r \sin\Delta\theta\right)
$$
$$
+ k_2\left(L\frac{di_{q'}}{dt} + R_s i_{q'} + \lambda_{pm}\omega_r \cos\Delta\theta\right)
\tag{6}
$$

The rate of change of the current is limited in solar applications, and hence by neglecting the contribution of the term $L\frac{di_{q'}}{dt}$, we have

$$
\omega = k_1\left(-\omega L i_{q'} - \lambda_{pm}\omega_r \sin\Delta\theta\right)
$$
$$
+ k_2\left(R_s i_{q'} + \lambda_{pm}\omega_r \cos\Delta\theta\right)
\tag{7}
$$

Solving for ω, we get

$$
\omega = \frac{k_2 R_s i_{q'}}{1 + k_1 L i_{q'}} + \frac{\sqrt{k_1^2 + k_2^2}}{1 + k_1 L i_{q'}}\lambda_{pm}\omega_r \sin\left(\Delta\theta + \tan^{-1}\left(\frac{k_2}{-k_1}\right)\right)
\tag{8}
$$

Substituting (8) in (1), we get

$$
\Delta\dot{\theta} = \omega_r - \omega
$$
$$
= \omega_r\left(1 - A\sin\left(\Delta\theta + \theta_c\right) - B\right)
\tag{9}
$$

979-8-3315-1612-3/25 $31.00 © 2025 IEEE

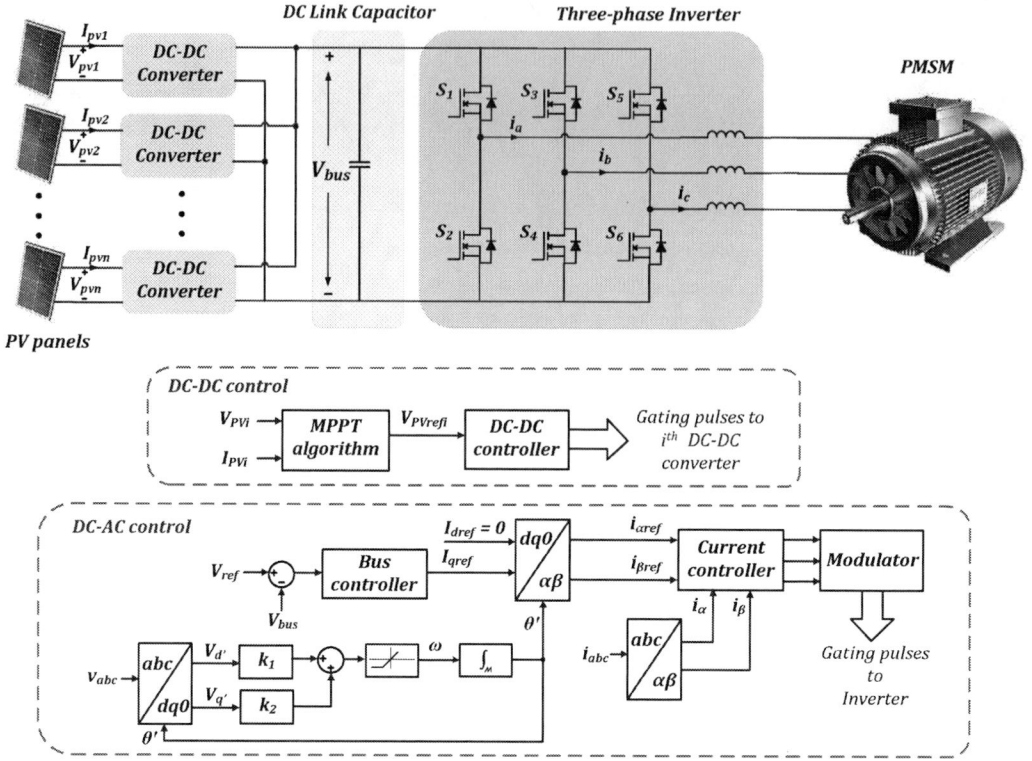

Fig. 5. Proposed control scheme for solar-powered PMSM drive.

where $A = \frac{\sqrt{k_1^2+k_2^2}}{1+k_1Li_{q'}}\lambda_{pm}$, $B = \frac{k_2R_si_{q'}}{\omega_r(1+k_1Li_{q'})}$, and $\theta_c = \tan^{-1}\left(\frac{k_2}{-k_1}\right)$. A unidirectional motor drive is required for applications such as water pumps. Hence, we assume operation with $\omega_r > 0$. From (9), the steady-state value of $\Delta\theta$ is found as

$$\Delta\theta^* = \sin^{-1}\left(\frac{1-B}{A}\right) - \theta_c. \tag{10}$$

We require the system to settle at $\Delta\theta = 0$ for MTPA control. This can be achieved by setting $\theta_c = \sin^{-1}\left(\frac{1-B}{A}\right)$. Fig. 4 provides a visual depiction of equation (9) for this choice of θ_c. In addition, we also require that this equilibrium point is stable. Assuming the system starts at point P, where $\Delta\dot{\theta}$ is positive, the control forces $\Delta\theta$ to increase until it reaches $\Delta\theta = 0$, at which point $\Delta\dot{\theta}$ becomes zero and the system remains at equilibrium. Similarly, starting at point Q, where $\Delta\dot{\theta}$ is negative, $\Delta\theta$ decreases until it reaches the stable equilibrium point at $\Delta\theta = 0$.

To achieve this stable control that attracts to the equilibrium point $\Delta\theta^* = 0$, we require the sinusoidal waveform intersect the $\Delta\theta = 0$ axis as shown in Fig. 4. This gives us two conditions: $\omega_r(1+A-B) > 0$ and $\omega_r(1-A-B) < 0$. Hence, for the controller to have a stable equilibrium at $\Delta\theta = 0$, we have the following conditions:

(i) $\theta_c = \sin^{-1}\left(\frac{1-B}{A}\right)$
(ii) $\omega_r(1+A-B) > 0$

(iii) $\omega_r(1-A-B) < 0$

Solving for these three conditions, we have

$$0 > k_1 > -\frac{1}{Li_{q'}} \tag{11}$$

$$k_2 = \frac{(1+k_1Li_{q'})}{\lambda_{pm}}\left(1 - \frac{i_{q'}R_s}{V_{q'}}\right) \tag{12}$$

Hence, by setting the drive frequency as $\omega = k_1V_{d'} + k_2V_{q'}$, with the choice of k_1 and k_2 according to (11) and (12) forces the controller to track the MTPA locus.

Fig. 5 shows the block diagram of the proposed control scheme for a solar microinverter-fed PMSM drive. The DC-DC converters are controlled to track the maximum power point (MPP) of individual PV panels. In addition, a power limitation function is implemented to limit the input power if the drive approaches the rated speed. The outputs of the DC-DC converters are tied to a DC link capacitor, which supplies power to a three-phase inverter. DC link voltage is regulated using a PI controller to directly set the stator current amplitude reference I_{qref}, while the frequency reference for the motor drive ω is derived to follow the MTPA trajectory. The phase reference for the stator current θ' is obtained from integrating the speed reference. The final stator current references in the stationary $\alpha\beta$ frame are given by $i_{\alpha ref} = I_{qref}\sin\theta'$ and $i_{\beta ref} = I_{qref}\cos\theta'$.

III. Stability Analysis of Proposed Scheme

The stability analysis of the motor drive under the proposed control can be carried out considering equations (1)-(3) with $\omega = k_1 V_d' + k_2 V_q'$, where k_1 and k_2 are chosen as in (11) and (12). In addition, the load torque is set as $T_L = T_0 + T_1 \left(\frac{2\omega_r}{P}\right)^2$ (pump load), and the total power input to the inverter is $P_{in} = \frac{3}{2} V_q' i_q'$. With $i_{d'}$ set to zero, the state variables are $\Delta\theta$, ω_r, and $i_{q'}$ and the dynamical equations governing the system are as follows.

$$\Delta\dot{\theta} = \omega_r + \frac{k_1 \lambda_{pm} \omega_r \sin\Delta\theta}{1 + k_1 L i_{q'}} - \frac{1}{\lambda_{pm}}\left(1 - \frac{3 i_{q'}^2 R_s}{2 P_{in}}\right) \quad (13)$$

$$\dot{\omega}_r = \frac{3 P^2 \lambda_{pm}}{8 J} i_{q'} \cos\Delta\theta - \left(\frac{T_0 P}{2J} + \frac{2 T_1}{PJ}\omega_r^2\right) - \frac{B}{J}\omega_r \quad (14)$$

$$\frac{di_{q'}}{dt} = \frac{1}{L}\left(\frac{2 P_{in}}{3 i_{q'}} - R_s i_{q'} - \lambda_{pm}\omega_r \cos\Delta\theta\right) \quad (15)$$

The eigenvalues of the system can be found by linearizing around the operating point $(\Delta\theta^*, \omega_r^*, i_{q'}^*)$ with $\Delta\theta^* = 0$ as

$$\lambda_{1,2} = \frac{1}{2}\left(-\left(\frac{2 P_{in}}{3 L (i_{q'}^*)^2} + \frac{R_s}{L} + \frac{B}{J} + \frac{4 T_1}{PJ}\omega_r^*\right)\right.$$

$$\left.\pm \sqrt{\left(\frac{2 P_{in}}{3 L (i_{q'}^*)^2} + \frac{R_s}{L} + \frac{B}{J} + \frac{4 T_1}{PJ}\omega_r^*\right)^2 - \frac{12 P^2 \lambda_{pm}^2}{8 J L}}\right)$$

$$(16)$$

$$\lambda_3 = \frac{k_1 \omega_r^* \lambda_{pm}}{1 + k_1 L i_{q'}^*}, \quad (17)$$

It can be observed that the eigen values $\lambda_{1,2}$ have negative real parts for $\omega_r^* > 0$. For the choice of k_1 such that $0 > k_1 > -\frac{1}{L i_{q'rated}}$, the eigen value λ_3 also becomes negative proving the stability of the proposed control scheme.

IV. Simulation Results

A 2.2 kW motor, with specifications as provided in the Appendix, was simulated with PSIM software to validate the efficacy of the proposed control scheme. The parameter $k_1 = -5$ was selected based on (11), and k_2 was calculated as per (12). A DC link voltage reference of 500 V and a bus capacitance of 8 μF were chosen, with the DC-AC stage control initiated when the voltage V_{bus} exceeded 450 V. Fig. 6 illustrates the startup of the drive at an input power of 500 W under the proposed control scheme. The drive frequency is set as $\omega = k_1 V_d + k_2 V_q$, and the frequency will initially be close to zero. To facilitate drive startup, a frequency limiter is utilized to set a minimum drive frequency ω_{min}. As rotor speed increases, motor voltage builds up, and drive frequency ramps up, aligning the dq and $d'q'$ reference frames. Conventional observer-based control requires a separate startup method, such as I-f startup, due to the non-observability phenomena at stand-still. In such methods, a constant current reference is utilized which can cause large oscillations in the DC link voltage, destabilizing the drive. In contrast, the

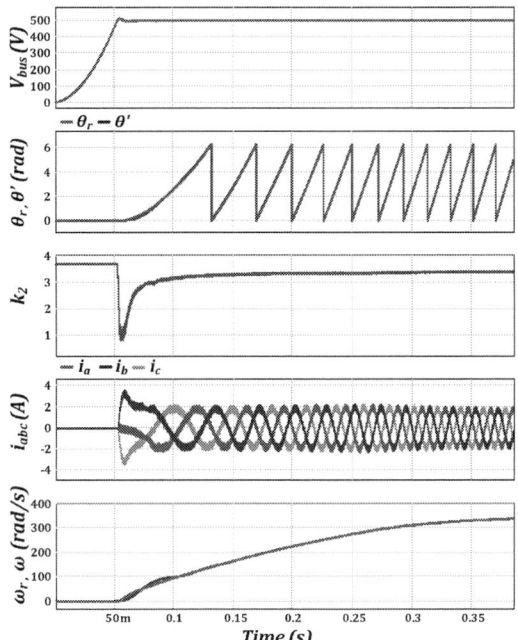

Fig. 6. Simulation of drive startup under proposed scheme.

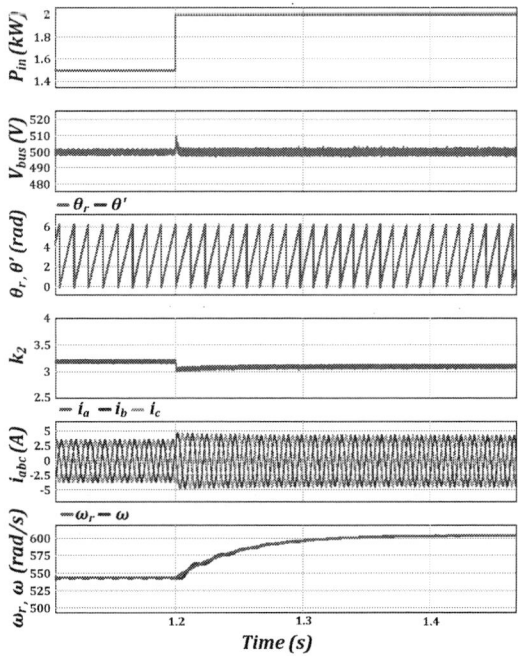

Fig. 7. Simulation waveforms under power transient of 500W.

proposed scheme achieves a smooth startup with minimal DC link fluctuation in a DC link, even with low stiffness, without the requirement of a dedicated startup routine.

Fig. 7 and Fig. 8 show the drive performance under input power and load transient respectively. In both scenarios, the proposed control scheme quickly stabilizes the DC link while modifying the drive frequency to follow the MTPA trajectory. Fig. 9 illustrates the variation of the coefficient

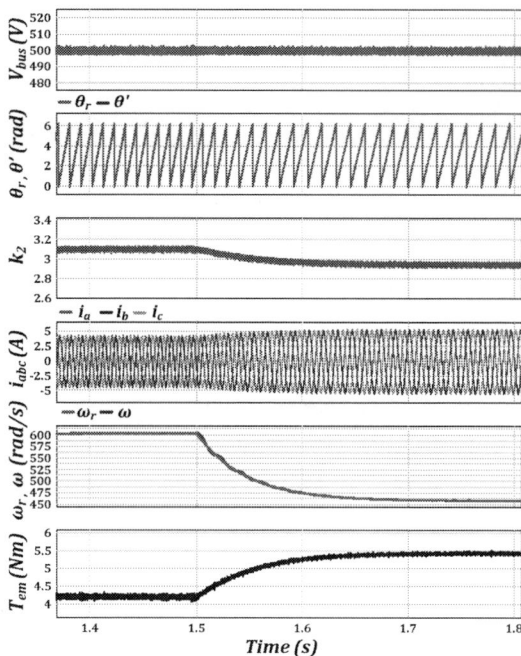

Fig. 8. Simulation waveforms under load transient of 1.5Nm.

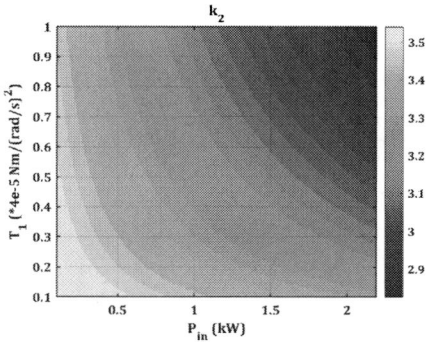

Fig. 9. Contour plot of k_2 with respect to input power P_{in} and varying torque profile $T_L = 0.5 + T_1\omega_m^2$.

k_2 with changes in input power and load torque profiles. Unlike conventional observer-based control with a cascaded control loop as shown in Fig. 2, the proposed scheme regulates the DC link by directly setting the current reference - an electrical state variable that has a fast dynamic behavior. This approach enables the use of bus controller with higher bandwidth. In addition, the proposed scheme also eliminates delays associated with the observer and hence demonstrates superior dynamic performance.

V. EXPERIMENTAL RESULTS

An experimental prototype of a solar microinverter was developed, and the proposed control strategy was implemented on an 8-pole 400 W Surface-Mounted Permanent Magnet Synchronous Motor (SPMSM) with specifications provided in the Appendix. The motor was coupled with a permanent magnet synchronous generator for testing purposes. An LLC

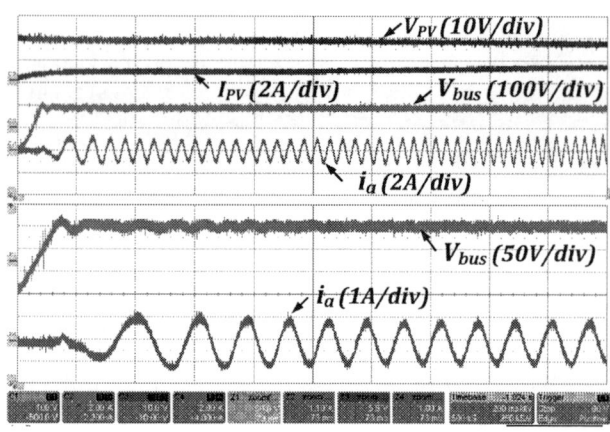

Fig. 10. Experimental waveform showing drive startup at irradiance of $100W/m^2$.

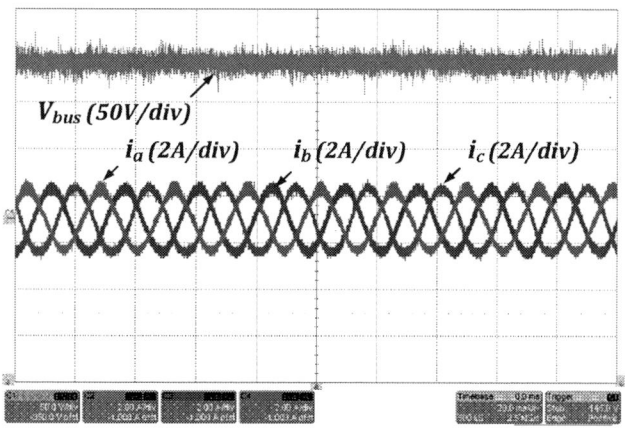

Fig. 11. Steady-state waveforms at irradiance of $100W/m^2$.

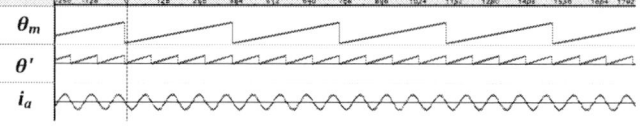

Fig. 12. Signaltap data at irradiance of $100W/m^2$.

resonant converter was utilized for the DC-DC stage and was powered by a TerraSAS PV simulator. A Perturb & Observe (P & O) algorithm was implemented for Maximum Power Point Tracking (MPPT). The entire control scheme was digitally implemented using a Field Programmable Gate Array (FPGA). The system incorporated an $8\,\mu$F DC link capacitor, with the DC link voltage regulated at 500 V. The inverter operation was enabled once the DC link voltage exceeded a threshold of 450 V, and the minimum drive frequency ω_{min} was set to 20 rad/s.

Fig. 10 shows the startup of the drive under a low irradiance

979-8-3315-1612-3/25 $31.00 © 2025 IEEE

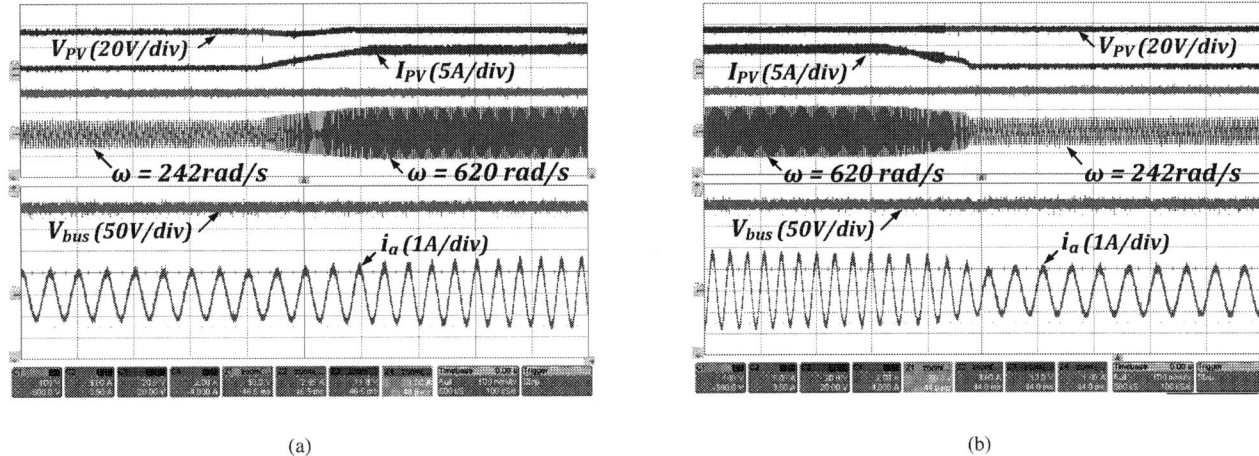

(a)   (b)

Fig. 13. Experimental waveforms under proposed scheme during input power transient due to (a) irradiance change from $100W/m^2$ to $500W/m^2$ (b) irradiance change from $500W/m^2$ to $100W/m^2$.

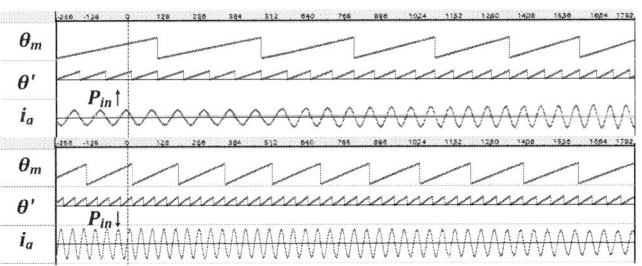

Fig. 14. Signatap data during input power transient.

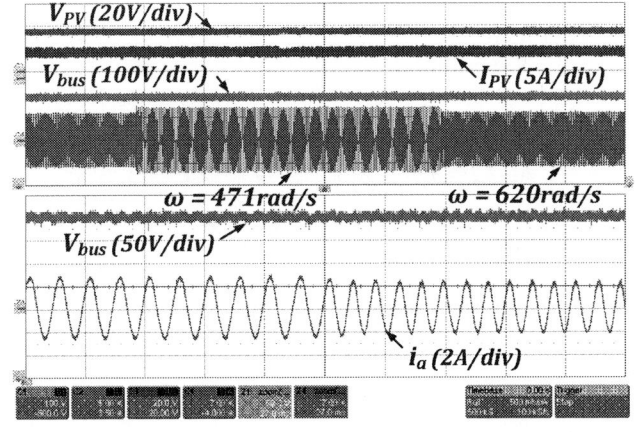

Fig. 15. Experimental waveforms during load transient of $0.35Nm$.

condition of $100\,W/m^2$. The PV voltage (V_{PV}) decreases, and the PV current (I_{PV}) increases as the MPPT algorithm moves the operating point towards the MPP (36V 1.4A). As the DC link capacitor charges and exceeds the minimum threshold, it triggers the control action of the DC-AC stage.

The enlarged view shows that the drive starts smoothly with minimal fluctuations in the DC link voltage. Fig. 11 displays the waveforms of the three-phase currents when the system reaches steady-state at a drive frequency of 242 rad/s. The encoder signals from the motor were fed to the FPGA to reconstruct the mechanical rotor angle $\theta_m = \frac{2}{P}\theta_r$. This was compared with the control angle θ' using a Signaltap logic analyzer as shown in Fig. 12. Since the motor has 8 poles, it exhibits four electrical cycles per mechanical cycle. It can be observed that the falling edges of both θ_m and θ' are aligned, verifying the MTPA tracking performance of the proposed controller.

Fig. 13 illustrates the drive performance under the proposed control scheme during input power transients caused by a change in irradiance condition of the PV panel, varying from 100 W/m^2 to 500 W/m^2 and vice-versa. The results indicate that the proposed scheme effectively regulates the DC link voltage while concurrently tracking the MTPA trajectory, as shown in Fig. 14. The transient response of the drive to a change in load torque is illustrated in Fig. 15. The experimental waveforms demonstrate the fast dynamic performance of the proposed control scheme, effectively stabilizing the DC link during transients despite its low stiffness. Additionally, the results validate the controller's ability to track the rotor dq frame by swiftly modulating the drive frequency in response to load or input power transients. Compared to conventional observer-based controllers, the proposed scheme features a simplified control structure that is easy to implement, exhibits a fast dynamic response, and eliminates the need for a separate startup routine.

It should be noted that the calculation of the gain coefficient k_2 is dependent on motor parameters such as resistance R_s and inductance L. Variations in these parameters during motor operation can lead the controller to deviate from the Maximum Torque Per Ampere (MTPA) trajectory. For example, a 40% reduction in inductance from the rated value can cause the controller to settle at $\Delta\theta = 0.056$ radians under rated load conditions. This results in a 0.16% reduction in torque, which is acceptable for applications such as water pumps.

VI. CONCLUSION

This paper introduced a novel control strategy for a solar-powered Permanent Magnet Synchronous Motor (PMSM) drive, addressing the challenges posed by solar power variability and limited energy storage in the DC link. The proposed scheme ensures robust control of motor drives with solar microinverters equipped with low DC link film capacitors, enhancing the power density and longevity of the system. The proposed approach directly modulates the drive frequency in response to power and load transients, maintaining stability and optimizing the torque output under varying solar irradiance. This method not only simplifies the implementation process but also eliminates the need for a dedicated startup process and enhances the dynamic performance of the motor drive. Both simulation and experimental results validate the effectiveness of the proposed scheme in following the MTPA trajectory while ensuring reliable operation without the complexities and limitations of traditional methods.

APPENDIX

Specifications of PMSM Used for Simulations

Rated Power (P_{rated}): 2.2 kW, Rated Voltage (V_{rated}): 230 V, Peak Magnetic Flux Density (λ_{pm}): 0.27 Wb, Stator Resistance (R_s): 1.8 Ω, Inductance (L): 5.3 mH.

Specifications of PMSM Used for Experiments

Rated Power (P_{rated}): 400 W, Rated Voltage (V_{rated}): 200 V, Peak Magnetic Flux Density (λ_{pm}): 0.094 Wb, Stator Resistance (R_s): 2.35 Ω, Inductance (L): 7 mH.

REFERENCES

[1] P. D. C. Perera, F. Blaabjerg, J. K. Pedersen, and P. Thogersen, "A sensorless, stable V/f control method for permanent-magnet synchronous motor drives," *IEEE Trans. Ind. Appl.*, vol. 39, no. 3, pp. 783–791, May-June 2003.

[2] Z. Tang, X. Li, S. Dusmez, and B. Akin, "A New V/f-Based Sensorless MTPA Control for IPMSM Drives," *IEEE Trans. Power Electron.*, vol. 31, no. 6, pp. 4400–4415, June 2016.

[3] L. Zhong, M. F. Rahman, W. Y. Hu, and K. W. Lim, "Analysis of direct torque control in permanent magnet synchronous motor drives," *IEEE Trans. Power Electron.*, vol. 12, no. 3, pp. 528–536, May 1997.

[4] F. Niu, B. Wang, A. S. Babel, K. Li, and E. G. Strangas, "Comparative evaluation of direct torque control strategies for permanent magnet synchronous machines," *IEEE Trans. Power Electron.*, vol. 31, no. 2, pp. 1408–1424, Feb. 2016.

[5] A. Consoli, G. Scarcella, and A. Testa, "Industry application of zero-speed sensorless control techniques for PM synchronous motors," *IEEE Trans. Ind. Appl.*, vol. 37, no. 2, pp. 513–521, Mar.-Apr. 2001.

[6] P. L. Xu and Z. Q. Zhu, "Carrier signal injection-based sensorless control for permanent-magnet synchronous machine drives considering machine parameter asymmetry," *IEEE Trans. Ind. Electron.*, vol. 63, no. 5, pp. 2813–2824, May 2016.

[7] G. Xie, K. Lu, S. K. Dwivedi, J. R. Rosholm, and F. Blaabjerg, "Minimum-voltage vector injection method for sensorless control of PMSM for low-speed operations," *IEEE Trans. Power Electron.*, vol. 31, no. 2, pp. 1785–1794, Feb. 2016.

[8] Z. Chen, M. Tomita, S. Doki, and S. Okuma, "An extended electromotive force model for sensorless control of interior permanent-magnet synchronous motors," *IEEE Trans. Ind. Electron.*, vol. 50, no. 2, pp. 288–295, Apr. 2003.

[9] Y. Zhao, Z. Zhang, W. Qiao, and L. Wu, "An extended flux model-based rotor position estimator for sensorless control of salient-pole permanent-magnet synchronous machines," *IEEE Trans. Power Electron.*, vol. 30, no. 8, pp. 4412–4422, Aug. 2015.

[10] Z. Qiao, T. Shi, Y. Wang, Y. Yan, C. Xia, and X. He, "New sliding-mode observer for position sensorless control of permanent-magnet synchronous motor," *IEEE Trans. Ind. Electron.*, vol. 60, no. 2, pp. 710–719, Feb. 2013.

[11] S. Bolognani, S. Calligaro, and R. Petrella, "Design issues and estimation errors analysis of back-EMF-based position and speed observer for SPM synchronous motors," *IEEE J. Emerg. Sel. Topics Power Electron.*, vol. 2, no. 2, pp. 159–170, June 2014.

[12] Z. Wang, K. Lu, and F. Blaabjerg, "A simple startup strategy based on current regulation for back-EMF-based sensorless control of PMSM," *IEEE Trans. Power Electron.*, vol. 27, no. 8, pp. 3817–3825, Aug. 2012.

[13] D. Chen, K. Lu, D. Wang, and M. Hinkkanen, "I-F control with zero D-axis current operation for surface-mounted permanent magnet synchronous machine drives," *IEEE Trans. Power Electron.*, vol. 38, no. 6, pp. 7504–7513, June 2023.

[14] B. Singh and S. Murshid, "A grid-interactive permanent-magnet synchronous motor-driven solar water-pumping system," *IEEE Trans. Ind. Appl.*, vol. 54, no. 5, pp. 5549–5561, Sept.-Oct. 2018.

[15] S. Murshid and B. Singh, "Double stage solar PV array fed water pump driven by permanent magnet synchronous motor," *IEEE Trans. Ind. Appl.*, vol. 57, no. 2, pp. 1736–1745, Mar.-Apr. 2021.

[16] E. T. Maddalena, C. G. d. S. Moraes, G. Bragança, L. G. Junior, R. B. Godoy, and J. O. P. Pinto, "A battery-less photovoltaic water-pumping system with low decoupling capacitance," *IEEE Trans. Ind. Appl.*, vol. 55, no. 3, pp. 2263–2271, May-June 2019.

[17] S. Meunier *et al.*, "Sensitivity analysis of photovoltaic pumping systems for domestic water supply," *IEEE Trans. Ind. Appl.*, vol. 56, no. 6, pp. 6734–6743, Nov.-Dec. 2020.

Design of a Compact Low-Loss MMC Double Submodule for MVDC and HVDC Applications

Ali Sharaf Addin, Rainer Marquardt, and Thomas Brückner

Laboratory for High Power Electronic Systems
Universität der Bundeswehr München
Neubiberg, Germany
e-mail: ali.sharaf@unibw.de

Abstract—**Double-Connected Submodules for Modular Multilevel Converters (MMC) have been a topic of interest in the past ten years due to their advantages in terms of reducing losses as well as footprint. Yet, there has been no practical implementation of them for use in Medium- or High-Voltage Direct-Current (MVDC, HVDC) applications due to the higher complexity compared to simple half-bridge or full-bridge submodules. A submodule topology that enables the reduction of semiconductor losses as well as capacitor size is the Double-Zero Submodule in Double Connection (DZDCSM). This paper presents a compact design of this topology based on 3.3-kV SiC MOSFET devices for a 2× 2 kV cell voltage rating. First measurement results on a full-scale prototype are presented.**

I. INTRODUCTION

The Modular Multilevel Converter (MMC) is the key component for HVDC transmission, connecting renewable power plants, and future meshed DC grids [1]–[3]. The characteristics of the MMC are governed by the types of submodules being utilized. Advanced types of submodules have been introduced in recent years that combine the benefits of low overall power losses with a bipolar output voltage, which is required for improved protection and current control, especially in case of DC faults [4]–[10]. One of the promising submodule topologies is the Double-Zero Submodule (DZSM) [1], which can replace the Full-Bridge Submodule (FBSM) featuring a switchable capacitor, see Fig. 1a. This allows connecting the upper and lower paths in parallel during the zero state and thus reduces the conduction losses compared to FBSM, significantly. If applied in a double connection (DZDCSM), as shown in Fig. 1b, the conduction losses, as well as the capacitor voltage ripple, can be reduced even further [11]–[13]. The DZDCSM features four terminal voltage levels: $+2V_C$ (connecting both capacitors in series), $\pm V_C$ (parallel connection of C_1 and C_2), and 0 V (paralleled bypass over the lower and upper branches). It has been shown that a fully SiC-equipped DZDCSM can operate with approximately 50 % less conduction losses than a series connection of two FBSM for the same installed SiC chip area [5], [13]. Until now, there has been no industrial implementation of the DZDCSM for MVDC or HVDC applications due to its seemingly complicated structure and the associated hardware and control design challenges [5].

This paper proposes a practical design concept for a compact DZDCSM and presents first measurement results to prove the concept. In the second section, a brief overview of the

TABLE I
ELECTRICAL PARAMETERS OF THE SUBMODULE PROTOTYPE

Parameter	Value
Nominal SM terminal voltages	0 V, ±2 kV, and 4 kV
Nominal arm current	1000/3 A (DC) | 380 A (AC)
Semiconductor type	3.3-kV 1-kA SiC MOSFET FF2000UXTR33T2M1
Capacitance $C_{x=\{1,2\}}$	2 mF
$\overline{V}_{C_{x=\{1,2\}}}$	2 kV ±15 %
$I_{C_{x=\{1,2\}}}$	150 A

electrical parameters of the submodule is provided. The third section highlights some key aspects of the practical design, followed by results of the practical tests. Finally, an evaluation of cost and efficiency is provided.

II. ELECTRICAL PARAMETERS

The submodule's voltage and current ratings have been chosen to enable an application in both MVDC and HVDC converter stations. Targeting applications with DC voltages of 20 kV and above, utilizing a semiconductor voltage rating of 3.3 kV leads to an MMC design with a minimum of seven or eight double submodules per converter arm. This appears to be adequate to gain a sufficient number of voltage levels and to ensure redundancy. Further assuming a power rating of 20 MW at 20 kV DC, leads to a DC current of 1 kA. Since a fully SiC-based DZDCSM offers the minimum conduction losses compared to a Silicon (Si) solution or a hybrid combination [13], it was decided to solely apply SiC MOSFET. The higher cost of SiC devices can be amortized by the loss reduction over time, which will be shown in the fifth section. The electrical parameters of the prototype are listed in Tab. I.

III. SUBMODULE DESIGN

To achieve a compact yet simple and reliable design of the DZDCSM with the parameters listed in Tab. I, a number of key points needs to be resolved, such as 1) realizing the submodule topology using commercial half-bridge power modules, 2) isolation coordination based on two submodule capacitors, 3)

979-8-3315-1612-3/25 $31.00 © 2025 IEEE

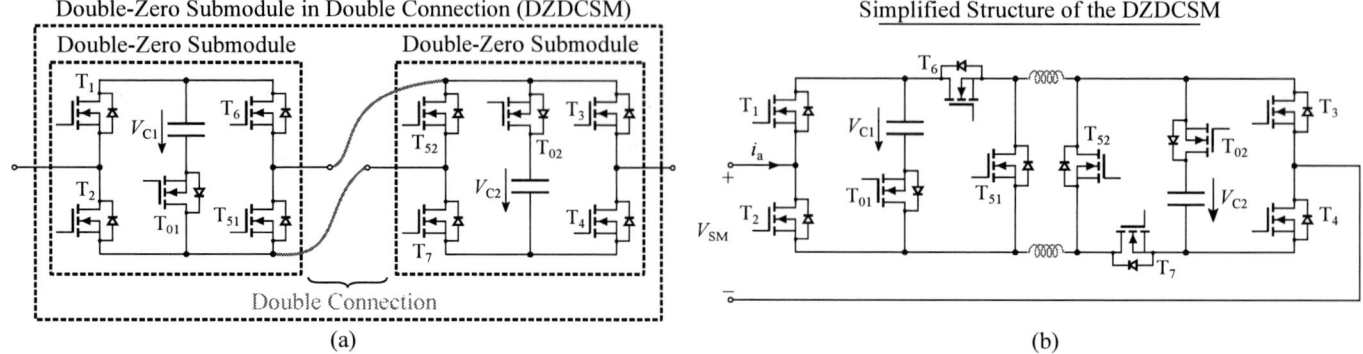

Fig. 1. Topology of the Double-Zero Submodule in Double Connection (DZDCSM) featuring SiC MOSFET

mechanical integration and minimizing the submodule volume. These points are addressed in the following.

A. Realization of the Submodule Topology

SiC switches of the required power rating are primarily available as half-bridge power modules in packages of the XHP (LV100, ...) family. These modules are optimized for application in standard two-level converters. The challenge arises in how to implement the submodule topology using these devices. Fig. 2 shows three possible configurations. They mainly differ in the way the capacitor switches (T_{01} and T_{02}) are realized. Configuration (a) utilizes one switch of a half bridge connected via the AC terminal to the upper branch as a capacitor switch. The second switch is connected to a resistor for discharging the capacitor in case of a submodule failure [12]. Since the power module's AC terminals are associated with higher inductances (highlighted in yellow in Fig. 2), they lead to an increased stray inductance in the commutation loops, which is very unfavorable for fast-switching SiC devices and is the case in configuration (a). The same problem also applies to configuration (b), where not only two, but all four commutation loops contain one AC terminal inductance. Please note that the double-connection parasitic inductances L_d for connecting both submodule parts are not involved in any commutation process and are not critical. So, having an AC terminal of a power module connected to that L_d is acceptable. To avoid AC terminals from being included in commutation loops as in (a) and (b), utilizing a half-bridge module with both switches connected in series as the capacitor switch might be a suitable solution. However, this would generate additional conduction losses. Therefore, a hard bypass across one of the two switches is implemented in configuration (c). The inductance of that bypass including the AC terminal (L_{sn}) together with the parallel body diode acts as a snubber, if the bypassed switch is kept in the off state. This will be further investigated in section IV. In design (c), the two discharging switches are realized using discrete single-chip semiconductors.

The DZDCSM switches can be divided into the groups $T_{1,4}$, $T_{2,3}$, $T_{51,52}$, $T_{6,7}$, and $T_{01,02}$. The group switches can be synchronously controlled and also generate the same losses due

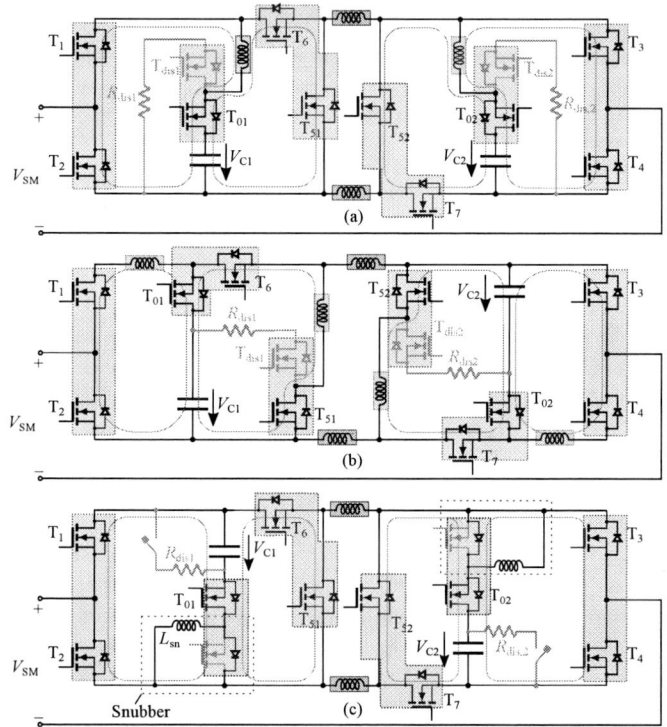

Fig. 2. Different options to realize the DZDCSM using half-bridge power modules highlighted in green

to the submodule symmetry. T_1 and T_4 are stressed the most — also for different power factors — because they are turned on for three switching states ($0\,\text{V}$, $+V_C$, $+2V_C$). A thermal design based on the submodule rating has shown that T_1 and T_4 require a parallel connection. Therefore, the submodule outer half bridges (T_1,T_2 and T_3,T_4) are implemented as a parallel connection of two XHP modules. Further information for the submodule operating point will be given in section V.

B. Isolation Coordination

An electrical potential within the submodule has to be chosen as a ground reference for the housings, heatsinks, and

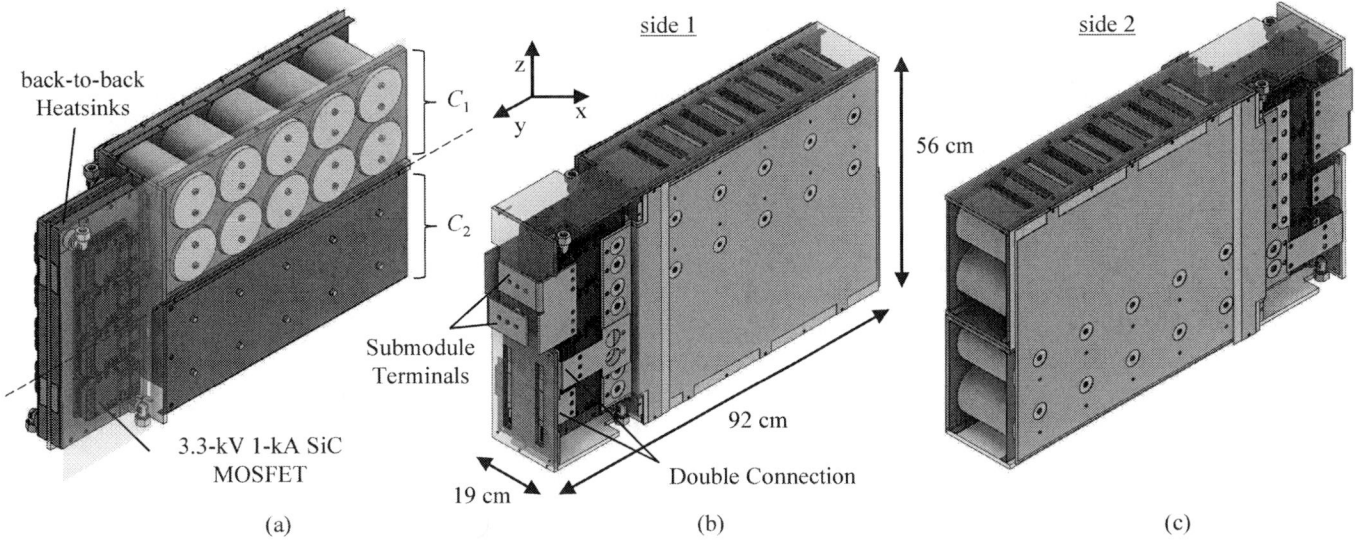

Fig. 3. CAD model of the DZDCSM prototype, (a): inner layout, (b): side-1, (c): side-2 (side-1 turned by 180° around the z-axis)

any electronic boards. In conventional submodules with one capacitor, the negative capacitor terminal is typically taken as a ground. Hence, the capacitor voltage is the reference value for determining the isolation voltage. As the DZDCSM contains two capacitors that can be inserted in series, having two grounds within it – linked to the negative capacitor terminals – would require designing clearance and creepage distances between differently grounded parts for an isolation voltage of a referenced nominal capacitor voltage. This is not preferable to achieve a compact design. So, it is better to have only one ground in the DZDCSM to avoid extra clearance and creepage distances. Furthermore, it is preferred that the potential difference from any electrical branch within the submodule to this ground point does not exceed one capacitor voltage considering any switching combinations of the switches within the submodule, including failure cases. These requirements are met by using the branch of the source of T_2 as a ground reference, solely. This makes the isolation requirement of the DZDCSM same as of conventional submodules.

C. Mechanical Integration

Smallest submodule dimensions are targeted to minimize the installation space of the converter stations. The developed structure of the DZDCSM is illustrated in Fig. 3 and a photograph of the realized prototype is shown in Fig. 4. The submodule consists of two sides designed symmetrically around the y-axis. Each side houses one DZSM comprising a bank of cylindrical film capacitors, half-bridge power modules, and a heatsink, which are highlighted in yellow in Fig. 3a. A common double-sided heatsink for both sides would be the preferable solution, but was not available with appropriate rating at the time of the construction. Both sides are double connected to form the DZDCSM, as shown in Fig. 3b. It should be noted that due to the double connection, the installed capacitance can be significantly reduced compared to that

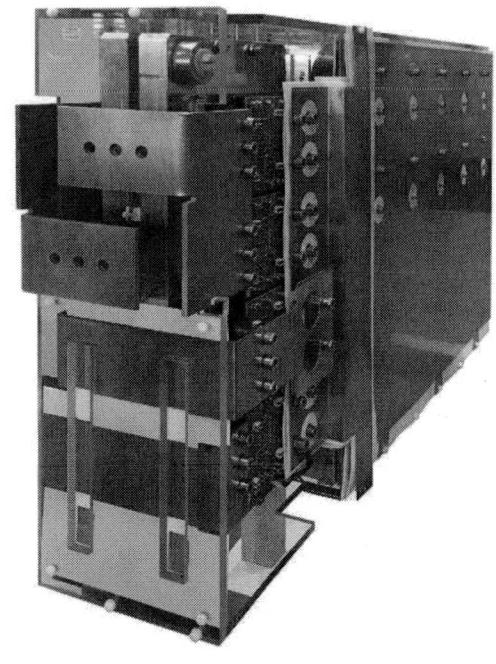

Fig. 4. Photograph of the DZDCSM prototype (dimensions: 92 cm, 19 cm, 56 cm, weight ≈ 80 kg)

of two equivalent FBSM [12], [13]. This provides a great benefit, since the capacitor occupies most of the submodule's volume. The chosen arrangement enables the usage of large busbars over both submodule side areas for connecting the capacitor banks with the semiconductors with a minimum stray inductance. For example, the busbar stray inductance of the first commutation loop of implementation (c) in Fig. 2 was estimated to be 13 nH using Ansys-Q3D. Adding 5 nH of the paralleled capacitors and 15 nH of the power modules [14]

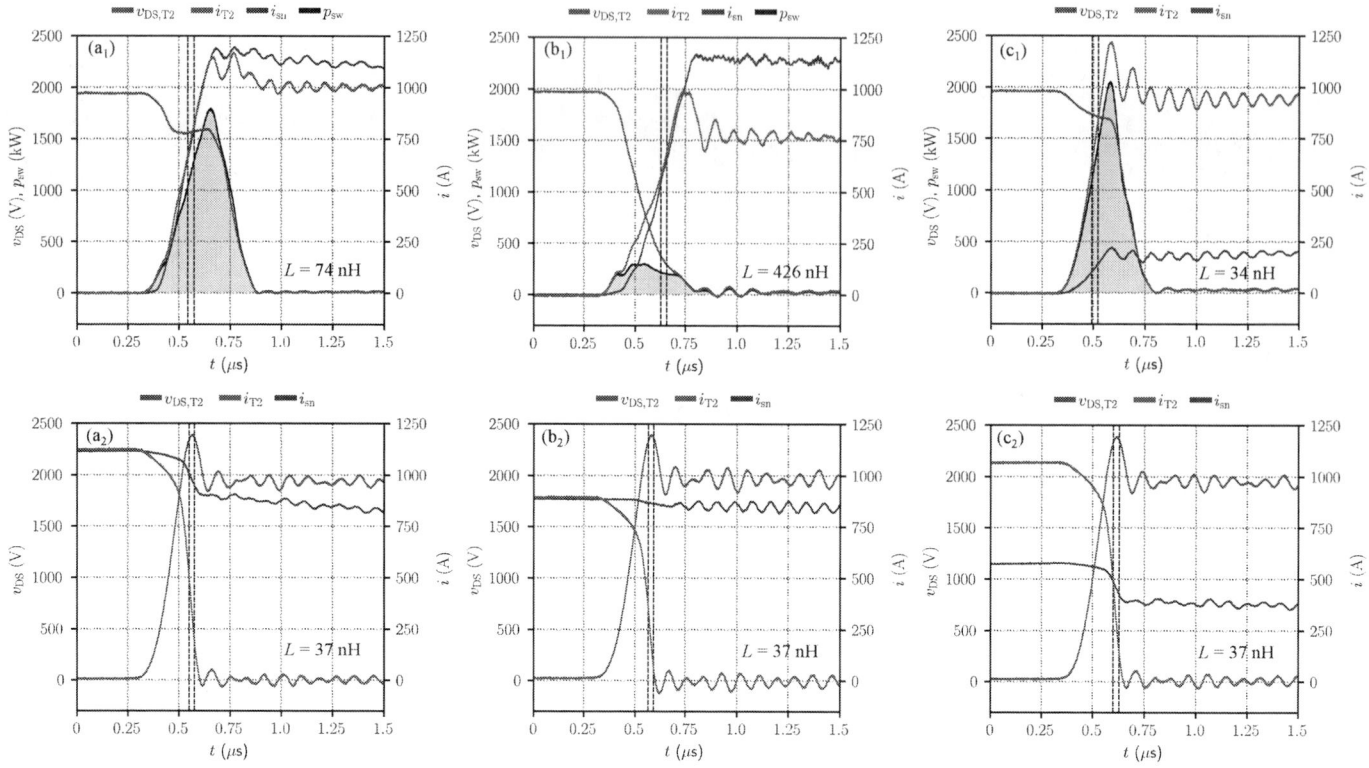

Fig. 5. Measured turn-on and -off transients of T_2 at $T_j = 25\,^\circ\text{C}$ for different scenarios: (a) active snubber case 1, a_1: 1000-A turn on, a_2: 1120-A turn off | (b) active snubber case 2, b_1: 750-A turn on, b_2: 890-A turn off | (c) inactive snubber, c_1: 940-A turn on, c_2: 1060-A turn off. The vertical dashed lines indicate the integration limits for calculating parasitic inductances

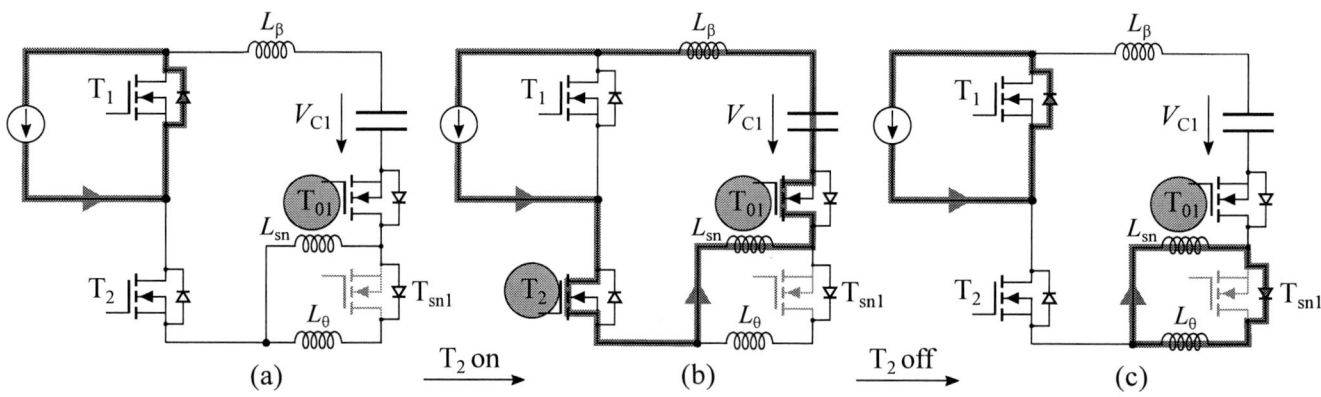

Fig. 6. Switching sequence for investigating the snubber function in a double-pulse test with active snubber (as of Fig. $5a_1$ and a_2), highlighted switches are in the on state

gives a total inductance of 33 nH. This design approach significantly improves space utilization while effectively minimizing stray inductance. Such low inductances would be difficult to achieve utilizing prismatic capacitors with isolated terminals.

IV. MEASUREMENT RESULTS

This section presents an analysis of the measurement results obtained from the prototype, focusing on two key aspects: First, evaluating the options for applying the capacitor switch (T_{01} and T_{02}) and their effect on the parasitic loop inductance. Second, analyzing the entire submodule operation by applying

a pulse sequence to investigate the current distribution, commutation behaviors, and also parasitic inductances within the submodule.

A. Snubber Function Investigation

As already mentioned, single-switch SiC MOSFET modules of the targeted power rating − needed for applying the capacitor switch − are currently not available on the market. Therefore, a half-bridge power module is used with one switch acting as a capacitor switch (T_{01}). In order to utilize the second switch (T_{sn1}) two options have been considered:

- *active snubber*: paralleling T_{sn1} with an electrical connection (L_{sn}) from the AC terminal to the DC+ terminal of the power module, with T_{sn1} always in the off state, or
- *inactive snubber*: same as *active snubber*, but with T_{sn1} always in the on state. This option is functionally equivalent to a single-switch power module.

Both scenarios eliminate the conduction losses of T_{sn1} due to the bypass. However, they differ in their impact on the loop inductance and the switching losses, which is the focus in the following. To make this investigation straightforward, only the first commutation loop in the submodule is considered by performing a double-pulse test on T_2 – applied as a parallel connection of two switches.

1) Active Snubber Scenario: The switching sequence for the double-pulse test is illustrated in Fig. 6, applying this scenario. During the turn-on process of T_2, the load current commutates from the body diode of T_1 to T_2 through L_{sn} and T_{01}, see Fig. 5a$_1$. At $t = 375$ ns, when the current starts increasing, the "red" current through T_2 (i_{T2}) becomes slightly higher than the "blue" current flowing through the snubber (i_{sn}), because i_{T2} also charges the output capacitance of T_{sn1}. After the charging process ends at $t = 440$ ns, i_{T2} and i_{sn} become equal. The voltage across L_{sn} during the di/dt phase appears across T_{sn1}, keeping the output capacitance charged. A substantial voltage drop in $v_{DS,T2}$ is observed, caused by the parasitic inductance ($L_\beta + L_{sn}$) in the loop. Integrating over the inductive voltage drop and dividing it by the current rise gives the turn-on inductance, which is 74 nH in this case. It is observed that i_{sn} rises to a higher value than the peak of i_{T2}. This results from the discharging current of the output capacitance of T_{sn1} during the negative dv/dt across L_{sn}, which is fed to i_{sn}. From its peak value, i_{sn} freewheels through the parallel body diode of T_{sn1} until it reaches the steady state of i_{T2}.

Fig. 5a$_2$ shows the turn-off process of T_2 for the *active snubber* scenario. As soon as the dv/dt phase begins, the snubber current starts commutating to the freewheeling path (from L_{sn} to L_θ). The drop observed in i_{T2} during this phase corresponds to the discharge of the output capacitance of T_1. When the drain-source voltage reaches V_{C1}, the di/dt phase begins, where – on the one hand – the current through i_{T2} commutates to the body diode of T_1 through L_β, and – on the other hand – the snubber current i_{sn} commutates to the freewheeling path T_{sn1} over L_θ completely. Both inductances L_β and L_θ contribute to the generated voltage overshoot seen in $v_{DS,T2}$. The calculated turn-off inductance is 37 nH. After ending the commutation, i_{sn} freewheels through the body diode of T_{sn1} until it reaches zero after about 15 μs.

In order to further investigate the active snubber function, the inductance L_{sn} was increased by replacing the copper plate for connecting the power module terminals (AC and DC+) with a cable. As can be seen from the turn-on process in Fig. 5b$_1$, the voltage generated by L_{sn} during the di/dt phase is high enough to enable a "break down" of $v_{DS,T2}$. This results in a 68-% reduction of the turn-on losses compared to a turn on with inactive snubber. The turn-off process in Fig. 5b$_2$ follows the same manner described in Fig. 5a$_2$ with a similar turn-off inductance. Taking the dissipated energy loss during the freewheeling phase of i_{sn} into consideration, the snubber offers 30 % loss reduction in total. A good design value for L_{sn} would lie within the range of the two variants shown above.

2) Inactive Snubber Scenario: The same turn-on and turn-off process described above is repeated, but with T_{sn1} always in the on-state. During the turn-on process (Fig. 5c$_1$), the current can flow through the paralleled L_{sn} and L_θ resulting in a lower turn-on inductance of 34 nH. The turn-off process, illustrated in Fig. 5c$_2$, is similar to that described in Fig. 5a$_2$ and results in a turn-off inductance of 37 nH. This is slightly higher than the 34 nH for the turn on because only L_θ is present during the turn off (without L_{sn} in parallel).

Overall, both options avoid extra conduction losses of T_{sn1}. The active snubber provides an extra feature for reducing the switching losses in the submodule.

B. Test of Submodule Using a Pulse Sequence

In the next measurement step, a pulse train emulating the typical switching pattern of the submodule was applied to evaluate the entire submodule's operation, to observe the current distribution in paralleled switches, capacitors, and bypass paths as well as to calculate commutation inductances in all four commutation loops.

Measurement Procedure: After pre-charging the capacitor banks to 2 kV each, a four-level pulse sequence (-2 kV$\rightarrow$ 0 V$\rightarrow +2$ kV$\rightarrow +4$ kV and vice versa) was applied to drive a positive current through a 3-mH inductor connected to the submodule terminals, see Fig. 7a. Not all switch currents in the submodule were measured due to the difficulty of positioning current sensors on certain switches. However, the set of switch currents measured provides sufficient information of the current distribution and commutation inductances in the submodule.

Current Distribution: Fig. 7b shows the currents in the paralleled switches $T_{2,1}$ and $T_{2,2}$, as well as $T_{4,1}$ and $T_{4,2}$. The currents are shared nearly ideally between the parallel-connected XHP-2 modules. For proper static sharing, the modules have been paired according to their on-state characteristics. Minimal deviations after the turn-on transients prove the good symmetry of the busbar design.

During the dead time at the transition from the negative state **(i)** to the bypass **(ii)** (turning off T_{01} and T_{02}), the body diodes of $T_{5,1}$ and $T_{5,2}$ start conducting, as can be seen in Fig. 7d. When turning T_1 and T_4 on, the current flows through the upper and lower bypass. It can be seen that the current in the lower bypass path (T_2, T_4 in Fig. 7b and T_7 in Fig. 7d) takes some time to balance with the upper bypass path (T_1, T_3, T_6, which were not measured). A slight increase in the lower bypass current is observed until it reaches its steady state along with the upper bypass, showing a well and fast balanced bypass paths. It should be noted that there is a small difference between the sum of switch currents and the inductor current in Fig. 7a, which is related to measurement errors. The

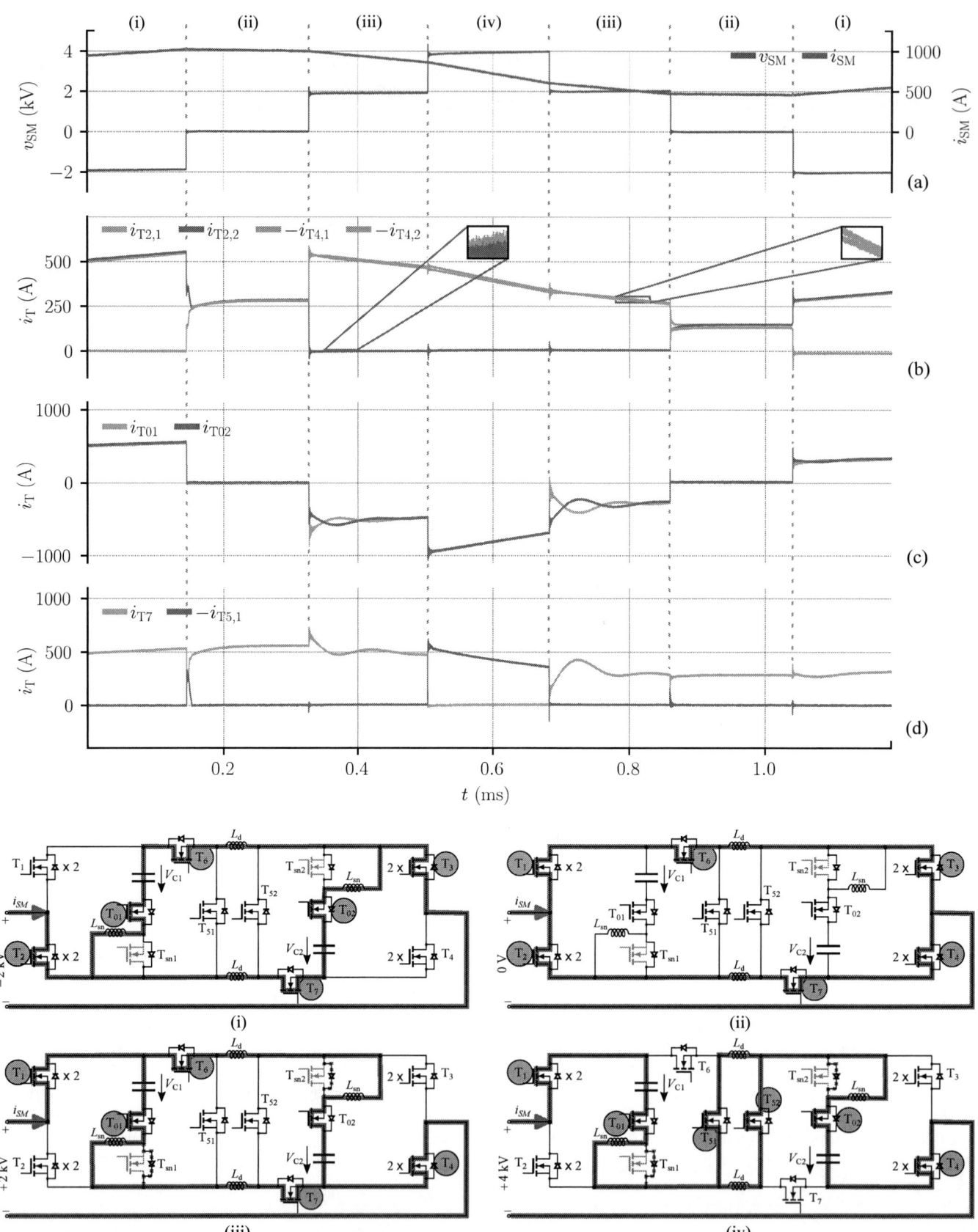

Fig. 7. Measured pulse sequence with a positive submodule current and associated switching states, highlighted switches are in the on state

TABLE II
Measured Turn-on Inductances During Pulse-Pattern Operation of the Submodule

Turned on switch	Condition	Active snubber	Inactive snubber
T_2	$i_{SM} > 0$, $+2\,kV \rightarrow 0\,V$	72 nH	32 nH
T_7	$i_{SM} > 0$, $+4\,kV \rightarrow +2\,kV$	70 nH	32 nH
T_4	$i_{SM} < 0$, $-2\,kV \rightarrow 0\,V$	78 nH	32 nH
$T_{5,1}$	$i_{SM} < 0$, $+2\,kV \rightarrow +4\,kV$	82 nH	42 nH

switches $T_{5,1}$ and $T_{5,2}$ can be also turned on during the bypass to further reduce the conduction losses.

During the transition from the bypass state (ii) to the positive state (iii) ($+2\,kV$, with T_2, T_3 turned off and T_{01}, T_{02} turned on), it is expected that the commutations would occur in the outer commutation loops without any change in the current at the middle of the submodule (T_6 and T_7). However, a small change in the current of T_7 is observed during this commutation, indicating that the two submodule parts are not fully decoupled and have unequal parasitic inductances. This results in unequal currents being commutated to both capacitors during this transition, which triggers an oscillation in the middle of the submodule, visible in the capacitor currents in Fig. 7c. The oscillation ends and settles within less than $200\,\mu s$, which is fast. This effect can be avoided by increasing the double-connection inductance L_d in order to further decouple both submodule parts.

This measurement was also conducted with a negative submodule current, but is not shown here as it provides no additional information due to the symmetry of the submodule.

Loop Inductances: To determine the parasitic inductances during regular operation, one switch per commutation loop was chosen (T_2, T_7, T_4, and $T_{5,1}$). T_2 and T_7 perform active turn-off and turn-on processes for a positive submodule current, while T_4, and $T_{5,1}$ for a negative. The summary of the measured turn-on inductances is listed in Tab. II for the active and inactive snubber options introduced in IV-A. Those for active snubber slightly differ from each other but they are in the range of the one measured in double-pulse testing (74 nH). The turn-off inductances for both options are expected to be slightly higher than the turn-on inductances of the inactive snubber option, as explained for the double-pulse testing.

Overall, the measured stray inductances lie in the range estimated and enable fast switching of the SiC devices despite the more complex commuation loops involving three switches.

V. Economical Evaluation of the DZDCSM

The implementation of the proposed submodule with $100\,\%$ SiC devices enforces higher semiconductor costs. This has to be compensated by lower energy costs during the life time. This section presents a brief efficiency and cost comparison of the DZDCSM vs. a conventional submodule with bidirectional voltage, viz. the FBSM. As a use case, a 20-kV-DC MMC with 20-MW rated power (and 30-MW overload) is assumed, equipped with either eight DZDCSM or 16 FBSM per converter arm, see operating point in Tab III. The inverter efficiency is crucial for the rated power, but for the thermal design the overload condition has to be considered. The FBSM is evaluated in two scenarios: one with Si-IGBT and another with SiC-MOSFET. The DZDCSM is applied solely with SiC-MOSFET, as already mentioned. The semiconductor devices considered in the design are:

- FZ2000R33HE4, single Si-IGBT in IHM-B package, 3.3 kV, 2 kA [15]
- FF2000UXTR33T2M1, half-bridge SiC-MOSFET in XHP-2 package, 3.3 kV, 1 kA [14]

The developed submodule prototype is designed to meet a junction temperature (T_j) limit of 135 °C under the overload condition, where its outer half bridges need parallel connections. For the Si-FBSM, no parallel connection per switch is required, while for the SiC-FBSM, each switch requires paralleling. In both scenarios $T_j \leq 135$ °C at $P = 30\,MW$ is met, as well.

The following comparison is performed at the nominal MMC operating point specified in Tab III ($P = 20\,MW$). Boundary conditions for this calculation consider:

- a submodule switching frequency of 150 Hz,
- on-state characteristics and switching losses for IGBT/Diode at a constant T_j of 125 °C,
- switching losses for SiC at a constant T_j of 125 °C,
- T_j dependency of $R_{DS,On}$ for SiC due to its significant impact, and
- a thermal chain of all submodules in one MMC arm connected in series.

Tab. IV summarizes the calculated results for Si-FBSM, SiC-FBSM, and SiC-DZDCSM. Since two FBSM are equivalent to one DZDCSM, the results are shown for two FBSM in series. A switching frequency (f_{sw}) of 150 Hz is assumed for each FBSM. This figure is in line with the present state-of-the-art industrial applications. The requirements for balancing the individual capacitor voltages are enforcing these minimum frequencies, if the capacitors are not significantly oversized. The DZDCSM features switching states with paralleled capacitors. This advantage enables to achieve a similar voltage ripple when operating with $f_{sw} = 150\,Hz$, too.

Taking the Si-FBSM as a reference, the SiC-FBSM achieves a total loss reduction of $59.8\,\%$, while the SiC-DZDCSM achieves a reduction of $76.37\,\%$. For increased switching frequencies, these results would shift even further in favor of the SiC devices.

To set the gain in efficiency in relation to the higher installation cost, estimates for module cost and loss capitalization are assumed. For SiC devices, a cost factor of three compared to Si devices seems to be a reasonable figure in the near future. Assuming a unit price of 800 € for the IGBT, and taking the current rating into account, the prospective cost per SiC module is estimated as:

$$3 \times 800\,€ \times \frac{1\,kA}{2\,kA} \times 2 = 2400\,€$$

979-8-3315-1612-3/25 $31.00 © 2025 IEEE

TABLE III
MMC Operating Point Used for Comparing DZDCSM vs. FBSM

Parameter	Value
V_{DC}	20 kV
$\overline{V}_C^{\Sigma}$ [a]	32 kV
P	20 MW (30 MW overload)
k [b]	1.25
r [c]	0.703
$\cos(\varphi_i)$	1

[a] sum of installed capacitor voltages per arm
[b] ratio of the peak fundamental AC voltage to the DC voltage generated by the arm
[c] submodule utilization factor

The assessment of losses is highly dependent on energy cost. Two scenarios are therefore considered: a medium loss capitalization rate of 6 €/W, and a high-end rate of 10 €/W. In the first scenario, the SiC-FBSM does not offer a cost advantage over the Si-FBSM, but it does in the second scenario. However, the SiC-DZDCSM promises to be cost-effective despite the higher SiC cost in both scenarios due to the significant reduction of the losses. Please note, that all submodules have been designed for an overload capability of 150 % (overload power $P = 30$ MW). This basic design requirement leads to an improved efficiency for all submodules at the nominal power $P = 20$ MW. The highest benefit, however, is found for the SiC-MOSFET devices due to their linear on-state characteristic.

VI. Conclusion

A compact and first-of-its-kind design of an advanced MMC submodule that reduces losses and footprint has been presented. Due to its electrical rating and physical structure, it is suitable for MVDC and HVDC applications. The challenges encountered during the design phase were addressed, and appropriate solutions proposed. Double-pulse tests and first measurements with pulse sequences on the full-scale prototype confirm its functional principles.

Acknowledgment

This research work has been carried out within the project DEFINE and is funded by dtec.bw – Digitalization and Technology Research Center of the Bundeswehr, which we gratefully acknowledge. dtec.bw is funded by the European Union – NextGenerationEU.

References

[1] R. Marquardt, "Modular multilevel converters: State of the art and future progress," *IEEE Power Electron. Mag.*, vol. 5, no. 4, 2018.

[2] M. A. Perez, S. Bernet, J. Rodriguez, S. Kouro, and R. Lizana, "Circuit topologies, modeling, control schemes, and applications of modular multilevel converters," *IEEE Trans. Power Electron.*, vol. 30, no. 1, 2015.

[3] A. Nami, J. Liang, F. Dijkhuizen, and G. D. Demetriades, "Modular multilevel converters for HVDC applications: Review on converter cells and functionalities," *IEEE Trans. Power Electron.*, vol. 30, no. 1, 2015.

TABLE IV
Comparison of Submodule Topologies for the Target Rating $P = 20$ MW

	2×FBSM		DZDCSM
semiconductor	Si IGBT FZ2000R33HE4	SiC MOSFET FF2000UXTR33T2M1	
assumed cost per module	800 €	2400 €	
no. of used modules	$8 = 2 \times T_{1,2,3,4}$	$8 = 4 \times T_{1,2,3,4}$	$8 = 2 \times T_{1,2,3,4} + 2 \times T_{5,01,02} + 1 \times T_{6,7}$
SM conduction losses	2,512 W	1,258 W	742 W
SM switching losses	746 W	52 W	28 W
SM total losses	3,258 W (100 %)	1,310 W (40.2 %)	770 W (23.63 %)
semiconductor efficiency	99.22 %	99.68 %	99.81 %
semiconductor cost	6,400 €	19,200 €	19,200 €
loss capitalization @ 6 €/W	19,548 €	7,860 €	4,620 €
total cost @ 6 €/W	**25,948 €**	**27,060 €**	**23,820 €**
loss capitalization @ 10 €/W	32,580 €	13,100 €	7,700 €
total cost @ 10 €/W	**38,980 €**	**32,300 €**	**26,900 €**

[4] S. Debnath, J. Qin, B. Bahrani, M. Saeedifard, and P. Barbosa, "Operation, control, and applications of the modular multilevel converter: A review," *IEEE Trans. Power Electron.*, vol. 30, no. 1, 2015.

[5] K. Jacobs, S. Heinig, D. Johannesson, S. Norrga, and H.-P. Nee, "Comparative evaluation of voltage source converters with silicon carbide semiconductor devices for high-voltage direct current transmission," *IEEE Trans. Power Electron.*, vol. 36, no. 8, 2021.

[6] L. Zhang *et al.*, "Modeling, control, and protection of modular multilevel converter-based multi-terminal HVDC systems: A review," *CSEEJ.Power Energy Syst.*, vol. 3, no. 4, 2017.

[7] X. Hu, Y. Zhu, J. Zhang, F. Deng, and Z. Chen, "Unipolar double-star submodule for modular multilevel converter with DC fault blocking capability," *IEEE Access*, vol. 7, 2019.

[8] J. Xu, P. Zhao, and C. Zhao, "Reliability analysis and redundancy configuration of MMC with hybrid submodule topologies," *IEEE Trans. Power Electron.*, vol. 31, no. 4, 2016.

[9] R. Li, L. Xu, D. Holliday, F. Page, S. J. Finney, and B. W. Williams, "Continuous operation of radial multiterminal HVDC systems under DC fault," *IEEE Trans. Power Del.*, vol. 31, no. 1, 2016.

[10] A. Schön and M.-M. Bakran, "Comparison of modular multilevel converter based HV DC-DC-converters," in *Proc. 18th Eur. Conf. Power Electron. Appl.*, 2016.

[11] C. Dahmen and R. Marquardt, "Power losses of advanced MMC submodule topologies using Si- and SiC-semiconductors," in *Proc. 21th Eur. Conf. Power Electron. Appl.*, 2019.

[12] A. Sharaf Addin, C. Dahmen, and T. Brückner, "Extended balancing and dimensioning of capacitors in MMC double submodules," in *Proc. 24th Eur. Conf. Power Electron. Appl.*, 2022.

[13] C. Dahmen and R. Marquardt, "Reduced capacitor size and on-state losses in advanced MMC submodule topologies," in *Proc. 22th Eur. Conf. Power Electron. Appl.*, 2020.

[14] Infineon, *FF2000UXTR33T2M1 SiC-MOSFET XHP-2 Module*, 2024, datasheet.

[15] ——, *FZ2000R33HE4 Si-IGBT IHM-B Module*, 2022, datasheet.

979-8-3315-1612-3/25 $31.00 © 2025 IEEE

A Series-type Dynamic Voltage Restorer Control Strategy to Cope with Voltage Swell

Jiazheng Zhang
School of Electrical and Information Engineering
Tianjin University
Tianjin,China
2022234388@tju.edu.cn

Hongyu Chen
School of Electrical and Information Engineering
Tianjin University
Tianjin,China
chy_2024234042@tju.edu.cn

Xi Chen
School of Electrical and Information Engineering
Tianjin University
Tianjin,China
2022234001@tju.edu.cn

Mingjun Bao
School of Electrical and Information Engineering
Tianjin University
Tianjin,China
baomingjun159@tju.edu.cn

Abstract—**In order to reduce the negative impact of grid swell on sensitive loads and improve the limitations of the dynamic voltage restorer (DVR) to cope with voltage swell, this paper introduces the composite control method on the basis of the proposed reactive power compensation calculation method and the closed-loop control strategy of the load voltage magnitude and the active output of the inverter, which solves the problem of power factor fluctuation of loads and guarantees the system's response time at the same time. Finally, the experiments are conducted with three single-phase back to back type DVRs as an example, and the results of the tests verify the feasibility and effectiveness of the proposed control.**

Keywords—Dynamic swell correction, voltage quality regulator, series-connect compensator.

I. INTRODUCTION

Power quality (PQ) is a growing concern as it affects many sensitive end-users including industrial and commercial electricity consumers. Studies have shown that voltage sags, transients and momentary interruptions account for 92% of PQ problems in distribution systems. In fact, voltage sags have always been a huge threat to the industry and even a 0.25 s voltage sag is enough to interrupt the manufacturing process and cause huge economic losses[1][2]. Voltage sags are generally categorized by their depth and duration. Typical sags can drop between 10% and 90% of the rated RMS voltage and last from 0.5 cycles to 1 minute.

The main topology of DVR is divided into series and parallel, the series type is mainly embodied in the DVR series connection between the grid and the load, when the grid voltage dips, the DVR is put into operation, by compensating for the part of the grid voltage drop, to protect the sensitive loads from the impact of the voltage dips. the energy required for DVR compensation can be directly derived from the energy storage unit, or through the grid rectification method to obtain the DVR, which is divided into the energy storage type DVR and back-to-back grid energy type DVR. The energy required for DVR compensation can be obtained directly from the energy storage unit or through the grid rectifier, and can be divided into energy storage type DVR and back-to-back grid energy extraction type DVR. Compared with the traditional UPS and UPQC, DVR as a part of the power compensation device effectively utilizes the residual voltage of the grid, which makes it cost-effective and efficient, and it is one of the most direct and economical choices in the field of voltage dips management at present.

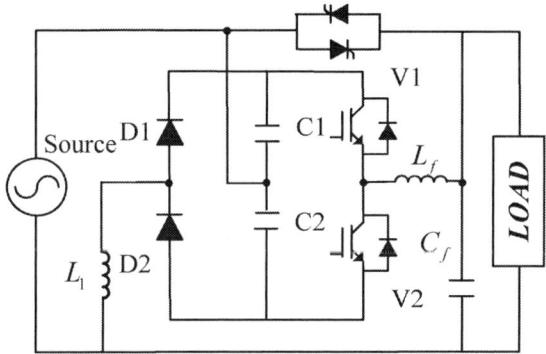

Fig. 1. Single-phase DySC configuration

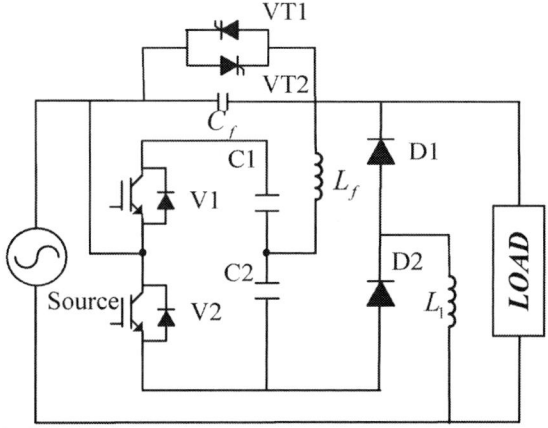

Fig. 2. SPB-AVQR topology

The common type of series dynamic voltage restorer (DVR) is the series DVR containing an isolation transformer, which is characterized by a simple topology, but at the same time it is also expensive because it contains a large capacity transformer, which makes it expensive, and in order to solve this problem,in [3] and [4], a transformerless topology is proposed as shown in Fig. 1 called Dynamic Sag Corrector (DySC), which has a low cost, small size, light weight and efficient sag mitigation system since it no longer requires a series transformer. But at the same time, because of its inability to compensate the deep sags occurring in the grid for a long period of time, an improved topology SPB-AVQR, shown in Fig. 2 , was proposed in [5] , which improves the device compensation capability by a unique shunt converter

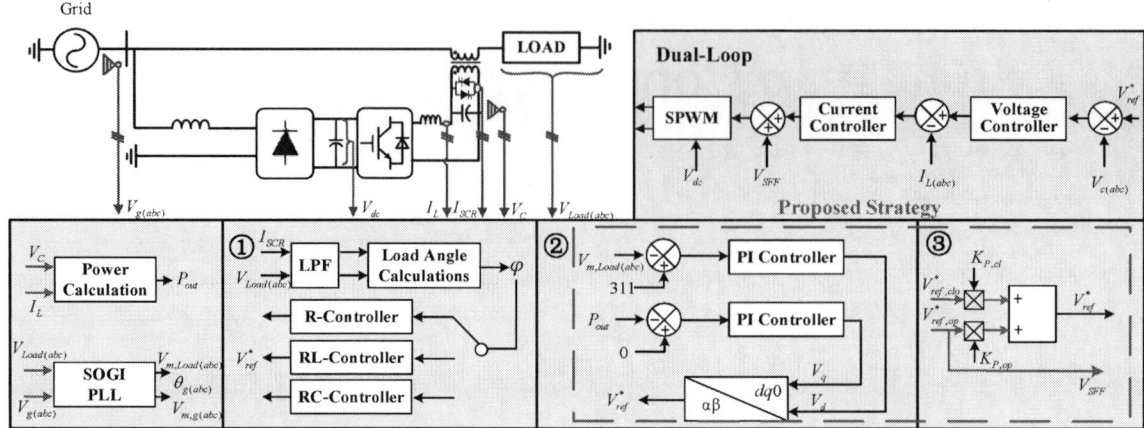

Fig.3 Proposed control strategy.

structure that acts as a parasitic boost circuit. However, this

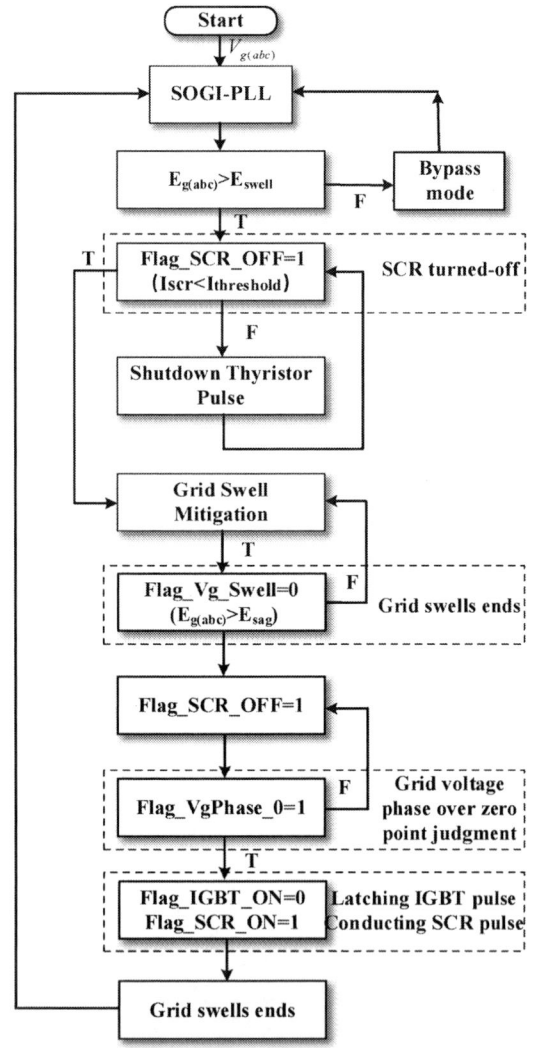

Fig.4. Control flow chart

topology itself brings the problem that when a deep sag occurs in the grid, in order to ensure the constant RMS value of the

output voltage of the device, it will lead to several times of the current flowing in its IGBTs, which puts great demands on the IGBT devices.

In traditional voltage swell mitigation, common strategies for both energy storage-based and back to back types DVRs include pre-sag compensation, in-phase compensation[7], and minimum energy compensation[8]. However, the first two strategies can cause the inverter to absorb active power, leading to DC voltage pump-up. The minimum energy compensation strategy is frequently used for addressing voltage swells but has limitations when dealing with fluctuating load power factors, causing the inverter to output additional active power.

This paper builds on the minimum energy compensation strategy by introducing an improved triple-loop control strategy and incorporating a regulation factor to develop a composite closed-loop control strategy. This approach allocates weights based on load fluctuations, using the output voltage reference value calculated by the minimum energy compensation strategy as a system feedforward reference value under appropriate weight. This is combined with the output voltage reference value calculated by the proposed triple-loop control under appropriate weight, resulting in a voltage reference value for a dual-loop voltage-current controller.

II. PROPOSED CONTROL STRATEGY

Fig. 3 shows the schematic diagram of the described series back to back type DVR and the novel voltage swell control strategy proposed in this paper. First, the power circuit is shown in the above figure, in which the grid side of the DVR transfers the grid power to the DC link through an uncontrolled rectifier circuit and is connected in series between the sensitive load and the grid through a controlled inverter circuit and a thyristor. Each inverter circuit consists of an h-bridge converter with an output LC filter, where Lf is the filter inductance and Cf is the filter capacitance.

The control algorithm of the system is shown in the bottom half of the figure. It consists of a dual-loop controller and optionally three outer-loop controllers.For simplicity of description, in this paper, the phase and magnitude of the grid and load voltages are obtained by SOGI-PLL. The standard RMS value of the grid is 220V and the magnitude is 311V.In the following part of this section, the control principle of the

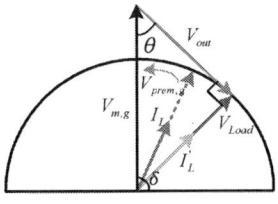
(a) Resistive Load

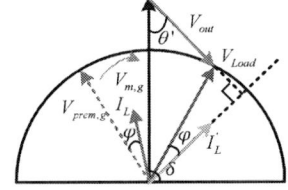
(b) Resistive-Inductive Loads

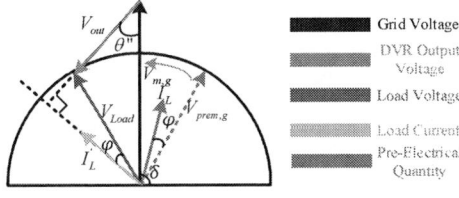
(c) Resistive-Capacitive Load

Fig.5: DVR Output Voltage Vector Diagram

novel triple-loop controller proposed in this paper is described in detail, respectively.

As shown in Fig.4 for this paper DVR swell suppression control flow chart, divided into eight stages first by the SOGI-PLL (Second Order Generalized Integrator-phase-locked loop) calculates the real-time grid amplitude, if $E_m > E_{swell}$ and the thyristor is not in the shutdown state (detecting the I_{scr}, if If $E_m > E_{swell}$ and the thyristor is not in the off state (detecting Iscr, if $I_{threshold}$ is less than three consecutive clock cycles, the thyristor is considered to be off), the system stops the thyristor pulses, and if the thyristor is in the off state, the system enters into the grid voltage transient rise suppression mode, and when the grid voltage is back to normal (Flag_Vg_Swell=0), the system will turn on the thyristor and block the IGBT pulses when the grid voltage is over the zero point in the phase.

a. Load angle direct calculation

As shown in Fig. 5, due to the short-term nature of the voltage swell, this method, under the premise of ignoring the sensitivity of the load to phase angle mutation, the system calculates the load phase angle after judging that the grid voltage transient rise occurs through the controller real-time preservation of the last cycle of Iscr and the load voltage V_{load} after the low-pass filtering, and when the controller determines that the load is a purely resistive load.

Where I_L is the load current before the grid swell. $V_{m,g}$, I_L' are the load current during the grid transient rise, and the grid voltage magnitude . θ is the angle between the DVR output voltage and the grid phase, δ is the grid phase angle during the grid transient rise, V_{load} is the load voltage, and V_{out} is the load voltage, φ is the load power factor angle.

It can be calculated through the following formula (1) to calculate the dual-loop controller voltage reference V_{ref}:

$$V_{ref}^* = \sin(\delta + \arcsin(\frac{V_{m,g}}{311}) - \pi) \times \sqrt{V_{m,g}^2 - 311^2} \quad (1)$$

As shown in Fig. 5(b), when the controller determines that the load is an inductive resistive load, the Vref of the dual-loop controller can be calculated by the following equations (2)(3).

$$V_{ref}^* = \sin(\delta + \arccos(\frac{V_{m,g}^2 + V_{m,out}^2 - 311^2}{2 \times V_{m,g} \times V_{m,out}}) - \pi) \times V_{m,out} \quad (2)$$

$$V_{m,out} = \sqrt{V_{m,g}^2 - (311 \times \cos\varphi)^2} - 311 \times \sin\varphi \quad (3)$$

As shown in Fig. 5(c), when the controller determines that the load is a resistive load, the Vref of the dual-loop controller can be calculated by the following equation (4):

$$V_{ref}^* = \sin(\delta - \arccos(\frac{V_{m,g}^2 + V_{m,out}^2 - 311^2}{2 \times V_{m,g} \times V_{m,out}}) + \pi) \times V_{m,out} \quad (4)$$

The key problem of this method is that, due to the uncontrolled rectifier circuit on the grid side of the equipment, if the load undergoes a phase angle change in the DVR operating mode, this control cannot be adjusted in real time and has poor robustness, which will lead to the problem of voltage swell that cannot be completely solved, and the problem of over-voltage on the DC link.

b. Triple-loop controller system

For the system to reach the problem of compensating voltage swell, this paper proposes a three-loop controller that can realize the suppression of voltage swell and DC link voltage over-voltage by controlling the output load voltage and the phase-shift control of the output power. The system output power P_{out} is obtained by measuring the inverter output voltage V_C and output current I_L, and the voltage reference value is obtained as:

$$\begin{cases} V_d^* = (311 - V_{m,LOAD(abc)}) \times (K_{P,V} + \dfrac{K_{i,V}}{s}) \\ V_q^* = (0 - P_{out(abc)}) \times (K_{P,P} + \dfrac{K_{i,P}}{s}) \end{cases} \quad (5)$$

$$V_{ref}^* = \begin{bmatrix} 1 & 0 \end{bmatrix} \times \left\{ \begin{bmatrix} \cos(\theta_{Load}) & -\sin(\theta_{Load}) \\ \sin(\theta_{Load}) & \cos(\theta_{Load}) \end{bmatrix} \times \begin{bmatrix} V_d^* \\ V_q^* \end{bmatrix} \right\} \quad (6)$$

where $V_{m,load(abc)}$ is the amplitude of the load voltage. $K_{P,v}$, $K_{i,v}$, $K_{P,P}$, $K_{i,P}$ are the load voltage controller and output power controller PI parameters, respectively, and is the load voltage phase.

Similarly, the key problem of this method is that, due to the negative feedback itself and the output power itself needs to go through a low-pass filtering link, this control, although realizes real-time adjustment according to the load switching, robust, but at the same time, it will lead to a slow response

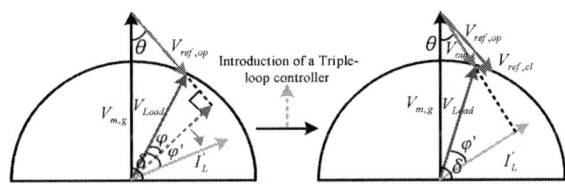

Fig.6. Vector Diagram of load power factor fluctuates

speed of the system, and can not quickly compensate for the voltage swell, and there is still a risk of making sensitive loads to fail to exist.

c. *Hybrid coordinated compensation strategy*

From the vector diagram of Fig. 6, when the load power factor fluctuates from φ to φ' ,the original above method A output voltage will not be able to be perpendicular to the load current, which makes the DVR active power not to be 0, which further results in a pumping up of the dc link voltage.With the introduction of method B on method A, the controller will calculate the appropriate $V_{ref,cl}$ in a closed loop.

Based on the above two methods, this paper further introduces the weighting factors $K_{p,op}$, $K_{p,cl}$, to obtain the voltage reference values as follows:

$$V_{ref}^* = V_{ref,op}^* \times K_{P,op} + V_{ref,cl}^* \times K_{P,cl} \quad (7)$$

Where $V_{ref,op}$, $V_{ref,cl}$ are the voltage reference values for the above proposed methods a and b respectively. Meanwhile, in order to increase the response speed of the system, $V_{ref,op}$ is added to the dual-loop controller as a feed-forward V_{sff}.

III. EXPERIMENTAL VERIFICATION

TABLE I.   PARAMETERS OF THE SYSTEM

System parameters			
Rated Grid Voltage	V_N =220VRMS/50Hz		
Load active power	P_L =10kW		
Circuit parameters			
Filter for Converter	L_l = 0.45mH; C_f= 40uF;		
DC Link Capacitor	C_{dc} = 5000uF;		
Switching Frequency	F_{sw}^L = 10kHz;		
Control parameters			
$K_{P,PR}$	10	K_p	30
$K_{r,PR}$	60	ω_c	5

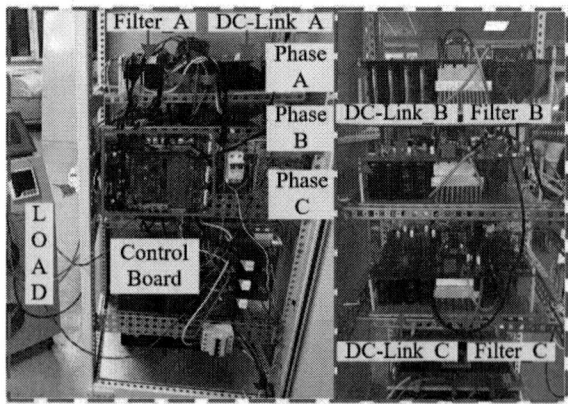

Fig.7. Experimental waveforms with 50% grid sag

In order to further confirm the feasibility of the proposed control and the characteristics such as fast response speed, this paper builds an experimental prototype with three single-phase back-to-back DVRs. The grid swell is realized

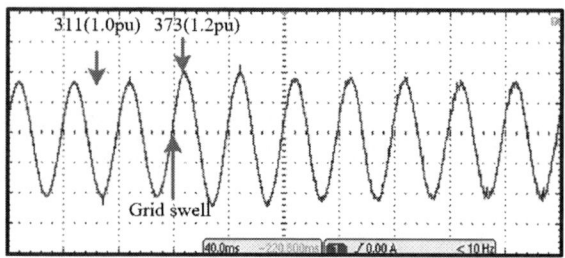

Fig.8. Experimental waveforms with 20% grid swell for method (b)

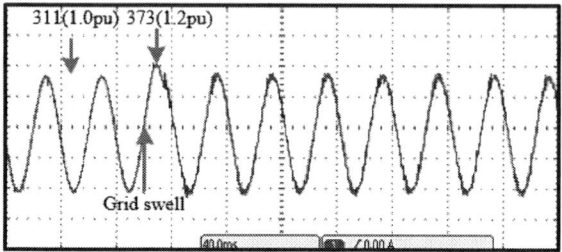

Fig.9. Experimental waveforms with 20% grid swell for method (c)

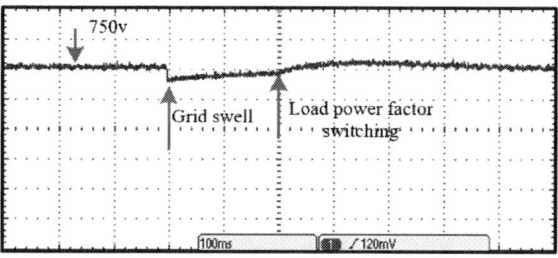

Fig.10. Experimental waveforms of DC link with 20% grid swell for method (d)

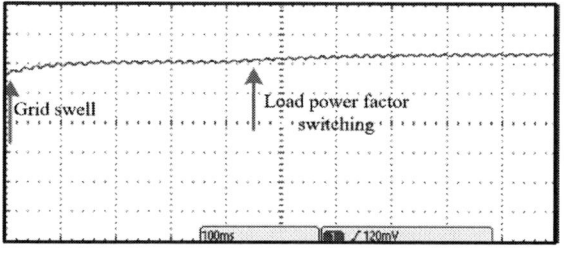

Fig.11. Experimental waveforms of DC link with 20% grid swell for method (a)

by the programmable power supply, and the experimental prototype is shown in Fig. 7.The parameters of the DVR are shown in Table 1.The experiment is divided into two stages. In the first stage, the grid is operated at 220V standard value. In stage 2, as shown in the experimental waveform Fig. 8, at the time point shown the power grid rises abruptly to 1.2 times the standard voltage, DVR puts into the three-loop controller proposed in this paper, and it can be seen from the waveform that there is a phenomenon of slow response in the problem of sudden rise. As shown in Fig. 9, the DVR is put into the composite controller proposed in this paper, and it can be observed that the response speed is obviously accelerated. As shown in Fig.10, the DC link voltage waveform under the existence of power factor jumping conditions, it can be seen that the proposed method compared with the traditional method shown in Fig.11. Fig.10, the proposed control method can be a good response to power factor fluctuations brought about by DC voltage pumping problems.

IV. CONCLUTION

In this paper, based on the reactive power compensation method in the minimum energy compensation strategy to suppress the voltage swell, applied to the general series-type energy storage-type and back-to-back DVRs. A three-loop controller is proposed for closed-loop control of the DVR output active power and load voltage, replacing the drawbacks of the traditional scheme that requires real-time sampling and calculation of the load power factor. On this basis, composite control and weight factor are introduced to further improve the response speed of the control. Experiments verify the feasibility of the proposed control.

REFERENCES

[1] M. F. McGranaghan, D. R. Mueller, and M. J. Samotyj, "Voltage sags in industrial systems," IEEE Trans. Ind. Appl., vol. 29, no. 2, pp. 397–403,Mar./Apr. 1993.

[2] A. Bendre, D. Divan, W. Kranz, and W. Brumsickle, "Equipment failures caused by power quality disturbances," in Proc. IEEE IAS Conf. Record, 2004, pp. 482–489. I. S. Jacobs and C. P. Bean, "Fine particles, thin films and exchange anisotropy," in Magnetism, vol. III, G. T. Rado and H. Suhl, Eds. New York: Academic, 1963, pp. 271–350.

[3] W. E. Brumsickle, R. S. Schneider, G. A. Luckjiff, D. M. Divan, and M. F. McGranaghan, "Dynamic sag correctors: Cost-effective industrial power line conditioning," IEEE Trans. Ind. Appl., vol. 37, no. 1, pp. 212–217, Jan./Feb. 2001.

[4] D. M. Divan, G. Luckjiff, R. Schneider, W. Brumsickle, and W. Kranz, "Dynamic voltage sag correction," U.S. Patent 6 118 676, Sep. 2000.

[5] Y. Lu, G. Xiao, B. Lei, X. Wu and S. Zhu, "A Transformerless Active Voltage Quality Regulator With the Parasitic Boost Circuit," in *IEEE Transactions on Power Electronics*, vol. 29, no. 4, pp. 1746-1756, April 2014.

[6] G. C. Xiao, Z. L. Hu, L. Zhang, and Z. A. Wang, "Variable Dc-bus controlstrategy for an active voltage quality regulator," in Proc. 24th Annu. IEEE Appl. Power Electron. Conf., 2009, pp. 1558–1563

[7] Molla E M, Kuo C C. Voltage sag enhancement of grid connected hybrid PV-wind power system using battery and SMES based dynamic voltage restorer[J]. IEEE Access, 2020, 8: 130003-130013.

[8] Naidu T A, Arya S R, Maurya R. Dynamic voltage restorer with quasi-newton filter-based control algorithm and optimized values of PI regulator gains[J]. IEEE Journal of Emerging and Selected Topics in Power Electronics, 2019, 7(4): 2476-2485.

[9] Danbumrungtrakul M, Saengsuwan T, Srithorn P. Evaluation of DVR capability enhancement-zero active power tracking technique[J]. IEEE Access, 2017, 5: 10285-10295.

Machine Learning Approach for Accurate Lithium-Ion Battery Temperature Prediction Using Electrochemical Features Independent of Battery SOC and SOH

Vincent Masabiar Tingbari
Department of Electrical Engineering
Chungnam National University
Daejeon, Republic of Korea
tingbari36@gmail.com

Oluwaseun Isaiah Ekuewa
Department of Electrical Engineering
Chungnam National University
Daejeon, Republic of Korea
ekuewao@gmail.com

Anshul Nagar
Department of Electrical Engineering
Chungnam National University
Daejeon, Republic of Korea
nagaranshul04@gmail.com

Asad Abbas
Department of Electrical Engineering
Chungnam National University
Daejeon, Republic of Korea
asadabbasturi.engr@gmail.com

Jamil Umar
Department of Electrical Engineering
Chungnam National University
Daejeon, Republic of Korea
umarjamil10000@gmail.com

Yuxin Zhang
Department of Electrical Engineering
Chungnam National University
Daejeon, Republic of Korea
zyx0605@naver.com

Woonki Na
Department of Electrical and Computer Engineering
California State University
Fresno, United States
wkna@csufrenso.edu

Jonghoon Kim
Department of Electrical Engineering
Chungnam National University
Daejeon, Republic of Korea
whdgns0422@cnu.ac.kr

Transportation electrification is a critical solution to reducing emissions and combating global climate change by shifting from fossil fuels to sustainable energy sources. However, the growing demand for fast charging introduces thermal challenges, increasing risks such as thermal runaway and battery degradation, especially in extreme temperatures. This study presents a machine-learning approach for predicting lithium-ion battery (LIB) temperature using electrochemical impedance spectroscopy (EIS) features with minimal sensitivity to the state-of-charge (SOC) and state-of-health (SOH). Through comprehensive analysis using the Pearson correlation coefficient (PCC), the study identifies EIS frequency ranges strongly correlated with temperature but remain unaffected by SOC or SOH variations. A support vector regression (SVR) model trained on these features achieves high accuracy, with RMSE and MAE values of 1.35°C and 0.81°C, respectively. The findings highlight the EIS magnitude at 6252 Hz as a reliable predictor of LIB temperature, offering significant advancements in thermal management for safer and more efficient battery operation.

Keywords—lithium-ion battery, temperature, electrochemical impedance, machine learning, support vector machine

I. INTRODUCTION

The electrification of transportation emerges as a highly promising solution for facilitating the transition from dependence on fossil fuels to more sustainable alternatives in response to the increasing emissions of greenhouse gases and the consequent alterations in global climate patterns [1], [2]. LIBs have attracted significant attention and have become the leading power source in aerospace and automotive applications due to their superior capacity and performance compared to

other rechargeable battery technologies. Heat is generated internally and builds up during the batteries' charging and discharging cycles. The demand for faster charging and discharging, particularly with high-capacity LIBs, can lead to potential thermal challenges or uneven heat distribution. Although electric mobility has gained significant market traction over the past decade, its broader adoption is still hindered by technological challenges, especially concerning the thermal management of LIBs. [3] and accelerated battery degradation due to high temperatures [4], [5]. When a LIB is subjected to high currents, its internal temperature can increase rapidly, potentially causing high-temperature gradients between the surface and the battery's core [6]. Since LIBs operate within a limited temperature range, an accurate thermal management system is essential to maintain safe, effective, and optimal temperature conditions. The rapid increase in battery temperature caused by the heightened heat generation during fast charging can drive batteries into hazardous conditions, even without internal short circuits [7], [8].

Addressing these temperature-related challenges requires efficient battery management systems (BMS), the most reliable operational solution for ensuring the safety, efficiency, and longevity of LIBs [9]. Given the substantial effect of temperature on multiple aspects of LIB performance, it is essential to monitor the thermal profile of batteries throughout the battery management process. For fast charging, effective monitoring of the thermal state allows for regulating the battery temperature within the optimal range through active thermal management. This ensures rapid charging occurs without compromising safety or reducing the battery's lifespan. Thermal

979-8-3315-1612-3/25 $31.00 © 2025 IEEE

data collected during battery operation allows the BMS to assess the energy and power capacity of LIBs under different performance conditions, aiding in the optimization of battery performance. Although significant efforts are made to understand thermal dynamics in the design phase of battery thermal management systems, comprehensive onboard monitoring of battery thermal profiles during operation has not yet received adequate attention. [10].

EIS measures impedance over a wide frequency range by assessing how the current responds to voltage perturbations. Often described as a "sensorless" technique, EIS does not require additional hardware, making it an efficient method for estimating thermal conditions. Another advantage of EIS is that it avoids the heat transfer delay commonly encountered with surface-mounted temperature probes. A fundamental relationship has been identified between battery impedance at specific frequencies and T_{int}. Despite its ability to provide real-time, non-invasive diagnostics, EIS has not yet been fully leveraged for comprehensive battery monitoring. Many studies have investigated methods for calculating the T_{int} of LIBs using various EIS features. However, the challenge with this method is that temperature is not the only factor affecting the battery's impedance; variables such as SOC and SOH also significantly influence impedance. Disentangling these coupled impedance responses, simultaneously impacted by multiple factors, remains a complex issue [11]. Consequently, it is crucial to address the selection of the optimal frequency range and the temperature-sensitive parameters involved.

A promising approach widely utilized for SOC or SOH estimation in battery state prediction is combining impedance data with machine learning (ML) techniques. However, temperature has been the focal point of discussion in only a few studies. McCarthy et al. [12] developed a model for online temperature estimation using a single-point stimulation frequency of 200 Hz. This concept was experimentally validated with data from cells subjected to over 100 cycles. To achieve accurate calculations, it was essential to calibrate each battery fitting model before testing. However, it is often impractical and tedious in practical applications to determine the fitting coefficients by pretesting each battery cell before integrating them into a battery pack. As a result, methods based on artificial intelligence appear to be promising solutions to these challenges, providing advantages in terms of scalability and generalizability. The study in [13] utilized an AC stimulation method to complete battery assembly while individually measuring the voltage of each cell to collect EIS data. Based on the EIS data, an artificial neural network (ANN) was employed to predict each cell's internal temperature (Tint) within a 3S2P battery module. However, this research did not account for the impact of SOH on temperature estimation, which limits its applicability throughout the battery's lifespan. Similarly, Ezahedi et al. [14] developed a framework that combined EIS measurements with Gaussian process regression for estimating the temperature of multiple high-capacity LIB cells using a single-point frequency. However, this study did not address the degradation effects as the battery approached the end of its life, a critical factor that could undermine the accuracy of battery state predictions. Furthermore, this method cannot be applied during the battery's operation. The study [14] presents a compilation of literature that centers on EIS-based temperature estimation; however, none of the research efforts predict the T_{int} using a singular feature [14]. Distinguishing coupled impedance responses concurrently affected by multiple factors remains a significant challenge. The primary contribution of this study is as follows: An in-depth analysis of the relationships between battery impedance spectra (ranging from 10 kHz to 100 mHz) and key parameters such as SOH, SOC, and internal temperature was conducted. This offers a deeper understanding of the electrochemical processes occurring within the battery, facilitating the advancement of cutting-edge battery technologies and materials. We employed the Pearson correlation coefficient to identify EIS features that remain consistently unaffected by variations in both SOH and SOC. This approach has not been extensively explored in the literature. Finally, support vector regression (SVR) was applied to predict the internal temperature, yielding high accuracy.

II. BATTERY EIS EXPERIMENT

Galvanostatic EIS (GEIS) was conducted on five 4.9 Ah INR21700-50E cylindrical batteries to study the relationship between EIS parameters and temperature changes. Using a BCS-815 battery cycling instrument, the batteries were tested across a temperature range of 0°C to 55°C with a 300mA rms sine wave current over a frequency range from 10 kHz to 100 mHz, and charging/discharging cycles were performed at 25°C up to 200 cycles. The test bench setup is shown in Fig. 1. All batteries underwent CC-CV charging and discharging cycles at 25°C, following these steps: (1) CC-CV charging at 2.45 A, (2) a two-hour rest, (3) CC-CV discharging, and (4) another two-hour rest, repeated for 40 cycles. After adjusting the temperature and allowing a six-hour thermal soak, a capacity test and an OCV test were conducted, followed by discharges in 5% increments with two-hour rests, repeating the EIS test until full discharge. This process was repeated at temperatures from 0°C to 55°C.

A. Battery internal temperature effect on EIS

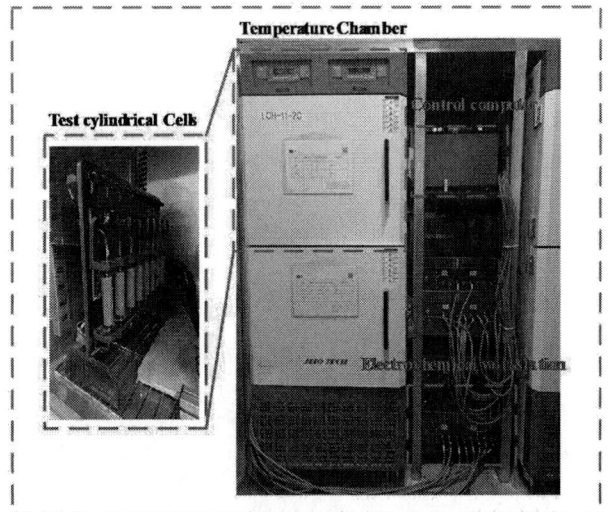

Fig. 1. Experimental platform

Fig. 2 presents the impedance spectra of two cells at various temperatures, illustrating the significant effect of temperature on the EIS spectra of LIBs. A clear inverse relationship is observed, where impedance in the semicircular region of the EIS spectra decreases as temperature rises. This shift is due to altered cell dynamics: at lower temperatures, reduced conductivity in the electrolyte and SEI layer limits lithium-ion transport, increasing electrode interface resistance. In contrast, higher temperatures enhance charge-transfer and diffusion-reaction processes, resulting in lower impedance. Notably, the ohmic resistance, indicated by the intersection of the EIS spectrum with the real axis, is largely influenced by the resistance at the clamp-interface.

B. Battery degradation effects on EIS

Battery impedance is closely linked to SOH, as illustrated by the EIS data in Fig. 3. With aging, the impedance spectra shift rightward along the real axis, indicating a gradual increase in ohmic resistance. The mid-to-high frequency range of the EIS spectra reveals two arcs that grow more distinct with battery aging, particularly the second arc. This shift results from ongoing electrolyte decomposition and SEI formation, which elevate battery resistance. The SEI layer, comprising inorganic and organic electrolyte decomposition products, forms on electrode surfaces and hinders Li-ion transport, further increasing resistance.

III. METHODOLOGY

A battery's electrochemical impedance provides insights into its internal electrochemistry, with distinct regions of the EIS spectrum corresponding to specific electrochemical processes [15]. As shown in Fig. 2, temperature changes inevitably impact EIS, as temperature influences these processes. Additionally, battery EIS varies with SOH and SOC, each of which uniquely affects internal electrochemical behavior [15], [16]. Battery impedance is affected by temperature, SOC, and SOH, and uncertainties in SOC and SOH can introduce errors in SOT estimates. To address this, a detailed EIS analysis is necessary to identify an ideal frequency range where specific impedance parameters—such as the imaginary part, real part, phase, and the intercept of the impedance spectrum with the real axis—are sensitive to temperature but remain unaffected by SOC and SOH. Establishing a link between battery temperature and impedance characteristics within this frequency range will help deepen the analysis of temperature effects on EIS. Since impedance metrics vary in sensitivity to temperature and other factors across frequencies, identifying the optimal frequency and impedance parameters is essential for accurate SOT estimation. The chosen features for SOT estimation must reliably reflect temperature variations in the battery without interference from SOC and SOH.

Furthermore, examining the relationship between the SOT estimation feature and the battery's T_{int} is crucial for enhancing machine learning model accuracy. An ML model is then trained using these optimal features, allowing it to map the nonlinear relationship between observed signals and SOT. The effectiveness of data-driven SOT estimations depends not only on the machine learning method but also on the design of the estimation model, which includes input features, model structures, and the quality and volume of training data.

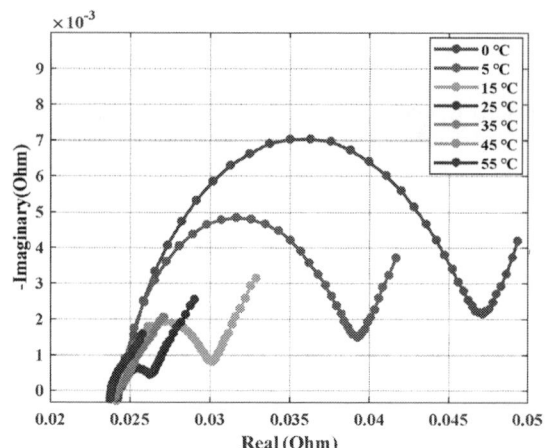

Fig. 2. Impedance spectra of two cells at varying temperatures

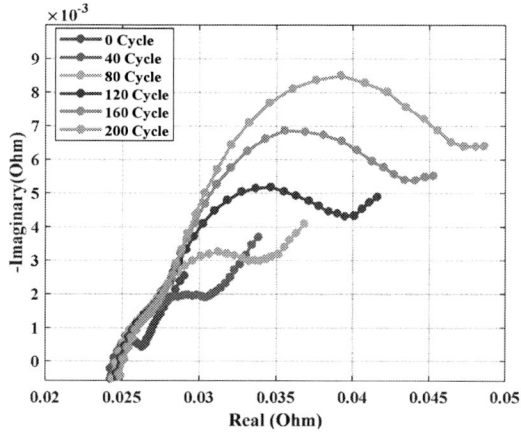

Fig. 3. Effect of battery aging on EIS with frequency range from 0.1 Hz to 10 kHz at 25°C

A. Feature extraction

For a high-accuracy model estimating battery temperature, the ML input data must be highly sensitive to temperature changes and minimally impacted by battery SOH and SOC. Fig. 4 illustrates the temperature sensitivity of EIS phase and magnitude, revealing that changes in these components are more significant at lower temperatures (0 to 25°C) and less pronounced at higher temperatures. Additionally, EIS responses vary by frequency: the magnitude component is highly sensitive to temperature in the lower and medium frequency ranges (100 mHz to 300 Hz), while the phase component exhibits greater sensitivity in the medium frequency range (10 Hz to 300 Hz).

Notably, temperature effects on impedance are frequency-dependent, with higher frequencies showing minimal temperature influence due to the dominance of resistive effects, as seen in Bode plots. A comprehensive correlation analysis will further examine the relationship between temperature and the EIS components.

B. Correlation analysis

979-8-3315-1612-3/25 $31.00 © 2025 IEEE

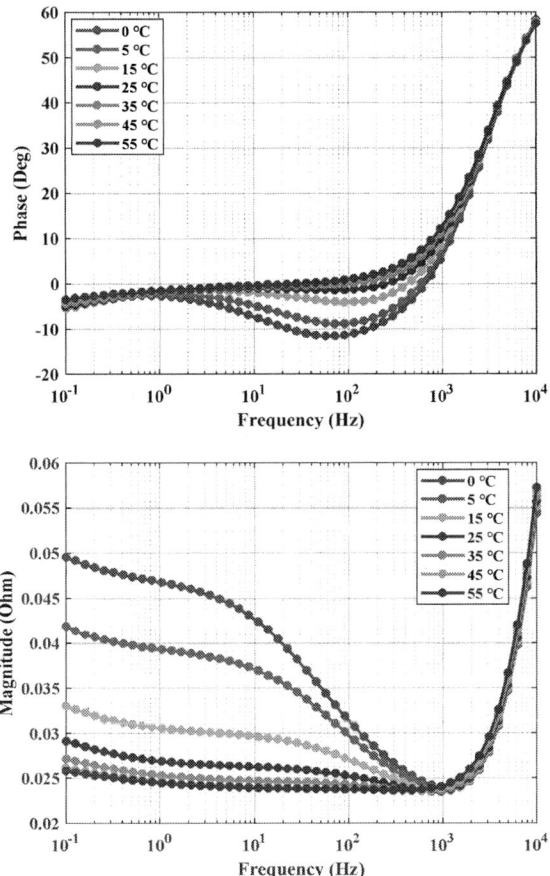

Fig. 4. Bode plot showing the variation of EIS features across a temperature range of 0 °C to 55 °C for a single cell.

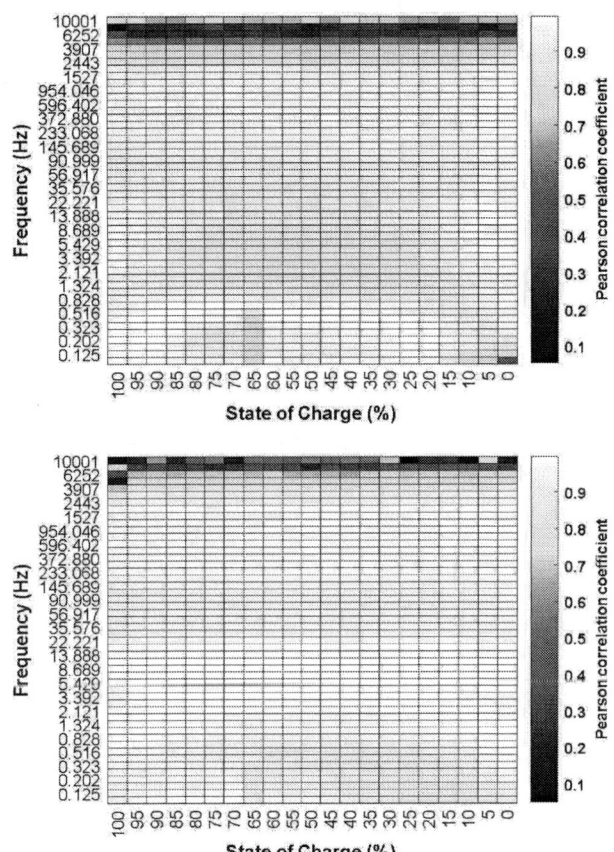

Fig. 5. PCC for the correlation between temperature and phase EIS (a) cell at 40 cycles, (b) cell at 200 cycles.

Pearson's Correlation Coefficient (PCC) analyzes the relationship between two variables. In the case of this paper, T_{int} and EIS to quantify the correlation between two variables, as given in Eq. (1).

$$R = \frac{\sum_{i=1}^{n}(x_i - \overline{x})(y_i - \overline{y})}{\sqrt{\sum_{i=1}^{n}(x_i - \overline{x})^2}\sqrt{\sum_{i=1}^{n}(y_i - \overline{y})^2}} \tag{1}$$

In Eq. x and y are random variables being analyzed. The terms x_i and y_i denote individual sample values indexed by i, with $\bar{x}$ and $\bar{y}$ as their respective means. The variable n represents the total number of data points. PCC values range from -1 to 1, where values close to 1 indicate a strong positive correlation, values near -1 represent a strong negative correlation, and values close to 0 suggest no correlation between the variables.

C. Temperature and EIS correlation analysis

Following that, Figs. 5 through 10 show that the effects of temperature on impedance vary with frequency, with high PCC values found in particular frequency ranges. In Fig. 5, the phase component shows high PCC values in the mid-frequency range, especially between 10 Hz and 100 Hz, with values exceeding 0.92 below 10 kHz at 100% SOC and 0.96 between 0.5 Hz and

10 kHz at 0% SOC. At medium SOC, temperature correlations decrease slightly but remain strong (PCC > 0.8) below 3 kHz. Weak correlations are observed at frequencies above 3 kHz across various aging levels. As shown in Fig. 6, the magnitude component of impedance behaves consistently across SOC and cycling stages. Frequencies above 1527 Hz have PCC values exceeding 0.95, and the range between 0.1 and 596.402 Hz has PCC values between 0.75 and 1 at all charge levels. Low PCC values (below 0.1) are observed between 596.402 and 954.046 Hz.

D. SOC and EIS correlation analysis

SOC significantly influences battery impedance, affecting both electrochemical and physical processes essential for battery management, performance, and diagnostics. Selecting frequencies where impedance is unaffected by SOC is crucial for accurate analysis. The phase impedance in Fig. 7 shows moderate to weak correlations from 100 Hz to 0.1 Hz, with PCC values between 0.6 and 0.9 for frequencies between 100 and 3907 Hz. Above 6252 Hz, the correlation weakens (PCC < 0.2). Similarly, the magnitude impedance in Fig.8 has frequencies above 3907 Hz showing very weak SOC correlations (PCC < 0.2), while frequencies between 2443 Hz and 2 Hz have

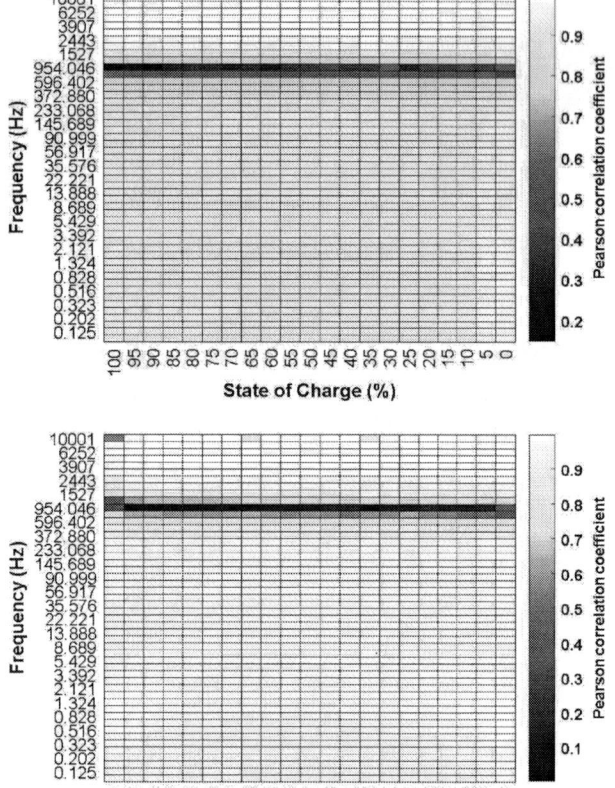

Fig. 6. PCC for the correlation between temperature and magnitude EIS (a) cell at 40 cycles (b) cell at 200 cycles.

magnitude components show a weak correlation with SOC, suggesting that the EIS response is mostly unaffected by SOC at these frequencies. However, moderate correlations at other frequencies imply some level of SOC influence. Regarding SOH, the magnitude component remains unaffected by SOH in frequencies above 6 kHz, unlike the phase component, which does not show independence from SOH across various temperatures and SOC levels. Based on PCC analysis, the frequency ranges between 6 kHz and 10 kHz for the magnitude component is found to be effective for predicting internal battery temperature, as it shows little sensitivity to both SOC

and SOH. Additional analysis, shown from Fig.5 - 10, pinpoints 6252 Hz as the most optimal frequency, balancing high sensitivity to temperature with minimal influence from SOC and SOH across all charge states.

G. Support vector regression

SVR is a machine learning technique for regression tasks that is founded on the ideas of Support Vector Machines (SVM). In contrast to conventional linear regression, SVR seeks to identify a function that maximizes the model's flatness while minimizing the error outside of a given tolerance margin, and approximating data inside this limit. SVR is especially well-suited for situations when accuracy with a narrow margin of error is crucial. The objective is to construct a function $f(x)$ that can predict the temperature with the least amount of error given a dataset $\{(x_i, y_i)\}, i = 1, 2, 3 \ldots \ldots n$, where y_i is the temperature value and x_i is the input feature vector (6252 Hz EIS-phase at various SOCs and cycles).

$$f(x) = w.\phi(x) + b \qquad (2)$$

The coefficient and bias that best fits the data with an error of ε are denoted by w and b. The SVR optimization problem can therefore be stated as follows:

$$\min_{w,b} \frac{1}{2} \|w\|^2 + C \sum_{i=1}^{n} \ell_c (f(x_i) - y_i) \qquad (3)$$

The trade-off between model complexity and the number of deviations greater than ε is managed by the regularization parameter C. Eq. 4 introduces the slack variables ξ_i and ξ_{i*}, which permit some points to reside outside the ε-tube.

$$\min_{w,b,\xi} \frac{1}{2} \|w\|^2 + C \sum_{i=1}^{n} (\xi_i + \xi_i^*) \qquad (4)$$

Eq.4 is subject to Eq.5

stronger correlations (PCC 0.5–0.9). Frequencies below 2 Hz show decreasing PCC as temperature decreases. Thus, the EIS phase, and magnitude exhibit very weak correlations with the SOC at frequencies above 6 kHz, making this frequency range ideal for predicting internal battery temperature without SOC influence.

E. SOH and EIS correlation analysis

Fig. 3 illustrates how battery aging affects impedance by showing how the EIS spectra fluctuate unevenly across frequencies as cycling continues. As a result, it is crucial to investigate how SOH affects EIS feature values. Only a few frequencies for the phase component are SOH-independent throughout SOC levels, as seen in Fig. 9. At high frequencies above 6 kHz, the magnitude component exhibits weak correlations that are constant across SOCs and temperatures, as seen in Fig. 10.

F. Optimal EIS feature summary for temperature estimation

The magnitude component of EIS correlate significantly with temperature in the high-frequency range (above 1500 Hz), however, the phase component indicates a weak correlation. At very high frequencies (above 3907 Hz), the phase and

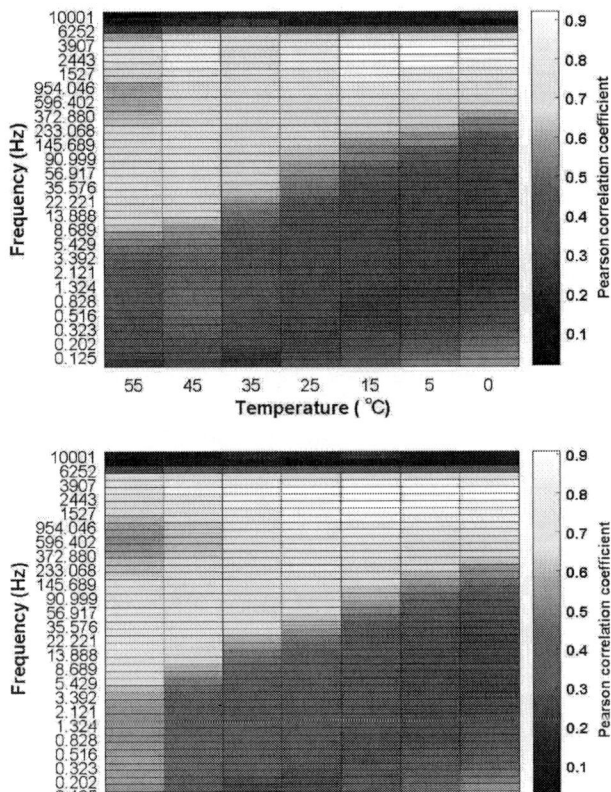

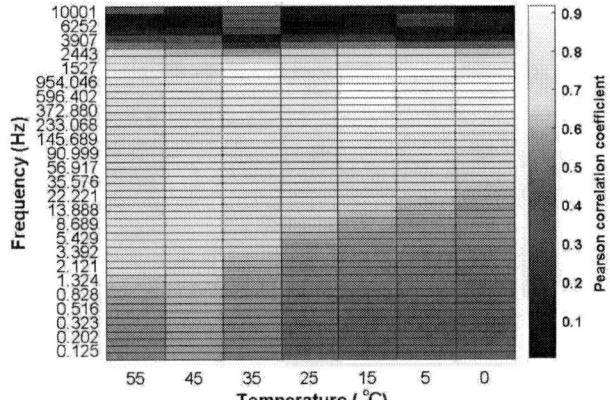

Fig. 8. PCC for the correlation between SOC and magnitude EIS

Fig. 7. PCC for the correlation between SOC and phase EIS (a) cell at 40 cycles, (b) cell at 200 cycles.

where αi and αi^* are Lagrange multipliers. The SVR model can, therefore, be expressed as

$$f\left(x\right)=\sum_{i}^{n}\left(\alpha_{i}-\alpha_{i}^{*}\right)K\left(x_{i},x\right)+b \qquad (8)$$

where $K(xi,x)=\phi(xi)\ \phi(xj)$ represents the kernel function. Radial basis function (RBF) is used and given by;

$$K_{RBF}\left(x_{i},x\right)=\exp\left(\left(-\left\|x_{i},x\right\|^{2}\right)\middle/2\sigma^{2}\right) \qquad (9)$$

where σ is the length scale of the kernel and $\|\cdot\|$ denotes the Euclidean distance

IV. RESULTS AND DISCUSSION

This section presents the outcomes of temperature prediction using the SVR model and offers a thorough evaluation of the models' functionality. The proposed method for estimating internal temperature involves three primary steps. First, impedance is measured across various SOCs and SOHs at different temperatures to obtain EIS data. In the second step, Pearson correlation analysis selects the most temperature-dependent features unaffected by SOC or SOH. Finally, an SVR model is trained and utilized to predict temperature. Magnitude data samples obtained from multiple cells across varying SOCs and SOHs are augmented and divided into training (80%) and testing (20%) sets. Fig. 10 compares the predicted and actual data distributions for the SVR model, considering the number of observations in the validation dataset. The predictions are fairly aligned with the real values. The SVR model predicts the temperature with RMSE and MAE evaluation metrics at 1.35 and 0.81 °C, respectively.

$$\begin{cases} y_{i}-w\cdot\phi(x)-b\leq\varepsilon+\xi_{i} \\ w\cdot\phi(x)+b-y_{i}\leq\varepsilon+\xi_{i}^{*} \\ \xi_{i},\xi_{i}^{*}\geq0 \end{cases} \qquad (5)$$

The problem can be stated in its dual form using the Lagrange multiplier method:

$$\begin{aligned} \max_{\alpha_{i}^{*},\alpha_{i}}=&-\frac{1}{2}\sum_{i,j}^{n}\left(\alpha_{i}^{*}-\alpha_{j}\right)\left(\alpha_{i}^{*}-\alpha_{j}\right)\phi\left(x_{i}\right)\phi\left(y_{j}\right)\\ &-\sum_{i}^{n}\alpha_{i}\left(y_{i}+\varepsilon\right)+\sum_{i}^{n}\alpha_{i}^{*}\left(y_{i}+\varepsilon\right) \end{aligned} \qquad (6)$$

Subject to

$$\begin{cases} \sum_{i}^{n}\left(\alpha_{i}-\alpha_{i}^{*}\right)=0 \\ 0\leq\alpha_{i},\alpha_{i}^{*}\leq C,\ \ i=1,2,\dots n \end{cases} \qquad (7)$$

V. CONCLUSION

This study demonstrates a machine learning method for predicting lithium-ion battery temperature using EIS features independent of SOC and SOH. Valuable insights into the electrochemical behavior of batteries across different SOCs and SOHs were gained, offering broader implications for battery

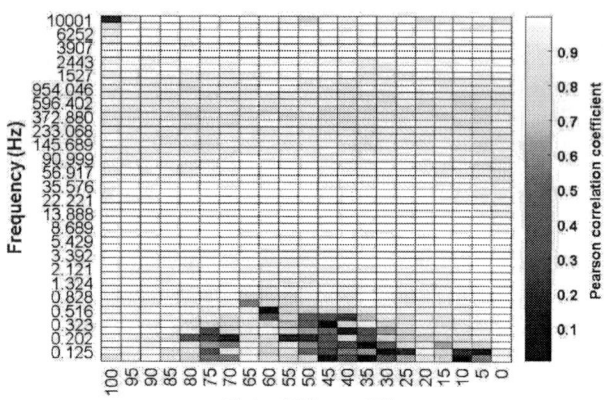

Fig. 9. PCC for the correlation between SOH and phase EIS at 25 °C

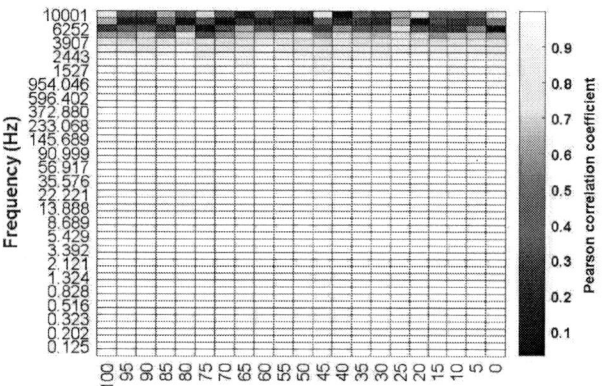

Fig. 10. PCC for the correlation between SOH and magnitude EIS at 25 °C

predictive accuracy, suggesting significant potential for enhancing battery thermal management in electric vehicles. These findings hold practical implications for improving battery safety, performance, and longevity, adding significant value to advancing energy storage technologies.

ACKNOWLEDGMENT

This research was supported by the 2024 government (Ministry of Trade, Industry and Energy) through the Korea Institute of Energy Technology Evaluation and Planning (Project No. RS-2024-00398346, Development of ESS Big Data-Based O&M and Asset Management Technology and Workforce Training) and the Korea Electric Power Research Institute (R23X005-03, development of cloud-based BMS element technology combining physical model and AI for ESS) and Korea Institute of Energy Technology Evaluation and Planning (KETEP) grant funded by the Korea government (MOTIE) (No. 20210501010020, Development on MMC-based ESS and Core Devices of High Voltage Hub-Station Grid-Connected with Renewable Energy Source).

REFERENCES

[1] B. Nykvist and O. Olsson, "The feasibility of heavy battery electric trucks," *Joule*, vol. 5, no. 4, pp. 901–913, Apr. 2021, doi: 10.1016/J.JOULE.2021.03.007.

[2] Z. P. Cano *et al.*, "Batteries and fuel cells for emerging electric vehicle markets," *Nat. Energy*, vol. 3, no. 4, pp. 279–289, 2018, doi: 10.1038/s41560-018-0108-1.

[3] X. Feng, M. Ouyang, X. Liu, L. Lu, Y. Xia, and X. He, "Thermal runaway mechanism of lithium ion battery for electric vehicles: A review," *Energy Storage Mater.*, vol. 10, pp. 246–267, Jan. 2018, doi: 10.1016/J.ENSM.2017.05.013.

[4] J. S. Edge *et al.*, "Lithium ion battery degradation: what you need to know," *Phys. Chem. Chem. Phys.*, vol. 23,

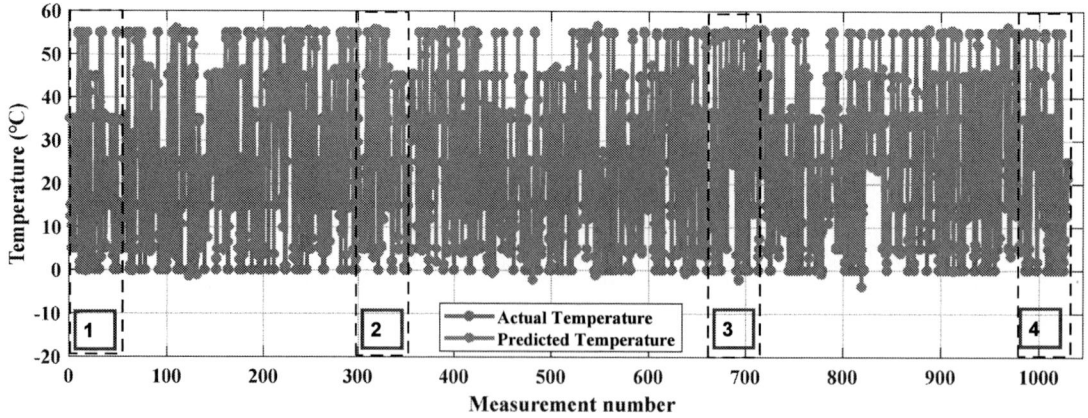

Fig. 11 Predicted and measured internal temperature by the SVR ML model

management systems and safety. By identifying frequencies where impedance is strongly temperature-dependent but unaffected by charge level or battery health, the approach offers reliable temperature estimation. The SVR model showed high

no. 14, pp. 8200–8221, 2021, doi: 10.1039/d1cp00359c.

[5] X. Lin, K. Khosravinia, X. Hu, J. Li, and W. Lu, "Lithium Plating Mechanism, Detection, and

Mitigation in Lithium-Ion Batteries," *Prog. Energy Combust. Sci.*, vol. 87, p. 100953, Nov. 2021, doi: 10.1016/J.PECS.2021.100953.

[6]   G. Xia, L. Cao, and G. Bi, "A review on battery thermal management in electric vehicle application," *J. Power Sources*, vol. 367, pp. 90–105, Nov. 2017, doi: 10.1016/J.JPOWSOUR.2017.09.046.

[7]   A. A. Adejare, F. E. Okemakinde, V. M. Tingbari, J. Lee, and J. Kim, "Comparative Analysis of Charging Protocol for Degradation Reduction and Remaining-Useful-Life Enhancement of a Lithium-Ion Battery," *Energy Technol.*, vol. 2400584, pp. 1–15, 2024, doi: 10.1002/ente.202400584.

[8]   W. Cai *et al.*, "A review on energy chemistry of fast-charging anodes," *Chem. Soc. Rev.*, vol. 49, no. 12, pp. 3806–3833, 2020, doi: 10.1039/C9CS00728H.

[9]   R. S. Longchamps, X.-G. Yang, and C.-Y. Wang, "Fundamental Insights into Battery Thermal Management and Safety," *ACS Energy Lett.*, vol. 7, no. 3, pp. 1103–1111, 2022, doi: 10.1021/acsenergylett.2c00077.

[10] Y. Zheng, Y. Che, X. Hu, X. Sui, D. I. Stroe, and R. Teodorescu, "Thermal state monitoring of lithium-ion batteries: Progress, challenges, and opportunities," *Prog. Energy Combust. Sci.*, vol. 100, p. 101120, Jan. 2024, doi: 10.1016/J.PECS.2023.101120.

[11] X. Du, J. Meng, Y. Amirat, F. Gao, and M. Benbouzid, "Exploring impedance spectrum for lithium-ion batteries diagnosis and prognosis: A comprehensive review," *J. Energy Chem.*, vol. 95, pp. 464–483, 2024, doi: https://doi.org/10.1016/j.jechem.2024.04.005.

[12] X. Hu, Y. Zheng, D. A. Howey, H. Perez, A. Foley, and M. Pecht, "Battery warm-up methodologies at subzero temperatures for automotive applications: Recent advances and perspectives," *Prog. Energy Combust. Sci.*, vol. 77, p. 100806, Mar. 2020, doi: 10.1016/J.PECS.2019.100806.

[13] M. Ströbel, V. Kumar, and K. P. Birke, "Temperature Estimation in Lithium-Ion Cells Assembled in Series-Parallel Circuits Using an Artificial Neural Network Based on Impedance Data," *Batteries*, vol. 9, no. 9, 2023, doi: 10.3390/batteries9090458.

[14] S. E. Ezahedi, M. Kharrich, and J. Kim, "Multi-cell sensorless internal temperature estimation based on electrochemical impedance spectroscopy with Gaussian process regression for lithium-ion batteries safety," *J. Energy Storage*, vol. 94, p. 112467, Jul. 2024, doi: 10.1016/J.EST.2024.112467.

[15] X. Wang *et al.*, "A review of modeling, acquisition, and application of lithium-ion battery impedance for onboard battery management," *eTransportation*, vol. 7, p. 100093, Feb. 2021, doi: 10.1016/J.ETRAN.2020.100093.

[16] J. G. Zhu, Z. C. Sun, X. Z. Wei, and H. F. Dai, "A new lithium-ion battery internal temperature on-line estimate method based on electrochemical impedance spectroscopy measurement," *J. Power Sources*, vol. 274, pp. 990–1004, Jan. 2015, doi: 10.1016/J.JPOWSOUR.2014.10.182.

A Battery Strings Circulating Current Blocking Method for Battery Energy Storage Systems

Haihong Long, Ziang Sun, Yucheng Fan, Xin Wu, Dehong Xu
College of Electrical Engineering
Zhejiang University
Hangzhou 310027, China
longhaihong@zju.edu.cn

Abstract—**Circulating current between paralleled battery strings within a Battery Energy Storage System (BESS) can significantly affect system efficiency, battery life, and safety. A circulating current blocking circuit and a corresponding control strategy for the BESS are proposed in this paper. It can block the circulating current during the charging mode, the discharging mode, the standby mode and the fault mode of the BESS. Simulation results validate the effectiveness of the proposed method.**

Keywords—parallel-connected battery strings, BESS, circulating current, soft-start

I. INTRODUCTION

Battery Energy Storage Systems typically obtains the required voltage by connecting battery cells in series, and increase current capacity by paralleling battery strings. Due to discrepancies of the characteristics of paralleled battery strings, the currents through them may differ significantly. Even in some case, some battery strings work at the charging mode while others work at the discharging mode. Circulation currents flows between battery strings. Circulation current between battery strings may lead to increased system losses, reduced battery life, and even causes thermal runaway [1]-[2].

There are two mainly categories of methods to suppress circulating currents between battery strings. One category is paralleling resistors across battery cell or modules to dissipate the energy of the battery strings with higher voltage with BMS [3]-[4]. Besides, it is not energy efficient, but also the string voltage equalizing will generate high of loss and heat, which may cause thermal runaway in cases. This approach significantly reduces system efficiency. Another category involves using power electronics converter to transfer energy from the high-voltage cell or module to the low-voltage cell or module [5]-[8]. Those approaches is ideal to equalize the string voltage. However, it is complex and costly.

A circulating current blocking circuit and control strategy for battery strings of BESS are proposed in this paper. By simply connecting a circulating current blocking circuit in series with each battery string branch, where each blocking circuit consists of two identical switches connected in series but the reversed direction, it can block the circulating currents between battery strings in charging, discharging, standby, and fault modes.

*This work was supported by National Natural Science Foundation of China under Grant 52037010.

This paper is organized as follows. The circulation current blocking circuit is introduced in the Section II. The operation mode and its control method are explained. The simulation results are given to verify the circulation current blocking circuit and control principle in the Section III. Finally, a summary is given in Section IV.

II. CIRCULATING CURRENT BLOCKING CIRCUIT

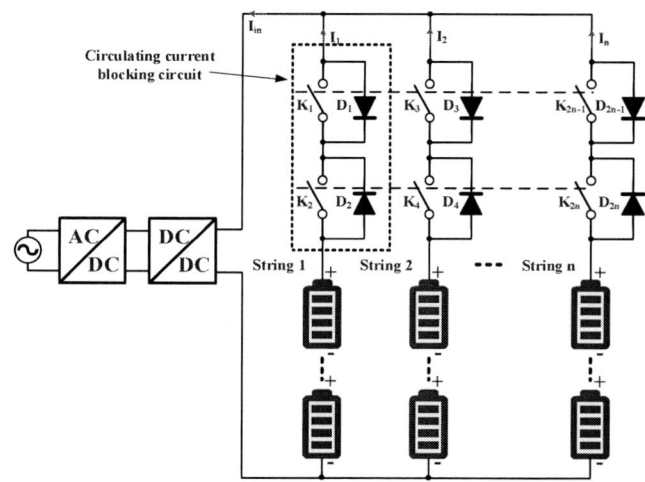

Figure 1: The schematic of the BESS and the proposed circuits

As shown in Figure 1, each battery string is composed of multiple battery cells connected in series to achieve the required DC voltage. Multiple battery strings are connected in parallel to achieve the required capacity of the BESS. Each battery string is connected in series with a circulating current blocking circuit, which consists of two identical units connected in reverse series. Each unit is composed of a switch with paralleled diode. The switch could be mechanical switch or power semiconductor device. By controlling the switches and utilizing the unidirectional conduction characteristic of the diodes, unidirectional current flow can be achieved during charging state and discharging state of the battery strings, effectively blocking the circulating currents between battery strings with different voltage.

A. Operation Mode Analysis

To explain the principle of the circulating current blocking, a simplified equivalent circuits are shown in Figure 2. The battery string is replaced by a voltage source V_n and internal resistance r_n. The working principle of the circulating blocking circuit is analyzed below in four different operating modes.

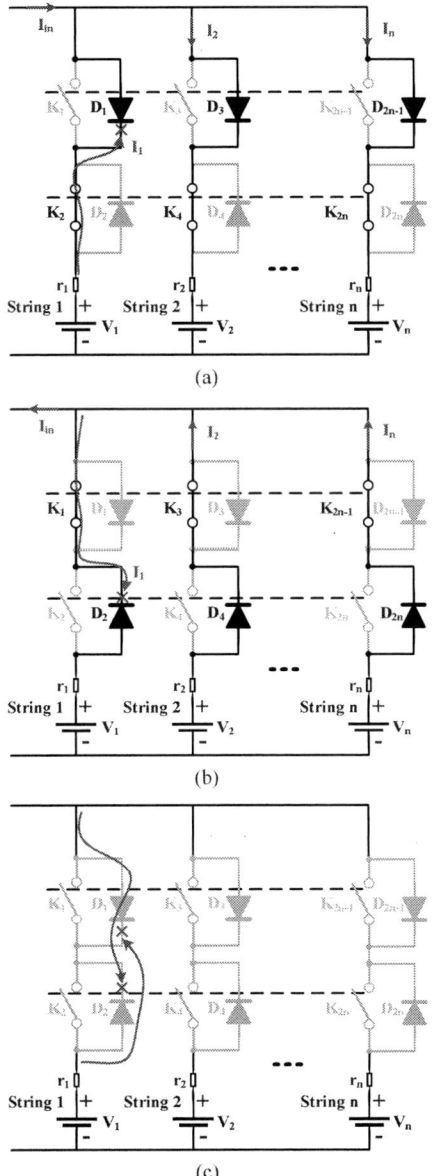

Figure 2: Operating mode equivalent circuits (a) Charging mode, (b) Discharging mode, (c) Standby mode or fault mode

(a) Charging mode: when the BESS absorbs the energy from the grid as shown in Figure 2(a), switches K_2, K_4, ..., K_{2n} turn on, and switches K_1, K_3, ..., K_{2n-1} turn off. Then, diodes D_1, D_3, ..., D_{2n-1} can block the potential circulating current among the battery strings.

(b) Discharging mode: When the BESS releases the energy to the grid as shown in Figure2 (b), switches K_1, K_3, ..., K_{2n-1} turn on, and switches K_2, K_4, ..., K_{2n} turn off. Then, diodes D_2, D_4, ...,

D_{2n} can block the potential circulating current among the battery strings.

(c) Standby mode/ Fault mode: when in a fault state or standby state, the BESS does not need to exchange energy. As shown in Figure 2(c), all the switches K_1, K_2, ..., K_{2n} turn off. Then, all the diodes D_1, D_2, ..., D_{2n} does not conduct. Thus, there is no current loop between battery strings.

B. Control Scheme

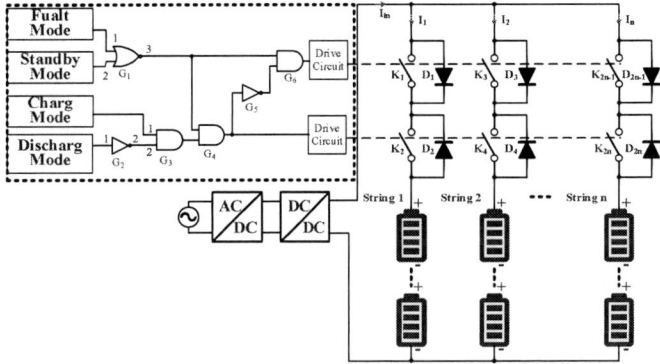

Figure 3: The Schematic diagram of the control structure of the proposed circuit

Figure 3 illustrates the control diagram of the circulating current blocking circuit. When the BESS operates in charging mode, input 1 of AND gate G_3 is at a high level, while the inputs of NOT gate G_2 and NOR gate G_1 are both at a low level. The logic circuit causes the output of AND gate G_6 to be low level and the output of AND gate G_4 to be high level, which respectively control the discharging switches K_1, K_3, ..., K_{2n-1} to turn off and the charging switches K_2, K_4, ..., K_{2n} to turn on. Correspondingly, when the BESS operates in discharging mode, input of NOT gate G_2 is at a high level, while the inputs of NOR gate G_1 and input 1 of AND gate G_3 are both at a low level. The logic circuit control switches K_1, K_3, ..., K_{2n-1} to turn on and switches K_2, K_4, ..., K_{2n} to turn off. When the BESS operates in standby mode, the input 2 of NOR gate G_1 is at a high level, and the inputs 1 of G_1, input of G_2, and input 1 of G_3 are all at a low level, controlling switches K_1, K_2, ..., K_{2n} to turn off. When the BESS operates in fault mode, the input 1 of NOR gate G_1 is at a high level, and the inputs of G_2, input 1 of G_3, and input 2 of G_1 are at a low level, controlling switches K_1, K_2, ..., K_{2n} to turn off.

To ensure the safe operation of the circulating current blocking circuit, it is very important to pay special attention to the potential significant voltage differences that may exist among battery strings during the start-up of charging mode and discharge mode. These differences can arise from the characteristics difference of these battery strings following a prolonged standby period. Failing to implement an appropriate soft-start strategy may lead to inrush current, which could potentially damage the battery strings or the components of the circulating current blocking circuit.

III. SIMULATION VERIFICATION

In this paper, simulation is used to verify the circulating current blocking circuit during the charging mode, discharging mode, standby mode and fault mode. The circuit and key parameters of the simulation are given in Figure 4 and Table I, respectively.

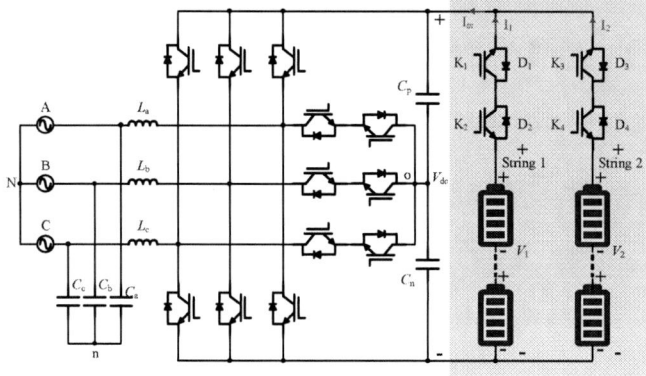

Figure 4. The simulation circuit of the BESS

TABLE I. THE KEY SIMULATION PARAMETERS OF THE BESS

Rate output power P_{rate}	200kW	Grid voltage V_g	380Vrms/50Hz
L_a, L_b, L_c	460 uH	C_a, C_b, C_c	25uF
C_p, C_n	11mF	Switching frequency f_s	10kHz
Battery string rate voltage	768 V	Maximum/minimum voltage of the string	864 V/648 V
Number of parallel battery strings	2	Inner resistance of battery string r	27 mΩ
Number of series battery cells	240	K1, K2, K3, K4	Infineon's IGBT FZ300R12KE3G

The BESS consists of two battery strings connected in parallel, and each battery string consists of 240 cells connected in series. The key parameters of the battery cell are listed in Table II.

TABLE II. KEY PARAMETERS OF THE BATTERY CELL

Rated Voltage	3.2 V	Capacity	280 Ah
Standard charging/discharging current	140 A/140 A	Maximum continue charging/discharging current I_{max}	280 A/280 A
Internal resistance r	0.11 mΩ	Maximum charging /discharging voltage	3.6V/ 2.7V

The BESS operates at the charging mode, and the total charging current is 250A as shown in Figure 5. At t >0, all the switches K₁ to K₄ are in turn on state, and the initial charging voltage of the battery string 1 and the battery string 2 are 775.2V and 762.3V, respectively. The string 2 current I_2 reaches charging current -275A while string 1 current I_1 becomes discharging current 25A. Approximately -150A circulation current flows between these two strings. Due to voltage difference of the battery strings, the higher voltage battery string 1 will charge the lower voltage battery string 2. At t = 1s, the

circulating current blocking circuit is activated. The higher voltage string 1 is blocked while the lower voltage string 2 is charged with total current I_{in} which is 250A.

With time going, the string 2 voltage will gradually increase, Once the voltage difference between the two strings becomes smaller, the string 1 will start to charge with the string 2. At t =8.1s, both strings will be charged together with a total charging current -250A.

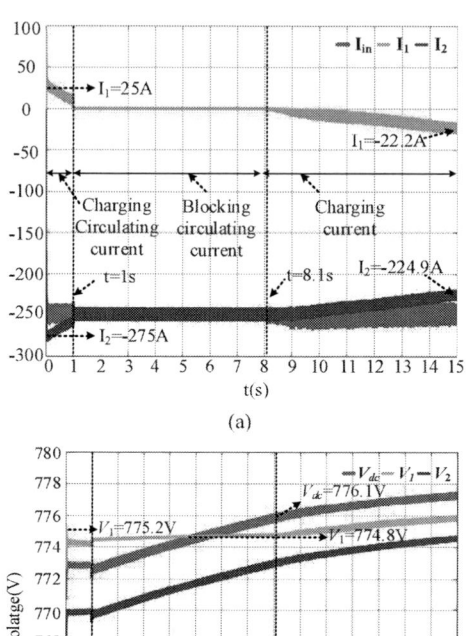

Figure 5. Simulation result of circulating current blocking in charging mode.

The BESS operates at the discharging mode, and the total charging current is 250A as shown in Figure 6. At t >0, all the switches K₁ to K₄ are in turn on state, and the initial discharging voltage of the battery string 1 and the battery string 2 are 814V and 800.5V, respectively. The string 1 discharging current I_1 reaches 297A while the string 2 operates at charging state with its current I_2 equal to -44A. Due to voltage difference of the battery strings, the higher voltage battery string 1 will charge the lower voltage battery string 2. 170.5A circulation current flows between these two strings. At t = 1s, the circulating current blocking circuit is activated. The lower voltage string 2 is blocked while the higher voltage string 1 is discharged with total current I_{in} which is 242A.

With time going, the string 1 voltage will gradually decrease. Once the voltage difference between the two strings becomes smaller, the string 2 will start to discharge with the string 1. At t =4.8s, both strings will discharge together with a total discharging current I_{in}.

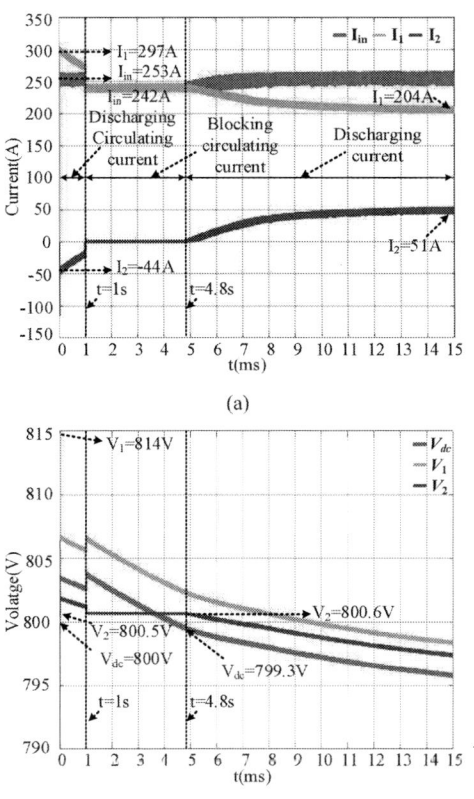

(a)

(b)

Figure 6. Simulation result of circulating current blocking in discharging mode

The BESS operates at the standby mode, and the total current I_{in} will be 0A as shown in Figure 7. At t >0, switches K_1 to K_4 are in turn on state. Due to voltage difference of the battery strings, the higher voltage battery string 1 will charge the lower voltage battery string 2. The string 2 current I_2 reach charging current -11.3A while string 1 current I_1 becomes discharging current 11.3A. About 11.3A circulation current flows between these two strings. At t =20ms, the circulating current blocking circuit is activated. The circulating current of the standby mode is cut off.

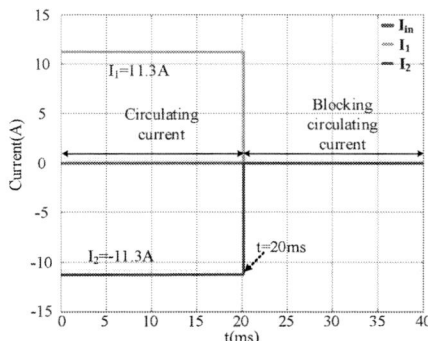

Figure 7. Simulation result of circulating current blocking in standby mode

As shown in Figure 8, during the discharging mode, the total discharging power is controlled to be 150kW. The maximum output power of the BESS is limited to 200kW. At t >0s, switches K_1 and K_3 are in turn on state, and switches K_2 and K_4 are in turn off state. Due to voltage difference of the battery

strings, the discharging current of the higher voltage battery string 1 I_1 is 118.4A, and the discharging current of the lower voltage battery string 2 I_2 is 97.7A. The total current I_{in}, which is the sum of I_1 and I_2, is 216.1A. due to the circulating current blocking circuit is activated, there is no circulating current. At t =20ms, an over load fault occurs, the discharging total current I_{in} of the BESS increases rapidly. At t=20.3ms, the output power exceeds 200kW, and the total current I_{in} is more than 300A. The circulating current blocking circuit operates in fault mode, switches K_1 to K_4 are turn off, which cuts off the fault current, preventing further increases in the battery string current that could potentially damage the battery strings and devices. Similarly, fault currents that occur during the charging mode can also be blocked.

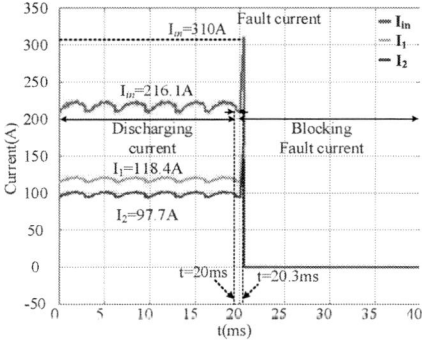

Figure 8. Simulation result of circulating current blocking in fault mode

IV. SUMMARY

The circulating currents of parallel-connected battery strings in BESS will affect the system efficiency, battery life and system safety. This paper proposes a circulating current blocking circuit and its control strategy, which can block the bidirectional circulating current and fault current between battery strings when the BESS is in charging mode, discharging mode, standby mode and fault mode. The proposed circuit and control are initially verified by the simulation. It is necessary to further verify the effective of the circulating current blocking circuit and control through experiments, which will be presented in the future.

REFERENCES

[1] Kun-Hsiang Lin, Li-Ren Yu, Chin-Sien Moo and Chun-Ying Juan, "Analysis on parallel operation of boost-type battery power modules," 2013 IEEE 10th International Conference on Power Electronics and Drive Systems (PEDS), Kitakyushu, Japan, 2013, pp. 809-813

[2] Xining Li, Lixing Lyu, Guangchao Geng, Quanyuan Jiang, Yu Zhao, Fuyuan Ma, Mengying Jin, "Power Allocation Strategy for Battery Energy Storage System Based on Cluster Switching," IEEE Transactions on Industrial Electronics, vol. 69, no. 4, pp. 3700-3710, April 2022.

[3] Vickey Vardwaj, Vishakha Vishakha, Vinay Kumar Jadoun, N. S. Jayalaksmi and Anshul Agarwal, "Various Methods Used for Battery Balancing in Electric Vehicles: A Comprehensive Review," 2020 International Conference on Power Electronics & IoT Applications in Renewable Energy and its Control (PARC), Mathura, India, 2020, pp. 208-213.

[4] Mohamed Daowd, Noshin Omar, Peter Van Den Bossche and Joeri Van Mierlo, "Passive and active battery balancing comparison based on MATLAB simulation," 2011 IEEE Vehicle Power and Propulsion Conference, Chicago, IL, USA, 2011, pp. 1-7.

[5] Yougui Guo, Peng Li, Ruijun Hou and Lie Xu, "Study of loop current suppression between multi-battery clusters in energy storage systems," 2024 IEEE 10th International Power Electronics and Motion Control Conference (IPEMC2024-ECCE Asia), Chengdu, China, 2024, pp. 3700-3705.

[6] Guangli Li, Jun Tian, Chaojin Feng, Qiao shi, Chao wang, chenhui yin, "Inter-cluster equalization regulation system for batteries," CN. Patent 217063321, July 26,2022

[7] Dehua shang, Wei Zhang, "Energy transfer type battery cluster parallel loop current control circuit,"CN.Patent 202210012803.3,April 12, 2022.

[8] Zhou Jianjie. "An energy storage system and a battery cluster equalization control method thereof," CN.Patent 202110249207. 2, Mar 28, 2021

A Hybrid Multilevel Converter-Based High-Gain Isolated DC/DC Converter for Grid-Tied Energy Storage Applications

Pengyu Fu, Yizhou Cong, Zhining Zhang, Jin Wang and Anant Agarwal

Department of Electrical and Computer Engineering
The Ohio State University
Columbus, OH, USA
fu.901@osu.edu

Abstract—A 7 kV/400 V isolated DC-DC converter utilizing 6.5 kV and 3.3 kV SiC MOSFETs is proposed for medium-voltage energy storage systems. The topology integrates a switched capacitor (SC) stage with modular multilevel converter (MMC) structure, achieving 12:1 voltage step-down before transformer isolation. The SC stage operates at 1 kHz with 6.5 kV devices, while MMC submodules switch at 20 kHz using 3.3 kV devices, enabling 40 kHz transformer operation through frequency multiplication. Capacitor voltage self-balancing is achieved through a novel switching scheme, and power flow is regulated through phase-shift control between primary and secondary voltages of transformer. Zero-voltage switching is realized for MMC submodules during voltage transitions. Simulation results validate the design, and experimental characterization of SiC devices demonstrates switching performance with dv/dt up to 40 V/ns.

Keywords—Medium-voltage silicon carbide, hybrid multilevel converter, high-gain isolated DC/DC converter, grid-tied energy storage

I. INTRODUCTION

Medium-voltage DC (MVDC) distribution networks are emerging as a solution for integrating renewable energy sources and energy storage systems (ESS), offering improved system resilience [1]. The key challenge lies in developing reliable and efficient power converters to interface ESS with MVDC networks while managing high voltage stress and maintaining high efficiency.

Recent advances in medium-voltage SiC MOSFETs have enabled various approaches to medium-voltage DC-DC conversion. Prior 2-level inverter based DC/DC converters have demonstrated 98% efficiency at 6 kV/400 V/40 kHz using 15 kV SiC devices [2] and 99% efficiency at 7 kV/400 V/48 kHz using 10 kV SiC devices [3]. However, reliability concerns persist regarding high-frequency operation of medium-voltage SiC devices [4] and transformer insulation stress [5]. Alternatively, a MMC-based 7 kV/400 V/30 kHz DC/DC converter using 3.3 kV SiC MOSFETs was demonstrated in [6].

This work proposes a hybrid multilevel converter (HMC) topology that combines the advantages of SC and MMC. The topology leverages newly available 6.5 kV and 3.3 kV SiC MOSFETs to optimize the trade-off between switching frequency, voltage stress, and system efficiency, while maintaining high reliability through reduced component stress.

II. TOPOLOGY AND OPERATION PRINCIPLES

A. Circuit Topology

The proposed converter topology is shown in Fig. 1. The key elements in the circuit include: medium-voltage stage, isolation stage and low-voltage stage. For the medium-voltage stage, it can be divided into SC section, decoupling section and MMC section. The SC section utilizes 6.5 kV SiC MOSFETs (S_{U1}, S_{U2}, S_{L1}, S_{L2}) interfacing with the V_{dc} input through split DC-link capacitors (C_1, C_2) rated at $V_{dc}/2$. S_{U1}/S_{L1} and S_{U2}/S_{L2} pairs operate in interleaved mode at f_{SC}. The decoupling section includes two series-connected capacitors C_3 and C_4, providing decoupling between SC and MMC sections. Each capacitor is rated at $V_{dc}/4$ and those two capacitors create a central split point N for transformer connection. In the MMC section, each arm contains two 3.3 kV SiC-based half-bridge submodules (selected based on device availability). Submodule SM_{U1} and SM_{U2} are in the upper arm, while submodule SM_{L1} and SM_{L2} are in the lower arm. In steady state, each submodules maintains voltage of $V_{dc}/6$.

The isolation stage features a medium-frequency transformer with its primary winding connected between the MMC leg midpoint P and the decoupling capacitors' midpoint N. The auxiliary inductor L_{aux}, integrated into the transformer leakage inductance, facilitates power flow control. The low-voltage stage employs a full H-bridge (S_{s1}-S_{s4}) operating at transformer frequency to interface with the battery. An output capacitor C_o provides ripple current filtering.

This configuration achieves 12:1 voltage step-down before transformer isolation while optimizing switching device stress and frequency. The 6.5 kV devices handle high-voltage/low-frequency switching in the SC section, while the 3.3 kV devices in MMC enable medium-frequency operation of transformer for reduced size.

B. Modulation Scheme

As shown in Fig.2, the proposed converter employs a novel modulation scheme that enables automatic submodule voltage self-balancing for MMC section while achieving desired voltage step-down ratio. The MMC section operates through

979-8-3315-1612-3/25 $31.00 © 2025 IEEE

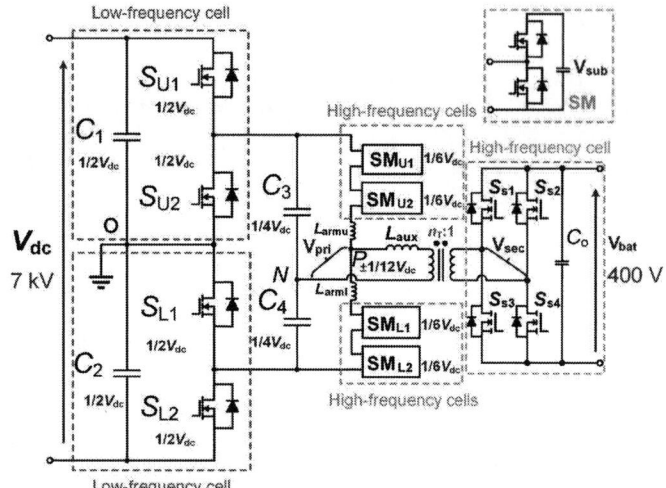

Fig. 1: Proposed 7 kV/400 V isolated DC/DC converter.

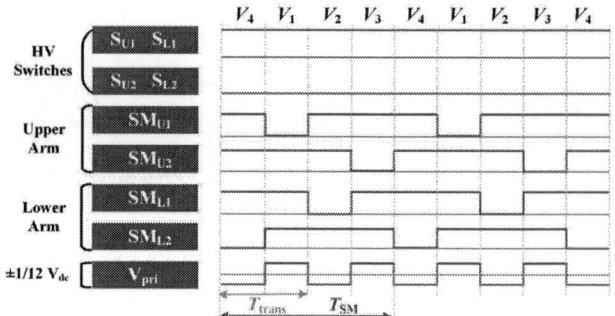

Fig. 2: Switching state placement for the switches on the medium-voltage side.

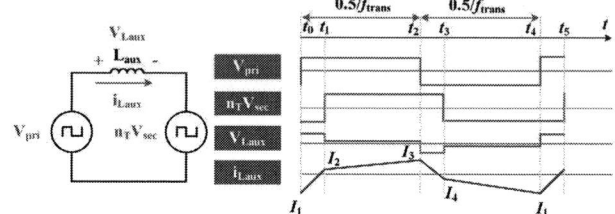

Fig. 3: Bidirectional power flow control principle.

four distinct switching states $\mathbf{V_1}$-$\mathbf{V_4}$, each occupying equal duration (1/4) of the submodule switching cycle T_{SM}. If 1 represents submodule insertion and 0 represents bypass state, then $\mathbf{V_1} = [0,1,1,1]^T$, $\mathbf{V_2} = [1,1,0,1]^T$, $\mathbf{V_3} = [1,0,1,1]^T$, $\mathbf{V_4} = [1,1,1,0]^T$.

The switching sequence maintains submodule capacitor voltages at $V_{dc}/6$ through a self-balancing mechanism described by:

$$\begin{bmatrix} \frac{V_{dc}}{2} \\ \frac{V_{dc}}{2} \\ \frac{V_{dc}}{2} \\ \frac{V_{dc}}{2} \end{bmatrix} = \begin{bmatrix} V_1 \\ V_2 \\ V_3 \\ V_4 \end{bmatrix}^T \begin{bmatrix} V_{CU1} \\ V_{CU2} \\ V_{CL1} \\ V_{CL1} \end{bmatrix} = \begin{bmatrix} 0 & 1 & 1 & 1 \\ 1 & 1 & 0 & 1 \\ 1 & 0 & 1 & 1 \\ 1 & 1 & 1 & 0 \end{bmatrix} \begin{bmatrix} V_{CU1} \\ V_{CU2} \\ V_{CL1} \\ V_{CL1} \end{bmatrix}. \quad (1)$$

This results in balanced submodule capacitor voltages:

$$V_{CU1} = V_{CU2} = V_{CL1} = V_{CL2} = \frac{V_{dc}}{6}. \quad (2)$$

The transformer primary voltage V_{pri} alternates between two levels based on the switching states. During $\mathbf{V_1}$ and $\mathbf{V_3}$:

$$V_{pri} = \frac{V_{dc}}{4} - \frac{V_{dc}}{6} = \frac{V_{dc}}{12}, \quad (3)$$

During $\mathbf{V_2}$ and $\mathbf{V_4}$:

$$V_{pri} = \frac{V_{dc}}{4} - \frac{2V_{dc}}{6} = -\frac{V_{dc}}{12}. \quad (4)$$

The modulation scheme creates a frequency multiplication effect, where the transformer operating frequency f_{trans} ($1/T_{trans}$) is twice the MMC submodule switching frequency f_{SM} ($1/T_{SM}$):

$$f_{trans} = 2f_{SM}. \quad (5)$$

To optimize overall converter efficiency, the modulation implements a multi-frequency domain strategy. The SC section using 6.5 kV devices operates at a lower frequency f_{SC} to minimize switching losses. The MMC section with 3.3 kV devices operates at medium frequency f_{SM}. The transformer and secondary H-bridge operate at f_{trans}.

C. Power Flow Control

The power flow control of the proposed converter follows the dual active bridge principle, as illustrated in Fig.3, where power transfer is controlled by the phase shift ϕ between primary and secondary square-wave voltages of transformer. The transformer primary voltage V_{pri} alternates between $\pm V_{dc}/12$ while the secondary voltage alternates between $\pm V_{bat}$. The phase shift angle ϕ is defined as:

$$\phi = \frac{\pi(t_1 - t_0)}{0.5/f_{trans}}, \quad (6)$$

where $t_1 - t_0$ represents the time delay between primary and secondary voltages. n_T is defined as the transformer turns ratio. r_v (=12 in this case study) is defined as the voltage ratio between V_{dc} and the primary voltage of transformer.

The auxiliary inductor current i_{Laux} can be derived for different intervals within a switching period T_{trans}. For $t_0 \leq t < t_1$:

$$i_{Laux}(t) = I_1 + \frac{V_{dc}/r_v + n_T V_{bat}}{L_{aux}}(t - t_0) = I_1 + K_1(t - t_0). \quad (7)$$

For $t_1 \leq t < t_2$:

$$i_{Laux}(t) = I_2 + \frac{V_{dc}/r_v - n_T V_{bat}}{L_{aux}}(t - t_1) = I_2 + K_2(t - t_1). \quad (8)$$

For $t_2 \leq t < t_3$:

$$i_{Laux}(t) = I_3 + \frac{-V_{dc}/r_v - n_T V_{bat}}{L_{aux}}(t - t_2) = I_3 + K_3(t - t_2). \quad (9)$$

For $t_3 \leq t < t_4$:

$$i_{Laux}(t) = I_4 + \frac{-V_{dc}/r_v + n_T V_{bat}}{L_{aux}}(t - t_3) = I_4 + K_4(t - t_3). \quad (10)$$

Based on the boundary conditions, the boundary currents can be calculated as:

$$I_1 = \frac{-n_T V_{bat} \phi}{2\pi f_{trans} L_{aux}}, \tag{11}$$

$$I_2 = I_1 + \frac{K_1 \phi}{2\pi f_{trans}}, \tag{12}$$

$$I_3 = I_2 + \frac{K_2(\pi - \phi)}{2\pi f_{trans}}, \tag{13}$$

$$I_4 = I_3 + \frac{K_3 \phi}{2\pi f_{trans}}. \tag{14}$$

Accordingly, the transferred power P_{tr} can be calculated as:

$$P_{tr} = \frac{V_{dc}}{r_v} \frac{n_T V_{bat}}{2\pi f_{trans} L_{aux}} \phi(1 - \frac{\phi}{\pi}). \tag{15}$$

III. CIRCUIT ANALYSIS AND SYSTEM DESIGN

A. Medium-Frequency Transformer Design

For transformer design, key circuit parameters including turns ratio n_T, apparent power S_{trans} and axillary inductor value L_{aux} need to be decided. Based on Eq. 15, the inductance value can be calculated as:

$$L_{aux} = \frac{n_T V_{bat} V_{dc}}{2\pi f_{trans} r_v P_{tr}} \phi(1 - \frac{\phi}{\pi}). \tag{16}$$

The transformer RMS current I_{rms} can be derived from the piecewise linear current, and the apparent power rating of the transformer can be calculated as:

$$S_{trans} = I_{rms} \frac{V_{dc}}{r_v}. \tag{17}$$

Fig. 4 presents a parametric analysis examining the relationship between transformer design parameters and operating characteristics. The analysis includes the impact of auxiliary inductance (20 μH - 150 μH) and transformer turns ratio (1.2 - 2.0) on phase shift angle, apparent power, and transformer RMS current. In the case study, the battery voltage range is 250 V to 450 V, with fixed transformer frequency of 40 kHz and power transfer of 10 kW. Through multi-objective optimization considering phase shift constraints ($10°$ - $45°$) and RMS current limitation (20 A), the optimal design parameters are determined as: auxiliary inductance of 69.9 μH and transformer turns ratio of 1.53. This combination results in a maximum phase shift angle of 30.0 degrees, minimum VA rating of 10.89 kVA, and RMS current of 18.7 A.

B. MMC Capacitor Design

The RMS current of submodule capacitor C_{SM} can be derived from the arm current and switching function:

$$i_{C_{SM}}(t) = S(t) \cdot i_{arm}(t). \tag{18}$$

The upper arm current consists of:

$$i_{up_{arm}}(t) = i_{z_{dc}} + \frac{1}{2}i_{Laux}(t), \tag{19}$$

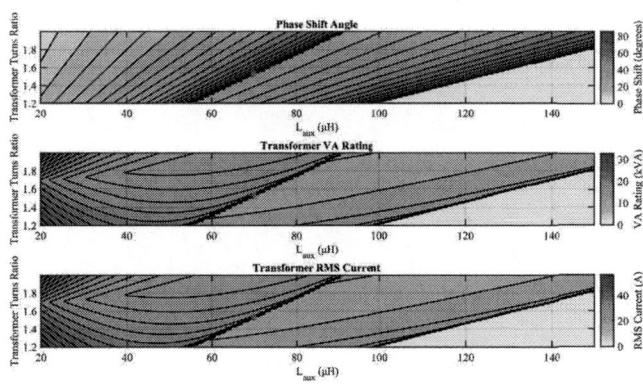

Fig. 4: Phase shift angle, apparent power and RMS current vs. auxiliary inductance at different turns ratios ($f_{trans} = 40$ kHz).

where $i_{z_{dc}} = P_{tr}/(V_{dc}/2)$ is the DC circulating current in MMC arms, and $i_{Laux}(t)$ is the transformer primary side current. The lower arm current is:

$$i_{lo_{arm}}(t) = i_{z_{dc}} - \frac{1}{2}i_{Laux}(t). \tag{20}$$

Given the switching sequence $[V_1, V_2, V_3, V_4]$ and the duty ratio $N_i/N = 3/4$, the RMS current of submodule capacitor is:

$$I_{C_{RMS}} = \sqrt{\frac{N_i}{N}} \sqrt{I_{z(dc)}^2 + \frac{1}{4}I_{Laux(rms)}^2}. \tag{21}$$

The capacitance can be determined from the peak-to-peak voltage ripple:

$$C_{SM} = \frac{\max|\int i_{C_{SM}}(t)dt|}{\Delta V_{SM}}. \tag{22}$$

The peak-to-peak voltage ripple can be calculated as:

$$\Delta V_{SM} = \max[v_{SM}(t)] - \min[v_{SM}(t)]. \tag{23}$$

For the given specifications in the case study: $V_{dc} = 7$ kV, $V_{SM} = 1.167$ kV, $P_{tr} = 10$ kW, $f_{trans} = 40$ kHz, $\Delta V_{SM}/V_{SM} = 5\%$, the calculated results for sizing MMC submodule capacitors are: $I_{C_{RMS}} = 7.45$ A, $C_{SM} = 3.5\mu F$, $\Delta V_{SM} = 51.97$ V(4.45%).

The DC-link capacitors C_3 and C_4 serve as critical decoupling components between the SC and MMC sections. Their design need account for both voltage ripple requirements and current handling capabilities. Given the symmetrical operation, the analysis focuses on C_3. The current through C_3 consists of half the transformer primary current:

$$i_{C3}(t) = i_{z_{ac}}(t) = \frac{1}{2}i_{Laux}(t). \tag{24}$$

The capacitor's current stress is characterized by its RMS value:

$$I_{C3(rms)} = \sqrt{\frac{1}{T_{trans}} \int_0^{T_{trans}} (\frac{1}{2}i_{Laux}(t))^2 dt}. \tag{25}$$

The steady-state voltage ripple across C_3 is defined by:

$$\Delta V_{C3} = \max[v_{C3}(t)] - \min[v_{C3}(t)]. \tag{26}$$

The relationship between capacitance and maximum charge variation:

$$C_3 = \frac{\max |\int i_{C3}(t)dt|}{\Delta V_{C3}}. \tag{27}$$

Using triangular current approximation, the charge variation during half of T_{trans} can be expressed as:

$$\Delta Q = \frac{1}{2} \cdot \frac{\Delta I_{pp}}{2} \cdot \frac{T_{trans}}{2} = \frac{\Delta I_{pp} T_{trans}}{8}, \tag{28}$$

where ΔI_{pp} represents peak-to-peak inductor current and $T_{trans} = 1/f_{trans}$ is the transformer period. The voltage ripple can then be expressed as:

$$\Delta v_{C3} = \frac{\Delta Q}{C_3} = \frac{\Delta I_{pp}}{8 f_{trans} C_3}. \tag{29}$$

Solving for the required capacitance:

$$C_3 = \frac{\Delta I_{pp}}{8 f_{trans} \Delta v_{C3}}. \tag{30}$$

For design specifications: $f_{trans} = 40$ kHz, $V_{C3,4} = V_{dc}/4 = 1.75$ kV, 2% voltage ripple requirement ($\Delta V_{C3,4} = 35$ V), the calculated peak-to-peak current ripple is $\Delta I_{pp} = 45.7$ A. The minimum required capacitance is $4.1\mu F$. Further calculation shows that required RMS current is 9.80 A and the actual voltage ripple is 46.9 V (2.6%) with $4.1\mu F$ capacitance.

C. Input and Output Capacitor Design

The switching functions for SC section with interleaved operation of S_{U1}/S_{L1} and S_{U2}/S_{L2} are:

$$S_{U1}(t) = \begin{cases} 1, & 0 \le t < \frac{T_{SCsw}}{2} \\ 0, & \frac{T_{SCsw}}{2} \le t < T_{SCsw} \end{cases}, \tag{31}$$

$$S_{U2}(t) = \begin{cases} 0, & 0 \le t < \frac{T_{SCsw}}{2} \\ 1, & \frac{T_{SCsw}}{2} \le t < T_{SCsw} \end{cases}. \tag{32}$$

The DC component into the MMC section comes from the DC source, and their difference needs to be handled by the capacitors C_1 and C_2. For interleaved operation:

$$i_{C1}(t) = \begin{cases} i_{dc_{in}} - i_{dc_{MMC}}, & 0 \le t < \frac{T_{SCsw}}{2} \\ i_{dc_{in}}, & \frac{T_{SCsw}}{2} \le t < T_{SCsw} \end{cases} \tag{33}$$

$$i_{C2}(t) = \begin{cases} i_{dc_{in}}, & 0 \le t < \frac{T_{SCsw}}{2} \\ i_{dc_{in}} - i_{dc_{MMC}}, & \frac{T_{SCsw}}{2} \le t < T_{SCsw} \end{cases}. \tag{34}$$

$i_{dc_{MMC}}$ is DC current into MMC section and $i_{dc_{in}}$ is the DC current at V_{dc} bus. Based on power balance, their relationship can be expressed as:

$$i_{dc_{MMC}} = 2 i_{dc_{in}}. \tag{35}$$

The voltage ripple now becomes:

$$\Delta v_{C1} = \frac{\max |\int i_{C1}(t)dt|}{C_1}. \tag{36}$$

Solving for the required capacitance:

$$C_1 = \frac{i_{dc_{in}} T_{SCsw}}{2 \Delta v_{C1}}. \tag{37}$$

The output capacitor C_o is designed to filter the high-frequency current ripple from the H-bridge. The current through the capacitor can be derived from the transformer secondary current and H-bridge switching function:

$$i_{sec}(t) = n_T i_{Laux}(t). \tag{38}$$

The rectified current after H-bridge:

$$i_{rect}(t) = \begin{cases} i_{sec}(t), & 0 \le t < \frac{T_{trans}}{2} \\ -i_{sec}(t), & \frac{T_{trans}}{2} \le t < T_{trans} \end{cases}. \tag{39}$$

The output capacitor current is the difference between rectified current and battery current:

$$i_{Co}(t) = i_{rect}(t) - I_{bat}, \tag{40}$$

where $I_{bat} = \frac{P_{tr}}{V_{bat}}$ is the battery current. For a given maximum output voltage ripple ΔV_o, the required capacitance can be calculated based on the peak-to-peak current ripple:

$$C_o = \frac{\Delta I_{Co}}{16 f_{trans} \Delta V_o}, \tag{41}$$

where $\Delta I_{Co} = \max(i_{rect}) - \min(i_{rect})$ is the peak-to-peak current ripple, and factor 16 comes from double frequency effect of H-bridge. The capacitor RMS current can be calculated as:

$$I_{Co(rms)} = \sqrt{\frac{1}{T_{trans}} \int_0^{T_{trans}} i_{Co}^2(t)dt}. \tag{42}$$

For the case study with $P_{tr} = 10$ kW, $V_{bat} = 400$ V, $f_{trans} = 40$ kHz, $\Delta V_o/V_o = 2\%$, $n_T = 1.53$, the calculated capacitance $C_o = 13.95\mu F$, the calculated RMS current $I_{Co(rms)} = 14.89$ A, and the calculated voltage ripple $\Delta V_o = 12.03$ V (3.01%).

D. Semiconductor Device Power Loss Analysis

The power losses in semiconductor devices are analyzed for three stages: MMC submodules, SC switches, and H-bridge. For each stage, both conduction and switching losses are considered based on operating conditions and device characteristics. For MMC stage with 3.3 kV SiC MOSFETs, The RMS arm current includes both DC and AC components:

$$I_{arm(rms)} = \sqrt{\left(\frac{P_{in}}{V_{dc}/2}\right)^2 + \frac{1}{4}I_{Laux(rms)}^2}. \tag{43}$$

The conduction loss for the switches in MMC submodules:

$$P_{cond(SM)} = R_{ds(on)}(T_j) I_{arm(rms)}^2. \tag{44}$$

For switching loss, considering the $[V_1, V_2, V_3, V_4]$ pattern, the proposed topology achieves natural zero-voltage switching (ZVS) ON for all MMC submodules during voltage transitions. When the transformer primary voltage transitions from $-V_{dc}/12$ to $V_{dc}/12$ ($\mathbf{V_4} \to \mathbf{V_1}$ or $\mathbf{V_2} \to \mathbf{V_3}$), the switch in the bypassed submodules in the upper arm achieves zero-voltage switching (ZVS) turn-ON because the anti-parallel diode of the switch conducts during the transition. When the transformer primary voltage transitions from $V_{dc}/12$ to $-V_{dc}/12$ ($\mathbf{V_1} \to \mathbf{V_2}$ or $\mathbf{V_3} \to \mathbf{V_4}$), the switch in the

979-8-3315-1612-3/25 $31.00 © 2025 IEEE

bypassed submodules in the lower arm achieves zero-voltage switching (ZVS) turn-ON. As a result, all switches in the MMC submodules naturally achieve ZVS turn-ON conditions during their respective switching transitions. Therefore, the submodule switching loss is decided by turn-off power loss:

$$P_{sw(SM)} = f_{SM} E_{off}(I_{off}, V_{SM}), \quad (45)$$

where the switching energy is scaled by voltage and current:

$$E_{off}(I, V) = E_{off(ref)} \frac{V}{V_{ref}} \frac{I}{I_{ref}}. \quad (46)$$

For SC stage with 6.5 kV SiC MOSFETs, switches S_{U1}, S_{U2}, S_{L1}, S_{L2} operate at 50% duty cycle:

$$P_{cond(SC)} = R_{ds(on)}(T_j)(\frac{P_{tr}}{V_{dc}/2})^2 D_{50\%}. \quad (47)$$

Switching losses at SC frequency:

$$P_{sw(SC)} = f_{SC}[E_{on}(I_{z(dc)}, \frac{V_{dc}}{2}) + E_{off}(I_{z(dc)}, \frac{V_{dc}}{2})]. \quad (48)$$

For H-bridge stage, the RMS current through H-bridge switches is:

$$I_{HB(rms)} = n_T \cdot I_{Laux(rms)}. \quad (49)$$

Conduction loss considering 50% duty cycle is:

$$P_{cond(HB)} = R_{ds(on)}(T_j)I_{HB(rms)}^2 D_{50\%}. \quad (50)$$

With ZVS operation of the switches, only turn-off losses occur:

$$P_{sw(HB)} = f_{trans}E_{off}(I_{HB(rms)}, V_{bat}). \quad (51)$$

Considering N submodules in MMC stage, the total semiconductor power loss can be calculated as:

$$P_{loss(total)} = N(P_{cond(SM)} + 2P_{sw(SM)})$$
$$+ 4(P_{cond(SC)} + P_{sw(SC)}). \quad (52)$$
$$+ 4(P_{cond(HB)} + P_{sw(HB)})$$

IV. EXPERIMENTAL AND SIMULATION RESULTS

A. Switching Performance of Medium-Voltage SiC Devices

A 10-kW converter prototype using both 6.5 kV and 3.3 kV SiC MOSFETs is under development. Both of the 6.5 kV and 3.3 kV SiC devices are engineering samples from GeneSiC. The typical $R_{ds(on)}$ of the 6.5 kV SiC devices is 160 mΩ. The typical $R_{ds(on)}$ of the 3.3 kV SiC devices is 60 mΩ. Initial characterization focused on evaluating the switching performance of both device types. The 3.3 kV SiC MOSFETs are tested under a blocking voltage of 2.4 kV, load current of 30 A, gate drive $V_{gs} = -5$ V/20 V and gate resistors $R_{g(on)} = 7.5\Omega$, $R_{g(off)} = 6.5\Omega$. The 6.5 kV SiC MOSFETs are tested under a blocking voltage of 3 kV, load current of 20 A and the same gate drive conditions.

The switching waveforms are shown in Fig. 5. For 3.3 kV devices, the turn-off transient in Fig. 5a exhibits a dv/dt of 47.2 V/ns, while the turn-on in Fig. 5b shows 24.6 V/ns. The 6.5 kV devices demonstrate switching speeds with dv/dt of 40 V/ns during turn-off (Fig. 5c) and 30.3 V/ns during turn-on

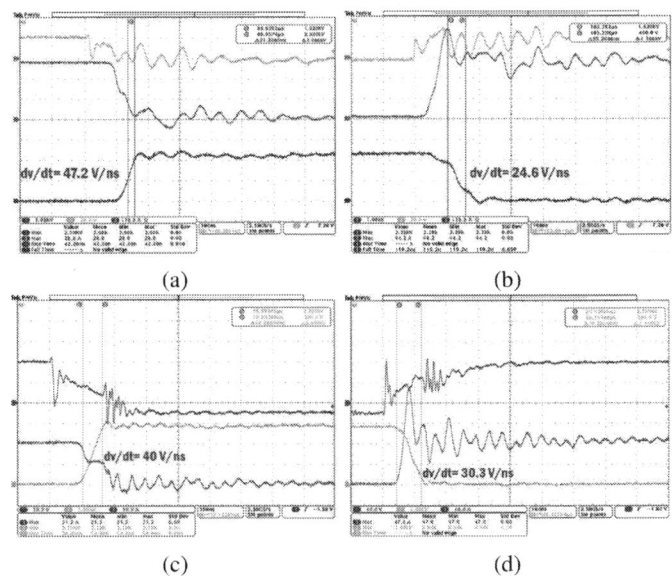

(a)　　　　　　　　(b)

(c)　　　　　　　　(d)

Fig. 5: Switching performance of 3.3 kV and 6.5 kV SiC MOSFETs. (a) turn-off transient of 3.3 kV SiC; (b) turn-on transient of 3.3 kV SiC; (c) turn-off transient of 6.5 kV SiC; (d) turn-on transient of 6.5 kV SiC.

(Fig. 5d) at 3 kV. The E_{on} and E_{off} of the 6.5 kV device at 3 kV/20 A are 2.97 mJ and 2.25 mJ, respectively.

Fig. 6 presents the switching energy characterization of the 3.3 kV SiC MOSFET at a 2 kV DC bus voltage under different drain currents. The switching performance was measured at room temperature with gate drive conditions of $V_{gs} = -5$ V/20 V and $R_{g(on)} = 7.5\Omega$, $R_{g(off)} = 6.5\Omega$. The device exhibits distinct behavior between turn-on and turn-off transitions. The turn-off energy (E_{off}) shows relatively weak current dependency, increasing gradually from 1.2 mJ at 10 A to 1.8 mJ at 30 A, with a tendency to saturate above 20 A. In contrast, the turn-on energy (E_{on}) demonstrates stronger current dependency, rising linearly from 2.8 mJ at 10 A to 6.0 mJ at 30 A. The total switching energy (E_{total}), dominated by turn-on losses, shows an approximately linear increase from 4.0 mJ to 7.8 mJ across the measured current range. This characterization confirms the device's suitability for medium frequency operation in the MMC stage, with turn-off losses being the primary consideration for efficiency optimization due to ZVS operation in MMC submodules.

At 10 kW rated power, with the $R_{ds(on)}$ and switching energy data, the power loss for the medium-voltage SiC devices can be calculated. For the MMC stage with 3.3 kV devices, the RMS arm current is 9.84 A, leading to conduction loss per submodule 5.81 W. With ZVS turn-on, only turn-off loss is considered where $E_{off} = 0.46$ mJ, resulting in switching loss 9.18 W per switch. For 4 submodules with 2 switches each, total MMC loss is $P_{loss_{MMC}} = 4 \cdot (5.81 + 9.18 \cdot 2) = 96.7$ W. For the SC stage with 6.5 kV devices, the switching energy at the operating voltage/current are $E_{off} = 0.375$ mJ and $E_{on} = 0.5$ mJ. With conduction loss $P_{cond(SC)} = R_{ds(on)}I_{z(dc)}^2 \cdot 0.5 = 0.65$ W, total SC loss for 4

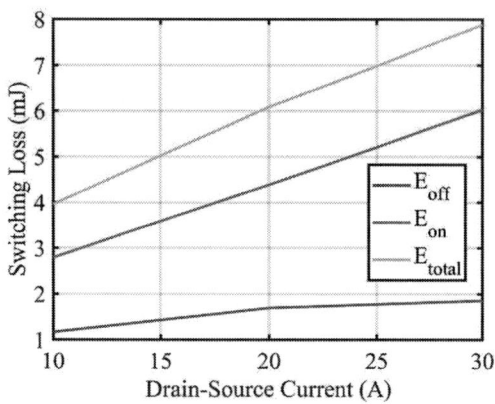

Fig. 6: Switching energy loss of the 3.3 kV SiC MOSFET at 2 kV DC bus voltage.

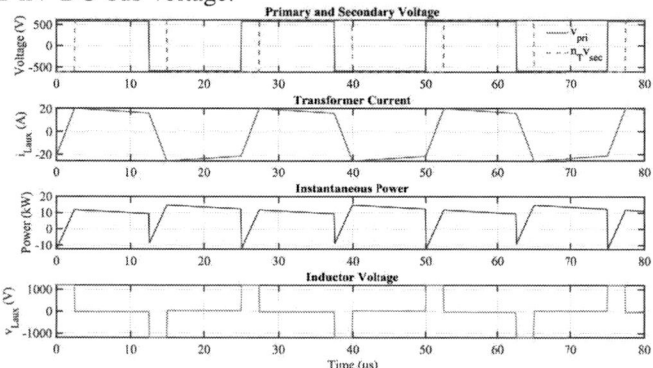

Fig. 7: Simulation results showing key waveforms of converter operation.

switches is $P_{loss_{SC}} = 4 \cdot (0.65 + 1 \text{ kHz} \cdot 0.875 \text{ mJ}) = 6.09 \text{ W}$.

B. Simulation Results of Converter Operation

The simulation results for the 10-kW medium-voltage DC-DC converter are presented in Figs. 7 and 8. Fig. 7 illustrates the key waveforms of the converter under steady-state operation. The transformer primary voltage (v_{pri}) exhibits a square-wave pattern of ± 583 V at 40 kHz, while the converted secondary voltage ($n_T v_{sec}$) switches between ± 612 V with a phase shift of 35°. The auxiliary inductor voltage (v_{Laux}) reaches peak values of ± 1195 V. The transformer current (i_{Laux}) demonstrates the characteristic trapezoidal waveform of dual-active bridge operation, with peak values of approximately ± 20 A. The instantaneous power waveform confirms stable power transfer of 10 kW from primary to secondary side, validating the proposed phase-shift control strategy. Fig. 8 demonstrates the zero-voltage switching (ZVS) operation of MMC submodules. The switching function (S_{SMU1}), submodule voltage (v_{SMU1}), and upper arm current (i_{arm}) waveforms are shown. During the turn-on transition (highlighted region), the negative arm current facilitates current through the anti-parellel diode, enabling ZVS operation of the switch.

V. CONCLUSION

This paper presents a HMC topology using 6.5 kV and 3.3 kV SiC MOSFETs for medium-voltage energy storage

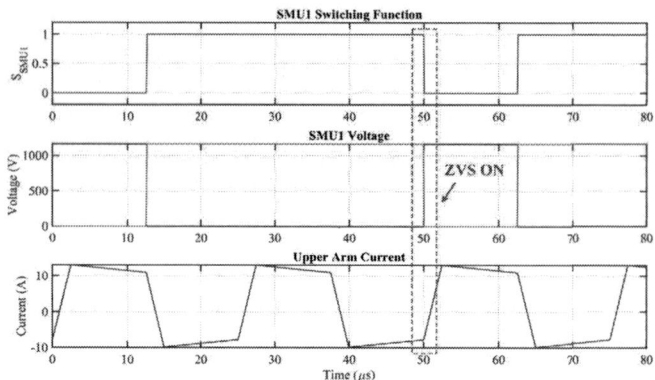

Fig. 8: ZVS ON of MMC submodules.

applications, achieving 7 kV/400 V conversion through integration of SC and MMC structures. Benefits of the solution include: 12:1 voltage step-down before the transformer and 2:1 frequency multiplication of transformer frequency to MMC submodule frequency; reduced component stresses with low switching frequency (1 kHz) of 6.5 kV switches and low voltage (583 V/400 V) and high frequency (40 kHz) operation of transformer; self-balancing of MMC submodule voltage; ZVS for MMC submodules during voltage transition. Initial experimental characterization of SiC devices validates their switching performance, with achieved dv/dt of 40 V/ns at turn-off. Simulation results demonstrate controlled power flow, ZVS operation and balanced voltage distribution in MMC submodules. Future work will focus on the experimental demonstration of the complete 10-kW converter prototype using the evaluated 6.5 kV and 3.3 kV SiC MOSFETs.

ACKNOWLEDGMENT

The authors gratefully acknowledge support for this work from the Energy Storage Program in the Office of Electricity Delivery and Energy Reliability at the U.S. Department of Energy.

REFERENCES

[1] B. Grainger and R. W. De Doncker, *Medium voltage DC system architectures*. IET, 2021, vol. 143.

[2] L. Wang, Q. Zhu, W. Yu, and A. Q. Huang, "A medium-voltage medium-frequency isolated DC–DC converter based on 15-kV SiC MOSFETs," *IEEE Journal of Emerging and Selected Topics in Power Electronics*, vol. 5, no. 1, pp. 100–109, 2016.

[3] D. Rothmund, T. Guillod, D. Bortis, and J. W. Kolar, "99% efficient 10 kV SiC-based 7 kV/400 V DC transformer for future data centers," *IEEE Journal of Emerging and Selected Topics in Power Electronics*, vol. 7, no. 2, pp. 753–767, 2018.

[4] A. B. Jørgensen, T. S. Aunsborg, S. Beczkowski, C. Uhrenfeldt, and S. Munk-Nielsen, "High-frequency resonant operation of an integrated medium-voltage SiC MOSFET power module," *IET Power Electronics*, vol. 13, no. 3, pp. 475–482, 2020.

[5] R. Agarwal, H. Li, Z. Guo, and P. Cheetham, "The effects of PWM with high dv/dt on partial discharge and lifetime of medium-frequency transformer for medium-voltage (MV) solid state transformer applications," *IEEE Transactions on Industrial Electronics*, vol. 70, no. 4, pp. 3857–3866, 2022.

[6] D. Xing, X. Li, Y. Zhang, Q. Cheng, Z. Zhang, B. Hu, A. Agarwal, J. Wang, and R. Guenther, "MMC-based high gain solid-state transformers for energy storage applications," in *2021 IEEE Applied Power Electronics Conference and Exposition (APEC)*. IEEE, 2021, pp. 1996–2002.

LCL Filter Parameter Selection Using Graphical Method for a 13.8 kV ac 1.1 MVA 7-level Flying Capacitor Grid-Connected Converter Utilizing Variable Switching Frequency

Arthur Mendes[1], David Nam, Mingze Gao, Timothy Thacker, Dong Dong, Rolando Burgos

Virginia Tech
Center for Power Electronics Systems
Blacksburg, Virginia USA
[1]arthurcm@vt.edu

Abstract—LCL filter is widely used in dc to ac power converters due to its ability to provide higher attenuation at high frequencies (HF), favorable inductor size trade-off, and enhanced system dynamics response. This is particularly advantageous in renewable energy and grid-connected applications which require compliance with power quality standards (e.g. IEEE Std 519 and IEEE Std 1547). Although the design procedure for LCL filters has been extensively discussed, existing methods often focus their consideration on 2-level or 3-level neutral point clamped converters, constant switching frequency, and passive damping. This paper presents an LCL filter design procedure using a graphical method for a 1.1 MVA 13.8 kV ac 22 kV dc 7-level flying capacitor multilevel (FCML) grid-connected converter based on 10 kV SiC MOSFETs. The method accounts for any number of levels, variable switching frequency, and active damping. This graphical method can help designers gain insight into filter performance and iterate to select the optimal filter parameters.

Index Terms—10 kV SiC MOSFET, flying capacitor multilevel converter (FCML), grid-connected, LCL filter, medium voltage converter, variable switching frequency.

I. INTRODUCTION

Grid-connected dc-ac pulse-width modulation (PWM) voltage source converters (VSC) require AC filters. Its main functions are to enable the connection between the power converter (a switched source) and the grid, and to attenuate switching frequency ripple and fundamental harmonics. Both attenuations are crucial for providing compliance with power quality standards, such as IEEE Std 519 for harmonic control on power systems.

The most common types of ac filters are L, LCL, LLCL, and 2-stage LCL [1]–[3]. Among these, the LCL filter often provides the most favorable trade-off between attenuation and size. The LCL filter achieves high attenuation after its resonant peak (60 dB/dec), hence requiring less overall inductance than the L and LLCL options. The latter is able to attenuate the resonant peak with an additional inductor in series with the capacitor, but it compromises high-frequency (HF) attenuation and may require a larger series inductance [4]. The 2-stage LCL filter achieves the highest HF attenuation but introduces

multiple resonant peaks, increased component count, and size, which can offset its benefits.

Even though the LCL filter and its design methods have been extensively discussed, there is no single general method that covers the full range of applications and all of the converter variations, such as topology, power and voltage ranges, modulation scheme, and switching frequency range. In [5]–[7], the traditional method of selecting LCL parameters is defined. However, limiting the filter capacitance value to a percentage ($\leq 5\%$) of the system base impedance to control the displacement power factor (DPF) can be an oversimplification in high-power grid-connected applications. The strict limits to the capacitance values could place the resonant frequency too close to the switching frequency, causing harmonic distortion and instability. Moreover, the converter controls can compensate for the displacement power factor. For example, in [8], a capacitor greater than 1 pu was designed for a 5 MW wind converter.

A design method for multi-megawatt grid-connected applications with low switching frequencies (e.g. $\leq 1.2\,kHz$) was developed in [9]. However, the design was targeted at a 3-level NPC converter; therefore, the effects of a flying capacitor (FC) voltage ripple, as well as active damping effects, on the parameter selection were not discussed.

In [2] the method to obtain the attenuation requirement plot was introduced for a 200 kVA 3-level NPC PV converter; however, the effects of variable switching frequency (VSF), flying capacitor ripple, and active damping on LCL filter parameter selection were not explored.

The volume, losses and temperature optimization method was presented in [10] for a low voltage three-phase PFC converter; hence, different DPF operations were not considered. Moreover, the filter attenuation was evaluated only at the switching frequency sideband, which is not ideal for the flying capacitor topology due to the extra harmonic distortion of its voltage ripple. In addition, the current distortion at lower frequencies can be enhanced by the filter resonant peak if proper damping is not achieved. A weight optimization design

---

979-8-3315-1612-3/25 $31.00 © 2025 IEEE

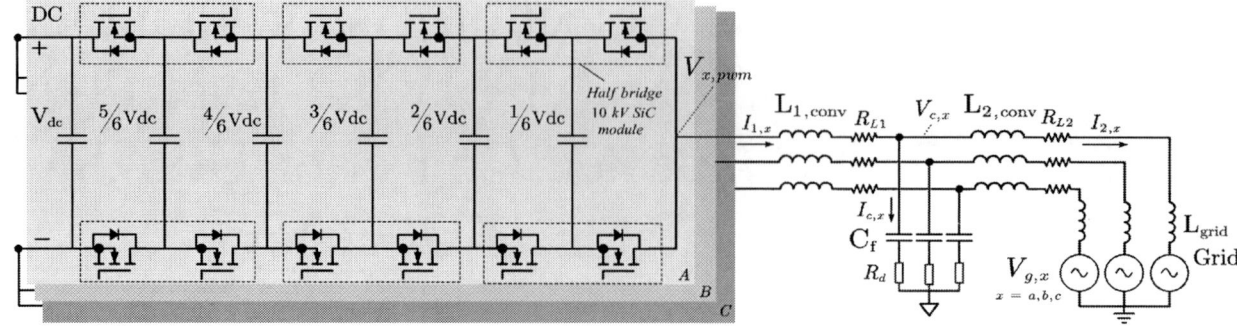

Fig. 1. 7-Level flying capacitor grid-connected converter and LCL scheme.

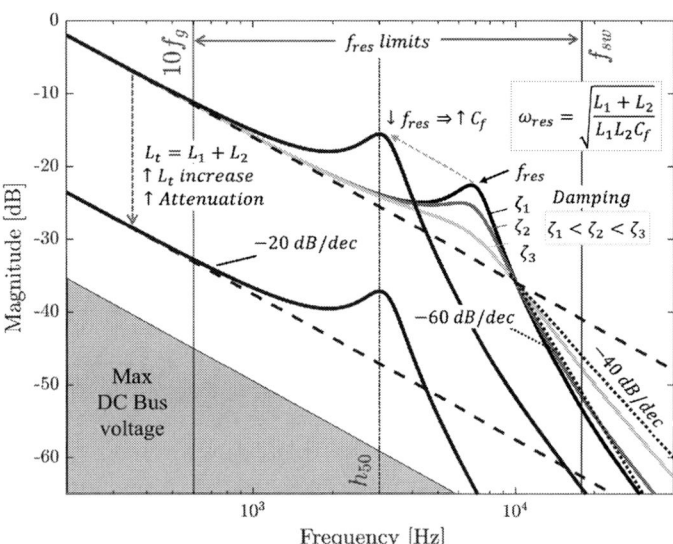

Fig. 2. Generic LCL filter transadmittance (Y_{21}) and design parameters.

method is presented by [11], but the FC voltage ripple effect on a multilevel converter is not explored, as this paper focused on a different topology.

A comprehensive review of the LCL design methods and a detailed evaluation of passive damping solutions (series R, shunt RC, series RLC, shunt RLC, trap + shunt RC, and two traps + shunt RC) was provided Beres et al [1]. The performance of the LCL filter was evaluated in terms of stability, losses, and size. This method focuses on cost-optimized LCL filters, meaning low series inductance and high filter capacitance, yielding low volume and cost, but requiring larger damping. The passive damping method was the focus of this study. One key aspect of the design procedure by Beres et al. is to calculate the filter attenuation requirement at the switching frequency, creating a design space based on the parameter values. However, for topologies like the flying capacitor, attenuation requirements must also be verified at intermediate frequencies because of its voltage ripple. The graphical method detailed in this work provides a valuable tool for visualizing filter performance across the full frequency range. In addition, this method allows for the verification of

the effectiveness of the damping strategy when a combination of passive and active damping is applied.

This paper aims to assist designers in selecting LCL filter parameters and to provide insights into the factors that influence filter performance through an attenuation requirement versus LCL filter transadmittance plot. This graphical method incorporates all relevant aspects of the converter, including topology, variable switching frequency, and active and passive damping. This work is organized as follows. Section II introduces the filter fundamentals and the converter specifications. Section III presents the graphical method and the procedure for parameter selection, and also demonstrates the factors that influence the graphical method (i.e. number of levels, converter ratings, FC ripple). Finally, Section IV details the final LCL parameter selection and simulation results for the 1.1 MVA FCML converter developed in [12] and illustrated in Fig. 1.

II. LCL FILTER FUNDAMENTALS AND CONVERTER DESIGN

Fig. 2 shows the transadmittance ($Y_{21} = I_g/V_{pwm}$) of the LCL filter. It is the main aspect of the filter design because it determines the filter attenuation performance with respect to the PWM voltage generated by the voltage source converter. The derivation of transadmittance has been studied extensively and will not be repeated here [1], [6], [9], [13]. The equations (1-5) are shown below, where R_d represents the damping resistor in series with the filter capacitor, and R_{L_1}, R_{L_2} are the inductors ESR.

Note that the highest attenuation of the filter (-60 dB/dec) occurs after the resonant peak, suggesting that a lower resonant frequency is preferred. However, achieving this requires an increased filter capacitance. Especially in medium voltage applications, it leads to a increase in volume, as many capacitors need to be connected in parallel to compensate for the many series-connected capacitive cells required to achieve a 22 kV blocking voltage capability. Alternatively, increasing the inductance will shift the curve downward, yielding an overall higher attenuation for the same frequency range plus a resonant peak occurring at a lower frequency, reducing the cost and size. Based on phasor analysis, the maximum filter inductance is defined by the converter maximum fundamental voltage amplitude, which is proportional to its maximum dc

$$Y_{21(LCL)} = \frac{(C_f R_d)s + 1}{as^3 + bs^2 + cs + d} \qquad (1)$$

$$a = L_1 L_2 C_f \qquad (2)$$

$$b = C_f(R_{L_2}L_1 + R_{L_1}L_2 + R_d(L_1 + L_2)) \qquad (3)$$

$$c = C_f(R_{L_1}R_{L_2} + R_d(R_{L_1} + R_{L_2})) \qquad (4)$$

$$d = R_{L_1} + R_{L_2} \qquad (5)$$

TABLE I
CURRENT HARMONIC LIMITS ACCORDING TO IEEE STD 519-2022

SCR[1]	< 20	20...50	50...100	100...1000	> 1000
$h < 11$[2]	4.0	7.0	10.0	12.0	15.0
$11 \leq h < 17$	2.0	3.5	4.5	5.5	7.0
$17 \leq h < 23$	1.5	2.5	4.0	5.0	6.0
$23 \leq h < 35$	0.6	1.0	1.5	2.0	2.5
$35 \leq h \leq 50$	0.3	0.5	0.7	1.0	1.4
TDD	5.0	8.0	12.0	15.0	20.0

I_{sc}: short circuit current, I_g: rated current, h: harmonic order.
[1] $SCR = I_{sc}/I_g$
[2] For $h \leq 6$, even harmonics are limited to 50% of the harmonic limits shown in the table.

bus voltage. The overshoot caused by the resonant peak can be detrimental to the stability of the system [1], [5] and the current harmonic spectrum. The overshoot can be limited by passive damping; however, a large damping resistor sacrifices the high-frequency (HF) attenuation, leading to HF harmonics and reduced converter efficiency. Hence, the proper selection of LCL parameters is essential for the power density, power quality, stability, and efficiency of the converter.

As described in [14], [15], this converter operates as a distributed energy resource and must comply with IEEE Std 1547. Therefore, its current harmonics must comply with Table I. A short circuit ratio (SCR) of 20 is considered, with harmonics beyond (50th) subject to the same requirements as h_{50}, representing the most stringent requirement.

III. GRAPHICAL METHOD FOR LCL PARAMETERS SELECTION PROCEDURE

The core of the graphical method consists in comparing the attenuation requirement curve $Y_{21(req)}$ with the transadmittance of the LCL filter $Y_{21(LCL)}$ (filter attenuation curve). If the LCL curve is smaller in magnitude than the requirement for the whole range of interest, the filter attenuation will achieve the harmonic current defined by the standard. On the other hand, if the filter transadmittance curve crosses the attenuation requirement curve, the current harmonics will exceed the standard limits at those frequencies.

A. Procedure

Fig. 3 presents the LCL parameter selection procedure based on the graphical method. First, the converter voltage spectrum must be obtained from time-domain simulations, which will embed all the converter operation characteristics into the graphical method, including the modulation scheme,

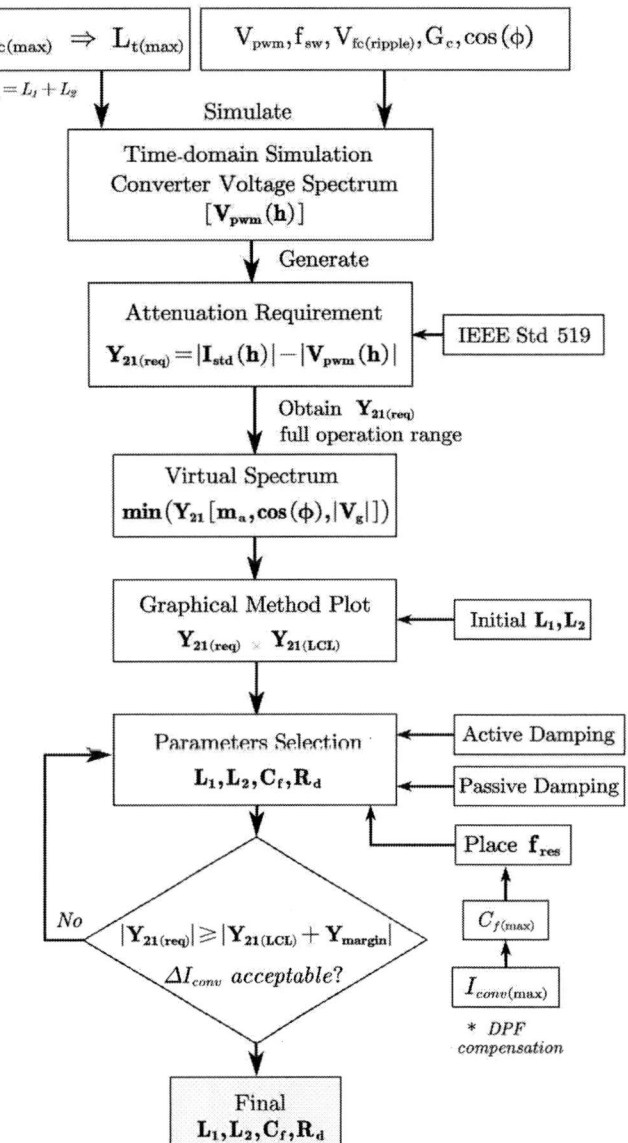

Fig. 3. LCL filter parameter selection procedure.

the FC voltage ripple, and active damping. In some multilevel topologies it is necessary to have a circulating current to balance the DC capacitors as in the case of NPC, or the flying capacitors in case of FCML or ANPC topologies. For that reason, the simulation must be connected to the grid. In this stage, an L-filter based on a percentage ($\leq 30\%$) of the calculated maximum inductance (see Section III-B) can be used to facilitate obtaining the converter voltage spectrum without requiring a well-rounded ac filter design beforehand.

The next step is to obtain the attenuation requirement by subtracting the converter voltage spectrum in each harmonic frequency from the harmonic current requirement defined by the standard, as shown in (6).

$$|Y_{21_{req}}(h)| = \frac{|I_{std}(h)|}{|V_{pwm}(h)|} = 20log_{10}(I_{std}) - 20log_{10}(V_{pwm}) \quad (6)$$

979-8-3315-1612-3/25 $31.00 © 2025 IEEE

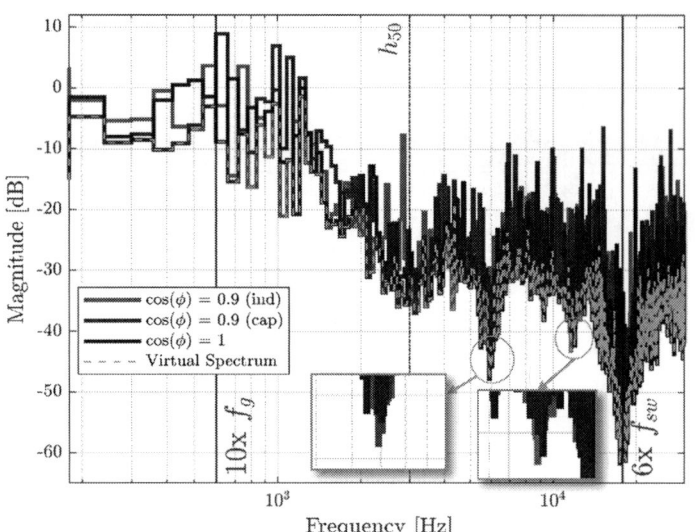

Fig. 4. Attenuation requirement accounting for different operation modes: virtual attenuation harmonic spectrum.

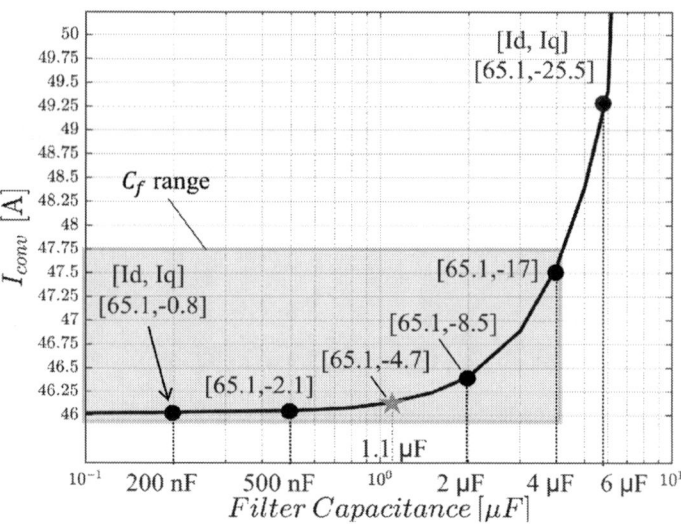

Fig. 5. 13.8 kVac 7-level converter rms current at unity power factor and rated active power (1.1 MW) as a function of the filter capacitance.

TABLE II
MAXIMUM FILTER CAPACITANCE

Method	Z_b	$C_{f(max)}$
Max Current (I_{rms})	173 Ω	4 μF
% Base Impedance (Z_b)	173 Ω	0.8 μF

To ensure that the filter design accommodates all the operating conditions, a virtual harmonic spectrum is generated based on the most stringent requirement (minimum magnitude) under various conditions, including different power factors and grid voltage levels. This is important because the power factor affects the FC voltage ripple [16], and consequently the current harmonics at the output. The output voltage harmonics of the converter is also affected by the modulation depth [17], which varies depending on the voltage levels of the grid and dc bus.

The latter is particularly relevant for some applications, for instance, solar inverters, as the dc bus voltage depends on the irradiation at the PV panels. Therefore, throughout the day the operation condition will vary substantially and the filter design must comply with the given requirement under all conditions. As shown in Fig. 4, the worst-case scenario for FC ripple occurs at the capacitive power factor, while at some other frequencies the inductive power factor imposes the most stringent requirement.

Finally, the virtual spectrum of the attenuation requirement is plotted and compared with the filter transadmittance [$Y_{21(req)}$ vs $Y_{21(LCL)}$]. To achieve this, the initial parameters of the LCL must be selected. As a reference, it is recommended to start with an L-filter transfer function that can marginally attenuate up to the switching frequency, further limiting the total inductance (L_t) selection range.

Based on the profile of the attenuation requirement curve, the resonant frequency f_{res} should be chosen. It must be sufficiently lower than the switching frequency sideband to allow for a significant reduction in total filter inductance. As Park et al. [11] the capacitance is minimized when L_1 is equal to L_2. The allowable range for the resonant frequency is $10f_g \leq f_{res} \leq f_{comb}$, to ensure that the converter bandwidth and low frequency harmonics remain unaffected by the resonant peak while maintaining the filter's ability to attenuate the switching frequency sideband [5], [6], [9],

[18]. An iterative approach can be used by the designer to achieve optimal parameter selection. The final step consists of verifying the filter parameters with respect to the maximum current ripple allowed, and the maximum current for filter capacitor power factor compensation. The first can be verified by simulation and iterated during the inductors loss design phase. A large current ripple means more core losses in the inductor on the converter side, affecting the overall efficiency of the converter. Lastly, the filter design must be verified for the maximum rms current it can provide to compensate for displacement caused by the filter capacitor selected. Fig. 5 presents the relationship between the filter capacitance values and the converter rms current considering unity power factor at rated power, this curve together with the converter thermal design defines the range for the filter capacitance. Note that for the converter under study, a capacitance of 4 uF requires 3.2% more power from the converter just to compensate for the filter capacitor, while 1.1 uF requires ten times lower current and can easily be acceptable by the converter thermal limit. In this application, applying the rule of thumb [5] [6], which limits the capacitor value to 5% of the base impedance, defined by $C_b = 0.05[1/(w_g Z_b)]$ and ($Z_b = V_{ll}^2/S_{conv}$), results in an conservative capacitance range, as demonstrated in Table II.

In general, considering the same split factor, a larger total inductance reduces the current ripple and the magnetic losses. However, it increases the volume, and, depending on the current values, a larger dc bus voltage is required. The latter increases the semiconductor switching losses and the voltage ratings of the power modules. A large filter capacitance would

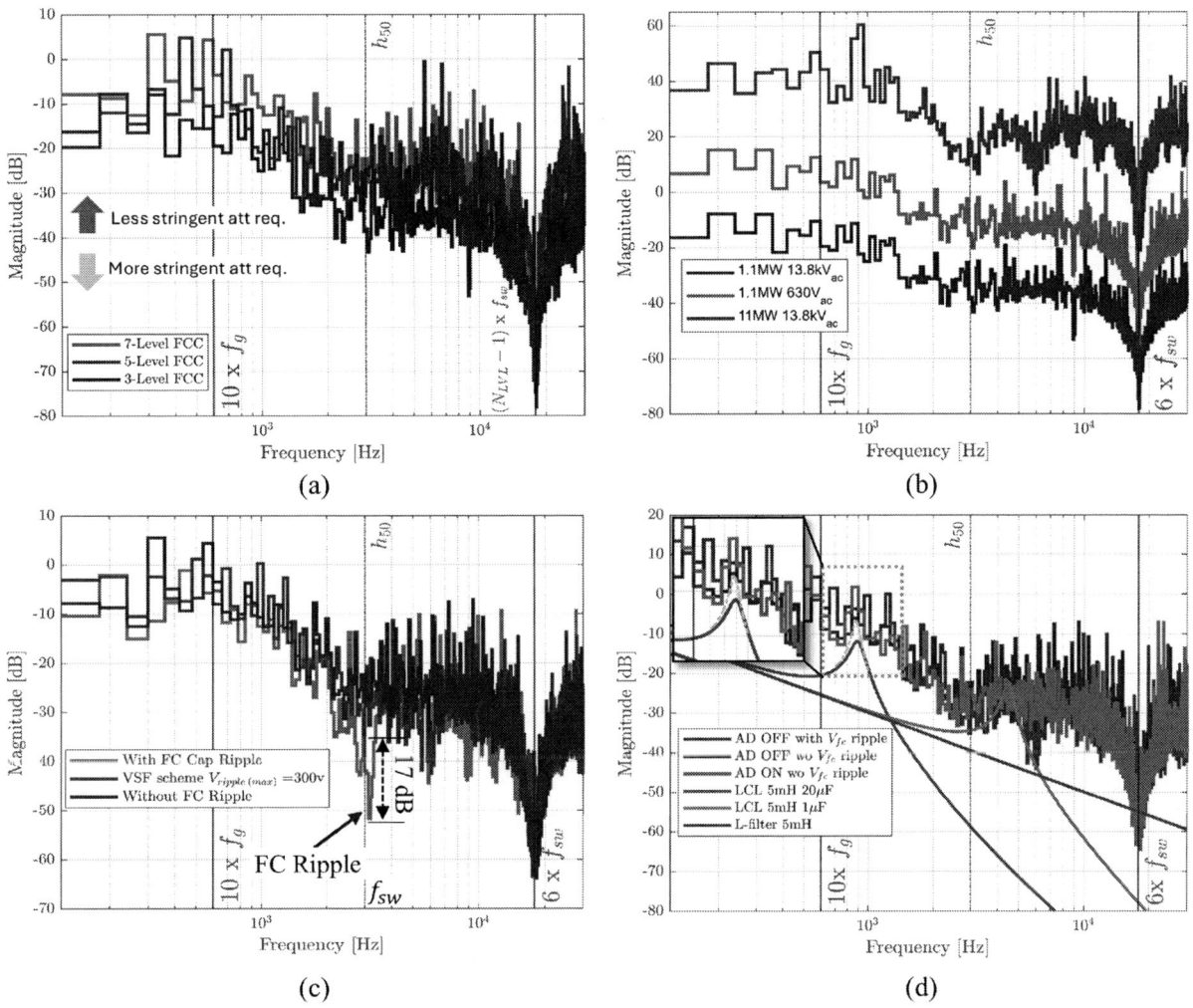

Fig. 6. Determinants of attenuation requirement profile (a) number of Levels (b) converter Rating (c) flying capacitor ripple and variable switching frequency (D) active damping

allow the inductance to be smaller but would require a larger current to compensate for its displacement, which would impact the semiconductor losses. In addition, a larger capacitor could need a larger damping, which would lower the overall converter efficiency.

B. Determinants of Attenuation Requirement $Y_{21(req)}$ Profile

Fig. 6 shows the main factors that change the profile of the attenuation requirement curve. As can be seen in Fig. 6(a), more levels shift the attenuation requirement curve upward for any modulation strategy, translating into a more relaxed filter requirement. As discussed in the previous section, Fig. 2, shifting the curve upward means a lower total filter inductance and, therefore, a smaller filter. Note that the output switching frequency is kept the same to have a fair comparison between the number of levels. On the other hand, if the cell switching frequency is kept the same, the output switching frequency will be lower for a converter with fewer levels. Consequently, having more levels on a PSPWM flying capacitor configuration

allows the filter resonant frequency to be pushed higher, yielding an even smaller filter capacitance.

Fig. 6(b) illustrates how the power, voltage and current ratings affect the filter parameter range. Two 1.1 MW converters are compared: a 13.8 kV 46 A converter investigated in this work and a commercial utility-scale solar PV inverter in the range of 630 V and 1 kA (e.g., ABB Ultra-1100, TMEIC PVL-L1000E, GE LV5). Note that the filter inductance values will be much higher for the medium voltage converter, even if the same amount of levels is considered, as the attenuation requirement curve of the 13.8 kV converter magnitude is smaller than the low voltage inverter. As demonstrated in the previous section and calculated by (6), a higher magnitude of converter voltage harmonics caused by the large dc bus voltage, combined with a lower magnitude of current harmonics, causes the attenuation requirement to be more stringent. Therefore, the required filter parameter values will be much higher for the MV converter. This does not directly translate to a larger volume, as inductors core and winding losses, voltage

TABLE III
MAXIMUM FILTER INDUCTANCE

—	Power	V_{dc}	V_{LL}	$L_{t(max)}$
LV Inverter	1.1 MVA	900 v - 1.3 kV	630 V	40 - 440 μH
MV Converter	1.1 MVA	22 kV	13.8 kV	5.4 mH
MV Converter	11 MVA	22 kV	13.8 kV	540 μH

TABLE IV
CONVERTER RATINGS

Parameter	Ratings
Power Rating	1.1 MVA
DC Voltage	22 kV
AC Line Voltage	13.8 kV
AC Current	46 A
Power Factor	$\pm 0.9 - 1.0$
Levels	7
Cell Voltage	3.7 kV
Flying Capacitors [DC to AC]	[1.6, 2.7, 3.2, 5.3, 8.0] μF
Power Module	10 kV SiC HB Module
Commutation Cell Switching Frequency (f_{sw})	3 kHz
AC Node Switching Frequency (f_{comb})	6× Cell f_{sw} = 18 kHz
FC Balancing Strategy	Variable Switching Frequency Cell [f_{sw}, $2 \times f_{sw}$]

winding distances, and wire/bar size are major contributing factors to the volume.

Similarly, for the same percentage of current ripple of the rated current, the much larger voltage and the low current ($< 50A$) will impose an extreme requirement on the MV converter, increasing the filter inductance. The LV inverter operates at higher current and is more susceptible to voltage drop in the filter inductors limitation, based on (7)-(8) below. As can be seen in Table III, the inductance is limited to the microhenry range while the MV converter is in the millihenry range. In contrast, the LV inverter operates at lower voltage and can benefit from the ease to obtain a higher capacitance. As expected, the attenuation requirement becomes less stringent as the medium voltage converter power is raised, for example, ten times its original value. The current harmonics and current ripple are not as small, but the dc bus voltage is kept the same, hence, the filter parameters can be relaxed. With zero sequence injection:

$$V_{ph(max)} = m_i \times \frac{V_{dc}}{2\sqrt{2}} \tag{7}$$

$$L_{t(max)} = \frac{V_{ph(max)} - V_{ph,grid}}{w_g I_{ph}} \tag{8}$$

The effect of the flying capacitor ripple on the attenuation requirement is shown in Fig. 6(c). Three cases are investigated, from highest to lowest ripple: without variable switching frequency (VSF); in blue, with VSF active at a threshold

voltage of 300 V; and in black, where the FC values are so large that the FC voltage ripple is eliminated. Note that a high FC voltage ripple imposes a more stringent (lower magnitude in the plot) around the cell switching frequency, which is ($N_{level} - 1$) times smaller than the output switching frequency. Therefore, the VSF scheme can significantly reduce the voltage ripple, easing the filter attenuation requirements and the switches peak blocking voltage.

Lastly, Fig. 6(d) shows the effects of active damping on the attenuation requirement plot. A bandpass filter is used to eliminate the frequencies around the resonant peak. In the attenuation requirement curve, these frequencies are presented as having a higher magnitude than the original curve. Using this feature, the designer can position the active damping range around the filter resonant peak to reduce the passive damping or to add a margin for component tolerances, which can be graphically verified from the $Y_{21(req)}$ vs $Y_{21(LCL)}$ plot.

IV. SIMULATION RESULTS

A simulation for the 7-level 13.8 kV converter is conducted to generate the inputs for the graphical method and, later, to verify the results of the selected LCL parameters.

A. Converter Design, Ratings and Application

The converter specifications can be found in Table IV. Its topology, modulation, power devices and ratings define the PWM voltage characteristics which, together with the application's current spectrum requirement, defines the attenuation needed from the LCL filter. The converter utilizes a 10 kV SiC MOSFET in a half-bridge module package. To keep the module voltage stress below 10 kV, a 7-level flying capacitor topology is used, providing a blocking voltage of 3.7 kV per switch and 7.4 kV per module. Each cell switches at 3 kHz, but the output switching frequency is proportional to the number of levels, which facilitates the filter design, as the resonant peak does not need to be placed at a lower frequency range to be able to attenuate the switching frequency harmonics. A variable switching frequency (VSF) scheme is designed and employed according to [16], to increase the power density and limit the flying capacitor voltage ripple. During the intervals of the fundamental frequency in which the FC capacitor voltage ripple reaches the threshold, the switching frequency is doubled, while the sampling frequency remains the same to avoid impacting the controls.

Fig. 7 shows the modulation signals for one phase-leg using discontinuous PWM (DPWM) and VSF. Note that the carrier, represented as a counter in Fig. 7(a), is reduced during the 220 to 224 ms interval, creating a higher number of transitions in the gate command shown in Fig. 7(b). After this interval, the phase-leg reference is clamped to achieve lower switching losses as per the DPWM scheme. In Fig. 7(c) the frequency can be seen toggling between the rated f_{sw} and $2 \times f_{sw}$.

Fig. 8(a) presents the grid phase voltage and current, as well as the converter PWM voltage at a capacitive power factor of 0.9. Fig. 8(b) shows the flying capacitor voltages and Fig. 8(c) shows the switching frequency. Note that the switching

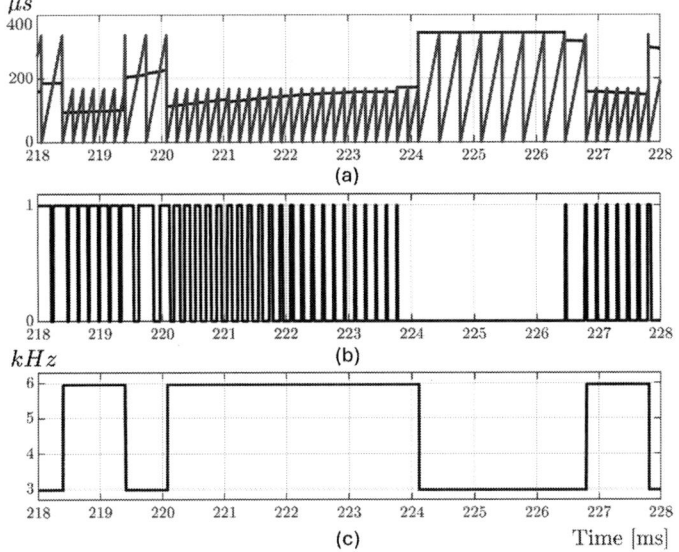

Fig. 7. Modulation signals for a single commutation cell (a) carrier and reference (b) gate command (c) cell switching frequency f_{sw}.

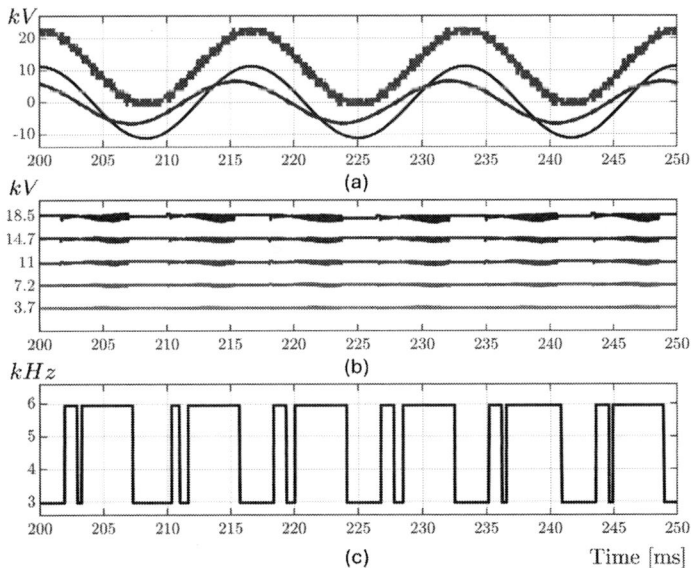

Fig. 8. Grid-connected 7-Level FC Converter Waveforms (a) grid V_{ph}, I_g and converter V_{pwm} (b) flying capacitor voltages (c) VSF trigger signal ($V_{th} = 500v$).

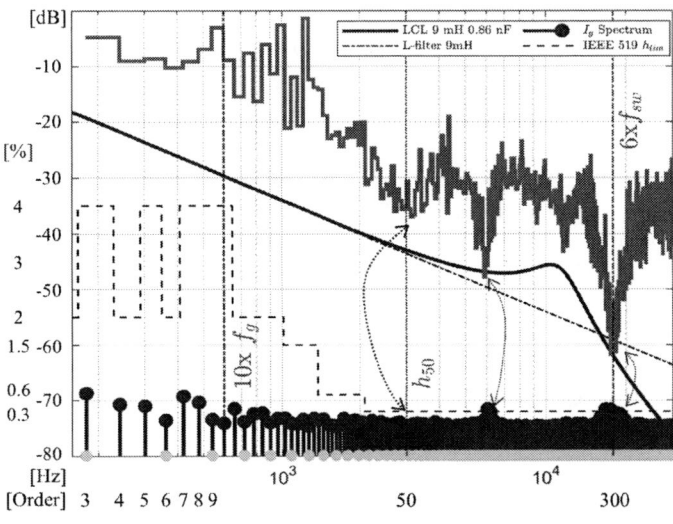

Fig. 9. Example of LCL parameters that does not fulfill the individual harmonic requirement: $L_t = 9\,mH$ and $C = 90\,nF$. Variable switching frequency ($V_{th} = 500v$). (Red) Attenuation requirement $Y_{21(req)}$ (black line) LCL transadmittance $Y_{21(LCL)}$ (stem curve) current harmonic spectrum (dashed line) individual current harmonics limits as per IEEE Std 519.

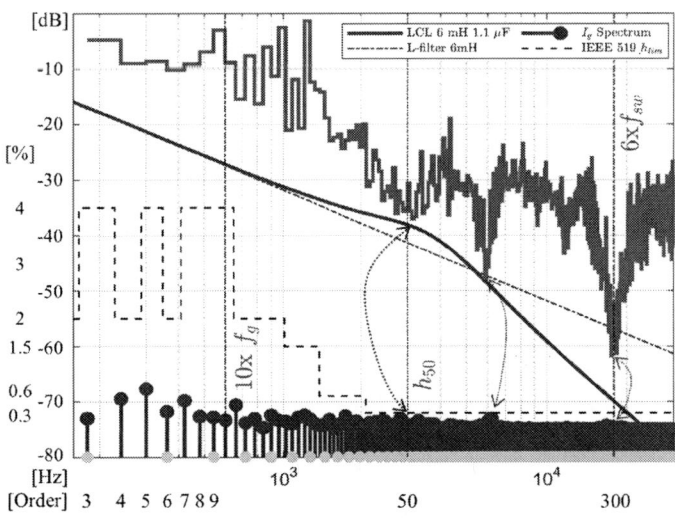

Fig. 10. Selected LCL parameters that fulfills the individual harmonic requirement: $L_t = 6\,mH$ and $C = 1.1\,\mu F$. Variable switching frequency ($V_{th} = 500v$). (red) Attenuation requirement $Y_{21(req)}$ (blue line) LCL transadmittance $Y_{21(LCL)}$ (stem curve) current harmonic spectrum (dashed line) individual current harmonics limits as per IEEE Std 519.

frequency increases as the FC voltage ripple increases and decreases as the ripple decreases.

The transadmittance $Y_{21(LCL)}$ of two sets of LCL parameters is plotted using the graphical method, and the current harmonic spectrum obtained from the simulation is also plotted. The first set of parameters, shown in Fig. 9, is selected such that the LCL transadmittance curve crosses the attenuation requirement curve, while the second set, shown in Fig.10, is designed to pass the filter requirements. The attenuation requirement is generated using the virtual attenuation spectrum, which means that all operation conditions are considered in

the frequency domain, as discussed in Section III and Fig. 4. However, the current spectrum of the grid obtained from simulation can only be generated for each operation condition at a time. In both cases, the capacitive power factor of 0.9 is selected because it imposes the most stringent requirement at the FC voltage ripple frequency.

Due to the resonant peak, it is not feasible to design the filter resonant frequency to fall between the FC ripple frequency and the switching frequency. As can be seen in Fig. 9, the LCL filter curve crosses the attenuation requirement. Even if a greater passive damping is used to mitigate the magnitude peak, the decrease of the HF attenuation after the resonant peak would hinder the filter transadmittance curve from meeting the attenuation requirement at the output switching frequency. Moreover, active damping is not useful in this frequency range because it is beyond the control bandwidth. Note that at frequencies where $Y_{21(LCL)}$ is greater than $Y_{21(req)}$, the individual harmonic exceeds the IEEE Std 519 limits, confirming the effectiveness of the graphical method. In Fig. 10 the FC voltage ripple frequency is close to the resonant frequency of the LCL filter, but it is not enough to cause the current harmonic spectrum to cross the individual harmonics ripple. Note that there is an amplification of the current harmonics when the LCL filter transadmittance gets closer to the attenuation requirement, i.e., at switching frequency range, but the limits are no exceeded.

V. SUMMARY AND CONCLUSION

A graphical method is presented to provide guidance for designers in selecting LCL filter parameters, which can be used for any ac filter topology. The attenuation requirement and filter transadmittance plots insights into the filter design; for example, the attenuation requirement is plotted based on virtual spectrum, guaranteeing compliance with the standard in all the operating conditions. The determinants of the attenuation requirement profile, such as the number of levels, power ratings, and FC voltage ripple, are useful resources to understand how to design a specific filter for any type of converter. The graphical method is verified in simulation by observing the current harmonic spectrum. At frequencies where the filter transadmittance has greater magnitude than the attenuation requirement, the grid current harmonic spectrum will not comply with the imposed requirement. The method could be adapted for any current harmonic spectrum requirement.

VI. ACKNOWLEDGMENT

This material is based on research supported in part by funding from the US Department of Energy (DOE), Office of Energy Efficiency and Renewable Energy (EERE), under award number DE-EE0009133 from the Office of Advanced Materials & Manufacturing Technologies Office (AMMTO).

REFERENCES

[1] R. N. Beres, X. Wang, M. Liserre, F. Blaabjerg, and C. L. Bak, "A Review of Passive Power Filters for Three-Phase Grid-Connected Voltage-Source Converters," *IEEE Journal of Emerging and Selected Topics in Power Electronics*, vol. 4, no. 1, pp. 54–69, Mar. 2016. [Online]. Available: http://ieeexplore.ieee.org/document/7350094/

[2] Y. Jiao and F. C. Lee, "LCL Filter Design and Inductor Current Ripple Analysis for a Three-Level NPC Grid Interface Converter," *IEEE Transactions on Power Electronics*, vol. 30, no. 9, pp. 4659–4668, Sep. 2015. [Online]. Available: http://ieeexplore.ieee.org/document/6918494/

[3] W. Wu, S. Feng, J. Ji, M. Huang, and F. Blaabjerg, "LLCL-filter based single-phase grid-tied aalborg inverter," in *2014 International Power Electronics and Application Conference and Exposition*. Shanghai, China: IEEE, Nov. 2014, pp. 658–663. [Online]. Available: https://ieeexplore.ieee.org/document/7037935

[4] M. Sanatkar-Chayjani and M. Monfared, "Design of *LCL* and *LLCL* filters for single-phase grid connected converters," *IET Power Electronics*, vol. 9, no. 9, pp. 1971–1978, Jul. 2016. [Online]. Available: https://onlinelibrary.wiley.com/doi/10.1049/iet-pel.2015.0922

[5] M. Liserre, F. Blaabjerg, and S. Hansen, "Design and Control of an LCL-Filter-Based Three-Phase Active Rectifier," *IEEE Transactions on Industry Applications*, vol. 41, no. 5, pp. 1281–1291, Sep. 2005. [Online]. Available: http://ieeexplore.ieee.org/document/1510826/

[6] A. Reznik, M. G. Simoes, A. Al-Durra, and S. M. Muyeen, "LCL Filter Design and Performance Analysis for Grid-Interconnected Systems," *IEEE Transactions on Industry Applications*, vol. 50, no. 2, pp. 1225–1232, Mar. 2014. [Online]. Available: http://ieeexplore.ieee.org/document/6571219/

[7] S. V. Araujo, A. Engler, B. Sahan, and F. L. M. Antunes, "LCL filter design for grid-connected NPC inverters in offshore wind turbines," in *2007 7th Internatonal Conference on Power Electronics*. Daegu, South Korea: IEEE, Oct. 2007, pp. 1133–1138. [Online]. Available: http://ieeexplore.ieee.org/document/4692556/

[8] R. Meyer and A. Mertens, "Design and optimization of LCL filters for grid-connected converters," in *2012 15th International Power Electronics and Motion Control Conference (EPE/PEMC)*. Novi Sad, Serbia: IEEE, Sep. 2012, pp. LS7a.1–1–LS7a.1–6. [Online]. Available: http://ieeexplore.ieee.org/document/6397501/

[9] A. A. Rockhill, M. Liserre, R. Teodorescu, and P. Rodriguez, "Grid-Filter Design for a Multimegawatt Medium-Voltage Voltage-Source Inverter," *IEEE Transactions on Industrial Electronics*, vol. 58, no. 4, pp. 1205–1217, Apr. 2011. [Online]. Available: http://ieeexplore.ieee.org/document/5601772/

[10] J. Muhlethaler, M. Schweizer, R. Blattmann, J. W. Kolar, and A. Ecklebe, "Optimal Design of LCL Harmonic Filters for Three-Phase PFC Rectifiers," *IEEE Transactions on Power Electronics*, vol. 28, no. 7, pp. 3114–3125, Jul. 2013. [Online]. Available: http://ieeexplore.ieee.org/document/6335486/

[11] K.-B. Park, F. D. Kieferndorf, U. Drofenik, S. Pettersson, and F. Canales, "Weight Minimization of LCL Filters for High-Power Converters: Impact of PWM Method on Power Loss and Power Density," *IEEE Transactions on Industry Applications*, vol. 53, no. 3, pp. 2282–2296, May 2017. [Online]. Available: http://ieeexplore.ieee.org/document/7831420/

[12] X. Lin, D. Nam, N. Yan, J. Stewart, A. Mendes, D. Dong, R. Burgos, and D. Boroyevich, "Design and Testing of a 13.8 kV/1.1MVA 3-Phase 5-Level Flying Capacitor Converter with 10 kV SiC MOSFETs," in *2023 IEEE Applied Power Electronics Conference and Exposition (APEC)*. Orlando, FL, USA: IEEE, Mar. 2023, pp. 158–164. [Online]. Available: https://ieeexplore.ieee.org/document/10131556/

[13] M. Liserre *, F. Blaabjerg, and A. Dell'Aquila, "Step-by-step design procedure for a grid-connected three-phase PWM voltage source converter," *International Journal of Electronics*, vol. 91, no. 8, pp. 445–460, Aug. 2004. [Online]. Available: http://www.tandfonline.com/doi/abs/10.1080/00207210412331306186

[14] B. Wang, R. Burgos, and B. Wen, "An Average Model for Three-Phase Five-level Flying Capacitor Converters with Phase-Shifted PWM," in *2022 IEEE Energy Conversion Congress and Exposition (ECCE)*. Detroit, MI, USA: IEEE, Oct. 2022, pp. 1–8. [Online]. Available: https://ieeexplore.ieee.org/document/9947563/

[15] N. N. Nandola, B. Wang, X. Wu, and R. Burgos, "Control of DC Microgrid based flexible cold-rolling steel mill Plant – an application of grid supporting rectifier," in *2023 IEEE 24th Workshop on Control and Modeling for Power Electronics (COMPEL)*. Ann Arbor, MI, USA: IEEE, Jun. 2023, pp. 1–7. [Online]. Available: https://ieeexplore.ieee.org/document/10221013/

[16] X. Lin, R. Burgos, and D. Dong, "Improved Variable Switching Frequency Control for Capacitor Voltage Ripple Regulation in Multilevel Flying Capacitor Converter," *IEEE Transactions on Power Electronics*, vol. 38, no. 5, pp. 5700–5705, May 2023. [Online]. Available: https://ieeexplore.ieee.org/document/10042070/

[17] D. G. Holmes and T. A. Lipo, *Pulse width modulation for power converters: principles and practice*. Hoboken, NJ: John Wiley, 2003.

[18] R. N. Beres, X. Wang, F. Blaabjerg, M. Liserre, and C. L. Bak, "Optimal Design of High-Order Passive-Damped Filters for Grid-Connected Applications," *IEEE Transactions on Power Electronics*, vol. 31, no. 3, pp. 2083–2098, Mar. 2016. [Online]. Available: http://ieeexplore.ieee.org/document/7118223/

Online Extraction of Electrochemical Impedance Spectroscopy Pattern Based on EV Load Profile and Short Time Fourier Transform for Diagnosis of Lithium-ion Battery Safety

Miyoung Lee
Department of Electrical Engineering
Chungnam National University
Daejeon, Republic of Korea
yoy0307@naver.com

Dongcheol Lee
Department of Electrical Engineering
Chungnam National University
Daejeon, Republic of Korea
cheol9883@naver.com

Youngmin Bae
Department of Electrical Engineering
Chungnam National University
Daejeon, Republic of Korea
qodudals1@gmail.com

Jongchan An
Department of Electrical Engineering
Chungnam National University
Daejeon, Republic of Korea
jongchanan94@gmail.com

Garam Yang
Department of Electrical Engineering
Chungnam National University
Daejeon, Republic of Korea
pp6573@naver.com

Woonki Na
Department of Electrical and Computer Engineering
California State University
Fresno, United States
wkna@csufresno.edu

Jonghoon Kim
Department of Electrical Engineering
Chungnam National University
Daejeon, Republic of Korea
whdgns0422@cnu.ac.kr

Batteries which convert chemical energy into electrical energy and play an important role in our daily lives and across various industries. The chemical reaction that occurs within a battery during operation of an electric vehicle (EV) depends on environmental conditions such as temperature, load, and humidity, and these variables greatly affect battery performance and life. However, there is challenging to analyze these electrochemical phenomena with only simple electric signals, such as direct current (DC) signal. For compensation, passive electrochemical impedance spectroscopy (PEIS) is derived by applying a signal processing algorithm that converts time series signals into frequency-domain signals in this study. Based on this, we propose to learn a siamese convolutional neural network (SCNN) algorithm on the converted PEIS pattern to the image and to effectively diagnose the state of the battery through similarity evaluation.

Keywords—Passive electrochemical impedance spectroscopy, Siamese convolution neural network, Signal processing, Similarity evaluation, Online fault diagnosis

I. INTRODUCTION

Recently, lithium-ion batteries have become widely available due to their high energy density, environmental friendliness, and long lifespan. As a result, they have become essential not only for electric vehicles (EVs) but also for renewable energy storage systems and portable electronics. However, lithium-ion batteries convert chemical energy generated by the redox reactions of lithium ions and electrons into electrical energy. The decomposition mechanism exhibited by the battery in this process varies depending on the operating

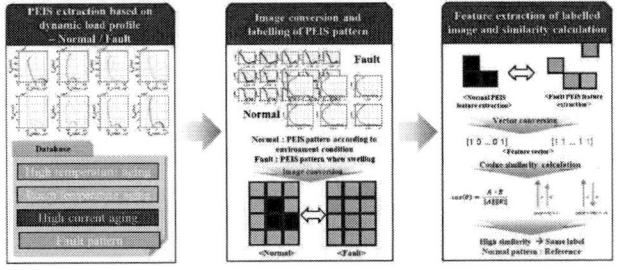

Fig. 1. Schematic diagram of safety diagnosis by extracting PEIS based on dynamic profile of lithium-ion battery and evaluating CNN similarity

conditions, especially temperature and load conditions [1]. These various degradation mechanisms can lead to rapid capacity loss under specific driving conditions and, in severe cases, may result in thermal runaway and fire hazards [2]. To prevent damage caused by sudden failures, various fault diagnosis techniques have been extensively studied. However, traditional fault diagnosis methods focus primarily on simple anomaly detection and rely heavily on electrical signals, such as current, voltage, and temperature, collected from the Battery Management System (BMS) [3]. These conventional approaches often employ threshold-based algorithms or basic statistical analyses, which are effective in detecting obvious anomalies. However, they have limitations in identifying subtle and complex degradation phenomena that occur in the early stages of failure. Moreover, these methods do not adequately consider the nonlinear characteristics of batteries, environmental factors, and variations in usage patterns, potentially masking early signs of degradation. Electrochemical impedance spectroscopy (EIS), which enables electrochemical analysis, can be applied to overcome the limitations of these electrical analyses. EIS is a technique that analyzes the response characteristics to AC signals applied at different frequencies, allowing for a more precise understanding of the internal state and degradation mechanisms of the battery. This provides the significant advantage of understanding the chemical state and changes in the battery, beyond simple electrical signal analysis. However, EIS requires specialized equipment such as frequency analyzers, which are costly and lack portability, posing limitations for real-time monitoring or large-scale applications [4]. To address these issues, a simpler and more cost-effective alternative that retains the advantages of EIS is required. Therefore, this study applied a signal processing algorithm that transforms time-domain characteristics into frequency-domain characteristics. This method is defined as passive electrochemical impedance spectroscopy (PEIS), and it was used to classify the environmental conditions of batteries exposed to various situations [5-6]. PEIS mimics the frequency response characteristics of EIS, enabling similar analysis results without complex equipment. This approach enables a more precise understanding of the battery's degradation state and improves the accuracy of fault prediction. The study demonstrated the potential of PEIS to precisely analyze the degradation state of batteries and efficiently classify battery characteristics under various environmental conditions. This can significantly enhance the fault diagnosis capabilities of BMS, thereby reducing the likelihood of unexpected failures and enhancing the overall reliability and safety of electric vehicles. The details are illustrated in Fig. 1. The PEIS patterns derived under various environmental conditions are classified into normal and abnormal categories. These classified PEIS patterns are then converted into images. The process of converting 1D vector signals into 2D images is implemented using various signal processing methodologies, such as FFT, spectrogram, and wavelet transform. However, these methods and others often require significant computational time and high computing power. In contrast, many algorithms used in computer graphics applications for image formation do not demand such high computational resources. Bresenham (1965) proposed a method for approximating n points in a given quadrant using simple arithmetic integer operations, such as addition, subtraction, and bit-shifting [7]. Here, n represents the number of samples, and this approach reduces computation time proportionally to n. The transformed image patterns are then used for fault diagnosis through deep learning. While deep learning methods yield impressive results in image analysis processes, their performance is heavily dependent on the quality of the training data. Some datasets may contain erroneous data, which can adversely affect the efficiency of the deep learning network. To address this issue, the Siamese neural network (SNN) was developed to improve feature extraction and comparison. SNN-based learning models typically use two or more sub-neural networks with identical weights and architectures to extract features from inputs. The extracted features are evaluated using a similarity criterion. The similarity criterion is based on the loss function of the SNN network, where the similarity score for a pair of inputs approaches '1' if they are similar, and '0' otherwise.

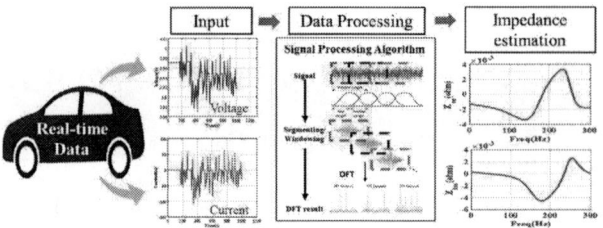

Fig. 2. A schematic flowchart of PEIS extraction based on signal processing and dynamic profile of lithium-ion battery.

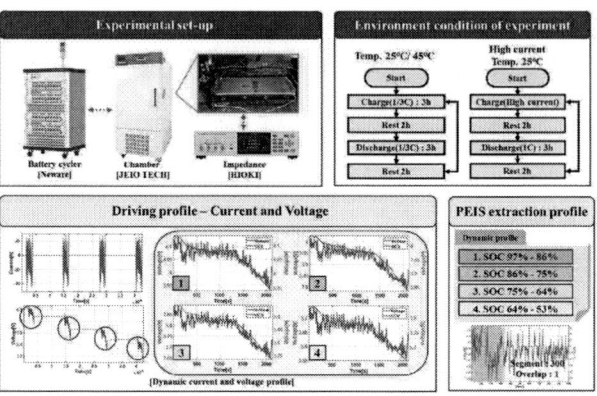

Fig. 3. A schematic flowchart of experimental set up for construction of dataset

This approach helps in enhancing the robustness of the fault diagnosis system, especially in cases where the dataset contains noise or anomalies. The ability to distinguish subtle differences between normal and faulty patterns increases diagnostic reliability, especially in complex operating environments. The remainder of this article is arranged as follows: Section 2 presents the methodology for deriving PEIS patterns based on STFT. Section 3 describes the experimental setup and validation of the PEIS fitting accuracy. Section 4 explains the safety diagnosis algorithm, and Section 5 concludes the article.

II. EXTRACTION OF ELECTROCHEMICAL IMPEDANCE SPECTROSCOPY BASED ON SHORT TIME FOURIER TRANSFORM

A. Signal processing algorithm

This study attempted to analyze the dynamic characteristics of the battery in the frequency domain by transforming data in the time domain into the frequency domain. The technique for analyzing the electrochemical characteristics of a lithium-ion battery in the frequency domain is called EIS, and in general, impedance is measured based on a voltage or current signal derived by applying a sinusoidal current or a voltage signal from 1 kHz to 0.01 Hz. There is an advantage of being able to non-destructively analyze the internal characteristics of the battery based on the amount of change in the anode/cathode/electrolyte depending on the frequency domain. Therefore, the analysis technique converted to the frequency domain by applying the signal processing technique was defined by PEIS to derive the pattern [8-11]. Signal processing algorithm is based on Euler equation, transforming time-series domain into frequency domain to interpret characteristics of frequency domain which are not contained in time-series domain. Fourier transform based on continuous time signal as shown in equation (1).

$$X(k) = \int_{-\infty}^{\infty} x(t) \cdot e^{jwt} dt \qquad (1)$$

In this study, to transform time-series data sampled at same time intervals into frequency domain, discrete time Fourier transforms (DFT) was applied equation (2) [12].

$$X(k) = \sum_{n=0}^{N-1} x[n] e^{(-jk\frac{2\pi}{N}n)} dt \qquad (2)$$

Also, a short time Fourier transform algorithm (STFT) was applied for transforming into real time frequency characteristics, and the corresponding equation is expressed in equation (3) [13].

$$X(h,k) = \sum_{n=0}^{N-1} x_{n+h} \omega_n e^{-jk\frac{2\pi}{N}n} \qquad (3)$$

In this study, the window function is used to divide the entire signal into sections for real-time conversion, with the Hamming window function chosen among various options. When performing STFT, parameters include sampling frequency, window size, and overlap, and the sampling frequency should be set to at least twice the maximum frequency held by the corresponding signal. The time series data for signal processing used in this study is an electric vehicle driving simulation profile, and 2128s of data measured at 1s intervals during one driving are collected. For the real-time extraction of EIS, the superposition degree was selected to be derived every 1s, and the window size was selected to represent electrochemical characteristics. Also, it was assumed that there is a frequency component of 300Hz in the corresponding region by compressing 300s as 1s. That is, it was assumed that there is data of 300Hz at 1Hz intervals, so the sampling frequency should be 600Hz or higher. In this study, 2000Hz was selected as the sampling frequency, and STFT was performed. In the case of STFT, 300 voltage and current data constitute one segment, and one segment is derived at 1s intervals. As a result of Fourier transform the voltage current, 1000 frequency components, the real part and the imaginary part, are derived for each segment, and each frequency interval is 7Hz, because f = (fs/sample). At this time, 11 frequencies are selected according to the EIS frequency selection criteria, enabling the Nyquist plot shown in Fig 2 to be obtained.

B. Optimization for PEIS extraction

When performing STFT, parameters include sampling frequency, window size, and overlap. The time-series data for signal processing converted in this study is an EV driving simulation pro-file, and 2128s of data measured at 1s intervals during one driving are collected. For the real-time extraction of EIS, the superposition degree was selected to be derived every

1s, and the window size was selected to represent electrochemical characteristics of 300s. 300s was selected, but it was assumed that there is a frequency component of 300Hz in the corresponding region by con-sidering 300s as 1s. Shortly, since the current signal is applied battery, there is electrochemical reaction such as polarization, charge transfer and mass transfer. That is, it was assumed that there is data of 300Hz at 1Hz intervals within preset segment in case 300s, so the sampling frequency should be 600Hz or higher. In this study, 2000Hz was selected as the sampling fre-quency, and STFT was performed. In the case of STFT in this way, 300 voltage and current data constitute one segment, and one segment is derived at 1s intervals. As a result of Fourier trans-forming the voltage current, 2000 frequency components symmetrically, the real part and the im-aginary part, are derived for each segment. 1,000 point is selected because, -1,000Hz to 1,000Hz is extracted, so only positive part is used. Real and imaginary impedance extraction is expressed (4)-(8) and shown in Fig. 3.

$$Z = \sqrt{Z_{Re}^2 + Z_{Im}^2} \qquad (4)$$

$$Z_{Re}(f) = \frac{\Re(v(f))}{\Re(i(f))} \qquad (5)$$

$$Z_{Re}(f) = \frac{\Im(v(f))}{\Im(i(f))} \qquad (6)$$

$$v(f) = \sum_{n=0}^{N-1} v_{n+h}(t)\omega_n e^{-jk\frac{2\pi}{N}n} \qquad (7)$$

$$i(f) = \sum_{n=0}^{N-1} i_{n+h}(t)\omega_n e^{-jk\frac{2\pi}{N}n} \qquad (8)$$

III. EXPERIMENTAL SET UP AND CONSTRUCTION OF DATASET FOR VALIDATION

A. Experimental set up and process for Electrochemical impedance spectroscopy

The experimental study was conducted using SKI commercial pouch-type lithium-ion battery E556. According to the data sheet, the nominal capacity of this battery is 55.6 Ah. The experiment was performed at a constant temperature of 25°C, maintained using a temperature and humidity chamber. Charging and discharging of the battery were performed using a battery cycler (Neware BTS4000), with a current. measurement range of 80A and an accuracy of approximately 0.1% of full scale. The time resolution is 100ms, and voltage and current data used to signal process algorithm are Fig. 2: EIS measurements were conducted using an impedance measurement device (HIOKI BT4560). The measurement frequency range is from 0.1 Hz to 1050 Hz. An experiment to verify the safety diagnosis of a PEIS-based lithium-ion battery consist of cycling under environmental conditions, a driving profile for PEIS extraction, and an EIS measurement test to verify the accuracy of PEIS fitting, as shown in Fig. 4. For cycling aging, all tests are conducted at 25°C. Under general conditions, both charging and discharging are performed at 1/3C rate. The driving profile is reduced by SOC 11% for one run, and after completion of the driving, the EIS test is conducted after 3 hours of rest for battery stabilization. This process is repeated 8 times from SOC 97% to 9%. Through these experiments, a dataset for PEIS extraction is constructed.

B. Verification data acquisition process for PEIS reliability based on signal processing

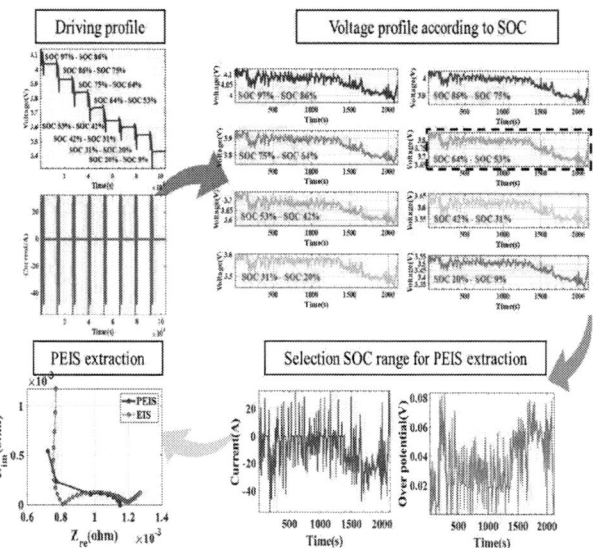

Fig. 4. PEIS extraction validation profile based on signal processing and dynamic profile

In this study, frequency components were derived from aperiodic dynamic signals such as EV driving profile based on signal processing algorithm. For securing reliability of EIS data extracted from driving profile, comparison based on real EIS data is performed. The STFT described in Section 2 was applied to the driving profile to extract PEIS through impedance calculation of the real and imaginary parts by frequency. When measuring EIS frequency selection criterion is that desired frequency area is set, and then 10 points are extracted at the point between frequencies where the frequency decreases by 10 digits, which is defined as 10 points per decade (Nd). In short, if frequency range is from 10kHz to 100mHz, a decade means 10kHz to 1kHz and 10 points of frequency will be selected within this frequency range. First, that frequency range is converted into a log scale, and the data which is converted into the log scale split into same 10 interval. And then, the same interval log scale is converted into exponential function and the converted data is selected as frequency for EIS measurement. For example, when frequency is in from 10kHz to 1kHz, the data will be converted in log scale, 4 to 3 and the data splits in 10 same intervals, like 3.9, 3.8, 3.7, ... and it is converted to exponential scale then it will be selected frequencies for EIS measurement. Since the time-series data segment is 300s data, time-series data which is desired to convert has maximum 300Hz frequency. Since sampling frequency fs is set as 2000 and the sample length N is 300, the frequency resolution is 7Hz, from equation (5) similar frequency was selected in the actual EIS measurement based on the corresponding frequency resolution. The selected frequency is expressed in Table I.

TABLE I.   FREQUENCY SIMILAR TO ACTUAL EIS

Frequency (Hz)			
206	200	193	186
180	173	166	160
153	146	140	133
126	120	113	106
100	93	80	60

The actual fitting results of EIS and PEIS are shown on Fig. 5. The EIS measured for each SOC area and the PEIS fitting extracted based on signal processing shows similar tendency. Fitting accuracy is under maximum error 5% by comparing real EIS and PEIS. When comparing the impedance error of the real part of the actual EIS and the extracted PEIS, the accuracy within the maximum error of 5% was secured. The performance of the algorithm was verified, and the PEIS extracted in this way was applied as the input data set of the safety diagnosis algorithm.

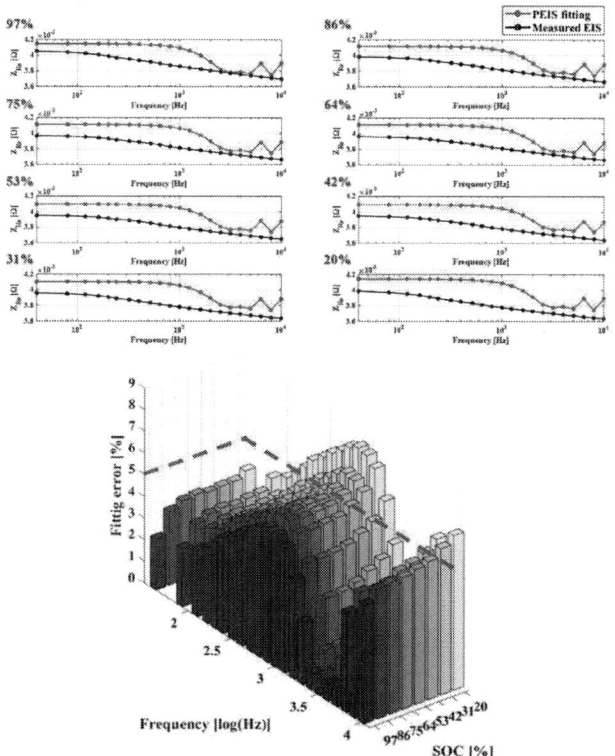

Fig. 5. Measured EIS and signal processing-based EIS fitting results according to the SOC range

IV. SAFETY DIAGNOSIS BASED ON CNN ALGORITHM AND SIMILARITY EVALUATION

In this paper, we propose a safety diagnosis algorithm for diagnosing abnormal conditions of electric vehicle batteries. To do this, we analyze the data of cells with defects in the driving data, extract defective PEIS patterns and normal PEIS patterns, and set the correct labels for them. The proposed algorithm converts these PEIS patterns into image format, inputs the converted images into a convolutional neural network (CNN), extracts key feature vectors, and evaluates the similarity based on these vectors. During the learning process of the algorithm, the CNN is trained to minimize the loss function using input data and the set labels. The input data uses PEIS patterns in the form of images generated through the Bresenham algorithm. The Bresenham algorithm converts the vector-type PEIS patterns into pixel-based image data, and thus matches the input format required for the image processing process of the CNN. During the learning process, the CNN extracts features from images through stages such as convolution, pooling, and flattening, and adjusts the weights of the network based on these features to optimize performance [14]. In the final stage, the CNN converts the PEIS patterns into vectors through the flattening process, and assigns the set labels based on these vectors. This algorithm can diagnose whether the battery condition is normal or defective by evaluating the similarity between the normal PEIS pattern and the input PEIS pattern when PEIS data is input in real time. The paper describes this similarity evaluation method in detail, and additionally introduces a similarity evaluation technique using Siamese neural network together with Bresenham algorithm. Siamese neural network helps to more accurately distinguish between normal and defective patterns by analyzing the similarity between PEIS patterns more precisely.

A. Bresenham algorithm for image conversion

The PEIS pattern imaging method based on the Bresenham algorithm is as follows. PEIS data includes information such as frequency, impedance magnitude, and phase, and is converted into two-dimensional coordinate data. Once the coordinate mapping is complete, Bresenham algorithm is used to construct a straight line using only integer additions and subtractions between two points. This creates an approximation of the straight line within the digital image, calculates the error at each point, and selects the next pixel based on this. For example, the method involves calculating the difference between the x and y coordinates of the starting point and the ending point, updating the error term at each iteration, and drawing a straight line incrementally point by point. In addition, the Bresenham algorithm dynamically adjusts the error calculation during the line drawing process, allowing it to maintain high accuracy while remaining computationally lightweight. The Bresenham algorithm's flow for converting PEIS data into two-dimensional coordinate data is illustrated in steps (9)-(13), and the detailed process is depicted in Fig 6.

$$x_{next} = x_2 \tag{9}$$

$$y_{next} = \begin{cases} y_1 \\ y_2 \end{cases} \tag{10}$$

$$y = mx + b \tag{11}$$

$$m = \frac{y_2 - y_1}{x_2 - x_1} \tag{12}$$

$$y = \frac{y_2 - y_1}{x_2 - x_1}(x - x_1) + y_1 \tag{13}$$

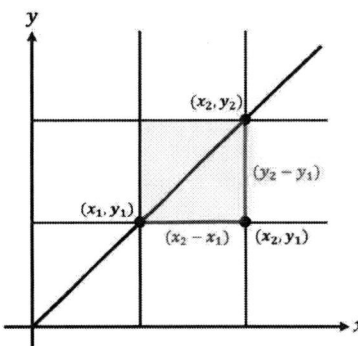

Fig. 6. Flowchart of Bresenham algorithm for converting to 2D coordinate data

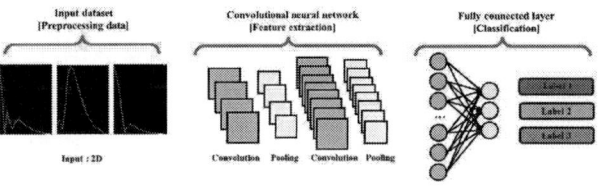

Fig. 7. CNN architecture using preprocessed data

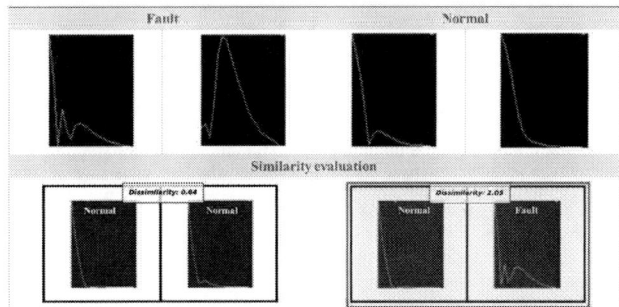

Fig. 8. Normal and fault EIS pattern diagnosis based on CNN algorithm and similarity evaluation

B. Similarity evaluation based on Siamese neural network

Based on the PEIS pattern images generated by the Bresenham algorithm, the CNN algorithm is employed to extract critical features from the images, and the similarity between these feature vectors is calculated to perform safety diagnosis. To achieve this, a Siamese Network with a CNN structure capable of processing two input datasets simultaneously is proposed. As illustrated in Figure 7, the Siamese Network processes each input data using identical CNN structures, extracts feature vectors for both inputs, and calculates their similarity. This similarity measure is then used to determine whether the data represents normal or faulty states based on a predefined threshold. In the diagnosis process, one of the normal images within the PEIS dataset is designated as the true label (or reference), and when a new image is input into the network, the similarity between the true label and the new input image is computed. This similarity score is then used to classify the input as either normal or faulty. During the learning phase, the network is trained to ensure that data with the same label are represented by feature vectors that are close to each other, while

$$Loss = Y_{similar} \cdot D^2(f(X_i), f(X_j)) +$$
$$(1 - Y_{similar}) \cdot \max(m - D^2(f(X_i), f(X_j)), 0)^2 \quad (14)$$

data with different labels are represented by feature vectors that are farther apart. This training process allows the network to effectively differentiate between normal and faulty images by learning the inherent structure and distribution of the data. In the final output stage of the CNN, feature vectors denoted as $f(X_i)$ and $f(X_j)$ are generated for each input image, and the similarity between the two feature vectors is computed. If the input images are similar, the output value approaches 1, and if the images are dissimilar, the output value approaches 0. This similarity-based evaluation enables precise fault diagnosis by quantifying the relationship between image pairs. To optimize this process, the loss function of the Siamese Network is designed to refine the distance relationship between the feature vectors. Specifically, the loss function, defined by equation (14), minimizes the difference for similar pairs while maximizing the difference for dissimilar pairs. Through this optimization process, the network effectively learns to identify and classify input data with high accuracy, making it a robust tool for diagnosing battery conditions and distinguishing between normal and faulty states with precision. This framework is particularly advantageous for applications where subtle differences in PEIS patterns need to be identified and leveraged for advanced diagnostics.

V. RESULTS AND ANALYSIS

To diagnose the failure and normal state of the electric vehicle battery, a PEIS-based image was extracted and converted into a feature vector to evaluate the similarity. If the distance between the feature vectors is 2 or more, the similarity is low, and if it is less than 2, the battery is diagnosed as normal, due to high similarity compared with normal labeled database in previous. The result is shown in Fig. 8, and when the dissimilarity is higher than predefined threshold, that pattern is diagnosed fault.

VI. CONCLUSION

In this paper, current and voltage data collected by applying dynamic profiles to lithium-ion batteries were transformed into the frequency domain using a STFT algorithm. The transformed time-series data, referred to as PEIS, were gathered under various operating conditions to construct a normal dataset. These PEIS patterns were subsequently used for feature extraction through a CNN algorithm [15], and the resulting feature vector was utilized to diagnose faults or normal states through similarity evaluation. This approach illustrates that even without actual measured EIS data, high diagnostic accuracy can be achieved, providing a novel and effective method for battery state diagnosis and prediction.

ACKNOWLEDGMENT

This research was supported by the 2024 government (Ministry of Trade, Industry and Energy) through the Korea Institute of Energy Technology Evaluation and Planning (Project No. RS-2024-00398346, Development of ESS Big Data-Based O&M and Asset Management Technology and Workforce Training). and by the National Institute of Information and Communications Technology Evaluation and Planning with financial resources from the government (Ministry of Science and ICT) in 2022 (No. 2022-1711152629, Functionality of Optimization Techniques in Machine Learning for SoC Estimation).

REFERENCES

[1] Xiong, Rui, et al. "Lithium-ion battery aging mechanisms and diagnosis method for automotive applications: Recent advances and perspectives." Renewable and Sustainable Energy Reviews 131 (2020): 110048.

[2] Ren, Dongsheng, et al. "A comparative investigation of aging effects on thermal runaway behavior of lithium-ion batteries." ETransportation 2 (2019): 100034.

[3] Shang, Yuzhao, et al. "Research progress in fault detection of battery systems: A review." Journal of Energy Storage 98 (2024): 113079.

[4] Koseoglou, Markos, et al. "A novel on-board electrochemical impedance spectroscopy system for real-time battery impedance estimation." IEEE Transactions on Power Electronics 36.9 (2021): 10776-10787.

[5] Tian, Jinpeng, et al. "Electrode ageing estimation and open circuit voltage reconstruction for lithium-ion batteries." Energy Storage Materials 37 (2021): 283-295.

[6] Mikheenkova, Anastasiia, et al. "Visualizing ageing-induced heterogeneity within large prismatic lithium-ion batteries for electric cars using diffraction radiography." Journal of Power Sources 599 (2024): 234190.

[7] Bresenham, Jack E. "Algorithm for computer control of a digital plotter." Seminal graphics: pioneering efforts that shaped the field. 1998. 1-6.

[8] Yang, Bowen, et al. "Research on online passive electrochemical impedance spectroscopy and its outlook in battery management." Applied Energy 363 (2024): 123046.

[9] Kuipers, Matthias, et al. "An algorithm for an online electrochemical impedance spectroscopy and battery parameter estimation: Development, verification and validation." Journal of Energy Storage 30 (2020): 101517.

[10] Liebhart, Bernhard, Lidiya Komsiyska, and Christian Endisch. "Passive impedance spectroscopy for monitoring lithium-ion battery cells during vehicle operation." Journal of Power Sources 449 (2020): 227297.

[11] Liebhart, Bernhard, Lidiya Komsiyska, and Christian Endisch. "Passive impedance spectroscopy for monitoring lithium-ion battery cells during vehicle operation." Journal of Power Sources 449 (2020): 227297.

[12] Zieliński, Tomasz P., and Tomasz P. Zieliński. "Discrete Fourier Transforms: DtFT and DFT." Starting Digital Signal Processing in Telecommunication Engineering: A Laboratory-based Course (2021): 65-92.

[13] Portnoff, Michael. "Time-frequency representation of digital signals and systems based on short-time Fourier analysis." IEEE Transactions on Acoustics, Speech, and Signal Processing 28.1 (1980): 55-69.

[14] Alzubaidi, Laith, et al. "Review of deep learning: concepts, CNN architectures, challenges, applications, future directions." Journal of big Data 8 (2021): 1-74.

[15] Zhang, Kai, et al. "Multi-fault detection and isolation for lithium-ion battery systems." IEEE Transactions on Power Electronics 37.1 (2021): 971-989

Enhanced Incremental Capacity Analysis for Evaluating Battery Degradation Mechanisms of Optimized Fast Charging Methods

Taehyeon Gong
Department of Electrical Engineering
Chungnam National University
Daejeon, Republic of Korea
xogus8948@naver.com

Jaehyeong Lee
Department of Electrical Engineering
Chungnam National University
Daejeon, Republic of Korea
wogud6136@naver.com

Sungjun Lee
Department of Electrical Engineering
Chungnam National University
Daejeon, Republic of Korea
lee990623@naver.com

Yura Kim
Department of Electrical Engineering
Chungnam National University
Daejeon, Republic of Korea
yura1120@naver.com

Bomyeong Ko
Department of Electrical Engineering
Chungnam National University
Daejeon, Republic of Korea
huskyholic@naver.com

Woonki Na
Department of Electrical and Computer Engineering
California State University
Fresno, United States
wkna@csufresno.edu

Sungjin Choi
Department of Electrical, Electronic and Computer Engineering,
University of Ulsan,
Ulsan, Republic of Korea
sjchoi@ulsan.ac.kr

Jonghoon Kim
Department of Electrical Engineering
Chungnam National University
Daejeon, Republic of Korea
whdgns0422@cnu.ac.kr

This paper presents a detailed analysis of the degradation modes associated with battery fast charging profiles. The analysis employs incremental capacity analysis (ICA) with advanced filtering techniques. The values for the loss of lithium inventory (LLI) and the loss of active material (LAM) were calculated based on the peaks extracted from the IC curve. Based on the ICA-based degradation mode analysis, the pulsed charging profile resulted in 0.004% less LLI and 0.01% less LAM compared to the constant current-constant voltage (CC-CV) charging profile. As a result, the degradation mode analysis demonstrated the performance of the pulse charging profile over CC-CV charging among fast charging profiles.

Keywords—Fast charging, Pulse charging, Incremental capacity analysis, Savitzky-Golay filter, Degradation mode analysis

I. INTRODUCTION

The growth of the electric vehicle (EV) market, the emergence of electric propulsion vessels, and the proliferation of portable electronic devices have collectively led to a significant surge in battery demand [1]. These changes drive technological innovation and address global energy challenges by reducing dependence on fossil fuels and encouraging the adoption of green energy solutions. The rapid advancement of battery technology, coupled with an emphasis on sustainability and efficiency, has rendered batteries a foundational component of contemporary energy systems. Batteries must provide high energy density and power output while supporting reliable operation over frequent and rapid charging cycles. These characteristics are paramount to enhancing user convenience and guaranteeing affordability, particularly in large-scale applications such as electric vehicles. For example, the ability to rapidly and efficiently charge an electric vehicle is directly correlated with its practicality. Similarly, portable electronic devices require rapid and reliable charging to meet user expectations for uninterrupted operation [2].

To fulfill these requirements, developing efficient and reliable rapid charging methodologies has become a significant target within the field of battery research. The objective of fast charging technology is not merely to reduce charging time; it also encompasses the mitigation of the degradation of battery performance that frequently occurs under fast charging conditions. Fast charging systems must be designed in a manner that minimizes these impacts while maximizing battery lifespan and efficient operation [3].

Conventional charging techniques, particularly constant current (CC) charging, have become widely accepted due to their straightforward implementation and reliable performance. In a CC charging system, the battery is charged by supplying a fixed current until it reaches a predetermined voltage threshold. This method is simple and effective for many applications, making it a popular choice in various industries. However, its simplicity comes with certain drawbacks that can negatively impact battery health over time. During the charging process, the continuous flow of current can result in the generation of significant heat within the battery cells. This heat can lead to thermal stress, which accelerates the aging process of the battery components. Additionally, CC charging can cause chemical imbalances, such as lithium plating on the anode and the degradation of the electrolyte, both of which compromise the battery's long-term performance and capacity retention.

These issues are particularly problematic for high demand applications, such as electric vehicles, which require fast and frequent charging cycles to meet operational needs. In such scenarios, the limitations of CC charging, including potential safety concerns and reduced battery lifespan, make it less suitable compared to more advanced charging techniques designed to mitigate these challenges [4].

To address these limitations, the constant current-constant voltage (CC-CV) charging method has become a widely adopted solution. The CC-CV method commences with a CC phase, whereby the battery is charged at a constant current until the target voltage is reached. Subsequently, the system transitions to a CV phase, wherein the voltage remains constant while the current gradually decreases [5]. This two-stage process mitigates the risk of overcharging and excessive heat buildup, rendering CC-CV an optimal choice for batteries in electric vehicles and other high-demand applications. Nevertheless, despite the notable advancements brought forth by CC-CV over conventional CC charging, certain challenges persist, particularly in instances of rapid charging, where issues such as heat accumulation and material deterioration may arise. Considering these considerations, researchers have investigated alternative approaches, such as pulse charging, which entails the application of pulsed currents interspersed with rest intervals. The intervals of rest permit the dissipation of heat and the restoration of ionic equilibrium within the battery, thereby reducing stress and degradation. This approach demonstrates the potential for enhancing battery health, optimizing performance, and prolonging the lifespan, particularly in applications necessitating rapid and frequent charging. The objective of this study is to compare the performance and degradation characteristics of CC-CV and pulse charging to determine the most effective charging strategy. The objective of the analysis is to identify the relative merits and shortcomings of each approach and to investigate the potential of pulse charging as a superior alternative for demanding battery applications [6].

To evaluate these methods, the study applies incremental capacity analysis (ICA), which is a highly effective tool for detecting detailed changes in the electrochemical properties of a battery. ICA offers valuable insights into degradation mechanisms, including lithium inventory loss (LLI) and loss of active material (LAM), which serve as crucial indicators of battery health [7]. However, ICA data is highly susceptible to noise, which can impede the interpretation of the results [8]. To address this issue, the moving average (MA) filter and Savitsky-Golay (SG) filter techniques are employed. These advanced filtering methods refine the data by reducing noise and increasing clarity, thereby facilitating the more accurate identification of key degradation patterns. In this paper, we undertake a comparative analysis of the CC-CV and pulse charging (PC) methods and apply ICA based on advanced filtering methods to provide comprehensive information for a deeper understanding and evaluation of battery health.

This paper is organized as follows: First, the optimal charging profile is identified using a comparative analysis of the capacity degradation rates associated with different

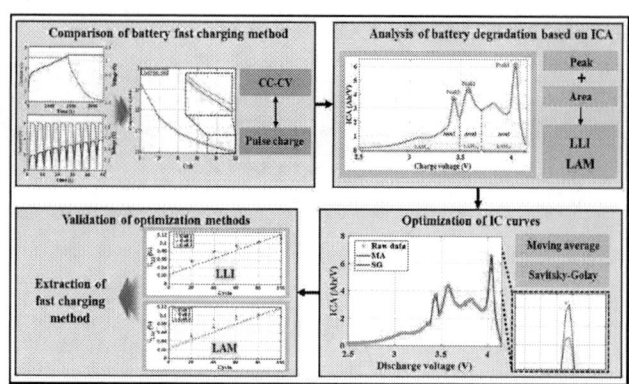

Fig. 1. Flowchart of the optimal IC curve extraction method for fast charging profile performance comparison

charging profiles. Next, an overview of the ICA method and advanced filtering techniques presents a proposed methodology for degradation mode analysis based on these techniques. It presents the results of a quantitative analysis of battery degradation modes based on the ICA method. Finally, it presents a summary of the findings presented in this paper.

II. EXPERIMENTAL SETUP AND METHODOLOGY

In this paper, two charging profiles were devised to compare the degradation patterns between CC-CV charging and pulse charging. This section describes the setup used in the experiments, the design of the charging profiles, and the capacity degradation results for each profile.

A. Experimental setup

The commercial cell adopted in this experiment is an LG HG2 18650 3000mAh NCM/graphite, exhibiting a charge cut-off voltage of 4.2V and a discharge cut-off voltage of 2.5V. An experimental platform was constructed to facilitate the charge and discharge experiments, and a constant temperature chamber was set at 40°C to accelerate the aging process of the battery. The experiment comprises two distinct components. The first is an aging experiment conducted based on battery profiles, while the second is a capacity experiment designed to extract parameters for the performance characteristics of the battery. The aging experiments comprised repeated cycles of full charge and discharge, enabling an accurate assessment of the extent of battery degradation. The analysis of the capacity measurement experiment serves to elucidate the underlying degradation mechanism [9]. In addition, the setup was optimized to ensure consistent testing conditions across all cells. This approach ensures that environmental and operational variances are minimized, improving the reliability of the results. The details of the experiment are as follows:

(1) Capacity test: It is essential to perform a CC discharge with 0.5C to a cut-off voltage of 2.5V. Two hours intervals should be permitted to stabilize the voltage. CC charge at 0.5C to a voltage of 4.2V should be applied, followed by CV charge at 4.2V until a current of 100mA is reached. Following the specified period of inactivity, CC discharge with 0.5C until a voltage of 2.5V is reached should be conducted to ascertain the discharge capacity.

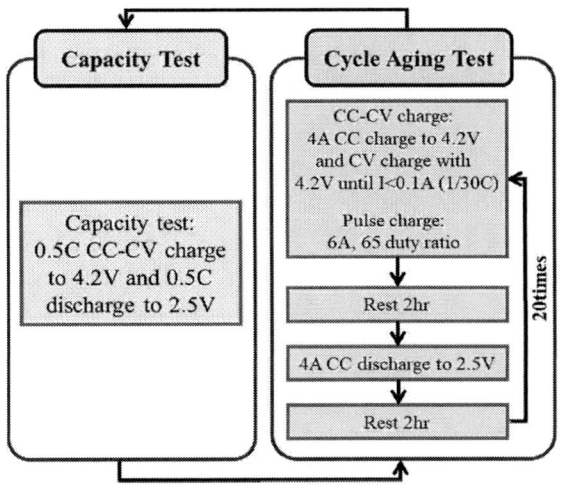

Fig. 2. Experimental procedure for applying fast charging profiles

(2) Aging test: The two designed charging profiles should be applied to each cell, and the aging test repeated 20 times. A detailed account of each profile will be provided in the subsequent session.

B. Design of charging profiles

In this experiment, a CC-CV profile was designed using the CC-CV charging method, and a pulse profile was created using the pulse charging method. If the battery is initially discharged, the CC-CV charging method will charge the battery with a constant current, which will gradually increase the battery voltage. Upon reaching a predetermined voltage level (4.2V), the battery voltage stabilizes while the charging current undergoes a gradual decline. Charging ceases when the charging current declines to a level below the cut-off point (100mA), which is typically less than 3% of the rated current. Consequently, CC-CV charging comprises two distinct phases: a constant current phase and a constant voltage phase. To guarantee the precision and dependability of the data, the corresponding CC-CV profiles were implemented in Cell 1 and Cell 2.

In this experiment, the standard PC mode was adopted. The standard PC mode represents a constant current charging method that incorporates periodic relaxation time [10]. The positive pulse current is maintained at a constant value, while the current is set to zero during the relaxation period. This process is performed repeatedly, enabling improvements in internal chemical changes and heat generation during battery operation. In this paper, the PC profile was configured based on the charging time of the CC-CV profile to equalize the experimental durations of both profiles. Accordingly, the PC charging profile was designed as specified in Table I and applied to Cell 3.

TABLE I.   PULSE CHARGING PROFILE SETTING VALUES

	Type	Mode	Range	Duty	Value
PC profile	GSTAT	Pulse-1	10A	65%	6A

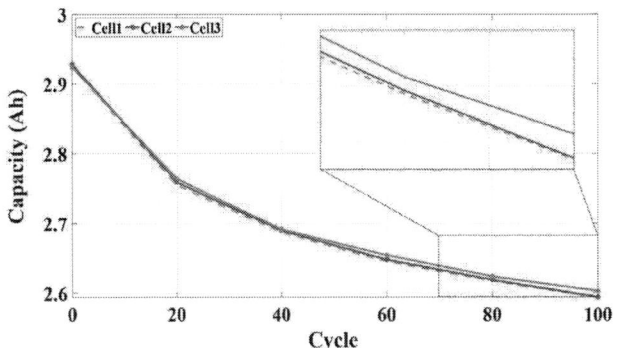

Fig. 3. A comparison of capacity reductions by fast charging profiles

C. Capacity reduction results by charging profiles

The aging and capacity tests were repeated at 20 cycles intervals for a total of 100 cycles, with data collected for each cycle. The capacity data for each cell over the 100 cycles is shown in Figure 2. The capacity test results show that the cells with the CC-CV profile exhibit a capacity degradation rate of approximately 11.6%. The cells with the PC profile showed a capacity degradation rate of approximately 11.0%.

These results suggest that among the fast charging profiles, the pulsed charging method can contribute to relatively mild battery degradation. In particular, the fact that the degradation rate remained in the 11% range despite repeated cycles of charge and discharge shows that this charging method is a favorable approach to preserve the long-term performance of the battery. This suggests that pulsed charging could be important in improving battery life and increasing the likelihood of continued use.

III.   ANALYSIS OF ICA WITH FILTERING TECHNIQUES

A. Incremental capacity analysis techniques

ICA is a highly effective technique for diagnosing the degradation state of lithium-ion batteries. By visualizing the voltage-capacity relationship within specific voltage ranges from the charge-discharge curves, ICA facilitates the analysis of battery degradation modes using direct current (DC) based signals. This method allows for insights into internal battery changes without the need for specialized equipment, making it highly advantageous for real-time implementation in battery management systems (BMS). Lithium-ion batteries' primary internal degradation mechanisms are typically categorized as the LAM and the LLI. LAM occurs due to structural changes or decomposition of cathode materials during charge and discharge cycles, reduction in energy capacity, and overall performance. LLI, otherwise, arises from irreversible lithium-ion losses during cycling, primarily caused by the formation and growth of the solid electrolyte interphase (SEI) layer, lithium metal plating, and electrolyte decomposition [11-12]. This loss diminishes the number of available lithium ions, leading to capacity fading and performance deterioration. ICA identifies such internal degradation modes by employing a derivative curve of voltage against capacity. By detecting subtle

979-8-3315-1612-3/25 $31.00 © 2025 IEEE

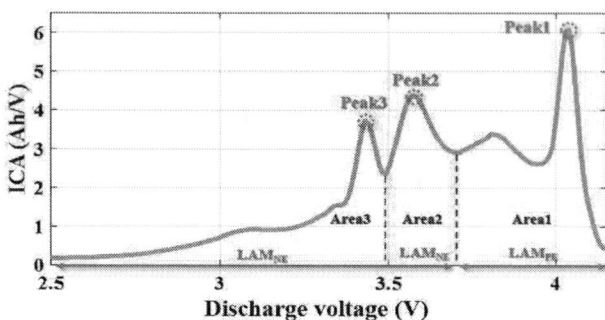

Fig. 4. Selection of peaks for incremental capacity analysis

changes in the voltage profile, ICA captures various physical and chemical transformations occurring within the battery. Each peak in the ICA curve reflects not a single degradation mechanism but rather a combination of factors. For instance, as illustrated in Fig. 4 and Table II, peak reduction and shifting are comprehensive outcomes of multiple degradation mechanisms, influenced by factors such as changes in electrolyte condition and structural alterations of active materials. Analyzing these shifting phenomena enables a more precise diagnosis of degradation causes, which can be instrumental in optimizing battery management strategies. Thus, ICA serves as a powerful tool for effectively diagnosing various degradation mechanisms of batteries through DC signal analysis. In this paper, ICA was applied to evaluate the degradation modes of lithium-ion batteries under different charging patterns. This research aims to identify the most effective charging strategies for preserving the state of health (SOH) of the battery by examining the impact of varying charging strategies on the battery degradation process. [13].

TABLE II. ANALYSIS OF IC PEAK-BASED
DEGRADATION MECHANISM

Degradation Mode	ICA Peak ①	ICA Peak ②	ICA Peak ③
LLI	↓	↓	←
LAM$_{PE}$	←	↓	↓
LAM$_{NE}$	↓	↓	↓

B. Filtering techniques for noise reduction in IC curve analysis

The original IC curve can be derived using equation (1). The IC curve is highly effective in detecting subtle changes in voltage signals, enabling the identification of critical physical and chemical changes occurring within the battery. However, the raw IC curve often contains excessive noise, which limits its applicability for detailed analysis. Therefore, noise removal through filtering algorithms is essential for accurate IC curve analysis.

$$\frac{dQ}{dV} = \frac{\Delta Q}{\Delta t} \times \frac{\Delta t}{\Delta V} = I \times \frac{\Delta t}{\Delta V} \quad (1)$$

To enhance robustness against data noise, the MA incremental capacity analysis (MA-ICA) method has been widely used [14]. Equation (2) demonstrates the MA-ICA method, which reduces noise by calculating the average value of specific data points and replacing each data point with its corresponding average. By averaging neighboring data points, this method minimizes unnecessary noise while preserving the original signal characteristics and maintaining the integrity of the data. This approach provides a more accurate and reliable analysis of battery performance, allowing for a clearer evaluation of the battery condition without distorting its essential characteristics or compromising analytical precision [15]. However, significant errors can occur if the number of data points used in MA-ICA is not appropriately chosen. To address these limitations, this paper proposes the use of the SG filter. The SG filter, compared to MA-ICA, better preserves the characteristics of peak signals and does not alter the data length. Based on the SG filtering algorithm, this paper proposes an enhanced filtering approach. The fundamental idea of the SG filtering algorithm is to apply a moving window to a specific segment of the data, fit a polynomial within the window, and substitute the central point with the smoothed value obtained from the polynomial. This process reduces noise while maintaining the overall shape of the data. The moving window refers to a range that includes the central data point and data points within a defined radius on either side. Equation (3) represents the fundamental formula of the SG filter [16]. This formula expresses the squared error between the measured data x_{i+j} at y_{i+j} and the polynomial approximation $P(x_{i+j})$. Here, i denotes the index of the current central data point, and j represents the relative position from the central point x_i. The range of j depends on the window size and spans from $-m$ to $+m$. This method effectively reduces noise by focusing more on the central data, thereby preserving and further enhancing the peak characteristics. Finally, the filtered value y_i is calculated using equation (4), where c_j represents the filter coefficients obtained through polynomial fitting. These coefficients play a critical role in reducing noise while preserving essential features of the data. This approach effectively reduces noise while maintaining the overall structure of the data. The SG filtering technique is particularly advantageous for analyzing IC curves as it preserves peak characteristics, making it more reliable for data analysis [17]. Fig. 5 and Fig. 6 present the IC curves and errors derived using MA-ICA and SG-ICA. Notably, SG-ICA significantly improves error compared to MA-ICA, particularly around approximately 4.1V. SG-ICA reduced the maximum error from 18% to 6%, reflecting an improvement of approximately 12%. These findings confirm the superior performance of SG-ICA over MA-ICA. In addition, the root mean square error (RMSE) results of SG-ICA and MA-ICA are shown in Table III.

$$\left(\frac{dQ}{dV}\right)_{MA-ICA} = \frac{d\left(\frac{1}{2m+1}\sum_{j=-m}^{m} Q_{i+j}\right)}{dV} \quad (2)$$

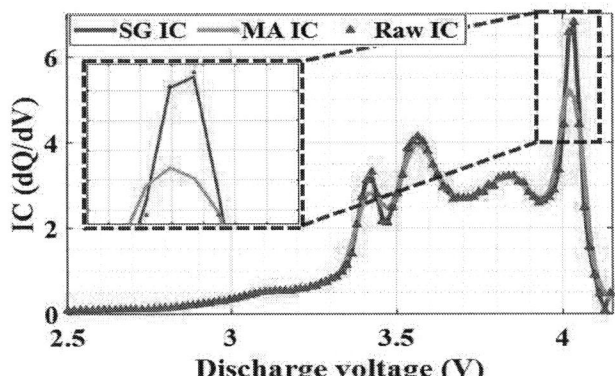

Fig. 5. Result of IC curves with the filtering techniques

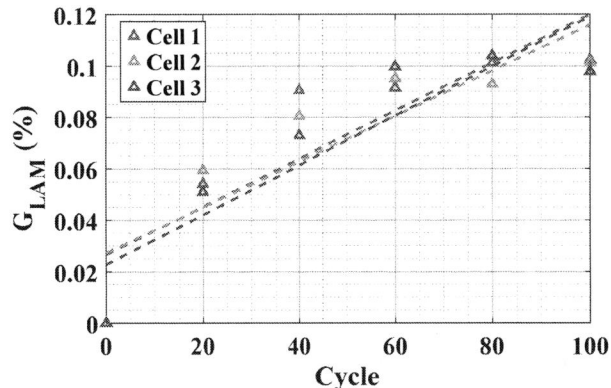

Fig. 7. Extraction of LAM ratio of each fast charging profile

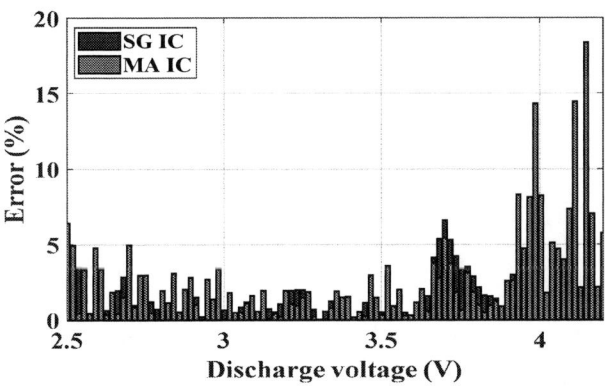

Fig. 6. Result of error with the filtering techniques

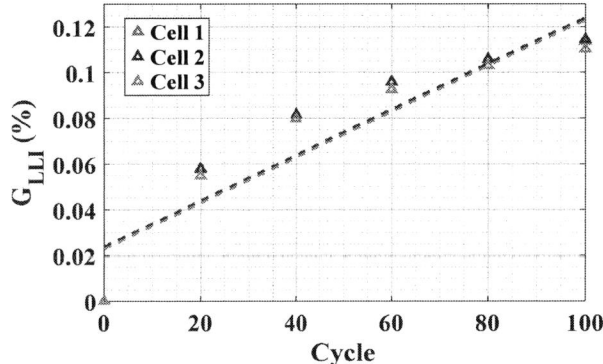

Fig. 8. Extraction of LLI ratio of each fast charging profile

The RMSEs of SG-ICA and MA-ICA are 2.03% and 4.21%, respectively, confirming that SG-ICA is superior to MA-ICA. Thus, in this study, SG-ICA was utilized to evaluate the degradation modes of batteries under fast charging profiles, demonstrating its effectiveness in improving the accuracy and reliability of battery performance assessments.

$$\sum_{j=-m}^{m} (y_{i+j} - P(x_{i+j}))^2 \quad (3)$$

$$\hat{y}_i = \sum_{j=-m}^{m} c_j y_{i+j} \quad (4)$$

TABLE III.   RMSE AND MAX ERRORS ON SG IC & MA IC

	RMSE	MAX error
SG IC	2.0273%	6.5805%
MA IC	4.2104%	18.3602%

IV. QUANTITATIVE ANALYSIS OF DEGRADATION MODES FOR ENHANCED RELIABILITY ASSESSMENT

A. Method of quantitative analysis of degradation modes

The analysis of battery degradation mechanisms under fast charging profiles was conducted using optimized IC curves processed with an SG filter. However, IC curves alone have limitations in quantitatively analyzing battery degradation and classifying specific degradation mechanisms. To overcome these limitations, quantitative assessments of internal chemical changes, such as LLI and LAM, were performed using equation (5) and equation (6). These equations provided quantitative values for performance degradation under various fast charging profiles, enabling a comparative analysis of performance loss across different charging strategies. Based on ICA derived outputs, such as the capacity change to voltage and the overall capacity, the equations identified regions with maximum occurrences, facilitating the determination of LLI and LAM related battery performance degradation. Consequently, these equations were applied to derive degradation modes across diverse fast charging profiles and to quantitatively analyze the impact of fast charging on battery degradation.

$$G_{LAM} = \frac{\left|\max(\frac{dQ}{dV})\right|_1 - \left|\max(\frac{dQ}{dV})\right|_n}{\left|\max(\frac{dQ}{dV})\right|_1} \quad (5)$$

$$G_{LLI} = \frac{\left|\max(Q)\right|_1 - \left|\max(Q)\right|_n}{\left|\max(Q)\right|_1} \quad (6)$$

B. Degradation mode analysis results

The quantitative analysis results of battery degradation under fast charging profiles are presented in Fig. 7 and Fig. 8. The findings indicate that performance degradation trends vary depending on the applied fast charging profiles. Specifically, for LLI, the conventional CC-CV profile resulted in approximately 0.114% degradation, whereas the pulse charging profile exhibited about 0.110% degradation, showing a difference of approximately 0.004% between the two patterns. This suggests that the PC profile mitigates lithium loss more effectively than the CC-CV profile.

Furthermore, the analysis of LAM revealed an even more pronounced difference. The CC-CV profile resulted in approximately 0.102% LAM, while the pulse charging profile exhibited only about 0.092%, reflecting a difference of approximately 0.01% between the two patterns. These quantitative results demonstrate that the application of the pulse charging profile significantly reduces battery degradation compared to the conventional CC-CV method. Therefore, the proposed profile's effectiveness in mitigating battery degradation has been validated, highlighting its superiority over traditional charging methods.

V. CONCLUSION

This study demonstrates that the PC profile outperforms the conventional CC-CV charging method in reducing battery degradation and maintaining battery health. To quantitatively assess battery performance degradation, the ICA technique was utilized to analyze internal chemical changes such as LLI and LAM. However, to improve the accuracy of deriving IC curves, a detailed analysis of degradation mechanisms using ICA enhanced with advanced filtering techniques was conducted, quantitatively confirming the advantages of pulse charging. Specifically, the application of MA-ICA and SG-ICA methods effectively minimizes noise interference, enabling precise interpretation of IC curves. The findings showed that the PC profile significantly mitigated LLI compared to the CC-CV profile, proving its potential as a superior strategy for high-performance battery systems. These results can contribute to the development of fast charging methods that enhance battery longevity and sustain performance in applications such as electric vehicles. To further refine and expand the applicability of this charging approach, future research should prioritize several key areas. First and foremost, it is of paramount importance to optimize the pulse charging profile to ensure reliability under a multitude of environmental and operational conditions, including but not limited to temperature fluctuations. Secondly, further advancement of signal filtering techniques, including MA-ICA and SG-ICA, is necessary to improve their compatibility with a wide range of complex battery systems. Exploring the internal electrochemical dynamics caused by pulse charging through modeling and simulations can provide valuable insights into the root causes of battery degradation. This research can be highly beneficial in the development of enhanced BMS to maximize battery performance, operational efficiency, and overall safety.

ACKNOWLEDGMENT (Heading 5)

This research was supported by the Korea Institute of Energy Technology Evaluation and Planning (No. RS-2024-00398346, Development of ESS Big Data-Based O&M and Asset Management Technology and Workforce Training) and the Korea Institute of Energy Technology Evaluation and Planning (No. 00426149, Development of Lithium Battery Packing and Enhanced Safety Control Technology with Liquid Cooling Application).

REFERENCES

[1] Xu, Chengjian, et al. "Future material demand for automotive lithium-based batteries."Communications Materials1.1 (2020): 99.

[2] Alhaider, Mohammed M., et al. "New temperature-compensated multi-step constant-current charging method for reliable operation of battery energy storage systems." *IEEE Access* 8 (2020): 27961-27972.

[3] Lin, Yiran, Bo Jiang, and Haifeng Dai. "Battery capacity estimation based on incremental capacity analysis considering charging current rate." World Electric Vehicle Journal 12.4 (2021): 224.

[4] Guo, Fei, et al. "Prediction of remaining useful life and state of health of lithium batteries based on time series feature and Savitzky-Golay filter combined with gated recurrent unit neural network." Energy 270 (2023): 126880.

[5] Stroe, Daniel-Ioan, and Erik Schaltz. "Lithium-ion battery state-of-health estimation using the incremental capacity analysis technique." IEEE Transactions on Industry Applications 56.1 (2019): 678-685.

[6] Kemény, M., P. Ondrejka, and M. Mikolasek. "Incremental Capacity Analysis for Prediction of Li-Ion Battery Degradation Mechanisms: Simulation Study."2020 13th International Conference on Advanced Semiconductor Devices And Microsystems (ASDAM). IEEE, 2020.

[7] Meda, Ujwal Shreenag, et al. "Solid Electrolyte Interphase (SEI), a boon or a bane for lithium batteries: A review on the recent advances." *Journal of Energy Storage* 47 (2022): 103564.

[8] Huang, Xinrong, et al. "A review of pulsed current technique for lithium-ion batteries." Energies 13.10 (2020): 2458.

[9] Xu, Chengjian, et al. "Future material demand for automotive lithium-based batteries."Communications Materials1.1 (2020): 99.

[10] Alhaider, Mohammed M., et al. "New temperature-compensated multi-step constant-current charging method for reliable operation of battery energy storage systems." *IEEE Access* 8 (2020): 27961-27972.

[11] Lin, Yiran, Bo Jiang, and Haifeng Dai. "Battery capacity estimation based on incremental capacity analysis considering charging current rate." World Electric Vehicle Journal 12.4 (2021): 224.

[12] Guo, Fei, et al. "Prediction of remaining useful life and state of health of lithium batteries based on time series feature and Savitzky-Golay filter combined with gated recurrent unit neural network." Energy 270 (2023): 126880.

[13] Stroe, Daniel-Ioan, and Erik Schaltz. "Lithium-ion battery state-of-health estimation using the incremental capacity analysis technique." IEEE Transactions on Industry Applications 56.1 (2019): 678-685.

[14] Kemény, M., P. Ondrejka, and M. Mikolasek. "Incremental Capacity Analysis for Prediction of Li-Ion Battery Degradation Mechanisms: Simulation Study."2020 13th International Conference on Advanced Semiconductor Devices And Microsystems (ASDAM). IEEE, 2020.

[15] Meda, Ujwal Shreenag, et al. "Solid Electrolyte Interphase (SEI), a boon or a bane for lithium batteries: A review on the recent advances." *Journal of Energy Storage* 47 (2022): 103564.

[16] Kemény, M., P. Ondrejka, and M. Mikolasek. "Incremental Capacity Analysis for Prediction of Li-Ion Battery Degradation Mechanisms: Simulation Study."2020 13th International Conference on Advanced Semiconductor Devices And Microsystems (ASDAM). IEEE, 2020.

[17] Meda, Ujwal Shreenag, et al. "Solid Electrolyte Interphase (SEI), a boon or a bane for lithium batteries: A review on the recent advances." *Journal of Energy Storage* 47 (2022): 103564.

Co-Estimation of SOC and SOT in Lithium-Ion Batteries Using an RLS-Based Heat Generation Model

Seongkyu Lee
Department of Electrical Engineering
Chungnam National University
Daejeon, Republic of Korea
kyu008520@gmail.com

Eunjin Kang
Department of Electrical Engineering
Chungnam National University
Daejeon, Republic of Korea
eunjini0716@gmail.com

Minhyeok Kim
Department of Electrical Engineering
Chungnam National University
Daejeon, Republic of Korea
rlaalsgur168@naver.com

Seunghyun Lee
Department of Electrical Engineering
Chungnam National University
Daejeon, Republic of Korea
tmdgus7072@gmail.com

Minwoo Song
Department of Electrical Engineering
Chungnam National University
Daejeon, Republic of Korea
bananakick4563@naver.com

Jaea Lee
Department of Electrical Engineering
Chungnam National University
Daejeon, Republic of Korea
jaeadorable@naver.com

Woonki Na
Department of Electrical and Computer Engineering
California State University
Fresno, United States
wkna@csufresno.edu

Jonghoon Kim
Department of Electrical Engineering
Chungnam National University
Daejeon, Republic of Korea
whdgns0422@cnu.ac.kr

Batteries are widely utilized in various applications, such as electric vehicles and energy storage systems, where they operate under diverse environmental conditions. High temperatures can decrease the stability of batteries and accelerate aging processes, while low temperatures reduce conductivity and diffusion rates, leading to diminished capacity and output. Therefore, maintaining a stable temperature during battery operation is critical for optimal performance and longevity. This paper focuses on analyzing battery heat generation prior to the implementation of thermal management strategies through a battery management system (BMS). To achieve this, the hybrid pulse power characterization (HPPC) test is conducted to assess the initial characteristics of the battery, including heat generation across different temperature ranges. Additionally, the recursive least squares (RLS) method is employed to inversely estimate the parameters of the battery model from voltage and current data, allowing for an accurate prediction of heat generation. Furthermore, the study integrates the state-of-charge (SOC) estimation into the analysis, enhancing the understanding of battery behavior under varying operational conditions. The designed battery heat generation model is subsequently validated by comparing the results with state-of-temperature (SOT) estimates and experimental data. This comprehensive approach not only improves the accuracy of thermal management but also contributes to more effective SOC estimation, ultimately leading to better battery performance and safety.

Keywords—Battery management system, Recursive least squares, Battery temperature, Lithium-ion battery

I. INTRODUCTION

One of the primary contributors to global warming is the emission of greenhouse gases, notably carbon dioxide, into the atmosphere. These emissions predominantly result from the combustion of fossil fuels in energy production and transportation. As the global threat of climate change intensifies, there is a growing impetus to transition towards eco-friendly energy sources and adopt transportation methods that can substantially reduce greenhouse gas emissions. Among these efforts, the shift from internal combustion engine vehicles to battery-powered electric vehicles is increasingly recognized as a pivotal strategy for mitigating the carbon footprint of the transportation sector.

Lithium-ion batteries have emerged as the most viable energy storage solution for electric vehicles due to their high energy density, extended driving range, rapid charging capability, and capacity to support high-performance driving. However, despite these advantages, lithium-ion batteries exhibit significant sensitivity to temperature variations, which can profoundly affect their performance, lifespan, and safety. For instance, at low temperatures, the electrochemical reactions within the battery slow down, leading to a reduction in available energy, thereby decreasing the vehicle's driving range. Conversely, at elevated temperatures, the rate of side reactions, such as the growth of the solid electrolyte interphase (SEI) layer and the decomposition of cathode materials, increases. These reactions expedite battery degradation, significantly shortening its operational lifespan and potentially compromising safety. Given this pronounced temperature sensitivity, it is imperative to maintain the operating temperature of lithium-ion batteries within an optimal range, typically between $288\ K$ and $308K$ [1]. Deviation from this temperature range not only diminishes the battery's efficiency and longevity but also introduces safety risks, such as thermal runaway. Consequently, the design of an effective battery cooling system is critical to ensuring the safe and reliable operation of lithium-ion batteries in electric vehicles.

979-8-3315-1612-3/25 $31.00 © 2025 IEEE

Precise prediction of battery heat generation is a crucial component of this design process. Inaccurate estimations can result in either excessive cooling or inadequate cooling, both of which can adversely affect the battery's stability and performance. A significant challenge in battery temperature management arises from the fact that the ratio of temperature sensors to batteries in a typical battery system is approximately 1:10 [2]. This ratio indicates that it is practically infeasible to obtain real-time temperature data for each individual battery within the system, as deploying sensors on every battery is not viable. Furthermore, the heat generation characteristics of batteries evolve as they age, further complicating the thermal management process [3]. Therefore, to obtain accurate temperature data for each battery, it is essential to develop and implement thermal models capable of estimating battery temperatures based on indirect measurements and other relevant parameters.

The objective of this study is to extract the initial battery parameter values using the hybrid pulse power characterization (HPPC) test. These initial parameters provide the foundational data necessary for understanding the thermal characteristics of the battery. Subsequently, the recursive least squares (RLS) algorithm is employed to continuously estimate battery parameters as the battery undergoes aging [4]. By integrating these estimated parameters into a comprehensive thermal model, this study aims to accurately predict the state-of-temperature (SOT) and state-of-charge (SOC) under various operating conditions [5]. The SOC estimation is performed using the Kalman Filter, which is widely recognized as one of the most effective methods for SOC estimation [6]. Predictive capability is crucial for optimizing battery cooling system design, ultimately enhancing the performance, safety, and lifespan of lithium-ion batteries.

II. BATTERY MODELLING

A. Experimental setup and environmental control

The battery cell specifications and experimental settings used in this study are shown in Table I and Fig. 1. To ensure accurate temperature measurements of the battery, it was necessary to minimize the influence of external conditions. Experiments were conducted within a chamber to maintain a constant ambient temperature around the battery. However, airflow within the chamber, introduced to maintain a uniform temperature, generated noise that could affect the precise measurement of battery heat generation. To examine the airflow dynamics within the chamber, the experiments were conducted as shown in Fig. 2. Eight measurement points were designated inside the chamber, as illustrated in Fig. 2(a). The airflow within the chamber originated from the outlet and exited through the inlet. Temperature and airflow velocity were measured when the chamber temperature decreased from 301K to 298K. The airflow velocity at each point is presented in Fig. 2(b). The analysis revealed an average airflow velocity of 0.9(m/s), which had a noticeable impact on the battery's temperature measurements. To minimize this effect, an inner chamber was constructed within the main chamber to reduce the airflow velocity Fig. 2(c). After installing the inner chamber Fig. 2(d), the airflow velocity near the center of the battery was measured When the chamber temperature decreased from $301K$ to $298K$.

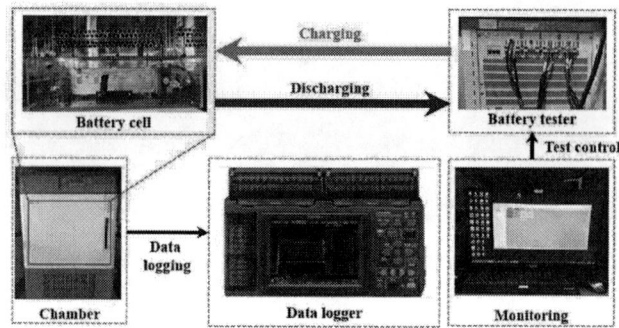

Fig. 1. Battery electrical characteristics test setup diagram

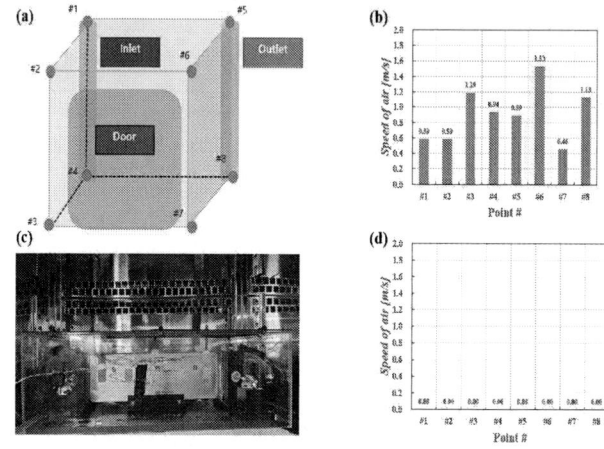

Fig. 2. Airflow inside the chamber (a) Measurement points (b) Air velocity (c) Set after installing the inner chamber (d) Air velocity after interior chamber installation

TABLE I. CELL SPECIFICATION

Item	Specification
Battery type	NMC
Voltage range	2.5~4.2V
Capacity	60.3 Ah

The airflow velocity around the battery was recorded as 0(m/s). The battery utilized in this study is a nickel-manganese-cobalt (NMC) pouch cell with a capacity of $60.3Ah$. Battery tests were conducted to extract the parameters of the electrical equivalent circuit model (EECM). The open circuit voltage (OCV) and HPPC test profiles are presented in Fig. 3. For the experiments, the battery was first discharged at a 1C- rate until the cutoff voltage of 2.5V was reached, followed by a rest period of 3 hours. Subsequently, the battery was charged using the constant-current constant-voltage (CC-CV) method at a 1C-rate rate until the cutoff current reached 0.02C-rate. The HPPC test was then performed iteratively to extract battery parameters, reducing the SOC by 5% increments. Each SOC level was followed by a rest period of 2 hours. Voltage and current changes were observed during the tests at various SOC levels using C-rates of 0.5, 1, 2, and 3C. The HPPC test procedures excluded C-rates that could trigger the upper or lower voltage limits.

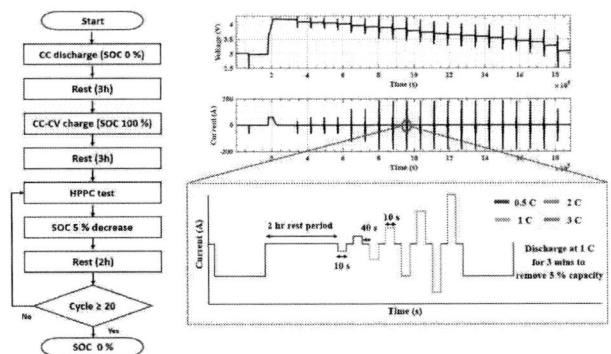

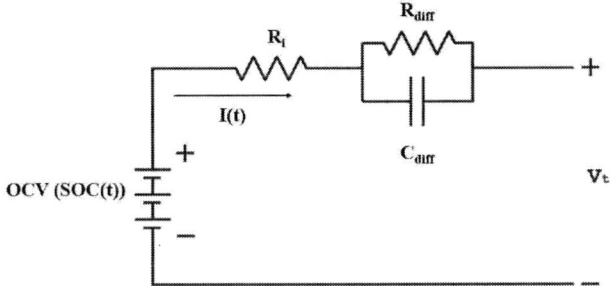

Fig. 3. Flowchart of the HPPC test performed at soc 5% intervals

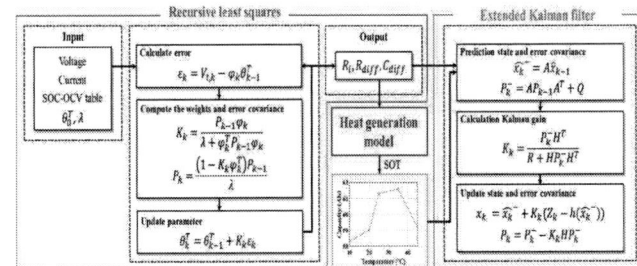

Fig. 5. Flowchart of co-estimation process for SOT and SOC

$$V_t = OCV - IR_i - R_{diff}\left(1 - e^{-\frac{t}{R_{diff}C_{diff}}}\right) \quad (1)$$

$$G(s) = \frac{R_i + \left(\frac{R_i}{R_{diff}C_{diff}} + \frac{T_s}{C_{diff}} - R_i\right)z^{-1}}{1 + \left(\frac{T_s}{R_{diff}C_{diff}} - 1\right)z^{-1}} \quad (2)$$

$$V_{t,k} = b_0 I_k - b_1 I_{k-1} + a_1(-OCV_{k-1} + V_{t,k-1}) + OCV \quad (3)$$

$$b_0 = R_i, \quad (4)$$

$$b_1 = \frac{R_i T_s}{R_{diff}C_{diff}} + \frac{T_s}{C_{diff}} - R_i \quad (5)$$

$$a_1 = \frac{T_s}{R_{diff}C_{diff}} - 1 \quad (6)$$

$$R_i = b_0, \ R_{diff} = \frac{b_1 - a_1 b_0}{1 + a_1}, C_{diff} = \frac{T_s}{b_1 - a_1 b_0} \quad (7)$$

$$Q_{total} = Q_{body} + Q_{tab,p} + Q_{tab,n} - Q_{loss} \quad (8)$$

$$Q_{body} = I^2 R_{total} - IT\frac{dOCV}{dT} \quad (9)$$

$$Q_{loss} = Q_{con} + Q_{rad} \quad (10)$$

$$Q_{con} = hA(T_b - T_a) \quad (11)$$

$$Q_{rad} = \varepsilon\sigma A(T_b{}^4 - T_a{}^4) \quad (12)$$

$$R_{total} = R_i + R_{diff} \quad (13)$$

Fig. 4 section

Fig. 4. Schematic diagram of the 1RC EECM

B. Battery electrical model

The EECM is widely utilized for SOC estimation due to its advantages, including a simplified structure, centralized parameters, ease of identification, and clear physical interpretation [7]. Among the commonly employed EECM, the n-RC model encompasses various configurations. Specifically, the 1RC model corresponds to the Thevenin model, and the 2RC model corresponds to the dual polarization (DP) model [8]. The battery pack is modeled using the EECM framework, where models with more than two RC circuits incorporate additional RC networks to analyze the transient characteristics of the charge and discharge profiles of the terminal voltage. In this study, the 1RC Thevenin model was adopted. The 1RC Thevenin model has been adopted, as consists of a series resistor circuit. In this model, R_0 represents the series resistance of the battery and R_1 represents the parallel resistance C_1 represents the diffusion capacitance of the battery electrode. Capacitance measures how fast lithium-ion can flow inside the battery electrode material under the influence of electrode shape and porosity factors. Parameters R_i, R_{diff}, and C_{diff} are equivalent parameters of the battery pack. equation (1) represents the voltage calculated using the EECM. Fig. 4 represents a schematic diagram of the EECM.

C. Recursive least squares based on SOT-SOC co-estimation

To achieve real-time terminal voltage estimation, it is necessary to derive the parameters of the EECM. In this paper, real-time model parameters are obtained using the RLS method. The flowchart of the RLS operation is shown in Fig. 5.

Real-time battery parameters were extracted using the RLS equation (2) to (7). equation (2) represents the transfer function in the EECM, while equation (3) defines the terminal voltage estimation based on the extracted parameters. equations (4), (5), (6), and (7) individually define the parameters, these extracted parameters were then utilized as inputs for the subsequent battery heating model. Lithium-ion batteries dissipate heat to the surroundings through convection, conduction, and radiation. The heat flow from the battery is expressed by equation (8). The heat generated within the battery is described by equation (9) following Bernardi's model, while equation (10) details the heat transfer to the external environment. The entropy coefficient is an important physical property that represents the thermal effects occurring during electrochemical reactions in a battery.

979-8-3315-1612-3/25 $31.00 © 2025 IEEE

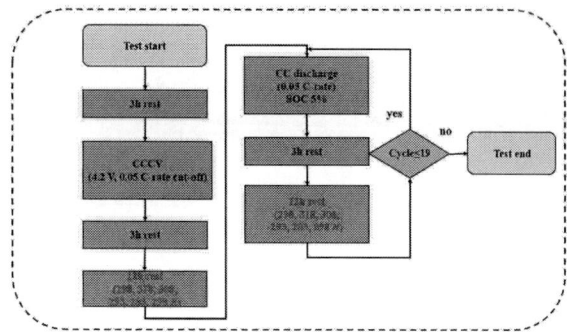

Fig. 6. Reversible heat coefficient test

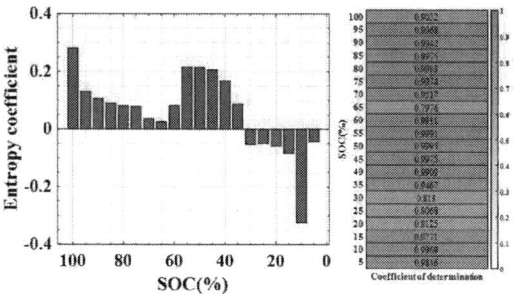

Fig. 7. Reversible heat coefficient test result

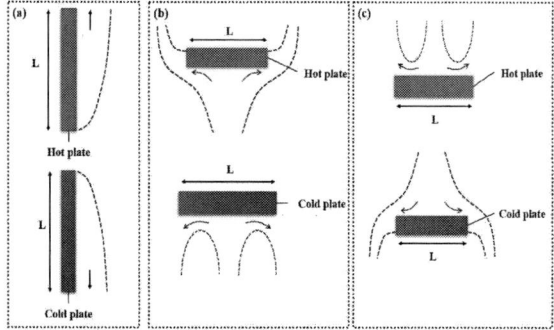

Fig. 8. Natural convection phenomena based on temperature and geometry

When electrochemical reactions such as charging and discharging occur within a battery, they are associated with changes in entropy. Specifically, positive entropy changes during charging and discharging indicate endothermic reactions, while negative entropy changes indicate exothermic reactions. To minimize conduction heat loss, the battery is isolated from the base surface, with heat losses considered only through convection and radiation, as shown in equations (11) and (12). The resistance value in equation (13) is calculated using parameters obtained through the RLS method. The reversible heat coefficient was determined from data collected at various temperatures (298K, 318K, 308K, 293K, 283K , 298K) during 0.05C-rate discharge. Data was collected after 12 hours stabilization period at each temperature, based on thermodynamic theory, and the calorimetric method, which employs a calorimeter [9]. Fig. 6 illustrates a schematic diagram of the experimental setup used to determine the reversible heating value of the battery.

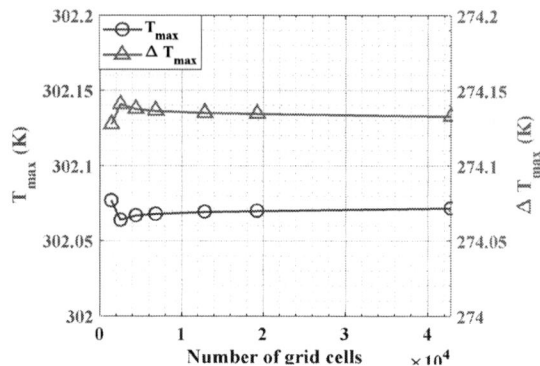

Fig. 9. Grid independence analysis

$$h = \frac{K}{L}(0.68 + \frac{0.67 Ra_L^{\frac{1}{4}}}{(1+(\frac{0.492k}{\mu C_p})^{\frac{9}{16}})^{\frac{4}{9}}} \quad (if\ Ra_L \leq 10^9) \quad (14)$$

$$h = \frac{K}{L}(0.825 + \frac{0.387 Ra_L^{\frac{1}{6}}}{(1+(\frac{0.492k}{\mu C_p})^{\frac{9}{16}})^{\frac{8}{27}}})^2 \quad (if\ Ra_L > 10^9) \quad (15)$$

$$Ra_L = Gr_L \Pr = \frac{g\beta(T_S - T\infty)L_C^3}{\nu}\Pr = \frac{g\beta(T_S - T\infty)L_C^3}{\nu\alpha} \quad (16)$$

$$Gr_L = \frac{g\beta(T_S - T\infty)L_C^3}{\nu^2} \quad (17)$$

$$\Pr = \frac{\nu}{\alpha} \quad (18)$$

During the 298K test, the first and last OCV were measured. For these two measurement points, values with a high entropy correlation coefficient were selected and used. Across the entire SOC range, the OCV extracted during the final 298K measurement exhibited a high correlation coefficient. Fig. 7 depicts the correlation between the calculated reversible heating coefficient of the battery and the battery parameters obtained during the reversible heating test. According to the SOC range, the coefficient was negative in the SOC ≤30% range and positive in the SOC >30% range. The entropy coefficient demonstrated a correlation coefficient of approximately 0.8 or higher across the entire SOC range. The heat loss coefficient (h) corresponds to the convective heat transfer coefficient, which typically ranges between 5 and 25 (W/m²·K) for natural convection. In this study, to derive a more accurate heat transfer coefficient, correlation coefficients were applied individually for each case by calculating the characteristic length based on the geometry of the batteries. Fig. 8 illustrates the characteristics of fluid flow and heat transfer around heat sources and cooling sources. Fig. 8(a) The vertical arrangement exhibits a clear upward and downward circulation of fluid due to natural convection, utilizing the chimney effect. convection, utilizing the chimney effect. Fig. 8(b) In the horizontal arrangement, convection occurs within the air layer heated by the upper hot plate, forming a horizontal flow. Finally, Fig. 8(c) the vertical separation arrangement demonstrates strong convective circulation between the heat source and the cooling source,

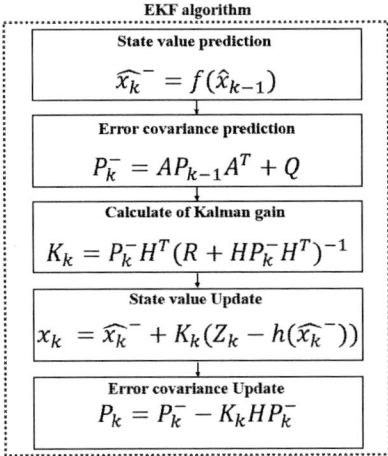

EKF algorithm

State value prediction

$$\widehat{x_k}^- = f(\hat{x}_{k-1})$$

Error covariance prediction

$$P_k^- = AP_{k-1}A^T + Q$$

Calculate of Kalman gain

$$K_k = P_k^- H^T (R + HP_k^- H^T)^{-1}$$

State value Update

$$x_k = \widehat{x_k}^- + K_k(Z_k - h(\widehat{x_k}^-))$$

Error covariance Update

$$P_k = P_k^- - K_k HP_k^-$$

Fig. 10. Schematic diagram EKF algorithm

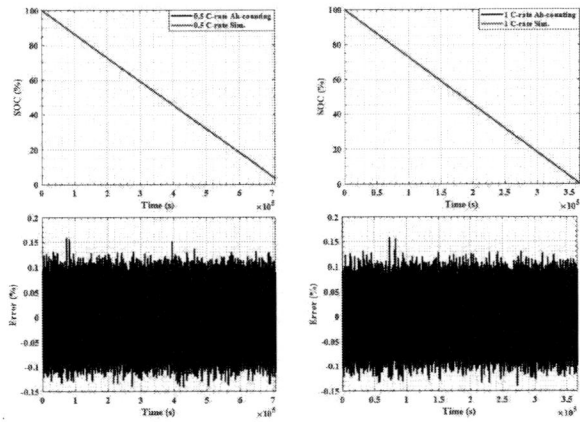

Fig. 11. SOC estimation results

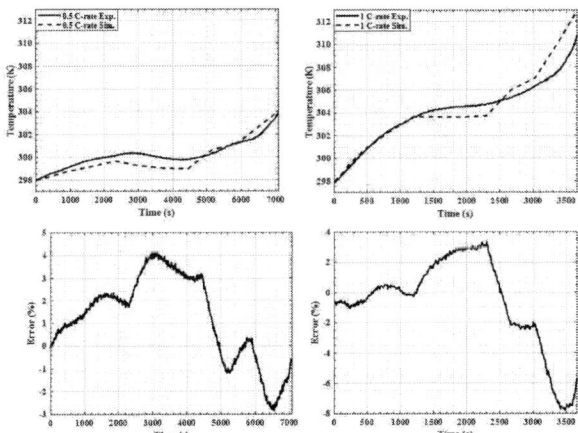

Fig. 12. SOT estimation results

with distinctly defined fluid flow paths. Influenced by the geometric structure of the battery, pouch batteries generate airflow in three different configurations driven by natural convection. equations (14) and (15) provide the formulas for calculating h under natural convection, with the specific equation depending on the value of Rayleigh number (Ra_L). Ra_L is defined as the product of the Grashof number(Gr_L) and the Prandtl number (Pr) as shown in equation (16). The Grashof number represents the relationship between buoyancy and viscosity within the fluid, while the Prandtl number describes the relationship between momentum diffusion and thermal diffusion. Both factors are critical in the thermal analysis of natural convection. However, directly measuring these parameters requires highly precise experimental conditions, making it extremely challenging in practice. Therefore, COMSOL, a widely used simulation software, was utilized for analysis. As shown in Fig. 9, a simulation convergence test was performed based on the number of mesh elements prior to conducting the battery analysis. It was confirmed that as the number of mesh elements increased, the simulation results converged to a constant value. For accuracy in the analysis, a mesh size of 42,789 was selected and applied to subsequent simulations. The method used for battery SOC estimation was the Extended Kalman Filter (EKF). The schematic diagram of the EKF process is presented in Fig. 10. EKF operates by updating the error covariance and adjusting the gain coefficients, allowing the state vector to gradually converge to the true value [10].

III. SOC-SOT ESTIMATION RESULTS

For model verification, the battery SOC-SOT was validated through a battery capacity test. The SOC and SOT of the battery were compared between the actual measured values and the estimated values under 0.5C-rate and 1C-rate discharge conditions. Fig. 11 shows a comparison of SOC under 0.5C-rate and 1C-rate conditions, while Fig. 12 presents a comparison of SOT under the same conditions. Detailed results are provided in Table II. As shown in Table II, the analysis at a 0.5C-rate revealed a maximum error of 4.13% and an average error of 1.95%, indicating that the predictions of battery heat generation are highly accurate when compared to actual measurements.

TABLE II.   ERRORS OF ESTIMATED SOT AND SOC

C-rate	SOT		SOC	
	Max. (%)	MAE (%)	Max. (%)	MAE (%)
0.5C-rate	4.1337	1.9545	0.1579	0.0241
1C-rate	7.8200	2.2092	0.1579	0.0252

Similarly, at a 1C-rate, the maximum error was 7.82%, and the average error was 2.21%. These results further validate the reliability of the proposed model.

IV. CONCLUSIONS

This study compares the experimentally measured core temperature of the battery under two current conditions (0.5C-rate and 1C-rate) with the actual temperature and the temperature estimated by the model. Future research will evaluate the consistency of parameter changes due to battery aging and the accuracy of parameters estimated by the RLS algorithm. The current data, without battery aging, demonstrates that temperature estimation remains within a maximum error of 7.82%. This indicates that the battery parameters extracted through the RLS algorithm accurately represent the actual internal parameters of the battery. This research is expected to contribute by providing fundamental data for assessing the performance of temperature estimation using the RLS algorithm. Future studies will include experiments on battery aging to verify whether the RLS algorithm can accurately estimate parameter variations caused by aging. Additionally, the accuracy of heat generation predictions considering battery aging will be evaluated.

ACKNOWLEDGMENT

This research was supported by the 2024 government (Ministry of Trade, Industry and Energy) through the Korea Institute of Energy Technology Evaluation and Planning (Project No. RS-2024-00398346, Development of ESS Big Data-Based O&M and Asset Management Technology and Workforce Training) and the Korea Evaluation Institute of Industrial Technology (Project No. 00404229, Development of Thermal Management Technology for Large-Capacity Batteries Over 80kWh with Direct Cooling Application).

REFERENCES

[1] Manuel Antonio Perez Estevez, Sandro Calligaro, Omar Bottesi, Carlo Caligiuri and Massimiliano Renzi, "An electro-thermal model and its electrical parameters estimation procedure in a lithium-ion battery cell," Energy, vol. 234, April 2021, 121296

[2] Xinfan Lin, Hector E. Perez, Jason B. Siegel and Anna G. Stefanopoulou, "Robust Estimation of Battery System Temperature Distribution Under Sparse Sensing and Uncertainty," IEEE Transactions on Control Systems Technology, vol. 28, May 2020, pp. 753–765.

[3] Q.L. Yue, C.X. He, M.C. Wu and T.S. Zhao, "Advances in thermal management systems for next-generation power batteries," International Journal of Heat and Mass Transfer, vol. 181, December 2021, 121853

[4] Depeng Kong, Shuhui Wang and Ping Ping, "A novel parameter adaptive method for state of charge estimation of aged lithium batteries," Journal of Energy Storage, vol. 44, December 2021, 103389

[5] Khadija Saqli, Houda Bouchareb, Nacer Kouider M'sirdi and Mohammed Oudghiri Bentaie, "Lithium-ion battery electro-thermal modelling and internal states co-estimation for electric vehicles," Journal of Energy Storage, vol. 63, July 2023, 107072

[6] Ines Baccouche, Bilal Manai, Hyoseong and Najoua Essoukri Ben, "SoC estimation of LFP battery based on EKF observer and a full polynomial Parameters-Model." 2020 IEEE 91st Vehicular Technology Conference, May 2020,

[7] Xin Lai, Yuejiu Zheng and Tao Sun, "A comparative study of different equivalent circuit models for estimating state-of-charge of lithium-ion batteries," Electrochimica Acta, vol. 259, January 2018, pp. 566–577

[8] Hongwen He, Rui Xiong, Hongqiang Guo and Shuchun Li, "Comparison study on the battery models used for the energy management of batteries in electric vehicles," Energy Conversion and Management, vol. 64, December 2012, pp. 113–121

[9] Ukmin Han, Hongseok Choi, Hyoseong Lee and Hoseong Lee, "Inverse Heat Transfer Analysis Method to Determine the Entropic Coefficient of Reversible Heat in Lithium-Ion Battery," International Journal of Energy Research, vol. 2023, February 2023,

[10] Xuan Zheng and Zhuqian Zhang, "State of charge estimation at different temperatures based on dynamic thermal model for lithium-ion batteries." Journal of Energy Storage, vol. 48, April 2022, 104011

Three-stage adaptive control strategy for stability improvement of grid-connected inverter in weak grid

Longxiang You
College of Electrical Engineering
Zhejiang University
Hangzhou, China
22310061@zju.edu.cn

Sicong Jin
College of Electrical Engineering
Zhejiang University
Hangzhou, China
12110035@zju.edu.cn

Xin Zhang
College of Electrical Engineering
Zhejiang University
Hangzhou, China
zhangxin_ieee@zju.edu.cn

Zuoshuai Wang
Wuhan Second Ship Design and Research
Institute
Wuhan, China
perfect403@qq.com

Sunqing Wang
China Ship Scientific Research Center
Wuxi, China
531875031@qq.com

Abstract—With the rapid development of power systems based on new energy, grid-connected inverter (GCI) is extensively utilized in new energy power system. However, the interaction between GCI and the grid impedance may easily induce system instability, especially in the case of weak grid. To improve the adaptability of GCI under different operating conditions, a three-stage adaptive control strategy based on voltage feedforward is proposed, which can switch three adaptive control strategies according to different operating conditions respectively. In each stage, SA-PSO algorithm is utilized to design feedforward path and control parameters in each operating condition sample, and then the polynomial fitting method is utilized to cover the continuous full operating range, which can be used to adaptively adjust the parameters. The proposed adaptive control strategy can ensure the better stability margin of GCI under different conditions. Finally, the OPAL-RT semi-physical experiment platform is used to verify the proposed strategy.

Keywords—grid-connected inverter, weak grid, small signal stability, adaptive control strategy

I. INTRODUCTION

Nowadays, with the ongoing development of distributed power generation systems, the percentage of new energy sources like solar and wind power in the grid is rising. The grid-connected inverters(GCI) have been widely used as an important interface device for new energy power system [1]. However, due to the unpredictable, irregularity, long-distance transmission and other features of the production of new energy power, the power grid gradually presents the characteristics of weak grid, which means the power grid impedance is constantly changing and cannot be ignored [2]. The interaction between grid-connected inverter and grid impedance may easily cause the oscillation instability problem, especially in the case of weak grid condition, which seriously threatens the stable operation of new energy system [3][4].

According to current studies, the impedance mismatch between grid-connected inverter and grid is one of the important reasons for the unstable operation of the inverter in weak grid condition. The impedance-based analysis method is commonly used to determine the cascaded stability between GCI and power grid, to be more specific, the stability is determined by whether the ratio of the grid impedance to the GCI impedance satisfies the (Generalized) Nyquist criterion [5].

When a grid-connected inverter is integrated into the grid, the grid impedance is often different for different point of common coupling(PCC). Moreover, with the addition, removal, and power changes of inverters in the power system, the grid impedance is also time-varying. These factors may lead to the instability of the grid connected inverter, even if the inverter is stable when operating at ideal grid. To solve this problem, the bandwidth of phase-locked loop(PLL) is reduced to improve the stability of grid-connected inverters in [6], while the performance of the inverter will be adversely affected by overly limited PLL bandwidth. On the other hand, a virtual impedance shaping method is proposed to ensure the output impedance of the inverter is passive in [7][8], while complex theoretical calculations are introduced. However, due to the conflict between the grid-connected inverter's cascaded stability and its own performance, the system dynamic performance is often deteriorated while improving the stability and robustness of the cascaded system. It is a problem worth researching how to achieve adaptive control of grid-connected inverters in accordance with different power grid intensities so as to improve the adaptability of inverters to different conditions.

This paper proposed a three-stage adaptive impedance reshaping strategy to improve the adaptability of GCI to different operating conditions. Different adaptive control strategy are adopted respectively, by dividing the unstable operating conditions into three levels: preliminary weak grid, weak grid and extremely weak grid. The simulated annealing-particle swarm optimization(SA-PSO) algorithm is utilized to adaptively optimize the relevant parameters in each operating condition, and then the polynomial fitting method is utilized to deploy the proposed adaptive strategy online. The proposed adaptive strategy can ensure the optimal stability margin of GCI under different power and grid impedance conditions, and the dynamic performance of the system is considered in some degree. Finally, the OPAL-RT experiment platform is used to verify the proposed adaptive strategy.

979-8-3315-1612-3/25 $31.00 © 2025 IEEE

II. SYSTEM MODELING AND STABILITY ANALYSIS

A. System Configuration

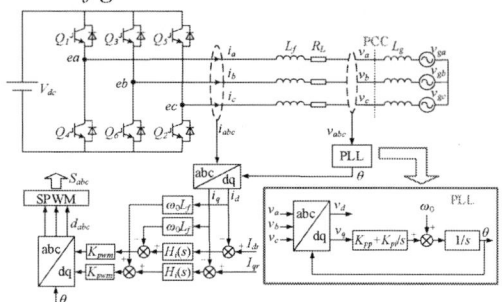

Fig. 1. Three-phase grid-connected inverter system.

The structure and control strategy of L-type three-phase grid-connected inverter are shown in Fig. 1. To identity the phase of grid, a Synchronous Reference Frame Phase-Locked Loop(SRF-PLL) is utilized. And the current is controlled by PI regulator under dq-frame. Where $H_i(s)=K_{ip}+K_{ii}/s$ is the current loop regulator, L_f and R_L are the output filter inductance and its parasitic resistance, respectively. L_g is the power grid side inductance. K_{pwm} is the modulation coefficient. ω_0 is the fundamental frequency of the grid. It should be noted that the grid impedance is considered as of pure inductance, which is utilized to simulate the worst circumstances.

B. Output Impedance Modeling

Due to the dynamics of the PLL, a phase angle difference exists between the dq-frame of the actual system and of the controller. The small signal model of grid-connected inverter system considering PLL dynamics is shown in Fig. 2 [9]. Hence, the dq-frame output impedance of GCI can be derived as:

$$Z_o = -\frac{\overset{\triangledown}{v}_s}{\overset{\triangledown}{i}_s} = (I - G_{PLL}^d V_{dc} - K_{pwm}V_{dc}G_{del}(G_{dec} - G_i)G_{PLL}^i)^{-1} \cdot$$

$$(Z_L - K_{pwm}V_{dc}G_{del}(G_{dec} - G_i)) \tag{1}$$

The corresponding matrix is defined as follows:

$$G_{PLL}^d = \begin{bmatrix} 0 & -D_q^s G_{PLL} \\ 0 & D_d^s G_{PLL} \end{bmatrix} \tag{2}$$

$$G_{PLL}^i = \begin{bmatrix} 0 & I_q^s G_{PLL} \\ 0 & -I_d^s G_{PLL} \end{bmatrix} \tag{3}$$

$$G_{del}(s) = \begin{bmatrix} \dfrac{1-0.75sT_s}{1+0.75sT_s} & 0 \\ 0 & \dfrac{1-0.75sT_s}{1+0.75sT_s} \end{bmatrix} \tag{4}$$

$$G_i(s) = \begin{bmatrix} H_i(s) & 0 \\ 0 & H_i(s) \end{bmatrix} \tag{5}$$

$$G_{dec}(s) = \begin{bmatrix} 0 & -\omega_0 L_f \\ \omega_0 L_f & 0 \end{bmatrix} \tag{6}$$

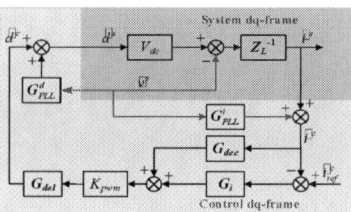

Fig. 2. Three-phase grid-connected inverter system.

$$Z_L(s) = \begin{bmatrix} sL_f + R_L & -\omega_0 L_f \\ \omega_0 L_f & sL_f + R_L \end{bmatrix} \tag{7}$$

$$I = \begin{bmatrix} 1 & 0 \\ 0 & 1 \end{bmatrix} \tag{8}$$

Where G_{del} is the control and sampling delays, T_s is sampling time, G_{PLL}^d and G_{PLL}^i are the conversion transfer function of duty cycle and current in different coordinate systems, respectively.

TABLE I shows the relevant parameters of the grid-connected inverter system. Fig. 3 shows the output impedance bode diagram of the grid-connected inverter. Since the dq-axis decoupling control approach is utilized, the stability of the system is mainly determined by the principal diagonal term. Furthermore, the negative impedance characteristic of $Z_{qq}(s)$ in low frequency band is one of the main reasons for the instability of grid-connected inverters in weak grid condition [10][11]. Hence, this paper only uses q-q axis impedance to carry out stability analysis and adaptive control design.

TABLE I.   SYSTEM PARAMETERS

Symbol	Description	Value
V_g	Grid voltage amplitude	311V
V_{dc}	DC bus voltage	800V
f_s	Switching frequency	10kHz
L_f	Inductance of the output filter	5mH
R_L	parasitic resistance of the filter	1mΩ
P	Active power	14kW
Q	Reactive power	0
K_{ip}	Proportional coefficient of current regulator	7
K_{ii}	Integral coefficient of current regulator	1000
K_{pp}	Proportional coefficient of PLL	1.4
K_{pi}	Integral coefficient of PLL	317
K_{pwm}	Modulation coefficient	1/800

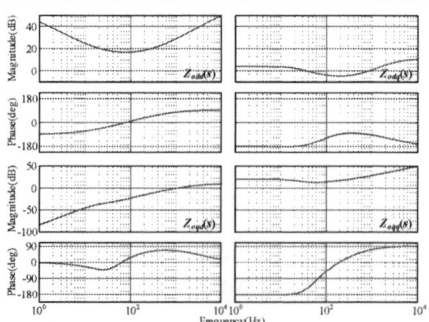

Fig. 3. The output impedance of grid-connected inverter.

C. Stability Analysis

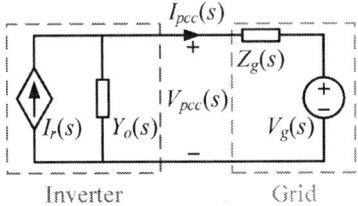

Fig. 4. Equivalent circuit of inverter-grid system.

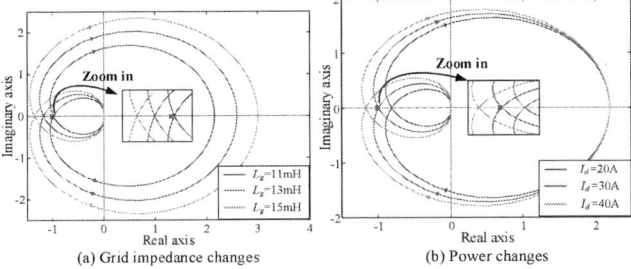

(a) Grid impedance changes (b) Power changes

Fig. 5. The impedance ratio Nyquist curve under different conditions.

Fig. 4 shows the equivalent circuit of inverter-grid system, where $Y_o(s)$ and $Z_g(s)$ represent the inverter output admittance and grid input impedance, respectively. Then, the stability of the cascaded system can be determined by whether the Nyquist curve of the impedance ratio $Y_o(s)Z_g(s)$ surrounds the point $(-1, j0)$ [5]. The Nyquist curve of impedance ratio under different grid impedance and power conditions is shown in Fig. 5, which indicates that the system is becoming instable with the weakening of the grid.

III. THREE-STAGE ADAPTIVE CONTROL STRATEGY

A. Three-stage control strategy description

Fig. 6 shows the implementation steps of the proposed three-stage adaptive control strategy. First, the possible operating

conditions of the inverter is determined and divided into four categories: strong grid(no instability problem), preliminary weak grid, common weak grid and extremely weak grid, which corresponds to the degree of system instability respectively.

In the preliminary weak grid stage, a second-order bandpass filter(BPF) is utilized to limit the influence range of voltage feedforward, which can minimize the influence on the original system so that the dynamic performance can be maintained [12] [13]. However, with the further weakening of the power grid(common weak grid stage), impedance reshaping in a small range is difficult to meet the stability requirements. Then, a voltage feedforward strategy with trap filter is further adopted. The center frequency ω_x of the trap filter is set to 600rad/s, so that the 5th and 7th harmonics in the abc coordinate system can be suppressed by the Park transformation. Furthermore, in extremely weak grid stage, the PLL parameters are introduced to help improve system stability.

B. Off-line design based on SA-PSO

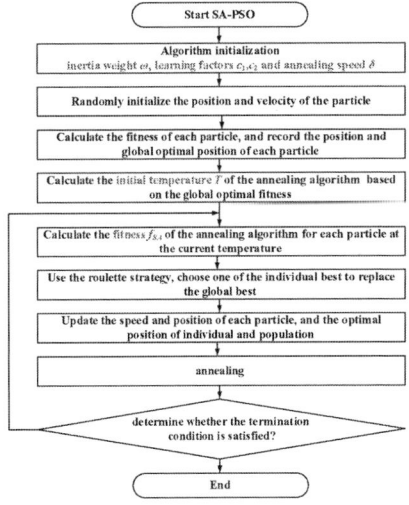

Fig. 7. The flow chart of SA-PSO algorithm.

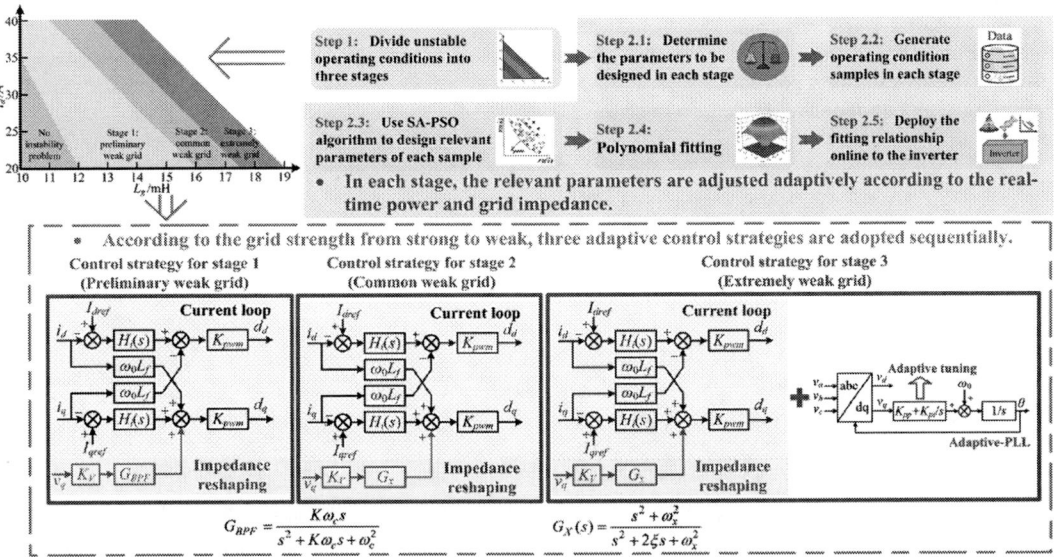

Fig. 6. Schematic diagram of three-stage adaptive control strategy under wide operating range.

In each stage, several sample points were selected, and the simulated annealing particle swarm optimization (SA-PSO) algorithm was used to design the feedforward path and PLL parameters. By introducing SA into PSO, the PSO algorithm can accept inferior solutions with a certain probability, which is conducive to breaking out of the local optimal solution and improving its global search capability [14]. The flow chart of SA-PSO algorithm is shown in Fig. 7.

Taking stage 1 adaptive control as an example, the parameters to be designed are K_v, K and f_c respectively. Based on experience, the parameter ranges are determined as follows: $K_v \in [0,2]$, $K \in [0,15]$, $f_c \in [f_{c\_init} - 2, f_{c\_init} + 2]$, where $\omega_c = 2\pi f_c$ is the initial center frequency of BPF determined by the impedance intersection frequency of the q-q axis. Relevant parameters can be designed in the same way in other stages.

Some crucial processes in SA-PSO algorithm are explained below:

1) Determination of annealing fitness f_{SA}: The annealing fitness function of each particle under the current temperature T is given as:

$$f_{SA}(P_{id}) = \frac{e^{-\frac{fitness(P_{id}) - fitness(P_{pd})}{T}}}{\sum_{i=1}^{n} e^{-\frac{fitness(P_{id}) - fitness(P_{pd})}{T}}} \quad (9)$$

Where P_{id} represents the position of the i-th particle, P_{pd} represents the global optimal position, and $fitness(P_{pd})$ represents its corresponding fitness function value.

2) Objective function:

It is worth noting that since the stability of the cascaded system and the dynamic performance of the system are often contradictory, the proposed strategy achieves a balance of dynamic performance by setting a reasonable cascaded stability margin. In order to comprehensively consider the cascaded stability and the dynamic performance of the system, 30° and 2dB is defined as the optimal phase margin and amplitude margin of the cascaded system. Furthermore, K and K_v are thought to be small as excellent in order to reduce the range of impact on the original system's impedance. Based on the above analysis, the fitness function can be defined as:

$$fitness = \alpha \left| G_{m\_c} - 2 \right| + \beta \left| P_{m\_c} - 30 \right| + \gamma |K| + \chi |K_v| \quad (10)$$

Where G_{m_c} and P_{m_c} represent the cascaded amplitude margin and phase margin of grid-connected inverter system, respectively. α, β, γ and χ are the weight coefficients, in this case $\alpha=6$, $\beta=4$, $\gamma=3$, $\chi=6$.

C. Online deployment based on polynomial fitting

Polynomial fitting of the offline design results is required in order to achieve continuous full operating condition coverage. The independent variables are determined to be grid impedance L_g and active current I_d, while the dependent variables are the center frequency f_c, passband width K, and feedforward coefficient K_v respectively. Taking the center frequency f_c as an example, in the fitting process, the highest power of L_g and I_d are

taken to the 5th and 3rd degrees respectively, and the polynomial is shown below:

$$
\begin{aligned}
f_c = f(L_g, I_d) =\ & p_{00} + p_{10}L_g + p_{01}I_d + p_{20}L_g^2 \\
& + p_{11}L_g I_d + p_{02}I_d^2 + p_{30}L_g^3 + p_{21}L_g^2 I_d + p_{12}L_g I_d^2 \\
& + p_{03}I_d^3 + p_{40}L_g^4 + p_{31}L_g^3 I_d + p_{22}L_g^2 I_d^2 + p_{13}L_g I_d^3 \\
& + p_{50}L_g^5 + p_{41}L_g^4 I_d + p_{32}L_g^3 I_d^2 + p_{23}L_g^2 I_d^3
\end{aligned} \quad (11)
$$

Fig. 8 to Fig. 10 show the fitting surfaces of adaptive adjustment parameters in each stage. According to theoretical study, impedance intersecting frequency decreases with increasing power grid impedance or active current. Hence, from the lightest working condition (L_g=10mH, I_d=20A) to the worst working condition, the center frequency f_c in Fig. 8(a) exhibits a trend of progressively decreasing. Moreover, the bandpass filter's action range and feedforward strength must be increased under harsher circumstances in order to maintain the same stability margin, in other words, K and K_v must be as high as feasible. The above analysis proves the rationality of fitting results, and Fig. 9 and Fig. 10 can be explained in the same way.

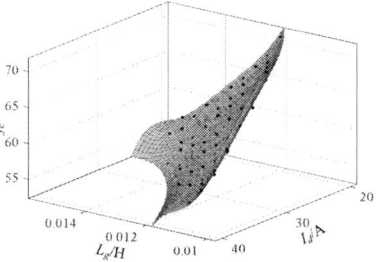

(a) f_c fitting surface

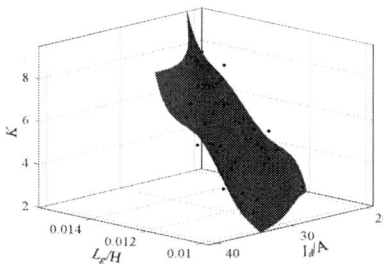

(b) K fitting surface

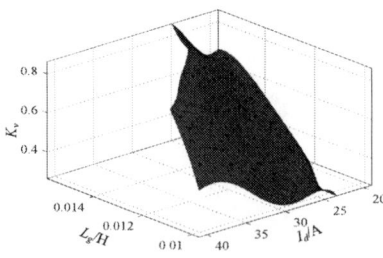

(c) K_v fitting surface

Fig. 8. The fitting surface of relevant parameters in stage 1.

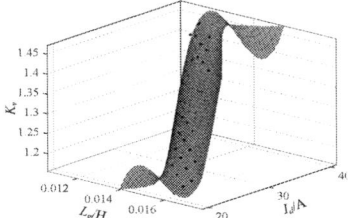

Fig. 9. The fitting surface of K_v in stage 2.

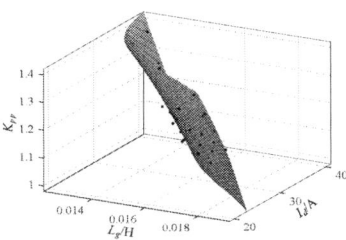

(a) K_{pp} fitting surface

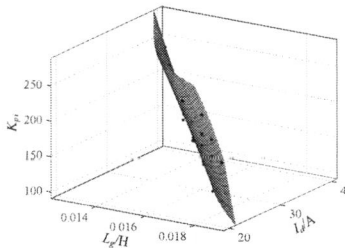

(b) K_{pi} fitting surface

Fig. 10. The fitting surface of relevant parameters in stage 3.

IV. EXPERIMENT VERIFICATION

Several typical operating conditions are constructed to verify the effectiveness of proposed adaptive control strategy. Experiments were carried out on the OPAL-RT semi-physical experimental platform as shown in Fig. 11.

Through comparative experiments, Fig. 12 to Fig. 17 show that the proposed strategy can ensure the stability of the system under different operating conditions. Moreover, the advantages of the proposed strategy in ensuring dynamic performance are proved by comparing with other stabilization strategy. Fig. 18 and Fig. 19 show the dynamic waveforms of the operating condition change under various control strategies , respectively, which indicates that the proposed strategy can ensure good dynamic settling time as much as possible.

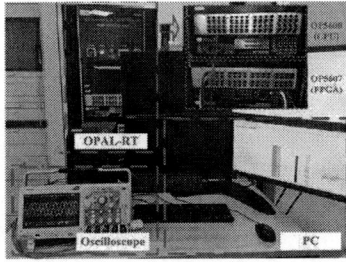

Fig. 11. Experimental platform.

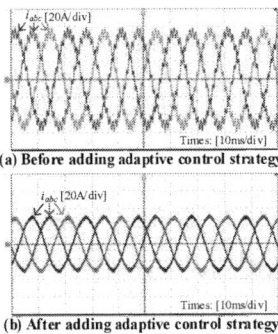

(a) Before adding adaptive control strategy

(b) After adding adaptive control strategy

Fig. 12. Three-phase current waveform when I_d=30A, L_g=12mH.

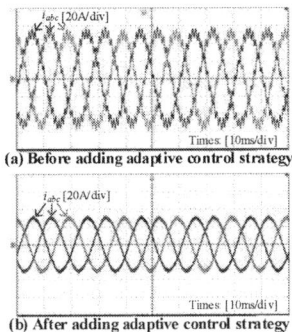

(a) Before adding adaptive control strategy

(b) After adding adaptive control strategy

Fig. 13. Three-phase current waveform when I_d=30A, L_g=14mH.

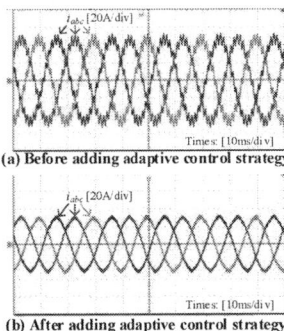

(a) Before adding adaptive control strategy

(b) After adding adaptive control strategy

Fig. 14. Three-phase current waveform when I_d=30A, L_g=16mH.

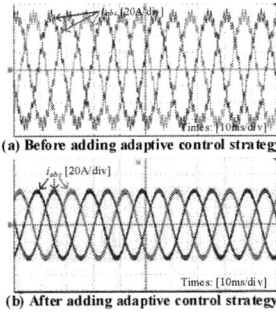

(a) Before adding adaptive control strategy

(b) After adding adaptive control strategy

Fig. 15. Three-phase current waveform when I_d=40A, L_g=10mH.

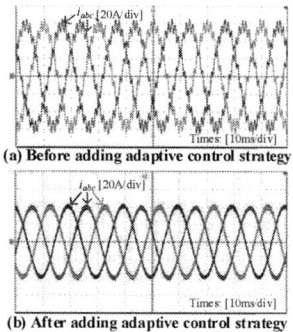

(a) Before adding adaptive control strategy

(b) After adding adaptive control strategy

Fig. 16. Three-phase current waveform when I_d=40A, L_g=12mH.

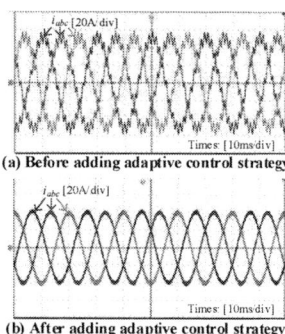

(a) Before adding adaptive control strategy

(b) After adding adaptive control strategy

Fig. 17. Three-phase current waveform when I_d=40A, L_g=14mH.

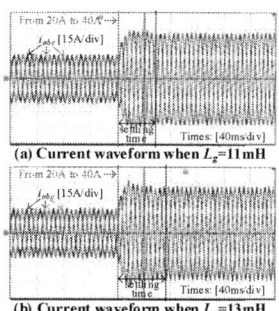

(a) Current waveform when L_g=11mH

(b) Current waveform when L_g=13mH

Fig. 18. Waveforms under power variation using proposed adaptive strategy.

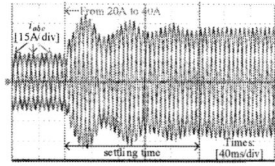

Fig. 19. Waveforms under power variation using controller parameter adjustment strategy(K_{pp}=1,K_{pi}=50).

V. CONCLUSION

This paper presents an adaptive control strategy for improving the stability of grid-connected inverter under variable operating conditions. The adaptive strategy effectively solves the problem that the traditional instability prevention process does not consider the condition fluctuation, and achieves the balance between stability and dynamic performance. Finally, the effectiveness of the proposed strategy is demonstrated through a number of comparative experiments under typical working conditions.

REFERENCES

[1] F. Blaabjerg, R. Teodorescu, M. Liserre and A. V. Timbus, "Overview of Control and Grid Synchronization for Distributed Power Generation Systems," *IEEE Transactions on Industrial Electronics*, vol. 53, no. 5, pp. 1398-1409, Oct. 2006.

[2] Z. Zhang, P. Wang, P. Jiang, F. Gao, L. Fu and Z. Liu, "Robust Control Method of Grid-Connected Inverters With Enhanced Current Quality While Connected to a Weak Power Grid," *IEEE Transactions on Power Electronics*, vol. 37, no. 6, pp. 7263-7274, June 2022.

[3] X. Wang, K. Qin, X. Ruan, D. Pan, Y. He and F. Liu, "A robust grid-voltage feedforward scheme to improve adaptability of grid-connected inverter to weak grid condition," *IEEE Transactions on Power Electronics*, vol. 36, no. 2, pp. 2384-2395, Feb. 2021.

[4] Z. Lin, X. Ruan, L. Wu, H. Zhang and W. Li, "Multi resonant component-based grid-voltage-weighted feedforward scheme for grid-connected inverter to suppress the injected grid current harmonics under weak grid," *IEEE Transactions on Power Electronics*, vol. 35, no. 9, pp. 9784-9793, Sept. 2020.

[5] J. Sun, "Impedance-based stability criterion for grid-connected inverters". *IEEE Transactions on Power Electronics*, 2011, vol. 26, no. 11, pp. 3075-3078.

[6] X. Wang, L. Harnefors and F. Blaabjerg, "Unified Impedance Model of Grid-Connected Voltage-Source Converters," *IEEE Transactions on Power Electronics*, vol. 33, no. 2, pp. 1775-1787, Feb. 2018.

[7] C. Wang, X. Wang, Y. He, D. Pan, H. Zhang, X. Ruan and X. Chen, "Passivity-Oriented Impedance Shaping for LCL-Filtered Grid-Connected Inverters," *IEEE Transactions on Industrial Electronics*, vol. 70, no. 9, pp. 9078-9090, Sept. 2023.

[8] A. Akhavan, J. C. Vasquez and J. M. Guerrero, "A simple method for passivity enhancement of current controlled grid-connected inverters," *IEEE Transactions on Power Electronics*, vol. 35, no. 8, pp. 7735-7741, Aug. 2020.

[9] B. Wen, D. Boroyevich, R. Burgos, P. Mattavelli and Z. Shen, "Analysis of d-q small-signal impedance of grid-tied inverters," *IEEE Transactions on Power Electronics*, vol. 31, no. 1, pp. 675-687, Jan. 2016.

[10] B. Wen, D. Dong, D. Boroyevich, R. Burgos, P. Mattavelli and Z. Shen, "Impedance-Based Analysis of Grid-Synchronization Stability for Three-Phase Paralleled Converters," *IEEE Transactions on Power Electronics*, vol. 31, no. 1, pp. 26-38, Jan. 2016.

[11] P. Yan, H. Wang, W. Sun, Y. Zhu and X. Ge, "A Method for Improving Stability of Grid-Connected Inverters Based on the Q-Axis Voltage Feedforward Control," *2020 IEEE 9th International Power Electronics and Motion Control Conference (IPEMC2020-ECCE Asia)*, Nanjing, China, 2020, pp. 3304-3309.

[12] Y. Hu, J. Xu, H. Qian, S. Bian and S. Xie, "Robustness and harmonics suppression of grid-connected inverters with different grid voltage feedforward compensations in weak grid," *2020 IEEE 29th International Symposium on Industrial Electronics (ISIE)*, 2020, pp. 779-784.

[13] X. Wu, X. Li, X. Yuan and Y. Geng, "Grid harmonics suppression scheme for LCL-type grid-connected inverters based on output admittance revision," *IEEE Transactions on Sustainable Energy*, vol. 6, no. 2, pp. 411-421, April 2015.

[14] S. Sudibyo, M. N. Murat and N. Aziz, "Simulated annealing-Particle Swarm Optimization (SA-PSO): Particle distribution study and application in Neural Wiener-based NMPC," *2015 10th Asian Control Conference (ASCC)*, Kota Kinabalu, Malaysia, 2015, pp. 1-6.

Degradation Analysis of Offshore Bifacial PV Modules under Multiple Climatic Stressors

Aidha Muhammad Ajmal, and Yongheng Yang

College of Electrical Engineering, Zhejiang University, Hangzhou 310058, China

Email: ama@zju.edu.cn, yoy@zju.edu.cn

Abstract—Offshore photovoltaic (PV) systems—especially those that use bifacial PV modules—offer tremendous promise for improving solar energy capture by capturing direct and reflected sunlight from the front and rear sides, respectively. Even so, the severe sea climate presents significant difficulties for solar systems as it expedites the degradation of PV modules and eventually lowers its operating efficiency. This paper presents a technique for predicting power degradation in offshore bifacial PV modules that considers salt stress and other climate factors. In this study, a cumulative damage model was tested to measure the combined impact of climatic stressors, including temperature fluctuations, relative humidity, ultraviolet radiation, and saltwater exposure, on offshore bifacial PV module's performance, in contrast to current and previous approaches that examine climatic effects on onshore monofacial PV modules separately. A genetic algorithm is adopted to optimize model parameters, achieving high predictive accuracy. Real-world data validates the model, revealing a forecast accuracy of less than 1.1%.

Index Terms—Offshore photovoltaic (PV) modules, bifacial PV module degradation, salinity stress, climatic stressors, damage model

I. INTRODUCTION

Photovoltaic (PV) systems reduce greenhouse gas emissions and dependency on fossil fuels, assisting the global transition to sustainable energy sources. Offshore PV systems have recently attracted much attention as a possible solution to land scarcity, particularly in densely populated places, by leveraging the abundantly available solar irradiation at sea [1]. In offshore applications, the water surface provides a suitable environment for the reflection of solar irradiation, making the bifacial PV modules ideal for offshore environments. These modules capture sunlight on both the front and rear sides, enhancing the efficiency of the PV system and increasing energy production [2].

Once the PV modules are installed in such systems, they become more susceptible to multiple climatic stresses like saltwater exposure, high relative humidity (RH), temperature fluctuations, and intense ultraviolet (UV) radiation, which contributes significantly to the decline in their performance and gradual degradation. In addition to the defects resulting from the manufacturing processes, which contribute to reducing the lifespan of these modules [3]. These climatic factors cause corrosion of metal components, delamination of protective layers, and degradation of solar cell materials [4]. Also, constant exposure to strong winds and waves exerts mechanical stress, potentially causing micro-cracks in PV modules and structural fatigue [5]. Moisture intrusion speeds up the degradation of PV modules, which is a particularly serious problem in offshore settings. The encapsulant characteristics, including the water vapor diffusion coefficient, water vapor transmission rate, and the flow of moisture between the glass and the solar cell, are significantly responsible for the rate of moisture infiltration. These elements work together over time to lower energy conversion efficiency and limit offshore PV systems' operating lifespan [6].

The infiltration of saltwater into the PV modules contributes to the corrosion of wires, frames, and supporting structures, which greatly affects the performance of electrical conduction, in addition to exposing the structures of the PV modules to damage. Also, salt deposits form a layer on the surfaces of the PV modules, which makes them receive non-uniform irradiation, causing the problem of partial shading. To mitigate these effects, offshore PV modules require specialized anti-corrosion materials, protective coatings, and regular maintenance to mitigate the impact of salinity and extend the system's operational lifespan. The impacts of environmental stressors have been studied separately using a variety of degradation rate models [7]. Most of these models are based on physical properties to predict the damage modes caused by these stresses [8]. For instance, the Arrhenius model is used to explain how temperature affects PV modules [9]. The majority of these degradation models concentrate on specific failure modes and are usually carried out in controlled indoor accelerated testing environments.

Despite advancements in materials and protective technologies, degradation remains a critical challenge, especially for offshore bifacial PV modules. The effects on individual climatic stressors have been extensively studied, but less is known about the combined impacts of different dynamic stresses, including salinity, temperature, RH, and UV radiation. To build more robust offshore PV systems, it is essential to comprehend how these stresses interact and lead to degradation. In order to forecast field degradation rates of offshore bifacial PV modules, this paper presents a cumulative damage model that bridges the gap between the individual physical degradation rate models, which were previously applied to conventional onshore PV modules. We will consider variables such as module temperature, temperature fluctuations, RH, UV, and the effects of salt accumulation. To refine the presented model, a genetic algorithm (GA) is applied to estimate the

979-8-3315-1612-3/25 $31.00 © 2025 IEEE

unknown model parameters, minimizing the error term and enhancing accuracy to predict the degradation rates and the lifespan of offshore bifacial PV modules. The presented model is validated by comparing predicted outcomes with actual measured data, achieving a forecast error within ±1.1%.

The rest of this paper is organized as follows: Section II presents the mathematical modeling based on the damage model, climatic stressors, and power model. Section III explains the system datasets and also validates the model, and discusses the results. Section IV concludes the paper.

II. MATHEMATICAL MODELING

A. Damage Model

To demonstrate that cumulative effects are suitable for specific degradation models, like chemical breakdown and wear-out processes, Nelson [10] created the cumulative degradation modeling method within the accelerated failure time framework. The authors in [11] expanded the cumulative damage model, taking into consideration the effects caused by different climatic factors. This method works especially well in settings where several stressors are present at the same time. The cumulative degradation model can be represented as

$$D_a(t) = D_p(t) + \varepsilon(t) \tag{1}$$

where $D_a(t)$ is the actual degradation of the PV module, $D_p(t)$ is the predicted degradation of the PV module, and $\varepsilon(t)$ is the error between actual and predicted values.

To comprehensively account for the effects of all environmental stressors, the cumulative function D_p caused by these stressors can be given as

$$D_p(t) = \int_0^t f(X(t), \alpha) \ dt \tag{2}$$

where $f(X(t), \alpha)$ represents the effects of the environmental stressors at time t, and $X(t)=[X_1(t), X_2(t), ..., X_n(t)]$, $(n=1 \ to \ m)$, represents n different climatic stressors affecting the PV module. Here, the parameter α corresponds to the degradation parameter associated with each environmental stressor. Accordingly, the cumulative degradation model in (1) can be rewritten as

$$D_a(t_{ij}) = D_p(t_{ij}, X_i(t_{ij})) + \varepsilon_i(t_{ij}) \tag{3}$$

$$D_p(t_{ij}, x_i(t_{ij})) = \alpha_0 + \int_0^t \prod_{l=1}^n f_l(x_{il}(i), \alpha_l) \ dt \tag{4}$$

in which α_0 is the initial power degradation value, α_i is the respective parameter for function f, n is the number of climatic stressors, and $\varepsilon_i(t_{ij})$ is the error term. This allows the integration of multiple dynamic climatic factors, enabling more accurate prediction of degradation over time, particularly in offshore environments where stressors interact in complex ways.

B. Climatic Stresses Models

To build a precise forecast model for the degradation of offshore bifacial PV modules, a detailed analysis of the primary climatic stressors is required. The majority of models for these

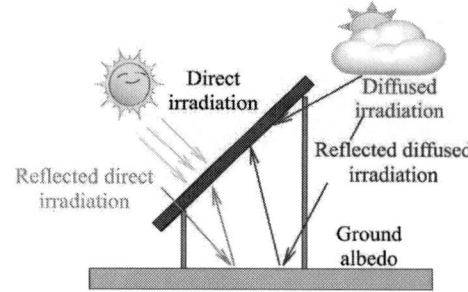

Fig. 1. Schematic diagram of irradiance acquisition for the bifacial PV module.

climatic stressors are based on physical and chemical methods. Nevertheless, chemical modeling is challenging because it requires the description of degrading properties, e.g., activation energy, using complex functions. As a result, physical modeling approaches are preferred. These models consider several significant factors, including module temperature, temperature variations, wind speed, salinity, RH, and UV exposure.

With the above, the Sandia temperature model is used to model the PV module temperature [12], which is heavily affected by wind speed and total irradiance:

$$T_m = T_{amb} + G_T(e^{a+bw_s}) \tag{5}$$

where T_m is the PV module temperature (°C), T_{amb} is the ambient temperature (°C), a and b are the parameters determined by the construction and materials of the module, w_s is the wind speed (m/s), and G_T is the total irradiance received from the front and rear sides of the bifacial PV module.

Fig. 1 shows how the total irradiance is distributed over the bifacial PV module. For the irradiance modeling of the bifacial PV module, weather data especially global horizontal irradiance (GHI) and diffuse horizontal irradiance (DHI) are required as the basic inputs of the model [13]. Usually, UV irradiation is considered to be 5% of the total irradiance [14]. The total irradiation of the bifacial PV module is then computed as

$$G_T = G_F + \varphi G_R \tag{6}$$

Here, G_F and G_R represent the irradiance received by the front and rear sides of the bifacial PV module, respectively, and φ is the bifaciality factor. These values can be further evaluated by

$$G_F = G_{F,b} + G_{F,d} + G_{F,grd} \tag{7}$$

$$G_F = \frac{S_{DNI}cos\theta}{sin\gamma} + \frac{S_{DIF}cos(1+cos\beta)}{2} + \frac{aS_{GHI}cos(1+cos\beta)}{2} \tag{8}$$

$$G_R = G_{R,b} + G_{R,d} + G_{R,grd} \approx G_{R,d} + G_{R,grd} \tag{9}$$

$$G_R = \frac{S_{DIF}cos(1+cos\beta)}{2} + \frac{aS_{GHI}cos(1-cos\beta)}{2} \tag{10}$$

where F and R refer to the front and rear sides, respectively, $G_{F,b}$, $G_{F,d}$, $G_{F,grd}$ is the beam irradiance, diffuse irradiance, and ground-reflected irradiance of the front side of the bifacial PV module, respectively, S_{GHI} is global horizontal irradiance, S_{DNI} is the direct horizontal irradiance, S_{DIF} is the diffused irradiance, α is the ground surface albedo, β is the tilt angle, γ the incidence angle of direct sunlight, and θ is the solar altitude angle.

To account for the thermal stressors, the effects of PV module temperature and cyclic temperatures are modeled using the Arrhenius [15] and Coffin-Manson models [16]. The Arrhenius model is expressed as

$$D_R(t) = A \times exp\left(-\frac{E_a}{k_B T}\right) \tag{11}$$

in which D_R is the degradation rate (% per year), E_a is the activation energy during the degradation process (eV), T is the PV module temperature (K), and k_B is the Boltzmann constant (8.62×10^{-5} eV/K). For cyclic temperature (N) stresses, the Coffin-Manson model is applied as

$$N = \frac{1}{\Delta T^\gamma} \tag{12}$$

where ΔT represents the module cyclic temperature (K), A, and γ are material-dependent parameters.

The effect of relative humidity (RH) is modeled using the Peck's model [17] as

$$D_R(T, RH) = A \times exp\left(-\frac{E_a}{k_B T}\right) \times RH^n \tag{13}$$

with RH being the relative humidity (%) and n being the unknown material-dependent parameter.

To capture the combined effects of multiple climatic stressors, the overall degradation is formulated based on the Norris-Landzberg model [18] as

$$D_R(T, \Delta T, UV, RH) = \alpha_1 exp\left(-\frac{\alpha_2}{k_B T}\right) \times \Delta T^{\alpha_3} \times UV^{\alpha_4} \times RH^{\alpha_5} \tag{14}$$

where $D_R(T, \Delta T, UV, RH)$ is the reaction rate, UV is the average ultraviolet irradiance (W/m²), and RH is the average relative humidity (%). The parameters α_1 to α_5 represent the frequency factor, activation energy, and the effects of cyclic temperature, UV radiation, and RH, respectively. These parameters are typically within the ranges specified in [19], [20].

Given the offshore environment, the effect of salinity concentration on module degradation is also considered. Following a similar functional form to the Peck's model with the Arrhenius term, the cumulative degradation model for offshore PV modules is then expressed as
Here, SA represents the salinity accumulation (mg/m²) and α_6 is the salinity impact parameter.

$$D_R(T, \Delta T, UV, RH, SA) = \\ \alpha_1 exp\left(-\frac{\alpha_2}{k_B T}\right) \times \Delta T^{\alpha_3} \times UV^{\alpha_4} \times RH^{\alpha_5} \times SA^{\alpha_6} \tag{15}$$

A more accurate prediction of the degradation rate is made possible by adding the salinity effects in the cumulative model, especially in offshore settings where salinity plays an important role. This approach guarantees that the combined effects of salt accumulation and other environmental stressors, like temperature, UV radiation, RH, and cyclic temperature fluctuations, are taken into consideration in the long-term performance and reliability research of bifacial PV modules [21].

C. Power Model

As the module power at the maximum power point is a widely employed metric by manufacturers for warranty, it has been chosen as the indicator for degradation. Therefore, degradation is characterized as the progressive decline in the maximum power of the PV modules over time. The main method of evaluating the degradation is to measure the initial power loss brought on by several climatic stressors, such as salt, temperature, RH, and UV exposure, which all contribute to a progressive decline in the generated power. The relation between the output power as a function of time can be found in [22] as

$$\frac{P(t)}{P_{ini}} = 1 - exp\left[-\left(\frac{B}{D_R t}\right)^\beta\right] \tag{16}$$

with $P(t)$ and P_{ini} being the PV module output power at time t and the initial output power, respectively; B being the material property; β being the shape parameter that describes the degradation behavior over time; D_R being the degradation rate constant for each degradation process.

While this constant-rate degradation model provides a basic understanding of power losses over time, it is less accurate for long-term forecasting. In real-world applications, the degradation rate often varies with time due to the cumulative effects of dynamic environmental conditions. For instance, degradation caused by salinity or UV exposure may accelerate as module materials weaken. However, for long-term degradation forecasting, a time dependent degradation rate produces more accurate predictions than using a constant degradation rate. Hence, the failure time (FT) model is suggested as

$$FT = \frac{B}{D_R\left(\left|ln(0.2)\right|\right)^{1/\beta}} \tag{17}$$

For the model in (17), the mean failure time function is derived by considering failure as a 20% loss of the initial power.

III. MODEL VALIDATION

A. System Datasets

To validate the effectiveness of the proposed model and due to the limited availability of historical data for bifacial PV modules in both onshore and offshore applications, we use the performance data of monofacial PV modules obtained from a power plant in Guangzhou, China [23]. The total annual average output power of 21 PV modules was used to determine their degradation rates after operating 16 years during the period from

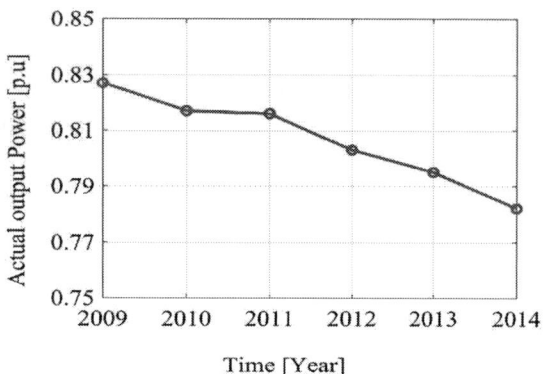

Fig. 2. Actual output power (per unit (p.u)) from 2009 to 2014.

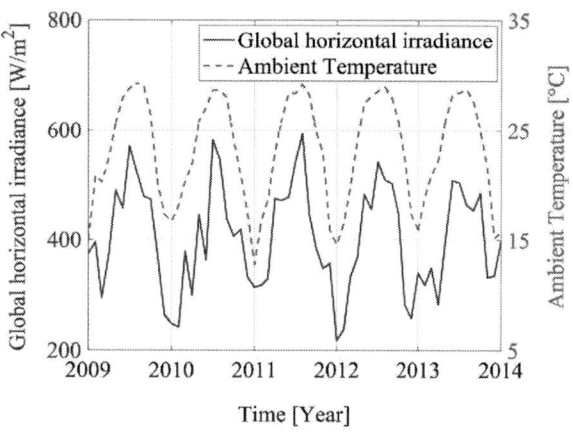

(a)

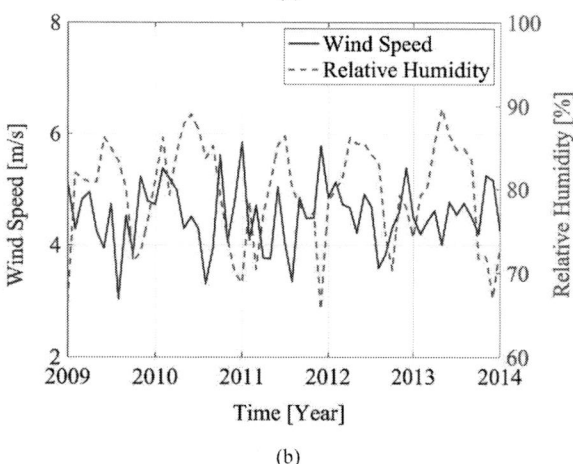

(b)

Fig. 3. Offshore monthly weather dataset from 2009 to 2014: (a) the monthly average global horizontal irradiance (GHI, W/m²) and the monthly ambient temperature (°C), and (b) the monthly relative humidity (%) and the monthly wind speed (m/s).

2009 to 2014 as presented in Fig. 2. In conjunction with this, monthly mean values of relevant climatic stressors, including temperature, RH, UV radiation, and wind speed, were collected from China meteorological administration database [24] for the same period, as illustrated in Fig. 3.

B. Parameters Evaluation

A genetic algorithm (GA) is employed to estimate the parameter values in the cumulative degradation model (based on the Peck model with the Arrhenius term) using the ordinary least squares (OLS) method [25]. The discretized data model is given as

$$D_p(t_{ij}, x_i(t_{ij})) = \alpha_0 + \sum_I^n \alpha_I exp\left(-\frac{\alpha_2}{k_B T}\right) \times \Delta T^{\alpha_3} \times UV^{\alpha_4} \times RH^{\alpha_5} \times SA^{\alpha_6}$$

(18)

The GA operates by generating a population consisting of sets of possible parameter values and applying genetic operations such as reproduction, crossover, and mutation. Through these operations, the algorithm iterates to identify the member with the smallest fitness value, where fitness is defined as the sum of squared losses. The OLS method minimizes the squared loss to optimize the unknown parameters of the developed model. In this case, the number of climatic stressors (n) is 5, and α_0 is set to 0 in this study because the performance data of PV modules is taken after working 16 years [23]. The fitted parameter values, along with their corresponding upper and lower confidence intervals (UCI and LCI), are presented in Table I.

TABLE I. PARAMETERS OF THE DEGRADATION MODEL

Parameter	α_1	α_2	α_3	α_4	α_5	α_6
Optimal value	0.80	0.36	1.01	0.5	0.8	0.1
UCI (+95%)	0.851	0.412	1.06	0.55	0.85	0.29
LCI (-95%)	0.753	0.314	0.96	0.45	0.75	0.01

To verify the correctness of the parameter values, the predicted annual degradation is compared with the actual degradation, as indicated in Table II. It is evident in Table II that the general accuracy of these predicted values is acceptable, given the limited degradation information except for the 2011 data point. There was no discernible variation in the climatic stressors in 2011. Therefore, it may be ruled out as an outlier even if the predicted values produced by the model differ too much from the actual value. The blue texts in Table II indicate the degradation estimates with the lowest relative deviation (RD). According to the corresponding RD in Table II, the mathematical model developed in this paper reduces the cumulative RD by 23.46%. This proves the model's benefit.

TABLE II. COMPARISON BETWEEN ACTUAL AND PREDICTED DEGRADATION RATES FROM 2009 TO 2013

Year	D_a (%/year)	D_p (%/year)	RD (%)	UCI (+95%)	LCI (-95%)
2009	1.203	1.19	1.08	1.21	1.14
2010	1.086	1.11	2.16	1.12	1.06
2011	0.532	1.08	103	1.10	1.06
2012	1.009	1.07	6.04	1.12	1.02
2013	1.629	1.11	31.86	1.16	1.06

C. Results and Discussion

Creating a comprehensive model encompassing various requirements to incorporate advancements in materials and designs is challenging. Additionally, validating such a model necessitates extensive datasets. In this section, the Sandia temperature model of (5) was used to estimate the temperature of the PV modules at the considered environmental conditions, and the irradiation models of (6) to (10). The module temperature (T_m), the maximum temperature (T_{max}), the minimum temperature (T_{min}), and total irradiance using these models are shown in Figs. 4 and 5, respectively, for bifacial and monofacial PV modules.

Using the parameters in Table I as well as the climate variables from the site (see Fig. 3), the model in (14) and (15) are applied to estimate average annual degradation rates for the bifacial and monofacial PV modules before and after salinity stress under offshore settings. Fig. 6 presents the differences in average monthly degradation rates of the bifacial and monofacial PV modules before and after adding salinity stress, respectively, where the relative differences in the predicted

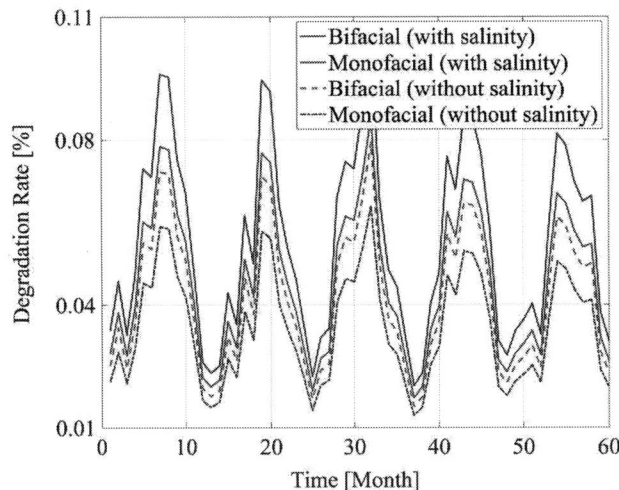

Fig. 6. Average monthly degradation rates of bifacial and monofacial PV modules with and without salinity effect.

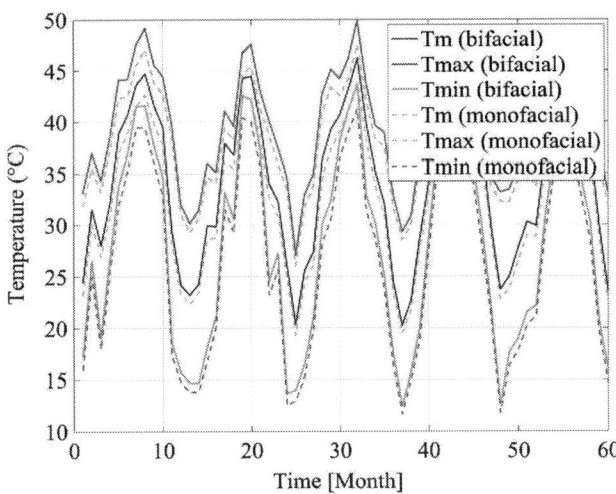

Fig. 4. The module temperature (T_m), maximum temperature (T_{max}), and minimum temperature (T_{min}) (°C) of the bifacial and monofacial PV modules using the Sandia model.

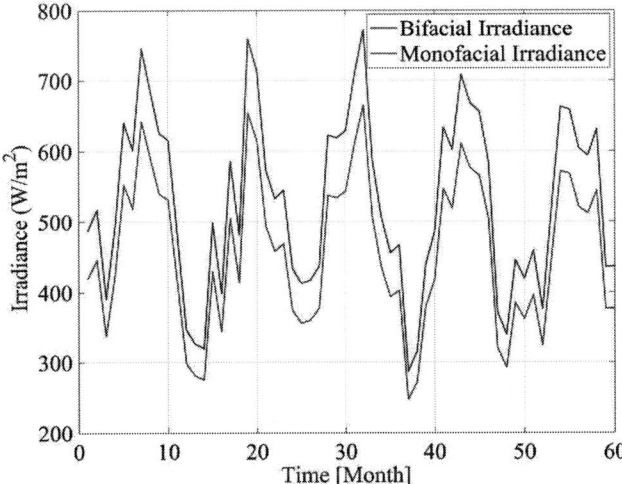

Fig. 5. Bifacial and monofacial irradiance (W/m²).

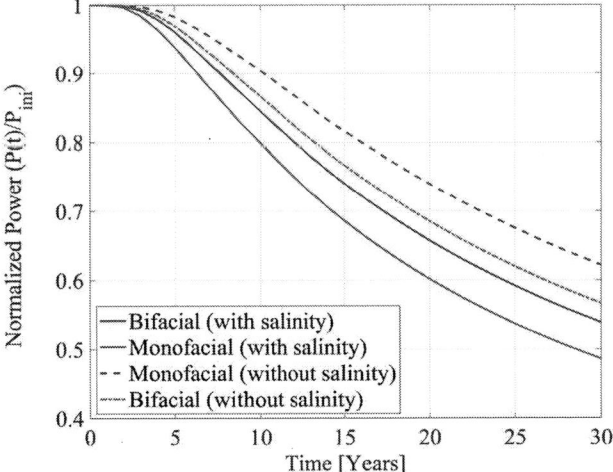

Fig. 7. Normalized power model of bifacial and monofacial PV modules with and without salinity effect.

degradation rates between the bifacial and monofacial PV modules are 0.75% before adding salinity stress conditions and 1.08% after adding salinity stress.

In this sense, a combination of environmental values contributes to the daily degradation rate. This is a way of analytically expressing the combined influence of environmental factors, but that is not what occurs physically. Using the PV module system data, the power model in (16) and the failure model in (17) are calibrated to extract the normalized power curve and predict the lifespan of the PV modules. The calibrated models are then applied to predict the long-term performance degradation by using a simple extrapolation mentioned as shown in Fig. 7.

Fig. 7 confirms that the degradation rate and power loss of bifacial PV modules are higher than those of monofacial modules due to their exposure to degradation on both the front and back sides, which increases the overall degradation rate by 2.1% compared with the monofacial PV modules, which expose to the degradation factors on one side only. Also, Fig. 7 shows that adding salinity stress to the PV modules in marine environments significantly increases the degradation rate of monofacial and bifacial PV modules by 1.2% and 2.23%, respectively. In addition, salinity stress contributes to expediting the expected failure time of the PV modules under offshore conditions by 2.5 years for the monofacial PV models and 3.3 years for the bifacial PV models.

IV. CONCLUSION

PV modules begin to suffer from gradual degradation and reduced lifespan as soon as they are installed in the field due to exposure to various external conditions. The salinity impact in offshore environments is a significant element that directly lowers the amount of solar energy that reaches the PV modules, along with weather-related effects. In this paper, the important elements of the degradation process of bifacial and monofacial PV modules were discussed under offshore settings. The damage model of offshore PV modules is essential for understanding the complex mechanisms that lead to degradation in marine environments. The performance and lifespan of solar PV panels are affected by several factors that contribute to the degradation process, including corrosion, salt deposition, UV exposure, and mechanical stress. A cumulative damage model with the salinity effect and other climatic stresses like temperature, RH, and UV has been tested to quantify these effects on the degradation rates of the PV modules. It has been tested for its ability to predict power output degradation using field data from a power plant in Guangzhou, China. When the measured and projected values are compared, the differences are found to be within ±1.1%. Significant additional loss has also been quantified as a result of the salt impact.

REFERENCE

[1] R. Cazzaniga and M. Rosa-Clot, "The booming of floating PV," *Solar Energy*, vol. 219, pp. 3-10, 2021.

[2] N. Kasahara, K. Yoshioka, and T. Saitoh, "Performance evaluation of bifacial photovoltaic modules for urban application," in *Proc. of 3rd World Conference on Photovoltaic Energy Conversion*, vol. 3, pp. 2455-2458, 2003.

[3] A. Kobi, C. M. F. Ke, P. A. Ndiaye, and V. Sambou, "Degradations of silicon photovoltaic modules: A literature review," *Solar Energy*, vol. 96, pp. 140-151, 2013.

[4] M. Aghaei, A. Fairbrother, A. Gok, S. Ahmad, S. Kazim, K. Lobato, and J. Kettle, "Review of degradation and failure phenomena in photovoltaic modules," *Renewable and Sustainable Energy Reviews*, vol. 159, pp. 112160, 2022.

[5] N. Bosco, T. J. Silverman, and S. Kurtz, "Climate specific thermomechanical fatigue of flat plate photovoltaic module solder joints," *Microelectronics Reliability*, vol. 62, pp.124–9 2016.

[6] D. J. Coyle, "Life prediction for CIGS solar modules part 1: modeling moisture ingress and degradation," *Progress in Photovoltaics: Research and Applications*, vol. 21, no. 2, pp. 156-172, 2013.

[7] A. Phinikarides, N. Kindyni, G. Makrides, and G. E. Georghiou, "Review of photovoltaic degradation rate methodologies," *Renewable and Sustainable Energy Reviews*, vol. 40, pp. 143-152, 2014.

[8] I. Kaaya, M. Koehl, A. P. Mehilli, S. de Cardona Mariano, and K. A. Weiss, "Modeling outdoor service lifetime prediction of PV modules: effects of combined climatic stressors on PV module power degradation," *IEEE Journal of Photovoltaics*, vol. 9, no. 4, pp. 1105-1112, 2019.

[9] O. Haillant, D. Dumbleton, and A. Zielnik, "An Arrhenius approach to estimating organic photovoltaic module weathering acceleration factors," *Solar Energy Materials and Solar Cells*, vol. 95, no. 7, pp. 1889-1895, 2011.

[10] W. Nelson, "Ch. 22. Prediction of field reliability of units, each under differing dynamic stresses, from accelerated test data," *Handbook of Statistics*, vol. 20, pp. 611-621, 2001.

[11] Y. Hong, Y. Duan, W. Q. Meeker, D. L. Stanley, and X. Gu, "Statistical methods for degradation data with dynamic covariates information and an application to outdoor weathering data," *Technometrics*, vol. 57, no. 2, pp. 180-193, 2015.

[12] J. A. Kratochvil, W. E. Boyson, and D. L. King, "Photovoltaic array performance model," Sandia National Laboratories (SNL), Albuquerque, NM, and Livermore, CA, USA, Rep., SAND 2004-353, 2004

[13] B. Sun, L. Lu, Y. Yuan, and P. Ocłoń, "Development and validation of a concise and anisotropic irradiance model for bifacial photovoltaic modules," *Renewable Energy*, vol. 209, pp. 442-452, 2023.

[14] K. O. Davis, M. P. Rodgers, G. Scardera, R. P. Brooker, H. Seigneur, N. Mohajeri, and W. V. Schoenfeld, "Manufacturing metrology for c-Si module reliability and durability Part II: Cell manufacturing," *Renewable and Sustainable Energy Reviews*, vol. 59, pp. 225-252, 2016.

[15] L. A. Escobar and W. Q. Meeker, "A review of accelerated test models," Statist. Sci., vol. 21, pp. 552–577, Nov. 2006.

[16] R. Pan, J. Kuitche, and G. TamizhMani, "Degradation analysis of solar photovoltaic modules: Influence of environmental factor," in *Proc. of Annual Reliability and Maintainability Symposium*, pp. 1–5, 2011.

[17] D. S. Peck, "Comprehensive model for humidity testing correlation," in *24th International Reliability Physics Symposium*, pp. 44-50, 1986.

[18] K. C. Norris, and A. H. Landzberg, "Reliability of controlled collapse interconnections," *IBM Journal of Research and Development*, vol. 13, no. 3, pp. 266-271, 1969.

[19] A. Sinha, H. Gopalakrishna, A. B. Subramaniyan, D. Jain, J. Oh, D. Jordan, and G. DamizhMani, "Prediction of climate-specific degradation rate for photovoltaic encapsulant discoloration," *IEEE Journal of Photovoltaic*. Vol. 10, no. 4, pp. 1093-1101, 2020.

[20] S. Pore, "Reliability of PV modules: Dependence on manufacturing quality and field climatic conditions," *Thesis, Arizona State University*, USA, 2017.

[21] C. G. Damo, C. Ozoegwu, C. Ogbonnaya, and C. Maduabuchi, "Effects of light, heat, and relative humidity on the accelerated testing of photovoltaic degradation using Arrhenius model," *Solar Energy*, vol. 250, pp. 335-346, 2023.

[22] I. Kaaya, J. A. Ascencio-Vásquez, M. Weiss M, and M. Topič, "Assessment of uncertainties and variations in PV modules degradation rates and lifetime predictions using physical models," *Solar Energy*, vol. 218, pp. 354-67, 2021.

[23] R. B. Yu, G. X. Liu, and H. Xu, "β distribution uniform expression based photovoltaic modules reliability evaluation with degradation data distribution," *Chinese Journal of Scientific Instrument*, vol. 36, no. 11, pp. 2586-2593, 2015.

[24] China Meteorological Administration. Available: http://www.escience.gov.

[25] A. B. Subramaniyan, R. Pan, J. Kuitche, and G. TamizhMani, "Quantification of environmental effects on PV module degradation: A physics-based data-driven modeling method," *IEEE Journal of Photovoltaics*, vol. 8, no. 5, pp. 1289-1296, 2018.

A Flexible Energy Management System for Solar Powered Electric-Bus Charging Stations

Supun Amarathunga, Pasan Gunawardena, Xiaoting Wang, Yunwei Li

Department of Electrical and Computer Engineering, University of Alberta, Edmonton, Canada
amarathu@ualberta.ca, pasan@ualberta.ca, xiaotin5@ualberta.ca, yunwei.li@ualberta.ca

Abstract—**This paper presents a flexible energy management system to manage an electric bus charging station incorporated with solar power, energy storage system and the main grid. To account for solar power uncertainties, random forest regression is employed to forecast solar irradiation. Robust scheduling frameworks are proposed to optimize charging schedules while minimizing operational costs, incorporating day-ahead pricing (DAP), grid-side dynamics, solar generation, and energy storage system (ESS). Two case studies demonstrate the scalability of the proposed energy management system for large-scale real-world electric bus fleets. Moreover, the energy management system effectively manages varying state-of-charge (SoC) levels, allocate chargers based on parking schedules, and utilize bidirectional power transfer for grid support and ancillary services.**

Index Terms—**Electric bus charging station, energy management system, energy storage system, solar, state of charge.**

I. INTRODUCTION

Energy management is critical for electric bus charging stations (EBCSs) due to high energy demands [1]. Recent methods for EBCSs include a cost-efficient energy management system (EMS) using mixed-integer linear programming (MILP) to maximize depot profits [2], while it assumes unrealistic power pricing. Another approach uses Gaussian distribution and Monte Carlo simulation for scheduling based on time-of-use (ToU) tariffs but has limited applicability [3].

More recently, Abdelshafy et al. [4] proposed an innovative EMS for battery and fuel cell bus charging stations, introducing a priority-based charging approach to minimize energy costs. However, it uses a rule-based system with annual-scale results. An integrated resource planning framework in [5] addresses investment and operational costs using a scenario-based chance-constrained optimization to handle uncertainties, though it focuses solely on daytime charging scenarios based on photovoltaic (PV) output. He et al. [6] proposed a network modeling framework to optimize EBCS charging at bus terminals for extended electric bus (EB) ranges, yet high power needs during short layovers limit its practicality. Liu et al. [7] proposed a three-layer stochastic EMS for transit center EBCS that minimizes operational costs while maintaining voltage quality. However, its use of three optimization algorithms adds significant computational burden for large EB fleets.

Another widely applied method, multi-objective optimization, minimizes costs and grid impact through improved simple optimization but requires significant computation for large fleets [8]. Li et al. [9] adopted joint optimization of charging schedules and charger deployment using an adaptive genetic algorithm (AGA) based on ToU tariffs but struggles with scalability. A stochastic EMS with renewable energy (RE) and bus-to-grid (B2G) capabilities employs a distributionally robust Markov decision process (DRMDP), but its heuristic solution remains computationally intensive for large-scale applications [10].

Two PEB charging scenarios, coordinated and uncoordinated, are analyzed for EBCS with and without ESS in [11], minimizing costs through a heuristic method but lacking renewable integration and availability-based scheduling. In [12], an EMS focuses on ESS sizing to reduce grid demand, aiding EBCS planning but not operational cost reduction, while also omitting renewable energy integration, which is crucial for minimizing GHG emissions. Qian et al. [13] proposed a smart EMS for EB battery swapping stations, but these stations require high capital investment and cannot be applied to standard EBCSs due to greater flexibility in their charging loads. A two-stage stochastic model in [14] schedules EV charging considering system constraints and uncertainties, but it lacks vehicle-specific schedules.

To address the above challenges, we propose a flexible EMS for a solar powered EBCS addressing computational burden which limits the application of EMS into large-scale bus fleets and uncertainties related with EBCS in real world scenarios. The proposed EMS formulates a MILP to generate day-ahead energy references for grid, ESS and load shedding under the uncertainties of solar generation, minimizing total operating cost. In the presented case study, parking schedules of 12 EBs are considered, and EB charging is scheduled considering not only the electricity price but also, availability of solar generation and other power system constraints.

The reminder of this paper is organized as follows. Section II briefly introduces random forest regression method to forecast solar irradiation and proposes two optimization-based EMSs for EBCSs. Section III presents a step by step procedure of the proposed EMSs and two case studies are provided in Section IV. Section V draws the conclusion.

II. PROPOSED EMS FOR ELECTRIC-BUS CHARGING STATIONS

This paper considers EBCS consisting of solar PV panels, ESS, EB load and grid power connection. The EBCS can import or export grid power and schedule load scheduling in emergencies. The goal is to minimize total operational costs considering penalties for potential load shedding while

979-8-3315-1612-3/25 $31.00 © 2025 IEEE

meeting physical and network constraints. Meanwhile, it aims to schedule EB loads in a more conservative way within the parking schedule to manage uncertainties (e.g., from solar irradiation) effectively.

A. Random Forest Regression

To model the uncertainty from renewable generations (e.g., solar PV panels), this paper adopts random forest regression [15] for forecasting solar irradiation, leveraging large-scale historical datasets (e.g., from Colorado park mountain [16]). Random forest regression ensembles predictions from multiple decision trees to enhance accuracy and robustness. By averaging the outputs of several decision trees trained on different subsets of the data, it effectively reduces the prediction variance [17]. Let $\mathbb{V}(X_i) = \sigma^2$ be the variance of the prediction (i.e., predicted solar irradiation) from an individual decision tree, where X_i denotes a single tree's prediction. Note that it is assumed that all decision trees have the same variance, solely to simplify the equation and demonstrate the relationship between the variances of ensemble mean and individual predictions. When ensembling n trees, the variance of the ensemble's mean prediction $\bar{X}$ will be:

$$\mathbb{V}\left(\bar{X}\right) = \mathbb{V}\left(\frac{1}{n}\sum_{i=1}^{n} X_i\right) = \frac{1}{n}\mathbb{V}\left(\sum_{i=1}^{n} X_i\right)$$
$$= \frac{\sigma^2}{n}\left(1 + (n-1)\rho\right) \qquad (1)$$

where the ensemble mean $\bar{X}$ is the average of predictions from all n decision trees and $\mathbb{V}\left(\bar{X}\right)$ is the variance of the ensemble mean predictions; ρ is the correlation coefficient between the prediction of any two trees. (1) implies that the ensemble variance $\mathbb{V}(\bar{X})$ diminishes as more trees are added (e.g., n increases) or as correlation between trees decreases. Particularly, random forest regression applies bagging [18] and random feature selection to produce a robust, stable model with reduced variance. See Fig. 1 for an overview of the random forest regression.

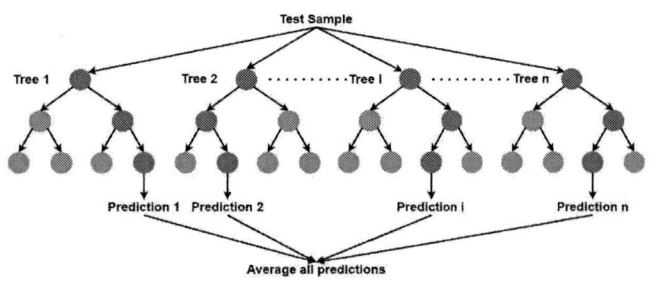

Fig. 1. Random forest regression method.

B. The Proposed EMS for EBCSs with Fixed EB Charging Load

This Subsection first considers EB charging using real world data [19]. Particularly, the number of EBs, available

charger capacity and count, morning and evening peak of bus transportation are considered in this approximation. In this case, we propose an EMS aiming to minimize the total daily operational cost, while allocating the maximum possible EB load. Furthermore, a penalty factor is integrated to deprioritize possible load shedding and no operational cost is considered for solar power. The objective function of EMS is as follows:

$$f = \sum_{t \in T} [P_t^{gp} C_t^{gp} - P_t^{gs} C_t^{gs} + P_t^s C_t^s] \Delta T \qquad (2)$$

where P_t^{gp} is the purchasing grid power, P_t^{gs} is the selling grid power, C_t^{gp} is the purchasing price of grid electricity by the customer, C_t^{gs} is the selling price of electricity to the grid, P_t^s is the load shedding, C_t^s is the penalty for load shedding at time interval t, and ΔT is the duration of the time interval.

Meanwhile, a set of constraints (from(3)-(10)) is considered in the proposed EMS. We consider the power balance constraint between power generation and consumption in EBCS:

$$P_t^{gp} + P_t^{\text{ESS d}} + P_t^{\text{PV}} + P_t^s = P_t^{gs} + P_t^{\text{ESS c}} + P_t^{\text{EB}} \qquad (3)$$

where $P_t^{\text{ESS d}}$ is the discharging power of the ESS; $P_t^{\text{ESS c}}$ is the charging power of the ESS; P_t^{PV} is the solar generation power, and P_t^{EB} is the charging load of EBs. The buying or selling power from/to the grid at time interval t is maintained within the maximum allowable grid power limit, $\bar{P}^g$.

$$0 < P_t^{gp} < \bar{P}^g X_t^g \qquad (4)$$

$$0 < P_t^{gs} < \bar{P}^g (1 - X_t^g) \qquad (5)$$

where X_t^g is binary variable that decides purchasing/ selling status of power from/to grid at time interval t.

Charging and discharging powers of ESS are constrained by the battery characteristics or the capacity of available chargers at EBCSs.

$$0 < P_t^{\text{ESS c}} < \bar{P}^{\text{ESS}} X_t^{\text{ESS}} \qquad (6)$$

$$0 < P_t^{\text{ESS d}} < \bar{P}^{\text{ESS}} (1 - X_t^{\text{ESS}}) \qquad (7)$$

where $\bar{P}^{\text{ESS}}$ is the maximum charging/ discharging power of ESS and X_t^{ESS} are binary variables that decide charging/ discharging status of ESS. State of charge (SoC) of ESS should be maintained within predefined safe limits to protect the state of health (SoH) of ESS.

$$\underline{\text{SoC}}_{\text{ESS}} < \text{SoC}_t^{\text{ESS}} < \overline{\text{SoC}}_{\text{ESS}} \qquad (8)$$

$$\text{SoC}_{t_0}^{\text{ESS}} = \text{SoC}_T^{\text{ESS}} \qquad (9)$$

where $\underline{\text{SoC}}_{\text{ESS}}$ and $\overline{\text{SoC}}_{\text{ESS}}$ are the lower and upper bounds of SoC level, respectively. Initial SoC of ESS and final SoC of ESS are maintained at same value (9). The SoC of ESS at time interval t, $\text{SoC}_t^{\text{ESS}}$ is updated based on charging/ discharging power of ESS as below.

$$\text{SoC}_t^{\text{ESS}} = \text{SoC}_{t-1}^{\text{ESS}} + \frac{\eta_c \cdot P_t^{\text{ESS c}}}{E_{\text{ESS}}} \times 100\% - \frac{P_t^{\text{ESS d}}}{\eta_d \cdot E_{\text{ESS}}} \times 100\% \qquad (10)$$

where η_c is the charging efficiency of ESS and η_d is the discharging efficiency and E_{ESS} denotes the capacity of ESS.

979-8-3315-1612-3/25 $31.00 © 2025 IEEE

C. The Proposed EMS for EBCSs with Flexible EB Charging Load

Two critical issues for EVs, including EBs, are limited driving range and long charging time. Transit services address these issues by identifying feasible routes and schedules for EB deployment, organizing them into operational groups known as EB blocks. Each block consists of multiple bus routes and shifts, typically covering morning or evening peak of transportation hours. Based on these assigned blocks, the parking times of EBs can be practically determined. In the second case, the proposed EMS is extended to address the scenarios where EB charging load is treated as a flexible load, with EB parking schedules used to schedule the charging of N EBs and define the charging load.

In this context, we define the daily parking schedule of the nth EB as $U_n = \left\{ \left(t_{1,a}^n, t_{1,d}^n \right), \left(t_{2,a}^n, t_{2,d}^n \right), \ldots, \left(t_{k,a}^n, t_{k,d}^n \right) \right\}$, where $t_{k,a}^n$ is the arrival time of the nth EB for the kth time to the EBCS in a day, and $t_{k,d}^n$ is the departure time. Let $\text{SoC}_{k,a}^n$ be the battery SoC of nth EB at k^{th} arrival to the EBCS. To improve the robustness of the proposed method, $\text{SoC}_{k,a}^n$ is assumed to be the minimum possible value $\text{SoC}_{\min}^n$ for each bus, upon arrival, despite of the block the EB covered. Moreover, EB batteries are fully charged during their parking time, i.e., the SoC of nth EB at the kth departure, $\text{SoC}_{k,d}^n$ reaches $\text{SoC}_{\max}^n$.

According to the above definitions, an extended EMS with a modified objective function compared to (2) is introduced:

$$f = \sum_{t \in T} \left(P_t^{gp} C_t^{gp} - P_t^{gs} C_t^{gs} + P_t^s C_t^s \right) \Delta T$$
$$+ \sum_{n=1}^{N} \sum_{i=1}^{k} \left(\text{SoC}_{\max}^n - \text{SoC}_{k,d}^n \right) C_p^{soc} \qquad (11)$$

which includes an additional penalty term to ensure that each EB is fully charged before its next departure from the EBCS. C_p^{soc} denotes high penalty cost applied for the deviation from the expected SoC level. Then, besides the constraints in (3)-(10), additional constraints (from (12)-(17)) associated with the EB charging and SoC are further considered. Clearly, each EB can be charged or discharged upon returning to the charging station, as described by:

$$P_t^n = \begin{cases} p_t^n, & t \in U_n \\ 0, & t \in T \setminus U_n \end{cases} \qquad (12)$$

with

$$0 < p_t^n < \bar{P}^c \qquad (13)$$

where P_t^n is the charging/ discharging power of nth EB during t^{th} time interval. For simplicity, we assume that each EB is connected to a charger as soon as it returns EBCS, having zero waiting time. p_t^n denotes the charging/discharging power of the nth EB at $t \in U_n$ and can be defined as above when B2G power transfer can not be facilitated. $\bar{P}^c$ is the maximum charging power of available EB chargers at EBCS and we assume that all EB chargers have the same maximum

charging capacity. EV chargers are typically categorized into three types: level 1 and level 2 (AC chargers) and level 3 (DC fast chargers), which are primarily used for opportunity charging [20]. When bidirectional chargers are used, constraint (13) can be updated as follows:

$$\underline{P}^c < p_t^n < \bar{P}^c \qquad (14)$$

where $\underline{P}^c$ is the maximum allowable discharging power of EB battery through bidirectional EB charger. The total EB load P_t^{EB} at t^{th} interval is calculated considering charging/discharging powers of all N EBs:

$$P_t^{\text{EB}} = \sum_{n \in N} P_t^n \qquad (15)$$

The total EB load served at any given time t depends on the grid constraints, available solar capacity, ESS SoC and time-based factors such as upcoming EB departures and whether it is daytime or nighttime. To meet these constraints, the total EB load is regulated within the allowable range from $\underline{P}^{\text{EB}}$ to $\bar{P}^{\text{EB}}$:

$$\underline{P}^{\text{EB}} < P_t^{\text{EB}} < \bar{P}^{\text{EB}} \qquad (16)$$

Furthermore, the SoC of nth EB battery at t, SoC_t^n is updated based on the charging/ discharging power during t^{th} time interval:

$$\text{SoC}_t^n = \text{SoC}_{t-1}^n + \frac{\eta \cdot p_t^n}{E_{EB}} \times 100\% \; ; \quad t \in U_n \qquad (17)$$

where E_{EB} denotes the capacity of the EB battery and SoC_{t-1}^n is defined by $\text{SoC}_{k,a}^n$ as previously defined. η is the charging efficiency of the EB batteries.

Remark 1. Note that a typical charging schedule charges EBs immediately upon their return to the EBCS with the objective function in (18), under the same constraints, is used for comparison with the proposed EMS (i.e., with objective function in (11), constraints (3)-(10), and (12)-(17)).

$$f = \sum_{t \in T} \left(P_t^{gp} C_t^{gp} - P_t^{gs} C_t^{gs} + P_t^s C_t^s \right) \Delta T$$
$$+ \sum_{n=1}^{N} \sum_{i=1}^{k} \sum_{t=t_{k,a}^n}^{t_{k,d}^n} \left(\text{SoC}_{\max}^n - \text{SoC}_{k,t}^n \right) C_p^{soc} \qquad (18)$$

III. THE PROPOSED FRAMEWORK FOR EMS OF EBCS

The proposed framework for EMS of EBCS is outlined step by step below.

Step 1. Input network data, and historical solar irradiation data. Forecast solar irradiation using random forest regression method. Pass the generated data to **Step 3**.

Step 2. Build the EMS framework based on the charging load of EBs in two cases. *Case 1:* EB charging load is approximated considering actual data of an EBCS [19]; *Case 2:* Parking schedule of EBs is used to identify the charging load and its flexibility to schedule charging.

Step 3. Solve the proposed EMS for EBCS (i.e., *Case 1:* from (2)-(10) and *Case 2:* from (3)-(17)) considering 24 hours time periods .

Step 4. Generate the results report including: the total energy cost for given EB charging load, the optimal power references for grid, ESS in hourly scale and SoC of ESS, the optimal charging schedule for given parking schedule and SoC variation of EB batteries.

IV. CASE STUDY

Two cases are studied based on how the charging load of EBs is defined. In *Case 1* the charging load of EBs is approximated from historical data. This case also compares the performance of the random forest regression method with the gradient boost regression (GBR) method for predicting solar irradiation [15]. In *Case 2*, the charging load of EBs is decided based on the optimal charging schedule proposed by EMS for a given parking schedule. However, solar power generation, hourly electricity prices, other system constraints are the same for both cases. The proposed EMS was implemented in Python language in Google Colab. The optimization problems in Section II-B and Section II-C were solved by coin-or branch and cut (CBC) open-source solver.

A. Case 1

This case considers an EBCS with $2000\,\mathrm{m}^2$ of solar panels (15% efficiency). Particularly, grid power importing/exporting limit, $\bar{P}^g$, is set to $400\,\mathrm{kW}$, and penalty (C_t^s) of 15 cents/kWh is applied for shedding load. Maximum charging/discharging power of ESS ($\bar{P}^{\mathrm{ESS}}$) is $120\,\mathrm{kW}$. Minimum ($\underline{\mathrm{SoC}}_{\mathrm{ESS}}$) and maximum ($\overline{\mathrm{SoC}}_{\mathrm{ESS}}$) SoC levels of ESS are maintained within 30% and 90%. The initial SoC of ESS is assumed to be $500\,\mathrm{kWh}$ out of $720\,\mathrm{kWh}$ capacity of ESS. Furthermore, a fleet of 7 EBs is considered and EBCS is equipped with 60 kW chargers.

Solar Irradiation Forecasting: We first forecast the solar irradiation (**Step 1**). Table I summarizes the solar irradiation prediction from random forest regression and gradient boost regression [15]. It is shown that random forest regression achieves higher prediction accuracy across root mean square error (RMSE), mean absolute error (MAE), and R-squared (R^2) metrics. Solar irradiation predictions using this method are compared with actual data for a typical day, as shown in Fig. 2.

TABLE I
PREDICTION ACCURACY OF SOLAR IRRADIATION

Performance metrics	Random Forest		Gradient Boost	
	5 min	*1 hour*	*5 min*	*1 hour*
RMSE	51.12	35.42	53.52	36.96
MAE	28.22	19.42	32.2	23.06
R^2	0.971	0.986	0.968	0.985

Then, the morning and evening peaks of bus transportation are considered when approximating the EB charging load as shown in Fig. 3. Table II shows hourly variation in grid electricity price. Once the network data and solar irradiation data are prepared, the proposed EMS is built (Section II-B : **Step 2** (2)-(10)).

Performance Evaluation of the Proposed EMS: Once the network data and solar irradiation data are prepared, solve

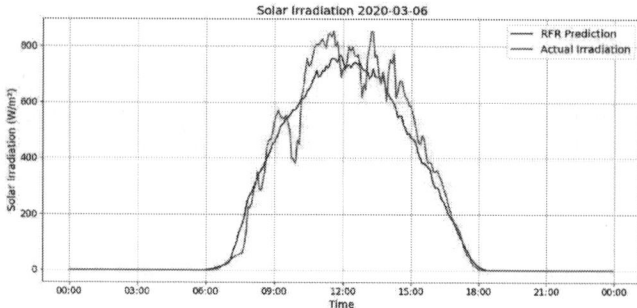

Fig. 2. Forecasted and actual solar irradiation.

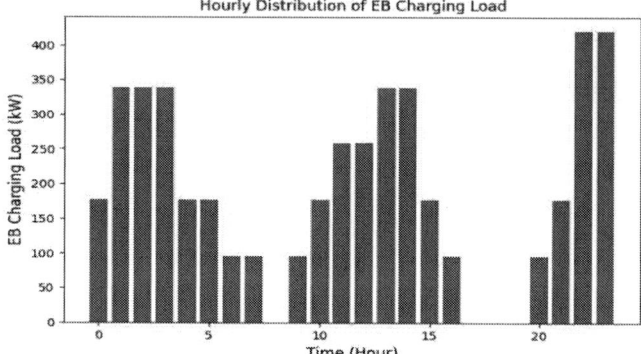

Fig. 3. Hourly EB charging load.

the built EMS (**Step 3**). It can be obtained that the minimum operational cost is CAD 125.89 per day under a grid power limit of 400 kW. When the grid power limit is reduced to 300 kW, EB load is shed at 22:00, increasing the minimum operational cost to CAD 140.61 per day, including load-shedding penalties. Since delays and outages in public transportation have significant social and economic impacts, proper charging scheduling is critical to minimize load shedding.

Proposed grid power references adapt to uncertain solar generation during the daytime, as shown in Fig. 4. For a 400 kW grid power limit, Fig. 5 presents the hourly grid power references (mean), ESS SoC, and electricity price variations.

TABLE II
HOURLY ELECTRICITY PRICE VARIATION

Hour	Electricity price (cents/kWh)	Hour	Electricity price (cents/kWh)
00:00	4.0	12:00	6.4
01:00	3.7	13:00	6.0
02:00	3.6	14:00	5.6
03:00	4.0	15:00	6.5
04:00	4.4	16:00	10.8
05:00	6.0	17:00	12.0
06:00	8.3	18:00	10.4
07:00	7.5	19:00	9.8
08:00	8.2	20:00	9.2
09:00	8.8	21:00	6.8
10:00	8.4	22:00	6.7
11:00	7.1	23:00	4.4

979-8-3315-1612-3/25 $31.00 © 2025 IEEE

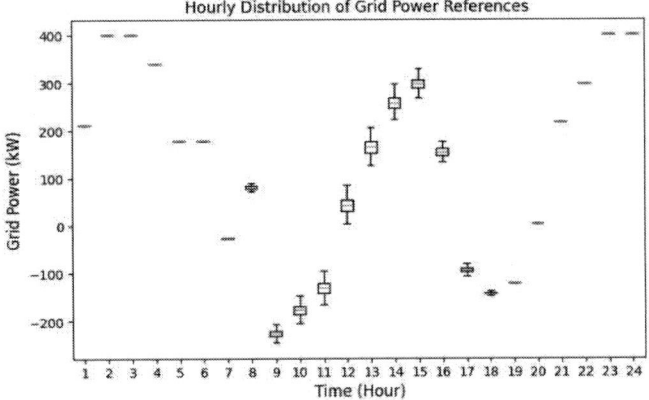

Fig. 4. Grid power references under uncertainty.

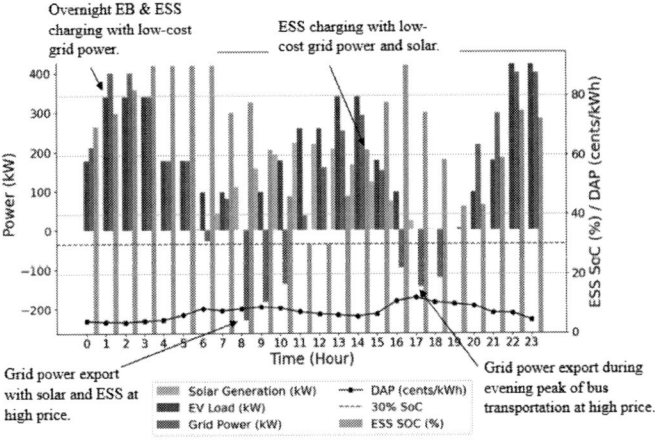

Fig. 5. Hourly references, SoC, and DAPs for approximate EB load

B. Case 2

This case considers parking schedule of 12 EBs, shown in Table III, to schedule EB charging, avoiding any possible load shedding carried out in *Case 1*. Aslo, each EB parks at the EBCS twice daily ($k = 2$) after completing its assigned EB blocks.

TABLE III
PARKING SCHEDULE OF EBS

Bus	*Parking duration 1*	*Parking duration 2*
A,G	00:00-06:00	14:00-18:00
B,H	06:00-14:00	18:00-24:00
C,I	09:00-13:00	21:00-03:00
D,J	03:00-09:00	14:00-21:00
E,K	06:00-12:00	20:00-24:00
F,L	00:00-16:00	12:00-20:00

The SoC of each returning EB, $SoC_{k,a}^n$, is set to a safe minimum (SoC_{min}^n) of 20%, while the SoC at the next departure, $SoC_{k,d}^n$, is maintained at the maximum (SoC_{max}^n) of 100% to ensure robustness against uncertainties. Other new system parameters for *Case 2* are listed in Table IV. All remaining parameters (e.g., electricity price, solar irradiation data, avail-

able solar generation, ESS capacity and initial energy storage) are same as *Case 1*.

TABLE IV
SYSTEM PARAMETERS FOR CASE 2

Parameter	*Value*
EB charger capacity	50 kW
EB battery capacity	180 kWh
Maximum possible EB charging load	300 kW
Grid power importing/ exporting limit	400 kW

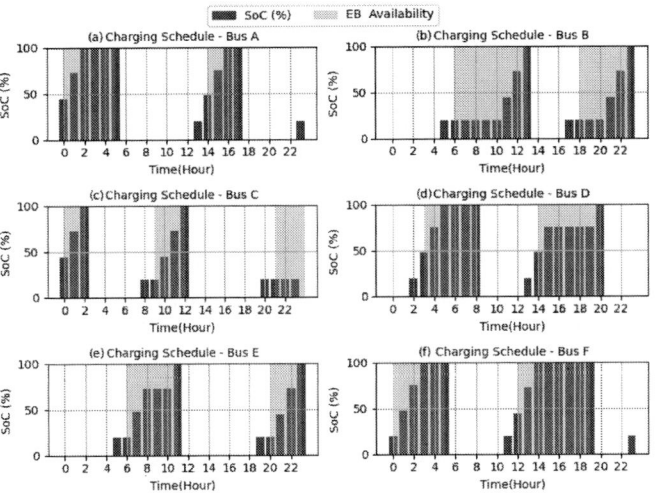

Fig. 6. Optimal charging schedule for EBs. The blue bars indicate the SoC of relevant EBs, the green highlights the EB parking time at EBCS, and their overlaps show the charging pattern. Each blue bar before the green highlighted parking time indicates the SoC of EB at its return to EBCS.

Fig. 6 illustrates the proposed optimal charging schedules for six selected EBs. For example, EB–E is parked at the EBCS from 06:00 to 12:00 and 20:00 to 24:00. However, its charging is delayed until after 06:00 to align with solar power availability and lower DAP in subsequent hours. A similar pattern is observed for EB–B, which, though parked from 06:00, begins charging at 11:00 when DAP drops. During the high DAP period (08:00–10:00), the EMS prioritizes exporting solar power and ESS energy to maximize profits. Conversely, EB–C is charged immediately despite high DAP due to its limited parking time before its scheduled departure at 12:00. Moreover, the SoC levels of EB–A, EB–D, and EB–F during their parking times indicate that bidirectional power transfer (B2G) could support ancillary grid services.

The total daily energy cost under the proposed EMS is CAD 50.91, with the EB load shown in Fig. 7. The EMS exports 1231.39 kWh to the grid over the day. In contrast, a typical charging schedule, which charges EBs immediately upon their return to the EBCS without considering these factors, results in a higher energy cost of CAD 83.07 and exports only 758.81 kWh to the grid. Hourly power references and ESS SoC variations for this typical schedule are shown in Fig. 8.

979-8-3315-1612-3/25 $31.00 © 2025 IEEE   3034

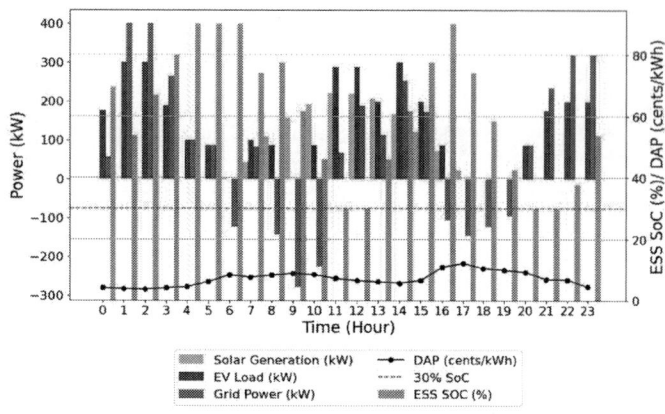

Fig. 7. Hourly references, SoC and DAPs under optimal charging

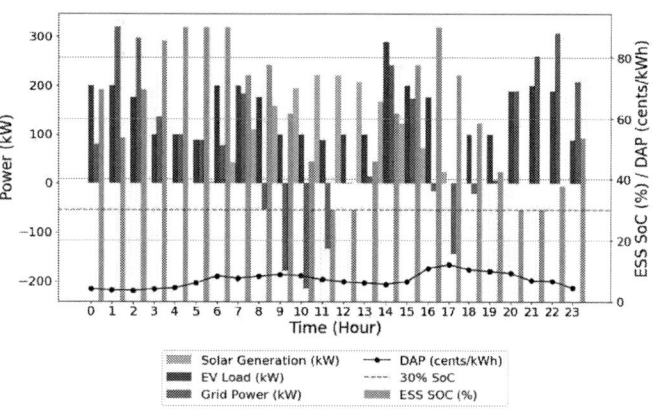

Fig. 8. Hourly references, SoC, and DAPs under immediate charging.

V. CONCLUSION

This paper proposes an adaptable and robust EMS for EBCSs considering fixed and flexible EB charging load. Specially, random forest regression is applied to address the solar power uncertainties. A robust scheduling framework is developed to optimize charging schedules, minimizing operational costs while integrating DAP, grid-side dynamics, solar generation, and ESS energy. Simulation studies in two cases demonstrate the scalability of proposed EMS for large-scale real-world EB fleets. Furthermore, the EMS can manage varying SoC levels, allocate chargers based on parking schedules, and incorporate B2G capabilities for grid support and ancillary services, offering a comprehensive solution for efficient and sustainable EBCS operations.

REFERENCES

[1] J.-M. Clairand, M. González-Roríguez, P. G. Terán, I. Cedenño, and G. Escrivá-Escrivá, "The impact of charging electric buses on the power grid," in *2020 IEEE Power Energy Society General Meeting (PESGM)*, 2020, pp. 1–5.

[2] S. M. Arif, T. T. Lie, B. C. Seet, and S. Ayyadi, "A novel and cost-efficient energy management system for plug-in electric bus charging depot owners," *Electric Power Systems Research*, vol. 199, p. 107413, 2021.

[3] R.-C. Leou and J.-J. Hung, "Optimal charging schedule planning and economic analysis for electric bus charging stations," *Energies*, vol. 10, no. 4, 2017.

[4] A. M. Abdelshafy, O. Samir, A. Elnozahy, and A. F. M. Ali, "Innovative energy management strategy of battery and fuel cell buses charging station," *Energy Conversion and Management*, vol. 315, p. 118815, 2024.

[5] Y. Cheng, W. Wang, Z. Ding, and Z. He, "Electric bus fast charging station resource planning considering load aggregation and renewable integration," *IET Renewable Power Generation*, vol. 13, no. 7, pp. 1132–1141, 2019.

[6] Y. He, Z. Liu, and Z. Song, "Optimal charging scheduling and management for a fast-charging battery electric bus system," *Transportation Research Part E: Logistics and Transportation Review*, vol. 142, p. 102056, 2020.

[7] Y. Liu and H. Liang, "A three-layer stochastic energy management approach for electric bus transit centers with pv and energy storage systems," *IEEE Transactions on Smart Grid*, vol. 12, no. 2, pp. 1346–1357, 2021.

[8] M. M. Hasan, N. Avramis, M. Ranta, A. Saez-de Ibarra, M. El Baghdadi, and O. Hegazy, "Multi-objective energy management and charging strategy for electric bus fleets in cities using various eco strategies," *Sustainability*, vol. 13, no. 14, 2021.

[9] X. Li, T. Wang, L. Li, F. Feng, W. Wang, and C. Cheng, "Joint optimization of regular charging electric bus transit network schedule and stationary charger deployment considering partial charging policy and time-of-use electricity prices," *Journal of Advanced Transportation*, vol. 2020, pp. 1–16, 2020.

[10] P. Zhuang and H. Liang, "Stochastic energy management of electric bus charging stations with renewable energy integration and b2g capabilities," *IEEE Transactions on Sustainable Energy*, vol. 12, no. 2, pp. 1206–1216, 2021.

[11] H. Chen, Z. Hu, H. Zhang, and H. Luo, "Coordinated charging and discharging strategies for plug-in electric bus fast charging station with energy storage system," *IET Generation, Transmission Distribution*, vol. 12, no. 9, pp. 2019–2028, 2018.

[12] I. Ojer, A. Berrueta, J. Pascual, P. Sanchis, and A. Ursúa, "Development of energy management strategies for the sizing of a fast charging station for electric buses," in *2020 IEEE International Conference on Environment and Electrical Engineering and 2020 IEEE Industrial and Commercial Power Systems Europe (EEEIC / ICPS Europe)*, 2020, pp. 1–6.

[13] Q. Dai, T. Cai, S. Duan, W. Zhang, and J. Zhao, "A smart energy management system for electric city bus battery swap station," in *2014 IEEE Conference and Expo Transportation Electrification Asia-Pacific (ITEC Asia-Pacific)*, 2014, pp. 1–4.

[14] F. Wu and R. Sioshansi, "A two-stage stochastic optimization model for scheduling electric vehicle charging loads to relieve distribution-system constraints," *Transportation Research Part B: Methodological*, vol. 102, pp. 55–82, 2017.

[15] A. Torres-Barrán, Álvaro Alonso, and J. R. Dorronsoro, "Regression tree ensembles for wind energy and solar radiation prediction," *Neurocomputing*, vol. 326–327, pp. 151–160, 2019.

[16] National Renewable Energy Laboratory (NREL), "Midc station redirector," https://www.nrel.gov/midc/spmd/, accessed: 2024-07-31.

[17] L. Breiman, "Random forests," *Machine Learning*, vol. 45, pp. 5–32, 2001.

[18] T. Hastie, R. Tibshirani, and J. Friedman, *The Elements of Statistical Learning: Data Mining, Inference, and Prediction*, 2nd ed. New York, NY, USA: Springer, 2009.

[19] MARCON, "Electric bus feasibility study for the city of edmonton," June 2016.

[20] U. D. of Energy. Alternative Fuels and A. V. D. Center, "Ev charger selection guide," jan 2018.

A Vienna Rectifier based Grid-Connected Powertrain for Hydrokinetic Turbine Systems

Peidong Li
School of EECS
Oregon State University
Corvallis, Oregon, USA
lipeid@oregonstate.edu

Md Tariquzzaman
School of EECS
Oregon State University
Corvallis, Oregon, USA
tariquzm@oregonstate.edu

Yue Cao
School of EECS
Oregon State University
Corvallis, Oregon, USA
Yue.Cao@oregonstate.edu

Abstract—**This work presents a novel hydrokinetic turbine (HKT) powertrain utilizing a Vienna rectifier for improved efficiency and reduced size. Unlike traditional HKT powertrain with a passive rectifier and DC/DC converters, the proposed powertrain integrates a Vienna rectifier as both a generator-side rectifier and step-up converter, enabling high-voltage DC bus for offshore applications. The dynamic current control strategy for the proposed powertrain is explained, which can achieve optimal power output by changing the tip speed ratio (TSR) across varying water velocities. Lastly, hardware experiments are conducted with the generator-motor setup and the converter system. Results demonstrate the system functionality to access the transient performance of the proposed powertrain.**

Index Terms—**hydrokinetic turbine, Vienna rectifier, powertrain**

I. INTRODUCTION

The hydrokinetic turbine (HKT) system is an emerging form of hydro-energy generation utilizing marine or riverine current with a direct hydro turbine in the waterbody. With reservoirs unnecessary, the HKT system is more flexible to install, making it applicable to a wider range of locations, and it also has minimal environmental impact [1].

The HKT system is similar to a wind turbine in that it needs to orient the turbine towards the water flow, which requires a yaw mechanism [2]. However, unlike a wind turbine that can pitch the blade to adjust the angle of attack, thus the input torque, many hydrokinetic turbines are fixed-pitch, which cannot tilt their blade to change the operating speed and torque [3]. Meanwhile, the HKT has a changeable working environment due to the nature of the water flow and limited mechanics of control, which requires a power electronics system as the powertrain to apply an adaptive control strategy [4]. The most common powertrain utilizes a passive rectifier on the gen-side and a boost converter [5]. This three-converter architecture is costly in size and not highly efficient. Due to environmental constraints, the HKT turbines can be installed a few hundred meters offshore. A high-voltage DC bus required by long-distance transmission is also difficult for traditional two-level converters. For optimal HKT performance, the underwater

This work was supported in part by the U.S. Department of Energy Advanced Research Projects Agency-Energy (ARPA-E) under Award DE-AR0001438 and in part by the U.S. Department of Energy Water Power Technologies Office under Award DE-EE0011381.

rectifier should also be compact to avoid disrupting water flow and affecting power output.

Various powertrains have been developed for wind turbines that have similar operating principles as HKTs. Common options include a diode rectifier with a boost DC/DC, a two-level back-to-back converter, and so on [6]. Among them, a Vienna rectifier-based powertrain is proposed since it has good power quality with reduced size [7]. However, applying the Vienna rectifier to HKTs requires control considerations on some of the HKTs' features. Compared with wind turbines, HKTs do not have the pitch control that can directly adjust the input torque. The input torque, which is also linked to the generator-side (gen-side) current, can be adjusted by controlling the turbine tip speed ratio (TSR) [8]. Therefore, to have maximum output power, the HKT's powertrain requires a dynamic current control at varying water velocities.

A Vienna rectifier-based powertrain for HKTs is thus proposed to address the above-mentioned issues. A control strategy for the powertrain to achieve optimal turbine speed is also explained. The content of this paper is organized into three sections. Section II introduces the system architecture. Section III explains the dynamic control for the proposed powertrain. The hardware evaluation is conducted in the last section.

II. SYSTEM ARCHITECTURE

As shown in Fig. 1, the proposed HKT system includes a direct-drive permanent magnet synchronous generator (PMSG) as the generator, and a powertrain consists of an offshore rectifier and an onshore inverter that connects to the rest of the system through a long-distance DC bus. A Vienna rectifier is used as a gen-side rectifier as well as a step-up converter. Both the generator and rectifier are installed underwater. The DC bus between the rectifier and the inverter can be shared by multiple offshore turbine units to form a modular system. The rectifier will work in the current control mode, enabling optimal turbine TSR, while the grid-following inverter regulates the DC bus voltage and feeds power to the grid.

The PMSG selection for the proposed system should prioritize the generator's functionality for direct driving at a lower speed since an underwater gearbox in the system is infeasible to repair and impacts long-term robustness. The generator

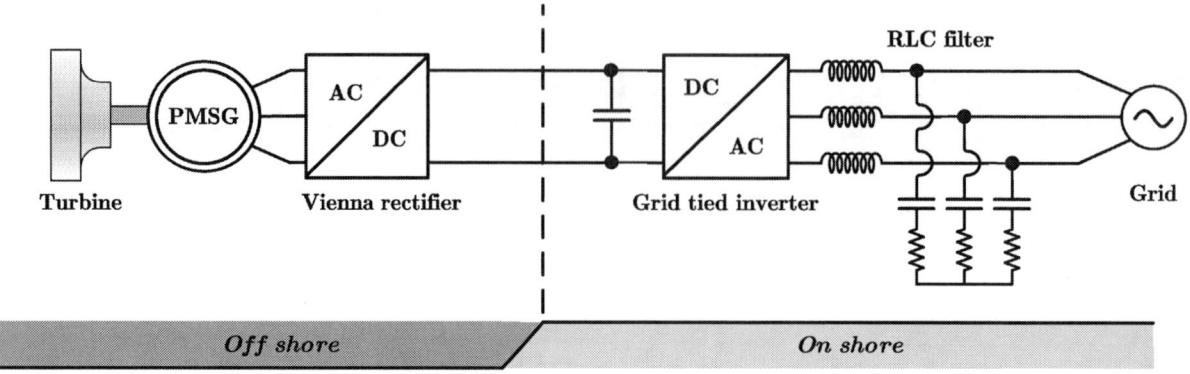

Fig. 1. System architecture.

should have high pole pairs to be able to operate at a low speed while at the same time being compact enough, in diameter, to minimize the power loss caused by the generator obstructing the turbine's own waterway.

Since the proposed system consists of many small power turbine units in a riverine or tidal application, each unit relies on its gen-side converter to perform control, which is attached underwater within the generator nacelle. In addition, the HKT units can span a few hundred meters off the shore, and long-distance power transmission can cause considerable losses compared with other parts of the system. Therefore, the powertrain will support a higher DC bus voltage.

An on-shore grid-side inverter will regulate the DC bus voltage and maintain a grid-tied connection. The grid-side inverter is located at the substation rather than having separate units for each turbine for the best efficiency. Since the inverter is on-shore, its maintenance and repair are easier and more feasible. Other passive filters on the grid side and DC side will also be on shore for convenient maintenance and to avoid large underwater loads.

III. POWERTRAIN DYNAMIC CONTROL

The HKT's working state can be modeled on its C_P (coefficient of performance), as shown in Fig. 3, which denotes the turbine's power capacity over TSR. The turbine generator's torque can be calculated based on the C_P, thus giving us the torque C_T curve. The C_P and C_T are defined as [9]:

$$C_P = \frac{2P}{\rho A V^3}, \quad C_T = \frac{2T}{\rho A V^2} \quad (1)$$

where P is the power, T is the torque, ρ is the water density, A is the cross sectional area of the turbine and V is the water velocity. Both curves typically have inversed-U shapes where the C_T curve has a leading peak. From point C to point A, the turbine will go from free-spinning to fully stopped by the braking mechanism, and the maximum power output point B happens between them, where TSR is optimal. The optimal TSR can be influenced by the turbine's design, but in practice, various factors such as wake effects [10] and biofouling [11] also play a significant role, making it difficult to determine the optimal TSR in advance. Therefore, it is not feasible to simply command the turbine to operate at the theoretically optimal TSR without considering real-world variables. It is necessary to estimate the optimal TSR for each turbine unit in real time.

The proposed dynamic current control is applied individually on each gen-side rectifier. The rectifier works in current control mode using sensed input voltage as the reference. A dynamic change is applied to the reference based on the turbine's output power trend. The operating range of the HKT under this control can be divided into three zones, as shown in Fig. 4. In Zone I, C_T increases with TSR, which means the turbine's input thrust decreases when it decelerates. Thus, if a load force larger than the static thrust T_0 is applied, the turbine will slow down until a full stop, making the generator side control unstable. Zone II and Zone III are separated by point B. In Zone II and III, C_T decreases with TSR, so the turbine will reach an equilibrium as long as the commanded load is less than T_{max} [12]. In practice, T_{max} cannot be precisely estimated, so control should be well-damped to avoid long overshoots in the current, which could drag the HKT unit into Zone I. In a direct-drive generator, the winding current amplitude is proportional to the load force on the turbine. Therefore, in the dynamic current control, the turbine will work in Zone III with gradual step changes to the current reference and monitor the output power level. If the output power rises at a step change, the next change should be in the same direction; if output power falls at a current reference change, flip the direction of the step change. This method is similar to the perturb-and-observe MPPT control. Since the water velocity is relatively stable, the dynamic reference will be fast enough to track the actual optimal TSR.

The control diagram of the proposed method is shown in Fig. 2. The three-phase current and voltage waveforms are sampled to calculate the input power. The power signal will go through a low-pass filter before using for dynamic power control since the motor's mechanical response is significantly slower than the power electronics transient. The bandwidth of the low-pass filter should be selected based on the turbine's mechanical inertia. Since the water velocity can be seen as

979-8-3315-1612-3/25 $31.00 © 2025 IEEE

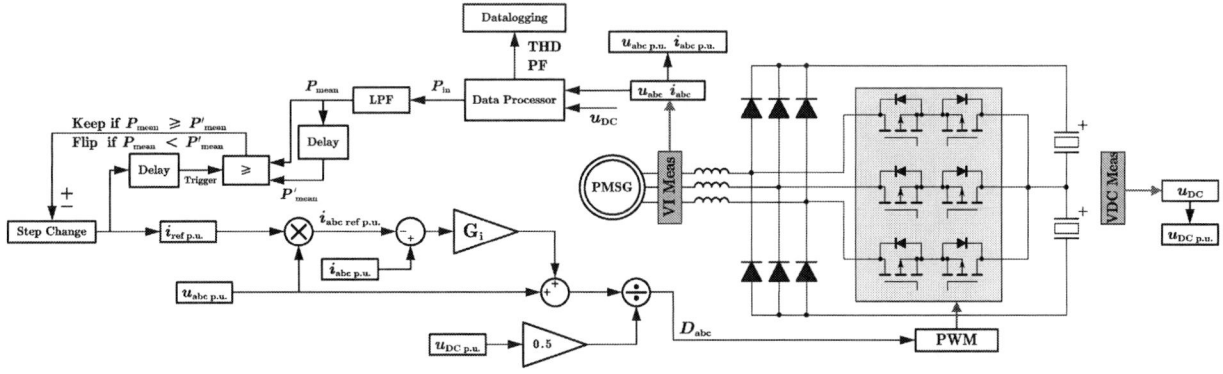

Fig. 2. Control diagram of proposed dynamic current control.

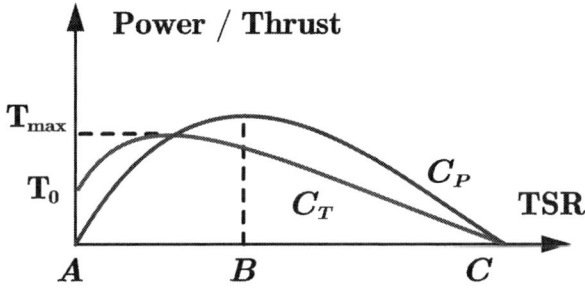

Fig. 3. HKT's C_T/C_P curve at a constant water speed.

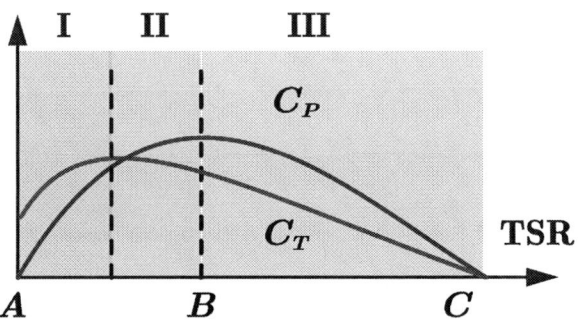

Fig. 4. HKT's working zones.

constant over a short period, the bandwidth can be rather low while still maintaining a good maximum power tracking performance. The direction of the current reference step change is determined based on the comparison of the power output change:

$$\Delta i_{\text{step p.u.}} = \Delta i'_{\text{step p.u.}} \times \text{sgn}(P_{\text{mean}} - P'_{\text{mean}})$$
$$i_{\text{ref p.u.}} = i'_{\text{ref p.u.}} + \Delta i_{\text{step p.u.}} \tag{2}$$

where $\Delta i_{\text{step p.u.}}$ is the step change value of the per unit current reference, P_{mean} is the average power output and $i_{\text{ref p.u.}}$ is the per unit current reference. Variables with the prime (′) sign refer to those in the previous control cycle. The dynamic current reference multiplies the input voltage to generate the current reference for each phase. The duty cycles of the device are calculated based on the current reference, three-phase voltage feedforward, and the DC bus voltage:

$$D_{\text{abc}} = \frac{(i_{\text{ref p.u.}} * u_{\text{abc p.u.}} - i_{\text{abc p.u.}})}{u_{\text{DC p.u.}}/2} \tag{3}$$

where D_{abc} are the duty cycles for all three phases, $u_{\text{abc p.u.}}$ and $i_{\text{abc p.u.}}$ are the three-phase per unit voltages and currents, and $u_{\text{DC p.u.}}$ is the per unit DC bus voltage.

IV. Performance Evaluation

A. Test Setup

The hardware setup is shown in Fig. 5. Initial tests are conducted on grid-tied mode with an NHR 9410 regenerative grid simulator to verify the functionality of the converter. Then, the Vienna rectifier is tested as the powertrain with the generator-motor test bench. The test bench has two PMSMs with similar parameters to emulate the actual turbine and generator. A high pole pair PMSG rated for low-speed application is selected with its key parameters listed in Table I. The generator has a waterproof nacelle which also allows direct convective cooling underwater. The powertrain is built with a back-to-back gen side Vienna rectifier and grid-tied inverter.

The Vienna rectifier uses six IDH16G120C5 diodes, and six IPP65R095C7 MOSFETs. Three hall effect CASR15-NP current transducers provide input line current data. The input AC and DC bus voltages are measured with a voltage divider with operational amplifiers. The analog sensing data are fed into the onboard ADC of the control card TMS320F28379D. The rectifier has a single-board design with a stacked aluminum PCB as the heatsink. The customizable inductor is used to facilitate testing in grid-tied mode and can be removed in later testing since generator windings have a large inductance. The power quality data from the AC and DC side of the rectifier are collected from a PA4000 power analyzer. The power data captured from the rectifier controller are only used for converter control since on-board measurement accuracy

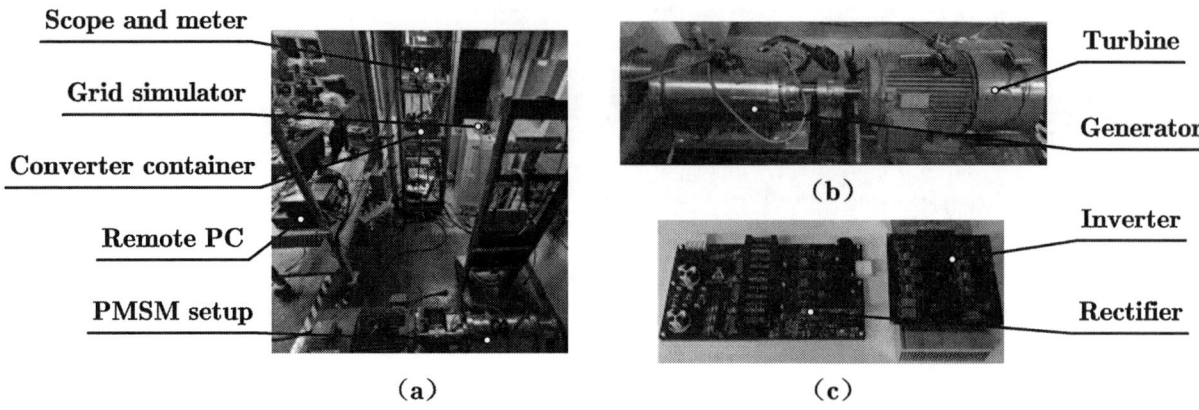

Fig. 5. Hardware setup: (a) system configuration, (b) generator-motor, and (c) powertrain converters.

TABLE I
GENERATOR PARAMETERS.

Parameter	Value
Stator resistance, R_s	3.711 Ω
d-axis magnetizing inductance, L_d	45.25 mH
q-axis magnetizing inductance, L_q	64.88 mH
Pole pairs, pp	12
Rated Power, P_{rated}	6 kW
Rated Speed, RPM_{rated}	150 RPM
Inertia, J	0.2564 kg m²

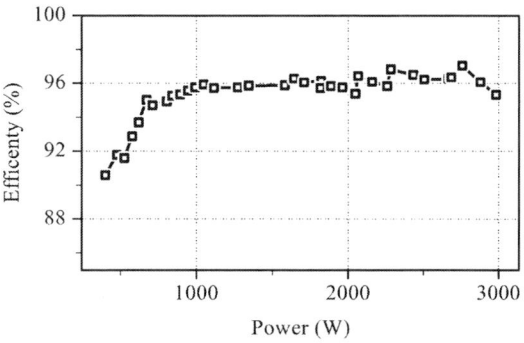

Fig. 6. Converter efficiency at grid-tied testing.

might suffer. The grid-side converter is a bi-directional DC power supply connected to a three-phase 480 VAC grid.

B. Rectifier Testing

The Vienna rectifier is first tested with a grid simulator to demonstrate its performance on grid-tied rectifying. The series inductor has an inductance of 1.3 mH, and the switching frequency is 20 kHz. The results of its efficiency and current total harmonic distortion (THD) data are shown in Fig. 6 and Fig. 7. The efficiency of the converter reaches about 96% at nominal working conditions. The current THD deteriorates as the power level increases. This might create an extra loss on the generator side but will not intervene in the grid side power quality as they are decoupled by a DC link. A current control strategy utilizing motor position information to generator PWM can help reduce the higher harmonics. For the actual turbine unit working at the optimal TSR, its power level is proportional to the water velocity and, thus, proportional to the generator voltage, which is contrary to a regular grid-tied rectifier working condition where the input voltage is rather constant. Therefore, the data at different power levels are obtained at a constant input current of 4 A but with different input voltages.

C. Dynamic Response

Further tests are conducted with the generator-motor setup to see the transient response of the system under the dynamic

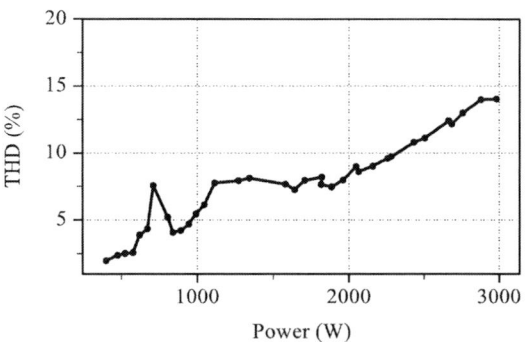

Fig. 7. Converter current THD at grid-tied testing.

current control. The motor is regulated by a separate variable-frequency drive (VFD) which works in speed mode at 125 RPM with an internal torque limit of 120 Nm. If the applied torque is greater than the turbine limit, the VFD will slow down to meet the torque limit until the motor fully stops. Given the friction torque induced by both machines and the coupling, the speed-torque response of the motor will have a

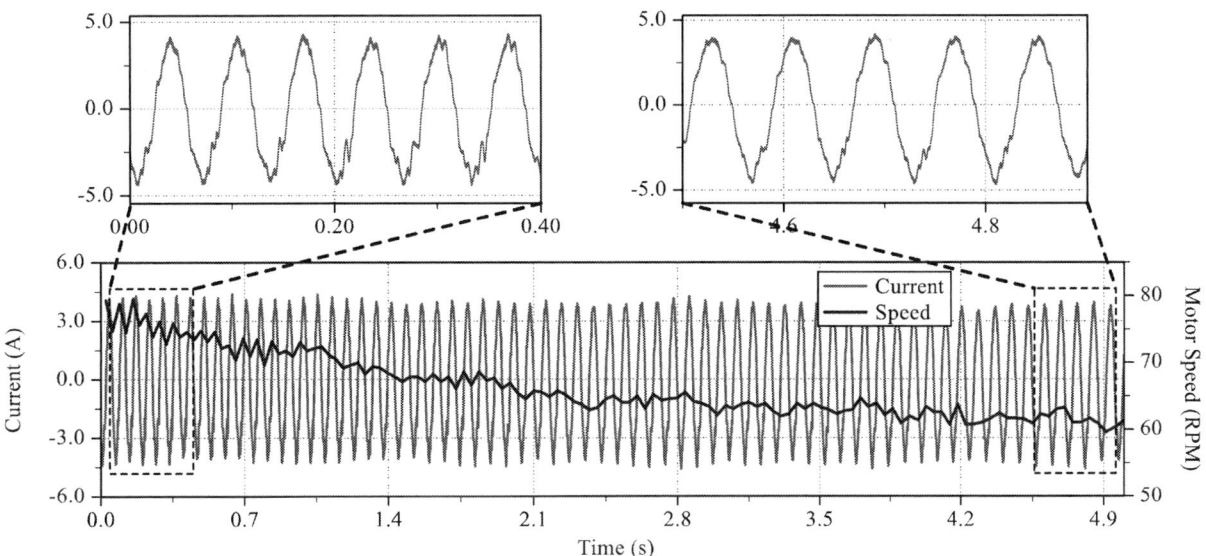

Fig. 8. Phase A current and generator speed at transient.

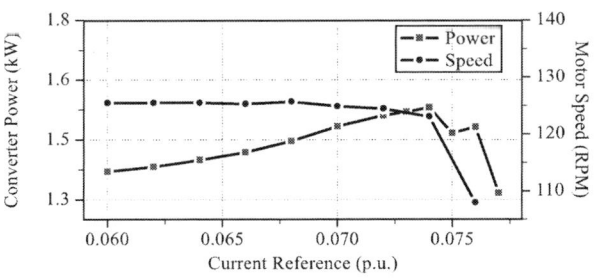

Fig. 9. Operating points near peak power.

linear region near the torque limit where its behavior is similar to the HKTs, as shown in Fig. 9. When the current reference is between 0.68 and 0.74, the setup sees an increasing power output with decreasing speed, and the power starts to drop fast after the optimal operation point. The dynamic current control will adjust the operating point to be near the optimal based on the power measurement. The detailed transient waveform is shown in Fig. 8, which demonstrates the transient for a step change in current reference at 0.2 s. The flickering motor speed causes harmonics in the current waveform which contributes to a higher THD compared with grid-tied rectifier tests. The motor speed starts to decrease slowly after the step change in control, and the entire transition takes about 5 seconds. The change should be kept small to prevent any overshoot that might damage the system.

V. CONCLUSION

This work demonstrated a Vienna rectifier based powertrain for HKT systems and a corresponding dynamic control strategy. The proposed control strategy can adjust the current reference to achieve variable speed control of the generator and track the optimal operating point. The novel powertrain simplifies the design by combining a gen-side rectifier and step-up converter with a Vienna rectifier, which increases the powertrain electrical efficiency and reduces the converter footprint. Hardware tests utilize a generator motor setup with two PMSMs to demonstrate the powertrain's performance at both grid-tied and generator tests and verify the functionality of the proposed dynamic current control to adjust turbine speed and power.

REFERENCES

[1] C. Niebuhr, M. van Dijk, V. Neary, and J. Bhagwan, "A review of hydrokinetic turbines and enhancement techniques for canal installations: Technology, applicability and potential," *Renewable and Sustainable Energy Reviews*, vol. 113, 2019.

[2] W. Tian, J. H. VanZwieten, P. Pyakurel, and Y. Li, "Influences of yaw angle and turbulence intensity on the performance of a 20 kw in-stream hydrokinetic turbine," *Energy*, vol. 111, pp. 104–116, 2016.

[3] Q. Sheng, S. S. Khalid, Z. Xiong, G. Sahib, and L. Zhang, "Cfd simulation of fixed and variable pitch vertical axis tidal turbine," *Journal of Marine Science and Application*, vol. 12, pp. 185–192, 2013.

[4] M. Tariquzzaman, P. Li, S. J. Barton, A. P. Thurlbeck, T. Kilgore, T. K. A. Brekken, and Y. Cao, "Multi-physics and multi-timescale modeling of hydrokinetic turbine energy conversion system," *IEEE Journal of Emerging and Selected Topics in Power Electronics*, pp. 1–1, 2024.

[5] R.-A. Chihaia, I. Vasile, G. Cîrciumaru, S. Nicolaie, E. Tudor, and C. Dumitru, "Improving the energy conversion efficiency for hydrokinetic turbines using mppt controller," *Applied Sciences*, vol. 10, no. 21, 2020.

[6] A. R. Nejad, J. Keller, Y. Guo, S. Sheng, H. Polinder, S. Watson, J. Dong, Z. Qin, A. Ebrahimi, R. Schelenz, F. Gutiérrez Guzmán, D. Cornel, R. Golafshan, G. Jacobs, B. Blockmans, J. Bosmans, B. Pluymers, J. Carroll, S. Koukoura, E. Hart, A. McDonald, A. Natarajan, J. Torsvik, F. K. Moghadam, P.-J. Daems, T. Verstraeten, C. Peeters, and J. Helsen, "Wind turbine drivetrains: state-of-the-art technologies and future development trends," *Wind Energy Science*, vol. 7, no. 1, pp. 387–411, 2022.

[7] M. Luqman, G. Yao, L. Zhou, and A. Lamichhane, "Analysis of variable speed wind energy conversion system with pmsg and vienna rectifier," in *Proc. IEEE Conference on Industrial Electronics and Applications (ICIEA)*, 2019, pp. 1296–1301.

[8] P. Li and Y. Cao, "A fault-tolerance mppt design and model analysis for hydrokinetic turbine systems," in *Proc. IEEE Energy Conversion Congress and Exposition (ECCE)*, 2023, pp. 252–257.

[9] J. Riglin, W. C. Schleicher, and A. Oztekin, "Diffuser Optimization for a Micro-Hydrokinetic Turbine," in *International Mechanical Engineering Congress and Exposition (IMECE)*, vol. 7, 2014.

[10] V. G. Nago, I. F. S. dos Santos, M. J. Gbedjinou, J. H. R. Mensah, G. L. Tiago Filho, R. G. R. Camacho, and R. M. Barros, "A literature review on wake dissipation length of hydrokinetic turbines as a guide for turbine array configuration," *Ocean Engineering*, vol. 259, 2022.

[11] J. M. Walker, K. A. Flack, E. E. Lust, M. P. Schultz, and L. Luznik, "Experimental and numerical studies of blade roughness and fouling on marine current turbine performance," *Renewable Energy*, vol. 66, pp. 257–267, 2014.

[12] L. Myers and A. Bahaj, "Power output performance characteristics of a horizontal axis marine current turbine," *Renewable Energy*, vol. 31, no. 2, pp. 197–208, 2006, marine Energy.

Condition Monitoring for DC-link Capacitors and PV arrays based on the Start-up Process of the PV System

Yongjie Liu, Ariya Sangwongwanich, Chen Liu, Xing Wei, Shuyu Ou
Tamás Kerekes, Jiahong Liu, Huai Wang
AAU Energy, Aalborg University, Aalborg DK-9220, Denmark
yoli@energy.aau.dk, ars@energy.aau.dk, chenl@energy.aau.dk, xwe@energy.aau.dk, so@energy.aau.dk
tak@energy.aau.dk, jliu@energy.aau.dk, hwa@energy.aau.dk

Abstract—This paper proposes a method to enable condition monitoring of DC-link capacitors and PV arrays simultaneously based on the start-up process in the PV system. First, the start-up and DC-link capacitor pre-charge process of the PV system based on PV arrays is analyzed. Then, the capacitance is estimated based on the charging process, which does not require additional hardware and complex algorithms, and at the same time is not affected by operation conditions, e.g., capacitor temperature and load variations. Moreover, the capacitor charging profile also includes information of the output characteristics from PV arrays. Thus, a digital twin model of PV arrays is built to extract series resistance (R_s) as health indicators for evaluating the degradation of PV arrays. Proof-of-concept experiments have been carried out to verify the feasibility of the proposed method.

Index Terms—condition monitoring, reliability, PV system, DC-link capacitors, PV arrays

I. INTRODUCTION

DC-link capacitors (DCC) and PV modules are essential components of the PV system. DCC play an important role in minimizing voltage fluctuations, mitigating harmonic disturbances, and managing transient power imbalances between the front and rear circuits [1]. Aluminum electrolytic capacitors (AECs) are widely applied in PV systems due to the combination of affordability and high energy density [2]. Meanwhile, PV modules are responsible for enabling the conversion of solar energy into usable power. [3].

However, they are susceptible to failure caused by harsh operating conditions in practical long-term operations (effects of electrical and thermal stresses). Thus, it is critical to implement health monitoring of DCCs and PV arrays to ensure the reliable performance of PV systems and avoid unscheduled shutdowns and catastrophic failures. With the degradation, the degradation of capacitors results in a gradual reduction in capacitance and an increase in equivalent series resistance (ESR) due to electrolyte evaporation and electrode corrosion [4]. The failure criteria for AECs is typically a 20% decrease in capacitance compared to the initial value [5]. For PV arrays, the equivalent series resistance (R_s) is increased with the degradation. In the past two decades, considerable research endeavours have focused on developing efficient methods for condition monitoring (CM) of DCCs and PV arrays in PV systems [6], [7].

Current techniques for CM of DCCs in grid-tied photovoltaic (PV) systems are generally categorized into two main approaches: methods based on capacitor voltage and current ripples, and those relying on externally injected signals [8]–[13]. Capacitor ripple-based methods require additional current sensors or calculate capacitor current indirectly to calculate capacitance. In [8], additional sensors are used to measure the DCC ripple voltage and current, and then the Goertzel algorithm is used to determine the ESR and capacitance. To eliminate the requirement for additional sensors, an alternative approach is to indirectly estimate the DC-link ripple current by analysing the output current of the PV array and calculating the input current of the PV inverter based on the three-phase AC current, the switching function, dead time, etc [10]. Moreover, ripple-based methods require high sampling rates, high resolution and high-frequency band-pass filter circuits to capture ripple current/voltage information and are susceptible to interference from operating conditions, measurement errors, and noise. External injected signal-based methods inject an external disturbance signal into the PV grid-connected inverter and then apply specific methods to analyze capacitor voltages and currents to estimate capacitance [11]–[13]. However, these methods require additional hardware and modification of the control strategy and may introduce new risks by interfering with the normal operation of the PV inverter. Moreover, the capacitances of the capacitors are also influenced by their core temperatures, which is also a challenge to obtain the core temperature in practical applications.

Compared to the above two methods, the CM method based on the charge/discharge process does not require additional hardware and high sampling frequency measurement while immunite to interference from operation conditions. It has been investigated in EV applications [14], [15]. However, a detailed discussion on the application of the PV system is absent in the literature. Thus, in this paper, pre-charging DCCs from the DC side of the PV inverter is analyzed, and corresponding condition monitoring technique for DCCs is realized. Moreover, existing literatures mainly focus on con-

979-8-3315-1612-3/25 $31.00 © 2025 IEEE

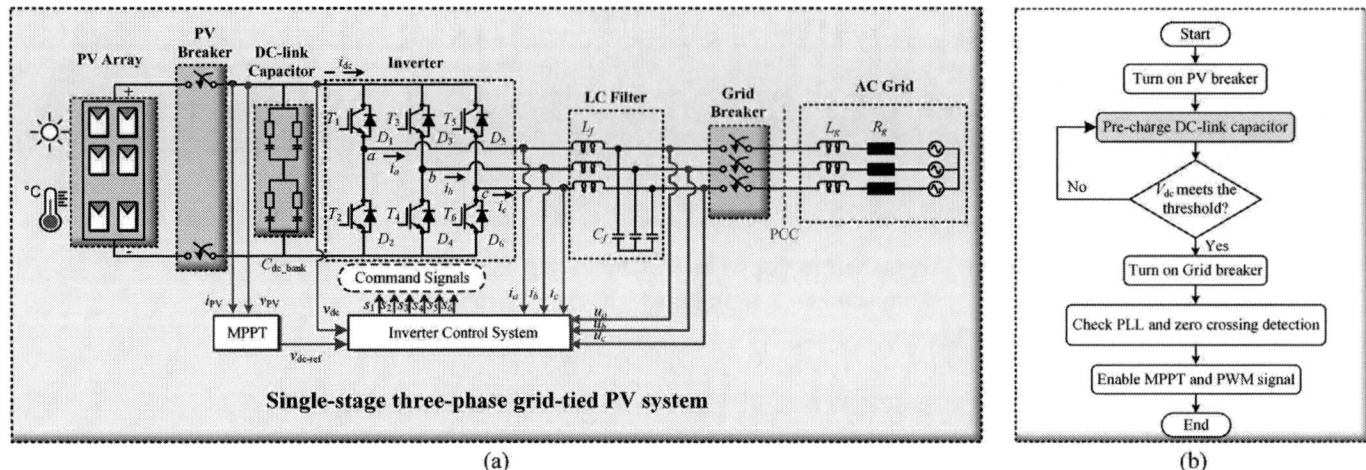

(a)

(b)

Fig. 1. Structure of the single-stage, three-phase grid-tied PV system. (a) System structure on pre-charge DCC from DC-side (PV arrays). (b) Start-up process of the grid-tied PV system charging from PV arrays. i_{PV} and v_{PV} denote the output current and voltage of the PV array during the pre-charge process of the capacitor, respectively. i_a, i_b and i_c are the measured three-phase AC currents. u_a, u_b, u_c are the grid voltages. v_{dc} represents the measured DC-link voltage. C_{dc_bank} represents the capacitance of the DCC.

dition monitoring for capacitors or PV arrays separately and propose the corresponding CM methods for each component.

In fact, the output characteristics of both in the PV system are not independent from each other, but rather correlated with each other. Therefore, this paper explores the utilization of the pre-charging process from the DC side to achieve the CM of the DCC and PV array simultaneously using a digital twin-based CM method.

II. START-UP PROCESS OF THE PV SYSTEM

The single-stage, three-phase, two-level grid-tied PV system illustrated in Fig. 1(a) consists of six IGBTs ($T_1 - T_6$) paired with antiparallel-connected diodes ($D_1 - D_6$). The output currents from the PV inverter are denoted as i_a, i_b and i_c, while u_a, u_b, u_c represent the corresponding grid voltages. The measured DC-link voltage is indicated by v_{dc}. The DCC bank with AECs is connected between the PV inverter and the PV array to limit the current ripple. The PV breaker or grid breaker can automatically connect/disconnect the PV array or AC grid. The start-up process of the PV system is crucial, especially for the pre-charge of DCCs to avoid large inrush currents that can prolong the service life of the power converter. Moreover, PV arrays can limit the pre-charge current to minimize the utilization of pre-charge resistors.

The start-up process of the PV inverter while charging the capacitor from the PV array is illustrated in Fig. 1(b). First, the grid breaker is off/disconnected from the grid. Then, the PV breaker is closed to pre-charge the DCC and the equivalent circuit is depicted in Fig. 2. The measured waveforms during the start-up of the PV system are illustrated in Fig. 3. When the PV breaker is closed, the DCC is short-circuited, and the output of the PV system operates at the short-circuit point (SC Point) ($V_{dc} = 0$). During the capacitor charging process, the voltage increases which makes the output of the PV panel gradually move toward the open-circuit point (OC

Point) ($V_{dc} = V_{oc}$). The I-V characteristic of the PV array is derived via the capacitor charging process, shown in Fig. 3(e). Moreover, it provides a potential for condition monitoring of the capacitor and PV array simultaneously based on the start-up charging process.

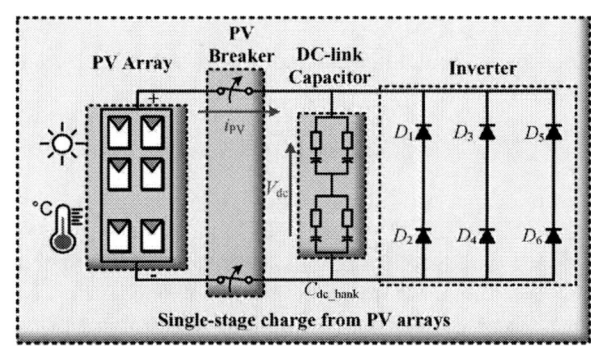

Fig. 2. Equivalent pre-charge circuit from PV arrays. C_{dc_bank} is the capacitance of the DCC. V_{dc} represents the DC bus voltage. i_{PV} represents the output current of PV arrays/DC side current. ($D_1 - D_6$) represent antiparallel connected diodes.

III. PROPOSED CM METHODS BASED ON PRE-CHARGE PROCESS

A. CM for DCCs

In this section, the CM method for the DCC based on the start-up process of the PV system is described. Based on Ohm's law, the instantaneous voltage across the DCC V_{dc} as shown in Fig. 2 at any time " t " can be expressed as:

$$V_{dc}(t) = ESR \times i_{PV}(t) + \frac{1}{C_{dc\_bank}} \int_0^t i_{PV}(t)dt \quad (1)$$

where C_{dc_bank} and ESR represent the capacitance and ESR of the DCC. i_{PV} is the output current of the PV array.

as follows:

$$V_{dc}(t_2) - V_{dc}(t_1) = \frac{1}{C_{dc\_bank}} \int_{t_1}^{t_2} i_{PV}(t)dt \qquad (3)$$

Finally, C_{dc_bank} can be estimated and expressed as follows:

$$C_{dc\_bank} = \frac{\int_{t_1}^{t_2} i_{PV}(t)dt}{V_{dc}(t_2) - V_{dc}(t_1)} \qquad (4)$$

Moreover, the output characteristics of the PV array can be obtained as described in Fig. 3(e), which makes it possible to further complete the CM of the PV array described in the next subsection.

B. CM for PV arrays

In this work, the single-diode model (SDM) is applied to model the I-V characteristic of PV cells / PV modules / PV strings / PV arrays. The SDM is described as follows:

$$I_{PV} = N_p I_{ph} - N_p I_d \left[\exp\left(\frac{q(V_{PV} + I_{PV}R_s N_s/N_p)}{AKT_c N_s} \right) - 1 \right]$$
$$- \frac{V_{PV} + I_{PV}R_s N_s/N_p}{R_{sh}N_s/N_p} \qquad (5)$$

where I_{ph} represents the photocurrent. I_d denotes the saturation current of diode. The series and shunt resistances are represented by R_s and R_{sh}, respectively. The Boltzmann constant K and the electron charge q are $K \approx 1.38064852 \times JK^{-1}$ and $q \approx 1.60217662 \times 10^{-19}C$, respectively. A is the ideality factor of the diode. V_{PV} and I_{PV} represent the output voltage and current. T_c is the temperature of solar cell. N_s and N_p represent the number of series-connected cells and parallel-connected substrings of the PV array, respectively.

As can be seen from (5), there are five parameters $x = [I_{ph}, I_d, A, R_s, R_{sh}]$ need to be estimated. The details of the TSM-240 PV module are presented in Table. I.

$$I_{ph} = \frac{G}{G_{STC}} \left[I_{ph,STC} + K_i I_{ph,STC}(T_c - T_{STC}) \right] \qquad (6)$$

$$I_d = I_{d,STC} \left(\frac{T}{T_{STC}} \right)^{3/A} \exp\left[\frac{qE_g}{AK} \left(\frac{1}{T_{STC}} - \frac{1}{T_c} \right) \right] \qquad (7)$$

The problem of model parameters estimation can be converted into an optimization problem, where the root mean square error (RMSE) is adopted as the objective function in (8) to minimize the error of measured and modeled I-V curves. Moreover, to ensure the stability of the parameter identification results to avoid coupling problems, A is fixed to the typical value of 1, and I_{ph} and I_d are solved via (6) and (7). The unknown parameters $[I_{ph,STC}, I_{d\,and\,STC}]$ under the standard test condition (STC) can be solved based on datasheet parameters [16]. G_{STC} and T_{STC} are the irradiance and temperature under STC. G is the measured irradiance. Thus, parameters R_s and R_{sh} mainly related to degradation are estimated using the grey wolf optimization (GWO) algorithm [17].

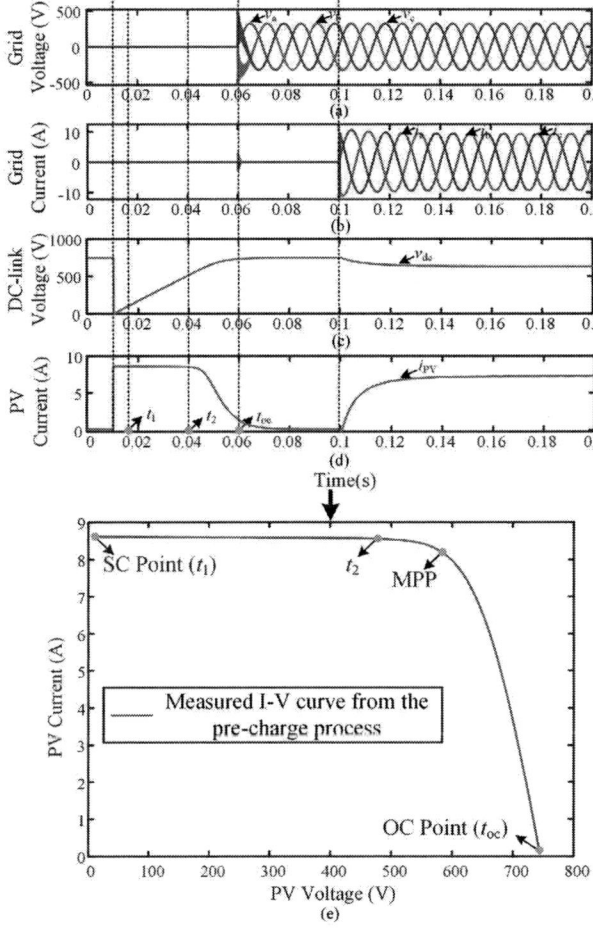

Fig. 3. Measured waveforms during the start-up process of the PV system (pre-charge the DCC from the PV array). (a) Measured three-phase grid voltages (v_a, v_b, v_c). (b) Measured three-phase grid currents (i_a, i_b, i_c). (c) Measured DC-link voltage V_{dc}. (d) Measured output current i_{PV} from the PV array. (e) Measured I-V curve of the PV array from the DCC charging process.

The DC-link voltage change during the charging process is calculated as follows:

$$V_{dc}(t_2) - V_{dc}(t_1) = ESR \times \{i_{PV}(t_2) - i_{PV}(t_1)\}$$
$$+ \frac{1}{C_{dc\_bank}} \int_{t_1}^{t_2} i_{PV}(t)dt \qquad (2)$$

From the short-circuit point to the maximum power point (MPP) of the PV array, the current changes are relatively small and almost constant ($i_{pv}(t_1) \approx i_{pv}(t_2)$). The waveforms of the start-up process including grid voltage, current, DC voltage and PV current are shown in Fig. 3. Moreover, the ESR of AECs is typically around 0.2 Ω, which produces a small voltage drop (less than 1 V) at low currents and can be neglected compared to the remarkable voltage changes during the charging process of the capacitor. Thus, the effect of ESR on the capacitance estimation can be eliminated and expressed

979-8-3315-1612-3/25 $31.00 © 2025 IEEE

TABLE I
SPECIFICATION OF THE TSM-240 PV MODULE

Parameters	Value
Maximum power ($P_{mpp,STC}$)	240 W
Voltage at maximum power point ($V_{mpp,STC}$)	29.7 V
Current at maximum power point ($I_{mpp,STC}$)	8.1 A
Open-circuit voltage ($V_{oc,STC}$)	37.3 V
Short-circuit current ($I_{sc,STC}$)	8.62 A
Temperature coefficient of current (K_i)	0.047 %/°C
Temperature coefficient of voltage (K_v)	-0.32 %/°C

The proposed CM method for the PV array is shown in Fig. 4. The objective function is outlined below:

$$ RMSE = \sqrt{\frac{1}{N_{points}} \sum_{i=1}^{N_{points}} \left(\frac{I_{model,i} - I_{meas,i}}{I_{meas,i}} \right)^2 } \quad (8) $$

where N_{points} represents the number of points on the I-V curve. $I_{model,i}$ and $I_{meas,i}$ denote the modeled and measured currents, respectively.

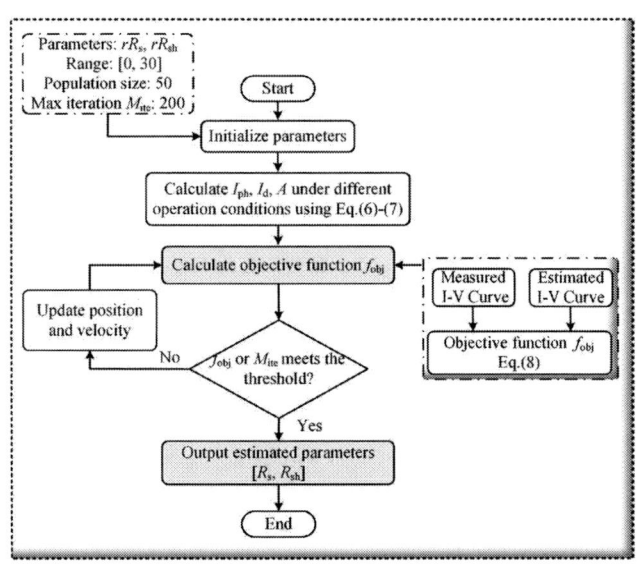

Fig. 4. Proposed digital twin method-based CM for PV arrays. f_{obj} is the objective function. R_s and R_{sh} are the estimated series and parallel resistances. I_{ph} and I_d are model parameters of the PV array for photocurrent and the saturation current of the diode, respectively.

IV. EXPERIMENTAL VERIFICATION AND ANALYSIS BASED ON THE HIL PLATFORM

A. Experimental results and analysis for capacitors CM

To evaluate the efficiency of the proposed method, a real-time hardware-in-the-loop (HIL) platform is implemented, as depicted in Fig. 5. The measured DC-link voltage (V_{dc}) and PV current (i_{PV}) during the start-up of the PV system are presented in Fig. 6. Then, the capacitance is calculated based on the CM method for the DCC described in Section III. A.

The capacitors are tested at different degradation levels from Case1 (healthy: 680μF) to Case5 (degradation: 544μF) with the capacitance decreased by 5% for each case in turn. The estimated results are provided in Fig.7. Estimated results are consistent with the actual degradation, which demonstrates the efficiency of the proposed method. The maximum error of estimated results is 0.06%.

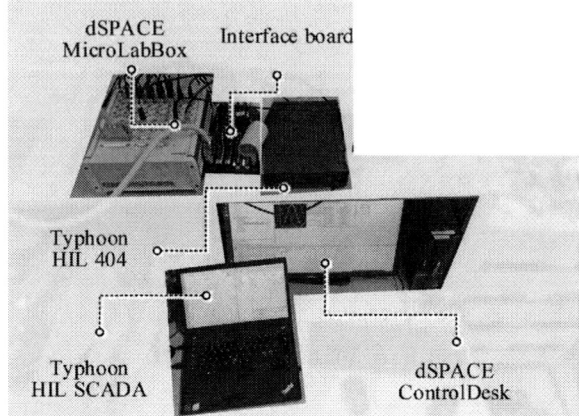

Fig. 5. A HIL-based experimental platform.

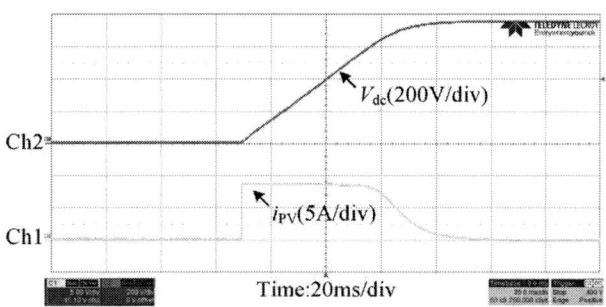

Fig. 6. Measured charging curves of DCCs (pre-charge from the PV arrays) based on the HIL platform. V_{dc} (200V/div) and i_{PV} (5A/div) are the measured voltage and the output current of the PV array, respectively.

B. Experimental results and analysis for PV arrays CM

The capacitor charging process is exploited to obtain the output characteristics of PV arrays simultaneously described in Section II and the CM of PV arrays is realized by the proposed method described in Section III.B. Experiments are tested at different degradation levels of PV arrays with accompanying increase of R_s from Case6 (healthy: $R_{s,STC}$), Case7 ($1.05*R_{s,STC}$), Case8 ($1.1*R_{s,STC}$), Case9 ($1.15*R_{s,STC}$), Case10 ($1.2*R_{s,STC}$), Case11 ($1.3*R_{s,STC}$), and experimental testing conditions of irradiance and temperature are 1000 W/m^2 and 25°C, respectively. The estimated results of R_s are displayed in Fig. 8 and are quite consistent with the actual degradation. The maximum error of estimated results is 0.88%, which further validates the efficiency of the proposed method.

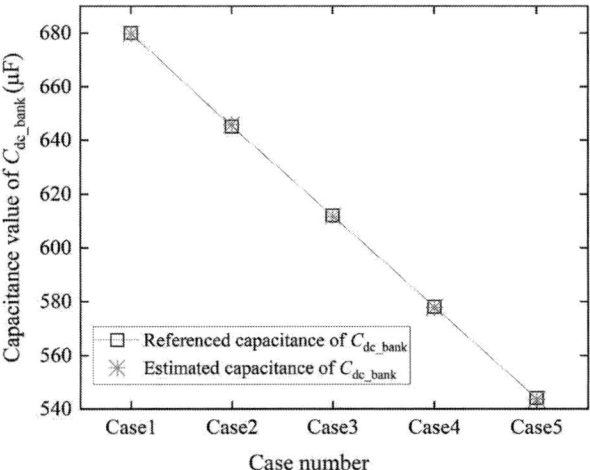

Fig. 7. Estimated results for the capacitance of C_{dc_link}. Case1 to Case5 represent the different degradation levels. The capacitance decreases with the degradation from Case1 to Case5 (Case1: 680μF, Case2: 645μF, Case3: 612μF, Case4: 578μF, Case4: 544μF)

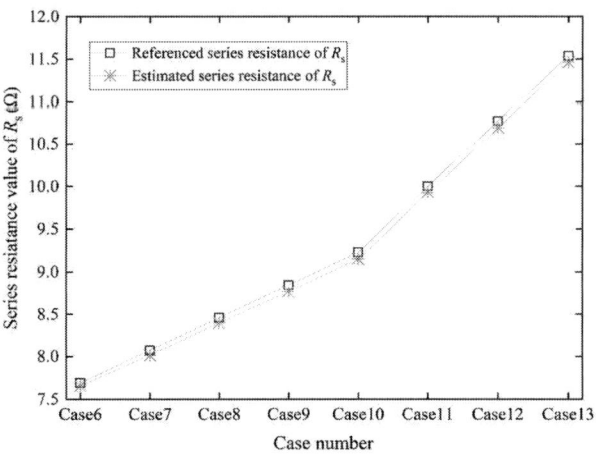

Fig. 8. Estimated results for the series resistance of R_s for PV arrays. Case6 to Case13 represent the different degradation levels of the PV array. The series resistance gradually increases with the degradation from Case6 to Case13 (Case6: 7.69Ω, *Case*7 : 8.08Ω, Case8: 8.46Ω, Case9: 8.85Ω, Case10: 9.23Ω, Case11: 9.99Ω, Case12: 10.69Ω, Case13: 11.54Ω)

V. EXPERIMENTAL VERIFICATION AND ANALYSIS BASED ON THE EXPERIMENTAL PLATFORM

A. Experimental verification and analysis for capacitors CM

To further verify the proposed method, the experimental setup for charging the DCC from the DC-side (PV array) is built as shown in Fig. 9. The PV simulator with high bandwidth (using linear amplify) is connected to the PV inverter via the PV breaker. The DCC of the PV inverter is composed of four aluminium electrolytic capacitors (EPCOS, 680uF, 400V, ±20%) arranged in series and parallel to increase the voltage and current capability. PV simulator software is used to remote control the output characteristics of the PV simulator. Then, the pre-charging of the DCC is accomplished

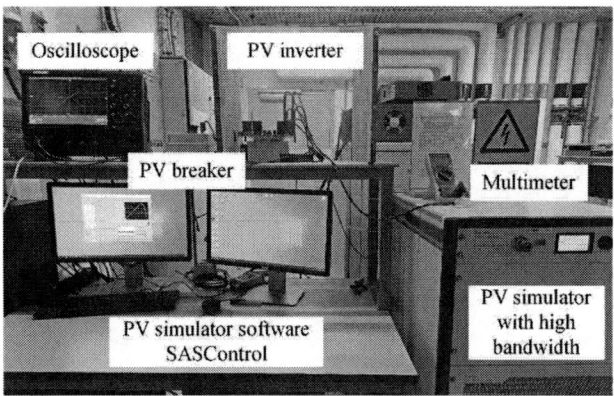

Fig. 9. Experimental platform of charging the DCC from the DC-side (PV array).

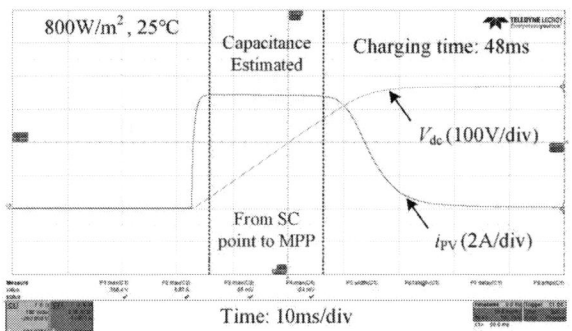

Fig. 10. Measured charging curves of DCCs (pre-charge DCCs from the PV arrays) based on the experimental platform. V_{dc} (100V/div) and i_{PV} (2A/div) are the measured DC-link voltage and the output current of the PV array, respectively.

by activating the PV breaker after the PV simulator is powered up and the waveforms of the pre-charging process are captured by an oscilloscope.

The measured charging curves are shown in Fig. 10. To ensure the accuracy of the capacitance estimation, currents and voltages of the DC-link voltage V_{dc} from 20V to 150V are extracted for the capacitance estimation. Because the PV current is almost constant ($i_{pv}(t_1) \approx i_{pv}(t_2)$) in this region, the effect of ESR can be neglected. Moreover, the capacitance of the capacitor is changed by connecting more capacitors in parallel to gradually decrease from the initial value of 665μF to 582μF, which emulates a decrease in the capacitance of the capacitor after degradation. The real capacitances are measured by the LCR meter at 100Hz and 23.7°C for 665μF, 645μF, 626μF, 602μF and 582μF, respectively. Then, capacitances are estimated under different irradiances as shown in Fig. 11. At low irradiation of 100 W/m², the maximum estimation error of capacitance is maximum between 2% and 3%, which is acceptable compared to the 20% drop in capacitance after the failure of the capacitor with the degradation.

B. Experimental results and analysis for PV arrays CM

The model parameters of PV arrays are identified based on the proposed method in Section II.B. The convergence curves

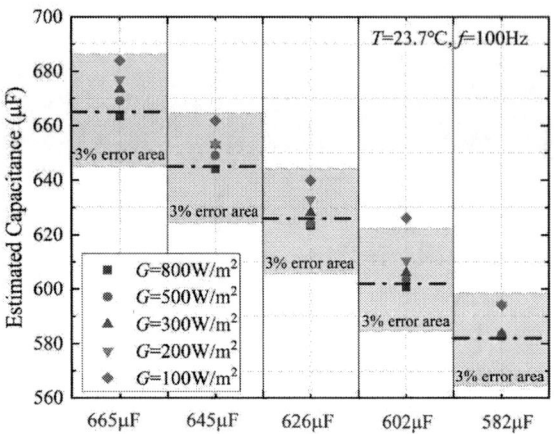

Fig. 11. Results of the estimated capacitance for different degradation levels under different irradiances.

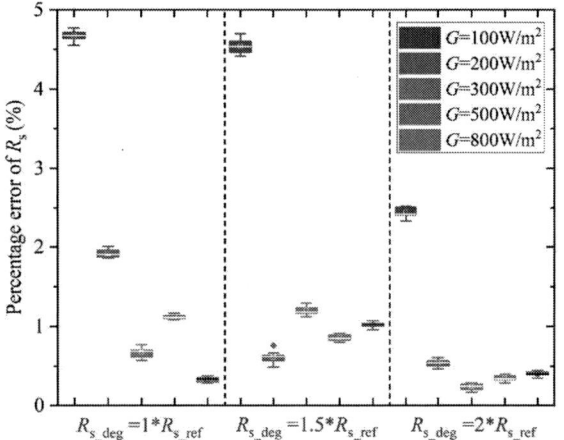

Fig. 12. Estimated results of model parameters for the PV array with different degradation states under different irradiances.

of the digital twin model at different irradiances increase gradually from 800 W/m² with RMSE error of 0.0065 A to 100 W/m² with RMSE error of 0.038 A, which can be seen from the convergence curve with a large error under low irradiation due to the measurement error. The estimation errors of R_s for different degradations and environmental conditions are analyzed as illustrated in Fig. 12, reveals that these errors are more pronounced under reduced irradiance levels. Moreover, the estimation accuracy is improved as the degradation increases and the maximum error is limited to 5%.

VI. CONCLUSIONS AND FUTURE WORK

In this paper, a method for concurrent CM of DCCs and PV arrays based on the start-up process in the PV system is presented. The advantages of the method are summarized as follows: 1) No extra hardware and sensor; 2) Without complex algorithms and modifications of the control. 3) Independent of operation conditions, capacitor temperature, and load variations. Experimental results of the HIL platform indicate that the maximum estimation error of the capacitance is 0.06%

and the estimation error of (R_s) the PV array is 0.88%. The experimental results of the experimental platform show that the estimation errors of the capacitance and the model parameters R_s of the PV arrays are limited to 3% and 5%.

REFERENCES

[1] H. Wang, H. Wang, G. Zhu, and F. Blaabjerg, "An overview of capacitive dc-links-topology derivation and scalability analysis," *IEEE Trans. Power Electron.*, vol. 35, no. 2, pp. 1805–1829, 2020.

[2] P. Sundararajan, M. H. M. Sathik, F. Sasongko, C. S. Tan, M. Tariq, and R. Simanjorang, "Online condition monitoring system for dc-link capacitor in industrial power converters," *IEEE Trans. Ind. Appl.*, vol. 54, no. 5, pp. 4775–4785, 2018.

[3] M. Meribout, "Sensor systems for solar plant monitoring," *IEEE Trans. Instrum. Meas.*, vol. 72, pp. 1–16, 2023.

[4] H. Wang and F. Blaabjerg, "Reliability of capacitors for dc-link applications in power electronic converters—an overview," *IEEE Trans. Ind. Appl.*, vol. 50, no. 5, pp. 3569–3578, 2014.

[5] Z. Zhao, H. Hu, Z. He, W. Lu, H. H.-C. Iu, F. Blaabjerg, and P. Davari, "A transient-modeling-based grey-box method for online monitoring of dc-link capacitors," *IEEE Transactions on Power Electronics*, vol. 38, no. 11, pp. 14 547–14 562, 2023.

[6] T. McGrew, V. Sysoeva, C.-H. Cheng, C. Miller, J. Scofield, and M. J. Scott, "Condition monitoring of dc-link capacitors using time–frequency analysis and machine learning classification of conducted emi," *IEEE Trans. Power Electron.*, vol. 37, no. 10, pp. 12 606–12 618, 2022.

[7] A. Triki-Lahiani, A. B.-B. Abdelghani, and I. Slama-Belkhodja, "Fault detection and monitoring systems for photovoltaic installations: A review," *Renewable Sustain. Energy Rev.*, vol. 82, pp. 2680–2692, 2018.

[8] P. Sundararajan, M. H. M. Sathik, F. Sasongko, C. S. Tan, J. Pou, F. Blaabjerg, and A. K. Gupta, "Condition monitoring of dc-link capacitors using goertzel algorithm for failure precursor parameter and temperature estimation," *IEEE Trans. Power Electron.*, vol. 35, no. 6, pp. 6386–6396, 2019.

[9] L. Ren, C. Gong, and Y. Zhao, "An online esr estimation method for output capacitor of boost converter," *IEEE Trans. Power Electron.*, vol. 34, no. 10, pp. 10 153–10 165, 2019.

[10] M. W. Ahmad, N. Agarwal, and S. Anand, "Online monitoring technique for aluminum electrolytic capacitor in solar pv-based dc system," *IEEE Trans. Ind. Electron.*, vol. 63, no. 11, pp. 7059–7066, 2016.

[11] Y. Wu and X. Du, "A ven condition monitoring method of dc-link capacitors for power converters," *IEEE Trans. Ind. Electron.*, vol. 66, no. 2, pp. 1296–1306, 2019.

[12] Y. Gupta, M. W. Ahmad, S. Narale, and S. Anand, "Health estimation of individual capacitors in a bank with reduced sensor requirements," *IEEE Trans. Ind. Electron.*, vol. 66, no. 9, pp. 7250–7259, 2018.

[13] P. Sun, C. Gong, X. Du, Q. Luo, H. Wang, and L. Zhou, "Online condition monitoring for both igbt module and dc-link capacitor of power converter based on short-circuit current simultaneously," *IEEE Trans. Ind. Electron.*, vol. 64, no. 5, pp. 3662–3671, 2017.

[14] T. F. Baumann, R. Murillo-Garcia, K. Papastergiou, and D. Peftitsis, "Discharge-based condition monitoring for electrolytic dc-link capacitors," *IEEE Trans. Power Electron.*, vol. 39, no. 12, pp. 16 622–16 637, 2024.

[15] H. Li, D. Xiang, X. Han, X. Zhong, and X. Yang, "High-accuracy capacitance monitoring of dc-link capacitor in vsi systems by lc resonance," *IEEE Trans. Power Electron.*, vol. 34, no. 12, pp. 12 200–12 211, 2019.

[16] T. Ma, W. Gu, L. Shen, and M. Li, "An improved and comprehensive mathematical model for solar photovoltaic modules under real operating conditions," *Sol. Energy*, vol. 184, pp. 292–304, 2021.

[17] S. Mohanty, B. Subudhi, and P. K. Ray, "A grey wolf-assisted perturb observe mppt algorithm for a pv system," *IEEE Trans. Energy Convers.*, vol. 32, no. 1, pp. 340–347, 2017.

Electrically and Thermally Efficient Reliable Power Converter Design for micro–Hydrokinetic Turbine

Md Tariquzzaman
School of EECS
Oregon State University
Corvallis, USA
tariquzm@oregonstate.edu

Peidong Li
School of EECS
Oregon State University
Corvallis, OR, USA
lipeid@oregonstate.edu

Yue Cao
School of EECS
Oregon State University
Corvallis, OR, USA
yue.cao@oregonstate.edu

Abstract—The wider adoption of hydrokinetic turbines (HKTs) is hindered by challenges such as high maintenance costs, low power conversion efficiency, and the reliability issues of power converters. This paper presents a novel power converter design optimized for 5 kW micro-HKT applications to address these challenges. The proposed design enhances electrical and thermal efficiency by utilizing Silicon Carbide (SiC) power MOSFETs. This study evaluates several converter topologies and semiconductor devices, comparing their conversion efficiency and case temperature through multi-physics HKT system average modeling. Based on these evaluations, a two-level SiC-based voltage source inverter (VSI) is selected and uniquely designed to enhance electrical efficiency, improve thermal management, and reduce the failure rate. The efficient power converter layout allows the DC bus capacitor to be placed onshore, with smaller parallel capacitors within the offshore HKT unit. This compact design is crucial for HKT applications, where space and weight are significant constraints. Hardware experimental results demonstrate that the developed SiC-based power converter achieves 98.4% efficiency at the rated 5 kW power output, with a maximum device case temperature of approximately 55.1°C under natural conduction cooling.

Index Terms—Hydrokinetic turbine, reliability, thermal, multiphysics, power converter, Aluminum substrate PCB

I. INTRODUCTION

Hydrokinetic turbines (HKTs) offer a viable alternative to solar and wind energy, particularly in regions with consistent water currents. These turbines can mitigate the intermittency of solar and wind power, thereby complementing overall power demand. Despite the potential, the power converter drivetrain's high maintenance costs and reliability concerns remain significant barriers to broader adoption [1].

The HKT drivetrain undergoes several energy conversions, including turbine-generator interactions, back-to-back power electronic conversions, and LCL filters, all contributing to energy losses. While its electrical topology and control mechanisms are similar to those in direct-drive wind turbine systems, the HKT electrical system exhibits unique thermal characteristics [2]. Heat transfer from the submerged power converter's enclosure to the surrounding water is primarily driven by

forced convection, which fluctuates with changes in hydro flow rates and temperatures [3].

The power converter is particularly susceptible to wear in submerged applications, directly impacting maintenance intervals and overall system availability [4]. Improving the reliability of the power converter subsystem is essential and achievable by minimizing component count and reducing thermal stress. Thus, choosing an appropriate converter circuit topology and designing an efficient layout is vital to meeting these objectives. Additionally, low-speed, high-torque micro-HKT applications require a high pole count permanent magnet synchronous generators (PMSGs), resulting in higher terminal voltages. Hence, DC bus voltage becomes higher typically between 700 V and 800 V [5]. Careful selection of insulation and thermal interface materials is necessary to enhance heat dissipation and insulation performance.

This paper presents the design of a two-level back-to-back voltage source inverter tailored for micro-HKT applications, optimized for high electrical and thermal efficiency, reliability, and a minimal component count. The power converter utilizes a stacked configuration with distinct power and control boards. The power stage board is built on an aluminum substrate single-layer PCB for superior thermal performance, minimizing the heat-sink enclosure footprint. In contrast, the control board is implemented on a four-layer FR4 substrate due to cost-effectiveness and low power traces.

The paper is organized as following. The micro-HKT system architecture is covered in Section II, followed by the power converter multiphysics average model, which is discussed in Section III. Section IV details the topology and device selection process. Section V addresses solutions to thermal challenges with a compact converter layout design. Hardware experiments validating the electrical and thermal efficiency of the power converter are provided in Section VI, and the paper concludes in Section VII.

II. HKT SYSTEM ARCHITECTURE

The architecture of the 5 kW micro-HKT system is illustrated in Fig. 1, featuring a back-to-back power converter with a filter to optimize power extraction from the HKT turbine and ensure robust grid integration. The HKT, PMSG, and generator-side inverter with microcontroller are housed

---

This work was supported in part by the U.S. Department of Energy Advanced Research Projects Agency-Energy (ARPA-E) under Award DE-AR0001438 and in part by the U.S. Department of Energy Water Power Technologies Office under Award DE-EE0011381.

979-8-3315-1612-3/25 $31.00 © 2025 IEEE

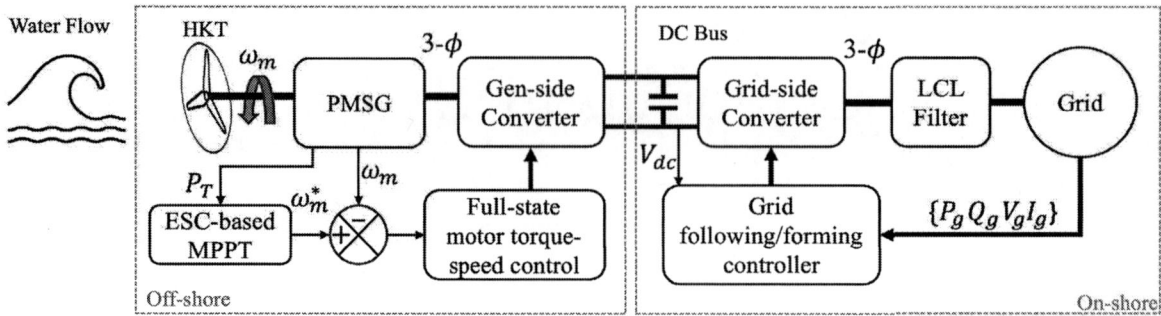

Fig. 1: Overview of HKT system level architecture.

offshore within a water-sealed nacelle. Maximum power can be achieved by deploying model-based maximum power point tracking (MPPT) control [5], hill-climbed MPPT control [6], or extremum-seeking control (ESC) [7]. In this paper, an ESC controller is adopted, which monitors the PMSG power output (P_T) and rotational speed (ω_m) and then adjusts the HKT-generator rotational speed reference (ω_m^*) to extract maximum power. Moreover, the ESC algorithm is particularly well-suited for HKT applications because the turbine's optimal power coefficient varies over time due to biofouling [8] and cavitation effects [9]. ESC can inherently track the optimal power trajectory despite the deterioration of the turbine characteristics.

The offshore generator-side power converter manages the torque and speed of the PMSG using full-state d-q reference frame-based field-oriented control. This offshore converter is essential for maintaining optimal turbine-generator operation, improving reliability crucial to ensure smooth HKT performance. The main DC bus capacitors are placed onshore, while smaller ones are placed offshore, thus reducing the offshore converter size. This configuration reduces the offshore footprint and component count, thereby enhancing reliability. The onshore grid-side inverter regulates the DC bus voltage (V_{dc}) and supplies power to grid-connected or islanded loads using grid-following or grid forming control. During islanded operation, active (P_g) and reactive (Q_g) power requirement set the operating frequency and voltage at any instant.

III. MULTIPHYSICS AVERAGE MODEL OF POWER CONVERTER

This study employs multiphysics average (MLPA) modeling to analyze critical subsystems, including the HKT, PMSG, and power converter, as outlined in [5]. The MLPA framework utilizes a first-principles, physics-based approach encompassing electrical, mechanical, and thermal domains. The analysis evaluates power losses and temperature across various subsystems under specific operating conditions, enabling informed design decisions.

This paper pays particular attention to power loss in semiconductor devices and capacitors. Insights from this analysis guide the topology and device selection, as discussed in Section IV. The primary contributors to power converter loss include switching and/or conduction losses in semiconductor

devices, diodes, and capacitors. Switching losses are calculated based on energy dissipation during the turn-on and turn-off processes, which depend on operating conditions (e.g., current and voltage) and MOSFET parasitic capacitance. The impact of gate-to-drain parasitic capacitance ($C_{gd's}$) and operating condition on voltage fall (tvf) and rise (tvr) time is calculated to quantify switching loss accurately, as mentioned in [10] using

$$t_{vf} = \frac{\frac{V_{DC}}{2}R_{gon}C_{gd1} + (\frac{V_{DC}}{2} - I_d R_{dson})R_{gon}C_{gd2}}{V_{drv} - V_m} \quad (1)$$

$$t_{vr} = \frac{\frac{V_{DC}}{2}R_{goff}C_{gd1} + (\frac{V_{DC}}{2} - I_d R_{dson})R_{goff}C_{gd2}}{V_m} \quad (2)$$

where R_{gon}, R_{goff}, and R_{dson} are gate turn-on, turn-off, and drain-to-source resistor, respectively; I_d, V_{drv}, V_{DC} and V_m are drain average current, gate drive voltage, DC bus and miller plateau voltage, respectively. Then, the switching loss per device (P_{sw}) at f_{sw} switching frequency is calculated using

$$P_{sw} = [V_{DC}I_d\frac{(tvf + tcr + tvr + tcf)}{2} + Q_{rr}V_{DC}]f_{sw} \quad (3)$$

where tcr, tcf, and Q_{rr} are the current rise, fall time, and diode reverse recovery charge, respectively, taken from the manufacturer datasheet.

The conduction loss in the MOSFETs (P_{cd}), body diode (P_d), and DC bus capacitor equivalent series resistance loss (P_c) are calculated as

$$P_{cd} = I_r^2 R_{dson}(Tp) \quad (4)$$

$$P_d = \frac{Q_{rr}V_{DC}f_{sw}}{4} \quad (5)$$

$$P_c = I_{cr}^2 R_{esr} \quad (6)$$

$$P_{ts} = P_{sw} + P_{cd} \quad (7)$$

where P_{ts} is the total loss dissipated in the semiconductor device. Using a thermal model, the calculated P_{ts} is used to measure the devices junction temperature stress, as shown in Fig. 2. Here, the semiconductor device average thermal model considers lumped thermal resistance across different interfaces such as junction to case(C_{jc}), thermal paste/grease (C_{tp}), heatsink (C_{hs}), and convective water cooling equivalent thermal

979-8-3315-1612-3/25 $31.00 © 2025 IEEE

resistance (C_{cwc}) to calculate junction (T_j) and case temperature (T_c) as

$$T_j = P_{ts}(C_{jc} + C_{tp} + C_{hs} + C_{cwc}) + T_a \qquad (8)$$

$$T_c = P_{ts}(C_{tp} + C_{hs} + C_{cwc}) + T_a \qquad (9)$$

where T_a is the ambient temperature outside the thermal boundary layer. The lumped convective water cooling thermal resistance C_{cwc} includes the thermal effort of the water flow condition. The detailed thermal modeling of the power converter for HKT application is discussed in [5].

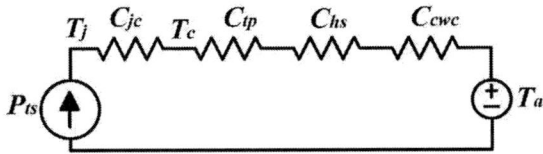

Fig. 2: Overview of HKT system level architecture.

IV. TOPOLOGY AND DEVICE SELECTION

The performance of a power converter is highly dependent on selecting the appropriate topology for the specific application. This decision must be guided by thoroughly understanding the system's requirements. This paper considers a 5 kW micro direct-drive HKT paired with a PMSG for electromechanical energy conversion. Direct-drive systems demand a compact PMSG with high torque and current, resulting in a higher machine constant and increased terminal voltage at the turbine-generator's rated speed operation.

The proposed methodology incorporates topology and device selection guided by the MLPA model. Fig. 3 presents the high-level design flow utilized in this process. The first step involves defining system requirements, such as DC bus voltage, power level, current, switching frequency, and harmonic limits, tailored for the micro-HKT application. A heuristic search is employed to identify the optimal DC bus voltage and switching frequency. The voltage range is constrained by the PMSG output and the maximum modulation index, while the frequency is selected based on filter size and IEEE 1547 THD standards. Subsequently, various feasible topologies for HKT applications are compared in terms of efficiency and reliability to identify the most suitable configuration. Among various possible gen-side converter topologies, front-end diode rectifiers followed by DC-DC converters or LLC resonant converters are adopted in [11]. However, this topology increases component count and extra diode conduction loss in overall conversion.

In contrast, an active power rectifier is preferred due to lower loss and the ability to control the PMSG-turbine speed. Considering 700 V DC bus voltage, a three-level VSI may be a compelling candidate but requires more components. However, with the availability of 1.2 kV SiC MOSFETs, converter topology is selected by comparing two- and three-level converter topology using the MLPA model.

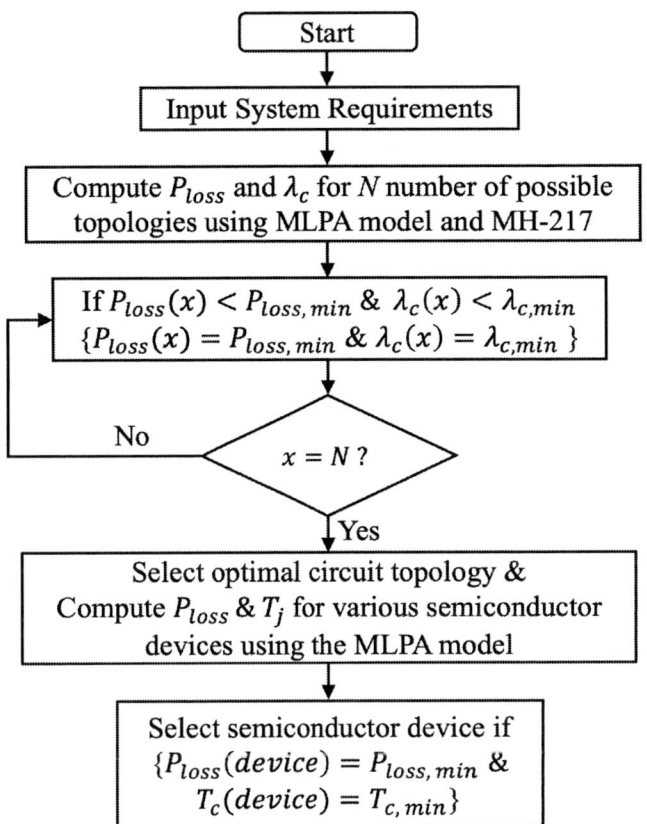

Fig. 3: High-level flow diagram for topology and device selection process.

Based on the MLPA model, the two-level converter topology achieves 0.87% higher efficiency. Moreover, according to the model in Military Handbook-217 [12], the converter topology failure rate (λ_c) is calculated based on the number of elements (N) with the reference failure rate (λ_{ref}), and number of components (k) using

$$\lambda_c = \sum_{i=1}^{k} \lambda_{ref}(i) N \qquad (10)$$

This analysis shows that the two-level VSI has a 50% lower total component failure rate due to having 50% fewer components.

After determining a two-level VSI as the optimal topology, device selection is performed based on a detailed power loss and thermal performance analysis. Since the two-level converter has fewer switches, it faces higher voltage stress and harmonics than the three-level VSI. At a 700-800 V DC bus, 1.2 kV rated semiconductor devices are considered to allow extra room for unwanted voltage spikes during turn-off transients and ensure safety. Thus, this paper selects four compelling semiconductor devices with 1.2kV breakdown voltage. With the selected semiconductor devices, simulation-based analysis is performed using the MLPA model. During the comparative analysis, operating conditions and design, such as power

979-8-3315-1612-3/25 $31.00 © 2025 IEEE

level, switching frequency, and thermal efforts, are assumed to be identical. Table I presents semiconductor devices' part numbers, packages, and the MLPA model's calculated peak efficiency, maximum case temperature ($T_{c,max}$), and energy generation. Energy generation is calculated over five years using water velocity profile data from [13]. Among the devices tested, NTBG020N120SC1 outperforms in terms of efficiency, thermal stress, and energy generation over five years from its counterparts and is selected to design the power converter board.

TABLE I: Simulation Based Comparative Study for Device Selection [5]

Manufacturer Part Number	Package	η_{peak} [%]	$T_{c,max}$ (^0C)	E_{gen} [MWh]
NTBG020N120SC1	D2PAK	98.86	40.78	22.55
NTH4L014N120M3P	D2PAK	98.42	46.18	21.84
C2M0025120D	TO-247	98.15	59.20	20.08
C3M0021120D	TO-247	97.43	60.94	17.73

V. CONVERTER LAYOUT DESIGN

Unlike wind turbines, HKTs usually operate in a direct-drive configuration, resulting in high current operation relative to the power level, posing challenges in designing a high-power layout without increasing the footprint. Traditional FR4 substrate PCBs exhibit poor thermal conductivity, increasing thermal stress on the power converters. Common solutions to improve heat dissipation and provide voltage insulation are the interfacing silicon thermal pad, metal core PCB, and direct bonded copper (DBC) [14]. Though DBC has the lowest thermal resistance compared to metal core PCB, it increases production costs and is particularly applicable in high-power applications. For the 5 kW micro-HKT application, an aluminum (Al) substrate PCB is utilized for the power stage board, as depicted in Fig. 4, to address thermal conductivity challenges. The design employs a 6061 Al material substrate, which provides exceptional thermal conductivity of 150 W/m·K, while the insulating layer achieves a thermal conductivity of 1.5 W/m·K. Detailed improvements in thermal performance are discussed in Section VI.

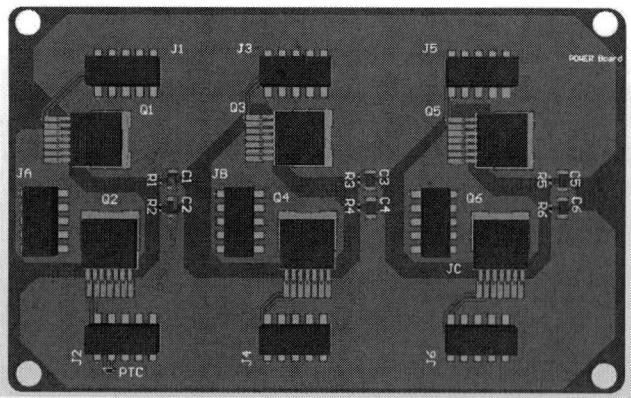

Fig. 4: Aluminum substrate single-layer power stage board.

The power stage board is equipped with SiC MOSFETs (Q1 to Q6), demonstrating higher efficiency under rated operating conditions, as shown in Table I. Smaller ceramic capacitors (C1 to C6) are placed across the drain and source of each leg to facilitate the connection of large DC bus capacitors located onshore. To address potential voltage imbalances caused by series-connected capacitors, parallel resistors are placed for voltage balancing. These smaller capacitors create a shorter commutation path, as shown in the red dotted line in Fig. 4, resulting in lower loop inductance. The board is tested with currents up to 20 A without overheating, thanks to the 2-oz copper, 0.150 mm dielectric, and 1.247 mm aluminum layers at the bottom, as shown in Fig. 5.

The control board is fabricated on a four-layer FR4 substrate PCB, stacked on the power stage board to minimize the overall gate loop and reduce footprint. Since the SiC MOSFETs are vulnerable to short circuit faults and quickly break down within 1 to 3 μs, conventional fuse protection will not prevent power converter damage. Thus, fast-acting desaturation (desat) or over-current protection at the microcontroller level is required. This control board incorporates desat protection, over-current protection, and device temperature monitoring features. The desat protection is realized using the new UCC21755Q1 TI gate driver integrated circuit. In contrast, the software includes over-current protection using the F28379D control card built-in comparator ADC pin without extra circuitry.

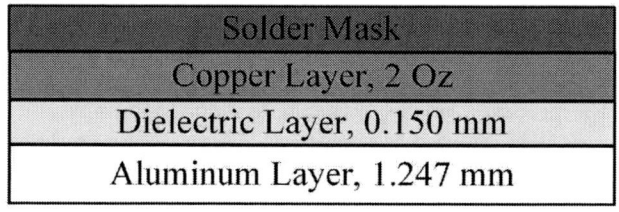

Fig. 5: Power stage board layer structure.

VI. HARDWARE EXPERIMENT RESULTS

A series of hardware experiments is conducted to validate the performance of the selected power converter design. Fig. 6 presents the experimental setup, which includes the generator-turbine system and the power converter hardware. In this configuration, a PMSM motor is used to emulate the behavior of an HKT system, acting as a prime mover with its torque controlled through external drives. The generator-side converter controls the water-sealed PMSG, ensuring precise regulation of its electrical output. A wide-angle view of the complete experimental setup is shown in Fig. 6(c) for a comprehensive understanding of the test arrangement.

The experimental results highlight the thermal performance and efficiency of the power converter employed in the micro-HKT system. A thermal image captured using a FLIR camera is shown in Fig. 7(a), revealing a maximum temperature of 55.1°C on the high-power section of the power converter's

979-8-3315-1612-3/25 $31.00 © 2025 IEEE

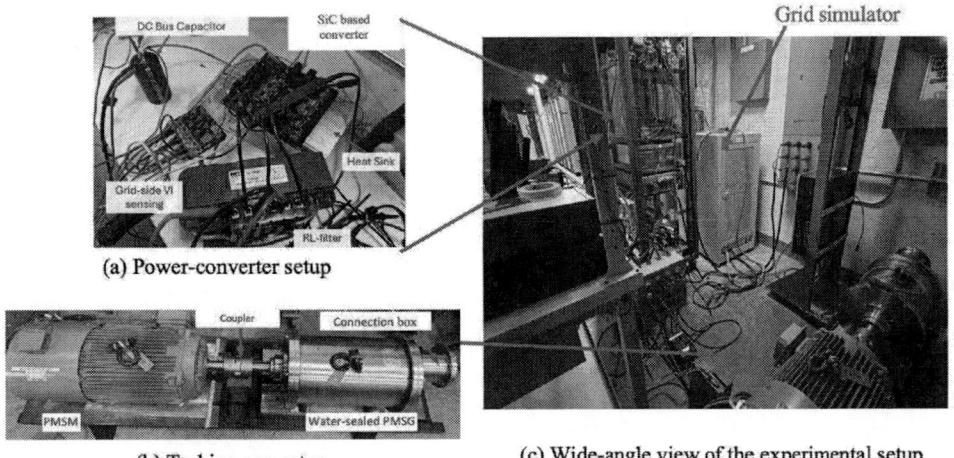

(a) Power-converter setup

(b) Turbine-gen setup

(c) Wide-angle view of the experimental setup

Fig. 6: Hardware experimental setup: (a) power converter, (b) turbine-generator, and (c) wide-angle view.

(a) PCB surface temperature using FLIR thermal camera

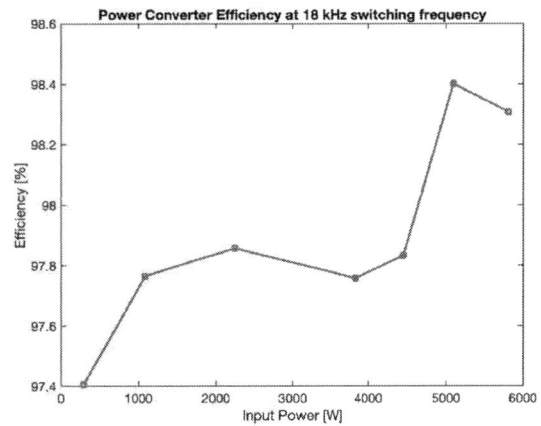

(b) Power converter efficiency over range of input power through hardware experiments

Fig. 7: Hardware results: (a) board surface temperature, and (b) power converter efficiency.

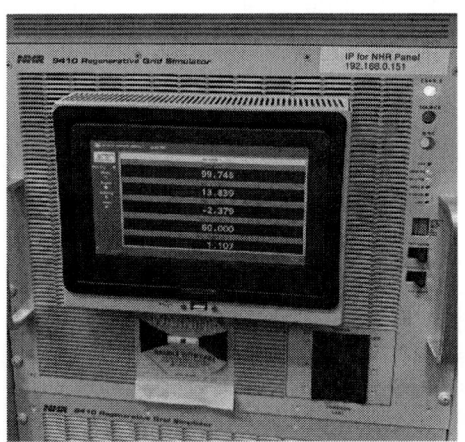

(a) NHR grid-simulator face plate showing power being fed to the grid (blue 'Sink' light and negative power shown on the display.)

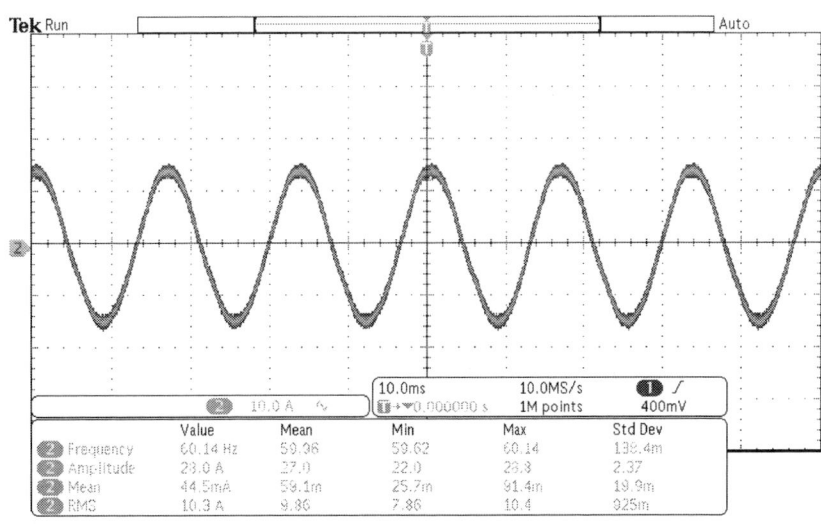

(b) Grid-side converter phase current

Fig. 8: Hardware experiment results: GFL operation of inverter and corresponding phase current waveforms

979-8-3315-1612-3/25 $31.00 © 2025 IEEE   3052

PCB during rated operation. This demonstrates the effectiveness of the thermal management system under natural conduction cooling. The design achieves efficient heat dissipation, even under high-power conditions. Furthermore, implementing natural water convective cooling as an extension to the current setup would significantly alleviate thermal stress on the converter, enhancing its operational reliability and extending its lifespan. These results emphasize the suitability of the selected converter topology for micro-HKT applications, providing robust performance under realistic operating conditions.

Fig. 7(b) illustrates the power converter's efficiency as a function of input power, measured at a switching frequency of 18 kHz. The efficiency curve demonstrates that the converter achieves a peak efficiency of approximately 98.4 % at the rated power level, highlighting its optimized design for high-performance operation. These results underscore the capability of the power converter to deliver high efficiency while effectively managing thermal constraints, particularly at its rated power output.

The grid-side converter transfers power to the grid simulator, leveraging voltage-sensing data to track the grid voltage vector accurately. This tracking is achieved using a synchronously rotating reference frame (SRF) phase-locked loop (PLL), which ensures precise phase alignment. This phase synchronization is critical for the d-q axis controller of the grid-side converter to regulate active and reactive power exchange with the grid effectively.

Fig. 8(a) depicts the power sinking operation of the grid-side converter under grid-following (GFL) mode, demonstrating stable power delivery to the grid simulator. Additionally, Fig. 8(b) presents the phase current waveform at a specific operating point, illustrating the converter's performance under load. Current harmonics are analyzed using a PA4000 power analyzer, revealing a total harmonic distortion of 3.8%. This value falls well within the IEEE 1547 standard for harmonic limits, ensuring compliance with grid integration requirements. During GFL operation, the system maintained a power factor of approximately 0.98 while feeding power up to 5.7 kW to the grid simulator. This demonstrates the converter's capability to efficiently deliver power with minimal distortion and a high degree of phase accuracy, making it suitable for high-performance renewable energy applications.

The experimental findings validate the suitability of the developed converter for the stringent performance and reliability requirements of the HKT system. The high efficiency and robust thermal management ensure minimal energy losses and enhanced durability, making it an ideal choice for micro-HKT applications.

VII. CONCLUSION

This paper presents a methodology for developing an aluminum substrate SiC-based VSI optimized for HKT applications, guided by an MLPA model for precise power loss, thermal behavior, and efficiency analysis. Utilizing a 6061 aluminum substrate enhances thermal management, while 1.2

kV SiC MOSFETs ensure high efficiency and robust performance. The 5 kW converter achieves over 98% peak efficiency, with thermal imaging and testing confirming operation within safe thermal limits. The compact, reliable design addresses key HKT challenges and demonstrates scalability for higher power ratings, offering a robust solution for renewable energy systems in marine environments.

REFERENCES

[1] N. D. Laws and B. P. Epps, "Hydrokinetic energy conversion: Technology, research, and outlook," *Renewable and Sustainable Energy Reviews*, vol. 57, pp. 1245–1259, 2016.

[2] A. P. Thurlbeck and Y. Cao, "Machine learning based condition monitoring for sic mosfets in hydrokinetic turbine systems," in *Proc. IEEE Energy Conversion Congress and Exposition (ECCE)*, 2022, pp. 1–7.

[3] T. Kilgore, M. Tariquzzaman, and Y. Cao, "Thermal management of sic mosfets within hydrokinetic applications," in *Proc. IEEE Energy Conversion Congress and Exposition (ECCE)*, 2022, pp. 1–8.

[4] S. Yang, A. Bryant, P. Mawby, D. Xiang, L. Ran, and P. Tavner, "An industry-based survey of reliability in power electronic converters," in *Proc. IEEE Energy Conversion Congress and Exposition*, 2009, pp. 3151–3157.

[5] M. Tariquzzaman, P. Li, S. J. Barton, A. P. Thurlbeck, T. Kilgore, T. K. A. Brekken, and Y. Cao, "Multi-physics and multi-timescale modeling of hydrokinetic turbine energy conversion system," *IEEE Journal of Emerging and Selected Topics in Power Electronics*, pp. 1–1, 2024.

[6] M. Liu, "Hydrokinetic turbine power converter and controller system design and implementation," Ph.D. dissertation, University of British Columbia, 2014.

[7] A. Ghaffari, M. Krstić, and S. Seshagiri, "Power optimization and control in wind energy conversion systems using extremum seeking," *IEEE Transactions on Control Systems Technology*, vol. 22, no. 5, pp. 1684–1695, 2014.

[8] J. M. Walker, K. A. Flack, E. E. Lust, M. P. Schultz, and L. Luznik, "Experimental and numerical studies of blade roughness and fouling on marine current turbine performance," *Renewable Energy*, vol. 66, pp. 257–267, 2014.

[9] D. Kumar and S. Sarkar, "A review on the technology, performance, design optimization, reliability, techno-economics and environmental impacts of hydrokinetic energy conversion systems," *Renewable and Sustainable Energy Reviews*, vol. 58, pp. 796–813, 2016.

[10] D. Graovac, M. Purschel, and A. Kiep, "Mosfet power losses calculation using the data-sheet parameters," *Infineon application note*, vol. 1, pp. 1–23, 2006.

[11] H. Dai, S. Acharya, D. Torrey, X. Yi, D. Jasinski, R. Thomas, M. H. Todorovic, and V. Rallabandi, "A hydrokinetic energy conversion system using underwater llc-type resonant converter," in *2023 IEEE Energy Conversion Congress and Exposition (ECCE)*. IEEE, 2023, pp. 3273–3278.

[12] U. S. D. of Defense, *Reliability Prediction of Electronic Equipment*. Department of Defense, 1982, vol. 217.

[13] Mississippi River at Baton Rouge, LA. United States Geographical Survey. [Online]. Available: https://waterdata.usgs.gov/monitoring-location/07374000/parameterCode=00065period=P7D

[14] Z. Guo and A. Q. Huang, "1.2 kv/400 a sic source turn-off mosfet intelligent power module," *IEEE Power Electronics Magazine*, vol. 11, no. 2, pp. 34–38, 2024.

979-8-3315-1612-3/25 $31.00 © 2025 IEEE

Comprehensive Evaluation of Cyber Attacks on Grid-Connected Smart Inverters

Rishabh Singla, Vishwam Raval, Hasan Ibrahim, Jaewon Kim, Prasad Enjeti, Narasimha Reddy
Power Electronics & Power Quality Laboratory, Dept of ECE, Texas A&M University,
College Station, TX 77845

Abstract—**This paper presents an in-depth study of cyberattacks on a grid-connected solar photovoltaic (PV) inverter, with a focus on denial-of-service (DoS) attacks. Four distinct scenarios are explored: (a) data flooding attacks, (b) man-in-the-middle (MITM) attacks, (c) compromised web server attacks, and (d) side-channel attacks. Preliminary laboratory experiments were conducted on a 3.8 kW grid-connected PV inverter powered by a PV emulator, illustrating the system's responses to each attack type. The results demonstrate the potential for DoS incidents and service interruptions. Future work will expand on this analysis and include a discussion of various mitigation strategies.**

Index Terms—**Smart inverters, Cyber-security, cyber-vulnerabilities, Denial-of-service (DoS), power stability, power grid security**

I. INTRODUCTION

The Solar Futures study by US Department of Energy projects a significant increase in the solar energy's portion of the United States' electricity supply, estimating it to reach 40% by 2035 and 45% by 2050 [1]. This growth places the solar industry at the forefront of the renewable energy transition, highlighting its vital role in shaping the nation's energy landscape. As the solar energy's share of the grid expands, attention turns to the integration challenges and opportunities it presents. Grid-tied photovoltaic (PV) inverters emerge as key players in ensuring the stability and reliability of energy infrastructure amid the sector's rapid evolution. Cybersecurity emerges as a critical aspect of this transformation as the industry confronts evolving threats like denial-of-service (DoS) and man-in-the-middle (MITM) attacks, along with emerging vulnerabilities such as side-channel attacks.

The crucial importance of robust cybersecurity protocols for cyber-physical systems is portrayed in [2], citing incidents involving the compromise of availability, integrity, and confidentiality. Examples include the ransomware assault on a major US fuel pipeline in 2021, resulting in the halt of crucial fuel distribution, the 2016 power outage in Ukraine, which was linked to compromised solar infrastructure [3], and a recent investigation examining the repercussions of warfare on critical infrastructure [4], highlighting the tangible threats posed by cyber-physical attacks on energy systems.

The industry's ongoing expansion, driven by the adoption of inverter-based Resources (IBRs) linked to distributed energy resources (DERs), necessitates a concerted effort to fortify cybersecurity frameworks against potential threats [5]. This paper presents a comprehensive analysis and experimental evaluation of network-based attack vectors in solar PV smart

inverters, enriching the dialogue on safeguarding critical infrastructure during the expansion of the renewable energy sector. It details four scenarios of cyber attacks: (a) data flooding attack, (b) man-in-the-middle attack, (c) compromised web server attack, and (d) side-channel attack.

Experimental results of the proposed attack vectors on a 3.8kW PV inverter powered by a solar PV emulator connected to the power grid in a laboratory setting are discussed in detail. Future work will include a more comprehensive analysis along with a discussion on several mitigation methods such as watermarking techniques [6] and side channel monitoring [7].

II. RELATED WORK

Exploiting vulnerabilities in the Modbus protocol, a widely used communication protocol for industrial automation systems, researchers demonstrated a stealth attack that injects false commands while remaining undetected by SCADA operators [8]. To evaluate cybersecurity strategies without compromising availability, a framework leveraging open-source tools and Docker containers was introduced [9]. Industry best practices for DERs at the device and distribution level were developed and validated through collaboration with a DER cybersecurity working group [10]. Comprehensive analysis of attack vectors targeting cyber-resilient smart inverters [3], as well as cyberphysical threats to IEC 61850 photovoltaic inverter systems [11], highlight the evolving cybersecurity landscape. Additionally, discussions on defending smart inverters in distribution systems against cyber threats included the implementation of an online Intrusion Detection System (IDS) to protect DER networks [12]. Despite these advancements, there remains a need for experimental validation of specific cyberattacks on modern smart inverters in realistic settings.

This work builds on these efforts by focusing on denial-of-service (DoS) attacks and experimentally analyzing their impact on a 3.8 kW grid-connected solar PV inverter powered by a PV emulator. This research provides insights into the system's responses to four distinct attack scenarios, including data flooding, man-in-the-middle (MITM), compromised web server, and side-channel attacks. By identifying vulnerabilities and proposing mitigation strategies, this study contributes to advancing the cybersecurity of renewable energy systems, ensuring their reliability and resilience.

979-8-3315-1612-3/25 $31.00 © 2025 IEEE

III. SYSTEM VULNERABILITY ASSESSMENT AND ATTACK SCENARIOS

Fig. 1 is an illustration of a commercial inverter deployed in a solar farm, a critical component of industrial scale renewable energy systems. These inverters are responsible for converting solar generated DC to grid compatible AC. In addition to their primary function, commercial inverters often support advanced capabilities such as net metering, which allows excess electricity to be sent back to the grid, and remote monitoring systems. Such functionalities are typically enabled through the integration of information technology (IT) and operational technology (OT), as highlighted in [13]. This connectivity, while beneficial for operational efficiency and scalability, introduces potential vulnerabilities that can be exploited. Cyber threats such as DoS attacks could exploit these connections to disrupt the inverter's communication with the grid, compromising energy generation and distribution at the utility level.

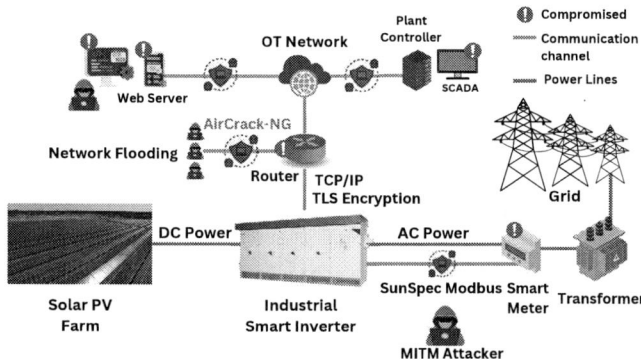

Fig. 1: A solar farm integrated with the power grid, exposing potential cybersecurity risks on an industrial scale smart inverter.

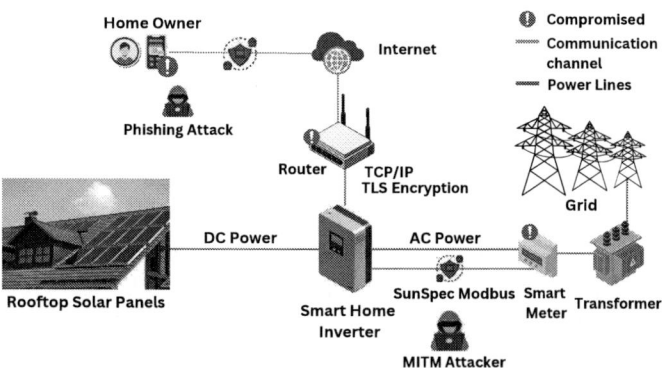

Fig. 2: A solar farm integrated with the power grid, exposing potential cybersecurity risks on an industrial scale smart inverter.

Fig. 2 depicts a smart home equipped with a residential inverter connected to the grid. These inverters, similar to their commercial counterparts, convert solar DC into grid-compatible AC but operate on a smaller scale. They are integral to the growing trend of DERs, which enable homeowners to not only reduce their reliance on the grid but also contribute to it through net metering. Additionally, modern residential inverters often come with smart features like app-based remote monitoring and management. However, this increased IT/OT integration in smart homes creates vulnerabilities. A targeted DoS attack on the residential inverter, for instance, could sever its communication with the grid, disrupting not just the homeowner's energy supply but also the broader stability of the local distribution network.

This section examines the vulnerabilities in these configurations, focusing on how DoS attacks can interfere with the functionality or communication of inverters. Such disruptions can prevent the effective integration of solar energy into the grid, potentially leading to energy interruptions and reduced system reliability. These figures above emphasize the interconnected nature of modern solar energy systems and the critical importance of robust cybersecurity measures to mitigate such risks.

A. Methodology

This research paper aims to assess the risks of grid disconnection and the exploitation of vulnerabilities of smart inverters within OT networks. Our approach involves conducting a comprehensive reconnaissance analysis to identify potential weaknesses. This includes examining open network ports, analyzing firmware, investigating the onboard microcontroller Joint Test Action Group (JTAG) interface, as well as physically accessing communication channels. The study also provides a detailed examination of the communication protocol between the smart meter and the inverter, with particular focus on the *SunSpec Modbus* protocol [14] and the web server responsible for adjusting the inverter's grid-side settings. Various vulnerabilities are analyzed, including those related to physical access, webserver security, and deficiencies in network protection measures.

B. Laboratory Setup

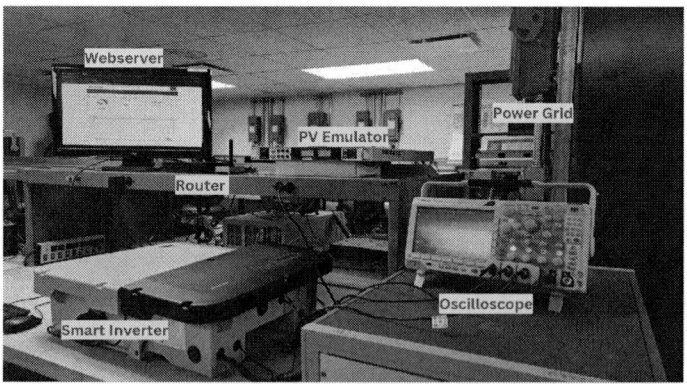

Fig. 3: A Smart PV Inverter, powered by a solar PV emulator, connected to the power grid and a local network via a router, enabling communication channel

Fig. 3 shows laboratory experimental arrangement comprising a photovoltaic (PV) emulator (5kW programmable dc power supply emulating a PV array), connected to a

3.8kW commercial smart PV inverter. Positioned between the inverter and the grid is a smart meter, which provides power measurements along with readings of voltage and current on the AC side. It communicates this data via the *SunSpec Modbus* protocol to the inverter. Telemetry data can be accessed through a web server situated on the smart inverter, operating on port 80 or 443. This access is provided within the OT network via SCADA monitoring and control for industrial applications, or through a web interface or mobile application via the internet for residential use.

C. Denial of Service Attack Scenarios and Test Results

A Denial of Service (DoS) attack is a an attempt to disrupt a target's normal operations, rendering it inaccessible to legitimate users. This is achieved by overwhelming the target with excessive traffic or exploiting vulnerabilities to force failures. Key methods include network flooding, which saturates bandwidth and blocks legitimate communication, and exploiting web server vulnerabilities, where weaknesses in software or configurations are targeted to cause crashes or resource exhaustion. More sophisticated methods involve man-in-the-middle (MITM) attacks, intercepting and manipulating communications, or side-channel attacks, which exploit indirect vulnerabilities like timing analysis or power consumption data. In critical systems such as grid-tied solar inverters, DoS attacks can disrupt grid connectivity, compromise energy distribution, and undermine the reliability of renewable energy systems.

1) OT Network Flooding Attack: Utilizing *Aircrack-NG*, a network penetration testing tool, malicious actors can monitor and infiltrate Wi-Fi networks by exploiting vulnerabilities in their configurations or encryption protocols. This tool, while designed for security testing, can also be leveraged for malicious purposes, including launching targeted attacks on critical infrastructure. A common method involves initiating network overloads by flooding the network with large data packets, such as ICMP, SYN, and HTTP packets. These floods specifically target TCP ports 80 and 443, which are widely used for web traffic and secure communications. In the case of smart inverters, which rely on these ports for remote monitoring and control, such attacks can cause significant disruptions. The excessive influx of traffic overwhelms the network, leading to service interruptions, performance degradation, and resource exhaustion.

$$N_{\text{total}} = \frac{B_{\text{avail}}}{P_s} \qquad (1)$$

As shown in Equation (1), the number of packets to be sent N_{total} can be calculated based on the available bandwidth B_{avail} and packet size P_s.

$$N_{\text{total}} >= R_{\text{server}} \qquad (2)$$

Equation (2) denotes the equilibrium point where the flood rate matches the server's transmission rate R_{server}, indicating a state of server saturation due to the flood of incoming packets.

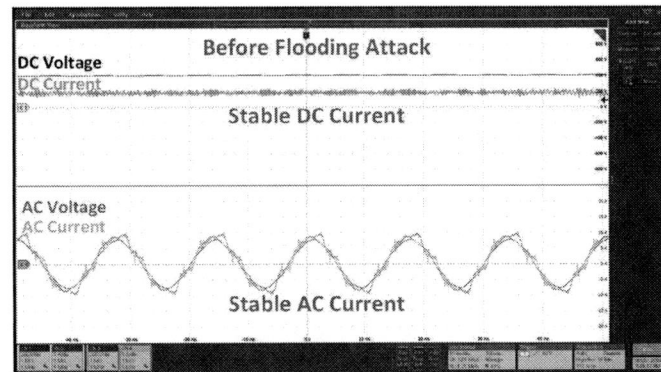

(a) Normal operation before the Data Flooding Attack (SYN) showing 1.4 kW power delivery to the grid with MPPT enabled.

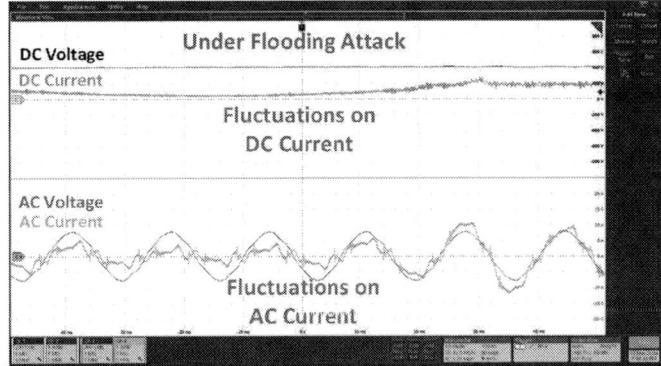

(b) Post-attack, illustrating sporadic AC/DC inverter currents. Data was obtained using the setup and oscilloscope shown in Fig. 3.

Fig. 4: (a) and (b) depict the system behavior before and after a Data Flooding Attack, respectively.

A comparison between sub-figures (a) and (b) in figure 4 demonstrates the inverter's response under flooding attack, where SYN Requests directed towards port 80 and 443 are sent to OT Network to disrupt the web server access. This led to the inverter's controller expending additional resources to process these requests, resulting in diminished efficiency and output current performance.

2) Man-in-the-Middle Replay Attack on Smart Meter Communication: If the communication channel of a smart meter is compromised, it becomes highly vulnerable to a MITM attack, posing a critical cybersecurity threat to the system. Attackers can intercept the RS-485 connection, a commonly used protocol in industrial communication, using tools like a serial port monitor to capture and manipulate the measurement packets exchanged between the smart meter and the controller. By replaying or altering these packets, the attackers introduce errors in the system, causing the smart meter to disconnect and forcing the inverter to restart. This disruption undermines the stability and efficiency of the solar power system.

Test results, as depicted in Figure 5, highlight the tangible impact of such an attack. The data shows a noticeable decrease in the power supplied to the grid, accompanied by a decline in the efficiency of the Maximum Power Point Tracking (MPPT) algorithm. This reduction not only affects energy output but

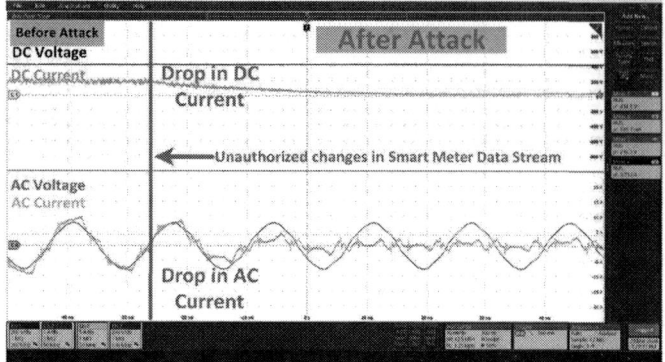

Fig. 5: Man-in-the-Middle attack causing disruption in inverter-smart meter data exchange, leading to fluctuations in AC output. Data was obtained using the setup and oscilloscope shown in Fig.3

also compromises the system's ability to maximize energy harvesting from solar panels. These findings emphasize the importance of secure communication channels and robust mitigation measures to safeguard smart energy systems against sophisticated cyber threats.

3) Webserver-based Attack: This section delves into the security vulnerabilities associated with attacks on inverters via compromised web servers, a critical point of access in modern grid connected systems. Attackers often exploit weaknesses in SCADA system monitoring devices or even end-user devices, which may be infected with malware such as keyloggers and spyware. These compromised devices can then be leveraged, through which the attackers can gain access to and infiltrate the web server controlling the inverters. This unauthorized access empowers them to manipulate login credentials, restrict access for legitimate users, reconfigure the grid-side settings, and potentially seize control of the inverter. By altering critical settings such as nominal grid voltage or regional grid standards, attackers can pose a sever threat to the stability and reliability of the power grid.

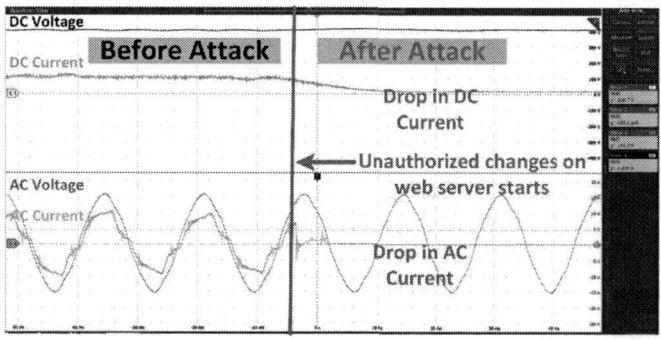

Fig. 6: Webserver-based DoS Attack disconnects the inverter from grid. Data was obtained using the setup and oscilloscope shown in Fig.3

Results of this test, presented in Figure 6, illustrate the impact of a web server attack on a commercial inverter. Wherein the attackers successfully gained administrative control through malware planted on the SCADA monitoring system or the end-user's device and changed grid configuration,

specifically nominal grid voltage and country grid standards. These unauthorized modifications resulted in the inverter's shutdown, directly interrupting the PV power generation to the grid.

4) Side Channel Attacks: A side channel attack exploits unintentional information leaks from a device, such as electromagnetic emissions, power consumption patters, or timing variations, to infer sensitive details about its operation. These attacks do not directly target the software or intended communication but instead target physical or environmental signals which are emitted during the device's operation, indirectly affecting the target through their manipulations. The presence of such vulnerabilities present significant risks to both performance and security of the system. An example

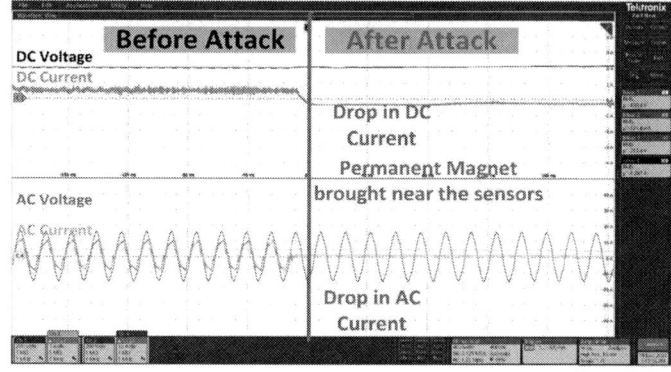

Fig. 7: Side-channel attack induced by placing a magnet near sensors, leading to a complete halt in AC power delivery from the inverter. Data was obtained using the setup and oscilloscope shown in Fig.3

of a side channel vulnerability for a grid tied inverter is the electromagnetic emissions from the current sensor. Attackers can use a strong magnet to distort the electromagnetic field used by the current sensor, causing inaccurate current readings. This interference degrades the inverters' performance and disrupts energy management by falsifying the data on power generation and distribution, as highlighted in [15]. Figure 7 illustrates the impact of such an attack, showing how exposure to a permanent magnet drives the DC current below zero, effectively halting the production of AC current.

D. Discussion

The experimental results demonstrate that grid-connected solar inverters are susceptible to various cyber-attacks. These findings highlight critical vulnerabilities in traffic management, communication security, system access controls, and physical security measures. To enhance the resilience of OT networks, we recommend advanced countermeasures such as rate limiting of network traffic, robust encryption, secure communication protocols, and physical security enhancements [16]–[18]. Additionally, leveraging dynamic watermarking [19] and machine learning-based anomaly detection [20]–[22] can provide early warning against cyber threats. Implementing a multi-layered defense strategy, including regular security audits, patch management, and redundancy mechanisms will ensure comprehensive protection and continuous operation of OT environments against evolving cyber threats [23].

IV. CONCLUSION

In conclusion, this paper presented an in-depth study of cyberattacks on a 3.8 kW solar PV inverter connected to the grid, highlighting the vulnerabilities of such systems in a laboratory environment. Four distinct attack scenarios were examined, ranging from a total Denial of Service (DoS) to sporadic interruptions in service, demonstrating the disruptive potential of these threats. These experimental results emphasize the urgency of addressing cybersecurity risks in grid-connected solar PV systems, which are critical to the growing renewable energy landscape. The future work will build upon these findings, offering a more comprehensive analysis and proposing mitigation strategies to guard against these types of attacks, ensuring the resilience and reliability of solar energy systems in the face of evolving cyber threats.

ACKNOWLEDGEMENTS

This work was funded in part by a grant from Qatar National Research Foundation grant. The authors would like to extend their sincere gratitude to Ben Loomis and Rajesh Kanungo from Tala Secure [24] for their technical expertise and support, which were invaluable to the successful completion of this research. Additionally, we thank Mohamed Zeid from Texas A&M University for his insightful feedback and assistance in the writing of this paper. Their collaboration and contributions were instrumental in advancing our work.

REFERENCES

[1] "Solar futures study," https://www.energy.gov/eere/solar/solar-futures-study, accessed: March 1, 2024.

[2] A. A. Wenli Duso, MengChu Zhou, "A survey of cyber attacks on cyber physical systems: Recent advances and challenges," p. 784, 2022. [Online]. Available: https://www.ieee-jas.net/en/article/doi/10.1109/JAS.2022.105548

[3] B. Ahn, T. Kim, S. Ahmad, S. Mazumder, J. Johnson, H. Mantooth, and C. Farnell, "An overview of cyber-resilient smart inverters based on practical attack models," *IEEE Transactions on Power Electronics*, vol. PP, pp. 1–18, 01 2023.

[4] R. Singla, S. Srinivasa, N. Reddy, J. M. Pedersen, E. Vasilomanolakis, and R. Bettati, "An analysis of war impact on ukrainian critical infrastructure through network measurements," in *2023 7th Network Traffic Measurement and Analysis Conference (TMA)*, 2023, pp. 1–10.

[5] "Ieee approved draft guide for cybersecurity of distributed energy resources interconnected with electric power systems," *IEEE P1547.3/D3.12, March 2023*, pp. 1–158, 2023.

[6] H. Ibrahim, J. Kim, P. Enjeti, P. R. Kumar, and L. Xie, "Detection of cyber attacks in grid-tied pv systems using dynamic watermarking," in *2022 IEEE Green Technologies Conference (GreenTech)*, 2022, pp. 57–61.

[7] P.-H. Huang, J. Kim, P. R. Kumar, J. Rajendran, and P. Enjeti, "Enhancing cybersecurity for industrial control systems: Innovations in protecting plc-dependent industrial infrastructures," *IEEE Internet of Things Journal*, vol. 11, no. 22, pp. 36 486–36 493, 2024.

[8] W. Alsabbagh, S. Amogbonjaye, D. Urrego, and P. Langendörfer, "A stealthy false command injection attack on modbus based scada systems," in *2023 IEEE 20th Consumer Communications Networking Conference (CCNC)*, 2023, pp. 1–9.

[9] E. Damián Gutiérrez Mlot, J. Saldana, and R. J. Rodríguez, "Towards a testbed for critical industrial systems: Sunspec protocol on der systems as a case study," in *2022 IEEE 27th International Conference on Emerging Technologies and Factory Automation (ETFA)*, 2022, pp. 1–4.

[10] R. S. de Carvalho and D. Saleem, "Recommended functionalities for improving cybersecurity of distributed energy resources," in *2019 Resilience Week (RWS)*, vol. 1, 2019, pp. 226–231.

[11] B. Kang, P. Maynard, K. McLaughlin, S. Sezer, F. Andrén, C. Seitl, F. Kupzog, and T. Strasser, "Investigating cyber-physical attacks against iec 61850 photovoltaic inverter installations," in *2015 IEEE 20th Conference on Emerging Technologies Factory Automation (ETFA)*, 2015, pp. 1–8.

[12] C. C. Sun, R. Zhu, and C. C. Liu, "Cyber attack and defense for smart inverters in a distribution system," *CIGRE Study Committee D2 Colloquium, Helsinki, Finland*. [Online]. Available: https://par.nsf.gov/biblio/10099566

[13] A. Ginter, *Engineering Grade OT security.* Abterra Technologies Inc. (October 16, 2023), 2023.

[14] "Sunspec modbus," accessed: March 3, 2024. [Online]. Available: https://sunspec.org/sunspec-modbus-specifications/

[15] A. Barua and M. A. A. Faruque, "Hall spoofing: A Non-Invasive DoS attack on Grid-Tied solar inverter," in *29th USENIX Security Symposium (USENIX Security 20)*. USENIX Association, Aug. 2020, pp. 1273–1290. [Online]. Available: https://www.usenix.org/conference/usenixsecurity20/presentation/barua

[16] J. Smith, "Network traffic shaping and rate limiting for enhancing cybersecurity," *Journal of Network Security*, 2017.

[17] W. Stallings, *Cryptography and Network Security: Principles and Practice.* Pearson, 2017.

[18] M. Johnson, "Physical security measures for high-security facilities," *Security Journal*, 2018.

[19] B. Satchidanandan and P. R. Kumar, "Dynamic watermarking: Active defense of networked cyber–physical systems," *Proceedings of the IEEE*, vol. 105, no. 2, pp. 219–240, 2016.

[20] B. Zhou, S. Liu, K. Hwang, and S. J. Lee, "Anomaly detection in cyber physical systems using recurrent neural networks," *IEEE Transactions on Information Forensics and Security*, vol. 14, no. 6, pp. 1457–1469, 2019.

[21] Y. Li and N. Ye, "Machine learning for anomaly detection: A systematic review," *IEEE Communications Surveys & Tutorials*, 2018.

[22] L. Khan and B. Thuraisingham, "A guide to behavioral analytics in cybersecurity," *Journal of Big Data*, 2016.

[23] J. H. Allen, S. Barnum, R. Ellison, G. McGraw, and N. R. Mead, "Software security engineering: A guide for project managers," 2008.

[24] (2024) Tala secure. [Online]. Available: https://www.talasecure.com/

Parallel Operation of Grid-Forming Converters based on Kuramoto Oscillators with Virtual Cable Emulation for Improved Power Sharing

Vikram Roy Chowdhury, Gab-Su Seo, and Barry Mather

Power Systems Engineering Center, National Renewable Energy Laboratory, Golden, CO 80401, USA

email: vikram.roychowdhury@nrel.gov, gabsu.seo@nrel.gov, barry.mather@nrel.gov

Abstract—**This paper introduces an advanced power control architecture designed to optimize the performance of grid-forming converters operating in parallel by leveraging a Kuramoto oscillator. Central to the proposed framework is the novel concept of emulated virtual cable impedance with a high resistance-to-inductance ratio, which significantly enhances power sharing and coordination among multiple converters, ensuring balanced, stable, and efficient operation of parallel-connected converters. The architecture is further strengthened by an adaptive dynamic control loop, which ensures system stability and resilience, even under fluctuating load conditions and challenging grid disturbances. This novel concept can not only improve parallel converter operation but also improve converter synchronization, response to nonideal grid conditions, and overall system reliability. The proposed system has been thoroughly evaluated through comprehensive evaluation in MATLAB/Simulink and PLECS simulation environments, with extensive test cases demonstrating its superiority in managing parallel converter operations and delivering consistent power in complex and demanding grid scenarios.**

Index Terms—**Grid-forming converters, emulated virtual cable, Kuramoto oscillators, Lyapunov energy function, parallel operation, power sharing.**

I. INTRODUCTION

Integrating multiple power sources, such as photovoltaic (PV) systems, wind energy, wave energy, and storage, to collectively serve a shared load naturally necessitates parallel operation of multiple converters [1]–[5]. This distributed power source approach can significantly enhance the reliability and efficiency of a power system by leveraging the complementary nature of different sources. However, the intermittency and variability of these sources pose significant challenges, requiring advanced control techniques [6]–[8]. Sophisticated control mechanisms are indispensable that can ensure grid stability, deliver a consistent and uninterrupted power supply against fluctuations in resource availability, and enhance the

This work was authored by the National Renewable Energy Laboratory, operated by Alliance for Sustainable Energy, LLC, for the U.S. Department of Energy (DOE) under Contract No. DE-AC36-08GO28308. Funding provided by U.S. Department of Energy Office of Energy Efficiency and Renewable Energy Water Power Technologies Office. The views expressed in the article do not necessarily represent the views of the DOE or the U.S. Government. The U.S. Government retains and the publisher, by accepting the article for publication, acknowledges that the U.S. Government retains a nonexclusive, paid-up, irrevocable, worldwide license to publish or reproduce the published form of this work, or allow others to do so, for U.S. Government purposes.

grid's adaptability to evolving energy demands. Such control systems play a critical role in achieving optimal power-sharing among diverse energy sources, facilitating seamless integration of renewable technologies, and ensuring system scalability to accommodate the growing penetration of renewables. Furthermore, they enhance grid resilience by mitigating the impact of disturbances, enabling real-time responses to dynamic operating conditions, and supporting the transition to a more sustainable energy paradigm. By aligning with ambitious sustainability objectives, these technologies enable the development of a robust, efficient, and carbon-neutral energy ecosystem. As the adoption of renewable energy accelerates globally, the continuous advancement of these control methodologies remains paramount in addressing future energy challenges and achieving a sustainable energy future.

In this paper, a novel methodology is introduced that utilizes the synchronization principles of Kuramoto-type oscillators to enhance the parallel operation of grid-forming (GFM) converters through the integration of emulated virtual cable impedance [5], [9]. Conventional approaches such as droop control, virtual synchronous machines (VSM), and virtual oscillator control (VOC) are widely recognized to exhibit Kuramoto-type oscillator behavior during synchronization [10]–[12]. This methodology extends the Kuramoto-type oscillator model to include voltage magnitude dynamics, addressing critical limitations in existing models. The incorporation of virtual cable impedance, characterized by a high resistance-to-inductance ratio, is employed in this work to improve stability margins, leveraging the well-established advantages of resistive droop for parallel operations [13]. Additionally, the inner-loop control is designed using a Lyapunov energy function-based architecture, renowned for its robustness in maintaining stability and control [14], [15]. The effectiveness of the proposed approach is validated through extensive modeling and simulations conducted in MATLAB/Simulink and PLECS, with various case studies demonstrating significant improvements in stability and performance across diverse operational scenarios. Section II provides an in-depth derivation of the Kuramoto-type oscillator model tailored for GFM converters, focusing on its dynamic behavior and synchronization principles. Section III elaborates on the design and implementation of the Lyapunov energy function-

979-8-3315-1612-3/25 $31.00 © 2025 IEEE

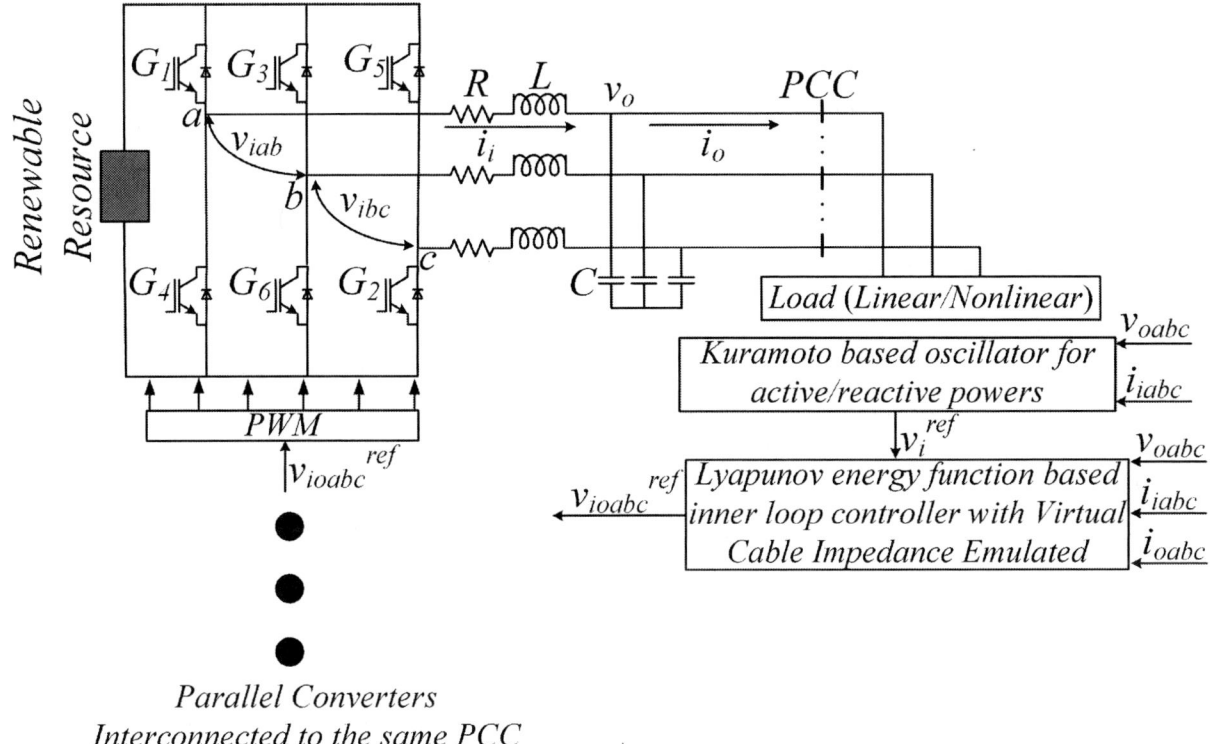

Fig. 1: Simplified block diagram of the proposed system for GFM Converters.

based inner loop architecture, incorporating emulated virtual cable impedance to enhance stability margins and control robustness. Section IV presents the simulation results and discussions, highlighting the performance improvements under varying operating conditions, and Section V concludes with key findings and potential directions for future work in advancing parallel converter operations.

II. KURAMOTO OSCILLATOR-BASED CONTROL ARCHITECTURE FOR GFM CONVERTERS

The proposed architecture is presented in Fig. 1. The dc bus of the GFM converter is connected to a power source, such as PV, wind, wave, or storage. The ac side is connected through an LC filter to the PCC, which can serve balanced or unbalanced and linear or nonlinear loads. The primary objective of a GFM converter is to regulate its output voltage and frequency, meeting regulations imposed to ensure the power quality of the system [16]. The general presentation of Kuramoto oscillators and their applications have been elaborated in [5], [9]. Kuramoto oscillators are generally used to analyze standard power control architecture like droop, VSM, or VOC, and it has not been used before as a simplified control architecture. In this work, the concept of the Kuramoto oscillator is utilized to generate a nonlinear oscillator architecture for advanced power-sharing performance.

A simplified per-phase circuit diagram from the inverter to the PCC is presented in Fig. 2. R_{vir} and L_{vir} are the emulated impedance of the virtual cable with the assumption $R_{vir} \gg$

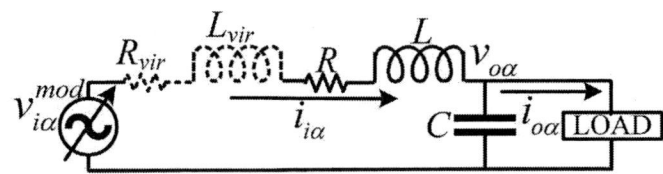

Fig. 2: Per-phase equivalent circuit of the inverter.

L_{vir}. The proposed Kuramoto type of oscillator utilized in this paper is presented as:

$$\begin{aligned} \omega &= \omega_{nom} - \frac{\psi V_i}{2(R+R_{vir})}\left[V_i - V_o \cos\left(\theta_i - \theta_o\right)\right], \\ V &= V_{nom} - \frac{\xi V_i V_o}{2(R+R_{vir})} \sin(\theta_i - \theta_o), \end{aligned} \tag{1}$$

where θ_i and θ_o are the load angles of the inverter and the PCC, respectively; V_i and V_o are the magnitude of the voltages of the inverter and the PCC, respectively; and ψ and ξ are user-defined constants equivalent of droop coefficients. Now it is observed from (1) that due to the emulation of the virtual impedance which is highly resistive, the inverter frequency relates to voltage magnitude, whereas the inverter voltage relates to the angle difference between the inverter and the PCC voltages. More details of active/reactive power relations under a resistive interfacing impedance architecture can be found in [17]. The next section briefly describes the Lyapunov energy function-based inner loop architecture for superior performance and power sharing.

III. LYAPUNOV ENERGY FUNCTION-BASED INNER LOOP CONTROLLER

Consider the per-phase equivalent circuit of Fig. 2. To reduce the number of equations, the overall implementation of this architecture is accomplished in the stationary two-phase $\alpha\beta$ reference frame. As presented in [14], [15], the Lyapunov energy function-based controller architecture is accomplished based on the dynamics of the interfacing impedance. However, as observed in Fig. 2, a virtual impedance is also taken into consideration for better stability and power-sharing performance [17]. Therefore, this virtual impedance is considered inside the inner loop control architecture to accomplish the overall control architecture. This type of control architecture is accomplished based on the reference, actual and error dynamics. The latter is utilized to generate the overall control law. Therefore, the error dynamics, chosen energy function, time derivative of the chosen energy function, and the overall control law are presented in (2) through (5):

$$(L + L_{vir})\frac{d\Delta i_{i\alpha}}{dt} = \begin{bmatrix} -(R + R_{vir})\Delta i_{i\alpha} \\ +\Delta m_\alpha \frac{V_{dc}}{2} - \Delta v_{o\alpha} \end{bmatrix}, \quad (2)$$
$$C\frac{d\Delta v_{o\alpha}}{dt} = \Delta i_{i\alpha},$$

$$U = \frac{1}{2}(L + L_{vir})\Delta i_{i\alpha}^2 + \frac{1}{2}C\Delta v_{o\alpha}^2, \quad (3)$$

$$\dot{U} = -(R + R_{vir})\Delta i_{i\alpha}^2 + \Delta m_\alpha \Delta i_{i\alpha}\frac{V_{dc}}{2}, \quad (4)$$

$$m_\alpha = m_\alpha^{ref} + \Delta m_\alpha = \frac{2}{V_{dc}}\begin{bmatrix} (L + L_{vir})\frac{di_{i\alpha}^{ref}}{dt} \\ +(R + R_{vir})i_{i\alpha}^{ref} \\ +v_{o\alpha}^{ref} + K_o\left(i_{i\alpha}^{ref} - i_{i\alpha}\right) \end{bmatrix}, \quad (5)$$

where V_{dc} is the dc bus voltage, $i_{i\alpha}$ and $i_{o\alpha}$ are the inverter side and load side currents, ref quantities indicate reference quantities and K_o is a user defined constant chosen to obtain required bandwidth for the controller. The overall control law is presented only for α axis. Similar architecture is also implemented in β axis and the overall architecture is realized. The voltage references for the capacitor denoted as $v_{o\alpha\beta}^{ref}$ is generated by the Kuramoto oscillator of (1) and used as an input for (5). The following section presents a detailed examination of key case studies, highlighting the proposed architecture's capabilities and performance. Through rigorous computer simulations, these case studies were thoroughly implemented and verified, providing valuable insights into the architecture's behavior and effectiveness in diverse situations.

IV. RESULTS AND DISCUSSIONS

The overall system is modeled in MATLAB/Simulink and PLECS and various case study results are presented for validation. Two parallel inverters are implemented in the simulation environment to run test cases for performance evaluation. The parameters of the simulation are elaborated in Table I. The result showing the performance during a step change in

TABLE I: PLANT AND COMPENSATOR PARAMETERS

Parameter	Value	Parameter	Value
$S_{inv,rated}$	5 KVA	V_{dc}	400 V
v_{PCC}	120 V(rms)	L	5.0 mH
R	1.0 Ω	C	20.0 μF
ψ_1	0.8	ψ_2	0.4
K_{o1}	100	K_{o2}	50
ϵ_1	0.2	ϵ_2	0.1
R_{vir1}	2.0 Ω	R_{vir2}	1.0 Ω
L_{vir1}	0.1 mH	L_{vir2}	0.05 mH
f_{sw}	20 kHz		

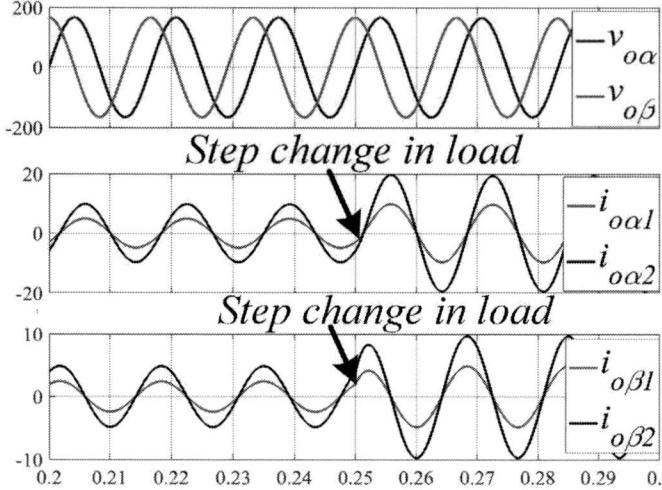

Fig. 3: Load voltage and current for the two three-phase inverters during a step change in unbalanced linear load.

linear unbalanced load is presented in Fig. 3. The power-sharing ratio between inverters 1&2 is set to be 2:1. Note that from this result, the proposed architecture achieves superior dynamic power-sharing performance during a step change. The next result shows the active and reactive powers during the same step change as shown in Fig. 4. The result confirms

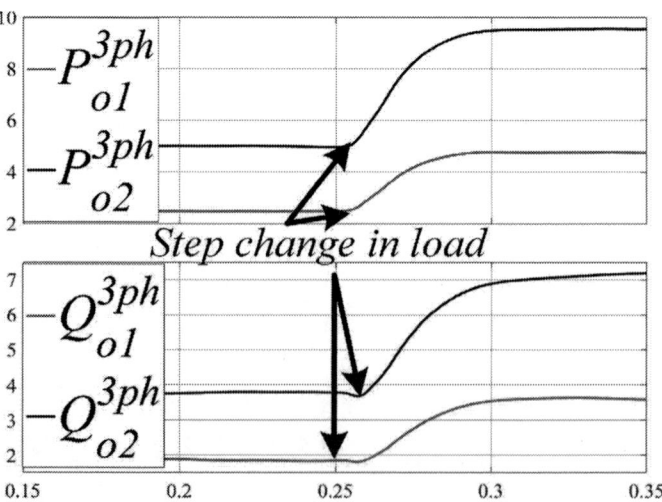

Fig. 4: Dynamic power-sharing performance of the two GFM converters during an unbalanced linear load transient.

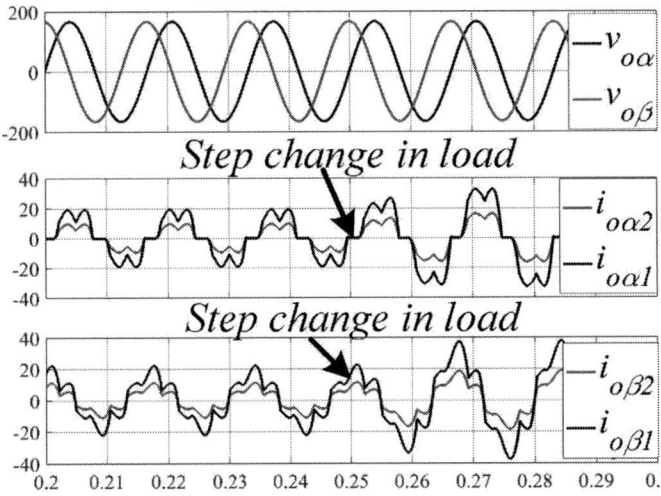

Fig. 5: Load voltage and current for the two three-phase inverters during a step change in nonlinear load.

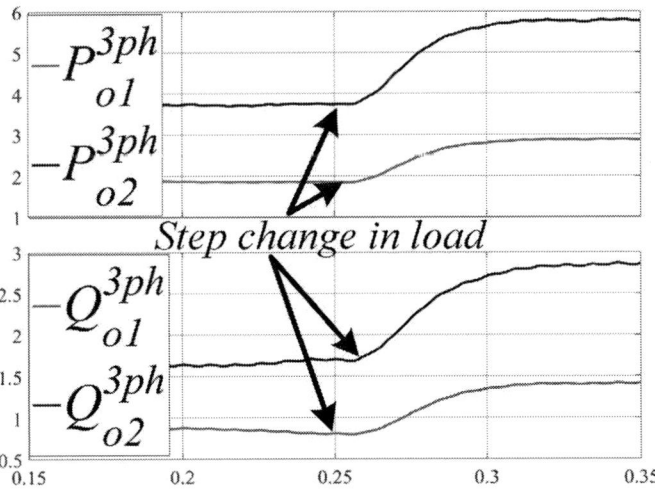

Fig. 6: Active and reactive power sharing performance of two GFM converters with the proposed control during a step change in nonlinear load.

a 2:1 power sharing between the two inverts, indicating the efficacy of the proposed approach. Fig. 5 illustrates the two inverters' parallel operation under a nonlinear (diode rectifier) load step change. As shown in the result, the invertersaccomplish superior dynamic performance as well as power-sharing performance with the proposed architecture. Finally, the result showing the active and reactive powers during the same nonlinear load transient is presented in Fig. 6. similar to Fig. 4, a successful accomplishment of power-sharing during nonlinear loading conditions underscores the potential of the Kuramoto oscillator-based control architecture with resistive virtual cable emulation.

V. CONCLUSION

This paper introduced a novel Kuramoto-based nonlinear oscillator architecture, designed to improve power sharing among multiple GFM converters. The proposed control architecture incorporates Lyapunov energy function-based inner loops to improve dynamic performance and power sharing. Through comprehensive case studies, we demonstrated the control architecture's effectiveness in maintaining the desired power-sharing among two GFM converters under unbalanced linear and nonlinear loading conditions. The simulation results also showcased the capability to regulate the PCC voltage within acceptable limits, even in the presence of nonlinear loads and allowable harmonic content. These findings underscore the potential of the proposed architecture to achieve superior performance metrics, particularly under non-ideal loading conditions. Future work will involve conducting additional case studies, stability analyses, and experimental validations, with a focus on exploring corner cases and refining the architecture's performance based on further study. By addressing the challenges in power sharing and voltage regulation using multiple GFM converters operating in parallel through a decentralized control concept, this work contributes to the development of more resilient and robust power systems.

REFERENCES

[1] M. P. Kazmierkowski and L. Malesani, "Current control techniques for three-phase voltage-source PWM converters: A survey," *IEEE Transactions on Industrial Electronics*, vol. 45, pp. 691–703, October 1998.

[2] S. B. Kjaer, J. K. Pedersen, and F. Blaabjerg, "A review of single phase grid-connected inverters for photovoltaic modules," *IEEE Transactions on Industry Application*, vol. 41, pp. 1292–1306, Oct 2005.

[3] M. Karimi-Ghartemani, "Linear and pseudo linear enhanced phased locked loop (ePLL) structures," *IEEE Transactions on Industrial Electronics*, vol. 61, pp. 1464–1474, Mar 2014.

[4] A. Timbus, M. Liserre, R. Teodorescu, P. Rodriguez, and F. Blaabjerg, "Evaluation of current controllers for distributed power generation systems," *IEEE Transactions on Power Electronics*, vol. 24, pp. 654–664, March 2009.

[5] N. Barrera Gallegos, M. Molinas, and V. Gasca Segura, "Synchronization properties of voltage source converters when seen as coupled oscillators based on the kuramoto model," in *MATHMOD 2018 Extended Abstract Volume, ARGESIM Report 55*. Vienna, Austria: ARGESIM, 2018, pp. 89–90.

[6] D. G. Holmes, T. A. Lipo, B. P. McGrath, and W. Y. Kong, "Optimized design of stationary frame three phase ac current regulators," *IEEE Transactions on Power Electronics*, vol. 24, pp. 2417–2426, November 2009.

[7] M. Malinowski, M. P. Kazmierkowski, S. Hansen, F. Blaabjerg, and G. D. Marques, "Virtual-flux-based direct power control of three-phase PWM rectifiers," *IEEE Transactions on Industry Applications*, vol. 37, pp. 1019–1027, Jul./Aug 2001.

[8] P. C. Loh and D. G. Holmes, "Analysis of multiloop strategies for LC/CL/LCL-filtered voltage-source and current-source inverters," *IEEE Transactions on Industry Applications*, vol. 41, pp. 644–654, Apr 2005.

[9] F. Dörfler and F. Bullo, "Synchronization and transient stability in power networks and non-uniform kuramoto oscillators," in *Proceedings of the 2010 American Control Conference*, 2010, pp. 930–937.

[10] M. C. Chandorkar, D. M. Divan, and R. Adapa, "Control of parallel connected inverters in standalone ac supply systems," *IEEE Trans. Ind Appl.*, vol. 29, no. 1, pp. 136–143, 1993.

[11] Q.-C. Zhong and G. Weiss, "Synchronverters: Inverters that mimic synchronous generators," *IEEE Transactions on Industrial Electronics*, vol. 58, pp. 1259–1267, April 2011.

[12] B. B. Johnson, S. V. Dhople, A. O. Hamadeh, and P. T. Krein, "Synchronization of parallel single-phase inverters with virtual oscillator control," *IEEE Transactions on Power Electronics*, vol. 29, no. 11, pp. 6124–6138, 2014.

[13] S. Mukherjee and V. R. Chowdhury, "Power-angle synchronization for grid-connected converter with fault ride-through capability for low-voltage grids," *IEEE Transactions on Energy Conversion*, vol. 33, pp. 970 – 979, September 2018.

[14] H. Komurcugil, N. Altin, S. Ozdemir, and I. Sefa, "Lyapunov-function and proportional-resonant-based control strategy for single-phase grid-connected VSI with LCL filter," *IEEE Transactions on Industrial Electronics*, vol. 63, pp. 2838–2849, May 2016.

[15] I. Sefa, S. Ozdemir, H. Komurcugil, and N. Altin, "An enhanced lyapunov-function based control scheme for three-phase grid-tied vsi with lcl filter," *IEEE Transactions on Sustainable Energy*, vol. 10, pp. 504–513, April 2019.

[16] UNIFI Consortium, "Specifications for grid-forming inverter-based resources: version 2," National Renewable Energy Lab.(NREL), Golden, CO (United States), Tech. Rep., 2024.

[17] V. R. Chowdhury and A. Singh, "Operation and control of a pv converter with enhanced stability based on virtual impedance emulation in a pseudo resistive grid," in *2023 IEEE Applied Power Electronics Conference and Exposition (APEC)*, 2023, pp. 3099–3104.

Enhancing Hydrogen Production in Hybrid Standalone Microgrids

Utkal Ranjan Muduli
APEC, EE Department
Khalifa University
Abu Dhabi, UAE
utkal.muduli@ku.ac.ae

Mohamed Shawky El Moursi
APEC, EE Department
Khalifa University
Abu Dhabi, UAE
mohamed.elmoursi@ku.ac.ae

Khalifa Al Hosani
APEC, EE Department
Khalifa University
Abu Dhabi, UAE
khalifa.halhosani@ku.ac.ae

Ahmed Al-Durra
APEC, EE Department
Khalifa University
Abu Dhabi, UAE
ahmed.aldurra@ku.ac.ae

Abstract—**Dynamic energy management plays a pivotal role in optimizing hydrogen production and ensuring power quality in hybrid standalone microgrids. This study investigates energy management in systems comprising photovoltaic panels, wind turbines, batteries, fuel cells, and electrolyzers, using the Mayfly Optimization Algorithm (MOA) for converter control. Boost converters are employed as maximum power point tracking devices for photovoltaic and wind systems, with all controllers integrated using MOA. Comparative analyses with Particle Swarm Optimization, Grey Wolf Optimization, and Genetic Algorithm reveal that MOA achieves superior performance in reducing operational costs and enhancing hydrogen production efficiency. Hardware-in-the-Loop simulations, conducted using two OPAL-RT devices, validate the proposed control methodology under real-time conditions. The approach demonstrates improved response times and maintains voltage stability under variable wind speeds, solar irradiance, and load conditions. By addressing sluggish hydrogen production dynamics, the study provides a cost-effective alternative to battery banks for high-power microgrid applications. This research contributes to advancing dynamic energy management in hybrid microgrids, supporting the integration of renewable energy into resilient and efficient systems.**

Index Terms—**Dynamic Energy Management, Microgrids, Hydrogen Production Optimization, Mayfly Optimization Algorithm, HIL Simulations.**

I. INTRODUCTION

As the global transition toward renewable energy accelerates, efficiently integrating diverse energy sources in microgrids becomes crucial for sustainable development. Hybrid standalone microgrids, combining photovoltaic (PV) panels, wind turbines, fuel cells, batteries, and electrolyzers, are emerging as promising solutions to provide reliable and clean energy, especially in remote areas lacking grid connectivity. These systems not only address energy scarcity but also contribute to reducing greenhouse gas emissions. Hybrid microgrids leverage multiple renewable energy sources, enhancing reliability by mitigating the intermittency of individual resources, while a common DC-link facilitates efficient energy management [1], [2]. Hydrogen storage, with its high energy density and long-term storage potential, provides advantages over traditional batteries, particularly for high-power applica-

tions, but its production faces challenges under fluctuating environmental conditions due to the sluggish dynamics of electrolyzers [3]–[5]. Addressing these issues requires innovative control strategies to optimize hydrogen production, improve power quality, and reduce operational costs in hybrid standalone microgrids [6].

Various studies have explored control strategies and optimization techniques to address these challenges. Ipsakis et al. [4] investigated power management in systems using renewable sources and hydrogen storage, emphasizing efficient energy utilization. Habib et al. [7] discussed optimal planning and energy management for PV-based standalone rural microgrids, highlighting the need for reliable power in remote areas. Saeed et al. [1] reviewed microgrid challenges, underscoring the complexities of integrating multiple renewable sources. Optimization algorithms like Particle Swarm Optimization (PSO) have been applied to microgrid cost optimization [8], while Genetic Algorithms (GA) have been used for sizing and control [9]. Nguyen et al. [10] introduced the Mayfly Optimization Algorithm (MOA), demonstrating effectiveness in complex search problems. Malla et al. [11] applied the Whale Optimization Algorithm for PV-based water pumping, indicating applicability in renewable contexts. Steward et al. [3] conducted a lifecycle cost analysis of hydrogen versus other storage technologies, highlighting economic feasibility. Further advancements include Enjeti and Choudhury's [12] development of a DC-side active filter for inverter power supplies to compensate unbalanced loads. Jiang and Fahimi [13] focused on active current sharing in fuel cell-battery hybrid systems, relevant to effective energy management. Memon and Kauhaniemi [14] explored hardware-in-the-loop testing for microgrid protection, demonstrating the utility of simulation tools like OPAL-RT devices. Gao et al. [5] modeled PV-wind-fuel cell-electrolyzer microgrids in real-time simulators, providing insights into system dynamics. Erdinc and Uzunoglu [6] emphasized detailed data utilization for performance evaluation in hybrid systems.

Recent advancements in microgrid management include coordinated control strategies and hybrid energy storage integration. In [15], a predictive control strategy manages wind-solar microgrids, extending storage lifespan and minimizing costs. A review of control strategies for DC microgrids with electric-hydrogen storage was provided in [16], focusing on efficiency and reliability. AI-based controllers for stabilizing

---

"This work was supported in part by Advanced Power and Energy Center (APEC), Khalifa Univeristy, Abu Dhabi, United Arab Emirates; and in part by the Advanced Technology Research Council ASPIRE Virtual Research Institute (VRI) Program, Abu Dhabi, United Arab Emirates, under Award VRI20-07."

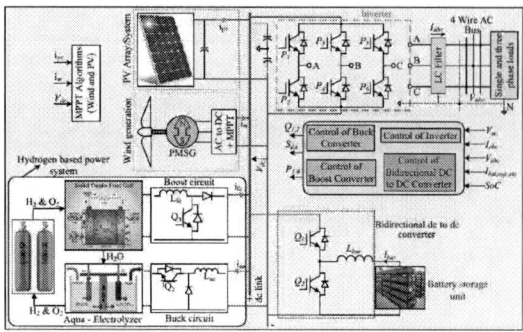

Fig. 1. Standalone Microgrid supplied by hybrid renewable power sources.

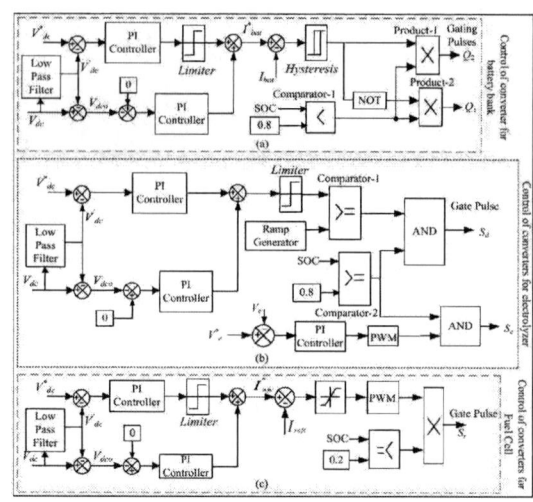

Fig. 2. Control schemes of (a) bidirectional circuit of battery unit, (b) buck circuit of electrolyzer, (c) boost circuit of fuel cell.

islanded microgrids using green hydrogen were explored in [17], improving frequency stability. The enhanced salp swarm algorithm was introduced in [18] for optimizing hybrid renewable systems, emphasizing cost reduction and reliability. A robust optimization framework for coordinating electric vehicles and hydrogen storage was developed in [19], enhancing flexibility under uncertainty. Feasibility and economic analysis of integrating hydrogen storage in hybrid systems were evaluated in [20], demonstrating environmental and economic benefits. Additional studies include a supervisory control system for MVDC microgrids using Z-source converters [21], and feasibility analysis of PV/wind-powered hydrogen production plants [22]. A planning model for standalone hydrogen-based microgrids was proposed in [23], improving computational efficiency. Real-time energy management methods for electric-hydrogen storage microgrids were implemented in [24], optimizing energy distribution.

The objectives of this study are threefold: (1) to develop a novel control methodology for converters in hybrid standalone microgrids by integrating the Mayfly Optimization Algorithm, (2) to enhance hydrogen production efficiency amid variable environmental conditions, and (3) to compare the performance of MOA with PSO, GWO, and GA in optimizing hydrogen production and reducing operational costs. The major contributions include: (a) demonstrating MOA's effectiveness in improving response times and hydrogen production, (b) validating the control strategy using hardware-in-the-loop simulations with OPAL-RT devices, and (c) providing a comprehensive comparative analysis of optimization algorithms in hybrid microgrids. By addressing previous limitations, this work advances dynamic energy management for hydrogen production and supports the shift toward efficient renewable energy systems.

II. MICROGRID CONTROL AND OPTIMIZATION

The standalone microgrid system illustrated in Fig. 1 integrates renewable energy sources, including solar photovoltaic (PV) panels, wind turbines, and hydrogen energy systems, to ensure sustainable power generation. Supporting components such as battery storage units, fuel cells, and an electrolyzer further enhance its reliability and efficiency. PV panels are directly connected to the DC link with maximum power point tracking (MPPT) functionality, eliminating the need for additional power circuits and improving energy conversion ef-

ficiency. Wind turbines utilize permanent magnet synchronous generators (PMSG), interfaced with rectifiers to supply a stable DC output to the common link. This hybrid architecture enables individual components to contribute efficiently to meeting varying load demands, enhancing the microgrid's stability and performance.

The hydrogen subsystem integrates an electrolyzer and a fuel cell into a closed-loop energy storage system, enabling efficient energy management in the microgrid. The electrolyzer utilizes surplus renewable energy to produce hydrogen, which is later converted back into electricity by the fuel cell during periods of low renewable energy availability. This arrangement addresses the intermittency of renewable sources and ensures a consistent power supply. A battery storage unit complements this setup by managing short-term energy fluctuations, maintaining system balance, and supporting reliable power delivery for varying load demands. Together, these components enhance the microgrid's sustainability and economic efficiency by providing a stable, integrated solution for renewable energy storage and distribution.

Control strategies are pivotal for optimizing energy flow and ensuring voltage stability across the microgrid. Bidirectional converters regulate the battery's charge-discharge cycles, while buck and boost circuits for the electrolyzer and fuel cell manage energy distribution during high renewable output or load changes, as depicted in Figures 2(b) and 2(c). The energy management system, illustrated in Fig. 3, coordinates these components and employs dq0-based inverter control (Fig. 4) to maintain balanced three-phase AC voltages, mitigating harmonic oscillations and mechanical wear. By integrating advanced controls and using HOMER software to determine optimal system capacities (Fig. 5), the microgrid achieves reduced operational costs and reliable performance, underscoring its feasibility for renewable energy applications.

A. Energy Balance and Hydrogen Production

To manage peak load under the assumption of no power generation from PV and wind sources, the fuel cell (FC) must be appropriately sized. The required size of the FC is estimated

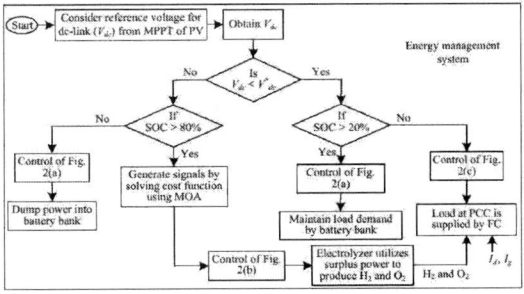

Fig. 3. Proposed energy management system.

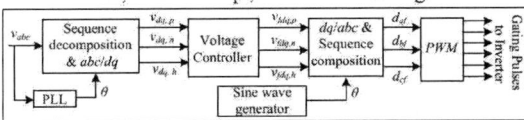

Fig. 4. Inverter control method based on $dq0$.

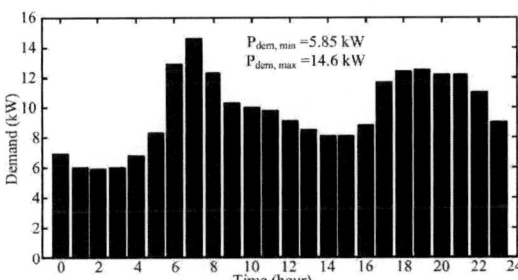

Fig. 5. Hourly based sample load profile.

to be 18.5 kW, with an additional 10% capacity allocated for optimal utilization. Similarly, the size of the electrolyzer is determined based on surplus power availability. Considering the electrolyzer's high cost, power rating of electrolyzer (W_{ele}) is considered only for 60% of the surplus power, as per the following calculation:

$$W_{ele} = (P_t - P_d) \times 60\% = 17.8 \text{ kW} \quad (1)$$

where P_t, P_d are the net power production and the minimum load demand in kW. The cost function accounts for various factors, including capital cost (C_{CAP}), maintenance and operational cost (C_{OM}), replacement cost (C_{REP}), salvage value (C_{SV}), electricity purchase (C_{GP}), electricity sale (C_{GS}), depreciation tax benefit (C_{DTB}), and hydrogen sale revenue (C_{HS}). The cost function (C) is formulated as:

$$\begin{aligned} C = C_{CAP} + C_{OM} + C_{REP} - C_{DTB} - C_{SV} \\ + C_{GP} - C_{GS} - C_{HS} \end{aligned} \quad (2)$$

The objective function (X) is calculated as:

$$\sum_{t=1}^{n} X = C_{mg} \times E_{mg}, \quad (3)$$

where E_{mg} represents the microgrid electricity. The cost per unit of electricity (C_{mg}) is directly derived from the total cost (C), ensuring that the cost function influences the objective function. If $E_{mg} < 0$, the system relies on grid electricity, and C_{mg} becomes the hourly variable import price, adjusting the

objective function accordingly. The energy balance equation ensures that the electricity demand ($E_{D(e)}$) is met by the sum of all available sources, including renewable generation, storage, and external grid supply. It is expressed as:

$$E_{D(e)} = E_{mg} + E_{WT} + E_{Hy} + E_{PV} + E_{FC(e)} - E_{EL(e)}. \quad (4)$$

where E_{WT} is the energy contribution from wind turbines. E_{Hy} and E_{PV} is the energy generated from hydrogen fuel cells and PV panels, respectively. $E_{FC(e)}$ is the electricity supplied by fuel cells, while $E_{EL(e)}$ is the electricity consumed by the electrolyzer for hydrogen production. The net microgrid electricity (E_{mg}) represents the shortfall or surplus after accounting for renewable generation and storage. A shortfall ($E_{mg} > 0$) requires grid electricity, influencing C_{GP}, while a surplus ($E_{mg} < 0$) allows for grid sales, impacting C_{GS}. Now, the hydrogen balance equation is:

$$E_{D(H_2)} = E_{EL(e)}\eta_{EL}^{e/H_2} \pm E_{inj(H_2)} - E_{FC(e)}\frac{1}{\eta_{FC}^{H_2/e}}. \quad (5)$$

Here, η_{EL}^{e/H_2} and $\eta_{FC}^{H_2/e}$ denote the efficiencies of the electrolyzer and fuel cell, respectively. The hydrogen production quantity is derived as:

$$Q = 80.7 \times \eta_F \times \frac{n_c I_{dc}}{ZF}, \quad (6)$$

where η_F is the Faraday efficiency, n_c is the number of cells, I_{dc} is the current, Z is the number of electrons, F is the Faraday constant, and A is the electrode area. The cell voltage required for the electrolyzer operation is:

$$V_{cell} = \frac{\Delta G}{ZF} + s \log\left(\frac{t_1 + \frac{t_2}{T} + \frac{t_3}{T^2}}{A} I + 1\right) + \frac{Tr_1 + r_2}{A} I \quad (7)$$

where ΔG is Gibbs free energy, T is the temperature, and s, t_1, t_2, t_3 are electrode overvoltage coefficients. Ohmic resistances are represented by r_1 and r_2. A step-down voltage converter is recommended to boost current, enhancing hydrogen production efficiency. The ideal voltage of the fuel cell is:

$$E = E^0 - \frac{RT}{2F} \ln\left(P_{H_2} \times P_{O_2}^{0.5}\right), \quad (8)$$

where P_{H_2} and P_{O_2} are the partial pressures of hydrogen and oxygen, respectively. Activation losses are given by:

$$V_{act} = \beta_1 + T \times [\beta_2 + \beta_3 \ln(CO_2) + \beta_4 \ln(i)]. \quad (9)$$

The voltage across Ohmic resistance is:

$$V_{ohmic} = -i \times R_{int}, \quad (10)$$

where $R_{int} = 1.605 \times 10^{-2} - 3.5 \times 10^{-5} T_{cell} + 8 \times 10^{-5} i$. These equations demonstrate how current density and resistive losses influence the overall voltage and efficiency of the fuel cell system. The voltage drop due to mass transport losses, also known as the loss of concentration, is mathematically expressed as:

$$V_{conc} = B \times \ln\left(1 - \frac{i}{i_{lim}}\right), \quad (12)$$

where B is a constant and i_{lim} denotes the limiting current. Consequently, the voltage across the fuel cell (FC) terminals can be determined as:

$$V_{\text{cell}} = E - (V_{\text{ohm}} + V_{\text{conc}} + V_{\text{act}}). \qquad (11)$$

B. Mayfly Optimization Algorithm Methodology

The optimization of the cost function in a microgrid is a key step in achieving operational and economic efficiency. The Mayfly Optimization Algorithm (MOA) is employed to minimize the cost function. The input variables to MOA include the following system parameters:

- Voltage Levels (V): Voltages across various components like PV panels, fuel cells, and electrolyzers.
- Current Flows (I): Currents generated or consumed by each system component.
- Energy Production Rates ($E_{\text{PV}}, E_{\text{WT}}, E_{\text{FC}}, E_{\text{EL}}$): Energy outputs from PV panels, wind turbines, fuel cells, and electrolyzers.
- Cost Components ($C_{\text{CAP}}, C_{\text{OM}}, C_{\text{REP}}, C_{\text{GP}}, C_{\text{GS}}, C_{\text{HS}}, C_{\text{DTB}}, C_{\text{SV}}$): These variables define the economic aspects of the microgrid system.

These input variables form the basis for evaluating the fitness function, which determines the quality of a solution at each iteration of the MOA. The Following determines the step-by-step guide for MOA implementation.

Step-1: Position Update for Male Mayflies: In MOA, the movement of male mayflies plays a crucial role in determining the minimum cost function. The position of a male mayfly, representing the cost function at a given iteration t, is updated as:

$$C_i^{t+1} = C_i^t + v_i^{t+1}, \qquad (12)$$

where C_i^t is the current position, and v_i^{t+1} is the updated velocity of the male mayfly.

Step-2: Velocity Update for Male Mayflies: The velocity function is defined as:

$$v_{kj}^{t+1} = v_{kj}^t g + e^{-\beta r_p^2} a_1 \left(\text{pbest}_{kj} - C_{kj}^t\right) \\ + e^{-\beta r_g^2} a_2 \left(\text{gbest}_j - C_{kj}^t\right), \qquad (13)$$

where k represents the mayfly number, j denotes the dimension, and t is the time step. Parameters a_1 and a_2 are positive attraction constants, β and g are visibility and gravitational coefficients, and r_p and r_g are Cartesian distances to the personal and global best positions, respectively. *Step-3:* Personal Best Update: The best solution for the kth mayfly, pbest_k, is updated as:

$$\text{pbest}_k = \begin{cases} C_k^{t+1}, & \text{if fitness}(C_k^{t+1}) < \text{fitness}(\text{pbest}_k) \\ \text{pbest}_k, & \text{otherwise.} \end{cases} \qquad (14)$$

where pbest_k is the personal best position of a mayfly based on its historical performance. gbest is the he global best solution among all mayflies at a given iteration. The movement of female mayflies adjusts their position to mate with males and improve the overall solution.

Step-4: Position Update for Female Mayflies: The position of a female mayfly is updated as:

$$z_i^{t+1} = z_i^t + u_i^{t+1}, \qquad (15)$$

where z_i^t is the current position, and u_i^{t+1} is the velocity of female mayfly.

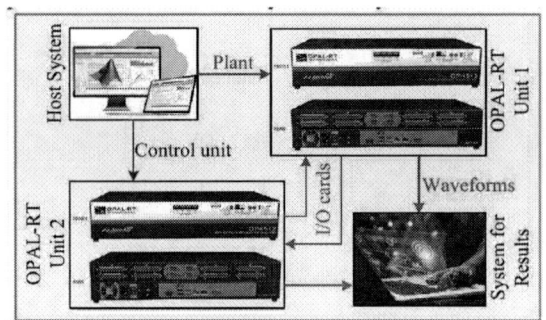

Fig. 6. Establishment of HIL.

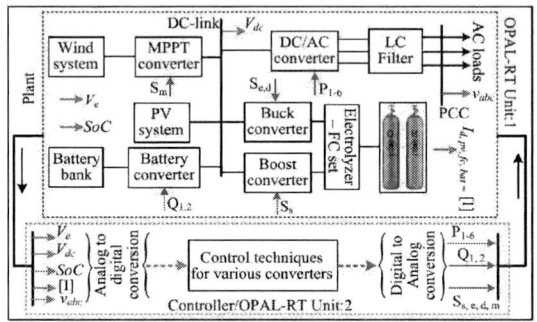

Fig. 7. HIL implementation of Fig. 1.

Step-5: Velocity Update for Female Mayflies: The velocity update rule depends on the fitness of the female compared to the male:

$$u_{kj}^{t+1} = \begin{cases} g u_{kj}^t + e^{-\beta r_{mf}^2} a_2 \left(C_{kj}^t - z_{kj}^t\right), & \text{if } f(z_k) > f(C_k) \\ u_{kj}^t g + flr, & \text{otherwise.} \end{cases} \qquad (16)$$

where $f(z_k)$ and $f(C_k)$ are the fitness functions.

Step-6: Offspring Generation: Offspring generation occurs by combining male and female mayflies' attributes:

$$\text{offspring}_1 = r_{of} \times \text{male} + (1 - r_{of}) \times \text{female}, \qquad (17)$$

$$\text{offspring}_2 = r_{of} \times \text{female} + (1 - r_{of}) \times \text{male}. \qquad (18)$$

The offspring are further refined using:

$$\text{offspring}_n' = \text{offspring}_n + k. \qquad (19)$$

If r_{of} is closer to 1, offspring solutions are more influenced by the male parent, promoting exploration of the search space. Similarly, if r_{of} is closer to 0, offspring solutions are more influenced by the female parent, enhancing local exploitation around the female's position. When $r_{of} \approx 0.5$, the offspring inherits equal contributions from both parents.

Step-7: The final cost function solution is updated as:

$$C^{(}t + 1) = C^{(}t) + \Delta X \times \text{offspring}_n'. \qquad (20)$$

This iterative procedure enables MOA to converge to the best solution, minimizing the cost function and optimizing the microgrid's performance.

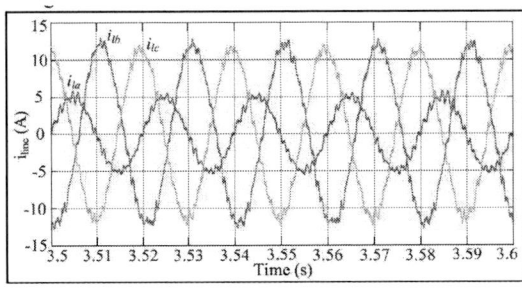

Fig. 8. 3-ph currents { Case- *A* }.

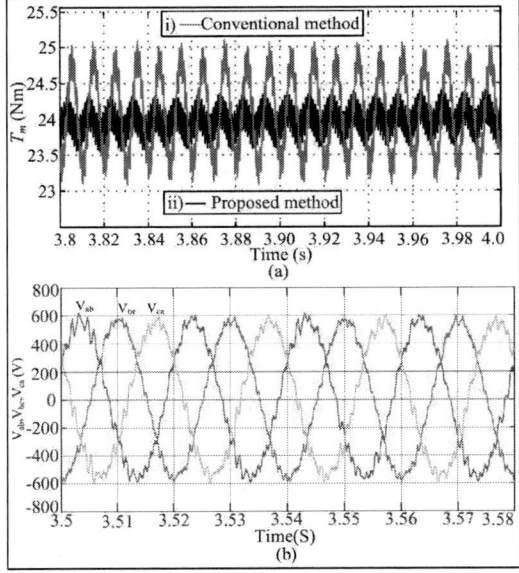

Fig. 9. Response of (a) electromagnetic torque, (b) lineline voltages { Case −*A*}.

III. RESULTS AND DISCUSSIONS

OPAL-RT modules are used to simulate real-time scenarios and test the performance of various systems in a controlled environment [15-16]. By integrating these modules into the HIL process, researchers and engineers can accurately replicate complex operating conditions and analyze the behavior of different components in a system. This allows for the identification of potential issues, optimization of system parameters, and validation of control algorithms before implementing them in real-world applications. The proposed methodology leverages the flexibility and scalability of OPAL-RT modules to create a comprehensive testing platform that can accommodate a wide range of systems, from power electronics and renewable energy systems to automotive and aerospace applications. By utilizing these modules, researchers can conduct detailed experiments, gather valuable data, and make informed decisions to improve the performance and reliability of their systems.

The HIL system is achieved through the utilization of two OPAL RT units. OPAL-RT unit 1 hosts a plant that includes PV, Wind, FC, Electrolyzer, battery, converters, loads etc. On the other hand, OPAL-RT unit 2 is responsible for implementing the proposed controllers. These two units are interconnected with I/O devices to facilitate data flow.

Another system is utilized to observe various waveforms. Fig. 6 illustrates the overall block representation of the HIL establishment. In this section, several results are presented using the HIL approach. Fig. 7 presents a comprehensive block diagram that depicts the HIL setup of the proposed method, including the relevant connecting configuration. We are evaluating different scenarios to demonstrate the efficiency and validate the functionality of the proposed control methodology.

A. Case-A: Operation with unbalanced loads

The Microgrid has been tested under the following unbalanced condition, as shown in Fig. 8. RMS currents of: A-Phase (i_{la}); B-Phase (i_{lb}); and C-Phase (i_{lc}) are 3.53 A ; 9.55 A ; and 8.45 A respectively. In Fig. 9(a), the response of the PMSG's torque under this unbalanced condition is shown with and without the proposed method of control algorithms. Therefore, it is feasible to minimize torque pulsations by utilizing the suggested control method. Nevertheless, the voltages at the PCC may exhibit imbalance as a result of unequal drops across the each phase of LC filter. However, the proposed method of inverter control is capable of making voltages as balanced through producing the accurate modulation indexes. Fig. 9(b) displays the balanced line voltages at PCC.

B. Case-B: Operation with Hydrogen source

In order to assess the system's performance while the electrolyzer is in operation, the SoC unit has been adjusted to approximately 80.0%, and till available of surplus power. The energy management strategy outlined in this document indicates that the electrolyzer starts to utilize excess power once the SoC exceeds 80.0%, as illustrated in Fig. 10(a). In Fig. 10(a), it can be seen that the SoC hits 80.0% at approximately 1.96 sec .

This scenario examines the sudden escalation in power demand from 7.5 kW to 15 kW , representing a 100% increase at 10.0 seconds, in order to evaluate the efficiency of the system when managed by the Fuel Cell. The energy management system's proposal entails the immediate response of the battery to fulfil the load demand, while the FC gradually begins supplying power due to its slow dynamics. As a result of this process, the SoC decreases to 20.0% at approximately 10.96 sec (see Fig. 10 (b)). The power from the FC begins to supply at 11.17 sec because of its sluggish characteristics and meets the load requirement at approximately 12.85 sec .

C. Case-C: Comparison among various optimization methods

The GWO, GA, PSO, and MOA are utilized to minimize the cost function, optimize hydrogen production, and improve the response time of a specified Microgrid. The response is presented in a per unit system, where the highest value is designated as ' 1 ' per unit, and relative values are applied to the other responses. The responses across different parameters are illustrated in Fig. 14. It can be seen from the Fig. that the cost of the Microgrid can be decreased with MOA in comparison to GWO, GA and PSO. The use of the proposed MOA leads to an enhancement in hydrogen production through timely decision-making in response to different variables. The maximum production achieved with MOA is designated as per

979-8-3315-1612-3/25 $31.00 © 2025 IEEE

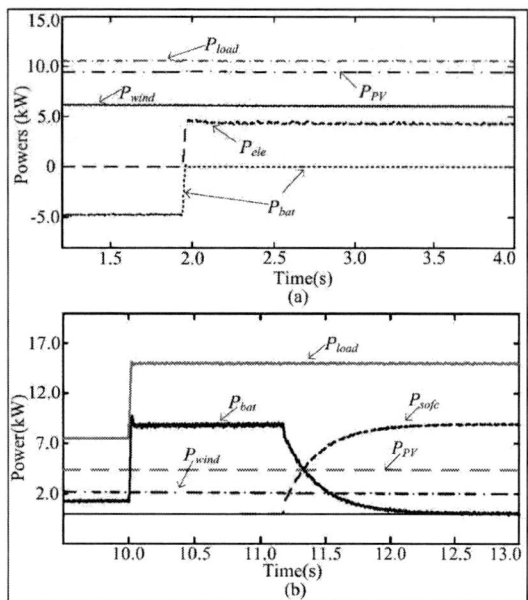

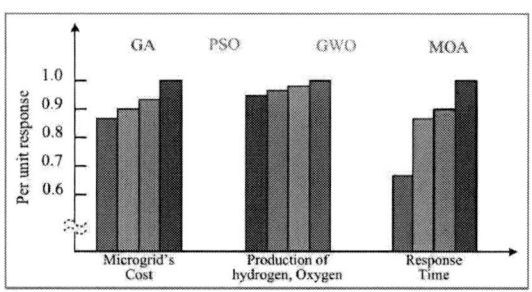

Fig. 10. Powers with operation of (a) Electrolyzer, (b) Fuel Cell { Case-B }.

Fig. 11. Comparison among optimization methods {CaseC}.

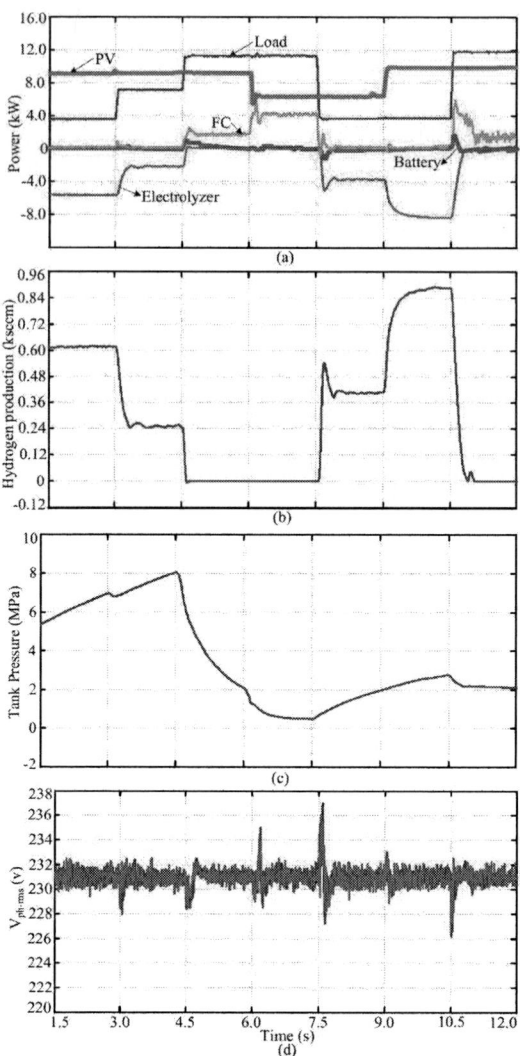

Fig. 12. Responses of (a) powers, (b) production of hydrogen, (c) tanks pressure, (d) RMS voltage at PCC { Case-C}

unit ' 1 ', while comparative values are allocated to alternative optimization methods. The time taken for responses by different optimization methods was finally presented, revealing that the MOA required a very short amount of time to reach a decision efficiently.

The production and utilization of hydrogen is depending on power generation and load. The response of various powers is considered in Fig. 15(a). Corresponding production and pressures in tanks of hydrogen are depicted in Figures 15(b) and (C) respectively. However, the proposed control methodology is stabilizing RMS voltage at PCC within limit to supply quality power to loads which is depicted in Fig. 15(d).

IV. CONCLUSION

This study advanced dynamic energy management in hybrid standalone microgrids to optimize hydrogen production and maintain power quality. The integration of the MOA into converter controls demonstrated improved efficiency, reduced costs, and faster response times compared to Particle Swarm Optimization, Grey Wolf Optimization, and Genetic Algorithm. The strategy maintained voltage stability despite fluctuations in load, wind speed, solar irradiance, and temperature.

Hardware-in-the-Loop simulations using OPAL-RT devices validated the approach under real-time conditions, showcasing its practicality and reliability. The method addresses sluggish hydrogen production dynamics, offering a cost-effective alternative to battery banks for high-power applications. By enhancing hydrogen production and economic efficiency, the approach contributes to the development of resilient and sustainable energy systems. Future research could explore the integration of additional energy systems or expand its application to larger-scale microgrids. This work supports the transition to more efficient renewable energy systems, emphasizing the role of advanced control strategies in achieving sustainability.

REFERENCES

[1] M. H. Saeed, W. Fangzong, B. A. Kalwar, and S. Iqbal, "A review on microgrids' challenges & perspectives," *IEEE Access*, vol. 9, pp. 166502–166517, 2021.

[2] C. Wang and M. H. Nehrir, "Power management of a stand-alone wind/photovoltaic/fuel cell energy system," *IEEE Transactions on Energy Conversion*, vol. 23, no. 3, pp. 957–967, 2008.

979-8-3315-1612-3/25 $31.00 © 2025 IEEE

[3] D. Steward, G. Saur, M. Penev, and T. Ramsden, "Lifecycle cost analysis of hydrogen versus other technologies for electrical energy storage," National Renewable Energy Laboratory (NREL), Tech. Rep., Nov 2009.

[4] D. Ipsakis, S. Voutetakis, P. Seferlis, F. Stergiopoulos, and C. Elmasides, "Power management strategies for a stand-alone power system using renewable energy sources and hydrogen storage," *International Journal of Hydrogen Energy*, vol. 34, no. 16, pp. 7081–7095, 2009.

[5] W. Gao, V. Zheglov, G. Wang, and S. M. Mahajan, "Pv-wind fuel cell-electrolyzer micro-grid modeling and control in real time digital simulator," in *International Conference on Clean Electrical Power*. IEEE, 2009, pp. 29–34.

[6] O. Erdinc and M. Uzunoglu, "The importance of detailed data utilization on the performance evaluation of a grid-independent hybrid renewable energy system," *International Journal of Hydrogen Energy*, vol. 36, no. 20, pp. 12664–12677, 2011.

[7] H. U. R. Habib *et al.*, "Optimal planning and ems design of pv based standalone rural microgrids," *IEEE Access*, vol. 9, pp. 32908–32930, 2021.

[8] S. Phommixay *et al.*, "Review on the cost optimization of microgrids via particle swarm optimization," *International Journal of Energy and Environmental Engineering*, vol. 11, no. 1, Mar 2020.

[9] E. Dursun and O. Kilic, "Comparative evaluation of different power management strategies of a stand-alone pv/wind/pemfc hybrid power system," *International Journal of Electrical Power and Energy Systems*, vol. 34, no. 1, pp. 81–89, 2012.

[10] B. M. Nguyen, T. Tran, T. Nguyen, and G. Nguyen, "Hybridization of galactic swarm and evolution whale optimization for global search problem," *IEEE Access*, vol. 8, pp. 74991–75010, 2020.

[11] S. G. Malla *et al.*, "Whale optimization algorithm for pv based water pumping system driven by bldc motor using sliding mode controller," *IEEE Journal of Emerging and Selected Topics in Power Electronics*, vol. 10, no. 4, pp. 4832–4844, 2022.

[12] P. Enjeti and S. Kim, "A new dc-side active filter for inverter power supplies compensates for unbalanced and nonlinear loads," in *IEEE Industry Applications Society Annual Meeting*, vol. 1. IEEE, 1991, pp. 1023–1031.

[13] W. Jiang and B. Fahimi, "Active current sharing and source management in fuel cell-battery hybrid power system," *IEEE Transactions on Industrial Electronics*, vol. 57, no. 2, pp. 752–761, 2010.

[14] A. A. Memon and K. Kauhaniemi, "Real-time hardware-in-the-loop testing of iec 61850 goose-based logically selective adaptive protection of ac microgrid," *IEEE Access*, vol. 9, pp. 154612–154639, 2021.

[15] M. B. Abdelghany, A. Al-Durra, and F. Gao, "A coordinated optimal operation of a grid-connected wind-solar microgrid incorporating hybrid energy storage management systems," *IEEE Transactions on Sustainable Energy*, vol. 15, no. 1, pp. 39–51, Jan 2024.

[16] W. Pei, X. Zhang, W. Deng, C. Tang, and L. Yao, "Review of operational control strategy for dc microgrids with electric-hydrogen hybrid storage systems," *CSEE Journal of Power and Energy Systems*, vol. 8, no. 2, pp. 329–346, March 2022.

[17] H. M. El Zoghby, A. S. Samir, A. F. Bendary, A. Hazem, H. S. Ramadan, M. M. Elmesalawy, and R. S. A. Afia, "Islanded microgrids frequency support using green hydrogen energy storage with ai-based controllers," *IEEE Access*, vol. 12, pp. 128129–128140, 2024.

[18] B. Modu, M. P. B. Abdullah, A. Alkassem, H. Z. A. Garni, and M. Alkabi, "Optimal design of a grid-independent solar-fuel cell-biomass energy system using an enhanced salp swarm algorithm considering rule-based energy management strategy," *IEEE Access*, vol. 12, pp. 23914–23929, 2024.

[19] M. M. Amiri, M. T. Ameli, M. R. Aghamohammadi, E. Bashooki, H. Ameli, and G. Strbac, "Day-ahead coordination for flexibility enhancement in hydrogen-based energy hubs in presence of evs, storages, and integrated demand response," *IEEE Access*, vol. 12, pp. 58395–58405, 2024.

[20] A.-R. Youssef, R. Abdelkareem, H. H. H. Mousa, and M. A. Ismeil, "Economic and technical evaluation of hydrogen storage in hybrid renewable systems with demand-side management: Upper egypt case study," *IEEE Access*, vol. 12, pp. 120250–120272, 2024.

[21] P. García-Triviño, L. de Oliveira-Assís, E. P. P. Soares-Ramos, R. Sarrias-Mena, C. A. García-Vázquez, and L. M. Fernández-Ramírez, "Supervisory control system for a grid-connected mvdc microgrid based on z-source converters with pv, battery storage, green hydrogen system and charging station of electric vehicles," *IEEE Transactions on Industry Applications*, vol. 59, no. 2, pp. 2650–2660, March 2023.

[22] K. Sayed, M. Khamies, A. G. Abokhalil, M. Aref, M. A. Mossa, M. Metab Almalki, and T. A. H. Alghamdi, "Feasibility study and economic analysis of pv/wind-powered hydrogen production plant," *IEEE Access*, vol. 12, pp. 76304–76318, 2024.

[23] X. Wu, B. Cao, B. Liu, and X. Wang, "A planning model of standalone hydrogen-based carbon-free microgrid through convex relaxation," *IEEE Transactions on Smart Grid*, vol. 14, no. 4, pp. 2668–2680, July 2023.

[24] Q. Li, X. Zou, Y. Pu, and W. Chen, "Real-time energy management method for electric-hydrogen hybrid energy storage microgrids based on dp-mpc," *CSEE Journal of Power and Energy Systems*, vol. 10, no. 1, pp. 324–336, January 2024.

LSTM-based Sub-Synchronous Oscillation Detection Scheme for Type 4 Wind Farm Interfaced with Weak AC Grid

Omar Abu-Rub [1], Muhammad F. Umar[2], Jana A. Sheikh Ali[3], Yazan Qiblawey[4], Abdulrahman Alassi[4], Maryam Saeedifard[1], Mohammad B. Shadmand[5]

[1]Department of Electrical & Computer Engineering, Georgia Tech, Atlanta, USA
[2]Department of Electrical and Computer Engineering, Texas A&M University at Qatar, Doha, Qatar
[3]Department of Electrical Engineering, Qatar University, Doha, Qatar
[4]Iberdrola Innovation Middle East, Doha, Qatar
[5]Department of Electrical & Computer Engineering, University of Illinois Chicago, Chicago, USA
omaraburub@gatech.edu, muhammad.umar@qatar.tamu.edu, jana.sheikhali@qu.edu.qa, yqiblawey@iberdrola.com, aalassi@iberdrola.com, maryam@ece.gatech.edu, shadmand@uic.edu

Abstract— Sub-synchronous oscillations (SSO) pose a significant threat to the stability of a power grid integrated with wind energy resources. Specifically in Type 4 wind farm's operation under weak grid conditions, SSO may be triggered because of the interaction of the fast dynamics of converter's control with the AC grid. The adverse effects of this type of SSO include sustained oscillations in the active power, point of common coupling (POC) voltage, and output current which can pose severe threat to the stability of the system and may lead to cascaded tripping across the network. Given these risks, it is crucial to ensure the effective and timely detection of SSO to mitigate potential instability and prevent potential cascading failures. Conventionally, SSO is identified by performing frequency scanning, impedance scanning approaches, developing and analyzing mathematical models (small signal and eigenvalue analysis). This paper presents a lightweight intelligent machine learning-based approach for swift detecting SSO in wind farms equipped with Type 4 WTG units. The proposed SSO detection scheme leverages a double layer LSTM (DL-LSTM) based neural network to devise swift, effective, and accurate SSO detection, addressing the shortcomings of previous approaches. Insightful case studies are presented in this paper to validate the effectiveness of the proposed ML based SSO detection scheme.

Keywords—low inertia power system, weak grid, sub synchronous oscillations, Type 4 wind energy system, machine learning

I. INTRODUCTION

The current power system is undergoing a transition to integrate a significant amount of renewable energy sources, including wind energy and solar energy. Several new challenges have surfaced due to the integration of renewable energy generation sources into the existing grid. Sub synchronous oscillations (SSO) is among the critical issues related to the stability of power system interfaced with wind energy systems. SSO occurs due to the significant exchange of energy between the turbine and generator at the natural frequency of the considered overall system under a disturbance [1, 2] . The natural frequency of the system is usually below the power system's synchronous frequency, which is 50 Hz or 60 Hz. Several instances of SSO have been documented, leading to

substantial damages to the power system and wind turbine generator (WTG) infrastructure. For instance, an SSO event occurred in Texas, USA, in 2009, due to the interaction between wind turbine generator (WTG) control and fixed series compensation, causing WTG to drop out from generating electricity and permanent damages to its structure. Similar incidents were observed in wind farms in the North China Power Grid [3, 4]. It was observed that WTG control contributes to unstable sub-synchronous power oscillations and permanent damages to the shaft of WTGs.

Timely detecting SSO in power systems integrated with wind energy is a challenging task due to several factors, including the complexity of interactions between wind turbines and the grid, the variability of wind conditions, and the dynamic nature of power system responses. SSO's characteristics such as incipient and low magnitude of sub-synchronous frequency makes it obscure for detecting schemes [5]. Moreover, SSO lowers the overall system damping ratio which makes it more prone to system instability and a small magnitude of oscillation can disrupt system stability and lead to cascaded tripping [6]. Usually the wind farms are located at the remote areas and integrated to the main grid via long distant transmission network. Due to large impedances imposed from long transmission network between the main grid and wind farms, the short circuit ratio (SCR) is negatively affected and the value of SCR at point of common coupling (POC) is greatly reduced. Furthermore, lower system inertia offered via WFs as compared to the synchronous generators makes the power system a potentially weak grid system . Resultantly, the sustained low-magnitude oscillations can escalate rapidly if triggered under a weak grid, with the amplitude increasing significantly within a few grid cycles. Therefore, detection requirements must account for the accurate detection of these oscillations within a few grid cycles.

Conventionally, in the existing literature, various SSO detection schemes were implemented for WFs integrated with the series compensation for long transmission lines [7]. These approaches include performing frequency scanning [8], using electromagnetic transients (EMT) mapping [9], eigenvalue

979-8-3315-1612-3/25 $31.00 © 2025 IEEE

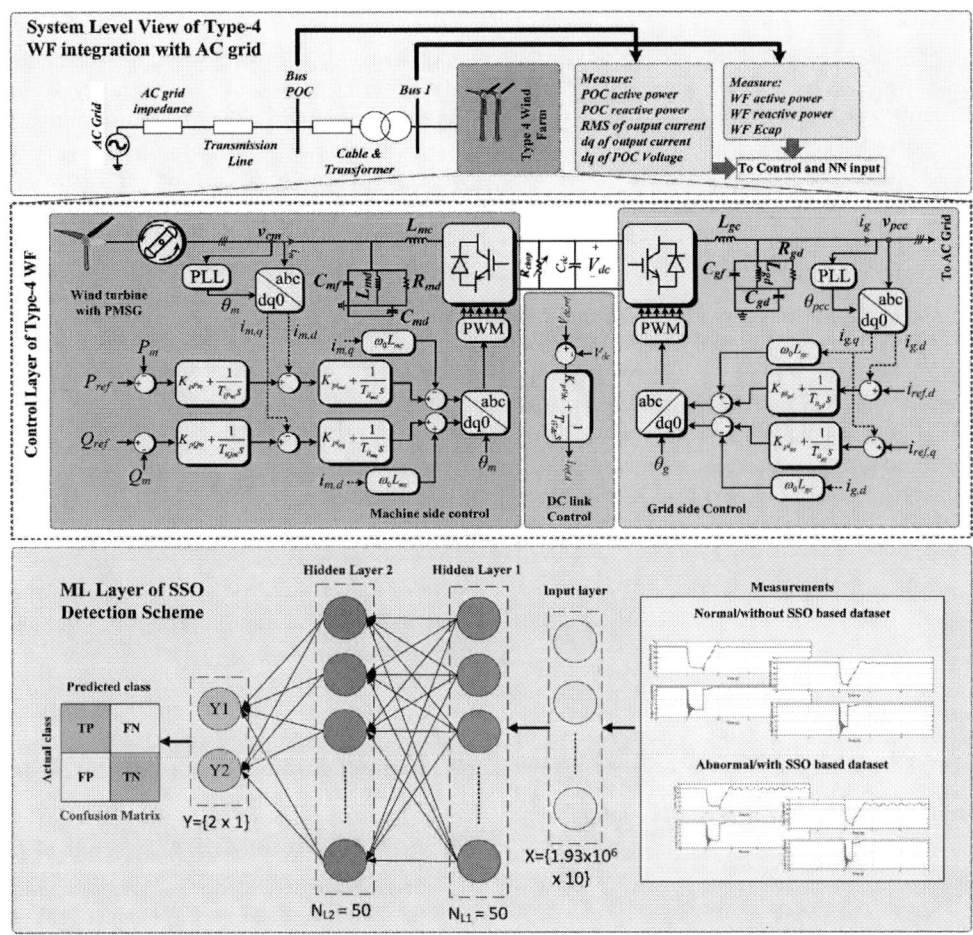

Fig.1. Overview of WF and its control integrated with AC grid and architecture of developed LSTM based ML scheme for SSO detection.

analysis [10], and impedance based small signal analysis [11]. In [12], WF's mathematical model was derived and then various properties of the SSO such as amplitude, frequency spectrum composition and shape envelope was investigated. The work presented in [13] leverages the periodogram data from the synchrophasors to determine the oscillations occurring due to SSO. This method is based on comparing a predefined threshold of frequencies originated due to the model oscillations with the magnitude of the frequencies triggered due to the SSO and if this threshold is crossed at that instant the SSO detection signal is made high. However, due to the high variability and intermittent nature of wind speed, it can affect the accuracy of the periodogram which may result in false alarm. In work [14], SSCI (sub synchronous control interaction) detection via using discrete time Fourier Transform (DTFT) is discussed. Nevertheless, the resolution of DTFT affects the determination of SSCI and several low frequency oscillations with small magnitude may get unnoticed that result in the loss of accuracy in detecting SSCI. The measured signal's envelope is utilized to filter out the synchronous frequency and leaving the sub-synchronous component of frequency. However, the authors suggest that the satisfactory performance is guaranteed for specific conditions and sub synchronous frequencies.

To overcome the shortcomings of the conventional approaches an ML based scheme is proposed for SSO detection

in Type 4 wind farms. This paper presents a light weight machine learning (ML) based SSO detection scheme that leverages a dual layer long short term memory (DL-LSTM) network. This approach automates a swift SSO detection within 3 grid cycles with very high accuracy and minimal false negative misclassification. Furthermore, applying the developed SSO detection scheme to Type 4 wind farms integrated with weak grids represents a unique and innovative application. As the type 4 wind turbine generator based wind farms are the new and progressing technology that features the benefits of the permanent magnet synchronous generator, so the instances of the SSO events are very rare in past. To the best of authors' knowledge, no such SSO detection scheme has been implemented in previous literature that addresses a new but very critical challenge in the modern-day power system due to several challenges such as lack of real world data for this type of SSO. Moreover, the previous scheme mainly encompasses the SSO triggering due to the interaction of series compensation with the wind turbine generator. But this work studies a new relationship between SSO triggering and weak grid conditions. Thus, this work uses an aggregated Type 4 wind farm integrated with weak AC grid to generate the SSO based events and create a comprehensive dataset for training and validation purposes of proposed DL-LSTM network. The remaining paper includes section II that discusses the derivation of dynamic model of control of Type 4 wind farm that will be used to generate the reliable dataset of the ML scheme. Section III explains the mechanism of dataset generation and the algorithm of the proposed DL-LSTM network is presented in section IV. The training and validation results of the proposed scheme are discussed in section V. Finally, section VI concludes the work and provides the future steps of the proposed scheme.

II. DYNAMIC MODELING OF BACK TO BACK CONVERTER CONTROL FOR TYPE 4 WF

The overall architecture of the network understudy is shown in Fig.1. It depicts a Type 4 aggregated wind farm interfaced with an AC grid which makes a weak AC network by applying modifications via changing the connected network impedances. The full control for the Type 4 aggregated wind farm is comprised of three parts, (i) machine side control, (ii) DC link capacitor voltage control, and (iii) grid side control. The dynamic relations of active and reactive power measurements are given by,

$$P_m = \frac{1}{2}(v_{cm,d}i_{m,d} + v_{cm,q}i_{m,q}) \tag{1}$$

$$Q_m = \frac{1}{2}(v_{cm,q}i_{m,d} - v_{cm,d}i_{m,q}) \tag{2}$$

where, $v_{cm,d}$ and $v_{cm,q}$ are the dq components of the voltage at machine side converter. $i_{m,d}$ and $i_{m,q}$ are the dq components of the machine side converter current. The measured active and reactive power is denoted by P_m and Q_m, respectively. The machine side converter's controller dynamic equations are given by,

$$i_{m,d}^* = K_{pPm}(P_{ref} - P_m) + \frac{1}{T_{iPm}}\int(P_{ref} - P_m)dt \tag{3}$$

$$i_{m,q}^* = K_{pQm}(Q_{ref} - Q_m) + \frac{1}{T_{iQm}}\int(Q_{ref} - Q_m)dt \tag{4}$$

$$v_{cm,d}^* = v_{cm,d} + K_{pi_{md}}(i_{m,d}^* - i_{m,d}) + \frac{1}{T_{ii_{md}}}\int(i_{m,d}^* - i_{m,d})dt \tag{5}$$
$$- \omega_0 L_{mc}i_{m,q}$$

$$v_{cm,q}^* = v_{cm,q} + K_{pi_{mq}}(i_{m,q}^* - i_{m,q}) + \frac{1}{T_{ii_{mq}}}\int(i_{m,q}^* - i_{m,q})dt \tag{6}$$
$$+ \omega_0 L_{mc}i_{m,d}$$

where, $i_{m,d}^*$ and $i_{m,q}^*$ are the decoupled machine side reference current in dq domain, K_{pPm}, T_{iPm}, K_{pQm}, and T_{iQm} are the proportional and integral (PI) controller gains related to regulating the reference active (P_{ref}) and reactive powers (Q_{ref}). The reference active and reactive powers are usually generated via maximum power point (MPPT) tracking algorithm or by the grid operator according to the grid requirements. $v_{cm,d}^*$ and $v_{cm,q}^*$ are the reference machine side converter voltage in dq domain. $K_{pi}md$, $T_{i}imd$, $K_{pi}mq$, and $T_{i}imq$ are the PI controller gains related to regulating the machine side current in dq domain. ω_0 is the nominal value of the rotational frequency and L_{mc} is the machine side filter's inductor. The controller to regulate the DC link voltage generates the reference output current for the grid side converter and it is given by,

$$i_{ref,d} = K_{pVdc}(V_{dcref} - V_{dc}) + \frac{1}{T_{iVdc}}\int(V_{dcref} - V_{dc})dt \tag{7}$$

where, K_{pVdc} and T_{iVdc} are the PI gains for DC link voltage controller. V_{dcref} and V_{dc} are the reference DC link voltage and measured DC link voltage, respectively. $I_{ref,d}$ is the generated reference current for the grid side converter control.

The grid side converter is operating in the grid-following or current controlled mode of operation. The direct axis reference current is usually generated via DC link control and the q-axis reference current is given by the grid operator according to the requirement of ancillary services from the converter. The relation of grid side converter control is given by,

$$v_{pcc,d}^* = v_{pcc,d} + K_{pi_{gd}}(i_{ref,d} - i_{g,d}) + \frac{1}{T_{ii_{gd}}}\int(i_{ref,d} - i_{g,d})dt \tag{8}$$
$$+ \omega_0 L_{gc}i_{g,q}$$

$$v_{pcc,q}^* = v_{pcc,q} + K_{pi_{gq}}(i_{ref,q} - i_{g,q}) + \frac{1}{T_{ii_{gq}}}\int(i_{ref,q} - i_{g,q})dt \tag{9}$$
$$- \omega_0 L_{gc}i_{g,d}$$

where, $v_{pcc,d}^*$, $v_{pcc,q}^*$, $v_{pcc,d}$ and $v_{pcc,q}$ are the reference voltages and point of common coupling voltages in dq domain, respectively. $i_{ref,d}$, $i_{ref,q}$, $i_{g,d}$ and $i_{g,q}$ are the reference current at grid side and measured grid current in dq domain, respectively. $K_{pi}gd$, $T_{i}igd$, $K_{pi}gq$, and $T_{i}igq$ are the PI gains to regulate the grid current. L_{gc} is the grid side filter's inductor. By leveraging the WF control dataset for the ML algorithm is generated and discussed in the following section of the paper.

III. DESCRIPTION OF DATASET GENERATION METHOD

Origination of SSO due to interaction between fast dynamics of the Type 4 control of WF and weak AC grid is a recent phenomenon and research on this issue is ongoing, so the availability of a comprehensive and large dataset on characteristics of this type of SSO is limited and not ample to compressively train and test a ML scheme. In this paper, a comprehensive data set is generated via test setup developed in PSCAD as shown in Fig. 1 that involves an aggregated Type 4 WF integrated with AC grid with different grid conditions. The measured values include active power, reactive power, RMS voltage, and the dq components for current and voltage measured at PCC and Bus 1, respectively (see Fig.1). Initially, a thorough controller sensitivity analysis is performed using the model developed in section II on the control of Type 4 WTG to identify the critical control parameters that can trigger SSO in the presence of a weak grid. Then, these parameters were varied in such a manner that can create induced SSO dataset.

A. Nnormal Dataset Samples

As deduced from the controller sensitivity analysis, the risk of SSO is decreased when gain of PLL at grid side is increased. Thus, while iteratively increasing the value of grid side PLL gain normal (SSO-free) dataset is generated and some of the results of this dataset are depicted in Fig. 2 and Fig. 3. Specifically, in the first subset of data, a short circuit fault is added as a disturbance at instant t= 2.5 sec for duration of 0.14 sec and in second subset of data, short circuit fault is added at instant t= 2.0 sec for duration of 0.25 sec. It can be seen as soon as the disturbance is removed the active power and d component

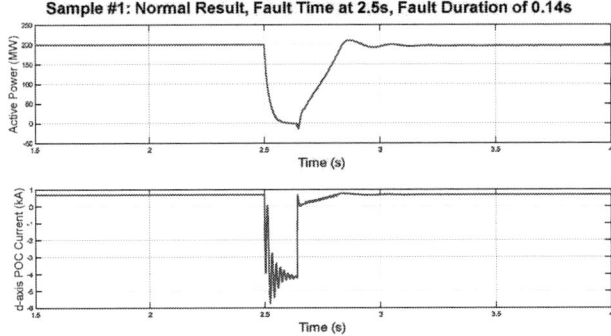

Fig. 2 Sample result of 1st subset of Normal dataset showing measured active power and d component of output current.

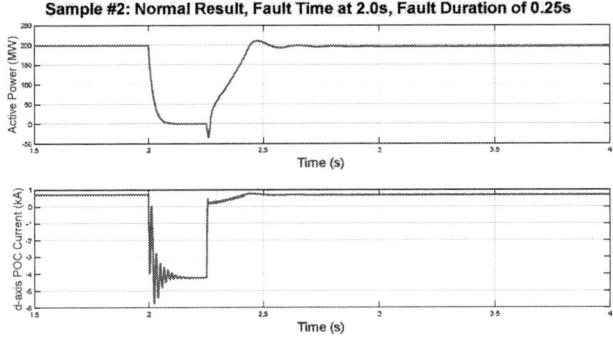

Fig. 3 Sample result of 2nd subset of Normal dataset showing measured active power and d component of output current.

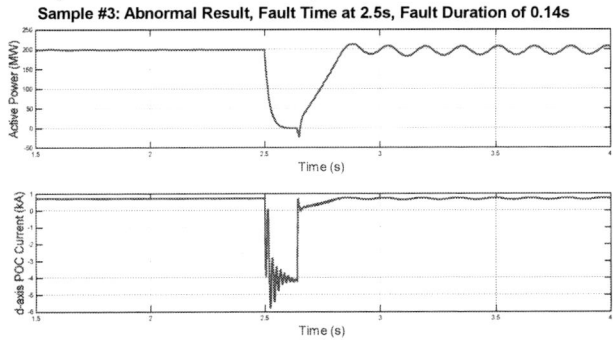

Fig. 4 Sample result of 1st subset of Abnormal dataset showing measured active power and d component of output current.

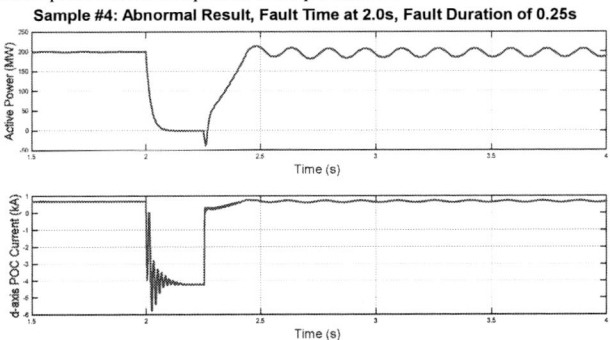

Fig. 5 Sample result of 2nd subset of Abnormal dataset showing measured active power and d component of output current.

of output current at PCC which denotes a SSO free and normal class of the dataset.

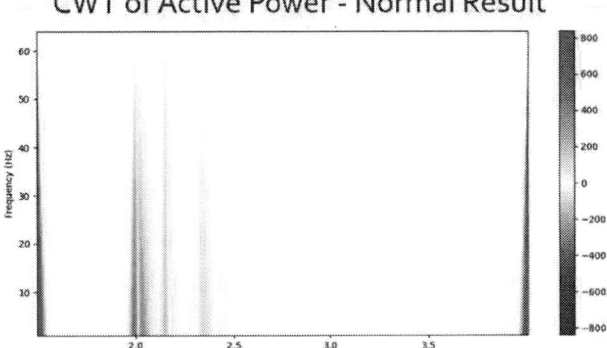

Fig. 6 CWT of active power of the normal dataset.

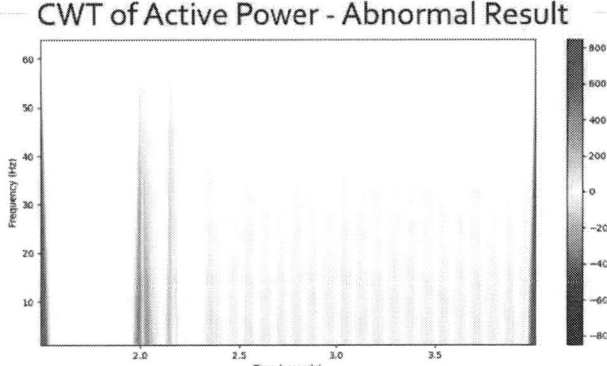

Fig. 7 CWT of active power of the abnormal dataset.

B. Abnormal Dataset Samples

To generate abnormal results or SSO based simulation results, the grid-side control parameter $K_{pi_{gd}}$ was increased and $T_{ii_{gd}}$ was decreased simultaneously. Deduced from the controller sensitivity analysis, the increase of $K_{pi_{gd}}$ and decrease of $T_{ii_{gd}}$ increases the risk of SSO, and therefore ensures that the results generated for this case are abnormal and contain induced SSO. Fig. 4 and Fig. 5 illustrate the 1st and 2nd subset of abnormal dataset. In both sample results as the fault is cleared, a low frequency and sustained oscillation is observed, and this refers to SSO in the output of the WF. The presence of SSO was confirmed via FFT analysis and this shows a low frequency component was present in the frequency spectrum of the active power and output current after the fault is cleared.

C. Verification of SSO Presence in Generated Dataset

The presence of SSO in the generated datasets is not only confirmed via simulation results and FFT analysis but the analytical method continuous wavelet transform (CWT) was also applied on the generated datasets. To do this, scalograms representing the signals in time-frequency representation were generated using CWT through Python programming. Fig. 6 shows the scalogram depicting the time-frequency representation of the SSO-free sample result. Fig. 7 displays the scalogram corresponding to sample result which contains SSO. The time-frequency representation of the abnormal result

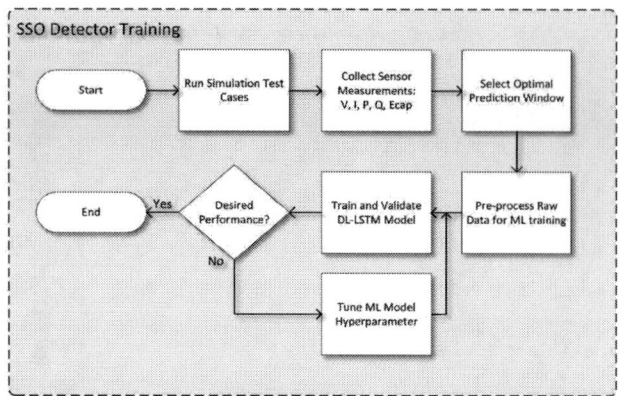

Fig. 8 Flowchart illustrating the training of the DL-LSTM based SSO detector

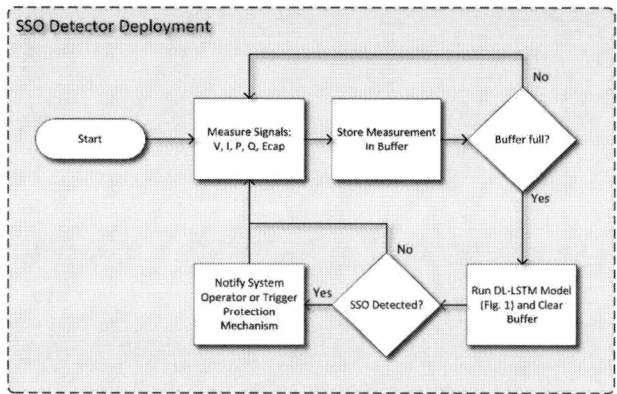

Fig. 9 Flowchart illustrating the deployment of the DL-LSTM based SSO detector

Table I: System parameters for understudy network

System Parameters	Value & Units
Rated power of one WTG unit	2 MVA
Number of WTG units	100
Nominal system frequency	60 Hz
Rated system power	200 MVA
Rated system AC voltage	33 kV
WTG's transformer ratio	33.0/0.69 kV
System's transformer ratio	230.0/33.0 kV
AC grid	230 kV

indicates a high frequency at the time of the fault at 2.0 seconds, and even after the fault is cleared, SSOs of around 40 Hz are sustained within the signal, confirming the presence of SSO.

IV. DL-LSTM DESCRIPTION

The SSO detection involves significant complexity and emphasizes the need for effective schemes, with machine learning demonstrating exceptional performance in addressing these challenges [15]. To achieve fast SSO detection within a prediction window of a few cycles, a lightweight yet accurate DL-LSTM machine learning model is implemented as seen in Fig. 1 and system specifications in Table I. The prediction latency is minimized by eliminating additional data processing and feature extraction computations, opting to use raw sensor measurements directly. The data is pre-processed by

Table II: Hyperparameter of the SSO detector

Parameters	Value
Number of Input Features	449
Hidden Layer 1	50 LSTM units
Hidden Layer 2	50 LSTM units
Optimizer	Adam optimizer
Batch Size	32
Epoch	50

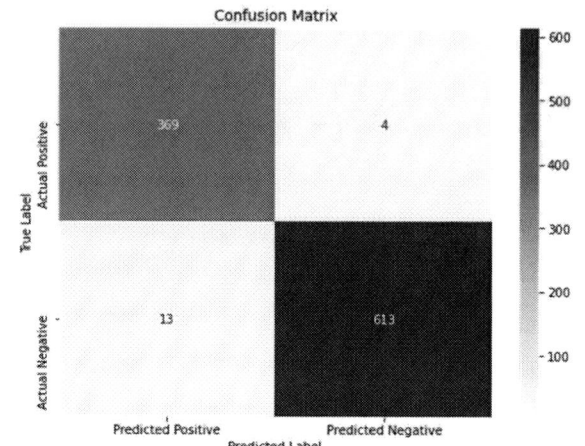

Fig. 10 Preliminary ML results

TABLE III: DL-LSTM PERFORMANCE

Metric	Value
Accuracy	0.98
False Negative Rate	0.01
Precision	0.97
F1 Score	0.99

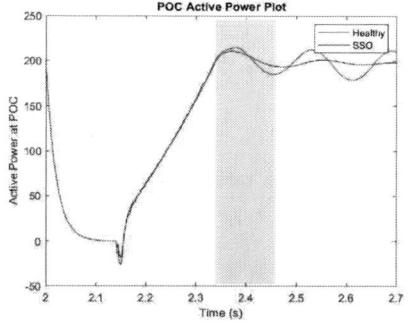

Fig. 11 Region of DL-LSTM misclassification

transforming, normalizing, and labeling it into a clean, usable format for training the ML model. The process of developing the DL-LSTM SSO detector is defined by the flowchart in Fig. 5 and Fig. 6. In SSO detection, false negatives occur when the system fails to identify a true SSO event, which is crucial to address as missing these events due to misdetection can lead to severe system instability, potential equipment damage, safety risk, and regulatory non-compliance. Therefore, a reliable SSO detector shall seek to minimize the number of false negatives and eliminate the false negative rate.

The first step in training the DL-LSTM SSO detector, illustrated in Fig. 8, involves data acquisition and collection from the PSCAD model mentioned in the previous section to generate cases with healthy and SSO simulation scenarios for the supervised machine learning implementation. The optimal prediction window size in this study was selected through a evaluation of multiple criteria, with emphasis on the mean prediction latency of the DL-LSTM model and ML performance metrics, particularly prioritizing minimal FNR and maximized classification accuracy. The methodology for hyperparameter optimization followed an iterative approach, with an ideal desired target performance of 0 FNR and 1.0 detection accuracy. The hyperparameters of the trained DL-LSTM model are summarized in Table II. The process of deploying the ML detector (Fig. 9) includes measuring the state space signals, storing them in a memory buffer dimensioned to the predetermined desired window size. For successful deployment, the sampling frequency of the signals must be identical to that of the trained ML model. Upon buffer saturation, the contents of the buffer are sent as input to the DL-LSTM trained model, where the detection outcomes are then transmitted to the system operator in real-time.

V. RESULTS AND DISCUSSION

The performance of DL-LSTM for a prediction window of 3 cycles and a training/testing split of 80/20 with 5-fold cross validation is illustrated by the confusion matrix in Fig. 10 and Table III. With an accuracy of 98% and false negative rate of only 0.01%, the results indicate the DL-LSTM model performs with high accuracy and low FNR as desired. Despite the high accuracy and a low FNR ratio, the DL-LSTM model's misclassifications strictly occur post fault/disturbance event, as shown and highlighted in Fig. 11. The post fault recovery is associated with a minor oscillatory characteristic resulting from the PI controllers regulating the system nominal values. The misclassification arises as the initial post-fault transients in normal and SSO operations share similar characteristics. It's important to note that this issue only occurs immediately after the fault/disturbance event and does not persist beyond that. The post disturbance misclassifications are expected to appear more frequently when reducing the window size and will need to be addressed when smaller window size is considered. Despite these minor misclassifications, the DL-LSTM model demonstrates robust performance, maintaining high accuracy and stability across various operational conditions.

VI. CONCLUSION AND FUTURE WORKS

The DL-LSTM model demonstrated high accuracy and a low false negative rate for real-time SSO detection within a 3-cycle prediction window, ensuring effective performance and reliability. This will allow the mitigation approach ample time to suppress the detected SSO in the Type 4 wind farms. The extension of this work discussed in [16] incorporate a multiclass detector to identify the control parameter likely causing SSO and provide recommendations for tuning the control loop to effectively mitigate it.

ACKNOWLEDGMENT

This publication is supported by Iberdrola S.A. as part of its innovation department research studies. Its contents are solely the responsibility of the authors and do not necessarily represent the official views of Iberdrola Group.

REFERENCES

[1] P. Catalán, Y. Wang, J. Arza, and Z. Chen, "A Comprehensive Overview of Power Converter Applied in High-Power Wind Turbine: Key Challenges and Potential Solutions," *IEEE Transactions on Power Electronics,* vol. 38, no. 5, pp. 6169-6195, 2023,

[2] J. Ma and Y. Shen, "Stability Assessment of DFIG Subsynchronous Oscillation Based on Energy Dissipation Intensity Analysis," *IEEE Transactions on Power Electronics,* vol. 35, no. 8, pp. 8074-8087, 2020,

[3] N. Verma, N. Kumar, S. Gupta, H. Malik, and F. P. G. Márquez, "Review of sub-synchronous interaction in wind integrated power systems: classification, challenges, and mitigation techniques," *Protection and Control of Modern Power Systems,* vol. 8, no. 2, pp. 1-26, 2023.

[4] L. Wang, X. Xie, Q. Jiang, H. Liu, Y. Li, and H. Liu, "Investigation of SSR in practical DFIG-based wind farms connected to a series-compensated power system," *IEEE Transactions on Power Systems,* vol. 30, no. 5, pp. 2772-2779, 2014.

[5] H. Du, J. Yan, M. Ghafouri, R. Zgheib, and M. Debbabi, "Modeling and Assessment of Cyber Attacks Targeting Converter-Driven Stability of Power Grids With PMSG-Based Wind Farms," *IEEE Transactions on Power Systems,* 2024.

[6] Y. Li, L. Fan, and Z. Miao, "Wind in weak grids: Low-frequency oscillations, subsynchronous oscillations, and torsional interactions," *IEEE Transactions on Power Systems,* vol. 35, no. 1, pp. 109-118, 2019.

[7] C. Wu, B. Hu, P. Cheng, H. Nian, and F. Blaabjerg, "Eliminating Frequency Coupling of DFIG System Using a Complex Vector PLL," in *2020 IEEE Applied Power Electronics Conference and Exposition (APEC),* 15-19 March 2020

[8] Z. Xiaoyu, *et al* , "Analysis and control on Sub-synchronous oscillation (SSO) of HVDC transmission for large-scale permanent magnet synchronous generators (PMSG)-based wind farm integration," *The Journal of Engineering,* vol. 2019, no. 16, pp. 2440-2444, 2019.

[9] H. A. Mohammadpour, J. Siegers, and E. Santi, "Controller design for TCSC using observed-state feedback method to damp SSR in DFIG-based wind farms," in *2015 IEEE Applied Power Electronics Conference and Exposition (APEC),* 15-19 March 2015 2015, pp. 2993-2998

[10] H. A. Mohammadpour, Y. J. Shin, and E. Santi, "SSR analysis of a DFIG-based wind farm interfaced with a gate-controlled series capacitor," in *2014 IEEE Applied Power Electronics Conference and Exposition - APEC 2014,* 16-20 March 2014 2014, pp. 3110-3117

[11] A. Ostadi, A. Yazdani, and R. K. Varma, "Modeling and Stability Analysis of a DFIG-Based Wind-Power Generator Interfaced With a Series-Compensated Line," *IEEE Transactions on Power Delivery,* vol. 24, no. 3, pp. 1504-1514, 2009, doi: 10.1109/TPWRD.2009.2013667.

[12] H. Ye, Y. Liu, P. Zhang, and Z. Du, "Analysis and detection of forced oscillation in power system," *IEEE Transactions on Power Systems,* vol. 32, no. 2, pp. 1149-1160, 2016.

[13] J. Follum and J. W. Pierre, "Detection of periodic forced oscillations in power systems," *IEEE Transactions on Power Systems,* vol. 31, no. 3, pp. 2423-2433, 2015.

[14] Z. Zhang, S. Liu, G. Zhu, and Z. Lu, "SSCI detection and protection in doubly fed generator based on DTFT," *The Journal of Engineering,* vol. 2017, no. 13, pp. 2104-2107, 2017.

[15] O. H. Abu-Rub, A. Y. Fard, M. F. Umar, M. Hosseinzadehtaher, and M. B. Shadmands, "Towards intelligent power electronics-dominated grid via machine learning techniques," *IEEE Power Electronics Magazine,* vol. 8, no. 1, pp. 28-38, 2021.

[16] Omar Abu-Rub, *et al*, "ML-Assisted Sub-synchronous Oscillation Detection and Localization in Type-4 Wind Farms under Weak Grid Conditions," in *50th Annual Conference of the IEEE Industrial Electronics Society (IECON),* Chicago, IL, USA, 2024: IEEE, pp. 1-4.

A Study of Module Design Method to Suppress the Oscillation Occurs Between Parallel-Connected Power Devices

Shinji Yato
System Solutions Engineering Headquarters
ROHM Co., Ltd.
Kyoto, Japan
shinji.yato@mnf.rohm.co.jp

Hiroto Sakai
Research and Development Center
ROHM Co., Ltd.
Kyoto, Japan
hiroto.sakai@rohm.co.jp

Hideo Araki
System Solutions Engineering Headquarters
ROHM Co., Ltd.
Kyoto, Japan
hideo.araki@dsn.rohm.co.jp

Shumei Shimosako
System Solutions Engineering Headquarters
ROHM Co., Ltd.
Kyoto, Japan
shumei.shimosako@rohm.co.jp

Abstract— **The use of power modules with multiple SiC-MOSFETs dies connected in parallel has increased significantly for traction inverter circuits in electric vehicles(xEVs). However, this configuration can cause problems with oscillation between the parallel-connected dies. This study proposes a module design method to suppress the oscillation. This design method is based on the approximate open-loop transfer function between the parallel-connected dies. The validity of the proposed design method was confirmed through simulations and experimental evaluations.**

Keywords—Power Module, Oscillation, Parallel Connected Dies, Parallel Switching, SiC MOSFET

I. INTRODUCTION

In recent years, taking advantages of the promising characteristics of high-voltage, low on-resistance, high-speed driving SiC-MOSFETs [1], the use of power modules with multiple SiC-MOSFET dies connected in parallel has increased significantly for traction inverter circuits in electric vehicles (xEVs). However, this configuration can cause problems with oscillation between the parallel-connected dies [2], "hereinafter referred to as oscillation". Therefore, preventing oscillation is one of the important issues in practical module design.

Fig.1 depicts the circuit that can be used to assess the stability of parallel-connected power devices [3][4]. Note that, L_{dd}, L_{gg}, and L_{ss} are inductances considering mutual inductance, not the partial inductances of point-to-point connections [5].

The open-loop characteristics from V_{gs_i} to V_{gs_o} in Fig.1 are examined by PLECS® (Plexim), and the bode diagram of parallel-connected power devices can be depicted as Fig.2 [4]. This characteristic is composed of "1st-order derivative," "2nd-order lag," "2nd-order lag," and "2nd-order lead" from low frequencies[4].

Therefore, this characteristic can be expressed by the following [4].

$$\frac{sK \cdot \omega_{p1}{}^2 \omega_{p2}{}^2}{\omega_z{}^2} \cdot \frac{G(s) = (s^2 + 2\zeta_z\omega_z s + \omega_z{}^2)}{(s^2 + 2\zeta_{p1}\omega_{p1}s + \omega_{p1}{}^2)(s^2 + 2\zeta_{p2}\omega_{p2}s + \omega_{p2}{}^2)} \quad (1)$$

Fig.2 depicts that the phase is below −180° between ω_{p2} and ω_z. From the Nyquist criterion, the system is unstable when the phase is below −180° at a gain of 0 dB [4].

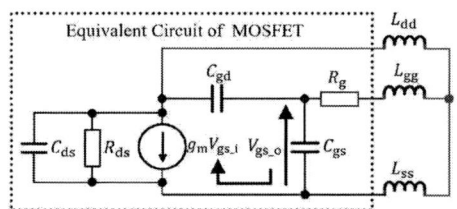

L_{dd}: Drain to Drain Inductance
L_{gg}: Gate to Gate Inductance
L_{ss}: Source to Source Inductance
R_g: Gate Resistance
R_{ds}: Drain – Source Resistance
C_{gd}: Gate – Drain Capacitance
C_{gs}: Gate – Source Capacitance
C_{ds}: Drain – Source Capacitance
g_m: Transconductance

Fig. 1. Equivalent circuit of parallel oscillation

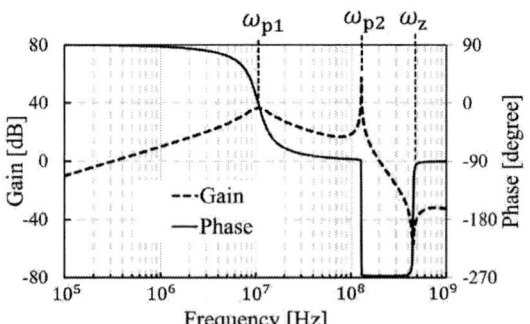

Fig. 2. Bode diagram of parallel-connected devices

979-8-3315-1612-3/25 $31.00 © 2025 IEEE

Based on to prevent oscillation, the distance between the second pole frequency ω_{p2} and the zero frequency ω_z in the approximate open-loop transfer function of the parallel-connected switching devices should be designed as close as possible [4]. Therefore, oscillation can be suppressed gradually by increasing the parameter ω_{p2}/ω_z which indicates the distance between ω_{p2} and ω_z. Nonetheless, this hypothesis has not been experimentally verified in the existing literature. Additionally, the module design criterion specifying the minimum value that ω_{p2}/ω_z should be designed to prevent oscillation is also unknown.

In this paper, a module design method is proposed to prevent oscillation by addressing the two issues mentioned above.

II. PARAMETERS THAT AFFECT ω_{p2}/ω_z

Under certain conditions, each coefficient of (1) can be expressed by the following approximate equations[4]. In this paper, the parallel operator "//" is defined as "$A//B = A \cdot B / (A+B)$."

$$K = g_m L_{ss} \qquad (2)$$

$$\omega_{p1} = \frac{1}{\sqrt{(L_{ss} + L_{gg})C_{gs}}} \quad , \quad \zeta_{p1} = \frac{R_g}{2}\sqrt{\frac{C_{gs}}{L_{ss} + L_{gg}}} \qquad (3)$$

$$\omega_{p2} = \frac{1}{\sqrt{(L_{dd} + L_{gg}//L_{ss})C_{ds}}} \quad , \quad \zeta_{p2} = \frac{1}{2R_{ds}}\sqrt{\frac{(L_{dd} + L_{gg}//L_{ss})}{C_{ds}}} \qquad (4)$$

$$\omega_z = \frac{1}{\sqrt{(L_{dd} + L_{gg})C_{gd}}} \quad , \quad \zeta_z = \frac{R_g}{2}\sqrt{\frac{C_{gd}}{L_{dd} + L_{gg}}} \qquad (5)$$

From (4) and (5), ω_{p2}/ω_z (the larger this is, the more the oscillation is suppressed) can be expressed simply by (6), which consist of two factors. The first factor is function of MOSFET's parasitic capacitances and the second factor is function of the module's parasitic inductances.

$$\frac{\omega_{p2}}{\omega_z} = \sqrt{\frac{C_{gd}}{C_{ds}}} \times \sqrt{\frac{L_{gg}/L_{ss}}{L_{dd}/L_{ss} + L_{dd}/L_{gg} + 1} + 1} \qquad (6)$$

The parasitic capacitances (C_{gs}, C_{ds} and C_{gd}) and inductances (L_{dd}, L_{gg} and L_{ss}) in parallel-connected MOSFETs are shown in Fig.3.

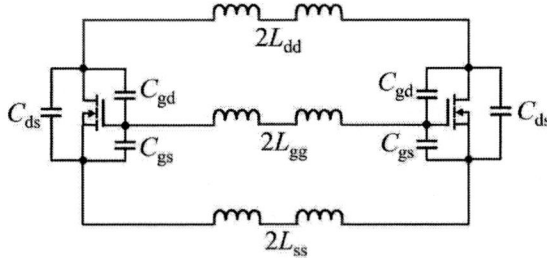

Fig. 3. Parasitic Inductances and Capacitances of MOSFET's in parallel

III. MODULE DESIGN METHOD TO SUPPRESS OSCILLATION

From (6), it can be observed that ω_{p2}/ω_z behaves as follows for each parasitic inductance change within the range in which the above-mentioned approximation is valid.

Fig.4 depicts ω_{p2}/ω_z behavior when L_{gg} changes. As L_{gg} decreases, ω_{p2}/ω_z converges to $\sqrt{C_{gd}/C_{ds}}$. As L_{gg} increases, ω_{p2}/ω_z increases.

Fig.5 depicts ω_{p2}/ω_z behavior when L_{ss} changes. As L_{ss} decreases, ω_{p2}/ω_z converges to $\sqrt{C_{gd}/C_{ds}}\sqrt{1 + L_{gg}/L_{dd}}$. As L_{ss} increases, ω_{p2}/ω_z converges to $\sqrt{C_{gd}/C_{ds}}$. No matter how small L_{ss} is, the maximum value of ω_{p2}/ω_z will always be limited to $\sqrt{C_{gd}/C_{ds}}\sqrt{1 + L_{gg}/L_{dd}}$.

Fig.6 depicts ω_{p2}/ω_z behavior for L_{dd} changes. As L_{dd} decreases, ω_{p2}/ω_z converges to $\sqrt{C_{gd}/C_{ds}}\sqrt{1 + L_{gg}/L_{ss}}$. As L_{dd} increases, ω_{p2}/ω_z converges to $\sqrt{C_{gd}/C_{ds}}$. No matter how small L_{dd} is, the maximum value of ω_{p2}/ω_z will always be limited to $\sqrt{C_{gd}/C_{ds}}\sqrt{1 + L_{gg}/L_{ss}}$.

As observed in Fig.6, the behavior of ω_{p2}/ω_z with respect to variations in L_{dd} is similar to that of L_{ss} (Fig.5). However, for most typical wire-bonded power modules, L_{dd} is significantly smaller than L_{ss}. Consequently, the variation in ω_{p2}/ω_z due to changes in L_{dd} is less pronounced compared to changes in L_{ss}. This implies that the parameter sensitivity of ω_{p2}/ω_z to L_{ss} is generally higher than that of L_{dd} in such modules.

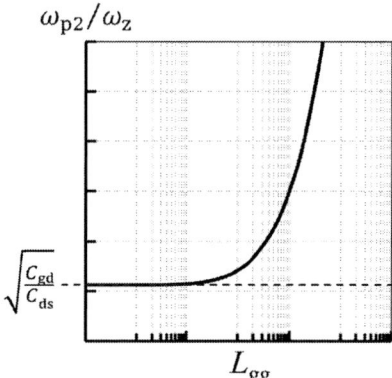

Fig. 4. ω_{p2}/ω_z behavior when L_{gg} changes

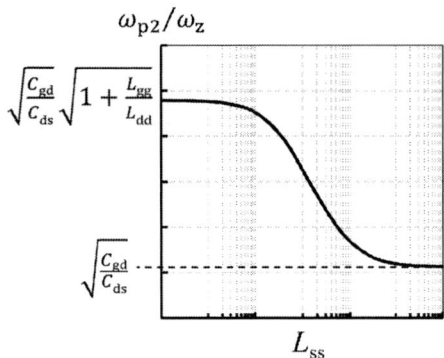

Fig. 5. ω_{p2}/ω_z behavior when L_{ss} changes

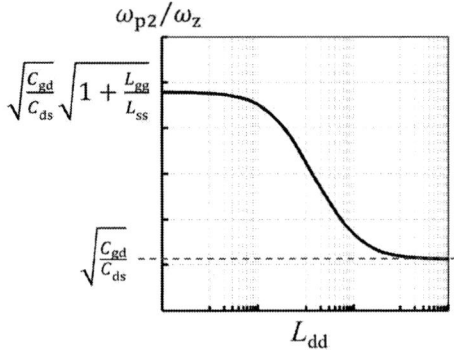

Fig. 6. ω_{p2}/ω_z behavior when L_{dd} changes

Therefore, it can be concluded that in order to increase ω_{p2}/ω_z and prevent oscillation, the following design procedure should be followed.

(i) Select a MOSFET with as high C_{gd}/C_{ds} ratio. This is crucial for suppressing oscillation, as ω_{p2}/ω_z is strongly influenced by this ratio. However, a balance must be struck, as a higher C_{gd} can contribute to self-turn-on[6] and increased switching losses. Careful optimization is necessary to achieve the desired trade-off.

(ii) Maximize L_{gg} and minimize L_{ss}, L_{dd} to increase ω_{p2}/ω_z.

(iii) Avoid excessively reducing L_{ss} and L_{dd}, as this can lead to reaching the upper limit of ω_{p2}/ω_z. Instead, consider increasing L_{gg} to further enhance ω_{p2}/ω_z.

IV. VERIFICATIONS

A. Verification by simulation

To verify this oscillation suppression module design method mentioned in section III, we performed simulations by using circuit simulator SIMetrix® (SIMetrix Technologies). Oscillations were replicated by using double pulse method [7] and were observed on the waveforms of gate-source voltage V_{gs} that can be seen primarily [8]. The simulation circuit is shown in Fig.7. In this verification, various combinations of parasitic inductances shown in TABLE I were simulated.

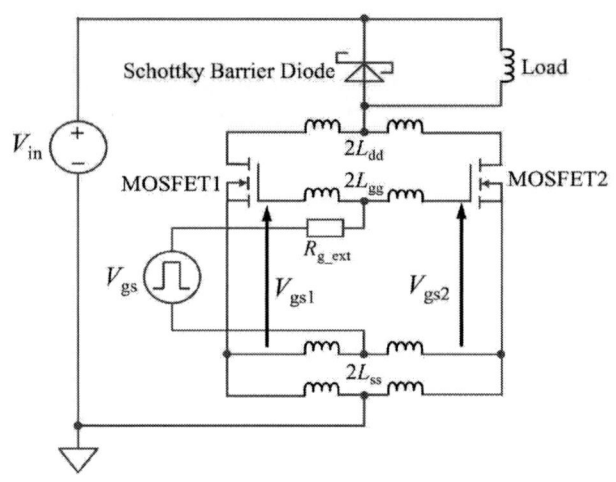

Fig. 7. Simulation Circuit in the time domain

TABLE I. SIMULATION PARAMETER CONDITIONS

V_{in} (V)	V_{gs} (V)	R_{g_ext} (Ω)	Load (μH)	C_{gd} (pF)	C_{ds} (pF)
500	0/18	5.6	200	10	199

	$2L_{dd}$ (nH)	$2L_{gg}$ (nH)	$2L_{ss}$ (nH)	ω_{p2}/ω_z
Cond. 1	0.8	22.3 to 32.0	13.8	0.35 to 0.40
Cond. 2	0.8	11.0 to 15.0	6.7	0.35 to 0.38
Cond. 3	0.8	38.0 to 54.0	26.7	0.34 to 0.38
Cond. 4	1.3	15.4	9.1 to 6.7	0.35 to 0.38
Cond. 5	1.3	10.0	5.6 to 4.5	0.34 to 0.36
Cond. 6	1.3	30.0	17.3 to 13.4	0.36 to 0.39

Fig.8 depicts the simulation results of oscillation waveforms in response to changes of ω_{p2}/ω_z for the V_{gs} turn-off in Cond.1. No-oscillation occurred at turn-on at these conditions. It can be seen that oscillation is suppressed as ω_{p2}/ω_z increases. Under other conditions, oscillations were suppressed as ω_{p2}/ω_z increases similarly.

From the magnified oscillation waveforms in Fig.8, V_{gs1} and V_{gs2} are being oscillated in counter phase each other since this is inter-die oscillation [8].

Oscillation amplitude ΔV_{gs} is defined as " $V_{gs1} - V_{gs2}$" and the maximum amplitude in an oscillation period ΔV_{gs_max} is used as the oscillation evaluation index.

Fig.9 depicts the gradual reduction of ΔV_{gs_max} due to ω_{p2}/ω_z increasing in Cond.1. Similarly under other conditions, ΔV_{gs_max} were also reduced gradually as ω_{p2}/ω_z increases.

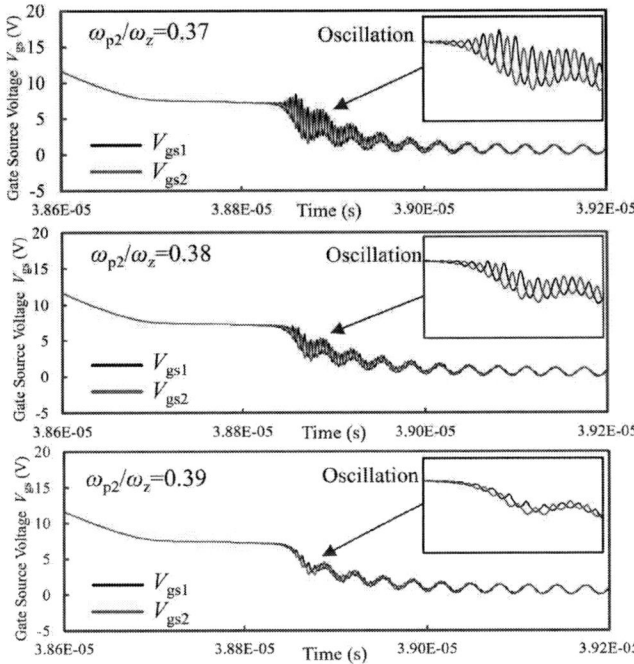

Fig. 8. Oscillation suppression due to ω_{p2}/ω_z increasing in Cond.1

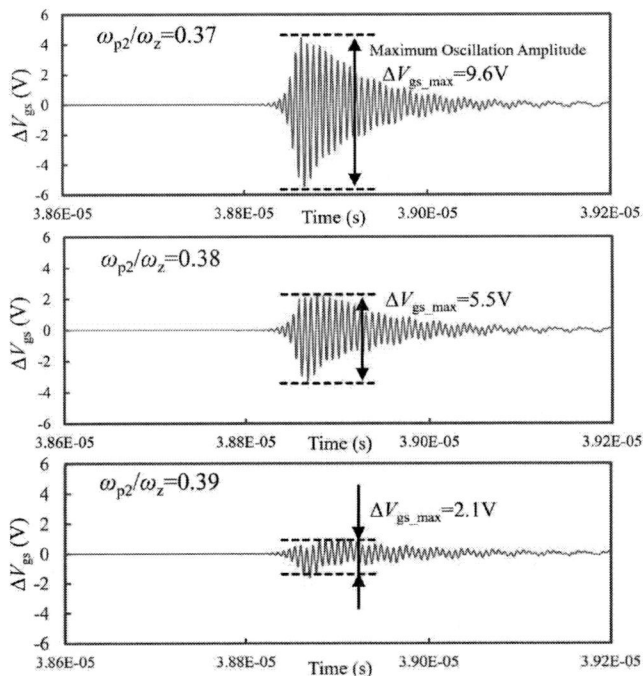

Fig. 9. Reduction of ΔV_{gs_max} due to ω_{p2}/ω_z increasing in Cond.1

Fig.10 depicts the simulation results of correlations between ΔV_{gs_max} and ω_{p2}/ω_z in Cond.1 to 6. Note that, ΔV_{gs_max} below 2V was considered to be 0V (No-oscillation).

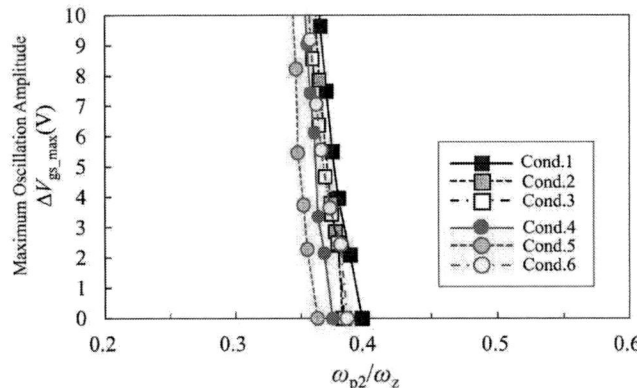

Fig. 10. ω_{p2}/ω_z dependence of ΔV_{gs_max}

In order to increase ω_{p2}/ω_z, whether L_{gg} was increased or L_{ss} was decreased, oscillations were suppressed gradually in all conditions, thereby it is confirmed the validity of the oscillation suppression module design method mentioned in section III.

It was also observed that the oscillation boundaries were concentrated nearby $\omega_{p2}/\omega_z=0.4$ and were completely suppressed when $\omega_{p2}/\omega_z \geq 0.4$, regardless of the parasitic inductance values. Therefore, it can be concluded that ω_{p2}/ω_z can be used as a module design criterion to prevent oscillation.

B. Verification by Experiment

Using the test circuit similar to the simulation shown in Fig.11, verification experiments were performed by changing L_{gg} and L_{ss} as in Cond.7 and 8 (TABLE II). These conditions correspond to Cond.1 and 4 respectively in the simulations mentioned above. In addition, each condition has been adjusted to avoid reaching the upper limit value of ω_{p2}/ω_z mentioned in section III.

Double pulse test method[7] was also used in the experiments as well as in the simulations. Note that, to properly observe the oscillation waveforms with small amplitude and fast frequency in V_{gs}, opto-isolated probes were used. To change L_{gg} and L_{ss}, the path length in the real test circuit was varied for each condition.

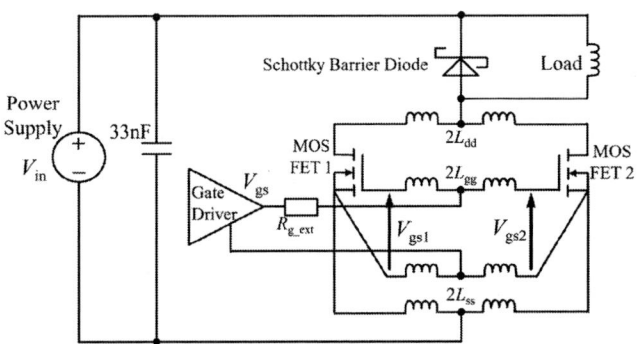

Fig. 11. Experimental Circuit

TABLE II. EXPERIMENTAL PARAMETER CONDITIONS

V_{in} (V)	V_{gs} (V)	R_{g_ext} (Ω)	Load (μH)	C_{gd} (pF)	C_{ds} (pF)
500	0/18	5.6	200	10	199

	$2L_{dd}$ (nH)	$2L_{gg}$ (nH)	$2L_{ss}$ (nH)	ω_{p2}/ω_z
Cond. 7	0.8	22.3 to 29.9	13.8	0.35 to 0.39
Cond. 8	1.3	15.4	8.9 to 7.2	0.35 to 0.37

Fig.12 depicts the experimental results of oscillation waveforms in response to changes of ω_{p2}/ω_z for the V_{gs} turn-off in Cond.7. Similar to the simulation, the following results (a) to (c) were obtained on the experiments. For Cond.8, similar results were also obtained as well.

(a) No-oscillation occurred at turn-on.

(b) Oscillations were suppressed gradually as ω_{p2}/ω_z increases.

(c) From the magnified oscillation waveforms in Fig.12, V_{gs1} and V_{gs2} are being oscillated in counter phase each other.

Fig.13 depicts the decrease in ΔV_{gs_max} as ω_{p2}/ω_z increases in Cond.7. These experimental results align with the simulation data. Similar trends were observed in Cond.8. However, in practical experiments, the maximum oscillation amplitudes did not reach the same levels of simulations. This is due to safety measures implemented to prevent device damage in case of excessive oscillation.

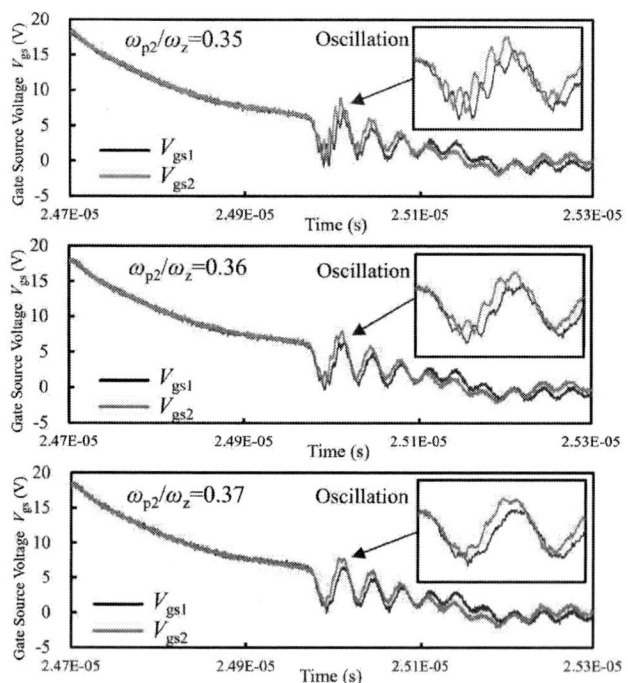

Fig. 12. Oscillation suppression due to ω_{p2}/ω_z increasing in Cond.7

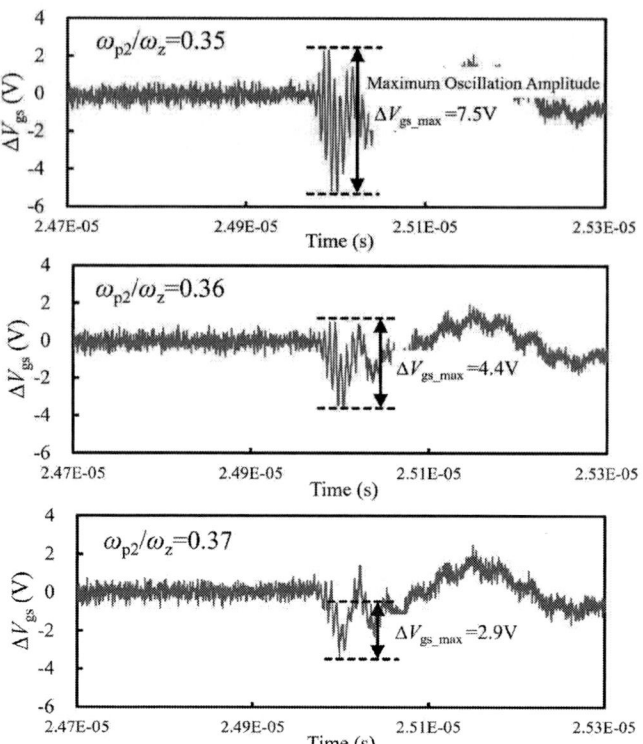

Fig. 13. Reduction of ΔV_{gs_max} due to ω_{p2}/ω_z increasing in Cond.7

Fig.14 depicts the experimental results of correlations between ΔV_{gs_max} and ω_{p2}/ω_z in Cond.7 and 8. In addition, the corresponded simulation results (Cond.1 and 4) are also drawn by doted lines in Fig.14 to compare the experimental results. The simulation results show good agreement with experimental results. Note that, ΔV_{gs_max} below 2V was considered to be 0V (No-oscillation) same as the simulations.

Similar to the simulations, the following results (d) and (e) were also obtained on the experiments.

(d) Oscillations were suppressed gradually as ω_{p2}/ω_z increases, whether L_{gg} was increased or L_{ss} was decreased.

(e) The oscillation boundaries were concentrated nearby $\omega_{p2}/\omega_z=0.4$ and oscillations were completely suppressed when $\omega_{p2}/\omega_z \geq 0.4$, regardless of the parasitic inductance values.

Therefore, it is also confirmed through experimental verifications that the oscillation suppression module design method mentioned in section III is valid, and ω_{p2}/ω_z can be used as a module design criterion to prevent oscillation.

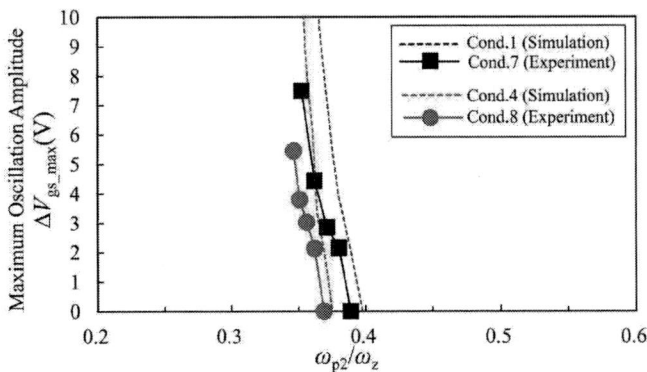

Fig. 14. ω_{p2}/ω_z dependence of ΔV_{gs_max} (Comparison of Experiments and Simulations)

V. CONCLUSION

This research proposes a design methodology to prevent oscillation in power modules by elucidating the relationship between the ω_{p2}/ω_z parameter and parasitic inductances. The effectiveness of this method has been verified through both simulation and experimental analysis under specific conditions.

While the criterion $\omega_{p2}/\omega_z \geq 0.4$ has proven effective in suppressing oscillations for the current conditions, further investigation is necessary to generalize its applicability. Additional tests are planned to explore the impact of factors such as increased parallel connections and variations in MOSFET characteristics on oscillation behavior. By expanding the scope of this research, we aim to develop a more robust and widely applicable design methodology for preventing oscillations in power modules.

ACKNOWLEDGMENT

The authors are grateful to our colleagues S.Kitagawa, M.Murata, Y.Okawauchi, R.Ishido and M.Noah for helpful discussions and comments on the manuscript.

REFERENCES

[1] Tsunenobu Kimoto, "Material science and device physics in SiC technology for high-voltage power devices," Japanese Journal of Applied Physics, Vol. 54, No. 4, 2015.

[2] Jhon G. Kassakian, David Lau, "An analysis and experimental verification of parasitic oscillations in paralleled power MOSFET's," IEEE Transaction on Electron Devices, ed 31, Vol. 7, pp.959-963, 1984.

[3] Katsuaki Saito, Tomoyuki Miyoshi, Daisuke Kawase, Seiichi Hayakawa, Toru Masuda, Yasushi Sasajima, "Simplified Model Analysis of Self-Excited Oscillation and Its Suppression in a High-Voltage Common Package for Si-IGBT and SiC-MOS," IEEE Transactions on Electron Devices, Vol. 65, No. 3, pp.1063-1071, 2018.

[4] Hiroto Sakai, Yuta Okawauchi, Shinji Yato, Hideo Araki, Takayuki Atago, Ken Nakahara, "Simplified Open-Loop Transfer Functions to Analyze Influential Parasitic Parameters for Oscillation Caused by Parallel Connected Transistors," 2023 35th International Symposium on Power Semiconductor Devices and ICs (ISPSD), pp.290-293, 2023.

[5] Igor Kasko, Sven E. Berberich, Matthias Spang, Stefan Oehling, "SiC MOS Power Module in Direct Pressed Die Technology and some Challenges for Implementation," 2020 32nd International Symposium on Power Semiconductor Devices and ICs (ISPSD), pp.364-367, 2020.

[6] Hirokatsu Umegam, Hiroki Ishibashi, Kimihiro Nanamori, Fumiya Hattori, Masayoshi Yamamoto, "Basic analysis of false turn-on phenomenon of power semiconductor devices with parasitic inductances," Electronics Letters, Vol. 52, No. 13, pp.1158-1159, 2016.

[7] Antonio Griffo, Jiabin Wang, Kalhana Colombage, Tamer Kamel, "Real-Time Measurement of Temperature Sensitive Electrical Parameters in SiC Power MOSFETs," IEEE Transactions on Industrial Electronics, Vol. 65, No.3, March 2018 pp.2663-2671.

[8] Florian Sawallich, Hans-Gunter Eckel, "Mitigating Inter-Chip Oscillation of paralleled SiC MOSFETs," 2023 25th European Conference on Power Electronics and Applications (EPE'23 ECCE Europe), pp.1-11, 2023.

A High-Efficient Hybrid Traction Inverter in Electric Vehicle Applications

Yousefreza Jafarian
Electrical and Computer Engineering
Queen's University
Kingston, Canada
19yjj@queensu.ca

Omid Salari
Electrical and Computer Engineering
Queen's University
Kingston, Canada
omid.salari@queensu.ca

Praveen Jain
Electrical and Computer Engineering
Queen's University
Kingston, Canada
praveen.jain@queensu.ca

Alireza Bakhshai
Electrical and Computer Engineering
Queen's University
Kingston, Canada
alireza.bakhshai@queensu.ca

Mohamed Z. Youssef
Electrical and Computer Engineering
Ontario Tech University
Oshawa, Canada
Mohamed.Youssef@ontariotechu.ca

Abstract—This paper introduces a new multisource inverter (MSI) configuration utilizing multiple energy sources, strategically developed for hybrid storage solutions within electric vehicle (EV) frameworks. The proposed configuration facilitates the integration of multiple DC sources while significantly reducing the number of semiconductor components. This approach optimizes cost, weight, complexity, and power density by eliminating magnetic elements from the circuit architecture. Through precise control of the MSI's switching states, the system can generate the desired AC output voltage and current from any combination of the connected DC sources, offering enhanced flexibility and efficiency in power management for EVs. The efficacy of the proposed topology is validated through different testing, employing both the European Urban Driving Cycle (EUDC) and experimental lab prototype. The EUDC simulation provides a standardized benchmark for assessing the topology's performance under typical urban driving conditions, while the experimental results offer empirical evidence of its practical viability and efficiency in real-world applications.

Keywords— DC/AC converters, electric vehicles (EVs), energy management system (EMS), hybrid storage systems (HSS), multi-source inverters (MSIs), permanent magnet synchronous machine (PMSM), space vector modulation (SVM).

I. INTRODUCTION

In recent decades, the electric vehicle (EV) industry has garnered significant attention, with rapid advancements in electrified powertrain technology. The performance of EVs is highly dependent on DC sources, traction inverter technology, and the efficiency of electric machines (EMs). While Li-ion batteries are the primary energy source for most commercial EVs, the electrified powertrain can be further enhanced by incorporating complementary DC sources such as Ultracapacitors (UCs) [1]. UCs offer higher power, faster charging and discharging capabilities, and virtually unlimited cycle life compared to Li-ion batteries, which primarily ensure long driving ranges [2], [3], [4]. There are two primary methods for integrating battery packs and UCs. The most basic configuration directly links the energy sources without intermediary power conversion devices, referred to as a passive connection. This approach, while straightforward, offers limited control over energy distribution between the DC sources and the load. This configuration features a straightforward design and an uncomplicated control circuit, resulting in reduced investment costs and a lighter overall weight. Despite these advantages, this structure cannot extract all the benefits of utilizing hybridizations of the sources due to the fixed output DC link voltage over the course of load variation [5]. An alternative approach for dynamic power flow management involves the implementation of high-capacity DC/DC conversion systems. This method offers active control over energy distribution within the electrical network. While this approach enhances system performance, it introduces additional complexities, including increased weight, volume, and cost, as well as reduced overall system efficiency [6], [7]. In response to these challenges, multi-source inverters (MSIs) have emerged as a promising solution for hybridizing DC sources while facilitating active power flow control [8]. MSIs offer a more streamlined and reliable approach, potentially mitigating the drawbacks associated with conventional DC/DC converter implementations. The concept of MSI was proposed in [8], focusing on connecting multiple DC sources to a single AC output using a one-step conversion process. By employing a T-type inverter topology, [8], [9] presented an MSI designed to integrate two separate DC sources. The proposed MSI in [8] operates in three modes depending on the required load condition. Both DC sources can be utilized individually, and the difference between the DC sources can serve as a third mode of operation. In [10] authors introduced a new configuration of MSI based on NPC topology to integrate two DC sources. The suggested configurations in [8], [9], [10] face two primary challenges: 1) This configuration lacks the capability to generate all potential combinations of the DC sources, particularly the boost feature that adds up the DC sources to increase the effective DC link voltage, and 2) they lack of extendibility feature, which limits its performance to only two DC sources. In [11], an MSI topology was proposed to enable all possible combinations of DC sources. While this design includes the add-up feature through the addition of two extra switches, it faces drawbacks such as a high number of switches and lacks of extendibility. Conversely, [12] introduced a

979-8-3315-1612-3/25 $31.00 © 2025 IEEE

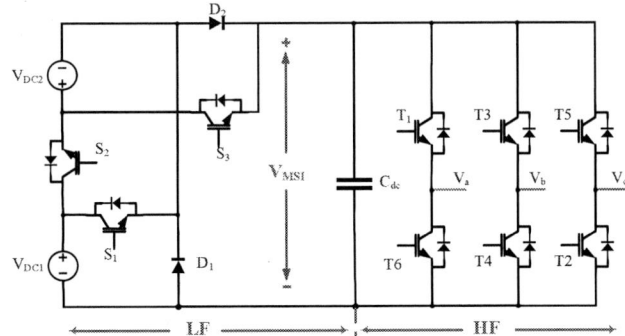

Fig. 1. The proposed extendible MSI architecture

modular MSI inverter that utilizes fewer switches to integrate DC sources more efficiently. They improved the life span of the battery packs by reducing the effective battery pack temperature. While the proposed topology addresses several previously identified shortcomings, the reliance on a large number of switches adversely impacts both the powertrain's efficiency and the overall system cost. Reference [13] as a continuation of the previous work in [12], introduces an optimized MSI topology that utilizes fewer number of switching elements. A flying capacitor-based topology was introduced in [14]. Although the topology in [14] has fairly a lower number of switching counts, it does not have the voltage boost combination. [15] introduced a new generalized MSI to harness the benefits of using MSI. Although they tried to introduce a generalized MSI, it suffers from high number of components and lower efficiency due to higher conduction loss.

In this paper, a new MSI topology is presented. The proposed MSI architecture, while maintaining the fundamental objective of previous MSI designs to connect multiple DC sources to a single AC load in a single-stage configuration, uses fewer components. The key benefit lies in its modular structure, which facilitates the generation of all possible combinations of DC source outputs. Notably, this topology achieves this versatility while employing the minimal number of active switches compared to existing expressed designs, representing a substantial improvement in efficiency and complexity reduction.

II. PROPOSED MSI TOPOLOGY

The proposed MSI topology consists of two main sections: low-frequency (LF) section and high-frequency (HF) section. The LF section generates all possible combinations of DC sources, while the HF part consists a two-level inverter, as illustrated in Fig. 1. With two DC sources, four possible combinations can occur during the generative mode, referred to as *Mode G1*, *Mode G2*, *Mode G3*, and *Mode G4*, and two regenerative modes, namely *Mode R1*, and *Mode R2*, which are detailed as follows. Assuming Vdc1> Vdc2, the DC link voltage is as follows: (1) *Mode G1*: (V_{dc2}). Only DC source two supplies the load, while DC source one does not contribute, (2) *Mode G2*: $V_{DC1} - V_{DC2}$. DC source one supplies the load and simultaneously charges DC source two, (3) *Mode G3*: (V_{dc1}). DC source one is the sole supplier of the load, and (4) *Mode G4*: $(V_{dc1} + V_{dc2})$. Both DC sources supply the load. The LF part switching state with the related DC link voltages are shown in

Table I. In a typical EV application, electric machines (EMs) are the load for the traction inverter. Then, for each torque-speed level required by the EM, the inverter can select a specific mode of operation, as shown in Fig.2.

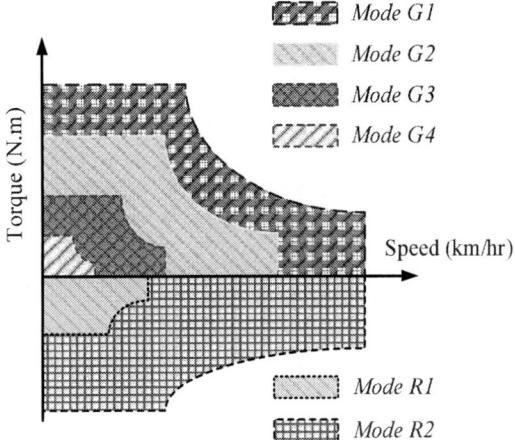

Fig. 2. Torque-speed characteristics for the proposed MSI

TABLE I: SWITCHING STATES IN ALL MODES OF OPERATIONS

M	Switching States						Line-to-Line Voltage
	LF part			HF part			
	S_1	S_2	S_3	T_1	T_3	T_5	V_{AB}
1	0	0	1	0	0	0	0
				P	0	0	V_{DC2}
				P	P	0	0
				0	P	0	$-V_{DC2}$
				0	P	P	$-V_{DC2}$
				0	0	P	0
				P	0	P	V_{DC2}
				P	P	P	0
2	0	1	0	0	0	0	0
				P	0	0	$(V_{DC1} - V_{DC2})$
				P	P	0	0
				0	P	0	$-(V_{DC1} - V_{DC2})$
				0	P	P	$-(V_{DC1} - V_{DC2})$
				0	0	P	0
				P	0	P	$(V_{DC1} - V_{DC2})$
				P	P	P	0
3	1	0	0	0	0	0	0
				P	0	0	V_{DC1}
				P	P	0	0
	0	1	1	0	P	0	$-V_{DC1}$
				0	P	P	$-V_{DC1}$
				0	0	P	0
				P	0	P	V_{DC1}
				P	P	P	0
4	1	0	1	0	0	0	0
				P	0	0	$(V_{DC1} + V_{DC2})$
				P	P	0	0
				0	P	0	$-(V_{DC1} + V_{DC2})$
				0	P	P	$-(V_{DC1} + V_{DC2})$
				0	0	P	0
				P	0	P	$(V_{DC1} + V_{DC2})$
				P	P	P	0

III. ADOPTED SPACE VECTOR (SVM) MODUALTION

The voltage of lines for the MSI output can be derived from the switching states presented in Table I. For a balanced three-phase system, the space vector voltage is computed using the following equation:

$$\vec{V}_{Space} = (V_{ab} + a.V_{bc} + a^2.V_{ca})\qquad(1)$$

where: $\vec{V}_{Space}$ represents the space vector voltage. V_{ab}, V_{bc}, and V_{ca} are the line-to-line voltages and $a = e^{j2\pi/3}$ is the complex operator. The factor 2/3 is used for power invariance in the transformation. The proposed MSI exhibits eight distinct switching combinations for each operational mode, resulting in eight voltage vectors: six non-zero vectors and two zero vectors. Through a comprehensive analysis of voltage vectors across all DC source combinations, four concentric hexagons are derived. By applying the SVM concept to this topology, all relevant equations can be adapted and utilized for the proposed MSI.

$$T_a V_1 + T_b V_2 + T_0 V_0 = (1/f_{Sw})V_{ref}\qquad(2)$$

where the three dwell times for the three nearest vectors are calculated by (2):

$$\begin{cases} T_a = M \times (1/f_{Sw}) \times \sin(\frac{\pi}{3} - \theta) \\ \quad T_b = M \times (1/f_{Sw}) \times \sin(\theta) \\ \quad T_0 = (1/f_{Sw}) - (T_a + T_b) \end{cases}\qquad(3)$$

where M represents the modulation index and it depends the MSI's mode of operation and is obtained by (4).

$$M = \sqrt{3}(\frac{V_{ref}}{V_{MSI}})\qquad(4)$$

where V_{MSI} represents the output voltage of LF part in the proposed MSI topology, as shown in Fig. 1. The magnitude of the reference voltage vector indicates the operational mode of the proposed MSI, and the angle between the voltage vector with the real axis (θ) shows the sector that the reference vector is located, and is calculated by (5).

$$\theta = \arctan(\frac{V_{ref-imag}}{V_{ref-real}})\qquad(5)$$

IV. COMPARISON ANALYSIS

A comprehensive comparative study of all popular MSIs is presented to demonstrate the advantages of the proposed MSI and to leverage the aforementioned benefits. To achieve a fair and reasonable conclusion regarding the effectiveness of the MSIs, the same criteria are applied to all MSI topologies. The first criterion, component counts, is presented in Table II.

TABLE II. COMPONENT COMPARISON STUDY

Topologies	N_{opr}	M_{max}	N_{HF-sw}	N_{LF-sw}	N_D	EXTN
[8]	3	0.75	12	0	0	NO
[11]	3	1.0	12	2	0	NO
[12]	4	1.0	6	6	0	YES
[13]	4	1.0	6	5	1	YES
[14]	3	0.75	10	0	0	YES
[16]	4	1.0	6	6	0	YES
Proposed	4	1.0	6	3	2	YES

where N_{opr}, M_{max}, N_{HF-sw}, N_{LF-sw}, N_d, and $EXTN$ represent the number of operations, maximum modulation index, number of switches in the HF section, number of switches in the LF

section, number of power diodes, and the extendibility feature, respectively.

Table III provides detailed information about the system. The loss comparison section can be divided into three sections, conduction loss in the LF section among those MSIs that utilized a two-level inverter in their outputs [12], [13], [14], [15], as shown in Fig. 3, conduction loss of the entire traction inverter between the proposed MSI and all existing MSIs, which is illustrated in Fig. 4, and Fig. 5 indicates the efficiency analysis between the proposed MSI and all existing MSIs in all four modes of operation.

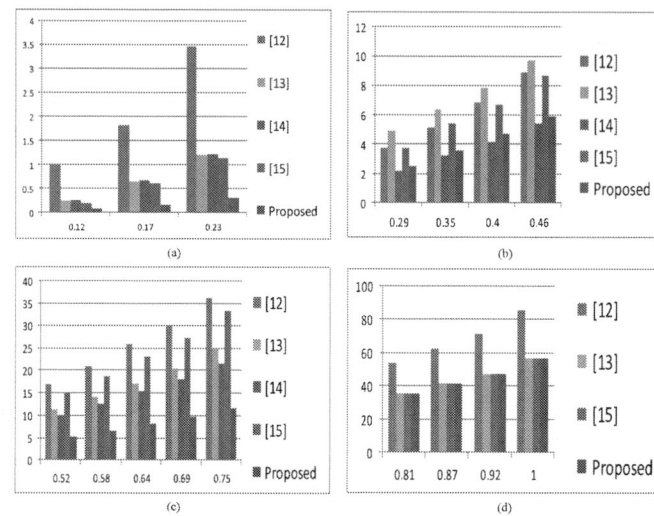

Fig. 3. Comparison of conduction losses in the LF section between the proposed MSI and those presented in [12], [13], [14], [15], categorized by operation modes: (a) Mode G1, (b) Mode G2, (c) Mode G3, and (d) Mode G4.

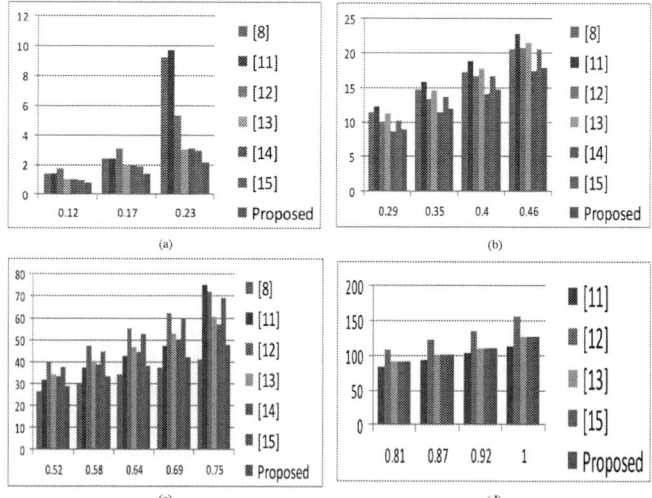

Fig. 4. Comparison of the total traction inverter conduction losses between the proposed MSI and all existing MSIs, evaluated for: (a) Mode G1, (b) Mode G2, (c) Mode G3, and (d) Mode G4.

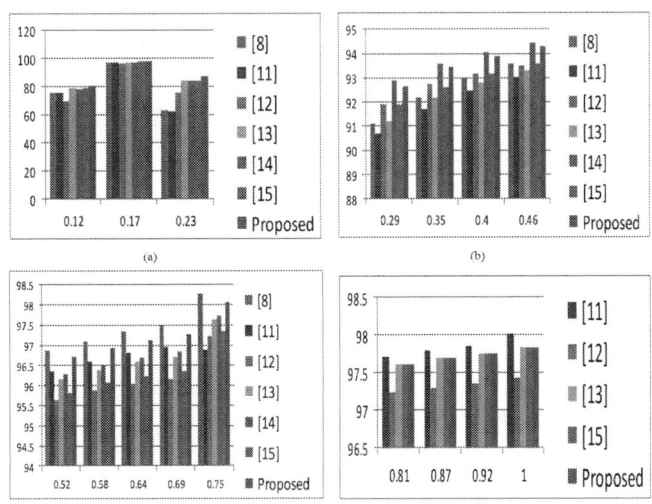

(a)

(b)

(c)

(d)

Fig. 5. Efficiency analysis comparing the proposed MSI with all existing MSIs across the following modes: (a) *Mode G1*, (b) *Mode G2*, (c) *Mode G3*, and (d) *Mode G4*.

TABLE III. THE SYSTEM PAREMETERS

Parameters	Value
Nominal Inverter's Power	8 kW
Voltage of DC Source One	300 V
Voltage of DC Source Two	100 V
Switching Frequency	10 kHz
Load	7+j11.3 Ω
Power Factor	80%

As shown in Fig. 3, the proposed MSI offers reduced conduction loss compared to MSIs utilizing a two-level inverter in their outputs. Regarding the entire traction inverter conduction loss, Fig. 4, although the MSI in [8] demonstrates lower loss in *Mode G3*, it lacks the capability to provide *Mode G4*. This is a significant advantage of MSIs that can produce a higher-level voltage ($V_{dc1} + V_{dc2}$). Although the MSI in [14] offers better efficiency in *Mode G2*, and the proposed MSI in [8] in *Mode G3*, these topologies are unable to produce every possible combination of DC sources. This limitation is a significant disadvantage compared to the proposed MSI.

V. SIMULATION AND EXPERIMENTAL RESULTS

A European standard drive cycle, the Extra Urban Driving Cycle (EUDC), is applied to a traction inverter equipped with the proposed Energy Management System (EMS) to control a permanent magnet synchronous motor (PMSM). Generally, the control strategy for a PMSM involves three hierarchical main control blocks: speed control, the PMSM controller (which includes maximum torque per ampere (MTPA), field weakening (FW), maximum torque per voltage (MTPV)), and the current controller, as depicted in Fig. 6. First, the measured shaft speed is compared with the reference speed, and a PI controller is used to eliminate the error signal. The output of the speed controller, along with the DC link voltage and the actual speed, serves as inputs to the PMSM control block. EV's operational conditions, the controller determines whether the motor should operate in the torque-constant region or the power-constant region. Subsequently, the PMSM control block generates the required currents on the direct (d) and quadrature (q) axes.

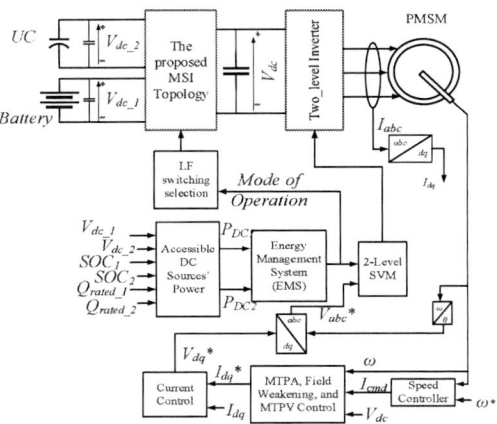

Fig. 6. The system structure used for testing the proposed topology under the EUD standard.

Then, these reference currents are compared with the actual measured currents and processed through the PI current controller block to minimize the error. The available power for both DC sources is calculated through the accessible DC sources' power block, and the outputs serve the proposed energy management system (EMS). Consequently, the mode of operation directs the LF switching selection and the two-level SVM blocks to control the switches HF sides. Fig. 7(a), and (b) illustrate the vehicle's speed reference and the load torque diagram for the EUDC standard, respectively. The corresponding PMSM reference and the actual shaft speed are shown in Fig. 7(c).

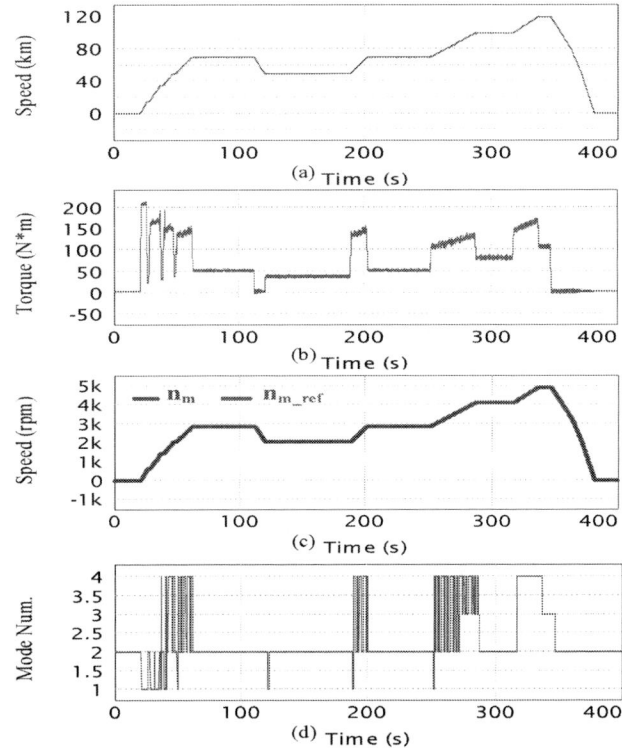

(a) Time (s)

(b) Time (s)

(c) Time (s)

(d) Time (s)

Fig. 7. Verification of proposed MSI for EUDC drive cycle: (a) Vehicle Speed, (b) Vehicle's required torque, (c) The reference and actual rotor speed, and (d) the proposed MSI's modes of operation.

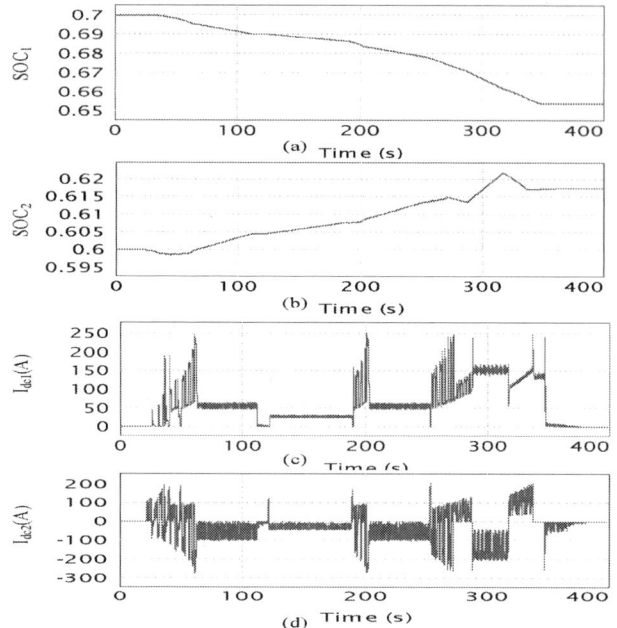

Fig. 8. Verification of proposed MSI for EUDC drive cycle: (a) SOC of the DC source One, (b) SOC of the DC source Two, (c) The output current of the DC source One., and (d) The output current of the DC source Two.

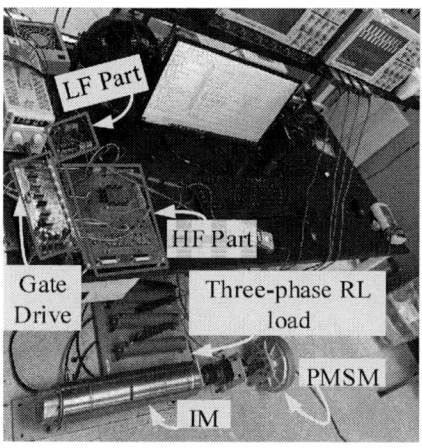

Fig. 9. The scale-down lab prototype.

As previously discussed, the proposed EMS block determines the mode of operation, with the results shown in Fig. 7(d). Based on the required power, the proposed MSI switches among all DC source combinations (from *Mode G1* to *Mode G4*). *Mode G2* is selected when the required power is low or when the SOC of the second DC source is below the maximum, allowing the main DC source to supply the load and simultaneously charge the second DC source. Both DC sources are batteries with different capacities and initial SOCs. The initial SOCs for the main and the second DC sources are 0.7 and 0.6, respectively, as shown in Fig. 8(a), and (b). The currents for the considering DC sources are depicted in Fig. 8(c), and (d). The parameters for the system under study are provided in Table IV and Table V. Fig. 9 illustrates the 1.2 kW laboratory prototype of the proposed MSI converter. The voltage rating of DC source One and DC source two are 300 Vdc, and 100 Vdc,

respectively. The functionality of the proposed MSI was further validated using a coupled IM and PMSM system, where the MSI-controlled PMSM drives the IM's shaft rotation. In this setup, the IM acts as a generator and is connected to a variable resistive-inductive (RL) load, compelling the proposed MSI to operate in different modes. The nominal shaft speed of the PMSM at the fundamental frequency of 120 Hz, given its 4 poles, is 3600 rpm. Experimental results of the coupled IM and PMSM system are depicted in Fig. 10(a)-(e). Fig. 10(a) presents the three-phase load currents and the PMSM line voltage, while Fig. 10(b)-(e) show zoomed-in views of Fig. 10(a) at various operational modes (*Mode G4*, *Mode G3*, *Mode G2*, and *Mode G1*). As seen in Fig. 10(b), when the modulation index is set to the 0.88, the proposed MSI operates in *Mode G4* and the stator frequency is around 104 Hz.

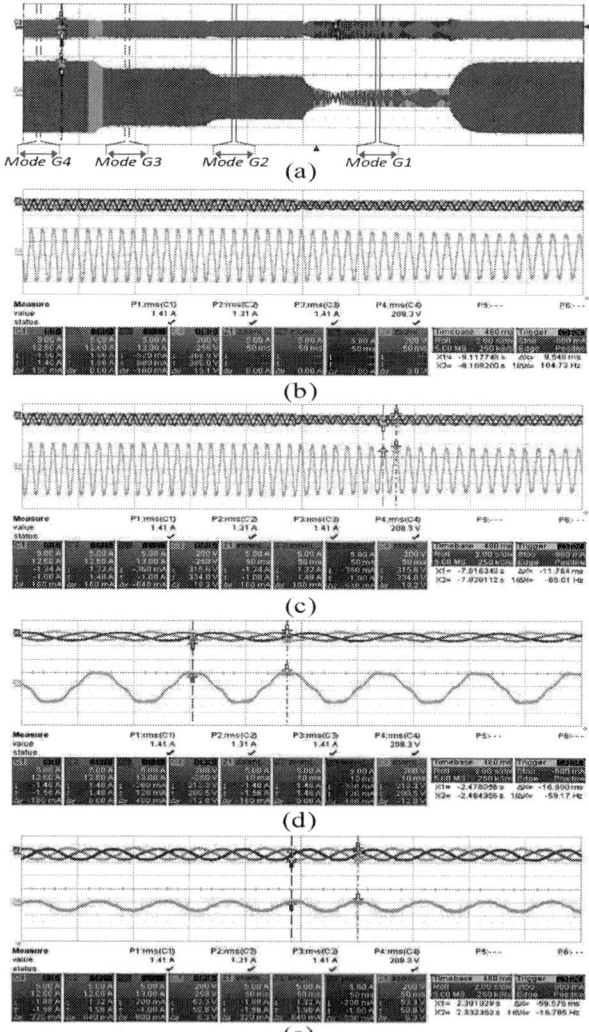

Fig. 10. Experimental results of coupled IM and PMSM system: (a) The three-phase load current and PMSM terminal voltage, (b) The zoomed-in view of (a) when the system operates in *Mode G4*, (c) The zoomed-in view of (a) when the system operates in *Mode G3*, (d) The zoomed-in view of (a) when the system operates in *Mode G2*, and (e) The zoomed-in view of (a) when the system operates in *Mode G1*.

979-8-3315-1612-3/25 $31.00 © 2025 IEEE

Fig. 10(b) clearly demonstrates the effectiveness of the control strategy, i.e., at higher frequencies, the proposed MSI operates in the highest mode of operation (*Mode G4*) where the MSI's output is equal to the summation of the DC sources' outputs. The corresponding speed in this mode is 3120 rpm. Fig. 10(c) illustrates the MSI outputs in *Mode G3*, when the modulation index is set to 0.63. In this mode, the stator frequency is 75.6 Hz, resulting in a shaft speed of 2268 rpm. Fig. 10(d) indicates the operation of *Mode G2*, where the modulation index is around 0.48 resulting the stator frequency be around 59.1 Hz. Consequently, the PMSM rotates at 1773 rpm. Finally, in Mode G1, the stator frequency is 16.7 Hz at modulation index equal to 0.14, and the PMSM rotates at 501 rpm, Fig. 10(e). It is worth mentioning that due to the trapezoidal back EMF of the PMSM motors, a trapezoidal voltage is expected across the PMSM line voltages.

TABLE IV. THE THREE-PHASE PMSM PARAMETERS

Stator resistance *(Ohm)*	0.02035
d-axis inductance *(H)*	0.0015
q-axis inductance *(H)*	0.005
Back EMF constant	147.95
Number of poles *(P)*	4
Moment of inertia	0.0206

TABLE V. VEHICLE PARAMETERS FOR SIMULATION RESULTS

Gear ratio *(S)*	5:1
Wheel's radius *(r)*	0.32
Vehicle mass *(kg)*	1140

VI. CONCLUSION

This article explores a new and an optimal MSI design that facilitates the utilization of UC banks in conjunction with battery packs to use of a single-stage integration approach, devoid of any magnetics, contributes to reductions in weight, and etc. A comparison with recently introduced MSI structures demonstrates that the proposed MSI topology provides a greater number of operating modes while utilizing fewer semiconductor switches. The performance of the proposed MSI topology is thoroughly investigated in all operating modes for active control of the hybrid storage system. Additionally, an EMS is introduced to control the PMSM behavior and control the power fellow between DC sources. The proposed MSI and EMS undergo rigorous evaluation using the EUDC. To verify the effectiveness of the new design, a scale-down model of the MSI was built and rigorously evaluated under various operating scenarios. This dual approach of simulation and experimental validation provides a robust assessment of the MSI's performance.

REFERENCES

[1] *Advanced Electric Drive Vehicles*. 2014. doi: 10.1201/9781315215570.

[2] F. Naseri, E. Farjah, and T. Ghanbari, "An efficient regenerative braking system based on battery/supercapacitor for electric, hybrid, and plug-in hybrid electric vehicles with BLDC motor," 2017. doi: 10.1109/TVT.2016.2611655.

[3] E. Chemali, M. Preindl, P. Malysz, and A. Emadi, "Electrochemical and Electrostatic Energy Storage and Management Systems for Electric Drive Vehicles: State-of-the-Art Review and Future Trends," 2016. doi: 10.1109/JESTPE.2016.2566583.

[4] P. K. Behera and M. Pattnaik, "Hybrid Energy Storage Integrated Wind Energy Fed DC Microgrid Power Distribution Control and Performance Assessment," *IEEE Trans Sustain Energy*, vol. 15, no. 3, 2024, doi: 10.1109/TSTE.2024.3354309.

[5] Y. J. Jelodar, O. Salari, M. Z. Youssef, J. Ebrahimi, and A. Bakhshai, "A Novel Control Scheme for Traction Inverters in Electric Vehicles With an Optimal Efficiency Across the Entire Speed Range," *IEEE Access*, vol. 12, pp. 25906–25916, 2024, doi: 10.1109/ACCESS.2024.3366934.

[6] H. F. Ahmed, H. Cha, S. H. Kim, D. H. Kim, and H. G. Kim, "Wide Load Range Efficiency Improvement of a High-Power-Density Bidirectional DC-DC Converter Using an MR Fluid-Gap Inductor," *IEEE Trans Ind Appl*, vol. 51, no. 4, 2015, doi: 10.1109/TIA.2014.2387485.

[7] Y. Shen, J. Xie, T. He, L. Yao, and Y. Xiao, "CEEMD-Fuzzy Control Energy Management of Hybrid Energy Storage Systems in Electric Vehicles," *IEEE Transactions on Energy Conversion*, vol. 39, no. 1, 2024, doi: 10.1109/TEC.2023.3306804.

[8] L. Dorn-Gomba, P. Magne, B. Danen, and A. Emadi, "On the Concept of the Multi-Source Inverter for Hybrid Electric Vehicle Powertrains," *IEEE Trans Power Electron*, vol. 33, no. 9, pp. 7376–7386, Sep. 2018, doi: 10.1109/TPEL.2017.2765247.

[9] L. Dorn-Gomba, E. Chemali, and A. Emadi, "A novel hybrid energy storage system using the multi-source inverter," in *Conference Proceedings - IEEE Applied Power Electronics Conference and Exposition - APEC*, 2018. doi: 10.1109/APEC.2018.8341086.

[10] E. Fedele, D. Iannuzzi, P. Tricoli, and A. Del Pizzo, "NPC-Based Multi-Source Inverters for Multimode DC Rail Traction Systems," *IEEE Transactions on Transportation Electrification*, vol. 9, no. 1, pp. 1289–1299, Mar. 2023, doi: 10.1109/TTE.2022.3175097.

[11] E. Chemali and A. Emadi, "On the concept of a novel Reconfigurable Multi-Source Inverter," in *2017 IEEE Transportation and Electrification Conference and Expo, ITEC 2017*, 2017. doi: 10.1109/ITEC.2017.7993356.

[12] O. Salari, K. H. Zaad, A. Bakhshai, and P. Jain, "Reconfigurable Hybrid Energy Storage System for an Electric Vehicle DC-AC Inverter," in *IEEE Transactions on Power Electronics*, Institute of Electrical and Electronics Engineers Inc., Dec. 2020, pp. 12846–12860. doi: 10.1109/TPEL.2020.2993783.

[13] J. Ebrahimi, O. Salari, S. Eren, K. Hashtrudi-Zaad, A. Bakhshai, and P. Jain, "Efficiency Improved Multi-Source Inverter for Hybrid Energy Storage Systems in Electric Vehicle Application," *IEEE Trans Power Electron*, vol. 37, no. 2, pp. 1982–1997, Feb. 2022, doi: 10.1109/TPEL.2021.3104759.

[14] J. Ebrahimi and S. Eren, "A Multi-Source DC/AC Converter for Integrated Hybrid Energy Storage Systems," *IEEE Transactions on Energy Conversion*, vol. 37, no. 4, pp. 2298–2309, Dec. 2022, doi: 10.1109/TEC.2022.3174518.

[15] M. A. Hosseinzadeh, M. Sarebanzadeh, C. F. Garcia, E. Babaei, J. Rodriguez, and R. Kennel, "A New Generalized Multisource Inverter for Electric Vehicles Controlled by Model Predictive," *IEEE Transactions on Industrial Electronics*, vol. 71, no. 9, pp. 10184–10197, Sep. 2024, doi: 10.1109/TIE.2023.3329248.

Dual-Use of Onboard Chargers to Achieve Controllable DC Bus Voltage for Electric Vehicles

Anuj Maheshwari*, Elie Libbos‡, Arijit Banerjee*
*Department of Electrical & Computer Engineering
University of Illinois at Urbana-Champaign, Urbana, IL, USA
Email: {anujm3, arijit}@ illinois.edu
‡Kilby Labs
Texas Instruments, Dallas, TX, USA
Email: e-libbos@ti.com

Abstract—The power flow architecture in electric vehicles must prioritize high efficiency and power density while consistently delivering the rated torque-speed envelope. In traditional setups, where the battery is directly connected to the traction inverter's dc link, the torque-speed envelope diminishes as the battery discharges. To address this limitation, adding a dc-dc converter between the battery and the traction inverter can help maintain torque-speed capabilities. However, this approach requires the dc-dc converter to handle the full traction power, making the system bulkier and less efficient. This paper introduces an alternative solution: leveraging the onboard charger to regulate the inverter dc-link during motoring. The analysis demonstrates that even with the onboard charger rated at only 15% of the traction motor's power, the system can maintain the rated torque-speed envelope down to a 20% battery state of charge (SOC). The paper explains the functionality of the proposed architecture, supplemented by efficiency color maps showing a 2% improvement across the torque-speed envelope. Experimental results comparing the conventional boost converter with the proposed architecture corroborate the analysis.

I. INTRODUCTION

Electric vehicle (EV) motors and their inverters are designed to deliver a rated torque-speed envelope across a wide operating condition [1], [2]. However, as the battery discharges, its terminal voltage decreases. References [3], [4] shows that at 20% state of charge (SOC), the battery voltage drops by 15% compared to its voltage at 100% SOC. This substantial reduction in the dc-bus voltage leads to the shrinking of the achievable torque-speed envelop when the battery is directly connected to the inverter. For example, Fig. 1 shows that at 20% SOC, the drive can maintain the rated torque only up to 84% of the base speed, and the available torque at maximum speed is reduced by 26% for a variable-pole induction motor drive (VPIM) [5], [6].

One solution to deliver the rated torque-speed envelop at reduced dc-bus voltage entails over-designing the motor and drive train [7], [8]. However, this approach introduces additional weight and reduces drive efficiency. Alternatively, another strategy involves integrating a dc-dc converter between the battery and the inverter, which can control the inverter's dc-link voltage to meet the torque-speed demand. A bidirectional and multilevel interleaved boost converters have been used to boost the dc-link voltage as the battery discharges [9], [10].

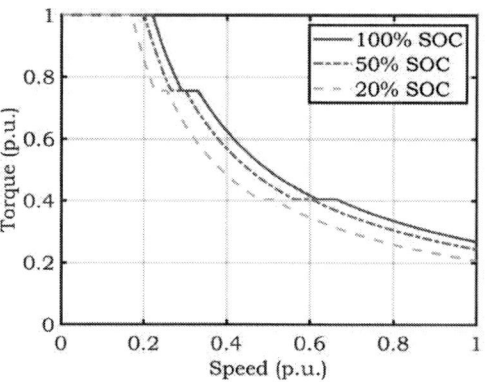

Fig. 1: Torque-speed envelop of EV motors shrinks with decreasing battery SOC for drive trains directly connected to the battery.

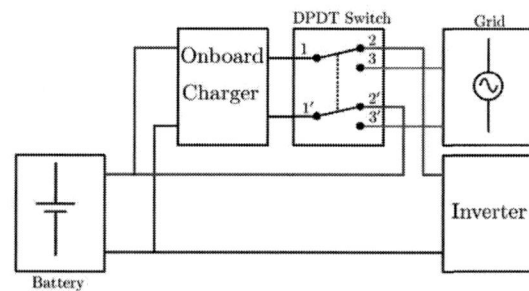

Fig. 2: Proposed architecture using the onboard charger as a partial power processing converter to regulate the inverter dc-bus during motoring operation. 1-1' connects to 2-2' during motoring and 3-3' during charging.

The ability to both reduce or increase the inverter's dc-link voltage has been shown to improve the drive efficiency in highly varying speed conditions typical of EV drive cycles [11]–[14]. References [15], [16] propose various converter architectures that can increase or decrease the dc-bus voltage relative to the battery voltage. However, these approaches entail processing the motor's entire power, resulting in a bulky and lossy dc-dc converter. Reference [17] uses a partial-power processing architecture with a dual port inverter. However, the

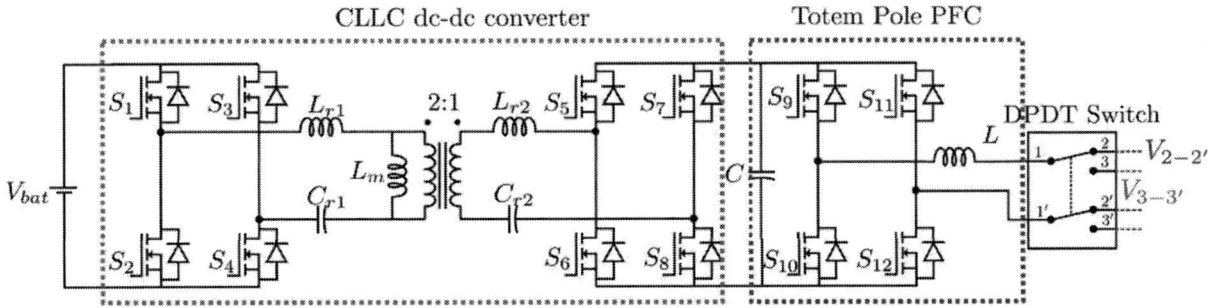

Fig. 3: The onboard charger consists of a totem-pole PFC followed by a CLLC converter. DPDT switch is connected to 2-2' for motoring mode and 3-3' for charging mode. This isolated and bidirectional architecture allows the charger to be used as a partial power processing converter in motoring mode.

partial converter is rated at 88% of the total power, resulting in marginal efficiency and power density benefits.

An onboard charger, which is typically a power-factor correction rectifier followed by an isolated dc-dc converter, is redundant during the motoring operation. This paper proposes a drive architecture that uses the onboard charger as a partial power processing converter to regulate the dc bus. Figure 2 presents the proposed architecture. The onboard charger is connected to a double-pole-double-throw (DPDT) switch. This switch can implemented either using a mechanical switch or an electronic switch. The switch terminal 1-1' is connected to 2-2', shown in blue, in motoring mode, and 3-3', shown in red, in charging mode, respectively. Eliminating a dedicated dc-dc converter leads to significant weight reduction from the EV drive architecture, improving the overall power density. Further, efficiency improvement of 2% is achieved over the entire torque-speed envelop due to the partial power processing capability of the proposed architecture. Section II explains the functionality of the proposed architecture. Section III compares the efficiency of the proposed architecture with a conventional boost converter over the entire torque-speed envelope. Section IV presents the experimental setup and provides preliminary results. Section V concludes the paper.

II. PROPOSED CONVERTER ARCHITECTURE

The onboard charger in the proposed architecture is chosen to be a totem pole power factor correction (PFC) rectifier circuit followed by a bidirectional CLLC converter, which is typical for Level 2 charging [18], [19], as shown in Fig. 3. The turns ratio of the CLLC stage transformer is set as 1:2 for an 800 V battery pack which boosts the standard 400 V dc-bus produced by PFC when operating in charging mode. In motoring mode, the CLLC converter is operated at the resonant frequency to maximize efficiency. Switches S_9, S_{10}, and the inductor L form the buck converter, which regulates the inverter dc-link voltage. The polarity of the voltage at terminal 1-1' is determined by switches S_{11} and S_{12}. When switch S_{11} is turned ON, the voltage at the terminal 2-2' is given by:

$$V_{2-2'} = D_b n V_{bat} - n V_{bat} = (D_b - 1) n V_{bat}$$
$$= -(1 - D_b) n V_{bat} \quad (1)$$

and when S12 is turned ON, the voltage at the terminal 2-2' is given by:

$$V_{2-2'} = D_b n V_{bat} \quad (2)$$

where n is CLLC stage transformer turns ratio and D_b is the duty ratio of the buck stage. The inverter dc-link voltage is given by:

$$V_{inverter} = V_{bat} + V_{2-2'} \quad (3)$$

With this configuration, the charger output $V_{2-2'}$ can regulate the inverter dc-link voltage irrespective of the battery SOC. As the charger is in series with the battery, the power it processes is proportional to its voltage $V_{2-2'}$. Since the battery voltage goes down by 15% at 20% SOC, the proposed architecture only processes 15% of the rated motor power to achieve the rated torque speed envelop. Moreover, as $V_{2-2'}$ can be both positive and negative, the inverter dc-link voltage can be either higher or lower than the battery voltage. The dc-link voltage reduction allows lower switching losses in the inverter. The optimum inverter dc-link voltage is decided by minimizing the total loss in the drive train, which will be presented in the following section.

III. OPTIMIZATION FRAMEWORK

This section presents the loss model that forms the basis for the optimization framework developed to compare the losses in the entire motor drive train using variable dc-bus voltage as a decision variable. The loss analysis for the entire drive train is described next using the per-slot model of VPIM described in [5].

A. Loss Analysis

The losses in the entire architecture are distributed among the motor, the inverter, and the front-end dc-dc converter. The

979-8-3315-1612-3/25 $31.00 © 2025 IEEE

stator and rotor copper loss equations are as follows:

$$P_{s,copper} = \frac{2}{Q_s} R_s (i_{ds}^2 + i_{qs}^2) \tag{4}$$

$$P_{r,copper} = \frac{2}{Q_s} R_r' \frac{i_{qs}^2}{(1+\sigma_r)^2} \tag{5}$$

where Q_s is the number of slots, R_s is the stator resistance and R_r' rotor resistance referred to the stator, i_{ds} and i_{qs} are currents in rotor flux reference frame, and σ_r accounts for rotor leakage ($\sigma_r L_m$ is the rotor leakage inductance). Core loss is modeled with a Steinmetz equation [20]:

$$P_{core} = C_h f B^\gamma + C_e f^2 B^2 = C_h' f i_{ds}^\gamma + C_e' f^2 i_{ds}^2 \tag{6}$$

where B is the peak flux density, γ is an empirical parameter, and C_h and C_e are hysteresis and eddy current coefficients respectively. C_h' and C_e' are modified coefficients to express core losses in terms of i_{ds}. The total motor loss is given by:

$$P_{motor} = P_{s,copper} + P_{r,copper} + P_{core} \tag{7}$$

The conduction loss in the inverter switch is given by:

$$P_{conduction,inv} = R_{ds,on} I_{leg,rms}^2 n_{inv} \tag{8}$$

where $R_{ds,on}$ is the on resistance of the MOSFET, n_{inv} is the number of inverter legs and $I_{leg,rms}$ is given by:

$$I_{leg,rms} = \frac{1}{\sqrt{2}} I_{leg,pk} = \frac{1}{\sqrt{2}} \frac{2C_p}{Q_s} \sqrt{i_{ds}^2 + i_{qs}^2} \tag{9}$$

where C_p is the number of slots in parallel. The switching loss in the inverter can be approximated by:

$$P_{switching,inv} = 2n_{inv} \frac{i_{sw,avg}}{I_{rated}} \frac{v_{sw}}{V_{rated}} f_{sw} E_{switching,rated} \tag{10}$$

where $i_{sw,avg}$ is the average current through the switch, v_{sw} is the voltage across the switch which is equal to inverter dc-bus voltage, I_{rated} and V_{rated} are the rated current and voltage of the switch, $E_{switching,rated}$ is the switching loss at the rated conditions obtained from the datasheet, and f_{sw} is the inverter switching frequency. $i_{sw,avg}$ is given by:

$$i_{sw,avg} = \frac{1}{\pi} I_{leg,pk} = \frac{1}{\pi} \frac{2C_p}{Q_s} \sqrt{i_{ds}^2 + i_{qs}^2} \tag{11}$$

The CLLC converter operates at resonant frequency and therefore has zero switching losses. The conduction loss in individual CLLC legs is given by:

$$P_{conduction} = R_{ds,on} I_{cllc,leg,rms}^2 \tag{12}$$

There are two legs on the primary and secondary sides of the CLLC converter. The rms current in the CLLC leg is given by:

$$I_{cllc,primary,rms} = \frac{\pi}{2\sqrt{2}} \frac{P_{charger}}{V_{battery}} \tag{13}$$

$$I_{cllc,secondary,rms} = nI_{cllc,primary,rms} \tag{14}$$

The copper in the transformer is given by:

$$P_{copper} = I_{cllc,primary,rms}^2 R_{primary} + I_{cllc,secondary,rms}^2 R_{secondary} \tag{15}$$

The core loss in the transformer is calculated using the Steinmetz equation. The conduction losses in the buck half-bridge are given by:

$$P_{conduction} = R_{ds,on} (I_{S9,rms}^2 + I_{S10,rms}^2) \tag{16}$$

where $I_{S9,rms}$ and $I_{S10,rms}$ is given by:

$$I_{S9,rms} = \frac{P_{inverter}}{V_{dc-bus}} \sqrt{D} \tag{17}$$

$$I_{S10,rms} = \frac{P_{inverter}}{V_{dc-bus}} \sqrt{1-D} \tag{18}$$

respectively. The switching loss is given by:

$$P_{switching,buck} = \frac{i_{sw,avg}}{I_{rated}} \frac{v_{sw}}{V_{rated}} E_{switching,rated} f_{sw} \tag{19}$$

where $i_{sw,avg}$ is given by:

$$i_{sw,avg} = \frac{P_{inveter}}{V_{dc-bus}} \tag{20}$$

where $P_{inveter}$ is given by:

$$P_{inveter} = \tau_e * \omega_m + P_{loss,motor} + P_{loss,inveter} \tag{21}$$

where τ_e and ω_m are the motor's output torque and mechanical speed. The total loss in the drive train for a given torque and mechanical speed is given by:

$$\text{Total loss} = f(i_{ds}, i_{qs}, p, v_{dc-bus}) = P_{loss,motor} + P_{loss,inveter} + P_{loss,dc-dc\ converter} \tag{22}$$

B. Optimization Problem Formulation

The optimization problem seeks a pole count and dc-bus voltage that minimizes the losses in the entire drive train for a given torque and mechanical speed. The decision variables for the optimization are d-axis current, q-axis current, pole count p, and dc-bus voltage:

$$x = [i_{ds}\ i_{qs}\ p\ v_{dc-bus}]^T \tag{23}$$

The optimal solution x^* minimizes the cost function f given by (22). The optimization problem is formulated as:

$$\min_x\ f(x) \tag{24}$$

subject to the following constraints:

1) equality constraint linking the mechanical speed to the electrical speed ω_e

$$\omega_e - \frac{p}{2}\omega_m - \frac{R_r' i_{qs}}{L_r' i_{ds}} = 0 \tag{25}$$

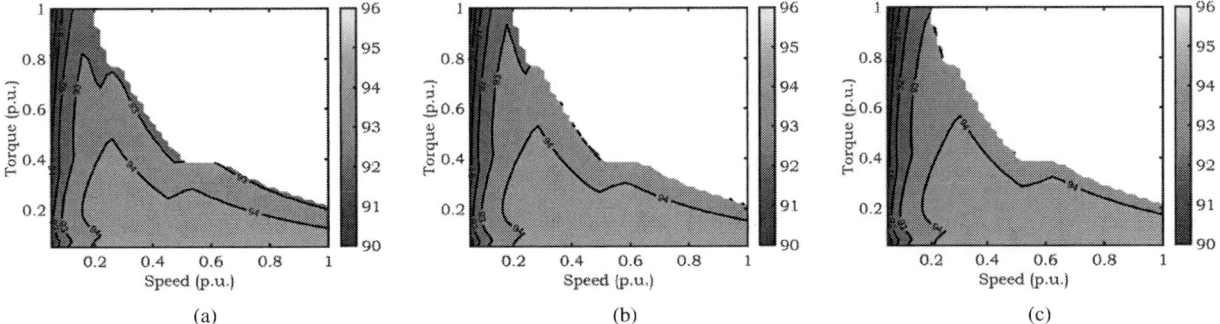

Fig. 4: Combined machine, inverter, and dc-dc converter efficiency color maps for the conventional interleaved boost dc-dc converter for different battery voltages of (a) 680 V (b) 740 V, and (c) 790 V

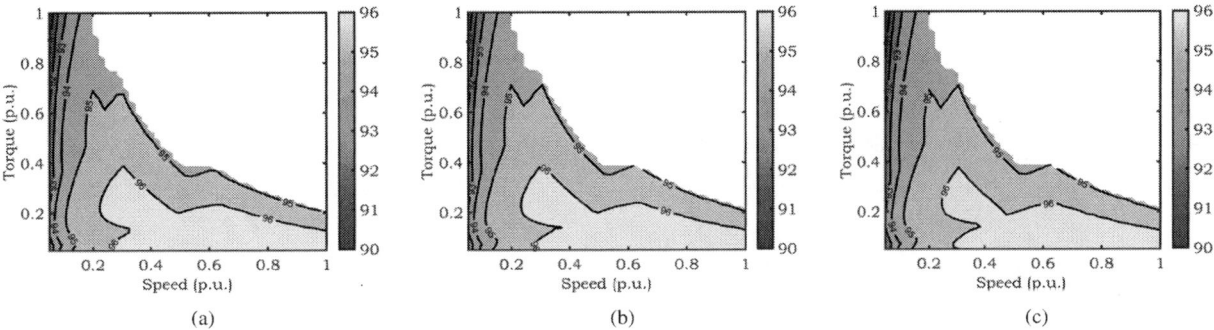

Fig. 5: Combined machine, inverter, and dc-dc converter efficiency contour maps for the proposed architecture for different battery voltages of (a) 680 V (b) 740 V, and (c) 790 V

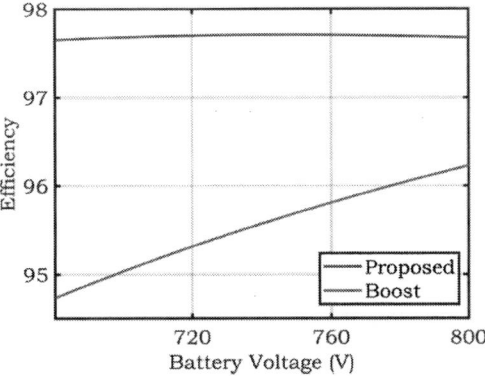

Fig. 6: Efficiency of the dc-dc stage at rated power for different battery voltage.

2) equality constraint to match the required torque

$$\tau_e - \frac{1}{p}\frac{C}{1+\sigma_r}i_{ds}i_{qs} = 0 \qquad (26)$$

where C is a constant depending upon machine dimensions.

3) current magnitude inequality constraint set by machine and drive limits

$$i_{ds}^2 + i_{qs}^2 \le (\frac{Q_s}{2}I_{rated})^2 \qquad (27)$$

4) voltage magnitude inequality constraint set by inverter dc-bus and pulse-width modulation (PWM) strategy

$$v_{ds}^2 + v_{qs}^2 \le (\frac{Q_s}{2}V_{rated})^2 \qquad (28)$$

where v_{ds} and v_{qs} in terms of i_{ds} and i_{qs} are given by:

$$v_{ds}^2 = (Rsi_{ds} - \omega_e kL_s i_{qs})^2 \qquad (29)$$

$$v_{qs}^2 = (Rsi_{qs} + \omega_e L_s i_{ds})^2 \qquad (30)$$

where k is given by:

$$k = 1 - \frac{L_m^2}{L_s L_r'} \qquad (31)$$

5) flux magnitude inequality constraint to enforce a flux limit

$$(L_s i_{ds})^2 + (kL_s i_{qs})^2 \le (\frac{Q_s}{2}\lambda_{rated})^2 \qquad (32)$$

6) maximum allowed dc-bus voltage set by switch voltage constraints

$$v_{dc-bus} \le V_{dc-bus,max} \qquad (33)$$

7) minimum possible dc-bus voltage set by the front-end dc-dc converter and the required dc-bus voltage to produce the torque subject to PWM strategy. The minimum voltage for conventional boost architecture for

979-8-3315-1612-3/25 $31.00 © 2025 IEEE
3092

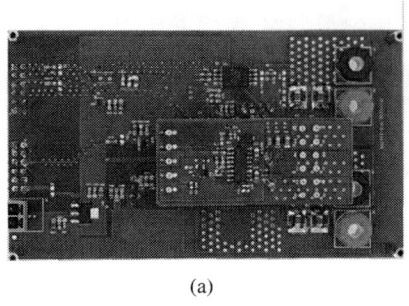

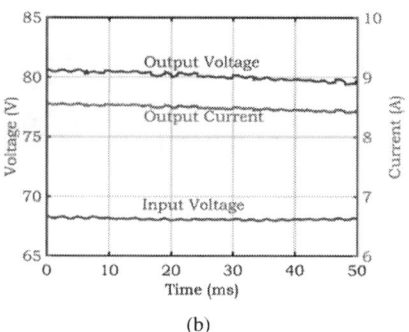

(a)

(b)

Fig. 7: (a) 80 V, 700 W prototype of the conventional boost switching at 100 kHz and (b) key experimental waveforms at 80 V, 700 W output

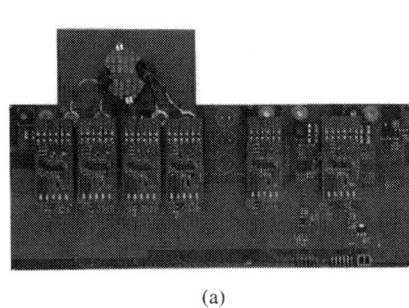

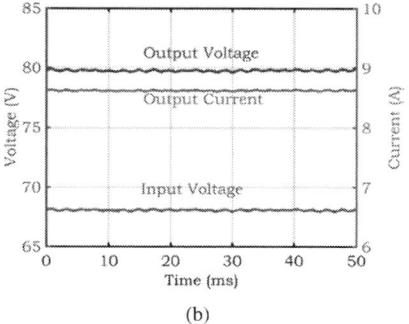

(a)

(b)

Fig. 8: (a) 80 V, 700 W prototype of the proposed architecture with CLLC stage switching at 450 kHz and buck stage switching at 100 kHz and (b) key experimental waveforms at 80 V, 700 W output

space vector modulation is given by:

$$v_{dc-bus} \geq max(\ C_s \sqrt{3(v_{ds}^2 + v_{qs}^2)}, V_{bat}\ ) \qquad (34)$$

The minimum voltage for the proposed architecture is given by:

$$v_{dc-bus} \geq max(\ C_s \sqrt{3(v_{ds}^2 + v_{qs}^2)},$$
$$V_{bat}(1-n)\ ) \qquad (35)$$

where C_s is the number of slots connected in series.

8) current constraint on the charger buck leg switches

$$i_{sw,rms} \leq I_{rated\_leg\_current} \qquad (36)$$

9) maximum allowed loss in the charger to ensure charger thermal limits are not violated

$$P_{loss,charger} \leq P_{max,charger} \qquad (37)$$

C. Optimization results

This section presents the efficiency improvements with the proposed architecture over the entire torque speed characteristics. A 158kW toroidal VPIM is used for machine parameters [21]. The base torque, base speed, and max speed of the machine are 302 Nm, 5000 rpm, and 22000 rpm respectively. A 22.5 kW rated onboard charger, which is 15% of the rated power of the motor, is used in this analysis. The efficiency

of the dc-dc converter stage for the conventional interleaved boost and the proposed architecture at rated power for different battery voltages is shown in Fig. 6. The proposed architecture achieves a 2.2x reduction in losses at rated power and 20% battery SOC compared to the interleaved boost converter. The efficiency improvements result from the reduced VA rating of the switch as the proposed architecture only processes 15% of the power processed by the conventional boost converter. Unlike boost, the proposed architecture can also reduce the dc-bus voltage to half of the battery voltage, allowing the converter to reduce the dc-bus voltage in low-torque low-speed regions. This dc-bus voltage reduction brings down the switching losses in the inverter improving the efficiency of the drive train. The efficiency of the entire drive train which includes switching and conduction losses in the front-end converter and inverter, and copper and core losses in the motor for the conventional boost and the proposed architecture for different battery voltages is shown in Fig. 4 and 5 respectively. The proposed architecture achieves 2% efficiency improvement on average which along with the reduction in weight due to the removal of the dc-dc converter will lead to improvements in the electric vehicle range.

IV. EXPERIMENTAL SETUP AND RESULTS

The scaled-down 80 V, 700 W GaN-based prototypes of the conventionally used boost converter and the proposed partial power processing converter are shown in Fig. 7 (a) and 8

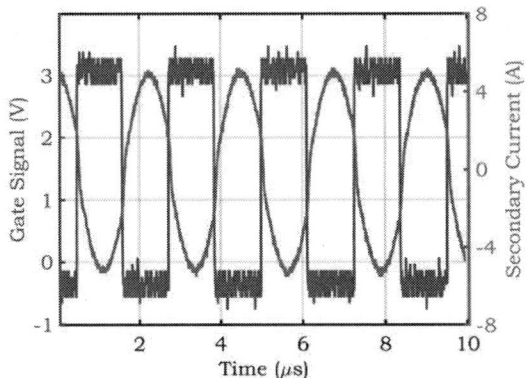

Fig. 9: Primary side switch gate signal (blue) and transformer secondary current (orange) for the CLLC stage. The CLLC stage works at the resonant frequency to maximize efficiency.

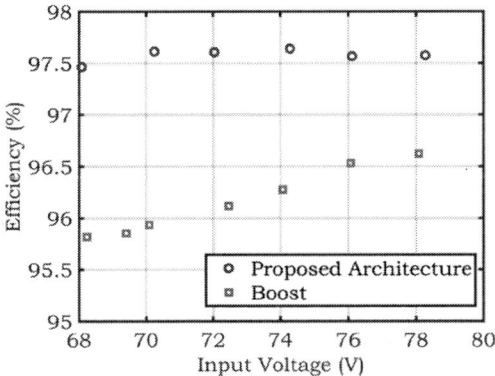

Fig. 10: Efficiency plots for the proposed architecture (blue circles) and the conventional boost converter (red squares) with changing input voltage at the rated output voltage of 80 V and output power of 700 W.

(a) respectively. A resistive load is used for the dc-dc stage to emulate the inverter and motor. The power and efficiency are measured using Keysight Integravision Power Analyzer PA2201A. Figure 7 (b) shows the key experimental waveforms for the boost converter at 68 V input and 80 V, 700 W output. Figure 8 (b) shows the key experimental waveforms for the proposed architecture for the same 68 V input and 80 V, 700 W output. However, in the proposed architecture, the converter only processes 105 W or 15% of the full power because of its partial-power processing capabilities. The reduction in processed power increases the drive efficiency for the proposed converter. Figure 9 shows the gate signal to the primary side switches of the CLLC stage and the secondary transformer current. The waveforms show that the CLLC stage operates close to resonant frequency to maximize efficiency. Figure 10 shows the efficiency comparison at rated output voltage and power with changing input voltage. The proposed architecture achieves higher efficiency compared to the conventional boost converter.

V. CONCLUSION AND FUTURE WORK

Drive architecture plays a significant role in the efficiency and power density of the EV power train. The dc-dc converter between the battery and the inverter is a large part of the power train architecture. This article presented a partial power processing architecture utilizing the onboard charger to eliminate the dc-dc converter, reducing the power train weight and increasing efficiency. The motor can meet its rated torque-speed requirement up to 20% SOC of the battery for a charger that is rated at just 15% of the motor power. The efficiency of the proposed architecture is compared with the conventionally used boost for 80 V, and 700 W prototypes and the proposed architecture achieved efficiency improvement of up to 1.75% at reduced battery SOC.

REFERENCES

[1] C. Liu, "Emerging electric machines and drives — an overview," *IEEE Transactions on Energy Conversion*, vol. 33, no. 4, pp. 2270–2280, 2018.

[2] U. D. P. U. D. D. Research, I. for Vehicle efficiency, and E. sustainability) *Electrical and electronics technical team roadmap - department of energy*, Oct 2017.

[3] W. Kim, P.-Y. Lee, J. Kim, and K.-S. Kim, "A robust state of charge estimation approach based on nonlinear battery cell model for lithium-ion batteries in electric vehicles," *IEEE Transactions on Vehicular Technology*, vol. 70, no. 6, pp. 5638–5647, 2021.

[4] C.-S. Huang, "An online condition-based parameter identification switching algorithm for lithium-ion batteries in electric vehicles," *IEEE Transactions on Vehicular Technology*, vol. 72, no. 2, pp. 1701–1709, 2023.

[5] E. Libbos, B. Ku, S. Agrawal, S. Tungare, A. Banerjee, and P. T. Krein, "Loss minimization and maximum torque-per-ampere operation for variable-pole induction machines," *IEEE Transactions on Transportation Electrification*, vol. 6, no. 3, pp. 1051–1064, 2020.

[6] E. Libbos, E. Krause, A. Banerjee, and P. T. Krein, "Inverter design considerations for variable-pole induction machines in electric vehicles," *IEEE Transactions on Power Electronics*, vol. 37, no. 11, pp. 13554–13565, 2022.

[7] N. Zhao and N. Schofield, "An induction machine design with parameter optimization for a 120-kw electric vehicle," *IEEE Transactions on Transportation Electrification*, vol. 6, no. 2, pp. 592–601, 2020.

[8] K. Field, "Mountain pass performance tests tesla model 3 power at different states of charge." https://tinyurl.com/3zd9bs2r, 2018. Accessed: 9 Oct. 2023.

[9] O. Hegazy, R. Barrero, J. Van Mierlo, P. Lataire, N. Omar, and T. Coosemans, "An advanced power electronics interface for electric vehicles applications," *IEEE Transactions on Power Electronics*, vol. 28, no. 12, pp. 5508–5521, 2013.

[10] V. Rathore, K. Rajashekara, P. Nayak, and A. Ray, "A high-gain multilevel dc–dc converter for interfacing electric vehicle battery and inverter," *IEEE Transactions on Industry Applications*, vol. 58, no. 5, pp. 6506–6518, 2022.

[11] T. Schoenen, M. S. Kunter, M. D. Hennen, and R. W. De Doncker, "Advantages of a variable dc-link voltage by using a dc-dc converter in hybrid-electric vehicles," in *2010 IEEE Vehicle Power and Propulsion Conference*, pp. 1–5, 2010.

[12] S. Sridharan and P. T. Krein, "Optimizing variable dc link voltage for an induction motor drive under dynamic conditions," in *2015 IEEE Transportation Electrification Conference and Expo (ITEC)*, pp. 1–6, 2015.

[13] K. K. Prabhakar, M. Ramesh, A. Dalal, C. U. Reddy, A. K. Singh, and P. Kumar, "Efficiency investigation for electric vehicle powertrain with variable dc-link bus voltage," in *IECON 2016 - 42nd Annual Conference of the IEEE Industrial Electronics Society*, pp. 1796–1801, 2016.

[14] H. Nisar and A. M. Bazzi, "Investigating the effect of inverter dc bus voltage on variable pole induction motor performance in electric vehicles," in *2024 IEEE Transportation Electrification Conference and Expo (ITEC)*, pp. 1–5, 2024.

[15] M. S. Khan, S. S. Nag, A. Das, and C. Yoon, "Analysis and control of an input-parallel output-series connected buck-boost dc–dc converter for electric vehicle powertrains," *IEEE Transactions on Transportation Electrification*, vol. 9, no. 2, pp. 2015–2025, 2023.

[16] M. S. Khan, S. S. Nag, A. Das, and C. Yoon, "A novel buck-boost type dc-dc converter topology for electric vehicle applications," in *2021 IEEE Energy Conversion Congress and Exposition (ECCE)*, pp. 1534–1539, 2021.

[17] D. Zhou, J. Wang, N. Hou, Y. Li, and J. Zou, "Dual-port inverters with internal dc–dc conversion for adjustable dc-link voltage operation of electric vehicles," *IEEE Transactions on Power Electronics*, vol. 36, no. 6, pp. 6917–6928, 2021.

[18] P. He, A. Mallik, A. Sankar, and A. Khaligh, "Design of a 1-mhz high-efficiency high-power-density bidirectional gan-based cllc converter for electric vehicles," *IEEE Transactions on Vehicular Technology*, vol. 68, no. 1, pp. 213–223, 2019.

[19] F. Jin, T. Yuan, A. Nabih, and Q. Li, "Efficient integrated magnetics with winding cancellation technique to reduce common-mode emi noise for a single phase cllc converter," *IEEE Transactions on Power Electronics*, pp. 1–16, 2024.

[20] C. Steinmetz, "On the law of hysteresis," *Proceedings of the IEEE*, vol. 72, no. 2, pp. 197–221, 1984.

[21] E. Libbos, E. Krause, A. Banerjee, and P. T. Krein, "Winding layout considerations for variable-pole induction motors in electric vehicles," *IEEE Transactions on Transportation Electrification*, pp. 1–1, 2023.

Isolated single-phase onboard chargers for BEV/PHEV using active power decoupling technology

Yosiki Amano
Department of Electrical and Mechanical Engineering
Nagoya Institute of Technology
Aichi, Japan
y.amano.808@stn.nitech.ac.jp

Keigo Nishimura
Department of Electrical and Mechanical Engineering
Nagoya Institute of Technology
Aichi, Japan
k.nishimura.846@stn.nitech.ac.jp

Hiroaki Matsumori
Department of Electrical and Mechanical Engineering
Nagoya Institute of Technology
Aichi, Japan
matsumori.hiroaki@nitech.ac.jp

Takashi Kosaka
Department of Electrical and Mechanical Engineering
Nagoya Institute of Technology
Aichi, Japan
kosaka.takashi@nitech.ac.jp

Kenichi Nagayoshi
Toyota Industries Corporation
Aichi, Japan
kenichi.nagayoshi@mail.toyota-shokki.co.jp

Kenichi Watanabe
Toyota Industries Corporation
Aichi, Japan
kenichi.watanabe.ac@mail.toyota-shokki.co.jp

Abstract— **This paper presents an isolated single-phase onboard charger used for BEV/PHEV using active power decoupling (APD) technology. The proposed circuit consists of an isolated SEPIC circuit and a totem pole PFC circuit and shares some power semiconductor devices to achieve smaller sizes, higher efficiency, and lower cost. Furthermore, by implementing APD control, the battery charging voltage ripple caused by twice frequency grid voltage is reduced. Therefore, the proposed circuit with APD control extends battery life. In this paper, the principle of APD control is explained, and the circuit operation for 1 kW output is verified through simulation and experiments.**

Keywords— *AC/DC converter, totem-pole PFC circuit, isolated SEPIC circuit, onboard chargers, active power decoupling (APD)*

I. INTRODUCTION

Onboard chargers used in BEV/PHEV are required to have functions such as power factor correction, isolation between the input side and output side, and output voltage control [1]. Figure 1 shows a typical single-phase onboard charger circuit with 12 semiconductor devices [2]. To reduce the number of semiconductor devices, we proposed the isolated onboard charger that integrates a Totem pole PFC circuit and an isolated SEPIC circuit as shown in Fig. 2. The proposed onboard charger consists of only six semiconductor devices, was well worked for power factor correction and low THD current in grid currents. However, the proposed converter remains battery charging current ripple of twice grid frequency component (i.e., 120 Hz). it affects battery life. Normally, to reduce the ripple component, an auxiliary circuit with the active power decoupling (APD) control is used [4]. However, an auxiliary circuit increases the total onboard charger cost [5-7]. Therefore, We propose APD control for Fig.2 which is realized without circuit modification since the

capacitor Cc works energy conversion and energy buffer by controlling switch Q_3. The principle of APD control is briefly explained, and then, the circuit operation for 1 kW output is verified through simulation and experiments.

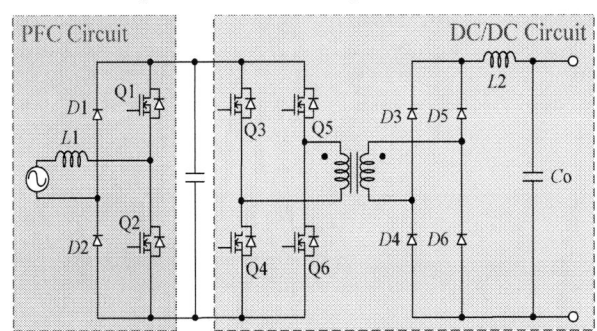

Fig. 1. Conventional onbord chareger.

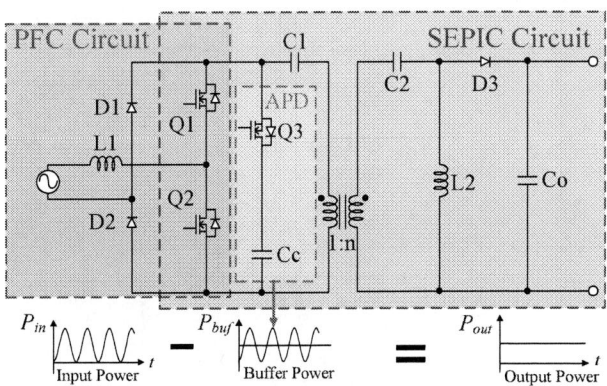

Fig. 2. Proposed onboard charger.

II. AC/DC CONVERSION WITH ACTIVE POWER DECOUPLING

A. The basic operation without active power decoupling

According to Ref. 3, when the input sinusoidal wave voltage is positive (half cycle of sinusoidal waveform), Q_1 is always ON, Q_2 performs PFC operation, and Q_3 turns OFF and ON according to the ON and OFF state of Q_2. This is to prevent Cc from short-circuiting when Q_1, Q_2, and Q_3 turn on simultaneously. Another half cycle of the sinusoidal waveform is also worked as opposite Q_1 and Q_2.

B. Active power decoupling control

The basic idea of the APD control is Q_1, Q_2, and Q_3 are operated independently to control capacitor Cc energy for power conversion and energy buffer for twice grid frequency component. In this case, APD control is achieved by adjusting the amount of current Q_{out} that passes through diode D3 within one switching cycle.

The control configuration is shown in Fig. 3. To prevent short-circuit by Cc energy when Q_1, Q_2, and Q_3 turn on simultaneously, logic gates are implemented. And the feedforward control to add the ideal duty cycle for output current amount control. Also, the control block contains, enhanced PLL control, input current control to realize PFC operation, and output voltage control to track the desired voltage. The operation waveform is shown in Fig. 4.

The amount of current supplied to the load side in a steady state is determined by the amount of current Q_{out} that passes through diode D_3 within one switching cycle. Figure 3 shows that i_{D3} begins to increase from 0 A when Q_2 turns off and decreases to 0 A when Q_3 turns off. Therefore, to derive the amount of current Q_{out} passing through diode D$_3$ in the steady state, let t_{inc} be the period from Q_2 to Q_3 and t_{dec} be the time from Q_3 to i_{D3} becoming 0.

$$Q_{\text{out}} = \frac{1}{2} i_{D3}(t_{inc}) \cdot (t_{inc} + t_{dec}) \tag{1}$$

Then, i_{D3} in each period can be expressed as follows. However, L_m is the excitation inductance on the primary, L_p is the leakage inductance on the primary, and n is the transformer turn ratio.

$$i_{D3}(t_{inc}) = \left[\frac{(L_m - L_p)(v_{cc} - v_{c1})}{nL_p(2L_m - L_p)} - \frac{L_m(v_{c2} - v_o)}{n^2 L_p(2L_m - L_p)} \right] \cdot t_{inc} \tag{2}$$

$$i_{D3}(t_{dec}) = -\frac{L_m - L_p}{nL_p(2L_m - L_p)} \cdot v_{c1} - \frac{L_m \cdot (v_{c2} - v_o)}{n^2 L_p(2L_m - L_p)} \tag{3}$$

After Q_3 switches from ON to OFF, i_{D3} becomes 0A after t_{dec}, so t_{dec} is obtained as

$$t_{dec} = \frac{v_{cc}}{v_{c1} + \dfrac{L_m}{n(L_m - L_p)} \cdot (v_{c2} + v_o)} \cdot t_{inc} \tag{4}$$

Substituting in (1), Q_{out} is obtained as follows.

$$Q_{out} = \frac{x(L_m - L_p)}{2nL_p(2L_m - L_p)} \cdot \frac{x + v_{c1} + \dfrac{L_m(v_{c2} + v_o)}{n(L_m - L_p)}}{v_{c1} + \dfrac{L_m(v_{c2} + v_o)}{n(L_m - L_p)}} \tag{5}$$

$$x = v_{cc} - v_1 - \frac{L_m}{n(L_m - L_p)} \cdot (v_{c2} + v_o) \tag{6}$$

The above shows that the larger x is, i.e., the larger the difference between v_{cc} and $v_1 + \frac{L_m}{n(L_m - L_p)} \cdot (v_{c2} + v_o)$ the larger Q_{out} will be. Conversely, the smaller this difference is, the smaller the current Q_{out} supplied to the load side becomes, and we consider that the output voltage ripple can be reduced. where V_{diff}^* is a constant representing the difference between v_{cc} and $v_1 + \frac{L_m}{n(L_m - L_p)} \cdot (v_{c2} + v_o)$.

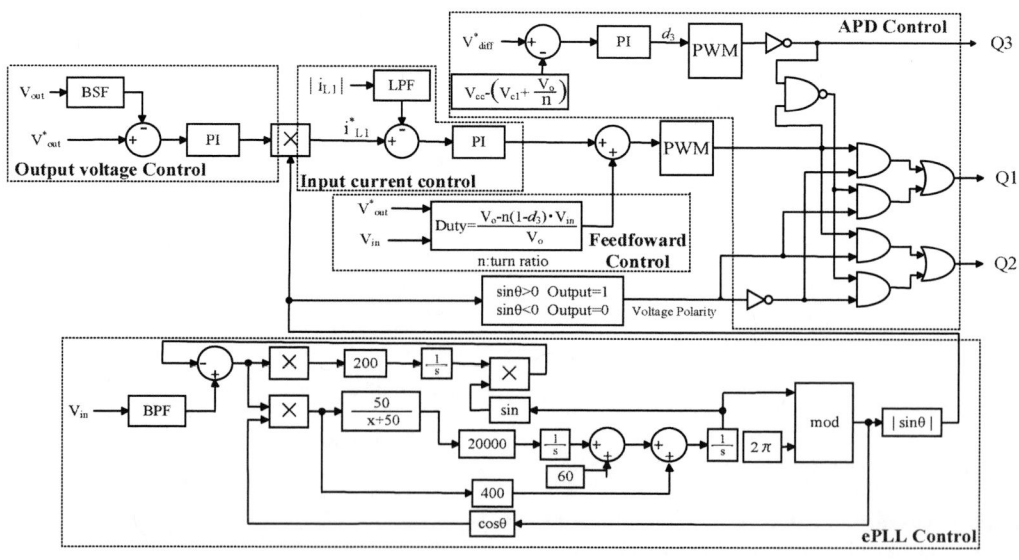

Fig. 3. Proposed Control block.

979-8-3315-1612-3/25 $31.00 © 2025 IEEE

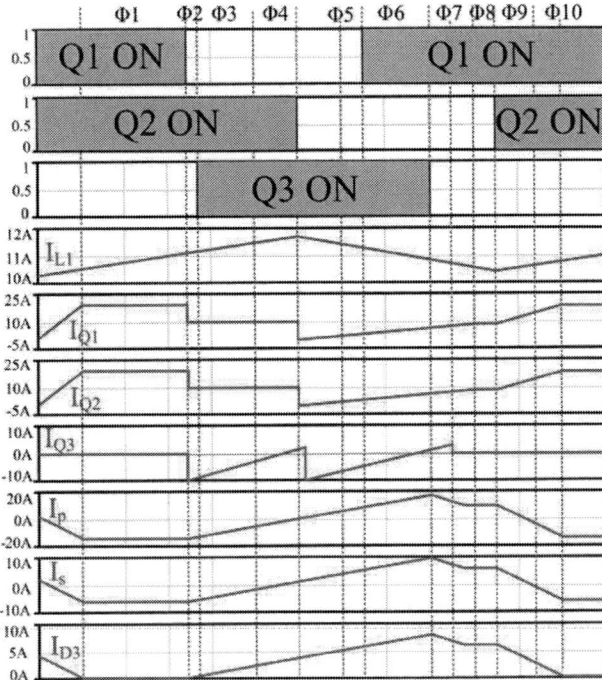

Fig. 4. Current/voltage waveform.

III. SIMULATION RESULTS

To validate the proposed APD control, the circuit simulation was conducted. The circuit parameters in the proposed converter were set as shown in Table 1. The input voltage is AC100 V_{rms}, and the output voltage is DC176 V (half battery voltage condition). In this condition, V_{diff}^* in Fig. 3 was set to 70 V.

The simulated input voltage and input current waveforms are shown in Fig. 5(a) and the output voltage waveforms are shown in Fig. 5(b). It can be seen that PFC operation is possible even with the APD control. Although the battery charge voltage ripple looks large at about 11 V, it can be reduced in the actual battery voltage condition (i.e. 350 V) because the amount of energy buffer power in capacitor Cc will be increased in a higher voltage condition [5].

IV. EXPERIMENT RESULTS

Figure. 6 shows the experimental setup. The circuit parameters are the same as the simulation conditions. Figure 7(a) shows the input voltage and current waveforms, and Figure 7(b) shows the output voltage waveforms. The experimental waveforms are well approximated by the simulated voltages and currents. Also, it can be confirmed that both input current and output voltage control are performed with a unity power factor (the power factor is 0.998), and the THD value of input current is 3.91 %, confirming the Japanese power utility standard (the THD is less than 5 %).

Table 1: The specification of the proposed converter.

L1, L2 (OC-250060-2/Micro Metal)	1 mH
Cc, C1, C2 (C4AQNBW5220M36J/KEMET)	22 µF
Co (EKHJ451VSN591MA54MA54M/ Nippon Chemi-Con)	1180 µF
Q1, Q2, Q3 (C3M0075120K/Cree)	Rated 1200 V
D1, D2, D3 (C4D20120A/Cree)	Rated 1200 V
Turn ratio of the transformer	1:2

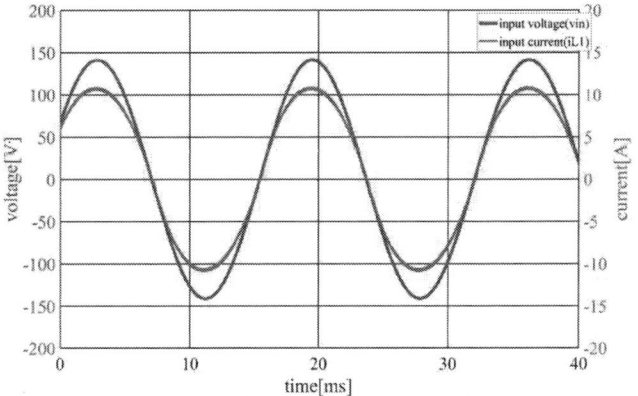

(a) Input voltage/current.

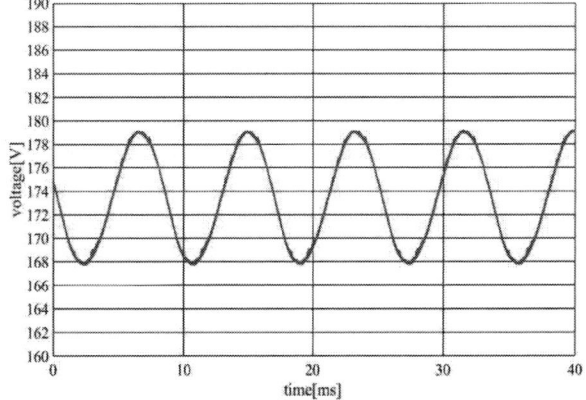

(b) Output voltage.

Fig. 5. Simulation result.

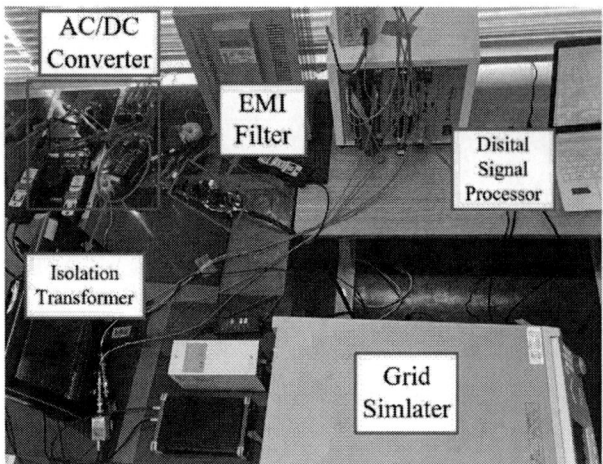

Fig. 6. Experimental configuration.

V. CONCLUSIONS

This paper proposed an active power decoupling control method to reduce output voltage ripple for the isolated onboard charger which integrates a totem pole PFC circuit and an isolated SEPIC circuit. The active power decoupling control was realized by adjusting the amount of current flowing on the output side which value is analyzed in the steady state condition. The proposed active power decoupling control was confirmed through simulations and experiments.

REFERENCES

[1] K. Ryota, H. Yusuke, and Y. Masaki, "Experimental Verification of Reducing Power Loss under Light Load Condition of a Bi-Directional Isolated DC/DC Converter for a Battery Charger-Discharger of Electric Vehicle," *IEEJ Journal of Industry Applications*, vol. 10, no. 3, pp. 371-383, May. 2021. J. Clerk Maxwell, A Treatise on Electricity and Magnetism, 3rd ed., vol. 2. Oxford: Clarendon, 1892, pp.68–73.

[2] D. Endo, H. Matsumori, T. Kosaka, K. Nagayoshi, and K. Watanabe, "Isolated AC/DC converter used in EV/PHEV battery charger from household AC outlet," *2022 IEEE Energy Conversion Congress and Exposition (ECCE)*, pp. 1-8, Nov. 2022.

[3] J. Yuan, L. Dorn-Gomba, A. Dorneles Cellegaro J. Reimers, and A. Emadi, "A Review of Bidirectional On-Board Chargers for Electric Vehicles," *IEEE Access*, vol. 9, pp. 51501-51518, Mar. 2021.

[4] Y. Liu, Y. Sun, and M. Su, "A Control Method for Bridgeless Cuk/Sepic PFC Rectifier to Achieve Power Decoupling," *IEEE Transactions on Industrial Electronics*, vol. 64, no. 9, pp. 7272-7276, Sep. 2017.

[5] T. Sakuraba, K. Kusaka, K.Orikawa, and J.Ito, "Requirements for Components of Single-Phase Inverter using Power Decoupling Technology toward High Power Density," *2016 18th European Conference on Power Electronics and Applications (EPE'16 ECCE Europe)*, pp. 1-10, Sep. 2016.

[6] H. Soonhwan, "Study on optimization and robust control of active power decoupling circuit based converters", PhD.THESIS, in Hokkaido University, 2019.

[7] H. Hu, N. Kutkut, I.Batarseh, and Z. John Shen, "Power Decoupling Techniques for Micro-inverters in PV Systems-a Review," *2010 IEEE Energy Conversion Congress and Exposition, Atlanta, GA, USA, 2010*, pp. 3235-3240, Nov. 2010.

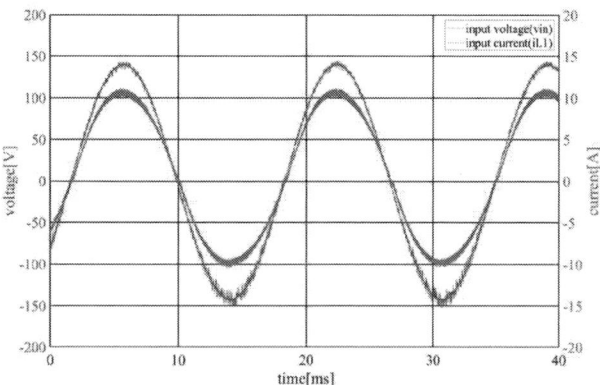

(a)   Input voltage/current.

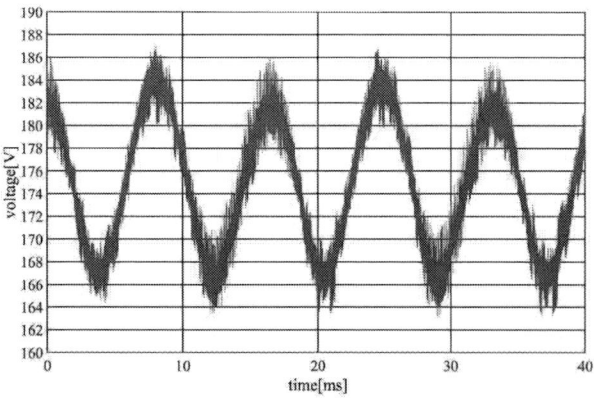

(b) Output voltage.

Fig. 7. Experimental results.

A Practical Use of xEVCap: The Modular and Standard DC-Link Capacitor Solution for the Main EV Powertrain Inverter

David Olalla
Product Development
TDK Electronics AG
Munich, Germany

Tomas Wagner
Mechanical Engineering
TDK Electronics SAU
Malaga, Spain

Fernando Rodriguez
Application Development
TDK Electronics SAU
Malaga, Spain

Alberto Espinar
Mechanical Engineering
TDK Electronics SAU
Malaga, Spain

Abstract— An innovative DC-link capacitor solution [1] for the main xEV powertrain Inverter has been presented to the market as of July 2024. The innovation is sustained by the conjunction of four pillars: modularity and scalability; design for application; design for manufacturing; and standardization (catalog product). After a brief review of these pillars already detailed in [1], this paper focuses on the practical study of xEVCap at the system level. The analysis is deployed in four major subject areas. The parasitics (ESL, ESR, ...) from the single capacitor element to the effective ones at the system level. The second area is the assembly and joint technologies with the busbar. The third area is the strategy for thermal decoupling between the capacitor and power semiconductors without a cooler. The paper concludes with the thermal integration with a cooler and how the various elements interact under different temperatures. Everything is supported by Ansys and CFD simulations and with the construction of a demonstrator for laboratory testing. This paper also presents the results of collaboration with recognized external partners, like power-semiconductor and busbar suppliers.

Keywords—xEVCap, DC-Link Capacitor, Traction Inverter, Busbar, cooling, ESR, ESL, WBG, SiC

I. Introduction: xEVCap as DC-Link Capacitor Solution

The DC-link capacitor acts as an energy buffer between input and output in the power converter. Its core mission is to keep the DC voltage stable, which means within the defined voltage limits by the system (also called smoothing function). Multiple capacitor technologies can be used as DC-link (Fig. 1-Up), and their suitability depends on voltage, power range, application, etc. [2]. Biaxially oriented polypropylene (BOPP) film capacitors are, as of today, the dominant technology solution used for the DC-link of the main traction inverter in xEV systems [4].

The DC-Link capacitor is one of the bulkiest and most customized components in xEV systems. If higher capacitance is required, for example to handle a higher I_{RMS}, the entire capacitor with its internal and external design must be modified. The Infineon evaluation kit "EVAL KIT HPD G1 SiC" [6] refers

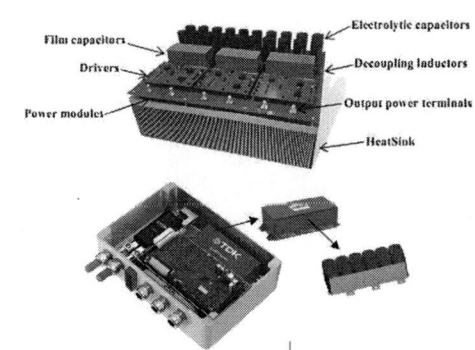

Fig. 1. **(Up)** Demonstrator of a 1500 V solar inverter by Infineon Technologies AG with Easy 3B module. TDK Electronics: Hybrid DC-link capacitor bank: High-frequency Film Capacitor (blue) + Aluminum Electrolytic (black) [3]. **(Bottom)** Reference design by Mankel-Engineering in collaboration with TDK and Infineon Technologies AG. Capacitor construction made with 6 parallel flat winding elements of polypropylene film capacitors.

to a TDK capacitor with 300 µF/855 V (B25655P8307K351). That solution would need external busbar adaptation to add DC terminals.

A. The first pillar: Modularity and Scalability

The bulk customized capacitor solution was divided into standardizable and modular capacitor units, as shown in Fig.2.

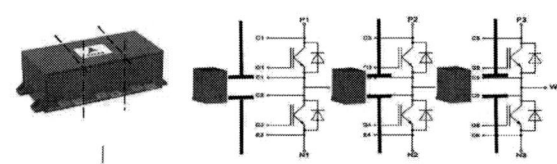

Fig. 2. The basic concept was to divide the bulk customized capacitor into standardizable units.

Every capacitor unit cannot provide the full capacitance and power required by the traction inverter. Still, paralleling them through a laminated busbar with a compensated inductance arrangement makes it possible. This solution can be easily scaled to any power level, power semiconductor supplier, model, system geometry, or cooling. DC terminals and other accessories can be incorporated into the busbar.

B. The second pillar: Design-for-Application

The target application spectrum was divided into Size 1 for low and mid-power and Size 2 for mid and high-power traction inverter systems The following table shows the extracted parameters of the DC-link capacitor linked to typical application parameters.

TABLE I.   DESIGN TO APPLICATION (SPLIT BY POWER) OF THE MODULAR DC-LINK CAPACITOR.

Power level	Size 1		Size 2	
	Low	Mid	Mid-High	High
C [µF]	$200 - 300$	$300 - 400$	$375 - 525$	$400 - 600$
$I_{c\,RMS}$ (max.) [A]	116	155	200	235
I_{dc} (max.) [A]	175	235	295	350
V_R [V]	850	850	850	850
P [kW]	150	200	250	300
$I_{ph.RMS}$ (max.) [A]	200	265	335	400

The whole DC-link capacitor was divided into individual units, paralleled, and connected through a busbar to the power semiconductors, as shown in Fig. 3.

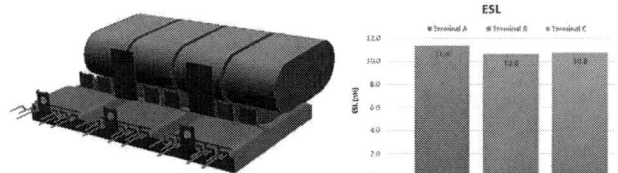

Fig. 3. Modular concept design for DC-link applications; standardized solution compatible with power semiconductor modules. Size 1 (400 µF).

C. The third pillar: Design-for-manufacturing

A design for manufacturing was developed, targeting standardization in terminal pitches and box sizes, similar to what is available for PCB components. A narrower width (W) will increase the I_{RMS} per capacitance, while a longer length or height (L or H) will increase the energy density. Raw materials design and production tooling are optimized to allow a scaled and automated mass manufacturing.

D. The fourth pillar: Standardization.

Two types of terminals, four voltage levels, and three different footprints are fully available and included in the web search and simulation tool CLARA (Capacitor Life And Rating Application). This tool simulates the components electrically and thermally under different operating conditions, also providing detailed data like mechanical and electrical, curves, 3D files, and complex Spice models valid in the time and frequency domain.

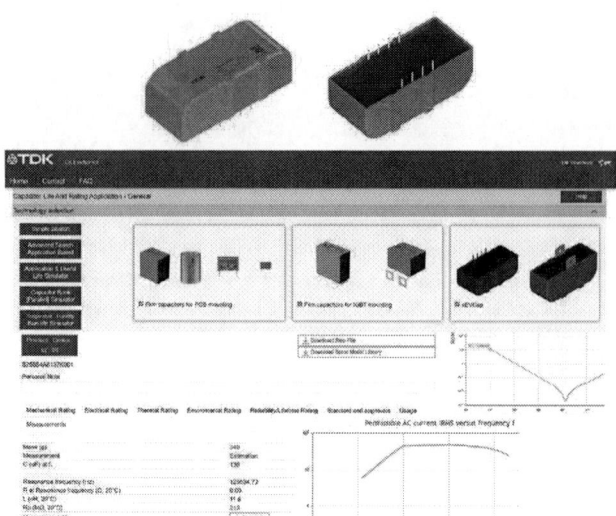

Fig. 4. xEVCap product details fully available in the web for all designers (https://www.tdk-electronics.tdk.com/en/2910862/design-support/design-tools/film-capacitors/clara)

This web tool provides comprehensive details necessary for designing DC-link capacitor solutions based on xEVCap. It includes curves of equivalent series resistance (ESR) and impedance (Z) versus frequency, as well as typical equivalent series inductance (ESL) values. However, it lacks indications of these critical parameters at the system level.

Note: For simplification, the term "DC-Link" is used to refer to the entire solution comprising multiple xEVCap capacitors connected in parallel via a laminated busbar, including the external connections but excluding other systems such as power semiconductor modules and the battery.

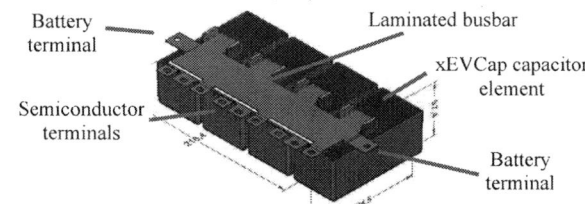

Fig. 5. "DC-Link" solution: Terms for following sections.

II. EVALUATION OF PARASITICS AT THE SYSTEM LEVEL

The primary parasitics of the DC-link, namely equivalent series resistance (ESR) and equivalent series inductance (ESL), significantly influence the overall system behavior. Accurately calculating these parameters during the design phase is challenging, particularly when using a component analyzer at higher frequencies. Instead, employing finite element modeling (FEM) electromagnetic software facilitates a virtual characterization of the entire DC-link solution (including capacitance, ESL, ESR, and impedance). This characterization enables the development of reliable SPICE models, which allow for a more precise simulation of the converter's operation [11].

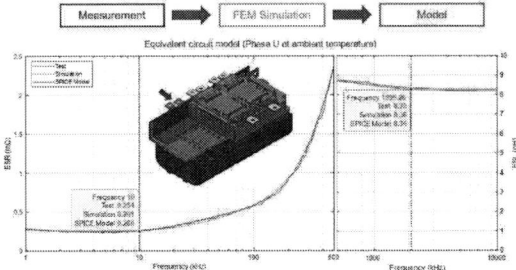

Fig. 6. Virtual characterization process.

A. ESR (Equivalent Series Resistance)

The product of current squared and ESR ($I^2 \times$ESR) quantifies the power losses generated by the current flowing through the DC-link capacitors. The ESR is not constant but varies significantly with frequency. This frequency dependency directly affects the thermal losses within the system. Two current spectrums with identical RMS values but different frequency ranges will result in different power losses. This also applies to the distribution of these losses (i.e., where they are generated).

Fig. 7 illustrates this concept: losses are calculated for the same automotive DC-link capacitor (with an identical ESR vs. frequency curve) and the same total capacitor current (in RMS). In one scenario, all current is applied at 10 kHz, while in the other, the current follows a typical automotive spectrum with a 10 kHz switching frequency at specified modulation parameters.

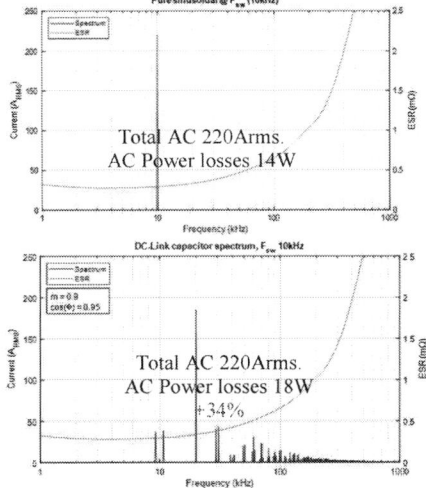

Fig. 7. Comparison of the power losses in a DC-link capacitor under single frequency 10kHz (up), and under a realistic frequency current spectrum (bottom)

There are five primary reasons why ESR increases with frequency: skin effect, inhomogeneous impedance, internal resonances, negative electromagnetic interaction, winding geometry and metal profile. xEVCap addresses the internal factors related to capacitor design. However, the design of the busbar and connections is crucial to mitigate inhomogeneous impedance, resonances, and negative electromagnetic interactions.

B. ESL (Equivalent Series Inductance)

To reduce the equivalent series inductance (ESL) of the DC-link, it is essential to understand the contribution of each component. It is a common misconception that the total ESL is simply the sum of the inductances of all its elements (capacitive elements, terminals, busbar, etc.). This assumption only applies to an ideal system without interactions between the elements. In reality, these interactions can either increase or decrease the system's ESL and must be evaluated on a case-by-case basis.

$$ESL_{DC-link} = \sum ESL_{parts} \pm \sum ESL_{interactions}$$

The following example illustrates a DC-link composed of five xEVCap capacitors (B25654A8806K001, 80 µF, 850 V). The ESL is simulated at 2 MHz, resulting in a total inductance of 8.3 nH from the central pair of terminals. If all parts of the system are considered independently, the inductances are 3.1 nH for semiconductor terminals, 3.7 nH for the busbar, and 1.1 nH (5.4 nH/5) for the five capacitors in parallel. The sum of these independent components is 7.9 nH, which is lower than the total system inductance of 8.3 nH. This discrepancy arises because the different elements interact, increasing the total inductance by 0.4 nH. Therefore, while the method of directly summing element inductances can provide a rough and quick estimation, it is not highly accurate.

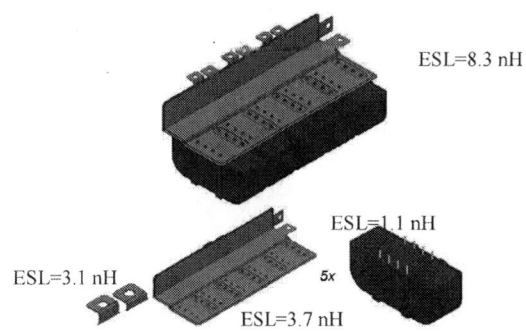

Fig. 8. The sum of component inductances (7.9 nH) is less than the total characterized inductance (8.3 nH) due to interaction among the individual elements.

There are various strategies to reduce the inductance of the DC-Link based on the three sources of ESL in the system:

- Optimize the connection to the power modules to reduce the inductance of the terminals.

- Maximize the overlap in the busbar.

- Increase the number of capacitive elements (xEVCap) connected in parallel. Fig. 9 (up) illustrates the improvement in ESL with an increased number of capacitive elements, together with the results of the double pulse test with 4 capacitive elements.

The total stray inductance measured in the double pulse test, as shown in Fig. 9 (bottom), comprises all components: capacitors, busbars, and semiconductors. This measurement matches the simulation results presented in Figure 8 for the DC-link, which includes capacitors, busbars, and terminals.

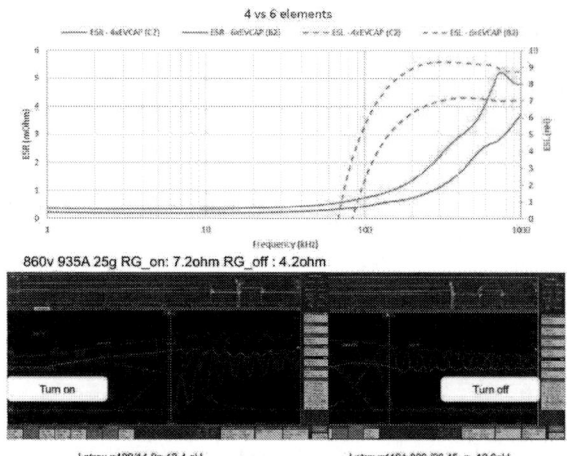

860v 935A 25g RG_on: 7.2ohm RG_off : 4.2ohm

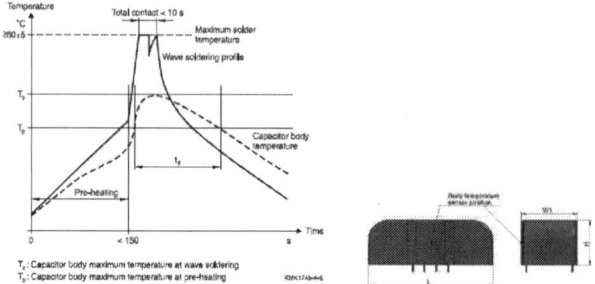

Fig. 11. Extract from mounting guidelines of B25654A* Datasheet [2].

Fig. 9. (Up). Simulation of ESR/ESL for a configuration 4 elements and 6 elements as per [1]. (Bottom) Double pulse test by ST Microelectronics (DSC Gemini 800 V/300 kW traction inverter). 4 x B25654A8137K002 (4x135 µF/850 V), showing a total stray inductance of 13 nH (all elements).

III. JOINT TECHNOLOGIES FROM ELEMENTS TO THE BUSBAR

The capacitor units should be connected in parallel via a laminated busbar, which then connects to the power semiconductors. While a printed circuit board (PCB) connection is another mounting option, it is not commonly used for the main power train inverter because a PCB can't handle high currents. Each capacitor features eight round copper terminals (four for each polarity) with a diameter of 1. 2mm, plated with pure tin. Proper mounting requires a soldering process to ensure a reliable finish.

The soldering process will subject capacitors to certain thermal stress. Therefore, it is crucial to adhere to the cautions

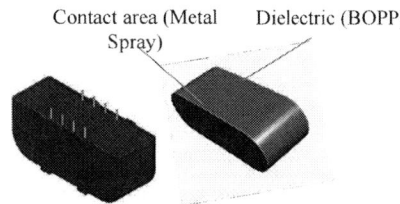

Fig. 10. xEVCap and its internal capacitive element.

specified in the datasheet [9]. Exceeding the recommended temperatures can cause internal damage, such as increased ESR due to issues in the electrode-film contact area, or decreased insulation resistance due to damage to the polypropylene dielectric. For capacitors mounted on PCBs using the wave soldering method, the maximum body temperature must be controlled within the specified limits [9]: During pre-heating (T_p) lower than +110 ° C; during soldering (T_s) lower than +120 ° C, and time of soldering (t_s) lower than 45 seconds.

A. The laminated busbar

Rogers Corporation has more than 40 years of experience designing and manufacturing ROLINX laminated busbars. A laminated busbar is a multilayer construction of conductors: copper or aluminum separated by thin dielectric materials,

laminated in one structure. Electrically, it is a multilayer circuit that distributes the current from the capacitors (buffer) to the power modules (IGBT) through specific connections. In this sense, it is equivalent to a multi-layer PCB, but with the advantage of using much thicker conductors. This allows it to carry much higher currents. Typical currents in a busbar ranges from a few dozen amps to several hundred, even kiloamps for the thickest busbar conductors.

There are specific solutions for the EV market, one of which has been specially developed to integrate film capacitors such as TDK xEVCap: ROLINX CapLink. This range of busbars displays unique design to solder capacitors into ROLINX laminated busbars. The combined assembly is lightweight, extremely low inductive, and offers high power density. These characteristics make it especially suitable for SiC devices.

TABLE II. ROLINX CAPLINK TYPICAL VALUES

Voltage rating [V]	400 to 1200
Peak voltage [V]	1600
I_{DC} or I_{AC} [A] (cont.)	60 to 500
Stray inductance [nH]	< 10 (depending on terminal design)
no. of capacitors	2 to 10+
Operating temperature [°C]	+105 (typ., continuous)
Relative humidity [%]	95 (at +55 °C)
Conductor material	Copper, Aluminum
Plating	Sn, Ni
Insulation CTI [V]	≤ 600

As they are fully custom made, electrical ratings can be extended and designed to fit as many capacitors as required by the application. For reference, 3 xEVCap type C can be accommodated into a 17 x 17cm CapLink with a busbar capable of carrying more than 400 A at 850 V.

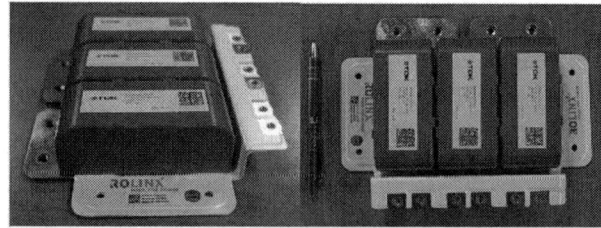

Fig. 12. Design example of three xEVCap (B25654A8137K001, 135 µF- with ROLINX CapLink. Collaboration between TDK and Rogers Corporation.

979-8-3315-1612-3/25 $31.00 © 2025 IEEE

B. Guidelines about soldering processes to laminated busbars

One of the approaches chosen by Rogers to integrate capacitors into a laminated busbar is a special busbar design that allows to concentrate heat around the soldering point, and perform a standard solder process: robot soldering, selective wave soldering, or wave soldering. Robot and selective wave soldering can be performed as described below. Wave soldering requires a protection mask to avoid damaging busbar insulation and to limit temperature rise during the process.

Rogers has experience in all above-mentioned processes, but the specific process performed for xEVCap is a manual solder process with iron and solder wire. Thanks to the special design of ROLINX CapLink around the soldering point, busbar to capacitor solder process follows the same steps as a standard pin soldering:

1. Place the capacitor(s) in position, with pins protruding through the soldering holes.
2. Touch the tip of the heated iron on the busbar conductor, and the capacitor pin simultaneously. Hold the position for 3-4 seconds to heat the conductor and the pin, until the solder melts and creates a soldering cone around the pin 3.
3. Remove the soldering iron and wait for the solder to cool down naturally.

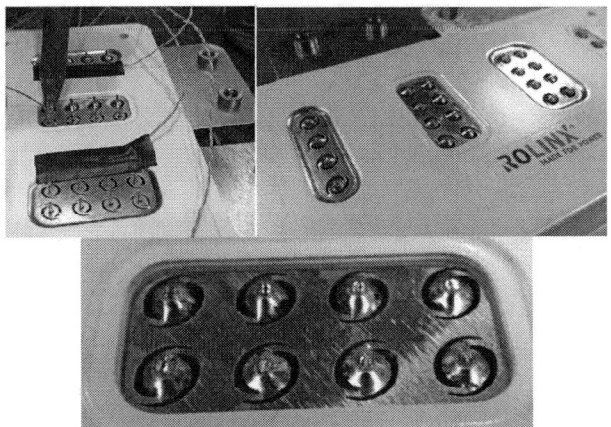

Fig. 13. Detail of manual soldering process and finished soldering points (by Rogers Corporation).

During test process, temperature was measured in the busbar and on the capacitors, as specified following figure 14. Measurement points were the following:

- Channel 1: Busbar surface, next to capacitor 1
- Channel 2: Capacitor 1
- Channel 3: Busbar surface, next to capacitor 2
- Channel 4: Capacitor 2
- Channel 5: Busbar surface, next to capacitor 3
- Channel 6: Capacitor 3

As it can be observed in figure 14, the temperature rises to a maximum of +40 °C on the busbar surface, and to less than +30 °C in the capacitors (which it is very low compared to the limits given in Fig. 11). High thermal conductivity of busbar allows to disperse heat fast, whereas the CapLink soldering design concentrates enough heat to melt solder and attain a proper joint.

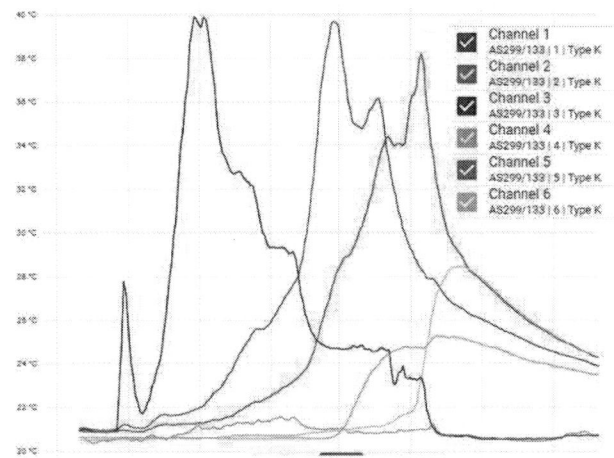

Fig. 14. Temperature monitoring during iron tip soldering. Very low temperature on busbars (channels 1, 3, and 5) and capacitors (channels 2, 4, and 6)

C. Trials with laser soldering process and TDK busbar design

Selective soldering methods are recommended to prevent damage to insulators while ensuring efficiency, cleanliness, and a high-quality soldering finish. In this experiment, instead of using the Rogers CapLink busbar, a custom-designed busbar connection for lead wires by TDK was utilized. A five-step laser soldering process is proposed in Fig. 15 to minimize heat input and solder material, while achieving an optimal solder joint through capillary action.

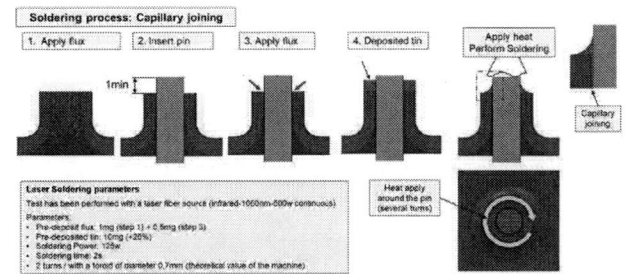

Fig. 15. Illustration of the experiment in 5 steps performed for the soldering of xEVCap lead wire terminals

The laser soldering equipment was specifically adapted for these trials with additional fixtures and custom programming. With two passes of the laser beam around the wire, there was no significant increase in temperature on the capacitor body (CH2 in Fig. 16). Only the area to be soldered (CH1) was heated.

The temperature was monitored in different areas. It was far below their maximum temperature ratings (Fig. 16).

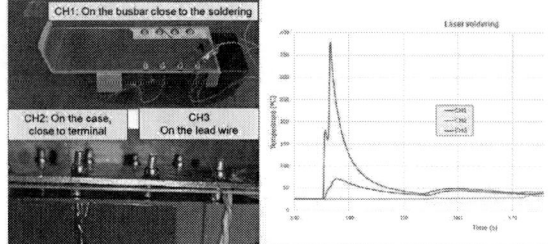

Fig. 16.   Temperature profiles in different areas. No relevant stress for the capacitor was observed.

The methodology involved trial and error until a stable soldering quality was achieved. Preliminary trials revealed issues such as gaps in the solder area or burnt tips and insulators, often due to manual tin-flux deposition control or excessive time and power. The primary challenge was that the laser equipment was not originally intended for this type of capillary soldering. The laser spot diameter had to be manually adjusted and was sometimes too wide during these trials.

For instance, reducing the power or spot diameter of the laser beam and slightly increasing the number of turns will promote a more homogeneous and progressive temperature distribution. This approach enhances capillary action and prevents burning of undesired areas, such as the tip and the insulator. Additionally, cutting the wire over the neck could contribute to tip burning due to chips. These aspects require further investigation and control to integrate this mounting process into mass production.

After several trials, the results were finally positive. A good soldering quality and a technically feasible process were achieved. This means a smooth soldering appearance, good wetting, and no cold solder. The tin material completely fills the gap between the pin and hole. There was no damage from the laser (due to excessive heating) on the plastic insulation or the pin, and no signs of burns.

Fig. 17.   xEVCap with TDK's busbar design. After laboratory trials of the soldering process. Detail of terminals and finishes

IV. THERMAL INTEGRATION AND MODELLING

Following the study of parasitics on the system level and the joint technologies with the busbar, the next step is to evaluate the system from a thermal perspective, considering other elements such as the cooler and semiconductors. Thermal systems are complex to model and evaluate due to the high interactions between all components. For the automotive DC-link capacitor, the key aspects to consider are the terminal temperatures, the contact with cooling surfaces, and the capacitor hotspot temperature in the dielectric. This value defines the lifetime of the component under specified mission profiles or allowed voltage operation.

A. Influence of temperature on the terminals

The temperature at the capacitor terminals is the primary external thermal boundary condition, significantly influencing thermal behavior, aside from ambient temperature. The DC-link solution includes all connections to the semiconductors and the battery. These connections are thermally affected by the current flow corresponding to each operation mode of the inverter, as well as by the temperatures at the semiconductor and battery connections. Depending on these temperatures, power will flow from or to the DC-link (busbar and xEVCap) through these terminals, thereby extracting or injecting power losses.

The terminals form a direct thermal path to the capacitive elements since the material used (copper) conducts heat very well and has a high cross-section. This significantly impacts the thermal behavior of the capacitor. The temperature of the dielectric (BOPP) will determine the maximum voltage and duration for which the capacitor can operate safely.

To estimate the temperature distribution in the capacitor modules, these boundary conditions must be well defined. User specifications usually include the maximum temperature at the terminals for a specific mission profile. It should be also noted that this temperature depends on the operation mode and is affected by the cooling. The effect of terminal temperatures can be analyzed by running a series of thermal simulations, keeping all parameters constant except for the terminal temperatures (Fig. 18).

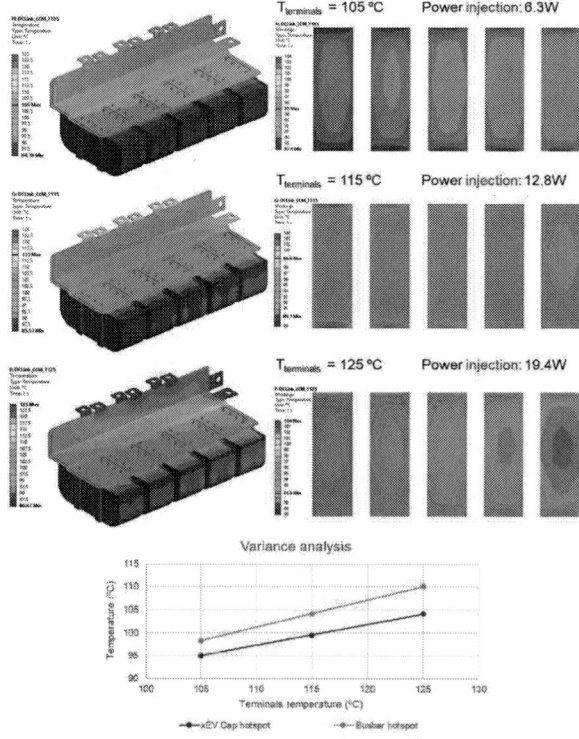

Fig. 18.   The importance of terminal temperature. Thermal simulation with three temperatures at the terminals: +105 °C, +115 °C, +125 °C. Cooperation between Infineon Technologies AG and TDK.

Fig. 18 shows the thermal results from simulations with capacitor losses of 35 W, applying DC battery current and AC current based on a given spectrum. The simulations were conducted with terminal temperatures set at +105 °C, +115 °C, and +125 °C, respectively. These results indicate that the system temperature is highly influenced by the terminal temperatures. Additionally, this boundary condition's effect is concentrated on the areas of the capacitor closest to the terminals. The xEVCap hotspot remains below these temperatures, as confirmed by subsequent real-world testing.

The terminal temperature is a key boundary condition affecting capacitor temperatures and, consequently, its lifetime under defined mission profiles. It directly influences the capacitor hotspot and depends on the instantaneous power of the inverter. Some users specify the power flow through the terminals, providing a more accurate definition of the boundary conditions for simulation.

B. Active cooling influence on thermal behavior

The other external condition that directly impacts the temperature distribution inside the capacitor is active cooling. For most automotive DC-link applications, this is necessary due to the high requirements in terms of current, and frequency and due to the inherent poor natural cooling. The most used approach is water cooling with a closed-loop circuit.

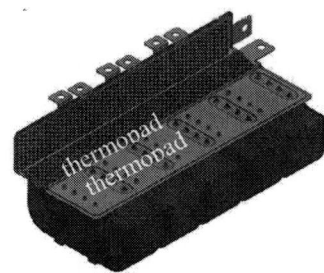

Fig. 19. DC-Link solution with thermopad in selected areas of the busbar to extract heat and decouple semiconductors and xEVCap thermally. Busbar design by Infineon Technologies AG

The cooling block enters in contact with the busbar (with the use of a thermopad) on the marked areas.

The positioning of the cooling surface aims two benefits:

• Heat extraction: cooling area on the busbar in the most effective position. The high thermal conductivity of copper helps to remove not only the losses from the busbar but also from the xEVCap through its terminals.
• Thermal decoupling: A cooling surface located between the semiconductor terminals and xEV capacitors thermally decouples DC-Link from the semiconductor modules (the biggest heat generator in the converter), reducing their influence on dielectric (BOPP) temperature.

Fig. 20. shows the temperature distribution in the DC-link solution with and without active cooling, and a heat exchange analysis showing the heat inputs and outputs of the complete system (power losses with boundary conditions).

When the system is not cooled (Figure 20, left), the maximum temperature increases. The heat generated by the capacitors is not adequately dissipated, leading to overheating of the capacitor dielectric. This heat becomes trapped between the semiconductors and capacitors. When cooling is applied (Fig. 20, right) the hottest spots of the system are the semiconductor terminals, and all the heat is extracted through the cooler. This is why this type of cooling is preferred for automotive applications.

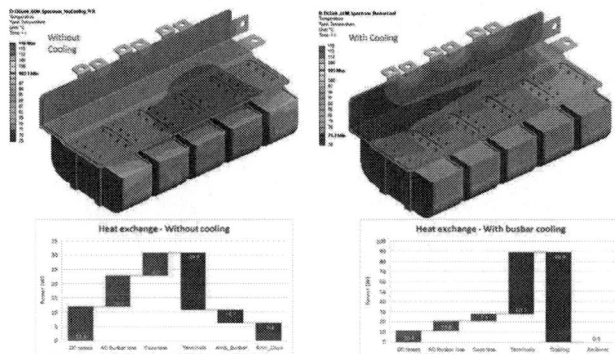

Fig. 20. Simulation of a DC-Link solution under two different scenarios: without cooling (left) / with cooling (right). Cooperation between Infineon Technologies AG and TDK

When possible, the active cooling surface should be over the copper surfaces and located between the semiconductor terminals and the capacitive elements to decouple the temperature of the capacitor dielectric from the temperature of the semiconductor terminals.

C. Thermal integration in the demonstrator

For this purpose, a demonstrator for laboratory testing has been developed according to the specifications required by automotive converters. It aims to demonstrate the thermal integration with a cooler and how the different elements interact under different heating/cooling conditions. The construction includes three xEVCap, (B25654A8806K001 80 µF/850 V; 56 A (RMS)), with a nominal power capability of approximately 150 kW. It is integrated into a system with a removable cooler and heater to simulate the losses of the power semiconductors.

Two different setups have been tested under pre-defined electrical conditions (168 A (RMS) at 20 kHz) at ambient temperature: One setup without external cooling or heating (Fig. 21 left), and the other setup with all the elements (heating and cooling, Fig. 21 right).

To simulate the power losses of the semiconductor (approx. 1500 W), nine cartridge heaters, each rated at 175 W, were placed over the cooler, replicating the positions of the semiconductors. Additionally, two more resistors were added at the connection with the busbar to reproduce the heat transferred by the semiconductor terminals, thereby maintaining the temperature at the semiconductor terminals.

To ensure good thermal contact and maximize the heat extracted by the cooler from the busbar, a thermal pad of 3,2 W/(m·K) and a thickness of 1.5 mm is used over the cooling area (Fig. 22).

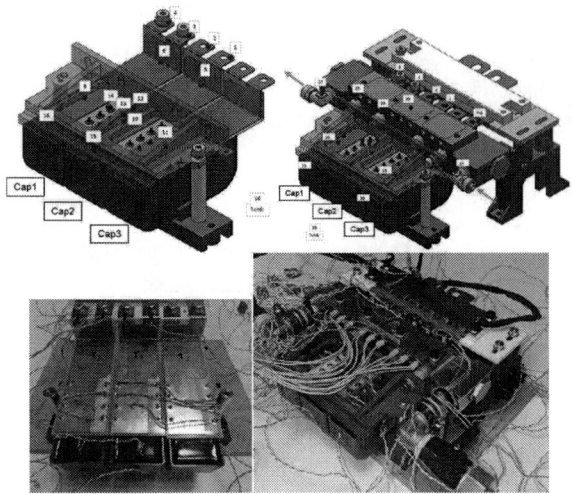

Fig. 21. Illustration of Demostrator setup for testing. One without a cooler (left) and the other with all the elements (heating and cooling) (right). Numbers indicate the position of the thermocouples.

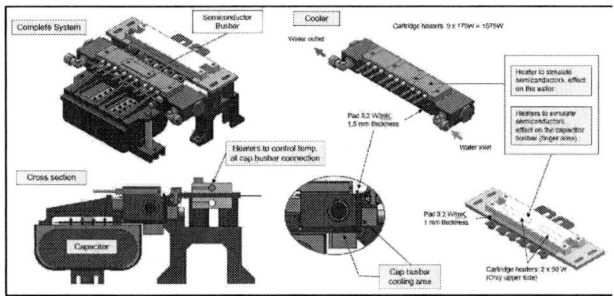

Fig. 22. Part description of the demonstrator.

A test matrix has been elaborated by varying the conditions of the water inlet temperature (+50 °C or +60 °C), the water flow rate (5 L/min or 8 L/min) and the temperature in the connection between the busbar and the semiconductors (ranging from +70 °C to +100 °C):

TABLE III.   TEST MATRIX FOR THERMAL TESTING

Test	Arms freq. (kHz)	Water Temp (°C)	Busbar (semicond.) Temp. (°C)	Water flow rate (L/min)	Active cooling
1a	20	without cooler	Not controlled	N/A	N
1b		50	70	8	Y
2a		50	80		Y
2b	20	60	80	8	Y
2c			90		Y
2d			100		Y
2e		60	80	5	Y

The test results (Fig. 23.) indicate a thermal decoupling between the capacitors and the heaters (which emulate semiconductor temperatures). The system effectively extracts heat from the capacitors, as the temperature of the busbar area in contact with the cooler is lower than the hottest point measured on the capacitors.

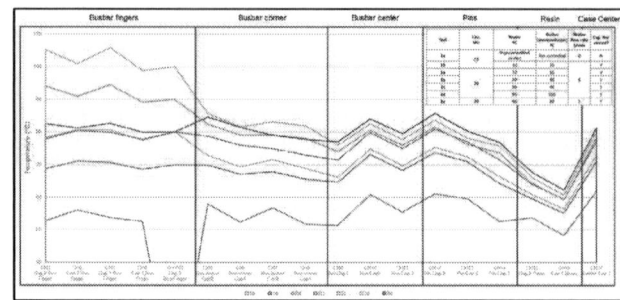

Fig. 23. Temperature results applying 168 A (RMS) of the test conditions defined in Table III. Channel location is indicated in Fig 21.

In the first setup (without cooler and no busbar heating, Fig. 21 left), the busbar temperature is about +55 °C, and the maximum temperature of +62.1 °C is observed in the case between capacitors (thermocouple 17). In the second setup (case 2E, with the whole integrated system with water temperature of about +60 to +64 °C and heater at +135 °C), the highest temperature is detected in the xEVCap terminals with a value of +85,7 °C.

The laser soldering connections between the xEVCap terminals and the busbar, have shown a good behaviour during the test, without significant overheating detected compared to the busbar plate (1,7 K in the worst case).

The conditions tested may not exactly match the specifications required by automotive converters, but these results will help designers better understand thermal distribution and optimize cooling strategies.

V. CONCLUSION

xEVCap offers an alternative DC-link solution for traction inverter power designers compared to traditional custom solutions. Custom solutions require high volume demand for industrialization, whereas xEVCap is an off-the-shelf solution. This approach is uncommon for traction inverter designers but is widely used in lower power electronics with printed circuit boards, product catalogs featuring multiple ratings, voltages, and dimensions. The same philosophy is applied to xEVCap, providing all necessary design information online.

Additional design aspects have been reviewed, including system-level parameters, busbar design, assembly technologies, and thermal strategies, focusing on the interaction between elements from a thermal perspective. These aspects are supported by advanced simulation tools, real-world testing in a demonstrator, and collaboration with leading companies in power semiconductors and busbars.

VI. FUTURE WORK

For the demonstrator, the next step is virtual characterization to develop accurate models of complete systems with thermal behavior under various power and cooling scenarios. Regarding the product, future investigations will focus on aspects such as miniaturization and integration with other components (e.g., EMC filter).

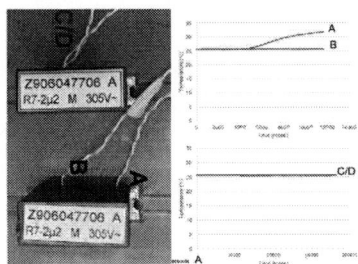

Fig. 24. Detail of temperature profile for laser soldering to an SMD film capacitor (X2- 2.2 μF/305 V (AC)). Laser soldering process applied to SMD film capacitors by Hamamatsu Photonics [12]

For joint technologies, mass production will always require process optimization, control, and cycle time reduction. For laser soldering, the tin wire should incorporate flux, and automation of preform insertion or auto-feeding the tin wire is essential. This laser soldering method could be used for SMD film capacitors, which would otherwise need larger and more expensive dielectrics and enclosures to be compatible with standard SMT reflow processes. Additionally, energy consumption, factory area, cleanliness, quality, process control, and yield are other factors that need to be investigated to support the incorporation of these new selective soldering processes for mounting electronic components.

ACKNOWLEDGMENT

Thanks to all contributors not explicitly mentioned as authors, but providing very valuable information for the paper and very useful learning lessons:

STMicroelectronics

Rogers Corporation

Infineon Technologies AG

Málaga (Spain) factory team

Nashik (India) factory team

REFERENCES

[1] David Olalla, Gayatri Kulkarni, Fernando Rodriguez, "A Modular DC-Link Capacitor Solution for the Main Powertrain Inverter of xEVs", PCIM 2024, June 2024.

[2] Shajjad Chowdhury, Emre Gurpinar, "Capacitor Technologies: Characterization, Selection, and Packaging for Next-Generation Power Electronics Applications" IEEE Transactions On Transportation Electrification, Vol. 8, No. 2, June 2022.

[3] Manuel Gomez, "Innovative film capacitor technologies for wide band-gap semiconductors," IEEE PSMA Capacitor Committee Workshop, April 2020.

[4] Yole Group, "Capacitors for Power Converters 2023", Market and Technology Report 2023.

[5] Wolfgang Rambow, Fabian Beck, Elvis Keli, Katharina Mankel, Mankel-Engineering, "Compact, Modular Inverter Manufactured Using Standard Components," Bodo's Power October 2022.

[6] Infineon Application Note, "Quickstart Manual for EVAL KIT HPD G1 SiC," www.infineon.com, 2022-12.

[7] IEC TS 63337 Ed. 1: "Basic qualification of DC-link film capacitors for automotive use - General requirements, test conditions and tests", February 2024.

[8] AEC Automotive Electronics Council, AEC-Q200 Rev E "Stress Test Qualification for Passive Components", March 2023

[9] TDK-Film capacitors – Power electronic capacitors- xEVCap Lead Wire Datasheet- B25654A*001, July 2024

[10] Rogers Corporation. "ROLINX® Laminated & Integrated Busbar Solutions. Design Rules", Version 03 / 2018

[11] Fernando Auñón, Fernando Rodríguez, Sergio Sepúlveda, David Olalla, "Film Capacitor Standard Series Digitalization: Electromagnetic & Thermal Modeling implementation in CLARA Web Tool", PCIM 2024, June 2024

[12] T-SMILS Laser heating system L15570 series | Hamamatsu Photonics [hamamatsu.com], Dec 2021

Optimized Bidirectional On-Board Charger Using a Novel Unfolder-DAB Topology

Héctor Sarnago, *Senior Member, IEEE*, Ignacio Álvarez, *Student Member, IEEE*, Pablo Briz, *Student Member, IEEE*,
and Óscar Lucía, *Fellow, IEEE*.
Department of Electronic Engineering and Communications
I3A, Universidad de Zaragoza
Zaragoza, Spain
hsarnago@unizar.es, olucia@unizar.es

Abstract— Electric vehicles rely on on-board charger converters to create a bidirectional power link between the battery and the grid when fast dc charging infrastructure is not available. Power density, weight, efficiency, and cost are among the key challenges for these converters, which must operate under a wide range of input and output voltages, as well as in single and multi-phase configurations, leading to costly and suboptimal designs. In contrast to classical designs where a dedicate rectifier branch should be included to operate in single-phase, the proposed paper features a novel unfolder-DAB structure to avoid extra components for single-phase operation. The proposed structure allows single-phase current rectification by using middle point connection over the primary of the DAB converter used to 1) provide isolation, and 2) adapt battery voltage variations according to its SOC.

As a result, and in contrast to classical implementations, all the power electronics are continuously used, independently of the grid configuration, yielding to a cost reduction and efficiency improvement.

Besides in the proposed bidirectional charger, an inherent dc-link voltage balance is performed, removing additional balancing resistors or extra measurements. The proposed converter has been successfully tested using a 11 kW OBC prototype.

Keywords—On-board charger, electric vehicle, dcx, dual active bridge.

I. INTRODUCTION

Nowadays, electric vehicle (EV) power train [1-4] greatly relies on modern power electronics solutions, being charging converters among the most complex and demanding ones [5-7]. On-board chargers (OBCs) [6, 8-11] are essential to ensure fast and efficient charge regardless the grid voltage, number of grid phases or the battery charge status. This is a complex operating scenario, where power conversion techniques are challenged to offer high performance, efficient and cost-effective solutions [12]. Fig. 1 shows the current architecture or EVs, where OBCs play a key role.

In the past, many different converter topologies have been proposed and analyzed [9, 13], including single and multi-phase solutions, unidirectional and bidirectional implementations [6, 10, 14-17], standard and totem-pole rectification strategies, and resonant [18], multi-level and dual-active-bridge dc-dc implementations [19-21]. However, there are still important limitations and challenges regarding simultaneous single- and multi-phase operation [22, 23], dc-link capacitor optimization [24] and semiconductor usage optimization both in terms of efficiency and cost. In this context, the overall power conversion cost plays a key role, since integrated converters with optimized optimization are mandatory in the current context.

In order to overcome these challenges, this paper proposes a new on-board charger topology featuring an unfolder-DAB [25] structure. The main benefits of this approach are that it enables single-phase operation without additional power electronics components, providing a substantial cost reduction, efficiency improvement, and a power dense implementation.

The remainder of this paper is organized as follows. Section II details the proposed power converter, including its topology and main waveforms. Section III presents the main implementation and experimental results. Finally, Section IV summarizes the main conclusions of this paper.

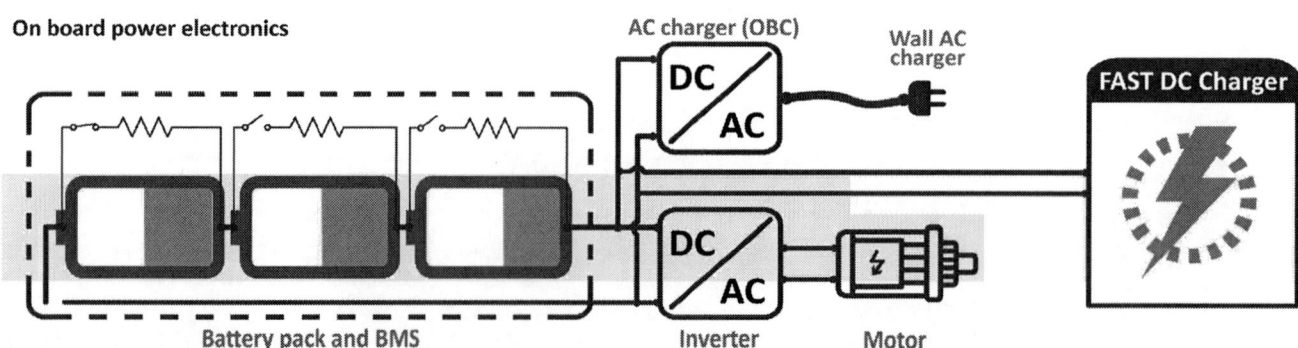

Fig. 1. EV architecture.

979-8-3315-1612-3/25 $31.00 © 2025 IEEE

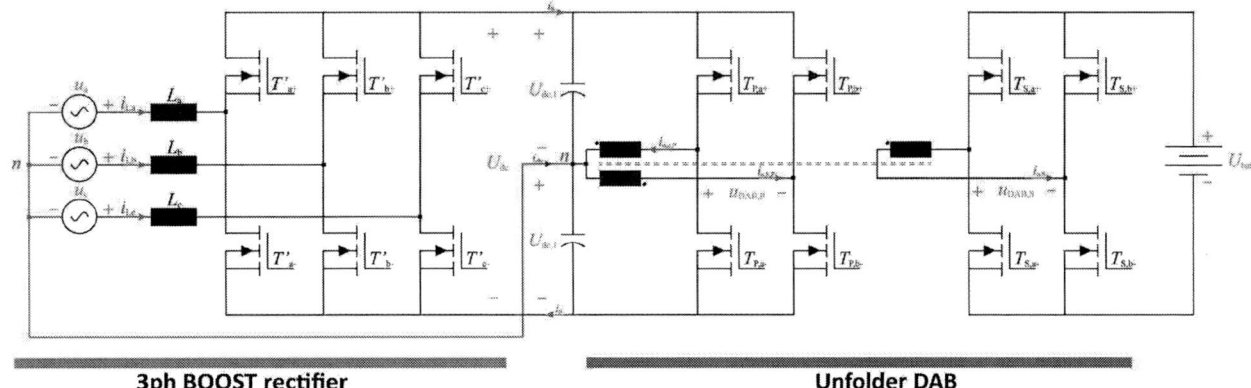

3ph BOOST rectifier   **Unfolder DAB**

Fig. 2. Proposed on-board charger topology.

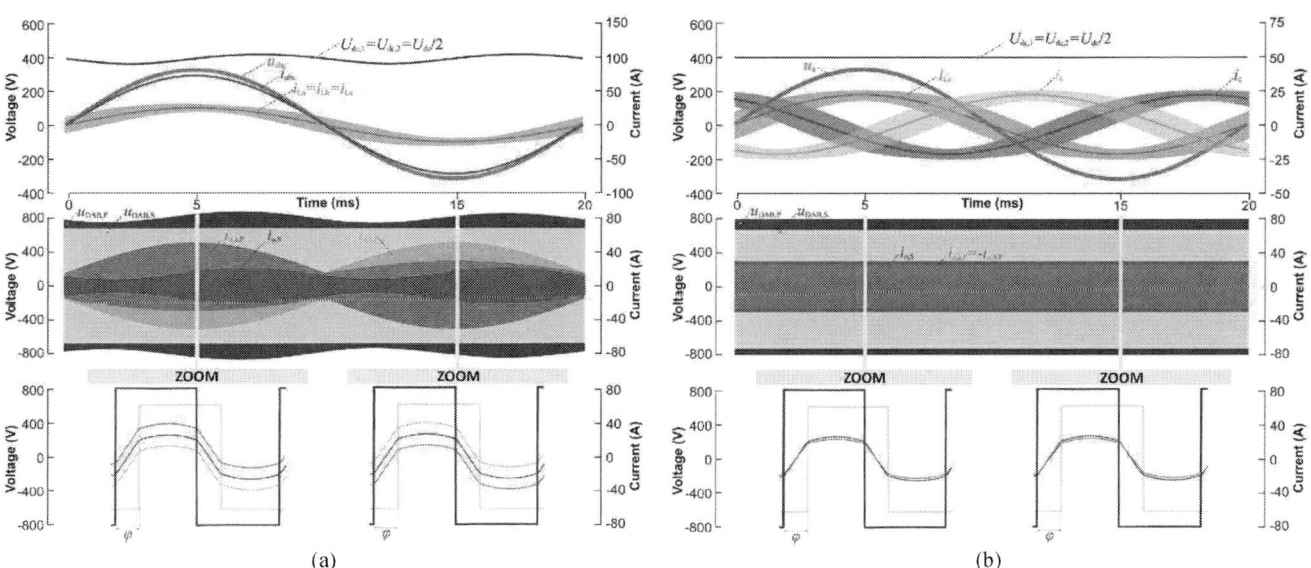

(a)   (b)

Fig. 3. Main waveforms of the proposed on-board charger topology: single-phase (a) and three-phase (b) operation.

II. PROPOSED POWER CONVERTER

The proposed power converter topology is depicted in Fig. 2. It is composed of a three-phase boost rectifier stage, composed by transistors $\{T'_{i+}, T'_{i-}\}$, and the boost inductor L_i, for each phase voltage, u_i, being $i=\{a, b, c\}$. To complete the design, an isolated DC-DC converter based on a DAB structure is proposed, composed by a primary full bridge $\{T_{P,a+}, T_{P,a-}, T_{P,b+}, T_{P,b-}\}$ the secondary bridge $\{T_{S,a+}, T_{S,a-}, T_{S,b+}, T_{S,b-}\}$, and the integrated transformer including sufficient leakage inductance to enable the proper control of the power flow in both directions. By ensuring a constant primary duty cycle of 50% and including a center tapped connection connected to both the mains neutral point, n, the split dc-link capacitors voltage results balanced, $U_{dc,1}=U_{dc,2}=U_{dc}/2$. This results in a novel integrated unfolder-DAB structure that enables the mains single-phase operation without additional rectifier branch, where half of the mains neutral current $\langle I_{o,b,P}\rangle_{Tsw}=-\langle I_{o,a,P}\rangle_{Tsw}=\langle I_{abc}\rangle_{Tsw}/2$, is split in each primary winding flowing in opposite directions, and therefore, not contributing to an increase of the magnetic field of the core.

As a consequence, the resulting duty cycles for the boost rectifier stages results, $D_i=(U_{dc}/2+u_i)/U_{dc}$, for each phase voltage, u_i, being $i=\{a, b, c\}$, while the minimum dc-link voltage result, $U_{dc,min}=2\sqrt{2}\ U_{abc,rms}$.

Fig. 3 shows the main waveforms of the proposed converter both in single phase operation (a) and three-phase (b). First, in case of single-phase operation the mains current flowing through the neutral point is rectified via the proposed unfolder DAB structure, equally split in the primary windings of the DAB transformer. Additionally, the inherent self-balance of the dc-link capacitors ensure a safe and reliable operation, compensating the mains power pulsation in case of single-phase operation. In the case of three-phase operation, no current is required to be flowing through the neutral point, resulting in the classical DAB operation of the DC-DC converter. As a result, the proposed converter exhibits important benefits in terms of multi-phase operation, power device utilization ration, dc-link capacitor design, and transformer complexity.

(a)   (b)

Fig. 4.   Proposed on-board charger topology experimental prototype: (a) top view and (b) bottom view of the power devices..

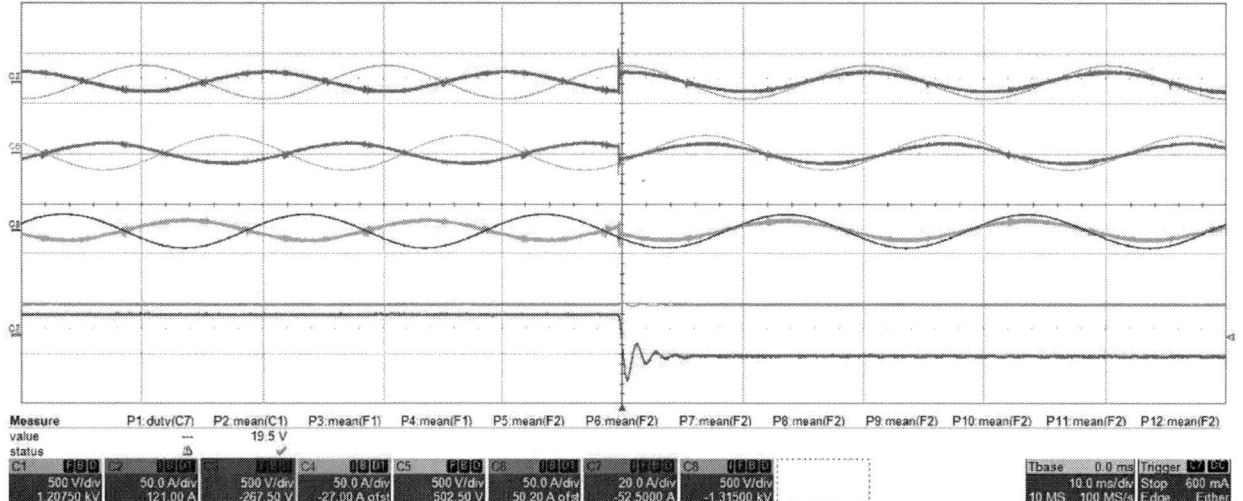

Fig. 5.   Main experimental waveforms at 11 kW for a bidirectional operation step in 3ph operation. From top to bottom: mains voltages (500V/div) and currents (50A/div), load current (20A/div) and battery voltage (500V/div).

III.   IMPLEMENTATION AND EXPERIMENTAL RESULTS

In order to prove the feasibility of the proposed power converter, an experimental prototype has been designed, implemented and tested. A 11 kW OBC has been designed, using 1200 SiC top-side-cooled devices from Infineon. 150 µH PFC boost inductors using High-Flux cores. The transformer has been implemented using a PM 62/49, N97 core. Finally, the converter is controlled using a Artix-7 100T Xilinx FPGA. Fig. 4 shows the top and bottom views of the implemented converter. Fig. 5 shows the top and bottom view if the designed and implemented prototype, including the electromechanical integration.

Fig. 5 shows the main experimental results of the proposed converter operating at 11 kW. In these experimental results, a bidirectional load step (from source to sink mode) is performed at maximum output power. These results validate the feasibility of the proposed converter.

IV.   CONCLUSIONS

This paper has proposed a novel single-stage multi-phase OBC topology. The proposed topology enables three-phase and single-phase operation with no additional elements required with optimized semiconductor usage. Besides, it achieves improved dc-link capacitor design as well as an optimized transformer implementation by means of a novel unfolder-DAB structure. The proposed converter has been tested using a 11-kW OBC prototype. Experimentation has proved the feasibility of this proposal in the complete output power operation range.

ACKNOWLEDGEMENT

This work was partly supported by Projects TED2021-129274B-I00, CNS2023-144980, and PDC2023-145837-I00 co-funded by MICIU/AEI/10.13039/501100011033, by "ERDF A way of making Europe", by the "European Union NextGenerationEU/PRTR", and by the DGA-FSE.

REFERENCES

[1] I. Aghabali, J. Bauman, P. J. Kollmeyer, Y. Wang, B. Bilgin, and A. Emadi, "800-V Electric Vehicle Powertrains: Review and Analysis of Benefits, Challenges, and Future Trends," *IEEE Transactions on Transportation Electrification,* vol. 7, no. 3, pp. 927-948, 2021, doi: 10.1109/TTE.2020.3044938.

[2] Z. Zhao, H. Hu, Z. He, H. H. C. Iu, P. Davari, and F. Blaabjerg, "Power Electronics-Based Safety Enhancement Technologies for Lithium-Ion Batteries: An Overview From Battery Management Perspective," *IEEE Transactions on Power Electronics,* vol. 38, no. 7, pp. 8922-8955, 2023, doi: 10.1109/TPEL.2023.3265278.

[3] M. Moradpour, P. Pirino, M. Losito, W.-T. Franke, A. Kumar, and G. Gatto, "Multi-Objective Optimization of the Gate Driver Parameters in a SiC-Based DC-DC Converter for Electric Vehicles," *Energies,* vol. 13, no. 14, p. 3720, 2020. [Online]. Available: https://www.mdpi.com/1996-1073/13/14/3720.

[4] N. Keshmiri, D. Wang, B. Agrawal, R. Hou, and A. Emadi, "Current Status and Future Trends of GaN HEMTs in Electrified Transportation," *IEEE Access,* vol. 8, pp. 70553-70571, 2020, doi: 10.1109/ACCESS.2020.2986972.

[5] M. Safayatullah, M. T. Elrais, S. Ghosh, R. Rezaii, and I. Batarseh, "A Comprehensive Review of Power Converter Topologies and Control Methods for Electric Vehicle Fast Charging Applications," *IEEE Access,* vol. 10, pp. 40753-40793, 2022, doi: 10.1109/ACCESS.2022.3166935.

[6] J. Yuan, L. Dorn-Gomba, A. D. Callegaro, J. Reimers, and A. Emadi, "A Review of Bidirectional On-Board Chargers for Electric Vehicles," *IEEE Access,* vol. 9, pp. 51501-51518, 2021, doi: 10.1109/ACCESS.2021.3069448.

[7] H. Tu, H. Feng, S. Srdic, and S. Lukic, "Extreme Fast Charging of Electric Vehicles: A Technology Overview," *IEEE Transactions on Transportation Electrification,* vol. 5, no. 4, pp. 861-878, 2019, doi: 10.1109/TTE.2019.2958709.

[8] A. Khaligh and M. D. Antonio, "Global Trends in High-Power On-Board Chargers for Electric Vehicles," *IEEE Transactions on Vehicular Technology,* vol. 68, no. 4, pp. 3306-3324, 2019, doi: 10.1109/TVT.2019.2897050.

[9] M. Yilmaz and P. T. Krein, "Review of Battery Charger Topologies, Charging Power Levels, and Infrastructure for Plug-In Electric and Hybrid Vehicles," *IEEE Transactions on Power Electronics,* vol. 28, no. 5, pp. 2151-2169, 2013, doi: 10.1109/TPEL.2012.2212917.

[10] H. Wouters and W. Martinez, "Bidirectional Onboard Chargers for Electric Vehicles: State-of-the-Art and Future Trends," *IEEE Transactions on Power Electronics,* vol. 39, no. 1, pp. 693-716, 2024, doi: 10.1109/TPEL.2023.3319996.

[11] M. Shao, Q. Zhang, Z. Wei, and H. Li, "A method for suppressing leakage current of non-isolated integrated on-board EV battery charger," *Chinese Journal of Electrical Engineering,* pp. 1-12, 2024, doi: 10.23919/CJEE.2024.000072.

[12] S. Lovati, "Differential power processing OBC for 400V/800V EV architecture," ed: Power Electronics News, 2023.

[13] H. Sarnago, O. Lucía, R. Jiménez, and P. Gaona, "Differential-power-processing on-board-charger for 400/800-V battery architectures using 650-V super junction MOSFETs," in *IEEE Applied Power Electronics Conference and Exposition*, 2021, pp. 564-568.

[14] H. Sarnago, I. Alvarez, and O. Lucia, "Bidirectional Isolated 400-12V DC-DC Converter With Improved Power Density and Full-Range Operation for EV Applications," in *PCIM Europe 2024; International Exhibition and Conference for Power Electronics, Intelligent Motion, Renewable Energy and Energy Management*, 11-13 June 2024 2024, pp. 2508-2512, doi: 10.30420/566262352.

[15] H. Sarnago, I. Alvarez, and O. Lucia, "22-kW Bidirectional Single-Stage Direct-Ac-Ac Power Conversion On-Board Charger with High-Power-Density Implementations," in *PCIM Europe 2024; International Exhibition and Conference for Power Electronics, Intelligent Motion, Renewable Energy and Energy Management*, 11-13 June 2024 2024, pp. 2443-2447, doi: 10.30420/566262344.

[16] Wolfspeed, "6.6 kW high power density bi-directional ev on-board charger," 2023. [Online]. Available: https://assets.wolfspeed.com/uploads/2020/12/crd-06600ff065n_application_note.pdf

[17] V. T. Tran, M. R. Islam, K. M. Muttaqi, and D. Sutanto, "An On-Board V2X Electric Vehicle Charger Based on Amorphous Alloy High-Frequency Magnetic-Link and SiC Power Devices," in *2019 IEEE Industry Applications Society Annual Meeting*, 29 Sept.-3 Oct. 2019 2019, pp. 1-6, doi: 10.1109/IAS.2019.8912310.

[18] C. Wei, D. Zhu, H. Xie, Y. Liu, and J. Shao, "A SiC-Based 22kW Bi-directional CLLC Resonant Converter with Flexible Voltage Gain Control Scheme for EV On-Board Charger," in *PCIM Europe digital days 2020; International Exhibition and Conference for Power Electronics, Intelligent Motion, Renewable Energy and Energy Management*, 7-8 July 2020 2020, pp. 1-7.

[19] S. Mukherjee, J. M. Ruiz, and P. Barbosa, "A High Power Density Wide Range DC–DC Converter for Universal Electric Vehicle Charging," *IEEE Transactions on Power Electronics,* vol. 38, no. 2, pp. 1998-2012, 2023, doi: 10.1109/TPEL.2022.3217092.

[20] I. Kim and J. W. Park, "Integration of DC-DC Converters for OBC, LDC, and TC in Electric Vehicles," *IEEE Transactions on Transportation Electrification,* pp. 1-1, 2023, doi: 10.1109/TTE.2023.3331033.

[21] I. Kougioulis, A. Pal, P. Wheeler, and M. R. Ahmed, "An Isolated Multiport DC–DC Converter for Integrated Electric Vehicle On-Board Charger," *IEEE Journal of Emerging and Selected Topics in Power Electronics,* vol. 11, no. 4, pp. 4178-4198, 2023, doi: 10.1109/JESTPE.2023.3276048.

[22] H. Sarnago and L. Ó, "Design and Experimental Verification of a Bidirectional EV On-Board Charger Featuring Multiphase Operation in Full Power/Voltage Ranges," *IEEE Open Journal of the Industrial Electronics Society,* vol. 5, pp. 458-467, 2024, doi: 10.1109/OJIES.2024.3406732.

[23] H. Sarnago, O. Lucía, S. Chhawchharia, D. Menzi, and J. W. Kolar, "Novel Bidirectional Universal 1-Phase/3-Phase-Input Unity Power Factor Differential AC/DC Converter," *Electronics Letters,* vol. 59, no. 13, pp. 1-4, 2023, doi: https://doi.org/10.1049/ell2.12857.

[24] H. L. Sarnago, O., "High Power Density On-Board Charger Featuring Power Pulsating Buffer," *IEEE Open Journal of Power Electronics,* vol. 5, pp. 162-170, 2024, doi: 10.1109/OJPEL.2024.3359271.

[25] H. Sarnago and O. Lucía, "Optimized EV ON-Board Charging Power Converter Using Hybrid DCX-Dab Topology," in *IEEE Applied Power Electronics Conference and Exposition,* 2024, pp. 1305-1309.

Critical Thermal Characterization of Next-Generation Solid-State Batteries for Automotive Battery Management Systems

Chandan Chetri *(Student Member, IEEE)*
Smart Transportation Electrification and Energy Research (STEER) Group
Department of Electrical, Computer, and Software Engineering
Ontario Tech University
Oshawa, Canada
chandan.chetri@ontariotechu.net

Sheldon Williamson *(Fellow, IEEE)*
Smart Transportation Electrification and Energy Research (STEER) Group
Department of Electrical, Computer, and Software Engineering
Ontario Tech University
Oshawa, Canada
sheldon.williamson@ontariotechu.ca

Abstract—**Solid-state batteries (SSBs) are at the forefront of next-generation energy storage technologies, promising enhanced safety, energy density, and longevity compared to traditional lithium-ion batteries. However, managing the thermal behavior of SSBs during charging is critical to ensuring their safe and reliable operation. This paper presents an in-depth analysis of the thermal characteristics of SSBs under various charging conditions, based on laboratory experiments conducted in controlled environments. The study specifically evaluates the thermal performance of SSBs in comparison to traditional lithium nickel cobalt aluminum oxide (NCA) batteries, with an emphasis on their potential application in e-mobility. The insights gained from this research are expected to inform the development of advanced thermal management strategies, contributing to the design of safer and more efficient solid-state battery technologies for e-mobility applications.**

Keywords—*solid state batteries, battery management systems, thermal characterization.*

I. INTRODUCTION

With increasing awareness about global sustainability, the focus of the transportation sector is shifting from conventional gasoline-based vehicles towards electric vehicles (EVs) and hybrid electric vehicles (HEVs). Lithium-ion batteries (LIBs) are widely used in these applications, forming the backbone of most EVs and HEVs due to their advantages, such as high energy density, power output, and rechargeability [1]. However, liquid electrolytes present in LIBs have limitations, especially regarding safety and thermal stability at extreme temperatures [2]. This becomes more pressing as EV technology demands higher performance. To tackle hindrances against the wider adoption of EVs and to eliminate range anxiety amongst users, much research is ongoing to develop higher energy density batteries. Solid-state batteries (SSBs) are emerging as a promising alternative, especially for e-mobility applications, due to their higher energy and power density [3]. For example, the energy density of Li-SSB was found to be 1143 Whl^{-1}, whereas that of a Li-NCA battery was only 635 Whl^{-1} [4]. With the increasing commercialization potential of solid-state electrolytes like lithium phosphorus oxynitride (LIPON),

lithium-ion superionic conductor (LISICON), and thio-LISICON, SSBs have attracted significant attention for their enhanced thermal safety compared to conventional LIBs [5].

Under low temperature conditions, the liquid electrolytes of traditional LIBs become viscous and freeze, which affects the migration of ions in the electrolyte and its diffusion between the anode and cathode [6], [7]. This reduces the ionic conductivity of the electrolytes and increases the internal impedance of the battery [8]. Moreover, fast charging under such conditions leads to lithium plating, a phenomena resulting from the uneven deposition of the Li-metal on the surface of the anode, leading to severe battery capacity loss [9], [10]. These unevenly deposited Li-metal forms dendritic structures and may even pierce the separator, leading to an internal short circuit and catastrophic failure such as thermal runaway of the entire battery pack [11]. During high temperature operation, the solid electrolyte interphase (SEI) forms between the electrode and the liquid electrolyte resulting into capacity loss of LIBs [12], [13]. Unlike the flammable liquid electrolytes in traditional LIBs, solid-state batteries (SSBs) have solid electrolytes which eliminates the freezing of the electrolytes at low temperatures and inhibits the formation of SEI layer at high temperatures, thereby offering better thermal stability than traditional LIBs [14], [15]. When LIBs are cycled at low temperatures, the rate of temperature rise of the battery increases due to increased internal impedances which results in accelerated battery degradation [16]. The understanding of the stability of SSBs for automotive application has not been researched much in literature. Thus, it is very crucial to understand the thermal behaviour of the SSB to help define the safe operating conditions and to prevent catastrophic damage.

This paper analyses the thermal behaviour of the Samsung 21700 lithium (NCA) nickel cobalt aluminium battery and the lithium-ion solid state battery. The charging and discharging tests have been conducted under a wide range of ambient temperatures. The results obtained shows the comparative analysis of the thermal characteristics of both the batteries and provide a practical insight of the benefits of using SSBs and

979-8-3315-1612-3/25 $31.00 © 2025 IEEE

NCA battery for automotive applications. The paper is divided into the following sections. Section II covers the experimental setup used. Section III discusses the detailed experimental results . Section IV discusses the conclusions drawn from the series of experiments., followed by references in the next section.

II. LAB EXPERIMENTAL SETUP AND EXPERIMENTS

The experimental setup used for this study is shown in Fig. 1. The experiments have been conducted inside the Associated Environmental System® thermal chamber, which is used for controlling the battery ambient temperature. To charge and discharge the battery, a bi-directional DC power supply from Elektro-Automatik (EA-PSB 10360-120 3U) with a rating of 360 V, ±120 A is used. This power supply is controlled using the EA Power supervisory controller for cycling the battery. To measure the temperature rise of the batteries, NTC Semitec104JT-050 thermistors are affixed to the surface of the battery using Kapton tape. The temperature readings are collected using the dSpace MicrolabBox® and MATLAB®. For the series of experiments, GF A674H2-S03 solid state battery and cylindrical Samsung INR21700-40T3 battery and have been used. The cathode of the solid state battery is made up of lithium cobalt oxide and the anode is made up of metallic lithium. The electrolytic medium used in this battery is lithium phosphorus oxynitride. The SSB has a rated capacity of 24 Ah and an energy of 87.6 Wh. The voltage rating ranges between 2.6 V and 4.2 V. The SSB is rated for a maximum of 24 A (1C) charge and 72 A (3C) discharge. The cathode of the cylindrical 21700 battery is made up of nickel cobalt aluminium oxide (NCA) and the anode is made up of graphite and silicon carbon nanocomposite. The NCA battery has a rated capacity of 4 Ah and an energy of 15.1 Wh. The voltage rating ranges between 2.5 V and 4.2 V and is rated for a maximum of 6 A (1.5C) charge and 35 A (8.75C) discharge. The specifications of both SSB and NCA battery is shown in Table I. The cells are tested at 25°C, 40°C, and 0°C ambient temperatures.

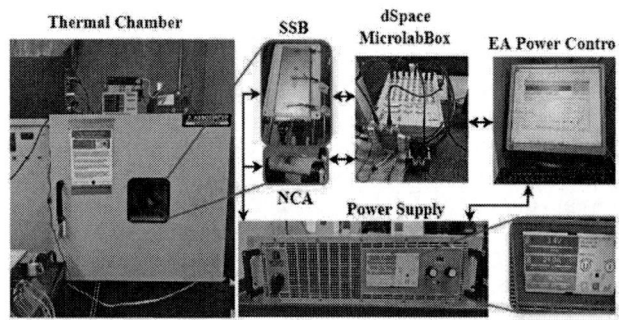

Fig.1 Experimental Setup

III. RESULTS AND DISCUSSIONS

The tests have been conducted on the commercially available Samsung 21700 NCA cells and GF A674H2-S03 solid state lithium batteries at temperatures of 0°C, 25°C and 40°C at 1C charging and discharging rates. The temperature range has been chosen to analyse the battery behaviour at extreme

temperatures of 0°C and 40°C compared to 25°C ambient temperature. The charging profile of both the batteries at 25°C ambient temperature are shown in Fig. 2 and Fig. 3. The discharge profile of both the batteries at 25°C ambient temperature are shown in Fig. 4 and Fig. 5. The voltage profiles of both the batteries at 1C rate at an ambient temperature of 25°C is shown in Fig. 2. It is observed that for both the batteries the voltage profile increases similarly. The temperature profile of the batteries during charging at 25°C ambient temperature is shown in Fig. 3. It is observed that the temperature profile of the SSB is different from that of the NCA cell. The SSB temperature profile shows a portion of flatted curve from nearly 15 minutes to 30 minutes during charging, and then increases more as the battery charges more, indicating that the SSB has an ability to maintain continuous temperature profile with no external cooling required. The temperature profile for the NCA battery shows gradual rise in the curve, with no periods of flattened portion. This indicates that the NCA battery does not cools down during charging, unlike the SSB. The maximum temperature attained for the SSB and NCA cells are 35.4°C and 32.9°C, respectively.

TABLE I.   BATTERY PARAMETERS

Parameters		Specifications	
		SSB	*NCA*
Model		GF A674H2-S03	Samsung INR21700-40T3
Chemistry	Cathode	Lithium Cobalt Oxide (LiCoO₂)	Lithium Nickel Cobalt Aluminium oxide (NCA)
	Anode	Lithium	Graphite + SCN
Nominal Capacity		24000 mAh	4000 mAh
Energy		87.6 Wh	15.1 Wh
Voltage Range		2.6 V - 4.2 V	2.5 V – 4.2 V
Rated Charge		24 A (1C)	6 A (1.5C)
Rated Discharge		72 A (3C)	35 A (8.75C)

The voltage profile of the batteries during discharging at ambient temperature of 25°C is shown in Fig 4. For both the batteries the voltage profiles are similar. The temperature profile during discharge at ambient temperature of 25°C is shown in Fig. 5. It is observed that the temperature rise of the SSB is significantly higher than the NCA cell throughout the charging process. This can be attributed to the higher discharge current of 24 A for the SSB, as compared to 4 A for the NCA cell during 1C discharge.

The voltage profiles of the batteries during 1C charging at 40°C is shown in Fig. 6. Similarly, as seen at 25°C ambient temperature, the voltage profile for both the batteries shows similar trend. The temperature profile of the batteries during 1C charging at 40°C ambient temperature is shown in Fig. 7. It is observed that the SSB shows a period of momentary cooling between 25 minutes and 30 minutes, without any exterior cooling. This behaviour can be attributed to the cooling area available for the SSB as compared to the NCA cell. However, the overall temperature rise for the SSB is significantly higher than the NCA cell. The maximum temperature rise of the SSB and NCA cell are 46.5°C and 42.9°C, respectively.

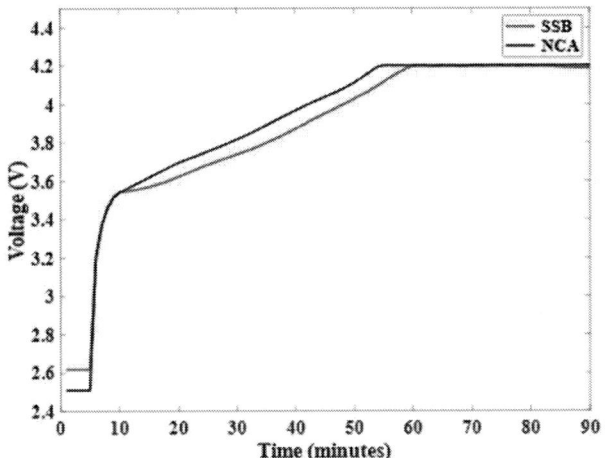

Fig. 2   Voltage profile during charging at 25°C ambient temperature

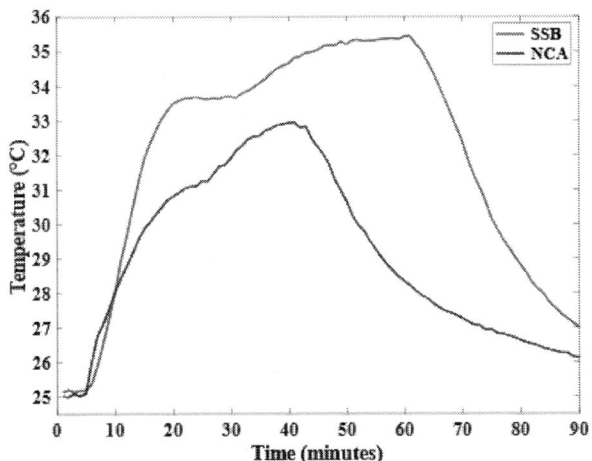

Fig. 3   Temperature profile during charging at 25°C ambient temperature

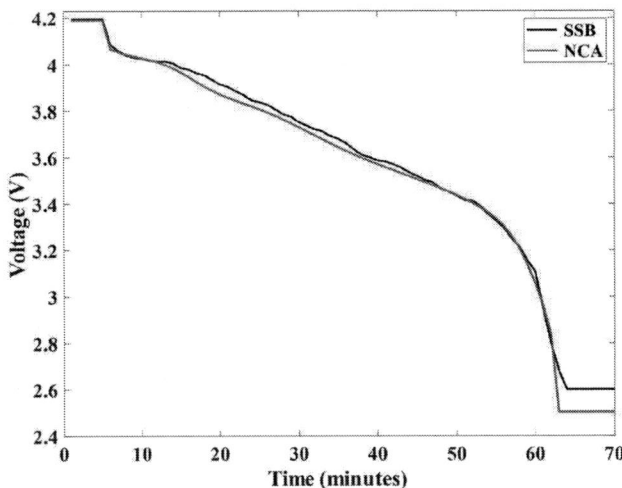

Fig. 4   Voltage profile during discharging at 25°C ambient temperature

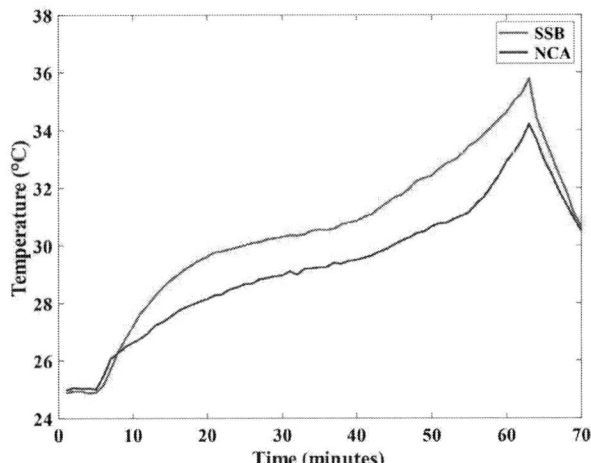

Fig. 5   Temperature profile during discharging at 25°C ambient temperature

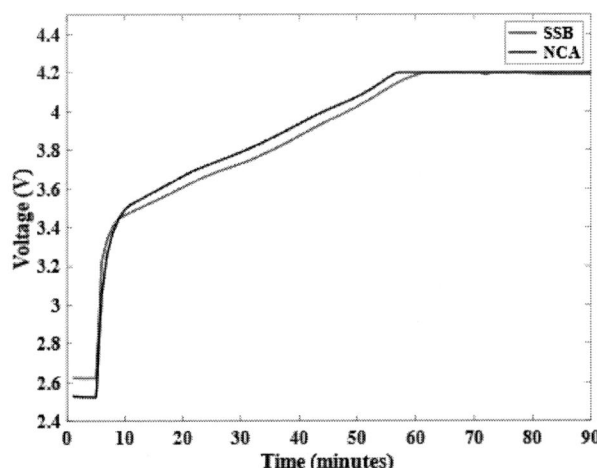

Fig. 6   Voltage profile during charging at 40°C ambient temperature

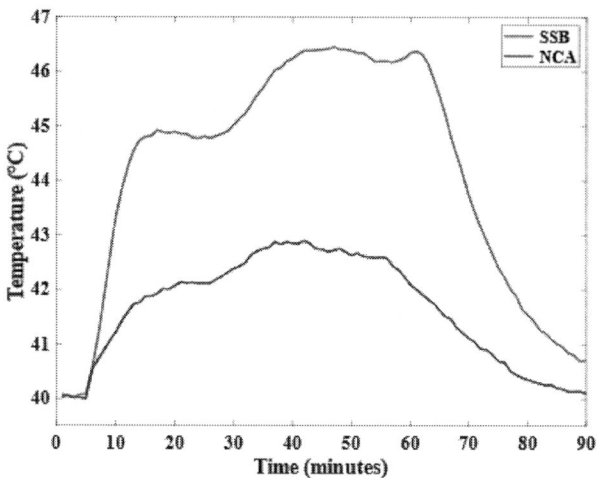

Fig. 7 Temperature profile during charging at 40°C ambient temperature

The voltage profile during 1C charging at 0°C ambient temperature is shown in Fig. 8. The voltage curve for SSB shows a momentary dip during the charging process at around 30% state of charge, whereas no such dip is observed for the NCA cell. At low temperatures, due to the phenomena of increased internal resistance and temporary lithium plating occurring in LIBs, this voltage dip is observed. The voltage dip for the NCA cell is observed at even lower temperatures [9]. The temperature profile during charging at 1C rate at ambient temperature of 0°C is shown in Fig. 9. The SSB shows a much flatter temperature curve as compared to the NCA cell. This can be attributed to higher cooling surface area available for the SSB as compared to the NCA cell. The maximum temperature rise of the SSB and NCA batteries are 18.7°C and 10.2°C, respectively.

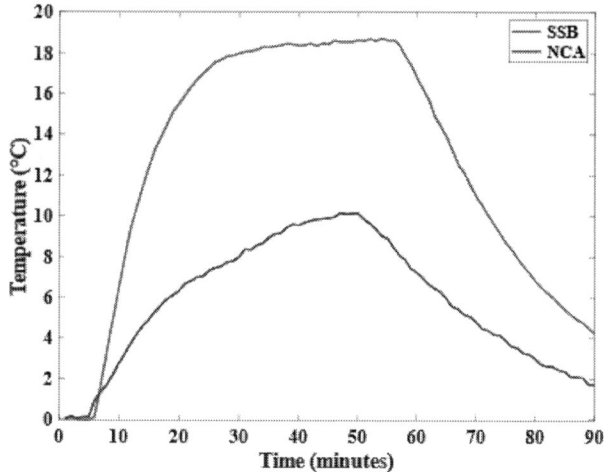

Fig. 9 Temperature profile during charging at 0°C ambient temperature

The summary of the battery charging profile comprising the current rates, maximum temperature attained and the maximum temperature rise are shown in Table II. Fig. 10 shows the maximum temperature rise of the SSB and NCA battery. At 0°C, the maximum temperature rise of the NCA and SSB are 10.2°C and 18.7°C, respectively. At 25°C, the maximum temperature rise of the NCA and SSB are 7.9°C and 10.4°C, respectively. At 40°C, the maximum temperature rise of the NCA and SSB are 6.5°C and 8.27°C, respectively. It is observed that the temperature rise of the SSB is always higher than the NCA battery. The temperature rise increases at lower temperatures compared to higher temperatures due to the increased internal impedances at lower temperature as compared to higher temperatures. Additionally, the SSB has a higher 1C charge current (24 A) compared to the NCA 1C charge current (4 A).

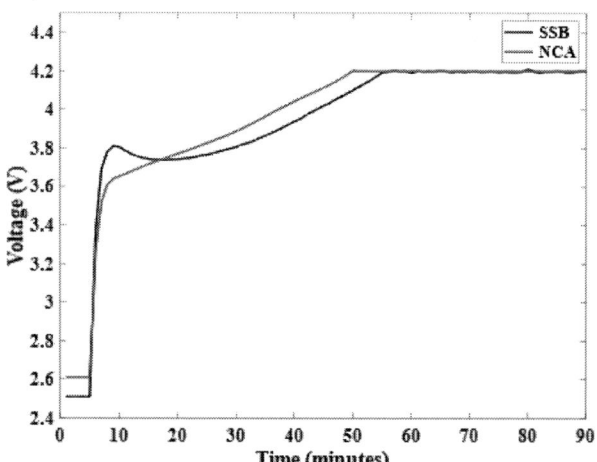

Fig. 8 Voltage profile during charging at 0°C ambient temperature

TABLE II. SUMMARY OF THE BATTERY TEMPERATURE DURING CHARGING AND DISCHARGING

Amb. Temp.	C-Rate(Charging/ Discharging)	Cell Type	Max. Temp. Attained (°C)	Max. Temp. Rise (°C)
25°C	Charging at 1C (24 A)	SSB	35.4	10.4
	Charging at 1C (4 A)	NCA 21700	32.9	7.9
	Discharging at 1C (24 A)	SSB	35.77	10.77
	Discharging at 1C (4 A)	NCA 21700	34.21	9.21
40°C	Charging at 1C (24 A)	SSB	48.27	8.27
	Charging at 1C (4 A)	NCA 21700	46.5	6.5
0°C	Charging at 1C (24 A)	SSB	18.7	18.7
	Charging at 1C (4 A)	NCA 21700	10.2	10.2

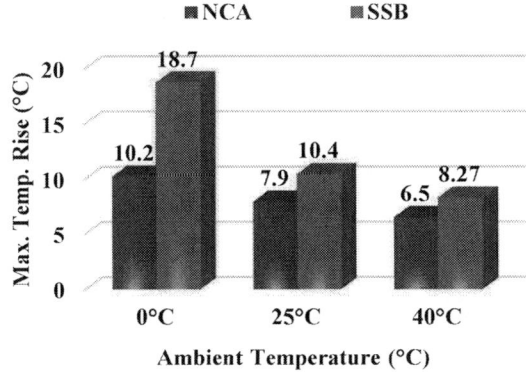

Fig. 10 Maximum temperature rise vs ambient temperature plot

IV. CONCLUSIONS

This paper discusses the comparison analysis of the temperature profile of the traditional lithium NCA battery with the solid state battery. During the 0°C, 25°C and 40°C experiments, the maximum temperature rise was obtained during 0°C experiment. This higher temperature rise during lower ambient temperature is attributed to the higher internal impedances of the batteries at lower temperatures. Additionally, due to the higher 1C charging current for the SSB (24 A) as compared to the NCA battery (4 A), the battery temperature for the SSB is consistently higher throughout. This higher charging current of the SSB can facilitate faster charging of the battery pack, which enhances user convenience. Though the SSB has higher temperature rise, the higher surface area available for the SSB facilitates better thermal cooling strategies compared of the cylindrical NCA battery. Additionally, the SSB temperature graphs shows period of flattened portion during charging experiments, which shows that the SSB can limit the rapid temperature rise compared to the NCA battery. The SSB also has lower temperature gradient compared to the NCA battery. The SSB also has higher capacity compared to the NCA battery. This will be beneficial for EVs as lesser number of SSB will be required in forming the battery pack of the same capacity, compared to NCA battery. This will help in designing less complex thermal management solutions compared to using the cylindrical NCA battery. These findings will be helpful in designing advanced automotive battery management systems incorporating he advantages of solid state batteries and efficient thermal management systems.

REFERENCES

[1] B. Diouf and R. Pode, "Potential of lithium-ion batteries in renewable energy," *Renew. Energy*, vol. 76, pp. 375–380, 2015, doi: 10.1016/j.renene.2014.11.058.

[2] Y. Zhang *et al.*, "Electrolyte Design for Lithium-Ion Batteries for Extreme Temperature Applications," *Adv. Mater.*, vol. 36, 2024, doi: 10.1002/adma.202308484.

[3] J. Janek and W. G. Zeier, "Challenges in speeding up solid-state battery development," *Nat. Energy*, vol. 8, pp. 230–240, 2023, doi: 10.1038/s41560-023-01208-9.

[4] J. Betz, G. Bieker, P. Meister, T. Placke, M. Winter, and R. Schmuch, "Theoretical versus Practical Energy: A Plea for More Transparency in the Energy Calculation of Different Rechargeable Battery Systems," *Adv. Energy Mater.*, vol. 9, 2019, doi: 10.1002/aenm.201900761.

[5] N. Sarfraz *et al.*, "Materials advancements in solid-state inorganic electrolytes for highly anticipated all solid Li-ion batteries," *Energy Storage Mater.*, vol. 71, Aug. 2024, doi: 10.1016/J.ENSM.2024.103619.

[6] X. Su *et al.*, "Liquid electrolytes for low-temperature lithium batteries: main limitations, current advances, and future perspectives," *Energy Storage Mater.*, vol. 56, pp. 642–663, 2023, doi: 10.1016/j.ensm.2023.01.044.

[7] A. Gupta and A. Manthiram, "Designing Advanced Lithium-Based Batteries for Low-Temperature Conditions," *Adv. Energy Mater.*, vol. 10, 2020, doi: 10.1002/aenm.202001972.

[8] X. Yang *et al.*, "Research progress on wide-temperature-range liquid electrolytes for lithium-ion batteries," *J. Power Sources*, vol. 624, Dec. 2024, doi: 10.1016/J.JPOWSOUR.2024.235563.

[9] C. Chetri and S. Williamson, "Effects of Fast Charging of EV Batteries at Low Temperatures Based on Temporary Lithium Plating and Temperature Gradients," in *2024 IEEE Energy Conversion Congress and Exposition (ECCE)*, 2024.

[10] X. Lin, K. Khosravinia, X. Hu, J. Li, and W. Lu, "Lithium Plating Mechanism, Detection, and Mitigation in Lithium-Ion Batteries," *Prog. Energy Combust. Sci.*, vol. 87, p. 100953, Nov. 2021, doi: 10.1016/J.PECS.2021.100953.

[11] Y. Qin *et al.*, "An ultra-fast charging strategy for lithium-ion battery at low temperature without lithium plating," *J. Energy Chem.*, vol. 72, pp. 442–452, 2022, doi: 10.1016/j.jechem.2022.05.010.

[12] T. Zhang *et al.*, "Designing composite solid-state electrolytes for high performance lithium ion or lithium metal batteries," *Chem. Sci.*, vol. 11, pp. 8686–8707, 2020, doi: 10.1039/d0sc03121f.

[13] M. T. F. Rodrigues, F. N. Sayed, H. Gullapalli, and P. M. Ajayan, "High-temperature solid electrolyte interphases (SEI) in graphite electrodes," *J. Power Sources*, vol. 381, pp. 107–115, 2018, doi: 10.1016/j.jpowsour.2018.01.070.

[14] A. Machín, C. Morant, and F. Márquez, "Advancements and Challenges in Solid-State Battery Technology: An In-Depth Review of Solid Electrolytes and Anode Innovations," *Batteries*, vol. 10, no. 1, 2024, doi: 10.3390/batteries10010029.

[15] L. Zhang *et al.*, "Recent advances in electrochemical impedance spectroscopy for solid-state batteries," *Energy Storage Mater.*, vol. 69, p. 103378, May 2024, doi: 10.1016/J.ENSM.2024.103378.

[16] C. Chetri, A. Samanta, and S. Williamson, "Critical Understanding of Temperature Gradient During Fast Charging of Lithium-ion Batteries at Low Temperatures," in *IECON 2023- 49th Annual Conference of the IEEE Industrial Electronics Society*, 2023, pp. 1–6. doi: 10.1109/IECON51785.2023.10312304.

979-8-3315-1612-3/25 $31.00 © 2025 IEEE

Nanocrystalline CMC Inductors for EV Charging: Trade Studies and Testing Standardization

Christopher Bracken
Mechanical Engineering and Materials Science
University of Pittsburgh
Pittsburgh, USA
csb80@pitt.edu

Mark A. Juds
Mechanical Engineering and Materials Science
University of Pittsburgh
Pittsburgh, United States
maj253@pitt.edu

Paul R. Ohodnicki, Jr.
Mechanical Engineering and Materials Science
University of Pittsburgh
Pittsburgh, United States
pro8@pitt.edu

Bharadwaj Reddy Andapally
CBMM
Flevoland, Netherlands
bharadwaj.andapally@cbmm.com

Jose Gato
Vacuumschemlze GmbH & Co.
KG Hanau, Germany
Jose.Gato@vacuumschmelze.com

Abstract—Recent advances in charging of electrical vehicles (EVs) places increasing demands on electromagnetic interference (EMI) mitigation, with common mode choke (CMC) inductors playing an important role. In this digest, we discuss a trade study of nanocrystalline and ferrite-core CMC inductors for specifications relevant for on-board EV charging, and we highlight a significant advantage of nanocrystalline solutions in terms of size and weight. Designs are pursued considering electrical and thermal performance requirements with both custom and commercially available cores. Conclusions of trade studies and presented designs are validated through both detailed characterization of commercial CMC inductors. Needs for standardization of testing of cores and inductors are also discussed, with testing to existing standards codified through data sheets made available publicly to designers and end-users.

Keywords—component, Common Mode Inductor, CMC, Common Mode Choke, Electromagnetic Interference, Electromagnetic Compliance, Electromagnetic Interference Mitigation

I. INTRODUCTION

Electromagnetic compatibility is a critical aspect for highly power dense electrical power conversion systems [1-3]. Designs of power electronic conversion systems to consider electromagnetic compatibility (EMC) requires integration of filtering devices which include common mode choke (CMC) inductors. The simplest form of a CMC inductor is a single core design with two separate coils, which provides both differential and common mode inductance to an electrical circuit (Fig. 2). The goal of a CMC/EMC inductor is to reduce common mode current [4]. Common mode current flows in the

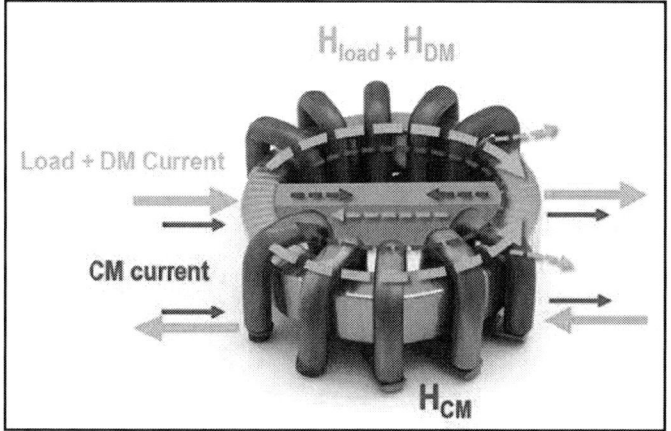

Figure 1: Example figures of common and differential mode currents and two-winding common mode choke (CMC) inductor schematics.

same direction of the source and return path completed in the circuit through ground, while differential current flows in opposite directions beginning at the source and ending in the return path [4]. CMC inductors play a critical role in a range of applications. Here we emphasize specifications relevant for on-board charging of electric vehicles.

II. DESIGN TRADE STUDY FOR COMMERICAL CMC INDUCTOR BENCHMARKING

Specifications were chosen from commercial CMC inductor products of Vacuumschmelze (VAC), for the purpose of custom designs to validate performance and compare performance, size, and weight based upon nanocrystalline cores as compared to ferrites. The specifications chosen were selected for consistency with commercial VAC CMC inductor #6127-X009 as illustrated in Table 1 [5]. Ferroxcube was chosen as the ferrite and benchmarked against two different nanocrystalline options, namely Magnetics, Inc. Finemet and VAC Vitroperm 500F. Assumed properties of the three different core options were taken from commercial

TABLE 1: COMMERCIAL INDUCTOR SPECIFICATIONS.

CMC INDUCTOR PART NUMBER	INNER DIAMETER (MM)*	OUTER DIAMETER (MM)*	HEIGHT (MM)*	MASS (G)*	INDUCTANCE (mH)	IMPEDANCE (Ω)	COMMON MODE CURRENT (mA)
6127-X009	12.37	28.23	13.52	27.91	3.3 at 10 kHz 0.8 at 100 kHz	216 at 10 kHz 780 at 100 kHz	45 mA at 10 kHz 93 mA at 100 kHz
6127-X016	10.98	24.63	12.80	28.32	9.8 at 10 kHz 2.3 at 100 kHz	696 at 10 kHz 2530 at 100 kHz	20 mA at 10 kHz 41 mA at 100 kHz
6127-X021	11.76	24.73	12.83	29.59	3.4 at 10 kHz 0.8 at 100 kHz	228 at 10 kHz 820 at 100 kHz	35 mA at 10 kHz 72 mA at 100 kHz

† - Refer to VAC Website [5] or the AMPED Website [6] for more detailed information on both components.
* - Dimensions are specifically of the box, not the core, as this information was unavailable.

datasheets [5]. An overview of the design process utilized to arrive at the recommended designs is presented in Figure 2, a simplified method that does not explicitly include EMI performance. The design process uses an iterative process, rather than a multi-objective simulation approach. To illustrate this procedure, a discussion of step one will be shown below. To begin with, define a magnetic core with a core flux area, A_{fe}, core flux path length, L_{fe}, a core air gap g on each side (typically a CMC has no air gap), and a core relative permeability μ_r based on the core material. An approximate effective core relative permeability can be calculated by:

$$\mu_{r-eff} = \frac{\mu_r}{1+2\left(\mu_r-1\right)\dfrac{g}{L_{fe}}}$$

(1)

Finding this value, the core inductance per square turn can be determined with also defining the number of core layers N_C as, for a given magnetic core manufactured off-the-shelf used for the design, the core inductance per square turn is defined as:

$$A_L = \frac{\mu_0 \mu_{r-eff} N_C A_{fe}}{L_{fe}}$$

(2)

By determining the core inductance per square turn, the number of turns each phase, N, that provides the desired common mode inductance can be calculated as:

$$N = \sqrt{\frac{L}{A_L}}$$

(3)

Note each phase here represents each separate coil, where one coil connects to one side of the circuit, and the second end to the other to complete the installation. A pictorial representation of the coil configurations is depicted in Figure 1.

Additionally, by defining the number of common mode phases, P_{cm}, the core window area A_{win}, and the wire diameter,

$$N_t = N P_{cm}$$

(3)

Note that the limitation of this quantity is the value needs to be less than $0.8\, A_{win}\, D_x^2$.

Finally, from these quantities, as well as noting the desired common mode current, I_{cm}, the core magnetic flux density can be calculated by

$$B = \frac{\mu_0 \mu_{r-eff} N I_{cm}}{L_{fe}}$$

(4)

Note the limitation for the CMC Inductor to operate correctly is to keep the core magnetic flux density less than the saturation magnetic flux density, B_{sat}.

Step two of the design procedure has a similar process with other quantities, such as winding capacitance, impedance, winding loss, and temperature rise of the core and windings, but these have not been included for brevity purposes.

A summary of the custom designs with their selected cores, performance metrics and size are presented in Table 2 along with the size of the commercial VAC #6127-X009 part. The distinct differences between the three designs are seen most effectively in the actual component size. Due in large part to the reduced relative permeability of the ferrite cores (~2,300) as compared to the nanocrystalline alloys (~95,000-108,000), the volumetric size and weight of the ferrite solution is substantially larger than the size of the nanocrystalline designs, ~10x by mass. In addition, the nanocrystalline core custom design performance metrics and size are in reasonable agreement with the commercially available VAC 6127-X009 indicating optimized commercial designs. Design trade studies were performed for the additional commercial VAC CMC inductors #6127-X016 and #6127-X021, with similar conclusions and hence details are not reported for brevity. The custom designs are for minimum mass. Note that turns can be reduced by increasing the number of cores.

TABLE 2: SUMMARY OF CUSTOM NANOCRYSTALLINE AND FERRITE INDUCTORS VS. SIZE OF VAC CMC INDUCTOR 6127-X009.

Component Core Technology	Component View at Core Level	Core Mass (kg)	Relative Permeability (μ_r)	Common Mode Inductance (mH)	Common Mode Current (mA-pk)	Rated Operating Current (A_{RMS})	Dimensions of Custom Design [LxWxH]	Dimensions of Standard Part Number (VAC 6127-X009) [LxWxH]
Ferroxcube Ferrite 3e94	*3 x 100 g – FerroxCube Tx51/32/19*	0.300	2,326	3.3 at 10 kHz	45	21	51 [mm] x 32 [mm] x 19 [mm]	25[mm] x 16[mm] x 10[mm]
Magnetic, Inc. Nanocrystalline	*2 x 17.6 g – Magnetics CMC025016010H*	0.035	108,000	3.3 at 10 kHz	45	21	25[mm] x 16[mm] x 10[mm]	25[mm] x 16[mm] x 10[mm]
Vitroperm 550F	*2 x 17.0 g – Allstar T60006-L2025-W380*	0.034	95,000	3.3 at 10 kHz	45	21	25[mm] x 16[mm] x 10[mm]	25[mm] x 16[mm] x 10[mm]

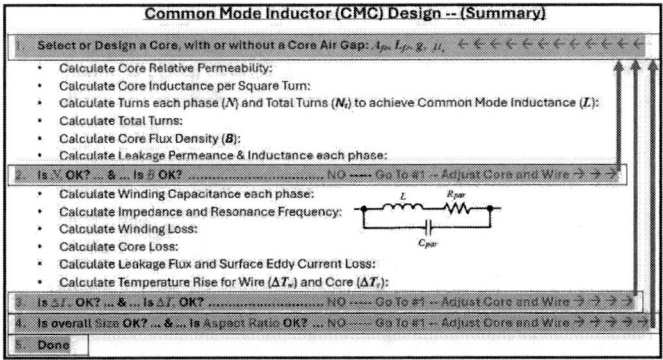

Figure 2: Proposed design process for CMC Inductors. Typically, a CMC has no core air gap (g).

D_x, the total number of turns can be calculated as:

III. DESIGN TRADE STUDY FOR EV ONBOARD CHARGER CMC INDUCTOR BENCHMARKING

As illustrated with a commercial CMC Inductor, the same procedure was applied for a specific application where the component would be used specifically for a custom inductor in an EV onboard charger application. The specifications were chosen using specifications from an innoelectric EV onboard charger, for the purpose of custom designs to validate performance and compare performance, size, and weight based upon nanocrystalline cores as compared to ferrites, as well as comparing to the trade study for the commercial CMC Inductor as illustrated in Table 3 [6]. Ferroxcube was chosen as the ferrite and benchmarked against two different nanocrystalline options, namely Magnetics, Inc. Finemet and VAC Vitroperm 500F. Assumed properties of the three different core options were taken from commercial datasheets [5]. The design process utilized to arrive at the custom designs is the same as presented in Figure 2, and a summary of the custom designs and their relative size and performance metrics are presented in Table 4. As with the commercial benchmark, the distinct differences between the three designs are seen most effectively in the component size.

Due mainly to reduced relative permeability (~2,200) of the ferrite core as compared to the nanocrystalline alloys (~95,000-108,000), the size and weight of the ferrite solution is substantially larger than the nanocrystalline designs, ~10x when measured by mass. Note that turns can be reduced by increasing the number of cores.

TABLE 3: SUMMARY OF EV CHARGER SPECIFICATIONS.

PRODUCT NAME	OBC42	OBC82
COMPONENT DESIGN	22 kW	
INPUT VOLTAGE (3-AC)	360 – 400 V	
INPUT VOLTAGE (1-AC)	103 – 253 V	
INPUT CURRENT (AC)	Max 32 A. (per phase)	
FREQUENCY (Hz)	50 – 60 (±1 %)	
EFFICIENCY (2 – 22 kW)	>94 %	
EFFICIENCY (2 – 10 kW)	>96 %	
OUTPUT VOLTAGE (DC)	210 – 510 V	400 – 850 V
OUTPUT CURRENT (DC)	Up to 65 A	Up to 45 A
DIMENSIONS (L X W X H)	610 x 391 x 120 mm	
WEIGHT (KG)	30	
OPERATING TEMPERARTURES (°C)	-45 to +65	

Component Core Technology	Component View at Core Level	Core Mass (kg)	Relative Permeability (μᵣ)	Common Mode Inductance (mH)	Common Mode Current (mAₚₖ)	Rated Operating Current (Aₚₖ)	Dimensions of Custom Design [LxWxH]
Ferroxcube Ferrite 3c94		0.518	2,181	1.4 at 10 kHz	160	32	80[mm] x 40[mm] x 15[mm]
				2 x 259 g – FerroxCube TX80/40/15			
Magnetic, Inc. Nanocrystalline		0.053	108,000	1.4 at 10 kHz	160	32	25[mm] x 16[mm] x 10[mm]
				3 x 17.6 g – Magnetics CMC025016010H			
Vitroperm 500F		0.051	95,000	1.4 at 10 kHz	160	32	25[mm] x 16[mm] x 10[mm]
				3 x 17.0 g – Allstar T60006-L2025-W380			

TABLE 4: SUMMARY OF CUSTOM NANOCRYSTALLINE AND FERRITE INDUCTORS FOR THE EV CHARGER SPECIFICATIONS.

IV. COMMON MODE CHOKE INDUCTOR TESTING AND STANDARDIZATION

In addition to component designs, validation through detailed testing is an important aspect of CMC inductor performance benchmarking. Establishment of standards in test requirements and procedures can play an important role in clarifying performance of components such as CMC inductors, as well as enabling nanocrystalline and ferrite cores. To aid in detailed component characterization and to validate the commercial specifications for VAC products, a detailed set of tests were developed according to existing standards relevant for on-board EV charging of CMC inductors and data was assembled into published datasheets [7]. Therefore, a key objective for work that seeks to generate relevant data sheets is to employ transparency with the measurement reporting seeking to set a standard for other manufacturers and third-party testing laboratories. This effort has the goal for setting a standard for other manufacturers and third-party testing laboratories for what should be reported, including the specific type and amount of data provided within. CISPR 17 is the most fundamental standard regarding measurements for CMC Inductors, and reported is insertion loss, S-parameters, and impedance [8]. IEC 61007 is also relevant for inductance and winding resistance [9]. Finally, IEC 60635 is also referenced for dimensional measurements (see clause three) [10]. Differential mode and current mode measurements were collected, with differential mode, shorting, one pair of windings. Plots and tables show the numerical and graphical depictions of the result. Test equipment lists are listed for what was used in the experiment. Test procedures are recorded for a sequential order and detailed equipment settings for how the test was conducted and how the results were obtained. Figures 3 and 4 show an example of how data is reported, with the insertion loss reported for the sample being discussed. The common mode inductance captured via the LCR Meter is shown in Figure 4. Also shown for illustration is the measurement of the winding resistance, as shown in Figure 5.

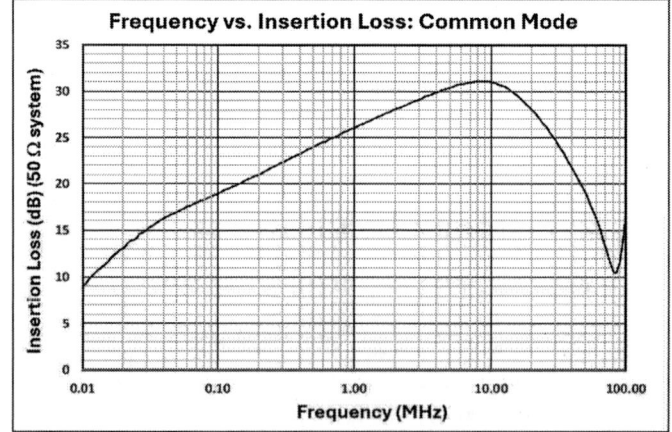

Figure 3: Insertion Loss vs. Frequency for a typical CMC.

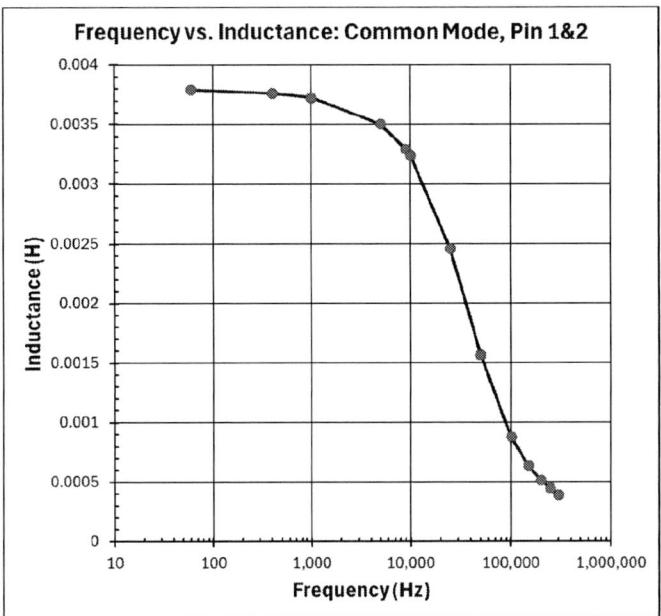

Figure 4: Inductance Measurements with LCR meter.

The winding resistance was captured using a DC Resistance Meter on one of the windings and waiting for the measurement to stabilize. Results resemble closely to what is reported from VAC, especially in the middle range of the frequency sweep where the instrumentation is most accurate. Figure 6 illustrates the typical test setup with two sets of equipment, with part (a) showing the Bode 100 Vector Network Analyzer that was employed, and (b) showing the LCR Meter that was used for specific frequencies.

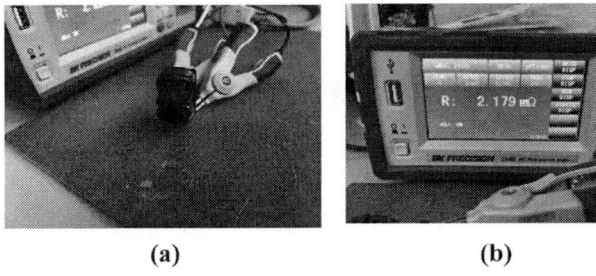

Figure 5: (a) Winding Resistance Setup: Core Layout and (b) Winding Resistance Reading.

Estimates of accuracy can also be derived. For example using insertion loss, the deviation between current measurements and the VAC datasheet is typically within 5% with exception of <~10kHz and >~100kHz for which the measurement instrumentation utilized here has relatively high errors. This was calculated by:

$$\%_{DEV} = \left(\frac{|Measured - Datasheet|}{Measured}\right)100\% \qquad (5)$$

Note the measurement deviation value was not published in the data sheets on the AMPED Website but could be in the future. This can also be calculated and compared by employing the values obtained in the datasheets published by the authors and VAC, and substituting these values into (5) above.

V. GAPS IN PRESENT STANDARDS AND DESIGN METHODS

Existing standards for CMC inductors capture basic functional performance requirements, but do not address the explicit EMI performance characteristics requiring additional standards to be applied when application relevant benchmarking is required. In addition, rational design of CMC inductors based upon nanocrystalline cores does not currently address all required EMI performance characteristics due in part to limited information available about the high frequency properties of nanocrystalline cores in the relevant frequency range. In contrast, ferrites exhibit low conductivity

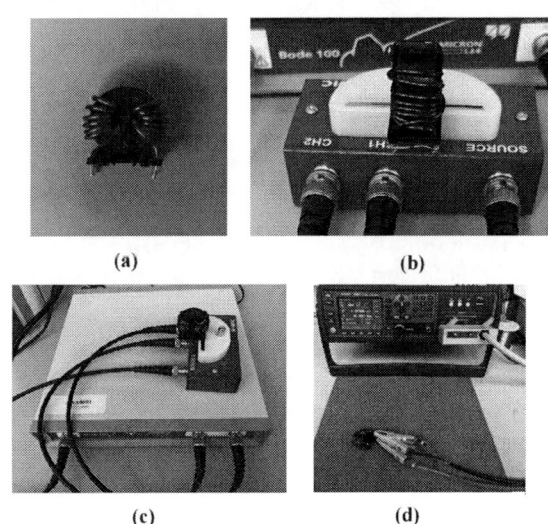

Figure 6: (a) 6127-X009 Sample, (b) Sample in Differential Mode. Impedance Setups with (c) Bode 100, (d) LCR Meter

and high frequency stability of permeability such that explicit measurements are not as critical in the frequency range of interest where EMI performance is desired.

Moving forward, several considerations must be considered. First, even though the standard used to conduct most tests regarding CMC Inductors, CISPR 17, is a relevant standard as shown within Table 1 of the standard [8], it is a broader standard not specific to CMC inductors even though they represent an important component in a typical EMI filter. Therefore, it may be advisable to specifically include additional detail within the CISPR 17 standard in future revisions, or alternatively to consider a need for one or more standards directly related to CMC inductors and related. The other factor worthy of note involves the design process when considering CMC and related inductors, which would benefit from explicitly including aspects of the high frequency EMI performance in the design process which may require additional measurements about high frequency core properties for nanocrystalline cores that is not typically available to be included. Simplified approaches such as the basic design

979-8-3315-1612-3/25 $31.00 © 2025 IEEE

process summarized in Figure 2 would benefit from expansion to include full requirements of the CMC inductor including high frequency EMI performance.

VI. CONCLUSIONS AND FUTURE WORK

CMC Inductors based upon nanocrystalline, and ferrite cores have been designed and investigated, showing significant size reductions for nanocrystalline based components in on-board EV charging applications. Detailed test results have been presented according to standard CISPR 17 and data sheets have been developed and published for public access [7], to support public access and transparency in measurements and reporting. At this moment, CISPR 17 has been a critical standard at the component level, and as illustrated in this digest, is most relevant for obtaining fundamental component information, including impedance, insertion loss, s-parameters, and more. More CMC Inductors have been tested, and more can be seen at the AMPED consortium website [7]. Additionally, an examination of the various factors that affect a CMC / EMC inductor design has been considered and recommendations are made for expanding upon the core properties and design considerations that are included for fully optimized designs including explicit measurements of EMI performance. Future work will be relevant to two areas of interest: 1) standards, and 2) prototyping / testing custom designs outlined within the trade study in this digest. Regarding standards, expansion of existing EMI standards to specifically consider component level performance considerations (e.g. CMC inductors) or alternatively new standards to describe requirements for the CMC and related inductors may be possible considerations moving forward. Prototyping and testing will also be performed of the custom designs, with subsequent datasheets to be assembled and published for the public. For more information, and to view these datasheets in full detail, these are viewable by visiting the AMPED website.

ACKNOWLEDGMENT

Acknowledged is Companhia Brasileria de Metalurgia e Mineração (CBMM) with funding support for the overall project work. Other generous contributions include Vacuumschmelze GmbH & Co. and Magnetics, Inc., donated cores and CMC Inductors for testing and prototyping.

REFERENCES

[1] C.R. Paul, *Introduction to Electromagnetic Compatibility*, John Wiley & Sons, Inc., Hoboken, 2006.

[2] H.W. Ott, *Electromagnetic Compatibility Engineering*, John Wiley & Sons, Inc., Hoboken, 2009.

[3] *"Intro to Noise (EMC): The Basics – Part 1: What is Noise? What is EMC? The Noise Problem is an Electronic Affliction in a Modern Society"* in TDK Tech-Mag, 2024. https://www.tdk.com/en/tech-mag/noise/01

[4] *"EMC Basics: Common Mode versus Differential Mode"* in All About Circuits, 2022. https://www.allaboutcircuits.com/industry-white-papers/emc-basics-common-mode-vs-differential-noise/

[5] *"Nanocrystalline Common Mode Chokes: Specification 6127-X009"* in Vacuumschmelze Automotive List of Standard Common Mode Chokes, 2024. https://vacuumschmelze.com/03_Documents/Datasheets%20-%20Drawings/Commom-Mode-Chokes/6127-X009.pdf

[6] *"On-Board Charger"* in Charging and Power Systems by innolectric, 2023. https://innolectric.ag/on-board-charger-2-2/?lang=en

[7] *Advanced Magnetics for Power and Energy Development Website. Standardization and Magnetic Testing.* https://pittamped.github.io/StandMag.html

[8] CISPR 17. Methods of measurement of the suppression characteristics of passive EMC filtering devices, 2011.

[9] IEC 61007. Transformers and inductors for use in electronic and telecommunication equipment – Measuring methods and test procedures, 2020.

[10] IEC 60635. Toroidal strip-wound cores made of magnetically soft material, 1978.

Predicting Efficiency of On-Board and Off-Board EV Charging Systems Using Machine Learning

1st Mohamed Yasko
Dept. of Electrical Engineering (ESAT)
KU Leuven - EnergyVille
Genk, Belgium
mohamed.yasko@kuleuven.be

2nd Fanghao Tian
Dept. of Electrical Engineering (ESAT)
KU Leuven - EnergyVille
Diepenbeek - Genk, Belgium
fanghao.tian@kuleuven.be

3rd Wilmar Martinez
Dept. of Electrical Engineering (ESAT)
KU Leuven - EnergyVille
Diepenbeek - Genk, Belgium
wilmar.martinez@kuleuven.be

4th Johan Driesen
Dept. of Electrical Engineering (ESAT)
KU Leuven - EnergyVille
Genk, Belgium
johan.driesen@kuleuven.be

Abstract—Prediction of charging energy and power profiles is crucial for optimal scheduling of different EV models. This paper proposes the idea of using advanced machine learning (ML) techniques to predict the operating efficiency of on-board (11 kW) and off-board (50 kW) EV charging systems. An experimental setup is developed to control the charging power, measure and collect datasets under several operating conditions using a real EV model. The collected dataset is used to train, validate, and compare ML techniques such as linear regression (LR), random forest (RF), artificial neural network (ANN) and conditional generative adversarial network (cGAN). The evaluation metrics used for this comparison are mean absolute error (MAE), mean square error (MSE) and the coefficient of determination (R^2). The results demonstrated that the RF and ANN performed better than the LR and cGAN models in both charging systems.

Index Terms—EV charging, efficiency, optimal scheduling, power profile, machine learning

I. INTRODUCTION

Electric vehicles (EVs) technologies will play an important role in the transport and energy sectors in the future. The high number of these EVs will contribute to the increase in the total energy and power demand. Therefore, it is clear that not controlling and optimally coordinating these demand curves can lead to power interruption, grid congestion, higher energy costs, inaccurate estimation of the required charging infrastructure, etc... [1].

EV charging systems are scheduled using smart algorithms that control charging power based on time flexibility and estimated energy requirements. Many charging scheduling strategies are proposed in the literature. These charging scheduling strategies are designed to meet different objectives and considerations, such as charging with a limited number of charging points [1], multiple charging station maximization [2], individual waiting time and operating cost optimization [3], etc. While the mathematical solutions presented in these previous studies might improve the scheduling in general, the proposed approaches need to consider the charger efficiency at

wide load operation, particularly at low power charging. This low charging power can increase power flexibility and hosting capacity but affect the charging duration, operating costs, total energy requirements, etc.

The low efficiency of chargers or power converters at light load operation is a well-known topic in power electronics [4], [5]. This charger behavior at a lower charging power profile must be considered in the charging scheduling strategy to optimize the charging efficiency. Few research works have investigated the idea of including charger efficiency in optimization scheduling. For instance, the authors in [6] demonstrated that the charging losses can be up to five times the size of the distribution system losses through three simulation scenarios based on assumed efficiency values. The charger's operation depends also on the battery conditions, particularly in off-board charging systems where the battery management systems control the charging power. Therefore, the charging control strategy is proposed to consider the battery's internal resistance using a simple battery model [7].

The most important gap observed in these previous studies is the assumed efficiency profile, which does not represent the efficiency values of the charger currently deployed in the on-board and off-board charging system industry. Additionally, the battery behavior using a modular off-board charging system needs to be better investigated and considered in the optimization schedule strategy.

The present paper proposes the idea of using machine learning (ML) technique to predict the operating efficiency of on-board and off-board charging based on the real-world charging datasets collected through an experimental work developed in this research as illustrated in Fig. 1. The traditional charging schedule approach can be improved by sending the DC power and the battery conditions in real-time or using ML models in the charging management system (CMS). Therefore, an optimal AC charging power profile that considers the operating efficiency and the battery conditions can be generated.

979-8-3315-1612-3/25 $31.00 © 2025 IEEE

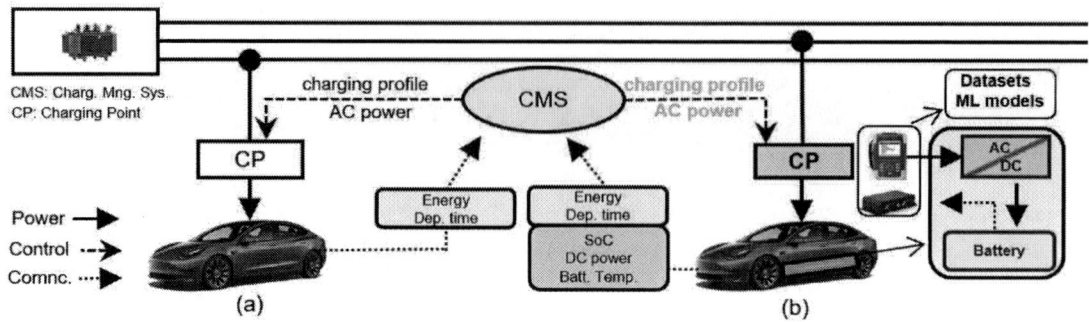

Fig. 1. (a) Conventional charging schedule system, (b) Proposed charging schedule system

TABLE I
CHARGING SYSTEMS AND TEST CONDITIONS PARAMETERS

Charging type	Battery energy (kWh)	Battery tech.	Power capacity (kW)	SoC (%)	Battery temperature (°C)
On-board	58	LFP	11	37-70	23-28
Off-board	58	LFP	50	30-100	21-46

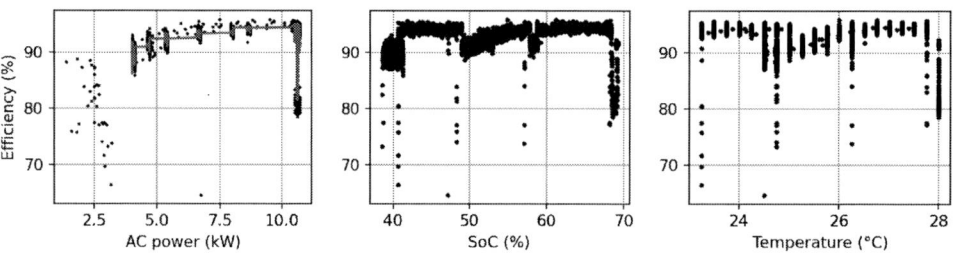

Fig. 2. Operating efficiency for on-board charging systems

ML models can potentially be used in EV charging systems where data access and communication among the charging point (CP), the power converter, and the battery still need to be fully mature. Previous studies on the application of ML are more focused on the prediction of the battery state of health [8], [9], the state of charge (SoC) [10], and the temperature [11]. ML is used at the charger or power converter level [12] to optimize the DAB converter's efficiency. In the present paper, similar ML techniques [13] such as linear regression (LR), random forest (RF), artificial neural network (ANN), and conditional generative adversarial network (cGAN) for regression are used to train and validate the models using the measurement datasets of real on-board and off-board charging systems.

Following this introductive part, methodology and data collection are presented in Section II. The ML techniques are discussed in Section III. Results and discussions are provided in Section IV. The conclusions of the research work are given in Section V.

II. METHODOLOGY AND DATA COLLECTION

As illustrated in Fig. 1., the proposed idea aims to include the operating efficiency of the charger by calculating it from the AC power measured from the CP and the DC power communicated from the CAN-bus of the EV model. As the

battery SoC and temperature can also be communicated from the same CAN-bus system, these parameters can be used to understand the battery behavior for better efficiency prediction over the wide operation range. In reality, the charger's efficiency depends mainly on the converter topology, the number of switching and passive components, the control strategies used to minimize switching and conduction losses, etc.

The charger of EV model manufacturers does not disclose these converter parameters. Therefore, a standard 11 kW on-board converter used in the widely adopted Tesla model Y [14] and a 50 kW off-board charger [15] are considered in this research work. The test conditions parameters and the EV model specifications for these two charging systems are provided in Table I. The collected charging dataset for the on-board charging systems, including the calculated efficiency in function of the input AC power, the battery SoC and temperature are illustrated in Fig. 2. A variable charging power profile was set to demonstrate the charger efficiency over a wide range. As the maximum charging power compared to the size of the battery is lower, the battery temperature is also lower due to insignificant heat generated in the battery during the charging process.

Similarly, the off-board charging system's operating efficiency in function of the input AC power, the battery SoC, and temperature are illustrated in Fig. 3. In this off-board charging

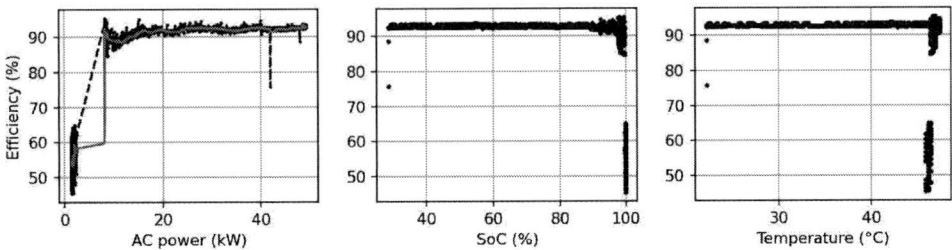

Fig. 3. Operating efficiency for off-board charging systems

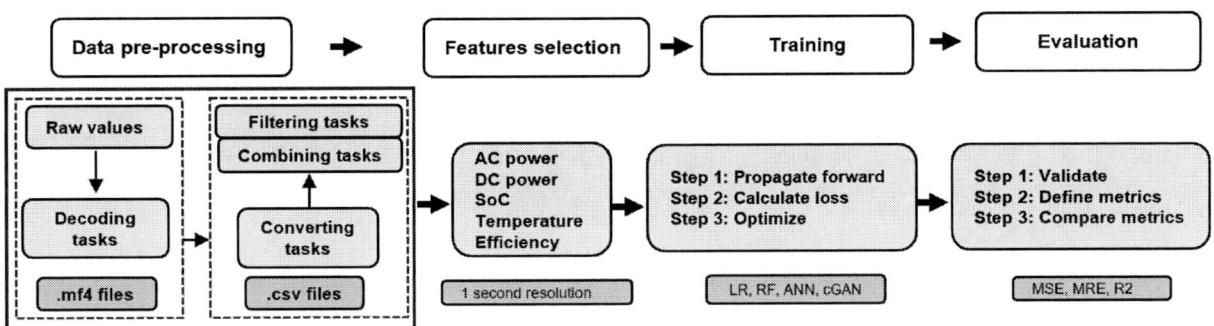

Fig. 4. Different steps followed in this research work

system, the higher power generates more heat in the battery. Therefore, the temperature increases during the charging process. In both charging scenarios, the lower efficiency is related to the lower charging power. The measurement method and the overall charging control system are explained in detail in [16]. These charging datasets illustrated in Fig. 2 and Fig. 3 are used to train ML techniques. Different steps followed in data processing, features selection, model training, validation, and evaluation illustrated in Fig. 4 are discussed in the following section.

III. ML MODELS

The ML techniques used in this research work include linear regression (LR), random forest (RF), artificial neural network (ANN), and conditional generative adversarial network (cGAN) [13]. A summary of the characteristics of these ML techniques is listed in Table II, and the architectures for RF, ANN, and cGAN are respectively shown in Fig. 5. The Python script developed in this work for these models can be found in [17]. A short explanation of these techniques is provided in the following subsections.

The most relevant variables that might impact efficiency are: AC power, SoC, and battery temperature. Before sending the data to models, several procedures must be followed. Firstly, the dataset from several sessions on the same EV is collected and combined. Secondly, a data cleaning process is implemented to clean the wrong data caused by the measurement fault and outliers caused by the time synchronization issue among measurement devices between the charger and the battery. Finally, an additional normalization is added for the ANN and cGAN model, where all the data are linearly scaled

from 0 to 1. In addition, all data are randomly divided into training (80%) and testing (20%) datasets.

A. LR

LR is a simple parametric model where the model fits the relationship between inputs and output by only linear operations. However, it is a simple method limited to modeling the dataset that intrusively has no linear relation.

B. RF

RF is a robust non-parametric ensemble learning method that constructs multiple decision trees during training and outputs the mean prediction of the individual trees, as shown in Fig 5. It effectively handles nonlinear relationships but faces the computational expensive problem when the dataset is too large.

C. ANN

ANN is a popular deep-learning model that can capture complex relationships between inputs and outputs. An ANN's fundamental structure is illustrated in Fig. 5. Nonlinear patterns are captured through the interconnections of neurons across multiple layers. However, this model lacks interpretability and relies heavily on the training data's size and quality.

D. cGAN

cGAN was originally developed to generate complex patterns, such as images, based on provided information. When applied to regression problems, it can effectively mitigate the overfitting issues, thereby enhancing the model's generalization ability. As shown in Fig. 5, cGAN incorporates input data

TABLE II
ML TECHNIQUES IN PREDICTING CHARGING AND BATTERY RELATED PARAMETERS

ML techniques	Type	Benefits	Drawbacks
LR	Parametric linear model	Simple, interpretable, fast	Limited to simple linear regression
RF	Ensemble learning	Relatively simple, non-linearity	Computationally expensive for large dataset
ANN	Deep learning model	Capable of complex pattern	Need large datasets, black-box
cGAN	Generative model	Deal imbalanced datasets, model complex distributions	Need large datasets, hard to train

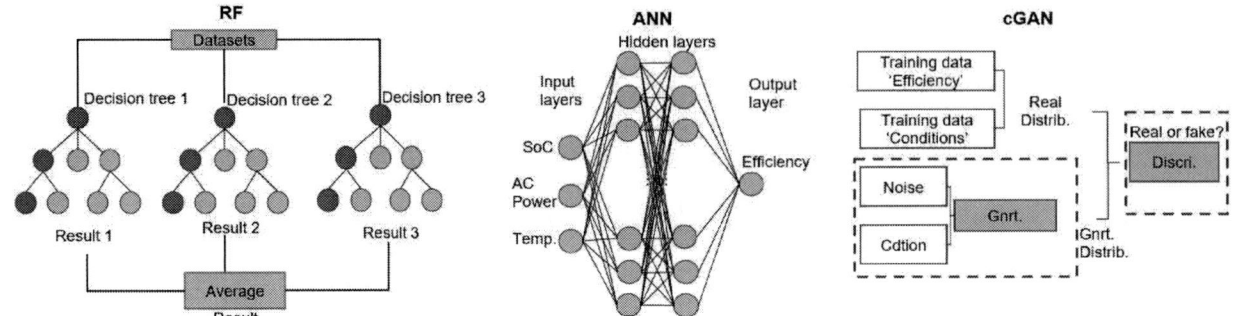

Fig. 5. ML architectures for RF, ANN and cGAN

as a condition, while an additional noise helps control and diversify the outputs. The generated results are continuously monitored by a discriminator, which ensures the model avoids overfitting while capturing the complex patterns of the data.

IV. RESULTS AND DISCUSSIONS

Both on-board and off-board charging sessions are applied to the four models described above. The cleaned datasets consist of 9,599 data points for on-board charging and 4,185 data points for off-board charging. As mentioned earlier, 20% of the data is reserved for testing purposes. The evaluation metrics include , mean absolute error (MAE), mean square error (MSE), and the coefficient of determination (R^2) defined in equations (1)-(3). MAE and MSE evaluate the model accuracy by measuring the absolute error or square error of the true result and the predicted result, while the R^2 assesses how well the predicted values explain the variance in the true values, indicating the proportion of variance captured by the model relative to the total variance.

$$\text{MAE} = \frac{1}{n} \sum_{j=1}^{n} |y_j - \hat{y}_j| \tag{1}$$

$$\text{MSE} = \frac{1}{n} \sum_{j=1}^{n} (y_j - \hat{y}_j)^2 \tag{2}$$

$$R^2 = 1 - \frac{\sum_{j=1}^{n}(y_j - \hat{y}_j)^2}{\sum_{j=1}^{n}(y_j - \bar{y})^2} \tag{3}$$

n is the number of data in the test dataset, and $\bar{y}$ stands for the average value of the output.

Table III shows the results of two testing datasets under the four ML models. Generally, the LR model fits the worst, indicating that the efficiency is a non-linear result of the

input conditions. Comparably, the other three ML models can catch the nonlinearity of the model. More specifically, RF and ANN have the best results, while the results from cGAN have slightly larger errors. This is because introducing a noise compromises the model's accuracy but increases its generality so the model can avoid over-fitting. However, the dataset adopted is from real charging sessions, where only a small range of data is collected, meaning most of the data comes from the steady charging period when AC power is high and stable. The lack of variety in the dataset is more likely to cause the under-fitting issue. Thus, the cGAN model performs poorly in this case.

Fig. 6 compares expected and true values from the four models. As seen, the LR model predicts very bad values, and the figure from cGAN spreads more widely, showing a larger error due to the uncertainties. The predicted values can follow the trend for the two models performing better, RF and ANN. However, the results from RF are spread equally to the two sides, meaning the results are more diverse. At the same time, ANN shows several similar predicted results given different actual values.

The RF model can generate multiple random samples, effectively mitigating the impact of these deviations. However, a notable limitation of the RF model is its inefficiency in handling large-scale datasets. As the volume of data collected from the measurements increases, ANN has the potential to outperform more to RF due to its scalability and capacity to model complex relationships. The efficiency values under similar conditions exhibit variability, particularly in on-board charging case, where the datasets include some measurement errors. These inconsistencies can confuse ML models (e.g. ANN), leading them to predict average values instead of capturing true patterns. Therefore, the quality of the dataset has a significant impact on the performance of the models.

979-8-3315-1612-3/25 $31.00 © 2025 IEEE   3127

TABLE III
EVALUATION OF THE PREDICTION RESULTS FOR THE USED ML MODELS

Metrics	On-board Charging				Off-board Charging			
	LR	RF	ANN	cGAN	LR	RF	ANN	cGAN
MAE	0.974	0.464	0.483	0.773	2.577	0.482	0.571	0.613
MSE	3.832	0.685	0.623	2.029	27.243	1.852	1.521	2.269
R^2	0.342	0.882	0.894	0.652	0.320	0.954	0.967	0.944

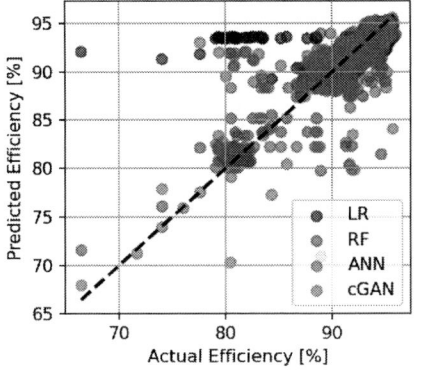 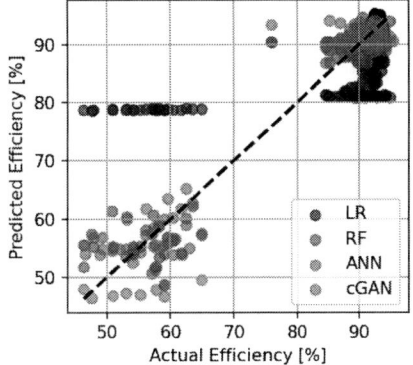

Fig. 6. Prediction results of the ML models in on-board (left) and off-board (right) charging systems

V. CONCLUSIONS

Incorporating charger efficiency and battery conditions into EV charging management systems can significantly enhance scheduling by reducing total energy requirements. This paper proposes a ML-based approach to predict the efficiency of on-board and off-board EV charging systems. The ML techniques were trained and validated using charging datasets collected from a real EV model. Performance metrics, including MAE, MSE, and R^2, were used to evaluate the models. The results indicate that RF and ANN outperformed LR and cGAN in both on-board and off-board charging scenarios.

Future work will focus on systematically collecting datasets across a wide range of conditions to improve data quality and the ML training process. These ML models can then be used as function approximators to define value functions within Q-learning frameworks, enabling integration of efficiency considerations into EV charging schedule optimization.

REFERENCES

[1] J. Liu et al., "Optimal EV Charging Scheduling by Considering the Limited Number of Chargers," in IEEE Transactions on Transportation Electrification, vol. 7, no. 3, pp. 1112-1122, Sept. 2021.

[2] C. B. Saner, A. Trivedi and D. Srinivasan, "A Cooperative Hierarchical Multi-Agent System for EV Charging Scheduling in Presence of Multiple Charging Stations," in IEEE Transactions on Smart Grid, vol. 13, no. 3, pp. 2218-2233, May 2022.

[3] M. Tan, Y. Ren, R. Pan, L. Wang and J. Chen, "Fair and Efficient Electric Vehicle Charging Scheduling Optimization Considering the Maximum Individual Waiting Time and Operating Cost," in IEEE Transactions on Vehicular Technology, vol. 72, no. 8, pp. 9808-9820, Aug. 2023.

[4] J. W. Kolar et al., "Extreme efficiency power electronics," 7th International Conference on Integrated Power Electronics Systems (CIPS), Nuremberg, Germany, 2012, pp. 1-22.

[5] Y. Jang, M. M. Jovanovic and D. L. Dillman, "Light-Load Efficiency Optimization Method," Twenty-Fourth Annual IEEE Applied Power Electronics Conference and Exposition, Washington, DC, USA, 2009, pp. 1138-1144.

[6] C. Crozier, M. Deakin, T. Morstyn and M. McCulloch, "Incorporating Charger Efficiency into Electric Vehicle Charging Optimization," IEEE PES Innovative Smart Grid Technologies Europe (ISGT-Europe), Bucharest, Romania, 2019, pp. 1-5.

[7] Y. Xu, "Optimal Distributed Charging Rate Control of Plug-In Electric Vehicles for Demand Management," in IEEE Transactions on Power Systems, vol. 30, no. 3, pp. 1536-1545, May 2015.

[8] X. Shu et al., "Ensemble Learning and Voltage Reconstruction Based State of Health Estimation for Lithium-Ion Batteries with Twenty Random Samplings," in IEEE Transactions on Power Electronics, vol. 38, no. 4, pp. 5538-5548, April 2023.

[9] J. Wu, H. Su, J. Meng and M. Lin, "State of Health Estimation for Lithium-Ion Battery via Recursive Feature Elimination on Partial Charging Curves," in IEEE Journal of Emerging and Selected Topics in Power Electronics, vol. 11, no. 1, pp. 131-142, Feb. 2023.

[10] M. A. Hannan et al., "SOC Estimation of Li-ion Batteries with Learning Rate-Optimized Deep Fully Convolutional Network," in IEEE Transactions on Power Electronics, vol. 36, no. 7, pp. 7349-7353, July 2021.

[11] D. Li et al., "Battery Thermal Runaway Fault Prognosis in Electric Vehicles Based on Abnormal Heat Generation and Deep Learning Algorithms," in IEEE Transactions on Power Electronics, vol. 37, no. 7, pp. 8513-8525, July 2022.

[12] Y. Tang et al., "Reinforcement Learning Based Efficiency Optimization Scheme for the DAB DC–DC Converter with Triple-Phase-Shift Modulation," in IEEE Transactions on Industrial Electronics, vol. 68, no. 8, pp. 7350-7361, Aug. 2021.

[13] K. Aggarwal, M. Kirchmeyer, P. Yadav, S. Keerthi and P. Gallinari, "Benchmarking Regression Methods: A comparison with CGAN," arXiv preprint arXiv:1905.12868, 2020. [Online]. Available: https://arxiv.org/abs/1905.12868

[14] EV Database, "Tesla Model Y" [Online]. Available: https://ev-database.org/car/1743/Tesla-Model-Y.

[15] ABB, "Terra 53 multi-standard DC charging station" [Online]. Available: https://library.e.abb.com/public/f704563d775748d0a22a7ff015931bd5/ABB_EVI_ProductLeaflet_Terra53_nd_web.pdf.

[16] M. Yasko, J. Driesen and W. Martinez, "Efficiency measurement and maximization for EV charging technologies,"IEEE Transportation Electrification Conference and Expo (ITEC), Chicago, IL, USA, 2024, pp. 1-5.

[17] F. Tian, M. Yasko, W. Martinez and J. Driesen EV charging efficiency prediction [Online]. Available: https://github.com/FanghaoT/EV_charging_efficiency_prediction_KUL.

High-Power and High-Speed Multi-Channel VCSEL Arrays with GaN Driver for Automotive LiDAR

Yifu Liu
JFS Laboratory
Wuhan, China
liuyifu@jfslab.com.cn

Sichao Li
JFS Laboratory
Wuhan, China
lisichao@jfslab.com.cn

Junlei He
JFS Laboratory
Wuhan, China
hejunlei@jfslab.com.cn

Changyu Hu
JFS Laboratory
Wuhan, China
huchangyu@jfslab.com.cn

Bill He
Vertilite Co., Ltd.
Shanghai, China
weiyi.he@vertilite.com

Karthik Krishnamurthy
JFS Laboratory
Wuhan, China
karthik@jfslab.com.cn

Andy Shen
JFS Laboratory
Wuhan, China
andyshen@jfslab.com.cn

Abstract—As the automotive industry races towards fully autonomous vehicles, diversity in image sensors plays a crucial role in enabling safety in autonomous driving. Apart from radar and camera, time-of-flight (ToF) LiDAR has been well developed as panacea in autonomous driving as it can provide high-resolution mapping of the vehicle's environment, long range detection and high frame real-time 3D representation (point cloud). High-resolution and long-range LiDAR requires high speed and high current driver for high power VCSEL and GaN HEMT turns out to be the perfect solution [8]-[15]. Compared with discrete single channel VCSEL, monolithic multi-channels VCSEL array is becoming increasingly important as it can provide wider field of view (FoV), high integration and excellent integrity. In this paper, high power density and high integration 8-channel VCSEL driven by GaN HEMTs are proposed. The GaN HEMT with a novel two-stage driving topology exhibits potential application in on-chip integration. Detailed modeling and circuits built of the 8 channel VCSEL are described and Ltspice simulation results are also demonstrated.

Keywords—LiDAR, Multi-channel VCSEL Array, GaN

I. INTRODUCTION

Nowadays, Radar and cameras have been widely adopted in autonomous driving [1]-[5]. As the current autonomous driving are becoming more and more intelligent, high resolution, long range and high frame sensors are the darling of the new era. However, there are still obvious shortages for Radar and cameras: the camera is hard to distinguish objects in poor vision environment (tunnel, heavy rain, foggy day, night…) and the Radar can't handle high moving speed situation and relatively long-range objects. In contrast, despite the fact that LiDAR sensor is relatively new in automotive applications, it is becoming widely adopted and assumed to be the next step towards full driverless capabilities since it is able to provide high resolution and real-time 3D vivid point cloud of surroundings under low visibility environment with long and wide detecting range [6][7]. That's to say, LiDAR perfectly answers the sensor requirements of intelligent autonomous driving.

The adoption of VCSEL in LiDAR arises from its superior properties in low power consumption, and circular light beam output [8]-[11]. Compared to single channel VCSEL, multiple lasers are connected in parallel to achieve much better FoV and high-long distance resolution. For common cathode type, the metal of VCSEL is connected to ground for better thermal dissipation and thus a high-side driver is required. Considering the high speed and short pulse power stage requirement, GaN FET with high side driver turns out to be the solution for VCSEL power stage in LiDAR as it has low Qg, low channel on resistance and good pulse-current capability.

The proposed novel GaN FET based 8-channel VCSEL driver for LiDAR is presented in First Section. The detailed modeling and related Ltspice simulation are demonstrated in Second Section. System layout design is described in Third Section.

II. SYSTEM TOPOLOGY

The proposed 8-channel GaN based VCSEL driver is shown in Fig. 1. Each single channel consists of a high-side GaN Power FET and a VCSEL Diode. These channels share the same DC bus and connected to the anode of the VCSEL array relatively. Therefore, every single GaN FET will be switching on a floating ground and relatively unique gate driving for each GaN FET is a must. The ground of the bus is exactly the common cathode of the VCSEL array, which provides good thermal dissipation. The cross section of co-package single channel VCSEL and GaN HEMT is shown in Fig. 2. The anode of VCSEL is wire-bonded to source pad of GaN FET and the cathode is connected to ground to help with thermals.

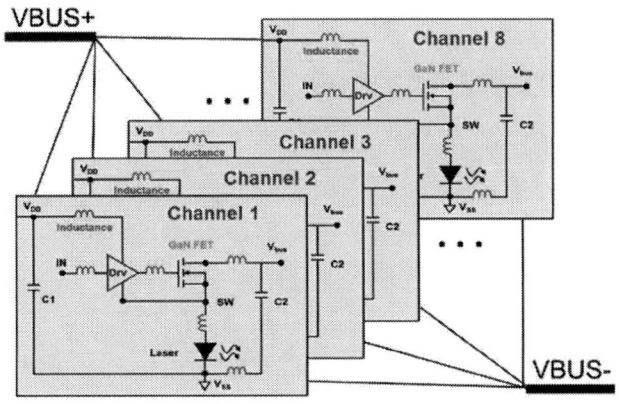

Fig. 1. Topology of proposed 8-channel VCSEL driver for LiDAR using GaN FET.

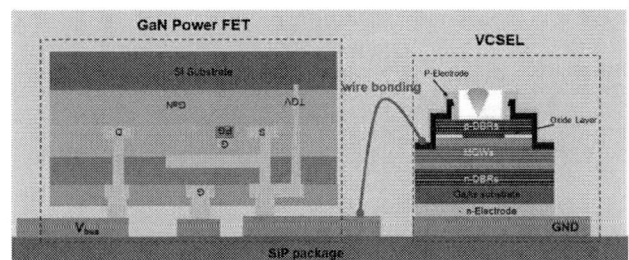

Fig. 2. Single channel co-package schematic.

To cooperate with this common cathode VCSEL driver structure, a novel two stage gate driving strategy is applied and demonstrated in Fig. 3. The first stage is bootstrap based half bridge. It is aimed to: (1) generate a level-shifted gate signal for high side FET; (2) provide a stable power supply on a floating ground. The high-side portion acts as the power supple of second stage. The low-side portion of the stage is used to quickly switch another FET to replenish the bootstrap cap. The second stage is simply a buffer stage. It receives the level-shifted gate signal from first stage and generates short and high current (5A – 7A) gate pulse for GaN FET. To ensure clean and short gate and power loop for LiDAR driver circuits, high-speed components with low propagation delay and small size are indispensable.

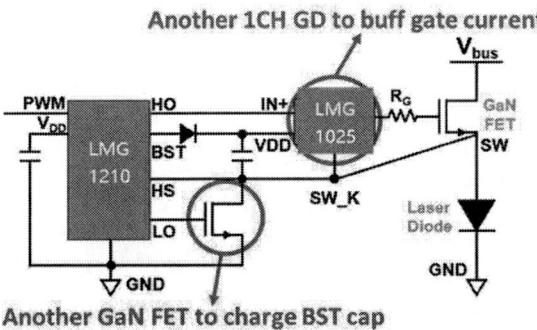

Fig. 3. 2-stage gate driving topology for high side GaN

As Fig. 3 shows, a COTS GaN FET is selected for high side power switch as it is capable of building high-speed, high-power density and high-efficiency power stage. Besides, 100V also leaves enough margin for safety concern due to *Vds* ringing. To cooperate with it, the second stage buff driver belongs to LMG1025 because of its independent 7A pull-up and 5A pull-down gate current capability. The LMG1210 plays the role of first stage gate driver to do the level shifting, pass the gate signal and power the LMG1025.

III. SIMULATION

Generally speaking, a VCSEL DC-IV characteristics modeling can be simplified into RLC model as shown in Fig. 4. To be as accurate as possible, the laser diode model is derived from the curve fitting using measurement data. Also, the bonding-wire inductance, series access resistance R_s, pad capacitance C_p, electro-plane capacitance C_m, junction resistance R_j are well estimated and extracted.

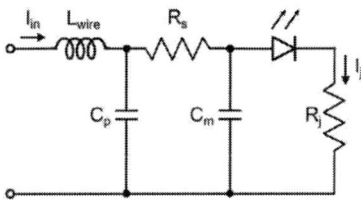

Fig. 4. VCSEL RLC model.

Further, the GaN FET is selected to build power stage, TI LMG1025 is chosen to be the buffer IC while LMG1210 to be the bootstrap half bridge IC. There models are also integrated into the Ltspice simulation and the results are shown in Fig. 5.

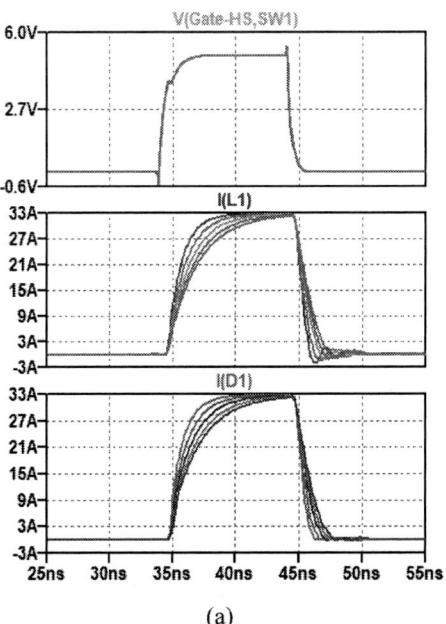

(a)

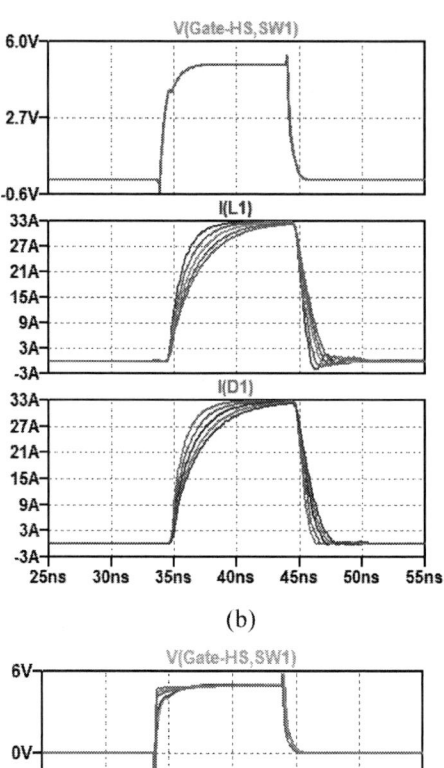

(b)

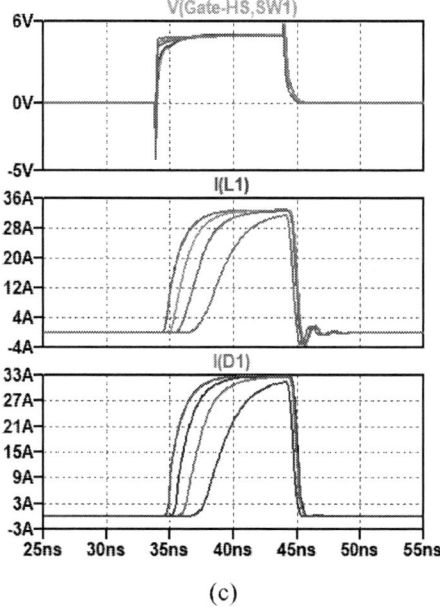

(c)

Fig. 5. VCSEL Ltspice simulation results: (a)Bonding-wire parasitic inductance sweeping at 0, 0.25nH, 0.5 nH, 0.75 nH and 1nH; (b) Power loop parasitic inductance sweeping at 0, 0.5nH, 1 nH, 1.5 nH and 2nH; (c) Gate resistor sweeping at 0, 0.5Ω, 1Ω, 2 Ω and 5Ω.

The current pulse at the output is 4.8ns/29.1A for a PWM input of 10ns. The VCSEL current shows 2.8ns rise time and 1ns fall time. Considering the co-package structure of the VCSEL channel, the inductance from wire-bonding need to be studied. By sweeping the bonding wire parasitic inductance as shown in Fig. 5(a), it is found that ringing exists in the extrinsic VCSEL current after the falling edge while no sign of ringing in intrinsic diode current. Additional inductance at the anode slows down the rise time and starts impacting the peak current for short pulses. Further, after adding the sweeping of power loop

parasitic in Fig. 5(b), the ringing becomes worse in both extrinsic VCSEL current and intrinsic diode current. That's to say, the bonding wire and power loop need to be designed as short and clean as possible to minimize the parasitic and the ringing issue. The sweeping of HS FET gate resistor is shown in Fig. 5(c), larger resistor makes the rise/fall time worse, but minimizes the oscillations/ringing on the extrinsic VCSEL current, caused by the LCR resonance.

IV. HARDWARE DEVELOPMENT

Well-designed system layout is very important to achieve high quality performance. 8 channels system layout after iterative revisions are presented in Fig. 6. In the layout design, components placement follows these rules: (1) Driver IC + GaN FET are placed as close as possible to minimize gate loop; (2) Carefully place the dc-link caps close to GaN FET + VCSEL in order to minimize the power loop; (3) Power traces and signal traces are separated to avoid any overlap between these two regions; (4) Mirror layout for each channel are essential to ensure the signal synchronization for multi-channel system.

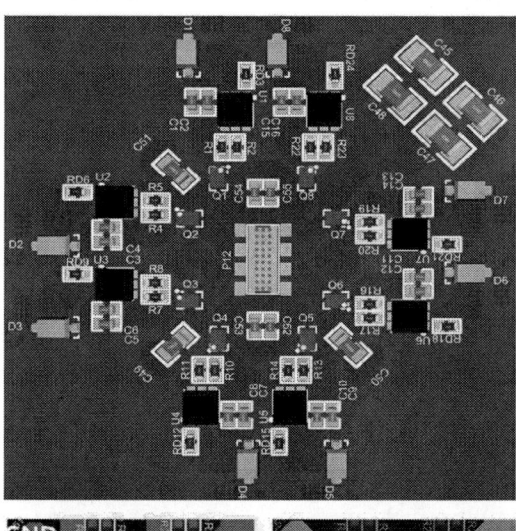

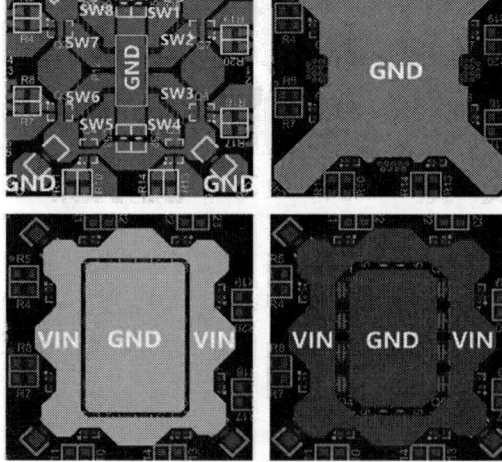

Fig. 6. 8 channels VCSEL driving system layout preview.

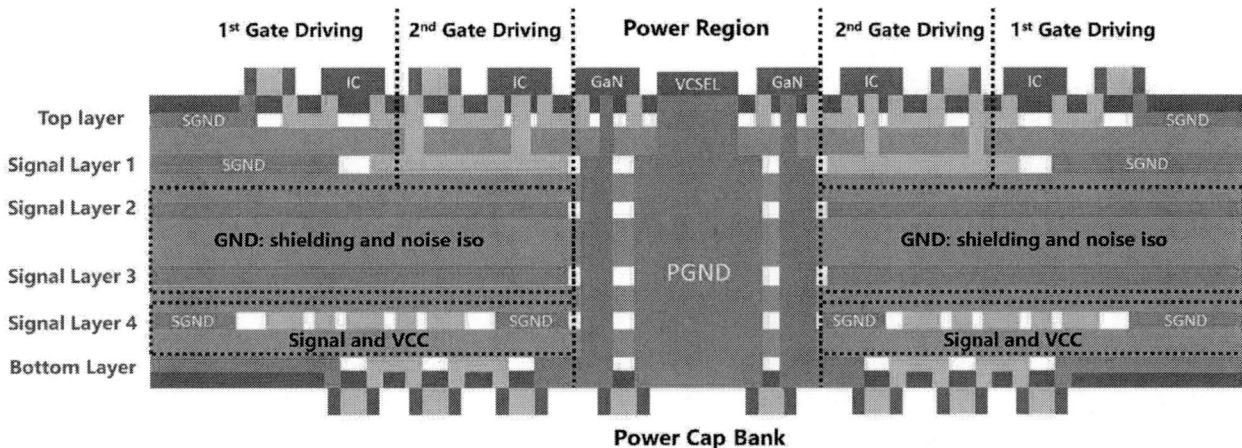

Fig.7. 8 channels VCSEL driving system layout preview.

A cross-section view of the designed 6-layers PCB structure is shown in Fig. 7. As the parasitic inductance deeply affects the system performance, it is very important to minimize it in both power and gate loop. To achieve it, following effective guidelines are to follow: (1) Pour ground area to either overlap or cling the power area to minimize noise; (2) Design the power and gate loop as short as possible and avoid direct overlap between them; (3) Place as many as possible Vias (fill with copper) for current path; (4) Set the Multi-layer bypass capacitors rank for noise filtering.

V. CONCLUSION

A novel high power and high speed 8 channel VCSEL driven by GaN HEMTs are proposed in this paper. Starting from one channel VCSEL, GaN power stage with two-stage gate driving strategy are elaborated. Detailed modeling of VCSEL is demonstrated. With proper selection of driver IC, an accurate Ltspice simulation of total eight channel VCSEL are conducted. Comprehensive simulation on the effect of parasitic on power loop and gate loop will be studied in full paper section. Further, a well-designed layout is presented, which focuses on parasitic minimization. The proposed 8 channel VCSEL driver will be first deployed on PCB and a well-designed PCB layout will be described in details. After that, the pulsing experiment will be conducted and related waveforms could be presented.

REFERENCES

[1] AD. K. Barton, "Radar system analysis and modeling," IEEE Aerosp. Electron. Syst. Mag., vol. 20, no. 4, pp. 23–25, Apr. 2005.

[2] J. Guerrero-Ibáñez, S. Zeadally, and J. Contreras-Castillo, "Sensor technologies for intelligent transportation systems," Sensors, vol. 18, no. 4, p. 1212, Apr. 2018

[3] E. Marti, M. A. de Miguel, F. Garcia, and J. Perez, "A review of sensor technologies for perception in automated driving," IEEE Intell. Transp. Syst. Mag., vol. 11, no. 4, pp. 94–108, Sep. 2019.

[4] B. S. Jahromi, T. Tulabandhula, and S. Cetin, "Real-time hybrid multi-sensor fusion framework for perception in autonomous vehicles," Sensors, vol. 19, no. 20, p. 4357, Oct. 2019.

[5] A. S. Mohammed, A. Amamou, F. K. Ayevide, S. Kelouwani, K. Agbossou, and N. Zioui, "The perception system of intelligent ground vehicles in all weather conditions: A systematic literature review," Sensors, vol. 20, no. 22, p. 6532, Nov. 2020.

[6] B. Behroozpour, P. A. M. Sandborn, M. C. Wu, and B. E. Boser, "LiDAR system architectures and circuits," IEEE Commun. Mag., vol. 55, no. 10, pp. 135–142, Oct. 2017.

[7] J. Lambert et al., "Performance analysis of 10 models of 3D LiDARs for automated driving," IEEE Access, vol. 8, pp. 131699–131722, 2020.

[8] S. T. S. Holmström, U. Baran, and H. Urey, "MEMS laser scanners: A review," J. Microelectromech. Syst., vol. 23, no. 2, pp. 259–275, Apr. 2014.

[9] H. W. Yoo et al., "MEMS-based LiDAR for autonomous driving," e i Elektrotechnik und Informationstechnik, vol. 135, no. 6, pp. 408–415, Oct. 2018.

[10] V. Milanovi´c, A. Kasturi, H. J. Kim, and F. Hu, "Iterative learning control algorithm for greatly increased bandwidth and linearity of MEMS mirrors in LiDAR and related imaging applications," in Proc. 17th MOEMS Miniaturized Syst., vol. 10545, W. Piyawattanametha, Y.-H. Park, and H. Zappe, Eds. Bellingham, WA, USA: SPIE, 2018, pp. 242–256

[11] B. L. Stann, J. F. Dammann, and M. M. Giza, "Progress on MEMSscanned ladar," in Proc. 21st Laser Radar Technol. Appl., vol. 9832, M. D. Turner and G. W. Kamerman, Eds. Bellingham, WA, USA: SPIE, 2016, pp. 197–205

[12] D. Wang, C. Watkins, and H. Xie, "MEMS mirrors for LiDAR: A review," Micromachines, vol. 11, no. 5, p. 456, Apr. 2020.

[13] M. J. Heck, "Highly integrated optical phased arrays: Photonic integrated circuits for optical beam shaping and beam steering," Nanophotonics, vol. 6, no. 1, p. 152, Jan. 2017.

[14] R. Fatemi, B. Abiri, A. Khachaturian, and A. Hajimiri, "High sensitivity active flat optics optical phased array receiver with a two-dimensional aperture," Opt. Exp., vol. 26, no. 23, pp. 29983–29999, Nov. 2018

[15] C. V. Poulton et al., "Long-range LiDAR and free-space data communication with high-performance optical phased arrays," IEEE J. Sel. Topics Quantum Electron., vol. 25, no. 5, pp. 1–8, Sep. 2019

Double Pulse Test Platform for Hybrid SiC-IGBT Switch Characterization and Optimal Gate Control Strategy for EV Traction Inverters

Rosario Attanasio
APMS Analog&Power
STMicroelectronics
Schaumburg (IL), USA
rosario.attanasio@st.com

Harsha Ademane
APMS Analog&Power
STMicroelectronics
Schaumburg (IL), USA
harsha.ademane@st.com

Ryan Satterlee
AMS Analog&Power
STMicroelectronics
Schaumburg (IL), USA
ryan.satterlee@st.com

Gianni Vitale
APMS Analog&Power
STMicroelectronics
Schaumburg (IL), USA
gianni.vitale@st.com

Abstract— **This paper evaluates the performance of a hybrid switch for Electric Vehicle (EV) traction inverters using a specialized Double Pulse Test Platform (DPTP) and automated characterization procedure. The Hybrid Switch is composed of Silicon (Si) Insulated Gate Bipolar Transistors (IGBTs) and Silicon Carbide (SiC) Power MOSFETs connected in parallel.**

The SiC/IGBT solution improves low-current performance due to its unique ON state characteristics and superior switching performance, which has garnered increasing interest from car manufacturers. By selecting the appropriate number of devices or silicon dice in parallel, the hybrid switch can achieve lower switching losses compared to an IGBT-based solution. However, the optimal selection of SiC and IGBT properties and gate control strategy is crucial. The hybrid switch requires precise experimental testing of the delay time between the turn-on and turn-off of SiC and IGBTs, which is time-consuming if done manually.

This paper introduces an automated method using a DPTP that significantly reduces the characterization time. Three inverter switch configurations are compared: a discrete hybrid switch with 4 parallel devices (1 SiC and 3 IGBTs) in a T-PAK package with a similar current rating, a full SiC solution, and a full IGBT solution with four devices in parallel in the same package. Experimental results demonstrate the benefits of the Si/SiC hybrid switch for EVs, and an optimal control strategy is proposed based on the selected hybrid switch configuration.

Keywords— *Hybrid SiC/IGBT Switch, Traction Inverter, SiC MOSFETs, IGBTs, Electric Vehicle.*

I. INTRODUCTION

Insulated Gate Bipolar Transistors and Silicon Carbide MOSFETs are key for modern EV traction systems. IGBTs are robust and cost-effective but have higher switching losses and slower speeds, reducing efficiency at high frequencies and low loads. SiC based inverters offer lower switching losses, higher efficiency, and faster switching but are costlier and often involve newer manufacturing processes. A hybrid SiC/IGBT solution combines SiC efficiency and speed with IGBTs robustness and cost-effectiveness, optimizing performance at low/partial loads

and reducing system cost while maintaining efficiency. The concept of using the parallel connection of MOSFET and IGBT switches was suggested in the past to improve inverter efficiency. The main idea was to exploit the conduction characteristic of the MOSFET at light load and the properties of the IGBT at medium/high load. This technique however could not be successfully implemented in hard switched PWM inverters because of the reverse recovery associated to the internal body diode of the MOSFET, which adds extra loss and offsets the gains of the lower conduction loss. For this reason, the hybrid concept was mainly used in soft switched inverters where different auxiliary circuits must be employed to achieve Zero Voltage Switching (ZVS) [1]. While these solutions demonstrated ultra-high efficiency of 99% with adaptive timing soft switching operation, their adoption has been limited by the higher number of components required to achieve ZVS.

The adoption and market availability of SiC MOSFETs compensate for the shortcomings of past hybrid solutions, mainly because of the almost negligible reverse recovery of the SiC body diode. Other advantages such as shorter switching time, smaller voltage drop, higher operating temperature, higher breakdown voltages and lower voltage distortion [2] result in operational advantages of the SiC solution versus Si-based ones, producing higher efficiency and power density in traction drives [3]. However, the cost of SiC MOSFETs has historically been higher than that of IGBTs, primarily due to the more complex and expensive manufacturing processes involved in producing SiC devices.

The current cost of SiC MOSFETs is 2-3 times higher than that of IGBTs of similar voltage and current rating and is projected to decrease by approximately 30-40% over the next five years as production volumes increase and manufacturing processes become more efficient [4]. This cost reduction could be driven by improvements in wafer quality, increased wafer sizes (transitioning from 4-inch to 6-inch and 8-inch wafers), and economies of scale as more manufacturers adopt SiC technology. Despite the promising advancements and potential cost reductions in SiC technology, several factors could lead to a scenario where the supply of SiC devices remains tight, and their prices do not become as competitive with IGBTs as

---

979-8-3315-1612-3/25 $31.00 © 2025 IEEE

anticipated. For example, quality and yield of SiC substrates from new suppliers, raw material shortages, demand surges driven by multiple applications benefitting from this technology, to name a few.

If the cost of SiC devices remains high due to supply constraints and manufacturing challenges, the adoption of hybrid SiC/IGBT switches presents a compelling solution. Hybrid switches leverage the advantages of both SiC MOSFETs and IGBTs. This approach allows to achieve a balance between performance and cost. The SiC MOSFETs, with their superior switching speeds and higher efficiency, handle low/medium load and high frequency operations and reduce overall switching losses. Meanwhile, the IGBTs manage the bulk of the current during low-frequency operations and sudden peak power demands. This combination not only enhances the overall efficiency and thermal performance of the power electronics system but also mitigates the high costs associated with using SiC devices exclusively. By optimizing the use of SiC components, hybrid switches can deliver significant performance improvements and energy savings, making them an attractive option for applications such as EVs and industrial motor drives.

The use of a hybrid switch requires careful selection of semiconductor electrical characteristics and precise gate control strategies, adding design complexity. The control strategies proposed in literature involve turning on the SiC device just before the IGBT and off shortly after. Optimal turn-on delay results in simultaneous activation, reducing turn-on losses by 73% compared to IGBT and 52% compared to SiC MOSFET. Optimal turn-off sequences the IGBT off before the SiC MOSFET, reducing turn-off losses by 61.4% [5][6]. The referenced analyses use a single IGBT and SiC device in parallel with a DPTP. Further studies analyze the impact of turn-off delay and SiC MOSFET dv/dt on IGBT current spike, proposing a parameter optimization method using a DPTP [7].

Turning on the SiC device before and off after the IGBT allows to achieve ZVS for the IGBTs but generates reliability concerns due the potential impact on SiC junction temperature since this device must carry the total system current for a short time.

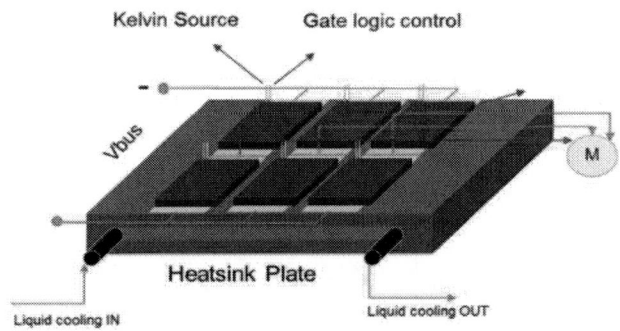

Fig. 1: Example of T-PAK based traction inverter

A method to control the junction temperature difference was proposed in [8] with the purpose of increasing reliability. A variable frequency current dependent control strategy is instead proposed in [9] and applied to a Full Bridge single phase inverter aiming at reducing power loss and improve reliability of previously proposed current dependent control strategies mainly based on information provided by the SiC MOSFET Safe Operating Area (SOA).

In summary, all the available research studies the control optimization of hybrid switches either based on theoretical considerations and/or experimental prototypes that use a small-scale power prototype or just two power devices, in standard through hole package, connected in parallel. Therefore, hybrid switch designs for automotive, aerospace, or renewable energy systems, where multiple devices or dice are connected in parallel, needs further research. For example, paralleling multiple IGBTs with SiC MOSFETs affects turn-on switching losses and high IGBT output capacitance can worsen performance if the SiC turns on first with non-optimal values of turn-on and turn-off gate resistors. Another observation is related to the reverse conduction mode of a hybrid switch where all the current flows across the Si diode packaged with the IGBT and having a significant lower forward voltage and reverse recovery charge compared to that of a SiC diode.

This paper introduces a DPTP for multiple parallel devices in the novel T-PAK package, an innovative solution for traction inverters [9]. An example of T-PAK based inverter is shown in Fig.1.

The T-PAK based inverter concept integrates power semiconductor devices into a compact, high-performance package, optimizing both thermal management and electrical performance. Advanced cooling solutions, such as liquid cooling or integrated heat sinks, are employed to dissipate heat efficiently and maintain optimal operating temperatures [10].

Additionally, sintering technology is used to attach the semiconductor dies to the substrate and the package to the heatsink baseplate, providing superior thermal and electrical conductivity compared to traditional soldering methods. This enhances the thermal performance and reliability of the inverter making it a highly attractive solution for modern power electronics applications.

The proposed DPTP is based on a bus bar concept that does not require sintering technology and allows for a simple assembly to use for lab characterization of any hybrid switch configuration. This platform can be used to analyze and characterize hybrid switches and extract optimal turn-on and turn-off delay times, by automatically measuring switching losses and parameters like dv/dt, di/dt, turn-off overvoltage, and IGBT turn-off current spike. This allows rapid identification of an optimized hybrid switch design once the optimal gate resistor values have been identified. The turn-on and turn-off switching energies (E_{on}, E_{off}) characterization results of a hybrid switch consisting of 3 IGBTs and 1 SiC FET selected to operate with a 400V bus and up to 1200A peak current are shown and compared to the results of similar solutions based either on SiC or IGBT technology. In addition the paper proposes a thermal model to estimate the junction temperature increase of the SiC device by means of a specific current dependent control strategy that overcomes the limitations of a simple SOA information based approach.

II. Double Pulse Test Platform for Hybrid Switches with T-Pak package

A. Target System Specifications

The proposed DPTP targets a high-efficiency EV traction inverter designed to deliver a peak power output of 480kW, ensuring high performance and efficiency for electric vehicle applications. The system operates with a DC bus voltage of 400Vdc and a peak current of 1200A. To achieve this power level, the inverter employs a hybrid configuration of power devices, consisting of one SiC MOSFET and three IGBTs connected in parallel. The specific devices used are the SCTHS250N65G3 SiC MOSFET and the STGST200G65DFAG IGBTs, all housed in the advanced ST-PAK package. This combination leverages the fast switching and low conduction losses of the SiC MOSFET with the cost-effective and robust performance of the IGBTs. The selected current ratio is 1 to 3, meaning that the SiC is selected to carry ¼ of the total current while the 3 IGBTs are selected to carry ¾ of the total current. The key system specifications are summarized in Table I.

TABLE I: TARGET SYSTEM SPECIFICATIONS

Parameter	Value	Units
Peak Power Output	480	kW
DC Bus Nominal Voltage	400	V
Peak Output Current	1200	A
Switching Frequency	10	kHz
Number of Devices in parallel per switch	4	
Current ratio	1:3	
IGBT V_{CEsat} (typ at Ic=200A, 25C)	1.52	V
Peak collector Current (Icp)	500	A
SiC FET R_{DSon} (Typ. @25C)	6.7	mΩ
SiC Pulsed Drain Current (I_{DM})	717	A

The 480kW peak power capability allows rapid acceleration and performance, making it ideal for high-end EV applications. The peak collector current and drain pulsed current are shown in the SOA of the two devices and limited by their maximum junction temperature.

B. Hybrid Switch Forward and Reverse Characteristics

The schematic of the hybrid switch is shown in Fig. 2. Each device is rated to carry ¼ of the total peak current. Such a choice allows to better exploit the use of the SiC FET and maximize the efficiency when considering standard driving cycles such as the Worldwide Light-Duty Vehicle Test Profile (WLTP).

The SiC FET conducts all the load current up a threshold that depends on the selected value of R_{DS_ON} and on the collector-emitter threshold voltage V_{TO} of the IGBT. If the voltage drop caused by the SiC FET drain current is lower than V_{TO} then the

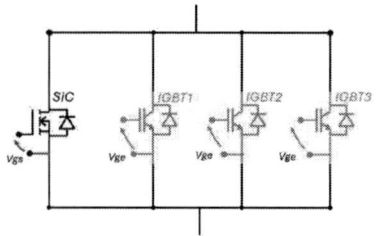

Fig. 2: Hybrid Switch configuration with 1 SiC FET and 3 IGBTs.

three IGBTs do not conduct any current, otherwise the load current is shared across the four devices.

The threshold current is defined as:

$$I_{TH} = \frac{V_{TO}}{R_{DS\_on}} \qquad (1)$$

Table II reports the on-resistance value and threshold voltage at 25 °C and 150 °C for the selected devices. The threshold current can vary significantly as the junction temperature of the power devices changes. Since the R_{DS_ON} of the SiC MOSFET is almost unaffected by temperature up to about 100C the major contribution to the I_{TH} variation comes from the V_{TO} variation with temperature.

TABLE II: THRESHOLD CURRENT VARIATION WITH TEMPERATURE.

Parameter	Typ. value @25°C	Typ. value at 150°C	Units
R_{DS_ON}	6.7	8.3	mΩ
V_{TO}	0.75	0.6	V
I_{TH}	111.9	72.2	A

The forward conduction characteristic of the hybrid switch is shown in Fig. 3 for a junction temperature of 150 °C. Two areas are identified: the first for a load current lower than the threshold current where the SiC FET only conducts, and the second for a load current higher than the threshold current where the SiC and the three IGBTs share the load current.

The current sharing is shown in Fig.4 where the SiC FET and single IGBT currents are plotted against the total load current for a junction temperature of 150 °C.

With the selected devices it is possible to achieve almost perfect current sharing at the maximum current of 1200A. This

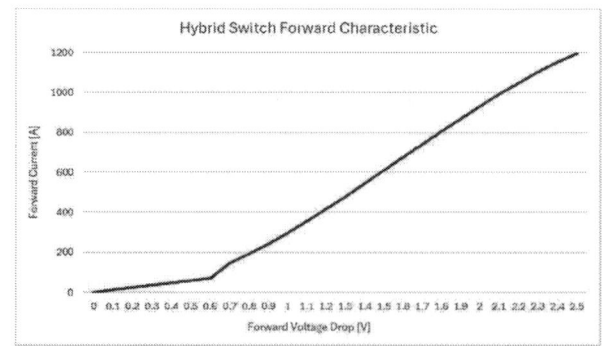

Fig.3: Hybrid Switch Forward Characteristics.

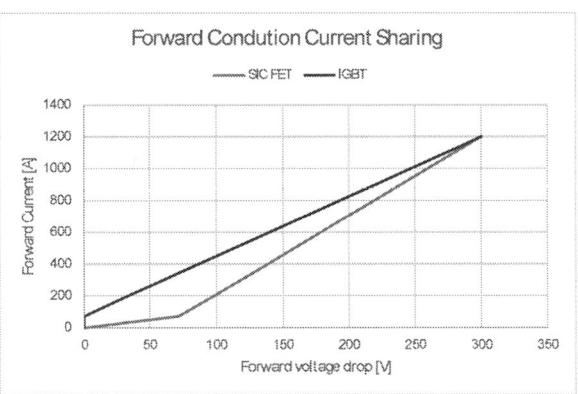

Fig.4: forward conduction current sharing of SiC and IGBTs.

load current is present mainly during maximum accelerations lasting for a limited time. Between 150A and 400A the SiC transistor conducts about 40% of the total load current. Between 400A and 1000A the SiC transistor conducts about 30% of the total load current and about 25% when the load current is above 1000A.

The current across the SiC and IGBTs can be calculated according to (3) and (4).

$$I_{SiC} = \frac{I_{load} \cdot \frac{R_{CE}}{3} + V_{TO}}{R_{DS\_on} + \frac{R_{CE}}{3}} \qquad (3)$$

$$I_{IGBT} = \frac{I_{load} \cdot R_{DS_{on}} - V_{TO}}{R_{DS\_on} + \frac{R_{CE}}{3}} \qquad (4)$$

where R_{CE} is the dynamic resistance of the IGBT, V_{TO} is the threshold voltage of the IGBT and R_{DS_ON} is the on-state resistance of the SiC transistor. Notice that these parameters are temperature dependent.

Similarly in Fig.5 and Fig.6 the reverse conduction characteristic of the hybrid switch and the current sharing during reverse conduction are shown. The SiC FET works in Synchronous Rectification (SR) and shares current with the Si diode packaged with the IGBT. If SR is not used the reverse conduction characteristic of the hybrid switch would simply be determined by the three parallel Si diodes in the IGBT packages.

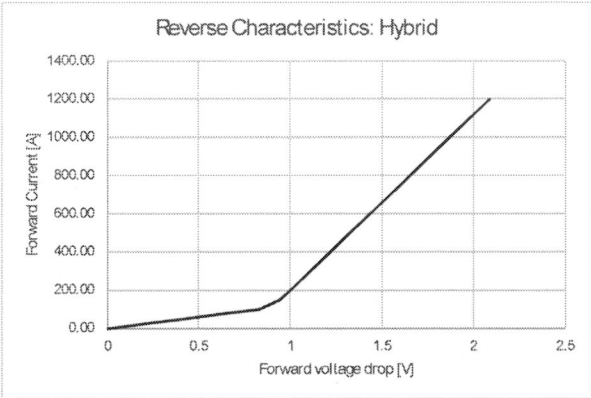

Fig.5: hybrid switch reverse conduction characteristic.

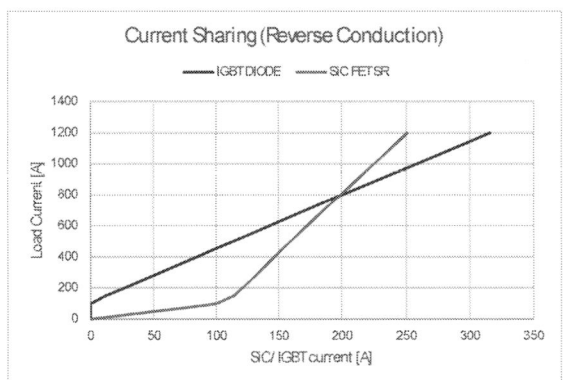

Fig.6: Reverse conduction current sharing.

C. Bus Bar Design

A DPTP is typically employed for the evaluation of dynamic characteristics of power transistors. The electrical schematic of the DPT is depicted in Fig. 7.

This circuit is the industry standard for dynamic characterization of semiconductor devices like IGBTs,

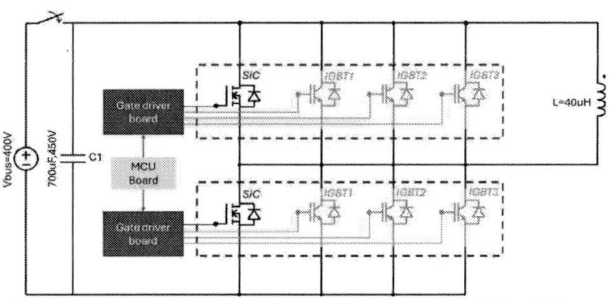

Fig.7: Schematic of the double pulse test platform for hybrid switches.

MOSFETs, and Wide Bandgap (WBG) devices such as SiC transistors. E_{on} and E_{off} energies from this test compare switching performance across devices. Hybrid switches have complex turn-on and turn-off processes due to sequential switching and additional losses. In fact, SiC and IGBTs can switch leading or lagging each other, with four control patterns shown in Fig. 8 (G1 for IGBT, G2 for SiC).

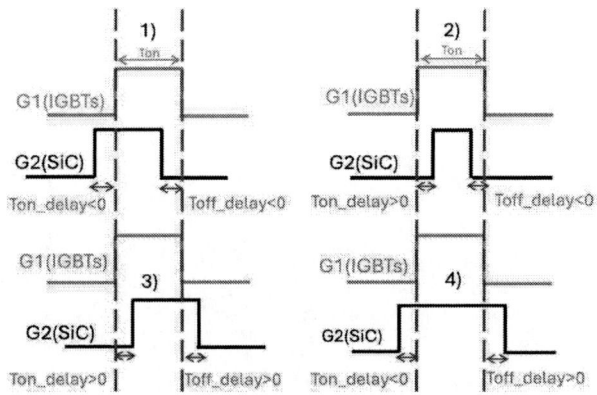

Fig.8: hybrid switch control patterns

979-8-3315-1612-3/25 $31.00 © 2025 IEEE

Delay times (t_{on_delay}, t_{off_delay}) are generated by a dedicated gate driver board controlled by a 32bit MCU producing gate driving signals delayed with tens of nanosecond resolution (e.g., 50 ns) with respect to each other. In this way it is possible to optimize the hybrid switch efficiency and traction inverter control strategy. The first technical challenge is designing a bus bar system with minimal parasitic inductance, precise T-PAK positioning, and accurate total current measurement of the hybrid switch. The bus bar design includes three main components: the top bus bar (green), insulating material (purple), and bottom bus bar (yellow), as shown in Fig. 9.

The insulator is sandwiched between conductor layers, with terminals for high voltage capacitor connection and current continuity. The bus bar length adjusts based on the number of parallel devices. The top bus bar has clearance grooves to prevent electrical contact with gate/source terminals and mounting holes for connecting the 40uH test inductor. The mid part connects high-side device sources to low-side device drains and the inductor, while the bottom part connects low-side device sources and the inductor for diode tests.

Positioning grooves on the top bus bar ensure precise T-PAK placement, and a center cut allows independent current measurement using Rogowski coils. The side links are removed after the assembly process is completed.

The parasitic inductance is estimated at 22nH, for a hybrid switch consisting of four T-PAK devices on the high-side and four on the low-side, using (5):

$$L_{par} = \frac{V_{DS_{peak}} - V_{BUS}}{{di_{OFF}}/{dt}} = \frac{507 - 400}{4.82} \approx 22nH \qquad (5)$$

where $V_{DS_{peak}}$ is the measured device overvoltage during a turn-off event with a switch current of 150A, bus voltage of 400V and 4.82A/ns. The total loop inductance includes the package inductance of the T-PAK devices and could be further reduced in an inverter implementation since the length of the current loop could be minimized by removing the central connection between high-side and low-side devices used in this test platform to measure the total hybrid switch current. The final DPTP assembly is shown in Fig. 10 where the two gate driver boards are mounted vertically to the hybrid switch.

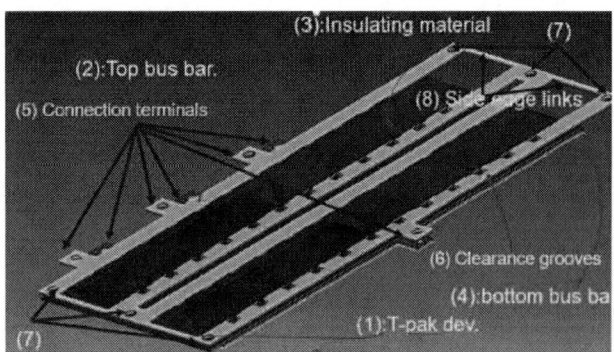

Fig.9: DPTP bus bar for T-PAK devices.

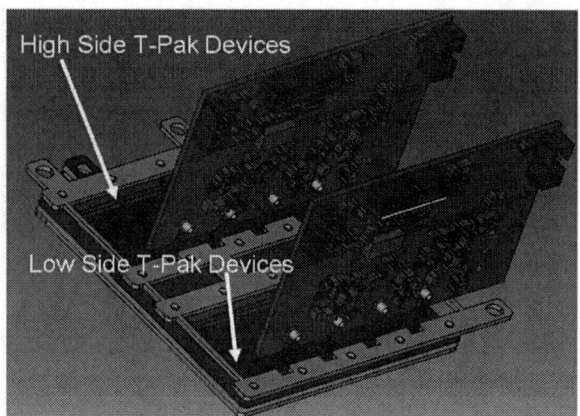

Fig. 10: DPTP bus bar assembled with gate driver boards.

III. AUTOMATED TEST PROCEDURE FOR HYBRID SWITCH CHARATERIZATION

An automated test procedure was used to efficiently test the three different inverter switch configurations. . For the hybrid switch, previous research showed characterization assuming a rather wide range of delay times between turn-on and turn-off of the SiC and IGBTs. Such a wide range is not particularly useful since not only does it add extra conduction losses for either the SiC or the three IGBTs (depending on the adopted control strategy) but can also impact reliability. For this reason the range of delay time was limited between -1.8us and +1.8us. Given that the delay time can be configured with a high-resolution granularity of few tens nanoseconds (ns) (for example 120ns), the resulting number of test points for each current level can be very high (30 test points for 120ns resolution). The automated test methodology is based on the utilization of a Comma-Separated Values (CSV) file, which is parsed using Python on a personal computer (PC). The critical parameters required to apply the correct switching pattern to the hybrid switch, such as gate turn-on order, gate turn-off order, delay on, and delay off, are transmitted to an STM32G4 MCU, which generates the PWM signals applied to the gate driver boards. Subsequently the modulation signals are applied to the power stage, and the resulting waveforms are captured by an oscilloscope. These waveforms, which include the SiC gate-source voltage, IGBT gate-emitter voltage, hybrid switch voltage, hybrid switch current, SiC FET current and IGBT current, are then downloaded to the PC for further analysis. The data acquired from the oscilloscope undergoes post-processing to determine the E_{on} and E_{off} energies for each test current as a function of the configured delay time. Additionally, the minimum turn-on energy (E_{on_min}) and minimum turn-off energy (E_{off_min}) can be computed from the stored data. If necessary, the resolution can be adjusted to refine the analysis or for further tuning of the system.

During the DPT the capacitor C1 is pre-charged at 400V and then discharged through the test inductor. As the test current increases with the duration of the first ON pulse, it is important to compensate for the discharge of C1. To achieve 400V during the first turn off and second turn on, the initial test voltage is

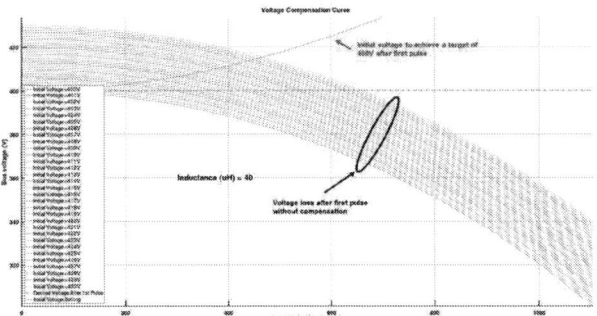

Fig.11: input voltage compensation as a function of the test current.

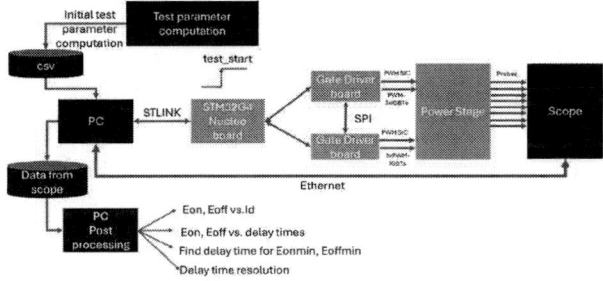

Fig.12: block diagram of the automated test procedure.

compensated as a function of the test current according to the curve shown in Fig.11.

This approach ensures precise and efficient characterization of the hybrid switch, optimizing the overall performance and reliability of the power electronic system. The overall concept is shown in the block diagram of Fig.12.

IV. EXPERIMENTAL RESULTS

The DPTP, shown in Fig.13, with software for automated test procedure was used to characterize and compare three use cases: a hybrid switch consisting of 1 SiC MOSFET and 3 IGBTs, a full SiC switch consisting of 4 SiC FETs and a full IGBT switch with 4 IGBTs. The single device specification and part numbers are reported in section II. In each case the high-side switch was matched to the low side one.

High voltage isolated probes were used to measure the DC voltage and the voltage across the low-side switch (or high-side for the diode characterization) while an optically isolated probe was used to monitor the gate of the SiC FET. The gate voltage

of the IGBTs was monitored with a standard non isolated probe. Rogowski coils were used to measure the total switch current and the single device current in the parallel configuration.

Fig. 13: DPTP for T-PAK Hardware.

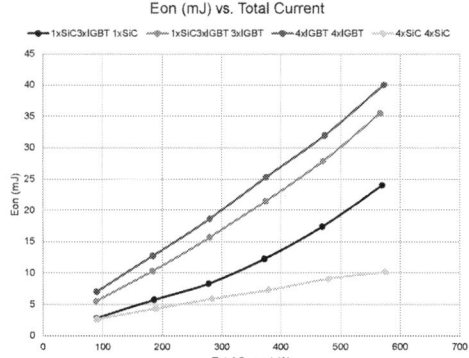

Fig. 14: hybrid switch turn-on energy loss Vs total test current

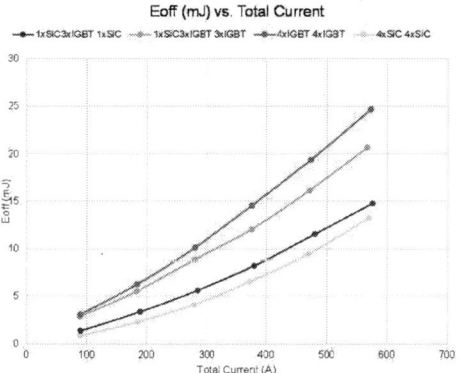

Fig. 15: hybrid switch turn-off energy Vs total test current

Fig. 14 and Fig. 15 show a comparison of E_{on} and E_{off} energies for the 3 configurations and test currents from 90A to 570A. The yellow curves are the switching energies for the full SiC configuration and the purple one is the full IGBT energy. In the middle, the light blue curve is the switching energy of the hybrid switch when the IGBTs commutation leads at turn-on and lags at turn-off the SiC one by 1.8us. The black curve is the hybrid switch energy loss when the SiC FET leads or lags the IGBTs commutation by 1.8us. There is a clear reduction in switching losses when this strategy is applied which is due to

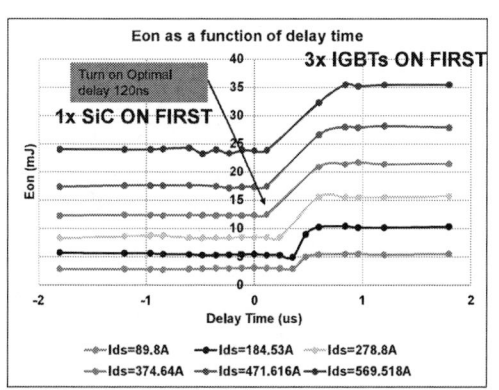

Fig. 16: E_{on} of the hybrid switch as a function of the delay time.

979-8-3315-1612-3/25 $31.00 © 2025 IEEE

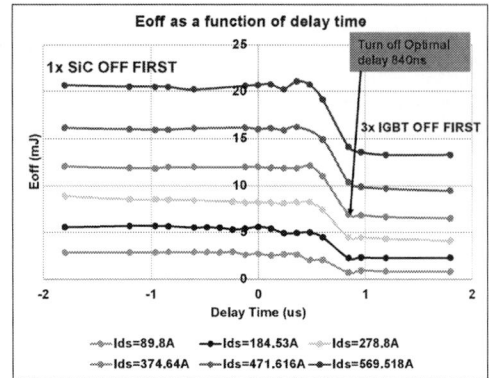

Fig. 17: E_{off} of the hybrid switch as a function of the delay *time*

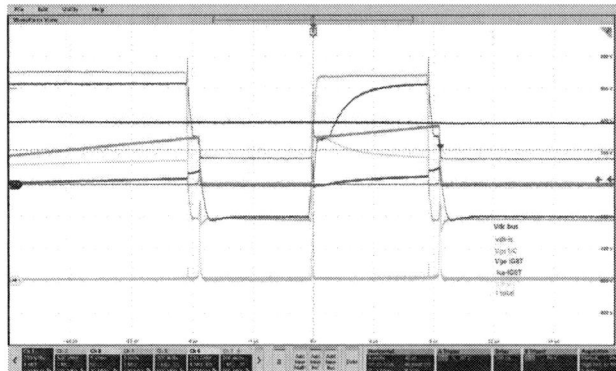

Fig. 18: turn-on waveforms with SiC PWM signal delay of 120ns.

ZVS of the IGBTs. Further improvements are possible with an optimized anticipation and delay of the SiC FET commutation.

Fig. 16 and Fig. 17 show the E_{on} and E_{off} as function of the applied delay time for different values of the test current. It is possible to notice that the minimum E_{on} is achieved when the PWM signal of the 3 IGBTs is applied 120ns before the PWM of the SiC device while the E_{off} is minimized when the SiC PWM signal is delayed by 840ns with respect to the IGBTs. Since IGBTs are much slower than SiC devices a small anticipation of 120ns results in a slightly delayed turn-on of the IGBTs with respect to the SiC FET. This is clear by looking at the details of the commutation in Fig. 18, acquired for a test current of 300A with a delay time of 120ns applied to the SiC FET. At turn-off it is necessary to wait for the IGBT current tail effect to be completed before applying the PWM signal to the SiC FET. It is worth noting that even though the effect of the optimized delay was evaluated up to 600A, the applicability in a real use case is limited both by the SiC FET SOA and by its junction temperature rise. Considering the diode reverse recovery causing a turn-on current spike of about 120A, it is possible to conclude that it is safe to turn the SiC FET on first up to about 300A so that the turn on current is well below the SOA limit and within the area limited by the maximum R_{DS_ON}. This assumption must be validated with a thermal model and then further proven with qualification tests. It is also recommended to verify through simulation that the junction temperature rise caused by the extra power dissipation in the SiC FET is within 50 °C to avoid impacts on lifetime. Above 300A and up to

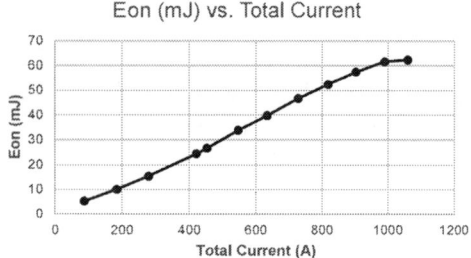

Fig.18: hybrid switch turn-on energy with IGBTs switching off after the SiC FET.

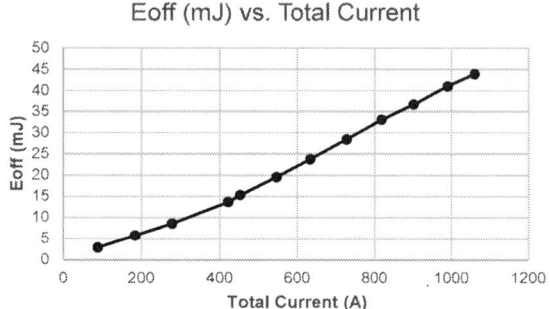

Fig. 19: hybrid switch turn-off energy with IGBTs switching off after the SiC FET.

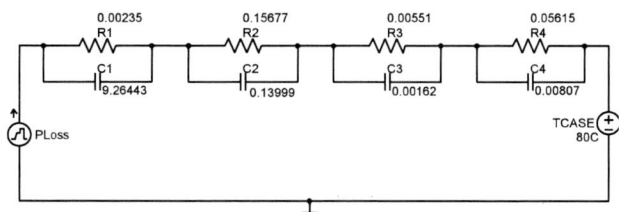

Fig. 20: hybrid switch thermal model.

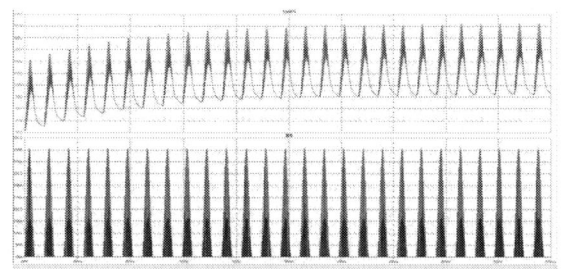

Fig. 21: thermal model results, SiC Tj (red) and Power loss generator (blue)

1200A it is necessary to change the commutation strategy and switch the three IGBTs with an anticipation of about 500ns.

Fig.18 and 19 show the E_{on} and E_{off} energies when this strategy is applied.

All the tests were conducted using a 10Ω turn-on gate resistor for both the SiC and IGBTs and a 10Ω turn-off resistor for the SiC FET and 5.1Ω turn-off resistor for the IGBTs. These values were selected to limit the drain source turn-off overvoltage and the reverse recovery turn-on current spike to

979-8-3315-1612-3/25 $31.00 © 2025 IEEE

safe levels. The thermal model used to verify the validity of the 300A switchover current is shown in Fig. 20, where the RC values are extracted by means of curve fitting of the single pulse transient thermal impedance function provided on the SiC FET datasheet.

The WLTP profile was used to identify the minimum RPM value at which this current occurs in order to calculate the number of cycles within an electrical period of the motor current (for a specific EV model). Then the instantaneous power loss generator model was created using the values of E_{on} and E_{off} extracted with the DPTP and adding the contribution of conduction losses. The resulting SiC power loss profile is shown in Fig.21 together with the resulting SiC junction temperature assuming a constant case temperature of 80 °C. The simulated junction temperature Tj indicates an average of 105°C and a maximum of 125°C. This results in a delta Tj of 35°C at 264 Hz. The delta Tj is below 50°C, confirming the suitability of this selection.

V. CONCLUSIONS

This paper demonstrates the advantages of a hybrid switch composed of three Silicon IGBTs and a SiC Power MOSFET for EV traction inverters. Compared to an IGBT solution, the hybrid switch significantly improves low-current performance and switching efficiency, which are crucial for EVs operating at low or partial loads. Using an automated double pulse test platform for T-PAK packaged devices, the characterization process is more efficient, allowing for the extraction of an optimal current-based control strategy that further enhances EV inverter efficiency. In conclusion, the hybrid switch is a viable and efficient alternative to traditional IGBT-based solutions and a cost competitive approach compared to a full SiC-based solutions. The automated characterization method is a valuable tool to quickly extract switching energies and optimal PWM delay times. Future work will detail the thermal model and the impact of the control strategy on standard driving range profiles such as the WLTP.

REFERENCES

[1] J. Lai *et al.*, "A hybrid-switch-based soft-switching inverter for ultra-high efficiency traction motor drives," *IEEE Trans. Ind. Appl.*, vol. 50, no. 3, pp. 1966–1973, May/Jun. 2014.17, no. 1, pp. 45- 55

[2] Xiaofeng Ding;Min Du;Chongwei Duan;Hong Guo;Rui Xiong;Jinquan Xu;Jiawei Cheng;Patrick Chi kwong Luk "Analytical and Experimental Evaluation of SiC-Inverter Nonlinearities for Traction Drives Used in Electric Vehicles" IEEE Transactions on Vehicular Technology Year: 2018,Volume: 67,Issue:1.

[3] K. Hamada, M. Nagao, M. Ajioka, and F. Kawai, "SiC—Emerging power device technology for next-generation electrically powered environmentally friendly vehicles," IEEE Trans. Electron Devices, vol. 62, no. 2, pp. 278–285, Feb. 2015.

[4] Yole Développement, "Power SiC 2021: Materials, Devices and Applications"

[5] Wang, Jun & Li, Zongjian & Jiang, Xi & Zeng, Cheng & GAE, John. (2018). "Gate Control Optimization of Si/SiC Hybrid Switch for Junction Temperature Balance and Power Loss Reduction". IEEE Transactions on Power Electronics, VOL. 34, NO. 2, February 2019

[6] H. Qin, R. Wang Q. Xun W. Chen, and S. XIE "Switching Time Delay Optimization for ``SiC/Si'' Hybrid Device in a Phase-Leg Configuration" IEEE Access 2169-3536 (ISSN) 693536 (eISSN) Vol. 9 p. 37542-37556 art. no 9339907 January 28, 2021,

[7] H. Qin, S. Xie, Q. Xun et al. "An optimized parameter design method of SiC/Si hybrid switch considering turn-off current spike" The 5th International Conference on Electrical Engineering and Green Energy CEEGE 2022, 8–11 June, 2022, Berlin, Germany.

[8] Z. Li et al., "Active gate delay time control of Si/SiC hybrid switch for junction temperature balance over a wide power range." IEEE Trans. Power Electron., vol. 35, no. 5, pp. 5354–5365, May 2020.

[9] Z. Peng et al., "A Variable-Frequency Current-Dependent Switching Strategy to Improve Tradeoff Between Efficiency and SiC MOSFET Overcurrent Stress in Si/SiC-Hybrid-Switch-Based Inverters," in IEEE Transactions on Power Electronics, vol. 36, no. 4, pp. 4877-4886, April 2021, doi: 10.1109/TPEL.2020.3026494.

[10] "Traction Inverter New Solutions" https://www.st.com/content/ccc/resource/sales_and_marketing/presentation/application_presentation/group0/83/a8/4e/89/ba/8b/46/6d/Traction-Inverter/files/ST-Traction-Inverter.pdf/jcr:content/translations/en.ST-Traction-Inverter.pdf

Critical Role of Individual Cell Temperature Monitoring in Mitigating Thermal Runaway and Reducing Accelerated Degradation in Lithium-ion Batteries

Mohit Sharma
Department of electrical, computer and software engineering
Ontario Tech University
Oshawa, Canada
mohit.sharma14@ontariotechu.net

Akash Samanta
Department of electrical, computer and software engineering
Ontario Tech University
Oshawa, Canada
akash.samanta@ontariotechu.net

William Locke
Department of electrical, computer and software engineering
Ontario Tech University
Oshawa, Canada
william.locke@ontariotechu.net

Sheldon Williamson
Department of electrical, computer and software engineering
Ontario Tech University
Oshawa, Canada
sheldon.williamson@ontariotechu.ca

Abstract—Thermal runaway in lithium-ion batteries (LIBs) is a critical risk, potentially leading to fires and explosions. This phenomenon can be triggered by overcharging, short-circuiting, physical damage, manufacturing defects, and overheating. Effective mitigation of thermal runaway relies on precise and close monitoring of individual cells within the battery pack, a task often complicated by the complexity and cost of implementing temperature sensors on each cell in automotive applications. This study demonstrates the importance of individual cell temperature monitoring by intentionally including an unhealthy cell (C#12) in a 14-cell LIB 21700 module equipped with an automotive-grade battery management system (BMS) from NXP®. Experimental results reveal that despite efficient BMS balancing, the weak cell exhibited a significantly higher temperature rise, with a final temperature difference of 6°C and a voltage difference of over 200 mV compared to healthy cells. These findings underscore the potential for thermal failure and runaway if individual cell temperatures are not closely monitored. Additionally, reliance on voltage-based control alone can lead to suboptimal battery pack utilization, as evidenced by a 5% capacity loss due to the weak cell. This research highlights the necessity of monitoring the temperature of each cell to prevent thermal runaway and ensure efficient battery performance.

Index Terms—transportation electrification, thermal management systems, battery management systems, and electric vehicles.

I. INTRODUCTION

Despite the fact that lithium-ion batteries (LIBs) are not the only energy storage systems available in the market, they are

This research is funded by Natural Sciences and Engineering Research Council of Canada (NSERC)-Idea-to-Innovation (I2I) Grants.
Mohit Sharma, Akash Samanta, William Locke, and Sheldon Williamson are with the Electrical, Computer, and Software Engineering, Ontario Tech University, ON, L1G 0C5, Canada.

widely preferred for automotive applications due to their high energy density, lightweight design, long cycle life, and ability to deliver high power and support fast charging. These advantages make LIBs ideal for electric vehicles (EVs), enabling greater range, performance, and efficiency [1]. While concerns about thermal runaway remain, advancements in battery management systems (BMS), thermal management technologies, and safer chemistries, such as lithium iron phosphate (LFP), have significantly enhanced their safety and reliability [2]. Additionally, the ongoing reduction in production costs and the expansion of charging infrastructure further solidify LIBs as the preferred choice for automotive energy storage solutions.

Effective BMS and thermal management require closely monitoring and carefully controlling heat generation in LIBs to maintain battery longevity and performance [3]. Excessive heat and high temperatures can trigger degradation processes, leading to reduced capacity and power loss [4]. The temperature rise within the LIB is primarily caused by electrochemical reactions, where heat from entropy changes is reversible during operation [5]. However, heat generated from charge transfer, ohmic losses, and mass transfer limitations is irreversible, resulting in a buildup of heat within the cell [6]. Monitoring temperature at the individual cell level is crucial for ensuring optimal safety and extending battery life.

Currently, temperature sensing is primarily performed at the module level rather than the cell level, which is suboptimal for battery thermal management systems (BTMS). Only a few EV manufacturers, such as Tesla, BMW, Nissan (Leaf), and Audi (e-tron series), employ BMS capable of monitoring temperatures at the cell level. In most commercial vehicles, the standard approach involves measuring temperature at various points on the surface or tab of LIB cells. However, in large-format LIBs, the surface temperature can differ significantly from the core temperature, which is more critical for performance and safety [7]. General Motors (GM) introduced the Ultium platform, which employs a wireless BMS (wBMS) capable of monitoring individual cell temperatures, although its commercial applications are yet to be realized.

Most of the commercially available BMS is designed to protect LIB cells from issues like over-voltage, under-voltage, overcurrent during charging and discharging, extreme temperatures, and cell balancing. However, significant temperature gradients under severe conditions can cause the separator to shrink or melt, potentially leading to an internal short circuit and uncontrolled temperature rise, known as thermal runaway. Often the thermal runaway starts from a single cell and propagates to the entire battery pack in a very time period as the cells are closely packed together within the battery pack. For instance, Tesla's Model S uses over 7,000 18650 LIB cells, and Mitsubishi's i-MiEV employs 88 large prismatic cells. Studies such as [8], [9] have developed techniques for estimating LIB cells' core and surface temperatures. Additionally, research like [10], [11] demonstrated cost-effective methods for accurate individual cell temperature measurement within battery modules.

Nonetheless, the importance of accurately monitoring individual cell temperatures to protect against thermal runaway and power loss using an automotive-grade BMS has not been fully demonstrated in the literature. This paper examines an automotive-grade BMS from NXP® and a 14-cell LIB module, including one unhealthy cell, to emphasize the significance of monitoring individual cell temperatures alongside cell voltages. The study also samples temperatures from various sections of the battery module to assess thermal safety and capacity loss, even under effective cell balancing by the BMS. Furthermore, this paper introduces a custom-built, highly precise, and accurate temperature monitoring system utilizing an STM32 microcontroller with a negative temperature coefficient (NTC) thermistor for real-time monitoring.

A. Major Contributions of the Paper

- This paper presents an efficient architecture for individual cell temperature monitoring, enabling the tracking of high-resolution temperature data over extended operational periods in automotive applications.
- It emphasizes the method for individually monitoring each cell's temperature and highlights the critical importance of such monitoring in preventing overheating and thermal runaway.
- The paper identifies the optimal location for installing temperature sensors when applying this monitoring system to a real electric vehicle battery module.
- It further discusses the potential applications of the battery pack after identifying weak cells, focusing on avoiding fire hazards and overheating—key safety considerations.

- Additionally, the paper demonstrates the practicality of the proposed individual cell temperature measurement system in mitigating overheating and preventing thermal runaway in automotive applications.
- Finally, the potential use of the recorded historical temperature data is also explored.

II. EXPERIMENTAL SETUP AND EXPERIMENT SCHEDULE

The experimental setup used in this research is illustrated in Fig. 1.

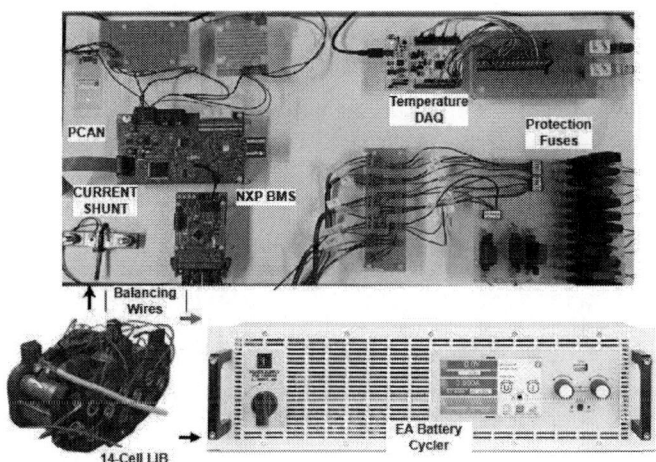

Fig. 1. Experimental setup for battery cycling and data collection.

It consists of 14 Samsung 21700 NCA cells, each with a capacity of 4 Ah, and an EA PSB 10360-120 battery cycler for charging and discharging the module. All 14 cells are connected in series. Intentionally, an unhealthy cell (C#12) is included in the battery module. The physical location of the week cell (C#12) is shown in Fig. 3. The setup also includes an automotive-grade NXP® BMS (RD33771-48VEVM) for the 14-cell battery module and an STM32 board connected to 14 NTC thermistors (103JT) for measuring individual cell temperatures.RD33771-48VEVM board has MC33771 series of battery cell controller ICs. These ICs are designed for LIB management in automotive and industrial applications, such as hybrid electric vehicles (HEVs), EVs, energy storage systems (ESS), and uninterruptible power supplies (UPS). They perform ADC conversions of differential cell voltages and current, as well as battery coulomb counting and temperature measurements. The MC33771C, for instance, supports 7 to 14 cells and features integrated 300 mA passive cell balancing with diagnostics. Both the NXP BMS and STM32 boards are connected to a common CAN bus, transmitting data with specific CAN IDs for monitoring, storage, and visualization. A dedicated Python script is used to decode all CAN data from the BMS and STM32 board. The STM32F446RE microcontroller features a 12-bit ADC that provides a resolution of 0.805 mV per step for a typical 3.3 V reference voltage, making it suitable for temperature-sensing applications with thermistors. The ADC achieves high accuracy, typically ±0.5

979-8-3315-1612-3/25 $31.00 © 2025 IEEE

LSB for integral non-linearity, and offers an effective number of bits (ENOB) exceeding 11 in practical applications. To maximize accuracy when measuring the resistance of a thermistor, which varies with temperature, it is essential to use a stable reference voltage, such as an external low-noise and temperature-stable voltage source, rather than relying on the internal reference. The thermistor is commonly used in a voltage divider circuit, and careful selection of the divider resistor ensures an optimal linearization range for the thermistor's response. Factors like ADC sampling time, which should be sufficiently long to minimize noise, and the impedance of the voltage divider must be considered to reduce input leakage effects. To further enhance accuracy, external noise should be minimized through shielding and bypass capacitors, while digital filtering, averaging of multiple ADC readings, and Steinhart-Hart equations or lookup tables can account for the thermistor's non-linear behavior. With proper design and calibration, the STM32F446RE ADC can achieve temperature measurement accuracy in the range of ±0.5°C to ±1°C, making it an excellent choice for thermistor-based temperature sensing in embedded systems. The setup includes safety fuses, which protect the BMS in the event of any unexpected incidents. The EA PSB 10360-120 power supply is also connected to the CAN bus, with a custom-built algorithm to send and receive commands from the centralized controller to charge and discharge the battery module. The entire battery pack operates according to scheduled instances of voltage and current, and the temperature data of the battery pack helps the algorithm determine the charging and discharging events. The inclusion of CAN (Controller Area Network) communication in the EA PSB 10360-120 power supply offers significant advantages in the experimental setup. CAN is a robust and reliable communication protocol designed for high-reliability applications, ensuring data integrity through error detection and correction mechanisms, even in noisy conditions. It supports real-time communication, making it ideal for precise and instantaneous monitoring or control of the power supply's parameters, such as voltage, current, and power. The details of the sensors, data acquisition systems, and software used are shown in Fig. 2. Here, an online Python-based real-time CAN data decoding architecture is used for real-time battery data collection and visualization.

For surface temperature measurement of the batteries, 103JT thermistors are used with the STM32 MCU board. All important battery parameters, such as the voltage of each of the 14 cells, stack voltage, and stack current, are decoded from the CAN bus data, stored on a local server, and displayed. The STM32 board provides the surface temperature data for the 14 cells. The system is a closed-loop setup where the BMS, temperature sensor module, and power supply are all connected to the same CAN bus, transmitting and decoding data in real time. This ensures that every event during the charging and discharging of the battery is monitored, and the decoded data is fed to a calibrated algorithm. This setup optimally protects the battery module from adverse conditions, including thermal runaway. The battery module is cycled at a

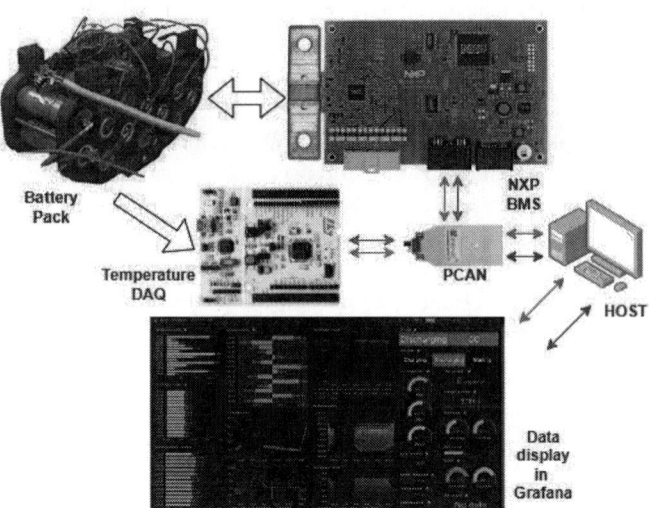

Fig. 2. Schematic layout of the data accusation, visualization, and control system.

Fig. 3. Physical location of cell#12 in the battery module.

1C charging rate, followed by a one-hour rest (cooling period), and then discharged at a 1C rate under controlled ambient temperature in a thermal chamber (Associated Environmental Systems®). All experiments are conducted at an ambient temperature of 25°C.

III. RESULTS AND DISCUSSION

Primarily, individual cell voltage, stack current, stack voltage, individual cell temperature, balancing current, and balancing voltage difference are collected to demonstrate the importance of closely monitoring individual cell temperature in an automotive battery pack. The plot of cell voltage and temperature in Cell#12 compared to the average voltage and temperature of the entire module are shown in Fig. 4 and Fig. 5 respectively. The individual cell temperature and balancing voltage are presented in Fig. 6 and Fig. 7 respectively.

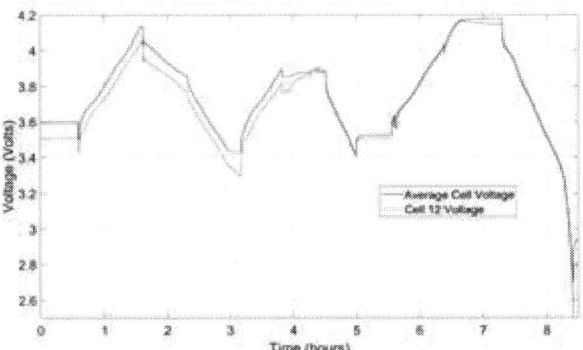

Fig. 4. Voltage of Cell#12 and Average voltage of module.

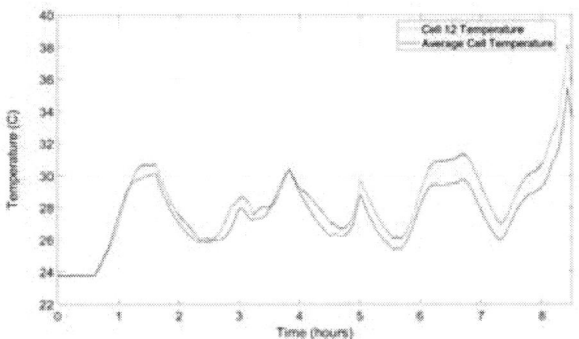

Fig. 5. Temperature of Cell#12 and Average temperature of module.

As shown in Fig. 6, although all the cells began charging at nearly the same temperature, by the end of discharging, cell#12 exhibited the highest temperature increase despite being subjected to the same charging and discharging current as the other cells. The temperature rise in Cell#12 was 4°C higher than that of the average temperature of the module and more than 6°C higher than that of the coolest cell in the module. It is also noticed that the peripheral cell#13 is getting heated up to the proximity of cell#12 which indicates the cascading effect that can lead to a thermal runaway condition. This highlights the drawback of relying on sample temperatures from the battery pack rather than monitoring individual cell temperatures. For example, degradation often varies among individual cells within a battery module or pack. As a result, thermal failure typically begins with a single cell and can quickly spread to the entire battery pack. If a temperature sensor is placed far from a weak or faulty cell, relying solely on sample temperature may not ensure an optimal level of thermal safety in the battery pack. Furthermore, even with effective cell balancing, the voltage difference in cell#12 compared to the average cell voltage is nearly 200 mV at the end of the discharge period, as shown in Fig. 7.

Fig. 4 highlights a sharp drop in the voltage of Cell#12 despite the same charging and discharging current in the battery module. Most automotive-grade BMS monitor and control the charging and discharging operations of the entire battery module based on the weakest cell in the module. Since the voltage of the weakest cell reaches the lower and upper cut-off thresholds more quickly than the healthier cells, controlling charging and discharging based on the voltage of the weakest cell leads to significant capacity loss for the entire battery pack. A capacity-based calculation using the conventional and highly accurate Coulomb counting method (under laboratory conditions) is used to estimate the capacity loss in this battery module. The study found that almost 5% of the 4 Ah total capacity of the 14-cell module is unutilized due to a single weak cell. This study involved a 14-cell module comprising all pristine cells, including the weak cell (C#12). With aging and the higher temperature rise in Cell#12, degradation will accelerate with continued battery cycling. Therefore, it is extremely important to monitor each individual cell's temperature, especially in automotive applications where LIBs are subjected to rapid charging and highly dynamic discharging conditions. The location of the temperature sensor on the cell is crucial to ensure accurate thermal management and to detect potential issues such as thermal runaway. In this study, the temperature sensors for 21700 LIB cells are considered at the central location of the surface (sidewall) [12]. The primary reason for selecting this location is that heat generated within the jelly roll transfers to the outer casing. Sidewall placement enables effective monitoring of the bulk temperature. The recommended placement is at the midpoint along the cylindrical length, as this is often where the thermal gradient is most pronounced. For higher accuracy, sensors can be placed at multiple locations along the sidewall (e.g., at 120° intervals around the circumference) to capture temperature uniformity. However, some additional considerations and recommendations for sensor placement are outlined below [13].

- Thermal Gradients: Temperature gradients can occur across the cell due to uneven current density or cooling conditions.
- Critical Regions: The regions most susceptible to overheating are the cell terminals (current collectors) and the center of the cell, where internal resistance generates more heat.
- Battery Configuration: In a pack, the placement should account for heat dissipation from adjacent cells and external cooling systems.
- Application-Specific Requirements: The application (e.g., automotive, aerospace) dictates the level of thermal sensitivity and the number of sensors used.

Several studies have highlighted that the optimal position for a temperature sensor is at the core of the cell, as this region often exhibits the highest internal temperature. The temperature difference between the surface and the core is typically in the range of 6°C to 10°C [14]. By accessing core temperature data, the BMS can gain a more accurate understanding of the cell's internal condition, significantly reducing response time in cases of overheating or thermal runaway [15]. However, accessing the core temperature directly is challenging. To address

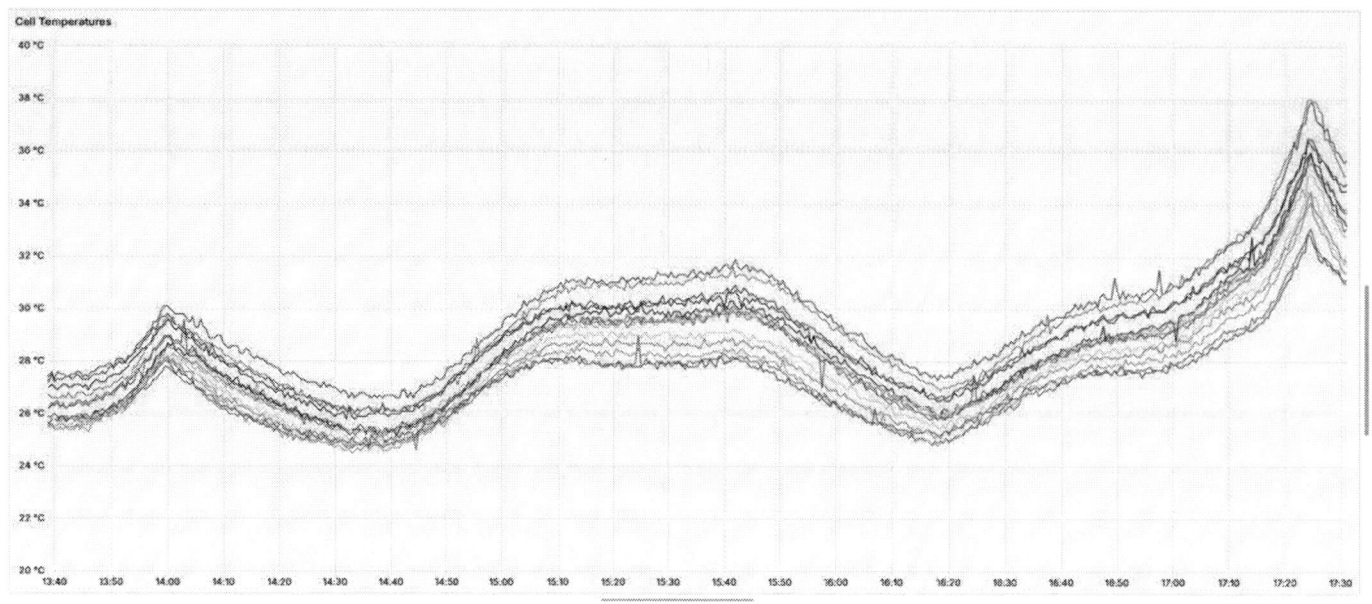

Fig. 6. Individual cell temperature of the 14-cell battery module.

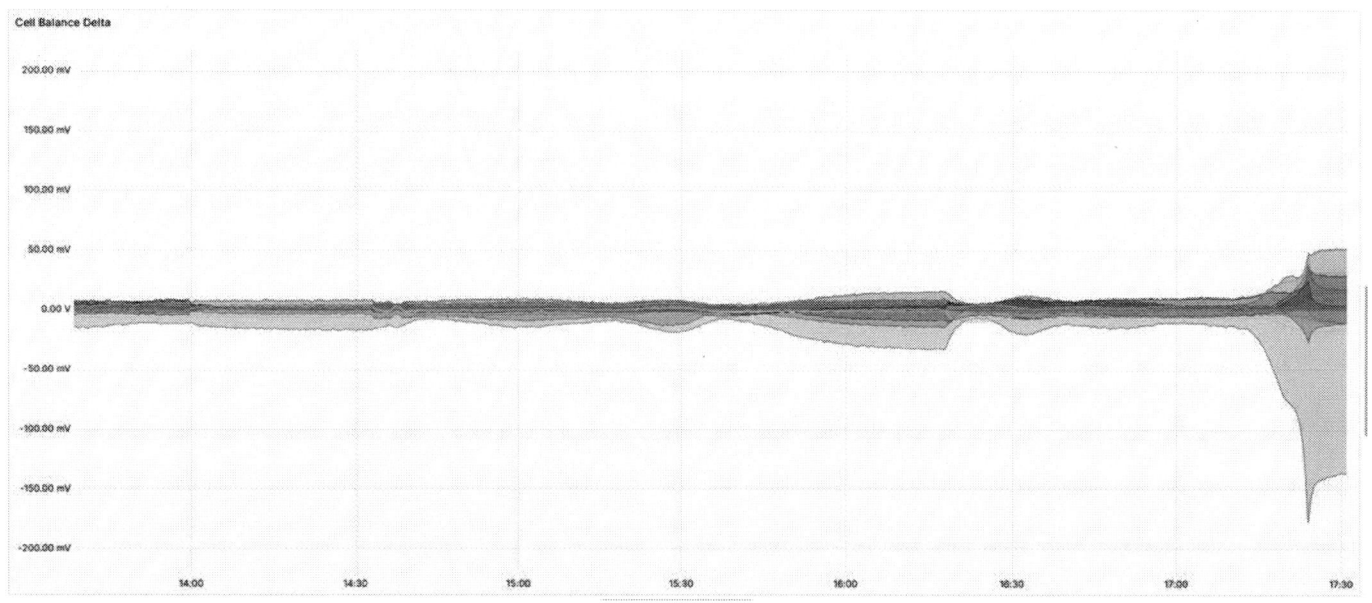

Fig. 7. Individual cell balancing profile of the 14-cell battery module.

this, researchers have proposed various estimation techniques to infer core temperature information for individual cells. Despite these efforts, the complexity of estimation methods, along with concerns about the accuracy and reliability of the inferred data, remain significant challenges. Thermal runaway threshold condition varies depending on the cell chemistry and design [16]. For common NMC (Nickel Manganese Cobalt) cells, this threshold is generally around 130°C, while for LFP (Lithium Iron Phosphate) cells, it is approximately 150°C due to their greater thermal stability [17]. Real-time testing has been conducted to evaluate a BMS's ability to respond to the

onset of thermal runaway and the reliability depends on the accurate information of the temperature data. Therefore, this paper demonstrated the importance of individual cell temperature monitoring and the capability of the developed architecture for temperature data monitoring and acquisition. Moreover, the idea of incorporating real-time testing results into future work is an excellent suggestion for further validating the BMS's responsiveness under thermal runaway conditions. A BMS developed considering the proposed architecture in this paper can efficiently identify an unhealthy cell within a pack. Once the unhealthy cell is identified, there could be many ways to

979-8-3315-1612-3/25 $31.00 © 2025 IEEE

handle to unhealthy cell. To handle an unhealthy or weak cell, where the temperature rises faster and higher compared to other cells, several measures can be implemented to prevent thermal runaway and further overheating. First, the BMS should isolate the affected cell by disconnecting it from the circuit, using safety mechanisms such as fuses or cell-balancing switches, to prevent further current flow. Additionally, thermal management strategies like targeted cooling (using liquid or air cooling systems) can be applied to the overheated cell to stabilize its temperature. The BMS should also redistribute the load across healthier cells to reduce stress on the weak cells. Continuous monitoring of the cell's parameters, such as voltage and impedance, is essential for early detection of anomalies. If the cell's condition worsens, it should be flagged for replacement during maintenance to ensure the overall safety and performance of the module. Furthermore, the historical recorded temperature data can play a crucial role in enhancing the safety, performance, and efficiency of LIBs in automotive applications. By analyzing historical temperature trends, the BMS can identify recurring patterns of thermal anomalies, aiding in the early detection of weak or deteriorating cells. This data also supports predictive modeling for better estimation of core temperatures, which are critical for preventing thermal runaway. Additionally, the long-term analysis of temperature fluctuations across operational cycles provides valuable insights into the battery's aging process, enabling more accurate state-of-health (SOH) assessments. Such data can also inform the development of advanced thermal management strategies and algorithms, ultimately improving the reliability and safety of the overall system.

CONCLUSIONS

This paper highlighted the importance of monitoring individual cell temperatures by focusing on a single weak cell within a 14-cell LIB module. This approach ensures optimal safety from thermal runaway, maximizes capacity utilization, and reduces battery degradation. The study showed that even with new, pristine cells, a weak cell could experience a temperature rise of up to 6°C during 0.5C charging and 1C discharging. Additionally, despite effective cell balancing, the voltage difference in the weak cell could reach up to 200 mV, resulting in over 5% underutilization of the 4 Ah battery module's capacity. The custom-built STM32 microcontroller-based temperature acquisition system proved effective, providing precise and accurate temperature measurements at a 10 Hz sampling rate, with a precision level of ±0.5% and an accuracy of 0.2%. Further, this study also highlighted that to handle an unhealthy or weak cell within a 14-cell module, where the temperature rises faster and higher compared to other cells, several measures can be implemented to prevent thermal runaway and further overheating. Overall, this demonstrates the practicality of the temperature measurement architecture for on-board BMS in automotive applications.

REFERENCES

[1] Y. Ding, Z. P. Cano, A. Yu, J. Lu, and Z. Chen, "Automotive li-ion batteries: current status and future perspectives," *Electrochemical Energy Reviews*, vol. 2, pp. 1–28, 2019.

[2] L. Kong, Y. Li, and W. Feng, "Strategies to solve lithium battery thermal runaway: from mechanism to modification," *Electrochemical Energy Reviews*, vol. 4, no. 4, pp. 633–679, 2021.

[3] K. Benabdelaziz, M. Maaroufi, and B. Ikken, "Degradation of lithium-ion batteries in electric vehicles at high temperatures: A case study," in *2018 6th International Renewable and Sustainable Energy Conference (IRSEC)*. IEEE, 2018, pp. 1–6.

[4] T. Gao, J. Bai, D. Ouyang, Z. Wang, W. Bai, N. Mao, and Y. Zhu, "Effect of aging temperature on thermal stability of lithium-ion batteries: Part a–high-temperature aging," *Renewable Energy*, vol. 203, pp. 592–600, 2023.

[5] L. Anekal, A. Samanta, and S. Williamson, "Rapid parameterization of lithium-ion batteries using frequency window identification technique for on-board charge control and battery management," in *2024 IEEE Applied Power Electronics Conference and Exposition (APEC)*. IEEE, 2024, pp. 1330–1337.

[6] L. B. Diaz, A. Hales, M. W. Marzook, Y. Patel, and G. Offer, "Measuring irreversible heat generation in lithium-ion batteries: An experimental methodology," *Journal of The Electrochemical Society*, vol. 169, no. 3, p. 030523, 2022.

[7] S. Surya, A. Samanta, V. Marcis, and S. Williamson, "Hybrid electrical circuit model and deep learning-based core temperature estimation of lithium-ion battery cell," *IEEE Transactions on Transportation Electrification*, vol. 8, no. 3, pp. 3816–3824, 2022.

[8] X. Du, J. Meng, J. Peng, Y. Zhang, T. Liu, and R. Teodorescu, "Sensorless temperature estimation of lithium-ion battery based on broadband impedance measurements," *IEEE Transactions on Power Electronics*, vol. 37, no. 9, pp. 10101–10105, 2022.

[9] Y. Ma, Y. Cui, H. Mou, J. Gao, and H. Chen, "Core temperature estimation of lithium-ion battery for evs using kalman filter," *Applied Thermal Engineering*, vol. 168, p. 114816, 2020.

[10] S. Novais, M. Nascimento, L. Grande, M. F. Domingues, P. Antunes, N. Alberto, C. Leitão, R. Oliveira, S. Koch, G. T. Kim *et al.*, "Internal and external temperature monitoring of a li-ion battery with fiber bragg grating sensors," *Sensors*, vol. 16, no. 9, p. 1394, 2016.

[11] K. M. Alcock, Á. González-Vila, M. Beg, F. Vedreño-Santos, Z. Cai, L. S. Alwis, and K. Goh, "Individual cell-level temperature monitoring of a lithium-ion battery pack," *Sensors*, vol. 23, no. 9, p. 4306, 2023.

[12] J. Patel, R. Patel, R. Saxena, and A. Nair, "Thermal analysis of high specific energy ncm-21700 li-ion battery cell under hybrid battery thermal management system for ev applications," *Journal of Energy Storage*, vol. 88, p. 111567, 2024.

[13] C. Chetri, A. Samanta, and S. Williamson, "Critical understanding of temperature gradient during fast charging of lithium-ion batteries at low temperatures," in *IECON 2023-49th Annual Conference of the IEEE Industrial Electronics Society*. IEEE, 2023, pp. 1–6.

[14] X. Zhang, H. Xiang, X. Xiong, Y. Wang, and Z. Chen, "Benchmarking core temperature forecasting for lithium-ion battery using typical recurrent neural networks," *Applied Thermal Engineering*, vol. 248, p. 123257, 2024.

[15] A. Samanta, C. Chetri, D. Karnehm, A. Neve, and S. Williamson, "Temporal sensitivity analysis of internal temperature informed charging algorithms and rapid thermal management system for e-mobility," in *2024 IEEE Energy Conversion Congress and Exposition (ECCE)*. IEEE, 2024.

[16] X. Feng, D. Ren, X. He, and M. Ouyang, "Mitigating thermal runaway of lithium-ion batteries," *Joule*, vol. 4, no. 4, pp. 743–770, 2020.

[17] S. Shahid and M. Agelin-Chaab, "Battery thermal management through simulation and experiment: Air cooling and enhancement," in *Handbook of Thermal Management Systems*. Elsevier, 2023, pp. 221–254.

979-8-3315-1612-3/25 $31.00 © 2025 IEEE

Loss-Optimized Design of a Triple Active Bridge DC-DC Converter for an Electric Vehicle Application

Sreejith Chakkalakkal[1], Kyle Kozielski[1], Wesam Taha[2], Yicheng Wang[2], Aniket Anand[2], and Ali Emadi[1]

[1]McMaster Automotive Resource Centre (MARC), McMaster University, Hamilton, ON, Canada
[2]R&D Americas, Schaeffler, Chatham, ON, Canada
Email: chakkas@mcmaster.ca

Abstract—This paper proposes a design optimization methodology for an isolated triple active bridge (TAB) DC-DC converter for an electric vehicle (EV) application. The optimization algorithm computes the optimal values of the high-frequency transformer leakage inductances to minimize converter losses and enhance efficiency across a broad range of operating conditions. The TAB is modeled using the generalized harmonic approximation (GHA) technique, which reduces computational complexity compared to time-domain analysis while maintaining high modeling accuracy across all converter operating zones. The TAB control variables are generated using particle swarm optimization (PSO) for the discrete operating points defined within the design search space. The TAB model, with the selected optimization strategy, is implemented in the PLECS Blockset software for an 800 V EV application with a maximum on-board charger (OBC) power of 9.6 kW and an auxiliary power module (APM) rated at 3.6 kW. The findings demonstrate that efficacy of the proposed design methodology effectively optimizes the leakage inductances of the three-port transformer, extends the zero-voltage switching (ZVS) range, and results in approximately a 23% reduction in converter losses under full-load conditions.

Index Terms—Converter design optimization, dc-dc converter, triple active bridge (TAB), particle swarm optimization (PSO), zero-voltage switching (ZVS).

I. INTRODUCTION

Triple active bridge (TAB) DC/DC converters provide an appealing solution for a wide range of modern applications requiring multiple power ports, sophisticated power flow control, and space-efficient design [1]–[3]. These converters are particularly advantageous for applications such as electric vehicles, renewable energy systems, and hybrid energy storage systems. Compared to traditional multi-stage topologies, TAB converters offer significant benefits, including reduced component count, lower costs, enhanced efficiency from fewer stages, improved power density, and more centralized and efficient control. However, designing a TAB DC-DC converter requires careful design considerations to minimize component stress, reduce system cost, and improve overall converter efficiency [4], [5].

In multi-port converters, including the dual active bridge (DAB) converter, selecting the appropriate transformer leakage inductance is crucial for managing component stress and optimizing converter efficiency. For traditional phase angle control, higher leakage inductance demands greater phase-shift angles for power transfer, leading to increased RMS and peak currents in the high-frequency transformer and semiconductor switches. Conversely, a lower leakage inductance limits the soft-switching range of the converter, reduces efficiency, and also makes thermal management more challenging.

Current stress and power losses in multi-port converters can be effectively managed by incorporating pulse width modulation (PWM) alongside phase angle control, as reported in various studies [6]–[10]. This additional control freedom regulates the volt-seconds applied across the high-frequency transformer and shapes the AC current waveforms of the converter to reduce device stress and minimize losses. Furthermore, introducing PWM in conjunction with phase-shift control can extend the zero voltage switching (ZVS) and improve both the efficiency and electromagnetic interference (EMI) performance of the converter. Since converter performance and efficiency heavily depend on the modulation strategy and control approach, design optimization should account for these factors in selecting transformer leakage inductances, semiconductor switches, and other passive components.

The leakage inductance selection and design optimization of high-frequency DC-DC converters have been extensively studied in previous literature. An efficiency-optimized design procedure for a DAB converter developed for EV applications is proposed in [11]. In this study, the converter is modeled in the time domain, and the optimization process includes selecting the transformer leakage inductances and turns ratio for a modulation strategy aimed at minimizing overall losses of the converter. The authors in [12] discusses the selection of minimum leakage inductance values for high-frequency DAB converter. The paper provides closed-form expressions to determine the minimum leakage inductance required to achieve ZVS during the converter operation.

A DAB design methodology is presented in [13] using an analytical model-based constrained optimization procedure. The research focuses on selecting the transformer turns ratios and leakage inductance values for triple-phase-shift modulation to achieve minimum RMS current opera-

979-8-3315-1612-3/25 $31.00 © 2025 IEEE

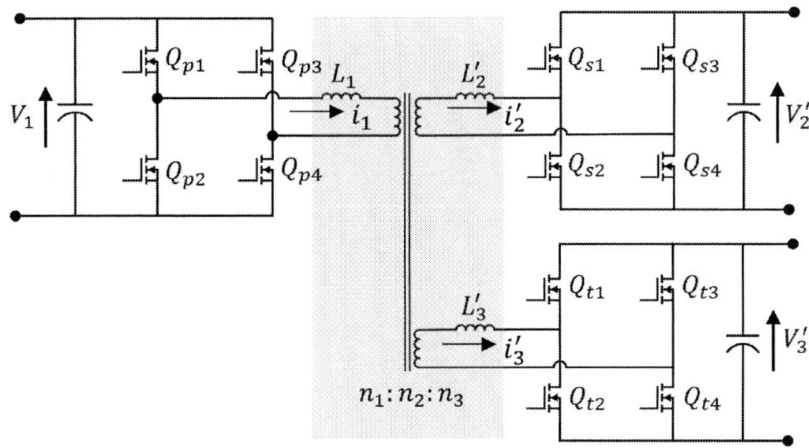

Fig. 1: Circuit of the TAB converter.

tion of the converter. An artificial intelligence (AI)-assisted design optimization of a high-frequency DAB with split inductors is proposed in [14]. Such topological variants are particularly beneficial for applications requiring an extended ZVS range, improved electromagnetic compatibility (EMC), and enhanced efficiency. [15] presents a numerical-based design optimization of a DAB for supercapacitor (SC) applications, targeting the selection of circuit parameters to minimize switching and conduction losses.

Design optimization studies for three-port converters are relatively limited in the literature, compared to those for DABs. A TAB design optimization based on fundamental harmonic approximation (FHA) is proposed in [16]. The TAB uses phase-angle control, and the design optimization focuses on selecting converter parameters to minimize reactive power and losses. While this approach is straightforward, it does not yield the optimal design due to model inaccuracies associated with FHA and the inherent limitations of using classical phase-shift control to optimize converter operation. In [17], a design optimization of a TAB with phase-shift control is proposed using a normalization approach. However, as with previous work, the authors do not consider the effects of multivariate modulation schemes and their impact on converter design optimization. The design process of a series-resonant three-port converter is discussed in [18] for phase-shift plus duty-cycle control. The converter is modeled using FHA, which is more suitable for resonant topologies due to the sinusoidal nature of the transformer current waveforms. To the best of the author's knowledge, none of the existing work addresses TAB design while considering its multivariable control flexibility and broader operating regions.

The present work proposes the optimal selection of transformer leakage inductances for a TAB converter in an electric vehicle OBC integrated with a low-voltage (LV) battery application. The individual leakage inductances of all three windings in the high-frequency transformer are optimized to minimize converter losses under the five-degree-of-freedom control strategy, also referred to as five variable control (FVC), as discussed in this paper. The TAB converter is modeled in the frequency domain using the generalized harmonic approximation (GHA) technique [19], [20], and the optimal control inputs for the specific operating points are generated using particle swarm optimization (PSO). The TAB losses and soft-switching ranges are analyzed for both the proposed and conventional design methodologies, and the results are compared. The loss-optimized TAB design is validated using PLECS Blockset analysis for a TPC converter designed for an 800 V EV application, with a maximum OBC power of 9.6 kW and an APM rated at 3.6 kW.

The remainder of the work is organized as follows. Section II explains the operational principle of the TAB converter and its modeling using GHA. Section III details the loss calculation of the MOSFETs and the high-frequency transformer used in the TAB. Sections IV and V discuss the proposed design optimization methodology and the PLECS Blockset simulation results, respectively. Finally, Section VI concludes the work.

II. OPERATING PRINCIPLE AND MODELING OF TAB

Figure 1 represents the schematic of a triple active bridge converter. V_1, V_2', and V_3' are the voltages across the TAB DC links, and i_1, i_2', and i_3' are the AC currents for ports 1, 2, and 3, respectively. The individual full-bridge modules are connected through the high-frequency transformer and series inductors, with the inductors either kept as discrete components or integrated within the transformer.

In its simplest form, the power flow among the various ports of a TAB is controlled by adjusting the phase angles between individual ports. However, for applications with wide DC voltage ranges, conventional phase-shift (PS) control leads to increased peak and RMS currents in the three-winding transformer, resulting in poor efficiency, especially under light-load conditions. Combined duty-cycle and phase-shift control significantly improves converter performance and reduces component stress [9], [21].

979-8-3315-1612-3/25 $31.00 © 2025 IEEE

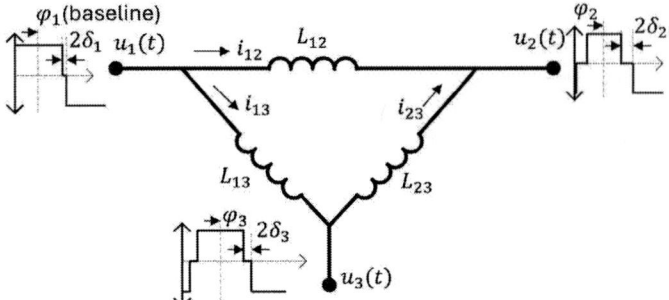

Fig. 2: Delta-equivalent circuit of the TAB converter.

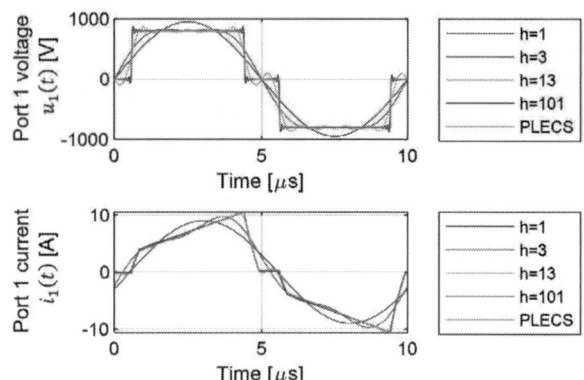

Fig. 3: Comparison of TAB GHA model with PLECS simulation results.

The delta equivalent representation of a TAB converter is shown in Fig. 2, referring to the primary side of the high-frequency transformer. The equivalent circuit is derived under the idealized approximation of the converter with negligible circuit parasitics and transformer magnetizing current. The detailed derivation of the equivalent circuit model is presented in [22], where the effective inductance between individual ports can be calculated using (1).

$$\begin{cases} L_{12} = L_1 + L_2 + \dfrac{L_1 L_2}{L_3} \\[2mm] L_{13} = L_1 + L_3 + \dfrac{L_1 L_3}{L_2} \\[2mm] L_{23} = L_3 + L_3 + \dfrac{L_2 L_3}{L_1} \end{cases} \quad (1)$$

Here, L_2 and L_3 are the transformer leakage inductances referred to the primary winding of the transformer. The delta-equivalent model of the TAB offers an effective and simplified way of modeling the three-port converter using analytical expressions.

Multiport converters can be modeled either in the time domain or in the frequency domain. In time-domain analysis, the converter voltage and current waveforms are represented using piecewise linear approximation techniques to derive expressions for the converter waveforms and other output quantities. However, this approach becomes computationally complex for converters employing duty-cycle control due to the loss of generality among various operating zones, which are distinguished by their phase angle and duty cycles [23], [24]. For a three-port converter, the analysis becomes even more complex, as it has higher number of operating zones compared to the DAB.

Frequency-domain modeling of the TAB provides a more generalized representation of the converter operation using closed-form analytical expressions. A loss-optimized TAB control methodology with PWM control is proposed in [25], utilizing first-order (fundamental) harmonic approximation (FHA) technique. The model accuracy of the TAB can be improved by considering higher-order harmonics approximation techniques, also referred to as generalized harmonic approximation (GHA) [19], [20], which will be used in the present converter optimization problem.

The instantaneous voltage expressions for the TAB converter can be obtained using the Fourier analysis of the converter voltage waveform, as shown in (2).

$$v_j(t) = \frac{4 V_j}{\pi} \sum_h \frac{1}{h} \cos(h\delta_j) \sin(h\omega t - \varphi_j) \quad (2)$$

Here, $h = 1, 3, 5, \ldots h_{max}$ represents the harmonic order, δ_j and φ_j represents the duty-cycle angle and phase-shift of the jth port, respectively, and $\omega = 2\pi f_{sw}$ is the switching frequency in rad/s.

The instantaneous current expressions between individual ports can be obtained from the delta-equivalent model of TAB shown in Fig. 2 by using (3).

$$i_{ij}(t) = \int_0^t \frac{v_i(t) - v_j(t)}{L_{ij}} \, dt \quad (3)$$

Figure 3 compares the TAB voltage and current obtained through GHA for various harmonic orders with the PLECS simulation results for a specific operating point. At lower harmonic orders ($h = 1$ and 3), the waveforms are inaccurate, leading to model uncertainties. [19] suggests an optimal harmonic order of 13 to achieve sufficient modeling accuracy without exceeding computational requirements.

The power flow between individual ports can be obtained using (4). The detailed expressions for the average power output and transformer RMS currents, as presented in [26], are utilized for the TAB optimization discussed in this paper.

$$P_{ij} = \frac{4}{\pi^3 f_{sw}} \sum_h \frac{1}{h^3} \left[\frac{V_i V_j}{L_{ij}} \cos(h\delta_i) \cos(h\delta_j) \sin\{h(\varphi_i - \varphi_j)\} \right] \quad (4)$$

III. CALCULATION OF TAB LOSSES

This work addresses power MOSFET losses and high-frequency transformer losses within the TAB design optimization framework, while excluding all other converter losses. Power MOSFETs are selected based on required

979-8-3315-1612-3/25 $31.00 © 2025 IEEE

TABLE I: MOSFETs selected for the various ports of the TAB.

MOSFET P/N	$V_{DS}[V]$	$I_D[A]$	$R_{DS}[m\Omega]$	$C_{oss}[pF]$
G3R40MT12K (Ports 1 & 2)	1200	49	40	81
IAUS260N10S5N019T (Port-3, two in parallel)	100	260	1.9	1386

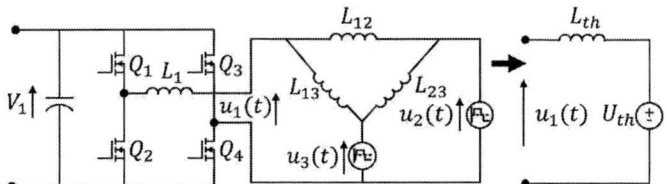

Fig. 5: TAB equivalent circuit representation for ZVS analysis.

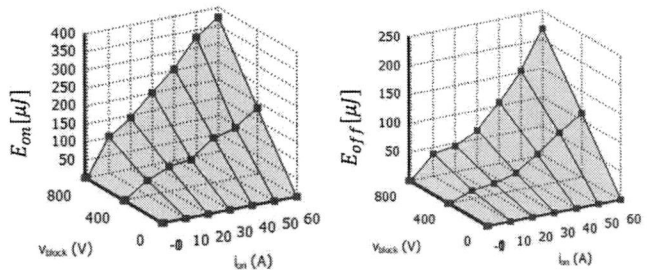

Fig. 4: Turn on & turn off energy of the SiC MOSFET G3R40MT12K obtained through DPT. $\{T_j : 65°C, V_{DC} : 800\,V, V_{GS} : -5\,V$ to $+15\,V$, and $R_G : 2.5\,\Omega\}$

device ratings and analyzed using the MOSFET figure-of-merit (FOM) as discussed in [27]. Transformer losses are modeled using short-circuit (SC) and open-circuit (OC) test results from the transformer prototype.

A. Calculation of MOSFET Losses

The MOSFETs selected for the TAB converter are listed in Table I. High-voltage SiC MOSFETs are used for ports 1 and 2, while port 3 uses two parallel Si MOSFETs. The total MOSFET losses consist of conduction and switching losses, as defined in (5):

$$P_{Tot} = P_{con} + P_{sw}, \tag{5}$$

The MOSFET conduction loss for a full bridge can be determined using (6):

$$P_{con} = \sum_{j=1}^{3} 2I_{j,rms}^2 R_{DS,j}, \tag{6}$$

where $I_{j,rms}$ is the RMS current of the jth port, and $R_{DS,j}$ is the effective on-state resistance of the MOSFETs used in the jth port.

The switching loss of the MOSFET is calculated using (7):

$$P_{sw} = f_{sw}\{E_{on}(1 - ZVS) + E_{off}\}, \quad ZVS \in \{1,0\} \tag{7}$$

where, E_{on} and E_{off} represent the MOSFET turn-on and turn-off energies, determined using the double-pulse test (DPT) results from on the MOSFET SPICE models, as illustrated in Fig. 4. To simplify the representation of the MOSFET switching energies, a regression model is used, expressing these energies as functions of the instantaneous voltage and current values obtained from the TAB GHA model discussed in Section II.

The ZVS term in (7) indicates whether the MOSFETs in a specific half-bridge satisfy the ZVS condition. For ZVS

to occur, the current polarity in a half-bridge must ensure that the MOSFET's body diode conducts before it is turned on. Additionally, the minimum current requirement is determined by the energy dissipated during the MOSFET's switching transitions across the Thevenin equivalent source, represented in Fig. 5. A detailed explanation of ZVS conditions and the minimum inductor current requirements for different switching intervals is provided in [19].

B. Transformer Loss Modeling

The prototype of the high-frequency transformer for the TAB converter and its equivalent circuit parameters are shown in Fig. 6. The TAB design uses a standalone high-frequency three-winding transformer, and the equivalent circuit parameters are determined using the SC and OC tests performed at the converter switching frequency for various voltage and current levels.

The transformer losses primarily consist of core losses and winding losses, which are considered in this optimization study as represented in (8).

$$P_{Tr} = P_{core} + P_{winding} \tag{8}$$

The core loss of the transformer is determined using the expression (9), where V_j is the RMS voltage applied across winding j, and $R_{C,j}$ is the core loss equivalent resistance of the three-winding transformer referred to the jth winding. The variation of R_C referred to the Port 3 winding of the transformer is shown in Fig. 6 with respect to the tertiary-side applied RMS voltage.

$$P_{core} = \frac{V_j^2}{R_{C,j}} \tag{9}$$

The total winding loss of the transformer is calculated using (10):

$$P_{winding} = \sum_{j=1}^{3} I_{j,rms}^2 R_{lj} \tag{10}$$

The transformer winding RMS currents for various ports ($I_{j,rms}$) can be calculated using the TAB GHA model. The phase winding resistance (R_{lj}) is the equivalent AC resistance measured at the converter switching frequency of $100\,kHz$.

As mentioned earlier, the transformer loss model discussed in this study is based on an empirical approach.

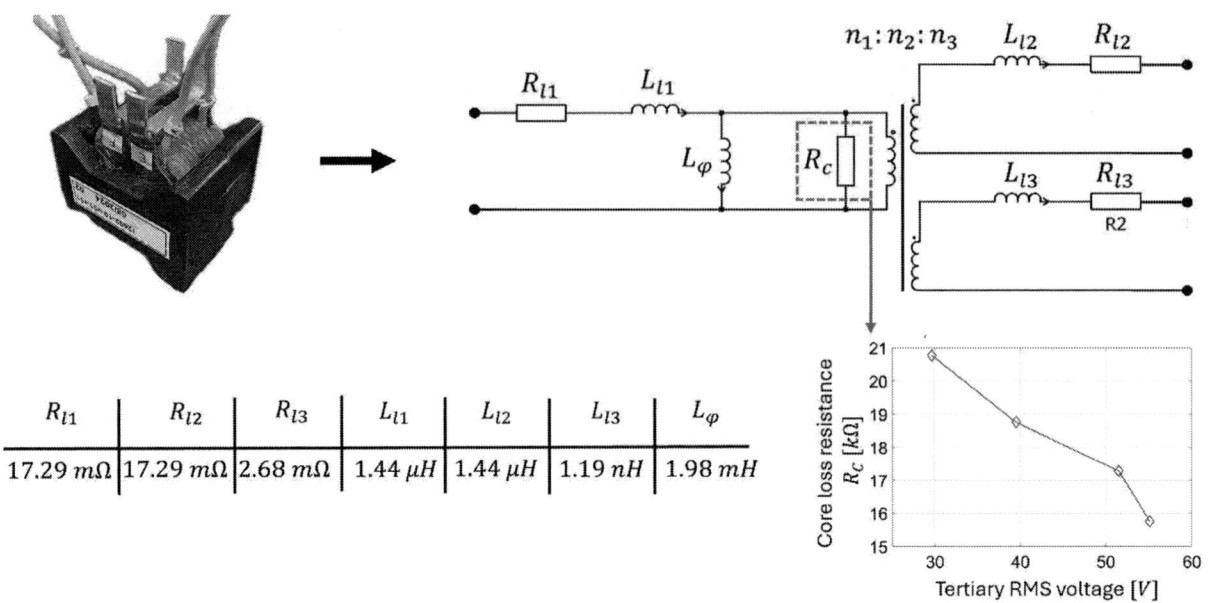

R_{l1}	R_{l2}	R_{l3}	L_{l1}	L_{l2}	L_{l3}	L_{φ}
$17.29\ m\Omega$	$17.29\ m\Omega$	$2.68\ m\Omega$	$1.44\ \mu H$	$1.44\ \mu H$	$1.19\ nH$	$1.98\ mH$

Fig. 6: Equivalent circuit representation and loss model of the high-frequency three-port transformer used in the TAB converter.

Designers may also incorporate an analytical loss model for the transformer or other magnetic components if the design targets component-level optimization to further enhance the converter performance.

IV. PROPOSED TAB DESIGN OPTIMIZATION

The proposed TAB design optimization selects the optimal values of transformer leakage inductors to minimize total converter losses (power semiconductor and transformer losses) across a wide range of operating conditions. The TAB uses FVC, where the power flow is controlled by varying the duty cycles of all three ports (δ_1, δ_2 and δ_3) and the phase angles of Port 2 and Port 3 with respect to Port 1 (φ_2 and φ_3). The analytical model of the TAB is derived using the GHA approach discussed in Section II for a selected harmonic order of 13.

Figure 7 illustrates the flowchart representation of the proposed TAB optimization process. The optimization algorithm averages the converter power loss across a wide range of operating conditions, characterized by different port voltages and power levels. The ranges for L_1, L_2', and L_3', represented as $L_{1,arr}$, $L_{2,arr}$, and $L_{3,arr}$ in the flowchart, are determined based on the calculated boundary values of the inductors capable of delivering the rated power for all values within the selected range of control variables and port voltages in the optimization problem. The voltage array ($V_{2,arr}$) represents the nominal and extreme points of the high-voltage battery, while the power array ($P_{x,arr}$) includes the output power levels for full-load and various partial-load conditions for the optimization study. The optimal

control variables for each operating point are computed using particle swarm optimization in MATLAB.

A. Particle Swarm Optimization (PSO) Problem

The objective function of the TAB optimization problem is defined using (11):

$$F = W_j J_o + W_c J_c \tag{11}$$

Here, J_o and J_c represent the individual cost functions corresponding to the total losses and the nonlinear equality constraints for the TAB converter. The calculations of J_o and J_c are defined in (12) and (13), respectively. $W_j = 5$ and $W_c = 2$ are the corresponding weighting factors selected empirically, using simulation studies, to achieve the required optimization performance.

$$J_o = P_{Tot} = P_{con} + P_{sw} + P_{Tr} \tag{12}$$

$$J_c = \sum_j \{P_{j(ref)} - P_j\}^2 \quad \text{where} \quad j = 2, 3. \tag{13}$$

A swarm size of 1000 and a maximum iteration of 500 have been chosen for the PSO problem to ensure better convergence and consistency in optimization. Additionally, the PS optimization problem involves five variables, corresponding to the FVC of the TAB, which determines the optimal duty cycles and phase angles for each operating point. The phase angle range is taken to be $-\frac{\pi}{2} \leq \varphi_j \leq \frac{\pi}{2}$, and the duty cycle angle range is $0 \leq \delta_j \leq \frac{\pi}{2}$ in the optimization process.

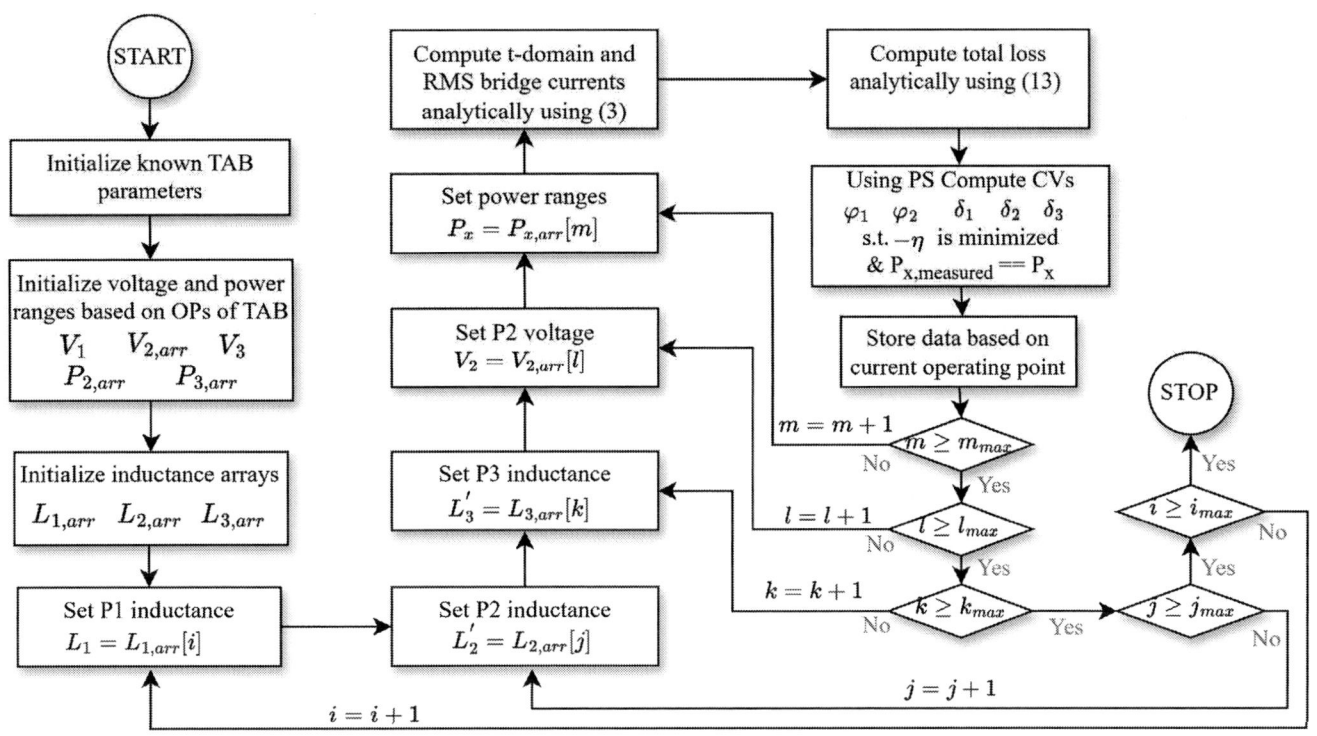

Fig. 7: Flowchart representation of the proposed TAB optimization scheme.

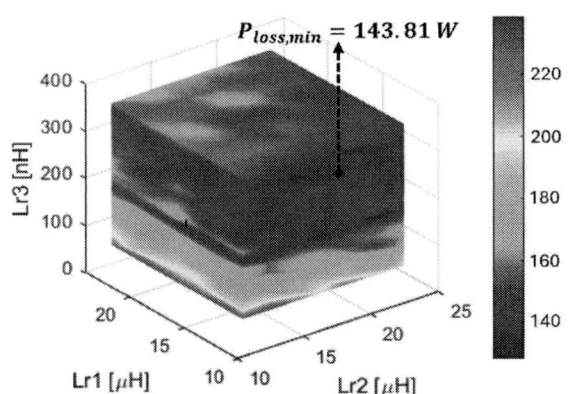

Fig. 8: TAB loss contour with respect to TAB leakage inductors.

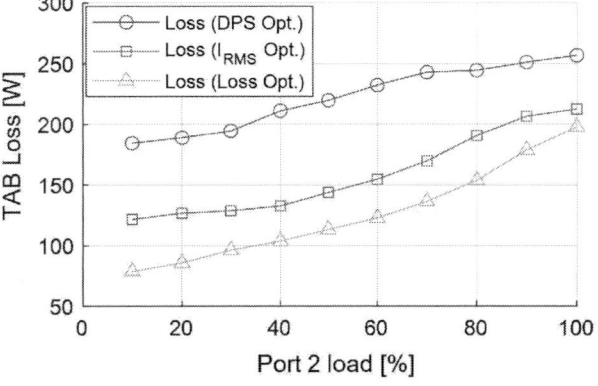

Fig. 9: MOSFET loss comparison for various TAB optimization techniques. $\{V_2 : 620V, P_3 : 1.8kW\}$

V. RESULTS & DISCUSSIONS

The TAB design optimization discussed in section IV and the circuit analysis of the optimally designed TAB are performed using PLECS Blockset software.

The maximum power rating of the TAB is selected to be 9.6 kW, targeted for an EV on-board charging application. The Port 1 voltage is fixed at 800 volts, while Port 2 voltage varies in the range of 620–840 V with respect to the state-of-charge (SOC) of the high-voltage EV battery. Port 3, also called the auxiliary power module (APM), is rated at 3.6 kW and a nominal voltage of 50 V. The turns ratio of the high-frequency transformer is selected to be $N_1 : N_2 : N_3 = 16 : 16 : 1$, based on the nominal voltage levels of the various ports. Hence, it is not part of the optimization process discussed in this study.

Fig. 8 illustrates the 3D contour of the total TAB converter losses obtained using the proposed optimization technique.

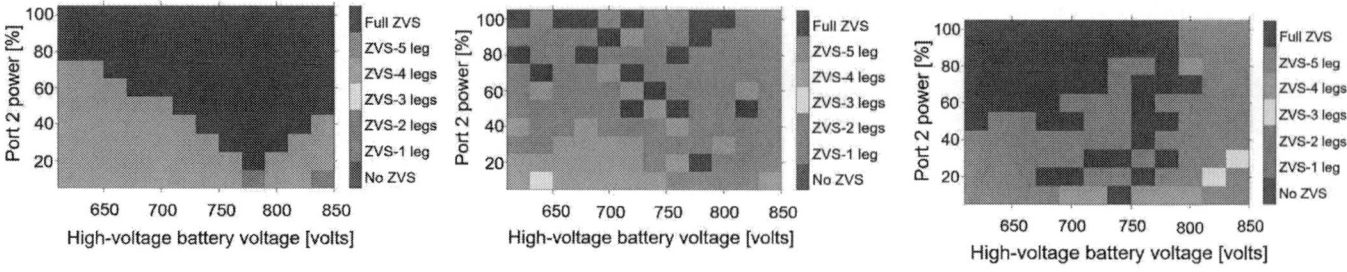

(a) Dual phase shift (DPS) modulation   (b) FVC with I_{RMS} minimization   (c) FVC with TAB loss minimization

Fig. 10: Comparison of ZVS performance of the TAB for various optimization techniques. $\{V_1 : 800\ V,\ V_3 : 50\ V,\ \&\ P_3 : 1.8kW\}$

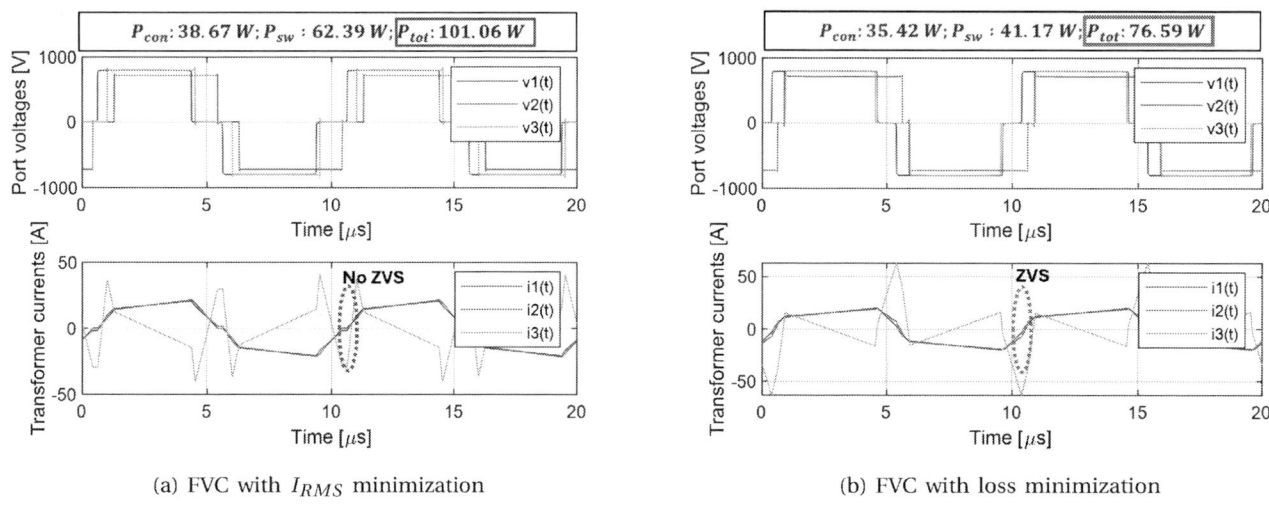

(a) FVC with I_{RMS} minimization   (b) FVC with loss minimization

Fig. 11: Comparison of TAB instantaneous waveforms. $\{V_{hv} : 720\ V; P_2 : 9.6\ kW; P_3 : 0\ kW\}$

The optimal leakage inductance values are $L_1 = 14\mu H$, $L_2^{'} = 22\mu H$, and $L_3^{'} = 166nH$, resulting in a minimum average TAB loss of 143.81 W.

The comparison of TAB losses for different optimization schemes is illustrated in Fig. 9 for various values of Port 2 power. The classical phase-shift control scheme, also known as dual-phase-shift modulation (DPS) in TAB converters, results in higher losses in the MOSFETs and transformer compared to other optimization techniques, primarily due to increased RMS and peak currents. Additionally, under light load conditions, the complete volt-seconds applied across the transformer cause higher core losses.

In contrast, the I_{RMS}-optimized control scheme reduces MOSFET conduction losses and transformer winding losses, improving converter efficiency, particularly under peak-load conditions. However, merely minimizing the RMS current does not guarantee optimal converter loss reduction, as losses depend significantly on the characteristics of the power MOSFETs and individual transformer windings.

Fig. 9 demonstrates that the proposed loss-optimized TAB design (FVC Opt.) achieves minimal overall TAB losses compared to other techniques, resulting in higher converter efficiency.

The soft-switching performance of the TAB for various designs is compared in Fig. 10. ZVS violations are significant for the DPS optimization scheme under light-load conditions, as depicted in Fig. 10a. The I_{RMS}-optimized control scheme focuses on reducing RMS current; however, it does not enhance ZVS conditions. In contrast, the proposed loss-optimized design extends the ZVS range across a wider operating range, as shown in Fig. 10c, resulting in reduced MOSFET losses and EMI concerns.

Figure 11 compares the instantaneous voltage and current waveforms of the TAB converter at a specific operating point for the I_{RMS}-optimized and loss-optimized designs. The proposed loss-optimized TAB design exhibits superior performance, achieving reduced losses by minimizing RMS and peak currents during switching events, as illustrated in Fig. 11b. Although the selected operating point is arbitrary, a consistent trend has been observed across various power and voltage levels, further validating the design.

VI. CONCLUSIONS

This paper presented a loss optimization design methodology for an isolated triple active bridge DC-DC converter in electric vehicle applications. The TAB is modeled using the generalized harmonic approximation technique, which improves modeling accuracy and reduces computational complexity compared to time-domain analysis techniques. The proposed optimization methodology selects the optimal values of high-frequency transformer leakage inductances to minimize converter losses over a wide range of operating conditions, leveraging the particle swarm optimization technique. Steady-state simulations in PLECS software demonstrate a significant reduction in MOSFET losses across all defined operating conditions. Additionally, the design extends the ZVS margin of the TAB, achieving approximately a 23% reduction in losses under full-load conditions at Port 2 while enhancing the converter's EMI performance. The proposed optimization methodology provides researchers and engineers with a comprehensive framework for designing high-efficiency, low-loss TAB converters, enabling improved performance and reliability in electric vehicle power systems.

REFERENCES

[1] H. Krishnaswami and N. Mohan, "Three-Port Series-Resonant DC–DC Converter to Interface Renewable Energy Sources With Bidirectional Load and Energy Storage Ports," *IEEE Transactions on Power Electronics*, vol. 24, no. 10, pp. 2289-2297, Oct. 2009, doi: 10.1109/TPEL.2009.2022756.

[2] H. Tao, A. Kotsopoulos, J. L. Duarte and M. A. M. Hendrix, "Multi-input bidirectional DC-DC converter combining DC-link and magnetic-coupling for fuel cell systems," *Fourtieth IAS Annual Meeting. Conference Record of the 2005 Industry Applications Conference*, 2005., Hong Kong, China, 2005, pp. 2021-2028 Vol. 3, doi: 10.1109/IAS.2005.1518725.

[3] Z. Wang, Q. Luo, Y. Wei, D. Mou, X. Lu and P. Sun, "Topology Analysis and Review of Three-Port DC–DC Converters," *IEEE Transactions on Power Electronics*, vol. 35, no. 11, pp. 11783-11800, Nov. 2020, doi: 10.1109/TPEL.2020.2985287.

[4] M. Michon, J. L. Duarte, M. Hendrix and M. G. Simoes, "A three-port bi-directional converter for hybrid fuel cell systems," *2004 IEEE 35th Annual Power Electronics Specialists Conference* (IEEE Cat. No.04CH37551), Aachen, Germany, 2004, pp. 4736-4742 Vol.6, doi: 10.1109/PESC.2004.1354836.

[5] P. Wang, X. Lu, W. Wang and D. Xu, "Hardware Decoupling and Autonomous Control of Series-Resonance-Based Three-Port Converters in DC Microgrids," *IEEE Transactions on Industry Applications*, vol. 55, no. 4, pp. 3901-3914, July-Aug. 2019, doi: 10.1109/TIA.2019.2906112.

[6] A. K. Jain and R. Ayyanar, "Pwm control of dual active bridge: Comprehensive analysis and experimental verification," *IEEE Transactions on Power Electronics*, vol. 26, no. 4, pp. 1215-1227, April 2011, doi: 10.1109/TPEL.2010.2070519.

[7] Dehong Xu, Chuanhong Zhao and Haifeng Fan, "A PWM plus phase-shift control bidirectional DC-DC converter," *IEEE Transactions on Power Electronics*, vol. 19, no. 3, pp. 666-675, May 2004, doi: 10.1109/TPEL.2004.826485.

[8] H. Bai and C. Mi, "Eliminate Reactive Power and Increase System Efficiency of Isolated Bidirectional Dual-Active-Bridge DC–DC Converters Using Novel Dual-Phase-Shift Control," *IEEE Transactions on Power Electronics*, vol. 23, no. 6, pp. 2905-2914, Nov. 2008, doi: 10.1109/TPEL.2008.2005103.

[9] H. Tao, A. Kotsopoulos, J. L. Duarte and M. A. M. Hendrix, "Transformer-Coupled Multiport ZVS Bidirectional DC–DC Converter With Wide Input Range," *IEEE Transactions on Power Electronics*, vol. 23, no. 2, pp. 771-781, March 2008, doi: 10.1109/TPEL.2007.915129.

[10] Haimin Tao, A. Kotsopoulos, J. L. Duarte and M. A. M. Hendrix, "Triple-half-bridge bidirectional converter controlled by phase shift and PWM," *Twenty-First Annual IEEE Applied Power Electronics Conference and Exposition*, 2006. APEC '06., Dallas, TX, USA, 2006, pp. 7 pp.-, doi: 10.1109/APEC.2006.1620700.

[11] F. Krismer and J. W. Kolar, "Efficiency-Optimized High-Current Dual Active Bridge Converter for Automotive Applications," *IEEE Transactions on Industrial Electronics*, vol. 59, no. 7, pp. 2745-2760, July 2012, doi: 10.1109/TIE.2011.2112312.

[12] S. Mukherjee, A. Kumar and S. Chakraborty, "Comparison of DAB and LLC DC–DC Converters in High-Step-Down Fixed-Conversion-Ratio (DCX) Applications," *IEEE Transactions on Power Electronics*, vol. 36, no. 4, pp. 4383-4398, April 2021, doi: 10.1109/TPEL.2020.3019796.

[13] D. Das and K. Basu, "Optimal Design of a Dual-Active-Bridge DC–DC Converter," in *IEEE Transactions on Industrial Electronics*, vol. 68, no. 12, pp. 12034-12045, Dec. 2021, doi: 10.1109/TIE.2020.3044781.

[14] C. Wang, T. -G. Zsurzsan and Z. Zhang, "Genetic Algorithm Assisted Parametric Design of Splitting Inductance in High Frequency GaN-Based Dual Active Bridge Converter," *IEEE Transactions on Industrial Electronics*, vol. 70, no. 1, pp. 522-531, Jan. 2023, doi: 10.1109/TIE.2021.3102398.

[15] A. Sengupta, T. Pereira and M. Liserre, "Design Optimization of Dual Active Bridge Converter for Supercapacitor Application," *IEEE Transactions on Power Electronics*, vol. 39, no. 9, pp. 11544-11557, Sept. 2024, doi: 10.1109/TPEL.2024.3401083.

[16] V. R. Kudaravalli, V. Uttam and V. M. Iyer, "A Design Methodology for Triple Active Bridge DC- DC Converter," *2022 IEEE International Conference on Power Electronics, Drives and Energy Systems (PEDES)*, Jaipur, India, 2022, pp. 1-7, doi: 10.1109/PEDES56012.2022.10080029.

[17] V. -L. Pham and K. Wada, "Design of Series Inductances in Triple Active Bridge Converter Using Normalization Procedure for Integrated EV and PV System," *2019 10th International Conference on Power Electronics and ECCE Asia* (ICPE 2019 - ECCE Asia), Busan, Korea (South), 2019, pp. 3027-3032, doi: 10.23919/ICPE2019-ECCEAsia42246.2019.8797142.

[18] N. Keshmiri, G. A. Mudiyanselage, S. Chakkalakkal, K. Kozielski, G. Pietrini and A. Emadi, "Design and Control Methodology of a Three-Port Resonant Converter for Electric Vehicles," *IEEE Open Journal of the Industrial Electronics Society*, vol. 3, pp. 650-662, 2022, doi: 10.1109/OJIES.2022.3215772

[19] P. Purgat, S. Bandyopadhyay, Z. Qin and P. Bauer, "Zero Voltage Switching Criteria of Triple Active Bridge Converter," *IEEE Transactions on Power Electronics*, vol. 36, no. 5, pp. 5425-5439, May 2021, doi: 10.1109/TPEL.2020.3027785.

[20] S. Dey, A. Mallik and A. Akturk, "Investigation of ZVS Criteria and Optimization of Switching Loss in a Triple Active Bridge Converter Using Penta-Phase-Shift Modulation," *IEEE Journal of Emerging and Selected Topics in Power Electronics*, vol. 10, no. 6, pp. 7014-7028, Dec. 2022, doi: 10.1109/JESTPE.2022.3191987.

[21] F. Krismer and J. W. Kolar, "Closed Form Solution for Minimum Conduction Loss Modulation of DAB Converters," *IEEE Transactions on Power Electronics*, vol. 27, no. 1, pp. 174-188, Jan. 2012, doi: 10.1109/TPEL.2011.2157976.

[22] A. Ganz, "A Simple, Exact Equivalent Circuit for the Three-Winding Transformer," *IRE Transactions on Component Parts*, vol. 9, no. 4, pp. 212-213, December 1962, doi: 10.1109/TCP.1962.1136764.

[23] Z. Qin, Y. Shen, P. C. Loh, H. Wang and F. Blaabjerg, "A Dual Active Bridge Converter With an Extended High-Efficiency Range by DC Blocking Capacitor Voltage Control," *IEEE Transactions on Power Electronics*, vol. 33, no. 7, pp. 5949-5966, July 2018, doi: 10.1109/TPEL.2017.2746518.

[24] Z. Guo, K. Sun, T. -F. Wu and C. Li, "An Improved Modulation Scheme of Current-Fed Bidirectional DC–DC Converters For Loss Reduction," *IEEE Transactions on Power Electronics*, vol. 33, no. 5, pp. 4441-4457, May 2018, doi: 10.1109/TPEL.2017.2719722.

[25] C. Zhao, S. D. Round and J. W. Kolar, "An Isolated Three-Port Bidirectional DC-DC Converter With Decoupled Power Flow Management," *IEEE Transactions on Power Electronics*, vol. 23, no. 5, pp. 2443-2453, Sept. 2008, doi: 10.1109/TPEL.2008.2002056.

[26] S. Dey and A. Mallik, "Multivariable-Modulation-Based Conduction Loss Minimization in a Triple-Active-Bridge Converter," *IEEE Transactions on Power Electronics*, vol. 37, no. 6, pp. 6599-6612, June 2022, doi: 10.1109/TPEL.2022.3141334.

[27] B. J. Baliga, *Advanced Power MOSFET Concepts*. Springer Science & Business Media, 2005.

A Magnetic-less DC/DC Converter with Pulse Charging for 800 V Powertrains from 400 V DC Fast Chargers

Duc Dung Le*, Shivam Chaturvedi*, Shahid Aziz Khan*, Mengqi Wang*, Mohamed Elshaer†
*Department of Electrical and Computer Engineering, University of Michigan-Dearborn, Dearborn, MI 48128 USA
†Ford Motor Company, Dearborn, MI 48124 USA
Email: dungle@umich.edu, shivamc@umich.edu, shahidkh@umich.edu, mengqiw@umich.edu, melshaer@ford.com

Abstract— In this letter, a magnetic-less DC/DC converter with a pulse-charging scheme is developed to achieve voltage compatibility between a DC fast-charging station and a high-voltage battery in an electric vehicle. It supports charging an 800 V battery system by typical fast-charging converters having an output voltage of 280 V to 420 V. The converter has a smaller component count than the existing ones, resulting in cost savings and loss reduction. In addition, the topology is operated at a low switching frequency to reduce the switching loss. Battery impedance has been examined by electrochemical impedance spectroscopy to determine a range of pulse frequencies, and the feasibility of the topology has been validated through an experiment with lead-acid batteries.

Keywords—800 V high-voltage EV battery, DC fast-charging station, electric vehicle (EV), magnetic-less DC/DC, pulse charging.

I. INTRODUCTION

Nowadays, one of the methods for enhancing the efficiency of the converter and motor in electric vehicles (EVs) is to increase the battery system voltage. There are many electric vehicles under development utilizing the high-voltage battery. Some of the EV automakers, including Porsche, Hyundai, and Kia, have configured their battery packs with a voltage level of 800 V instead of 400 V. Therefore, the current conducting through to the inverter and motor is reduced by 1/2, resulting in lower I2R losses and smaller wiring sizes [1].

However, most public fast-charging stations are 400 V DC chargers, which are inadequate to support the high charging voltage of 800 V. Thus, an auxiliary DC/DC converter is necessary to obtain compatibility between those fast chargers and the high-voltage EV battery system. In 2019, a 150kW fast-charging booster was introduced into the Porsche Taycan, which has provided a solution to charge the 800 V powertrain at a 400 V fast charger [2]. This boost topology consists of two half-bridges, one resonant tank, and two clamping capacitors. Power losses of the resonant tank, which mainly include AC copper and core losses, are significant for the total power loss of the topology. The converter has also been developed to boost the 300 V battery voltage to the 600 V DC bus voltage in hybrid electric vehicles [3].

A three-phase interleaved boost converter is one of the most popular step-up DC/DC converters, but its inductors are lossy and bulky components [4]. In 2021, the Hyundai Ioniq 5 was built with the 800 V Electric Global Modular Platform, where

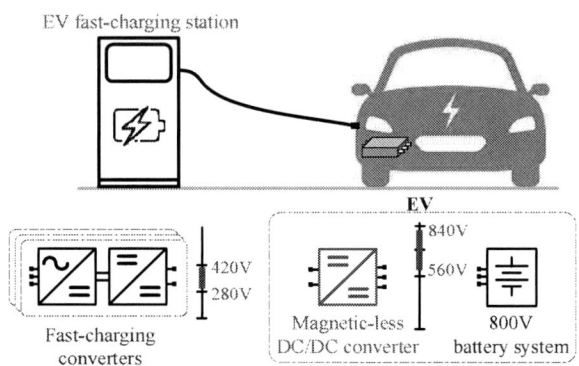

Fig. 1. Implementation of magnetic-less DC/DC converter in EV to charge the 800 V battery system at 400 V DC fast-charging station.

the inverter and the motor are reconfigured as a three-level interleaved boost converter to charge the high-voltage battery from a 400 V DC fast charger [5]. However, it is inefficient at light loads and the device voltage stress is equal to that of a high-voltage battery. In [6–7], a three-level boost converter has been shown to have half the voltage stresses on devices compared with those of the interleaved boost converter, but the inductor is still significant in the total power loss.

To step up the output voltage, a magnetic component is used the most, which is one of the most lossy and bulky components in the converter. Therefore, it is desirable to develop magnetic-less converters to reduce size, weight, and loss. In [8], a multilevel DC/DC conversion system that is composed of multiple magnetic-less converters has been studied for hybrid electric vehicles. In [9], a DC/DC converter has also been suggested to integrate with a 400 V on-board charger to charge an 800 V battery, where upper and lower batteries are charged alternately. Nevertheless, the DC/DC converter must be turned off before charging each module in every cycle to avoid high voltage stresses on the components in the on-board charger. In addition, the pulse frequency is very low, which does not meet the minimum battery impedance.

To improve the charging performance, a pulse charging method has been proposed [10], [11]. The short-term relaxation brings many favorable merits to the battery such as the elimination of polarization, slowing down of capacity decay,

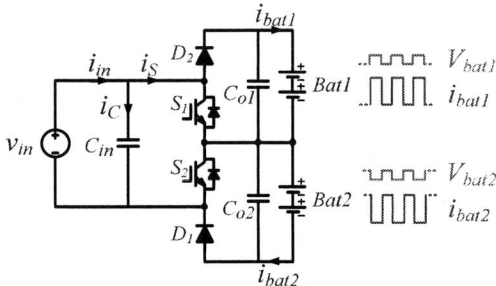

Fig. 2. Circuit configuration of magnetic-less converter.

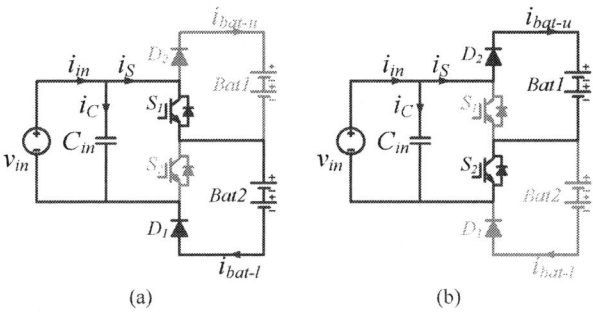

(a) (b)

Fig. 3. Current paths to charge batteries. (a) S_1 is turned ON and lower batteries are charged. (b) S_2 is turned ON and upper batteries are charged.

and acceleration of ion diffusion [10]. Therefore, pulse charging can be considered as a promising fast charging method [11].

In this letter, a magnetic-less DC/DC converter with pulse charging has been proposed to charge an 800 V battery system at a 400 V DC fast charger, as shown in Fig. 1. The converter has been developed with a smaller component count, leading to cost savings and loss reduction. An input capacitor is designed to reduce the voltage stresses for components in the front-end converters at deadtime. The converter efficiency is also improved by running at low switching frequency. The effectiveness of the converter has been verified by experimental results.

II. TOPOLOGY AND OPERATING PRINCIPLE

A. Converter Configuration and Operating Principle

The circuit configuration of the magnetic-less DC/DC converter is illustrated in Fig. 2. It includes two pulse-controlled switches (S_1 and S_2), which are used to control charging frequency; two diodes (D_1 and D_2); and input and output capacitors (C_{in}, C_{o1} and C_{o2}). Its input is connected to the 400 V fast-charging converter, and v_{in} is the input voltage. At the output side, the high-voltage battery is separated into upper and lower modules, Bat1 and Bat2.

Fig. 3 demonstrates an operating principle of the converter. During a switching period, there are two charging modes, which charge the upper and lower batteries alternately. Firstly, the lower battery, Bat2, is charged by turning ON S_1, the current flow path shown in Fig. 3(a). Secondly, the upper battery, Bat1, is charged by the current flowing through S_2 and D_2, shown in Fig. 3(b). It is noted that the maximum voltage stresses on

TABLE I
COMPARISON OF COMPONENT COUNT AND EFFICIENCY

Topologies	Proposed converter	Three-level boost converter [6], [7]	Switched tank converter [2], [3]	Three-phase interleaved boost converter [4], [5]
Switch	2 (650V/62.5A)	2 (650V/62.5A)	2 (650V/99A)	6 (1200V/31A)
Diode	2 (650V/62.5A)	2 (650V/62.5A)	2 (650V/99A)	-
Capacitor	3	2	3	2
Inductor	-	1	1	3
Switching frequency	Low	High	High	High
Efficiency	99.3%	98.9% [6], 97.3% [7]	98.3% [2], 97.4% [3]	98% [4]

Fig. 4. Input capacitor voltage ripple according to switching states of S_1 and S_2.

semiconductor devices depends on v_{in}. Therefore, 650 V rated power devices are used for the proposed converter.

B. Comparison Studies

A comparison with the existing topologies is listed in Table I. The RMS current values of power switches are measured from the simulation result of each converter, with a power rating of 50 kW. The proposed converter has the lowest number of passive components. Also, it can eliminate the need for thermal management for an inductive component in the converter design. Therefore, the efficiency of the proposed converter shows significant improvement. In addition, its control is not complicated, with S_1 and S_2 being switched complementarily with 50% of the duty cycle. In the case of unbalanced SOCs between the upper and lower batteries, the balancing performance can be achieved effortlessly by regulating the duty cycle.

III. INPUT CAPACITOR DESIGN AND PULSE FREQUENCY DETERMINATION

A. Converter Configuration and Operating Principle

C_{in} is mandatory for reducing voltage stresses on components in the front-end converters at the deadtime. The filter capacitor current, i_C, is given as:

$$i_C = i_{in} - i_S, \tag{1}$$

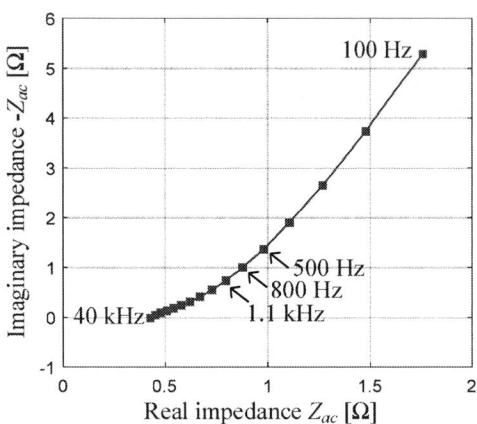

Fig. 5. Nyquist plot for lead-acid battery.

where i_{in} and i_S are the input current and converter current. They are composed of the average and ripple components as:

$$\overline{I}_C + \tilde{i}_C = \overline{I}_{in} + \tilde{i}_{in} - \overline{I}_S - \tilde{i}_S . \qquad (2)$$

The average current through the filter capacitor is zero, and thus $\overline{I}_{in} = \overline{I}_S$. If the filter capacitance is large enough, the input current ripple is zero ($\tilde{i}_{in} = 0$). $\tilde{i}_C$ is derived as:

$$\tilde{i}_C = -\tilde{i}_S = \overline{I}_S - i_S . \qquad (3)$$

The filter capacitor voltage ripple, $\tilde{v}_{Cin}$, is given as:

$$\tilde{v}_{C_{in}} = \frac{1}{C_{in}} \int_0^t \tilde{i}_C dt = \frac{1}{C_{in}} \int_0^t \left(\overline{I}_S - i_S \right) dt . \qquad (4)$$

Fig. 4 illustrates the waveforms of $\tilde{v}_{Cin}$ and switching states; the stray inductance and equivalent series resistance are negligible. $\tilde{v}_{Cin}$ is derived as:

$$\tilde{v}_{Cin} = \begin{cases} (\overline{I}_S - i_{bat1})\mathrm{T}_1 & t_0 \leq t \leq t_1 \\ \overline{I}_S t_d & t_1 \leq t \leq t_2 \\ (\overline{I}_S - i_{bat2})\mathrm{T}_3 & t_2 \leq t \leq t_3 \\ \overline{I}_S t_d & t_3 \leq t \leq t_4 \end{cases}, \qquad (5)$$

where t_d is the deadtime. From (5), the maximum voltage ripple can be calculated as:

$$\Delta v_{Cin\_\max} = \frac{\overline{I}_S t_d}{C_{in}}, \qquad (6)$$

Therefore, the minimum filter capacitance is achieved as:

$$C_{in} \geq \frac{\overline{I}_S t_d}{\Delta v_{Cin\_\max}} . \qquad (7)$$

TABLE II
PARAMETERS USED FOR EXPERIMENT

Parameters	Symbol	Value
Power	P	25 kW
Input voltage	v_{in}	241 ~ 400 V
Output voltage	v_o	471 ~ 800 V
Input capacitance	C_{in}	100 μF
Output capacitance	C_{o1}, C_{o2}	330 μF
Switching frequency	f_{sw}	1100 Hz
Dead time	t_d	2 μS

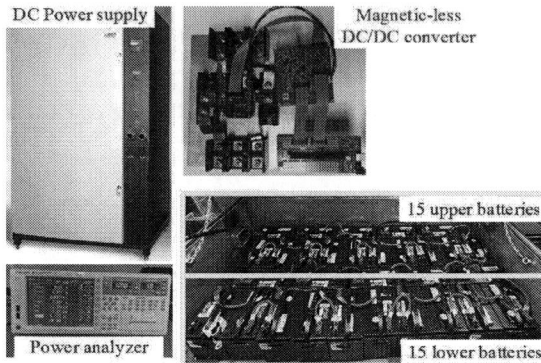

Fig. 6. Experimental setup.

The input capacitance with 100 μF has been selected in this letter.

B. Pulse Frequency Determination

To reduce the switching loss, S_1 and S_2 can be operated with a low switching frequency. However, to reduce the energy loss in the battery charging process [10], the pulse frequency should be determined according to the AC impedance of the battery. For this letter, a series of NAPA Marine lead-acid batteries (12 V) has been used for the upper and lower batteries. Fig. 5 shows the Nyquist plot (impedance spectrum) obtained by electrochemical impedance spectroscopy (Bio-Logic VMP3). As can be seen from the plot, the ohmic resistance of Z_{ac} becomes minimum at 40 kHz, the boundary point between the middle and high frequencies. However, 40 kHz is relatively high for pulse charging and the zero current condition is too short to distribute the ions evenly.

In Fig. 2, the converter current, i_S, is expressed as:

$$i_S(t) = \frac{V_{diff}}{\sqrt{\dfrac{L}{C_o} - \left(\dfrac{R_{ds\_on}}{2} \right)^2}} \sin\left(\sqrt{\dfrac{1}{LC_o} - \left(\dfrac{R_{ds\_on}}{2L} \right)^2}\, t \right) e^{-\frac{R_{ds\_on}}{2L}t}$$
$$+ I_{bat} \qquad (8)$$

where V_{diff} is the voltage difference between the input voltage and battery voltage, L is the stray inductance, and R_{ds_on} is the drain-source on resistance. The di/dt of i_{bat} is limited by L, which is taken into consideration in the pulse frequency determination. In the experimental setup, L is 1.2 μH. To obtain

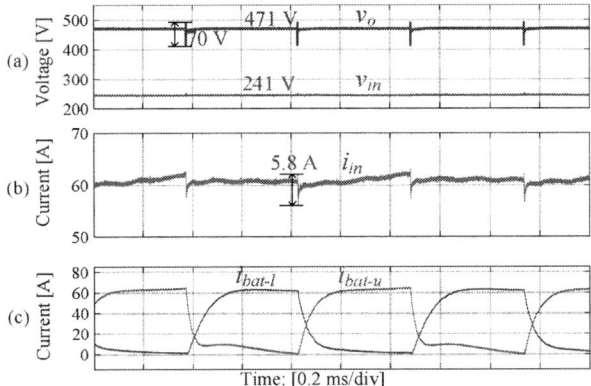

Fig. 7. Performance of converter at 12.5 kW. (a) Input and output voltages. (b) Input current. (c) Upper and lower battery's currents.

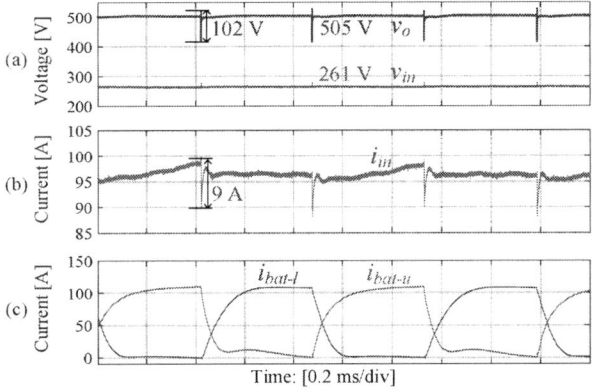

Fig. 8. Performance of converter at 25 kW. (a) Input and output voltages. (b) Input current. (c) Upper and lower battery's currents.

a steady state of pulse amplitude at 110 A (25 kW), the pulse frequency of 1.1 kHz is selected. In addition, the pulse frequency for Li-ion batteries is usually from 1 kHz to 1.5kHz [10]–[12]. Also, the switching loss of the converter is low.

IV. EXPERIMENTAL RESULTS

An experimental study with 30 lead-acid batteries has been conducted to verify the proposed topology. Fig. 6 shows the hardware setup of the system. The battery is connected to the converter from 165 cm away. The parameters for the experiment are given in Table II.

Fig. 7 shows the steady-state operation at 12.5 kW. Fig. 7(a) shows the input and output voltages of the converter, where the average value of v_{in} is 241 V. As can be seen, the input voltage does not change abruptly at the dead time, and the input voltage ripple is about 8 V. Therefore, it can avoid high voltage stresses for components in the front-end converters. The output voltage (v_o) is about 471 V, which is lower than twice that of v_{in} owing to voltage drops at the drain-source on-state resistance and bus bars. In Fig. 7(b), the input current (i_{in}) is maintained at 61 A with 5.8 A of fluctuation. Fig. 7(c) illustrates the currents of the upper and lower batteries. *Bat1* and *Bat2* are charged alternately with a pulse frequency of 1.1 kHz and 0.5 of the duty cycle. The pulse amplitude of the charging current is 63 A.

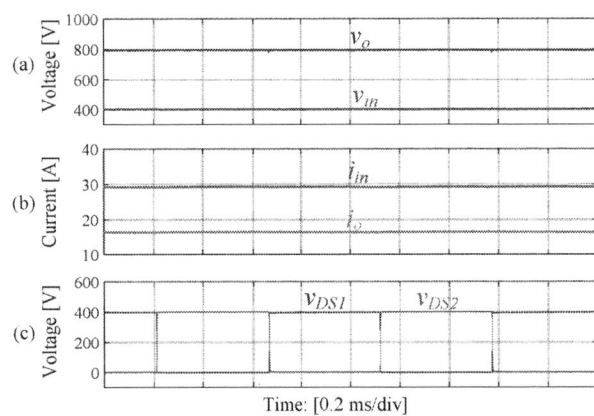

Fig. 9. Performance of converter with resistive load at $v_i = 400$ V. (a) Input and output voltages. (b) Input and output currents. (c) Drain-source voltages of S_1 and S_2.

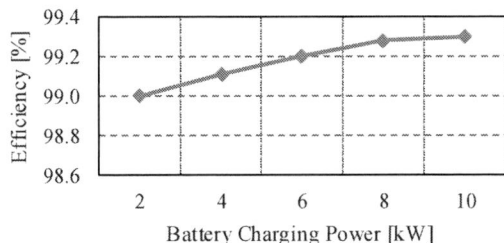

Fig. 10. Measured efficiency of proposed converter.

Fig. 8 shows the steady-state operation at 25 kW, where v_{in} is equal to 261 V and the pulse amplitude is increased. During operation, if the SOCs of the upper and lower batteries are different, they can be balanced by adjusting the duty cycles. Since the number of lead-acid batteries is not enough to validate the effectiveness of the converter at 800 V output, an operation with resistive load has been performed. Fig. 9 shows the performance of the converter with a resistive load when the input voltage is equal to 400 V. In Fig. 9(a), v_o is maintained well at 800 V. Fig. 9(c) demonstrates the drain-source voltages of switches, where they are switched complementarily with 50% of the duty cycle.

The converter efficiency is measured at different powers, as illustrated in Fig. 10. With low switching frequency, the conduction losses of the semiconductor devices are major contributions. The efficiency reaches to about 99.3% at 10 kW.

V. CONCLUSIONS

In this letter, a magnetic-less converter with a pulse charging scheme has been proposed to achieve voltage compatibility between the 400 V DC fast charger and the 800 V high-voltage EV battery. It is composed of semiconductor devices with low voltage stress, and it has a smaller passive component count than the existing ones. In addition, the converter has been operated at a low switching frequency to minimize the switching loss. The input capacitor has been designed and the pulse frequency range is determined according to the AC impedance of the battery. A prototype has been built and tested

979-8-3315-1612-3/25 $31.00 © 2025 IEEE

at 25 kW with lead-acid batteries, and a high efficiency of 99.3% is obtained.

REFERENCES

[1] I. Aghabali, J. Bauman, P. J. Kollmeyer, Y. Wang, B. Bilgin and A. Emadi, "800-V Electric Vehicle Powertrains: Review and Analysis of Benefits, Challenges, and Future Trends," IEEE Trans. Transport. Electrific., vol. 7, no. 3, pp. 927-948, Sept. 2021.

[2] High Power Charge Pump with Inductive Elements, by Enzo Illiano. (2017, Aug 30). WO/2018/046370 [Online]. Available: https://patents.google.com/patent/WO2018046370A1/en.

[3] Z. Ni, Y. Li, C. Liu, M. Wei and D. Cao, "A 100-kW SiC Switched Tank Converter for Transportation Electrification," in IEEE Trans. Power Electron, vol. 35, no. 6, pp. 5770-5784, June 2020.

[4] J. Zhang, J. -S. Lai, R. -Y. Kim and W. Yu, "High-Power Density Design of a Soft-Switching High-Power Bidirectional dc–dc Converter," in IEEE Trans. Power Electron, vol. 22, no. 4, pp. 1145-1153, July 2007.

[5] Electric Vehicle, by Y. -W. Lee, et al. (2019, Apr 03). EP3461679A2 [Online]. Available: https://patents.google.com/patent/EP3461679A2/en.

[6] Youn, H.-S, Yun, D.-H, Lee, W.-S and Lee, I.-O. "Study on Boost Converters with High Power-Density for Hydrogen-Fuel-Cell Hybrid Railway System" Electronics, vol. 9, no. 5, pp. 771, May 2020.

[7] M. Lee and J. -S. Lai, "Fixed-Frequency Hybrid Conduction Mode Control for Three-Level Boost PFC Converter," in IEEE Trans. Power Electron, vol. 36, no. 7, pp. 8334-8346, July 2021.

[8] M. Shen, F. Z. Peng and L. M. Tolbert, "Multilevel DC–DC Power Conversion System with Multiple DC Sources," in IEEE Trans. Power Electron, vol. 23, no. 1, pp. 420-426, Jan. 2008.

[9] J. -Y. Kim, B. -S. Lee, D. -H. Kwon, D. -W. Lee and J. -K. Kim, "Low Voltage Charging Technique for Electric Vehicles With 800 V Battery," in IEEE Trans. Ind. Electron, vol. 69, no. 8, pp. 7890-7896, Aug. 2022.

[10] L. Jiang, et al., "Optimal Charging Strategy with Complementary Pulse Current Control of Lithium-Ion Battery for Electric Vehicles," in IEEE Trans. Transport. Electrific., vol. 8, no. 1, pp. 62-71, March 2022.

[11] D. R. R. Kannan and M. H. Weatherspoon, "The Effect of Pulse Charging on Commercial Lithium Nickel Cobalt Oxide (NMC) Cathode Lithium-ion Batteries", J. Power Sources, vol. 479, Dec. 2020.

[12] Chen, Liang-Rui, et al. "Sinusoidal-ripple-current Charging Strategy and Optimal Charging Frequency Study for Li-ion Batteries." IEEE Trans. Ind. Electron., vol.60, no.1, pp.88-97, Jan. 2013.

Boosting Charger Efficiency: A GaN-Based Flyback Converter with Energy Recycling

Ahmad Nabizadah, *Student Member, IEEE*, Majid Ghasemi Korrani, *Student Member, IEEE*, and Babak Fahimi, *Fellow, IEEE*

Renewable Energy and Vehicular Technology (REVT) lab

Department of Electrical Engineering, University of Texas at Dallas

Richardson, TX, USA

Email: Majid.GhasemiKorrani@utdallas.edu

Abstract—In this paper, design and implementation of a high-efficiency flyback converter tailored for mobile phone chargers is presented. The proposed converter has been evolved through three generations, each addressing specific power loss challenges. The first-generation design utilized a standard flyback converter with an RCD clamp snubber, resulting in an efficiency of 76%. The second generation introduced synchronous rectification, minimizing output rectifier losses. The third generation, which is the focal point of this paper, further enhanced efficiency by recycling leakage energy using a GaN switch and a recycling snubber circuit. This design achieved an experimental efficiency of 88.6% at 10W output power, 4.6% higher than the previous generation that does not use leakage energy recycling. Additionally, it outperformed Samsung chargers by up to 2% and Apple chargers by up to 12.6%. The converter is simulated and implemented to validate its performance and efficiency and to verify the efficacy of the proposed concepts. Both simulation and experimental results confirm the effectiveness of the proposed design.

Keywords—flyback converter, Mobile charger, Leakage energy, Gan switches, Power conversion.

I. INTRODUCTION

Mobile phone chargers are essential in today's society, powering billions of devices that keep us connected. The prevalence of mobile phones in both personal and professional settings highlights the necessity for reliable and efficient charging solutions. In 2023, the global cell phone charger market revenue reached $18.5 billion. By 2032, it is expected to grow to $41.3 billion, with wired chargers generating $21.3 billion and wireless chargers contributing $20.0 billion. With over 7 billion chargers in use, the need for energy-efficient chargers has become critical. Enhancing the efficiency of mobile phone chargers by even 1% can lead to significant energy savings globally, reducing electricity consumption and their environmental impact. [1].

This improvement extends battery life, enhances device performance, and increases user satisfaction. Therefore, the drive for higher efficiency in mobile phone chargers is not merely a technical challenge but a crucial step towards sustainable energy use and environmental conservation [2]-[8]. In the 3C (Computer, Communication, and Consumer Electronics) market, a variety of converters are employed, with the flyback converter being a standout choice due to its inherent advantages [8]-[13]. These include a simple structure, cost-effectiveness, support for multiple outputs, and electrical isolation capabilities [14]. The flyback converter is highly suitable for low-power applications, such as mobile phones, power delivery devices, and tablet computers. Its straightforward design and ability to provide multiple output voltages make it an ideal solution for these applications, further emphasizing the importance of optimizing efficiency in this family of converters [14]-[18].

Flyback converters require an air gap in the transformer core to store energy. Increasing this air gap enables the transformer to store more energy, but it reduces permeability and increases leakage inductance [18]-[23]. Consequently, conventional flyback converters often need a snubber circuit to mitigate voltage spikes caused by this increased leakage inductance [6]. While effective in suppressing voltage spikes, snubber circuits can reduce converter efficiency and cause power switch losses due to hard switching, ultimately leading to poor overall efficiency. There are multiple methods to recover this energy and boost overall efficiency. In [20]-[24], the active clamp flyback converter was used to recycle leakage inductance energy and to achieve soft switching for both primary and auxiliary switches, enhancing full-load efficiency. However, under light load conditions and in the presence of variations in leakage inductance and snubber capacitor, the secondary-side rectifier conduction angle is affected, which leads to efficiency reduction. Additionally, using the two active switches in this method increases cost [25].

The major losses in smartphone chargers are divided into switching losses and clamp losses[26]. Various studies have reduced the aforementioned losses and enhanced converter efficiency using timing generators and pulse frequency modulation (PFM) [27].The mentioned methods primarily address the switching frequency losses, which are not dominant in smartphone charger applications. However, the major loss in the low-power flyback chargers is the clamp loss, which needs more attention [28]. Accordingly, considering both types of losses, proposing smartphone chargers with higher efficiency is a research direction that attracts significant attention due to the widespread use of these chargers globally [29]-[30].

In the present paper, three generations of the flyback converter were designed and implemented for charging applications. In the first generation, the goal was to identify the main sources of power losses. A standard flyback with an RCD clamp snubber was implemented, revealing that most losses occurred in the clamp circuit and the diode rectifier on the secondary side, resulting in an efficiency of 76%. The second-generation design focused on reducing the highest loss, which was the output rectifier loss due to the diode's forward voltage drop, even when using Schottky diodes. To address this, the diode was replaced with a low R_{ds-on} MOSFET, and synchronous rectification was employed. The third version, which is the focus of this paper, aimed to reduce clamp losses

979-8-3315-1612-3/25 $31.00 © 2025 IEEE

by recycling the leakage energy. By designing a flyback converter based on GaN switches and using a recycling snubber, the efficiency of the proposed charger increased to 88.6%. Thus, the proposed charger offers more efficiency than current market chargers.

II. PRINCIPAL OF OPERATION

A. Conventional flyback converter

The flyback converter offers the advantage of achieving high voltage gains by adjusting the transformer's turn ratio. This enables direct connection of a battery or multiple parallel batteries to the DC bus, eliminating the need for series connections and reducing imbalances in battery state-of-charge. The flyback converter has a straightforward design and provides galvanic isolation between the battery and the load, making it well-suited for charger systems. With just one switch, it can operate with high efficiency. Conventional flyback converters are commonly used in phone chargers. In this converter, an RCD snubber is essential to protect the switch from voltage spikes. However, a significant drawback is power loss due to leakage energy dissipated in the resistor, which lowers overall efficiency. Furthermore, the voltage drop across the output diode is significant, as it carries the primary circuit current, resulting in substantial power loss. Addressing this diode loss and other design limitations could further enhance the flyback converter's efficiency.

B. Principal operation of the flyback converter with recycling snubber

Fig. 1(a) illustrates the proposed flyback converter with a leakage energy recycling mechanism. The input stage comprises an EMI filter and a full-bridge rectifier, which charges the bulk capacitor, C_{bulk}. The snubber capacitor, C_2, stores the transformer's leakage energy. Inductor L_2 then channels this stored energy from C_2 back into the bulk capacitor and magnetizing inductor L_m. The operation of this circuit can be divided into three stages:

Stage1: Stage 1 begins when the primary side FET, S_1, is turned on, as illustrated in Fig. 1(b). Current flows from the bulk capacitor (C_{bulk}) through S_1 and into the transformer, where energy is stored in the magnetizing inductor. Concurrently, the previously stored leakage energy in C_2 is transferred to L_2, as indicated by the blue current path. Equation (1) describes the current oscillation within this LC circuit, where L_2 is the snubber inductor, C_2 is the snubber capacitor, and V_{C0} and I_{L0} represent the initial values of the capacitor voltage and inductor current, respectively. Once all the energy from the capacitor is transferred to L_2, the inductor releases energy back into the snubber capacitor, charging it with the opposite polarity. After C_2 is fully charged to this opposite polarity and the inductor current reaches zero, diode D_1 turns off to prevent any further energy transfer from C_2 to the inductor.

$$I_{leak}(t) = V_{c0}\sqrt{\frac{C_2}{L_2}}\sin\left(\frac{t}{\sqrt{L_2 C_2}}\right) + I_{L0}\cos\left(\frac{t}{\sqrt{L_2 C_2}}\right) \quad (1)$$

Stage 2: This stage is illustrated in Fig. 1(c). At this stage, the primary side switch S_1 is turned off, and the current through the switch becomes zero. However, the oppositely charged snubber

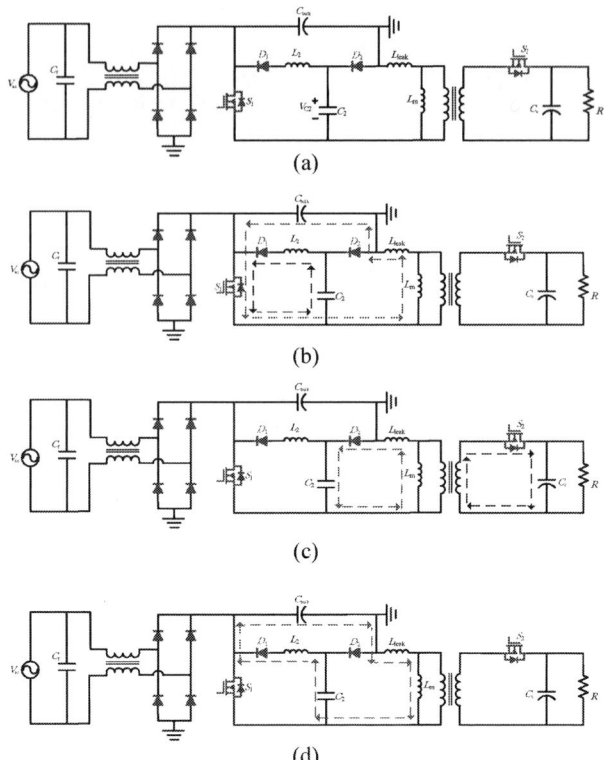

Figure 1: Modes of operation in the proposed flyback with recycling snubber. (a) proposed flyback converter topology (b) state 1 of operation S1 is on (c) state 2 of operation S_1 is off and S2 is on (d) state 3 of operation

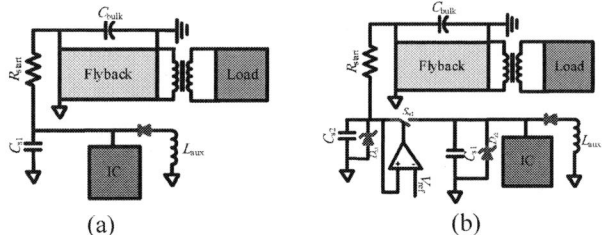

Figure 2: Start-up circuit of the flyback converter. (a) conventional start-up circuit (b) proposed start-up circuit

capacitor C_2 will keep the magnetizing current flowing until all of its energy, including any stored leakage energy, is transferred to the magnetizing inductance while its voltage reaches zero. Then, the magnetizing inductance will begin to charge the snubber capacitor back in the positive polarity. When the snubber capacitor voltage reaches the reflected output voltage, S_2's body diode begins to conduct, and the magnetizing energy is being transferred to the secondary side. Simultaneously, the leakage energy of the current cycle will transfer to the snubber capacitor, causing its voltage to rise above the reflected output voltage. Equation (2) represents the current passing through the leakage inductance (L_l) and snubber capacitor (C_2). Time t_0 in equation (3) represents the time when the current reaches zero, the capacitor voltage stabilizes, and diode D_2 turns off.

$$I_{leak}(t) = V_{C0}\sqrt{\frac{C_2}{L_l}}\sin\left(\frac{t}{\sqrt{L_lC_2}}\right) + I_{L0}\cos\left(\frac{t}{\sqrt{L_lC_2}}\right) \qquad (2)$$

Furthermore, the leakage current in equation (2) can be substituted into equation (4) to calculate the energy transferred from the leakage inductance to the snubber capacitor. This calculation of leakage energy is important for selecting the appropriate value for the snubber capacitor, as will be described in Section II. Once all the leakage energy is transferred to the snubber capacitor, the leakage current will become zero. Additionally, when the body diode of S_2 begins to conduct, synchronous rectification starts, allowing S_2 to turn on and improve rectification efficiency.

$$t_0 = \frac{1}{\sqrt{LC}}tan^{-1}\left(\frac{I_{L0}}{V_{c0}}\sqrt{\frac{L_l}{C_2}}\right) \qquad (3)$$

$$E = \int_0^{t0}\frac{1}{2}LI_{leak}^2 \qquad (4)$$

Stage 3: Stage 3 begins when the magnetizing current reaches zero, causing the converter to enter discontinuous mode. In this mode, the reflected output voltage drops to zero. This allows the snubber capacitor to recycle its energy back to the bulk capacitor until its voltage decreases to match that of the bulk capacitor. The leakage recycling path is illustrated in Fig. 1(d) by the red current path ($C_2 \rightarrow L_2 \rightarrow D_1 \rightarrow C_{bulk} \rightarrow L_m \rightarrow C_2$). As a result, a portion of the leakage energy from the previous cycle is transferred to the bulk capacitor, where it will be utilized in subsequent cycles. The amount of leakage energy recycled during this stage depends on the duration for which the converter remains in discontinuous mode before switch S_1 turns on. If the converter stays in discontinuous mode for an extended period, nearly all the leakage energy will be recycled at this stage. Conversely, if the discontinuous period is brief, only a portion of the energy will be recycled, and the remaining energy will be returned to the magnetizing inductance once S_1 turns on, as explained in Stages 1 to 2.

C. Second Primary Ground and Start-up circuit

Unlike conventional flyback circuits, the proposed topology features the switch in a high-side configuration. Although high-side switching can be done with high-side gate drivers, it is more expensive and less efficient due to its higher quiescent current than low-side gate drivers. Using a high-side primary gate driver can lead to reduced efficiency in a converter where efficiency is critical. Therefore, the proposed converter uses a low-side gate driver to drive the high-side switch to minimize costs while maintaining high efficiency. To enable the low-side driver to control the high-side switch, the source node of the switch (Fig. 1(a).) is designated as a second floating ground. All primary-side circuits reference this ground node. The transformer auxiliary winding is used to supply power to the ICs relative to this second ground node. However, initially, no switching occurs, resulting in zero voltage across the auxiliary winding. A startup circuit is implemented to provide the necessary initial current to the ICs until the converter enters continuous switching and the auxiliary voltage reaches a

sufficient level. The total startup current required for the circuit is approximately 1 mA. To supply this current, a large resistor, R_{start}, can be connected from the C_{bulk} node to the VCC capacitor of the ICs (C_{s1}), as illustrated in Fig. 2(a). Under direct current (DC), the bulk capacitor ground and the second floating ground (the source node of the switch) can be considered as being short-circuited since the only component between them is the magnetizing inductance of the transformer. As a result, C_{s1} will charge from C_{bulk}, allowing the circuit to start up. However, there is a significant issue with this simple startup method. Assuming a bulk voltage (V_{bulk}) of 170V, the resistor R_{start} needs to be 170 kΩ to supply the 1 mA startup current, as determined by equation (5). This presents a problem because the 1 mA current will continue to flow through the resistor even after the converter has reached a steady state, leading to a power loss of 170 mW. This loss can significantly reduce efficiency in a 10W converter, especially during low-power operation. Although using a very large R_{start} can decrease these losses, it will not provide sufficient current for startup.

$$R_{start} = \frac{V_{bulk}}{I_{start}} = 170k \qquad (5)$$

To enhance efficiency, the startup circuit depicted in Fig. 2(b) is proposed. In this configuration, a very large resistor, R_{start}, valued at 2.2 MΩ, is connected to the capacitor C_{s2}, which is a 6.8 μF, 50V ceramic capacitor. Due to the large resistance value, the power loss is negligible, approximately 10 mW. During the startup process, C_{s2} begins charging very slowly, drawing around 77 μA of current. The switch S_{w1} remains off to prevent the ICs from draining current from C_{s2} while it is charging. Once the voltage across C_{s2} reaches 20V, S_{w1} is activated, allowing a large current to flow from C_{s2} to C_{s1}. The capacitor C_{s1} has the same capacitance as C_{s2}, enabling it to charge rapidly to 10V. This capacitance is sufficient to maintain the Vcc voltage until the startup phase is complete and the auxiliary winding takes over. Consequently, this method leads to a minimal power loss of 13 mW across R_{start}.

III. SIMULATION RESULTS

A. Simulation of the proposed flyback in MATLAB Simulink

Fig. 3 shows the proposed converter's waveforms simulated in MATLAB. The waveforms are for the parameters of Table 1. This section will describe one cycle's leakage recycling and the primary side operation. The leakage recycling occurs in two phases.

Phase 1: At time t_1, the primary side switch (S_1) is off, and the voltage across the snubber capacitor (V_{C2}) is equal to the reflected voltage of 120V, with a stored energy of 7.2 μJ. Between t_1 and t_2, all the leakage energy from the current cycle is transferred to the snubber capacitor, causing its voltage to increase to 240V, corresponding to a stored energy of 28.8 μJ. This means that 21.6 uJ has been added to the capacitor's stored energy due to the leakage inductance. By the end of the switching cycle, all 21.6 μJ of leakage energy must be recycled, and the snubber capacitor must return to its initial stored energy of 7.2 μJ.

979-8-3315-1612-3/25 $31.00 © 2025 IEEE

The overshoot in the switch is 120V, which equals the increase of the snubber capacitor's voltage above the reflected voltage.

Between t_2 and t_3, the switch remains off while the magnetizing current decreases linearly, transferring magnetizing energy to the secondary side. At t_3, the magnetizing current reaches zero, and the converter enters discontinuous mode. The reflected voltage approaches zero, while the voltage across the switch approaches the bulk capacitor voltage of 170V. Since the reflected output voltage has reached zero, this allows the snubber capacitor (C_2) to recycle its energy back to the bulk

Table 1: Simulation Parameters

Parameters	Value
Input Voltage	170VDC
Snubber inductor L1	470 uH
Snubber Capacitor C2	1 nF
Switching Frequency	75 kHz
Transformer Turn Ration	22:1
Output Voltage/Current	5.4 V, 2A

capacitor (C_{bulk}) through the current loop shown in Fig. 1(d). As a result, the voltage across the snubber capacitor decreases to 125V, with a stored energy of 7.81 µJ. This indicates that out of a total leakage energy of 21.6 µJ, 20.99 µJ has been recycled, leaving 0.61 µJ remaining. The reason for the high amount of energy recycling is that the converter remained in the discontinuous mode long enough for the snubber capacitor to recycle most of the leakage energy. In situations where the discontinuous period is shorter, significantly smaller amount of energy will be recycled during this phase, and the remainder will be recycled in the subsequent phase. This concludes the first phase of energy recycling.

Phase 2: At t_4, the switch turns on, causing the voltage across the capacitor to decrease as its energy is transferred to the snubber inductor (L_2) through the current loop shown in Fig 1(b). Simultaneously, the magnetizing current begins to rise linearly, and the switch current is the sum of the snubber circuit current and the magnetizing current. By t_5, the snubber capacitor voltage reaches zero, and all energy is transferred to the snubber inductor. The snubber inductor keeps the current flowing and charges the capacitor with the opposite polarity. At t_6, all the energy from the snubber inductor is transferred back to the snubber capacitor, charging it to -125V. The energy in the capacitor returns to 7.8 µJ. Between t_6 and t_7, the current due to the recycling circuit is zero, and the current flowing through the switch consists solely of the magnetizing current, which increases linearly until the switch is turned off at t_7. When this occurs, the switch current drops to zero. However, the oppositely charged snubber capacitor keeps the magnetizing current flowing by transferring its energy (7.8 µJ) back to the magnetizing inductance, causing its voltage to reach zero (The current loop is shown in Fig. 1(c)). Following this, the magnetizing energy recharges the capacitor up to the reflected output voltage of approximately 120V, with a stored energy of 7.2 µJ at t_8. The energy difference of 0.61 µJ is recycled back to the magnetizing inductor. Although this amount of recycled energy is small, in situations where the discontinuous period is

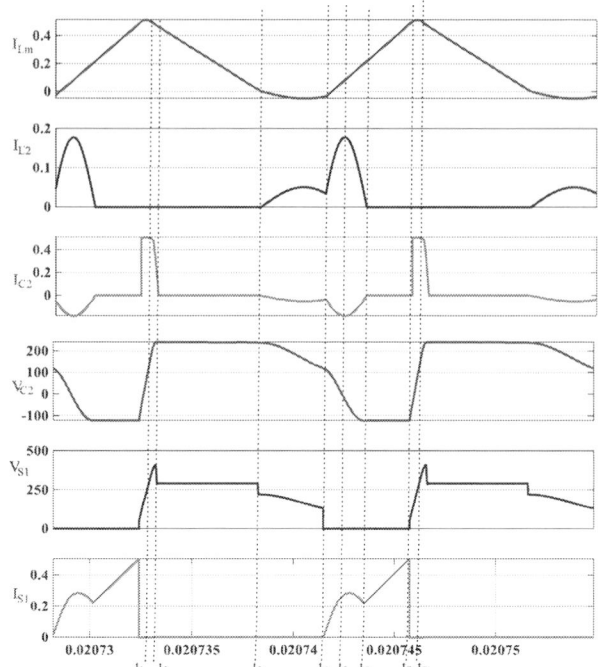

Figure 3: Simulation waveforms of the proposed converter (a) Magnetizing current (b) Snubber inductor current (c) Snubber capacitor current (d) Snubber capacitor voltage (e) Primary switch drain-source voltage (f) Primary switch drain-source current.

small, a larger amount of the energy is returned to the magnetizing inductor during this phase. This concludes the second phase of recycling. At this stage, the energy is returned to the initial state with a stored energy of 7.2 µJ, having recycled all the leakage energy of the previous cycle, and the loop continues.

B.   Snubber circuit design

To select the appropriate snubber capacitor, it is essential to understand the energy stored in the leakage inductor. The maximum amount of energy in the leakage inductor occurs just before the primary switch turns off, as this is when the magnetizing current reaches its peak. The energy stored in the leakage inductance is expressed by equation 6.

$$E_{Lk} = \frac{1}{2} L_{LK} I_{LKPK}^2 \tag{6}$$

is used. The peak current at the leakage inductance (I_{lkpk}) is equal to the peak magnetizing current and is given by equation (7). The leakage inductance (L_{lk}) can be measured or obtained from the manufacturer's datasheet. In this converter, an off-the-shelf transformer with a maximum leakage inductance of 40 µH.

$$I_{Lkpk} = \frac{V_{DC} D}{(L_M + L_{LK}) f} \tag{7}$$

$$E_{Lk} = \frac{1}{2} L_{LK} I_{LKPK}^2$$

$$I_{Lkpk} = \frac{V_{DC}D}{(L_M + L_{LK})f}$$

Where D is the duty cycle, and f is the switching frequency. With D = 0.34, Vdc = 170V, Lm = 1310µH, and f = 75kHz, the peak leakage current is calculated to be 0.571A. The leakage energy is determined using equation 8, resulting in a value of 6.52µJ. This energy is approximate, and for precise leakage energy calculations, equations 2-4 can be used. This energy will be transferred to the snubber capacitor once the snubber capacitor voltage reaches the reflected output voltage. Consequently, the final stored energy in the snubber capacitor will be the previously stored energy plus the leakage energy, expressed as follows:

$$E_{C_f} = E_{C_i} + E_{LK} \qquad (8)$$

$$E_{lk} = Ec_f - E_{C_i} = \frac{1}{2} C((nVo + \Delta V)^2 - (nVo)^2) \qquad (9)$$

Where C represents the snubber capacitor, nVo denotes the reflected output voltage, and ΔV indicates the change in the voltage across the snubber capacitor due to the leakage energy, which is equal to the overshoot voltage across S_1 (i.e., V_{DS}). Therefore, ΔV is a design parameter that can be selected based on the desired overshoot for the switch. By solving for C, the following equation is obtained:

$$C_{snubber} = \frac{2E_{LK}}{2\Delta V \cdot nVo + \Delta V^2} \qquad (10)$$

Choosing a higher overshoot voltage (ΔV) results in a smaller snubber capacitance and a significant improvement in efficiency at low power. This is because larger capacitance allows more energy to cycle through the leakage recycling circuit, leading to increased losses. Therefore, for this design, a voltage overshoot of 110V is selected, corresponding to a 37% overshoot across the primary switch. The reflected output voltage (nVo) is 122V, and the leakage energy (E_{LK}), calculated previously, is 6.52µJ. Plugging these results into equation 10 yields a snubber capacitance of approximately 335pF. This capacitance is an estimate, and the optimal value is determined experimentally.

Size considerations limit the choice of snubber inductor. A very small inductance can lead to a high RMS current through the switch, resulting in significant energy dissipation across the FET. Conversely, a very large inductance requires more space, is costly, and may not be able to transfer all the leakage energy from the snubber capacitor in time, leading to excessive. overshoot. Generally, a snubber inductance that is 2 to 3 times smaller than the magnetizing inductance is sufficient. For this design, an inductance of 470 µH has been selected, which is 2.8 times smaller than the magnetizing inductance of 1310 µH.

Figure 4: Proposed flyback prototype PCB board

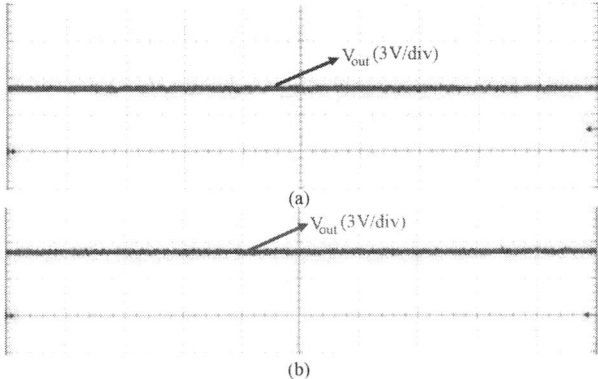

Figure 5: Output voltage of flyback converter. (a) Light load operation (b) Full load operation

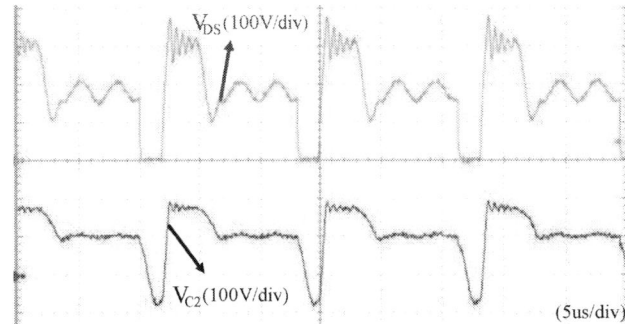

Figure 6: Drain-source voltage of switches (blue) and snubber capacitor voltage under light load condition

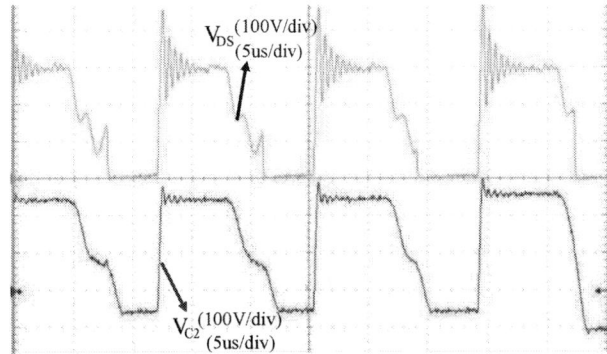

Figure 7: Drain-source voltage of switches (blue) and snubber capacitor voltage under Full load condition

IV. Experimental Results

A prototype designed for charger applications has been developed and implemented using the parameters outlined in Table 2 to evaluate the feasibility of the proposed converter. The experimental setup, shown in Fig. 4, uses the UCC3813 controller IC, utilizing peak current mode control. A 650V GaN switch from Nexperia is used on the primary side to reduce conduction and switching losses. Experimental results are presented under light and full load operations. Fig. 5(a) and Fig. 5(b) illustrate the output voltage under both light load and full load conditions, which is maintained at 5.4V. Fig. 6 demonstrates the voltage across the switch and snubber capacitor during light load conditions. As shown in Fig. 6, when the switch is turned on, the capacitor charges with reverse polarity, as indicated by the simulation. When the switch is turned off, the capacitor recharges with positive polarity, storing the leakage energy from the current cycle. Once all the leakage energy has been transferred to the snubber capacitor, its voltage stabilizes. Additionally, as the converter transitions into discontinuous mode and the reflected output voltage drops to zero, the voltage of the snubber capacitor decreases as the stored energy is recycled back into the bulk capacitor. Fig. 7 illustrates the snubber capacitor alongside the drain-source voltage during full load operation. In a fully loaded scenario, the duration of discontinuous mode is shorter compared to that during light load operation. However, this discontinuous period is still long enough to recycle almost all the leakage energy. Similar behavior is seen across the snubber capacitor voltage under full load conditions.

Table 2: System Parameters

Parameters	Value
Input Voltage	120-240VAC
Snubber inductor L1	470 uH
Snubber Capacitor C2	330 pF
Switching Frequency	75 kHz
Transformer Turn Ration	140:6
Output Voltage/Current	5.3 V, 2A,0.47A

B. Power loss analysis

The efficiency of the proposed flyback converter was evaluated through tests that measured the input and output voltage and current. These measurements are illustrated in Fig. 8 and Fig. 9 for full and light load conditions. Equation (11) defines the efficiency of the converter for each load during one cycle of input voltage.

$$P = \frac{\int_0^{T_s}(V_{out}(t)I_{out}(t))dt}{\int_0^{T_s}(V_{in}(t)I_{in}(t))dt} \tag{11}$$

The converter was developed and tested for all three generations. The first generation was a standard flyback converter. The second generation incorporated synchronous rectification to reduce power loss. The third generation added a recycling snubber and modifications for startup to eliminate clamp losses.

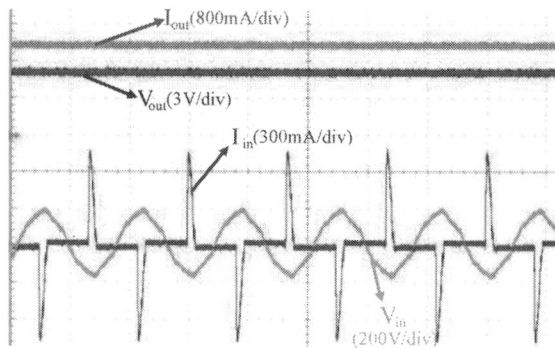

Figure 8: Output and Input voltage and current in light load condition

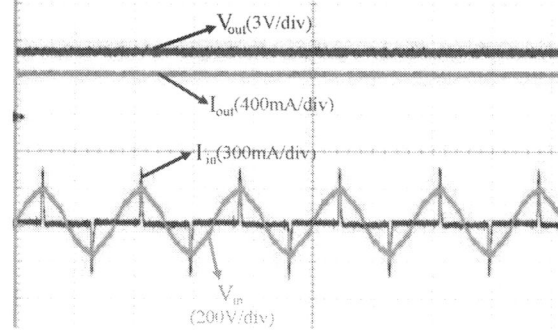

Figure 9: Output and Input voltage and current in light load condition

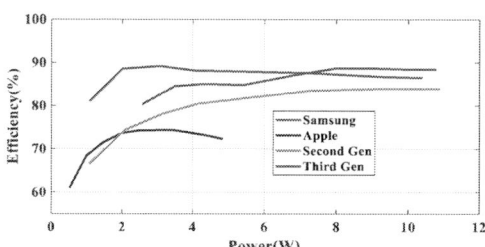

Figure 10: Efficiency comparison with market chargers

As shown in Fig. 10, this design achieved experimental efficiency of 88.6% at a 10W output, which is 4.6% higher than the previous generation that lacked leakage recycling. Additionally, this converter outperformed Samsung chargers (EP-TA200) by up to 2% and Apple chargers (A1385) by up to 12.6%. Furthermore, the performance at light loads stay competitive and given the smaller amount of transferred power during light load charging, the overall efficiency of the proposed system over the entire charging cycle is expected to be superior to those of commercially available alternatives.

V. Conclusion

This paper introduces a high-efficiency flyback converter for mobile phone chargers, achieving significant improvements in energy efficiency through three generations of design. The final version integrates GaN switches and a recycling snubber circuit, enhancing efficiency to 88.6% at full power, outperforming commercially available chargers. The results validate the effectiveness of the proposed design and highlight its potential for reducing global energy consumption in low-power applications. This research contributes to global efforts to reduce energy consumption and environmental impact, aligning with the broader goals of sustainable development.

REFERENCES

[1] https://market.us,

[2] S. Xu, S. Xu, Q. Qian, C. Wang, S. Lu and W. Sun, "Sample-Data Modeling for Active Clamp Flyback Converter in Critical Conduction Mode with PCM and ZVS control at Variable Switching frequency," *2020 IEEE Applied Power Electronics Conference and Exposition (APEC)*, New Orleans, LA, USA, 2020.

[3] K. Elissa, "Title of paper if known," unpublished. X. Huang, J. Feng, W. Du, F. C. Lee and Q. Li, "Design consideration of MHz active clamp flyback converter with GaN devices for low power adapter application," *2016 IEEE Applied Power Electronics Conference and Exposition (APEC)*, Long Beach, CA, USA, 2016,

[4] A. Abramovitz, C. -S. Liao and K. Smedley, "State-Plane Analysis of Regenerative Snubber for Flyback Converters," in *IEEE Transactions on Power Electronics*, vol. 28, no. 11, pp. 5323-5332, Nov. 2013,

[5] M. Li, Z. Ouyang and M. A. E. Andersen, "Analysis and Optimal Design of High-Frequency and High-Efficiency Asymmetrical Half-Bridge Flyback Converters," in *IEEE Transactions on Industrial Electronics*, vol. 67, no. 10, pp. 8312-8321, Oct. 2020,

[6] J. Park *et al.*, "Quasi-Resonant (QR) Controller With Adaptive Switching Frequency Reduction Scheme for Flyback Converter," in *IEEE Transactions on Industrial Electronics*, vol. 63, no. 6, pp. 3571-3581, June 2016,

[7] M. Ghasemi, A. Honarbakhsh, M. Saradarzadeh and M. Hamzeh, "Ultra-Wide Voltage Range Control of DC-DC Full-Bridge Converter with Hysteresis Controller," *2022 13th Power Electronics, Drive Systems, and Technologies Conference (PEDSTC)*, Tehran, Iran, Islamic Republic of, 2022.

[8] J. Park *et al.*, "Quasi-Resonant (QR) Controller With Adaptive Switching Frequency Reduction Scheme for Flyback Converter," in *IEEE Transactions on Industrial Electronics*, vol. 63, no. 6, pp. 3571-3581, June 2016,

[9] Y. -T. Yau and T. -L. Hung, "A Flyback Converter With Novel Active Dissipative Snubber," in *IEEE Access*, vol. 10, pp. 108145-108158, 2022,

[10] C. Wang, D. Sun, X. Zhang, W. Gu and S. Gui, "A Constant Current Digital Control Method for Primary-Side Regulation Active-Clamp Flyback Converter in CCM Mode," *2021 IEEE Energy Conversion Congress and Exposition (ECCE)*, Vancouver, BC, Canada, 2021,

[11] C. Wang, D. Sun, X. Zhang, J. Hu, W. Gu and S. Gui, "A Constant Current Digital Control Method for Primary-Side Regulation Active-Clamp Flyback Converter," in *IEEE Transactions on Power Electronics*, vol. 36, no. 6, pp. 7307-7318, June 2021,

[12] H. Luo, T. Zang, S. Chen and B. Zhou, "An Adaptive Off-Time Controlled DCM Flyback PFC Converter With Unity Power Factor and High Efficiency," in *IEEE Access*,

[13] J. Park *et al.*, "Quasi-Resonant (QR) Controller With Adaptive Switching Frequency Reduction Scheme for Flyback Converter," in *IEEE Transactions on Industrial Electronics*, vol. 63, no. 6, pp. 3571-3581, June 2016,

[14] X. Wu, Z. Wang and J. Zhang, "Design Considerations for Dual-Output Quasi-Resonant Flyback LED Driver With Current-Sharing Transformer," in *IEEE Transactions on Power Electronics*, vol. 28, no. 10, pp. 4820-4830, Oct. 2013,

[15] M. Bahrami-Fard, M. Ghasemi Korrani, M. Rastegar, P. Balsara and B. Fahimi, "Error Compensation Strategy in Encoderless Surface Mounted

PMSM Drives," 2024 IEEE Energy Conversion Congress and Exposition (ECCE), Phoenix, AZ, USA, 2024.

[16] R. -L. Lin and S. -H. Hsu, "Design and Implementation of Self-Oscillating Flyback Converter With Efficiency Enhancement Mechanisms," in *IEEE Transactions on Industrial Electronics*, vol. 62, no. 11, pp. 6955-6964, Nov. 2015,

[17] S. -W. Lee and H. -L. Do, "A Single-Switch AC–DC LED Driver Based on a Boost-Flyback PFC Converter With Lossless Snubber," in *IEEE Transactions on Power Electronics*, vol. 32, no. 2, pp. 1375-1384, Feb. 2017.

[18] C. Wang, D. Sun, X. Zhang, J. Hu, W. Gu and S. Gui, "A Constant Current Digital Control Method for Primary-Side Regulation Active-Clamp Flyback Converter," in *IEEE Transactions on Power Electronics*, vol. 36, no. 6, pp. 7307-7318, June 2021.

[19] M. Bahrami-Fard, B. M. Mosammam, M. Hassan Ghaderi, D. D. Patel, P. Balsara and B. Fahimi, "An Effective Cooling System for High Torque Electric Motors Using Microchannels and Two-phase Coolants," *2024 IEEE Transportation Electrification Conference and Expo (ITEC)*, Chicago, IL, USA, 2024.

[20] C. -Y. Tang, W. -Z. Lin and Y. -C. Tan, "An Active Clamp Flyback Converter With High Precision Primary-Side Regulation Strategy," in *IEEE Transactions on Power Electronics*, vol. 37, no. 9, pp. 10281-10289, Sept. 2022.

[21] P. -J. Liu and J. -W. Lee, "A Bidirectional Voltage-Mode Controlled Converter for Power Distribution in a Low-Power Dual-Output Flyback Converter," in *IEEE Transactions on Industrial Electronics*, vol. 71, no. 11, pp. 14021-14032, Nov. 2024.

[22] T. N. T. Tran, H. -Y. Xu and J. -M. Wang, "Development of Active-Clamp Flyback Converter for Improving Light-Load Efficiency," in *IEEE Journal of Emerging and Selected Topics in Power Electronics*, vol. 12, no. 3, pp. 2456-2469, June 2024.

[23] M. Bahrami-Fard, M. Ghasemi Korrani, M. Rastegar, P. Balsara and B. Fahimi, "Adaptive PLL-based Sensorless Control for CSI-Fed PMSM Drives Used in Submersible Pumps," IECON 2024- 50th Annual Conference of the IEEE Industrial Electronics Society, Chicago, IL, USA, 2024.

[24] S. Xu, Q. Qian, T. Tao, S. Lu and W. Sun, "Small Signal Modeling and Control Loop Design of Critical Conduction Mode Active Clamp Flyback Converter," in *IEEE Transactions on Power Electronics*, vol. 36, no. 6, pp. 7250-7263, June 2021

[25] A. Sarkar, N. Deshmukh and S. Anand, "Modified PWM Scheme to Reduce Reverse Conduction Loss in GaN-Based Independently Controlled Multiple Output Flyback Converter," in *IEEE Transactions on Power Electronics*, vol. 37, no. 11, pp. 12968-12972, Nov. 2022.

[26] Z. Ma, S. Wang, H. Sheng and S. Lakshmikanthan, "Modeling, Analysis and Mitigation of Radiated EMI Due to PCB Ground Impedance in a 65 W High-Density Active-Clamp Flyback Converter," in *IEEE Transactions on Industrial Electronics*, vol. 70, no. 12, pp. 12267-12277, Dec. 2023

[27] X. Liu, Y. Feng, M. Leng, M. Li and X. Zhang, "Single-Stage Integrated Buck-Flyback PFC Converter With Low Total Harmonic Distortion Control," in *IEEE Journal of Emerging and Selected Topics in Power Electronics*, vol. 12, no. 2, pp. 1782-1792, April 2024

[28] T. Meng, H. Ben, C. Li and F. Wu, "Investigation of the Integrated-Transformer Winding Architectures in the Input-Series Flyback Auxiliary Power Supply Considering the Stray Capacitances," in *IEEE Transactions on Power Electronics*, vol. 36, no. 2, pp. 1790-1803, Feb. 2021.

[29] Y. -E. Wu and Y. -T. Ke, "A Novel Bidirectional Isolated DC-DC Converter With High Voltage Gain and Wide Input Voltage," in *IEEE Transactions on Power Electronics*, vol. 36, no. 7, pp. 7973-7985, July 2021

[30] Y. Chen, R. Wang, X. Liu and Y. Kang, "Gate-Drive Power Supply With Decayed Negative Voltage to Solve Crosstalk Problem of GaN Synchronous Buck Converter," in *IEEE Transactions on Power Electronics*, vol. 36, no. 1, pp. 6-11, Jan. 2021,

979-8-3315-1612-3/25 $31.00 © 2025 IEEE

A Hybrid Three-Level Buck Converter with Flying Supercapacitor for High Load Current Surge Capability using Peak Current Mode Control

Finlay Lodge
Department of Electronic and
Electrical Engineering
University of Strathclyde
Glasgow, United Kingdom
finlay.lodge.2018@uni.strath.ac.uk

Rafael Peña-Alzola
Department of Electronic and
Electrical Engineering
University of Strathclyde
Glasgow, United Kingdom
rafael.pena-alzola@strath.ac.uk

Martin MacFadyen
Department of Electronic and
Electrical Engineering
University of Strathclyde
Glasgow, United Kingdom
martin.macfadyen@strath.ac.uk

Patrick Norman
Department of Electronic and
Electrical Engineering
University of Strathclyde
Glasgow, United Kingdom
patrick.norman@strath.ac.uk

Mark Sweet
Rolls-Royce plc
Derby, United Kingdom
mark.sweet@rolls-royce.com

Graeme Burt
Department of Electronic and
Electrical Engineering
University of Strathclyde
Glasgow, United Kingdom
graeme.burt@strath.ac.uk

Abstract—In aircraft power systems, many non-propulsive loads require large bursts of current, such as electrohydrostatic actuators for positioning control surfaces. Hybridized energy storage systems utilizing batteries and supercapacitors prove an attractive option for supplying power systems with highly variable loads; however, such systems require many more components, increasing mass and volume. In this paper the use of a hybrid three-level buck converter is proposed, featuring an integrated flying supercapacitor for combined energy storage and multilevel operation, capable of providing large pulses of power. In addition to using Peak Current Mode control, a new algorithm is implemented to manage current-limiting and level selection, facilitating high load current surge capability. Advantages include reduced component count, short-circuit protection, and reduced battery stress, inductor current ripple and voltage ripple. This paper explains the operation and functionality of the proposed topology and discusses the proposed control algorithm. A model, developed in MATLAB/Simulink, is used to demonstrate the behavior of the converter and validate the proposals.

Index Terms—Three level buck converter (3LBC), flying capacitor, peak current mode (PCM).

I. INTRODUCTION

Significant advances are being made in the electrification of aircraft, including the use of electrohydrostatic actuators (EHAs) [1] and electric brake system [2] in newer Boeing 787 models, and the replacement of hydraulic actuators with EHAs for Airbus A380 and A350 passenger aircraft [3] [4], reducing the weight of the A380 by 450 kg [5]. The peak power consumption of typical actuators is expected to reach ten times the nominal power draw [6], highlighting the need for power systems capable of handling large, intermittent surges in load current. Multilevel converters such as the three-level buck converter (3LBC) are becoming increasingly popular due to

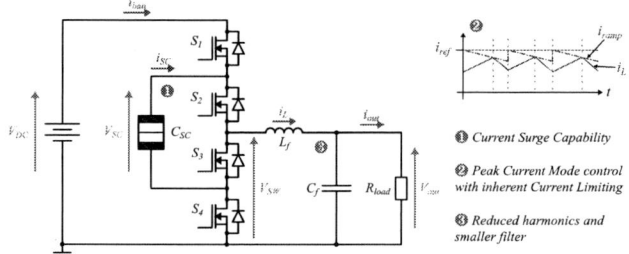

Fig. 1: Three-level buck topology with flying supercapacitor.

their lower harmonic content, reduced dv/dt and voltage stress, and smaller filter size [7]. These topologies are well suited to aerospace applications, where systems must be lightweight, reliable and well-protected [8].

Peak Current Mode (PCM) control is an attractive control method for the three-level buck converter, enabling fast control and current-limiting with simpler dynamics than voltage-mode control. The balancing of the FC voltage and the loss of the inductor current ripple which acts as a carrier signal for the PWM have been identified in literature as key sources of instability in the three-level flying capacitor PCM-controlled buck. Several papers [9] [10] [11] [12] have identified that ramp-compensated Valley Current Mode (VCM) control facilitates FC voltage balancing, while stability conditions exist for PCM. Stability analyses have been undertaken to derive stability characteristics [13] and discrete-time large signal models [14] for ramp-compensated and uncompensated PCM and VCM. Several control strategies have been proposed to overcome the FC voltage runaway and zero-ripple point issues, including

979-8-3315-1612-3/25 $31.00 © 2025 IEEE

sensorless stabilization using the duty cycle mismatch between switch groups S_1/S_4 and S_2/S_3 [15], digital predictive current-mode control for PCM, VCM and average current-mode controllers [16], combined PCM-VCM and trapezoidal modulation around the zero-ripple point [17], and PCM with peak offsetting of the around the zero-ripple point [18].

The structure of the paper is as follows: Section II presents an overview of the three-level buck converter and comments on the design and sizing of the flying supercapacitor module. Section III details the stability criteria for conventional PCM control required to ensure FC voltage balancing. In section IV, simulation results are presented to verify the proposed functionality. Section VI concludes this paper and recommends directions for future research.

II. THREE-LEVEL BUCK CONVERTER

A. Proposed Topology

The proposed topology, shown in Fig. 1, consists of a three-level buck converter with flying supercapacitor (FSC) for hybrid energy storage. The three-level operation arises from the use of a capacitor which is balanced at half of the DC-link voltage during steady-state operation. The three-level nature arises from the consideration of the capacitor as a constant DC source which can either be placed in anti-series with the supply or connected directly to the switching node, both of which result in a switching node voltage V_{SW} of half the DC voltage. The use of the middle level in the modulation strategy reduces the voltage swing at the switching node by half during normal operation, as the converter can be modulated between upper and middle levels when the desired output voltage is greater than half the DC voltage, and modulated between middle and lower levels when the desired output voltage is below half the DC voltage. This allows the filter inductor and output capacitor to be reduced by a factor of four and eight respectively [10] [13].

By integrating energy storage with the flying capacitor functionality, the converter can supply surges in load which exceed the current rating of the battery if the output voltage is lower than the capacitor voltage. This paper proposes the use of a supercapacitor to facilitate temporary energy storage and three-level operation due to its superior specific energy when compared with conventional capacitors. Additionally, in the event of a battery failure the proposed topology can quickly reconfigure to supply the load in buck configuration using the remaining energy in the supercapacitor for emergency operation or to sustain power until the fault is cleared.

In instances where the supercapacitor voltage differs significantly from the nominal value – typically after supplying a load surge – the middle level voltage will increase significantly above half the DC voltage. This must be accounted for to ensure stable modulation as the output voltage must fall between the respective switch-node voltages of the middle and upper/lower levels being modulated. In conventional implementations of the 3LBC topology, the modulation issues that arise from significant deviation of the capacitor voltage are not typically encountered as the voltage is tightly regulated at half the nominal supply voltage. However, for load surge capability this is a not-infrequent occurrence and must be considered when developing the control algorithm.

B. Design Considerations

Batteries, while offering high energy density, are limited in their discharge rates by cell electrochemistry. In systems where peak loads may be significantly higher than nominal power draw, oversizing the battery may lead to issues with mass and volume. Conventional capacitors demonstrate superior specific power, but possess limited energy storage capability and discharge quickly. Supercapacitors bridge this gap, delivering higher specific power than batteries while achieving energy densities 10-100 times greater than conventional capacitors. This significant reduction in volume makes supercapacitors particularly attractive for temporary bulk energy storage in tandem with batteries to supply nominal loads.

Integrating the supercapacitor within the topology as a flying capacitor allows the voltage balancing to be achieved while also providing bulk energy storage. Traditional architectures typically interface supercapacitors with the DC-link bus through a dedicated DC-DC converter. The structure proposed in this paper reduces the semiconductor count, though it necessitates a larger FSC module comprising series-connected supercapacitors to accommodate their very low voltage rating.

All supercapacitors possess inherent variations and operation over a long lifetime may result in parameter drift as the components degrade. Over time, this may result in sufficient variation in capacitance that some series-connected supercapacitors may exceed their rated voltage. A large FSC module would require passive or active voltage balancing to prolong the lifetime of the supercapacitors and ensure safe operation of the module. Despite this drawback, the significantly higher energy density of supercapacitors makes this arrangement more viable than regular capacitors for combining bulk energy storage and facilitation of the current surge behavior proposed in this paper. The topology is best suited to applications where either supercapacitor energy can be recycled i.e. discharged back into the source, or where load surges will occur often enough between start-up and shutdown to justify the additional energy storage.

The supercapacitor is sized based on the specifications for average load during surge operation and the time duration for which the converter must be able to supply the load surge. It is recommended that the supercapacitor does not discharge below half its nominal voltage as this may increase the voltage swing at V_{SW} under certain conditions, increasing the inductor current ripple and necessitating a larger filter design. The proposed limit allows for the release of 75% of the stored energy while ensuring the ripple remains tolerable.

The energy that is available to be discharged by the supercapacitor module can be defined in terms of the capacitance, C_{FSC}, the nominal DC voltage, V_{DC}, and the discharge voltage, $V_{FSC,min}$.

979-8-3315-1612-3/25 $31.00 © 2025 IEEE   3168

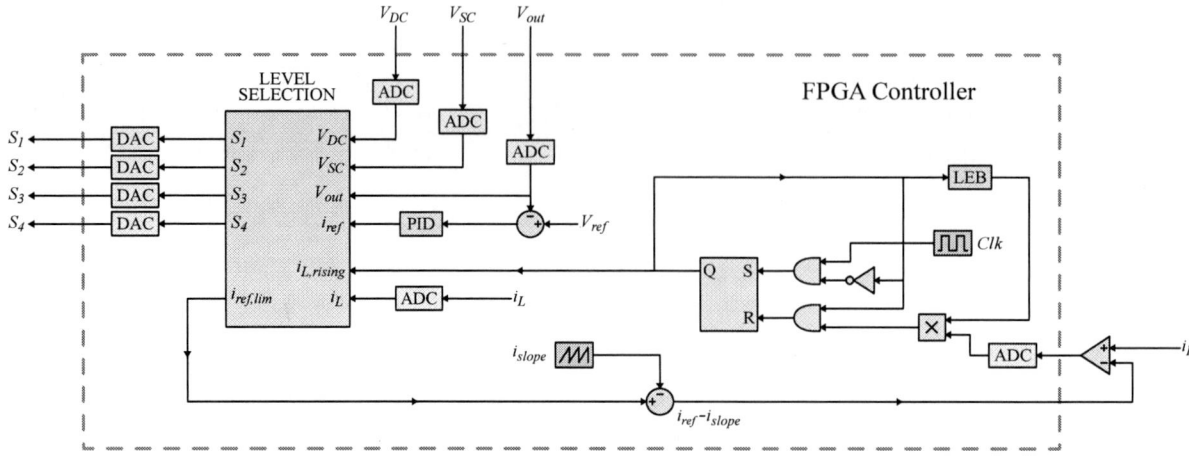

Fig. 2: Control diagram, including PCM control, leading-edge blanking and level selection function block.

$$\Delta E_{FSC} = \frac{1}{2} C_{FSC} \left(\frac{1}{4} V_{DC}{}^2 - V_{FSC,min}{}^2 \right) \quad (1)$$

Accordingly, the maximum ideal time for which a specified average load surge can be sustained can be calculated.

$$T_{surge,max} = \frac{\overline{P_{surge}}}{\Delta E_{FSC}} \quad (2)$$

From this, the minimum capacitance required to sustain a specified average load surge over a given time can be derived using the following equation.

$$C_{FSC,min} = \frac{\overline{P_{surge}}}{\frac{1}{2} T_{surge,max} \left(\frac{1}{4} V_{DC}{}^2 - V_{FSC,min}{}^2 \right)} \quad (3)$$

III. PEAK CURRENT MODE CONTROL

Peak Current Mode (PCM) control has emerged as a popular control strategy for commercial power supplies, owing to its fast control and inherent current-limiting capability. While this control methodology has been comprehensively analyzed in the literature for two-level converters, its implementation in multilevel converters presents significant challenges. When applied to a three-level buck converter, PCM demonstrates instability unless the inductor current ripple substantially exceeds the load current. This stability constraint severely impacts the practical implementation of PCM-controlled 3LBC systems designed for wide output voltage ranges, necessitating either a restricted output voltage range or acceptance of excessive current ripple. The converter is only stable if the following criterion is met:

$$\frac{\Delta I_L}{I_{out}} > r(M) \quad (4)$$

where $r(M)$ is a function of the voltage conversion ratio, M:

$$r(M) = \begin{cases} \frac{2(0.5 - M)}{M} & M < 0.5 \\ \frac{2(M - 0.5)}{1 - M} & M > 0.5 \end{cases} \quad (5)$$

and M is defined as the ratio of the output voltage to the input voltage.

$$M = \frac{V_{out}}{V_{DC}} \quad (6)$$

In a conventional implementation of PCM for a 3LBC, operation relies on two complementary switching pairs: S_1/S_4 (denoted as A/A') and S_2/S_3 (denoted as B/B'). Under stable conditions, both switching pairs maintain equal duty cycles, facilitating balanced charging and discharging of the flying capacitor during each switching cycle.

$$D_A = D_B = D \quad (7)$$

However, when the stability criterion is not satisfied, duty cycle mismatches occur, leading to voltage imbalance. As with two-level topologies, ramp compensation is also necessary to eliminate further instability due to subharmonic oscillations when $M > 0.5$.

To address this limitation, this paper proposes a methodology that dynamically adjusts the FC voltage based on its measured value normalized to the DC-link voltage. This approach employs a function block to determine optimal switching states and govern modulation. The algorithm proposed in this digest combines the advantages of conventional PCM control with the ability to maintain a balanced FC voltage independently of the voltage conversion ratio and stability criterion.

IV. PROPOSED ALGORITHM

A. Summary of Proposed Control Algorithm

The algorithm proposed in this digest regulates the FC voltage by applying PCM to modulate between the three levels, and using a novel function algorithm to configure the middle state and set the current limit based on the measured

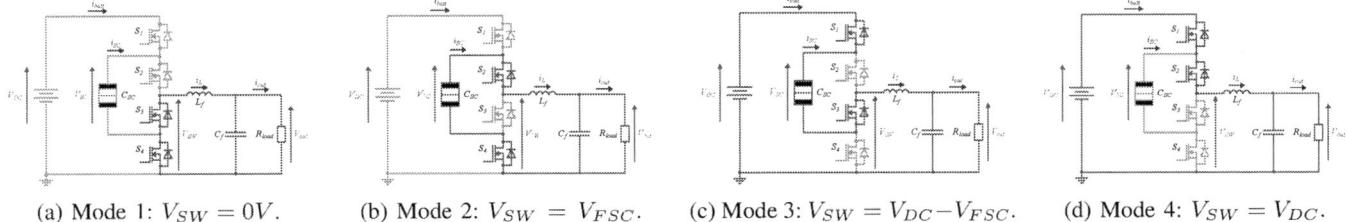

(a) Mode 1: $V_{SW} = 0V$.   (b) Mode 2: $V_{SW} = V_{FSC}$.   (c) Mode 3: $V_{SW} = V_{DC} - V_{FSC}$.   (d) Mode 4: $V_{SW} = V_{DC}$.

Fig. 3: 3LBC modes of operation. Mode 2 discharges and Mode 3 charges the FSC.

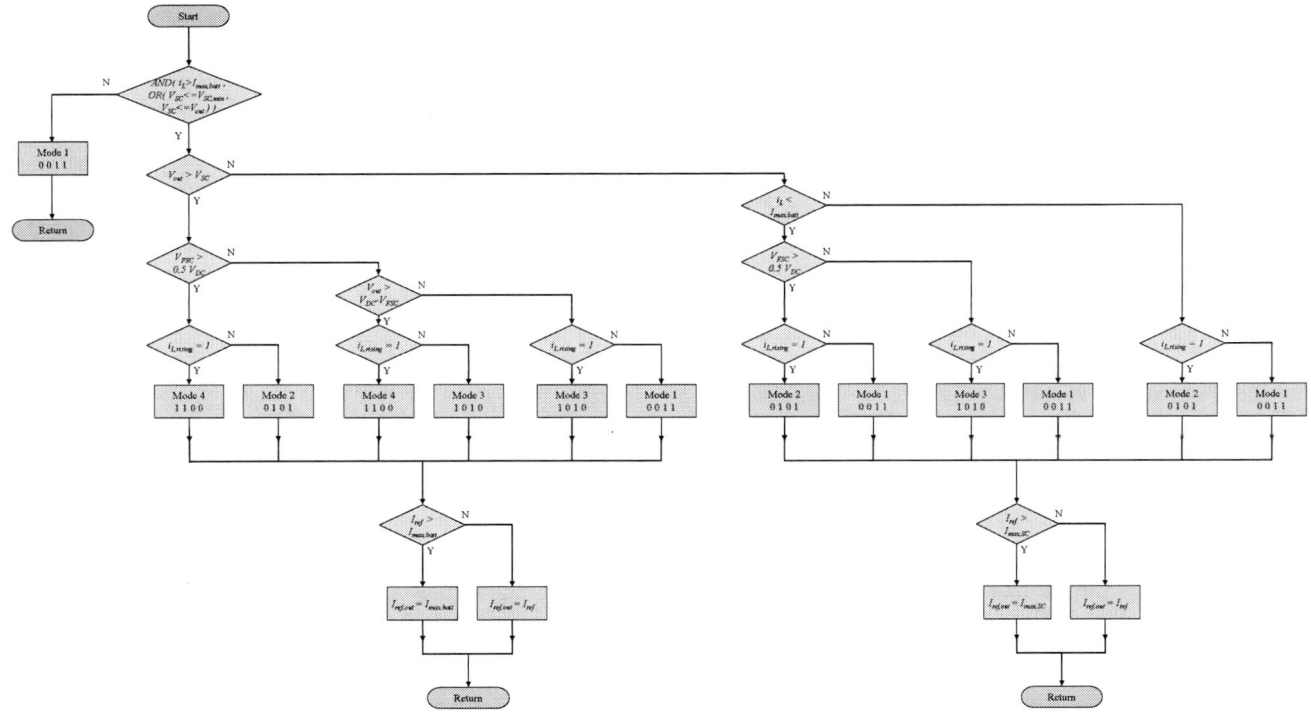

Fig. 4: Flowchart of function block procedure to configure switching states and manage current-limiting.

FC voltage and normalized output voltage. The function block updates many times over one switching cycle, ensuring fast reconfiguration and current-limiting. Although the voltage measurement is subject to noise, the size of the supercapacitor ensures that the voltage change over one switching cycle is sufficiently small that the impact of noise on the FC voltage ripple is negligible.

Fig. 2 shows a typical FPGA-based implementation of the PCM controller with leading-edge blanking and the function block required for level selection and current limiting. The voltage reference is set by the user, and the current reference is generated from the voltage error by a PI controller tuned using the Symmetrical Optimum criterion, the output of which is then input to the function block. The function limits the current reference if the inductor current exceeds the relevant current rating for the envelope of operation, before feeding the reference back to the PCM control. Conventional PCM control uses the latch output as the gate signal to control switching. In the proposed control scheme, the latch signal is used as an input for the function block to indicate whether the inductor

current should be rising or falling. The function block sets the switch states accordingly to increase or decrease the inductor current and facilitate power transfer.

B. Modes of Operation of Converter

The converter has four primary modes of operation, dictated by the switch configuration. These modes are shown in Fig. 3 and are used to describe the operation of the algorithm. Mode 1 and Mode 4 represent the lower and upper voltage levels respectively. Modes 2 and 3 are used to facilitate the middle voltage level: Mode 2 is configured to discharge the FSC while Mode 3 is connected to charge the FSC.

C. Function Block Behavior

A flowchart of the detailed function block behavior is shown in Fig. 4. The function first determines whether the supercapacitor voltage is below both the lower FSC voltage limit or the output voltage, and if the inductor current is less than the battery current limit. The supercapacitor cannot supply below the specified limits and must be charged up by configuring Mode 3. If the inductor current exceeds the battery current limit, charging cannot take place and Mode 1 is configured until the FSC can be recharged.

The function then determines if the output voltage is greater than the FSC voltage. If this is true, the inductor current must be limited to $I_{batt,max}$ and the modulated modes must be determined for proper balancing of the FSC. This is described in Section IV-C1 as Case A.

If the output voltage is lower than the FSC voltage, the converter may supply load surges. If the measured inductor current is below the battery current limit, the converter operates normally, described in Section IV-C2 as Case B.

If the measured inductor current exceeds the battery current limit, the converter is supplying a load surge, and the supercapacitor must be used to supply the load until either the surge passes, or the supercapacitor voltage drops below the output voltage or lower discharge limit. This is described in Section IV-C3 as Case C.

1) Case A: If the FSC voltage is greater than half of the DC-link voltage, the FSC must be discharged. This means the middle level is configured as Mode 2. Thus, the converter configures Mode 4 if the inductor current is rising or Mode 2 if the inductor current is falling.

If the FSC voltage is less than half of the DC-link voltage, the FSC must be charged. This means that the middle level is configured as Mode 3. If during Mode 3 the voltage at V_{SW} is greater than V_{out}, the converter modulates between upper and middle levels, i.e. configures Mode 4 if the inductor current is rising or Mode 3 if the inductor current is falling. Else if during Mode 3 the voltage at V_{SW} is greater than V_{out}, the converter modulates between middle and lower levels, i.e. configures Mode 3 if the inductor current is rising or Mode 0 if the inductor current is falling.

Finally, the current limiting is applied. If the current reference exceeds the rated battery current, the current is limited to $I_{batt,max}$ before feeding back to the PCM control.

2) Case B: If the output voltage is below the FSC voltage then the converter modulates between middle and lower levels. As the battery current limit is not exceeded, both Mode 2 and Mode 3 are available to use. Therefore, if the FSC voltage is greater than half of the DC-link voltage, the converter configures Mode 2 if the inductor current is rising or Mode 1 if the inductor current is falling. Else, if the FSC voltage is less than half of the DC-link voltage, the converter configures Mode 3 if the inductor current is rising or Mode 1 if the inductor current is falling.

If the current reference exceeds the rated battery current, the inductor current is limited to $I_{SC,max}$ before feeding back to the PCM control. However, if the current begins to rise above $I_{batt,max}$, the function will move to Mode C to supply the load surge while ensuring that the battery is protected.

3) Case C (Load Surge): As the inductor current exceeds the rated battery current, the converter can only operate in Mode 1 or Mode 2. Thus, the converter configures Mode 2 if the inductor current is rising or Mode 1 if the inductor current

is falling. If the FSC is discharged as low as the output voltage or the lower FSC voltage limit, this will be detected at the beginning of the function as described earlier, and Mode 1 will be configured to protect the battery and converter.

If the current reference exceeds the rated FSC current, the inductor current is limited to $I_{max,FSC}$ before feeding back to the PCM control.

V. SIMULATIONS AND RESULTS

Simulations were performed to validate the proposed topology and control strategy prior to future hardware implementation. The FSC voltage was set to 50 V to align with the planned experimental setup utilising commercially-available Maxwell BMOD0058 supercapacitor modules (16 V, 58 F). These modules can be connected in series to form a 64 V, 14.5 F FSC module and operated at a maximum of 50 V to provide a safety margin. The simlation parameters were scaled accordingly, with the DC-link voltage set to 100 V and the nominal FSC voltage to 50 V to match the anticipated hardware specifications. A full list of specifications for the simulations is provided in Table I.

TABLE I: Model parameters for 3LBC simulations.

Parameter	Symbol	Value
DC-Link Voltage	V_{DC}	100 V
FSC Voltage	V_{FSC}	50 V
Battery Current Rating	$I_{batt,max}$	3 A
FSC Current Rating	$I_{FSC,max}$	8 A
FSC Capacitance	C_{FSC}	14.5 F
Filter Inductor	L	9 mH
Switching Frequency	f_{sw}	10 kHz
Function Sampling Frequency	f_s	1.28 MHz

A model was constructed in MATLAB/Simulink to validate the current-limiting behavior and load surge behavior. The function was implemented using a MATLAB Function block within the Simulink environment.

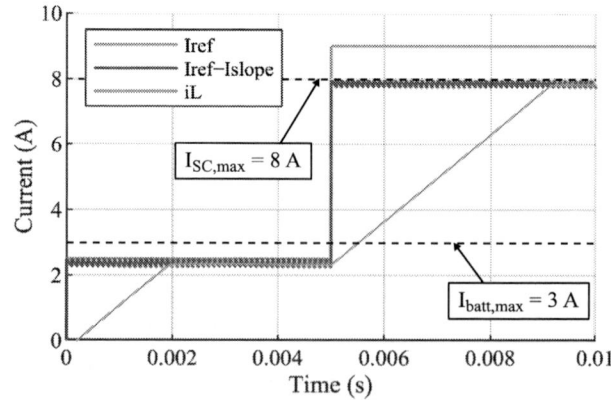

Fig. 5: Inductor current limiting during load surge.

The current-limiting behavior was tested using a load step to significantly increase the current draw at 0.005 seconds.

The results in Fig. 5 show the modulation and current-limiting of the peak inductor current. Initially, the current reference is below the rated battery current and the algorithm modulates Mode 2/3 and Mode 1 to maintain a constant output voltage and balance the FSC. It can be observed that the current-limited reference uses ramp compensation to eliminate subharmonic instability. When a surge in load current occurs and the current reference exceeds the maximum FSC current, I_{ref} is limited by the function block, and the inductor current follows the current-limited, ramp-compensated reference.

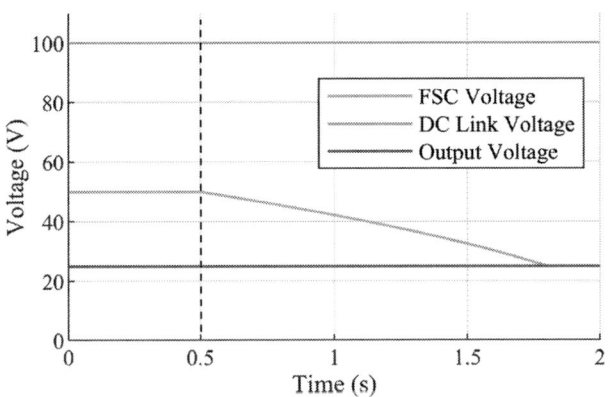

Fig. 6: FSC voltage discharge during load surge.

The results in Fig. 6 show the full discharging of the FSC. The FSC capacitance has been significantly reduced to improve simulation time, however is still sufficient to validate the behavior. When the current increases at 0.5 seconds, the reference current exceeds the rated battery current. The controller then modulates between Mode 1 and Mode 2 to supply the load until the FSC discharges as low as the output voltage, demonstrating the load surge behavior of the converter. As the FSC can no longer transfer power to the output, the converter is held in Mode 1 until the inductor current falls below the battery current limit.

The results in both simulations match the predicted behavior and validate the proposed functionality of the 3LBC converter and control scheme.

VI. Conclusions

In this paper, a hybrid three-level buck converter topology incorporating a flying supercapacitor has been presented, demonstrating significant advantages for aerospace power systems which are required to handle periodic surges in load current. A sizing procedure for the supercapacitor has been presented, along with important considerations regarding the FSC module structure and voltage balancing requirements of series-connected supercapacitors. The proposed algorithm combines Peak Current Mode control with a novel function block for level selection and current limiting, successfully managing current protection, FSC voltage balancing, and load surge operation while eliminating the FSC voltage balancing instability. Simulation results obtained through MATLAB/Simulink

modeling validate the proposed topology and control strategy, demonstrating stable operation during load surges and current-limiting behavior. Recommended future work includes hardware validation of the control using a prototype converter, and expansion of the control algorithm to include soft-start capability. Additionally, future research will investigate the use of duty cycle information for FSC voltage balancing and modulation during load surges, eliminating the need for an isolated sensor to measure the flying supercapacitor voltage.

References

[1] C. Wang, I.-S. Fan, and S. King, "Failures mapping for aircraft electrical actuation system health management," PHM Society European Conference, vol. 7, pp. 509–520, 06 2022.

[2] J.-C. Mare, "Review and analysis of the reasons delaying the entry into service of power-by-wire actuators for high-power safety-critical applications," Actuators, vol. 10, p. 233, 09 2021.

[3] D. van den Bossche and Airbus, "The A380 flight control electrohydrostatic actuators, achievements and lessons learnt," 25th International Congress of the Aeronautical Sciences, 01 2006.

[4] Q. Haitao, Y. Fu, X. Qi, and Y. Lang, "Architecture optimization of more electric aircraft actuation system," Chinese Journal of Aeronautics (CJA), vol. 24, pp. 506–513, 08 2011.

[5] I. Chakraborty, D. Mavris, M. Emeneth, and A. Schnegans, "A methodology for vehicle and mission level comparison of more electric aircraft subsystem solutions - application to the flight control actuation system," Proceedings of the Institution of Mechanical Engineers Part G Journal of Aerospace engineering, vol. 229, 06 2014.

[6] Z. Cao, H. Liu, Y. Zhou, and C. Zhang, "Energy optimization charateristic analysis of more electric aircraft flight control system," in 2018 IEEE 2nd International Electrical and Energy Conference (CIEEC), 2018, pp. 227–231.

[7] G. P. Adam, O. Anaya-Lara, G. M. Burt, S. J. Finney, B. W. Williams, and J. R. McDonald, "Comparison between flying capacitor and modular multilevel inverter," in 2009 35th Annual Conference of IEEE Industrial Electronics (IECON 2009), 2009.

[8] G. Buticchi, P. Wheeler, and D. Boroyevich, "The more-electric aircraft and beyond," Proceedings of the IEEE, vol. 111, no. 4, pp. 356–370, 2023.

[9] B. M. Cassidy, D. S. Ha, and Q. Li, "Constant on-time 3-level buck converter for low power applications," in 2015 IEEE Energy Conversion Congress and Exposition (ECCE), 2015, pp. 1434–1441.

[10] D. Reusch, F. C. Lee, and M. Xu, "Three level buck converter with control and soft startup," in 2009 IEEE Energy Conversion Congress and Exposition, 2009, pp. 31–35.

[11] J. S. Rentmeister and J. T. Stauth, "A 48v:2v flying capacitor multilevel converter using current-limit control for flying capacitor balance," in 2017 IEEE Applied Power Electronics Conference and Exposition (APEC), 2017, pp. 367–372.

[12] J. S. Rentmeister, C. Schaef, B. X. Foo, and J. T. Stauth, "A flying capacitor multilevel converter with sampled valley-current detection for multi-mode operation and capacitor voltage balancing," in 2016 IEEE Energy Conversion Congress and Exposition (ECCE), 2016, pp. 1–8.

[13] E. Abdelhamid, G. Bonanno, L. Corradini, P. Mattavelli, and M. Agostinelli, "Stability properties of the 3-level flying capacitor buck converter under peak or valley current-programmed-control," in 2018 IEEE 19th Workshop on Control and Modeling for Power Electronics (COMPEL), 2018, pp. 1–8.

[14] G. R. Chilukuri, P. Majumder, and S. Kapat, "Closed-loop stability analysis of digitally current mode controlled three-level buck converter using a simplified discrete-time modeling framework," in 2023 IEEE Applied Power Electronics Conference and Exposition (APEC), 2023, pp. 1200–1206.

[15] E. Abdelhamid, L. Corradini, P. Mattavelli, G. Bonanno, and M. Agostinelli, "Sensorless stabilization technique for peak current mode controlled three-level flying-capacitor converters," IEEE Transactions on Power Electronics, vol. 35, no. 3, pp. 3208–3220, 2020.

[16] G. Bonanno and L. Corradini, "Digital predictive peak current-mode control for three-level buck converters," in 2019 20th Workshop on Control and Modeling for Power Electronics (COMPEL), 2019, pp. 1–7.

979-8-3315-1612-3/25 $31.00 © 2025 IEEE

[17] L. Lu, A. Prodić, G. Calabrese, G. Frattini, and M. Granato, "Current programmed mode control of multi-level flying capacitor converter near zero-ripple current region," in 2019 IEEE Applied Power Electronics Conference and Exposition (APEC), 2019, pp. 3064–3070.

[18] L. Lu, S. M. Ahsanuzzaman, A. Prodić, G. Calabrese, G. Frattini, and M. Granato, "Peak offsetting based cpm controller for multi-level flying capacitor converters," in 2018 IEEE Applied Power Electronics Conference and Exposition (APEC), 2018, pp. 3102–3107.

Supercritical Carbon Dioxide (sCO_2)-Cooled Current Source Inverter-based Integrated Motor Drive for MW-scale Electric Aviation Applications

Hang Dai[*] John Yagielski[*] Thomas Jahns[#] Kum-Kang Huh[*] Vandana Rallabandi[+] Libing Wang[*] Tarak Saha[*]
Wenda Feng[#] Bulent Sarlioglu[#]

[*]GE Aerospace Research, [#]University of Wisconsin-Madison, [+]Oak Ridge National Laboratory
Email: [*]{hang.dai, john.yagielski, huhk, libing.wang, tarak.saha}@geaerospace.com, [#]jahns@engr.wisc.edu,
[+]rallabandivp@ornl.gov, [#]wfeng33@wisc.edu, [#]sarlioglu@wisc.edu,

Abstract— **Wide-bandgap enabled current-source inverter (CSI) is a promising power converter topology that has cleaner output voltage waveform, higher fault tolerance, and higher temperature operation capability than conventional voltage-source inverters (VSI). This paper presents a WBG-enabled CSI for a 2-MW integrated motor drive (IMD) for electric aviation application. The CSI system consists of two series-connected CSI that forms as a unit, and three such units are connected in parallel to drive a specially designed surface permanent magnet synchronous motor (PMSM) that has six individual groups of three-phase windings. Furthermore, the CSI's dc-link inductor and power modules are cooled with supercritical carbon dioxide (sCO_2), and also cools the integrated PMSM's stator windings. The proposed CSI-based MW-scale IMD achieves an estimated 97.3% efficiency and 18.1 kW/kg power density. Its high-power density and promising fault-tolerance characteristics make it a competitive solution for MW-scale electric aviation applications.**

Keywords—current source inverter, electric aviation, integrated motor drive, sCO_2

I. INTRODUCTION

High power density and fault-tolerant motor drive is key to more electric aircraft [1]. In recent years, the development of integrated motor drive (IMD), which integrates power electronics drive(s) with an electric motor is proven, in both industry and academia, to significantly boost the propulsion system's power density [2].

However, due to the tight integration, the drive unit will experience higher ambient temperature than conventional standalone drive that is separated from the motor through a lengthy cable. Specifically, electric machines usually could operate with hot spot temperature of 180 °C or higher [3], and this temperature is higher than the maximum temperature limits of many power converter's key components (e.g., power semiconductor devices and capacitors). As a result, thermal

management is usually a significant challenge for IMD development. As shown in [4], a single water cooling jacket is designed to cool both the power converter and the motor housing. Nevertheless, heat generated in the machine needs to flow through multiple layers (e.g., wire insulation, slot liner, etc.) before reaching the cooling jacket, resulting in large thermal resistance and reduced cooling effectiveness [5]. To improve the cooling performance, there are efforts of designing hollow conductor machine windings with direct cooling and has achieved significant winding current density improvement (increased from 4A/mm^2 to 14 A/mm^2 [6]). Despite these eminent works, the hollow conductors are usually cooled using fluids. To circulate the fluid through the windings of the electric machine, significant pressure drop would occur, and a large pump is needed.

In addition to the cooling challenge, the conventionally adopted VSI topology for the motor drive has several limitations for use in IMD applications [7]. Specifically, VSI requires the use of dc-link capacitors that are usually temperature limited [8]. Despite years of development, the commercially available high temperature dc-link capacitors are usually limited to 125 °C and have significantly reduced lifespan as compared to operating at temperatures below 80 °C [9]. In addition, for the dominant two-level VSI topology, when driving permanent magnet synchronous machines (PMSMs), there are deleterious short-circuit fault and uncontrolled generation fault conditions [7]. Furthermore, the fast dv/dt at the output of the VSI results in significant overvoltage, machine winding stress and common-mode (CM) electromagnetic interference (EMI) for the motor drive system [7].

To tackle the discussed challenges, this paper presents a MW-scale IMD using CSI topology that has high-temperature capable and rugged dc-link inductor and with more benign fault performance [10], [11]. The CSI topology also has better CM-

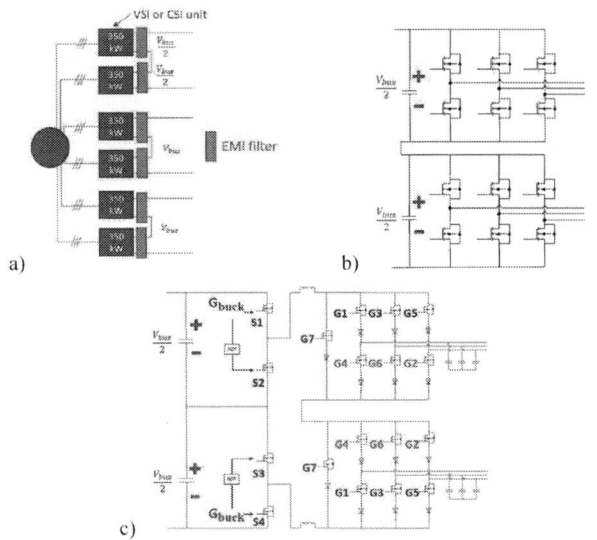

Fig. 1 a) 2MW motor drive circuit diagram; b) Two-level VSI topology; c) Proposed CSI topology

EMI performance than VSI [12], [13]. To effectively cool the high-power CSI's dc-link inductor and machine winding, a hollow conductor-type inductor using nanocrystalline material is designed and manufactured, along with a multi-phase PMSM using hollow conductor stator winding. sCO$_2$-based coolant that has high specific heat capacity and lower fluid density than oil is adopted to cool the CSI's dc-link inductor, SiC power module cold plate and electric machine's winding.

II. Comparison of VSI and CSI for MW-Scale Electric Aviation Application

A. 2 MW IMD Design

The proposed 2-MW system is shown in Fig. 1a, with dc bus voltage in the range of 2 to 2.6 kV [14]. To relieve the voltage stress and to add modularity, two converters are connected in series to form a unit, and three such units are connected in parallel. For the electric machine, it is designed to have six groups of independent three-phase windings. Note that each converter is designed to deliver 350 kW output power to the machine, given the efficiency of the machine being 94%.

For the converter, the conventional two-level VSI and a newly proposed dual buck front-end CSI [15] are comparatively evaluated, with topologies shown in Fig. 1b and c, respectively. For the VSI, it has two conventional two-level VSI connected in series, and each converter is driving a three-phase winding of the machine. Because of the sharing of the voltage, each VSI capacitor and power semiconductor switch only need to block half of the dc bus voltage, providing more component selection freedom.

Table I - Summary of the Parameters of Single-buck Front-end CSI

Switch Freq	Input Voltage	dc-link current	Fundamental frequency	dc-link inductor	Output filter capacitor (per phase)
20 kHz	2.0 - 2.6 kVdc	369 A	1500 Hz	54 uH	20 uF

In comparison, the newly proposed CSI has a dual-buck front-end converter that interfaces with high voltage dc bus. It has two symmetrically placed H7-CSI topologies, with each driving a three-phase PMSM winding. Similar to the series connected VSI, the front-end capacitor and switch in the proposed CSI also only need to block half of the dc bus voltage. However, it is noticeable that the CSI has dc-link inductor and requires the use of reverse voltage blocking switch, which is realized by connecting a SiC-Schottky diode in series with the SiC-MOSFET. The detailed operation of the proposed CSI has been discussion in [15].

Key parameters of the CSI are shown in Table I. VSI is designed to have the same switching frequency, motor fundamental frequency and dc-link input voltage.

B. Efficiency Comparison

For IMD, efficiency is an important metric as it directly relates to the cooling requirement. To maintain a fair comparison of the two converters' efficiency, equivalent operating conditions are maintained for the two converters to have the same dc bus voltage, switching frequency, fundamental frequency, machine power factor, device junction temperature, etc. A GE-built 1.7 kV SiC module [16] has been used for both converters. For the SiC module, its conduction loss and switching loss are extracted through experiment testing from half-bridge setups. Additionally, 1.7kV SiC Schottky diode by Navitas [17] is connected in series with the SiC module to realize the CSI's switch. The simulated device junction temperatures are shown in Fig. 2a, and the efficiencies are shown in Fig. 2b, for the VSI and CSI.

It is seen that the CSI has lower efficiency. This is mainly due to the use of extra front-end stage, dc-link inductors and

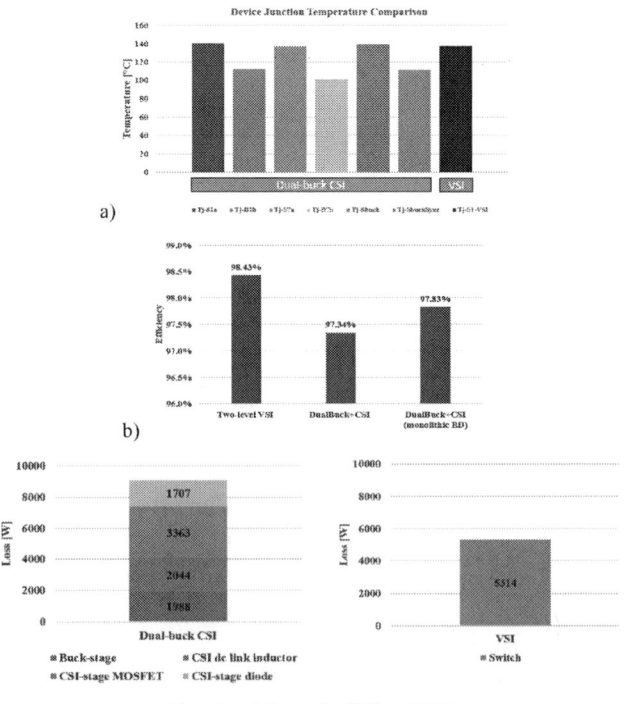

c) loss breakdown for CSI and VSI

Fig. 2 Comparison of predicted motor drive efficiency of VSI and CSI

hybrid reverse voltage blocking switches. Specifically, the key loss breakdown of VSI and CSI are shown in Fig. 2c. It is seen that the front-end buck stage incurs 22% of the loss. For CSI in motor drive applications, the buck stage is usually needed to avoid inrush current during machine start-up process and to enables buck-boost operation to achieve improved motor drive system performance [18]. In recent years, there are research of single-stage CSI that avoids the use of front-end stage, which holds promise for avoiding the extra front-end loss [19].

In addition, the CSI stage has 19% of loss coming from the conduction loss of the extra series-connected diodes. Recently, there are already reports of using monolithic bidirectional switch for CSI that eliminates the extra diode voltage drop [20], [21]. Assuming the use of BD switch for the CSI stage, the efficiency of the CSI is increased to 97.83%. It is worth mentioning that CSI using BD switch requires special gate signal placement, and the PWM modulation strategies proposed in [22] could be applied in the proposed CSI topology.

Moreover, the dc-link inductor constitutes another 22% of the total loss. As will be shown later, significant losses come from the inductor winding due to the high current density enabled by the sCO_2 cooling to reduce the inductor weight.

C. Other Performance Comparison

Although the efficiency for the CSI is inferior to that of the VSI, some of the CSI's benefits discussed below offer opportunities to offset at least a portion of the disadvantages. Like all complex systems, a fair engineering evaluation of a candidate architecture must be based on the aggregated assessment of all its key features and metrics. Several of these features discussed below highlight potential system advantages of the CSI.

Specifically, CSI has the benefits in terms of reduced voltage stress for the machine, higher fault tolerance especially when driving permanent magnet machines, lower CM EMI, voltage boost capability for increased machine voltage and reduced machine mass [18]. In this section, the output waveform quality and CM-EMI will be focused.

The simulated output line-line voltage and phase current waveforms of the two converters are shown in Fig. 3. As seen, the VSI has high dv/dt voltage. Despite IMD's short connecting cable length from the converter to the machine terminal, there are still significant overvoltage issue happening for the internal windings [23]. In comparison, CSI has close-to-sinusoidal output voltage waveforms, which is promising for improving the lifespan of the machine winding [24]. The THDs of the voltage and current waveforms of the VSI are about 60% and 6%, respectively. In comparison, the THDs of the voltage and current waveforms of the CSI are about 9% and 1%, which clearly show significant benefits of CSI. This includes reduced aging of the motor insulation system, with impact approaching a 50% reduction.

Another key metric for the motor drive unit is the CM-EMI. For aviation application, the power converter needs to meet DO-160G EMI standards. Conventionally, the VSI-based power converter for aviation application has EMI filter that constitutes nearly 30% of the total converter volume. To evaluate the EMI performance, the simulation setup in Fig. 4a

is built. The load side has 20nF per phase parasitic capacitance being modeled to represent machine winding's coupling capacitance to the ground. The dc-side CM (common-mode) EMI is defined in (1) and compared in Fig. 4b. For the dual-buck CSI topology, realistic switch delay of 20ns has been considered, with more details in [15]. As seen, the CSI topology has significantly lower (more than 20 dB lower) CM-EMI as compared to VSI topology, reducing the required CM-EMI filter weight and volume.

$$I_{bundle} = I_{dcP} + I_{dcN} \qquad (1)$$

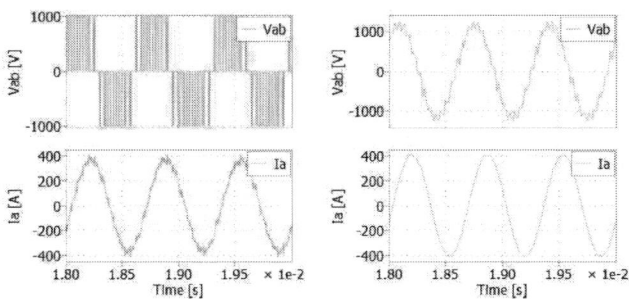

Fig. 3 Simulated motor drive module line-line voltage and phase current waveforms at 2 kV dc bus voltage operation condition: a) VSI; and b) CSI

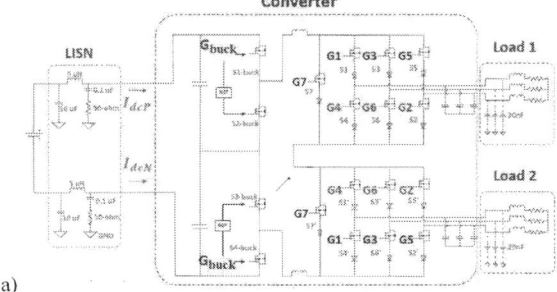

a)

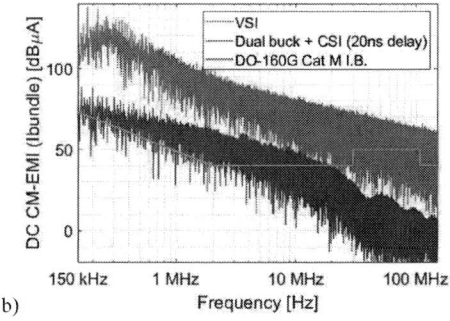

b)

Fig. 4 CM-EMI comparison of series-connected VSI and dualbuck+H7CSI configurations without EMI filters (20 ns gating mismatch is assumed for CSI's dual-buck stage and CSI stages): a) simulation setup; b) CM-EMI comparison

III. SCO_2-COOLED CSI-IMD

Despite several promising advantages of the CSI topology, there are challenges that need to be resolved, including the weight and loss management of the dc-link inductor, and

979-8-3315-1612-3/25 $31.00 © 2025 IEEE

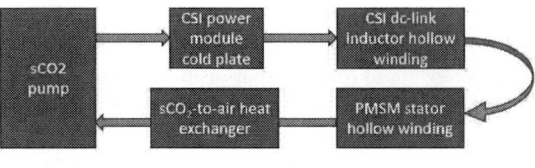

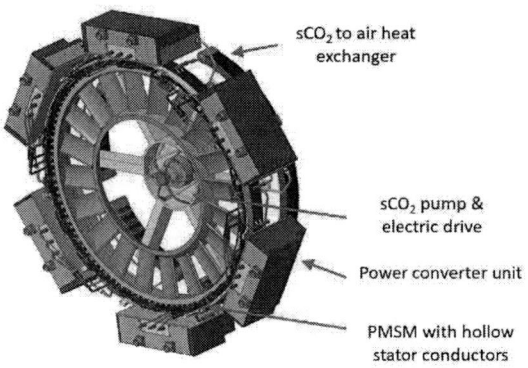

Fig. 5 Cooling of the CSI IMD: a) cooling loop; b) 3D CAD model

thermal cooling of the whole converter. In this section, the design and testing of the dc-link inductor and the sCO_2 cooling system will be discussed.

A. Cooling Design for the CSI-IMD

As highlighted earlier, recent efforts to design IMD with a common coolant for both the inverter and the motor have been presented [4]. However, traditional coolants (water, oil, etc.) usually have high fluid density. In passing the coolant through the power converter and the machine winding, significant pressure drop occurs, and a bulky thermal management system is needed. In contrast, the proposed CSI-IMD adopts sCO_2 coolant that has lower fluid density and gas-like properties, yielding a pumping power. Because of this, the CSI-IMD is designed to have the cooling loop as shown in Fig. 5. As seen, sCO_2 is passed through passages to cool the power module's heatsink, the dc-link inductor, and then the PMSM's hollow-conductor windings.

It is worth mentioning that cryogenic cooling of the inverter is another commonly used cooling methods, and several studies revealed its benefits especially for CSI [25].

B. Thermal Estimation of the CSI-IMD Power Module

An additively manufactured heatsink (as shown in Fig. 6) is used to cool the power module. Based on FEA simulation, the thermal resistances from the power module to the sCO_2 coolant are extracted. The thermal resistance is used in PLECS circuit for predicting the device junction temperatures, with results shown in Fig. 2. The rated sCO_2 coolant inlet temperature is 90°C. As seen, the devices operate with junction temperature lower than their maximum temperature limit at the rated operating condition.

Fig. 6 sCO_2-cooled cold plate for CSI IMD power module

C. Design of CSI-IMD dc-link Inductor Cooled by sCO_2

As described in [7], conventionally, the dc-link inductor takes a significant proportion of the CSI weight and volume. One method to reduce the inductor weight is to increase switching frequency. As shown in (2), the increased switching frequency (f_{sw}) directly results in the reduction of the required dc-link inductance for the same current ripple (Δi_{dc-pp}). In addition to employing SiC-MOSFET to increase the switching frequency, the proposed CSI-IMD's dc-link inductor uses nanocrystalline-based core with distributed airgaps, and its winding is using hollow copper tube that flows sCO_2 coolant inside.

The designed CSI dc-link inductor is shown in Fig. 7a. To achieve the CSI's required dc-link current of 370A, two parallel windings are adopted. To tolerate high dc-link current and to avoid core saturation, a total of eight air gaps are distributed across the magnetic flux loop. FEA simulation has been used to simulate the magnetic and loss performance of the designed inductor. The peak magnetic flux density is 1.48 T as shown in Fig. 7b, and the estimated total loss is 2044 W, with loss breakdowns shown in Table II.

$$L_{bus} = \frac{\sqrt{3}V_{ph-in}}{4f_{sw}\Delta i_{dc-pp}} \qquad (2)$$

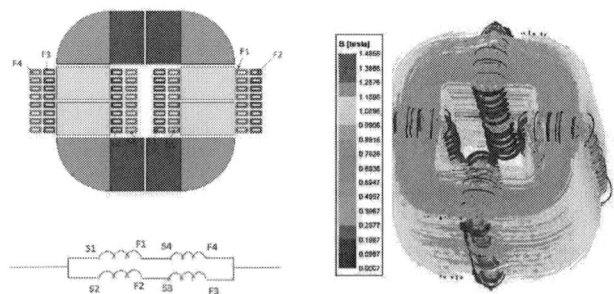

a) Designed CSI dc-link inductor   b) FEA simulation of CSI inductor

Fig. 7 CSI dc-link inductor design

979-8-3315-1612-3/25 $31.00 © 2025 IEEE

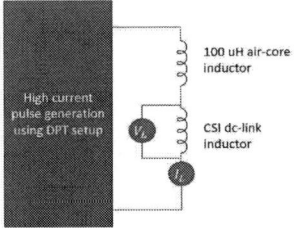

a) High bias current generation

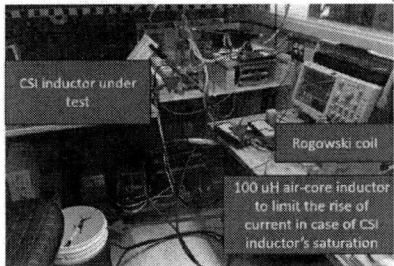

b) Picture of the test setup

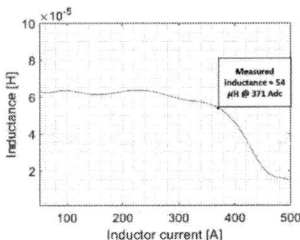

c) Extracted dc-link inductance under bias current

Fig. 8 Experiment evaluation of the CSI dc-link inductor

TABLE II - LOSS AND INDUCTANCE SIMULATION OF THE DESIGNED CSI DC-LINK INDUCTOR

DC loss @ 370A [W]	822
AC loss 370A dc and 20kHz ripple [W]	1065
Core loss @ 20kHz [W]	25
Air gap loss [W]	132
Core weight [kg]	2.08
Copper weight [kg]	0.45
Inductance @ 370Adc [uH]	53

a) sCO$_2$ pump

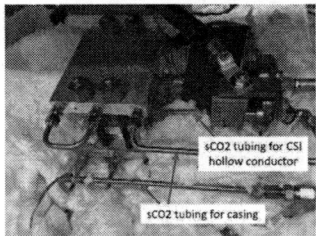

b) CSI dc-link inductor under testing

Fig. 9: CSI's dc-link inductor under sCO$_2$ cooling

The CSI's inductor has been manufactured and experimentally evaluated, including measurement of the inductance under the rated 370A dc link bias current. To obtain said inductance, the inductor was tested, as shown in Fig. 8. The pulse generation circuit provides high current to the inductor. A 100uH air-core inductor is placed in series to limit the current rise rate in case of CSI inductor's saturation. The voltage and current across the inductor have been recorded, and the inductance is extracted according to (3), where V_L is the voltage across the inductor, R_{dc} is the inductor's winding resistance. The obtained inductance is determined in Fig. 8c. As seen, under the rated current of 370A, the dc-link inductor has inductance of 54 uH, less than 2% difference as compared to the design value in Table II.

$$L = \frac{V_L - I_L \times R_L}{\frac{di_L}{dt}} \quad (3)$$

D. Inductor sCO$_2$ Cooling Testing

The inductor has also been tested using the sCO$_2$ test rig in Fig. 9. Due to test equipment limitations, only a dc current of 420Adc is applied to the windings, which generated approximately 1020 watts of dc loss. With 0.014 kg/min sCO$_2$ flow rate, the recorded temperature rise of the inductor is 26.0 deg K.

E. Power Density Estimation

Mass-based power density (kW/kg) is a critical metric for power converters in aerospace applications, making estimation of the power converter mass an important topic.

A detailed CAD model was developed specifically for the proposed converter, resulting in a set of mass estimates for all of the power converter components. The inductor mass is based on the actual mass of the prototype dc-link inductor shown in Fig. 9. The total estimated power converter masses for 2-MW six-converter architecture are presented in Fig. 10.

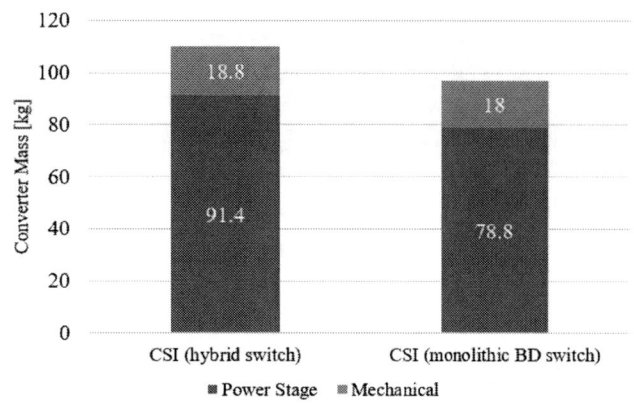

Fig. 10 Estimated weight of the 2-MW CSI-IMD power converter

These include the same two CSI converter configurations defined in the preceding efficiency discussion. Among the two CSI power converter configurations, the version using the existing GE SiC MOSFETs with series-connected Schottky diodes results in a higher estimated mass of 110.2 kg, corresponding to 18.1 kW/kg power density. By introducing monolithic RVB switches that eliminate the diodes, the CSI power converter mass is reduced to 96.8 kg.

IV. CONCLUSIONS

This paper has presented a 2 MW-scale CSI-based IMD for electric aircraft applications. The 2-MW system is divided into six power converter modules, with two modules in series to form a unit, and three such units are connected in parallel. The proposed CSI adopts symmetric buck converters and three-phase CSIs. Such topology helps relieve the voltage/current stress of the converters and achieves significant common-mode EMI reduction.

Comparative study of the proposed CSI and conventional VSI is performed. It has found the proposed CSI has 1.1% lower efficiency than VSI due to additional loss of the RB switch, dc-link inductors and front-end stage. Despite lower efficiency, proposed CSI-IMD has more than 20dB lower CM-EMI, much cleaner output voltage and current waveforms, higher temperature operation capability, higher fault tolerance and better boost capability to extend the PMSM's operating regions.

To minimize the CSI weight, compact dc-link inductor using nanocrystalline core with distributed airgaps is designed and experimentally evaluated. Furthermore, sCO_2 coolant that has high specific heat capacity and low fluid density is designed to cool the converter. The effectiveness of the sCO_2 cooling system has been verified using simulation and experiment testing.

In the future, system-level testing integrating power converter, electric machine and cooling system is desired to evaluate the whole CSI-IMD's performance.

ACKNOWLEDGMENT

The authors gratefully acknowledge the support from U.S. Advanced Research Project Agency-Energy (ARPA-E) under Project Number DE-AR0001353. The views and opinions expressed in this paper are those of the authors and do not necessarily reflect those of the U.S. ARPA-E.

REFERENCES

[1] S. Wu, C. Tian, W. Zhao, J. Zhou, and X. Zhang, "Design and Analysis of an Integrated Modular Motor Drive for More Electric Aircraft," IEEE Trans. Transp. Electrification, vol. 6, no. 4, pp. 1412–1420, Dec. 2020, doi: 10.1109/TTE.2020.2992901.

[2] S. Pickering, P. Wheeler, F. Thovex, and K. Bradley, "Thermal Design of an Integrated Motor Drive," in IECON 2006 - 32nd Annual Conference on IEEE Industrial Electronics, Nov. 2006, pp. 4794–4799. doi: 10.1109/IECON.2006.348109.

[3] T. M. Jahns and H. Dai, "The past, present, and future of power electronics integration technology in motor drives," CPSS Trans. Power Electron. Appl., vol. 2, no. 3, pp. 197–216, Sep. 2017, doi: 10.24295/CPSSTPEA.2017.00019.

[4] A. Tenconi, F. Profumo, S. E. Bauer, and M. D. Hennen, "Temperatures Evaluation in an Integrated Motor Drive for Traction Applications," IEEE Trans. Ind. Electron., vol. 55, no. 10, pp. 3619–3626, Oct. 2008, doi: 10.1109/TIE.2008.2003099.

[5] X. Chen, J. Wang, A. Griffo, and A. Spagnolo, "Thermal Modeling of Hollow Conductors for Direct Cooling of Electrical Machines," IEEE Trans. Ind. Electron., vol. 67, no. 2, pp. 895–905, Feb. 2020, doi: 10.1109/TIE.2019.2899542.

[6] P. Lindh et al., "Direct Liquid Cooling Method Verified With an Axial-Flux Permanent-Magnet Traction Machine Prototype," IEEE Trans. Ind. Electron., vol. 64, no. 8, pp. 6086–6095, Aug. 2017, doi: 10.1109/TIE.2017.2681975.

[7] Y. Zhang, "Investigation of Approaches for Improving the Performance and Fault Tolerance of Permanent Magnet Synchronous Machine Drives Using Current-Source Inverters," University of Wisconsin-Madison, 2016.

[8] J. Wang, Y. Li, and Y. Han, "Integrated Modular Motor Drive Design With GaN Power FETs," IEEE Trans. Ind. Appl., vol. 51, no. 4, pp. 3198–3207, Jul. 2015, doi: 10.1109/TIA.2015.2413380.

[9] KEMET, "New C4AK Series High Temperature, Long Life DC-Link Film Capacitors." Accessed: Nov. 20, 2024. [Online]. Available: https://www.kemet.com/en/us/technical-resources/new-c4ak-series-high-temperature-long-life-dc-link-film-capacitors.html

[10] L. Zheng, X. Han, R. P. Kandula, and D. Divan, "DC-Link Current Minimization Control for Current Source Converter-Based Solid-State Transformer," IEEE Trans. Power Electron., vol. 37, no. 10, pp. 11865–11875, Oct. 2022, doi: 10.1109/TPEL.2022.3178433.

[11] X. Guo, "Three-Phase CH7 Inverter With a New Space Vector Modulation to Reduce Leakage Current for Transformerless Photovoltaic Systems," IEEE J. Emerg. Sel. Top. Power Electron., vol. 5, no. 2, pp. 708–712, Jun. 2017, doi: 10.1109/JESTPE.2017.2662015.

[12] D. Zhang, M. Leibl, J. Muhlethaler, J. Huber, and J. W. Kolar, "Analytical Modeling and Comparison of EMI Pre-Filter Noise Emissions of Three-Phase Voltage and Current DC-Link Converters," IEEE Trans. Power Electron., pp. 1–16, 2024, doi: 10.1109/TPEL.2024.3427826.

[13] Y. Xu, Z. Wang, P. Liu, and J. He, "A Soft-Switching Current-Source-Inverter-Fed Motor Drive With Reduced Common-Mode Voltage," IEEE Trans. Ind. Electron., vol. 68, no. 4, pp. 3012–3021, Apr. 2021, doi: 10.1109/TIE.2020.2978691.

[14] D. Zhang, J. He, and D. Pan, "A Megawatt-Scale Medium-Voltage High-Efficiency High Power Density 'SiC+Si' Hybrid Three-Level ANPC Inverter for Aircraft Hybrid-Electric Propulsion Systems," IEEE Trans. Ind. Appl., vol. 55, no. 6, pp. 5971–5980, Nov. 2019, doi: 10.1109/TIA.2019.2933513.

[15] H. Dai, K. K. Huh, C. Li, T. M. Jahns, B. Sarlioglu, and J. Yagielski, "Symmetric Dual-Buck Front-End Current-Source Converter with Low Common-Mode EMI," presented at the IEEE Workshop on Wide Bandgap Power Devices and Applications (WiPDA), Dayton, US, 2024.

[16] GE Aerospace, "GE Aerospace Electrical Power Silicon-Carbide." Accessed: Dec. 16, 2024. [Online]. Available: https://www.geaerospace.com/systems/electrical-power

[17] Navitas, "Silicon Carbide: The Facts." Accessed: Dec. 16, 2024. [Online]. Available: https://navitassemi.com/silicon-carbide-the-facts/

[18] F. Chen, W. Feng, H. Ding, S. Lee, T. M. Jahns, and B. Sarlioglu, "Comprehensive Efficiency Analysis of Current Source Inverter Based SPM Machine Drive System for Traction Applications," in 2020 IEEE Energy Conversion Congress and Exposition (ECCE), Oct. 2020, pp. 3002–3009. doi: 10.1109/ECCE44975.2020.9236270.

[19] D. Benatti, G. Migliazza, E. Carfagna, F. Immovilli, and E. Lorenzani, "Novel Single-Stage Current Source Inverter: Extension to Low-Speed Region in Motor Drive Applications," IEEE Trans. Ind. Electron., vol. 71, no. 9, pp. 10335–10345, Sep. 2024, doi: 10.1109/TIE.2023.3335461.

[20] S. Narasimhan, A. Kanale, S. Bhattacharya, and J. B. Baliga, "Performance Evaluation of 3.3 kV SiC MOSFET and Schottky Diode Based Reverse Voltage Blocking Switch for Medium Voltage Current Source Inverter Application," IEEE Access, vol. 11, pp. 89277–89289, 2023, doi: 10.1109/ACCESS.2023.3302916.

[21] M. Guacci et al., "Three-phase two-third-PWM buck-boost current source inverter system employing dual-gate monolithic bidirectional GaN e-FETs," CPSS Trans. Power Electron. Appl., vol. 4, no. 4, pp. 339–354, Dec. 2019, doi: 10.24295/CPSSTPEA.2019.00032.

[22] H. Dai, R. A. Torres, J. Gossmann, W. Lee, T. M. Jahns, and B. Sarlioglu, "A Seven-Switch Current-Source Inverter Using Wide Bandgap Dual-Gate Bidirectional Switches," IEEE Trans. Ind. Appl., vol. 58, no. 3, pp. 3721–3737, May 2022, doi: 10.1109/TIA.2022.3149461.

[23] M. T. Fard, J. He, L. Wei, R. Ilka, B. Mirafzal, and F. Fateh, "Mitigation of Motor Reflected Overvoltage Fed by SiC Drives - A New Solution based on Smart Coils," IEEE Trans. Power Electron., pp. 1–10, 2024, doi: 10.1109/TPEL.2024.3502352.

[24] ABB, "Technical Guide No. 102: Effects of AC Drives on Motor Insulation." [Online]. Available: https://library.e.abb.com/public/fec1a7b62d273351c12571b60056a0fd/voltstress.pdf

[25] M. Ul Hassan, Y. Wu, A. Imran Emon, Z. Yuan, and F. Luo, "Design Considerations and Their Optimization for a Two-Level GaN-Based Current Source Inverter," IEEE J. Emerg. Sel. Top. Power Electron., vol. 12, no. 3, pp. 3173–3187, Jun. 2024, doi: 10.1109/JESTPE.2024.3391005.

The Challenge of Thermal Runaway in Soft Magnetic Materials for Inductive Power Transfer

Yibo Wang
Department of Electrical Engineering
City University of Hong Kong
Hong Kong SAR, China
yibwang2-c@my.cityu.edu.hk

Ben Zhang
Department of Electrical Engineering
City University of Hong Kong
Hong Kong SAR, China
zhang.ben@cityu.edu.hk

Weisheng Guo
Department of Electrical Engineering
City University of Hong Kong
Hong Kong SAR, China
weishguo@cityu.edu.hk

Tianlu Ma
Department of Electrical Engineering
City University of Hong Kong
Hong Kong SAR, China
tianlu.ma@my.cityu.edu.hk

Sheng Ren
Department of Electrical Engineering
City University of Hong Kong
Hong Kong SAR, China
shengren3-c@my.cityu.edu.hk

C. Q. Jiang
Department of Electrical Engineering
City University of Hong Kong
Hong Kong SAR, China
chjiang@cityu.edu.hk

Abstract—**This paper investigates the thermal runaway effect in soft magnetic materials (SMCs). The SMCs are essential for modern power electronics and are used in applications such as inductors, transformers, and wireless power transfer systems. Despite their importance, the thermal runaway of the SMCs poses a significant challenge to the reliability and safety of these applications, particularly with inductive power transfer (IPT). To address this challenge, the paper compares two materials: a benchmark MnZn ferrite and a Fe-based nanocrystalline material. The core losses of these materials were measured inside a thermal chamber to determine their relationship with different ambient temperatures. The findings reveal that both materials exhibit a valley temperature point where the core loss density is minimized. However, this valley temperature varies with different flux densities and shows distinct behaviors for ferrites and nanocrystalline materials. Furthermore, the thermal runaway effect is mathematically explained based on thermal radiation and convection calculations of an example IPT system. The analysis can be used to estimate and prevent thermal runaway during the pre-design stage.**

Keywords—*Thermal management, inductive power transfer (IPT), soft magnetic materials*

I. Introduction (*Heading 1*)

Soft magnetic materials (SMC) are widely adopted in IPT applications to enhance magnetic coupling and reduce leakage flux [1], [2], [3], [4]. MnZn ferrite is the most commonly used soft magnetic material in IPT systems. However, its properties, such as core loss, permeability, and saturation flux density, can vary significantly with temperature.

High-power wireless charging systems are designed for passenger vehicles, industrial vehicles, and electric ships, where ambient temperatures range from -20 to 55 °C. The internal magnetic material, which generates both loss and heat, can exceed 100 °C. These temperature variations affect the operating conditions of IPT systems by altering the magnetic properties of ferrites, potentially degrading performance over time. Additionally, magnetic loss produces heat. If this heat isn't dissipated quickly, the system may reach its temperature limit, leading to potential thermal defects or standards violations. As ferrite loss starts to increase exponentially beyond a certain temperature, the system risks thermal runaway, posing a danger to operations

and public safety, and reducing the market acceptability of IPT technology.

Some literature has identified the thermal runaway feature of ferrites. The high thermal instability of the ferrite materials can result in thermal stress which can cause materials to crack [5] and even cause fire hazards [6]. The study in [7] has pointed out the critical temperature for the thermal runaway of a magnetic material. Nevertheless, the influences of the flux densities are not analyzed. The thermal instability feature of existing power ferrites does not draw sufficient attention as this problem is not severe in converter applications. One reason is that active cooling in IPT chargers is costly, especially for vehicle assemblies. Academic research often focuses on instantaneous performance without addressing long-term operation. Insights from industry patents, such as [8] from Tesla, Inc., reveal innovations like a temperature sensor array covering the entire coil and ferrite areas, indicating concerns about the thermal performance of ferrite.

In addition to MnZn ferrites, the Fe-based nanocrystalline materials has gained increasing popularity for IPT applications in recent years. The thermal performance of the Finemet nanocrystalline material and ferrite are compared in [9], [10] and the core loss are analyzed in [11], [12].

The current literature on thermal analysis of magnetic core materials for IPT systems is lacking, particularly concerning the critical thermal runaway effect. This is essential for practical IPT applications. As demand for higher power density increases, the core material's flux density is raised, complicating magnetic design due to the interrelated factors of flux density, temperature, and core loss. Predicting performance during pre-design becomes challenging, leading to potential repetitive design corrections. Uncontrolled thermal runaway can cause overheating, fire hazards, material degradation, and potential system failures, reducing reliability. These issues impede the market growth of inductive power transfer, despite its benefits.

This paper aims to thoroughly analyze the thermal behaviors and investigate the thermal runaway effect of magnetic materials used in IPT systems. The objective is to provide a theoretical and mathematical foundation for analyzing thermal stability and to offer a comprehensive comparison of existing materials.

979-8-3315-1612-3/25 $31.00 © 2025 IEEE

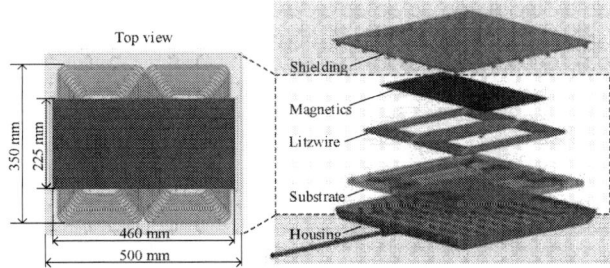

Fig. 1. Assembly of typical high-power inductive power transfer systems and the components for thermal analysis.

II. REVIEW OF HEAT GENERATION AND DISSIPATION IN IPT MAGNETIC COUPLERS

A. Loss generation of the soft magnetic materials

In Fig. 1, the typical assembly structure of an IPT system is shown. The system's overall losses come from components like Litz wire windings, aluminum shielding, and soft magnetic materials. To focus on the thermal behavior and performance of magnetic materials, the system is simplified by excluding the housing and aluminum shielding, with the system facing upwards as vehicle assembly. The typical core loss density for magnetic materials is calculated using the empirical Steinmetz equation, which includes temperature coefficients, as shown in equation (1).

$$P_{core} = C_m \cdot f^{\alpha} \cdot B^{\beta} \cdot (C_2 \cdot T_c^2 + C_1 \cdot T_c^2 + C_0) \qquad (1)$$

where P_{core} is the power density in kW/m³, C_m is the loss coefficient, f is the excitation frequency, B is the flux density inside the core material, and T_c is the core temperature. C_2, C_1, and C_0 are the temperature coefficients for the T_c^2, T_c^1, and T_c^0 respectively. These coefficients are usually regarded as constant when estimating the core loss. However, for thermal stability analysis, it is important to take into account the valley point of the core loss. This valley point, denoted as T_v, is defined as the temperature at which core loss is minimized, as shown in (2).

$$T_v = -\frac{C_1}{2 \cdot C_2} \qquad (2)$$

Constant temperature coefficients lead to a fixed valley point, which contradicts the observed changes in core loss behavior at different flux densities. To resolve this, this paper employs varying temperature coefficients as a function of flux density:

$$\begin{cases} C_0 = f_0(B) \\ C_1 = f_1(B) \\ C_2 = f_2(B) \end{cases} \qquad (3)$$

The detailed measurement results at different flux densities and temperatures are presented in Section III.

B. Thermal equilibrium of magnetic cores in IPT

The heat dissipation power P_{diss} to the ambient environment for magnetic cores includes the free convection power P_{conv}, the radiation power P_{rad}, and the conduction power P_{cond}. The first two dissipate heat directly into the air, while the last involves heat dissipation through contact with the litz wire windings and the housing. The mathematical

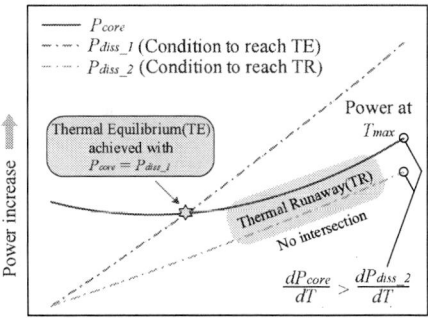

Fig. 2 Mathematical interpretation of thermal equilibrium and thermal runaway effects of magnetic materials.

interpretation can be expressed in (4).

$$\begin{cases} P_{diss} = P_{conv} + P_{rad} + P_{cond} \\ P_{conv} = h \cdot A \cdot (T_c - T_{amb}) \\ P_{rad} = \sigma \cdot \varepsilon \cdot A \cdot ((T_c + 273)^4 - (T_{amb} + 273)^4) \\ P_{cond} = \dfrac{k \cdot A \cdot (T_c - T_{amb})}{d} \end{cases} \qquad (4)$$

where h is the convection heat transfer coefficient, A is the convection and radiation surface area of the magnetic cores, σ is the Stefan-Boltzmann constant and ε is the emissivity. T_c and T_{amb} represent the core and ambient temperatures, respectively. k and d are the thermal conductivity and thickness of the Litz wires and housing compound, respectively. It is evident that P_{rad} depends solely on the temperatures once the geometry is fixed, whereas P_{conv} is related to the heat transfer. The heat transfer coefficient can be calculated as:

$$h = \frac{k_{air} \cdot Nu}{L} \qquad (5)$$

where k_{air} is the thermal conductivity of air, Nu is the Nusselt number and L is the characteristic length. The Nusselt number can be calculated according to the thermal parameters [13]. For the magnetic cores, the thermal equilibrium can be reached once the following condition is fulfilled:

$$P_{core} \cdot V_c = P_{diss} \qquad (6)$$

where V_c is the core volume.

Fig. 2 illustrates an example of a mathematical representation of thermal runaway and thermal equilibrium states. Thermal equilibrium is achieved when the loss curve intersects with the dissipation curve. If they do not intersect, thermal runaway occurs. In the state of thermal runaway, the increases of core loss and temperature interact as positive feedback with both values grows exponentially.

C. Magnetic materials for IPT

In this paper, two magnetic materials are analyzed and used in the IPT experiments: MnZn ferrite N95 and Fe-based Nanocrystalline Ribbon Cores (NRC). These materials are among the most efficient magnetic cores available on the market. Their wide availability ensures the reproducibility of the analysis.

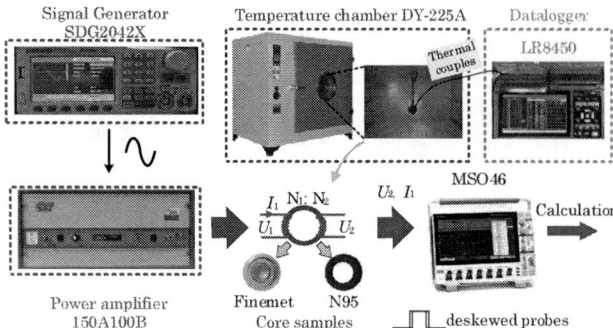

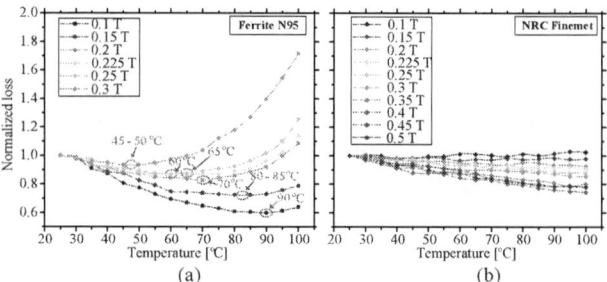

Fig. 4. Normalized core loss results at different temperatures. (a) Ferrite N95. (b) NRC Finemet.

Fig. 3 Core loss measurement setups for Finemet and N95 materials under different flux densities and temperatures.

III. CORE LOSS ANALYSIS OF SOFT MAGNETIC MATERIALS AT VARIOUS TEMPERATURES

A. Measurement setups

In Figure 3, the measurement setup for core loss analysis is illustrated. A SDG2042X signal generator produces sinusoidal signals to control the 150A100B power amplifier, which generates sinusoidal excitation for the core samples. These samples are toroidal and configured with a conventional two-winding setup, with the secondary winding left open. The secondary voltage U_2 and primary current I_1 are measured using the MSO46 digital oscilloscope. Current and voltage probes are deskewed to correct for inherent time delays.

To examine behavior under varying ambient temperatures, ferrite, and Finemet core samples are placed inside a DY-225A temperature chamber. Thermocouples attached to the core samples provide real-time temperature readings, recorded by the LR8450 datalogger to monitor actual core temperatures. The collected voltage and current signals are processed to calculate core loss and flux densities. Since the evaluated IPT operates at an 85 kHz nominal condition, measurements of the toroidal materials are also conducted with excitation fixed at 85 kHz to focus on thermal performance.

B. Measurement with different conditions

The ambient temperature ranges from 25 °C to 100 °C in 5 °C steps. For N95, flux densities are measured up to 0.3 T; beyond this, saturation causes waveform distortion and errors in core loss measurements. Finemet NRC is tested up to 0.5 T.

Fig. 4 shows core loss data for N95 and Finemet, highlighting differences. For ferrite, the core loss vs. temperature curve is quadratic, with the valley temperature shifting as flux density changes. Core loss increases more at higher flux densities. Finemet shows stable core loss with minimal change up to 0.35 T; core loss decreases with higher temperatures.

Fig. 5(a) and 5(b) illustrate normalized core loss, showing the temperature valley point. Normalization uses core loss at 25 °C for each flux density. For ferrite, the valley temperature shifts: 90 °C at 0.1 T, 80-85 °C at 0.15 T, 70 °C at 0.2 T, 65 °C at 0.225 T, 60 °C at 0.25 T, and 45-50 °C at 0.3 T. Valley shape varies with flux density. At 0.1 T, core loss drops by 40% at the valley, but only 10% at 0.3 T.

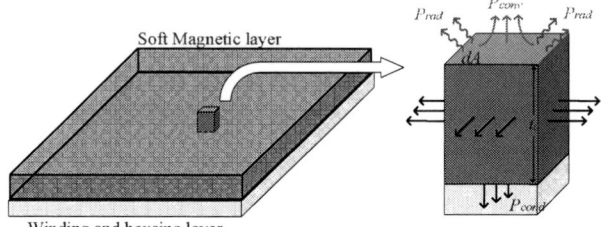

Fig. 5. The infinitesimal thermal model of the magnetic layers in IPT.

For Finemet, core loss remains constant at 0.1 T and 0.15 T, with less than 5% variation. At higher flux densities, core loss decreases steadily across temperatures, lowest at 100 °C. Reduction from initial loss increases from 8% at 0.2 T to 27% at 0.45 T, then slightly drops to 22% at 0.5 T.

IV. THERMAL STABILITY ANALYSIS OF THE MAGNETIC CORES IN IPT WITH NRC AND FERRITE

A. Simplification of the thermal stability evaluation.

The thermal modeling of an IPT system is complex and varies with assembly, making thermal stability predictions challenging. In Fig. 2, thermal runaway starts at the hottest spot with a positive gradient, causing its temperature to rise faster than elsewhere. Thus, focusing on this spot is key for assessing stability. To simplify the analysis, tests are done without aluminum shielding, with the magnetic layer facing up, assuming uniform temperatures across the thickness.

The hottest spot is modeled as shown in Fig. 5. It's considered an infinitesimal cuboid, with side faces touching adjacent material, the top face exposed to air for natural convection, and the bottom face contacting winding layers. Heat transfer from the sides is ignored due to uniform temperature in adjacent areas.

Therefore, the heat dissipation power can be simplified with free convection P_{conv}, radiation from the top surface P_{rad}, and heat conduction from the bottom surfaces P_{cond}. The total dissipation power is normalized with the core loss at 25 °C at each B_m flux density condition.

$$P_{diss\_n} = \frac{P_{conv} + P_{rad} + P_{cond}}{dA \cdot t_c \cdot P_{core@25°C,B_m}} \quad (7)$$

where t_c is the thickness, the dA component in the denominator can be cancelled out with the respective part inside the dissipation power.

The estimated heat transfer parameters are in Table I, based on relevant literature. Litz wire's thermal conductivity

TABLE I.   ESTIMATED PARAMETERS FOR THERMAL STABILITY
ANALYSIS.

Parameters	Value
Radial thermal conductivity of Litzwire	1.5 W/m°C
Thermal conductivity of PVC housing	0.17 W/m°C
Thermal conductivity of air	0.026 W/m°C
Emissivity of Ferrite N95	0.85
Emissivity of Finemet NRC	0.6

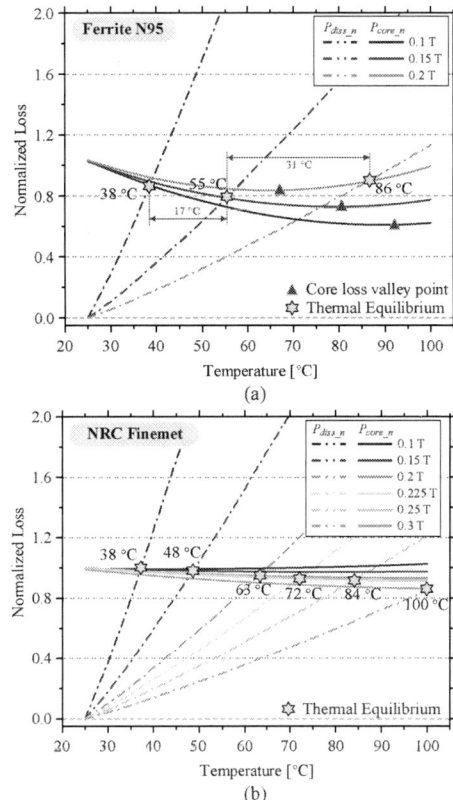

Fig. 6. Thermal stability prediction for Finemet and N95 materials under different flux densities.

is highly anisotropic. In magnetic layers, heat flux passes through insulations, impregnations, and copper wires to the PVC housing. Heat resistance is mainly due to the insulation and impregnation, which have much lower conductivity than copper. The average value is 1.5 W/m°C [14]. MnZn ferrite emissivity ranges from 0.7 to 0.9 [15], while Finemet varies from 0.06 to 0.79 [16]. In this study, emissivity is calibrated using an infrared camera and thermocouples at room temperature, assuming constant values.

B. Prediction of thermal stability

Fig. 6 shows the thermal stability prediction based on the core loss measurement data in Section III and the thermal analysis. For the ferrite materials in Fig. 6(a), at a flux density up to 0.2 T, thermal equilibrium is expected at intersection temperatures of 38 °C for 0.1 T, 55 °C for 0.15 T, and 86 °C for 0.2 T. Notably, at 0.2 T, the equilibrium temperature exceeds the core loss valley point, suggesting that stabilization beyond the minimum loss temperature is possible if the rise in core loss with temperature is less than the heat dissipation power. Higher flux density of ferrite materials will lead to thermal runaway.

The thermal stability of NRC Finemet is shown in Fig. 6(b), highlighting its ability to stabilize up to 0.3 T. Compared to N95, Finemet materials have a relatively lower core loss, leading to a lower expected stabilizing temperature than ferrite materials. Additionally, the increase in stabilizing temperature is less significant in Finemet than in N95. For instance, when the flux density rises from 0.1 T to 0.15 T and then to 0.2 T, the stabilizing temperature of N95 increases by 17 °C and 31 °C, respectively. In contrast, for Finemet, the increases are 10 °C and 15 °C. This difference is due to the smaller increase in core loss with respect to flux density in Finemet.

V. EXPERIMENTAL VERIFICATIONS.

A. Experiment setups

The experimental setup is shown in Fig. 7. The winding and housing are made from the materials listed in Table I. The magnetic material covers an area of 460 mm × 225 mm × 4 mm. For the Finemet setup, two edge shields made of nanocrystalline flake ribbon (NFR) materials are added to address the issue of concentrated flux density on the edges and achieve a more uniform distribution . Both NFR and NRC use Finemet material compositions. For the N95 setup,

six core plate blocks are spliced into a single core layer with minimal air gaps.

Based on the flux density distribution from finite element method (FEM) simulations, the hottest spots on the magnetic cores are identified. Four K-type thermocouples are attached to these locations on the N95 and Finemet cores to measure the maximum temperatures. The temperature readings are recorded using a Hioki LR8450 data logger.

The IPT evaluation platform includes primary and secondary magnetic couplers, excited by an inverter linked to a DC power supply. The secondary pad connects to a rectifier, with the rectified DC current fed back to the input side. This setup enables small power draws from the grid while achieving large power flow during experiments. A Fotric P5-L25-L46 thermal camera is used to capture the thermal distribution and verify the flux density distribution from FEM simulations.

B. System parameters

The system parameters are summarized in Table II. The primary pad utilizes ferrite N95 for both setups and is equipped with active cooling to ensure stable operation. The secondary pad serves as the observation and measurement platform. Compared with N95, the Finemet material's high permeability results in higher self and mutual inductances, as well as coupling factors. A series-series compensation topology is employed, with the tuned resonant frequency closely matched to the operating frequency, thereby ensuring a sinusoidal excitation is generated. The thermal testing is conducted at a room temperature of 23 °C.

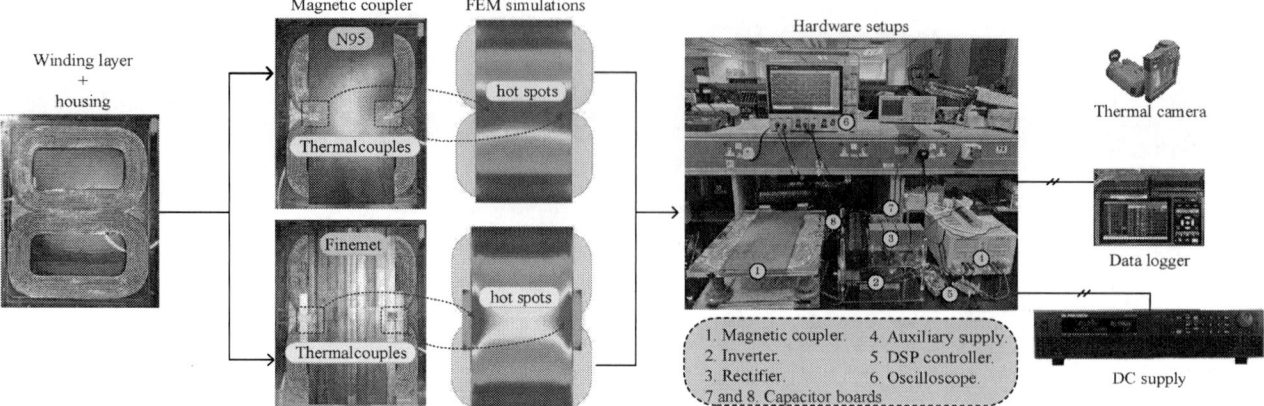

Fig. 7. Configuration process for experimental hardware setup.

TABLE II. ESTIMATED PARAMETERS FOR THERMAL STABILITY ANALYSIS.

Parameters	Ferrite N95	NRC Finemet
Air gap	120 mm	120 mm
Primary inductance	140.1 μH	140.5 μH
Secondary inductance	140.3 μH	145.1 μH
Mutual inductance	36.73 μH	38.69 μH
Coupling factor	0.262	0.271
Switching frequency	85.5 kHz	85.5 kHz
Room temperature	~23 °C	~23 °C

C. Investigation of thermal runaway effect

The thermal distribution of the ferrite and NRC materials is initially examined. As shown in Fig. 8, a comparison under a 15 A output current reveals that both N95 and Finemet show some thermal inhomogeneity. Hotspots appear near the ferrite material's splicing gaps due to localized flux density concentration. Over time, the temperature near the middle edge rises sharply, matching the flux density concentration from FEM simulations. The temperature quickly reaches the overtemperature protection (OTP) threshold of 100 °C in under 10 minutes.

In contrast, the Finemet setup shows a more gradual temperature increase, reaching a maximum of 65 °C after 30 minutes and 67.5 °C after 1 hour, without dramatic rises.

Besides the risk of overheating, there is a significant threat from the high thermal stress caused by the rapid temperature increase during the thermal runaway state. Fig. 15 illustrates the condition of the N95 plates following a short test with 0.3 T. The extremely high rate of temperature change generates substantial thermal stress near the thermal runaway area. The thermal expansion coefficient of ferrite is around $12 \diamond 10^{-6}$/K [17], which is considered relatively low. Despite this, thermal shocks are observed in two different samples during separate tests, leading to the sudden cracking of the plates. This highlights the continuous challenge of ensuring the thermal stability of ferrite materials for the reliability and safety of IPT systems.

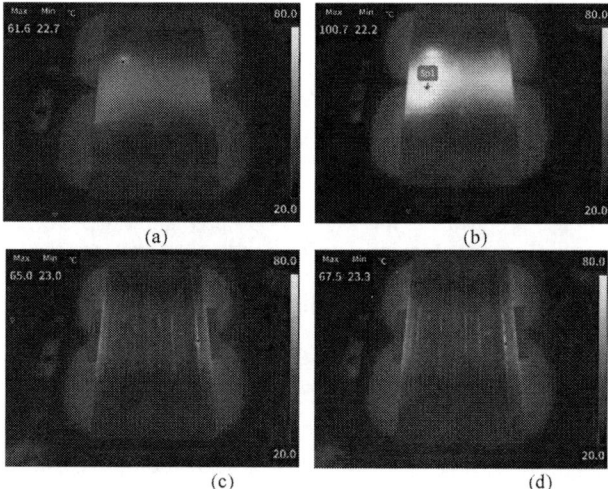

Fig. 8. Comparison of thermal distribution in ferrite N95 and NRC Finemet materials at 15 A with varying operational times: (a) N95 after 30 minutes, (b) N95 at OTP after 40 minutes, (c) Finemet after 30 minutes, and (d) Finemet after 1 hour.

The nanocrystalline material exhibits an even lower thermal expansion, typically below $12 \diamond 10^{-6}$/K, with some negative values reported at around 80 °C [18]. This suggests that the thermal stress generated is expected to be lower than in MnZn ferrites. Using Fe-based nanocrystalline material in IPT systems shows superior thermal stability. However, the maximum operating temperature is limited to 155°C due to the epoxy resin used in the lamination process.

D. Conclusion.

This study examines the thermal stability of soft magnetic materials in inductive power transfer (IPT) applications, focusing on system reliability and safety. Two materials are analyzed: MnZn ferrite N95 and Fe-based nanocrystalline ribbon core (NRC) Finemet.

Initial experiments in a temperature chamber show that ferrite exhibits fluctuating core loss behavior, with loss dependent on flux density. In contrast, NRC consistently shows decreased core loss as temperature rises. Ferrite's primary loss is hysteresis, linked to magnetocrystalline anisotropy, while NRC's loss is from eddy currents, which decrease with temperature due to increased resistivity. A simplified thermal model assesses their thermal stability.

Further IPT platform experiments confirm that ferrite easily enters thermal runaway (TR) at flux densities over 0.2 T, while NRC reaches thermal equilibrium (TE) regardless of flux density. At 15 A, N95's core loss spikes over 74.4 W, reducing efficiency by 1.27% and triggering an overtemperature shutdown. These findings highlight challenges in IPT systems, with Fe-based nanocrystalline alloys offering a promising alternative.

ACKNOWLEDGMENT

This work was supported in part by the Research Grants Council, Hong Kong SAR, China, under CRF-YCRG C1002-23Y, and in part by a grant from the Science Technology and Innovation Committee of Shenzhen Municipality, China, under project SGDX2021082310400 3034. Also, the authors are with Joint Laboratory of Energy Saving and Intelligent Maintenance for Modern Transportations.

REFERENCES

[1] D. Wang et al., "Modern Advances in Magnetic Materials of Wireless Power Transfer Systems: A Review and New Perspectives," *Nanomaterials*, vol. 12, no. 20, Art. no. 20, Jan. 2022.

[2] B. S. Gu, T. Dharmakeerthi, S. Kim, M. J. O'Sullivan, and G. A. Covic, "Optimised Magnetic Core Layer in Inductive Power Transfer Pad for Electric Vehicle Charging," *IEEE Transactions on Power Electronics*, pp. 1–11, 2023.

[3] S. Kim et al., "Thermal Evaluation of an Inductive Power Transfer Pad for Charging Electric Vehicles," *IEEE Transactions on Industrial Electronics*, vol. 69, no. 1, pp. 314–322, Jan. 2022.

[4] Y. Gu, J. Wang, Z. Liang, and Z. Zhang, "Flexible Constant-Power Range Extension of Self-Oscillating System for Wireless In-Flight Charging of Drones," *IEEE Transactions on Power Electronics*, vol. 39, no. 11, pp. 15342–15355, Nov. 2024.

[5] P. A. J. Lawton, F. J. Lin, and G. A. Covic, "Magnetic Design Considerations for High-Power Wireless Charging Systems," *IEEE Transactions on Power Electronics*, vol. 37, no. 8, pp. 9972–9982, Aug. 2022.

[6] S. Niu, H. Yu, S. Niu, and L. Jian, "Power Loss Analysis and Thermal Assessment on Wireless Electric Vehicle Charging Technology: The Over-temperature Risk of Ground Assembly Needs Attention," *Applied Energy*, vol. 275, p. 115344, Oct. 2020.

[7] F. Farahmand, F. P. Dawson, and J. D. Lavers, "Critical Temperature for Thermal Runaway in a Magnetic Material," *IEEE Trans. Magn.*, vol. 44, no. 11, pp. 4513–4516, Nov. 2008.

[8] X. Zhang, S. O'MEARA, and X. Fan, "Temperature Sensors and Applications in Wireless Charging," WO2024182405A1, Sep. 06, 2024 Accessed: Nov. 14, 2024. [Online]. Available: https://patents.google.com/patent/WO2024182405A1/en?oq=WO+20 24%2f182405

[9] M. Yang et al., "Application-Oriented Characterization and Analysis of Core Materials Under Medium-Frequency Condition," *IEEE Transactions on Power Electronics*, vol. 38, no. 9, pp. 11245–11259, Sep. 2023.

[10] D. E. Gaona, C. Q. Jiang, and T. Long, "Highly Efficient 11.1-kW Wireless Power Transfer Utilizing Nanocrystalline Ribbon Cores," *IEEE Transactions on Power Electronics*, vol. 36, no. 9, pp. 9955–9969, 2021.

[11] Y. Wang, C. Q. Jiang, L. Mo, X. Wang, W. Guo, and B. Zhang, "Design and Analysis of A Multi-segment Multi-permeability Core for EV Wireless Charging with Enhanced Efficiency and Thermal Performances," *Applied Energy*, vol. 375, p. 124181, Dec. 2024.

[12] Y. Wang, C. Q. Jiang, L. Mo, J. Xiang, Z. Luo, and W. Miao, "Magnetic Analysis of Permeability Configurable Nanocrystalline Flake Ribbons for Medium Frequency Energy Conversion Applications," *Materials Today Sustainability*, vol. 27, p. 100795, Sep. 2024.

[13] J. Ma, Z. Li, Y. Liu, M. Ban, and W. Song, "Thermal Analysis and Optimization of the Magnetic Coupler for Wireless Charging System," *IEEE Transactions on Power Electronics*, vol. 38, no. 12, pp. 16269–16280, Dec. 2023.

[14] X. Liu, D. Gerada, Z. Xu, M. Corfield, C. Gerada, and H. Yu, "Effective Thermal Conductivity Calculation and Measurement of Litz Wire Based on the Porous Metal Materials Structure," *IEEE Transactions on Industrial Electronics*, vol. 67, no. 4, pp. 2667–2677, Apr. 2020.

[15] P. Thakur, D. Chahar, S. Taneja, N. Bhalla, and A. Thakur, "A Review on MnZn Ferrites: Synthesis, Characterization and Applications," *Ceramics International*, vol. 46, no. 10, Part B, pp. 15740–15763, Jul. 2020.

[16] A. Talaat et al., "Radio-Frequency Rapid Thermal Processing Enabling Spatial Phase Transformation and Nanocrystallization of Soft Magnetic Amorphous Alloys," *Advanced Engineering Materials*, vol. 24, no. 10, p. 2200208, 2022.

[17] TDK Corp. Material Characteristics. Accessed: Aug. 24, 2023. [Online]. Available: https://product.tdk.com/system/files/dam/doc/product/ferrite/ferrite/ferrite-core/catalog/ferrite_mnzn_material_characteristics_en.pdf.

[18] C. Su, Y. Niu, H. Li, G. Huang, and X. Wang, "Thermal Expansion Behavior of Amorphous Fe-Si-B Alloys," *Journal of Iron and Steel Research, International*, vol. 18, no. 6, pp. 74–78, Jun. 2011.

A Capacitively Coupled Alternative Electric Field Control for Freeze-Free based High Quality Food Preservation

1st Jaeyong Cho
Life Solution Team
Samsung Research
Seoul, Republic of Korea
jy7954.cho@samsung.com

2nd Junhyeong Park
Life Solution Team
Samsung Research
Seoul, Republic of Korea
jhyeong.park@samsung.com

3rd Sung-Bum Park
Life Solution Team
Samsung Research
Seoul, Republic of Korea
sb21.park@samsung.com

4th Daehyun Kim
Life Solution Team
Samsung Research
Seoul, Republic of Korea
daehyun2.kim@samsung.com

5th Jinsoo Choi
Life Solution Team
Samsung Research
Seoul, Republic of Korea
jins27.choi@samsung.com

Abstract— **Freezing has been extensively used as the most common method in food storage application. However, it causes unavoidable poor quality of perishable foods such as meat and fishes by texture damage from ice nucleation of internal moisture. As a potential solution, lossless supercooling preservation technology using alternative electric field is newly proposed which has two parallel copper plate electrodes with 13.56 MHz and 50 Ω-matched impedance using CLC T-network on foods for deep penetration depth and maximum power transfer, separately. For 150 g of beef with 50 W, the supercooling state successfully last over 3 days with the meat temperature of -3°C at the ambient temperature of -10°C (the temperature to prevent bacterial growth) without phase transition from water to ice. Among several storage methods, the supercooled sample has the outstanding storage performance as the lowest drip loss of 0.9%, 7.8 times lower than the freezed sample and the lowest color change ratio of 8.1%, 2.9 times lower than the freezed one. These findings provide a solution for long-term and fresh food storage to meet the strong demand of consumers.**

Keywords— *Supercooling, Food preservation, Alternative electric field, ISM band, Impedance matching*

I. INTRODUCTION

Freezing is a common method used to preserve fresh food for long period such as fruits, fishes, and especially meat by suppressing microbial activity related to food poisoning and spoilage [1-3]. However, it can lead to irreversible deterioration in taste and quality due to texture damage from ice crystal formation of water molecules inside fresh foods [4,5]. To overcome the limitation of the previous food preservation, freeze-free technologies using supercooling phenomenon have been researched as a way to avoid ice crystallization and extend

the shelf life of food with high quality, suggesting an alternative to the food shortages and serious environmental damage caused by food waste [6-8]. Supercooling is the process of lowering the temperature of an object such as foods below its freezing point without the phase transition from water molecules to ice [9,10].

In Figure 1, the performance indexes of supercooling preservation technology are generally the degree of supercooling which means the temperature difference between the freezing point and supercooled temperature, and the duration period of supercooled state [11,12]. As supercooling is a physicochemical phenomenon related to nucleation process, there are physical and chemical external factors, which are cooling intensity of a freezer, chemical additives in the target material. The supercooled state is highly unstable from the external factors and ice nucleation can be easily triggered by a stochastic process.

To control the unstable supercooling state, a number of studies have been executed using electromagnetic treatment using electric field or magnetic field during cooling and freezing to suppress ice nucleation of water molecules [13-15]. In their study, Mok et al. (2017) utilized a combined treatment of pulsed electric field and oscillating magnetic field to achieve a supercooling state (-6.5℃) for chicken breasts at an ambient temperature of -7℃. However, further investigation is necessary considering the requirement of both fields simultaneously and the fact that the supercooled duration lasted only 12 hours. Moreover, the samples were artificially managed into a cubic block. In another study by Anese et al. (2012), they applied radiofrequency wave of 27.12 MHz to aid in the freezing process, resulting in reduced drip loss and improved tissue quality of the treated meat, which can be attributed to the formation of smaller ice crystals. Nonetheless, the system was not supposed to realize and maintain the supercooling state without undergoing freezing.

In this paper, a supercooling preservation system using alternative frequency electric field is newly proposed to prevent ice nucleation through control of water molecules inside foods without direct contact to foods in a commercial freezer. To design the supercooling system practical from an industrial point of view, the alternative electric field with 13.56 MHz, one of ISM band is applied to food between two parallel copper electrodes [16,17]. Furthermore, 50 Ω load impedance is implemented with CLC T-matching topology and LC resonance tank technique as the industrially promised value in terms of high power transfer and low signal distortion [18,19]. It is proven that the external electric field during freezing process can successfully suppress ice nucleation based on the classical nucleation theory and thermodynamic law [20,21]. Last, the evaluation of food storage quality is conducted comparing supercooled beef with the refrigerated and frozen ones to verify the superiority of supercooling-based storage.

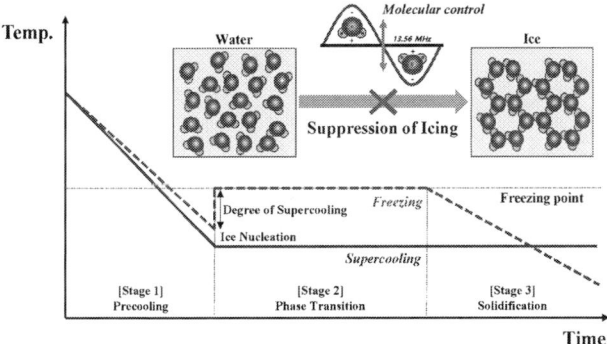

Fig. 1. Temperature profile of food during freezing process.

II. THEORY

A. Ice Nucleation Theory

The classical nucleation theory was applied to examine the external electric field effects on ice nucleation in foods [22]. In liquid-solid equilibrium system, the total change in the free energy ΔG_T during homogeneous nucleation of a spherical ice nucleus of radius r is given by

$$\Delta G_T = \frac{4}{3}\pi r^3 (\Delta g_v + \Delta W_E) + 4\pi r^2 \gamma \tag{1}$$

, where Δg_v is the difference in free energy per unit volume between the liquid and solid phases, ΔW_E is the free energy change due to the electric field, and γ is the interfacial free energy.

Under the uniform electric field E, ΔW_E is determined as in Equation 2, where v_c is volume per molecule, x is a number of molecules in unit volume, and ε_0, ε_l, and ε_s are the relative permittivity of vacuum, liquid phase, and the solid phase, respectively.

$$\Delta W_E = \frac{3\varepsilon_0 \varepsilon_s v_c x E^2}{8\pi}\left(\frac{1 - \varepsilon_s/\varepsilon_l}{2 + \varepsilon_s/\varepsilon_l}\right) \tag{2}$$

Then, ΔG_T depending on r can be sketched in Figure 2, where ΔG_T^*, which is determined by differentiating (1), is the free energy barrier to form a larger nuclei as in Equation 3.

$$\Delta G_T^* = \frac{16\pi\gamma^3}{3(\Delta g_v + \Delta W_E)^2} \tag{3}$$

It can be identified that the ice nucleus creation is impeded by the electric field in the case of $\varepsilon_l > \varepsilon_s$ by increasing ΔG_T^* with positive value of ΔW_E, where Δg_v is a negative value. As explained, it is possible to preserve the foods as the supercooling state under freezing point by applying electric field, with an operating frequency that satisfy $\varepsilon_l > \varepsilon_s$ (foods with high water content).

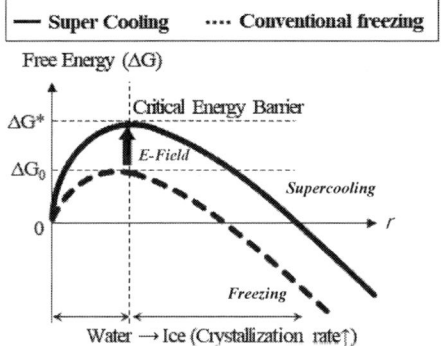

Fig. 2. Gibbs free energy for water molecule nucleation.

B. Thermodynamics-based Supercooling Theory

The environment in which food is stored in a refrigerator is assumed to be a closed system (where mass does not pass through the boundary and only energy moves), and when interpreted using the first law of thermodynamics, which means the law of energy conservation, the relationship between heat and work is the same as expressed in Equation 4 [23]. Q is the amount of heat supplied to the system from outside and $\triangle$U is the change of energy inside the system, and W is the work done by the system to the outside.

Considering sub-zero food storage environment inside a refrigerator, Q is negative value during cooling and freezing process, $\triangle$U is the change in the internal energy of the food which related to the food temperature, and W is the work done by AEF (Alternative Electric Field) to the system. To summarize, the energy balance degree between Q from the indoor temperature of a freezer and W from AEF results in the food temperature as below.

$$Q = \Delta U + W_{AEF} \tag{4}$$

Additionally, the amount of heat (Q) is also interpreted by multiplying the specific heat (c), the object mass (m) and the change of object temperature (ΔT)

$$Q = c \cdot m \cdot \Delta T \tag{5}$$

Meanwhile, in order to apply work to the system in the form of heat, dielectric heating based on AEF mainly is utilized from radio frequency (RF) band of 3 kHz to 300 MHz and even 2.45 GHz (microwave oven), where electromagnetic wave energy penetrates into the material and causes vibration of polarized water molecules inside the object [24,25]. It is the high-speed and volumetric heating method through friction, rotation, and alignment of water molecules according to the continuously changed direction of the electric field according to Equation 6,

$$P = 2\pi f \cdot \varepsilon_0 \cdot \varepsilon_r \cdot \tan\delta \cdot |E|^2 \qquad (6)$$

, where f is the frequency of applied alternative electric field, ε_0 is the vacuum permittivity, ε_r is the dielectric constant, $\tan\delta$ is the loss tangent related to dielectric heating index as an object's intrinsic characteristics, and $|E|$ is the magnitude of electric field.

To be commercialized, electronic devices must adhere to communication protocols using ISM band such as 13.56, 27.12 MHz, 2.45 GHz and etc, designated frequency band for industry, science and medical fields around the world [26]. In terms of penetration depth which means volumetric heating index by AEF, the higher the frequency, the shallower the penetration depth (δ) into the object. For example, in the case of 2.45 GHz microwave oven, it is easy to influence only exposed surface of the object because of short penetration depth, meaning that the electric field could not affect water molecules near the center of object such as foods [27]. Thus, in this paper, 13.56 MHz was selected for the AEF-based supercooling system which has the deepest penetration depth among ISM band considered by

$$\delta = \sqrt{\frac{1}{\pi f \sigma \mu}} \qquad (7)$$

, where f is the frequency of alternative electric field, σ is the conductivity and μ is the permeability of the object.

By combining Equation 4-6, the final temperature of object during supercooling can be assumed by Equation 9. It means the object temperature could be controlled by both storage temperature and AEF intensity as well as preventing ice nucleation.

$$\Delta U = Q - \int (2\pi f \cdot \varepsilon_0 \cdot \varepsilon_r \cdot \tan\delta \cdot |E|^2)\, dt \qquad (8)$$

$$= cm(T_2 - T_1) - \int (2\pi f \cdot \varepsilon_0 \cdot \varepsilon_r \cdot \tan\delta \cdot |E|^2)\, dt$$

$$\therefore T_2 = T_1 + \frac{\Delta U + \int (2\pi f \cdot \varepsilon_0 \cdot \varepsilon_r \cdot \tan\delta \cdot |E|^2)dt}{cm} \qquad (9)$$

Before confirming supercooling phenomenon of foods using AEF, the pretest was conducted to figure out supercooling performance according to indoor temperature of a freezer related to the amount of heat (Q). Figure 3(b) describes the temperature-time curve of 150 ml pure water, the similar condition of 150 g of beef. In -12.5°C ambient temperature, the phase transition of the water takes place immediately without supercooling state. The supercooling state temporarily appeared at -8.0°C and the state last longer at the higher temperature of -6.5°C. In -5°C condition which is the highest ambient temperature, the supercooling state was stably maintained over 3.5 days. From the results, it can be assumed that the supercooling performance is inversely proportional to the gap between air temperature and cooled food temperature. To confirm the supercooling state, the physical impact was applied to the water and the state was

immediately broken and the ice nucleation was progressed in Figure 3(c).

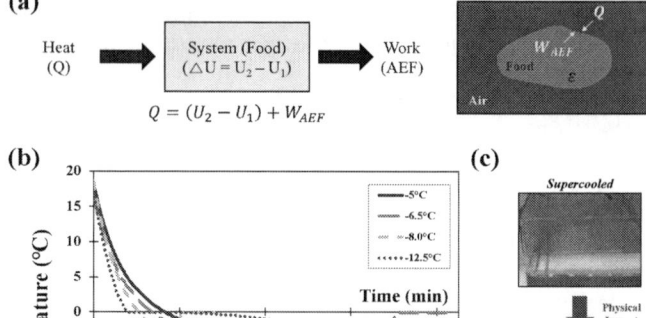

Fig. 3. (a) Relationship between heat and work in a closed system (The first law of thermodynamics). (b) Supercooling performance of water according to ambient temperature inside a freezer. (c) Phase transition of water to ice after physical impact to prove supercooling state.

III. RESULTS AND DISCUSSION

A. Supercooling Storage System by 50 Ω-based Electric Field

Considering the experimental results of supercooled water, the AEF-based system was constructed for the supercooling storage experiment in the commercial freezer (FS-120F, Lassele, South Korea). The temperature was set to -10°C during the experiment because -10°C is the temperature of preventing bacteria growth. To ensure uniform electric field to the beef, two parallel copper plates were selected with the size of 30(W) × 20(L) × 0.2(H) cm³ with 6 cm gap. Figure 4(a) shows the experimental setup for analysing food impedance and building the LC resonance tank for 13.56 MHz utilizing the vector network analyzer (MS4647B, Anritsu, China). The 150 g of beef between the electrodes have very low capacitance of 14 pF and the inductance value was calculated by Equation 10 to compensate the capacitance so that only real impedance value remains for 50 Ω impedance matching. Along the equation, the compensation inductor of 9.84 μH at 13.56 MHz was placed in series to delete reactance of the capacitor as resonance tank technique [28].

$$f = \frac{1}{2\pi\sqrt{LC}} \qquad (10)$$

Figure 4(b) describes the experimental setup of the AEF-based supercooling system and Figure 4(c) shows the circuit schematic of the system. For applying 13.56 MHz alternative electric field to stored foods, the driving circuits are configured from grid source to PFC, DC/DC Converter, Power Amplifier, Impedance Matching and Electrodes mentioned in detail on Section 4. In the lab scale, the signal generator (E4432B, Keysight, USA) and the power amplifier (2204, EMPOWER, USA) were utilized. To make the system optimized with regard to maximum power transfer and minimum signal distortion, all the units such as cables are set up with 50 Ω impedance. Load impedance with resonated inductor is 12 Ω and CLC T-matching network is used to increase the load to 50 Ω as indicated in Figure 4(d).

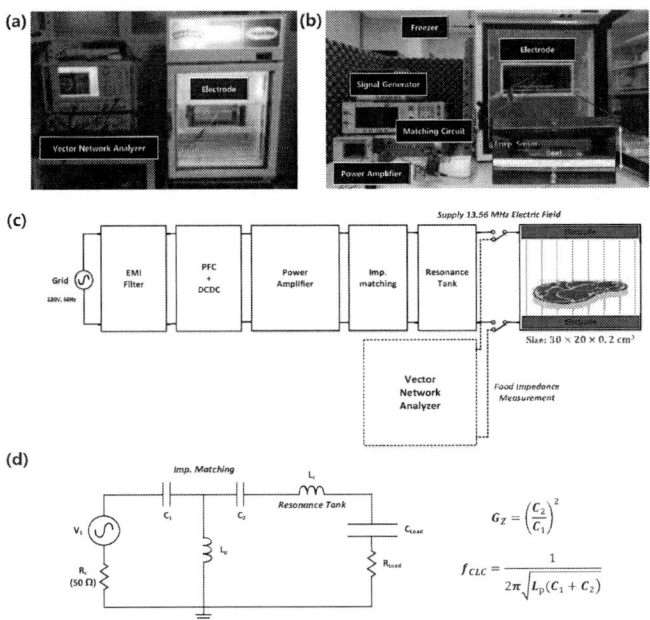

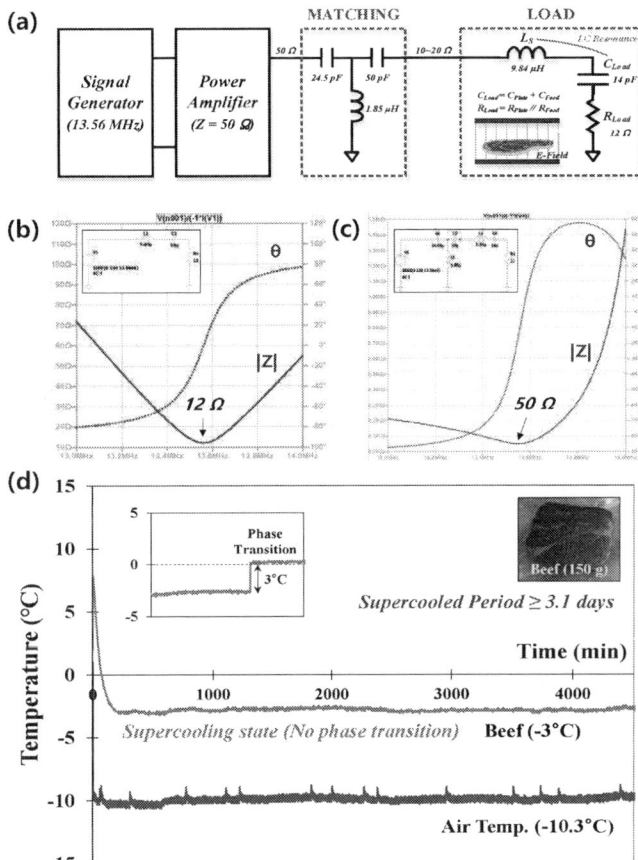

Fig. 4. (a) Experimental setup for food impedance analysis using vector network analyzer. (b) Experiment setup for AEF-based supercooling system. (c) Schematic of AEF system including measurement of food impedance and supply of 13.56 MHz electric field. (d) Equivalent circuit of the AEF system with CLC impedance matching network (T-network).

Figure 5(a) describes the values of the lumped components such as capacitors and inductors to be used for 50 Ω impedance matching at 13.56 MHz. Without the impedance matching part in the circuit simulation tool, LTspice, the load impedance has 12 Ω impedance at 13.56 MHz and the mismatched impedance between the source impedance of 50 Ω and the load impedance of 12 Ω results in power refraction, meaning of low efficiency and circuit instability in Figure 5(b). When CLC T-matching network was applied in front of the load, the load impedance looked like 50 Ω on the source side.

Using the determined components of the circuit, the system was constructed and experimented for supercooling. To compare with other storage methods, beef samples (brisket, 150 g) were divided into three cases and stored in the freezers for 3 days, separately; one was located between the electrodes with 13.56 MHz alternative electric field at -10°C, while the remainders were stored without electric field at 4°C (Refrigerated) and -19°C (Freezed), separately. To measure the temperature of the samples and indoor air in high electric field, fiber optic temperature sensors (FOM-H201-8, OMEGA Engineering, USA) are used. The electric field between the electrodes was controlled as 40 kV/m at the 50 W condition.

Figure 5(d) shows the temperatue-time curve of the successfully supercooled sample which was well-controlled as about -3°C at the ambient temperature of -10°C over 3 days. When the supercooling state was broken at the last moment, the freezing point of the beef was confirmed as about 0°C according to the sensor. Thus, the supercooling degree and the duration period is about 3°C and 3 days at -10.3°C, individually. In the

case of the measured air temperature, it often shows peak point due to limitation of temperature control in the commercial freezers.

Fig. 5. (a) 13.56 MHz AEF circuit with 50 Ω impedance matching referring measured food impedance. (b) Impedance curve of the AEF circuit with resonance tank. (c) Impedance curve of the AEF circuit with resonance tank and CLC T-matching network. (d) Temperature-time curve of supercooled beef at -10°C. The inlet describes the depth of supercooling through confirming the freezing point of about 0°C.

B. Food Storage Performance Evaluation

To compare the storage performance for three cases, the evaluation was conducted after 3 days storage based on two criteria: drip loss (water loss caused by texture damage) and chromaticity (surface color difference) [29, 30]. Drip loss was estimated as a percentage of loss weight ($w_0 - w$) divided by initial weight (w_0) in Equation 11, which is the commonly used method. The initial and final weight values of the samples were measured before and after each treatment.

$$\text{Drip loss (\%)} = \frac{w_0 - w}{w_0} \times 100 \qquad (11)$$

The surface color of the samples was measured by the colorimeter (NR100, 3NH Technology, China). To evaluate the color changes uniformly across the sample surfaces, the five points of each sample were selected and averaged before and after storage. Furthermore, in the paper, total color difference is defined as an average percentage of total colors' difference (L*:

lightness, a*: red/greenness, b*: yellow/blueness). In Figure 6 and Table 1, the supercooled sample has the higher preservation quality with lower drip loss and chromaticity difference than other two methods.

$$\text{Total } Color\ Change\ (\%) = \frac{\Delta E}{E} \times 100(\%)$$

$$= \frac{\sqrt{(\Delta L^*)^2 + (\Delta a^*)^2 + (\Delta b^*)^2}}{\sqrt{L_0^{*2} + a_0^{*2} + b_0^{*2}}} \times 100 \quad (12)$$

Figure 6 shows the beef samples after treatment and Table 1 shows the quantitatively measured evaluation results of storage performance following the two criteria (drip loss and chromaticity). In terms of drip loss, the supercooled sample shows the lowest drip loss of 0.9% after 3 days storage, which is better than the refrigerated one and is about 7.8 times lower than the freezed sample of 7.1%. When it comes to chromaticity, the supercooled sample has the lowest total color change of 8.1%, about 2.9 times lower than the freezed one of 23.6%. Therefore, the evaluation results prove superior food preservation performance of AEF-based supercooling system as a potential solution for fresh and long-term storage of perishable foods.

Supercooled	Refrigerated	Freezed
	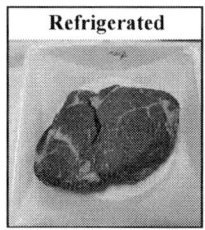	

Fig. 6. The samples after 3 days of treatments (Supercooled, Refrigerated and Freezed).

TABLE I.   EVALUATION RESULTS OF TREATED SAMPLES.

Criteria	Index	Supercooled	Refrigerated	Freezed
Drip loss (150 g)	Weight	150.2 / 148.9	150.4 / 148.6	140.0 / 130.1
	Drip loss	**0.9%**	1.2%	7.1%
Chrom-aticity	L* (Light)	38.40 / 36.66	42.45 / 40.20	39.90 / 37.46
	a* (R-G)	21.24 / 18.16	22.84 / 13.87	19.39 / 9.27
	b* (Y-B)	11.99 / 11.01	14.29 / 12.18	11.88 / 8.88
	ΔE (Ratio)	**3.67 (<u>8.1%</u>)**	9.62 (18.9%)	10.86 (23.6%)

IV. USER SCENARIO

Assuming that the supercooling compartment is a part of a commercial refrigerator (or freezer) except a stand-alone device, two parallel electrodes (copper, etc.) are placed above and below the compartment for non-freezing at sub-zero temperature as shown in Figure 7a. In several food storage cases such as refrigerating, freezing, and supercooling inside a refrigerator, it is essential for the refrigerator to have control functions to simultaneously measure and control both the temperature values of the food and the indoor air.

In order to control the indoor temperature of a refrigerator, it is necessary to use an EMI (Electro Magnetic Interference) filter to filter out EMI noise by receiving grid power, PFC (Power Factor Correction) circuit to increase power factor of electronic devices and convert AC voltage into DC voltage, DC/DC converter to convert DC voltage to the desired DC voltage and 3-phase inverter to drive the compressor motor to control ambient temperature inside the refrigerator.

To independently control the food temperature inside the refrigerator, 13.56 MHz alternative electric field is used for water molecules of food to be rotated and thereby fractionized. It consists of a power amplifier for signal & power amplification, an impedance matching circuit to improve transmission efficiency by compensating for the impedance of the load into 50 Ω, and electrodes that generate an alternating electric field. Additionally, temperature sensors such as an optical fiber type or IR (Infrared Radiation) camera are required to measure both the indoor air temperature and the food temperature.

Based on the control system above, elaborate temperature measurement and control are required for non-freezing storage in sub-zero environment. For general freezing storage at -19°C, the three states occur: cooling, phase transition from water to ice, and freezing without supercooling state in Figure 7(b). As indoor temperature gradually increases by the freezing point of food, supercooling state temporarily appears before entering phase transition period as shown in Figure 7(c). When controlling the gap between food and indoor temperatures to be close in Figure 7(d), the supercooling performance dramatically increases in terms of retention period but, it needs higher food storage temperature than previous storage condition (-19°C) which could induce bacterial and mold growth. Therefore, Figure 7(e) shows the potential solution using AEF which can make supercooling state and retain the state stably even at previous low storage temperature. The precise control of the system is necessary to prevent corner cases such as fluctuation of indoor temperature, when putting in or taking out foods and etc.

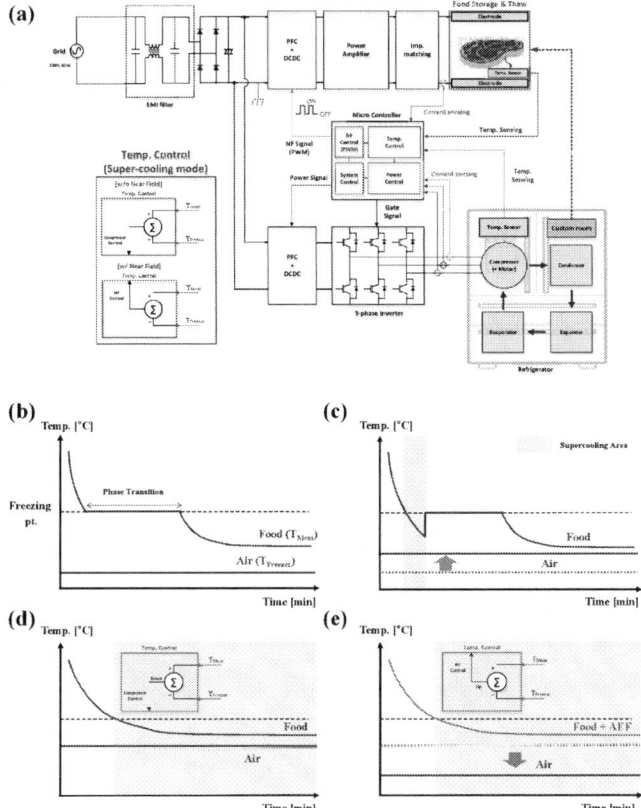

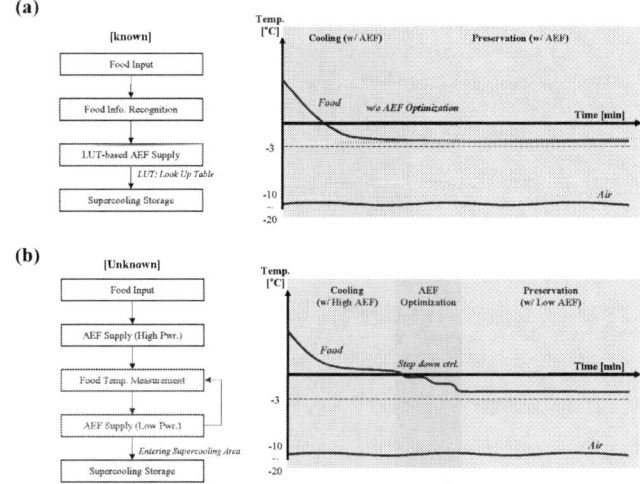

Fig. 8. (a) Supercooling storage algorithm in two cases (a) when a user knows food conditions such as size, mass, type and (b) when a user doesn't know food conditions.

V. CONCLUSIONS

The freeze-free food preservation system using 13.56 MHz alternative electric field based on the ice nucleation theory and the thermodynamics was developed for long-term and fresh storage of perishable food. To enhance power transfer efficiency to the food, the load was matched with 50 Ω impedance using CLC T-matching network and resonance tank. AEF of 50 W was applied to 150 g of beef at about -10°C indoor temperature of a freezer and the sample successfully entered into supercooling state and maintained the state. With the supercooled sample, the food quality evaluation after 3 days is implemented compared to previous food preservation methods (refrigerated and freezed) in terms of drip loss and chromaticity. Finally, the supercooled sample shows the lowest drip loss of 0.9%, which is about 7.8 times lower than the freezed sample of 7.1%. It also has the lowest color change of 8.1%, about 2.9 times lower than the freezed one of 23.6%. The results prove high food preservation quality of AEF-based supercooling system as a high potential solution for fresh and long-term storage of perishable foods. For application to a real environment, the AEF-added refrigerator control algorithm was suggested as a user scenario. Further studies are needed to demonstrate coverage of various food species and storage conditions with the AEF supercooling system.

Fig. 7. (a) Integrated schematic of AEF-added refrigerator control system. (b) Temperature-time curve of food in a general refrigerator (-19°C). (c) Occurrence of supercooling state according to higher storage temperature. (d) Enhancement of supercooling performance with lower gap between food and indoor air temperature. (e) Controllable supercooling state using AEF below lower temperature.

As user scenarios, supercooling storage process are divided into two cases: when food information is supplied and when food information is unknown. In the case where food information is supplied in advance from users, preset intensity of AEF is applied based on pre-entered value in the system according to the food information such as look-up table in Figure 8(a). On the other hand, when food information is unknown, step-by-step AEF intensity lowering algorithm is performed and simultaneously the built-in temperature sensor is used to track the food temperature and guide it to gradually enter the supercooling area as shown in Figure 8(b).

REFERENCES

[1] Speck, M. L., and B. Ray, "Effects of freezing and storage on microorganisms in frozen foods: a review," Journal of food protection, vol. 40, no. 5, pp. 333-336, 1977.

[2] Qian, S., Li, X., Wang, H., Sun, Z., Zhang, C., Guan, W., & Blecker, C, "Effect of sub‐freezing storage (-6, -9 and -12° C) on quality and shelf life of beef," International Journal of Food Science & Technology, vol. 53, no. 9, pp. 2129-2140, 2018.

[3] Gould, G. W., and N. J. Russell, "Major, new, and emerging food-poisoning and food-spoilage microorganisms." Food preservatives, pp. 1-13, 2003.

979-8-3315-1612-3/25 $31.00 © 2025 IEEE

[4] Jia, G., Chen, Y., Sun, A., & Orlien, V., "Control of ice crystal nucleation and growth during the food freezing process," Comprehensive Reviews in Food Science and Food Safety, vol. 21, no. 3, pp. 2433-2454, 2022.

[5] Tan, Mingtang, Jun Mei, and Jing Xie, "The formation and control of ice crystal and its impact on the quality of frozen aquatic products: A review," Crystals, vol. 11, no. 1, pp. 68, 2021.

[6] Fukuma, Y., Yamane, A., Itoh, T., Tsukamasa, Y., and Ando, M., "Application of supercooling to long-term storage of fish meat," Fisheries science, vol. 78, pp. 451-461, 2012.

[7] You, Y., Li, M., Kang, T., Ko, Y., Kim, S., Lee, S. H., and Jun, S., "Application of supercooling for the enhanced shelf life of Asparagus," Foods, vol. 10, no. 10, pp. 2361, 2021.

[8] Park, D. H., Lee, S., Lee, J., Kim, E. J., Jo, Y. J., Kim, H., Choi, M. J. and Hong, G. P., "Stepwise cooling mediated feasible supercooling preservation to extend freshness of mackerel fillets," LWT, vol. 152, pp. 112389, 2021.

[9] Stonehouse, G. G., and J. A. Evans, "The use of supercooling for fresh foods: A review," Journal of Food Engineering, vol. 14, pp. 74-79, 2015.

[10] Lilley, D., Lau, J., Dames, C., Kaur, S., and Prasher, R, "Impact of size and thermal gradient on supercooling of phase change materials for thermal energy storage," Applied Energy, vol. 290, pp. 116635, 2021.

[11] Osuga, R., Koide, S., Sakurai, M., Orikasa, T., & Uemura, M., "Quality and microbial evaluation of fresh-cut apples during 10 days of supercooled storage," Food Control, vol. 126, pp. 108014, 2021.

[12] Lu, Wan, and Tassou, S. A., "Characterization and experimental investigation of phase change materials for chilled food refrigerated cabinet applications," Applied Energy, vol. 12, pp. 1376-1382, 2013.

[13] Lin, H., Xu, Y., Guan, W., Zhao, S., Li, X., Zhang, C., and Liu, J., "The importance of supercooled stability for food during supercooling preservation: a review of mechanisms, influencing factors, and control methods," Critical Reviews in Food Science and Nutrition, pp. 1-15, 2023.

[14] Mok, J. H., Her, J. Y., Kang, T., Hoptowit, R., and Jun, S., "Effects of pulsed electric field (PEF) and oscillating magnetic field (OMF) combination technology on the extension of supercooling for chicken breasts," Journal of Food Engineering, vol. 196, pp. 27-35, 2017.

[15] Anese, M., Manzocco, L., Panozzo, A., Beraldo, P., Foschia, M., and Nicoli, M. C., "Effect of radiofrequency assisted freezing on meat microstructure and quality," Food Research International, vol. 46, no. 1, pp. 50-54, 2012.

[16] Naik, M., Lavanya, D., Thangaraju, S., Modupalli, N., and Natarajan, V., Dielectric Heating: Recent Trends and Applications in Food Processing, 1st ed., Food Processing and Preservation Technology, 2022, pp. 3-35.

[17] Ferdous, M. S., Koupaie, E. H., Eskicioglu, C., and Johnson, T., "An experimental 13.56 MHz radio frequency heating system for efficient thermal pretreatment of wastewater sludge," Progress In Electromagnetics Research B, vol. 79, pp.83-101, 2017.

[18] Twiname, R. P., Thrimawithana, D. J., Madawala, U. K., and Baguley, C. A., "A dual-active bridge topology with a tuned CLC network," IEEE Transactions on Power Electronics, vol. 30, no. 12, pp. 6543-6550, 2014.

[19] Jonokuchi, H., Nakashima, O., Hiwatari, D., and Hirayama, H., "Zero-current switching with LC resonant tank circuit and capacitor isolation DC-DC converter," 22nd IEEE.European Conference on Power Electronics and Applications, pp. P-1, 2020.

[20] Wang, C., Wu, J., Wang, H., and Zhang, Z., "Classical nucleation theory of ice nucleation: Second-order corrections to thermodynamic parameters," The Journal of Chemical Physics, vol. 154, no. 23, pp. 234503, 2021.

[21] Yan, G., Liu, Y., Qian, S., and Yu, J., "Theoretical study on a vapor compression refrigeration system with cold storage for freezer applications," Applied Thermal Engineering, vol. 160, pp. 114091, 2019.

[22] Yan, J. Y., and G. N. Patey, "Heterogeneous ice nucleation induced by electric fields," The Journal of Physical Chemistry Letters, vol. 2, no. 20, pp. 2555-2559, 2011.

[23] de Cindio, B., "Thermodynamic properties of food materials," Engineering principles of unit operations in food processing. Woodhead Publishing, Chapter 4, pp.65-106, 2021.

[24] Chandrasekaran, S., Ramanathan, S., and Basak, T., "Microwave food processing—A review," Food research international, vol. 52, no. 1, pp. 243-261, 2013.

[25] Bengtsson, N. E., and Ohlsson, T., "Microwave heating in the food industry," Proceedings of the IEEE, vol. 62, no. 1, pp. 44-55, 1974.

[26] Kumbhar, A., "Overview of ISM bands and Software-defined Radio Experimentation," Wireless Personal Communications, vol. 97, no. 3, pp. 3743-3756, 2017.

[27] Ramaswamy, H., and Tang, J., "Microwave and radio frequency heating," Food Science and Technology International, vol. 14, no.5, pp. 423-427, 2008.

[28] Sarnago, H., Lucia, O., and Burdio, J. M., "A versatile resonant tank identification methodology for induction heating systems," IEEE Transactions on Power Electronics, vol. 33, no. 3, pp. 1897-1901, 2017.

[29] Qiao, J., Wang, N., Ngadi, M. O., Gunenc, A., Monroy, M., Gariepy, C., and Prasher, S. O., "Prediction of drip-loss, pH, and color for pork using a hyperspectral imaging technique," Meat science, vol. 76, no. 1, pp. 1-8, 2007.

[30] Zhu, S., Ramaswamy, H. S., and Simpson, B. K., "Effect of high-pressure versus conventional thawing on color, drip loss and texture of Atlantic salmon frozen by different methods," LWT-Food Science and Technology, vol. 37, no. 3, pp. 291-299, 2004.

The Characteristics of the Long Length Primary Loop and the Power Supply for the SCMaglev's DWPT System

Keisuke Yamamoto
Central Japan Railway Company
2-1-85 Konan, Minato-ku,
Tokyo,108-8204, Japan
keisuke.yamamoto@jr-central.co.jp

Jun Enomoto
Central Japan Railway Company
2-1-85 Konan, Minato-ku,
Tokyo,108-8204, Japan
jun.enomoto@jr-central.co.jp

Shunsaku Koga
Central Japan Railway Company
2-1-85 Konan, Minato-ku,
Tokyo,108-8204, Japan
shunsaku.koga@jr-central.co.jp

Junichi Kitano
Central Japan Railway Company
2-1-85 Konan, Minato-ku,
Tokyo,108-8204, Japan
j.kitano@jr-central.co.jp

Abstract—Superconducting Maglev (SCMaglev) operates at 500 km/h, and the commercial line is currently under construction. The SCMaglev adopted a dynamic wireless power transfer (DWPT) system to supply power to on-board. The current DWPT system has been under operation in the Yamanashi Maglev Line since 2014. The DWPT system has distinctive features, long length primary loop (coil) extending over a few kilometers and a relatively lower working frequency 9.8 kHz. This paper describes the configuration of the DWPT system and compares it to other DWPT systems. Simplified methods are proposed to calculate inductance, stray capacitance, and impedance-frequency characteristics, considering both lumped and distributed circuit of the long-length loop. The calculated values are close to the measured values. The impedance changes even during power supply, and field verification clarified these changes. The reactance of the primary loop is adjusted to accommodate these changes. These theoretical and experimental verifications were conducted using actual primary loop and SCMaglev's vehicle during the operations supplying 300 kW and running at 500 km/h.

Keywords—*Superconducting Maglev, dynamic wireless power transfer, lumped circuit, distributed circuit, electromagnetic induction*

I. INTRODUCTION

The superconducting Maglev (SCMaglev) is an ultra-high speed railway and a mass transit system. SCMaglev is currently under operation in the Yamanashi Maglev Line (YML), and under construction for the commercial operation in Japan [1]. The commercial line is expected to achieve two key objectives. First, aims to create a mega region by connecting largest cities, Tokyo, Nagoya, and Osaka. Second, seeks to provide a transportation alternative to conventional high-speed railway, Tokaido Shinkansen, strengthening resilience against various types of disasters.

The power for the propulsion and levitation is generated from magnetic forces acting between the coils on the side walls and the superconducting coils on the cars. On the other hand, the power primarily for the passenger services was previously supplied by on-board fuel generator. Since majority of the commercial line will be built in tunnels in mountainous areas

and in deep subterranean in urban areas, the power supply method must be no environmental impact, to be more precise, no smoke emissions. The contact power supply from overhead wires as used in the conventional railways is not feasible due to the ultra-high speed. For these reasons, a dynamic wireless power transfer (DWPT) system has been developed. Fundamental development began in 2008, and based on the outcomes, the current DWPT system covering approximately 16 km on the YML, was established in 2014. The SCMaglev's operation utilizing the DWPT system has continued steadily up to present day.

This paper consists of four parts. Chapter 2 explains configuration and distinctive features of the DWPT system, and compares its specifications to the those of other DWPT systems developed worldwide. Chapter 3 and 4 describe simplified methods to calculate the electric characteristics of the primary loop, which is the most significant component of the system. Chapter 3, deals with the inductance and stray capacitance, while Chapter 4 explores impedance-frequency (Z-F) characteristics. Chapter 5 presents impedance changes that occur during power supply, discusses reactance adjustment to ensure the stable supply, and provides data obtained during the supply of 300 kW while running at 500 km/h.

II. FEATURES OF THE SCMAGLEV'S DWPT SYSTEM AND COMPARISON TO OTHER DWPT SYSTEMS

A. System Configulation

Fig. 1 shows the schematic configuration of the SCMaglev's DWPT system in the commercial line. The rated output power of the ground-side inverter is 9.8 kHz, 350 A, 6 kV, and 2.1 MVA. The output current is controlled constant. The rated output power of each on-board converter and of total of dozens converters are 20 kW (DC) and approximately 1 MW. On-board converter controls the receiving power constant and power factor close to 1. Both sides' controls work independently without communication. The power capacity of the ground-side is larger than the total of the on-board load, which accommodates the load variations and transmission losses.

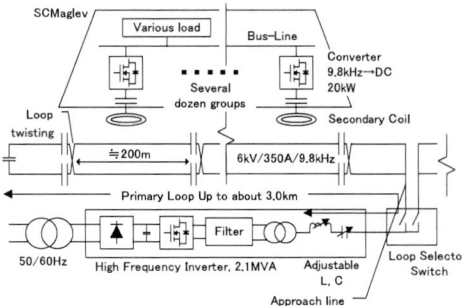

Fig. 1. Schematic configuation of the SCMaglev's DWPT system (in the commericial line).

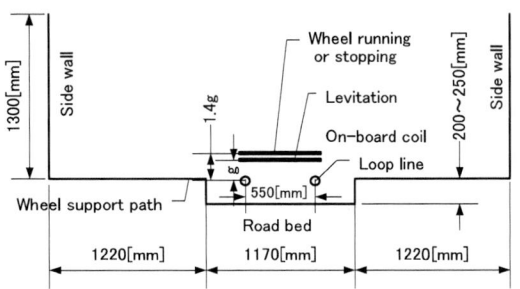

Fig. 2. Spatial relationship between primary loop, secondary coil, and surface of guideway.

Fig. 3. Appearance of the loop, loop lines and series capasitor.

TABLE I. COMPARATIVE OVERVIEW OF THE MAIN SPECIFICATIONS OF OTHER DEVELOPED DWPT SYSTEMS

Name	K-AGT [2]	ORNL [3]	Utah Univ. [4]	SCMaglev
Country	Korea	USA	USA	Japan
Application	Railway	Car	Bus	Railway
Rated power	1 MW	20 kW	25 kW	1 MW commercial line
Frequency	60 kHz	22 kHz	20 kHz	9.8 kHz
Gaap length	50 mm	162 mm	150 mm	around 200 ㎜
Length or No. of pri. coil	128 m	2 Size not listed	R350 mm×2	STD. 2.5 km Max. 3.0km
No of sec. coil	4	1	2	Dozens

This DWPT system uses the electromagnetic induction method with lower working frequency, rather than the magnetic resonance method with higher frequency commonly utilized in EV's DWPT systems. The combination of lower frequency and the electromagnetic induction method enables long-distance power transmission and provides a wide tolerance to impedance variations as described in Chapter 5.

Ground-side supply posts are situated approximately every 5 km and in close proximity to the Maglev line. The output of the high frequency inverter, the main equipment in the post, is connected to the primary loop. The primary loop consists of multiple equipment, including adjustable reactor and capacitor, loop selector switch, approach line, and the loop. The loop is composed of loop lines and series capacitors and laid on roadbed in the guideway where SCMaglev runs. The standard length of the loop is 2.5 km with the maximum length of approximately 3.0 km. The loop is twisted approximately every 200 m in order to mitigate the induction effects on nearby parallel installations.

On-board side, the total length of the 16-car SCMaglev is 400 m in the commercial line. Several dozen secondary coils are installed at the bottom of the cars. These are connected to series capacitors and converters. Each output of the converters is connected to the bus-line, which supplies power to the various on-board components.

Fig. 2 illustrates the spatial relationship between the primary loop, the secondary coil, and the surface of the guideway. The gap length between the loop and the coil is around 200 mm and varies depending on the driving mode, stopping, running with wheels or levitating. The ratio of the gap length for these driving modes is approximately 1.4, with levitating is closer.

The details of each equipment of the primary loop are described below, the main focus of this paper. The adjustable reactor and capacitor are used to adjust the input reactance of the primary loop to keep the input voltage below the rated value while maintaining the input current constant. The loop selector switch allows the system to switch from one inverter to either of two primary loops based on the car's position and speed. The approach line is a twisted line laid from the supply post to the guideway to prevent electromagnetic radiation. The loop lines consist of two reciprocating wires supported at the height of the wheel support paths. The series capacitors are placed at approximately 200-meter intervals with the loop twisted in the capacitor box. Fig. 3 displays the photos of the loop lines and series capacitor. The details about on-board side, refer to [5].

B. Comparison to other DWPT systems

This chapter compares several developed or implemented DWPT systems to the SCMaglev's one. Table 1 shows the comparative overview of the specifications.

K-AGT applied the large captative DWPT system for high-speed railway [2]. The rated power, working frequency, and the gap length are 1 MW, 60 kHz, and 50 mm respectively. The length of the primary coil is 128 m. Oak Ridge National Lab. (ORNL) developed the 20 kW system with 22 kHz and gap length 162 mm for car [3]. Utah State University, 25 kW, 20 kHz, and 150 mm for bus [4]. Systems for automobiles are designed for point-to-point supply. While those for railways are designed for line-to-point supply.

Addressing the ground-side and comparing above DWPT systems to the SCMaglev's system, the distinctive features are narrow down to the long length of the primary loop and the lower frequency. The features were determined based on conditions, such as electromagnetic compatibility and system rationality. In the beginning of the development, there was no literature or knowledge on electric characteristics with similar features. Therefore, various verifications were conducted to clarify the characteristics.

III. INDUCTANCE AND STRAY CAPACITANCE OF PRIMARY LOOP

This chapter discusses the theoretical and field verification results of the inductance and stray capacitance of the primary loop which determine the resonance frequency and specifications of the adjustable reactor and capacitor. There were three challenges associated with the verifications.

a. How to handle conditions close to the ground and formulate the ground earth return in mathematical term for the inductance calculation.

b. How to account for the ground shape of the guideway as surface of ground to calculate the stray capacitance.

c. Developing simplified methods without using dedicated software such as the Finite Element Method (FEM).

A. Thereotical Calculation of Inductance

Fig. 4 (a) presents a simplified diagram of the positional relationship of the loop, ground surface, and earth return used in the inductance calculation. Note that the shape of the ground surface differs from that in Fig. 2; it is modified to a flat plane. In the simplified calculation does not require the accurate surface shape. The location of the earth return, sufficiently deep below the ground surface, allows for this simplification. In Fig. 4 (a), p stands for the complex penetration depth [6] of the earth return current in the ground. Equation (1) defines the depth. Angular frequency $\omega = 2\pi \cdot 9.8 \times 10^3$ [rad/s], relative permeability $\mu_r = 1$, and conductivity $\sigma = 0.01$ [s/m]. σ referred to the value of railway structures dealing with inductive disturbance.

$$p = \frac{1}{\sqrt{j\omega\mu_0\sigma}} = 50.84 - j50.84 \text{ [m]} \tag{1}$$

The working impedance represented in (2). Here, μ_0, $d = 550$ [mm], and $r_{lo} = 26$ [mm] mean permeability, distance between two loop lines, and outer radius of the conductor in the line respectively. The approximation is based on the relative magnitude of the individual terms ($Z_i \ll Z_s - Z_m$).

$$Z_w = Z_i + Z_s - Z_m \cong Z_s - Z_m = \frac{j\omega\mu_0}{2\pi} \ln\frac{d}{r_{lo}} \tag{2}$$

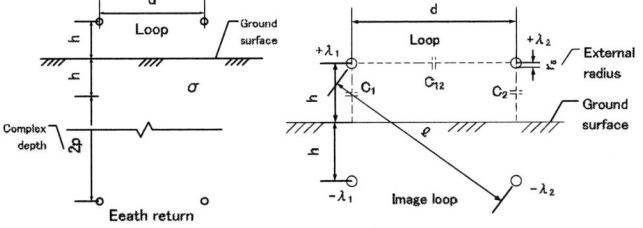

(a) For inductance calculation   (b) For capacitance calculation

Fig. 4. Spatial relationship between primary loop and ground surface for inductance and stray capacitance calculation.

The self-impedance Z_s is expressed in (3) and the mutual-impedance Z_m is in (4). Here, $h = 200 \sim 250$ [mm] denotes height of the loop line from the ground surface. These approximations are based on $h, d \ll p$.

$$Z_s = \frac{j\omega\mu_0}{2\pi} \ln\frac{\sqrt{(2h+2p)^2}}{r_{lo}} \cong \frac{j\omega\mu_0}{2\pi} \ln\frac{2p}{r_{lo}} = R_s + j106.18 \text{ [}\Omega\text{/km]} \tag{3}$$

$$Z_m = \frac{j\omega\mu_0}{2\pi} \ln\frac{\sqrt{(2h+2p)^2+d^2}}{d} \cong \frac{j\omega\mu_0}{2\pi} \ln\frac{2p}{d}$$
$$= R_m + j60.06 \text{ [}\Omega\text{/km]} \tag{4}$$

Therefore, the working inductance L_w is given in (5). For more details, refer to literature [7].

$$L_w \cong \frac{Z_w}{\omega} = 0.75 \text{ [}\mu\text{H/m]} \tag{5}$$

B. Thereotical Calculation of Stray Capacitance

Fig. 4 (b) illustrates the positional relationship used in the stray capacitance's calculation. The surface of the guideway considered as ground surface was converted to a flat plane, it is similar simplification used in the inductance calculation. This simplification is based on the distance from the loop line to the surface of each part of the guideway close to the line. The distance to the roadbed surface directly below is 200 to 250 mm and to the side surface of the wheel support path is 300 mm, Since the roadbed is closer and electromagnetic coupling is stronger, this approximate conversion is valid.

Equation (6) represents the relationship between potential v, potential coefficient p, and electric charge λ. ε_0 means permittivity and other symbols refer to Fig. 4 (b). The inverse matrix of p yields the matrix of coefficients of capacitance and induction q in (7).

$$v = \begin{bmatrix} V_1 \\ V_2 \end{bmatrix} = \frac{1}{2\pi\varepsilon_0} \begin{bmatrix} \ln\frac{2h-r_s}{r_s} & \ln\frac{l-r_s}{d-r_s} \\ \ln\frac{l-r_s}{d-r_s} & \ln\frac{2h-r_s}{r_s} \end{bmatrix} \begin{bmatrix} \lambda_1 \\ \lambda_2 \end{bmatrix} = p\lambda \tag{6}$$

$$q = p^{-1} = 2\pi\varepsilon_0 \begin{bmatrix} \ln\frac{2h-r_s}{r_s} & \ln\frac{l-r_s}{d-r_s} \\ \ln\frac{l-r_s}{d-r_s} & \ln\frac{2h-r_s}{r_s} \end{bmatrix}^{-1} \tag{7}$$

The stray capacitances are calculated with elements of the matrix q. The combined stray capacitance C_c is determined as (8). Stray capacitance C_1, C_2 to the ground and C_{12} between two lines are derived in (9). The range of the value is determined by the variation in the height of the lines. In the literature [7], more accurate calculation method is proposed by conformal mapping and transformation using elliptic functions.

$$C_c = \frac{C_1}{2} + C_{12} = 9.70 \sim 9.25 \text{ [pF/m]} \tag{8}$$

$$C_1 = C_2 = q_{11} + q_{12} = q_{21} + q_{22}, \quad C_{12} = q_{12} = q_{21} \tag{9}$$

C. Field Verification

Filed verifications were conducted to verify the inductance and stray capacitance. Two 200 m loops were installed on the guideway. Through the measurements, values of 0.77 [$\mu H/m$] (same value for both loops) for inductance and 9.85 and 10.24 [$\mu F/m$] for stray capacitance were obtained. The differences from the calculated values were 3 % for inductance and 2 to 10 % for capacitance. The larger difference in capacitance is due to the equation not accounting for the sag of the loop lines and the slight variations in the heights of the lines depending on their locations. However, the calculated values is practically sufficient. The stray capacitance is considerably small compared to the capacitance of series capacitance, which is in the order of 1 μF. The verifications indicated that the stray capacitance does not significantly affect the series resonance adjustment of the primary loop. Therefore, the specifications of the adjustable equipment can be design based of the inductance and its accuracy. Both values influence the Z-F characteristics discussed in the next chapter.

IV. THE FREQUENCY CHARACTERISTICS OF THE PRIMARY LOOP

The primary loop is designed to occur series resonance near the working frequency by treating it as a lumped circuit. The resonance frequency is mainly determined by the inductance of the loop, the values of adjustable reactor and capacitance, and the capacitance of the series capacitors. Additionally, distributed circuit resonance occurs due to the long length and working frequency. This distributed circuit resonance arises from the inductance and the stray capacitance of the loop, as discussed in Chapter 3.

In the initial stage of the development, it was necessary to establish design guidelines including both the lumped and distributed circuit resonance. Particularly, the method was required for longer length loop, as the resonance frequency of distributed circuit shifts towards that of the lumped circuit. Furthermore, a simplified method to calculate the Z-F characteristics was needed. The frequency range targeted by the method is within ± 5 kHz of working frequency, as the range affects the characteristics of power supply.

A. Simplified Caluculation Method of the Z-F Characteristics

This section explains a simplified method for calculating the Z-F characteristics of long length equipment such as the approach line and loop. When one derives equations including both lumped and distributed circuit characteristics, combining F-matrix and distributed circuit theory. For the loop, (10) is obtained according to conventional theories. Each F-matrix, $\boldsymbol{F}_{lp}$ $\boldsymbol{F}_{lpn}$, $\boldsymbol{F}_{cs}$, $\boldsymbol{F}_{cst}$ stands for overall loop, the n-th section of the loop separated by series capacitors, the series capacitor, and the terminal capacitor, respectively. The symbols, γ_{lp}, L_{csn}, Z_{0lp}, C_s, and C_{st} represent propagation constant, the length of the n-th loop, characteristic impedance, propagation constant, series capacitor's capacitance, and the terminal capacitor's capacitance.

$$\boldsymbol{F}_{lp} = \boldsymbol{F}_{lp1}\boldsymbol{F}_{cs}\boldsymbol{F}_{lp2}\boldsymbol{F}_{cs} \cdots \boldsymbol{F}_{lpn}\boldsymbol{F}_{cst} \qquad (10)$$

$$\boldsymbol{F}_{lpn} = \begin{bmatrix} \cosh\gamma_{lp}L_{csn} & Z_{0lp}\sinh\gamma_{lp}L_{csn} \\ \dfrac{1}{Z_{0lp}}\sinh\gamma_{lp}L_{csn} & \cosh\gamma_{lp}L_{csn} \end{bmatrix}$$

$$\boldsymbol{F}_{cs} = \begin{bmatrix} 1 & -j\dfrac{1}{\omega C_s} \\ 0 & 1 \end{bmatrix}, \quad \boldsymbol{F}_{cst} = \begin{bmatrix} 1 & -j\dfrac{1}{\omega C_{st}} \\ 0 & 1 \end{bmatrix}$$

For the approximation to simplify the equation, assuming the use of average length $L_{cs_{ave}}$ of each loop section and of average capacitance $C_{s_{ave}}$ of C_s and C_{st}, (12). $\cosh\gamma_{lp}L_{csn}$, $Z_{0lp}\sinh\gamma_{lp}L_{csn}$, and $\dfrac{1}{Z_{0lp}}\sinh\gamma_{lp}L_{csn}$ are each approximated by 1, lumped circuit impedance of series resistance and inductance of the section, and lumped circuit admittance of parallel stray capacitance (c_{lps}) in turn. $\boldsymbol{F}_{lpa}$ in (13) expresses the approximated and series connected F-matrix of one loop section and one series/terminal capacitor. For the following explanation, uses B_{lpa} and C_{lpa} as elements of $\boldsymbol{F}_{lpa}$ (14).

$$L_{cs_{ave}} = \frac{L_{lp}}{N}, \quad C_{s_{ave}} = \frac{\sum_{k=1}^{N-1} C_{sk} + C_{st}}{N} \qquad (12)$$

$$\boldsymbol{F}_{lpa} = \boldsymbol{F}_{lpn}\boldsymbol{F}_{csn} \cong \begin{bmatrix} 1 & (r_{lp} + j\omega l_{lp})L_{cs_{ave}} - j\dfrac{1}{\omega C_{s_{ave}}} \\ j\omega c_{lps}L_{cs_{ave}} & 1 \end{bmatrix}$$
$$(13)$$

$$\boldsymbol{F}_{lpa} = \begin{bmatrix} 1 & B_{lpa} \\ C_{lpa} & 1 \end{bmatrix} \qquad (14)$$

The approximated overall loop's F-matrix $\boldsymbol{F}'_{lp}$ is shown in (15). This approximation is made possible by the value of C_{lpa}. C_{lpa} takes values between $j(0.7 \sim 2.0) \times 10^{-4}$ in the required frequency range, and obtained terms multiplied by C_{lpa}^2 in the process of matrix multiplication are sufficiently small and able to neglect. These approximations reveal that the initial equation has been simplified. The number of series connections can be treated as a coefficient. Therefore, neither matrix multiplication nor power calculation for N is required.

$$\boldsymbol{F}'_{lp} = \boldsymbol{F}_{lpa}\boldsymbol{F}_{lpa} \dots \boldsymbol{F}_{lpa} \cong \begin{bmatrix} 1 & B_{lpa} \\ C_{lpa} & 1 \end{bmatrix}^N$$

$$\cong \begin{bmatrix} 1 + \dfrac{1}{2}N(N-1)B_{lpa}C_{lpa} & NB_{lpa} + \dfrac{1}{6}N(N-1)(N+1)B_{lpa}^2 C_{lpa} \\ NC_{lpa} & 1 + \dfrac{1}{2}N(N-1)B_{lpa}C_{lpa} \end{bmatrix}$$
$$(15)$$

Similar to the simplification of the loop's equation, the simplified equation of the approach line $\boldsymbol{F}'_{ap}(s)$ can also be derived in (16). Note that there is no series capacitor in the approach line.

$$\boldsymbol{F}'_{ap}(s) = \begin{bmatrix} 1 & (r_{ap} + sl_{ap})L_{ap} \\ sc_{ap}L_{ap} & 1 \end{bmatrix} \qquad (16)$$

The combined F-matrix of the loop and approach line F'_{aplp} and input impedance of the combined circuit z_{aplp} are shown in (17). The subscripts of the symbols indicate the element number of the matrix. As a side note, this equation can be used even for the transient response when converts $j\omega$ to s.

$$F'_{aplp} = F'_{ap}F'_{lp}, \quad z_{aplp} = \frac{F'_{aplp(1,2)}}{F'_{aplp(2,2)}} \quad (17)$$

B. Comparison of Calculated and Measured Z-F Char.

Field verifications was conducted to verify the characteristics of two examples, one with an approach line at 25m and a loop at 2420 m, and another with at 25m and 3070 m. The loop length in the latter case is nearly the maximum length in the commercial line. Fig. 5 compares the measured Z-F characteristics to the calculated ones. The parameters used in the calculation are provided in Table Ⅱ. The inductance and stray capacitance differ from those in Chapter 3. The difference in the inductance is due to the difference of conductor's shape of the loop line. The difference of the stray capacitance is due to the measured sight, heights of the loop lines from the roadbed is closer than the that in Chapter 2.

The calculated Z-F characteristics by the proposed method closely matched the measured ones in both cases. As a reference, calculated characteristics as lumped and distributed circuit are shown in the Figs. In the required frequency range, the lumped ones deviate from the measured values as the frequency increase

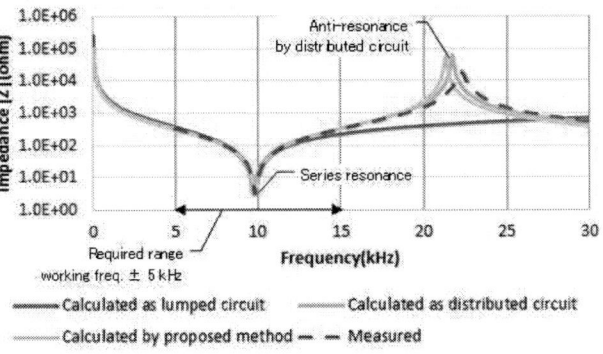

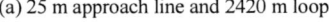

(a) 25 m approach line and 2420 m loop

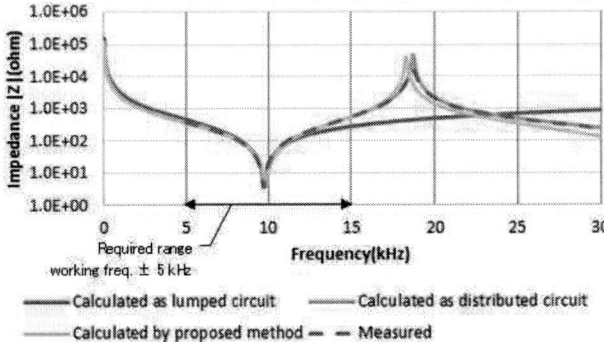

(b) 25 m approach line and 3070 m loop

Fig. 5. Comparison of measured and calculated Z-F char. of primary loop.

or decrease from working frequency. These results demonstrate the need to consider the distributed circuit.

V. CHANGES OF IMPEDANCE CHARACTERISTICS AND POWER SUPPLY DURING ULTRA-HIGH SPEED RUNNING

The Z-F characteristics change even during the power supply. The impedance at the working frequency changes by three factors. One is understandably the resistance increase with the entry of the cars. The power factor of the receiving power is controlled at close to 1, making on-board load and loss predominantly resistive to the ground-side. Two is decrease in capacitance of series capacitor owing to the temperature rise due to energization. This factor is designable based on the temperature characteristics of the capacitors and their quantity. Three is decrease of the reactance as a result of shielding the magnetic flux around the loop by the cars' body. The third factor could not be determined until field experiment was conducted.

The voltage equation (18) represents the view from the input of the primary loop, illustrating the elements of each impedance. Here, V_{in}, I_{in}, z_{aplp}, x_{adj}, Δx_c, Δx_s, and P_{ob} are input voltage and current, input impedance described in Chapter 4, reactance of the adjustment components, reactance decreases by the energized capacitors and shielding, and the on-board load and loss.

$$V_{in} = \left(z_{aplp} + jx_{adj} + j\Delta x_c + j\Delta x_s\right)I_{in} + \frac{P_{ob}}{I_{in}} \quad (18)$$

The changes of input reactance were verified in the field. Fig. 6 provides the results when the vehicle, 5-car vehicle, with 300 kW load, and running at 500 km/h, entries the loop. The period ⅲ in the Fig. represents the time from when the front of the vehicle enters until the rear finishes the entry. The reactance decreased approximately 0.5 ohm, converting to reactive power

TABLE II. PARAMETERS IN CULUCLATIONS OF Z-F CHAR.

(a) Loop's parameters

Parameters	Values
Inductance l_{lp}	0.84*2 [μH/m]
Stray capacitance c_{lp}	10.6 [pF/m]
Series capacitance C_s, C_{st}	1.47/2 or 2.94/2 [μF]
Length L_{lp}	2420 [m], 3070 [m]
No. of sections N	13, 16
Char. impedance Z_{0lp}	$398 - j1.51$[ohm]
Propagation constant γ_{lp}	$1.51 \times 10^{-6} + j3.98 \times 10^{-4}$

(a) Approach line's parameters

Parameters	Values
Inductance l_{ap}	0.12*2 [μH/m]
Stray capacitance c_{ap}	280 [pF/m]
Length L_{ap}	25 [m]
Char. impedance Z_{0ap}	$28.7 - j0.22$ [ohm]
Propagation constant γ_{ap}	$5.93 \times 10^{-6} - j7.56 \times 10^{-4}$

based on 350 A, rated current of the loop, it equates 60 kvar. In the commercial line, the value is necessary to convert this value by the composition of the cars, 16-car. Temperature rise of capacitors also causes the change indicating in the period ⅰ in the Fig. The reactance reduced by 0.5 ohm (excluding the reduction by shielding) for about 10 seconds of the energization.

Under these conditions, the power is supplied stably. On the ground-side, the input current is maintained constant, and on the on-board side, the output power is sustained constant if there is no variation of on-board load. Both are controlled by PID controls. The input impedance adjustment also contributes to stability. As mentioned in Chapter 2, this system uses the electromagnetic induction method, eliminating need to align the resonance frequency closely to the working frequency. The input reactance of the primary loop is adjusted higher, specifically to a positive reactance of several ohms, in anticipation of the reactance decrease. It ensures the output voltage of the inverter keep not to exceed the rated value even with the higher reactance and load resistance. Other measures, such as using of reactive power compensator, were not adopted considering costs increase.

Additionally, Fig. 7 represents progressions on both the ground and on-board side when changes the input current setpoint of the primary loop from 350 A to 300 A. Since the gap length between the loop and coils shortens during levitation, as mentioned in Fig. 2, there is room to reduce the current. Changing the setpoint means resistance, as observed from the ground-side, increases while the vehicle maintains the receiving power constant. Despite the current change, the input current of the primary loop maintained constantly. On the on-board side, the current increases after the current of primary loop change to ensure the receiving power constant. Inspite of the impedance change due to various changes and variations, the current on the ground-side and the receiving power on the on-board side maintained constant. These are indication of stable power supply.

VI. CONCLUSIONS

This paper described the characteristics of the primary loop of the SCMaglev's DWPT systems. This system has the distinctive features, the loop length is on the order of kilometers

and working frequency is approximately 10 kHz. The system configuration and specifications are introduced and compared to those of the other DPWT systems. Simplified methods are explained to calculate the inductance, stray capacitance, and impedance-frequency characteristics, including both lumped and distributed circuit characteristics of the primary loop, and they yield values close to the measured ones. The reactance decreases due to magnetic flux shielding by the vehicle's body was clarified. Based on the characteristics of the primary loop, reactance decrease, and low working frequency, the input reactance of the primary loop is adjusted several ohms positive to accommodate impedance changes and variations. They were unveiled through the theoretical and experimental verifications using actual primary loop and SCMaglev's vehicle, which received 300 kW power supply and was running at 500 km/h.

REFERENCES

[1] M. Uno, "Chuo Shinkansen Project using Superconducting Maglev System," Japan Railway & Transport Review, No. 68, Oct. 2016.

[2] J. H. Kim, B.-S. Lee J.-H. Lee, S.-H. Lee, C.-B. Park, S.-M. Jung, S. -G. Lee, K.-P. Yi, and J. Baek, "Development of 1-MW Inductive Power Transfer System for a High-Speed Train," IEEE Trans. on Industrial Electronics, Vol. 62, No. 10, pp.6242–6250, Oct. 2015.

[3] O. C. Onar, M. Chinthavali, S. L. Campbell, L. E. Seiber, C. P. White, and V. P. Galigekere, "Modeling, Simulation, and Experimental Verification of a 20-kW Series-Series Wireless Power Transfer Sytem for a Toyota RAV4 Electric Vehicle," 2018 IEEE Transportation Electrification Conference and Expo (ITEC), Long Beach, CA, USA, 2018, pp. 874–880.

[4] R. Tavakoli and Z. Pantic, "Analysis, Design, and Demonstration of a 25-kW Dynamic Wireless Charging System for Roadway Electric Vehicles," IEEE Journal of Emerging and Selected Topics in Power Electronics. Vol. 6, No. 3, Sep, 2018.

[5] D. Shimode, T. Murai, and T. Sawada, "Design of ferrite cores of Inductive Power Collection Coils for Movig Vehicles," Proceedings of the 2014 International Power Electronics Conference (IPEC-Hiroshima 2014), pp. 1122–1127, May. 2014.

[6] A. Deri, G. Tevan, S. Semlyen, and A. Castanheira, "The Complex Ground Return Plane A Simplified Model for Homogeneous and Multi-Layer Earth Return," IEEE Trans. on Power Apparatus and Systems, Vol PAS-100, No. 8, Aug. 1981.

[7] K. Yamamoto, J. Enomoto, S. Koga, and J. Kitano, "Theoretical Consideration of the Long-length Primary Loop used in SCMaglev's DWPT System," IEEJ Journal of Industry Applications, Vol. 13 pp. 580–586, 2024.

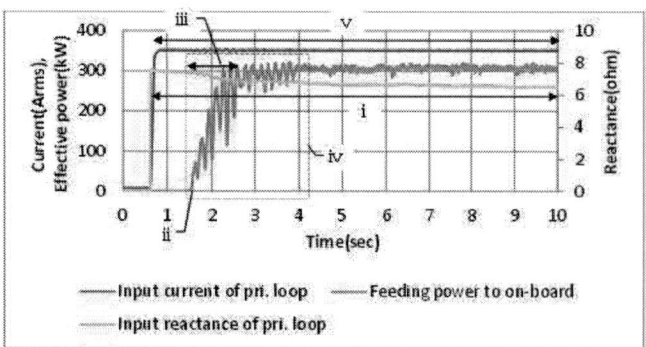

ⅰ Input reactance decreases due to temperature increase of the series capacitors.
ⅱ The front of SCMaglev entered loop.
ⅲ Input reactance decreases due to magnetic flux shielding by Maglev's body.
ⅳ Feeding power fluctuates due to converters' start up and power control.
ⅴ Maintain input current of pri. loop despite changes in reactance and power.

Fig. 6   Reactance change and input current of the pri. loop, and feeding power during vehicle running at 500 km/h entries to the loop.

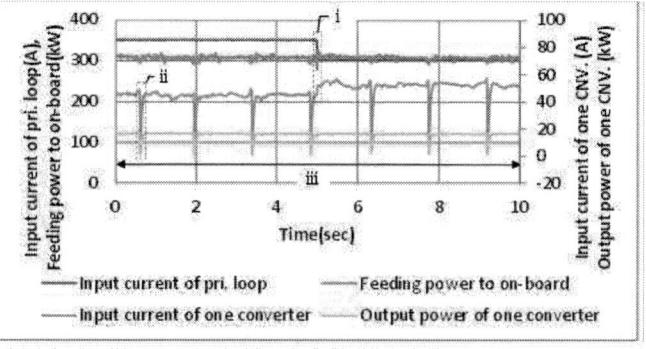

ⅰ Input of pri loop current setpoint adjusted 350A to 300A,
 on-board input current changes immediately to maintain constant power.
ⅱ Dips occurs periodically due to passing over the loop twistings.
ⅲ Input current of on-board varies constantly due to vehicle's oscillation.
 Total feeding power maintained nearly constant despite changes and variations.

Fig. 7. Progressions of current and power on ground and on-board side when the pri. loop's current changes during the vehicle running at 500 km/h.

A Wireless EV Charging System with a Double-Sided LCC Network Using Variable Switching Frequency and DC-link Voltage Control

Chae-Lyn Kim
Department of Electrical and Computer Engineering
Sungkyunkwan University
Suwon, South Korea
kcl0218@g.skku.edu

Hyeonu Jo
Department of Electrical and Computer Engineering
Sungkyunkwan University
Suwon, South Korea
jhw981@g.skku.edu

Ju-A Lee
Department of Electrical and Computer Engineering
Sungkyunkwan University
Suwon, South Korea
ljten88@g.skku.edu

Dong Hyeon Sim
Department of Electrical and Computer Engineering
Sungkyunkwan University
Suwon, South Korea
sdh5419@skku.edu

Byoung Kuk Lee[†]
Department of Electrical and Computer Engineering
Sungkyunkwan University
Suwon, South Korea
bkleeskku@skku.edu

Abstract— **This paper proposes a design method and control scheme for an 11 kW 2-stage wireless charging system using a double-sided LCC (DLCC) network, considering a wide misalignment and battery voltage range. The proposed design method and control scheme fully utilize the resonance characteristics of the compensation network to achieve the battery charging profile only through ground assembly (GA)-side control within a limited frequency range. Based on the formula analysis of the DLCC network, the design process is described first. Subsequently, a control algorithm to satisfy the battery charging profile is proposed based on the design results. Finally, a loss analysis is performed, and a prototype is constructed to verify the proposed method through experiments.**

Keywords— *2-stage wireless charging system, Inductive power transfer (IPT), Double-sided LCC (DLCC) compensation network, Variable switching frequency, Variable DC-link voltage*

I. INTRODUCTION

Wireless charging systems for electric vehicles (EVs) are commonly configured in a 3-stage configuration combining power factor correction (PFC), inductive power transfer (IPT), and the battery management (BM) stage, as shown in Fig.1. However, since the BM converter reduces the power density of the vehicle-side circuit, research is being conducted on a 2-stage system without the BM converter. In the existing 3-stage wireless charging system, the IPT converter only functions as a power transfer unit between the ground assembly (GA) and vehicle assembly (VA) pads, with its output determined by the coupling coefficient (k) between the pads. Consequently, the BM converter is responsible for output control to meet the battery charging profile requirements. However, in a 2-stage system, the IPT converter takes the role of the BM converter, therefore it should fulfill the battery charging profile. To satisfy the charging profile in the IPT converter, there are two parameters available to regulate the output voltage: the DC-link

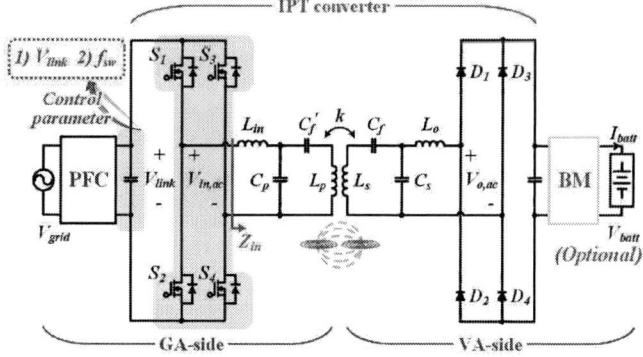

Fig. 1. Conceptual diagram of wireless charging system.

voltage (V_{link}) and the switching frequency (f_{sw}) of the GA-side. Nevertheless, DC-link voltage is limited with the device rating, and the range of switching frequency is also restricted to $79 - 90$ kHz according to the wireless charging standard SAE J2954 [1].

Additionally, as the output capacity of the system increases, the widened battery voltage leads to frequency bifurcation. Fig. 2 illustrates an example of the output voltage (V_{batt}) and the phase (φ_{in}) of the input impedance of the IPT converter in a 2-stage system according to the coupling conditions. In Fig. 2, the output voltage can be regulated by controlling switching frequency. However, as the battery voltage increases, frequency bifurcation causes φ_{in} to invert under specific frequency conditions. This phase inversion causes the IPT converter to operate in the zero current switching (ZCS) region, resulting in high current stress on the MOSFET. Furthermore, as the k varies, the parameters of the IPT pad also change, causing the ϕ_{in} to fluctuate depending on misalignment conditions between the GA and VA pads.

979-8-3315-1612-3/25 $31.00 © 2025 IEEE

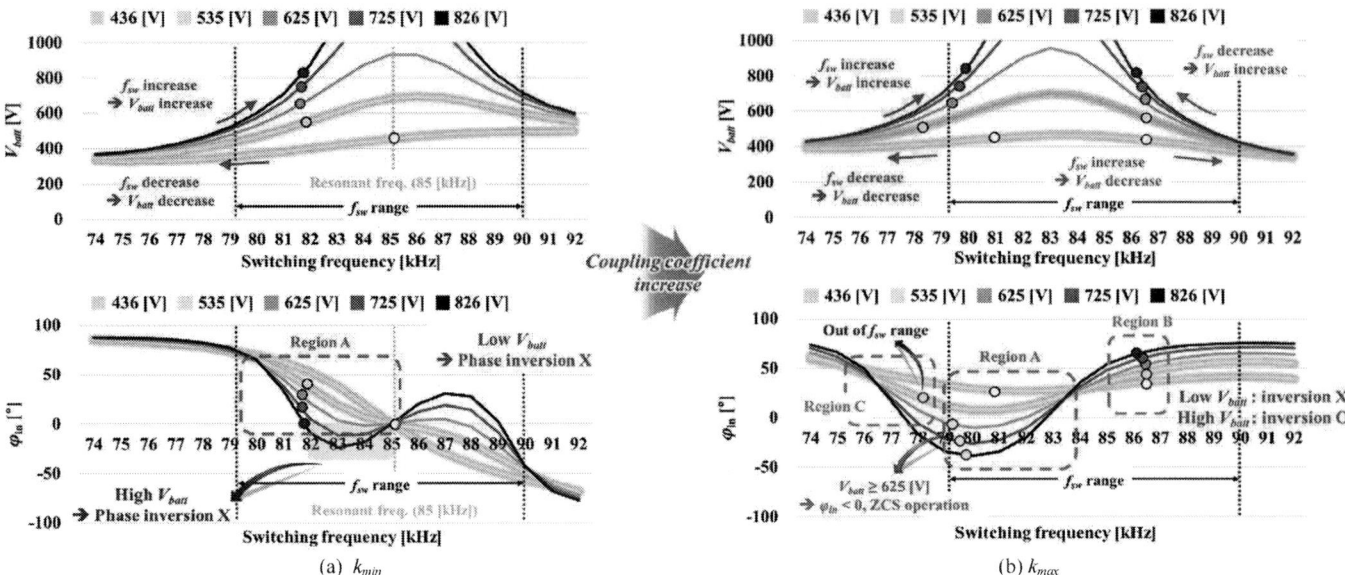

Fig. 2. Characteristic of output voltage and input impedance phase according to coupling coefficient change.

Therefore, when considering all misalignment and a wide battery voltage range, the available switching frequency range is significantly reduced, making it challenging to achieve the desired output while avoiding ZCS operation. In summary, the IPT converter for a 2-stage wireless charging system should comply with the following requirements: (1) The battery charging profile should be satisfied within the limited switching frequency range, and (2) ZCS operations should be avoided under all output voltage and coupling conditions.

In previous research, the LCC-S compensation network was applied to a 2-stage system with a battery voltage of $240 - 410$ V and a power rating of 3.3 kW, with the VA-side configured to operate as a bridgeless rectifier [2][3]. According to the study, the IPT converter operates with fixed f_{sw} near the zero-phase angle frequency, and the battery voltage is regulated by V_{link} control and the bridgeless rectifier. However, applying this control method to an 800 V battery presents challenges in achieving a wide output voltage range through V_{link} control of the IPT converter and bridgeless rectifier. Additionally, the LCC-S network contains a single capacitor on the VA-side, making it sensitive to impedance changes caused by wide load fluctuations. Furthermore, hard switching losses from the bridgeless rectifier lead to efficiency degradation, and the control becomes more complex as it requires synchronization between the GA and VA.

Therefore, to reduce additional losses and simplify system control, this paper focuses on utilizing the GA-side of the IPT converter to satisfy the battery charging profile while maintaining zero voltage switching (ZVS) operation. For this, a double-sided (DLCC) compensation network is more suitable than the LCC-S compensation network, by higher tolerance for the fluctuation of coupling and load conditions [4][5]. Although research on IPT converters with DLCC networks has been conducted, most studies have either focused on designs with fixed coupling coefficients [6][7] or considered only horizontal misalignment. Additionally, since most of these systems operate at a fixed frequency [6]-[9], they utilize only a limited portion of

the resonance characteristics of the compensation network. Furthermore, these systems are typically verified under specific operating conditions, which may not fully account for the entire battery charging profile.

Therefore, to overcome these limitations, this paper proposes a design and control strategy of an IPT converter that can satisfy the battery charging profile only f_{sw} and V_{link} control, in consideration of wide misalignment and battery voltage conditions. By applying the proposed design method and control scheme, the resonance characteristics of the compensation network are fully utilized, enabling the achievement of the battery charging profile through frequency control. This approach ensures ZVS operation across the entire range while simplifying the overall control process. To propose the design method of the IPT converter, the design process of DLCC network is explained through formula analysis first. Subsequently, a design method is proposed that meets output conditions within the limited frequency range while enhancing efficiency. Based on the proposed design, a control method that can satisfy the battery charge profile using f_{sw} and V_{link} control is proposed. Finally, a loss analysis compares the efficiency with conventional design, and a prototype is constructed to verify the proposed design method through experiments.

II. DESIGN PROCESS FOR DLCC COMPENSATION NETWORK

This section provides an overview of the design process for the DLCC compensation network. The system specifications for the initial design of the compensation network are summarized in Table I. The DLCC network is designed based on (1) – (3), ensuring that the reactance is zero at the resonance frequency.

$$\omega_o = \frac{1}{\sqrt{L_o C_s}} = \frac{1}{\sqrt{L_{in} C_p}} \tag{1}$$

$$j\omega_o L_o = j\omega L_s + \frac{1}{j\omega_o C_f} \tag{2}$$

979-8-3315-1612-3/25 $31.00 © 2025 IEEE

TABLE I. DESIGN SPECIFICATION OF IPT CONVERTER

Parameter		Value
DC-link voltage (V_{link})		$\leq$ 800 V
Switching frequency (f_{sw})		79 – 90 kHz
Max. charging current (I_{batt})		25 A
Rated power (P_{out})		11 kW
Battery voltage (V_{batt})		436 – 826 V
Coupling coefficient (k)		0.098 – 0.21
Self-inductance of IPT pads	L_p (k_{min} / k_{max})	165.19 / 154.63 uH
	L_s (k_{min} / k_{max})	157.14 / 166.57 uH

TABLE II. SELECTED PARAMETERS OF THE PROPOSED DLCC NETWORK

Parameter	Conventional	Proposed
Input inductance (L_{in})	27 uH	26.8 uH
GA-side parallel capacitor (C_p)	129.849 nF	122.051 nF
GA-side series capacitor (C_f')	25.371 nF	23.636 nF
Output inductance (L_o)	27 uH	26.8 uH
VA-side parallel capacitor (C_s)	129.849 nF	122.051 nF
VA-side series capacitor (C_f)	26.941 nF	25.097 nF

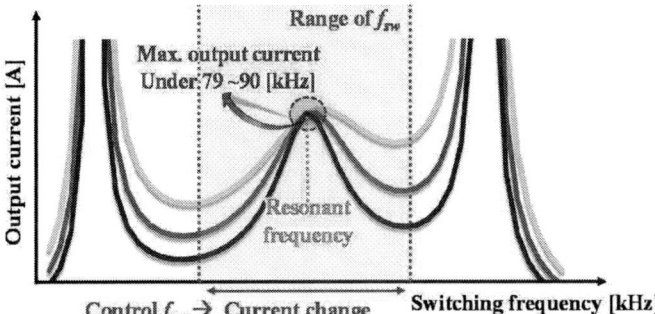

Fig. 3. Output current characteristic of double-sided LCC network.

$$j\omega_o L_{in} = j\omega_o L_p + \frac{1}{j\omega_o C_f'} \qquad (3)$$

The output power of the IPT converter is expressed through input impedance of the DLCC network, as shown in (4).

$$P_o = V_{o,ac} I_{o,ac} = \left\{ \frac{k\sqrt{L_p L_s}}{\omega L_{in} L_o} \right\} V_{in,ac} V_{o,ac} \qquad (4)$$

The pad parameters (L_p, L_s) and the k are determined by the pad manufacturer, while the parameters of the compensation network can be derived based on the selected $V_{in,ac}$ and $V_{o,ac}$. The product of the compensation inductors (L_{in}, L_o) is determined by the selection of input and output voltages, with the inductors designed to meet all output voltage conditions. To ensure reliable power transfer across all k conditions, the wireless charging system is designed under the minimum k (k_{min}). The input and output voltages are then selected considering the variation in k and the voltage range, as discussed in the following paragraph.

The output power of the DLCC network is proportional to the input voltage and the k, as shown in (4), guiding the input voltage selection at the design point. The k of the wireless charging system, as shown in Table I, fluctuates up to twice that of k_{min}. Therefore, to achieve the same load output under k_{max}, only half of the input voltage required at k_{min} is needed. A higher input voltage minimizes the resonance current under the same load condition, reducing conduction losses. Considering efficiency, the highest feasible input voltage is selected under the k_{min} condition. Consequently, the V_{link} voltage at the design point is set to 800 V, considering the allowable V_{link} range.

The output voltage at the design point is determined based on the output characteristics of the DLCC network, as shown in

Fig. 3. Since the DLCC network delivers maximum current at resonance, it is configured to supply 25 A at 436 V under resonance conditions, corresponding to the maximum charging current for an 11 kW load. Therefore, 436 V, which provides maximum current under peak load conditions, is selected as the output voltage at the design point. Consequently, as the charging current decreases due to increased battery voltage, charging current control is managed through switching frequency adjustment.

By selecting the input and output voltages based on the worst-case condition of the IPT converter as the design point, the charging profile can be satisfied through f_{sw} adjustments. Substituting the input and output voltages and design conditions into (4) determines the product of L_{in} and L_o, and the compensation inductances are selected through ratio adjustment. After specifying the compensation inductances, the capacitances of the compensation network can be derived using (2), (3).

III. PROPOSES DESIGN OF DLCC NETWORK AND CONTROL METHOD FOR IPT CONVERTER

A. Conventional design method

In general, as recommended by SAE J2954, a compensation network is designed with a resonance frequency of 85 kHz. The output characteristics are shown in Fig. 2, and Table II shows the design results. Under the k_{min} condition, the target output voltage is met within the switching frequency range, with operation in region A enabling ZVS. In region A, when the battery voltage exceeds the target voltage, the switching frequency is decreased. And when the battery voltage is below the target voltage, the switching frequency is increased to achieve the required voltage, as shown in Fig. 2(a).

However, at k_{max} as shown in Fig. 2(b), ZCS operation occurs under specific voltages ($V_{batt} \geq 625$ V) due to parameter fluctuations from the increased k, making it necessary to operate in region B to achieve ZVS. In this case, to reduce the battery voltage under the k_{max} condition, the frequency control direction is reversed, increasing the switching frequency. Accordingly, since the control direction of the wireless charging system depends on the k condition, precise k measurement under all misalignment conditions is required before charging begins.

Additionally, in region B, φ_{in} significantly increases to 30–70° as shown in Fig. 2(b), which causes a delay in the input current of the compensation network, leading to higher switching loss and reactive power components due to the turn-off current in the inverter on the VA side. This reduces the overall system efficiency. To address this, a compensation network design is needed to minimize the input impedance phase delay, maintain a consistent control direction within the

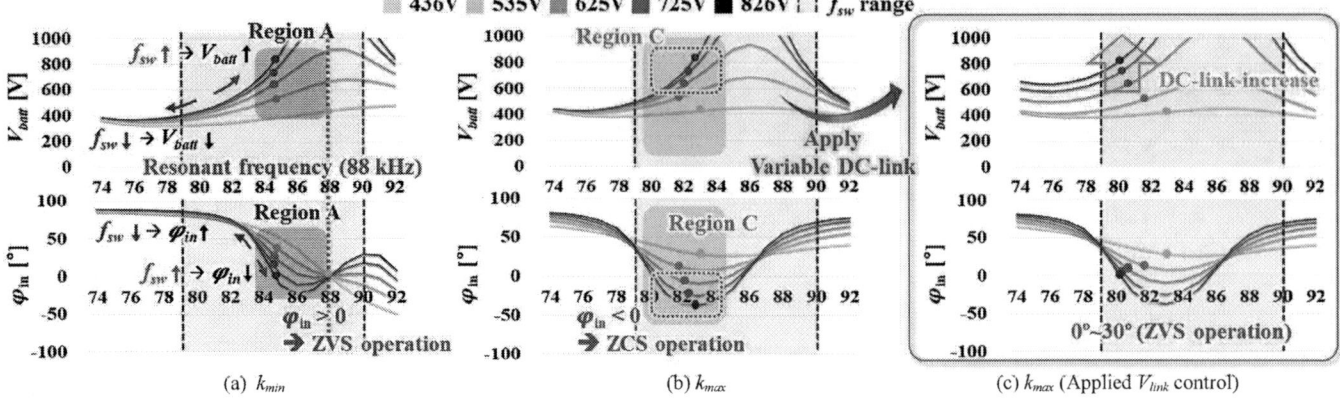

Fig. 4. Characteristic curve of double-sided LCC network applying proposed method.

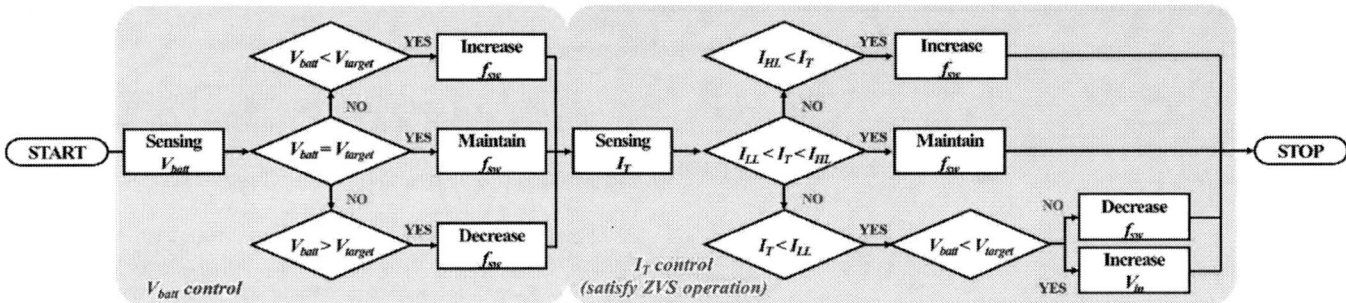

Fig. 5. Control scheme for the IPT converter.

switching frequency range, and allow ZVS operation.

B. Proposed design method

This section proposes a compensation network design approach to mitigate excessive phase delay in input impedance caused by an increased k, maintaining a constant control direction regardless of k. Region C in Fig. 2(b), is utilized, and the resonance frequency is adjusted from 85 kHz to 88 kHz to remain within the allowable frequency range. Table II presents the DLCC network design results, and output characteristic from applying this design shown in Fig. 4.

To achieve ZVS operation within the frequency range and meet battery voltage requirements, the IPT converter operates in region A under k_{min} conditions, as shown in Fig. 4(a). In this area, if the output voltage exceeds the target voltage, f_{sw} is reduced, and if below, f_{sw} is increased. This design ensures a consistent frequency control direction across varying k values, with region C applied under k_{max} conditions. Consequently, the phase of the input impedance improves from 30°–70° to 0°–30°. However, when k increases to k_{max}, as shown in Fig. 4(b), ZCS operation may still occur if the battery voltage exceeds 625 V. To address this, additional V_{link} voltage control is applied, as illustrated in Fig. 4(c). An increase in input voltage does not affect the phase of the input impedance, enabling ZVS operation while maintaining a consistent control direction within the limited frequency range, regardless of k.

C. Control method of Proposed design method

Fig. 5 illustrates the control scheme of the IPT converter

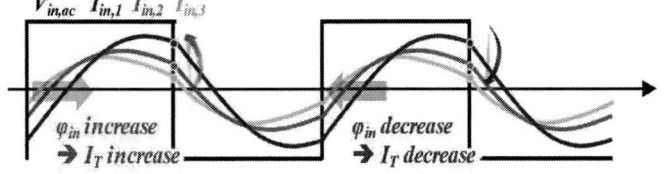

Fig. 6. Example of input current (I_{in}) waveform (φ_{in} : $I_{in,1} > I_{in,2} > I_{in,3}$).

using the proposed compensation network design. Since the IPT converter is designed to operate in region A regardless of the k, the same frequency control can be applied based on the output voltage condition. Therefore, after sensing the battery voltage, the direction of f_{sw} control is determined by comparing it with the target output voltage. When f_{sw} changes for battery voltage control, φ_{in} changes accordingly, as shown in Fig. 4. This phase change affects the turn-off current (I_T) of the input current (I_{in}), as illustrated in Fig. 6.

Therefore, to prevent ZCS operation that may occur when f_{sw} increases, a phase control method is applied. This control is achieved by detecting the I_T and maintaining its magnitude within between the upper limit (I_{HL}) and the lower limit (I_{LL}). When these two control methods are applied to actual battery charging, under k_{min} conditions, ZVS operation and the battery charging profile can be achieved using f_{sw} and I_T control. Furthermore, for k_{max}, as the battery voltage increases above 625 V, V_{link} voltage control is applied in addition to f_{sw} control to maintain ZVS operation.

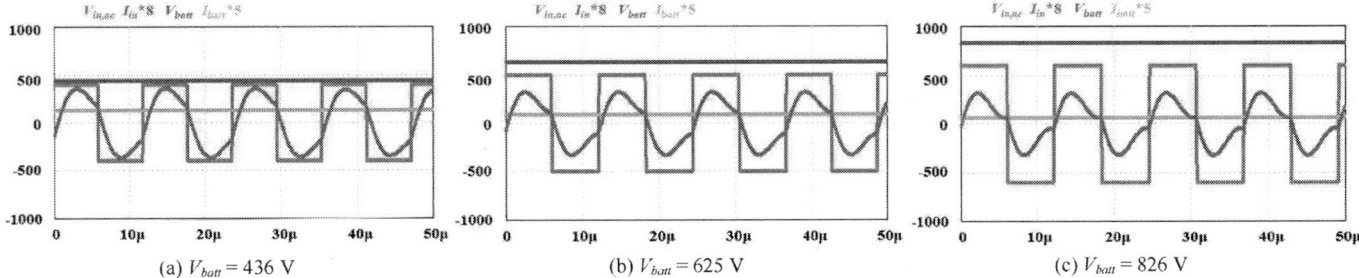

(a) V_{batt} = 436 V   (b) V_{batt} = 625 V   (c) V_{batt} = 826 V

Fig. 7. PSIM simulation waveform under maximum coupling coefficient.

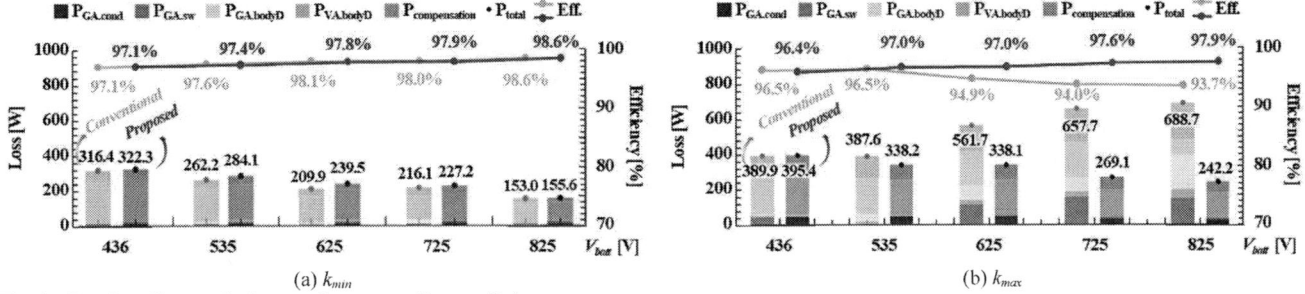

(a) k_{min} (b) k_{max}

Fig. 8. Results of loss analysis according to coupling coefficient.

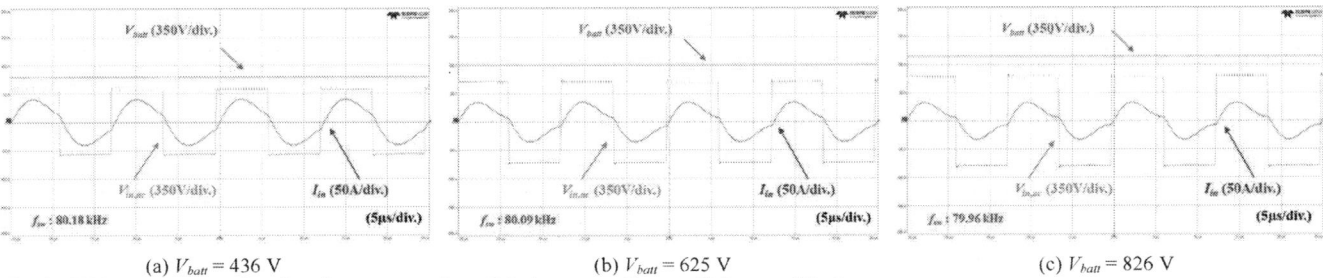

(a) V_{batt} = 436 V   (b) V_{batt} = 625 V   (c) V_{batt} = 826 V

Fig. 9. IPT converter waveform based on output voltage (Maximum coupling coefficient condition).

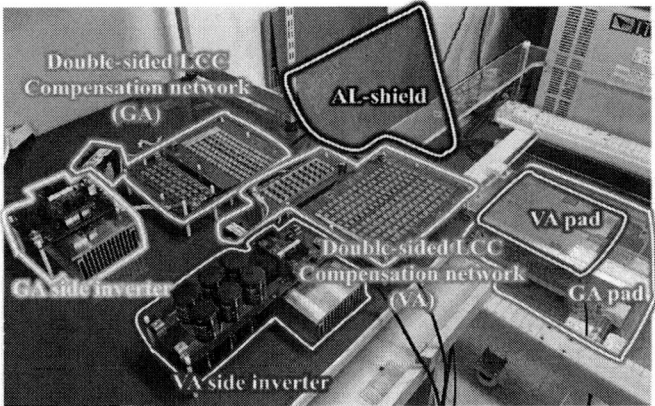

Fig. 10. Prototype and experimental setup.

IV. VALIDATION OF PROPOSED DESIGN AND CONTROL METHOD

A. Results of simulation-based loss analysis

With the proposed design and control method applied, PSIM simulations are conducted to verify system operation. The simulation waveform is examined under the k_{max} condition, where operational differences occur. Under this condition, V_{link} control is used to achieve ZVS operation across all operating areas, as shown in Fig. 7, resulting in an increase in input voltage as the output voltage increase.

Subsequently, the effectiveness of the proposed method is confirmed through simulation-based loss analysis. Fig. 8 illustrates the results of a loss analysis conducted to verify the advantages of the proposed design method compared to the design method with resonance frequency suggested by SAE J2954. In the k_{min} condition, there is no significant difference in operation between the two design methods other than the change of the resonance frequency from 85 kHz to 88 kHz. Consequently, the difference in efficiency is minimal, with an average difference of 0.13% and a maximum difference of 0.27%.

In contrast, under the k_{max} condition, applying the proposed design method significantly reduces the inductive component of the input impedance. This reduction in the I_T significantly decreases the switching loss on the GA side. Additionally, as the control increases V_{link}, the resonance current decreases when the output voltage increases, reducing body diode and compensation network losses on the GA side. As a result, the proposed design method achieved an efficiency improvement of 1.81% on average, with gains of up to 3.7%.

The loss analysis showed that while the efficiency difference is minor under k_{min}, the efficiency of the proposed design method under k_{max} is significantly improved. Since most wireless charging occurs under high coupling conditions (at k_{max}), the proposed design and control method, which provides a substantial efficiency boost at maximum coupling, is advantageous for the high-efficiency operation of wireless charging systems.

B. Experimental verification

Fig. 9 shows the experimental waveform for the 11 kW 2-stage wireless charging system with the proposed design and control method applied, while the prototype and experimental setup are shown in Fig. 10. As there is no operational difference under the k_{min} condition, the experiment is conducted under the k_{max} condition, where significant operational differences are observed. The experiment results, as shown in Fig. 9, indicate that frequency control was applied to meet the battery output voltage requirement. Additionally, to enable ZVS operation across all output voltage conditions, V_{link} control was applied alongside f_{sw} control. Consequently, $V_{in,ac}$ voltages increased under conditions of 625 V or higher, achieving the target output voltage. Finally, stable operation is confirmed at 10.9 kW, the maximum load capacity supported in the experimental setup.

V. CONCLUSION

This paper proposes a design and control method for an 11 kW 2-stage wireless charging system utilizing a DLCC network. The output characteristics of the compensation network are analyzed, comparing the conventional design method with the proposed method. Subsequently, a control strategy is introduced to ensure that the wireless charging system satisfies the battery output voltage requirement while achieving high efficiency within a limited frequency range. After verifying the operation through PSIM simulations, a loss comparison analysis based on simulation results is conducted. Under the k_{max} condition, where wireless charging predominantly occurs, the proposed method demonstrates an average efficiency improvement of 1.81% and a maximum efficiency improvement of 3.75%, making it advantageous for high-efficiency operation. Finally, experimental verification was performed under a 10.9 kW load condition.

ACKNOWLEDGMENT

This work was supported by the Announcement of Materials/Parts Technology Development Program (20024898, Development of 600kW Battery Emulator for dynamometer system) funded By the Ministry of Trade, Industry & Energy(MOTIE, Korea)

This work was supported by Korea Institute of Energy Technology Evaluation and Planning(KETEP) grant funded by the korea government(MOTIE)(RS-2024-00422103, EV Smart Charging Platform Innovation Research Center)

REFERENCES

[1] Wireless power transfer for light-duty plug-in/electric vehicles and alignment methodology, SAE International, Warrendale, PA, USA, Tech. Rep. J2954, Nov. 27, 2017.

[2] M. Kim, D. Joo, S. Ann, and B. K. Lee. "Two-Stage Inductive Power Transfer Charger for Electric Vehicles," *Trans. Korean Inst. Power Electron*, 22(2), 134-139, 2017.

[3] S. Ann and B. K. Lee, "Analysis of Impedance Tuning Control and Synchronous Switching Technique for a Semibridgeless Active Rectifier in Inductive Power Transfer Systems for Electric Vehicles," *IEEE Trans. Power Electron.*, vol. 36, no. 8, pp. 8786-8798, Aug. 2021.

[4] K. Shi, T. Feng, J. Jiang, P. Wang, Z. Meng, and C. Tang, "A Highly Magnetic Integrated Method of LCC-Compensated IPT System with Excellent Misalignment Tolerance." *IEEE Trans. Power Electron.*, vol. 38, no. 12, pp. 16256-16268, Dec. 2023.

[5] T. Mishima and C. Lai, "Zero-Phase-Angle Load-Independent and Adaptable Dual-Side LCC Inductive Wireless Power Transfer System," *IEEE Trans. Transp. Electrif.*, vol. 10, no. 2, pp. 3492-3503, June 2024.

[6] Y. Chen, H. Zhang, C. Shin, C. Jo, S. Park and D. H. Kim, "An Efficiency Optimization-Based Asymmetric Tuning Method of Double-Sided LCC Compensated WPT System for Electric Vehicles," *IEEE Trans. Power Electron.*, vol. 35, no. 11, pp. 11475-11487, Nov. 2020.

[7] N. Fu, J. Deng, Z. Wang and D. Chen, "An LCC–LCC Compensated WPT System With Switch-Controlled Capacitor for Improving Efficiency at Wide Output Voltages," *IEEE Trans. Power Electron.*, vol. 38, no. 7, pp. 9183-9194, July 2023.

[8] S. Li, W. Li, J. Deng, T. D. Nguyen and C. C. Mi, "A Double-Sided LCC Compensation Network and Its Tuning Method for Wireless Power Transfer," *IEEE Trans. on Veh. Technology*, vol. 64, no. 6, pp. 2261-2273, June 2015.

[9] V. -T. Nguyen, V. -B. Vu, G. Gohil and B. Fahimi, "Coil-to-Coil Efficiency Optimization of Double-Sided LCC Topology for Electric Vehicle Inductive Chargers," *IEEE Trans. Ind. Electronics*, vol. 69, no. 11, pp. 11242-11252, Nov. 2022.

Class E/EF Inductive Power Transfer to Achieve Stable Output under Variable Low Coupling

Yifan Zhao[1], Mowei Lu[1], Heyuan Li[2], Zhenbin Zhang[3], Minfan Fu[2] and Stefan M. Goetz[1*]

[1] *Department of Engineering, University of Cambridge, Cambridge, United Kingdom.*
[2] *School of Information Science and Technology, ShanghaiTech University, Shanghai, China*
[3] *School of Electrical Engineering, Shandong University, Jinan, China.*
Email: smg84@cam.ac.uk

Abstract—**This paper develops an inductive power transfer (IPT) system with stable output power based on a Class E/EF inverter. Load-independent design of Class E/EF inverter has recently attracted widespread interest. However, applying this design to IPT systems has proven challenging when the coupling coefficient is weak. To solve this issue, this paper uses an expanded impedance model and substitutes the secondary side's perfect resonance with a detuned design. Therefore, the system can maintain stable output even under a low coupling coefficient. A 400 kHz experimental prototype validates these findings. The experimental results indicate that the output power fluctuation remains within 15% as the coupling coefficient varies from 0.04 to 0.07. The peak power efficiency achieving 91%.**

Index Terms—**Class E/EF inverter, Stable output, extended impedance method**

I. INTRODUCTION

In recent years, inductive wireless power transfer technology has conquered various fields and demonstrated its versatility and applicability across diverse sectors. Unlike conventional charging methods that offer robust wired connections and generally reliable power transfer [1]–[19], inductive power transfer (IPT) systems are inherently sensitive to coupling, which means that any variations in the coupling coefficient between the transmitter and the receiver can cause significant fluctuation in the output power. The power fluctuation can impact the efficiency and stability of the power transfer process. Consequently, the establishment of an IPT system that is robust against coupling variations, capable of maintaining consistent and reliable power transfer despite changes in coupling conditions, is crucial for advancing inductive power transfer technology and ensuring its widespread adoption and success in various applications [20]–[27].

The single-switch Class E/EF inverter has garnered widespread application across various industries and applications due to its exceptional cost-effectiveness and high operational efficiency [28]–[31]. However, the intrinsic load sensitivity of resonant converters poses significant challenges when they feed wireless power transfer systems. Accordingly, a load-independent design of the Class E/EF inverter seems to constitute a promising solution to overcome the challenge [32]. Nevertheless, the implementation of a load-independent Class

This work was supported by National Natural Science Foundation of China under Grant 52477013 and Lingang Laboratory under Grant NO. LG-GG-202402-06-10.

E/EF inverter is constrained by a minimum resistance requirement [32]. This means that the inverter must be connected to a load with a resistance that is above a certain threshold to ensure stable and efficient operation. When deployed within an IPT system, based on the reflected impedance model, there is an inherent lower bound on the coupling coefficient that must be met to maintain stable power. Given that the majority of IPT systems operate under weak coupling, this seemingly optimal scheme harbors substantial limitations.

This study comprehensively analyses and solves the critical failure of load-independent design in Class E/EF driven IPT systems when they are operating under weak coupling conditions. In response, we propose an innovative solution by substituting the ideal resonance on the secondary side with a detuned design. This detuned design incorporates an expanded impedance model [33], [34] to further explore the potential. This paper found that when the reflection impedance shifts from resistive to capacitive, it can effectively counteract the destabilizing effects of weak coupling. Moreover, this paper further elucidate theoretically why an inductive nature is more favorable for stable output in IPT systems.

II. TOPOLOGY OF CLASS-E/EF INVERTER BASED IPT SYSTEM

Fig 1(a) depicts the topology of a basic IPT system driven by a Class E inverter. V_{dc} is the input voltage. S is the switch, whose duty cycle is D and frequency is f_s. The TX coil's self inductance L_{tx} would resonate with C_0 at frequency f_s with additional reactance X. Please note that L_{tx} will resonate with C_0 directly, which is different from a traditional resonate tank. In contrast to the traditional Class E inverter, L_1 serves as a resonant inductor in place of a choke. The resonance frequency between L_1 and C_1 could be normalized with respect to w_s; a frequency factor q is then defined as

$$X = \omega_s L_{tx} - 1/(\omega_s C_0), \quad (1)$$

$$q = 1/\left(\omega_s \sqrt{L_1 C_1}\right). \quad (2)$$

Meanwhile, Fig. 1(b) illustrates the topology of the IPT system driven by the Class EF inverter, where the input inductor L_f serves as an RF choke. C_1 is a shunt capacitor that absorbs the switch junction capacitance. At f_s, L_0 and

C_0 resonate with the additional reactance X_0. The shunt tank L_2 and C_2 would resonate at f_2, which can be described as

$$X_0 = \omega_s L_{tx} - 1/(\omega_s, C_0) \tag{3}$$

$$f_2 = 1/\left(2\pi\sqrt{L_2 C_2}\right). \tag{4}$$

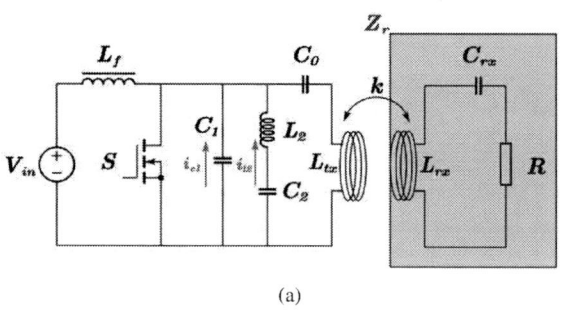

(a)

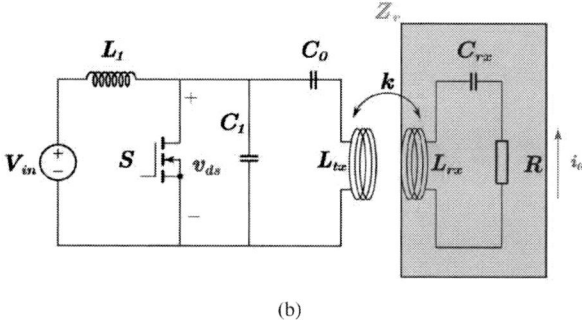

(b)

Fig. 1: Circuit model. (a) Basic IPT system driven by class EF inverter. (b) Basic IPT system driven by class E inverter.

III. DESIGN METHODOLOGY

A. Load-independent Class-E inverter

When the quality factor Q is high, the output current i_0 and voltage across the switch v_{ds} can obtained as follows with β and p as given:

$$i_o(\omega t) = I_m \sin(\omega t + \phi) \tag{5}$$

$$\frac{\nu_{DS}(\omega t)}{V_{in}} = \frac{I_m}{\omega C_1 V_{in}} \int_{2\pi D}^{\omega t} \frac{i_{C_1}(\tau)}{I_m} d\tau = q^2 p \beta(\omega t) \tag{6}$$

$$\beta(\omega t) = \int_{2\pi D}^{\omega t} \frac{i_{C_1}(\omega t)}{I_m} d\omega t, \quad p = \frac{\omega L_1 I_m}{V_{in}} \tag{7}$$

The switch voltage is filtered by the total impedance of jX and Z_r. Consequently, upon applying the Fourier transform to v_{ds}, its sine and cosine components are respectively applied to the resistive and reactive parts of the entire branch, i.e.,

$$\begin{aligned}
\frac{\nu_{Z_r}}{V_{in}} &= \frac{1}{\pi} \int_{2\pi D}^{2\pi} \frac{\nu_{ds}(\omega t)}{V_{in}} \sin(\omega t + \phi) d\omega t \\
&= \frac{q^2 p}{\pi} \int_{2\pi D}^{2\pi} \beta(\omega t) \sin(\omega t + \phi) \, d\omega t = \frac{q^2 p}{\pi} \psi_1,
\end{aligned} \tag{8}$$

$$\begin{aligned}
\frac{\nu_{jX}}{V_{in}} &= \frac{1}{\pi} \int_{2\pi D}^{2\pi} \frac{\nu_{ds}(\omega t)}{V_{in}} \cos(\omega t + \phi) d\omega t \\
&= \frac{q^2 p}{\pi} \int_{2\pi D}^{2\pi} \beta(\omega t) \cos(\omega t + \phi) d\omega t = \frac{q^2 p}{\pi} \psi_2.
\end{aligned} \tag{9}$$

There are two design objectives: Constant ZVS and constant output, which can be derived as

$$\frac{\partial}{\partial p}\left(\frac{q^2 p}{\pi}\psi_1\right) = \frac{q^2}{\pi}\frac{\partial p\psi_1}{\partial p} = 0 \text{ over range of } p, \tag{10}$$

$$\beta(2\pi) = 0 \text{ over } \mathbb{D}_p. \tag{11}$$

Therefore, the design variable can be obtained as

$$\frac{\nu_{jX}}{V_{in}} = \frac{q^2 p}{\pi}\psi_2(q,p,\phi,D) = \frac{i_m X}{V_{in}} = p\frac{X}{\omega L_1}, \tag{12}$$

$$\frac{X(q,\phi,D)}{\omega L_1} = \frac{q^2}{\pi}\psi_2(q,\phi,D) \tag{13}$$

The solution of above equation is $q=1.2915$, $\frac{X(q,\phi,D)}{\omega L_1} = 0.2663$.

B. Load-independent Class-EF inverter

Beginning with the series $L_2 C_2$ network, its current can obtained as

$$\frac{i_{L_2}}{I_{in}}(\omega t) = A_2 \cos(q_2 \omega t) + B_2, \tag{14}$$

$$\sin(q_2 \omega t) - \frac{q_2^2 p}{q_2^2 - 1}\sin(\omega t + \phi) + \frac{1}{k+1}, \tag{15}$$

$$k = \frac{C_1}{C_2}, q_2 = \frac{1}{\omega}\sqrt{\frac{C_1 + C_2}{L_2 C_1 C_2}} = q_1\sqrt{\frac{k+1}{k}}, \tag{16}$$

$$p = \frac{C_2}{C_1 + C_2}\frac{I_m}{I_{in}} = \frac{1}{k+1}\frac{I_m}{I_{in}}, \tag{17}$$

where k, q_2, p can be obtained from (16) and (17).

Meanwhile, the current flow into C_1 and the drain voltage follow

$$\frac{i_{C_1}}{I_{in}}(\omega t) = 1 - p(k+1)\sin(\omega t + \phi) - \frac{i_{L_2}}{I_{in}}(\omega t), \tag{18}$$

$$\frac{\nu_{ds}(\omega t)}{V_{in}} = 2\pi\frac{\beta(\omega t)}{\alpha}, \tag{19}$$

$$\beta(\omega t) = \int_{2\pi D}^{\omega t} \frac{i_{C_1}}{I_{in}}(\tau) d\tau, \tag{20}$$

$$\alpha = \int_{2\pi D}^{2\pi} \beta(\omega t) d\omega t. \tag{21}$$

Similar to the Class-E design, the Fourier transform yields

$$\begin{aligned}
\frac{\nu_{Z_r}}{V_{in}} &= \frac{2}{\alpha}\int_{2\pi D}^{2\pi} \beta(\omega t)\sin(\omega t + \phi) d\omega t = \frac{2}{\alpha}\psi_1, \\
\frac{\nu_{jX}}{V_{in}} &= \frac{2}{\alpha}\int_{2\pi D}^{2\pi} \beta(\omega t)\cos(\omega t + \phi) d\omega t = \frac{2}{\alpha}\psi_2.
\end{aligned} \tag{22}$$

There are also two design objectives: Constant ZVS and constant output power per

$$\frac{\partial}{\partial p}\left(p\frac{\psi_1(p)}{\alpha(p)}\right) = 0 \quad \text{over } \mathbb{D}_p, \tag{23}$$

$$\beta(2\pi) = 0 \quad \text{over } \mathbb{D}_p. \tag{24}$$

Therefore, the design variable can be obtained as

$$\frac{1}{\omega R_L C_1} = \frac{\pi p^2 (k+1)^2}{\alpha(p)}, \tag{25}$$

$$\omega X C_1 = \frac{1}{\pi p(k+1)}\psi_2(q_1, k, \phi, D), \tag{26}$$

$$I_m = 2\frac{\psi_1(p)}{\alpha(p)}\frac{V_{\text{in}}}{R}. \tag{27}$$

The solution of above equation is $q=1.3$, $\frac{X(q,\phi,D)}{\omega L_1} = 0.3533$.

C. Detuned Compensation Design

As mentioned before, this design has the minimum coupling-variation requirement. In scenarios with low coupling coefficients, the detuned secondary design allows for the full release of design variables.

1) Expanded impedance model: To employ the expanded impedance model, the initial step involves representing circuit elements through their respective impedance matrices. Define the critical harmonic order range as $[-N, N]$, where harmonics outside this range are disregarded. It is crucial to set this range to optimize computational efficiency without compromising design accuracy. Consequently, all impedance expressions form square matrices of dimension $(2N + 1)$. Notably, resistors, inductors, and capacitors are represented by diagonal matrices.

The matrix form of a resistor follows

$$\mathbf{Z_R} = \begin{bmatrix} R & \cdots & 0 & \cdots & 0 \\ \vdots & \ddots & \ddots & \ddots & \vdots \\ 0 & \ddots & R & \ddots & 0 \\ \vdots & & & & \vdots \\ \vdots & \ddots & \ddots & \ddots & \vdots \\ 0 & \cdots & 0 & \cdots & R \end{bmatrix}, \tag{28}$$

the matrix form of inductors equivalently

$$\mathbf{Z_L} = \begin{bmatrix} -jN\omega L & \cdots & 0 & \cdots & 0 \\ \vdots & \ddots & \ddots & \ddots & \vdots \\ 0 & \ddots & 0 & \ddots & 0 \\ \vdots & \ddots & \ddots & \ddots & \vdots \\ 0 & \cdots & 0 & \cdots & jN\omega L \end{bmatrix}. \tag{29}$$

The matrix form of capacitors follows

$$\mathbf{Z_C} = \begin{bmatrix} -jN\omega C & \cdots & 0 & \cdots & 0 \\ \vdots & \ddots & \ddots & \ddots & \vdots \\ 0 & \ddots & 0 & \ddots & 0 \\ \vdots & \ddots & \ddots & \ddots & \vdots \\ 0 & \cdots & 0 & \cdots & jN\omega C \end{bmatrix}^{-1}. \tag{30}$$

The transistor is modeled as a resistance that varies with time, featuring a significant OFF-state resistance R_{off} and a minor ON-state resistance R_{on}. This applies to all harmonic orders p within the range $[N, N]$ per

$$R_{S,p} = \begin{cases} R_{\text{ON}}D + R_{\text{OFF}}(1-D), & p = 0 \\ R_{\text{ON}} - R_{\text{OFF}}\frac{\sin(p\pi D)}{p\pi}e^{-jp\pi D}, & p \neq 0 \end{cases}. \tag{31}$$

Note that D is the duty cycle of the driving signal. The impedance expression of the active switch is

$$\mathbf{Z_S} = \begin{bmatrix} R_{S,0} & \cdots & R_{S,-N} & \cdots & R_{S,-2N} \\ \vdots & \ddots & \ddots & \ddots & \vdots \\ R_{S,N} & \ddots & R_{S,0} & \ddots & R_{S,-N} \\ \vdots & \ddots & \ddots & \ddots & \vdots \\ R_{S,2N} & \cdots & R_{S,N} & \cdots & R_{S,0} \end{bmatrix}. \tag{32}$$

The dc input voltage can be expressed as

$$\mathbf{V}_{in} = \begin{bmatrix} 0 & \cdots & 0 & V_{\text{in}} & 0 & \cdots & 0 \end{bmatrix}^{\text{T}}. \tag{33}$$

Besides simulating the steady-state waveforms, the supplied power and load power at steady state can also be derived as

$$\begin{aligned} \bar{P}_{\text{in}} &= \frac{1}{T}\int_T |V_{\text{DC}}(t)i_0(t)|\,\mathrm{d}t \\ &= \sum_{k=-\infty}^{+\infty} V_{\text{DC},k}I_{0,k} \\ &= \mathbf{V}_{\text{DC}}^{\text{T}}\mathbf{Y}_{\text{Class-E}}\mathbf{V}_{\text{DC}}, \\ \bar{P}_{\text{out}} &= \mathbf{V}_{\text{C}}^{\text{T}}\mathbf{Y}_{\text{load}}\mathbf{V}_{\text{C}}. \end{aligned} \tag{34}$$

2) Analysis Results: As depicted in Fig. 2, this work uses an expanded impedance model. For the specific topology, determined values, and given range of k, we can calculate the power drop ratio β, and by select minimum β, can get the candidate points, which are listed in TABLE I.

TABLE I: Parameters and constraint condtions of design.

Parameters	Value	Parameters	Value
L_1	10 μH	V_{in}	30 V
L_{tx}	140 μH	L_{rx}	50 μH
R_L	12.5 Ω	C_0	1.15 nF
C_1	9.49 nF	C_{rx}	3.3 nF
Q_{tx}	350	Q_{rx}	251

The selected design point reveals an inductive characteristic on the secondary side. This occurs because an inductive secondary side results in a capacitive impedance when reflected to

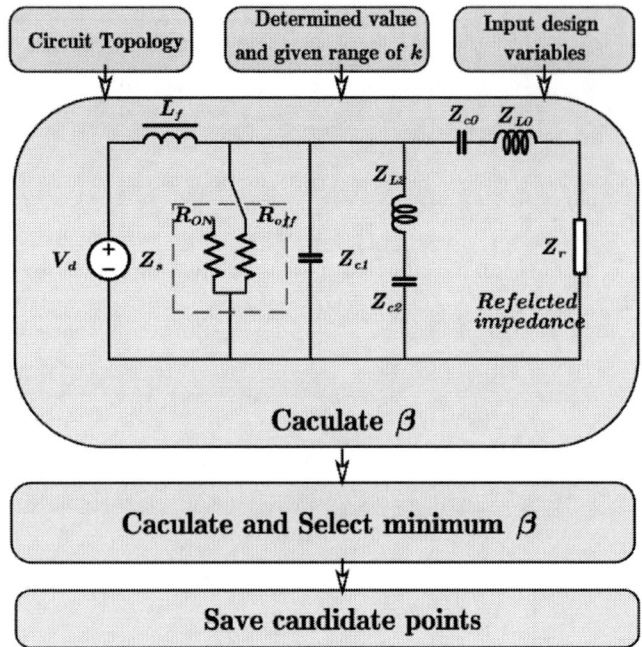

Fig. 2: Design flow chat

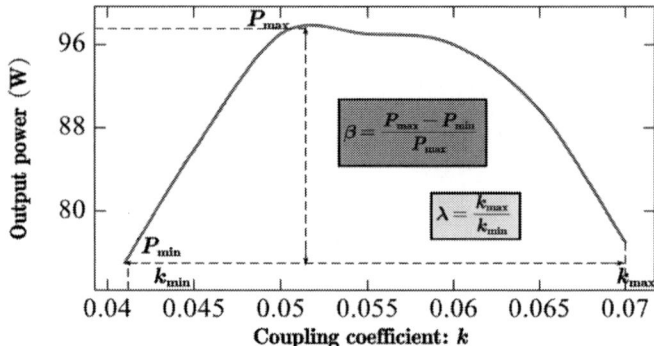

Fig. 3: k-dependent power under proposed design.

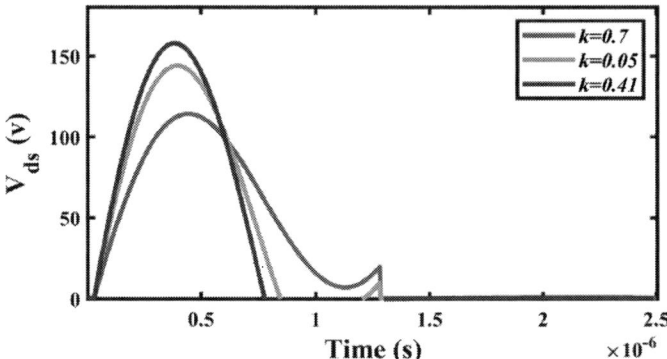

Fig. 4: v_{ds} under different k.

the primary side. Within a load-independent design framework, even in scenarios where low coupling leads to performance degradation, the voltage at the switching node remains relatively constant. Consequently, if the reflected impedance to the primary side is capacitive, it can compensate for a portion of the transmitter coil's impedance and secure a larger share of the voltage, thereby facilitating the maintenance of a stable output.

IV. SIMULATION RESULTS

The point selection model does not account for the effects of parasitic resistances and the nonlinear junction capacitance. Therefore, additional validation through simulation is necessary to confirm the model's effectiveness. Using the parameters of Table I, the quality factors of the inductive components are taken into consideration, and the switch is modeled with a Spice model with the GaN transistor GS66508B.

Fig. 3 illustrates the relationship between output power and varying k. It demonstrates that for $k < 0.05$, P_0 increases with k, whereas for $k > 0.05$, P_0 decreases with k. In this context, β is 20%.

Fig. 4 presents the waveform of the voltage across the switch v_{ds}. It can be observed that under this design, ZVS only holds true for certain coupling coefficients. However, this does not necessarily imply an ineffective design, as the core objective of this paper is to achieve stable output, and constant ZVS is not the primary design goal. On the other hand, the implementation of ZVS aims to achieve high efficiency. Section V describes the experimental verification to demonstrate that this design, even with partial loss of ZVS, can still maintain high efficiency.

V. EXPERIMENTAL RESULTS

An experimental IPT system, designed as previously described, is depicted in Fig. 7. Both the TX and RX coils are crafted from Litz wire, with specifications of 0.1 mm by 150 strands and a diameter of 1.7 mm. All system parameters are detailed in Table I. The circuit uses high-precision, high-quality-factor NP0/C0G ceramic capacitors. The system employs a GaN-based transistor (GS66508B). It is important to note that a constant switch junction capacitance of 200 pF is taken into account and is to be incorporated into C_1. Fig. 5 demonstrates an experimental IPT system.

Fig. 5 graphs typical waveforms for different coupling conditions. It furthermore displays the measured output power and efficiency. The power fluctuation rate β is 15% which successfully aligning with the predictions made by the model-based design. It can also be observed that as the coupling coefficient increases, the efficiency gradually improves. Although partial ZVS is lost, an efficiency above 85% is still maintained, which verifies the feasibility of the design.

VI. CONCLUSIONS

This work solves the shortcomings of load-independent Class E/EF designs when subjected to low coupling variations. A detuned design on the secondary side can liberate global variables. Analysis with an expanded impedance model identifies the optimal design points. Experimental results indicate that under low coupling conditions, the power fluctuation rate

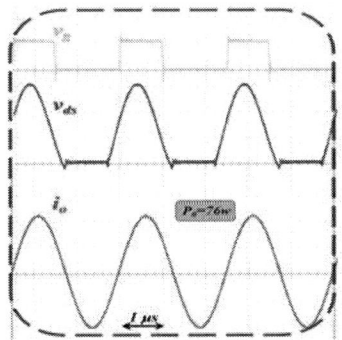 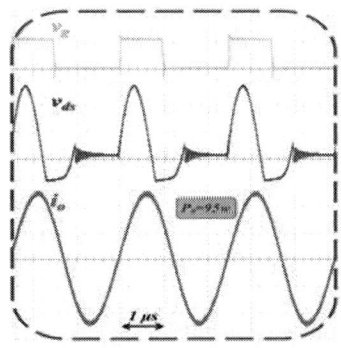 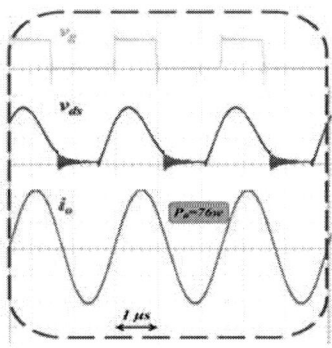

Fig. 5: Waveform at different k: v_g (5V/div), v_{ds} (50V/div), i_0 (2A/div). (a) k = 0.04. (b) k = 0.05. (c) k = 0.07.

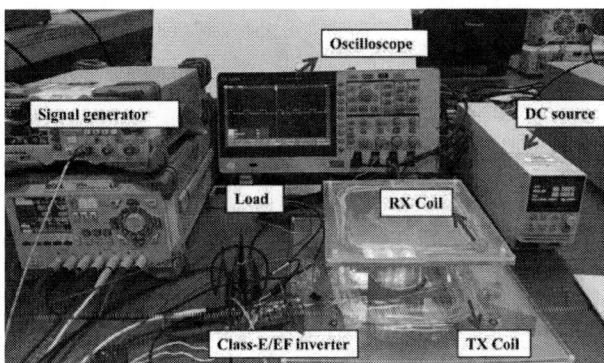

Fig. 6: Experimental setup.

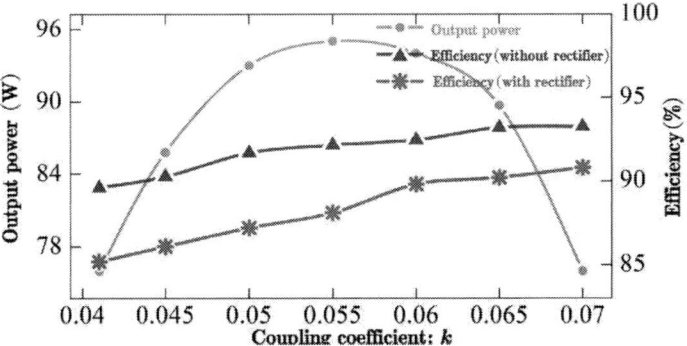

Fig. 7: Output power and efficiency.

remains at 15%, with efficiency maintained within the range of 86–91% as the coupling coefficient k varies from 0.04 to 0.07.

REFERENCES

[1] H. Wang, N. Tashakor, W. Jiang, W. Liu, C. Q. Jiang, and S. M. Goetz, "Hacking encrypted frequency-varying wireless power: Cyber-security of dynamic charging," *IEEE Transactions on Energy Conversion*, vol. 39, no. 3, pp. 1947–1957, 2024.

[2] S. Liu, Y. Wu, L. Zhou, R. Mai, Z. He, and S. M. Goetz, "A two-dimensional misalignment-tolerant ipt system based on three-arm voltage doubler rectifier," in *2022 IEEE Energy Conversion Congress and Exposition (ECCE)*, 2022, pp. 1–7.

[3] H. Wang, K. T. Chau, W. Liu, and S. M. Goetz, "Design and control of wireless permanent-magnet brushless dc motors," *IEEE Transactions on Energy Conversion*, vol. 38, no. 4, pp. 2969–2979, 2023.

[4] X. Yang, J. Li, K. Wang, U. Kim, Z. Zhang, and K.-B. Park, "A weighting factor design approach for fcs-mpc techniques based on pso and k-means algorithm," in *2022 IEEE Energy Conversion Congress and Exposition (ECCE)*, 2022, pp. 1–8.

[5] Z. Wu, H. Han, J. Lin, S. Xie, Y. Sun, Z. Tang, and F. Blaabjerg, "Admittance-based stability analysis of resistance-emulating controlled grid-connected voltage source rectifiers," *IEEE Transactions on Industrial Electronics*, vol. 70, no. 10, pp. 10 076–10 088, 2023.

[6] S. Ni, C. Li, and Z. Zheng, "Control strategy of a hybrid sic-si traction inverter for direct-drive multiphase pmsms in marine propulsion," *IEEE Transactions on Power Electronics*, vol. 39, no. 12, pp. 16 400–16 414, 2024.

[7] Z. Wei, H. Wang, Y. Lu, D. Shu, G. Ning, and M. Fu, "Bidirectional constant current string-to-cell battery equalizer based on l2c3 resonant topology," *IEEE Transactions on Power Electronics*, vol. 38, no. 1, pp. 666–677, 2023.

[8] M. Lu, M. Qin, W. Mu, J. Fang, and S. M. Goetz, "A hybrid gallium-nitride–silicon direct-injection universal power flow and quality control circuit with reduced magnetics," *IEEE Transactions on Industrial Electronics*, vol. 71, no. 11, pp. 14 161–14 174, 2024.

[9] M. Lu, M. Qin, J. Kacetl, E. Suresh, T. Long, and S. M. Goetz, "A novel direct-injection universal power flow and quality control circuit," *IEEE Journal of Emerging and Selected Topics in Power Electronics*, vol. 11, no. 6, pp. 6028–6041, 2023.

[10] X. Huang, Y. Kong, Z. Ouyang, W. Chen, and S. Lin, "Analysis and comparison of push–pull class-e inverters with magnetic integration for megahertz wireless power transfer," *IEEE Transactions on Power Electronics*, vol. 35, no. 1, pp. 565–577, 2020.

[11] H. Sekiya, T. Tokano, W. Zhu, Y. Komiyama, and K. Nguyen, "Design procedure of load-independent class-e wpt systems and its application in robot arm," *IEEE Transactions on Industrial Electronics*, vol. 70, no. 10, pp. 10 014–10 023, 2023.

[12] C. H. Lee, G. Jung, K. A. Hosani, B. Song, D.-k. Seo, and D. Cho, "Wireless power transfer system for an autonomous electric vehicle," in *2020 IEEE Wireless Power Transfer Conference (WPTC)*, 2020, pp. 467–470.

[13] Z. Yue, Q. Zhang, Z. Yang, R. Bian, D. Zhao, and B.-Z. Wang, "Wall-meshed cavity resonator-based wireless power transfer without blocking wireless communications with outside world," *IEEE Transactions on Industrial Electronics*, vol. 69, no. 7, pp. 7481–7490, 2022.

[14] M. Venkatesan, R. Narayanamoorthi, K. M. AboRas, and A. Emara, "Efficient bidirectional wireless power transfer system control using dual phase shift pwm technique for electric vehicle applications," *IEEE Access*, vol. 12, pp. 27 739–27 755, 2024.

[15] T. Mishima and C.-M. Lai, "Zero-phase-angle load-independent and -adaptable dual-side lcc inductive wireless power transfer system," *IEEE Transactions on Transportation Electrification*, vol. 10, no. 2, pp. 3492–3503, 2024.

[16] D.-W. Seo, J.-H. Lee, and H.-S. Lee, "Optimal coupling to achieve maximum output power in a wpt system," *IEEE Transactions on Power Electronics*, vol. 31, no. 6, pp. 3994–3998, 2016.

[17] T. Fujita, T. Yasuda, and H. Akagi, "A dynamic wireless power transfer system applicable to a stationary system," *IEEE Transactions on Industry Applications*, vol. 53, no. 4, pp. 3748–3757, 2017.

[18] F. Lu, H. Zhang, H. Hofmann, and C. C. Mi, "An inductive and capacitive combined wireless power transfer system with lc-compensated topology," *IEEE Transactions on Power Electronics*, vol. 31, no. 12, pp. 8471–8482, 2016.

[19] D. Ahn and P. P. Mercier, "Wireless power transfer with concurrent 200-khz and 6.78-mhz operation in a single-transmitter device," *IEEE Transactions on Power Electronics*, vol. 31, no. 7, pp. 5018–5029, 2016.

[20] L. Gu and J. Rivas-Davila, "1.7 kW 6.78 MHz wireless power transfer with air-core coils at 95.7

[21] L. Gu, W. Liang, and J. R. Davila, "Design of very-high-frequency synchronous resonant dc-dc converter for variable load operation," in *2017 IEEE Energy Conversion Congress and Exposition (ECCE)*, 2017, pp. 3447–3454.

[22] X. Tian, W. Liu, K. T. Chau, and S. M. Goetz, "Omnidirectional magnetic resonant extender design for underwater wireless charging system," *IEEE Journal of Emerging and Selected Topics in Power Electronics*, vol. 12, no. 4, pp. 3325–3333, 2024.

[23] M. Liu and M. Chen, "Dual-band wireless power transfer with reactance steering network and reconfigurable receivers," *IEEE Transactions on Power Electronics*, vol. 35, no. 1, pp. 496–507, 2020.

[24] X. Tian, J. Zhang, H. Wang, and S. M. Goetz, "Design and analysis of automatic modulation and demodulation strategy in wireless power and drive transfer system," *IEEE Transactions on Industrial Electronics*, 2024.

[25] P. Zhao, J. Liang, H. Wang, and M. Fu, "Detuned lcc/ss compensation for stable-output inductive power transfer system under ultra-wide coupling variation," *IEEE Transactions on Power Electronics*, 2023.

[26] H. Feng, T. Cai, S. Duan, X. Zhang, H. Hu, and J. Niu, "A dual-side-detuned series–series compensated resonant converter for wide charging region in a wireless power transfer system," *IEEE Trans. Ind. Electron.*, vol. 65, no. 3, pp. 2177–2188, 2018.

[27] H. Feng, A. Dayerizadeh, and S. M. Lukic, "A coupling-insensitive x-type ipt system for high position tolerance," *IEEE Trans. Ind. Electron.*, vol. 68, no. 8, pp. 6917–6926, 2021.

[28] J. Xu, Z. Tong, and J. Rivas-Davila, "1 kW MHz wideband class e power amplifier," *IEEE Open Journal of Power Electronics*, vol. 3, pp. 84–92, 2022.

[29] K. Surakitbovorn and J. M. Rivas-Davila, "A simple method to combine the output power from multiple class-e power amplifiers," *IEEE Journal of Emerging and Selected Topics in Power Electronics*, vol. 10, no. 2, pp. 2245–2253, 2022.

[30] Z. Tong and J. M. Rivas-Davila, "Wideband push-pull class e amplifier for rf power delivery," in *2023 IEEE 24th Workshop on Control and Modeling for Power Electronics (COMPEL)*, 2023, pp. 1–7.

[31] S. Liu, M. Liu, S. Yang, C. Ma, and X. Zhu, "A novel design methodology for high-efficiency current-mode and voltage-mode class-e power amplifiers in wireless power transfer systems," *IEEE Transactions on Power Electronics*, vol. 32, no. 6, pp. 4514–4523, 2017.

[32] S. Aldhaher, D. C. Yates, and P. D. Mitcheson, "Load-independent class e/ef inverters and rectifiers for mhz-switching applications," *IEEE Transactions on Power Electronics*, vol. 33, no. 10, pp. 8270–8287, 2018.

[33] Y. Jiang, H. Li, Y. Liu, J. Liang, H. Wang, and M. Fu, "Multiconstraint design of single-switch resonant converters based on extended impedance method," *IEEE Journal of Emerging and Selected Topics in Power Electronics*, vol. 11, no. 2, pp. 1901–1912, 2023.

[34] J. Liang and W.-H. Liao, "Steady-state simulation and optimization of class-e power amplifiers with extended impedance method," *IEEE Transactions on Circuits and Systems I: Regular Papers*, vol. 58, no. 6, pp. 1433–1445, 2011.

A Motorized Air-Core Variable Inductance Winding Structure

Xindong Li
Department of EEE
The University of Manchester
Manchester, United Kingdom
xindong.li@postgrad.manchester.ac.uk

Sampath Jayalath
Department of Electrical Engineering
University of Cape Town
Cape Town, South Africa
sampath.jayalath@uct.ac.za

Cheng Zhang
Department of EEE
The University of Manchester
Manchester, United Kingdom
cheng.zhang@manchester.ac.uk

Abstract—Air-core windings, which are free from core losses and flux saturation, are widely used in power conversion applications such as wireless power transfer, resonance converters, and induction heating. They are typically paired with compensating capacitors to form resonant tanks handling ac currents. However, the component tolerances and the winding misalignment will cause discrepancies between the circuit's optimal operating frequency and the designated frequency. Real-time adjustment of the inductance values in applications will resolve this problem. This paper introduces a motorized planar linkage structure that adjusts the winding's equivalent aperture, allowing for continuous inductance adjustment. This adjustment can be made during operation with a feedback system controlling the inductance to maintain the optimal operating point. An example of a misaligned wireless power transfer system is introduced in this paper to demonstrate the benefit of the variable inductance air-core winding, achieving optimal performance.

Index Terms—Variable shape, variable inductance, air-core windings, wireless power transfer

I. INTRODUCTION

Air-core windings are widely used in power conversion applications, such as wireless power transfer systems, resonance convertors, and induction heating [1]–[4]. This is because they do not use magnetic cores to enhance magnetic fields, which experience core losses and flux saturation issues [5]. In addition, air-core windings are lighter and cheaper than cored ones.

Air-core windings are often connected to compensating capacitors to form resonant tanks that withhold ac current. In practice, the component tolerances, including both the inductance and capacitance, and/or misalignment of the windings can cause the optimal frequency to deviate from the designated switching frequency. To address this problem, one approach is to change the operating frequency to the optimal value [6]–[9]. However, in some applications, changing the frequency to fit the optimal operation point is not an option as the frequency is regulated, e.g. 85 kHz in SAE J2954 [10], [11], and the *industrial, scientific and medical* (ISM) band of 6.78 MHz with a bandwidth of 30 kHz. Another approach is to change the capacitance and/or the inductance of the resonant tank [12], [13]. To do this in real time is difficult because currently, common variable capacitors and variable inductors are typically designed for low power, signal-oriented electronic applications, but not for high power scenarios. Some previous

works can adjust the inductance values by altering the core [14], [15]. However, in scenarios where cores are not preferred, e.g. avoiding hysteresis losses and saturation issues or needing to be light-weighted, this method cannot be used.

While extensive research has been conducted on high power applications with inductors having variable inductance, there is a notable lack of studies addressing windings having variable shapes. In [16], a variable inductor is used in the impedance matching dc-dc converter for a PV panel to reduce overall inductor size and get a higher current. In [17], variable inductor is used in the power factor compensation for an ac-dc converter, where various kinds of core air gaps are tested to get different L-i characteristics. In the two applications above, variable inductance is achieved by taking advantage of non-linearity of L-i characteristic curves of saturated cores, which has the aforementioned disadvantages. In [18] the equivalent inductance values of air-core winding sets can be adjusted to get higher efficiency with a dual-layer structure on both two sides. However, this method still adopts fixed windings and cannot accommodate component tolerance issues directly. In a review of tuneable inductors [19], some other methods are also introduced to variable inductance, such as changing the number of turns or changing the air-gap length of the core. All these methods increase the complexity of the mechanical designs or have the disadvantages of having cores.

To resolve these issues, a motorized planar linkage structure (PLS) is proposed in this paper. This structure can adjust the shape of the air-core winding on-the-fly to optimize the overall circuit's performance. This adjustment can be either manually or automatically controlled. It will directly modify the self-inductance of the winding. During the adjustment, the winding's length remains constant whilst the windings are neither loosened nor tightened throughout the variation. The proposed structure is demonstrated in a wireless power transfer system, where the optimal operating point is maintained by adjusting the inductance of the air-core winding. The proposed structure is also applicable to other applications, such as resonance converters and induction heating systems.

979-8-3315-1612-3/25 $31.00 © 2025 IEEE

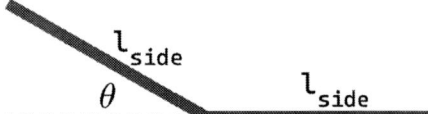

Fig. 1: Illustration of an angled limb

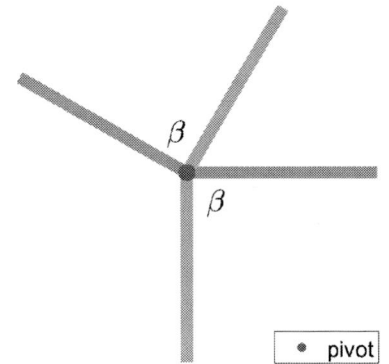

Fig. 2: Illustration of a scissors structure

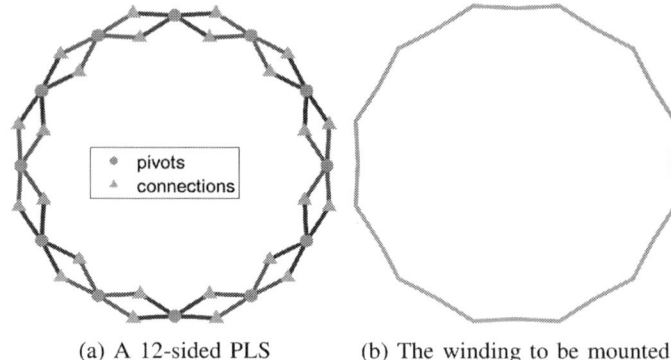

(a) A 12-sided PLS (b) The winding to be mounted

Fig. 3: Illustrations of a 12-sided PLS and a one-turn air-core winding to be mounted on it

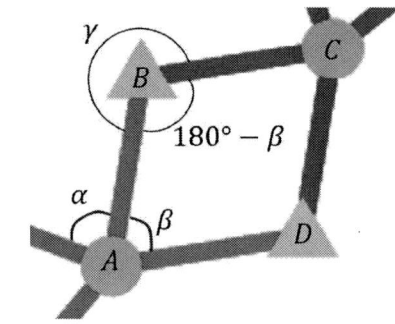

Fig. 4: A local zoom of the example PLS in Fig .3

II. METHODOLOGY

A. The planar linkage structure

The foundation of the work of this paper is a *planar linkage structure* (PLS) that can change the shape of the winding mounted on it whilst maintaining the constant winding length. The key component that forms the PLS is an angled limb shown in Fig. 1. The two segments of the limb are of the same length denoted by l_{side}. The exterior angle is denoted as angle θ. By locating the pivoting vertices of two identical angled limbs together, they form a structure that looks like a pair of scissors, as is shown in Fig. 2. The two limbs can rotate freely around the pivot. The angular misalignment between the two limbs is denoted as angle β.

The proposed PLS is then formed by concatenating multiple identical scissors structures. The number of scissors structures is defined as the side number. The example PLS in this paper is 12-sided. Fig. 3(a) shows how the 12 scissors structures are connected. They are coloured alternating red and blue to be distinguishable. When β increases, the pivots and connection points shrink to the center, as well as the area of the PLS gets smaller. This process is named folding, and the inverse process is named unfolding. To keep the length of the winding on the PLS constant during folding and unfolding, the winding is mounted along the exterior perimeter of the PLS by fixing the wire with pivots and exterior connection points. Fig. 3(b) shows the appearance of a one-turn winding when it is mounted to the PLS.

For a N_0-sided PLS, the necessary and sufficient condition for the angled limbs to form it is that θ must be equal to $\frac{360°}{N_0}$, as proved by the following: The interior perimeter of a N_0-sided PLS is a N_0-pointed star, which is also a $2N_0$-sided simple polygon. Its convex vertex angles are denoted by α, whilst its concave vertex angles are denoted by γ, as is shown in Fig. 4, a local zoom of the example PLS. Because all angled limbs are identical, the lengths of the 4 sides of quadrilateral ABCD are all equal to l_{side}, so quadrilateral ABCD is a diamond. Therefore, we have

$$\alpha = 180° - \theta - \beta, \tag{1}$$

and

$$\begin{aligned}\gamma &= 360° - (180° - \beta)\\ &= 180° + \beta\end{aligned} \tag{2}$$

The sum of the interior angles of the N_0-pointed star is $2(N_0 - 2) \times 180°$. So, we have

$$2(N_0 - 2) \times 180° = N_0(\alpha + \gamma) \Rightarrow \theta = \frac{360°}{N_0}. \tag{3}$$

For the example 12-sided PLS, θ is 30°.

Fig. 5 shows the folding process of the 12-sided PLS. The folding ratio is measured by the angle between the bisector of one angled limb and the initial position of the bisector when the PLS is fully unfolded. The folding ratio angle is denoted by ϕ and is measured in gradian (grad) but not degree (°) for the convenience of future discussions (100 grad = 90°). The mathematical minimum folding ratio of a PLS is 0 grad when the two angled limbs of the same scissors structure coincide completely, as is shown in Fig. 5(a). When all interior connection points arrive at the center point, the PLS is at its mathematical maximum folding ratio, as shown in Fig .5(e).

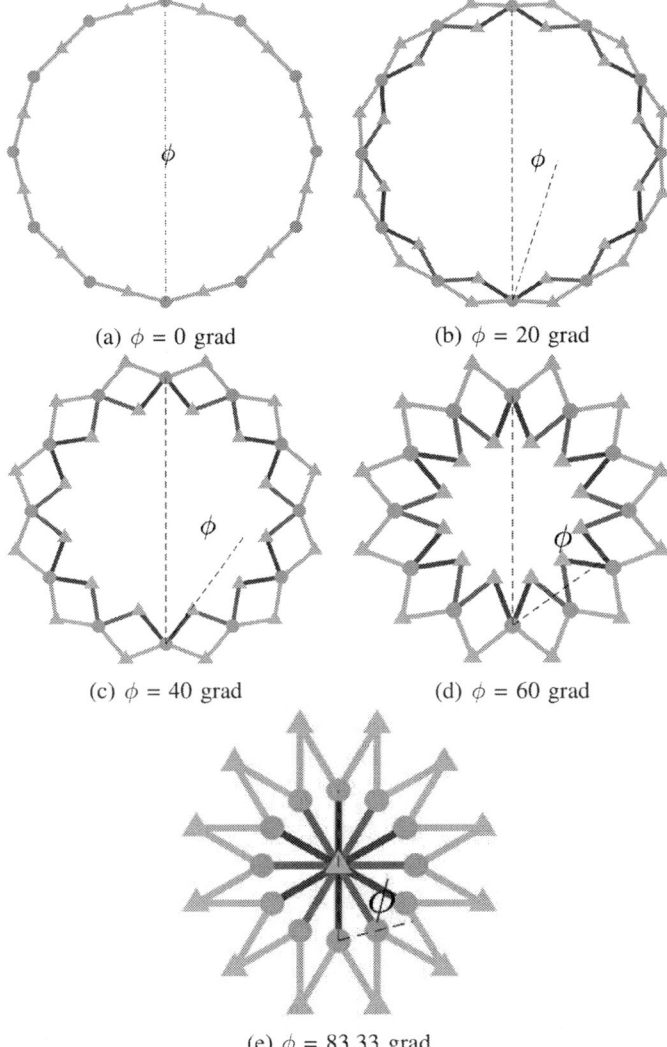

(a) $\phi = 0$ grad

(b) $\phi = 20$ grad

(c) $\phi = 40$ grad

(d) $\phi = 60$ grad

(e) $\phi = 83.33$ grad

Fig. 5: Illustrations of the example 12-sided PLS at different folding ratios during folding process, with winding mounted

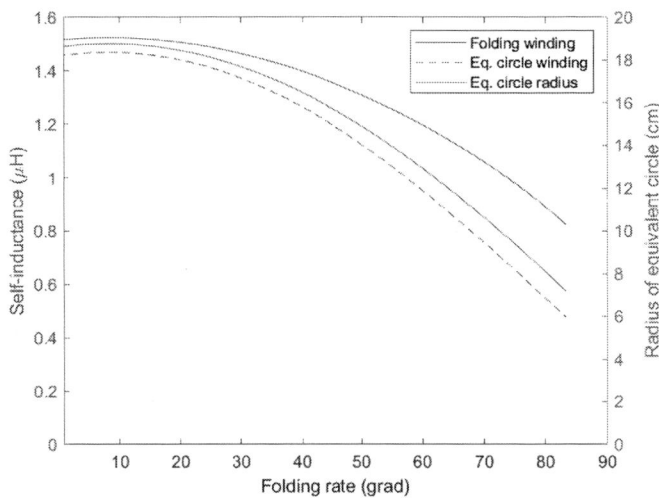

Fig. 6: Calculated self-inductance of a one-turn winding mounted on a 12-sided PLS while folding, $l_{side} = 5$ cm, whilst compared to its equivalent circular loop winding

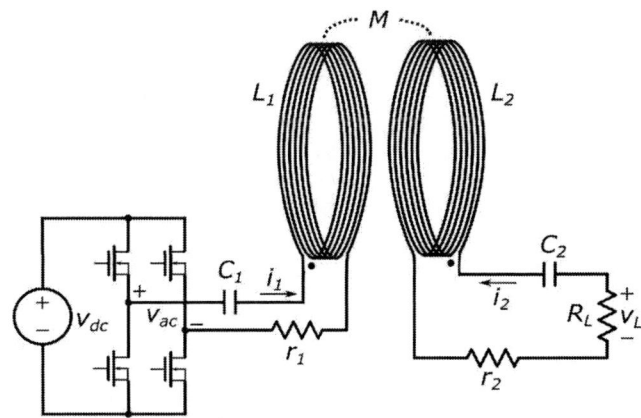

Fig. 7: Circuit topology of the feasibility verification experiment wireless power transfer system

According to Fig .5(e), we can derive the equation to calculate the mathematical maximum folding ratio

$$
\begin{aligned}
\phi_{max}^{math} &= \frac{1}{2}(180° - \theta) \\
&= 90° - \frac{\theta}{2} = 100 \ \text{grad} - \frac{\theta}{2}.
\end{aligned}
\tag{4}
$$

For the example 12-sided PLS, ϕ_{max}^{math} is 83.33 grad. For real-world PLSes, mathematical minimum and maximum folding ratios are unachievable due to mechanical limits, i.e. the width and thickness of the limbs.

The self-inductance of the air-core winding mounted on the PLS varies as it reshapes, and the value can be numerically evaluated by segmentation methods [20]. Fig. 6 shows the self-inductance of a single-turn winding, the cross-sectional radius of which is 0.4 mm, changes along the folding process when $N_0 = 12$ and $l_{side} = 5$ cm. Fig. 6 also compares the winding

self-inductance to the self-inductance of its equivalent circular loop of the same cross-sectional radius. The radius of the equivalent circle is the average of the distances from the center to the pivots and from the center to the exterior connection points. It is shown that both winding inductance and equivalent circle inductance increase before folding ratio reaches around 10 grad. After that, both two inductances start to decrease. The higher the folding ratio, the faster the decrement. Winding self-inductance is always higher than the equivalent circle self-inductance during the folding process. When the variable air-core winding is deployed in an application of various of windings, the mutual inductance of it between other windings will also change depending on the shapes and positions of the other windings.

B. Experiment methodology

The variable self and mutual inductances give us an opportunity to have one more option of adjustment when we need

979-8-3315-1612-3/25 $31.00 © 2025 IEEE

to improve the performance of the applications. To verify the feasibility of this theory, The variable winding is used as the transmitting coil in a series-series inductive wireless power transfer system. The schematic of the system is shown in Fig. 7. By applying KVL at both sides of the system, we can establish the following equations

$$\begin{cases} Z_1 I_1 + Z_m I_2 = V_1 \\ Z_m I_1 + Z_2 I_2 = 0 \end{cases}, \qquad (5)$$

where

$$\begin{cases} Z_1 = j\omega L_1 + \frac{1}{j\omega C_1} + r_1 \\ Z_2 = j\omega L_2 + \frac{1}{j\omega C_2} + r_2 + R_L \\ Z_m = j\omega M \end{cases}. \qquad (6)$$

In (5) and (6), V_1 denotes the phasor form of the fundamental component of v_{ac}, ω denotes the operating angular frequency. M denotes the mutual inductance of the two windings. $I_1, I_2, L_1, L_2, C_1, C_2, r_1,$ and r_2 denote the current phasors, winding self-inductances, compensating capacitances, and winding internal resistances of the two sides, respectively. R_L denotes load resistance. We can find I_1 and I_2 by solving (5) and the result is

$$\begin{cases} I_1 = \frac{Z_2}{Z_1 Z_2 - Z_m^2} V_1 \\ I_2 = -\frac{Z_m}{Z_1 Z_2 - Z_m^2} V_1 \end{cases}. \qquad (7)$$

When dealing with wireless power transfer system, we usually care about the efficiency of the system. Efficiency is the ratio between the power consumed by the load between the total power generated by the power source. If we consider the dc-ac converter as part of the power source, then power loss is caused by r_1 and r_2. From this point of view, we have

$$\eta = \frac{P_L}{P_L + P_{r_1} + P_{r_2}}. \qquad (8)$$

where η demotes efficiency, P_L, P_{r_1}, and P_{r_2} denote the power consumed by R_L, r_1, and r_2, respectively. By substituting (6) and (7) into (8), we have

$$\begin{aligned} \eta &= \frac{I_2 \bar{I}_2 R_L}{I_2 \bar{I}_2 R_L + I_1 \bar{I}_1 r_1 + I_2 \bar{I}_2 r_2} \\ &= \frac{(\omega M)^2 R_L}{\left[(r_2 + R_L)^2 + (\omega L_2 - \frac{1}{\omega C_2})^2 \right] r_1 + (\omega M)^2 (r_2 + R_L)}. \end{aligned} \qquad (9)$$

According to (9), efficiency η is a function of L_2 and M. Therefore, it can be increased by adjusting them. In the experiment bellow, how efficiency increases with changing M will be shown.

III. EXPERIMENTAL SETUP AND RESULTS

A. The prototype PLS

Fig. 8 shows the 3-D model of the prototype PLS to be used in the experiment. Parts in same color are identical. Fig. 9 is a

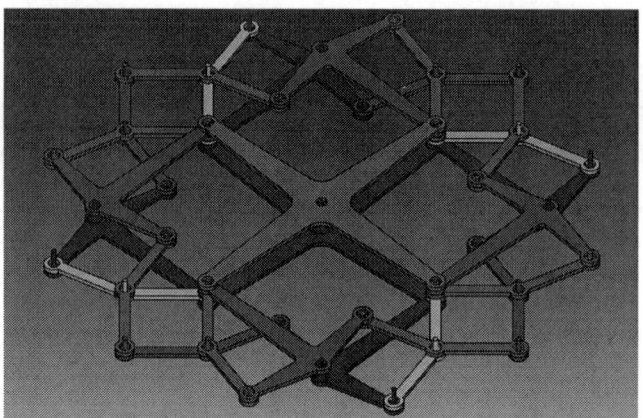

Fig. 8: 3-D model of the prototype PLS in CAD software

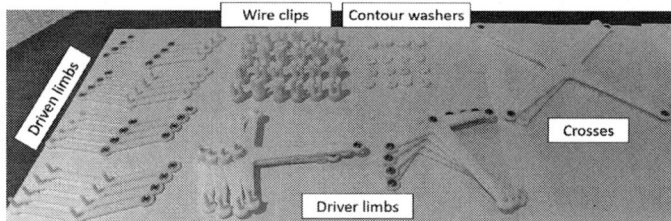

Fig. 9: Photograph of unassembled 3-D printed parts of the prototype PLS

photograph of unassembled 3-D printed parts. To implement the folding functionality of the PLS, there are two types of angled limbs: driven limbs and driver limbs. The driven limbs are in the shape of simple angled rods, whilst driver limbs are in a shape that looks like the letter 'T'. To operate the driver limbs, two more cross shaped parts are mounted in the center of the PLS. The cross at the bottom is the base cross, which is fixed on the surface where the PLS is deployed. The cross at the top is the driver cross, which is driven by a motor and rotates around the center to operate the driver limbs, so that the whole PLS can fold or unfold. The motor is controlled manually by a user in the experiment. Due to the thickness of the parts, the pivots and exterior connections of the scissors structure are at different heights. To solve this problem, various washers of different thicknesses are applied at the positions mentioned above to elevate the wire clips to the same height so that the air-core winding mounted on it can be planar. Bearings are also applied at all connection points in the PLS to smooth the folding and unfolding. The selected l_{side} is 5 cm.

Fig. 10 shows a group of snapshots of the prototype PLS with winding mounted during the folding process. The litz wire winding has 7 turns and is tightly, vertically arranged by the wire clips. Through the transparent acrylics base board, we can see the motor underneath it, which drives the top driver cross to fold and unfold the whole PLS. The available folding ratio of the prototype is from 8.2 grad to 46.7 grad due to mechanical structure limits. Fig. 11 is a photograph of the receiving winding. The receiving winding is in the shape

(a) ϕ = 8.2 grad

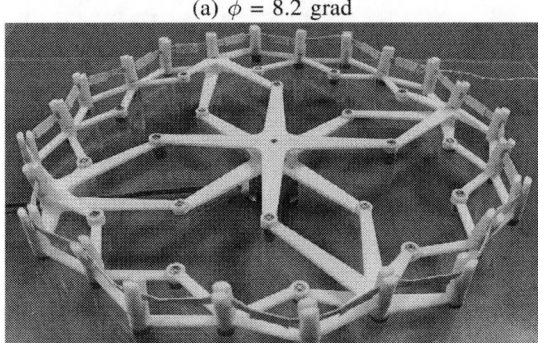

(b) ϕ = 30 grad

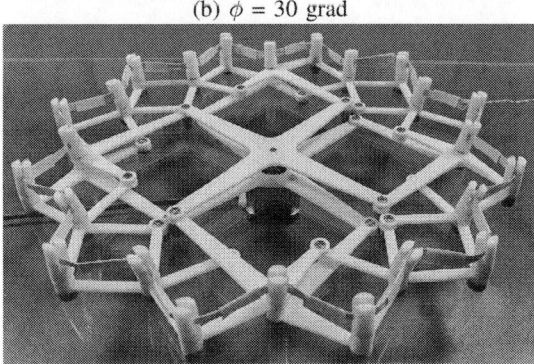

(c) ϕ = 46.7 grad

Fig. 10: Snapshots of the prototype PLS at different folding ratios with winding mounted

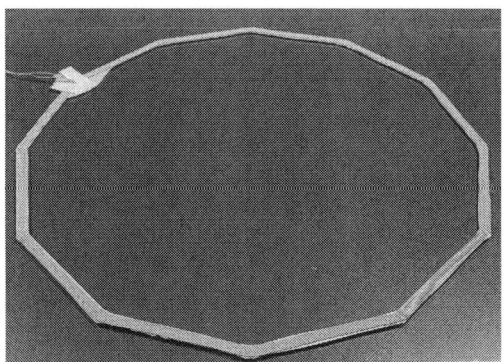

Fig. 11: Photograph of the receiving winding

(a) 0 cm misalignment

(b) 10 cm misalignment

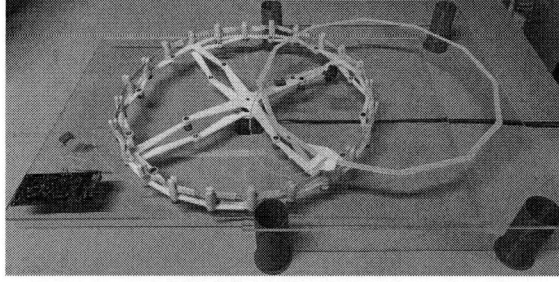

(c) 20 cm misalignment

Fig. 12: Snapshots of the winding pair at different misalignments, with ϕ = 8.2 grad

of a regular dodecagon (twelve-sided regular polygon). The distance from its center to its interior vertexes is 18.9 cm. The receiving winding is elevated by 4 3-D printed supports to maintain a constant vertical distance between the PLS winding. To test the efficiency improvement ability of the prototype PLS, the wireless power transfer experiment is run at three different misalignments, as is shown in Fig .12. The parameters of the experiment components are collected in TABLE I. The manufacturer IDs of the key components are shown in TABLE II. In the experiments, the motor is manually controlled by the user.

B. Experiment results

The measured values of the self-inductance of the prototype PLS winding and the mutual inductance between the winding

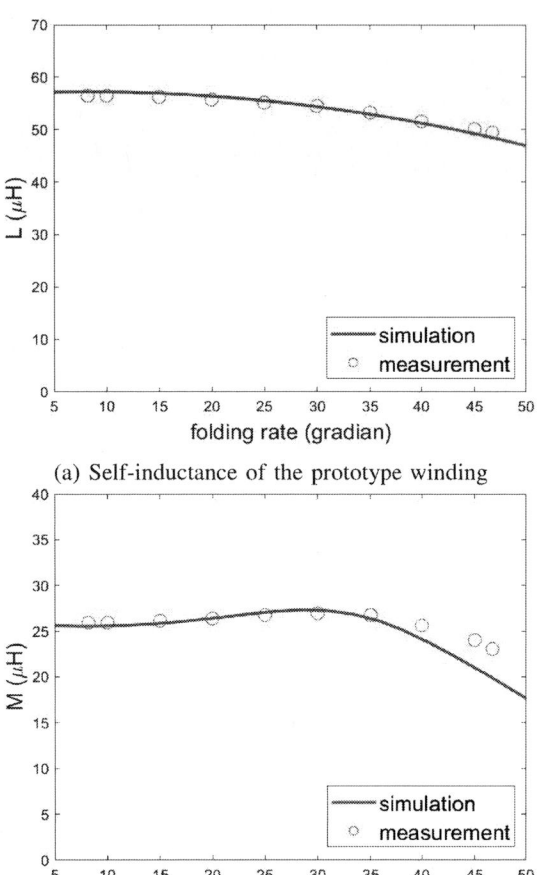

(a) Self-inductance of the prototype winding

(b) Mutual inductance of the winding pair at 0 cm misalignment

Fig. 13: Measured inductances during the folding process

TABLE I: Component parameters

Parameter	Symbol	Value	Unit
Self-inductance of L_2	L_2	50.62	μH
Capacitance of C_1	C_1	10	nF
Capacitance of C_2	C_2	10	nF
Resistance of r_1	r_1	0.780	Ω
Resistance of r_2	r_2	0.786	Ω
Resistance of R_L	R_L	120	Ω
ac voltage frequency	f	700.0	kHz
Distance between windings	$d_{winding}$	2.5	cm
Litz wire cross-sectional radius	r_{litz}	0.4	mm

pairs during the folding process are respectively shown in Fig. 13. The measurements are also compared to software simulation and we can see that the measured values match the simulation well. According to Fig .13(a), self-inductance decreases when the PLS is folding. The larger the folding ratio, the faster the decrement, which matches what we expected in Fig. 6. According to Fig .13(b), the mutual inductance remains constant at the beginning of the folding process. Then, it starts to increase slowly when the folding ratio is about 15 grad. After reaching the peak value at folding ratio of about 30 grad, it starts and keeps dropping fast until the end of the

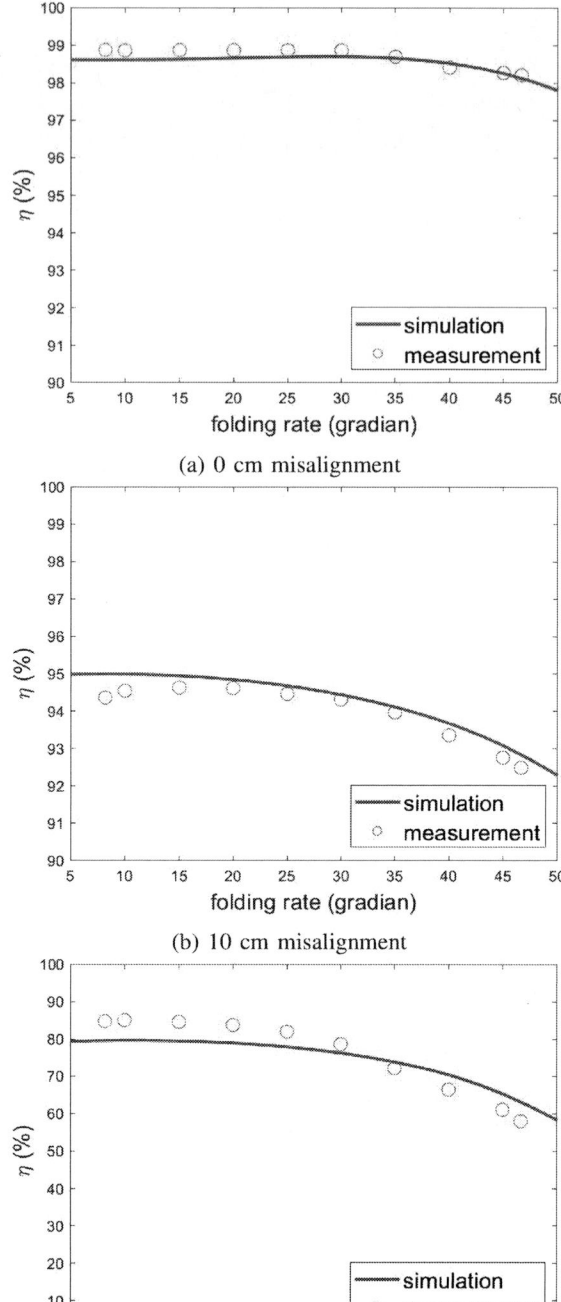

(a) 0 cm misalignment

(b) 10 cm misalignment

(c) 20 cm misalignment

Fig. 14: Measured efficiency variations at different misalignments

folding process. The reason for the variation trends of the two curves is that, at the beginning of the folding process, the winding's shape almost does not change, so the inductances are almost constant. When it reaches a specific folding ratio (about 30 grad in this example), the angle between the winding

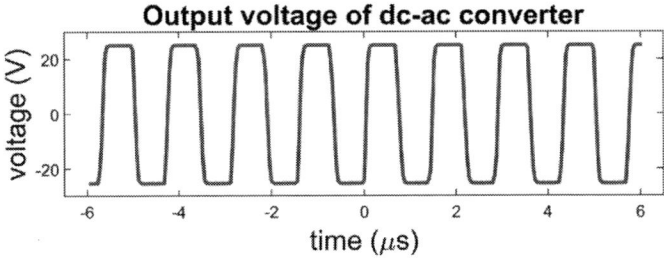

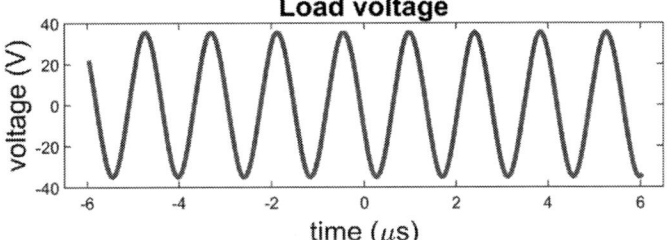

Fig. 15: Measured voltage waveforms

TABLE II: Experiment setup and components

Component	MID
DC power source	Aim-TTi CPX200D
Oscilloscope	Tektronix MSO58LP
Switch MOSFETs	Vishay SiRA02DP
Switching signal generation	Analog Devices AD9837
Stepper Motor	NEMA 17

segments on the limb sides start to get smaller quickly, which means the winding segments are getting more parallel to their neighbouring segments. Therefore, the magnetic field generated by them cancels more between each other, leading to both decreasing self-inductance and decreasing mutual inductance.

To measure and plot the efficiency curve versus folding ratio at different misalignments, The prototype winding is tested under three misalignments: 0 cm, which represents no misalignment, 10 cm, which represents a small misalignment, and 20 cm which represents a large misalignment that almost equals half of the average winding radius. All measurements are made at $P_L = 5$ W. v_{dc} of Fig .7 is adjusted to maintain the load power constant. The measured waveforms of v_{ac} and v_L of 10 cm misalignment at 8.2 grad folding ratio are shown in Fig. 15. The variations of efficiency at different misalignments are shown in Fig. 14. According to Fig. 14(a), when there is no misalignment, the dodecagon winding inductive power transfer system can operate at almost 100% efficiency. Under this situation, folding the PLS just changes the efficiency a little bit, in a small range of 1% width. We can see that the trend of the curve and measurements in Fig. 14(a) are similar to the Fig. 13(b)'s, which means that this small change is caused by varying mutual inductance. When it comes to Fig. 14(b), The maximum available efficiency drops to about 94.6% due to the increasing misalignment. The starting efficiency when the PLS is fully unfolded, is about 94.2%. After reaching the peak value, it decreases with the increasing folding ratio. The ending efficiency when the PLS is fully folded is about

92.5%. According to Fig. 14(c), the whole efficiency curve drops lower than when the misalignment was 10 cm. Because the increment around $\phi = 10$ grad is so small compared to the decrement afterwards, we can consider the curve as monotonic decreasing. The efficiency changing range is much larger than the previous two situations: from 58% to 88%. From this result, we can conclude that, under the large misalignment scenarios of inductive wireless power transfer applications, efficiency can be significantly improved by unfolding the PLS and the winding on it. The increment can be up to 30%.

IV. CONCLUSION AND FUTURE WORK

A. Conclusion

In this paper, a planar linkage structure is proposed. The structure is able to change the shape of the air-core winding mounted on its exterior contour by folding or unfolding. The inductance of the winding is changed while maintaining constant length. This kind of variable inductance winding can be used in many power conversion applications to improve operating performance such as efficiency when adjustment of voltage or frequency is unachievable. The PLS is made by connecting many identical, specially designed scissors structures. For the example 12-sided PLS researched, the mathematical minimum inductance can be down to half of the mathematical maximum inductance. The available inductance range of real-world windings is narrower due to mechanical limits.

A prototype PLS is built with 12 scissors structures, 5 cm side length, with a 7-turn air-core winding mounted on its exterior contour. The available folding ratio of the prototype is from 8.2 grad to 46.7 grad. The windings is tested in a series-series inductive wireless power transfer system to verify its ability to increase efficiency under scenarios where large misalignment exists. The change of efficiency during the folding process is measured at 3 different misalignments: 0 cm , 10 cm and 20 cm. The experiment result shows that when the misalignment is high, even equal to half of the average radius of the PLS winding, it can increase the efficiency of the wireless power transfer system by 30%, which is a significant improvement.

B. Future work

The future work will be mainly focused on applying automatic controls to the motor. For example, in wireless power transfer systems, the controller on the transmitter side senses the primary winding voltage and current amplitudes and phases and may apply perturb and observe method to track the optimal operating points of the fold, subject to misalignment and load perturbations.

REFERENCES

[1] Z. Zhang, H. Pang, A. Georgiadis, and C. Cecati, "Wireless power transfer—an overview," *IEEE Transactions on Industrial Electronics*, vol. 66, no. 2, pp. 1044–1058, 2019.
[2] S. Y. R. Hui, W. Zhong, and C. K. Lee, "A critical review of recent progress in mid-range wireless power transfer," *IEEE Transactions on Power Electronics*, vol. 29, no. 9, pp. 4500–4511, 2014.

[3] G. K. Y. Ho, Y. Fang, and B. M. H. Pong, "A multiphysics design and optimization method for air-core planar transformers in high-frequency llc resonant converters," *IEEE Transactions on Industrial Electronics*, vol. 67, no. 2, pp. 1605–1614, 2020.

[4] M. A. Elazm, A. Ragheb, A. Elsafty, and M. Teamah, "Computational analysis for the effect of the taper angle and helical pitch on the heat transfer characteristics of the helical cone coils," *Archive of Mechanical Engineering*, pp. 361–375, 2012.

[5] K. Orikawa, S. Kanno, and S. Ogasawara, "A winding structure of air-core planar inductors for reducing high-frequency eddy currents," *IEEE Transactions on Industry Applications*, vol. 58, no. 6, pp. 7572–7580, 2022.

[6] W. Li, W. Mei, Q. Yuan, Y. Song, Z. Dongye, and L. Diao, "Detuned resonant capacitors selection for improved misalignment tolerance of lcc-s compensated wireless power transfer system," *IEEE Access*, vol. 10, pp. 49 474–49 484, 2022.

[7] Y.-S. Lai and M.-H. Yu, "Online autotuning technique of switching frequency for resonant converter considering resonant components tolerance and variation," *IEEE Journal of Emerging and Selected Topics in Power Electronics*, vol. 6, no. 4, pp. 2315–2324, 2018.

[8] Z. Zhang, S. Zheng, S. Luo, D. Xu, P. T. Krein, and H. Ma, "An inductive power transfer charging system with a multiband frequency tracking control for misalignment tolerance," *IEEE Transactions on Power Electronics*, vol. 37, no. 9, pp. 11 342–11 355, 2022.

[9] L. Tian, F. Yang, B. Cai, S. Li, K. Liu, and H. Zhao, "High misalignment tolerance in efficiency of wpt system with movable intermediate coil and adjustable frequency," *IEEE Access*, vol. 9, pp. 139 527–139 535, 2021.

[10] M. A. Houran, X. Yang, W. Chen, and M. Samizadeh, "Wireless power transfer: Critical review of related standards," in *2018 International Power Electronics Conference (IPEC-Niigata 2018 -ECCE Asia)*, 2018, pp. 1062–1066.

[11] T. Samanchuen, K. Jirasereeamornkul, C. Ekkaravarodome, and T. Singhavilai, "A review of wireless power transfer for electric vehicles: Technologies and standards," in *2019 4th Technology Innovation Management and Engineering Science International Conference (TIMES-iCON)*, 2019, pp. 1–5.

[12] H. Zeng and F. Z. Peng, "Non-linear capacitor based variable capacitor for self-tuning resonant converter in wireless power transfer," in *2018 IEEE Applied Power Electronics Conference and Exposition (APEC)*, 2018, pp. 1375–1379.

[13] C.-M. Lai, C.-H. Hsu, Y.-C. Lin, M.-Z. Chen, and T.-Z. Zheng, "A high-efficiency wireless power transfer system with variable capacitor technique for compensating coil misalignment," in *2024 IEEE Wireless Power Technology Conference and Expo (WPTCE)*, 2024, pp. 688–692.

[14] J. M. Alonso, M. Perdigão, M. A. Dalla Costa, S. Zhang, and Y. Wang, "Variable inductor modeling revisited: The analytical approach," in *2017 IEEE Energy Conversion Congress and Exposition (ECCE)*, 2017, pp. 895–902.

[15] Z. He, G. Zhang, Z. Chen, and S. S. Yu, "A review of variable-inductor-based power converters for eco-friendly applications: Fundamentals, configurations, and applications," *Chinese Journal of Electrical Engineering*, vol. 9, no. 3, pp. 50–71, 2023.

[16] L. Zhang, W. G. Hurley, and W. Wölfle, "A new approach to achieve maximum power point tracking for pv system with a variable inductor," in *The 2nd International Symposium on Power Electronics for Distributed Generation Systems*, 2010, pp. 948–952.

[17] W. Wolfle, W. Hurley, and S. Lambert, "Quasi-active power factor correction: the role of variable inductance," in *2001 IEEE 32nd Annual Power Electronics Specialists Conference (IEEE Cat. No.01CH37230)*, vol. 4, 2001, pp. 2078–2083 vol. 4.

[18] A. Fereshtian and J. Ghalibafan, "Impedance matching and efficiency improvement of a dual-band wireless power transfer system using variable inductance and coupling method," *AEU - International Journal of Electronics and Communications*, vol. 116, p. 153085, 2020. [Online]. Available: https://www.sciencedirect.com/science/article/pii/S1434841118331364

[19] R. E. A. Azim, "A review of tunable inductors for power electronics: Techniques and applications," *e-Prime - Advances in Electrical Engineering, Electronics and Energy*, vol. 9, p. 100655, 2024. [Online]. Available: https://www.sciencedirect.com/science/article/pii/S2772671124002353

[20] C. Zhang, X. Chen, K. Chen, and D. Lin, "Iptvisual: Visualisation of the spatial energy flows in inductive power transfer systems with arbitrary

winding shapes," *World Electric Vehicle Journal*, vol. 13, no. 4, 2022. [Online]. Available: https://www.mdpi.com/2032-6653/13/4/63

Wireless Power Transfer System with Automatic Tuning Capability in Metallic Environment

Renjie Zhang
School of Electrical Engineering
State Key Laboratory of EIPE
Xi'an Jiaotong University
Xi'an, China
zrj123456@stu.xjtu.edu.cn

Yue Wu
School of Electrical Engineering
State Key Laboratory of EIPE
Xi'an Jiaotong University
Xi'an, China
dywuyue@stu.xjtu.edu.cn

Delin Zhao
School of Electrical and
Electronic Engineering
Nanyang Technological University
Singapore, Singapore
zhao0473@e.ntu.edu.sg

Yaohua Li
School of Electrical and
Electronic Engineering
Nanyang Technological University
Singapore, Singapore
yaohua001@e.ntu.edu.sg

Yongbin Jiang
School of Electrical and
Electronic Engineering
Nanyang Technological University
Singapore, Singapore
yongbin.jiang@ntu.edu.sg

Yi Tang
School of Electrical and
Electronic Engineering
Nanyang Technological University
Singapore, Singapore
yitang@ntu.edu.sg

Huan Yuan
School of Electrical Engineering
State Key Laboratory of EIPE
Xi'an Jiaotong University
Xi'an, China
huanyuan@xjtu.edu.cn

Xiaohua Wang
School of Electrical Engineering
State Key Laboratory of EIPE
Xi'an Jiaotong University
Xi'an, China
xhw@mail.xjtu.edu.cn

Mingzhe Rong
School of Electrical Engineering
State Key Laboratory of EIPE
Xi'an Jiaotong University
Xi'an, China
mzrong@xjtu.edu.cn

Abstract—**Wireless power transfer (WPT) systems utilize alternating magnetic fields to transfer energy. However, metal objects within these magnetic fields generate eddy currents, which increase power losses, damage the system, and reduce power transfer efficiency and capabilities. This paper proposes an automatic tuning assist circuit (ATAC) to mitigate the impacts of metal objects on WPT systems. First, an LCC-S compensated WPT system is adopted to theoretically analyze the metal plate's effects on the receiver side. Moreover, the equivalent model of the metal-affected LCC-S circuit is established, and the ATAC is proposed to eliminate the coil self-inductance variations caused by the metal plate. Finally, comprehensive experiments are conducted to validate the performances and effectiveness of the ATAC when a metal plate is added to the receiver side of the WPT system. Experimental results indicate that the proposed ATAC can significantly improve the peak power transfer efficiency of the WPT system from 66% to 75%, together with an increment of peak transmitted power of over 210%. The proposed ATAC eliminates the need for additional magnetic shielding materials and provides a more compact and environmentally friendly WPT system design idea.**

Keywords—Wireless power transfer, metal environment, metal eddy current, automatic tuning, LCC-S topology

I. INTRODUCTION

With the rapid advancement of WPT technology across various fields, it has been extensively applied in consumer electronics, electric vehicles, spacecraft, and underwater detectors for power supply purposes [1-4]. Significant research efforts focus on WPT technology in diverse media environment, including atmospheric and space environment, underwater settings, and metallic environment. These studies, tailored to different application scenarios, provide a robust theoretical foundation for deploying WPT systems across various domains [5-8]. Metallic environment have a pronounced impact on WPT systems, and metal near the receiver is unavoidable in scenarios such as consumer electronics, automotive systems, and spacecraft [9-12].

Two primary approaches are employed to mitigate the adverse effects of metallic environment on WPT systems. The first involves using metal foreign object detection technology; when metallic foreign objects are detected, the operation of the WPT system is halted, and power transfer resumes only after the metal is removed manually or mechanically. This method is typically applied in high-power, high-efficiency applications to prevent the negative impacts caused by metal interference. The second approach seeks to reduce the influence of metallic environment by incorporating magnetic shielding materials or altering the circuit topology [13,14]. While this method cannot entirely eliminate the effects of metallic environment, it enables relatively efficient power transfer in many low-power scenarios. In the design of numerous products, WPT technology is often adopted as the energy transfer method to minimize physical interfaces while meeting the power supply requirements of devices [15]. However, devices commonly contain substantial amounts of metal, and the prevailing method to mitigate the impact of metallic environment on WPT systems is by adding magnetic shielding materials. Using such materials significantly increases the system's size and weight, posing a critical challenge in enhancing power transfer efficiency and maximum power output while maintaining a lightweight design.

Several studies have explored compensation topologies to enhance the performance of WPT systems. The LCC-S compensation topology, in particular, has been extensively investigated for its ability to provide a stable transmission current and improve system robustness against parameter variations [16]–[19]. Guo and Cui [16] analyzed the optimal configuration of an LCC-S-type WPT system, demonstrating improved efficiency and load adaptability. Tian et al. [17] optimized parameters for dynamic wireless power transfer in electric vehicles using the LCC-S topology, achieving better power transfer over varying distances. Ge [18] applied the LCC-S compensated resonant topology to lithium battery charging systems, highlighting its effectiveness in maintaining resonance under load variations. Zheng et al. [19] developed a small-signal model for inductive power transfer systems using LCC-S compensation, facilitating enhanced system design and control.

The aforementioned studies only concern about improving WPT system performances under ideal conditions with few considerations of the challenges posed by metallic environment. Metal-induced eddy currents can significantly degrade power transfer efficiency and output power, and the standard LCC-S topology lacks mechanisms to compensate for these effects.

Other researchers have focused on mitigating interference in WPT systems through automatic compensation methods. Ishihara et al. [20]–[22] proposed automatic active compensation techniques to counteract cross-coupling and impedance mismatches in multiple-receiver resonant inductive coupling WPT systems. Additionally, Zhang et al. [23] presented a self-tuning WPT system with constant voltage output under resonance frequency shifts, addressing efficiency maintenance when environmental factors cause frequency deviations. However, these methods are primarily designed for multi-receiver scenarios or frequency shifts and do not specifically target the issues caused by metal-induced eddy currents near the receiving coil.

To address these limitations, this paper proposes a novel approach that integrates an ATAC with the LCC-S compensation topology to counteract the detrimental effects of a metallic plate near the receiving coil in a WPT system. The ATAC injects reactive power into the receiving circuit, effectively neutralizing the additional reactance introduced by metal-induced eddy currents. This integration allows the system to maintain resonance and achieve efficient power transfer without relying on bulky magnetic shielding materials like ferrite cores. Consequently, the proposed method substantially improves the power transfer capability and efficiency of the WPT system in metallic environment while reducing material usage and overall system weight. This innovation offers a practical solution for deploying efficient and lightweight WPT systems in applications where proximity to metal is unavoidable.

The paper is organized as follows: Section 1 presents the equivalent model of coils in metallic environment and analyzes the equivalent circuit of the WPT system under these conditions. A WPT system with automatic tuning capabilities is introduced, and its tuning mechanism is theoretically derived. Section 2 details experiments verifying the proposed system's output consistency in the presence of a metal plate. Finally, Section 3 concludes the paper based on the experimental results.

II. CIRCUIT MODEL ANALYSIS

A. Analysis of the Effect of a Metal Plate on a Single-Turn Coil

The three-dimensional structural diagram of the WPT system is presented in Fig. 1, which has a single aluminum metal plate at the receiver. The system includes the transmitter circuit, the receiver circuit, the transmitting coil $Coil_1$, the receiving coil $Coil_2$, and the aluminum metal plate, which is placed directly above the receiving coil.

To better analyze the effect of the metal plate on the coil, we start by calculating its impacts on a single-turn coil, as shown in Fig. 2. According to Fig. 2, a single-turn coil is perpendicular to the Z axis with its center coincides with the Z axis. The coil has a radius a and carries a unit current $\dot{I}$ with angular frequency ω. A metal plate with a

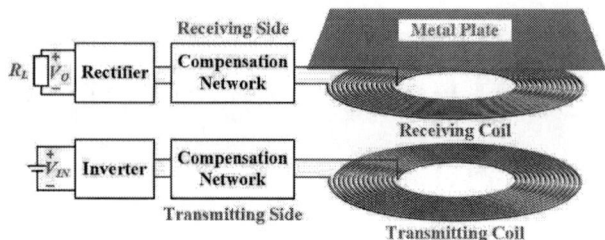

Fig. 1 WPT system with metal plate at the receiver side

thickness of h is parallel placed above the coil at a distance of d. The volume resistivity of the metal plate is ρ_v, and its surface resistivity can be expressed as $\rho_s = \rho_v/h$.

According to the proof by Geselowitz D. B. et al. in [14], the electromagnetic field generated by the eddy currents in the metal plate is approximately equivalent to the mirror-image electromagnetic field generated by an imaging coil $Coil_1$. A_ϕ and A_ϕ' are the magnetic vector potentials generated by the unit current in coil1 and the eddy currents in the metal plate, respectively. In cylindrical coordinates, the angle ϕ is irrelevant to the electromagnetic field of this system. Therefore, the coordinate system can be adopted as cylindrical coordinates, and the coordinates (r, z) of $Coil_1$ in the coordinate system can be expressed as (a, d).

The electromagnetic field generated by the eddy currents in the metal plate near the metal plate is approximately equivalent to the electromagnetic field produced by a mirror-image coil, which can be calculated by:

$$A_\phi'(r,z) = -A_\phi(r,-z) + j\frac{2\rho_S}{\mu\omega}B_r(r,-z) \tag{1}$$

The component $-A\phi(r,-z)$ is comparable to the magnetic vector potential generated by an equivalent coil $Coil_1'$ with a unit current of the same magnitude. $j2\rho_S/\mu\omega B_r(r,-z)$ is the magnetic flux density B generated by $Coil_1$ in the r-direction, whose amplitude is proportional to $2\rho_S/\mu\omega$.

Therefore, the impedance change ΔZ induced on $Coil_1$ by the electromagnetic field generated by the eddy currents in the metal plate can be expressed as:

$$\begin{aligned}
\Delta Z &= \frac{\Delta \dot{V}}{\dot{I}} \\
&= \frac{-j\omega(2\pi a)A_\phi'(a,d)}{\dot{I}} \\
&= j\omega\Delta L + \Delta R \\
&= j\omega\left[2\pi a A_\phi(a,-d)\right] + \frac{4\pi a\rho_S}{\mu}B_r(a,d)
\end{aligned} \tag{2}$$

Where ΔV, ΔL, and ΔR are the voltage, inductance, and resistance changes of $Coil_1$, which are induced by the electromagnetic field of the eddy currents, respectively. For a single-turn coil located at $z = d$, the magnetic vector potential A_ϕ and B_r produced by a unit current can be

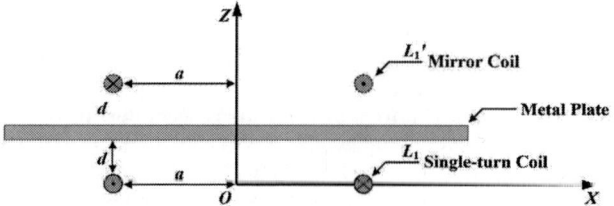

Fig. 2 Cross-sectional analysis of a non-ferromagnetic

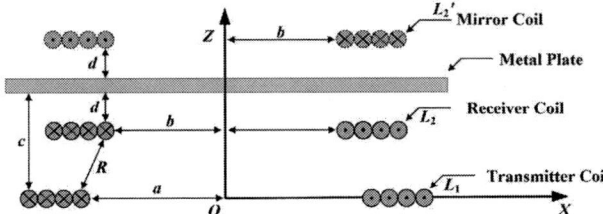

Fig. 3 Cross-sectional analysis of two multi-turn coils in the presence of a metal plate.

expressed as:

$$
\begin{cases}
A_\phi(r,z) = \dfrac{\mu}{2\pi}\sqrt{\dfrac{a}{r}}G(k) \\[2mm]
B_r(r,z) = \dfrac{\mu}{2\pi}\dfrac{-(z-d)}{r\sqrt{(a+r)^2+(z-d)^2}} \\[2mm]
\qquad \cdot\left[-K(k)+\dfrac{a^2+r^2+(z-d)^2}{(a-r)^2+(z-d)^2}E(k)\right]
\end{cases}
\tag{3}
$$

where:

$$
\begin{cases}
G(k) = \left(\dfrac{2}{k}-k\right)K(k)-\dfrac{2}{k}E(k) \\[2mm]
k = \sqrt{\dfrac{4ar}{(a+r)^2+(z-d)^2}} \\[2mm]
K(k) = \int_0^{\frac{\pi}{2}}\dfrac{d\psi}{\sqrt{1-k^2\sin^2\psi}} \\[2mm]
E(k) = \int_0^{\frac{\pi}{2}}\sqrt{1-k^2\sin^2\psi}\,d\psi
\end{cases}
\tag{4}
$$

In (4), $K(k)$ is the first kind elliptic integral, and $E(k)$ is the second kind elliptic integral. Thus, we can calculate the ΔL and ΔR as:

$$
\begin{cases}
\Delta R = \dfrac{2\rho_s d}{\sqrt{a^2+d^2}}\left[-K(k)+\dfrac{a^2+2d^2}{2d^2}E(k)\right] \\[2mm]
\Delta L = a\mu G(k) \\[2mm]
k = \dfrac{a}{\sqrt{a^2+d^2}}
\end{cases}
\tag{5}
$$

B. Analysis of the Effect of a Metal Plate on Multi-Turn Coils

Fig. 3 shows a cross-sectional view of two coaxial multi-turn coils with the presence of a metal plate. When the metal plate is not considered, and $Coil_1$ and $Coil_2$ are placed in the infinite free space, the mutual inductance between two coaxial parallel coils can be calculated with Neumann's formula:

$$
\begin{aligned}
M &= \frac{\mu_0 N_1 N_2}{4\pi}\int_{L_2}\int_{L_1}\frac{d\vec{l_1}\cdot d\vec{l_2}}{R} \\
&= \frac{\mu_0 N_1 N_2 ab}{4\pi}\cdot \\
&\quad \int_0^{2\pi}\int_0^{2\pi}\frac{\cos(\varphi_1-\varphi_2)\,d\varphi_1 d\varphi_2}{\sqrt{a^2+b^2+(c-d)^2-2ab\cos(\varphi_1-\varphi_2)}}
\end{aligned}
\tag{6}
$$

In practical calculations, the integration in (6) can be transformed into an elliptic integral for easier computation, as shown below:

$$
M = \mu N_1 N_2 \sqrt{ab}\,G\left(\sqrt{\frac{4ab}{(a+b)^2+(c-d)^2}}\right)
\tag{7}
$$

It can be seen from (7) that in free space, without considering the physical dimensions of the coils, the mutual inductance between two coaxial parallel coils only depends on the number of turns, coil radius, and the distance between the coils.

When considering the metal plate, the additional magnetic flux generated by eddy currents causes changes in mutual inductance. Therefore, the change in mutual inductance between coils near the metal plate can be evaluated through the mutual inductance between coils $Coil_1$ and $Coil_2$. Using the image method and Neumann's formula, the changed ΔM can be obtained:

$$
\Delta M = -\mu N_1 N_2 \sqrt{ab}\,G\left(\sqrt{\frac{4ab}{(a+b)^2+(c+d)^2}}\right)
\tag{8}
$$

By applying the superposition theorem, when the changes in self-inductance and self-resistance of a single-turn coil are known, the changes in self-inductance and self-resistance of a multi-turn coil can be derived. Here, ΔL_i and ΔR_i can be obtained from (5):

$$
\begin{cases}
\Delta L = \displaystyle\sum_{i=1}^{N}\Delta L_i + 2\sum_{i\neq j}^{N}\Delta M_{ij} \\[2mm]
\Delta R \approx \displaystyle\sum_{i=1}^{N}\Delta R_i
\end{cases}
\tag{9}
$$

C. Equivalent Circuit Model

Based on the previous derivation, the equivalent model of the WPT system with a metal plate is shown in Fig. 4. The mutual inductances among the transmitting coil, receiving coil, and metal plate are considered, and the circuit model is designed based on the LCC-S topology.

In the circuit, V_{in} represents the DC input voltage. The full-bridge inverter, consisting of MOSFETs Q_1 to Q_4, converts V_{in} into an AC voltage V_1. The transmitting circuit includes the resonant capacitor C_1 and transmitting coil L_1. Similarly, the receiving circuit comprises the resonant capacitor C_2 and receiving coil L_2, with mutual inductance M_{12} between L_1 and L_2. The rectifier circuit uses diodes D_1 to D_4 to convert the AC voltage back to DC, and the smoothing capacitor C_L provides a stable DC voltage V_O to the load R_L.

The metal plate is represented in Fig. 4 by an equivalent circuit consisting of an inductance L_3 in series with a resistance R_3, based on Loos's equivalent model [24]. There are mutual inductances M_{13} and M_{23} between L_3 and the transmitting coil L_1 and receiving coil L_2, respectively.

The metal eddy current circuit impedance is reflected onto the transmitting and receiving circuits to simplify the analysis. The reflected impedance on the transmitting circuit is given by (10):

$$
\begin{cases}
Z_{24} = \dfrac{(\omega M_{24})^2}{Z_4} = \dfrac{(\omega M_{24})^2}{R_4+j\omega L_4} \\[3mm]
\quad = \dfrac{(\omega M_{24})^2}{R_4^2+(j\omega L_4)^2}R_4 - j\dfrac{(\omega M_{24})^2}{R_4^2+(j\omega L_4)^2}\omega L_4 \\[3mm]
Z_{34} = \dfrac{(\omega M_{34})^2}{Z_4} = \dfrac{(\omega M_{34})^2}{R_4+j\omega L_4} \\[3mm]
\quad = \dfrac{(\omega M_{34})^2}{R_4^2+(j\omega L_4)^2}R_4 - j\dfrac{(\omega M_{34})^2}{R_4^2+(j\omega L_4)^2}\omega L_4
\end{cases}
\tag{10}
$$

and on the receiving circuit by:

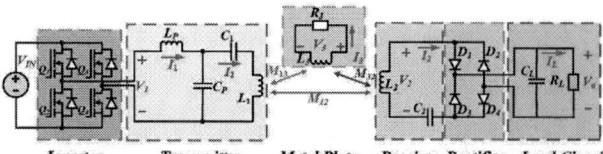

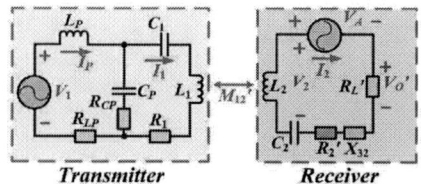

Fig.6 Simplified equivalent circuit diagram

Fig. 4 Circuit diagram of a WPT system with metal plates

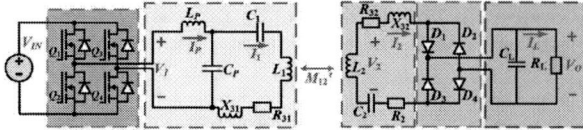

Fig. 5 System circuit diagram for simplified single metal plate

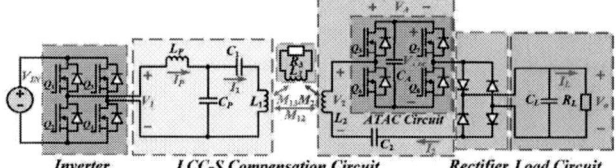

Fig. 7 The proposed automatic tuning wireless power transfer system.

$$Z_{34} = \frac{(\omega M_{34})^2}{Z_4} = \frac{(\omega M_{34})^2}{R_4 + j\omega L_4}$$

$$= \frac{(\omega M_{34})^2}{R_4^2 + (j\omega L_4)^2} R_4 - j\frac{(\omega M_{34})^2}{R_4^2 + (j\omega L_4)^2} \omega L_4 \quad (11)$$

The equivalent circuit diagram with the reflected impedances is shown in Fig. 5. In Fig. 5, R_{31} and X_{31} are the reflected resistance and reactance on the transmitting circuit, and R_{32} and X_{32} are those on the receiving circuit. Their expressions are provided in (12):

$$\begin{cases} R_{42} = \frac{(\omega M_{24})^2}{R_4^2 + (j\omega L_4)^2} R_4 \\ R_{43} = \frac{(\omega M_{34})^2}{R_4^2 + (j\omega L_4)^2} R_4 \\ X_{42} = -j\frac{(\omega M_{24})^2}{R_4^2 + (j\omega L_4)^2} \omega L_4 \\ X_{43} = -j\frac{(\omega M_{34})^2}{R_4^2 + (j\omega L_4)^2} \omega L_4 \end{cases} \quad (12)$$

Due to the eddy currents in the metal plate, the internal resistance of the transmitting and receiving coils increases, the self-inductance decreases, and the mutual inductance M_{12} changes to M_{12}'. As a result, the system's resonance condition is affected since the resonant capacitors remain unchanged.

Resonant wireless power transfer systems are sensitive to changes in resonant frequency. The transmitted power and transmission efficiency will decrease if the system is not adjusted to maintain resonance.

D. Automatic Tuning Circuit

Since the distance between the metal plate and the transmitting coil L_1 discussed in this paper is relatively large and much greater than the distance to L_2, the mutual inductance M_{13} is extremely small. Therefore, M_{13} is neglected in this paper, and only the effect of M_{23} is considered. Based on this, an equivalent circuit simplified from Fig. 5 is obtained, as shown in Fig. 6. The inverter circuit is equivalent to an AC power source V_1 in this figure. The parameters R_2', L_2', and R_L' in the receiving circuit can be expressed by (13).

$$\begin{cases} R_3' = R_3 + \frac{(\omega M_{34})^2}{R_4^2 + (j\omega L_4)^2} R_4 \\ L_3' = L_3 - \frac{(\omega M_{34})^2}{R_4^2 + (j\omega L_4)^2} L_4 \\ R_L' = \frac{\pi^2}{8} R_L \end{cases} \quad (13)$$

This paper proposes a self-tuning wireless power transfer system to address the resonance shift caused by eddy currents in the metal plate. The circuit is shown in Fig. 7. In the figure, the topology of the transmitting end is changed to an LCC topology. To keep I_1 in a constant current state, L_P, C_P, C_1, and L_1 need to be configured to satisfy:

$$\begin{cases} j\omega L_P + \frac{1}{j\omega C_P} = 0 \\ j\omega L_1 + \frac{1}{j\omega C_1} + \frac{1}{j\omega C_P} = 0 \end{cases} \quad (14)$$

As shown in Fig. 7, the receiving circuit includes an automatic tuning assist circuit within the red dashed box, along with the traditional series topology. This circuit consists of capacitor C_A and four GaNFETs Q_5 to Q_8. V_{A_DC} represents the DC voltage across C_A, and V_A is the output voltage of the ATAC circuit. This circuit generates a voltage that is 90° out of phase with the circuit current, forming an equivalent reactance to counteract X_{32} caused by metal eddy currents.

The operation timing diagram of the proposed WPT system is shown in Fig. 8. To enable the self-tuning WPT system work, drive signals of Q_1 to Q_8 need to be set in advance, as shown in Fig. 8. The drive signals of Q_5 to Q_8 have a fixed phase difference θ with respect to the drive signals of Q_1 to Q_4 in the inverter. By adjusting θ, the current I_2 at the receiving side can be corrected under the interference of metal eddy currents. The characteristics of the ATAC determine this; when the circuit operates in a steady state, the phase difference between the current I_2 in the receiving circuit and the drive signals of Q_5 and Q_8 is 90°. When the ATAC is not operating, metal eddy currents cause the receiving circuit current I_2', as shown in Fig. 8, to lag in phase and decrease in amplitude, reducing the system's power and efficiency.

In the initial state with V_{A_DC}, the receiving current is I_2'. When Q_5 and Q_8 are turned on, if I_2' is positive, C_A charges

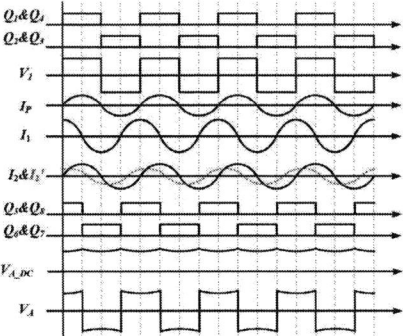

Fig. 8 The working sequence diagram of the proposed circuit

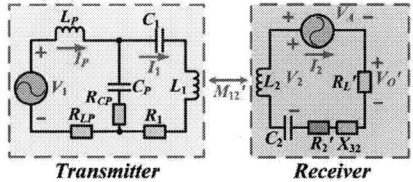

Fig. 9 Simplified circuit diagram of proposed system

through these transistors; C_A discharges if I_2' is negative. Similarly, Q_6 and Q_7 control charging and discharging based on the direction of I_2'. The voltage across C_A changes according to the charging and discharging durations, determined by the proportion of the cycle where I_2' is positive or negative.

As V_{A_DC} increases, an inverted voltage V_A is generated. Due to its 90° phase difference with I_2, this voltage offsets part of the circuit's reactance, causing I_2' to approach I_2. When the phase difference between I_2 and the drive signals of Q_5 and Q_8 is exactly 90°, V_{A_DC} stabilizes, and the receiving current settles at I_2. The assist circuit effectively acts as an automatically adjustable reactor. Consequently, a simplified equivalent schematic diagram can be derived from Figs. 6 and 7, as shown in Fig. 9.

Since the transmitting circuit is less affected by metal eddy currents, the effects of R_{LP}, R_{CP}, and R_1 are neglected when calculating the transmitting current I_1. In the receiving circuit, due to the significant influence of metal eddy currents on the receiving coil, the equivalent resistance R_2' and X_{32} need to be considered in the equivalent circuit. When the ATAC is not operating, V_1, I_1, V_2, and V_O in Fig. 9 can be obtained as below, where ω is the angular frequency of the system.

$$
\begin{cases}
I_1 = \dfrac{V_1}{j\omega L_P} \\[2mm]
V_2 = M_{12}' \dfrac{V_1}{L_P} \\[2mm]
V_O' = M_{12}' \dfrac{V_1 R_L'}{L_P(R_2' + R_L' + X_{32})}
\end{cases}
\tag{15}
$$

where V_O' is the equivalent voltage of the equivalent resistance after the rectifier circuit and load resistance. When the ATAC is operating, the output voltage of the equivalent load will increase and can be re-expressed as:

$$
V_O' = M_{12}' \frac{V_1 R_L'}{L_P(R_2' + R_L')}
\tag{16}
$$

By comparing (15) and (16), it can be concluded that although the output voltage is still affected by the mutual inductance M_{12}' and R_2', the most significant influence from X_{32} is eliminated by the adopted ATAC and the whole WPT system will be maintained in the resonant state. Although the self-tuning WPT system will still lose energy due to the thermal effects caused by metal eddy currents, the resonant network will not be affected by the metal due to the operating ATAC. The current and voltage in the receiving

circuit are maintained in phase, which means the reactive power will be reduced compared to the original circuit, thus improving both the output power and efficiency. The following formulas calculate the system output power P_{OUT} and overall efficiency η:

$$
P_{OUT} = \frac{\left(M_{12}' V_1\right)^2 R_L'}{\left[L_P(R_2' + R_L')\right]^2}
\tag{17}
$$

$$
\eta = \frac{I_2^2 R_L'}{I_P^2 R_{LP} + (I_P - I_1)^2 R_{CP} + I_1^2 R_1 + I_2^2(R_2' + R_L')}
\tag{18}
$$

III. EXPERIMENTAL VERIFICATION

Fig. 10 presents the experimental platform setup, which includes a signal generator, a driver power supply, an electronic load, an oscilloscope, an ammeter, and a DC voltage source.

In the experiment, coaxial transmitting coil L_1 and receiving coil L_2 are placed on two parallel acrylic plates 8cm apart, and a square aluminum metal plate is placed parallel 2cm above the receiving coil. The metal plate has a side length of 30cm and a thickness of 3mm. The DC voltage source provides constant input voltage V_{IN} during the experiment. The DC input power is measured using an inline ammeter.

According to the parameters in Table 1, comparative experiments are conducted to compare the operating states of the LCC-S topology circuit alone and the proposed circuit under the influence of metal. Figure 11 shows the oscilloscope waveforms of V_1, I_1, I_2, and V_A for both the standard LCC-S topology and the proposed circuit in a metal environment. The specific parameters in the experiment are shown in Table 1.

By comparing Fig. 11(a) and 11(b), we observe that the transmitting currents I1 in both circuits are almost identical under the same input and load conditions. Their phases lag 90° behind V_1, consistent with the expression of V_1 and I_1 in

Fig. 10 Experimental diagram of automatic tuning system in metal environment

TABLE I Experimental Parameter Settings

Name	Parameter	Name	Parameter
L_P	30.21 μH	C_3	25.13 nF
L_1	98.80 μH	f_P	100 kHz
L_2	100.86 μH	f_A	100 kHz
C_P	84.97 nF	θ	90°
C_1	36.79 nF	R_L	10–100 Ω
C_A	220 μF	V_{IN}	50 V

(15). However, the currents I_2 in the receiving circuits of the two systems are significantly different under the same condition. The phase difference between the current I_2 and I_1 of the LCC-S topology circuit alone is only 10° under the influence of the metal. At the same time, the self-tuning WPT system corrects the phase of current I_2 back to the normal 90° phase difference. This change not only increases the magnitude of I_2 but also corrects its power factor, effectively improving the power and efficiency of the system, which aligns with the conclusions from theoretical analysis.

The transmission efficiency of the WPT system was compared with and without ATAC by varying the load resistance R_L while keeping other conditions constant. The results are shown in Fig. 12.

As shown in Fig. 12(a) when the load is larger than 50 Ω in a metal environment, the output power difference between the self-tuning system and the LCC-S topology is negligible. This occurs because R_L is much larger than R_2' and X_{32}, as explained by (15) and (16), leading to similar output powers.

As the load resistance decreases, the proposed system's output power becomes significantly higher than that of the LCC-S topology. The ATAC eliminates the effect of X_{32}, allowing the load to receive a higher voltage and increasing output power as demonstrated in (17). Consequently, the peak output power increases from 8.9 W to 19.1 W, leading

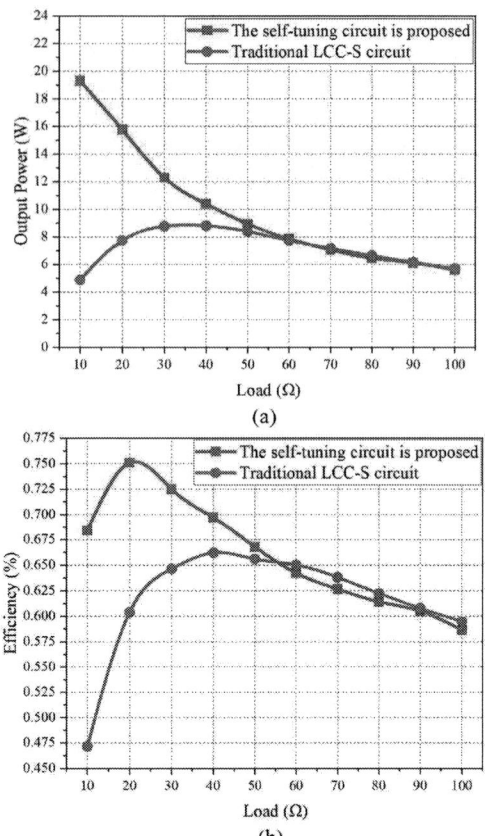

Fig. 12 The power and efficiency diagrams of the two circuits in a metal environment. (a) Power diagram. (b) Efficiency diagram.

to a 114.6% increment over the LCC-S topology.

Similarly, Fig. 12(b) indicates negligible efficiency differences for load resistances above 50 Ω. However, as the load resistance decreases, the efficiencies of both systems first increase and then decrease. The self-tuning system achieves a maximum efficiency of 75% at 20 Ω, while the peak efficiency of the LCC-S topology is only 67.5%.

These results demonstrate that the automatic tuning circuit enhances both output power and efficiency in a metal environment. By implementing maximum power point tracking and optimal efficiency tracking technologies, the system can operate efficiently under these conditions.

IV. CONCLUSION

This paper introduces an automatic tuning assist circuit to reduce the impacts of the metal plate on WPT systems and minimize the induced system parameter variations. An LCC-S compensated WPT system is adopted as an example to analyze the effects of the metal plate on the receiver side, and its corresponding equivalent model is established. A new ATAC is proposed to eliminate the effects of the metal plate on the coil self-inductance. Compared to the traditional LCC-S topology, the ATAC can improve maximum output power by 114.6% and increase transfer efficiency from 67.5% to 75%. Under the same load condition, the ATAC can achieve a peak efficiency increment of 23.5%. This system is promising for scenarios with unavoidable metal interference and compact design requirements, maintaining

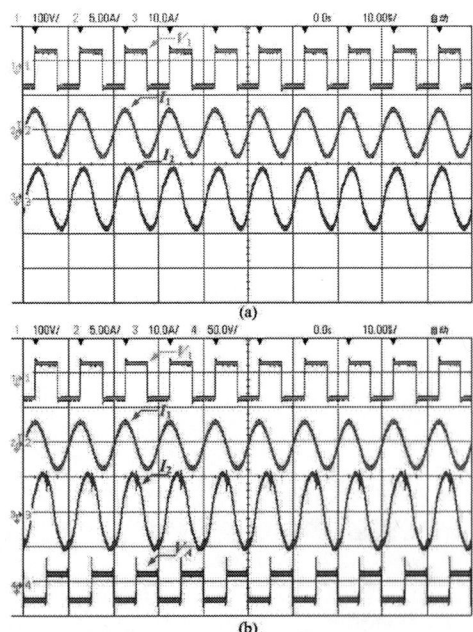

Fig. 11 The working waveforms of the two circuits with metal plate. (a) LCC-S circuit. (b) The proposed automatic tuning circuit.

979-8-3315-1612-3/25 $31.00 © 2025 IEEE

high power transfer capability and efficiency without additional magnetic shielding.

REFERENCES

[1] Y. Zhang, B. Luo, H. Cheng, *et al.*, "A multi-loads capacitive power transfer system for railway intelligent monitoring systems based on single relay plate," *International Journal of Circuit Theory and Applications*, 2023.

[2] C. Liang, X. Wang, R. Zhang, P. Zhao, C. Zhao, H. Yuan, A. Yang, J. Chu, and M. Rong, "An anti-offset CPT system with multiple pickups for mobile desktop application," *IEEE Transactions on Power Electronics*, 2023.

[3] W. Liu, B. Luo, X. He, *et al.*, "Analysis of compensation topology with constant-voltage/current output for multiple loads capacitive power transfer system," *CSEE Journal of Power and Energy Systems*, pp. 1–12, 2023.

[4] Q. Yang, X. Zhang, and P. Zhang, "Intelligent wireless power transmission cloud network for electric vehicles," *Transactions of China Electrotechnical Society*, vol. 38, no. 1, pp. 1–12, 2023.

[5] Y. Zhang, Z. Shen, W. Pan, *et al.*, "Constant current and constant voltage charging of wireless power transfer system based on three-coil structure," *IEEE Transactions on Industrial Electronics*, vol. 70, no. 1, pp. 1066–1070, 2023.

[6] Z. Zhu, Y. Wu, X. Wang, et al., "Maximum efficiency tracking of a wireless power transfer system with 3-D coupling capability using a planar transmitter coil configuration," *IEEE Transactions on Power Electronics*, vol. 39, no. 8, pp. 10594–10604, 2024.

[7] Y. Wu, Y. Jiang, Y. Li, H. Yuan, X. Wang, and Y. Tang, "Precise parameterized modeling of coil inductance in wireless power transfer systems," *IEEE Transactions on Power Electronics*, vol. 39, no. 9, pp. 11746–11757, 2024.

[8] C. Jiao, X. Yang, J. Yang, *et al.*, "Coupling-independent constant-voltage output LCC/S compensation inductive power transfer system based on multi-objective optimization theory," *Transactions of China Electrotechnical Society*, pp. 1–15, 2023.

[9] H. Zhang, H. Wang, L. Ding, *et al.*, "The analysis of magnetic emission based on inductive power transfer through metal barriers," *Computer Simulation*, vol. 37, no. 7, pp. 253–259, 2020.

[10] A. Yu, "Research on wireless energy and information cooperative transmission in metal environment," M.S. thesis, Chongqing University, 2019.

[11] Y. Xiang, "Calculation of the magnetic force between parallel coaxial current-carrying circle coils," *Journal of Chongqing University*, no. 6, pp. 51–54, 1997.

[12] H. Yang, Z. Zhang, Z. Kuang, *et al.*, "Research on piezoelectric direct current wireless power transfer system," *Audio Engineering*, vol. 40, no. 10, pp. 33–35, 2016.

[13] W. Zhong, F. Xiang, and C. Hu, "Metal object detection with detection coils perpendicular to power coils for wireless power transfer systems," *IEEE Transactions on Power Electronics*, vol. 38, no. 9, pp. 10530–10534, 2023.

[14] D. B. Geselowitz, Q. T. N. Hoang, and R. P. Gaumond, "The effects of metals on a transcutaneous energy transmission system," *IEEE Transactions on Biomedical Engineering*, vol. 39, no. 9, pp. 928–934, 1992.

[15] Y. Jiang, L. Wang, J. Fang, et al., "A joint control with variable ZVS angles for dynamic efficiency optimization in wireless power transfer system," *IEEE Transactions on Power Electronics*, vol. 35, no. 10, pp. 11064–11081, 2020.

[16] Y. Guo and N. Cui, "Research on optimal configuration and characteristics based on LCC-S type wireless power transfer system," *Transactions of China Electrotechnical Society*, vol. 34, no. 18, pp. 3723–3731, 2019.

[17] Y. Tian, Z. Zhu, J. Tian, *et al.*, "Parameters optimization of electric vehicles dynamic wireless power transfer system based on LCC-S compensation topology," *Journal of Mechanical Engineering*, vol. 57, no. 14, pp. 150–159, 2021.

[18] S. Ge, "WPT system of lithium battery based on LCC-S compensated resonant topology," *Journal of Changzhou Institute of Technology*, vol. 36, no. 2, pp. 28–32, 2023.

[19] G. Zheng, K. Zhao, H. Wang, *et al.*, "Small-signal model for inductive power transfer systems using LCC-S compensation," *Transactions of China Electrotechnical Society*, vol. 37, no. 21, pp. 5369–5376, 2022.

[20] M. Ishihara, K. Fujiki, K. Umetani, *et al.*, "Automatic active compensation method of cross-coupling in multiple-receiver resonant inductive coupling wireless power transfer systems," in *Proc. IEEE Energy Conversion Congress and Exposition (ECCE)*, IEEE, 2019.

[21] M. Ishihara, K. Fujiki, K. Umetani, *et al.*, "Autonomous system concept of multiple-receiver inductive coupling wireless power transfer for output power stabilization against cross-interference among receivers and resonance frequency tolerance," *IEEE Transactions on Industry Applications*, vol. 57, no. 4, pp. 3898–3910, 2021.

[22] M. Ishihara, K. Umetani, and E. Hiraki, "Impedance matching to maximize induced current in repeater of resonant inductive coupling wireless power transfer systems," in *Proc. IEEE Energy Conversion Congress and Exposition (ECCE)*, IEEE, 2018.

[23] R. Zhang, H. Yuan, M. Rong, *et al.*, "Self-Tuning WPT System With Constant Voltage Output Under Resonance Frequency Shift," *IEEE Transactions on Power Electronics*, vol. 39, no. 1, pp. 1713–1722, 2024.

[24] Y. Le Bihan, "Study on the transformer equivalent circuit of eddy current nondestructive evaluation," *NDT & E International*, vol. 36, no. 5, pp. 297–302, 2003.

Design of Wireless Power Transmitters for Enhanced Transmission Distance and Output Power

Kaiyuan Wang
School of Electrical and Electronic Engineering
Nanyang Technological University
Singapore
kaiyuan002@e.ntu.edu.sg

Shuang Zhao
School of Electrical Engineering and Automation
Hefei University of Technology
Hefei, Anhui, China
shuang.zhao@hfut.edu.cn

Shuye Shang
School of Engineering
University of California, Merced
Merced, CA, USA
shuyeshang@ucmerced.edu

Eric Ka-wai Cheng
School of Engineering
University of California, Merced
Merced, CA, USA
ericcheng@ucmerced.edu

Siew-Chong Tan
Department of Electrical and Electronic Engineering
University of Hong Kong
Hong Kong, China
sctan@eee.hku.hk

Yun Yang
School of Electrical and Electronic Engineering
Nanyang Technological University
Singapore
yun.yang@ntu.edu.sg

Abstract—This paper proposes a design method for the transmitter in wireless power transfer (WPT) systems that significantly enhances both output power and power transfer distance. The transmitter comprises a source resonator being compensated at the resonant frequency and an amplifier resonator being compensated above the resonant frequency. With the proposed designs, the transmission distance and the output power of typical WPT systems can be significantly improved as compared to the conventional WPT systems without the amplifier resonators and the WPT systems with ferrites. Practical results confirm that the output power of the WPT system using the proposed transmitter increases approximately 20.76 times compared to a conventional transmitter operating at the same charging distance. Moreover, with equal output power, the charging distance of the proposed transmitter can be extended by about 2.75 times compared to the conventional counterpart.

Keywords—wireless transmitter design, amplifier resonator, enhanced output power, enhanced power transfer distance

I. INTRODUCTION

Inductive-type wireless power transfer (WPT) technology is increasingly recognized as a viable alternative or complement to heavy and bulky batteries in emerging power devices, such as portable tools, robotics, and electric vehicles (EVs). Transmission distance remains a critical concern in WPT systems due to leakage flux. As the distance between two coupled coils increases, the output power of the WPT system decreases significantly. Consequently, WPT systems are typically used in applications where the transfer distance is shorter than the coil dimension [1].

To maintain sufficient power supply over relatively long galvanic gaps, various types of power amplifier coils, known as relay coils, have been proposed. In [2], a four-coil WPT system is described, where two relay coils are placed alongside the source and receiver coils. This configuration achieves approximately 40% transmission efficiency and delivers 60 W of output power at a resonance frequency of 10 MHz, with a transmission distance of 2 m. Subsequent studies have highlighted the critical role of the additional relay coils in impedance matching within four-coil WPT systems [3]-[4]. In [3], a tuning method is used to maximize power transfer efficiency. In [4], detailed explanations of impedance matching for four-coil WPT systems are provided, showing that transmission efficiency can be enhanced by adjusting the distance between the transmitter coil and its adjacent relay or the receiver coil and its corresponding relay.

Similar to [4], the distance between the relay coil and the receiver coil is adjusted to achieve the optimal mutual inductance for enhancing efficiency [5]. In [6], a formulation for power transfer efficiency based on reflected load theory and coupled-mode theory is presented. In [7], two relay coils are positioned on the transmitter side, boosting the apparent coupling coefficient at the operating frequency and achieving higher system efficiency compared to conventional four-coil systems.

It is noted in [8] that three-coil WPT systems can achieve higher output power than four-coil systems. Thus, three-coil WPT systems have been explored for use in implantable devices [9]-[11]. Comparisons between the efficiency of two-coil and three-coil WPT systems are discussed in [12]-[14]. In [12], it is shown that three-coil WPT systems are more energy-efficient than two-coil systems, as the relay coil in the three-coil system shifts current stress from the transmitter and generates a large relay current to maximize magnetic coupling with the receiver. In [13], the concept of a critical coupling coefficient is introduced as the threshold at which two- and three-coil systems achieve the same transmission efficiency. The three-coil system exhibits higher efficiency when the critical coupling coefficient is lower than this threshold. In [14], the three-coil WPT system is demonstrated to reduce current stress and electromagnetic field emissions more effectively under misalignment compared to two-coil systems. Beyond three and four-coil systems, multiple relay coils can be strategically placed between the source and receiver coils to form a domino WPT system [15].

The WPT systems discussed above typically use planar relay designs with capacitors compensated at resonant frequencies. In this paper, we analyze a system where the relay coil is compensated off the resonant frequency. The relay coil is positioned beneath the transmitting coil to avoid complexity of the receiver. Two relay coils with different winding structures are proposed, and the magnetic field distribution of the proposed transmitter design is simulated. The proposed transmitter design achieves significant improvements in both transmission distance and output power compared to traditional WPT systems.

II. DESCRIPTIONS OF THE PROPOSED DESIGN

A. Coil structure

The proposed two transmitter coil structures are depicted as shown in Fig. 1. Both transmitter coils comprise a source coil and an amplifier coil. The source coil and the amplifier coil are not connected. The source coil can be a traditional transmitter coil with the planar structure being placed on the top surface. The amplifier coil can be either planar coil or conical coil. For the planar coil as shown in Fig. 1(a), the amplifier coil is wound outside the source coil. For the conical coil as shown in Fig. 1(b), the amplifier coil is wound around the side surface of the truncated cone to form a conical structure. The three-dimensional views of the amplifier coils are shown in Fig. 2. For the planar one, r_1 and r_2 are the inner and outer radius of the coils, respectively. For the conical one, r_1 and r_2 are the radius of the upper and lower surfaces of the truncated cone. r_1 is also the outer radius of the source coil. h denotes the height of the truncated cone.

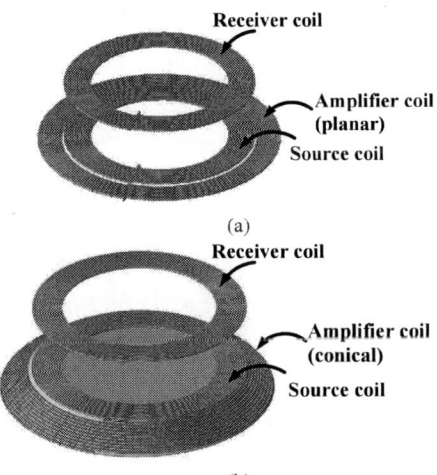

(a)

(b)

Fig. 1. Coil winding structures of the proposed WPT system with the (a) planar amplifier coil and (b) conical amplifier coil.

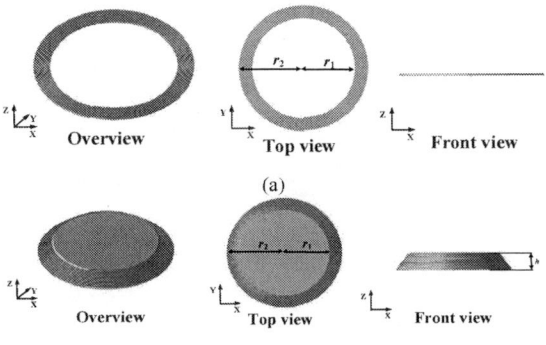

(a)

(b)

Fig. 2. Three-dimensional views of the proposed (a) planar amplifier coil and (b) conical amplifier coil.

B. Resonator circuit design and analysis

The equivalent circuit diagram of the proposed WPT system is shown in Fig. 3. The LCC-compensated topology is used for the source and receiver resonator, and the series-compensated topology is adopted for the amplifier resonator. M_1 is the mutual inductance between the source coil L_s and the receiver coil L_r. M_2 is the mutual inductance between L_s and the amplifier coil L_a. M_3 is the mutual inductance between L_r and L_a.

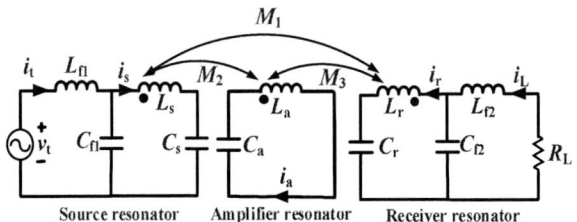

Fig. 3. Equivalent circuit of the proposed WPT system.

In the source resonator, L_s is compensated by the LCC components C_s, L_{f1}, and C_{f1}. The parameters are designed to satisfy:

$$\frac{1}{\sqrt{L_{f1}C_{f1}}} = \frac{1}{\sqrt{(L_s - L_{f1})C_s}} = \omega_o \quad (1)$$

where $\omega_o = 2\pi f_o$. f_o is the resonant frequency. v_t is the equivalent voltage source of the system. The compensation circuit design of the receiver resonator is similar to that of the source resonator. The receiver coil L_r is compensated by the LCC components C_r, L_{f2}, and C_{f2}. The parameters are designed to satisfy

$$\frac{1}{\sqrt{L_{f2}C_{f2}}} = \frac{1}{\sqrt{(L_r - L_{f2})C_r}} = \omega_o \quad (2)$$

R_L is the equivalent load resistance of the system. Based on (1) and (2), the excitation current (i.e., i_s) of the source coil can be regarded as constant, which can be expressed as

$$i_s = \frac{v_t}{j\omega_o L_{f1}} \quad (3)$$

To simplify the analysis without losing generality, L_{f1} and L_{f2} are designed to be the same as $L_{f1} = L_{f2} = L_f$. Based on the Kirchhoff's law, the load current can be derived as

$$i_L = -\frac{M_1 i_s + M_3 i_a}{L_f} \quad (4)$$

The current of the amplifier coil can be further derived as

$$i_a = \frac{\omega_o M_3 R_L}{j\omega_o L_f X_a} i_L - \frac{\omega_o M_2}{X_a} i_s \quad (5)$$

where $X_a = \omega_0 L_a - 1/(\omega_0 C_a)$ is the equivalent impedance of the amplifier resonator. By substituting (5) into (4), the root mean square (RMS) value of i_L can be derived as

$$I_L = \frac{L_f|M_1 X_a - \omega_o M_2 M_3|I_s}{\sqrt{L_f^4 X_a^2 + R_L^2 M_3^4}} \quad (6)$$

If the amplifier coil is not used, the RMS value of i_L can be derived as

$$I'_L = \frac{M_1 I_s}{L_f} \quad (7)$$

979-8-3315-1612-3/25 $31.00 © 2025 IEEE   3228

Then, the ratio (i.e., α) of the output current with and without using the amplifier resonator can be derived based on (6) and (7) as

$$\alpha = \frac{I_L}{I_L'} = \begin{cases} \dfrac{X_a - \beta}{\sqrt{\gamma + X_a^2}}, & X_a > \beta \\[3mm] \dfrac{\beta - X_a}{\sqrt{\gamma + X_a^2}}, & X_a < \beta \end{cases} \tag{8}$$

where $\beta = \omega_o M_2 M_3 / M_1$ and $\gamma = M_3^4 R_L^2 / L_f^4$. To further analyze the relationship between α and X_a, partial derivative of α with respect to X_a for $X_a > \beta$ can be derived as

$$\partial(\alpha, X_a) = \frac{(\gamma + \beta X_a)}{(\gamma + X_a^2)^{\frac{3}{2}}} > 0 \tag{9}$$

According to (9), for $X_a > \beta$, α will increase when X_a is increased. Due to $\lim\limits_{X_a \to +\infty} \alpha = 1$, α is always lower than 1 for $X_a > \beta$. Therefore, the amplifier resonator will reduce the output current as well as the output power of the system when $X_a > \beta$. Based on (8), the partial derivative of α with respect to X_a for $X_a < \beta$ can be derived as

$$\partial(\alpha, X_a) = \frac{-(\gamma + \beta X_a)}{(\gamma + X_a^2)^{\frac{3}{2}}} \tag{10}$$

According to (10), $\partial(\alpha, X_a)$ is positive for $X_a < -\gamma/\beta$ while it is negative for $X_a > -\gamma/\beta$. Due to $\lim\limits_{X_a \to -\infty} \alpha = 1$, α is always higher than 1 for $X_a < -\gamma/\beta$. Thus, the amplifier would enhance the output power of the system when $X_a < -\gamma/\beta$. By solving $\partial(\alpha, X_a) = 0$, the maximal α can be obtained when X_a satisfies

$$X_a = \frac{-\gamma}{\beta} = \frac{-M_1 M_3^3 R_L^2}{\omega_o M_2 L_f^4} \tag{11}$$

The corresponding maximal α can be obtained as

$$\alpha_{\max} = \sqrt{1 + \frac{L_f^4 M_2^2 \omega_o^2}{M_1^2 M_3^2 R_L^2}} \tag{12}$$

To investigate the relationship between the resonant frequency of the amplifier resonator and the source resonator, the equivalent impedance of the amplifier resonator can be determined as

$$X_a = \omega_o L_a \left(1 - \frac{\omega_a^2}{\omega_o^2}\right) \tag{13}$$

where ω_a is the resonant angle frequency of the amplifier resonator. According to (11), if ω_a is designed to be greater than ω_o (i.e., $\omega_a > \omega_o$), X_a is negative such that the output power of the system has the potential to be increased by including the amplifier resonator. According to (8), the output characteristics is influenced by several parameters, including load resistance and mutual inductances. To

illustrate the effective range of the amplifier, Fig. 4 shows the relationship between the output current ratio and the resonant frequency of the amplifier under different load and coupling conditions. For simplification, it is assumed that the coupling coefficient between the source and the receiver (i.e., k_1) and between the amplifier and the receiver (i.e., k_3) are the same. The coupling between the source and the amplifier (i.e., k_2) is strong, as the two coils are closely placed. Fig. 4(a) represents the WPT system with relatively strong coupling between the source and receiver. Clearly, there is a range where the ratio of the output current is greater than 1, with the maximum ratio reaching approximately 3 for $R_L = 5$ and $\omega_a/\omega_o = 1.01$. When the main coupling is reduced to 0.2, the effective range increases significantly, as shown in Fig. 4(b). Therefore, the proposed transmitter design is better suited for long-distance wireless charging applications.

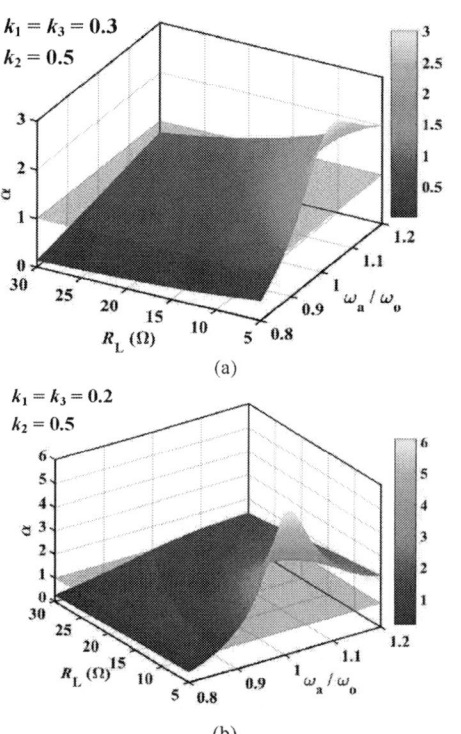

Fig. 4. Relationship between α and ω_a/ω_o under different load and coupling conditions. (a) $k_1 = 0.3$ and (b) $k_1 = 0.2$.

III. SIMULATION OF THE MAGNETIC FIELD DISTRIBUTION

Simulations are conducted using a combination of Ansys Simplorer and Maxwell 2023R1. Four models were constructed in Maxwell, as illustrated in Fig. 5. Model A depicts the conventional WPT system. I In Model B, a ferrite layer is placed beneath the source coil. Models C and D are developed using the system with two different amplifier designs: one with a conical structure and the other with a planar structure. Both amplifiers have identical dimensions and the same number of turns. The distance between the source and receiver is fixed at 8 cm, and the circuit structure shown in Fig. 3 is built in Simplorer. The RMS value of v_t is set as 5 V. The operating frequency is set as 85 kHz. The key parameters of the models in Maxwell, along with the compensated network in Simplorer, are listed in Table 1. The magnetic flux densities are illustrated in Fig. 6. Clearly, the magnetic flux densities in Models B to D are stronger than in Model A. The magnetic flux distribution in Model C is similar to that of Model A, exhibiting strong magnetic flux density at the edges of the coils. In contrast, the magnetic flux densities in Models B and D are more concentrated in the center.

TABLE I. MAIN PARAMETERS IN SIMULATION

Description	Value	Description	Value
Source and receiver coils of Model A in Maxwell			
Outer radius (cm)	10	Radius of each turn (cm)	1
Number of turns	12	Inductance (μH)	48.6
Ferrite of Model B in Maxwell			
length*width*height (mm)	300*300*2	Relative permeability	1500
Amplifier coils of Model C and D in Maxwell			
Outer radius r_2 (cm)	15	Inner radius r_1 (cm)	10
Height h (cm)	3	Number of turns	10
Radius of each turn (cm)	1	Inductance (μH)	52 (Model C) 47 (Model D)
Circuit parameters in Simplorer			
L_{f1} and L_{f2} (μH)	17.5	C_s and C_r (nF)	112.7
C_{f1} and C_{f2} (nF)	200	R_L (Ω)	2

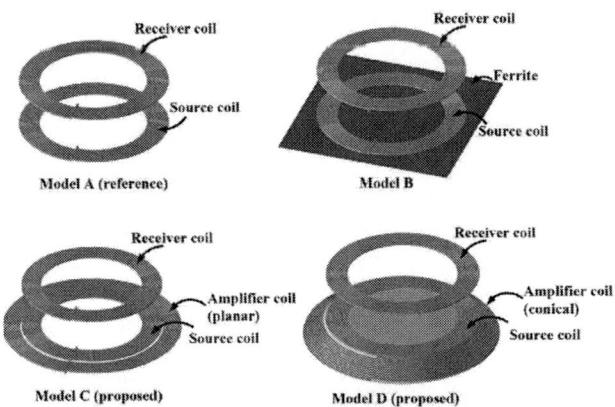

Fig. 5. Coil structures in simulation.

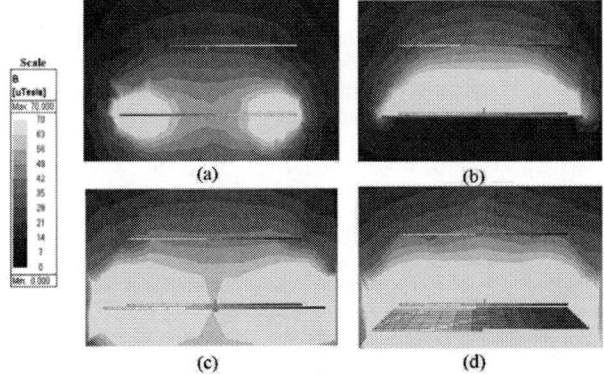

Fig. 6. Simulated magnetic field distribution for (a) Model A, (b) Model B, (c) Model C, and (d) Model D.

IV. EXPERIMENTAL VERIFICATIONS

Experiments are conducted on an LCC-LCC compensated WPT system using the invented prototypes, as shown in Fig. 7. Comparisons are made among the WPT system without amplifier coils and ferrites (i.e., Model A), the WPT system with ferrites (i.e., Model B), the WPT system with the proposed planar amplifier coil (i.e., Model C), and the WPT system with the proposed conical amplifier coil (i.e., Model D). The dimensions of the ferrites and amplifier coils are provided in Fig. 8. The ferrite material (PC 95) has a thickness of 2 mm. The source coil and the conical amplifier are mounted on a truncated cone model fabricated using 3D printing technology with polylactic acid (PLA) material. The other parameters of the WPT system are listed in Table 2. The input DC voltage is set to 5 V. The operating frequency, as well as the resonant frequencies of the source and receiver coils, are set at 85 kHz. The resonant frequencies of the amplifier resonators are 86 kHz.

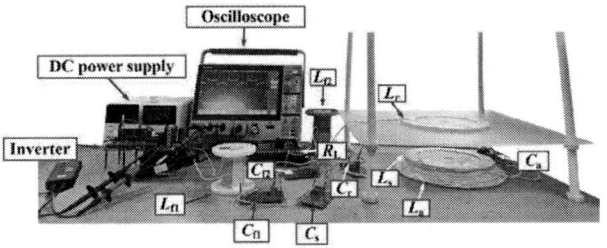

Fig. 7. Photograph of the experimental setup.

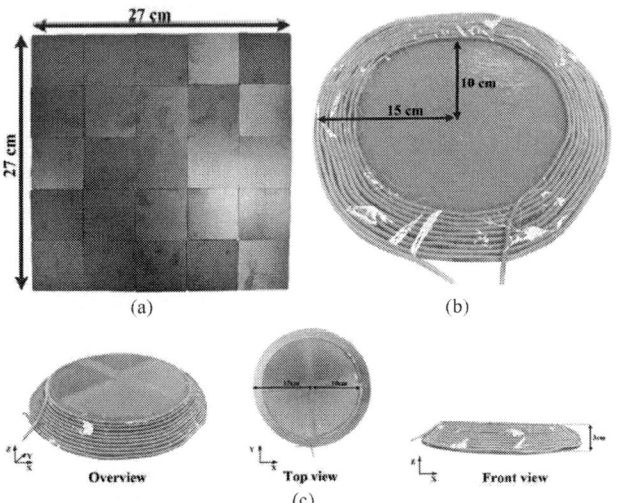

Fig. 8. Dimensions of the (a) Ferrites, (b) planar amplifier coil, and (c) conical amplifier coil used in the experiment.

TABLE II. MAIN PARAMETERS OF THE DOUBLE-SIDE LCC-COMPENSATED WPT SYSTEM IN EXPERIMENT

Symbol	Value	Symbol	Value	Symbol	Value
L_{f1} (μH)	15.3	L_{f2} (μH)	15.7	C_{f1} (nF)	230.3
C_{f2} (nF)	223.9	L_s (μH)	48.8	L_r (μH)	49.1
C_s (nF)	104.2	C_r (nF)	105.4	R_L (Ω)	1-8
L_a for model D (μH)	41.8	L_a for model C (μH)	39.2		

The comparisons of experimental output power and efficiency are shown in Figs. 9 and 10, respectively. Fig. 9 illustrates that the average output power of the WPT systems across load resistances ranging from 1 Ω to 8 Ω are 0.41 W, 0.93 W, 6.83 W, and 8.51 W. The output power of the WPT systems with the proposed transmitter coil designs incorporating amplifier coils is enhanced by approximately 16.66 times and 20.76 times compared to the conventional system without amplifier coils. As shown in Fig. 10, the use of ferrite achieves the highest efficiency among all designs.

The average efficiency across load resistances from 1 Ω to 8 Ω for Models A to D is measured as 67.4%, 76.1%, 50.4%, and 56.9%, respectively.

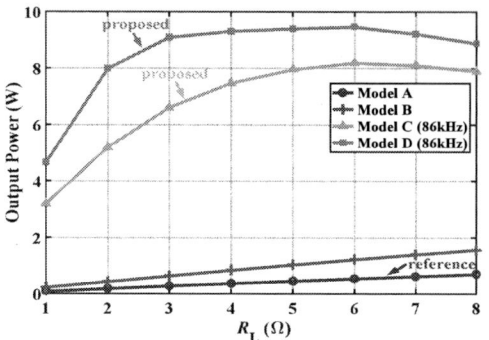

Fig. 9. Comparisons of the output power of the four models in experiment.

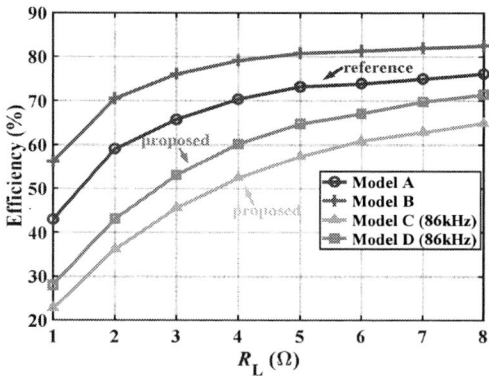

Fig. 10. Comparisons of the system efficiency (dc-dc) of the four models in experiment.

To achieve the same output power of 0.38 W with R_L = 4 Ω, the charging distance among the four models are 8 cm, 9.7 cm, 21.5 cm, and 22 cm, respectively. The comparisons are exhibited in Fig. 11. The charging distance of the proposed designs can be extended about 2.69 times and 2.75 times, respectively.

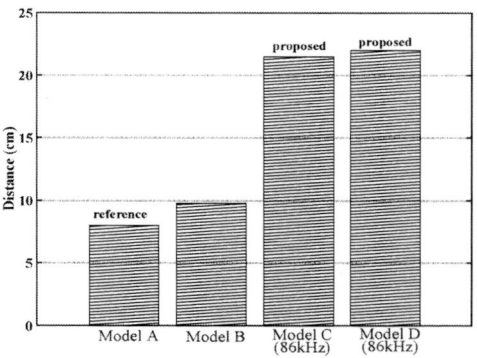

Fig. 11. Comparisons of the galvanic gaps of the WPT systems with the same output power.

V. Conclusion

This paper introduces a design method for wireless transmitters that incorporates a source resonator and an amplifier resonator with different resonant frequencies, significantly increasing power transfer distance and output power compared to conventional two-coil WPT systems. The relationship between the output current ratio and the resonant frequency of the amplifier under different load and coupling conditions is presented, and the magnetic field distribution of the proposed transmitter is simulated. The proposed transmitter design is better suited for long-distance WPT applications, as the effective power enhancement area is more significant under weak coupling conditions.

Acknowledgment

The authors would like to thank the financial supports from the A*Star MTC Young Individual Research Grant (YIRG) M23M7c0115 and the Ministry of Education (MoE) Academic Research Fund (AcRF) Tier-1 under Grant RG134/23.

References

[1] A. Setjoadi, and N. Sharpe, "Wireless power transfer using weakly coupled magnetostatic resonators," in Proc. *Energy Convers. Congr. Expo.*, 2010, pp. 4179 – 4186.

[2] A. Kurs, A. Karalis, R. Moffatt, J. D. Joannopoulos, P. Fisher, and M. Soljacic, "Wireless power transfer via strongly coupled magnetic resonances," *Science*, vol. 317, no. 5834, pp. 83-86, Jul. 2007.

[3] J. Chen, T. H. Chu, C. L. Lin, and Z. C. Jou, "A study of loosely coupled coils for wireless power transfer," *IEEE Trans. Circuits and Syst.—Part II: Express Briefs*, vol. 57, no. 7, pp. 536-540, Jul. 2010.

[4] S. Cheon, Y. H. Kim, S. Y. Kang, M. L. Lee, J. M. Lee, and T. Zyung, "Circuit-model-based analysis of a wireless energy-transfer system via coupled magnetic resonances," *IEEE Trans. Ind. Electron.*, vol. 58, no. 7, pp. 2906-2914, Jul. 2011.

[5] K. Wang, Y. Yang and E. K. -W. Cheng, "Exploration of four coils magnetically coupled resonant inductive power transfer system with efficiency optimization under unsymmetrical power wiring structure," in *Annual Conference of the IEEE Industrial Electronics Society (IECON)*, pp. 1-5, 2023.

[6] M. Kiani and M. Ghovanloo, "The circuit theory behind coupled-mode magnetic resonance-based wireless power transmission," *IEEE Trans. Circuits Syst.—Part I*, vol. 59, no. 8, pp. 1-10, Aug. 2012.

[7] S. Moon and G. -W. Moon, "Wireless power transfer system with an asymmetric four-coil resonator for electric vehicle battery chargers," *IEEE Trans. Power Electron.*, vol. 31, no. 10, pp. 6844-6854, Oct. 2016.

[8] M. Kiani, U. -M. Jow, and M. Ghovanloo, "Design and optimization of a 3-coil inductive link for efficient wireless power transmission," *IEEE Trans. Biomed. Circuits Syst.*, vol. 5, no. 6, pp. 579-591, Dec. 2011.

[9] S. H. Kang, J. H. Choi, F. J. Harackiewicz, and C. W. Jung, "Magnetic resonant three-coil WPT system between off/in-body for remote energy harvest," *IEEE Microw. Wireless Compon. Lett.*, vol. 26, no. 9, pp. 741-743, Sep. 2016

[10] P. J. Abatti, C. M. de Miranda, M. A. P. da Silva, and S. F. Pichorim, "Analysis and optimization of three-coil wireless power transfer systems," *IET Power Electron.*, vol. 11, no. 1, pp. 68-72, 2018.

[11] A. K. RamRakhyani and G. Lazzi, "On the design of efficient multi-coil telemetry system for biomedical implants," *IEEE Trans. Biomed. Circuits Syst.*, vol. 7, no. 1, pp. 11–23, Feb. 2013.

[12] W. X. Zhong, C. Zhang, X. Liu, and S. Y. R. Hui, "A methodology for making a three-coil wireless power transfer system more energy efficient than a two-coil counterpart for extended transfer distance," *IEEE Trans. Power Electron.*, vol. 30, no. 2, pp. 933–942, Feb. 2015.

[13] D. W. Seo, "Comparative analysis of two- and three-coil WPT systems based on transmission efficiency," *IEEE Access*, vol. 7, pp. 151962-151970, 2019.

[14] J. Zhang, X. Yuan, C. Wang and Y. He, "Comparative analysis of two-coil and three-coil structures for wireless power transfer," *IEEE Trans. Power Electron.*, vol. 32, no. 1, pp. 341-352, Jan. 2017.

[15] X. Hou, Y. Su, Z. Zuo, X. Dai, and Y. Fei, "A novel analysis method based on quadratic eigenvalue problem for multirelay magnetic coupling wireless power transfer," *IEEE Trans. Power Electron.*, vol. 36, no. 9, pp. 9907-9917, Sept. 2021.

Optimization of Wireless Power Transfer Waveforms and In-Vivo Receivers for Implantable Medical Devices

Hanbing Liu
Department of Electrical and Computer Engineering
University of Maryland
College Park, USA
Email: lhb42@umd.edu

Xin Zan
Department of Electrical and Computer Engineering
University of Maryland
College Park, USA
Email: xinzan@umd.edu

Abstract—**Providing high power levels of several watts, achieving high-efficiency in-vivo receivers, and reaching implantable size simultaneously within the specific absorption rate (SAR) and heat dissipation limits are challenging for wireless power transfer (WPT) in implantable medical devices (IMDs). To overcome these challenges, waveforms and receivers together with operating frequencies should be optimized. In this paper, we propose a solution that optimizes WPT waveforms with different peak-to-average power ratios for different loads, controlled by duty cycles, quality factors, and phase shifts between the transmitter and receiver. With a diameter of 65 mm, three-turn, four-layer PCB coils operating at 6.78 MHz, and a coupling coefficient of 0.255, the optimized system can achieve a maximum power of 5.9 W at a load of 20 Ω, and 6.4 W at a load of 25 Ω under SAR and heat dissipation limits, reaching the end-to-end efficiency of 76.4% and 77.4%, respectively. Compared to conventional operating conditions, the optimized system increases the power level by 19.6% and 7.0% at load resistances of 20 Ω and 25 Ω, respectively.**

Index Terms—**wireless power transfer, implantable medical devices, waveform and rectifier optimization, peak-to-average power ratio, SAR, heat dissipation, low quality factor, current-mode class D**

I. INTRODUCTION

Wireless power transfer (WPT) is a promising solution for wirelessly powering implantable medical devices (IMDs), as it eliminates the need of percutaneous power wires or the need for additional surgeries to replace in-vivo batteries, thereby reducing further risks to patients. While the devices are implanted in the human body, safety considerations always come first. Specific absorption rate (SAR) and heat dissipation caused by power transfer between transmitter and receiver coils restrict the power levels. To satisfy the power levels of some high power IMDs under safety regulations, higher power and higher efficiency can be achieved by optimizing WPT waveforms and in-vivo receivers together with operating frequencies in WPT systems.

Different IMDs require different power levels and sizes [1]–[8]. Devices like neurostimulators, cardiac pacemakers, and capsule endoscopy typically need less than 100 mW [8], allowing for small implant coils and diverse circuit topologies

[1]–[4]. A transferred power of 170 mW can be achieved by using an Rx coil with a diameter of 35 mm with a Class-E inverter and a Class-DE rectifier [1]. A dual-band EnerCage system with two multi-coil inductive links is designed to power the small IMD ($2.5 \times 2.5 \times 1.5$ mm^3) serving as the receiver at the power level of 2.7 mW [2]. A receiver coil with a diameter of 1 mm enables power transfer efficiency and power delivered to the load at 2.4% and 1.3 mW with the coil segmentation method [3]. By using a current-mode class D (CMCD) inverter and passive full-bridge rectifiers with the Rx coils having a diameter of 1 cm, 5.4 mW is delivered to each receiver that is arbitrarily distributed over a large area [4].

In contrast, devices like implantable pulse generators (IPGs), left ventricular assist devices (LVADs), and artificial pancreas systems demand higher power, up to several watts. Based on a generic transcutaneous transformer model, a remote power supply using a resonant topology for artificial hearts is analyzed and designed for easy controllability and high efficiency with an operating frequency range of 250-320 kHz [5]. Although the system can transfer 12-60 W power, the receiver of the system may be large for the implant. To achieve high efficiency, a 0.8 MHz self-driven synchronous rectifier circuit with minimized volume is developed, which can transfer 30 W power with an efficiency above 95% at a coil separation distance of 20 mm and a coil diameter of 70 mm [6]. A distinct sandwiched structure is used in both the transmitter and receiver coils to increase the coupling coefficient, thus improving the output power up to 5 W at 160 kHz [7]. While some high power wireless systems exist and perform well in terms of power and efficiency [5]–[7], all of them do not adequately address SAR and heat dissipation requirements, which are critical for IMDs. Some of them are too large for implantation owing to relatively low operating frequencies of several hundred kHz, compared to miniaturized systems operating at tens of MHz [9].

In this paper, we investigate a method to optimize the WPT waveforms and in-vivo receivers for IMDs with different load resistances to push the power to several watts under the limits of SAR and heat dissipation. To illustrate the method,

979-8-3315-1612-3/25 $31.00 © 2025 IEEE

CMCD topology shown in Fig. 1(a) is used as both transmitter and receiver for an example, because it can easily reach zero-voltage switching (ZVS) at a wide resistive load range and does not have high-side switches [10]. The symmetric CMCD WPT system is analyzed under both high and low quality factors on the receiver side with high quality factors on the transmitter side. Different waveforms across WPT coils are also achieved by varying waveforms' duty cycles and phase shifts [11] between the transmitter and receiver together with Q_p and Q_s, as shown in Fig. 1(b), resulting in efficiency and power improvement. The operating frequency is selected as 6.78 MHz to have a tradeoff balance between the in-vivo coil size and the power level under the SAR and heat dissipation limits. Our system tends to get the highest output power for a given resistive load under the SAR and heat dissipation limits. The optimized system achieves 5.9 W output power with 76.4 % end-to-end efficiency at 20 Ω load resistance with a receiver duty cycle of 0.6; 6.4 W output power with 77.4 % end-to-end efficiency at 25 Ω load with a receiver duty cycle of 0.8. Compared to conventional operation with the receiver duty cycle of 1, the optimized system increases the output power by 19.6 % and 7.0 % at the loads of 20 Ω and 25 Ω, respectively.

II. WAVEFORM AND RECEIVER OPTIMIZATION

To illustrate the optimization of waveforms, a CMCD rectifier is used as an example for the in-vivo receiver. To get the highest output power of the WPT system under SAR, heat dissipation, and size limits, the voltage across the secondary side coil $v_s(t)$ and output load voltage V_{out} are derived under different quality factors and duty cycles. We also optimize the phase shifts between the transmitter and receiver, namely ϕ_1 and ϕ_3. The selection of the operating frequency is also considered.

A. Theoretical Derivations of Receiver Coil Voltage

1) Any Quality Factor Operation: Low quality factor operation results in a low VA rating for devices and components and a reduced circulating current in the resonant tank, thus, decreasing heat dissipation and enhancing system efficiency. We present a general method to derive the secondary coil voltage $v_s(t)$ and output voltage V_{out}, applicable to both high quality factors Q_s and low Q_s operation of the CMCD rectifier on the receiver side.

Under the assumption of high quality factors Q_p on the primary side, the different harmonics of the primary coil voltage are all in phase. The primary side and WPT coils of the symmetric CMCD WPT system can be reflected to the secondary side as a superposition of several voltage sources [10], [12] $v_1, v_3, v_5...$, where v_1, v_3, and v_5 are the 1st, 3rd, and 5th harmonics of the equivalent voltage input to the receiver, and an effective inductor $L_{eff,s}$, as shown in Fig. 2(a) [12]. $L_{eff,s} = (1 - k^2)L_s$, where k is the coupling coefficient and L_s is the secondary coil inductance. Using Norton's Theorem, the voltage source equivalent system can be converted into the

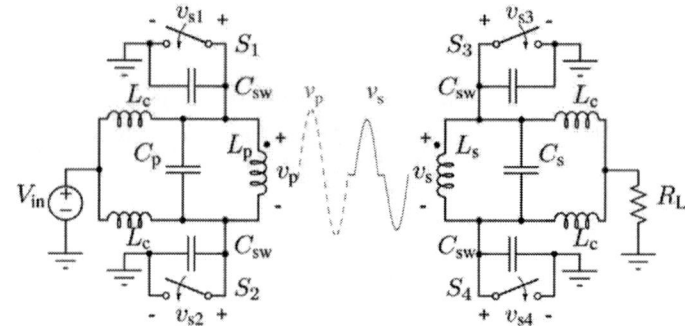

(a) A symmetric CMCD WPT system is used to demonstrate the optimization of WPT waveforms and in-vivo receivers.

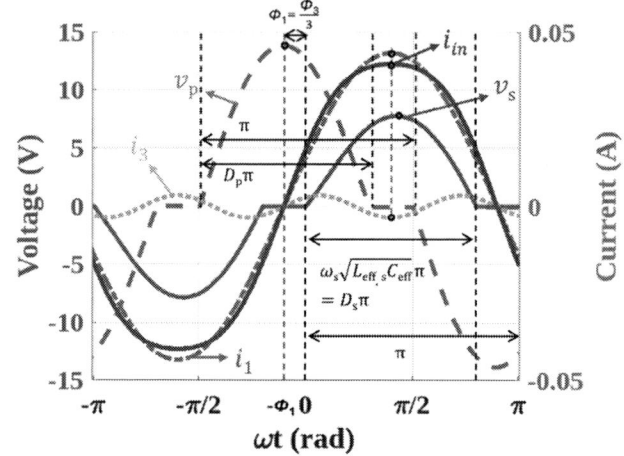

(b) Phase shift $\phi_1 = \frac{\phi_3}{3}$ is between the primary coil voltage v_p and secondary coil voltage v_s. i_{in} is the equivalent input current source of the receiver side, i_1 is the fundamental component of i_{in}, and i_3 is the third harmonic of i_{in}. The peaks are represented by black circles.

Fig. 1. The symmetric CMCD WPT system and its optimized operation for IMDs.

current source equivalent system, as shown in Fig. 2(b). The relationship between i_1, i_3, i_5 and v_1, v_3, v_5 can be given as:

$$I_n = \frac{V_n}{jn\omega L_{eff,s}}, \quad n = 1, 3, 5... \tag{1}$$

where I_n and V_n are the phasors of i_n and v_n, respectively. i_1, i_3, and i_5 are the 1st, 3rd, and 5th harmonics of the equivalent current input to the receiver, i_{in}, as shown in Fig. 2(b).

When S_3 is off, S_4 is on, using KCL on node x in Fig. 2(b),

$$i_{in} + i_L = i_C + i_{Csw} + I_1, \tag{2}$$

where i_L is the current in the resonant inductor $L_{eff,s}$, i_C is the current in the resonant capacitor C_s, i_{Csw} is the current in the parasitic switch capacitor C_{sw} of the switch S_3, and I_1 is the current in the choke inductor L_c. With large L_c, I_1 is constant. Plug $i_C = C_s \frac{dv_s}{dt}$ and $i_{Csw} = C_{sw} \frac{dv_s}{dt}$ into (2), where v_s is also the voltage across the receiver resonant tank as shown in Fig. 2(b), the following equation can be derived,

$$(C_s + C_{sw}) \frac{dv_s}{dt} = i_{in} + i_L - I_1. \tag{3}$$

979-8-3315-1612-3/25 $31.00 © 2025 IEEE   3233

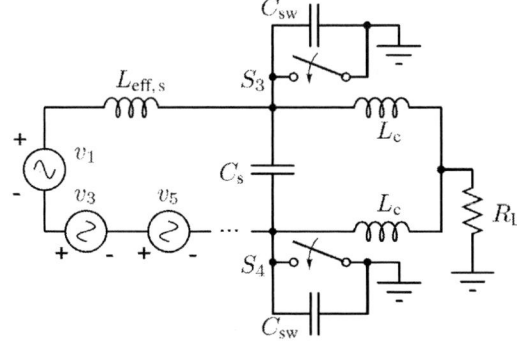

(a) The voltage source equivalent circuit of the in-vivo receiver.

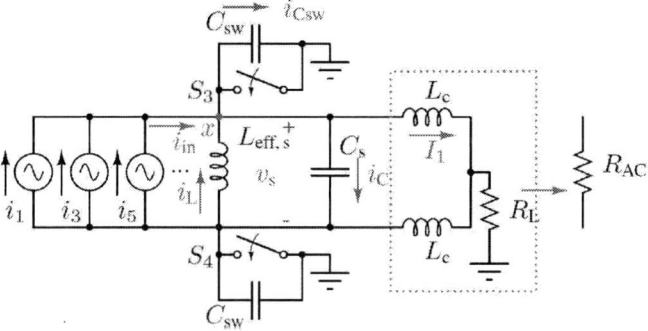

(b) The current source equivalent circuit of the in-vivo receiver.

Fig. 2. The equivalent circuits of the in-vivo receiver. (a) The primary side and WPT coils are equivalently represented as the superposition of several voltage sources, v_1, v_3, v_5... in series with the effective inductor $L_{\mathrm{eff,s}}$ [10], [12]. (b) The primary side and WPT coils are equivalently represented as the superposition of several current sources, i_1, i_3, i_5... in parallel with the effective inductor $L_{\mathrm{eff,s}}$. Two chokes L_c together with the output resistive load R_L are equivalent to a resistive load R_{AC} using power balance.

C_s and C_{sw} can be combined as an effective capacitor C_{eff}. In order to get the expression of v_s, Laplace transformation of (3) is

$$V_s(s) = \frac{I_{\mathrm{in}}(s) + I_L(s) - I_1(s)}{C_{\mathrm{eff}}s} + \frac{v_0(t=0)}{s}. \quad (4)$$

Since the initial starting point is set as shown in Fig. 1(b), $v_s(0) = 0$ owing to ZVS switching and CMCD rectifier operation.

To simplify the derivation, only the 1st and 3rd harmonics are considered, which is reasonable and leads to a small deviation compared to the analysis with full harmonics since the amplitudes of the higher harmonics are relatively small compared to the amplitudes of the 1st and 3rd harmonics. Thus,

$$i_{\mathrm{in}} = \frac{ka}{\omega_s L_{\mathrm{eff,s}}} \sin\left(\omega_s t + \phi_1\right) + \frac{kb}{3\omega_s L_{\mathrm{eff,s}}} \sin\left(3\omega_s t + \phi_3\right), \quad (5)$$

where ω_s is the switching frequency, a and b are the peak values of the 1st and 3rd harmonics of the primary coil voltage v_p and they are linearly proportional to input voltage V_{in}. The duty cycle of the primary side D_p determines the ratio between a and b [12]. When $D_p = 1$, $b = 0$. ϕ_1 is the phase shift between i_1 and v_s, as shown in Fig. 1(b), which is also the phase shift between the peak value of v_p and the start rising

point of v_s when Q_p is high. Under high Q_p assumption, $\phi_3 = 3\phi_1$ and i_1, i_3, and i_{in} all have zero cross points at $\omega t = -\phi_1 = -\frac{\phi_3}{3}$. The relationship between v_s and i_L is given by:

$$v_s = L_{\mathrm{eff,s}} \frac{di_L}{dt}. \quad (6)$$

Do Laplace transformation of (5), (6), and I_1, and plug them into (4), we have

$$V_s(s) = \left\{ \frac{ka}{\omega_s L_{\mathrm{eff,s}}} \frac{s\sin\phi_1 + \omega_s\cos\phi_1}{s(s^2 + \omega_s^2)C_{\mathrm{eff}}} \right.$$
$$+ \frac{kb}{3\omega_s L_{\mathrm{eff,s}}} \frac{s\sin\phi_3 + 3\omega_s\cos\phi_3}{s[s^2 + (3\omega_s)^2]C_{\mathrm{eff}}}$$
$$\left. + \frac{i_L(0) - I_1}{s^2 C_{\mathrm{eff}}} \right\} \Big/ \left(1 + \frac{1}{s^2 C_{\mathrm{eff}} L_{\mathrm{eff,s}}}\right). \quad (7)$$

After doing a reverse Laplace transformation for each term and simplifying them by defining $D_s = \sqrt{L_{\mathrm{eff,s}} C_{\mathrm{eff}}}\omega_s$, we have

$$v_s = A + B + (C + D)\sin\left(\frac{\omega_s t}{D_s}\right) - E\cos\left(\frac{\omega_s t}{D_s}\right)$$

$$A = \frac{ka}{1 - D_s^2}\cos(\omega_s t + \phi_1);$$

$$B = \frac{kb}{1 - (3D_s)^2}\cos(3\omega_s t + \phi_3);$$

$$C = \frac{k}{D_s}\left\{\frac{a\sin\phi_1}{1 - D_s^2} + \frac{b\sin\phi_3}{3[1 - (3D_s)^2]}\right\}; \quad (8)$$

$$D = \sqrt{\frac{L_{\mathrm{eff,s}}}{C_{\mathrm{eff}}}}(i_L(0) - I_1);$$

$$E = \left\{\frac{ka\cos\phi_1}{1 - D_s^2} + \frac{kb\cos\phi_3}{1 - (3D_s)^2}\right\}.$$

Two equations are needed to calculate the two unknowns $i_L(0)$ and I_1. The first equation uses the fact that the average voltage across L_c is zero:

$$\frac{\int_0^{\sqrt{L_{\mathrm{eff,s}} C_{\mathrm{eff}}}\pi} v_s dt}{2\frac{\pi}{\omega_s}} = V_{\mathrm{out}} = 2I_1 R_L, \quad (9)$$

where V_{out} is the dc output voltage across the output resistive load R_L. The second equation uses the fact that the peak energy stored in the capacitor and inductor, as shown in Fig. 3, are the same:

$$\frac{1}{2}L_{\mathrm{eff,s}} i_L(0)^2 = \frac{1}{2}C_{\mathrm{eff}} V_{s,\mathrm{pk}}^2, \quad (10)$$

where $i_L(0)$ is the current in the effective inductor $L_{\mathrm{eff,s}}$ when $\omega t = 0$ while $V_{s,\mathrm{pk}}$ is the peak voltage across the resonant capacitor C_s. This equation is strictly valid when Q_s is high. When Q_s decreases, the energy stored in the two chokes are different at inductor peak energy point and capacitor peak energy point, which will cause a small difference between the peak energy in the capacitor and inductor. However, (10) can still serve as a reasonable approximation, which has been verified in simulation, because I_1 is assumed to be constant without any ripples.

979-8-3315-1612-3/25 $31.00 © 2025 IEEE

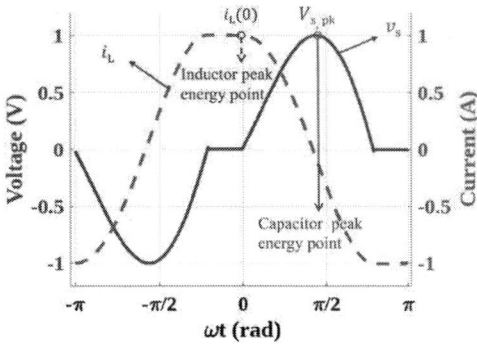

Fig. 3. Times when energy stored in the resonant capacitor and coil inductor reach the peaks. v_s is the secondary coil voltage. i_L is the secondary coil current. Waveforms come from the LTspice simulation when $Q_s = 3.86$.

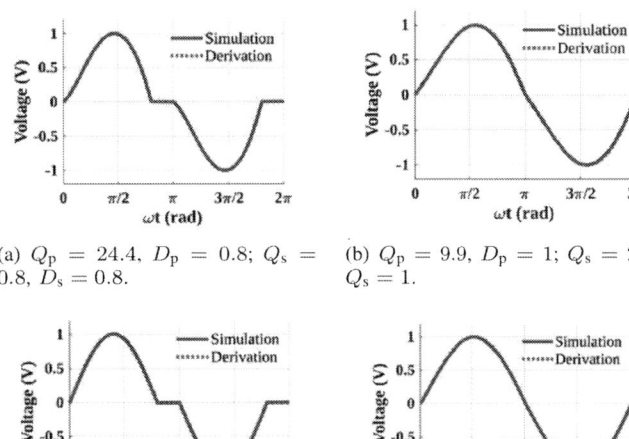

(a) $Q_p = 24.4$, $D_p = 0.8$; $Q_s = 0.8$, $D_s = 0.8$.

(b) $Q_p = 9.9$, $D_p = 1$; $Q_s = 2.4$, $Q_s = 1$.

(c) $Q_p = 8.4$, $D_p = 0.8$; $Q_s = 11.8$, $D_s = 0.8$.

(d) $Q_p = 10.3$, $D_p = 1$; $Q_s = 9.6$, $D_s = 1$.

Fig. 4. Secondary coil voltages v_s under different quality factors and duty cycles. v_s with relatively low Q_s in Figs. 4(a) and 4(b) verifies the theoretical derivation in (8) through LTspice simulation. v_s with relatively high Q_s in Figs. 4(c) and 4(d) verifies the theoretical derivation in (16) through LTspice simulation. Q_p and Q_s are the primary side and secondary side quality factors, respectively. D_p and D_s are the primary side and secondary side duty cycles, respectively. Voltages are normalized to the corresponding peaks.

To find the $V_{s,pk}$, let $\frac{dv_s}{dt} = 0$. There should be a ξ in $(0, \sqrt{L_{eff,s}C_{eff}}\pi)$, which makes $\frac{dv_s(\xi)}{dt} = 0$. So, $V_{s,pk} = v_s(\xi)$. Two equations with the two unknowns are formulated, which can be solved numerically. Thus, the v_s can be expressed analytically.

It should be noted that when $t = 0$ and $t = \sqrt{L_{eff,s}C_{eff}}\pi$, v_s in (8) is always zero as long as

$$\phi_1 = \frac{1 - \sqrt{L_{eff,s}C_{eff}}\omega_s}{2}\pi \text{ and } \phi_3 = 3\frac{1 - \sqrt{L_{eff,s}C_{eff}}\omega_s}{2}\pi. \tag{11}$$

In other words, ZVS can always be achieved at these points regardless of the loads R_L.

As shown in Fig. 1(b), the waveform v_s is positive during $t = 0$ to $t = \sqrt{L_{eff,s}C_{eff}}\pi$, and the length of half cycle is $\frac{\pi}{\omega_s}$. Thus, the duty cycle of the secondary side, namely D_s, is

defined as

$$D_s = \frac{\sqrt{L_{eff,s}C_{eff}}\pi}{\frac{\pi}{\omega_s}} = \omega_s\sqrt{L_{eff,s}C_{eff}}, \tag{12}$$

namely, $\omega_s^2 L_{eff,s}C_{eff} = D_s^2$. With the defination of D_s, (11) can be expressed as:

$$\phi_1 = \frac{1 - D_s}{2}\pi \text{ and } \phi_3 = 3\frac{1 - D_s}{2}\pi. \tag{13}$$

Analytical derivation of v_s with ZVS under low Q_s in (8) is verified through LTspice simulation, as shown in Figs. 4(a) and 4(b). The expression of Q_s is defined in Appendix. A, which is independent of D_s.

For the conventional operation where $D_p = D_s = 1$, which means that the two switches on each side operate complementarily, with each switch turning on for 50% of the switching period. In this case, with $\phi_1 = \frac{1-D_s}{2}\pi$ in (13) and $b = 0$, (8) can be simplified by calculating the limit when $D_s \to 1$, resulting in

$$v_s = \left[\frac{ka\omega_s t}{2} + \sqrt{\frac{L_{eff,s}}{C_{eff}}}(i_L(0) - I_1)\right]\sin(\omega_s t). \tag{14}$$

2) High Quality Factor Operation: When the CMCD rectifier operates at a high Q_s, the derivation is simplified as a special case of any quality factor as derived above. Assuming the the receiver coil voltage v_s is purely sinusoidal, which means the peak of v_s happens at exactly the middle of the half cycle, namely at $\xi = \sqrt{L_{eff,s}C_{eff}}\pi/2 = \frac{D_s\pi}{2\omega_s}$. Plug ξ into (8),

$$V_{s,pk} = \frac{k}{D_s}\left\{\frac{a\sin\phi_1}{1-D_s^2} + \frac{b\sin\phi_3}{3[1-(3D_s)^2]}\right\} + \sqrt{\frac{L_{eff,s}}{C_{eff}}}(i_L(0) - I_1). \tag{15}$$

With high Q_s assumption, the two unknowns in (9) and (10) can be calculated as $I_1 = \frac{V_{out}}{2R_L} = \frac{V_{s,pk}D_s}{2\pi R_L}$ and $i_L(0) = V_{s,pk}\sqrt{\frac{C_{eff}}{L_{eff,s}}}$. Then, $v_s(t)$ in (8) is

$$v_s = A + B + F\sin\left(\frac{\omega_s t}{D_s}\right) - E\cos\left(\frac{\omega_s t}{D_s}\right)$$

A, B, and E are from (8); $\tag{16}$

$$F = \frac{2\pi R_L}{\omega_s D_s L_{eff,s}}\left\{\frac{a\sin\phi_1}{1-D_s^2} + \frac{b\sin\phi_3}{3[1-(3D_s)^2]}\right\},$$

and

$$V_{out} = \frac{2R_L k}{\omega_s}\frac{1}{L_{eff,s}}\left\{\frac{a\sin\phi_1}{1-D_s^2} + \frac{b\sin\phi_3}{3[1-(3D_s)^2]}\right\}. \tag{17}$$

Intuitively, from (16), when R_L and corresponding Q_s increases, the coefficient F of $\sin(\frac{\omega_s t}{D_s})$ will dominate v_s and v_s will become more sinusoidal, validating the high Q_s assumption. Analytical derivation of v_s with high Q_s in (16) is verified through LTspice simulation, as shown in Figs. 4(c) and 4(d).

For the conventional operation with $D_p = D_s = 1$ and high Q_s, LC tank is tuned at $\omega_s = \frac{1}{\sqrt{L_{eff,s}C_{eff}}}$ and $F'\sin(\frac{\omega_s t}{D_s})$ dominates in (16). In this case, with $\phi_1 = \frac{1-D_s}{2}\pi$ in (13), and

979-8-3315-1612-3/25 $31.00 © 2025 IEEE

$b = 0$, (16) can be simplified by calculating the limit when $D_s \to 1$, resulting in

$$v_s = \frac{ka\pi^2 R_L \sin(\omega_s t)}{2\omega_s L_{\text{eff, s}}}. \tag{18}$$

Thus, $V_{s,\text{pk}} = \frac{ka\pi^2 R_L}{2\omega_s L_{\text{eff, s}}}$ and $V_{\text{out}} = \frac{V_{s,\text{pk}}}{\pi} = \frac{ka\pi R_L}{2\omega_s L_{\text{eff, s}}}$. These results match the derivations of conventional operation in [10].

B. Optimization of Phase Shift

To achieve the highest V_{out} across a given R_L and v_p, thus highest output power P_{out}, ϕ_1 and ϕ_3 in Section II-A needs to be optimized. The optimization is performed under the high Q_s assumption.

The current source equivalent model in Fig. 2(b) is redrawn in Fig. 5, the current entering the tank i_{tank} is the segmented i_{in} during the green region

$$i_{\text{tank}} = \begin{cases} i_{\text{in}}, & t \in [0, \frac{\pi D_s}{\omega_s}] \cup [\frac{\pi}{\omega_s}, \frac{\pi}{\omega_s} + \frac{\pi D_s}{\omega_s}]. \\ 0, & t \in [\frac{\pi D_s}{\omega_s}, \frac{\pi}{\omega_s}] \cup [\frac{\pi}{\omega_s} + \frac{\pi D_s}{\omega_s}, \frac{2\pi}{\omega_s}]. \end{cases} \tag{19}$$

When the two switches are on, the secondary coil voltage v_s is zero, as shown in the grey region in Fig. 5. During this period, i_{in} flows through the two switches without entering the resonant tank, so the energy stored in the inductor and capacitor remains unchanged when $v_s = 0$.

The segments of i_{tank} and v_s when $v_s = 0$ are eliminated just for this analysis, resulting in pseudo i'_{tank} and v'_s. Since the resonant tank is tuned at $\frac{\omega_s}{D_s} = \frac{1}{\sqrt{L_{\text{eff, s}} C_{\text{eff}}}}$, the pseudo current $i'_{\text{tank, 1st}}$ in R_{AC}, namely $\frac{v'_s}{R_{\text{AC}}}$, is purely sinusoidal with a pseudo frequency of $\frac{\omega_s}{D_s}$ when Q_s is large, whose fundamental component amplitude $I'_{\text{tank, 1st}}$ can be represented by the fundamental harmonic of i'_{tank},

$$I'_{\text{tank, 1st}} = \frac{ka}{\pi D_s L_{\text{eff, s}}} \left[\sin(D_s \pi + \phi_1) + \sin \phi_1 \right] \left[\frac{2 D_s}{\omega_s(1 - D_s^2)} \right]$$
$$+ \frac{kb}{3\pi D_s L_{\text{eff, s}}} \left[\sin(3 D_s \pi + \phi_3) + \sin \phi_3 \right] \left\{ \frac{2 D_s}{\omega_s[1 - (3 D_s)^2]} \right\}. \tag{20}$$

When ϕ_1 and ϕ_3 are tuned to make the peak of v'_s aligned with the peak of i'_{tank}, which is the same phase shift to guarantee ZVS operation in (13), we have $\phi_1 = \frac{1-D_s}{2}\pi$ and $\phi_3 = 3\frac{1-D_s}{2}\pi$. Assuming ϕ_1 and ϕ_3 are both shifted by α, ϕ_1 and ϕ_3 become:

$$\begin{cases} \phi_1 = \frac{1-D_s}{2}\pi + \alpha, \\ \phi_3 = 3\frac{1-D_s}{2}\pi + \alpha. \end{cases} \tag{21}$$

Plug (21) into (20)

$$I'_{\text{tank, 1st}} = \frac{ka}{\pi D_s L_{\text{eff, s}}} \left[2\cos\left(\frac{\pi D_s}{2}\right) \cos \alpha \right] \left[\frac{2 D_s}{\omega_s(1 - D_s^2)} \right]$$
$$+ \frac{kb}{3\pi D_s L_{\text{eff, s}}} \left[2\cos\left(\frac{3\pi D_s}{2}\right) \cos \alpha \right] \left\{ \frac{2 D_s}{\omega_s[1 - (3 D_s)^2]} \right\}, \tag{22}$$

which reaches the maximum point when $\cos \alpha = 1$. Thus, $\alpha = 0$.

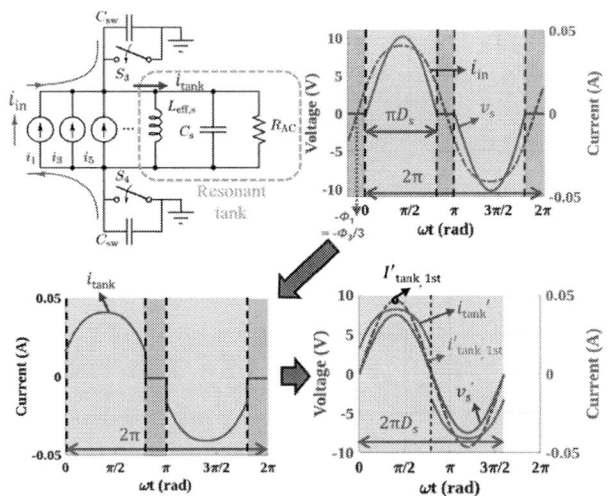

Fig. 5. Segmented waveforms for phase shift optimization. The current source equivalent circuit of the in-vivo receiver with an equivalent resistive load R_{AC}. The waveforms of the input current i_{in} in the current source equivalent in-vivo receiver, the current i_{tank} flow into the resonant tank, and the pseudo tank input current i'_{tank} and the fundamental component of pseudo output current $i'_{\text{tank, 1st}}$ in R_{AC} for analysis.

Considering power balance, we can calculate the equivalent $R_{\text{AC}} = \frac{\pi^2 R_L}{2 D_s}$. Then, the output voltage can be expressed as:

$$V_{\text{out}} = V_{s,\text{pk}} \frac{D_s}{\pi} = I'_{\text{tank, 1st}} R_{\text{AC}} \frac{D_s}{\pi}$$
$$= \frac{2 R_L k}{\omega_s} \frac{1}{L_{\text{eff, s}}} \left\{ \frac{a \sin(\frac{1-D_s}{2}\pi)}{1 - D_s^2} + \frac{b \sin(3\frac{1-D_s}{2}\pi)}{3[1 - (3 D_s)^2]} \right\}. \tag{23}$$

As shown above, V_{out} reaches the maximum value when ϕ_1 and ϕ_3 are tuned to make the peak of v_s aligned with the peak of i_{in}, which is also aligned with the peaks of i_{tank} and $i_{\text{tank, 1st}}$, namely $\phi_1 = \frac{1-D_s}{2}\pi$ and $\phi_3 = 3\frac{1-D_s}{2}\pi$. So, V_{out} in (17) can be expressed as the same as (23).

When $Q_s = 2.4$, which is the lowest Q_s in this paper, ϕ_1 and ϕ_3 in (13) can still be regarded as optimal based on LTspice simulation.

C. SAR and Power Dissipation Analysis

Safety is the top priority when designing implantable medical devices. By using the integral form of Faraday's Law and the relationship between magnetic flux ϕ_B and inductor coil current i_L, the relationship between electric field E and coil voltage v_s can be derived as: $\oint_{\partial \Sigma} \mathbf{E} \cdot d\mathbf{l} = -L_s \frac{di_L}{dt} = -v_s$, where $\partial \Sigma$ is a closed loop along the edge of the coil with a diameter of d. $|\mathbf{E}|$ is the same on the close loop. Consider the root mean square (RMS) value of v_s and E, the integral form can be reduced to $V_{s,\text{rms}}^2 = E_{\text{rms}}^2 \cdot (\pi d)^2$. Given the SAR definition [13]–[15],

$$\text{SAR} = \frac{\sigma_{\text{skin}} |E_{\text{rms}}|^2}{\rho_{\text{skin}}}, \tag{24}$$

where σ_{skin} is the electrical conductivity of the skin, ρ_{skin} is the density of the skin. To calculate SAR along the edge of

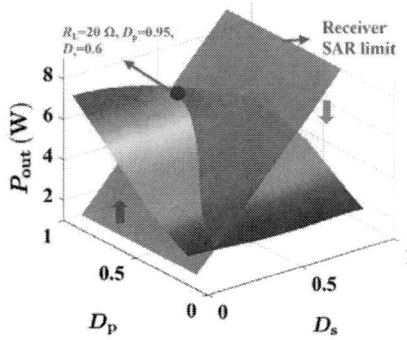

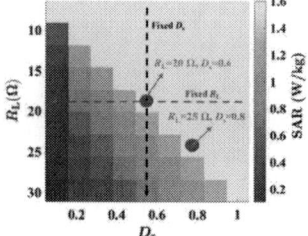
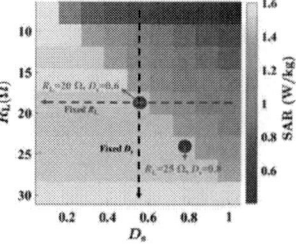

(a) SAR values of transmitters. (b) SAR values of receivers.

Fig. 8. SAR values of transmitters and receivers for each operating point in Fig. 7. The black line and red line indicates the optimal point with $R_L = 20\,\Omega$ and $D_s = 0.6$.

Fig. 6. The 3D plot of P_{out} as a function of the primary side duty cycle D_p and the secondary side duty cycle D_s for a fixed transmitter SAR of 1.6 W/kg, without considering a receiver SAR limit. The system operates with $Q_p = 6.9$, $Q_s = 2.4$, and a load resistance of $R_L = 20\,\Omega$. A guide of how to maximize P_{out} under different duty cycles for a given load without receiver SAR limit. The receiver SAR limit is shown as the grey surface. The red arrow indicates operating points are above the receiver SAR limit, and the green arrow indicates operating points are below the receiver SAR limit. The data points are calculated from (17).

The 3D plot of P_{out} under different D_s and R_L within the receiver SAR and heat dissipation limits is shown in Fig. 7 for $D_p = 0.95$ and transmitter SAR of 1.6 W/kg. These data points come from LTspice simulation. When R_L increases at a given duty cycle D_s, as shown in the black line in Fig. 7, P_{out} initially rises and reaches a peak when the receiver SAR=1.6 W/kg. If R_L continues to increase, the input dc voltage must be reduced to avoid exceeding the receiver SAR limit, which in turn decreases the P_{out}. Similarly, for a given R_L, it is always preferable to reduce D_s to achieve higher P_{out} until receiver SAR reaches the limit, as shown in the red line in Fig. 7. The SAR values of the transmitters and receivers for each operating point in Fig. 7 is shown in Figs. 8(a) and 8(b), respectively. When the system operates at the two optimal pink points, both transmitter SAR and receiver SAR are at the limit of 1.6 W/kg.

The grey surfaces in Figs. 6 and 7 are the receiver SAR limits of each operating point in terms of power, where receiver SAR exceeds the limit above the grey surfaces and receiver SAR is within the limit below the grey surfaces. The pink points are the optimal operating points at $R_L = 20\,\Omega$ and $R_L = 25\,\Omega$, which are also the optimal operating points for hardware implementation in Section III.

D. Consideration of Operating Frequency and Coil

To achieve high power and high efficiency WPT system for high power IMDs under SAR, heat dissipation, and size limits, the operating frequency and coil should be taken into account.

The coil diameter d is selected as 65 mm, which is the similar size used by a commercial company Resonant Link [17] for high power IMDs. Since σ_{skin} is a monotonically increasing function of frequency in the range of 1-100 MHz [16], $\sigma_{\text{skin}} = f(\omega_s)$, from (25),

$$\text{SAR} \propto \frac{f(\omega_s)V_{\text{rms}}^2}{d^2}. \tag{26}$$

Because a and b are both proportional to V_{in} [12], based on (17), V_{out} is proportional to V_{in}, R_L and reversely proportional to ω_s, $L_{\text{eff, s}}$. Then,

$$V_{\text{out}} \propto \frac{V_{\text{in}}R_L}{\omega_s L_{\text{eff, s}}}. \tag{27}$$

When the coil is single turn, approximately $L_{\text{eff, s}} \propto d$ [18]. Given $V_{\text{in}} = \frac{\sqrt{2D_p}}{\pi}V_{\text{p, rms}}$, $V_{\text{out}} = \frac{\sqrt{2D_s}}{\pi}V_{\text{s, rms}}$, fixed SAR, d,

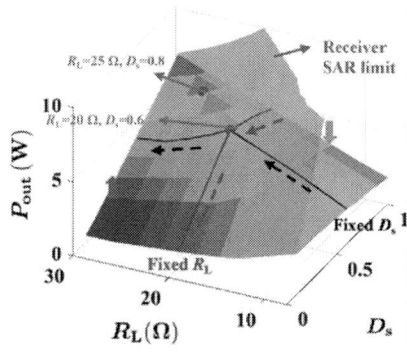

Fig. 7. P_{out} at different D_s and R_L within receiver SAR and heat dissipation limits. The receiver SAR limit is shown as the grey surface. D_p is optimized as 0.95 with the transmitter below the SAR limit. The black line and red line indicates the optimal point with $R_L = 20\,\omega$ and $D_s = 0.6$. Data points come from LTspice simulation.

the coil, where $|\mathbf{E}|$ is the maximum in the WPT coils, it is expressed as

$$\text{SAR} = \frac{\sigma_{\text{skin}}|V_{\text{s, rms}}|^2}{(\pi d)^2 \rho_{\text{skin}}}. \tag{25}$$

At the operating frequency of 6.78 MHz, $\sigma_{\text{skin}} = 0.07$ S/m, [16], $\rho_{\text{skin}} = 1100$ kg/m^3 [13], [14], and the coil diameter $d = 65$ mm. Based on the calculation, the maximum RMS voltage of v_p and v_s in this paper is 32 V, which leads to the SAR limit of 1.6 W/kg [13], [14].

Thermal analysis indicates that the surface power density of the system must be less than 40 mW/cm^2 to ensure the temperature rise of the skin does not exceed 2 °C [13], [14]. This requirement limits the receiver loss to 1.3 W.

The 3D plot of output power P_{out} without the receiver SAR limit under different D_p and D_s with $Q_p = 6.9$ and $Q_s = 2.4$ is shown in Fig. 6, calculated from (17), where the transmitter SAR is fixed at the limit of 1.6 W/kg. Larger D_p gives larger P_{out} for a given R_L and D_s as long as the receiver is within the SAR limit. Higher output power can be achieved but it exceeds the receiver SAR limit.

979-8-3315-1612-3/25 $31.00 © 2025 IEEE

D_p and D_s; $V_{out} \propto \frac{1}{\sqrt{f(\omega_s)}}$ and $R_L \propto \omega_s$ can be derived from (26) and (27), respectively. So

$$P_{out} \propto \frac{1}{f(\omega_s)\omega_s}. \tag{28}$$

Thus, the operating frequency of 6.78 MHz is selected among the ISM bands, because it has higher output power than that of the other ISM operating frequencies considering the SAR limit.

Heat dissipation generated by the receiver exceeds the limit of 1.3 W when the transmitter and receiver coil are both single turn at 6.78 MHz. To meet the heat dissipation limit, the coil is made into three turns to decrease the RMS current in the coil, thus, reducing the coil loss. Based on the LTspice simulation, a three-turn 65 mm coil operating at 6.78 MHz can achieve an output power of 6.6 W, compared to 4.8 W at 13.56 MHz and 2.1 W at 27.12 MHz with a single turn. The simulation from LTspice compares the output power at different operating frequencies with fixed coil diameter at SAR= 1.6 W/kg and conventional operation at $D_p = D_s = 1$, as shown in Table I. P_{out} of different operating frequencies in Table I is maximum under the given loads R_L, SAR, and heat dissipation limits. Lower D_s should be considered when R_L is smaller.

TABLE I
COMPARISON OF OPTIMIZED SYSTEMS AT DIFFERENT OPERATING FREQUENCIES AND NUMBER OF TURNS UNDER THE SAME TRANSMITTER AND RECEIVER SAR LIMITS AND COIL SIZE.

Operating frequency(MHz)	6.78	13.56	27.12
Turns of coil	3	1	1
$R_L (\Omega)$	30.5	9.7	19.5
P_{out}(W)	6.6	4.8	2.1

III. HARDWARE IMPLEMENTATION AND RESULTS

A. Component Selection and Implementation

EPC 2203 was chosen as switch because of small $R_{ds,on}$ and 80 V voltage rating. The gate driver utilized the architecture in [19] with logic inverters SN74LVC2G04 for driving capability and a wideband RF transformer WBC8-1L for signal isolation. The WPT coils, which are identical on both transmitter and receiver, were fabricated as four-layer FR4 PCB inductors with 1-oz copper for inner layers and 2-oz copper for outer layers. The choke inductors were the conical BCL-272JL from Coilcraft, which were selected for small stray capacitance and a high impedance across a wide bandwidth. The explicit resonant capacitors were from VJ HIFREQ Series ceramic capacitors from Vishay, which were selected for high quality factors at 6.78 MHz and high self-resonant frequencies well above 6.78 MHz. Each D_s operation needed different C_s for ZVS. The components are listed in Table II.

The identical PCB layout for both transmitter and receiver is shown in Fig. 9. To minimize the receiver size, the other part of the CMCD rectifier is encircled by the receiver coil. It is calculated that the encircled circuit with a diameter of 28 mm only accounts for 10% magnetic flux [20] of total over the whole 65 mm coil at $z = 0$ plane, which will not affect

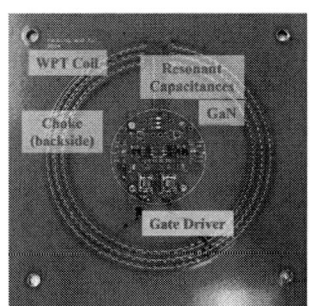

Fig. 9. The PCB layout for both transmitter and receiver of the symmetric CMCD WPT system for IMDs.

TABLE II
COMPONENT SELECTION AND PARAMETER DESIGN

Components & Parameters	Values
Choke Inductor	BCL-272JL 2.8 μH
GaN HEMT	EPC 2203
Gate Driver	SN74LVC2G04 & WBC8-1L
WPT PCB Coils	0.95 μH (3 turns, $d_{average} = 6.5$ cm)
Resonant Capacitor ($D_p = D_s = 0.95$)	520 pF
Transfer Distance	1.7 cm
Coupling Coefficient	0.255

the encircled circuit operation and performance much. If less flux is needed over the encircled circuit, field cancellation can be used [21].

B. Implementation Results

A symmetric CMCD WPT system can be designed under the guideline in [10]. To avoid the open circuit of the current source, the maximum D_p and D_s are both set to be 0.95 for overlap time. Typical waveforms of the transmitter tuned with $D_p = 0.95$ are shown in Fig. 10, where SAR = 1.6 W/kg, maintaining the same for all implementation results. Larger D_p gives larger P_{out} for a given R_L and D_s as long as the receiver is within the SAR limit.

Typical waveforms of the two optimal points of the tuned transmitter and receiver are shown in Figs. 11 and 12 when $R_L = 20\,\Omega$ and $R_L = 25\,\Omega$, respectively. Both examples are designed at the upper bound of SAR limit 1.6 W/kg for both the transmitter and receiver for maximum P_{out}. At $R_L = 20\,\Omega$, the system is optimized with $D_p = 0.95$, $D_s = 0.6$ while $Q_p = 6.9$, $Q_s = 2.4$. The system has $P_{in} = 7.672$ W, $P_{out} = 5.858$ W, $P_{gate} = 0.147$ W, $\eta_{dc\text{-}dc, \text{ w/o gating loss}} = 76.4\%$, and $\eta_{total} = 74.8\%$. At $R_L = 25\,\Omega$, the system is optimized with

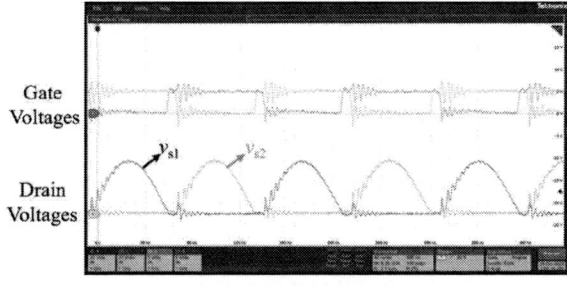

Fig. 10. Typical waveforms of the tuned unloaded CMCD transmitter when $V_{in} = 14$ V and $D_p = 0.95$. The transmitter is operating at the upper bound of SAR limit, namely SAR = 1.6 W/kg.

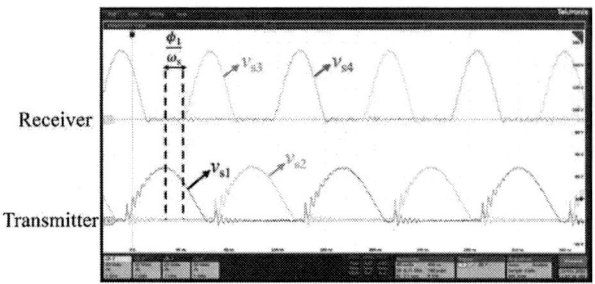

Fig. 11. Waveforms of v_p and v_s under $D_p = 0.95$ and $D_s = 0.6$ at $R_L = 20\,\Omega$ with $Q_p = 6.9$ and $Q_s = 2.4$. Both transmitter and receiver operate at the upper bound of SAR limit, namely SAR = 1.6 W/kg. The system has $P_{in} = 7.672$ W, $P_{out} = 5.858$ W, $P_{gate} = 0.147$ W, $\eta_{dc\text{-}dc,\ w/o\ gating\ loss} = 76.4\%$, and $\eta_{total} = 74.8\%$.

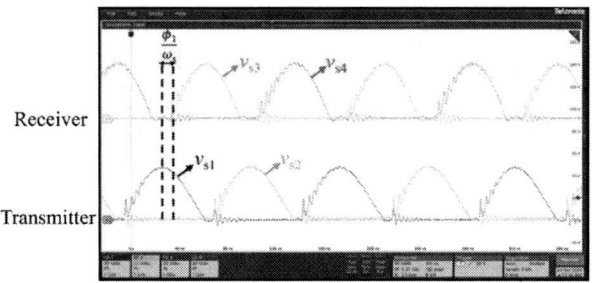

Fig. 12. Waveforms of v_p and v_s under $D_p = 0.95$ and $D_s = 0.8$ at $R_L = 25\,\Omega$ with $Q_p = 5.5$ and $Q_s = 3.1$. Both transmitter and receiver operate at the upper bound of SAR limit, namely SAR = 1.6 W/kg. The system has $P_{in} = 8.316$ W, $P_{out} = 6.433$ W, $P_{gate} = 0.147$ W, $\eta_{dc\text{-}dc,\ w/o\ gating\ loss} = 77.4\%$, and $\eta_{total} = 76.0\%$.

$D_p = 0.95$, $D_s = 0.8$ while $Q_p = 5.5$, $Q_s = 3.1$. The system has input power $P_{in} = 8.316$ W, $P_{out} = 6.433$ W, total gating loss $P_{gate} = 0.147$ W, $\eta_{dc\text{-}dc,\ w/o\ gating\ loss} = 77.4\%$, and $\eta_{total} = 76.0\%$. Power and loss are measured with corresponding dc currents and voltages through 6.5-digit multi-meters (34461A from Keysight). The waveforms are obtained using four 1 GHz probes (TPP1000 from Tektronix) with a 1 GHz, 6.25 GS/s oscilloscope (MSO44B from Tektronix).

C. Comparison with Different D_s for a given R_L

At $R_L = 25\,\Omega$, $D_s = 0.8$ and 0.95, while at $R_L = 20\,\Omega$, $D_s = 0.6$, 0.7, 0.8, and 0.95. D_p is fixed at 0.95 for all conditions to compare the different operating waveforms, output power, and efficiency under the SAR and heat dissipation limits. The experiment does not go below $D_s = 0.6$ at $R_L = 20\,\Omega$ and $D_s = 0.8$ at $R_L = 25\,\Omega$ to avoid exceeding the SAR limit. The experimental waveforms of v_{s1} in transmitter under $D_p = 0.95$ with $Q_p = 6.9$ are shown in Fig. 13(a), while the waveforms of v_{s3} in receiver under different D_s at $R_L = 20\,\Omega$ with $Q_s = 2.4$ are shown in Fig. 13(b) with the optimized phase shift in (13). All switch voltage waveforms achieve ZVS.

P_{out} of the system operating at $D_p = 0.95$ under different D_s and R_L are shown in Fig. 14. The derivation and simulation results come from Figs. 6 and 7. The maximized P_{out} under the SAR limit is with $D_s = 0.6$ at $R_L = 20\,\Omega$ and $D_s = 0.8$ at $R_L = 25\,\Omega$, as the boxes shown in Fig. 14. Under the same R_L and D_p, P_{out} will be larger at smaller D_s. Under the same

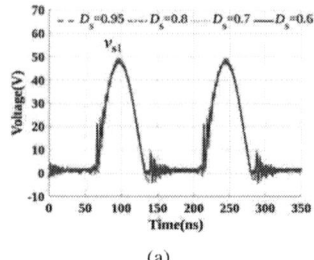 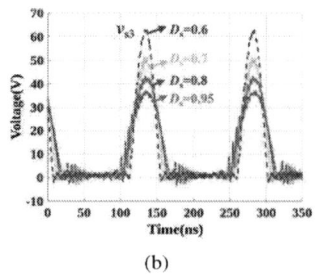

(a)   (b)

Fig. 13. Waveforms of v_{s1} and v_{s3} under the same $D_p = 0.95$ and different D_s at $R_L = 20\,\Omega$. All switch voltage waveforms achieve ZVS. (a) Waveforms of v_{s1}. The transmitter operates at the upper bound of SAR limit. (b) Waveforms of v_{s3}. The receiver with $D_s = 0.6$ operates at the upper bound of SAR limit and the other D_s operations are below SAR limit.

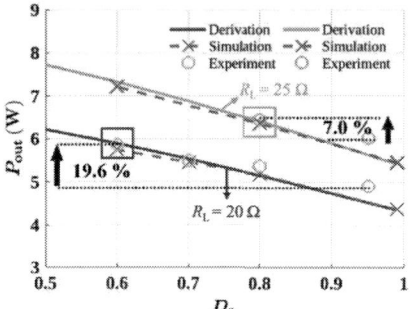

Fig. 14. The comparison of the experimental P_{out} with the derivation and simulation ones under $R_L = 20\,\Omega$ and $R_L = 25\,\Omega$. $D_p = 0.95$. The systems under the six experimental points are all operating within the SAR limit. The optimized system increases the power level by 19.6% and 7.0% at load resistances of $20\,\Omega$ and $25\,\Omega$, respectively, compared to conventional operation.

D_p and D_s, P_{out} will be larger at larger R_L. The efficiency at $D_p = 0.95$ under different D_s and R_L is shown in Fig. 15. It is worth nothing that the system gets both higher output power and efficiency at the two optimal points, as the boxes shown in Figs. 14 and 15. Compared to the conventional operation when $D_p = D_s = 0.95$, the optimized system increases the power level by 19.6% and 7.0% at $R_L = 20\,\Omega$ and $R_L = 25\,\Omega$, respectively, as shown in Fig. 14.

IV. CONCLUSION

Optimizing the waveforms and in-vivo receivers, in terms of quality factors, duty cycles, and phase shifts, for WPT at different loads in the range of several watts under the limits

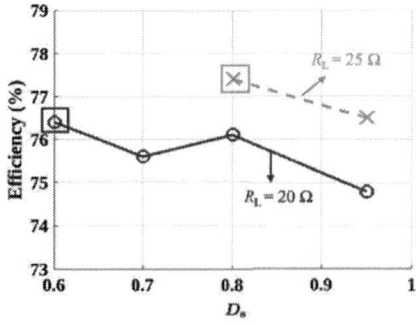

Fig. 15. The dc-dc without gating loss efficiency $\eta_{dc\text{-}dc,\ w/o\ gating\ loss}$ under different D_s and R_L at $D_p = 0.95$.

979-8-3315-1612-3/25 $31.00 © 2025 IEEE   3239

of SAR, heat dissipation, and coil size provides higher output power for high power IMDs. Compared to the conventional operation with high quality factor and maximum duty cycle, the proposed method reduces heat dissipation and improves efficiency by lowering the in-vivo receiver quality factor while increasing the output power by operating at a duty cycle less than 1. Compared to conventional operation, the optimized system increases the power level by 19.6% and 7.0% at load resistances of $20\,\Omega$ and $25\,\Omega$, respectively.

REFERENCES

[1] T. P. G. van Nunen, R. M. C. Mestrom, and H. J. Visser, "Wireless power transfer to biomedical implants using a class-E inverter and a class-DE rectifier," *IEEE Journal of Electromagnetics, RF and Microwaves in Medicine and Biology*, vol. 7, no. 3, pp. 202–209, 2023.

[2] Y. Jia, S. A. Mirbozorgi, P. Zhang, O. T. Inan, W. Li, and M. Ghovanloo, "A dual-band wireless power transmission system for evaluating mm-sized implants," *IEEE Transactions on Biomedical Circuits and Systems*, vol. 13, no. 4, pp. 595–607, 2019.

[3] S. A. Mirbozorgi, P. Yeon, and M. Ghovanloo, "Robust wireless power transmission to mm-sized free-floating distributed implants," *IEEE Transactions on Biomedical Circuits and Systems*, vol. 11, no. 3, pp. 692–702, 2017.

[4] B. Lee, D. Ahn, and M. Ghovanloo, "Three-phase time-multiplexed planar power transmission to distributed implants," *IEEE Journal of Emerging and Selected Topics in Power Electronics*, vol. 4, no. 1, pp. 263–272, 2016.

[5] Q. Chen, S. C. Wong, C. K. Tse, and X. Ruan, "Analysis, design, and control of a transcutaneous power regulator for artificial hearts," *IEEE Transactions on Biomedical Circuits and Systems*, vol. 3, no. 1, pp. 23–31, 2009.

[6] O. Knecht, R. Bosshard, and J. W. Kolar, "High-efficiency transcutaneous energy transfer for implantable mechanical heart support systems," *IEEE Transactions on Power Electronics*, vol. 30, no. 11, pp. 6221–6236, 2015.

[7] C. Liu, C. Jiang, J. Song, and K. T. Chau, "An effective sandwiched wireless power transfer system for charging implantable cardiac pacemaker," *IEEE Transactions on Industrial Electronics*, vol. 66, no. 5, pp. 4108–4117, 2019.

[8] Y. Zhou, C. Liu, and Y. Huang, "Wireless power transfer for implanted medical application: A review," *Energies*, vol. 13, no. 11, p. 2837, 2020.

[9] D. J. Perreault, J. Hu, J. M. Rivas, Y. Han, O. Leitermann, R. C. Pilawa-Podgurski, A. Sagneri, and C. R. Sullivan, "Opportunities and challenges in very high frequency power conversion," in *2009 Twenty-Fourth Annual IEEE Applied Power Electronics Conference and Exposition*, pp. 1–14, 2009.

[10] X. Zan and A.-T. Avestruz, "100 MHz symmetric current-mode class D wireless power transfer," *IEEE Journal of Emerging and Selected Topics in Power Electronics*, vol. 11, no. 4, pp. 4508–4525, 2023.

[11] X. Zan and A.-T. Avestruz, "27.12 MHz bi-directional wireless power transfer using current-mode class D converters with phase-shift power modulation," in *2018 IEEE PELS Workshop on Emerging Technologies: Wireless Power Transfer (Wow)*, pp. 1–6, 2018.

[12] X. Zan and A.-T. Avestruz, "Wireless power transfer for implantable medical devices using piecewise resonance to achieve high peak-to-average power ratio," in *2017 IEEE 18th Workshop on Control and Modeling for Power Electronics (COMPEL)*, pp. 1–8, 2017.

[13] A. Ahlbom, U. Bergqvist, J. Bernhardt, J. Cesarini, M. Grandolfo, M. Hietanen, A. Mckinlay, M. Repacholi, D. H. Sliney, J. A. Stolwijk, *et al.*, "Guidelines for limiting exposure to time-varying electric, magnetic, and electromagnetic fields (up to 300 GHz)," *Health physics*, vol. 74, no. 4, pp. 494–521, 1998.

[14] I. of Electrical and E. Engineers, *IEEE standard for safety levels with respect to human exposure to radio frequency electromagnetic fields, 3 kHz to 300 GHz*. IEEE, 2006.

[15] X. Shi, B. H. Waters, and J. R. Smith, "SAR distribution for a strongly coupled resonant wireless power transfer system," in *2015 IEEE Wireless Power Transfer Conference (WPTC)*, pp. 1–4, IEEE, 2015.

[16] S. Gabriel, R. W. Lau, and C. Gabriel, "The dielectric properties of biological tissues: III. parametric models for the dielectric spectrum of tissues," *Physics in Medicine Biology*, vol. 41, p. 2271, nov 1996.

[17] R. Link, "Wireless charging platform for high power implantable medical devices." Online, 2024. Available: https://cdn.prod.website-files.com/6356a48dff70259fddface01/66979dd283a47cee53b82b69_Resonant%20Link_Datasheet_LVAD%20%26%20TAH%20Charger%20-%20Updated.pdf.

[18] F. W. Grover, *Inductance calculations: working formulas and tables*. Courier Corporation, 2004.

[19] X. Zan and A.-T. Avestruz, "Isolated ultrafast gate driver with variable duty cycle for pulse and VHF power electronics," *IEEE Transactions on Power Electronics*, vol. 35, no. 12, pp. 12678–12685, 2020.

[20] J. M. Griffith and G. W. Pan, "Time harmonic fields produced by circular current loops," *IEEE Transactions on Magnetics*, vol. 47, no. 8, pp. 2029–2033, 2011.

[21] X. Zan and A.-T. Avestruz, "Field cancellation for circuits encircled by VHF wireless power transfer coils," *IEEE Transactions on Power Electronics*, vol. 38, no. 1, pp. 46–52, 2023.

APPENDIX

A. Definition of Quality Factor

By definition, the quality factor is defined as:

$$Q = \frac{\text{Energy stored in the tank}}{T \cdot \text{Energy dissipated}} \cdot 2\pi, \tag{29}$$

where $T = \frac{2\pi}{\omega_s}$. With the CMCD rectifier parameter in Fig. 2(b), Q_s can be expressed as:

$$Q_s = \frac{\frac{1}{2}L_{\text{eff, s}}I_{\text{L, pk}}^2}{T \cdot P_{\text{out}}} \cdot 2\pi, \tag{30}$$

where $I_{\text{L, pk}}$ is the peak current in the inductor. Under the high Q_s assumption, $V_{\text{out}} = \frac{D_s V_{\text{s, pk}}}{\pi}$. Given $|V_{\text{s, pk}}| = \frac{\omega_s}{D_s}L_{\text{eff, s}}I_{\text{L, pk}}$,

$$P_{\text{out}} = \frac{D_s^2 V_{\text{s, pk}}^2}{\pi^2 R_L} = \frac{D_s^2 I_{\text{L, pk}}^2 (\frac{\omega_s}{D_s})^2 L_{\text{eff, s}}^2}{\pi^2 R_L}. \tag{31}$$

Finally,

$$Q_s = \frac{\pi^2 R_L}{2\omega_s L_{\text{eff, s}}}. \tag{32}$$

979-8-3315-1612-3/25 $31.00 © 2025 IEEE

Comparison of Compact Power Amplifier Designs for High Frequency Resonant Wireless Power Transfer Systems at 6.78 MHz using High-Q Resonators

Manuel Rueß[©], Kilian Müller, Mathias C. J. Weiser[©], Ingmar Kallfass[©]

Institute of Robust Power Semiconductor Systems
University of Stuttgart
Stuttgart, Germany
Mail: manuel.ruess@ilh.uni-stuttgart.de

Abstract—**This paper presents a comparison of highly compact power amplifiers for 6.78 MHz, 100 W wireless power transfer systems based on high-Q resonators. The use of multilayer self-resonant structures in the transmission path significantly improves quality factors while maintaining a very compact form factor. This imposes specific requirements on the power amplifier to achieve high efficiencies in a small package. In contrast to the efficient but complex design of a Class Φ_2 topology, this work investigates highly compact Class D amplifier designs. Alongside a Current-Mode Class D topology, a ZVS Voltage-Mode Class D/DE topology is introduced, which achieves 93% efficiency at 60 W output power with a highly reduced form factor. Thanks to an additional series resonant circuit at the output, this topology can also be employed for wireless power transfer systems based on high-Q resonators. The size-optimized designs utilize GaN transistors with integrated drivers. At a coupling distance of 2 cm, a coupling factor of 0.2, and a maximum input voltage of 80 V, these amplifiers were tested in a WPT system, achieving up to 70 W transmitted power at 6.78 MHz.**

Index Terms—**WPT, Power Amplifier, GaN, Class D, resonant, power density**

I. INTRODUCTION

Wireless Power Transfer (WPT) via resonant inductive coupling is increasingly gaining importance in fields such as electro-mobility, medical implants, robotics, and wearable devices [1]–[4]. In particular, for medical implants and wearable devices, a high power density at low voltage levels is preferred to significantly reduce form factor and improve safety. By increasing the transmission frequency to the MHz range, near-field WPT systems can be improved in terms of efficiency, transmitted power, size, and transmission range [1], [5], [6]. The system efficiency of such WPTs is closely linked to the transformer quality factor (Q-factors) and the coupling coefficient, as illustrated in Figure 1 [7]. Therefore a high-Q

This work is funded by the German Research Foundation (DFG) as part of the 3D-CeraGaN project (No.: 462828009). The author would like to thank the Ministry of Science, Research and Arts of the Federal State of Baden-Württemberg for the financial support of the project within the Innovation Campus Future Mobility (ICM), project LAB16.

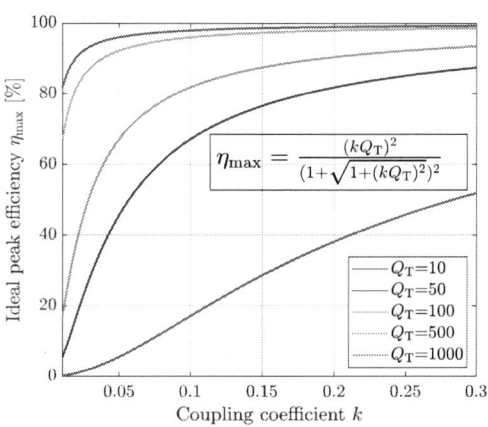

Figure 1. Ideal peak efficiency of WPT systems depending on the coupling coefficient k and the quality factor of the resonant tank Q_T. With larger distances between transformers, high Q-factors are required to achieve high efficiencies. [7]

transformer enables high efficiencies at increased transmission range.

In [8], [9], a concept of a multilayer self-resonant structure (MSRS) is demonstrated achieving maximum Q-factors at small form factors. These resonators combine coil and compensation capacitance in one component and represent a parallel compensated system. The state of the art (SotA) in WPT systems, as depicted in Figure 2, achieves efficiencies of over 90 % at 6.78 MHz. At these frequencies, the use of MSRS systems has proven advantageous for achieving maximum DC-to-DC efficiencies at very high power densities [5], [10]. In addition to a high Q-factor transmission system, a suitable power amplifier (PA) is required for operating a MSRS based WPT system. In [5], [10], Class Φ power amplifiers are used to transmit power in the lower kW range with efficiencies up to 95 %. However, Class Φ amplifiers are highly complex in their design and implementation. In comparison, Class D PAs provide a very compact and straightforward alternative

979-8-3315-1612-3/25 $31.00 © 2025 IEEE

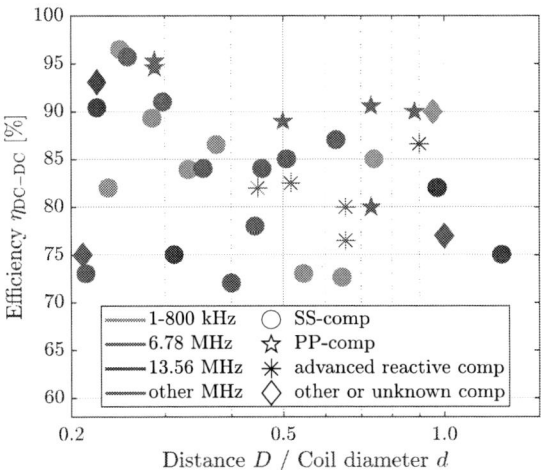

Figure 2. SotA in efficiency of resonant inductive WPT systems as a function of the transmission distance D normalized to the coil diameter d [1]–[6], [9], [10], [14], [16]–[19], [28], [29].

for realizing WPT systems based on high-Q MSRS [11], [12].

This work investigates the implementation of WPT systems based on MSRS and focuses on highly compact PA topologies. For this purpose, different versions of PA topologies, including a series stacked (SS) Class Φ_2 push-pull power amplifier with a T-network (PPT), Current-Mode Class D (CM Class D), and Zero Voltage Switching (ZVS) Voltage-Mode Class D/DE (VM Class D/DE), are analyzed in detail. These topologies are evaluated in a WPT system analyzing their handling and tuning of the resonant frequency, power density, and efficiency. The implementations are designed with a maximum input voltage of $V_{DC,in} = 80$ V and power ratings of up to $P_{DC,in} = 100$ W. For a highly compact PA, power transistors with integrated drivers are used. Among the topologies examined, the ZVS VM Class D/DE topology presented in this paper combines compact design with high efficiency, outperforming both the CM Class D and SS Class Φ_2 PPT configurations. First, the MSRS is briefly introduced. Subsequently, the designs of the amplifiers are presented. The performance of the amplifiers will be evaluated, including a 6.78 MHz WPT system to demonstrate and discuss their compatibility with MSRS-based systems.

II. High-Q MSRS Design for 6.78 MHz

The concept of MSRS was initially proposed by Sullivan et al. in [8] and further modified by Stein et al. in [13]. The MSRS is capable of handling high power densities while simultaneously providing a very high Q-factor normalized to its size. The quality factor can be up to six times higher than that of a conventional coil of the same size [14]. This ensures outstanding efficiency in power-dense MHz WPT systems [3], [5], [10]. The concept employs a stack-up comprising thin, C-shaped conductive foil layers and thin, circular dielectric film layers. The aforementioned layers are stacked in an alternating sequence, forming internal capacitances resonating with the inductance of the structure at a specific frequency while

ensuring equal current sharing across the conductive layers. The resonant frequency of the resulting LC-tank can be tuned by varying the number of layers and their physical dimensions. The modified MSRS by Stein et al. employs copper layers laminated on a thin and cost-effective polyimide (PI) substrate, resulting in a simplified manufacturing process of the MSRS. However, this approach has the disadvantage of necessitating twice as many foil layers to achieve a given resonant frequency in comparison to the initial MSRS design. This is due to the modification of the MSRS, which has resulted in the elimination of the low-quality dielectric PI layers in terms of their contribution to the internal capacitance of the resonator.

In this work, a MSRS based on [13] is utilized for a resonant/transmission frequency of 6.78 MHz and a core diameter of $d = 75$ mm. The prototype, shown in Figure 3, uses 12 μm copper, 50 μm PI, and 25 μm poly-tetrafluoroethylene (PTFE) layers. With an outer diameter of 59 mm and an inner diameter of 28.5 mm, the inductance of the stack-up is determined to be $L = 126$ nH. For a resonant frequency of 6.78 MHz, a capacitance of $C = 4.37$ nF is required, which can be realized using 12 layer sections according to the theory from [13]. Therefore, a layer section consists of two copper layers and one PTFE layer, as illustrated in Figure 3. As described above, the PI layers only act as mechanical support for the copper layers and contribute neither to the resonant capacity nor significantly to the losses of the structure [10]. However, impurities in the layer stack, such as dust, can reduce the capacity of the structure by increasing the distance between the Flex-PCB and PTFE layers. Therefore, 15 layers are used to achieve the desired resonant frequency. At resonant a quality factor of $Q_{6.78} = 1343$ is measured via Z-parameter using a vector network analyzer (VNA), as shown in Figure 3. The parallel compensated MSRS, necessitates the use of a sinusoidal voltage for driving purposes. This indicates that the

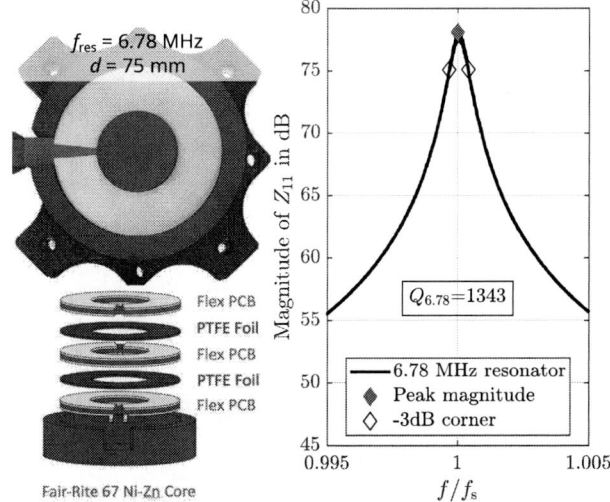

Figure 3. MSRS setup for a resonant frequency of 6.78 MHz. The used concept of the resonator stack and the assembled prototype is shown on the left. On the right Z-Parameter measurement and the resulting Q-factor is presented.

979-8-3315-1612-3/25 $31.00 © 2025 IEEE

Table I
COMPARATIVE TABLE OF PA TOPOLOGIES FOR WPT SYSTEMS USING [11], [15], [20], [21], [23]–[26]

Parameter	ZVS Class D	Class DE	CM Class D	Class E push-pull	Class Φ_2 PPT	SS Class Φ_2 PPT
Device peak voltage	$V_{\text{DC,in}}$	$V_{\text{DC,in}}$	$\pi\,V_{\text{DC,in}}$	$3.6\,V_{\text{DC,in}}$	$2.1\,V_{\text{DC,in}}$	$1.05\,V_{\text{DC,in}}$
Device RMS current	$\frac{\pi}{2}\,I_{\text{DC,in}}$	$1.9\,I_{\text{DC,in}}$	$\frac{1}{\sqrt{2}}\,I_{\text{DC,in}}$	$0.77\,I_{\text{DC,in}}$	$0.95\,I_{\text{DC,in}}$	$1.9\,I_{\text{DC,in}}$
Voltage gain DC-AC	$0.55\,V_{\text{DC,in}}$	N/A	$2.22\,V_{\text{DC,in}}$	$1.414\,V_{\text{DC,in}}$	$2.43\,V_{\text{DC,in}}$	$1.215\,V_{\text{DC,in}}$
Active device utilization	0.32	0.26	0.23	0.18	0.25	0.25
Number of switches	2	2	2	2	2	2
Duty cycle	0.5	0.33	0.5	0.5	0.3-0.35	0.3-0.35
WPT Compensation	SS	PP	PP	SS	SS,PP	SS, PP

MSRS is incompatible with Voltage-Mode PAs and rectifiers, which are commonly proposed in high-performance WPT systems (typically featuring a series compensation).

III. SPACE-OPTIMIZED POWER AMPLIFIERS FOR PP-COMPENSATED RESONATORS

Table I provides an overview of various PA topologies used in WPT systems. In Voltage-Mode PAs, the output voltage waveform is a (quasi-) rectangular wave, while the output current waveform is sinusoidal. Consequently, topologies such as ZVS Class D and Class E cannot be directly used with an MSRS. However, in some cases, additional LC series resonant circuits must be implemented to enable operation (e.g., in DE and Φ_2 topologies), resulting in higher losses and increasing form factor. Current-Mode PAs, such as the CM Class D topology, can directly drive a parallel-compensated transformer system, for example.

In this work, the power amplifiers, shown in Figure 4 are investigated for a MSRS use case and a highly compact design. The parameters listed in Table II were used for input power levels up to $P_{\text{DC,in}} = 100\,\text{W}$. To achieve such power levels, always three EPC21701 transistors connected in parallel are used as switch cell to increase the current-carrying capability. Figure 5 illustrates the circuit diagrams of the PA topologies investigated in this work. In addition to the SS Class Φ_2 PPT [15] and the CM Class D [12], a combination of ZVS Class D and Class DE is presented. It is referred to ZVS Class D/DE in the following

The amplifiers are compared in their efficiency both through simulation and measurement. Figure 6 shows the DC-AC efficiencies as a function of output power. While for the SS Class Φ_2 PPT and ZVS Class D/DE amplifiers, a real AC load Z_L can be used, this is not possible for the CM Class D amplifier. In this case, Z_L includes not only the load but also the resonant circuit of the WPT system. Therefore, only simulated results are presented for the CM Class D amplifier, where the needed parallel RLC resonant tank is modeled ideally with a real part corresponding to the connected load. Real transistor models and inductors with a series resistance of $100\,\text{m}\Omega$ were used to compare the different loss behavior.

A. CM Class D

The CM Class D topology is characterized by a significantly reduced number of required components, especially when combined with the MSRS. This makes it ideal for very compact WPT systems. However, the ZVS operation in this

Table II
PARAMETERS OF THE DEMONSTRATED PA DESIGNS AND TEST SETUP

	SS Class Φ_2 PPT	ZVS Class D/DE	CM Class D
$\mathbf{V_{\text{DC,in}}}$	15 V to 35 V	15 V to 65 V	10 V to 20 V
Duty cycle	0.35	0.35	0.51
Active device		EPC21701 80 V	
Volume	$29.45\,\text{cm}^3$	$19.65\,\text{cm}^3$	$18.75\,\text{cm}^3$
Load		first: $6\,\Omega$ analog second: tuned 6.78 MHz MSRS System 126 nH, 4.37 nF, $k = 0.2$, small signal $Q \approx 1300$ variable electronic DC load	

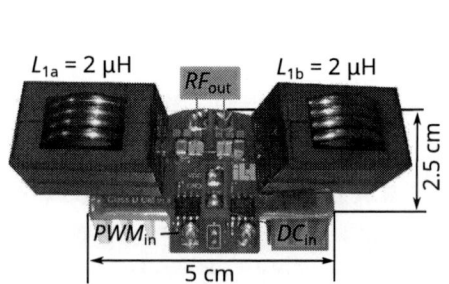
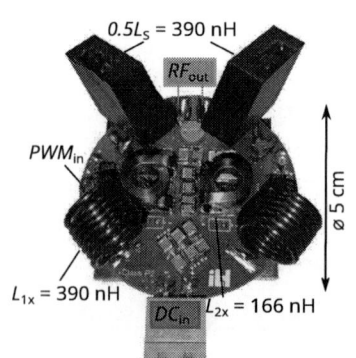
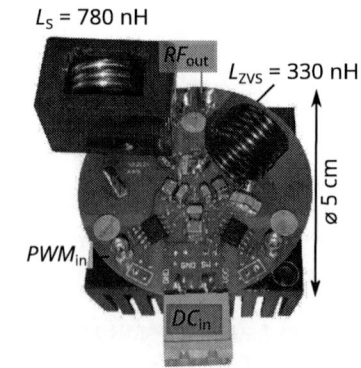

Figure 4. Illustration of the realized PA prototypes based on the EPC21701 laser driver. CM Class D (left), SS Class Φ_2 PPT (mid), and ZVS Class D/DE (left). In order to achieve better cooling and a more compact design, the transistors were placed on the bottom of the PCB.

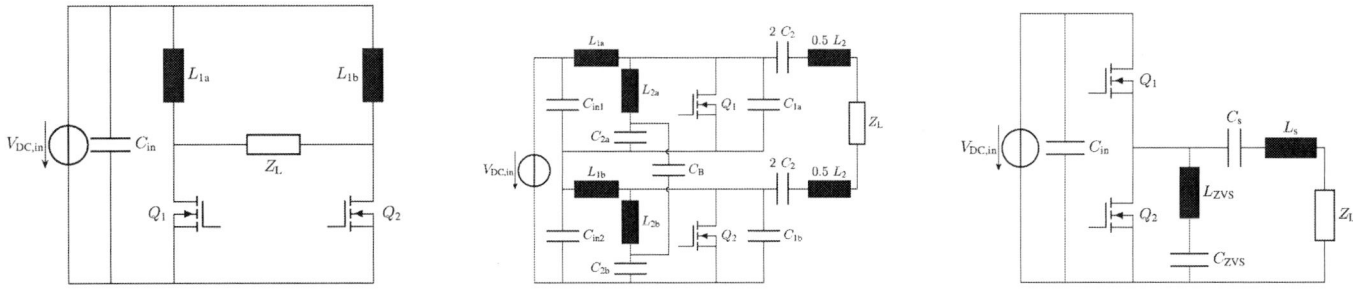

Figure 5. Simplified circuit diagrams of the PA topologies used, with resonators, rectifier and load resistor combined into the impedance Z_L. CM Class D (left), the SS Class Φ_2 PPT (mid) and the proposed ZVS Class D/DE (right).

topology depends on the tuning of the resonators as well as the matching between the PA and the resonator. This leads to a much more complex tuning process between the amplifier and the transmission system, as two degrees of freedom must be satisfied: the optimal ZVS of the amplifier and the optimal frequency matching of both resonators. In simulation, the CM Class D achieves the highest efficiency among the PAs, reaching 97 % due to the used ideal LC resonant tank. When the losses in the coils are considered negligible, the primary losses in this topology are conduction losses occurring in the transistors. However, in a WPT system, even small deviations from the resonant frequency lead to increased switching losses, as ZVS operation no longer occurs.

B. SS Class Φ_2 PPT

In contrast, the SS Class Φ_2 PPT, with its series resonant circuit comprising C_s and L_s, allows the switching frequency to be tuned to the resonant frequency independently of the load Z_L. This topology is therefore load-independent and operates in ZVS, achieved via a LC path in shunt to the resistor using L_{2x} and C_{2x}. The downside of this topology for compact designs is the increased number of LC tanks. While these enable an efficient PA operation, they also result in a much more complex design and increased volume. At lower voltages and the same output power, high losses occur in the inductors due to the increased current. Therefore, the SS Class Φ_2 PPT variant achieved an efficiency of only 88 % in simulation and 86 % in measurements. For comparison, a SS Class Φ_2 PPT High Voltage (HV) based on 650 V transistors is also shown. This version can operate at higher input voltages and, due to the reduced current for the same power, achieves a significant increase in efficiency, reaching up to 95 %. However, the inductors of the HV version have higher quality factors due to their larger size, resulting in a 4.6 times larger PA volume compared to the SS Class Φ_2 PPT developed in this work.

C. ZVS Class D/DE

The presented ZVS Class D/DE topology is based on the implementation presented in [11] and has been enhanced with a resonant reactance network to efficiently drive a parallel-compensated transmission system. Similar to the SS Class Φ_2 PPT variant, the ZVS Class D/DE exhibits a low voltage stress on the transistors, which is three times lower compared to the

CM Class D, enabling the use of active components with lower voltage ratings, which often exhibit better dynamic properties. In addition to its compact design, the ZVS Class D/DE variant stands out with load-independent ZVS, using L_{ZVS} and C_{ZVS}, switching behavior and achieved efficiencies of 95 % in simulation and 93 % in measurements. During real tests, a maximum input voltage of 50 V was applied to protect the transistors with a breakdown voltage (V_{BDS}) of 80 V. This also highlights that the output power of this variant is significantly more limited compared to the other topologies, as can be observed from its lower voltage gain.

The measurements, presented in Figure 6 clearly show that the number of used inductors in a topology, especially when carrying high currents, is critical to the efficiency of the PA. All measured efficiencies remain below the simulated maximum values, with this effect increasing at higher current levels in the coils as the loss mechanisms become more relevant and the coil Q-factors reduce due to the increasing real impedance component. The efficiency of the amplifiers studied here is mainly determined by the losses of the coils and thus by their Q-factors. Since, the Q-factor of inductors is scaled down with decreasing size, the choice of them for PA design is always a compromise between the compactness of the system and its efficiency. Particularly at high frequencies, it is difficult to achieve a high current carrying capacity at low resistance, a high quality factor and a small footprint at the same time. In this study, therefore, either air-core coils or coils with a special NiZn ferrite core were used, which exhibit low losses at MHz frequencies. In all the PAs used, ZVS ensures that the switching losses in the active components of the amplifier are minimized. ZVS is achieved by different methods in the different topologies considered. In ZVS Class D/DE and SS Class Φ_2 PPT this minimizes losses in the amplifier and enables very efficient operation over a wide range of operating points. Additionally the ZVS operation of the PA and the resonant frequency of the MSRS can be optimized independently from each other. This is a great advantage when tuning for most efficient operation of the WPT system.

In summary, the CM Class D amplifier achieves the highest efficiency, exceeding 97 %, but only under optimal tuning between the PA and the resonator. The SS Class Φ_2 PPT, with an efficiency of 88 %, is significantly less efficient under the

979-8-3315-1612-3/25 $31.00 © 2025 IEEE

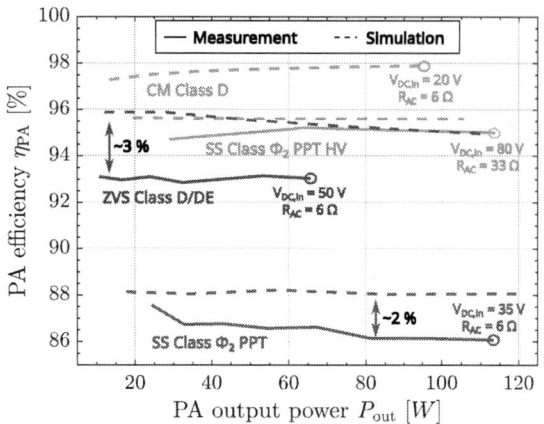

Figure 6. Simulated and measured PA efficiencies up to 80 V Drain-Source voltages using an AC-load which was adjusted to reach 100 W output power per PA.

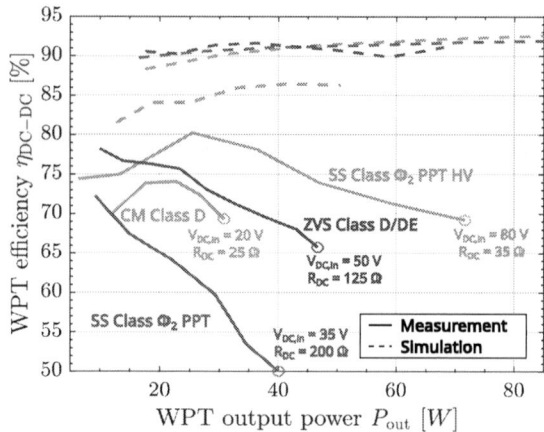

Figure 8. Simulated and measured efficiencies over the transmitted power in the WPT system using the proposed PA topologies.

given requirements due to high currents and increased losses in the inductors, though it could perform better in higher voltage classes. With an efficiency of 93 % and the ability to tune the PA and resonator separately to the resonant frequency, the ZVS Class D/DE provides the best overall performance. In terms of volume, the CM Class D has a total volume of approximately $18.75\,\mathrm{cm}^3$, the SS Class Φ_2 PPT about $29.45\,\mathrm{cm}^3$, and the ZVS Class D/DE around $19.65\,\mathrm{cm}^3$, assuming nearly identical coil sizes. As a result, the ZVS Class D/DE with its two inductors is also one of the most compact designs (Fig. 10).

IV. POWER AMPLIFIER EVALUATION IN A 6.78 MHz WPT SYSTEM BASED ON MSRS

The three PA topologies presented are now analyzed in more detail in a WPT system. The experimental WPT system for this analysis is based on a send (TX) and a receive (RX) resonator according to II and is shown in Figure 7. In addition to the two MSRS building the transmission system, the setup includes the PA, in this case the CM Class D, which is supplied by a DC voltage source, a simple diode rectifier and a DC load on the receiver side. The maximum voltage across the transistors, together with the given voltage gain of the PA topology (Table I), limits the maximum achievable power

over a given load resistance range. This means that the input voltages and/or load resistance values must be adjusted for each PA under test to achieve comparable power levels.

The high quality factor of the MSRS enables high WPT efficiency, but brings the disadvantage that precise tuning of the resonators is required to transmit the desired power. When using MSRS with Q values above 1000, even small deviations between the resonant frequencies result in a greatly increased impedance of the transmission system, which reduces the transmitted power significantly. Therefore, both MSRS were preset to their desired resonant frequencies using a VNA. The TX resonator was tuned to exactly 6.78 MHz, while the RX resonator was tuned to approximately 7 MHz to account for the additional capacitance of the rectifier diodes, which lowers the RX MSRS resonant frequency.

In Figure 8, the simulated and measured efficiencies of the WPT systems are presented. The efficiency is measured between the amplifier's input power $P_{\mathrm{DC,in}}$ and the output power $P_{\mathrm{DC,out}}$ at the load. The simulation includes models for the transistors, diodes, and the MSRS system. The measurements at a switching frequency of 6.78 MHz demonstrate

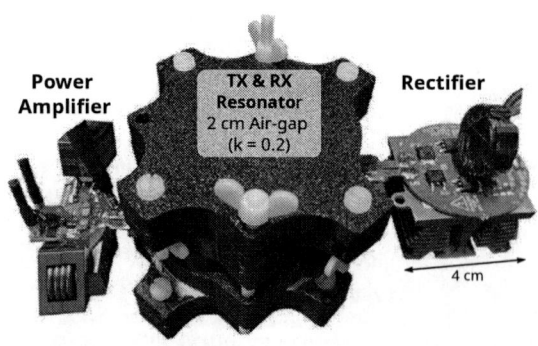

Figure 7. Realization of the 6.78 MHz WPT system, which was used for the evaluation of the power amplifiers. Includes TX and RX MSRS and rectifier.

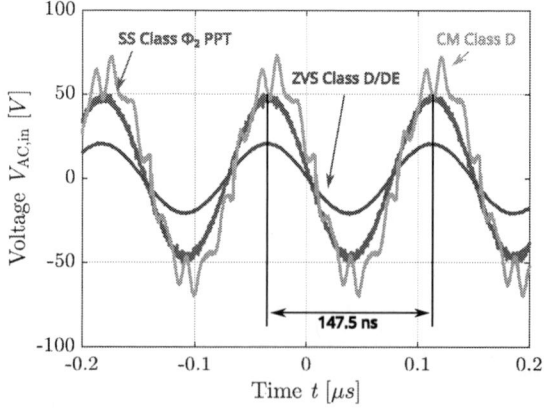

Figure 9. AC voltage measured at the input of the MSRS for the three investigated PA topologies.

979-8-3315-1612-3/25 $31.00 © 2025 IEEE

efficiencies ranging between 70 % and 80 %. The deviation from the simulated values can be attributed to losses in the resonator structure due to a mismatched resonant frequency. Specifically, the measurements indicate significant efficiency reductions at higher transmitted power levels, caused by defects or impurities in the resonator that lead to increased losses. Regardless of the losses in the resonator and rectifier, a clear trend between the amplifier topologies can be observed. The SS Class Φ_2 PPT topology is noticeably disadvantaged due to the specified low voltage and high current levels. At higher voltages and reduced currents, efficiency improvements of 15 % to 20 % could theoretically be achieved but are not feasible with the design based on the $EPC21701$. The CM Class D topology demonstrates its maximum efficiency at an input voltage of 15 V, under optimal tuning between the PA and resonator. Deviations in input voltage or load resistance, however, result in the loss of ZVS behavior, leading to reduced efficiency. Figure 9 shows the voltage at the input of the resonator for the three topologies. With CM Class D, it is clear that the additional parasitic capacitance in combination with the line inductance between the PA and resonator generates an additional resonant frequency lowering both, the PA efficiency and the transmission efficiency. This can be minimized by appropriate tuning the TX MSRS capacitance for a specific operating point. However, if the input voltage or the load changes, the system is de-tuned. This illustrates the difficulties in tuning the WPT system with such a topology.

The proposed ZVS Class D/DE topology stands out with the highest efficiency compared to the other setups. Its advantage lies in the reduced number of inductive components and the optimal tuning of the PA achieved through the additional series resonant network. Compared to the SS Class Φ_2 PPT topology, an efficiency improvement of up to 10 % was observed over a 50 W output power range. Similarly, a 3 % improvement in efficiency was achieved compared to the CM Class D topology.

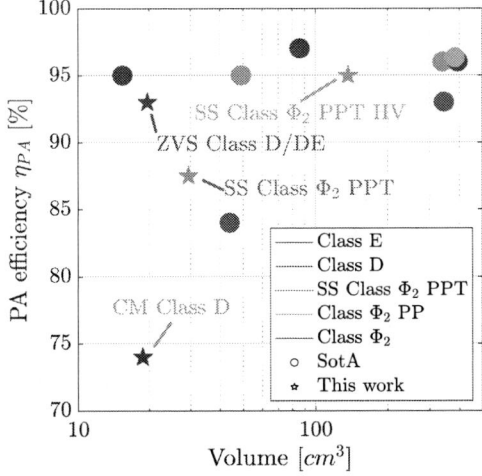

Figure 10. The PAs presented in this work compared with the SotA of PAs used in WPT systems in terms of efficiency and volume [5], [11], [15]–[19].

V. CONCLUSION AND FUTURE WORK

In this work, the amplifier topologies SS Class Φ_2 PPT, CM Class D, and ZVS Class D/DE were compared and investigated in detail for their use in a 6.78 MHz MSRS WPT system. The MSRS enables the realization of WPT systems with very high Q-factors at MHz frequencies. However, amplifiers and the MSRS must be well-tuned to fully utilize the efficiency advantage by the high Q-factors. To further improve a compact and efficient WPT design, PAs benefit from a reduced number of inductors, which often have the highest losses in such designs. The measured WPT efficiencies in this work were mostly influenced by high losses in the resonator and rectifier set up used for testing purposes. With an optimized design, significantly higher efficiencies are possible. However, the focus of this work was on the design of a very compact and efficient PA that is capable of operating a MSRS WPT system.

The very simple and compact CM Class D topology, with a volume of $18.75\,cm^3$, promises an exceptionally compact overall system. This is achieved through the integration of the transmission resonator into the PA design, resulting in a minimal number of components. However, the efficiency of the PA and WPT transmission depends directly on the quality factor of the MSRS and the tuning between the PA and the resonator. The compact design of the CM Class D thus comes with significantly higher tuning requirements for the WPT system, which are challenging to implement in practice. In comparison, the SS Class Φ_2 PPT offers the ability to independently tune the MSRS and PA to the transmission frequency. However, this topology is heavily loss-prone due to the high number of inductors operating at relatively high currents. Under the conditions specified in this work, the SS Class Φ_2 PPT achieves an efficiency of only 88%. With a volume of $29.45\,cm^3$, this PA is nearly twice as large as the CM Class D due to the higher number of components.

The demonstrated modified ZVS Class D/DE variant presented in this work impresses with its compact design and a volume of $19.65\,cm^3$ in combination with its high efficiency of 93%. Thanks to the additional series resonant circuit at the output of the power amplifier, the resonant frequency can be tuned independently of the transmission system, similar to the SS Class Φ_2 PPT topology. This enables very precise adjustment to the resonator and consequently also high efficiency over a wide output range. Compared to the SotA in PA designs for WPT systems, shown in Figure 10, it performs among the best with an efficiency of 93% in relation to the total volume of the PA designs. The ZVS Class D/DE variant thus offers optimal conditions for realizing a highly compact WPT system based on MSRS technology at a transmission frequency of 6.78 MHz.

By further improving the rectifier design, WPT systems in the MHz range in particular can benefit significantly. Future work will focus on a compact and efficient rectifier design to further improve the efficiency and size of WPT systems.

979-8-3315-1612-3/25 $31.00 © 2025 IEEE

ACKNOWLEDGMENT

The authors would like to express their gratitude to Fair-Rite and Böhme for providing valuable resources. The support offered was highly beneficial and significantly contributed to the successful completion of this research project.

REFERENCES

[1] K. Surakitbovorn and J. Rivas-Davilla, "Design of a GaN-Based Wireless Power Transfer System at 13.56 MHz to Replace Conventional Wired Connection in a Vehicle," 2018 International Power Electronics Conference (IPEC-Niigata 2018 -ECCE Asia), Niigata, Japan, 2018, pp. 3848-3854, doi: 10.23919/IPEC.2018.8507473.

[2] H. Takanashi, Y. Sato, Y. Kaneko, S. Abe and T. Yasuda, "A large air gap 3 kW wireless power transfer system for electric vehicles," 2012 IEEE Energy Conversion Congress and Exposition (ECCE), Raleigh, NC, USA, 2012, pp. 269-274, doi: 10.1109/ECCE.2012.6342813.

[3] R. Qin, J. Li and D. Costinett, "A 6.6-kW High-Frequency Wireless Power Transfer System for Electric Vehicle Charging Using Multilayer Nonuniform Self-Resonant Coil at MHz," in IEEE Transactions on Power Electronics, vol. 37, no. 4, pp. 4842-4856, April 2022, doi: 10.1109/TPEL.2021.3120734.

[4] O. Knecht, R. Bosshard and J. W. Kolar, "High-Efficiency Transcutaneous Energy Transfer for Implantable Mechanical Heart Support Systems," in IEEE Transactions on Power Electronics, vol. 30, no. 11, pp. 6221-6236, Nov. 2015, doi: 10.1109/TPEL.2015.2396194.

[5] Gu, Lei, Grayson Zulauf, Aaron Stein, Phyo Aung Kyaw, Tuofei Chen, und Juan Manuel Rivas Davila. „6.78-MHz Wireless Power Transfer With Self-Resonant Coils at 95% DC–DC Efficiency". IEEE Transactions on Power Electronics 36, Nr. 3 (März 2021): 2456–60. https://doi.org/10.1109/TPEL.2020.3014042.

[6] Ulmer, Sabrina, Klaus-Dieter Kächele, und Kathrin Kocher. Class Phi2 Amplifier Using GaN HEMTs at 13.56MHz with Tuned Transformer for Wireless Power Transfer. DE: VDE VERLAG GMBH, 2022. https://doi.org/10.30420/565822239 .

[7] E. Waffenschmidt and T. Staring, "Limitation of inductive power transfer for consumer applications," 2009 13th European Conference on Power Electronics and Applications, Barcelona, Spain, 2009, pp. 1-10.

[8] Sullivan, Charles R., und Lotfi Beghou. „Design methodology for a high-Q self-resonant coil for medical and wireless-power applications". In 2013 IEEE 14th Workshop on Control and Modeling for Power Electronics (COMPEL), 1–8. Salt Lake City, UT, USA: IEEE, 2013. https://doi.org/10.1109/COMPEL.2013.6626460.

[9] Stein, Aaron L. F., Phyo Aung Kyaw, Jesse Feldman-Stein, und Charles R. Sullivan. „Thin self-resonant structures with a high-Q for wireless power transfer". In 2018 IEEE Applied Power Electronics Conference and Exposition (APEC), 1044–51. San Antonio, TX, USA: IEEE, 2018. https://doi.org/10.1109/APEC.2018.8341144.

[10] Stein, Aaron L. F., Phyo Aung Kyaw, und Charles R. Sullivan. „Wireless Power Transfer Utilizing a High-Q Self-Resonant Structure". IEEE Transactions on Power Electronics 34, Nr. 7 (Juli 2019): 6722–35. https://doi.org/10.1109/TPEL.2018.2874878.

[11] De Rooij, Michael A. „The ZVS voltage-mode Class D amplifier, an eGaN FET-enabled topology for highly resonant wireless energy transfer", 1608–13. Charlotte, NC, USA: IEEE, 2015. https://doi.org/10.1109/APEC.2015.7104562.

[12] H. Kobayashi, J. Hinrichs and P. M. Asbeck, "Current mode Class D power amplifiers for high efficiency RF applications," 2001 IEEE MTT-S International Microwave Sympsoium Digest (Cat. No.01CH37157), Phoenix, AZ, USA, 2001, pp. 939-942 vol.2, doi: 10.1109/MWSYM.2001.967047.

[13] Stein, Aaron L. F., Phyo Aung Kyaw, und Charles R. Sullivan. „High-Q self-resonant structure for wireless power transfer". In 2017 IEEE Applied Power Electronics Conference and Exposition (APEC), 3723–29. Tampa, FL, USA: IEEE, 2017. https://doi.org/10.1109/APEC.2017.7931234.

[14] Stein, Aaron L.F., Phyo Aung Kyaw, und Charles R. Sullivan. „The Feasibility of Self-Resonant Structures in Wireless Power Transfer Applications". In 2018 IEEE PELS Workshop on Emerging Technologies: Wireless Power Transfer (Wow), 1–6. Montréal, QC: IEEE, 2018. https://doi.org/10.1109/WoW.2018.8450931.

[15] Gu, Lei, Grayson Zulauf, Zhemin Zhang, Sombuddha Chakraborty, und Juan Rivas-Davila. „Push–Pull Class Φ_2 RF Power Amplifier". IEEE Transactions on Power Electronics 35, Nr. 10 (Oktober 2020): 10515–31. https://doi.org/10.1109/TPEL.2020.2981312.

[16] Gu, Lei, und Juan Rivas-Davila. „1.7 kW 6.78 MHz Wireless Power Transfer with Air-Core Coils at 95.7% DC-DC Efficiency". In 2021 IEEE Wireless Power Transfer Conference (WPTC), 1–4. San Diego, CA, USA: IEEE, 2021. https://doi.org/10.1109/WPTC51349.2021.9458037.

[17] Choi, Jungwon, Daisuke Tsukiyama, und Juan Rivas. „Comparison of SiC and eGaN devices in a 6.78 MHz 2.2 kW resonant inverter for wireless power transfer". In 2016 IEEE Energy Conversion Congress and Exposition (ECCE), 1–6. Milwaukee, WI, USA: IEEE, 2016. https://doi.org/10.1109/ECCE.2016.7854938.

[18] Zulauf, Grayson, und Juan M. Rivas-Davila. „Single-Turn Air-Core Coils for High-Frequency Inductive Wireless Power Transfer". IEEE Transactions on Power Electronics 35, Nr. 3 (März 2020): 2917–32. https://doi.org/10.1109/TPEL.2019.2932178.

[19] Liu, Ming, Minfan Fu, und Chengbin Ma. „Parameter Design for a 6.78-MHz Wireless Power Transfer System Based on Analytical Derivation of Class E Current-Driven Rectifier". IEEE Transactions on Power Electronics 31, Nr. 6 (Juni 2016): 4280–91. https://doi.org/10.1109/TPEL.2015.2472565.

[20] Glaser, John S., und Juan M. Rivas. „A 500 W push-pull dc-dc power converter with a 30 MHz switching frequency". In 2010 Twenty-Fifth Annual IEEE Applied Power Electronics Conference and Exposition (APEC), 654–61. Palm Springs, CA, USA: IEEE, 2010. https://doi.org/10.1109/APEC.2010.5433602.

[21] Kaczmarczyk, Z.; Jurczak, W.: A Push–Pull Class E Inverter With Improved Efficiency. In: IEEE Transactions on Industrial Electronics 55 (2008), April, Nr. 4, 1871–1874. http://dx.doi.org/10.1109/TIE.2007.907665. – DOI 10.1109/TIE.2007.907665. – ISSN 0278–0046

[22] Rivas, Juan M., Yehui Han, Olivia Leitermann, Anthony D. Sagneri, und David J. Perreault. „A High-Frequency Resonant Inverter Topology With Low-Voltage Stress". IEEE Transactions on Power Electronics 23, Nr. 4 (Juli 2008): 1759–71. https://doi.org/10.1109/TPEL.2008.924616.

[23] Eroglu, Abdullah: Introduction to RF power amplifier design and simulation. Boca Raton: CRC Press, Taylor & Francis Group, CRC Press is an imprint of the Taylor & Francis Group, an informa Business, 2016. – ISBN 978–1–4822–3164–9

[24] Walker, John L. B. (Hrsg.): Handbook of RF and microwave power amplifiers. Cambridge, New York : Cambridge University Press, 2012 (The Cambridge RF and microwave engineering series). – ISBN 978–0–521–76010–2

[25] Sekiya, Hiroo, Xiuqin Wei, Tomoharu Nagashima, und Marian K. Kazimierczuk. „Steady-State Analysis and Design of Class DE Inverter at Any Duty Ratio". IEEE Transactions on Power Electronics 30, Nr. 7 (Juli 2015): 3685–94. https://doi.org/10.1109/TPEL.2014.2339355.

[26] Koizumi, H., T. Suetsugu, M. Fujii, K. Shinoda, S. Mori, und K. Iked. „Class DE high-efficiency tuned power amplifier". IEEE Transactions on Circuits and Systems I: Fundamental Theory and Applications 43, Nr. 1 (Januar 1996): 51–60. https://doi.org/10.1109/81.481461.

[27] Kyaw, Phyo Aung, Aaron L. F. Stein, und Charles R. Sullivan. „Power Handling Capability of Self-Resonant Structures for Wireless Power Transfer". In 2018 IEEE PELS Workshop on Emerging Technologies: Wireless Power Transfer (Wow), 1–6. Montréal, QC, Canada: IEEE, 2018. https://doi.org/10.1109/WoW.2018.8450908.

[28] Kwan, Christopher H., Juan M. Arteaga, David C. Yates, und Paul D. Mitcheson. „Design and Construction of a 100 W Wireless Charger for an E-Scooter at 6.78 MHz". In 2019 IEEE PELS Workshop on Emerging Technologies: Wireless Power Transfer (WoW), 186–90. London, United Kingdom: IEEE, 2019. https://doi.org/10.1109/WoW45936.2019.9030648.

[29] Kwan, Christopher H., Juan M. Arteaga, Samer Aldhaher, David C. Yates, und Paul D. Mitcheson. „A 600W 6.78MHz Wireless Charger for an Electric Scooter". In 2020 IEEE PELS Workshop on Emerging Technologies: Wireless Power Transfer (WoW), 278–82. Seoul, Korea (South): IEEE, 2020. https://doi.org/10.1109/WoW47795.2020.9291303.

Analysis and Design of Capacitive Coupling Wireless Power Transfer System Using Load-Independent Class-EF Inverter

Takumi Kobayashi†, Yutaro Komiyama†, Akihiro Konishi†, Hiroaki Ota‡,
Yuki Ito‡, Taichi Mishima‡, Takeshi Uematsu‡, Kien Nguyen†, Hiroo Sekiya†
† Graduate School of Science and Engineering, Chiba University, Chiba, Japan
‡ OMRON Corporation, Kyoto, Japan
Emal: takumi.kobayashi.0426@chiba-u.jp, sekiya@faculty.chiba-u.jp

Abstract—This study proposes an analysis and design of a capacitive coupling wireless power transfer (WPT) system using a load-independent (LI) class-EF inverter. The proposed system consists of a LI class-EF inverter, two pairs of coupling capacitors, an LCCL filter, and a class-D rectifier. Conventional capacitive coupling WPT system using a LI class-E inverter increases circuit cost by introducing high-voltage switches due to the high voltage stress of the switches, which is a bottleneck for achieving higher power. Another problem is that the input-output voltage ratio is uniquely determined once the switch time ratio and coupling capacitor specifications are determined, thus limiting design freedom. The proposed system uses a class-EF inverter to reduce the voltage stress on the switches and an LCCL filter to convert the LI constant-current (CC) output characteristic of the class-EF inverter to a constant-voltage (CV) characteristic, thereby achieving zero-voltage-switching (ZVS) and constant output voltage regardless of load variations. The output voltage can also be adjusted arbitrarily depending on the design of the LCCL filter. This study analytically demonstrates the design method of the proposed circuit and shows the effectiveness of the proposed circuit through actual device verification.

Keywords—Class-EF inverter, Capacitive Coupling, Load-independent, Wireless Power Transfer

I. INTRODUCTION

The development of wide-bandgap semiconductor devices has enabled higher-frequency circuits and accelerated the development of high-frequency wireless power transfer (WPT) systems. Applications of high-frequency WPT systems include implantable devices [1], automatic guided vehicles [2], and industrial robot arms [3].

One of the typical WPT methods is a capacitive coupling type. The capacitive coupling WPT system transmits power through the electric field between flat plates. Since the electric field responsible for power transmission is mainly limited to the area between the plates, the system is not easily affected by surrounding metallic materials and can be expected to be highly safe in metallic ambient environments.

In view of the application of WPT systems, it is required to maintain high power-delivery efficiency and constant output voltage against load variations. In general, load variations are feedback from the receiving side to the transmission side, and the inverter is controlled to maintain a constant output

voltage. However, in a WPT system, wireless communication is required for feedback from the power receiver to the power transmitter, so simpler measures are required.

As one approach, a capacitive coupling WPT system using a load-independent class-E inverter was proposed as shown in Fig. 1 [4]. A LI class-E inverter is a particular mode of operation of a class-E inverter designed with specific circuit parameters. This system can maintain zero-voltage-switching (ZVS) and constant output voltage without feedback control. However, the maximum switch voltage of a class-E inverter is approximately 3.5 times the input voltage, resulting in high voltage stress, which increases the circuit cost by introducing high voltage switches and becomes a bottleneck toward a higher power. Another issue is the narrow design freedom that the duty ratio of the switch and the capacitance of the coupling capacitors determine the input-output voltage gain.

This study proposes an analysis and design of a capacitive coupling WPT system using the LI class-EF inverter. As shown in Fig. 2, the proposed system uses the LI design for the class-EF inverter because the LI class-EF inverter achieves the ZVS without feedback control regardless of load variations and maintains the constant amplitude of the output current. Furthermore, by using a class-EF inverter, the maximum switch voltage is limited to 2.5 times the input voltage. However, assuming applications of WPT, a constant-voltage (CV) output is desirable, so it is necessary to convert the constant-current (CC) output on the inverter. Therefore, the proposed circuit uses an LCCL filter to convert the CC output on the transmission side into a CV output on the receiver side. The output voltage can be designed arbitrarily by the circuit constants of the LCCL filter. Furthermore, we conducted experiments on actual devices and confirmed that the proposed circuit achieves LI operation, demonstrating the effectiveness of the proposed circuit.

II. PROPOSED CIRCUIT

A. Circuit Configuration and Basic Operations

Fig. 3 shows the circuit configuration of the proposed capacitive coupling WPT system using a LI class-EF inverter. The proposed circuit consists of a LI class-EF inverter, two

979-8-3315-1612-3/25 $31.00 © 2025 IEEE

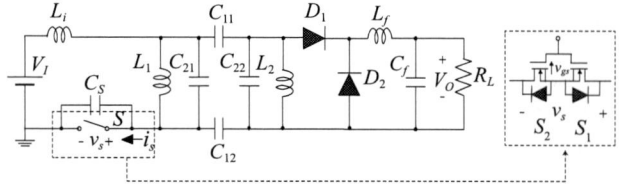

Fig. 1. Capacitive coupling WPT system with LI class-E inverter.

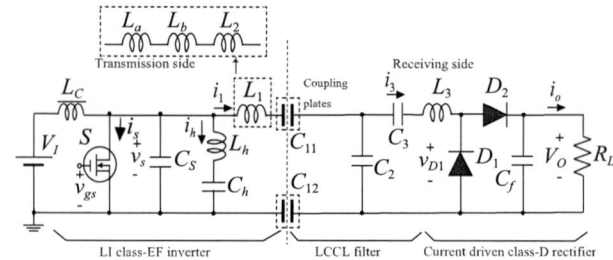

Fig. 3. Proposed circuit topology.

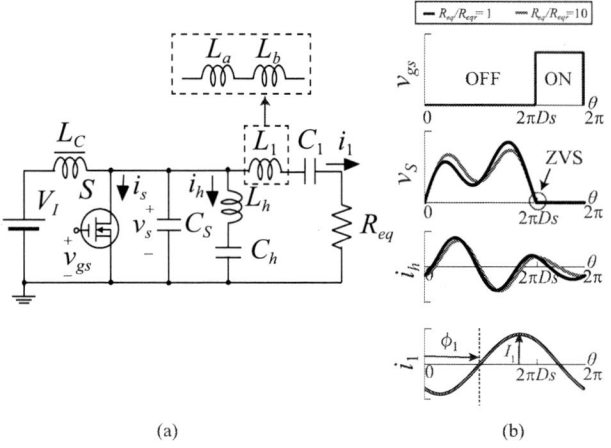

Fig. 2. LI class-EF Inverter. (a) Circuit topology. (b) Operating waveform example.

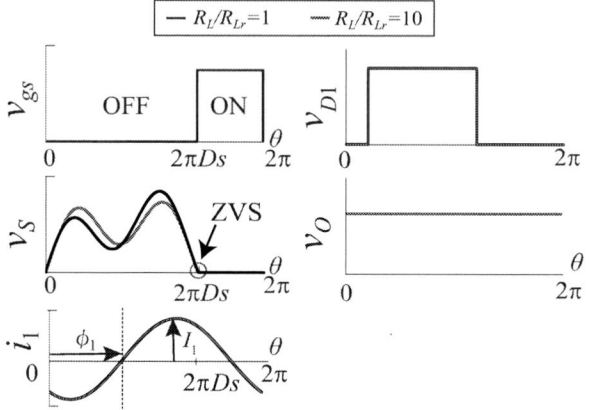

Fig. 4. Operating waveform example of proposed circuit.

pairs of coupling capacitors C_{11} and C_{12}, an LCCL filter, and a class-D rectifier. L_1 of the proposed system is divided into L_{1-2} of the resonant inductance of the LI class-EF inverter and L_2 of the LCCL filter. Furthermore, L_{1-2} of the resonant inductance of the class-EF inverter is divided into L_a and L_b, where L_a is the resonant component, and L_b is the extra inductance for the phase shift.

Fig. 4 shows an example of the operating waveform of the proposed system. In the proposed system, the switch always achieves ZVS at turn-on regardless of load variations due to the LI operation of the class-EF inverter. Harmonic currents of appropriate amplitude and phase generated by the higher-order resonant filter $L_h - C_h$ flows into the shunt capacitor C_s with low impedance during the switch-off period, reducing the peak switch voltage v_s. When the class-EF inverter achieves LI operation, the output current i_1, whose amplitude and phase are constant regardless of the load, flows into the LCCL filter. The current i_3 passing through the LCCL filter flows into the class-D rectifier, and the diode voltage v_{D1} of the class-D rectifier is a square wave due to the smoothing capacitor C_f. By using a sufficiently large smoothing capacitor C_f, the output voltage V_o of the rectifier becomes a DC voltage without ripple.

Fig. 5 shows the equivalent circuit transformation procedure. The flow of the equivalent circuit transformation is as follows. Fig. 5(a) is transformed to Fig. 5(b) by converting the class-D rectifier to the pure resistance R_i [5]. Next, to replace Fig. 5(b) with Fig. 5(c), the LCCL filter is converted to an

impedance viewed from the inverter. Here, the LCCL filter must satisfy the following conditions (1) and (2) so that the reactance component of the input impedance is constant with respect to load variations. Conditions (1) and (2) are expressed as

$$1 - \omega^2 L_2 C_2 = 0, \tag{1}$$

$$\omega(L_3 - L_2) - \frac{1}{\omega C_3} = 0. \tag{2}$$

In this case, the equivalent impedance is expressed as

$$R_{eq} = \frac{1}{\omega^2 C_2^2 R_i} = \frac{\omega^2 L_2^2}{R_i} = \frac{\pi^2 \omega^2 L_2^2}{2R_L}, \tag{3}$$

$$X_{eq} = 0, \tag{4}$$

where only the resistive component remains. By the above equivalent transformations, the proposed circuit can be equivalently attributed to the same topology as the LI class-EF inverter. Therefore, the LI class-EF inverter of the proposed circuit achieves ZVS and constant output current regardless of load variations. The output voltage V_o is expressed as

$$V_o = \frac{I_1}{\sqrt{2}} \sqrt{R_{eq} R_L} = \frac{\pi \omega L_2}{2} I_1, \tag{5}$$

where I_1 is the amplitude of the output current i_1 of the inverter; the LI design of the class-EF inverter [6] ensures that I_1 is constant regardless of load variations, and consequently $V\phi$ is also constant. The output voltage can be designed

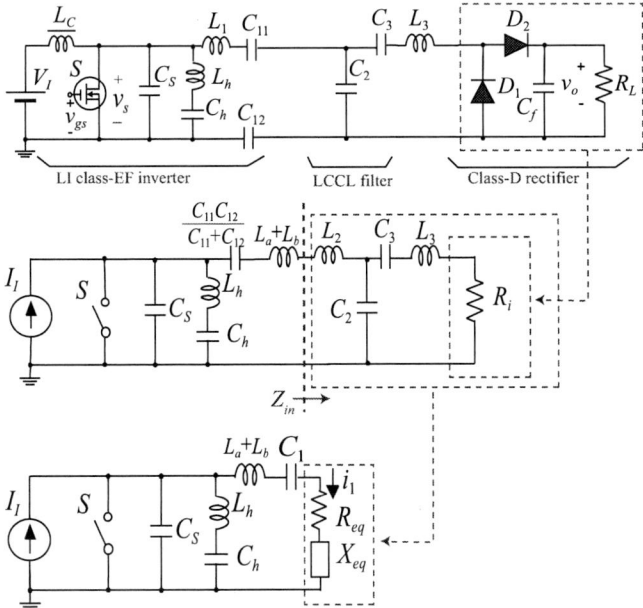

Fig. 5. Equivalent circuit transformation of the proposed system.

arbitrarily according to the inductance L_2. By satisfying the above design conditions, the proposed capacitive-coupling WPT system can achieve both the LI operation and freedom output-voltage design.

III. CIRCUIT EXPERIMENT

A. Specifications and Circuit Design

In this design, as experimental specifications, input voltage $V_I = 80$ V, operating frequency $f = 6.78$ MHz, rated output voltage $V_{Or} = 16.8$ V, rated load resistance $R_{Lr} = 14.7$ Ω, and off-duty ratio $D_S = 0.7$ are given. The range of variations of the load resistance is $R_{Lr} \leq R_L \leq 10R_{Lr}$. The two coupling capacitors used are made of a 30 cm×30 cm×2 mm, aluminum plate A6061 (eMetals) [8] and Kapton film 3-1966-05 (AS ONE) with relative permittivity ($\varepsilon_r = 3.4$) [9], bonded with copper tape and soldered to the conductors. The capacitance of the coupling capacitor is 1.0 nF when the distance between the electrode plates is set to 1 mm. L_3 has degrees of freedom, and the L_3 created is 327 nH.

From the experimental specifications and [6], the important parameters for the LI design of class-EF inverter are obtained as

$$\phi_1 \approx -2.20, \tag{6}$$

$$\omega_h = \frac{1}{\omega\sqrt{L_h C_h}} \approx 1.67, \tag{7}$$

$$\gamma = \frac{C_h}{C_s} \approx 0.764, \tag{8}$$

and

$$I_1 \approx 15.3 f C_s V_I = 1.59. \tag{9}$$

where ϕ_1 is the phase difference between the output current and turn-off instant of the drive signal, ω_h is the ratio of the harmonic resonance frequency to the resonance frequency, and γ is the ratio of the shunt capacitor C_s to C_h. From the experimental specifications and (8), C_h is obtained as

$$C_h = 0.764 C_s = 146 \text{ pF}. \tag{10}$$

By substituting (10) into (7), L_h is obtained as

$$L_h = \frac{1}{\omega^2 \omega_h^2 C_h} = 1.36 \ \mu\text{H}. \tag{11}$$

From the value of the coupling capacitor C_1, L_a is obtained as

$$L_a = \frac{1}{\omega^2 C_1} = 1.10 \ \mu\text{H}. \tag{12}$$

From the LI conditions in (6)-(8), and (27) of [7], L_b is obtained as

$$L_b = \frac{1}{\pi \omega I_1} \int_0^{2\pi} v_s(\theta)\cos(\theta + \phi)\, d\theta$$
$$= 986 \text{ nH}. \tag{13}$$

By substituting V_{or} and (9) into (5), L_2 is obtained as

$$L_2 = \frac{2V_o}{\pi \omega I_1} = 158 \text{ nH}. \tag{14}$$

From (12)-(14), L_1 is obtained as

$$L_1 = L_a + L_b + L_2 = 2.25 \ \mu\text{H}. \tag{15}$$

By substituting (14) into (1), C_2 is obtained as

$$C_2 = \frac{1}{\omega^2 L_2} = 3.48 \text{ nF}. \tag{16}$$

By substituting L_3 and (14) into (2), C_3 is obtained as

$$C_3 = \frac{1}{\omega^2(L_3 - L_2)} = 3.33 \text{ nF}. \tag{17}$$

Table I shows the circuit parameters of the proposed circuit. Note that the measured C_s does not include the parasitic capacitance of the switches. In the experiment, GS66502B (GaN System) is used as the switching device, STPS5H100 (STMicroelectronics) as the rectifier diode, and UCC27512 (Texas Instruments) as the gate driver. A photograph of the implemented circuit is shown in Fig. 6. The size of the substrate created is 4.5 cm × 6.5 cm for the transmitter side and 4.0 cm × 4.0 cm for the receiver side.

B. Waveform Evaluation

Fig. 7 shows the operating waveforms of the proposed circuit for $R_L/R_{Lr} = 1$ and $R_L/R_{Lr} = 10$. The figure shows that the switch voltage achieves ZVS regardless of the load variations. The peak switch voltage is 200 V, which is 2.5 times lower than the input voltage of 80 V. Furthermore, the amplitude and phase of the current i_1 flowing through L_1 are constant regardless of load variations, resulting in a constant output voltage V_o. Therefore, it is confirmed that the proposed circuit is robust to load variations.

TABLE I
ELEMENT VALUE OF THE PROPOSED CIRCUIT

Symbol	Analytical	Measured
L_h	1.36 μH	1.36 μH(ESR : 171 mΩ)
L_1	2.25 μH	2.25 μH(ESR : 258 mΩ)
L_3	325 nH	327 nH(ESR : 60 mΩ)
C_s	191 pF	80 pF
C_h	146 pF	146 pF
C_{11}	1.00 nF	0.999 nF(ESR : 533 mΩ)
C_{12}	1.00 nF	1.01 nF(ESR : 640 mΩ)
C_2	3.48 nF	3.48 nF
C_3	3.33 nF	3.33 nF

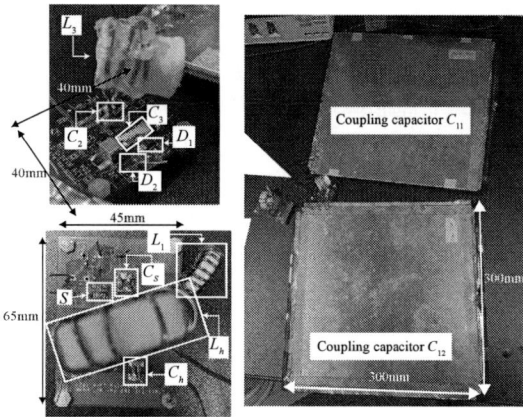

Fig. 6. Experimental circuit.

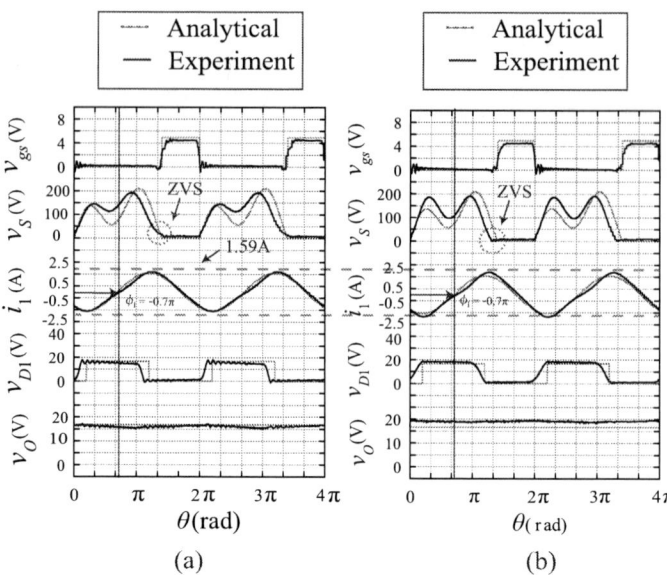

Fig. 7. Operating waveform (solid line: experimental waveform, dashed line: analytical waveform) (a) $R_L/R_{Lr} = 1$ (b) $R_L/R_{Lr} = 10$.

C. Circuit Characteristics

Fig. 8 shows the analytical and experimental output voltage V_o and power-delivery efficiency η for load variations. From the Fig. 8, the output voltage V_o is constant with respect to the load variations. In addition, the power-delivery efficiency η is equivalent to the analysis, and the maximum power-delivery efficiency is 81.2 % under rated condition $R_L/R_{Lr} = 1$. The power-delivery efficiency decreases as the load resistance increases, but this is because the equivalent load resistance in the class-EF inverter becomes smaller as the load variations increase, and the power loss in the ESR of the circuit elements becomes dominant.

A comparison of the conventional and proposed systems is shown in Table II, where I_m in the table is the amplitude of the constant output current of the inverter of the conventional system, and C_m is the same as C_1 in Fig. 5 of this paper. Table II shows that the proposed system has a smaller maximum switch voltage than the conventional system. The equation for the output voltage V_o shows that the proposed system can be adjusted using the L_2 of the LCCL filter, whereas the conventional system depends on the off-time ratio of the switch and the capacitance of the coupling capacitor, indicating a wide range of design options. It can also be seen that the proposed system is as efficient as the conventional system, although the power and other factors are different.

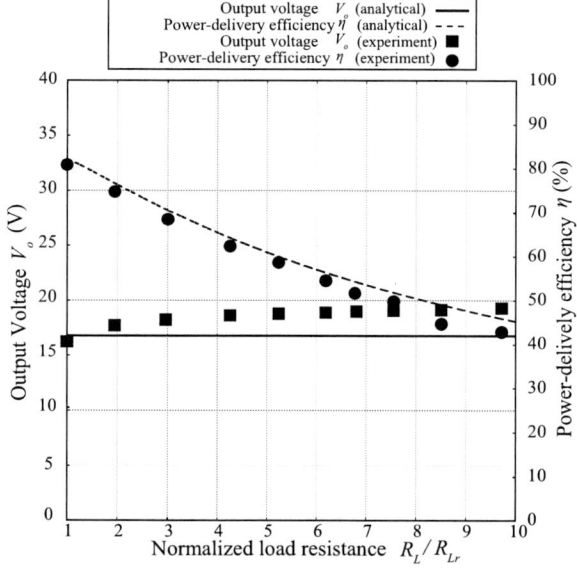

Fig. 8. Characteristics of the proposed system with respect to load variations.

IV. CONCLUSION

This study presents an analysis and design method for a capacitive coupling LI WPT system that can be designed for any output voltage independent of the load resistance by applying an LCCL filter. The proposed system achieves ZVS independent of load variations and constant output voltage without additional control by using a LI design. In the experiment, the experimental circuit always achieved the designed output voltage and ZVS independent of the load variations. As a result, the proposed system achieved LI operation and high

TABLE II
COMPARISON OF CONVENTIONAL AND PROPOSED SYSTEMS.

	[4]	Proposed
Number of passive components	11	13
The operating frequency f	1.00MHz	6.78MHz
Typical ON-duty ratio	0.5	0.3
Soft switching type	ZVS	ZVS
Inverter output characteristic	CC	CC
Analytical switch-voltage stress V_{Smax}/V_I at the rated condition	3.5	2.5
Formula for output voltage V_o	$\frac{2I_m}{\pi\omega C_m}$	$\frac{\pi\omega L_2}{2}I_1$
Maximum power-delivery efficiency η in the experiment	81.1 %	81.2 %

power conversion efficiency regardless of the MHz drive. This demonstrates the effectiveness of the proposed design method.

REFERENCES

[1] Chang, Fu-Wen and Hsieh, Ping-Hsuan, "A 13.56-MHz Wireless Power Transfer Transmitter with Impedance Compression Network for Biomedical Applications," *2020 IEEE International Symposium on Circuits and Systems (ISCAS)*, pp.1-5, 2020

[2] M. Sugino and T. Masamura, "The wireless power transfer systems using the Class E push-pull inverter for industrial robots," *2017 IEEE Wireless Power Transfer Conference (WPTC)*, Taipei, Taiwan, 2017, pp. 1-3,

[3] H. Sekiya, K. Tokano, W. Zhu, Y. Komiyama and K. Nguyen, "Design Procedure of Load-Independent Class-E WPT Systems and Its Application in Robot Arm," *IEEE Transactions on Industrial Electronics*, vol. 70, no. 10, pp. 10014-10023, 2023.

[4] K. Shinde and H. Koizumi, "Capacitive Power Transfer System Using a Load-Independent Class E Zero Voltage Switching Parallel Resonant Inverter and a Class D Voltage-Driven Rectifier," *IEEE Transactions on Circuits and Systems II: Express Briefs*

[5] M. T. Outeiro, G. Buja and D. Czarkowski, "Resonant Power Converters: An Overview with Multiple Elements in the Resonant Tank Network," *IEEE Industrial Electronics Magazine*, vol. 10, no. 2, pp. 21-45, June 2016

[6] Y. Komiyama , A. Komanaka, H. Ota, Y. Ito, T. Mishima, T. Uematsu, A. Konishi, W. Zhu, K. Nguyen, and H. Sekiya, "Analysis and design of high frequency WPT system using load-independent inverter with robustness against load variations and coil misalignment," *IEEE Access*, pp. 1-1, 2024.

[7] Hanxiao Wang, Yutaro Komiyama, Xiuqin Wei, Akihiro Konishi, Kien Nguyen, Hiroo Sekiya, "Analysis and design of high-frequency multiple-output wireless power transfer system with load-independent class-E/F inverter," *international journal of circuit theory and applications*, Jan. 2024.

[8] "A6061," eMetals, https://www.e-metals.net/product/201457/, June 2024

[9] "3-1966-05 Polyimide film Kapton（R）," AS ONE https://jp.misumi-ec.com/vona2/detail/223007899167/?HissuCode=3-1966-05, June 2024

Design and Optimization of a 600 W Wireless Drone Charger for High Gravimetric Power Density

Arka Basu *Student Member, IEEE* and Daniel Costinett, *Senior Member, IEEE*
Department of Electrical Engineering and Computer Science
The University of Tennessee, Knoxville
Knoxville, TN, USA
abasu1@vols.utk.edu

Abstract—**This article presents a systematic approach to reduce the total weight of the receiver components in a wireless charger for drones. An overall design and optimization methodology, based on loss, thermal and weight models, incorporating all stages of the on-board receiver, is presented to maximize the gravimetric power density of the wireless charger. A 600 W prototype leveraging GaN technology is built to validate the system modeling and design. The receiver achieves a gravimetric power density of 11.8 W/g not considering the weight of controls and sensing.**

Index Terms—**Wireless power transfer (WPT), power density, synchronous rectifier, swiched-capacitor converter.**

I. INTRODUCTION

Wireless charging of small unmanned aerial vehicles (UAV), or drones, has received significant attention since this technology renders full autonomy to UAV fleets [1]. These systems are designed to achieve high gravimetric power density of the on-board receiver by reducing the weight of the receiver (Rx) coil and converter, and simultaneously increasing power output to maximize drone flight time and minimize recharging time. In this application designs which achieve below maximum efficiency may be preferable if they result in weight savings greater than the additional thermal management weight required. The offboard transmitter components, conversely, are not heavily weight-constrained and are designed to help reduce the weight of the on-board receiver.

Power density optimization related to weight, area and volume of magnetic [2]–[4] or capacitive couplers [5] are abundant in wireless power transfer (WPT) literature. Other applications, such as consumer electronics and data center applications, focus on high power density of converter stages [6], [7]. For drone applications, [8] introduces an overall modeling, design and optimization methodology, based on loss, thermal and weight models, incorporating all stages of the on-board receiver, to reduce the weight of the wireless charger. The design in [8] uses a buck dc-dc converter, where the output inductor accounts for 40% of the converter weight. To

---

Research was sponsored by the Army Research Laboratory and was accomplished under Cooperative Agreement Number W911NF2220007. The views and conclusions contained in this document are those of the authors and should not be interpreted as representing the official policies, either expressed or implied, of the Army Research Office or the U.S. Government. The U.S. Government is authorized to reproduce and distribute reprints for Government purposes notwithstanding any copyright notation herein.

---

further improve the power density of the receiver, this article considers inductor-less topologies for the dc-dc stage of the receiver. Also, for efficient cooling at a higher output power, thermal management systems consisting of fan and heat sink are considered for the receiver converters along with the PCB cooling solution adopted in [8].

Section II presents the loss, thermal and weight models of the different stages of the receiver. Using these models the various stages of the receiver are designed and optimized simultaneously to maximize the gravimetric power density of the on-board receiver in Section III. Section IV discusses the experimental results. Finally, Section V gives the conclusions.

II. SYSTEM STRUCTURE AND MODEL

The schematic of the system being designed is shown in Fig. 1. The Rx coil with compensating capacitors, a full-bridge (FB) diode rectifier (DR) or active rectifier (AR), and an optional pure switched-capacitor (SC) converter constitute the receiver. The basic 2:1 SC, series-parallel (SP) and ladder are considered as candidate topologies in this article to achieve high power density. Fig. 2 shows the schematics of the 2:1 SC, 3:1 SP and 3:1 ladder converters, respectively. The transmitter includes a FB inverter and a transmitter (Tx) coil with matching capacitors. The analysis and design in this article are based on the specifications in Table I.

A. Electrical Model of Receiver

1) Loss Model of Rectifier: The AR is desired to operate at near-resistive load condition, with zero-voltage switching

TABLE I
SYSTEM SPECIFICATIONS

Parameter	Specification
V_{out}	22.2 V
P_{out}	600 W
Coil temperature rise	80° C
Power stage temperature rise	50° C

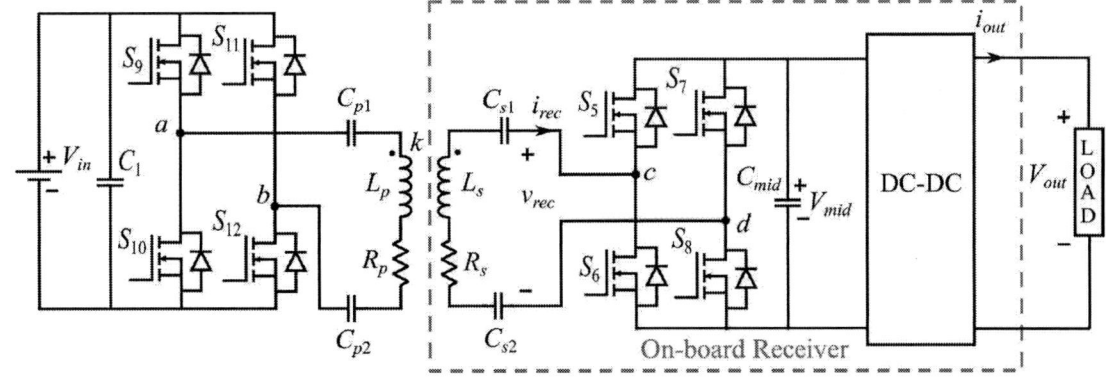

Fig. 1. Schematic of the designed system, composed of inverter, coils, active rectifier, and optional dc-dc converter.

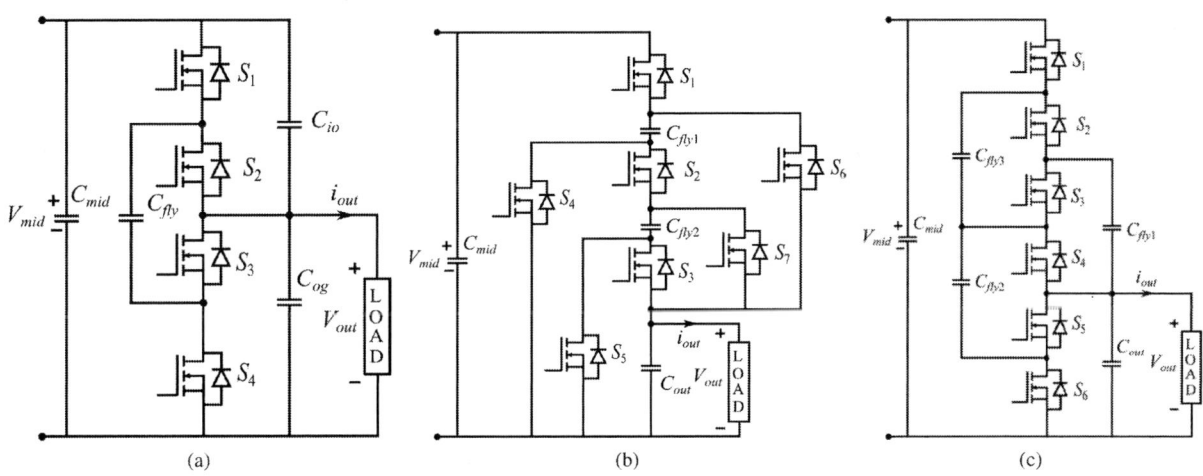

Fig. 2. Circuit schematics: (a) 2:1 SC converter, (b) 3:1 SP [11], and (c) 3:1 Ladder [11].

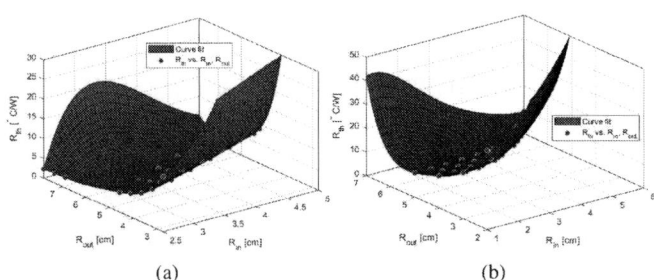

Fig. 3. Curve fitting of R_{th} for different gauge coils, obtained from Icepak simulations: (a) 14 AWG, (b) 22 AWG.

(ZVS) achieved through nominal phase-shift. The minimum dead time to turn on all the FETs at zero voltage is [9]

$$dt_{rec} = \frac{1}{\omega_{s,wpt}} \cos^{-1}\left(1 - \frac{C_{eq,Q,rec}V_{mid}\omega_{s,wpt}}{I_{rec}}\right) \quad (1)$$

where the nonlinear C_{oss} of the FET is approximated by the charge equivalent linear capacitance $C_{eq,Q,rec}$ [10], $I_{rec} = \pi P_{out}(2V_{mid})^{-1}$ is the peak rectifier input current, and $\omega_{s,wpt}$ is the rectifier switching frequency. Due to ZVS of rectifier FETs, only gate charge loss is considered in terms of switching loss.

The AR conduction loss is

$$P_{cond,AR} = R_{ds,AR}I_{rec}^2 \quad (2)$$

where $R'_{ds,AR}$ is the FET conduction resistance when it is on. Conduction loss for the DR is

$$P_{cond,DR} = 4(V_0 I_{d,avg} + I_{d,rms}^2 R_d)$$
$$= 4\left(V_0\frac{I_{rec}}{\pi} + \frac{I_{rec}^2}{8}R_d\right) \quad (3)$$

where V_0 is the diode threshold volatge and R_d is the dynamic ON-resistance of the diode.

2) Loss Model of dc-dc Stage: The circuit models of the various dc-dc switched capacitor topologies are built in PLECS to construct discrete time state space models of the converters in MATLAB. Augmented state space models are then used to solve the steady-state average efficiency as detailed in [12]–[16]. Circuit models include losses due to transistor on-resistance, C_{oss} capacitance, and flying capacitor ESR.

3) Loss Model for Rx coil: Litz wire of different gauges is considered to model the Rx coil. The length of the winding, with negligible pitch between turns, is

$$l_{rx} = \pi N_{rx}(d_{out,rx} + d_{in,rx})/2 \quad (4)$$

979-8-3315-1612-3/25 $31.00 © 2025 IEEE

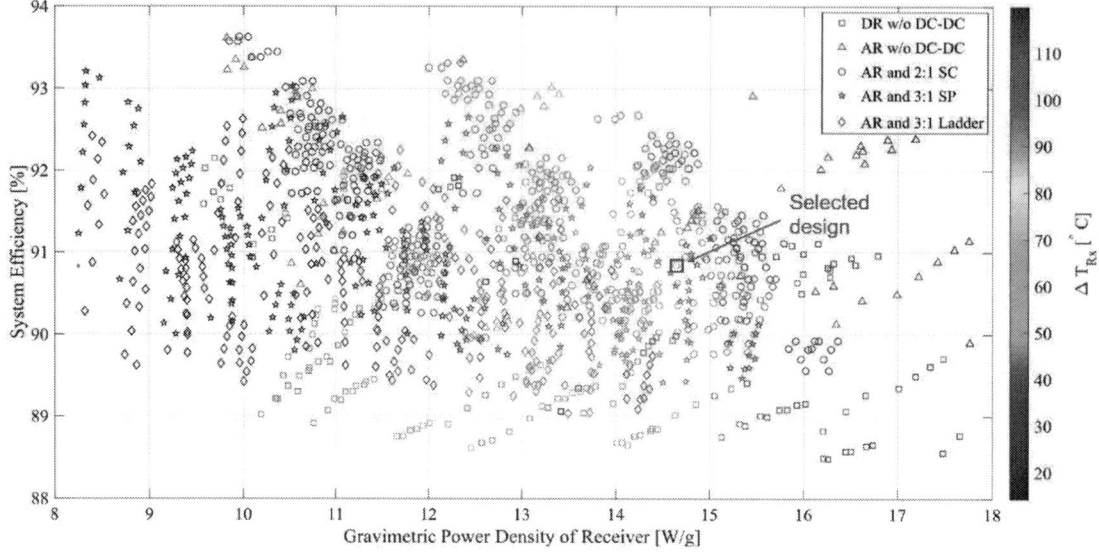

Fig. 4. System efficiency vs receiver gravimetric power density trade-off for the evaluated designs at 600 W.

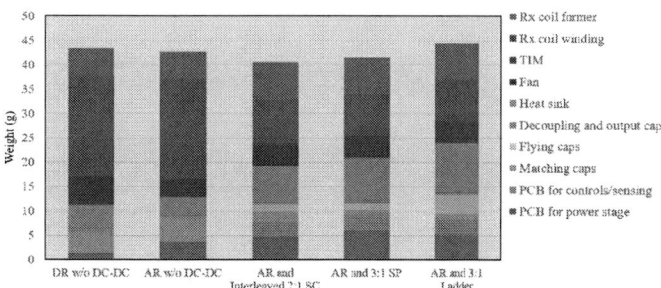

Fig. 5. Receiver weight breakdown for each optimized topology at 600 W.

TABLE II
DETAILS OF THE SELECTED DESIGN

Component	Part number	Parameter	Value
$S_1 - S_4$	EPC2067	$f_{s,SC}$	500 kHz
$S_5 - S_8$	EPC2071	$f_{s,wpt}$	105 kHz
Heat sink	CoolInnovations 3-121202UBFA	C_{fly}	43 μF
Fan	ASB0305HP-00	$C_{s1} = C_{s2}$	2.17 μF
		Rx coil	3 turns, $d_{in,rx}$ = 9 cm, 100/38 AWG Litz

where $d_{in,rx}$ and $d_{out,rx}$ are the inner and outer diameters of the coil. Then the coil DC resistance is

$$R_{dc} = \rho_{cu} \frac{l_{rx}}{a_{w,rx}} \tag{5}$$

where $a_{w,rx}$ is the bare copper area of the wire and ρ_{cu} is the copper resistivity.

To account for the losses due to skin effect and proximity effect, the AC resistance factor [17] of the coil is

$$F_R = \frac{R_{ac}}{R_{dc}} = 1 + \frac{k_{rx}\pi^3(n_{rx}N_{rx})^2 d_s^6}{3 \cdot 192\delta_{rx}^4 b_{rx}^2}. \tag{6}$$

where n_{rx} and N_{rx} are the number of strands and number of turns, d_s is the strand diameter, δ_{rx} and b_{rx} are the skin depth and winding breadth of the coil, and k_{rx} is a factor for Rx coil magnetic field distribution. Then the power loss in the coil is

$$P_{rx} = \frac{1}{2}I_{rec}^2 R_{dc} F_R \tag{7}$$

B. Receiver Thermal Model

1) Thermal Model for Rectifier and SC Stage: In [8], large regions of copper on the receiver PCB are used to dissipate the converter power loss through natural convection. Ellison's correlation model [18] is used to compute the required copper plane area. Apart from this PCB cooling solution, heat sinks under natural and forced convection are also considered for thermal management of the converters on the on-board receiver. A single heat sink, and for the air-cooled designs a single fan along with the heat sink is considered for cooling both the rectifier and SC stages. The temperature rise of the entire power stage consisting of the rectifier and SC stages is

$$\Delta T_{ps} = (P_{SC} + P_{rec})(R_{\theta cs} + R_{\theta sa}) \tag{8}$$

where P_{SC} and P_{rec} are the total losses of the SC and rectifier stage, respectively. $R_{\theta cs} = d_{TIM}/(k_{TIM}A_{TIM})$ is the thermal resistance of the TIM, where d_{TIM}, A_{TIM} and k_{TIM} are the thickness, surface area and thermal conductivity of the TIM, respectively. $R_{\theta sa}$ is the thermal resistance of the heat sink obtained from the data sheet. For the naturally cooled heat sinks $R_{\theta sa}$ value corresponding to that condition is selected. For the forced cooling cases $R_{\theta sa}$ changes with the air flow rate, with higher air flow resulting in a lower value of $R_{\theta sa}$.

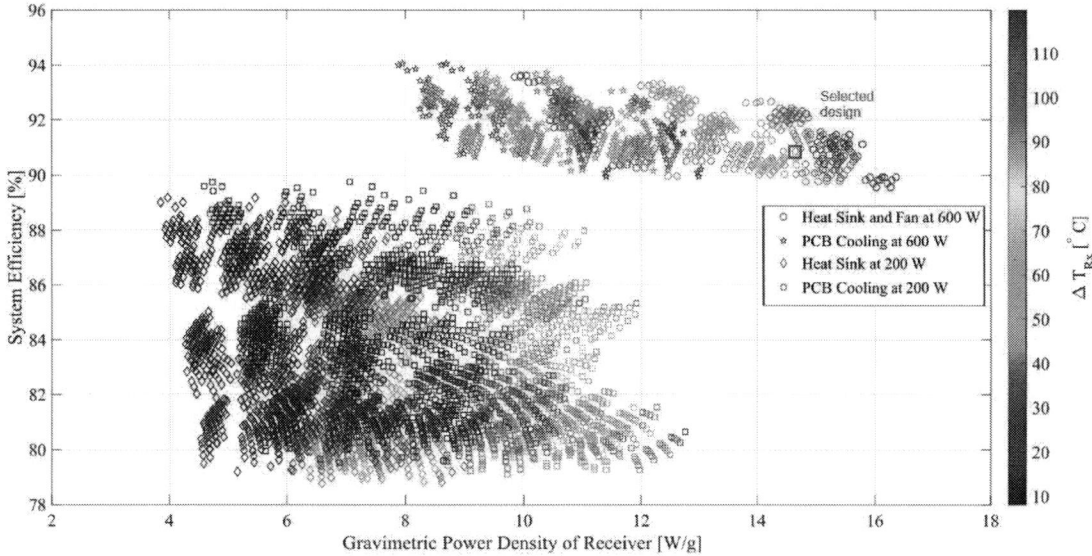

Fig. 6. System efficiency vs receiver gravimetric power density trade-off for AR plus 2:1 SC at different powers and different cooling schemes.

2) *Thermal Model of Rx Coil:* The Rx coil thermal resistances (R_{th}) are generated by curve-fitting data obtained from running thermal simulations in Ansys Icepak as described in [8]. In this work, the model is improved from [8] by incorporating R_{th} values for all the considered equivalent wire sizes. In the simulations the coils are approximated as toroids having a rectangular cross-section. The equation for the curve fit is

$$R_{th} = \sum_{m=0}^{5} \sum_{n=0}^{5} P_{mn} R_{in}^m R_{out}^n \tag{9}$$

where R_{in} and R_{out} are the inner and outer radii of the Rx coil, and P_{mn} are the fitted coefficients. Two examples of the curve fit for two different equivalent wire gauges are shown in Fig. 3.

The resulting empirical model for the Rx coil temperature rise is

$$\begin{aligned}
&a\Delta T_{rx}^2 + b\Delta T_{rx} + c = 0 \\
&a = 3 \cdot 768 b_{rx}^2 \rho_{cu,amb}^2 (0.5 I_{rec}^2 R_{th} R_{dc,amb} \alpha_{cu}^2 - \alpha_{cu}), \\
&b = 3 \cdot 768 b_{rx}^2 \rho_{cu,amb}^2 (I_{rec}^2 R_{th} R_{dc,amb} \alpha_{cu} - 1), \\
&c = 0.5 I_{rec}^2 R_{th} R_{dc,amb} (3 \cdot 768 b_{rx}^2 \rho_{cu,amb}^2 + \\
&\qquad \pi^3 \omega_{s,wpt}^2 \mu^2 n_{rx}^2 d_s^6 k_{rx})
\end{aligned} \tag{10}$$

where $R_{dc,amb}$ is the dc resistance of the coil at ambient, $\rho_{cu,amb}$ and α_{cu} are the copper resistivity at ambient temperature and the temperature coefficient of resistance for copper.

C. Weight Models

To model the PCB weight the PCB area is approximated first. A PCB layout is designed with the switches, decoupling, output, half of the matching and 0603 flying capacitors (if any) on the top layer, with the power stage being top-cooled for the heat sink designs. The bottom layer contains the bigger size flying capacitors, remaining half of the compensating capacitors, and the gate drive, auxiliary components and power supplies to improve electrical performance by reducing parasitics. An example design of the top layer PCB layout is illustrated though the AR plus 2:1 SC case. The AR FETs are placed in a column with switch node and ground traces separating them. The four SC FETs are placed beside the AR FETs separated by the V_{mid} trace. S_1 and S_2 are placed in one row, and S_3 and S_4 are placed in a second row with their switch node traces and 0603 flying capacitors separating the two rows. For the interleaved design, the second phase is placed beside the first phase separated by 2 mm. The smallest heat sink area is calculated so that it covers the AR and SC FETs allowing for a 3 mm clarence on all sides. The output, matching and rectifier decoupling capcitors with ground traces to complete the power loop are placed beyond the heat sink. Thus the total area of the PCB top layer is calculated. The total area of the PCB bottom layer is approximated as three times the bottom layer component area to facilitate routing and hand-soldering. The greater of these two areas is considered as the PCB area, which changes with the power stage size and complexity. For the PCB cooled designs, the bigger of the copper plane area required to dissipate the converter losses through natural convection and the PCB bottom layer area is considered as the PCB area. The weight of the receiver PCB is modeled for a 31 mil thick PCB with FR4 dielectric, and four 1-oz copper layers, two of them having fill factors of 70% and the other two of 30%. The weights of heat sink, fan and capacitors are obtained from their respective datasheets.

The thermal interface material (TIM) weight is modeled as

$$m_{TIM} = \rho_{TIM} d_{TIM} A_{TIM} \tag{11}$$

where ρ_{TIM} is the density of the TIM. The Rx coil winding mass is modeled as the sum of the copper weight and the insulation weight [8].

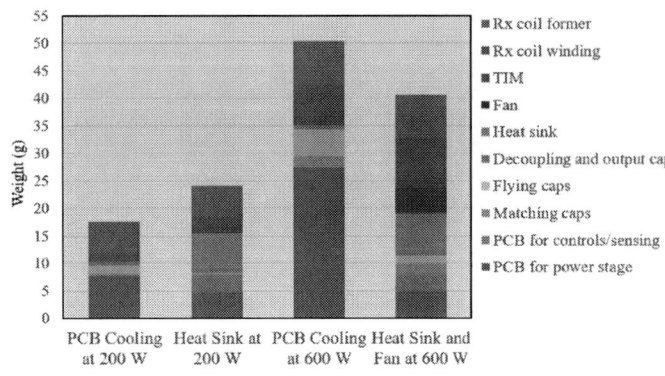

Fig. 7. Receiver weight breakdown for AR plus 2:1 SC at 200 and 600 W.

TABLE III
PROTOTYPE COMPONENTS AND PARAMETERS

Component	Part number	Specification
$S_1 - S_4$	EPC2067	40 V, 1.55 mΩ
$S_5 - S_8$	EPC2071	100 V, 2.2 mΩ
C_{mid}	GRM32ER71J106KA12L	63 V, 10 μF × 6
C_{fly}	GRM32ER71J106KA12L; GRM188R6YA106MA73D	63 V, 10 μF × 12; 35 V, 10 μF × 12
$C_{io} = C_{og}$	GRM188R6YA106MA73D	35 V, 10 μF × 6
C_{s1} and C_{s2}	C2220C944K5GLCAUTO; C2220C274J5GACTU	50 V, 0.94 μF × 2; 50 V, 0.27 μF
Heat sink	CoolInnovations 3-121202UBFA	1.23 in × 1.23 in × 0.2 in
TIM	TG-A1780-40-40-1.0	17.8 W/m·K
Fan	ASB0305HP-00	5.1 CFM
Gate driver	UCC27282-Q1	–
Digital isolator	ISO6720FBDR	–
Rx coil	–	3 turns, $d_{in,rx}$ = 9 cm, 100/38 AWG Litz

III. SYSTEM-WIDE DESIGN AND OPTIMIZATION

The goal of this work is to optimize the system-level design space combining all stages of the receiver for maximizing gravimetric power density of the on-board receiver. Thus, low-Q receiver coils with a small number of turns may show superior performance despite being suboptimal in efficiency-constrained applications.

Because both the transmitter and receiver coils are being designed, V_{mid} is not fixed and is a design variable, allowing switched capacitor converters of varying conversion ratios while keeping the same output voltage and power. Higher voltage and lower current may benefit the Rx coil and rectifier, allowing reduced conduction loss and weight of the thermal system, while lower voltage may be advantageous for the dc-dc converter weight and loss. Hence, this analysis considers receiver designs with and without dc-dc switched capacitor stages between the rectifier and load.

For the Rx coil design, Litz wires of 38 AWG strands, and equivalent gauges from AWG 10 to AWG 26 are considered

for AC loss reduction. $d_{in,rx}$ range over 3 cm to 13 cm, and 1 to 9 turns are considered as part of the optimization. For the DR, Schottky diodes of peak reverse voltages from 40 V to 100 V and average rectified currents from 15 A to 100 A are considered. For the AR and SC stages, EPC eGaN FETs of drain-to-source voltages from 40 V to 100 V and continuous drain currents from 29 A to 90 A are considered. For the SC stage, 35 V to 100 V and 10 μF to 47 μF Murata capacitors are considered to implement the flying capacitance. AR switching frequency $f_{s,wpt}$ and SC converter switching frequency $f_{s,SC}$ are swept from 50 kHz to 1 MHz and 100 kHz to 5 MHz, respectively. Up to two phases of the SC converters are considered for optimizing the dc-dc stage. CoolInnovation fan sinks of footprints from 1 in × 1 in to 2.46 in × 1.78 in, and BGA heat sinks of footprints from 0.54 in × 0.54 in to 2.46 in × 1.23 in are considered as part of the optimization. DC fans of dimensions from 20 mm × 20 mm to 40 mm × 40 mm and air flow rate from 1.12 CFM to 7 CFM are considered for the forced cooled designs. EPC recommended T-Global A1780 is considered as TIM.

The weights of the Rx coil, receiver stage PCB, heat sink, fan, TIM, flying, decoupling, output and matching capacitors are calculated in the design and optimization. The trade-off between system efficiency and gravimetric power density of the on-board receiver for the considered designs is shown in Fig. 4. The two phase interleaved designs for the 3:1 SC converters were not included in Fig. 4 because they were worse for all possible cases. The heat sink, fan, TIM and flying capacitor are selected such that the total receiver weight is minimized. Also, the Rx coil designs are iterated for various V_{in} and the combination yielding minimum total weight for the Rx coil and compensating capacitor is selected.

From the weight breakdown of the receiver for each optimized topology detailed in Fig. 5, it is clear that the total receiver weight goes down from rectifier-only designs to a 2:1 interleaved SC stage, but the weight again goes up as we move to a 3:1 SC stage. The Rx coil weight goes down as V_{mid} increases, but the power stage weight goes up due to increased weight of the larger PCB area and thermal management required to deal with the increased dc-dc converter losses, resulting in minimum total receiver weight for the AR plus 2:1 interleaved SC design. Table II reports the details of the selected design.

Apart from being lower power density, the DR without dc-dc designs also lack any controls on the receiver side. For safe battery charging, this design would need additional wireless controls and communication to allow charging control from the transmitter side. Phase shift regulation in the AR-based topologies [19] allows local receiver-side charging control.

The effectiveness of the PCB cooling solution at lower powers can be seen from Fig. 6. While at 600 W the air-cooled AR plus 2:1 SC designs have better power density compared to the PCB cooled designs, the receivers with PCB cooling have better gravimetric power density than the heat sink designs at 200 W. From the weight breakdown of the lightest 2:1 SC designs under different cooling mechanisms and different

Fig. 8. Photographs of the 600 W receiver prototype: (a) power stage, (b) Rx coil.

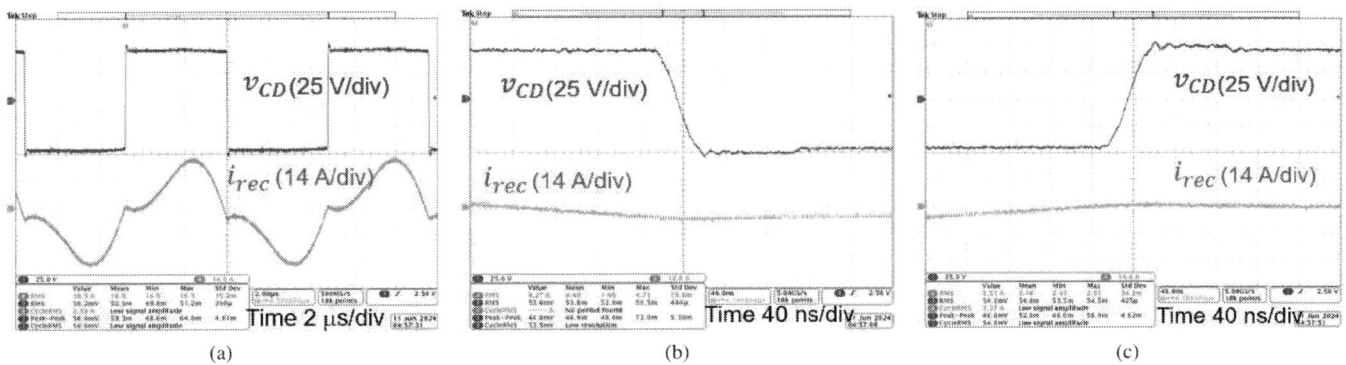

Fig. 9. i_{rec} and v_{cd} of the AR at full power: (a) two full periods, (b) zoomed-in view of falling edge of v_{CD}, and (c) zoomed-in view of rising edge of v_{CD}.

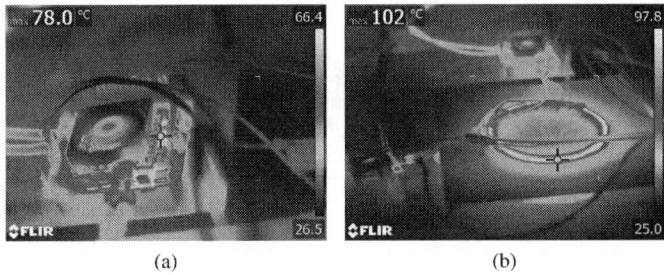

Fig. 10. Thermal photographs of prototype at full power: (a) receiver power stage, (b) Tx and Rx coils.

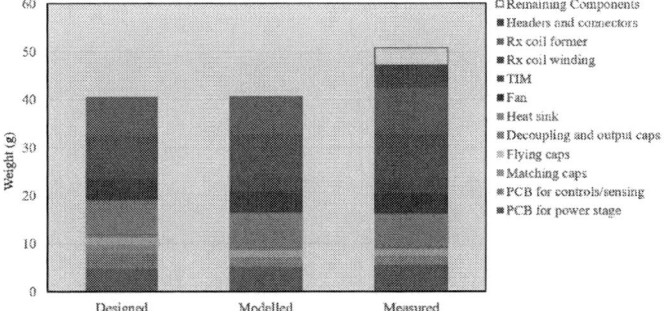

Fig. 11. Weight distribution of the 600 W receiver prototype.

powers shown in Fig. 7, it is clear that at 600 W the PCB weight dominates due the higher losses. So, the total weight of the receiver is more for the PCB cooled designs. But at 200 W due to lower losses, the PCB weight for the PCB cooled design is less that the PCB plus thermal management weight for the heat sink design, resulting in lower all-over receiver weight. At 200 W the optimized single phase designs have better power density for both thermal management schemes. The optimized interleaved designs turn out to be lighter weight for both cooling mechanisms at 600 W.

IV. EXPERIMENTAL RESULTS

A 600 W prototype of the optimized AR plus interleaved 2:1 SC based receiver is built to validate the system modeling and design. Table III lists the prototype components and parameters. A photograph of the receiver power stage and Rx coil is given in Fig. 8. $S_{11} - S_{41}$ and $S_{12} - S_{42}$ denote the two phases of the interleaved 2:1 SC converter in Fig. 8. The tests have been conducted using an electronic load at the output. The waveforms focusing on the falling and rising edges

979-8-3315-1612-3/25 $31.00 © 2025 IEEE

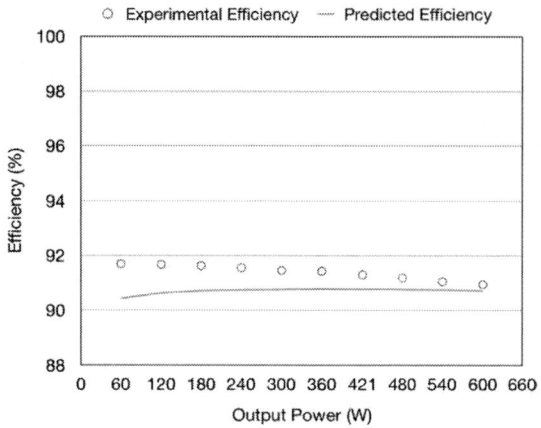

Fig. 12. Measured and predicted system efficiencies of the prototype.

of v_{CD} in Fig. 9(b-c) show that the AR FETs achieve ZVS. Thermal images of the power stage with the optimized thermal management, and of Rx and Tx coils under natural convection conditions at full load are depicted in Fig. 10.

The weight breakdown of the receiver is detailed in Fig. 11. The "measured" column reports the weights measured from the fabricated prototype. The "modeled" column takes into account the discrepancies between the optimized design and the constructed prototype. These two columns do not include the designed PCB weight for sensing and controls since the prototype was tested using a launchpad. The slight discrepancy in the Rx coil weight is attributed to the leads of the prototype Rx coil not considered in the design and optimization. The weight of the TIM is doubled since a 1 mm thick TIM is used in the hardware, rather than the 0.5 mm thick TIM considered in the design. Finally, power stage PCB weight is slightly higher due to the additional PCB area to accommodate the connectors and headers used in the prototype. The on-board receiver prototype has achieved a gravimetric power density of 11.8 W/g. The predicted and measured dc-dc efficiency of the entire system from inverter input to SC output are shown in Fig. 12 using a fixed load resistance. The prototype achieves 90.9% efficiency at full power.

V. CONCLUSION

A system-wide design and optimization method is presented to maximize the gravimetric power density of the on-board receiver in wireless charging of drones. After analyzing different receiver structures, such as active rectifier or diode rectifier directly connected to the load, and SC converters of different conversion ratios, the active rectifier plus interleaved 2:1 switched capacitor design is found to achieve maximum gravimetric power density. The selected design was prototyped for experimental validation. Measurements verify the modeled weights of the Rx coil, receiver stage PCB, heat sink, fan, TIM, flying and matching capacitors. Tests up to 600 W have been conducted with the optimized cooling solution for the power stage and natural convection for the Rx coil. The Power stage and Rx coil thermal performances are consistent with expec-

tations. The on-board receiver achieves a gravimetric power density of 11.8 W/g not considering weight of controls and sensing. The prototype achieves 90.9% efficiency at full load. It is evident that the system losses exhibit good agreement with the loss/thermal models. Also, the efficacy of the PCB cooling solution for improving power density at lower powers is demonstrated. Analysis shows that the 2:1 SC design to achieve better power density at 200 W with the PCB cooling scheme compared to the more traditional heat sink option.

REFERENCES

[1] Y. Shao, N. Kang, H. Zhang, R. Ma, M. Liu, and C. Ma, "A lightweight and robust drone MHz WPT system via novel coil design and impedance matching," *IEEE Trans. on Ind. Appl.*

[2] M. Abdelraziq, S. Paul, F. Bartels, and Z. Pantic, "Optimization of efficiency and receiver-coil mass in an autonomous 700-W S-S IPT system for UAV applications," in *Proc. IEEE Appl. Power Electron. Conf. Expo. (APEC)*, Orlando, FL, USA, Mar. 2023, pp. 803–810.

[3] C. Cai, S. Wu, L. Jiang, Z. Zhang, and S. Yang, "A 500-w wireless charging system with lightweight pick-up for unmanned aerial vehicles," *IEEE Trans. on Power Electron.*, vol. 35, no. 8, pp. 7721–7724, Aug. 2020.

[4] S. Wu, C. Cai, L. Jiang, J. Li, and S. Yang, "Unmanned aerial vehicle wireless charging system with orthogonal magnetic structure and position correction aid device," *IEEE Trans. on Power Electron.*, vol. 36, no. 7, pp. 7564–7575, Jul. 2020.

[5] S. Sinha, A. Kumar, B. Regensburger, and K. K. Afridi, "A new design approach to mitigating the effect of parasitics in capacitive wireless power transfer systems for electric vehicle charging," *IEEE Trans. on Transp. Electrification*, vol. 5, no. 4, pp. 1040–1059, Dec. 2019.

[6] Z. Ye, Y. Lei and R. C. N. Pilawa-Podgurski, "The Cascaded Resonant Converter: A Hybrid Switched-Capacitor Topology With High Power Density and Efficiency," in IEEE Transactions on Power Electronics, vol. 35, no. 5, pp. 4946-4958, May 2020.

[7] C. Gammeter, F. Krismer and J. W. Kolar, "Weight Optimization of a Cooling System Composed of Fan and Extruded-Fin Heat Sink," in IEEE Transactions on Industry Appl., vol. 51, no. 1, pp. 509-520, Jan.-Feb. 2015.

[8] A. Basu, K. Froehle and D. Costinett, "Design and Optimization of a High Gravimetric Power Density Receiver for Wireless Charging of Drones," 2023 IEEE 24th Workshop on Control and Modeling for Power Electronics (COMPEL), Ann Arbor, MI, USA, 2023, pp. 1-8.

[9] J. Li, R. Qin, J. Sun, and D. Costinett, "Systematic design of a 100-W 6.78-MHz wireless charging station covering multiple devices and a large charging area," *IEEE Trans. on Power Electron.*, vol. 37, no. 4, pp. 4877–4889, Apr. 2022.

[10] D. Costinett, D. Maksimovic, and R. Zane, "Circuit-oriented treatment of nonlinear capacitances in switched-mode power supplies," *IEEE Trans. Power Electron.*, vol. 30, no. 2, pp. 985–995, Feb. 2015.

[11] M. D. Seeman and S. R. Sanders, "Analysis and Optimization of Switched-Capacitor DC–DC Converters," in IEEE Transactions on Power Electronics, vol. 23, no. 2, pp. 841-851, March 2008.

[12] J. A. Baxter and D. J. Costinett, "Broad-Scale Converter Optimization Utilizing Discrete Time State-Space Modeling," 2022 IEEE Design Methodologies Conference (DMC), Bath, United Kingdom, 2022, pp. 1-6.

[13] J. A. Baxter and D. J. Costinett, "Converter Analysis Using Discrete Time State-Space Modeling," 2019 20th Workshop on Control and Modeling for Power Electronics (COMPEL), Toronto, ON, Canada, 2019, pp. 1-8.

[14] J. A. Baxter and D. J. Costinett, "Steady-State Convergence of Discrete Time State-Space Modeling with State-Dependent Switching," 2020 IEEE 21st Workshop on Control and Modeling for Power Electronics (COMPEL), Aalborg, Denmark, 2020, pp. 1-8.

[15] J. A. Baxter and D. J. Costinett, "Power Converter and Discrete Device Optimization Utilizing Discrete Time State-Space Modeling," 2023 IEEE 24th Workshop on Control and Modeling for Power Electronics (COMPEL), Ann Arbor, MI, USA, 2023.

[16] A. Kumar, J. Lu and K. K. Afridi, "Enhanced-accuracy augmented state-space approach to steady-state modeling of resonant converters," 2015 IEEE 16th Workshop on Control and Modeling for Power Electronics (COMPEL), Vancouver, BC, Canada, 2015, pp. 1-6.

[17] C. R. Sullivan, "Optimal Choice for Number of Strands in a Litz-Wire Transformer Winding," *IEEE Trans. on Power Electron.*, vol. 14, no. 2, pp. 283–291, Mar. 1999.

[18] G. N. Ellison, *Thermal Computations for Electronic Equipment.* Malabar, FL: R. E. Krieger Publishing Company, 1989.

[19] S. Cochran and D. Costinett, "Discrete Time Synchronization Modeling for Active Rectifiers in Wireless Power Transfer Systems," 2019 20th Workshop on Control and Modeling for Power Electronics (COMPEL), Toronto, ON, Canada, 2019, pp. 1-8.

Stabilization Method for DC-bus Oscillation in Dynamic Wireless Power Transfer Systems

Yuki Ochiai, Keisuke Kusaka

Dept. of Electrical, Electronics, and Information Engineering

Nagaoka University of Technology

Nagaoka, Japan

Email: s201067@stn.nagaokaut.ac.jp, kusaka@vos.nagaokaut.ac.jp

Abstract— **Dynamic wireless power transfer systems will have a DC bus, which supplies power to inverters with transmission coils intermittently placed over long distances on the road surface. As the length of the DC bus increases, the wiring inductance also increases. This effect affects inverters that are located far from the power source of the DC bus, requiring them to connect through this increased wiring inductance. When the inverter performs current control on the output side, it may cause negative impedance, resulting in the DC-bus voltage oscillation. This paper proposes a control method to prevent instability of the DC bus voltage by utilizing positive feed-forward (PFF) control. The experimental results with the small-scale prototype show that the proposed control effectively prevents instability.**

Keywords— *DC power supply, Feed-forward control, Dynamic wireless power transfer, Magnetic resonance coupling, Electric vehicle*

I. INTRODUCTION

In recent years, the transition from gasoline and diesel vehicles to electric vehicles (EVs) has been actively promoted to reduce greenhouse gas emissions. However, the current adoption rate of EVs remains low. This is primarily due to several challenges, such as the high cost of EVs caused by expensive batteries, the long charging times, and the limited driving range. To address these issues, wireless power transfer (WPT) for EVs during motion has been extensively studied [1]-[3]. Implementing in-motion WPT technology can extend the driving range even with smaller onboard batteries. Consequently, the smaller battery capacity reduces the overall cost of EVs. Furthermore, the ability to charge vehicles while driving eliminates the need to stop to recharge. These advantages suggest that the practical adoption of in-motion WPT could significantly accelerate the widespread use of EVs.

A system configuration for a dynamic WPT has been proposed, as shown in Fig. 1 [4]. In this configuration, multiple primary coils are embedded intermittently along the roadway. Each transmission coil is connected to an inverter, which supplies power to the coil. The inverters are connected to a DC bus that runs along the roadway. The length of the DC bus may be long since the WPT system is expected to be applied to highways. As the DC bus length increases, the wiring inductance also becomes significant. Consequently, inverters far from the AC-DC converter that supplies power to the DC bus are connected via non-negligible wiring inductance. In such a system, if output power is controlled on the inverter side, negative resistance caused by the inverter's control may lead to oscillation in the DC-link capacitor voltage [7].

This paper aims to clarify the instability conditions of the WPT system. First, a simplified model is constructed to represent the system under conditions where the wiring inductance is sufficiently large and instability occurs. Then, a control method is proposed to prevent instability by applying a positive feedforward (PFF) control strategy [8] to the simplified model.

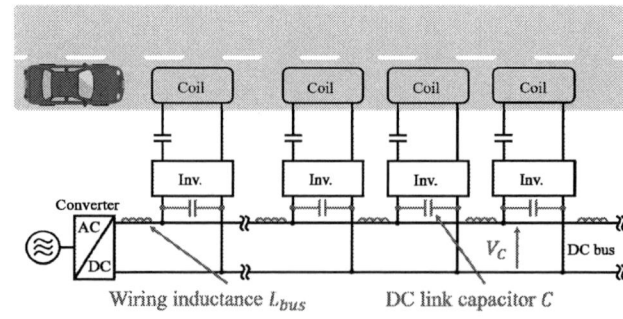

Fig. 1. System structure of dynamic charging for electric vehicles.

II. DC BUS VOLTAGE INSTABILITY CONDITIONS

Using a simplified model, this chapter derives the instability conditions for inverters connected via the DC bus in the in-motion WPT system. In the in-motion wireless power transfer (WPT) system shown in Fig. 1, power is supplied to each inverter through the DC bus from an AC-DC converter that serves as the DC power source. When each inverter performs output power control, negative resistance is introduced by the feedback control of the inverter. The negative resistance, combined with the wiring inductance of the DC bus, may cause an oscillation in the DC-link voltage. The next paragraph will explain the instability conditions for inverters connected via the DC bus using a simplified model.

The instability conditions are developed from the simplified model of the dynamic WPT. Fig. 2 illustrates the simplified system used to derive the instability conditions of the WPT system. In this system, V_g represents the voltage source supplying power to the DC bus, while L_{bus} and R_{bus} denote the wiring inductance and resistance of the DC bus, respectively. The "Constant power load" in Fig. 2 represents a power transmission unit far from the DC bus voltage source. This unit behaves as a constant power load (CPL) due to feedback control. Fig. 3 shows the current-voltage (I-V) characteristics of the CPL. The slope of the tangent at an operating point, dv_{in} / di_b, is equivalent to a resistance-like parameter and is denoted as r'. At the operating point v_{in}, r' is expressed as follows:

$$r' = \frac{dv_{in}}{di_b} = -\frac{v_{in}^2}{P} \tag{1}$$

In this paragraph, the stability conditions are derived. The negative resistance r' may cancel the resistive component R_{bus} of the wiring under certain conditions. This is known to cause oscillations between the wiring inductor L_{bus} and the input capacitor C_{s1}.

Assume that in Fig. 2, the supply voltage V_g momentarily changes by Δv_g to $V_g + \Delta v_g$. It causes the change in the current flowing through the inductor and the load voltage. The current flowing in the inductor changes by Δi_{in} from I_{in} to $I_{in} + \Delta i_{in}$. Also, the load voltage changes by Δv_{in} from V_{in} to $V_{in} + \Delta v_{in}$. The following equation is developed from these changes.

$$\begin{bmatrix} \Delta \dot{v}_{in} \\ \Delta \dot{i}_{in} \end{bmatrix} = \begin{bmatrix} \dfrac{P}{C_{s1}V_{in}^2} & \dfrac{1}{C_{s1}} \\ -\dfrac{1}{L_{bus}} & -\dfrac{R_{bus}}{L_{bus}} \end{bmatrix} \begin{bmatrix} \Delta v_{in} \\ \Delta i_{in} \end{bmatrix} + \begin{bmatrix} 0 \\ \dfrac{1}{L_{bus}} \end{bmatrix} \Delta v_g \tag{2}$$

$$A = \begin{bmatrix} \dfrac{P}{C_{s1}V_{in}^2} & \dfrac{1}{C_{s1}} \\ -\dfrac{1}{L_{bus}} & -\dfrac{R_{bus}}{L_{bus}} \end{bmatrix} \tag{3}$$

The characteristic equation $\Delta(s)$ of this system is expressed as Eq. (4), where I is the identity matrix. Applying the Routh-Hurwitz stability criterion to Eq. (4), the stability conditions can be derived as follows:

$$\Box(s) = \det[sI - A] \tag{4}$$

$$-\frac{P}{C_{s1}V_{in}^2} + \frac{R_{bus}}{L_{bus}} > 0 \tag{6}$$

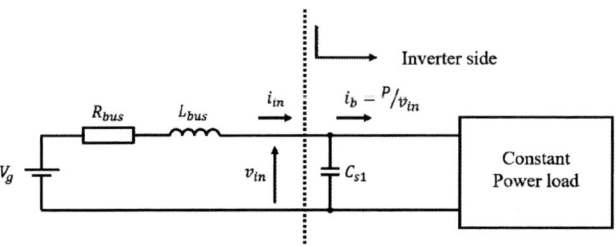

Fig. 2. A simple system concerning instability caused by constant power loads.

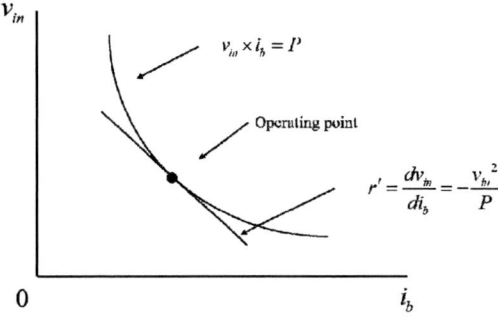

Fig. 3. Negative impedance behavior of constant power load.

$$\Delta(s) = \det \begin{bmatrix} s - \dfrac{P}{C_{s1}V_{in}^2} & -\dfrac{1}{C_{s1}} \\ \dfrac{1}{L_{bus}} & s + \dfrac{R_{bus}}{L_{bus}} \end{bmatrix} = s^2 + \left(-\dfrac{P}{C_{s1}V_{in}^2} + \dfrac{R_{bus}}{L_{bus}} \right)s - \dfrac{PR_{bus}}{C_{s1}L_{bus}V_{in}^2} + \dfrac{1}{C_{s1}L_{bus}} \tag{5}$$

979-8-3315-1612-3/25 $31.00 © 2025 IEEE

$$\alpha = 2\cos^{-1}\left(\frac{\pi}{2\sqrt{2}}\frac{E_{ie}}{V_g}\right) \quad (9)$$

$$-\frac{PR_{bus}}{C_{s1}L_{bus}V_{in}^{2}}+\frac{1}{C_{s1}L_{bus}}>0 \quad (7)$$

Simplifying Eq. (7) results in:

$$-\frac{PR_{bus}}{V_{in}^{2}}+1>0 \quad (8)$$

Thus, an oscillation in the DC bus voltage will occur if both Eqs. (6) and (8) are not satisfied.

III. PROPOSED CONTROL METHOD

A. System Overview

Fig. 4 shows the circuit diagram of the wireless power transfer (WPT) system, including the DC bus. In this circuit, wiring inductance and resistance components are inserted between the DC bus power source and the inverter to simulate a power transmission unit far from the DC power source. Fig. 5 presents the control block diagram of the inverter. In this control scheme, since the switching frequency is high, a constant current is supplied to the primary coil by controlling the envelope i_{ie}, which connects the peak values of the primary sinusoidal current.

As shown in Fig. 6, a peak detection circuit is used to detect the current envelope i_{ie}. The voltage V_{il} represents the signal from the current sensor, and V_{ie} is the output signal of the peak detection circuit. This circuit utilizes the forward characteristics of the diode; when V_{il} exceeds the voltage of capacitor C_p, the capacitor is charged. Then, the circuit outputs the peak value of V_{il} at all times. Using this circuit, the envelope is detected from the primary current of the WPT system, and the phase shift α of the inverter output is adjusted via feedback control. Specifically, Eq. (9) in Fig. 5 describes the conversion of the PI controller's output into the phase shift α.

However, this feedback control lets the inverter behave as a negative resistance, leading to oscillations in the DC bus voltage. Positive feed-forward (PFF) control is introduced into the control block to address this issue. PFF control detects the fluctuation component of the DC-link voltage V_c using a high-pass filter. The fluctuation component is then multiplied by a gain K_F and added to the error in the feedback loop to mitigate instability.

B. Derivation of plant models and design of PI controllers

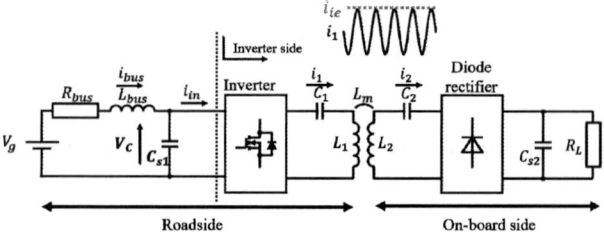

Fig. 4. Circuit configuration emulating a DC bus and unit.

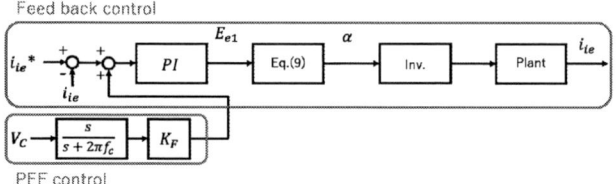

Fig. 5. Control block diagram.

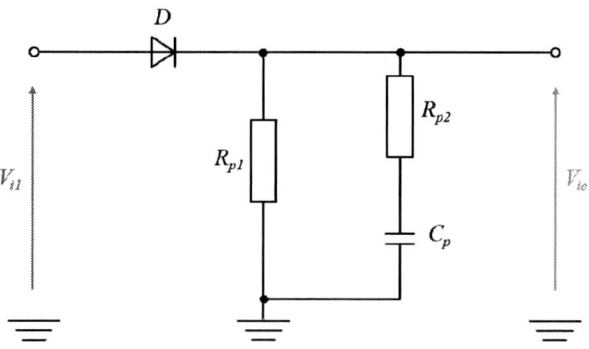

Fig. 6. Peak detection circuit.

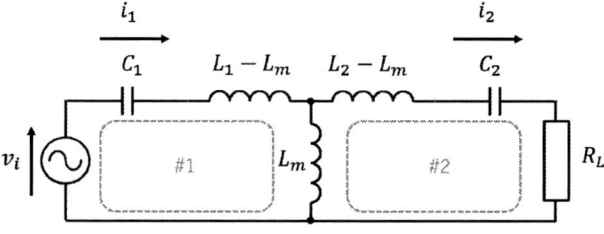

Fig. 7. T-type equivalent circuit.

To design the control system, the transfer function of the plant for control of the envelope of the primary current in the series-series (S/S) resonant circuit is derived in this section. Fig. 7 shows the equivalent T-type circuit of the S/S resonant circuit, where the internal resistances of the primary and secondary coils are neglected for simplicity.

Eq. (11) and (12) are obtained from Fig. 7, defining two current loops I_1 and I_2. Here, the following approximate equation is obtained. In equation (12), $1 / \omega_0$ appears each time a partial integral is performed. This time, we simplify the equation by ignoring the third terms because the third term is proportional to the inverse cube of the angular frequency ω.

$$v_1 = \frac{1}{C_1} \int i_1(t)\,dt + (L_1 - L_M)\frac{di_1(t)}{dt} + L_M\left(\frac{di_1(t)}{dt} - \frac{di_2(t)}{dt}\right) \tag{10}$$

$$L_M\left(\frac{di_1(t)}{dt} - \frac{di_2(t)}{dt}\right) = (L_2 - L_M)\frac{di_2(t)}{dt} + \frac{1}{C_2}\int i_2(t)\,dt + R_L i_2(t) \tag{11}$$

$$
\begin{aligned}
\frac{1}{C_1}\int i_1(t)\,dt &= \frac{1}{C_1}\int I_1\sin(\omega_o t)\,dt \\
&= \frac{-1}{\omega_o C_1}\left(I_1\cos(\omega_o t) - \int \frac{dI_1(t)}{dt}\cos(\omega_o t)\,dt\right) \\
&= \frac{-1}{\omega_o C_1}\left\{I_1\cos(\omega_o t) - \frac{1}{\omega_o}\left(\frac{dI_1(t)}{dt}\sin(\omega_o t) - \int \frac{d^2 I_1(t)}{dt^2}\sin(\omega_o t)\,dt\right)\right\} \\
&= \frac{-1}{\omega_o C_1}\left[I_1\cos(\omega_o t) - \frac{1}{\omega_o}\left\{\frac{dI_1(t)}{dt}\sin(\omega_o t) + \frac{1}{\omega_o}\left(\frac{d^2 I_1(t)}{dt^2}\cos(\omega_o t) - \int \frac{d^3 I_1(t)}{dt^3}\cos(\omega_o t)\,dt\right)\right\}\right] \\
&= \frac{-1}{\omega_o C_1}I_1\cos(\omega_o t) + \frac{1}{\omega_o^2 C_1}\frac{dI_1(t)}{dt}\sin(\omega_o t) - \frac{1}{\omega_o^3 C_1}\left(\frac{d^2 I_1(t)}{dt^2}\cos(\omega_o t) - \int \frac{d^3 I_1(t)}{dt^3}\cos(\omega_o t)\,dt\right) \\
&\approx \frac{-1}{\omega_o C_1}I_1\cos(\omega_o t) + \frac{1}{\omega_o^2 C_1}\frac{dI_1(t)}{dt}\sin(\omega_o t)
\end{aligned}
\tag{12}
$$

$$
\begin{aligned}
L_M\frac{di_2(t)}{dt} &= L_M\left(\frac{di_2(t)}{dt}\cos(\omega_o t) - \omega_o I_2(t)\sin(\omega_o t)\right) \\
&\approx \omega_o L_M I_2\sin(\omega_o t)
\end{aligned}
\tag{13}
$$

$$V_1(t) = 2L_1\frac{dI_1(t)}{dt} + \omega_o L_M I_2(t) \tag{14}$$

$$\omega_o L_M I_1(t) = 2L_2\frac{dI_2(t)}{dt} + R_L I_2(t) \tag{15}$$

By eliminating $I_2(t)$ from Eqs. (14) and (15), the following equation is Eq. (16).

Taking the Laplace transform of Eq. (16), the envelope model from V_1 to I_1 is expressed as Eq. (17):

$$G(s) = \frac{\dfrac{1}{2L_1}s + \dfrac{R_L}{4L_1 L_2}}{s^2 + \dfrac{R_L}{2L_2}s + \dfrac{(\omega L_M)^2}{4L_1 L_2}} \tag{17}$$

Here,

$$L_M = k\sqrt{L_1 L_2} \tag{18}$$

allowing for the following rearrangement:

$$P(s) = \frac{\dfrac{1}{2L_1}s + \dfrac{R_L}{4L_1 L_2}}{s^2 + \dfrac{R_L}{2L_2}s + \dfrac{\omega_o^2 k^2}{4}} \tag{19}$$

$$R_L V_1 + 2L_2\frac{dV_1(t)}{dt} = 4L_1 L_2\frac{d^2 I_1(t)}{dt^2} + 2L_1 R_L\frac{dI_1(t)}{dt} + (\omega_o L_M)^2 I_1(t) \tag{16}$$

Assuming $L_1 = L_2 = L$, the equation can be further simplified to:

$$P(s) = \frac{2Ls + R_L}{4L^2 s^2 + 2LR_L s + L^2\omega_o^2 k^2} \tag{20}$$

Eq. (20) represents the plant's transfer function used in this study.

Using this plant's transfer function, the proportional gain K_p and integral gain K_i were determined via pole placement. The resulting values are $K_p = 0.50$ and $K_i = 5500$.

C. PFF controller settings

This section describes the gain settings of the PFF controller. The gain K_F of the PFF controller in Fig. 5 is determined from the disturbance transfer function. The disturbance transfer function from the detected DC link voltage V_c to the plant output I_{ie} depends on the gain K_F of the PFF controller. To suppress oscillations in the DC bus voltage, it is desirable for the disturbance transfer function to remain below 0dB. To satisfy this condition, K_F must be set to a value less than 1. If K_F is negative, the phase is inverted, which amplifies the oscillations of V_c. Therefore, K_F must satisfy $0 < K_F < 1$.

979-8-3315-1612-3/25 $31.00 © 2025 IEEE

which may limit the power transmission capacity of the DWPT system. Additionally, since the output signal of the PFF controller is added to the input of the PI controller, a larger output signal from the PFF controller increases the error relative to the current reference value.

To address these issues, K_F should be set to a small value within the range $0 < K_F < 1$, ensuring that the maximum V_c during instability is kept below a desired threshold.

IV. EXPERIMENTAL RESULTS

The stability limits and the effects of PFF control were experimentally verified using a mini-model. Fig. 8 shows the appearance of the experimental setup. The mini-model circuit configuration is depicted in Fig. 9, and the circuit parameters are listed in Table 1. The resonance circuit and load parameters were designed for maximum efficiency based on the required maximum power and the DC bus power supply voltage [12].

First, the current reference was step-changed between stable and unstable conditions to confirm the DC bus oscillation conditions in this system. Fig. 9 shows the DC bus voltage when the reference value was step-changed from 0.42 to 0.50p.u. According to the stability conditions derived from Eq. (6) and Eq. (8), the system exceeded the stability limit as the primary current increased from 0.42 to 0.50p.u., confirming the stability conditions through experimental validation.

Fig. 10 shows the DC link voltage when PFF control was applied during operation. In this experiment, PFF control was applied to stabilize the system after the DC link voltage began oscillating, with the reference value set at 0.67 p.u. From Fig. 10, it was confirmed that applying PFF control successfully avoided instability. The oscillation frequency of the DC link voltage was approximately 98 Hz, which matched the resonant frequency of the DC bus wiring inductance and the DC link capacitor.

Finally, Fig. 11 illustrates the relationship between K_F, the maximum DC link voltage V_{c_max}, and the rise time T_r. As K_F increases, V_{c_max} decreases, but the rise time becomes longer, reducing the chargeable power. Therefore, K_F should be set to a small value within the range that suppresses V_c to an acceptable maximum DC link voltage while ensuring adequate rise time.

V. CONCLUSIONS

This study introduced a control method that integrates positive feed-forward (PFF) control with current envelope control to prevent oscillations in the DC link voltage of a

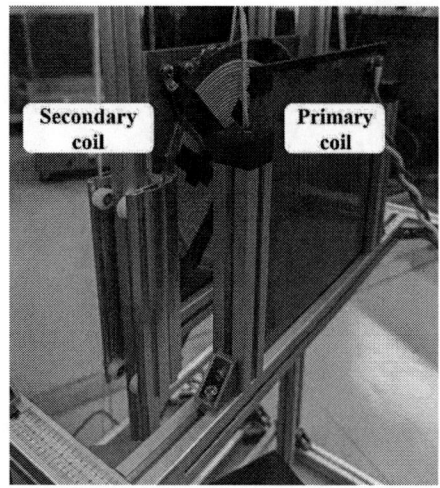

(a) Coils

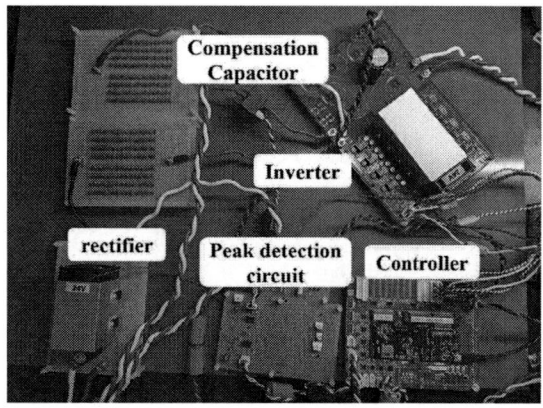

(b) Circuit

Fig. 8. Experimental setup.

Table1. Parameters of scaled-model.

Parameter	Symbol	Value
DC bus voltage	V_s	70.0 V
Resonant frequency	f_o	85.0 kHz
Load resistance	R_L	6.67 Ω
Transmitting coil	L_1, L_2	63.2 μH
ESR	R_1, R_2	0.320 Ω
Compensation Capacitor	C_1, C_2	55.4 nF
Wire resistance	R_{bus}	0.1 Ω
Wire inductor	L_{bus}	5.5 mH
Primary DC link capacitor	C_{s1}	470 μF
Secondary DC link capacitor	C_{s2}	120 μF

wireless power transfer system with an extended DC bus. Experimental validation demonstrated that the application of PFF control effectively mitigates instability. Furthermore, it was found that increasing the PFF control gain led to a longer rise time and a decrease in the maximum DC link voltage.

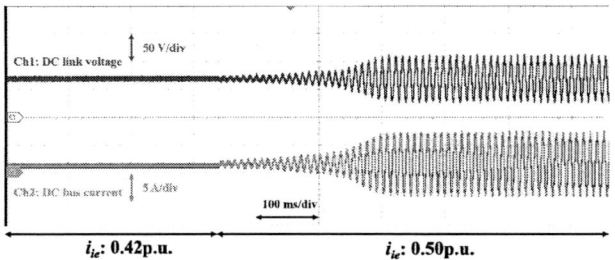

Fig. 9. Oscillation waveform at stability limit by measurement.

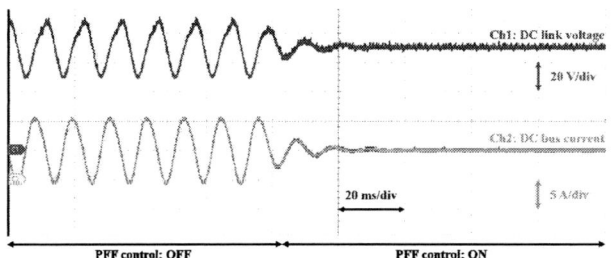

Fig. 10. Oscillation comparison with PFF control.

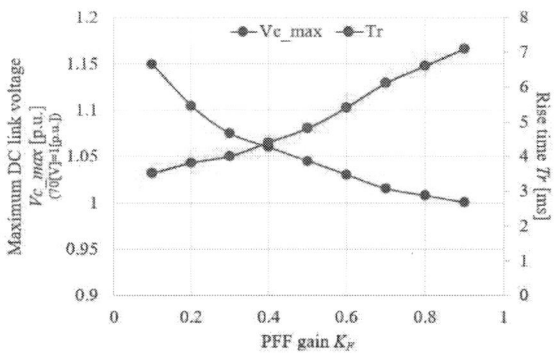

Fig. 11. Relationship between KF and Vc_max, Tr.

REFERENCES

[1] Zhang, Zhen, et al. "Wireless power transfer—An overview." *IEEE transactions on industrial electronics* 66.2 (2018): 1044-1058.

[2] Mahesh, Aganti, Bharatiraja Chokkalingam, and Lucian Mihet-Popa. "Inductive wireless power transfer charging for electric vehicles–a review." *IEEE access* 9 (2021): 137667-137713.

[3] Feng, Hao, et al. "An LCC-compensated resonant converter optimized for robust reaction to large coupling variation in dynamic wireless power transfer." *IEEE Transactions on Industrial Electronics* 63.10 (2016): 6591-6601.

[4] Daita Kobayashi, Takehiro Imura, Yoichi Hori, "Real-time Maximum Efficiency Control in Dynamic Wireless Power Transfer System" *IEEJ Transaction on Industry Applications* Vol.136 No.6 pp.425-432 (2016).

[5] Zhang, Xin, et al. "Adaptive active capacitor converter for improving stability of cascaded DC power supply system." *IEEE Transactions on Power Electronics* 28.4 (2012): 1807-1816.

[6] Zhang, Xin, Qing-Chang Zhong, and Wen-Long Ming. "Stabilization of cascaded DC/DC converters via adaptive series-virtual-impedance control of the load converter." *IEEE Transactions on Power Electronics* 31.9 (2016): 6057-6063.

[7] K. Miura, H. Watanabe, J. -i. Itoh, T. Kiribuchi and H. Tokusaki, "Damping Controller Integrated into Output Current Control Loop and Design for Multiple Servo Drive Systems Connected to Common DC-Bus Line," 2023 IEEE Applied Power Electronics Conference and Exposition (APEC), Orlando, FL, USA, 2023, pp. 2367-2374, doi: 10.1109/APEC43580.2023.10131476.

[8] Jeung, Yoon-Cheul, et al. "Design of passivity-based damping controller for suppressing power oscillations in DC microgrids." *IEEE Transactions on Power Electronics* 36.4 (2020): 4016-4028.

[9] S. Chen and T. Czaszejko, "Partial discharge test circuit as a spark-gap transmitter," in IEEE Electrical Insulation Magazine, vol. 27, no. 3, pp. 36-44, May-June 2011.

[10] Daisuke GUNJI, Takehiro IMURA, Hiroshi FUJIMOTO. " Secondary Voltage Envelope Model and Application to Control System Design on Wireless Power Transfer using Magnetic Resonance Coupling." THE INSTITUTE OF ELECTRONICS, INFORMATION AND COMMUNICATION ENGINEERS (2014): 45-50.

[11] T. Hamada, T. Fujita and H. Fujimoto, "Fast Start-Up Control of Both-Side Current Without Overshoot Focusing on Rectification Timing for Dynamic Wireless Power Transfer Systems," in IEEE Journal of Emerging and Selected Topics in Industrial Electronics, vol. 5, no. 3, pp. 1039-1047, July 2024, doi: 10.1109/JESTIE.2023.3288476.

[12] Bosshard, Roman, et al. "Modeling and η-α Pareto Optimization of Inductive Power Transfer Coils for Electric Vehicles." IEEE Journal of Emerging and Selected Topics in Power Electronics 3.1 (2014):

Unveiling Aliasing Effect on Resonant Pole Locations in Wireless Battery Chargers

Anwesha Mukhopadhyay and Daniel Costinett
Department of Electrical Engineering and Computer Science,
The University of Tennessee, Knoxville
Knoxville, TN, USA
amukhopa@utk.edu, dcostine@utk.edu

Abstract—**Resonance frequency bifurcation is a common phenomenon in resonant tank circuits used in wireless battery chargers, leading to a drift between the transmitter and receiver frequencies. Studying this drift helps to define the acceptable operating frequency range to ensure efficient power transfer and maintain soft switching conditions. While this understanding aids in improving steady-state performance and efficiency, the control dynamics depend on the small-signal model and the resonant pole locations of the system under control. State-of-the-art literature, based on the discrete-time (DT) model, indicates that the plant resonant poles show up at frequencies much below the coil resonance frequencies. However, the qualitative origin of the poles at the lower frequency range is not apparent. It is also unclear how the resonant poles and the tank resonance frequencies are correlated. In this work, the origin of the resonant poles is investigated and a method is suggested for the quick approximations of the pole locations without numerically solving the discrete-time model. The analysis leads to understanding the relation between the pole locations and the sampling and tank resonance frequencies. A series-series LC resonant converter-based battery charger is considered for the study and experimental validation. The derived conclusions are verified with the results available in the literature. Though the analysis evolves based on a full-bridge converter, the formulation lends itself to complex resonant converter topologies.**

Index Terms—**Wireless battery charger, series resonant converter, tank resonance frequencies, discrete-time modeling, aliasing, resonant poles, maximum control bandwidth**

I. INTRODUCTION

Wireless chargers are used for charging batteries in a diverse range of applications, including medical devices, consumer electronics, small unmanned or heavy electric vehicles, etc. [1]–[5]. Resonant converters have emerged as a popular topology for transmitting a few watts to several kilowatts of power utilizing inductive coupling between transmitter and receiver coils [6]. This work focuses on a 200 W battery charger, designed for charging drone batteries. To maximize the flight time and enable rapid charging of the drone, the receiver must be lightweight while having the capability of handling high power. Gravimetric power density optimization of the drone

This research was sponsored by the Army Research Laboratory and was accomplished under Cooperative Agreement Number W911NF2220007. The views and conclusions contained in this document are those of the authors and should not be interpreted as representing the official policies, either expressed or implied, of the Army Research Office or the U.S. Government. The U.S. Government is authorized to reproduce and distribute reprints for Government purposes notwithstanding any copyright notation herein.

[7] favors a series-series resonant converter with an active rectifier at the receiver (Rx) end. An active rectifier, unlike a diode bridge rectifier, allows for soft-switching, leading to reduced losses and better thermal management. Output regulation without any additional power stage is another advantage that leads to achieve high efficiency in a light-weight design.

However, soft-switching or output regulation in constant current (CC) or voltage (CV) charging requires synchronous switching of the rectifier with respect to the transmitter (Tx) at a specified angle, decided by the control. The air gap between the Tx and Rx converters makes their coherent operation challenging in the absence of any communication between them [8]. Therefore, the Rx-end rectifier is synchronized with the Tx-end inverter using local sensing and control [9]. The synchronization requires high bandwidth control to prevent the Rx from falling out of step with the Tx during the transient. However, the control bandwidth is often limited by the frequencies of the dominant resonant poles of the plant model to ensure sufficient phase and gain margins. A precise knowledge of the resonant poles is, thus, critical for maximizing the control bandwidth.

Modeling of resonant converters has been extensively studied in literature [10]–[13]. The state-space averaging does not work for resonant converters as the key state variables do not have any DC component and the variables exhibit significant ripple, far from being small. Extended describing function-based method [12] and discrete-time analysis methods [14], [15] emerged as promising techniques to accurately predict the high frequency dynamics up to the switching frequency. The describing function approach [16] uses fundamental frequency approximation, which means only fundamental frequency components of the tank variables are considered. However, in low-Q tank, the harmonics of the switching frequency are also significant. The discrete-time (DT) model is more general and accurate, applicable to all converters without any special design requirement to be satisfied [14]. It captures the inherently sampled nature of the practical system and can predict the small-signal behavior up to the switching frequency. The computation rigor and involved numerical solutions are often seen as one of the limitations of DT modeling [13]. However, a Matlab toolbox [17], based on DT modeling, enables all types of power converter design and exact analysis while making it adoptable for practical

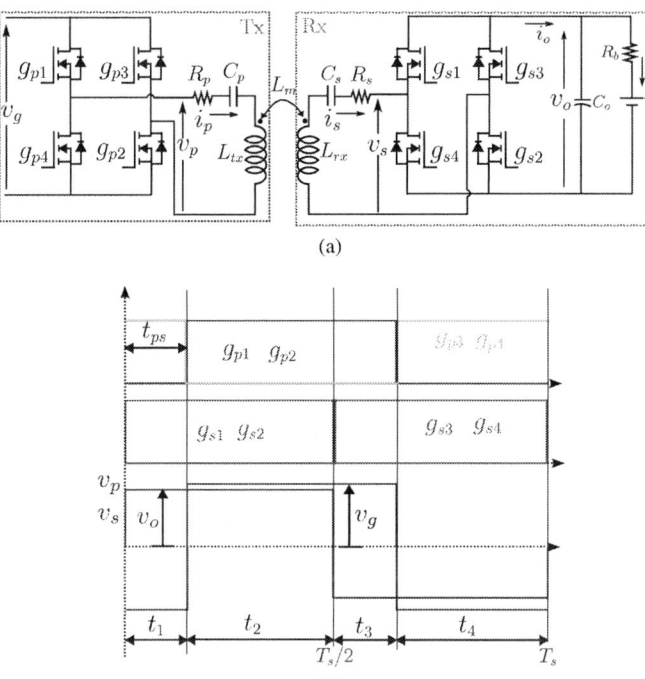

(a)

(b)

Fig. 1. (a) Circuit schematic of a wireless battery charger and (b) its switching pattern.

use without computational complexities. This work utilizes the tool to generate accurate dynamic models for verifying the proposed analysis and conclusion across different designs [7], [9], [18] of series-series LC resonant converters.

Fig. 1(a) depicts the considered system where the power stage consists of a series-series LC resonance converter. The compensating capacitors C_p and C_s are designed to resonate with the Tx and Rx coil inductances L_{tx} and L_{rx}, respectively, at equal or nearly equal frequencies. The switching pulses in Fig. 1(b) depict four subintervals (t_1-t_4) that decide the phase-shift ϕ_{ps} ($=2\pi t_{ps}/T_s$) between the waveforms of the Tx and Rx coil voltages, v_p and v_s, respectively. Fig. 2 depicts the tank equivalent circuit [9], where $L_p = L_{tx} - L_m$, $L_s = L_{rx} - L_m$, and $L_m = k\sqrt{L_{tx}L_{rx}}$ with k being the coupling coefficient between Tx and Rx coils.

Even if the compensating capacitors are selected such that $L_{tx}C_p = L_{rx}C_s$, the resonance frequency splits into multiple resonance frequencies if there is some change in the circuit parameters, load resistance, or degree of coupling [19], [20]. This phenomenon, commonly called bifurcation, can result in loss of zero voltage switching (ZVS) if not taken care

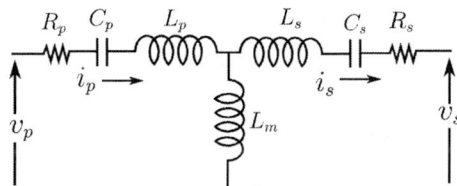

Fig. 2. Equivalent circuit of Tx and Rx tank.

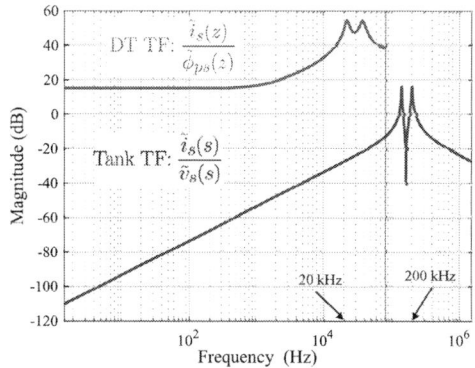

Fig. 3. Frequency responses of tank TF and DTTF are overlaid to highlight the difference between the tank resonance frequencies and resonant pole locations.

of by suitable control efforts. As seen in [19], [21], the bifurcated resonance frequencies lie in the vicinity of the designed resonance frequency ($\omega_r = \omega_{rTx} = \omega_{rRx}$). However, the discrete-time control-to-output transfer function of series-series (SS) resonant converters in [9], [22] predicts the presence of resonant pole pairs at frequencies nearly an order of magnitude below that of the tank resonance frequencies.

Fig. 3 depicts the frequency responses of a resonant tank circuit, similar to the one shown in Fig. 1(a). The transfer gain, relating the Rx coil current (i_s) and rectifier input voltage (v_s), shows that the resonance frequency of the Tx and Rx tank is bifurcated to frequencies in the range of 150 - 200 kHz. In the same plot, the frequency response of the DT plant model ($G_{\phi_{ps}i_s}(z)$) between ϕ_{ps} and i_s is laid over. The resonant poles of $G_{\phi_{ps}i_s}(z)$ are seen in the range of 20 - 40 kHz, which is approximately one decade lower than the frequencies of the resonant tank.

While the quantitative accuracy of the discrete-time (DT) model of the SS resonant converter has been experimentally verified in literature [9], the qualitative origin of the resonant poles at a frequency much lower than both the transmitter and receiver tanks' resonance frequencies is not apparent. However, the locations of these resonant poles are critical in the control loop design which aims to maximize the control bandwidth. In the present context of the wireless battery charger system for drones, the synchronization of the Rx-end active rectifier with the Tx-end inverter requires high bandwidth control. Therefore, knowledge of the resonant pole locations and a quick prediction helps to decide the stable control loop design criteria and limits. With this background, the present paper aims to investigate the following

- The resonant pole locations predicted by the DT model through a simplified approach.
- The correlation between the bifurcated tank resonance frequencies and the frequencies of the resonant pole pairs.

The next section provides a brief review of the DT modeling which forms the basis of the subsequent derivations that lead to the prediction of the resonant poles in Section III. A correlation between the tank resonance frequencies and the locations of the resonant pole pair is highlighted and graphically

979-8-3315-1612-3/25 $31.00 © 2025 IEEE

validated in Section IV. Section V presents the simulation and experimental results, followed by the conclusion of this work.

II. REVIEW OF DISCRETE-TIME MODELING

Discrete-time modeling is developed based on the solution of piecewise linear equivalent circuits pertaining to different sub intervals in a switching cycle T_s. Thus, it can capture switching frequency dynamics facilitating an accurate reconstruction of the variables from the model. Referring to Fig. 1(b), four subintervals exist, which means four linear equivalent circuits. However, the constant output voltage (v_o) assumption allows to obtain a time-invariant representation of the equivalent circuit, with v_p and v_s as depicted in Fig. 1(b). This allows us to concentrate on the resonant poles only, ignoring the output port dynamics, dictated by C_o and R_b. The assumption does not impact the controller design and structure as long as $\omega_{pr} < 1/R_b C_o$, where ω_{pr} are the frequencies corresponding to the resonant pole pair. In other words, it assumes the resonant pole pairs are the principal dynamic limiting the control bandwidth, usually true for battery charging applications.

A. State-Space Formulation

The state equations corresponding to the equivalent circuit of Fig. 2 is given by (1).

$$\dot{X} = \mathbf{A}X + \mathbf{B}U \tag{1}$$

where, $X = [i_p \; i_s \; v_{cp} \; v_{cs}]^T$ and $U = [v_p \; v_s]^T$ are state and input vectors. The state matrix $(\mathbf{A})_{4\times4}$ and the input matrix $(\mathbf{B})_{4\times2}$ are given by (2) and (3) with $\Sigma L_{sq} \triangleq L_{tx}L_{rx} - L_m^2$.

$$\mathbf{A} =$$

$$\begin{bmatrix} -\dfrac{R_p L_{rx}}{\Sigma L_{sq}} & \dfrac{-R_s L_m}{\Sigma L_{sq}} & \dfrac{-L_{rx}}{\Sigma L_{sq}} & \dfrac{-L_m}{\Sigma L_{sq}} \\ \dfrac{-R_p L_m}{\Sigma L_{sq}} & \dfrac{-R_s L_{tx}}{\Sigma L_{sq}} & \dfrac{-L_m}{\Sigma L_{sq}} & \dfrac{-L_{tx}}{\Sigma L_{sq}} \\ \dfrac{1}{C_P} & 0 & 0 & 0 \\ 0 & \dfrac{1}{C_s} & 0 & 0 \end{bmatrix} \tag{2}$$

$$\mathbf{B} = \begin{bmatrix} \dfrac{L_{rx}}{\Sigma L_{sq}} & \dfrac{L_m}{\Sigma L_{sq}} & 0 & 0 \\ -\dfrac{L_m}{\Sigma L_{sq}} & \dfrac{L_{tx}}{\Sigma L_{sq}} & 0 & 0 \end{bmatrix}^T \tag{3}$$

If the initial state at $t = t_0$ is defined as $X = x_0$ and a small interval (t_i) is considered within which the input vector $U_i(t) = [v_p \; v_s]^T$ remains constant, (1) is solved for any instant $t \in t_i$ as

$$X(t) = e^{\mathbf{A}(t-t_0)}X_0 + \mathbf{A}^{-1}(e^{\mathbf{A}(t-t_0)} - \mathbf{I})\mathbf{B}U \tag{4}$$

B. Steady-State Analysis

At steady-state, the states become periodic over the switching cycle T_s, yielding,

$$X_{ss} =$$

$$(\mathbf{I} - \prod_{i=1}^{4} e^{\mathbf{A}t_i})^{-1} \left[\sum_{i=1}^{4} \prod_{j=i+1}^{4} e^{\mathbf{A}t_j} \mathbf{A}^{-1} (e^{\mathbf{A}t_i} - \mathbf{I})\mathbf{B}U_i \right] \tag{5}$$

where $\mathbf{I}$ is a 4×4 identity matrix. If the state variables have half-wave anti-symmetry (as in this example), (5) can be simplified considering only two subintervals lying within the half of the switching period ($T_s/2$), as derived in [9].

C. Small-Signal Model

The discrete-time model, given by (4) is a large-signal linear model with respect to input vector U, but non-linear function of the switching times (t_i, t_j) that contain the control phase (ϕ_{ps}) information, due to the presence of the exponential term. Therefore, a linear model is obtained considering a small-signal perturbation ($\tilde{\phi}_{ps}$) in the phase-shift ϕ_{ps} at a steady-state operating point corresponding to ϕ_{pso}, such that $\tilde{\phi}_{ps} << \phi_{pso}$. Fig. 4 depicts the equivalent perturbation $\tilde{t}_{ps} = \dfrac{\tilde{\phi}_{ps}T_s}{2\pi}$ in the time-shift, obtained by perturbing the active rectifier gate pulses at the Rx-end. The resulting sub-intervals are designated as t_1', t_2', t_3', and t_4'. Following the steps elaborated in [23], the Z-transfer function vector (4×1) between $\tilde{\phi}_{ps}$ to any variable $\tilde{x}$ is

$$G_{\phi_{ps}x}(z) = \dfrac{\tilde{x}(z)}{\tilde{\phi}_{ps}(z)} = (z\mathbf{I} - \mathbf{F})^{-1}\mathbf{G} \tag{6}$$

where, $\mathbf{F} = e^{\mathbf{A}T_s}$ is the natural response matrix and

$$\mathbf{G} = \left(e^{\mathbf{A}(t_3+t_4)}e^{\mathbf{A}t_2}\mathbf{B}(U_1 - U_2) + e^{\mathbf{A}t_4}\mathbf{B}(U_3 - U_4) \right) \dfrac{T_s}{2\pi} \tag{7}$$

is the forced response matrix. Alternatively, the numerical solution and frequency response of any of these transfer functions of interest can be found utilizing Switched Mode Power Supply Toolbox [17].

In this formulation, the DT model inherently models a sampled system, having the sampling rate T_s. The sampled

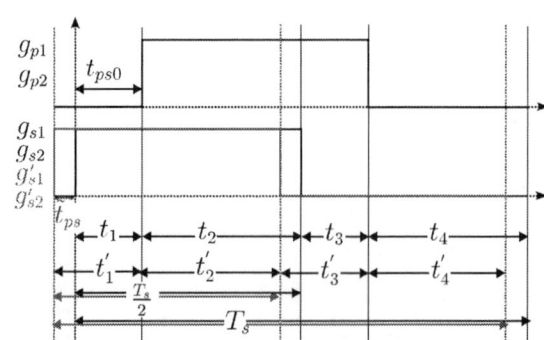

Fig. 4. Switching pulses for Tx and Rx bridges at steady-state (black) and with small-signal perturbation $\tilde{t}_{ps}$ in time-shift between them (red).

TABLE I
COEFFICIENTS OF (8) TO FIND THE EIGENVALUES OF **A**

Coefficient	Expression	Coefficient	Expression
k_1	$\dfrac{R_p(L_m + L_s) + R_s(L_p + L_m)}{L_m L_s + L_s L_p + L_m L_p}$	k_2	$\dfrac{R_p R_s C_p C_s + C_p(L_p + L_m) + C_s(L_s + L_m)}{C_p C_s(L_m L_s + L_s L_p + L_m L_p)}$
k_3	$\dfrac{R_p C_p + R_s C_s}{C_p C_s(L_m L_s + L_s L_p + L_m L_p)}$	k_4	$\dfrac{1}{C_p C_s(L_m L_s + L_s L_p + L_m L_p)}$

nature of the system is inherent, resulting from the sampling process that occurs in the analog-to-digital (A/D) converter and/or modulator at a sampling frequency f_s [24]. The phase-frequency detector, essential for PLL implementations for synchronizing the Tx and Rx, also results in an inherent sampling per period [9]. Regardless of the implementation or control loop being considered, the state transition matrix remains $\mathbf{F} = e^{\mathbf{A}T_s}$, which models the natural response of the ac waveforms in Fig. 2, sampled once per switching period.

III. ORIGIN OF RESONANT POLES

The pole locations in a linear system are predicted from the eigenvalues which are also the roots of the characteristic equation that considers all the state variables. Based on the number of energy storage elements, the system under consideration (Fig. 1) has five state variables: i_p, i_s, v_{cp}, v_{cs}, and v_o. In [22], a full-order model of the LC tank is considered with all five state variables. However, as mentioned in Section II, a voltage-stiff output port assumption reduces the order of the system to four ac-only states while focusing on the resonant pole pairs.

A. Analytical Derivation of Resonant Pole Locations

Though the eigenvalues of the natural response matrix $\mathbf{F}$ give the pole locations, obtaining the analytical expression of eig($\mathbf{F}$) is not straightforward. Hence, let us first consider the eigenvalues (λ), obtained from the solution of

$$\lambda^4 + k_1\lambda^3 + k_2\lambda^2 + k_3\lambda + k_4 = 0 \tag{8}$$

corresponding to the of the state matrix **A** of the LC tank. The expressions for k_1-k_4 are given in Table I. The solution of λ requires to solve a quartic equation. Therefore, a closed-form solution can only be obtained using a symbolic mathematical solver, while compromising on the design-oriented analytical insights. However, the following assumptions can reduce the equation to a biquadratic one, improving the comprehensibility of the solution-

(i) Tx and Rx tank circuits are lossless with $R_p, R_s = 0$.

(ii) The resonance frequencies of Tx ($\omega_{r_{Tx}}$) and Rx tank ($\omega_{r_{Rx}}$) are equal ($= \omega_r$), making $L_{tx}C_p = L_{rx}C_s$.

These assumptions do not affect the pole locations significantly, as the coils are typically designed with a high-quality factor (Q), having small R_p and R_s. Thus, with assumption (i), (8) reduces to

$$\lambda^4 + k_2'\lambda^2 + k_4 = 0 \tag{9}$$

where

$$k_2' = \frac{L_{tx}C_p + L_{rx}C_s}{\Sigma L_{sq}C_p C_s} \tag{10}$$

If the resonance frequencies of Tx ($\omega_{r_{Tx}}$) and Rx tank ($\omega_{r_{Rx}}$) are assumed equal ($= \omega_r$), the eigenvalues of matrix **A** are found as in (11), indicating dependence on the resonance frequency and coefficient of coupling (k).

$$\lambda_1, \lambda_2 = \pm j\omega_r \frac{1}{\sqrt{1-k}} \quad ; \quad \lambda_3, \lambda_4 = \pm j\omega_r \frac{1}{\sqrt{1+k}} \tag{11}$$

To find the eigenvalues of the natural response matrix **F**, the following identity is used, which is true for any square matrix [25].

$$\text{eig}(e^X) = e^{\text{eig}(X)} \tag{12}$$

Thus, the approximate locations of the resonance poles of the plant transfer function are

$$\Lambda_i = e^{\lambda_i T_s} : i = \{1, 2, 3, 4\} \tag{13}$$

The exact pole predictions are achieved if (8) is solved numerically. It is to be noted that (13) gives the pole locations in the z-domain. These can be equated to $re^{j\omega_{pri}T_s}$, to find ω_{pri} s, which are the frequencies of the poles in rad/s.

B. Prediction Accuracy and Approximation Error

For verification of the prediction and error introduced due to the above-mentioned assumptions, three different designs are considered [7], [9], as detailed in Table II. For design I, switching frequency f_s is equal to the the Tx and Rx tank resonance frequencies. However, for the other two designs $\omega_{rTx} \neq \omega_{rRx}$ and switching frequency is also different. Both high-Q and low-Q tank circuits are considered. The discrete-time transfer function (DTTF) obtained using [17] is considered the basis for comparison. Fig. 5 depicts the discrete-time frequency responses for the three designs under consideration, considering v_o as one of the states. Although this paper does not treat v_o as a state, use of the toolbox [17] allows for a full-order model with all states included. The close match between the pole frequencies in Fig. 5 and the predicted values in Table II (from (13)) indicates that model order reduction does not affect the resonant pole locations and, consequently, the bandwidth limit.

Table II shows that the approximate analysis predicts the pole locations with high accuracy (error< 1.5%) when ω_{rTx} and ω_{rRx} are close to each other. However, for design III, where the difference between these two frequencies are large and Q_{tx} is small, the error in the approximate analysis

979-8-3315-1612-3/25 $31.00 © 2025 IEEE

TABLE II
RESONANT POLE FREQUENCIES: EXACT AND PREDICTED WITH APPROXIMATION

Converter Details (Se-Se LC resonant)	Tank Specification	Critical Parameters	Resonant Pole Frequency (kHz)		Error (%)
			DT model	Approx (13)	
I. P=200 W $f_s = 165$ kHz [7]	L_{tx}, L_{rx} = 23.49, 2.8 μH, k=0.34 C_p=39.61 nF, C_s=331.66 nF R_p, R_s = 0.146, 0.076 Ω	$2\pi f_s = \omega_{rTx} = \omega_{rRx}$ $Q_{Tx} = 166.345$ $Q_{Rx} = 38.03$	22.29 37.91	22.21 37.67	0.61 0.33
II. P=600 W $f_s = 105$ kHz [18]	L_{tx}, L_{rx} = 71.87, 2.45 μH, k=0.37 C_p= 34 nF, C_s=1.08 μF R_p, R_s=0.198, 0.039 Ω	$\frac{\omega_{rTx}}{\omega_{rRx}} = 0.997$ $Q_{Tx} = 227.42$ $Q_{Rx} = 36.07$	16.09 26.28	16.03 25.93	0.39 1.33
III. P=20 W $f_s = 150$ kHz [9]	L_{tx}, L_{rx}=10.75, 12.1 μH, k=0.5 C_p= 235.9 nF, C_s=175 nF R_p, R_s=0.155, 0.39 Ω	$\frac{\omega_{rTx}}{\omega_{rRx}} = 0.916$ $Q_{Tx} = 43.64$ $Q_{Rx} = 21.32$	1.48 64.89	2.0 64.55	-35.7 0.53

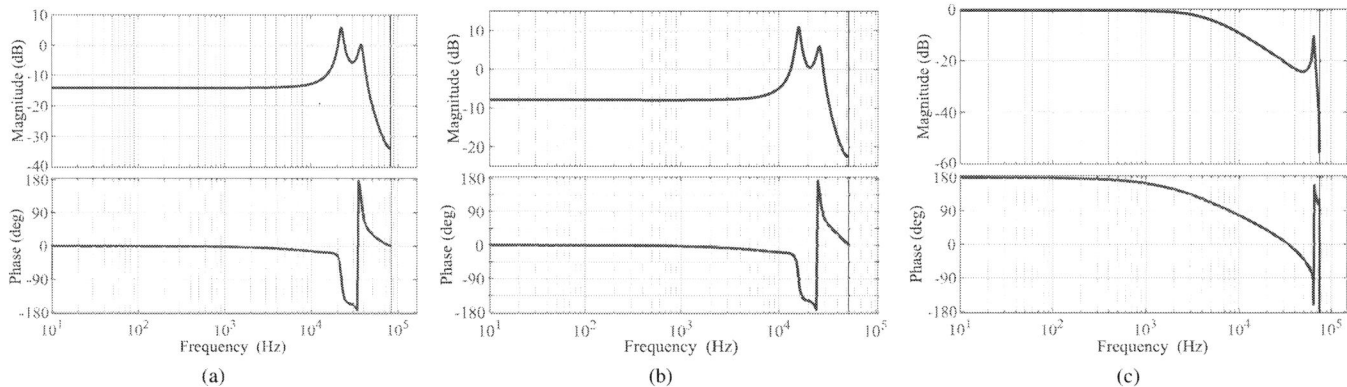

Fig. 5. Discrete-time (DT) transfer function for $\tilde{v}_o(z)/\tilde{\phi}_{ps}(z)$ for (a) design I, (b) design II, and (c) design III.

increases significantly (35.7%). For $\omega_{rTx} \neq \omega_{rRx}$, the pole locations are calculated based on (a) $\omega_r = \omega_{rTx}$, (b) $\omega_r = \omega_{rRx}$, and (c) $\omega_r = \sqrt{\omega_{rTx}\omega_{rRx}}$. Table II presents results for case (c), which yields the smallest error. Although the accuracy of the approximate pole prediction decreases with a larger mismatch between ω_{rTx} and ω_{rRx} and lower Q, the simplified expression in (13) relates the pole locations to the tank resonance frequencies, providing a quick ballpark approximation.

IV. SAMPLING IMPACT ON THE POLE FREQUENCIES

A. Impact of Sampling and Frequency Aliasing

The z-domain pole locations and corresponding frequencies are further studied to understand the sampling effect on the pole locations. Without resorting to the calculation of the exponential of the eigenvalues, the eigenvalues λ can directly be mapped into the frequencies of the resonant poles.

In a practical system, which is inherently a sampled-data system due to the per-period sampling actions within the A/D converter, modulator, or phase frequency detector, sampling artifacts are expected in the natural response. Since

the discrete-time (DT) model accounts for these effects, it accurately quantifies the sampling impacts. Let's consider the 200 W design (design I)- Table III gives the eigenvalues of the equivalent tank circuit. The frequencies corresponding to the eigenvalues are referred to as f_{rL} and f_{rH}. As the sampling frequency is equal to the switching frequency f_s in this formulation, both the eigenvalues exceed the Nyquist frequency, $f_s/2$= 82.5 kHz. Therefore, in any response, the frequency components corresponding to both f_{rL} and f_{rH} will show up at their aliased frequencies, which are below the Nyquist frequency. The aliased frequencies [26] are

$$|f_{rL} - f_s| = 22.2 \text{ kHz}$$
$$|f_{rH} - f_s| = 37.6 \text{ kHz} \quad (14)$$

It is seen that these frequencies are the same as those of the resonant poles, tabulated in Table II, obtained from an exact discrete-time model [22] and approximate predictions using (11)-(13). It is to be noted that the resonant poles are located at frequencies ($\approx$ 22.2 kHz and 37.6 kHz) much lower (7.5 times) than the Tx and Rx tank resonant and switching frequencies (165 kHz). Similarly, for design II and III, following (14), the

979-8-3315-1612-3/25 $31.00 © 2025 IEEE

TABLE III
EIGENVALUES AND FREQUENCY OF $\mathbf{A}$

Design	$\lambda = \mathrm{eig}(\mathbf{A})$ ($\times 10^5$ rad/sec)	Frequency (kHz) $\left(\frac{\mathrm{Imag}(\lambda)}{2\pi}\right)$	Aliased Frequency (kHz)
I.	$-0.064 \pm 8.98j$	$f_{rL} = 142.9$	22.2
	$-0.13 \pm 12.74j$	$f_{rH} = 202.8$	37.6
II.	$-0.042 \pm 5.60j$	$f_{rL} = 89.3$	15.8
	$-0.089 \pm 8.25j$	$f_{rH} = 131.3$	26.3
III.	$-0.241 \pm 9.34j$	$f_{rH} = 148.6$	1.41
	$-0.072 \pm 5.35j$	$f_{rL} = 85.14$	64.9

aliased frequencies are found, as provided in Table III. A close match between the aliased frequencies (Table III) and resonant pole frequencies (Table II) are observed in all three designs considered.

B. Graphical Interpretation

Fig. 6 depicts the magnitude response of the tank $G_{v_{cs}v_s}(j\omega) = \dfrac{\tilde{v}_{cs}(j\omega)}{\tilde{v}_s(j\omega)}$ for the design I, which relates the state v_{cs} to input v_s. This is obtained from the state-space model of the equivalent tank circuit, depicted in Fig. 2. Irrespective of the choice of state (i_p, i_s, v_{cp}, v_{cs}) and input (v_p, v_s) pair, the eigenvalues and hence the pole locations remain unchanged. The solid green trace indicates the resonance at the frequencies corresponding to the eigenvalues (λ) of $\mathbf{A}$. The frequency response is frequency shifted by $\pm n f_s : n = \{1, 2\}$ to obtain $G_{v_{cs}v_s}(\pm j\omega \mp n\omega_s)$ (dotted traces) and overlaid on the $G_{v_{cs}v_s}(j\omega)$. The vertical solid brown lines highlight the sampling frequency f_s and Nyquist frequency f_N. When any pole is located above f_N, its aliased frequency in the range $0 < f < f_N$ appears in the frequency response of the discrete-time transfer function (DTTF) of the converter that relates the state variables to the control input $\tilde{\phi}_{ps}$. This is verified in Fig. 7.

The frequency response of discrete-time transfer function (DTTF) $G_{v_{cs}\phi_{ps}}(z)$ is shown in Fig. 7 with the resonant pole frequencies labeled (ω_{pr1}, ω_{pr2}). Any other transfer function relating other states to the control input will not alter the resonant poles, as confirmed by comparing Fig. 7 and Fig. 5(a). The converter and tank parameters are as provided in Table II for design I. Among the aliased frequency responses $G_{v_{cs}v_s}(\pm j\omega \mp n\omega_s)$ of the tank, shown in Fig. 6, those lie in the range of $0 \leq f < f_N$ (highlighted in Fig. 6) are overlaid on the DTTF $G_{v_{cs}\phi_{ps}}(z)$. It shows that the resonant pole frequencies of the DTTF coincide with the aliased frequencies $f_s - f_{rL}$ and $f_{rH} - f_s$, as expected.

This aliasing effect causes the resonant frequencies of the tank, which are generally designed close to f_s, to show up in the control loops at much lower frequencies, complicating wide-bandwidth closed-loop control design. To this point, the analysis has assumed a sampling rate equal to the converter switching frequency f_s. A higher sampling rate could mitigate the in-band aliasing effect [24], but its feasibility

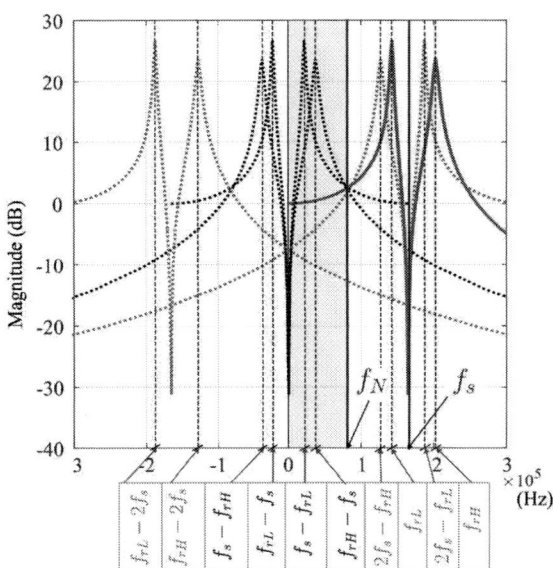

Fig. 6. Frequency response of tank TF $G_{v_{cs}v_s}(j\omega)$ (green) and its aliases $G_{v_{cs}v_s}(\pm j\omega \mp n\omega_s)$ (dotted) obtained by shifting $G_{v_{cs}v_s}(j\omega)$ by $\pm n f_s$ along both positive and negative frequency axes.

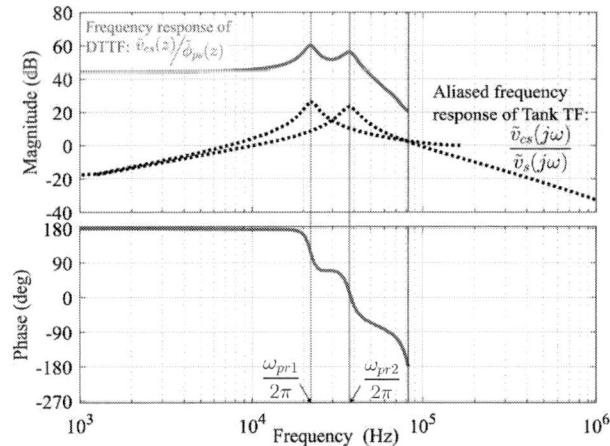

Fig. 7. Resonant pole frequencies from the DTTF frequency response and aliased frequency response of the tank which lies in $[0, f_N]$.

depends on the waveforms being sampled. For instance, phase measurements in the synchronization loop [9] are inherently sampled at f_s due to zero crossing detector (ZCD) and phase frequency detector (PFD) circuitry. Waveforms in the tank can be sampled at a higher rate but will cause the aliasing of switching ripple onto DC, requiring additional design consideration. Waveforms on the DC-side, including v_o and i_o can be sampled at a higher rate but will still exhibit some degree of low-frequency aliasing due to the rectification of the AC waveforms.

V. EXPERIMENTAL VALIDATION

Fig. 8 shows the experimental setup for the 200 W prototype (design I). Two different loads are considered: (A) R load and (B) Battery load.

Fig. 8. Experimental setup depicting Tx and Rx converters and coils.

A. R Load

Fig. 9(a) depicts experimentally obtained typical waveforms of the Tx and Rx coil voltages and currents when v_s leads v_p

by an angle $\phi_{ps} = 75°$. To find the resonant pole locations experimentally, a resistive load of 8 Ω is considered. This makes $v_b = 0$ and $R_b = 8$ Ω in Fig. 1(a). Following a step change in the phase ϕ_{ps}, the change in output voltage v_o and load current i_b are captured in Fig. 9(b). The captured data is averaged over the switching cycle T_s to filter out the switching frequency component and the change in the load current $\tilde{i}_b$ is depicted in Fig. 9(c). A periodic oscillation causing local minima is observed at every 7th sample. The frequency of this oscillation, $\frac{1}{7T_s} \approx 23.5$ kHz, closely matches the dominant resonance frequency, obtained from the proposed eigenvalue analysis with better than 5% accuracy.

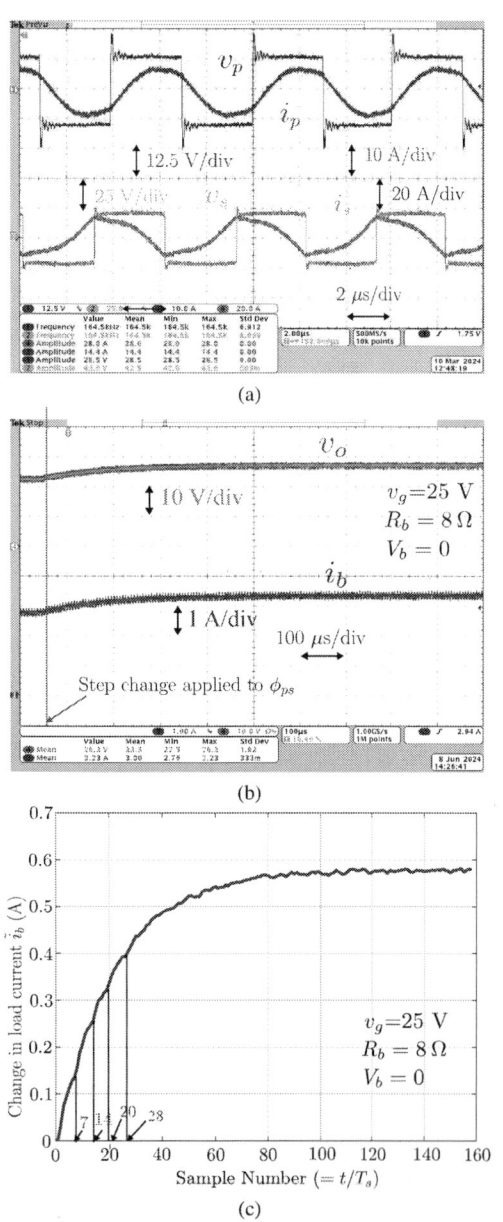

Fig. 9. Experimental results with R load depicting (a) different key variables at steady-state, (b) dynamics of output voltage v_o and load current i_b, and (c) resonance frequency found from experimental results.

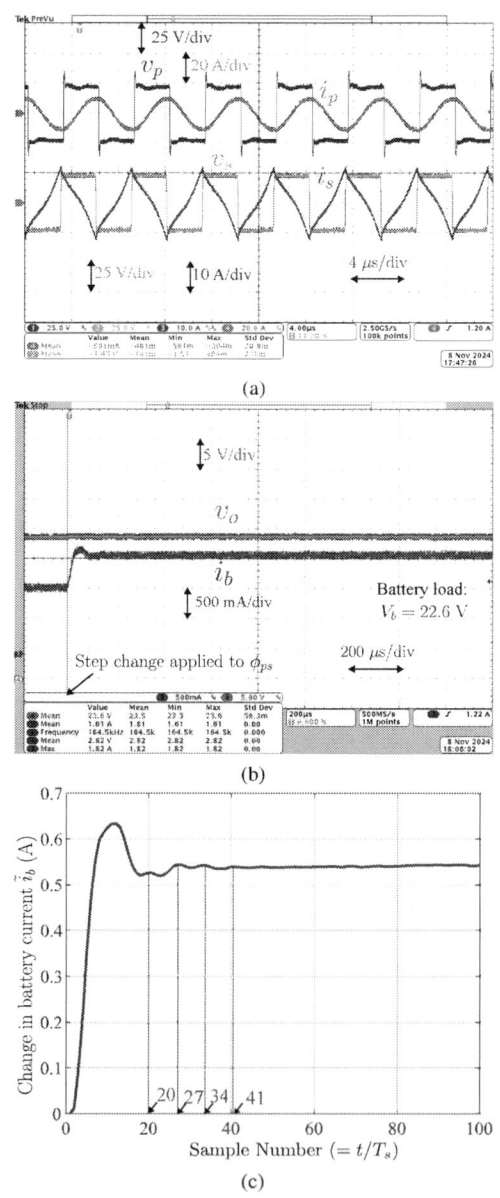

Fig. 10. Experimental results with battery load, depicting (a) different key variables at steady-state, (b) dynamics of output voltage v_o and battery current i_b, and (c) resonance frequency found from experimental results.

B. Battery Load

Fig. 10(a) depicts steady-state experimental waveforms when the R-load is replaced by a battery ($V_b = 22.6$ V). The waveforms are captured for $V_g = 25$ V and v_s leads v_p by an angle $\phi_{ps} = 22.5°$. As the battery is nearly in full charged condition, ϕ_{ps} is kept small to avoid overcharging of the battery. Following a step change of $7.5°$ in the phase ϕ_{ps}, the change in v_o and i_b are captured in Fig. 10(b). The captured data is averaged over the switching period T_s to filter out the switching frequency component in Fig. 10(c). The change from R load to battery load results in qualitative change in the response. However, a periodic oscillation during transient is observed, which remains periodic over 7 switching period (≈ 23.5 kHz). This indicates the change in the output port dynamics do not alter the resonant pole locations, which is decided by the tank parameters and the sampling frequency. With R load, a first-order open-loop transient response, dictated by the pole at $1/R_bC_o$, is observed. However, for battery load, due to the voltage-stiff output port, $\omega_{pr} < 1/R_bC_o$, as discussed in Section II. Regardless of the nature of the load, the experimental results confirm the resonant pole locations obtained from the DT model as well as the poles predicted from eigenvalue analysis considering the effect of frequency aliasing. The resonant pole frequency, showing up in the dynamic response, sets the limit for achievable maximum bandwidth.

VI. Conclusion

This paper identifies the origin of resonant poles in a series-series wireless power transfer system and correlates the pole locations with the sampling frequency and tank resonant frequencies. The analytical derivation, based on finding the eigenvalues of the natural response matrix, provides an intuitive understanding of the pole locations of the discrete-time model. The analytical results for three different designs are validated against their respective discrete-time (DT) models having T_s sampling rate. DT transfer functions (DTTFs) corresponding to different states (v_{cs}, v_o) are considered to demonstrate that the analysis is agnostic to the choice of states. A close match is found between the predicted and actual pole frequencies. For the design I (200 W system), the resonant pole locations are experimentally verified for both resistive and battery loads. The findings are valuable for designing stable control systems with multiple loops while maximizing control bandwidth.

References

[1] X. Lu, P. Wang, D. Niyato, D. I. Kim, and Z. Han, "Wireless charging technologies: Fundamentals, standards, and network applications," *IEEE Communications Surveys & Tutorials*, vol. 18, no. 2, pp. 1413–1452, 2016.

[2] S. Aldhaher, P. D. Mitcheson, J. M. Arteaga, G. Kkelis, and D. C. Yates, "Light-weight wireless power transfer for mid-air charging of drones," in *2017 11th European Conference on Antennas and Propagation (EUCAP)*, 2017, pp. 336–340.

[3] H. Feng, R. Tavakoli, O. C. Onar, and Z. Pantic, "Advances in high-power wireless charging systems: Overview and design considerations," *IEEE Trans. Transport. Electrific.*, vol. 6, no. 3, pp. 886–919, 2020.

[4] E. Nadi and H. Zhang, "Design and analysis of a detuned series-series ipt system with solenoid coil structure for drone charging applications," in *2024 IEEE Transport. Electrific. Conf. Expo. (ITEC)*, 2024, pp. 1–4.

[5] X. Wang, M. Leng, X. Zhang, Q. Tian, X. Zhou, B. Guo, and H. Ma, "Multioutput wireless charger for drone swarms with reduced switch requirements and independent regulation capability," *IEEE Trans. Ind. Electron.*, vol. 71, no. 5, pp. 4883–4895, 2024.

[6] A. Sagar, A. Kashyap, M. A. Nasab, S. Padmanaban, M. Bertoluzzo, A. Kumar, and F. Blaabjerg, "A comprehensive review of the recent development of wireless power transfer technologies for electric vehicle charging systems," *IEEE Access*, vol. 11, pp. 83 703–83 751, 2023.

[7] A. Basu, K. Froehle, and Costinett, D., "Design and optimization of a high gravimetric power density receiver for wireless charging of drones," in *2023 IEEE 24th Workshop Control Model. Power Electron. (COMPEL)*, 2023, pp. 1–8.

[8] X. Li, D. Zhou, F. Li, J. Zou, and X. Liu, "A high-power multi-rectifier wpt system with a hybrid modulation for wide-range zvs operation," *IEEE Trans. Transport. Electrific.*, pp. 1–1, 2024.

[9] S. Cochran and D. Costinett, "Discrete time synchronization modeling for active rectifiers in wireless power transfer systems," in *2019 20th Workshop Control Model. Power Electron. (COMPEL)*, 2019, pp. 1–8.

[10] V. Vorperian and S. Cuk, "Small signal analysis of resonant converters," in *1983 IEEE Power Electron. Specialists Conf.*, 1983, pp. 269–282.

[11] J. Sun and H. Grotstollen, "Averaged modeling and analysis of resonant converters," in *Proceedings of IEEE Power Electron. Specialist Conf. - PESC '93*, 1993, pp. 707–713.

[12] E. Yang, F. Lee, and M. Jovanovic, "Small-signal modeling of series and parallel resonant converters," in *[Proceedings] APEC '92 Seventh Annual Appl. Power Electron. Conf. Expo.*, 1992, pp. 785–792.

[13] S. Tian, F. C. Lee, and Q. Li, "A simplified equivalent circuit model of series resonant converter," *IEEE Trans. Power Electron.*, vol. 31, no. 5, pp. 3922–3931, 2016.

[14] D. J. Packard, "Discrete modeling and analysis of switching regulators," PhD thesis, California Institute of Technology, May 1976, available at https://resolver.caltech.edu/CaltechETD:etd-01082008-110208.

[15] M. Elbuluk, G. Verghese, and J. Kassakian, "Sampled-data modeling and digital control of resonant converters," *IEEE Trans. Power Electron.*, vol. 3, no. 3, pp. 344–354, 1988.

[16] E. X.-Q. Yang, "Extended describing function method for small-signal modeling of resonant and multi-resonant converters," PhD thesis, Virginia Polytechnique Institute and State University, Feb 1994, available at https://vtechworks.lib.vt.edu/server/api/core/bitstreams/682bb00b-a10f-4d72-9bc9-645c027ea9ce/content.

[17] J. A. Baxter and D. Costinett. (2023) Switched mode power supply toolbox, v0.2.x. [Online]. Available: https://github.com/costinet/AURA

[18] A. Basu and D. J. Costinett, "Design and optimization of a switched capacitor converter based 600 W wireless drone charger for high gravimetric power density," in *2025 IEEE Appl. Power Electron. Conf. Expo. (APEC)*, 2025.

[19] K. Aditya and S. S. Williamson, "Design guidelines to avoid bifurcation in a series–series compensated inductive power transfer system," *IEEE Trans. Ind. Electron.*, vol. 66, no. 5, pp. 3973–3982, 2019.

[20] J.-W. Lee, D.-G. Woo, S.-H. Ryu, B.-K. Lee, and H.-J. Kim, "Practical bifurcation criteria considering coil losses and compensation topologies in inductive power transfer systems," in *2015 IEEE International Telecommunications Energy Conference (INTELEC)*, 2015, pp. 1–6.

[21] A. Kunwar, "Design and implementation of an inductive power transfer system for wireless charging of future electric transportation," PhD thesis, University of Ontario Institute of Technology, August 2016, available at https://www.scribd.com/document/439092981/Aditya-Kunwar.

[22] A. Mukhopadhyay, K. Froehle, A. Basu, and D. Costinett, "CC-CV control of wireless battery charger based on discrete-time small-signal model," in *2024 IEEE Workshop Control Model. Power Electron.(COMPEL)*, 2024, pp. 1–8.

[23] D. Costinett, R. Zane, and D. Maksimovic, "Discrete time modeling of output disturbances in the dual active bridge converter," in *2014 IEEE Appl. Power Electron. Conf. Expo. - APEC 2014*, 2014, pp. 1171–1177.

[24] L. Corradini, D. Maksimović, P. Mattavelli, and R. Zane, *Digital Control of High-Frequency Switched-Mode Power Converters*, 2nd ed. John Wiley Sons, Ltd, 2015.

[25] C. D. Meyer, *Matrix Analysis and Applied Linear Algebra*, 2nd ed. Society for Industrial and Applied Mathematics (SIAM), 2000.

[26] J. Proakis and D. Manolakis, *Digital Signal Processing: Principles, Algorithms and Applications*, 3rd ed. Prentice-Hall International, 1996.

Integrated Hybrid Inductive and Capacitive Power Transfer System with Asymmetrical PCB Self-Resonator

Yao Wang[1], Zhen Sun[1], Xiangrong Zhang[1], Yun Yang[1*] and Shu Yuen Ron Hui[2]

1. Nanyang Technological University, School of Electrical and Electronic Engineering, 50 Nanyang Avenue, Singapore, 639798
2. The University of Hong Kong, Department of Electrical and Electronic Engineering, Pokfulam, Hong Kong
{yao.wang, yun.yang}@ntu.edu.sg, {sunz0028, zhan0649}@e.ntu.edu.sg, ronhui@eee.hku.hk

Abstract—This paper presents the design of a hybrid wireless power transfer (WPT) system with asymmetrical printed-circuit-board (PCB) based self-resonators, integrating both inductive and capacitive power transfer (IPT and CPT) channels. The PCB-based self-resonant coupler consists of four PCB-coil plates with two different sizes, which work as the coupled inductance for IPT as well as the capacitive plates for CPT. With a typical stacked four-plate configuration, both inductive and capacitive mutual couplings are achieved between transmitter and receiver, contributing to a self-resonant hybrid WPT with an integrated coupler and no external compensation components. Detailed theoretical analysis and system modeling are provided and a 250W hybrid WPT system is implemented with an asymmetrical coupler consisting of 210-mm and 140-mm PCB-coil plates. With transfer distances of 12mm and 37mm, the implemented hybrid WPT system works at the self-resonant frequencies of 3.19MHz and 3.57MHz, which respectively demonstrate a peak DC-DC efficiency of 87.3% with 155.7W at 12mm and 86.7% with 237.5W at 37mm, validating the effectiveness of the designed hybrid WPT system.

Keywords—Wireless power transfer, hybrid inductive and capacitive power transfer, PCB coils, self-resonator.

I. INTRODUCTION

Wireless power transfer (WPT) technology is revolutionizing the modern energy delivery pattern by cutting off troublesome metal cables [1]-[3], which has shown great potential in electric vehicles (EVs) [4]-[5], implanted medical devices [6]-[7], and consumer electronics [8]-[10]. Inductive power transfer (IPT) and capacitive power transfer (CPT) are two kinds of mainstream wireless power transfer (WPT) technologies. Combining the merits of both IPT and CPT systems, recently, the hybrid WPT system has attracted researchers' attention due to its potential for reduced components, higher power transfer capability, and improved misalignment tolerance [11]-[16].

Three conventional hybrid WPT system structures are provided in Fig.1, which are constructed by adding additional capacitive coupling to an IPT system [12], adding additional inductive coupling to a CPT system [13], or implementing independent IPT and CPT links [14]. However, the main drawback of conventional hybrid WPT systems is the usage of two sets of separate inductive and capacitive couplers, which makes the whole system bulky and takes up large space in practical applications. Although some attempts have been made to mitigate this issue by compactly configuring the inductive and capacitive couplers, such as stacking the

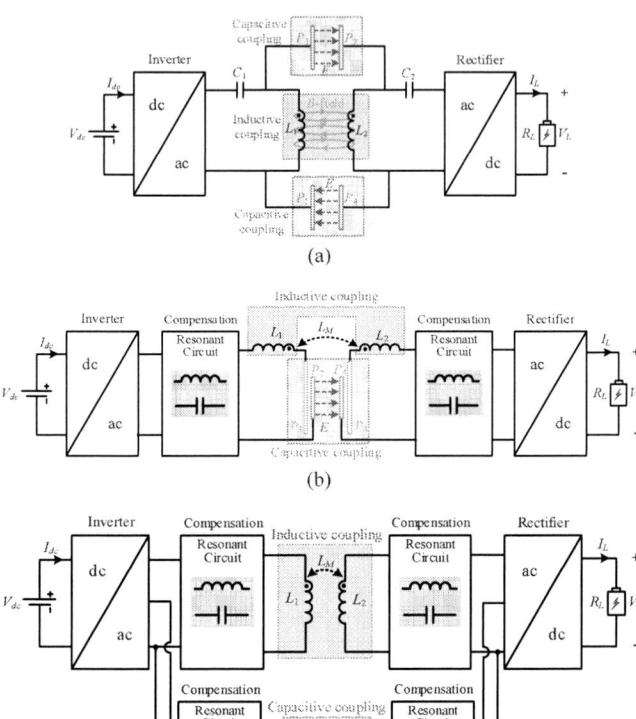

Fig. 1. Conventional hybrid WPT systems constructed by (a) adding an additional capacitive link to a CPT system [12]; (b) adding an additional inductive link to an IPT system [13]. (c) constructing independent IPT and CPT links [14].

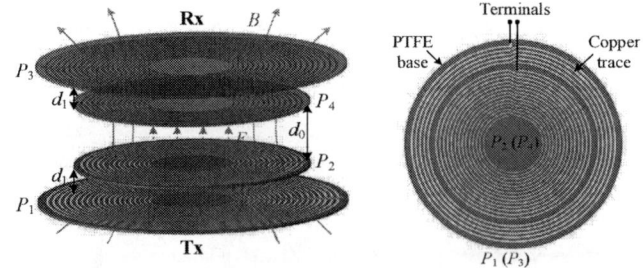

Fig. 2. Proposed hybrid inductive and capacitive coupler based on asymmetrical PCB self-resonator.

inductive and capacitive couplers together [15] and embedding the inductive coupler into the capacitive coupler [16]. However, the root concern is not fully addressed due to the failure to integrate the IPT and CPT links into one coupler. Recently, printed-circuit-board (PCB) based self-resonators have been proposed to design WPT systems with high compactness [17]-[21], in which the PCB coil inductance can resonate with its parasitic capacitance. These PCB self-resonators have shown compatibility to work in either IPT or CPT mode [21], which provides the potential to achieve hybrid inductive and capacitive power transfer with an integrated coupler.

This paper proposed an integrated hybrid inductive and capacitive power transfer system based on asymmetrical PCB self-resonators. As shown in Fig.2, the proposed asymmetrical self-resonant coupler consists of four PCB-coil plates with different sizes. With a typical stacked four-plate configuration, coupling capacitance for CPT is achieved, meantime, the mutual inductance between the coil inductance of the transmitter (Tx) and receiver (Rx) provides the channel for IPT, contributing to hybrid inductive and capacitive power transfer with an integrated coupler. In the proposed PCB-based self-resonant coupler, the inductance of copper trace works as the compensation for coupling capacitance, and vice versa, which achieves a self-resonant structure without using any external compensation, realizing high compactness, simplicity, cost-effectiveness, and ease of mass production. To validate the proposed design, a 250W hybrid WPT system is implemented with an asymmetrical self-resonant coupler consisting of 210-mm and 140-mm PCB-coil plates, which demonstrate the power transfer capability of 155.7W at 12mm with a DC-DC efficiency of 87.3%, and 237.5W at 37mm with a DC-DC efficiency of 86.7%.

II. PROPOSED HYBRID WPT SYSTEM WITH ASYMMETRICAL PCB-BASED SELF-RESONANT COUPLER

A. Hybrid WPT System Design

The structure of the proposed PCB-based self-resonant coupler is demonstrated in Fig.2, which consists of four PCB coil plates, P_1, P_2, P_3, and P_4, manufactured by printing spiral circular copper trace with a thickness of 1oz (35μm) on the polytetrafluoroethylene (PTFE) board. P_1 and P_2 are transmitter (Tx) while P_3 and P_4 are receiver (Rx). Transfer distance is represented by d_0 which is changeable while the gap d_1 in Tx and Rx are fixed. P_2 and P_4 are smaller than P_1 and P_3, and a stacked configuration is designed with P_2 and P_4 inside and P_1 and P_3 outside.

Based on the proposed asymmetrical PCB self-resonator, the developed hybrid WPT system is provided in Fig.3. A push-pull current-source inverter consisting of inductors L_{dc1} and L_{dc2} and switches S_1 and S_2 is used to generate a sinusoidal excitation voltage v_1, and i_1 is the input current. L_r is the resonant inductor compensating the drain-source capacitances C_{ds1} and C_{ds2} of switches. i_2 and v_2 are the AC current and voltage at the output side and a full-bridge rectifier is used to achieve DC output for the load.

B. Modeling of the Hybrid Couplings

The demonstration of the inductive and capacitive

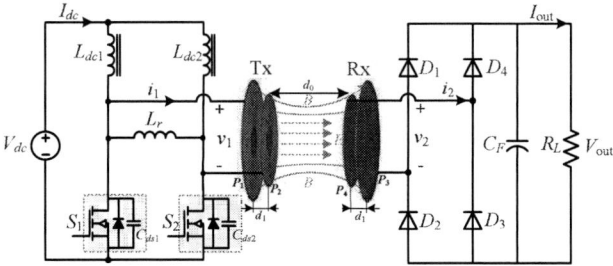

Fig.3. Proposed hybrid wireless power transfer system

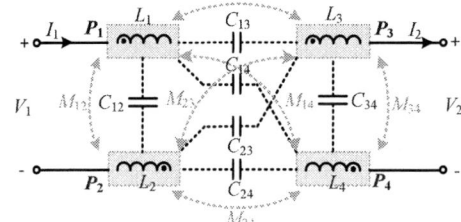

Fig.4 Inductive and capacitive couplings within the proposed self-resonant coupler

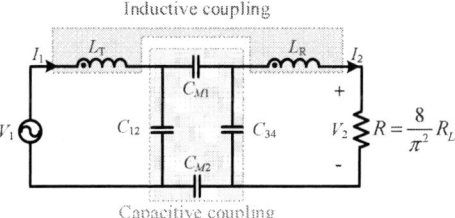

Fig.5. Equivalent circuit of the proposed hybrid WPT system.

couplings within the self-resonant coupler is provided in Fig.4. Coupling capacitances C_{13} between P_1 and P_3, and C_{24} between P_2 and P_4 mainly comprise the channel for capacitive power transfer. C_{12} and C_{34} are the parasitic capacitance between P_1 and P_2, P_3 and P_4, which are mainly used to compensate the coil inductance. C_{14} and C_{23} are the cross-coupling capacitances. L_1, L_2, L_3, and L_4 are respectively the inductances of four PCB-coil plates. The inductive mutual couplings M_{12} and M_{34} can respectively enhance the self-inductances of transmitter and receiver, while the inductive couplings M_{13}, M_{14}, M_{23}, and M_{24} between transmitter and receiver comprise the channel for IPT.

Based on the presented hybrid WPT system design, the simplified equivalent circuit model is provided in Fig.5. V_1 (RMS) is the AC input voltage, and R is the equivalent AC load resistance. According to [21], V_1 and R are given as:

$$V_1 \approx 2.22V_{dc}, R = 8R_L / \pi^2 \qquad (1)$$

L_T and L_R are respectively the equivalent self-inductance of Tx and Rx, M_{TR} is the equivalent mutual inductance, and k_L is the inductive coupling coefficient, specified below.

$$\begin{cases} L_T = L_1 + L_2 + 2M_{12}, \quad L_R = L_3 + L_4 + 2M_{34} \\ M_{TR} = M_{13} + M_{24} + M_{14} + M_{23} \\ k_L = M_{TR} / \sqrt{L_T \cdot L_R} \end{cases} \qquad (2)$$

979-8-3315-1612-3/25 $31.00 © 2025 IEEE

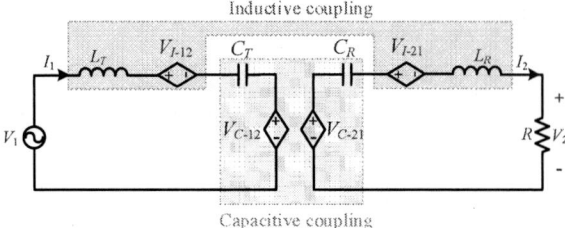

Fig.6. Behavioral voltage source-based equivalent circuit.

C_{M1} and C_{M2} are the equivalent coupling capacitances in forward and backward paths. The self-capacitances C_1, C_2, mutual capacitance C_M, and the capacitive coupling coefficient k_C are defined below.

$$
\begin{cases}
C_M = \dfrac{C_{M1} \cdot C_{M2}}{C_{M1} + C_{M2}} = \dfrac{C_{13} \cdot C_{24} - C_{23} \cdot C_{14}}{C_{13} + C_{14} + C_{23} + C_{24}} \\[2mm]
C_1 = C_{12} + C_M, \quad C_2 = C_{34} + C_M \\[2mm]
k_C = \dfrac{C_M}{\sqrt{C_1 \cdot C_2}}
\end{cases}
\tag{3}
$$

Based on the z-parameter model of two-port network theory [3], the inductive coupling can be equivalent to two current-controlled behavioral voltage sources $V_{I\text{-}12}$ and $V_{I\text{-}21}$, and capacitive couplings can be equivalent to $V_{C\text{-}12}$, and $V_{C\text{-}21}$, given below:

$$
\begin{cases}
V_{I\text{-}12} = I_2 \cdot j\omega M_{TR} \\[1mm]
V_{I\text{-}21} = I_1 \cdot j\omega M_{TR} \\[1mm]
V_{C\text{-}12} = -I_2 \cdot \dfrac{1}{j\omega C_M (1/k_C^2 - 1)} \\[2mm]
V_{C\text{-}21} = I_1 \cdot \dfrac{1}{j\omega C_M (1/k_C^2 - 1)}
\end{cases}
\tag{4}
$$

With four behavioral voltage sources $V_{I\text{-}12}$, $V_{I\text{-}21}$, $V_{C\text{-}12}$, and $V_{C\text{-}21}$ introduced, the circuit model is further simplified into Fig.6. According to [3], the capacitance C_T and C_R are defined in (5).

$$
\begin{cases}
C_T = C_1 \cdot (1 - k_C^2) = C_{12} + \dfrac{C_{34} C_M}{C_{34} + C_M} \\[2mm]
C_R = C_2 \cdot (1 - k_C^2) = C_{34} + \dfrac{C_{12} C_M}{C_{12} + C_M}
\end{cases}
\tag{5}
$$

C. Circuit Working Principle

According to Fig.6, it is straightforward to derive the zero-phase-angle (ZPA) operating frequency f, provided in (6).

$$
f = \frac{1}{2\pi\sqrt{L_T C_T}} = \frac{1}{2\pi\sqrt{L_R C_R}}
\tag{6}
$$

According to Fig.6, when working at ZPA frequency, the input voltage V_1 and the behavioral-voltage sources $V_{I\text{-}12}$ and $V_{C\text{-}12}$ satisfy the relationship below:

$$
\begin{cases}
V_1 = V_{C\text{-}12} + V_{I\text{-}12} = I_2 \cdot j\omega M_{TR} + j\dfrac{I_2}{\omega C_M (1/k_C^2 - 1)} \\[3mm]
V_2 = V_{C\text{-}21} - V_{I\text{-}21} = I_1 \cdot \dfrac{1}{j\omega C_M (1/k_C^2 - 1)} - I_1 \cdot j\omega M_{TR}
\end{cases}
\tag{7}
$$

According to (7), the output current I_2 (RMS) is derived as:

$$
\begin{cases}
I_1 = j\dfrac{V_2}{\omega M_{TR} + \dfrac{1}{\omega C_M (1/k_C^2 - 1)}} \\[5mm]
I_2 = \dfrac{V_1}{j\left(\omega M_{TR} + \dfrac{1}{\omega C_M (1/k_C^2 - 1)}\right)}
\end{cases}
\tag{8}
$$

With ZPA working frequency, input current I_1 and the induced voltages $V_{I\text{-}12}$ and $V_{C\text{-}12}$ are in phase. The transferred power via inductive and capacitive links are respectively represented by P_{IPT} and P_{CPT}, given below:

$$
\begin{cases}
P_{IPT} = I_1 \cdot V_{I\text{-}12} \\[1mm]
P_{CPT} = I_1 \cdot V_{C\text{-}12}
\end{cases}
\tag{9}
$$

The power transfer ratio between IPT and CPT links are derived in (10), which are respectively proportional to the inductive and capacitive coupling coefficients k_L and k_C.

$$
Ratio = \frac{P_{IPT}}{P_{CPT}} = \frac{|V_{I\text{-}12}|}{|V_{C\text{-}12}|} = \frac{k_L}{k_C}
\tag{10}
$$

III. HARDWARE IMPLEMENTATION

A. Implementation of the Self-Resonant Coupler

Fig.7 demonstrates the implemented PCB-based self-resonant coupler, and the main specifications are provided in Table I. PCB coil plates P_1 and P_3 have an outer diameter of 210mm with 20-turn planar spiral circular coils and the plates P_2 and P_4 have an outer diameter of 145mm with 14-turn coils. To achieve a high Q-factor of the self-resonant coupler, polytetrafluoroethylene (PTFE) is used in the PCB base instead of conventional FR-4 material. The thickness of the printed copper trace is 1oz (35μm). The distance d_0 in Tx or Rx is set at 3.5mm, which achieves a self-resonant frequency of Tx and Rx of 3.845MHz.

B. Implementation of the Hybrid WPT System

Fig.8 presents the implemented hardware of the hybrid WPT system. A DC voltage of 80V is used at the input side and a push-pull current-source inverter is designed based on two gallium nitride (GaN) FETs GS66508T and controlled by DSP TMS320F28335, aiming at achieving quasi-sine AC excitation. The rectifier is designed based on four silicon carbide (SiC) diodes C4D02120A. An electronic load with a resistance range of 100 Ω~1000 Ω is used.

With a relatively long transfer distance, e.g., 100mm, the coupling capacitance between Tx and Rx for the CPT link is very weak and the system will mainly work in IPT mode, which has been tested in previous literature [20]. With a short

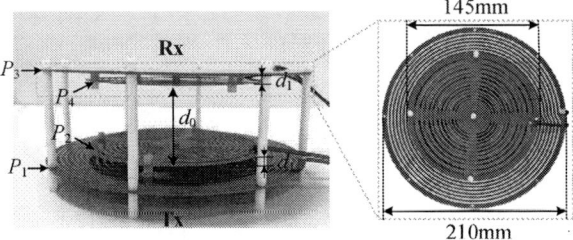

Fig.7 Implementation of the asymmetrical PCB self-resonant coupler.

TABLE I. MAIN SPECIFICATIONS OF THE PCB-BASED SELF-COUPLER.

Parameter	Value	
	P_1 & P_3	P_2 & P_4
Diameter of PCB D_{PCB}	210mm	145mm
Diameter of innermost coil D_{in}	60mm	18mm
Diameter of Outermost coil D_{out}	200mm	130mm
Coil pitch	2.482mm	2.2mm
Coil width of the innermost turn	0.794mm	1.013mm
Coil width of the outermost turn	1.626mm	3.5mm
Number of coil turns	20	14
Material of PCB base	Polytetrafluoroethylene (PTFE)	
Thickness of PCB base	1.56mm	
Thickness of copper trace	1oz (35μm)	
Distance d_1	3.5 mm	
Self-resonant frequency of Tx/ Rx	3.845MHz	

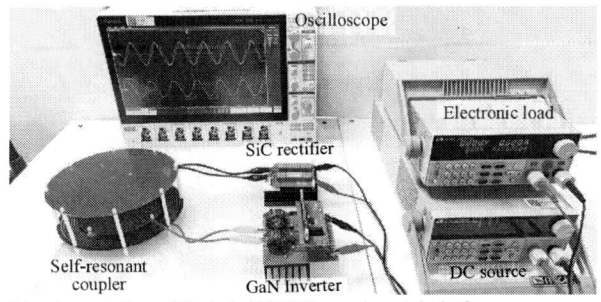

Fig.8 Implementation of the hybrid WPT experimental platform.

TABLE II. SPECIFICATIONS OF THE IMPLEMENTED HYBRID WPT SYSTEM.

Parameter	Value
V_{dc}	80V (DC)
Transfer distance d_0	37mm/12mm
Lateral misalignment	0~50mm
Plate gap d_1	3.5mm
Switching frequency f_s	3.57MHz/3.19MHz
GaN FET	GS66508T
Rectifier diode	C4D02120A
Magnetic ring	fair-rite 67
R_L	100Ω~1000 Ω

distance (e.g., 37mm and 12mm), the system will work in hybrid WPT mode, in which the ZPA operating frequencies are respectively set at 3.57MHz and 3.19MHz.

IV. EXPERIMENTAL VALIDATION

A. Experimental Waveforms

Fig.9 (a) and Fig.9 (b) respectively show the experimental waveforms of the hybrid WPT system at the transfer distances of 12mm and 37mm with an input DC voltage of 80V and operating frequencies of 3.19MHz and 3.57MHz, respectively. It shows that the input voltage v_1 has low harmonics with

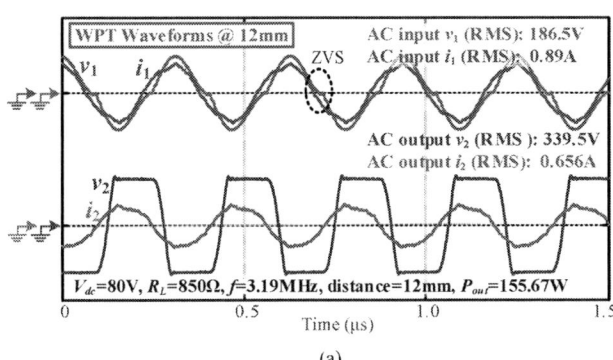

(a)

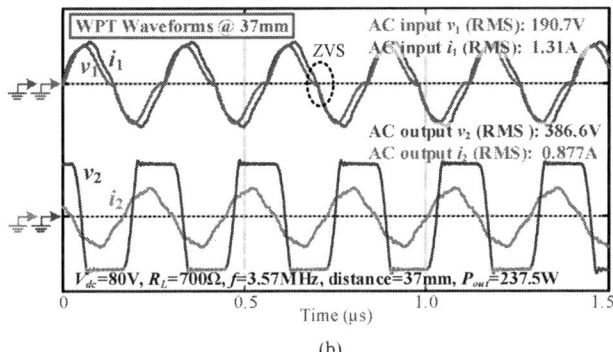

(b)

Fig.9 Experimental waveforms at (a) 12mm and (b) 37mm.

quasi-sine waveforms and zero-voltage-switching (ZVS) is achieved. The input current i_1 is in phase with input voltage v_1, contributing to the ZPA working condition. The AC output voltage v_2 is in square-wave and the AC output current i_2 slightly leads v_2 due to the parasitic capacitance of rectifier diodes. The impact of the rectifier parasitic capacitance becomes significant with a large load resistance R_L (light-load condition).

B. System Performance Versus Load Variation

With an input DC voltage of 80V, Fig.10 demonstrates the DC output current I_{out} versus load resistance R_L at different transfer distances. It shows that with a wide load variation from 100Ω to 1000Ω, the output DC current I_{out} respectively shows a drop of 24.0% at 37mm and 39.5% at 12mm. The parasitic resistance of the PCB coils accounts for the current fluctuation versus load variation. Meantime, with a high operating frequency of multi-MHz, the parasitic capacitance of the rectifier can't be ignored, which will cause the detuning of the whole system when the load changes, leading to the drops of I_{out}.

Fig.11 shows the transferred DC power versus load resistance at distances of 12mm and 37mm. With a distance of 12mm and a load range of 100Ω~1000Ω, the output power first increased from 39.8W at R_L=100Ω to the peak of 160.5W at R_L=700Ω and then slightly drops to 145.7W at 1000Ω. With a distance of 37mm, the DC output power P_{out} monotonously increases from 43.5W at R_L=100Ω to 258W at R_L=1000Ω. It shows that the output power shows significant nonlinearity at the load range of high resistance of R_L, which is mainly attributed to the impact of rectifier diode capacitances.

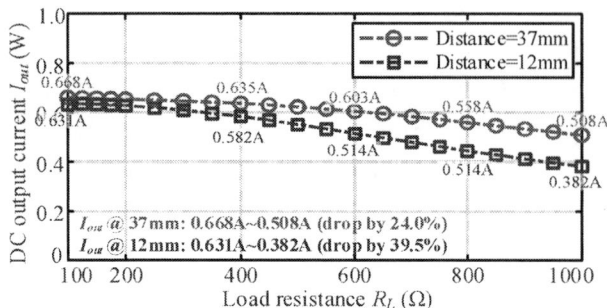

Fig.10 Output DC current I_{out} versus load resistance R_L.

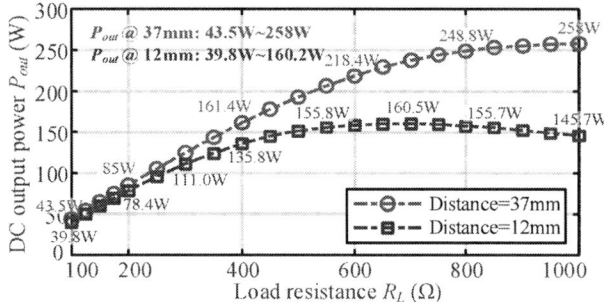

Fig.11 Output power P_{out} versus load resistance R_L.

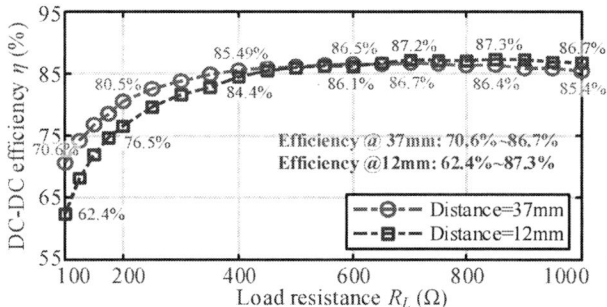

Fig.12 DC-DC efficiency η versus load resistance R_L.

Fig.12 shows the DC-DC efficiency versus load resistance at 12mm and 37mm of the implemented hybrid WPT system. At a transfer distance of 12mm, the system efficiency increases from 62.4% at $R_L=100\Omega$ to the maximum of 87.3% at $R_L=850\Omega$ before it slightly drops to 86.7% at $R_L=1000\Omega$. The output power achieves 155.7W at the maximum efficiency point. With a transfer distance of 37mm, the system efficiency ranges from 70.6% to 86.7% and the maximum efficiency of 86.7% is achieved at $R_L=700 \Omega$ with a DC output power of 237.5W.

C. Misalignment Tolerance

Fig.13 shows the system output power versus lateral misalignment from 0 to 50mm with fixed input DC voltage and load resistance. With a transfer distance of 12mm, the system power remains almost unchanged within a misalignment of 10mm, and then demonstrates significant drops versus the further increased misalignment and the system power reduces from 160.5W at aligned case to 28.1W at 50-mm misalignment. In comparison, with a transfer distance of 37mm, the system power is also stable between

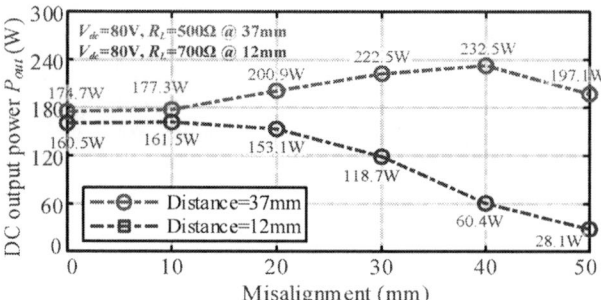

Fig.13 Output power P_{out} versus misalignment.

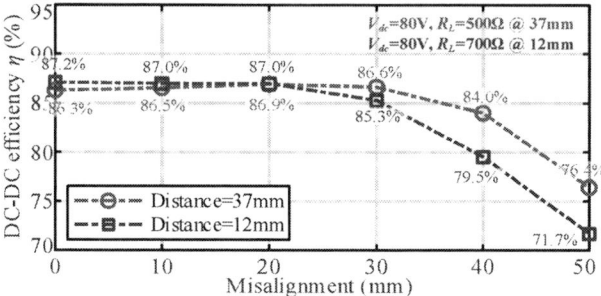

Fig.14 DC-DC efficiency η versus misalignment.

174.7W~177.3W within 10-mm misalignment and then increases with misalignment, achieving the peak of 232.5W at a misalignment of 40mm before falling back to 197.1W at a misalignment of 50mm.

Fig.14 compares the DC-DC efficiency versus misalignment when working at different transfer distances. At a transfer distance of 12mm, the system efficiency first remains unchanged at over 87% with misalignment of 0~20mm, and then drops rapidly when misalignment exceeds 20mm, which achieves 71.7% at a maximum misalignment of 50mm. At the case of 37-mm transfer distance, the implemented WPT system demonstrates higher misalignment tolerance. System efficiency remains stable with misalignment of 0~30mm and starts to drop within 30mm~50mm, which reduces to 76.4% at misalignment of 50mm.

When the misalignment occurs, the coupling inductance and capacitance between Tx and Rx will be reduced, affecting the system output power and efficiency. Meantime, the changed coupling condition will also cause the detuning of the implemented hybrid WPT system, which further impacts the system output power and deteriorates the system efficiency. Overall, with a misalignment within 30mm, the system power and efficiency do not demonstrate significant drops.

V. CONCLUSIONS

This paper presents an integrated hybrid inductive and capacitive power transfer system design with an asymmetrical PCB-based self-resonant coupler. The theoretical analysis and modeling are provided, and the hardware of the PCB-based self-resonant coupler is demonstrated. A 250W hybrid WPT system is implemented, and the experimental results validate the effectiveness of the proposed design. With transfer distances of 12mm and 37mm, the maximum DC-DC efficiency respectively achieves 87.3% at 155.7W and 86.7% at 237.5W. The misalignment tolerance is also tested, showing acceptable power and efficiency performance within 30-mm misalignment. Future work will focus on the optimization of the self-resonant coupler oriented for efficiency maximization.

REFERENCES

[1] Z. Zhang, H. Pang, A. Georgiadis and C. Cecati, "Wireless power transfer—an overview," *IEEE Trans. Ind. Electron.*, vol. 66, no. 2, pp. 1044-1058, Feb. 2019.

[2] J. Dai and D. C. Ludois, "A survey of wireless power transfer and a critical comparison of inductive and capacitive coupling for small gap applications," *IEEE Trans. Power Electron.*, vol. 30, no. 11, pp. 6017–6029, Nov. 2015.

[3] Y. Wang, H. Zhang, Y. Cao and F. Lu, "Remaining Opportunities in Capacitive Power Transfer Based on Duality With Inductive Power Transfer," IEEE Trans. Transp. Electrif, vol. 9, no. 2, pp. 2902-2915, June 2023.

[4] S. Li, S. Lu and C. C. Mi, "Revolution of electric vehicle charging technologies accelerated by wide bandgap devices," *Proceedings of the IEEE*, vol. 109, no. 6, pp. 985-1003, June 2021.

[5] D. Patil, M. K. McDonough, J. M. Miller, B. Fahimi and P. T. Balsara, "Wireless power transfer for vehicular applications: overview and challenges," *IEEE Trans. Transp. Electrific.*, vol. 4, no. 1, pp. 3-37, March 2018.

[6] H. J. Kim, H. Hirayama, et al., "Review of near-field wireless power and communication for biomedical applications," *IEEE Access*, vol. 5, pp. 21264-21285, Oct. 2017.

[7] K. Agarwal, R. Jegadeesan, et al., "Wireless power transfer strategies for implantable bioelectronics," *IEEE Rev. Biomed. Eng.*, vol. 10, pp. 136-161, 2017

[8] X. Liu, "Qi Standard Wireless Power Transfer Technology Development Toward Spatial Freedom," *IEEE Circuits Syst. Mag.*, vol. 15, no. 2, pp. 32-39, Secondquarter 2015.

[9] Qi Specification: [online] https://www.wirelesspowerconsortium.com/knowledge-base/specifications/download-the-qi-specifications/

[10] N. S. Jeong and F. Carobolante, "Wireless Charging of a Metal-Body Device," *IEEE Trans. Microw. Theory Tech.*, vol. 65, no. 4, pp. 1077-1086, April 2017

[11] X. Li, C. Tang, X. Dai, P. Deng and Y. Su, "An inductive and capacitive combined parallel transmission of power and data for wireless power transfer systems," *IEEE Trans. Power Electron.*, vol. 33, no. 6, pp. 4980-4991, June 2018.

[12] Luo, B., Long, T., Mai, R., Dai, R., He, Z. and Li, W., "Analysis and design of hybrid inductive and capacitive wireless power transfer for high-power applications," *IET Power Electronics*, vol.11, no. 2263-2270, 2018.

[13] F. Lu, H. Zhang, H. Hofmann and C. C. Mi, "An inductive and capacitive combined wireless power transfer system with LC-compensated topology," *IEEE Trans. Power Electron.*, vol. 31, no. 12, pp. 8471-8482, Dec. 2016.

[14] D. Vincent, P. S. Huynh and S. S. Williamson, "A link-independent hybrid inductive and capacitive wireless power transfer system for autonomous mobility," *IEEE J. Emerg. and Sel. Topics in Ind. Electron.*, vol. 3, no. 2, pp. 211-218, April 2022.

[15] X. Zhang, G. Li, T. Chen, F. Wang, Q. Yang and W. Xu, "A high-efficiency hybrid wireless power transfer system with low plate voltage stresses," *IEEE Trans. Power Electron.*, vol. 39, no. 8, pp. 10546-10557, Aug. 2024

[16] X. Gao et al., "Design and analysis of a new hybrid wireless power transfer system with a space-saving coupler structure," *IEEE Trans. Power Electron.*, vol. 36, no. 5, pp. 5069-5081, May 2021

[17] K. Chen and Z. Zhao, "Analysis of the double-layer printed spiral coil for wireless power transfer," *IEEE J. Emerg. Sel. Topics Power Electron.*, vol. 1, no. 2, pp. 114-121, Jun. 2013.

[18] J. Qu, L. He, N. Tang, et al., "Wireless power transfer using domino-resonator for 110-kV power grid online monitoring equipment," *IEEE Trans. Power Electron.*, vol. 35, no. 11, pp. 11380-11390, Nov. 2020.

[19] Z. Yı, M. Lı, B. Muneer, G. He and X. -X. Yang, "Self-Resonant antisymmetric planar coil for compact inductive power transfer system avoiding compensation circuits," *IEEE Trans. Power Electron.*, vol. 36, no. 5, pp. 5121-5134, May 2021.

[20] Q. Wang, M. A. Saket, A. Troy and M. Ordonez, "A self-compensated planar coil for resonant wireless power transfer systems," *IEEE Trans. Power Electron.*, vol. 36, no. 1, pp. 674-682, Jan. 2021.

[21] Y. Wang, K. Wang, K. Li, Y. Yang and S. Y. R. Hui, "Multi-MHz inductive and capacitive power transfer systems with PCB-based self-resonators," *IEEE Trans. Power Electron.*, vol. 39, no. 10, pp. 14077-14090, Oct. 2024.

High frequency noise reduction method of the class E power amplifier

Kyungmin Lee

Samsung Research

Samsung Electronics

Seoul, South Korea

km1019.lee@samsung.com

Sungku Yeo

Samsung Research

Samsung Electronics

Seoul, South Korea

sungku.yeo@samsung.com

Abstract— **Low-temperature dryers using the radio frequency (RF) electric field utilize dielectric heating to overcome disadvantages of conventional heat type dryer like shrinkage and damage of fabric. In order to increase the drying effect, a high frequency RF electric field of 13.56MHz is generated in the kW level. In several kW power circuits, 13.56 MHz is a relatively high frequency, resulting in a lot of noise. A method of reducing high-frequency noise by using harvesting in a low-temperature dryer is proposed. It utilizes the harvesting energy generated in the dryer; the magnetic energy of the motor and thermal energy of the drying unit. For verification of the proposed methods, the multi-channel power amplifier was connected to the real dryer load and used in the experiment.**

Keywords— Energy harvesting, class E power amplifier.

I. INTRODUCTION

Electric dryers are widely used because they not only dry fabrics quickly but also contain functions such as sterilization. Conventional dryers use heating coils to dry fabrics through high-temperature hot air. However, these dryers have the disadvantage of shrinking and damaging fabrics due to high temperatures. Recently, a dryer using a heat pump has been developed instead of a heat method [1]-[3]. However, since the dryer using the heat pump also operates at a temperature of 60 to 70 degrees, it cannot be completely free from damage to the fabric due to high temperature.

The dryer using the dielectric heating phenomenon operates at a temperature of 40 to 50 degrees, so it can be completely free from fabric damage [4]-[6]. Low-temperature dryers using dielectric heating use power amplifiers to apply high frequency electric field of tens of MHz to fabrics. From the power circuit standpoint, the frequency of tens of MHz and the power of kW are not general solutions. Ordinary power circuits are driven at several kW power and tens of kHz frequencies, and communication circuits are driven at hundreds of W power and tens of MHz frequencies.

II. PROPOSED CIRCUIT AND OPERATIONAL PRINCIPLE

The high frequency noise reduction circuit using energy harvesting of the low temperature dryer is proposed. The power unit of the dielectric heating low temperature dryer consists of an electromagnetic interference (EMI) filter, a power factor correction circuit (PFC), a DCDC converter, a class E power amplifier, an impedance matching circuit, and an electrode. Since the PFC and DCDC converters operate at frequencies of tens of kHz, the high frequency noise of ~MHz does not occur. Since the class E power amplifier operates at ~MHz and kW levels, the high frequency noise occurs a lot [7]-[10]. In general, the class E power amplifier of a low-temperature dryer is operated at frequencies of 6.78MHz, 13.56MHz, and 27MHz, and the proposed circuit is operated at 13.56MHz.

The proposed circuit uses two main methods for noise cancellation (Fig. 1). The first is ground separation. The circuit consists of a power unit having a large amount of power flows and a signal unit for controlling the power unit. Noise occurs in the power unit that has a switching element, and is delivered to the signal unit; it causes many signal circuit malfunctions. In the proposed circuit, the power part and the signal part are separated. Because the ground is separated, an isolated gate driver is used to drive the class E power amplifier in the signal unit. The second is energy harvesting. Because the signal part is isolated from the power part, there needs driving power source to drive the signal generation circuits. In this circuit, the power for the signal part is generated through the energy harvesting in the dryer.

The signal unit consists of a part for energy harvesting, a power conversion circuit, a 13.56MHz source generator, and an isolated gate driver. The 13.56MHz source generator can be configured by a microcontroller or crystal. In addition, the AC-DC converter converts AC power collected through harvesting into DC level power. When the circuit is configured in this way, the power unit and the signal unit are completely separated. That is, even if the noise due to load change, external injection, and PCB layout parasitic components occurs at the power unit, the signal unit is not affected and a correct gate signal can be applied to the power unit.

The harvesting energy can be collected in various ways in the dryer. The first is to collect magnetic energy using a motor. The dryer has a tumble motor for rotating the drying unit and an intake and exhaust motor for discharging moisture to the outside. Because the motor converter the electric energy to rotational energy using the magnetic elements, there is always a leakage magnetic field around the motor. In the proposed circuit, a harvesting unit of a magnetic core and a wire is placed around the motor to collect the magnetic energy (Fig. 2(a)). By placing the magnetic harvesters in several places around the motor and merging the collected energy, the harvesting energy can be stored in the energy storage element through the ac dc converter. This energy is used to drive the signal part of the circuit.

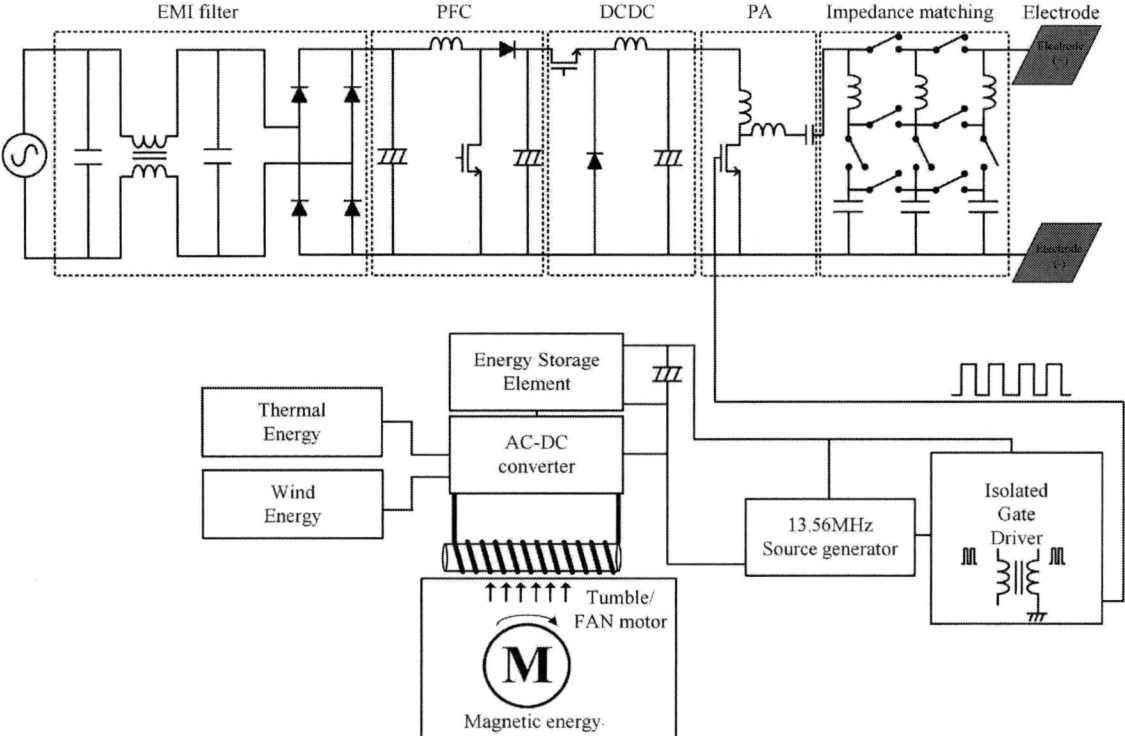

Fig. 1. The proposed circuit of high frequency noise reduction method of the low temperature dryer using magnetic energy harvesting

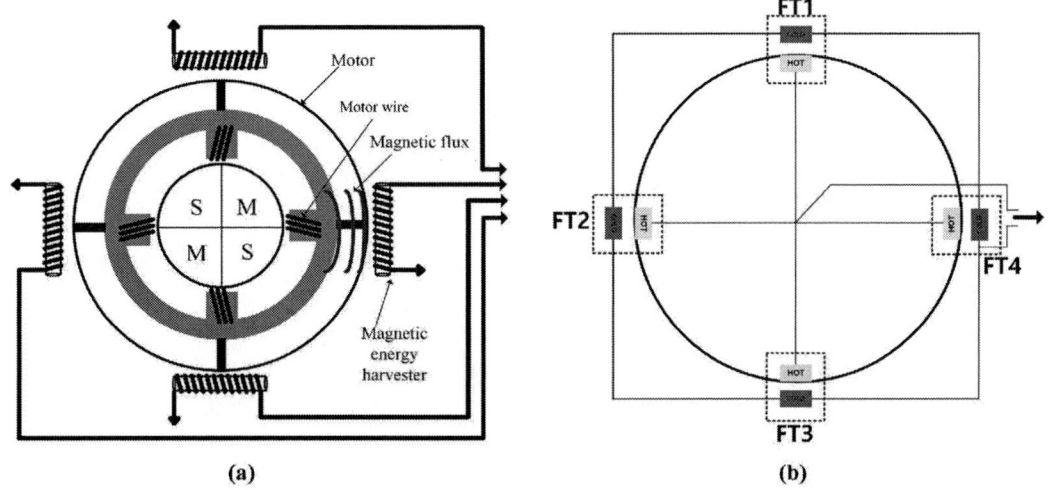

Fig. 2. The structure of the proposed harvester; (a) magnetic energy harvesting using the motor, (b) thermal harvesting using the peltier.

The second energy harvesting source is thermal energy (Fig. 2(b)). The dryer has a drum in which the drying material spins at high temperature. The energy harvesting using the temperature difference between the inside and outside of the drum is possible. A device called a peltier is used for thermal energy harvesting. Peltier is a device that collects thermal energy using temperature differences. Peltier electrodes can be placed inside and outside the drum, respectively, and the harvesting energy can be collected in the energy storage element.

III. EXPERIMENTAL RESULTS

A multi-channel class E power amplifier was fabricated to verify the proposed circuit with the prototype of the dielectric heating dryer (Fig. 3(a)). The circuit was driven at 13.56MHz and 1kW, and the operation was compared under the ideal load condition and the actual load condition. Under ideal conditions, the drain voltage and gate voltage operated normally (Fig.3(b)), and it was confirmed that noise occurred under the actual 3kG dryer load (Fig.3(c)).

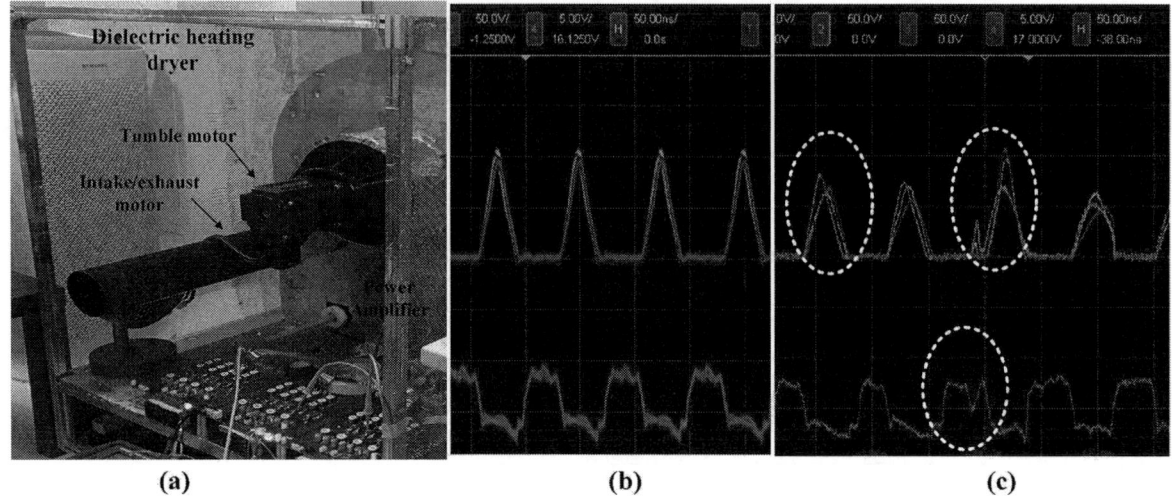

Fig. 3. Experimental results; (a) Prototype of the dielectric heating dryer, (b) drain and gate voltage without any noise, and (c) drain and gate voltage waveform with any noise.

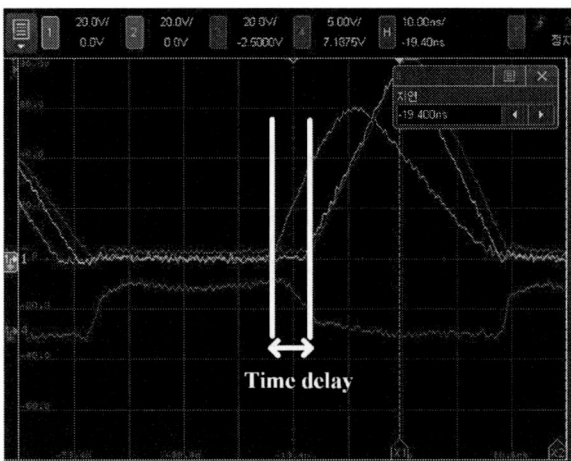

Fig. 4. Detail drain and gate waveform with noise.

As shown in fig. 3(a), the dielectric heating dryer has two motors, a tumble motor and an intake and exhaust motor. Recently, as the demand for dryers has increased, it is necessary to use a motor with a capacity of at least 300 W because more than 20 kg of drying clothes need to be dried. To prevent circuit noise issues caused by near fields, a universal type motor is used for the motor. Since the universal motor is directly connected to the grid power source of 85~265 VAC, it can be free from noise compared to a DC motor.

The drain and gate voltage waveforms including noise were measured (Fig. 4). In a Class E power amplifier, the drain voltage rises as a resonant waveform when the main switch is turned off. The rise in drain voltage started at different points for each channel with time delay in Fig. 5. This means that the gate drivers of each channel were turned off at different timings, indicating that either the 13.56 MHz source generator or the gate driver was malfunctioning due to high frequency noise. This malfunction caused by the noise can be eliminated by separating the grounds with the proposed harvesting circuit.

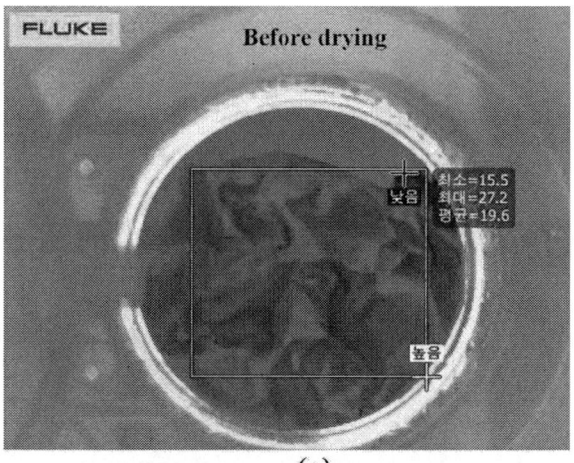

(a)

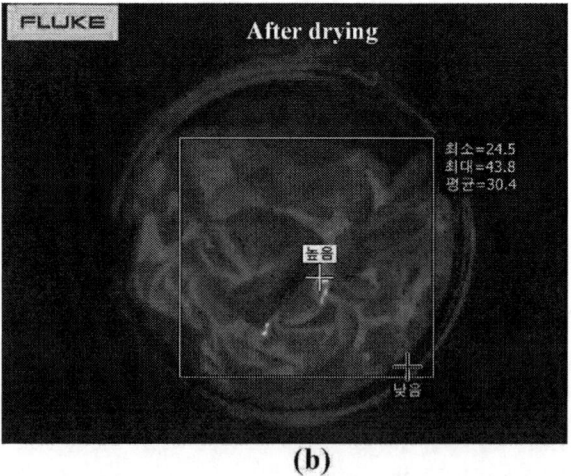

(b)

Fig. 5. The temperature of dryer (a) before drying and (b) during drying

To confirm the possibility of thermal energy harvesting, the temperature inside the drum before and during drying was measured (Fig. 5). Before drying, the maximum temperature was 27.2°C (Fig. 5(a)), while during drying, the temperature reached 43.8°C (Fig. 5(b)). For the purpose of low-temperature dryers aimed at preventing fabric damage, the interior of the drum is maintained at around 50°C through internal temperature sensors and control circuits. By utilizing peltier elements, it is possible to harvest this thermal energy generated within the drum during the drying process. By monitoring the amount of harvested energy and the power consumption of the control circuit, it is also possible to control the internal temperature of the drum for optimal harvesting.

The temperature of drum inside has a tradeoff between fabric damage and the generation of thermal energy. As shown in fig. 5, the temperature inside the drum is formed at around 40°C. This temperature can be changed depending on applied power, drum size, and electrode placement. For the clothes with less fabric damage, more thermal energy can be collected by increasing the temperature.

IV. CONCLUSION

The high frequency noise reduction circuit for the low temperature dryer using energy harvesting is proposed. The proposed circuit separates the ground of the power part and the signal part. In addition, harvesting energy generated from the dryer is used to supply power to the signal unit. In this paper, the magnetic energy harvesting using the magnetic field of dryer motor and thermal energy harvesting using the difference of temperature between inside and outside of the drum was used. By using the class E power amplifier of 13.56MHz and 1kW, the environment in which noise occurs was verified.

REFERENCES

[1] J.M. Chang, J.H. Wu, and Y.S. Wu, "Energy Efficient Shower System with Waste Heat Recovery" in 2020 *3rd IEEE International Conference on Knowledge Innovation and Invention (ICKII),* Aug. 2020.

[2] G. Zambonin, F. Altinier, L. Corso, M. Sessolo, A. Beghi, and G. A. Susto, "Soft Sensors for Estimating Laundry Weight in Household Heat Pump Tumble Dryers" in 2018 IEEE *14th International Conference on Automation Science and Engineering (CASE),* Aug. 2018.

[3] K. Wang, T. Jiang, and J. Shi, "Thermal analysis and design of solar heat pump drying wheat system" in 2023 IEEE *2nd International Conference on Electrical Engineering, Big Data and Algorithms (EEBDA),* Feb. 2023.

[4] F. Ahmad, A. Jorgensen, and S. Munk-Nielsen, "Modeling and Operation of Series-Parallel Resonant Load in Industrial RF Dielectric Heating Application" *IEEE Transactions on Industry Applications, Early Access,* Mar. 2023.

[5] I. Kano, and Y.Takahashi, "Effect of Electric Field Generated by Microsized Electrode on Pool Boiling" *IEEE Transactions on Industry Applications,* vol. 49, no. 6, pp. 2382-2387, Jun. 2013.

[6] G. C. Abdin, F. Freschi, M. Mitolo, A. Poggio, and M. Tartaglia, "District Heating Safety Issues: Interactions Between Grounding Systems and Thermal Installations" *IEEE Transactions on Industry Applications,* vol. 52, no. 3, pp. 2040-2045, Dec. 2015.

[7] J. Lesage and J. Penn, "Comparison of Additive Noise of DAC Technologies for Low Noise Microwave Frequency Synthesizers" in 2022 IEEE *Joint Conference of the European Frequency and Time Forum and IEEE International Frequency Control Symposium (EFTF/IFCS),* Apr. 2022.

[8] Y. Zhan, Z. Yin, J. Liu, R. Zhang, and X. Sun, "IPMSM Sensorless Control Using High-Frequency Voltage Injection Method With Random Switching Frequency for Audible Noise Improvement" *IEEE Transactions on Industrial Electronics,* vol. 67, no. 7, pp. 6019-6030, Aug. 2019.

[9] N. Gajanur, M. Greidanus, S. Mazumder, and M. Abbaszada "Impact and Mitigation of High-Frequency Side-Channel Noise Intrusion on the Low-Frequency Performance of an Inverter" *IEEE Transactions on Industrial Electronics,* vol. 37, no. 10, pp. 11481-11485, Oct. 2022.

[10] S. Park, Y. Sohn, and G. Cho, "SiC-Based 4 MHz 10 kW ZVS Inverter With Fast Resonance Frequency Tracking Control for High-Density Plasma Generators" *IEEE Transactions on Industrial Electronics,* vol. 35, no. 3, pp. 3266-3275, Mar. 2020.

Single-Stage Three-Phase Buck-Matrix Rectifier with Series-Parallel Connected Transformers for High-Power 48 V Data Center Power Supplies

Yuki Ishikura
Murata Manufacturing Co., Ltd.
Kyoto, Japan
y_ishikura@murata.com

Chinmay Bhagat
Murata Manufacturing Co., Ltd.
Kyoto, Japan
chinmay.bhagat@murata.com

Abstract—This paper proposes a single-stage three-phase buck-matrix rectifier for high power data center power supplies. The proposed converter is composed with series-parallel connected transformers to reduce magnetizing current and current stress in synchronous rectifier FETs. An active regenerative snubber is used to reduce snubber power losses and voltage oscillation in synchronous rectifier FETs, enabling use of low-voltage FETs, which further reduces the conduction loss. In addition, the proposed converter realizes ZVS operation with single-stage power conversion. Circuit configuration and features of the proposed converter are presented along with theoretical discussions and simulation analysis of the parallel-connected rectifier bridges. Operation of the proposed converter is demonstrated using a prototype rated for 8 kW, three-phase 400 Vrms AC input and 50 V DC output. Experimental results show that the proposed converter can realize unity power factor and THD of 4.7% at rated power, achieving a maximum power conversion efficiency of 97.7% and a rated power conversion efficiency of 97.3%.

Keywords—Single-Stage, Buck-Matrix Rectifier, ZVS, Series-parallel connected transformers, Active regenerative snubber

I. INTRODUCTION

With the rapid development of cloud computing and artificial intelligence (AI), the power consumption of high performance central processing units (CPUs) and graphic processing units (GPUs) is increasing rapidly in datacenter applications [1][2].

To accommodate this high-power demand, 48 V power bus architecture has been introduced for data center power supplies to replace the standard 12 V bus architecture [3][4]. Therefore, high-power, high-density, and high-efficiency AC-DC power supply units (PSUs) compatible with 48 V power bus architecture are required [5].

In general, three-phase isolated AC-DC converters are employed when high power, in order of over tens of kilowatts is required [6]. Three-phase isolated AC-DC conversion can be realized using two stage circuit configuration, with Vienna or six-switch boost rectifier followed by an isolated DC-DC converter [6][7]. Despite the simple structure of these boost rectifiers, the power density of three-phase isolated AC-DC converter is low compared with single-phase isolated AC-DC

converters, as these boost rectifiers are usually hard switched and converter cannot operate at high switching frequency.

Alternatively, three-phase buck-matrix AC-DC rectifier is considered a promising single-stage candidate topology for HVDC (High Voltage Direct Current) system [8]-[12]. High-efficiency and high-density can be expected because of nonexistence of the intermediate DC-link capacitor bank and the second stage isolated DC–DC converter, and owing to zero-voltage switching (ZVS) operation. However, there are few studies applying the three-phase buck-matrix rectifier to high-power 48 V power bus architecture.

Although, large voltage oscillation occurs in synchronous rectifier FETs of the three-phase buck-matrix rectifier, as it realizes ZVS operation in the same manner as the phase-shift full-bridge (PSFB) DC-DC converter [10][13]. The maximum voltage generated by this voltage oscillation is huge, requiring FETs with high withstand voltage, in turn increasing conduction loss in the synchronous rectifier circuit. To withstand voltage oscillation, several passive snubbers [14][15], clamping circuits [16][17], and active regenerating snubbers [18][19] have been proposed for the PSFB DC-DC converter. The passive snubbers utilize RC snubber or RCD snubber, which generate huge power loss and have limited voltage clamping capability. This would result in increased cooling requirements for passive snubbers in high-power converters, eventually decreasing the power density. In the clamping circuits, the winding structure of the transformer is complicated, and the conduction loss in the clamping circuit cannot be ignored for high-efficiency applications. On the other hand, the active regenerating snubbers are a good solution to reduce power loss and large voltage oscillation in synchronous rectifier FETs. However, there are few studies applying active regenerating snubbers to the three-phase buck-matrix rectifier for obtaining high-efficiency and their effectiveness is unknown.

This paper proposes a single-stage three-phase buck-matrix rectifier with series-parallel connected transformers for high power data center power supplies. The proposed converter can reduce magnetizing current of transformers and current stress in synchronous rectifier FETs. An active regenerative snubber is also used to reduce snubber power losses and voltage oscillation in synchronous rectifier FETs. Detailed analysis of the parallel-connected synchronous rectifier circuits is

979-8-3315-1612-3/25 $31.00 © 2025 IEEE

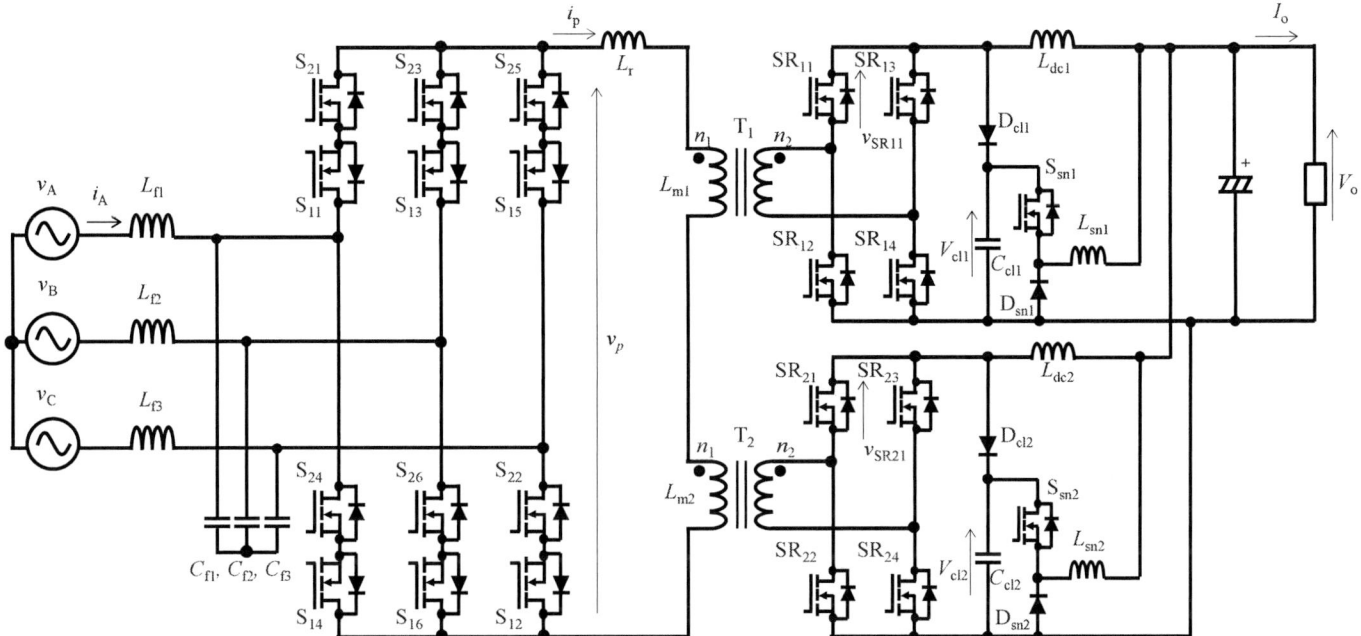

Fig. 1. Single-stage three-phase buck-matrix rectifier with series-parallel connected transformers.

discussed. To verify the operation of the proposed converter, a 50 V DC output and 8 kW prototype is fabricated and tested.

II. CONFIGURATION OF THE PROPOSED CONVERTER

Fig. 1 shows the single-stage three-phase buck-matrix rectifier with series-parallel connected transformers. It consists of input filters, a matrix bridge at primary, followed by a high-frequency transformer circuit, full-bridge rectifiers at secondary, output filters, and active regenerative snubbers. The matrix bridge comprises 12 FETs S_{ij} (i=1-2, j=1-6). The high-frequency transformer circuit consists of a resonant inductor L_r and two transformers T_1, T_2. The primary windings of transformers T_1 and T_2 are connected in series, and secondary windings are connected to each full-bridge rectifier circuit. The outputs of two full-bridge rectifier circuits are connected in parallel through dc inductors L_{dc1}, L_{dc2}. By connecting full bridge rectifier circuits in parallel, it is possible to support high output current. Since same current flows through primary windings, the current in secondary rectifier circuits is automatically balanced and equalized [20][21]. This configuration reduces conduction loss in the matrix bridge by reducing magnetizing current and current stress in the synchronous rectifier FETs. In general, the current-doubler rectifier is suitable for low-voltage and high-current application [22]. On the contrary, to reduce conduction loss by using FETs with withstand voltage of less than or equal to 150 V, full-bridge rectifier is adapted in the proposed converter. Furthermore, by using full-bridge rectifier, the excitation frequency of the dc inductors L_{dc1}, L_{dc2} is twice the switching frequency of the converter, leading to reduced inductance of the dc inductors L_{dc1}, L_{dc2}.

The proposed converter uses a resonant inductor L_r to charge and discharge the parasitic capacitance of the primary FETs to achieve ZVS operation [10]. Similar to that in the PSFB DC-DC converter, parasitic capacitance of transformer windings and secondary-side rectifier FETs resonate with the resonant inductance L_r causing voltage oscillations [13][14]. High-voltage secondary-side rectifier FETs are required as the maximum value of this voltage oscillation is more than twice secondary-side output voltage in the transformer. In order to reduce the maximum value of this voltage oscillation, the active regenerative snubbers comprising a clamping circuit and a buck converter on each full-bridge synchronous rectifier is employed [18]. Voltage oscillation energy is stored in the clamp capacitors C_{cl1}, C_{cl2} through the clamp diodes D_{cl1}, D_{cl2}, and is regenerated to output of the proposed converter by the buck converter. This suppresses maximum voltage of oscillation and reduces power loss in snubbers as compared to passive snubbers, allowing use of low voltage FETs with low on-resistance in the synchronous rectifier FETs.

III. OPERATING ANALYSIS OF THE SECONDARY SIDE RECTIFIER CIRCUIT

In the proposed configuration, two transformers transmit the same amount of power since primary windings are connected in series and current in transformers is the same. The current ripple of dc inductor Δi_{dc1} and Δi_{dc2} are expressed as follows:

$$\Delta i_{dc1} = \frac{\frac{n_2}{n_1}(\frac{L_{m1}}{L_{m1}+L_{m2}})v_p(t) - V_o}{L_{dc1}} \, mT_s \tag{1}$$

$$\Delta i_{dc2} = \frac{\frac{n_2}{n_1}(\frac{L_{m2}}{L_{m1}+L_{m2}})v_p(t) - V_o}{L_{dc2}} \, mT_s \tag{2}$$

Where, number of windings (n_1 and n_2), magnetic inductance (L_{m1} and L_{m2}), primary side transformer voltage ($v_p(t)$), output voltage (V_o), dc inductor (L_{dc1} and L_{dc2}), modulation ratio (m), and switching period (T_s) are the circuit parameters.

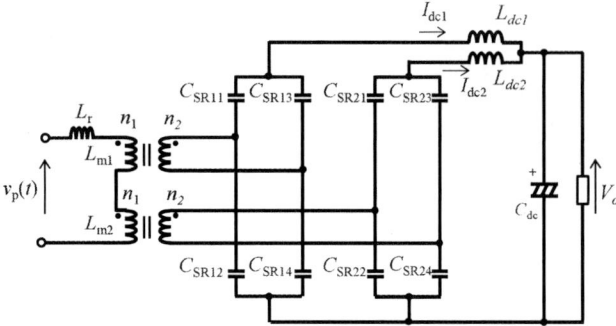

Fig. 2. An equivalent circuit in the synchronous rectifier.

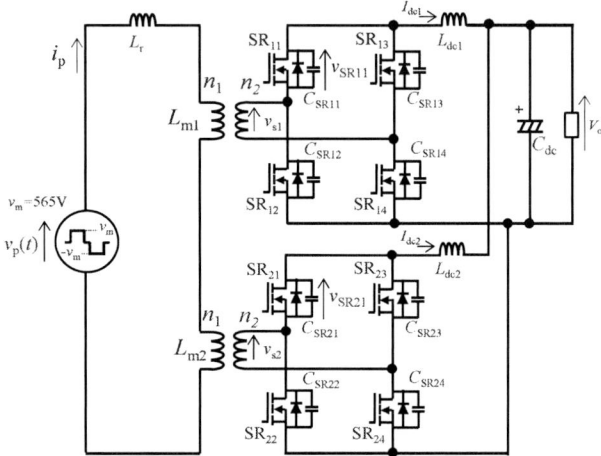

Fig. 3. Simulation model.

Table I: Circuit parameters in simulation

Resonant inductor	L_r	4 µH
DC inductor	L_{dc1}, L_{dc2}	5 µH
Winding ratio	n_1, n_2	4:1
Switching frequency	f_s	70 kHz
Parasitic capacitance	C_{SRij} (i=1-2, j=1-4)	5000 pF
DC capacitor	C_{dc}	16.9 mF

However, if there is difference in the magnetizing inductances L_{m1}, L_{m2}, instantaneous power transmitted by each transformer becomes non-uniform, and from equation (1) and (2), current ripple of dc inductor becomes different.

An equivalent circuit of the synchronous rectifier in the proposed converter shown in Fig. 2. It consists of the resonant inductor L_r, two transformers (L_{m1} and L_{m2}), parasitic capacitance of FETs C_{SRij} (i=1-2, j=1-4) and dc inductors (L_{dc1} and L_{dc2}) . In this equivalent circuit, two resonance phenomena occur during switching operation. First is a low frequency resonance phenomenon which occurs between the parasitic capacitance of the synchronous rectifier FETs and DC inductances due to the influence of difference in two dc inductor current ripples. The resonant frequency is expressed as follows:

$$f_{r1} = \frac{1}{2\pi\sqrt{C_{SR}(L_{dc1} + L_{dc2})}} \qquad (3)$$

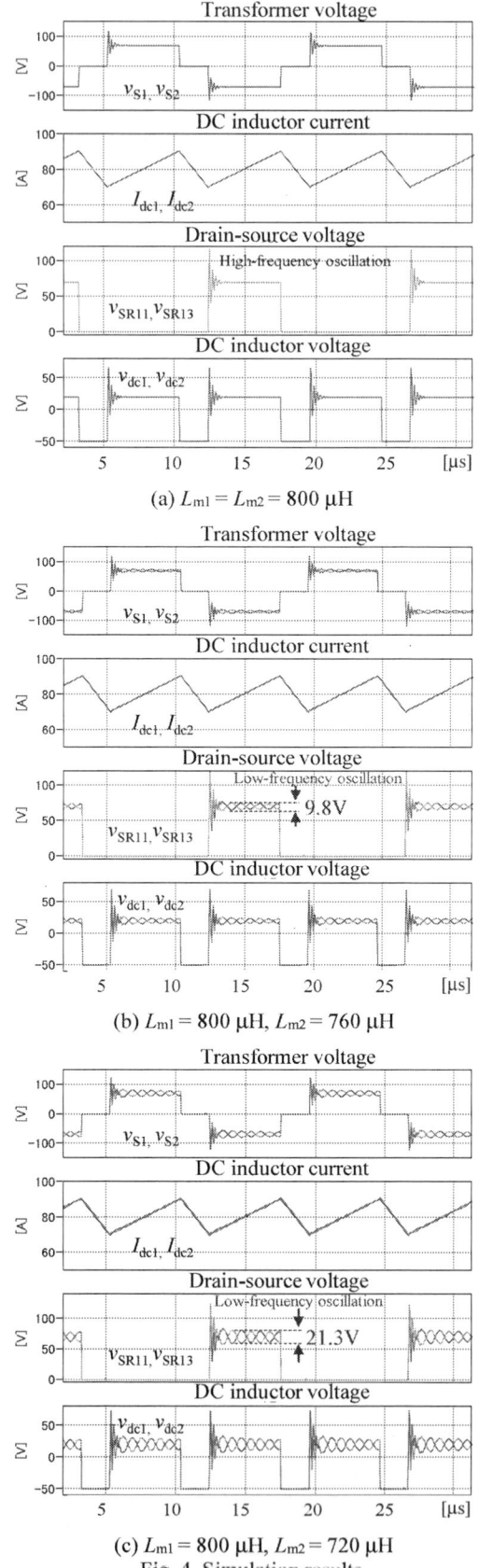

(a) $L_{m1} = L_{m2} = 800$ µH

(b) $L_{m1} = 800$ µH, $L_{m2} = 760$ µH

(c) $L_{m1} = 800$ µH, $L_{m2} = 720$ µH

Fig. 4. Simulation results.

Second is a high frequency resonance phenomenon which occurs between the parasitic capacitance of the synchronous rectifier FETs and the resonant inductance. The resonant frequency is expressed as follows:

$$f_{\mathrm{rh}} = \frac{1}{2\pi\sqrt{4C_{\mathrm{SR}}(\frac{n_2}{n_1})^2 L_{\mathrm{r}}}} \tag{4}$$

To clarify the influence of the magnetizing inductance L_{m1}, L_{m2} to the secondary rectifier circuits, behavior of parallel-connected full-bridge rectifier circuits is analyzed using the circuit simulator PLECS®. Fig.3 and table I show the simulation circuit model and circuit parameters. The simulation analysis is performed under following conditions: input voltage $v_p(t)$ is a square wave consisting of ±565 V and 0 V, output voltage $V_o = 50$ V, output current $I_o = 160$ A. Using the simulation circuit shown in Fig.3 and these analysis conditions, the switching operation in the rectifier circuit is analyzed when the two magnetizing inductances are changed.

Fig. 4(a)-(c) show analysis results in case of changing the magnetizing inductance L_{m2}. When the magnetizing inductance L_{m1}, L_{m2} are the same at 800μH shown in Fig.4(a), only high frequency voltage oscillation, shown in equation (4), occurs in the secondary-side output of transformers (v_{S1}, v_{S2}) drain-source voltages (v_{SR11}, v_{SR13}) and dc inductor voltages (v_{dc1}, v_{dc2}). This high-frequency voltage oscillation require higher voltage FETs and increase conduction losses in the rectifier circuit. To suppress maximum voltage of high-frequency oscillation, the proposed converter utilizes the active regenerative snubber. When the magnetizing inductance are different which $L_{m1} = 800$μH and $L_{m2} = 760$μH shown in Fig. 4(b), not only high-frequency voltage oscillation, but also low-frequency voltage oscillation occurs. This low-frequency voltage oscillation, shown in equation (3), occurs in the secondary-side output of transformers (v_{S1}, v_{S2}) drain-source voltages (v_{SR11}, v_{SR13}) and dc inductor voltages (v_{dc1}, v_{dc2}). When the difference in the magnetizing inductances becomes higher, with $L_{m1} = 800$μH and $L_{m2} = 720$μH shown in Fig. 4(c), low-frequency voltage oscillation ripple tends to increase. Low-frequency voltage ripple in Fig.4(c) is twice as large as compared with Fig.4(b). This leads to increase conduction loss in synchronous rectifier FETs, iron loss in dc inductors and transformers, and electromagnetic interference. To suppress low-frequency voltage oscillation, it is necessary to make the magnetizing inductances L_{m1}, L_{m2} of the two transformers connected in series as uniform as possible.

IV. EXPERIMENTAL RESULTS

To demonstrate the validity and effectiveness of the proposed converter, a prototype is fabricated and tested. The experimental evaluation is conducted under conditions tabulated in Table II. In this prototype, to allow large current in the synchronous rectifier circuit, five FETs are connected in parallel. Small RC passive snubber ($R_{sn} = 22$Ω, $C_{sn} = 4700$pF) is also added to each synchronous rectifier FETs. Active regenerative snubbers operate with fixed on-time PWM

Table II: Experimental test condition

Input ac voltage	v_{in}	400 Vrms, 60Hz
Output dc voltage	V_o	50 V
Output Current	I_o	30-160 A
Switching frequency	f_s	70 kHz
Resonant inductor	L_r	4 μH
DC inductor	L_{dc1}, L_{dc2}	5.5 μH
DC capacitor	C_{dc}	16.9 mF
Winding ratio	n_1, n_2	4:1
Magnetizing inductance	L_{m1}, L_{m2}	805 uH, 822 uH
Primary side SiC-FET	S_{ij} (i=1-2, j=1-6)	1200 V, 18 mΩ
Secondary side Si-FET	SR_{ij} (i=1-2, j=1-4)	135 V, 3.7 mΩ
Dead time	T_{dead}	200 ns
Passive snubbers	R_{sn}, C_{sn}	22 Ω, 4700 pF
Clamp capacitors	C_{cl1}, C_{cl2}	7.8 μF
Snubber inductor	L_{sn1}, L_{sn2}	68 μH
Snubber frequency	f_{sn}	60 kHz
Fixed on time	T_{on}	5 μs

switching and regenerate stored energy in the clamp capacitors to the dc output.

Fig. 5(a) shows the experimental waveforms of input line to line voltage v_{ab} and input phase current i_a when the output current I_o is 160 A. The phase of the input phase current i_a is delayed by 30° with respect to the input AC line to line voltage v_{ab}. This indicate that the phase of the input current i_a is almost the same phase as the input AC phase voltage v_a, and unity power factor (over 0.99) is achieved. Although total harmonic distortion (THD) in the input current i_a is 4.7% at rated power, small current distortion every 60° of input phase current waveform is observed. This is due to swapping of switching sequence during sector transition. This current distortion can be reduced by performing correction control for the duty cycle loss [10].

Fig. 5(b) shows the experimental waveforms of transformer voltage v_p and transformer current i_p when the output current I_o is 160 A. Transformer voltage v_p is sequentially changed from high voltage to low voltage to zero voltage. To realize ZVS operation in all FETs, 6 segment PWM switching sequence [10] is adopted for the matrix bridge, which can be verified from the primary side transformer voltage v_p, which changes in 6 steps.

Fig. 6(a) shows the drain-source voltage of the synchronous rectifier FETs v_{SR11}, v_{SR21} and clamping capacitor voltage V_{cl1}, V_{cl2} waveforms. Clamping capacitor voltage V_{cl1}, V_{cl2} are almost constant values, 87 Vrms and 85 Vrms, respectively. Clamping capacitors are charged every switching operation, and the energy stored in the clamping capacitor is regenerated to the output-side. In this test condition, total regenerated power in two active snubbers is around 200 W. Furthermore, although low-frequency voltage oscillation occurs in v_{SR11} and v_{SR21}, the amplitude is suppressed to less than 5 V, as the difference in magnetizing inductances (L_{m1}, L_{m2}) is small.

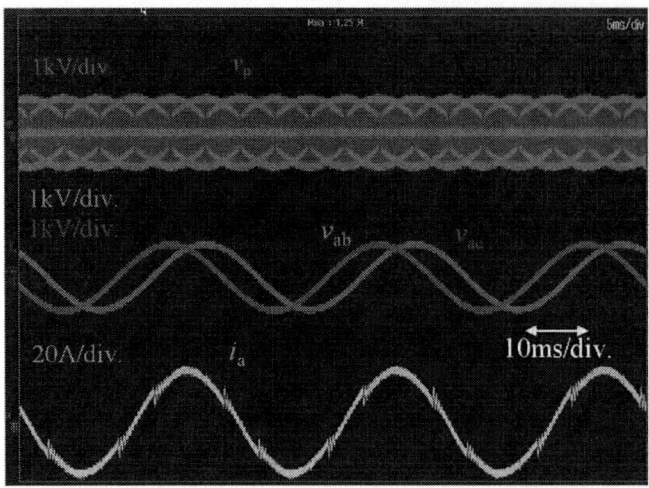

(a) Input line to line voltage and input current

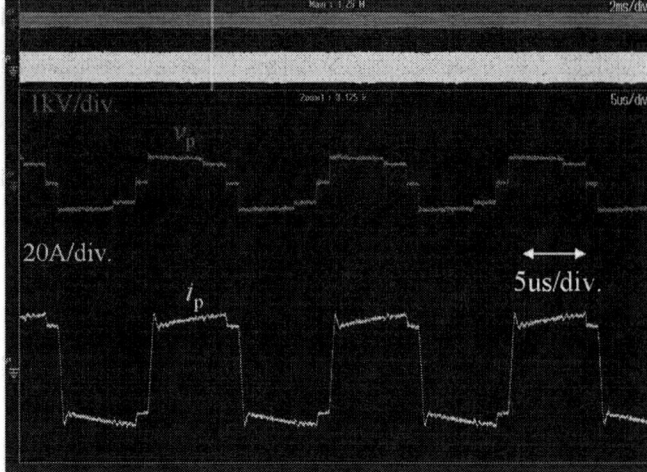

(b) Transformer voltage and transformer current

Fig. 5. Operating waveforms in primary side.

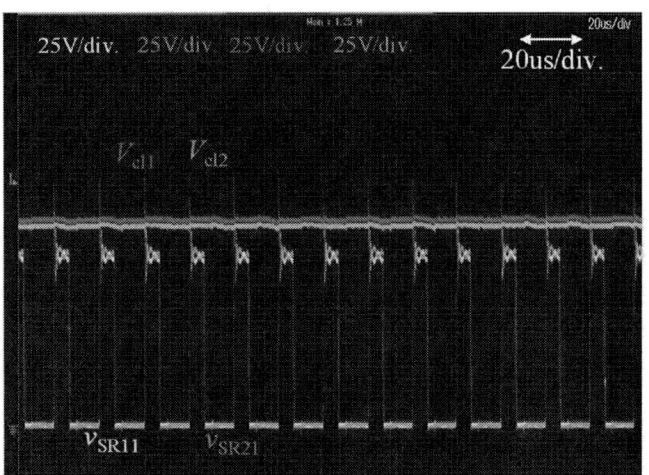

(a) Drain-source voltage in FETs and clamping capacitor voltages

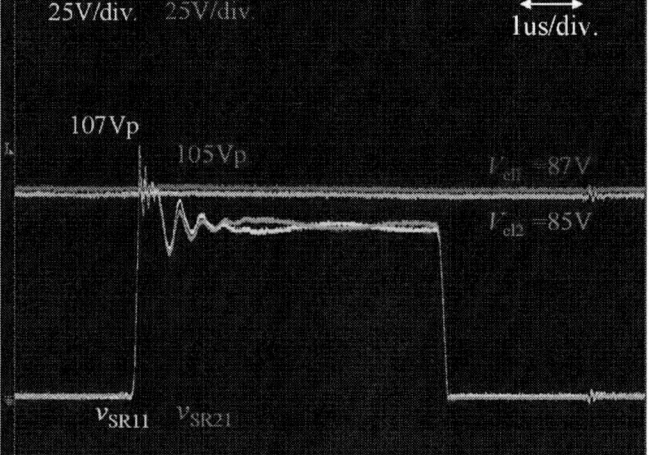

(b) Drain-source voltage in FETs

Fig. 6. Operating waveforms in secondary side.

Fig.6(b) shows zoomed-in drain-source voltage waveforms of the synchronous rectifier FETs. These switching waveforms are clamped at the clamping capacitor voltages V_{cl1}, V_{cl2} and the maximum drain-source voltage v_{SR11}, v_{SR21} are 107 V and 105 V, respectively. These maximum voltages are around 1.5 times the theoretical secondary transformer voltage. By using active regenerating snubbers, it is possible to reduce the maximum value of high-frequency voltage oscillation in the synchronous rectifier FETs. This leads to adoption of low-voltage FETs as the synchronous rectifier FETs, which reduces conduction loss and enable higher efficiency.

However, due to high-frequency voltage oscillation, the maximum drain-source voltage v_{SR11}, v_{SR21} exceed the clamping capacitor voltage by around 20 V. This is due to the parasitic inductance present in the clamping circuit, which prevents charging the surge energy into the clamp capacitors. The maximum value of voltage oscillation can be reduced by reducing the parasitic inductance in the printed circuit board

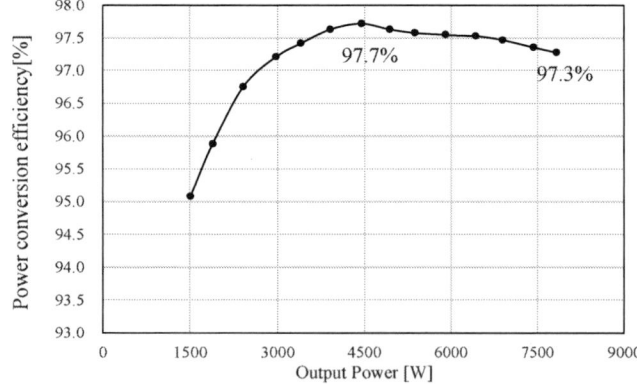

Fig. 7. Power conversion efficiency.

(PCB) pattern and equivalent stray inductance (ESL) in the clamping capacitor.

979-8-3315-1612-3/25 $31.00 © 2025 IEEE   3289

Fig. 7 shows the power conversion efficiency characteristic obtained with the proposed converter when the output current is varied from 30 A to 160 A. Maximum power conversion efficiency of 97.7% is obtained, along with power conversion efficiency of 97.3% at rated load.

V. CONCLUSION

This paper presented a single-stage three-phase buck-matrix rectifier for high power data center power supplies. The proposed converter can reduce magnetizing current of transformers and current stress in synchronous rectifier FETs. The active regenerative snubber is also used to reduce snubber power losses and voltage oscillation in the synchronous rectifier FETs. The configuration of the proposed converter and the mechanism of voltage oscillation in parallel-connected rectifier bridges was discussed and clarified using simulation analysis. In order to confirm the validity and the effectiveness of the proposed converter, the prototype rated for 8 kW, three-phase 400 Vrms AC input and 50 V DC output was fabricated and tested. Experimental results show that the proposed converter realized unity power factor and THD of 4.7% with the single-stage power conversion and ZVS operation. By reducing the difference in magnetizing inductances of the two transformers, the amplitude of the low-frequency voltage ringing is reduced. Also, active regenerative snubbers suppress maximum drain-source voltage, reducing the voltage to 1.5 times the theoretical secondary transformer voltage. Furthermore, the proposed converter achieved a maximum power conversion efficiency of 97.7% and rated power conversion efficiency of 97.3%.

REFERENCES

[1] R. Y. Sun, S. Webb, Y. -F. Liu and P. C. Sen, "Optimization and Design of a 48-to-12 V 35 A Split-Phase Dickson Switched-Capacitor Converter," *in Proc. IEEE Applied Power Electronics Conference and Exposition*, pp. 1900-1907, 2021.

[2] M. Qiu, X. Liu, and D. Cao, " A New 8x Matrix Autotransformer Switched Capacitor DC-DC Converter for Datacenter Application," *in Proc. IEEE Energy Conversion Congress and Exposition*, pp. 6526-6528, 2023.

[3] M. Ahmed, C. Fei, F. Lee and Q. Li, "48-V Voltage Regulator Module With PCB Winding Matrix Transformer for Future Data Centers," *IEEE Trans. on Industrial Electronics*, vol. 64, no. 12, pp. 9302–9310, 2017.

[4] X. Li and S. Jiang. Google 48v power architecture. 2017 IEEE Applied Power Electronics Conference and Exposition, Tampa, FL, 2017.

[5] A. Nabih, F. Jin and Q. Li, "Efficient Integrated Transformer–Inductor With High PCB Utilization and Optimized Core," *IEEE Trans. on Industrial Electronics*, vol. 71, no. 6, pp. 5653–5662, 2024.

[6] J. W. Kolar and T. Friedli, "The Essence of Three-Phase PFC Rectifier Systems - Part I," IEEE Transactions on Power Electronics, vol. 28, no. 1, pp. 176–198, Jan 2013.

[7] L. Huber, M. Kumar, and M. M. Jovanovic´, "Performance comparison of three-step and six-step PWM in average-current-controlled three-phase six-switch boost PFC rectifier," IEEE Trans. Power Electron., vol. 31, no. 10, pp. 7264–7272, Oct. 2016.

[8] V. Vlatkovi´c and D. Borojevi´c, "Digital-signal-processor-based control of three- phase space vector modulated converters," *IEEE Transactions on Industrial Electronics*, vol. 41, no. 3, pp. 326–336, Jun. 1994.

[9] V. Vlatkovic, D. Borojevic, and F. Lee, "A zero-voltage switched, three-phase isolated PWM buck rectifier," *IEEE Trans. on Power Electronics*, vol. 10, no. 2, pp. 148–157, 1995.

[10] J. Afsharian, D. David Xu, B. Gong, and Z. Yang, "Space vector demonstration and analysis of zero-voltage switching transitions in three-phase isolated PWM rectifier," *in Proc. IEEE Energy Conversion Congress and Exposition*, 2015, pp. 2477–2484.

[11] J. Afsharian, D. Xu, B. Gong, and Z. Yang, "Reduced duty-cycle loss and output inductor current ripple in a ZVS switched three-phase isolated PWM rectifier," *in Proc. IEEE Applied Power Electronics Conference and Exposition*, Mar. 2016, pp. 33–37.

[12] J. Afsharian, D. Xu, B. Wu, B. Gong, and Z. Yang, "The Optimal PWM Modulation and Commutation Scheme for a Three-Phase Isolated Buck Matrix-Type Rectifier," *IEEE Tran. on Power Electronics*, vol. 33, no. 1, pp. 110–124, 2018.

[13] X. Zhang, C. Li, H. Yang and Y.Guan, "Control strategy analysis and loop design of full-bridge phase-shift soft-switching DC-DC converter," in Proc. International Power Electronics and Motion Control Conference, July.2016.

[14] S. Lin and C. Chen, "Analysis and design for RCD clamped snubber used in output rectifier of phase-shift full bridge-bridge ZVS converters," *IEEE Transactions on Industrial Electronics*, vol.45, no.2, Apr.1998, pp.358–359.

[15] X. Wu, X. Xie, J. Zhang, R. Zhao, and Z. Qian, "Soft switched full bridge dc–dc converter with reduced circulating loss and filter requirement," *IEEE Trans. on Power Electronics*, vol. 22, no. 5, pp. 1949–1955, 2007.

[16] M. Lee, D. Choi, J. Bae and J. Chae, "A New Secondary Clamp Diode for Phase-Shift Full-Bridge Converter," *in Proc.* International Power Electronics Conference (IPEC-Himeji ECCE Asia), 2022, pp. 1596–1600.

[17] C. Y. Lim, Y. Jeong, and G. W. Moon, "Phase-Shifted Full-Bridge DC-DC Converter With High Efficiency and High Power Density Using Center-Tapped Clamp Circuit for Battery Charging in Electric Vehicles," IEEE Trans. Power Electron., vol. 34, no. 11, pp. 10945-10959, Nov. 2019.

[18] M. Cacciato and A. Consoli, "New regenerative active snubber circuit for ZVS phase shift Full Bridge converter," *in Proc. IEEE Applied Power Electronics Conference and Exposition*, 2011, pp. 1507–1511.

[19] M. Mahapatra, A. Pal, and K. Basu, "Analysis and Design of Active Snubber of A Step-up Phase Shifted Full Bridge DC-DC Converter Considering Parasitics," *in Proc. IEEE Energy Conversion Congress and Exposition*, 2020, pp. 5447–5451.

[20] Y. Shem, W. Zhao, Z. Chen, and C. Cai, "Full-Bridge LLC Resonant Converter with Series-Parallel Connected Transformers for Electric Vehicle On-Board Charger", IEEE Access., vol.6, pp.13490-13500, Mar. 2023.

[21] D. H. Sim, C. Kim, H.Jo, Ju. Lee and B. K. Lee, "Optimized Dual Transformer Wiring Method for High Efficiency Operation of a 25kW LLC Converter," *in Proc. IEEE Applied Power Electronics Conference and Exposition*, 2024, pp. 2006–2010.

[22] X. Lou and Q. Li, " 300A Single-stage 48V Voltage Regulator with Multiphase Current Doubler Rectifier and Integrated Transformer," *in Proc. IEEE Applied Power Electronics Conference and Exposition*, pp. 1004-1010, 2022.

Sector Transition PWM Modulation Scheme for a Three-Phase Isolated Buck-Matrix Rectifier

Chinmay Bhagat
Murata Manufacturing Co., Ltd.
Kyoto, Japan
chinmay.bhagat@murata.com

Yuki Ishikura
Murata Manufacturing Co., Ltd.
Kyoto, Japan
y_ishikura@murata.com

Abstract— This work proposes a PWM modulation scheme pertaining to sector transitions for three-phase isolated buck-matrix rectifier. PWM modulation schemes realizing UPF (Unity Power Factor) operation and ZVS (Zero-Volt Switching) turn-on at matrix bridge by dividing input AC voltage into sectors are available in existing literature, but issues related to sector transitions are not discussed. The proposed modulation scheme ensures that excessive current or voltage stresses are not experienced by matrix bridge switches during sector transitions. Moreover, synchronous rectification for switches with forward-biased diodes can be performed at matrix bridge using the proposed modulation scheme, not compromising on power conversion efficiency. Analysis of stress causing mechanism at sector transition boundaries is presented along with simulation results. The proposed modulation scheme is verified experimentally and peak efficiency of 97.7%, and a rated-power efficiency of 97.3% is obtained on an 8 kW, 400 Vrms AC input/50 V DC output converter prototype, while eliminating current stress due to AC source short-circuit and reducing peak drain-source voltage at matrix bridge by 10% during sector transitions.

Keywords—buck-matrix rectifier, modulation, sector, ZVS

I. INTRODUCTION

Proliferation of AI (Artificial Intelligence) and ML (Machine Learning) requires high-performance datacenters to meet ensuing challenging computational needs, with a greater number of more powerful GPUs [1][2]. With power consumption per GPU said to double every 6 months, server rack power density is bound to increase. Owing to this, and a need for greener datacenters, the power delivery architecture of hyperscalar datacenters is poised to undergo fundamental design changes. In this development trend, 48 V power bus architecture has been introduced for data center power supplies to replace the standard 12 V bus architecture [3][4]. Moreover, the concept of disaggregated power rack, with one power rack envisioned have power rating of up to 1 MW, is now being proposed for future datacenters [5]. All these developments demand more compact, efficient, and high-power front-end power supplies to power the server racks [6].

Conventional datacenter front-end power supplies are single-phase two-stage designs [7]. Single-phase power supplies require balancing of the loads connected to the three phases, as unbalanced load operation can cause energy losses in distribution cables and transformers as well as inefficient operation of the central UPS [8]. In general, three-phase PFC power supplies are better suited for the changing requirements in datacenter application and are shown to attain higher power densities [9]. A three-phase power supply solution for datacenter application would typically have two stages: a three-phase PFC stage followed by an isolated DC-DC conversion stage. The DC link of two stage design can have a bulk capacitor to fulfill the holdup requirements.

However, with increasing power density of power supply units (PSUs), traditional hold-up requirements will move towards system side, as solutions with higher energy density - like Lithium-ion capacitors or their combination with Lithium-ion batteries will be needed to meet the requirement of keeping the system operational in the event of disturbances in the AC grid. This will further serve the purpose of mitigating system reactive power requirements and decoupling from the intermittency of renewable energy resources [8][10]. Matrix-based single-stage three-phase PFC rectifiers, utilizing matrix bridge for HF (high-frequency) AC generation, are suitable candidates for this application as they offer higher power density and better power conversion efficiency with single stage AC-DC power conversion without an intermediate DC link [11]. However, matrix-based PFC rectifiers are not explored much for practical applications due to need for complex control strategies to operate them. Matrix-based single stage PFC rectifiers can be divided into 4 major categories [12]. The quasi single-stage buck and boost derived PFC rectifiers and, single-stage buck and boost derived PFC rectifiers. This work focuses on single-stage buck matrix rectifiers, which could be a good choice for datacenter power supplies requiring buck operation from AC grid to the DC output.

Buck-matrix rectifier was proposed in [13][14] along with modulation scheme based on the phase-shifted full-bridge DC-DC converter [15], enabling ZVS, UPF operation and output voltage regulation in single-stage isolated topology. Many works have proposed modulation techniques for the converter since then. [16] proposed SVPWM based modulation while demonstrating direct AC-to-AC voltage conversion. Absolute value logic SPWM modulation scheme was reported in [17]. Both these works implement 6-segment PWM modulation. Another work [18] has presented detailed analysis of a 6 segment PWM scheme for the buck-matrix rectifier. Apart from these, several 8-segment PWM schemes were reported. [19] presented a digital controller based SVM modulation technique. Modulation scheme proposed in [13] along with the novel circuit topology was an 8-segment scheme. [20] has presented an analysis of switching transitions and soft commutation for another 8-segment PWM scheme. [21] has proposed a new 8-segment PWM scheme realizing ZVS operation for all 12 matrix bridge switches. All these previously reported works have focused on the basic converter operation, realizing ZVS at the matrix bridge and UPF operation, optimizing circuit components, while generating stable DC output voltage in the single-stage topology. Other works have proposed modulation schemes for operation under abnormal conditions of the AC grid, to operate the buck-matrix rectifier as required of the PSUs in the datacenter application [22]. Although a lot of work has been done on modulation techniques for buck-matrix rectifier - appropriately selecting input AC phases for realizing UPF operation while dividing them into sectors, issues involved during sector transition are not documented.

This work focuses on developing modulation scheme for buck-matrix rectifier focusing on addressing issues during AC voltage sector transitions. First, the basic operation of buck-matrix rectifier is described along with the role of modulation scheme in realizing ZVS and UPF operation. Then, issues in existing modulation scheme during sector transition are elucidated, followed by the proposed modulation scheme solving these issues. Also, new sector definition for incorporating the sector-transition modulation scheme is proposed.

II. CONVERTER OPERATION PRINCIPLE

Fig. 1 shows three-phase buck-matrix rectifier. The converter has LC filter at input AC side for commutation at matrix bridge and at output for filtering rectified DC voltage. Three-phase, line-frequency AC voltage V_{ac} is converted to single-phase HF AC by the matrix bridge at primary, isolated using an HF transformer and rectified to DC voltage using a HF rectifier stage at secondary. Leakage inductance L_r is used to realize ZVS turn-on of matrix bridge switches. AC input voltage V_{ac} is divided into 12 sectors as shown in Fig. 2, such that the relative voltage relationship within a sector remains unchanged. Depending on sector, matrix bridge is virtually divided into two group of two-phase bridges, bridge x and bridge y. Type A PWM modulation scheme [11] is adopted for buck-matrix rectifier. Using this modulation strategy ensures high-to-low (H2L) transition of primary voltage v_p for all sectors of V_{ac}, enabling ZVS turn-on of matrix bridge switches. It also has the merit of reducing DC inductor current ripple and duty cycle loss as

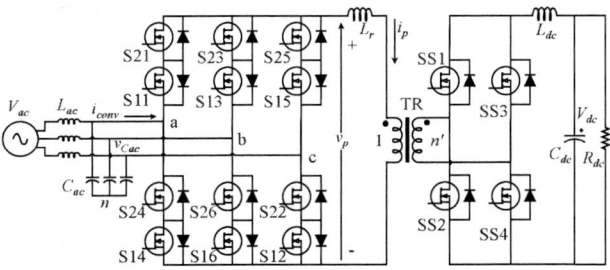

Fig. 1: Three-phase single-stage buck-matrix rectifier

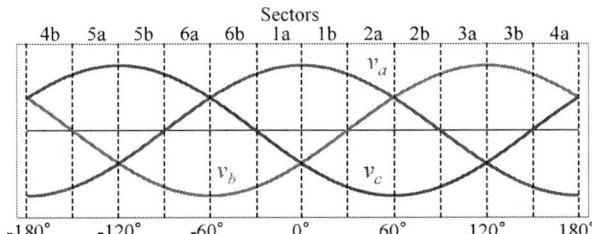

Fig. 2: Sectors as defined in existing literature

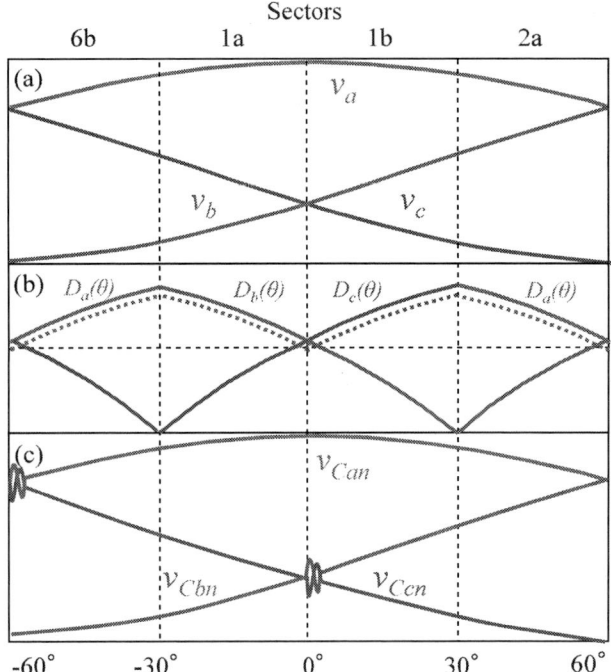

Fig. 3: Problem with a2b sector transition:
(a) 3-phase AC voltages, (b) Duties of matrix bridge switches [dashed: effective duty, solid: asserted duty], (c) Input LC filter capacitor voltages

compared to other reported modulation strategies. UPF operation can be realized simultaneously by adjusting on-duties of bridge x and bridge y proportional to AC input voltage V_{ac}. Additionally, to improve power conversion efficiency and power density, switches with forward-biased diodes are turned-on for synchronous rectification in primary and secondary.

III. ISSUES WITH EXISTING MODULATION SCHEME

Issue during transition from sector 'a' to sector 'b', referred as 'a2b sector transition' in the following text, in sector 1 as an example is discussed in the following text. Similar analysis can be performed for other sectors. Ideally, Phase B voltage v_b is minimum in sector 1a and Phase C voltage v_c is minimum in sector 1b, both being equal at a2b sector transition. Matrix bridge legs connected to phase A and B comprise bridge x, whereas legs connected to phase A and C comprise bridge y. While assuring flux cancellation in transformer core every switching cycle, I_x+, I_y+, I_0, I_x-, I_y-, I_0 vector sequence is asserted in sector 1a, whereas changed vector sequence I_y+, I_x+, I_0, I_y-, I_x-, I_0 is used in sector 1b, to ensure high-to-low transition of transformer primary voltage v_p. Duty cycle loss ΔD occurs twice every switching cycle when an active vector (I_x or I_y) is asserted after zero vector (I_0), as shown in Fig. 5. Power cannot be transferred to secondary during duty cycle loss period, when polarity of primary current i_p reverses. This results in reduction of effective duty of the active vector experiencing the duty cycle loss, as shown in Fig. 3.

As the vector sequence is changed for sectors 1a and 1b, vector experiencing reduction in duty cycle changes from I_x in sector 1a to I_y in sector 1b, resulting in step change in effective duty of

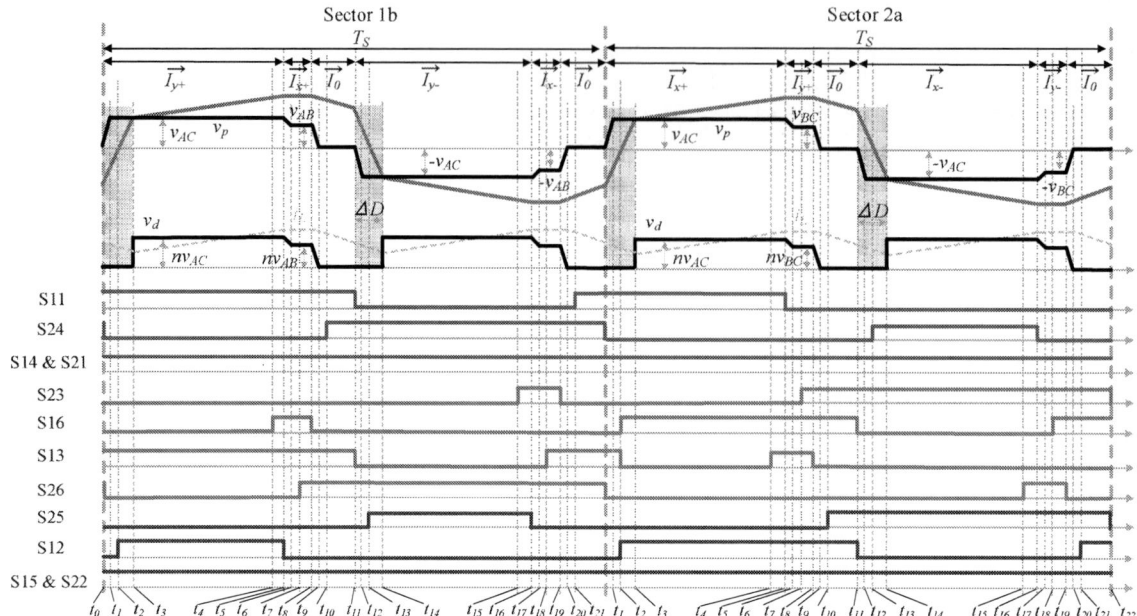

Fig. 4: PWM control signals during b2a sector transition from sector 1 to 2 using existing modulation scheme

active vector during a2b sector transition. This leads to sudden change in input phase currents of phase B and phase C, eventually manifesting in the form of voltage oscillations at the capacitor of input LC filter. As a result, the underlying assumption of input AC voltage magnitude relationship being unchanged within a sector is rendered void, leading to short-circuit of AC source and high current stress at matrix bridge switches. Stable power conversion operation cannot be realized in such scenario as the power semiconductor switches of the matrix bridge will eventually fail due to excessive thermal stresses caused by short-circuits during a2b sector transitions.

On the other hand, an analysis of matrix bridge circuit during transition from sector 'b' to sector 'a', referred as 'b2a sector transition' in the following text, during transition from sector 1 to 2, reveals the issue pertaining to it. As shown in Fig. 4, the modulation scheme is transitioned from sector 1b to sector 2a at time instant t_2 in sector 2a. The vector I_y of sector 1b utilizes the same matrix bridge switches of phase A and C as vector I_x in sector 2a. Although, the state of phase B switches in sector 1b and 2a differs, and hence along with turning-on of switch S12, switch S16 is also turned-on and S13 is turned-off at the same instant. This results in hard switching and leads to voltage surge on matrix bridge switches while transitioning from I_0 to I_x of the following sector.

IV. PROPOSED SECTOR DEFINITION AND PWM MODULATION SCHEME

This paper proposes new sector definition as shown in Fig. 5. Transition sectors are added to the sector definition of Fig. 2, for all sector transitions, resulting in a total of 24 sectors. In Fig. 5, '$xabt$ (x=1-6)' depicts areas of a2b scheme implementation in sector x, whereas 'xyt (y=x+1, $x,y \in$ [1-6])' denotes areas of b2a scheme implementation while transitioning from sector x to y.

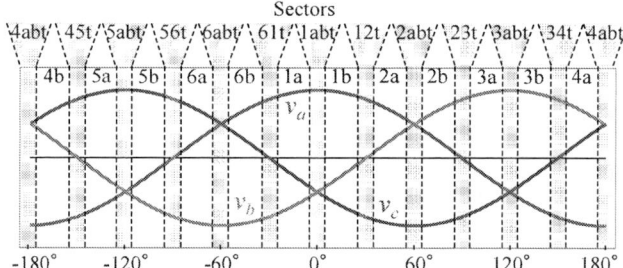

Fig. 5: Proposed sector definition incorporating transition sectors

Fig. 6 shows proposed sector transition modulation scheme for sector 1. Modulation scheme for sector 1abt, shown in Fig. 6(a), implements complementary PWM switching control for diagonally placed matrix bridge switches (S13 & S25, S26 & S12) in Fig. 1, of upper and lower arms, respectively, in legs connected to phase voltages with equal voltage amplitude. With this modification, synchronous rectification is disabled in some of the matrix bridge switches during a2b sector transition. This ensures that the AC source doesn't get shorted even if the AC voltage magnitude relationship changes due to change in vector sequence.

Proposed modulation scheme for sector 12t is shown in Fig. 6(b). 4-segment PWM is implemented in sector 12t. Switches of phase B (S23, S16, S13, S26) with narrow PWM pulses are turned-off, resulting in a 4-step voltage v_p at transformer primary. Due to this, the proposed modulation scheme avoids switching of two different switches at b2a transition instant as discussed in previous section, with possibility of asserting two different active current vectors simultaneously. Hence, hard switching is avoided, resulting in reduced drain-source voltage surge on matrix bridge switches at sector transition instants. The duration of sector 12t and other b2a transition sectors is kept

979-8-3315-1612-3/25 $31.00 © 2025 IEEE

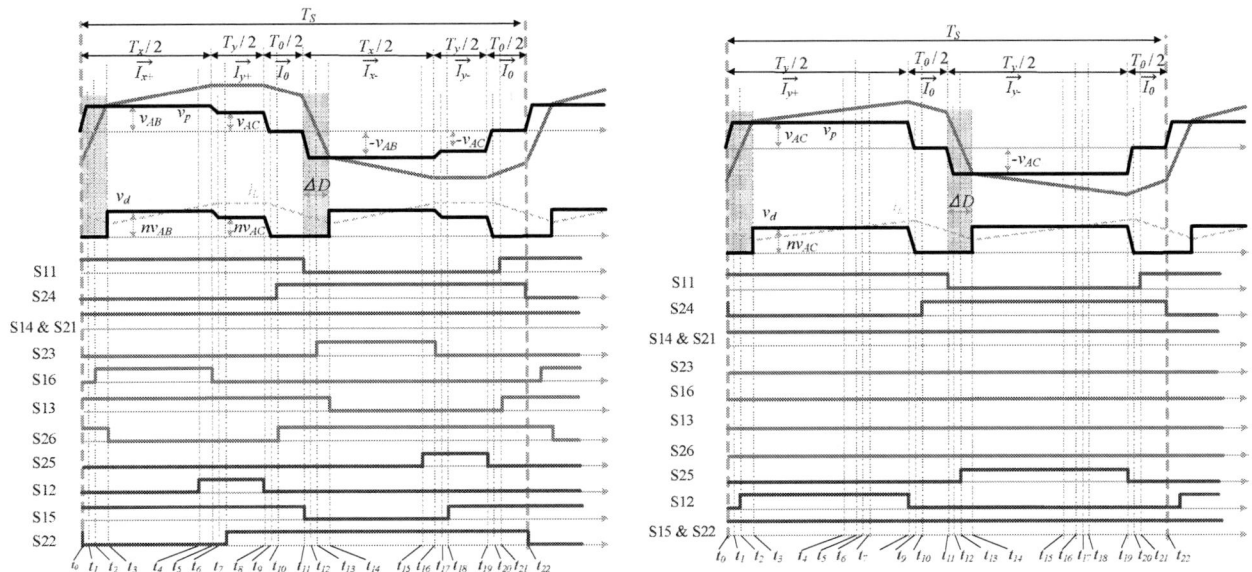

(a) Proposed modulation scheme for sector 1abt

(b) Proposed modulation scheme for sector 12t

Fig. 6: Proposed modulation scheme for sector 1

small to avoid input AC current distortion during sector transitions.

V. SIMULATION AND EXPERIMENTAL RESULTS

A buck-matrix rectifier prototype is built, and simulation and experimental studies are carried out to verify the validity and efficacy of the proposed modulation scheme. Evaluation conditions for simulation and experiments are given in Table I. Block diagram of prototype converter is shown in Fig. 7. This circuit configuration is adopted to handle high current in the secondary side while reducing transformer magnetizing current to reduce conduction losses in the primary.

Simulation results with the existing PWM modulation scheme is shown in Fig. 8(a). Short-circuit of AC source during a2b sector transition, causing high current stress in the matrix

Table I: Simulation and experimental evaluation conditions

Input AC voltage	V_{ac}	400 Vrms, 60 Hz
Output DC voltage	V_{dc}	50 V
Output Current	I_o	30 - 160 A
Switching frequency	f_s	70 kHz
Resonant inductor	L_r	4 µH
DC inductor	$L1_{dc}, L2_{dc}$	5.5 µH
DC capacitor	C_{dc}	16.9 mF
Winding ratio	n	0.25
Primary side SiC-FET	S_{ij} (i=1-2, j=1-6)	1200 V, 18 mΩ
Secondary side Si-FET	SR_{ij} (i=1-2, j=1-4)	135 V, 3.7 mΩ
Dead time	T_{dead}	200 ns

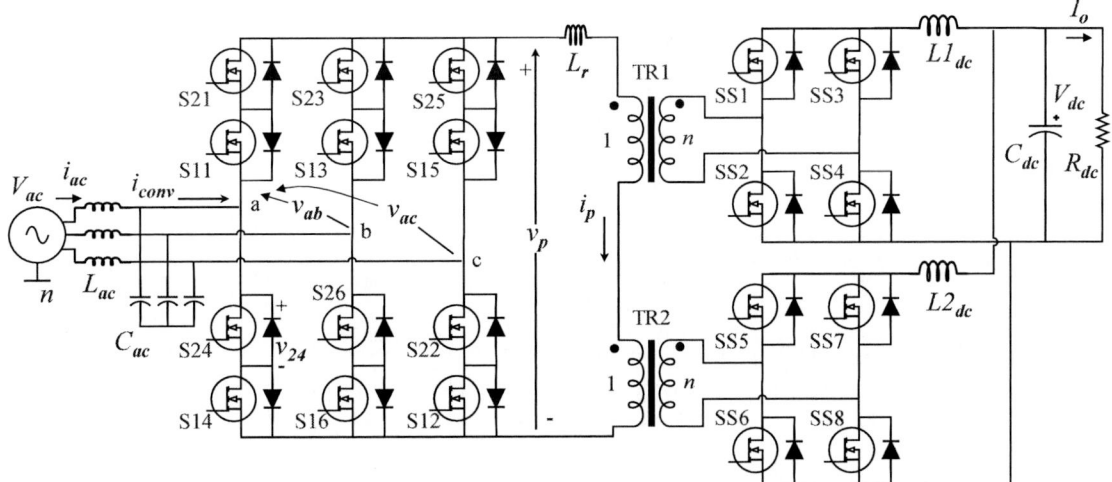

Fig. 7: Circuit diagram for prototype 8 kW buck-matrix rectifier

bridge switches, can be observed from the converter input current i_{conv} waveform. Whereas, it can be concluded from the simulation result of Fig. 8(b) that the proposed modulation scheme eliminates this short-circuit.

Results of experiments on the prototype taken at input AC voltage of 270 Vrms are shown in Figs. 9(a) and 9(b). Both the results of Fig. 9(a) and Fig. 9(b) are taken with proposed modulation scheme for a2b transition implemented, as a short-circuit of AC source at matrix bridge can lead to a catastrophic failure. Elimination of AC source short-circuit, and hence current stress at matrix bridge, can be observed from both the experimental results. Whereas, as can be seen from Fig. 9(a), existing PWM modulation scheme results in large spike at drain-

source voltage of matrix bridge switch S24 (v_{24}) at the transition boundary of sector 1 and 2. Similar observations are made for matrix bridge switches during other b2a sector transitions. Experimental results with proposed modulation scheme are shown in Fig. 9(b). Reduction of voltage spike in v_{24} is evident. Measurement of drain-source voltage of matrix bridge switches with respect to V_{ac} confirm reduction in peak drain-source voltage by 10% across all b2a sector transitions, confirming effectiveness of the proposed scheme.

The experimental result of Fig. 10(a) and 10(b) are taken at experimental conditions mentioned in Table I. Fig. 10(a) shows the input line-to-line voltages v_{ab} and v_{ac} waveforms along with input phase a current i_a. It can be observed that the phase a

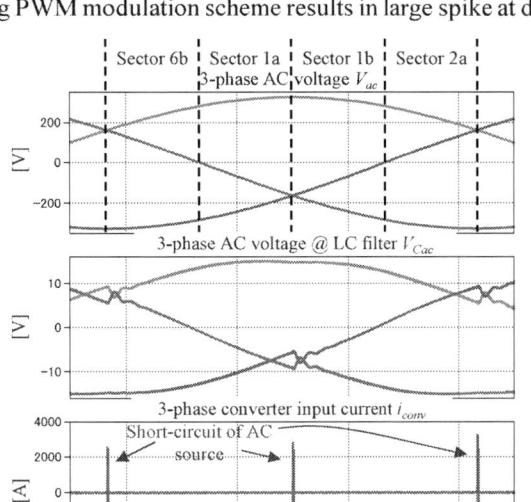

(a) With existing modulation scheme

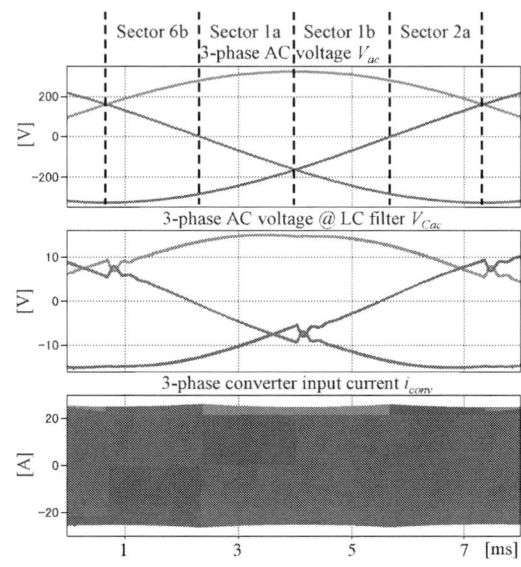

(b) With proposed modulation scheme for a2b transition

Fig. 8: Simulation results on the prototype buck-matrix rectifier circuit

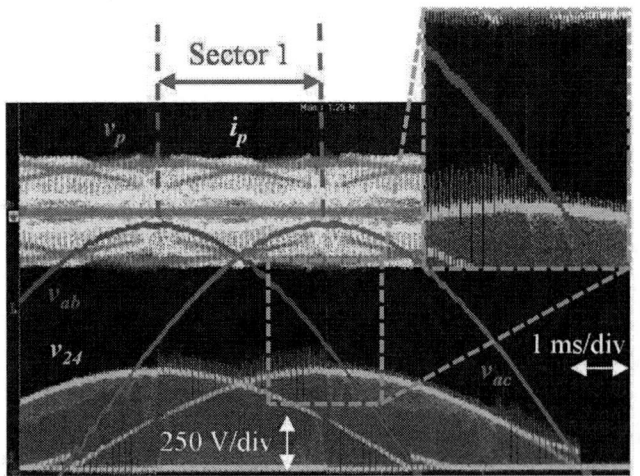

(a) High voltage stress on S24 (v_{24}) during transition from Sector 1 to 2 (b2a transition) using existing modulation scheme

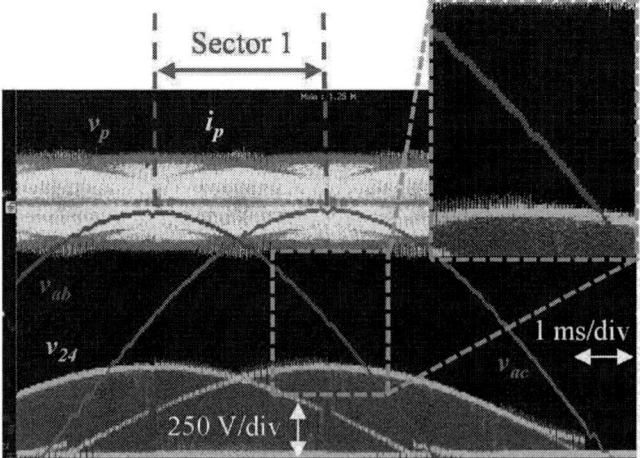

(b) Reduction of voltage stress on matrix bridge switches after adopting proposed modulation scheme

Fig. 9: Experimental results showing waveforms during transition from sector 1 to 2

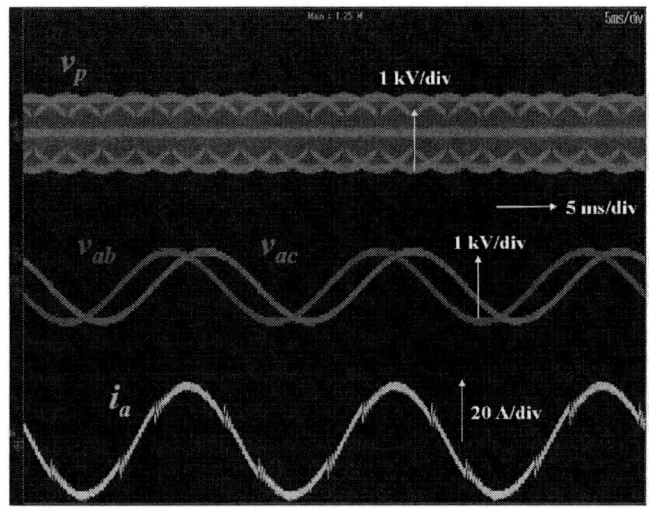
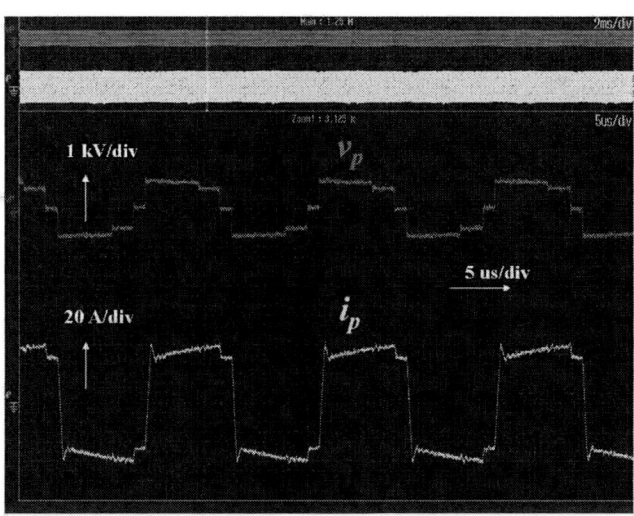

(a) Transformer input voltage, AC input voltage and current

(b) Transformer input voltage and current

Fig. 10: Experimental results showing waveforms at rated input and output conditions

current i_a lags the line voltage v_{ab} by approximately 30°. Thus, it can be concluded that the converter realizes UPF operation. Also, the input current THD is measured to be 4.7% at the rated load condition. This is due to changed vector sequence in 'a' and 'b' sectors, as explained in section III, causing distortion in input phase currents every 60°. This oscillation can be minimized to improve the THD by compensating for the duty cycle loss as reported in [11]. Fig. 10(a) also shows the transformer input voltage v_p, the envelope of which has 360 Hz (6 times the line frequency) ripple. This is due to line-to-line AC voltages being asserted to the transformer primary which change every 60° with changing sectors.

Experimental result of Fig. 10(b) shows zoomed-in transformer input voltage v_p and transformer primary current i_p waveforms. 6-step voltage waveform can be observed with high-to-low voltage transitions realizing ZVS turn-on [11] at the matrix bridge switches.

Fig. 11 shows the power conversion efficiency of the experimental prototype from 1.5 kW output power to the rated load of 8 kW. Peak efficiency of 97.7% and rated power conversion efficiency of 97.3% is realized, showing that high power conversion efficiencies can be achieved using the

proposed modulation scheme, not compromising on power conversion efficiency and power density of the buck-matrix rectifier.

VI. CONCLUSION

This work underlined issues in existing PWM modulation schemes for buck-matrix rectifier during input AC voltage sector transitions. Mechanism of high current and voltage stresses during sector transitions were discussed. Improved PWM modulation scheme along with new sector definition for incorporation of new scheme were proposed, that improves converter reliability. Furthermore, simulation results on the prototype converter circuit were presented, showing the effectiveness of proposed modulation scheme. Finally, the proposed modulation scheme was implemented on an 8 kW converter prototype. Stable converter operation, eliminating current stress due to AC source short-circuit, and reduction of peak drain-source voltage at matrix bridge by 10% up to rated load were verified in experiments. Peak efficiency of 97.7%, and efficiency of 97.3% along with input current THD of 4.7% at full load were obtained during experiments, showing the proposed modulation scheme can realize high power conversion efficiency and maintain low input current harmonics while enabling ZVS and UPF operation.

REFERENCES

[1] R. Y. Sun, S. Webb, Y. -F. Liu and P. C. Sen, "Optimization and Design of a 48-to-12 V 35 A Split-Phase Dickson Switched-Capacitor Converter," *in Proc. IEEE Applied Power Electronics Conference and Exposition*, pp. 1900-1907, 2021.

[2] M. Qiu, X. Liu, and D. Cao, "A New 8x Matrix Autotransformer Switched Capacitor DC-DC Converter for Datacenter Application," *in Proc. IEEE Energy Conversion Congress and Exposition*, 2023.

[3] M. Ahmed, C. Fei, F. Lee and Q. Li, "48-V Voltage Regulator Module With PCB Winding Matrix Transformer for Future Data Centers," *IEEE Trans. on Industrial Electronics*, 2017.

[4] X. Li and S. Jiang. Google 48V power architecture. 2017 IEEE Applied Power Electronics Conference and Exposition, Tampa, FL, 2017.

[5] J. Adrian, L. Olariu, B. Sok, "Mt Diablo - Disaggregated Power Fueling the Next Wave of AI Platforms," in Azure Infrastructure Blog, *Link*: Mt

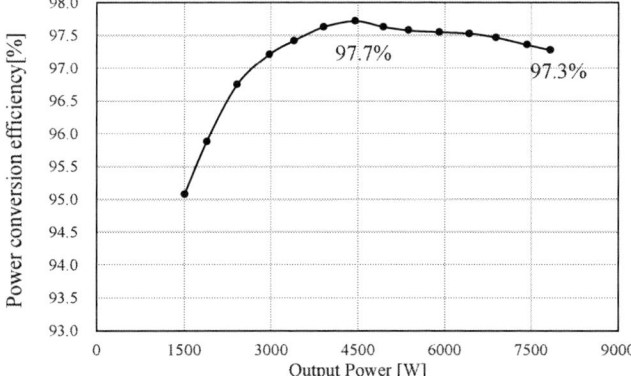

Fig. 11: Power conversion efficiency

Diablo - Disaggregated Power Fueling the Next Wave of AI Platforms (microsoft.com)

[6] A. Nabih, F. Jin and Q. Li, "Efficient Integrated Transformer–Inductor With High PCB Utilization and Optimized Core," *IEEE Trans. on Industrial Electronics*, vol. 71, no. 6, pp. 5653–5662, 2024.

[7] A. Pratt, P. Kumar and T. V. Aldridge, "Evaluation of 400V DC distribution in telco and data centers to improve energy efficiency," INTELEC 07 - 29th International Telecommunications Energy Conference, 2007.

[8] S. Zhao, N. Khan, S. Nagarajan and O. Trescases, "Lithium-Ion-Capacitor-Based Distributed UPS Architecture for Reactive Power Mitigation and Phase Balancing in Datacenters," in IEEE Transactions on Power Electronics, Aug. 2019.

[9] T. Friedli, M. Hartmann and J. W. Kolar, "The Essence of Three-Phase PFC Rectifier Systems-Part II," in IEEE Transactions on Power Electronics, Feb. 2014.

[10] W. Lin, J. Lin, Z. Peng, H. Huang, W. Lin, K. Li, "A systematic review of green-aware management techniques for sustainable data center," *Sustainable Computing: Informatics and Systems*, 2024.

[11] J. Afsharian, D. Xu, B. Wu, B. Gong and Z. Yang, "The Optimal PWM Modulation and Commutation Scheme for a Three-Phase Isolated Buck Matrix-Type Rectifier," in IEEE Transactions on Power Electronics, Jan. 2018.

[12] J. W. Kolar, U. Drofenik and F. C. Zach, "VIENNA rectifier II-a novel single-stage high-frequency isolated three-phase PWM rectifier system," in IEEE Transactions on Industrial Electronics, Aug. 1999.

[13] V. Vlatkovic, D. Borojevic and F. C. Lee, "A zero-voltage switched, three-phase isolated PWM buck rectifier," in IEEE Transactions on Power Electronics, March 1995.

[14] V. Vlatkovic, D. Borojevic, X. Zhuang and F. C. Lee, "Analysis and design of a zero-voltage switched, three-phase PWM rectifier with power factor correction," PESC '92 Record. 23rd Annual IEEE Power Electronics Specialists Conference, Toledo, Spain, 1992.

[15] J. A. Sabate, V. Vlatkovic, R. B. Ridley, F. C. Lee and B. H. Cho, "Design considerations for high-voltage high-power full-bridge zero-voltage-switched PWM converter," Fifth Annual Proceedings on Applied Power Electronics Conference and Exposition, Los Angeles, CA, USA, 1990.

[16] S. Ratanapanachote, Han Ju Cha and P. N. Enjeti, "A digitally controlled switch mode power supply based on matrix converter," in IEEE Transactions on Power Electronics, Jan. 2006.

[17] Z. Yan, K. Zhang, J. Li and W. Wu, "A Novel Absolute Value Logic SPWM Control Strategy Based on De-Re-Coupling Idea for High Frequency Link Matrix Rectifier," in IEEE Transactions on Industrial Informatics, May 2013.

[18] J. Afsharian, D. David Xu, B. Gong and Z. Yang, "Space vector demonstration and analysis of zero-voltage switching transitions in three-phase isolated PWM rectifier," 2015 IEEE Energy Conversion Congress and Exposition (ECCE), Montreal, QC, Canada, 2015.

[19] V. Vlatkovic and D. Borojevic, "Digital-signal-processor-based control of three-phase space vector modulated converters," in IEEE Transactions on Industrial Electronics, June 1994.

[20] R. Garcia-Gil, J. M. Espi, E. J. Dede and E. Sanchis-Kilders, "A bidirectional and isolated three-phase rectifier with soft-switching operation," in IEEE Transactions on Industrial Electronics, June 2005.

[21] X. Lin, Y. Li, J. Afsharian and D. Xu, "An Improved PWM Scheme to Achieve Zero-Voltage Switching for All Devices in Three-Phase Isolated Matrix Rectifier," 2018 International Power Electronics Conference (IPEC-Niigata 2018 -ECCE Asia), Niigata, Japan, 2018.

[22] J. Afsharian, D. Xu, B. Wu, B. Gong and Z. Yang, "A New PWM and Commutation Scheme for One Phase Loss Operation of Three-Phase Isolated Buck Matrix-Type Rectifier," in IEEE Transactions on Power Electronics, Nov. 2018.

Adaptive Capacitance Circuit for Optimal Dynamic Impedance Matching in Variable Reluctance Energy Harvesting Applications

Alejandro Redondo, Fernando Pérez, Sofía García, Gabriel Mujica and Airán Francés

Centro de Electrónica Industrial
Universidad Politécnica de Madrid
Madrid, Spain
Email: alejandro.redondo.ayala@upm.es, fernando.ptorrero@upm.es, sofia.garcia.ordonez@alumnos.upm.es,
gabriel.mujica@upm.es, airan.frances@upm.es

Abstract—**Energy Harvesting (EH) is considered one of the topics that will enable the mass deployment of autonomous Internet of Things (IoT) devices. This paper is focused on Variable Reluctance Energy Harvesters (VREH) which gather energy from the rotation. As these systems feature an inductive output impedance and generate a sinusoidal voltage with variable frequency, it is necessary to perform capacitive impedance matching to maximize power transfer for each working frequency. In this paper, an adaptive capacitance circuit optimization methodology is designed and implemented for a VREH application with the aid of genetic algorithms (GA). This enables to determine in the design phase, for different number of capacitors in parallel, their value and which combinations to use at each frequency to reach the power closest to the maximum possible one. The system is tested in an experimental setup, obtaining output power levels close to the ideal ones over a wide range of frequencies.**

Index Terms—**Dynamic Impedance Matching, Capacitor Array, Genetic Algorithms, Variable Reluctance Energy Harvesting**

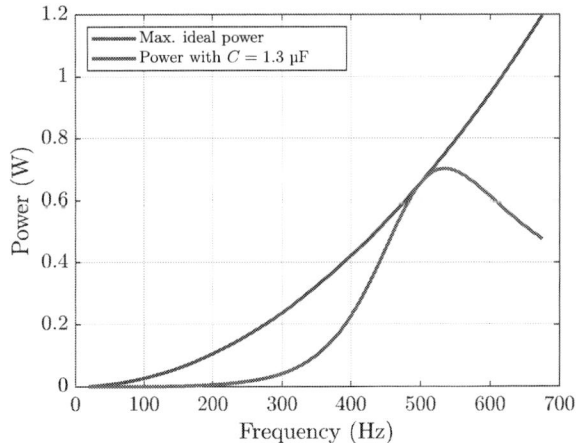

Fig. 1. Extracted power with one capacitor.

I. INTRODUCTION

Deployment of Wireless Sensor Networks (WSN) in order to collect data for predictive maintenance, environment monitoring, or asset tracking, among others applications, has experienced great growth in recent years as part of the Internet of Things (IoT) paradigm. These devices are usually battery powered, and their life expectancy is expected to be ten years or more [1]. However, they can be located in remote, difficult to access, or hazardous locations, so frequent battery replacement is not feasible. Energy Harvesting (EH) addresses this power gap by powering devices with energy from their environment [2]. Harvesters are classified according to their energy source: piezoelectric [3], solar [4], thermoelectric, and so on [5].

Variable Reluctance Energy Harvesters (VREH) obtain energy from rotation. They have recently gained attention due to their promising results, robustness, and low complexity even at low rotational speeds [6]. They can be modeled as an alternating sinusoidal source whose frequency and peak voltage increase linearly with the rotational speed, in series with a resistance and an inductance. Due to the inductive behavior, there will be a phase mismatch between the voltage

and current, so the power delivered to the load will not be the maximum.

To perform impedance matching in order to obtain the maximum power transfer, it is necessary to add a series capacitor that cancels the effect of the inductor. Its value depends on the inductance and the frequency of the generated voltage [7], as shown in:

$$C = \frac{1}{(2\pi f)^2 L}. \qquad (1)$$

If only one capacitor is used, the system is only efficient at the resonance frequency (Fig. 1), so to cover all frequencies range, a solution in which the capacitance varies dynamically is needed.

Other applications that require impedance matching are Radio Frecuency (RF) energy harvesters [8], [9] or RF plasma systems [10], [11] among others. RF EH systems are designed to operate around a particular frequency and an impedance matching network is used to adjust the bandwidth of matching [8]. Capacitor arrays are also used in RF plasma systems

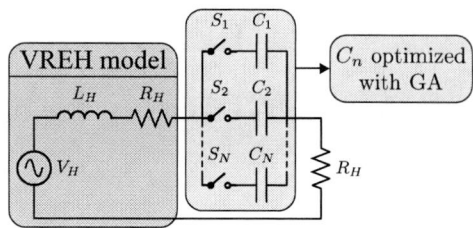

Fig. 2. System model and proposed solution.

TABLE I
SUMMARY OF VREH SPECIFICATIONS

Parameter	Value	Description
L_H	77.5 mH	Equivalent inductance
R_H	60 Ω	Equivalent resistance
f_H	20 − 675 Hz	Voltage frequency range
V_H^{RMS}	0.5 − 16.95 V	Generated RMS range
C_H	817 μF − 717.35 nF	Matching capacitor range

[10], [11] and RF EH systems [9], but they work with capacitances in the order of nanofarads and picofarads. In VREH application, as frequencies are low, capacitances in the order of microfarads are required so commercial embedded variable capacitors, which are in the order of picofarads, do not suit the requirements.

In previous VREH works implementing experimental setups, two scenarios can be distinguished. In the first one, the inductive behavior is neglected, as in [12]–[18], because the rotational speed of the harvester is low and the resistive effect is dominant. In the second one, a single capacitor is added [6], [7], [19]–[21]. However, it is selected for one nominal condition that is fixed, whereas the case in which the speed of the VREH could change dynamically is not regarded.

This study focuses on VREH in smart bearing applications, such as for truck wheels, where variable speed is to be accounted for. To extract maximum power, ideally, a matching capacitor would be needed for each of the working frequencies. However, in a real application this is not possible due to space and cost constraints.

In this article, the impedance matching for VREH in variable speed conditions is addressed with a capacitor array. It contains N parallel capacitors that can be combined with switches to provide $(2^N - 1)$ capacitances (Fig. 2).

Notice that selecting the optimum number of capacitors, their specific values within the given range, and the best combination at every frequency is not straightforward. Therefore, an optimization methodology based on genetic algorithms (GA) to obtain the optimal number and value of the capacitors is proposed.

The remainder of this paper is organized as follows. Section II introduces the VREH model. In Section III the design of the genetic algorithm and the obtained capacitor values used in the experimental phase are explained. Section IV includes a description of the experimental setup and the main results. Finally, Section V concludes the paper.

II. SYSTEM MODEL

The design of the harvester is beyond the scope of this paper and the electrical model shown in Fig. 2 is considered, with a given inductance and resistance of values $L_H = 77.5$ mH and $R_H = 60$ Ω respectively.

The harvester's output voltage value and frequency are determined by the speed of the vehicle depending on the design of the VREH. In this application their values are as follows:

$$f = 7.5 \cdot v, \quad (2)$$

$$V_{\mathrm{RMS}} = f \cdot 0.0355 \cdot \frac{1}{\sqrt{2}}, \quad (3)$$

where v is the speed of the vehicle in km/h, f is the frequency in Hz and V_{RMS} in V is the RMS value of the generated voltage.

In the speed range of the vehicle (up to 90 km/h) the generated RMS voltage reach 16.95 V at 675 Hz which correspond to the maximum speed of the truck. The specifications of the harvester are summarized in Table I, which includes the range of capacitors needed according to (1).

III. OPTIMAL DYNAMIC IMPEDANCE MATCHING CIRCUIT

A. Design of the genetic algorithm

To optimize the number of capacitors, their values, and which combination of capacitors to use at each frequency, a methodology based on genetic algorithms has been developed. The genetic approach allows for searching in a large space of potential solutions, thus being suitable for this optimization problem. Genetic algorithms work in several steps. First, a random population of possible solutions, encoded as chromosomes, are selected. Then, they are evaluated with a fitness function, which gives an indication of their performance. Based on the evaluation, the best ones are selected, and they are subjected to mutation and crossover processes. These new solutions are evaluated, and the process is repeated until termination constraints are met.

In the proposed methodology, each possible solution is made up of the N individual values of the capacitors, represented in the chromosome with their corresponding integer values of resonant frequencies (1).

The maximum power that could be obtained,

$$P_{max}(f) = \frac{V_{\mathrm{RMS}}^2(f)}{4 \cdot R_H}, \quad (4)$$

is considered to be that which could be extracted if at each frequency the inductance were canceled by a matching capacitor (1) in series with the VREH, and the load were a resistor R_H. For each chromosome, i.e., for each of the possible solutions, there are $2^N - 1$ possible combinations of the capacitors in parallel. Each combination provides the following power,

$$P_i(f) = \frac{V_{\mathrm{RMS}}^2(f) R_H}{(2R_H)^2 + (2\pi f L_H - \frac{1}{2\pi f C_i})^2}, \quad (5)$$

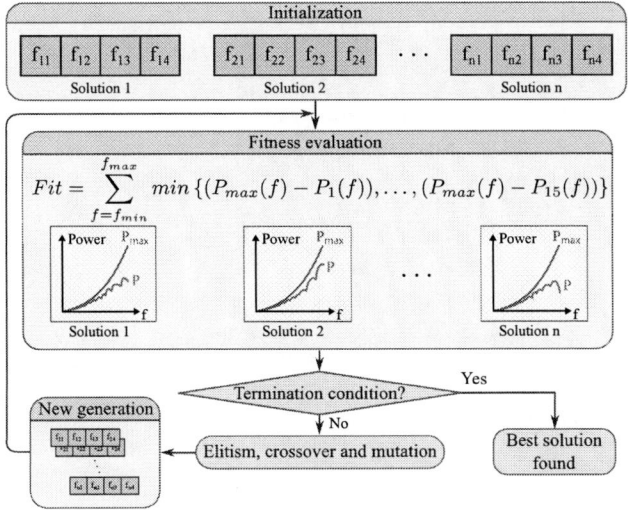

Fig. 3. Scheme of the optimization algorithm.

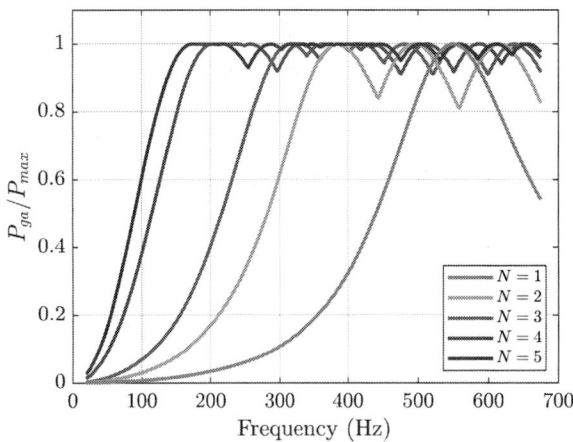

Fig. 4. Ratio of the extractable power with the capacitors given by the genetic algorithm with respect to the maximum extractable power for different numbers of capacitors.

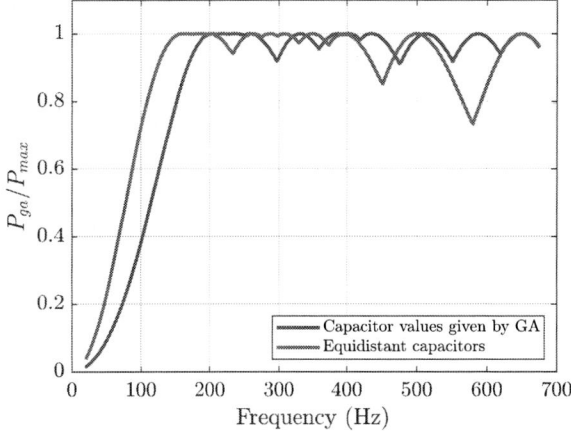

Fig. 5. Ratio of extracted power with 4 capacitors values given by GA and its combinations versus using 4 equidistant capacitors and their combinations.

where i indicates each of the possible combinations. Note that in these power calculations the load is considered to be a resistance of the same value as the harvester's one, that is, using the simplified model of Fig. 2. In addition, although in the experimental application the output is DC and, therefore, a rectification and DC conditioning stage is necessary, their effects on the power have been neglected in the design of the genetic algorithm.

The fitness function, i.e.,

$$Fit = \sum_{f=f_{min}}^{f_{max}} min\Big\{ \big\{P_{max}(f) - P_i(f)\big\} \\ : i = 1, 2, \ldots, 2^N - 1 \Big\}, \quad (6)$$

in a loop that runs through all the frequencies, calculates the maximum obtainable power at each frequency and the actual power obtained with each capacitor combination, choosing from all of them the one that gives the highest power. Then, the differences between the maximum and actual power for every frequency are added, which is the value of the fitness function that the algorithm minimizes. The algorithm is summarized in Fig. 3.

The fitness function minimizes the difference between the maximum extractable power and that which would be obtained for a certain capacity value. Therefore, the differences at high frequencies are penalized more, since at those frequencies greater power is obtained and thus where more is contributed to the sum in (6). Consequently, this genetic algorithm design will give solutions that optimizes better at high frequencies, which is advantageous in this application since in general there will be long periods at speeds higher than 50 km/h.

At the end of the execution, the algorithm indicates the value of each of the N capacitors to be used and which combination of capacitors to select at each of the working frequencies.

B. Results from the Genetic Algorithm

Denote $P_{ga}(f)$ as the power at frequency f using the capacitors combination given by the genetic algorithm and calculated according to (5).

Fig. 4 shows the ratio of the $P_{ga}(f)$ with the maximum extractable power for different number of capacitors (N). It is observed that the higher the number of capacitors, the better the power output at low frequencies. Also the range of frequencies in which the power output is closer to the ideal one increases.

Taking as a reference the fitness value for a single capacitor, when increasing to two capacitors the enhancement increment is 34.8%, at three capacitors 16.02%, at four capacitors 7.5% and 4% for five devices.

It should be also noted that increasing the number of capacitors enlarges the size of the solution, the losses associated with the elements, the complexity and the cost. That is why

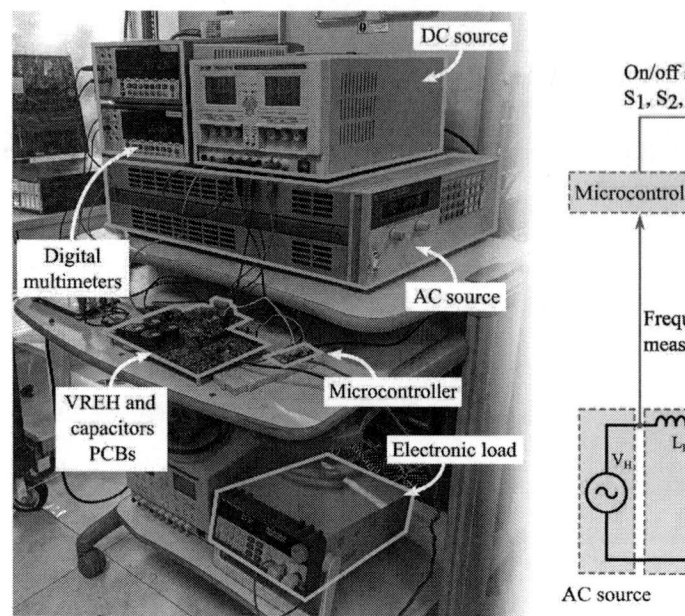

Fig. 6. Experimental setup.

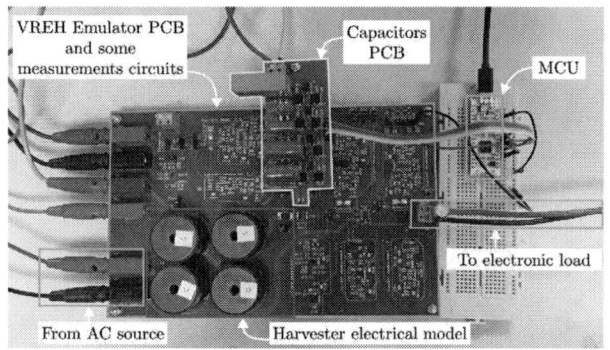

Fig. 7. Harvester and capacitor PCB detail.

TABLE II
SPECIFICATIONS OF THE MAIN PCB ELEMENTS

Description	Element ID
Harvester Coils	Murata 1422604C
Harvester resistance	TE Connectivity 354022RJT
Rectifier diodes	Toshiba 08F30
MOSFET switches	Infineon BSC120N12LSGATMA1
MOSFET isolated drivers	Skyworks SI8751AB-ISR
C1, 0.75 µF film capacitor	Panasonic ECWF4754RJL
C2, 0.91 µF film capacitor	Panasonic ECWF4914HL
C3, 1.20 µF film capacitor	Panasonic ECWF2125JA
C4, 4.50 µF film capacitor	KEMET C4AEGBU4450A1WJ

in the remainder of this paper a solution with 4 capacitors and their combinations will be considered.

For the case of four capacitors, the resulting optimal values for the specified VREH are 0.77 µF, 0.94 µF, 1.23 µF, and 4.55 µF.

To highlight the effectiveness of the genetic algorithm to find an optimal solution, the comparative result using the capacitors given by the genetic algorithm versus taking four equidistant capacitors and their combinations is shown in Fig. 5. The equidistant ones are those matching equispaced frequencies at 200 Hz, 350 Hz, 500 Hz, and 650 Hz, corresponding to 8.17 µF, 2.67 µF, 1.31 µF, and 0.77 µF respectively. As in the case of the capacitors given by the genetic algorithms, the combinations have been chosen in such a way that at each frequency the maximum possible power is obtained. The initial equispacing value of 650 Hz has been chosen to give more weight to high frequencies.

The results show that in the case of genetic algorithms,

the power obtained over 167 Hz is not less than 90% of the maximum possible one, while in the case of equispaced capacitors it drops to 73% at 580 Hz. Therefore, for the whole frequency range, using the optimized values given by GA lead to more power extraction than using equidistant values especially at high frequencies.

IV. EXPERIMENTAL RESULTS

To test the proposed solution, the setup shown in Fig. 6 was used. The VREH was emulated with a custom PCB that implements its electrical model, a diode rectifier, and additional measurement circuits, including a sensor to measure the harvester's frequency. An AC source provides the sinusoidal voltage of the harvester, and an electronic load is connected to the output of the rectifier.

To facilitate prototyping, the array of capacitors has been implemented on an additional PCB, which is connected to the VREH. In order to avoid variation on the capacitance due to the effect of the voltage in the capacitor's terminals, film capacitors have been chosen. The commercially available ones

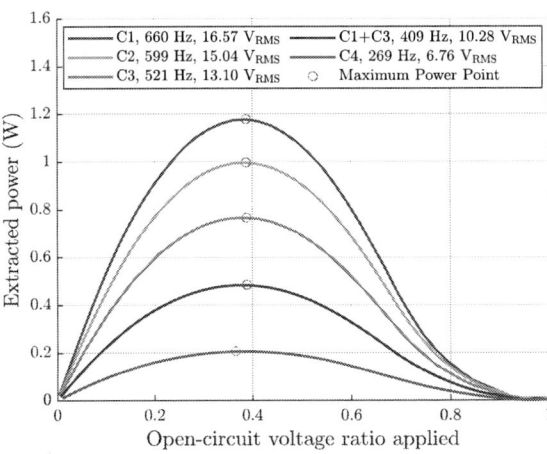

Fig. 8. MPPT experimental obtained curves. Each of which is for the specified activated capacitor, and the AC source set to the frequency and RMS value indicated in the legend.

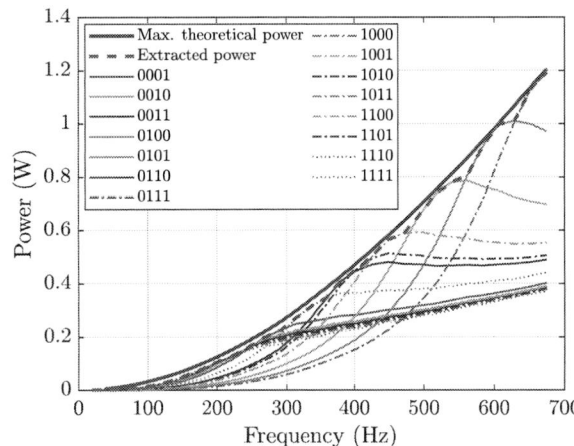

Fig. 9. Power obtained experimentally. For the curves showing the power with each combination of capacitors, it is indicated in the legend with a 1 which ones are active as [C1, C2, C3, C4].

that are closest to those given by the genetic algorithm have been selected. The values installed in the capacitors PCB are 0.75 µF, 0.91 µF, 1.2 µF, and 4.5 µF. The harvester emulator PCB with the capacitors PCB attached to it, is shown in Fig. 7 and their main elements are listed in Table II. Each switch consists of two MOSFETs (Infineon's BSC120N12LS) with their sources connected to block AC currents, and a single isolated driver (Skyworks's SI8751AB-ISR), used to control them. A microcontroller (STMicroelectronics's ST32L031K6), based on the frequency measurements, provides the enable signals to the switches according to the optimal results given by the optimization algorithm which are stored in a lookup table.

Between the harvester and the load, usually, a power management unit (PMU) is placed to ensure proper power extraction and energy dispatch to the devices connected to it.

Assuming ideal rectification and perfect impedance matching, according to the maximum power transfer theorem, the highest power could be extracted if a resistor with the value of the resistance of the harvester is placed after the rectifier. Nonetheless, the PMU rarely will has such characteristic, consequently systems that find the maximum power point are normally used by means of techniques such as maximum power point tracking (MPPT).

To ensure the highest possible power extraction, in this paper a fraction of the open circuit voltage of the harvester is applied after the rectifier with an electronic load. It emulates the operation of a power management unit (PMU) using that method to find the maximum power point.

Experiments have been conducted by performing voltage sweeps for a given frequency and its matching capacitor activated. In them, different values of voltage at the load have been tested in order to find which one allows to extract the highest power from the harvester. These tests are shown in Fig. 8. It was found that the voltage that maximizes the extracted power, is about 40% of the open circuit voltage,

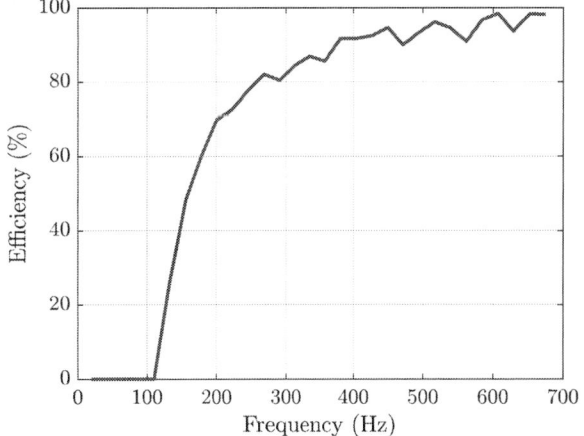

Fig. 10. Experimental obtained efficiency.

which is the peak value of (3).

The proposed system has been validated by measuring the extracted power for every combination of capacitors over the entire range of frequencies of the harvester. The results are shown in Fig. 9. The thick blue line shows the maximum power that could be extracted if perfect impedance matching was performed at every frequency, and the discontinuous red line shows the obtained power with the proposed solution. The rest of the curves show the power obtained with each of the 15 combinations of capacitors.

Regarding losses, these mainly come from the rectifier, the MOSFETs and their drivers. The MOSFETs are only activated at certain times depending on the voltage frequency. In this application it is expected that the speed variations are not too fast, so the switching losses are expected to be negligible. Therefore, only the conduction losses due to the resistance between the drain and the source are considered. As for the driver, the losses considered are those due to the driver power supply, which, according to the datasheet typical supply current value, are 4.95 mW for each driver constant over the entire operating range.

Since the current and voltage measurements are taken at the output of the harvester, before the electronic load, the losses attributable to the MOSFETs and diodes are already included in the measurements.

Fig. 10 shows the efficiency (with respect to the maximum power that could be obtained) including driving losses. It is over 80% for frequencies above 250 Hz and over 90% for frequencies higher than 370 Hz. Notice that for low frequencies, since the static consumption of the drivers is higher than the power obtained, the efficiency has been set to 0.

V. CONCLUSIONS AND FUTURE WORK

This paper presents an adaptive impedance matching solution for VREH based on a capacitor array optimized with genetic algorithms. The proposed circuit can dynamically cancel out the inductive impedance of the harvester with the optimal capacitor combination in applications where the frequency is highly variable, such as vehicle wheels. The efficiency over the entire frequency range is significantly improved when compared to solutions that use a fixed capacitor. Future work includes enhancement of the algorithm to consider DC signal conditioning and rectification circuit losses as well as taking into account space constraints.

ACKNOWLEDGMENT

This work is supported by the LoLiPoP IoT project, Grant PCI2023-143407, funded by MICIU/AEI /10.13039/501100011033 and cofunded by European Comission. LoLiPoP IoT has received funding from the Chips Joint Undertaking (Chips JU) under grant agreement No 101112286. The JU receives support from the European Union's Horizon Europe research and innovation programme and accordingly from the participating countries.

REFERENCES

[1] J. Portilla, G. Mujica, J.-S. Lee, and T. Riesgo, "The extreme edge at the bottom of the internet of things: A review," *IEEE Sensors Journal*, vol. 19, no. 9, pp. 3179–3190, 2019.

[2] F. Akhtar and M. H. Rehmani, "Energy replenishment using renewable and traditional energy resources for sustainable wireless sensor networks: A review," *Renewable and Sustainable Energy Reviews*, vol. 45, pp. 769–784, 2015.

[3] M. Shirvanimoghaddam *et al.*, "Towards a green and self-powered internet of things using piezoelectric energy harvesting," *IEEE Access*, vol. 7, pp. 94 533–94 556, 2019.

[4] H. Zhang, X. Liu, M. Kedia, and R. S. Balog, "Photovoltaic hybrid power harvesting system for emergency applications," in *2013 IEEE 39th Photovoltaic Specialists Conference (PVSC)*, 2013, pp. 2902–2907.

[5] T. Sanislav, G. D. Mois, S. Zeadally, and S. C. Folea, "Energy harvesting techniques for internet of things (iot)," *IEEE Access*, vol. 9, pp. 39 530–39 549, 2021.

[6] Y. Xu, S. Bader, M. Magno, P. Mayer, and B. Oelmann, "System implementation trade-offs for low-speed rotational variable reluctance energy harvesters," *Sensors*, vol. 21, no. 18, 2021, Art. no. 6317.

[7] Y. Xu, S. Bader, and B. Oelmann, "Design, modeling and optimization of an m-shaped variable reluctance energy harvester for rotating applications," *Energy Conversion and Management*, vol. 195, pp. 1280–1294, Sep. 2019.

[8] I. D. Bougas, M. S. Papadopoulou, A. D. Boursianis, K. Kokkinidis, and S. K. Goudos, "State-of-the-art techniques in RF energy harvesting circuits," *Telecom*, vol. 2, no. 4, pp. 369–389, 2021.

[9] D. Khan *et al.*, "A high-efficient wireless power receiver for hybrid energy-harvesting sources," *IEEE Transactions on Power Electronics*, vol. 36, no. 10, pp. 11 148–11 162, Oct. 2021.

[10] J. Min and Y. Suh, "Hybrid variable capacitor in impedance matcher for RF plasma process," in *2021 IEEE 12th Energy Conversion Congress & Exposition - Asia (ECCE-Asia)*, Singapore, Singapore, May 2021, pp. 2410–2414.

[11] J. Min, B. Chae, Y. Suh, J. Kim, and H. Kim, "Next-generation variable capacitors to reduce capacitance variable time using SiC MOSFETs and p-i-n diodes in 13.56-MHz RF plasma systems," *IEEE Journal of Emerging and Selected Topics in Power Electronics*, vol. 10, no. 2, pp. 1353–1362, Apr. 2022.

[12] M. Kroener, N. Moll, S. K. T. Ravindran, P. Mehne, and P. Woias, "Characterization of a variable reluctance harvester," *Journal of Physics: Conference Series*, vol. 557, no. 1, nov 2014, Art. no. 012035.

[13] Y. Xu, Y. Zhang, S. Bader, B. Oelmann, and J. Cao, "Three-phase variable reluctance energy harvesting," *Energy Conversion and Management: X*, vol. 14, 2022, Art. no. 100211.

[14] Y. Zhang *et al.*, "Enhanced variable reluctance energy harvesting for self-powered monitoring," *Applied Energy*, vol. 321, 2022, Art. no. 119402.

[15] Y. Xu, S. Bader, and B. Oelmann, "A survey on variable reluctance energy harvesters in low-speed rotating applications," *IEEE Sensors Journal*, vol. 18, no. 8, pp. 3426–3435, Apr. 2018.

[16] Y. Zhang, X. Wu, Y. Lei, J. Cao, and W.-H. Liao, "Self-powered wireless condition monitoring for rotating machinery," *IEEE Internet of Things Journal*, vol. 11, no. 2, pp. 3095–3107, Jan. 2024.

[17] Y. Zhang, H. Zhu, Y. Xu, J. Cao, S. Bader, and B. Oelmann, "Theoretical modeling and experimental verification of rotational variable reluctance energy harvesters," *Energy Conversion and Management*, vol. 233, 2021, Art. no. 113906.

[18] M. Kroener, S. K. T. Ravindran, and P. Woias, "Variable reluctance harvester for applications in railroad monitoring," *Journal of Physics: Conference Series*, vol. 476, no. 1, Dec. 2013, Art. no. 012091.

[19] Y. Gong *et al.*, "A variable reluctance based rotational electromagnetic harvester for the high-speed smart bearing," *Smart Materials and Structures*, vol. 31, no. 4, Mar. 2022, Art. no. 045023.

[20] Y. Gong, S. Wang, Z. Xie, and W. Huang, "Parameter study of the variable reluctance energy harvester for smart railway axle box bearing," in *2020 International Conference on Sensing, Measurement & Data Analytics in the era of Artificial Intelligence (ICSMD)*, Xi'an, China, Oct. 2020, pp. 464–469.

[21] Y. Gong *et al.*, "Self-powered wireless sensor node for smart railway axle box bearing via a variable reluctance energy harvesting system," *IEEE Transactions on Instrumentation and Measurement*, vol. 70, pp. 1–11, 2021, Art. no. 9003111.

Gallium Nitride (GaN) Based Topology Comparison for Low Power Battery Charging Applications

Jai Aditya Chaudhary
AMS Analog&Power STMicroelectronics
Schaumburg,USA
jaiaditya.chaudhary@st.com

Rosario Attanasio
AMS Analog&Power STMicroelectronics
Schaumburg, USA
rosario.attanasio@st.com

Gianni Vitale
AMS Analog&Power STMicroelectronics
Schaumburg, USA
gianni.vitale@st.com

Abstract— **This paper presents the design and implementation of a GaN-based Power Converter for low-power battery chargers. It compares the performance of the Asymmetrical Half Bridge Flyback (AHB-FB) with 2-Switch Flyback (2SW-FB) and resonant LCC topologies, all using the same 600V/150mΩ GaN Half Bridge (HB) System in Package (SiP). High Voltage (HV) GaN High Electron Mobility Transistors (HEMT) SiP technology enables a simpler implementation and smaller form factor compared to standard, discrete silicon solutions. Additionally, GaN-SiP facilitates soft switching over a wider operational range with minimal circulating current, enhancing system efficiency.**

The AHB-FB, LCC, and 2SW-FB converters use dedicated analog controllers to achieve Zero Voltage Switching (ZVS) for the AHB and LCC, and Quasi-Resonant (QR) Operation for the 2SW-FB.

The paper outlines the experimental results for a battery charger designed around these three topologies: AHB-FB, LCC, and 2SW-FB. Key factors for comparison include efficiency for performance, Bill Of Materials (BOM) for cost, and size for power density. The analysis highlights the advantages of Flyback topologies, particularly in terms of reduced components and cost. The findings emphasize the benefits of GaN-based Flyback configurations in these power applications.

Keywords— Gallium Nitride (GaN) System in Package (SIP), Battery Charger, 2-Switch Flyback (2SW-FB), Asymmetrical Half Bridge Flyback (AHB-FB), Zero Voltage Switching, Zero Current Switching, Synchronous Rectification (SR).

I. INTRODUCTION

The advancements in battery technology have driven the growth of various battery-powered applications, enhancing convenience and usability. With improvements in batteries, the demand for faster, more reliable, and efficient charging solutions has increased. Manufacturers are exploring simpler, smaller, and cost-effective power topologies to achieve this. Power tool batteries for example consist of 3 to 5 cells in series, resulting in 9 to 21 V. An ideal fast battery charger can fully charge and maintain batteries with different cell configurations, delivering up to 10 Amps or 200 W. It is a power converter that delivers a constant current controlled by an internal regulator loop at the voltage imposed by the connected battery.

Resonant topologies like the LCC mentioned in [1], offering wider operating range and peak efficiencies of 94.5%, utilize an integrated transformer [2] achieving low tolerance on inductance values that saves PCB space and additional costs. The AHB-FB converter is an attractive topology for its high peak efficiency [3]. Like the LCC, in the AHB-FB, the primary switches operate in ZVS and the secondary side rectifier in Zero Current Switching (ZCS) conditions [4]. A 2SW-FB with GaN switches controlled in QR mode is simple, well understood, and easy to implement [5]. It efficiently reuses the energy stored in the leakage inductance, unlike a traditional Flyback where this energy is dissipated by a resistor, making it suitable for 200W applications.

This paper presents the design of two Flyback converters, an 2SW-FB and AHB-FB based on GaN SiP [6], that operate over a wide output voltage range and compares their performance with similarly sized LCC converters for battery charging applications. GaN SiP technology in power conversion minimizes heatsink size and eliminates cooling fans, reducing costs, and improving reliability. The practical implementation of these converters is demonstrated by 200W prototypes developed with GaN SiP and Synchronous Rectification (SR).

The main electrical specifications and design parameters are summarized in Table I. The range of output regulation allows the battery charger to be used for charging a variety of batteries with different cell configurations. This flexibility is crucial for accommodating the diverse requirements of various battery types and ensuring efficient and reliable charging across a wide range of applications.

TABLE I. ELECTRICAL SPECIFICATIONS AND DESIGN PARAMETERS

Spec/Parameter	Symbol	Value	Unit
Input Voltage	V_{in}	90-132 / 180-277	V RMS
Switching Frequency	F_{sw}	65-250	kHz
Output Voltage Range	V_{out}	8-22	V
Output Current (Max.)	I_{out}	10	A
PWM Control Input	V_{ICTL}	0-3.3	V

II. PROPOSED DESIGN

A. Topology Description of a battery Charger

The proposed battery charger block diagram featuring an input Electromagnetic Interference (EMI) filter, followed by a full wave rectifier diode bridge with a large capacitance to limit the voltage ripple is shown in fig.1. The input capacitance also helps mitigating high-frequency switching-related voltage ripple, HV GaN SiP with 650V/150mΩ GaN transistors in Half Bridge configuration and dedicated gate driver drives the primary side of the high frequency transformer. The controller on the primary side generates two PWM control signals and with a suitable dead time to avoid cross conduction in the HB. The system under consideration incorporates synchronous rectification. A Synchronous Rectifier (SR) typically features a low on-state resistance (R_{DS_ON}) Metal-Oxide-Semiconductor Field-Effect Transistor (MOSFET). The output current and voltage of the battery charger are regulated using a Constant Current/Constant Voltage (CC/CV) control circuit, which includes a voltage reference generator, two operational amplifiers for the feedback loop, and a compensation network. This circuit is powered by an auxiliary 12V voltage generated from the output DC voltage and referenced to the negative output terminal. The CC/CV feedback signal is then transmitted to the analog controller through an optocoupler. When this circuit is used in a battery charger an additional digital controller is required. This digital controller is necessary to identify the battery maximum charging current and implement a specific, optimized charging profile.

The critical design choice that determines the performance, cost, and form factor of a battery charger is the selection of the power converter topology. This choice affects the efficiency, range of operation, complexity, and overall effectiveness of the charger.

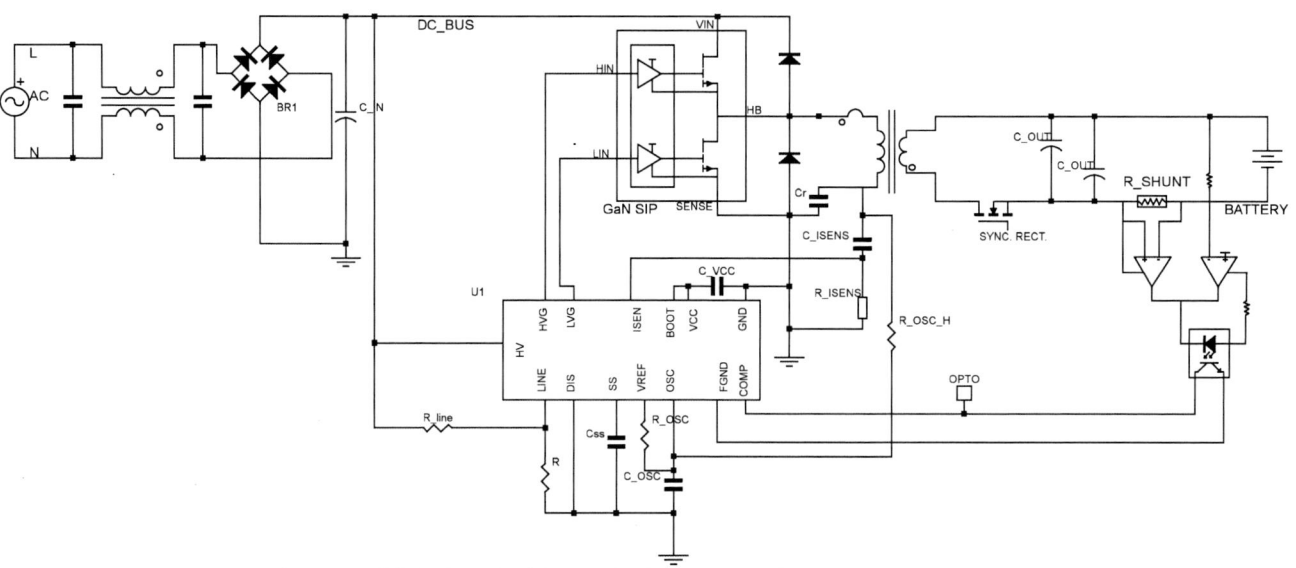

Fig. 1. Simplified schematic of the complete battery charger based on AHB-FB topology.

B. Design Considerations

The primary design considerations for the battery charger include selecting an appropriate power topology that performs well across varying output terminal voltages and input AC voltage ranges. The key criteria for comparing converters fall into three main categories: performance, cost, and size. Performance can be measured by peak efficiency, average efficiency, and the ability to maintain stable output despite variations in input voltage and load conditions. An ideal topology maintains high efficiency through changes in load and line conditions. Cost encompasses both the electronic bill of materials (BoM) and the human effort required to design the converter, with lower costs benefiting manufacturers and consumers by making the product more competitive. Size pertains to the form factor, with a smaller form factor leading to more compact and portable designs, which are highly valued in applications like power tools and portable electronics. Balancing these factors ensures a high-performing, cost-effective, and compact battery charger.

C. 2 Switch Flyback with GaN HEMT

The block diagram of the GaN based 2SW-Flyback converter is shown in Fig. 2.

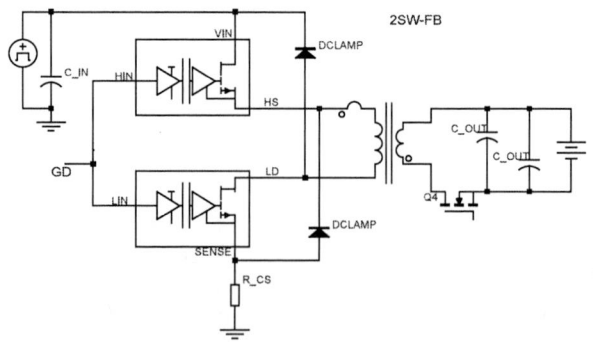

Fig. 2. Block Diagram of 2SW Flyback topology.

979-8-3315-1612-3/25 $31.00 © 2025 IEEE

On the primary side, the two GaN switches are operated simultaneously using two gate drivers, each referenced to the source pin of the GaN devices. This topology needs a specific gate driver circuit as opposed to the widely known HB gate drivers. During the conduction time (T_{on}), the transformer stores energy as magnetic flux, resulting from the current drawn from the power source. During the OFF time (T_{off}) of the switches the transformer is demagnetized, the stored energy is transferred to the load at the converter output terminals through a secondary side rectifier composed of a low R_{DS_ON} Silicon MOSFET driven by an SR controller. Unlike a traditional Flyback converter, the 2SW-FB topology uses two diodes (D_{CLAMP}) to return the energy stored in the leakage inductance of the Flyback transformer to the input capacitor after the switches are turned off.

The primary side gate controller is an analog peak current mode controller with transformer demagnetization detection, enabling Quasi-Resonant(QR) switching over a wide range. This mode, also known as Critical Conduction Mode or Transition Mode (TM), has been widely discussed [6].

The controller keeps the switches on for the duration of T_{on} until the transformer's primary current (I_{p_pk}) reaches a set level defined by the feedback loop. This current is sensed through the CS pin using a current sense resistor(R_CS) placed in series with the low-side switch. The voltage across the current sense resistor is continuously monitored to enable pulse-by-pulse current limitation, preventing over-dissipation in the primary switches during output short circuit and overload conditions. The Zero Current Detection (ZCD) pin detects transformer demagnetization using an auxiliary winding on the primary side of the transformer.

The fundamental equations for a flyback converter are as follows:

$$T_{on} = L_f \frac{I_{p\_pk}}{Vin_{dc}} \quad (1)$$

$$V_R = (V_{out} + V_F) \cdot \frac{N_{pri}}{N_{sec}} \quad (5)$$

$$T_{dmag} = L_f \frac{I_{s\_pk}}{V_{out}} \quad (2)$$

$$P_{sw\_on} = \frac{1}{2} \cdot V_{sw}^2 \cdot C_d \cdot F_{sw} \quad (6)$$

$$I_{in} = \frac{1}{2} \cdot I_{p\_pk} \cdot \frac{T_{on}}{T_{sw}} \quad (3)$$

$$P_{sw\_off} = \frac{1}{2} \cdot V_{in} \cdot I_{p_{pk}} \cdot T_{c\_off} \cdot F_{sw} \quad (7)$$

$$T_V = \frac{1}{2 \cdot f_r} \quad (4)$$

$$f_T = \frac{1}{2 \cdot P_{IN} \cdot L_P \cdot \left(\frac{1}{V_{IN}} + \frac{1}{V_R}\right)^2} \quad (8)$$

The reflected voltage (V_R) shown in (5), helps define the important design trade-offs, including the RMS and peak values of the primary current and secondary currents. In a 2SW-FB converter, V_R should be less than the DC link voltage supplied to the converter; otherwise, the clamping diodes D_{CLAMP} will continue to conduct during the T_{off} of the primary side switches.

Equation (6) represents the power loss(P_{sw_on}) due to the turn-on commutation, where C_d is the equivalent drain capacitance of the switch, and V_{SW} is the valley of the drain voltage of the low-side switch. Equation (6) gives the switching loss at turn-off due to the crossing of drain current and drain in the crossover time T_{c_off}. The power loss related to conduction (P_{COND}) can be estimated by calculating the RMS values of primary current and the R_{DS_ON} of the selected GaN device. Increasing V_R reduces the duty cycle (D), peak secondary current, and the valley of the drain voltage, thereby reducing the turn-on related power loss in each primary switch, in (6)

In (8) f_T is the theoretical Switching frequency of the converter considering TM operation. The time T_V in (4) is the time duration after demagnetization of the transformer until the valley of the drain voltage. During T_V, the Primary inductance (L_P) and the drain capacitance resonate at the resonant frequency (f_r). T_V is a finite duration and hence it contributes to the actual switching frequency F_{SW} of the converter.

$$F_{SW} = \frac{2 \cdot f_T}{1 + \frac{f_T}{f_r} + \sqrt{1 + \frac{f_T}{f_r}}} \quad (9)$$

The relationship in (8) and (9) between F_{SW} and primary inductance helps in choosing the typical frequency of operation as a trade-off between switching losses and the size of the magnetic component.

D. AHB-FB with GaN HEMT

The block diagram of the Asymmetrical Half Bridge Flyback (AHB-FB) converter is shown in Fig. 3.

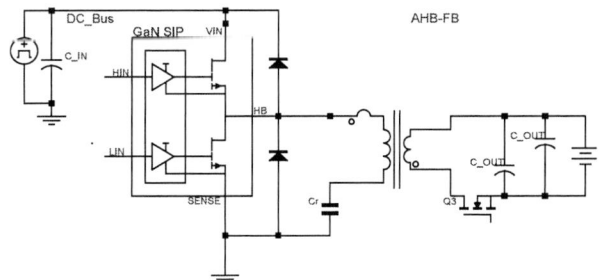

Fig. 3. Block Diagram of converter with AHB-FB topology.

On the primary side, the two GaN switches are driven by the HB gate diver integrated into the GaN SiP operated with complementary control using a PWM controller. In steady state, during the conduction time of the high side switch (T_{on_HS}), the current flows from the input capacitor through the high side switch into the series combination of transformer primary and resonant capacitor Cr and into the ground return.

Once the HS switch is turned off, during dead time T_{d_HL}, the HB node capacitance C_{HB} is discharged. The low side switch is turned-on in zero voltage conditions, eliminating the turn-on switching loss .The leakage inductance (Lr) and the resonant capacitor Cr resonate, and the secondary winding conducts for time $T_{r\_sec} = \pi\sqrt{Lr \cdot Cr}$. The secondary side rectifier, which is a low R_{DS_ON} Silicon MOSFET, is turned on under Zero Current Switching (ZCS) [8][9].

After the transformer demagnetization, the LS switch is kept on until the current in the primary side reverses its direction and has a negative value. The current I_{ZCD} is required to charge the HB node capacitance during the dead time T_{d_LH} to create zero voltage turn-on conditions for the HS switch.

979-8-3315-1612-3/25 $31.00 © 2025 IEEE

$$I_{ZCD} = C_{HB} \cdot \frac{V_{DC}}{T_{d\_LH}} \quad (10) \qquad C_{HB} = 2 \cdot C_{oss} + C_{STRAY} \quad (11)$$

Where, C_{HB} is the total node capacitance, V_{DC} is the voltage across C_{HB}.

The analog controller for AHB-FB is an integrated current mode PWM controller with high voltage floating structure for HS gate driver. The switching frequency is set externally with the passive components connected to the oscillator section of the device. The PWM block of the devices applies the duty cycle to the high side switch, based on the secondary side feedback. The high side device is kept on for T_{ON_HS} until the current (I_{HS_pk}) reaches a level set by secondary side feedback. The controller also imposes a limit on the maximum duty cycle of the HS switch which is $D_{max\_HS} = 50\%$. and applies a dead time to avoid cross conduction in the HB.

In steady state operation the average voltage V_{Cr} at Cr can be approximated as in (12) using the volt-second balance principle.

$$V_{Cr} = V_{in} \cdot D \quad (12)$$

where D is the Duty cycle of HS Switch. During the conduction of secondary rectifier, the voltage at Cr is applied to the primary of the transformer and the output voltage is

$$V_{OUT} = V_{Cr} \cdot \frac{N_s}{N_p} = \frac{V_{Cr}}{N} \quad (13)$$

$$V_{OUT} = V_{in} \cdot \frac{D}{N} \quad (14)$$

The above equations are defined for the average model of the topology when it is assumed that the resonant capacitor is a constant voltage source, and the magnetizing inductance (L_m) is much higher than the leakage inductance (L_s).

The peak and the valley of the magnetizing current (I_m) ripple is defined by (15) and (16) respectively

$$I_{m\_peak} = I_{m\_avg} + \frac{\Delta I_m}{2} \quad (15)$$

$$I_{m\_valley} = I_{m\_avg} - \frac{\Delta I_m}{2} \quad (16)$$

$$\Delta I_m = \frac{V_{in} - V_{cr}}{L_m} \cdot T_{on\_HS} = N \cdot V_{out} \cdot \frac{(1-D)}{F_{sw}} \quad (17)$$

In above equation, the average current in the converter primary is $I_{m\_avg} = I_{out}/N$. Combining with (15),(16),(17) the switching frequency (F_{sw}) can be calculated as:

$$F_{sw} = \frac{N^2 \cdot V_{out} \cdot (1-D)}{2 \cdot I_{out} \cdot L_m} \quad (18)$$

To maintain ZVS for wide output voltage, the On-time of the LS switch should be increased as the output voltage decreases since it takes longer for the magnetizing current I_m to return to a negative value. In other words, the frequency must be reduced to maintain ZVS. This can be easily achieved by using an additional resistor R_{OSC_H} between the OSC pin and resonant capacitor terminal.

The total conduction losses in primary side are related to the conduction loss in the HS GaN HEMT P_{cond_HS} and LS GaN HEMT P_{cond_LS}. A portion of the total conduction power

loss is the result of the I_{zcd} current, which is mandatory to achieve ZVS in each switching cycle.

The C_{OSS} of GaN HEMT is four times lower than that of Silicon Junction MOSFET with similar R_{DS_ON} and hence the need of I_{ZCD} gets significantly lowered.

The design details of LCC converter have been widely discussed in [1].

E. Transfomer Size, Key components and pracital aspects.

In a Flyback converter, the transformer stores the energy as magnetic flux. For ferrite materials, the flux density (B) is practically limited to 0.3T. The physical size of the transformer is directly proportional to the energy it stores, the saturation current and the inductance value. The operation of a 2SW-FB converter is like that of a single-switch Flyback converter and so is the transformer size. Conversely, in an AHB-FB converter, the energy is stored both as charge in the resonant capacitor and as magnetic flux in the transformer according to (19) and (20).

$$E_L = \frac{1}{2} \cdot L_m \cdot I_{m\_peak}^2 \quad (19)$$

$$E_C = \frac{(I_{m\_peak} \cdot T_{on\_HS})^2}{8 \cdot C_r} \quad (20)$$

$$I_{m\_peak} = \frac{(V_{in} - V_{cr})}{L_m} \cdot T_{on\_HS} \quad (21)$$

For simplicity , it is assumed that the current at the beginning of turn-on instance of the HS GaN (T_{on_HS}) is zero and the voltage at Cr remains constant.

In the AHB-FB, the transformer needs to store less energy. A significant amount is also stored in the resonant capacitor. For a similar range of switching frequencies , the 2SW-FB magnetizing inductance is180uH while for the AHB is 80uH. This smaller inductance and energy storage requirement result in a smaller core compared to the 2SW-FB converter. The LCC transformer requires higher inductance, physical separation between windings to integrate significant leakage inductance and needs two symmetrically wound windings on the secondary side, which poses a size constraint on the transformer's footprint. Fig. 4 has the pictures of transformers used in experimental boards for each topology.

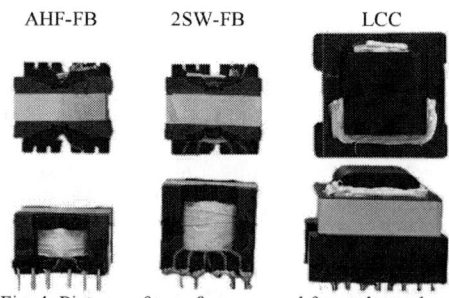

AHF-FB   2SW-FB   LCC

Fig. 4. Pictures of transformers used for each topology.

Experimentally, the RMS current in the primary of the AHB-FB converter is higher than that in the 2SW-FB converter, 1.9 A and 1.29 A respectively. For the LCC converter, the RMS current in the primary was measured to be 1.85 A, which is comparable to that in the AHB-FB converter.

The RMS current in the secondary of the 2SW-FB converter is slightly higher, at 11.1 A, compared to 10.67 A in the AHB-FB converter. In contrast, the RMS current in the secondary windings of the LCC converter is much lower. Notably, for a rating of 10 A output current, the RMS current in the SR MOSFET for both the AHB and 2SW-FB converters is significantly higher than that in the LCC converter with the full-wave rectifier configuration.

While the difference in copper losses between the AHB-FB and 2SW-FB converters is minimal, there is a three-fold difference in core losses due to the size, with the AHB-FB transformer being smaller and more efficient. Table II summarizes the number and size of key component used in each power topology in consideration.

TABLE II. KEY COMPONETS IN EACH POWER TOPOLOGY.

Particulars	2SW-FB	AHB	LCC
GaN HEMT	2	2	2
Resonant Capacitor	0	1	2 to 3
SR MOSFET	1	1	2
Transformer Size	Medium PQ3232	Smallest PQ3221	Large EE42/42/15
Clamping diodes	2	1	0

All the 3 topologies use two primary GaN HEMT and both the Flyback converters uses one less secondary side rectifier as compared to LCC topology which uses 3 resonant capacitors. AHB_FB also needs one resonant capacitor.

Additionally, the 2SW-FB needs a very low forward voltage drop diode in parallel with the SR MOSFET to improve efficiency. This is because there is a short delay before the SR MOSFET gate signal is applied by the controller to starts its conduction. The peak of the current during this time can be as high as 25Amps in certain conditions. This is not a problem with AHB-FB since the current shape resembles that of a half sinusoid and begins with zero value. However, for optimized performance, the AHB-FB needs a SR MOSFET with very low C_{OSS}. The C_{OSS} of SR MOSFET contributes to the current dip effect that increases the RMS values of current in transformer. The effect is discussed in [10]. The current dip effect limits from using the lowest R_{DS_ON} MOSFET for the required breakdown voltage.

For the AHB-FB and the LCC, the presence of a resonant capacitor connected to power ground allows the use of a capacitive divider circuit for primary side current sensing This type of sensing circuit eliminates the power loss in current sense resistance (R_CS) used in the 2SW-FB, as shown in Fig. 2.

The characterization parameters of the GaN SiP for AHF-FB and LCC and discrete GaN used in 2SW-FB converter are summarized in Table III. Fig. 5 shows the PCBs used for the comparison. The HB GaN SiP is placed on 40x30mm Daughter Board(DB) PCB modules along with the circuit necessary for the gate driver bias voltages. For the 2SW-FB, discrete GaN devices were used. This requires the use of a galvanically isolated gate driver for closely matched operation required to drive for the HS and LS switches

synchronously. Both the daughter boards discussed were designed with a two layers structure and 75μm copper thickness.

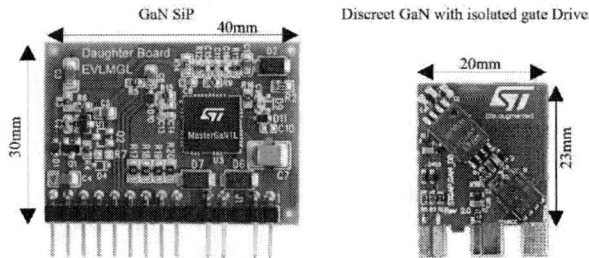

Fig. 5. Daughter board PCBs for HB GaN SiP and GaN HEMT.

TABLE III. CHARACTERIZATION PARAMETERS OF GaN

Parameter	MASTERGAN1L	SGT120R65AL
V_{BR_DS}	650V	650V
Static R_{DS_ON} @ 25°C	150mΩ	75mΩ
Static R_{DS_ON} @ 125°C	330mΩ	185
C_{OSS_TR}	50pF	83pF
Q_G	2nC	3nC
V_{GS_TH}	1.7V	1.8V

III. EXPERIMENTAL RESULTS

A. Prototype board Description

The battery chargers based 2SW-FB, AHB-FB discussed in the previous section were implemented with constant current and constant voltage feedback to validate the concept and evaluate the performance. The performance data for LCC was extracted from an optimally designed prototype board that was used for an earlier study[1]. A picture of the GaN based 2SW-FB and AHB-FB is shown in Fig. 6.

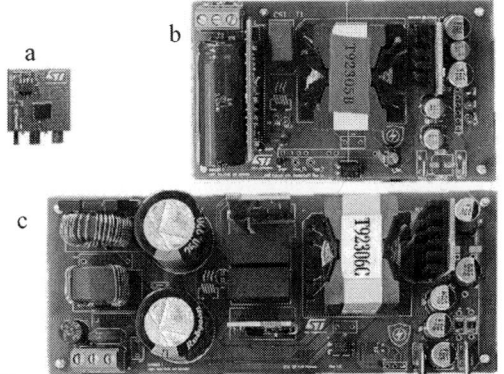

Fig. 6. Picture of the SR DB(a), AHB-FB(b), 2SW-FB with input filter and rectifier(c).

The HV GaN daughter boards are mounted vertically on the main power board. The R_{TH_J-A} in case of GaN SiP was measured at 40°C/W and 50°C/W in case of discrete GaN. The EMI and rectification components are placed on the left side of the main board. The controller and all the additional circuitries are mounted on the bottom of the main PCB for each converter. In AHB-FB, the transformer is based on a PQ3221 geometry core while the 2SW-FB has a PQ3232 to ensure a compact design.

979-8-3315-1612-3/25 $31.00 © 2025 IEEE

TABLE IV. DESIGN PARAMETERS OF EACH CONVERTER

Particulars	2SW-FB	AHB	LCC
Turns Ratio Np/Ns	7:01	5:01	7.5:1
Primary Inductance, Lp	180uH	80uH	660uH
Leakage Inductance, Ls	3% of Lp	3% of Lp	160uH
Transformer Core	PQ3232	PQ3221	EE42/42/15
Current Sense Resistor	220mΩ	-	-
Resonant Capacitor, Cr	-	680nF	47nF

B. Key Waveforms, Plots and Thermal Results

Fig. 7 shows the switching waveforms of the transformer primary current (yellow trace) together with the gate control signal of the LS device (pink trace) and Drain Source voltage(V_{DS}) of LS GaN switch (blue trace) of the AHB-FB and 2SW-FB converter operating with an input of $V_{DC} = 325V$ and output of 22V at 10A. V_{DS} was captured with capacitive coupling only to see ZVS and Valley Switching without the probe touching the electrical node.

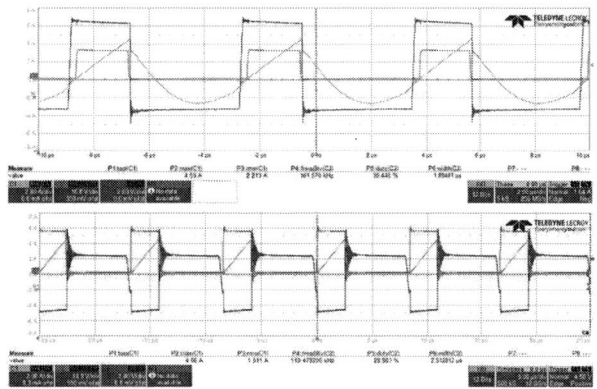

Fig. 7. Primary Side Switching waveforms of AHB-FB(top) and 2SW-FB(bottom).

In Fig. 8 blue trace is the current in the secondary winding, the pink trance is the gate signal of the SR MOSFET applied by the rectification controller.

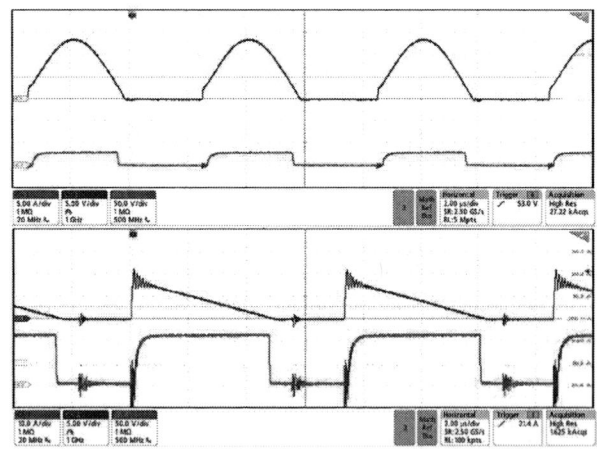

Fig. 8. Secondary Side Switching waveforms of AHB-FB(top) and 2SW-FB(bottom).

It is worth observing how the current shape differs in both the Flyback topologies, being half sinusoidal for AHB and a saw tooth for the 2SW-FB. V_{DS} was captured with capacitive

coupling only to see ZVS and Valley Switching without the probe touching the electrical node.

Fig. 9 has the plot of efficiency versus the output voltage for different changing currents for the AHB-FB supplied by a constant DC source set at 325Volts. The same was plotted for the 2SW-FB and shown in Fig. 10. It is to be noted that circuit bias supply consumption was not included in the power measured for this comparison. The efficiency numbers were captured for all the three converters with same R_{DS_ON} GaN HEMT on the primary side.

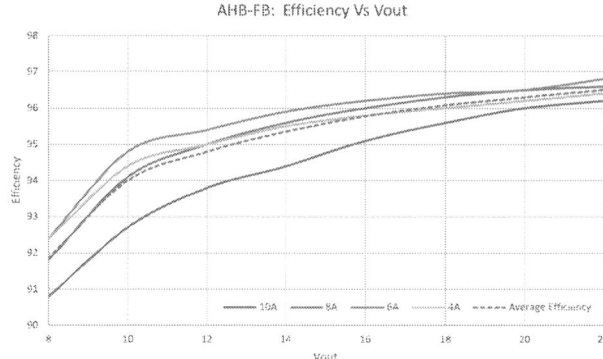

Fig. 9. Efficiency of AHB at different output voltages and current.

The green dotted trace is the average efficiency of AHB plotted for each output voltage.

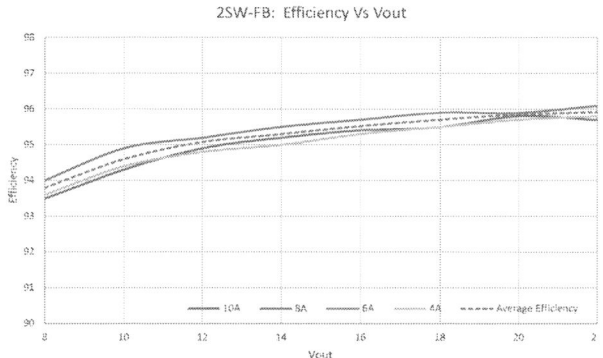

Fig. 10. Efficiency of 2SW-FB at different output voltages and current.

The peak efficiency in case of AHB-FB is 96.8% and occurs at 22V/8A and 96.1% for 2SW-FB running in the same load conditions. The average efficiency of the 2SW-FB is 95.2% which is slightly higher than that of AHB-FB measured at 95.0%. The LCC lacks behind through the operating range where the peak is about 94% and average is 88.0%.

Fig. 11 presents a bar graph illustrating the average and peak efficiency of each converter at four different output voltage levels. While the performance of all three converters is comparable at maximum output voltage, the LCC topology begins to lag in efficiency as the output voltage decreases. The average and peak efficiency of the LCC topology continue to decline with decreasing output voltage. This decline is due to increased switching losses resulting from a higher switching frequency required to reduce the output voltage. In contrast, the QR 2SW-FB and ZVS AHB-FB converters experience a reduction in switching frequency as the output voltage decreases, maintaining better efficiency.

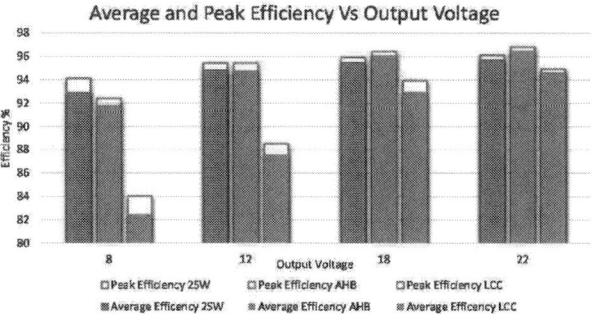

Fig. 11. The average and peak efficiency for each GaN based topology for different output voltage.

Table V contains the measured values of current frequency and duty cycle of each converter running at same input and output conditions.

TABLE V. MEASURED CURRENTS AND OPERATING CONDITIONS

Parameter	2SW-FB	AHB	LCC
I_{p_pk} (A)	3.86	4	1.8
I_{p_rms} (A)	1.29	1.9	1.75
I_{s_pk} (A)	23.4	17.5	12.5
I_{s_rms} (A)	11.1	10.6	8.8
F_{sw} (Khz)	137	169	93
Duty, D (%)	30	29	50

The 2 pie charts in Fig. 12 are the estimated loss distribution for the AHB and 2SW-FB Flyback converter. The losses were estimated for the same operating conditions from the measured parameter.

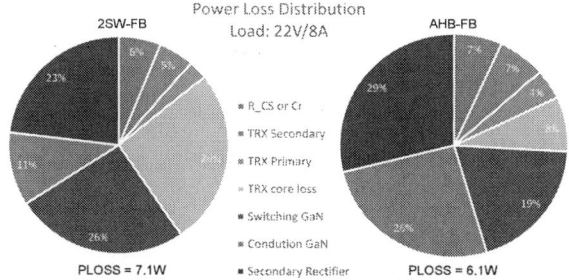

Fig. 12. Power Loss Distribution for $V_{DC} = 325V$

The thermal images in Fig 13. were captured for the converter running at same load condition. It can be observed that SR MOSFET in AHB is significantly hotter because of higher R_{DS_ON} related losses than that of 2SW-FB where the losses are also shared by a diode placed on the bottom of the main board. The transformer in case of AHB-FB is cooler than in 2SW-FB thanks to smaller core volume and lower flux swing. The temperature of the GaN switch is comparable in both AHB-FB and 2SW-FB configurations. Despite the higher losses in 2SW-FB, the presence of two packages allows for more effective power dissipation.

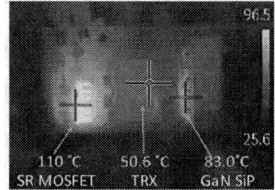 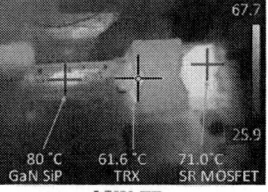

AHB-FB   2SW-FB

Fig. 13. Thermal map of AHB-FB and 2SW-FB with same GaN SiP on primary side.

The key criteria for comparing converters fall into three main categories: Performance, Cost, and Size. Fig. 16 is the radar chart for visualizing the performance and design comparison of GaN-based converter topologies.

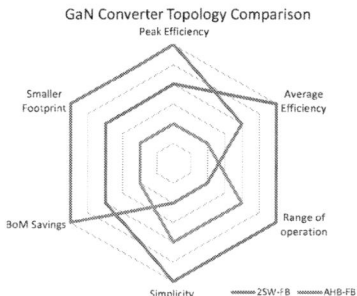

Fig. 16. Radar Chart for comparison of the 3 topologies

To summarize Fig 16, the blue line represents the 2SW-FB topology, which is the easiest to design and optimize, offering the highest average efficiency and a good balance in other criteria. The orange line represents the AHB-FB topology, which achieves the highest peak efficiency and offers a compact design, but optimization requires a digital PWM controller and high design effort. The gray line represents the LCC topology, which benefits from widely available resonant controllers but falls behind in efficiency, cost, and footprint compared to the flyback topologies.

IV. CONCLUSION

The paper outlines the design, implementation and comparison of 3 topologies for a low power battery charging solution using GaN technology. It discusses the main design parameters that are highly valued by manufacturers and consumers of such systems. The study proposes three topologies: Asymmetrical Half Bridge Flyback (AHB-FB), Resonant LCC, and 2-Switch Flyback (2SW-FB) that enable power delivery over a wide range of load conditions.

The study concludes that GaN-based power converters, particularly the AHB-FB and 2SW-FB topologies, offer significant advantages in terms of efficiency, size, and cost for low-power battery charging applications. The use of GaN HEMTs enhances performance, reduces losses, and supports the development of more efficient, cost-effective, and compact power converters. The findings underscore the potential of GaN technology to revolutionize power conversion in battery charging applications, making it a valuable choice for manufacturers and consumers alike.

REFERENCES

[1] R. Attanasio, H. Ademane, J. Hyslop and G. Vitale, "A GaN Based LCC Converter for Lithium-Ion Battery Chargers," 2023 IEEE Applied Power Electronics Conference and Exposition (APEC), Orlando, FL, USA, 2023, pp. 2489-2495, doi:10.1109/APEC43580.2023.10131630.

[2] S. De Simone, C. Adragna and C. Spini, "Design guideline for magnetic integration in LLC resonant converters," 2008 International Symposium on Power Electronics, Electrical Drives, Automation and Motion, Ischia, Italy, 2008, pp. 950-957, doi: 10.1109/SPEEDHAM.2008.4581225.

[3] L. Huber and M. M. Jovanović, "Analysis, design, and performance evaluation of asymmetrical half-bridge flyback converter for universal-line-voltage-range applications," 2017 IEEE Applied Power Electronics Conference and Exposition (APEC), Tampa, FL, USA, 2017, pp. 2481-2487, doi:10.1109/APEC.2017.7931047.

[4] Li-Ming Wu and Chen-Yin Pong, "A half bridge flyback converter with ZVS and ZCS operations," 2007 7th Internatonal Conference on Power Electronics, Daegu, Korea (South), 2007, pp. 876-882, doi: 10.1109/ICPE.2007.4692511.

[5] "170W high input voltage two switch flyback based on L6565 and 1500V K5 MOSFETs," STMicroelectronics, AN5287.

[6] "600 V half-bridge enhancement mode GaN HEMT with high voltage driver," MASTERGAN1L Datasheet, STMicroelectronics, DS14355.

[7] Y. Panov and M. M. Jovanovic, "Adaptive off-time control for variable-frequency, soft-switched flyback converter at light loads," in *IEEE Transactions on Power Electronics*, vol. 17, no. 4, pp. 596-603, July 2002, doi: 10.1109/TPEL.2002.800958.

[8] Han Li, Wenjun Zhou, Shiping Zhou and Xiao Yi, "Analysis and design of high frequency asymmetrical half bridge flyback converter," 2008 International Conference on Electrical Machines and Systems, Wuhan, China, 2008, pp. 1902-1904.

[9] G. Spiazzi and S. Buso, "The Asymmetrical Half-Bridge Flyback Converter: a Reexamination," 2020 IEEE Energy Conversion Congress and Exposition (ECCE), Detroit, MI, USA, 2020, pp. 405-411, doi:10.1109/ECCE44975.2020.9235370.

[10] M. Li, Z. Ouyang and M. A. E. Andersen, "Analysis and Optimal Design of High-Frequency and High-Efficiency Asymmetrical Half-Bridge Flyback Converters," in IEEE Transactions on Industrial Electronics, vol. 67, no. 10, pp. 8312-8321, Oct. 2020, doi:10.1109/TIE.2019.2950845.

Server motherboard power performance study under immersion cooling environment

Meng Wang
Sales and Marketing Group
Intel Corporation
Shanghai, PRC
meng.wang@intel.com

Haiyan Wang
R&D Center
Nettrix Information Industry Corporation
Beijing, PRC
grace.wang@nettrix.com.cn

Pavan Kumar
Data Center Power Solutions
Intel Corporation
Hillsboro, OR, USA
pavan.kumar@intel.com

Haijin Zhang
Data Center Power Solutions
Intel Corporation
Shanghai, PRC
haijin.zhang@intel.com

Xiang Li
Sales and Marketing Group
Intel Corporation
Shanghai, PRC
xiang.j.li@intel.com

Fengwei Bian
Sales and Marketing Group
Intel Corporation
Shanghai, PRC
fengwei.bian@intel.com

Jianting Deng
R&D Center
Nettrix Information Industry Corporation
Beijing, PRC
justin.deng@nettrix.com.cn

Jiaqi Zhu
R&D Center
Nettrix Information Industry Corporation
Beijing, PRC
jackie.zhu@nettrix.com.cn

Yiming Lei
R&D Center
Nettrix Information Industry Corporation
Beijing, PRC
francis.lei@nettrix.com.cn

Abstract—**This paper provided a holistic analytical and experimental power performance impact study on a single-phase immersion cooling two socket server system, included board and system level performance comparison analysis between air and immersion cooling. With proven benefit in this study, cost down opportunities are analyzed to show case that design optimization can be achieved for immersion cooling environment-based server motherboard power design, which will greatly help total cost of ownership (TCO) of datacenter and improve product competitiveness.**

Keywords—datacenter, server motherboard, single-phase immersion cooling, CPU, voltage regulator, cost down.

I. Introduction

With more energy consumption demand from datacenter to meet high computing performance needs, such as AI, cloud computing, autonomous driving, cloud storage for managing business etc., high power server system trends to apply immersion cooling technology to improve end to end efficiency and reduce carbon footprint.

Current immersion cooling server motherboard power design in industry is generally leveraged from traditional air-cooling design without optimization and cost ineffective, corresponding validation process and performance data are leveraged as well, industry does not have very detailed and specialized research in immersion cooling environment based sever motherboard power performance and validation process.

To address above challenges, this paper investigated power performance impact through lab experiments and power simulation based on an Intel eagle stream (EGS) platform two socket 350W CPU server system under single-phase[1] immersion cooling environment. This study included motherboard and closed chassis server system level power performance comparison between air and immersion cooling, corresponding design optimization opportunities, such as cost saving are discussed as well. This paper is intended as a proof of concept to show the potential benefits of reduced temperature on the power performance due to immersion cooling. It is a first step in establishing further study of the VR behavior and potential benefits under immersion cooling environments.

It is shown from result that, compared with air cooling, server motherboard in immersion cooling can achieve 4.7% reduction of dc resistance of CPU core power delivery path (Rpath); 0.68% increasement of CPU core power rail VCCIN voltage regulator (VR) efficiency and 15W+ input power saving per one CPU socket full load stressing; server system can achieve up to 106W input power saving; which can greatly improve datacenter energy efficiency. Meanwhile, thermal study indicates most components in sever system are less than 50°C, which will significantly drive sever motherboard power design optimization. By capacitor cost down investigation, it can help cost saving around $20 per board, which greatly help total cost of ownership (TCO) of datacenter and improve product competitiveness.

II. Experiment and Result

This section describes the experimental setup methodology to evaluate power performance impact in single phase[1] immersion cooing environment with desired result provided. Test units are Intel eagle stream platform-based server motherboard and close chassis system, they will be put into immersion cooling tank to explore crucial performance items as follows: CPU core power rail VCCIN VR performance test, Rpath, efficiency, input power saving and thermal. Test data will be compared and analyzed with that in air cooling to demonstrate real benefits brought in immersion cooling.

Single phase immersion cooling tank coolant take use of Noah3000, a type of fluorinated cooling liquids, liquid temperature in the tank is around 30°C.

A. Motherboard level power performance comparison

An Intel eagle stream platform based two socket server motherboard is used for testing the performance difference between immersion cooling and air cooling as Fig1. Shown.

This is only board level performance comparison without processor in socket. Put an Intel Gen5 VRTT[2] in one CPU socket to execute CPU VR power performance test, no other

979-8-3315-1612-3/25 $31.00 © 2025 IEEE

external device installed in motherboard, experiment is based on same motherboard, VRTT, power supplier unit (PSU), which are leveraged using in both air and immersion cooling validation, that means an apple to apple comparison.

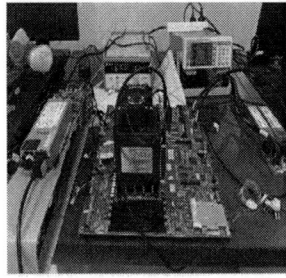

Fig. 1. Air cooling(left) and Immersion cooling (right) motherboard level test setup.

1) CPU VCCIN VR performance

In general, Intel server platform CPU VR test plan is used as a fundamental document for server original design manufacturer(ODM) /original equipment manufacturer (OEM) to evaluate VR performance. Here we take use of eagle stream platform CPU VR test plan[3] and follow intel required test procedure in this document. The specific testing process is as follows, firstly taking use of Gen5 VRTT and run Intel VR test software to validate motherboard one socket side VCCIN VR test to check VR performance in air cooling, then submerging the same motherboard in a dielectric fluid in a tank, carefully connecting test point with external test instruments outside the tank, and run test software again for the same socket side VR to get performance data under immersion cooling condition.

It is shown from test results that all test items passed Intel power spec, VR output voltage met CPU Vmin and Vmax voltage requirement, even pass spec with more margin under immersion cooling test condition. Among these results, it is worth highlighting the following performance advantages.

a) VR current Iout report error rate in immersion cooling is lower than motherboard in air cooling condition from light load to heavy load shown as Fig.2. This will correspond to an improvement in CPU turbo performance accuracy, and higher power CPUs will benefit more.

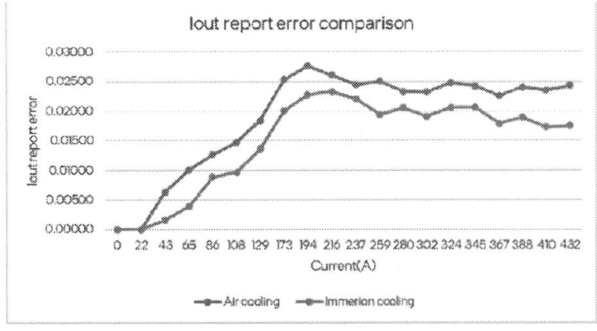

Fig.2. VCCIN VR Iout accuracy comparison

b) VR transient response performance is improved.
VR output with 1.83V, load current dynamically changed from 60A to 430A, 10% to 50% duty cycle, 305hz to 1Mhz frequency sweeping, with these test condition , it is shown

from validation result that both Vmin and Vmax pass Intel spec with additional margin, such as Vos(overshoot) improves 6 mv compared with air cooling as Fig.3 shown, which means more design margined can be obtained.

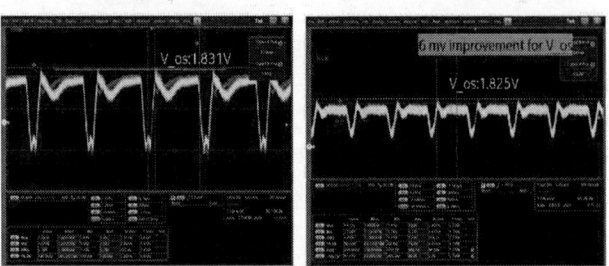

Fig.3. VCCIN transient performance comparison

2) CPU VCCIN Rpath, efficiency and system input power saving

VCCIN Rpath usually refers to the impedance path of CPU core power rail including print circuit board (PCB), package and socket, low Rpath will uplift CPU performance and bring more PCB power loss saving.

In this board level study, we measured VCCIN PCB Rpath from VR output to CPU remotes sense under different workload generated by Gen5 VRTT. Meanwhile, VR efficiency is tested, and board input power information is recorded by input power meter connected with motherboard PSU side.

Compared with air cooling, at VCCIN thermal design current (TDC[2]) level workload, the experimental results show that VCCIN Rpath at remote sense point can reduce 4.71%, VR efficiency increase 0.68%, system input power saving reduce 15.2W.

3) VCCIN thermal

To prove how much benefits brought from thermal side, this study measured temperatures at different locations in server motherboard, thermo-couples were placed throughout top side and bottom side of VCCIN VR power delivery area, included VR phase area, power stage, inductor, PCB interface and CPU socket cavity side and etc. shown as Fig.4.

As we can see from the result, at VCCIN TDC level workload, among the 11 locations, temperatures can reduce up to 54.1°C at bottom power stage area in immersion cooling environment and overall temperature is lower than 50°C, mostly is 35°C around.

Location	100%TDC		
	Air cooling	Immersion cooling	Improvement
1.Top- Phase1 power stage	74.8℃	33.5℃	41.3℃
2.Top- Phase1 inductor	67.8℃	34.8℃	33℃
3.Top- Phase5 power stage	95.8℃	46.9℃	48.9℃
4.Top- Vccin power corridor area	71.9℃	35.7℃	36.2℃
5.Bot- Between phase4&5	97.1℃	43℃	54.1℃
6.Bot- Phase1 input MLCC decoupling	69.2℃	33.3℃	35.9℃
7.Bot- Phase output spcap	80.5℃	35.6℃	44.9℃
8.Bot- Vccin power corridor area	77.3℃	36.5℃	40.8℃
9.Bot- Vccin edge cap	66.3℃	39.1℃	27.2℃
10.Bot- Vccin cavity cap	49.8℃	32.2℃	17.6℃
11.Bot- Vccin cavity cap	49℃	33.5℃	15.4℃

Fig.4. VCCIN thermal result at TDC level and test setup

These low temperatures data further support previous section test result mentioned low power loss with additional efficiency improvement benefits due to good heat dissipation, and it means to have potential opportunities for VR design optimization, such as cost down. This will be discussed in the following cost down analysis section.

B. Closed chassis system level performance comparision

To make the experimental conclusions more convincing and informative, this section analyze actual product level performance comparison between immersion cooling and air cooling in a closed chassis fully functional system with processors and other devices installed.

Two Intel eagle stream platform-based server closed chassis systems with air-cooling design and immersion cooling design are being studied shown in Fig5. Two systems leveraged same CPU, memory, hard drive, and other devices. Chassis, fan, power supply unit (PSU) and motherboard cannot be leveraged due to different design in these two systems and difficult operations to implement. Using Intel software stressing CPU workload to 100% thermal design power (TDP) 305W level.

Fig.5. Air cooling and Immersion cooling system level test setup

1) VCCIN Rpath, efficiency and system input power saving

Rpath is measured from VR output to far end point as Fig.6 right side bottom picture, the results show a 0.36% reduction, and system input power can achieve 106.1W saving in immersion cooling. The main reason of minor improvement is, air cooling system has extra fan which drives these data to be very closed, and special heatsink design helps CPU VR thermal dissipation a lot as Fig.6 right side top picture shown. Therefore, it is necessary to consider the impact of system fan design on Rpath performance when conducting such comparative experiments.

2) VCCIN thermal

Test Location	PTU Stress with 100% TDP	
	Air cooling	Immersion cooling
1.Top- Phase4 power stage	55.3°C	46.4°C
2.Top- Phase4 inductor	52.0°C	42.0°C
3.Top- Phase5 power stage	59.4°C	44.3°C
4.Top- Vccin power corridor area	55.0°C	49.9°C
5.Bot- Between phase4&5	57.4°C	45.4°C
6.Bot- Between phase5&6	56.5°C	45.5°C
7.Bot- Vccin power corridor area	56.8°C	40.2°C
8.Bot- Vccin edge cap PC457	57.1°C	37.1°C
9.Bot-Vccin cavity cap PC433	67.9°C	41.3°C
10.Bot- Vccin cavity cap PC278	65.0°C	35.0°C

Fig.6. VCCIN thermal result at CPU TDP level and test setup

Similar to board level test, this study measured temperatures at various locations in close chassis server system by placing several thermo-couples throughout top side and bottom side of VCCIN VR power delivery area. Among those locations, temperatures can reduce up to 30°C and major components temperature are less than 50°C in immersion cooling.

C. PI Simulation

In addition to experiment results, CPU VCCIN DC Rpath and AC impedance simulation were conducted based on above EGS sever motherboard, which proved similar power performance improvement trend in immersion cooling as below Fig.7 shown.

Under immersion liquid cooling and air cooling environment, temperature is known to be the main influencing factor for motherboard performance, different temperatures are chosen to simulate these two environments. Using 90°C as worse case temperature for air cooling condition, 60°C for immersion cooling condition. All components and board models are built base on this two thermal setting.

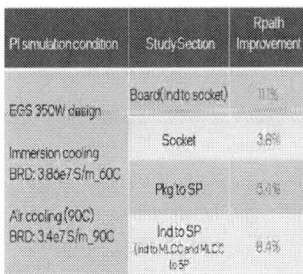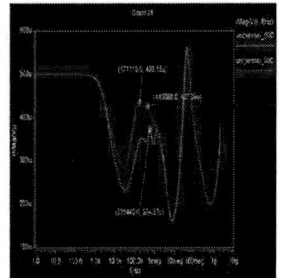

Fig.7. VCCIN DC Rpath and AC impedance simulation under different temperature.

D. Cost down opportunity overview

As demonstrated by above benefits in immersion cooling environment, which bring many potential opportunities for server motherboard power design to be optimized, such as cost down. The following aspects can be considered,

1) Multilayer ceramic capacitor type selection

There are many types of multilayer ceramic capacitor (MLCC) are designed into server motherboard, such as X7R, X5R, X6S,COG,X5R and etc. One of the indicators for selecting these capacitors is temperature characteristics, their temperature range usually is as follows, X7R(55°C ~+125°C), COG(-55°C ~+125°C),X6S(-55°C ~+105°C) , X5R(-55°C ~+85°C).

As above VCCIN thermal result in this paper indicated, whether from motherboard level or system level, most components are less than 50°C in immersion cooling at CPU TDP level, consequently, many MLCC capacitor with X7R, COG,X6S can be potentially replaced with X5R.

It is roughly estimated for the EGS platform motherboard this paper studied; it can achieve around $20 cost reduction if whole board MLCC are replaced with X5R capacitors.

2) Cap removal due to gain more design margin.

In the previous board level VCCIN VR performance study section, it demonstrated extra voltage margin benefits for transient response in immersion cooling, such as several mv improvement for Vmin,Vos and etc. From power design perspective, which can potentially help drive the reduction of number or capacity of VR decoupling and even CPU package cap.

3) Power stage heatsink removal

As we can see from thermal evaluation data, VR phase 5 power stage, the component with the highest temperature in VR all phases, from light load to high load, its temperature is far lower than air cooling test result as Fig.8 shown , so there is no need to add heatsink to cool the power stage, it can be removed in immersion cooling.

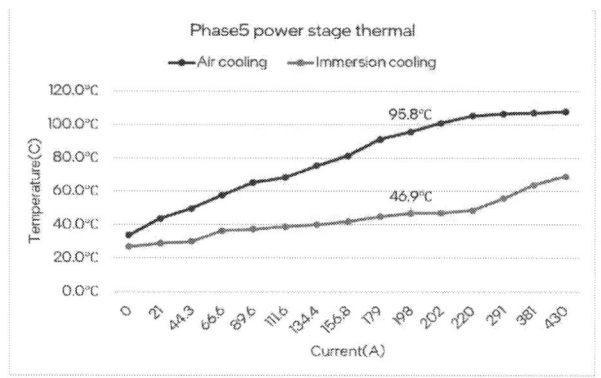

Fig.8.Board level VCCIN VR phase 5 power stage thermal measurement

4) VR phase reduction.

According to test result, VR efficiency is 90.49% at 198A in air-cooling, 90.64% at 220A in immersion cooling, that means VR in immersion cooling can afford additional 22A with similar efficiency, which is almost one more phase capability, that will potentially drive to save one VR phase for power design in immersion cooling environment.

III. Conclusion and Future work

To know the power performance impact in immersion cooling environment, you must have concrete experimental data to demonstrate, while it is very hard to execute power test in immersion cooling tank due to its environment and operation complexity. Use power test tool VRTT for validating VR performance in liquids is also a pioneering work. In addition, it is first time to do component level thermal analysis under immersion cooling environment.

This paper overcome above challenges and deeply researched power performance differences and impact between immersion cooling and air cooling from board level and close chassis system level, conduct and provide constructive suggestions to optimize power design based on proven benefits. This can be a research reference for server industry ODM/OEM to do similar study based on their specific products and optimize their power design in the future.

In addition to the power performance impact study in this paper covered, component level reliability, lifetime impact in immersion cooling is also one technical aspect that needs to be considered. Detailed analysis and experimental results to ensure that the benefits accrued are mainly due to the improvement in the power parameters and not due to tool/instrumentation benefits in immersion cooling needs to

be established. Next step, this paper will expand the research findings of this section to give a full picture view of power performance impact to further optimize immersion cooling server power design in the future.

Acknowledgment

We would like to thank Horthense D. Tamdem for providing many constructive suggestions on this paper, thank Kevin J. Whale and Dave Merritt for test related study sharing and guidance on VRTT setup in immersion cooling, thank Dawei Hu's inputs based on previous study, thank Dongrui Xue and Yuehong Fan for thermal measurement related guidance. And we also would like to thank Kun He, Zhengguo Fan, Lili Deng, Ho Wang, Gary Gao for the fully support on this proof of concept(POC)project research study.

References

[1] Suchismita Sarangi, Eric D. McAfee, Drew G. Damm and Jessica Gullbrand, "Single-phase immersion cooling performance in intel servers with immersion influenced heatsink," 38th Semiconductor Thermal Measurement, Modeling & Management Symposium, 2022.

[2] Intel, "Eagle Stream Platform Design Guide",2023

[3] Intel, "Eagle Stream Platform CPU Power Delivery Test Plan",2023

Practical PCB Design Considerations for GaN HEMTs Based Isolated DC-DC Converter

Gaureej Gauttam, Student Member, IEEE
Electrical and Computer Engineering
University of Houston
Houston, United States
ggauttam@uh.edu

Harish S. Krishnamoorthy, Senior Member, IEEE
Electrical and Computer Engineering
University of Houston
Houston, United States
hskrishn@uh.edu

Sai Sushma Pasupuleti, Student Member, IEEE
Electrical and Computer Engineering
University of Houston
Houston, United States
spasupu3@uh.edu

Abstract— This paper proposes improved PCB design techniques for converters using Gallium Nitride High Electron Mobility Transistors (GaN HEMTs). An isolated DC-DC converter designed for high-temperature (>150 ℃) applications- such as defense, oil and gas, etc.- is considered as the test case for validating the proposed concepts. The primary focus of this paper is to address the challenge of reducing voltage spikes in the drain-to-source voltage (V_{ds}) by minimizing parasitic elements, thereby improving voltage quality in high power density GaN HEMT designs. The proposed 'closed-form' layout for a half-bridge configuration effectively decreases the parasitic inductances, a significant stride in enhancing voltage stability and overall device performance.

Ansys (Q3D analysis) simulations, for the conventional designs and the proposed closed-form layout are done and results are compared. These simulation results are validated through hardware experiments, providing comprehensive confirmation of the design's effectiveness. It is shown that the proposed closed-form layout of GaN HEMT presents a substantial reduction in the V_{ds} spikes.

Keywords— GaN layout, dV/dt, Parasitic, half-bridge, high-temperature.

I. INTRODUCTION (*HEADING 1*)

In wideband gap devices, Gallium Nitride (GaN) high electron mobility transistors (HEMTs) are recognized for their compact size, high power capabilities, fast switching speeds, and low on-resistance (R_{dsON}), making them highly effective in high power density applications [1]. These attributes are particularly valuable in defense and military systems, where reliability and performance under extreme conditions are paramount. Despite their superior performance compared to previous generations of silicon (Si) and silicon carbide (SiC) devices, GaN HEMTs still present challenges related to high di/dt, causing spikes in the drain-to-source voltage (V_{ds}). This necessitates meticulous layout design to ensure operational consistency and device damages which are even more important in extreme environment and critical applications.

GaN HEMTs have been extensively tested for high-temperature (HT) applications up to 125 °C. However, there is significant potential for further research and development in HT

applications exceeding 150 °C, which could simplify the deployment of GaN HEMTs in extreme environments. Such advancements are crucial not only for military systems but also for subsurface characterization, where they could enhance the reliability of downhole drilling, geothermal energy extraction, and carbon capture, utilization, and storage (CCUS). For HT PCB layouts, various packaging options are available, with the quad flat package (QFP) 32-pin with kovar-based hermetic package being one popular option (shown in Fig. 1).

Fig. 1. GaN HEMT device (QFP 32pin Hermetic Package).

A layout design focused on minimizing stray inductances and capacitances to enhance the switching characteristics of a SiC MOSFET is discussed in [2]. In [3] and [4], a vertical power loop design for GaN devices from GaN Systems is presented, aimed at reducing switching noise and voltage overshoot. This design achieves a V_{ds} overshoot reduction of up to 17.2% for a 68V overshoot at 400V DC input. However, it faces challenges related to charge imbalance, reliability, and high common-source inductance. To address parasitic inductance, [5] introduces a vertical lattice loop structure for GaN HEMTs that increases magnetic flux cancellation, resulting in a 25% V_{ds} overshoot at 200V operation. Additionally, [6] examines the impact of PCB layout on high-frequency GaN-based converters, highlighting a design for e-GaN FETs from EPC that reduces parasitic inductance by 40%.

High-speed switching in GaN devices introduces two primary types of noise: one caused by high voltage changes (dV/dt) and the other by rapid current changes (dI/dt). Both types of noise can severely impact the performance and reliability of GaN devices, especially in high-temperature (HT) environments.

The dV/dt noise predominantly affects the gate driver circuit and can lead to gate driver failure. This issue arises when the switching paths overlap with the high-power paths in the PCB layout, creating unwanted interference. To mitigate this, the PCB layout must be designed to eliminate any overlap between these paths. Additionally, employing isolators with high common-mode transient immunity (CMTI) on the gate driver

---

The information, data, or work presented herein was funded in part by the Advanced Research Projects Agency-Energy (ARPA-E), U.S. Department of Energy, under Award Number DE-AR0001582

side significantly reduces susceptibility to dV/dt-induced failures.

On the other hand, dI/dt noise affects the main power path and can result in device shoot-through, potentially damaging the GaN devices. Mitigating dI/dt-related spikes in drain-to-source voltage (V_{ds}) at high temperatures and enhancing heat dissipation are crucial challenges, particularly for defense, military, and oil and gas applications where reliability in extreme environments is paramount.

This paper addresses these challenges by presenting a detailed layout design and optimization for GaN HEMTs in HT applications up to 175 °C. The design utilizes 32-pin Kovar-based hermetic packaging and incorporates the Tagore Technology GaN FET, 'TP44200NH,' which features GaN-on-GaN technology with an integrated driver. This FET was extensively tested in hardware experiments to validate the proposed methods.

A closed-form layout design for a half-bridge configuration is proposed to minimize parasitic inductances and reduce V_{ds} voltage spikes and overshoot, ensuring reliability under high-stress conditions. By adopting a modular approach, this design can be replicated across multiple stages or legs of an isolated DC-DC converter. Key design features include careful placement of bypass capacitors to address high dI/dt noise, optimized power loop placement to minimize via-induced inductance, and high-power trace isolation to prevent dV/dt noise coupling.

To enhance thermal management, the design explores strategic placement of thermal vias, pads, and copper pours. These elements facilitate efficient heat dissipation, critical for maintaining device performance and longevity in demanding environments. Furthermore, the paper compares the parasitic effects and V_{ds} voltage spikes observed in lateral, vertical, and proposed layouts. Simulation and experimental results validate the advantages of the proposed layout in minimizing noise and thermal issues, making it a robust solution for HT GaN applications in mission-critical fields.

The most commonly used topologies for high-temperature (HT) applications, such as downhole drilling, geothermal energy extraction, CCUS, and other extreme environment scenarios, typically employ isolated DC-DC converter designs, including half-bridge or full-bridge configurations. These topologies are often developed to power amplifiers, extending their utility to critical applications such as MRI machines in the healthcare sector [7]. Consequently, this paper uses the half-bridge topology as a case study, with all comparisons and analyses conducted based on this configuration.

The organization of this paper is as follows: Section II provides a detailed discussion on the factors affecting the performance of HT GaN HEMTs. Section III presents a comparison of the proposed layout topology with widely used pre-existing topologies, analyzed using ANSYS Q3D. Section IV highlights the experimental results, demonstrating the benefits of the proposed layout. Finally, Section V concludes the work and outlines potential future directions for this application.

II. FACTORS AFFECTING HT GAN HEMTs PERFORMANCE

This section will cover the mathematical and theoretical analysis for both parasitic inductance and heat dissipation techniques.

A. Parasitic Inductance

The parasitic inductances can cause a high V_{ds} voltage spike during the turn ON and turn OFF transient of the GaN FETs. The V_{ds} voltage spike can be estimated as per the equation (1):

$$V_L = L_p \frac{dI}{dt} \qquad (1)$$

where the V_L is the voltage spike due to the parasitic inductance L_p, the current I rising from zero to the rated value during the transition time t.

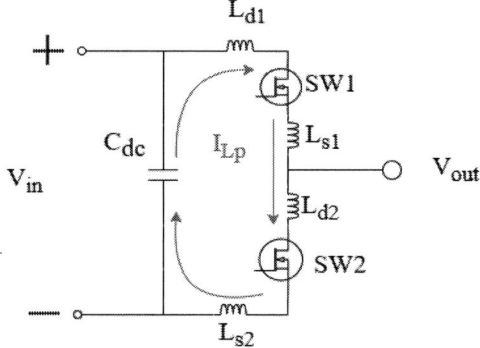

Fig. 2. Parasitic inductance in the power loop of a half bridge converter.

In a half-bridge configuration, all possible parasitic inductances in the power loop are illustrated in Fig. 2. Here, where L_{d1}, L_{s1} represent the drain and source-side parasitic, while L_{d2}, L_{s2} correspond to those of switch 2 (SW2). The change in current during switching (dI) can vary from a few milliamps to several amps, depending on the application. For this study, a current change of 1–2 A is considered.

Despite the relatively small dI of 1–2 A, GaN FETs exhibit an exceptionally short transition time (dt) of only a few nanoseconds, resulting in significant voltage spikes. These spikes are directly proportional to the parasitic inductances in the power loop, as described by Equation (1). Therefore, minimizing parasitic inductance in the layout design is critical to ensuring reliable operation.

The total parasitic inductance of the power loop (L_p) can be calculated using [8]:

$$L_p \cong \mu_0 \mu_r \frac{d}{w} l \left(\frac{0.27}{1 - 0.74 e^{-0.45(d/w)}} \right) \qquad (2)$$

Here, l and w represent the length and width of the trace wire, which determine the physical dimensions of the conductor within the PCB. These dimensions directly influence the inductance and resistance of the trace. The parameter d refers to the distance between two adjacent current-carrying paths, which impacts the magnetic coupling and mutual inductance between the traces. A larger d reduces coupling, while a smaller d increases it, potentially causing unwanted interference. μ_0 is the permeability of free space and μ_r is the relative permeability of material.

979-8-3315-1612-3/25 $31.00 © 2025 IEEE

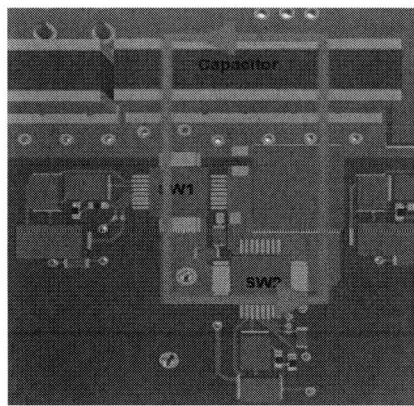

Fig. 3. Proposed closed-form layout for HT GaN HEMTs with hermetic package.

To address this, the proposed layout, shown in Fig. 3, reduces parasitic inductance by shortening the overall path length (l), as described in Equation (2). Such spikes can significantly impact the performance and reliability of GaN-based systems, especially in high-temperature and high-frequency applications. Additionally, the layout mitigates inductance introduced by vias, which typically contribute a few nanohenries. By placing the power loop entirely on a single PCB layer, the design avoids the need for vertical current transitions between layers, which can otherwise increase loop inductance and degrade system performance. This approach ensures a more streamlined and efficient current flow, directly improving switching characteristics and reducing electromagnetic interference (EMI).

Furthermore, a strategically placed small bypass capacitor near the FETs addresses high-frequency noise caused by rapid current changes (dI/dt) during device switching. This capacitor acts as a local energy reservoir, absorbing transient currents and suppressing oscillations that can otherwise propagate through the circuit, potentially causing gate driver failures or signal distortions.

The simulation section will provide a detailed analysis of the optimized layout design and parasitic reduction using ANSYS Q3D for the half-bridge configuration.

B. Heat Dissipation

In GaN-based power converters, effective heat dissipation is critical for ensuring reliable operation and maintaining device performance, especially under high-power and high-temperature conditions. Conventional cooling methods, such as external fan cooling and large heat sinks, are often employed to manage thermal stress. These are typically supplemented by thermal pads and vias integrated into the PCB design to further facilitate heat dissipation.

Experimental findings have revealed that thermal vias should be exclusively dedicated to heat dissipation rather than carrying electrical current. When current flows through these vias, it can introduce additional resistive heating, thereby negating their intended thermal management benefits and increasing the overall temperature of the device.

An effective enhancement involves the application of a copper pour on both sides of the thermal vias, as depicted in Fig. 4. This approach increases the surface area available for heat conduction, allowing for more efficient thermal transfer to ambient or connected heat sinks. Additionally, maintaining an uninterrupted connection between the copper pour and the vias improves heat flow, ensuring that hotspots are minimized. Integrating these strategies into a new layout design resulted in a temperature reduction of approximately 10 °C under 1 A input current conditions.

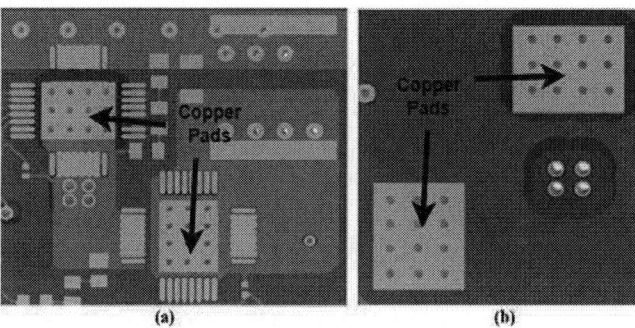

Fig. 4. PCB layout with copper pour and no internal layer connection (a) Bottom layer. (b) Top Layer.

III. SIMULATION AND COMPARISON

Ansys Q3D simulations were conducted to calculate parasitic inductances for various layout configurations, enabling a detailed comparison of their effectiveness in minimizing parasitic effects. Figure 4 illustrates the layout configurations analyzed during these calculations, each designed to emphasize specific power loop arrangements.

The analysis revealed that the conventional lateral power loop layout (Fig. 4(a)) exhibited a parasitic inductance of 17.24 nH, while the conventional vertical power loop layout (Fig. 4(b)) showed a slightly higher inductance of 21.24 nH. The previously optimized mixed-layout configuration, combining elements of both lateral and vertical arrangements (Fig. 4(c)), demonstrated an improved inductance of 13.44 nH.

In stark contrast, the proposed layout configuration, shown in Fig. 3, achieved a significantly reduced parasitic inductance of 7.38 nH. This represents a notable improvement, nearly halving the inductance of the previously optimized design.

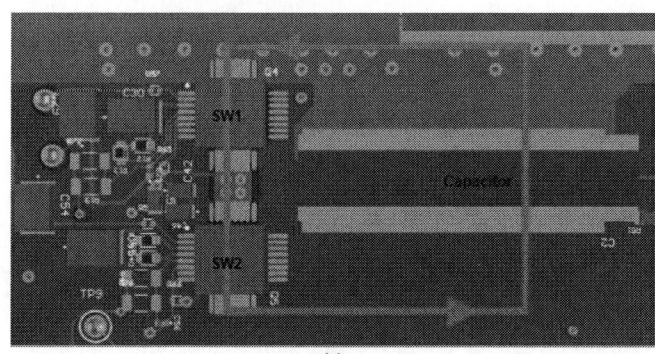

(a)

979-8-3315-1612-3/25 $31.00 © 2025 IEEE

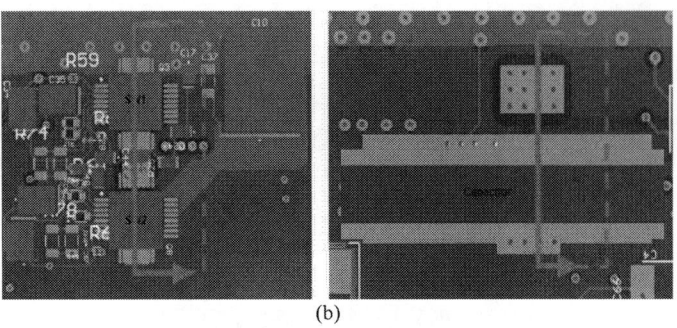

(b)

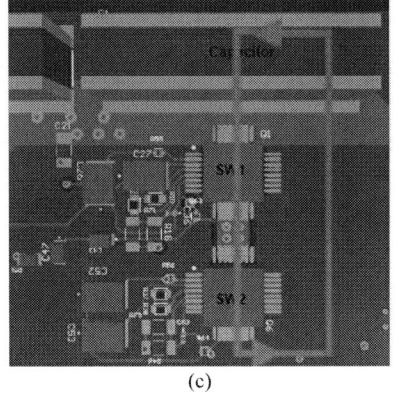

(c)

Fig. 4. Layout design and power loop for different configuration. (a) Lateral power loop (b) Vertical power loop (c) Mixed power loop.

IV. EXPERIMENTAL RESULTS

The hardware testing results of the proposed closed-form layout, compared with the conventional optimal layout, are presented in Fig. 6 (a) and Fig. 6 (b). These tests revealed a significant improvement in reducing Vds voltage spikes. Specifically, the conventional layout exhibited a spike magnitude of 70% at an operating voltage of 20 V, while the proposed layout reduced this spike to just 20%.

Furthermore, the superiority of the proposed layout became even more pronounced at higher operating voltages. When tested at 120 V under high-temperature (HT) conditions of 175 °C, the spike reduction was further minimized to only 5%, as shown in Fig. 6 (c). This underscores the scalability of the proposed layout's benefits at elevated stress conditions, which are typical in real-world high-power applications.

To ensure reliability, all hardware tests were performed at a sustained high temperature of 175 °C for a minimum duration of 20 minutes under continuous operation. This testing demonstrates the practical advantages of the proposed layout and its potential for broader adoption in high-performance power electronic converters.

As shown in Fig. 7(a), the temperature rise of the upper GaN switch ($SW1$) is observed to be 51.1 °C under a load of 1 A when there is no copper pour and the vias are conducting current. In contrast, as shown in Fig. 7(b), the temperature of the same GaN FET reduces to approximately 41.8 °C for the same load when proper thermal dissipation techniques are applied. The improved heat dissipation is achieved through the careful addition of

thermal vias and the application of a copper pour beneath the FET and on the opposite side of the PCB. These optimizations effectively reduce the temperature rise of the FET by about 10 °C, highlighting the importance of efficient thermal management in GaN-based converters.

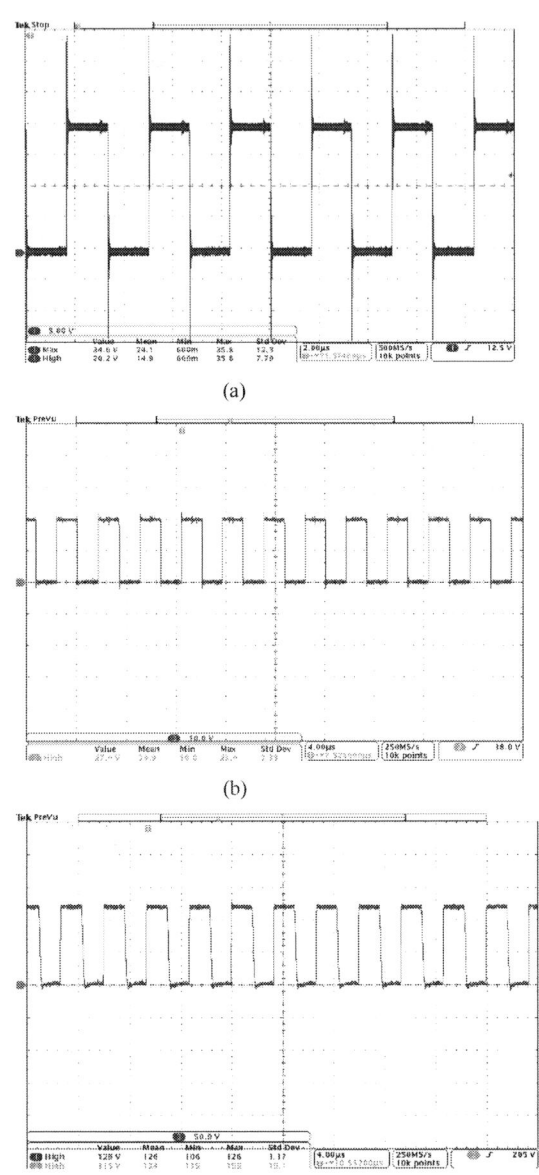

(a)

(b)

(c)

Fig. 6. (a) V$_{ds}$ for conventional layout FET V$_{ds}$ for input voltage of 20 V (V$_{ds}$ spike of 70 % of the FET voltage), (b) V$_{ds}$ for the proposed layout FET V$_{ds}$ for an input voltage of 20 V (V$_{ds}$ spike of ~10 % of the voltage), (c) V$_{ds}$ for the proposed layout FET V$_{ds}$ for an input voltage of 120 V (V$_{ds}$ spike of ~5 % of the voltage).

979-8-3315-1612-3/25 $31.00 © 2025 IEEE   3319

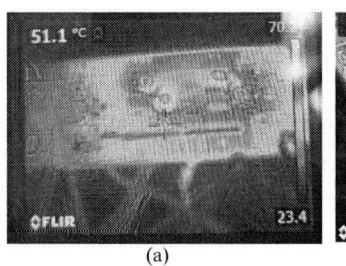 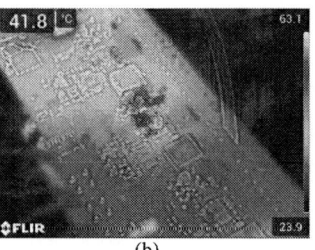

(a)   (b)

Fig. 7. (a) Temperature of the GaN FET for initial case. (b) Temperature of GaN FET with improved thermal padding.

V. Conclusion And Future Work

The proposed closed-form layout for GaN HEMTs-based isolated DC-DC converters demonstrates a substantial improvement in reducing V_{ds} voltage spikes and enhancing thermal dissipation. These features make it particularly suitable for high-temperature (HT) and critical applications in sectors such as defense, military, oil and gas, health sector and other industries where reliability under extreme environmental conditions is paramount.

Temperature measurements under identical current conditions validate the effectiveness of the thermal management techniques. Additionally, both simulation results and hardware tests verified a significant reduction in parasitic inductance and V_{ds} voltage spikes compared to prior GaN-based designs. These advancements directly contribute to improved system performance, increased device reliability, and extended operational lifespan.

Rigorous testing was conducted at a high ambient temperature of 175 °C on a specially designed PCB optimized for HT applications. The proposed layout consistently maintained robustness and stability under these extreme conditions, validating its suitability for demanding environments.

References

[1] A. I. Emon, Mustafeez-ul-Hassan, A. B. Mirza, J. Kaplun, S. S. Vala and F. Luo, "A Review of High-Speed GaN Power Modules: State of the Art, Challenges, and Solutions," in IEEE Journal of Emerging and Selected Topics in Power Electronics, vol. 11, no. 3, pp. 2707-2729, June 2023, doi: 10.1109/JESTPE.2022.3232265.

[2] Z. Chen, D. Boroyevich, R. Burgos and F. Wang, "Characterization and modeling of 1.2 kv, 20 A SiC MOSFETs," 2009 IEEE Energy Conversion Congress and Exposition, San Jose, CA, USA, 2009, pp. 1480-1487, doi: 10.1109/ECCE.2009.5316106.

[3] N. Haryani, X. Zhang, R. Burgos and D. Boroyevich, "Static and dynamic characterization of GaN HEMT with low inductance vertical phase leg design for high frequency high power applications," 2016 IEEE Applied Power Electronics Conference and Exposition (APEC), Long Beach, CA, USA, 2016, pp. 1024-1031, doi: 10.1109/APEC.2016.7467996.

[4] X. Zhang, Z. Shen, N. Haryani, D. Boroyevich and R. Burgos, "Ultra-low inductance vertical phase leg design with EMI noise propagation control for enhancement mode GaN transistors," 2016 IEEE Applied Power Electronics Conference and Exposition (APEC), Long Beach, CA, USA, 2016, pp. 1561-1568, doi: 10.1109/APEC.2016.7468075.

[5] S. -S. Yang, J. -H. Soh and R. -Y. Kim, "Parasitic Inductance Reduction Design Method of Vertical Lattice Loop Structure for Stable Driving of GaN HEMT," 2019 IEEE 4th International Future Energy Electronics Conference (IFEEC), Singapore, 2019, pp. 1-8, doi: 10.1109/IFEEC47410.2019.9014921.

[6] D. Reusch and J. Strydom, "Understanding the Effect of PCB Layout on Circuit Performance in a High-Frequency Gallium-Nitride-Based Point of Load Converter," in IEEE Transactions on Power Electronics, vol. 29, no. 4, pp. 2008-2015, April 2014, doi: 10.1109/TPEL.2013.2266103.

[7] G. Gauttam, H. S. Krishnamoorthy, A. K. Deka and S. Chen, "Miniaturized Pulsed Power Supply for Magnetic Resonance Imaging (MRI) Application," 2023 IEEE Energy Conversion Congress and Exposition (ECCE), Nashville, TN, USA, 2023, pp. 6417-6422, doi: 10.1109/ECCE53617.2023.10362679.

[8] B. Sun, Z. Zhang and M. A. E. Andersen, "Research of PCB Parasitic Inductance in the GaN Transistor Power Loop," 2019 IEEE Workshop on Wide Bandgap Power Devices and Applications in Asia (WiPDA Asia), Taipei, Taiwan, 2019, pp. 1-5, doi: 10.1109/WiPDAAsia.2019.8760312.

979-8-3315-1612-3/25 $31.00 © 2025 IEEE

Data-Driven Characterization and Forecasting of Metal-Oxide Varistor Degradation in DC Circuit Breakers

Zhi Jin (Justin) Zhang, Yang Liu, Lukas Graber, Maryam Saeedifard

School of Electrical and Computer Engineering
Georgia Institute of Technology
Atlanta, USA
{justin.zhang, yliu1024}@gatech.edu, {lukas.graber, maryam}@ece.gatech.edu

Abstract—**Energy absorption devices play a critical role in semiconductor-based dc circuit breakers (DCCBs), where they generate a counter-voltage to reverse fault current direction and dissipate fault energy during the interruption process. The metal-oxide varistor (MOV) is the mainstream candidate for this role. However, since MOVs are primarily designed for electrical surge protection, their degradation patterns under dc fault interruption conditions, characterized by exposure to unique DCCB voltage and current waveforms, have not been thoroughly studied. Understanding and predicting MOV degradation (an indicator of MOV lifetime) is crucial for ensuring the reliability and longevity of DCCBs. Therefore, there is a need to (i) systematically test and characterize the degradation of MOVs during dc fault interruptions and (ii) develop forecasting methods to estimate their long-term degradation. In this work, MOV degradation data is collected using an experimental DCCB testbed designed for accelerated degradation testing. Based on this data, statistical analysis and machine learning techniques are then employed to analyze the data, providing insight into DCCB MOV degradation trends and enabling the forecasting of their long-term performance. This work lays the foundation for enhancing the operational reliability and life-cycle management of DCCBs.**

Index Terms—**DC circuit breaker, metal-oxide varistor, accelerated degradation test, machine learning, data-driven, forecasting, neural networks.**

I. INTRODUCTION

Medium- and high-voltage ($\geq$ 1 kV) semiconductor-based dc circuit breakers (DCCBs), a key component of dc power systems, can be generally categorized into solid-state circuit breakers (SSCBs) and hybrid circuit breakers (HCBs), as depicted in Fig. 1 [1], [2]. SSCBs and HCBs use the solid-state switch path and the mechanical disconnect switch path to conduct nominal system current, respectively. Despite this distinction, both follow a similar fault interruption process. During a fault, the current is redirected into the solid-state switches, which are subsequently turned off to commutate the current into the metal-oxide varistors (MOVs). The MOVs insert a counter-voltage to reverse the rate of rise of the fault current and dissipate the fault energy [2]. While MOVs are crucial for fault interruption, they experience energy pulses and thermal cycles far more severe than those encountered in their conventional applications, such as electrical surge

protection [3], [4]. Consequently, MOVs have been identified as a lifespan bottleneck in DCCBs [3]–[5], highlighting the need for detailed characterization, modeling, and forecasting of their degradation patterns. These patterns are directly linked to MOV lifetimes and play a critical role in (i) enhancing DCCB reliability and (ii) optimizing DCCB production and operation.

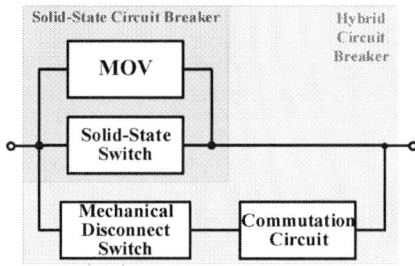

Fig. 1. General structures of solid-state and hybrid circuit breakers. A commutation circuit is required only in hybrid circuit breakers and can be placed in the path with either the mechanical disconnect switch or the solid-state switch [1].

To address these challenges, this work employs an accelerated degradation testing (ADT) platform to investigate MOV degradation under DCCB operating conditions. The degradation data collected from the ADT is analyzed using statistical methods and machine learning techniques to understand and forecast long-term degradation trends. The results reveal variability in degradation patterns among MOVs of the same manufacturer and model, likely due to manufacturing imperfections or variations. A data transformation is applied to standardize these patterns, resulting in a normal distribution that improves the performance of data-driven algorithms for forecasting. Lastly, this study implements two machine learning models – one based on artificial neural networks (ANNs) and the other on recurrent neural networks (RNNs) – both of which achieve satisfactory and comparable forecasting performance. The degradation/lifetime forecasting models developed here, which have not been previously reported in the technical literature, offer significant benefits:

- **Predictive Maintenance:** Forecasting MOV lifetimes

enables proactive maintenance and replacement scheduling, ensuring safe and reliable fault interruptions. This approach is analogous to remaining useful life estimation for electric vehicle batteries [6], [7], industrial chemical plant assets [8], [9], and power transformers [10].

- **Data-Driven Modeling:** Understanding MOV degradation mechanisms in DCCB applications is vital for selecting and manufacturing MOVs that meet reliability and lifespan requirements. Deciphering this mechanism requires a large amount of ADT results. However, ADT is time-consuming. The machine learning model, which captures different degradation patterns, can significantly reduce the number of ADT experiments, leading to cost reduction, time saving, and improved efficiency for DCCB MOV manufacturers and researchers.

The remainder of this paper is organized as follows. Section II describes the ADT for MOV characterization. Section III presents and analyzes the obtained MOV degradation patterns. Section IV details the machine learning models developed for MOV degradation prediction. Section V concludes the paper.

II. THE DCCB ADT SETUP

The ADT aims to record the degradation process of DCCB MOVs. This section describes the ADT setup, procedure, and the rationale for using MOV clamping voltage as a degradation/lifetime indicator.

A. Setup Description

A custom ADT testbed is designed and built specifically for DCCB MOVs [11]. The schematic of the setup is shown in Fig. 2(a), whose pictures are depicted in Fig. 2(b) (i.e., the power circuitry) and Fig. 2(c) (i.e., the control and data acquisition station). The testbed operates with a dc source voltage of 200 V (supplied by an 1.65 mF capacitor bank). The current-limiting inductor is 230 μH. The DCCB in this setup utilizes Semikron SEMiX404GB17E4s IGBTs as the solid-state switches. Samples from the Panasonic ERZE08A201 MOV model are used for degradation testing. The ADT logic is implemented and executed using a National Instruments cRIO-9082 FPGA controller.

B. ADT Procedure

As shown in Fig. 2(a), the testbed can host up to five MOV samples (i.e., MOV$_1$ to MOV$_5$), each connected to a dedicated IGBT switch (i.e., S$_1$ to S$_5$) for sequential insertion into the fault circuit. The sequential scheme enhances ADT efficiency by testing one MOV sample per fault event, allowing repetitive pulsing and stress for all samples without simultaneous testing. The detailed ADT procedure follows these steps:

1) A fault is initiated by turning on all IGBTs.
2) **MOV$_1$** is inserted to interrupt the fault and is then switched out of the circuit.
3) The fault is re-initiated, and **MOV$_2$** is inserted.
4) This process repeats through **MOV$_5$**, after which the cycle restarts with **MOV$_1$**.

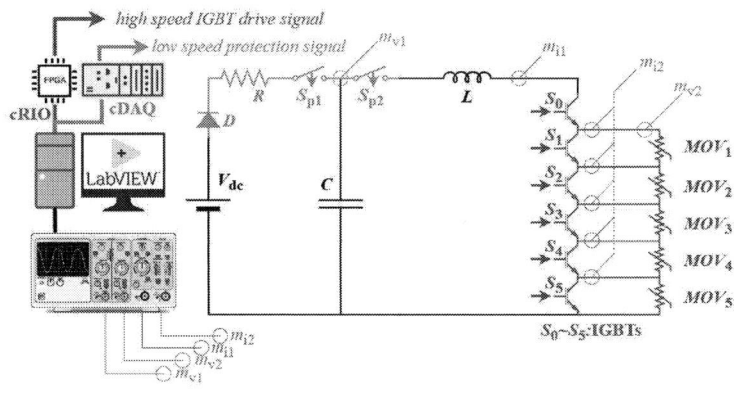

(a)

(b)

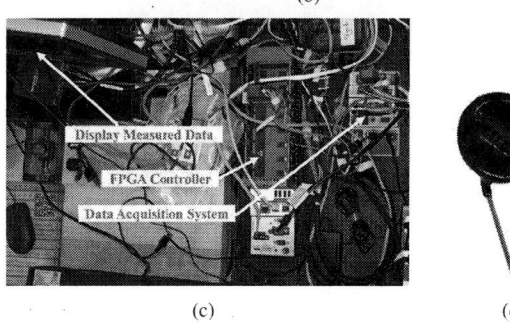

(c)   (d)

Fig. 2. Illustrations of (a) the schematic of the ADT setup, (b) the DCCB test circuit, (c) the control and data acquisition station, and (d) the Panasonic ERZE08A201 MOV under test.

In this setup, the fault interruption time is approximately 300 μs. For every test or shot, an MOV sample experiences a peak fault current of around 110 A, and dissipates approximately 4 J of energy. To simulate real-world conditions where fault events occur sporadically, a dwell time of 3.456 sec. is imposed for each MOV insertion event. This results in a cooling period of approximately 17.28 sec. per MOV sample (5 × 3.456 sec.).

Waveforms of MOV voltage, current, and energy during a fault interruption event/shot are illustrated in Fig. 3. The complete ADT procedure is graphically represented in Fig. 4.

C. MOV Clamping Voltage as Degradation/Lifetime Indicator

Generally, a device's lifetime is determined by its service time until failure, which can be categorized into two types: (i) catastrophic failure (device can no longer serve), and (ii) performance failure (certain performance of the device has deviated from designed/expected value) [12], [13]. The

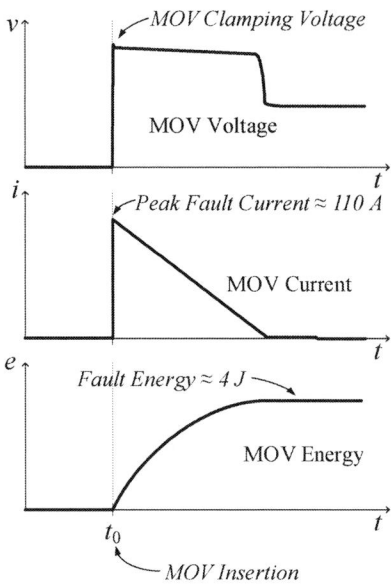

Fig. 3. Illustrative MOV voltage, current, and energy waveforms during one fault interruption event.

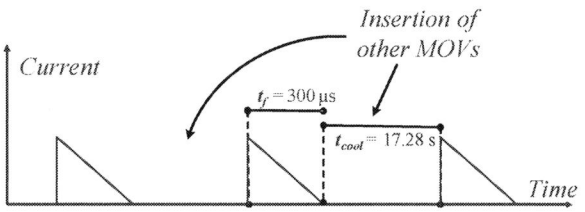

Fig. 4. Graphical illustration of the ADT procedure that performs repetitive fault interruptions. The purple triangle represents the current waveform of a single MOV. Between each test of this MOV, other four MOVs are sequentially tested in-turn to maximize ADT efficiency.

occurrence of either type of failure will end the device lifetime.

The focus of this work is on performance failure, as it exhibits a long-term pattern. Based on the standards and technical literature [14]–[16], an MOV reaches its end-of-life when its clamping voltage deviates by more than 10% from the initial value. Therefore, tracking the MOV clamping voltage over repeated fault interruptions provides a direct measure of its degradation pattern and lifetime. As a result, the clamping voltage value (indicated in Fig. 3) is recorded throughout ADT to monitor and evaluate MOV performance.

III. DCCB MOV DEGRADATION CHARACTERIZATION

A. Experimental Results

In this work, 20 MOV samples are tested, with their clamping voltages recorded over multiple fault interruption events (denoted by the shots). Figure 5 illustrates the clamping voltage degradation patterns of those MOVs, with each colored curve representing an individual sample. As a reference, the nominal clamping voltage of the Panasonic ERZE08A201 MOV is 340 V [17]. Therefore, the empirical data reveals significant variability in clamping voltage profiles among samples from the same product line. These samples also deviate from the nominal voltage value.

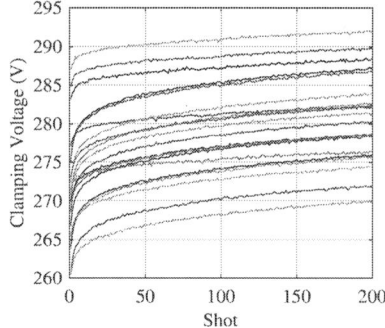

Fig. 5. MOV clamping voltage degradation patterns over the number of fault interruption events/shots. Each colored curve represents the degradation profile of one MOV sample, with a total of 20 samples/curves.

To simplify data processing and interpretation, baseline normalization was applied to the empirical data from Fig. 5. Each degradation curve was divided by its initial value, yielding the normalized voltages shown in Fig. 6(a). The data shows that degradation patterns vary significantly across MOV samples. Even within the same product model, some samples exhibit much faster degradation than others. The histogram of clamping voltages in Fig. 6(b) confirms that the distribution is non-normal, complicating the ability to predict degradation behavior.

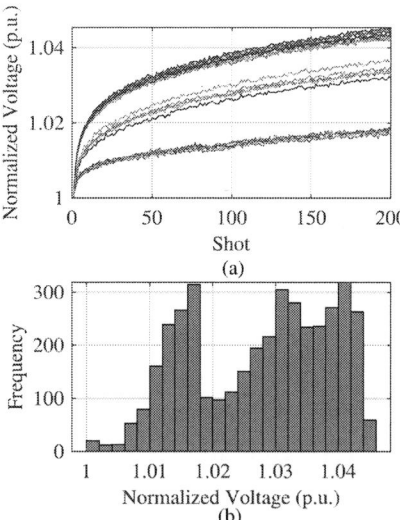

Fig. 6. Waveforms for (a) baseline normalized MOV clamping voltage degradation patterns over the number of fault interruption events/shots and (b) their corresponding histogram.

Moreover, the data exhibit non-stationarity, meaning the structure of the data changes over time [18]. As shown in Figs. 5 and 6, the clamping voltages are non-stationary and demonstrate an increasing trend with the number of shots. Non-stationarity introduces additional challenges for modeling and regression [18].

To address these complexities, a data transformation is employed to render the degradation trends more predictable and easier to model. Despite the variation in the trends, their overall shapes remain consistent: The voltages experience a steep initial rise, followed by a gradual increase over time. Based on this observation, a differencing operation is performed to standardize the trends. That is, the clamping voltage (V) from the current shot (s) is subtracted from the voltage at the subsequent shot ($s + 1$):

$$\Delta V(s) = V(s + 1) - V(s). \tag{1}$$

The transformed data, i.e., the clamping voltage differences ΔV, is depicted in Fig. 7(a). Except for the initial decrease that corresponds to the steep initial voltage rise (as shown in the zoomed-in overlay on Fig. 7(a)), the differencing operation effectively makes the degradation patterns stationary. The transformed data exhibits a skew-normal distribution with a non-zero mean and a rightward tail when plotted as a histogram (Fig. 7(b)). The non-zero mean with a small magnitude reflects the gradual increase in clamping voltage over time, whereas the tail represents the sharp voltage rise. The normal distribution suggests that, despite the seemingly varying degradation patterns across MOV samples as shown in Fig. 5, the underlying degradation mechanism (i.e., the voltage change as a function of fault interruption events) is likely identical across MOV samples.

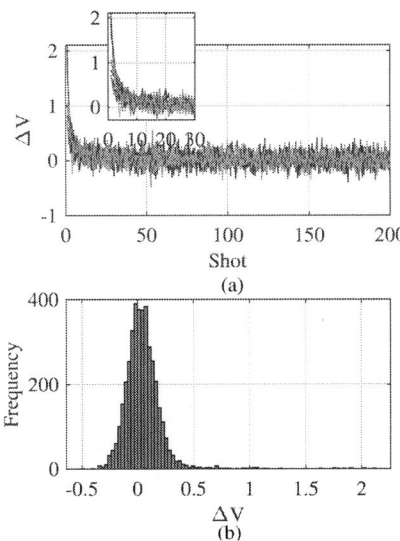

Fig. 7. Waveforms for (a) MOV clamping voltage differences (ΔV) over the number of fault interruption events/shots (with a zoomed-in view of shots #1 to #30) and (b) their corresponding histogram.

B. Physical Explanation of the Degradation Mechanism

MOV degradation is primarily attributed to the migration of charge carriers, which accumulate at the grain boundaries of the zinc oxide material. This accumulation deteriorates the double Schottky barrier at these boundaries [19], [20].

If left unaddressed, this deterioration can result in localized current density concentrations, overheating of the MOV material, and eventual thermal runaway [19], [20]. To prevent such failures, it is essential to implement degradation and lifetime forecasting algorithms. These algorithms, presented in the next section, facilitate proactive maintenance and efficient scheduling for MOV replacement.

IV. DCCB MOV Degradation Forecasting

As mentioned previously, degradation forecasting plays a crucial role in determining the remaining useful life and reliability of critical devices. It also helps to reduce the time and resources spent on ADT by estimating the degradation process rather than conducting further experiments. However, the technical literature has yet to address degradation forecasting for DCCB MOVs, a key requirement for the reliable manufacturing and operation of the DCCBs. One primary challenge is the lack of available degradation data for these MOVs. Therefore, leveraging the empirical degradation data obtained from this work (presented in Section III-A), degradation forecasting algorithms are developed.

The DCCB MOV degradation forecasting problem is illustrated in Fig. 8. The task is to predict MOV clamping voltages for the next N DCCB shots, based solely on the clamping voltage data from the previous M shots, where N is preferably significantly greater than M (i.e., $N \gg M$). Due to the complex multi-physics nature of the degradation process, data-driven approaches are preferred, as they have shown promising performance in other degradation/lifetime prediction applications [6]–[10], [21], [22].

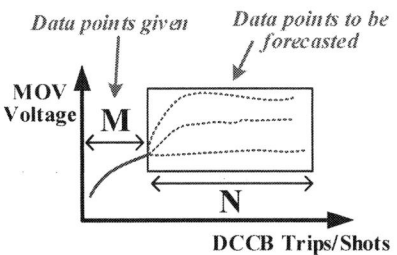

Fig. 8. Graphical illustration of the DCCB MOV degradation/lifetime forecasting problem.

It is proposed that the transformed dataset (i.e., the MOV clamping voltage differences, ΔV), rather than the original dataset (i.e., the MOV clamping voltages, V), be used as the target of data-driven forecasting. The transformed dataset exhibits stationarity and normal distribution features, making it more suitable for prediction. After predicting the voltage differences, they will be added back to the initial M data points to obtain the predicted clamping voltages for the next N shots. This is referred to as an univariate time-series forecasting problem [23]. To address this problem, three data-driven models are implemented and compared.

A. Curve of Best Fit

A conventional data regression technique, known as the curve of best fit, is implemented to establish a baseline forecasting performance. The curve is derived using data shown in Fig. 7(a). The least squares regression method is

979-8-3315-1612-3/25 $31.00 © 2025 IEEE

applied to obtain the parameters of the curve equation, where the sum of the squared differences between the curve and the data points is minimized. The resulting curve equation is:

$$\Delta V(s) = 2.0998e^{-0.3638s} + 0.0111, \qquad (2)$$

where the MOV clamping voltage differences ΔV is a function of the number of shots (s).

The result of the curve fitting is shown in Fig. 9(a). MOV #15 and #20 are randomly chosen as test samples for degradation forecasting, with the results depicted in Figs. 9(b) and 9(c), respectively. The root mean square error (RMSE) for samples #15 and #20 are 2.85 V and 0.94 V, respectively. From the results, it is evident that this conventional data regression technique is insufficient for degradation forecasting, as (i) noticeable errors exist for both MOV test samples and (ii) the performance is inconsistent among samples. These issues highlight the need for more sophisticated data-driven methods that can provide accurate predictions across all MOV samples.

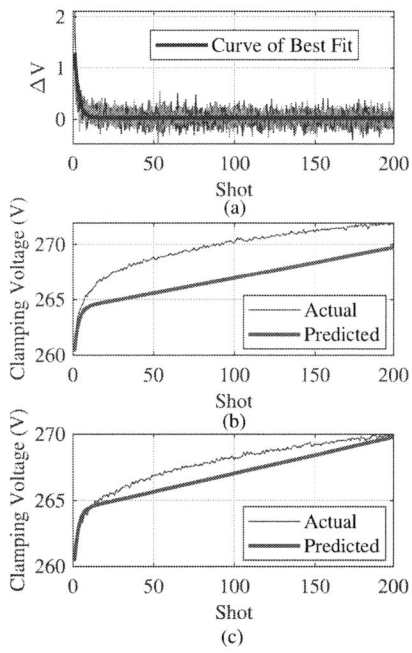

Fig. 9. Waveforms for (a) the curve of best fit for degradation forecasting, (b) results of degradation forecasting on MOV sample #15 and (c) MOV sample #20.

B. Long Short-Term Memory (LSTM)

A machine learning model, i.e., the long short-term memory (LSTM), is implemented for degradation forecasting. The LSTM is an improved version of the conventional recurrent neural network (RNN) to address problems such as vanishing gradients and is well-established for sequential data regression and forecasting [24].

Based on grid search and a 10-fold cross-validation, an LSTM model with an input dimension of 20, an output

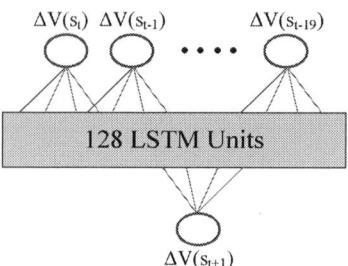

Fig. 10. The LSTM architecture implemented in this work.

dimension of 1, and a single hidden layer of 128 LSTM units is selected. In other words, the LSTM uses ΔV values from the past 20 shots (i.e., s_{t-19} to s_t) to predict the value ΔV of the next shot (i.e., s_{t+1}). The architecture of the LSTM model is depicted in Fig. 10. Accordingly, the training process of the LSTM involves slicing the degradation data into sequences of length 20 as input and predicting 1 step ahead as output. The loss function used during training is represented by the mean squared error (MSE) for regression, and the Adam optimizer is employed to minimize the MSE during training.

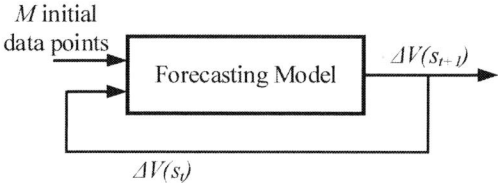

Fig. 11. The recursive multi-step forecasting strategy.

As shown in Fig. 8, long-term/multi-step forecasting capability is needed. As a result, a closed-loop/recursive structure is implemented, as illustrated in Fig. 11. In this structure, the present prediction is used as the input (i.e., the lagged observation) to forecast the value at the next shot. To elaborate, the LSTM initially receives empirical data from the first 20 shots (i.e., 1st to 20th shots, representing $M = 20$ data points given). The model then predicts the value at 21st shot. Subsequently, this predicted value is used in place of past observations (i.e., 2nd to 21st shots) to forecast the value of the next shot (i.e., 22nd shot), and the whole process repeats itself for the N data points, where $N \gg M$. Note that since the LSTM is trained off-line, it does not need to be re-tuned or re-trained when applied to new MOV samples, as long as the training data includes appropriate inter-sample deviations.

For comparison purposes, MOV #15 and #20 are again selected as the test samples, while the remaining 18 MOV samples are used for training. Their voltage forecasting results are presented in Figs. 12(a) and 12(b), respectively. The initial 20 empirical data points are used as the input to the LSTM, so the predicted voltages start at shot #21. This corresponds to $M = 20$ and $N = 180$. Compared with the results from Fig. 9, the predicted voltages using the LSTM are closer to the real values, with RMSE of 0.21 V and 0.18 V for samples

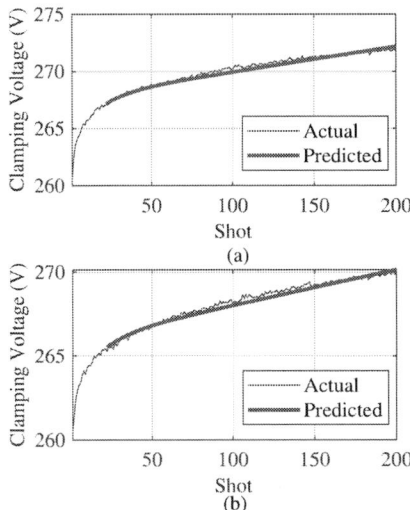

Fig. 12. Results of degradation forecasting on (a) MOV sample #15 and (b) MOV sample #20 using the LSTM.

#15 and #20, respectively. This demonstrates the advantage of the machine learning methods over the conventional curve fitting techniques for degradation forecasting.

C. Artificial Neural Network (ANN)

Recent advancements in long-term time-series data forecasting suggest that RNN-based solutions (e.g., transformers) may not outperform the conventional artificial neural network models in certain applications [25]. This observation is particularly important because ANNs require less time and resources for training and implementation compared to RNNs. Therefore, this is investigated by comparing an ANN model with the aforementioned LSTM model (a version of RNN) in the MOV degradation forecasting application.

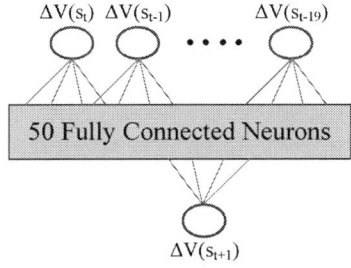

Fig. 13. The ANN architecture implemented in this work.

The ANN architecture resembles that of the LSTM, with the same input and output dimensions and a hidden layer of 50 fully connected neurons, as illustrated in Fig. 13. The training, multi-step forecasting, and testing procedures are identical to those of the LSTM model. The testing results for MOV samples #15 and #20 are shown in Figs. 14(a) and 14(b), respectively. They have RMSE of 0.2 V (sample #15) and 0.17 V (sample #20). Comparing the results from Fig. 12 and

Fig. 14 for the same MOV samples, both ANN and RNN provide comparable and satisfactory long-term forecasting performance. This is promising, as ANNs tend to be less complex than RNNs, which makes them advantageous for large-scale forecasting applications where training time and resource usage are important considerations.

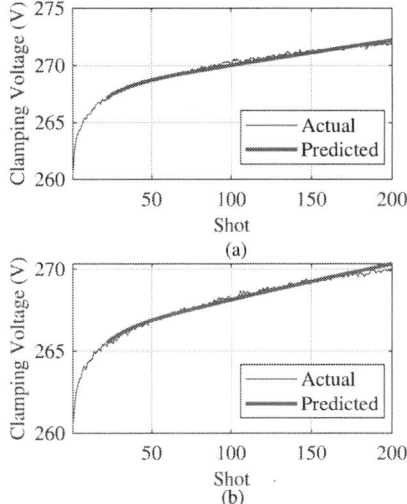

Fig. 14. Results of degradation forecasting on (a) MOV sample #15 and (b) MOV sample #20 using the ANN.

V. CONCLUSIONS AND FUTURE WORK

The study of MOV degradation patterns and their prediction is critical for the efficient and reliable production and operation of dc circuit breakers. However, such investigations are scarcely addressed in the technical literature. In this work, a DCCB MOV degradation testing platform is developed and utilized to obtain empirical degradation data. Using this data, data-driven forecasting methodologies are implemented and compared. Two machine learning models, namely LSTM and ANN, are able to accurately predict the long-term degradation profiles of various MOV samples in a recursive fashion, given only a small amount of input data.

Future work will analyze the MOV degradation mechanism and failure modes from both statistical and physical perspectives. Particularly, degradation patterns and failures of MOVs that differ in shape and size will be compared. For example, the square-cube law suggests that thermally-induced failure modes are underrepresented in smaller MOVs compared to their larger counterparts. Therefore, it is worth investigating if there is a common underlying failure mechanism for different types of MOVs.

In addition, physics-informed data-driven techniques will be explored to not only improve the forecasting results, but also predict catastrophic failure scenarios. Some physical features that can be considered include the change of barrier heights due to charge migration, coefficient of thermal expansion, and stress-strain curves.

ACKNOWLEDGMENT

Mr. Yang Liu and Dr. Lukas Graber were supported by the Advanced Research Projects Agency-Energy (ARPA-E), U.S. Department of Energy, Grant DE-AR0001524 in the SF6-FREE program.

REFERENCES

[1] Y. He *et al.*, "A survey of hybrid circuit breakers: Component-level insights to system-wide integration," *IEEE Open Journal of Power Electronics*, vol. 5, pp. 513–533, 2024.

[2] R. Rodrigues, Y. Du, A. Antoniazzi, and P. Cairoli, "A review of solid-state circuit breakers," *IEEE Transactions on Power Electronics*, vol. 36, no. 1, pp. 364–377, 2020.

[3] Z. J. Zhang, M. Bosworth *et al.*, "Lifetime-based selection procedures for DC circuit breaker varistors," *IEEE Transactions on Power Electronics*, vol. 37, no. 11, pp. 13 525–13 537, 2022.

[4] C. Xu, A. Rockhill *et al.*, "Evaluation tests of metal oxide varistors for DC circuit breakers," *IEEE Open Access Journal of Power and Energy*, vol. 9, pp. 254–264, 2022.

[5] A. P. Yu, B. X. Guo, C. L. Wang, D. Z. Zhou, E. G. Wang, and F. Z. Zhang, "Lifetime estimation for hybrid HVDC breakers," *International Journal of Electrical Power & Energy Systems*, vol. 120, p. 106035, 2020.

[6] S. A. Hasib *et al.*, "A comprehensive review of available battery datasets, RUL prediction approaches, and advanced battery management," *IEEE Access*, vol. 9, pp. 86 166–86 193, 2021.

[7] N. Yang, H. Hofmann, J. Sun, and Z. Song, "Remaining useful life prediction of Lithium-ion batteries with limited degradation history using random forest," *IEEE Transactions on Transportation Electrification*, vol. 10, no. 3, pp. 5049–5060, 2024.

[8] M. Bogojeski, S. Sauer, F. Horn, and K.-R. Müller, "Forecasting industrial aging processes with machine learning methods," *Computers & Chemical Engineering*, vol. 144, p. 107123, 2021.

[9] Z. Li, J. Zhou, H. Nassif, D. Coit, and J. Bae, "Fusing physics-inferred information from stochastic model with machine learning approaches for degradation prediction," *Reliability Engineering & System Safety*, vol. 232, p. 109078, 2023.

[10] L. Cheim, L. Lin, and A. Dagnino, "Probabilistic transformer fault tree analysis using Bayesian networks," in *IEEE PES T&D Conference and Exposition*, 2014, pp. 1–5.

[11] Y. Liu *et al.*, "Designing a platform to evaluate metal oxide varistors for DC circuit breaker applications," in *IEEE Electrical Insulation Conference (EIC)*, 2024, pp. 5–8.

[12] K. M. Blache and A. B. Shrivastava, "Defining failure of manufacturing machinery and equipment," in *IEEE Proceedings of annual reliability and maintainability symposium (RAMS)*, 1994, pp. 69–75.

[13] W. B. Nelson, *Accelerated testing: statistical models, test plans, and data analysis*. John Wiley & Sons, 2009.

[14] "Varistors for use in electronic equipment - Part 1: Generic specification," in *IEC 61051-1*, 2018.

[15] L. Muremi, P. Bokoro, and W. Doorsamy, "Electrical degradation characteristics of low voltage Zinc Oxide varistor subjected to AC switching surges," in *IEEE 5th International Conference on Condition Assessment Techniques in Electrical Systems (CATCON)*, 2021, pp. 195–200.

[16] M. Reinhard, V. Hinrichsen, B. Richter, and F. Greuter, "Energy handling capability of high-voltage metal-oxide surge arresters Part 2: Results of a research test program," in *Cigré Session*, 2008.

[17] Panasonic, "Varistors (ZNR Surge Absorber), Series E, Type D," 2020.

[18] C. Cheng, A. Sa-Ngasoongsong, O. Beyca, T. Le, H. Yang, Z. Kong, and S. T. Bukkapatnam, "Time series forecasting for nonlinear and non-stationary processes: a review and comparative study," *IISE Transactions*, vol. 47, no. 10, pp. 1053–1071, 2015.

[19] K. Chuong, Y. Liu, L. Graber, and L. M. Garten, "Improving the microstructure of ZnO-based metal oxide varistors using cold sintering," in *IEEE Electrical Insulation Conference (EIC)*, 2024, pp. 63–66.

[20] J. He, C. Cheng, and J. Hu, "Electrical degradation of double-Schottky barrier in ZnO varistors," *AIP Advances*, vol. 6, no. 3, 2016.

[21] C. Yuhang and H. Bing, "Prediction of bearing degradation trend based on LSTM," in *IEEE Symposium Series on Computational Intelligence (SSCI)*, 2019, pp. 1035–1040.

[22] E. Pieri, A. Kyprianou, A. Phinikarides, G. Makrides, and G. E. Georghiou, "Forecasting degradation rates of different photovoltaic systems using robust principal component analysis and ARIMA," *IET Renewable Power Generation*, vol. 11, no. 10, pp. 1245–1252, 2017.

[23] G. E. Box, G. M. Jenkins, G. C. Reinsel, and G. M. Ljung, *Time series analysis: forecasting and control*. John Wiley & Sons, 2015.

[24] J. Schmidhuber and S. Hochreiter, "Long short-term memory," *Neural Computation MIT-Press*, vol. 9, no. 8, pp. 1735–1780, 1997.

[25] A. Zeng, M. Chen, L. Zhang, and Q. Xu, "Are transformers effective for time series forecasting?" in *Proceedings of the AAAI Conference on Artificial Intelligence*, vol. 37, no. 9, 2023, pp. 11 121–11 128.

A Thyristor-Based Fault Current Bypass Solid-State Circuit Breaker for DC Microgrid Applications

Jiale Zhou
Dept. of Electrical & Computer Engineering
University of North Carolina at Charlotte
Charlotte, NC, USA
jzhou20@charlotte.edu

Xiuhu Sun
Dept. of Electrical & Computer Engineering
University of North Carolina at Charlotte
Charlotte, NC, USA
xsun17@charlotte.edu

Qiang Mu
Dept. of Electrical & Computer Engineering
University of North Carolina at Charlotte
Charlotte, NC, USA
qmu1@charlotte.edu

Tiefu Zhao
ECE Department
BATT CAVE Research Center
University of North Carolina at Charlotte
Charlotte, NC, USA
Tiefu.Zhao@charlotte.edu

Abstract—Thyristor-based solid-state circuit breaker (SSCB) has high efficiency as a protection device for DC microgrid distribution. However, the passive turn-off and microsecond-level turn-off time make thyristor-based SSCBs more suitable for high-inertia DC systems. Large inductance in DC systems extends the fault clearing time. This paper proposes a thyristor-based fault current bypass SSCB (TFCB-SSCB) for DC microgrids. The proposed topology, compared to conventional thyristor-based SSCBs, not only offers high efficiency but also achieves rapid isolation of the source and fault side, with the only additional cost being a switch with a low breakdown voltage. Initially, the superiority of the proposed topology is analyzed in comparison to the prior-art thyristor-based topology. Next, the operating principles and key design considerations are presented for the proposed topology. Finally, the effectiveness of the proposed topology is verified through simulation results on an 800 V/ 165 A TFCB-SSCB model and experimental results on a 100 V/68 A TFCB-SSCB prototype. The fault isolation time of the proposed topology is five times faster than that of the traditional topology in simulations and twice as fast in experiments.

Keywords—DC circuit breaker, Solid-state circuit breaker, Fault current bypass, DC Microgrid, Thyristor

I. INTRODUCTION

DC microgrids are increasingly achieving high penetration due to their low impedance and high efficiency characteristics [1-5]. A DC microgrid can interface with the utility through multilevel inverters or rectifiers [6-11], and connect to loads via DC-DC converters [12-15]. Solid-state circuit breaker (SSCB), serving as a protective device within DC microgrid systems, plays an indispensable role and has consistently garnered attention from both industry and academia, as shown in Fig. 1 [16-20]. It can be seen that the DC circuit breakers are located on the DC bus, ahead of different converters. Compared with fully controlled power semiconductors, thyristor-based SSCB has high efficiency as a protection device for DC microgrid distribution [21]. However, the passive turn-off and microsecond-level turn-off time make thyristor-based SSCBs more suitable for high-inertia DC systems [22]. Large inductance in DC systems extends the fault clearing time. Long time of isolating the source and fault side can increase the risk to maintenance personnel and adversely affect reclosing

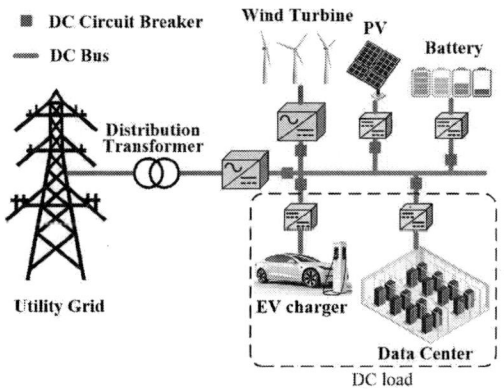

Fig. 1. DC circuit breakers in DC microgrid.

operations. Fully controlled fault current bypass-based SSCBs have been reported in [17], [23], while thyristor-based fault current bypass SSCB (TFCB-SSCB) has not been well explored.

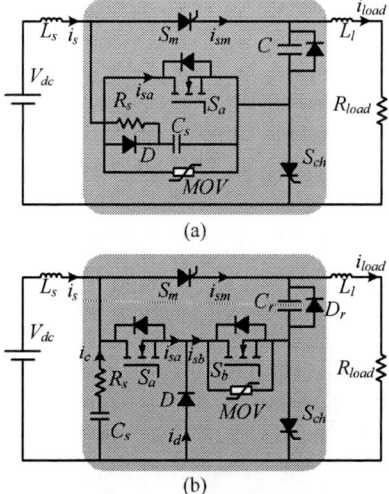

Fig. 2. Two thyristor-based SSCB topologies. (a) Topology in [22]; (b) Proposed TFCB-SSCB.

Literature [26] proposed a thyristor-based fault current bypass SSCB to achieve fast fault isolation. However, the MOV paralleled to the auxiliary switch continues to dissipate energy after the auxiliary switch turns off, leading to an extended the fault isolation time.

To better illustrate the contributions of the proposed topology, the topology in Fig. 2(a) is used for comparison with the proposed topology. It can be seen that the clamping voltage component metal oxide varistor (MOV) will be connected between the source and the fault side during the fault clearing time. This indicates that there remains an electrical connection even when the semiconductor switch is turned off, resulting in an extended fault isolation time. In systems with large inductive loads, the fault clearing time is further prolonged, potentially causing a drop in the DC bus voltage and impacting other loads connected to the DC bus. To address this issue, this paper proposes the TFCB-SSCB topology.

II. PROPOSED TFCB-SSCB

A. Circuit Topology

Fig. 2(b) illustrates the proposed TFCB-SSCB topology, which includes two thyristors: one is used as the main switch, S_m, while the other, S_{ch} serves as the pre-charge switch. Additionally, the commutation circuit consists of two fully controlled switches (S_a and S_b), a diode (D_a), and a resonant capacitor C_a. To mitigate the charging current when C_a is charged during initial operation, S_{ch} can be connected in series with a resistor. The MOV is placed in parallel with S_b, which can be selected a low breakdown voltage. The circuit also includes

a snubber circuit comprising snubber resistance (R_s) and snubber capacitance (C_s), as well as a current isolation diode (D). The proposed TFCB-SSCB not only offers high efficiency but also enables the rapid isolation of the source from the fault side, reducing the impact of the fault side on the DC bus, which also connects to other loads.

B. Operating Principle

Fig. 3 and Fig. 4 illustrate the operating modes and electrical waveforms of the TFCB-SSCB topology during fault current interruption, respectively. i_s, i_{sm}, i_{sa}, i_{sb}, i_{load}, and i_d represent the source current, main switch current, auxiliary switch S_a current, auxiliary switch S_b current, load current, and diode current, respectively. I_{th} represents the threshold value for turning on the auxiliary switch S_a, while I_n represents the normal current. v_{sm}, v_{sa}, and v_{sb} represent the voltages across the main switch, auxiliary switch S_a, and S_b, respectively. V_{dc} and V_{clamp} denote the DC link voltage and the clamping voltage of the MOV, respectively.

Mode 1 & Mode 2 (Before t_0): Before normal operation, S_b is turned on to bidirectionally bypass the MOV. During the charging operation in *Mode 2*, S_m is in the ON state, and both the snubber capacitor and the resonant capacitor are charged.

Mode 3 (Before t_0): The proposed TFCB-SSCB operates in normal operation, and $i_s = i_{sm} = i_{load} = I_n$.

Mode 4 ($t_0 \leq t < t_1$): A short-circuit fault occurs at $t=t_0$. S_a turns on when the fault current reaches I_{th} at $t=t_1$.

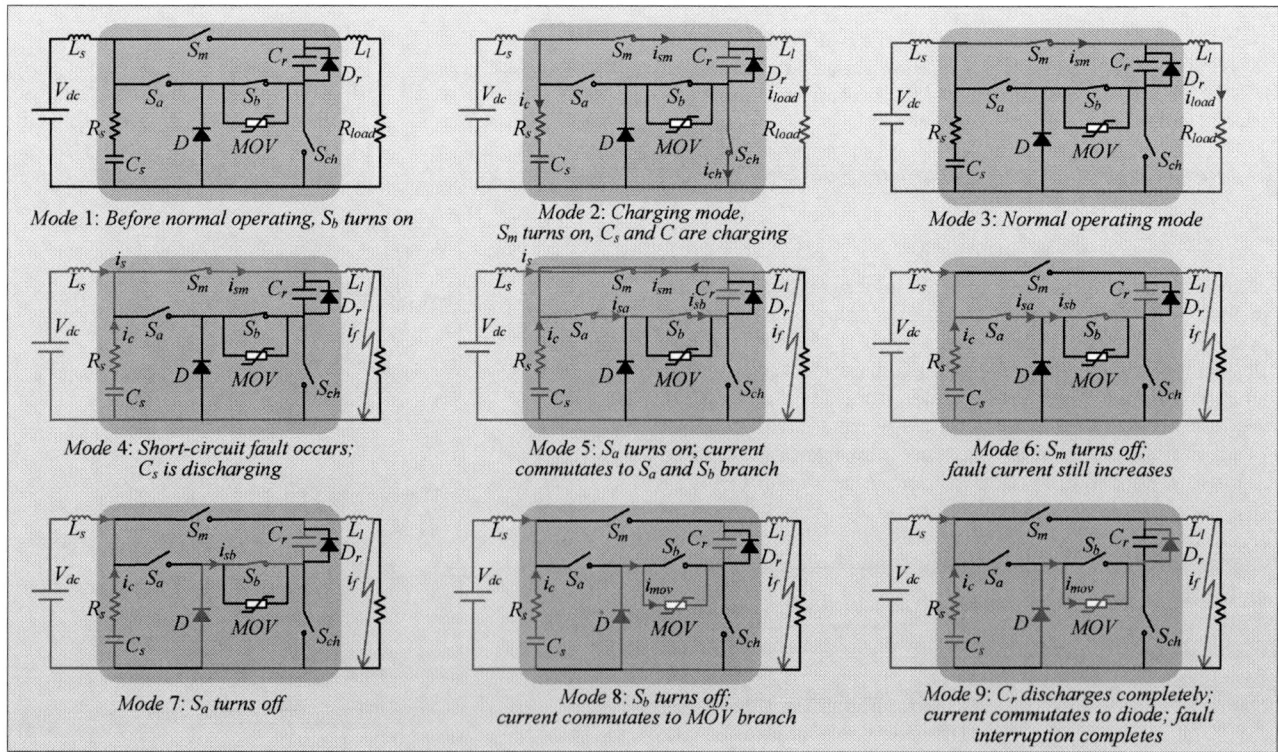

Fig. 3. Operating modes of the proposed TFCB-SSCB during fault current interruption.

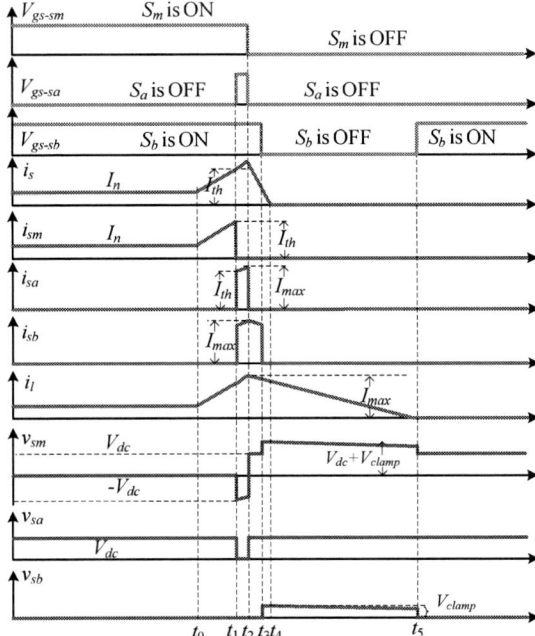

Fig. 4. Electrical waveforms of the proposed TFCB-SSCB.

Mode 5 ($t_1 \leq t < t_2$): The auxiliary switch S_a turns on and current reaches I_{th} at $t=t_1$, and S_m completely turns off. Consequently, i_{sm} drops to zero and cannot increase further. Then, S_a turns off at $t=t_2$, isolating the source side from the fault location.

Mode 6 ($t_2 \leq t < t_3$): The fault current flows through S_b, and v_{sm} reaches V_{dc}. Then, S_b turns off at $t=t_3$. Afterward, v_{sm} rises $V_{dc}+V_{clamp}$, while v_{sb} is clamped at V_{clamp}.

Mode 7 ($t_3 \leq t < t_4$): The source current decreases to zero.

Mode 8 ($t_4 \leq t < t_5$): The load current decreases to zero through the MOV, indicating that the fault has been fully cleared.

C. Design Considerations

To reduce the current in S_m to zero, two criteria need to be satisfied: $i_{sm}=0$ and $v_{sm}<0$ during the interval t_2-t_1, which is greater than t_q. It means that the resonant capacitor should meet the following equation.

$$I_{max} \cdot t_q \cdot \alpha < V_{dc} \cdot C_r \qquad (1)$$

Where I_{max} is the maximum current when S_a turns off, t_q is the turn-off time of the thyristor, and α is a redundancy factor ranging between 1 and 3 [22].

The voltage rating of S_m should exceed $V_{dc}+V_{clamp}$. The voltage ratings of D_r, D, S_a, and S_{ch} should be greater than V_{dc}, and the voltage rating of S_b should be greater than V_{clamp}. For the snubber design, the criteria are elaborated in [24-25], [27-31].

III. SIMULATION RESULTS

In order to verify the effectiveness of the proposed topology, a simulation with 800 VDC, 50 A rated current, and 165 A

protection current is conducted in the MATLAB/Simulink environment. Table I lists the simulation parameters.

Fig. 5 and Fig. 6 show a comparison of the simulation results between the topology in [22] and the proposed topology. It can be seen from Fig. 5(a) and Fig. 6(a) that the fault isolation time is 5 times faster than that of the topology in [22] by using the proposed topology. Fig. 6(b) illustrates that the maximum voltage across S_m is around 1000 V, which equals $V_{dc}+V_{clamp}$. The time interval is set to 200 μs. S_m is turned off when S_a turns on, and S_m does not turn on again. The voltage across S_a exhibits a voltage spike due to the system inductance and snubber circuit, which should be smaller than the voltage rating of S_a. In Fig. 5(f), the load current decreases to zero under $V_{clamp}-V_{dc}$, whereas the load current in Fig. 6(f) decreases to zero under V_{clamp}.

IV. EXPERIMENTAL RESULTS

In order to further verify the effectiveness of the proposed topology, a prototype with 100 VDC and 68 A protection current is built, as shown in Fig. 7. S_3 is a circuit breaker that controls the DC power supply, which charges the resonant capacitor, as shown in Fig. 7(a). The experimental parameters are listed in Table II. The MOV-20D820K from Bourns is selected as the voltage clamping component.

TABLE I - SIMULATION PARAMETERS

Parameters	Values	
	Topology in [22]	Proposed
DC source voltage (V)	800	800
Normal load current (A)	50	50
Fault current (A)	165	165
MOV clamping voltage (V)	1000	200
System inductance (μH)	20	20
Load inductance (mH)	5	5
Resonant capacitance (μF)	50	50
Snubber capacitance (μF)	-	33
Snubber resistance (Ω)	-	0.5

TABLE II - EXPERIMENTAL PARAMETERS

Parameters	Value or type
DC source voltage (V)	100
Fault current (A)	27
Interruption current (A)	68
MOV clamping voltage (V)	135
Load inductance (μH)	282
Resonant capacitor (μF)	60
Resonant inductance (μH)	2
Snubber capacitance (μF)	60
Snubber resistance (Ω)	0.5
S_m and S_{ch}	SK055MTP
S_a and S_b	IKQ75N120CT2XKSA1
D and D_r	C4D20120

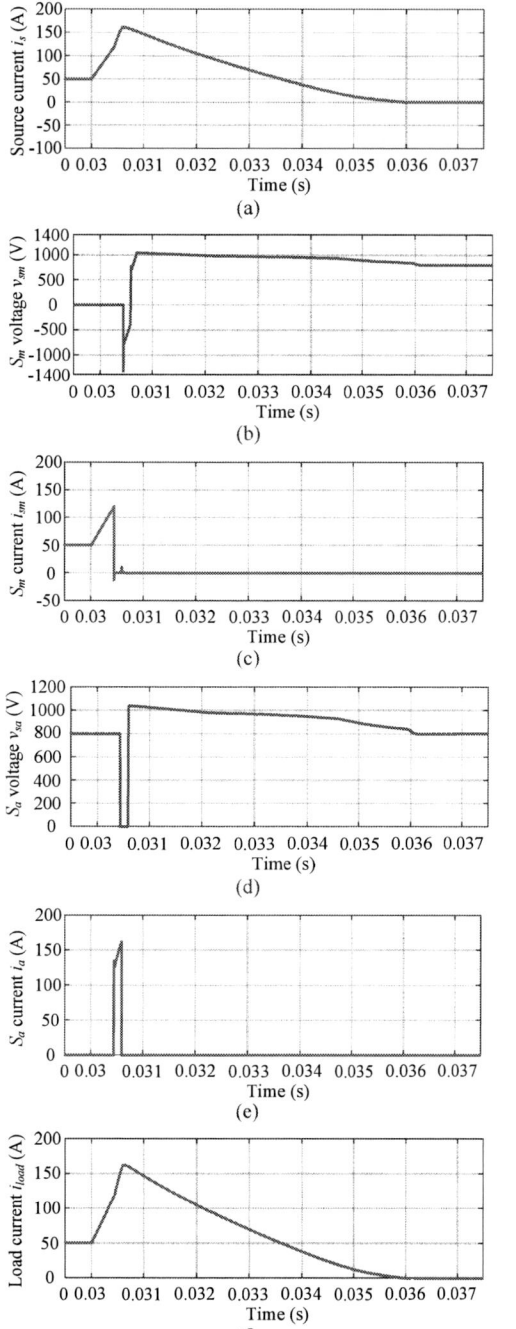

Figure 5: Topology in [22]. (a) Source current (b) Main switch voltage, (c) Main switch current, (d) S_a voltage, (e) S_a current, (f) Load current.

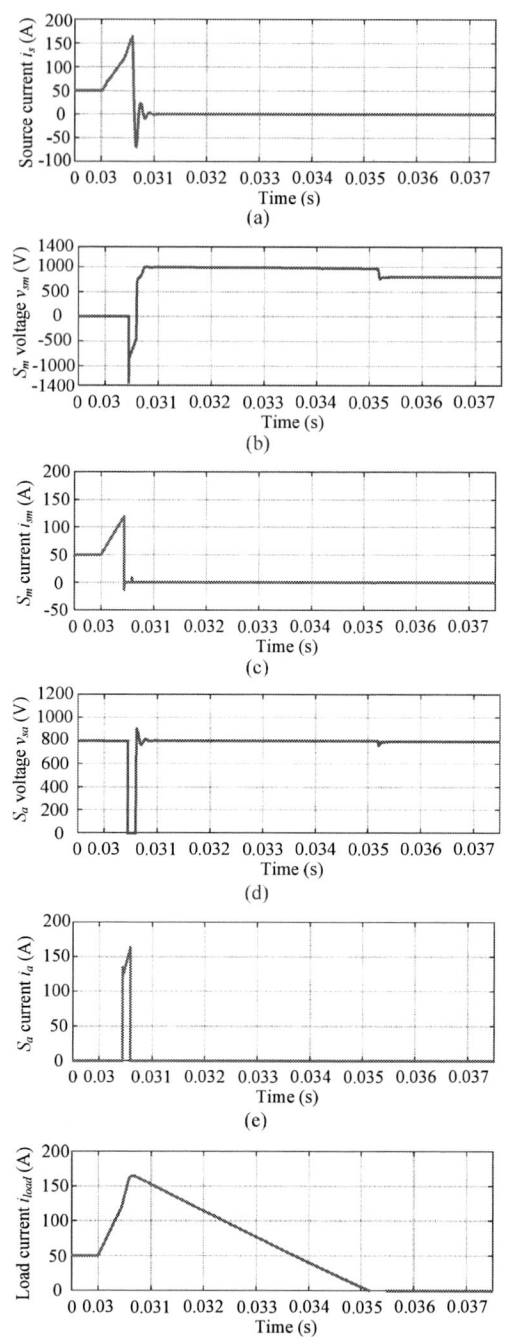

Fig. 6. Proposed TFCB-SSCB. (a) Source current (b) Main switch voltage, (c) Main switch current, (d) S_a voltage, (e) S_a current, (f) Load current.

Fig. 8 illustrates the experimental results of the topology in [22]. It can be seen that the main switch is interrupted at 31.5 A, and i_{sm} equals i_s. Then, the current is commutated to the S_a branch, with the maximum interruption current reaching 80.4 A. When S_a turns off, the energy is dissipated in the MOV. The total isolation time is approximately 500 μs.

Fig. 9 illustrates the experimental results of the proposed topology. It can be seen that the main switch is interrupted at

31.5 A. It is noted that $i_s < i_{sm}$, which is due to the discharge of the the snubber capacitor during the fault. Then, the current is subsequently commutated to the S_a and S_b branches. The rate of change of the current in these branches is higher than that of the S_m current. That's because the resonant capacitor discharges from an initial voltage of 100V. It is noted that the time interval from when S_a turns on to when it turns off is 120 μs, which is significantly longer than the turn-off time of the SK055MTP.

979-8-3315-1612-3/25 $31.00 © 2025 IEEE   3331

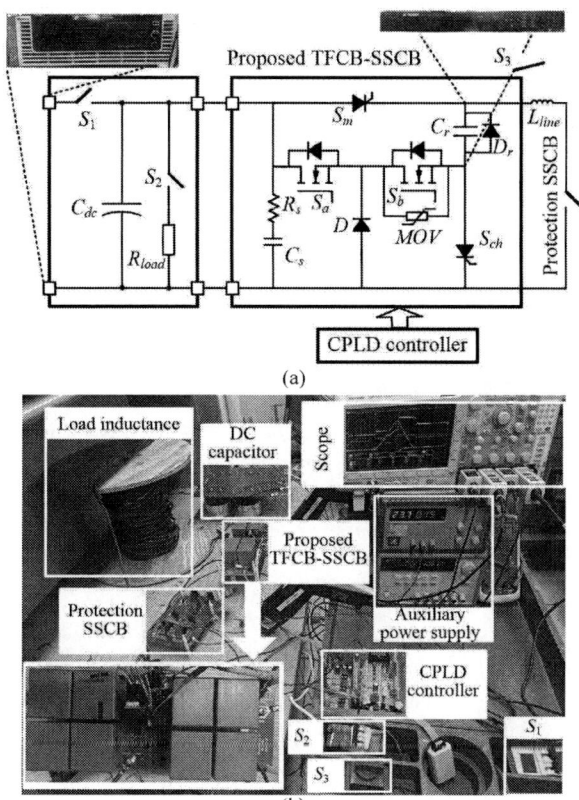

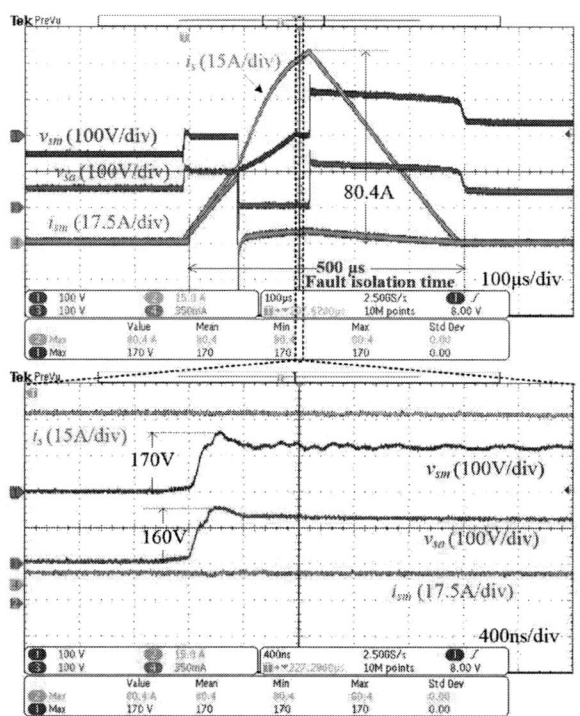

Fig. 7. Test platform. (a) Test diagram, (b) Prototype.

This interval ensures that the thyristor is fully turned off. The maximum interruption current is 68.1 A, which is smaller that

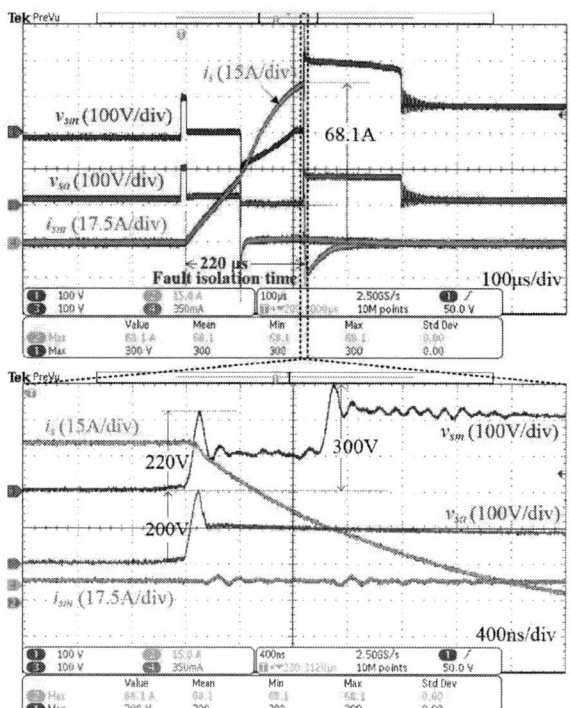

Fig. 9. Experiment waveforms of the proposed TFCB-SSCB

that of the topology in [22]. The isolation time between the source side and fault side is approximately 220 μs, excluding the MOV energy dissipation time. This represents a 50% reduction compared to the traditional topology.

When S_a turns off, the voltage spike caused by the snubber circuit is approximately 200 V. This voltage spike is not superimposed with the voltage caused by the MOV because there is a 1 μs delay between the turn-off times of S_a and S_b. This will be beneficial for increasing the DC bus voltage in the future.

V. CONCLUSIONS

This paper proposes a thyristor-based fault current bypass SSCB (TFCB-SSCB) for DC microgrids. The topology analysis, work principles, and key design considerations are introduced. Next, simulation results validate that the fault isolation time is 5 times faster than that of the traditional topology by using the proposed topology. The experimental results demonstrate that the fault isolation time is 220 μs, representing a 50% reduction compared to the traditional topology. Future work will aim to increase the DC bus voltage. Future work will focus on increasing the DC bus voltage to verify the higher voltage utilization ratio of the proposed topology.

REFERENCES

[1] L. Qi, P. Cairoli, Z. Pan, C. Tschida, Z. Wang, V. R. Ramanan, L. Raciti, and A. Antoniazzi, "Solid-state circuit breaker protection for DC shipboard power systems: breaker design, protection scheme, validation testing," IEEE Trans. Ind. Appl., vol. 56, no. 2, pp. 952-960, March-April 2020.

[2] Q. Huang, H. Chen, X. Xiang, C. Li, W. Li and X. He, "Islanding Detection With Positive Feedback of Selected Frequency for DC

Fig. 8. Experimental waveforms of the topology in [22]

Microgrid Systems," in IEEE Transactions on Power Electronics, vol. 36, no. 10, pp. 11800-11817, Oct. 2021.

[3] X. Yang, Y. Lyu, K. Wang, U. Kim, Z. Zhang and K. -B. Park, "A Computationally Efficient FCS-MPC Imitator for Grid-Tied Three-Level NPC Power Converters Based on Sequential Artificial Neural Network," 2022 IEEE Energy Conversion Congress and Exposition (ECCE), Detroit, MI, USA, 2022, pp. 1-6.

[4] J. Teng, X. Sun, X. Liu, W. Zhao and X. Li, "Power Mismatches Elimination Strategy for MMC-Based Photovoltaic System and Lightweight Design," in IEEE Transactions on Power Electronics, vol. 38, no. 9, pp. 11614-11629, Sept. 2023.

[5] J. Tian, C. Zhuo, F. Wang and H. Deng, "Dual-Side Asymmetric Duty Modulation Based on Accurate Soft-Switching Characteristics Modeling for DAB-Based DC Microgrid," in IEEE Journal of Emerging and Selected Topics in Power Electronics, vol. 12, no. 3, pp. 3146-3160, June 2024.

[6] J. Zhou, X. Sun, T. Zhao, Q. Mu and X. Guo, "An Improved One-Dimensional Space Vector Modulation for Three-Phase Five-Level Cascaded H-Bridge Inverters," 2023 IEEE Energy Conversion Congress and Exposition (ECCE), Nashville, TN, USA, 2023, pp. 3349-3354.

[7] H. Cao, G. Zhu, F. Diao and Y. Zhao, "Novel Power Decoupling Methods for Three-Port Triple-Active-Bridge Converters," 2022 IEEE Applied Power Electronics Conference and Exposition (APEC), 2022, pp. 1833-1837.

[8] M. F. Uddin, U. Asif, F. Mukhtar and O. Imtiaz, "Single-Phase Transformer-less Dynamic Power Flow Control of Transmission Network using Cascaded Converter based SSSC," 2022 Global Energy Conference (GEC), Batman, Turkey, 2022, pp. 32-37.

[9] J. Teng, X. Sun, M. Zhang, W. Zhao and X. Li, "Low-Capacitance CHB-Based SST Based on Resonant Push–Pull Decoupling Channel," in IEEE Transactions on Industrial Electronics, vol. 71, no. 3, pp. 2477-2488, March 2024.

[10] J. Teng, X. Sun, Z. Bu, W. Zhao and X. Li, "Optimization Scheme Based on High-Frequency Link Interconnection of Submodules," in IEEE Transactions on Power Electronics, vol. 36, no. 12, pp. 13645-13659, Dec. 2021.

[11] M. Gao, V. Mitrovic, and R. Burgos, "Overview of FPGA-based Synchronization Techniques in Distributed Control Systems for Multilevel Power Converters," IECON 2024- 50th Annual Conference of the IEEE Industrial Electronics Society, Chicago, IL, USA, 2024.

[12] J. Tian, F. Wang, F. Zhuo, Y. Wang, H. Wang and Y. Li, "A Zero-Backflow-Power EPS Control Scheme With Multiobjective Coupled-Relationship Optimization in DAB-Based Converter," in IEEE Journal of Emerging and Selected Topics in Power Electronics, vol. 10, no. 4, pp. 4128-4145, Aug. 2022.

[13] Q. Huang, Y. Li, Z. Ma, Y. Yang, Y. Lai and S. Wang, "RLC Balance Technique of Transformer to Reduce CM EMI for Isolated DC-DC Converters," 2023 IEEE Energy Conversion Congress and Exposition (ECCE), Nashville, TN, USA, 2023, pp. 2945-2952.

[14] A. H. Ismail, H. Cao, A. Al-Hmoud, Z. Ma, X. Du and Y. Zhao, "A 3.3 kV Silicon Carbide MOSFET Based Building Block for Medium-Voltage Ultra-Fast DC Chargers," 2023 IEEE Applied Power Electronics Conference and Exposition (APEC), Orlando, FL, USA, 2023, pp. 1693-1700.

[15] H. Cao, N. Lin, P. Darvish, Y. Yang, Z. Wang and Y. Zhao, "Enhanced Triple Phase Shift Modulation Strategy for ANPC-DAB Converter to Extend Soft Switching Range," 2024 IEEE Applied Power Electronics Conference and Exposition (APEC), Long Beach, CA, USA, 2024, pp. 445-452.

[16] S. Zhao, R. Kheirollahi, Y. Wang, H. Zhang and F. Lu, "Implementing Symmetrical Structure in MOV-RCD Snubber-Based DC Solid-State Circuit Breakers," in IEEE Transactions on Power Electronics, vol. 37, no. 5, pp. 6051-6061, May 2022.

[17] R. Kheirollahi, S. Zhao and F. Lu, "Fault Current Bypass-Based LVDC Solid-State Circuit Breakers," in IEEE Transactions on Power Electronics, vol. 37, no. 1, pp. 7-13, Jan. 2022.

[18] S. Zhao, C. Xu, L. Ravi, Z. Dong and P. Cairoli, "Review and Analysis of Voltage Clamping Circuits With Low Overvoltage Ratios for DC Circuit Breakers," in IEEE Open Journal of the Industrial Electronics Society, vol. 5, pp. 651-662, 2024.

[19] A. H. Ismail et al., "High-Density High-Power Converter using 3.3-kV All-Silicon Carbide Modules," 2023 IEEE Energy Conversion Congress and Exposition (ECCE), Nashville, TN, USA, 2023, pp. 1831-1835.

[20] H. Liu, J. Zhou, T. Zhao and X. Xu, "Si IGBT and SiC MOSFET Hybrid Switch-Based Solid State Circuit Breaker for DC Applications," 2022 IEEE Energy Conversion Congress and Exposition (ECCE), Detroit, MI, USA, 2022, pp. 1-6.

[21] X. Song, P. Cairoli, Y. Du and A. Antoniazzi, "A Review of Thyristor Based DC Solid-State Circuit Breakers," in IEEE Open Journal of Power Electronics, vol. 2, pp. 659-672, 2021.

[22] R. Kheirollahi, S. Zhao, Y. Wang, H. Zhang and F. Lu, "Proactive Thyristor-Based DC Solid-State Circuit Breaker," 2023 IEEE Applied Power Electronics Conference and Exposition (APEC), Orlando, FL, USA, 2023, pp. 1616-1621.

[23] T. Pang, M. F. Rahman and M. D. Manjrekar, "A Ground Clamped Solid-State Circuit Breaker for DC Distribution Systems," 2020 IEEE Energy Conversion Congress and Exposition (ECCE), Detroit, MI, USA, 2020, pp. 989-994

[24] J. Zhou, H. Liu, T. Zhao, X. Xu and Y. Wang, "SiC Bidirectional Solid-State Circuit Breaker with Soft-Start Function for Motor Control Center," 2023 IEEE Applied Power Electronics Conference and Exposition (APEC), Orlando, FL, USA, 2023, pp. 2307-2312.

[25] J. Zhou, T. Zhao and Y. Wang, "Lifetime Extension for Solid-State Circuit Breakers in Motor Control Center Applications," 2023 IEEE Energy Conversion Congress and Exposition (ECCE), Nashville, TN, USA, 2023, pp. 6411-6416.

[26] R. Kheirollahi, S. Zhao, H. Zhang and F. Lu, "Fault Current Bypass-Based DC SSCB Using TIM-Pack Switch," in IEEE Transactions on Industrial Electronics, vol. 70, no. 4, pp. 4300-4304, April 2023.

[27] T. Pang and M. D. Manjrekar, "A Surgeless Diode-Clamped Multilevel Solid-State Circuit Breaker for Medium-Voltage DC Distribution Systems," in IEEE Transactions on Industrial Electronics, vol. 69, no. 7, pp. 7329-7339, July 2022.

[28] X. Song, Y. Du and P. Cairoli, "Survey and Experimental Evaluation of Voltage Clamping Components for Solid State Circuit Breakers," 2021 IEEE Applied Power Electronics Conference and Exposition (APEC), Phoenix, AZ, USA, 2021, pp. 401-406.

[29] F. Liu, W. Liu, X. Zha, H. Yang, and K. Feng, "Solid-state circuit breaker snubber design for transient overvoltage suppression at bus fault interruption in low-voltage DC microgrid," *IEEE Trans. Power Electron.*, vol. 32, no. 4, pp. 3007–3021, Apr. 2017.

[30] Z. J. Shen, G. Sabui, Z. Miao, and Z. Shuai, "Wide-bandgap solid-state circuit breakers for DC power systems: device and circuit considerations," *IEEE Trans. Electron Devices*, vol. 62, pp. 294–300, Jan. 2015.

[31] J. Magnusson, R. Saers, L. Liljestrand, and G. Engdahl, "Separation of the Energy Absorption and Overvoltage Protection in Solid-State Breakers by the Use of Parallel Varistors," *IEEE Trans. Power Electron.*, vol. 29, no. 6, pp. 2715-2722, June 2014.

979-8-3315-1612-3/25 $31.00 © 2025 IEEE

Single-Stage Three-Phase AC-AC Isolated Inertialess Converter (IIC) for Industrial Drives

Brad Houska[*], Decheng Yan[*], Aniruddh Marellapudi[*], Satish Belkhode[**], Joseph Benzaquen[*], and Deepak Divan[*]

Georgia Institute of Technology, Atlanta, United States
**Indian Institute of Technology, Roorkee, India*
bhouska3@gatech.edu

Abstract—**Industrial drive systems require large, heavy, and costly drive filter and transformer subsystems in order to mitigate EMI, harmonics, and other converter interactions. This paper proposes a bidirectional, single-stage, three-phase AC-AC matrix-type converter with high frequency isolation to consolidate the variable speed drive, drive transformer, and drive filter subsystems to lower size, cost, and weight. Previously, this topology has been experimentally characterized using PWM and SVM-based modulation at low power, but results have shown problems with THD, efficiency, and transformer saturation. This paper expands on previous work by providing a more thorough explanation of the SVM-based modulation and discussing and validating a power factor control for the topology. The SVM-based modulation is validated in simulation with parasitics and a four-step commutation, with an input and output current THD of 2.7% and 0.2%. In addition, the impact of modulation, parasitics, and commutation on transformer saturation is analyzed, which is an important challenge that has not been previously discussed for this topology.**

Index Terms—**Matrix converters, high frequency link, transformer, single stage, three-phase, AC-AC, isolated, industrial motor drive, bidirectional switch**

I. INTRODUCTION

With the sophistication of industrial processes, the demand for industrial motors with speed, torque and position control is expected to significantly grow. The gold standard for motor control is the variable speed drive (VSD) with active front end (AFE), which improves controllability and energy efficiency [1] but introduces system-level problems from increased EMI, common mode currents, harmonics, and circulating currents between VSDs [2]. To counteract issues introduced by VSDs, drive systems contain 1) VSD transformer(s), 2) VSD input filter(s), and 3) VSD output filter(s). These subsystems are external to the VSD but comprise the majority of drive system size and cost. For instance, the VSD transformer comprises up to 50% of system size and 70% of system weight [3]. In addition, EMI filters can comprise up to 30% of system volume [4].

System-level costs and size can be dramatically reduced by including high frequency (HF) isolation and filtering within the AFE and VSD, as depicted in Fig. 1. Therefore, a 3ϕ-3ϕ architecture with HF isolation and reduced EMI is needed. The most common 3ϕ-3ϕ architecture with isolation contains three stages: a voltage source rectifier (VSR), rectifier DC link, isolated HF DC/DC stage, inverter DC link, and voltage

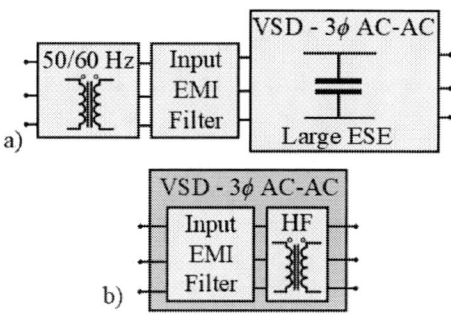

Fig. 1. (a) Traditional drive system with separate VSD transformer, input filter and VSD (b) proposed integration of subsystems within the VSD converter

source inverter (VSI). This solution suffers from size, cost and efficiency penalties due to its three stages and has large conducted EMI due to the VSR and VSI. Another candidate topology is a single stage soft-switching 3ϕ-3ϕ flyback-based converter called the S4T [5]. The S4T has full ZVS, controlled switching dv/dt, and a current-source converter type behavior which reduces conducted EMI [6]. However, the S4T's flyback-type transformer and output CL filter become bulky and expensive at higher power levels.

The candidate topology discussed in this paper is a single-stage, bidirectional 3ϕ-3ϕ converter with high frequency isolation that can reduce size, weight and cost compared to the S4T and 3-stage AC-AC converter. The converter is designed and controlled so that minimal energy is stored within the transformer magnetizing flux Φ_m and leakage flux Φ_{lkg}. For this reason, the converter is called the Isolated Inertialess Converter (IIC). There is no intermediate energy storage element (ESE) or output filter, which keeps converter size small when scaling to higher power levels, unlike the S4T flyback transformer, which stores energy in L_m. In addition, the high frequency transformer lowers input-side common-mode (CM) EMI, which reduces the EMI filtering requirements of the input-side filter. In addition, the power devices have a high utilization factor, unlike the flyback-style operation of the S4T.

Two research groups have performed preliminary low-power hardware builds and testing of the 3ϕ-3ϕ IIC [7] [8]. A carrier-based modulation was tested in [7], but experienced poor THD due to low switching frequency and a lack of synchronization

979-8-3315-1612-3/25 $31.00 © 2025 IEEE

between rectifier and inverter bridges. Improved current THD and unity power factor was achieved with a virtual indirect SVM-based modulation with deadtime compensation in [8], but input current waveforms had low order harmonics, and the converter had a low efficiency of 91.4%.

This paper builds on previous work by first proposing the 3ϕ-3ϕ IIC as a novel solution to consolidate and reduce size of industrial drive subsystems. Then a thorough mathematical description and validation are provided for the SVM-based modulation and input power factor control (PFC) in Section II. Lastly, implementation challenges are analyzed in Section III, including transformer saturation and transformer leakage management, and a built hardware prototype is discussed and compared to the two previous builds of the 3ϕ-3ϕ IIC.

II. TOPOLOGY AND PRINCIPLES OF OPERATION

A. DC to AC

The 3ϕ-3ϕ IIC can be understood by first examining the DC-AC IIC variant [9] of Fig. 2. The DC bridge (S_{a+}, S_{a-}, S_{b+} and S_{b-}) transfers power by applying HF voltage pulses of alternating polarity and equal volt-seconds across V_{pn} so that the transformer magnetizing flux Φ_m is kept close to zero. The voltage pulses across V_{pn} are reflected to the transformer secondary and applied to the secondary-side load. If the desired output voltage V_{AB} is positive, then the DC bridge and inverter synchronize switching patterns to output $+V_{DC}$. To synthesize a negative output voltage of $-V_{DC}$ across V_{AB}, the DC bridge and inverter are anti-synchronized.

The IIC can use multiple types of modulation. The duration of each voltage pulse can be fixed, in a discrete pulse modulation scheme such as $\Sigma\Delta$ modulation [9], as shown Fig. 2, or the pulse width can vary according to a PWM duty cycle.

When commutating between positive and negative half cycles, the DC bridge uses a zero vector S_0, either from $[S_{a+}S_{b+}]$ or $[S_{a-}\ S_{b-}]$ held high. During the DC bridge zero vector, V_{pn} and the transformer secondary V_{PN} are held at $0V$, if transformer parasitics are ignored. While the transformer secondary voltage is held at $0V$, the secondary-side inverter soft switches to its next vector.

The load inductor holds current stiff at I_L. Therefore, while switching the inverter bridge from the positive to negative vector, the transformer leakage L_{lkg} experiences a large $\frac{dI}{dt}$. To mitigate the leakage-induced voltage spike, a clamping circuit ($cc1$) is placed across the transformer secondary. To minimize clamp losses, small L_{lkg} is desired. Therefore, a coaxial winding transformer (CWT), which inherently has low L_{lkg} [10], is used for both the DC/AC IIC, and the 3ϕ-3ϕ IIC.

B. 3ϕ-3ϕ AC IIC Modulation

The 3ϕ-3ϕ IIC schematic, control diagram, and cycle-level switching waveforms using SVM are shown in Fig. 3. Example SVM vectors for the rectifier and inverter in one switching cycle are shown in Fig. 4. The 3ϕ-3ϕ IIC is similar to the DC-AC IIC, in that the damped input CL filter impresses a stiff voltage and the output inductive filter impresses a stiff current. One difference is that a blocking cap C_b is used to

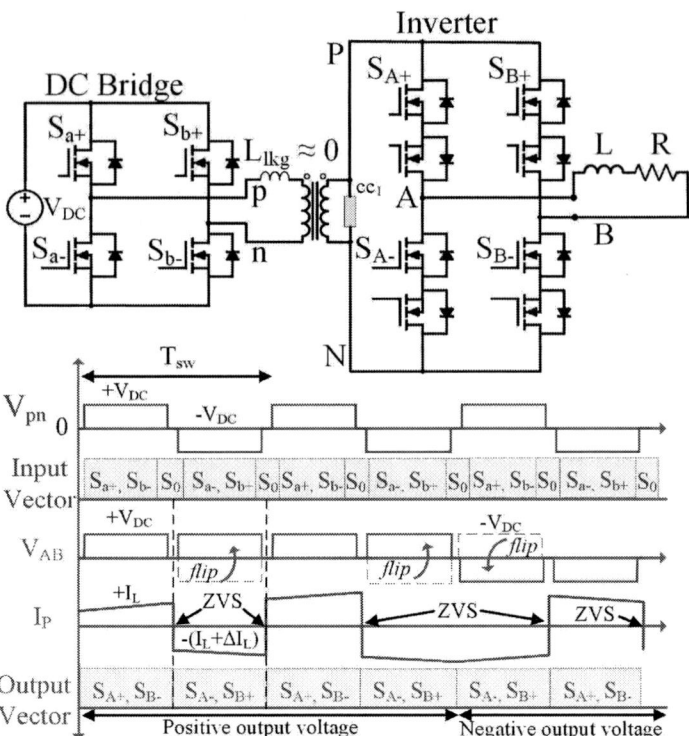

Fig. 2. Schematic of DC-1ϕAC IIC and idealized switching waveforms [9]

prevent volt-second mismatch from saturating the transformer. Volt-second error accrues over time due to switch turn-on and turn-off timing mismatch, CL filter voltage ripple, and other parasitic effects. The blocking capacitor requires current-second balance of transformer current I_p to prevent buildup of V_{C_b}.

Objectives for modulation and control include: 1) input PFC, 2) motor control, 3) bidirectional power flow, 4) 3ϕ sinusoidal, low-THD input current and output current, and 5) transformer volt-second and current-second balance to minimize magnetizing flux Φ_m and blocking cap C_b charge.

The 3ϕ-3ϕ IIC can adapt modulation schemes from the indirect matrix converter [11] by splitting each switching cycle into positive and negative half cycles applied across the transformer. A version of matrix-type SVM [12] is adapted to the 3ϕ-3ϕ IIC because of its good dynamic response, low sampling frequency, and wide application range [13].

In matrix-type SVM [12], the rectifier utilizes input bridge duty cycles $D_{i,1}$ and $D_{i,2}$ and line-to-line voltages $V_{1,ll}$ and $V_{2,ll}$ to synthesize a cycle-averaged constant voltage across the transformer $\overline{V_{pn}}$, also known as a virtual DC link voltage, shown in (1) and (2), where m_i is input modulation index, θ_i^* is input SVM reference angle, $V_{i,pk}$ is input voltage phase amplitude, and ϕ_i is rectifier bridge phase shift.

$$D_{i1} = m_i sin(\frac{\pi}{3} - \theta_i^*), \quad D_{i2} = m_i sin(\theta_i^*) \quad (1)$$

$$\overline{V_{pn}} = D_{i1}V_{1,ll} + D_{i2}V_{2,ll} = \frac{3}{2}V_{i,pk}m_i cos(\phi_i) = const \quad (2)$$

979-8-3315-1612-3/25 $31.00 © 2025 IEEE

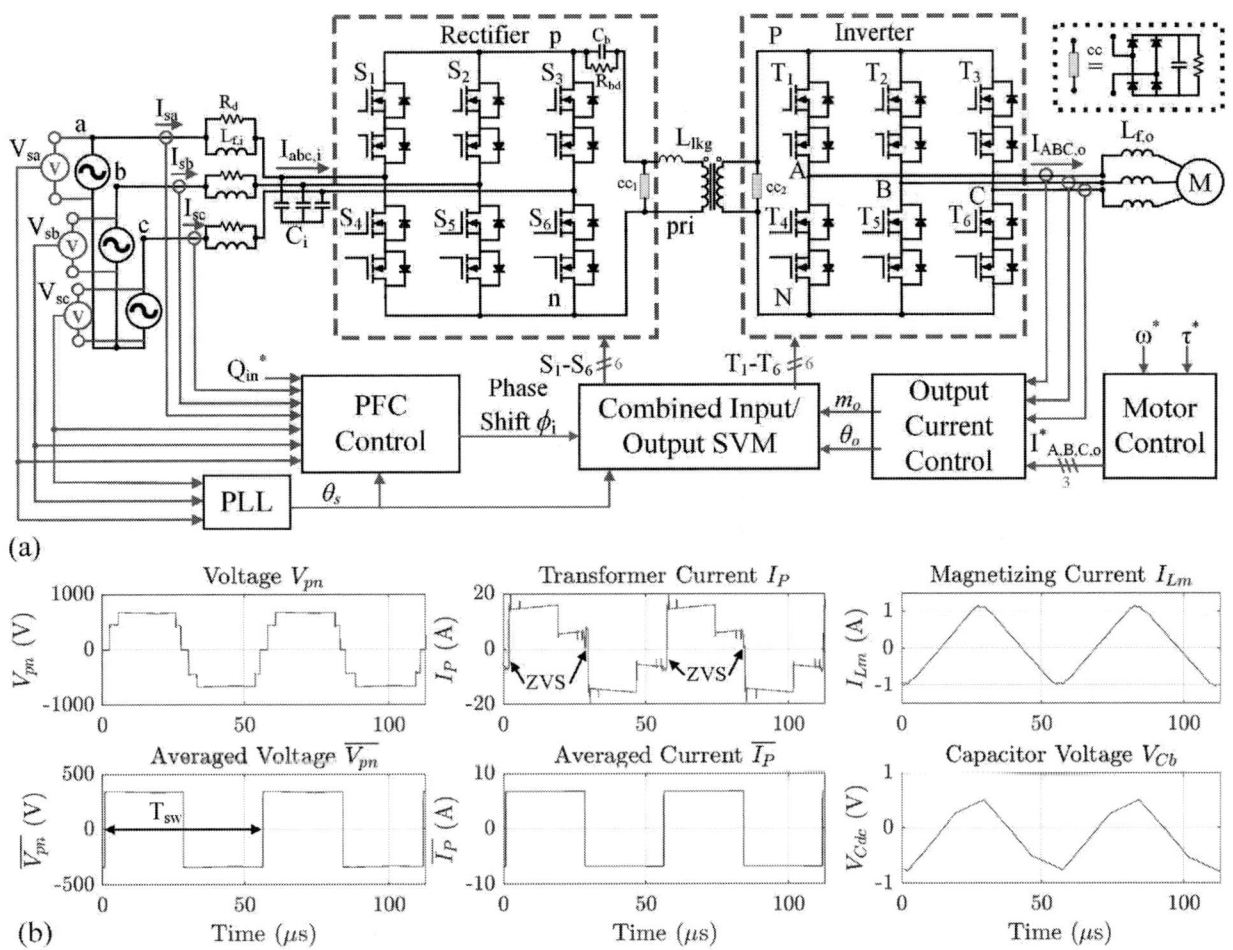

Fig. 3. (a) 3ϕ-3ϕ IIC schematic and high-level control diagram and (b) simulated switching cycle waveforms

The inverter then multiplies the cycle-averaged virtual DC link voltage $\overline{V_{pn}}$ by output bridge duty cycles $D_{o,1}$ and $D_{o,2}$ to synthesize a reference output voltage $\overline{V_o}^*$ in the $\alpha\beta$ frame in (3) and (4), where θ_o^* is output SVM reference angle, n_{npri} and n_{sec} are transformer primary and secondary turns, and V_{ok} is the kth SVM vector in the $\alpha\beta$ fame.

$$D_{o1} = m_o sin(\frac{\pi}{3} - \theta_o^*), \quad D_{o2} = m_o sin(\theta_o^*) \quad (3)$$

$$\overline{V_o}^* = D_{o1}V_{o1} + D_{o2}V_{o2}, \quad V_{ok} = \frac{2}{3}(\frac{n_{sec}}{n_{pri}})\overline{V_{pn}}e^{\frac{(k-1)\pi}{3}i} \quad (4)$$

Input and output SVM reference angles θ_i^* and θ_o^* are calculated in (5), in which they and are rotated back to the 1st SVM segment by taking into account input and output SVM segment number n_i and n_o. The input reference angle θ_i^* is derived from source voltage PLL angle θ_s and rectifier phase shift ϕ_i from the PFC. The output reference angle θ_o^* is derived from output phase reference ϕ_o^*.

$$\theta_i^* = \theta_s + \phi_i + \frac{\pi}{6} - \frac{\pi}{3}(n_i - 1), \quad \theta_o^* = \phi_o^* - \frac{\pi}{3}(n_o - 1) \quad (5)$$

Equations (6) and (7) show that in the combined rectifer-inverter SVM, the input duty cycles with modulation index m_i are multiplied by the output duty cycles with modulation index m_o. Therefore, m_i can be set to 1, while only m_o is adjusted. To output a voltage with magnitude $V_{o,pk}^*$, the output modulation index m_o is set to $\frac{\sqrt{3}V_{o,pk}^*}{V_{pn}}$.

$$T_1 = D_{i1}D_{o1}\frac{T_{sw}}{2}, \quad T_2 = D_{i2}D_{o1}\frac{T_{sw}}{2} \quad (6)$$

$$T_3 = D_{i1}D_{o2}\frac{T_{sw}}{2}, \quad T_4 = D_{i2}D_{o2}\frac{T_{sw}}{2} \quad (7)$$

$$T_0 = \frac{T_{sw}}{2} - T_1 - T_2 - T_3 - T_4 \quad (8)$$

Similarly to the cycle-averaged voltage $\overline{V_{pn}}$, the inverter uses output line currents and output duty cycles to synthesize a cycle-averaged constant current $\overline{I_P}$, also known as a virtual DC link current, as shown in (9) [12]. The cycle-averaged current $\overline{I_P}$ generated by the inverter is used by the rectifier vectors I_{i1} and I_{i2} with input duty cycles to synthesize the desired 3ϕ rectifier reference current $\overline{I_i}^*$ in the $\alpha\beta$ frame. The kth SVM rectifier vector in the $\alpha\beta$ frame is denoted by I_{ik}.

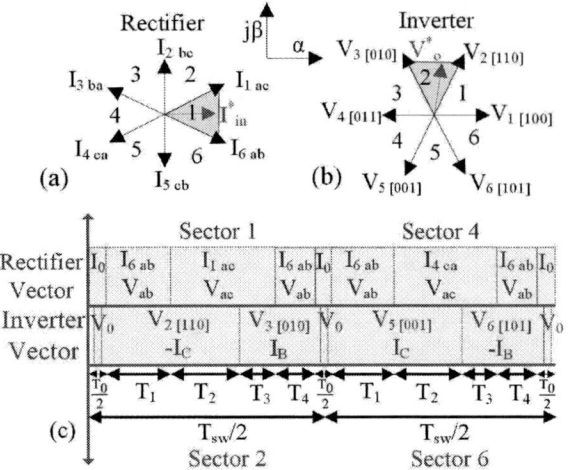

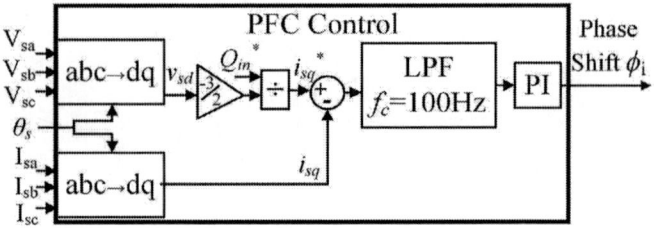

Fig. 5. 3ϕ-3ϕ IIC input power factor control block diagram

Fig. 4. Example rectifier SVM diagram with current reference vector I_i^* (a), example inverter SVM diagram with output voltage reference vector V_o^* (b), (c) vectors over one switching cycle

$$\overline{I_P} = D_{o1}I_{1,ph} + D_{o2}I_{2,ph} = \frac{3}{2}I_{o,pk}m_o cos(\phi_L) = const \quad (9)$$

$$\overline{I_i^*} = D_{i1}I_{i1} + D_{i2}I_{i2}, \quad I_{ik} = \frac{2}{3}\left(\frac{n_{pri}}{n_{sec}}\right)\overline{I_P}e^{\left(\frac{(k-1)\pi}{3} + \frac{\pi}{6}\right)i} \quad (10)$$

C. Commutation

A current-bidirectional commutation [14] was chosen for both the rectifier and inverter, in which the input voltage polarity is known, and the output current polarity does not need to be known. This was chosen because motor loads often have output current phase angles ϕ_L greater than $\frac{\pi}{6}$. Since an inverter SVM segment spans $\frac{\pi}{3}$, if the load angle is greater than $\frac{\pi}{6}$, then inverter vectors within the same SVM segment can have either a positive or negative output current. Current-bidirectional commutation permits safe commutation near the output current zero crossing within an SVM segment. In addition, the input SVM line-line voltage vector can safely be assumed to be positive, since the input current phase angle ϕ_i is very small while achieving PFC.

The inverter-side commutation is partially soft switching, as shown in Fig. 3(b) and Fig. 4. The soft switching method is identical to that discussed for the DC-AC IIC in Section II-A. During the zero voltage vector of the rectifier (I_0), the voltage V_{pn} is zero while the inverter switches between vectors. One difference between the DC-AC IIC and the 3ϕ-3ϕ IIC is that the blocking capacitor voltage V_{Cb} will be reflected to the transformer secondary voltage V_{PN}, which will induce a small amount of hard switching loss.

D. Control

Matrix converter control can be adopted for the 3ϕ-3ϕ IIC almost without modification, due to the lack of energy storage, identical three phase input CL filter and identical output inductive filter. An additional requirement is that the

effect of control dynamics on the magnetizing flux Φ_m and blocking cap voltage V_{Cb} must also be considered.

A block diagram of the PFC is shown in Fig. 5. The inputs to the PFC are source voltages $V_{abc,s}$ and currents $I_{abc,s}$, source voltage phase angle θ_s, and input reactive power reference Q_{in}^*. In the control, Q_{in}^* and the d-axis component of measured source voltage V_{sd} is used to generate a q-axis current reference i_{sq}^*. The error in i_{sq} is passed through a 1st order low pass filter (LPF) with cutoff frequency of 100 Hz to prevent control overreaction of ϕ_i to harmonic currents. Then, the filtered i_{sq} error signal is passed through a PI block to generate a rectifier phase shift ϕ_i. Equations (1), (5), and (10) show that adjusting ϕ_i will modify the input rectifier reference angle θ_i^*, which adjusts the rectifier duty cycles D_{i1} and D_{i2} applied to the virtual DC link current I_P until $I_{s,abc}$ is in phase with $V_{s,abc}$.

Fig. 6 shows simulated 3ϕ-3ϕ IIC waveforms for a 3ϕ RL load with output resistance R_o. Parameters are listed in Table I. Parasitics such as transformer secondary-side capacitance C_{xfmr} and leakage inductance L_{lkg} were included. Input PFC, output V/f control from 0 Hz to 60 Hz, and transformer volt-second balance were verified. No control was implemented for output current. The simulation implements four-step current-bidirectional commutation [14] with a commutation step duration (T_{comm}) of 200 ns and a minimum vector time ($T_{min\,vec}$) of 1 μs. The input and output current THD at full load are 2.7% and 0.2% respectively.

TABLE I
SIMULATION AND HARDWARE PARAMETERS

Parameter	Value	Parameter	Value
$V_{in,ll-rms}$	480 V	$V_{o,ll-rms}$	395 V
f_{in}	60 Hz	f_o	60 Hz
L_m	7.6 mH	C_b	250 μF
$n_p : n_s$	3:3	L_{lkg}	670 nH
R_{bd}	2.8 Ω	$C_{xfmr,\,sec}$	1 nF
$T_{min\,vec}$	1 μs	$T_{comm\,step}$	200 ns
f_{sw}	18 kHz	R_d	21 Ω
$L_{f,i}$	1.25 mH	C_i	7 μF
$L_{f,o}$	28.6 mH	R_o	18.2 Ω
$k_{P,PFC}$	0.025	$k_{I,PFC}$	0.5
$T_{ramp,\,V/f}$	0.2 s	Commutation	4-step

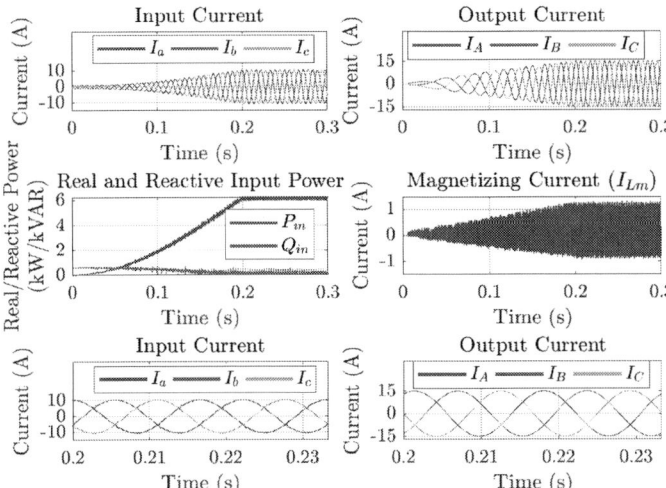

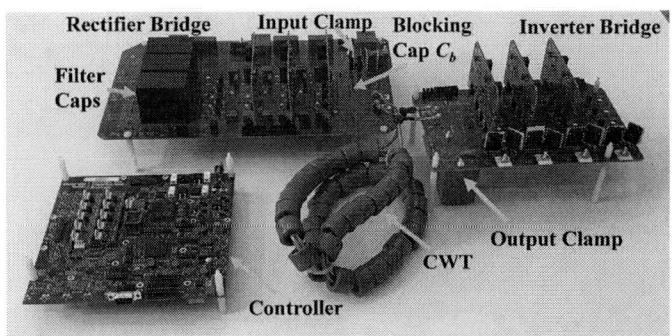

Fig. 7. Photo of 3ϕ-3ϕ IIC hardware setup

Fig. 6. Simulated input current, output current, real and reactive power, and magnetizing current demonstrating output V/f control, input PFC, and low THD current with four-step commutation

III. HARDWARE AND IMPLEMENTATION-RELATED CHALLENGES

A. Comparison to previous work on the 3ϕ-3ϕ IIC

Two research groups have performed preliminary low-power hardware builds and testing of the 3ϕ-3ϕ IIC [7] [15]. Both groups encountered hardware challenges and did not continue to work on the topology after preliminary prototypes. Therefore, it is important to learn from challenges from previous implementations of the 3ϕ-3ϕ IIC.

The hardware prototype in this work is shown in Fig. 7. Several changes have been made from previous hardware builds of the 3ϕ-3ϕ IIC, shown in Table II.

A major concern from previous experimental results was the low reported efficiency (91.7%) [8], due to hard switching, using Si devices, and having transformer leakage-induced losses from an L_{lkg} of 3.18 μH. The efficiency concern was addressed by choosing a commutation scheme with partial ZVS of the inverter as discussed in Section II-C, by using a low-L_{lkg} CWT [10], and by using G3R45MT17K SiC devices to lower conduction and switching loss.

The usage of SiC devices enables switching frequency to be increased. In order to keep the size of the CWT sufficiently small (below 1 meter in length) with low losses and L_{lkg} below

TABLE II
COMPARISON OF HARDWARE BUILDS OF THE SAME TOPOLOGY AS THE 3ϕ-3ϕ IIC

Specification	Group 1 [7]	Group 2 [15] [8]	Proposed Work
f_{sw} (Input)	1 kHz	10 kHz	18 kHz
Rectifier Modulation	Mod. PWM	SVM	SVM
f_{sw} (output)	5 kHz	10 kHz	18 kHz
Inverter Modulation	PWM	SVM	SVM
Soft Switching	No	No	Partial ZVS
Power Device	IGBT	IGBT	SiC MOSFET
Transformer	Standard	Standard	CWT

1 μF, a switching frequency of at least 18 kHz is needed. One competing factor to decrease f_{sw} is that the minimum vector duration $T_{vec,\,min}$ is desired to be less than 5% of the duration of half of a switching cycle $\frac{T_{sw}}{2}$. Due to the 4-step commutation, the minimum vector duration was chosen to be $5 * T_{comm\,step}$, yielding a minimum vector time of 1μs.

B. Transformer Saturation

No previous work has focused on transformer saturation, which is a major hardware challenge of the 3ϕ-3ϕ IIC. Over time, a mismatch in transformer volt-seconds due to inexact semiconductor turn-on/turn-off times, capacitor voltage ripple, and other parasitic effects will lead to transformer saturation, if not properly managed. For this reason, the blocking capacitor C_b was used. Saturation can be analyzed with the equivalent circuit of Fig. 8. A high pass filter is formed by L_m and C_b to remove DC and low-frequency voltage components generated by mismatch in volt-seconds from the supply or current-seconds from the load inductors.

The choice of L_m, C_{dc} and damping resistor R_{bd} are important. First, L_m is picked with a value large enough so that ΔI_{Lm} does not impact the input current THD. However, there is a direct correlation between L_m and L_{lkg}, so L_m was chosen small enough so that L_{lkg} was less than 1 μH. Next, a conservatively large value of C_b was chosen to minimize V_{Cb} in the event of volt-second or current-second mismatch. Then the damping resistance R_{bd} was chosen to critically damp resonance between C_{dc} and L_m. Lowering the value of R_{bd} damps LC resonance at the price of increased losses and larger steady-state magnetizing current I_{Lm}. Fig. 9 shows the simulated values of I_{Lm}, V_{Cb}, and P_{Rbd}, using the same simulation parameters from Table I and Fig. 6. It can be concluded that for the simulated case with critical damping,

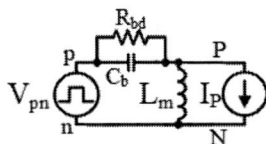

Fig. 8. Equivalent circuit to model I_{Lm} and V_{Cb}

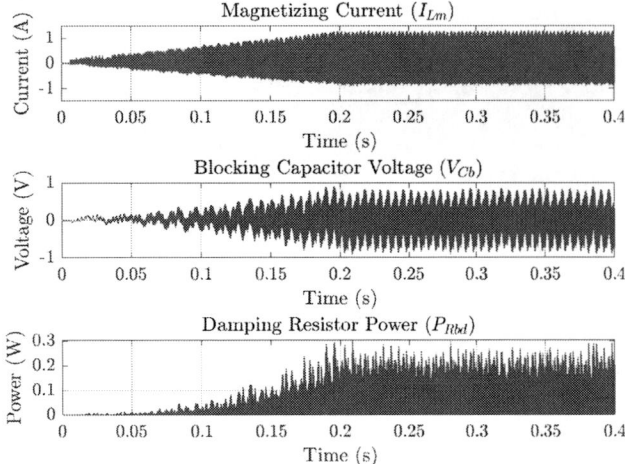

Fig. 9. Simulated waveforms for I_{Lm}, V_{Cb}, and P_{Rbd}

there are minimal losses within the damping resistor R_{bd}, while the resonance between L_m and C_b is removed.

One concern was that the dead-time commutation step with duration $T_{comm\,step}$ during the four-step commutation could lead to volt-second and current-second imbalance and transformer saturation. The simulated waveforms of Fig. 6 show that dead-time does not directly impact saturation. This is because the volt-seconds and current-seconds applied during commutation dead-time during the positive half cycle cancels out with commutation dead-time during the negative half cycle.

C. Leakage Management

Primary and secondary clamping circuits $cc1$ and $cc2$ clamp the L_{lkg}-induced voltage spike which occurs when the inverter changes vectors, as shown in Fig. 3. The transformer secondary bridge clamp $cc2$ is also tied to output phase terminals A, B, and C to absorb output filter inductance energy $L_{f,o}$ in the case of an inverter open circuit. Therefore, $cc2$ has a larger clamp capacitor, which is visible in Fig. (7).

IV. CONCLUSION AND FUTURE WORK

In this paper, a bidirectional, isolated single-stage matrix-type 3ϕ-3ϕ converter named the 3ϕ-3ϕ IIC has been proposed to integrate the VSD, VSD transformer, and filtering subsystems for industrial motor drive systems to reduce size, cost, and weight. The 3ϕ-3ϕ IIC's modulation and PFC have been explained and validated in simulation with a four-step commutation and relevant parasitics. Previous hardware builds of the 3ϕ-3ϕ IIC were compared to the built prototype of this paper, and the issue of transformer saturation was analyzed.

In future work, hardware results will fully validate the modulation and control of the 3ϕ-3ϕ IIC, and efficiency will be characterized.

ACKNOWLEDGEMENTS

The authors are grateful for the support provided by the Center for Distributed Energy (CDE) at Georgia Tech for this work.

REFERENCES

[1] ABB. (2021) White paper: Achieving the paris agreement the vital role of high-efficiency motors and drives in reducing energy consumption. ABB.

[2] G. L. Skibinski, R. M. Tallam, M. Pande, R. J. Kerkman, and D. W. Schlegel, "System design of adjustable speed drives, part 1: Equipment and load interactions," *IEEE Industry Applications Magazine*, vol. 18, no. 4, pp. 47–60, 2012.

[3] H. Abu-Rub, S. Bayhan, S. Moinoddin, M. Malinowski, and J. Guzinski, "Medium-voltage drives: Challenges and existing technology," *IEEE Power Electronics Magazine*, vol. 3, no. 2, pp. 29–41, 2016.

[4] B. Narayanasamy and F. Luo, "A survey of active emi filters for conducted emi noise reduction in power electronic converters," *IEEE Transactions on Electromagnetic Compatibility*, vol. 61, no. 6, pp. 2040–2049, 2019.

[5] L. Zheng, R. P. Kandula, K. Kandasamy, and D. Divan, "Single-stage soft-switching solid-state transformer for bidirectional motor drives," in *2017 IEEE Energy Conversion Congress and Exposition (ECCE)*, 2017, pp. 2498–2505.

[6] H. Dai, R. A. Torres, T. M. Jahns, and B. Sarlioglu, "Comparative study of conducted common-mode emi in wbg-enabled dc-fed three-phase current-source inverter," *IEEE Journal of Emerging and Selected Topics in Power Electronics*, vol. 10, no. 6, pp. 7188–7204, 2022.

[7] H. Cha and P. Enjeti, "A three-phase ac/ac high-frequency link matrix converter for vscf applications," in *IEEE 34th Annual Conference on Power Electronics Specialist, 2003. PESC '03.*, vol. 4, 2003, pp. 1971–1976 vol.4.

[8] K. Koiwa, J.-i. Itoh, and M. Shioda, "Improvement of waveform for high frequency ac-linked matrix converter with svm based on virtual indirect control," in *2015 IEEE Applied Power Electronics Conference and Exposition (APEC)*, 2015, pp. 3359–3366.

[9] S. Belkhode, N. Prabhu, J. Benzaquen, and D. Divan, "Single-stage bidirectional inertia-less isolated dc/ac converter," in *2024 IEEE Applied Power Electronics Conference and Exposition (APEC)*, 2024, pp. 348–353.

[10] M. S. Rauls, D. W. Novotny, D. M. Divan, R. R. Bacon, and R. W. Gascoigne, "Multiturn high-frequency coaxial winding power transformers," *IEEE Transactions on Industry Applications*, vol. 31, no. 1, pp. 112–118, 1995.

[11] L. Wei and T. Lipo, "A novel matrix converter topology with simple commutation," in *Conference Record of the 2001 IEEE Industry Applications Conference. 36th IAS Annual Meeting (Cat. No.01CH37248)*, vol. 3, 2001, pp. 1749–1754 vol.3.

[12] L. Huber and D. Borojevic, "Space vector modulated three-phase to three-phase matrix converter with input power factor correction," *IEEE Transactions on Industry Applications*, vol. 31, no. 6, pp. 1234–1246, 1995.

[13] J. Zhang, L. Li, and D. G. Dorrell, "Control and applications of direct matrix converters: A review," *Chinese Journal of Electrical Engineering*, vol. 4, no. 2, pp. 18–27, 2018.

[14] R. Narwal, S. Rawat, A. Kanale, T.-H. Cheng, A. Agarwal, S. Bhattacharya, B. J. Baliga, and D. C. Hopkins, "Analysis and characterization of four-quadrant switches based commutation cell," in *2023 IEEE Applied Power Electronics Conference and Exposition (APEC)*, 2023, pp. 209–216.

[15] K. Inoue, M. Shioda, M. Katade, A. Goto, S. Morishita, J. Itoh, and K. Koiwa, "Space vector modulation based on virtual indirect control for high frequency ac-linked matrix converter," in *2014 International Power Electronics Conference (IPEC-Hiroshima 2014 - ECCE ASIA)*, 2014, pp. 130–137.

AUTHOR INDEX

Abarzadeh, Mostafa 1261
Abbas, Asad 2973
Abotaleb, Youssef 1850
Abrams, Kerry J. 1781
Abramson, Rose A. 291, 2805
Abu-Rub, Omar 3071
Abu-Zaher, Mustafa 2327
Acero, Jesús 2468
Addin, Ali Sharaf 2960
Adeli, Mohammad Hassan 1489
Ademane, Harsha 3133
Adisurja, Ananda Tjakra 1255
Adragna, Claudio 958
Afrasiabi, Seyedeh Nazanin 1279
Afridi, Khurram K. 1640, 1646
Agarwal, Anant 2986
Ahammed, Md Tanvir 2220
Ahmad, Faheem 175
Aider, Youssef 1026
Aiello, Natale 738
Ajmal, Aidha Muhammad 3024
Akamatsu, Keiji 1728
Akter, Tanzila 1844, 2407
Akuta, Hector 761
Alam, Md Didarul 1746
Alam, Muhammad Muneeb 1051, 2569
Alassi, Abdulrahman 3071
Alathamneh, Mohammad 1953
Al-Durra, Ahmed 622, 2871, 3064
Alenezi, Ali 1217
Alexander, Mark 2162
Aleyasin, Seyed Hossein 1408
Ali, Abdelrahman 429
Ali, Jana A. Sheikh 3071
Ali, Kawsar 1529
Alkhatib, Mohamed 1940
Allen, Mark G. 1791
Allgeier, Jan 1919
Allioua, Abdelmoumin 2125
Alou, Pedro 1197
Al-Smadi, Mohammad K. 2779, 2840
Altin, Necmi 1489
Álvarez, Ignacio 3109
Aly, Mokhtar 746, 895, 2290, 2327
Alzahrani, Ahmad 1230
Alzate, Cesar 401
Amano, Yoshiki 3096
Amarathunga, Supun 3030

Amirabadi, Mahshid 1465, 1983
Amitkumar, K. S. 1279
Amler, Adrian 1759, 1767
Amor, Yacine Ayachi 1781
An, Jongchan 3000
Anand, Aniket 1096, 3147
Anand, Sandeep 69
Anantha, Neeraj 1121
Andapally, Bharadwaj Reddy 3119
Andersen, Michael A. E. 246
Anderson, Blake 1850
Ando, Masato 2681
Anekal, Latha 1224
Anjum, Waseah 1148
Antoszczuk, Pablo Daniel 479
Anurag, Anup 9, 442, 1318
Ao, Chengkang 171
Arai, Takamasa 2821
Araki, Hideo 3077
Aravind, G. 1610, 2785
Arduini, Douglas 1159
Asadi, Peyman 2162
Asel, Thaddeus J. 2419
Ashikaga, Toru 2284
Asllani, Besar 2051
Atkinson, Joshua 401
Attanasio, Rosario 3133, 3304
Attukadavil, Jenson Joseph C. 1481
Atwimah, Samuel K. 185, 207
Aunsborg, Thore Stig 175
Avenas, Yvan 1396, 2562
Aygun, Deniz 195
Azzopardi, Stéphane 2718
Bader, Samuel 1681
Bae, Jung-Soo 2228
Bae, Youngmin 3000
Baek, Jaeil 491
Bagci, F. Selin 880
Bahrami-Fard, Milad 930
Bak, Yeongsu 1734
Bakhshai, Alireza 3083
Balakrishnan, Manu 1286
Balamurali, Aiswarya 1096
Balda, Juan C. 27
Balen, Gleisson 1935
Balutto, Mattia 479
Banaie, Amin 1184
Banerjee, Arijit 3089

Bansal, Divyanshu 1610, 2785
Bantemits, Georgios.. 479
Bao, Mingjun 2628, 2968
Bao, Xiaokun 1143
Barbosa, Peter M. 2002
Barbosa, Peter 9, 442, 1318, 2082, 2296
Barik, Tapas 1184
Barros, Stayner Nóbrega.. 689
Barzegarkhoo, Reza 90
Basu, Arka 3253
Basu, Shibaji 464
Batard, Christophe 1076
Bau, Plinio 195
Bauer, Pavol 609
Bavi, Danial 385
Bazzi, Ali 2332, 2510
Beckemeyer, Randy 2082
Beig, Abdul R. 2647
Beinarys, Rytis 2009
Belanger, Matthew.. 2833
Belikov, Juri 1622
Belkhode, Satish 164, 3334
Benson, Mikayla 2413
Bergveld, H.J. 1451
Bertolini, Alessandro 2640
Beura, Kalpana 1940
Beushausen, Steffen.. 2589
Bezerra, Pedro A.M. 2361
Bhagat, Chinmay 3285, 3291
Bhambay, Rajul.. 2920
Bhattacharya, Subhashish 370, 552, 1347, 1866
Bhuse, Tejas.. 1, 54
Biadene, D. 2014
Bian, Fengwei 3312
Bien, Franklin 1629
Blaabjerg, Frede............................ 696, 912, 1501
Blanco, Cristian 1935
Blaquière, Jean-Marc.. 2718
Blij, Nils Hans Van Der 479
Boby, Mathews 1279
Boisseau, Sébastien.. 828
Boisson, Guillaume Piquet...................................... 1396
Bojoi, Radu 472, 1408
Bolaños, Robert E. 1190
Boles, Jessica D. 1012
Bonanno, Giovanni 1666
Borowy, Bogdan S. 1326
Boroyevich, Dushan 2228
Bosch, Michael 2387
Boutet, Jérôme.. 828
Bracken, Christopher 544, 3119
Bradford, Paul 2393

Brandão, Danilo I. 1355
Brandão, Dener A. de L.. 1355
Briz, P. 147
Briz, Pablo.. 3109
Brown, Alyssa 231
Brown III, Buck F. 1153
Brückner, Thomas 2960
Bruyere, Paul.. 2562
Bu, Jiankang.. 854
Bugade, Vikas 821
Burdío, José-Miguel.. 2468
Burgos, Rolando.................... 111, 409, 1495, 1551, 2992
Burnett, Hunter.. 401
Burt, Graeme.. 3167
Buttay, Cyril.. 2051
Cairnie, Mark.. 2228, 2616
Calabretta, Michele.. 1070
Cammarata, Federica... 252
Campbell, Steven 2797
Cao, Hanqing.. 1810
Cao, Hui 27
Cao, Yue 3036, 3048
Carretero, Claudio 2468
Castro, Alejandro 1427
Catanoso, Matthew.. 1791
Cattani, Alberto.. 2640
Cazzaniga, Daniele.. 958
Cazzitti, Sacha J. 1512
Cerutti, Stefano 738
Cervera, Pedro Alou 788
Cervone, Andrea.. 1305
Chae, Jongyoon.. 1899
Chagas, Rafael Bogo Portal 2446
Chakkalakkal, Sreejith 3147
Chakraborty, Shiladri 69
Chambon, Clément.. 828
Chamorro, Luis Ruiz 788
Chandrasekhar, Nurani...................................... 2889
Chang, Che-Wei 1495, 1551, 1564
Chang, Chuan-En 664
Chang, Jun-Yang 16
Chang, Yi-Chun.. 2143
Chareyron, Mathilde 828
Chatterjee, Bhaskar 1919
Chatterjee, Kallol 821
Chaturvedi, Shivam...................................... 2624, 3155
Chaturvedi, V... 1451
Chaudhary, Jai Aditya 3304
Chavarria, Jose 491
Chavez, Fredo 385
Cheema, Muhammad Ali Masood...................................... 768
Chellamuthu, Anand.. 1286

Chen, Cai 2343, 2369, 2426
Chen, Ching-Jan 664, 2131, 2143, 2725, 2735, 2741
Chen, Chun-Yen 938
Chen, Eric 1274
Chen, Guozhu 2059
Chen, Hao 906
Chen, Hongyu 2968
Chen, Hua 518
Chen, Hung-Chi 2687
Chen, Jiahong 1114
Chen, Jiann-Fuh 2043
Chen, Kai-Hui 887
Chen, Kevin J. 1047
Chen, Minjie 139, 349, 510, 566, 1274, 1693, 1882, 2438
Chen, Qiling 2375
Chen, Shih-Gang 938
Chen, Tianxiao 2361
Chen, Ting 2846
Chen, Wanjun 192
Chen, Xi 2628, 2968
Chen, Xingyu 1537, 1741
Chen, Yilun 2535
Cheng, Eric Ka-Wai 3227
Cheng, Jinpeng 1832
Cheng, Kuang-Yao 2157
Cheng, Lin 966, 1687
Cheng, Qi 2236
Cheng, Tzu-Ping 2900
Cheng, Yan 1047
Cheng, Yun-Keng 887
Cheshire, Audrey 682
Chetri, Chandan 3114
Chiu, Huang-Jen 389
Chiu, Jui-Yang 16, 900
Cho, Jaeyong 3187
Choi, Beomseok 491
Choi, Dongho 2311
Choi, Dongmin 1899
Choi, Jinsoo 3187
Choi, Jungwon 1874
Choi, Seokwon 2268
Choi, Seungdeog 943, 951, 1026, 1420, 1858
Choi, Sunghyuk 1659
Choi, Sungjin 3006
Choksi, Kushan 2582
Choo, Vin Loong 1576
Choong, Yin Quen 505
Chowdhury, Vikram Roy 645, 761, 1465, 3059
Chuang, Cheng-Ta 664, 2725
Chung, Henry Shu-Hung 98, 1507, 1582
Ciabattoni, Matteo 1646
Ciardo, S. Yuri 252

Clark, Landon.. 919
Cobos, Álvaro 1427
Cobos, José A. 1427
Coday, Samantha 971, 2249
Collings, William M. 185
Cong, Yizhou... 2986
Contreras-Barrios, René 629
Coomans, Bart 195
Corradini, Luca 1, 54, 334, 2764
Costa, Levy F. 1334, 1341
Costinett, Daniel... 3253, 3267
Cox, James 538
Cronin, Jared 2865
Croston, José Andrés Aguilar 2051
Crovetti, Paolo Stefano 738
Cruz, Alfonso 860
Cruz, Mario F. 670
Cui, Hongchang 202
Cui, Wen Tao 1108
Cui, Yujia 2932
Curbow, Austin 1167
D'Amato, Davide 689
Da-Cunha-Alves, Wendell 429
Dai, Hang 3174
Dang, Yongliang 278, 2482
Dannehl, Kai 1774
Dardeer, Mostafa 906
Darvish, Peyman 2453
Das, Shuvangkar Chandra 1184
Datta, Kishalay 1715
Datta, Promit 586
Davari, S. Alireza 2290
De, Vivek 518, 1681
Deboi, Brian 1167
Deboy, Gerald 1444, 2260
Defaz, Samuel 2576, 2582
Dekka, Apparao 2647
Delmar, Aria 1242
Deneke, Niklas 848
Deng, Jianting 3312
Deniz, Erkan 1489
Deppe, Conner 2393
Derbey, Alexis 2562
Desai, Nachiket 1681
Descamps, Anne-Sophie 1076
Deshpande, Ankit Vivek 1459
Dev, Archit 2920
DeVoto, Douglas 1824
Diao, Naizhe 757
Dieckerhoff, Sibylle 2361
DiMarino, Christina 586, 1836, 2228, 2616
Ding, Peiyang 2375

Ding, Wenlong	2713
Divan, Deepak	164, 3334
Do, Huong	491, 1681
Dobakhshari, Sina Salehi	1673
Dominguez, Miguel Alvarez	1640
Dong, Minhai	2075
Dong,	111, 1495, 1551, 1564, 2992
Driesen, Johan	3124
Driussi, Francesco	479
Dryden, Daniel M.	2419
Du, Bangli	436, 2752
Duan, Bin	2713
Dujic, Drazen	266, 1063, 1305
Dutta, Soham	711
Dworakowski, Piotr	2051
Eguchi, Shinichiro	2828
Ekuewa, Oluwaseun Isaiah	2973
Elasser, Youssef	510, 566, 2438
Elezab, Ahmed	670
El-Fouly, Tarek H.M.	622
Ellis, Nathan M.	2276
Ellis, Philip	1781
El-Refaie, Ayman M.	1551
El-Refaie, Ayman	1230, 1495
El-Saadany, Ehab F.	622
Elsanabary, Ahmed	746
Elshaer, Mohamed	3155
Emadi, Ali	670, 3147
Endo, Shun	2681
Eni, Emanuel	2746
Enjeti, Prasad	727, 1217, 1459, 3054
Enomoto, Jun	3194
Enslin, Johan	1153
Espinar, Alberto	3100
Espinoza, Angel	214
Estrin, Julia	132
Etta, Dheeraj	1640, 1646
Evzelman, Michael	594
Expósito, Alberto Delgado	788, 1803
Fahimi, Babak	930, 3160
Fahmy, Youssef A.	272
Falkenberg, Niklas	2772
Fan, Junchong	1203
Fan, Yucheng	2981
Farantatos, Evangelos	1184
Farivar, Glen G.	1927
Fassi, Youssof	828
Fein, Martin	2348
Feng, Hao	1832
Feng, Kaiyuan	2894
Feng, Wenda	3174
Fernandes, Arnold	1311

Fernandes, Baylon G.	1481
Ferrari, Maximiliano	637
Figueroa, Alejandro	1427
Filho, Braz de J.C.	1355, 1615
Fiore, Michele	1070
Flannery, John	285
Flaten, Paul	682
Forouzesh, Mojtaba	1673, 1892
Forsyth, Andrew J.	1512
Foster, Geoffrey M.	207
Fox, Aidan P.	185
Fox, Matthew	1791
Francés, Airán	868, 3298
Francois, Thomas W.	1311
Frank, Simon	2348
Freeman, Andrew	1242
Fu, Minfan	809, 2846, 3206
Fu, Pengyu	1203, 2986
Fujisaki, Keisuke	1797
Fujita, Jun	1383
Funaki, Tsuyoshi	2813
Funatani, Kenji	2654
Funatsu, Shohei	1237
Furukawa, Akihiko	1383
Gaafar, Mahmoud A.	775, 906, 2327
Gajare, Siddhesh	214
Galamb, Andrew	2527
Gallage, Nirashi Polwaththa	874
Gangadhar, Pratheesh	2920
Gao, Alex	2149
Gao, Ju	171, 225
Gao, Mingze	2992
Gao, Xiang	2846
Gao, Xiaoguang	2070
Gao, Yuan	524, 1034
Gao, Yuntian	278, 2482
Garcia, Enrique	538
García, Pablo	1935
Garcia, Ricardo	214
García, Sofía	3298
García-Espinosa, Antoni	1774
Garza-Arias, Enrique	1459
Gasparini, Alessandro	2640
Gato, Jose	3119
Gautam, Sushanta	185
Gauthier, Jean-Yves	2051
Gauttam, Gaureej	3316
Geboers, Tim	436
Gennaro, Francesco	738
Georgescu, Sorin	180
Georgiev, Daniel G.	185, 207
Gessner, Joerg	1889

Ghanayem, Haneen 1953
Ghartemani, Masoud Karimi 943
Ghitelman, Kolman Puterman 2101
Ghosh, Mohendro Kumar 1326
Ghosh, Prosenjit 2541
Ghosh, Subarto Kumar 1420
Giardine, Francesca 151
Gil, Pablo M. 1701
Ginot, Nicolas 1076
Giuffrida, Simone 472
Gockel, Hendrik 2125
Goetz, Stefan M. 1754, 2846, 3206
Goicoechea, Javier 1427
Gomez-Rivera, Luis F. 1774
Gong, Jiakun 219
Gong, Minxiang 518
Gong, Taehyeon 3006
Gong, Xiaowu 1114
Gonzalez, Reynaldo S. 1190
Gonzalez-Castaño, Catalina 629
Goodrich, Dakota 719
Goto, Akiko 1569
Gouy, Louison 1076
Graber, Lukas 860, 3321
Grainger, Brandon 544, 1326
Green, Andrew J. 2419
Griepentrog, Gerd 2125
Grigoryan, Davit 566, 1882, 2438
Groon, Fabian 90
Guan, Quanxue 895
Guenther, Robert 1203
Guichon, Jean-Michel 2562
Guillod, Thomas 1816
Gunawardena, Pasan 3030
Guo, Heng 2713
Guo, Jiacheng 2375
Guo, Weisheng 3181
Guo, Xiaoqiang 757
Guo, Zhengchen 2703
Guo, Zhongyin 2070
Gurudiwan, Shubhangi 719, 2194
Guthrie, Travis 2162
Gutierrez, Harold 1159
Ha, Jung-Ik 457, 1659, 2268, 2937
Habibolahi, Zahra Sadat 2202
Haddadi, Aboutaleb 1184
Hajisadeghian, Hossein 1666
Halawa, Ali 1473
Hamani, Rachid 1889
Hameed, Aamna Nasir 1673
Hameed, Asad 1972
Han, Yi 103

Hanhart, Michael 2757
Hanna, Rachelle 1396
Hansen, Frederik Lillebæk 2380
Hanson, Alex J. 231, 1121, 2521
Hanson, Alex 77, 2857
Hao, Weijia 2109
Harbi, Ibrahim 895, 2290
Haryani, Nidhi 442
Hasan, Abu Shahir Md Khalid 1844, 2407
Hasan, Md Zakir 1026
Hasan, Syed Imam 1294, 2698
Hassan, Alaaeldien 2327
Hassan, Najam Ul 834
Hassan, Nazmul 1746
Hata, Katsuhiro 1084, 1102, 2284, 2551
Hayashi, Tetsuya 423
He, Bill 3129
He, Binghui 1673
He, JiangBiao 919, 1368
He, Jiayin 171, 225
He, Junlei 3129
He, Xinlong 2066
Heckel, Thomas 1759
Hedenik, Marina 1519
Hedeshi, Hamid Montazeri 2202
Hegde, Anantha 1728
Heinen, Stefan 2757
Heiries, Vincent 828
Heldwein, Marcelo Lobo 2446
Hemming, Samuel 670
Heo, Go Woon 1723
Herbert, Edward 2495
Hernandez, Arturo Sanchez 530
Herzer, Stefan 1286
Higashiyama, Koji 1728
Hiller, Marc 1919, 2348
Hiraki, Eiji 321, 2654
Hiraoka, Toshio 285
Hirase, Yuko 1946
Hisamochi, Hirofumi 1414
Hobart, Karl D. 185, 207
Hoene, Eckart 2361
Hokmabad, Hossein Nourollahi 1622
Hong, Kang 2096
Hontz, Micheal R. 207
Horibe, Masahiro 2821
Hornbuckle, Malachi 363, 2241
Horowitz, Logan 151, 2276
Hosani, Khalifa Al 1940, 2871, 3064
Hossain, Md Maksudul 2407
Hossain, Mohammad Safayet 1184
Hou, Ting 2375

Hou, Zhengming	1913, 2851
Houska, Brad	3334
Howell, Brandon	2162
Hsieh, Chun-Yu	2735
Hsieh, Hsin-Che	815
Hsu, Jun-Ming	938
Hu, Borong	1439, 2597
Hu, Changsheng	2894
Hu, Changyu	3129
Hu, Jhih-Cheng	2692
Hu, Jiangang	2932
Hu, Shoudong	2764
Huang, Alex Q.	1786
Huang, Cheng	505, 1946
Huang, Hao-Ran	664, 2131
Huang, Ming-Shi	938, 2692
Huang, PengHao	1217
Huang, Peng-Hao	727
Huang, Qinghui	1173, 2603
Huber, J.	2014
Huber, Jonas	1318
Huber, Laszlo	442
Hudgins, Jerry L.	2877
Hudgins, Jerry	1850
Huh, Kum-Kang	3174
Hui, Shu Yuen Ron	3275
Hung, Chien-Chih	16, 900
Hung, Yu-Ting	2735
Huo, Zhenguo	2660
Husain, Iqbal	1746
Husev, Oleksandr	1622, 2173
Hussain, Amir	1990
Hwang, Yun Seong	733
Iannuzzo, Francesco	738, 1070
Ibáñez-Muñoz, Esteban	629
Ibrahim, Ahmed	775
Ibrahim, Eltaib Abdeen D.	775
Ibrahim, Hasan	727, 3054
Ibrahim, Mohamed	670
Ide, Tomoya	1946
Ikriannikov, Alexandr	2149
Iliæ, Milan	2764
Ilka, Reza	1368
Imaeda, Yuta	2431
Imaoka, Jun	2431
Imperiali, Luc	1318
Inokuchi, Seiichiro	2356
Inoue, Shuntaro	782
Irie, Yusuke	2828
Ishido, Ryosuke	1797
Ishihara, Masataka	321, 2654
Ishikura, Yuki	3285, 3291

Ishizuka, Yoichi	2828
Ishraq, Naveed	34, 1135
Islam, Md Khurshedul	943, 951
Islam, Md Majharul	2407
Islam, Nasherul	2059
Islam, Sarwar	1824
Ismail, Ahmed H.	2453
Isobe, Takanori	1946
Ito, Yuki	3248
Itoh, Jun-Ichi	21, 48, 2913
Ivimey, Arjun	464
Iwabuchi, Akio	2828
Iwamoto, Motomitsu	1108
Iyer, Rahul K.	157
Iyer, Vignesh	2764
Jacobs, Alan G.	207
Jafarian, Yousefreza	3083
Jahns, Thomas	3174
Jain, Akshat	658
Jain, Praveen	464, 616, 2953, 3083
Jalakas, Tanel	1622
Jalalabadi, Esmaeil	416
Janabi, Ameer	2597
Jayalath, Sampath	3212
Jeong, Seogyong	834
Jeong, Won Hyo	2937
Jerez, Raiphy	2249
Jha, Kunal	1519
Ji, Shengchang	278, 2482
Ji, Shiqi	2857
Ji, Yichao	966, 1687
Ji, Yingfeng	2889
Jia, Xiaoting	1564
Jiang, C.Q.	795, 3181
Jiang, W.L.	1451
Jiang, Wei	2343, 2426
Jiang, Xi	1114
Jiang, Yang	978
Jiang, Yongbin	3220
Jiao, Dong	1913, 2851
Jiao, Yang	416
Jin, Feng	429
Jin, Liyang	1020, 1564
Jin, Sicong	3018
Jin, Zhiyang	860
Jing, Mengmeng	2713
Jo, Hyeonu	3200
Jo, Hyunkyeong	1629
Jochmans, Thomas	258
Johnson, Brian	711
Johnson, Ken	2510
Jørgensen, Asger Bjørn	357, 1034

Jørgensen, Jannick Kjær	175, 357
Joshi, Kishor	943
Juds, Mark A.	1326, 3119
Jung, Jee-Hoon	834
Jung, Jun-Hyung	689
Jurkov, Alexander	124, 132
Kabashima, Takamune	1728
Kachura, Avram	449, 1905
Kai, Toshihiro	423
Kalathy, Abirami	616, 2953
Kallfass, Ingmar	2387, 3241
Kamalapur, Aakash	2228
Kamran,	252
Kanakri, Haitham	2029
Kanathipan, Kajanan	768
Kandeel, Youssef	285
Kang, Byeong-Woo	2948
Kang, Doug	180
Kang, Eunjin	3012
Kang, Gyeong-Gu	566, 2438
Kang, Seung Hyun	733
Kang, Yong	2066, 2343, 2369, 2426
Kano, Yuko	782
Kanungo, Gautam Dey	821
Kar, Narayan C.	1096
Karanth, Shashank	2746
Karimi-Ghartemani, Masoud	1858
Kataoka, Soya	1237
Katsura, Kenshiro	1299
Kaufmann, Maik	1286
Kawahara, Chihiro	2356
Kawamoto, Keisuke	1569
Kawano, Akihiro	1977
Kelkar, Kapil	1519
Kennel, Ralph	895
Kerekes, Tamás	738, 3042
Khaburi, Davood Arab	895
Khadka, Purushottam	1040, 2400
Khalid, Saad	2569
Khalife, Khalil	479
Khan, Faisal	1824
Khan, N.	1451
Khan, Nisar Ahmed	2569
Khan, Shahid Aziz	2624, 3155
Khandelwal, Sourabh	385
Khandla, Dhaval	2920
Khanna, Mudit	854
Khanna, Raghav	185, 207
Khatua, Mausamjeet	1681
Khorasani, Ramin Rahimzadeh	2101
Kim, Byeong-Il	1734
Kim, Chae-Lyn	3200

Kim, Daehyun	3187
Kim, Dong Hwan	1723
Kim, Dongmin	1899
Kim, Han-Gyu	951
Kim, Hongrae	1746
Kim, Hyeon Soo	733
Kim, Jae-Seong	925
Kim, Jaewon	727, 3054
Kim, Jeonghun	761
Kim, Jonghoon	2973, 3000, 3006, 3012
Kim, Jong-Hun	834
Kim, Joon-Seok	1734
Kim, Jungho	1629
Kim, Katherine A.	880
Kim, Minhyeok	3012
Kim, Min-Sik	834
Kim, Myeong-Ho	834
Kim, Namwon	703
Kim, Sung-Oh	2943
Kim, Yura	3006
Kimball, Jonathan W.	1311
Kimpara, Renata	703
Kirtley, James L.	2474
Kisacikoglu, Mithat John	1602
Kishikawa, Ryoko	2821
Kishimoto, Sumiaki	285
Kitano, Junichi	3194
Klidbari, Mohammadreza Khodaparast	2202
Klymenko, Mariia	1590
Knapp, Jeffrey	854
Knappstein, Lukas	2772
Knoll, Jack	2228, 2616
Ko, Bomyeong	3006
Kobayashi, Takumi	3248
Koch, Dominik	2387
Koehler, Andrew D.	185, 207
Koga, Shunsaku	3194
Koga, Takahiro	2828
Kokkonda, Raj Kumar	1347
Kolar, J.W.	2014
Kolar, Johann W.	1318
Kolli, Nithin	1347
Komiyama, Yutaro	3248
Komo, Hideo	2356
Kondo, Hiroki	1102
Kondo, Ryota	2813
Kong, Jiaze	2167
Kong, Jie	2380
Kong, Rui	696
Konishi, Akihiro	3248
Koppolu, Manoj	2920
Korrani, Majid Ghasemi	930, 3160

Kosaka, Takashi 1237, 3096
Koseoglou, Sokratis 479
Kotani, Junichi 579
Kouro, Samir 775, 2327
Kozak, Joseph P. 1211
Kozielski, Kyle 3147
Kragl, Robert 1051
Krishnamoorthy, Harish S. 3316
Krishnamurthy, Harish 1681
Krishnamurthy, Karthik 3129
Krishnan, Sahana 151, 291, 2805
Kritprajun, Paychuda 1184
Ku, Han 900
Kubulus, Pawel Piotr 1034
Kularatna, Nihal 378, 874
Kularatna-Abeywardana, Dulsha 874
Kulasekaran, Siddharth 491
Kumar, Misha 2002, 2082
Kumar, Pavan 530, 3312
Kusaka, Keisuke 3261
Kusunoki, Shigeru 2551
Kutrolli, Uiliam 2332
Kwak, Jin Woong 2541
Kwon, Hyukjae 566, 2438
Kwon, Man Jae 733
Ladhar, Manraj Singh 2322
Laha, Arpan 616, 2953
Lahuerta, Óscar 2468
Lai, Jih-Sheng 815, 1058, 1913, 2851
Lai, Rixin 2138
Lai, Yanwen 1173, 2603
Laird, Ian 2088
Lam, John 768, 2022
Lamar, Diego G. 1701, 1959
Lawniczak, Celine 1129
Lawson, Wayne 1403, 1781
Lazzarin, Telles Brunelli 342
Le, Duc Dung 3155
Le, DucDung 2624
Le, Hoang 2647
Le, Thanh-Long 2718
Leary, Alex M. 2516
Lee, Bonyoung 1629
Lee, Byoung Kuk 733, 1723, 3200
Lee, Byunghun 834
Lee, Chen-Chan 1058
Lee, Dongcheol 3000
Lee, Dong-Choon 1267
Lee, Dongsu 457
Lee, Eun Woo 2311
Lee, Hoi 2236
Lee, Jaea 3012

Lee, Jaehyeong 3006
Lee, Ju-A 3200
Lee, Jun Young 2311
Lee, June-Seok 1734, 2311
Lee, Justin 2138
Lee, Juwon 457
Lee, Kahyun 2937
Lee, Kangbeen 2413
Lee, Kevin 327, 1261, 2907
Lee, Kyo-Beum 925, 2317, 2943, 2948
Lee, Kyungmin 2547, 3281
Lee, Miyoung 3000
Lee, Po-Chang 900
Lee, Seongkyu 3012
Lee, Seunghyun 3012
Lee, Sungjun 3006
Lee, Taewoo 1659
Lee, Ting-Lun 2143
Lee, Wen-Hsuan 2043
Lee, Woongkul 1473, 2413
Lee, Yun-Jin 2317
Lehman, Brad 761, 1465, 1983
Lehmeier, Thomas 1767
Lei, Weihao 1143
Lei, Yiming 3312
Leslie, Alec 401
Leyrer, Thomas 2920
Li, Bing 2932
Li, Chun-I 2741
Li, Duo 307
Li, Haoran 510, 566, 1882, 2438
Li, Heyuan 809, 2846, 3206
Li, Hui 1248, 2075
Li, Jiajun 1590
Li, Lingyun 524
Li, Peidong 3036, 3048
Li, Pengwei 2332
Li, Qiang 202, 299, 429, 498, 1433, 1537, 1557,
.... 1741, 2228, 2488
Li, Ruqi 1159
Li, Sichao 3129
Li, Tien-Sheng 111
Li, Xiang 3312
Li, Xiaoling 1824
Li, Xindong 3212
Li, Xinze 1143
Li, Xuewen 751
Li, Yang 2576
Li, Yanqiao 1590
Li, Yaohua 3220
Li, Yi 1153
Li, Yilei 2035

Li, Yiming 1173
Li, Yuan 761, 1465
Li, Yunwei 3030
Li, Zehui 485
Li, Zhenchao 1305
Lian, Zhina 1090
Liang, Gaowen 1927
Liang, Jingyuan 1108
Liang, Katherine 363, 2241
Liang, Tsorng-Juu 887
Liang, Yaogan 1084
Liao, Hong-Xuan 2692
Liao, Hsuan 2043
Liao, Kuo Fu 2043
Liao, Mian 139, 349, 1882
Libbos, Elie 3089
Lim, Gyu Cheol 2937
Lim, Je-Yeong 1723
Lim, Jong-Hun 1723
Lin, Fanfan 1143
Lin, Jesse 1211
Lin, Jinshu 2075
Lin, Lei 2535
Lin, Qing 409
Lin, River 1159
Lin, Wei-Ren 258
Linares, Daniel Ríos 1375
Liserre, Marco 90, 118, 689, 1148
Liske, Andreas 2348
Liu, Baihan 2343, 2426
Liu, Caifeng 2066
Liu, Chen 3042
Liu, Chien-Lung 2692
Liu, Ching-Yao 1058
Liu, Christopher 1403
Liu, Chun-Hung 1026
Liu, Gao 357, 1034
Liu, Hanbing 3232
Liu, Haoyang 2361
Liu, Hong 2634
Liu, Hongru 2675
Liu, Hualong 1363, 1597
Liu, Jia 751
Liu, Jiahong 3042
Liu, Jiaxin 2343, 2369, 2426
Liu, Jinjie 1114
Liu, Jinjun 751
Liu, Kevin 1274
Liu, Liming 1746
Liu, Ming 2521
Liu, Sijia 2369, 2426
Liu, Wen-Chin B. 315

Liu, Wentao 1090
Liu, Xiaosen 1544, 2556, 2675
Liu, Xiaoshan 429
Liu, Y. 1451
Liu, Yan-Fei 1673, 1892
Liu, Yang 2675, 3321
Liu, Yifu 3129
Liu, Yongjie 1501, 3042
Liu, Yu-Chen 2179
Liu, Yunting 1179
Liu, Zeguo 1687
Liu, Zengyang 2488
Liu, Zhan 2521
Liu, Zhanlei 278, 2482
Liu, Ziheng 171, 225
Locher, Fabrice 2495
Locke, William 3141
Lodge, Finlay 3167
Logi, Sean 880
Long, Haihong 651, 2981
Long, Teng 1439, 2597
Loparo, Kenneth A. 2698
Lope, Ignacio 2468
López, Abraham 1959
Lopez-Torres, Carlos 1774
Lu, Che-Yu 1967, 2900
Lu, Fengwang 98
Lu, Guo-Quan 586, 2228
Lu, Lucas 416
Lu, Mowei 1754, 2846, 3206
Lu, Wei 2117
Luan, Shaokang 1034
Lucía, O. 147
Lucía, Óscar 3109
Luckett, Benjamin 919
Luise, Claudio 2640
Lukic, Srdjan 2527
Lumod, Phen 1159
Luo, Fang 2576, 2582
Luo, Tianming 2035
Lv, Jianwei 2343, 2369, 2426
Ma, D. Brian 2541
Ma, Dingkun 2375
Ma, Guangji 2070
Ma, Hangxiao 978
Ma, Tianlu 3181
Ma, Zhedong 1173, 2603
Ma, Zhiyuan 1786
Ma, Zhuxuan 27
Maaz, Syed Mohammad 1267
Mabuchi, Yuuichi 2681
MacFadyen, Martin 3167

Madadi, Mehrnaz	370
Maddela, Avinash	1715
Maekawa, Sari	2924
Mahbub, S. Tahmid	157
Maheshwari, Anuj	3089
Maji, Sounak	1640
Major, Joshua	1824
Mak, Pui-In	978
Maksimoviæ, Dragan	1, 54, 334, 682, 2764
Malannino, Claudia	252
Mallik, Ayan	34, 1135
Mallik, Ranajay	658
Mandrile, Fabio	472
Manjrekar, Madhav	2883
Mannan, Tahmid Ibne	1420, 1858
Manos, Konstantinos	1274, 1693
Mansour, Mahmoud	719
Mantooth, H. Alan	1844, 2407
Manzoni, Stefano	958
Marcault, Emmanuel	1396
Marellapudi, Aniruddh	164, 3334
Marianne, Julien	828
Marin, Brandon	491
Marquardt, Rainer	2960
Martin, Alexander	1211
Martin, Sébastien	828
Martin, Trent	1, 54
Martinez, Wilmar	238, 258, 436, 2167, 2752, 3124
Martinez-Limia, Alberto	1051
Martins, João R.R.O.	1889
Martins, Rui P.	978
März, Martin	1759, 1767
Mather, Barry	645, 3059
Mathieu, Frédéric	2495
Mathúna, Cian Ó.	285
Matiushkin, Oleksandr	1622, 2173
Matsumori, Hiroaki	1237, 3096
Matsumoto, Hirokazu	579
Matsumoto, Yohei	2681
Matsuo, Takayoshi	2932
Mattavelli, P.	2014
Mattavelli, Paolo	2667
Maureira-Riquelme, Ángel	629
Mauromicale, Giuseppe	1070
Mavencamp, Dan	2157
Mazariegos, Pablo	1427
Mazzer, Simone	1444, 2254, 2260
McDonald, Brent	42
McGrew, Tyler	1557
Mekhilef, Saad	746
Mendes, Arthur	2992
Mercier, Patrick P.	315

Metwly, Mohamed Y.	919
Meyer, Stefan	1034
Miao, Honglei	2059
Michelis, Stefano	479
Milivojeviæ, Nikola	1, 54
Min, Hao	1090
Min, Hyungki	1629
Min, Run	2109
Minato, Yuichiro	2913
Mirafzal, Behrooz	2461
Mirkoviæ, Nikola	788
Mishima, Taichi	3248
Mishra, Santanu K.	2213
Mitcheson, Paul D.	1653
Mitrovic, Vladimir	409
Mitsui, Koji	1299
Miyamae, Masaki	2681
Miyanjou, Kazuki	1977
Miyazaki, Tatsuya	1797
Mo, Liping	795
Mo, Xianghao	1375
Mohammad, Mostak	1635
Mohammadi, Sajjad	2474
Mohseni, Parham	2173
Moniruzzaman, Md	943, 951, 1420
Montejano, Misael	637
Monticone, Francesco	1646
Montoya-Acevedo, Diego	629
Moon, Gun-Woo	1899
Moon, Jinyeong	1473, 2413
Moorthy, Radha Sree Krishna	2797
Morris, Lauryn	1311
Moschopoulos, Gerry	1972
Motoori, Shuichiro	1977
Motto, Eric	2356
Mou, Di	2857
Mou, Shin	2419
Mounesi, Reza	1791
Moursi, Mohamed Shawky El	2871, 3064
Mousavi, Mahdi S.	2290
Mu, Qiang	1388, 2790, 3328
Mu, Wei	2597
Mu, Xuchu	978
Muduli, Utkal Ranjan	1940, 2871, 3064
Mueller, Lukas	538
Muenz, Ulrich	1184
Mühlethaler, Jonas	2495
Mujica, Gabriel	868, 3298
Mukhopadhyay, Anwesha	3267
Mukunoki, Yasushige	2356
Müller, Kilian	3241
Mulumudi, Guru Abhilash	1135

Munk-Nielsen, Stig	175, 357, 1034
Murakami, Haruhiko	1569
Muravleva, Ekaterina	1850
Murillo-Yarce, Duberney	1959
Murray, Samantha K.	1905
Murukesan, Karthick	180
Muscat, Isaac	449
Musolino, Francesco	738
Mustakin, Zaheen	1388, 2790
Na, Woonki	2973, 3000, 3006, 3012
Nabila, Kashfia Tajmim	2877
Nabizadah, Ahmad	3160
Nag, Kumar Joy	990, 997
Nagahara, Teruaki	1569
Nagai, Yoshiyuki	423
Nagano, Masanori	285
Nagar, Anshul	2973
Nagasawa, Shinobu	2610
Nagayoshi, Kenichi	1102, 3096
Nakagaki, Akito	2654
Nakagawa, Shigeki	1797
Nakamura, Hirokazu	1728
Nakamura, Keisuke	1237
Nakano, Satoshi	2551
Nakashima, Junichi	2356
Nakata, Yosuke	1383
Nakata, Yuki	21, 2913
Nam, David	2992
Namadmalan, Alireza	2474
Namba, Akira	1797
Namburi, Krishna	1294
Naradhipa, Adhistira M.	498
Narasimhan, Sneha	1866
Narumanchi, Sreekant	1824
Nasiri, Adel	1489, 1791
Nassaji, Abolfazl	2290
Nassar, Rajaie	586, 2228
Nations, Mark	552
Naval, Sourav	1012
Navarro-Rodríguez, Ángel	1935
Neal, Adam T.	2419
Nelms, R.M.	1953, 2703
Nelson, Blake	395, 1167
Nelson, Tolen M.	207
Nelson, Tolen	185
Ng, Wai Tung	983, 1108
Ngo, Khai D. T.	2228
Ngo, Khai	586
Ngo, Minh	111
Nguyen, Allen T.	840
Nguyen, Calvin	1274
Nguyen, Duy T.	231, 1121

Nguyen, Hien	1, 54
Nguyen, Kien	3248
Nguyen, Tung-Tan	389
Ni, Chuan	2117
Nielsen, Morten Rahr	357
Nikmaram, Behnam	2290
Ning, Guangdong	809
Ning, Guangfu	2096
Ning, Shangxian	2660
Nishijima, Kimihiro	1977
Nishimura, Keigo	3096
Nishio, Haruhiko	1108
Nishizawa, Shin-Ichi	2551
Nitta, Honami	1797
Noesges, Brenton A.	2419
Noguchi, Koichiro	1569
Noh, Young-Seok	518
Norman, Patrick	3167
Notake, Koki	1299, 1414
Núñez, Guillermo	1197
Nuzzo, Jeremy	2387
O'Driscoll, Seamus	285, 2009
Oberdieck, Karl	1051, 2589
Oboreh-Snapps, Oroghene	1311
Ochiai, Yuki	3261
Ohi, Toshi	2821
Ohno, Takashi	21
Ohodnicki, Paul R.	544, 2516, 3119
Ohodnicki, Paul	370, 1326
Okamoto, Takahiro	321
Olalla, David	3100
Olimmah, Marshal	395
Onar, Omer C.	1635
Onishi, Hiroyuki	2431
Onuma, Naoto	2681
Opificius, Julian	401
Orabi, Mohamed	775, 906, 2327
Orlando, Tailan	342
Orr, Allison	1211
Oruganti, V.S.R.Varaprasad	801
Ota, Hiroaki	3248
Ou, Shuyu	1501, 3042
Ouyang, Ziwei	246, 252, 1810
Pahlevani, Majid	616, 2953
Pakala, Sriharsh	505
Palani, Praveenkumar	62
Pallantla, Manikanta	2708
Palmal, Manas	1874
Pan, Ci	192
Pan, Qishan	2207
Panja, Pijush Kanti	821
Paplham, Tyler W.	2516

Parashar, Sanket...1347
Paredes-Camacho, Alejandro.............................1774
Park, Junhyeong...3187
Park, Sung-Bum...3187
Parkhideh, Babak.....................................1388, 2790
Parreiras, Thiago M. 1355, 1615
Pasupuleti, Sai Sushma.....................................3316
Patle, Nagesh...2805
Paul, Sayan 1, 54, 334
Paulino, Glaucio H...1274
Pavone, Mario Giuseppe......................................738
Peña-Alzola, Rafael 1935, 3167
Peng, Fang Z..761
Peng, Hongjie 171, 225
Peng, Xiaochuan...1090
Penof, David..1519
Pereira, Joao..637
Pereira, Lucas 1388, 2790
Pereira, Thiago Antonio 118
Peretz, Mor Mordechai.......................................594
Pérez, Fernando 868, 3298
Pérez, Sara 1197
Perez-Farre, Quirc...1774
Perreault, David J..132
Perreault, David...124
Petriæ, Ivan Z. 157
Petriæ, Ivan..2764
Petrillo, Gaia...266
Petucco, Andrea..2667
Pfost, Martin 573, 1129, 1576, 2772
Philippe, Antoine 1396
Phukan, Ripun 2082, 2296
Phung, Thanh Hai 195
Picot-Digoix, Mathis..2718
Piel, Joshua J..2419
Pietrini, Giorgio 670
Pigott, J. 1451
Pilawa-Podgurski, Robert C. N................... 151, 157, 291, 558,
....2276, 2805
Pillonnet, Gaël 315
Pirson, Nicolas..258
Pizzuto, Matteo...1096
Plum, Thomas...1919
Pong, Man-Hay..389
Pool-Mazun, Erick...1459
Popoviæ, Zoya...682
Porras, David A..27
Porter, Matthew 1020, 1564
Pou, Josep..1927
Pourjafar, Saeid...2173
Prabhakar, Siva..69
Pradhan, Rachit...670

Prakash, Surya...1940
Preindl, Matthias.......................................272, 1255
Prodiæ, Aleksandar 307, 990, 997
Punjabi, Shobhana...1159
Qahouq, Jaber A. Abu..................................2779, 2840
Qi, Nianzun...357, 1034
Qian, Ting 2117
Qian, Yijie 524
Qiblawey, Yazan...3071
Qin, Yuan...1564
Qin, Zian 609
Qiu, Tian...1040, 2400
Queiroz, Samuel S....................................1334, 1341
Quenette, Vincent..1889
Rabenold, Elizabeth...2249
Radhakrishnan, Kaladhar.............................491, 1681
Radici, Christian.......................1403, 1512, 1781
Rafiq, Aamir... 395
Rahman, Md Rashedur...................................943, 951
Rahman, Mohammad Dehan....................1844, 2407
Rahouma, Ahmed..................................... 27
Rajagopal, Narayanan 1836
Rajpurohit, Chirayu...2764
Raju, Soniya.. 378
Rallabandi, Vandana..................................1635, 3174
Ram, Achala..2920
Ramasubramanian, Deepak...............................1184
Ramirez, Juan..1211
Ramkumar, S..2708
Ramos, Gabriel V. 1355, 1615
Ramos, Regina 1197, 1375
Ran, Li...1832
Rana, Dilip...1040
Rana, Mandeep S. 2213
Rao, Yifan...1274
Rashid, Syed Saeed.....................................1640, 1646
Rathore, Vikas Kumar 594
Raval, Vishwam...727, 3054
Ravichandran, Krishnan.....................................1681
Rawat, Shubham..1347
Raychowdhury, Arijit 518
Reddy, Narsimha...3054
Redondo, Alejandro......................................868, 3298
Reinotas, Jurgis...1754
Ren, Linhao...2343
Ren, Sheng 3181
Ren, Xufu...1439
Restrepo, Carlos 629
Rettner, Cornelius...1759
Richardeau, Frédéric..2718
Rikiishi, Yasuhiro...2284
Ripamonti, Giacomo 479

Ristic-Smith, Aleksandar 1529
Rivas-Davila, Juan 363, 2241
Rizkalla, Maher 2029
Rizzolatti, Roberto 1444, 2254, 2260
Roberts, Gianluca 307
Rodgers, Aidan 1242
Rodriguez, Ezequiel Ramos 1927
Rodriguez, Fernando 3100
Rodriguez, José 746, 895, 2290, 2327
Rodríguez, Juan 1701, 1959
Rogers, Daniel 1529
Rogers, Michael 1569, 2356
Ronanki, Deepak 2647
Rong, Mingzhe 3220
Rong, Zhenshuai 1439
Rosa, Bruno M.G. 1653
Round, Simon 1803
Roy, Soham 1121
Rubinic, Jaksa 416
Rueß, Manuel 3241
Ruiz, Juan M. 2002
Ruiz, Juan 442, 2082, 2296
Ruoff, Dominik Alexander 2589
Ruppert, Daniel 1759
Russo, Andrea 252
Ruszczyk, Adam 1803
Sa, Satyam 103, 449
Saberi, Sajad 2840
Sadasivan, Arya 2461
Sadilek, Tomas 401
Sado, Kerry 2833
Saeedifard, Maryam 2051, 3071, 3321
Saelens, Jonathan 1311
Saggini, Stefano 479, 1444, 2260
Saha, Subrata 1237
Saha, Tarak 3174
Sahoo, Subham 696, 912, 1501
Sai, Ranajit 285
Saiga, Kazuma 2551
Saito, Shoji 1569
Saito, Wataru 2551
Sakai, Hiroto 3077
Salari, Omid 3083
Salehi, Maryam 2883
Samanta, Akash 3141
Sambo, Haifah B. 291
Sandoval, Rolando 1459
Sangwongwanich, Ariya 738, 1501, 3042
Sanjakdar, Omar 1396, 2562
Santi, Enrico 2833, 2865
Santos, Ion Leandro Dos 342
Santos Jr., Euzeli Cipriano Dos 2029

Sanusi, Bima Nugraha 246, 1810, 2035
Saraf, Pushkar 77
Sarajian, Ali 895
Sarda, Radhika 62, 1927
Sarlioglu, Bulent 3174
Sarnago, H. 147
Sarnago, Héctor 3109
Sarofim, Seif 449
Sati, Shraf Eldin 622
Sato, Yuji 1383
Sato, Yuki 579
Satterlee, Ryan 3133
Satyamsetti, Vijayakrishna 1403
Sauter, Bailey 2764
Sayed-Ahmed, Ahmed 2932
Sba, Baher Abu 2453
Sbabo, Paolo 2667
Schanen, Jean-Luc 2562
Scheideler, William 1590
Scherer, Yohannes Amilcar Tekle 342
Sebastián, Javier 1959
Sebata, Kohei 1977
Sekiya, Hiroo 3248
Selvarasu, Uthandi 761, 1465
Sen, Paresh C. 1892
Sen, Tanuj 139, 349, 1882
Sengstock, Jonathan 1242
Sengupta, Arkadeb 90, 118, 1148
Seo, Gab-Su 602, 645, 3059
Seo, Seoktae 1629
Sethupandi, Abishek 62
Seugnet, Léo 2718
Shadmand, Mohammad B. 3071
Shafei, Ahmad El 1326
Shah, Shreyas B. 670
Shahbazi, Reza 1179
Shahsavar, Tala Hemmati 1622
Shang, Shuye 3227
Shao, Hang 2138
Shao, Linbo 1020, 1564
Sharma, Mohit 3141
Shen, Andy 3129
Shen, Xiaobing 436, 2167
Shi, Guannan 1564
Shillaber, Luke 2597
Shimada, Takae 2681
Shimosako, Shumei 3077
Shin, Se-Un 834
Shivdikar, Saumil 2400
Shoji, Tomokazu 2821
Shrestha, Niranjan 801
Shu, Wenze 524

Siddiquee, Ashraf	1294, 1602
Silveira, Hector Bessa	342
Sim, Dong Hyeon	3200
Sim, Si Yuan	505
Singh, Anurag	1, 54, 334
Singh, Prashant	1026
Singla, Rishabh	3054
Siraj, Ahmed	1040, 2400
Sitta, Alessandro	1070
Smith, John	637
Solecki, Alex	1242
Solomentsev, Michael	77
Son, Gibong	1741
Song, Chen	2075
Song, Keqi	1507
Song, Minwoo	3012
Song, Qihao	202
Song, Xiaoqing	1844, 2407
Song, Yubo	696
Song, Zhihao	327
Sönmez, Ertuðrul	1051
Soundararajan, Soundhariya G.	238
Souri, Naser	2202
Sowers, Elizabeth A.	2419
Sozer, Yilmaz	1294, 1602, 2698
Spiazzi, Giorgio	2667
Spieler, Matthias	1495, 1551
Sridhar, Sundaramoorthy	299
Sriram, Vaisambhayana B.	62, 1927
Srivastava, Shubham	2213
Starke, Michael	637, 703
Stauth, Jason T.	1590, 1715
Steiner, Mark	2356
Stella, Fausto	1408
Steyaert, Bernard	1255
Steyn-Ross, Alistair	378, 874
Stillwell, Andrew	1242
Stokowski, Nicole	1242
Strache, Sebastian	2569
Strathman, Sophia A.	1311
Streit, Jochen	1051
Strezelecki, Ryszard	2173
Stricula, Justin	401
Sturdivant, Maurice	544
Su, Gui-Jia	1635
Su, Mei	2096
Sugie, Hisashi	2730
Sui, Qingcheng	436, 2752
Sukita, Yohei	1102
Sullivan, Charles R.	840, 1816
Sun, Bosheng	1990
Sun, Kai	2488
Sun, Lingwei	1108
Sun, Peiyuan	2375
Sun, Ruize	192
Sun, Weifeng	524
Sun, Xiuhu	3328
Sun, Zhen	3275
Sun, Ziang	2981
Sund, Jade	971
Sune, Joseph Benzaquen	164, 3334
Suntharalingam, Piranavan	670
Suzuki, Asamira	1728
Swaminathan, Madhavan	2101
Sweet, Mark	3167
Swoboda, Philipp	2348
Syed, Hadiuzzaman	1051
Szczublewski, Austin M.	185
Tadakuma, Toshiya	2610
Taguchi, Koichi	2356
Taha, Wesam	3147
Tajima, Shin	782
Takahashi, Keita	2610
Takahashi, Yoshiaki	1414
Takamiya, Makoto	1084, 1102, 2551
Takamura, Yota	1797
Takayama, Naoki	2681
Takeda, Ryo	2821
Takeuchi, Kosuke	21
Takeuchi, Toshiro	2828
Takishima, Kenta	423
Takizawa, Sota	2924
Tan, Matthew	1882
Tan, Siew-Chong	3227
Tanaka, Kenichiro	579
Tanaka, Ryota	423
Tanaka, Shinsaku	2284
Tanaka, Toshiyuki	2828
Tang, Ho-Tin	1507, 1582
Tang, Wenyuan	1363, 1597
Tang, Yi	3220
Tant, Mike	1824
Tariquzzaman, Md.	3036, 3048
Tarutani, Masayoshi	1383
Tatetsu, Riku	1977
Tayebi, Milad	854
Teng, Fei	2527
Teng, Yiyina	757
Terauchi, Naoya	285
Terzija, Vladimir	757
Thacker, Thimothy	2992
Then, Han Wui	1681
Thevar, Madasamy Palavesha	62
Thike, Rajendra	1279

Thirumoorthi, Sathya Rupan.................................. 1866
Thurlbeck, Alastair P... 1602
Tian, Fanghao 2167, 3124
Tian, Jiachen... 2327
Tian, Xiaoyang... 1754
Tingbari, Vincent Masabiar................................ 2973
Tomey, Hala.. 1211
Tomioka, Shohei...579
Tong, Junhong... 1786
Tong, Qiaoling... 2109
Torres, Javier.. 1197
Torres, Renato Amorim...................................... 1495
Touhami, Mustapha 1012
Tran, Ngoc Ho 2569
Trescases, O. 1451
Trescases, Olivier.................... 103, 449, 676, 1905
Tripathi, Anshuman................................... 62, 1927
Tsai, Chieh-Ju 664, 2131, 2143, 2725, 2735, 2741
Tschanz, James 1681
Tseng, Chien-Hao 2725
Tsou, Ming-Chang...887
Tsuchida, Takayuki...285
Tuzizila, Jeremie 401
Uchida, Yasuo... 48
Uddarraju, Praneeth 1311
Uddin, Muhammad Fasih 2453
Uegaki, Shin 1383
Uematsu, Takeshi... 3248
Ulrich, Burkhard 1707, 2303, 2502
Umanand, L. 1610, 2785
Umar, Jamil... 2973
Umar, Muhammad F. 3071
Umetani, Kazuhiro...................................... 321, 2654
Ursino, Mario................................ 1444, 2254, 2260
Uzum, Alper.. 1294, 2698
Vagnon, Eric 2562
Vanderwegen, Wout...238
Varadarajan, Kamal...180
Vasiæ, Miroslav 788, 1375, 1803
Vedula, Inder.. 1, 54
Vergès, Gaël 1905
Vico, Enrico 1408
Vines, Peter........................... 1403, 1512, 1781
Vinnac, Sébastien.. 2718
Vitale, Gianni.. 3133, 3304
Vohl, Kenny 2757
Wagner, Tomas... 3100
Walters, Andrew...951
Wang, Cheng Feng 103, 449
Wang, Daming 2369
Wang, Haiyan 3312
Wang, Haoyu 485, 983, 1544, 2207, 2556, 2675, 2857

Wang, Hongjie...719, 2393
Wang, Huai.............................912, 2380, 3042
Wang, Jin.............................1203, 2986
Wang, Jinyan.............................171, 225
Wang, Jun.............................538, 1850, 2088
Wang, Kaiyuan.............................3227
Wang, Kejia 505
Wang, Kun 2088
Wang, Kunrong.............................2162
Wang, Laili.............................2375
Wang, Lei.............................2162
Wang, Liang 983
Wang, Libing.............................3174
Wang, Lichong 757
Wang, Linguo.............................2070
Wang, Lisheng.............................1248
Wang, Liwei.............................1153
Wang, Maojun 225
Wang, Meng.............................3312
Wang, Mengqi.............................2624, 3155
Wang, Pinhe.............................246, 2035
Wang, Qiong 498
Wang, Rudy.............................9, 1318
Wang, Rui.............................1063
Wang, Shaozhe.............................1459
Wang, Shukai.............................566, 1882, 2438
Wang, Shumeng.............................761, 1465
Wang, Shuo 1173, 2603
Wang, Sunqing.............................3018
Wang, Wei.............................2692
Wang, Xiao 219
Wang, Xiaohua.............................3220
Wang, Xiaosheng 795
Wang, Xiaoting.............................3030
Wang, Xiaoyu 416
Wang, Xinlin 809
Wang, Xiongfei.............................1615
Wang, Xuan.............................1791
Wang, Xuliang.............................1544, 2556, 2675
Wang, Yan.............................1544, 2556, 2675
Wang, Yao.............................3275
Wang, Yibo.............................795, 3181
Wang, Yicheng.............................3147
Wang, Yiju.............................1368
Wang, Yulei 219
Wang, Yunxin.............................2675
Wang, Zijian.............................1537
Wang, Ziyao 485
Wang, Zuoshuai.............................3018
Watabe, Kiyoto.............................2551
Watanabe, Hiroki.............................21, 48
Watanabe, Kenichi 1102, 3096

Wehr, Erik	2757
Wei, Anran	1983
Wei, Bo	327
Wei, Jinxiao	1832
Wei, Xing	3042
Wei, Xuanjing	1403
Wei, Yuxin	2713
Weihs, Leon	2757
Weiser, Mathias C.J.	2387, 3241
Weng, Sheldon	1681
Wens, Mike	195
Wheeler, Patrick	895
Wicht, Bernhard	848
Wick, Lukas	2380
Williamson, Sheldon	801, 1224, 2322, 3114, 3141
Wilson, Marcus	378
Winkler, Joseph	848
Wipprecht, Lukas	2303
Wojewoda, Leigh	491
Wong, Andy	2833
Wouters, Hans	238, 258, 2167
Wright, Jason	401
Wu, Alan	307
Wu, Chih-Chiang	2687
Wu, Hsiang-Kai	2687
Wu, Shang-Syun	2179
Wu, Taotao	1090
Wu, Teng	2634, 2660
Wu, Tsai-Fu	16, 900
Wu, Xin	651, 2981
Wu, Xinke	1995
Wu, Yang	139, 349
Wu, Yanqing	1995
Wu, Yingzhe	1248
Wu, Yue	1114, 3220
Wu, Yuxuan	2582
Wunderlich, Andrew	2865
Wunderlich, Ralf	2757
Xi, Zichen	1020
Xia, Xiaoyi	2022
Xiang, Zhangwei	429
Xiao, Junjie	609
Xie, Biyun	919
Xu, Dehong	651, 2894, 2981
Xu, Guo	2096
Xu, Haoran	2109
Xu, Huangsheng	2207
Xu, Limei	2075
Xu, Shen	524
Xu, Wentao	1012
Xu, Wenzhe	2426
Xu, Xinmiao	1433
Xu, Yun	1114
Xu, Ziyang	2521
Xue, Hui	2117
Xue, Yuxiang	1248
Yabuta, Shigenori	2821
Yagielski, John	3174
Yamaguchi, Koji	1299, 1414
Yamamoto, Keisuke	3194
Yamamoto, Masayoshi	2431
Yamanaka, Kimito	1797
Yan, Decheng	3334
Yan, Yiyang	2343, 2426
Yan, Zhaoheng	1114
Yan, Zhixing	357, 1034
Yang, Bowen	602
Yang, Garam	3000
Yang, Hélène T.W. Ma	983
Yang, Juchen	1203
Yang, Liu	2369
Yang, Qichen	860
Yang, Qiuzhe	1537
Yang, Robert	180
Yang, Xin	1020, 1564
Yang, Xingyu	1953
Yang, Xinliang	409
Yang, Yirui	1173, 2603
Yang, Yongheng	3024
Yang, Yun	3227, 3275
Yang, Zineng	1020
Yao, Wenxi	327
Yao, Yuzhou	1203
Yasko, Mohamed	3124
Yato, Shinji	3077
Ye, Liang	285
Ye, Zhengyu	2117
Yeo, Howe Li	62
Yeo, Sungku	2547, 3281
Yi, Lifang	2413
Yi, Zheyuan	2488
Yin, Shan	1248, 2075
Yin, Tianxiang	2535
Yoneyama, Rei	2356
Yoshimoto, Kantaro	423
Yoshimura, Yuto	2654
You, Longxiang	3018
Youssef, Mohamed Z.	3083
Yu, Hao	2343, 2426
Yu, Jingshu	1681
Yu, Ruiyang	854
Yu, Sheng-Han	2131
Yu, Sheng-Yang	42, 1990
Yu, Wensong	2220

Yu, Xiang	1892
Yuan, Hao	2117
Yuan, Huan	3220
Yuan, Jiaqi	670
Yuan, Jingyi	966
Yuan, Song	1114
Yuan, Tianlong	429
Yuan, Tianshu	2375
Yun, Dam	1659, 2268
Zaabi, Omar Al	1940
Zade, Aditya	719, 1004, 2194
Zaitsu, Toshiyuki	1977, 2654
Zaizen, Shohei	2551
Zaman, Mohammad Shawkat	676
Zan, Xin	3232
Zane, Regan	719, 1004, 2194
Zeineldin, Hatem H.	622
Zekorn, Tobias	2757
Zeng, Hank	2138
Zeng, Jia-En	1967
Zeng, Wenliang	510
Zeng, Zheng	219
Zhan, Cao	278, 2482
Zhang, Ben	3181
Zhang, Bing	2070
Zhang, Bo	192
Zhang, Bohua	573
Zhang, Boran	2675
Zhang, Boyi	1850, 2296
Zhang, Cheng	1512, 3212
Zhang, Chenghui	2713
Zhang, Chi	9
Zhang, Desheng	2109
Zhang, Fuxing	2059
Zhang, Haijin	3312
Zhang, Hely	1286
Zhang, Heng	2369
Zhang, Hengbin	1248
Zhang, Hong	2375
Zhang, Honglang	2075
Zhang, Jiazheng	2628, 2968
Zhang, Jincheng	978
Zhang, Jinfeng	1439
Zhang, Li	2535
Zhang, Qingzheng	2894
Zhang, Renjie	3220
Zhang, Shengke	214
Zhang, Shiqi	757
Zhang, Tianyi	2066
Zhang, Weihang	978
Zhang, Xiangrong	3275
Zhang, Xin	505, 1143, 3018

Zhang, Xiong	2343, 2426
Zhang, Yi	2380
Zhang, Yichi	2380
Zhang, Yifan	2343, 2369, 2426
Zhang, Yifu	2746
Zhang, Yingjie	2603
Zhang, Yuanxin	2857
Zhang, Yuhao	202, 1020, 1564
Zhang, Yuxin	2973
Zhang, Zhe	2907
Zhang, Zhenbin	2846, 3206
Zhang, Zheyu	1040, 1153, 2400
Zhang, Zhi Jin	3321
Zhang, Zhining	1203
Zhang, Zichen	1850
Zhao, Delin	3220
Zhao, Fangzhou	1615
Zhao, Hongbo	357, 1034
Zhao, Shuang	2369, 3227
Zhao, Shuofeng	1824
Zhao, Tiefu	1388, 2790, 3328
Zhao, Tuo	1274
Zhao, Wending	1995
Zhao, Yifan	2521, 2846, 3206
Zhao, Yue	27, 2453
Zheng, Zexiang	2343, 2369
Zhou, Daniel H.	1693
Zhou, Daniel	566, 1274, 2438
Zhou, Dao	2380
Zhou, Fei	2541
Zhou, Feng	2624
Zhou, Jiale	1388, 2790, 3328
Zhou, Kunxiao	809
Zhou, Lufan	1803
Zhou, Mingde	2207
Zhou, Wenqi	2589
Zhou, Xigen	2157
Zhou, Yan	1767
Zhou, Yi	651
Zhou, Yuan	2535
Zhou, Yuebin	2369
Zhou, Zongjie	1047
Zhu, Jiaqi	3312
Zhu, Jinli	761, 1465
Zhu, Junjie	2070
Zhu, Lingyu	278, 2482
Zhu, Liyan	1020
Zhu, Yicheng	558, 2276
Zhu, Zhenhai	1995
Zhuo, Fang	2327
Zolfi, Pouya	1230
Zou, Huanghaohe	1786

Zou, Jiaao..2066
Zou, Jiarui... 558, 2276, 2805
Zou, Mingrui...219
Zou, Xudong...2066
Zou, Xuecheng...2109
Zufferli, Kevin 1444, 2260
Zuo, Yu 258, 436, 2167, 2752
Zuo, Zhiling..2070
Zynger-Capaverde, Betina 2562

IEEE
445 Hoes Lane
Piscataway, NJ 08854-4141

ISBN 979-8-3315-1612-3